# DIAGNOSTIC IMAGING IN INTERNAL MEDICINE

Ronald L. Eisenberg, M.D.

PROFESSOR AND CHAIRMAN
DEPARTMENT OF RADIOLOGY
LOUISIANA STATE UNIVERSITY SCHOOL OF MEDICINE
SHREVEPORT, LOUISIANA

**McGraw-Hill Book Company**

New York St. Louis San Francisco Auckland Bogotá Guatemala Hamburg
Johannesburg Lisbon London Madrid Mexico Montreal New Delhi Panama
Paris San Juan São Paulo Singapore Sydney Tokyo Toronto

Notice

Medicine is an ever-changing science. As new research and clinical experience broaden our knowledge, changes in treatment and drug therapy are required. The editors and the publisher of this work have checked with sources believed to be reliable in their efforts to provide drug dosage schedules that are complete and in accord with the standards accepted at the time of publication. However, readers are advised to check the product information sheet included in the package of each drug they plan to administer to be certain that the information contained in these schedules is accurate and that changes have not been made in the recommended dose or in the contraindications for administration. This recommendation is of particular importance in connection with new or infrequently used drugs.

**DIAGNOSTIC IMAGING IN INTERNAL MEDICINE**



1234567890 HAL HAL 898765

ISBN 0-07-019262-6

This book was set in Optima Roman by General Graphic Services, Inc.; the editors were Robert E. McGrath and T. Fiore Lavery; the production supervisor was Avé McCracken; the designer was Russell H. Till.
Halliday Lithograph Corporation was printer and binder.

**Library of Congress Cataloging in Publication Data**

Eisenberg, Ronald L.
Diagnostic imaging in internal medicine.

"Written as a radiographic complement to Harrison's Principles of internal medicine, 10th edition"—Pref.
Includes bibliographies and index.
1. Diagnosis, Radioscopic. 2. Imaging systems in medicine. 3. Internal medicine. I. Harrison, Tinsley Randolph, date. II. Title. [DNLM: 1. Internal Medicine. 2. Nuclear Magnetic Resonance—diagnostic use. 3. Radiography. 4. Ultrasonics—diagnostic use. WN 200 E36d]
RC78.E54 1985 616.07'57 84-28899
ISBN 0-07-019262-6

To Zina, Avlana, and Cherina

# Contents

## Chapter Seventy-Two: Miscellaneous Disorders of the Central Nervous System 935



## Chapter Seventy-Three: Diseases of the Spine and Spinal Cord 941



# PART THIRTEEN: MISCELLANEOUS DISORDERS

## Chapter Seventy-Four: Granulomatous Diseases of Unknown Etiology 951



## Chapter Seventy-Five: Diseases of the Spleen and Lymph Nodes 957



## Chapter Seventy-Six: Diseases Due to Environmental Hazards and Chemical and Physical Agents 965



## Chapter Seventy-Seven: Cutaneous Diseases 977



## Chapter Seventy-Eight: Miscellaneous Congenital Disorders 981



## Appendix: Procedures in Diagnostic Imaging 987

# Contributors

**Lanning W. Houston, M.D.,** Assistant Professor of Radiology, Department of Radiology, University of Wisconsin in Madison, participated substantially in the writing of Part Twelve "Disorders of the Nervous System" and provided almost all the illustrations for this part.

**R. Brooke Jeffrey, M.D.,** Associate Professor of Radiology, Department of Radiology, University of California in San Francisco, participated substantially in the writing of Chapter Fifty-Two "Tumors of the Liver" and Chapter Fifty-Five "Diseases of the Pancreas" and provided many of the illustrations for these chapters.

**Peter C. Meyers, M.D.,** Assistant Professor of Radiology, Department of Radiology, Louisiana State University in Shreveport, provided a large number of illustrations from his extensive teaching file.

**Edward A. Sickles, M.D.,** Associate Professor of Radiology, Department of Radiology, University of California in San Francisco, wrote and provided the illustrations for the section on breast cancer in Chapter Five.

# Foreword

*Diagnostic Imaging in Internal Medicine* is a suitable companion to the respected and prestigious Harrison's *Principles of Internal Medicine*, 10th edition. This extensively illustrated, meticulous text serves as an excellent guide for the interpretation and understanding of the diagnostic imaging procedures which today are indispensable in the modern practice of medicine.

Benefiting from rapid technological advances, radiology has made the current diagnostic imaging approaches more efficient and much less invasive. Today, many of the diagnostic procedures are performed on an out-patient basis, thus shortening or totally avoiding hospital stay. Not only have procedures become less invasive but they are significantly more sensitive in detecting tissue abnormalities. The diagnoses made by the new imaging procedures, however, are still not specific in most instances but depend on the correlation of imaging findings with clinical information. Without clinical and laboratory data, diagnostic imaging is not at its best and the information it provides is not optimized.

This book also describes the new developments in interventional radiology. These include imaging-guided needle biopsies for histologic diagnosis and therapeutic measures such as drainage, opening of closed, normally existing channels, or occlusion of pathologic communications. The wide scope of this book only emphasizes the intimate connections between clinical medicine and diagnostic imaging and that the great advances which have occurred in both have been enhanced by their mutual interdependence.

Alexander R. Margulis, M.D.
Professor and Chairman
Department of Radiology
University of California Medical Center
San Francisco, California

# Preface

During the past few years, the rapid development of new sophisticated imaging modalities has made radiology the fastest growing specialty in medicine. Ultrasound, computed tomography (CT), digital subtraction angiography, and magnetic resonance imaging (nuclear magnetic resonance) have greatly expanded the capabilities of relatively noninvasive diagnostic imaging. New interventive techniques, such as balloon arterial dilatation, insertion of nephrostomy tubes, biliary drainage, and abscess drainage and needle biopsy under ultrasound or CT guidance have become effective diagnostic and therapeutic alternatives to the high morbidity of surgery.

More than ever before, internists and family practitioners must be aware of the extensive role of diagnostic imaging modalities in their daily practice. This book was written as an imaging complement to Harrison's *Principles of Internal Medicine*, 10th edition, and the organization and subheadings of this book follow that classic text as much as is practical. Nevertheless, the detailed written description of major imaging findings for many clinical disorders, as well as the wealth of representative illustrations that include the newer imaging techniques (ultrasound, CT, digital subtraction angiography, magnetic resonance imaging), can be correlated with any of the major internal medicine textbooks. A carefully developed reference list offers the serious reader additional sources in which to pursue subjects in more depth than is possible within the space limitations of this book.

The radiologist in residency or practice will find this book a useful guide for planning the proper imaging approach to specific clinical problems. It also can serve as a general overview of the spectrum of imaging findings for a wide variety of disorders in internal medicine for both the radiologist and referring physician.

As stringent controls over medical practice and payments are being instituted by the government and third party payers to attempt to curb the skyrocketing costs of medical care, it has become increasingly important for physicians to formulate a rational imaging approach to diagnosis rather than the previous practice of ordering a wealth of unnecessary examinations. Whenever possible, this book suggests the best single test for initially evaluating a given clinical problem before embarking on a description of the findings of other diagnostic modalities. The illustrations are also selected to reflect this concept. For example, in the area of neuroradiology where CT has revolutionized the diagnostic approach, the vast majority of illustrations are CT scans. Although the findings on angiography, radionuclide brain scans, and skull films are mentioned in the text, there are few illustrations since these techniques have much less bearing on current radiologic practice.

I want to express my thanks to Betty DiGrazia for the

many hours she spent in the arduous task of typing and retyping the manuscript. I gratefully appreciate the efforts of James Kendrick of George Washington University, Eugenia Kolsanoff in San Francisco, and the medical communications department at the Louisiana State University School of Medicine in Shreveport for skillfully photographing the many illustrations. Thanks also should go to the many physicians who have kindly allowed me to use their published material as illustrations in this book. Finally, I acknowledge the encouragement and support of Joseph Brehm and Terry Fiore Lavery, editing supervisor, who have made the immense technical problems of preparing a book as painless as possible.

RONALD L. EISENBERG

# PART ONE

# BIOLOGICAL CONSIDERATIONS IN THE APPROACH TO CLINICAL MEDICINE

## Chapter One
# Clinical Immunology

### IMMUNODEFICIENCY DISEASES

Immunodeficiency diseases may be congenital, spontaneously acquired, or iatrogenic. These disorders may primarily involve humoral or cell-mediated immunity, or they may reflect the complete impairment of both immunologic systems. Patients with immunodeficiency diseases are especially prone to develop infections caused by opportunistic organisms that are of little consequence in patients with an intact immune system. *Candida albicans, Pneumocystis carinii,* and *Giardia lamblia* are among those organisms that most frequently infect the respiratory and gastrointestinal tracts of immunodeficient patients.

#### Di George's Syndrome

Di George's syndrome is the classic example of the impairment of cell-mediated immunity. The hallmark of Di George's syndrome is the congenital absence of the thymus gland, though carefully performed autopsies may reveal a tiny, histologically normal thymus in an ectopic location. Associated abnormalities include hypocalcemic tetany due to aplasia of the parathyroid glands, congenital cardiac defects (especially involving the great vessels), and malformations of the face, ears, and nose. Frontal chest radiographs demonstrate a narrow mediastinum and the absence of a thymic shadow; on lateral projections, the retrosternal space is clear and relatively lucent.[1]

#### Agammaglobulinemia

The classic example of impaired humoral immunity is congenital agammaglobulinemia, an inherited disorder that is transmitted as an X-linked recessive trait and thus is seen exclusively in males. Severe sinus and pulmonary infections in these children begin late in the first year of life, after passive immunity has worn off. Radiographically, there is a virtual absence of the normal shadow of adenoid tissue in the posterior nasopharynx (Neuhauser's sign) and thus a large pharyngeal airway. Because the draining lymph nodes do not tend to enlarge in the normal way in response to local infection, patients with congenital agammaglobulinemia usually demonstrate extensive pulmonary parenchymal disease (recurrent and poorly resolving pneumonias) with a striking absence of hilar lymph node enlargement. Unlike the situation with Di George's syndrome, the thymic shadow is normal.

Acquired agammaglobulinemia may be associated with malnutrition, protein-losing enteropathy, and intestinal lymphangiectasia. It may also result from lymphoreti-

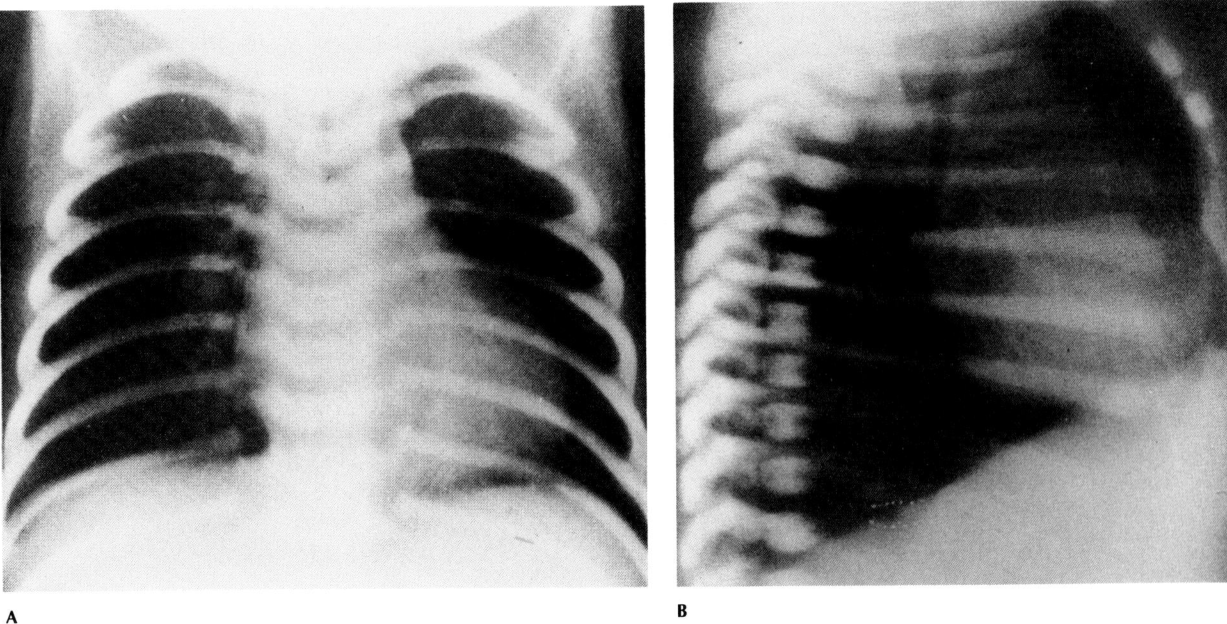


**Figure 1.1. Di George's syndrome.** (A) Frontal and (B) lateral views of the chest. The thymus is not visible. The retrosternal area is lucent, and the anterior border of the heart is sharply defined. (From Kirkpatrick, Capitanio, and Pereira,[1] with permission from the publisher.)

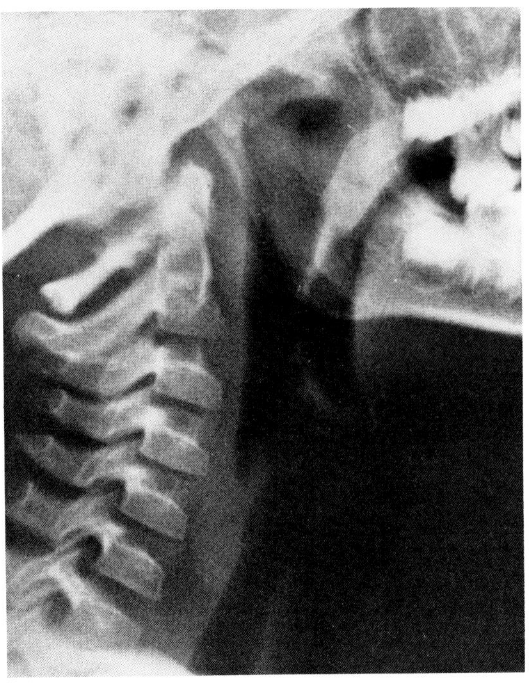


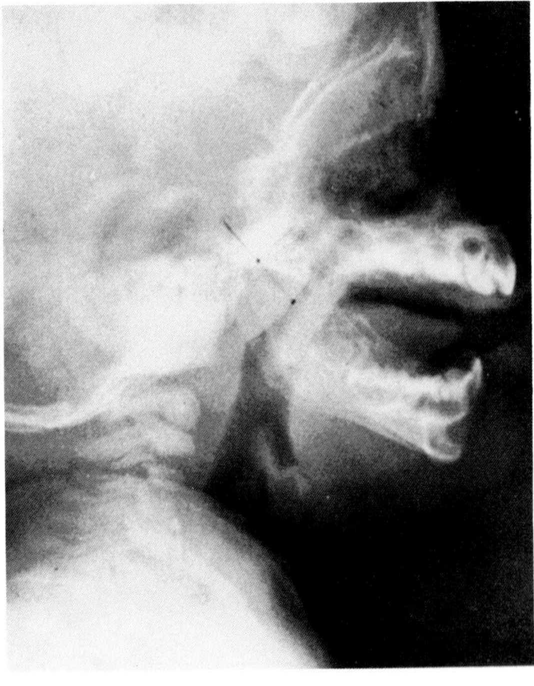


**Figure 1.2. Congenital hypogammaglobulinemia.** (A) There is almost no lymphoid tissue in the pharynx of this 8-year-old boy. (B) A radiograph of the neck of a normal infant demonstrates a normal amount of lymphoid tissue in the posterior nasopharynx. (From Kirkpatrick, Capitanio, and Pereira,[1] with permission from the publisher.)

cular malignancy and from its treatment with radiation, antilymphocyte serum, or cytotoxic drugs. As with the congenital form, sinus and pulmonary infections dominate the clinical picture, and chest radiographs may demonstrate recurrent and poorly resolving pneumonias, bronchiectasis, atelectasis, and empyema. Unlike the congenital type, acquired agammaglobulinemia often results in hyperplastic lymphoid tissue and is frequently accompanied by enlargement of the mediastinal lymph nodes. In the gastrointestinal tract, hyperplastic lymph

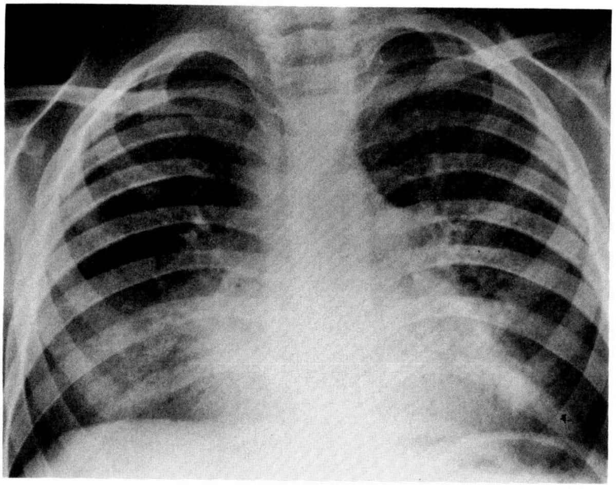
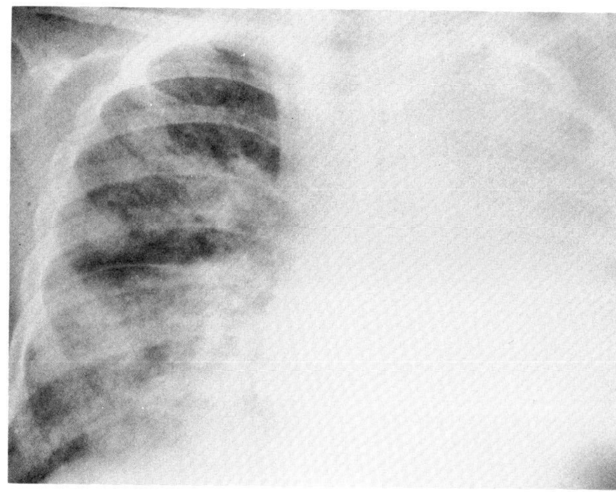

**Figure 1.3. Hypogammaglobulinemia with recurrent pneumonia.** (A) Bilateral basilar pneumonia involves the right middle and lower lobes and the left lower lobe. (B) This film taken 2 years later, 1 day before the patient's death, shows a diffuse right-sided infiltrate with complete filling of the left hemithorax by a massive pleural effusion.

follicles may appear as innumerable tiny polypoid masses uniformly distributed throughout the involved segment of bowel. This nodular lymphoid hyperplasia primarily affects the jejunum but can also involve the remainder of the small bowel and the colon. Most patients with nodular lymphoid hyperplasia due to acquired agammaglobulinemia have diarrhea and malabsorption; associated infection by *Giardia lamblia* may cause the underlying intestinal fold pattern to be irregularly thickened and grossly distorted. It must be noted, however, that in children and young adults the presence of multiple small, symmetric nodules of lymphoid hyperplasia in the distal small bowel is normal.[1–4]

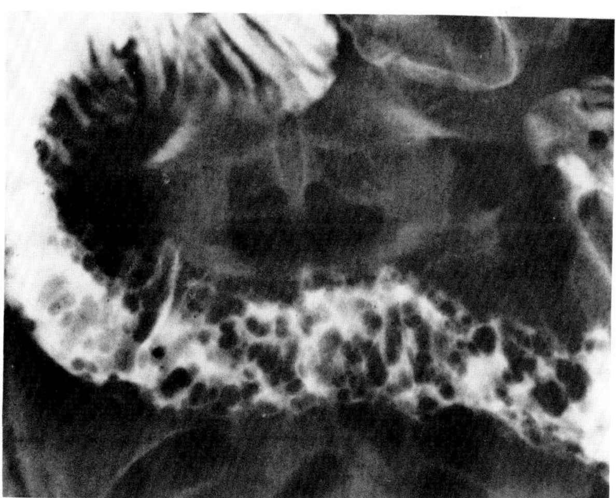

**Figure 1.4. Nodular lymphoid hyperplasia.** Multiple filling defects in the small bowel suggest polypoid masses. (From Eisenberg,[35] with permission from the publisher.)

## Swiss Type Agammaglobulinemia

In Swiss type agammaglobulinemia there is a congenital deficiency of both humoral and cell-mediated immunity. Children with this rapidly fatal disorder have a small thymus and virtually absent lymphoid tissue and are susceptible to severe skin, respiratory, and gastrointestinal infections.

## Acquired Immunodeficiency Syndrome (AIDS)

The acquired immunodeficiency syndrome (AIDS), which primarily affects young homosexual males, is characterized by a profound and sustained impairment of cellular immunity that results in recurrent or sequential disseminated opportunistic infections and a particularly aggressive form of Kaposi's sarcoma (see the section on "Kaposi's Sarcoma" in Chapter 5). AIDS has also been reported in intravenous drug abusers, hemophiliacs, and a group of heterosexual, non-drug-abusing Haitians.

AIDS predominantly involves the lungs, gastrointestinal tract, and central nervous system. Pulmonary infections, often due to *Pneumocystis carinii* pneumonia, are characterized by a sudden onset, a rapid progression to diffuse lung involvement, and a marked delay in resolution. Gastrointestinal manifestations include a variety of sexually transmitted diseases involving the rectum and the colon, infectious processes (such as shigellosis, amebiasis, candidiasis, and giardiasis), and alimentary tract dissemination of Kaposi's sarcoma. Central nervous system disorders include atypical brain abscesses and meningeal inflammation, most commonly due to toxoplasmosis and cryptococcosis.[5–9]

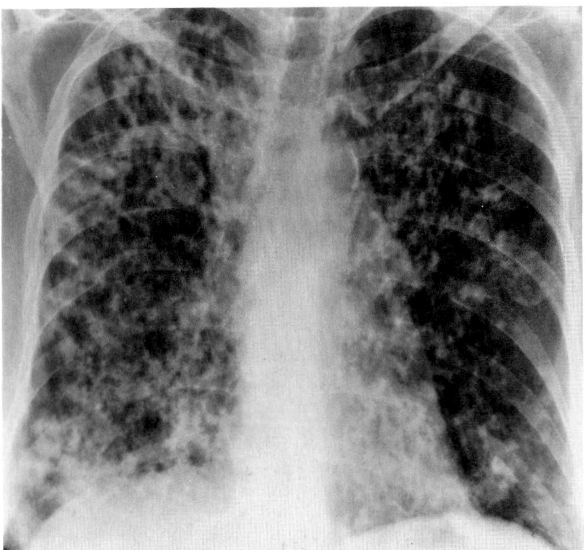

**Figure 1.5. Acquired immunodeficiency syndrome.** Diffuse bilateral pulmonary infiltrates due to superimposed *Pneumocystis carinii* pneumonia.

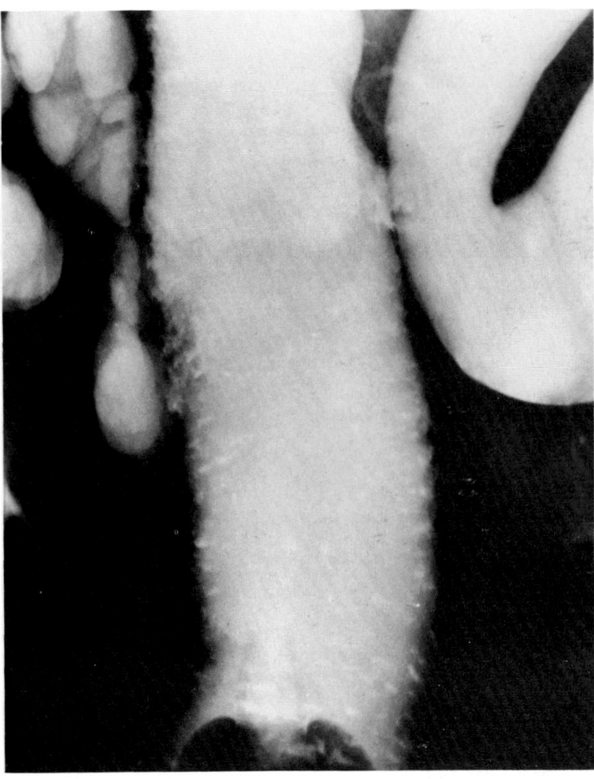

**Figure 1.6. Acquired immunodeficiency syndrome.** Severe ulcerating colitis in the rectosigmoid region. Superficial ulcers and inflammatory edema produce a serrated outer margin of the barium-filled colon.

## PLASMA CELL NEOPLASMS AND RELATED DISORDERS

### Multiple Myeloma

Multiple myeloma is a disseminated malignancy of plasma cells that may be associated with bone destruction, bone marrow failure, hypercalcemia, renal failure, and recurrent infections. The disease primarily affects persons between 40 and 70 years of age. Typical laboratory findings include an abnormal spike of monoclonal immunoglobulin and the presence of Bence Jones protein in the urine. Up to 20 percent of patients with multiple myeloma develop amyloidosis (see the section on amyloidosis in this chapter).

The classic radiographic appearance of multiple myeloma is multiple punched-out osteolytic lesions scattered throughout the skeletal system and best seen on lateral views of the skull. Because the bone destruction is due to the proliferation of plasma cells distributed throughout the bone marrow, the flat bones containing red marrow (vertebrae, skull, ribs, pelvis) are primarily affected. The appearance may be indistinguishable from that of a metastatic carcinoma, though the lytic defects in multiple myeloma tend to be more discrete and uniform in size. The sharply circumscribed lytic lesions eventually tend to coalesce, destroying large segments of bone. The destructive lesion may break through the cortex and periosteum to form a soft tissue mass. This most commonly involves a rib and rarely occurs in metastatic disease. Pathologic fractures are common, especially in the ribs, vertebrae, and long bones. Solitary or diffuse areas of sclerosis, simulating osteoblastic metastases, are rarely seen.

Extensive plasma cell proliferation in the bone marrow with no tendency to form discrete tumor masses may produce generalized skeletal deossification simulating postmenopausal osteoporosis. In the spine, decreased bone density and destructive changes in multiple myeloma are usually limited to the vertebral bodies, sparing the pedicles (lacking red marrow) which are frequently destroyed by metastatic disease. The severe loss of bone substance in the spine often results in multiple vertebral compression fractures. Because multiple myeloma causes little or no stimulation of new bone formation, radionuclide bone scans may be normal even with extensive skeletal infiltration.

Mild or severe renal failure develops in about 20 percent of patients with multiple myeloma. This may be related to tubular damage from large quantities of filtered monoclonal proteins or from hypercalcemia, uric acid nephropathy, or amyloidosis. In the past, dehydration induced by the preparation for some radiographic procedures has been implicated in the precipitation of renal failure. However, many now feel that if dehydration is avoided, a contrast study such as excretory urography or computed tomography (CT) is not contraindicated in the absence of renal failure.

Because of immunodeficiencies associated with abnormal globulins, pneumonia is a common complication of multiple myeloma and is the most frequent cause of death.[10,11]

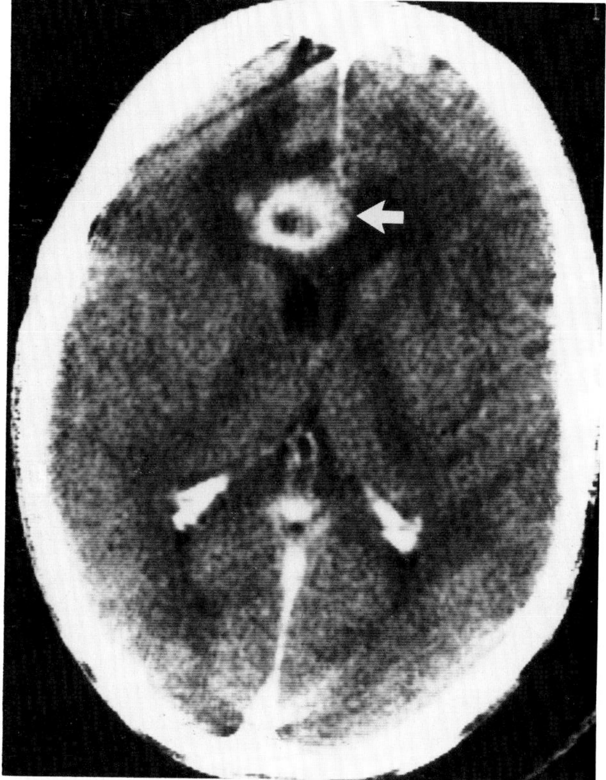
A

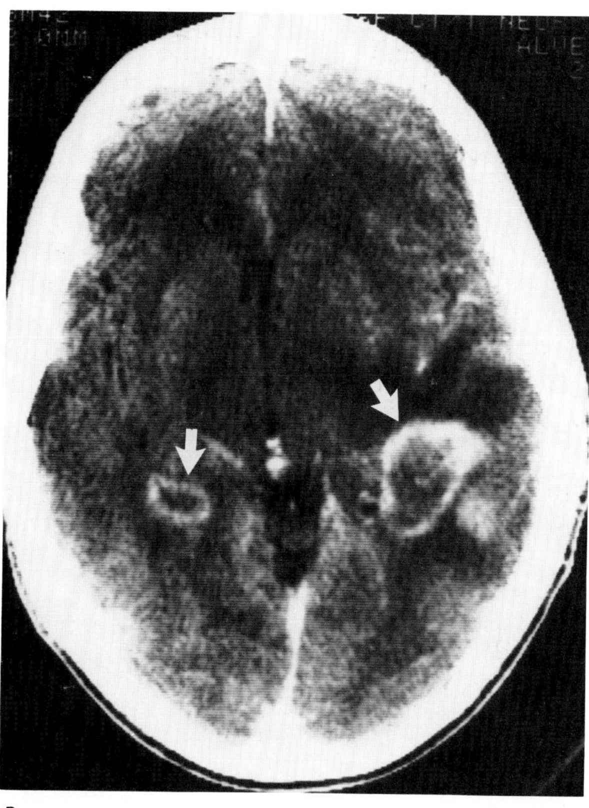
B

**Figure 1.7. Central nervous system involvement in acquired immunodeficiency syndrome.** (A) A CT scan shows a candidal abscess as a cystic lesion with a thick zone of enhancement (arrow) near the genu of the corpus callosum. (B) A CT scan in another patient demonstrates multiple brain abscesses (toxoplasmosis), which appear as lucent lesions with rings of enhancement. (From Kelly and Brant-Zawadzki,[6] with permission from the publisher.)

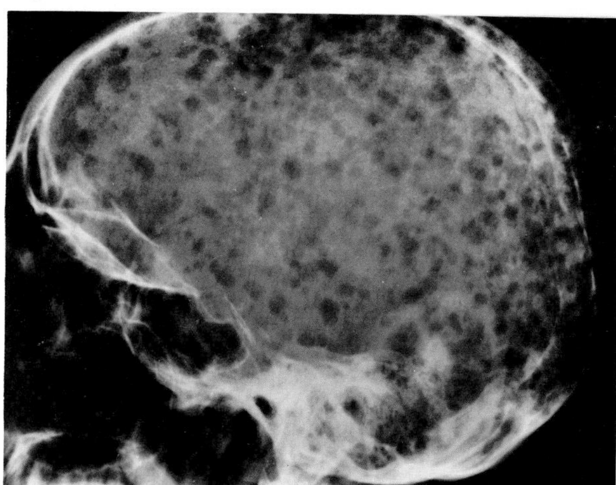

**Figure 1.8. Multiple myeloma.** Diffuse punched-out osteolytic lesions scattered throughout the skull.

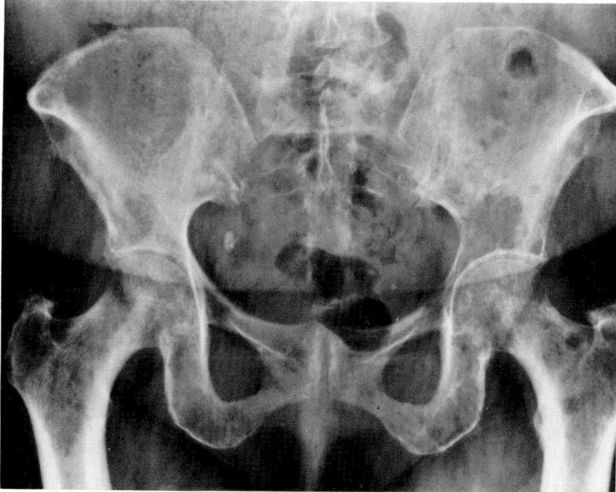

**Figure 1.9. Multiple myeloma.** Diffuse osteolytic lesions involving the pelvis and proximal femurs.

## Localized Myeloma (Solitary Plasmacytoma)

Infrequently, a single plasma cell tumor presents as an apparent solitary destructive bone lesion with no evidence of the major disease complications usually associated with multiple myeloma. A solitary plasmacytoma causes a central area of destruction, usually in the shaft of a long bone. The tumor may be highly destructive, expanding or ballooning the bone before it breaks through the cortex. Residual streaks of bone can produce a tra-

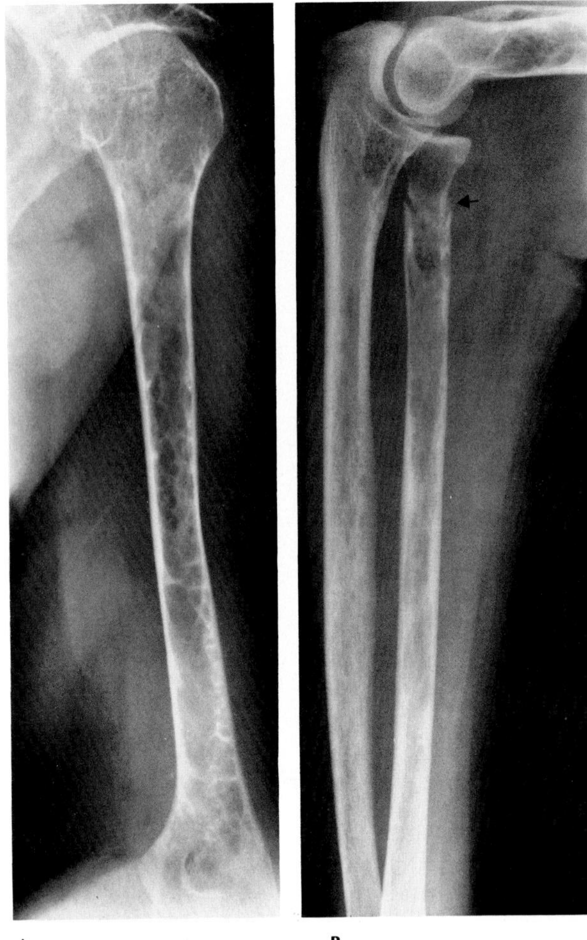


**Figure 1.10. Multiple myeloma.** (A) Generalized deossification of the humerus with thinning of the cortex. (B) A similar process in the proximal radius is associated with a pathologic fracture (arrow).

beculated or "soap bubbly" radiographic appearance, especially in the pelvis. In the spine, an affected vertebral body can collapse or be completely destroyed; a surrounding soft tissue mass is often present. Although plasmacytomas may remain solitary for long periods, widespread destructive lesions usually appear within 1 or 2 years as the condition develops into typical multiple myeloma.

## Extramedullary Myeloma

Abnormal plasma cell proliferation occasionally occurs outside the bone marrow. These soft tissue, or extramedullary, plasmacytomas may be single or multiple and may arise at any site in the body. The vast majority occur in the head and neck, primarily in the nasal cavity, paranasal sinuses, and upper airway. Rarely, an extramedullary myeloma may involve the gastrointestinal tract and present as a nonspecific filling defect or as an infiltrating, constricting lesion. Most patients with an extramedullary myeloma respond favorably to treatment, and, even with local recurrences, patients may have long survival rates.[12–14]

## Waldenström's Macroglobulinemia

Waldenström's macroglobulinemia is a plasma cell dyscrasia involving those cells that synthesize macroglobulins (IgM). Large amounts of a monoclonal IgM are found in the sera of patients with the disorder. It has an insidious onset in late adult life and is characterized clinically by anemia, bleeding, hepatosplenomegaly, and the hyperviscosity syndrome.

Chronic pleural effusion and interstitial pulmonary infiltrates are the major findings on chest radiographs. Hyperviscosity of the blood can lead to cardiomegaly and high-output heart failure. Mediastinal masses may be due to extramedullary hematopoiesis or to infiltration by neoplastic cells. Immunologic disturbances often lead to the development of recurrent pneumonia.

Malignant lymphoplasmacytic cells are deposited in many tissues, particularly the reticuloendothelial structures. Although gastrointestinal symptoms are unusual, malabsorption can occur. The lacteals and the lamina propria of the small bowel villi are filled with a proteinaceous material that is mostly IgM. As the villi become greatly distended and even visible to the naked eye, a sandlike radiographic pattern is produced. Pleural and pericardial effusions not associated with congestive heart failure may occur because of secondary involvement by the abnormal cells.

The major skeletal abnormality is generalized osteoporosis with a widening of marrow spaces and a loss of trabeculae. Solitary or multiple punched-out lesions may occur, though much less frequently than in multiple myeloma.[15]

## Heavy Chain Diseases

Heavy chain diseases are rare disorders of immunoglobulin peptide synthesis that produce lymphoma-like clinical syndromes with lymphadenopathy, hepatosplenomegaly, and recurrent infections.

The most common is alpha chain disease, in which an extensive plasmacytic infiltrate throughout the small bowel mucosa results in diarrhea and a severe malabsorption syndrome. The disorder primarily affects young adults living in the Mediterranean region. Infiltration of the lamina propria by mononuclear cells (predominantly plasma cells) causes distorted villous architecture and a radiographic pattern of coarsely thickened, irregular mucosal folds. A diffuse pattern of small nodules is occasionally seen.

The other heavy chain diseases (gamma and mu) simulate diffuse lymphoma and chronic lymphocytic leukemia, respectively. Widely scattered osteolytic bone lesions in these conditions may be indistinguishable from those of multiple myeloma.

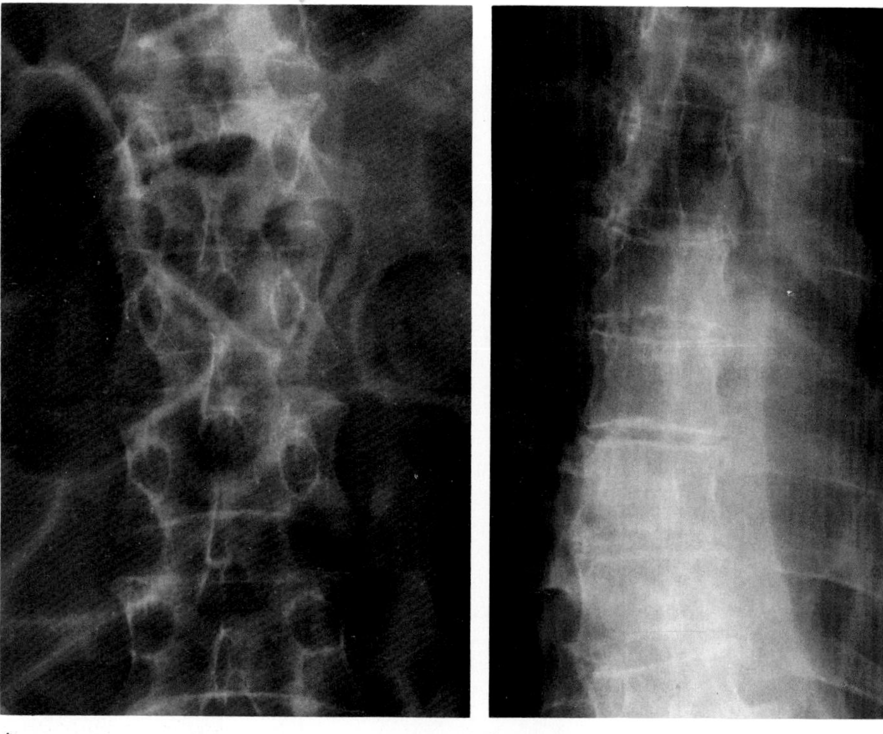

A B

**Figure 1.11. Pedicle sign in multiple myeloma.** (A) Destructive changes and a loss of height of several vertebral bodies have occurred in this patient with diffuse myeloma of the spine. The pedicles, however, retain their normal appearance. (B) In this patient with metastatic oat cell carcinoma, multiple pedicles have been destroyed. The vertebral bodies demonstrate relatively little radiographic involvement. (From Eisenberg,[36] with permission from the publisher.)

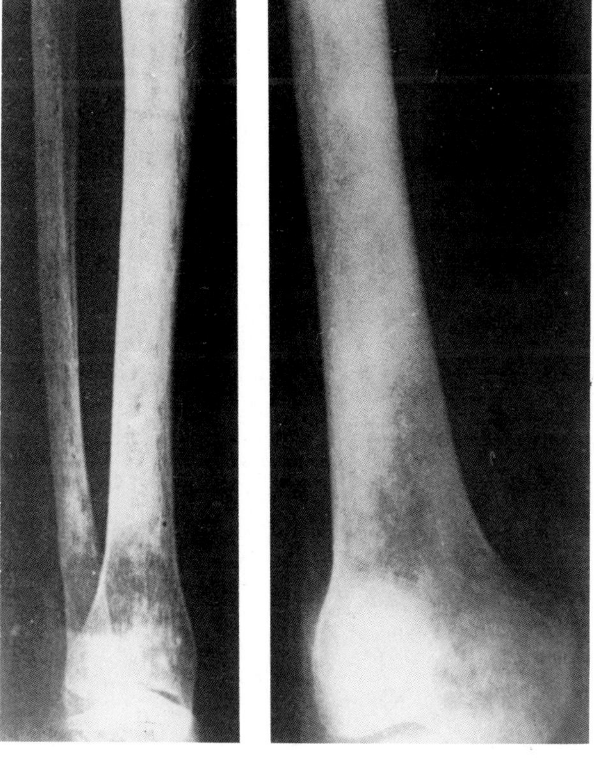

A B

**Figure 1.12. Sclerotic myeloma.** Views of (A) the leg and (B) the femur demonstrate diffuse and nodular sclerosis. Cortical thickening of the tibia encroaches on the medullary canal. Similar changes were evident in the pelvis. (From Meszaros,[37] with permission from the publisher.)

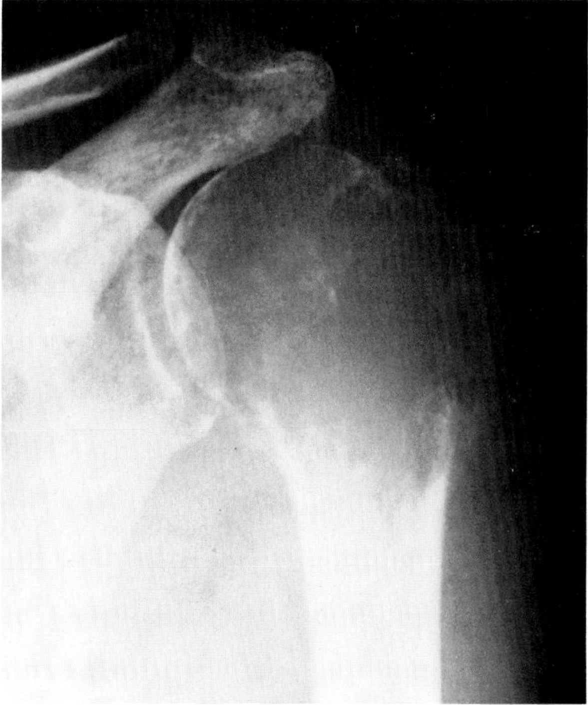

**Figure 1.13. Solitary plasmacytoma of the humeral head.** The highly destructive lesion has expanded the bone and broken through the cortex.

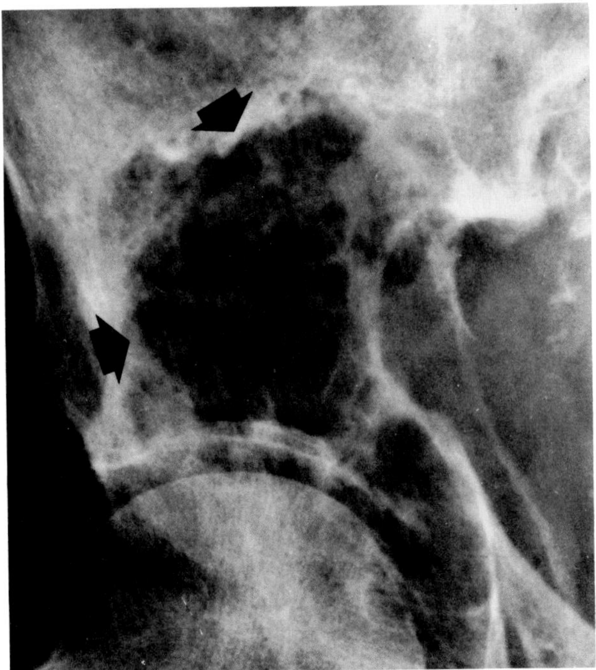

**Figure 1.14. Solitary plasmacytoma of the ilium (arrows).** Some residual streaks of bone remain in this osteolytic lesion, producing a soap bubbly appearance.

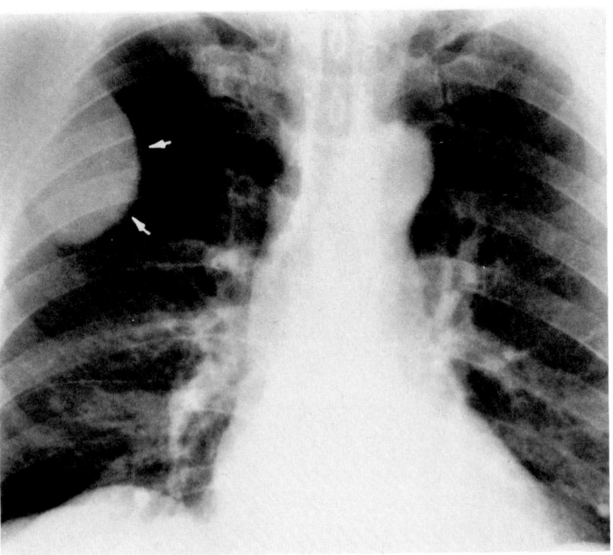

**Figure 1.15. Extramedullary myeloma.** A large extrapleural mass (arrows) containing a proliferation of plasma cells.

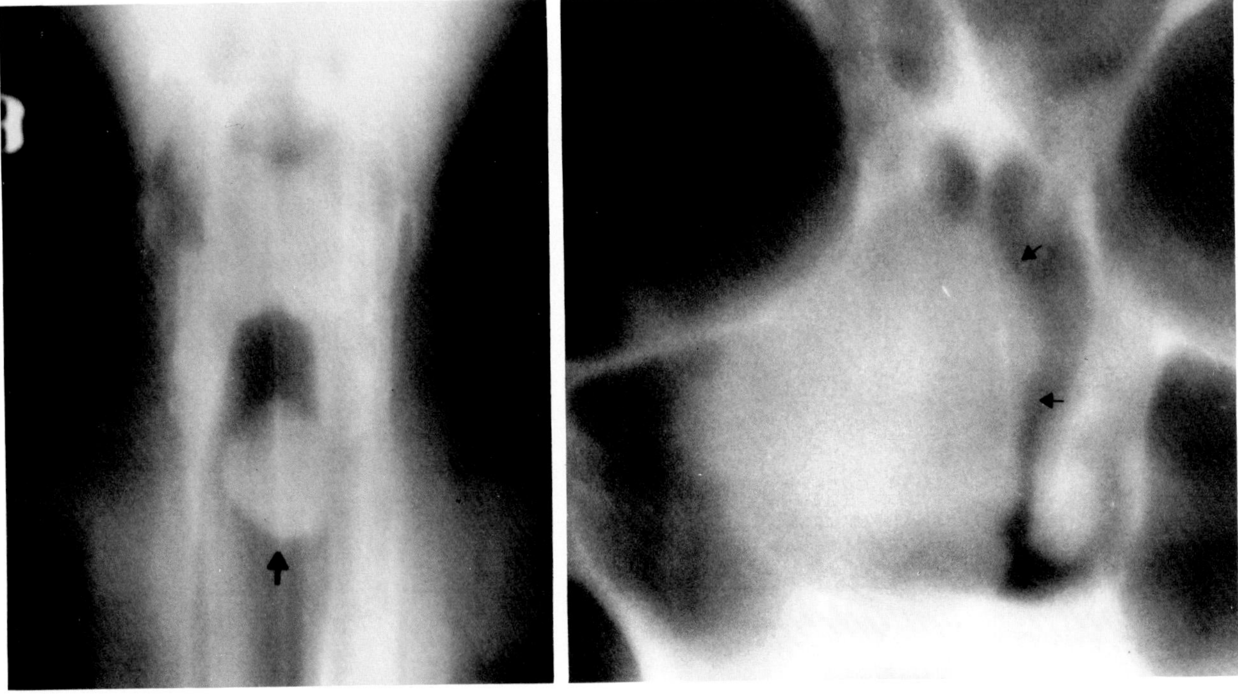

A B

**Figure 1.16. Extramedullary myeloma.** (A) Proliferation of plasma cells forms a mass in the trachea (arrow). (B) A large plasmacytoma of the right nasal cavity (arrows) causes destruction of the medial wall of the right maxillary antrum. (A, from Gromer and Duvall,[14] with permission from the publisher; B, from Schabel, Rogers, Rittenberg, et al,[12] with permission from the publisher.)

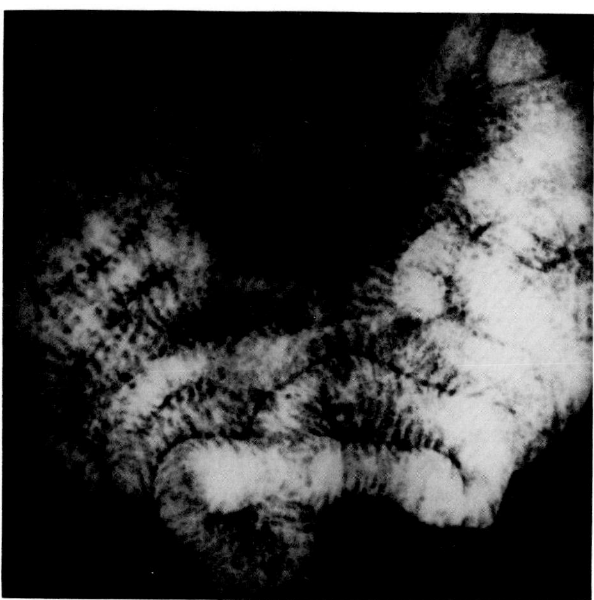

**Figure 1.17. Waldenström's macroglobulinemia.** Greatly distended small bowel villi produce a sandlike pattern.

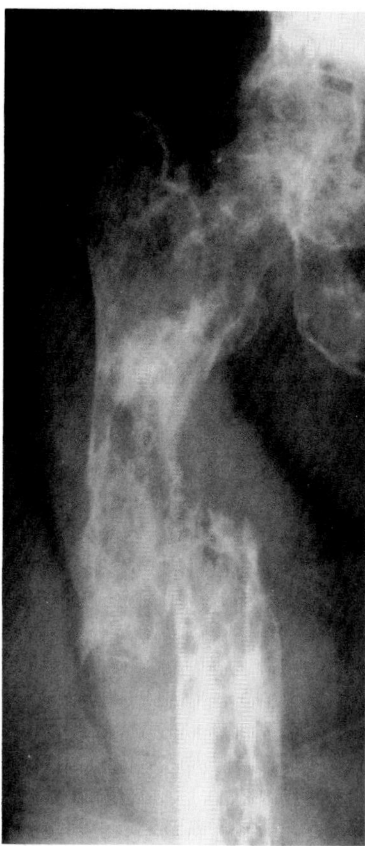

**Figure 1.18. Heavy chain disease.** The diffuse, destructive bone lesions have led to a pathologic fracture of the midshaft of the femur.

## NEUTROPHIL DYSFUNCTION DISEASES

### Chronic Granulomatous Disease

Chronic granulomatous disease is an inherited, usually fatal disorder that predominantly affects males and is characterized by severe recurrent infections of the skin, lymph nodes, liver, and bones. The infections are caused chiefly by staphylococci and certain gram-negative bacteria. The underlying disorder is the inability of morphologically normal neutrophilic leukocytes to kill these bacteria, even after successful phagocytosis. The persistence of low-virulence bacteria within the leukocytes incites a granulomatous response. In the lung there is also extensive fibrous thickening of alveolar walls, and dense infiltration by lymphocytes and plasma cells.

The major radiographic finding in chronic granulomatous disease is recurrent pneumonia that typically resolves incompletely or regresses into abscess formation. Hilar lymph node enlargement is usually striking. Multiple areas of parenchymal consolidation may coalesce into large, relatively discrete granulomatous masses, occasionally with calcification. Extensive fibrosis and dense infiltration by lymphocytes and plasma cells may produce a local or generalized reticulonodular or honeycomb pattern.

A common complication of chronic granulomatous disease is an osteomyelitis that primarily involves the small bones of the hands and feet. Diffuse granulomatous abscesses often cause enlargement of the liver and spleen, both of which may contain speckled calcification. Cervical and abdominal lymph nodes also can calcify.[16,17]

## AMYLOIDOSIS

Amyloidosis is a disorder in which multiple major organs are damaged by deposition of the amorphous fibrous protein amyloid in extracellular tissues. Primary amyloidosis occurs with no evidence of pre-existing or coexisting disease. Amyloidosis is more commonly secondary to such conditions as multiple myeloma, chronic infectious diseases (osteomyelitis, tuberculosis, leprosy), chronic inflammatory diseases (rheumatoid arthritis, ankylosing spondylitis), or aging. Heredofamilial amyloidosis is a group of inherited disorders in which the deposition of amyloid is associated with familial Mediterranean fever and a variety of neuropathic, renal, cardiovascular, and other syndromes.

Radiographic manifestations of amyloidosis depend on which of the organ systems are involved with the often ubiquitous deposition of amyloid. In renal amyloidosis there is initially bilateral, symmetric renal enlargement with normal collecting systems. Late in the disease the kidneys may be reduced in size and a gradual reduction in renal function may lead to renal failure. The nephrotic syndrome and renal vein thrombosis are often

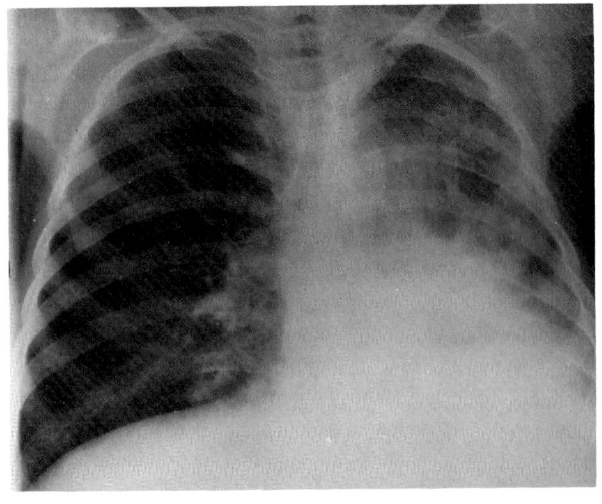
A

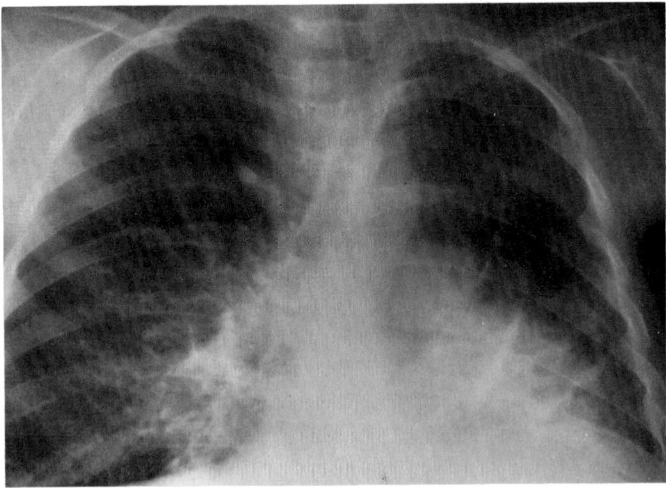
B

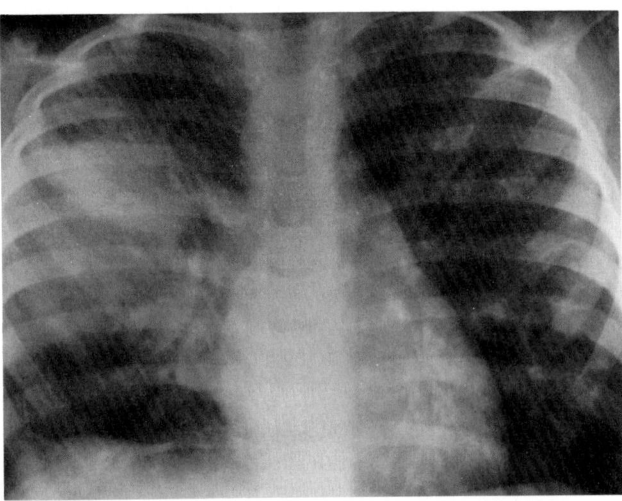
C

**Figure 1.19. Recurrent pneumonia in chronic granulomatous disease.** (A) The initial examination demonstrates a diffuse patchy infiltrate in the left lung with an associated pleural effusion. (B) Several months later there are bilateral lower lobe infiltrates. (C) Six months after the initial film, a large coalescent right midlung mass is associated with a pleural effusion. *Pseudomonas* was cultured from the sputum on all three occasions.

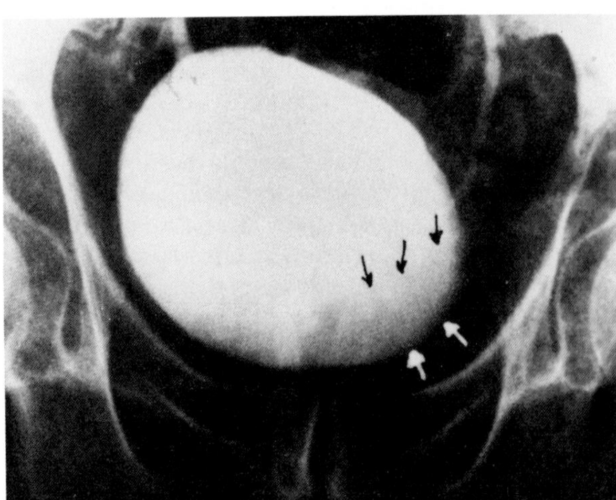

**Figure 1.20. Amyloidosis of the bladder.** There is an irregular lobulated mass (arrows) near the bladder neck. The upper tracts were normal. (From Strong, Kelsey, and Hoch,[18] with permission from the publisher.)

associated abnormalities. Localized accumulation of amyloid may appear as a mass in the bladder or cause a partially obstructing, stricture-like defect in the ureter.

Amyloid often infiltrates the gastrointestinal tract and can produce a broad spectrum of radiographic findings. Massive deposition of amyloid in the muscular layers of the esophagus can cause a failure of relaxation of the lower esophageal sphincter and an achalasia pattern. Distorted peristalsis can produce a dilated esophagus with multiple tertiary contractions. Amyloid infiltration in the stomach can cause thickening and rigidity of the gastric wall, especially in the antrum, leading to luminal narrowing and the radiographic pattern of linitis plastica. Localized deposition of amyloid can cause thickening of rugal folds or discrete, often-ulcerated gastric filling defects. Involvement of the small intestine occurs in a majority of cases of generalized amyloidosis and may lead to a protein-losing enteropathy and malabsorption. Deposition of amyloid in the muscular layer impairs the peristaltic activity. As the bowel wall thickens, the lumen

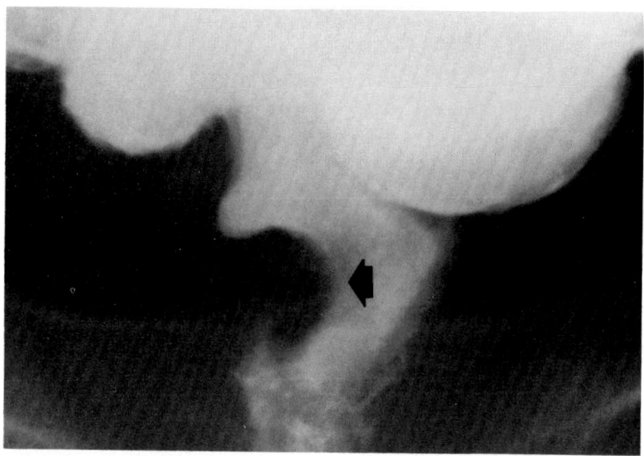

A

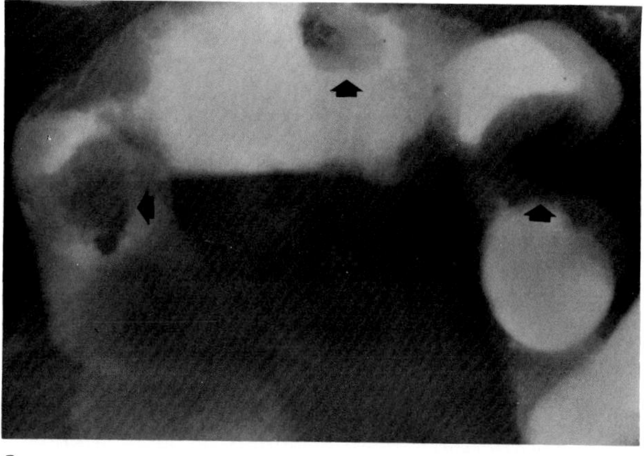

B

**Figure 1.21. Amyloidosis of the colon.** (A) Rectal amyloidosis presents as a localized filling defect (arrow). (B) Multiple discrete filling defects (arrows) represent areas of localized deposition of amyloid in the sigmoid colon. (From Eisenberg,[35] with permission from the publisher.)

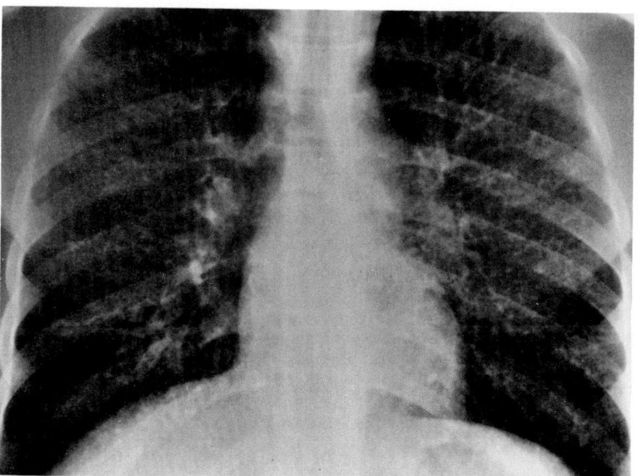

**Figure 1.22. Pulmonary amyloidosis.** A diffuse reticulonodular pattern similar to that of other causes of interstitial fibrosis.

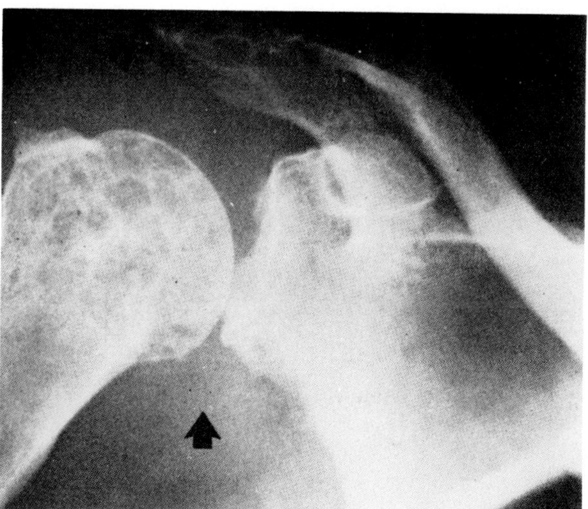

**Figure 1.23. Skeletal amyloidosis.** There is increased soft tissue density surrounding the left shoulder joint (arrows) and lytic changes in the head of the humerus. (From Grossman and Hensley,[22] with permission from the publisher.)

becomes increasingly narrow, and complete obstruction can occur. Occlusion of small blood vessels can cause ischemic enteritis with ulceration, intestinal infarction, and hemorrhage. Radiographically, amyloidosis is characterized by sharply demarcated thickening of folds throughout the small bowel. The folds can be symmetric and uniform or can be grossly irregular with nodularity and tumor-like defects. Concomitant thickening of the mesentery causes separation of small bowel loops. In the colon, amyloidosis may present as single or multiple filling defects or may produce a pattern of ulcerations, effacement of haustral markings, and narrowing and rigidity (especially in the rectum and sigmoid) that simulates the radiographic appearance of chronic ulcerative colitis. Hepatosplenomegaly frequently occurs.

There are three major patterns of amyloid deposition in the chest, differing markedly in their anatomic sites and the effects they produce. In the tracheobronchial form, the deposition of amyloid in the walls of the tracheobronchial tree produces tumor-like masses that may obstruct the bronchial lumen and lead to atelectasis, obstructive pneumonitis, or chronic bronchiectasis. In the nodular parenchymal form, single or, more often, multiple nodules of amyloid develop within the lung parenchyma. These slow-growing masses can become calcified or can cavitate. In the diffuse alveolar septal form, the deposition of amyloid in the interstitial tissues produces a reticulonodular pattern similar to that of interstitial fibrosis from other causes. Enlargement of hilar and mediastinal nodes may be massive and associated with dense calcification. The deposition of amyloid in the heart can cause cardiomegaly, decreased mobility of the ventricular wall, and eventually congestive heart failure.

Amyloid infiltration of the periarticular and articular structures causes prominent soft tissue swelling. The invasion of contiguous bones may cause multiple small erosions. The replacement of bone by large deposits of amyloid may appear radiographically as localized, well-demarcated areas of bone destruction that most frequently involve the proximal femur and the upper part of the humerus. Diffuse infiltration of the marrow causes generalized demineralization with collapse of vertebral bodies, though this appearance is usually a manifestation of underlying myeloma.[18–22]

## URTICARIA

The characteristic mucosal pattern of large, round or polygonal raised plaques in a grossly dilated bowel may reflect an allergic reaction of the colonic mucosa to medication. This "colonic urticaria" predominantly involves the right colon, can be seen without concomitant cutaneous lesions, and regresses once the offending medication is withdrawn. A pattern similar to that of colonic urticaria has been reported in several other conditions (herpes zoster, ischemia, obstructing carcinoma, cecal volvulus); the common denominator seems to be submucosal edema.[23,24]

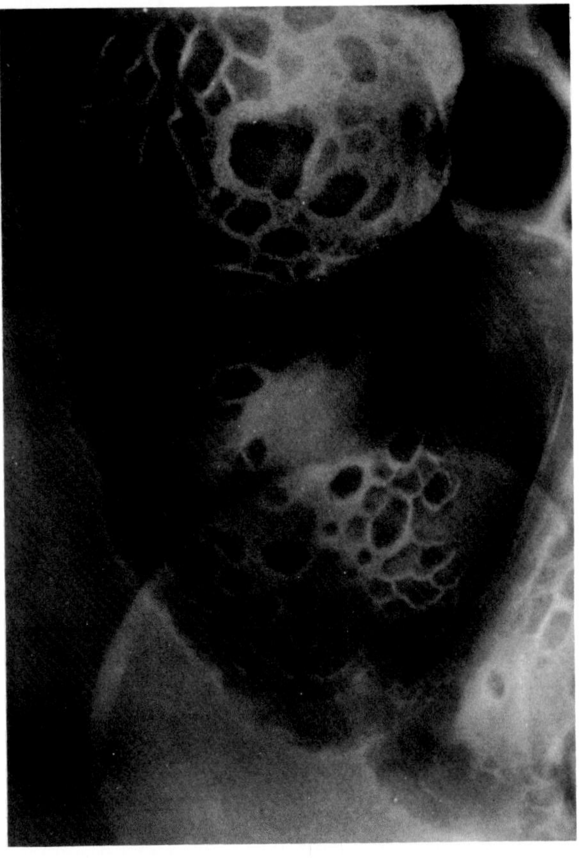

**Figure 1.24. Colonic urticaria.** Large, polygonal raised plaques are visible in a dilated cecum and ascending colon. (From Eisenberg,[35] with permission from the publisher.)

## ANGIONEUROTIC EDEMA

Angioneurotic edema is characterized by edematous, frequently localized, swellings of the skin, mucous membranes, or viscera. In this inherited disorder (a dominant trait with irregular penetrance), visceral manifestations may be the first or the only symptoms, and the patient may initially be thought to have an acute surgical abdomen. Clinical attacks are periodic and occur at irregular intervals. The most serious complication is laryngeal edema, which is associated with a significant mortality rate.

The radiographic findings in angioneurotic edema are present only when the patient is in visceral crisis; they rapidly revert to normal as the attack subsides. Angioneurotic edema most commonly causes regular thickening of small bowel folds, which tends to be more localized than the generalized thickening due to intestinal edema seen in patients with hypoproteinemia. Adjacent loops of small bowel are often separated because of mural and mesenteric thickening. Thumbprinting in the small bowel or the colon may simulate intestinal ischemia or intramural bleeding. The complete reversibility, especially when combined with a characteristic family history and evidence of a previous angioneurotic reaction, should allow differentiation of this condition from other causes of thumbprinting or regular thickening of small bowel folds.[25]

## ALLERGIC RHINITIS (HAY FEVER)

Allergic rhinitis is characterized by sneezing, rhinorrhea, obstruction of the nasal passages, conjunctival and pharyngeal itching, and lacrimation. During acute attacks there may be edematous thickening of the mucous membranes in the paranasal sinuses. The paranasal sinuses may be clouded or completely opaque, often with air-fluid levels. Mucosal polyps, protrusions containing edematous fluid with variable amounts of eosinophilic infiltration, are often found in the nasal cavity and paranasal sinuses of patients with recurrent allergic rhinitis. Polyps in the maxillary sinus produce single or multiple, smooth soft tissue densities that project from the sinus wall. Although mucosal polyps also occur in the other paranasal sinuses, they are more difficult to demonstrate radiographically in these locations.

## VASCULITIS

### Periarteritis Nodosa (Polyarteritis Nodosa)

Periarteritis nodosa is a connective tissue disorder in which a necrotizing inflammation involves all layers of the walls of arterioles and small arteries. The lesions of

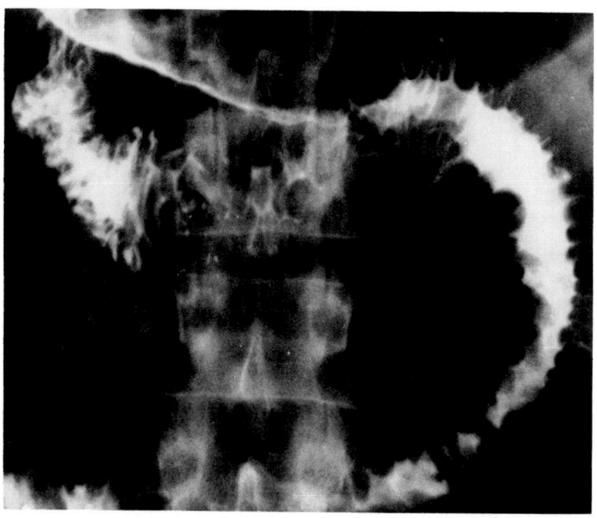

A

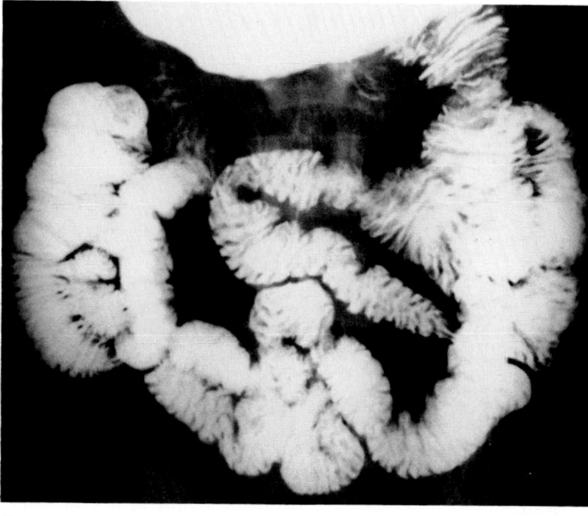

B

**Figure 1.25. Angioneurotic edema.** (A) Regular thickening of folds with thumbprinting in a portion of the proximal jejunum during an acute attack. (B) A repeat small bowel barium examination 36 hours after (A) demonstrates a dramatic return to a normal bowel pattern. (From Pearson, Buchignani, Shimkin, et al,[25] with permission from the publisher.)

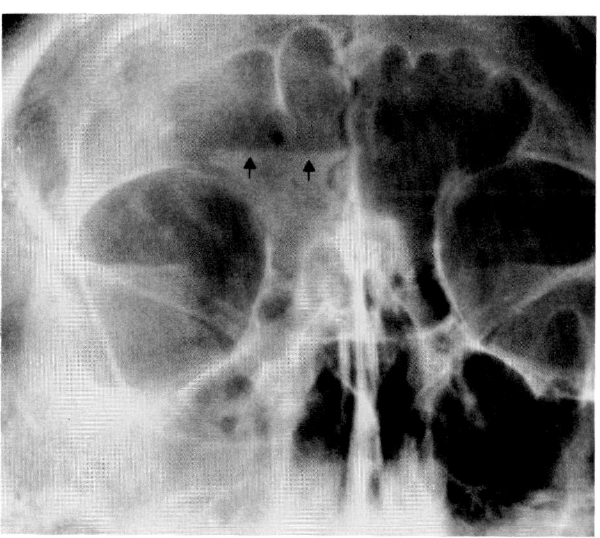

A

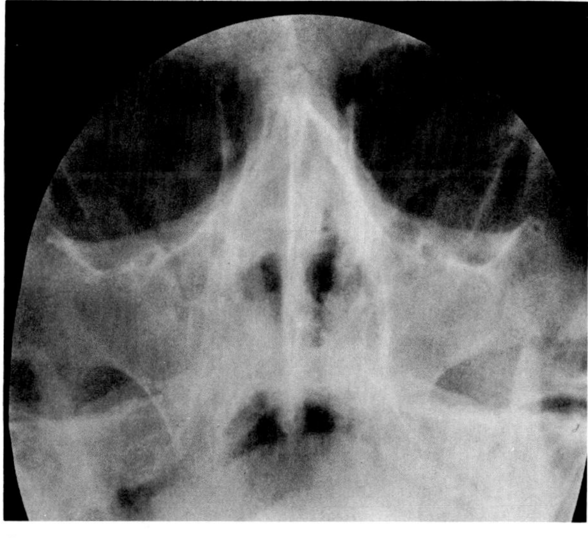

B

**Figure 1.26. Allergic rhinitis.** (A) There is mucosal thickening in the right maxillary and ethmoid sinuses with an air-fluid level (arrows) in the right frontal sinus. (B) There is diffuse, bilateral opacification of the maxillary and ethmoid sinuses.

periarteritis nodosa are widespread throughout the body, most commonly involving the arteries of the heart, mesentery, kidneys, muscles, and nerves. The extent and location of the lesions dictate the type and severity of the clinical symptoms. The disease usually begins in adulthood and has a male predominance.

The radiographic hallmark of periarteritis nodosa is the presence of multiple small aneurysms that most often arise at the bifurcation of arteries. Arteriography is essential to demonstrate the characteristic aneurysms in small arteries of the kidneys, mesentery, liver, and pancreas during the acute phase of the disease. In late stages, narrowing and thrombosis of the arteries in these areas may lead to ischemic necrosis.

In the gastrointestinal tract, mesenteric vasculitis can compromise the blood supply to a segment of the intestine and cause ulcerative, ischemic, or hemorrhagic changes in the bowel wall. In severe cases, massive bleeding, multiple infarctions, and perforation may occur. Acute abdominal symptoms may mimic a surgical abdomen and result in an unavoidable, but unnecessary, laparotomy. Ischemia and inflammation can also produce irregular narrowing of the gastric antrum.

More than half the patients with periarteritis nodosa

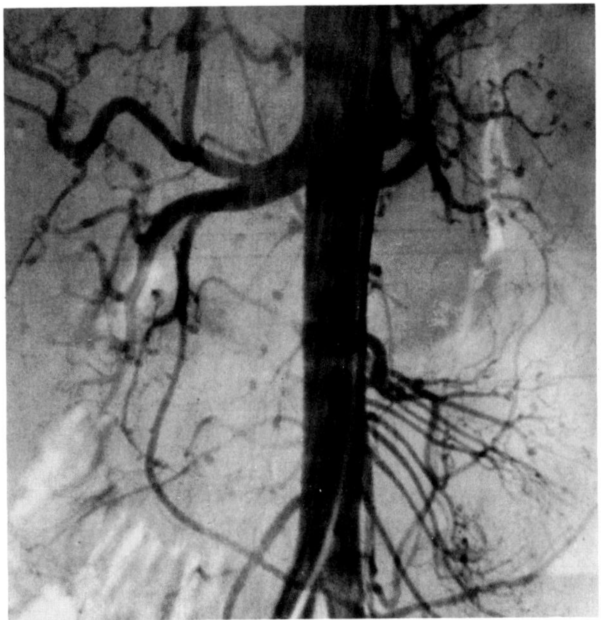

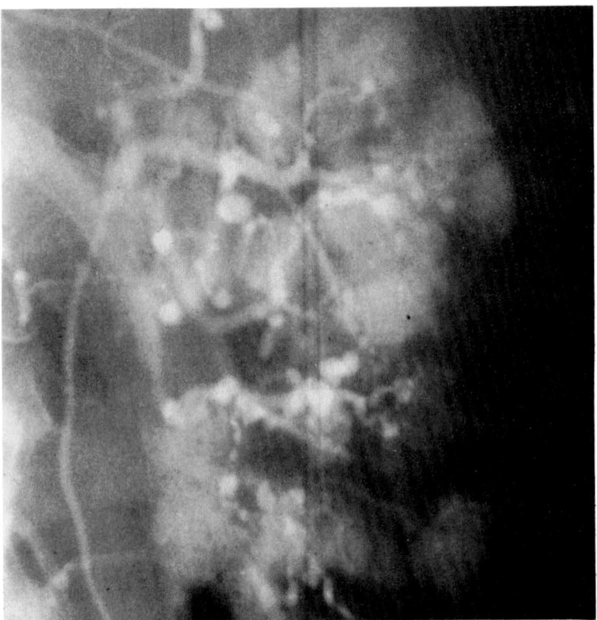

A B

**Figure 1.27. Periarteritis nodosa.** (A) An aortogram demonstrates innumerable small aneurysms arising from vessels throughout the abdomen. Note that the aneurysm formation spares the aorta and its major branches. (B) A renal arteriogram demonstrates multiple small aneurysms. Note the extremely irregular margin of the kidney reflecting multiple infarctions from rupture of some of these intrarenal aneurysms.

have renal disease. Hypertension and hematuria are common. Arteriography typically demonstrates multiple aneurysms of the segmental, interlobar, arcuate, and interlobular arteries. Rupture of these intrarenal aneurysms may cause renal infarcts or perinephric or parenchymal hematomas.

Myocardial ischemia or infarction due to involvement of coronary arteries can lead to cardiac enlargement and congestive heart failure. Pericarditis and pleuritis, with or without effusions, are common.

It is difficult to estimate the frequency with which polyarteritis nodosa involves the lungs, since necrotizing inflammation and granuloma formation that are accompanied by lung involvement and eosinophilia should be classified as allergic granulomatosis (see the following section).[26–28]

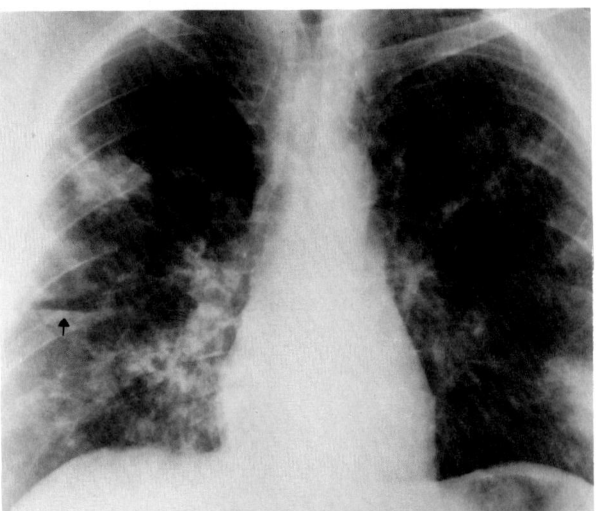

**Figure 1.28. Allergic granulomatosis.** A frontal chest radiograph demonstrates multiple pulmonary infiltrates in a nonsegmental distribution. A cavity is manifest by an air-fluid level in the right chest (arrow). The chest radiograph returned to normal following high-dose steroid therapy. (From Degesys, Mintzer, and Vrla,[38] with permission from the publisher.)

## Allergic Granulomatosis

Allergic granulomatosis is a necrotizing vasculitis that is separated from periarteritis nodosa because of prominent eosinophilia, the presence of perivascular granulomas, lung involvement, and the clinical association with bronchial asthma. Except for the pulmonary manifestations, the organ involvement is similar to that of periarteritis nodosa.

A characteristic appearance on chest radiographs is fleeting, patchy consolidations with a nonsegmental distribution. There is often a changing radiographic picture on serial films, with progression and regression of the infiltrates reflecting the appearance of new lesions and the healing of old ones. Cavitation of confluent densities may simulate the appearance of Wegener's granulomatosis. Clearly defined or hazy parenchymal nodules may simulate metastases or miliary tuberculosis. Large perihilar densities (batwing appearance) may simulate pulmonary edema.[29,37]

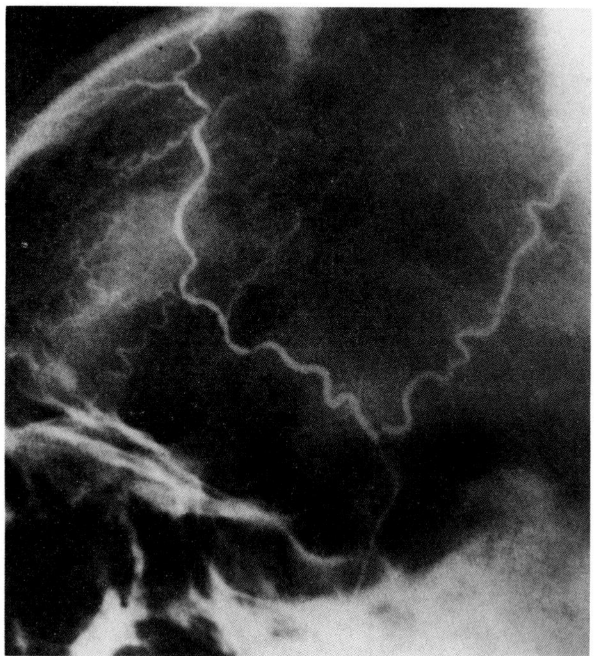

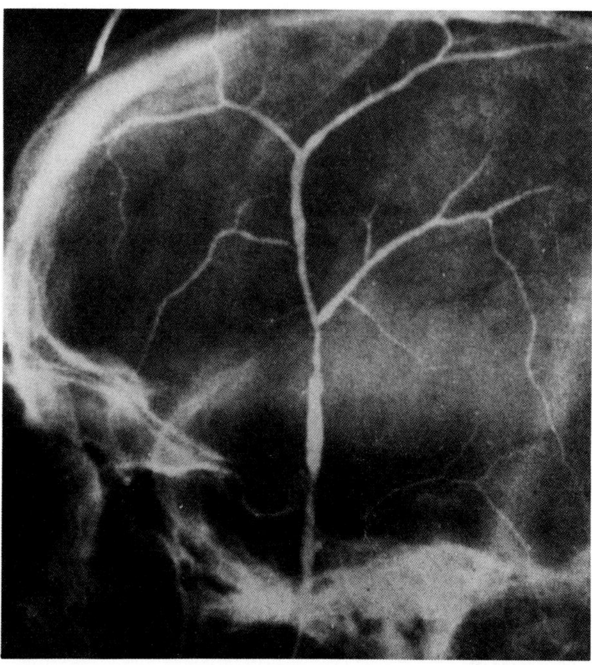

A B

**Figure 1.29. Giant cell arteritis involving the superficial temporal artery.** (A) An external carotid arteriogram demonstrates numerous contracted and slightly dilated segments in branches of the superficial temporal artery. (B) In another patient, multiple dilated and contracted segments are seen in both the main stem and the branches of the superficial temporal artery. (From Gillanders,[30] with permission from the publisher.)

## Wegener's Granulomatosis

This condition is discussed in the section "Granulomatous Diseases of Unknown Etiology" in Chapter 66.

## Giant Cell Arteritis (Temporal Arteritis)

Giant cell arteritis is a subacute inflammatory disease of large- and medium-sized arteries in elderly persons. The condition was initially termed *temporal arteritis* because the temporal arteries were most strikingly involved in the earliest recorded cases. However, temporal arteritis is merely a type of giant cell arteritis, and an identical pathologic process can involve the intracranial and extracranial carotid and vertebral arteries, the aorta and its branches, and any systemic artery. Aortic aneurysms and dissections, mesenteric ischemia, myocardial infarction, and claudication of the lower extremities can all be due to giant cell arteritis.

When the disease involves the superficial temporal arteries, external carotid arteriography demonstrates focal areas of irregularity, narrowing, and dilatation.

Giant cell arteritis responds dramatically to treatment with corticosteroids.[30]

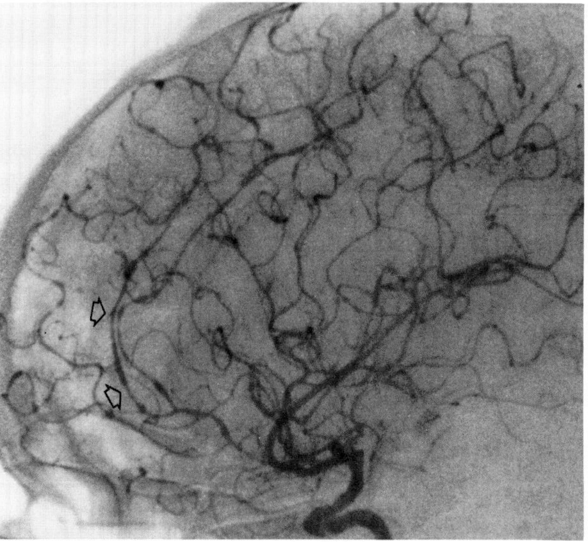

**Figure 1.30. Giant cell arteritis diffusely involving the intracranial vessels.** Multiple areas of vascular beading and narrowing (arrows), most marked in the anterior cerebral artery and its branches.

## SYSTEMIC LUPUS ERYTHEMATOSUS

Systemic lupus erythematosus is a connective tissue disorder that primarily involves young or middle-aged women and most likely represents an immune-complex disease. The presentation and the course of the disease are highly variable. Characteristic laboratory findings include antibodies to nuclear antigens, elevation of immunoglobulin levels, and elevation of the erythrocyte sedimentation rate. Cutaneous manifestations are common and include the characteristic butterfly rash, discoid lupus, alopecia, and photosensitivity.

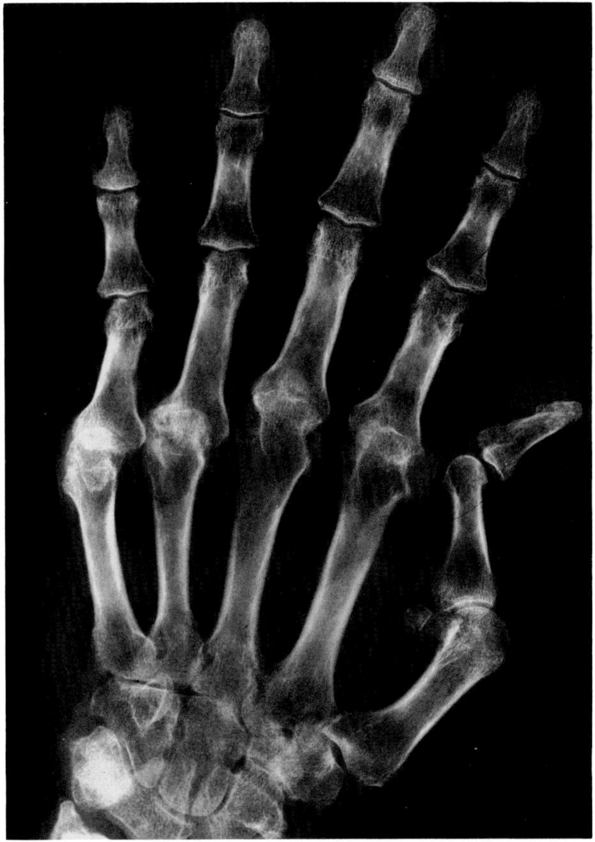

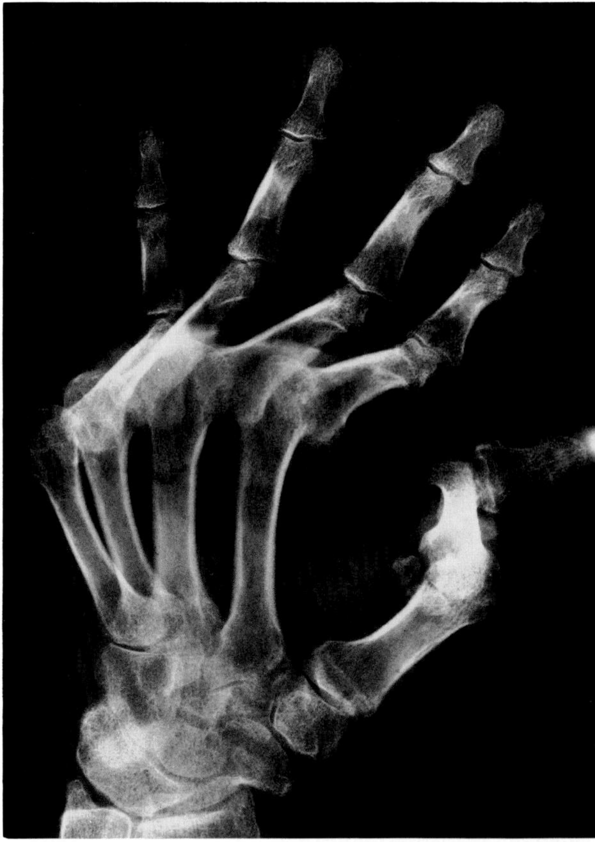

**Figure 1.31. Systemic lupus erythematosus.** Frontal and oblique views of the hand show subluxation of the phalanges at the metacarpal articulations, and hyperextension deformities of the proximal interphalangeal joints. Note the absence of erosive changes. (From Brown and Forrester,[39] with permission from the publisher.)

Polyarthritis and arthralgia are the most frequent clinical complaints in patients with systemic lupus erythematosus. A characteristic finding is subluxations and malalignment of joints in the absence of erosions. Typical deformities include ulnar deviation at the metacarpophalangeal joints and hyperextension and hyperflexion deformities (boutonniere, swan neck) at the interphalangeal joints. Ischemic necrosis in the femoral or humeral heads may be due to either the primary disease process or steroid therapy.

Cardiopulmonary abnormalities frequently develop in patients with systemic lupus erythematosus. Pleural effusions, usually bilateral and small, but occasionally massive, occur in about one-half of the patients. Although the effusions may be asymptomatic, most represent a pleuritis accompanied by elevation of the diaphragm and basilar atelectasis. Enlargement of the cardiac silhouette is generally the result of pericarditis and pericardial effusion. Primary myocardial involvement can also cause cardiomegaly and pulmonary edema. The pulmonary changes are nonspecific and commonly consist of poorly defined, patchy areas of increased density, usually in the lung bases and situated peripherally. These most likely represent pneumonia due to secondary infection.

Although kidney involvement, often leading to renal failure, is one of the most serious manifestations of systemic lupus erythematosus, no specific urographic findings are seen. Enlargement of the liver, spleen, and lymph nodes occurs in about one-fourth of the patients. Although gastrointestinal symptoms are frequent, the radiographic findings are minimal. Decreased motility can cause dilatation of the esophagus and small bowel. A necrotizing vasculitis can result in massive bleeding, multiple infarctions, and perforation.

Many of the radiographic manifestations of systemic lupus erythematosus tend to disappear during spontaneous remissions or following steroid therapy.

A syndrome similar to systemic lupus erythematosus may be induced by certain drugs, especially procainamide, hydralazine, and isoniazid.[31–34,39]

## REFERENCES

1. Kirkpatrick JA, Capitanio MA, Pereira RM: Immunologic abnormalities: Roentgen observations. *Radiol Clin North Am* 10:245–259, 1972.
2. Margulis AR, Feinberg SB, Lester RG, et al: Roentgen manifestations of congenital agammaglobulinemia. *Radiology* 69:354–359, 1957.

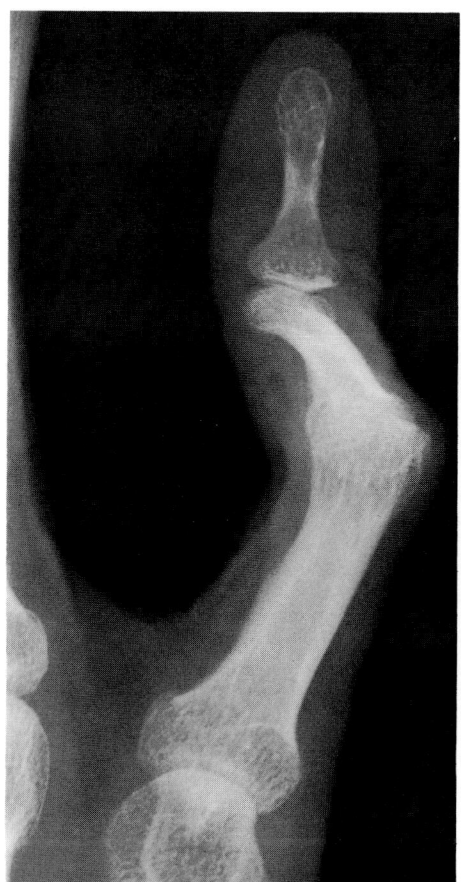

A

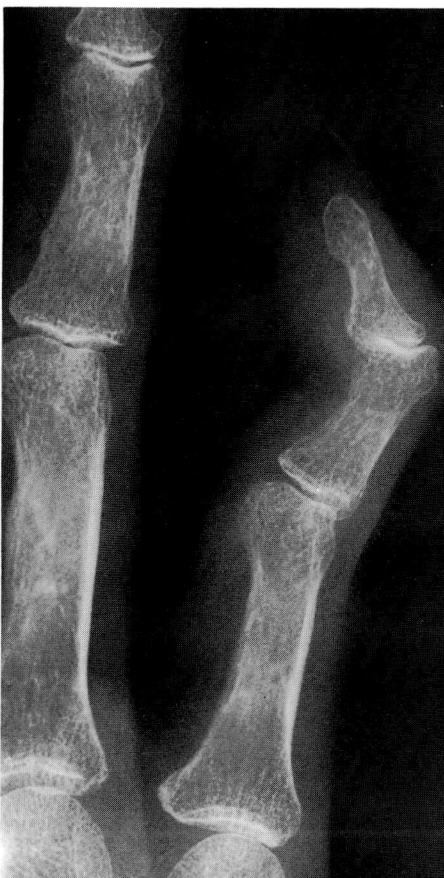

B

**Figure 1.32. Systemic lupus erythematosus.** (A) Flexion of the proximal interphalangeal joint and hyperextension of the distal interphalangeal joint result in a boutonniere deformity. (B) Hyperextension of the proximal interphalangeal joints and flexion of the distal interphalangeal joints produce a swan neck deformity. (From Forrester, Brown, and Nesson,[40] with permission from the publisher.)

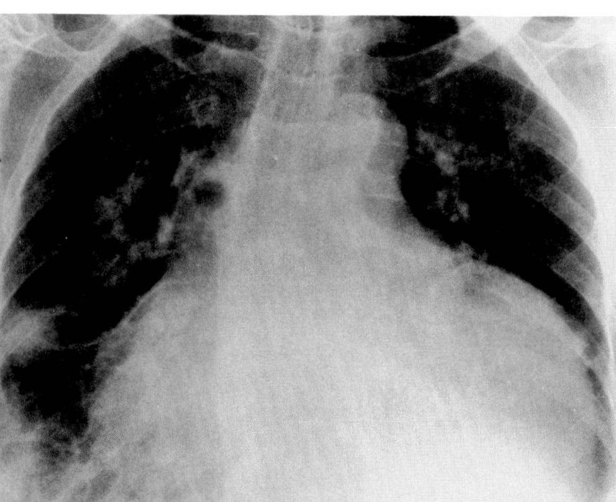

**Figure 1.33. Systemic lupus erythematosus.** A frontal chest film demonstrates massive cardiomegaly due to a combination of pericarditis and pericardial effusion. There also are bilateral pleural effusions, more marked on the right, with some streaks of basilar atelectasis.

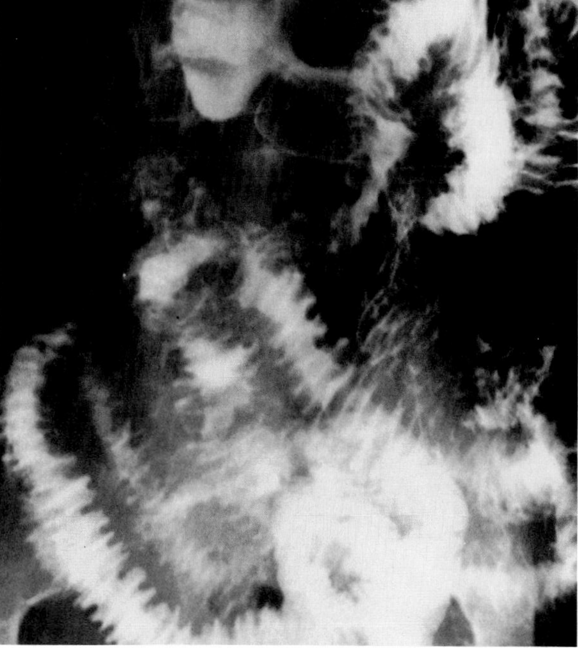

**Figure 1.34. Systemic lupus erythematosus.** A necrotizing vasculitis with hemorrhage into the bowel wall produces regular thickening of small bowel folds. (From Eisenberg,[35] with permission from the publisher.)

3. Ament ME, Rubin CE: Relation of giardiasis to abnormal intestinal structure and function in gastrointestinal immunodeficiency syndromes. *Gastroenterology* 62:216–226, 1972.
4. Ajdukiewicz AB, Youngs GR, Bouchier IAD; Nodular lymphoid hyperplasia with hypogammaglobulinemia. *Gut* 13:589–595, 1972.
5. Whelan MA, Kricheff II, Handler M, et al: Acquired immunodeficiency syndrome: Cerebral computed tomographic manifestations. *Radiology* 149:477–484, 1983.
6. Kelly WM, Brant-Zawadzki MB: Acquired immunodeficiency sydrome: Neuroradiologic findings. *Radiology* 149:485–491, 1983.
7. Hill CA, Harle TS, Mansell PWA: The prodome, Kaposi's sarcoma, and infections associated with acquired immunodeficiency syndrome. *Radiology* 149:393–399, 1983.
8. McCauley DI, Naidich DP, Leitman BS, et al: Radiographic patterns of opportunistic lung infections and Kaposi's sarcoma in homosexual men. *AJR* 139:653–658, 1981.
9. Sider L, Mintzer RA, Mendelson EB, et al: Radiographic findings of infectious proctitis in homosexual men. *AJR* 139: 667–671, 1982.
10. Baltzer G, Jacob H, Esselborn H: Contrast media and renal function in multiple myeloma. *Fortschr Roentgenstr* 129: 208–211, 1978.
11. Jacobson HG, Poppel MM, Shapiro JH, et al: The vertebral pedicle sign. *AJR* 80:817–821, 1958.
12. Schabel SI, Rogers CI, Rittenberg GM, et al: Extramedullary plasmacytoma. *Radiology* 128:625–628, 1978.
13. Wiltshaw E: The natural history of extramedullary plasmacytoma and its relation to solitary myeloma of bone and myelomatosis. *Medicine* 55:217–238, 1976.
14. Gromer RC, Duvall AJ: Plasmacytoma of the head and neck. *J Laryngol Otol* 87:861–872, 1973.
15. Khilnani MT, Keller RJ, Cuttner J: Macroglobulinemia and steatorrhea: Roentgen and pathologic findings in the intestinal tract. *Radiol Clin North Am* 7:43–55, 1969.
16. Gold RH, Douglas SD, Preger L, et al: Roentgenographic features of the neutrophil dysfunction syndromes. *Radiology* 92:1045–1054, 1969.
17. Caldicott WJH, Baehner RL: Chronic granulomatous disease of childhood. *AJR* 103:133–139, 1968.
18. Strong GH, Kelsey D, Hoch W: Primary amyloid disease of the bladder. *J Urol* 122:463–466, 1974.
19. Pear B: Radiographic manifestations of amyloidosis. *AJR* 111:821–832, 1971.
20. Brown J: Pulmonary amyloidosis. *Clin Radiol* 15:358–367, 1964.
21. Legge DA, Carlson HC, Wollaeger EE: Roentgenologic appearance of systemic amyloidosis involving the gastrointestinal tract. *AJR* 110:406–412, 1970.
22. Grossman RE, Hensley GT: Bone lesions in primary amyloidosis. *AJR* 101:872–875, 1967.
23. Seaman WB, Clements JL: Urticaria of the colon: A nonspecific pattern of submucosal edema. *AJR* 138:545–547, 1982
24. Berk RN, Millman SJ: Urticaria of the colon. *Radiology* 99:539–540, 1971.
25. Pearson KD, Buchignani JS, Shimkin PM, et al: Hereditary angioneurotic edema of the gastrointestinal tract. *AJR* 116:256–261, 1972.
26. Nicks AJ, Hughes F: Polyarteritis nodosa "mimicking" eosinophilic gastroenteritis. *Radiology* 116:53–54, 1975.
27. Han SY, Jander HP, Laws HL: Polyarteritis nodosa causing severe gastrointestinal bleeding. *Gastrointest Radiol* 1:285–287, 1976.
28. Capps JH, Klein RM: Polyarteritis nodosa as a cause of perirenal and retroperitoneal hemorrhage. *Radiology* 94:143–146, 1970.
29. Sokolov RA, Rachmaninoff N, Kaine HD: Allergic granulomatosis. *Am J Med* 32:131–141, 1962.
30. Gillanders LA: Temporal arteriography. *Clin Radiol* 20:47–50, 1969.
31. Lubowitz R, Schumacher HR: Articular manifestations of systemic lupus erythematosus. *Ann Intern Med* 74:911–921, 1974.
32. Noonan CD, Odone DT, Engleman EP, et al: Roentgenographic manifestations of joint disease in systemic lupus erythematosus. *Radiology* 80:837–843, 1963.
33. Greenberg JH, Lutcher CL: Drug induced systemic lupus erythematosus. *JAMA* 222:191–193, 1972.
34. Levin DC: Proper interpretation of pulmonary roentgen changes and systemic lupus erythematosus. *AJR* 111:510–517, 1971.
35. Eisenberg RL: *Gastrointestinal Radiology: A Pattern Approach*. Philadelphia, JB Lippincott, 1983.
36. Eisenberg RL: *Atlas of Signs in Radiology*. Philadelphia, JB Lippincott, 1984.
37. Meszaros WT: The many facets of multiple myeloma. *Semin Roentgenol* 9:219–228, 1974.
38. Degesys GE, Mintzer RA, Vrla RF: Allergic granulomatosis. *AJR* 135:1281–1287, 1980.
39. Brown JC, Forrester DM: Arthritis. In Eisenberg RL, Amberg JR (eds): *Critical Diagnostic Pathways in Radiology: An Algorithmic Approach*. Philadelphia, JB Lippincott, 1981, pp 280–298.
40. Forrester DM, Brown JC, Nesson JW: *The Radiology of Joint Disease*. Philadelphia, WB Saunders, 1978.

**Chapter Two**

# Nutritional Disorders

## KWASHIORKOR

Kwashiorkor is a disease of severe protein malnutrition that primarily affects the autonomic nervous system and the gastrointestinal tract. Disordered motility produces dilated, hypotonic loops of small bowel, especially in the jejunum. Hypoalbuminemia leads to diffuse edema and ascites and a markedly protuberant abdomen. Fatty replacement of liver tissue may cause the right upper quadrant to appear excessively lucent on plain abdominal radiographs. Damage to pancreatic acinar cells can lead to pancreatic lithiasis and the radiographic small bowel pattern seen with malabsorption disorders. Retarded bone growth with thinned cortices is often associated. Computed tomography (CT) demonstrates varying degrees of cerebral atrophy with ventricular dilatation and widening of the Sylvian fissure.[1–4]

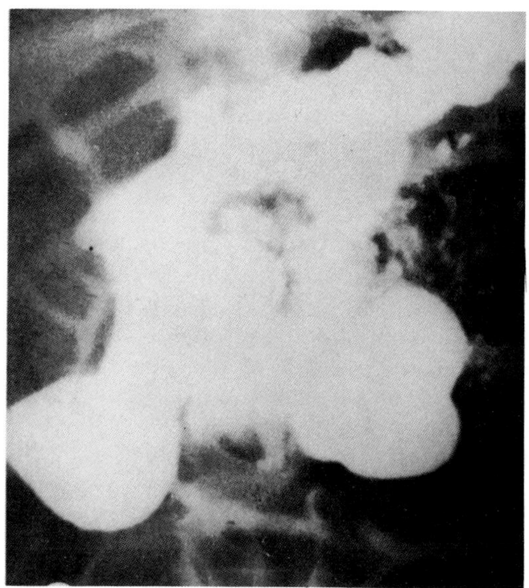

**Figure 2.1. Kwashiorkor.** A frontal view of the abdomen 2 hours after the ingestion of barium shows dilated, hypotonic loops of duodenum and jejunum; the appearance of the loops remained unchanged several hours later. (From Kowalski,[1] with permission from the publisher.)

## OBESITY

Obesity refers to an excess of adipose tissue that develops when the caloric intake consistently exceeds the caloric expenditure. It may be related simply to personal habits

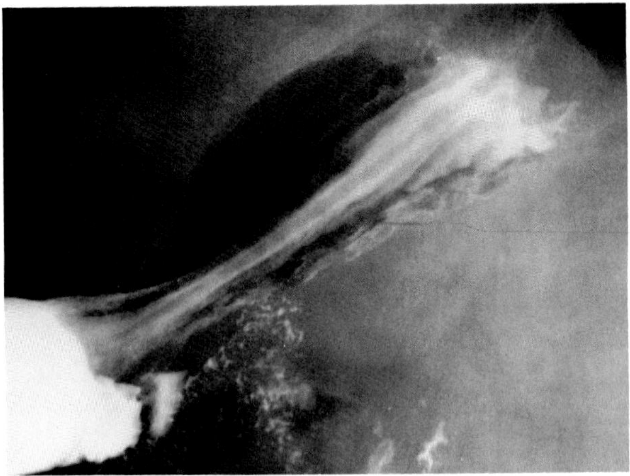

**Figure 2.2. Obesity.** Enlargement of the retrogastric space with no evidence of a discrete mass impressing the posterior margin of the stomach.

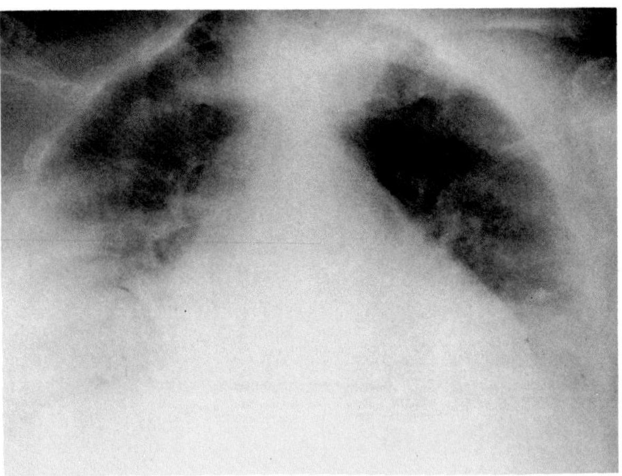

**Figure 2.3. Pickwickian syndrome.** Profound obesity has led to severe hypoventilation and secondary polycythemia; this causes marked cardiomegaly and an engorgement of the pulmonary vessels.

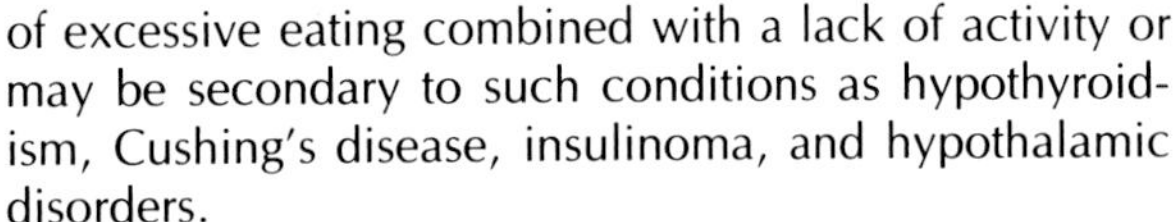

of excessive eating combined with a lack of activity or may be secondary to such conditions as hypothyroidism, Cushing's disease, insulinoma, and hypothalamic disorders.

Excess adipose tissue can cause displacement of normal abdominal structures, producing such radiographic patterns as widening of the retrogastric and retrorectal spaces. An extreme increase in the intra-abdominal volume causes diffuse elevation of the diaphragm with a relatively transverse position of the heart (simulating cardiomegaly), prominence of pulmonary markings, and atelectatic changes at the bases of the lungs. In the most severe form of obesity (Pickwickian syndrome), profound hypoventilation can result in hypoxia, retention of carbon dioxide, secondary polycythemia, and pulmonary hypertension with cor pulmonale. An excessive deposition of fatty tissue can also appear radiographically as widening of the mediastinum and prominence of the pericardial fat pads.

There are three major surgical procedures for morbid obesity. The jejunoileal bypass involves an end-to-end anastomosis of the proximal jejunum to the distal terminal ileum. Gastric operations include gastroplasty, in which a small upper gastric remnant is connected to a larger lower gastric pouch by a narrow channel, and gastric bypass, in which a small upper gastric pouch is directly connected to the small bowel. The presence of surgical clips and sutures combined with either gastric narrowing or an abnormal pathway for the flow of contrast is radiographic evidence of these surgical procedures for morbid obesity.[5,6,11]

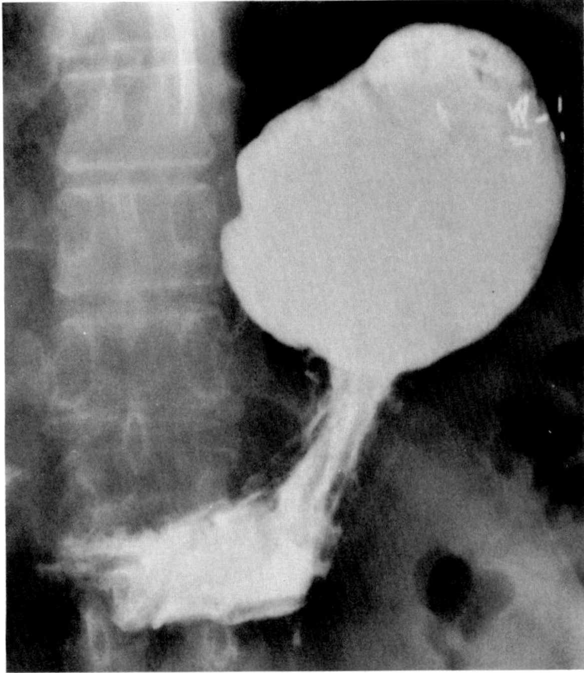

**Figure 2.4. Gastroplasty surgery for weight reduction.** The presence of surgical clips and sutures in combination with distal gastric narrowing should suggest the proper diagnosis. (From Eisenberg,[11] with permission from the publisher.)

## THIAMINE DEFICIENCY (BERIBERI)

Thiamine deficiency primarily involves the cardiovascular and nervous systems. Initially, peripheral vasodilatation causes increased cardiac output that produces a generalized enlargement of the cardiac silhouette and increased pulmonary vascular markings. With progressive disease, biventricular myocardial damage leads to congestive heart failure. Clinical and radiographic improvement may be dramatic following thiamine administration.

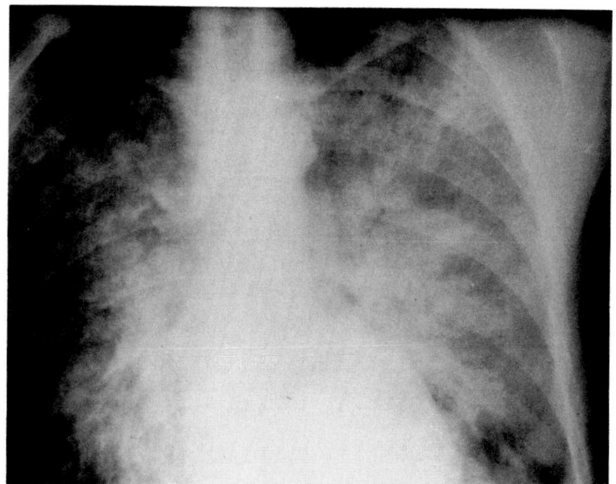

**Figure 2.5. Thiamine deficiency (beriberi).** Diffuse pulmonary edema due to severe congestive heart failure.

Noninflammatory degeneration of myelin sheaths due to thiamine deficiency produces encephalopathy and peripheral neuropathy.[7]

## VITAMIN C DEFICIENCY (SCURVY)

The deficiency of ascorbic acid (vitamin C) in scurvy leads to an inability of the supporting tissues to produce and maintain intercellular substances (collagen, osteoid, dentin) and vascular endothelium. In children, disordered chondroblastic and osteoblastic activity causes bone changes that are most prevalent where growth is normally most rapid (sternal end of the ribs; distal end of the femur, radius, and ulna; proximal end of the humerus; and both ends of the tibia and fibula).

In infantile scurvy, radiographs show generalized osteoporosis ("ground-glass" appearance) with blurring or disappearance of trabecular markings, and severe cortical thinning. Widening and increased density of the zone of provisional calcification produce the characteristic "white line" of scurvy. A relatively lucent osteoporotic zone forms on the diaphyseal side of the white line and often gives the appearance of a double epiphysis. This porotic zone is easily fractured, permitting the dense zone to become impacted on the shaft and jut laterally beyond it, thus giving rise to characteristic marginal spur formation (Pelken's spur). The epiphyseal ossification centers are demineralized and surrounded by a dense, sharply demarcated ring of calcification (Wimberger's sign of scurvy). If epiphyseal dislocations have not occurred, the appearance of the skeletal structures usually returns to normal following appropriate therapy.

Subperiosteal hemorrhage often occurs along the shafts of the long bones. Calcification of the elevated periosteum and hematoma is a radiographic sign of healing.[8,9]

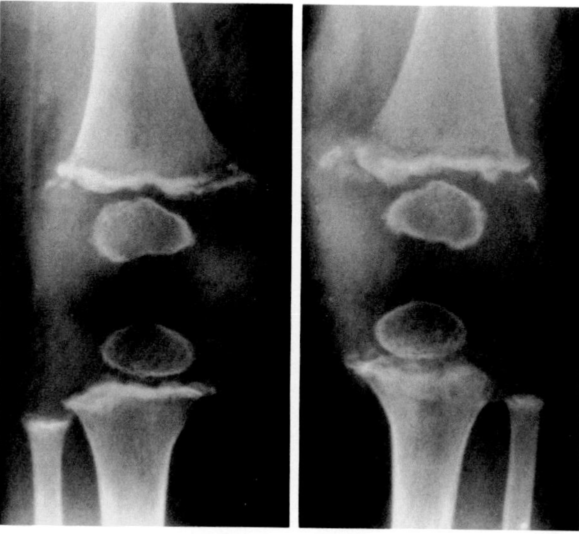

**Figure 2.6. Scurvy.** Frontal views of both knees demonstrate widening and increased density of the zone of provisional calcification, producing the characteristic white line of scurvy. Note also the submetaphyseal zone of lucency and the characteristic marginal spur formation (Pelken's spur). The epiphyseal ossification centers are surrounded by a dense, sharply demarcated ring of calcification (Wimberger's sign).

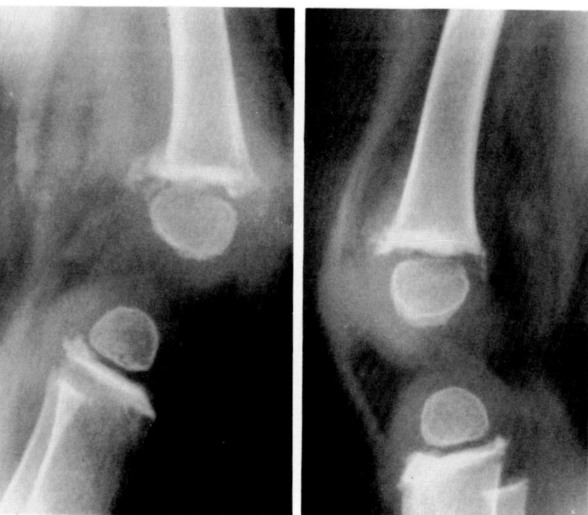

**Figure 2.7. Scurvy.** Lateral views of both knees again demonstrate the characteristic white lines, Pelken's spurs, and Wimberger's sign of scurvy.

## HYPERVITAMINOSIS A

Chronic excessive intake of vitamin A produces a syndrome characterized by bone and joint pain, hair loss, pruritis, anorexia, dryness and fissures of the lips, hepatosplenomegaly, and yellow tinting of the skin.

Radiographic changes are most commonly seen in children between the ages of 1 and 3. There is solid or

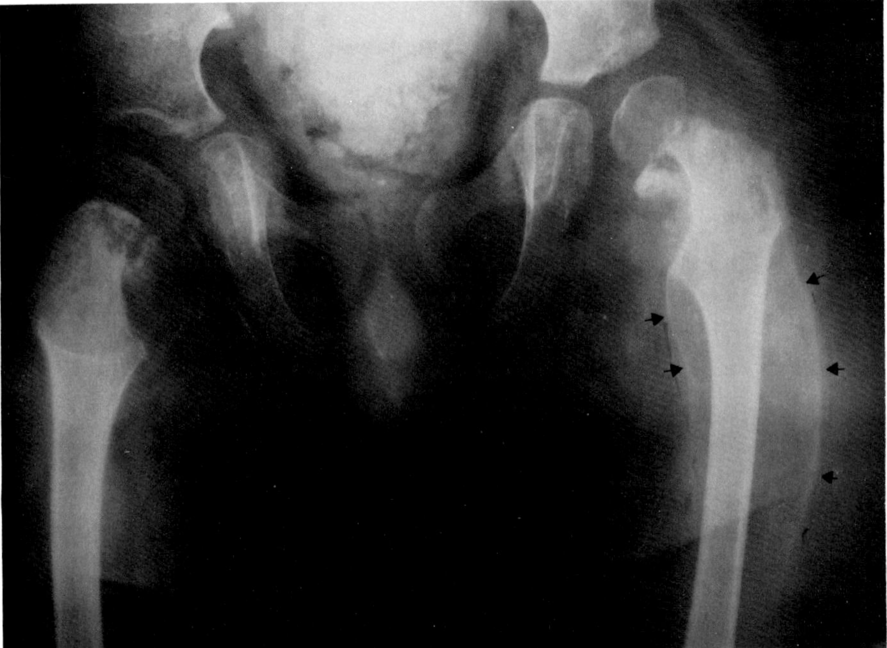

**Figure 2.8. Scurvy.** There is a large calcifying subperiosteal hematoma of the femoral shaft (arrows). (From Stoker and Murray,[9] with permission from the publisher.)

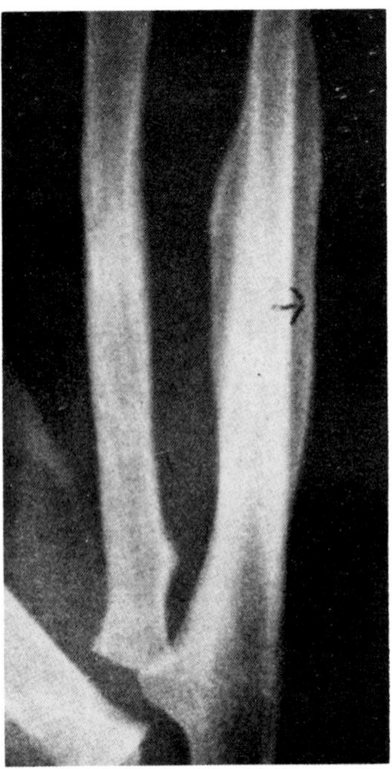

**Figure 2.9. Hypervitaminosis A.** Long, wavy cortical hyperostosis of the ulna (arrow). (From Caffey,[10] with permission from the publisher.)

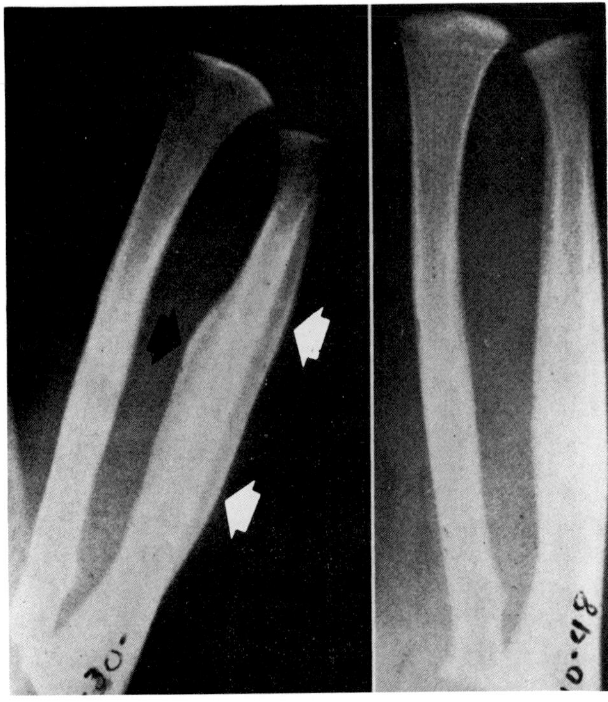

A B

**Figure 2.10. Hypervitaminosis A.** (A) Thin, wavy, shell-like cortical thickening (arrows) during the active phase of poisoning. (B) Four months after stoppage of the vitamin concentrate, the hyperostosis is shrunken, smooth, and more sclerotic. (From Caffey,[10] with permission from the publisher.)

lamellated periosteal new bone formation, which is greatest near the center of the shaft and tapers toward the end.

The periosteal thickening in hypervitaminosis A can be differentiated from infantile cortical hyperostosis (Caffey's disease) because it rarely involves the mandible.[9]

## REFERENCES


1. Kowalski R: Roentgenologic studies of the alimentary tract in kwashiorkor. *AJR* 100:100–112, 1967.
2. Adams P, Berridge FR: Effects of kwashiorkor on cortical and trabecular bone. *Arch Dis Child* 44:705–709, 1969.
3. El-Tatawy S, Badrawi N, El-Bislawy A: Cerebral atrophy in infants with protein energy malnutrition. *Am J Neurorad* 4:434–436, 1983.
4. Mathur KC, Pant HC, Mathur GP, et al: Radiological appearance of the skeleton in malnourished children. *Ind J Radiol* 33:40–44, 1979.
5. Burwell CS, Robin ED, Whaley RD, et al: Extreme obesity associated with alveolar hypoventilation—a Pickwickian syndrome. *Am J Med* 21:811–818, 1956.
6. Hammond DI, Freeman JB: The radiology of gastroplasty for morbid obesity. *J Can Assoc Radiol* 33:21–24, 1982.
7. Jones JH: Beriberi heart disease. *Circulation* 19:275–283, 1959.
8. McCann F: The incidence and value of radiologic signs of scurvy. *Br J Radiol* 35:683–686, 1962.
9. Stoker DJ, Murray RO: Skeletal changes in hemophilia and other bleeding disorders. *Semin Roentgenol* 9:185–193, 174.
10. Caffey J: Chronic poisoning due to excess of vitamin A. *Pediatrics* 5:672–688, 1950.
11. Eisenberg RL: *Gastrointestinal Radiology: A Pattern Approach.* Philadelphia, JB Lippincott, 1983.

## Chapter Three
# Metabolic Disorders

## INHERITED DISORDERS OF AMINO ACID METABOLISM

### Phenylketonuria

Phenylketonuria is an inborn error of metabolism in which an enzyme deficiency results in the impaired conversion of phenylalanine to tyrosine. Untreated patients with this condition usually suffer profound mental retardation, hyperactivity, and seizures, all related to brain atrophy that can be demonstrated as dilatation of the ventricles and sulci on computed tomography (CT). Hypopigmentation (underproduction of melanin) and eczema are common.

The most common skeletal manifestation of phenylketonuria is cupping and beaking of the distal metaphyses of the radius and ulna. Fine opaque spicules extend vertically from the metaphysis into the epiphyseal cartilage. With continued growth, the original spicules are incorporated into the osseous metaphysis to form vertical striations.[1,2,36]

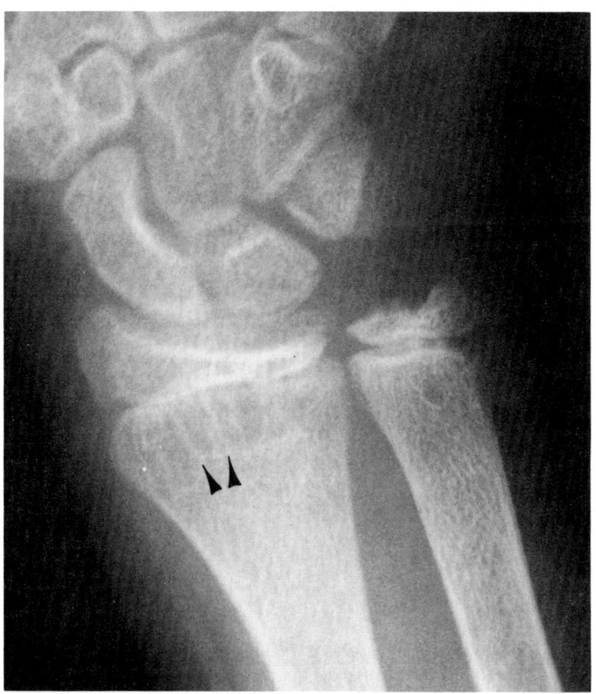

**Figure 3.1. Phenylketonuria.** Vertical metaphyseal spicules (arrowheads) in a mentally retarded 12-year-old girl with untreated phenylketonuria. (From Woodring and Rosenbaum,[36] with permission from the publisher.)

### Homocystinuria

Homocystinuria is an inborn error of methionine metabolism that causes a defect in the structure of collagen or elastin and an appearance similar to that of Marfan's syndrome.

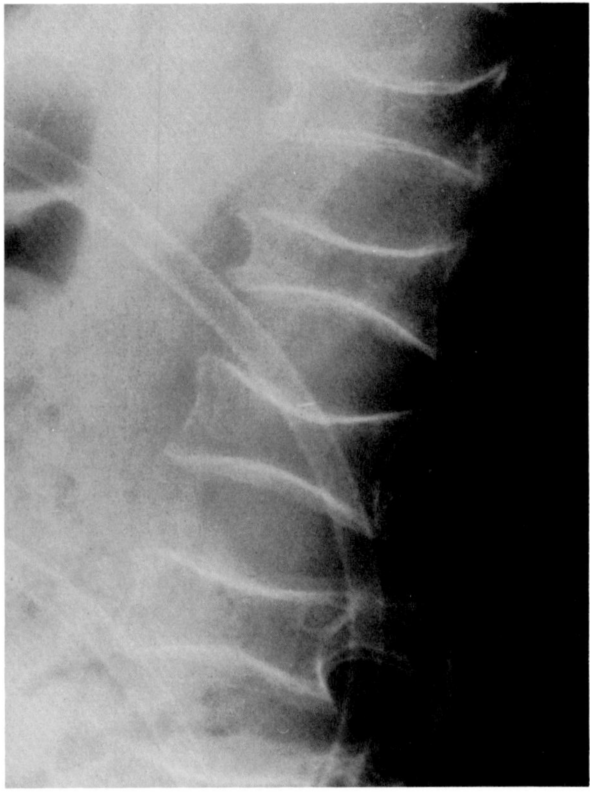

**Figure 3.2. Homocystinuria.** Striking osteoporosis of the spine is associated with biconcave deformities of the vertebral bodies. (From Thomas and Carson,[37] with permission from the publisher.)

The most frequent and striking radiographic feature of homocystinuria is osteoporosis of the spine (extremely rare in Marfan's syndrome), which is often associated with biconcave deformities of the vertebral bodies. Osteoporosis with cortical thinning is also common in the long bones. There is usually widening of the metaphyses and enlargement of the ossification centers of the long bones, most noticeably at the knees. Unlike Marfan's syndrome, homocystinuria does not usually result in arachnodactyly and scoliosis.

Patients with homocystinuria have a tendency to develop arterial and venous thrombosis, and premature occlusive vascular disease is the major cause of death. Because relatively minor damage to the vascular endothelium can incite a life-threatening vascular complication, angiographic procedures should be avoided.[3,4,37]

## STORAGE DISEASES OF AMINO ACID METABOLISM

### Cystinosis

Cystinosis is a metabolic disease characterized by the abnormal accumulation of cystine in body tissues. In children, cystine accumulation in the kidney causes renal insufficiency that may lead to vitamin D–resistant rickets and growth retardation. Unlike cystinuria, cystinosis does not cause the urinary concentration of cystine to be particularly elevated.

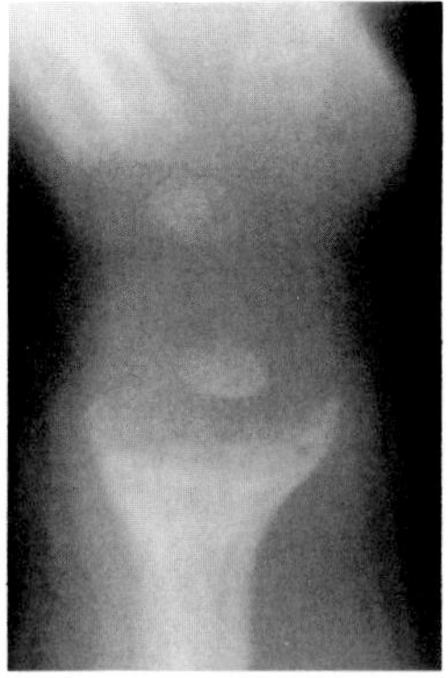

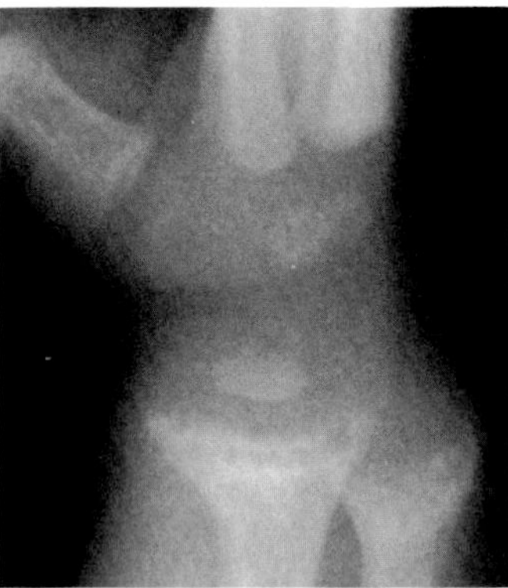

**Figure 3.3. Cystinosis.** Two views of the wrist demonstrate the typical changes of rickets, including cupping and fraying of the metaphyses with loss of the normal sharp metaphyseal line and blurring of the ossification centers.

In the adult (benign) form of cystinosis, cystine is deposited in the cornea but not in the kidney, and no specific radiographic abnormalities are produced.

### Hyperoxaluria

Primary hyperoxaluria is an inborn error of metabolism in which chronic excessive urinary excretion of oxalic acid leads to deposition of calcium oxalate in the kidney. The resulting nephrocalcinosis and nephrolithiasis lead to recurrent urinary tract obstruction and infection, hypertension, renal failure, and early death. The radiographic appearance is indistinguishable from that of renal tubular acidosis or other causes of nephrocalcinosis.

Oxalate crystals can also deposit in extrarenal tissues. In the long bones this may cause extensive demineralization or discrete cystic lucencies with sclerotic margins that primarily involve the metaphyseal and subperiosteal areas.

Severe renal disease may cause secondary hyperparathyroidism, which then dominates the radiographic picture.

Calcium oxalate deposits are far more common in secondary oxalosis, which may be related to overabsorption of dietary oxalate in patients with bacterial overgrowth syndromes, chronic disease of the pancreas and biliary tract, decreased small bowel absorptive surface (ileal resection, jejunoileal bypass), or Crohn's disease.[5–7]

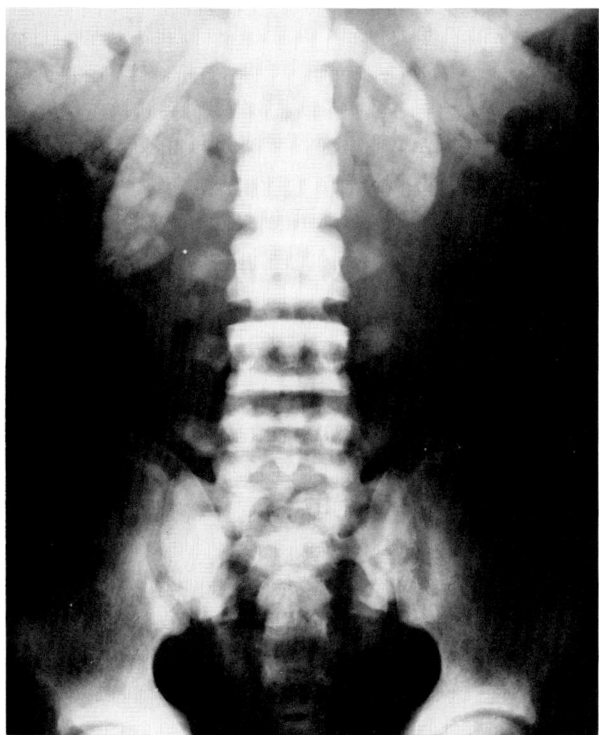

**Figure 3.4. Primary calcium oxalosis.** There are diffuse, mottled renal parenchymal calcifications. Other evidence of the disease includes a "rugger-jersey" spine and sclerotic bands in the iliac crest and acetabula. (From Carsen and Radkowski,[6] with permission from the publisher.)

## Alkaptonuria and Ochronosis

Alkaptonuria is a rare inborn error of metabolism in which an enzyme deficiency leads to an abnormal accumulation of homogentisic acid in the blood and urine. The urine is either very dark on voiding or becomes black after standing or being alkalinized. The disorder often goes unrecognized until middle life, when deposition of the black pigment of oxidized homogentisic acid in cartilage and other connective tissue produces a distinctive form of degenerative arthritis (ochronosis).

In ochronosis the pathognomonic radiographic finding is dense laminated calcification of multiple intervertebral disks that begins in the lumbar spine and may extend to involve the dorsal and cervical regions. The intervertebral disk spaces are narrowed, the vertebral bodies are osteoporotic, and limitation of motion is common. A severe degenerative type of arthritis may also develop in the peripheral joints, especially the shoulders, hips, and knees. Since osteoarthritis is rare in the shoulder, degenerative changes in this joint in a younger patient should suggest ochronosis. Cartilage degeneration produces joint space narrowing, marginal osteophytes, and subchondral sclerosis. Calcifications may form in the synovial membrane and periarticular tissues, and episodes of acute arthritis may occur.[8–10]

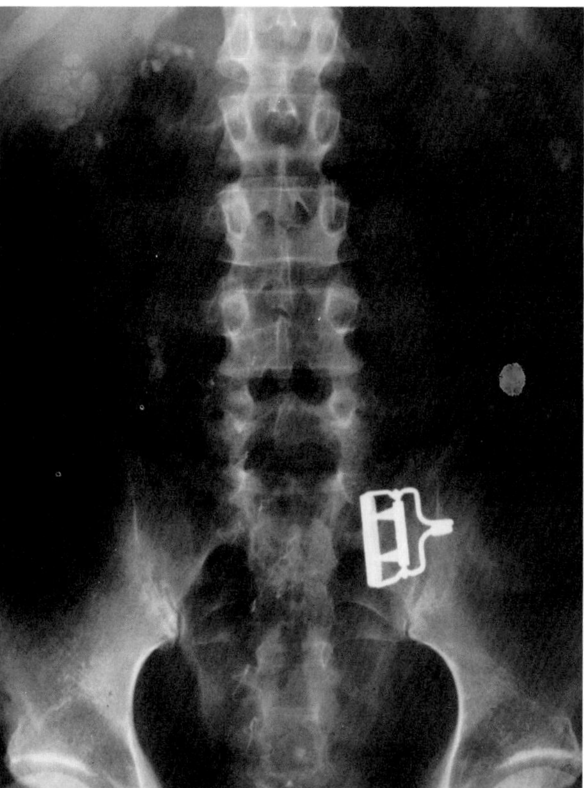

**Figure 3.5. Secondary oxaluria associated with Crohn's disease.** Multiple calcifications are evident in both kidneys, both ureters, and the bladder. Calcifications are also present in the gallbladder and the cystic duct. (From Chikos and McDonald,[5] with permission from the publisher.)

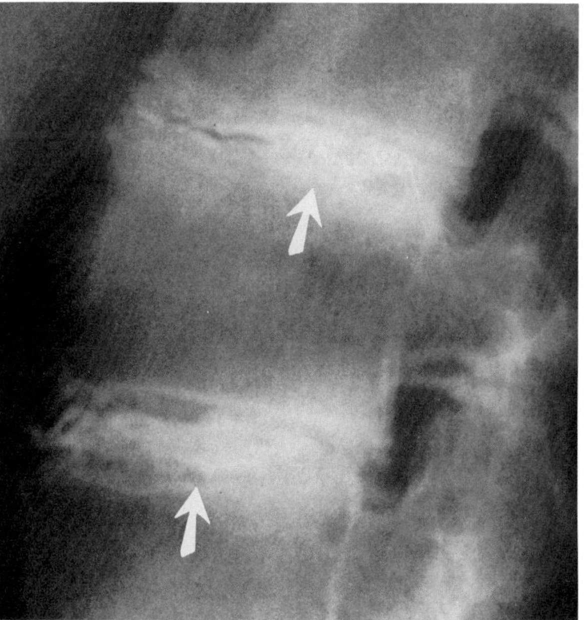

**Figure 3.6. Ochronosis.** There is calcification of the nucleus pulposus of several intervertebral disk spaces (arrows). (From Eisenberg,[38] with permission from the publisher.)

## DISORDERS OF AMINO ACID TRANSPORT

### Cystinuria

Cystinuria, the most common inborn error of amino acid transport, is characterized by impaired tubular absorption and excessive urinary excretion of the dibasic amino acids (lysine, arginine, ornithine, and cystine). Because cystine is the least soluble of the naturally occurring amino acids, its overexcretion predisposes to the formation of renal, ureteral, and bladder calculi. Urolithiasis can cause hematuria, flank pain, renal colic, obstructive uropathy, and infection and may lead to progressive renal insufficiency.

Pure cystine stones are not radiopaque and can only be demonstrated on urograms, where they appear as filling defects in the urinary tract. Stones containing the calcium salts of cystine are radiopaque and can be detected on plain abdominal radiographs.[11]

## DISORDERS OF PURINE METABOLISM

### Gout

Gout refers to a diverse group of conditions characterized by an increase in the serum urate concentration that leads to the deposition of uric acid crystals in the joints, cartilage, and kidney. Several inherited enzyme defects can cause overproduction of uric acid (primary gout). In secondary gout, hyperuricemia can be due to an overproduction of uric acid because of increased turnover of nucleic acids (myeloproliferative and lymphoproliferative disorders, myeloma, polycythemia, hemolytic anemia, metastatic carcinoma), drugs (chemotherapy, thiazides), or a decrease in the excretion of uric acid because of renal failure.

The primary manifestation of acute gout is an exquisitely painful arthritis that initially attacks a single joint, primarily the first metatarsophalangeal. Radiographic changes develop late in the disease and only after repeated attacks. Therefore, negative radiographs do not exclude gout and thus are of no aid in early diagnostic evaluation.

Deposition of nonopaque urate crystals in the joint synovial membrane and on the surface of articular cartilage incites an inflammatory reaction that produces the

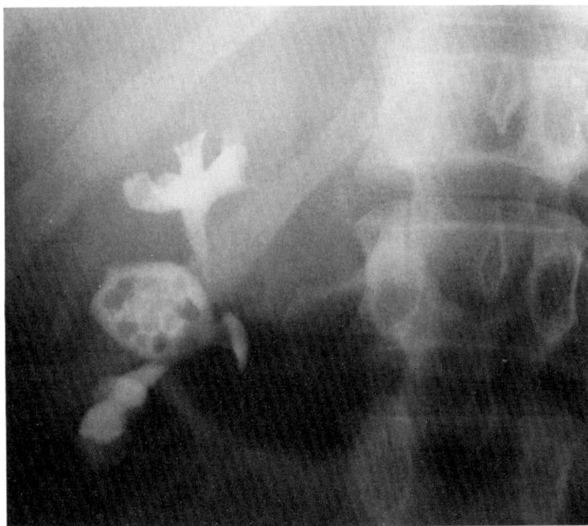

**Figure 3.7. Cystinuria.** An excretory urogram demonstrates multiple radiolucent cystine stones in the renal pelvis.

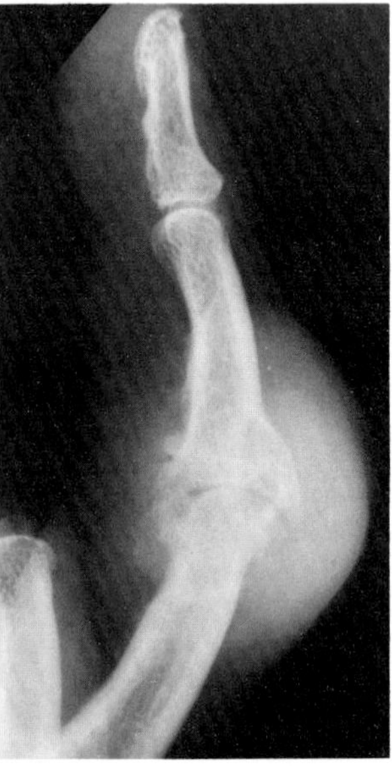

**Figure 3.8. Gout.** Severe joint effusion and periarticular swelling about the proximal interphalangeal joint of a finger. Note the associated erosion of articular cartilage.

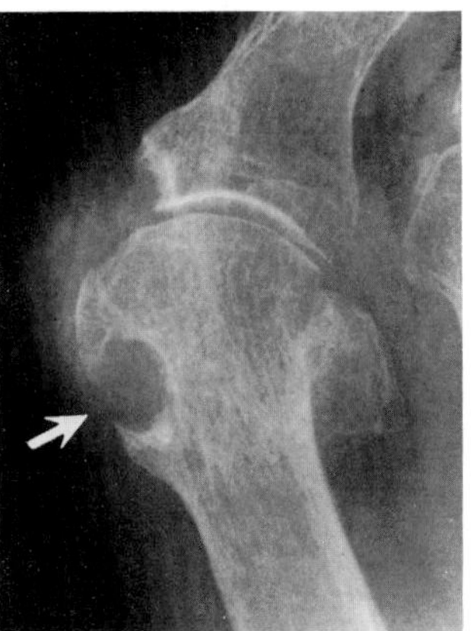

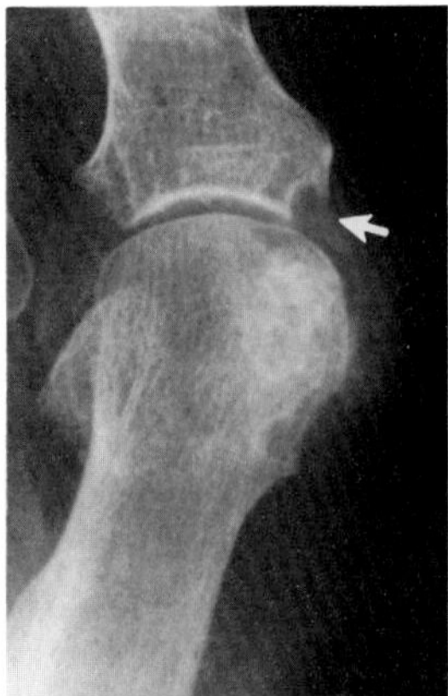

**Figure 3.9. Gout.** Two examples of typical "rat-bite" erosions about the first metatarsophalangeal joint (arrows). The cystlike lesions have thin sclerotic margins and characteristic overhanging edges.

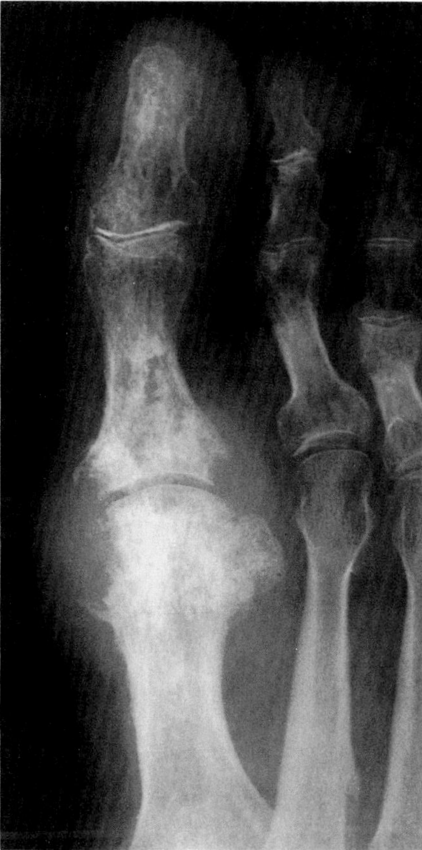

**Figure 3.10. Gout.** Diffuse joint effusion and periarticular swelling about the first metatarsophalangeal joint, with erosions and reactive sclerosis.

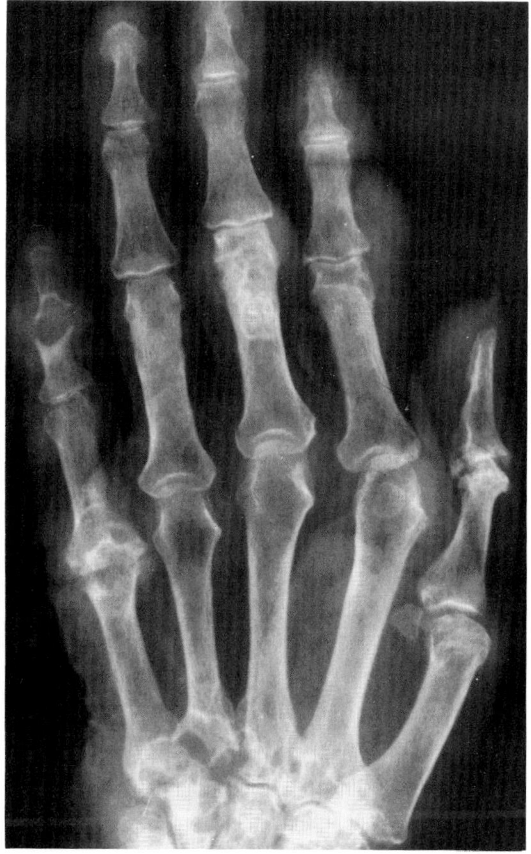

**Figure 3.11. Gout.** Diffuse deposition of urate crystals in the periarticular tissues of the hand produces multiple large, lumpy soft tissue swellings representing gouty tophi. Note the erosive changes that especially involve the carpal bones and the distal interphalangeal and metacarpophalangeal joints of the fifth digits.

earliest radiographic sign of joint effusion and periarticular swelling. Aggregated deposits of urate crystals (tophi) form along the margins of the articular cortex and erode the underlying bone, producing small, sharply marginated, punched-out defects at the joint margins of the small bones of the hand and foot. These erosions often have the appearance of cystlike lesions with thin sclerotic margins and characteristic overhanging edges ("rat bite"). In advanced disease, severe destructive lesions are associated with joint space narrowing and even fibrous ankylosis. Because patients are relatively free of symptoms between acute exacerbations of the disease, the bone density remains relatively normal, and diffuse osteoporosis is not part of the radiographic appearance.

Continued deposition of urate crystals in the periarticular tissues causes the development of characteristic large, lumpy soft tissue swellings representing gouty tophi. Tophi classically occur at the first metatarsophalangeal joint. Other common sites include the ear, the olecranon bursa, and the insertion of the Achilles tendon. Indeed, bilateral enlargement of the olecranon bursae, often with erosion or spur formation, is pathognomonic of gout. Tophi consisting only of sodium urate are of soft tissue density. Deposition of calcium in these urate collections causes the tophaceous periarticular masses to become radiopaque.

Calcification in the articular cartilage and the menisci of the knee (chondrocalcinosis) resembling pseudogout is a common finding in patients with gout. Ischemic necrosis of the femoral heads, medullary bone infarcts, irregular erosions of the sacroiliac joints, and large popliteal cysts resembling those in rheumatoid arthritis can also occur.

Although some renal dysfunction occurs in almost all patients with gouty arthritis, no radiographic abnormalities in the urinary tract can be demonstrated unless uric acid stones are formed. Pure uric acid stones are not radiopaque and can only be demonstrated on urograms, where they appear as filling defects in the pelvocalyceal system or ureter. Stones containing the calcium salts of uric acid are radiopaque and can be detected on plain abdominal radiographs.[12,13]

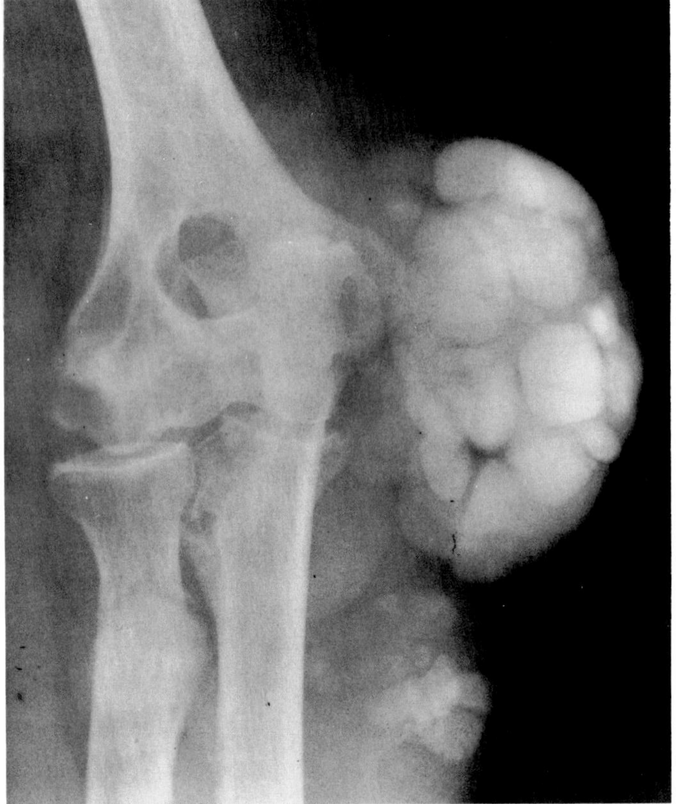

A

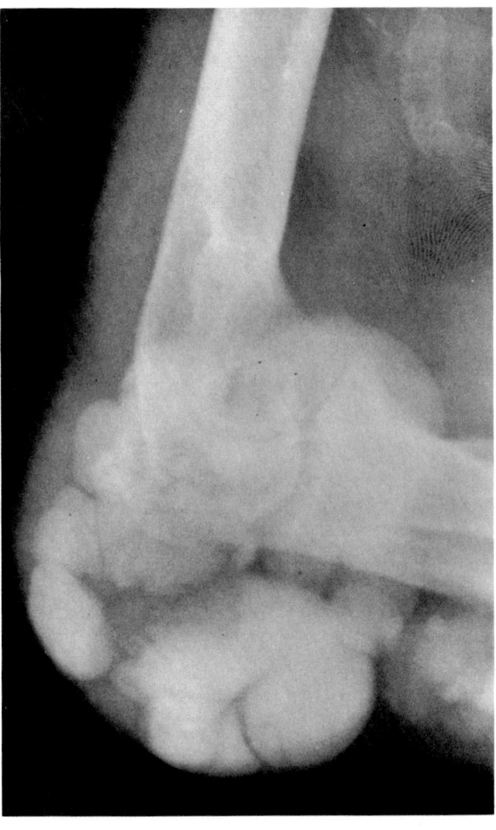

B

**Figure 3.12. Gout.** (A) Frontal and (B) lateral views demonstrate a tophaceous lesion of long duration with massive deposition of calcium about the elbow.

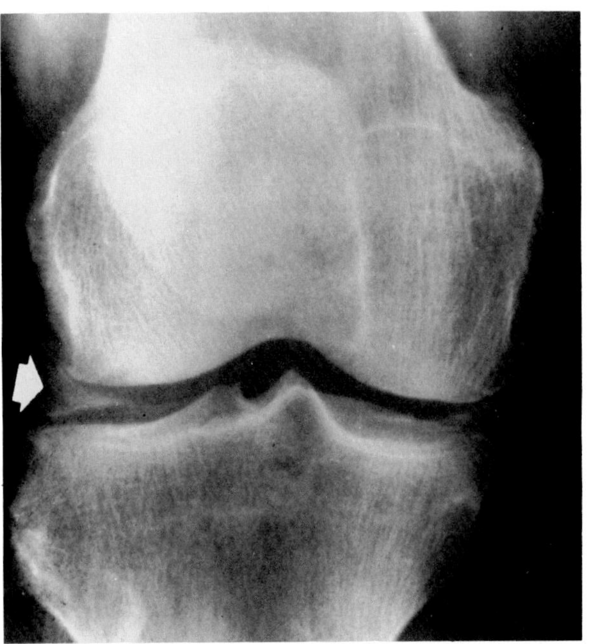

**Figure 3.13. Gout.** Chondrocalcinosis about the knee (arrow) simulates pseudogout.

### Lesch-Nyhan Syndrome

The Lesch-Nyhan syndrome is a rare inherited disorder of purine metabolism in which hyperuricemia and a profound overproduction of uric acid are associated with mental and growth retardation, abnormal aggressive behavior, and self-mutilation by biting of the fingers and lips. In children, radiographic findings include delayed skeletal maturation, valgus deformity and subluxation of the hips, and urinary tract calculi. Older patients may demonstrate typical radiographic signs of gout.[14,15]

### Xanthinuria

Xanthinuria is a rare hereditary disorder of purine metabolism in which there is a decreased concentration of uric acid in the serum and urine and an increase in the urinary excretion of xanthine and other purine derivatives. About one-third of patients with this condition have urinary xanthine stones, which are usually nonopaque.[16]

## HEMOCHROMATOSIS

Hemochromatosis is an iron-storage disorder in which the deposition of iron in parenchymal cells causes tissue

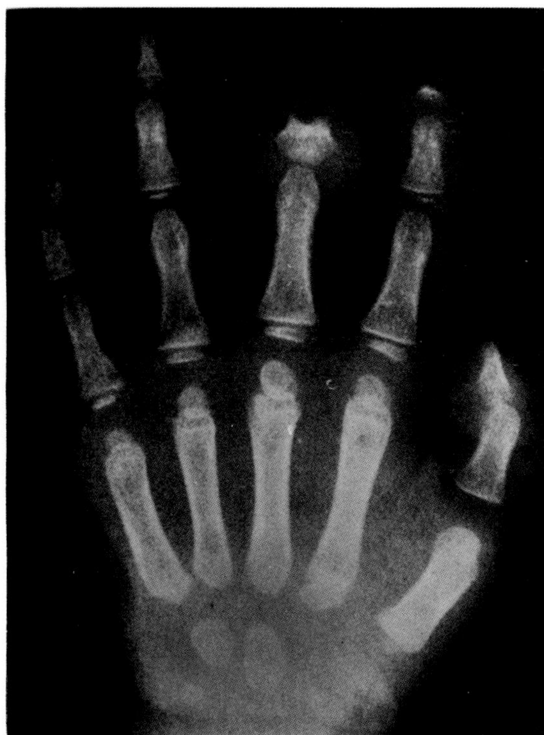

**Figure 3.14. Lesch-Nyhan syndrome.** A film of the hand demonstrates amputation of the index and middle fingers from a self-inflicted bite. Although the child is 5 years old, the bone age is that of a 3-year-old child. (From Becker and Wallin,[15] with permission from the publisher.)

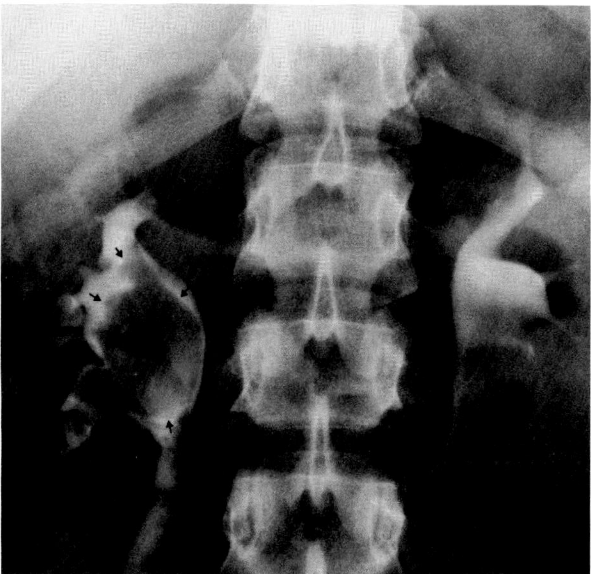

**Figure 3.15. Xanthinuria.** A large lucent stone (arrows) almost fills the right pelvocalyceal system.

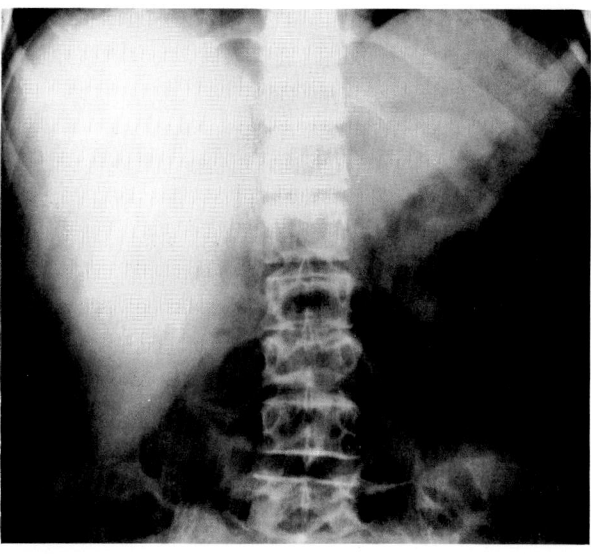

**Figure 3.16. Hemochromatosis.** An abdominal radiograph demonstrates an extremely dense liver shadow in the right upper quadrant caused by the parenchymal deposition of iron. (From Smith and Quattromani,[17] with permission from the publisher.)

damage and functional insufficiency in various organs, especially the liver, pancreas, spleen, kidneys, heart, and endocrine glands. Idiopathic hemochromatosis is an inherited disorder in which there is an inappropriate increase in the absorption of iron from the gastrointestinal tract. Hemochromatosis is more commonly secondary and is found in patients with anemia due to abnormal erythropoiesis (thalassemia), alcoholics with chronic liver disease, and persons with chronic excessive iron ingestion. The clinical spectrum in hemochromatosis includes skin pigmentation (hence the name bronze diabetes), diabetes mellitus, liver and cardiac dysfunction, arthropathy, and hypogonadism.

Enlargement and fibrosis of the liver are almost always present in patients with hemochromatosis. Portal hypertension, esophageal varices, and ascites may occur as late findings. Plain abdominal radiographs may demonstrate a very dense liver shadow in the right upper quadrant caused by the parenchymal deposition of iron. Hepatocellular carcinoma develops in about one-third of the patients.

About half the patients with hemochromatosis develop an arthropathy that most commonly involves the small joints of the hands, especially the second and third metacarpophalangeal joints. The major features include subarticular cysts and erosions, joint space narrowing, osteophytes, irregularity and sclerosis of the articular surface, subluxation, and flattening and widening of the metacarpal heads. In larger joints, especially the knees and hips, joint space narrowing and marginal osteophyte formation simulate osteoarthritis. Diffuse osteoporosis of the spine may be associated with vertebral collapse.

Acute brief attacks of synovitis simulating pseudogout may occur in patients with hemochromatosis. Chondrocalcinosis, not necessarily related to the symptomatic

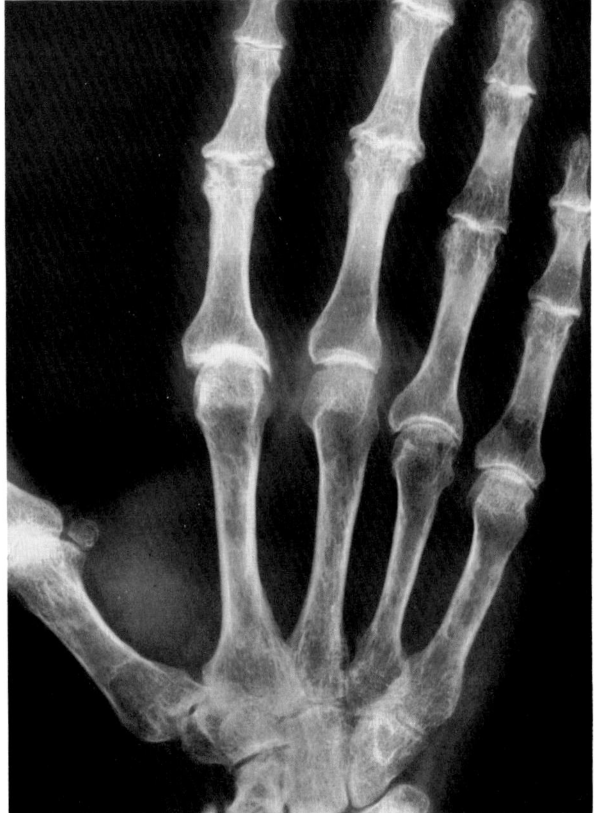

**Figure 3.17. Hemochromatosis.** A hand film demonstrates diffuse joint space narrowing with scattered erosions, osteophytes, and articular sclerosis.

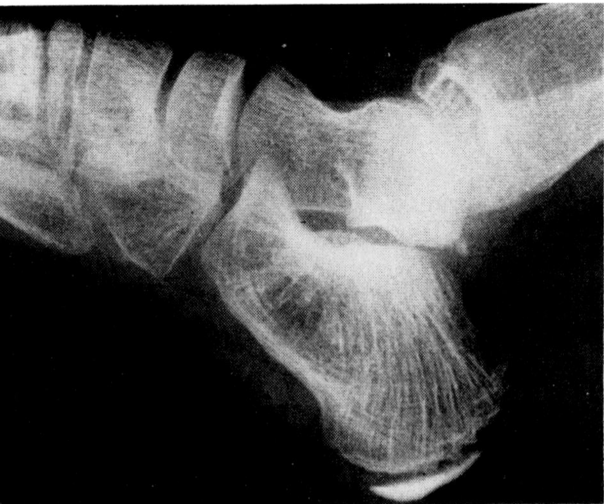

**Figure 3.18. Wilson's disease.** A lateral view of the ankle and foot demonstrates marked demineralization, thinning of the cortex, and coarsening of the trabecular pattern, all best seen in the os calcis. (From Mindelzun, Elkin, Scheinberg, et al,[20] with permission from the publisher.)

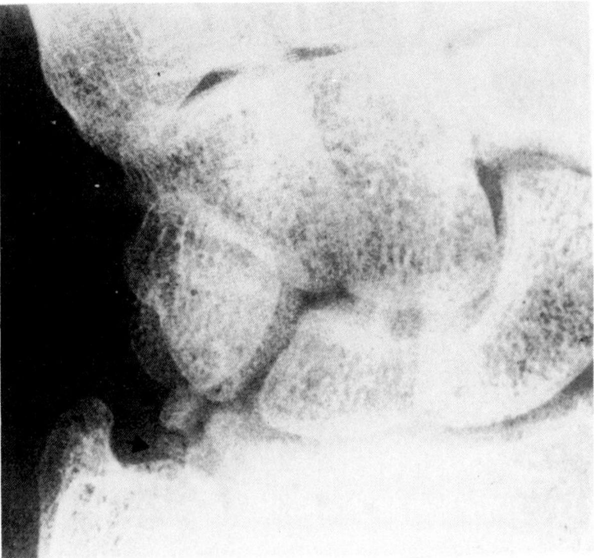

**Figure 3.19. Wilson's disease.** Characteristic ossicles (arrows) are seen in the region of the triangular fibrocartilage of the wrist. (From Dalinka, Reginato, and Golden,[39] with permission from the publisher.)

joints, occurs in more than half the patients with arthropathy. The knee is the joint most often involved, though calcification may also appear in the triangular cartilage of the wrist and about the shoulder, elbow, hip, and symphysis pubis.

Deposition of iron in the heart causes diffuse cardiomegaly and may lead to congestive heart failure.[17–19]

## WILSON'S DISEASE

Wilson's disease is a rare familial disorder in which impaired hepatic excretion of copper results in toxic accumulation of the metal in liver, brain, and other organs. Clinically, patients with Wilson's disease present with cirrhosis of the liver, often simulating acute hepatitis, neurologic or psychiatric disturbances, and a characteristic pigmentation of the cornea (Kayser-Fleischer ring).

About half the patients with Wilson's disease demonstrate skeletal changes, especially bone demineralization with cortical thinning and coarsened trabeculation. Various bone and joint abnormalities have been reported, especially subarticular cyst formation and fragmentation of subchondral bone that primarily occur in the hands and wrists.

Increased levels of copper in the kidney can rarely produce renal tubular dysfunction and lead to renal rickets or adult osteomalacia.[20,21,39]

## PORPHYRIA

Acute intermittent porphyria is a familial metabolic disease characterized by repeated attacks of severe, colicky

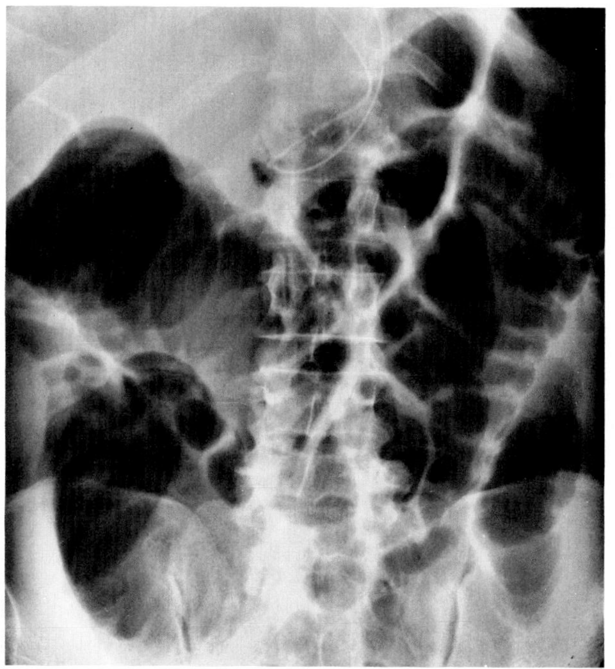

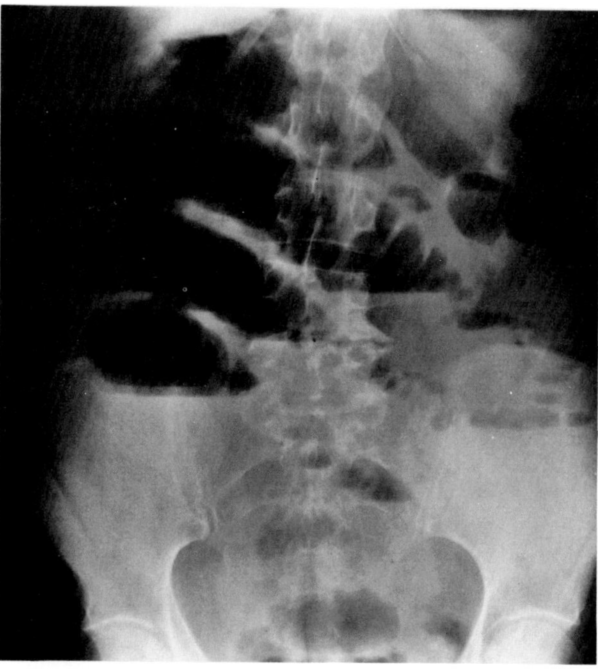

A B

**Figure 3.20. Porphyria.** (A) Supine and (B) upright views of the abdomen demonstrate adynamic ileus simulating mechanical obstruction in a patient with acute intermittent porphyria. (From Eisenberg,[38] with permission from the publisher.)

abdominal pain in assocation with obstipation. Clinical and radiographic symptoms and signs often lead to the erroneous diagnosis of bowel obstruction. Although many patients are operated upon for this reason, no organic obstruction is found. The diagnosis is usually made from the chance observation that the urine becomes dark on exposure to light or because of the development of characteristic neurologic symptoms, which lead to a search for the presence of abnormal porphyrins in the urine and feces.[22,38]

## GLYCOGEN STORAGE DISEASES

The glycogen storage diseases (GSDs) are a group of genetic disorders involving the pathways for the storage of carbohydrates as glycogen and for its utilization to maintain blood sugar and to provide energy. An excess amount of normal or abnormal glycogen infiltrates multiple organs; other metabolic changes, especially hyperuricemia and gout, are usually associated.

### Von Gierke's Disease

Von Gierke's disease (GSD type Ia) is the most common type of glycogen storage disorder. Radiographic abnormalities in this condition include marked hepatomegaly, delayed skeletal maturation, a loss of bone density, and a lack of modeling. Patients who survive into adulthood may develop the uric acid stones and skeletal changes of gout.[23,40]

### Pompe's Disease

Pompe's disease (GSD type II) is a generalized glycogen storage disorder that primarily involves the central nervous system, heart, and muscles. The bones are not affected in this condition.

Excess deposition of glycogen in the heart causes massive cardiac enlargement with a globular configuration simulating endocardial fibroelastosis. The pulmonary vasculature remains normal until heart failure develops.[24]

## LIPID STORAGE DISEASES

### Gaucher's Disease

Gaucher's disease is an inborn error of metabolism characterized by the accumulation of abnormal quantities of complex lipids in the reticuloendothelial cells of the spleen, liver, and bone marrow. There are two clinical forms of the disease. In the fulminating infantile type, there is rapid enlargement of the liver and spleen, causing a protuberant abdomen, and severe central nervous system and respiratory abnormalities that lead to rapid deterioration and death within the first 2 years.The more common adult form has a variable severity and age of onset with no evidence of neurologic or pulmonary involvement.

The most striking radiographic changes of Gaucher's disease occur in the skeletal system. Infiltration of the bone marrow with abnormal lipid-containing cells causes

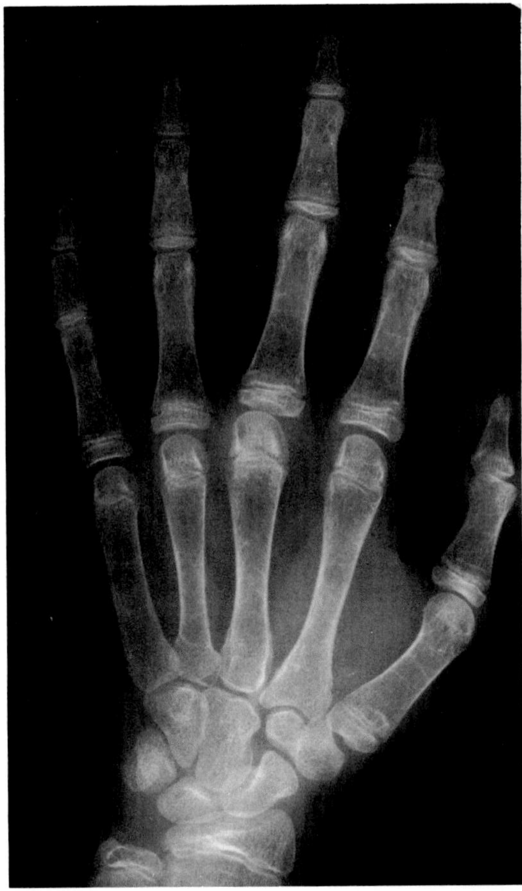

**Figure 3.21. Von Gierke's disease.** A view of the left hand demonstrates osteopenia, abnormal modeling, and numerous growth arrest lines in the distal radius. Although the patient had a chronologic age of 22 years, the osseous maturity of the left hand corresponds to the standard of an 11-year-old boy. (From Miller, Stanley, and Gates,[40] with permission from the publisher.)

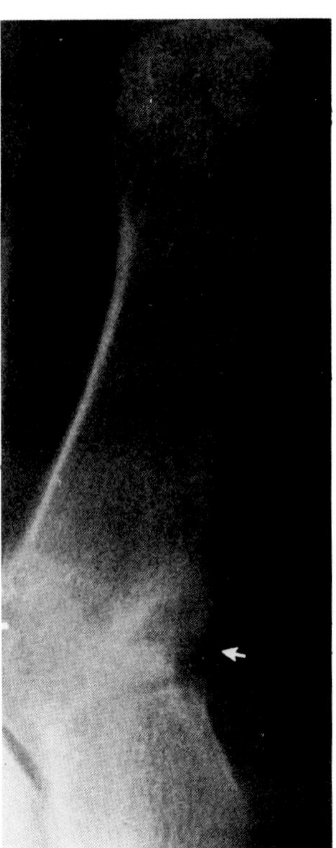

**Figure 3.22. Von Gierke's disease.** There are erosive and cystic changes (arrow) at the base of the first metatarsal. Clinical gout started 5 years previously at the age of 26. (From Preger, Sanders, Gold, et al,[23] with permission from the publisher.)

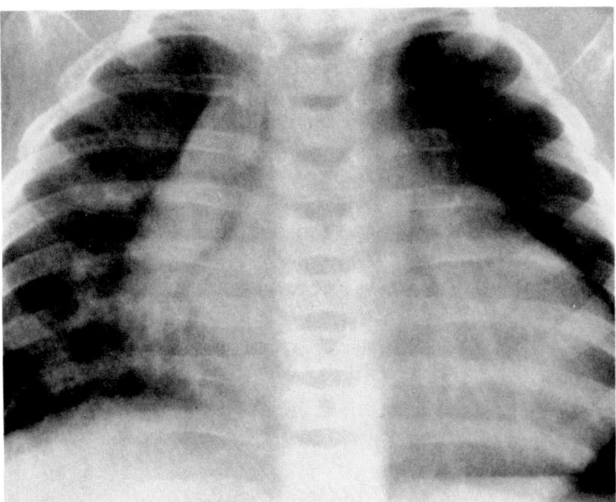

**Figure 3.23. Pompe's disease.** Massive cardiac enlargement with a globular configuration due to excess deposition of glycogen in the heart.

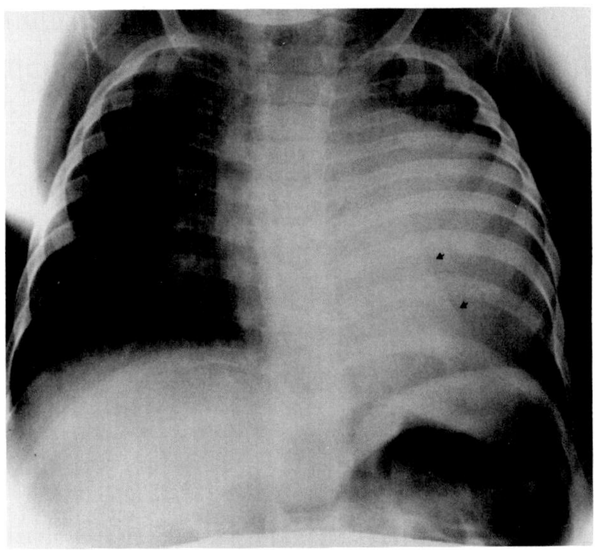

**Figure 3.24. Pompe's disease.** A grossly enlarged heart with a collapsed left lower lobe (arrows). (From Nihill, Wilson, and Hugh-Jones,[24] with permission from the publisher.)

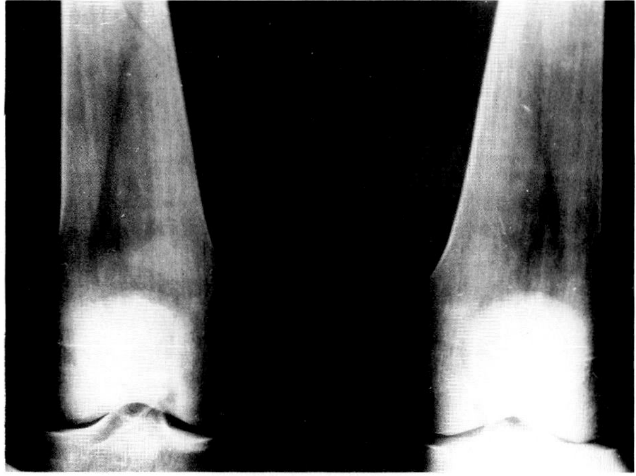

**Figure 3.25. Gaucher's disease.** The distal ends of the femurs show typical underconstriction and cortical thinning (Erlenmeyer flask appearance). (From Levin,[26] with permission from the publisher.)

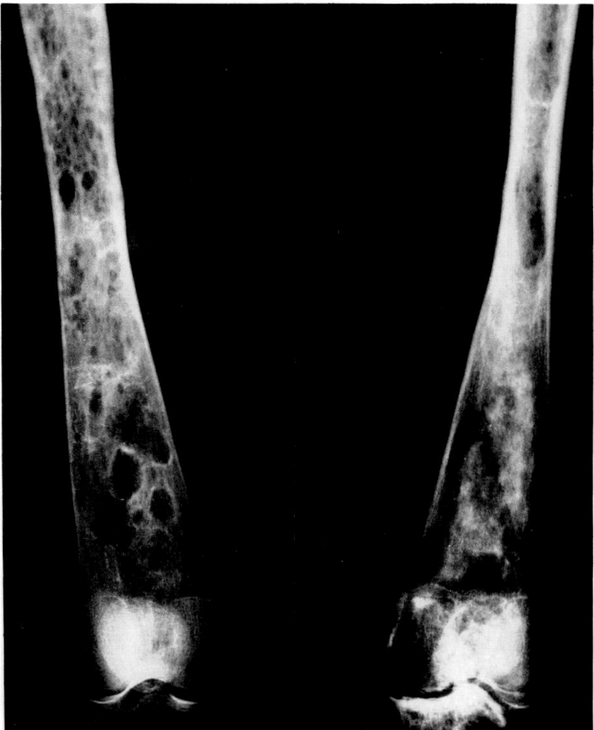

**Figure 3.26. Gaucher's disease.** Extensive replacement of normal bone by lipid-containing cells produces a cystlike appearance. In some areas the cortex is markedly thinned, and there is irregularity of the gross bone contour. The femurs are appreciably underconstricted. (From Levin,[26] with permission from the publisher.)

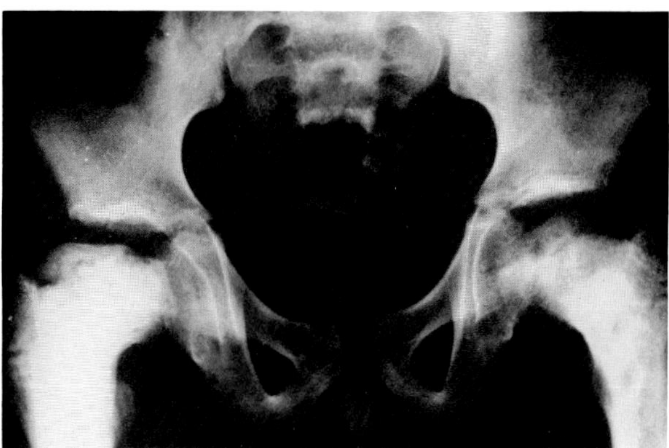

**Figure 3.27. Gaucher's disease.** Bilateral aseptic necrosis of the femoral heads.

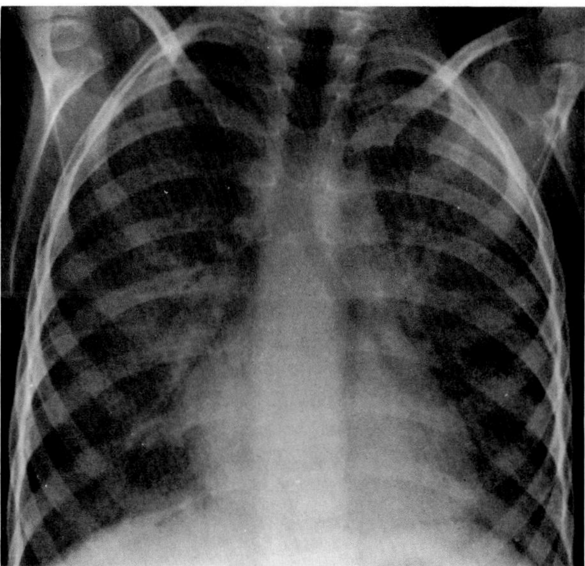

**Figure 3.28. Gaucher's disease.** Diffuse interstitial infiltrates throughout the lungs.

a loss of bone density with expansion and cortical thinning of the long bones, especially the femur. Deossification commonly results in pathologic fractures of the long bones and compression of the vertebral bodies. Focal collections of abnormal cells may cause lucent defects sharply circumscribed by coarsened trabeculae, simulating metastases or myeloma. Secondary bone repair may cause patchy or diffuse areas of sclerosis. Marrow infiltration in the distal femur causes abnormal modeling and flaring and the characteristic (but nonspecific) Erlenmeyer flask deformity.

Aseptic necrosis is a common complication, especially in the femoral heads, but also often involving the humeral head and the bones of the wrists and ankles. Bone infarcts frequently develop in the metaphyses of the long bones.

The spleen is usually markedly enlarged; hepatomegaly is common. Diffuse pulmonary infiltrates are occasionally found in the acute form of the disease, though much less commonly than in Niemann-Pick disease.[25,26]

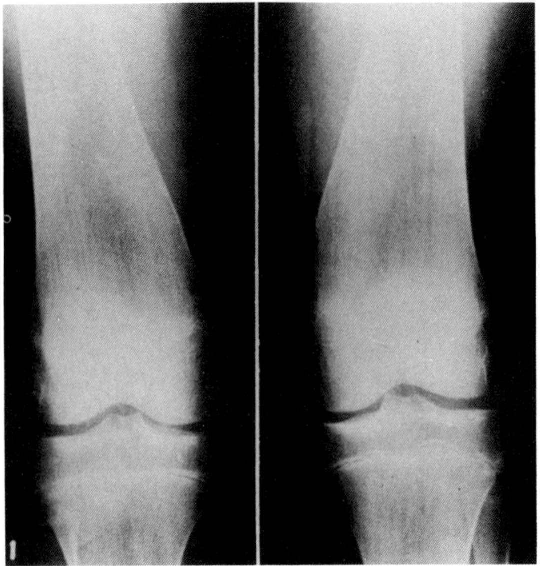

**Figure 3.29. Niemann-Pick disease.** Thin cortices and a lack of normal modeling of the distal femurs simulate the pattern in Gaucher's disease. (From Lachman, Crocker, Schulman, et al,[27] with permission from the publisher.)

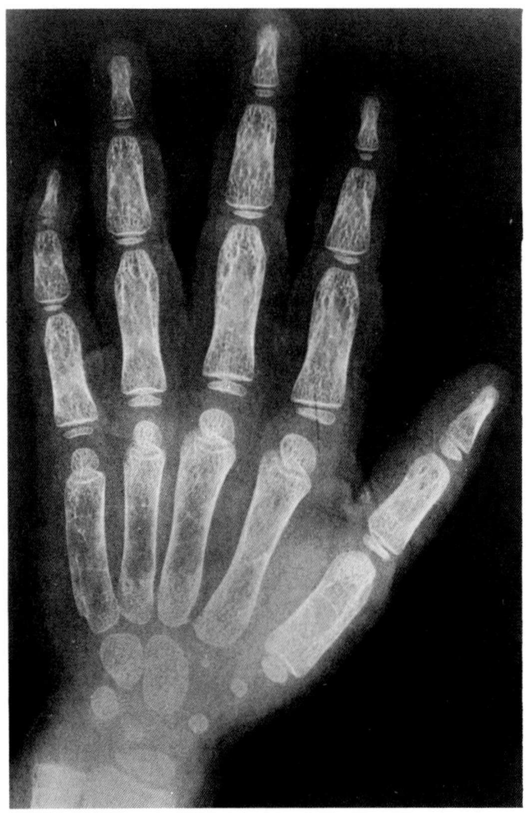

**Figure 3.30. Niemann-Pick disease.** Diffuse widening of the metacarpals and phalanges with thin cortices and coarsened trabeculae. (From Lachman, Crocker, Schulman, et al,[27] with permission from the publisher.)

### Niemann-Pick Disease

Niemann-Pick disease is an inborn error of lipid metabolism in which there is an abnormal deposition of sphingomyelin. The most common form of the disease begins in infancy and is rapidly fatal, with severe central nervous system and pulmonary involvement and marked hepatosplenomegaly. A cherry-red macular spot is often seen. Less common forms of the disease are milder and more slowly progressive.

Radiographic changes are often seen in patients who survive until late childhood or adolescence. Most patients demonstrate diffuse nodular pulmonary infiltrates with linear strands that produce a honeycomb effect or a coarse reticulonodular pattern on chest radiographs. The skeletal abnormalities closely resemble those of Gaucher's disease, with abnormal modeling, cortical thickening, and coarsened trabeculae. However, the presence of nodular interstitial pulmonary infiltrates and an early age of onset suggest Niemann-Pick disease.[27]

## MUCOPOLYSACCHARIDOSIS

The mucopolysaccharidoses are genetically determined disorders of mucopolysaccharide metabolism that result in a broad spectrum of skeletal, visceral, and mental abnormalities.

### Hurler's Syndrome (Gargoylism)

The clinical characteristics of Hurler's syndrome include a large, bulging head with eyes wide apart, a sunken bridge of the nose, an everted lip, a protruding tongue,

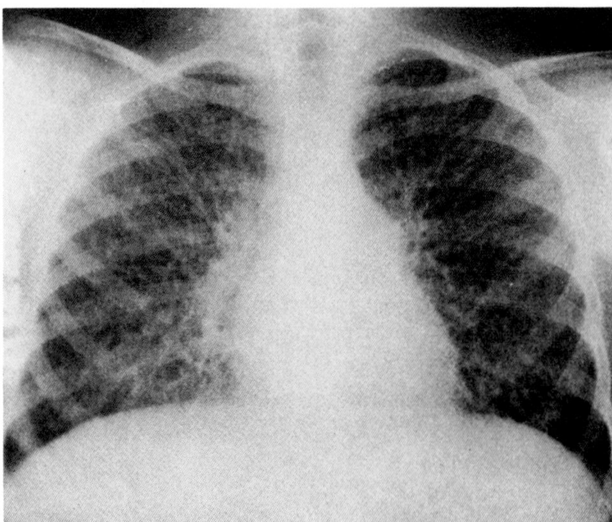

**Figure 3.31. Niemann-Pick disease.** Coarse reticulondular infiltrates diffusely involve both lungs, producing a honeycomb effect. (From Lachman, Crocker, Schulman, et al,[27] with permission from the publisher.)

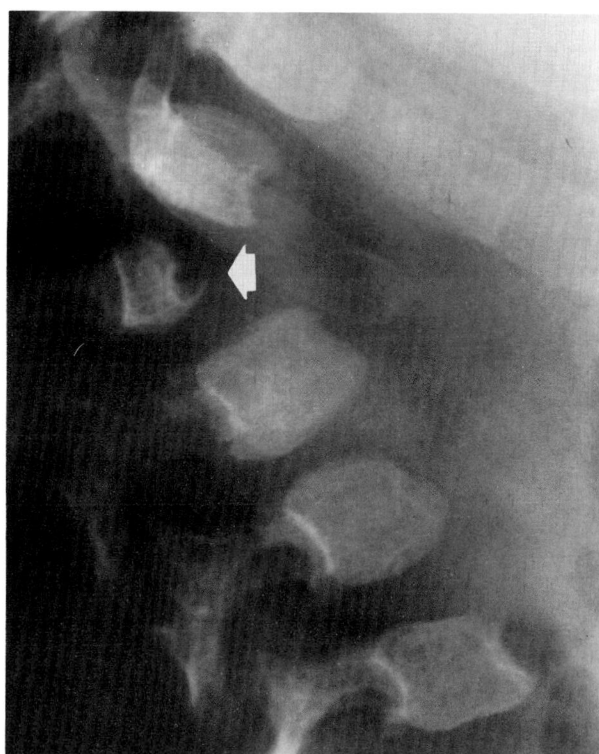

**Figure 3.32. Hurler's syndrome.** A lateral view of the lumbar spine demonstrates hypoplasia and posterior displacement of the second lumbar vertebra (arrow), producing an accentuated kyphosis.

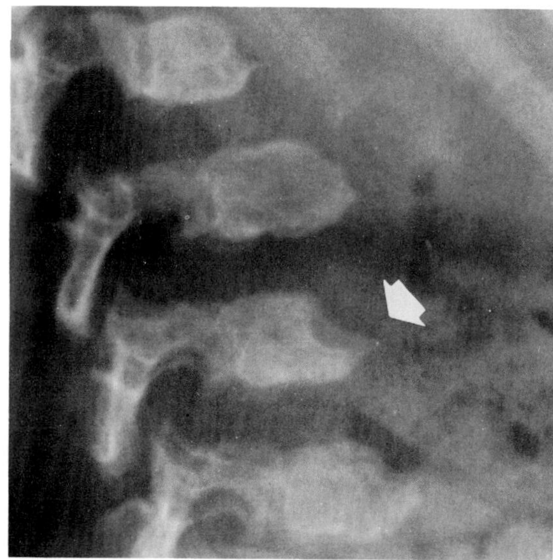

**Figure 3.33. Hurler's syndrome.** Typical inferior beaking (arrow) of the anterior margin of a vetebral body.

and widely separated and poorly formed teeth, all of which produce a heavy and ugly facies that has been likened to that of a gargoyle. The patient with Hurler's syndrome usually develops characteristic clouding of the cornea and has prominent hepatosplenomegaly, short stature with restricted joint motion and flexion deformities, and progressive mental retardation.

The most distinctive radiographic changes in Hurler's syndrome involve the vertebral bodies. The centra are oval and either increased or normal in height with inferior beaking of the anterior vertebral margin, in contrast to the flattened vertebral bodies (vertebra plana) with central beaking of Morquio's syndrome. The centrum of the second lumbar vertebra is usually hypoplastic and displaced posteriorly, giving rise to an accentuated kyphosis or gibbus deformity. Changes in the long bones, most marked in the upper extremities, include swelling in the central portions of the bones (due to cortical thickening or widening of the medullary canal) with tapering of one or both ends. Similar central widening of the ribs with tapering at the vertebral end produces a canoe paddle appearance with a decrease in the intercostal space.

The skull is abnormally large and usually dolichocephalic. A characteristic finding is the J-shaped sella, a shallow, elongated sella with a long anterior recess extending under the anterior clinoid processes. The sinuses and mastoids are usually poorly developed.

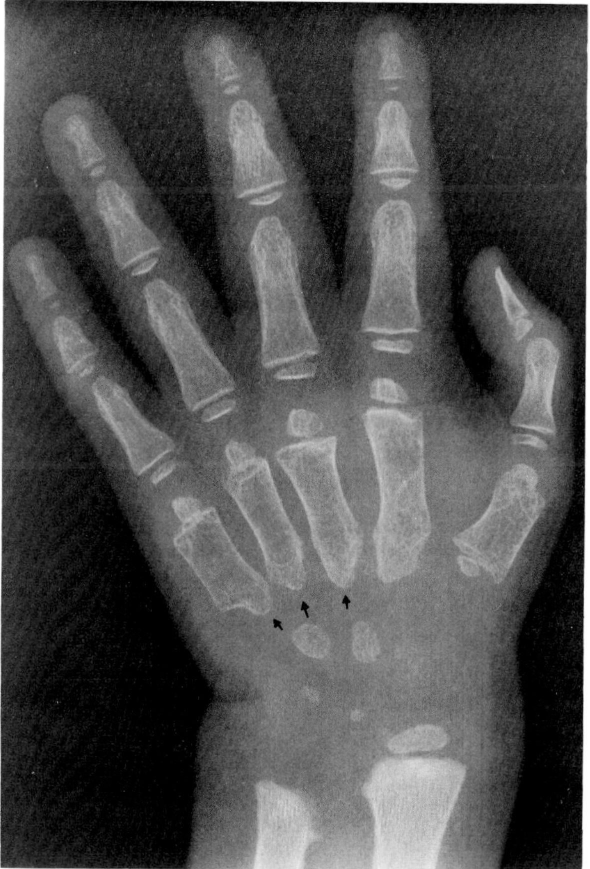

**Figure 3.34. Hurler's syndrome.** A view of the hand demonstrates widening of the metacarpals, which show thinning of the cortices and regular tapering at the proximal ends. The epiphyseal plates of the distal radius and ulna are abnormally angled toward each other. Tapering of the distal ends of the phalanges produces a typical bullet shape.

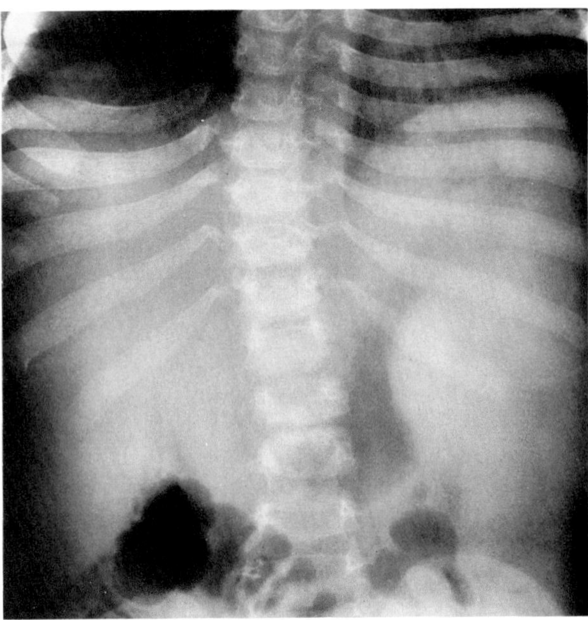

**Figure 3.35. Hurler's syndrome.** Central widening of the ribs with tapering at the vertebral ends produces a canoe paddle appearance with a decrease in the width of the intercostal space. Note also the prominent enlargement of the liver and spleen.

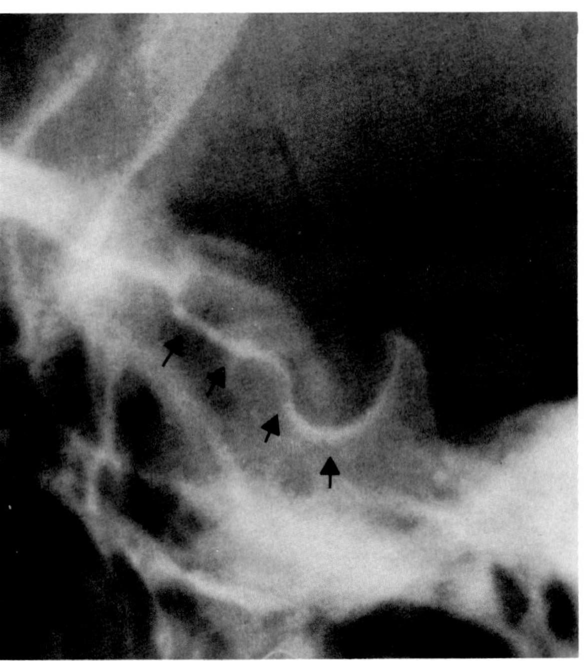

**Figure 3.36. Hurler's syndrome.** The sella is shallow and elongated (arrows) with a long anterior recess extending under the anterior clinoid process (J-shaped sella).

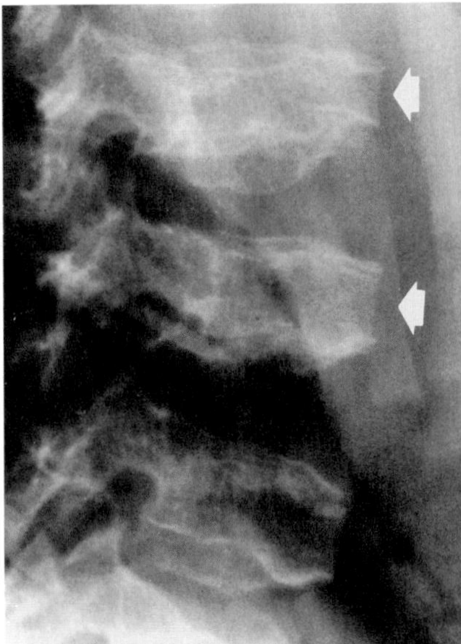

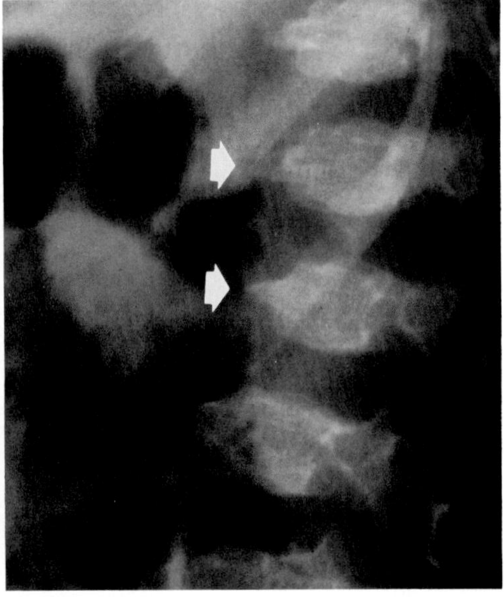

**Figure 3.37. Morquio's syndrome.** Two examples of universal flattening of the vertebral bodies with central anterior beaking (arrows).

The other mucopolysaccharidoses, other than Morquio's syndrome, tend to have radiographic findings similar to those of Hurler's syndrome, though they are less severe and tend to occur at a later age.[28,29]

## Morquio's Syndrome

Morquio's syndrome is a congenital disturbance of mucopolysaccharide metabolism characterized by striking dwarfism, kyphosis, deformities of the hands and feet, and diffuse corneal opacification. Unlike patients with Hurler's syndrome, patients with Morquio's syndrome are not mentally retarded, and the skull and facial bones are not involved. Many patients have marked hypoplasia of the odontoid process that can cause cervical dislocation and some degree of spinal cord compression. Aortic regurgitation is often associated.

The most characteristic skeletal change in Morquio's syndrome is universal flattening of the vertebral bodies (vertebra plana). The superior and inferior surfaces of the dorsolumbar vertebral bodies are irregular and appear to approximate as a central anterior beak, unlike the normal or enlarged vertebrae with inferior beaking in Hurler's syndrome. One vertebra, usually the first or second lumbar, is hypoplastic and posteriorly placed, resulting in a sharp, angular kyphosis.

Epiphyseal ossification centers may be multiple and are often irregular. They are late in appearing but mature normally. Although the long bones may demonstrate tapering, this is not as marked as in Hurler's syndrome. Fragmentation, flaring, and flattening of the femoral heads combined with irregular deformity of the acetabula often result in subluxation at the hip. The femoral neck is thick and short, and varus or valgus deformity may develop.[30,31]

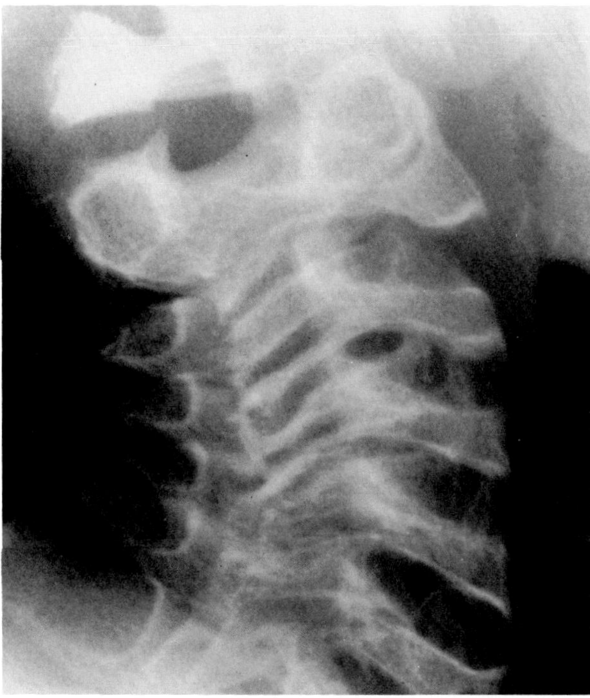

**Figure 3.38. Morquio's syndrome.** Generalized flattening of the cervical vertebral bodies.

## DISORDERS OF CONNECTIVE TISSUES

### Ehlers-Danlos Syndrome

The Ehlers-Danlos syndrome comprises a group of inherited generalized disorders of connective tissue whose major features include fragile and hyperextensible skin, easy bruising, and loose-jointedness. The most characteristic radiographic abnormality is calcification of fatty nodules in the subcutaneous tissues of the extremities. The nodules range from 2 to 10 mm and appear as central lucent zones with ringlike calcification, simulating phleboliths. They must be differentiated from calcified subcutaneous parasites, especially cysticerci, which tend to be aligned along muscular and fascial planes rather than randomly distributed in the soft tissues.

Tissue hyperelasticity leads to such nonspecific musculoskeletal abnormalities as scoliosis, deformities of the thoracic cage, hypermobility of the joints, and subluxations. Congenital anomalies such as spina bifida, spondylolisthesis, and radial-ulnar synostosis are often associated. Stenoses of the pulmonary arteries and aneurysms of the great vessels can occur. Gastrointestinal abnormalities include megaduodenum and multiple duodenal and jejunal diverticula that can result in malabsorption.[32,41]

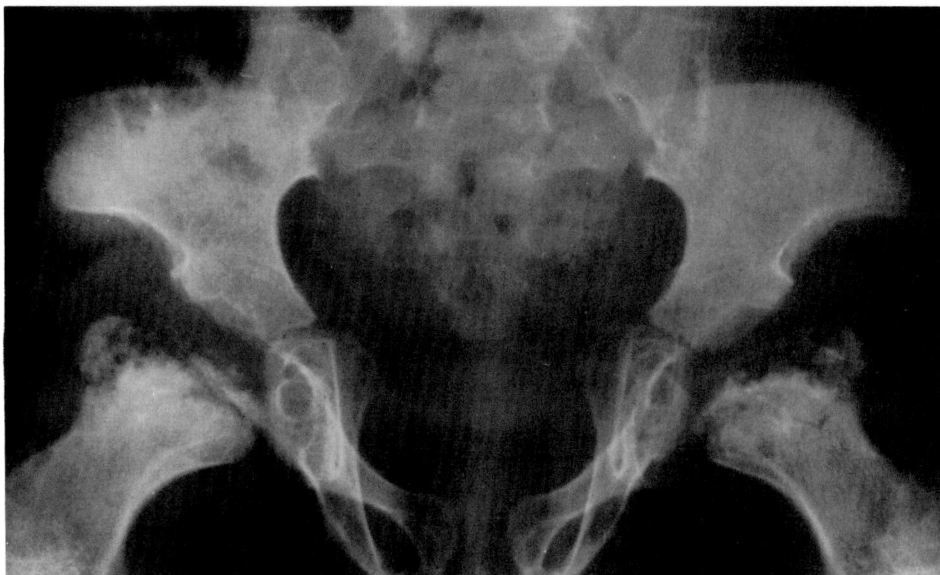

**Figure 3.39. Morquio's syndrome.** Flattening and irregularity of the femoral capital epiphyses is associated with widening and irregularity of the acetabula.

## Marfan's Syndrome

Marfan's syndrome is an inherited generalized disorder of connective tissue with ocular, skeletal, and cardiovascular manifestations. Most patients are tall and slender and appear emaciated because of a decrease in subcutaneous fat. A typical feature of Marfan's syndrome is bilateral dislocation of the lens of the eye due to the weakness and redundancy of its supporting tissues. A laxity of ligaments and other supporting structures of the joints leads to loose-jointedness or double-jointedness, sclerosis, recurrent dislocations, and pes planus (flatfoot). Sternal displacement is common and may be inward (pectus excavatum, or funnel chest) or upward (pectus carinatum, or pigeon breast).

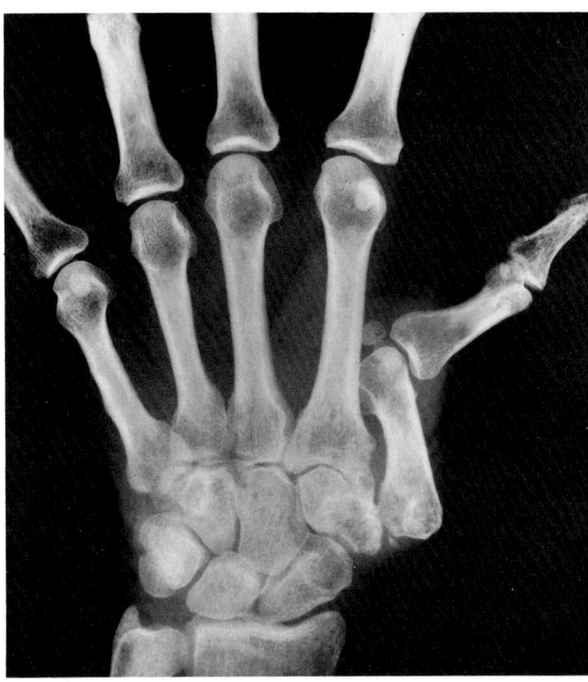

**Figure 3.40. Ehlers-Danlos syndrome.** Subluxations of the first carpal-metacarpal and metacarpophalangeal joints due to hyperelasticity of the connective tissue.

The major radiographic abnormality is elongation and thinning of the tubular bones, most pronounced in the hands and feet. A characteristic feature is arachnodactyly, which refers to the excessively long and slender metacarpals and phalanges. Unlike homocystinuria, which is similar clinically, Marfan's syndrome does not cause osteoporosis.

Almost all patients with Marfan's syndrome have abnormalities in the cardiovascular system. Medial necrosis of the wall of the aorta causes progressive dilatation of the proximal portion of the ascending aorta. On plain chest radiographs this is evidenced by a bulging of the upper right portion of the cardiac silhouette and an unusual prominence of the pulmonary outflow tract as it is displaced by the dilated aorta. Dissecting aneurysm is a serious complication that may kill the patient in early life. Aneurysmal dilatation of the pulmonary artery, descending thoracic aorta, and abdominal aorta also occurs. Aortic insufficiency often leads to left ventricular enlargement. A "floppy" mitral valve may lead to regurgitation and left atrial enlargement.

Relatively infrequent pulmonary manifestations of Marfan's syndrome include diffuse pulmonary emphysema, and apical emphysematous bullae that may lead to pneumothorax. Rarely, prominence of the interstitial structures causes a honeycomb appearance of the lungs.[33,42]

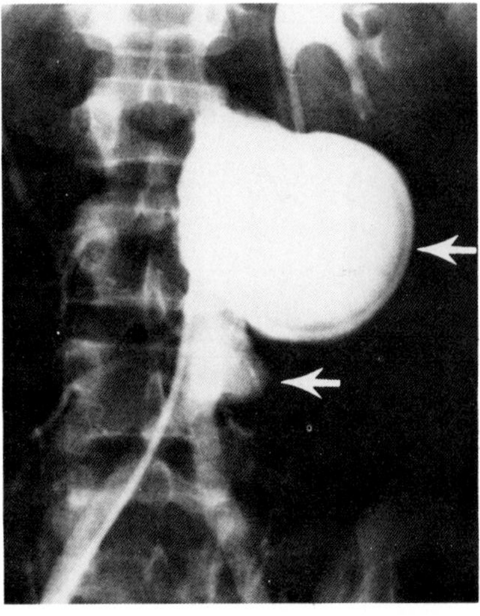

A

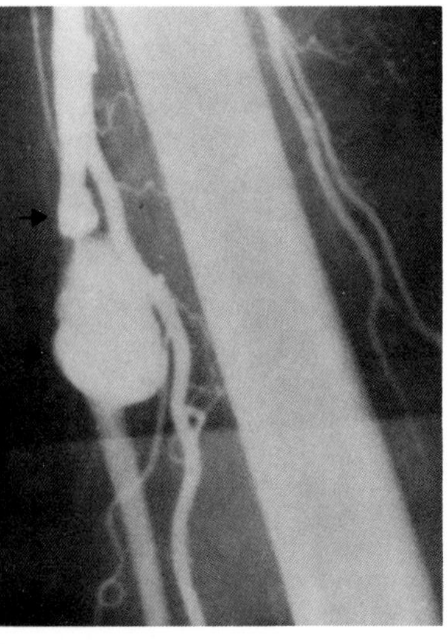

B

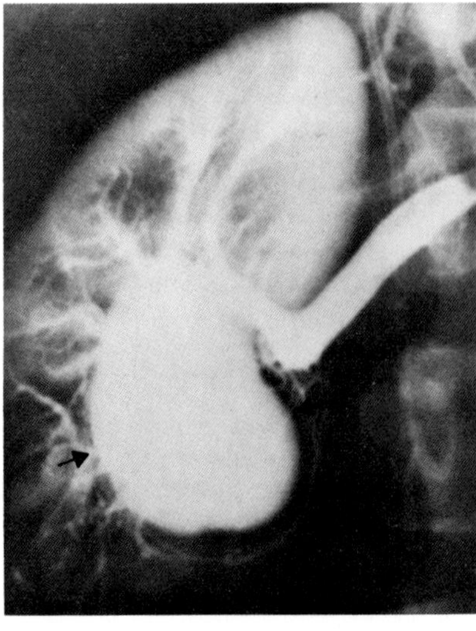

C

**Figure 3.41. Ehlers-Danlos syndrome.** Multiple aneurysms in the Ehlers-Danlos syndrome. (A) Two separate saccular aneurysms in the abdominal aorta. (B) Two saccular aneurysms in the left brachial artery. (C) A large right renal artery aneurysm. (From Mirza, Smith, and Lim,[41] with permission from the publisher.)

## Osteogenesis Imperfecta

This condition is discussed in the section "Other Dysplasias" in Chapter 61.

## Pseudoxanthoma Elasticum

Pseudoxanthoma elasticum is a hereditary systemic disorder characterized by widespread degeneration of elastic fibers that results in cutaneous, ocular, and vascular manifestations in children and young adults. The name derives from the characteristic yellowish xanthoma-like skin lesions found in virtually all patients. Angioid streaks in the fundi, and retinal hemorrhage and scarring, can cause significant visual loss.

Most patients with pseudoxanthoma elasticum have some combination of peripheral, cardiac, mesenteric, or cerebrovascular abnormalities. Intermittent claudication, early-onset hypertension or coronary artery disease, and severe upper gastrointestinal bleeding are common manifestations of the disease. Arteriography may demonstrate occlusive vascular changes and angiomatous malformations, highly suggestive findings in a young patient with recurrent abdominal pain or repeated gastrointestinal bleeding.

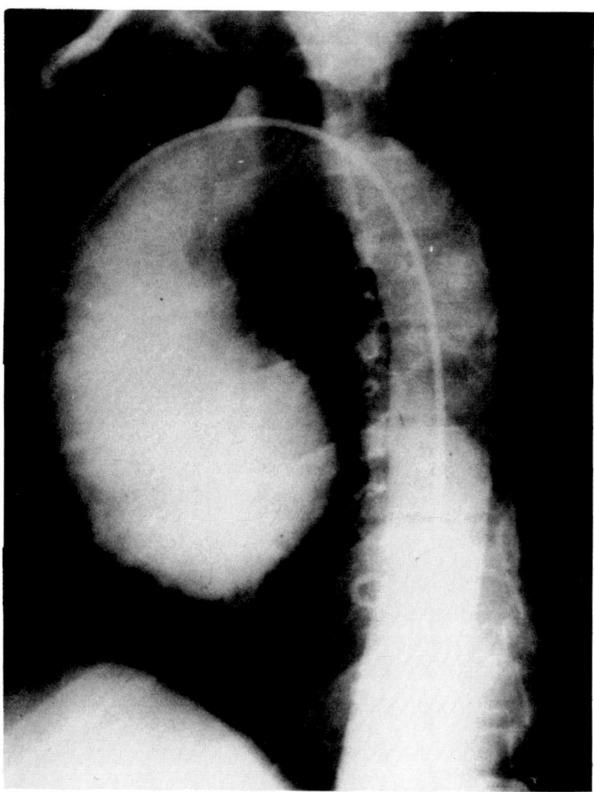

**Figure 3.43. Marfan's syndrome.** An arteriogram shows enormous dilatation of the aneurysmal ascending aorta. (From Ovenfors and Godwin,[42] with permission from the publisher.)

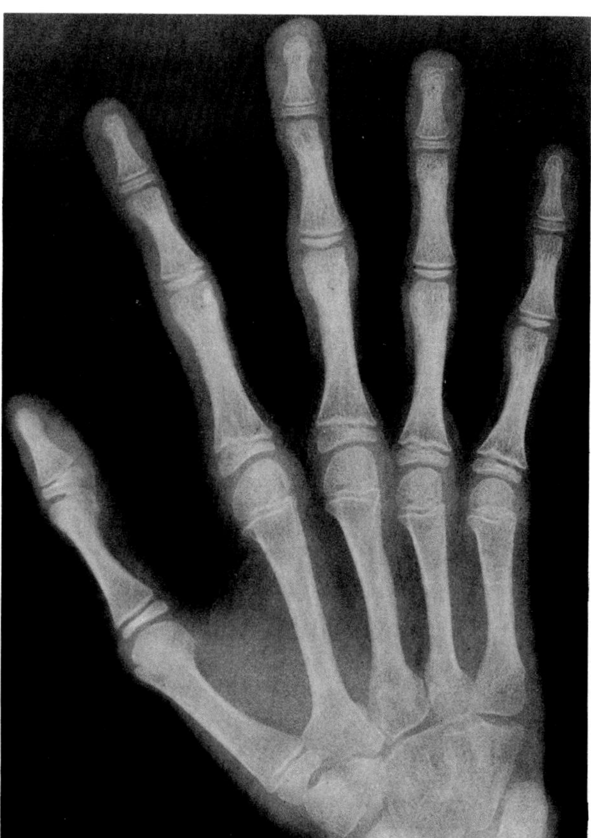

**Figure 3.42. Arachnodactyly in Marfan's syndrome.** The metacarpals and phalanges are unusually long and slender.

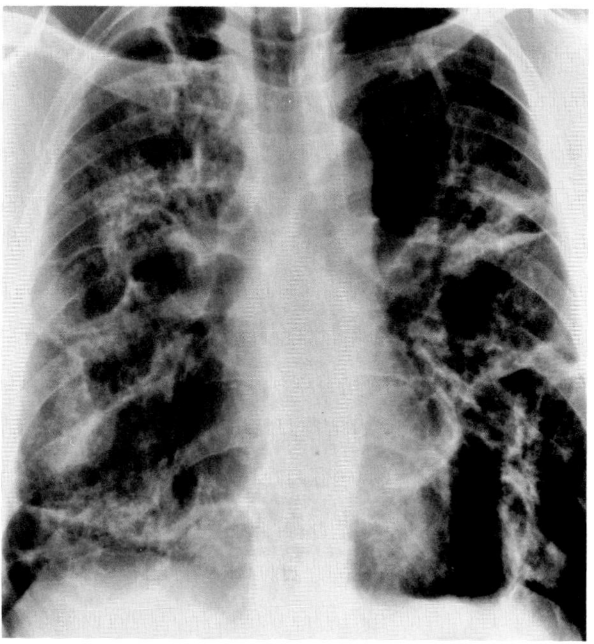

**Figure 3.44. Marfan's syndrome.** Severe pulmonary emphysema and interstitial fibrosis with diffuse bullous formation.

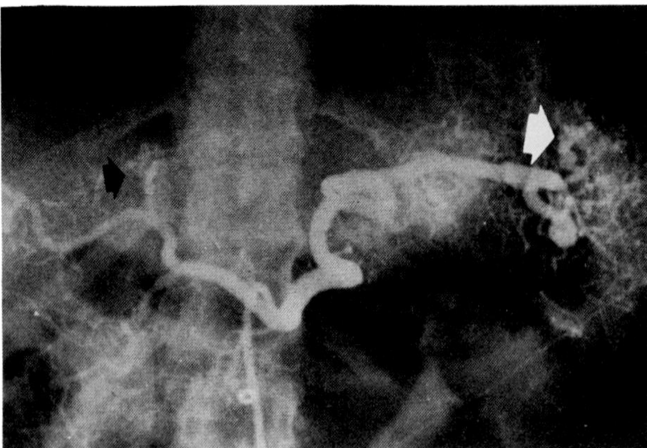

**Figure 3.45. Pseudoxanthoma elasticum.** A selective celiac arteriogram in a patient with abdominal pain and gastrointestinal bleeding demonstrates angiomatous malformations in the spleen and liver (arrows). (From Bardsley and Koehler,[35] with permission from the publisher.)

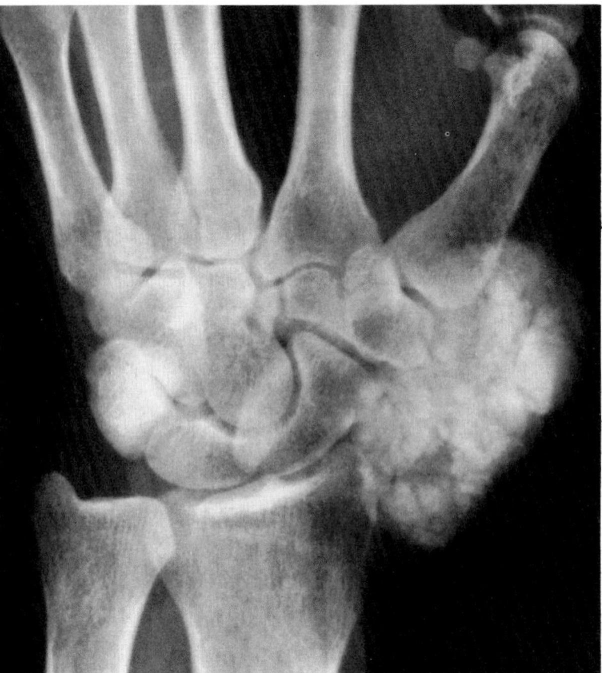

**Figure 3.46. Pseudoxanthoma elasticum.** Extensive calcification in soft tissues on the radial side of the wrist.

Arterial calcification, both intimal and medial, is commonly present. Unusual forms of calcification in this condition include linear calcification of the dermal layer and extensive collections of amorphous soft tissue or periarticular deposits.[34,35]

## REFERENCES


1. Murdock MM, Holman GH: Roentgenologic bone changes in phenylketonuria. *Am J Dis Child* 107:523–527, 1964.
2. Feinberg SB, Fisch RO: Roentgen findings in growing long bones in phenylketonuria. *Radiology* 78:394–398, 1962.
3. Morreels CL, Fletcher BD, Weilbaecher RG, et al: The roentgenographic features of homocystinuria. *Radiology* 90: 1150–1158, 1968.
4. Brill PW, Mitty HA, Gaull GE: Homocystinuria due to cystathionine synthase deficiency: Clinical-roentgenographic correlations. *AJR* 121:45–54, 1974.
5. Chikos PM, McDonald GB: Regional enteritis complicated by nephrocalcinosis and nephrolithiasis. *Radiology* 121:75–76, 1976.
6. Carsen GM, Radkowski MA: Calcium oxalosis: A case report. *Radiology* 113:165–166, 1974.
7. Weber AL: Primary hyperoxaluria. Roentgenographic, clinical and pathological findings. *AJR* 100:155–161, 1967.
8. Ortiz AC, Neal EG: Alkaptonuria and ochronotic arthritis. *Clin Orthop* 25:147–155, 1962.
9. Lasker FJ, Sargison KD: Ochronotic arthropathy. *J Bone Joint Surg* 52B:653–666, 1970.
10. Ward FR, Engelbrecht PJ: Alkaptonuria and ochronosis. *Clin Radiol* 14:170–174, 1963.
11. Crawhall JC, Watts RWE: Cystinuria. *Am J Med* 45:736–755, 1968.
12. Watt I, Middlemiss H: The radiology of gout. *Clin Radiol* 26:27–36, 1975.
13. Martel W: The overhanging margin of bone. A roentgenologic manifestation of gout. *Radiology* 91:755–756, 1968.
14. Nyhan WC: Clinical features of the Lesch-Nyhan syndrome. *Arch Intern Med* 130:186–192, 1972.
15. Becker MH, Wallin JK: Congenital hyperuricosuria. *Radiol Clin North Am* 6:239–243, 1968.
16. Seegmiller JE: Xanthine stone formation. *Am J Med* 45:780–783, 1968.
17. Smith WL, Quattromani F: Radiodense liver in transfusion hemochromatosis. *AJR* 128:316–317, 1977.
18. Jensen PE: Hemochromatosis: A disease often silent but not invisible. *AJR* 126:343–351, 1976.
19. Twersky J: Joint changes in idiopathic hemochromatosis. *AJR* 124:139–144, 1975.
20. Mindelzun R, Elkin M, Scheinberg I, et al: Skeletal changes in Wilson's disease. *Radiology* 94:127–132, 1970.
21. Aksoy M, Camli N, Dincol K, et al: Osseous changes in Wilson's disease. *Radiology* 102:505–510, 1972.
22. Furste W, Ayres PR: Acute intermittent porphyria. *Arch Surg* 72:426–430, 1956.
23. Preger L, Sanders GW, Gold RH, et al: Roentgenographic skeletal changes in the glycogen storage diseases. *AJR* 107:840–847, 1969.
24. Nihill MR, Wilson DS, Hugh-Jones K: Generalized glycogenosis type II (Pompe's disease). *Arch Dis Child* 45:122–129, 1970.
25. Greenfield GB: Bone changes in chronic adult Gaucher's disease. *AJR* 110:800–807, 1970.
26. Levin B: Gaucher's disease. *AJR* 85:685–696, 1961.
27. Lachman R, Crocker A, Schulman J, et al: Radiologic findings in Niemann-Pick disease. *Radiology* 108:659–664, 1973.
28. Horrigan DW, Baker DH: Gargoylism: Review of roentgen skull changes with description of new findings. *AJR* 86:473–477, 1961.
29. Melhem R, Dorst JP, Scott CI, et al: Roentgen findings in mucolipidosis III (pseudo-Hurler polydystrophy). *Radiology* 106:153–160, 1973.
30. Schenk EA, Haggerty J: Morquio's disease. A radiologic and morphologic study. *Pediatrics* 34:839–850, 1964.
31. Langer LO, Carey LS: The roentgenographic features of the KS mucopolysaccharidosis of Morquio. *AJR* 97:1–20, 1966.
32. Beighton P, Thomas ML: Radiology of the Ehler–Danlos syndrome. *Clin Radiol* 20:354–361, 1969.

33. Brenton DP, Dow CJ: Homocystinuria and Marfan's syndrome. *J Bone Joint Surg* 54B:277–298, 1972.
34. James AE, Eaton B, Blazek JV, et al: Roentgen findings in pseudoxanthoma elasticum. *AJR* 106:642–647, 1969.
35. Bardsley JL, Koehler PR: Pseudoxanthoma elasticum: Angiographic manifestation in abdominal vessels. *Radiology* 93: 559–562, 1969.
36. Woodring JH, Rosenbaum HD: Bone changes in phenylketonuria reassessed. *AJR* 137:241–243, 1981.
37. Thomas PS, Carson NAJ: Homocystinuria. *Ann Radiol* 21:95–104, 1978.
38. Eisenberg RL: *Gastrointestinal Radiology: A Pattern Approach*. Philadelphia, JB Lippincott, 1983.
39. Dalinka MK, Reginato AJ, Golden DA: Calcium deposition diseases. *Semin Roentgenol* 17:39–48, 1982.
40. Miller JH, Stanley P, Gates GF: Radiology of glycogen storage diseases. *AJR* 132:379–387, 1979.
41. Mirza FH, Smith PL, Lim WN: Multiple aneurysms in a patient with Ehlers-Danlos syndrome. *AJR* 132:993–995, 1979.
42. Ovenfors CO, Godwin JD: Aortic aneurysms and dissections. In Eisenberg RL, Amberg JR (eds): *Critical Diagnostic Pathways in Radiology: An Algorithmic Approach*. Philadelphia, JB Lippincott, 1981.

## Chapter Four
# Endocrine Diseases

## DISEASES OF THE PITUITARY

### Hypopituitarism

Because the pituitary gland controls the level of secretion of gonadal and thyroid hormones as well as the production of growth hormone, decreased function of the pituitary causes profound generalized disturbances in bone growth and maturation. In children, hypopituitarism typically leads to the Lévi-Lorain type of dwarfism, in which the delayed appearance of epiphyseal centers causes the failure of bones to grow normally in length or width. This results in a person who is small in stature and sexually immature though well-proportioned and of normal mentality. In spite of the short stature of these patients, epiphyseal closure is greatly delayed, sometimes up to the fifth decade. The lack of marginal epiphyses of the vertebrae may result in a relative platyspondlysis. In many patients, there is a delay in the eruption of the teeth, which tend to become impacted because their size is not affected. The arrest in the growth of the skeleton occurs during childhood, when the cranial vault is proportionately greater in relation to the facial bones than in the adult. Since this discrepancy remains into adulthood in hypopituitary dwarfism, the relatively large skull may be mistakenly thought to result from hydrocephaly. Although the sella turcica is often small in childhood hypopituitarism, an underlying pituitary tumor can cause sellar enlargement.

Hypopituitarism occurring after adolescence usually results in few radiologic findings. The heart and kidneys are often small, and calcification or ossification may develop in the articular cartilages.[1,2]

### Hyperpituitarism

Hyperpituitarism results from an excess of growth hormone produced by an eosinophilic adenoma or by generalized hyperplasia of the eosinophilic cells of the anterior lobe of the pituitary gland. The development of this condition before enchondral bone growth has ceased results in gigantism; hyperpituitarism beginning after bone growth has stopped produces acromegaly.

Generalized overgrowth of all the body tissues is the underlying abnormality in acromegaly. Proliferation of cartilage may cause joint space widening, especially of the metacarpophalangeal and hip joints. The slight increase in length of each of the seven articular cartilages for each digit leads to perceptible lengthening of the fingers. Overgrowth of the tips of the distal phalanges produces thick bony tufts with pointed lateral margins. The concomitant hypertrophy of the soft tissues produces the characteristic square, spade-shaped hand of acromegaly. Degenerative changes develop early and are associated with prominent hypertrophic spurring; unlike typical osteoarthritis, acromegaly results in joint spaces that remain normal or are even widened. The tubular

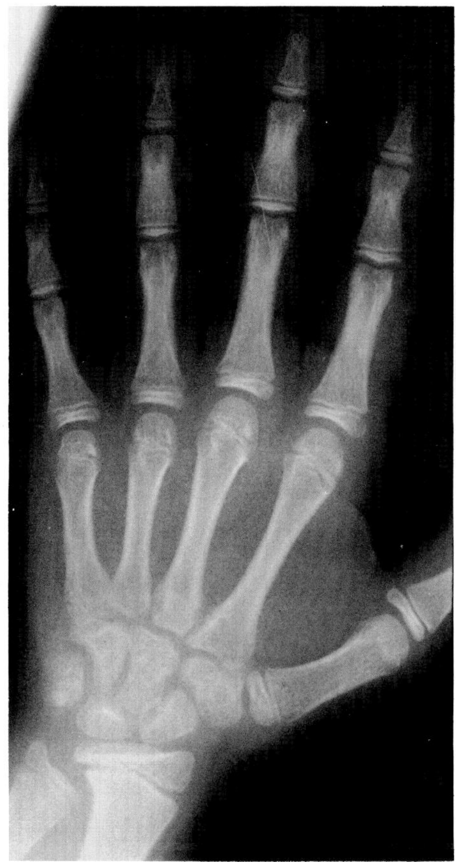

A

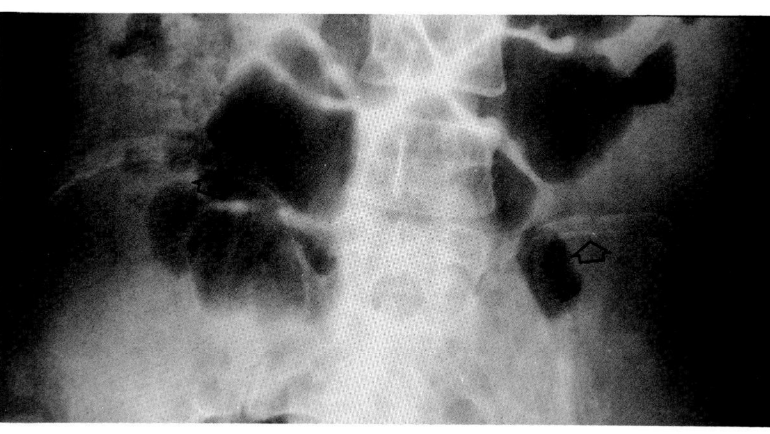

B

**Figure 4.1. Retarded bone age and delayed epiphyseal closure in hypopituitarism.** (A) A frontal view of the hand of a 15-year-old girl shows a radiographic bone age of $9\frac{1}{2}$ years. (B) A frontal view of the pelvis of a 39-year-old man shows persistence of the iliac crest ossification centers (arrows), which normally fuse by 30 years of age.

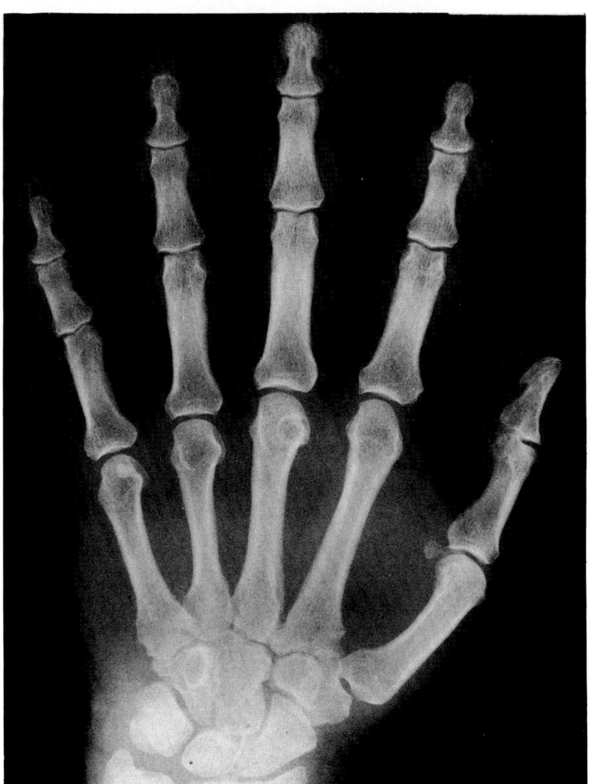

**Figure 4.2. Acromegaly.** A frontal view of the hand shows widening of the metacarpophalangeal joints, thickening of the soft tissues of the fingers, and overgrowth of the tufts of the distal phalanges (arrows).

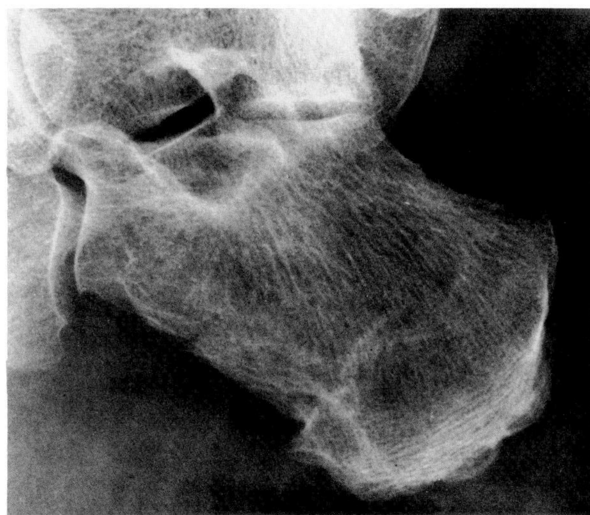

**Figure 4.3. Acromegaly.** Prominent thickening of the heel pad (32 mm on the original radiograph).

bones have a coarsened trabecular pattern; cystic changes often occur, especially in the carpal bones and femoral trochanters.

Thickening of the heel pads (the soft tissue inferior to the plantar aspect of the calcaneus) to greater than 23 mm is highly suggestive of acromegaly. However, a similar appearance may also be seen in patients with obesity, myxedema, thyroid acropachy, or generalized edema.

The bones of the skull become thickened and have increased density, often with obliteration of the diploic space. This bone thickening is especially prominent in the frontal and occipital regions, leading to characteristic frontal bossing as well as enlargement of the occipital protuberance. The paranasal sinuses (especially the frontal) become enlarged, and the mastoids are usually overpneumatized. Lengthening of the mandible and an increased mandibular angle produce the prognathous jaw, one of the typical clinical features of acromegaly. The radiographic appearance of enlargement and pressure erosion of the sella turcica due to a pituitary adenoma is discussed in the following section.

In the spine, hypertrophy of cartilage causes an increased width of the intervertebral disk spaces. Appositional bone growth results in an increase in the size of the vertebral bodies, best seen on lateral projections. Hypertrophy of soft tissues may produce an increased concavity of the posterior aspect of the vertebral bodies (scalloping); this is most prominent in the lumbar spine.

The increased growth of bone at the costochondral junctions lengthens and thickens the ribs. In the pelvis, prominent spurring may cause typical bony beaking of the symphysis pubis.

Extraskeletal manifestations of acromegaly include visceral enlargement, especially of the heart and kidney, enlargement of the tongue (which may narrow the pharyngeal airway and cause characteristic outward tilting of the teeth), and calcification of cartilage in the pinna of the ear.

Gigantism is manifested as an excessively large skeleton. If hypersecretion of growth hormone continues after epiphyseal closure, the soft tissue and bony changes of acromegaly are superimposed.[2-4]

## Pituitary Adenoma

Pituitary adenomas, almost all of which arise in the anterior lobe, constitute more than 10 percent of all intracranial tumors. Symptoms of pituitary tumors may result from a mass effect causing compression of parasellar structures or from an alteration in pituitary trophic hormone production (increased levels with secreting adenomas; decreased levels caused by the compression of the pituitary gland by a nonsecreting tumor).

Thin-section computed tomography (CT) is the examination of choice in evaluating a patient with a suspected pituitary tumor. After the intravenous administration of contrast material, large pituitary tumors are typically homogeneous and hyperdense with respect to surrounding brain tissue. Areas of necrosis or cyst formation within the tumor cause internal areas of low density. Most pituitary microadenomas are of lower density than the normal pituitary gland. CT can also demonstrate adjacent bone erosion, the extension of the tumor be-

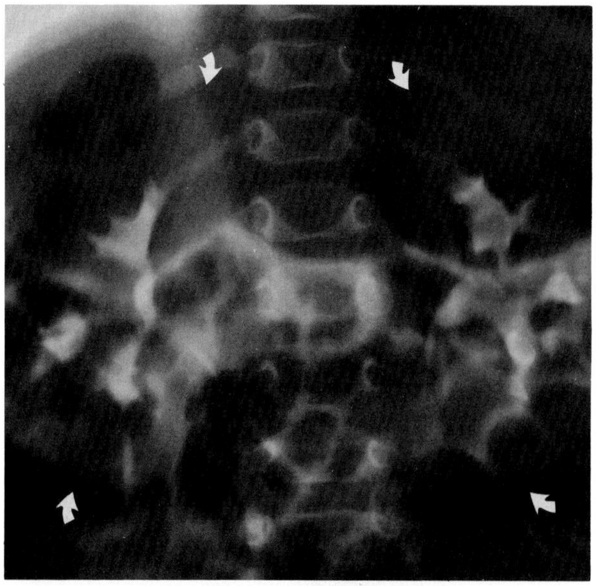

**Figure 4.4. Gigantism.** Striking bilateral enlargement of well-functioning kidneys. The arrows point to the superior and inferior margins of the kidneys, both of which are about five vertebral bodies in length.

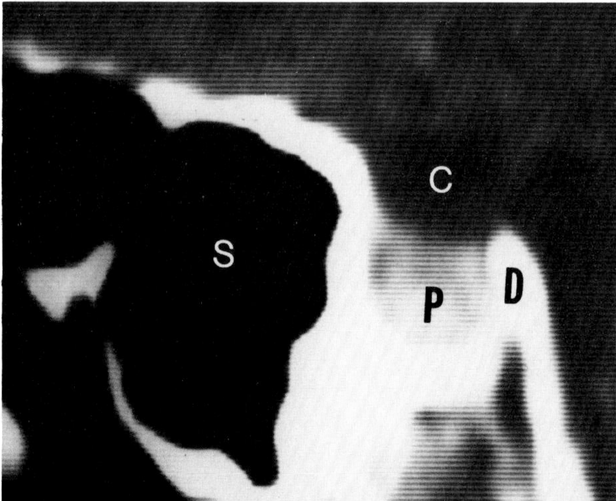


**Figure 4.5. Pituitary adenoma.** A sagittal CT reformation shows a homogeneous normal pituitary gland (P) in a 35-year-old woman with Cushing's disease. The air-filled sphenoid sinus (S) is anterior, the dorsum sellae (D) is posterior, and the cerebrospinal fluid–filled suprasellar cistern (C) is superior. At surgery, however, a 2-mm adenoma was removed. This case stresses the fact that a normal CT of the sella does not exclude a small functioning pituitary tumor.

yond the confines of the sella, and the impression of nearby structures such as the third ventricle, optic nerves, or optic chiasm.

Although plain skull radiographs can show enlargement of the sella turcica, erosion of the dorsum sellae, and a double floor resulting from the unequal downward growth of the mass, this imaging modality now is of value only in the incidental detection of sellar enlargement on films taken for other purposes.

Thin-section pluridirectional tomography was until recently the best procedure for demonstrating pituitary microadenomas. These tumors, which are less than 1 cm in diameter, cause hypersecretion of pituitary hormones but are too small to compress nearby structures. Focal areas of expansion or bone thinning producing asymmetry, usually at the anteroinferior aspect of the sellar floor, were considered indicative of the presence and location of the tumor. However, the recent reports of a significant number of false-positive and false-negative results with this technique and the high accuracy of high-resolution CT have virtually eliminated the use of pluridirectional tomography in cases of suspected pituitary microadenomas.

**Figure 4.6. Pituitary prolactinoma.** A coronal CT reformation shows a low-density microadenoma (arrows) within the pituitary gland in a 29-year-old woman with amenorrhea, galactorrhea, and prolactinemia. As in this patient, most microadenomas are of lower density than the pituitary gland.

Cerebral arteriography and digital subtraction angiography can reliably demonstrate arterial abnormalities that can complicate pituitary surgery. Therefore, these imaging modalities are often used in the preoperative evaluation of pituitary adenoma to detect incidental aneurysms or congenital abnormalities of the internal carotid artery or to show arterial encasement by the tumor.[5-8]

## Empty Sella Syndrome

The downward extension of the subarachnoid space through a widened or absent diaphragm of the sella produces sellar enlargement that may suggest an underlying pituitary tumor. Pulsations of cerebrospinal fluid cause remodeling and symmetric expansion of the sella. Occasionally, an empty sella may be associated with the spontaneous or post-therapy involution of a pituitary adenoma. In such cases, the sellar enlargement is greater and often asymmetric.

The empty sella usually is an incidental finding on

**Figure 4.7. Nonfunctioning pituitary macroadenoma.** A sagittal CT reformation in a 60-year-old man with a progressive bitemporal visual field defect shows a pituitary tumor (arrows) growing superiorly into the suprasellar cistern. The visual defects were due to compression of the optic chiasm. Although the adenoma caused a significant mass effect, it did not cause any abnormality in the endocrine function of the gland.

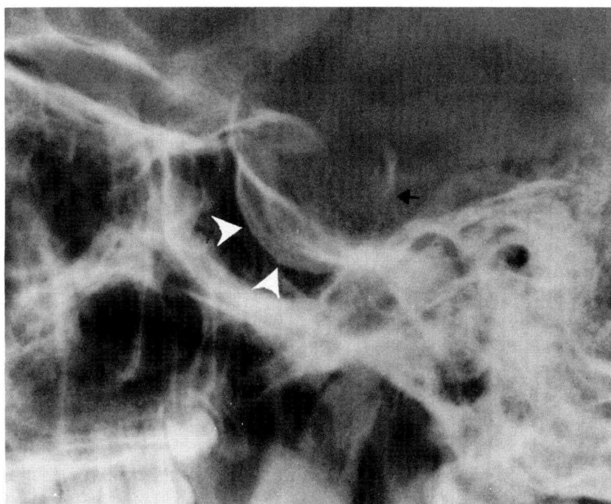

**Figure 4.8. Chromophobe adenoma causing hypopituitarism.** A plain lateral radiograph of the skull shows ballooning of the sella turcica with downward displacement of the floor (arrowheads) into the posterior portion of the sphenoid sinus. Note also the thinning and erosion of the dorsum sellae (arrow) by the intrasellar tumor.

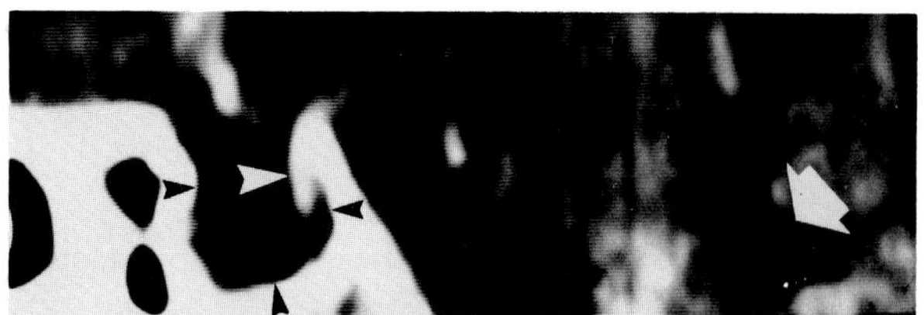

**Figure 4.9. Empty sella.** A sagittal CT reformation shows the sella filled with cerebrospinal fluid (black arrows) that is of the identical low density as the fluid within the fourth ventricle (large white arrow). The enhancing structure in the posterior part of the sella (white arrowhead) represents the pituitary stalk (infundibulum).

plain skull radiographs or CT. Infrequently, prolapse of the optic chiasm produces visual impairment. Plain films show lobular enlargement of the sella with thinning and bowing of the dorsum, an appearance that mimics that of an intrasellar tumor. Although in the past pneumoencephalography was required to demonstrate the presence of subarachnoid air within the empty sella, the diagnosis can now be made easily by the CT demonstration of a rounded low-density area occupying the sella. CT following the cisternal administration of metrizamide may occasionally be required to determine whether the hypodense region within the sella is due to an extension of the subarachnoid cistern or a necrotic tumor.[9–12]

### Diabetes Insipidus

The impaired renal conservation of water in diabetes insipidus results from low blood levels of antidiuretic hormone; this reflects deficient vasopressin release by the neurohypophysis in response to normal physiologic stimuli. Severe polyuria can lead to massive dilatation of the renal pelves, calyces, and ureters, and this probably represents a compensatory alteration to accommodate the huge volume of excreted urine.[13,14]

## DISEASES OF THE THYROID

### Radioactive Iodine Scanning

Scanning following the administration of radioactive iodine is the major imaging modality for demonstrating both functioning and nonfunctioning thyroid tissue. This technique is used to localize palpable nodules, to determine the function of nodules, to detect nonpalpable lesions (especially in patients with a history of neck irradiation), and to evaluate the extent of residual tissue after surgical or radioisotopic thyroid ablation. Radionuclide scanning may be combined with uptake, stimulation, or suppression techniques to better characterize thyroid lesions.

The normal thyroid gland tends to be butterfly-shaped with an isthmus connecting the lower portions of the two lobes. The distribution of radioactivity is uniform throughout the normal gland. Diffuse thyroid enlargement without hyperthyroidism most frequently represents a multinodular goiter or Hashimoto's thyroiditis. In the patient with hyperthyroidism, diffuse thyroid enlargement suggests Graves' disease.

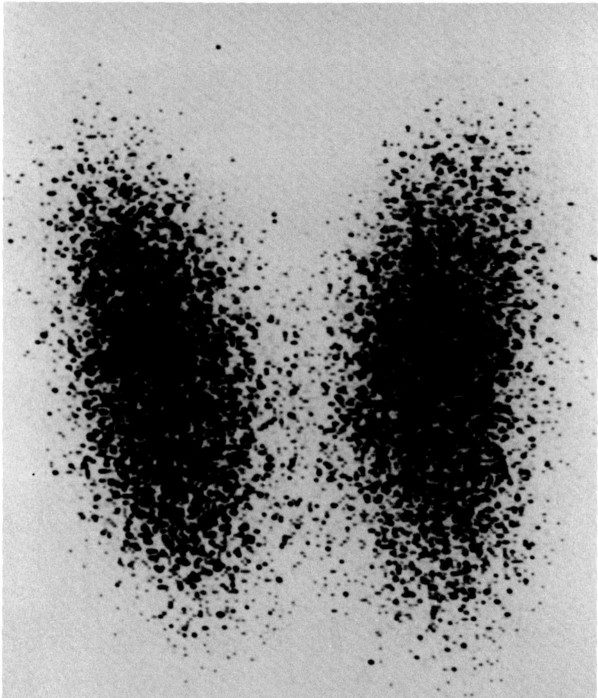

**Figure 4.10. Normal radioactive iodine scan.** Uniform distribution of isotope activity throughout the gland.

Masses within the thyroid appear as hyperfunctioning ("hot") or poorly functioning ("cold") nodules. A hot nodule demonstrates increased radionuclide uptake compared with surrounding thyroid tissue. These hot nodules usually represent autonomously functioning thyroid tissue and are rarely malignant, although malignancy is sometimes reported elsewhere in the same gland. Large hyperfunctioning nodules may completely suppress the remaining thyroid tissue so that only the nodule itself is visualized. An autonomous nodule may eventually develop central hemorrhage or cystic change and evolve into a nonfunctioning (cold) nodule.

Cold thyroid nodules contain less isotope per unit tissue mass than adjacent normal thyroid tissue. Most cold nodules represent poorly functioning adenomas. Thyroid cysts, carcinoma, thyroiditis, and involutional areas of antecedent hyperplasia can also produce this appearance. In young patients, the absence of uptake in the region of a solitary cold thyroid nodule is associated with a 10 to 25 percent probability that the nodule is malignant. Although ultrasound can distinguish a thyroid cyst from the other causes of cold nodules, the modality is of little value in further characterizing noncystic thyroid masses.[15–17]

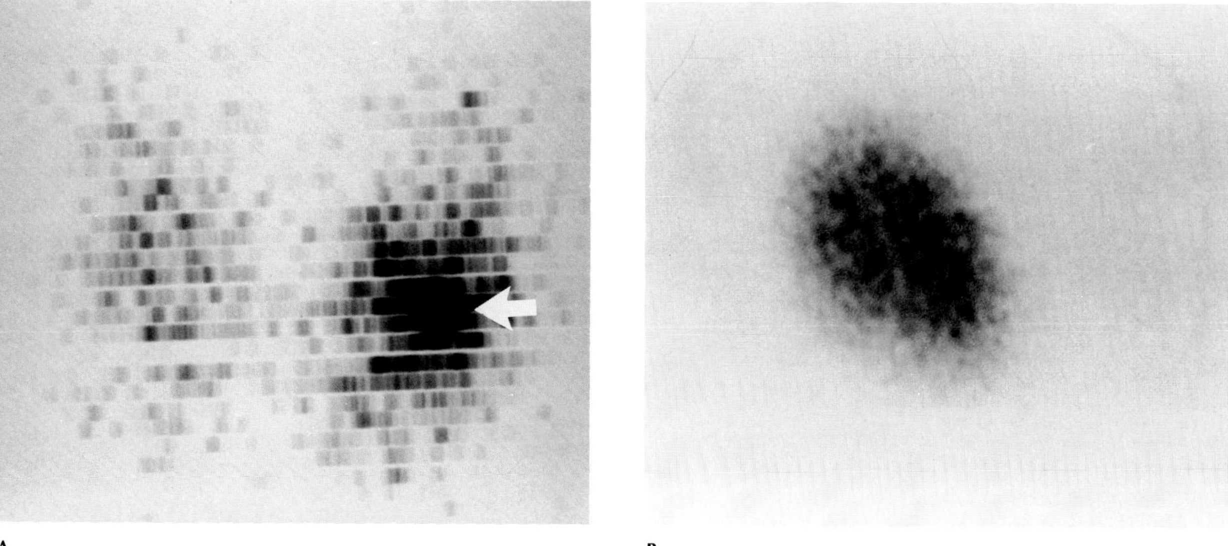

Figure 4.11. **Hyperfunctioning (hot) nodules.** (A) The localized area of increased radionuclide uptake on the left (arrow) has partially suppressed radionuclide uptake in the remainder of the left lobe and in the entire right lobe of the gland. (B) In this patient, the large hot nodule on the right has completely suppressed radionuclide uptake in the nonvisualized left lobe of the thyroid.

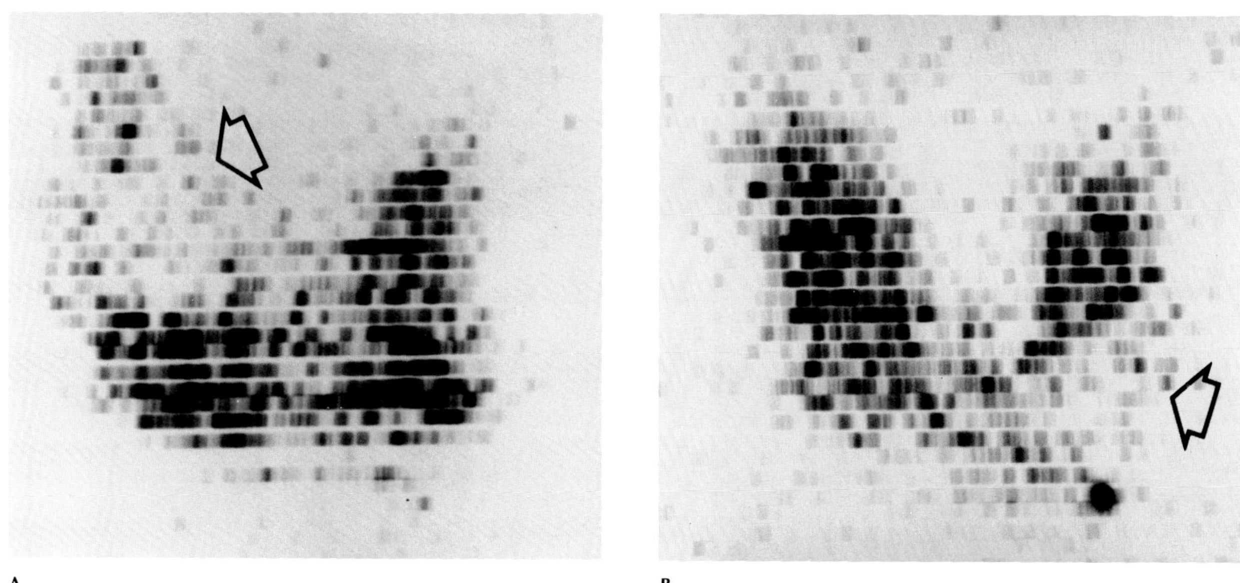

Figure 4.12. **Nonfunctioning (cold) nodules.** (A) Carcinoma in the right lobe of the thyroid (arrow). (B) Adenoma in the left lobe of the thyroid (arrow).

## Hypothyroidism

Hypothyroidism can result from any structural or functional abnormality that leads to an insufficient synthesis of thyroid hormone. Hypothyroidism dating from birth (cretinism) results in multiple developmental abnormalities. When hypothyroidism begins in adulthood, the accumulation of hydrophilic mucopolysaccharides in the ground substance of the dermis and other tissues leads to thickening of the facial features and doughy induration of the skin (myxedema).

Children with cretinism typically have a short stature; coarse features with a protruding tongue, a broad, flat nose, and widely set eyes; sparse hair; dry skin; and a protuberant abdomen with an umbilical hernia. The major radiographic abnormalities include a delay in the appearance and subsequent growth of ossification centers, epiphyseal dysgenesis (fragmented epiphyses with multiple ossification foci), and retarded bone age. Skull changes are common and include an increase in the thickness of the cranial vault, underpneumatization of the sinuses and mastoids, widened sutures with delayed

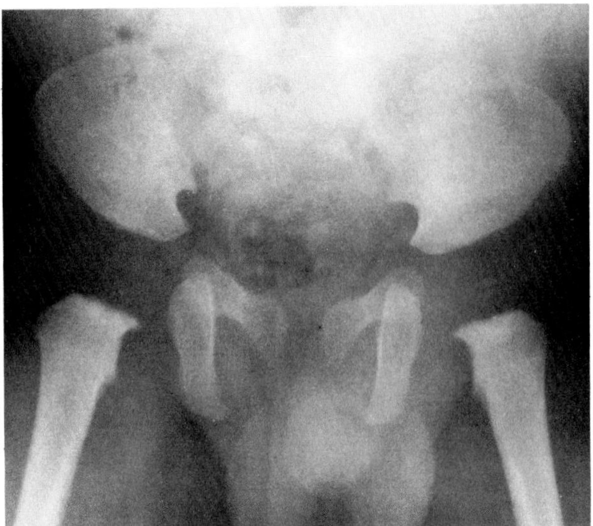

**Figure 4.13. Cretinism.** A 17-month-old child with a complete absence of ossification centers about the hip. No metabolic disorder other than hypothyroidism causes so severe a delay in enchondral ossification.

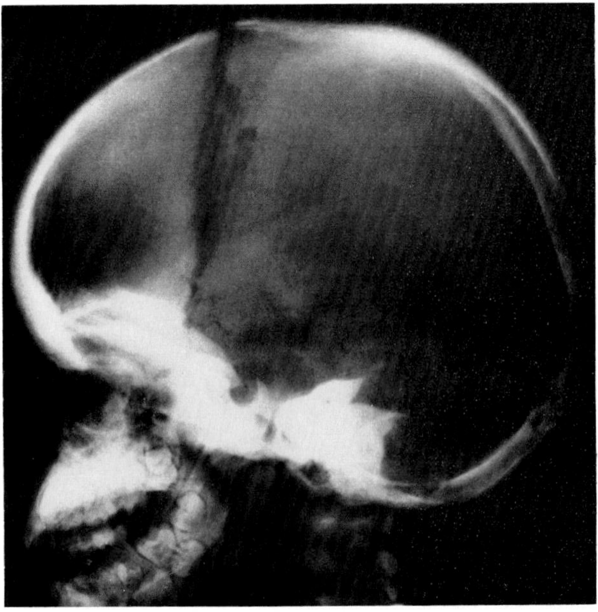

**Figure 4.14. Cretinism.** A lateral view of the skull shows an increased density at the base, small, underdeveloped sinuses, and hypoplasia of the teeth with delayed eruption. The retardation of facial maturation makes the face appear small relative to the size of the calvarium.

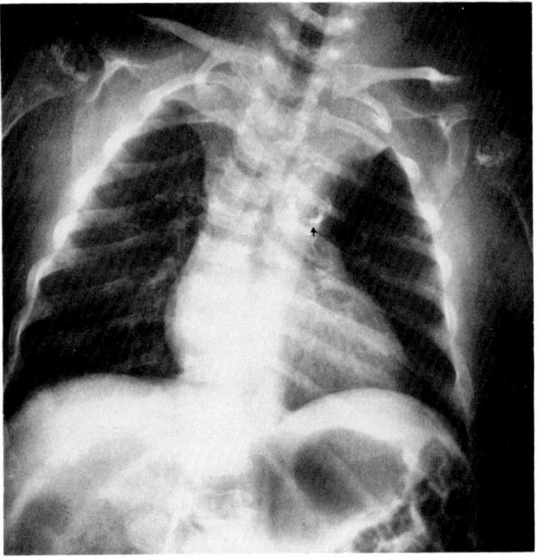

A

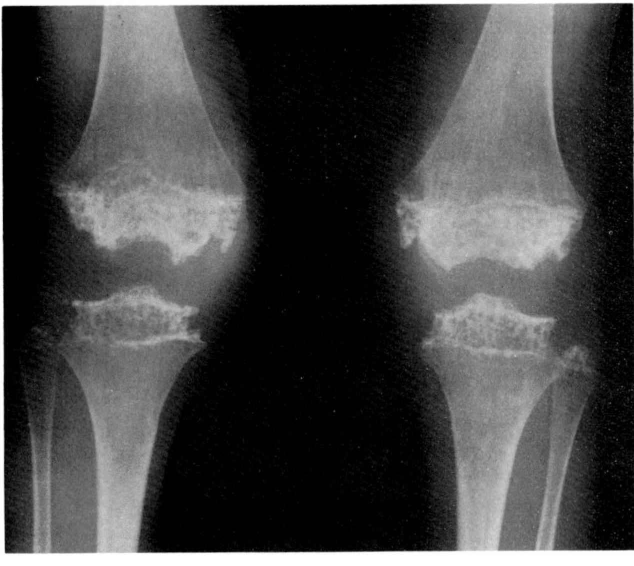

B

**Figure 4.15. Cretinism.** (A) A frontal view of the chest demonstrates scoliosis, thinning of the ribs, fragmentation of the epiphyseal centers of the humeral head with spotty calcification, and premature calcification of the aorta (arrow). (B) The epiphyses about the knee have uneven density and irregular margins.

closure, and a delay in the development and eruption of the teeth. In many adult cretins, little or no residual skeletal deformity persists, even though insufficient or no treatment was given.

Adult hypothyroidism has an insidious onset with nonspecific symptoms including lethargy, constipation, cold intolerance, slowing of intellectual and motor activity, and weight gain in spite of a decreased appetite. Dry skin, stiff aching muscles, and a deepening voice with hoarseness often occur. Radiographically, the heart is typically enlarged because of pericardial effusion. Soft tissue thickening is often seen on films of the extremities, and adynamic ileus is a common finding on abdominal views.[18,19]

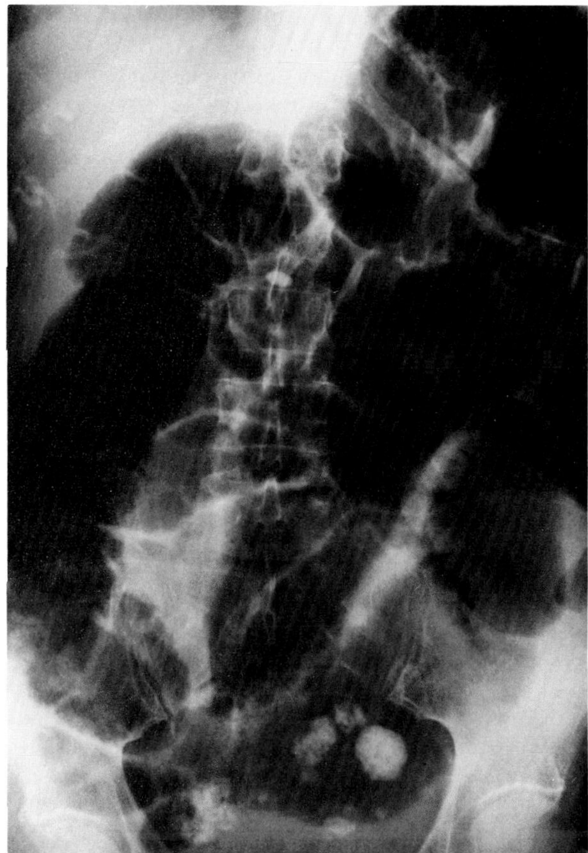

**Figure 4.16. Hypothyroidism.** Adynamic ileus with diffuse dilatation of gas-filled loops of large and small bowel.

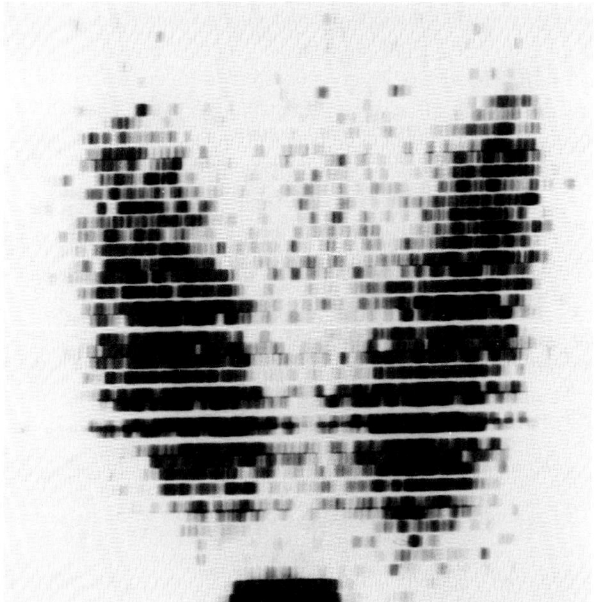

**Figure 4.17. Hyperthyroidism.** A radioactive iodine scan in a patient with Graves' disease demonstrates diffuse enlargement of the thyroid gland with prominence of the pyramidal lobe and a homogeneous increase in isotope uptake.

## Hyperthyroidism

Hyperthyroidism results from the excessive production of thyroid hormone, either from the entire gland (Graves' disease) or from one or more functioning adenomas (Plummer's syndrome). Graves' disease is a relatively common disorder that most often develops in the third and fourth decades and has a strong female predominance. The major clinical symptoms include nervousness, emotional lability, an inability to sleep, tremors, excessive sweating, and heat intolerance. Weight loss is frequent, usually despite an increased appetite.

Radiographically, radioactive iodine scans in patients with Graves' disease typically demonstrate diffuse enlargement of the thyroid gland, often with prominence of the pyramidal lobe. The radioiodine uptake is diffusely and homogeneously increased. Thyrotoxicosis can cause high-output cardiac failure along with generalized cardiomegaly and pulmonary congestion. In children, hyperthyroidism accelerates the appearance and growth of secondary ossification centers and is often associated with osteoporosis. Unilateral or bilateral exophthalmos secondary to Graves' disease can be demonstrated by CT as thickening of the extraocular muscles.[20]

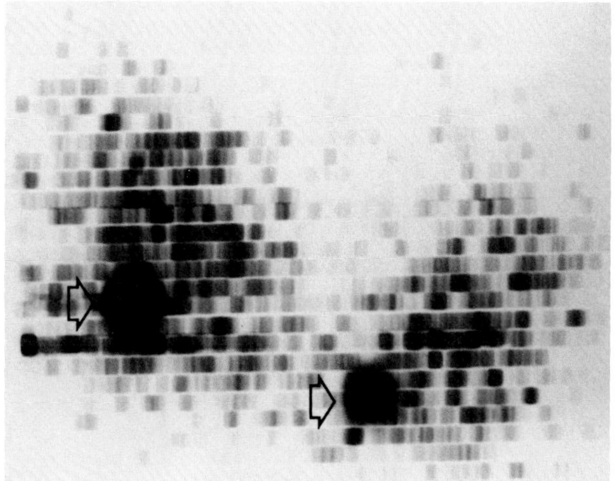

**Figure 4.18. Hyperthyroidism.** A radioactive iodine scan in a patient with Plummer's syndrome demonstrates two small hot nodules (arrows) in a normal-sized gland.

#### THYROID ACROPACHY

Thyroid acropachy is a rare complication of hyperthyroid disease characterized by progressive exophthalmos, relatively asymptomatic swelling of the hands and feet, clubbing of the digits, pretibial myxedema, and periosteal new bone formation. Almost all cases of thyroid acropachy develop following therapy for the primary hyperthyroid condition; thus most patients are either euthyroid or hypothyroid when symptoms develop. Radiographically, there is diffuse soft tissue swelling along with clubbing of the digits. Spiculated periosteal new bone formation tends to involve the midportions of the diaphyses of the phalanges and metacarpals, causing a fusiform contour that is often asymmetric. Multiple small radiolucencies within the irregular periosteal new bone may produce a bubbly or lacy appearance.[21,22]

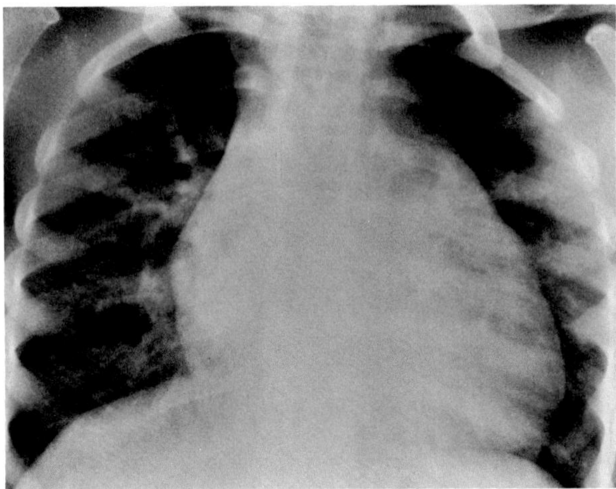

**Figure 4.19. Hyperthyroidism.** The generalized enlargement of the heart and engorged pulmonary vascularity reflect high-output cardiac failure.

## Goiter

A goiter is an enlargement of the thyroid gland that does not result from an inflammatory or neoplastic process and is not initially associated with thyrotoxicosis or myxedema. Simple (nontoxic) goiter results when one or more factors impair the capacity of the thyroid gland in the basal state to secrete the quantities of active hormones necessary to meet the needs of the peripheral tissues. This eventually results in a sufficient increase in both the functioning thyroid mass and the cellular activity to overcome the mild or moderate impairment of hormone synthesis, permitting the patient to remain metabolically normal though goitrous. On a radioactive iodine scan, a nontoxic goiter usually appears as symmetric or asymmetric enlargement of the thyroid gland. Plain radiographs and esophograms often show the enlarged thyroid gland impressing or displacing the trachea and esophagus.

Toxic multinodular goiter is a not infrequent consequence of long-standing nontoxic goiter. In this condition, one or more areas of the gland become independent from thyrotropic hormone (TSH) stimulation. Radioactive iodine scans most commonly show accumulation of iodine diffusely but in patchy foci throughout the gland. Another pattern consists of iodine accumulation in one or more discrete nodules within the gland, with the remainder of the gland being essentially nonfunctional.[23]

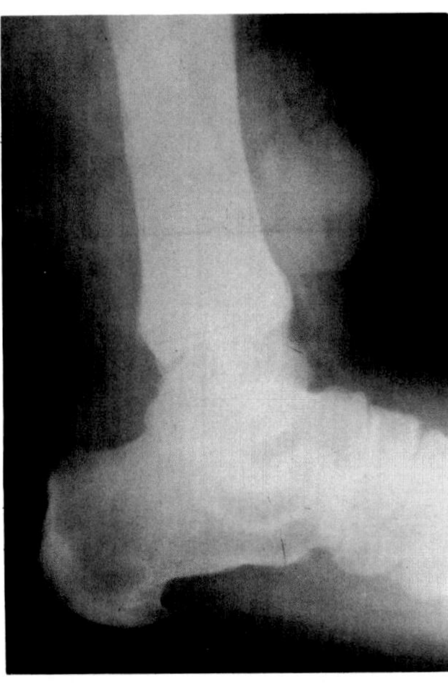

A

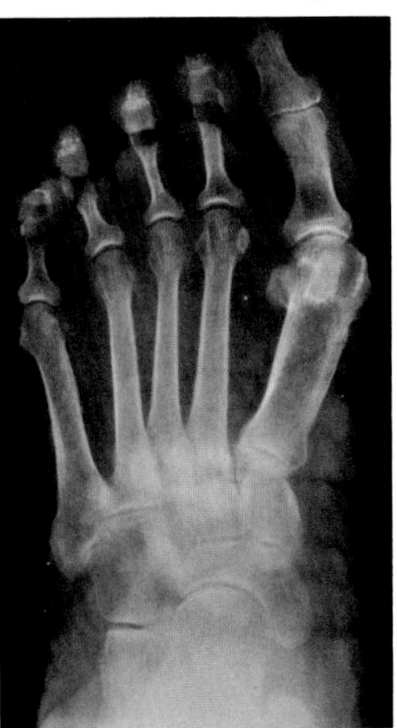

B

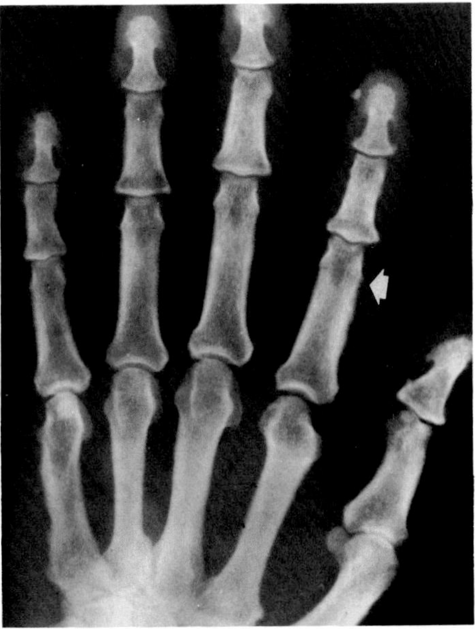

C

**Figure 4.20. Thyroid acropachy.** Views of the ankle and lower leg (A) and the foot (B) show diffuse soft tissue swelling. (C) A frontal view of the hand shows spiculated periosteal new bone formation, seen best on the radial aspect of the proximal phalanx of the second digit (arrow).

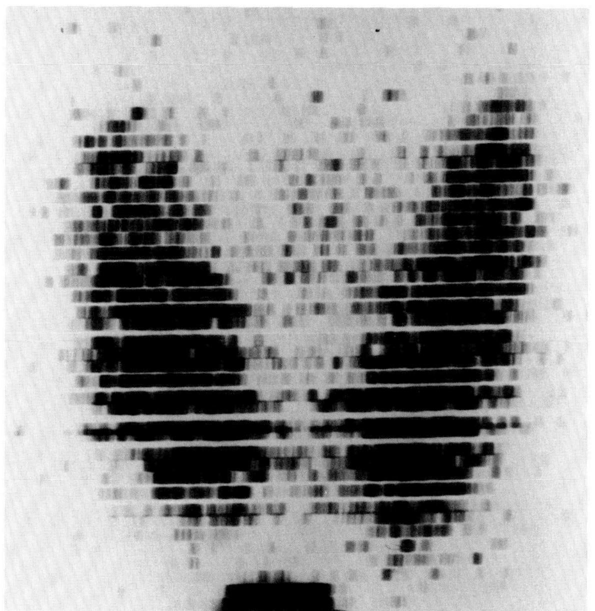

**Figure 4.21. Simple (nontoxic) goiter.** A radioactive iodine scan shows a homogeneous increase in the uptake of isotope in a diffusely enlarged thyroid gland.

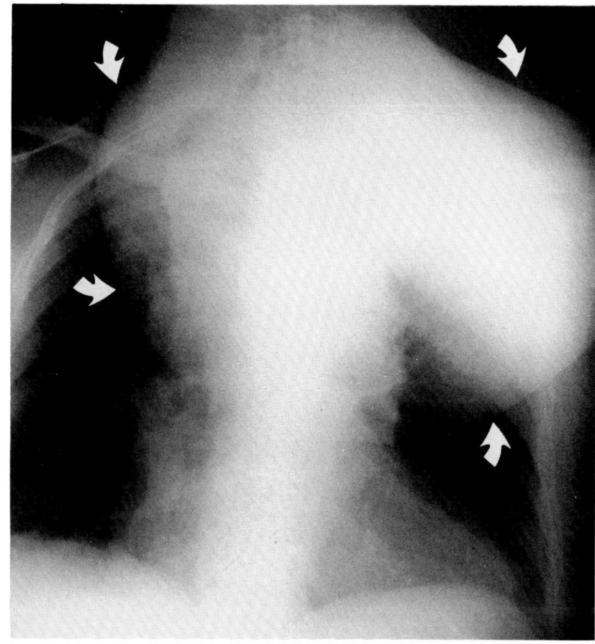

A

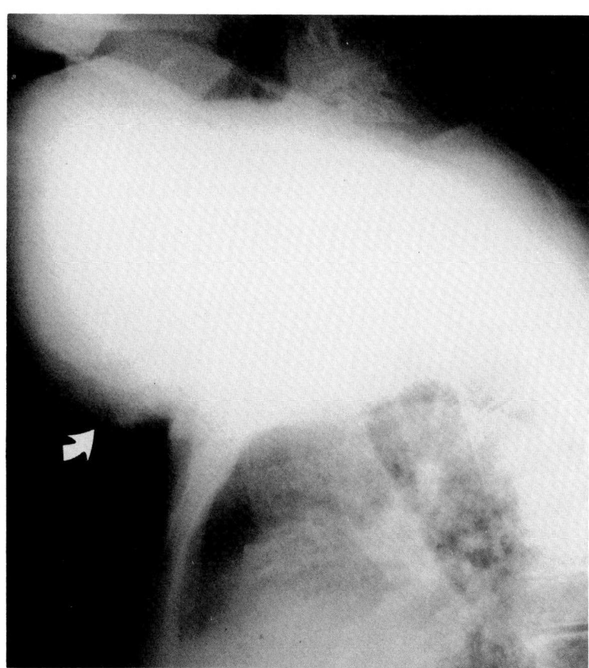

B

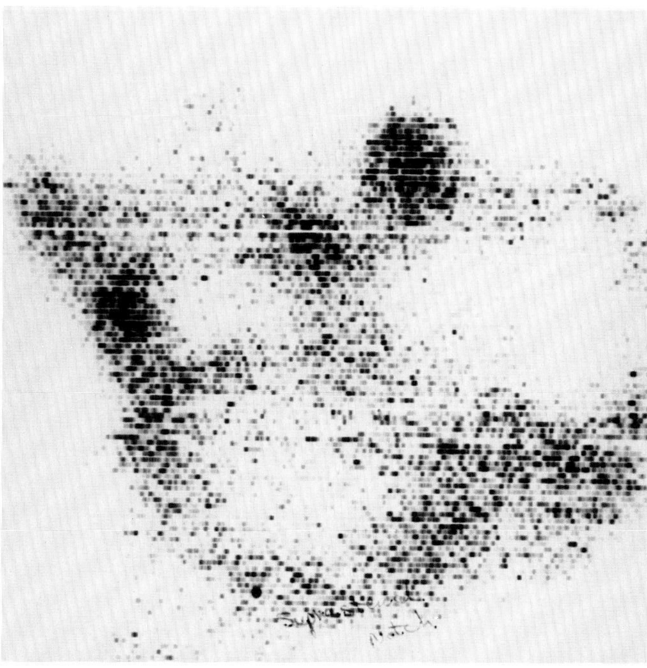

C

**Figure 4.22. Goiter.** (A) Frontal and (B) lateral views of the chest demonstrate a huge soft tissue mass (arrows). (C) The patchy uptake of radioactive iodine throughout the mass proves that the lesion represents massive enlargement of the thyroid gland.

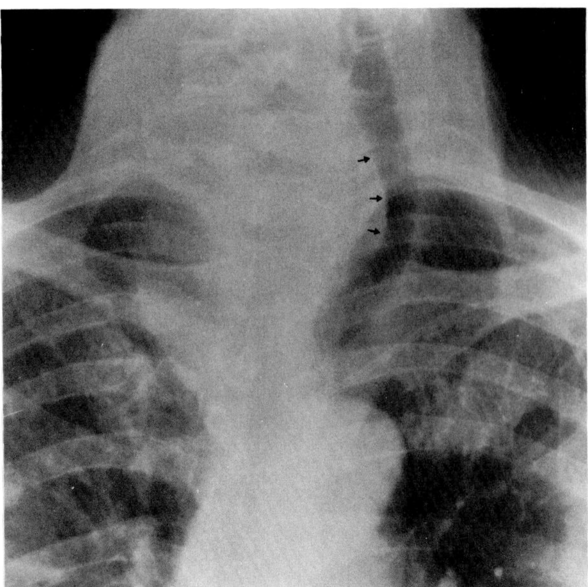

**Figure 4.23. Goiter.** The markedly enlarged thyroid appears as a soft tissue mass impressing the trachea (arrows) and displacing it to the left.

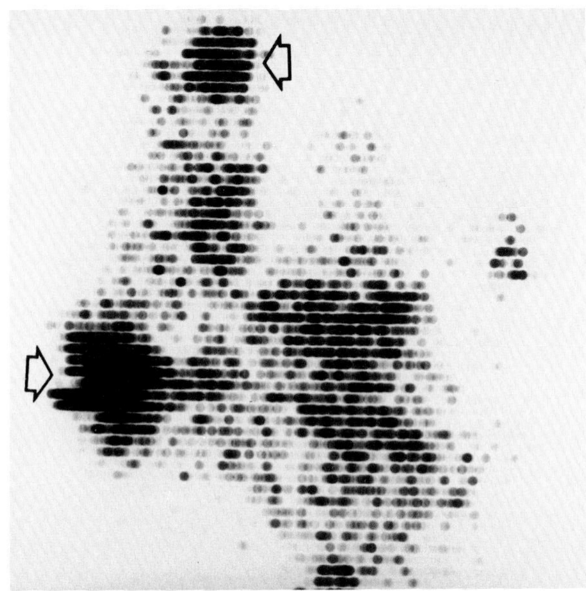

**Figure 4.24. Toxic multinodular goiter.** A radioactive iodine scan shows patchy uptake of isotope with several discrete hot nodules (arrows) within the enlarged gland.

## Thyroid Adenoma

Thyroid adenomas are encapsulated tumors that vary greatly in size and usually compress contiguous tissue. They may be located within the neck, where they tend to cause deviation or compression of the trachea, or they may extend substernally and appear as masses in the superior portion of the anterior mediastinum. Calcification not infrequently develops within the mass.

The size, location, and functional activity of a thyroid adenoma are best demonstrated by radioactive iodine scanning. A thyroid adenoma can appear as a hot or a cold nodule, depending on its functional capacity. Ultrasound demonstrates a thyroid adenoma as a nonspecific echogenic mass (especially if calcified) that is indistinguishable from thyroid carcinoma. [24,25,124]

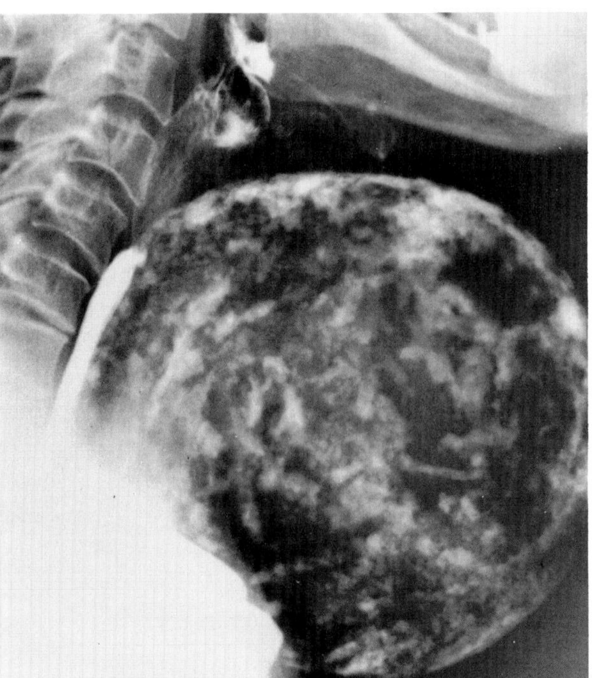

**Figure 4.25. Thyroid adenoma.** A lateral view of the neck shows a huge calcified thyroid mass.

## Thyroid Carcinoma

There are three major types of thyroid carcinoma: papillary, follicular, and medullary. Papillary carcinoma, the most common type, has a bimodal frequency with peaks occurring in adolescence and young adulthood and again in later life. The tumor is usually slow-growing, and it typically spreads to regional lymph nodes, where it may remain indolent for many years. Metastases to the lungs often cause only mild, nonspecific thickening of bronchovascular markings, though the metastases may also appear as miliary and nodular densities that predominantly involve the lower lobes.

Histologically, follicular carcinoma closely mimics normal thyroid tissue. The tumor usually undergoes early hematogenous spread, especially to the lung and bone. Skeletal metastases (which may be the initial presentation) primarily involve the skull, pelvis, and upper extremities and tend to produce entirely lytic, expansile destruction that extends into the soft tissues and is associated with little or no periosteal reaction.

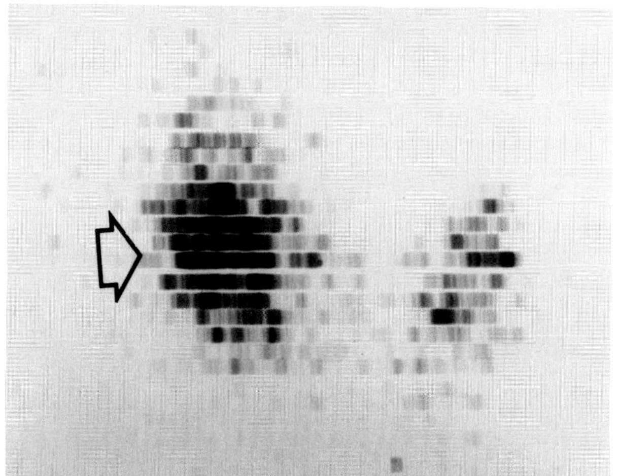

**Figure 4.26. Thyroid adenoma.** A radioactive iodine scan demonstrates a hot nodule (arrow) in the right lobe; the hot nodule suppresses the level of isotope uptake in the remainder of the thyroid gland.

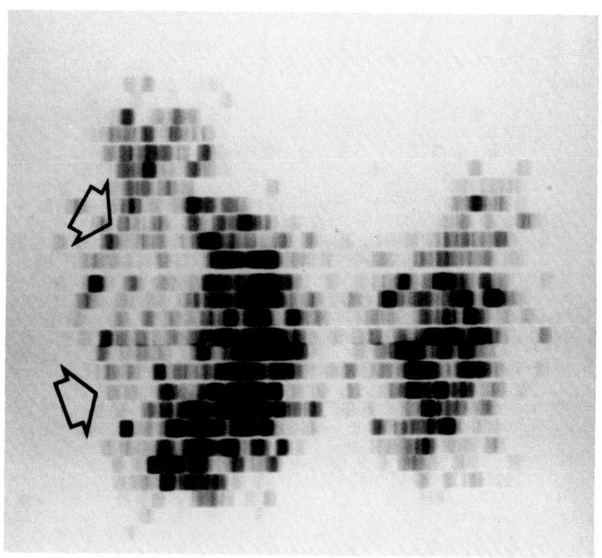

**Figure 4.27. Thyroid adenoma.** A radioactive iodine scan demonstrates a cold nodule (arrow) in the right lobe of the gland.

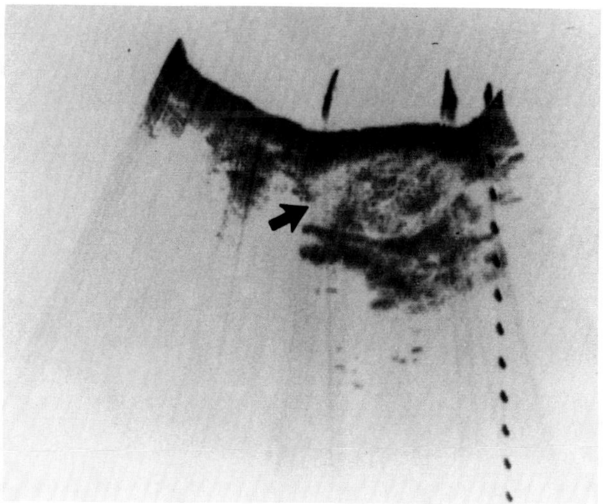

**Figure 4.28. Thyroid adenoma.** A longitudinal sonogram demonstrates a nonspecific echogenic mass (arrow). (From Gooding,[125] with permission from the publisher.)

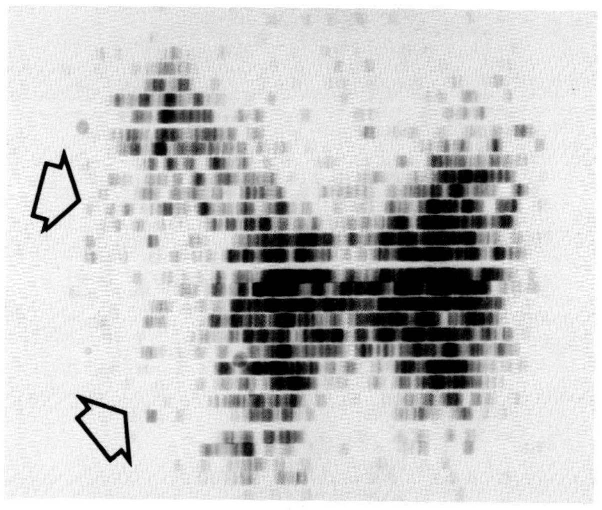

**Figure 4.29. Thyroid carcinoma.** A radioactive iodine scan shows a solitary cold nodule (arrows) that corresponded to the patient's palpable mass.

Medullary carcinoma is the least common type of thyroid malignancy. At least 10 percent of the cases are familial, most often appearing as a component of the multiple endocrine neoplasia syndrome. Dense, amorphous calcifications can often be seen within the tumor.

On radioactive iodine scans, thyroid carcinoma usually appears as a solitary cold nodule that corresponds to a palpable mass. A nodule that is functioning (hot) can essentially be excluded from a diagnosis of thyroid carcinoma. Ultrasound typically demonstrates thyroid carcinoma as an echogenic mass that is indistinguishable from a benign neoplasm.

There is a substantially increased risk of thyroid cancer in persons with a history of therapeutic neck irradiation in childhood. In the past, such irradiation was used for benign disease such as enlargement of the tonsils, adenoids, and thymus; middle ear disease; and a variety of cutaneous disorders, including acne.[26,27,124]

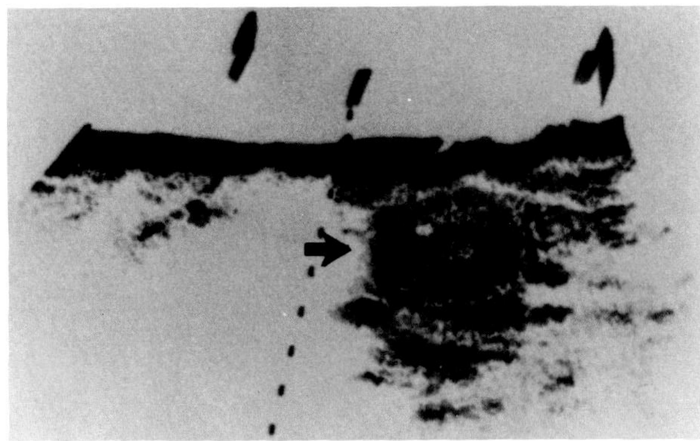

**Figure 4.30. Thyroid carcinoma.** A longitudinal sonogram shows an echogenic mass (arrow) indistinguishable from a benign neoplasm. (From Gooding,[125] with permission from the publisher.)

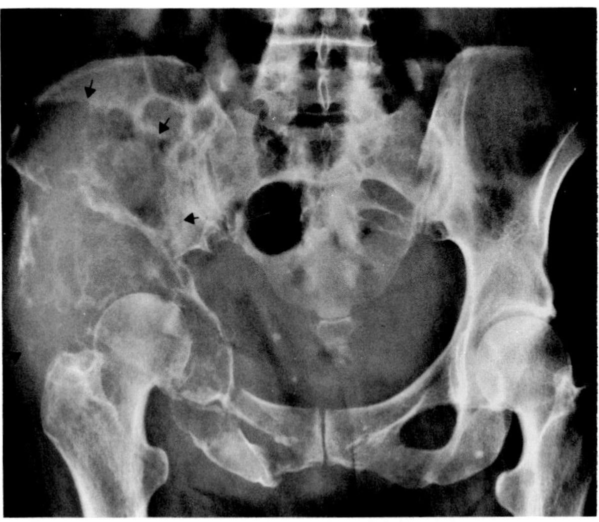

**Figure 4.31. Metastatic thyroid carcinoma.** A large area of entirely lytic, expansile destruction (arrows) involves the left ilium.

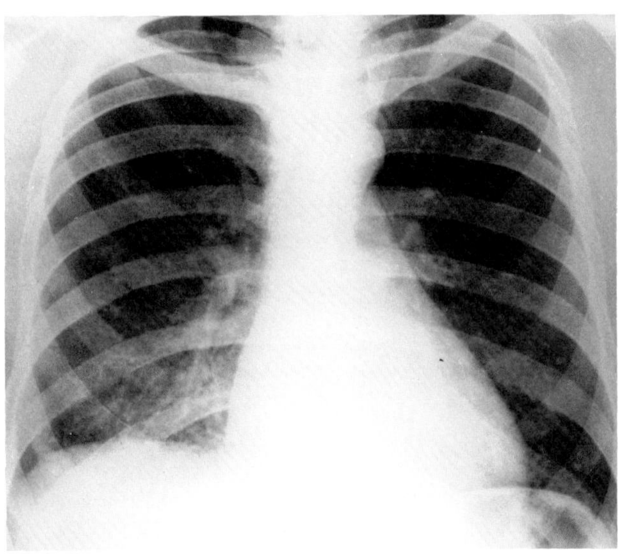

A

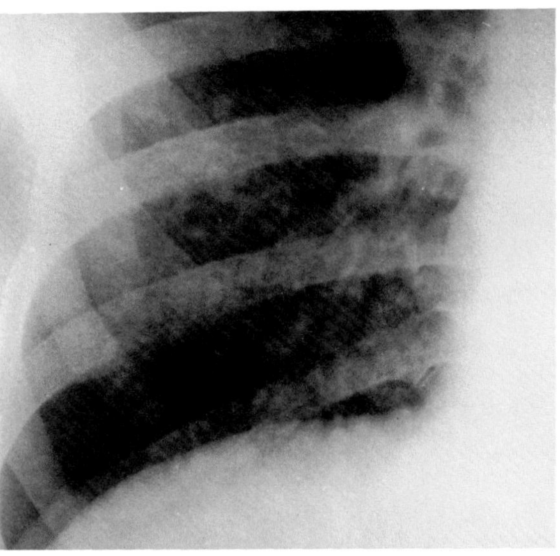

B

**Figure 4.32. Metastatic thyroid carcinoma.** (A) A frontal chest radiograph and (B) a coned view of the right lower lung demonstrate a diffuse pattern of miliary and larger nodular pulmonary densities.

## Thyroiditis

Hashimoto's thyroiditis is the most common form of chronic thyroiditis. Most frequently occurring in women of middle age, Hashimoto's thyroiditis appears to reflect an autoimmune disorder. The radioactive iodine scan typically shows inhomogenous activity in an enlarged thyroid gland, which often has scalloped margins. The progressive replacement of thyroid parenchyma by lymphocytes or fibrous tissue may lead to frank hypothyroidism and the decreased uptake of radioactive iodine.

Subacute (de Quervain's) thyroiditis is a granulomatous process that appears to be viral in origin and usually develops following an upper respiratory infection. The condition may be indolent, with smoldering symptoms of referred, rather than local, pain; less commonly, the onset is acute, with severe pain and tenderness over all or a portion of the gland. Subacute thyroiditis typically affects young to middle-aged women and is characterized by a marked depression in the uptake of radioactive iodine. If left untreated, the disorder eventually subsides and normal thyroid function returns.

Pyogenic thyroiditis and chronic fibrosing (Reidel's) thyroiditis are rare disorders. Pyogenic thyroiditis is usually preceded by a pyogenic infection elsewhere and is characterized by tenderness and swelling of the thyroid

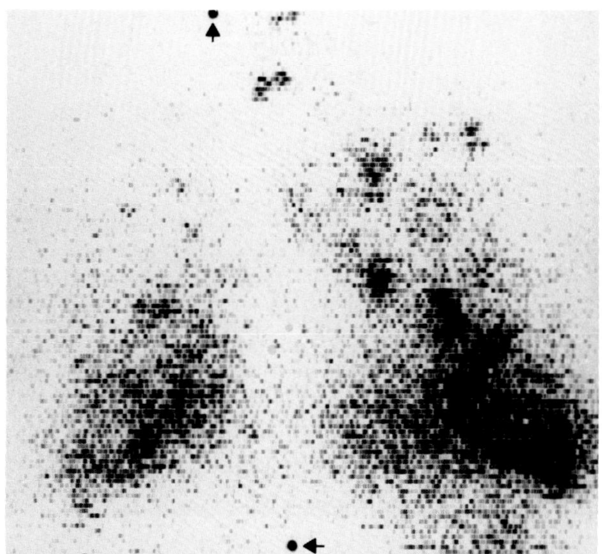

**Figure 4.33. Metastatic thyroid carcinoma.** A radioactive iodine scan over the chest shows multiple areas of patchy isotope uptake; these represent diffuse metastases to the lungs. The round markers (arrows) at the top and bottom of the scan represent the suprasternal notch and xiphoid process, respectively.

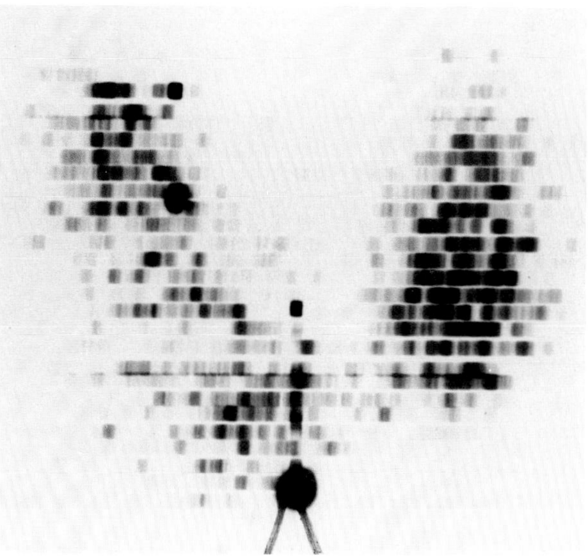

**Figure 4.34. Hashimoto's thyroiditis.** A radioactive iodine scan shows patchy inhomogeneous activity in an enlarged thyroid gland.

gland, redness and warmth of the overlying skin, and constitutional signs of infection. In Reidel's thyroiditis, which may be associated with mediastinal and retroperitoneal fibrosis, intense fibrosis of the thyroid gland and surrounding structures leads to induration of the tissues of the neck.[128]

## DISEASES OF THE ADRENAL CORTEX

### Cushing's Syndrome

In Cushing's syndrome, the excess production of steroids with glucocorticoid activity may be due to generalized bilateral adrenocortical hyperplasia or may be secondary to a functioning adrenal or even nonadrenal tumor.

In the past, generalized adrenal enlargement in Cushing's syndrome has been best demonstrated by venography, which also shows a generalized convexity of the adrenal contour and an increased separation of the intraglandular branches. More recently, CT and ultrasound have been used in the evaluation of the patient with Cushing's syndrome. CT demonstrates thickening of the wings of the adrenal gland, which appears to have a stellate or Y-shaped configuration in cross section.

Benign and malignant tumors of the adrenal cortex are less common causes of Cushing's syndrome than is nontumorous adrenal hyperfunction. Although as a general rule the larger the adrenal cortical tumor and the more abrupt the onset of clinical symptoms and signs the more likely the tumor is to be malignant, the differentiation between adenoma and carcinoma may be impossible at the time of histologic examination; only the clinical course may define the nature of the tumor. Excretory urography with nephrotomography is usually used initially to evaluate the patient with clinical and laboratory evidence of Cushing's syndrome due to a tumor. Plain films may demonstrate calcification or displacement of the abundant perirenal fat around the periphery of the mass. Left-sided masses tend to rotate the upper pole of the kidney laterally so that the axis of the kidney no longer parallels the psoas margin. Larger left-sided masses may displace the kidney downward or indent the posterior aspect of the stomach. On the right, much of the adrenal gland is truly suprarenal in location. Thus, a right-sided mass commonly indents the superior aspect

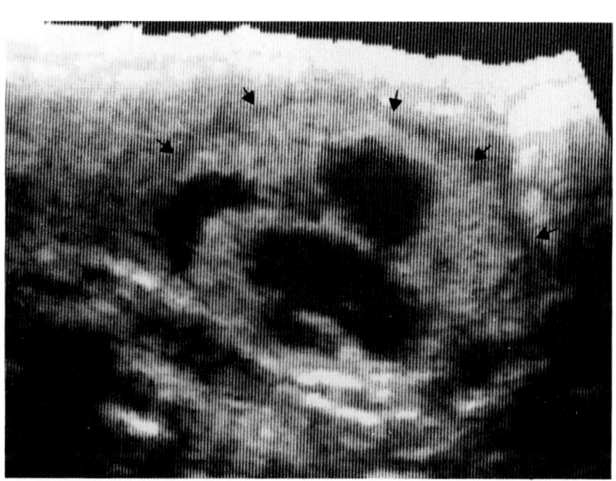

**Figure 4.35. Hashimoto's thyroiditis.** Ultrasound demonstrates a complex mass (arrows) containing large cystic areas.

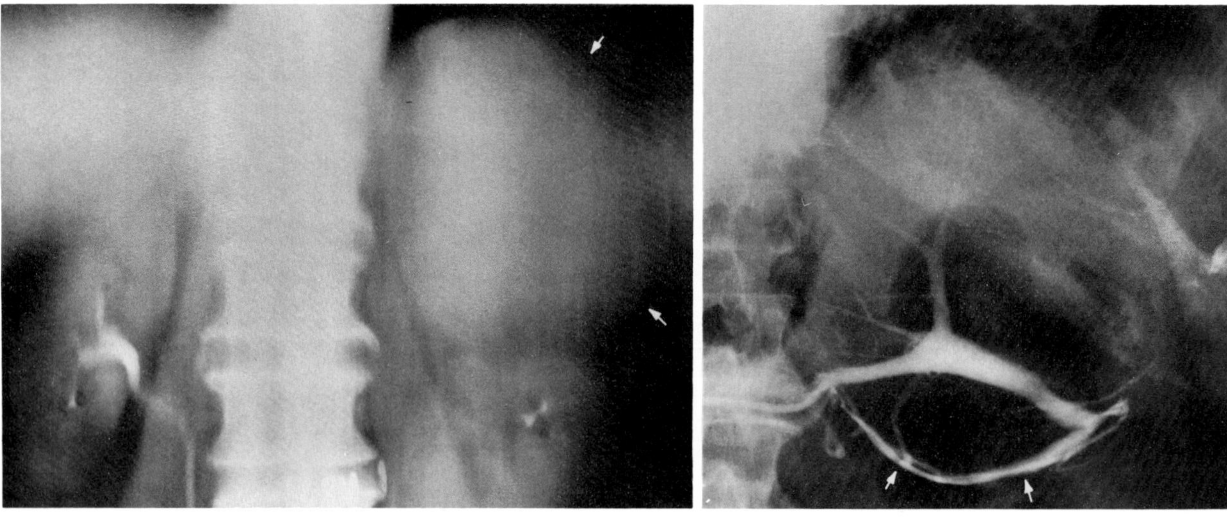


**Figure 4.36. Large adrenal adenoma causing Cushing's syndrome.** (A) A nephrotomogram demonstrates a huge suprarenal mass (arrows) causing indentation and downward displacement of the left kidney. (B) Venography shows a circumferential vein (arrows) defining the inferior extent of the mass.

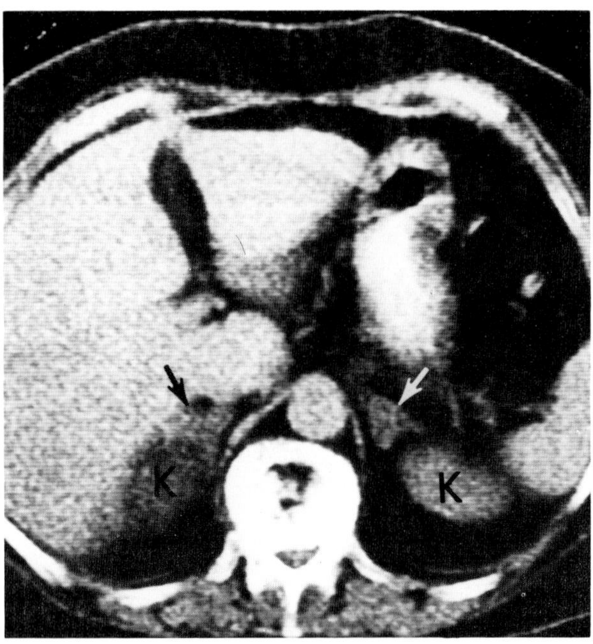


**Figure 4.37. Cushing's syndrome due to adrenal hyperplasia.** Although the adrenal glands are enlarged (arrows), their normal configuration is maintained. (From Weyman and Glazer,[116] with permission from the publisher.)

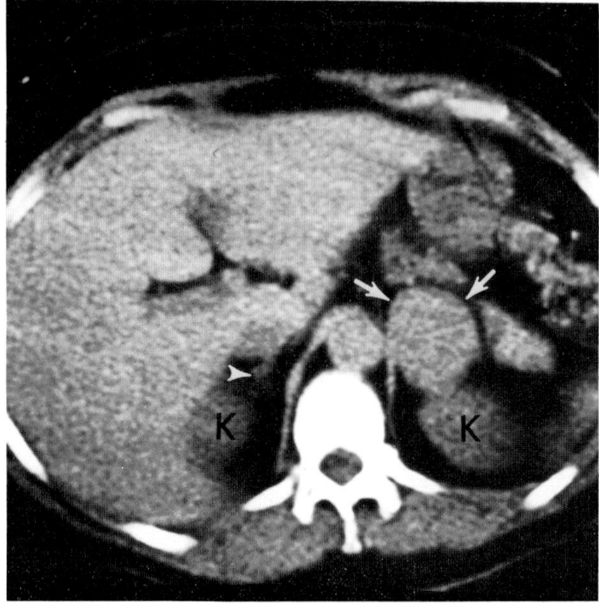


**Figure 4.38. Cushing's syndrome due to cortical adenoma.** A 4-cm mass in the left adrenal gland (arrows) is seen posterior to the tail of the pancreas and anterior to the kidney (K). The arrowhead points to the normal right adrenal gland. (From Weyman and Glazer,[116] with permission from the publisher.)

of the right kidney, eventually displacing the kidney downward.

CT or ultrasound is usually the next diagnostic step; the former is often of more value, since the abundance of retroperitoneal fat may preclude an optimal ultrasound examination. Because adrenal tumors causing Cushing's syndrome are usually 3 cm or larger by the time they are clinically apparent, almost all can be demonstrated by CT. However, the tumor occasionally may have such a low attenuation number that it is difficult to separate from retroperitoneal fat. Ultrasound typically demonstrates an adrenal tumor as a poorly echogenic, relatively homogeneous mass; the tumor may have a stronger echo pattern if necrosis develops. Although the bulk of the

**Figure 4.39. Cushing's syndrome due to adrenal carcinoma.** A longitudinal prone sonogram shows a sonolucent mass (M) that is largely extrarenal although the kidney (K) is compressed. (From Mitty and Yeh,[28] with permission from the publisher.)

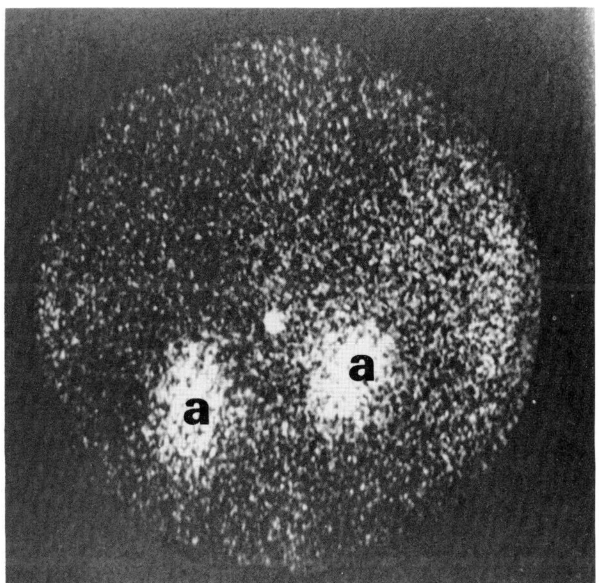


**Figure 4.40. Cushing's syndrome due to bilateral adrenocortical hyperfunction.** A radiocholesterol scan obtained 6 days after the injection of the isotope demonstrates bilateral dense uptake of the radionuclide in the adrenal glands (a). (From Chatal, Charbonnel, and Guihard,[117] with permission from the publisher.)

adrenal tumor is usually seen beyond the expected renal outline, it may be difficult to sonographically separate an adrenal mass from an upper pole renal tumor.

Adrenal venography has been widely employed to demonstrate adrenal masses. Unlike arteriography, in which several vessels supplying the gland must be injected, venography requires that only one vein be injected on each side to opacify the adrenal. In addition, venography also permits the aspiration of blood samples for assessing the level of adrenal hormones. Because the cortical adenomas of Cushing's syndrome are generally well-circumscribed, a circumferential vein can often be

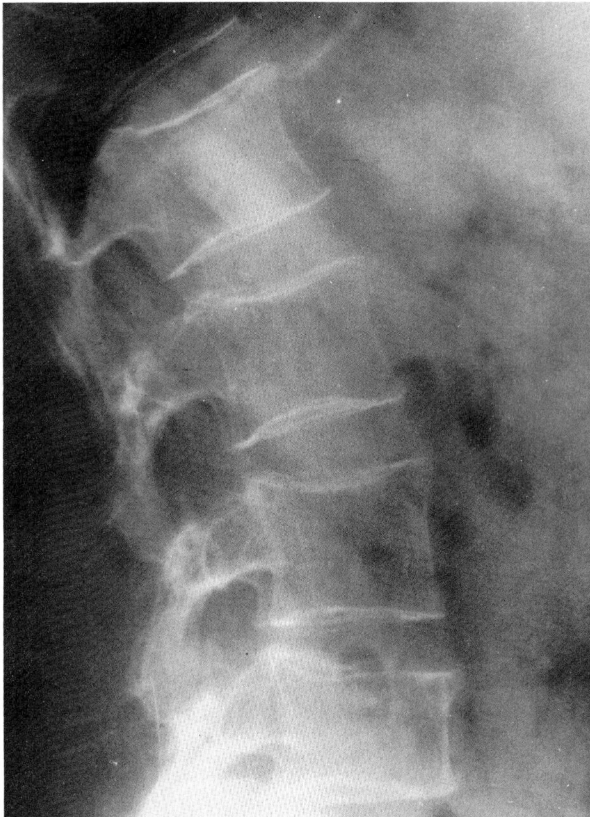

**Figure 4.41. Diffuse osteoporosis in Cushing's syndrome due to adrenal hyperplasia.** There is marked demineralization and an almost complete loss of trabeculae in the lumbar spine. The vertebral end plates are mildly concave, and the intervertebral disk spaces are slightly widened. There is compression of the superior end plate of L4. (From Reynolds and Karo,[118] with permission from the publisher.)

seen defining the extent of the mass. The presence of well-developed accessory draining veins, or the presence of tumor nodules in the adrenal vein, is a venographic sign of malignancy.

In some centers, radiocholesterol scanning has been employed in the evaluation of patients with Cushing's syndrome. Bilateral hyperfunction results in higher-than-normal radioactivity in both adrenal glands. Adenomas cause intense uptake on the side of the tumor along with markedly diminished or no activity on the suppressed, contralateral side. Adrenal cortical carcinomas demonstrate little uptake of the isotope. Because cortisol produced by the tumor suppresses the contralateral gland, no adrenal uptake of isotope may be seen. The major drawback to the radionuclide method is the need to perform multiple scans over several days while the isotope is being incorporated into the adrenal steroids that are being synthesized.

Cushing's syndrome produces radiographic changes in multiple systems. Diffuse osteoporosis causes generalized skeletal demineralization that may lead to the

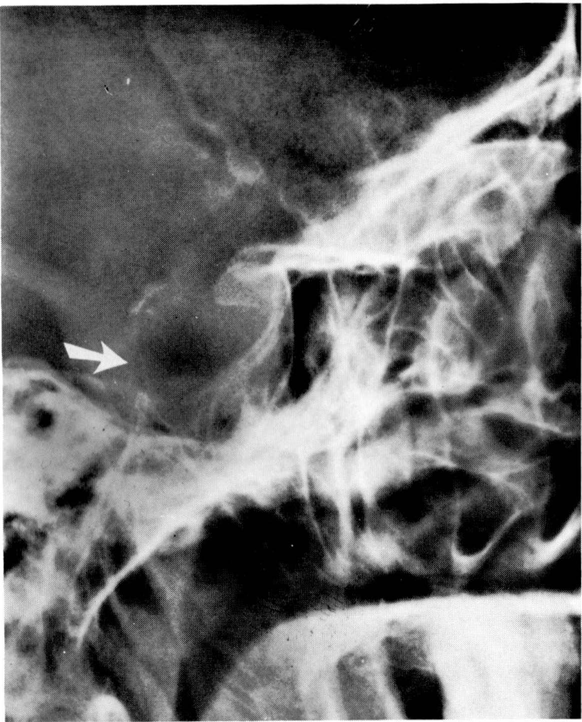

A

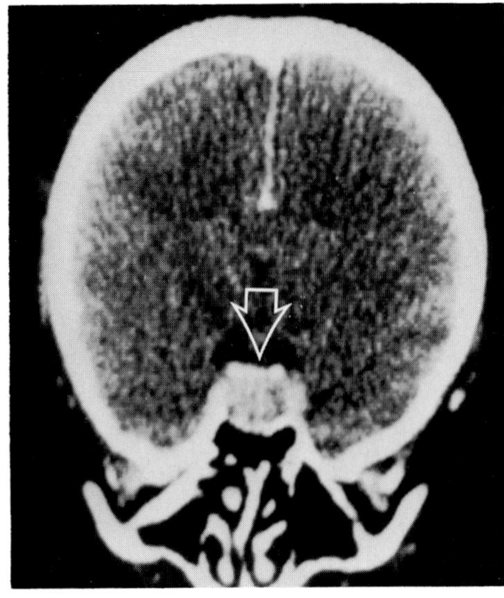

B

**Figure 4.42. Nelson's syndrome.** (A) A lateral film of the sella turcica shows general enlargement of the pituitary fossa with marked thinning of the dorsum sellae (arrow). (B) A CT scan demonstrates the vascular tumor filling the enlarged sella (arrow). (From Mitty and Yeh,[28] with permission from the publisher.)

collapse of vertebral bodies, spontaneous fractures (which usually heal with extensive callus formation), and aseptic necrosis of the head of the femur or humerus. Widening of the mediastinum due to excessive fat deposition sometimes develops in Cushing's syndrome and can be confirmed by CT. Hypercalcuria caused by the predominantly catabolic effect of elevated levels of cortisol can lead to nephrocalcinosis or renal calculi.

Imaging of the sella turcica by conventional or computed tomography is important in the routine assessment of the patient with Cushing's syndrome. A majority of patients with nontumorous adrenal hyperfunction are found at surgery to have an intrasellar lesion. It is important to emphasize, however, that small pituitary microadenomas may be present in asymptomatic patients. Following adrenal surgery, a pituitary adenoma develops in up to one-third of the patients and produces progressive sellar enlargement (Nelson's syndrome). For this reason, yearly follow-up sellar tomograms may be indicated after adrenalectomy.

Nonpituitary tumors producing adrenocorticotropic hormone (ACTH) may cause adrenal hyperfunction and Cushing's syndrome. The most common sites of origin are the lung, thymus, and pancreas; about half of these tumors can be demonstrated on chest radiographs.[28–35, 116–119]

## Adrenal Carcinoma

About half of adrenal carcinomas are functioning tumors that cause Cushing's syndrome, virilization, feminization, or aldosteronism. The tumors grow rapidly and are usually large masses at the time of clinical presentation.

Excretory urography demonstrates an adrenal carcinoma as a nonspecific suprarenal mass that is often lobulated and tends to displace the kidney without evidence of invasion. Ultrasound shows the tumor as a complex mass that may be difficult to separate from an upper pole renal tumor. CT demonstrates an adrenal carcinoma as a large mass that often contains low-density areas resulting from necrosis or prior hemorrhage. Because lymphatic and hepatic metastases are common at the time of clinical presentation, CT scans at multiple abdominal levels are necessary to define the extent of the primary tumor and to detect metastases prior to attempts at resection. Extension of the tumor into the renal vein and vena cava can also be detected by CT, especially following the injection of intravenous contrast medium.

Angiography demonstrates an adrenal carcinoma as a vascular lesion that is often associated with arteriovenous lakes, neovascularity, and early venous filling. The angiographic appearance often mimics that of a vascular pheochromocytoma.[28,36–39,119]

## Metastases to the Adrenals

The adrenal gland is one of the most common sites of metastatic disease. The primary tumors that most frequently metastasize to the adrenal gland are carcinomas of the lung and breast; carcinomas of the kidney, ovary, and gastrointestinal tract, as well as melanomas, also often metastasize to the adrenals. Metastatic enlarge-

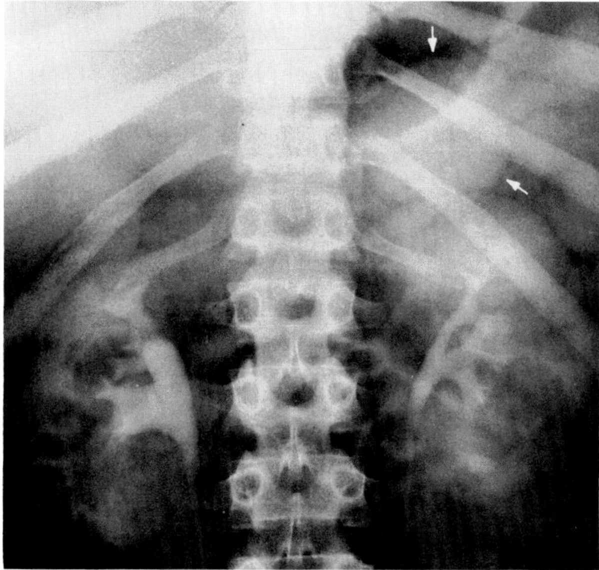

**Figure 4.43. Adrenal carcinoma.** An excretory urogram demonstrates a nonspecific 5-cm mass (arrows) superior to the left kidney.

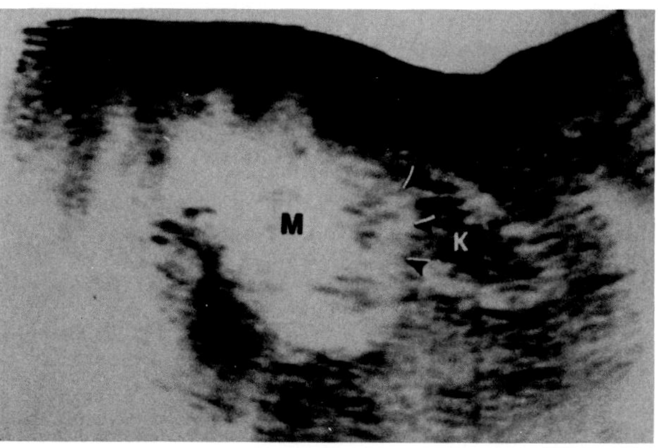


**Figure 4.44. Adrenal carcinoma.** A posterior longitudinal sonogram demonstrates a faint border (arrowheads) between a large suprarenal mass (M) and the kidney (K). Note the areas of strong echoes within the mass. (From Yeh, Mitty, Rose, et al,[119] with permission from the publisher.)

ment of the adrenal glands can cause downward displacement of the kidney with flattening of the upper pole. Ultrasound demonstrates adrenal metastases as solid lesions with rare focal cystic areas due to necrosis. CT shows these lesions as soft tissue masses that vary considerably in size and are frequently bilateral. However, the ultrasound and CT patterns are indistinguishable from those of primary malignancies of the gland. Therefore, in the face of a known primary tumor elsewhere, it is usually assumed that an adrenal mass is metastatic. If necessary, a needle biopsy under ultrasound guidance may be of value to determine whether the adrenal lesion is primary or metastatic.[28,40,41]

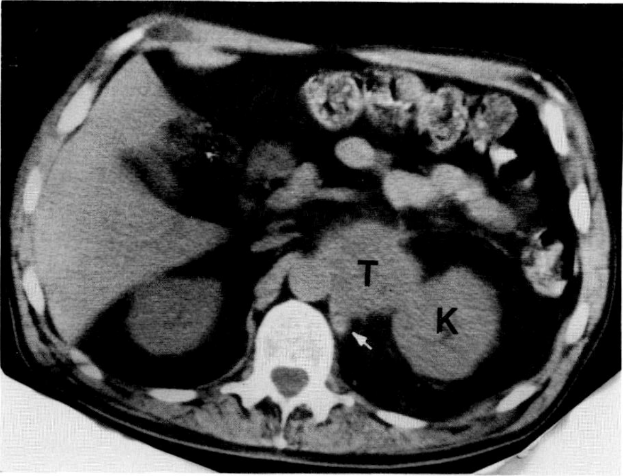


**Figure 4.45. Adrenal carcinoma.** A CT scan shows a large soft tissue tumor (T) invading the anteromedial aspect of the left kidney (K) and left crus of the diaphragm (arrow). (Courtesy of Nolan Karstaedt, M.D., and Neil Wolfman, M.D.)

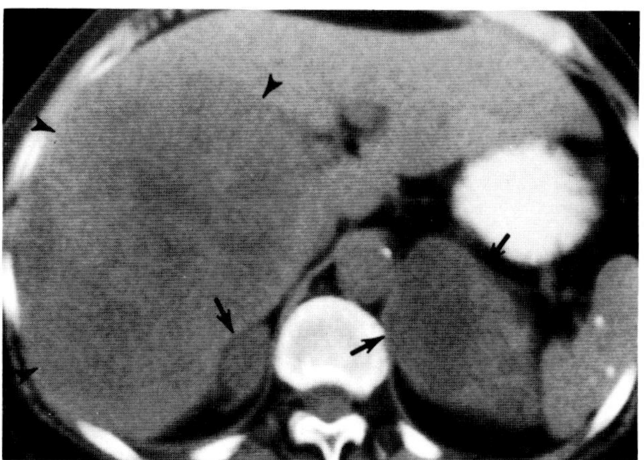

**Figure 4.46. Metastases to the adrenals.** Bilateral adrenal metastases (arrows) in a patient with colonic carcinoma. A large liver metastasis (arrowheads) is also present. (From Weyman and Glazer,[116] with permission from the publisher.)

## Aldosteronism

Primary aldosteronism, which presents as hypertension and hypokalemia with inappropriate kaliuresis, may be due to an adrenal cortical adenoma (Conn's syndrome) or to bilateral hyperplasia of the zona glomerulosa. The clinical manifestations of primary aldosteronism can be cured by the resection of an adenoma, but are little affected by the removal of both glands in the patient with bilateral hyperplasia. Therefore, the role of radiographic imaging is to demonstrate the location of adenomas that may be otherwise difficult to detect during exploratory surgery.

Because aldosteronomas are often very small (unlike the large cortical adenomas in patients with Cushing's

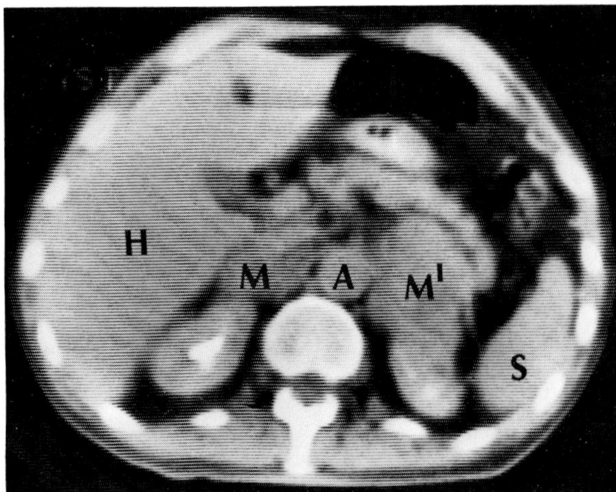


**Figure 4.47. Metastases to the adrenals.** Bilateral adrenal metastases in a patient with carcinoma of the lung. A CT scan clearly demonstrates the right (M) and left (M') adrenal metastases; H = liver; S = spleen; A = aorta. (From Mitty and Yeh,[28] with permission from the publisher.)

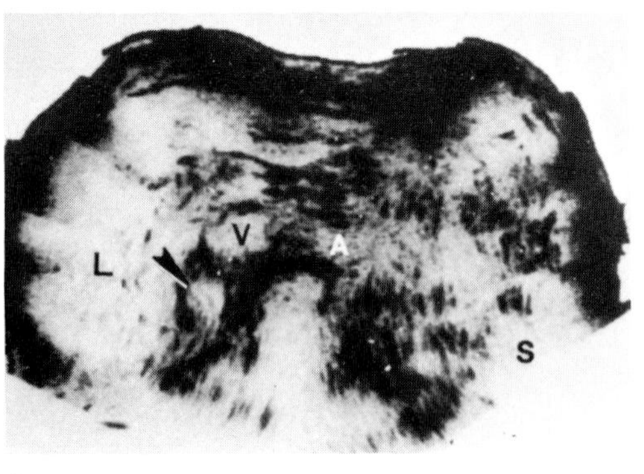


A

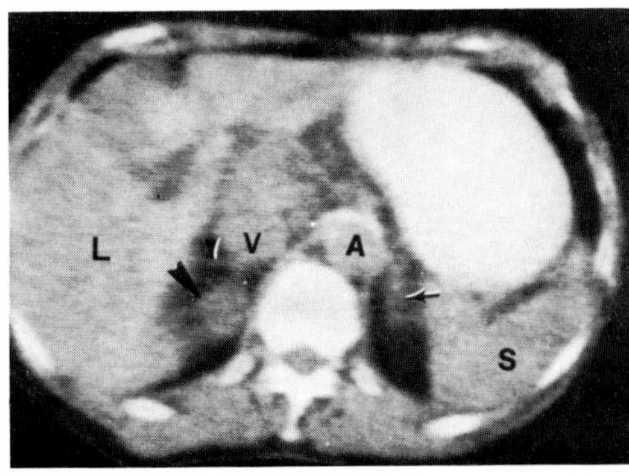


B

**Figure 4.48. Aldosteronism.** (A) A transverse sonogram demonstrates a 2-cm macronodule (arrowhead) in the posterior aspect of the right adrenal gland. L = liver; V = inferior vena cava; A = aorta; S = spleen. (B) The corresponding CT scan also shows the nodule (large arrowhead). Note the normal right lateral wing (small arrowhead) and the thickened left gland (arrow). (From Mitty and Yeh,[28] with permission from the publisher.)

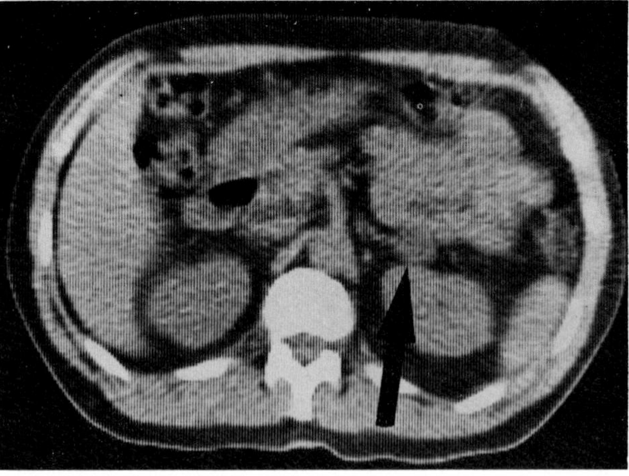

**Figure 4.49. Aldosteronism.** A small mass (arrow) anterior to the left kidney. (From Burko, Smith, Kirschner, et al,[122] with permission from the publisher.)

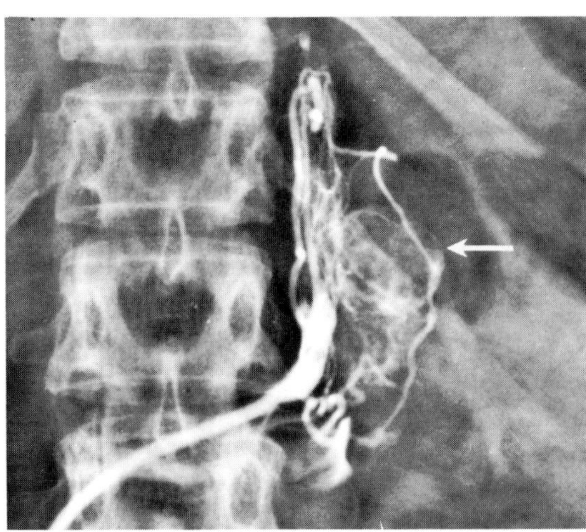

**Figure 4.50. Aldosteronism.** A selective venogram clearly defines a 2-cm adenoma. (From Mitty and Yeh,[28] with permission from the publisher.)

syndrome), excretory urography with tomography has limited usefulness. The major value of urography is to exclude intrinsic renal disease, which is a common cause of secondary hyperaldosteronism.

Adrenal venography with biochemical assay of a sample of adrenal blood has been the major method for localizing aldosteronomas. Filling the venous system within the mass may produce a pattern of vessels resembling the spokes of a wheel. Smaller tumors may be identified only by the displacement of vessels about their periphery; with larger lesions, circumferential veins may define the extent of the mass. To obtain meaningful results, it is essential that samples of venous blood for biochemical assay be obtained before performing contrast venography.

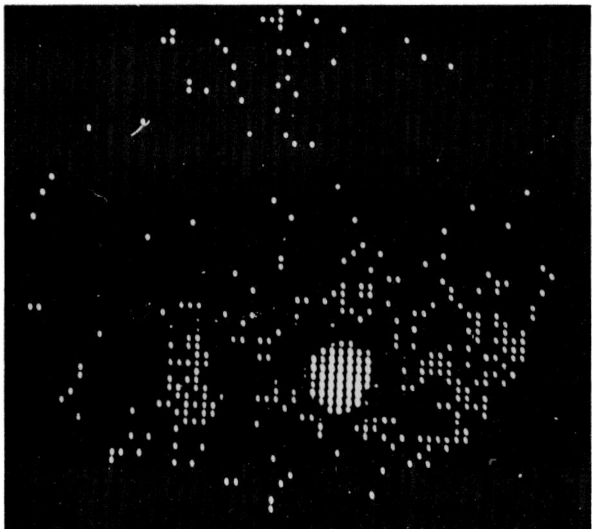

**Figure 4.51. Aldosteronism.** A $^{131}$I-19-iodocholesterol dexamethasone suppression scan in the posterior position obtained 6 days after the administration of the isotope clearly defines the right-sided tumor. (From Chatal, Charbonnel, and Guihard,[117] with permission from the publisher.)

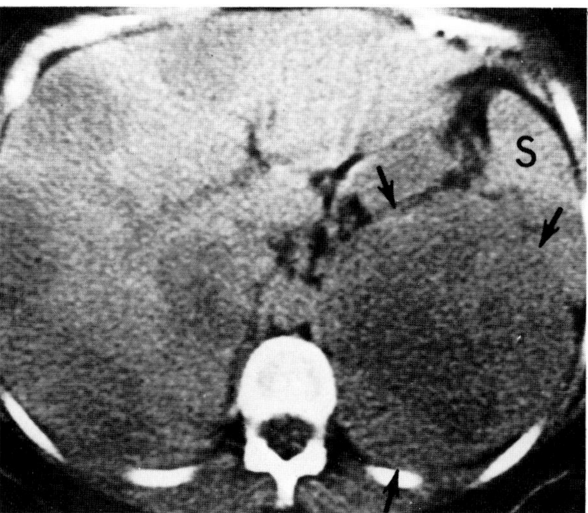


**Figure 4.52. Adrenogenital syndrome due to adrenal carcinoma.** A CT scan shows a large mass in the left upper quadrant (arrows) in the region of the adrenal gland; the mass displaces the spleen (S) anteriorly. Multiple round metastases are present in the liver. (From Karstaedt, Sagel, Stanley, et al,[120] with permission from the publisher.)

CT has to a large extent replaced venography as the major imaging modality for detecting aldosteronomas. This noninvasive technique demonstrates the small adrenal cortical adenoma as a contour abnormality of the gland. Occasionally, tumors with low attenuation coefficients are difficult to distinguish from retroperitoneal fat. Ultrasound is less effective because the tumors are often too small to be detected with this technique.

As with other forms of adrenal disease, radioisotope scanning with iodocholesterol derivatives may have diagnostic value but is infrequently performed because of the prolonged period of study that is necessary and the consequent relatively high cost. Dexamethasone suppression has been used as an important means of differentiating an adenoma from hyperplasia, since dexamethasone decreases the uptake of iodocholesterol in normal adrenal tissue but does not change the uptake in an adenoma; thus the adenoma becomes more easily recognized.

Radiographically detectable causes of secondary hyperaldosteronism include renin-secreting tumors, renal artery stenosis, malignant hypertension, and bilateral chronic renal disease.[28,42–46,117]

## Adrenogenital Syndrome

The adrenogenital syndrome is caused by the excessive secretion of androgenically active substances by the adrenal gland. In the congenital form, a specific enzyme deficiency that prevents the conversion of cortisol precursors causes continuous ACTH stimulation and bilateral hyperplasia. The elevated levels of androgens result in accelerated skeletal maturation along with premature epiphyseal fusion that may lead to dwarfism.

Most cases of acquired adrenogenital syndrome are due to adrenocortical tumors. These adrenal masses, which are usually radiographically indistinguishable from the cortical tumors associated with Cushing's syndrome, can often be detected by adrenal venography, CT, or ultrasound. An adrenal carcinoma occasionally has virilizing effects, though it more commonly appears as a palpable mass rather than as a functionally active neoplasm.[28]

## Adrenocortical Insufficiency

The clinical manifestations of adrenocortical insufficiency vary from those of a chronic, insidious disorder (easy fatigability, anorexia, weakness, weight loss, increased melanin pigmentation) to an acute collapse with hypotension, rapid pulse, vomiting, and diarrhea.

The most common cause of adrenal insufficiency is probably iatrogenic and secondary to exogenous steroid administration. Primary adrenocortical insufficiency (Addison's disease) results from progressive adrenocortical destruction, which must involve more than 90 percent of the glands before clinical signs of adrenal insufficiency appear. In the past, Addison's disease was usually attributed to tuberculosis; at present, most cases reflect idiopathic atrophy, probably on an autoimmune basis. In areas in which the disease is endemic, histoplasmosis is an occasional cause of adrenal insufficiency. Acute inflammatory disease causes a generalized enlargement of the adrenals that can be demonstrated by a variety of

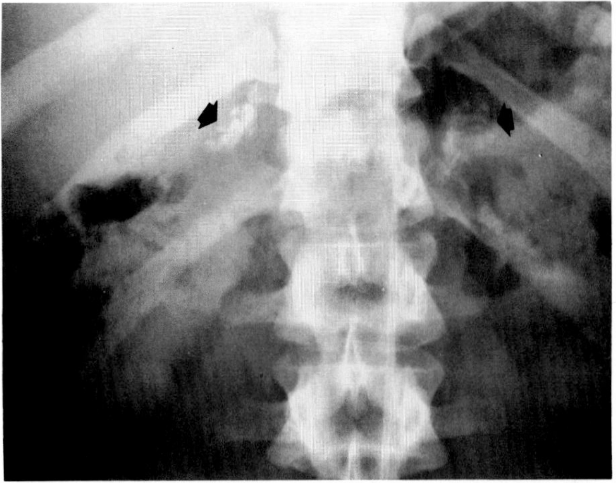

**Figure 4.53. Adrenocortical insufficiency (Addison's disease).** There are bilateral adrenal calcifications (arrows) in a patient with tuberculous adrenal disease.

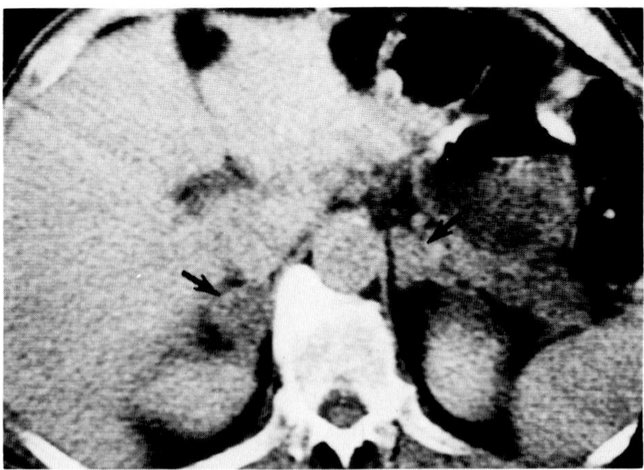

**Figure 4.54. Adrenocortical insufficiency due to disseminated histoplasmosis.** A CT scan demonstrates bilateral adrenal enlargement (arrows). (From Karstaedt, Sagel, Stanley, et al,[120] with permission from the publisher.)

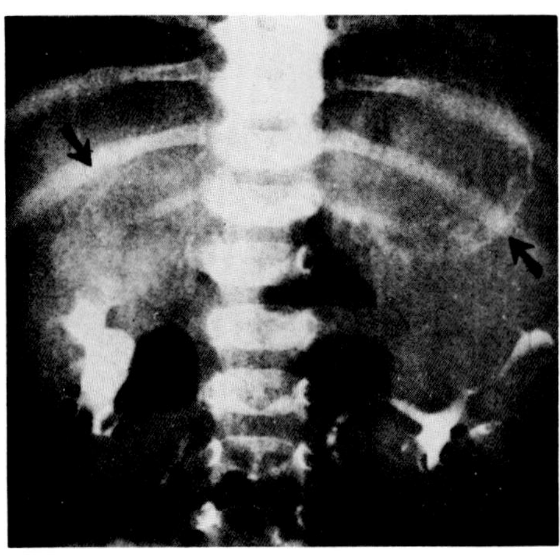

A

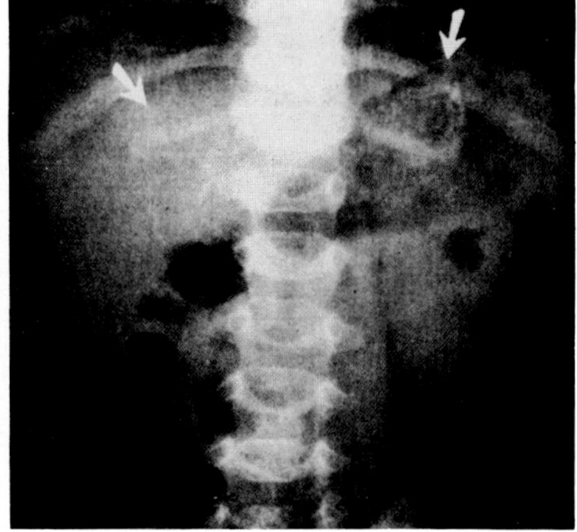

B

**Figure 4.55. Neonatal adrenal hemorrhage.** (A) An excretory urogram in a 2-month-old infant demonstrates calcification surrounding bilateral adrenal masses (arrows). (B) Three months later, the masses have reduced in size and the calcification has assumed a triangular shape (arrows). (From Rose, Berdon, and Sullivan,[51] with permission from the publisher.)

imaging techniques. Focal enlargement with angiographic evidence of distorted vasculature may simulate a primary adrenal tumor. Adrenal calcification develops in up to 25 percent of patients with tuberculous adrenal disease. However, this calcification is of little diagnostic value since asymptomatic adrenal calcification may be seen in normal persons. Other radiographic findings occasionally seen in patients with adrenal insufficiency include a small heart and calcification of the cartilage of the ear.

A severe form of acute adrenal insufficiency is the Waterhouse-Friderichsen syndrome, in which adrenal hemorrhage is usually associated with overwhelming septicemia, especially due to meningococcus. Patients with this syndrome often have enlarged adrenals because of the associated bleeding, though the acute course of the disease usually precludes extensive radiographic evaluation.

In infants, acute adrenal insufficiency can result from bilateral hemorrhage (see the following section).[28,47,48]

## Neonatal Adrenal Hemorrhage

Neonatal adrenal hemorrhage most often occurs in infants born to diabetic mothers and in infants with an abnormal obstetric history (prematurity, the use of forceps, breech delivery). During the total body opacification phase of high-dose excretory urography, compression and displacement of adrenal tissue by the central hemorrhage produce a dense vascular rim surrounding a lucent avascular suprarenal mass. The vascular rim subsequently calcifies, and this permits neonatal adrenal hemorrhage to be distinguished from a hydronephrotic upper pole renal duplication, which may have a similar urographic appearance. The peripheral, usually triangular, calcification of neonatal adrenal hemorrhage must be differentiated from the diffuse and granular calcification of a neuroblastoma.

Ultrasound initially demonstrates neonatal adrenal hemorrhage as a complex mass in which areas of echoes

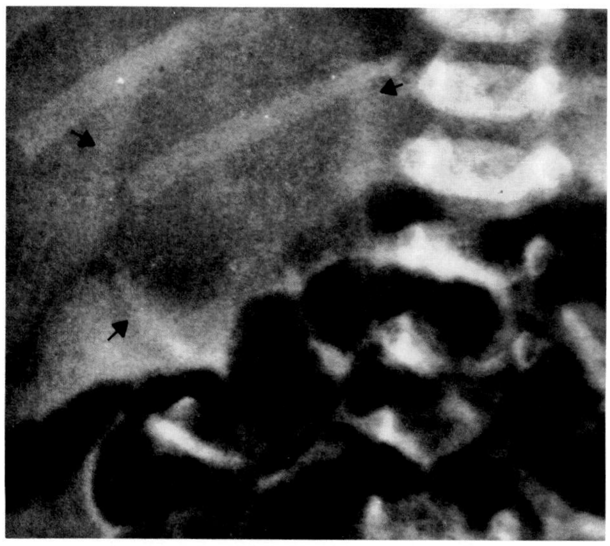

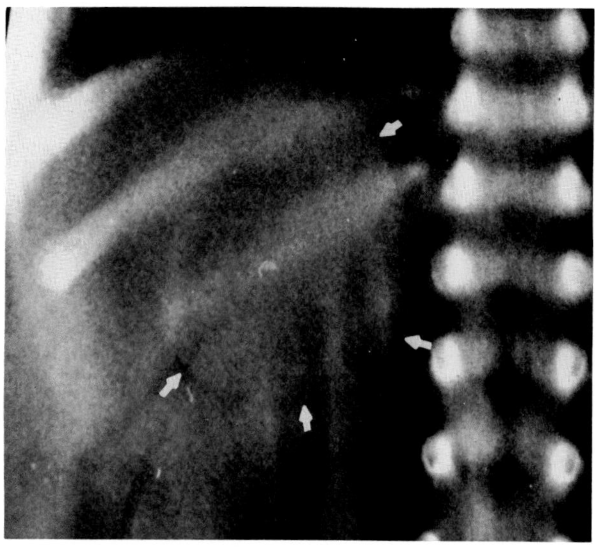

A B

**Figure 4.56. Neonatal adrenal hemorrhage.** (A) An excretory urogram in a 6-day-old infant demonstrates a lucent mass with a dense vascular rim, flattening and depressing the right kidney. (B) Three weeks later, a plain film tomogram of the right upper quadrant shows a calcified rim (arrows) in the identical position as the vascular rim on the previous excretory urogram. (From Brill, Krasna, and Aaron,[49] with permission from the publisher.)

(due to clotting) are mixed with echo-free regions. As the echogenic clot lyses, the lesion becomes predominantly sonolucent.[28,49–51]

## DISEASES OF THE ADRENAL MEDULLA AND SYMPATHETIC NERVOUS SYSTEM

### Pheochromocytoma

A pheochromocytoma is a tumor of chromaffin tissue that most commonly arises in the adrenal medulla. The tumor produces vasopressor substances (epinephrine and norepinephrine) that can cause an uncommon but curable form of hypertension. About 10 percent of pheochromocytomas are extra-adrenal in origin. About 10 percent of the patients have bilateral tumors, and a similar percentage of pheochromocytomas are malignant.

Because in almost all patients the diagnosis of a pheochromocytoma can be made with biochemical tests, radiographic imaging serves as a confirmatory study and for the localization of the tumor. Plain films may demonstrate calcification with a curvilinear eggshell or stellate configuration. Excretory urography, even with nephrotomography, may be of limited value since the kidney is often not displaced even when an adrenal pheochromocytoma is large. This modality may be of value, however, in detecting extra-adrenal tumors causing either lateral displacement of the ureters or a filling defect in the bladder.

CT and ultrasound are very useful in the localization of pheochromocytomas. The cross-sectional images not only detail the extent of an adrenal lesion but also define

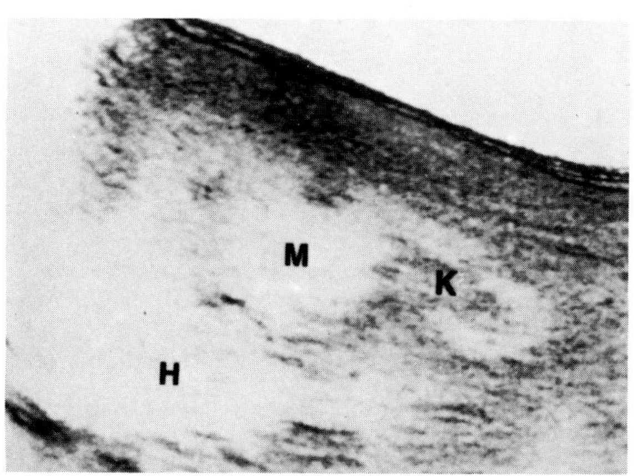


**Figure 4.57. Pheochromocytoma.** A posterior longitudinal sonogram shows the right adrenal mass (M) anterosuperior to the right kidney (K) and posterior to the liver (H). (From Mitty and Yeh,[28] with permission from the publisher.)

the status of adjacent structures and can demonstrate bilateral or multiple pheochromocytomas, extra-adrenal tumors, and metastases. Pheochromocytomas generally appear as round, oval, or pear-shaped masses. On ultrasound, they tend to be slightly less echogenic than liver and kidney parenchyma (because of their relatively homogeneous cellular structure), though focal necrosis can cause marked echogenicity. On CT, pheochromocytomas often have an attenuation value less than that of liver or renal parenchyma and may simulate thick-walled cystic lesions.

Although adrenal venography is a highly accurate method of localizing adrenal pheochromocytomas and

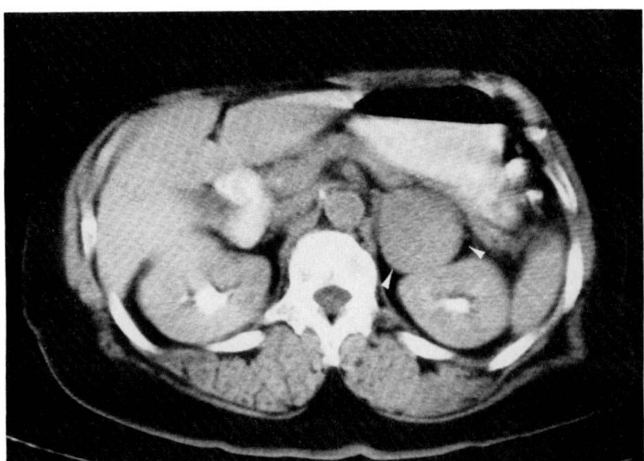

**Figure 4.58. Pheochromocytoma.** A CT scan demonstrates a large pear-shaped mass (arrows) anterior to the left kidney.

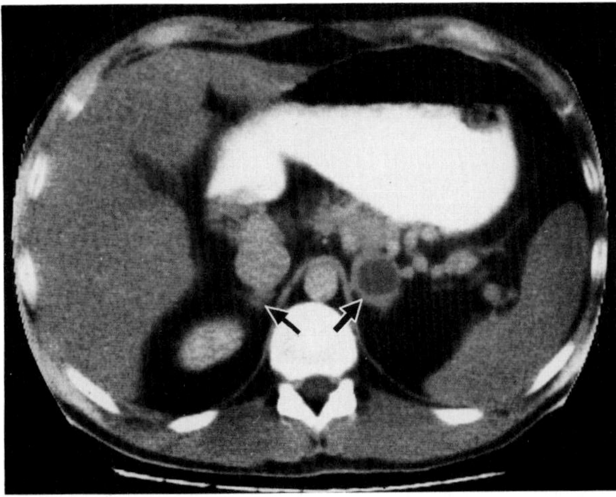

**Figure 4.59. Bilateral adrenal pheochromocytomas (arrows).** A CT scan following the administration of oral and intravenous contrast material shows a left adrenal lesion with peripheral enhancement and a low-density center; the appearance simulates a thick-walled cystic lesion. The patient also had medullary thyroid carcinoma (multiple endocrine neoplasia type II). (From Welch, Sheedy, van Heerden, et al,[60] with permission from the publisher.)

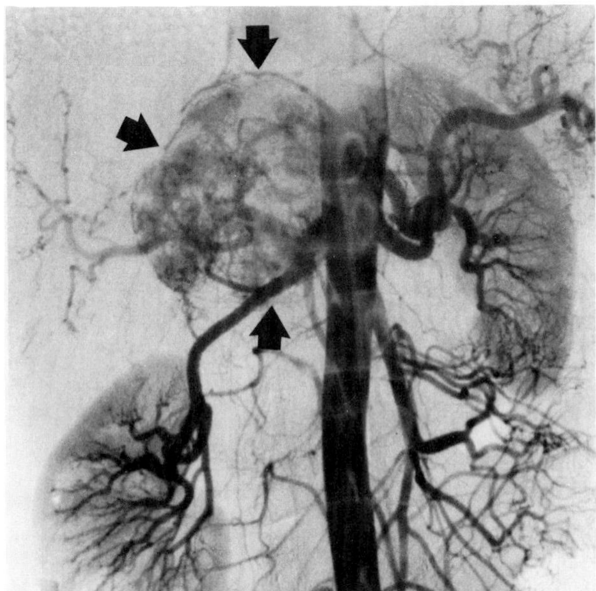

**Figure 4.60. Pheochromocytoma.** Arteriography demonstrates a large hypervascular right adrenal mass (arrows) displacing the right kidney inferiorly. (From Burko, Smith, Kirchner, et al,[122] with permission from the publisher.)

permits the drawing of blood samples for the quantification of catecholamine levels, the diagnostic role of this modality has been greatly reduced by CT and ultrasound. Because in patients with pheochromocytomas the arterial injection of contrast causes a marked elevation in blood pressure that must be controlled by $\alpha$-adrenergic blocking agents, and in view of the need for multiple injections (because several vessels feed each adrenal gland), selective arteriography is much less frequently performed.

Most extra-adrenal pheochromocytomas arise in the abdomen; a few are found in the chest or neck. The tumor may be located anywhere along the sympathetic nervous system, in the organ of Zuckerkandl, in chemoreceptor tissues such as the carotid body or the glomus jugulare, in the wall of the urinary bladder, or even in the kidney or ureter. Masses may displace the ureter or kidney or may appear as a filling defect in the bladder. CT and ultrasound are highly accurate in demonstrating small extra-adrenal pheochromocytomas not detectable on excretory urograms.

A higher-than-normal incidence of pheochromocytoma occurs in patients with neurofibromatosis, von Hippel-Lindau disease, and the multiple endocrine neoplasia syndromes.[28,52–60,119]

## Neuroblastoma

Neuroblastoma, a tumor of adrenal medullary origin, is the second most common malignancy in children. About 10 percent of these tumors arise outside the adrenal gland, primarily in sympathetic ganglia in the neck, chest, abdomen, or pelvis. The tumor is highly malignant and tends to attain great size before detection.

Calcification is common in neuroblastoma (occurring in about 50 percent of the cases), in contrast to the relatively infrequent calcification in Wilms's tumor, from which neuroblastoma must be differentiated. Calcification in a neuroblastoma has a fine granular or stippled appearance. Occasionally, there may be a single mass of amorphous calcification. Calcification can also develop in metastases of neuroblastoma in paravertebral lymph nodes and liver.

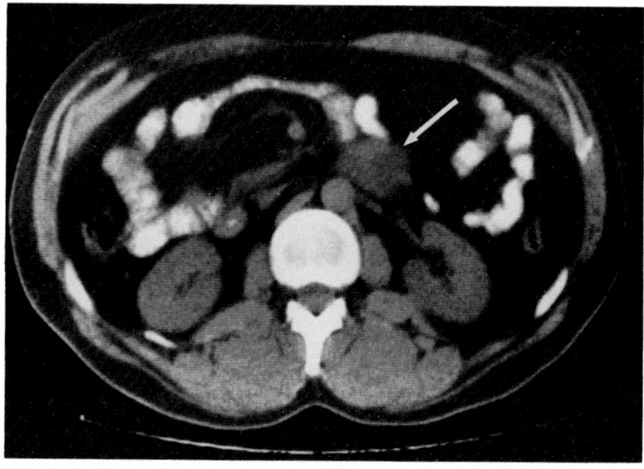

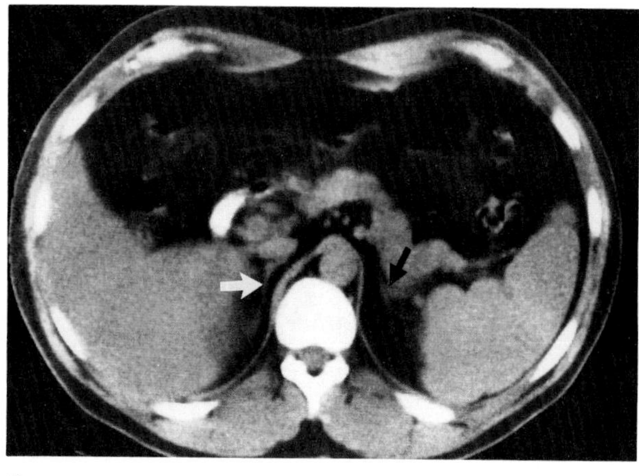

A B

**Figure 4.61. Ectopic pheochromocytoma.** (A) A CT scan shows a soft tissue mass (arrow) adjacent to the aorta, in front of the left renal vein. (B) A scan taken at a higher level demonstrates that both the right and the left adrenal glands are normal (arrows). (From Welch, Sheedy, van Heerden, et al,[60] with permission from the publisher.)

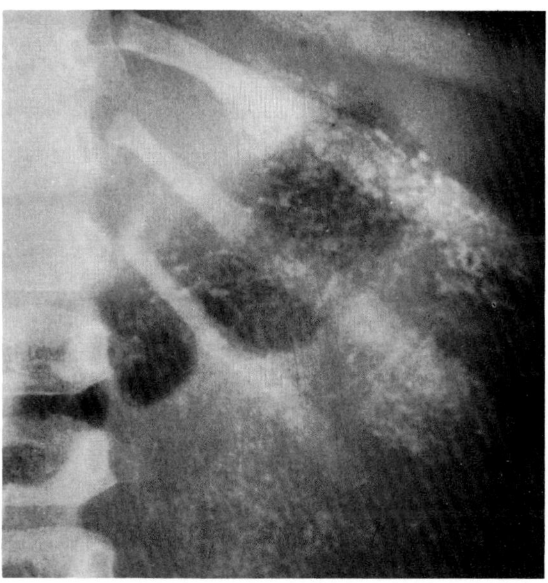

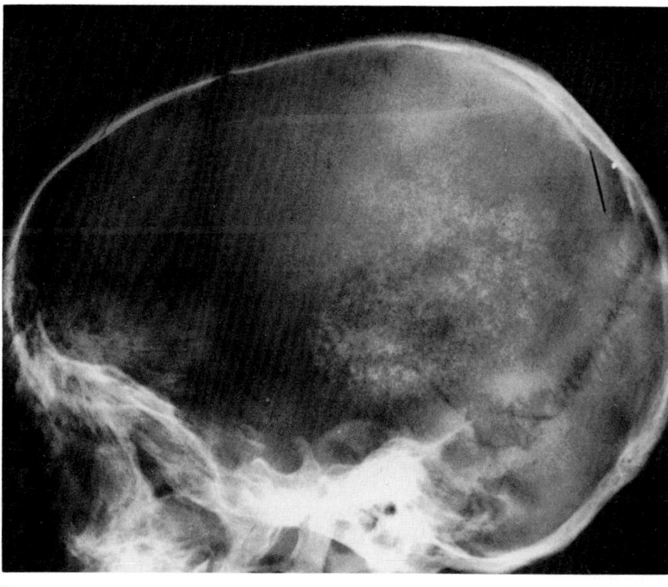

A B

**Figure 4.62. Neuroblastoma metastatic to bone.** (A) A plain film of the left upper abdomen shows diffuse granular calcification within a large neuroblastoma. (B) A lateral view of the skull shows similar calcific deposits within a metastatic lesion in the calvarium. Note the sutural widening consistent with increased intracranial pressure.

Excretory urography usually demonstrates downward and lateral renal displacement by the tumor mass. Neuroblastoma tends to cause the entire kidney and its collecting system to be displaced as a unit, unlike Wilms's tumor, which has an intrarenal origin and thus tends to distort and widen the pelvocalyceal system.

Because of its nonionizing character, ultrasound is a superb modality for evaluating abdominal masses in children. A neuroblastoma appears as a solid or semisolid mass that may be partly cystic but is separate from the kidney. The tumor is often diffusely highly echogenic, probably because of necrosis and hemorrhage. CT can show evidence of tumor spread to lymph nodes and the sympathetic chain, widening of the paravertebral stripe, and metastases to the liver and chest. This modality can also be used to assess the response to treatment.

Metastases to bone, liver, and lungs are common in neuroblastoma. Metastases to the skull typically cause spreading of cranial sutures because of plaques of tumor tissue growing along the surface of the brain. Bone destruction leads to a granular pattern of osteoporosis that is often associated with thin whisker-like calcifications coursing outward and inward from the tables of the skull. Metastases in long tubular bones most commonly involve the medial surfaces of the proximal metaphysis of the humerus and the distal metaphysis of the femur. Metastases are often multiple and relatively symmetric and present a permeative destructive pattern. Periosteal reaction is common and may appear as a layer of calcification parallel to the cortex or as perpendicular spicules similar to those found in the skull. Patchy areas of sclerosis may occur.

Neuroblastomas arising in the chest appear as posterior mediastinal masses. Metastases to the chest most commonly cause the asymmetric enlargement of mediastinal nodes. Metastases to the pulmonary parenchyma are infrequent.[28,61–64,82]

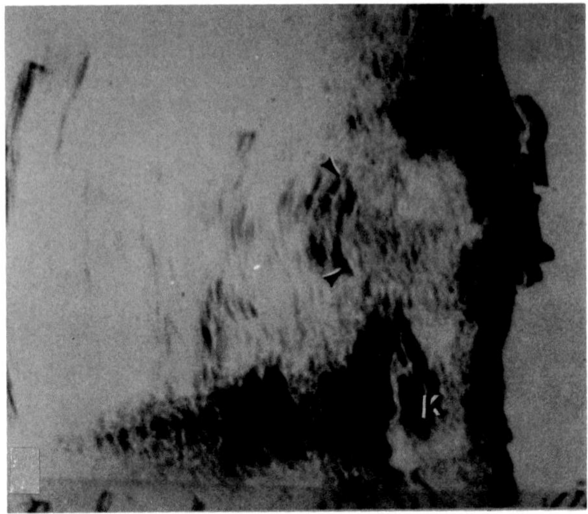

**Figure 4.63. Neuroblastoma.** A posterior longitudinal ultrasound scan demonstrates a suprarenal mass (arrowheads) that appears more echogenic than most adrenal tumors. The examination was performed with the patient in a sitting position, since the patient could not lie prone because of a markedly protuberant abdomen. (From Yeh, Mitty, Rose, et al,[119] with permission from the publisher.)

## Carotid Body Tumors

The carotid body is a chemoreceptor that appears to alter vasomotor activity and the respiratory rate in response to changes in carbon dioxide and oxygen tensions as well as to changes in blood pH. Tumors of the carotid body (chemodectoma) are rare and histologically benign, though they are reported to metastasize in about 5 percent of the cases. The tumors are situated in the neck at or near the bifurcation of the common carotid artery. Most are asymptomatic; cough, dysphagia, and hoarseness occasionally occur.

Carotid arteriography demonstrates a vast network of fine vessels in a highly vascular mass near the bifurcation of the common carotid artery. Large lesions may displace the internal and external carotid arteries. Because about 5 percent of carotid body tumors are bilateral, arteriography should be performed on both sides of the neck.

Ultrasound and CT can be used to detect carotid body tumors. On ultrasound, the lesion appears as a weakly echogenic, well-circumscribed solid mass. On CT, especially dynamic scanning with multiple images after

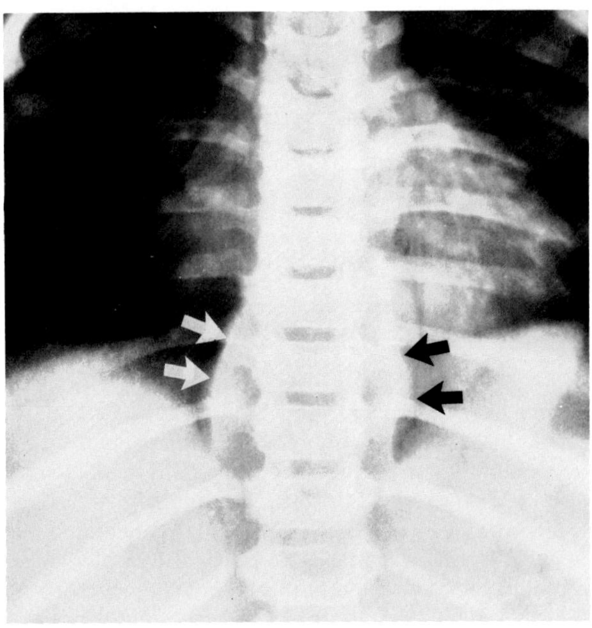

A

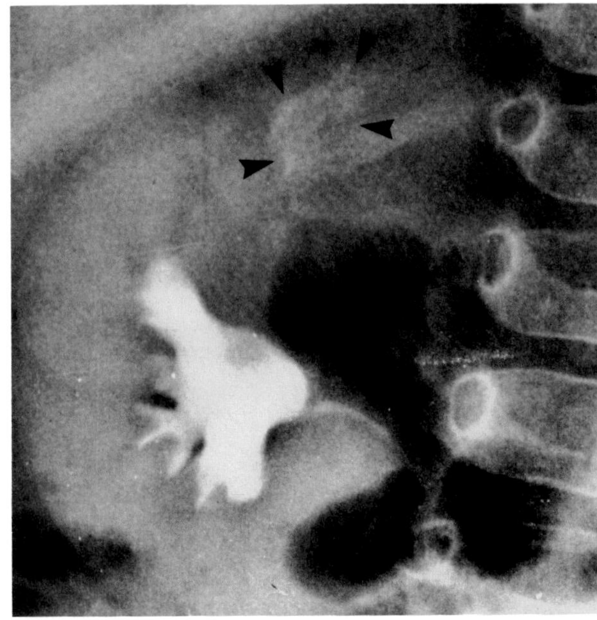

B

**Figure 4.64. Neuroblastoma.** (A) Bilateral widening of the paravertebral stripes (arrows) represents metastatic deposits. (B) An excretory urogram demonstrates the calcified tumor causing downward and lateral displacement of the upper pole of the right kidney. (From Friedland, Filly, Goris, et al,[126] with permission from the publisher.)

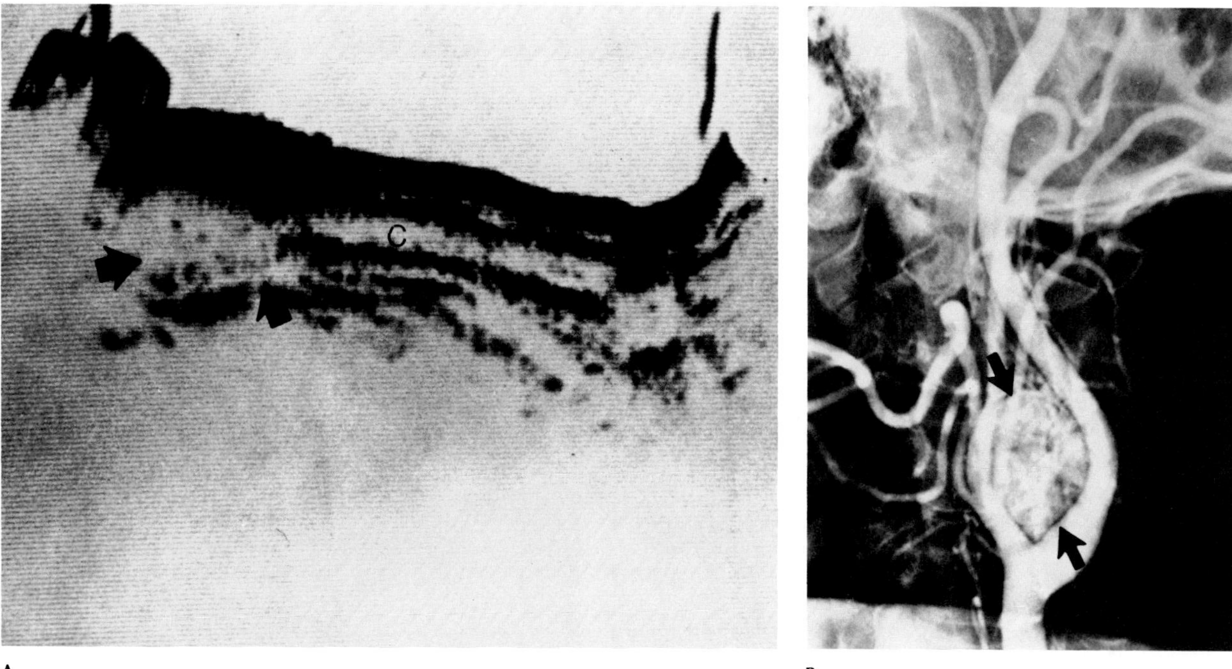


**Figure 4.65. Carotid body tumor.** (A) A longitudinal sonogram shows a solid mass (arrow) at the carotid bifurcation (C). Some artifactual echoes are seen within the carotid artery. (B) An arteriogram demonstrates characteristic tumor hypervascularity (arrows). (From Gooding,[67] with permission from the publisher.)

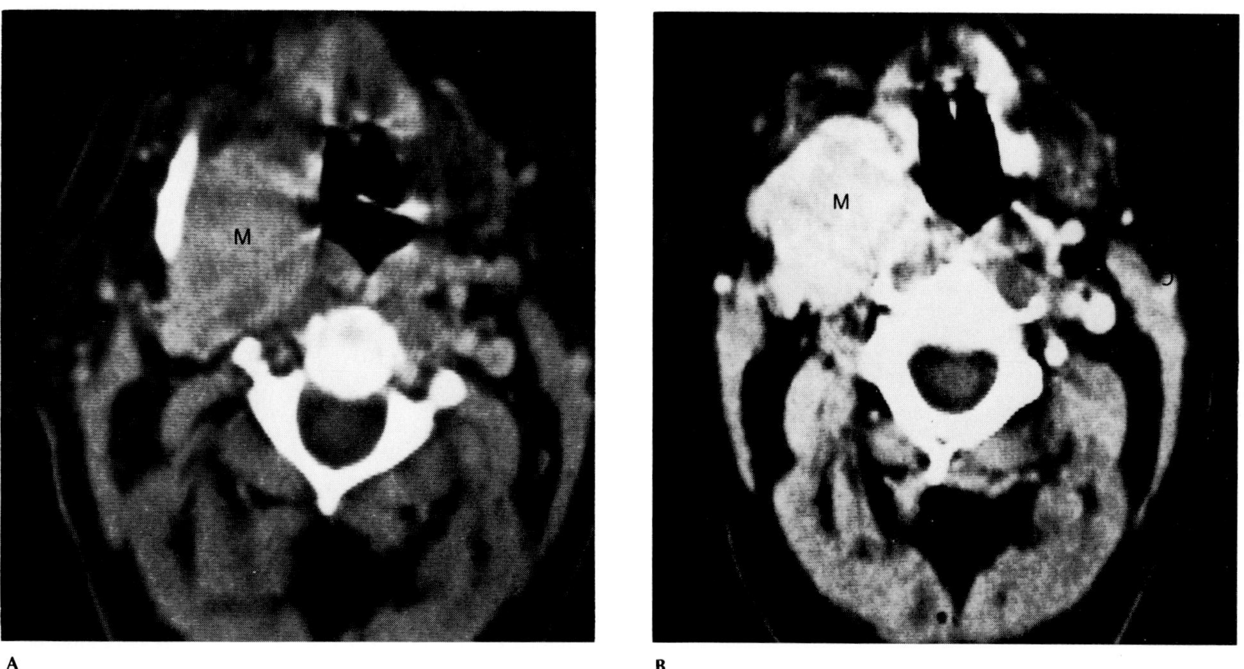


**Figure 4.66. Carotid body tumor.** CT scans before (A) and after (B) the infusion of constrast material show the characteristic slightly lobulated contour and the intense, homogeneous enhancement of the right submandibular tumor mass. (From Mancuso,[69] with permission from the publisher.)

the rapid intravenous injection of contrast material, a carotid body tumor appears as a markedly enhancing mass that is clearly separate from adjacent muscles and vascular structures.[65–69]

## DIABETES MELLITUS

Patients with diabetes mellitus are prone to develop late complications involving multiple organ systems. Atherosclerotic disease and subsequent ischemia involving the coronary, extracerebral, and peripheral circulations occur earlier and are more extensive in diabetics, especially those who smoke. There is a high incidence of peripheral vascular calcification as a result of the often-associated Mönckeberg's sclerosis, in which calcium deposition follows the degeneration of smooth muscle cells in the media of medium-sized muscular arteries. Men with diabetes may demonstrate characteristic calcification of the vas deferens; the calcification appears as bilaterally symmetric parallel tubular densities that run medially and caudally to enter the medial aspect of the seminal vesicles at the base of the prostate.

Diabetics have an increased susceptibility to infection; this especially affects the feet and may lead to severe osteomyelitis that produces bone destruction without periosteal reaction. Diabetic neuropathy with the loss of proprioception or deep pain sensation may lead to repeated trauma on an unstable joint. Degeneration of cartilage, recurrent fractures and fragmentation of subchondral bone, soft tissue debris, and marked proliferation of adjacent bone can lead to total disorganization of the joint (Charcot's joint) (see the section "Neuropathic Joint Disease" in Chapter 64). Vascular disease with diminished blood supply can lead to gas gangrene, in which bubbles or streaks of gas develop in the subcutaneous or deeper tissues.

Diabetic neuropathy often causes radiographically evident abnormalities in the gastrointestinal tract. In the esophagus there are decreased primary peristalsis, tertiary contraction waves, and delayed emptying, especially in the recumbent position. Delayed gastric emptying and decreased peristalsis occur in up to one-third of diabetics; acute ketoacidosis can cause atony and severe gastric dilatation. Reduced duodenal peristalsis can lead to relative narrowing of the transverse duodenum with proximal dilatation (superior mesenteric artery syndrome). Dilatation of the small bowel with normal folds, occasionally associated with severe diarrhea and malabsorption, has been reported in diabetics, especially when the disease is complicated by hypokalemia. Severe gaseous distension of the large bowel without an organic obstruction (colonic ileus) may cause constipation and present a radiographic appearance sim-

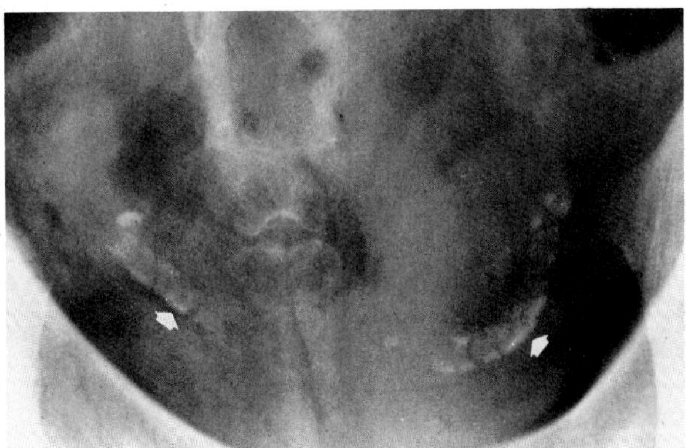

**Figure 4.67. Diabetes mellitus.** Bilaterally symmetric parallel tubular densities in the pelvis (arrows) represent the typical calcification of the vas deferens.

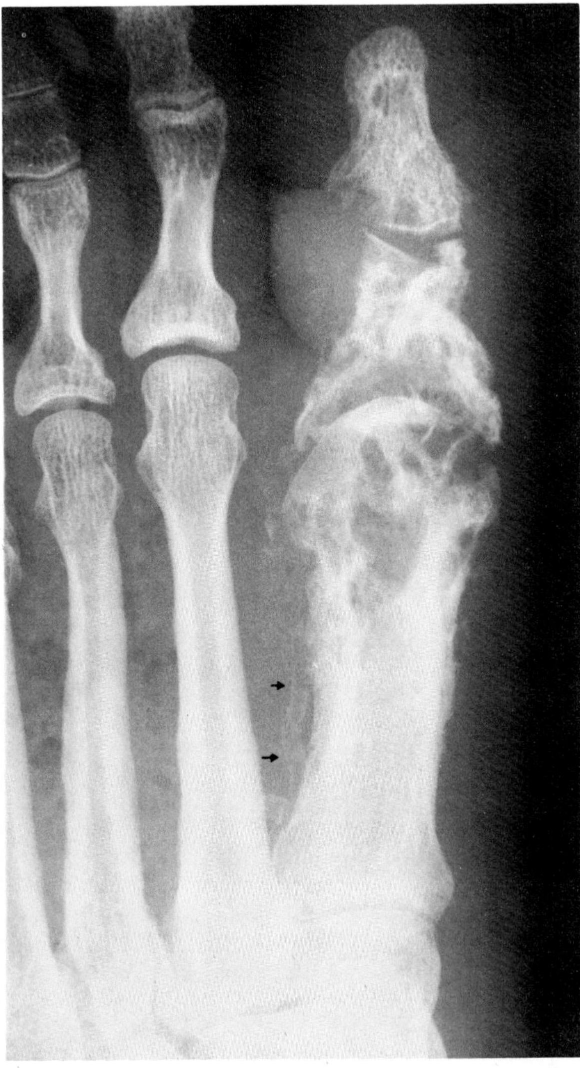

**Figure 4.68. Diabetic osteomyelitis.** Severe bone destruction of the distal first metatarsal and the proximal phalanges of the great toe. There are areas of sclerosis and periosteal reaction. Note the typical calcification of peripheral vessels (arrows) and the gas bubbles within the soft tissues.

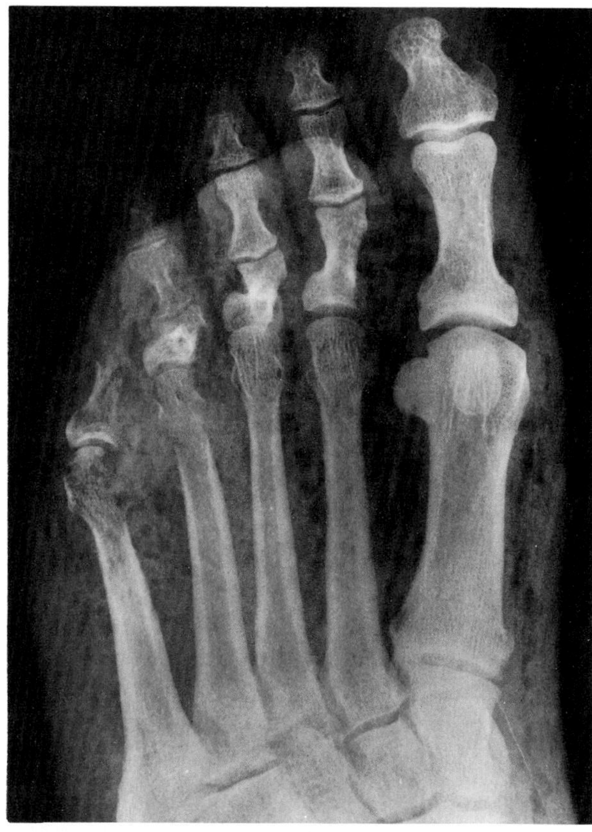

**Figure 4.69. Diabetic gangrene.** Diffuse destruction of the phalanges and the metatarsal head of the fifth digit. Note the large amount of gas within the soft tissues of the foot.

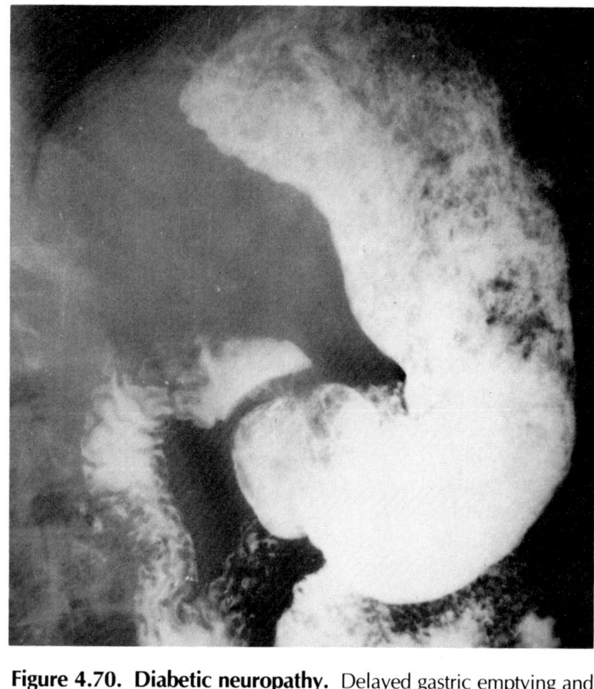

**Figure 4.70. Diabetic neuropathy.** Delayed gastric emptying and decreased peristalsis lead to a dilated, atonic stomach with a substantial solid gastric residue. There is no evidence of gastric outlet obstruction. (From Gramm, Reuter, and Costello,[72] with permission from the publisher.)

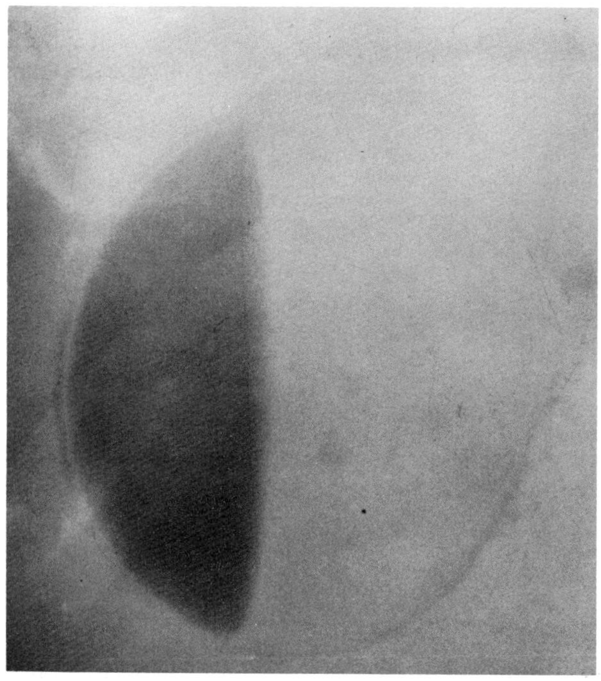

A

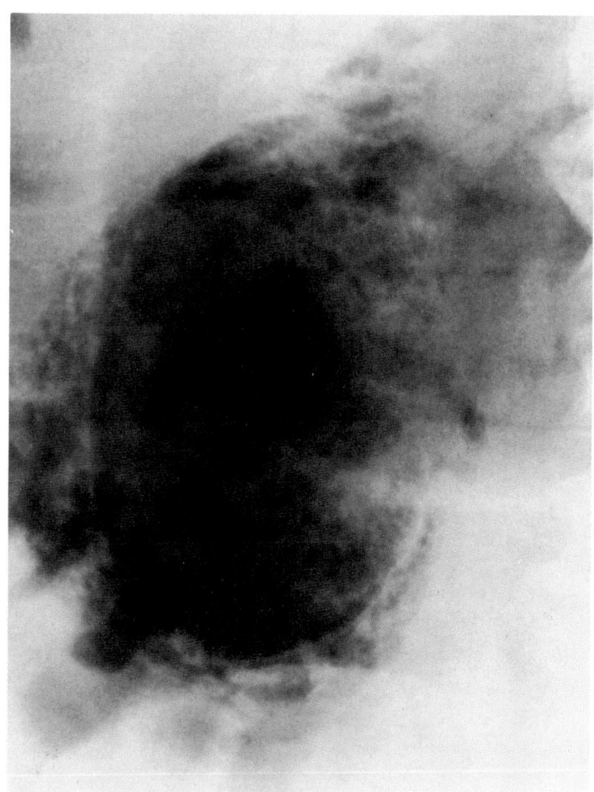

B

**Figure 4.71. Gas-forming visceral infections in diabetes.** (A) Emphysematous cholecystisis appears as gas within both the lumen and the wall of the gallbladder. (B) Emphysematous pyelonephritis produces gas in and around an enlarged, nonfunctioning kidney. (A, from Eisenberg,[71] with permission from the publisher; B, From Davidson,[121] with permission from the publisher.)

ulating that of mechanical colonic obstruction. Diabetic neuropathy can also lead to gallbladder enlargement with bile stasis and an increased incidence of gallstones. Emphysematous cholecystitis with gas in the lumen and wall of the gallbladder is a severe complication that occurs almost exclusively in diabetics.

Renal disease is a common complication and a leading cause of death in diabetics. Acute and chronic pyelonephritis, renal papillary necrosis, and cystitis often occur. Nodular sclerosis of the renal glomeruli and arterioles (Kimmelstiel-Wilson syndrome) may occur in chronic diabetics and may lead to the nephrotic syndrome or chronic renal failure. Diabetic neuropathy can also cause dilatation and atony of the bladder with incomplete emptying.

Mucormycosis infection, which primarily involves the sinuses and lungs, occurs virtually only in uncontrolled diabetics (see the section "Mucormycosis" in Chapter 14).[70–72]

## HORMONE-SECRETING TUMORS OF THE PANCREAS

### Insulinoma

Although islet cell tumors of the pancreas may cause severe endocrine disorders, they are usually small, and their preoperative localization presents a diagnostic challenge. Ultrasound is often the initial imaging modality since this noninvasive procedure can detect small tumors in the head or body of the pancreas if there is no overlying bowel gas. CT with bolus injection of contrast material and rapid sequential scanning can demonstrate the transient increase in contrast enhancement (tumor blush) that is characteristic of these tumors, which may be impossible to differentiate from surrounding pancreatic tissue on noncontrast scans. In the past, selective celiac and mesenteric arteriography has been used for the preoperative demonstration of islet cell tumors, though the overall detection rate was less than 50 percent. At arteriography, the tumors show a nonspecific capillary blush (tumor stain) along with displacement of adjacent small vessels.

### Other Hormone-Secreting Tumors of the Pancreas

Tumors of the pancreatic islets can synthesize a variety of hormones other than insulin. Ulcerogenic islet cell tumors (gastrinomas) produce the Zollinger-Ellison syndrome, which is characterized by intractable ulcer symptoms, hypersecretion of gastric acid, and diarrhea. (See the section "Zollinger-Ellison syndrome" in Chapter 44.) Diarrheogenic islet cell tumors produce the WDHA syndrome; the acronym stands for *w*atery *d*iarrhea, *h*ypokalemia, and *a*chlorhydria, which are major features of the clinical picture. ACTH production by pancreatic islet cell tumors causes Cushing's syndrome; the release of serotonin by pancreatic tumors may cause the carcinoid syndrome. Glucagonomas, many of which are malignant and metastasizing, cause a clinical syndrome characterized by a distinctive skin lesion, weight loss, and anemia.

Because functional islet cell tumors usually become apparent by their hormonal effects rather than by the

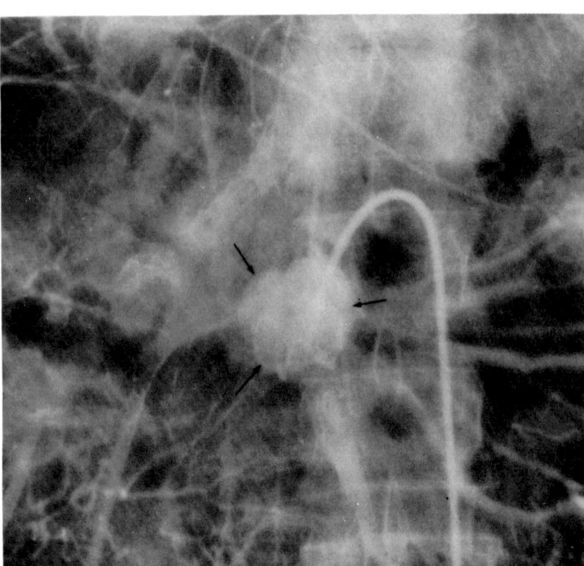

**Figure 4.72. Insulinoma.** A film obtained during the late arterial phase of a superior mesenteric artery injection demonstrates an intense tumor blush (arrows) in the pancreatic bed. (Courtesy of Robert Gray, M.D.)

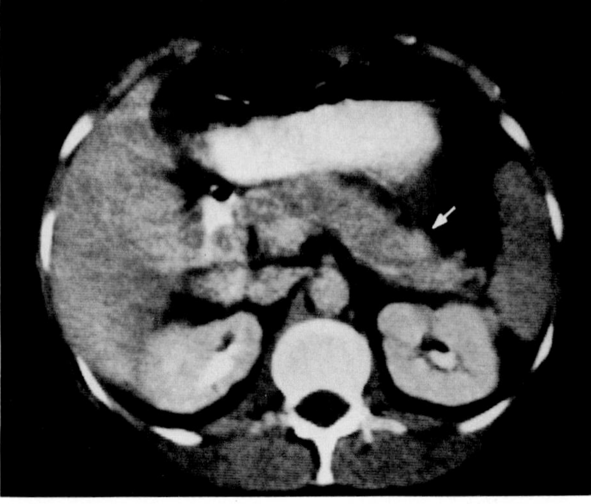

**Figure 4.73. Gastrinoma.** A CT scan with contrast material reveals a 1-cm enhancing lesion (arrow) within the normal contour of the pancreatic tail. Multiple 0.5- to 2-cm pancreatic gastrinomas were found at surgery in this patient with clinical evidence of the Zollinger-Ellison syndrome. (From Stark, Moss, Goldberg, et al,[76] with permission from the publisher.)

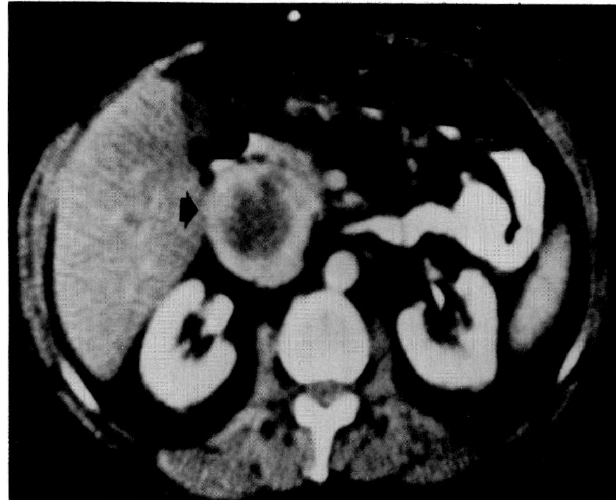

**Figure 4.74. Resectable 6-cm gastrinoma arising from the head of the pancreas.** A CT scan shows peripheral contrast enhancement (vascular blush) about the mass (arrowhead). The remainder of the pancreas, the retroperitoneum, and the liver were normal. (From Stark, Moss, Goldberg, et al,[76] with permission from the publisher.)

consequences of tumor bulk, they are often small and difficult to detect. CT demonstrates the primary pancreatic tumor in less than half of the cases. Many pancreatic islet cell tumors are malignant, and hepatic metastases may be seen on CT examination. Pancreatic arteriography may show intense tumor staining in patients with vascular lesions.[73–77]

## DISORDERS OF THE TESTIS

### Testicular Tumors

Testicular tumors are the most common neoplasms in men between the ages of 20 and 35. Almost all testicular tumors are malignant and tend to metastasize to the lymphatics that follow the course of the testicular arteries and veins and drain into para-aortic lymph nodes at the level of the kidneys.

The homogeneous, medium-level echogenicity of the normal testis on ultrasound examination provides an excellent background for the detection of testicular disease. A localized testicular tumor appears as a circumscribed hypoechogenic, hyperechogenic, or mixed echogenic nidus in an otherwise uniform testicular echo structure. Although this is a somewhat nonspecific pattern that can also be seen with orchitis, hemorrhage, infarction, abscess, or chronic torsion, the accuracy of ultrasound in detecting testicular neoplasms is high. Ultrasound is of special value in detecting an occult malignant testicular neoplasm in patients with germ cell tumor metastases who have testicles that appear normal on palpation.

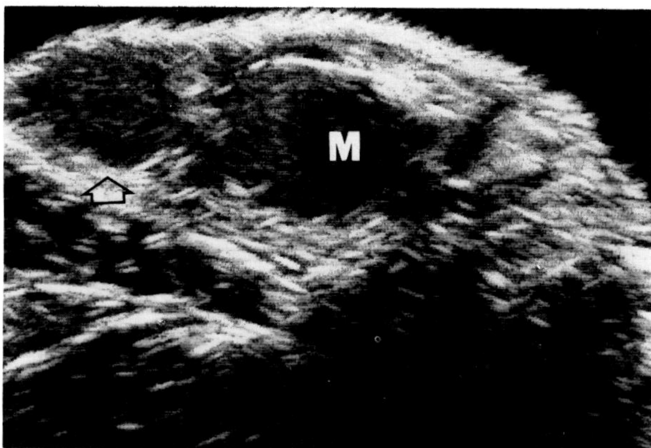


**Figure 4.75. Testicular carcinoma.** A transverse sonogram shows a predominantly sonolucent mass (M) within an enlarged left testis. The right testis (arrow) is normal. (Courtesy of Robert Mevorach, M.D.)

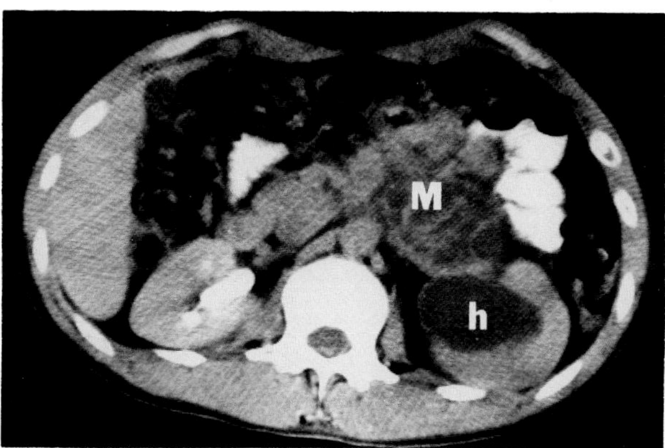


**Figure 4.76. Metastatic testicular seminoma.** A CT scan through the level of the kidneys shows diffuse nodal metastases containing characteristic low-attenuation areas. Extrinsic pressure on the lower left ureter has caused severe hydronephrosis with dilatation of the renal pelvis (h).

Because testicular tumors first metastasize to lymph nodes at the renal hilar level, a nodal group not normally opacified with bipedal lymphography, CT is generally considered the preferred method for the initial staging of patients with known testicular neoplasms. In addition to demonstrating metastatic nodal enlargement, CT can also delineate the exact extent of the tumor mass and detect any spread to the lung or liver. Because lymphography can detect neoplastic replacement in normal-sized lymph nodes, it may be performed in patients in whom the CT scan is equivocal or negative.[78–81]

### Testicular Torsion and Epididymitis

Testicular torsion refers to the twisting of the gonad on its pedicle that leads to the compromise of circulation

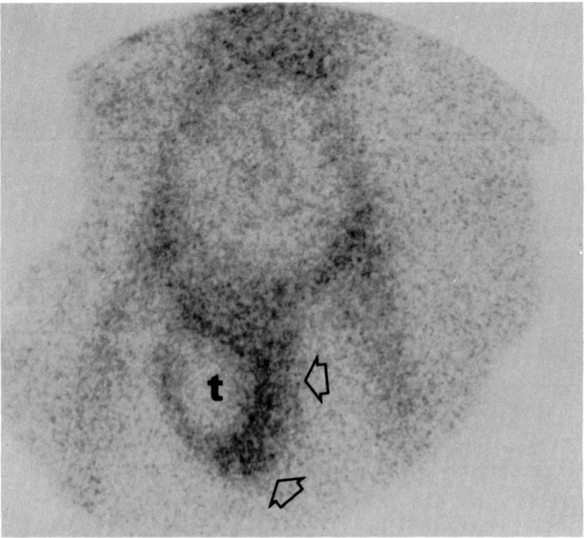


**Figure 4.77. Testicular torsion.** Severely diminished arterial perfusion causes the testicle to appear as a rounded, cold area (t) on a radionuclide scan. The surrounding rim of increased activity represents the blood supply to the scrotal sac (arrows).

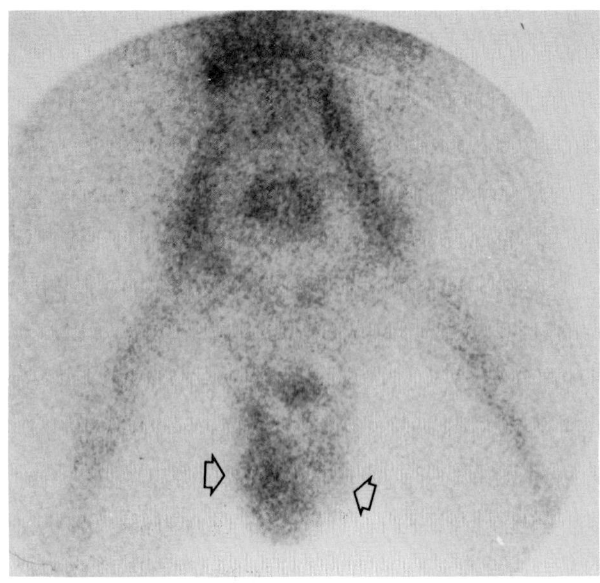

**Figure 4.78. Epididymitis.** A radionuclide scan shows high isotope uptake in the region of the testicle (arrows) due to increased blood flow.

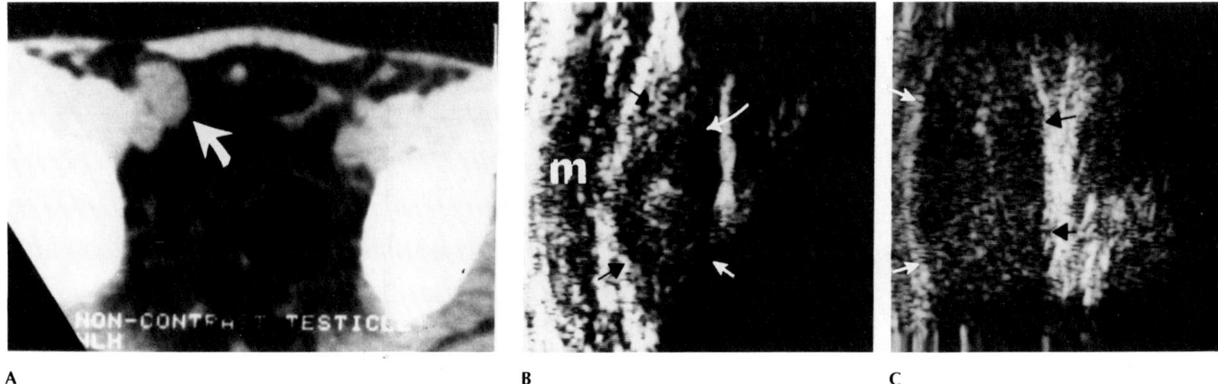


A B C

**Figure 4.79. Intra-abdominal undescended testis in a 12-year-old boy.** (A) A CT scan at the midpelvis shows the undescended testis (arrow) on the surface of the iliac vessels deep to the anterior abdominal wall. The attenuation value of the testis is a little less than that of the surrounding soft tissues. (B) A real-time ultrasound scan of the right lower abdomen demonstrates the testis as an oval structure (arrows) deep to the abdominal wall muscle layer (M). (C) A real-time ultrasound image of the normally descended testis in the left scrotum (arrows). The undescended testis is approximately half the size of the scrotal testis. (From Wolverson, Houttuin, Heiberg, et al,[123] with permission from the publisher.)

and the sudden onset of severe scrotal pain. Although primarily a clinical diagnosis, the scrotal pain and swelling of testicular torsion may be difficult to distinguish from that due to epididymitis. In such cases, Doppler ultrasound or radionuclide studies are of value.

Doppler ultrasound demonstrates the presence of intratesticular arterial pulsations. In testicular torsion, the arterial perfusion is diminished or absent, while in epididymitis there is increased blood flow. Similarly, the radionuclide angiogram shows that the isotope activity on the twisted side is either slightly decreased or at the normal, barely perceptible level. On the uninvolved side, the perfusion should be normal; when compared with the decreased activity on the involved side, it appears increased. Static nuclear scans demonstrate a rounded, cold area replacing the testicle in patients with torsion, but a hot area in those with epididymitis.

On gray-scale ultrasound, acute and subacute torsion result in enlargement and decreased echogenicity of the affected testis. There usually is associated skin thickening and an alteration of the long axis of the testis from cephalocaudal to more horizontal. Chronic torsion produces a small homogeneously echo-poor testicle with a large uniformly echogenic epididymis. Acute epididymitis

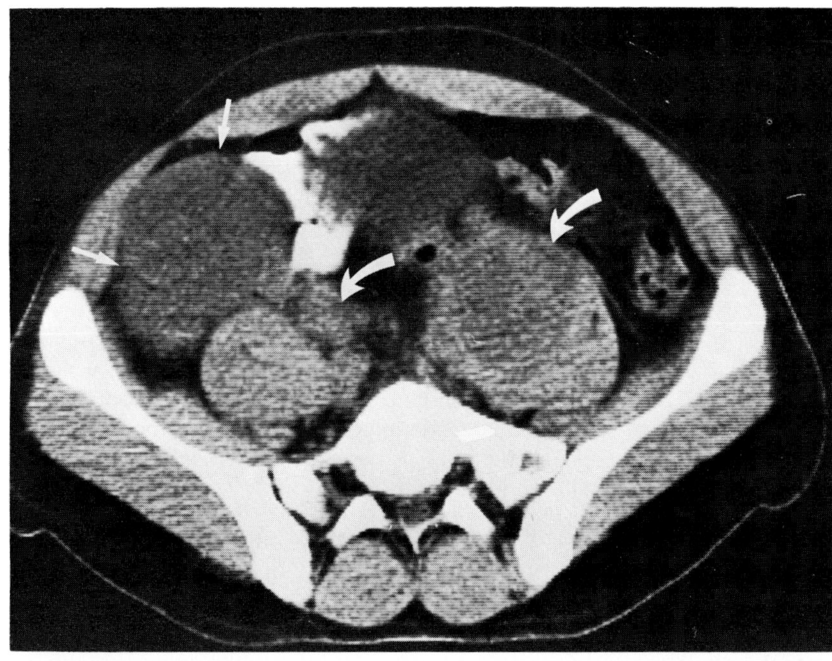

A

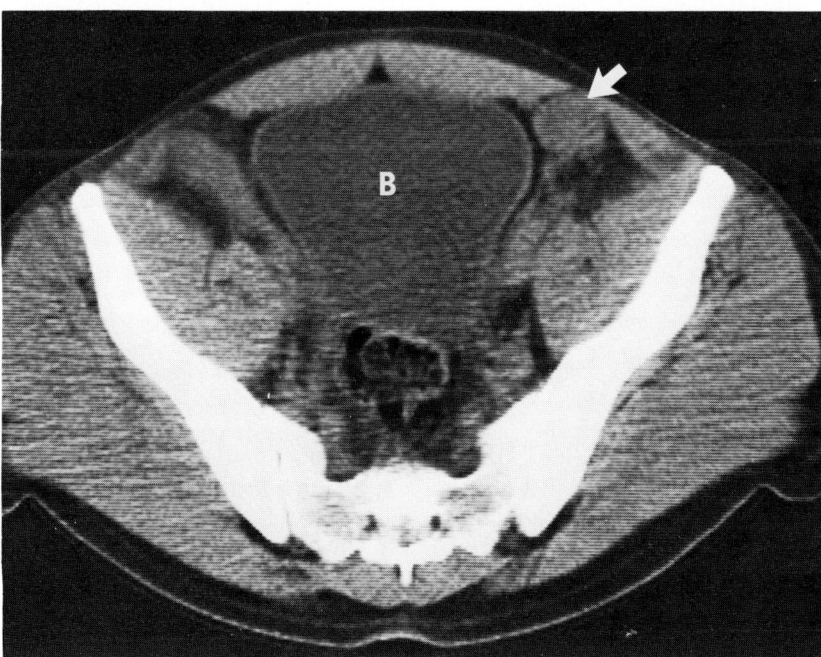


B

**Figure 4.80. Bilateral undescended testes.** (A) The undescended right testis is enlarged by a mass (straight arrows) representing an embryonal cell carcinoma. The tumor has metastasized to lymph nodes (curved arrows), which are enlarged. (B) The nontumorous intra-abdominal left testis (arrow) appears as a smaller, rounded structure adjacent to the bladder (B). (From Jeffrey,[127] with permission from the publisher.)

causes enlargement of the inflamed epididymis; nonhomogeneous enlargement of the testicle, with areas of decreased echogenicity; a marked increase in peritesticular fluid; and peritesticular soft tissue–skin thickening.[83–86]

## Undescended Testis (Cryptorchism)

Near the end of gestation, the testis normally migrates from its intra-abdominal position through the inguinal canal into the scrotal sac. If one of the testicles cannot be palpated within the scrotum, it is important to determine whether this represents an absent testis or an ectopic position of the testis. The rate of malignancy is up to 40 times higher in the undescended (intra-abdominal) than in the descended testis. Because of this extremely high rate of malignancy, the diagnosis of undescended (intra-abdominal) testis usually leads to orchiopexy in patients younger than 10 years of age and orchiectomy in those seen after puberty.

In the absence of a palpable testicle, ultrasound is usually employed as a screening technique. The modality carries no radiation risk and is highly accurate in demonstrating the majority of undescended testicles that are located in the inguinal canal. The undescended testicle appears as an oval mass with uniformly distributed medium-level echoes that is situated in the expected course of testicular descent. However, sonography is not successful in detecting ectopic testicles in the pelvis or abdomen. If ultrasound fails to demonstrate an undescended testis, CT is indicated.[87,88]

## DISORDERS OF THE FEMALE REPRODUCTIVE TRACT

### Female Infertility

The major radiographic procedure for evaluating infertile women is hysterosalpingography, in which the uterine cavity and the fallopian tubes are opacified following the injection of contrast material into the uterus. In the normal woman with patent fallopian tubes, contrast material extravasating into the pelvic peritoneal cavity outlines the peritoneal surfaces and often loops of bowel within the pelvis. Developmental anomalies or fibrosis from pelvic inflammatory disease may cause occlusion of one or both of the fallopian tubes, and in such cases there is no evidence of the contrast material reaching the peritoneal cavity. In addition to assessing tubal patency, hysterosalpingography can also demonstrate uterine abnormalities contributing to infertility, such as intrauterine fibroids, severe uterine flexion or retroversion, and other congenital and acquired malformations.[89]

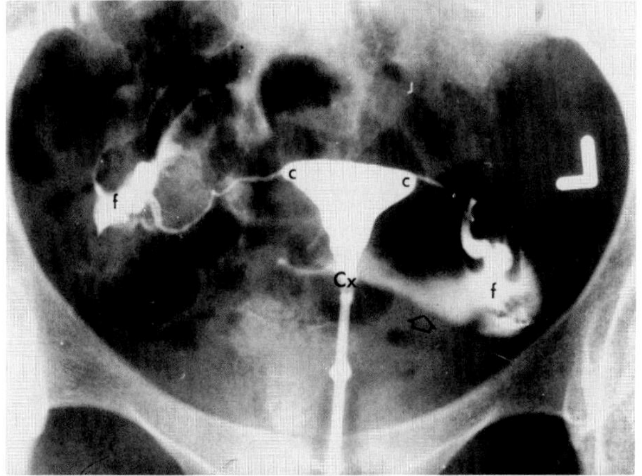


**Figure 4.81. Normal hysterosalpingogram.** The arrows point to the bilateral spill of contrast material into the peritoneal cavity. Cx = internal cervical os; c = cornua of the uterus; Rt and Lt = right and left fallopian tubes; f = fimbriated portion. (From Yune, Klatte, Cleary, et al,[89] with permission from the publisher.)

### Trophoblastic Disease

Trophoblastic disease refers to a spectrum of pregnancy-related disorders ranging from benign hydatidiform mole to the more malignant and frequently metastatic choriocarcinoma. The clinical features of hydatidiform mole include abnormal vaginal bleeding, a uterus that appears large for dates, and abnormally high levels of human chorionic gonadotropin. On ultrasound the lesion typically appears as a large, soft tissue mass of placental (trophoblastic) tissue filling the uterine cavity and containing low-to-moderate-amplitude echoes; numerous small cystic fluid–containing spaces are scattered throughout. There is no evidence of a developing fetus. Multiple larger sonolucent areas, which represent areas of degeneration or internal hemorrhage, may be seen within the molar tissue.

About half of choriocarcinomas follow pregnancies complicated by hydatidiform mole. The remainder occur after spontaneous abortion, ectopic pregnancy, or normal deliveries. The ultrasound appearance of choriocarcinoma is similar to that of benign hydatidiform mole and usually consists of a large complex mass in the expected position of the uterus. Highly echogenic tissue is seen interspersed with multiple cystic areas which probably represent hemorrhage. Ultrasound may also demonstrate metastatic deposits in the liver. Choriocarcinoma tends to metastasize to the lung, where it typically produces multiple large masses that rapidly regress once appropriate chemotherapy is instituted.[90]

### Ovarian Cysts and Tumors

#### DERMOID CYST (TERATOMA) OF THE OVARY

Dermoid cysts, the most common type of germ cell tumors, contain skin, hair, teeth, and fatty elements, all of which typically arise from ectodermal tissue. About half of all ovarian dermoid cysts contain some calcification. This is usually in the form of a partially or completely formed tooth; less frequently, the wall of the cyst is partially calcified. The characteristic calcification combined with the relative radiolucency of the lipid material within the lesion is pathognomonic of an ovarian dermoid cyst.

The most common ultrasound appearance of a dermoid cyst is a complex, predominantly solid mass containing high-level echoes arising from hair and/or calcification within the mass. The highly echogenic nature of these masses may make it difficult to completely delineate the mass or to distinguish it from surrounding gas-containing loops of bowel. As with other ovarian tumors, the more irregular and solid the internal components of the mass, the more likely that it is malignant.[91–94]

#### CYSTADENOMA AND CYSTADENOCARCINOMA OF THE OVARY

Cystadenomas and cystadenocarcinomas of the ovary often produce calcifications that can be detected on plain

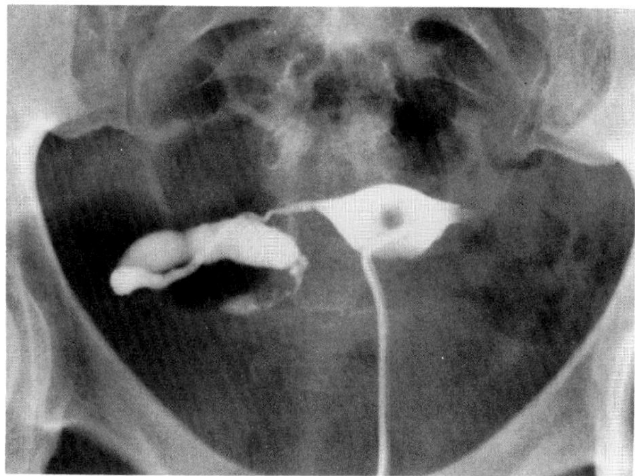
A

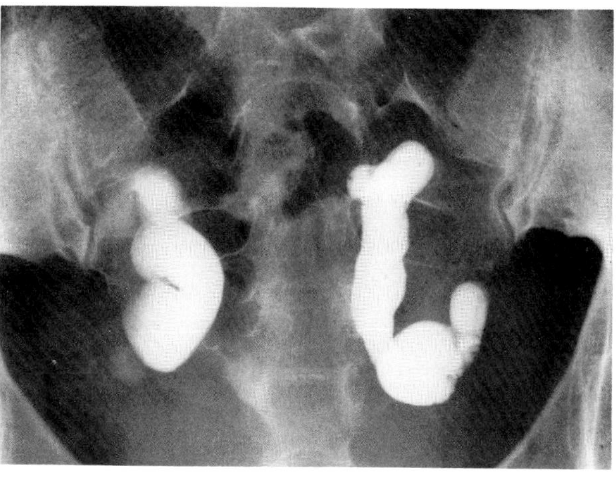
B

**Figure 4.82. Hydrosalpinx.** (A) Unilateral and (B) bilateral gross dilatation of the fallopian tubes without evidence of free spill of contrast material into the peritoneal cavity.

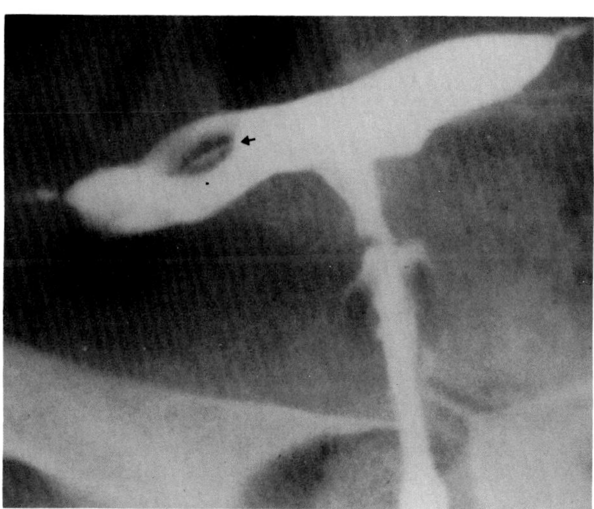

**Figure 4.83. Uterine polyp.** Filling the uterine cavity with a small volume of contrast material outlines a solitary filling defect on the right side of the fundal portion. (From Yune, Klatte, Cleary, et al,[89] with permission from the publisher.)

abdominal radiographs. Psammomatous bodies are depositions of calcium carbonate located in the fibrous stroma of papillary mucinous cystadenomas and cystadenocarcinomas. These psammomatous calcifications appear as scattered, fine amorphous shadows that are barely denser than the normal soft tissues and can therefore be easily missed unless they are extensive. Serosal and omental implants of serous cystadenocarcinoma can appear as diffuse, ill-defined collections of granular amorphous calcification. Less frequently, metastases are seen as sharply circumscribed masses of fairly homogeneous density with occasional rim calcification. Calcified metastatic deposits along the lateral abdominal wall adjacent to the peritoneal fat stripe are characteristic. Because ovarian metastases tend to be distributed along the course

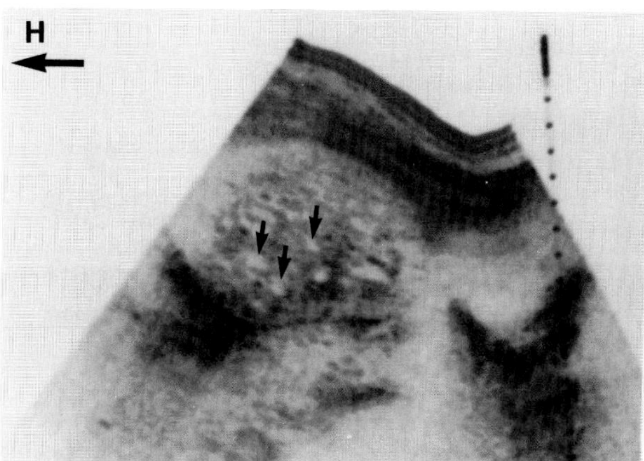


**Figure 4.84. Hydatidiform mole.** A longitudinal sonogram in a patient in the second trimester demonstrates a large, moderately echogenic mass filling the central uterine cavity. Note the numerous small cystic spaces (arrows) that represent the markedly hydropic chorionic villi. (From Callen,[90] with permission from the publisher.)

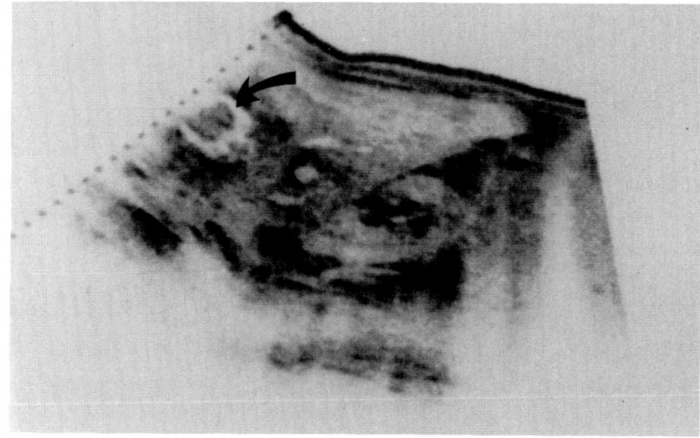

**Figure 4.85. Choriocarcinoma.** A longitudinal sonogram shows a necrotic metastasis with hemorrhage (arrow) in the liver. The patient died of massive hemoperitoneum from bleeding hepatic metastases. (From Callen,[90] with permission from the publisher.)

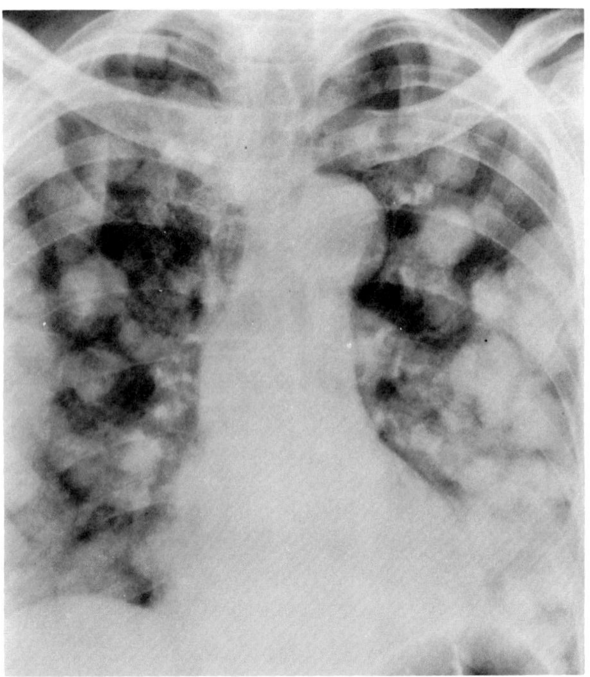

**Figure 4.86. Choriocarcinoma.** Multiple large metastatic deposits scattered diffusely throughout the lungs.

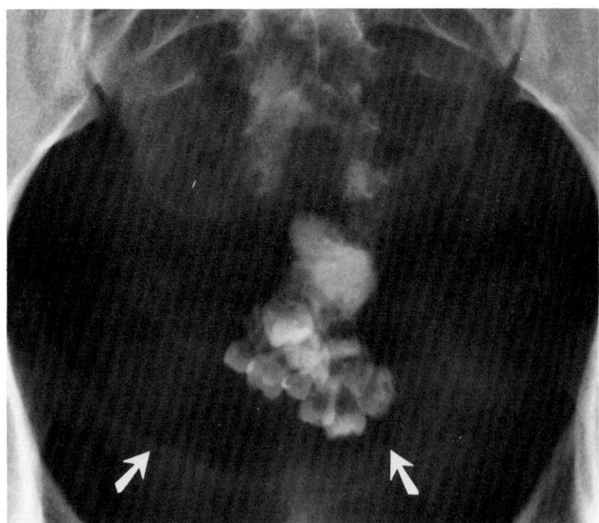

**Figure 4.87. Dermoid cyst containing multiple well-formed teeth.** Note the relative lucency of the mass (arrows), which is composed largely of fatty tissue. (From Eisenberg,[71] with permission from the publisher.)

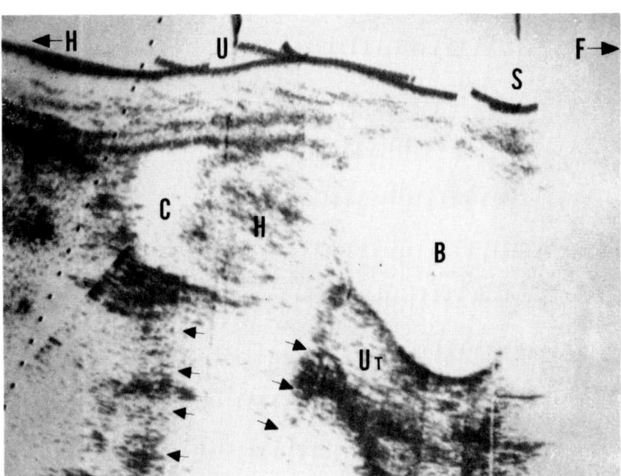


**Figure 4.88. Dermoid cyst.** A longitudinal sonogram demonstrates a mass containing both fluid and hair (H). Note the characteristic acoustic shadow (between arrows) cast by the hair within the lesion. C = cyst; B = bladder; Ut = uterus; S = symphysis pubis; U = umbilicus; H = head; F = feet. (From Burke, Smith, Kirchner, et al, [122] with permission from the publisher.)

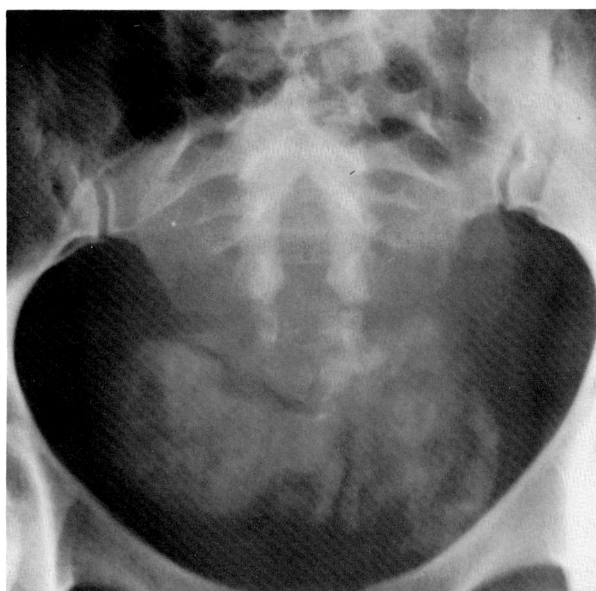

**Figure 4.89. Psammomatous calcification.** Diffuse, ill-defined collections of granular amorphous calcification are visible within this cystadenocarcinoma of the ovary.

of the mesentery of the colon, these vague and diffuse calcifications can initially be mistaken for feces or previously ingested barium.

On ultrasound examination, cystadenomas and cystadenocarcinomas typically appear as large cystic masses with internal septa. The number and the arrangement of the internal septa do not appear to correlate with whether the mass is benign or malignant. However, the more solid and irregular the areas within the mass, the more likely it represents a malignant tumor. In addition, the association of ascites with an ovarian mass strongly suggests underlying malignancy. Hepatic metastases associated with ovarian carcinoma usually appear as relatively hypoechoic masses within the liver. Peritoneal

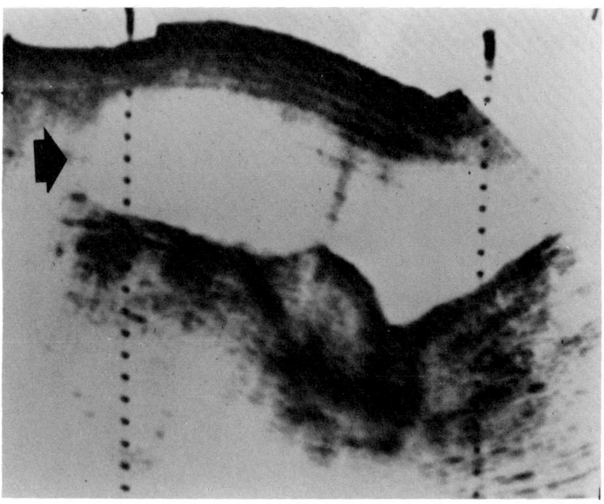

A

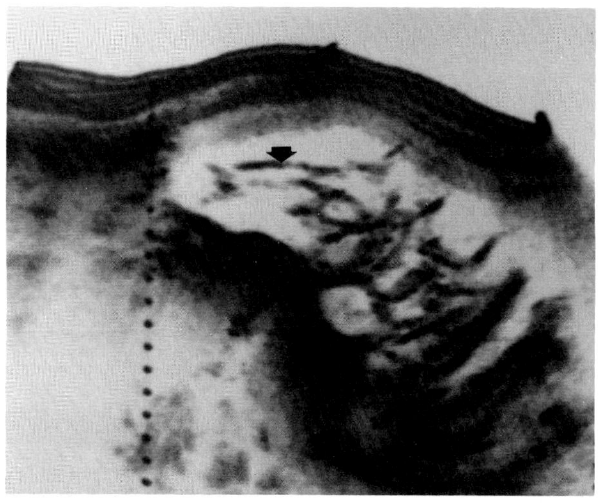

B

**Figure 4.90. Cystadenoma.** (A) A longitudinal sonogram in an asymptomatic girl demonstrates a serous cystadenoma as a large sonolucent, completely cystic mass. (B) A longitudinal sonogram demonstrates a complex, predominantly cystic mass containing several thin and well-defined septations (arrow), an appearance characteristic of a mucinous cystadenoma. (From Fleischer, Wentz, Jones, et al,[92] with permission from the publisher.)

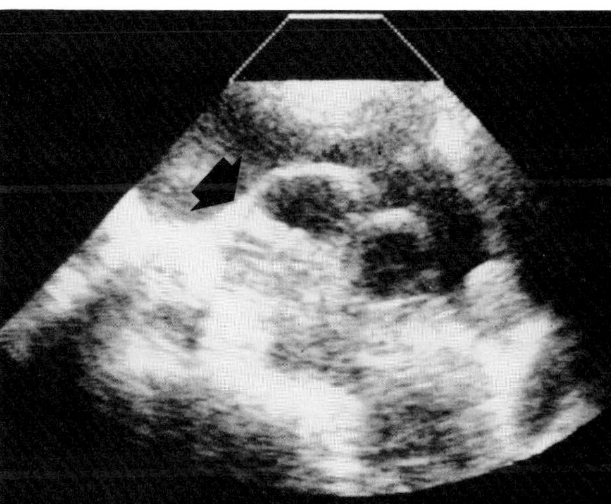

**Figure 4.91. Mucinous cystadenocarcinoma.** A sonogram demonstrates a predominantly solid mass with some cystic components and associated ascites. (From Fleischer, Wentz, Jones, et al,[92] with permission from the publisher.)

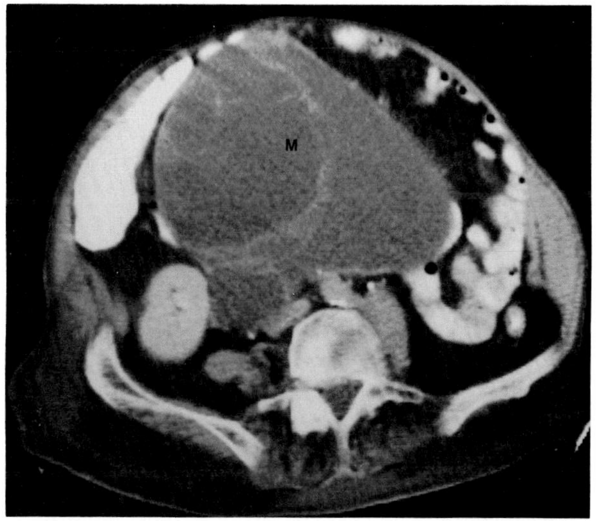


**Figure 4.92. Mucinous cystadenoma.** A CT scan demonstrates a large, low-attenuation mass (M) that contains multiple septa. (From Gross, Moss, Mihara, et al,[97] with permission from the publisher.)

metastases larger than 2 cm can be detected when surrounded by ascites.

CT demonstrates benign cystadenomas as low-density lesions that typically have a thin, smooth wall and relatively homogeneous contents. Higher CT attenuation values may be produced by hemorrhage or infection within the cyst; multiple septa are typically seen in a mucinous cystadenoma. Depending on the degree of differentiation, cystadenocarcinomas have a CT appearance that ranges from completely cystic to predominantly solid.

Ovarian carcinomas usually spread by implanting widely on the omental and peritoneal surfaces. This can produce the characteristic CT appearance of an "omental cake," an irregular sheet of nodular soft tissue densities beneath the anterior abdominal wall. Compared with sonography, CT is superior in detecting tumor adherence to bowel, ureteral involvement, and retroperitoneal adenopathy.[92–93,95–97]

#### MEIGS'S SYNDROME

Meigs's syndrome consists of the association of ascites and non-neoplastic pleural effusion with an ovarian fibroma. The ascites and the pleural effusion may be unilateral or bilateral, and they disappear rapidly following

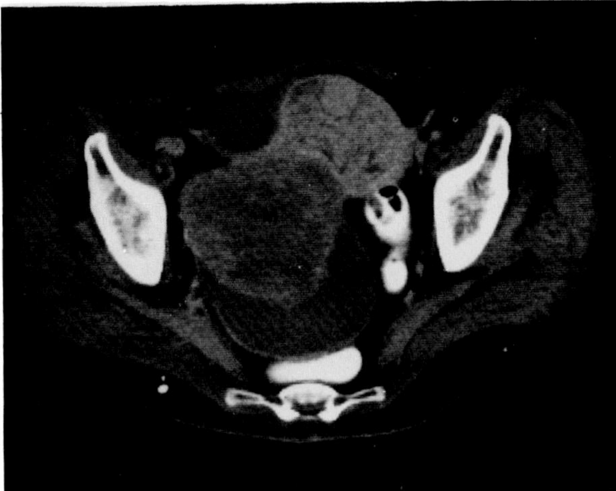

**Figure 4.93. Cystadenocarcinoma.** A CT scan demonstrates a large mass (M) with prominent solid and cystic components. (From Gross, Moss, Mihara, et al,[97] with permission from the publisher.)

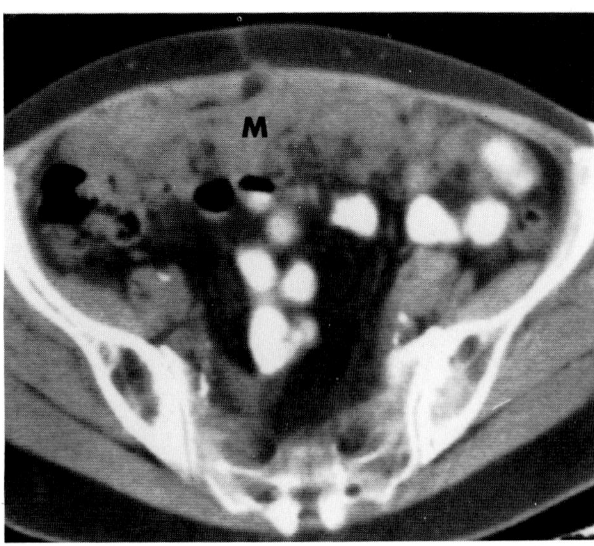


**Figure 4.94. Omental metastases (M) secondary to serous cystadenocarcinoma of the ovary.** Note the posterior displacement of adjacent contrast-filled bowel loops. (From Lee and Balfe,[102] with permission from the publisher.)

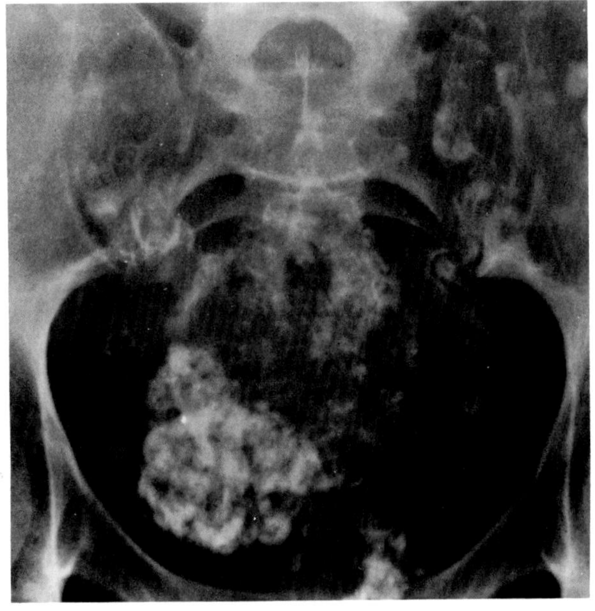

A

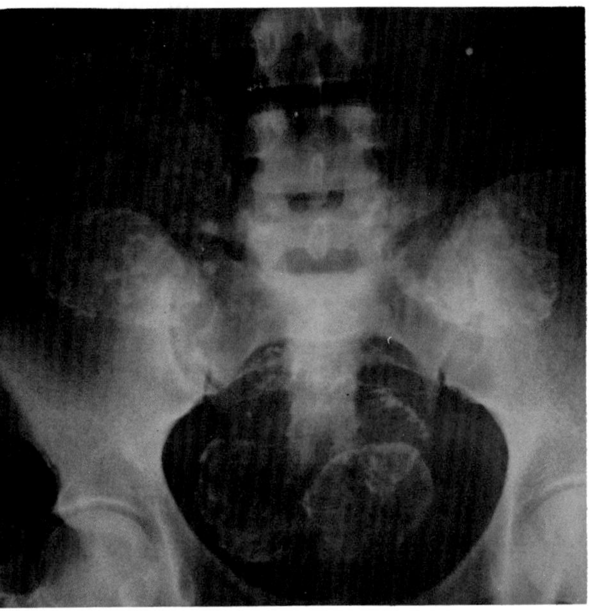

B

**Figure 4.95. Calcified uterine fibroid (leiomyoma).** (A) Characteristic stippled or whorled appearance of the calcifications. (B) In this patient, the calcification extends far out of the confines of the pelvis. (From Eisenberg,[71] with permission from the publisher.)

the surgical removal of the diseased ovary. It must be remembered that an identical radiographic appearance of ascites and pleural effusion in a patient with an ovarian mass can also be produced by a malignant ovarian tumor with diffuse metastases.[98]

## Tumors of the Uterus

### LEIOMYOMA (FIBROID) OF THE UTERUS

Uterine fibroids are by far the most common calcified lesions of the female genital tract. These frequently multiple tumors have a characteristic mottled, mulberry, or popcorn type of calcification and appear on plain abdominal radiographs as smooth or lobulated nodules with a stippled or whorled appearance. A very large calcified fibroid occasionally occupies the entire pelvis or even extends out of the pelvis to lie in the lower abdomen.

During excretory urography, persistent uterine opacification is often seen in patients with an underlying uterine leiomyoma. The tumor typically presses on the fundus of the bladder, causing a lobulated extrinsic impression that differs from the smooth impression usually seen with

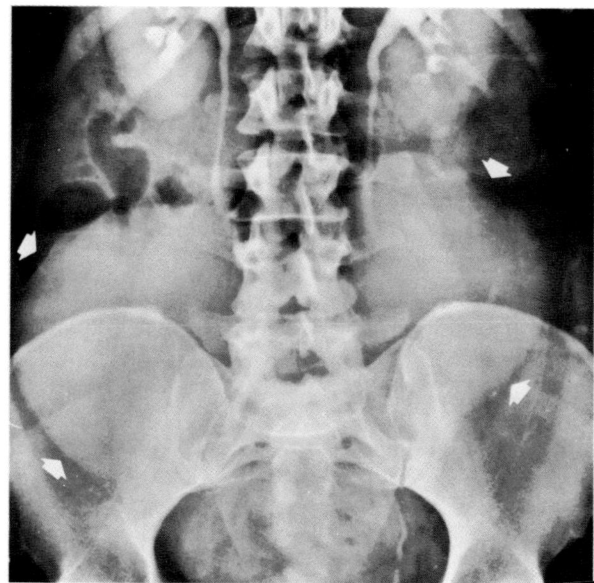

**Figure 4.96. Uterine fibroid.** An excretory urogram demonstrates persistent dense opacification of a huge uterine leiomyoma (arrows). (From Eisenberg,[71] with permission from the publisher.)

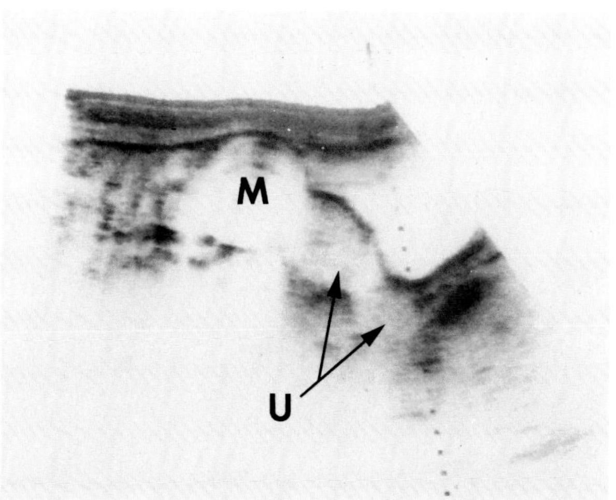


**Figure 4.97. Uterine fibroid.** A longitudinal sonogram demonstrates a pedunculated leiomyoma as a hypoechoic mass (M) projecting from the fundus of the uterus (U). Decreased sound transmission through the mass indicates its solid nature. (From Gross and Callen,[99] with permission from the publisher.)

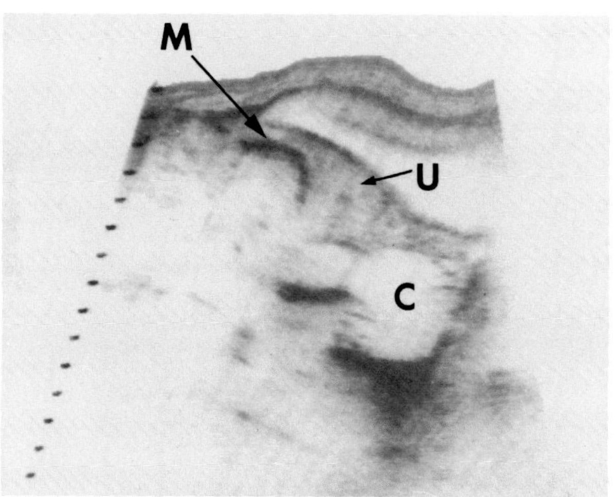


**Figure 4.98. Calcified uterine fibroid.** A longitudinal sonogram shows the calcification as a high-amplitude echo (M) within the uterus (U) with resultant acoustic shadowing. Incidentally noted is a cystic adnexal mass (C). (From Gross and Callen,[99] with permission from the publisher.)

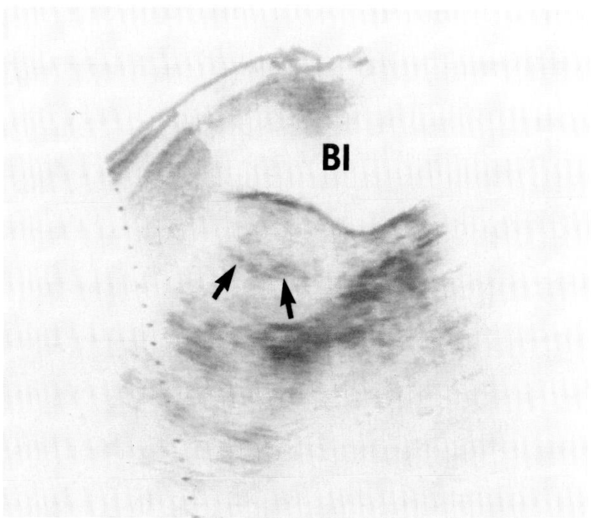


**Figure 4.99. Endometrial carcinoma.** A longitudinal sonogram shows the uterus to be enlarged and bulbous. There are clusters of high-amplitude echoes (arrows) in the region of the central cavity echo. Bl = bladder. (From Gross and Callen,[99] with permission from the publisher.)

ovarian cysts. Extension of a leiomyoma into the parametrium may cause medial displacement of the pelvic ureter or ureteral compression leading to hydronephrosis.

The classic ultrasound appearance of a uterine leiomyoma is a hypoechoic, solid contour-deforming mass in an enlarged, inhomogeneous uterus. Fatty degeneration and calcification cause focal increased echogenicity; the calcification may result in acoustic shadowing. Degeneration or necrosis may cause decreased echogenicity and increased through transmission, sometimes causing the leiomyoma to simulate a cystlike mass.[97,99–101]

#### CARCINOMA OF THE ENDOMETRIUM

Adenocarcinoma of the endometrium is the predominant neoplasm of the uterine body and is the most common invasive gynecologic neoplasm. Most patients present clinically with postmenopausal bleeding. Excretory urography may demonstrate an enlarged uterus impressing or invading the posterior wall and fundus of the bladder. The typical ultrasound appearance of endometrial carcinoma is an enlarged uterus with irregular areas of low-level echoes and bizarre clusters of high-intensity echoes. Unless evidence of local invasion can

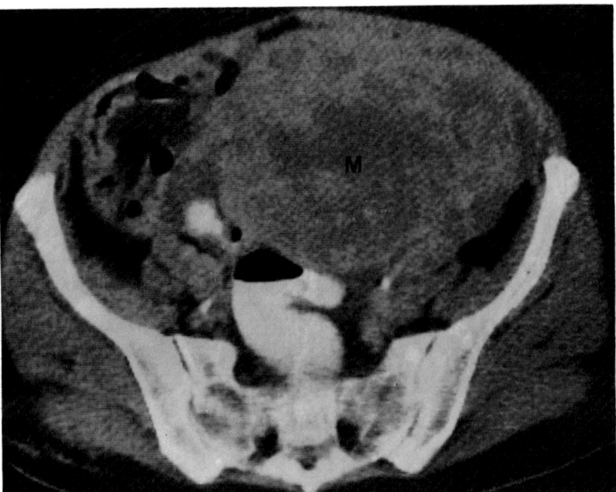


**Figure 4.100. Endometrial carcinoma.** A CT scan demonstrates a large, inhomogeneous uterine mass (M). (From Gross, Moss, Mihara, et al,[97] with permission from the publisher.)

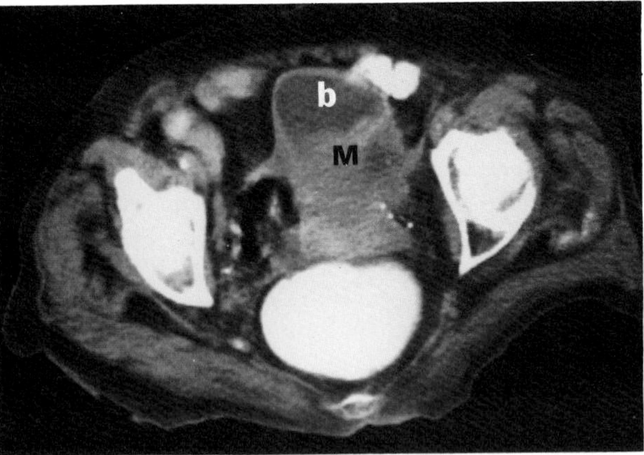


**Figure 4.101. Bladder invasion by endometrial carcinoma.** A CT scan shows a mass (M) obliterating the fat planes between the bladder (B) and the uterus. Urine within the bladder outlines thickening of the posterior bladder wall. (From Gross, Moss, Mihara, et al,[97] with permission from the publisher.)

be demonstrated, the ultrasound findings are indistinguishable from those of leiomyomas, which often occur in patients with endometrial carcinoma. CT demonstrates focal or diffuse enlargement of the body of the uterus. This modality is especially useful in detecting clinically unsuspected omental and nodal metastases in patients with advanced disease, as well as in evaluating patients with suspected neoplastic recurrence and in following the response to chemotherapy or radiation treatment.[97,99,100,102]

#### CARCINOMA OF THE CERVIX

Carcinoma of the cervix is the second most common form of cancer in women. At the time of initial staging, one-third of the patients have unilateral or bilateral hydronephrosis that can be demonstrated by excretory urography or ultrasound. Indeed, the most common cause of death in patients with carcinoma of the cervix is impairment of renal function due to ureteral obstruction. Spread to the ureters is most commonly by direct extension, though metastases through lymphatic vessels can occur. Extension of the tumor to the bladder may cause an irregular filling defect; direct infiltration of perirectal tissues may produce an irregular narrowing of the rectosigmoid and a widening of the retrorectal space.

Ultrasound usually demonstrates a cervical carcinoma as a solid retrovesical mass, though it appears indistinguishable from a benign cervical myoma. This modality is of value in staging cervical carcinoma, since it may detect parametrial or paracervical thickening of soft tissue, involvement of the pelvic sidewalls, extension into the bladder, and pelvic adenopathy. Nevertheless, CT is more accurate in detecting pelvic sidewall invasion and therefore is usually the initial staging procedure in patients in whom there is a clinical suspicion of advanced disease. CT can demonstrate nodal enlargement due to metastases, and involvement of the rectum and bladder. This modality is also the procedure of choice for monitoring tumor response to treatment and for assessing suspected recurrence.[97,99,100,102]

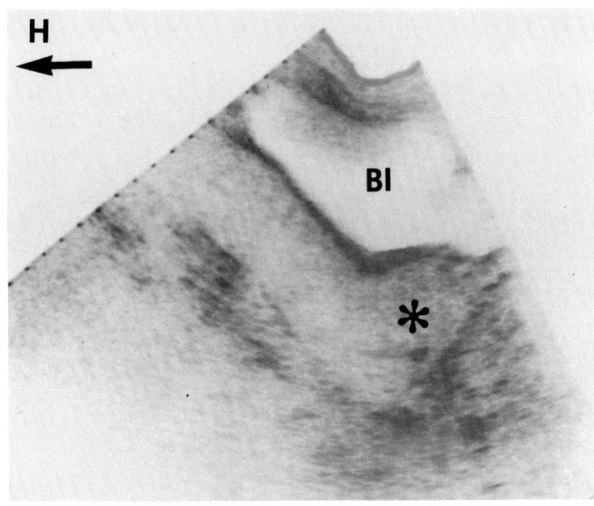


**Figure 4.102. Carcinoma of the cervix.** A sonogram demonstrates a solid, echogenic retrovesical mass that is indistinguishable from a cervical myoma. (From Gross and Callen,[99] with permission from the publisher.)

#### ENDOMETRIOSIS

Endometriosis is the presence of heterotopic foci of endometrium in an extrauterine location. Although tissues next to the uterus (ovaries, uterine ligaments, rectovaginal septum, pelvic peritoneum) are most frequently involved in endometriosis, the gastrointestinal and urinary tracts can also be affected.

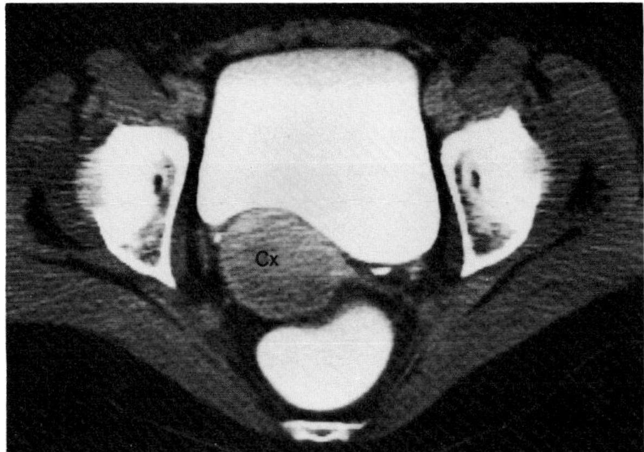


**Figure 4.103. Carcinoma of the cervix.** A CT scan demonstrates inhomogeneity of the enlarged cervix (Cx). (From Gross, Moss, Mihara, et al,[97] with permission from the publisher.)

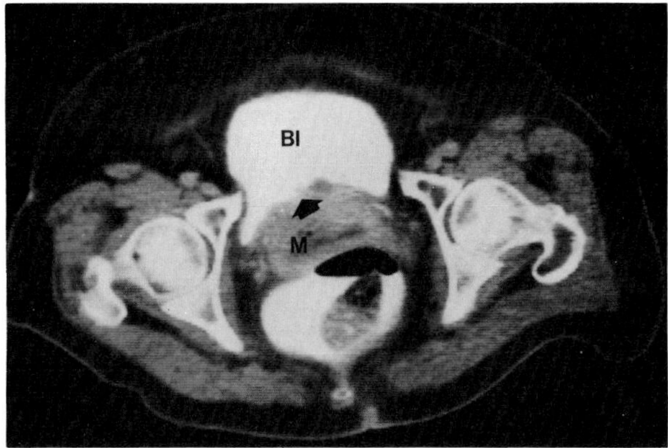


**Figure 4.104. Bladder invasion by carcinoma of the cervix.** A CT scan shows irregularity (arrow) of the posterior margin of the contrast-filled urinary bladder (Bl) with an adjacent inhomogeneous cervical mass (M). (From Gross and Callen,[99] with permission from the publisher.)

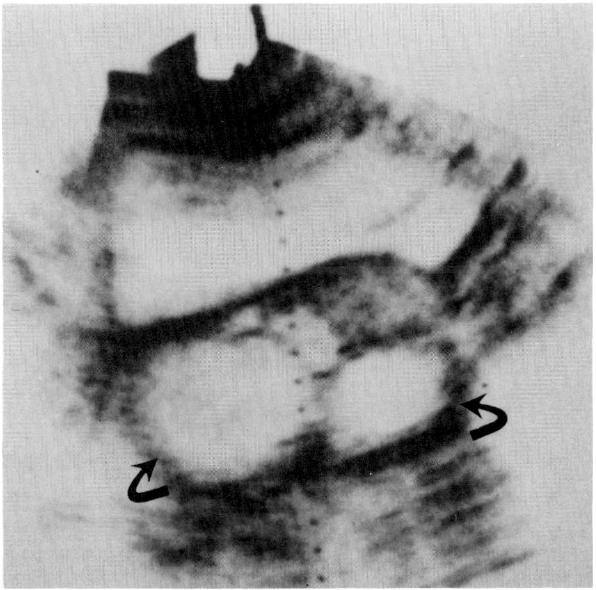

**Figure 4.105. Endometriosis.** A sonogram demonstrates several sonolucent masses simulating multiple follicular cysts arising from the ovary. (From Fleischer, Wentz, Jones, et al,[92] with permission from the publisher.).

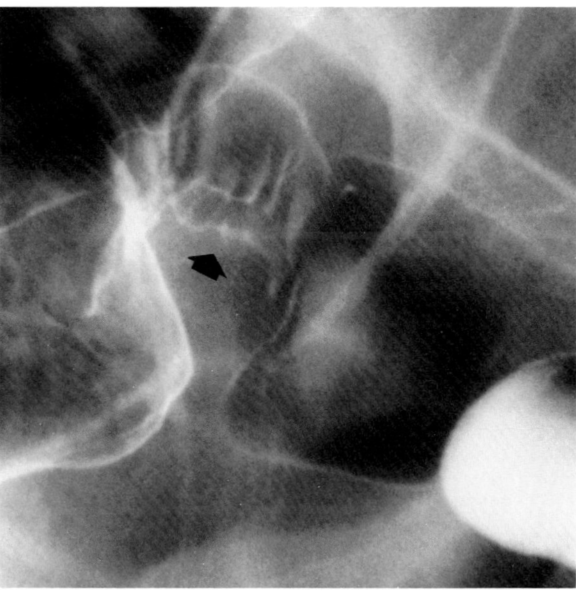

**Figure 4.106. Endometriosis.** There is an intramural mass in the proximal sigmoid colon near the rectosigmoid junction (arrow). The sharply defined, eccentric defect simulates a flat saddle cancer. Pleating of the adjacent mucosa is due to secondary fibrosis. (From Eisenberg,[71] with permission from the publisher.)

On ultrasound, endometriosis is a great mimicker that can produce a spectrum of appearances ranging from a nearly completely sonolucent mass to a highly echogenic solid structure. In predominantly cystic masses, some thickness or irregularity of the wall can usually be demonstrated. The variable amount of internal solid components in an endometrioma is related to clot formation and retraction, fibrosis, and liquefaction.

Endometriosis involves primarily those parts of the bowel that are situated in the pelvis, especially the rectosigmoid colon. An isolated endometrioma typically appears as an intramural defect, often with pleating of the adjacent mucosa due to secondary fibrosis. The sharply defined, eccentric defect may simulate a flat saddle cancer. However, in contrast to the mucosal pattern in a primary colonic malignancy, the mucosal pattern underlying and adjacent to an endometrioma usually remains intact. Endometriosis can also appear as a constricting lesion simulating annular carcinoma. Radiographic findings favoring a diagnosis of endometriosis are an intact mucosa, a long lesion with tapered margins, and the absence of ulceration within the mass. The repeated shedding of endometrial tissue and blood into the peritoneal cavity can lead to the development of

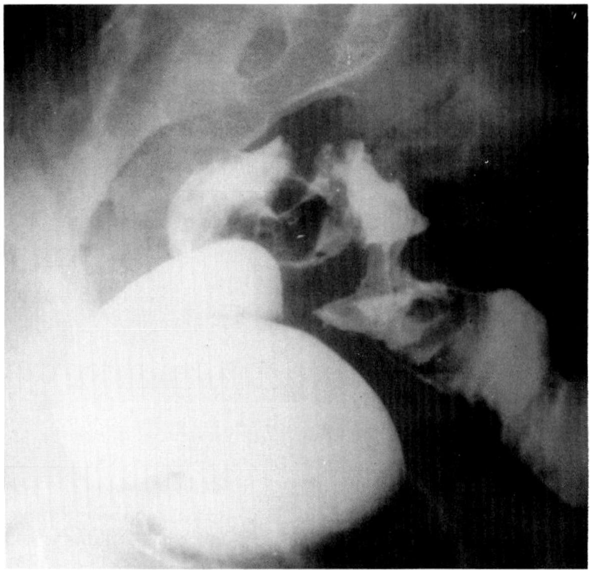

**Figure 4.107. Endometriosis.** Stenotic areas and multiple polypoid lesions are evident in a 7-cm segment of sigmoid colon. The mucosa is intact. The 43-year-old patient had had a previous episode of partial intestinal obstruction. (From Spjut and Perkins,[104] with permission from the publisher.)

dense adhesive bands, causing extrinsic obstruction of the bowel.

Endometriosis involving the urinary tract most commonly produces ureteral obstruction below the level of the pelvic brim. The condition mimics a ureteral tumor and may appear as an intraluminal mass or stricture. A smooth, rounded or multilobular filling defect in the bladder may occasionally be seen.[103–105]

## DISORDERS OF SEXUAL DIFFERENTIATION

### Gonadal Dysgenesis (Turner's Syndrome)

Gonadal dysgenesis is characterized by primary amenorrhea, sexual infantilism, short stature, multiple congenital anomalies, and bilateral streak gonads in phenotypic females with an absence or defect of an X chromosome. Clinically, there is often lymphedema of the hands and feet (due to lymphatic hypoplasia), webbing of the neck, a shieldlike chest with widely spaced nipples, cubitus valgus, and a hearing defect.

A characteristic, but nonspecific, skeletal abnormality is shortening of the fourth metacarpal and sometimes also the fifth metacarpal. This produces a positive metacarpal sign, in which a line drawn tangential to the distal ends of the heads of the fifth and fourth metacarpals passes through the head of the third metacarpal (indicating the disproportionate shortening of the fourth and fifth metacarpals), rather than extending distal to the head of the third metacarpal as in a normal person. The short fourth metacarpal can also be demonstrated by the phalangeal sign, in which the total length of the distal and proximal phalanges exceeds by 3 mm or more the length of the fourth metacarpal; in contrast, in normal persons the total length of the distal/proximal phalanges of the fourth finger is approximately equal to the length of the fourth metacarpal.

In the knee, a typical finding is enlargement of the medial femoral condyle with flattening or depression of the opposing tibial plateau. Hypogonadism may lead to generalized osteoporosis and delayed skeletal maturation. Deformities of the pelvis, clavicles, and ribs may also occur.

Various urinary tract anomalies, especially horseshoe kidney and other types of malrotation, are often seen in patients with gonadal dysgenesis. Coarctation of the aorta is the most common cardiovascular anomaly; because this condition most often affects men, its appearance in a female should suggest the possibility of underlying gonadal dysgenesis.

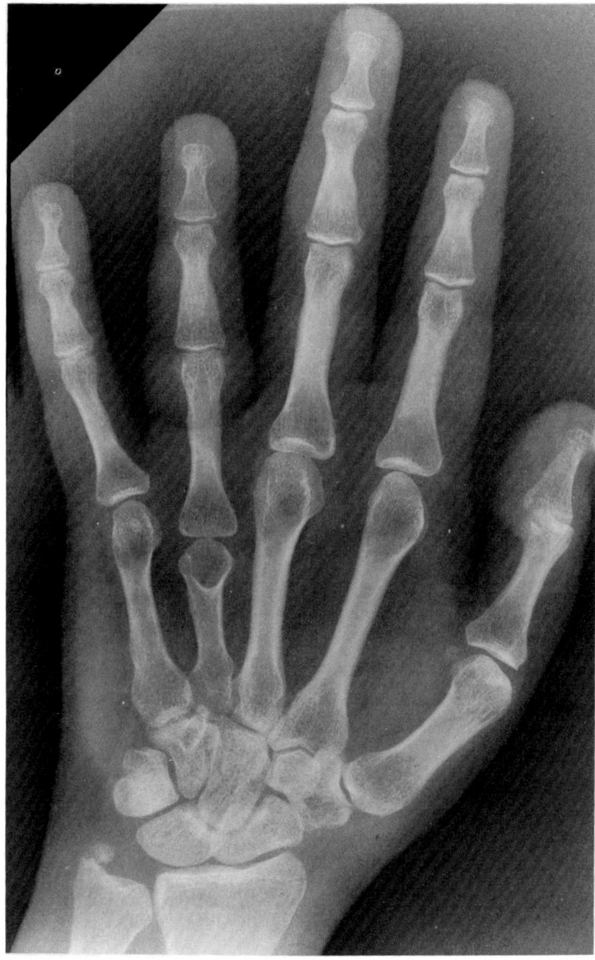

**Figure 4.108. Gonadal dysgenesis.** A frontal view of the hand shows a short fourth metacarpal with a resulting positive metacarpal sign.

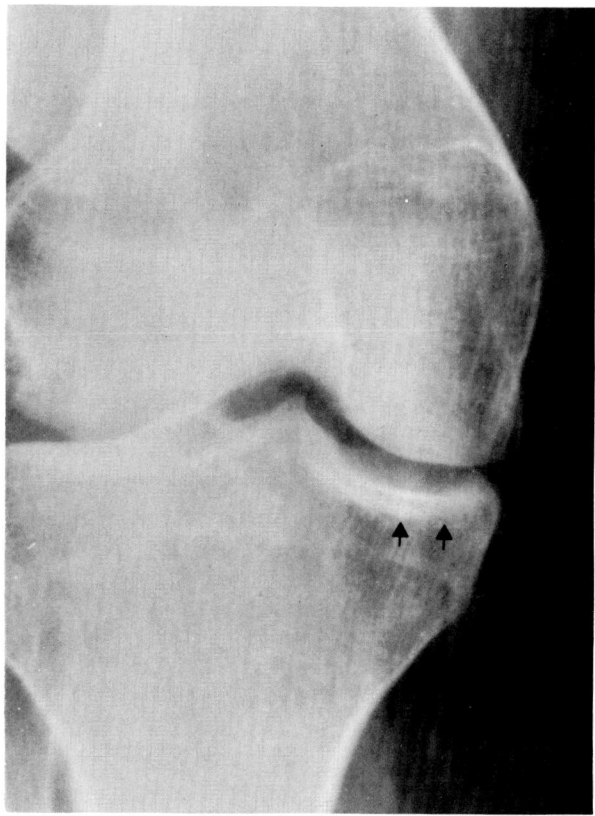

**Figure 4.109. Gonadal dysgenesis.** A frontal view of the knee shows characteristic enlargement of the medial femoral condyle with depression of the opposing tibial plateau (arrow).

The small dysgenetic or streak gonads usually cannot be demonstrated by ultrasound. The uterus remains prepubertal in size and configuration unless exogenous estrogen is given.[106–109]

### Klinefelter's Syndrome

Klinefelter's syndrome is a disorder in phenotypic males that is characterized by small firm testes, azoospermia, gynecomastia, and elevated levels of plasma gonadotropins. The fundamental defect is the presence in the male of two or more X chromosomes. Radiographically, the skeletal changes are both less common and less marked than in patients with Turner's syndrome, its female counterpart. A positive metacarpal sign and phalangeal preponderance are seen in fewer than 25 percent of the patients. The ungual tufts of the distal phalanges often have a peculiar square shape. Shortening or curving (clinodactyly) of the fifth finger is frequently seen. Hypogonadism may lead to delayed epiphyseal fusion and retarded bone maturation.[110]

## DISEASES OF THE PINEAL GLAND

### Pineal Calcification

The pineal gland contains sufficient calcium to be radiographically detectable on up to 75 percent of skull

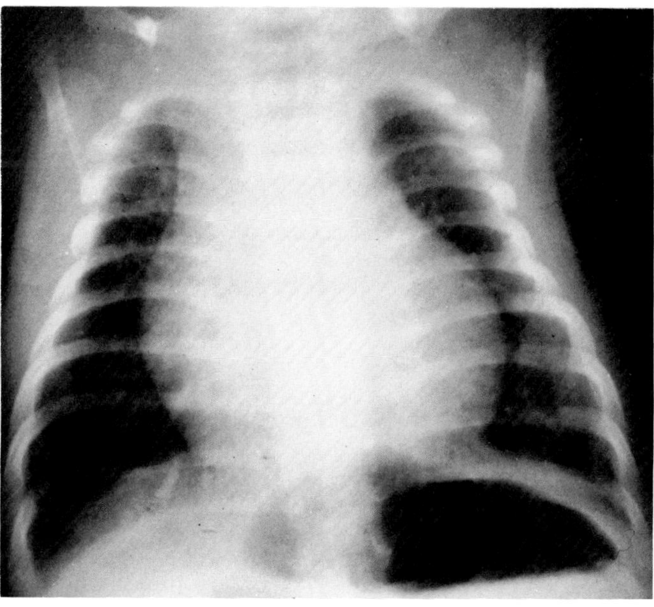

A

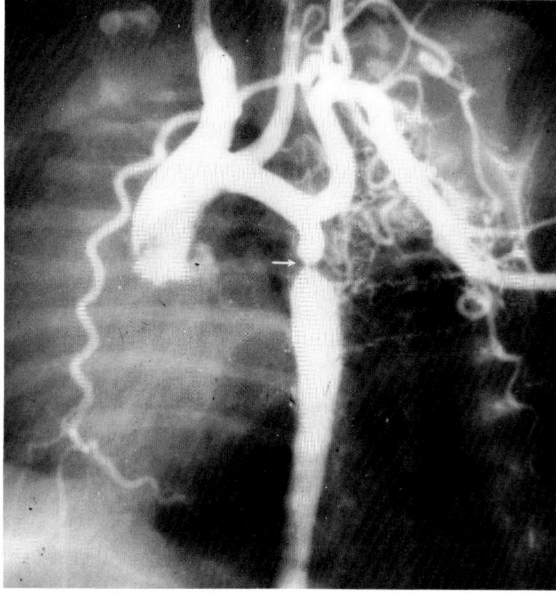

B

**Figure 4.110. Gonadal dysgenesis.** (A) A frontal view of the chest in a 1-month-old girl with shortness of breath demonstrates globular enlargement of the heart. (B) An aortogram shows an area of narrowing of the aorta (arrow), just distal to the origin of the left subclavian artery, with rich collateral circulation. A small ductus, which was not demonstrated by the aortogram, entered the aorta distal to the coarctation. (From Cooley and Schreiber,[124] with permission from the publisher.)

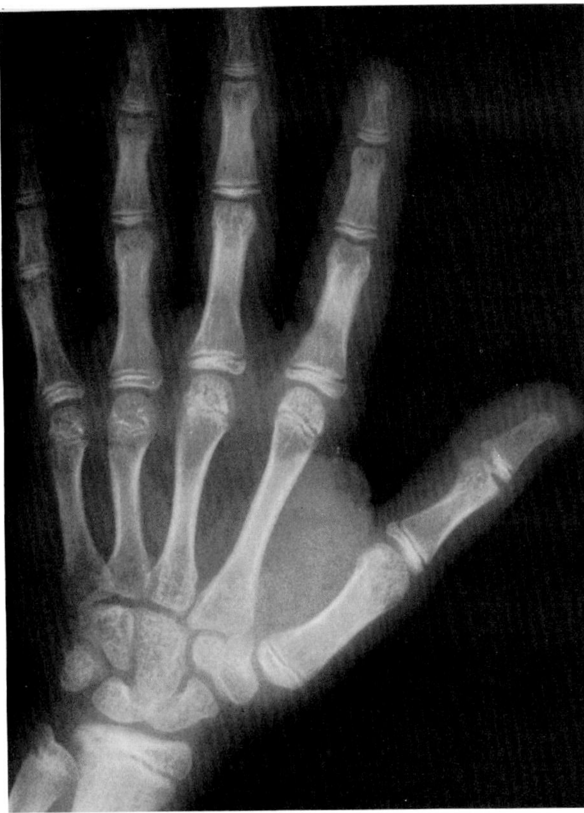

**Figure 4.111. Klinefelter's syndrome.** A disproportionate shortening of the fourth and fifth metacarpals (positive metacarpal sign) is combined with a positive phalangeal sign, which refers to the disproportionate length of the phalanges of the fourth finger (proximal + distal = 4.4 cm) compared with that of the fourth metacarpal (3.9 cm).

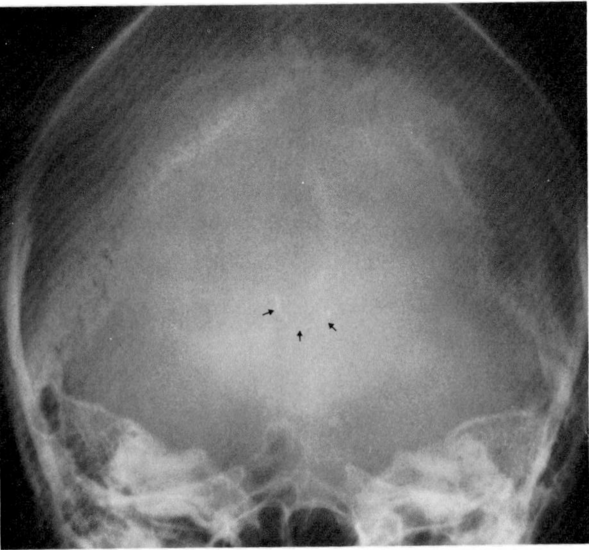

**Figure 4.112. Calcification in a pinealoma.** A large (1.5cm) central calcification in the region of the pineal gland suggests an underlying pineal tumor.

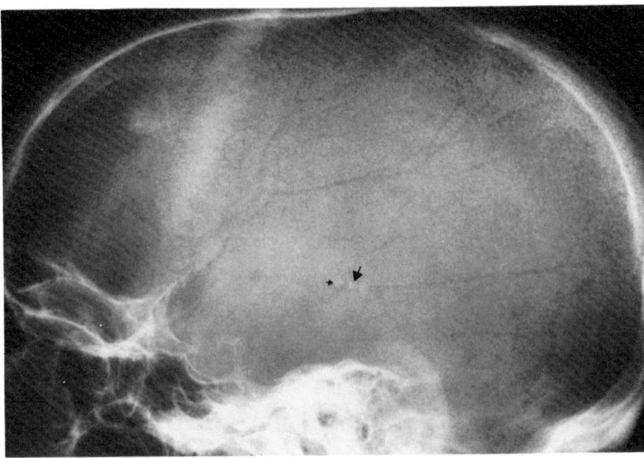

**Figure 4.113. Pineal and habenular calcification.** A lateral view of the skull shows calcification in the pineal gland (large arrow). Calcification in the habenular commissure (small arrow) lies a few millimeters anterior to the pineal gland and has a typical C-shaped configuration that opens posteriorly.

films in adults. Pineal gland calcification becomes increasingly frequent with advancing age and is rarely noted in children. Indeed, the demonstration of pineal calcification in a young child suggests an underlying neoplasm. Pineal calcification may appear as an irregular, amorphous cluster of densities or as a solitary mass. Because the gland is a midline structure, lateral displacement of a calcified gland may be a valuable clue to the presence of an intracranial lesion. A tumor usually displaces the pineal gland to the opposite side; an atrophic lesion occasionally shifts the gland toward the affected side. To allow for some degree of skull asymmetry, a variation of pineal position of 1 to 2 mm from the midline is generally considered to be within the normal range.

A calcified pineal gland is usually less than 1 cm in diameter. A larger calcification in this region suggests a pineal tumor or a calcified arteriovenous malformation.

Pineal calcification must be differentiated from calcification in the habenular commissure, which lies about 5 mm anterior to the pineal gland and has a characteristic C-shaped configuration that opens posteriorly.

## Pineal Tumors

The most common tumors of the pineal region are germinomas and teratomas, both of which occur predominantly in males under 25 years of age and may be associated with precocious puberty.

On CT, germinomas appear as hyperdense or isodense masses that tend to deform or displace the posterior aspect of the third ventricle and often obliterate the quadrigeminal cistern. Punctate calcification can often be detected within the mass. Intense enhancement of the tumor occurs following the injection of contrast material. Teratomas in the pineal region typically appear as

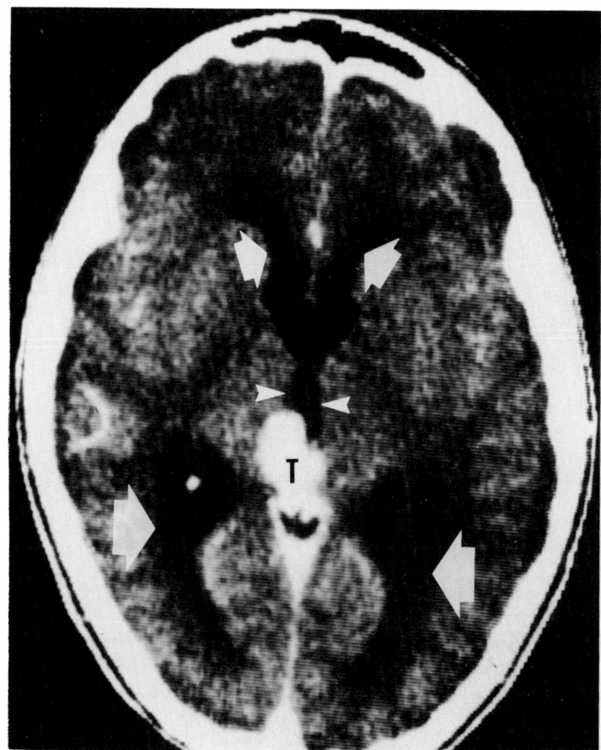


**Figure 4.114. Pineal germinoma.** A CT scan after the administration of contrast material shows an enhancing tumor (T) in the pineal region of a young girl with paralysis of upward gaze, headaches, and nausea (Perinaud's syndrome). The minimal dilatation of the third ventricle (arrowheads) and lateral ventricles (arrows) indicates mild hydrocephalus, which developed because of obstruction of the posterior portion of the third ventricle by the tumor.

hypodense masses with internal calcification. Occasionally, other formed elements (e.g., teeth) can occur. Contrast enhancement is usually much less pronounced than with germinomas. Large pineal tumors may cause obstructive hydrocephalus with ventricular dilatation.

A small number of tumors with the histologic appearance of pinealomas develop elsewhere in the brain at some distance from the normal pineal gland. These "ectopic pinealomas" generally occur in the anterior aspect of the third ventricle or within the suprasellar cistern. They may produce a clinical triad of bitemporal hemianopsia, hypopituitarism, and diabetes insipidus that simulates a craniopharyngioma.[111–114]

## MULTIPLE ENDOCRINE NEOPLASIA SYNDROMES

Some patients have multiple tumors (or hyperplasia) of various endocrine glands simultaneously or in different stages of life. In multiple endocrine neoplasia type I, the lesions occur in the anterior pituitary, parathyroids, pancreatic islet cells, and adrenal cortex. Occasionally, adenomatous goiters in the thyroid glands may occur. In Sipple's syndrome (multiple endocrine neoplasia type II), familial pheochromocytoma is associated with medullary thyroid carcinoma. The pheochromocytomas are often bilateral or multiple and frequently develop in an extra-adrenal location. Characteristic dense calcifications are sometimes seen in medullary carcinoma of the thyroid, in contrast to the psammomatous calcifications in thyroid carcinomas of the papillary type. Hyperparathyroidism may develop, due either to a parathyroid adenoma or to compensatory hyperparathyroidism secondary to hypocalcemia resulting from the excess thyrocalcitonin produced by the medullary thyroid carcinoma.

In multiple endocrine neoplasia type III, pheochromocytoma is associated with medullary thyroid carcinoma, mucosal neuroma, and sometimes a marfanoid habitus. The diffuse intestinal ganglial neuromatosis in this condition may produce abnormal haustral markings, thickened mucosal folds, diverticula, and dilatation of the colon; the resulting appearance may occasionally be misdiagnosed as ulcerative colitis, Crohn's disease, or congenital megacolon.

(Illustrations of the various tumors constituting the multiple endocrine neoplasia syndromes can be found in the appropriate sections.)[28,57,115]

### REFERENCES


1. Greenblatt RB, Nieburg HW: Some endocrinologic aspects of retarded growth and dwarfism. *Med Clin N Am* 31:712–730, 1947.
2. LeMay M: The radiologic diagnosis of pituitary disease. *Radiol Clin N Am* 5:303–315, 1967.
3. Edeiken J: *Roentgen Diagnosis of Disease of Bone*. Baltimore, Williams & Wilkins, 1981.
4. Greenfield GB: *Radiology of Bone Diseases*. Philadelphia, JB Lippincott, 1980.
5. Raji MR, Kishore PRS, Becker DP: Pituitary microadenoma: A radiological-surgical correlative study. *Radiology* 139:95–99, 1981.
6. Robertson WB, Newton TH: Radiologic assessment of pituitary microadenomas. *AJR* 131:489–492, 1978.
7. Levine HG, Kleefield J, Rao KCVG: The base of the skull, in Lee SH, Rao KCVG (eds): *Cranial Computed Tomography*. New York, McGraw-Hill, 1983.
8. Hilal SK: Angiography of juxtasellar masses. *Semin Roentgenol* 6:75–93, 1971.
9. Rozario R, Hammerschlag SB, Post KD, et al: Diagnosis of empty sella with CT scan. *Neuroradiology* 13:85–88, 1977.
10. Hall K, McAllister VL: Metrizamide cisternography in pituitary and juxtapituitary lesions. *Radiology* 134:101–108, 1980.
11. Kaufman B: The "empty" sella turcica–A manifestation of the intrasellar subarachnoid space. *Radiology* 90:931–941, 1968.
12. Shore IN, DeCherney AH, Stein KM, et al: The empty sella syndrome. *JAMA* 227:69–70, 1974.
13. Wheeler JS, Adelson WJ: Pituitary diabetes insipidus associated with progressive urinary tract dilatation. *J Urol* 92:64–67, 1964.
14. Jalowitz PA, Randall RV, Greene LP: Dilatation of the urinary tract associated with pituitary and nephrogenic diabetes insipidus. *J Urol* 103:327–329, 1970.

15. Miller JM, Hamburger JI: The thyroid scintigram: I. The hot nodule. *Radiology* 84:66–74, 1965.
16. Miller JM, Hamburger JI, Mellinger RC: The thyroid scintigram: II. The cold nodule. *Radiology* 85:702–710, 1965.
17. Chilcote WS: Gray scale ultrasonography of the thyroid. *Radiology* 120:381–383, 1976.
18. Rybak M: Skeletal dysplasia and bone maturation in hypothyroidism of children. *Ann Radiol* 13:243–249, 1970.
19. Wietersen FK, Ablow RM: The radiological aspects of thyroid disease. *Radiol Clin North Am* 5:255–266, 1967.
20. Enzman D, Marshall WH, Rosenthal AR, et al: Computed tomography in Graves' opthalmopathy. *Radiology* 118: 615–620, 1976.
21. Scanlon GT, Clemett AR: Thyroid acropachy. *Radiology* 83:1039–1042, 1964.
22. McCarthy J, Twersley J, Lion M: Thyroid acropachy. *J Can Assoc Radiol* 26:199–202, 1976.
23. Renda F, Holmes RA, North WA, et al: Characteristics of thyroid scans in normal persons, hyperthyroidism, and nodular goiter. *J Nucl Med* 9:156–159, 1968.
24. Margolin FP, Winfield J, Steinbach HL: Patterns of thyroid calcification. *Invest Radiol* 2:208–214, 1967.
25. Blum M, Goldman AB, Herskovic A, et al: Clinical application of thyroid echography. *N Engl J Med* 287:1164–1169, 1972.
26. Pearson KD, Wells SA, Keiser HR: Familial medullary carcinoma of the thyroid, adrenal pheochromocytoma and parathyroid hyperplasia. *Radiology* 107:249–256, 1973.
27. Arnold J, Pinksy S, Ryo UY, et al: $^{99m}$Tc pertechnetate thyroid scintigraphy in patients predisposed to thyroid neoplasms by prior radiotherapy to the head and neck. *Radiology* 115:653–657, 1975.
28. Mitty HA, Yeh HC: *Radiology of the Adrenals with Sonography and CT*, Philadelphia, WB Saunders, 1982.
29. Nelson DH, Meakin JW, Thorn GW: ACTH-producing tumors following adrenalectomy for Cushing's syndrome. *Ann Intern Med* 52:560–569, 1960.
30. Robertson WD, Newton TH: Radiologic assessment of pituitary microadenomas. *AJR* 131:49–55, 1978.
31. Moses DC, Schteingart DE, Sturman MF, et al: Efficacy of radiocholesterol imaging of the adrenal glands in Cushing's syndrome. *Surg Gynecol Obstet* 139:201–204, 1974.
32. Wahner HW, Northcutt RC, Salassa RM: Adrenal scanning: Usefulness in adrenal hyperfunction. *Clin Nucl Med* 2:253–264, 1977.
33. Lecky JW, Wolfman NT, Modic CW: Current concepts of adrenal angiography. *Radiol Clin North Am* 14:309–352, 1976.
34. Sample WF, Sarti DA: Computed tomography and gray scale ultrasonography of the adrenal gland: A comparative study. *Radiology* 128:377–383, 1978.
35. Dunnick NR, Schaner EG, Doppman JL, et al: Computed tomography in adrenal tumors. *AJR* 132:43–46, 1979.
36. Dunnick NR, Doppman JL, Geelhood GW: Intravenous extension of endocrine tumors. *AJR* 135:471–476, 1980.
37. King DR, Lack EE: Adrenal cortical carcinoma. A clinical and pathological study of 49 cases. *Cancer* 44:239–244, 1979.
38. Bernardino ME, Goldstein HM, Green B: Gray scale ultrasonography of adrenal neoplasms. *AJR* 130:741–744, 1978.
39. Colapinto RF, Steed BL: Arteriography of adrenal tumors. *Radiology* 100:343–350, 1971.
40. Cedermark BJ, Ohlsen H: Computed tomography in the diagnosis of metastases of the adrenal glands. *Surg Gynecol Obstet* 152:13–16, 1981.
41. Vas W, Zylak CJ, Mather D, et al: The value of abdominal computed tomography in the pretreatment assessment of small cell carcinoma of the lung. *Radiology* 138:417–418, 1981.
42. Conn JW, Cohen EL, Herwig KR: The dexamethasone-modified adrenal scintiscan in hyporeninemic aldosteronism (tumors vs. hyperplasia). A comparison with adrenal venography and adrenal venous aldosterone levels. *J Lab Clin Med* 88:841–856, 1976.
43. Dunnick NR, Doppman JL, Mills SR, et al: Preoperative diagnosis and localization of aldosteronomas by measurement of corticosteroids in adrenal venous blood. *Radiology* 133:331–333, 1979.
44. Ganguly A, Pratt JH, Yune HY, et al: Computerized tomographic scanning and primary aldosteronism. *N Engl J Med* 301:558, 1979.
45. Prosser PR, Sutherland CM, Scullin DR: Localization of adrenal aldosterone adenoma by computerized tomography. *N Engl J Med* 300:1278–1279, 1979.
46. Davidson JK, Morley P, Hurley GD, et al: Adrenal venography and ultrasound in the investigation of the adrenal gland: An analysis of 58 cases. *Br J Radiol* 48:435–450, 1975.
47. Jarvis JL, Jenkins D, Sosman MC, et al: Roentgenologic observations in Addison's disease: A review of 120 cases. *Radiology* 62:16–29, 1954.
48. Crispell KR, Parson W, Hamlin J, et al: Addison's disease associated with histoplasmosis: Report of four cases and review of literature. *Am J Med* 20:23–29, 1956.
49. Brill PW, Krasna IH, Aaron H: An early rim sign in neonatal adrenal hemorrhage. *AJR* 127:289–291, 1976.
50. Mineau DE, Koehler PR: Ultrasound diagnosis of neonatal adrenal hemorrhage. *AJR* 132:443–444, 1979.
51. Rose J, Berdon WE, Sullivan T, et al: Prolonged jaundice as presenting sign of massive adrenal hemorrhage in the newborn. *Radiology* 98:263–272, 1971.
52. Tisnado J, Amendola MA, Konerding KF, et al: Computed tomography versus angiography in the localization of pheochromocytoma. *J Comput Assist Tomogr* 4:853–859, 1980.
53. Stewart BH, Bravo EL, Meaney TF: A new simplified approach to the diagnosis of pheochromocytoma. *J Urol* 122:579–581, 1979.
54. Pickering RS, Hartman GW, Weeks RE, et al: Excretory urographic localization of adrenal cortical tumors and pheochromocytomas. *Radiology* 114:345–349, 1975.
55. Laursen K, Damgaard-Pedersen K: CT for pheochromocytoma diagnosis. *AJR* 134:277–280, 1980.
56. Muhvhoff G, Pohle D, Sack H: The radiological diagnosis of pheochromocytoma with special reference to angiography and the specific preparation required. *Radiology* 116:759–767, 1975.
57. Cho KJ, Freier DP, McCormick TL, et al: Adrenal medullary disease in multiple endocrine neoplasia type II. *AJR* 134:23–29, 1980.
58. Fill WL, Lamcell JM, Polk NO: The radiographic manifestations of Von Hippel-Lindau disease. *Radiology* 133:289–295, 1979.
59. Fries JG, Chamberlin JA: Extra-adrenal pheochromocytoma: Literature review and report of a cervical pheochromocytoma. *Surgery* 63:268–279, 1968.
60. Welch TJ, Sheedy PF, van Heerden JA, et al: Pheochromocytoma: Value of computed tomography. *Radiology* 148:501–503, 1983.
61. Juhl JH: *Essentials of Roentgen Interpretation*. Philadelphia, Harper & Row, 1981.
62. Barrett AF: Sympaticoblastoma (neuroblastoma). *Clin Radiol* 14:34–41, 1963.
63. Young LW, Rubin P, Hanson RE: The extra-adrenal neuroblastoma. *AJR* 108:75–80, 1970.
64. Kincaid OW, Hodgson JR, Dockerty MB: Neuroblastoma: A roentgenologic and pathologic study. *AJR* 78:420–436, 1957.
65. Gehweiler JA, Bender WR: Carotid arteriography in the diagnosis and management of tumors of the carotid body. *AJR* 104:893–898, 1958.
66. Laster DW, Citrin CM, Thyng FJ: Bilateral carotid body tumors. *AJR* 115:143–147, 1972.
67. Gooding GAW: Gray scale ultrasound detection of carotid body tumors. *Radiology* 132:409–410, 1979.
68. Shugar MA, Mafee MF: Diagnosis of carotid body tumors by dynamic computerized tomography. *Head Neck Surg* 4:518–521, 1982.
69. Mancuso AA: Computed tomography of the neck. In Moss

AA, Gamsu G, Genant HK (eds): *Computed Tomography of the Body*. Philadelphia, WB Saunders, 1983.
70. Beck RE; Roentgen findings in the complications of diabetes mellitus. *AJR* 82:887–896, 1959.
71. Eisenberg RL: *Gastrointestinal Radiology: A Pattern Approach*, Philadelphia, JB Lippincott, 1983.
72. Gramm HF, Reuter K, Costello P: The radiologic manifestations of diabetic gastric neuropathy and its differential diagnosis. *Gastrointest Radiol* 3:151–155, 1978.
73. Robins JM, Bookstein JJ, Oberman HA, et al: Selective angiography in localizing islet-cell tumors of the pancreas. *Radiology* 106:525–528, 1973.
74. Gunther RW, Klose KJ, Ruckert K, et al: Islet-cell tumors: Detection of small lesions with computed tomography and ultrasound. *Radiology* 148:485–488, 1983.
75. Dunnick NR, Doppman JL, Mills SR, et al: Computed tomographic detection of nonbeta pancreatic islet-cell tumors. *Radiology* 135:117–120, 1980.
76. Stark DD, Moss AA, Goldberg HI, et al: CT of pancreatic islet-cell tumors. *Radiology* 150:491–494, 1983.
77. Gray RK, Rosch J, Grollman JH: Angiography in diagnosis of islet-cell tumors. *Radiology* 97:39–44, 1970.
78. Hricak H, Filly RA: Sonography of the scrotum. *Invest Radiol* 18:112–121, 1983.
79. Arger PH, Mulhern CB, Coleman BG, et al: Prospective analysis of the value of scrotal ultrasound. *Radiology* 141:763–766, 1981.
80. Glazer HS, Lee JKT, Leland MG, et al: Sonographic detection of occult testicular neoplasms. *AJR* 138:673–675, 1982.
81. Wilson PC, Day DL, Valvo JR, et al: Scrotal ultrasound with an Octoson. *RadioGraphics* 2:24–38, 1982.
82. Boldt DW, Reilly BJ: Computed tomography of abdominal mass lesions in children: Initial experience. *Radiology* 124:371–378, 1977.
83. Thompson IM, Latourette H, Chadwick S, et al: Diagnosis of testicular torsion using Doppler ultrasonic flow meter. *Urology* 6:706–707, 1975.
84. Holder LE, Martire JR, Holmes ER, et al: Testicular radionuclide angiography and static imaging: Anatomy scintigraphic interpretation, and clinical indications. *Radiology* 125:739–752, 1977.
85. Pederson JR, Holm HH, Hald T: Torsion of the testis diagnosed by ultrasound. *J Urol* 113:66–68, 1975.
86. Bird K, Rosenfield AT, Taylor KJW: Ultrasonography in testicular torsion. *Radiology* 147:527–534, 1983.
87. Madrazo BL, Klugo RC, Parks JA, et al: Ultrasonography demonstration of undescended testes. *Radiology* 133:181–183, 1979.
88. Lee JKT, McClennan BL, Stanley RJ, et al: Utility of computed tomography in the localization of the undescended testes. *Radiology* 135:121–125, 1980.
89. Yune HY, Klatte EC, Cleary RE, et al: Hysterosalpingography in infertility. *AJR* 121:642–651, 1974.
90. Callen PW: Ultrasonography in evaluation of gestational trophoblastic disease, in Callen PW (ed): *Ultrasonography in Obstetrics and Gynecology*. Philadelphia, WB Saunders, 1983.
91. Guttman PH: In search of the illusive benign cystic ovarian teratoma: Application of the ultrasound "tip of the iceberg" sign. *J Clin Ultrasound* 5:403–406, 1977.
92. Fleischer AC, Wentz AC, Jones HW, et al: Ultrasound evaluation of the ovary, in Callen PW (ed): *Ultrasonography in Obstetrics and Gynecology*. Philadelphia, WB Saunders, 1983.
93. Meire H, Farrant P, Guha T: Distinction of benign from malignant ovarian cysts by ultrasound. *Br J Obstet Gynaecol* 85:893–899, 1978.
94. Sandler M, Silver T, Karo J: Gray-scale ultrasonic features of ovarian teratomas. *Radiology* 131:705–709, 1979.
95. Teplick JG, Haskin ME, Alavi A: Calcified intraperitoneal metastases from ovarian carcinoma. *AJR* 127:1003–1006, 1976.
96. Moncada R, Cooper JA, Garces M, et al: Calcified metastases from malignant ovarian neoplasm: A review of the literature. *Radiology* 113:31–35, 1974.
97. Gross BH, Moss AA, Mihara K, et al: Computed tomography of gynecologic diseases. *AJR* 141:765–773, 1983.
98. Meigs JV, Cass JW: Hydrothorax and ascites in association with fibroma of the ovary. *Am J Obstet Gynecol* 53:249–254, 1937.
99. Gross BH, Callen PW: Ultrasound of the uterus, in Callen PW (ed): *Ultrasonography in Obstetrics and Gynecology*. Philadelphia, WB Saunders, 1983.
100. Walsh JW, Brewer WH, Schneider V: Ultrasound diagnosis in diseases of the uterine corpus and cervix. *Semin Ultrasound* 1:30–40, 1980.
101. Birnholz JC: Uterine opacification during excretory urography: Definition of a previously unreported sign. *Radiology* 105:303–307, 1972.
102. Lee JKT, Balfe DM: Pelvis, in Lee JKT, Sagel SS, Stanley RJ (eds): *Computed Body Tomography*. New York, Raven Press, 1983.
103. Sample WF: Pelvic inflammatory disease and endometriosis, in Sanders RC, James AE (eds): *Ultrasonography in Obstetrics and Gynecology*. New York, Appleton Century Crofts, 1980.
104. Spjut HJ, Perkins DE: Endometriosis of the sigmoid colon and rectum. *AJR* 82:1070–1075, 1959.
105. Fleischer AC, James AE: *Introduction to Diagnostic Ultrasonography*. New York, John Wiley & Sons, 1980.
106. Kosowicz J: The roentgen appearance of the hand and wrist in gonadal dysgenesis. *AJR* 93:354–361, 1965.
107. Reveno JS, Palubinskas AJ: Congenital renal abnormalities in gonadal dysgenesis. *Radiology* 86:49–51, 1966.
108. Nora JJ, Torres FG, Sinha AK, et al: Characteristic cardiovascular anomalies of XO Turner syndrome, XX and XY phenotype and XO/XX Turner mosaic: *Am J Cardiol* 25:639–641, 1970.
109. Grimes CK, Rosenbaum DM, Kirkpatrick JA: Pediatric gynecologic radiology. *Semin Roentgenol* 17:284–301, 1982.
110. Ohsawa T, Furuse M, Kikuchi Y, et al: Roentgenographic manifestations of Klinefelter's syndrome. *AJR* 112:178–184, 1971.
111. Zimmerman RA, Bilaniuk LT, Wood JH, et al: Computed tomography of pineal, parapineal, and histologically related tumors. *Radiology* 137:669–677, 1980.
112. Futrell NN, Osborne AG, Cheson BD: Pineal tumors: CT-Pathological spectrum. *Am J Neuroradiol* 2:415–420, 1981.
113. Tod PA, Porter AJ, Jamieson KG: Pineal tumors. *AJR* 120:19–26, 1974.
114. Cole H: Tumors in the region of the pineal. *Clin Radiol* 22:110–117, 1971.
115. Sipple JH: The association of pheochromocytoma with carcinoma of the thyroid. *Am J Med* 31:163–166, 1961.
116. Weyman PJ, Glazer HS: Adrenals, in Lee JKT, Sagel SS, Stanley RS (eds): *Computed Body Tomography*. New York, Raven Press, 1983.
117. Chatel JF, Charbonnel B, Guihard D: Radionuclide imaging of the adrenal glands. *Clin Nucl Med* 3:71–82, 1978.
118. Reynolds WA, Karo JJ: Radiologic diagnosis of metabolic bone disease. *Orthop Clin North Am* 3:521–532, 1972.
119. Yeh HC, Mitty HA, Rose J, et al: Ultrasonography of adrenal masses. *Radiology* 127:467–482, 1978.
120. Karstaedt N, Sagel SS, Stanley RJ, et al: Computed tomography of the adrenal gland. *Radiology* 129:723–730, 1978.
121. Davidson AJ: *Radiologic Diagnosis of Renal Parenchymal Disease*. Philadelphia, WB Saunders, 1977.
122. Burko H, Smith CW, Kirchner SG, et al: Hypertension, in Eisenberg RL, Amberg JR (eds): *Critical Diagnostic Pathways in Radiology: An Algorithmic Approach*. Philadelphia, JB Lippincott, 1981.
123. Wolverson MK, Houttuin E, Heiberg E, et al: Comparison of computed tomography with high-resolution real-time ultrasound in the localization of the impalpable undescended testis. *Radiology* 146:133–140, 1983.

124. Cooley RN, Schreiber MH: *Radiology of the Heart and Great Vessels*. Baltimore, Williams & Wilkins, 1978.
125. Gooding GAW: Neck masses, in Eisenberg RL, Amberg JR (eds): *Critical Diagnostic Pathways in Radiology*. Philadelphia, JB Lippincott, 1981.
126. Friedland GW, Filly R, Goris ML, Gross D, Kempson RL, Korobkin M, Thurber BD, Walter J: *Uroradiology: An Integrated Approach*. New York, Churchill Livingstone, 1983.
127. Jeffrey RB: Computed tomography of lymphovascular structures and retroperitoneal soft tissues, in Moss AA, Gamsu G, Genant HK: *Computed Tomography of the Body*. Philadelphia, WB Saunders, 1983.
128. Matin P: *Clinical Nuclear Medicine*. Garden City, N.Y., Examination Publishing Co., 1981.

## Chapter Five
# Neoplasia

## BREAST CANCER

It is estimated that 1 of every 11 American women will develop breast cancer at some time in her life. Current surgical and radiation therapy techniques provide highly effective treatment, but only if the cancers are detected when localized to the breast itself. Unfortunately, most breast tumors are discovered accidentally rather than in the course of regular survey examinations. By this time, the majority have spread either to regional lymph nodes or systemically, accounting for the current high mortality (about 50 percent) that makes breast cancer the leading cause of cancer death in women.

Periodic careful physical examination of the breast, done either by a trained health professional or by the patient herself, will discover cancers that are smaller and more likely to be localized. Even-smaller, nonpalpable, and potentially more curable lesions can be detected by mammography, a radiographic examination that is by far the most effective breast diagnostic procedure. Indeed, routine mammography combined with physical examination is the only approach currently available that promises to significantly reduce breast cancer mortality.

The two major radiographic techniques for diagnosing breast cancer are screen-film mammography and xeromammography. Screen-film imaging uses a specially designed x-ray screen that permits the proper exposure of film by many fewer x-rays than would otherwise be necessary. This produces a conventional black-and-white image at a very low radiation dose. Xeromammography is an adaptation of the standard xerographic photocopying process; the blue-and-white x-ray images are made on paper rather than on film. Although many radiologists have strong personal preferences for one or the other technique, both screen-film mammography and xeromammography, when performed properly, produce images of essentially equal diagnostic quality. The major differences between the two techniques are that xeromammography can be performed using a general-purpose x-ray machine, while screen-film imaging requires somewhat more expensive equipment; and the radiation dose of screen-film mammography is three to five times lower than that of xeromammography.

Almost all breast cancers present mammographically as a tumor mass or clustered calcifications, or both. Either feature, when clearly demonstrated, is so suspicious of malignancy as to require prompt biopsy whether the lesion is palpable or not. The typical malignant tumor mass is poorly defined, has irregular margins, and demonstrates numerous fine linear strands or spicules radiating out from the mass. This appearance is characteristic but not diagnostic of malignancy and is in stark contrast to the typical mammographic picture of a benign mass, which has well-defined, smooth margins and a round, oval, or gently lobulated contour. Unfortunately, many masses demonstrate mammographic features in-

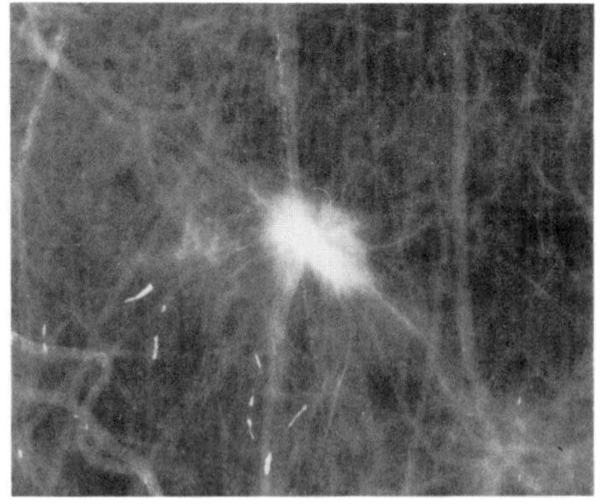

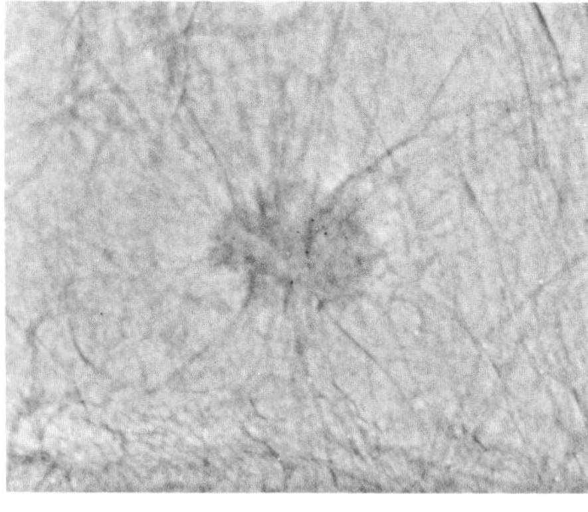

A B

**Figure 5.1. Small breast cancers.** (A) A screen-film mammogram and (B) a xeromammogram demonstrate poorly defined malignant masses with irregular margins and numerous fine strands (spicules) radiating from them.

**Figure 5.2. Radiographic features of benign breast masses.** (A) A screen-film mammogram demonstrates a smooth, round fibroadenoma with clearly defined margins. (B) A xeromammogram demonstrates the smooth contour and sharply defined margins of a benign cyst.

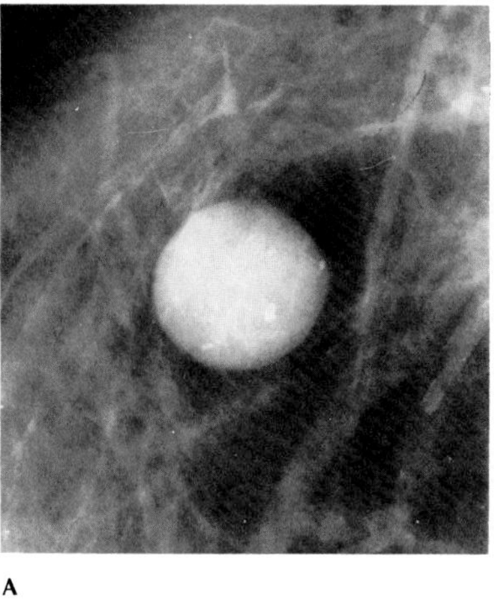

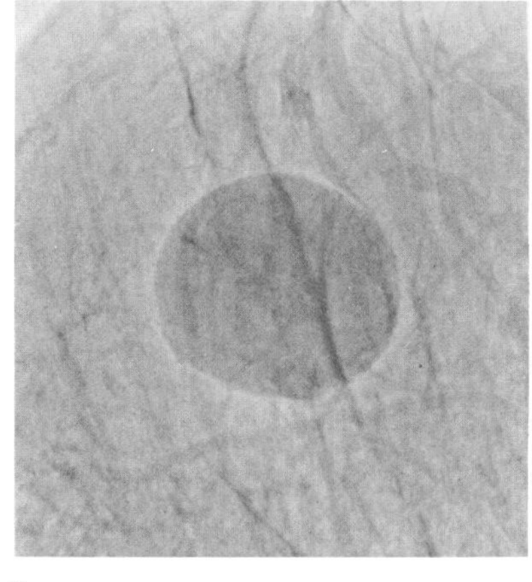

A B

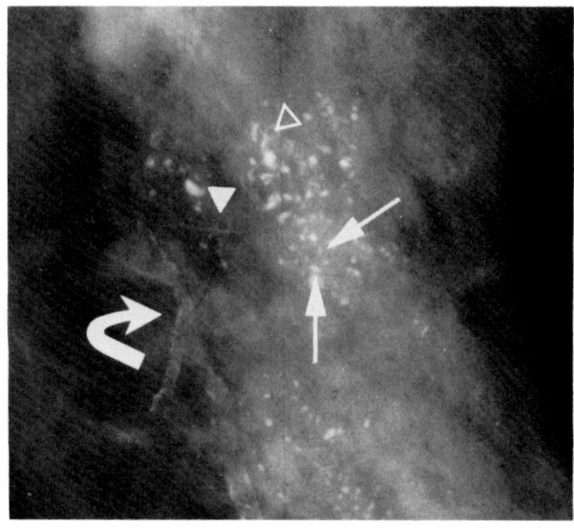

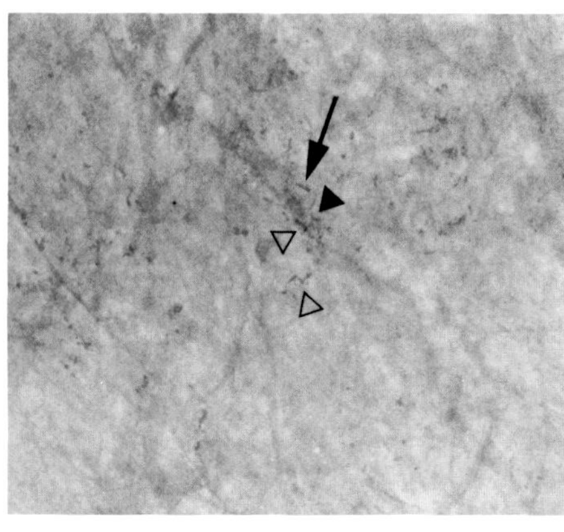

A B

**Figure 5.3. Nonpalpable breast cancers presenting as clusters of calcifications.** (A) A screen-film mammogram and (B) a xeromammogram show numerous tiny calcific particles in one segment of one breast, with the linear (arrows), curvilinear (closed arrowheads), and branching (open arrowheads) forms characteristic of malignancy. On the screen-film mammogram, a benign calcification in the wall of an artery is easily recognized by its large size and tubular distribution (curved arrow).

termediate between these two extremes or are not imaged with sufficient clarity to permit specific mammographic diagnosis. Further complicating this situation, many nonpalpable breast cancers do not even exhibit the mammographic features of a mass, either because the tumor is too small to be distinguished from adjacent dense breast parenchyma or because it is confined to the basement membrane of one or several contiguous ducts and therefore does not form a discrete mass.

The second major mammographic presentation of breast cancer is as clustered calcifications. These calcifications are typically numerous, very small, and lo-

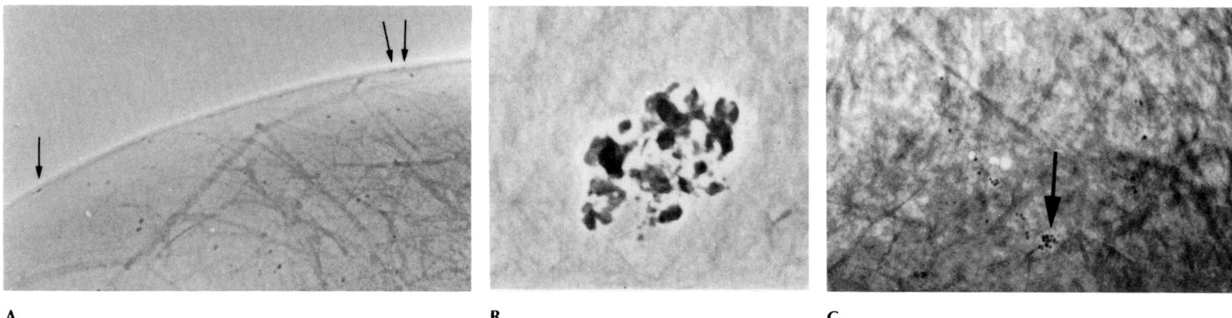

A B C

**Figure 5.4. Typical benign breast calcifications seen on xeromammograms.** (A) Skin calcifications are round to oval and, when imaged tangentially, can be localized to the skin (arrows). (B) The calcifications in a degenerating fibroadenoma are large and amorphous and often have a "popcorn" appearance. (C) The calcifications of adenosis are as small as those of carcinoma, but are round to oval, are usually scattered througout both breasts, and when clustered are grouped very close together (arrow).

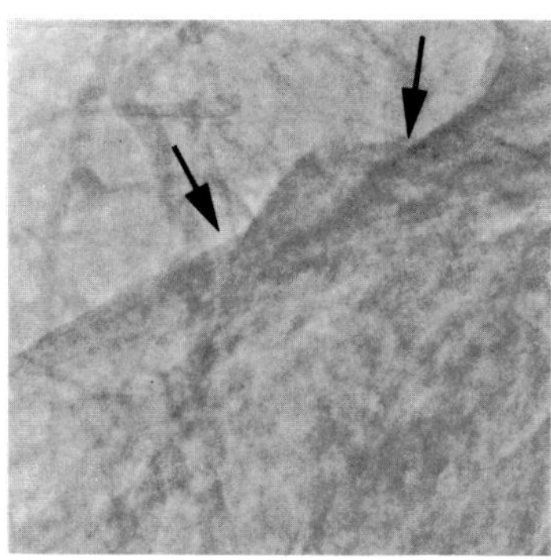

A

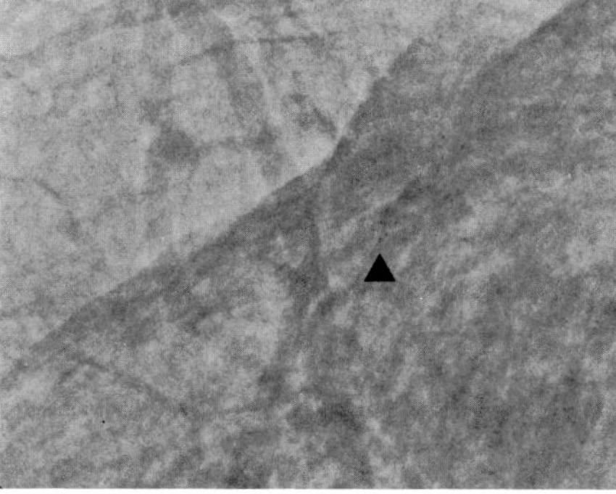

B

**Figure 5.5. The value of magnification mammography in carcinoma of the breast.** (A) A conventional mammogram shows minor distortion of the normal breast architecture, as evidenced by slight parenchymal retraction (arrows). This finding does not raise sufficient suspicion of malignancy to require a biopsy. (B) A magnification mammogram shows tiny clustered calcifications, with at least one rod form (arrowhead). These findings led to a biopsy, which revealed infiltrating duct carcinoma.

calized to one segment of the breast, and they demonstrate a wide variety of shapes, including fine linear, curvilinear, and branching forms. Only about half of breast cancers present mammographically as clusters of calcifications. Fortunately, typical calcification is seen in many of the nonpalpable intraductal cancers, which often do not form mass lesions and might otherwise escape detection. Benign calcifications are very common; the vast majority are easily recognized as benign because of their size, location, distribution, and shape. However, if benign calcifications are small and clustered or if the shapes of individual calcific particles cannot be seen clearly, it may be difficult if not impossible to distinguish them from malignant lesions.

Magnification imaging and grid-assisted imaging are technical enhancements to standard mammography that have been developed over the last few years. Magnification is accomplished by separating the film or xerographic plate from the breast, thereby causing the mammographic shadow image to enlarge. However, this technique is successful only when performed with equipment designed to emit x-rays from a very tiny source. When proper equipment is used, magnification mammography can substantially increase diagnostic accuracy by detecting a small cancer that would otherwise have been overlooked or, more commonly, by converting equivocal mammographic interpretations to definitively malignant or benign diagnoses. Grid-assisted imaging utilizes a moving x-ray grid to enhance the contrast of screen-film mammography images. Like magnification imaging, this technique may permit demonstration of subtle mammographic features that are either inapparent or poorly defined using conventional techniques. Both magnification and grid-assisted mammography impart about twice the standard radiation dose to the breast. Therefore, they usually are used only when conventional mammograms are of suboptimal quality or produce equivocal interpretations.

Even if the mammographic features are characteristic of cancer, confirmation by either aspiration cytology or biopsy is required before proceeding to definitive cancer treatment. This is necessary since a benign lesion, such as fat necrosis or postbiopsy scarring, can rarely mimic the mammographic appearance of a malignancy. If a benign cyst is a likely possibility, this diagnosis can be established or excluded by aspiration or by ultrasound examination (see the following section). Aspiration has the advantage of being both diagnostic and therapeutic. It is also less expensive but is feasible only for masses that are palpable. If a solid lesion is demonstrated, further tissue diagnosis is necessary using either aspiration cytology or open biopsy. Either approach is satisfactory for palpable masses. Nonpalpable lesions require open biopsy with the aid of mammographic localization and specimen radiography to ensure removal of the suspicious lesion.

## Other Breast Imaging Procedures

Ultrasound can differentiate cystic from solid masses and thus substantially reduce the number of biopsies done for benign cysts. Unfortunately, ultrasound is considerably less accurate in distinguishing benign from malignant masses. It is especially ineffective in detecting nonpalpable cancers, particularly those that present with calcifications alone. Therefore, it cannot substitute for mammography as a standard examination and plays only an adjunctive role.

Thermography is a noninvasive imaging test that meaures the heat emitted from the skin of the breast. It is reasonably successful in demonstrating large, palpable cancers but detects only about half of the nonpalpable cancers that can be identified by mammography. Although some investigators have suggested that thermography is useful in predicting the risk of future breast cancer, this has yet to be confirmed in a prospective controlled trial. Currently, there is no clear indication for thermography in the evaluation of breast disease.

Transillumination (diaphanography) attempts to image cancer by the demonstration of a focal area of decreased light transmission through the breast. The technique is particularly sensitive in documenting the presence of hematomas. However, compared with mammography, transillumination is relatively ineffective in detecting small and nonpalpable cancers, especially those located deep within the breast.

Magnetic resonance imaging is a new technique that has been successful in demonstrating several large, palpable breast cancers. However, the number of patients examined with this technique thus far, especially patients with nonpalpable cancer, is too small to determine its potential usefulness.

## Indications for Breast Imaging Procedures

Breast imaging is useful to the extent that it provides diagnostic information not available by physical examination. Mammography is by far the most successful of the breast imaging techniques and is the only imaging procedure that should be considered for cancer detection. It is widely used for "diagnostic" purposes to help establish or exclude the possibility of cancer in a woman with a palpable breast lesion. However, a more important contribution of mammography in this situation is its ability to permit the physician to evaluate the rest of both breasts for nonpalpable cancer. Indeed, mammography should be performed preoperatively on all women over the age of 30 who are scheduled to undergo biopsy for a lesion of unknown cause. Other, less common indications include a bloody nipple discharge, focal skin dimpling or nipple retraction of recent onset, unexplained axillary adenopathy, and the search for a primary tumor site in a woman with documented metastatic adenocarcinoma.

The most important use of mammography is the screening of asymptomatic women for early breast cancer. Only by identifying cancers before they present with clinical signs and symptoms can we expect to find the most potentially curable lesions. For women aged 50 and above, it is clear that annual screening with mammography and physical examination will result in a significant reduction in death from breast cancer. There is also very strong indirect evidence that the same degree of mortality reduction extends to women aged 40 to 49, although proof must await completion of the randomized controlled studies now under way. Several national medical organizations have recently published guidelines for the use of screening mammography. The American Cancer Society recommends a baseline examination before the age of 40, repeat studies at 1- to 2-year intervals until the age of 50, and yearly examinations thereafter.

Several major obstacles have limited the implementation of effective mammographic screening. There is a public perception that the radiation dose used in mammography induces as many cancers as it might help to cure. In reality, current mammographic techniques now are so low in radiation dose that the benefit exceeds the risk many times over. Two other obstacles are the lack of sufficient radiographic equipment, x-ray technologists, and properly trained radiologists to study the tens of millions of women who should be screened each year, and the high cost of the examination. Indeed, only by offering readily available, cost-effective mammography examinations can we expect to have a major impact on breast cancer mortality.[1–14]

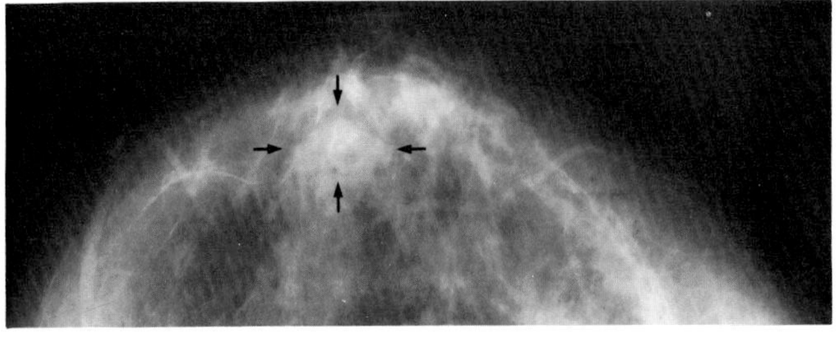

A

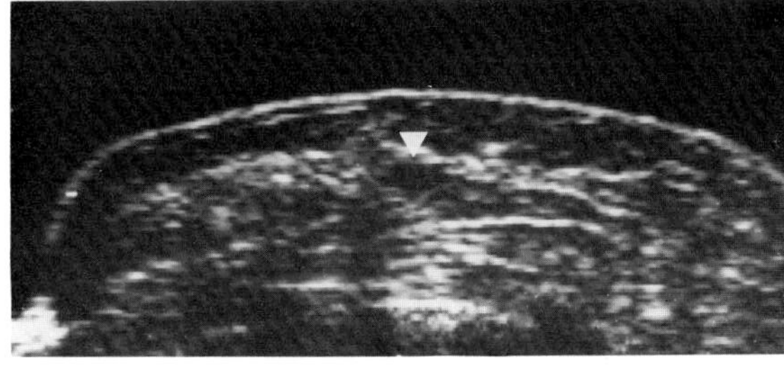

B

**Figure 5.6. The value of sonography in demonstrating a breast cyst.** (A) A mammogram shows a somewhat ill-defined noncalcified mass (arrows). Although the mass is probably benign, a malignancy cannot be excluded. (B) A sonogram documents that the mass is a simple cyst (arrowhead). Because such lesions are invariably benign, biopsy can be averted.

## CARCINOMA OF THE PROSTATE

Carcinoma of the prostate is the second most common malignancy in men. The disease is rare before the age of 50, and the incidence increases with advancing age. The tumor can be slow-growing and asymptomatic for long periods or can behave aggressively with extensive metastases.

Carcinoma of the prostate is best detected by palpation of a hard, nodular, and irregular mass on routine rectal examination. Radiographically, carcinoma of the prostate often elevates and impresses the floor of the contrast-filled bladder. Unlike the contour in benign prostatic hypertrophy, the contour in carcinoma usually is more irregular. Bladder neck obstruction, trigonal infiltration, or invasive obstruction of the ureters above the bladder may produce obstruction of the upper urinary tract.

Carcinoma of the prostate may spread by direct extension, the lymphatics, or the bloodstream. The spread of carcinoma of the prostate to the rectum can produce a large, smooth, concave pressure defect, a fungating ulcerated mass simulating primary rectal carcinoma, or a long, asymmetric annular stricture. Concomitant widening of the retrorectal space is often seen. Ultrasound

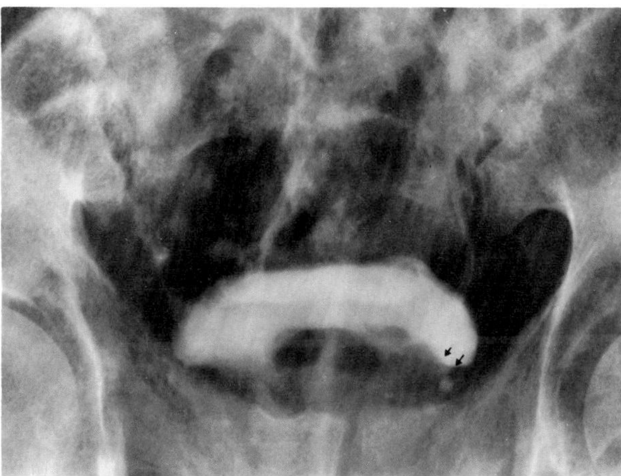

**Figure 5.7. Carcinoma of the prostate.** An excretory urogram demonstrates elevation of and impression on the floor of the contrast-filled bladder. The impression is somewhat irregular, especially at the lateral margins (arrows). Note the blastic mestastases in the sacrum and the medial aspects of the ilia.

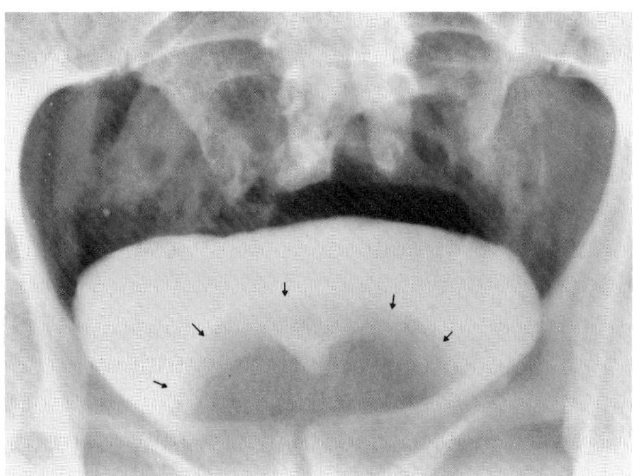

**Figure 5.8. Benign prostatic hypertrophy.** An excretory urogram shows a huge mass (arrows) causing elevation and a smooth impression on the contrast-filled bladder.

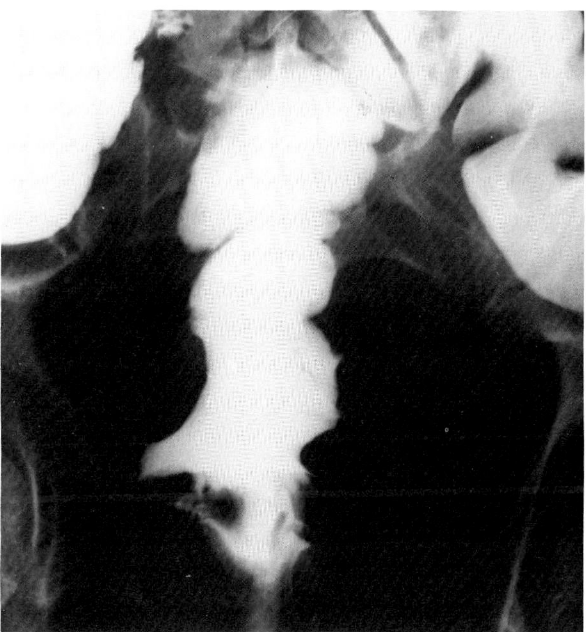

**Figure 5.9. Carcinoma of the prostate involving the colon.** The rectum, which has scalloped margins, shows lack of distension and extrinsic infiltration of the bowel wall. (From Gengler, Baer, and Finby,[49] with permission from the publisher.)

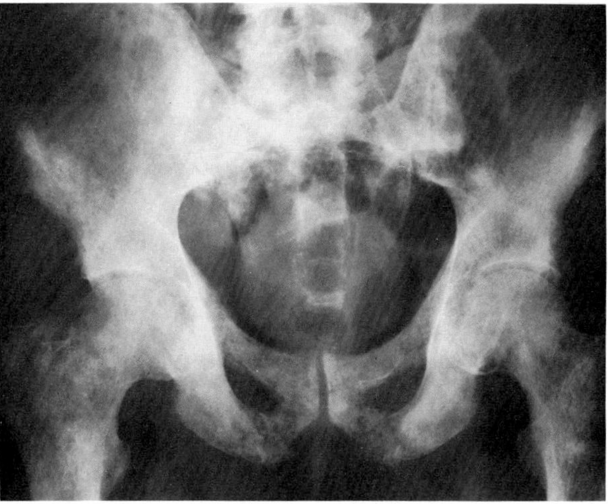

**Figure 5.10. Metastatic carcinoma of the prostate.** The pelvis and proximal femurs are riddled with diffuse osteoblastic lesions.

examination of the pelvis may document the degree of extension of the tumor into the bladder and the seminal vesicles. Computed tomography (CT) may aid in defining the extent of the tumor and in detecting metastases in enlarged lymph nodes.

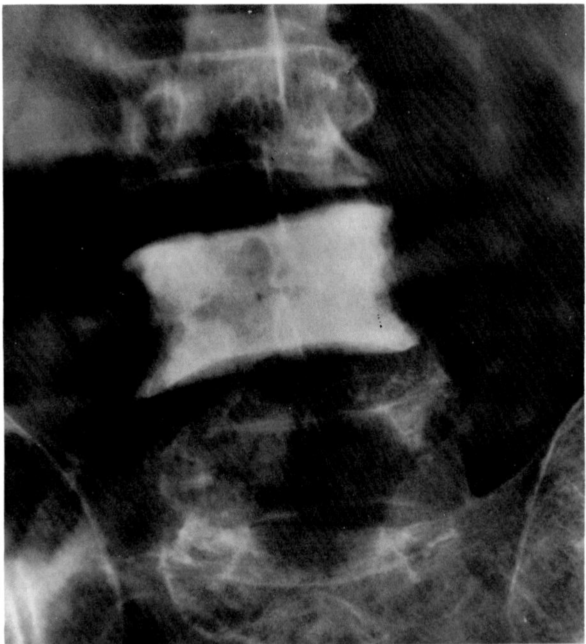

**Figure 5.11. Metastatic carcinoma of the prostate.** Diffuse sclerosis involves an entire vertebral body (ivory vertebra).

The most common hematogenous metastases are to bone. They primarily involve the pelvis, lumbar spine, femurs, thoracic spine, and ribs. These lesions are most commonly osteoblastic and appear as multiple rounded foci of sclerotic density or, occasionally, diffuse sclerosis involving an entire bone ("ivory vertebra"). Patients with bony metastases usually have strikingly elevated levels of serum acid phosphatase. Because significant bone destruction or bone reaction must occur before a lesion can be detected on plain radiographs, the radionuclide bone scan is the best screening technique for detecting asymptomatic skeletal metastases in patients with carcinoma of the prostate. However, since the radionuclide scan is much less specific than sensitive and may show increased uptake in multiple disorders of bone, conventional radiography of the affected site should be performed when a positive scan is obtained.

Other major sites for visceral metastases are the lungs, pleura, liver, and adrenal glands.[15–19,49]

## ACUTE LEUKEMIA

Acute leukemia is a neoplastic proliferation of white blood cells. In childhood leukemia, radiographically detectable skeletal involvement occurs in up to 70 percent of cases as a result of the infiltration of leukemic cells in the marrow. The earliest radiographic sign of disease is usually a transverse radiolucent band at the metaphyseal ends of the long bones, most commonly about the knees, ankles, and wrists. Although in infancy this appearance is nonspecific and also occurs with malnutrition or systemic disease, the presence of these transverse lucent

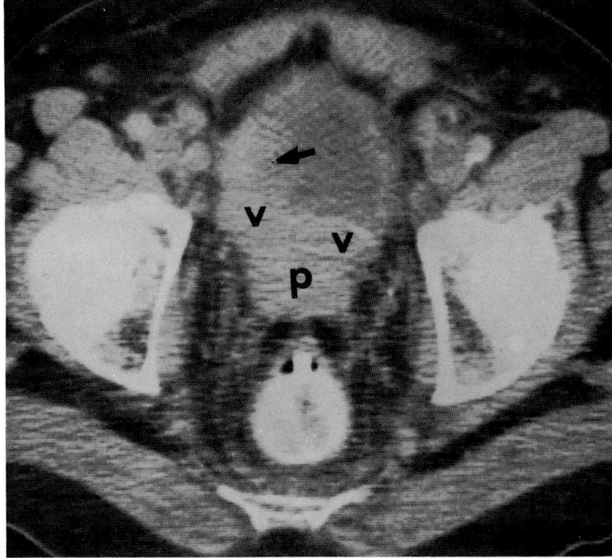


A

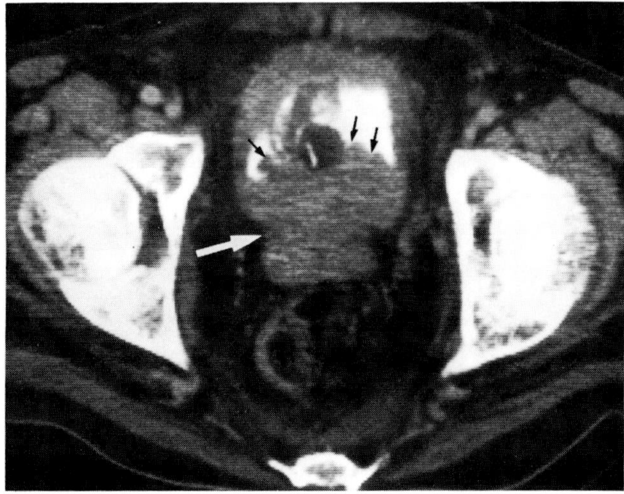

B

**Figure 5.12. Metastatic carcinoma of the prostate.** (A) A CT scan shows prostatic carcinoma (p) invading the wall of the bladder (arrow) and seminal vesicles (v). (B) In another patient, a CT scan shows prostatic carcinoma invading the bladder (black arrows) and seminal vesicles. The sharp angle between the seminal vesicles and the prostate is lost (white arrow). (From Thoeni,[19] with permission from the publisher.)

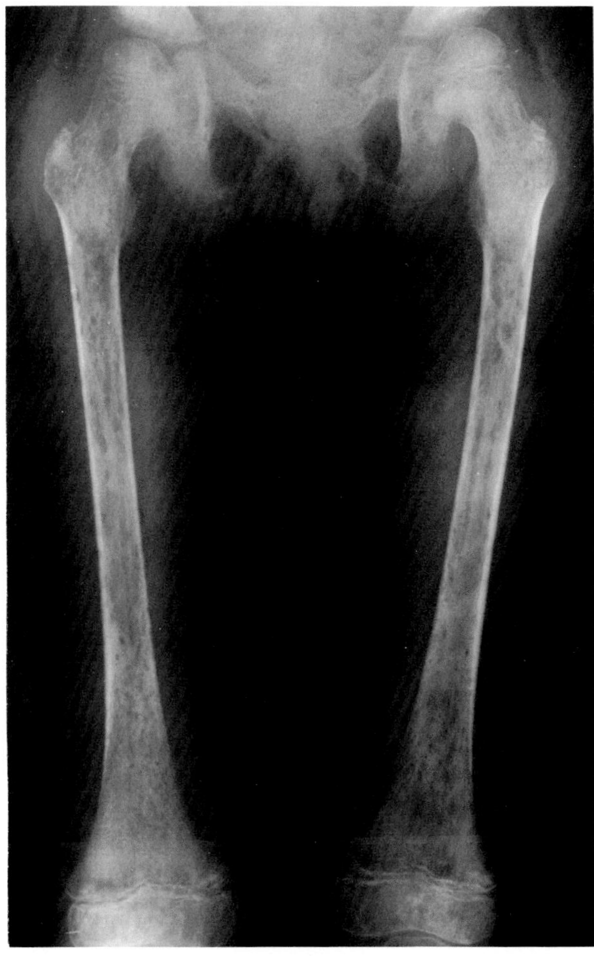

**Figure 5.13. Acute leukemia.** Proliferation of neoplastic cells in the marrow has caused extensive destruction of bone in both femurs.

metaphyseal bands after the age of 2 strongly suggests acute leukemia.

As the proliferation of neoplastic cells in the marrow becomes more extensive, actual destruction of bone may occur. This may cause patchy lytic lesions, a permeative moth-eaten appearance, or diffuse destruction with cortical erosion. A reactive response to proliferating leukemic cells can cause patchy or uniform osteosclerosis; subperiosteal proliferation incites periosteal new bone formation.

In both children and, occasionally, adults with acute leukemia, diffuse skeletal demineralization may result from both leukemic infiltration of the bone marrow and alteration of protein and mineral metabolism. This is best demonstrated in the spine and may result in vertebral compression fractures.

Parenchymal leukemic infiltration or secondary edema and hemorrhage can cause bilateral enlargement of the kidneys. Associated stretching of the pelvocalyceal system may simulate polycystic kidney disease.

Enlargement of mediastinal and hilar lymph nodes is the most common abnormality on chest radiographs. Diffuse bilateral reticular changes may simulate lymphangitic spread of carcinoma. The nonspecific pulmonary infiltrates seen in patients with acute leukemia are usually due to hemorrhage or secondary infection.

Enlargement of the spleen and, less frequently, the liver is common in acute leukemia. However, the degree of splenomegaly is usually much less than that in chronic leukemia.[20–23]

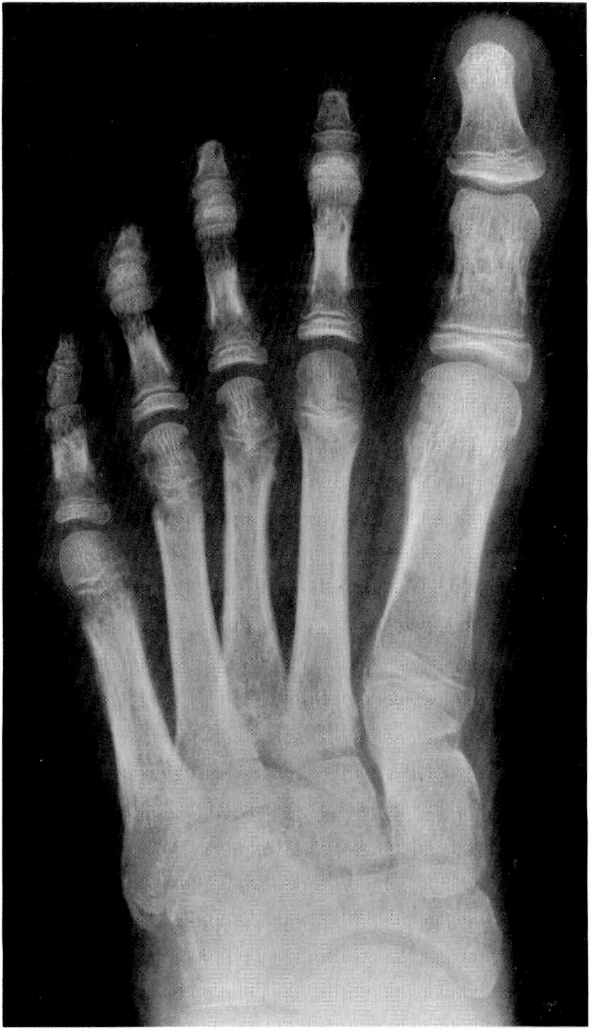

**Figure 5.14. Acute leukemia.** In addition to radiolucent metaphyseal bands, there is frank bone destruction with cortical erosion involving many of the metatarsals and proximal phalanges.

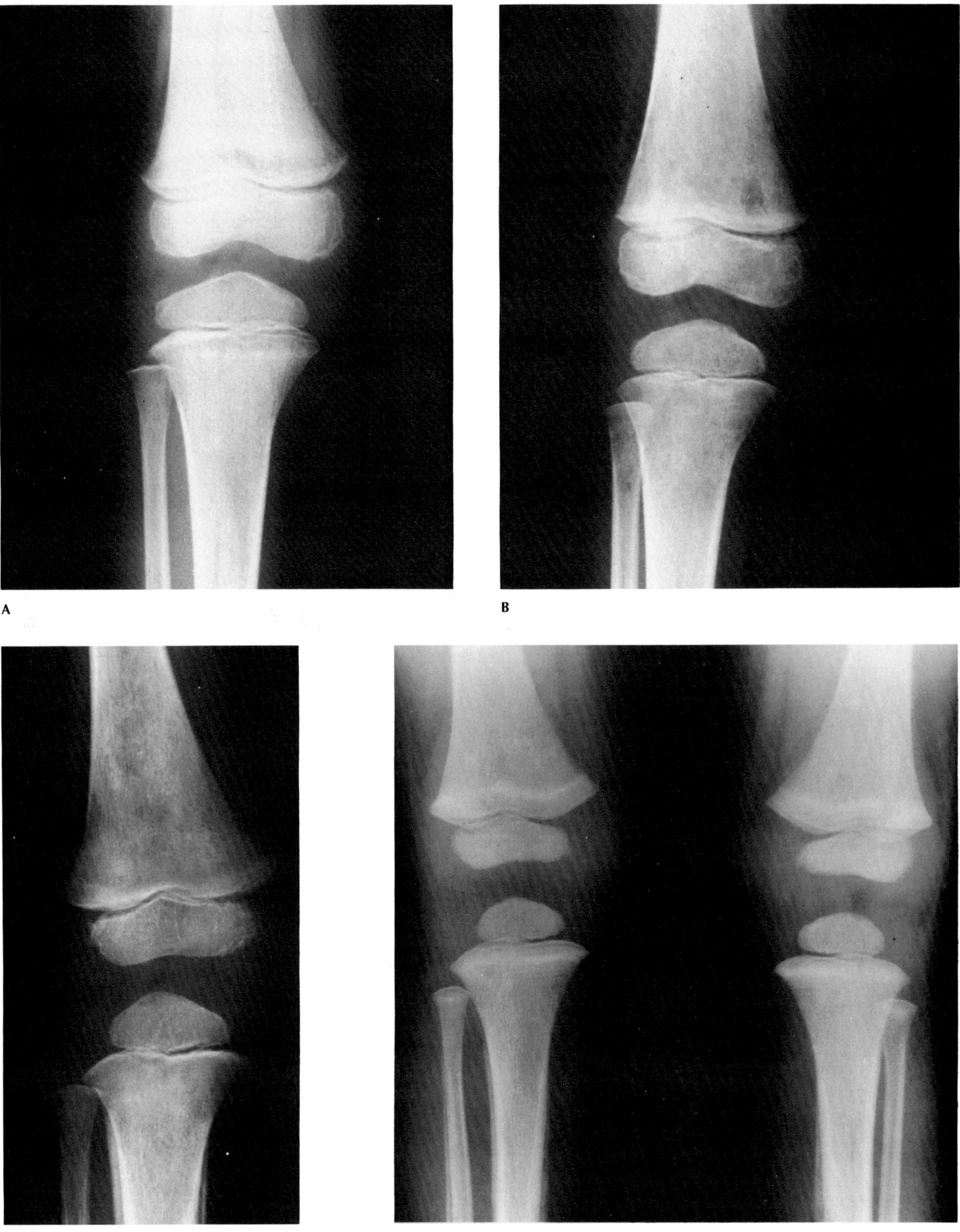


**Figure 5.15. Bone changes of acute leukemia in four children.** (A) Radiolucent metaphyseal bands. (B) Osteolytic foci and metaphyseal cortical destruction. (C) Periosteal elevation. (D) Bands of metaphyseal sclerosis with periosteal response. (A, B, and C from Pear,[20] with permission from the publisher.)

## CHRONIC LEUKEMIA

The radiographic abnormalities in chronic leukemia are often similar to those in the acute disease, though their frequency and degree may vary. Skeletal changes are much less common and are usually limited to generalized demineralization in the flat bones, where active marrow persists in adulthood. Demonstration of focal areas of destruction, or periosteal new bone formation, suggests transformation into an acute phase of the disease.

Hilar and mediastinal adenopathy are common, especially in chronic lymphocytic leukemia. As in the acute form of the disease, pulmonary infiltrates usually reflect hemorrhage or secondary infection, often by opportunistic agents. Pleural effusion, most commonly unilateral, may be secondary to infection, heart failure, or lymphatic obstruction as well as actual leukemic involvement of the pleura. Congestive heart failure commonly results from the associated severe anemia.

Marked splenomegaly is an almost-constant finding in patients with chronic leukemia. Leukemic infiltration of the gastrointestinal tract can produce single or multiple intraluminal filling defects or appear as an infiltrative process that may be indistinguishable from carcinoma. Renal infiltration can cause bilateral enlargement of the kidneys. In chronic lymphocytic leukemia, enlargement of retroperitoneal or mesenteric lymph nodes can cause displacement or obstruction of structures in the genitourinary or gastrointestinal tracts.[20,22,24,25]

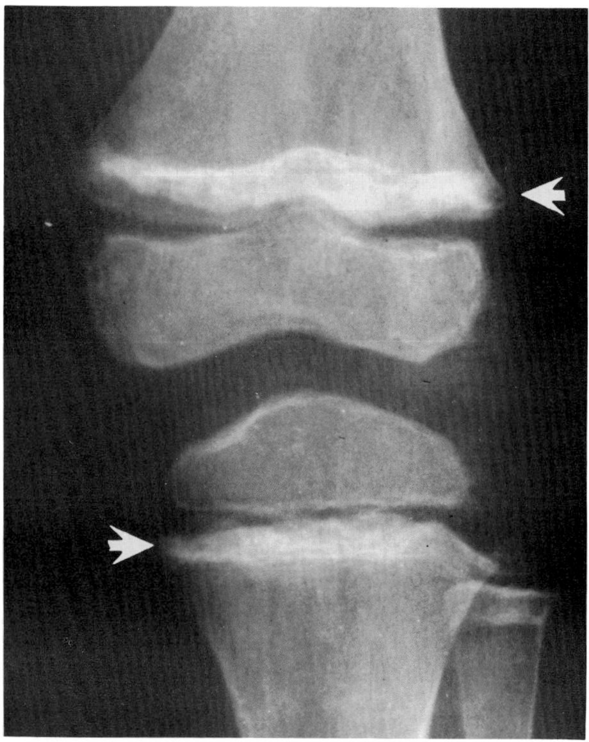

**Figure 5.16. Chronic leukemia.** Following therapy with methotrexate, dense, irregular sclerosis has developed about the metaphyses of the distal femur and proximal tibia (arrows).

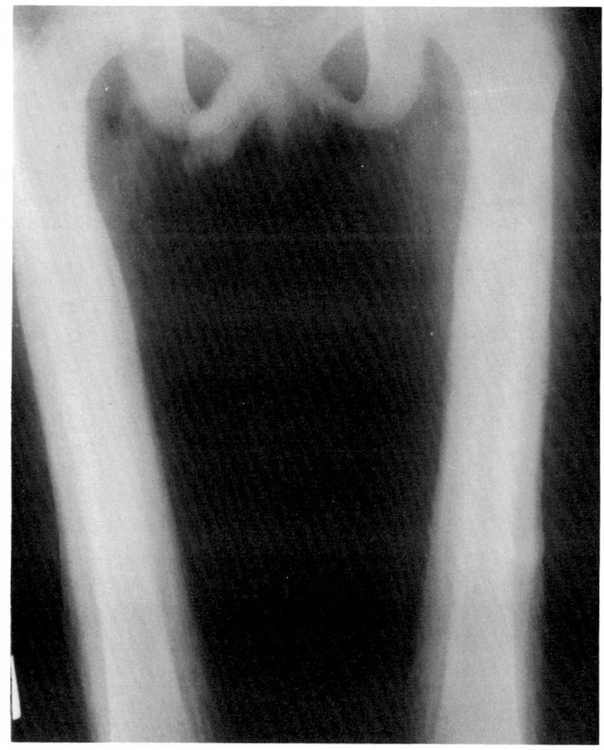

A

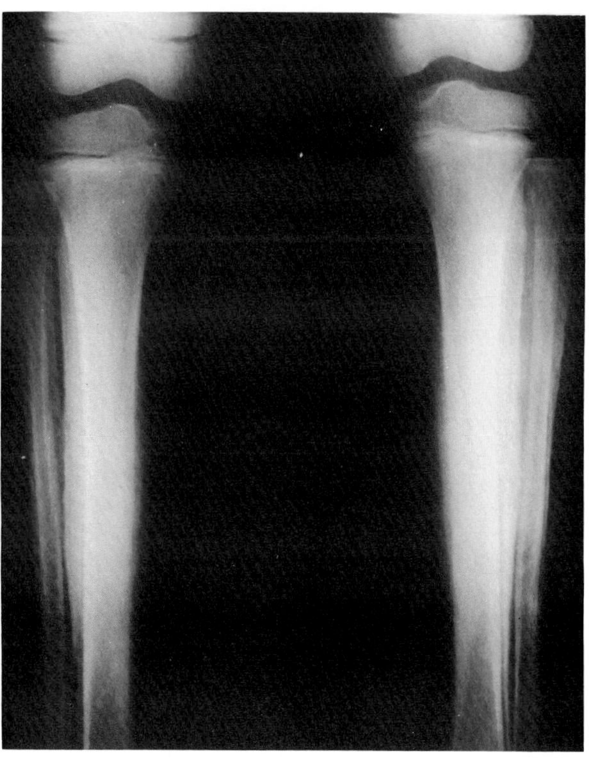

B

**Figure 5.17. Chronic leukemia.** Pronounced periosteal new bone formation cloaking (A) the femurs and (B) the tibias and fibulas.

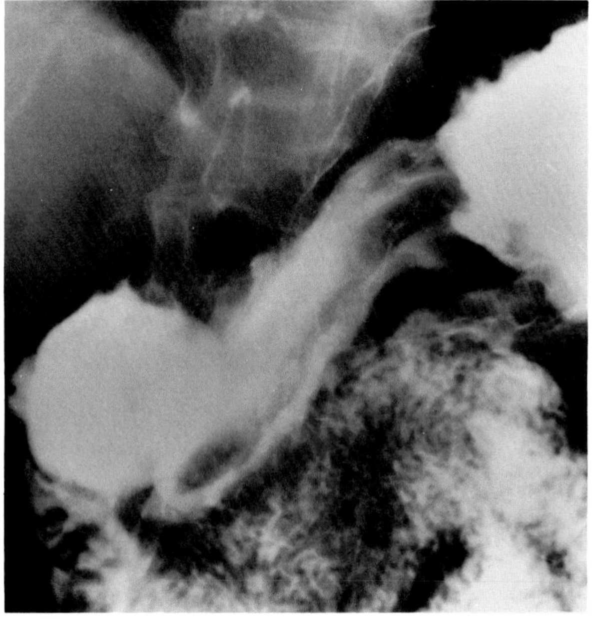

**Figure 5.18. Chronic leukemia.** Infiltration of the wall of the stomach causes diffuse thickening of the gastric folds. (From Eisenberg,[29] with permission from the publisher.)

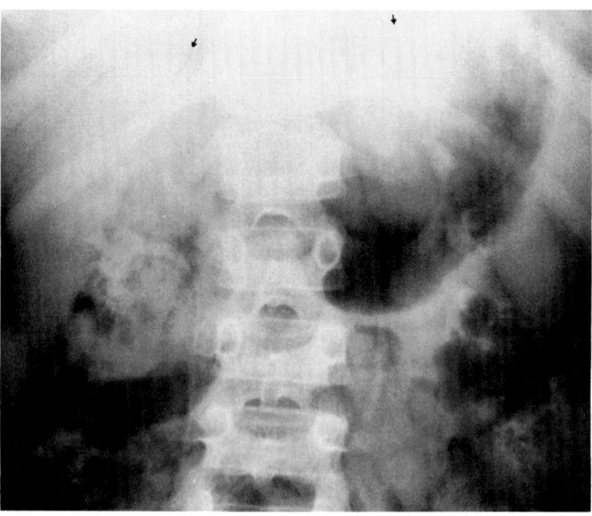

**Figure 5.19. Chronic leukemia.** Diffuse infiltration of leukemic cells has caused bilateral enlargement of the kidneys (arrows).

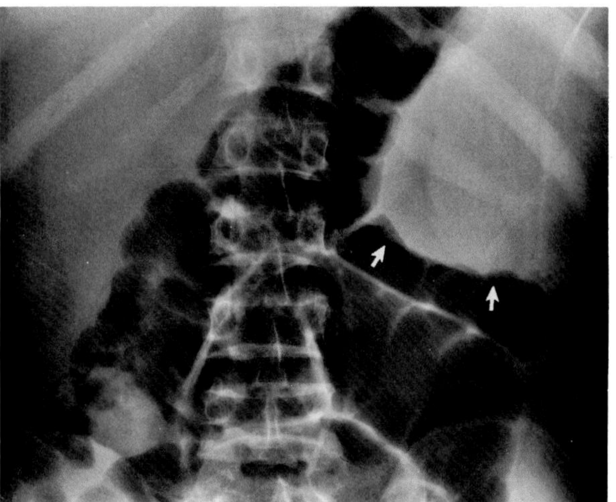

**Figure 5.20. Chronic leukemia.** Marked enlargement of the spleen (arrows).

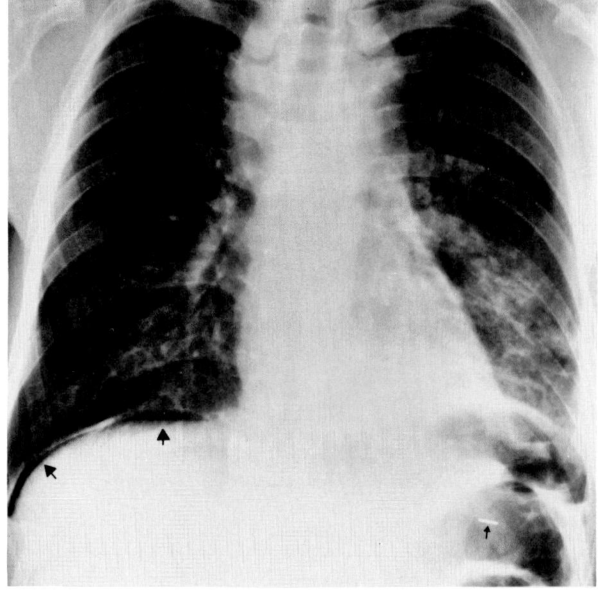

**Figure 5.21. Lymphoma.** A chest radiograph demonstrates diffuse widening of the upper portion of the mediastinum due to lymphadenopathy. There is an ill-defined lymphomatous parenchymal infiltrate at the left base. The metallic clip overlying the region of the spleen (small arrow) and the small amount of free intraperitoneal gas seen under the right hemidiaphragm (large arrows) are evidence of a recent exploratory laparotomy and splenectomy for staging of the lymphoma.

## LYMPHOMA

Lymphoma arises in the lymph nodes or in the lymphoid tissues of parenchymal organs such as the gastrointestinal tract, lung, or skin. Ninety percent of cases of Hodgkin's disease originate in the lymph nodes; 10 percent are of extranodal origin. In contrast, parenchymal organs are more often involved in non-Hodgkin's lymphomas; about 40 percent of these lymphomas are of extranodal origin.

Mediastinal lymph node enlargement is the most common radiographic finding in lymphoma. It is seen on initial chest radiographs of about half of the patients with Hodgkin's disease and one-third of those with non-Hodgkin's lymphoma. Mediastinal lymph node enlargement is usually asymmetric but bilateral. Involvement of anterior mediastinal and retrosternal nodes is common, a major factor in the differential diagnosis from sarcoidosis, which rarely produces radiographically visible enlargement of nodes in the anterior compartment. Calcification may develop in intrathoracic lymph nodes after mediastinal irradiation.

Pleuropulmonary involvement usually occurs by direct extension from mediastinal nodes along the lymphatics of the bronchovascular sheaths. Radiographically, this may appear as a coarse reticulonodular pattern, as solitary or multiple ill-defined nodules, or as patchy areas of parenchymal infiltrate that may coalesce to form a large homogeneous nonsegmental mass. Cavitation of parenchymal lesions may develop. At times, it may be difficult to distinguish a superimposed infection following radiation therapy or chemotherapy from the continued spread of lymphomatous tissue. Pleural effusion occurs in up to one-third of patients with thoracic lymphoma; extension of the tumor to the pericardium can cause pericardial effusion.

Primary pulmonary non-Hodgkin's lymphoma is a rare condition that presents as a homogeneous mass that may occupy an entire lung. Unlike a bronchogenic carcinoma, the lymphomatous mass rarely obstructs the bronchial tree by either intraluminal invasion or extrabronchial compression; thus an air bronchogram can almost invariably be identified. When most or all of a segment or lobe is involved, primary pulmonary lymphoma may simulate acute pneumonia.

About 5 to 10 percent of patients with lymphoma have involvement of the gastrointestinal tract, primarily the stomach and small bowel. Gastric lymphoma often presents as a large, bulky polypoid mass, usually irregular and ulcerated, that may be indistinguishable from a carcinoma. A multiplicity of malignant ulcers or an aneurysmal appearance of a single huge ulcer (the diameter of which exceeds that of the adjacent gastric lumen) is characteristic of lymphoma. Other findings suggestive of lymphoma include relative flexibility of the gastric wall, enlargement of the spleen, and associated prominence of retrogastric and other regional lymph nodes that cause extrinsic impressions on the barium-filled stomach.

Invasion of the gastric wall by an infiltrative type of lymphoma (especially the Hodgkin's type) can cause a severe desmoplastic reaction and a radiographic appearance that mimics that of scirrhous carcinoma. However, unlike scirrhous carcinoma, which results in rigid-

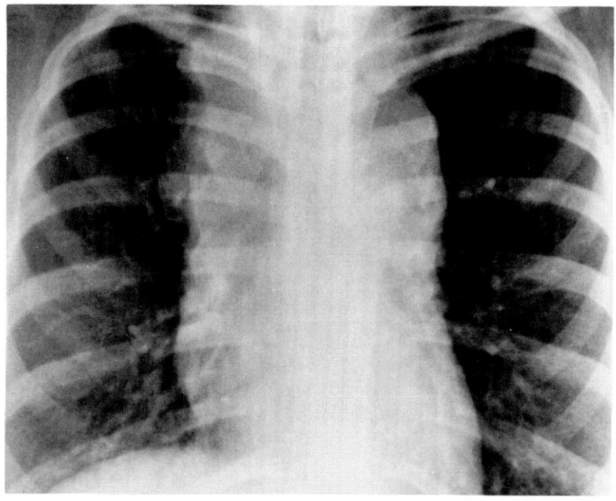
A

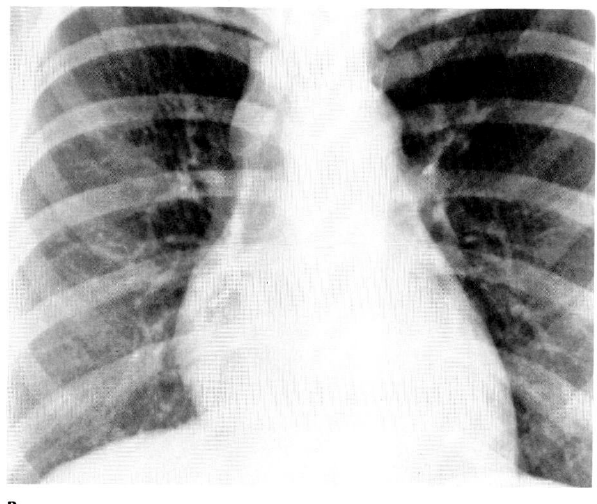
B

**Figure 5.22. Lymphoma.** (A) An initial chest film demonstrates marked widening of the upper half of the mediastinum due to pronounced lymphadenopathy. (B) Following chemotherapy there is a dramatic decrease in the width of the upper mediastinum.

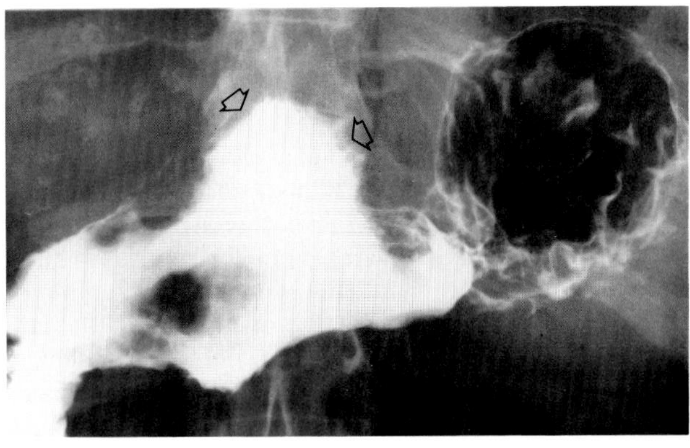

A

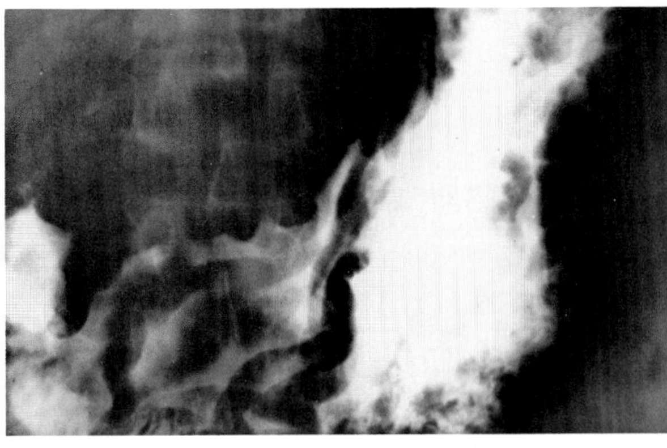

B

**Figure 5.23. Gastric lymphoma.** (A) A huge malignant ulcer (arrows), the diameter of which exceeds that of the adjacent gastric lumen. (B) Severe thickening of the stomach folds, most prominent in the antrum. (From Eisenberg,[29] with permission from the publisher.)

**Figure 5.24. Small bowel lymphoma.** There is generalized irregular thickening of the small bowel folds. Localized segmental circumferential infiltration by the tumor causes a constricting napkin-ring lesion (arrow). (From Eisenberg,[29] with permission from the publisher.)

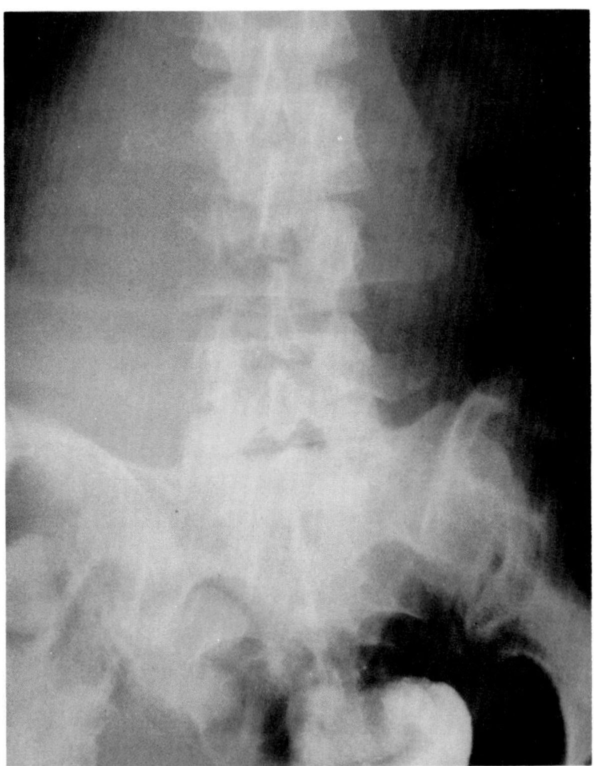

**Figure 5.25. Lymphoma.** There is erosion of the right lateral aspect of the third and fourth lumbar vertebrae with complete destruction of the right transverse process of L4. Note the large adjacent soft tissue mass and the striking lateral displacement of the psoas margin, both due to extensive lymphadenopathy.

ity and fixation, Hodgkin's disease often results in preservation of residual peristalsis and flexibility of the stomach wall.

Thickening, distortion, and nodularity of rugal folds simulating Ménétrier's disease is another pattern of gastric lymphoma. If the enlarged rugal folds predominantly involve the distal portion of the stomach and the lesser curvature, especially in combination with splenomegaly or lymph node enlargement, lymphoma is more likely. However, if the process stops at the incisura and spares the lesser curvature, if there is no ulceration or no true rigidity, or if excess mucus can be demonstrated, Ménétrier's disease is the probable diagnosis.

Lymphoma can produce virtually any pattern of abnormality in the small bowel. The disease may be lo-

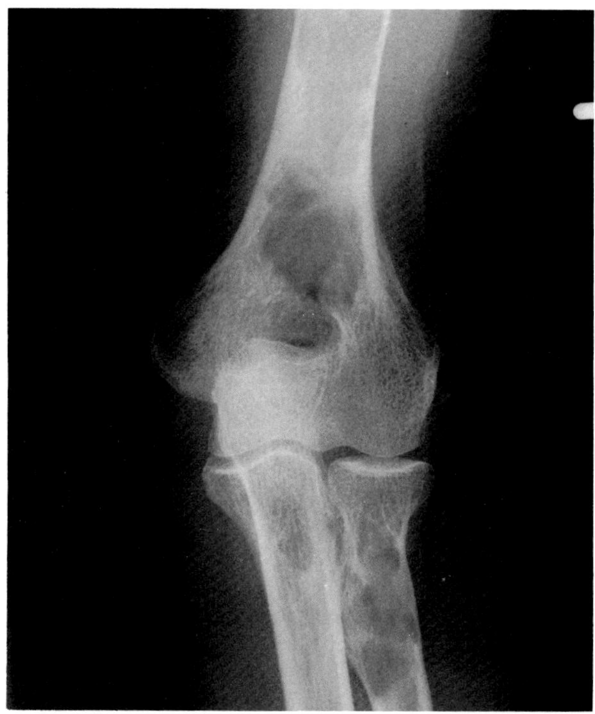

**Figure 5.26. Lymphoma.** Multiple lytic defects in the distal humerus and proximal radius and ulna simulate metastatic disease. (From Pear,[20] with permission from the publisher.)

calized to a single intestinal segment, be multifocal, or cause diffuse involvement. A classic radiographic appearance is infiltration of the bowel wall with narrowing of the lumen and irregular thickening or obliteration of the mucosal folds. If the tumor produces large masses in the bowel that necrose and cavitate, the central core may slough into the bowel lumen and produce aneurysmal dilatation of the bowel wall. Lymphoma may present as a discrete polypoid mass, often large and bulky with irregular ulcerations, or as multiple intraluminal or intramural filling defects simulating metastatic disease. Diffuse submucosal infiltration combined with mesenteric involvement causes separation of small bowel loops.

In the colon, lymphoma can occasionally produce an area of localized narrowing or a lobulated polypoid mass that is radiographically indistinguishable from carcinoma. Unlike carcinoma, localized lymphoma tends to be unusually bulky and to extend over a longer segment of the colon. Diffuse thickening of folds can produce multiple irregular filling defects that are either scattered throughout the colon (simulating multiple polyposis) or limited to a short segment.

Skeletal involvement can be demonstrated in about 15 percent of patients with lymphoma. The spine is most frequently affected; the pelvis, sternum, ribs, long bones, and skull are less frequently involved. Direct extension

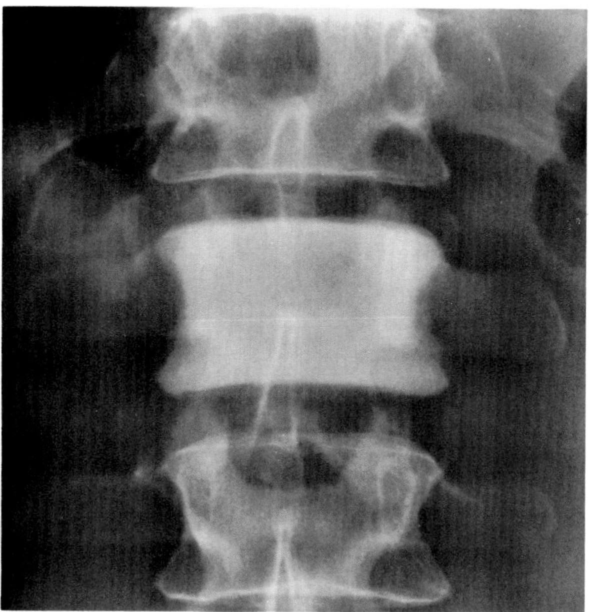

**Figure 5.27. Lymphoma.** Dense sclerosis of a vertebral body (ivory vertebra).

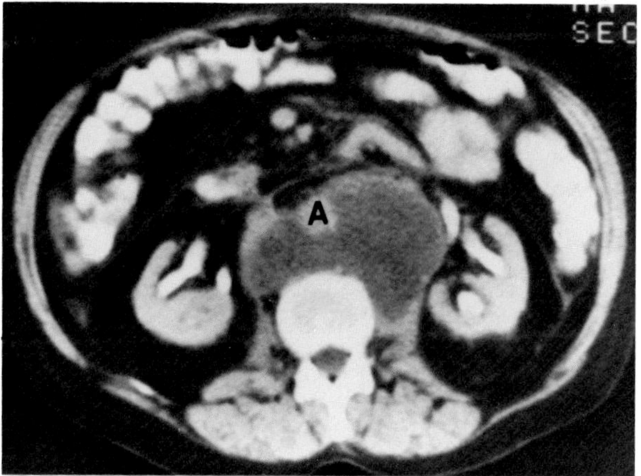


**Figure 5.28. Lymphoma.** A CT scan demonstrates anterior displacement of the abdominal aorta (A) away from the spine due to lymphomatous involvement of retroaortic and para-aortic nodes.

from adjacent lymph nodes causes bone erosion, especially of the anterior surfaces of the upper lumbar and lower thoracic spine. Paravertebral soft tissue masses may occur. The hematogenous spread of lymphoma produces a mottled pattern of destruction and sclerosis that may simulate metastic disease. Dense vertebral sclerosis (ivory vertebra) may develop in Hodgkin's disease.

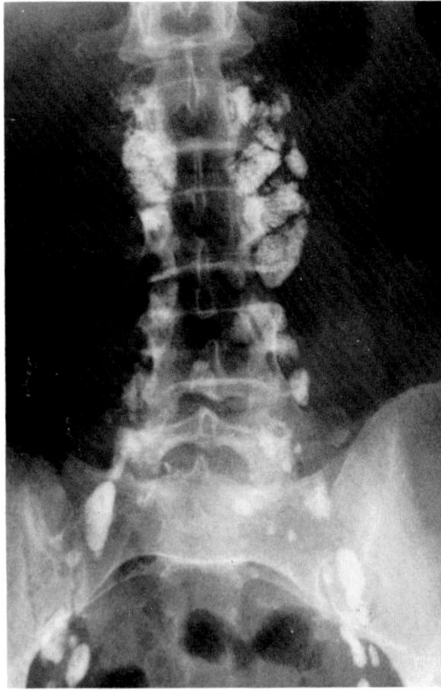
A

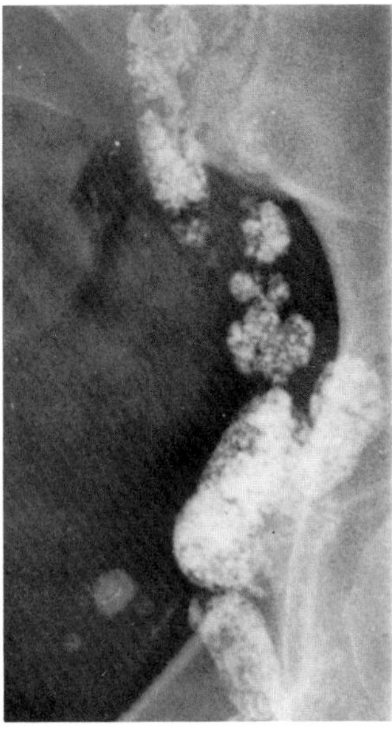
B

**Figure 5.29. Lymphoma.** (A) A staging lymphogram demonstrates tumor involvement of para-aortic and paracaval lymph nodes. (B) In another patient, a lymphogram demonstrates foamy, enlarged pelvic nodes compatible with metastases. (From Marglin and Castellino,[26] with permission from the publisher.)

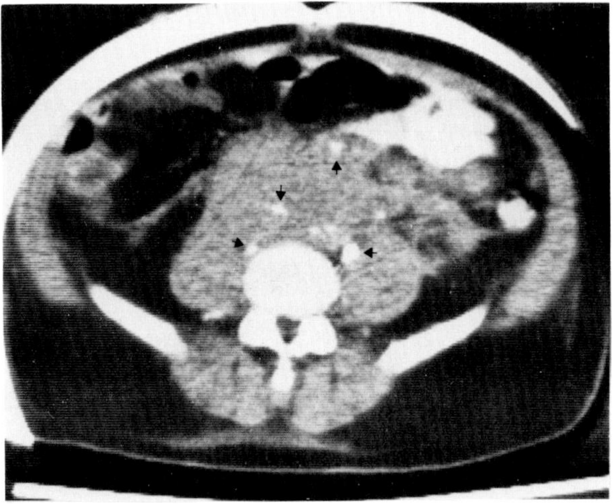
A

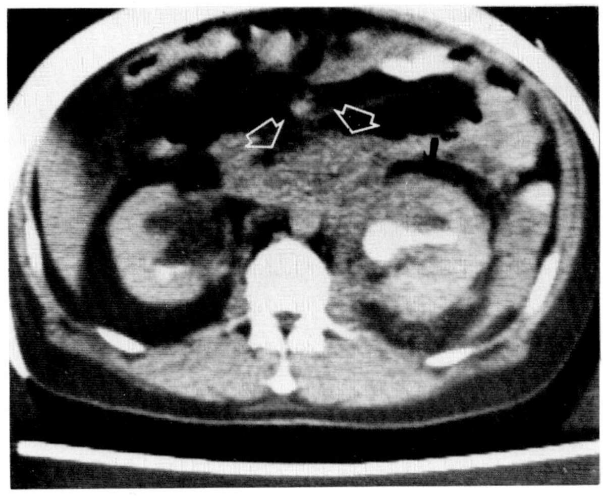
B

**Figure 5.30. Lymphoma.** (A) A CT scan through the upper portion of the iliac crest at approximately the L3–L4 level shows a large lobulated retroperitoneal mass. Although contrast material is seen within many lymph nodes (arrows) that were abnormal at lymphography, the extent of disease is greater than that demonstrated by the opacified nodes. (B) A CT scan more cephalad through the kidneys shows the more superior extension of the retroperitoneal mass (open arrows), which was not opacified during lymphography. In addition, this scan demonstrates the unsuspected finding of an extension of the mass into the anterior aspect of the left kidney (closed arrows), as well as the previously known right hydronephrosis. (From Marglin and Castellino,[26] with permission from the publisher.)

Diffuse lymphomatous infiltration may cause renal enlargement with distortion, elongation, and compression of the calyces. Single or multiple renal nodules or perirenal masses may displace or distort the kidney. Diffuse retroperitoneal lymphoma can displace the kidneys or ureters and obliterate one or both psoas margins.

Once the diagnosis of lymphoma is made, it is essential to determine the status of the abdominal and pelvic lymph nodes. This is necessary for both the initial staging and treatment planning and for assessing the efficacy of treatment and detecting tumor recurrence.

There is controversy about the best radiographic approach for the staging of a patient with known lymphoma. Although lymphography has long been the most accurate examination for demonstrating lymphomatous involvement of abdominal and pelvic nodes, noninvasive cross-sectional imaging techniques (CT and ultrasound) are now considered by many to be the diagnostic methods of choice. Unlike CT and ultrasound, which rely primarily on an increase in node size as a criterion for determining tumor involvement, lymphography can detect microscopic tumor foci, and alterations in architecture within normal-sized nodes. It also can distinguish large nodes that contain tumor from similarly enlarged nodes that demonstrate only benign reactive changes. Following formal staging procedures, postlymphography abdominal radiographs can provide an inexpensive and accurate means of assessing therapeutic efficacy and of detecting relapse. However, the lymph nodes in the upper para-aortic, retrocrural, renal hilus, splenic hilus, porta hepatis, and mesenteric areas cannot be adequately examined by lymphography and require CT or ultrasound. CT is also useful in delineating the exact extent of the tumor mass. It often shows that the abnormality demonstrated with lymphography is only the tip of the iceberg and that more extensive disease, possibly important in planning radiation treatment, is actually present. CT can also be used to follow the response to treatment, especially since there may be insufficient residual contrast on postlymphography abdominal films to make a diagnosis.

In practice, CT is generally the first procedure employed in staging patients with lymphoma, especially those with non-Hodgkin's lymphoma that tends to produce bulky masses in the mesenteric and high retrocrural areas where the contrast material used in lymphography does not reach. An abnormal CT scan eliminates the need for the more invasive lymphography; a normal CT scan obtained at 2-cm intervals can exclude retroperitoneal adenopathy with high confidence. Lymphography is of value primarily in Hodgkin's disease, which infrequently involves the mesenteric nodes, often does not produce bulky masses, and may cause alterations of internal architecture only (which cannot be detected with CT) in normal-sized nodes. Lymphography is also indicated when CT is equivocal because of either a lack of fat or gross motion artifacts.[26–34]

## BURKITT'S LYMPHOMA

Burkitt's lymphoma is a common childhood tumor of the reticuloendothelial system. The disease predominantly affects children in Central Africa and New Guinea and is characterized by swelling and bony lesions of the mandible and maxilla. A tumor histologically indistinguishable from the African variety is increasingly being reported in North American children. Most of these children,

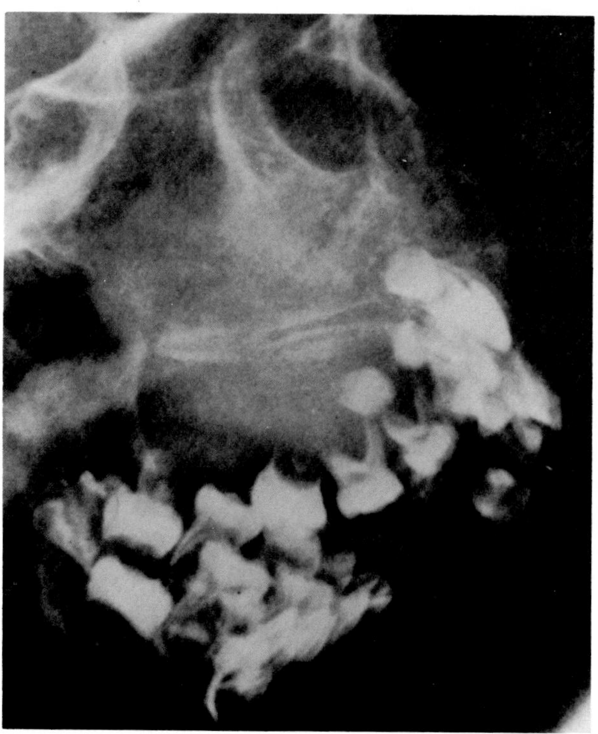

A

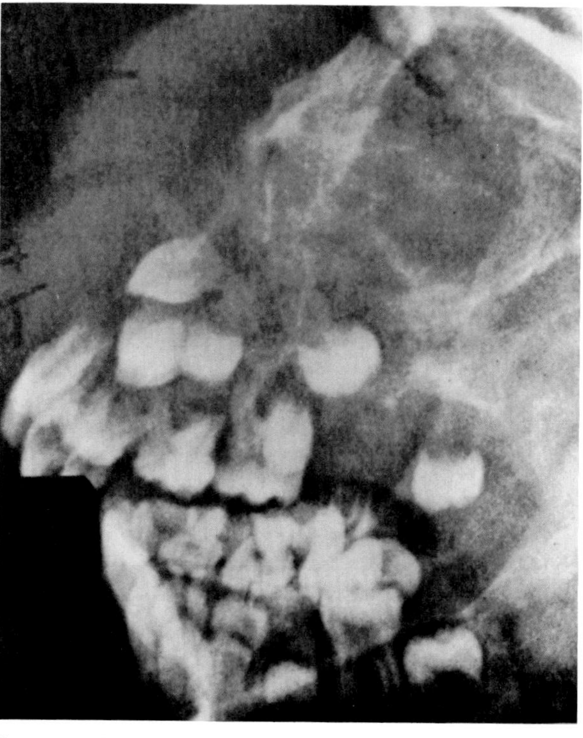

B

**Figure 5.31. Burkitt's lymphoma.** Two views of a patient with advanced tumors of the maxillary areas show soft tissue masses, dental displacement, loss of the lamina dura, and an abnormal medullary bone pattern. (From Whittaker,[37] with permission from the publisher.)

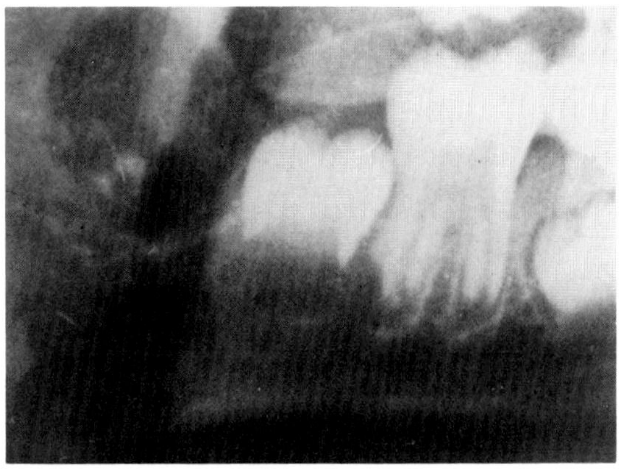

A

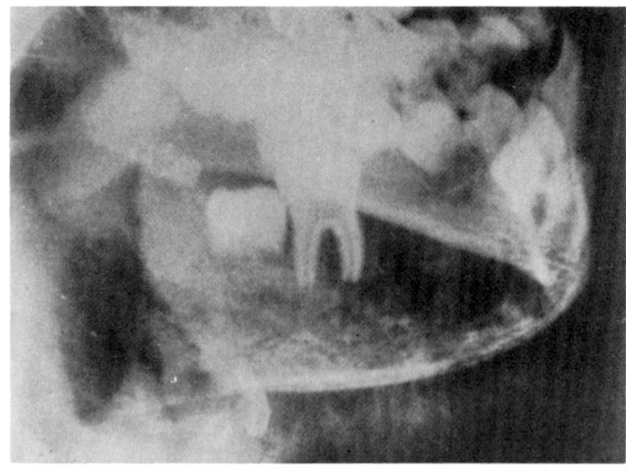

B

**Figure 5.32. Burkitt's lymphoma.** (A) There is a focal loss of the lamina dura around the germinal follicles and roots of several teeth. There is also a loss of trabecular structure. (B) Permeative cortical destruction with a loss of teeth and with periosteal new bone formation. (From Cockshott and Middlemiss,[50] with permission from the publisher.)

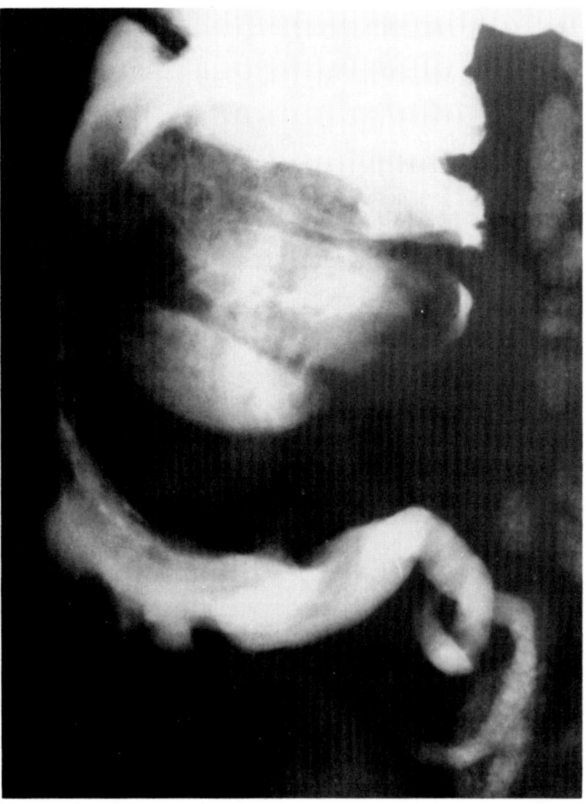

**Figure 5.33. Burkitt's lymphoma.** A huge mass fills essentially the entire cecum. (From Eisenberg,[29] with permission from the publisher.)

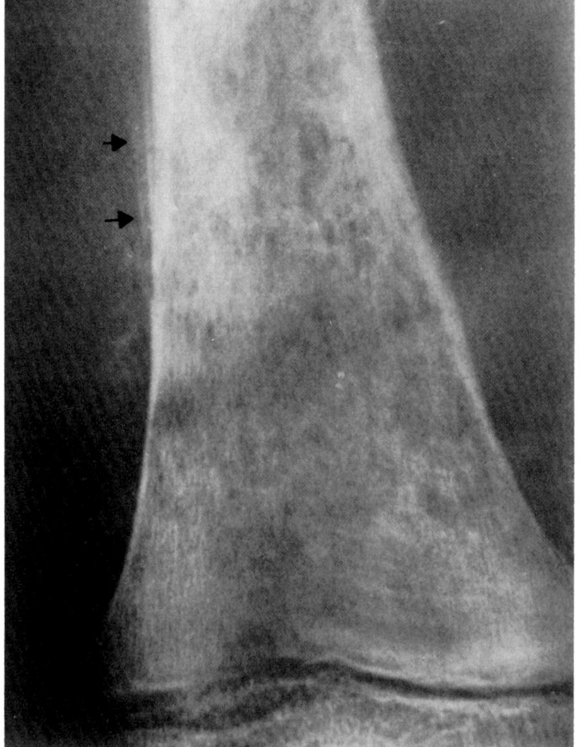

**Figure 5.34. Burkitt's lymphoma.** Erosive, permeative destruction of the distal femur with lamellated periosteal new bone formation (arrow). (From Cockshott and Middlemiss,[50] with permission from the publisher.)

unlike their African counterparts, have demonstrated some involvement of the gastrointestinal tract, predominantly masses in the ileocecal area that often cause intussusception or obstruction.

Radiographically, the African form of Burkitt's lymphoma typically causes a progressive lytic destructive lesion in the mandible or maxilla. Associated soft tissue masses and periosteal reaction are often seen. Although radiographically evident bone lesions are infrequent in American Burkitt's lymphoma, bone marrow involvement is far more common than in the African variety. African Burkitt's lymphoma may cause discrete renal masses, in contrast to the diffuse renal infiltration occasionally seen in the American form. Extrarenal retroperitoneal lymphomatous masses may cause displacement or obstruction of the ureter. Mediastinal adenopathy and pulmonary parenchymal nodules are infrequent; pleural effusion is seen in about two-thirds of patients with American Burkitt's lymphoma but is rare in the African form, in which pleural fluid is usually secondary to cardiac involvement.

Central nervous system disease is seen in up to 20 percent of patients with African Burkitt's lymphoma and may cause paraplegia secondary to spinal cord compression by an intradural or extradural mass.[35–37,50]

## CARCINOID TUMOR

Carcinoid tumors are slowly growing tumors arising from silver-staining (argentaffin) cells found in the crypts of Lieberkühn. Carcinoids most commonly arise in the appendix. These carcinoids are usually single lesions that show much less tendency to metastasize than carcinoids arising elsewhere in the gastrointestinal tract.

The second most common site of carcinoids is the small bowel, especially the ileum. Small bowel carcinoids occasionally present as small, sharply defined submucosal lesions. More commonly, infiltration of the mesentery and an intense desmoplastic response result in the characteristic appearance of diffuse luminal narrowing, and the separation, fixation, and abrupt angulation of intestinal loops.

Rectal carcinoids are usually small, solitary, and asymptomatic and present as polypoid protrusions into the lumen. Rarely, they appear as infiltrating or annular lesions indistinguishable from adenocarcinoma. Rectal carcinoids develop metastases in about 10 percent of cases but rarely give rise to the carcinoid syndrome.

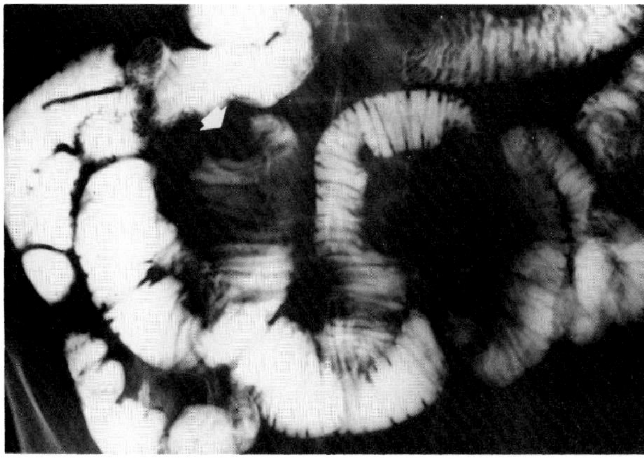

**Figure 5.35. Carcinoid tumor.** A polypoid filling defect (arrow) in the small bowel.

**Figure 5.36. Carcinoid tumor.** Separation, fixation, and angulation of intestinal loops with diffuse luminal narrowing are caused by the intense desmoplastic response incited by the tumor.

**Figure 5.37. Carcinoid tumor.** An intense desmoplastic reaction incited by the tumor causes kinking and angulation of the bowel and separation of small bowel loops in the midabdomen. (From Eisenberg,[29] with permission from the publisher.)

The likelihood of metastases from carcinoid tumors of the small bowel and the colon is directly related to the size of the primary lesion. Metastases are rare from primary tumors under 1 cm; they develop from about half of the tumors between 1 and 2 cm and from about 90 percent of primary lesions larger than 2 cm.

Carcinoid tumors tend to metastasize to the liver and may involve this organ extensively. Bone metastases, which often are osteoblastic, can occur. Deposition of fibrous tissue on the endocardium of the valvular cusps and cardiac chambers often leads to tricuspid insufficiency and stenosis of the pulmonary valve.

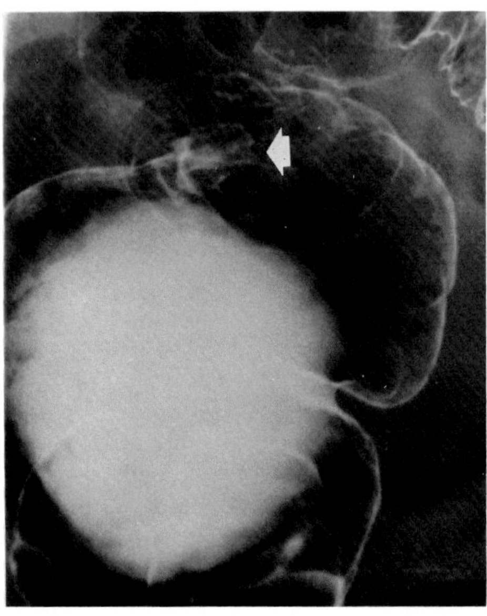

**Figure 5.38. Carcinoid tumor.** The submucosal mass in the rectum presents as a sessile polyp protruding into the lumen (arrow). (From Eisenberg,[29] with permission from the publisher.)

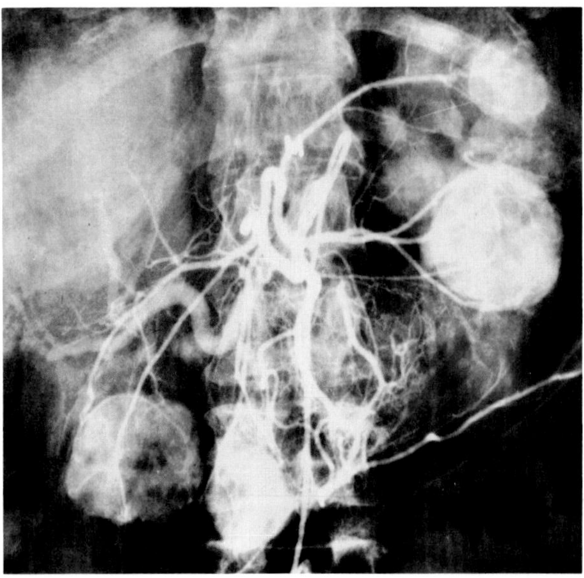

**Figure 5.39. Carcinoid syndrome.** There are multiple large, extremely vascular metastases from a carcinoid tumor of the small bowel.

Benign bronchial carcinoids constitute the vast majority of bronchial adenomas (see the section "Tumors of the Lung" in Chapter 33). Carcinoid tumors can also arise in the stomach, pancreas, and thyroid, and from ovarian and testicular teratomas.

The characteristic carcinoid syndrome consists of skin flushing, diarrhea, cyanosis, and bronchoconstriction. The clinical symptoms are due to circulating serotonin produced by the carcinoid tumor. Because serotonin released into the portal venous system is inactivated in the liver, the carcinoid syndrome is seen almost exclusively in patients with liver metastases, in whom serotonin is released directly into the systemic circulation without being inactivated. These liver metastases typically appear as multiple large, extremely vascular lesions on hepatic arteriograms. The small bowel is the primary tumor site in almost all patients with the carcinoid syndrome; however, only a minority of small bowel lesions present with this endocrine syndrome.[38–41]

## CUTANEOUS MANIFESTATIONS OF INTERNAL MALIGNANCY

### Kaposi's Sarcoma

Kaposi's sarcoma is a systemic disease that characteristically affects the skin and causes an ulcerated hemorrhagic dermatitis. The skin lesions are sensitive to irradiation and have a low-grade malignant course. The typical nodules or pigmented patches initially involve the extremities and are frequently associated with intense lymphedema, which causes the limbs to become firm, thick, and heavily pachydermatous. The disease is most common in middle-aged or elderly men from countries of eastern Europe, northern Italy, and parts of Africa. In recent years, Kaposi's sarcoma has been described with increasing frequency in homosexual men and in conjunction with the acquired immunodeficiency syndrome (AIDS).

Most patients with multiple skin nodules inevitably develop visceral disease as well, though this may not be manifest at the time of the original diagnosis. Metastases to the small bowel are relatively common and consist of multiple reddish or bluish-red nodules that intrude into the lumen of the bowel. Similar lesions can develop throughout the gastrointestinal tract. Central ulceration of the metastases causes gastrointestinal bleeding and a characteristic radiographic appearance of multiple "bull's-eye" lesions simulating metastatic melanoma.

Pulmonary abnormalities in patients with Kaposi's sarcoma may represent metastatic infiltrates or, more commonly, superinfection by opportunistic organisms such as *Pneumocystis carinii*. Destructive foci in bone may be caused by erosion from an external soft tissue deposit or from an intramedullary lesion that expands the bone to produce a cystlike mass.[42–45,50]

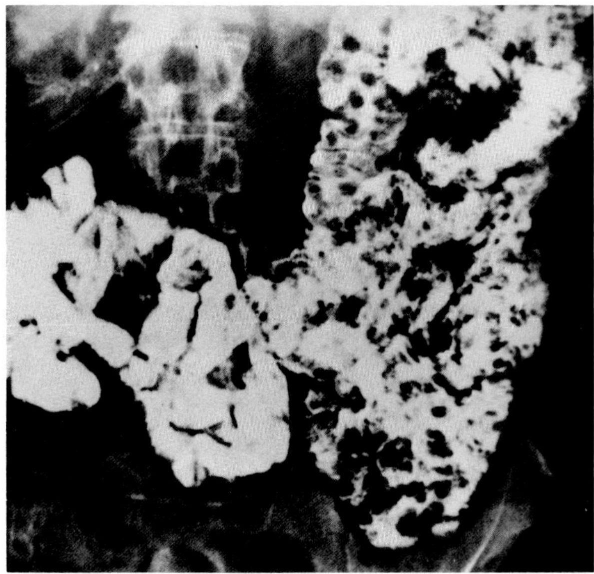

**Figure 5.40. Kaposi's sarcoma.** A small bowel study shows multiple intramural nodules (predominantly involving the jejunum) that distort the mucosal pattern and produce contour defects as well as intraluminal lucencies. (From Bryk, Farman, Dallemand, et al,[45] with permission from the publisher.)

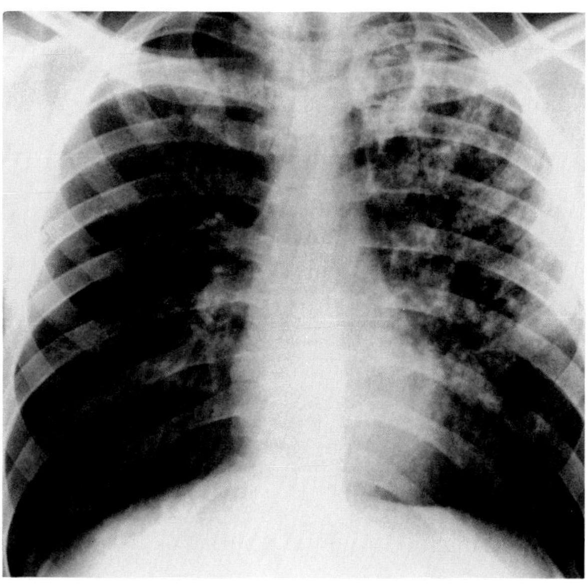

**Figure 5.41. Kaposi's sarcoma.** Patchy interstitial infiltrate and air space consolidation involving much of the left lung reflect a combination of Kaposi's sarcoma of lung parenchyma and an opportunistic fungal infection.

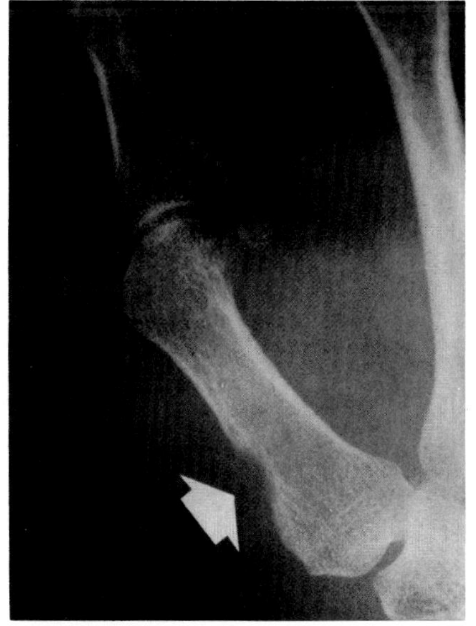

A

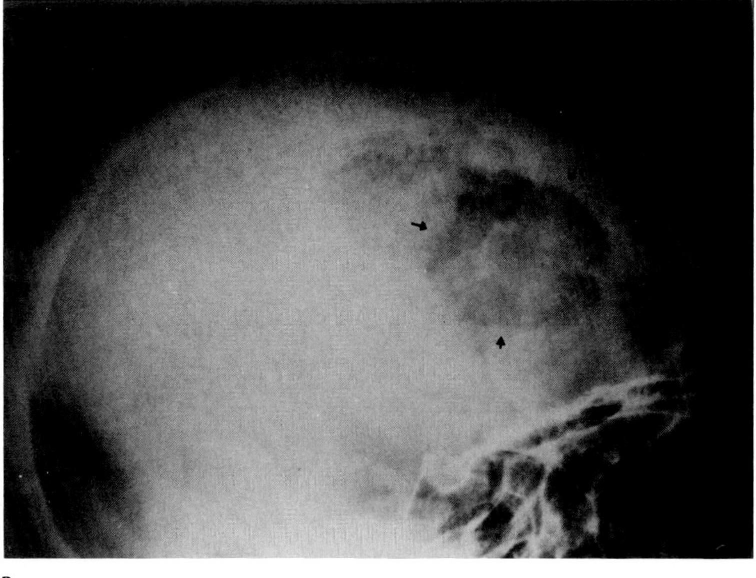

B

**Figure 5.42. Kaposi's sarcoma involving bone.** (A) Erosive marginal cortical bone destruction (arrow). Note the soft tissue prominence due to superficial Kaposi's deposits. (B) An erosive defect in the skull (arrows). (From Cockshott and Middlemiss,[50] with permission from the publisher.)

## Acanthosis Nigricans

Acanthosis nigricans is a premalignant skin disorder characterized by papillomatosis, pigmentation, and hyperkeratosis. When it involves the esophagus, multiple verrucous proliferations throughout the mucosa (similar to the skin changes) may produce the radiographic appearance of finely nodular filling defects simulating candidiasis or superficial spreading eosphageal carcinoma. Acanthosis nigricans has been associated with a high incidence of malignant tumors, especially adenocarcinomas of the stomach.[46]

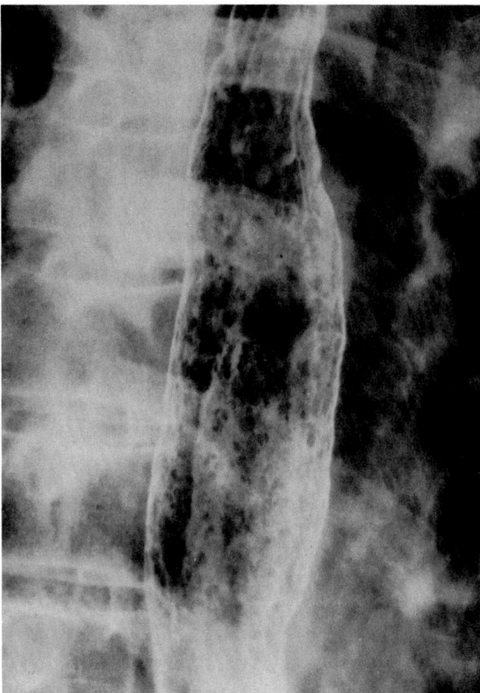

**Figure 5.43. Acanthosis nigricans.** Innumerable finely nodular lesions are seen throughout the esophagus without evidence of serration of the esophageal margins. (From Itai, Kogure, Okuyama, et al,[46] with permission from the publisher.)

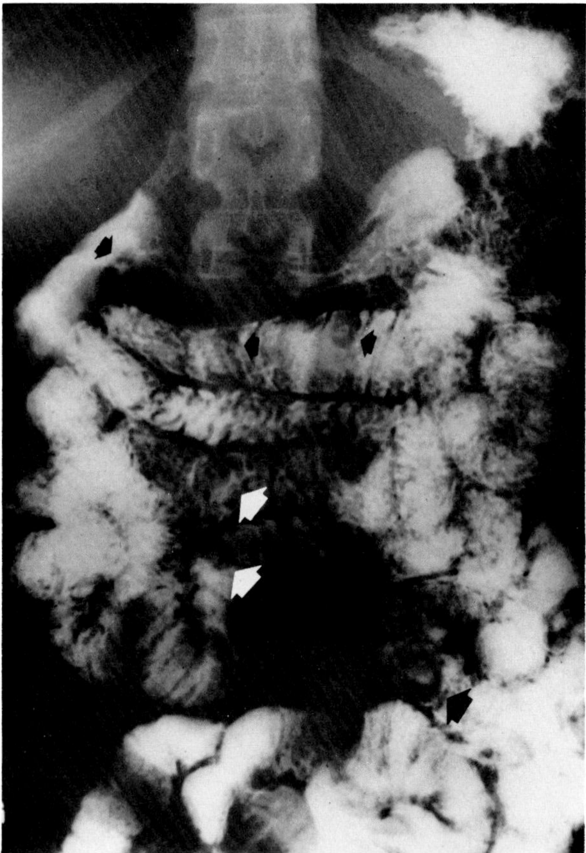

**Figure 5.44. Metastatic melanoma.** Multiple nodular filling defects (arrows) varying from 6 mm to 2 cm in diameter are present throughout the small bowel. Central ulceration can be identified in most of the lesions; in some the central ulcer is large in relation to the size of the mass, causing a bull's-eye appearance. At least 10 separate, discrete nodular lesions could be identified on the complete series of original radiographs. (From Cavanagh, Buchignani, and Rulon,[51] with permission from the publisher.)

## MALIGNANT MELANOMA

Malignant melanoma metastasizes widely and frequently involves the gastrointestinal tract, usually sparing the large bowel. Metastases of this primary skin malignancy can be well-circumscribed, round or oval nodules, plaques, or sessile or pedunculated polypoid masses. As the metastasis outgrows its blood supply, central necrosis and ulceration produce a dense, barium-filled central crater surrounded by the sharply marginated nodular mass (bull's-eye, or target, lesion). Metastases in the form of an enlarging pedunculated mass projecting into the bowel lumen can lead to intussusception. Gastrointestinal metastases can be the first clinical manifestation of metastatic melanoma; at times it can be impossible to identify the primary tumor site.

Metastatic melanoma can also produce multiple nodules in the lungs, destructive lesions in bone with neither new bone formation nor reactive sclerosis, intraluminal filling defects in the pelvocalyceal collecting system of the urinary tract, and masses in the adrenal gland.[47,48,51]

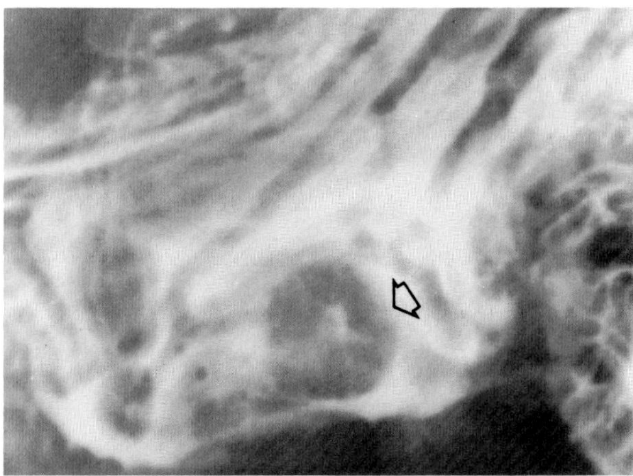

**Figure 5.45. Metastatic melanoma.** A large, ulcerated filling defect with a bull's-eye appearance (arrow). Several smaller, nodular lesions can also be seen. (From Eisenberg,[29] with permission from the publisher.)

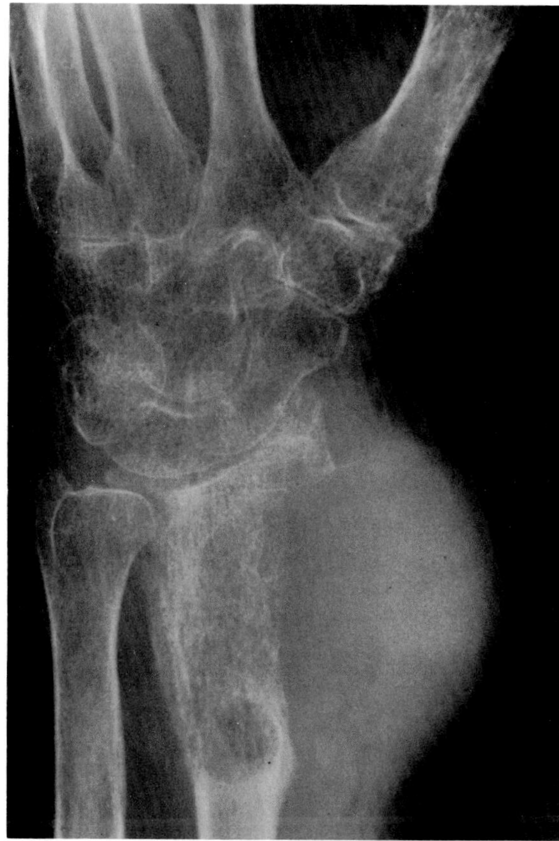

**Figure 5.46 Malignant melanoma.** There is a large soft tissue mass with erosion of the distal radius. The patient initially had a lesion excised from this location 13 years previously and had suffered multiple subsequent recurrences.

## REFERENCES


1. Moskowitz M: Minimal breast cancer redux. *Radiol Clin N Amer* 21:93–113, 1983.
2. Sickles EA: The relative merits of xeromammography and screen-film mammography, in Margulis AR, Gooding CA (eds): *Diagnostic Radiology 1981*. San Francisco, University of California Press, 1981, pp 395–402.
3. Martin JE: *Atlas of Mammography. Histologic and Mammographic Correlations*. Baltimore, Williams & Wilkins, 1982.
4. Tabar L, Dean PB: *Teaching Atlas of Mammography*. New York, Thieme-Stratton, 1983.
5. Sickles EA: Microfocal spot magnification mammography using xerographic and screen-film recording systems. *Radiology* 131:599–607, 1979.
6. McSweeney MB, Sprawls P, Egan RL: Mammographic grids, in Feig SA, McLelland R (eds): *Breast Carcinoma: Current Diagnosis and Treatment*. New York, Masson, 1983, pp 169–176.
7. Sickles EA, Filly RA, Callen PW: Sonography for the detection of benign breast lesions. *Radiology*, in press.
8. Sickles EA, Filly RA, Callen PW: Breast cancer detection with sonography and mammography: Comparison using state-of-the-art equipment. *AJR* 140:843–845, 1983.
9. Dodd GD: Heat-sensing devices and breast cancer detection, in Feig SA, McLelland R (eds). *Breast Carcinoma: Current Diagnosis and Treatment*. New York, Masson, 1983, pp 207–225.
10. Sickles EA: Breast cancer detection with transillumination and mammography. *AJR*, in press.
11. Shapiro S: Evidence on screening for breast cancer from a randomized trial. *Cancer* 39:2772–2782, 1977.
12. Feig SA: Mammographic screening: Benefit and risk, in Feig SA, McLelland R (eds): *Breast Carcinoma: Current Diagnosis and Treatment*. New York, Masson, 1983, pp 351–363.
13. American Cancer Society: Guidelines for the cancer-related checkup. Recommendations and rationale. *CA* 30:224–227, 1980.
14. American Cancer Society: Mammography guidelines 1983: Background statement and update of cancer-related checkup guidelines for breast cancer detection in asymptomatic women age 40 to 49. *CA* 33:255–257, 1983.
15. Shaffer DL, Pendergrass HP: Comparison of enzyme, clinical, radiographic, and radionuclide methods of detecting bone metastases from carcinoma of the prostate. *Radiology* 121:431–434, 1976.
16. Dennis JM: The solitary dense vertebral body. *Radiology* 77:618–621, 1961.
17. Meyers MA, Oliphant M, Teixidor H: Metastatic carcinoma simulating inflammatory colitis. *AJR* 123:74–83, 1975.
18. Klein LA: Prostatic carcinoma. *N Engl J Med* 300:824–833, 1979.
19. Thoeni RF: Computed tomography of the pelvis, in Moss AA, Gamsu G, Genant HK (eds): *Computed Tomography of the Body*. Philadelphia, WB Saunders, 1983, pp 987–1053.
20. Pear BL: Skeletal manifestations of the lymphomas and leukemias. *Semin Roentgenol* 9:229–240, 1974.
21. Nixon GW, Gwinn JL: The roentgen manifestations of leukemia in infancy. *Radiology* 107:603–607, 1973.
22. Klatte EC, Yardley J, Smith EB, et al: Pulmonary complications of leukemia. *AJR* 89:598–609, 1963.
23. Gowdy JF, Neuhauser EBD: Roentgen diagnosis of diffuse leukemic infiltration of the kidneys in children. *AJR* 60:13–21, 1948.
24. Green RA, Nichols NJ: Pulmonary involvement in leukemia. *Am Rev Respir Dis* 80:833–844, 1959.
25. Limberakis AJ, Mossler JA, Roberts L, et al: Leukemic infiltration of the colon. *AJR* 131:725–728, 1978.
26. Marglin S, Castellino R: Lymphographic accuracy in 632 consecutive previously untreated cases of Hodgkin's disease and non-Hodgkin's lymphoma. *Radiology* 140:351–353, 1981.
27. Lee JKT, Stanley RJ, Sagel SS, et al: Limitations of the post-lymphangiogram plain abdominal radiograph as an indicator of recurrent lymphoma: Comparison to computed tomography. *Radiology* 134:155–158, 1980.
28. Glazer HS, Lee JKT, Balfe DM: Non-Hodgkin's lymphoma: Computed tomographic demonstration of unusual extranodal involvement. *Radiology* 149:211–217, 1983.
29. Eisenberg RL: *Gastrointestinal Radiology: A Pattern Approach*. Philadelphia, JB Lippincott, 1983.
30. Filly R, Blank N, Castellino RA; Radiographic distribution of intrathoracic disease in previously untreated patients with Hodgkin's disease and non-Hodgkin's lymphoma. *Radiology* 120:277–281, 1976.
31. Fraser RG, Pare JAP: *Diagnosis of Diseases of the Chest*. Philadelphia, WB Saunders, 1979.
32. Fisher AMH, Kendall B, Van Leuven BD: Hodgkin's disease. A radiological survey. *Clin Radiol* 13:115–127, 1962.
33. Baron MG, Whitehouse WM: Primary lymphosarcoma of the lung. *AJR* 85:294–308, 1961.
34. Lee JKT, Stanley RJ, Sagel SS, et al: Accuracy of computed tomography in detecting intra–abdominal and pelvic adenopathy in lymphoma. *AJR* 131:311–315, 1978.
35. Alford BA, Coccia PF, L'Heureux PR: Roentgenographic features of American Burkitt's lymphoma. *Radiology* 124:763–770, 1977.
36. Cockshott WP: Radiological aspects of Burkitt's tumour. *Br J Radiol* 38:172–180, 1965.

37. Whittaker LR: Burkitt's lymphoma. *Clin Radiol* 24:339–346, 1973.
38. Balthazar EJ: Carcinoid tumors of the alimentary tract: Radiographic diagnosis. *Gastrointest Radiol* 3:47–56, 1978.
39. Peavy PW, Rogers JV, Clements JL, et al: Unusual osteoblastic metastases from carcinoid tumors. *Radiology* 107:327–330, 1973.
40. Ureles AL: Diagnosis and treatment of malignant carcinoid syndrome. *JAMA* 229:1346–1348, 1974.
41. Bancks NH, Goldstein HM, Dodd GD: The roentgenologic spectrum of small intestinal carcinoid tumors. *AJR* 123:274–280, 1975.
42. Brown RKJ, Huberman RP, Vanley G: Pulmonary features of Kaposi's sarcoma. *AJR* 139:659–660, 1982.
43. Rose HS, Balthazar EJ, Megibow AJ, et al: Alimentary tract involvement in Kaposi's sarcoma. *AJR* 139:661–666, 1982.
44. McCauley DI, Naidich DP, Leitman BS, et al: Radiographic patterns of opportunistic lung infections and Kaposi's sarcoma in homosexual men. *AJR* 139:653–658, 1982.
45. Bryk D, Farman J, Dallemand S, et al: Kaposi's sarcoma of the intestinal tract: Roentgen manifestations. *Gastrointest Radiol* 3:425–430, 1978.
46. Itai Y, Kogure T, Okuyama Y, et al: Diffuse finely nodular lesions of the esophagus. *AJR* 128:563–566, 1977.
47. Goldstein HM, Begdoun MT, Dodd GD: Radiologic spectrum of melanoma metastatic to the gastrointestinal tract. *AJR* 129:605–612, 1977.
48. Oddson TA, Rice RP, Seigler HF, et al: The spectrum of small bowel melanoma. *Gastrointest Radiol* 3:419–423, 1978.
49. Gengler L, Baer J. Finby N: Rectal and sigmoid involvement secondary to carcinoma of the prostate. *AJR* 125:910–917, 1975.
50. Cockshott WP, Middlemiss H: *Clinical Radiology in the Tropics*. Edinburgh, Churchill Livingstone, 1979.
51. Cavanagh RC, Buchignani JS, Rulon DB: RPC of the month from the AFIP. *Radiology* 101:195–200, 1971.

# PART TWO
# INFECTIOUS DISEASES

## Chapter Six
# General Overview

## LOCALIZED INFECTIONS AND ABSCESSES

### Abdominal Abscesses

Even with the availability of broad-spectrum antibiotics and sophisticated surgical techniques, the morbidity and mortality of abdominal abscesses remain high. The newer imaging modalities of computed tomography (CT), ultrasound, and radionuclide scanning have greatly improved the ability to accurately detect intra-abdominal abscesses. In selected cases, percutaneous drainage of abdominal abscesses is indicated as a safe and effective alternative to surgical intervention.

Plain abdominal radiographs and contrast studies of the gastrointestinal tract are often abnormal in patients with intra-abdominal abscesses. Ectopic gas or gas-fluid levels, soft tissue masses, pleural fluid or elevation of the diaphragm, and focal dilatation of adjacent bowel loops are all important diagnostic clues. However, the nonspecific nature of these findings, and the inability to precisely define the extent of the abscess, significantly limit the overall usefulness of conventional radiographic techniques.

The classic ultrasound appearance of an abscess is a sonolucent collection surrounded by thick, irregular walls and clearly separable from the normal structures of the abdomen, pelvis, and retroperitoneum. In practice, however, a spectrum of patterns ranging from purely cystic to purely solid may occur. The thick, purulent contents of an abscess often produce a pattern of diffuse, weak echoes. If there are septations or clumps of necrotic debris within the central cavity, coarse and irregular echoes are seen. The internal layering of different components of the abscess may produce the bandlike pattern of a fluid-fluid level. Gas collections, either large or in the form of microbubbles, reflect incident sound waves almost completely and cause distal shadowing. Both the irregularity of the abscess walls and the nature of the abscess fluid itself result in poorer definition of the wall-fluid interface than might be expected. Similarly, the through transmission of sound that is characteristic of simple cysts may not be present in abscesses. Although this wide variety of patterns may appear confusing, ultrasound can detect more than 90 percent of abdominal abscesses. Ultrasound is especially valuable for detecting abscesses in the right upper quadrant or pelvis; pancreatic abscesses and those involving the left upper quadrant may be obscured by overlying bowel gas from an accompanying ileus. Because the examination requires close contact between the transducer and the skin, ultrasound may be difficult to employ in postsurgical patients with recent incisions, wound dressings, drainage tubes, superficial infections, or stomas.

The CT appearance of an abdominal abscess varies with its age and maturity. Before a mature abscess cavity forms, a phlegmon may alter the normal organ contours and obliterate the adjacent soft tissue planes. These findings, in the appropriate clinical setting, may permit the very early detection of a focal infective process. A well-formed abscess appears as a soft tissue or low-density mass that neither conforms to the contour of the normal parenchymal organs nor lies within the confines of the bowel. Abscesses commonly appear homogeneous, although septations and necrotic debris may present a pattern of varying attenuation values. The thick, irregular wall of a mature abscess can often be seen. The most specific CT finding of abdominal abscess, seen in almost half of the cases, is the presence of gas within the central cavity. Intravenous contrast agents may be used to advantage in the detection of abscesses by CT. After contrast material is administered, the hypervascular wall of the abscess is enhanced while the central cavity remains unchanged. The thickening of fascial planes and the abnormal density of adjacent intraperitoneal or extraperitoneal fat indicate extension of the inflammatory process. The relation of abdominal abscesses to adjacent bowel loops can be readily determined by the oral or rectal administration of dilute contrast material to opacify the gastrointestinal tract.

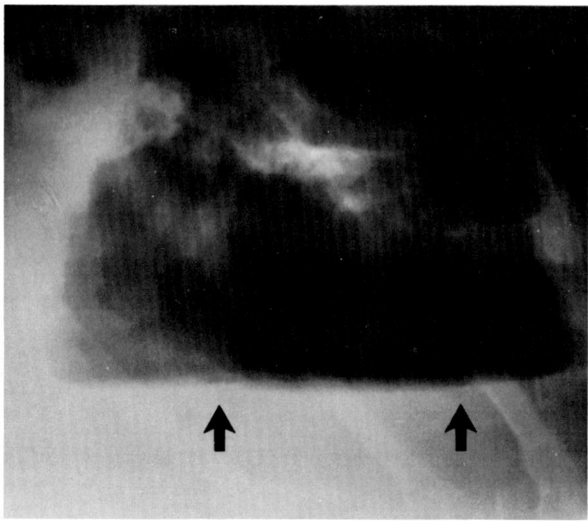

**Figure 6.1. Hepatic abscess.** A lateral decubitus view demonstrates a gas-fluid level (arrows) in this abscess, which contains a large amount of soft tissue necrotic debris.

Although CT has a high sensitivity in abscess detection, the findings are somewhat nonspecific, and a similar pattern can be produced by old hematomas, tumors with central necrosis, and complicated cysts.

There are two basic approaches to the detection of abdominal abscesses by radionuclide scanning. If agents that accumulate in normal parenchyma are used, negative defects (cold spots) indicate displacement of the normal tissue by a pathologic process. However, techniques using radionuclides that selectively accumulate at sites of inflammation have largely supplanted parenchymal imaging agents in abscess detection. The most widely used technique employs gallium 67 citrate or indium 111–labeled leukocytes. The migration of white blood cells to sites of infection results in an increased local concentration of radioactivity. A major disadvantage of gallium scanning is the accumulation of the radio-

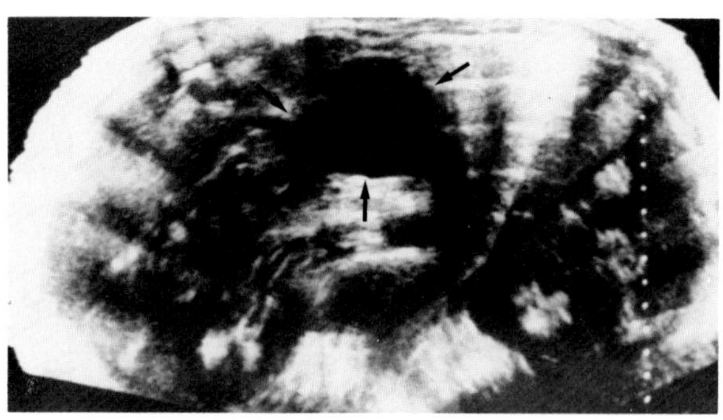

A

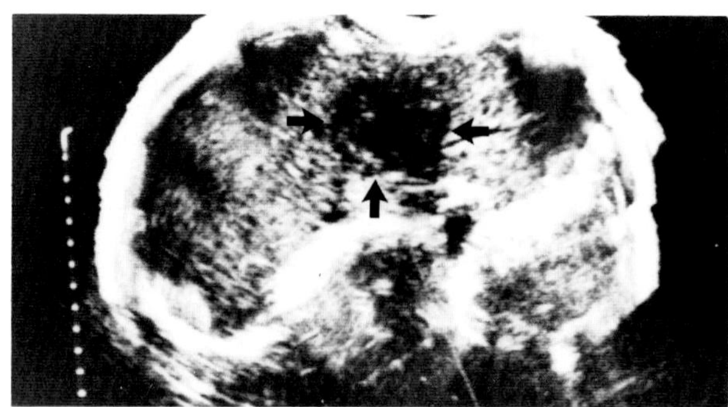

B

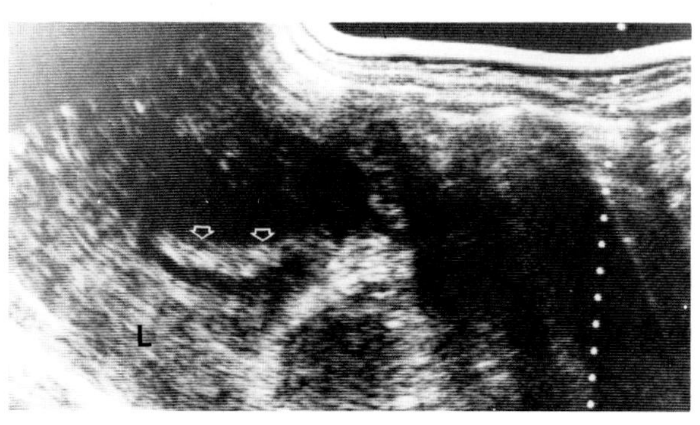


C

**Figure 6.2. Ultrasound of abdominal abscesses.** (A) A transverse sonogram through the pancreas demonstrates a large sonolucent mass with good posterior acoustic enhancement (arrows) that represents a cystic-appearing pancreatic abscess. Note the smooth walls and the absence of internal echoes, both typical of cystic lesions. (B) A transverse sonogram demonstrates an amebic liver abscess (arrows) as an irregular mass, with extensive echoes within the lesion and poor posterior acoustic enhancement. (C) A parasagittal sonogram through the liver (L) in a patient with a liver abscess demonstrates layering of echogenic debris (open arrows) in the posterior portion of a large sonolucent collection. (From Kressel and McLean,[1] with permission from the publisher.)

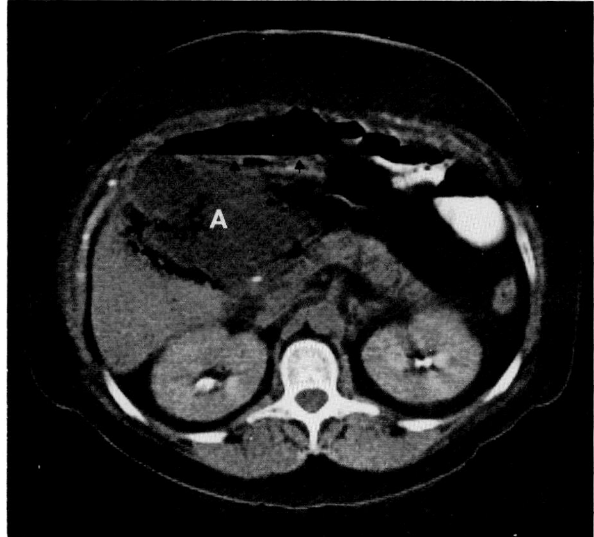


**Figure 6.3. Subhepatic abscess.** The well-formed abscess (A) appears as a low-density mass that neither conforms to the contour of the normal parenchymal organs nor lies within the confines of the bowel. Note the gas-fluid level within the abscess (arrows).

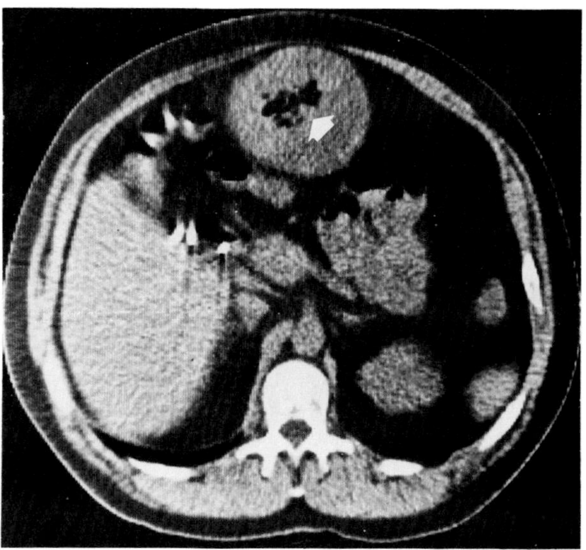

**Figure 6.4. Abdominal abscess.** There is gas (arrow) within the central cavity of the abscess.

nuclide in normal structures (colon, liver, spleen); this may obscure abscesses and require scanning over a 3-day period until background levels have sufficiently decreased. This delay may present significant management problems in acutely ill patients in whom rapid intervention is essential. Despite its sensitivity in detecting inflammatory processes, gallium uptake is nonspecific for abscesses and may occur in a variety of neoplasms (hepatomas, lymphomas). Indium 111 is only available for investigational use at this time.

#### PERCUTANEOUS DRAINAGE OF ABDOMINAL ABSCESSES

The major criteria for patient selection for percutaneous drainage of abdominal abscesses include a well-defined, unilocular abscess, a safe access route for catheter insertion, and the availability of immediate surgical backup if the procedure is not successful or if there are complications. CT is used to provide the detailed anatomic map necessary for planning the appropriate route; ultrasound (or CT) is then used to guide the needle into the abscess cavity. Complete resolution of the abscess can be achieved in more than 90 percent of cases. The complication rates of percutaneous abscess drainage are low (less than the operative morbidity) and include transient bacteremia and either perforation of a viscus or contamination of the pleural space during insertion of the catheter.[1]

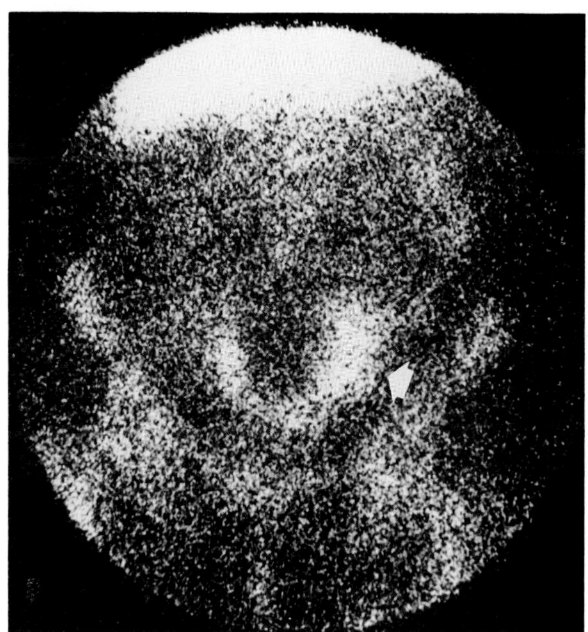

**Figure 6.5. Pelvic abscess.** A radionuclide scan ($^{111}$In-WBC) demonstrates an accumulation of the isotope within the abscess (arrow). (From Kressel and McLean,[1] with permission from the publisher.)

### Subphrenic Abscesses

A subphrenic abscess most commonly develops as a complication of abdominal surgery or closed blunt trauma, or as a sequela of generalized peritonitis caused by a perforation in the gastrointestinal or the biliary tract. The earliest radiographic findings associated with subphrenic inflammation are elevation and restricted motion of the hemidiaphragm on the affected side. An inflammatory pleural reaction at the base of the lung produces a nonpurulent (sympathetic) pleural effusion. By displacing the fundus of the stomach downward, a left subphrenic abscess produces abnormal widening of the distance be-

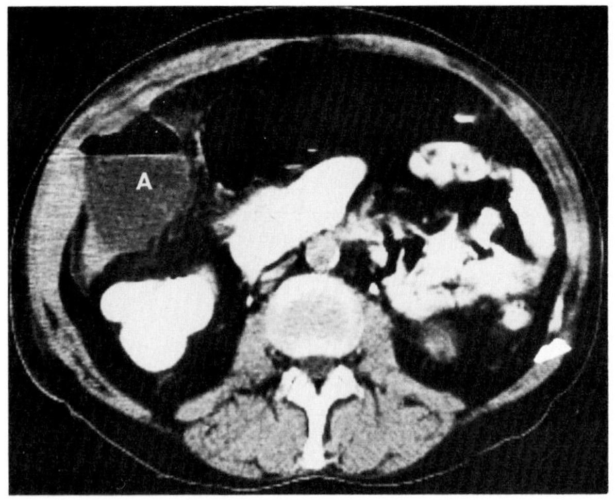


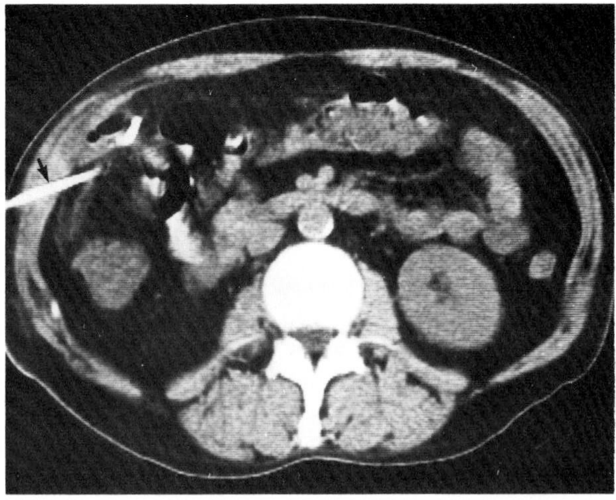

A B

**Figure 6.6. Percutaneous drainage of abdominal abscess.** (A) A large right paracolic gutter abscess with a gas-fluid level is secondary to a perforated duodenal ulcer. Note the swelling of the adjacent abdominal muscles. (B) The abscess cavity has been completely evacuated via the drainage catheter (arrow).

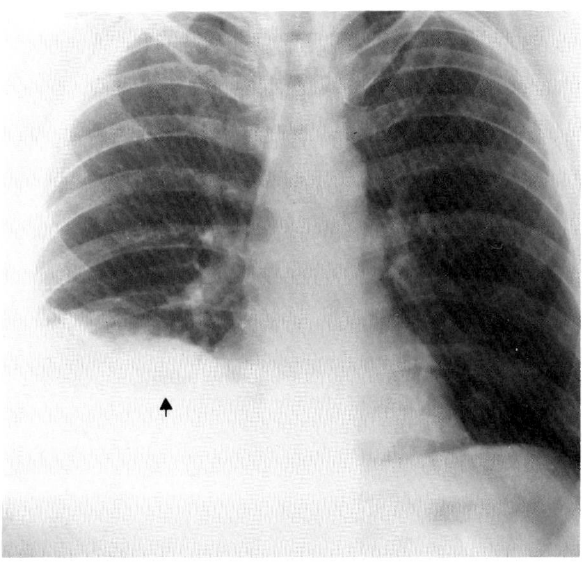

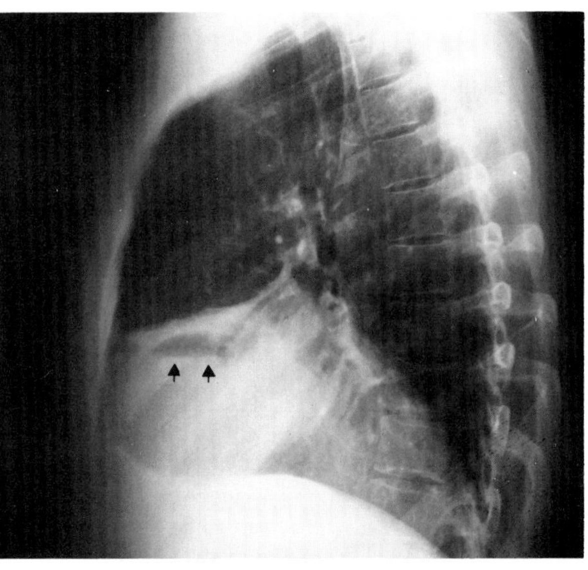

A B

**Figure 6.7. Right subphrenic abscess.** (A) Frontal and (B) right lateral views demonstrate elevation of the right hemidiaphragm, a small right pleural effusion, and gas in the abscess cavity (arrow). (From Connell, Stephens, Carlson, et al,[12] with permission from the publisher.)

tween the air-filled stomach and the diaphragm and causes a pressure defect on the contrast-filled fundus. Extraluminal gas can be identified on plain abdominal radiographs in more than two-thirds of patients with subphrenic abscesses. Gas bubbles intermixed with necrotic material or pus within the abscess produce a typical pattern of mottled radiolucency. A gas-fluid level can often be seen on upright or decubitus views. Although this appearance can be confused with that of gas accumulations in the bowel, the constancy of position of the gas shadows in multiple projections and on serial radiographs clearly indicates that they are outside the lumen of the bowel. Plain-film identification of gas-containing abscesses is more difficult in the left subphrenic region than in the right because the stomach and splenic flexure normally contain gas.

CT and ultrasound are of value in detecting subphrenic abscesses. On the left, CT is the examination of choice because large amounts of overlying gas and technical difficulties often limit the use of ultrasound.[1,12]

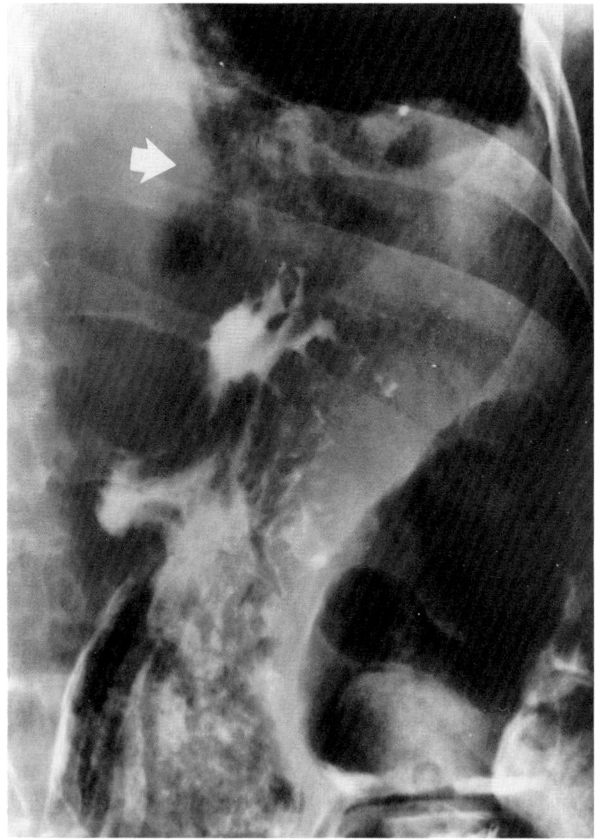

**Figure 6.8. Left subphrenic abscess.** The characteristic mottled radiolucent appearance of the abscess (arrow), which is located above the fundus of the stomach, is due to gas bubbles intermixed with necrotic material and pus.

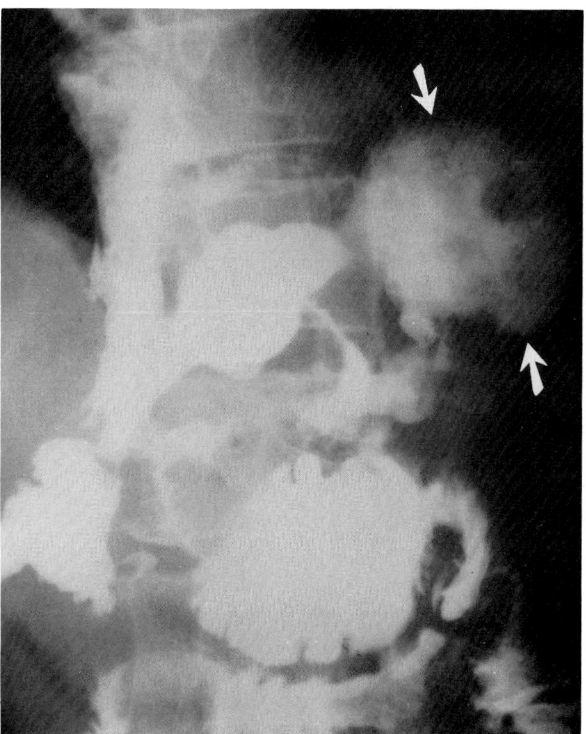

**Figure 6.9. Left subphrenic abscess.** After the oral administration of barium, contrast material is seen to enter the large left subphrenic abscess (arrows).

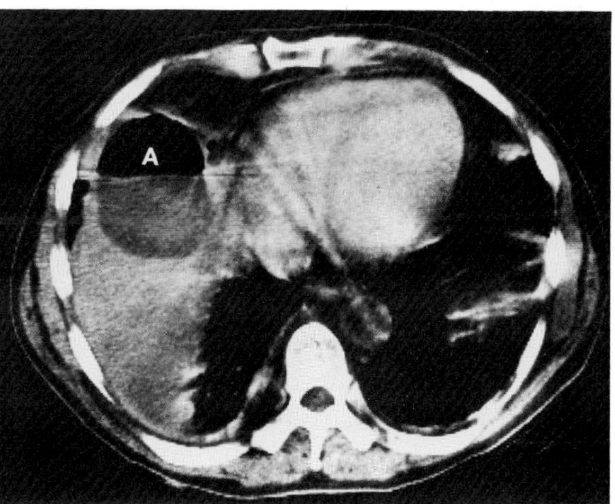


**Figure 6.10. Right subphrenic abscess.** A prominent gas-fluid level is seen within the mass.

## Lesser Sac Abscesses

In a patient with a lesser sac abscess, an upright film demonstrates a gas-fluid level in the left upper abdomen; this level extends slightly over the midline but does not reach the diaphragm. The stomach is displaced anteriorly, and the colon inferiorly. The lesser sac is frequently inaccessible to ultrasound, because of the spine posteriorly and gas in the stomach and colon anteriorly. Thus CT is the study that can most consistently image abscesses in the lesser sac. However, if the patient has a paucity of intra-abdominal fat, CT scans are difficult to interpret, and in such cases the diagnosis of lesser sac abscess may require radionuclide scanning with gallium 67 citrate or indium 111–tagged white blood cells.[1,13]

## Renal Abscesses

Renal infection causing abscess formation can be due to antecedent urinary tract disease (urinary tract infection, obstructive uropathy, trauma, instrumentation) or to direct or hematogenous spread of extraurinary infection. A renal abscess usually does not spread to the contralateral side because the medial fascia surrounding the kidney is closed and the spine and the great vessels act as a natural deterrent. The inflammatory process can extend around the entire kidney, though it is usually most

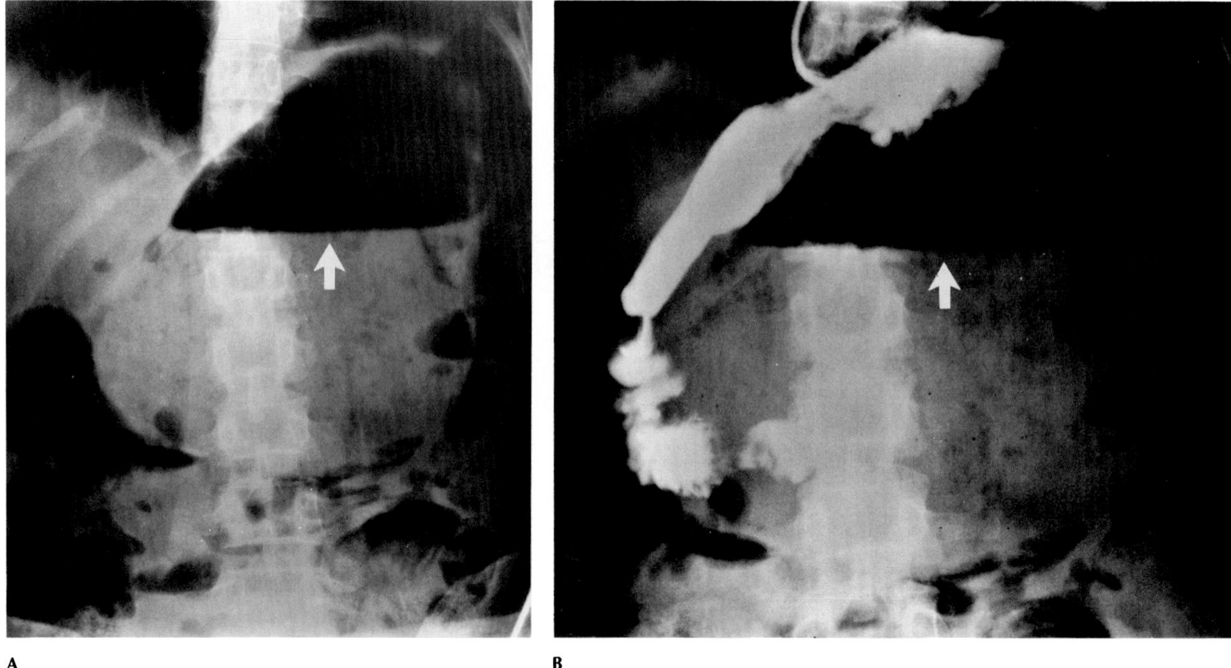


**Figure 6.11. Lesser sac abscess.** (A) A plain abdominal radiograph and (B) a film from a barium study reveal a huge abscess cavity with a prominent gas-fluid level (arrows). (From Eisenberg,[13] with permission from the publisher.)

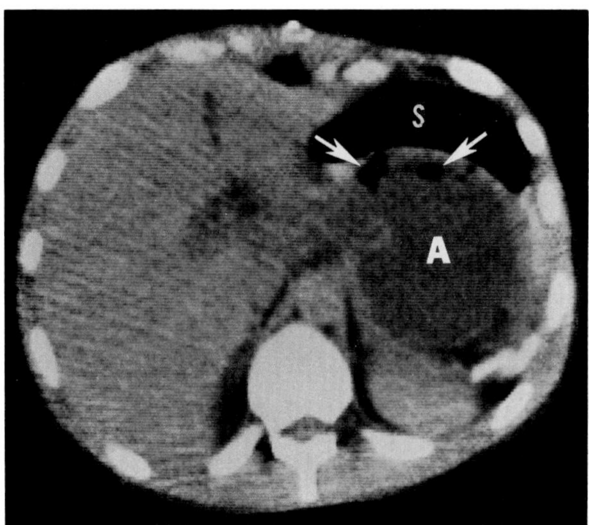


**Figure 6.12. Lesser sac abscess.** A CT scan demonstrates a large hypodense abscess (A) posterior to the stomach (S). Note the multiple gas bubbles (arrows) within the abscess cavity.

pronounced on the dorsal and superior aspects where the renal fascia is open and the surrounding tissues offer little resistance.

The radiographic demonstration of extraluminal gas around or inside the kidney is nearly pathognomonic of a renal or perirenal abscess. Perirenal infection can be diffuse and poorly defined or can present as a localized mass. Because the exudate usually localizes in the dorsolateral perirenal fat near the lower pole, the kidney tends to be displaced anteriorly. Displacement of the descending duodenum and the ascending or descending colon is common, as is obliteration of the upper half of the psoas muscle shadow.

Nephrotomography may demonsrate a poorly demarcated lucent area that simulates a cyst but has an irregular contour and often a thickened wall. A single large intrarenal abscess may mimic a neoplasm by producing a localized bulge of the renal margin and distorting adjacent calyces.

Ultrasound and CT are very sensitive in demonstrating intrarenal abscesses as well as the perinephric abscesses that are commonly associated with intraparenchymal infection. Perinephric abscesses cause thickening of Gerota's fascia, a subtle early change that is consistently demonstrated by CT. Radionuclide scanning with isotopically labeled agents that selectively accumulate at sites of inflammation (gallium 67 citrate, indium 111–labeled leukocytes) can detect a renal abscess by demonstrating an increased concentration of radioactivity in the region of the kidney.[2–6,14]

## INVASIVE ENTERIC PATHOGENS

### *Campylobacter jejuni*

*Campylobacter jejuni* has recently been recognized as a common human enteric pathogen and may be the most common cause of specific infectious colitis. Patients with this disease typically present with an acute onset of diarrhea, abdominal pain, fever, and constitutional symp-

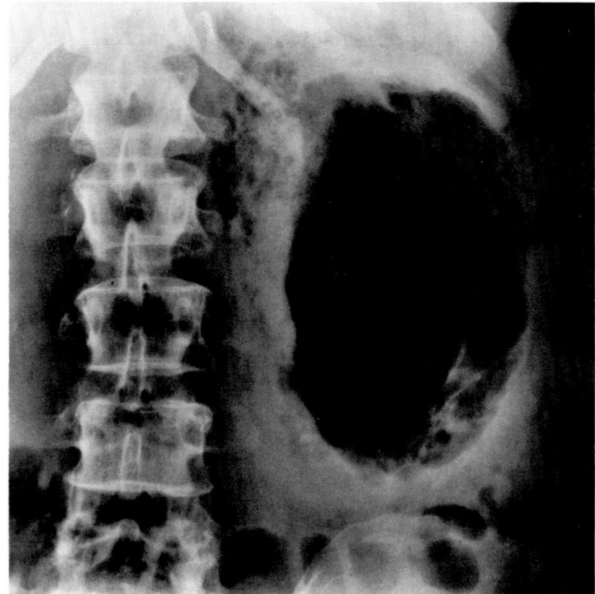

A

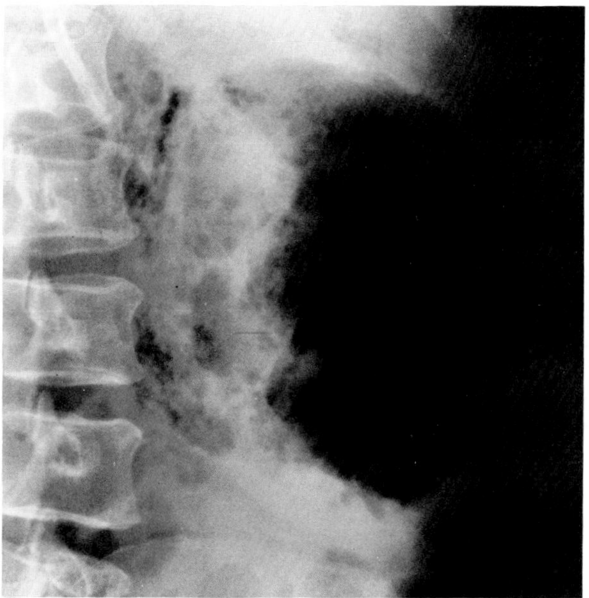

B

**Figure 6.13. Renal abscess.** (A) Frontal and (B) oblique views show large amounts of extraluminal gas within and around the left kidney.

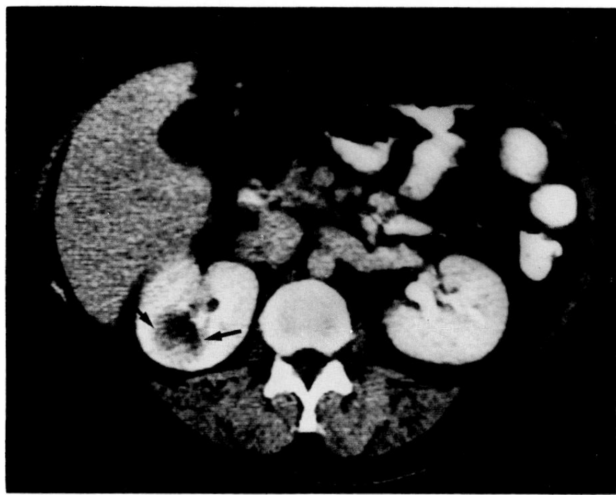

**Figure 6.14. Renal abscess.** A contrast-enhanced CT scan through both kidneys demonstrates a discrete low-density area (arrows) which proved to be an abscess on diagnostic needle aspiration.

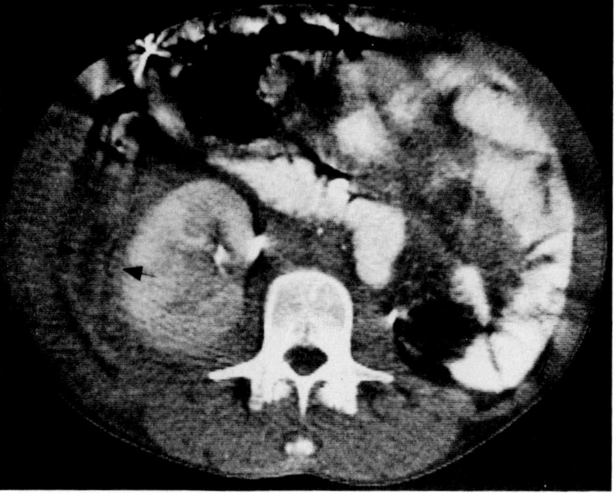

**Figure 6.15. Perinephric abscess.** A CT scan demonstrates a candidal perinephric abscess as a low-density area (arrows) adjacent to the right kidney. (From Hoddick, Jeffrey, Goldberg, et al,[14] with permission from the publisher.)

toms. Because the radiographic pattern is indistinguishable from that of ulcerative colitis, *C. jejuni* infection should be ruled out with appropriate cultures and immunologic studies before the diagnosis of ulcerative colitis is made in a patient presenting with a single episode of acute colitis.[7–9]

## Shigellosis

This is discussed in the section "*Shigella* Infections" in Chapter 9.

## Salmonellosis

This is discussed in the section "*Salmonella* Infections" in Chapter 9.

## *Yersinia enterocolitica*

This is discussed in the section "*Yersinia* Infections" in Chapter 9.

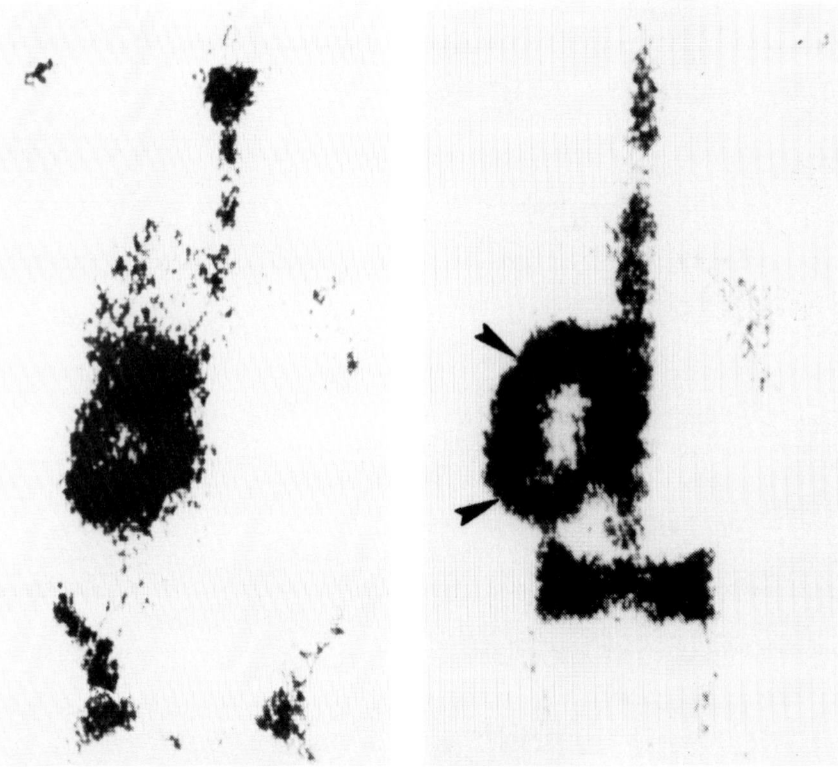


**Figure 6.16. Renal abscess.** (A) Anterior and (B) posterior views from a gallium 67 radionuclide scan show an intense accumulation of the isotope around the right kidney (arrows), with a central area of reduced activity. The abscess was surgically drained of over a liter of exudate. (From Baum, Vincent, Lyons, et al,[6] with permission from the publisher.)

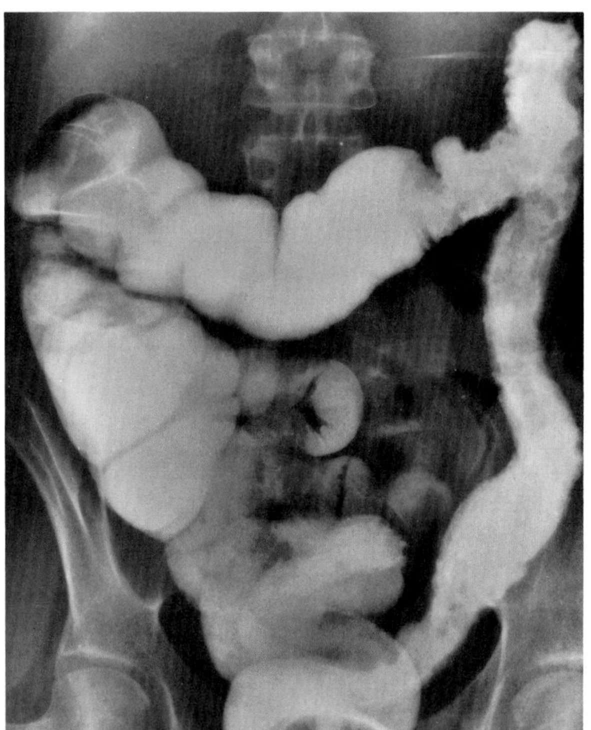

**Figure 6.17. Colitis caused by *Campylobacter jejuni*.** The radiographic pattern is indistinguishable from ulcerative colitis. Cultures and immunologic studies were necessary for the proper diagnosis. (From Eisenberg,[13] with permission from the publisher.)

## PELVIC INFLAMMATORY DISEASE

Pelvic inflammatory disease refers to ascending infection of the uterus, fallopian tubes, and broad ligament. It is usually the result of venereal disease (especially gonorrhea) in women of childbearing age, though it can also develop postpartum or postabortion or be a complication of intrauterine devices. If pelvic inflammatory disease is not promptly and adequately treated, adhesions may obstruct the fallopian tube and lead to a pyosalpinx. Concomitant ovarian infections can produce tubo-ovarian abscesses, which are usually bilateral. The spill of purulent material from the fallopian tubes, or the rupture of a pyosalpinx, can lead to peritonitis and the formation of a pelvic abscess.

Ultrasound is the diagnostic procedure of choice for detecting pelvic inflammatory disease and pelvic abscesses. The fluid-filled urinary bladder provides an excellent acoustic window for examining the supravesical and paravesical spaces and the pouch of Douglas; the distended bladder also displaces confusing loops of small bowel out of the pelvis.

The earliest ultrasound findings in acute pelvic inflammatory disease are nonspecific and include a slight enlargement of the uterus and adnexal structures. The uterus demonstrates decreased echogenicity and indistinct borders. Pyosalpinx and tubo-ovarian abscess typically present as tubular adnexal masses that are sonolucent and compatible with fluid collections. However,

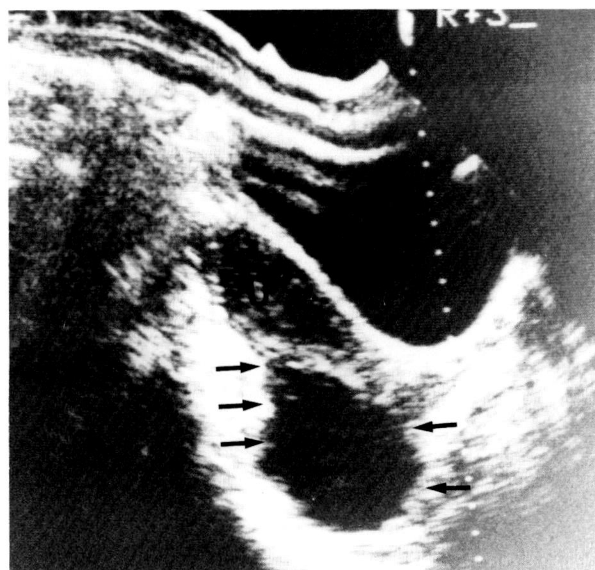

**Figure 6.18. Pelvic abscess.** A sagittal sonogram through the midpelvis of a 22-year-old woman illustrates a large collection (arrows) posterior to the uterus (U). Note the irregular walls of the collection and the good through sound transmission. Weak echoes are noted within the sonolucent region. (From Kressel and McLean,[1] with permission from the publisher.)

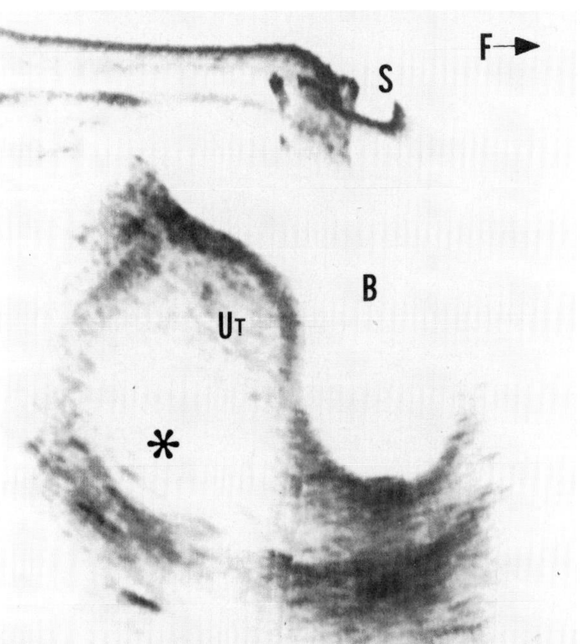


**Figure 6.19. Pelvic inflammatory disease.** A sagittal sonogram shows obscuration of the posterior aspect of the uterus (Ut) by a large sonolucent mass (asterisk) in the cul-de-sac. B = bladder; S = symphysis pubis; F = feet. (From Laing,[15] with permission from the publisher.)

abscesses may also have thick and irregular, or "shaggy," walls or may contain echoes or fluid levels representing the layering of purulent debris. Free pelvic fluid may indicate peritonitis. If the free fluid becomes loculated, a peritoneal abscess may form. This is especially common within the cul-de-sac, the most dependent point of the peritoneal space in the supine patient. Some abscesses have a very echogenic appearance due to small gas bubbles created by gas-forming organisms.

The residua of acute infection, or the subacute reinfection of previous acute pelvic inflammatory disease, can lead to a broad spectrum of often-confusing ultrasound findings. In chronic abscesses, fibrosis and scarring may produce solid-appearing pelvic masses that may be difficult to distinguish from neoplasms. Adhesions may lead to the fixation of bowel loops and omentum in the pelvis, and this may be mistaken for pelvic cysts or masses. A hydrosalpinx, in which the purulent exudate of a pyosalpinx is replaced by serous fluid, may retain a tubular appearance, permitting its identification as a fallopian tube, or may become markedly distended and be similar to other large, simple adnexal cysts. As active infection subsides, the uterus regains its sharp borders and the echogenicity of the uterus and endometrial cavity returns to normal.

Plain abdominal or pelvic radiographs are of little value in detecting pelvic inflammatory disease and pelvic abscesses. Abnormal gas collections may be masked by fecal material in the rectum and in loops of small bowel. Asymmetry of soft tissue contours, such as the

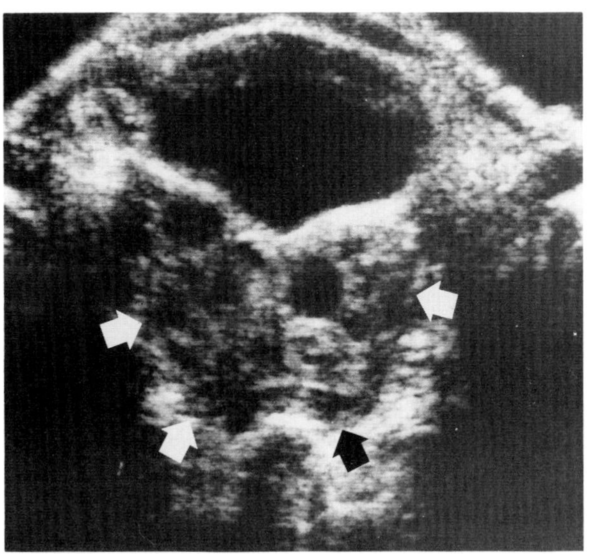

**Figure 6.20. Chronic pelvic inflammatory disease.** A transverse sonogram demonstrates large, complex cystic and echogenic masses (arrows). (From Berland, Lawson, and Foley,[10] with permission from the publisher.)

obturator internus muscle, is frequently present but nonspecific.

There is a significantly increased incidence of pelvic infection in patients with an intrauterine device (IUD). Unlike most forms of pelvic inflammatory disease, the

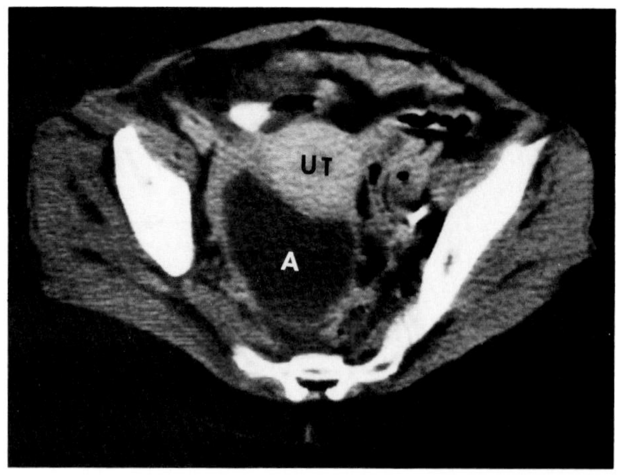


A

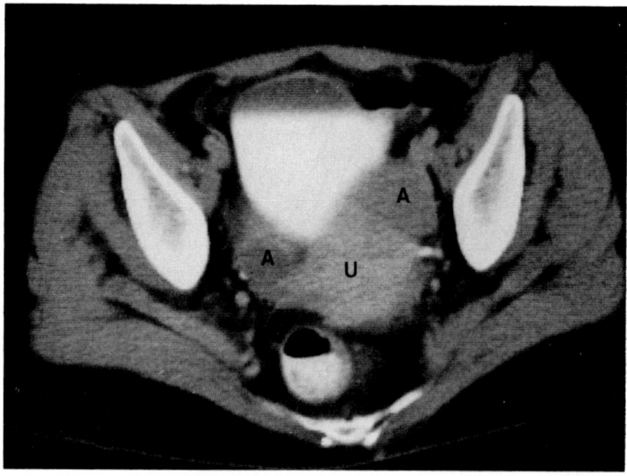


B

**Figure 6.21. Pelvic inflammatory disease.** CT of tubo-ovarian abscesses. (A) A large low-density mass (A) posterior to the uterus (UT). (B) Bilateral adnexal masses (A) anterior and lateral to the uterus (U). (B, from Gross, Moss, Mihara, et al,[16] with permission from the publisher.)

tubo-ovarian abscesses associated with IUDs are more often unilateral. The IUD itself is usually visible as an intensely echogenic structure in the central uterine cavity.

CT demonstrates a pelvic abscess as a fluid collection with a thickened wall and a low-density center. The presence of multiple low-density areas representing gas within the mass is virtually diagnostic for an abscess.[10,11,15,16]

## REFERENCES


1. Kressel HY, McLean GK: Abdominal abscess, in Eisenberg RL, Amberg JR (eds): *Critical Diagnostic Pathways in Radiology: An Algorithmic Approach.* Philadelphia, JB Lippincott, 1981, 177–195.
2. Rabinowitz JG, Kinkhabwala MN, Robinson T: Acute renal carbuncle: The roentgenographic clarification of a medical enigma. *AJR* 116:740–748, 1972.
3. Mendez G, Isikoff MB, Morillo G: The role of computed tomography in the diagnosis of renal and perirenal abscesses. *J Urol* 122:582–586, 1979.
4. Koehler PR: The roentgen diagnosis of renal inflammatory masses: Special emphasis on angiographic changes. *Radiology* 112:257–266, 1974.
5. Goldman SM, Minkin SD, Naraval FD, et al: Renal carbuncle: The use of ultrasound in the diagnosis and treatment. *J Urol* 118:525–528, 1977.
6. Baum S, Vincent NR, Lyons KP, et al: *Atlas of Nuclear Medicine Imaging.* New York, Appleton-Century-Crofts, 1981.
7. Blaser MJ, Parsons RB, Wang WLL: Acute colitis caused by *Campylobacter fetus jejuni. Gastroenterology* 78:448–453, 1980.
8. Brodey PA, Fertig S, Aron JM: *Campylobacter enterocolitis:* Radiographic features. *AJR* 139:1199–1201, 1982.
9. Kollitz JPM, Davis GB, Berk RN: *Campylobacter colitis:* A common infectious form of acute colitis. *Gastrointest Radiol* 6:227–229, 1981.
10. Berland LL, Lawson TL, Foley WD: Sonographic evaluation of pelvic infections, in Callen PW (ed): *Ultrasonography in Obstetrics and Gynecology.* Philadelphia, WB Saunders, 1983.
11. Uhrich PC, Sanders RC: Ultrasonic characteristics of pelvic inflammatory masses. *J Clin Ultrasound* 4:199–204, 1966.
12. Connell TR, Stephens DH, Carlson HC, et al: Upper abdominal abscess. *AJR* 134:759–765, 1980.
13. Eisenberg RL: *Gastrointestinal Radiology: A Pattern Approach.* Philadelphia, JB Lippincott, 1983.
14. Hoddick W, Jeffrey RB, Goldberg HI, et al: CT and sonography of renal and perirenal infections. *AJR* 140:517–523, 1983.
15. Laing FC: Pelvic masses, in Eisenberg RL, Amberg JR (eds): *Critical Diagnostic Pathways in Radiology: An Algorithmic Approach.* Philadelphia, JB Lippincott, 1981, pp 244–259.
16. Gross BH, Moss AA, Mihara K, et al: Computed tomography of gynecologic diseases. *AJR* 141:765–773, 1983.

**Chapter Seven**

# Gram-Positive Cocci

## PNEUMOCOCCAL INFECTIONS

Pneumococcal pneumonia is an acute pulmonary infection caused by *Streptococcus* (formerly *Diplococcus*) *pneumoniae*. Because the infectious organisms reach the lung when upper respiratory secretions are aspirated during sleep, the lower lobes and the right middle lobe are most commonly involved. Pneumococcal pneumonia may occur in otherwise healthy people but is much more common in vagrants, alcoholics, and other compromised hosts. The usual clinical presentation of acute pneumococcal pneumonia is abrupt, with fever, shaking chills, cough, slight expectoration, and intense pleural pain.

Although pneumococcal pneumonia is often considered synonymous with "lobar pneumonia," a complete lobe is infrequently involved in pneumococcal pneumonia. The typical radiographic appearance is a homogeneous consolidation of lung parenchyma that begins in the peripheral air spaces of the lung and thus almost invariably abuts against a visceral pleural surface. A characteristic finding is visualization of air within intrapulmonary bronchi (air bronchogram sign) coursing through the virtually airless zone of surrounding lung parenchyma. An air bronchogram is almost always present, and its absence should cast doubt upon the diagnosis. Although the physical findings in the lung soon return to normal after appropriate therapy, radiographic evidence of residual pulmonary consolidation may persist for several months, especially in patients with alcoholism or chronic obstructive airway disease.

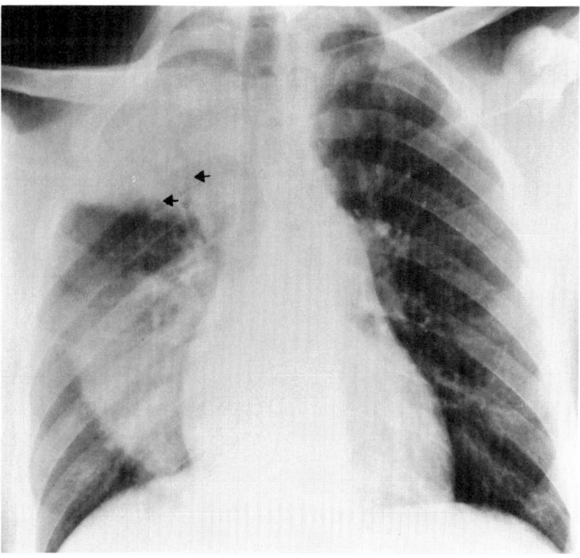

**Figure 7.1. Pneumococcal pneumonia.** Homogeneous consolidation of the right upper lobe and the medial and posterior segments of the right lower lobe. Note the associated air bronchograms (arrows).

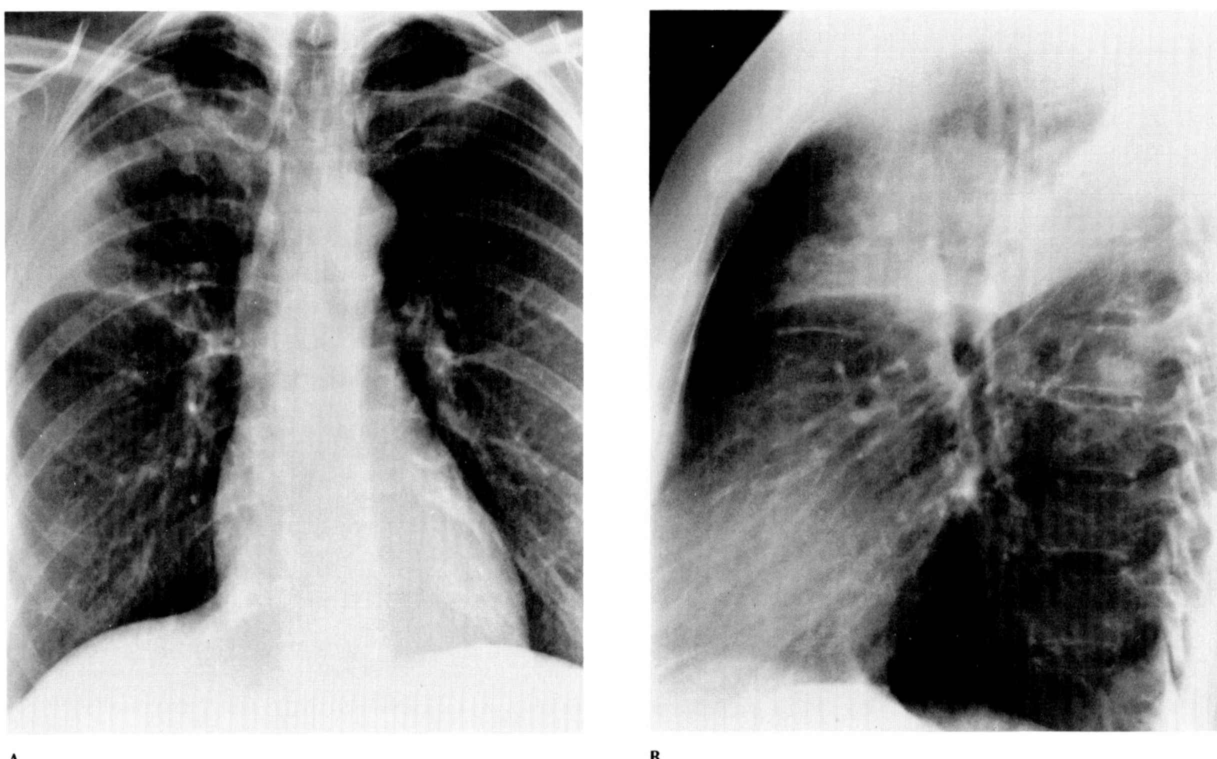

A B

**Figure 7.2. Pneumococcal pneumonia.** (A) Frontal and (B) lateral views of the chest show homogeneous air space consolidation of the anterior and posterior segments of the right upper lobe. The pneumonia involves the peripheral air spaces of the lung and abuts against visceral pleural surfaces.

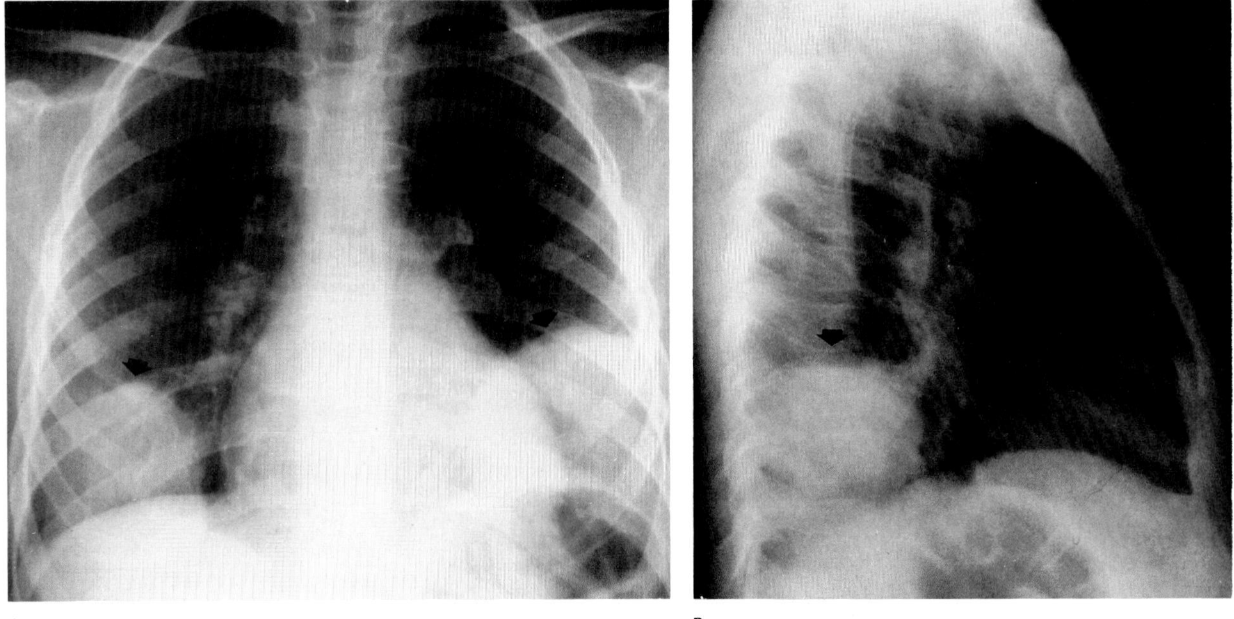

A B

**Figure 7.3. "Spherical" pneumonia.** (A) Frontal and (B) lateral views of the chest show a rounded soft tissue density in the posterolateral aspects of both lower lobes (arrows) with mild bilateral hilar prominence. (From Rose and Ward,[1] with permission from the publisher.)

A rare, but important, form of acute pneumococcal pneumonia in children is the so-called round or spherical pneumonia, in which a well-circumscribed spherical consolidation on both frontal and lateral views simulates a pulmonary or mediastinal mass. The appropriate diagnosis can be made on the basis of the clinical presentation (a mild respiratory infection followed by a sudden, acute febrile illness) and the rapid response to appropriate antibiotic therapy.

In contrast to its frequency before the advent of antibiotics, radiographically demonstrable pleural effusion is now uncommon in patients with pneumococcal pneumonia. Similarly, empyema and lung abscess are rare complications if the initial infection is properly treated.

Generalized distension of gas-filled loops of small and large bowel (adynamic ileus) is often present in severely ill patients. Acute gastric dilatation is a rarer and more serious gastrointestinal complication. Other rare complications include pericarditis, bacterial endocarditis, arthritis, meningitis and brain abscess, and peritonitis.[1,2]

## STAPHYLOCOCCAL INFECTIONS

Staphylococci usually produce relatively harmless superficial suppurative infections but may also cause severe infections of the lungs, heart, bones and joints, kidneys, and gastrointestinal tract.

Acute staphylococcal pneumonia occurs most frequently in children, especially during the first year of life. The onset is abrupt, with high fever, tachypnea, and

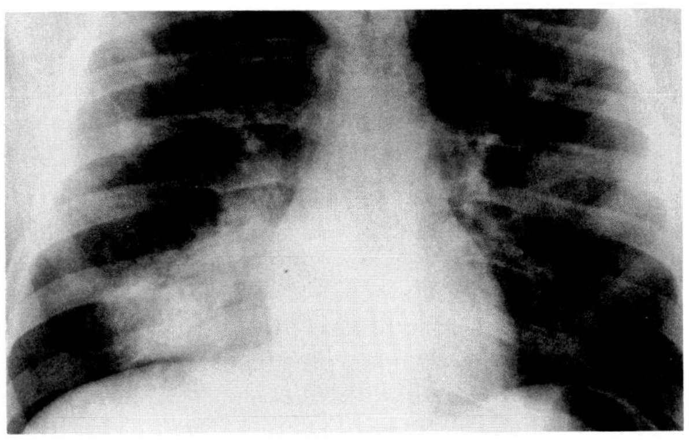

A

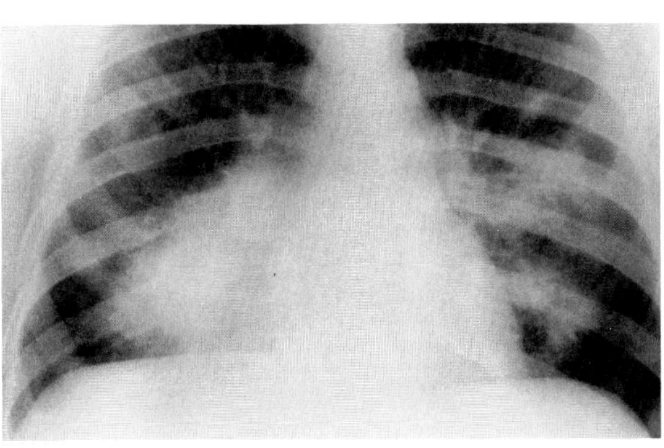

B

**Figure 7.4. Staphylococcal bronchopneumonia.** (A) An initial film shows an area of bronchopneumonia at the right base. (B) A film obtained 5 days later shows progression of the right basilar pneumonia and the development of patchy air space consolidation at the left base.

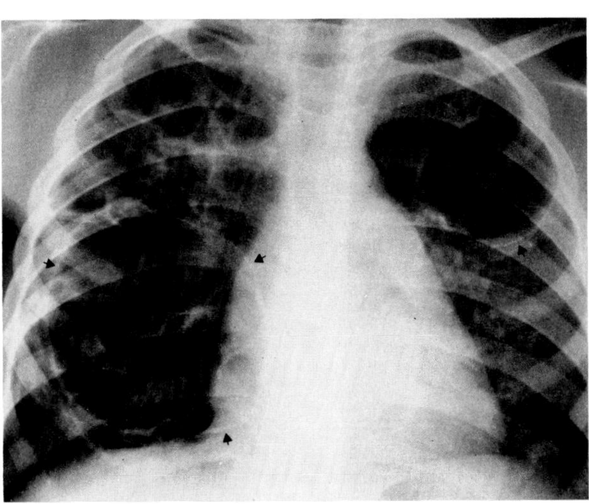

**Figure 7.5. Staphylococcal pneumonia.** Residual thin-walled cystic spaces (pneumatoceles) in the pulmonary parenchyma many years after a childhood staphylococcal pneumonia.

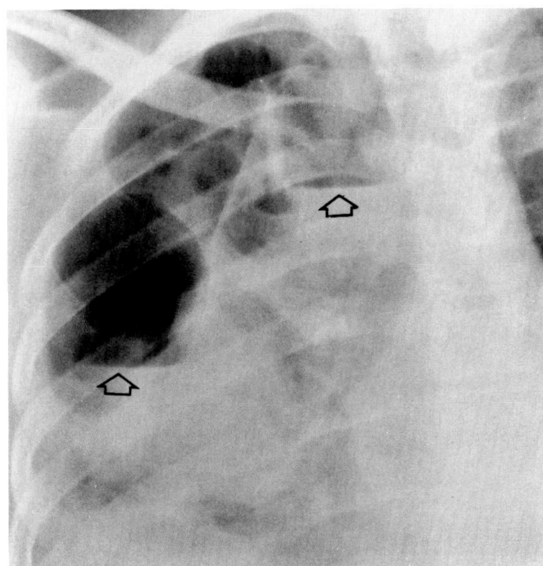

**Figure 7.6. Staphylococcal pneumonia.** Diffuse air space consolidation, multiple lung abscesses with air-fluid levels, and a large pleural effusion.

cyanosis; prompt treatment is necessary to save the patient's life. Extensive alveolar infiltrates develop rapidly, usually involving a whole lobe or even several lobes. Because an acute inflammatory exudate fills the airways, segmental collapse and a loss of volume may accompany the consolidations. For the same reason, air bronchograms are infrequent, and their presence should suggest an alternative diagnosis. A characteristic radiographic finding in childhood staphylococcal pneumonia is the development of pneumatoceles, thin-walled cystic spaces in the parenchyma. Pneumatoceles often contain fluid levels and occasionally are large enough to fill a hemithorax. The rupture of a pneumatocele leads to pneumothorax. Most pneumatoceles disappear spontaneously within 6 weeks, though some may persist for several months. Pleural effusion (or empyema) often occurs.

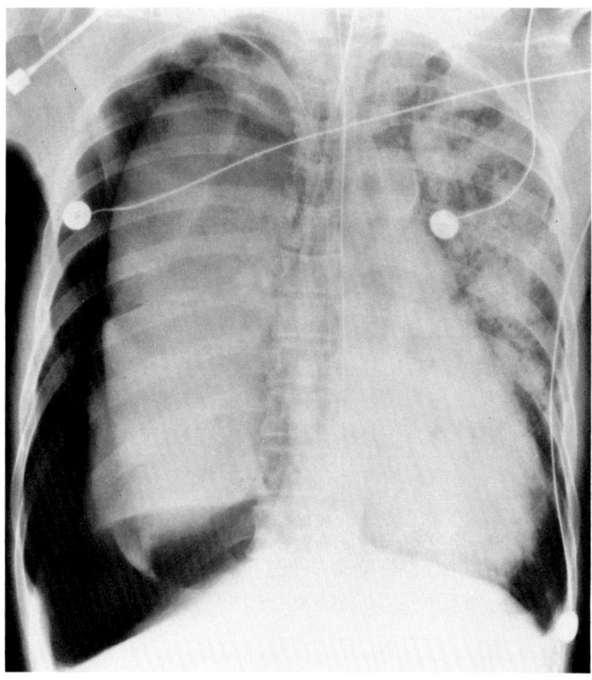

**Figure 7.7. Staphylococcal pneumonia with pneumothorax.** There is consolidation of the left upper lobe and entire right lung, with a moderate right pneumothorax. The extensive consolidation prevents further collapse of the right lung. The pneumothorax was due to the rupture of a pneumatocele, though no pneumataocele can be identified on this film.

Staphylococcal pneumonia in adults usually develops in hospitalized patients whose resistance has been lowered by disease or a recent operation, or as a complication of a viral respiratory infection (measles, epidemic influenza). The patchy infiltrates are most commonly bilateral and often lead to single or multiple lung abscesses, fluid-containing cavities that typically have an irregular, shaggy inner wall. Pleural effusion or empyema is common. Pneumatoceles rarely develop in adults.

*Staphylococcus aureus* is the most common cause of osteomyelitis (discussed in the section "Osteomyelitis" in Chapter 62.) The organism is also the most frequent cause of pyogenic arthritis, in which the infection is secondary to either osteomyelitis of an adjacent bone or systemic disease. The earliest radiographic sign of staphylococcal arthritis is soft tissue swelling, which is most easily demonstrated in the hip and knee. Rapid cartilage destruction causes early joint space narrowing, a late finding in tuberculous arthritis. Hyperemia and disuse atrophy lead to periarticular demineralization. Cortical destruction can rarely be detected in less than 1 week after the onset of symptoms, and thus erosive bone changes are of little value in the early diagnosis of staphylococcal arthritis. The destruction of the entire articular cartilage results in bony ankylosis.

Postantibiotic staphylococcal enteritis occurs after a course of orally administered broad-spectrum antibiotics, usually tetracycline. The disease is most commonly found among hospitalized patients and is caused by antibiotic-resistant strains that enter the gastrointestinal tract by the nasopharyngeal route and grow profusely in the intestine once the population of normal intestinal flora has been significantly reduced by oral antibiotics. Mild staphylococcal enteritis subsides rapidly once the antibiotic to which the organism is resistant is discontinued and the normal intestinal flora is allowed to return. In

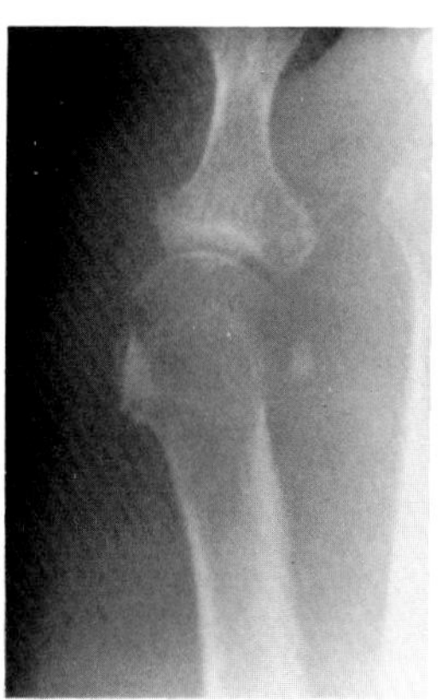

A

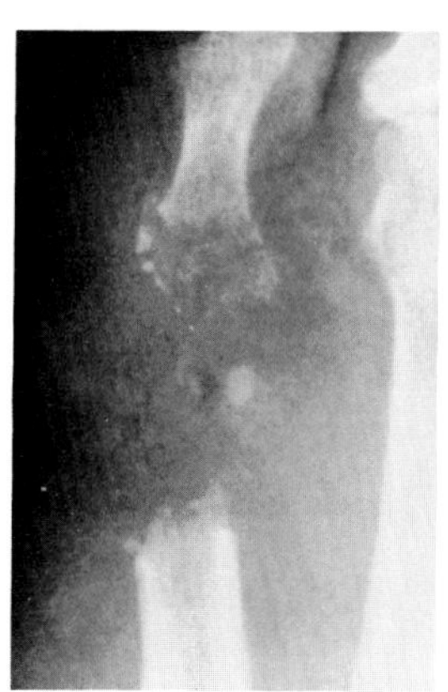

B

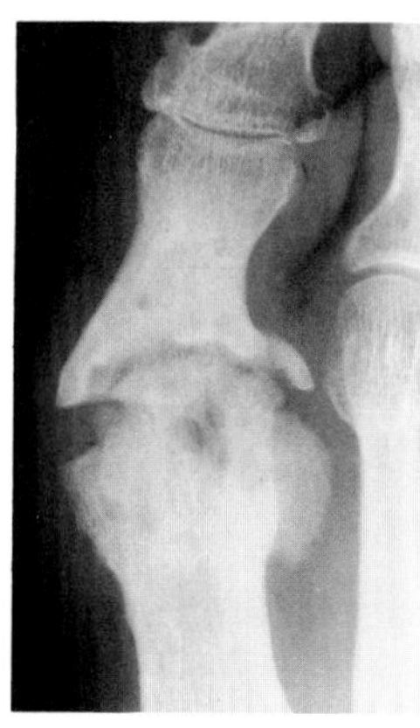

C

**Figure 7.8. Staphylococcal osteomyelitis.** (A) An initial film of the first metatarsophalangeal joint shows soft tissue swelling and periarticular demineralization due to hyperemia. (B) Several weeks later there is severe bony destruction about the metatarsophalangeal joint. (C) In another patient, irregular bony destruction with reactive sclerosis about the first metatarsophalangeal joint represents chronic staphylococcal osteomyelitis.

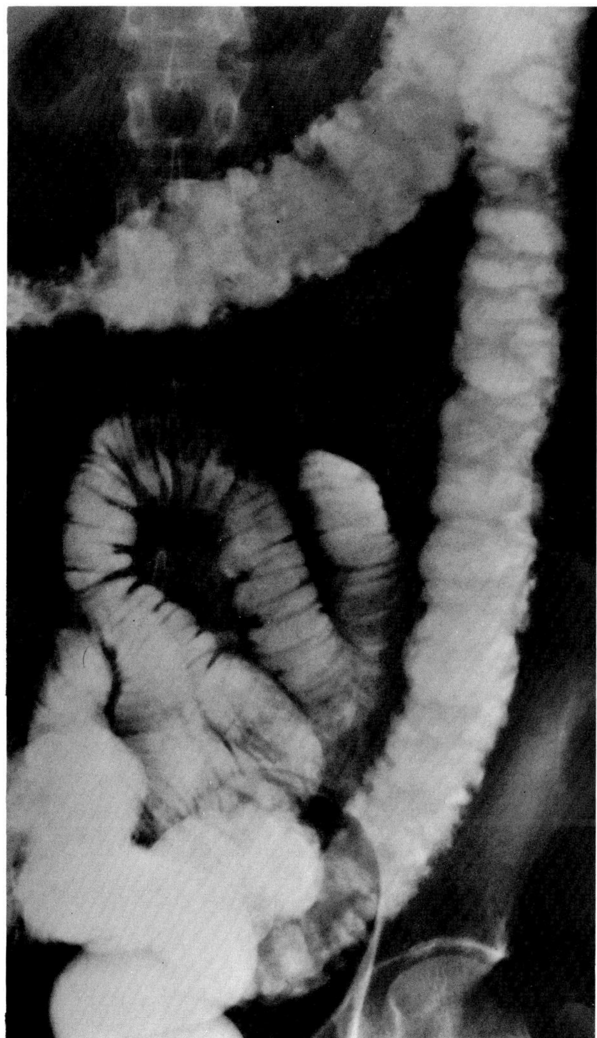

**Figure 7.9. Staphylococcal colitis.** Severe ulcerating colitis that developed after a course of orally administered broad-spectrum antibiotics. (From Eisenberg,[10] with permission from the publisher.)

severe enteritis it may be necessary to administer an antibiotic to which the *Staphylococcus* is sensitive. Although barium enema examinations are rarely performed in patients with signs and symptoms of staphylococcal enteritis, they can demonstrate the characteristic features of a generalized ulcerating colitis.[3–7]

## STREPTOCOCCAL INFECTIONS

Streptococci are among the most common bacterial pathogens of humans and are responsible for a diverse spectrum of diseases. Most infections involve the skin, pharynx, and lymph nodes and present no radiographic abnormalities.

Streptococcal pneumonia is uncommon and usually follows viral infections such as measles, pertussis, and epidemic influenza. The onset of acute streptococcal pneumonia is generally abrupt, with pleural pain, shaking chills, fever, and purulent, often blood-tinged, sputum. The radiographic manifestations are indistinguishable from those of acute staphylococcal pneumonia and consist of homogeneous or patchy consolidation in a segmental distribution, with a lower lobe predominance and often some loss of volume. Unlike staphylococcal infection, streptococcal pneumonia rarely causes the development of pneumatoceles. Before the advent of antibiotics, the early and rapid accumulation of empyema fluid was a characteristic feature of streptococcal pneumonia. Complications of streptococcal pneumonia include extension to the pericardium (purulent pericarditis), mediastinitis, pneumothorax, and bronchiectasis.

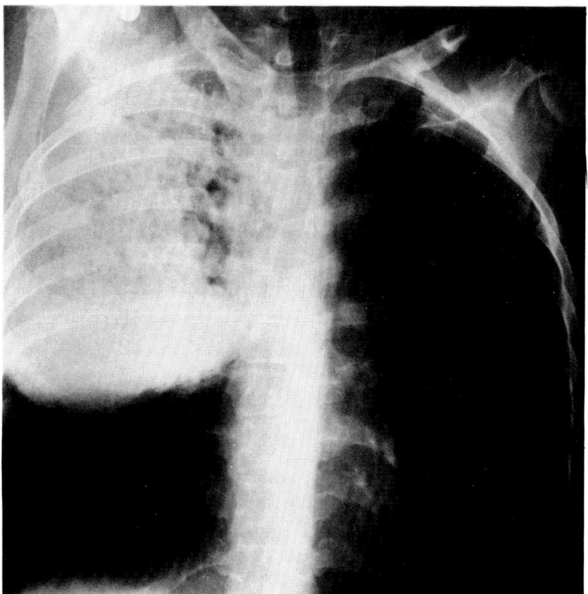

**Figure 7.10. Streptococcal pneumonia and empyema.** A large mottled opacity over the right upper lung represents an extensive empyema that obscures the underlying parenchymal pneumonia. The patchy air densities within the empyema indicate communication with the bronchial tree. Before the advent of antibiotics, the early and rapid accumulation of empyema fluid was a characteristic feature of streptococcal pneumonia.

Rheumatic fever and glomerulonephritis, severe complications of group B streptococcal infections, are discussed in other sections (see the sections "Rheumatic Fever" in Chapter 22 and "Glomerulonephritis" in Chapter 37).[8,9]

### REFERENCES


1. Rose RW, Ward BH: Spherical pneumonias in children simulating pulmonary and mediastinal masses. *Radiology* 106: 179–182, 1973.
2. Stephen JJ, Johanson WG, Pierce AK: The radiographic resolution of Streptococcus pneumoniae pneumonia. *N Engl J Med* 293:798–801, 1975.

3. Ceruti E, Contreras J, Neira M: Staphylococcal pneumonia in childhood. *Am J Dis Child* 122:386–391, 1971.
4. Meyers HI, Jacobsen G: Staphylococcal pneumonia in children and adults. *Radiology* 72:665–671, 1959.
5. Wiita RM, Cartwright RR, Davis JG: Staphylococcal pneumonia in adults: A review of 102 cases: *AJR* 86:1083–1087, 1961.
6. Huxtable KA, Tucker AS, Wedgwood RJ: Staphylococcal pneumonia in childhood: Long-term follow-up. *Am J Dis Child* 108:262–265, 1964.
7. Argen ST, Wilson CH, Wood P, et al: Suppurative arthritis. Clinical features of 42 cases. *Arch Intern Med* 17:661–666, 1966.
8. Burnmeister RW, Overholt EC: Pneumonia caused by hemolytic streptococcus. *Arch Intern Med* 111:367–374, 1963.
9. Kevy SV, Lowe BA: Streptococcal pneumonia and empyema in childhood. *N Engl J Med* 264:738–743, 1961.
10. Eisenberg, RL: *Gastrointestinal Radiology: A Pattern Approach*. Philadelphia, JB Lippincott, 1983.

## Chapter Eight
# Gram-Negative Cocci

## MENINGOCOCCAL INFECTIONS

*Neisseria meningitidis* (meningococcus) is a gram-negative coccus that primarily attacks young children or young adults in institutions or military service. The organism most commonly causes meningitis, the radiographic findings of which are discussed in the section "Acute Bacterial Meningitis" in Chapter 70. Up to 20 percent of patients with generalized meningococcal infection develop fulminant meningococcemia (Waterhouse-Friderichsen syndrome), which is associated with vasomotor collapse and shock. Arthritis is a common complication of systemic infection, though permanent joint changes are rare. Although it is unusual, meningococcus may cause a primary pneumonia that presents a radiographic pattern of alveolar consolidation that may be bilateral and associated with pleural effusion.[1,2]

## GONOCOCCAL INFECTIONS

Gonorrhea is a venereal disease caused by *Neisseria gonorrhoeae*. The systemic manifestations with radiographic abnormalities include proctitis and arthritis.

The symptoms associated with gonorrheal proctitis are similar to those of other forms of ulcerative proctitis and include rectal burning and itching, purulent anal discharge, and blood and mucus in the stools. Barium enema examination is normal in most patients with gonorrheal proctitis. Infrequently, mucosal edema and ulceration confined to the rectum can be demonstrated. Gonorrheal proctitis responds to specific antibiotic therapy.

The changes in gonorrheal arthritis are indistinguishable from those of tuberculous arthritis, though the time

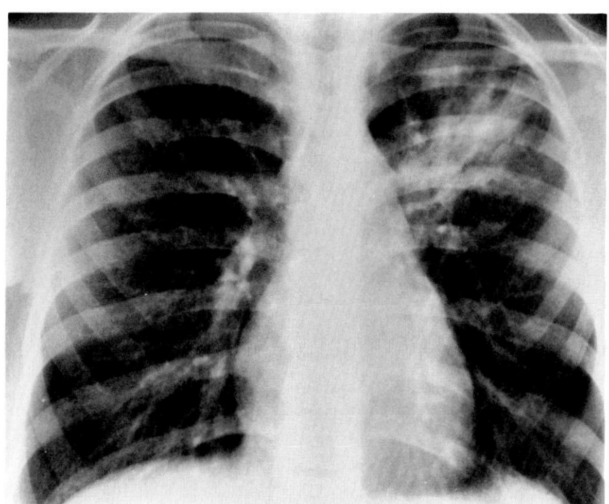

**Figure 8.1. Meningococcal pneumonia.** Nonspecific alveolar consolidation in the right upper lobe.

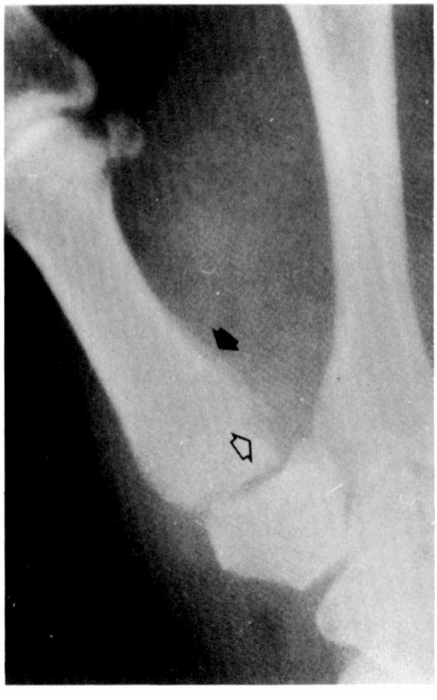

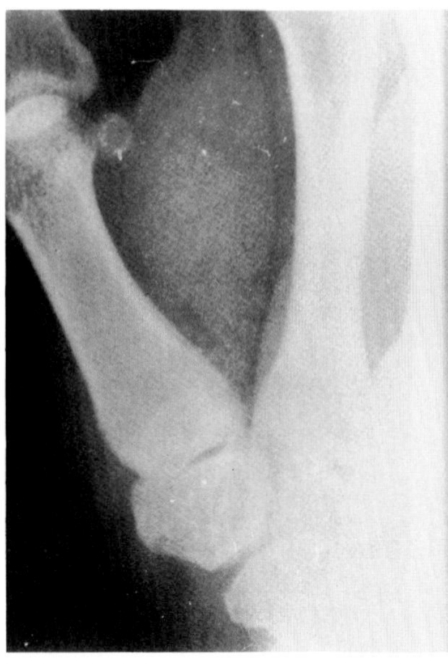

A B

**Figure 8.2. Gonococcal arthritis.** (A) After 2 months of arthritic symptoms, there is an osteolytic defect at the base of the first metacarpal with associated periosteal reaction. (B) A film obtained 3 months later, after a course of penicillin therapy, shows healing of the bone defect. (From Keiser, Ruben, Wolinsky, et al,[3] with permission from the publisher.)

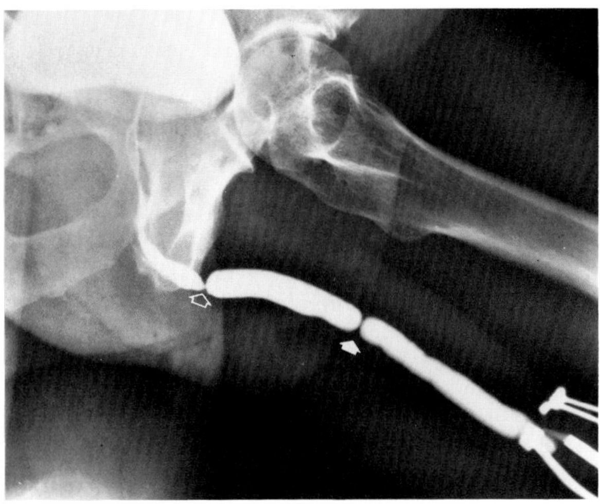

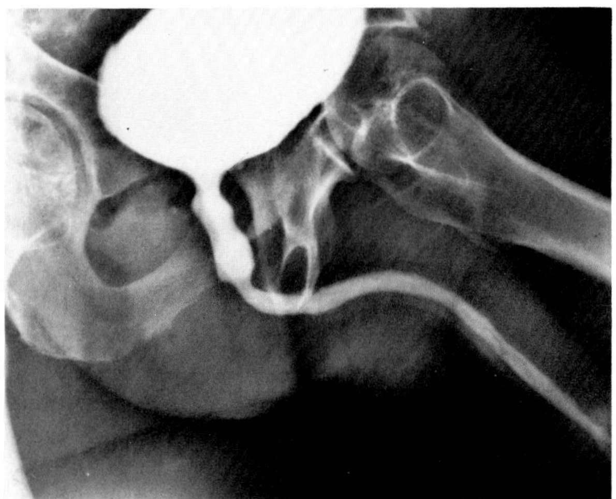

A B

**Figure 8.3. Gonococcal urethral stricture.** (A) An initial retrograde urethrogram shows diffuse stricture of the bulbar urethra and high-grade stenoses in the anterior (closed arrow) and posterior (open arrow) portions of the urethra. (B) Following balloon dilatation, a voiding urethrogram shows marked improvement in the appearance of the urethra. (From Russinovich, Lloyd, Griggs, et al,[6] with permission from the publisher.)

course is usually much more rapid. Periarticular soft tissue swelling may initially affect multiple joints, especially the knees, wrists, elbows, and ankles. Progressive disease may lead to articular erosion and joint space narrowing. Acute infection simulates other types of suppurative arthritis; the chronic changes resemble those of tuberculosis.

Urethral stricture is a complication of urethritis that can be demonstrated on urethrograms. Radiographic manifestations of gonococcal pelvic inflammatory disease are described in the section "Pelvic Inflammatory Disease" in Chapter 6.[3–6]

## REFERENCES


1. Schaad UB: Arthritis due to *Neisseria meningitidis*. *Rev Infect Dis* 2:880–888, 1980.
2. Irwin RS, Woelk WK, Coudon WL: Primary meningococcal pneumonia. *Ann Intern Med* 82:493–498, 1975.
3. Keiser H, Ruben FL, Wolinsky E, et al: Clinical forms of gonococcal arthritis. *N Engl J Med* 279:234–240, 1968.
4. Owen RL, Hill JL: Rectal and pharyngeal gonorrhea in homosexual men. *JAMA* 220:1315–1318, 1972.
5. Kibukamusoke JW: Gonorrhea and urethral stricture. *Br J Vener Dis* 41:135–136, 1965.
6. Russinovich NAE, Lloyd LK, Griggs WP, et al: Balloon dilatation of urethral strictures. *Urol Radiol* 2:33–37, 1980.

**Chapter Nine**

# Enteric Gram-Negative Bacilli

## INFECTIONS DUE TO ENTEROBACTERIACEAE

The Enterobacteriaceae, enteric gram-negative bacilli, cause a group of diseases that can produce a broad spectrum of clinical and radiographic findings.

### *Escherichia coli*

*Escherichia coli* accounts for more than 75 percent of urinary tract infections (pyelonephritis, pyelitis, cystitis, asymptomatic bacteriuria). The organism can usually be cultured from a perforated or inflamed appendix or from abscesses secondary to perforated diverticula, peptic ulcers, or mesenteric infarction. Gallbladder involvement can lead to the development of gas in the lumen and wall of the organ (emphysematous cholecystitis; see the sections "Emphysematous Cholecystitis" in Chapter 53 and "Diabetes Mellitus" in Chapter 4). The organism may produce abscesses anywhere in the body. Subcutaneous infections may result in gas in the soft tissues, a pattern similar to that of clostridial infection. Pulmonary consolidation is an unusual complication. When it does occur, pleural effusion is frequent and cavitation and massive pulmonary gangrene are rare. Extensive bacteremia can cause prolonged hypotension and gram-negative shock.[1]

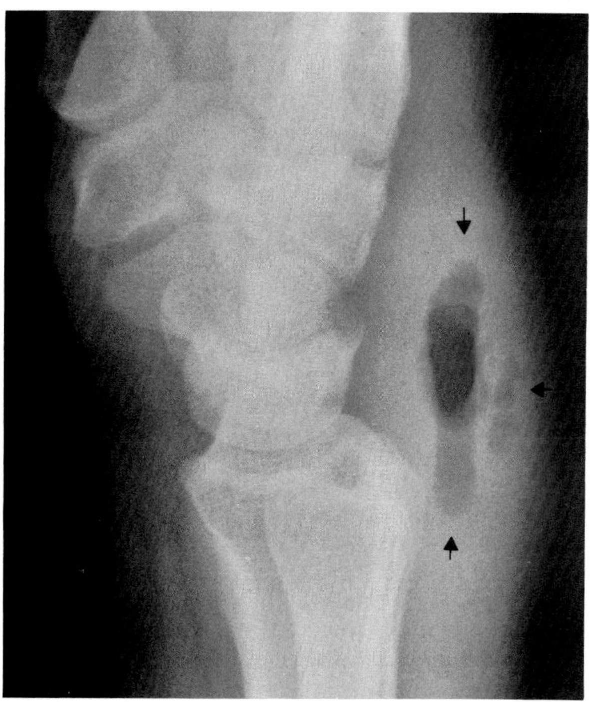

**Figure 9.1.** ***Escherichia coli.*** Gas within a large soft tissue abscess that appeared clinically as a red, tender fluctuant mass over the dorsum of the wrist.

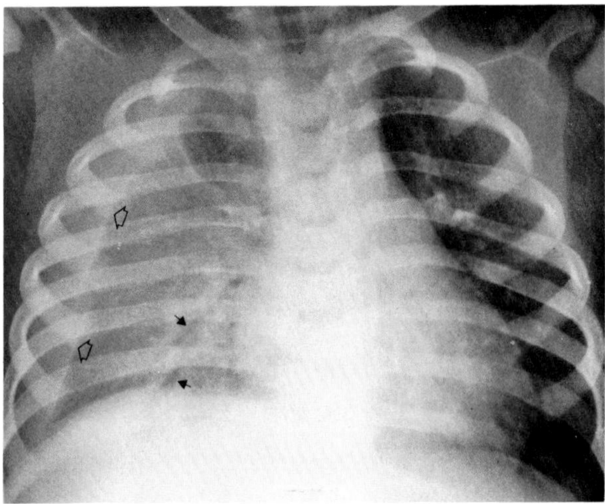

**Figure 9.2. *Escherichia coli* pneumonia.** Extensive pulmonary consolidation involving much of the right lung. Note the air bronchograms (closed arrows) and large pleural effusion (open arrows).

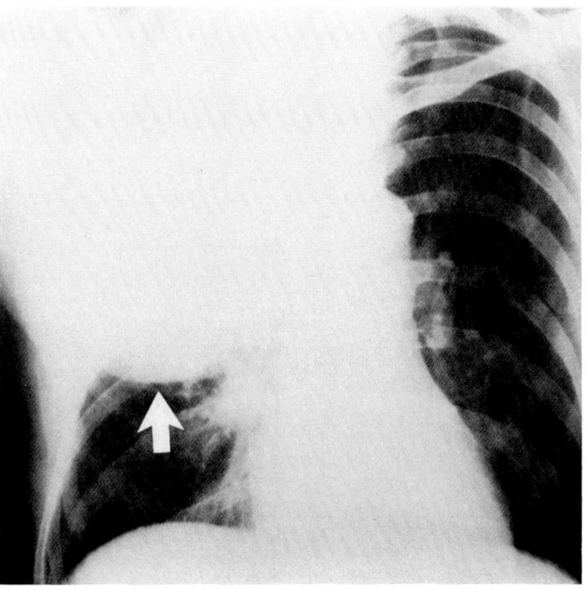

**Figure 9.3. *Klebsiella* pneumonia.** Downward bulging of the minor fissure (arrow) due to enlargement of the right upper lobe with inflammatory exudate. (From Felson, Rosenberg, and Hamburger,[2] with permission from the publisher.)

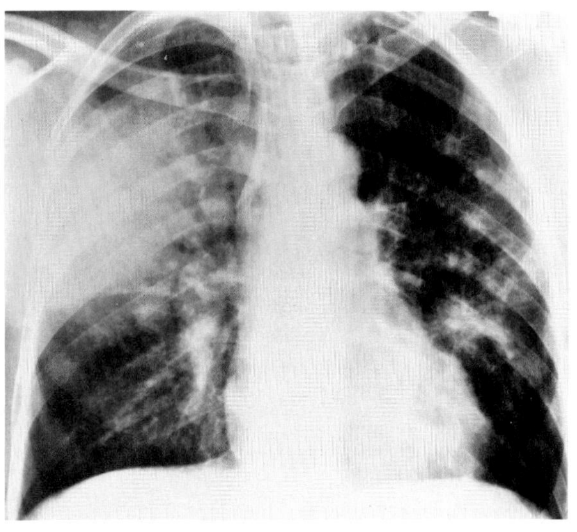

A

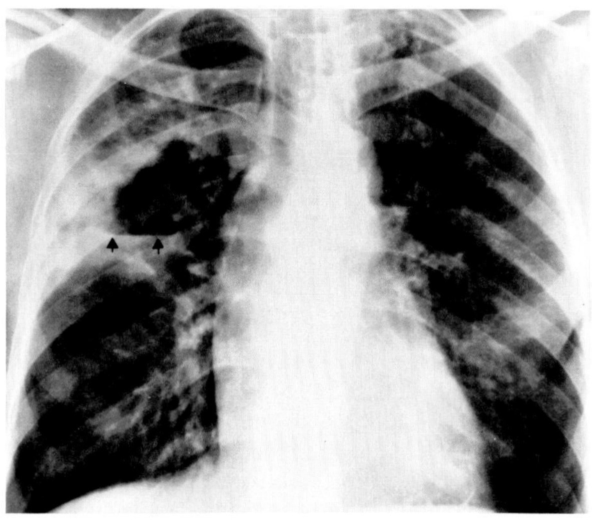

B

**Figure 9.4. *Klebsiella* pneumonia.** (A) Air space consolidation involving much of the right upper lobe. (B) Progression of the necrotizing infection produces a large abscess cavity with an air-fluid level (arrow).

## *Klebsiella*

*Klebsiella* (Friedländer's bacillus) produces an acute alveolar pneumonia that is most common in men over 40 years of age and often develops in alcoholics. Diabetes mellitus and chronic bronchopulmonary disease also predispose to infection by this organism. *Klebsiella* usually produces a homogeneous parenchymal consolidation containing air bronchograms, a pattern that simulates that of pneumococcal pneumonia. Although the pneumonia is most often located in the right upper lobe, lower lobe and bilateral involvement are not infrequent. Because the infiltrate usually abuts against an adjacent fissure, the border of the consolidation tends to be sharp and distinct. The disease typically induces such a large inflammatory exudate that the volume of the affected lobe or portion of a lobe is greatly increased, with a resultant characteristic bulging of an adjacent interlobar fissure. Unlike acute pneumococcal pneumonia, *Klebsiella* pneumonia causes frequent and rapid cavitation,

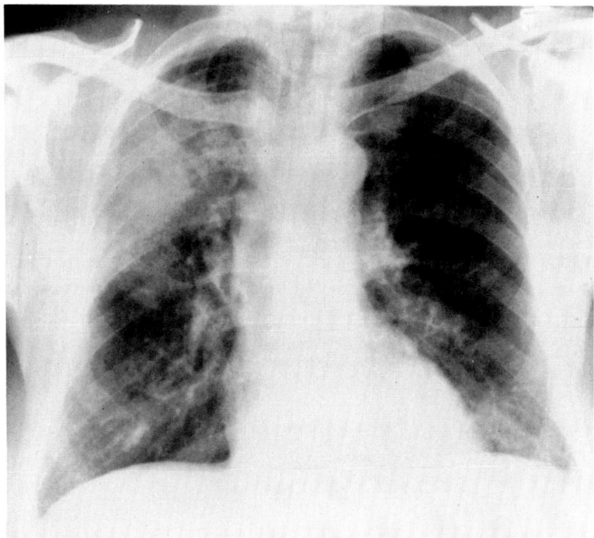

**Figure 9.5.** ***Serratia* pneumonia.** Early focal bronchopneumonia in the posterior segment of the right upper lobe. (From Balikian, Herman, and Godleski,[30] with permission from the publisher.)

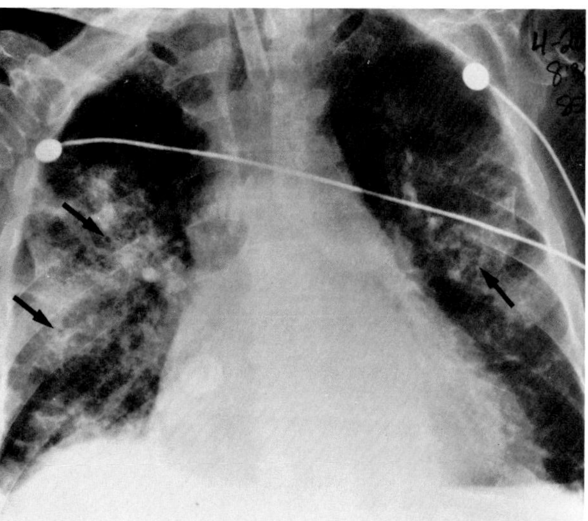

**Figure 9.6.** ***Enterobacter* pneumonia.** Bilateral diffuse bronchopneumonia associated with multiple small abscesses (arrows), which proved to be thick-walled and measured 0.5 to 1.5 cm at necropsy. Organizing pneumonia, fibrosis, and severe tracheobronchitis were also seen. (From Balikian, Herman, and Godleski,[30] with permission from the publisher.)

and there is a much greater incidence of pleural effusion and empyema. Massive lung gangrene, the separation of large masses of necrotic lung into an abscess cavity, can occur but is rare.

*Klebsiella* pneumonia may occasionally progress to a chronic necrotizing infection, with multiple areas of cavitation and fibrosis simulating tuberculosis.[2]

## *Serratia* and *Enterobacter*

*Serratia marcescens* is a common saprophyte of soil, water, and sewage. Infections due to the organism are usually hospital-acquired and involve elderly, debilitated patients, most of whom have been receiving antibiotic therapy. The source of infection is often contaminated solutions in respiratory therapy devices. *Serratia* typically produces a bronchopneumonia without abscess formation, but often with pleural effusion.

*Enterobacter* pneumonia also generally occurs in elderly hospitalized patients with major medical or surgical illnesses. The pneumonia presents a mixed alveolar-interstitial pattern with predominant lower lobe involvement. Small cavities develop in about half of the patients; minimal or moderate effusion is present in the same proportion.[3,4,30]

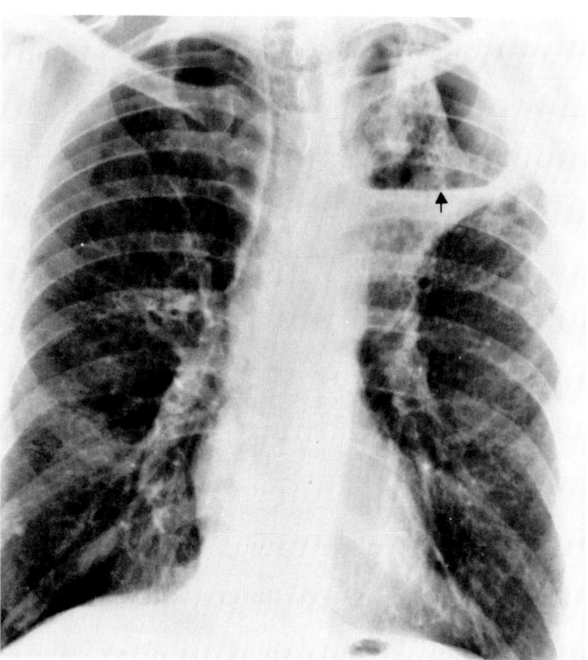

**Figure 9.7.** ***Proteus* pneumonia.** A large, thick-walled left upper lobe abscess with an air-fluid level (arrow) and an associated infiltrate.

## *Proteus*

*Proteus* is a ubiquitous organism that is rarely a primary invader and produces disease in locations previously infected by other organisms. It frequently contaminates surgical wounds, burns, and decubitus ulcers. Infections of the ears and mastoid sinuses can have intracranial extension, leading to lateral sinus thrombosis, meningitis, brain abscess, and bacteremia. *Proteus* organisms are a common cause of urinary tract infections, espe-

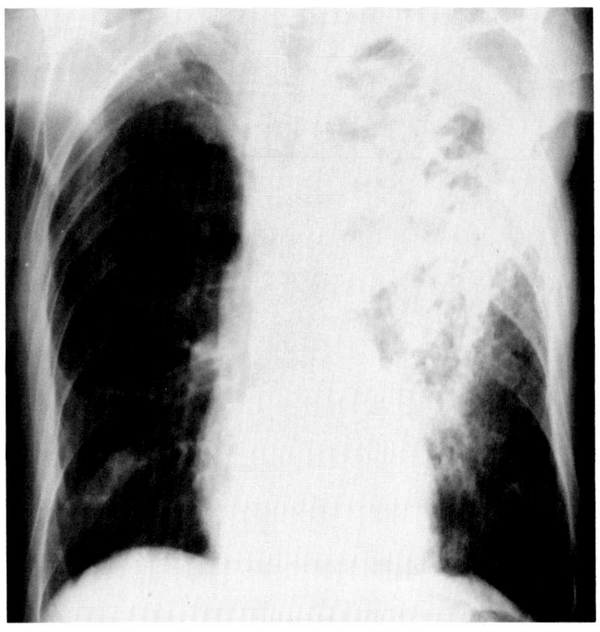

A

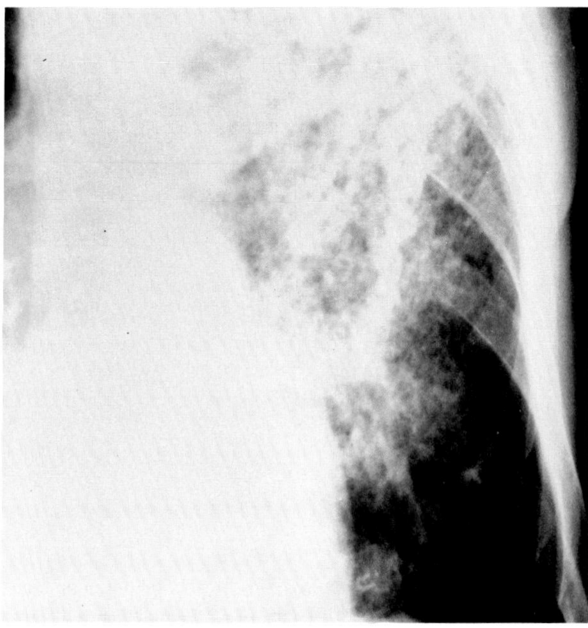

B

**Figure 9.8.** ***Pseudomonas*** **pneumonia.** (A) A frontal chest film demonstrates diffuse nonsegmental air space consolidation involving much of the left lung. (B) A coned view of the left lung better shows the hazy alveolar nodular infiltrate with multiple small lucent areas consistent with cavitation. (From Unger, Rose, and Unger,[4] with permission from the publisher.)

cially in patients who have had an obstructive uropathy, a history of instrumentation of the bladder, or repeated courses of chemotherapy. The organism may involve an anatomically normal urinary tract in patients with diabetes mellitus. *Proteus* pneumonia usually presents as an alveolar consolidation that predominantly involves the lower lobes; it may be associated with multiple abscesses, pleural effusion, and a decrease in lung volume.[4,5]

## *PSEUDOMONAS* INFECTIONS

*Pseudomonas aeruginosa* is a gram-negative bacillus that may produce a severe infection in patients with local tissue damage or diminished resistance. The disease is usually found in hospitalized patients receiving steroids, antibiotics, or cytotoxic or immunosuppressive drugs. It especially affects premature infants, patients with burns, and geriatric patients with debilitating diseases. Hospital equipment and "sterile" solutions may harbor the organism. Patients with tracheostomies are particularly at risk, and a common source of infection is heavily contaminated nebulizers attached to artificial ventilators.

*Pseudomonas* infection may initially present a pattern of widespread patchy or nodular shadows that are widely distributed throughout both lungs and that may coalesce. Numerous tiny translucencies are often seen within areas of consolidation. Rather than representing true micro-

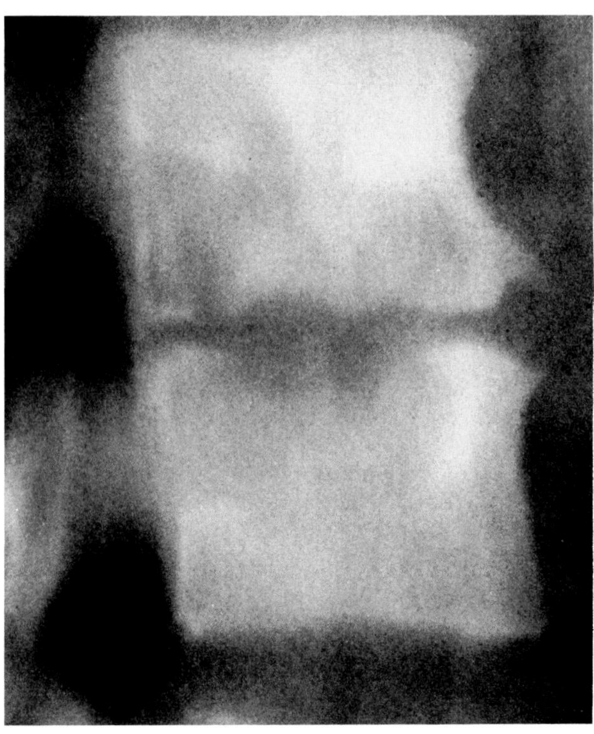

**Figure 9.9.** ***Pseudomonas*** **osteomyelitis.** A tomogram shows the destructive process in the second and third lumbar vertebrae, irregular narrowing of the intervertebral disk space, and reactive sclerosis.

abscesses, they probably reflect areas of air-containing parenchyma unaffected by the inflammatory process. Although the disease has a lower lobe predominance, extensive bilateral parenchymal consolidation is common. Large excavating abscesses up to 10 cm in diameter may rapidly develop. Small pleural effusions may occur, though they are not a prominent part of the radiographic appearance.

*Pseudomonas* is an important cause of osteomyelitis in drug addicts and in patients who are immunosuppressed or suffer from debilitating disease. The disease primarily involves the spine, causing destructive changes at the contiguous vertebral margins and some sclerosis and narrowing of the intervertebral disk space.

*Pseudomonas* organisms may also cause urinary tract infection (especially in patients with an obstructive uropathy who have been subjected to repeated urethral manipulations or urologic surgery), epidemic diarrhea of infancy, iatrogenic meningitis, and bacterial endocarditis following open-heart surgery.[4,6]

## MELIOIDOSIS

Melioidosis is an endemic disease in southeast Asia that is caused by *Pseudomonas pseudomallei*. Acute infections are characterized by multiple small abscesses in the lungs. In subacute disease the lung abscesses tend to be more extensive, and widespread extrapulmonary disease most commonly causes subcutaneous and deep abscesses, draining sinuses, osteomyelitis, septic arthritis, and nodules or purulent abscesses in the lungs, liver, spleen, kidneys, and brain.

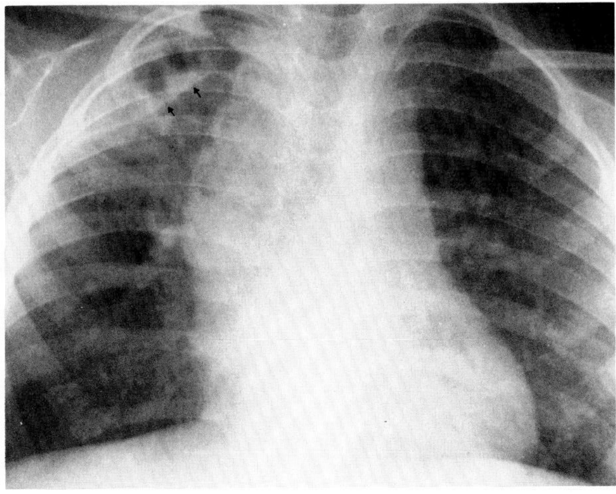

**Figure 9.10. Melioidosis.** Melioidosis in a young Laotian causes cavitation in the right upper lobe (arrows) with volume loss. The large amount of right-sided mediastinal adenopathy is unusual.

The radiographic appearance of pulmonary infection in melioidosis is often indistinguishable from that of tuberculosis or other granulomatous diseases. In the acute phase there is usually a pattern of generalized, irregular, small nodular densities (simulating miliary tuberculosis), which may rapidly cavitate or coalesce into a segmental or lobar consolidation. As in tuberculosis, the upper lobes are most commonly involved. However, unlike tuberculosis, melioidosis is not usually accompanied by hilar adenopathy and pleural effusion.

Osteomyelitis may result from the direct extension of a subcutaneous infection or from hematogenous dissemination. The long bones, pelvis, sternum, and rib cage are most commonly affected. Diffuse osteoporosis and cortical destruction result in a patchy, moth-eaten appearance with little sclerosis or periosteal reaction. Draining sinuses may develop from the infected bone.

Abscesses in the liver, spleen, kidneys, or brain may be both clinically and radiographically indistinguishable from those caused by other infectious processes. The location and extent of the lesions can be demonstrated by ultrasound, computed tomography, or radionuclide scanning.

Melioidosis should be a differential consideration in any patient from an endemic area who has a febrile illness and a radiographic pattern of tuberculosis but from whom tubercle bacilli cannot be isolated.[7–9]

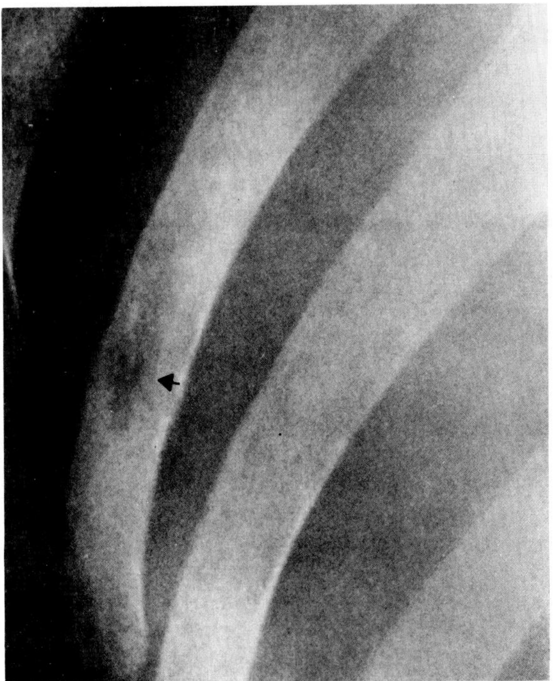

**Figure 9.11. Melioidosis.** A coned view shows a focal area of osteomyelitis (arrow) in the right eighth rib resulting from the direct extension of a subcutaneous infection. (From Jones and Ross,[9] with permission from the publisher.)

# *SALMONELLA* INFECTIONS

## Typhoid Fever

Typhoid fever is an acute systemic illness caused by *Salmonella typhosa*. The disease is transmitted by food and water contaminated by human feces containing the bacteria. Once in the gastrointestinal tract, the organisms are phagocytized by lymphoid tissue, particularly in the Peyer's patches of the terminal ileum. The organisms multiply there and produce raised plaques, which appear radiographically as irregular thickening and nodularity of mucosal folds that are usually limited to the terminal ileum but can involve the colon. Necrosis of the overlying mucosa causes ulceration. In severe disease, cecal narrowing and irregularity produce the coned cecum pattern. After treatment, the ileum (and cecum) usually returns to normal, though healing with fibrosis and stricture can occur.

Typhoid fever must be distinguished radiographically from Crohn's disease of the terminal ileum. Ileal involvement in typhoid fever is symmetric; skip areas and fistulas do not occur. In addition, most patients with typhoid fever have clinical and radiographic evidence of splenomegaly.

Extensive involvement of the muscular and serosal layers of the bowel may occasionally lead to perforation of the ileum with evidence of free intraperitoneal gas on abdominal radiographs obtained in the upright or decubitus positions. Colonic disease can rarely lead to toxic megacolon.[10]

## Other *Salmonella* Infections

*Salmonella* gastroenteritis (food poisoning) occurs after the ingestion of heavily contaminated food and mainly affects the terminal ileum, though colon changes can also be seen. The symptoms and signs of salmonellosis vary from acute gastroenteritis to severe septicemia. In typical food poisoning there is a sudden onset of fever, nausea, vomiting, and diarrhea after a very short incubation period, often as little as 12 hours. In most cases the disease is self-limited and there is complete recovery in 4 to 5 days.

Patients with suspected *Salmonella* gastroenteritis seldom undergo contrast examination, since the symptoms

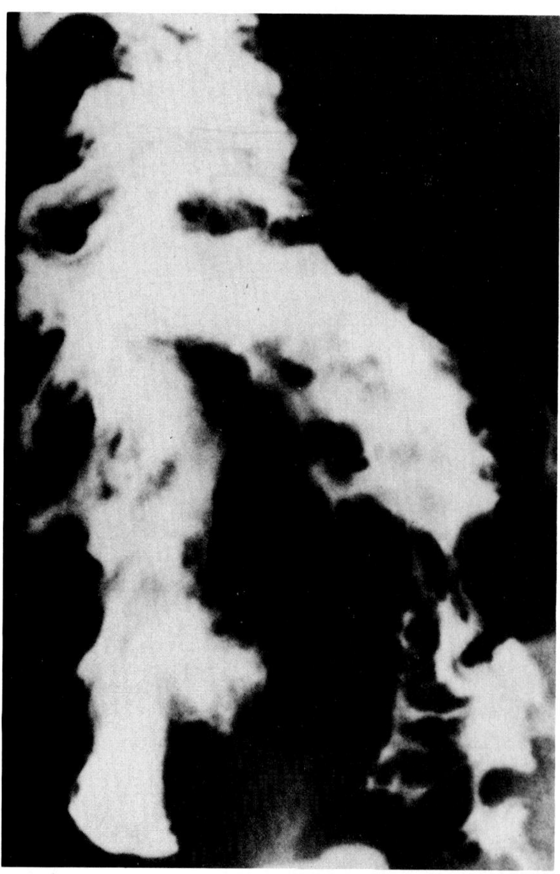

A

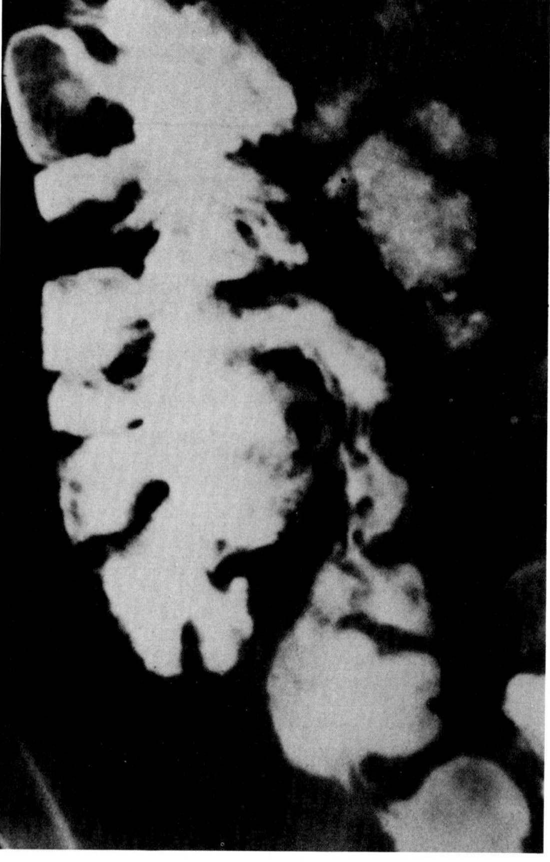

B

**Figure 9.12. Typhoid fever.** (A) Nodularity and irregularity of the terminal ileum with deformity of the cecum. (B) After therapy the ileum and cecum return to normal. (From Francis and Berk,[10] with permission from the publisher.)

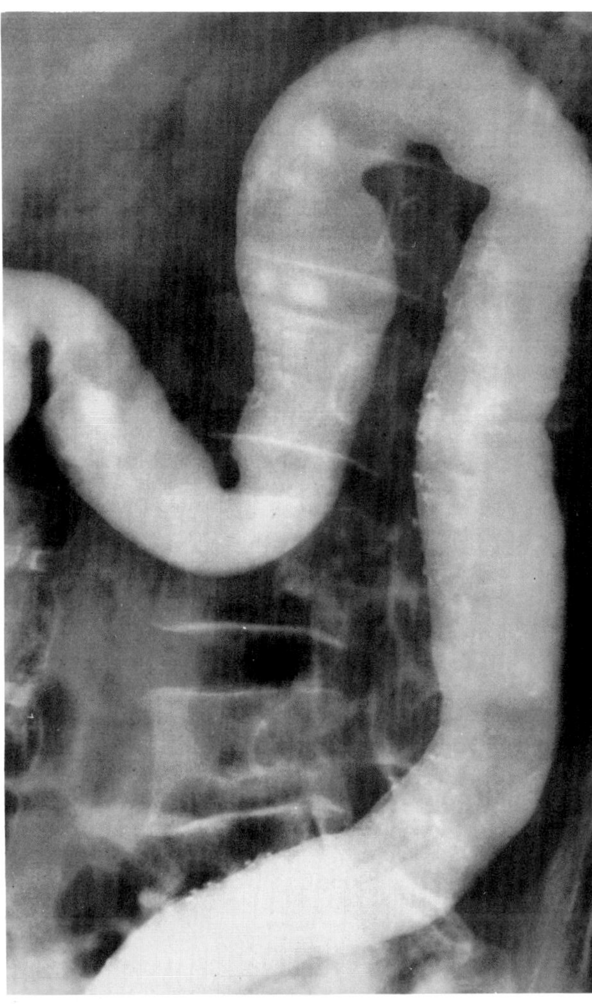

**Figure 9.13. Salmonellosis.** Diffuse fine ulcerations simulate ulcerative colitis. (From Eisenberg,[31] with permission from the publisher.)

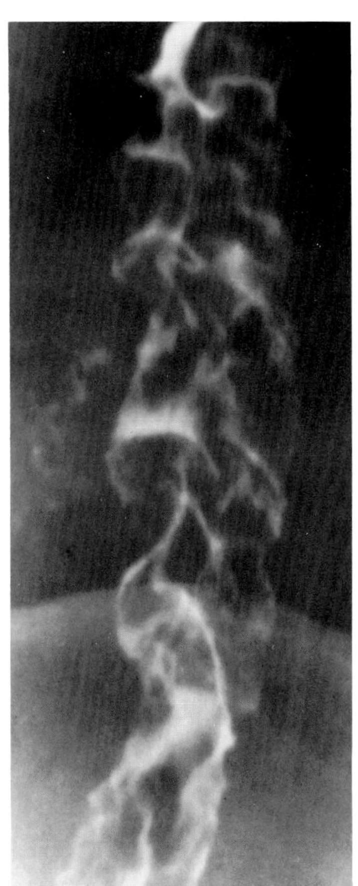

**Figure 9.14. Salmonellosis.** A postevacuation film from a barium enema examination demonstrates ulceration and irregular thickening of the mucosal folds in the descending colon. (From Eisenberg,[31] with permission from the publisher.)

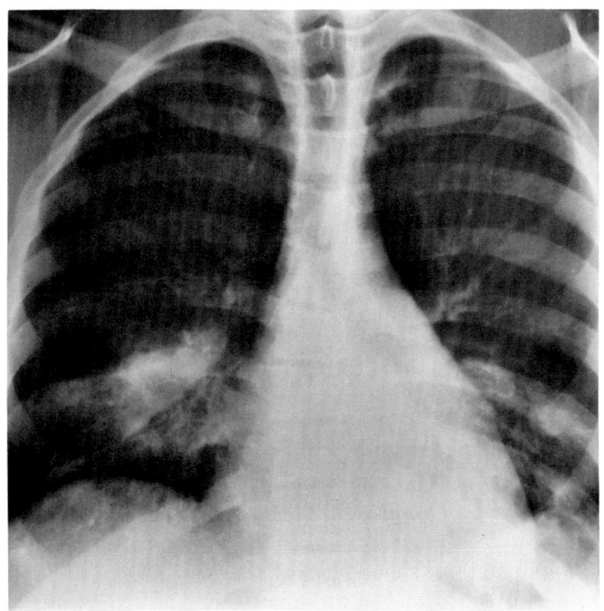

A

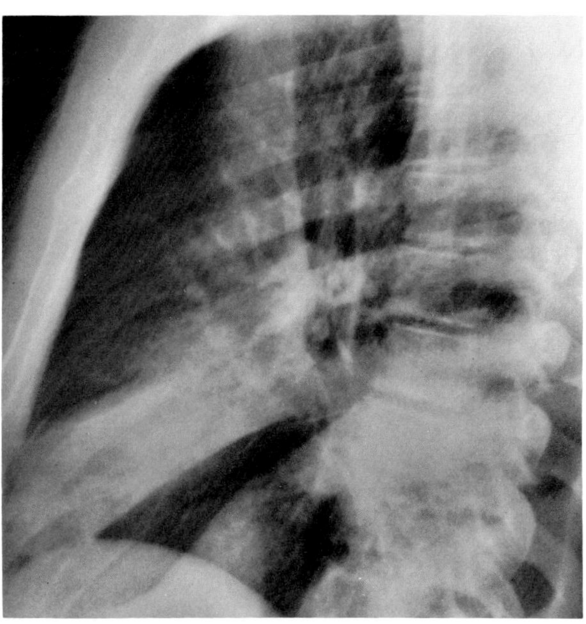

B

**Figure 9.15. Salmonella pneumonia.** (A) Frontal and (B) lateral views of the chest demonstrate right middle lobe and lingular pneumonia. A left pleural effusion is also present.

are acute and suggest the underlying self-limited condition. When performed, however, contrast studies demonstrate the involved portions of the small and large bowel to be atonic with edematous and nodular mucosa. Diffuse fine ulcerations often occur. Radiographic differentiation from shigellosis is frequently impossible in the colon, though concomitant involvement of the terminal ileum strongly suggests salmonellosis as the correct diagnosis.

Although salmonellae are most commonly pathogens in the gastrointestinal tract, the organisms may also cause bronchopneumonia (secondary to aspiration), or a diffuse miliary pattern of infection secondary to bacteremia. Pulmonary involvement most often occurs in patients with disseminated malignant disease or in those who are immunosuppressed. The alveolar infiltrates are often diffuse and bilateral; cavitation, pleural effusion, and empyema are common.

In disseminated disease, *Salmonella* organisms may lodge in the bone marrow, especially at sites of thromboses or infarcts. *Salmonella* is usually the causative organism of the acute osteomyelitis that is a frequent complication in patients with sickle cell disease.[11]

## *SHIGELLA* INFECTIONS

Shigellosis (bacillary dysentery) is an acute, self-limited infection of the intestinal tract characterized by diarrhea, fever, and abdominal pain. Unlike salmonellosis, which mainly affects the terminal ileum, shigellosis predominantly involves the colon. Shigellae enter the bowel in food and drink that have been contaminated by infected fecal material. The organisms penetrate the colonic mucosa and grow rapidly, liberating exotoxins and endotoxins that cause inflammation of the colon and rectum. Mucosal edema and submucosal infiltration are associated with an outpouring of mucoid and blood-streaked exudate, which fills the lumen of the gut. As the necrotic tissue sloughs, shallow and ragged ulcers remain, partially or completely encircling the bowel.

In acute bacillary dysentery, a barium enema examination generally cannot be tolerated. If such an examination is performed, severe spasm of the colon may prevent complete filling. When a barium enema examination is successful, the radiographic appearance is related to the severity and stage of the colon involvement. The acute, severe form is a pancolitis characterized by deep "collar button" ulcers, intense spasm, and mucosal edema. In less severe disease, superficial ulcerations with coarse, nodular edematous folds can involve the entire colon or be segmental, primarily affecting the rectum, sigmoid, and, less commonly, the descending portion. Nonspecific findings, such as spasm, haustral distortion, and excess fluid, usually accompany the mucosal changes. Fistulous tracks may develop.

Because it may be impossible to distinguish between severe shigellosis and acute ulcerative or Crohn's colitis, bacteriologic investigation is often required for a specific diagnosis. It is critical that the precise causative agent be identified, since steroid therapy, often used to treat noninfectious forms of colitis, is obviously contraindicated in shigellosis.[12,13]

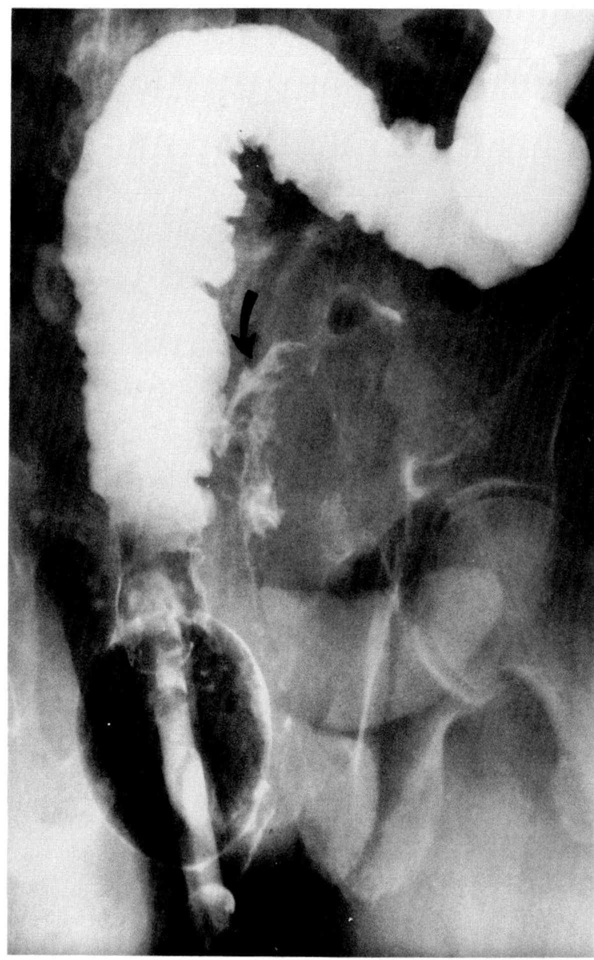

**Figure 9.16. Shigellosis.** Mucosal edema and ulceration primarily involve the rectosigmoid. Note the fistulous tract (arrow). (From Eisenberg,[31] with permission from the publisher.)

## *HEMOPHILUS* INFECTIONS

### *Hemophilus influenzae*

*Hemophilus influenzae* is a common inhabitant of the upper respiratory tract in healthy persons. Serious *Hemophilus influenzae* infections primarily affect children under the age of 4 and older patients who have undergone antibiotic therapy or who suffer from diseases that increase their general susceptibility to infection (e.g.,

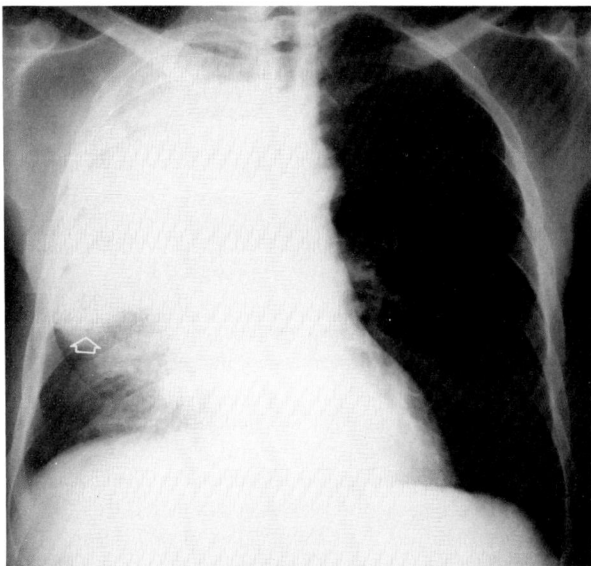

**Figure 9.17. *Hemophilus influenzae* pneumonia.** Acute lobar consolidation with downward bulging of the minor fissure due to enlargement of the right upper lobe. (From Francis and Francis,[18] with permission from the publisher.)

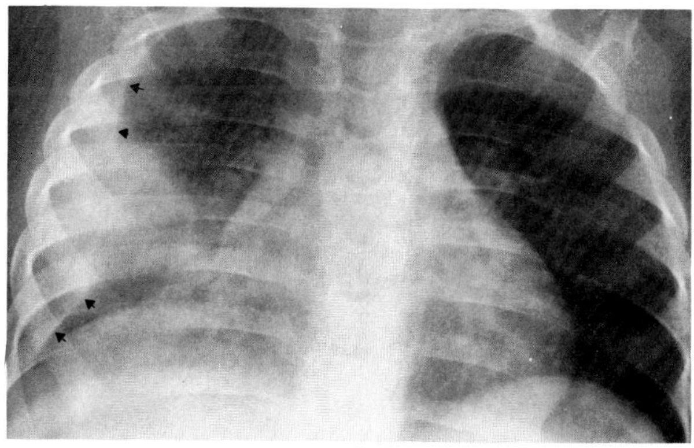

**Figure 9.18. *Hemophilus influenzae* pneumonia.** In addition to the ill-defined right lower lung consolidation, note the extensive degree of pleural thickening and/or fibrinous exudate (arrows) that appears out of proportion to the associated parenchymal infiltrate. (From Vinik, Altman, and Parks,[14] with permission from the publisher.)

sickle cell disease, alcoholism, immunodeficiency disorders, chronic pulmonary disease). *Hemophilus influenzae* pneumonia may present as an acute, often bilateral, bronchopneumonia or as a unilateral lobar or segmental consolidation simulating pneumococcal disease. A characteristic finding is an extensive degree of pleural involvement that often appears out of proportion to the associated parenchymal infiltrate. There is usually marked pleural thickening and/or fibrinous exudate with a relatively small amount of free pleural fluid. Unlike staphylococcal pneumonia in children, *Hemophilus influenzae* pneumonia is not accompanied by expansion to adjacent lobes and pneumatocele formation. In adults, *Hemophilus influenzae* can cause chronic bronchitis and, infrequently, a diffuse miliary bronchopneumonia without significant pleural reaction.

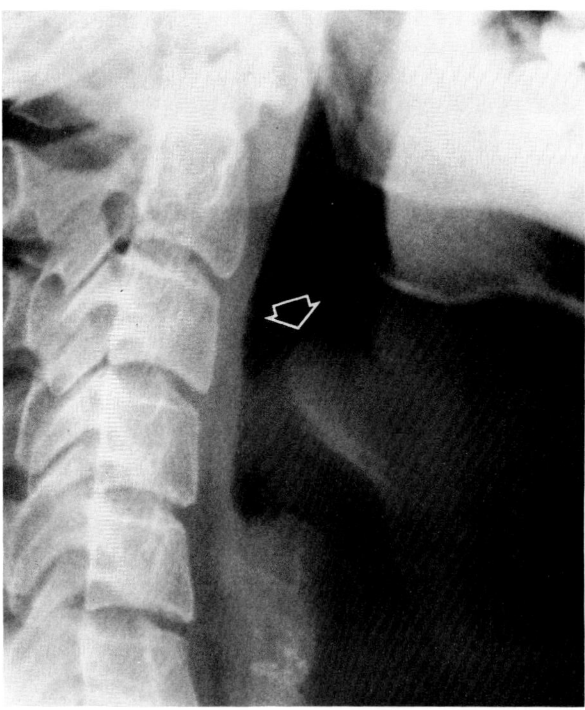

**Figure 9.19. *Hemophilus influenzae* epiglottitis.** A soft tissue lateral view of the neck demonstrates rounded thickening (arrow) of the normally narrow epiglottic shadow.

*Hemophilus influenzae* is the leading cause of epiglottitis, a potentially lethal disease that often requires airway intubation. On soft tissue lateral views of the neck, acute epiglottitis results in a rounded thickening of the normally narrow epiglottic shadow, giving it the configuration and approximate size of an adult thumb.

*Hemophilus influenzae* is a common cause of childhood bacterial meningitis and the second leading cause of childhood otitis media. It occasionally leads to endocarditis, pericarditis, pyogenic arthritis, and brain abscess.[14–16]

## *Hemophilus pertussis* (Whooping Cough)

Although whooping cough is often considered to have been largely eradicated by immunization, immunity is apparently not lifelong, and pertussis has become a not uncommon cause of bronchitis in adults. Acute pertussis most frequently affects children under 2 years of age. Chest radiographs demonstrate various combinations of atelectasis, segmental pneumonia (usually in the lower lobes or middle lobe), and hilar lymph node enlargement. An increase in perihilar markings, or a coalescence of air space consolidation contiguous to the heart, may obscure the normally sharp cardiac borders and produce a shaggy heart contour. Although the shaggy heart was initially considered pathognomonic of pertussis, an identical appearance is more commonly seen with viral pulmonary infection and also can develop in patients with asbestosis.[17–19]

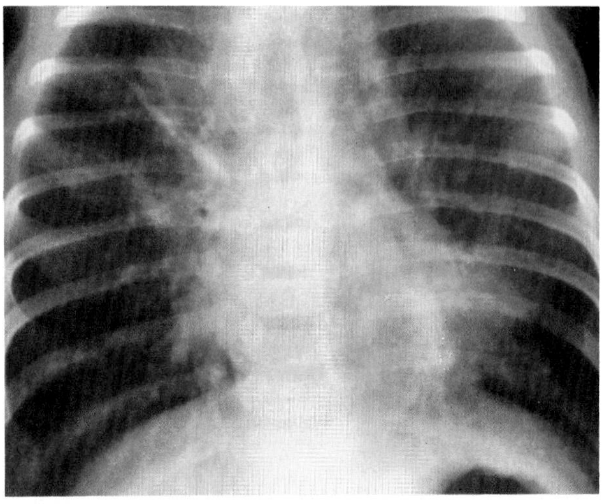

**Figure 9.20. *Hemophilus pertussis* (whooping cough).** Bilateral central parenchymal infiltrates and linear areas of atelectasis obscure the normally sharp cardiac borders to produce the shaggy heart contour.

## BRUCELLOSIS

Brucellosis is primarily a disease of animals (cattle, swine, goats, sheep) that is transmitted to humans by the ingestion of infected dairy products or meat, or through direct contact with animals, their carcasses, or their excreta. The most characteristic radiographic findings are seen in the spine, especially the lower thoracic and lumbar vertebral bodies. The earliest changes occur at the periphery of the vertebral bodies and appear as a loss of cortical definition or frank erosions of the anterior and superior margins. As the disease progresses, there is reactive sclerosis and the proliferation of new bone, with hypertrophic spur formation anteriorly and laterally. A loss of height of the intervertebral disk space is an early finding. The development of a paraspinal abscess causes swelling of the paravertebral soft tissues. In the less common central type of vertebral lesion, lytic destruction in the vertebral body leads to vertebral collapse with varying degrees of wedging. The overall radiographic appearance in brucellosis closely simulates that of tuberculous spondylitis.

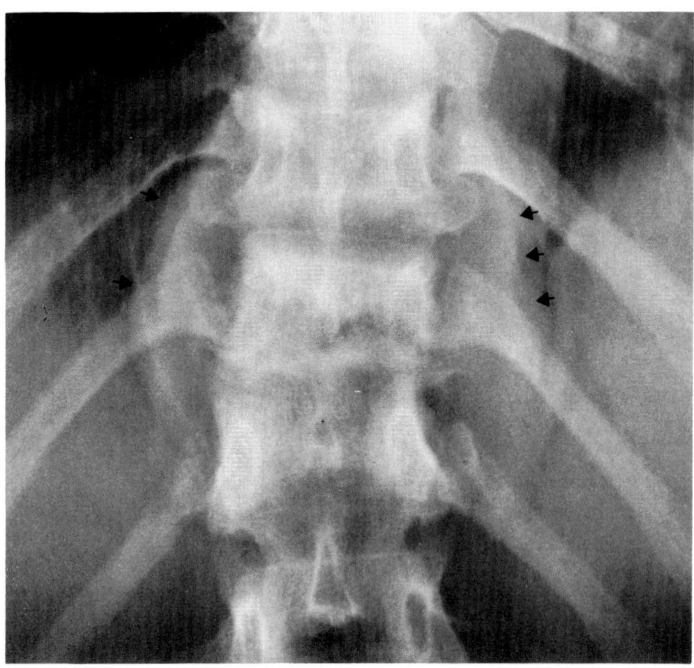

A

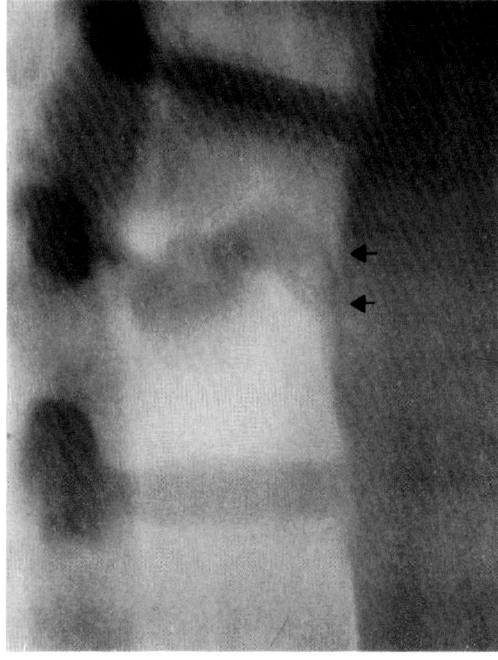

B

**Figure 9.21. Brucellosis.** (A) A frontal plain film of the lower thoracic spine demonstrates loss of height of the T11 and T12 vertebral bodies. The inferior end plate of T11 and the superior end plate of T12 are destroyed. Swelling of the paravertebral soft tissues (arrows) is also seen. (B) A lateral tomogram of the lower thoracic spine demonstrates cortical destruction with sclerosis of the inferior end plate of T11 and the superior end plate of T12. There is a mild degree of anterior wedging. The overall radiographic appearance is indistinguishable from that of tuberculosis spondylitis.

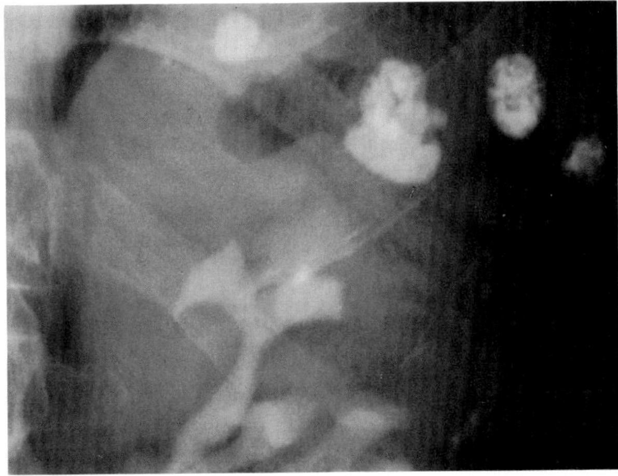

**Figure 9.22. Chronic brucellosis.** Multiple calcified granulomas in the spleen. (From Eisenberg,[31] with permission from the publisher.)

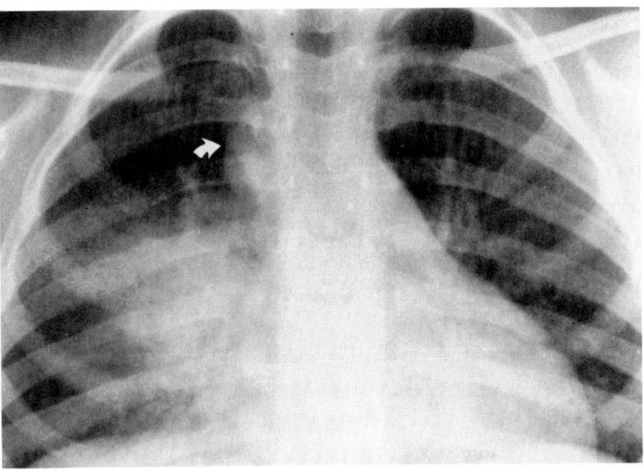

**Figure 9.23. Tularemia pneumonia.** Air space consolidation involving the right middle lobe and a portion of the right upper lobe. Note the right paratracheal nodal enlargement (arrow).

Multiple calcified granulomas and chronic abscesses of the spleen and, less frequently, the liver may be demonstrated in chronic brucellosis. Unlike the lesions in histoplasmosis and tuberculosis, the lesions in chronic brucellosis tend to be still active and suppurating even in the presence of calcification. The calcified nodules in chronic brucellosis are larger (about 1 to 3 cm in diameter) and consist of a flocculent center in a radiolucent area that is surrounded by a laminated calcified rim.

Uncommon radiographic manifestations of brucellosis include a destructive arthritis, osteomyelitis of the long bones, nonspecific pneumonia with pleural effusion, and hydrocephalus due to postinflammatory adhesive arachnoiditis.[20,21]

## TULAREMIA

Tularemia is an infectious disease caused by the gram-negative bacillus *Francisella tularensis,* which is usually transmitted to humans from infected animals (rodents, small mammals) or insect vectors. In most patients, the disease consists of a primary ulcerative cutaneous lesion and pronounced regional lymph node enlargement (ulceroglandular form). Hematogenous spread from a distant site, or the inhalation of organisms (chiefly by laboratory workers), may produce the pulmonary form of the disease.

Radiographically, tularemia causes patchy infiltrates, which may occur in any lobe and may be bilateral and/or multilobar. Ipsilateral hilar adenopathy and pleural effusion occur in about half of the cases. Abscess formation and cavitation are rare. Although ovoid consolidations with unilateral adenopathy in a patient in an area in which the disease is endemic suggest the diagnosis, the radiographic findings are usually nonspecific and simulate those of other inflammatory and neoplastic disorders.[22–24]

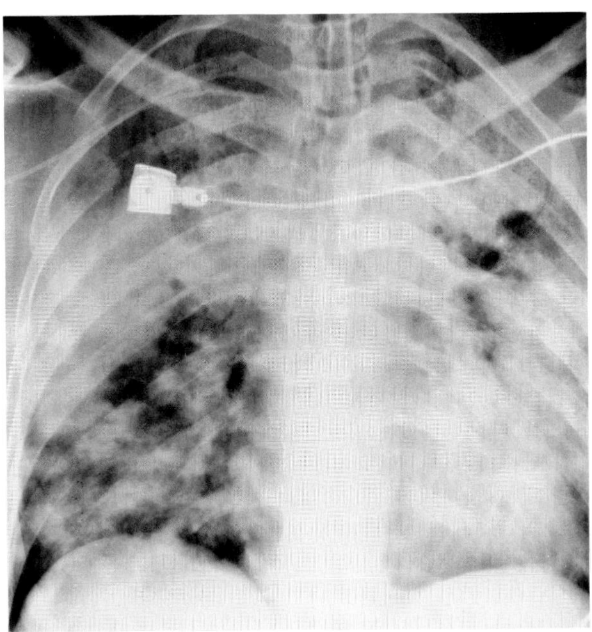

**Figure 9.24. Plague pneumonia.** Diffuse air space consolidation involves both lungs.

## *YERSINIA* INFECTIONS

### Plague

The pneumonic type of plague *(Yersinia pestis)* causes severe pulmonary consolidation, necrosis, and hemorrhage and is usually fatal. Patchy segmental infiltration or dense lobar consolidation simulates pneumococcal pneumonia. Severe edematous swelling may cause bowing of an adjacent pleural fissure, simulating *Klebsiella* pneumonia. At times the radiographic findings in the thorax may be limited to enlargement of the hilar and

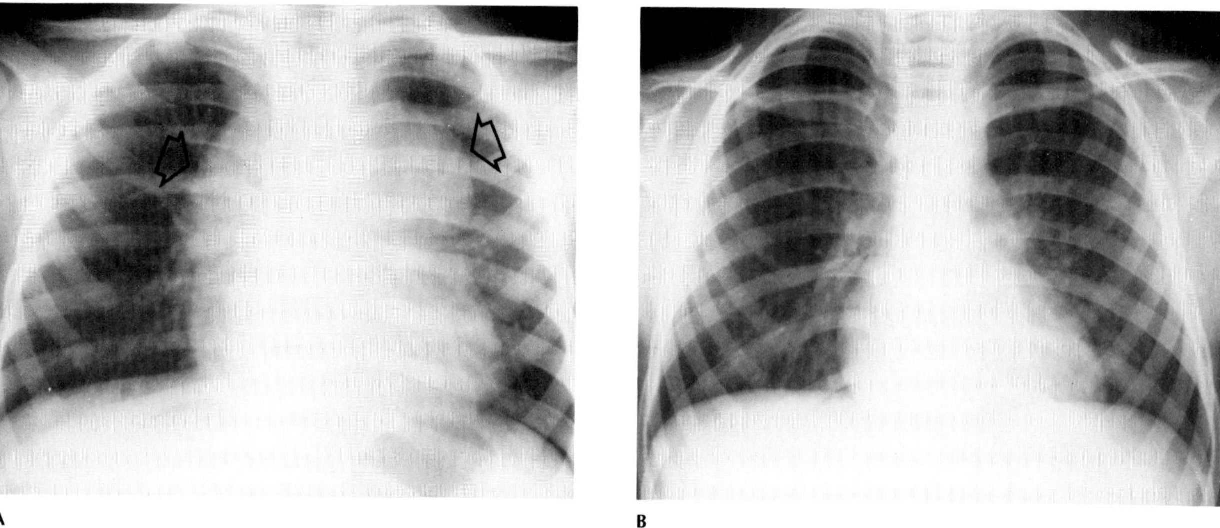


**Figure 9.25. Bubonic plague.** (A) An initial film demonstrates massive enlargement of the mediastinal lymph nodes. (B) Following chloramphenicol therapy, a repeat chest film demonstrates complete clearing of the lymphadenopathy. (From Sites and Poland,[25] with permission from the publisher.)

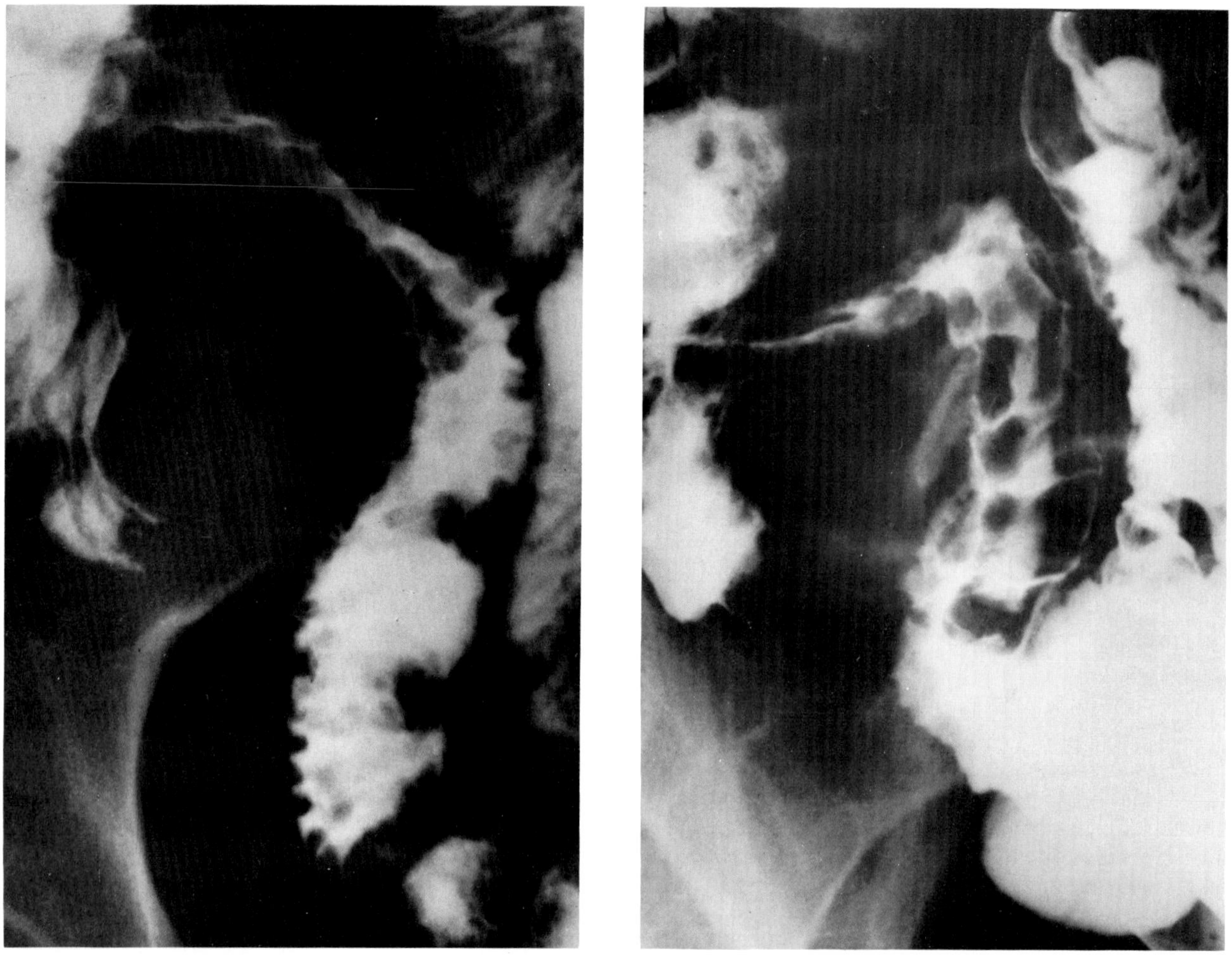

A B

**Figure 9.26. *Yersinia enterocolitica.*** (A) Numerous small nodules, marked edema, and moderate narrowing of the lumen combine to give the terminal ileum an appearance of irregularly thickened folds. (B) Nodular pattern. (From Ekberg, Sjostrom, and Brahme,[29] with permission from the publisher.)

paratracheal lymph nodes. If the patient is not treated early with appropriate antibiotics, the inflammatory process usually spreads rapidly to other lobes and leads to a fatal outcome.[25,26]

## *Yersinia enterocolitica*

*Yersinia enterocolitica* is a gram-negative bacillus that is a member of the same genus as the organism causing plague. In children, *Yersinia enterocolitica* infections most frequently cause acute enteritis with fever and diarrhea; in adolescents and adults, acute terminal ileitis or acute mesenteric adenitis simulating appendicitis more commonly occurs. *Yersinia enterocolitica* usually causes a focal disease involving short segments of the terminal ileum, though it can also affect the colon and rectum.

The most common radiographic pattern is coarse, irregular thickening of small bowel mucosal folds. Nodular filling defects and ulceration are also seen. Densely packed nodules surrounded by deep ulcerations can produce a pattern resembling the cobblestoning appearance of Crohn's disease. During the healing stage of the disease, the mucosal thickening decreases and tiny 1- to 2-mm filling defects appear, producing a granular pattern (follicular ileitis) that can persist for many months.

*Yersinia enterocolitica* can also cause multiple small colonic ulcerations similar to those seen in Crohn's colitis; cecal narrowing and irregularity can produce the pattern of a coned cecum.[27–29]

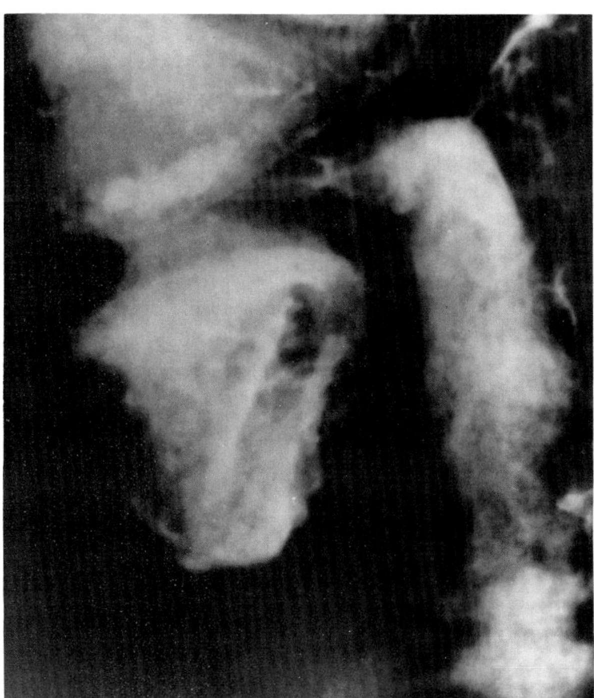

**Figure 9.27. *Yersinia enterocolitica.*** Conical narrowing and irregular margins of the cecum are present, with mild inflammatory changes in the terminal ileum. (From Eisenberg,[31] with permission from the publisher.)

**REFERENCES**


1. Gutman E, Pongdee O, Park YS: Massive pulmonary gangrene. *Radiology* 107:293–294, 1973.
2. Felson B, Rosenberg LS, Hamburger M: Roentgen findings in acute Friedlander's pneumonia. *Radiology* 53:559–565, 1949.
3. Meltz DJ, Grieco MH: Characteristics of *Serratia marcescens* pneumonia. *Arch Intern Med* 132:359–364, 1973.
4. Unger JD, Rose HD, Unger GF: Gram-negative pneumonia. *Radiology* 107:283–291, 1973.
5. Tillotson JR, Lerner AM: Characteristics of pneumonia caused by *Bacillus proteus*. *Ann Intern Med* 68:287–294, 1968.
6. Renner RR, Coccaro AP, Heitzman ER, et al: *Pseudomonas:* A prototype of hospital-based infection. *Radiology* 105: 555–562, 1972.
7. Sweet RS, Wilson ES, Chandler BF: Melioidosis manifested by cavitary lung disease. *AJR* 103:543–547, 1968.
8. James AE, Dixon GP, Johnson H: Melioidosis: A correlation of the radiologic and pathologic findings. *Radiology* 89:230–235, 1967.
9. Jones GP, Ross JA: Melioidosis. *Br J Radiol* 36:415–417, 1963.
10. Francis RS, Berk RN: Typhoid fever. *Radiology* 112:583–585, 1974.
11. Slomic AM, Rousseau PL *Salmonella* colitis. *Ann Radiol* 19:431–437, 1976.
12. Farman J, Rabinowitz JG, Meyers MA: Roentgenology of infectious colitis. *AJR* 119:375–381, 1973.
13. Christie AB: Bacillary dysentery. *Br Med J* 2:285–288, 1968.
14. Vinik M, Altman DH, Parks RE: Experience with *Hemophilus influenzae* pneumonia. *Radiology* 86:701–706, 1966.
15. Francis JB, Francis PB: Bulging (sagging) fissure sign in *Hemophilus influenzae* lobar pneumonia. *South Med J* 71:1452–1453, 1978.
16. Pearlberg J, Haggar AM, Saravolatz L, et al: *Hemophilus influenzae* pneumonia in the adult. *Radiology* 151:23–26, 1984.
17. Barnhard HJ, Kniker WT: Roentgenologic findings in pertussis: With particular emphasis on the "shaggy heart" sign. *AJR* 445–450, 1960.
18. Fawcitt J, Parry HE: Lung changes in pertussis and measles in childhood. A review of 1894 cases with a follow-up study of the pulmonary complications. *Br J Radiol* 30:76–82, 1957.
19. Morse SJ: Pertussis in adults. *Ann Intern Med* 68:953–954, 1968.
20. Buchanan TM, Faber LC, Feldman RA: Brucellosis in the United States, 1960–1972: An abattoir-associated disease. *Medicine* 53:403–413, 1974.
21. Williams E: Brucellosis. *Br Med J* 1:791–793, 1973.
22. Ivie JM: Roentgenological observations on pleuropulmonary tularemia. *AJR* 14:466–472, 1955.
23. Overholt EL, Tigertt WD: Roentgenographic manifestations of pulmonary tularemia. *Radiology* 74:758–765, 1960.
24. Dennis JM, Boudreau RP: Pleuropulmonary tularemia. *Radiology* 68:25–30, 1957.
25. Sites VR, Poland JD: Mediastinal lymphadenopathy in bubonic plague. *AJR* 116:567–570, 1972.
26. Douglas RF: Chest roentgenology in early diagnosis of infectious disease (plague pneumonia). *Milit Med* 128:104–106, 1963.
27. Lachman R, Soong J, Wishon G, et al: *Yersinia* colitis. *Gastrointest Radiol* 2:133–135, 1977.
28. Atkinson GO, Gay BB, Ball TI, et al: *Yersinia enterocolitica* colitis in infants: Radiographic changes. *Radiology* 148:113–116, 1983.
29. Ekberg O, Sjostrom B, Brahme F: Radiological findings in *Yersinia* ileitis. *Radiology* 123:15–19, 1977.
30. Balikian JP, Herman PG, Godleski JJ: Serratia pneumonia. *Radiology* 137:309–317, 1980.
31. Eisenberg RL: *Gastrointestinal Radiology: A Pattern Approach*. Philadelphia, JB Lippincott, 1983.

**Chapter Ten**

# Other Aerobic Bacteria

## LEGIONNAIRES' DISEASE

Legionnaires' disease is an acute bacterial respiratory infection that may cause a fulminant, often fatal, pneumonia. The most frequent initial radiographic abnormality is a patchy or fluffy alveolar infiltrate in a single lobe. Well-circumscribed nodular infiltrates can also occur. By the time of maximal involvement, the pneumonia has generally progressed to the adjacent lobes and the contralateral side. Small pleural effusions are common; cavitation and hilar adenopathy are unusual. Most patients with Legionnaires' disease respond well to erythromycin, though the radiographic resolution often lags behind the clinical response.[1,2]

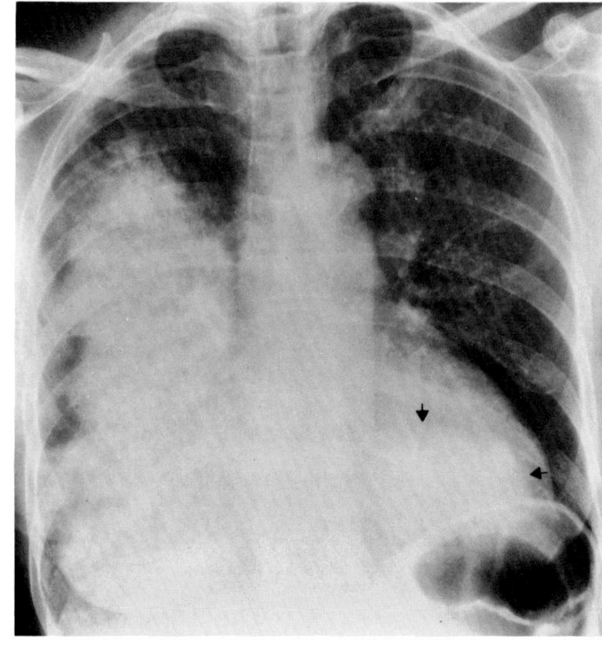

**Figure 10.1. Legionnaires' disease.** Extensive consolidation of much of the right lung, with a smaller area of infiltrate (arrows) at the left base.

## PITTSBURGH PNEUMONIA AGENT

The Pittsburgh pneumonia agent *(Legionella micdadei)* is a recently described human pathogen that primarily affects patients receiving intensive immunosuppressive therapy, especially renal transplant recipients being treated for episodes of allograft rejection. A variety of radiographic findings have been observed in the few cases

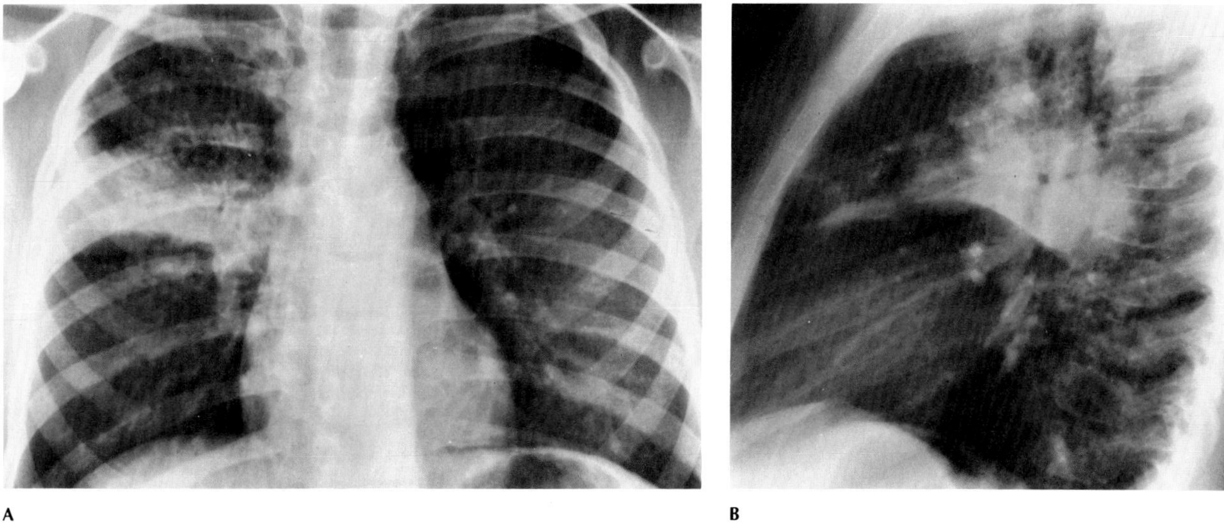

A B

**Figure 10.2. Legionnaires' disease.** (A) Frontal and (B) lateral views of the chest demonstrate a patchy alveolar infiltrate in the right upper lobe.

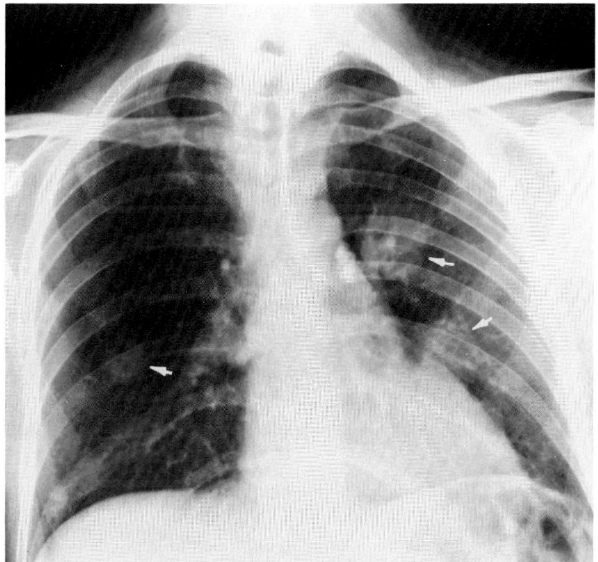

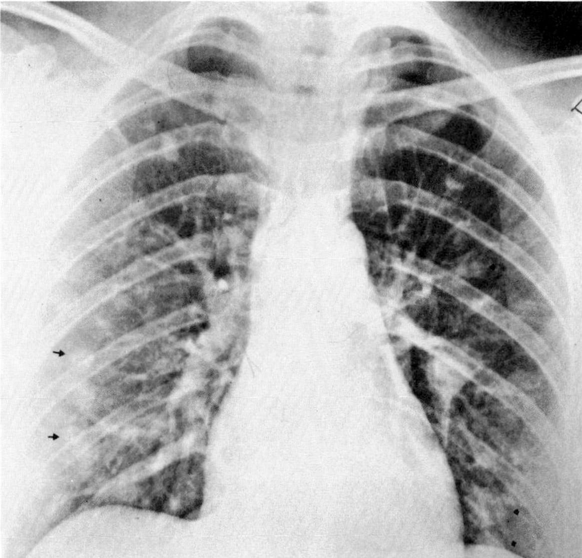

**Figure 10.3. Pittsburgh pneumonia agent.** Two patients with multiple nodular pulmonary infiltrates (arrows) in both lungs following renal transplantation and 2 days after completion of methylprednisolone pulse therapy for rejection. (From Pope, Armstrong, Thompson, et al,[3] with permission from the publisher.)

reported. There may be single or multiple nodules and segmental infiltrates involving part or all of one or more lobes, with a tendency for rapid growth of the pulmonary densities. Following erythromycin therapy, definite radiographic improvement usually is seen.[3,4]

## LISTERIOSIS

Listeriosis is a rare bacterial disease that primarily occurs as an intrauterine infection with a high neonatal mortality rate, or as a disease of the newborn. Although it most commonly produces meningitis, listeriosis of the newborn can involve the lungs and produce air space consolidations, interstitial fibrosis, or a diffuse miliary pattern of coarse, irregular granular densities distributed throughout both lungs. In adults, listeriosis has been reported in association with malignant disease, especially of the reticuloendothelial system.[5–7]

## ANTHRAX

Anthrax is a bacterial disease of cattle, sheep, and goats that primarily affects humans who inhale spores from infected animals or their products (e.g., wool, hides). In

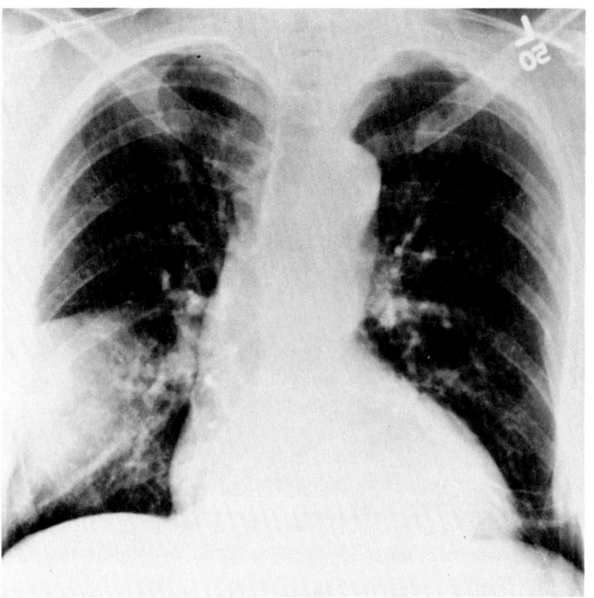

**Figure 10.4. Pittsburgh pneumonia agent.** An air space consolidation involves most of the right middle lobe. This film was obtained 22 days after a renal transplant, 5 days after azathioprine was discontinued, and during tapering of prednisone. (From Pope, Armstrong, Thompson, et al,[3] with permission from the publisher.)

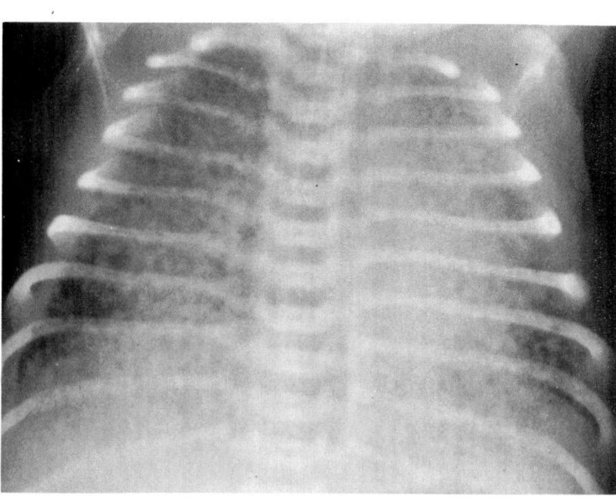

**Figure 10.5. Listeriosis.** A diffuse miliary pattern of coarse, irregular granular densities is distributed throughout both lungs.

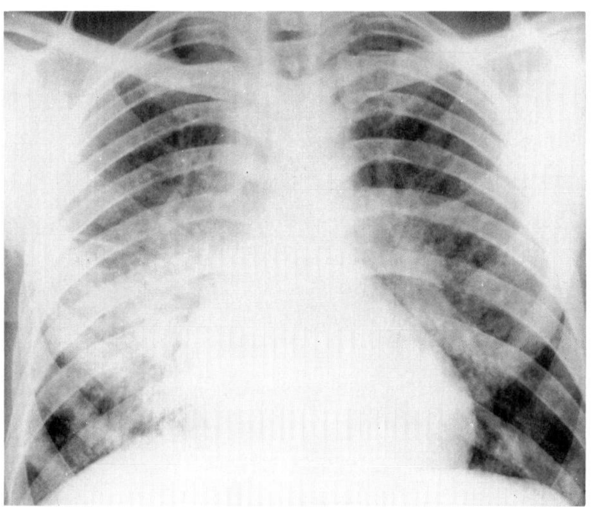

A

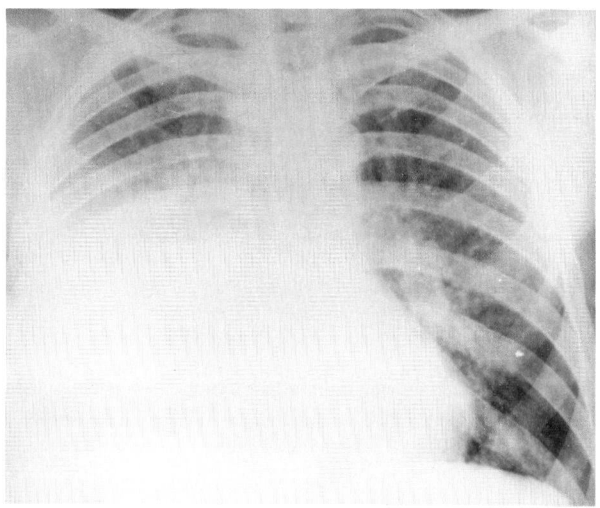

B

**Figure 10.6. Anthrax.** (A) Mediastinal enlargement and diffuse parenchymal nodularity, with more significant involvement of the right lower lobe. (B) Examination 4 months later shows continued mediastinal enlargement, with right pleural effusion and diffuse consolidation of the right lower lobe. (From Brachman, Pagano, and Albrink,[8] with permission from the publisher.)

disseminated disease the characteristic radiographic finding is mediastinal widening resulting from lymph node enlargement and a hemorrhagic mediastinitis. Patchy parenchymal infiltrates may develop, and pleural effusion is common.[8,9]

## DONOVANOSIS (GRANULOMA INGUINALE)

Donovanosis is a chronic ulcerative venereal disease involving the skin and lymphatics of the genital or perianal areas. Fatal disseminated disease, involving the

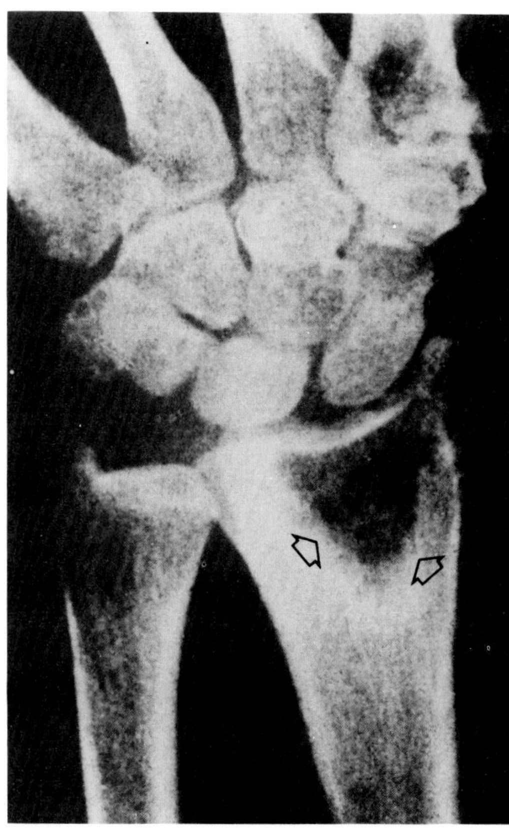

**Figure 10.7. Donovanosis.** A destructive focus in the distal radius (arrows) of a teenage girl in the West Indies. (From Cockshott and Middlemiss,[11] with permission from the publisher.)

bones, joints, or liver, has been rarely reported after years of chronic local infection. Multiple osteolytic areas without sclerotic reaction may produce an appearance resembling that of metastases or multiple myeloma. The flat bones and the distal ends of the long bones are most commonly involved.[10]

## REFERENCES


1. Reed JC, McLelland R, Nelson P: Legionnaires' disease. *AJR* 131:892–894, 1978.
2. Kirby BD, Peck H, Meyer RD: Radiographic features of Legionnaires' disease. *Chest* 76:562–565, 1979.
3. Pope TL, Armstrong P, Thompson R, et al: Pittsburgh pneumonia agent: Chest film manifestations. *AJR* 138:237–241, 1982.
4. Muder RR, Reddy SC, Yu VL, et al: Pneumonia caused by Pittsburgh pneumonia agent: Radiologic manifestations. *Radiology* 150:633–637, 1984.
5. Willich E: A new pulmonary manifestation of listeriosis in newborn babies: The interstitial pneumopathy. *Ann Radiol* 10:285–288, 1967.
6. Louria DB, Hensle T, Armstrong D, et al: Listeriosis complicating malignant disease: A new association. *Ann Intern Med* 67:261–281, 1967.
7. Moore PH, Brogdon BG: Granulomatosis infantiseptica (listeriosis). *Radiology* 79:415–419, 1962.
8. Brachman PS, Pagano JS, Albrink WS: Two cases of fatal inhalation anthrax. *N Engl J Med* 265:203–208, 1961.
9. Plotkin SA, Brachman PS: Inhalation anthrax. *Am J Med* 29:992–1001, 1960.
10. Kirkpatrick DJ: Donovanosis (granuloma inguinale): A rare cause of osteolytic bone lesions. *Clin Radiol* 21:101–105, 1970.
11. Cockshott WP, Middlemiss H: *Clinical Radiology in the Tropics*. Edinburgh, Churchill Livingstone, 1979.

## Chapter Eleven
# Anaerobic Bacteria

### TETANUS

Tetanus is an acute, often fatal, disease caused by an exotoxin produced in a wound by *Clostridium tetani*. It is characterized by a generalized increased rigidity and by convulsive spasms of skeletal muscles. Most radiographic abnormalities in tetanus reflect either complications of the disease or overly vigorous therapy. Severe compression fractures of the spine are common, are usually multiple, and primarily involve the midthoracic vertebrae. Although the degree of compression may be substantial, the fractures rarely cause pain and do not affect the prognosis or lead to neurologic sequelae. Stripping and elevation of the periosteum during muscle spasm may lead to the development of myositis ossificans, especially around the hips and along the vertebrae.

A toxic myocarditis may result in cardiomegaly and pulmonary edema. Nonspecific pneumonia is a common late complication of tetanus and a major cause of death.[1,2]

### OTHER CLOSTRIDIAL INFECTIONS

#### Gas Gangrene

Bubbles or streaks of gas in the subcutaneous or deeper tissues may be seen in gas gangrene, a clostridial infection that develops in anoxic, devitalized tissues in which the arterial circulation has been compromised by trauma, constricting tourniquets or casts, or obliterative arterial disease. At an early stage it may be difficult to distinguish gas formed by such anaerobic bacteria from air that has been introduced through a penetrating injury. Serial examinations demonstrate a steady decrease in externally introduced air, in contrast to an increasing amount of gas and an extension of gas from the known site of soft tissue injury in a patient with gas gangrene. The gas may remain superficial and produce subcutaneous bubbles in a netlike structure or may infiltrate into the deeper tissues and outline muscle bundles. The disease is rapidly fatal unless appropriate treatment is begun promptly.

#### Antibiotic-Associated Colitis

*Clostridium difficile* has been demonstrated in the stools of a high percentage of patients with pseudomembranous colitis arising after antibiotic therapy. The bacterium elaborates a cytotoxic substance that destroys human cells in culture and produces a severe enterocolitis when injected into the cecum of animals. The clinical symptoms can be identified within 1 day to 1 month after the initiation of the antibiotic therapy. Most patients recover uneventfully after the withdrawal of the offending antibiotic and the institution of adequate fluid and electrolyte replacement. (See the section "Pseudomembranous Colitis" in Chapter 47 for radiographic findings.)

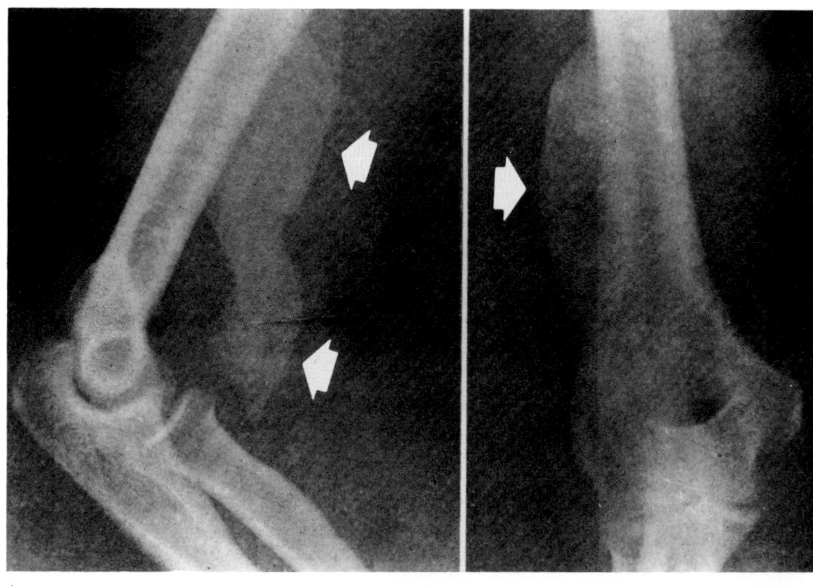


**Figure 11.1. Tetanus.** (A) Lateral and (B) frontal views of the elbow demonstrate myositis ossificans (arrows) over the flexor surface of the elbow about the anterolateral aspect of the distal humerus. (From Femi-Parse and Olowu,[2] with permission from the publisher.)

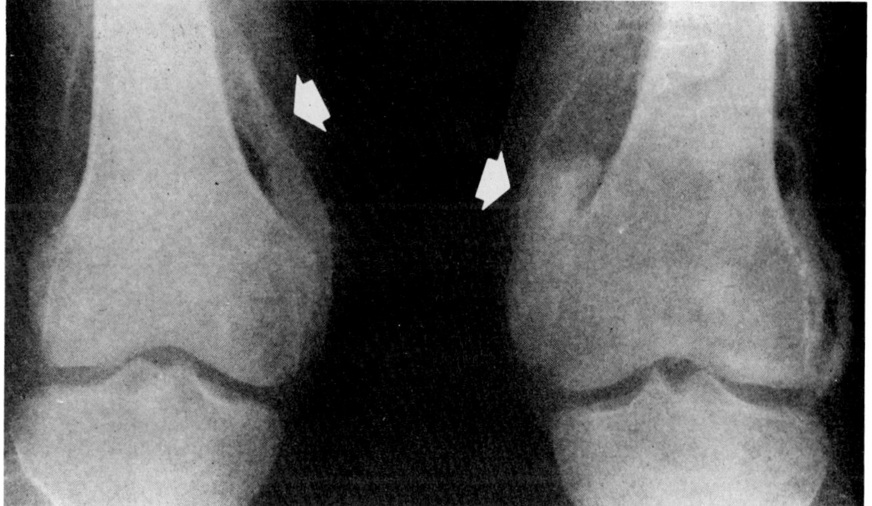

**Figure 11.2. Tetanus.** Plaques of bony masses are attached to the medial and lateral condyles of the distal femurs. (From Femi-Parse and Olowu,[2] with permission from the publisher.)

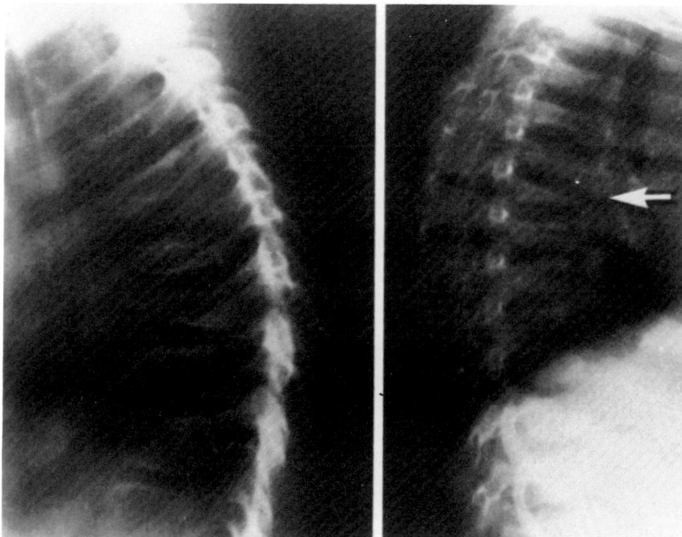

**Figure 11.3. Tetanus.** Typical compression fractures of the upper dorsal spine. (From Cockshott and Middlemiss,[6] with permission from the publisher.)

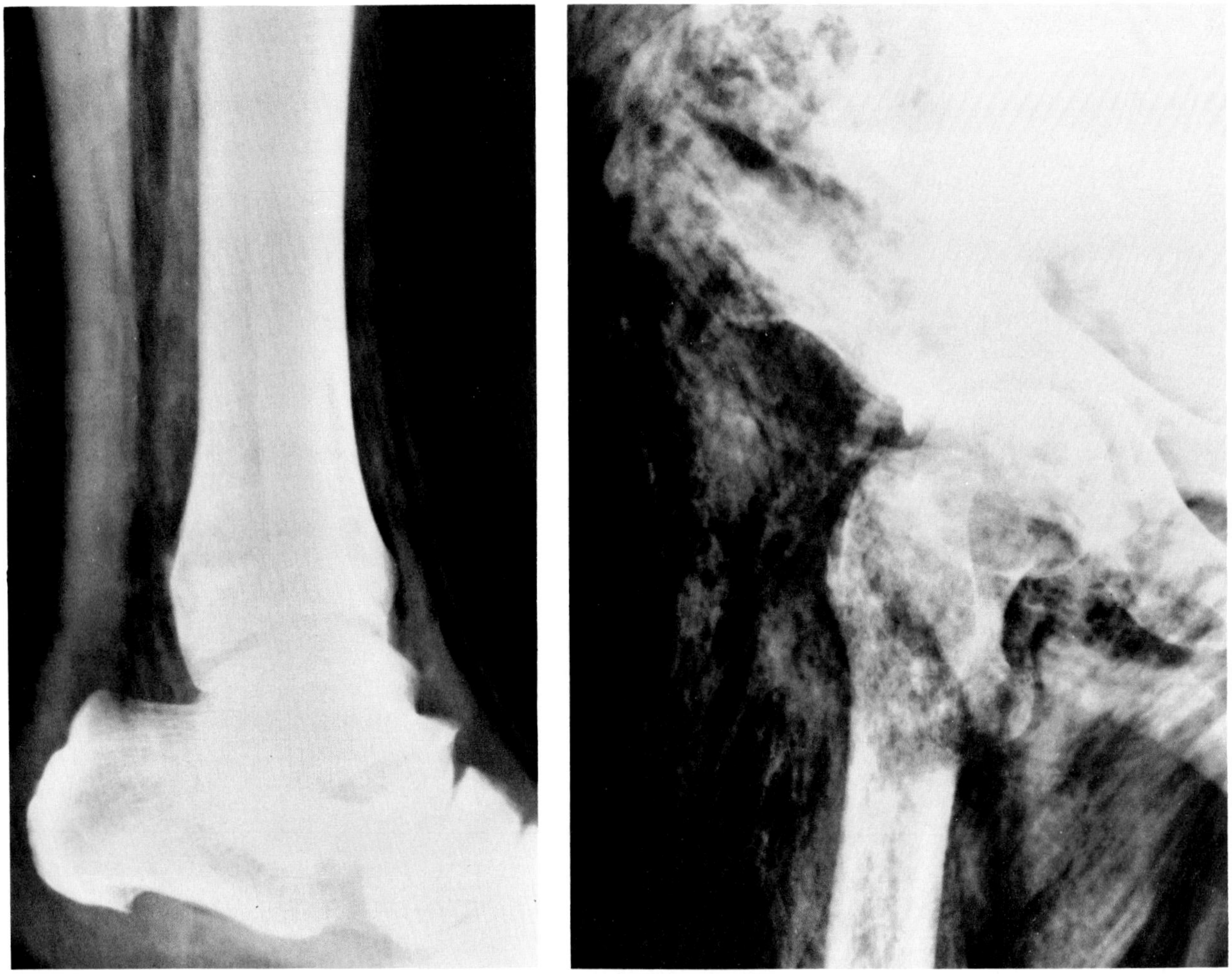

A B

**Figure 11.4. Gas gangrene.** Films of (A) the lower leg and (B) the upper leg and pelvis demonstrate large amounts of air in subcutaneous and deeper tissues in a diabetic who had suffered an injury to the thigh. The air clearly outlines multiple muscle bundles.

## Food Poisoning

Enteritis necroticans is a fatal hemorrhagic necrosis of the small bowel caused by *Clostridium perfringens*. The radiographic abnormalities in this condition include thickening and ulceration of the bowel wall with segmental stenoses and fistulization. Bowel perforation can lead to extraluminal gas-fluid collections. Most cases of *Clostridium perfringens* food poisoning, however, are self-limited, and recovery is rapid.

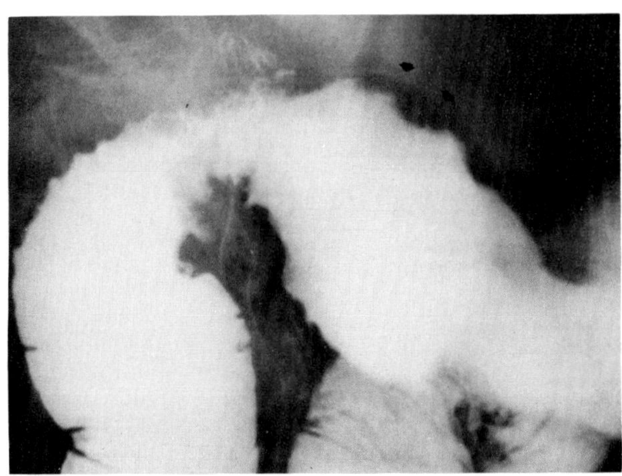

**Figure 11.5. Phlegmonous emphysematous gastritis.** There is severe, irregular ulceration of the distal stomach with air in the wall (arrows). (From Eisenberg,[7] with permission from the publisher.)

## Miscellaneous Clostridial Infections

Clostridial infection in other areas can cause intramural gas, which can be detected on plain abdominal radiographs. *Clostridium welchii* is one of the causes of phlegmonous gastritis, an often fatal condition in which bubbles of gas may be seen in the wall of the stomach.

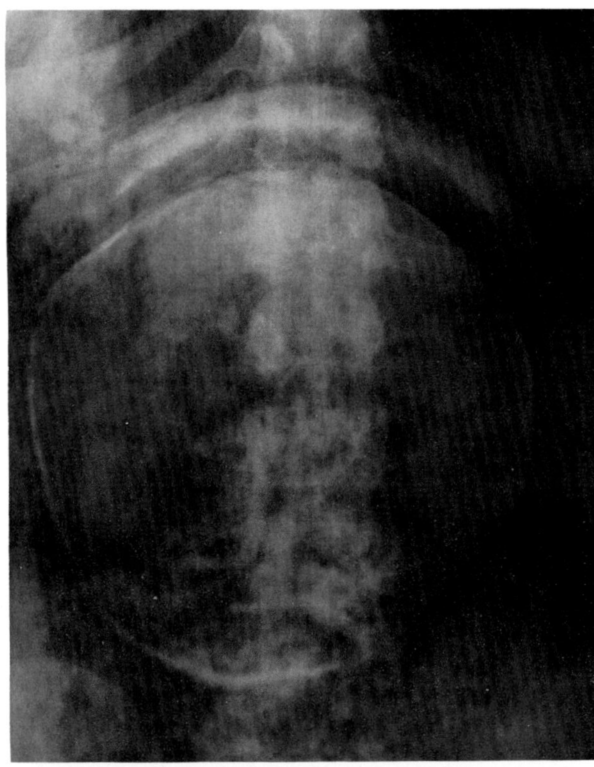

**Figure 11.6. Gas gangrene of the uterus.** Gas is evident in the wall of the uterus and in the surrounding soft tissues. (From Eisenberg,[7] with permission from the publisher.)

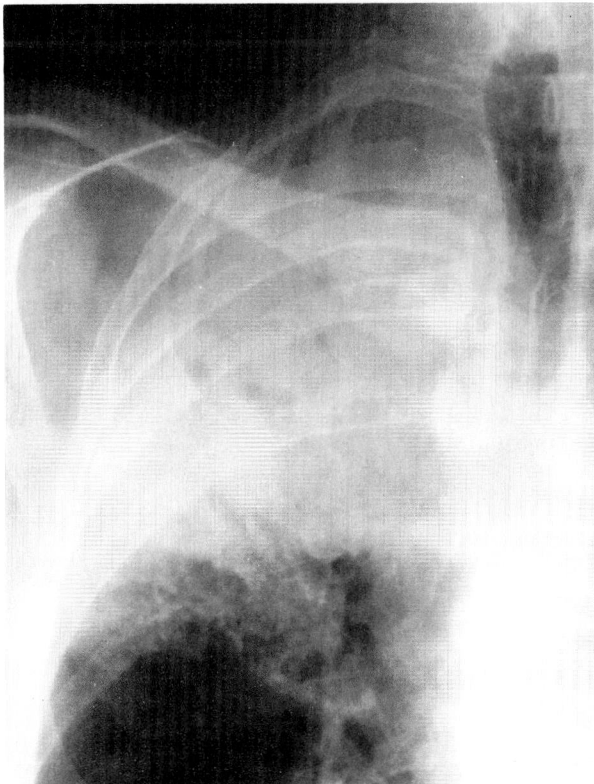

A

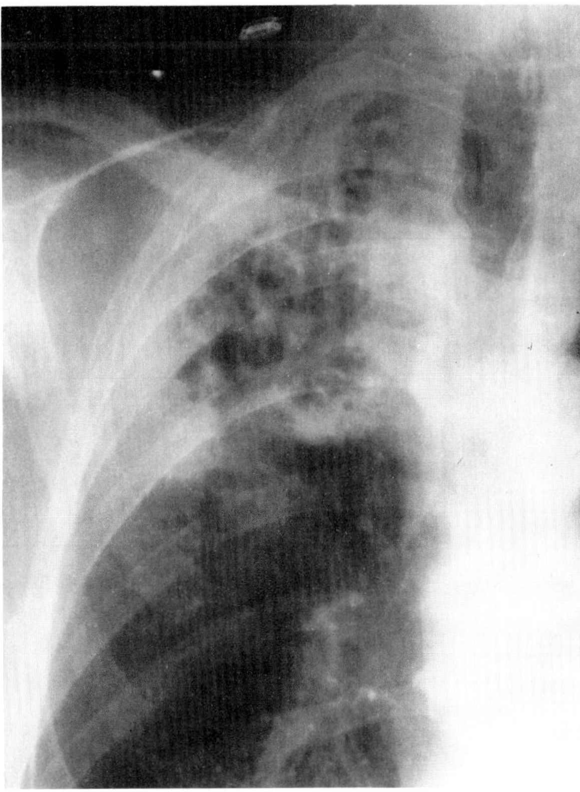

B

**Figure 11.7. Necrotizing clostridial pneumonia.** (A) An initial film demonstrates a large consolidation in the right upper lobe, with small areas of lung necrosis. (B) One week later, much more of the lung is necrotic. (From Landay, Christensen, Bynum, et al,[5] with permission from the publisher.)

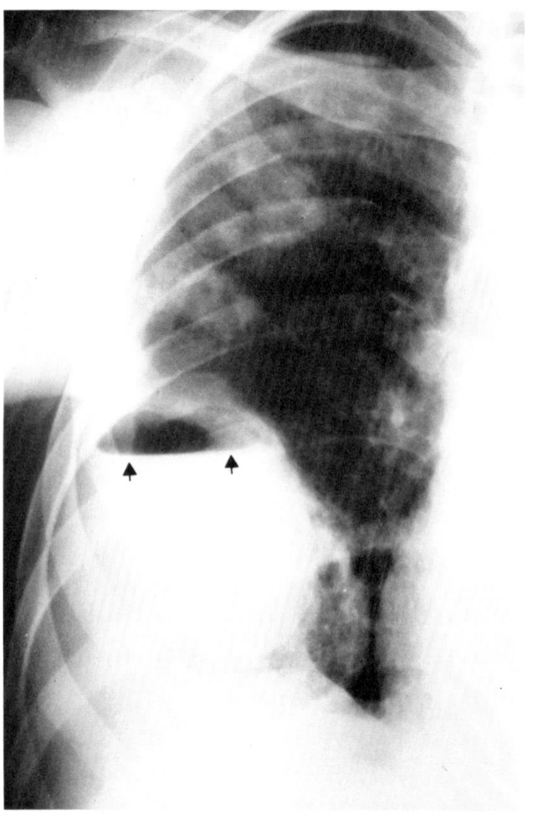

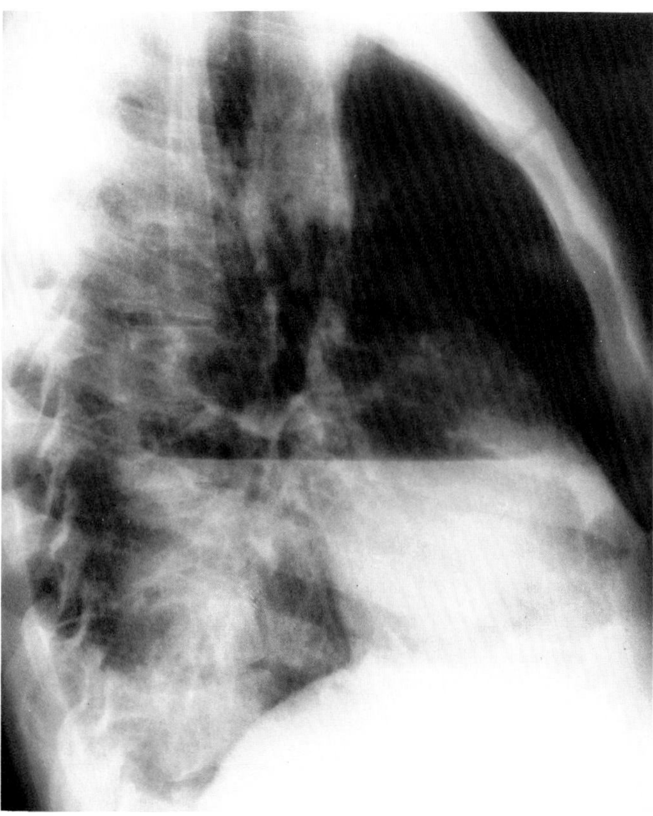

A B

**Figure 11.8. Clostridial pyopneumothorax.** (A) Frontal and (B) lateral views demonstrate a large empyema with an air-fluid level (arrows) representing a bronchopleural fistula. The patchy infiltrate in the right upper lung represents endobronchial spread. (From Landay, Christensen, Bynum, et al,[5] with permission from the publisher.)

Clostridial infection can complicate chronic cholecystitis and produce air within the lumen and wall of the gallbladder (see the section "Emphysematous Cholecystitis" in Chapter 53). Gas in the wall of the uterus and in pelvic soft tissues may develop in clostridial postabortal and puerperal sepsis. Pulmonary infection most commonly appears as a noncavitary pneumonia, which often develops necrosis despite antimicrobial therapy. Clostridial involvement of the pleura can rarely result in empyema or pyopneumothorax, with the gas formed by the infecting organisms.[3–5]

## REFERENCES


1. Evans KT, Walley RV: Radiological changes in tetanus. *Br J Radiol* 36:729–731, 1963.
2. Femi-Pearse D, Olowu AO: Myositis ossificans: A complication of tetanus. *Clin Radiol* 22:89–92, 1971.
3. Matousek V: Radiographic manifestations of clostridial infection encountered at Queen Elizabeth Hospital, Adeleide. *Australas Radiol* 13:387–398, 1969.
4. Goldberg NH, Rifkind D: Clostridial empyema. *Arch Intern Med* 115:421–425, 1965.
5. Landay MJ, Christensen EE, Bynum LJ, et al: Anaerobic pleural and pulmonary infections. *AJR* 134:233–240, 1980.
6. Cockshott WP, Middlemiss H: *Clinical Radiology in the Tropics*. Edinburgh, Churchill Livingstone, 1979.
7. Eisenberg RL: *Gastrointestinal Radiology: A Pattern Approach*. Philadelphia, JB Lippincott, 1983.

## Chapter Twelve
# Mycobacterial Diseases

## TUBERCULOSIS

### Primary Tuberculosis

Primary pulmonary tuberculosis has traditionally been considered a disease of children and young adults. However, with the dramatic decrease in the prevalence of tuberculosis (especially in children and young adults), primary pulmonary disease can develop at any age.

There are four basic radiographic patterns of pulmonary tuberculosis: lobar or segmental infection, lymphadenopathy, pleural effusion, and miliary involvement. These patterns may occur singly or in combination; because tuberculosis is a continuing process, the radiographic appearance varies with the stage of the disease.

The tuberculous infiltrate presents as a lobar or segmental air space consolidation that is usually homogeneous, dense, and well-defined. Primary tuberculous pneumonia may affect any lobe; it is essential to remember that the diagnosis cannot be excluded because the infection is not in the upper lobe. The pneumonia of primary tuberculosis typically clears slowly (usually from the periphery toward the hilum), unlike the relatively rapid clearing of an acute bacterial infection. Associated enlargement of the hilar or mediastinal lymph nodes is very common in a primary tuberculous infection. A combination of a focal parenchymal lesion and enlarged hilar or mediastinal nodes produces the classic primary complex, an appearance strongly suggestive of primary tuberculosis.

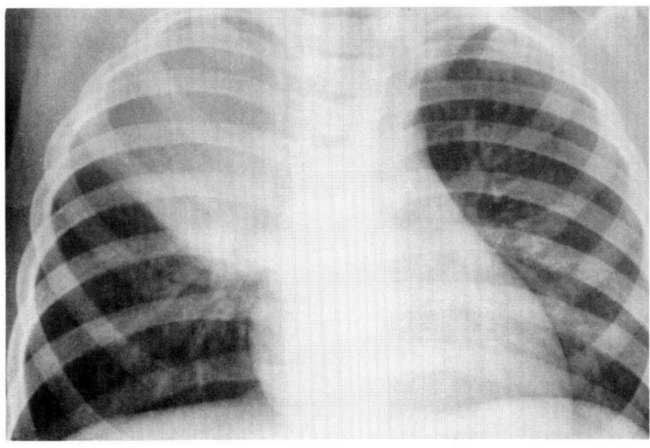

**Figure 12.1. Primary tuberculosis.** Consolidation of the right upper lobe.

A primary tuberculous infection may involve the pleura, especially in adults. Parenchymal disease and lymphadenopathy may be absent. Most primary tuberculous pleural effusions are unilateral and clear rapidly with treatment. Chronic effusion can lead to thickening of the pleura, which may become calcified and produce hazy, ill-defined densities overlying the hemithorax.

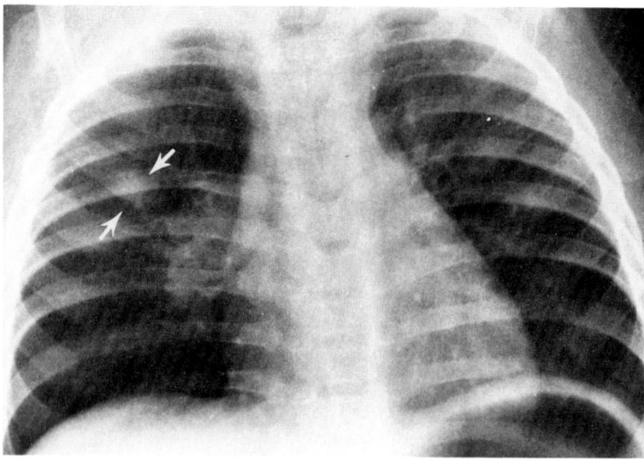

**Figure 12.2. Primary tuberculosis.** The combination of a focal parenchymal lesion (arrows) and enlarged right hilar lymph nodes produces the classic primary complex.

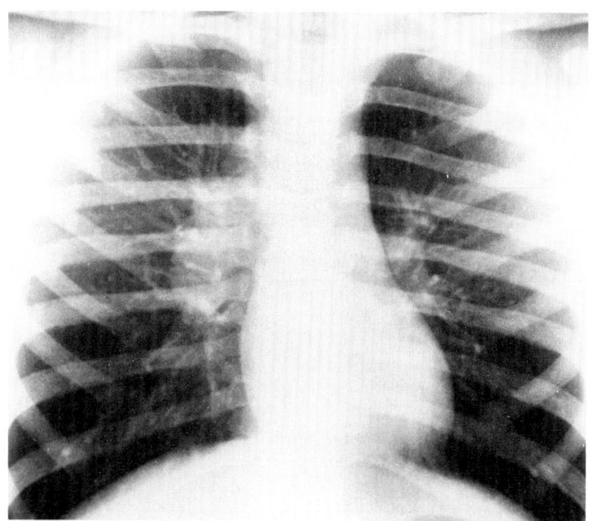

**Figure 12.3. Primary tuberculosis.** Enlargement of right hilar nodes without a discrete parenchymal infiltrate.

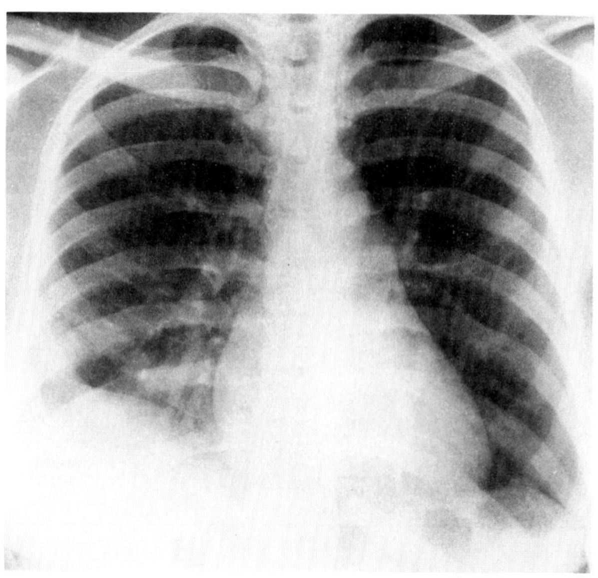

**Figure 12.4. Primary tuberculosis.** Unilateral right tuberculous pleural effusion without parenchymal or lymph node involvement.

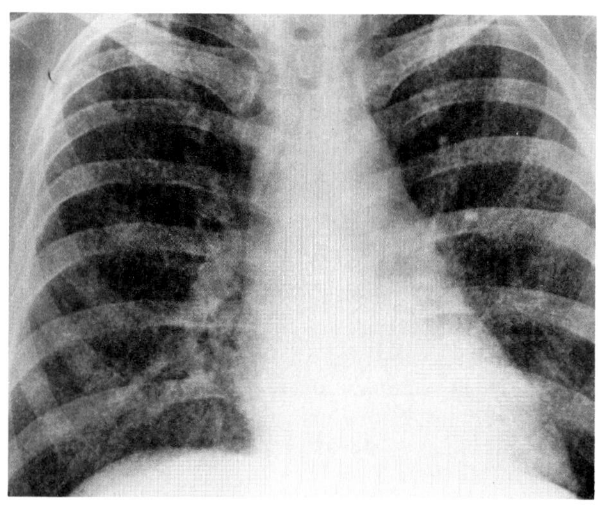

**Figure 12.5. Primary tuberculosis.** Miliary pattern.

Miliary tuberculosis represents hemotogenous dissemination of the disease. Innumerable fine discrete nodules distributed uniformly throughout both lungs produce a uniform granularity that may initially appear as a mild thickening of the interstitial markings. Pleural effusion and lymphadenopathy may occur. As the infection progresses, the miliary nodules coalesce and may cavitate. In the early stages of miliary disease, the chest radiograph may be normal even though the patient is ill with fever and hepatosplenomegaly. After proper treatment, miliary tuberculosis usually clears completely. Although multiple small calcifications may develop, this pattern is much more likely to be a sequela of histoplasmosis, coccidioidomycosis, or varicella pneumonia.

Tuberculous pneumonia that does not progress beyond the cellular exudate stage may resolve completely and leave a normal lung. However, if necrosis and caseation develop, some fibrous scarring occurs. Calcification may develop within both the parenchymal and the nodal lesions, and this may be the only residue of primary tuberculous infection on subsequent films. Even after treatment has begun, the involved lymph nodes may continue to enlarge and produce bronchial impression or erosion and obstructive atelectasis.

Primary pulmonary tuberculosis may respond poorly to therapy and continue to progress, especially in patients with an immunodeficiency or diabetes and in those receiving steriod therapy. The pneumonia may break down into multiple necrotic cavities or a single large abscess filled with caseous material. Rupture of the lesion permits spread of the infection to other lobes or segments, particularly in the opposite lung. Extension to the pleural surface can lead to pneumothorax and empyema.[1,2]

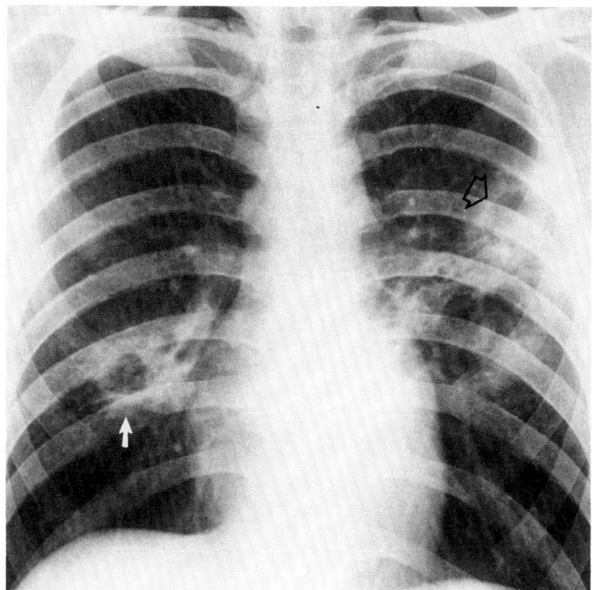

**Figure 12.6. Secondary tuberculosis.** Bilateral cavitary lesions with relatively thick walls (arrows).

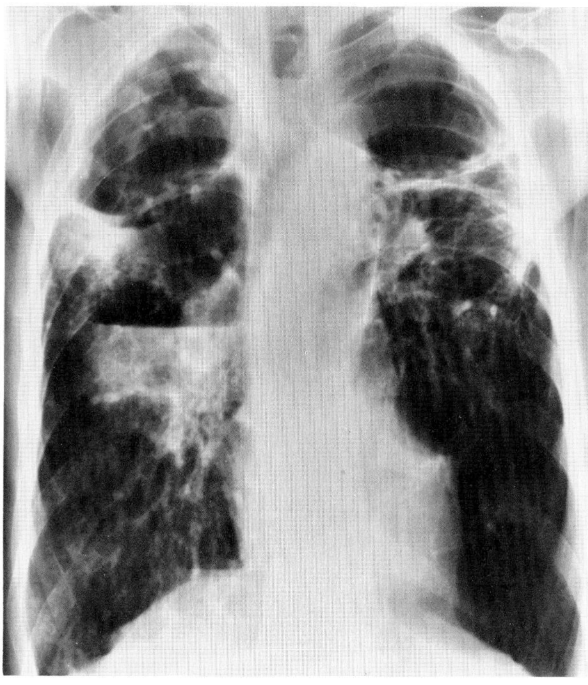

**Figure 12.7. Secondary tuberculosis.** Multiple large cavities with air-fluid levels within both upper lobes. Note the chronic fibrotic changes and upward retraction of the hila.

## Secondary (Reactivation) Tuberculosis

Secondary tuberculosis most commonly affects the upper lobes, especially the apical and posterior segments. The isolated involvement of the anterior segment of an upper lobe is rare. Bilateral, though often asymmetric, upper lobe disease is very common and almost diagnostic of reactivation tuberculosis. Because an apical lesion may be obscured by overlying clavicle or ribs, an apical lordotic view is often of value. Tuberculous involvement of the lower lobe is much less common. The superior segment of a lower lobe is the usual site; isolated basilar involvement is rare.

Secondary tuberculosis initially presents as a nonspecific hazy, poorly marginated alveolar infiltrate with a few nodules and a few fine dense lines. The infiltrate often radiates outward from the hilum or occurs in the periphery of the lung. As the disease progresses, additional ill-defined localized lesions are associated with pulmonary scarring, which distorts the normal vascular and mediastinal structures and eventually leads to retraction of the adjacent intralobar fissure and displacement of the hilum.

Because necrosis and liquefaction are part of the histopathologic process of secondary tuberculosis, cavitation is common. There is a wide variation in the appearance of tuberculous cavities, which may be thick- or thin-walled, large or small, single or multiple. Bilateral cavitation within an ill-defined, amorphous parenchymal reaction is likely to be tuberculous in origin. A thick wall (more than 2 mm) or an ill-defined inner wall with surrounding haze strongly suggests active disease. Fluid is often present within tuberculous cavities,

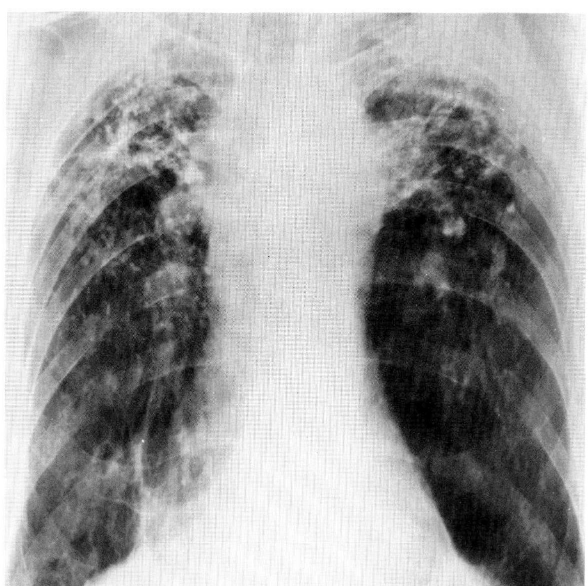

**Figure 12.8. Secondary tuberculosis.** Bilateral fibrocalcific changes at the apices. There is upward retraction of the hila.

though its presence or absence is unrelated to the prognosis. Because the presence of cavitation may influence patient management, conventional tomography is often used to better evaluate the tuberculous lesion. Tuberculous cavities are often difficult to distinguish from bullae. As a general rule, bullae have a uniformly thin, well-defined wall with no perifocal haze and tend to vary little in size over a prolonged period.

Secondary pulmonary tuberculosis heals slowly. Serial radiographs demonstrate decreased parenchymal infiltrates, progressive obliteration of cavities, fibrosis, and, often, calcification. Fibrous stranding and subsequent contraction of the scars causes loss of volume of the involved segment or lobe and a decrease in the size of the hemithorax. The trachea and other mediastinal structures are retracted to the involved side; in upper lobe disease, the hilum is elevated.

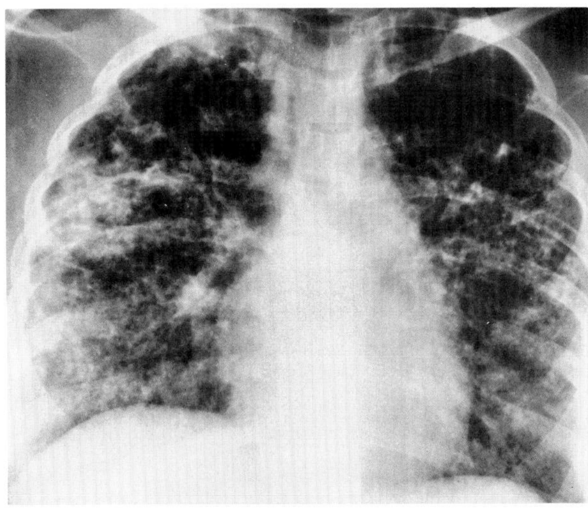

**Figure 12.9. Secondary tuberculosis.** Diffuse interstitial fibrosis pattern.

The development of pleural fluid in secondary tuberculosis, which usually reflects an empyema caused by the direct spread of the disease into the pleural cavity, differs from the benign pleural effusion that often occurs in primary disease. Pleural calcification frequently develops. A spontaneous air-fluid level in the pleural space indicates the development of a bronchopleural fistula. Other complications of secondary pulmonary tuberculosis include pneumothorax, bronchostenosis, broncholithiasis, pneumothorax, and the direct spread of infection to the chest wall.

It is difficult to radiographically determine the activity of secondary tuberculosis. An unchanged appearance of fibrosis and calcification on serial films is usually considered evidence of "healing" of the tuberculous process. Nevertheless, even densely calcified lesions can contain central areas of necrosis in which viable organisms can still be found even after long periods of apparent inactivity. Of course, new cavitation or an increasing amount of pulmonary infiltrate indicates active disease.[1,2]

## Tuberculoma

A tuberculoma is a sharply circumscribed parenchymal nodule, often containing viable tubercle bacilli, that can develop in either primary or secondary disease. Although the residual localized caseation may remain unchanged for a long period or permanently, a tuberculoma is potentially dangerous since it may break down at any time and lead to dissemination of the disease. Radiographically, tuberculomas appear as single or multiple pulmonary nodules, usually 1 to 3 cm in diameter; they can occur in any part of the lung but are more common in the periphery and in the upper lobes. The nodule is usually indistinguishable from a malignant mass and may be well-defined or have a vague, hazy, and indistinct

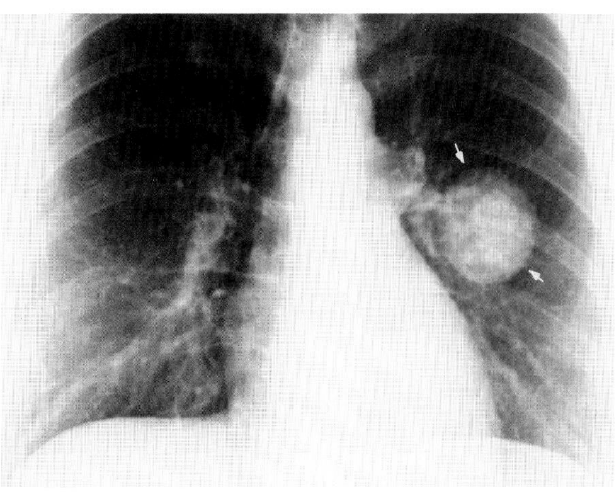

A

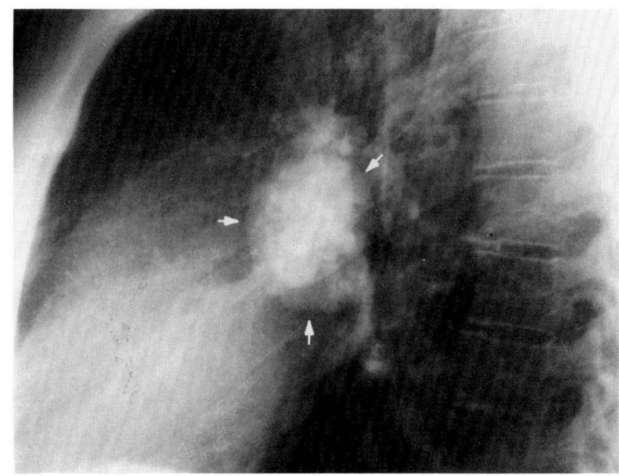

B

**Figure 12.10. Tuberculoma.** (A) Frontal and (B) lateral views of the chest show a large left-lung soft tissue mass (arrows) containing dense central calcification.

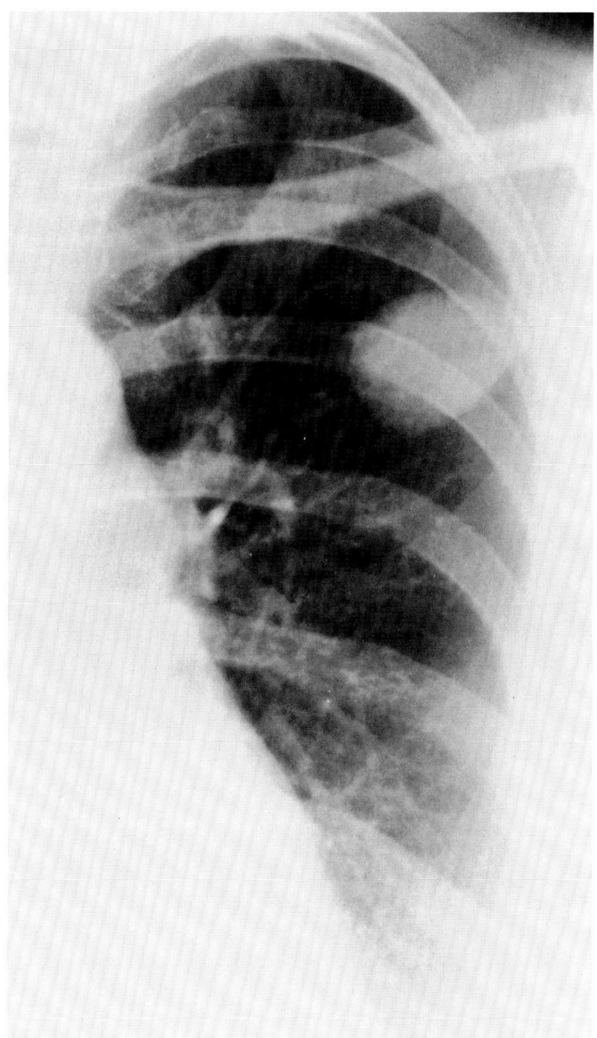

**Figure 12.11. Tuberculoma.** A single smooth and well-defined pulmonary nodule in the left upper lobe. In the absence of a central nidus of calcification, this appearance is indistinguishable from that of a neoplasm.

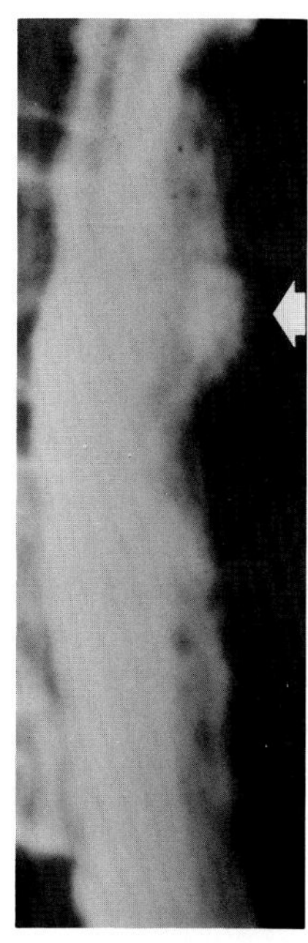

**Figure 12.12. Tuberculosis of the esophagus.** A large midesophageal ulcer (arrow) with surrounding inflammatory edema mimics carcinoma. Note the generalized irregularity, which represents diffuse ulcerative disease. (From Eisenberg,[7] with permission from the publisher.)

border. A central nidus of calcification (which may only be detectable on tomograms) strongly suggests that the lesion represents a tuberculoma; the lack of calcification, however, is of no diagnostic value.[1,2]

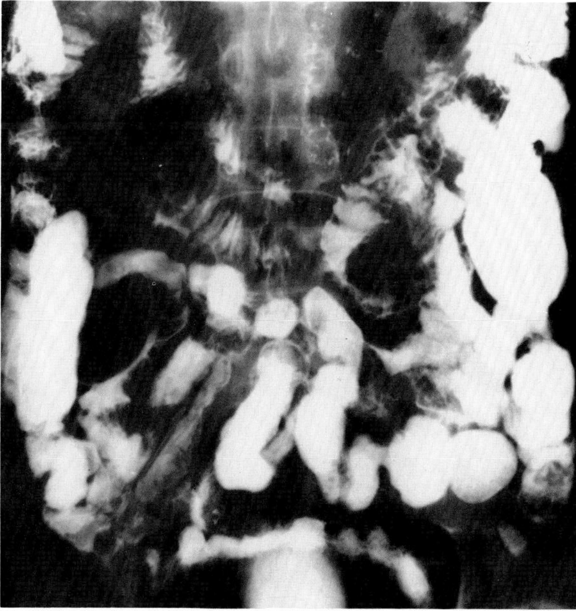

**Figure 12.13. Tuberculosis of the small bowel.** The irregular thickening of folds, segmental narrowing, and separation of bowel loops produce a pattern indistinguishable from that of Crohn's disease.

## Gastrointestinal Tuberculosis

Tuberculosis of the esophagus is rare and almost invariably secondary to terminal disease in the lungs. It can result from the swallowing of infected sputum, direct extension from laryngeal or pharyngeal lesions, or contiguous extension from caseous hilar lymph nodes or infected vertebrae; it can also be part of disseminated miliary disease.

The most common manifestation of tuberculosis of the esophagus is single or multiple ulcers. An intense fibrotic response often causes a narrowing of the esophageal lumen. Numerous miliary granulomas in the mucosal layer occasionally give the appearance of multiple nodules. Sinuses and fistulous tracts are common; their presence in a patient with severe pulmonary tuberculosis who has a midesophageal ulcer or a stricture in the region of the tracheal bifurcation should suggest a tuberculous cause.

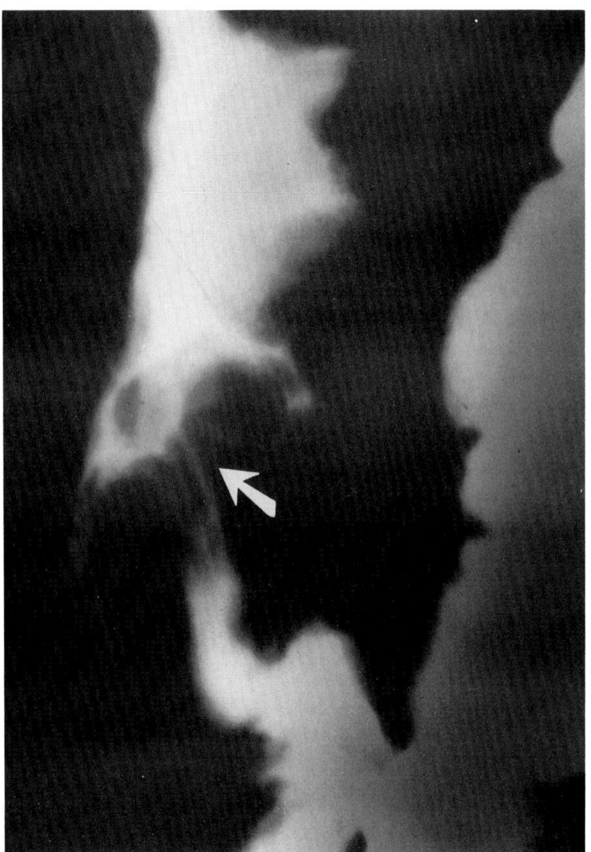

**Figure 12.14. Tuberculosis of the region of the ileocecal valve.** The valve (arrow) has an inverted umbrella-like appearance as a result of spasm. Note the severe narrowing and almost complete obliteration of the cecum. (From Eisenberg,[7] with permission from the publisher.)

In gastric tuberculosis both diffuse inflammation and fibrotic healing cause rigidity, narrowing, and even obstruction of the distal stomach (linitis plastica pattern) that mimic scirrhous carcinoma or lymphoma. Gastric ulcerations and fistulas between the antrum and the small bowel can simulate the radiographic appearance of involvement of the stomach by Crohn's disease.

Tuberculosis of the small bowel can produce a radiographic pattern indistinguishable from Crohn's disease, though the disease tends to be more localized than Crohn's disease and predominantly affects the ileocecal region. Tuberculosis typically causes generalized irregularity and distortion of small bowel folds, as well as separation of small bowel loops due to granulomatous infiltration of the bowel wall and mesentery. Fibrosis leads to rigidity of the bowel wall and narrowing of the lumen.

The healing of acute tuberculous inflammation can cause both shortening and narrowing of the terminal ileum and the characteristic pattern of a coned cecum. Thickening, straightening, and rigidity of the ileocecal valve can make the terminal ileum appear to empty directly into the stenotic ascending colon, and this causes nonopacification of the fibrotic, contracted cecum (Stierlin's sign). In some patients, tuberculous involvement of the ileocecal region may be difficult to distinguish from amebiasis. In tuberculosis, concomitant ileal changes (mucosal ulceration and nodularity, mural rigidity, luminal narrowing, fistulas) are often seen; in amebiasis, in contrast, the terminal ileum is not involved.

Colonic tuberculosis predominantly involves the ascending and the transverse colon, almost invariably in continuity with cecal disease. Segmental colonic involvement can occur, primarily in the sigmoid region. In the early stages of the disease, superficial or deep mucosal ulcerations can sometimes be identified radiographically, though intense spasm and irritability often prevent adequate filling of the involved portion of colon. In chronic stages, reparative healing with fibrosis leads to distortion, rigidity, and narrowing of the lumen. Segmental involvement can produce an annular ulcerating lesion mimicking colonic carcinoma. Indeed, tuberculosis should be considered a strong diagnostic possibility whenever segmental colonic narrowing develops in a young adult who originates from an area in which the disease is endemic, especially if there is chest radiograph or skin test evidence of active or chronic tuberculosis.[3–7]

## Lymph Nodes

The most common site of lymph node involvement in tuberculosis is the hilum draining the pulmonary site of the initial infection. Noncalcified nodes appear as smooth or lobulated soft tissue densities that are often unilateral and indistinguishable from adenopathy or masses with other causes. With healing, the nodes become smaller and often demonstrate amorphous or concentric calcification.

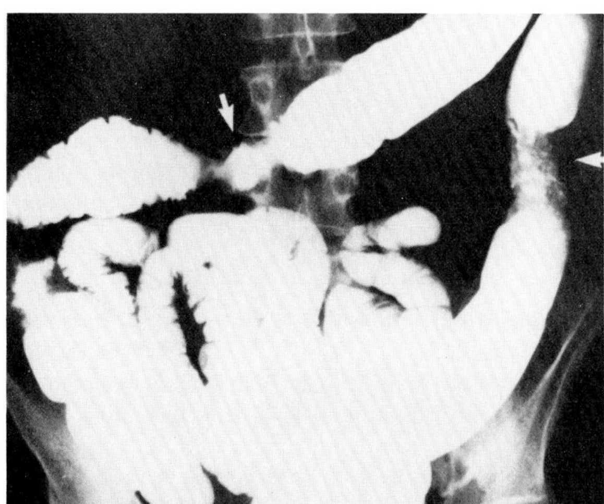

**Figure 12.15. Tuberculosis of the colon.** There are two sharply demarcated areas of segmental colonic inflammation and narrowing. (From Carrera, Young, and Lewicki,[6] with permission from the publisher.)

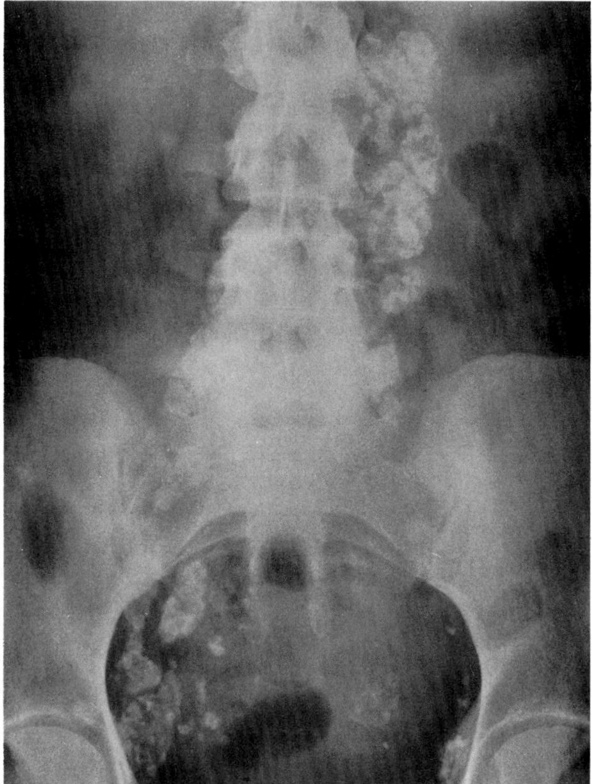

**Figure 12.16. Tuberculous lymph nodes.** Diffuse calcification of lymph nodes lying along the course of the aorta and the iliac arteries.

Tuberculous infection can result in mottled calcification of mesenteric lymph nodes, though this appearance more commonly is the result of histoplasmosis. In the past, cervical adenitis (scrofula) due to tuberculosis often produced large soft tissue masses of nodes matted together. With healing, these cervical lymph nodes tended to become densely calcified in a typically mottled or stippled pattern.

## Genitourinary Tuberculosis

The hematogenous spread of tuberculosis may lead to the development of small granulomas scattered in the cortical portion of the kidneys, usually in a juxtaglomerular location. Initially, the excretory urogram is normal in spite of positive urine cultures for tuberculous organisms. The spread of infection to the renal pyramid causes an ulcerative, destructive process in the tips of the papillae. The resulting irregularity of the involved calyces, which is usually the first urographic sign of tuberculosis, is indistinguishable from papillary necrosis from other causes. As the papillary lesions progress, they enlarge and form irregular cavities that may communicate with the collecting system. Fibrosis and stricture formation in the pelvocalyceal system cause narrowing or obstruction of the infundibulum to the affected calyx. Cortical scarring and parenchymal atrophy may simulate the appearance of chronic bacterial pyelonephritis. Intrarenal tuberculous granulomas that do not communicate with the collecting system may mimic neoplastic masses.

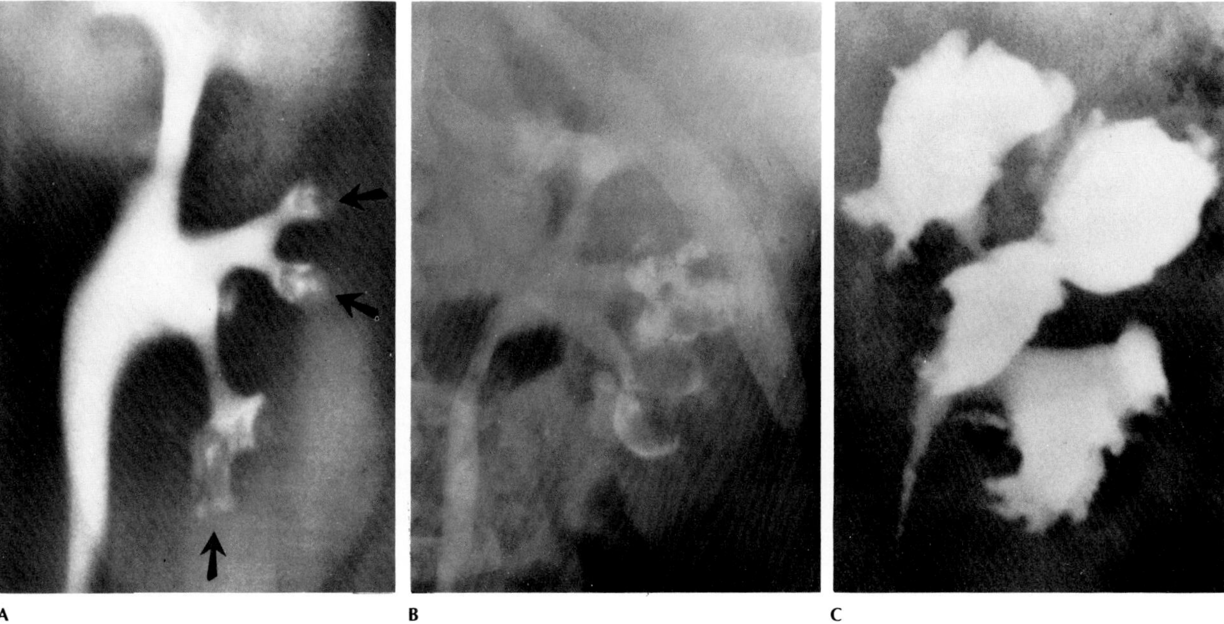


**Figure 12.17. Stages of papillary destruction by active tuberculosis.** (A) Early. (B) Intermediate. (C) Advanced. (From Tonkin and Witten,[8] with permission from the publisher.)

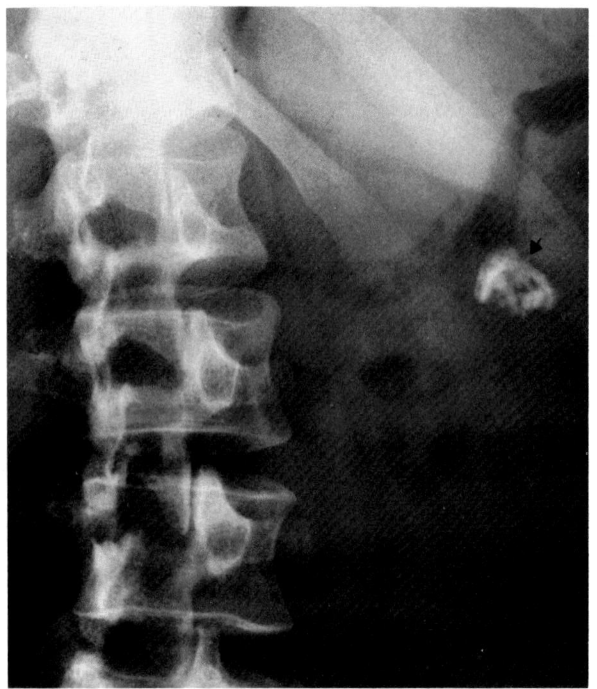

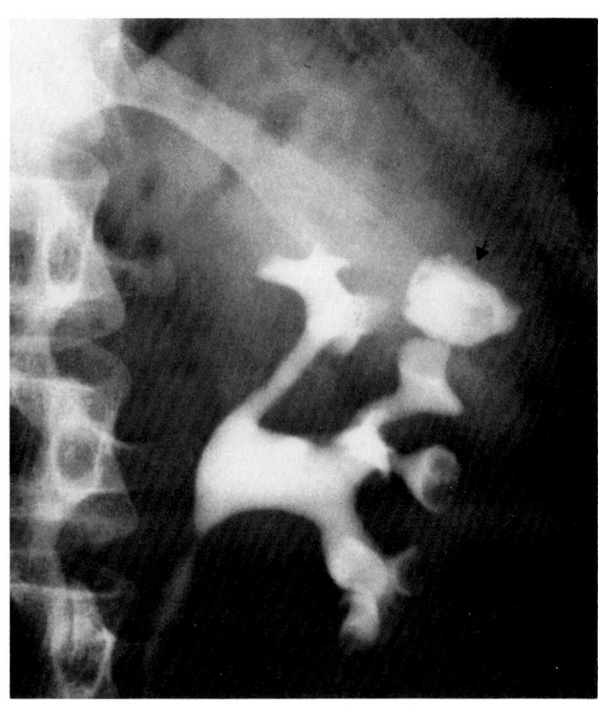

A B

**Figure 12.18. Renal tuberculosis.** (A) A plain abdominal film demonstrates calcification (arrow) in the region of the left kidney. (B) A film from an excretory urogram shows that the calcification is located within an irregular cavity (arrow) that communicates with the collecting structures and represents an area of parenchymal necrosis with erosion into an adjacent calyx.

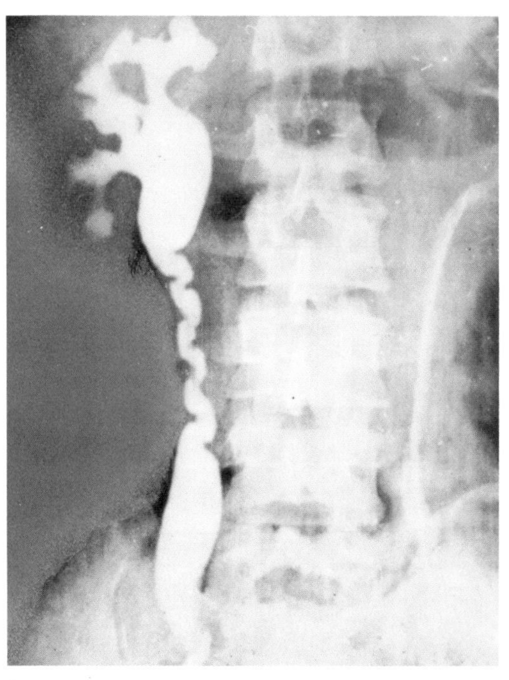

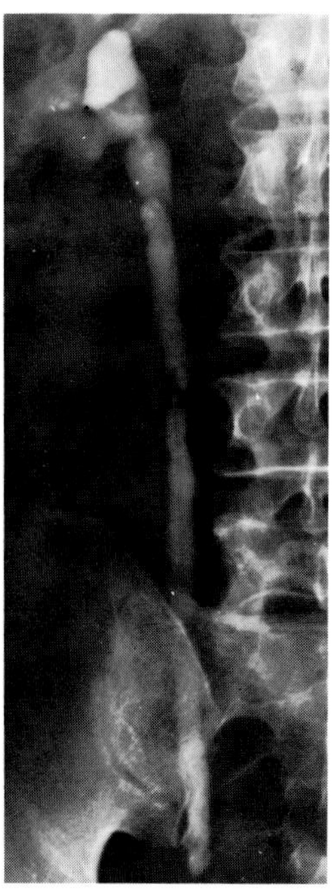

A B

**Figure 12.19. Ureteral tuberculosis.** (A) Segmental areas of dilatation and constriction produce a corkscrew, or beaded, pattern. (B) Thickening and fixation of the ureter (pipestem ureter). (A, from Ney and Friedenberg,[15] with permission from the publisher; B, from Witten, Myers, and Utz,[16] with permission from the publisher.)

In the ureter, tuberculosis typically causes multiple ulcerations of short or long segments. These ulcerations result in a ragged, irregular appearance of the ureteral wall. As the disease heals, there are usually multiple areas in which ureteral strictures alternate with dilated segments, and this produces a beaded, or corkscrew, appearance. In advanced cases, the wall of the ureter may become thickened and fixed with no peristalsis; this results in a pipestem ureter that runs a straight course toward the bladder.

Tuberculous involvement of the urinary bladder may produce mural irregularities simulating carcinoma or,

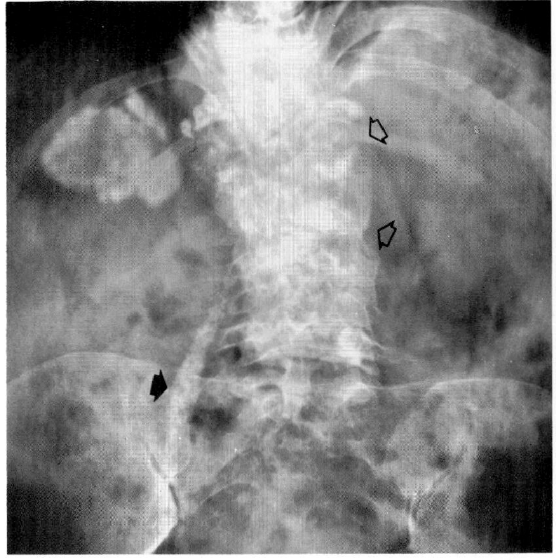


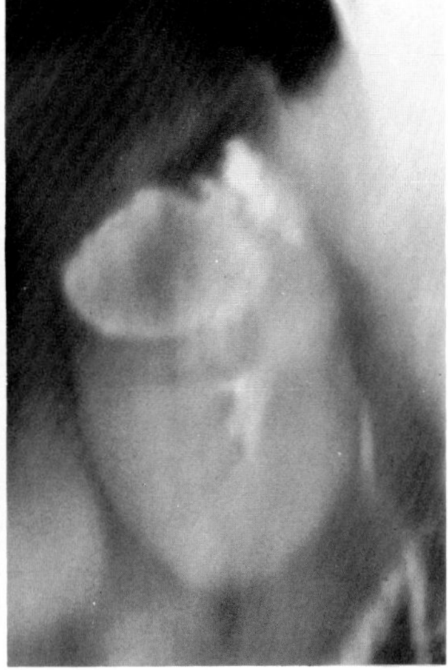


**Figure 12.20. Calcified tuberculoma.** (A) A plain film and (B) a nephrotomogram demonstrate a large calcified tuberculoma involving the upper pole of the right kidney. Note the diffuse destructive changes in the dorsolumbar spine (open arrows) and the calcified right psoas abscess (solid arrow). (From Tonkin and Witten,[8] with permission from the publisher.)

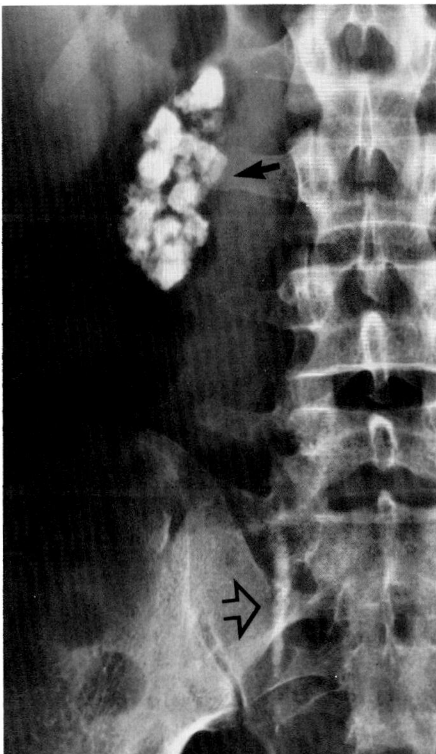

**Figure 12.21. Tuberculous autonephrectomy.** A plain film shows coarse irregular calcification that retains a reniform shape (black arrow). Note also the tuberculous calcification of the distal right ureter (open arrow). (From Tonkin and Witten,[8] with permission from the publisher.)

more commonly, may produce a smooth, contracted bladder with a thickened wall.

Plain abdominal radiographs may demonstrate calcification of the renal parenchyma. Flecks of calcification in multiple tuberculous granulomas can present as nephrocalcinosis. With progressive disease, gross amorphous and irregular calcifications can develop. Eventually, the entire nonfunctioning renal parenchyma may be replaced by massive calcification (autonephrectomy). Calcification of the ureter and bladder is infrequent and generally detectable only after extensive tuberculous changes are already evident in the kidney.

Tuberculosis can lead to irregular calcification of the vas deferens and the seminal vesicles, presenting a pattern indistinguishable from the calcification of these same structures that is more commonly seen in diabetic patients. Pelvic calcifications in the midline near the pubic symphysis may be caused by tuberculous prostatitis.[8]

## Skeletal System

Tuberculous osteomyelitis most commonly involves the thoracic and lumbar spine. The infection tends to begin in the anterior part of the vertebral body, adjacent to the intervertebral disk. Irregular, poorly marginated bone destruction within the vertebral body is often associated with a paravertebral abscess, an accumulation of purulent material that produces a fusiform soft tissue mass about the vertebra. The shadow of a paravertebral abscess is best seen in the dorsal spine; in the lumbar region, the only evidence of abscess formation may be outward bulging of the psoas muscle shadow. The spread of tuberculous osteomyelitis causes a narrowing of the adjacent intervertebral disk and the extension of infection and bone destruction across the disk to involve the contiguous vertebral body. Unlike pyogenic infection, tuberculous osteomyelitis is rarely associated with periosteal reaction or bone sclerosis. In the untreated patient, progressive vertebral collapse and anterior wedging lead to a characteristic sharp kyphotic angulation and gibbus deformity. Healed lesions may demonstrate mottled calcific deposits in a paravertebral abscess and moderate recalcification and sclerosis of the affected bones.

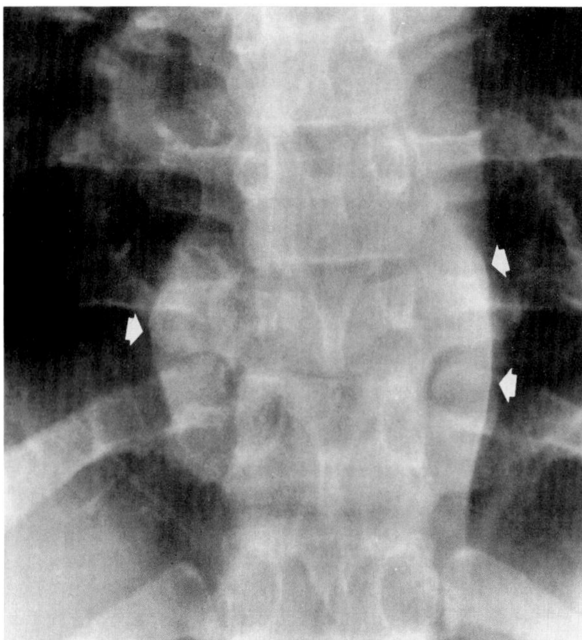

**Figure 12.22. Tuberculous osteomyelitis of the spine.** There is poorly marginated destruction along with loss of the superior and inferior end plates of the T9 vertebral body. Note the large paravertebral abscess that produces a fusiform soft tissue mass about the vertebrae (arrows).

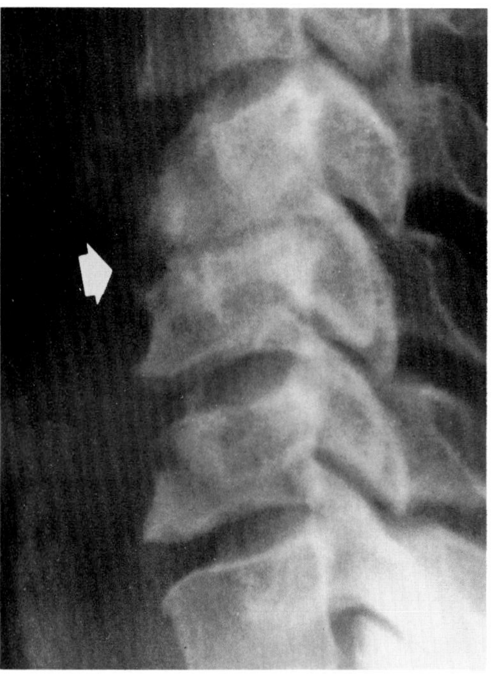

**Figure 12.23. Tuberculous osteomyelitis of the cervical spine.** Narrowing of the intervertebral disk space is accompanied by diffuse bone destruction involving the adjacent vertebrae. Note the lack of sclerotic reaction.

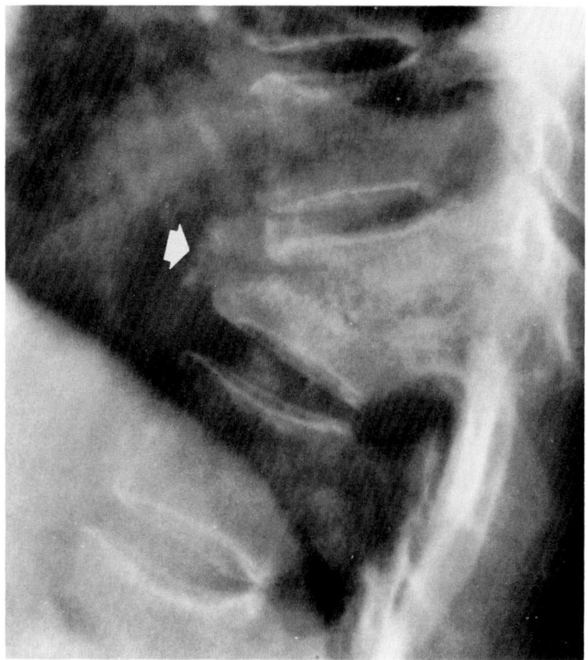

A

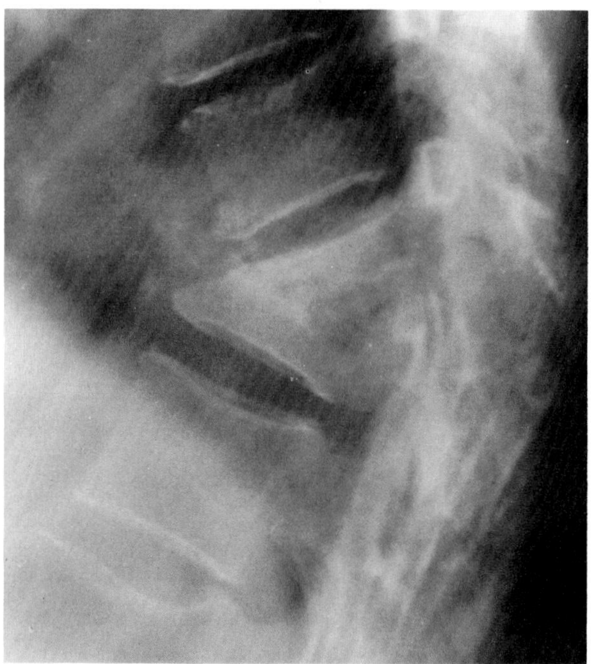

B

**Figure 12.24. Tuberculous osteomyelitis of the dorsal spine.** (A) An initial film demonstrates vertebral collapse and anterior wedging of adjacent midthoracic vertebrae (arrow). The residual intervertebral disk space can barely be seen. (B) Several months later there is virtual fusion of the collapsed vertebral bodies, producing a characteristic sharp kyphotic angulation (gibbus deformity).

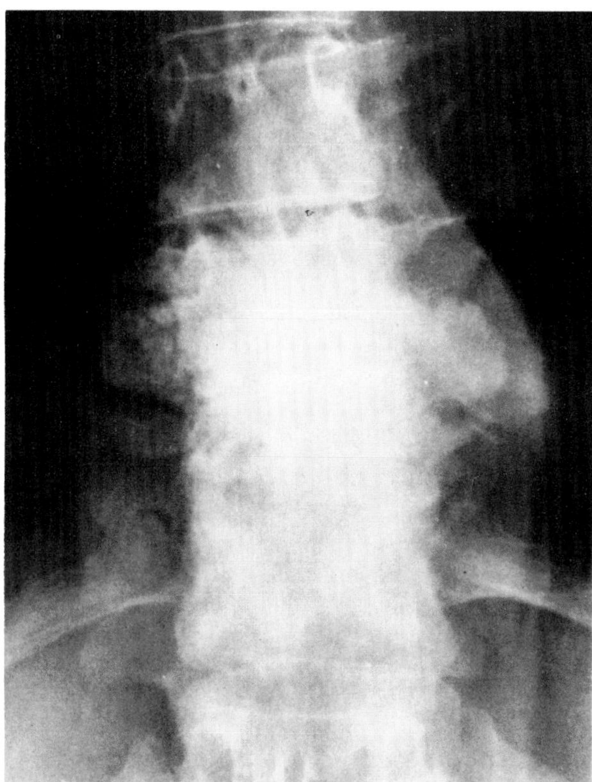

**Figure 12.25. Tuberculous osteomyelitis.** Dense calcification within the large paraspinal abscess virtually obscures the underlying vertebral destruction.

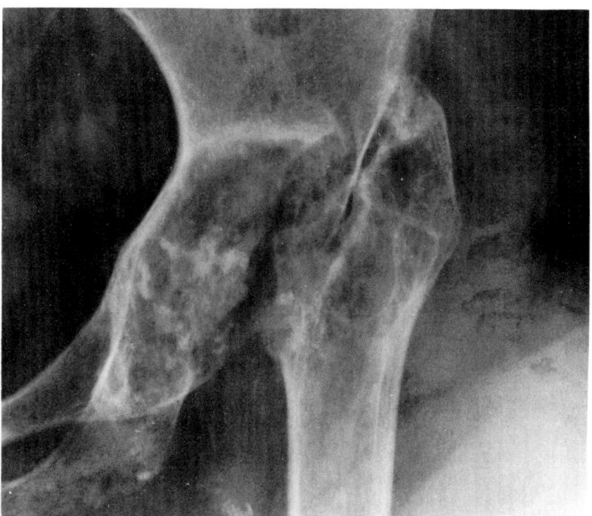

**Figure 12.26. Tuberculosis of the hip.** There has been essentially complete destruction of the femoral head with superior displacement of the residual femur. Calcific debris within the cold abscess has almost replaced the hip joint.

Tuberculosis can produce a low-grade chronic infection of the long bones that appears radiographically as a generally destructive lytic process with minimal or no periosteal reaction. The spectrum of radiographic appearances is wide, varying from localized, well-circumscribed, expansile lesions to diffuse, uniform, honeycomb-like areas of destruction that are often associated with pathologic fractures. Chronic draining sinuses may develop, especially in children.

Involvement of the bones of the hands and feet occurs rarely, primarily in children. This infective dactylitis (spina ventosa) produces characteristic expansion of a phalanx with irregular destruction of the bone architecture; periosteal reaction is usually absent. This expanded appearance must be differentiated from syphilitic dactylitis, which is also seen in children but in which the bone is thickened as a result of periosteal calcification that forms a dense shell around the lesion.

Tuberculous arthritis preferentially involves the large and medium-sized joints, especially the hip and knee. The sacroiliac, shoulder, elbow, and ankle joints are less frequently affected. The earliest radiographic evidence of tuberculous arthritis is periarticular swelling caused by a combination of synovial hypertrophy and joint effusion. Chronic synovitis leads to hyperemia and periarticular osteoporosis along with blurring and even loss

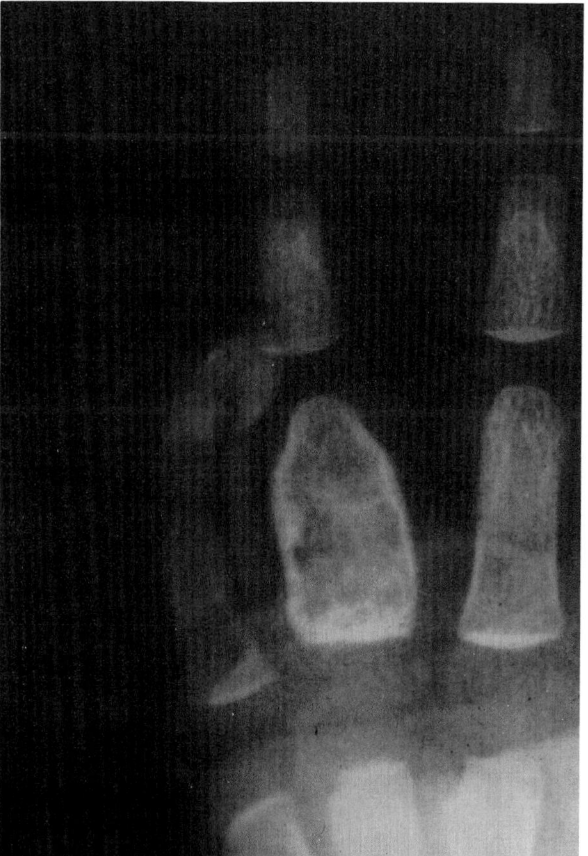

**Figure 12.27. Tuberculous dactylitis.** Typical expansion of a phalanx along with irregular destruction of bone architecture. Note the absence of periosteal reaction, which differentiates this appearance from that of syphilitic dactylitis. (From Chapman, Murray, and Stoker,[9] with permission from the publisher.)

of the normally sharp subarticular outline. Secondary marginal erosions may mimic those of rheumatoid arthritis. As the disease progresses, there is often evidence of destruction of articular cartilage and subarticular bone. In the weight-bearing joints, there is a tendency for the preservation of cartilage at points of maximum weight-bearing. Therefore, radiographically evident narrowing of the joint space is often a relatively late finding. Gross disorganization of the joint structures along with the disappearance of articular cartilage, the ragged destruction of the bone ends, and sequestra formation may eventually develop, though little reactive sclerosis occurs.

Tuberculous arthritis must be differentiated from pyogenic infection and rheumatoid arthritis. In chronic pyogenic arthritis, the joint space is diffusely narrowed without the tendency, seen in tuberculosis, for the preservation of articular cartilage at points of maximum weight-bearing. Unlike tuberculosis, pyogenic disease often results in sclerotic reaction and hypertrophic spurs along the edges of the bone, while marginal erosions are unusual. Bony ankylosis, which frequently develops after the destruction of articular cartilage in chronic pyogenic infection, is rarely seen with tuberculosis. Unlike rheumatoid arthritis, in which diffuse joint involvement is characteristic, tuberculous arthritis rarely involves more than a single joint at any one time.[9]

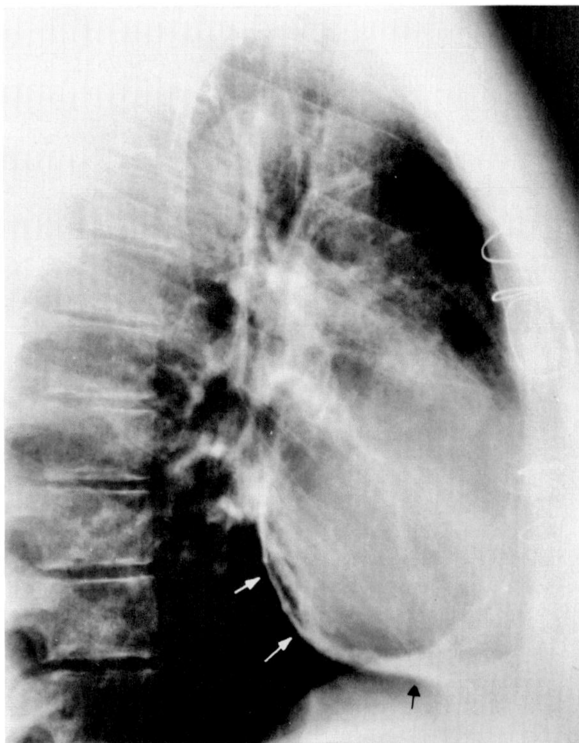

**Figure 12.28. Tuberculous pericarditis.** A lateral view of the chest demonstrates dense plaques of pericardial calcification (arrows) in a patient with chronic constrictive pericarditis due to tuberculosis.

## Tuberculous Pericarditis

Tuberculous pericarditis is usually the result of the spread of infection to the pericardium from mediastinal lymph nodes or from contiguous segments of lung. Radiographically, pericarditis due to tuberculosis produces a generalized enlargement of the cardiac silhouette, often with calcification, that is indistinguishable from that of pericarditis due to other causes. Ultrasound, radionuclide scanning, or computed tomography may be used to assess the thickness of the pericardium and confirm the presence of pericardial fluid.[1]

## Tuberculous Peritonitis

The spread of tuberculosis to the peritoneum incites the production of a large amount of ascitic fluid. This is easily detectable on plain abdominal radiographs as a general abdominal haziness (ground-glass appearance) and an increased soft tissue density that shifts to the pelvis on upright films. On barium examination, ascitic fluid causes the separation of small bowel loops.[1]

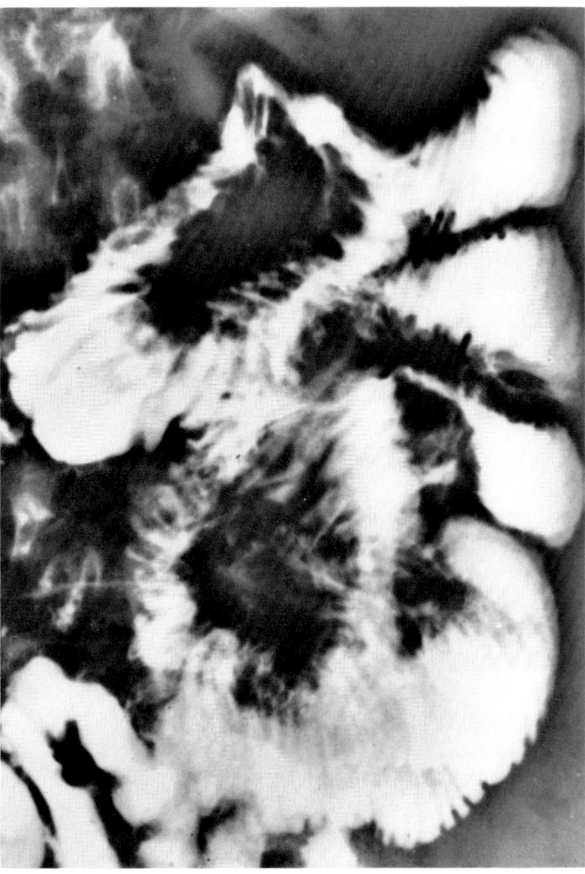

**Figure 12.29. Tuberculous peritonitis and enteritis.** There is separation of bowel loops, irregular narrowing, fold thickening, and angulation.

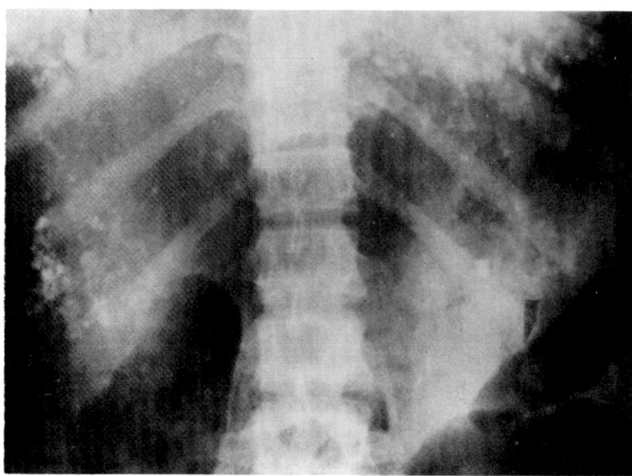

**Figure 12.30. Tuberculosis in the liver and spleen.** Diffuse small calcifications represent healed foci of tuberculosis in the liver and spleen. (From Darlak, Moskowitz, and Kattan,[17] with permission from the publisher.)

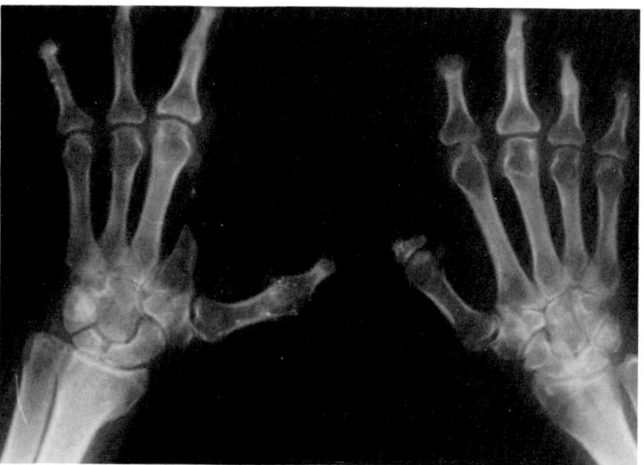

**Figure 12.31. Leprosy.** Views of the hands show severe phalangeal resorption and some examples of typical pencil-like configurations. (Courtesy of Robert R. Jacobson, Pearl Mills, and Tanya Thomassie.)

## Liver and Spleen

Healed foci of granulomatous disease secondary to tuberculosis (and, more frequently, histoplasmosis) are the most common intrahepatic calcifications. These calcifications tend to be small (1 to 3 cm), multiple, dense, and discrete and to be scattered throughout the liver. Tuberculosis sometimes can produce moderately large, solidly calcified granulomas or nodular, popcorn-like, or even laminated calcifications.[7]

## LEPROSY

Leprosy (Hansen's disease) is a chronic granulomatous mycobacterial infection that involves superficial tissues, especially the skin, peripheral nerves, and nasal mucosa. The typical radiographic abnormalities in the hands and feet reflect severe neurotrophic changes due to an insensitivity to pain that allows repeated trauma and infection to go untreated.

The classic appearance of leprosy is the slowly progressive resorption of bone. In the hands, this process begins at the terminal tufts of the distal phalanges and extends proximally until the proximal phalanges are destroyed. In the feet, the destructive changes usually begin at the metatarsophalangeal joints and proceed in both directions. The concentric resorption of the metatarsals produces a typical "pencil-line" tapering of the distal ends. Complete resorption of the metatarsals leaves the toes apparantly dangling in space. Acute osteomyelitis in the fingers and feet is frequent and presents the same radiographic appearance as other infections. Because of the insensitivity to pain, the bone changes are usually advanced on initial examinations and may show extensive destruction, periostitis, fragmentation, and sequestration. Neurotrophic changes may cause complete joint disorganization of the Charcot type. Infrequently, the granulomatous lesions of leprosy produce focal areas of bone destruction, which tend to develop a sclerotic rim during the healing phase.

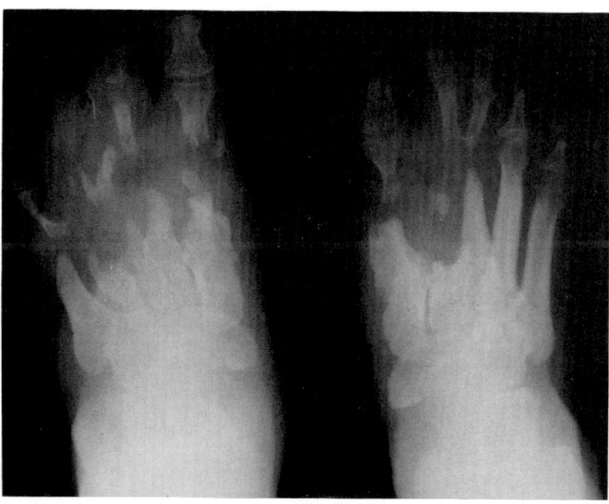

**Figure 12.32. Leprosy.** Views of the feet show the most severe changes at the metatarsophalangeal joints along with marked bone destruction and pencil-like resorption. (Courtesy of Robert R. Jacobson, Pearl Mills, and Tanya Thomassie.)

An uncommon, though virtually pathognomonic, radiographic sign of leprosy is the calcification of nerves in the distal extremities.[10,11]

## OTHER MYCOBACTERIAL INFECTIONS

Atypical mycobacteria may produce pulmonary infections that are clinically and radiographically indistinguishable from tuberculosis. The lesions often form thin-walled cavities with minimal surrounding parenchymal

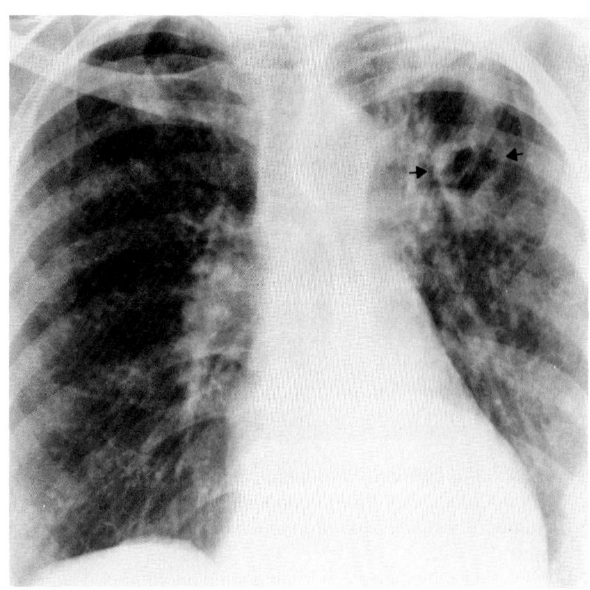

**Figure 12.33. Atypical mycobacterial infection.** Cavitary lesion (arrows) in the left upper lobe. The wall of the cavity is mildly irregular, and there is minimal parenchymal disease.

disease. Pleural effusion and hilar adenopathy are much less common than in tuberculosis. Patients with an atypical mycobacterial infection have a negative tuberculin test and do not respond to antituberculous therapy.

In children, atypical mycobacteria can rarely produce osteomyelitis, which can vary from an indolent, chronic local bone lesion to widespread dissemination with a fulminant course.[12–14]

## REFERENCES


1. Palmer PES: Pulmonary tuberculosis—Usual and unusual radiographic presentations. *Semin Roentgenol* 14:204–243, 1979.
2. Juhl JH: *Essentials of Roentgen Interpretation*, Philadelphia, JB Lippincott, 1981.
3. Balthazar EJ, Bryk D: Segmental tuberculosis of the colon: Radiographic features in seven cases. *Gastrointest Radiol* 5:75–80, 1980.
4. Ito J, Kobayashi S, Kasugai P: Tuberculosis of the esophagus. *Am J Gastroenterol* 65:454–456, 1976.
5. Thoeni, RF, Margulis AR: Gastrointestinal tuberculosis. *Semin Roentgenol* 14:283–294, 1979.
6. Carrera GS, Young S, Lewicki AM: Intestinal tuberculosis. *Gastrointest Radiol* 1:147–155, 1976.
7. Eisenberg RL: *Gastrointestinal Radiology: A Pattern Approach*. Philadelphia, JB Lippincott, 1983.
8. Tonkin AK, Witten DM: Genitourinary tuberculosis. *Semin Roentgenol* 14:305–318, 1979.
9. Chapman M, Murray RO, Stoker BJ: Tuberculosis of the bones and joints. *Semin Roentgenol* 14:266–282, 1979.
10. Enna CD, Jacobsen RR, Rausch RO: Bone changes in leprosy. *Radiology* 100:295–306, 1971.
11. Newman H, Casey B, DuBois JJ, et al: Roentgen features of leprosy in children. *AJR* 114:402–410, 1972.
12. Anderson DIT, Grech P, Townshend RH, et al: Pulmonary lesions due to opportunist mycobacteria. *Clin Radiol* 26:461–469, 1975.
13. Curry FJ: Atypical acid-fast mycobacteria. *N Engl J Med* 272:415–417, 1965.
14. Keller RH, Runyon EH: Mycobacterial diseases. *AJR* 92:528–539, 1964.
15. Ney C, Friedenberg RM: *Radiographic Atlas of the Genitourinary System*. Philadelphia, JB Lippincott, 1981.
16. Witten DM, Myers GH, Utz DC: *Emmett's Clinical Urography*. Philadelphia, WB Saunders, 1977.
17. Darlak JJ, Moskowitz M, Kattan KR: Calcifications in the liver. *Radiol Clin North Am* 18:209–219, 1980.

# Chapter Thirteen
# Spirochetal Diseases

## SYPHILIS

Syphilis is a chronic sexually transmitted systemic infection caused by the spirochete *Treponema pallidum*. In most instances, radiographic abnormalities are manifestations of the tertiary stage of the disease.

Cardiovascular syphilis primarily involves the ascending aorta, which may become aneurysmally dilated and often demonstrates linear calcification of the wall. Because calcification of the ascending aorta is relatively unusual in atherosclerotic disease, the finding has in the past been considered pathognomonic of syphilis. However, tertiary syphilis is now so rare that it can be assumed that most patients with calcification in the ascending aorta actually have advanced atherosclerotic disease. Syphilitic aortitis often involves the aortic valvular ring and produces aortic regurgitation with enlargement of the left ventricle. Stenosis of the coronary ostia is common and may be demonstrated by coronary arteriography.

Syphilitic involvement of the skeletal system most commonly produces radiographic findings of chronic osteomyelitis, which usually affects the long bones and skull. The destruction of bone incites a prominent periosteal reaction, and dense sclerosis is the most outstanding feature. Unlike pyogenic osteomyelitis, syphilitic osteomyelitis rarely causes sequestra. Gumma formation causes an ill-defined lytic area that is surrounded by an

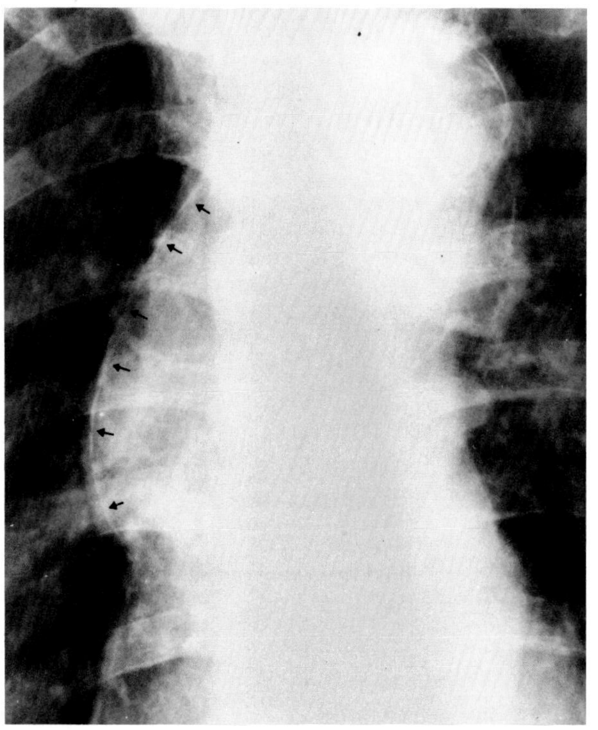

**Figure 13.1. Syphilitic aortitis.** There is aneurysmal dilatation of the ascending aorta with extensive linear calcification of the wall (arrows). Some calcification is also seen in the distal aortic arch.

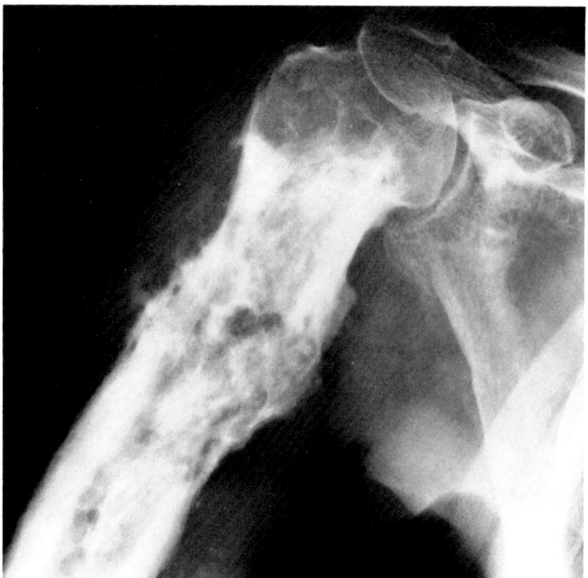

**Figure 13.2. Syphilitic osteomyelitis.** A view of the proximal humerus demonstrates diffuse lytic bone destruction with reactive sclerosis and periosteal new bone formation.

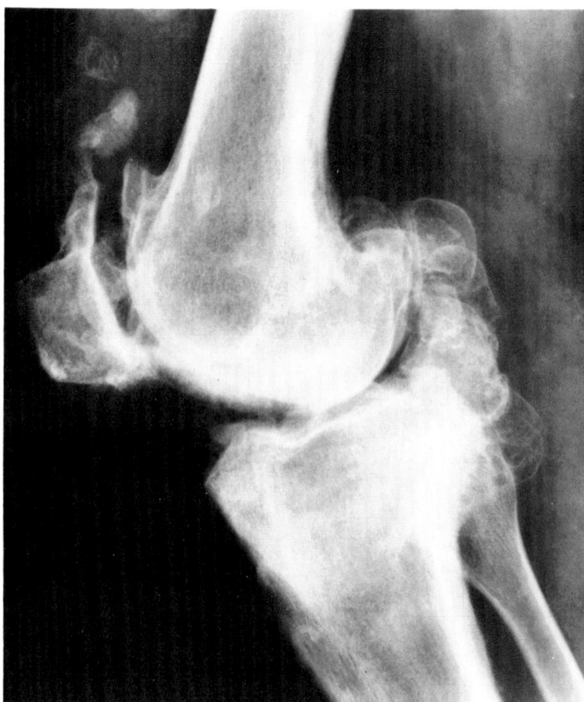

**Figure 13.3. Neurotrophic joint disease (Charcot's joint) in syphilis.** A lateral view of the knee shows disorganization of the joint with bone erosions, reactive sclerosis, soft tissue and ligamentous calcifications, and free joint bodies.

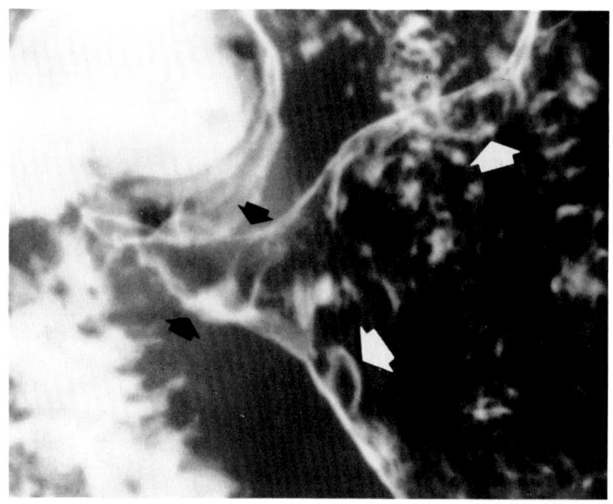

**Figure 13.4. Syphilis of the stomach.** Diffuse thickening of the gastric wall results in narrowing of the antrum (black arrows) and scattered gummatous polyps (white arrows).

extensive radiodense proliferation of the surrounding bone. Syphilis is a major cause of neuropathic joint disease (Charcot's joint), in which bone resorption and total disorganization of the joint are associated with calcific and bony debris. (See the section "Neuropathic Joint Disease" in Chapter 64.)

Tertiary syphilis involving the stomach and other portions of the gastrointestinal tract is now exceedingly rare. Although discrete, nodular, gumma-like lesions can occur, a more common appearance is diffuse thickening of the gastric wall that results in mural rigidity and narrowing of the lumen indistinguishable from that caused by scirrhous carcinoma.[1,9]

## Congenital Syphilis

The radiographic abnormalities in congenital syphilis primarily involve the skeletal system and may be divided into two types. In infantile syphilis, the bone changes are present at birth or in early infancy and are often accompanied by signs of secondary syphilis. In late congenital syphilis, radiographic abnormalities appear in later childhood, adolescence, or early adult life and are associated with characteristics of tertiary disease.

The earliest sign of infantile syphilis is widening of the epiphyseal plate. This is usually accompanied by a transverse band of decreased density across the metaphysis, a nonspecific radiographic sign that also occurs in other systemic childhood diseases. The transverse bands of lucency and the widened epiphyseal plate may simulate the signs of scurvy, though the characteristic ringed

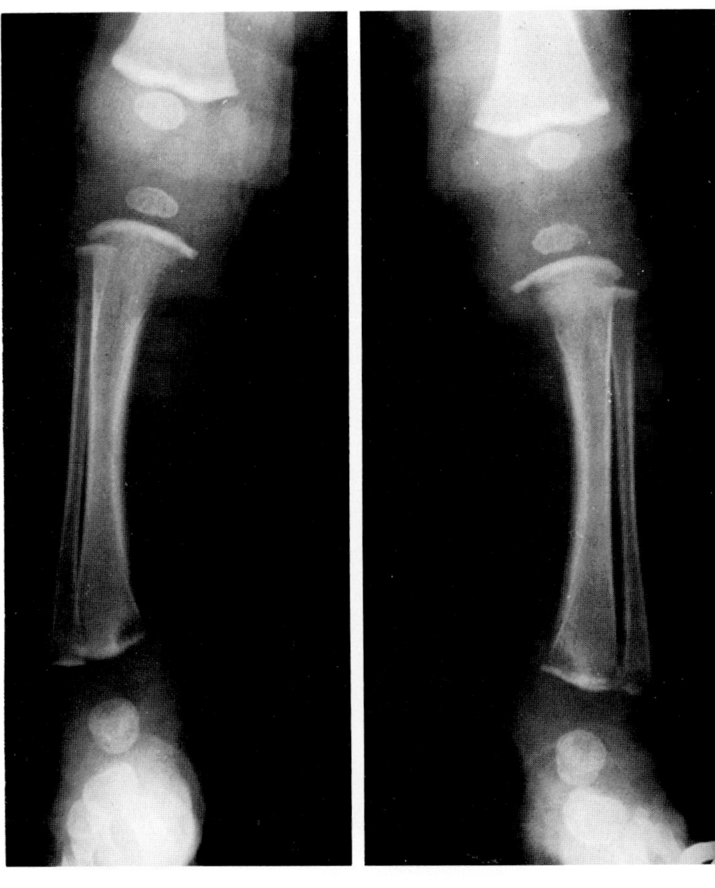

**Figure 13.5. Congenital syphilis.** Symmetric destruction at the medial borders of the tibias (Wimberger's sign), a finding pathognomonic of syphilis. Note the destructive areas at the distal ends of the tibias and fibulas and the increased density at the juxtaepiphyseal areas of bone. (From Edeiken,[7] with permission from the publisher.)

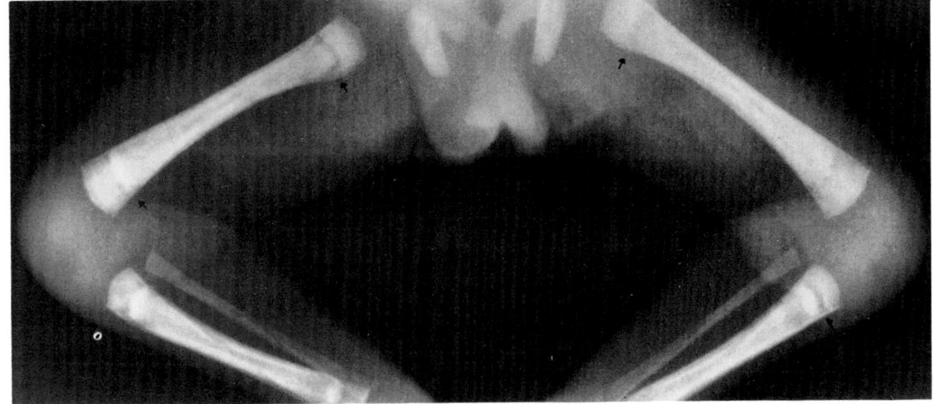

**Figure 13.6. Congenital syphilis.** Transverse bands of lucency (arrows) in the diaphysis of the femurs and tibias.

epiphyses and ground-glass demineralization of scurvy are not seen in congenital syphilis. A pathognomonic sign of syphilis is the symmetric destruction of the medial corners of the proximal tibial metaphyses (Wimberger's sign of syphilis). Irregular serration and fraying of the metaphysis may simulate early rickets. In the diaphysis, focal areas of cortical destruction cause patchy osteolytic lesions. These usually incite an intense periosteal reaction, so that solid or lamellated periosteal new bone formation is generally the dominant radiographic finding.

The most common radiographic appearance of late congenital syphilis is cortical thickening and increased density of the shafts of the long bones; these signs reflect periosteal reaction to underlying gummas. Dense cortical thickening of the anterior surface of the upper tibia produces the classic saber shin appearance.[2,3,9]

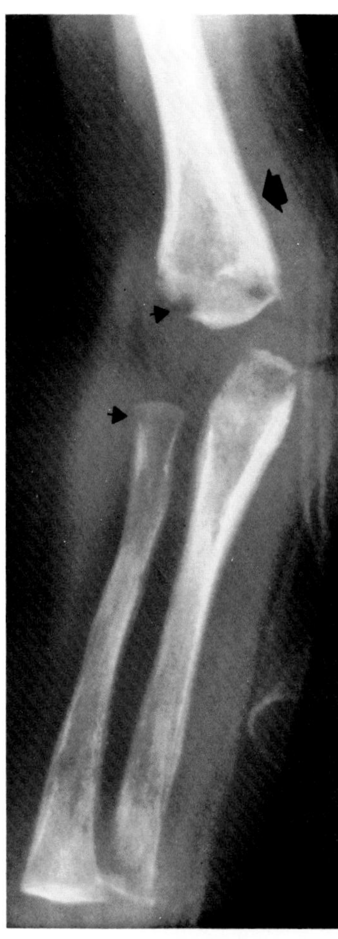

**Figure 13.7. Congenital syphilis.** A radiograph of the arm demonstrates transverse bands of decreased density across the metaphyses (small arrows) along with patchy areas of bone destruction in the diaphyses. There is solid periosteal new bone formation (large arrow), which is best seen about the distal humerus.

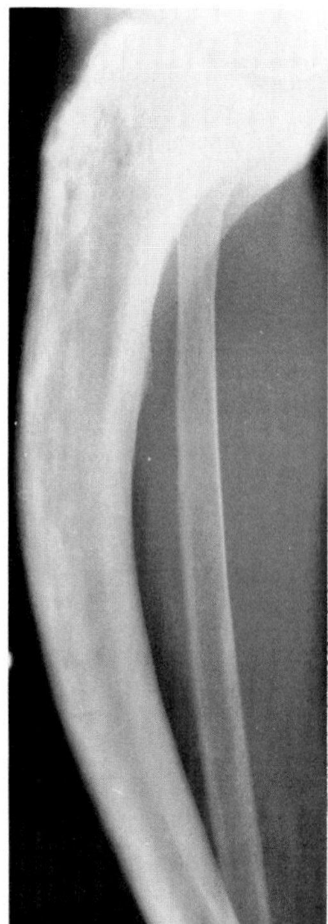

**Figure 13.9. Congenital syphilis.** Saber shin appearance.

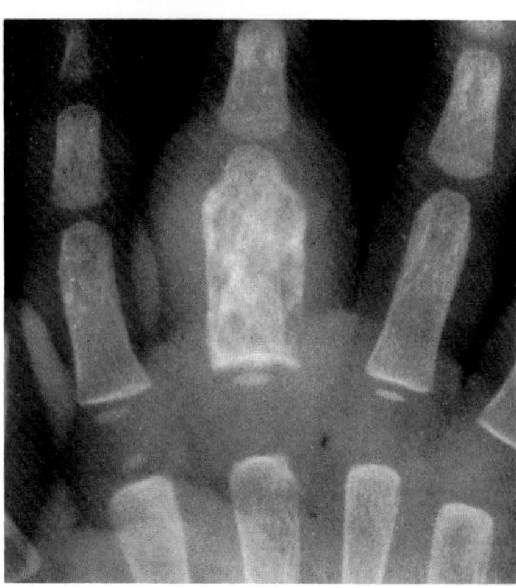

**Figure 13.8. Congenital syphilitic dactylitis.** Typical destructive expansion of a phalanx with periosteal calcification forming a dense shell around the lesion.

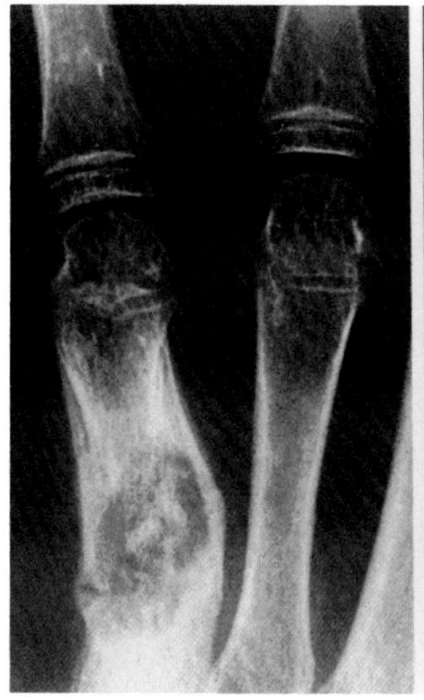

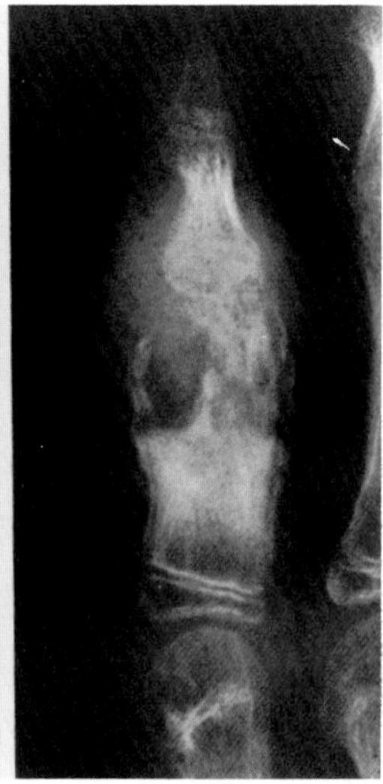

**Figure 13.10. Yaws dactylitis.** Examples of cortical and medullary granulomas along with intense periosteal new bone formation. (From Cockshott and Middlemiss,[10] with permission from the publisher.)

## YAWS

Yaws is a tropical infection caused by *Treponema pertenue*. It is characterized by an initial skin lesion followed by relapsing, eventually destructive lesions of the bones and joints that may be indistinguishable from those of syphilis. The bones of the forearms and hands are most often involved. Gumma formation in bone causes lytic destruction and surrounding sclerosis and incites an exuberant periosteal new bone formation. Shortening or absence of the phalangeal diaphyses in the hands has been reported. Granulomatous soft tissue nodules may develop in chronic disease. Destruction of the nose, maxilla, palate, and pharynx may lead to rhinopharyngitis mutilans.[4,5,10]

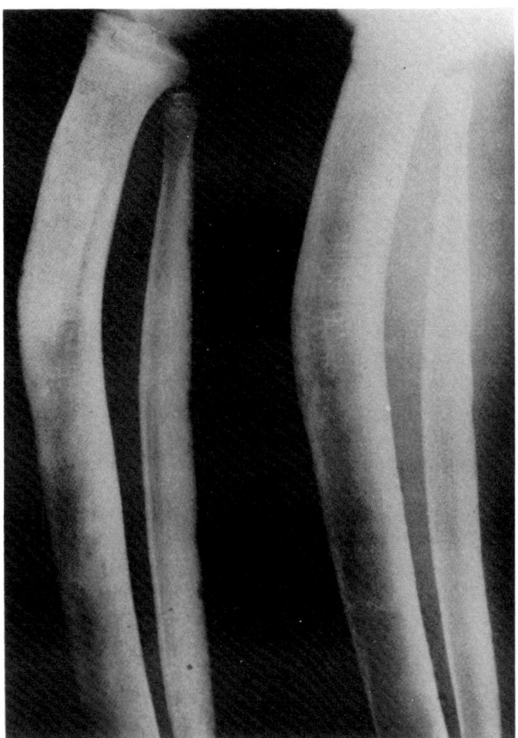

**Figure 13.11. Yaws.** Bilateral saber shin deformities of the tibia simulate the appearance in syphilis. (From Cockshott and Middlemiss,[10] with permission from the publisher.)

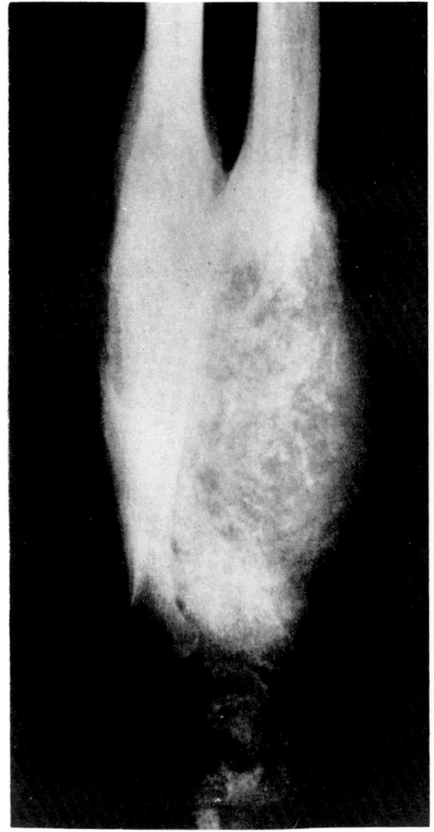

A

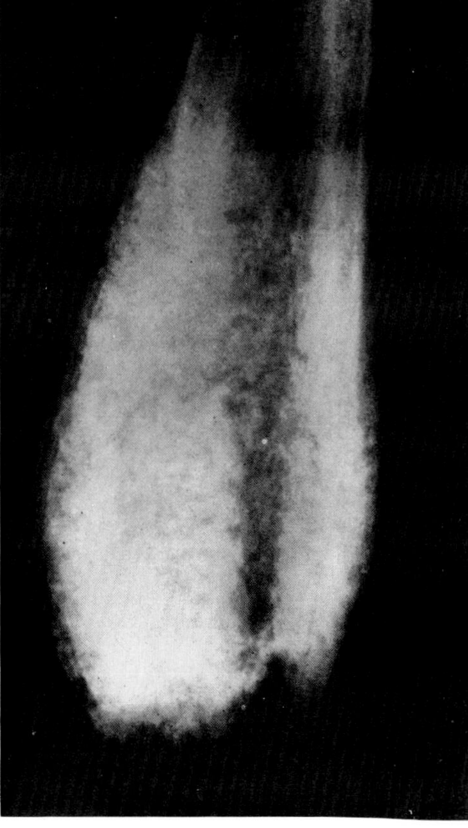

B

**Figure 13.12. Yaws.** (A) A view of the right forearm shows an expanding inflammatory process with surrounding sclerosis. (B) In another patient, massive patchy new bone formation affects both bones of the forearm. Strands of new bone extend in the line of the interosseous ligament. (From Cockshott and Davies,[5] with permission from the publisher.)

## LEPTOSPIROSIS

The radiographic abnormalities in leptospirosis are usually confined to the lungs. Direct action of the spirochete toxin on pulmonary capillaries causes hemorrhage and edema without inflammation; this produces a pattern of peripheral nonsegmental, patchy, or diffuse air space consolidation without pleural effusion or hilar enlargement.[6,7]

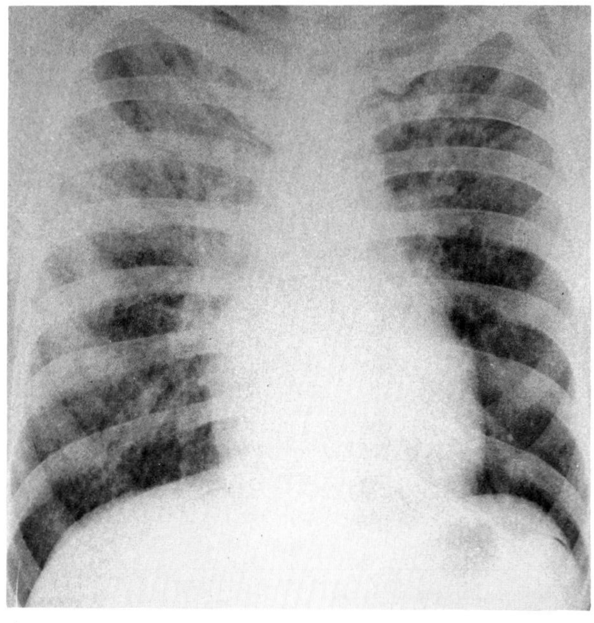

A

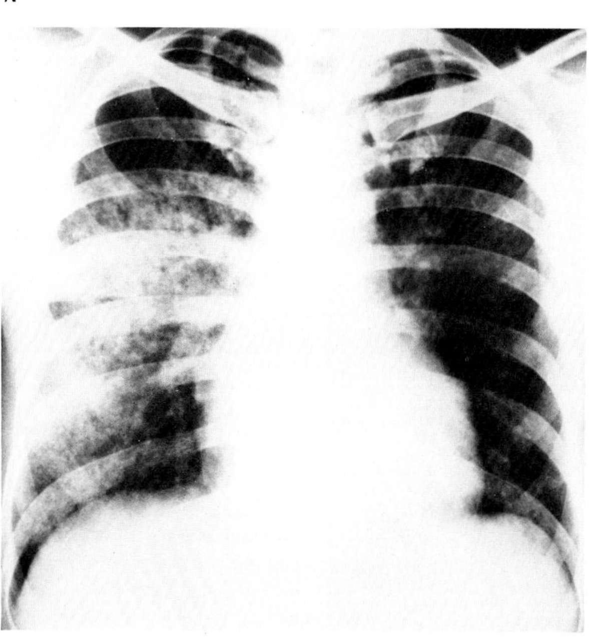

B

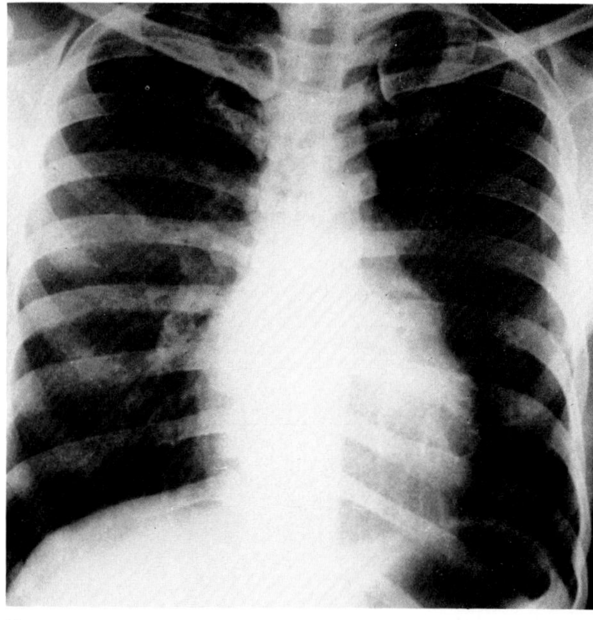

C

**Figure 13.13. Leptospirosis.** (A) An initial film taken the day before hospital admission demonstrates extensive opacification bilaterally, more marked on the right. (B) On the day of admission, a chest radiograph demonstrates patchy consolidation on the right and minimal changes on the left. (C) Following 3 days of penicillin therapy, there is rapid resolution of the pulmonary opacities. (From Poh and Soh,[7] with permission from the publisher.)

## TROPICAL ULCER

Tropical ulcers, caused by Vincent's types of fusiform bacilli and spirochetes, are extremely common throughout much of Africa and are often associated with lesions in the underlying bone (ulcer osteoma). The chronic ulcers most often affect children and young adults and are usually located in the mid or lower leg.

The earliest radiographic finding is fusiform periosteal reaction localized to the bone beneath the ulcer. This periosteal new bone blends with the cortex and produces a sclerotic cortex, which can become over 1 in thick and give rise to the classic ivory osteoma. An even more common sequela results from osteoporosis distal to the ulcer, followed by bulbous expansion of the medullary cavity toward the ulcer and thinning of the cortex in this region (cancellous ulcer osteoma). A serious complication of tropical ulcers is malignant transformation, which occurs in about 2 percent of the patients and should be suspected if the ulcer begins to increase in size or if radiographs demonstrate cortical destruction and a soft tissue mass.[8]

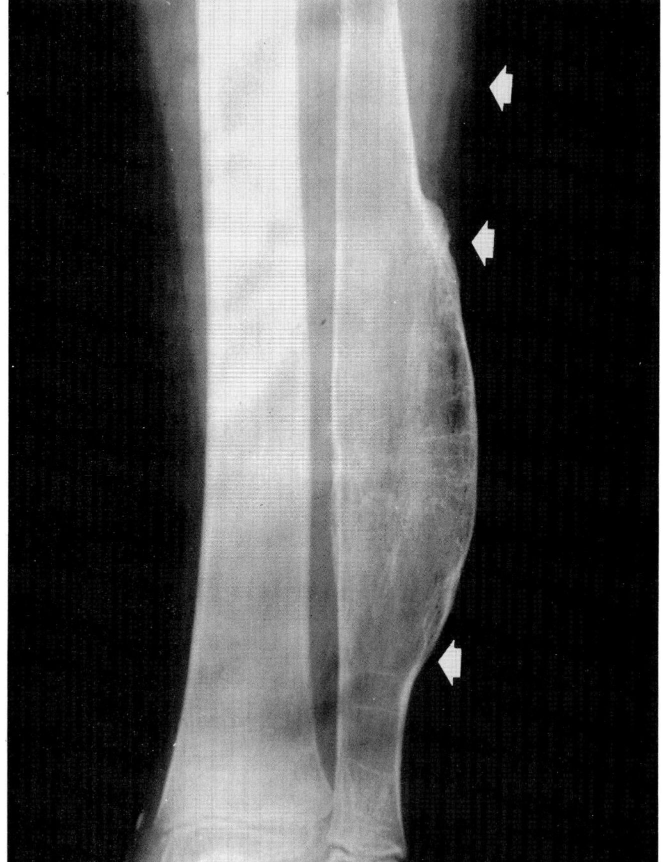

A

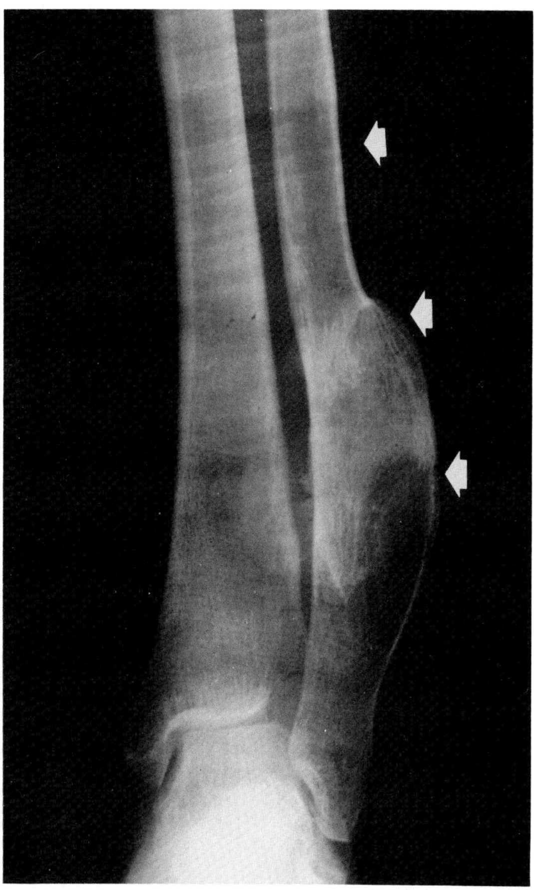

B

**Figure 13.14. Cancellous ulcer osteoma.** Two examples in children, both of whom had had their ulcers for more than 5 years. Note the loss of overlying soft tissue (arrows are at the skin surface) and the distal osteoporosis. (From Kolawole and Bohrer,[8] with permission from the publisher.)

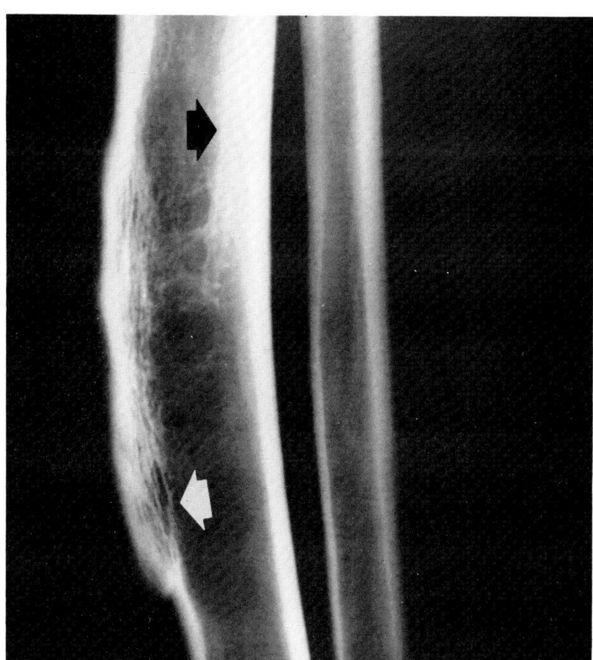

**Figure 13.15. Ivory osteoma of tropical ulcer.** There is cortical thickening of the tibia on the side opposite the ulcer (black arrow), which had been present for 1 year. Medullary resorption is starting at the inner margin of the osteoma, and the solid cortex is beginning to show a trabecular pattern (white arrow). (From Kolawole and Bohrer,[8] with permission from the publisher.)

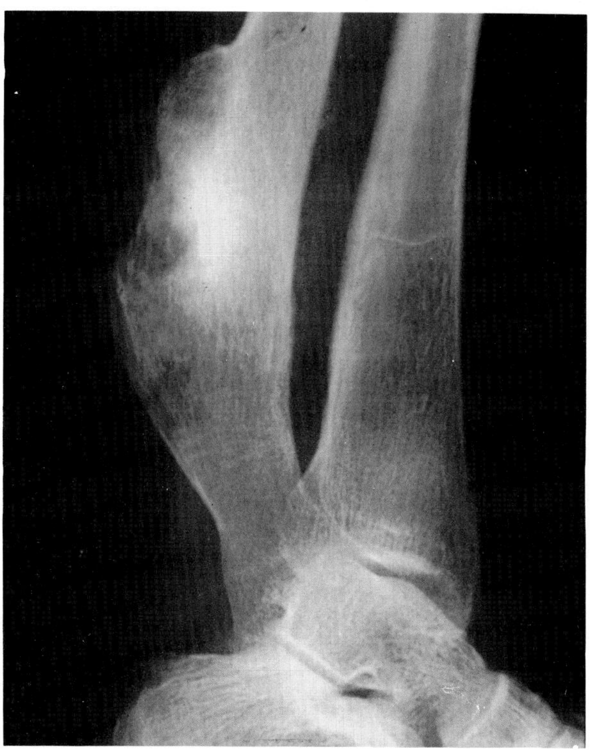

**Figure 13.16. Malignant cancellous ulcer osteoma.** The ulcer had been present for 20 years. There is swelling of the soft tissue, a discrete soft tissue mass, and destruction of the cortex of the osteoma. (From Kolawole and Bohrer,[8] with permission from the publisher.)

## REFERENCES


1. Johns D: Syphilitic disorders of the spine. *J Bone Joint Surg* 52B:724–731, 1970.
2. Cremin BJ, Fisher RM: The lesions of congenital syphillis. *Br J Radiol* 43:333–341, 1970.
3. Coblentz DR, Cimini R, Mikiti VG, et al.: Roentgenographic diagnosis of congenital syphilis in the newborn. *JAMA* 212:1061–1064, 1970.
4. Riseborough AW, Joske RA, Vaughan BF: Hand deformities due to yaws in Western Australian aborigines. *Clin Radiol* 12:109–113, 1961.
5. Cockshott WP, Davies AGM; Tumoural gummatous yaws. *J Bone Joint Surg* 42B:785–791, 1960.
6. Health CW, Alexander AD, Galton MM: Leptospirosis in the United States. *N Engl J Med* 273:915–922, 1965.
7. Poh SC, Soh CS: Lung manifestations in leptospirosis. *Thorax* 25:751–755, 1970.
8. Kolawole TM, Bohrer SP: Ulcer osteoma: Bone response to tropical ulcer. *AJR* 109:611–618, 1970.
9. Edeiken J: *Roentgen Diagnosis of Diseases of Bone*. Baltimore, Williams & Wilkins, 1981.
10. Cockshott WP, Middlemiss H: *Clinical Radiology in the Tropics*. Edinburgh, Churchill Livingstone, 1979.

## Chapter Fourteen
# Higher Bacteria and Fungi

## ACTINOMYCOSIS

Actinomycosis is an indolent suppurative infection, caused by gram-positive higher bacteria, that produces radiographic abnormalities in the lungs, pleura, bones, and gastrointestinal tract.

The typical pulmonary pattern of actinomycosis is an acute nonsegmental air space consolidation, which may resemble pneumonia or a tumor mass. With appropriate therapy, most cases resolve without complication. If therapy is not instituted, the lesion may cavitate and develop into a lung abscess. Extension of the infection

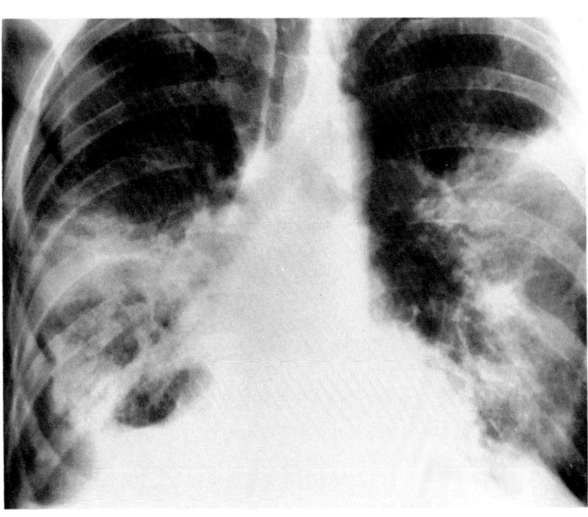

A

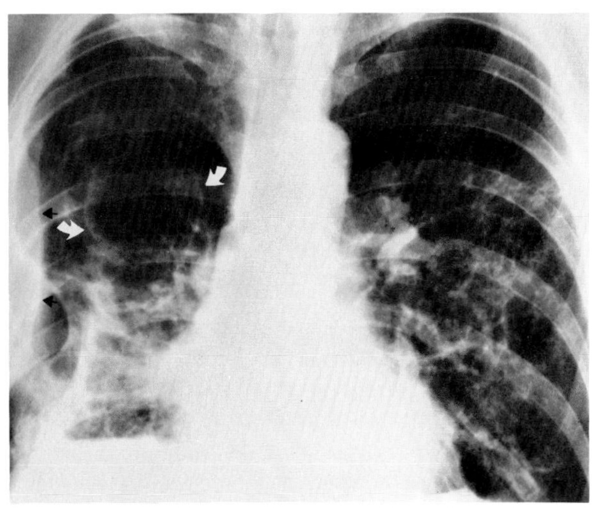

B

**Figure 14.1. Actinomycosis.** (A) Bilateral nonsegmental air space consolidations. (B) Because appropriate treatment was not instituted, the patient developed a large cavity (white arrows), and infection extended into the pleura to produce an empyema (black arrows). The right fifth posterior rib was partially resected during surgery, which revealed thickening of the pleura circumferentially around the right lower lobe. Sulfur granules were detected on pleural biopsy.

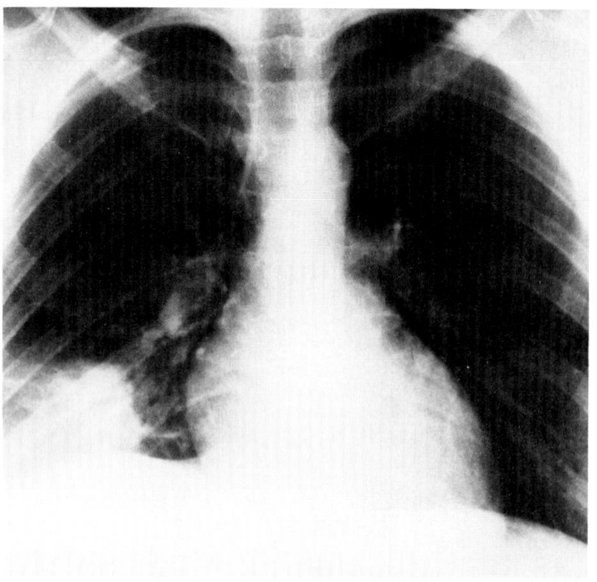
A

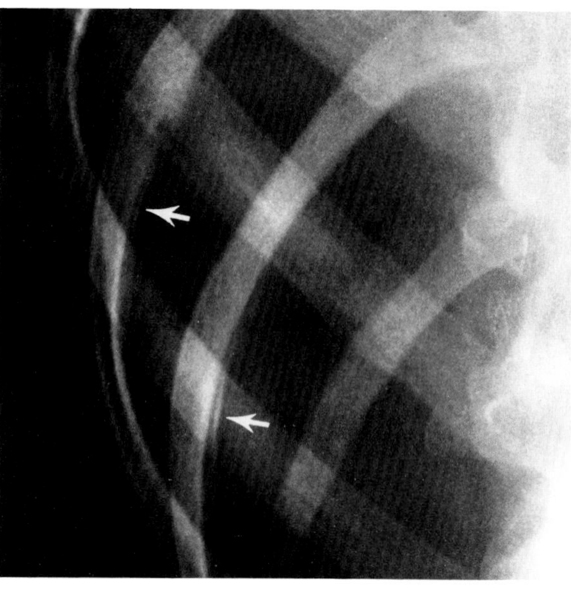
B

**Figure 14.2. Actinomycosis.** (A) A plain chest radiograph demonstrates right lower lobe air space consolidation. (B) A coned view of the lower right ribs demonstrates periosteal reaction (arrows) involving the tenth and eleventh ribs.

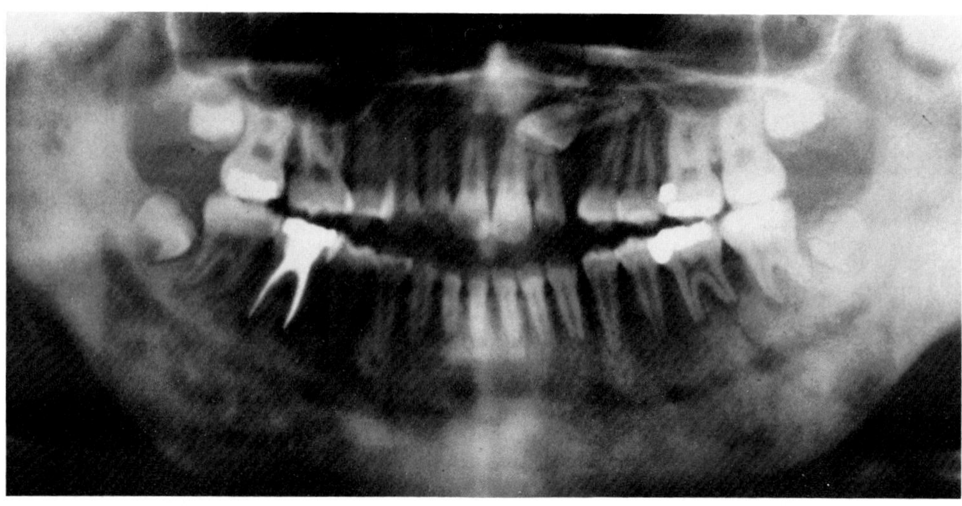

**Figure 14.3. Cervicofacial actinomycosis.** Diffuse lytic changes are seen throughout the mandible. This patient manifested an unusually large amount of periosteal involvement and bone sclerosis.

into the pleura produces an empyema, which classically leads to osteomyelitis of the ribs and the formation of a sinus tract. Before the advent of antibiotics, rib destruction and sinus tracts in the chest wall commonly developed; this advanced stage is now rarely seen.

Actinomycosis involving bone is usually secondary to infection of adjacent soft tissues. Because cervicofacial actinomyosis is the most common form of the disease, the mandible is most frequently involved; it demonstrates bone destruction, usually without substantial new bone formation. When the thoracic wall is affected, destruction of the ribs is usually associated with thick periosteal new bone formation, an appearance that is very suggestive of actinomycosis. Hematogenous spread can occasionally result in multiple destructive lesions with some sclerosis, a pattern indistinguishable from that of chronic osteomyelitis caused by other organisms.

Actinomycosis of the gastrointestinal tract involves predominantly the ileocecal region. Extensive fistula formation and a palpable soft tissue mass indenting the distal ileum and cecum produce a nonspecific radiographic pattern simulating that of Crohn's disease or periappendiceal abscess. The diagnosis of actinomycosis is suggested by the presence of a sinus tract extending to the overlying skin.

Pelvic actinomycosis, once rare, is now being seen in women with an implanted intrauterine contraceptive device (IUD). Ultrasound may demonstrate a complex adnexal mass simulating pelvic inflammatory disease or a tumor.[1–6]

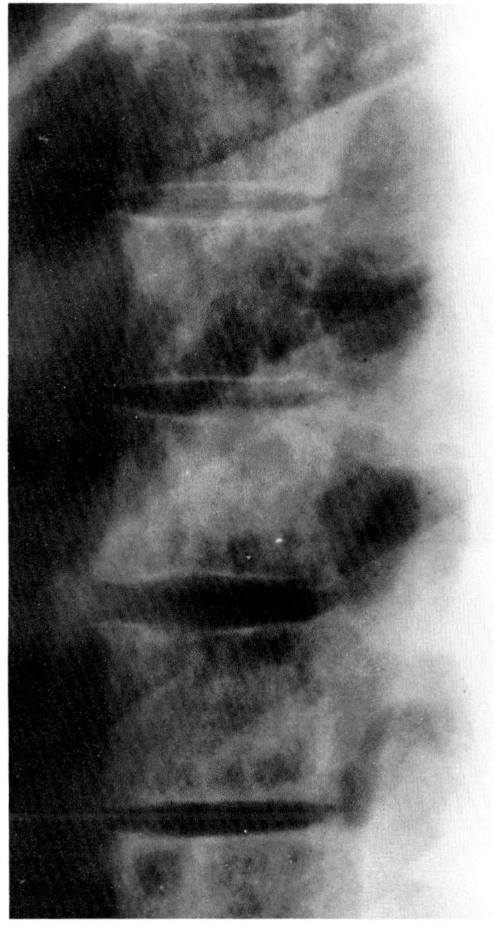

**Figure 14.4. Actinomycosis.** Diffuse osteolytic destruction involving multiple vertebral bodies represents hematogenous spread from a primary pulmonary focus of infection.

## NOCARDIOSIS

Nocardiosis is a suppurative infection caused by a ubiquitous gram-positive higher bacterium. The organism is usually an opportunistic invader that infects patients with serious pre-existing conditions, such as steroid or immunosuppressive therapy or underlying malignancy.

Nocardiosis initially involves the lungs, where it produces single or multiple areas of nonspecific pneumonia. Like actinomycosis, the pulmonary infiltrate of nocardiosis often cavitates and, if not appropriately treated, extends to produce pleural empyema, rib destruction, and chest-wall abscesses.

Hematogenous dissemination to the brain and subcutaneous tissue is frequent in nocardiosis. The brain lesions are typically multiple abscesses that can rupture into the ventricle and result in purulent meningitis.[7,8,37,43]

## CRYPTOCOCCOSIS (TORULOSIS)

Cryptococcosis is a granulomatous disease caused by the yeastlike fungus *Cryptococcus neoformans* that lives in soil, particularly that contaminated by pigeon droppings. Although primary cryptococcal infections occur, the organism is most frequently an opportunistic invader that infects debilitated patients with chronic reticuloendothelial system disease (leukemia, lymphoma) or patients undergoing steroid or antibiotic therapy.

The major pulmonary manifestation of cryptococcosis is a single, fairly well circumscribed mass that is usually in the periphery of the lung and is often pleural-based. This radiographic pattern may be indistinguishable from carcinoma or other chronic pulmonary granulomas.

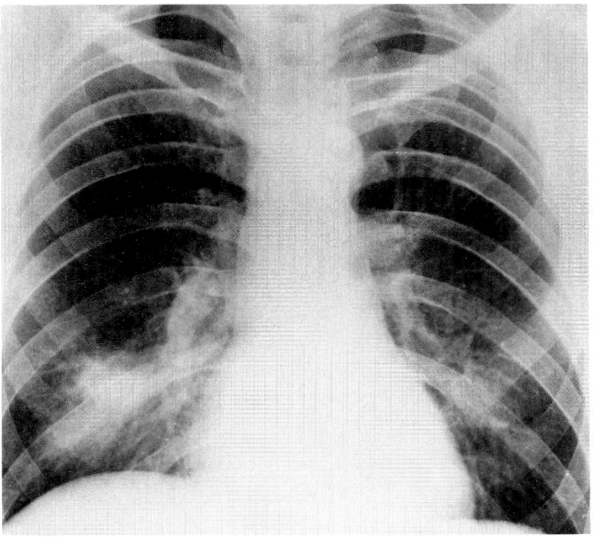

A

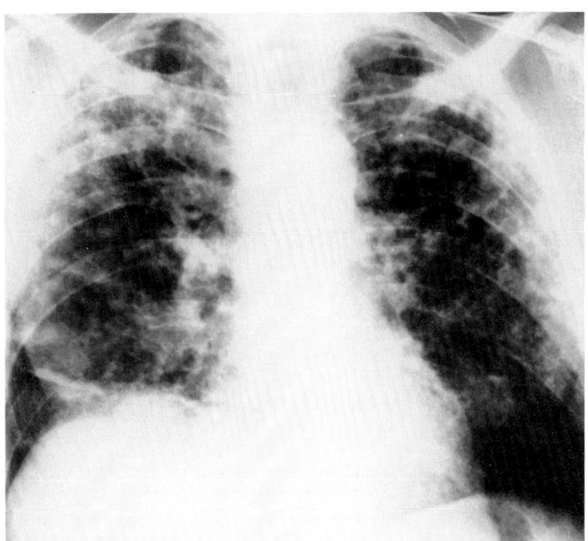

B

**Figure 14.5. Nocardiosis.** (A) An initial chest radiograph demonstrates an area of nonspecific pneumonia in the right lower lobe. (B) Without appropriate therapy, the infection spread to involve both lungs diffusely with a patchy infiltrate and multiple small cavities.

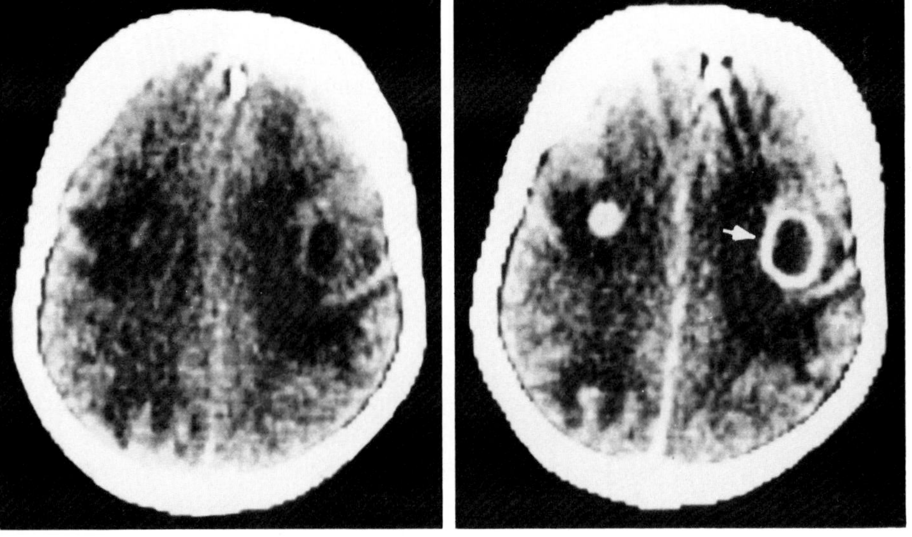


**Figure 14.6. Nocardiosis complicating chronic lymphocytic leukemia.** (A) Precontrast and (B) postcontrast CT scans demonstrate three lesions with prominent surrounding edema. The right hemisphere lesion (arrow) with ring enhancement had a well-developed abscess capsule at surgery. The lesions resolved completely after appropriate antibiotic therapy. (From Enzmann, Brant-Zawadzki, and Britt,[37] with permission from the publisher.)

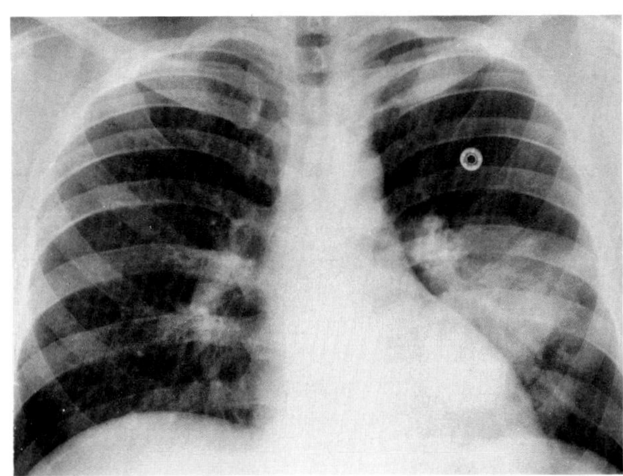

**Figure 14.7. Cryptococcosis.** A single, fairly well circumscribed, masslike consolidation in the superior segment of the left lower lobe.

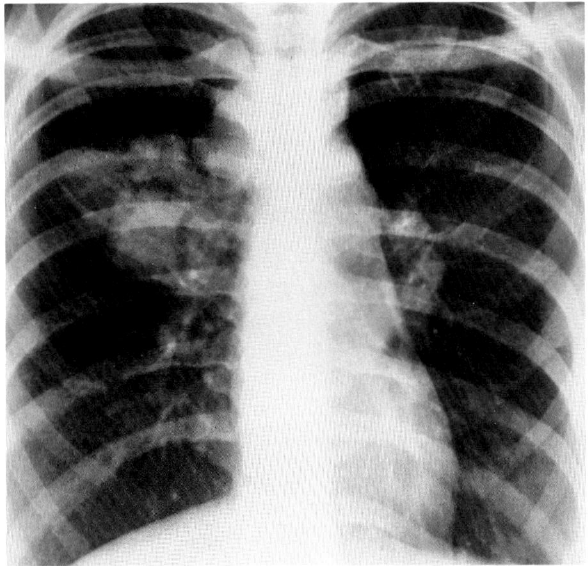


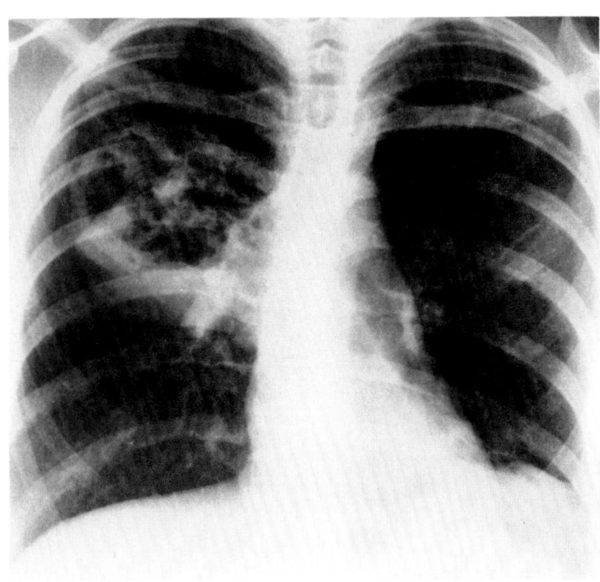


**Figure 14.8. Cryptococcosis.** (A) An initial chest film demonstrates an air space consolidation in the right upper lung. (B) With progression of the infection, the right upper lung pneumonia has cavitated (arrows) and a left lower lobe air space consolidation (arrows) has developed.

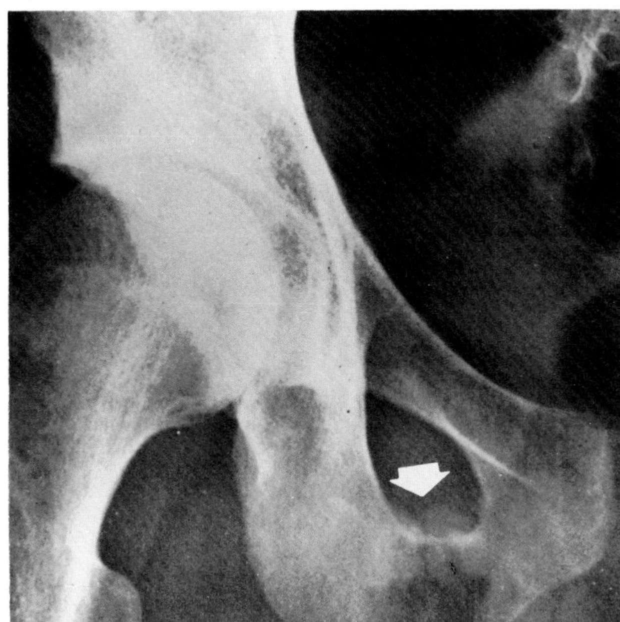

A

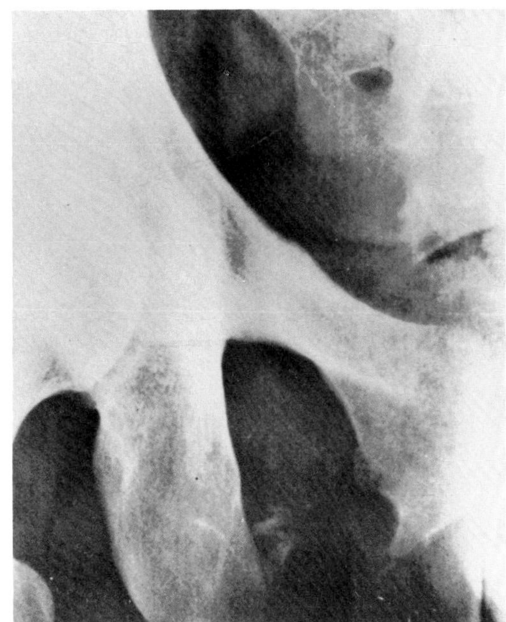

B

**Figure 14.9. Cryptococcosis.** (A) There is a well-circumscribed cystic-appearing lesion at the ischiopubic junction (arrow). Faint periosteal reaction is present along the upper margin, and a small cortical erosion is seen at the inferior edge. (B) Six months after bone curettage and drainage of a gluteal abscess, there is no evidence of healing. (From Rosen and Jacobson,[10] with permission from the publisher.)

Cavitation is relatively uncommon compared with its frequency in the other mycoses. An ill-defined infiltrate may simulate pneumonia, and widely disseminated disease may produce a miliary pattern of diffuse nodules.

The uncommon skeletal involvement in cryptococcosis most frequently affects the pelvis, ribs, and skull. It appears as single or multiple, well-defined lytic lesions associated with minimal sclerosis or periosteal reaction.

Focal cryptococcal brain abscesses may be demonstrated by computed tomography (CT) as sharply demarcated dense masses with central areas of lesser density.[9,10]

## BLASTOMYCOSIS

Blastomycosis is a granulomatous disease caused by the yeastlike fungus *Blastomyces dermatitidis* that is found in the soil and is inhaled into the respiratory tract. The pulmonary manifestations of blastomycosis are variable and nonspecific. In acute disease, patchy areas of air space consolidation simulate pneumonia. Cavitation and miliary nodules infrequently occur. In chronic disease, fibronodular changes in the upper lobe, sometimes with cavitation, can simulate tuberculosis. Blastomycosis may appear as a solitary pulmonary mass which, when associated with unilateral lymph node enlargement, may closely mimic a bronchogenic carcinoma.

The skeletal system is involved in about half of the patients with disseminated disease. The radiographic appearance is that of a nonspecific chronic osteomyelitis. Often multiple, the lesions demonstrate marked destruction with or without surrounding bone sclerosis. Secondary soft tissue abscesses and draining sinuses are common. In the spine, vertebral body destruction and an associated paraspinal abscess simulate tuberculosis.[10–13,38]

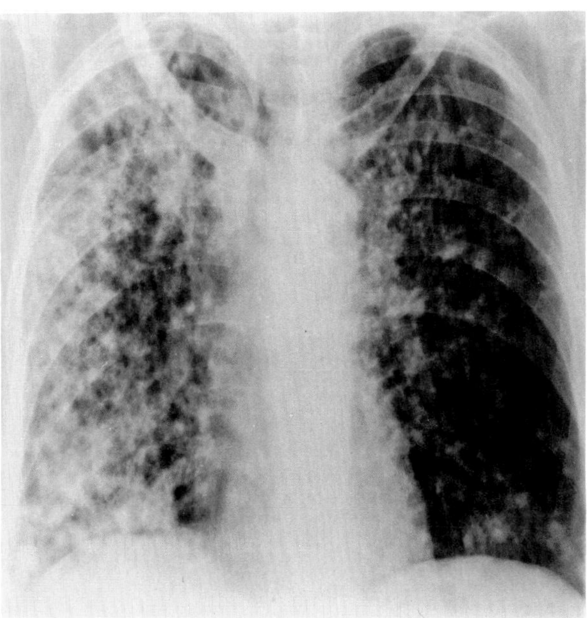

**Figure 14.10. Blastomycosis.** Diffuse patchy areas of air space consolidation that are most prominent in the right lung. In several areas the pattern simulates that of large miliary nodules.

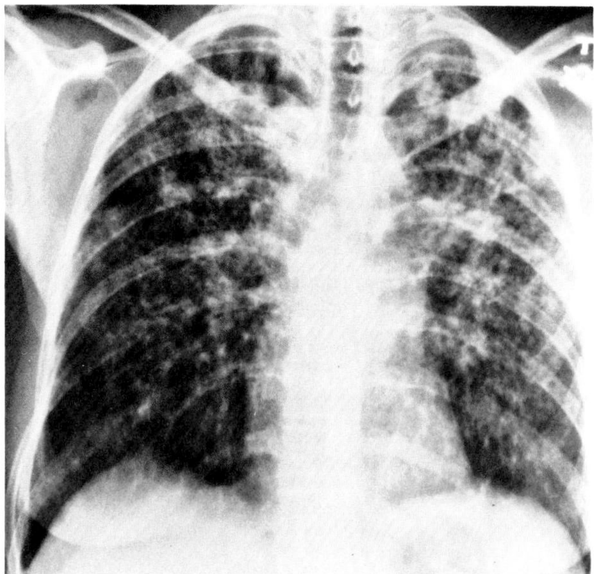

**Figure 14.11. Blastomycosis.** Diffuse interstitial disease with upper lobe predominance. Note the volume loss in the upper lobe and the overdistension of the lower lobes along with the formation of bullae at the bases. (From Halvorson, Duncan, Merten, et al,[38] with permission from the publisher.)

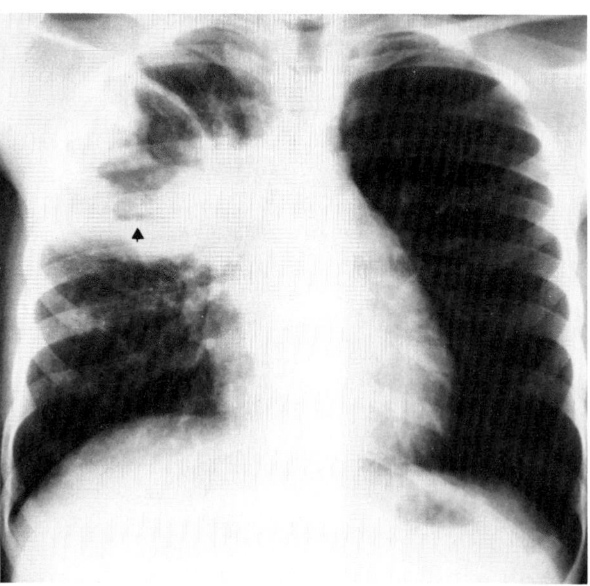

**Figure 14.12. Blastomycosis.** A right upper lobe cavity with thick walls and a faintly visible air-fluid level (arrow). There is an associated soft tissue mass along the lateral wall of the cavity. (From Halvorson, Duncan, Merten, et al,[38] with permission from the publisher.)

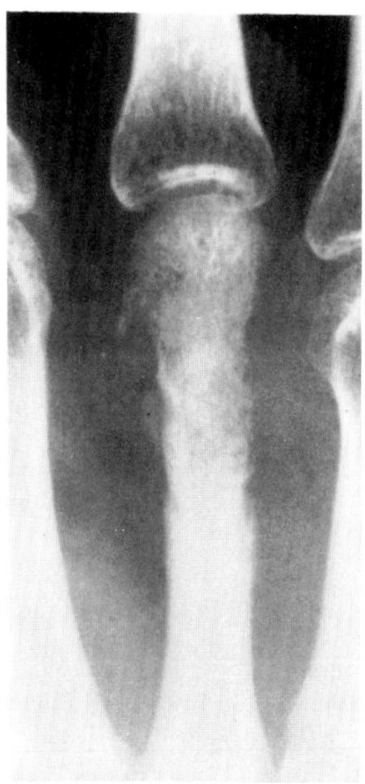

**Figure 14.13. Blastomycosis.** Osteomyelitis with irregular destruction involving the distal half of a metacarpal.

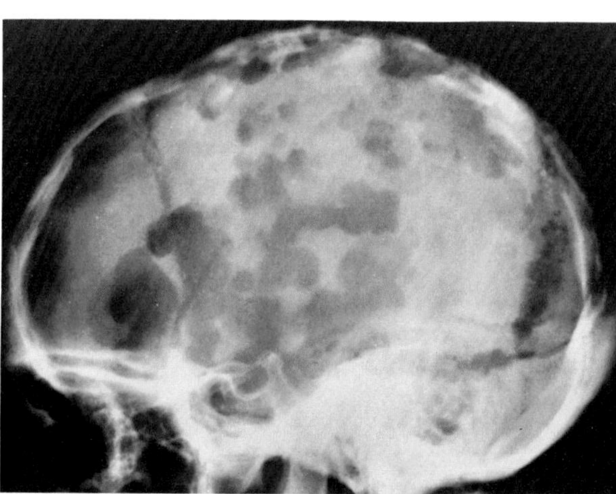

**Figure 14.14. Blastomycosis.** Diffuse areas of osteolytic destruction affect most of the calvarium.

## HISTOPLASMOSIS

Histoplasmosis, caused by the fungus *Histoplasma capsulatum,* is a common disease that often produces a radiographic appearance simulating that of tuberculosis.

The primary form of histoplasmosis is usually relatively benign and often passes unnoticed. Chest radiographs may demonstrate single or multiple areas of pulmonary infiltration that are most often in the lower lung and are often associated with hilar lymph node enlargement. Although this pattern simulates the primary complex of tuberculosis, pleural effusion rarely occurs with histoplasmosis. In children, striking hilar adenopathy, which may cause bronchial compression, may develop without radiographic evidence of parenchymal disease. Hilar lymph node calcification is common in adults. A common manifestation of pulmonary histoplasmosis is a solitary, sharply described, granulomatous nodule (his-

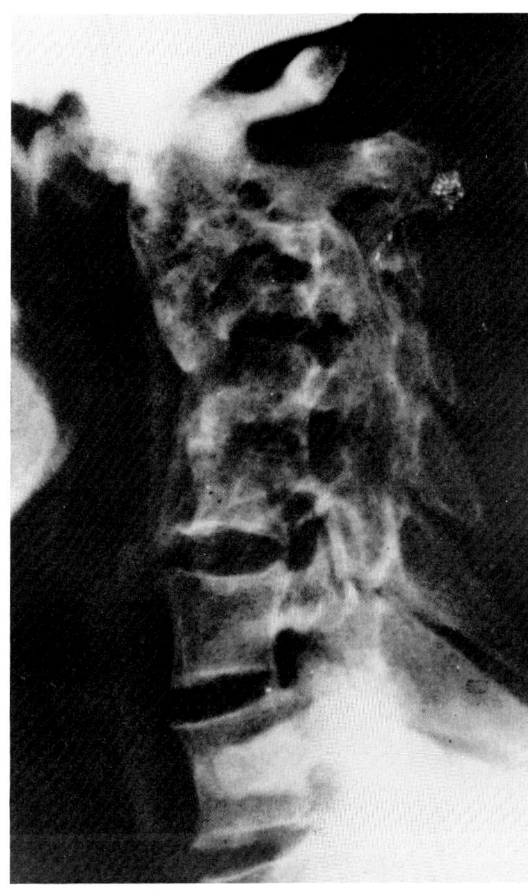

**Figure 14.15. Blastomycosis.** Nonspecific chronic osteomyelitis involves the upper cervical spine. There is narrowing and partial obliteration of contiguous intervertebral disk spaces, and partial fusion involving the second, third, and fourth vertebral bodies and their neural arches. (From Rosen and Jacobson,[10] with permission from the publisher.)

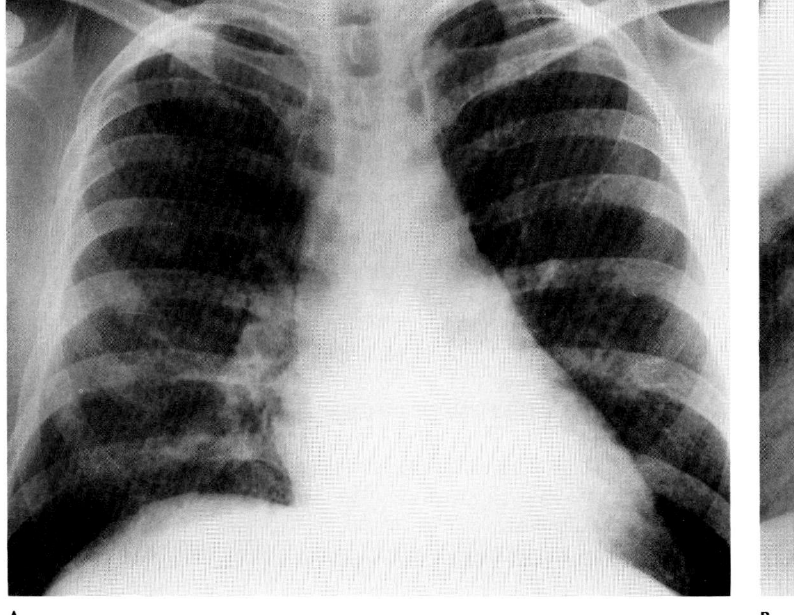

A

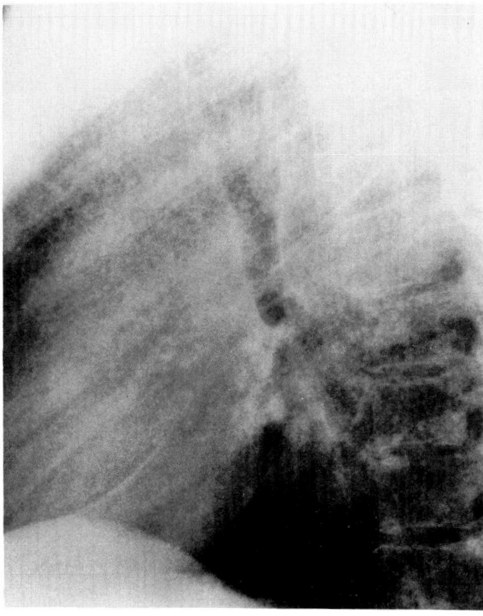

B

**Figure 14.16. Miliary histoplasmosis.** (A) Frontal and (B) lateral views of the chest demonstrate multiple soft tissue nodules scattered throughout both lungs in a pattern simulating miliary tuberculosis.

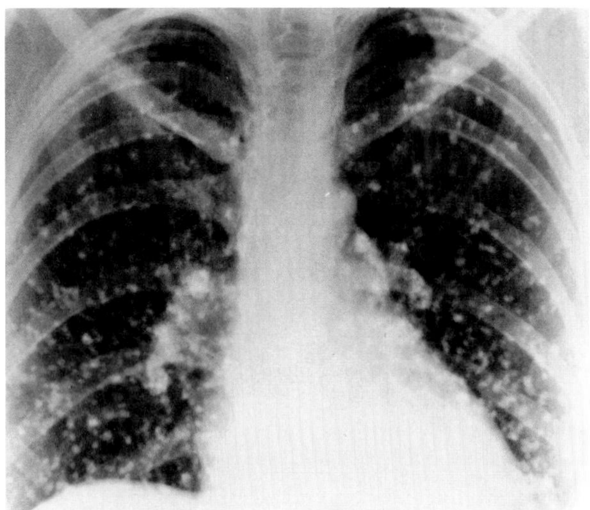

**Figure 14.17. Histoplasmosis.** Diffuse calcifications within the lungs produce a snowball pattern.

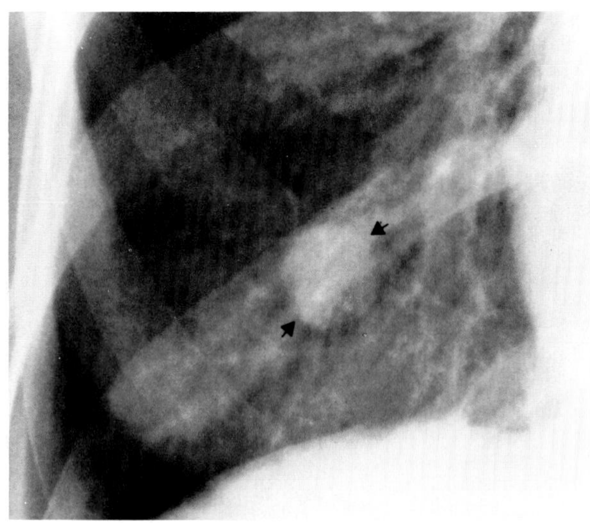

**Figure 14.18. Histoplasmoma.** There is a solitary, sharply circumscribed granulomatous nodule in the right lower lobe.

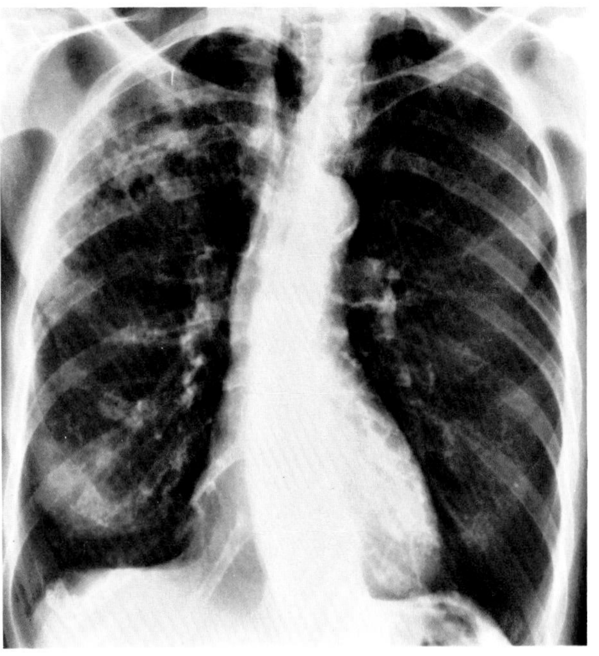

A

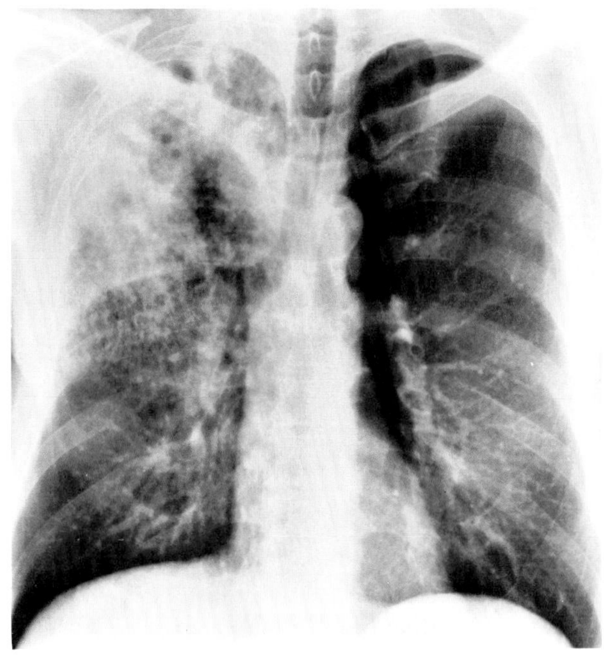

B

**Figure 14.19. Chronic histoplasmosis.** (A) An initial film demonstrates an ill-defined area of parenchymal consolidation in the right upper lobe. (B) One week later, there is dramatic extension of the infiltrate, which now involves most of the upper half of the right lung.

toplasmoma), which is usually less than 3 cm in diameter and is most often in a lower lobe. Central, rounded calcification within the mass (target lesion) is virtually pathognomonic of this disease. Multiple soft tissue nodules scattered throughout both lungs may simulate miliary tuberculosis. These shadows may clear completely or may fibrose and persist, often appearing on subsequent chest radiographs as widely disseminated punctate calcifications.

The more chronic form of histoplasmosis is characterized by zones of parenchymal consolidation, often large and with a loss of lung volume, that usually develop in an upper lobe. Cavitation is common, and the radiographic appearance closely simulates reinfection tuberculosis.

Histoplasmosis can incite progressive fibrosis in the mediastinum. This can cause obstruction of the superior vena cava, pulmonary arteries, and pulmonary veins, as well as severe narrowing of the esophagus.

Diffuse calcification in the liver, spleen, and lymph

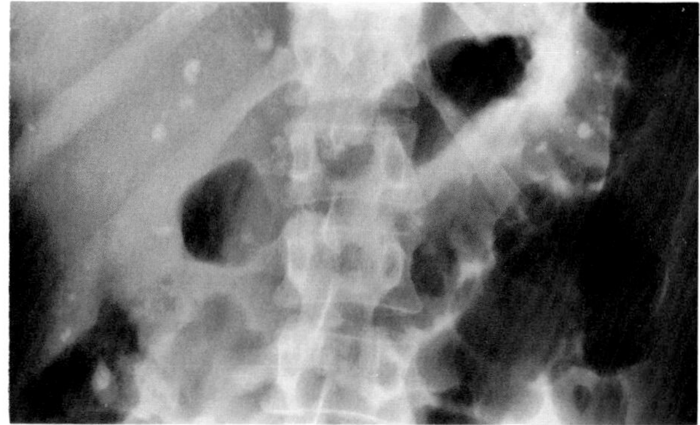

**Figure 14.20. Histoplasmosis.** Scattered foci of calcification within the liver and spleen represent healed histoplasmosis.

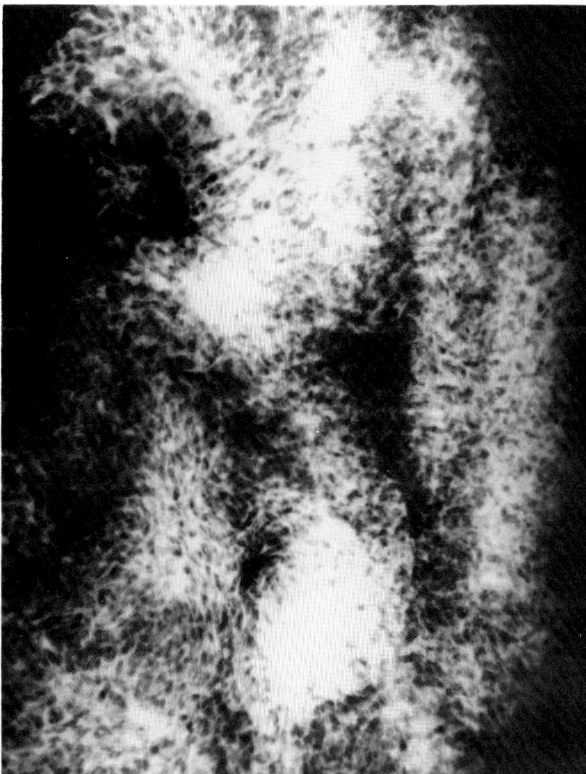

**Figure 14.21. Histoplasmosis.** Innumerable filling defects diffusely blanket the small bowel with a sandlike covering superimposed on irregular distorted folds. (From Eisenberg,[39] with permission from the publisher.)

nodes is virtually diagnostic of histoplasmosis, especially in areas in which the disease is endemic (e.g., the Mississippi and Ohio valleys of the United States). These calcifications tend to be small, multiple, dense, and discrete, though occasionally they may appear as moderately large, solidly calcified granulomas. Disseminated disease rarely affects the gastrointestinal tract. The infiltration of enormous numbers of *Histoplasma*-laden macrophages into the lamina propria, accompanied by intense villous edema, causes irregularly thickened and distorted small bowel folds. When seen on end, the folds appear as innumerable filling defects that seem to diffusely blanket the small bowel with a sandlike covering. Other rare complications of histoplasmosis include pericarditis, chronic meningitis, and Addison's disease from adrenal involvement.[14–16]

## COCCIDIOIDOMYCOSIS

Coccidioidomycosis is a granulomatous infection caused by the fungus *Coccidioides immitis* that is endemic in

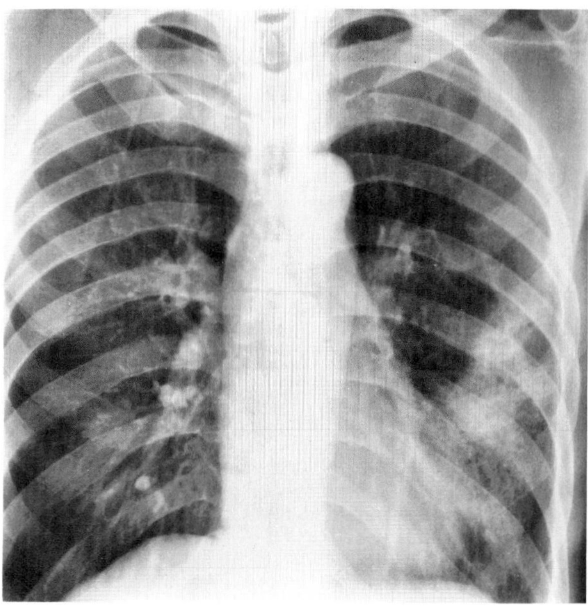

**Figure 14.22. Coccidioidomycosis pneumonia.** An ill-defined area of patchy infiltrate in the left lower lung.

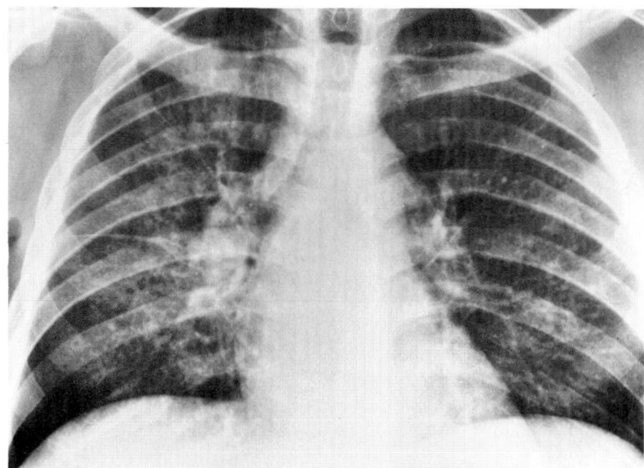

**Figure 14.23. Coccidioidomycosis pneumonia.** A diffuse pattern of fine nodules throughout the lungs simulates miliary tuberculosis in this patient with disseminated disease.

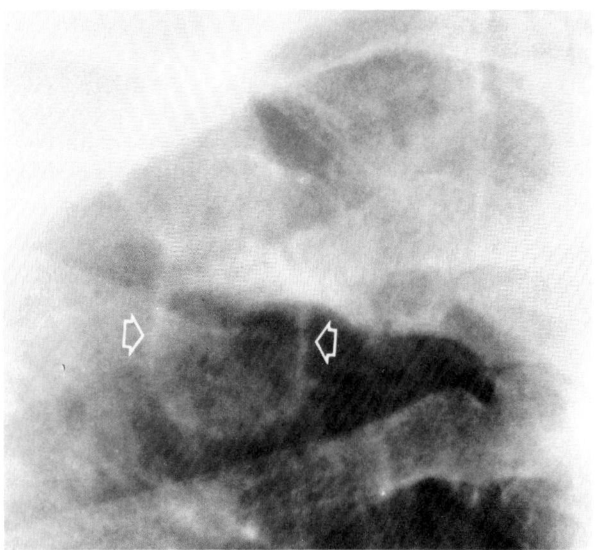

**Figure 14.24. Coccidioidomycosis.** A typical thin-walled cavity (arrows) without surrounding reaction.

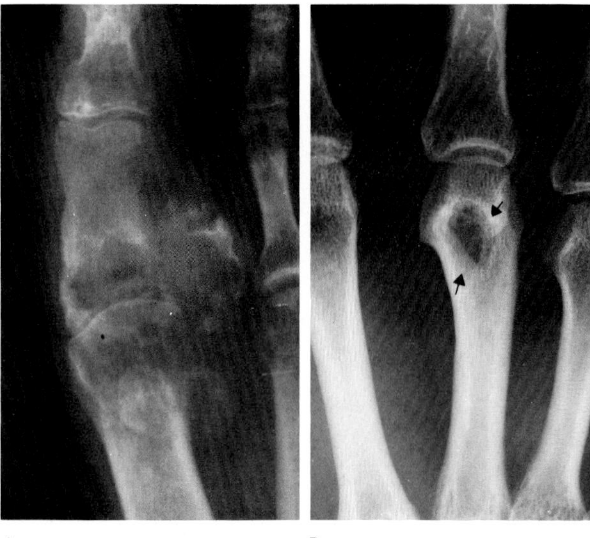

**Figure 14.25. Coccidioidomycosis.** (A) Diffuse destruction of the first metatarsal and proximal and distal phalanges of the great toe. There is complete disruption of the cortical margin of the proximal phalanx along with an associated soft tissue mass (focal granuloma). (B) In another patient, there is a typical, well-marginated, punched-out lytic defect in the head of the third metacarpal (arrows). (From McGahan, Graves, Palmer, et al,[40] with permission from the publisher.)

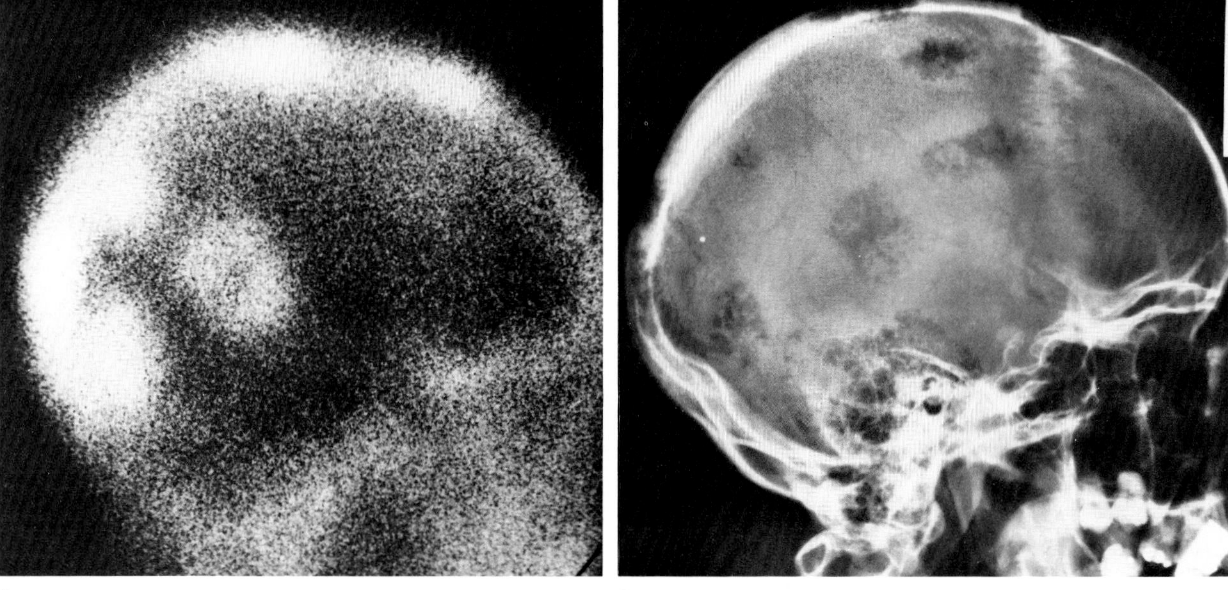

**Figure 14.26. Coccidioidomycosis of the skull.** (A) A radionuclide bone scan demonstrates several areas of increased uptake of isotope. (B) Skull radiograph shows multiple circumscribed lytic defects within the calvarium. (From McGahan, Graves, Palmer, et al,[40] with permission from the publisher.)

the southwestern United States, especially in the San Joaquin valley of California.

Pulmonary involvement usually begins as a fleeting area of patchy pneumonia that is often accompanied by ipsilateral hilar adenopathy and, less frequently, by pleural effusion. Unilateral or bilateral hilar and mediastinal adenopathy occasionally develop even in the absence of parenchymal disease. Disseminated coccidioidomycosis produces a diffuse nodular pattern simulating miliary tuberculosis.

Although the primary disease usually clears completely, residual active parenchymal infection may persist. Well-defined nodules of varying sizes may occur singly or in groups. These nodules infrequently calcify and may excavate to form thin- or thick-walled cavities. Thin-walled cavities without surrounding reaction are suggestive of coccidioidomycosis. The thick-walled cavities are nonspecific and simulate those of tuberculosis. The pulmonary inflammation may persist as a chronic pneumonia or may progress to fibronodular disease.

Skeletal involvement occurs in almost one-fourth of the cases of disseminated coccidioidomycosis. The first evidence of bone involvement may be soft tissue swelling due to an overlying abscess, especially near the sternoclavicular junction and about the small bones of the hand. Bone lesions are usually multiple and involve primarily the ends of the long bones and the bony prominences. Bone destruction initially produces sharply marginated lytic lesions. The presence of a sclerotic margin indicates a chronic infectious process. Vertebral destruction and an associated paraspinal mass, representing an abscess, may produce a radiographic appearance indistinguishable from that of tuberculosis.[17–19,43]

## CANDIDIASIS

*Candida* is a common yeastlike fungus that is found in the mouth, throat, sputum, and feces of many normal children and adults. Acute infection by *Candida* develops when there is an immunologic imbalance between the host and the normal body flora. Candidiasis occurs most frequently in patients with leukemia, lymphoma, diabetes, or other chronic debilitating diseases and in patients receiving prolonged steroid, immunosuppressive, or antibiotic therapy. Radiographic abnormalities most commonly involve the esophagus, lung, and kidney.

Candidiasis of the esophagus may have an insidious onset or may cause acute intense odynophagia. It is important to remember that esophageal candidiasis can occur with minimal or absent thrush (oral involvement). If the condition is promptly treated, complete recovery can be rapid. However, if the organism enters the circulation through ruptured esophageal vessels, systemic invasion can lead to dissemination of *Candida* to other organs and often to death.

Candidiasis involves primarily the distal third of the esophagus. Initially there is only abnormal motility and a slightly dilated, virtually atonic esophagus. The earliest morphologic changes are small, marginal filling defects (simulating tiny air bubbles) with fine serrations along the outer border. The classic radiographic appearance in more advanced esophageal candidiasis is a shaggy marginal contour caused by deep ulcerations and sloughing of the mucosa. Nodular, edematous thickening of the intervening mucosa and the presence of pseudomembranous plaques produce an irregular cobblestone pattern. Narrowing and rigidity extending over a long segment of the esophagus may develop during the healing phase.

Pulmonary candidiasis produces a nonspecific radiographic pattern and almost always reflects hematoge-

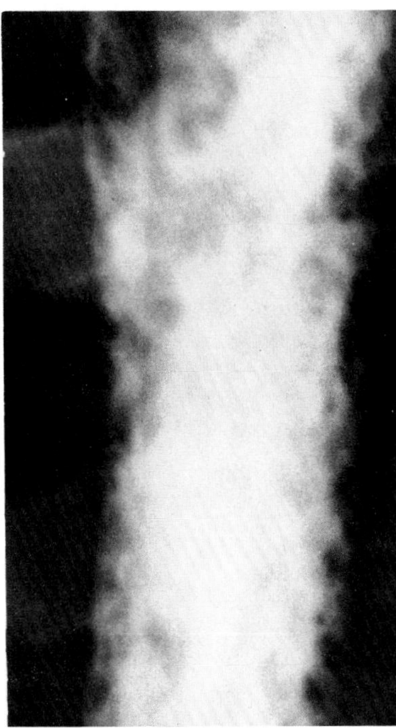

**Figure 14.27. Esophageal candidiasis.** Multiple round and oval nodular filling defects are evident throughout the barium-filled esophagus. (From Eisenberg,[39] with permission from the publisher.)

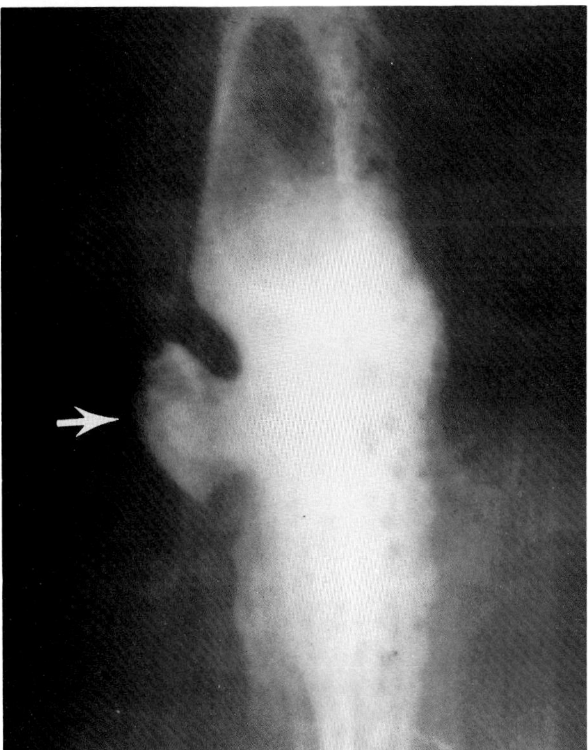

**Figure 14.28. Esophageal candidiasis.** A nodular cobblestone pattern is seen in combination with a large, discrete ulceration (arrow). (From Eisenberg,[39] with permission from the publisher.)

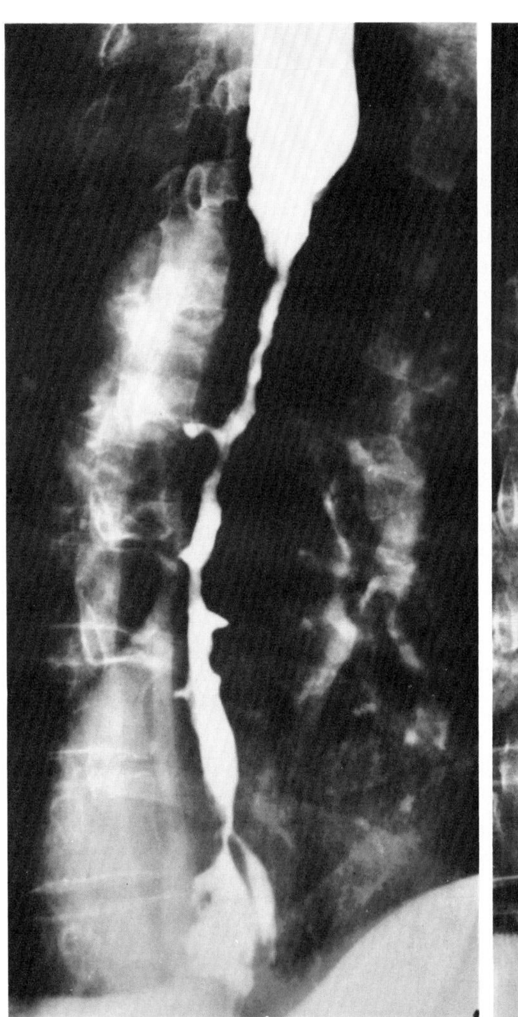

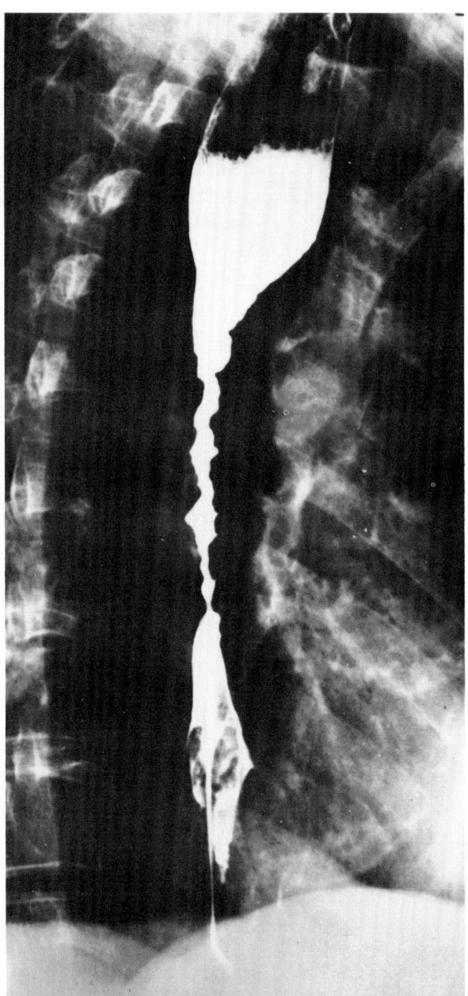



**Figure 14.29. Esophageal candidiasis.** (A) There are diffuse transverse ulcerations and irregular esophageal narrowing. (B) In another patient, there is irregular narrowing of the distal two-thirds of the esophagus with multiple transverse ulcerations. (From Ott and Gelfand,[23] with permission from the publisher.)

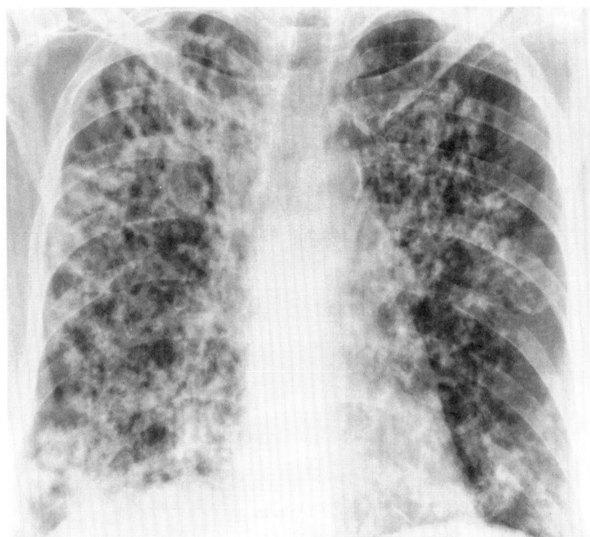

**Figure 14.30. Pulmonary candidiasis.** A diffuse pattern of ill-defined nodules throughout both lungs.

nous dissemination. The most common appearances are patchy, segmental, homogeneous air space consolidation and a diffuse pattern of ill-defined miliary nodules. Cavitation and hilar adenopathy may occur.

Renal candidiasis also results from hematogenous spread of the infection. It often appears on excretory urograms as shaggy, irregular filling defects (fungus balls) in the renal pelvis, which often extend into the infundibula and upper ureter. Candidiasis can result in acute papillary necrosis or chronic pyelonephritis, often with a marked decrease in renal function. The development of fungus balls in the bladder usually follows the catheterization or instrumentation of a patient who has diabetes mellitus or who is receiving broad-spectrum antibiotics.[20–24]

## ASPERGILLOSIS

Aspergillosis is almost always a secondary infection in which the fungus colonizes a damaged bronchial tree,

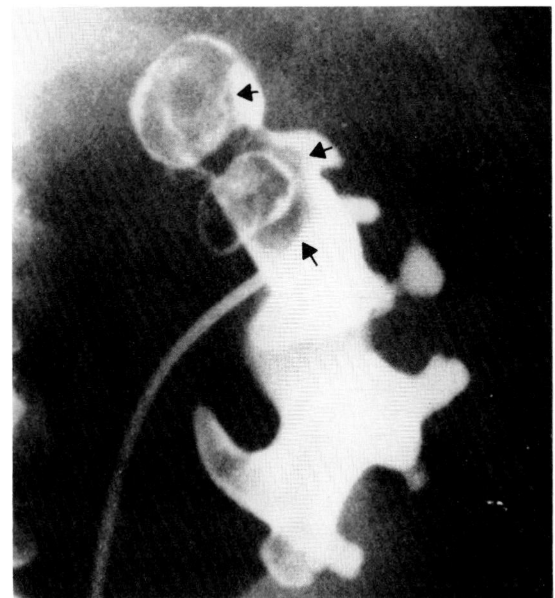

**Figure 14.31. Renal candidiasis.** A retrograde pyelogram demonstrates a large filling defect (arrows) in the left renal pelvis. (From Boldus, Brown, and Culp,[21] with permission from the publisher.)

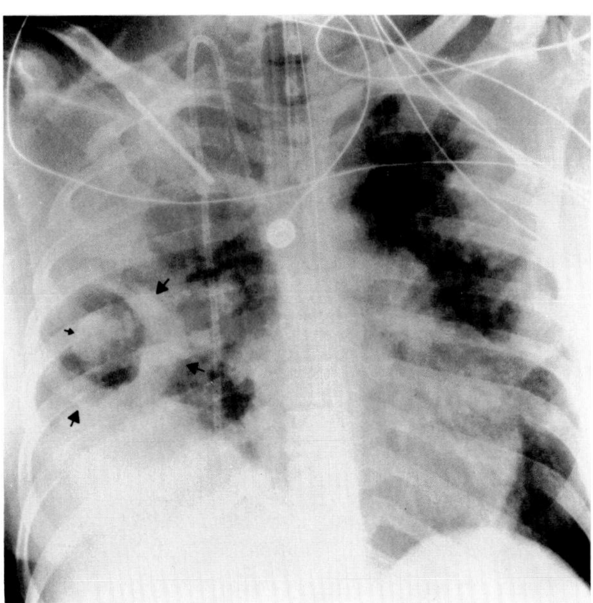

**Figure 14.32. Aspergillosis.** In the right lower lobe there is a large, thick-walled cavity (large arrows) containing an intracavitary fungus ball (small arrow).

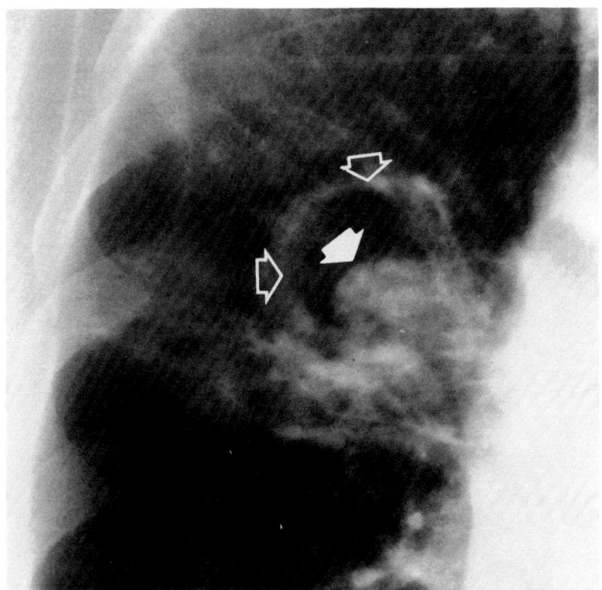

**Figure 14.33. Aspergillosis.** A mycetoma (solid arrow) appears as a homogeneous rounded mass that is separated from the thick wall of the cavity by a crescent-shaped air space (open arrows).

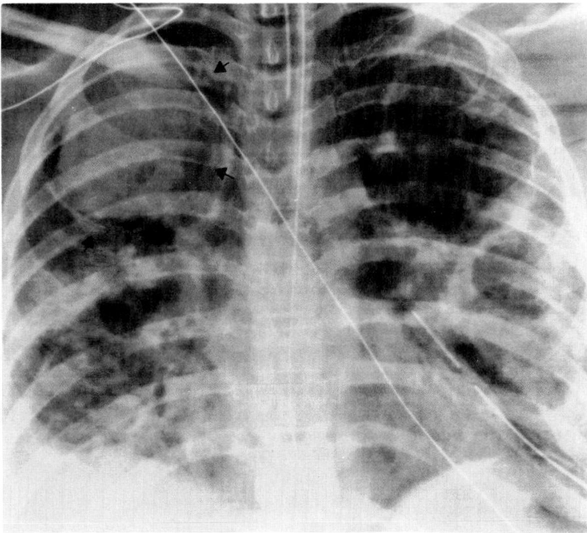

**Figure 14.34. Aspergillosis.** Multiple cavities of varying sizes are superimposed on a diffuse pulmonary infiltrate. A fungus ball almost fills the large cavity in the right upper lobe (arrows). A right pleural effusion is also seen in this patient with chronic lymphocytic leukemia.

pulmonary cyst, or cavity of a patient with underlying lung disease. It may also be an opportunistic invader in patients with debilitating disease and those on antibiotic or immunosuppressive therapy.

The radiographic hallmark of secondary aspergillosis is a pulmonary mycetoma, a conglomeration of intertwined fungal hyphae matted together with fibrin, mucus, and cellular debris within a pulmonary cavity. The fungus is almost always a pure saprophyte, and the underlying cavity can usually be ascribed to other causes, such as tuberculosis, bronchiectasis, sarcoidosis, histoplasmosis, a bronchial cyst, a chronic bacterial abscess, or a carcinoma. The mycetoma appears as a solid homogeneous rounded mass within a spherical or ovoid cavity; the fungal mass is separated from the wall of the cavity by a crescent-shaped air space. Air-fluid levels are not seen within the cavities. The intracavitary fungus ball occurs much more commonly in the upper than in the lower lobes, presumably because it so frequently occupies a tuberculous cavity. The cavities are usually thin-walled and are often contiguous to a pleural surface, which may be thickened. The mycetoma may move within the cavity when the patient changes position. Calcification of the mycelial mass can appear as scattered small nodules, a fine peripheral rim, or an extensive process involving most of the fungus ball.

In patients with severe debilitating diseases, such as reticuloendothelial malignancies, and those on steroid, immunosuppressive, or antibiotic therapy, secondary aspergillosis may cause single or multiple areas of pulmonary consolidation with poorly defined margins. Cavitation may occur. Miliary spread throughout the lungs is very rare.

Primary infection with *Aspergillus* is extremely rare and usually presents the radiographic pattern of a nonspecific pneumonia that may progress to abscess formation.

Allergic bronchial aspergillosis is a distinct clinical entity that probably represents a hypersensitivity reaction, in which the fungal infection develops in adult asthmatics and is associated with eosinophilia. The major radiographic findings relate to mucous plugs of aspergilli and eosinophils in segmental bronchi. These plugs appear as characteristic homogeneous, bandlike densities that resemble gloved fingers and usually involve the upper lobes and the more central, rather than peripheral, bronchi. Thickening of bronchial walls causes parallel or slightly tapering line shadows ("tramlines") that appear as hairline ring shadows when seen on end. Patchy or homogeneous air space consolidation may occur. These infiltrates are usually transient but may persist unchanged for a prolonged period.[25–27]

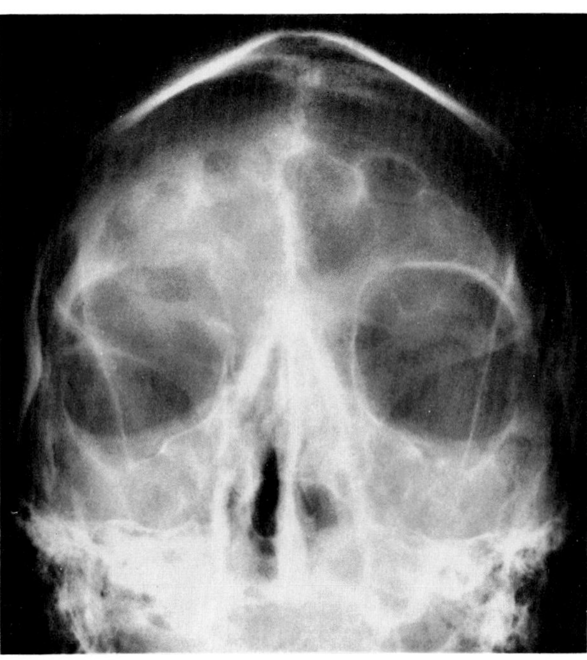

**Figure 14.35. Pansinusitis with osteomyelitis in mucormycosis.** There is destruction of the roof of the right orbit with a loss of the mucoperiosteal line of the right frontal sinus.

## MUCORMYCOSIS

The fungus that causes mucormycosis is generally not pathogenic in humans, and severe disseminated infection usually occurs in patients with diabetes or an underlying malignancy (leukemia, lymphoma). The disease usually originates in the nose and paranasal sinuses, where it produces mucosal thickening and clouding of one or more of the sinuses. There often is an associated air-fluid level. Extension of the infection can cause destruction of the walls of the sinus and an appearance that simulates that of a malignant neoplasm. Carotid arteriography may demonstrate invasion or obstruction of the carotid siphon, and CT may show extension of the infection into the substance of the brain.

Pulmonary mucormycosis is a progressive severe pneumonia that is widespread and confluent, and often cavitates. It may occasionally produce a solitary pulmonary nodule.[28–30,41]

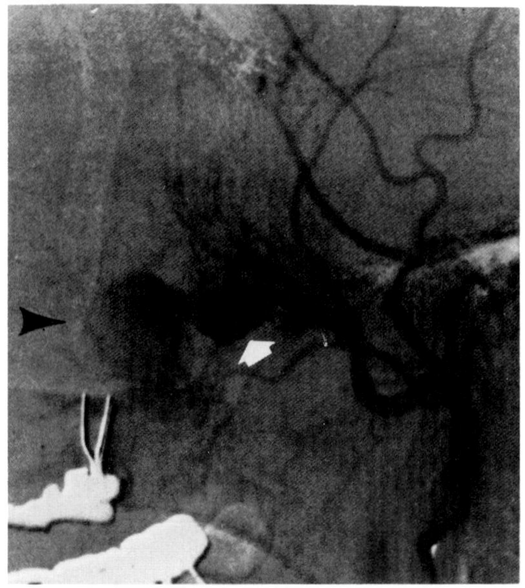

**Figure 14.36. Mucormycosis.** An external carotid arteriogram shows a pseudoaneurysm (white arrow) arising from the sphenopalatine artery and extravasation of contrast material into the left maxillary sinus (arrowhead). Clinically, the patient had severe upper nasal and oral hemorrhaging at the time of the arteriogram.

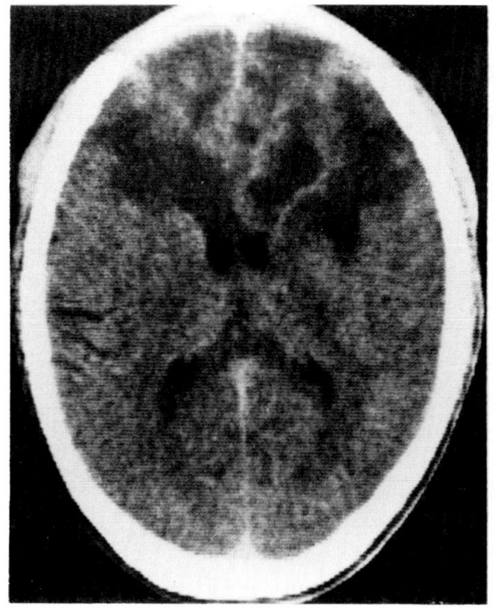


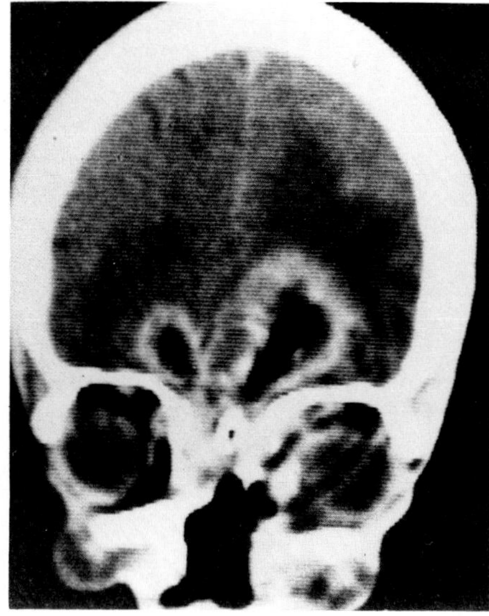


**Figure 14.37. Mucormycosis.** (A) A CT scan with contrast enhancement shows multiple diffuse inflammatory masses involving both frontal lobes and causing pressure on the frontal horns. (B) An enhanced coronal CT scan obtained 3 weeks later shows bilateral frontal lobe abscesses and associated cerebritis. The air within the right-sided abscess relates to an abnormal communication between the abscess and the right ethmoid sinus. (From Lazo, Wilner, and Metes,[41] with permission from the publisher.)

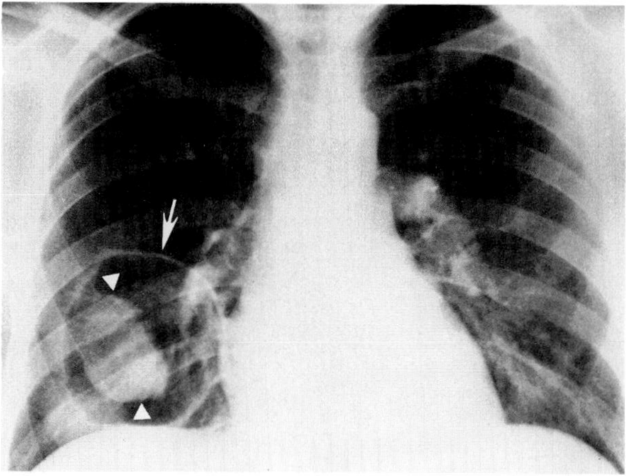

**Figure 14.38. Fungus ball in pulmonary mucormycosis.** A frontal chest radiograph demonstrates a large, thin-walled cavity (arrows). Within the cavity is a smooth, elliptical, homogeneous mass (arrowheads) representing the fungus ball.

**Figure 14.39. Sporotrichosis.** There is a severe destructive arthritis that has led to chronic osteomyelitis of the talus and distal tibia with virtual ankylosis across the tibiotalar joint.

## SPOROTRICHOSIS

Sporotrichosis is a chronic infection caused by the fungus *Sporothrix schenckii* that is usually limited to the skin and the draining lymphatics. In rare instances, disseminated disease can involve the skeletal system and the lungs.

The disease tends to involve the knee, the elbow, the wrist, and the small joints of the hands and feet and produces an extensive destructive arthritis with large-joint effusions. The radiographic appearance often simulates rheumatoid or tuberculous arthritis.

The pulmonary lesions of sporotrichosis produce a variety of nonspecific radiographic patterns, which include cavitary nodular masses, fibronodular infiltrates, and chronic pneumonia. Hilar lymph node enlargement occurs in a high percentage of the cases and may cause bronchial obstruction. The infection may spread through the pleura into the chest wall, creating a sinus tract.[31–33,42]

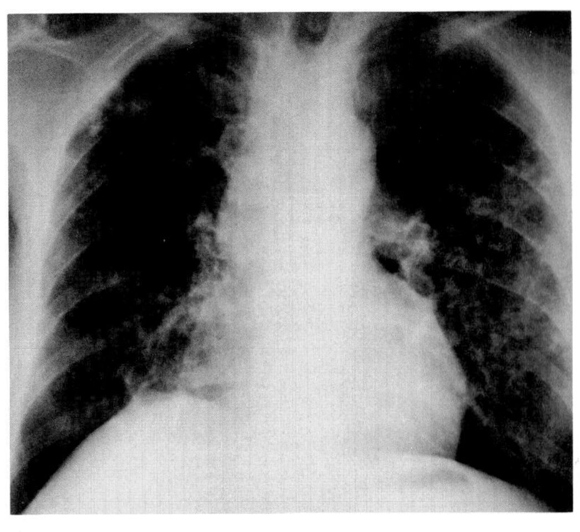

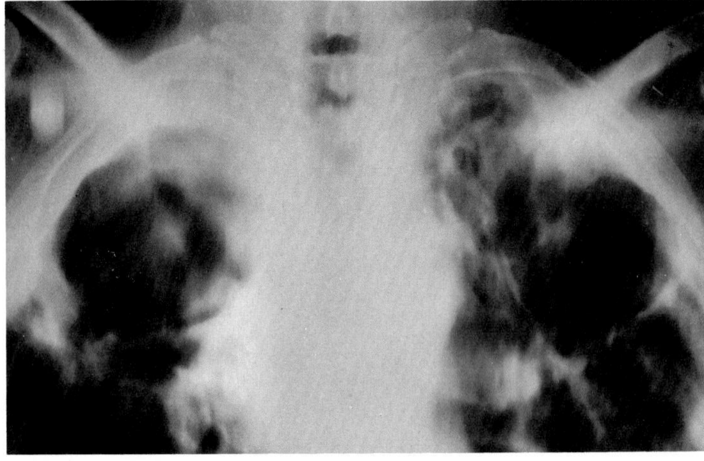

A B

**Figure 14.40. Sporotrichosis.** (A) A frontal chest film demonstrates diffuse interstitial lung disease and a right upper lobe cavity. (B) A frontal tomogram obtained 4 years later demonstrates marked progression with extensive bilateral upper lobe cavities. (From Naimark and Tiu,[42] with permission from the publisher.)

## PARACOCCIDIOIDOMYCOSIS (SOUTH AMERICAN BLASTOMYCOSIS)

Paracoccidioidomycosis is a fungal infection caused by *Paracoccidiodes brasiliensis* that is common in Brazil and occasionally occurs elsewhere in South and Central America. The organism probably enters the body by inhalation and often causes a nonspecific bilateral patchy pneumonia. Disseminated disease may affect the gastrointestinal tract and typically causes severe narrowing, rigidity, and mucosal irregularity of the cecum and terminal ileum; this produces a coned cecum appearance simulating amebiasis or Crohn's disease.[34,35]

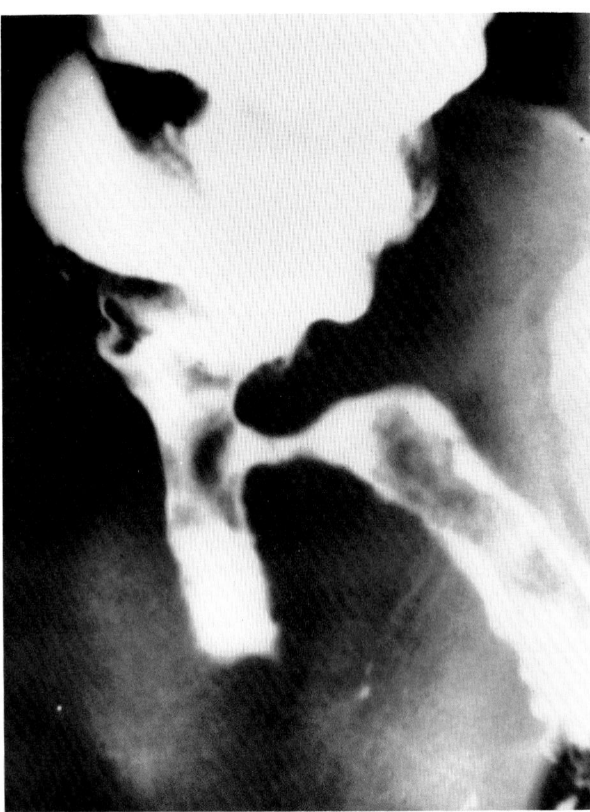

**Figured 14.41. Paracoccidioidomycosis.** There is severe narrowing and rigidity of the terminal ileum and cecum (coned cecum). (From Avritchir and Perroni,[34] with permission from the publisher.)

## MYCETOMA

Mycetoma is a chronic fungal infection in which soil organisms enter the skin through minor trauma. The most common site of infection is the tarsometarsal area of the foot (madura foot). The major radiographic findings include deep soft tissue swelling, often with superficial nodules, and diffuse, irregular bone destruction. Areas of endosteal sclerosis are common, and this produces a mixed lytic-sclerotic appearance. Sinus tracts are frequent, and there may be a palisade or a bristled periosteal reaction.[36]

### REFERENCES

1. Flynn MW, Felson B: The roentgen manifestations of thoracic actinomycosis. *AJR* 110:707–716, 1970.
2. Frank P, Strickland B: Pulmonary actinomycosis. *Br J Radiol* 47:373–378, 1974.
3. Schwarz J, Baum GL: Actinomycosis. *Semin Roentgenol* 5:58–71, 1970.
4. Smith DL, Lockwood WR: Disseminated actinomycosis. *Chest* 67:242–244, 1975.
5. Varadarajan MJ: Actinomycosis of bone. *Punjab Med J* 10:321–325, 1961.

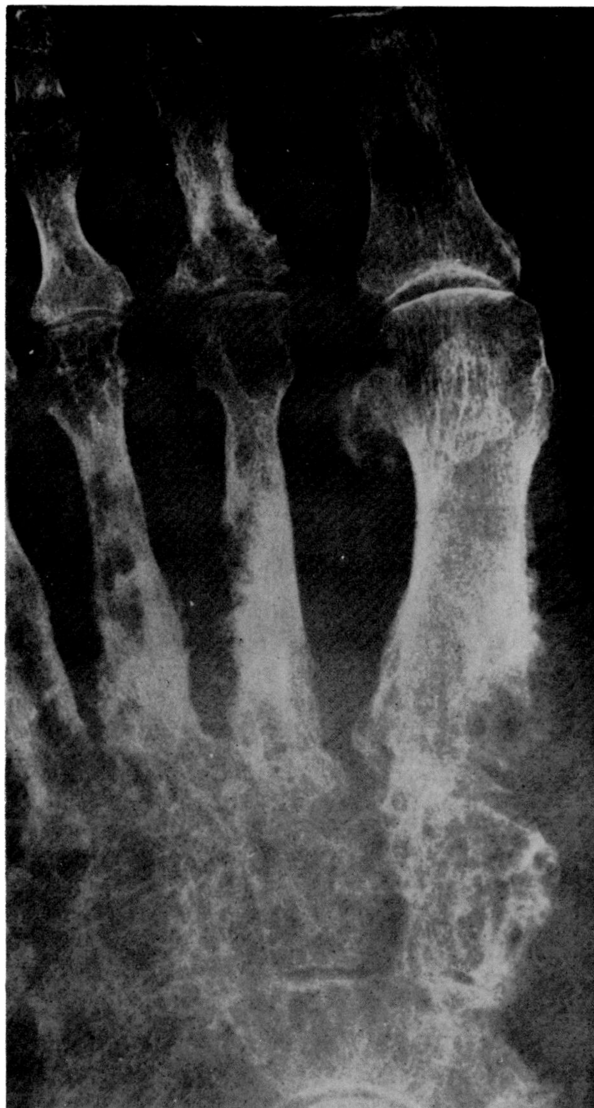

**Figured 14.42. Madura foot.** Diffuse, confluent, and predominantly osteolytic lesions involve the entire foot of an itinerant laborer in Alabama who often could not afford shoes. Minimal periosteal reaction (arrows) is identified along the medial aspect of the first metatarsal and along the lateral aspect of the fourth metatarsal. The tarsal and tarsometatarsal joints are destroyed, and the soft tissue swelling is pronounced. (From Rosen and Jacobson,[10] with permission from the publisher.)


6. Weese WC, Smith IM: A study of 57 cases of actinomycosis over a 36 year period. *Arch Intern Med* 135:1562–1568, 1975.
7. Grossman CB, Bragg DG, Armstrong D: Roentgen manifestations of pulmonary nocardiosis. *Radiology* 96:325–330, 1970.
8. Hathaway BM, Mason KN: Nocardiosis. A study of fourteen cases. *Am J Med 32:903–909, 1962.*
9. Gordonson J, Birnbaum W, Jacobson G, et al: Pulmonary cryptococcosis. *Radiology* 112:557–561, 1974.
10. Rosen RS, Jacobson G: Fungus disease of bone. *Semin Roentgenol* 1:370–391, 1966.
11. Witorsch P, Utz JP: North American blastomycosis: A study of 40 patients. *Medicine* 47:169–200, 1968.
12. Hawley C, Felson B: Roentgen aspects of intrathoracic blastomycosis. *AJR* 75:751–757, 1956.
13. Joyce PF: A rare clinical presentation of blastomycosis. *Skeletal Radiol* 2:239–242, 1977.
14. Forrest JV: Common fungal diseases of the lungs. Histoplasmosis. *Radiol Clin North Am* 11:163–176, 1970.
15. Christoforidis AJ: Radiologic manifestations of histoplasmosis. *AJR* 109:478–490, 1970.
16. Bank S, Trey C, Gans I, et al: Histoplasmosis of the small bowel with "giant" intestinal villi and secondary protein-losing enteropathy. *Am J Med* 39:492–501, 1965.
17. Greendyke WH, Resnick DL, Harvey WC: Primary coccidioidomycosis. *AJR* 109:491–499, 1970.
18. Dalinka MK, Dinninberg S: Roentgenographic features of osseous coccidioidomycosis. *J Bone Joint Surg* 53:1157–1162, 1971.
19. Schwarz J, Baum GL: Coccidioidomycosis. *Semin Roentgenol* 5:29–37, 1970.
20. Clark RE, Minagi H, Palubinskas AJ: Renal candidiasis. *Radiology* 101:567–572, 1971.
21. Boldus RA, Brown RC, Culp DA: Fungus balls in the renal pelvis. *Radiology* 102:555–557, 1972.
22. Lewicki AM, Moore JP: Esophageal monoliasis: A review of common and less frequent characteristics. *AJR* 125:218–225, 1975.
23. Ott DJ, Gelfand DW: Esophageal stricture secondary to candidiasis. *Gastrointest Radiol* 2:323–325, 1978.
24. Greer AE: *Disseminating Fungus Diseases of the Lung,* Springfield, Ill, Charles C Thomas, 1962.
25. McCarthy DS, Simon G, Hargreave FE: Radiological appearances in allergic bronchopulmonary aspergillosis. *Clin Radiol* 21:366–375, 1970.
26. Freundlich IM, Israel HL: Pulmonary aspergillosis. *Clin Radiol* 24:248–253, 1973.
27. Curtis AM, Smith GJW, Ravin CE: Air crescent sign of invasive aspergillosis. *Radiology* 133:17–21, 1979.
28. Bartrum RJ, Watneck M, Herman PG: Roentgenographic findings in pulmonary mucormycosis. *AJR* 117:810–815, 1973.
29. Gabriele OF: Mucormycosis. *AJR* 83:227–235, 1960.
30. McBride RA, Corson JM, Gammin GJ: Mucormycosis. *Am J Med* 28:832–846, 1960.
31. Winter TQ, Pearson KV: Systemic sporotrichosis. *Radiology* 104:579–584, 1972.
32. Ridgeway NA, Whitcombe FC, Erickson EE, et al: Primary pulmonary sporotrichosis. *Am J Med* 32:153–160, 1962.
33. Comstock C, Wolson AH: Roentgenology of sporotrichosis. *AJR* 125:651–655, 1975.
34. Avritchir Y, Perroni AA: Radiological manifestations of small intestinal South American blastomycosis. *Radiology* 127: 607–609, 1978.
35. Schwarz J, Baum GL: Paracoccidioidomycosis (South American blastomycosis). *Semin Roentgenol* 5:69–75, 1970.
36. Davies AGM: Bone changes in madura foot (mycetoma). *Radiology* 70:841–847, 1958.
37. Enzmann DR, Brant-Zawadzki M, Britt RH: CT of central nervous system infections in immunocompromised patients. *AJNR* 1:239–243, 1980.
38. Halvorson RA, Duncan JD, Merten DF, et al: Pulmonary blastomycosis. *Radiology* 150:1–5, 1984.
39. Eisenberg RL: *Gastrointestinal Radiology: A Pattern Approach.* Philadelphia, JB Lippincott, 1983.
40. McGahan JP, Graves DS, Palmer PES, et al: Classic and contemporary imaging of coccidioidomycosis. *AJR* 136:393–404, 1981.
41. Lazo A, Wilner HI, Metes JJ: Craniofacial mucormycosis. *Radiology* 139:623–627, 1981.
42. Naimark A, Tiu S: Primary pulmonary sporotrichosis. *J Canad Assoc Radiol* 30:129–130, 1979.
43. Fraser RG, Pare JAP: *Diagnosis of Diseases of the Chest.* Philadelphia, WB Saunders, 1979.

# Chapter Fifteen
# Rickettsial Infections

About half of the patients with Q fever develop pneumonia, which typically produces a dense, homogenenous, segmental, or lobar consolidation simulating pneumococcal disease. The pulmonary infiltrates predominantly affect the lower lobes and may be bilateral. Pleural effusion occurs in about one-third of the cases; hilar involvement and small focal lesions are rare.

Pulmonary involvement has occasionally been reported in patients with other rickettsial diseases such as Rocky Mountain spotted fever and severe cases of ty-

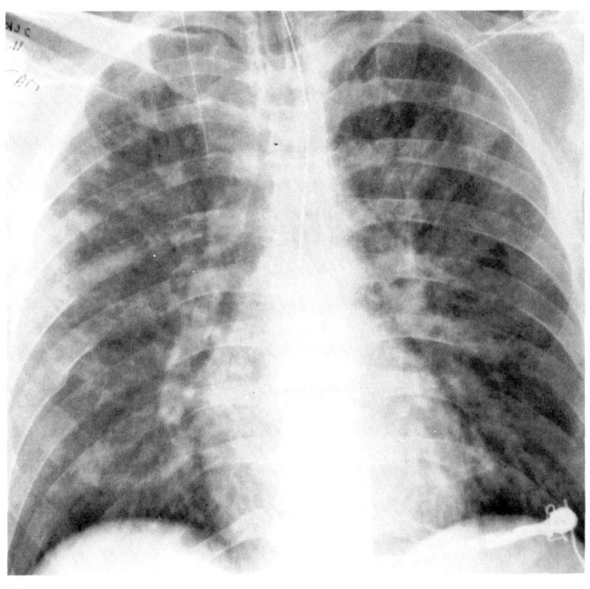

A

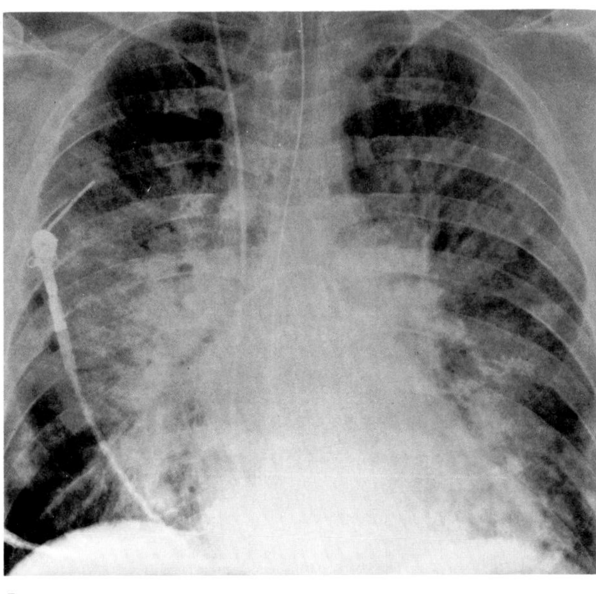

B

**Figure 15.1. Rocky Mountain spotted fever.** (A) Diffuse interstitial pulmonary infiltrates progressing to (B) diffuse end air space consolidation 1 day later. The patient subsequently died of the infection. (From Martin, Choplin, Shertzer, et al,[4] with permission from the publisher.)

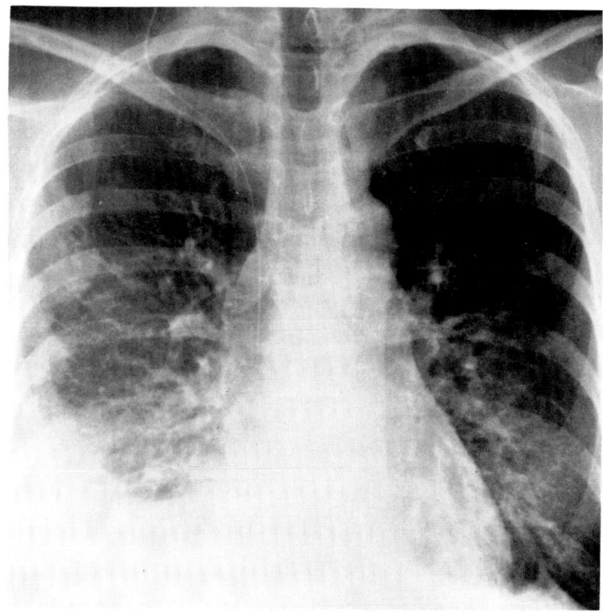

**Figure 15.2. Rocky Mountain spotted fever.** Bibasilar end air space consolidation with right pleural effusion. The patient recovered. (From Martin, Choplin, Shertzer, et al,[4] with permission from the publisher.)

phus. The nonspecific alveolar and interstitial infiltrates that are usually scattered throughout the chest may reflect either focal rickettsial involvement or secondary bacterial superinfection.[1–4]

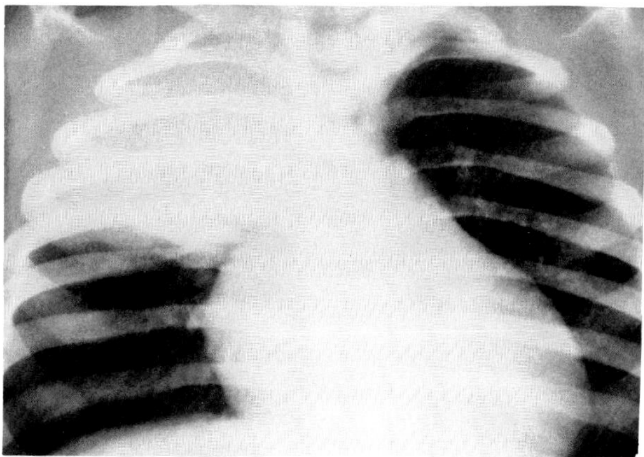

**Figure 15.3. Q fever.** Right upper lobe air space consolidation simulating pneumococcal pneumonia.

## REFERENCES


1. Johnson JE, Kadull PJ: Laboratory-acquired Q fever: A report of 50 cases. *Am J Med* 41:391–403, 1966.
2. Jacobson G, Denlinger RB, Carter RA: Roentgen manifestations of Q fever. *Radiology* 53:739–749, 1949.
3. Marrie TJ, Haldane EV, Noble MA, et al: Q fever in Maritime Canada. *Can Med Assoc J* 126:1295–1300, 1982.
4. Martin W, Choplin RH, Shertzer ME: The chest radiograph in Rocky Mountain spotted fever. *AJR* 139:889–893, 1982.

## Chapter Sixteen
# Mycoplasma and Chlamydia

### *MYCOPLASMA PNEUMONIAE* INFECTION

Infection with *Mycoplasma pneumoniae* causes a radiographic appearance that is indistinguishable from that of many viral pneumonias. Most infections are mild, though the radiographic signs are more extensive than might be suspected from the physical examination. The infection is characterized by intrafamily spread, and children 5 to 10 years old, especially boys, have the greatest incidence of disease.

Initially, acute interstitial inflammation appears as a fine or coarse reticular pattern. This is followed by patchy air space consolidation that is usually segmental and predominantly involves the lower lobes. Bilateral and multilobular involvement are common. Pleural effusion is infrequent.[1–3]

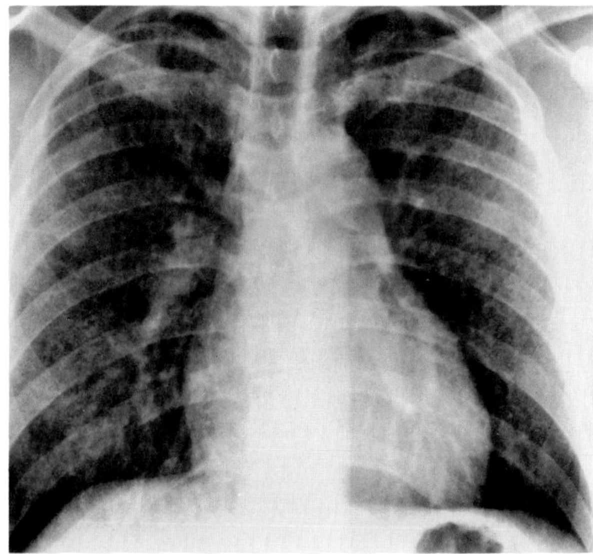

**Figure 16.1. *Mycoplasma pneumoniae* infection throughout both lungs.** A diffuse fine reticular pattern represents acute interstitial inflammation. The radiographic appearance is indistinguishable from that of most viral pneumonias.

### LYMPHOGRANULOMA VENEREUM

Lymphogranuloma venereum is a sexually transmitted chlamydial infection that is especially common in the tropics. The rectum is the first and usually the only por-

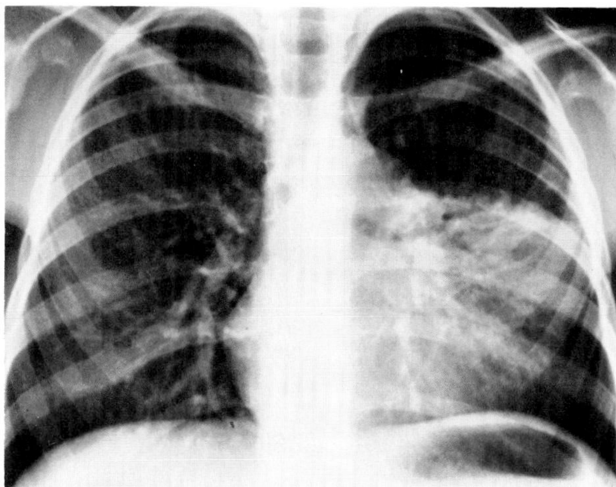

**Figure 16.2. *Mycoplasma pneumoniae* infection.** Patchy air space consolidation involves both segments of the lingula.

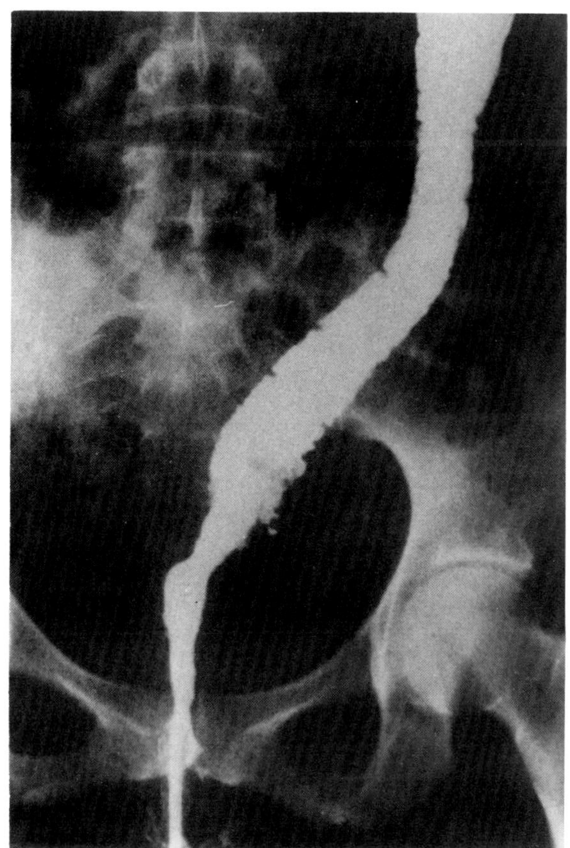

**Figure 16.4. Lymphogranuloma venereum.** There is a long stricture of the rectum and lower sigmoid with shortening and straightening of the more proximal colon. Mucosal ulceration is also seen. (From Cockshott and Middlemiss,[6] with permission from the publisher.)

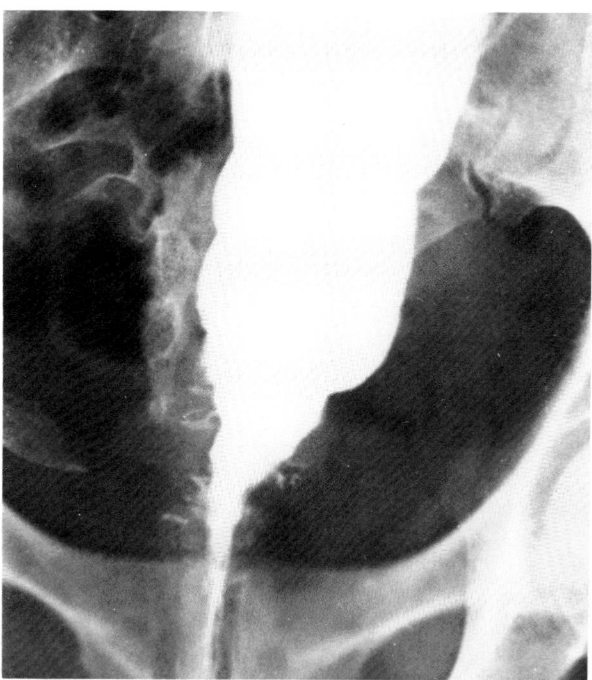

**Figure 16.3. Lymphogranuloma venereum.** There is a long rectal stricture with multiple deep ulcers. (From Dreyfuss and Janower,[5] with permission from the publisher.)

tion of the colon involved. Invasion and blockage of the rectal lymphatics, combined with secondary infection, lead to rectal edema with cellular infiltrate in the submucosa and muscularis. In the early stages of the disease, the bowel is spastic and irritable and has boggy and edematous mucosa and multiple shaggy ulcers. Fistulas and sinus tracts of varying length are frequently present.

The hallmark of lymphogranuloma venereum is the development of a rectal stricture in the chronic stage of the disease. These strictures are usually long and tubular, beginning just above the anus and varying in radiographic appearance from a short, isolated narrowing to a stenotic segment up to 25 cm long. The mucosa is irregular, with multiple deep ulcers; the lumen can be so narrow that it resembles a thin string. The portion of normal colon proximal to the stricture is usually dilated, has a loss of haustration, and gradually tapers in a smooth, conical fashion. In a patient with a rectal stricture, the demonstration of fistulas and sinus tracts communicating with perirectal abscess cavities, the lower vagina, or perianal skin should suggest a diagnosis of lymphogranuloma venereum.[4–6]

## PSITTACOSIS

Psittacosis is a chlamydial infectious disease of birds that is most often transmitted to humans by members of the

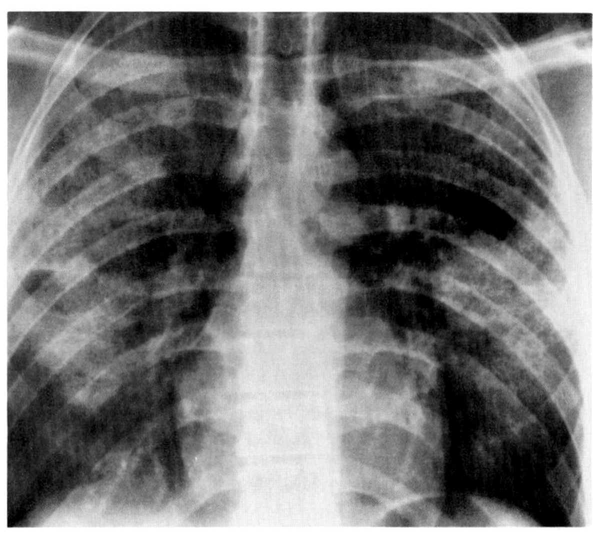

**Figure 16.5. Psittacosis.** The bilateral patchy pulmonary infiltrates developed in the owner of a pet store.

parrot family. The radiographic changes in the lungs are nonspecific and, as in most nonbacterial pneumonias, are more extensive than the physical signs of the disease would suggest. Psittacosis usually appears as a multifocal, patchy infiltrate that is often bilateral and has a tendency to change rapidly in appearance and distribution. The disease may also appear as a lobar or wedge-shaped consolidation or may present a diffuse reticular or miliary pattern. Resolution of the radiographic abnormalities is often delayed for many weeks after a clinical cure has occurred.[7,8]

## REFERENCES


1. Stenstrom R, Jansson E, von Essen R: *Mycoplasma* pneumonias. *Acta Radiol [Diagn]* 12:833–841, 1972.
2. Rosmus HH, Pare JAP, Mason AM, et al: Roentgenographic patterns of acute *Mycoplasma* and viral pneumonitis. *J Can Assoc Radiol* 19:74–79, 1968.
3. Putman CE, Curtis A McB, Simeone JF, et al: *Mycoplasma* pneumonia. Clinical and roentgenographic patterns. *AJR* 124: 417–422, 1975.
4. Annamuthodo H, Marryatt J: Barium studies in intestinal lymphogranuloma venereum. *Br J Radiol* 34:53–57, 1961.
5. Dreyfuss JR, Janower ML: *Radiology of the Colon*. Baltimore, Williams & Wilkins, 1980.
6. Cockshott WP, Middlemiss H: *Clinical Radiology in the Tropics*. Edinburgh, Churchill Livingstone, 1979.
7. Schaffner W, Brutz DJ, Duncan GW, et al: The clinical spectrum of endemic psittacosis. *Arch Intern Med* 119:433–443, 1967.
8. Stenstrom R, Jansson E, Wager O: Ornithosis pneumonia with special reference to roentgenological lung findings. *Acta Med Scand* 171:349–356, 1962.

# Chapter Seventeen
# Viral Diseases

## VIRAL PNEUMONIA

Pneumonia can be caused by many types of respiratory viruses. In many patients, there is a disparity between the severity of the respiratory symptoms and the relative paucity of radiographic changes. Conversely, radiographic changes often persist for days or weeks after clinical recovery. The most common appearance is a peribronchial infiltrate that reflects edema and mononuclear cell infiltration of the interstitial tissues. This may extend to produce a diffuse reticular pattern throughout the lungs. Nodular or patchy alveolar infiltrates can also occur. Viral pneumonia has a predilection for the lower lobes. Lobular, segmental, and lobar opacifications may occur. Pleural effusion is rare in adults but occurs occasionally in children.[1–3]

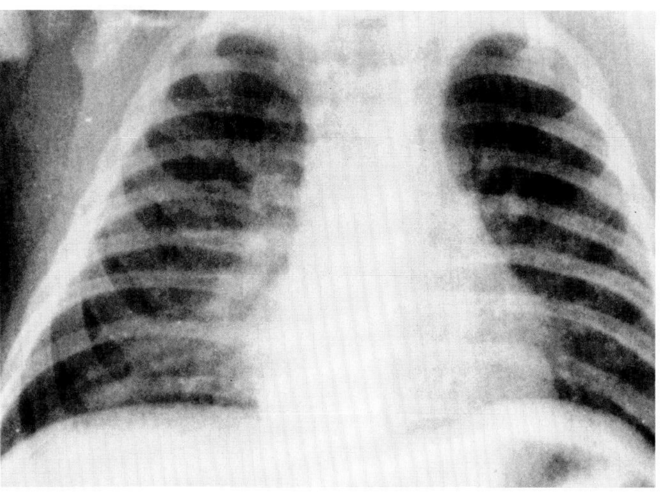

A

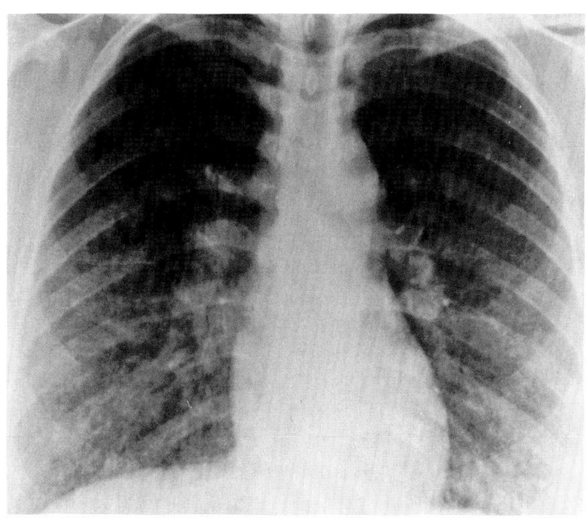

B

**Figure 17.1. Typical viral pneumonia.** Diffuse interstitial infiltrates with perihilar haze in (A) a child and (B) an adult.

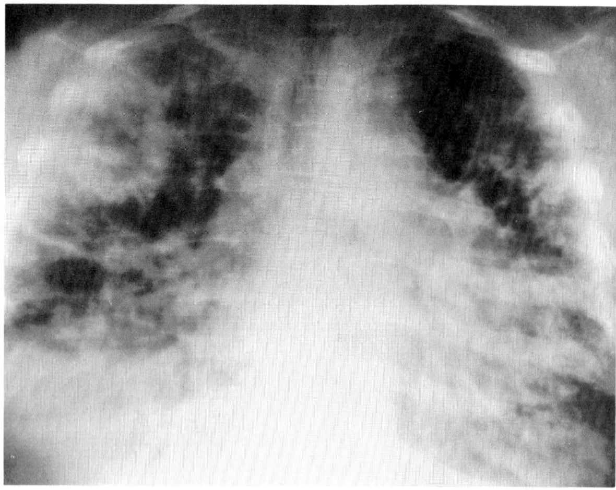

**Figure 17.2. Viral pneumonia.** A diffuse peribronchial infiltrate with associated air space consolidation obscures the heart borders (shaggy heart sign) in this patient with severe viral pneumonia. A patchy alveolar infiltrate in the right upper lung is also seen.

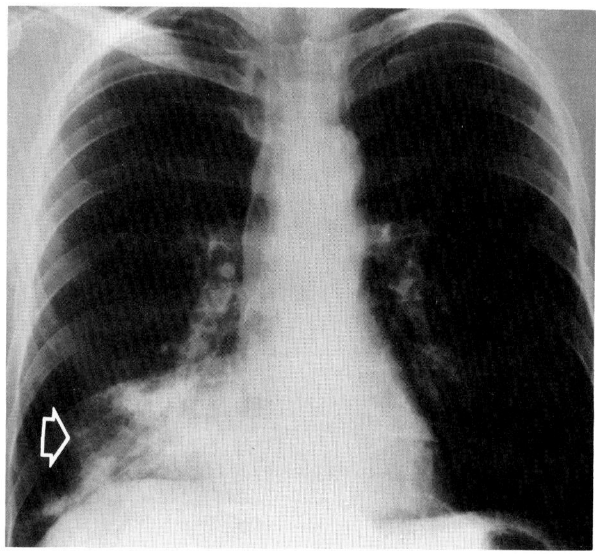

**Figure 17.3. Influenza pneumonia.** Relatively homogeneous air space consolidation at the left base (arrow). Although the appearance simulates a bacterial lobar pneumonia, this infiltrate has irregular borders unlike the sharp margins seen with lobar consolidation.

## INFLUENZA

Influenza is a highly contagious acute respiratory infection characterized by the sudden onset of headache, myalgia, fever, and prostration. Pneumonia is an uncommon, but dreaded, complication of influenza. Although often localized and of mild to moderate severity, it is sometimes fatal. In mild cases, a patchy or homogeneous segmental infiltrate is indistinguishable from other viral pneumonias. The disease most commonly involves the lower lobes and is often bilateral. In severe epidemics

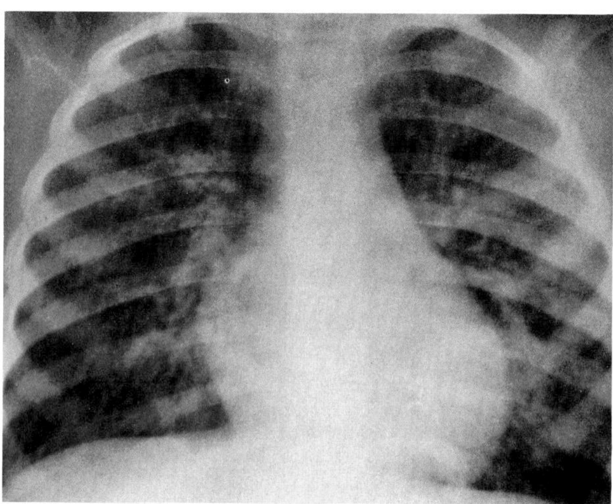

**Figure 17.4. Measles pneumonia.** A diffuse, reticular interstitial infiltrate predominantly involves the lower lungs.

of the past, the pneumonia was often of the interstitial type with a coarsened bronchovascular pattern extending out from the hila and relatively sparing the periphery. Most fatalities in influenza are due to secondary bacterial pneumonia, especially due to staphylococci, that develops within 2 weeks after the initial viral infection.[4–6,33]

## MEASLES (RUBEOLA)

Primary pneumonia caused by measles virus produces a diffuse reticular pattern that has a predilection for the bases and reflects the interstitial location of the disease. Involvement of the reticuloendothelial system results in prominent hilar and mediastinal adenopathy. Segmental consolidation and atelectasis frequently develop, usually because of superimposed bacterial infection. A rare pulmonary complication is interstitial giant cell pneumonia, a fulminant and often fatal condition that is seen most often in children suffering from severe systemic disease (leukemia, congenital immunodeficiency, malnutrition). This pneumonia may occur in the absence of the typical measles exanthem.[7–9]

## RUBELLA (GERMAN MEASLES)

Rubella is usually a benign febrile exanthem, but when it occurs in pregnant women during the first trimester, transplacental passage of the virus may lead to fetal infection which results in serious malformations. Among the abnormalities reported in congenital rubella are congenital heart disease, eye lesions, deafness, hepatosplenomegaly, thrombocytopenic purpura, and mental retardation.

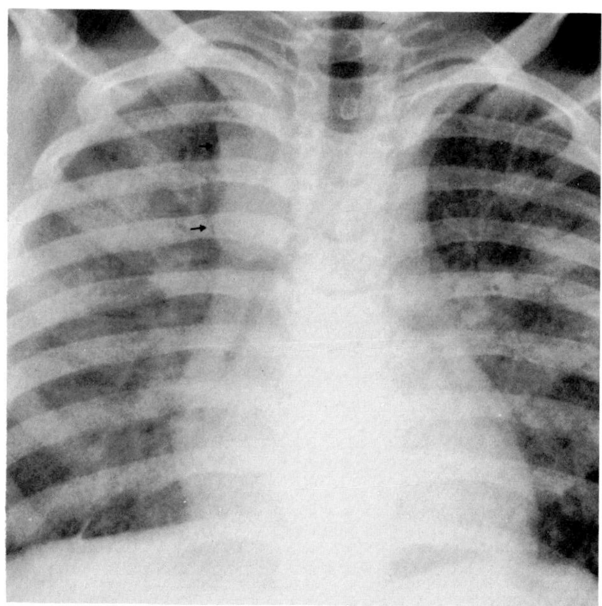

**Figure 17.5. Measles pneumonia.** A diffuse, reticular interstitial infiltrate with a focal area of consolidation in the right upper lobe. Note the striking right hilar and mediastinal adenopathy (arrows).

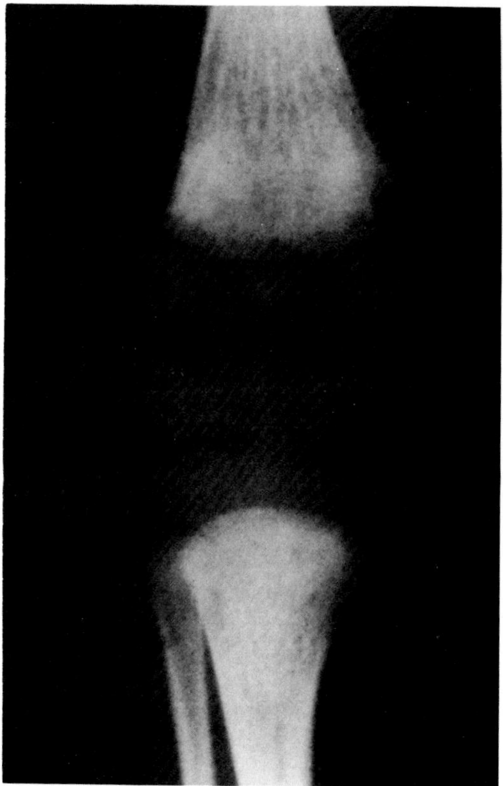

**Figure 17.6. Rubella.** A radiograph of the knee in a 1-day-old girl with a maternal history of rubella demonstrates alternating lucent and sclerotic longitudinal striations extending perpendicular to the epiphyseal plate and parallel to the long axis of the bone (celery stick sign). (From Singleton, Rudolph, Rosenberg, et al,[10] with permission from the publisher.)

Radiographically, congenital rubella primarily affects the skeletal system. The changes are seen at birth and involve the metaphyses of the long bones, especially at the knee. The zones of provisional calcification are absent or poorly defined, and the metaphyses are irregular with coarse trabeculation. Unlike congenital syphilis, rubella causes no periosteal reaction. The most striking characteristic is alternating lucent and sclerotic longitudinal striations that extend perpendicular to the epiphyseal plate and parallel to the long axis of the bone. This "celery stick" pattern fades into normal-appearing bone in the shafts. Although similar bone changes may occur in cytomegalic inclusion disease, which also occurs in newborns, the typical brain calcifications seen on skull radiographs of patients with this disease are rarely seen in rubella.

The osseous manifestations of rubella improve rapidly in those infants who thrive; the zones of provisional calcification become sharp and regular, and the metaphyseal trabeculations become normal.

Polyarthritis involving the small joints of the hands may be a complication of rubella, especially in young women. The swelling of the wrists, fingers, and knees is transient and usually disappears within 2 weeks.[10,11]

## SMALLPOX (VARIOLA)

At this time, smallpox is considered to no longer exist in nature, and the infrequent radiographic manifestations are primarily of historical interest. Smallpox osteomyelitis is a rare complication in children and causes metaphyseal destruction that rapidly spreads to involve the shaft and incite a periosteal reaction. Although the bone lesions may completely regress, they usually result in premature epiphyseal fusion and subsequent severe deformity. Infection of an adjacent joint may lead to ankylosis.[12,32]

## CHICKENPOX (VARICELLA)

Varicella pneumonia is rare in children but develops in about 15 percent of adults with chickenpox. It is invariably associated with skin lesions and appears 1 to 6 days after the onset of the rash. The degree of pulmonary involvement correlates with the severity of the rash; patients may be virtually asymptomatic or may develop life-threatening disease.

Varicella pneumonia charcteristically produces an extensive, bilateral fluffy nodular infiltrate that tends to coalesce near the hilum and lung bases. Hilar lymph node enlargement is often present but may be difficult to appreciate because of contiguous consolidation in the perihilar parenchyma. Although the infiltrates are transitory, with radiographic findings changing from day to

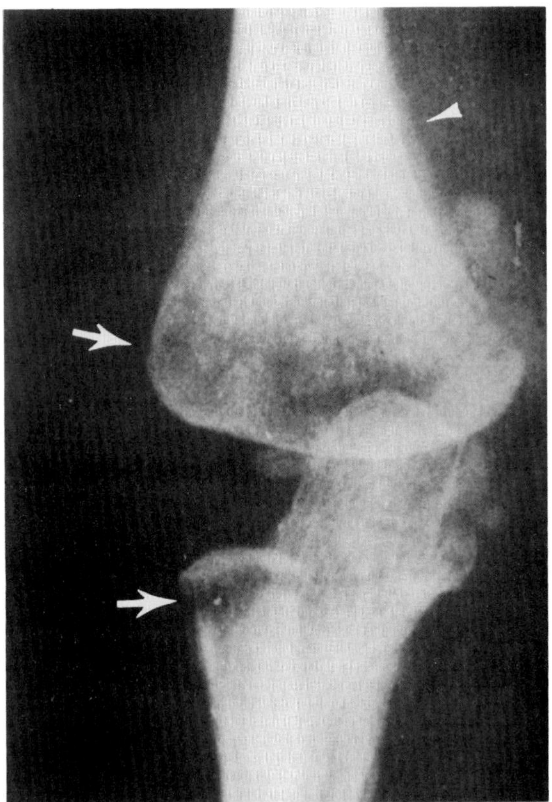

**Figure 17.7. Smallpox osteomyelitis.** Two weeks after the onset of the disease, the elbow joint is disorganized, and a large effusion has disrupted articular relations. Note the lytic destructive changes in the metaphyses (arrows) and the periosteal reaction (arrowhead) at the lower end of the humerus. (From Cockshott and Middlemiss,[32] with permission from the publisher.)

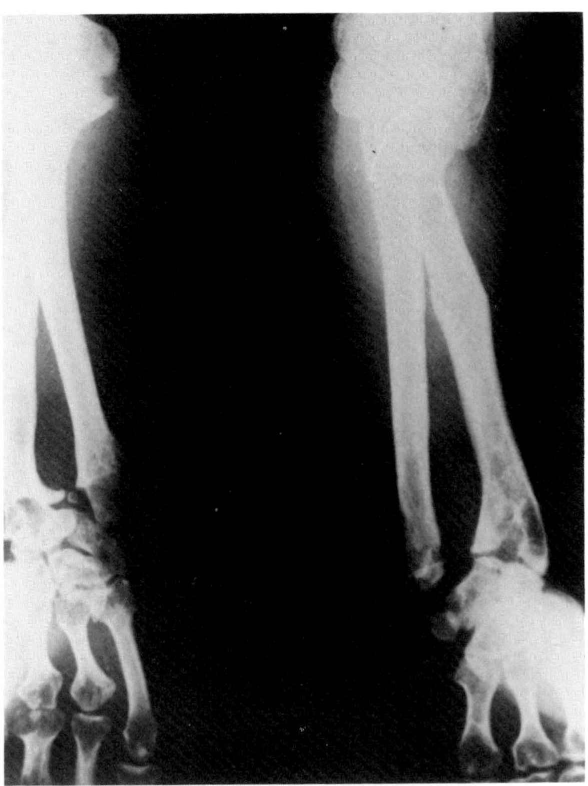

**Figure 17.8. Smallpox.** Telescoped forearms and shortened metacarpals secondary to smallpox osteomyelitis. (From Cockshott and Middlemiss,[32] with permission from the publisher.)

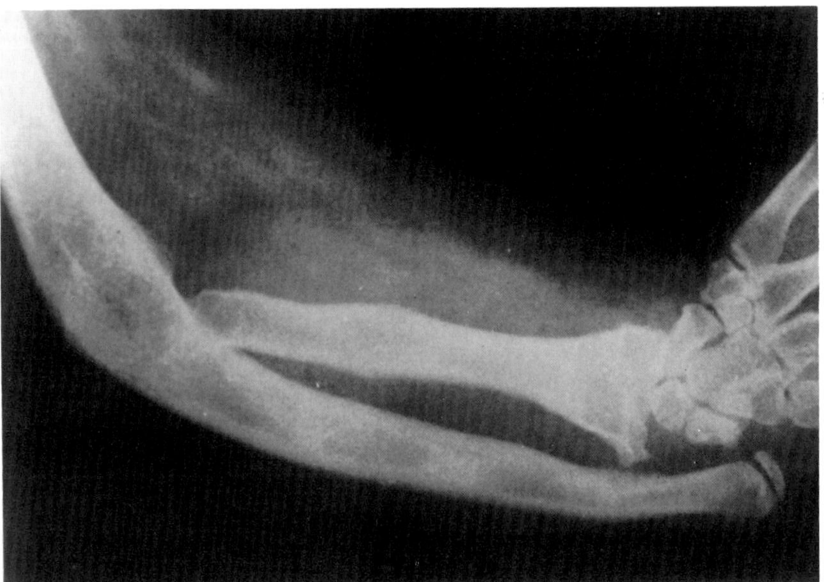

**Figure 17.9. Smallpox.** An ankylosed elbow, a shortened radius, and dislocation of the distal radioulnar joint after childhood smallpox. (From Cockshott and Middlemiss,[32] with permission from the publisher.)

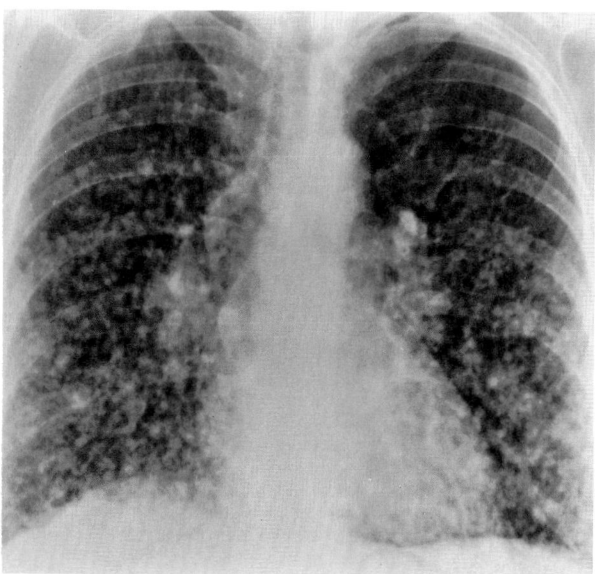

**Figure 17.10. Chickenpox pneumonia.** Bilateral, coarse miliary infiltrates distributed diffusely throughout both lungs.

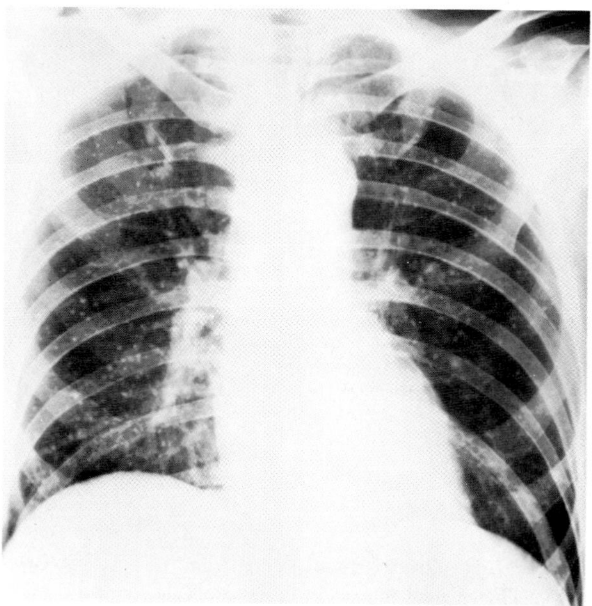

**Figure 17.11. Healed chickenpox pneumonia.** Multiple, tiny calcific shadows measuring 1 to 2 mm in diameter are scattered widely and uniformly throughout both lungs. This 42-year-old asymptomatic man had had florid chickenpox with acute pneumonia 15 years earlier. (From Fraser and Pare,[33] with permission from the publisher.)

day, complete clearing may be slow. In patients with fatal disease, the pulmonary involvement may be virtually total, with little or no visible aerated lung.

The classic appearance of healed varicella pneumonia is the presence of tiny miliary calcifications, scattered widely throughout both lungs, which develop several years following the acute infection. This radiographic pattern is indistinguishable from that of healed histoplasmosis.[13–15]

## HERPES ZOSTER

Herpes zoster is an exanthematous neurocutaneous disorder secondary to reactivation of or reinfection by a large pox virus. Rarely, the disease causes small ulcerations in a narrowed portion of the colon, a radiographic pattern similar to that of a segmental ulcerating colitis. This pattern corresponds to the ulcerative cutaneous changes that sometimes follow the more characteristic vesicular skin lesions. The short length of the colonic lesion and the typical clinical history and skin lesions should suggest the correct diagnosis. The virus can also produce colonic mucosal blebs that appear as multiple small, discrete polygonal filling defects with sharp angular margins simulating colonic urticaria. These blebs correspond morphologically and temporally to the vesicular phase of the cutaneous lesion and are segmentally arrayed in a corresponding or noncorresponding dermatome.

A rare complication of herpes zoster is unilateral diaphragmatic paralysis.[16,17]

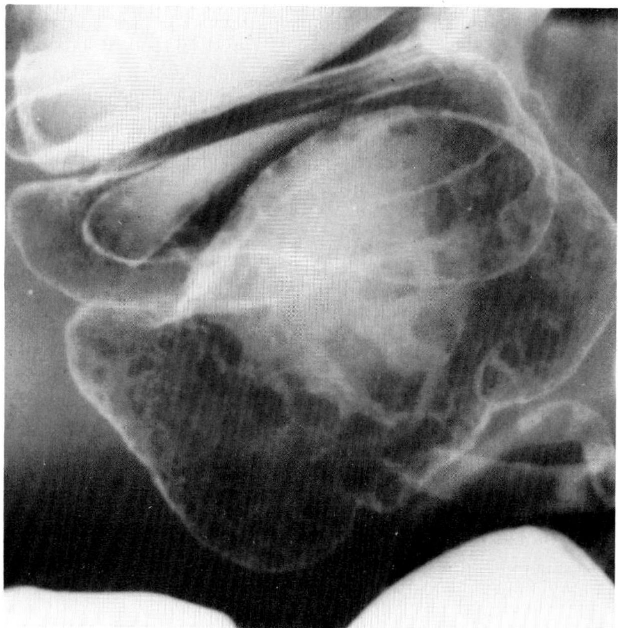

**Figure 17.12. Herpes zoster.** Polygonal filling defects with sharp angular margins are evident in the cecum.

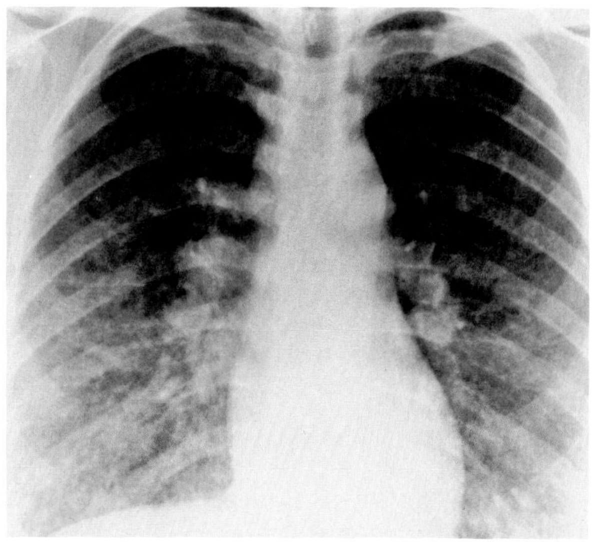

**Figure 17.13. Coxsackievirus pneumonia.** There is a generalized, nonspecific increase in the bronchovascular markings, a typical finding in viral pneumonia.

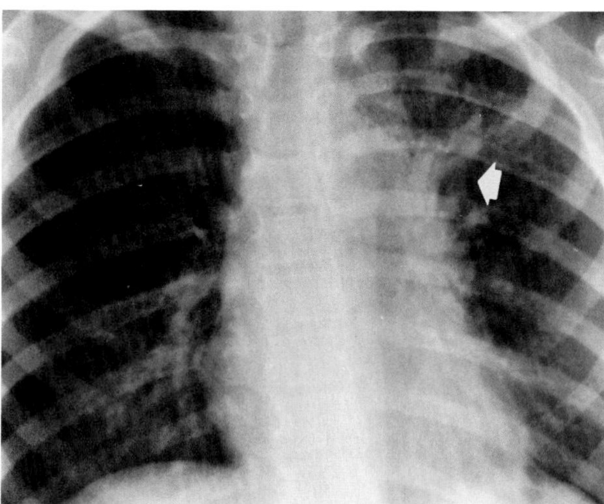

**Figure 17.14. Echovirus pneumonia.** Nonspecific, patchy air space consolidation (arrow) in the posterior portion of the left upper lobe obscures the aortic knob.

## COXSACKIEVIRUS AND ECHOVIRUS

Coxsackievirus and echovirus are common enteric viruses that produce a broad spectrum of nonspecific clinical signs and symptoms including upper and lower respiratory disease, pleuritis, myopericarditis, aseptic meningitis, rashes, and diarrhea. The radiographic findings are usually confined to the chest. Patchy parenchymal consolidations, bilateral hilar lymph node enlargement and a nonspecific increase in bronchovascular markings have been described in the pneumonia caused by these organisms. Enlargement of the cardiac silhouette can be produced by myocarditis, pericarditis, or both. Echocardiography can demonstrate the presence of a pericardial effusion. Pleural effusion frequently develops.[18–20,37]

## POLIOMYELITIS

The radiographic manifestations of poliomyelitis reflect muscular weakness and paralysis due to necrosis and a loss of motor nerve cells in the spinal cord, brain, or cranial nerves. Decreased peripheral muscle tone causes a loss of muscular soft tissue volume and secondary bone atrophy; this produces thinning and demineralization of the bones of the extremities. Severe scoliosis of the thoracolumbar spine can develop. In patients with severely paralyzed extremities, calcification and ossification may form in periarticular soft tissues and may appear to bridge across the joint. This finding is most common about the hip.

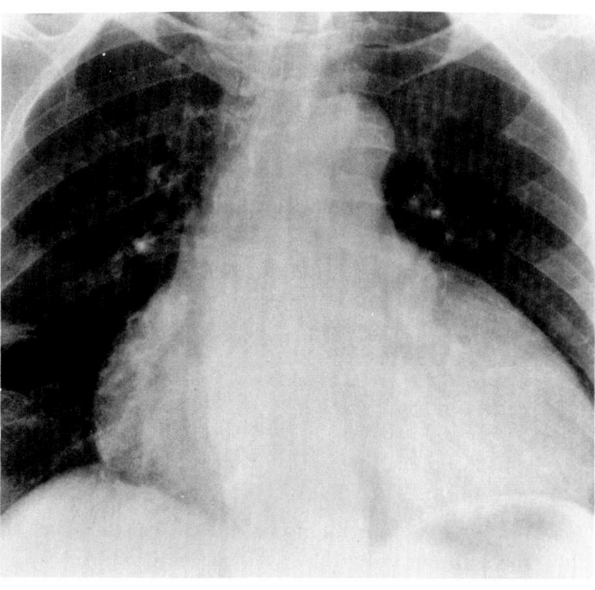

**Figure 17.15. Coxsackievirus pericardial disease.** Globular enlargement of the cardiac silhouette reflects a combination of pericarditis and pericardial effusion. There are small pleural effusions bilaterally.

Involvement of the phrenic nerve can cause paralysis of the diaphragm. Weakness of the accessory muscles of respiration may cause the patient to require lifelong mechanically assisted ventilation. Radiographic changes associated with the chronic use of a respirator include erosive rib lesions, presumably secondary to the continued pressure of the scapula against the posterior aspect of the ribs, and discal calcification in the thoracolumbar spine.

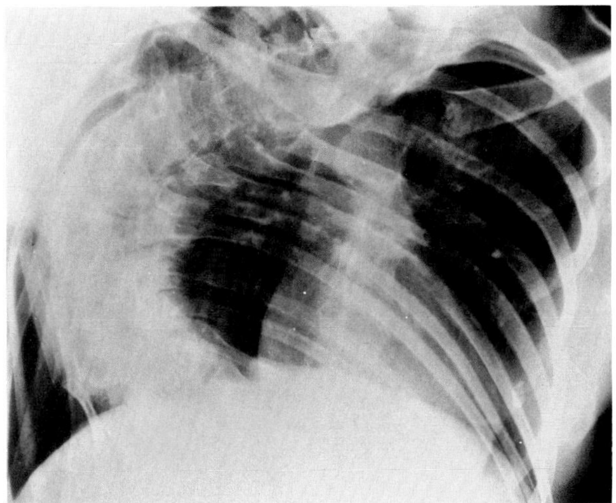

Figure 17.16. **Poliomyelitis.** Severe scoliosis of the thoracic spine.

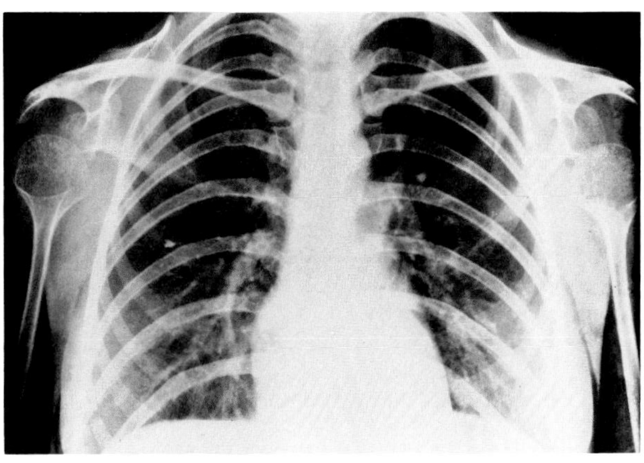

**Figure 17.17. Poliomyelitis.** Severe thinning of the humeri and ribs bilaterally. Note the bilateral shoulder dislocations.

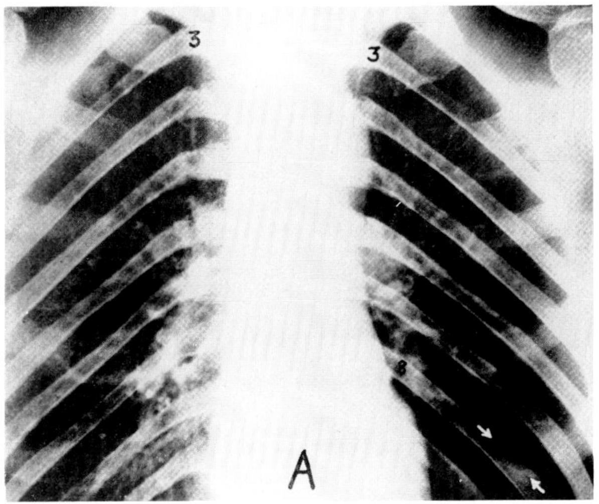


A

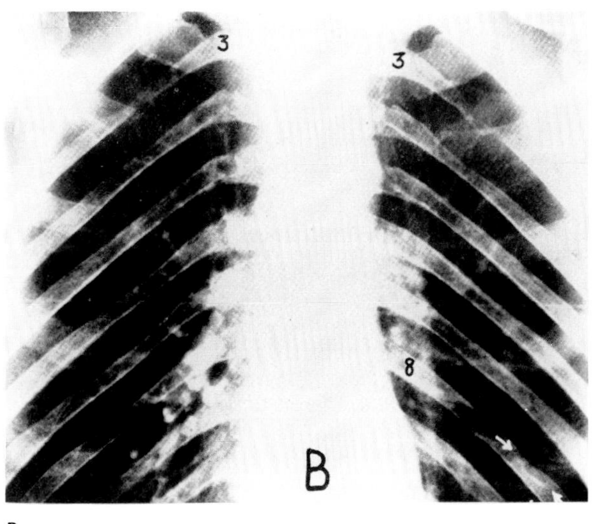


B

**Figure 17.18. Rib lesions associated with chronic use of a respirator in poliomyelitis.** (A) Seventeen months after the onset of severe paralytic poliomyelitis, there is minimal cortical thinning and an indentation of the third ribs posteriorly. Arrows indicate a localized shallow erosion of the eighth rib immediately opposite the inferior angle of the scapula on the left. (B) Fifteen months later, there has been progression of the third rib involvement bilaterally. The erosion of the eighth rib on the left (arrows) is also more pronounced at this time. (From Bernstein, Loeser, and Manning,[22] with permission from the publisher.)

Although there are no direct effects of polio virus on the lung, weakness of the thoracic musculature and retention of bronchial secretions can lead to recurrent pneumonia and focal atelectasis.

No radiographic changes are usually seen during acute polio virus infection.[21-23]

## HERPES SIMPLEX VIRUS

Herpes simplex infestation of the esophagus can produce a clinical and radiographic pattern indistinguishable from *Candida* esophagitis. Herpetic esophagitis predominantly affects patients with disseminated malignancy, debilitating illness, or abnormal immune systems (steroids, chemotherapy, radiation). Although the radiographic appearance of ulcers and multiple diffuse, plaquelike filling defects resembles that of candidiasis, the presence of discrete ulcers on an otherwise normal background mucosa is strongly suggestive of herpetic esophagitis. Because the viral inflammation is usually self-limited, a response to antifungal therapy cannot be considered proof that the characteristic radiographic changes are the result of *Candida* infection. Esophagoscopy with biopsy and cytologic examination is required for the diagnosis of herpetic esophagitis.

Infrequently, a stricture of the esophagus may develop during the healing phase of the disease.[24,25]

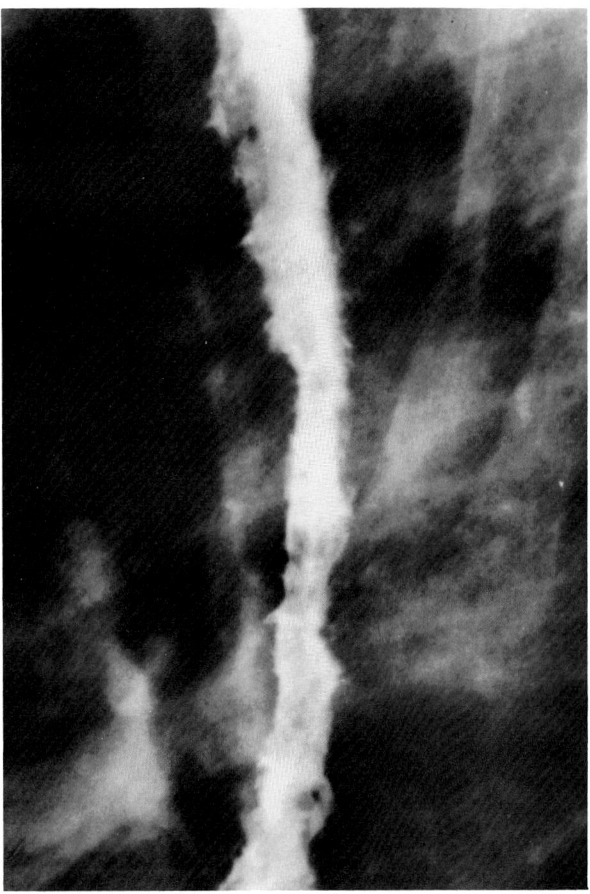

**Figure 17.19. Herpes simplex esophagitis.** The diffuse irregularity and ulceration of the esophagus are indistinguishable from *Candida* esophagitis. (From Eisenberg,[34] with permission from the publisher.)

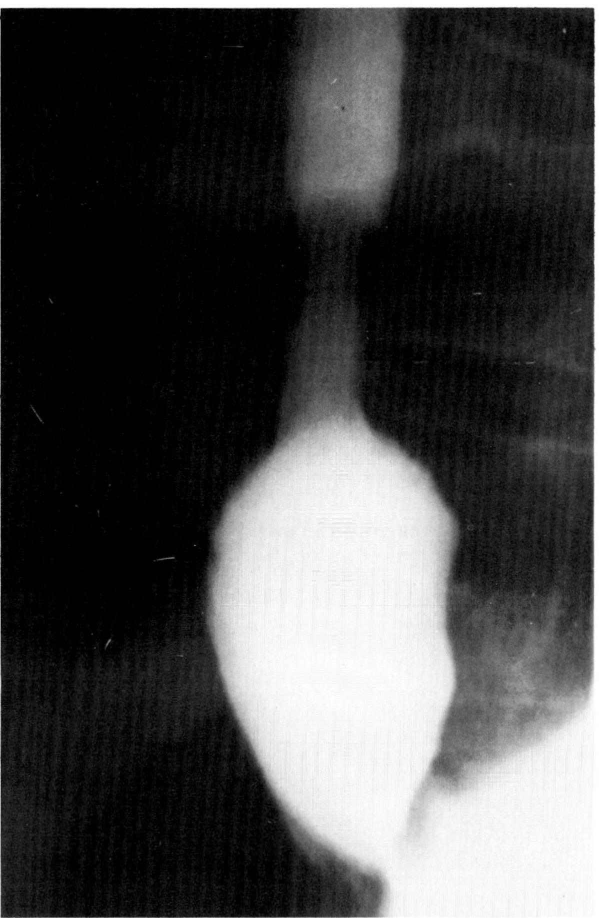

**Figure 17.20. Herpes simplex esophagitis.** A short, smooth stricture in the distal esophagus. (From Eisenberg,[34] with permission from the publisher.)

## CYTOMEGALOVIRUS INFECTION (CYTOMEGALIC INCLUSION DISEASE)

Cytomegalic inclusion disease is a viral disorder that predominantly affects neonates and small infants and produces a clinical syndrome which includes jaundice, hepatosplenomegaly, purpura, and respiratory distress. The infection is most commonly transmitted from mother to fetus in utero. Most children who survive the acute illness have microcephaly with severe mental retardation. Cytomegalic inclusion disease in the newborn and young infant characteristically produces multiple intracranial calcifications that have a periventricular distribution and appear stippled or curvilinear. Similar calcifications can be seen in toxoplasmosis and tuberous sclerosis. In addition, in cytomegalic disease there is usually marked atrophy of the brain with dilatation of the ventricles.

Infants with cytomegalic inclusion disease may demonstrate a nonspecific interstitial pneumonia with progressive dyspnea and cyanosis. Moderate splenomegaly and hepatomegaly are frequently seen, especially in patients with jaundice. Although diffuse bony sclerosis occasionally may be demonstrable at birth, this soon progresses to demineralization and atrophy. Metaphyseal changes similar to those seen in congenital rubella have been reported.

In adults, cytomegalic inclusion disease may produce a respiratory infection that primarily involves patients with underlying reticuloendothelial disease or immunologic deficiencies, or those receiving immunosuppressive therapy (especially after renal transplantation). Chest radiographs demonstrate the rapid development of diffuse bilateral alveolar infiltrates along with nodular densities that are particularly numerous in the outer third of the lungs. This nonspecific pattern must be differentiated from that of *Pneumocystis carinii* pneumonia, which also commonly affects immunosuppressed patients.

Cytomegalovirus-induced colonic ulcers are the most important cause of severe lower gastrointestinal bleeding in immunosuppressed transplant recipients. Early diagnosis and prompt surgical intervention are essential in

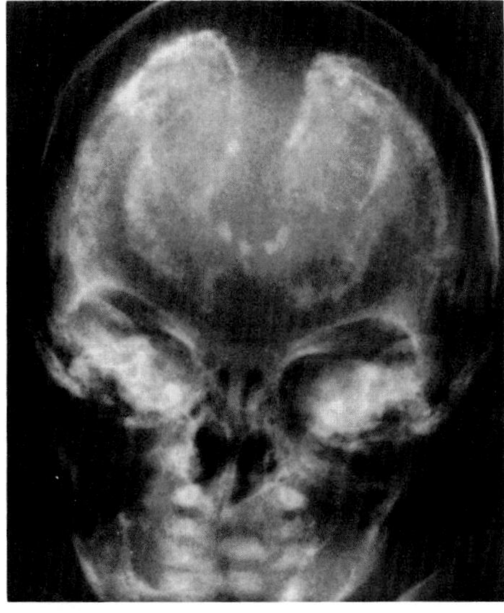

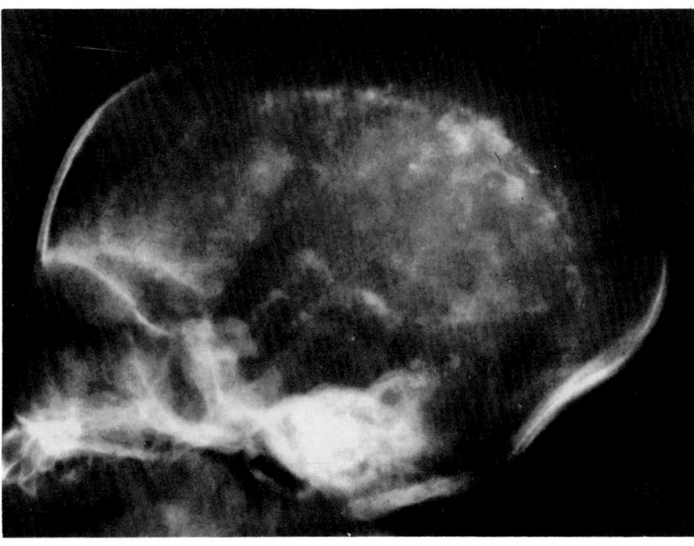

A B

**Figure 17.21. Cytomegalovirus.** (A) Frontal and (B) lateral views of the skull of a young infant demonstrate multiple intracranial calcifications with a typical periventricular distribution.

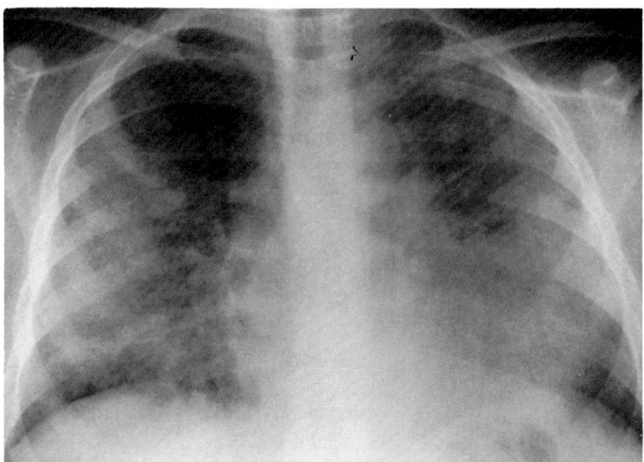

**Figure 17.22. Cytomegalovirus pneumonia.** Diffuse bilateral infiltrates, with a peripheral predominance, that developed rapidly in a patient treated with steroids following renal transplantation.

the management of this often-fatal complication. Cytomegalovirus usually involves the cecum, causing mucosal ulceration with severe local inflammation. Prominent local edema due to associated vasculitis may produce luminal narrowing, thumbprinting, or even tumor-like defects.[26–29]

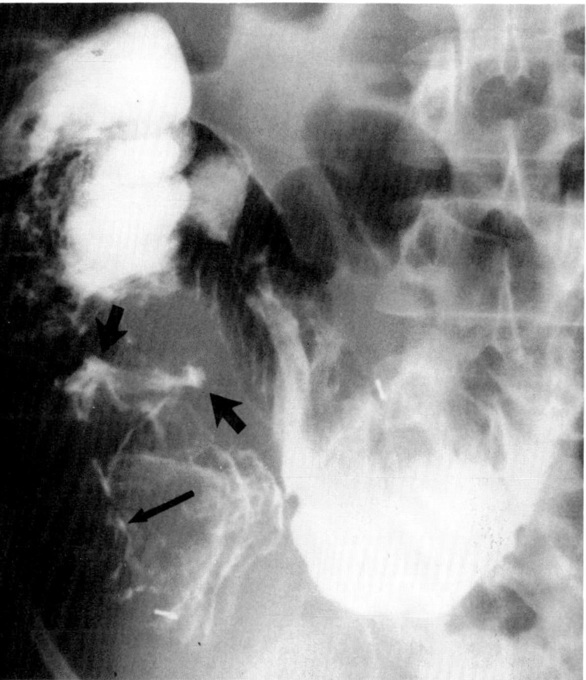

**Figure 17.23. Cytomegalovirus-induced colonic ulceration.** A postevacuation film demonstrates a markedly edematous cecum and ascending colon with mucosal irregularity and ulcers in a patient who had undergone a renal transplant. (From Cho, Tisnado, Liu, et al,[27] with permission from the publisher.)

## INFECTIOUS MONONUCLEOSIS

Generalized lymphadenopathy and splenomegaly are characteristic findings in infectious mononucleosis. Hilar lymph node enlargement, usually bilateral, can be demonstrated in about 15 percent of the cases. Pneumonia is a rare complication that can appear as a diffuse reticular pattern indicating interstitial disease or as a patchy, nonspecific air space consolidation.

Extensive marrow involvement in infectious mononucleosis may rarely cause transient periosteal reaction and pain in the limbs.[30,31]

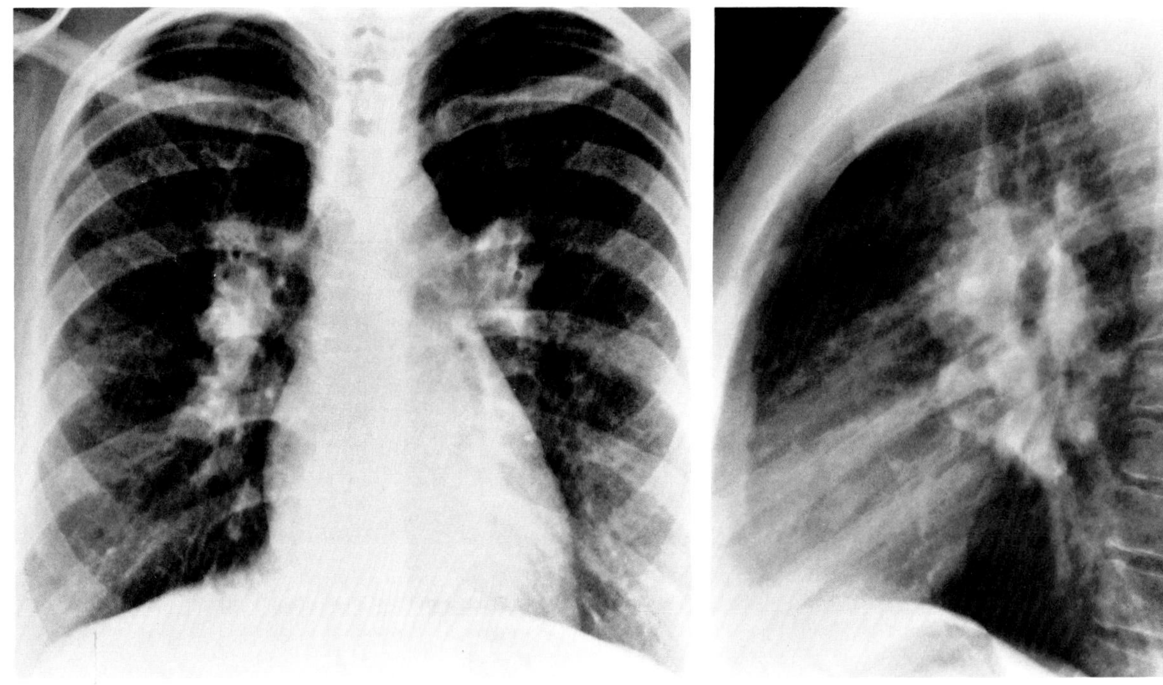

A B

**Figure 17.24. Infectious mononucleosis.** (A) Frontal and (B) lateral views of the chest demonstrate marked bilateral hilar adenopathy.

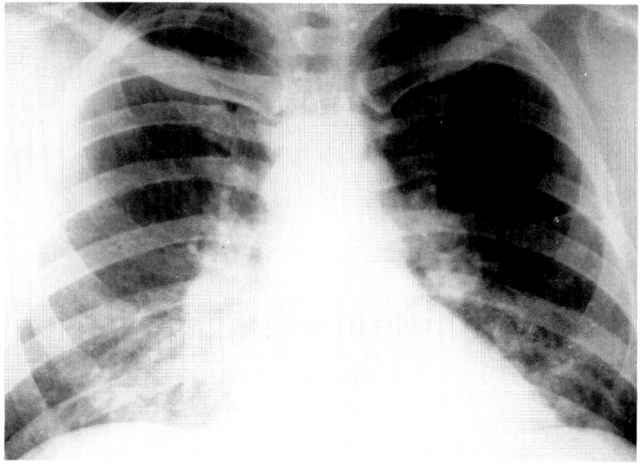

**Figure 17.25. Infectious mononucleosis.** A frontal chest radiograph shows bilateral hilar adenopathy and a bilateral fine interstitial pattern with predominantly basal distribution. (From Lander and Palayew,[31] with permission from the publisher.)

## REFERENCES

1. Conte P, Heitzman ER, Markarian B: Viral pneumonia: Roentgen pathological correlations. *Radiology*. 95:267–272, 1970.
2. Forbes JA, Bennett NM, Gray NJ: Epidemic bronchiolitis caused by a respiratory syncytial virus. Clinical aspects. *Med J Aust* 2:933–939, 1961.
3. Scanlon GT, Unger JD: The radiology of bacterial and viral pneumonias. *Radiol Clin North Am* 11:317–338, 1973.
4. Fry J: Influenza A (Asian) 1957. Clinical and epidemiological features in a general practice. *Br Med J* 1:259–263, 1959.
5. Galloway RW, Miller RS: Lung changes in the recent influenza epidemic. *Br J Radiol* 32:28–31, 1959.
6. Soto PJ, Broun GO, Wyatt JP: Asian influenzal pneumonitis. A structural and virologic analysis. *Am J Med* 27:18–25, 1959.
7. Quinn JL: Measles pneumonia in an adult. *AJR* 91:560–563, 1964.
8. Fawcitt J, Parry HE: Lung changes in pertussis and measles in childhood. A review of 1894 cases with a follow–up study of the pulmonary complications. *Br J Radiol* 30:76–82, 1957.
9. Sokolowski JW, Cordray DR, Canton EF, et al: Giant cell interstitial pneumonia. *Am Rev Respir Dis* 105:417–420, 1972.
10. Singleton EB, Rudolph AJ, Rosenberg HE, et al: The roentgenographic manifestations of the rubella syndrome in newborn infants. *AJR* 97:82–91, 1966.
11. Rabinowitz JG, Wolf BS, Greenberg EI, et al: Osseous changes in rubella embryopathy (congenital rubella syndrome). *Radiology* 85:494–500, 1965.
12. Cockshott WP, McGregor M: The natural history of osteomyelitis variolosa. *Clin Radiol* 10:57–66, 1959.
13. Southard ME: Roentgen findings in chickenpox pneumonia. Review of the literature and report of five cases. *AJR* 76:553–539, 1956.
14. Raider L: Calcification in chickenpox pneumonia. *Chest* 60:504–507, 1971.
15. Burton GC, Sayer WJ, Lillington GA: Varicella pneumonitis in adults: Frequency of sudden death. *Dis Chest* 50:179–185, 1966.
16. Seaman WB, Clements JL: Urticaria of the colon: A nonspecific pattern of submucosal edema. *AJR* 138:545–547, 1982.
17. Menuck LS, Brahme F, Amberg J, et al: Colonic changes of herpes zoster. *AJR* 127:273–276, 1976.
18. Lerner AM, Klein JO, Levin HS, et al: Infections due to Coxsackie virus group A type 9, in Boston, 1959, with special reference to exanthems and pneumonia. *N Engl J Med* 263:1265–1272, 1960.
19. Cramblett HG, Rosen L, Parrott RH, et al: Respiratory illness in six infants infected with a newly recognized ECHO virus. *Pediatrics* 21:168–177, 1958.

20. Munk J, Lederer KT: Radiologic observations of 33 cases of primary interstitial myocarditis. *Clin Radiol* 9:195–201, 1958.
21. Gilmartin D: Cartilage calcification and rib erosion in chronic respiratory poliomyelitis. *Clin Radiol* 17:115–120, 1966.
22. Bernstein C, Loesner W, Manning L: Erosive rib lesions in paralytic poliomyelitis. *Radiology* 70:368–372, 1958.
23. Hooper FM: A case of soft tissue ossification following poliomyelitis. *J Coll Radiol Aust* 7:198–200, 1963.
24. Levine MS, Laufer I, Kressel HY, et al: Herpes esophagitis. *AJR* 136:863–866, 1981.
25. Meyers C, Durkin MG, Love L: Radiographic findings in herpetic esophagitis. *Radiology* 119:21–22, 1976.
26. Allen JH, Riley HD: Generalized cytomegalic inclusion disease. *Radiology* 71:246–262, 1958.
27. Cho FR, Tisnado J, Liu CI, et al: Bleeding cytomegalovirus ulcer of the colon: Barium enema and angiography. *AJR* 136:1213–1215, 1981.
28. Graham CB, Thal A, Wassum CS: Rubella-like bone changes in congenital cytomegalic inclusion disease. *Radiology* 94:39–43, 1970.
29. Goodman N, Daves ML, Rifkind D: Pulmonary roentgen findings following renal transplantations. *Radiology* 89:621–625, 1967.
30. Burrows FGO: Transient periosteal reaction in illness diagnosed as infectious mononucleosis. *Radiology* 98:291–292, 1971.
31. Lander P, Palayew MJ: Infectious mononucleosis: A review of chest roentgenographic manifestations. *J Can Assoc Radiol* 25:303–306, 1974.
32. Cockshott WP, Middlemiss H: *Clinical Radiology in the Tropics*. Edinburgh, Churchill Livingstone, 1979.
33. Fraser RG, Pare JAP: *Diagnosis of Disease of the Chest*. Philadelphia, WB Saunders, 1979.
34. Eisenberg RL: *Gastrointestinal Radiology: A Pattern Approach*. Philadelphia, JB Lippincott, 1983.

## Chapter Eighteen
# Protozoans and Helminths

### AMEBIASIS

Amebiasis is a widespread primary infection of the colon, caused by the protozoan *Entamoeba histolytica*, which is acquired by the ingestion of food or water contaminated by amebic cysts. The amebas tend to settle in areas of stasis and thus primarily affect the cecum and, to a lesser extent, the rectosigmoid and the hepatic and splenic flexures. The patient is asymptomatic (in a carrier state) until the protozoan actually invades the wall of the colon. Penetration of the bowel wall by the organism incites an inflammatory reaction that leads to a broad clinical spectrum varying from mild abdominal discomfort to an acute illness with severe diarrhea, cramping abdominal pain, and blood and mucus in the stools.

In the early stages of amebic colitis, barium studies demonstrate superficial ulcerations superimposed on a pattern of mucosal edema or nodularity, along with spasm and a loss of the normal haustral pattern. The cecum is the primary site of involvement in up to 90 percent of the patients with clinical disease. Chronic inflammation and fibrosis can concentrically narrow the lumen of the cecum, which eventually assumes a characteristic cone-shaped configuration. The ileocecal valve is almost invariably thickened, rigid, and fixed in an open position. However, in contrast to Crohn's disease and tuberculosis, in which terminal ileum involvement is the rule, in amebiasis the terminal ileum is usually normal. An additional differential point is the dramatic change in radiographic appearance seen in amebiasis within 2 weeks of the institution of the antiamebic therapy.

Scattered areas of segmental colonic involvement can occur in amebiasis, and generalized severe colitis sometimes develops. Deep penetrating ulcers, pseudopolyps, cobblestoning, and thumbprinting may be demonstrated. Multiple skip lesions are frequent, unlike the continuous disease in ulcerative colitis. A definitive diagnosis of colonic amebiasis requires demonstration of the parasite in the stool or rectal biopsy. Complications of amebic colitis include toxic megacolon, perforation, sinuses, fistulas, and pericolic abscesses.

Although patients with acute amebiasis typically have segmental spasm and irritability, chronic benign strictures are uncommon complications if the patients are treated properly. When stenoses occur, they are often multiple and long, tapering gradually at both ends into adjacent bowel of normal caliber and mucosal pattern.

An ameboma is a focal, hyperplastic granuloma caused by secondary bacterial infection of an amebic abscess in the bowel wall. Amebomas are most common in the cecum and ascending colon, and about half are multiple. Radiographically, amebomas are characterized by eccentric or concentric thickening of the entire circumference of the bowel wall. They can appear as discrete luminal masses or as annular, nondistensible lesions with irregular mucosa that simulate colonic carcinoma. Fac-

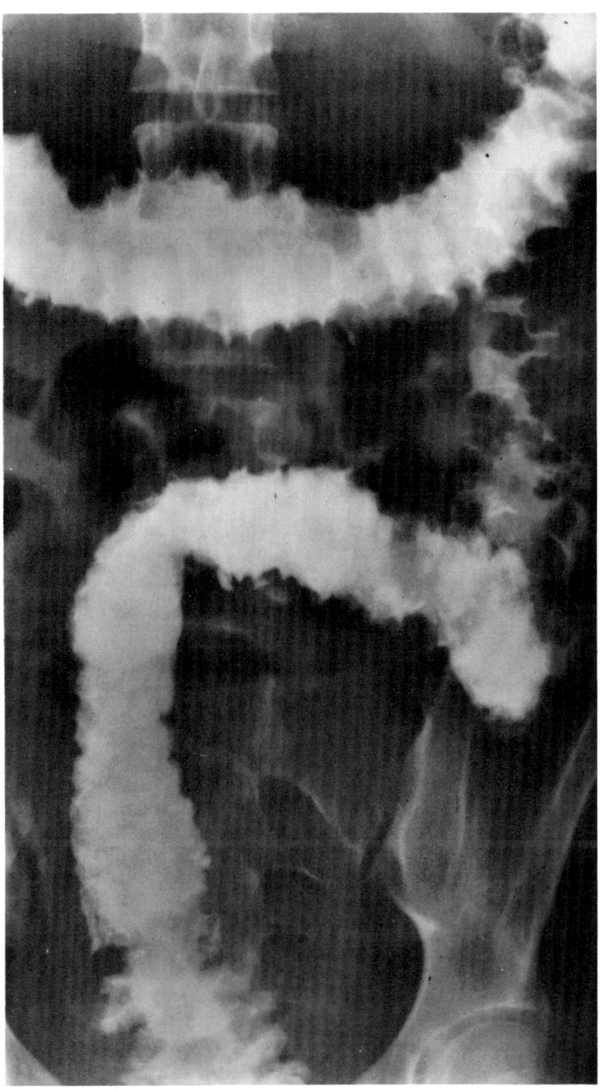

**Figure 18.1. Amebic colitis.** Diffuse ulceration and mucosal edema mimic ulcerative colitis. (From Eisenberg,[69] with permission from the publisher.)

tors favoring the diagnosis of ameboma rather than a malignant lesion include multiplicity, greater length, and pliability of the lesion; evidence of amebiasis elsewhere in the colon; and rapid improvement with antiamebic therapy.

Hepatic abscesses are the most frequent extracolonic complication of amebiasis, occurring in about one-third of all patients with amebic dysentery. About two-thirds of the amebic abscesses are solitary; the remainder are multiple and may coalesce into a single, large liver abscess. The lesions are most often located in the posterior portion of the right lobe of the liver, since this region receives most of the blood draining the right colon because of the "streaming" effect in portal vein flow. Most abscesses enlarge upward, producing a "bulge" in the right hemidiaphragm. There is often restricted diaphragmatic motion and associated small pleural effusion and basilar atelectasis. Amebic abscesses in the liver can be accurately localized by ultrasound, computed tomography (CT), or radionuclide scans.

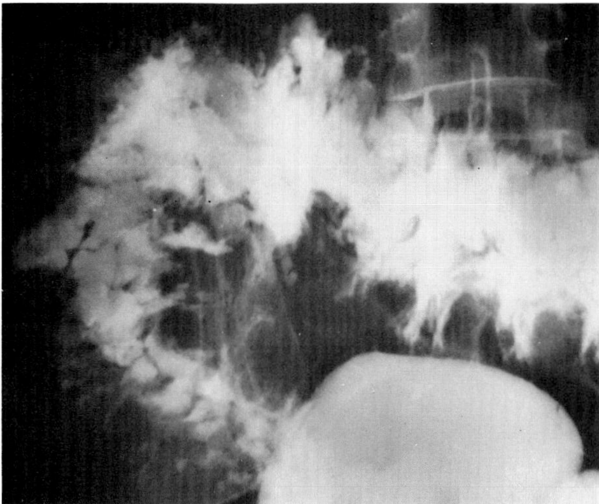

**Figure 18.2. Amebic colitis.** Deep, penetrating ulcers produce a bizarre appearance. (From Eisenberg,[69] with permission from the publisher.)

Gas in a hepatic amebic abscess may result from secondary infection or from a fistulous communication with the biliary tract or bowel. Mural calcification in a chronic amebic abscess is generally associated with secondary infection.

Pleuropulmonary involvement usually arises from the direct extension of the hepatic infection through the right hemidiaphragm, though occasionally it may be of hematogenous origin. The earliest radiographic abnormality is haziness and loss of definition of the outline of the hemidiaphragm, along with the development of a small triangular "cloud" whose apex extends toward the hilum and whose base follows the usually elevated diaphragmatic contour. A more clearly defined infiltrate may develop, usually with obliteration of the costophrenic and cardiophrenic angles. An extensive pleural effusion may obscure the underlying lung involvement. Metastatic abscesses from bronchogenic or hematogenous spread may develop in any part of the lung. These abscesses may be thin- or thick-walled and are clinically and radiographically indistinguishable from other lung abscesses.

Extension of an abscess from the left lobe of the liver to the pericardium is the most dangerous complication of amebic hepatic abscess. This may cause an acute purulent infection that may result in constrictive pericarditis. Rupture of a liver abscess or perforation of a colonic ulcer may lead to peritonitis.[1–4,69,70]

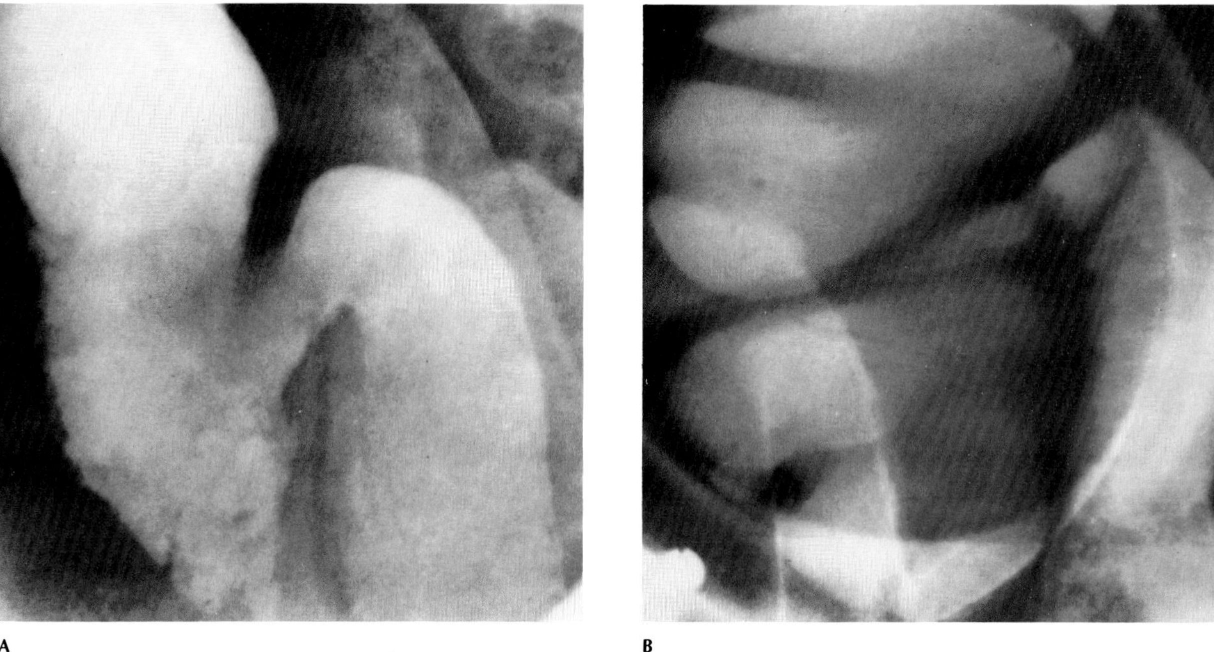


**Figure 18.3. Cecal amebiasis.** (A) The small, shallow ulcers produce an irregular cecal margin and a finely granular mucosa. The terminal ileum is not involved. (B) After a course of antiamebic therapy, the cecum and the ileocecal valve appear normal. (From Eisenberg,[69] with permission from the publisher.)

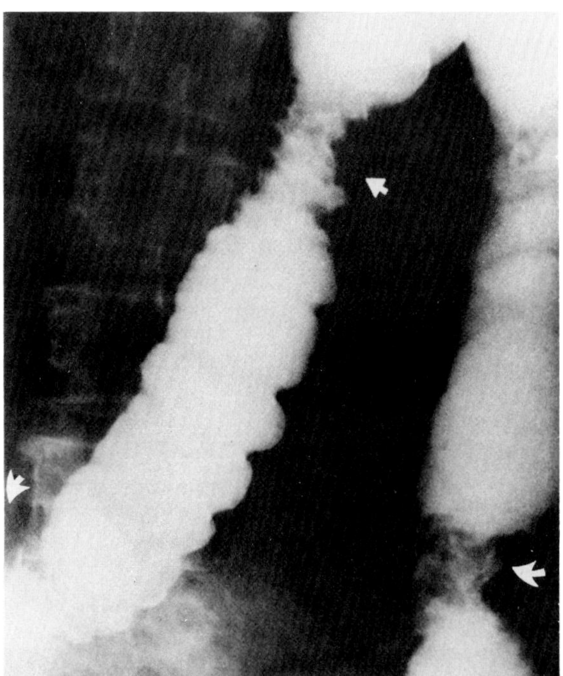

**Figure 18.4. Ameboma.** Three separate constricting lesions (arrows) in the colon. The multiplicity of the process favors an inflammatory rather than a malignant cause. (From Eisenberg,[69] with permission from the publisher.)

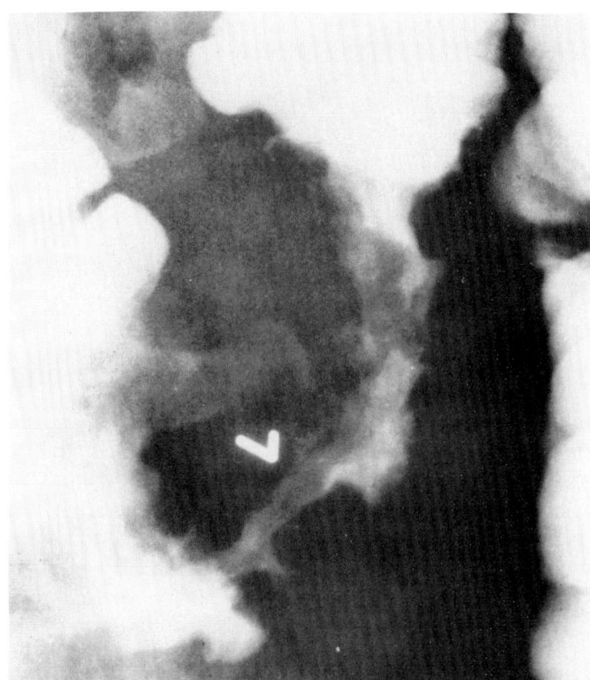

**Figure 18.5. Ameboma.** There is an irregular constricting lesion in the transverse colon. The relatively long area of involvement tends to favor an inflammatory cause. (From Eisenberg,[69] with permission from the publisher.)

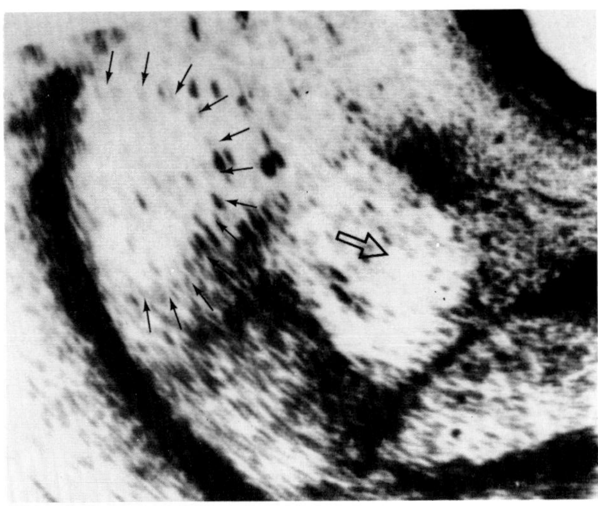

**Figure 18.6. Hepatic amebic abscesses.** A sagittal ultrasound scan shows two abscesses in the right lobe, both with low-level, homogeneous fine echoes. One has a small, central, echo-free area (open arrow); the other is homogeneous (small arrows). (From Ralls, Meyers, and Lapin,[70] with permission from the publisher.)

## MALARIA

Malaria is a widespread protozoan disease transmitted to humans by the bite of *Anopheles* mosquitoes. The radiographic findings in malaria are nonspecific. Enlargement of the spleen is common. Splenic infarction and infection can result in a splenic abscess, which on erect abdominal films appears as an air-fluid level that resembles the fundus of the stomach but is situated more laterally. The severe and often fatal pulmonary edema that may complicate *Plasmodium falciparum* infections produces a nonspecific, diffuse alveolar pattern without cardiac enlargement. Bronchitis and pneumonia may also occur. In blackwater fever, acute renal failure and uremia may be associated with enlargement of the kidneys.[5,6]

## KALA AZAR

Kala azar (black fever) is a widespread infection caused by a protozoan of the genus *Leishmania*, which is characterized by skin lesions, hepatosplenomegaly, lymphadenopathy, and pancytopenia. The radiographic hallmark is a striking enlargement of the spleen, which may become so large that it fills the abdomen and reaches the pelvis, producing a general abdominal haze suggesting ascites.[72]

## CHAGAS' DISEASE

Chagas' disease is due to infection by the protozoan *Trypanosoma cruzi* and develops from a bite by an infected reduviid bug and the resultant contamination of the punctured skin by the insect's feces. These bloodsucking insects usually acquire the trypanosomes by feeding on the armadillo, the chief host for the organism.

Acute or chronic myocardiopathy in Chagas' disease often produces cardiac enlargement, which usually in-

A

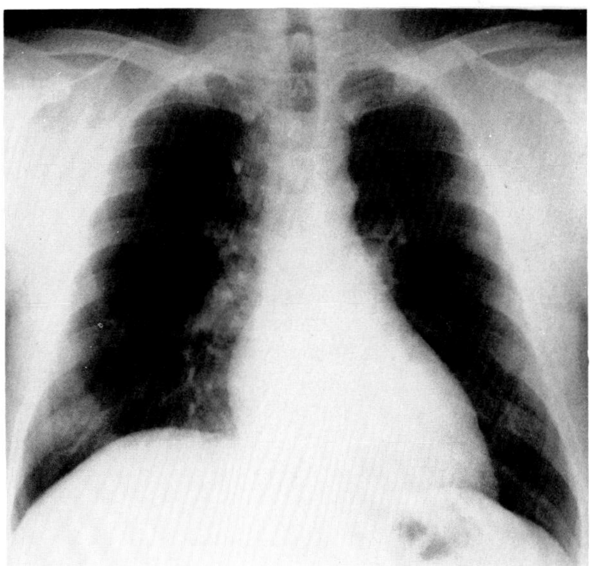

B

**Figure 18.7. Acute malaria.** (A) An admission chest film shows interstitial pulmonary edema manifested by perihilar and perivascular haziness, interlacing septal lines, and thickened interlobar fissures. (B) A discharge film shows complete resolution of the interstitial edema. (From Godard and Hansen,[5] with permission from the publisher.)

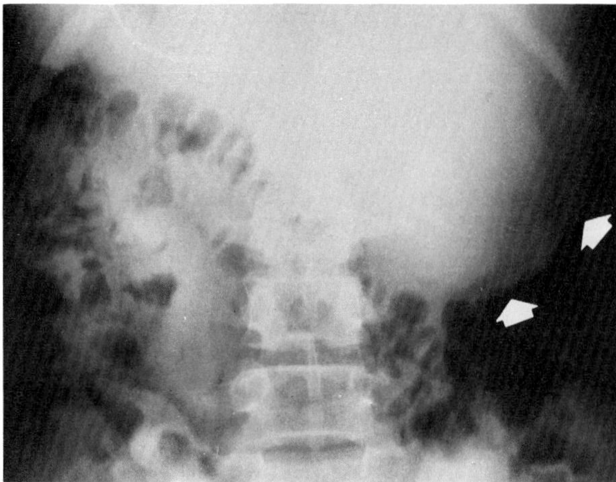

**Figure 18.8. Kala azar.** There is massive enlargement of the spleen (arrows), which fills the entire left upper quadrant of the abdomen.

volves all chambers but may predominantly affect the left side. The pulmonary vasculature usually remains normal, even in patients with congestive failure, since concomitant failure of the right ventricle prevents overloading of blood in the lungs.

In the esophagus, increased muscular contraction initially causes esophageal spasm, which may result in dysphagia. The destruction of ganglion cells in the myenteric plexuses, probably due to a neurotoxin, eventually produces a radiographic pattern identical to that of achalasia. This pattern includes the failure of relaxation of the lower esophageal sphincter, a characteristic tapering of the distal esophageal segment, and a severe dilatation of the proximal esophagus.

The destruction of the colonic myenteric plexuses in Chagas' disease causes striking elongation and dilatation, especially of the rectosigmoid and the descending colon. The radiographic appearance is similar to that caused by the developmental aganglionosis of the distal colon in patients with Hirschsprung's disease (see the section "Aganglionic Megacolon" in Chapter 48 for further information). Patients with megacolon due to Chagas' disease have severe chronic obstipation, and plain abdominal radiographs demonstrate a massive amount of retained feces throughout a grossly dilated colon, which virtually fills the entire abdomen. Fecal impaction and sigmoid volvulus are frequent complications.

In advanced Chagas' disease, the ureters, small bowel, bronchi, gallbladder, and biliary ducts can occasionally be markedly dilated.[7,8,72]

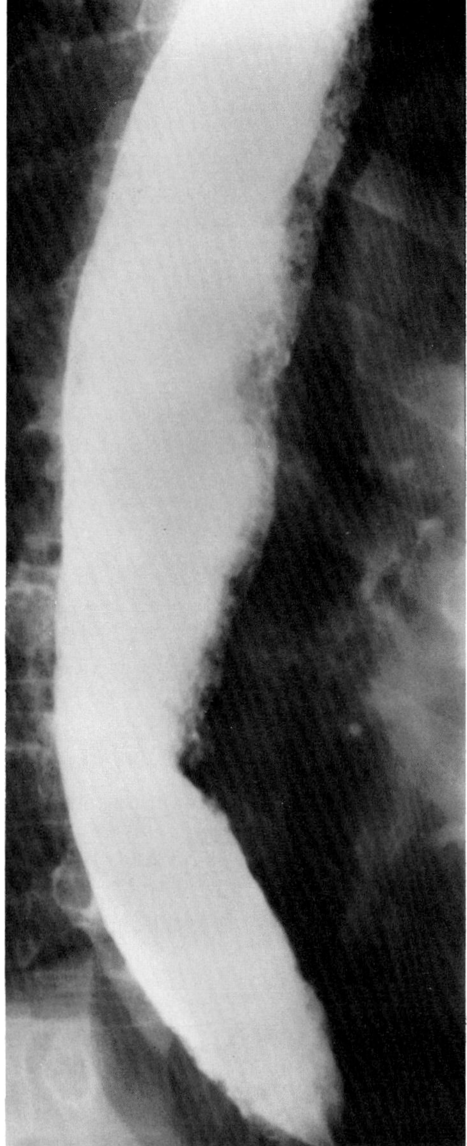

**Figure 18.9. Chagas' disease of the esophagus.** There is esophageal dilatation and aperistalsis with a large amount of residual food, a pattern simulating achalasia.

## TOXOPLASMOSIS

Congenital toxoplasmosis is the most common cause of scattered intracranial calcifications in the neonate. The bilateral calcifications vary greatly in appearance. In general, calcifications in the cerebral cortex tend to appear as punctate flecks or slightly larger, discrete nodules; those in the basal ganglia or thalamus are usually more striated or curvilinear; meningeal calcifications are more plaquelike. The calcifications are scattered diffusely throughout the brain parenchyma, most often in the parietal region, but are also found in specific cerebral structures (choroid plexus, ependyma of the ventricles). The diffuse intracranial calcifications in toxoplasmosis may be indistinguishable from those in cytomegalic inclusion disease.

In up to 80 percent of the patients, obstruction of the aqueduct or one of the foramina by toxoplasmic gran-

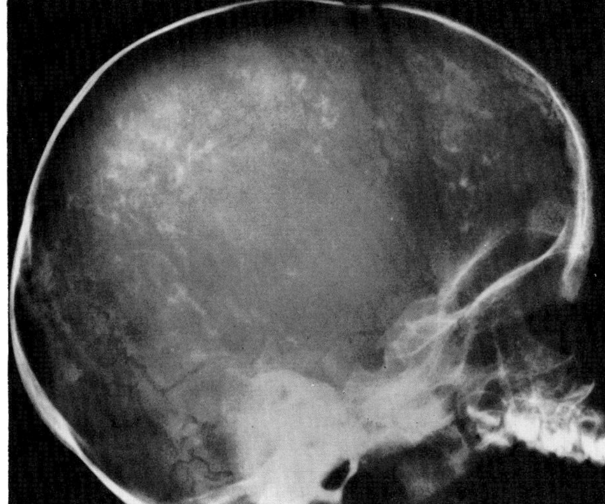

**Figure 18.10. Toxoplasmosis.** A lateral skull film demonstrates diffuse flecks, plaques, and nodules of calcification scattered throughout the brain parenchyma and meninges.

ulomas causes hydrocephalus, which appears radiographically as a large head with enlargement of the fontanelles, widening of the sutures, and striking convolutional impressions. Postinflammatory scarring with cerebral atrophy may result in microcephaly.

Acquired toxoplasmosis infection in adults primarily involves the lungs. Chest radiographs demonstrate a nonspecific interstitial pneumonitis along with accentuation of hilar and perihilar markings due to pulmonary congestion. Air space consolidation and hilar adenopathy may develop in advanced disease. Intracranial calcifications, which are extremely common in the congenital form of the disease, do not occur in the acquired variety.

Toxoplasmosis infection is especially virulent in immunocompromised patients, especially those receiving immunosuppressive therapy for lymphoproliferative malignancies or following transplantation. Central nervous system involvement is common and may lead to the development of a brain abscess that can be demonstrated with CT.[9–12,71]

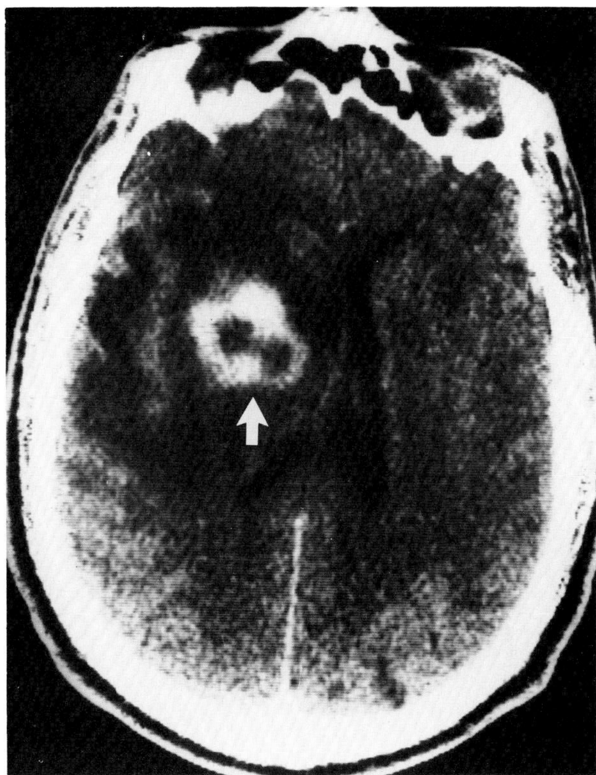

**Figure 18.11. Toxoplasmosis.** A CT scan following the injection of contrast material demonstrates a multicystic lesion (arrow) with irregularly thick margins of enhancement and considerable mass effect; this represents a toxoplasmic brain abscess in a patient with acquired immunodeficiency syndrome. (From Kelly and Brant-Zawadzki,[71] with permission from the publisher.)

## *PNEUMOCYSTIS CARINII* PNEUMONIA

*Pneumocystis carinii* pneumonia is a protozoan disease of the compromised host that occurs in two settings: an epidemic form among premature or debilitated infants, and a sporadic illness among children or adults who are immunosuppressed (especially patients treated for lymphoproliferative diseases or those who have received renal transplants). The typical early radiographic finding is a hazy, perihilar granular infiltrate that spreads to the periphery and appears predominantly interstitial. In later stages, the pattern progresses to patchy areas of air space

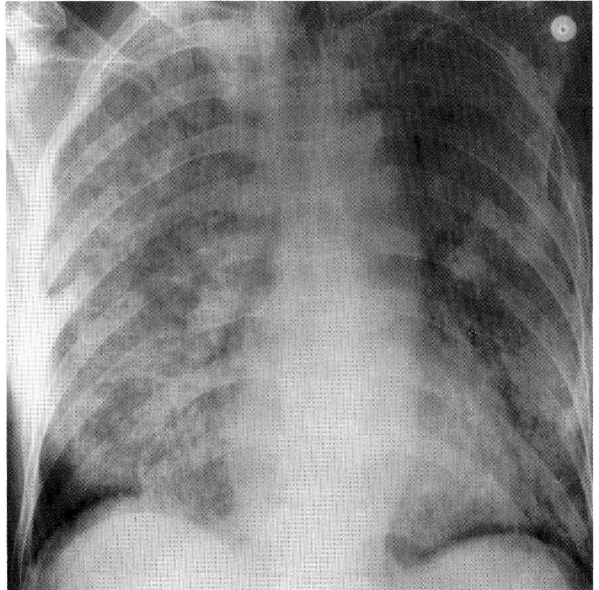

**Figure 18.12. *Pneumocystis carinii* pneumonia.** A hazy, granular infiltrate involves both lungs in a patient receiving steroid therapy following renal transplantation.

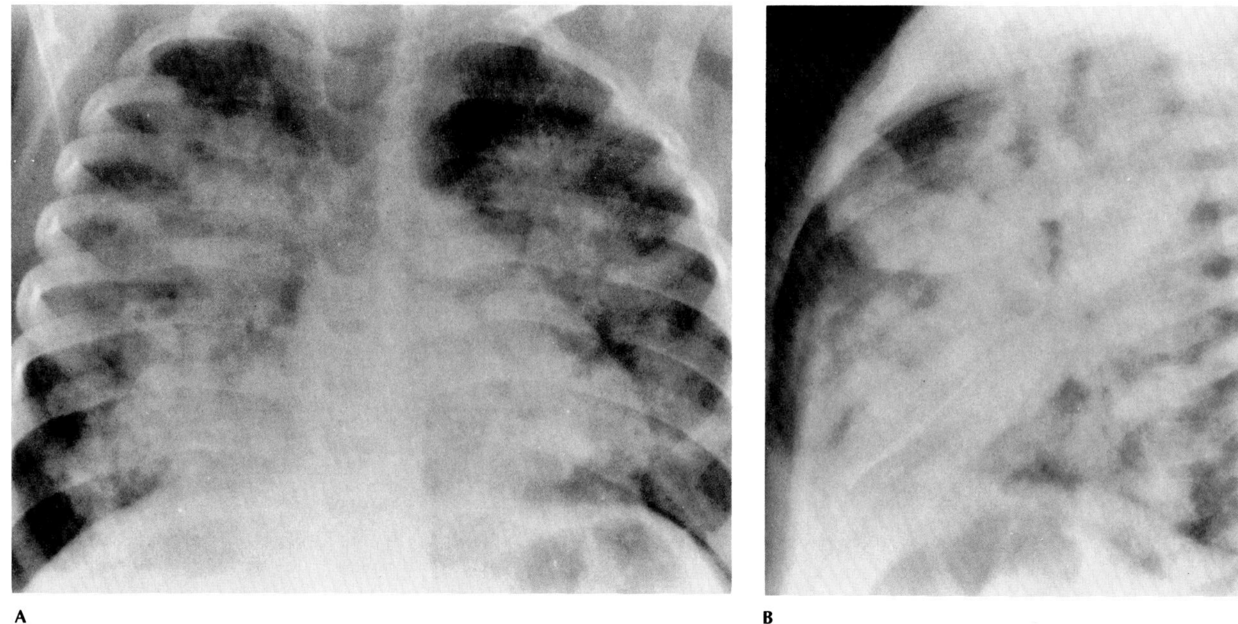

A B

**Figure 18.13. *Pneumocystis carinii* pneumonia.** (A) Frontal and (B) lateral views of the chest demonstrate a patchy air space consolidation with air bronchograms that predominantly involves the central portions of the lungs and closely simulates pulmonary edema. This child was undergoing immunosuppressive therapy for leukemia.

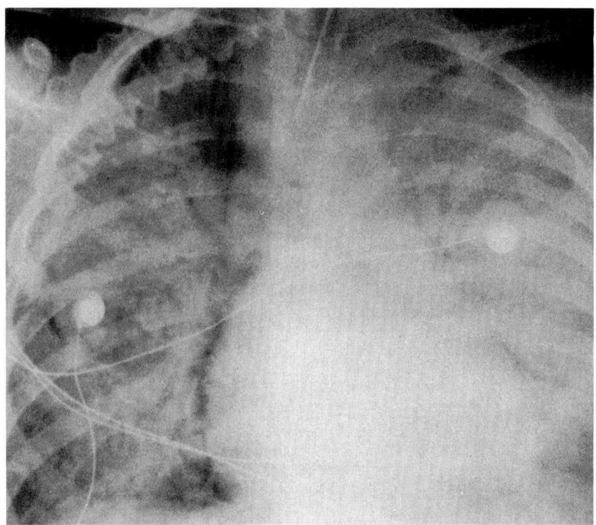

**Figure 18.14. *Pneumocystis carinii* pneumonia.** Severe, bilateral air space consolidation with air bronchograms. The patient was undergoing immunosuppressive therapy for lymphoma and died shortly after this radiograph was obtained.

consolidation with air bronchograms, indicating the alveolar nature of the process. The peripheral lungs are clear, though there may be focal areas of atelectasis or emphysema due to compression of alveoli and alveolar ducts by the interstitial involvement. Massive consolidation with virtually airless lungs may be a terminal appearance. Hilar adenopathy and significant pleural effusions are rare and should suggest an alternative diagnosis.

The radiographic appearance may closely resemble pulmonary edema, bacterial pneumonia, or the pneumonia caused by cytomegalovirus, which also tends to attack renal transplant recipients. Because *Pneumocystis carinii* cannot be cultured and the disease it causes is usually fatal if untreated, an open-lung biopsy is often necessary if a sputum examination reveals no organisms in a patient suspected of having this disease.[13–15]

## GIARDIASIS

*Giardia lamblia* is a protozoan parasite harbored by millions of asymptomatic persons throughout the world. Clinically significant infestations occur predominantly in children, postgastrectomy patients, and travelers to areas in which the disease is endemic (e.g., Leningrad, India, and the Rocky Mountains of Colorado). The infection is apparently acquired through drinking water, and the clinical symptoms range from a mild gastroenteritis to a severe, protracted illness with profuse diarrhea, cramping, malabsorption, and weight loss. There is a striking incidence of giardiasis in patients with agammaglobulinemia, in whom a generally lowered resistance permits the usually innocuous *Giardia* organisms to flourish in the small bowel.

Radiographic abnormalities in giardiasis are usually limited to the duodenum and jejunum. Diffuse mucosal inflammation and widespread infiltration by inflammatory cells result in irregularly tortuous small bowel folds that present a distinctly nodular appearance when seen

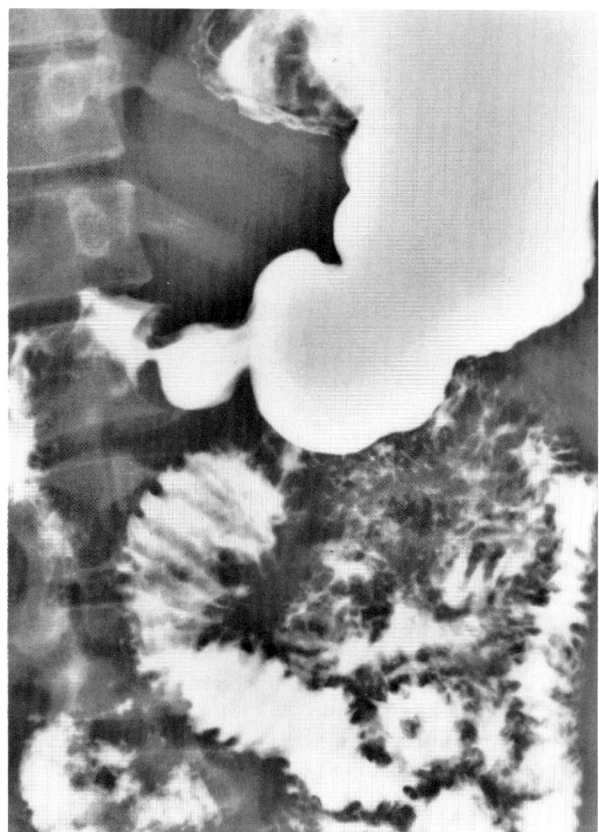

**Figure 18.15. Giardiasis.** The irregular fold thickening is most prominent in the proximal small bowel. (From Eisenberg,[69] with permission from the publisher.)

on end. Increased secretions cause blurring of small bowel folds, which may simulate the malabsorption pattern. Hypermotility frequently causes a rapid transit time, and the lumen of the bowel is often narrowed because of spasm. When giardiasis complicates an immunodeficiency state, the irregularly thickened folds can be superimposed on a pattern of multiple, tiny filling defects characteristic of nodular lymphoid hyperplasia.

An unequivocal diagnosis of *Giardia lamblia* infestation requires demonstration of the characteristic cysts in formed feces or the trophozoites in diarrheal stools, duodenal secretions, or jejunal biopsies. The eradication of the parasite after treatment with quinacrine or metronidazole results in a return of the small bowel pattern to normal.[16,17,69]

## TRICHINOSIS

Trichinosis in humans is contracted by the ingestion of undercooked meat, usually pork, containing the encysted larvae of *Trichinella spiralis*. From the gastrointestinal tract, the larvae enter skeletal muscle, where they grow and again begin to encyst. Although calcification of the encysted larvae is common pathologically, their small size makes them difficult to detect radiographically. Larvae carried to sites other than skeletal muscles do not encyst but disintegrate, often stimulating a granulomatous inflammatory reaction. This process can lead to an intense myocarditis or the development of a vasculitis involving small arterioles and capillaries of the brain and meninges. Pulmonary involvement can cause consolidations in the lungs.

## FILARIASIS

Filariasis is a group of disorders caused by infection with threadlike nematodes that invade the lymphatics and

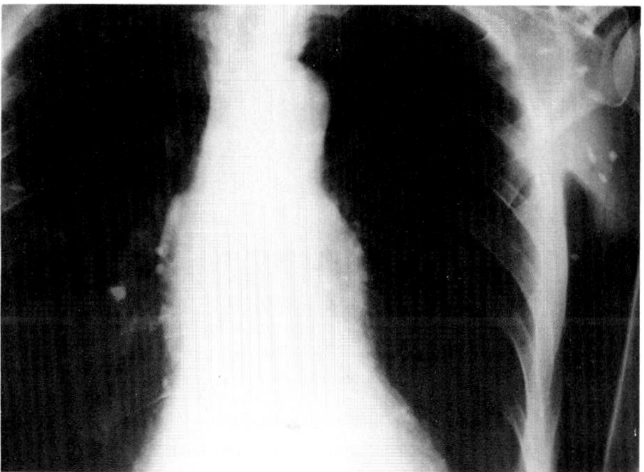

**Figure 18.16. Trichinosis.** There is calcification of multiple encysted larvae, best seen in the soft tissues about the left shoulder.

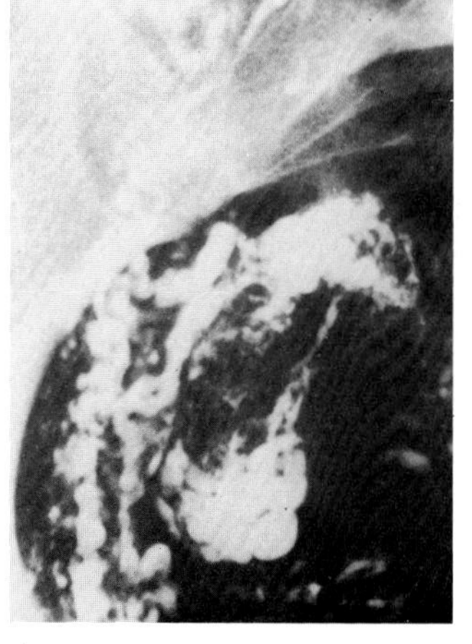

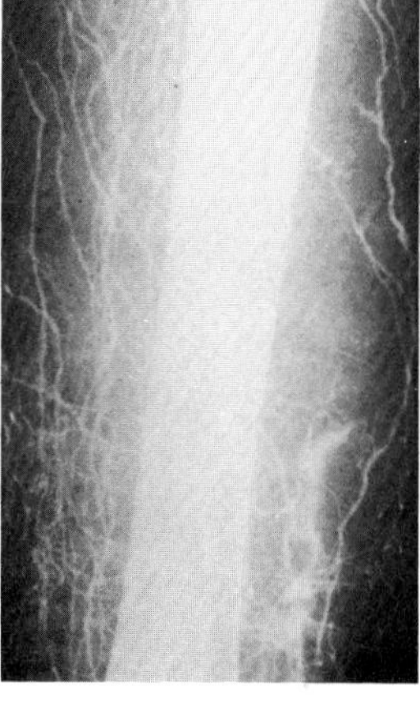

A B

**Figure 18.17. Filariasis.** (A) A lymphogram demonstrates large, varicose, redundant lymphatic channels due to filarial obstruction. The patient also had chyluria. (B) Gross filarial leg edema with increased and dilated channels and some dermal backflow. (From Cockshott and Middlemiss,[72] with permission from the publisher.)

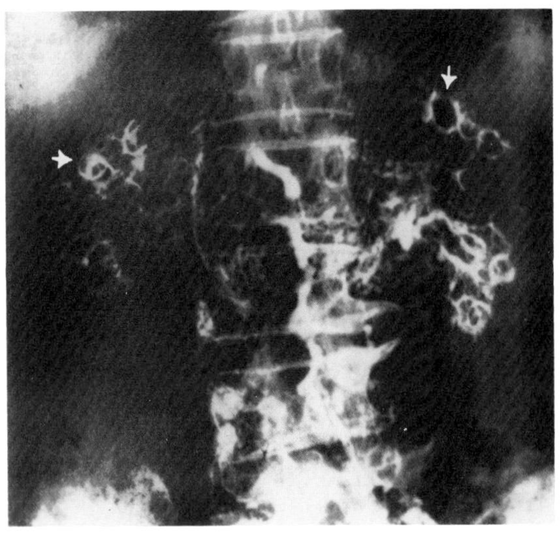
A

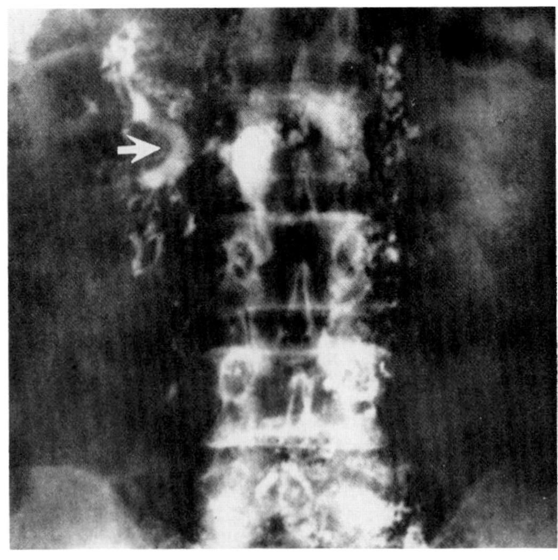
B

**Figure 18.18. Filariasis.** (A) A lymphogram shows bilateral filling of the renal lymphatics by reflux from abnormal para-aortic nodes. Many of the renal lymphatic channels are dilated and tortuous, and some appear as circular densities surrounding the bases of renal papillae (arrows). (B) In another patient, the right renal lymphatics are tortuous and dilated. Lymphatopelvic fistulization, which produces chyluria, permits opacification of the renal pelvis (arrow). (From Akisada and Tani,[19] with permission from the publisher.)

connective tissues and produce reactions ranging from acute inflammation to chronic scarring. Occlusion of inguinal and pelvic lymph nodes by the worms, and secondary inflammation, cause severe edema, thickening, and fibrosis of the soft tissues and can lead to elephantiasis. Lymphography is technically difficult to perform in these patients, though it may be of value in excluding congenital types of lymphedema due to aplastic or hypoplastic lymphatics. The para-aortic nodes are initially enlarged and contain mottled filling defects. The nodes later become small and granular and eventually fail to fill on lymphograms. The small superficial lymphatic channels are dilated and tortuous and more numerous than normal (dermal backflow).

The dilated lymphatics may rupture into surrounding tissues. If a fistulous connection develops between the renal lymphatics and the renal pelvis, lymphography demonstrates retrograde filling of dilated, tortuous renal lymphatics via the para-aortic nodes. The small, abnormal renal lymphatics empty slowly into the renal collecting system and produce chyluria.

The thoracic duct often demonstrates segmental dilatation and tortuosity, and this produces an accordion-like or beaded appearance. Complete obstruction of the thoracic duct is rare.

In chronic filariasis with elephantiasis, the bones of the extremity may demonstrate marked periosteal reaction and irregular cortical thickening. This is the result of the soft tissue reaction, rather than a direct effect on the bone.[18,19,72]

### Tropical Pulmonary Eosinophilia

Tropical pulmonary eosinophilia is an aberrant type of filariasis in which there is an eosinophilic inflammatory reaction that leads to granuloma formation and fibrosis. Radiographically, the pulmonary involvement is diffuse and symmetric. Multiple small nodules with indistinct outlines produce a pattern that often appears as a generalized increase in lung markings. Patchy areas of consolidation that change rapidly simulate the transient, shifting infiltrates of Loeffler's pneumonia. The response to appropriate therapy is dramatic and usually leads to complete clearing.[20,21]

### Loiasis

Humans are infected by the larvae of the filaria *Loa loa* via the bite of the mango fly. The mature worm incites focal allergic inflammation in the eye and subcutaneous tissues. Radiographically, the worm can be identified only when it is dead and calcified. The calcified worm may be curled up into a coil or appear as a thin, cotton-like thread of density. The calcifications are often difficult to visualize and are best seen in the thinner portions of the hand or foot.[22,72]

### Dirofilariasis

Dirofilariasis is a disease caused by the dog heartworm, which occasionally is transmitted to humans by several types of mosquitoes. The trapping and subsequent death of a worm in a small arteriole of the lung causes focal pulmonary infarction, which appears radiographically as a solitary, spherical, well-circumscribed nodule 1 to 2 cm in diameter. Bilateral pulmonary nodules may also occur. The appearance is indistinguishable from that of a malignant coin lesion, and a biopsy specimen is required for diagnosis.[23]

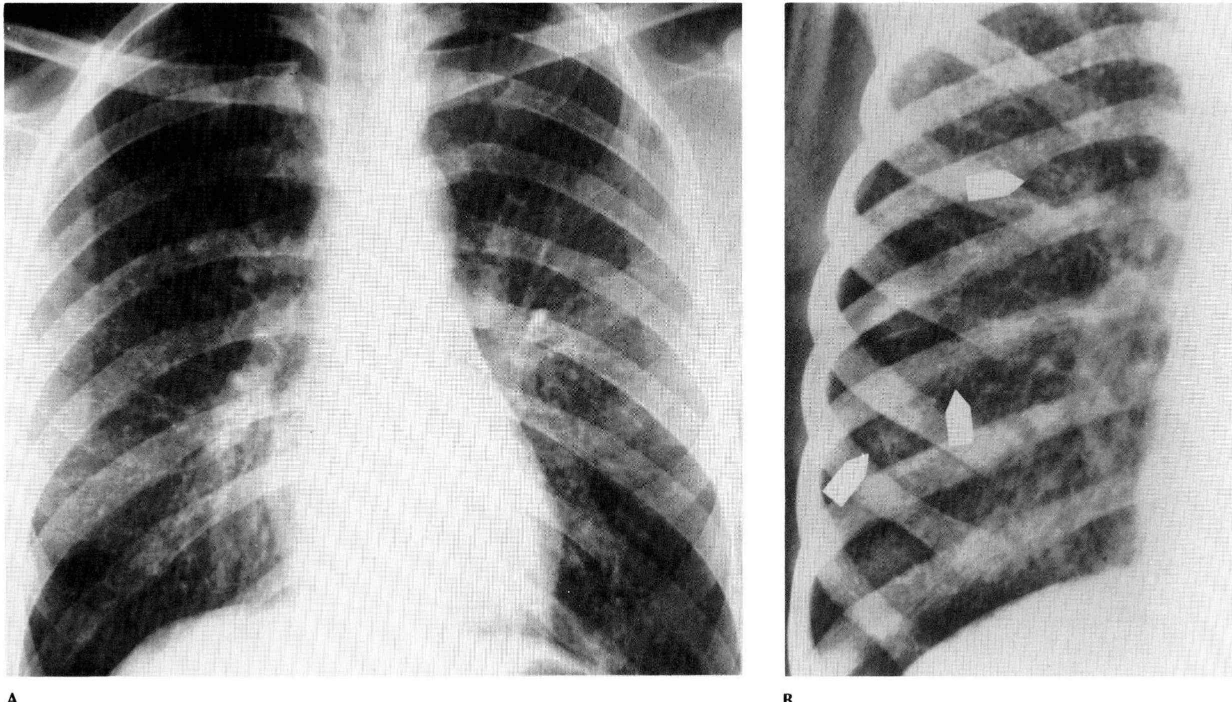

**Figure 18.19. Tropical pulmonary eosinophilia.** (A) Multiple small nodules with indistinct outlines produce a pattern of generalized increase in lung markings. (B) A coned view of the right lung in another patient better demonstrates the punctate opacities in most areas (a few groups are indicated by arrows). There is also hilar prominence and haze, increased striation, and a thickened horizontal fissure. (From Herlinger,[27] with permission from the publisher).

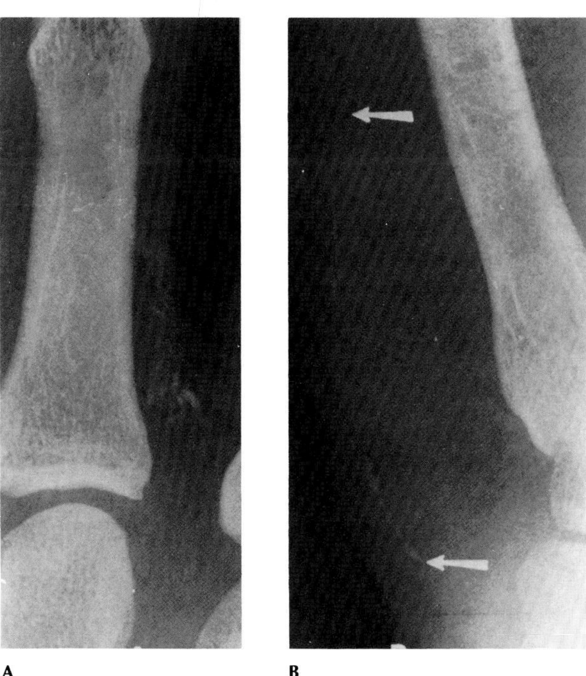

**Figure 18.20. Loiasis.** Calcified worms seen in (A) convoluted and (B) extended form. (From Cockshott and Middlemiss,[72] with permission from the publisher.)

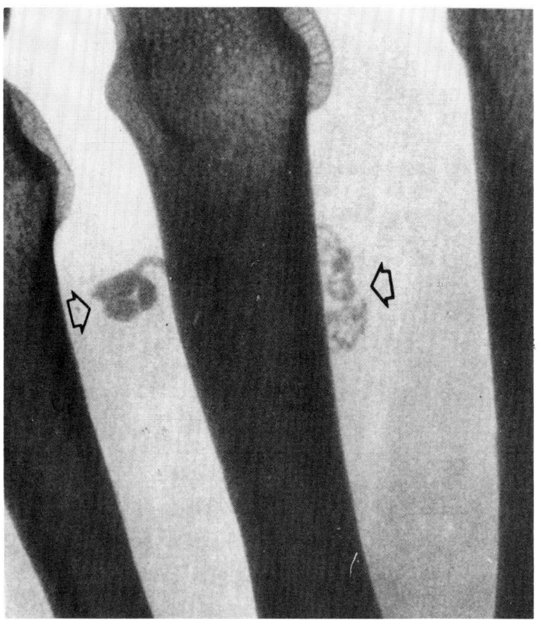

**Figure 18.21. Loiasis.** Calcified worms (arrows). (From Williams,[23] with permission from the publisher.)

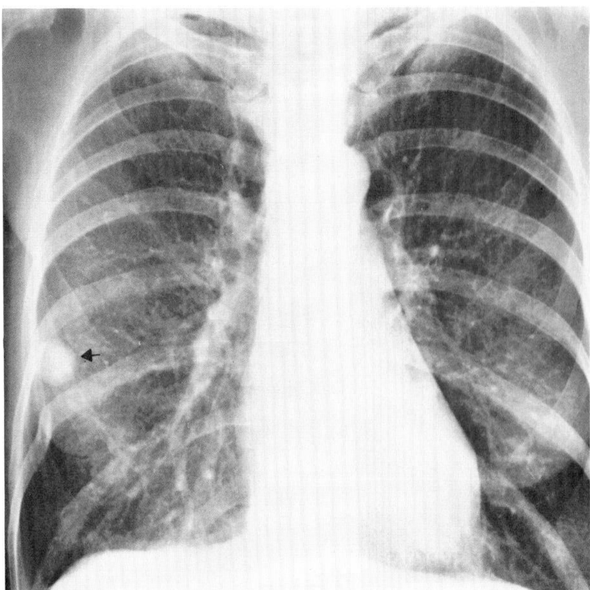

**Figure 18.22. Dirofilariasis.** A well-circumscribed solitary pulmonary nodule (arrow) that is indistinguishable from a malignant coin lesion.

## SCHISTOSOMIASIS (BILHARZIASIS)

Schistosomiasis is a widespread infectious disease caused by a blood fluke that inhabits the portal venous system of the liver. After partially maturing in the body of a suitable snail host, the *Schistosoma* organism emerges into the water and enters the human host by penetrating the unbroken skin or buccal membrane. Those larvae that make their way to the liver develop into adult worms, which can then migrate against the flow of portal venous blood into small venules where the females deposit their eggs. Eggs lying close to the mucosal surface may rupture into the lumen of the bladder or gut before being carried to the outside in the urine or feces. Pulmonary infestation is caused by organisms that reach the peripheral venules and are carried to the right side of the heart and then to the pulmonary capillaries. The irritative effect of ova that pass through or lodge in the walls of the bowel, bladder, or pulmonary capillaries stimulates an inflammatory response that includes granuloma formation, obliterative vasculitis, and progressive fibrosis.

Colon involvement is usually due to *Schistosoma mansoni* or *Schistosoma japonicum*. Because the adult worm has a predilection for entering the inferior mesenteric vein, the sigmoid colon and rectum are most commonly involved. The most characteristic radiographic appearance is multiple filling defects (usually 1 to 2 cm large) due to the development of polypoid granulomas. These masses are friable and vascular, and they bleed easily with the passage of feces; this explains the frequently bloody stools associated with severe infestation. Individual masses may become so large that they

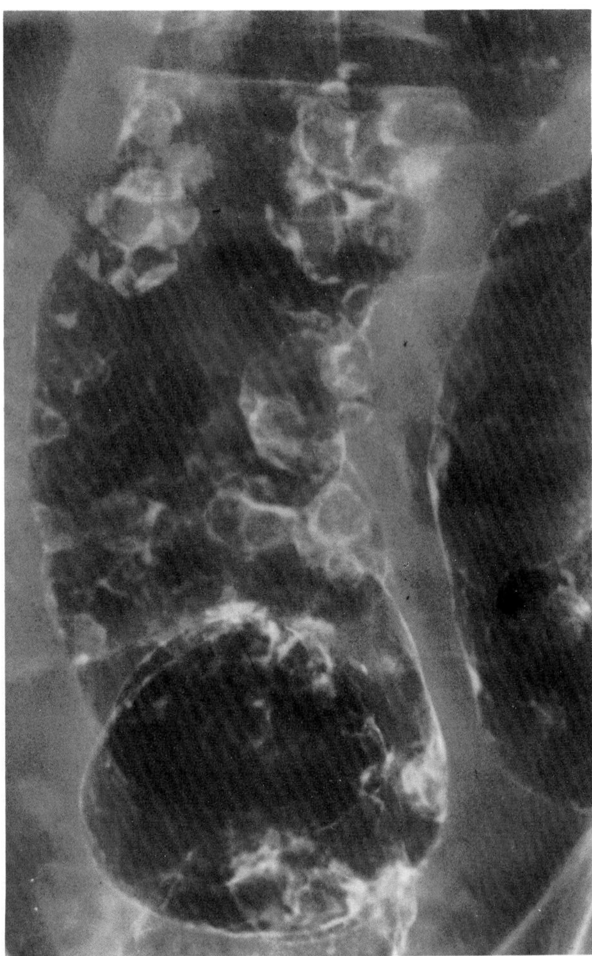

**Figure 18.23. Schistosomiasis.** Multiple filling defects in the colon represent polypoid granulomas. (From Eisenberg,[69] with permission from the publisher.)

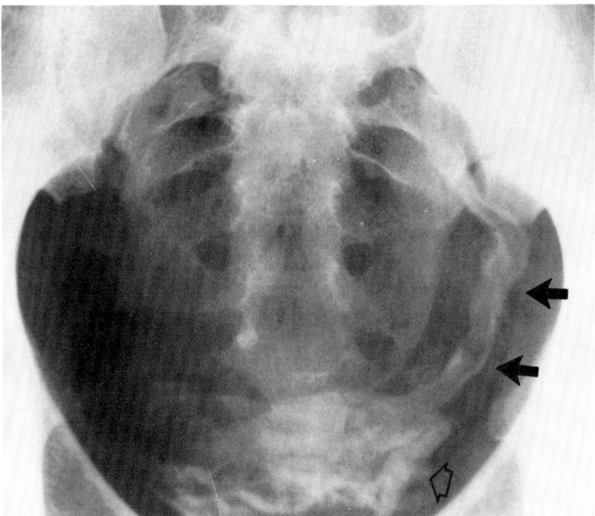

**Figure 18.24. Schistosomiasis.** There is calcification of the distal ureter (solid arrows) and the bladder (open arrow). (From Eisenberg,[69] with permission from the publisher.)

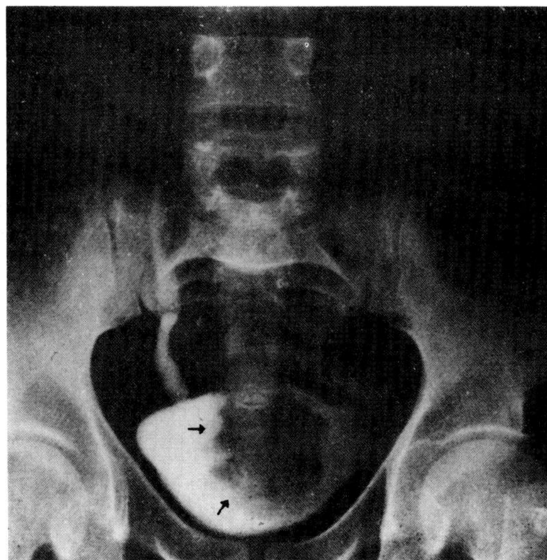

**Figure 18.25. Schistosomiasis.** An excretory urogram shows a huge filling defect on the left side of the bladder (arrows); this represents a bilharzial carcinoma. (From Al-Ghorab,[27] with permission from the publisher.)

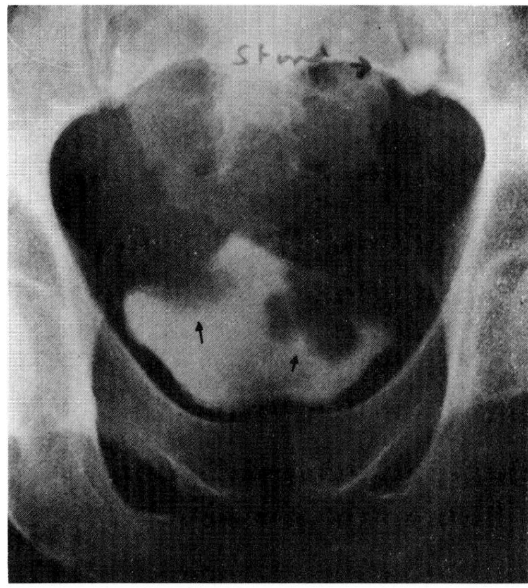

**Figure 18.26. Schistosomiasis.** A cystogram shows two symmetrically placed benign bilharzial polyps (arrows) of the bladder; the polyps simulate malignant lesions. (From Al-Ghorab,[27] with permission from the publisher.)

cause obstruction or intussusception. Tiny ulcerations or mural spiculations may simulate ulcerative colitis and appear as a diffuse granular pattern on double-contrast enemas. Spasm, disturbed motility, and a loss of the haustral pattern are common. Reactive fibrosis and extensive pericolonic infiltration can narrow the colon and produce a radiographic pattern indistinguishable from that of Crohn's disease or colonic malignancy.

*Schistosoma haematobium* primarily involves the urinary tract, especially the wall of the bladder and lower ureters. Submucosal edema causes the bladder outline to appear hazy and indistinct on contrast studies. Granulomatous polyps projecting into the lumen can simulate malignancy. Segmental fibrosis causes decreased contractility, and the bladder can become small and deformed. Distortion of the trigone leads to ureterovesical reflux along with aperistalsis and dilatation of the distal ureters.

Calcification of the bladder is a classic radiographic finding in schistosomiasis, seen in about 50 percent of the patients with urinary tract disease. Initially, this calcification is most apparent and extensive at the base of the bladder, where it forms a linear, opaque shadow parallel to the upper border of the pubic bone. With further calcium deposition, the linear density encircles the entire bladder. Somewhat less frequently, schistosomiasis causes ureteral calcification, which appears as two roughly parallel dense lines separated by the caliber of the ureter. The calcification is heaviest in the pelvic portion of the ureter and gradually decreases as it approaches the kidneys.

Squamous cell carcinoma of the bladder is a frequent late complication of schistosomiasis in Egypt, though not in other areas. This tumor alters the appearance of the calcified bladder by disrupting the continuity of the homogeneous line of calcification in the area of neoplastic infiltration.

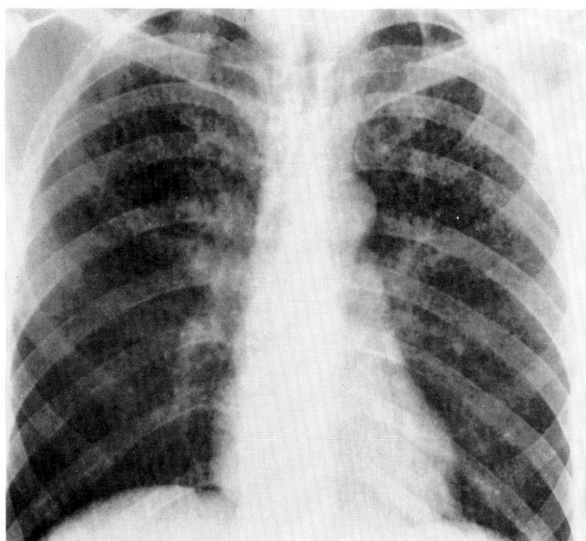

**Figure 18.27. Schistosomiasis.** Perivascular granulomas produce small nodular and linear densities that are distributed diffusely throughout the lungs in a miliary pattern simulating tuberculosis.

Schistosomiasis may cause two forms of pulmonary disease. A transient pneumonia simulating Loeffler's syndrome can occur coincidentally with the passage of the larvae through the pulmonary capillaries. Dissemination of ova throughout the pulmonary vascular tree produces a more serious form of pulmonary disease that often leads to an obliterative arteritis, interstitial fibrosis, pulmonary hypertension, and cor pulmonale. Radiographically, perivascular granulomas produce small nodular and linear densities that are diffusely distributed throughout the lungs in a miliary pattern simulating tuberculosis. Pulmonary hypertension appears as progressive dilatation of the main pulmonary artery and its branches with rapid tapering toward the periphery. Aneurysmal dilatation of the main pulmonary arteries can occur.

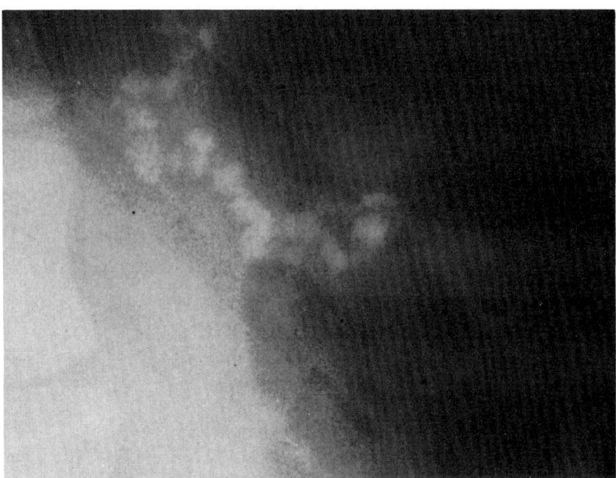

**Figure 18.28. Dracunculiasis.** A calcified guinea worm in the right lobe of the liver. (From Darlak, Moskowitz, and Kattan,[30] with permission from the publisher.)

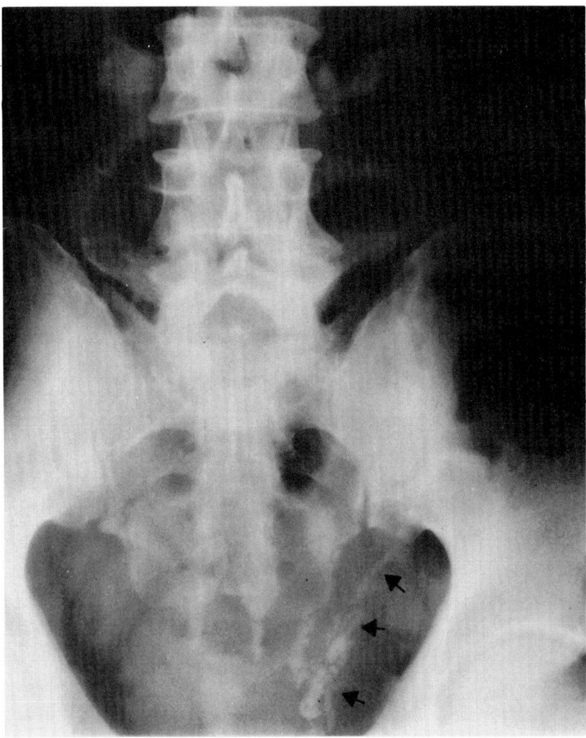

**Figure 18.29. Dracunculiasis.** A long serpiginous calcified worm (arrows) within the pelvis.

The deposition of eggs in periportal connective tissue can incite an inflammatory reaction that leads to multiple granulomas in the liver and subsequent reactive fibrosis. This causes liver enlargement, presinusoidal portal hypertension, and often massive splenomegaly. The development of esophageal varices is an indication of chronic, irreversible liver disease. Cholangiography may demonstrate irregular narrowing and tortuosity of interhepatic bile ducts simulating the pattern in cirrhosis.[24–29]

## DRACUNCULIASIS (GUINEA WORM)

Dracunculiasis is acquired by humans through the contamination of water with the free-swimming larvae of the guinea worm. The parasite typically settles in the soft tissues of the limbs, especially the lower extremities. The live worm cannot be identified radiographically. After its death, however, the guinea worm often calcifies and may appear as a long, stringlike, serpiginous or curvilinear opacification, which is often coiled and may be several feet long. The calcification is frequently segmented and beaded, because muscle movement breaks up the necrotic worm.[18,30]

## INTESTINAL NEMATODES

### Trichuriasis (Whipworm)

*Trichuris trichiura* (whipworm) is a ubiquitous parasite that is primarily found in warm, moist tropics. Humans become infested by ingesting whipworm ova from contaminated soil or vegetables. The ova hatch into larvae that migrate down to the cecum, where they develop

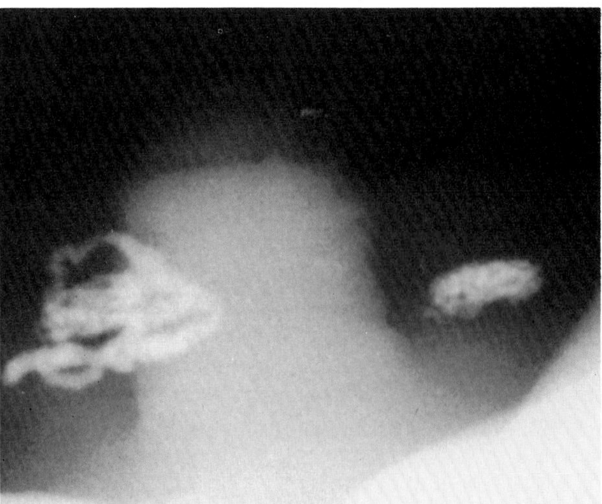

**Figure 18.30. Dracunculiasis.** Long, serpiginous calcified worms coiled within the scrotum.

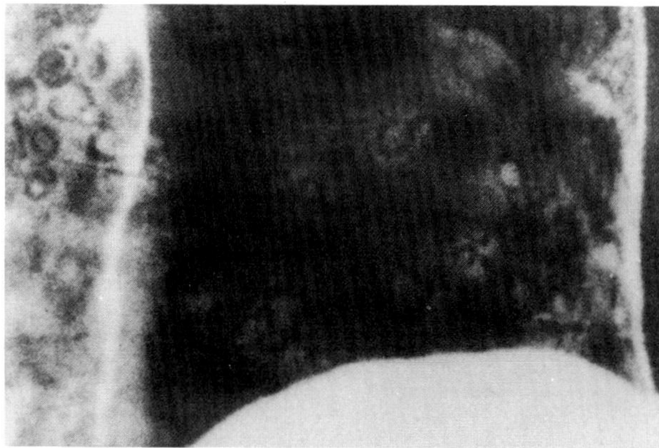

**Figure 18.31. Trichuriasis.** A close-up film of the ascending colon from a double-contrast enema demonstrates innumerable coiled worms in a patient harboring a heavy load of *Trichuris trichiura*. The black bar marker is 1 cm long. (From Cockshott and Middlemiss,[72] with permission from the publisher.)

into adult worms. The whipworms become attached to the colon and lie imbedded between intestinal villi, coiled upon themselves with their blunt caudal portions projecting into the bowel lumen. Mild infections are usually asymptomatic, but heavy infestation, especially in children, can result in chronic diarrhea, abdominal pain, dehydration, weight loss, rectal prolapse, eosinophilia, and microcytic anemia.

Radiographically, trichuriasis causes a diffuse, granular mucosal pattern throughout the colon along with considerable flocculation of barium due to abundant mucus secretions surrounding the tiny whipworms. In addition, multiple filling defects are produced by the individual outlines of the innumerable worms, which are attached to the mucosa with their posterior portions either tightly coiled or unfurled in a whiplike configuration.[31,32,72,74]

## Ascariasis

Infection by *Ascaris lumbricoides* is characterized by an early, pulmonary phase related to larval migration, and a later, prolonged intestinal phase. After the host has ingested contaminated vegetables, water, or soil, the larvae of the worm hatch in the small bowel. Following a complicated migration through the bowel wall, the lung, and the tracheobronchial tree, the larvae again enter the alimentary tract and reach the small bowel, where they mature into adult worms. If the worms are not numerous, the mere presence of adult worms in the small bowel is usually asymptomatic or associated with vague and nonspecific abdominal complaints. The complications of ascariasis include intestinal obstruction, peritonitis (if the worms penetrate the bowel), biliary colic (if the worms enter the bile duct), and hemoptysis (as the worms pass through the lungs en route to the bowel).

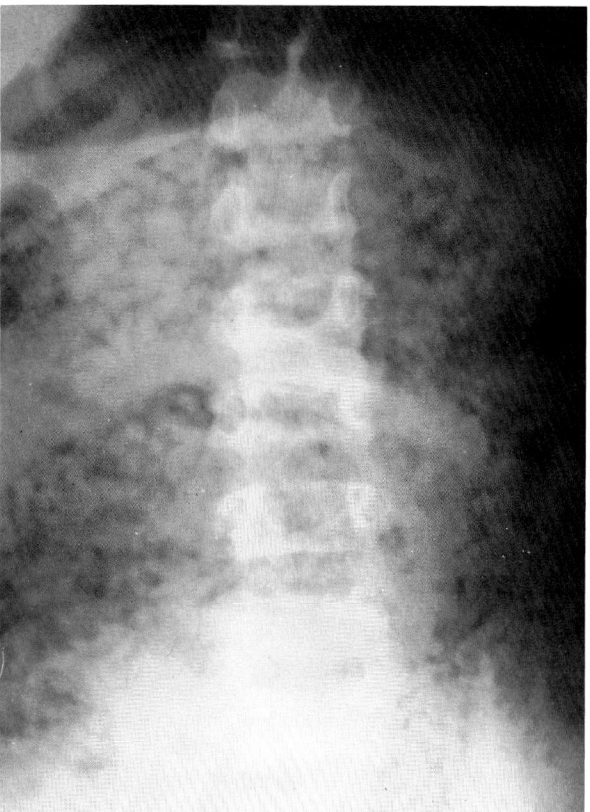

**Figure 18.32. Ascariasis.** A plain abdominal radiograph demonstrates enormous numbers of impacted worms as convoluted masses outlined by gas (Medusa locks sign). (From Ellman, Wynne, and Freeman,[35] with permission from the publisher.)

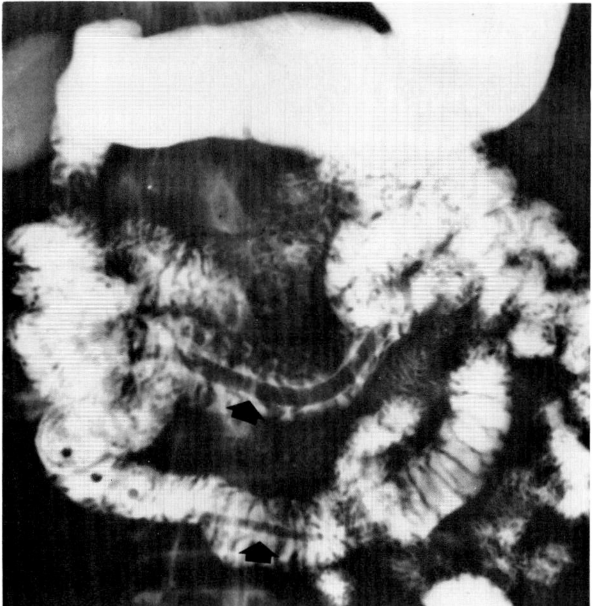

**Figure 18.33. Ascariasis.** The worms appear as elongated, radiolucent filling defects (arrows) within the small bowel. (From Eisenberg,[69] with permission from the publisher.)

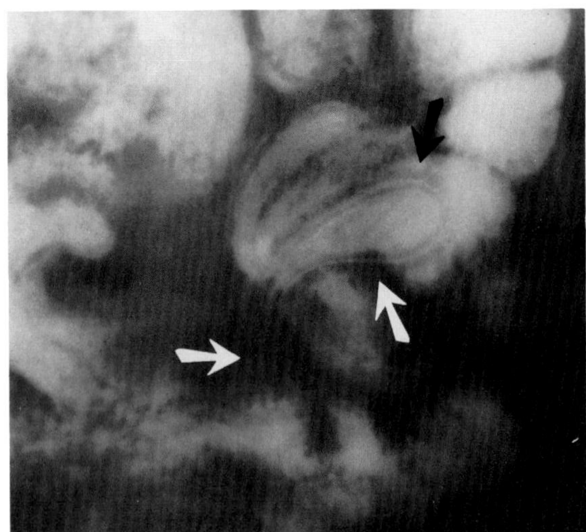

**Figure 18.34. Ascariasis.** Barium can be seen in the intestinal tract of the worm. (From Weissberg and Berk,[34] with permission from the publisher.)

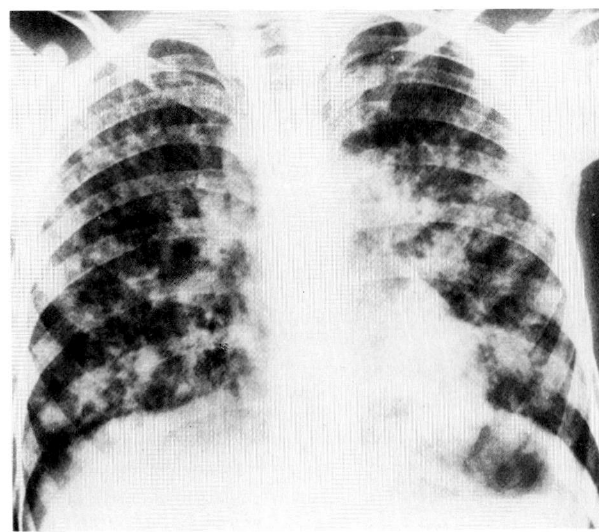

**Figure 18.35. Ascariasis.** Extensive pulmonary infiltrates due to the presence of *Ascaris* larvae in the lungs. (From Cockshott and Middlemiss,[72] with permission from the publisher.)

On plain abdominal radiographs, the trapping of intestinal gas between masses of worms can give rise to an appearance suggesting coiled locks of hair (Medusa locks). With the patient erect, the worm bolus can appear as an irregular nodularity that distorts the gas-fluid levels in the bowel ("hump" sign).

On barium studies, the worms appear as elongated, radiolucent filling defects that are most often found in the distal jejunum. If the patient has been fasting for 12 hours or longer, an ascaris often ingests barium, and the gastrointestinal tract of the worm then appears as a thin, longitudinal opaque density bisecting the length of the lucent filling defect. Masses of coiled worms that have clumped together may produce one or more rounded intraluminal filling defects in the small bowel.

The migration of *Ascaris lumbricoides* from the intestines to the biliary tract can result in partial obstruction complicated by cholangitis, cholecystitis, and liver abscess. Biliary ascariasis can also cause pancreatitis, either by the mechanical effect of the worms in the common duct or by the migration of the worms into the main pancreatic duct. As in the intestine, worms in the biliary system produce characteristic long, linear filling defects. At times the worms coil in the bile duct and appear as more discrete masses. Disruption of the normal architecture of the sphincter of Oddi may permit reflux of intestinal contents into the common bile duct, which causes gas in the biliary tree.

Migrating *Ascaris* larvae in the lungs may produce patchy areas of homogeneous consolidation, which are often concentrated in the perihilar regions and may be variable and migratory. With more severe involvement, the shadows tend to coalesce and assume a lobular pattern. The death of a migrating larva within the lung may result in granuloma formation and the appearance of a solitary pulmonary nodule.[33–36,72]

## Anisakiasis

*Anisakis* is an ascaris-like nematode that spends its larval stage within tiny crustaceans that are eaten by such saltwater fish as herring, cod, and mackerel. The disease primarily involves persons from Japan, Holland, or Scandinavia who are in the habit of eating raw, slightly salted, or vinegar-pickled fish. The larvae penetrate the mucosa of the gastrointestinal tract and cause acute, cramping abdominal pain within a few hours of ingestion. The disease is usually self-limited, because the worms cannot grow in humans and die within a few weeks.

In the stomach, mucosal edema causes a localized or generalized pattern of coarse, broad gastric folds. A definitive radiographic diagnosis requires the demonstration of a threadlike filling defect, about 30 mm in length, that represents the larva itself. The worms can appear serpiginous, circular, or ringlike and can change their shape during an examination. Thickening of the bowel wall of the ascending colon or terminal ileum may produce the coned cecum pattern.[37,38,73]

## Hookworm Disease

Hookworm infection in humans results from the presence of adult *Ancylostoma duodenale* and/or *Necator americanus* within the small intestine, usually the jejunum and proximal ileum. Although many infected persons are asymptomatic, untreated patients who have a heavy infestation may develop malabsorption, severe anemia, and malnutrition.

Most patients with hookworms show no radiographic abnormalities on barium examination of the upper gastrointestinal tract. Nonspecific small bowel changes in-

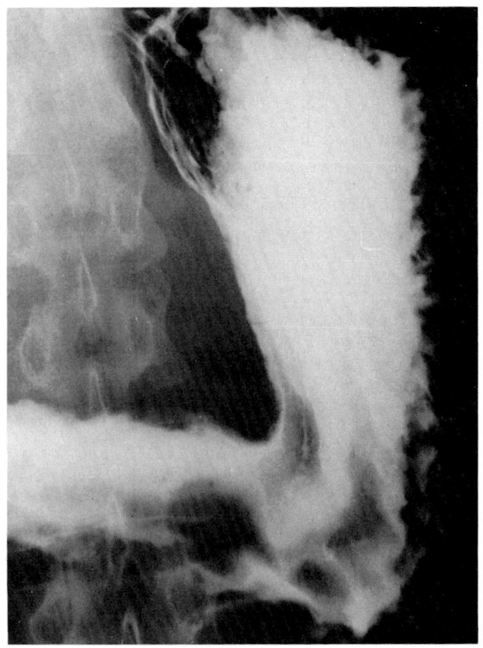

A

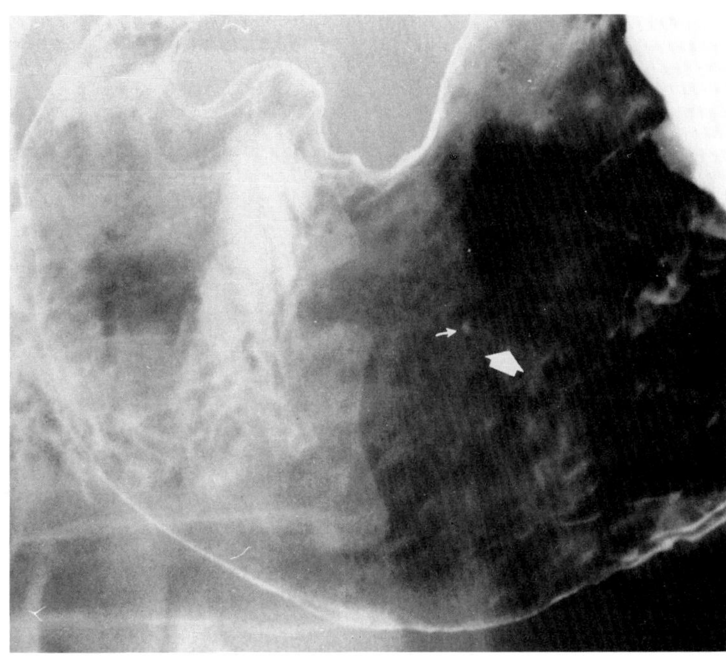

B

**Figure 18.36. Anisakiasis.** (A) Generalized fold thickening represents mucosal edema in a patient who developed abdominal pain and nausea 8 hours after eating raw mackerel. (B) A double-contrast study shows a small collection of barium at the site at which the parasite penetrated the gastric mucosa (small arrow) and reveals the thin outline of the larva itself (large arrow). (From Nakata, Takeda, and Nakayama,[73] with permission from the publisher.)

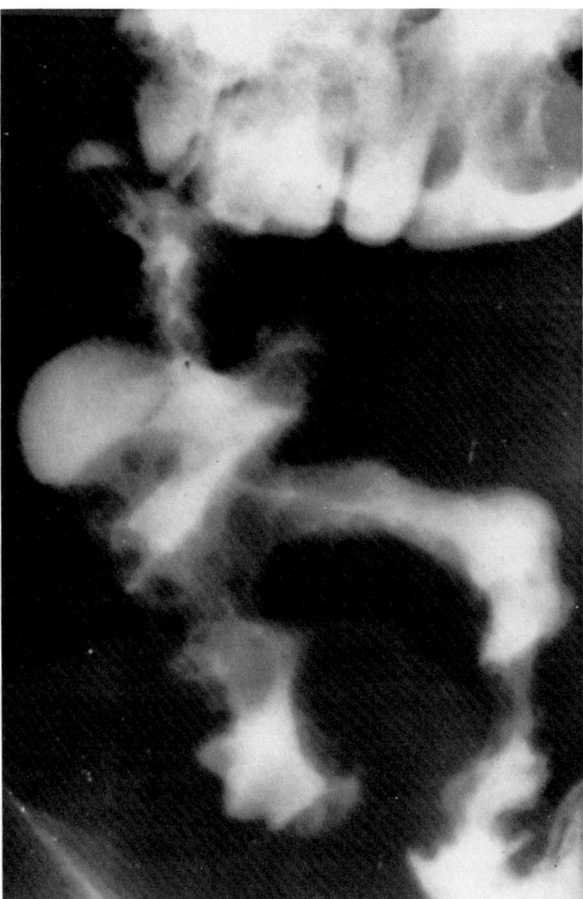

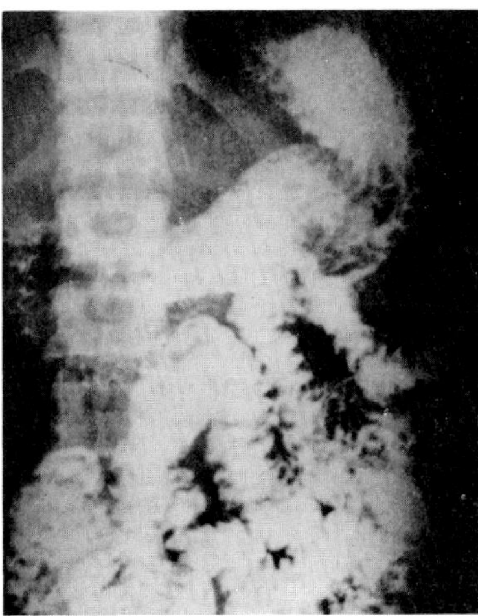

**Figure 18.38. Hookworm disease.** A small bowel study shows prominent thickening of mucosal folds in the jejunum. (From Chuttani, Puri, and Misra,[39] with permission from the publisher.)

**Figure 18.37. Anisakiasis.** Severe inflammatory changes are seen in the cecum, ascending colon, and ileocecal valve in a patient who developed severe abdominal pain after eating raw fish. (From Eisenberg,[69] with permission from the publisher.)

clude mucosal fold thickening and distortion, narrowing of intestinal loops, and occasionally one or more rounded intraluminal filling defects due to masses of coiled worms clumped together.

The most common radiographic finding in patients with chronic severe hookworm disease is mild or moderate generalized cardiac enlargement due to the patient's profound anemia and hypoproteinemia.[39–41]

## Cutaneous Larva Migrans (Creeping Eruption)

The larvae of the dog and cat hookworm (*Ancylostoma braziliense*) can penetrate and migrate under the skin, causing intense itching. In about half of the patients, the larvae reach the lung and cause transient, migratory pulmonary infiltrates associated with an increased number of eosinophils in the blood and sputum. The pulmonary lesions begin about 1 week after the skin eruption and usually resolve within several months.[42]

## Strongyloidiasis

*Strongyloides stercoralis* is a roundworm that exists in warm, moist climates where there is frequent fecal contamination of the soil. When females of the parasite are swallowed, they invade the mucosa and produce an infection that predominantly involves the proximal small bowel but can affect any part of the gastrointestinal tract from the stomach to the anus. Mild intestinal disease is

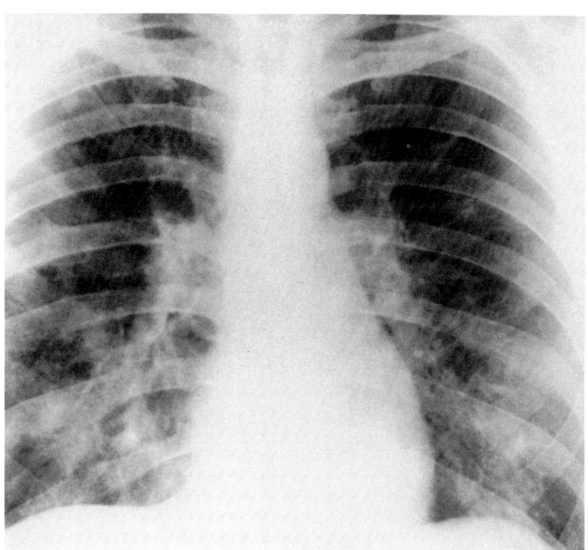

**Figure 18.39. Cutaneous larva migrans.** A frontal chest radiograph demonstrates multiple small irregular areas of air space consolidation widely scattered throughout both lungs. (From Kalmon,[42] with permission from the publisher.)

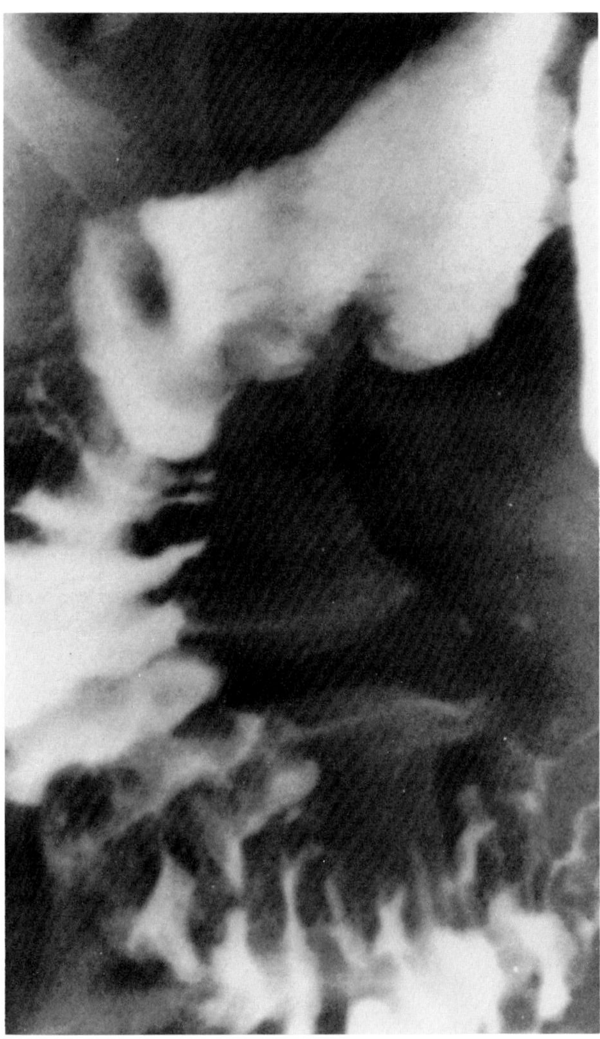

**Figure 18.40. Strongyloidiasis.** Irregular, at times nodular, thickening of folds throughout the duodenal sweep. (From Eisenberg,[69] with permission from the publisher.)

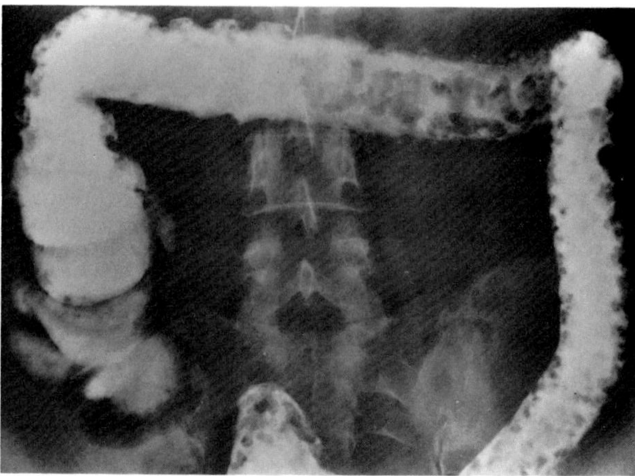

**Figure 18.41. Strongyloidiasis.** There is a diffuse ulcerating colitis with deep and shallow ulcers and pronounced mucosal edema. (From Eisenberg,[69] with permission from the publisher.)

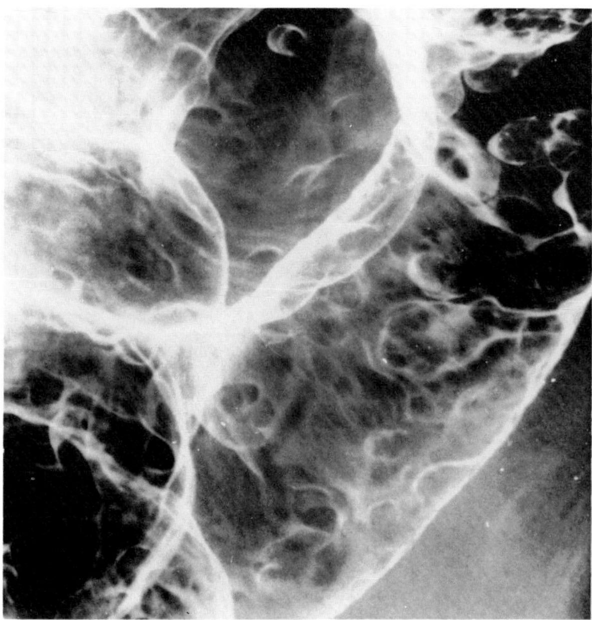

**Figure 18.42. Strongyloidiasis.** Multiple hyperplastic pseudopolyps simulate ulcerative or Crohn's colitis. (From Eisenberg,[69] with permission from the publisher.)

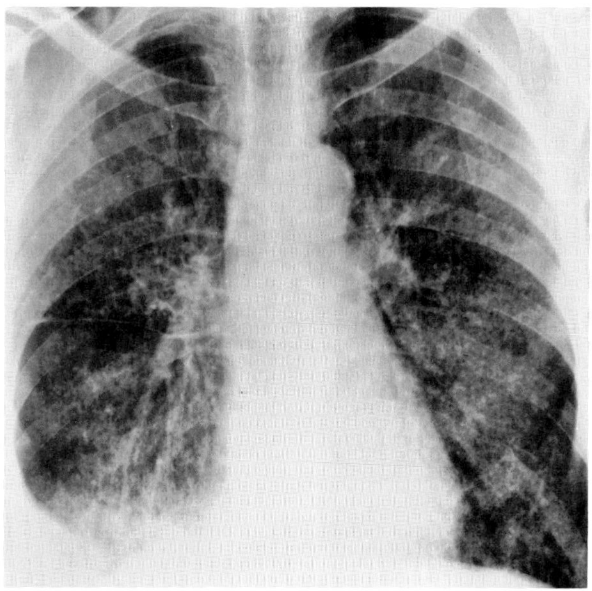

**Figure 18.43. Strongyloidiasis.** A chest radiograph during the stage of larval migration shows a pattern of miliary nodules diffusely distributed throughout both lungs. There also is a large right pleural effusion.

often asymptomatic, but severe symptoms of abdominal pain, nausea, vomiting, weight loss, and fever can occur. Severe diarrhea and steatorrhea can clinically mimic acute tropical sprue.

Radiographically, strongyloidiasis most commonly produces irritability and irregular thickening of the mucosal folds of the duodenum and proximal jejunum. Hyperperistalsis and increased secretions cause blurring of the mucosal fold pattern. Ulceration and luminal stenosis, often over a long segment, can simulate Crohn's disease.

Severe colitis, characterized by both small and large ulcers, mucosal edema, pseudopolyposis, and the loss of haustral markings, is an unusual manifestation of strongyloidiasis. Colonic infection is often associated with overwhelming sepsis, hemorrhage, and death, though healing with stricture formation can occur.

Strongyloidiasis involving the stomach can produce nodular intramural filling defects secondary to granuloma formation. In advanced disease, severe inflammatory changes and diffuse fibrosis can cause mural rigidity and the linitis plastica pattern. At times, narrowing of the gastric outlet can be so severe as to cause delayed gastric emptying.

Although the chest radiograph is normal in most patients with strongyloidiasis, during the stage of larval migration there may be fine miliary nodules or ill-defined patchy areas of air space consolidation similar to those seen in Loeffler's syndrome. In immunosuppressed patients, an overwhelming *Strongyloides* infection of the lungs may cause a radiographic pattern simulating severe bilateral pulmonary edema.[43–45]

## OTHER TREMATODES OR FLUKES

### Paragonimiasis

Paragonimiasis is a chronic infection of the lung caused by a trematode. The infection is acquired by eating raw, or poorly cooked, infected crab or crayfish. Many patients with a heavy infestation are asymptomatic. Others have chronic bronchitis and bronchiectasis and present with cough, pain, hemoptysis, and brownish sputum.

The manifestations of paragonimiasis on chest radiographs depend on the stage and severity of the disease. The migrating larvae may cause an exudative pneumonia that primarily involves the bases and produces a patchy air space consolidation with a fluffy cotton-wool appearance. The infiltrate resolves within several weeks, leaving residual nodules or cysts that usually lie close to a pleural surface and contain the mature worms imbedded in fibrous tissue. Localized pleural thickening is often seen. The most characteristic radiographic feature is the "ring shadow," composed of a thin-walled cyst with a prominent crescent-shaped opacity along one side of its border. Another typical finding is a track, or burrow, that runs an irregular tortuous course in the pulmonary parenchyma and communicates with the adjacent cyst. Conventional tomography is often required to demonstrate these burrows. With the death of the parasite, fibrosis and calcification occur; this presents a radiographic pattern closely resembling tuberculosis.

Cerebral paragonimiasis often causes intracranial calcifications that appear as amorphous, punctate densities, ill-defined small nodules, or aggregates of round or oval

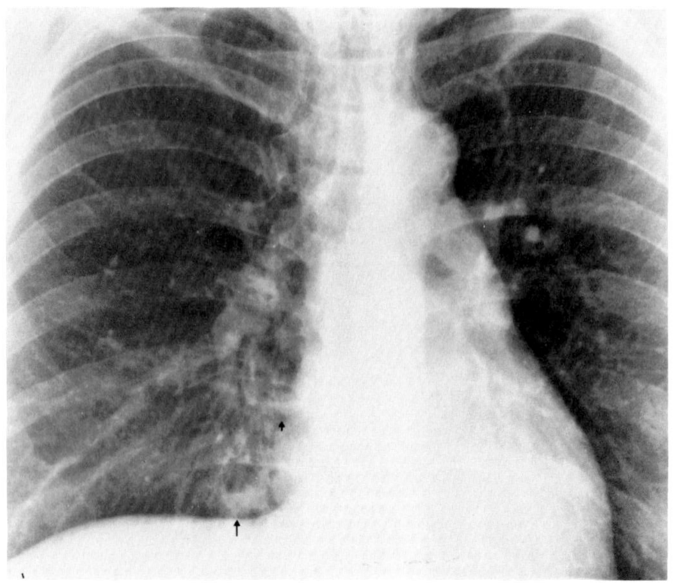

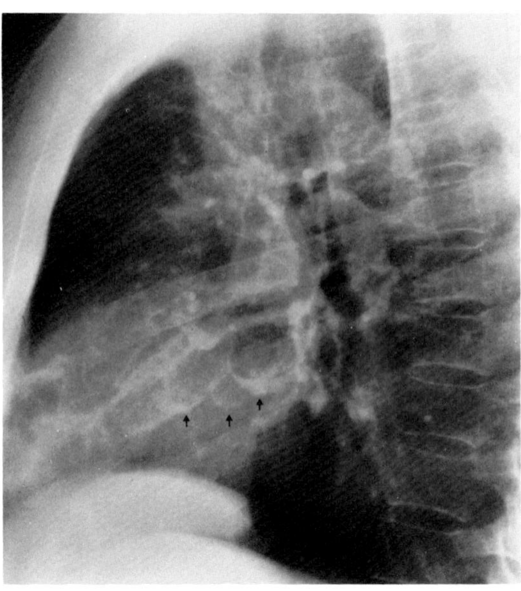

A B

**Figure 18.44. Paragonimiasis.** (A) Frontal and (B) lateral chest radiographs demonstrate multiple cysts (arrows) within the right middle lobe. The cysts are thin-walled and most have a prominent crescent-shaped opacity along one side of their borders, the characteristic ring shadow of paragonimiasis.

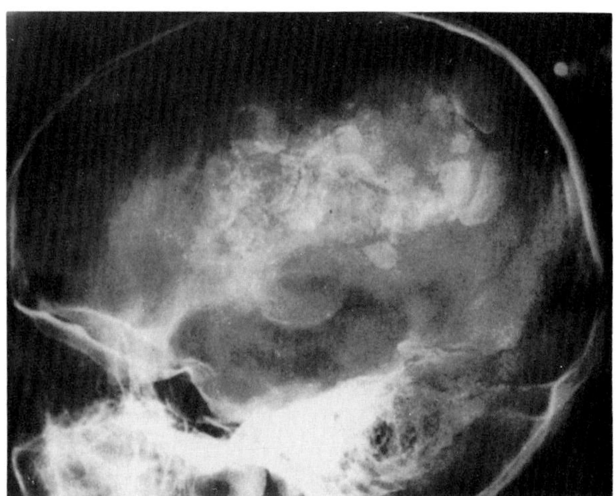

**Figure 18.45. Paragonimiasis.** Characteristic soap-bubble appearance of calcification in the parietal area and posterior part of the frontal lobe. The dorsum sellae is not visible, a result of increased intracranial pressure. (From Oh,[47] with permission from the publisher.)

cysts that have peripheral areas of increased density ("soap bubbles"). Individual cysts are often small, but when they are grouped together, a cluster may measure up to 10 cm in diameter. The calcifications are almost always unilateral and usually occur in the parietal and occipital lobes. There is evidence of increased intracranial pressure in about one-third of the patients with cerebral involvement. Large cysts or abscesses may appear as avascular space-occupying lesions on cerebral angiography or CT. Subcortical atrophy and arachnoiditis often develop.[46–49]

## Clonorchiasis

*Clonorchis sinensis* is a parasitic liver fluke that is acquired by the ingestion of raw freshwater fish. The parasite migrates from the duodenum into the biliary tree, where it may live for many years. The adult worms usually reside in the small intrahepatic bile ducts, where they incite an inflammatory reaction that predisposes to cholangitis, liver abscess, hepatic duct stones (with ova or adult flukes forming the nidus), or even common duct obstruction. At cholangiography, the flukes appear as typical linear filling defects when seen in profile; when viewed en face, the worms produce round filling defects simulating calculi.

A higher-than-normal incidence of carcinoma of the intrahepatic bile ducts has been associated with *Clonorchis* infestation.[50,69]

## Fascioliasis

In sheep-growing areas, *Fasciola hepatica* infects persons who ingest pond water or watercress contaminated with the metacercarial form of the liver fluke life cycle. The adult worms reside in the small intrahepatic bile ducts, where they produce epithelial hyperplasia and periductal fibrosis. The resultant biliary stasis favors bacterial infection and leads to a radiographic pattern identical to that of *Clonorchis* infection with cholangitis (alternating strictures and dilatation of the biliary tree), liver abscess, hepatic duct stones, and even common duct obstruction.[51]

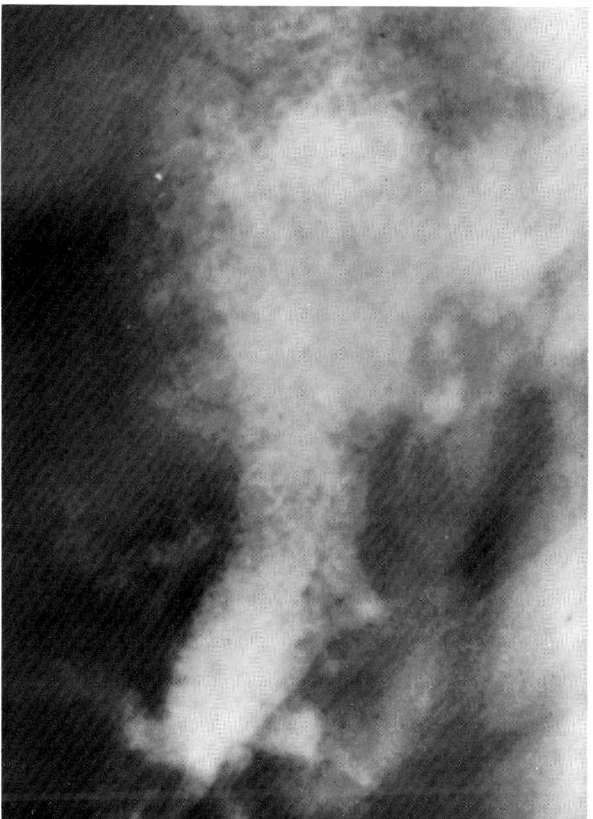

**Figure 18.46. Clonorchiasis.** There are multiple filling defects in the biliary system. Many of these filling defects represent coexistent calculi, which are often seen in this condition. (From Eisenberg,[69] with permission from the publisher.)

## CESTODE (TAPEWORM) INFECTIONS

### Taeniasis

Humans are the only definitive hosts of the beef and pork tapeworms (*Taenia saginata* and *Taenia solium*). After the larvae in undercooked, or raw, infected meat are released by the digestive juices, they evaginate, attach to the gut mucosa, and grow to become adult worms living in the intestinal lumen. Although the parasite is widespread, it is infrequently demonstrated radiographically on barium examinations. The tapeworm typically appears as an unusually long linear filling defect in the lower jejunum or ileum.[52,53]

### Cysticercosis

Cysticercosis refers to the invasion of human tissue by the larval form of a tapeworm, usually *Taenia solium* (pork tapeworm). The larvae develop in the subcutaneous tissues, muscle, and viscera, and, most significantly, in the eye and brain. Although there is little tissue reaction while the organism is viable, the dead larvae invoke a marked tissue response that produces muscular pain, weakness, fever, and eosinophilia. Within a few years, calcification develops.

The classic radiographic appearance of cysticercosis is multiple linear or oval calcifications in the soft tissues. These calcified cysts often have a noncalcified central area and almost always have their long axes in the plane

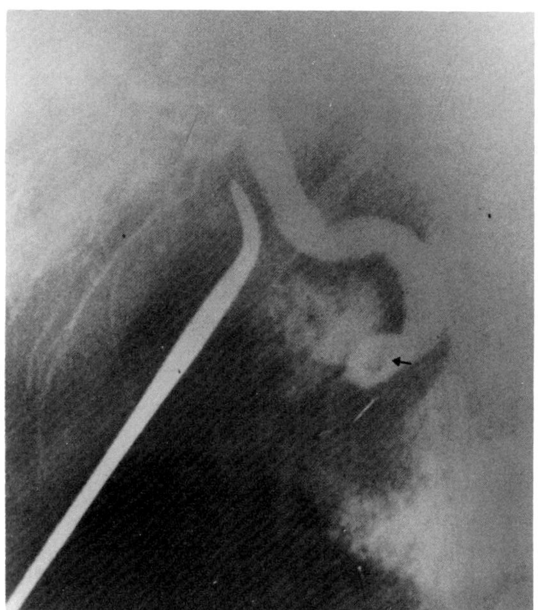

**Figure 18.47. Fascioliasis.** A cholangiogram shows a filling defect (arrow) near the ampulla; the defect represents a worm coiled in the distal portion of the bile duct. (From Belgraier,[51] with permission from the publisher.)

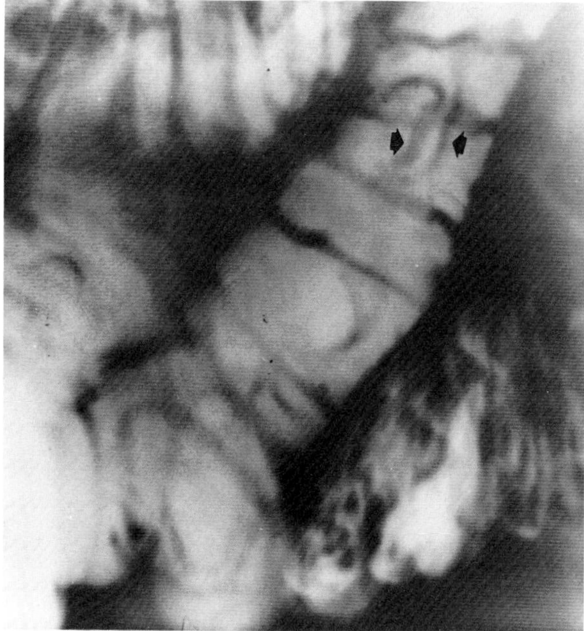

**Figure 18.48. *Taenia solium* (tapeworm).** Multiple small filling defects (arrows) are visible in this segment of small bowel. (From Eisenberg,[69] with permission from the publisher.)

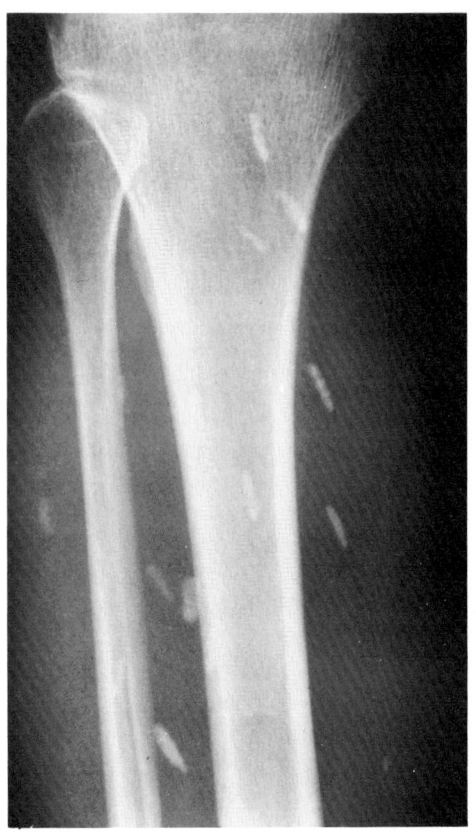

A

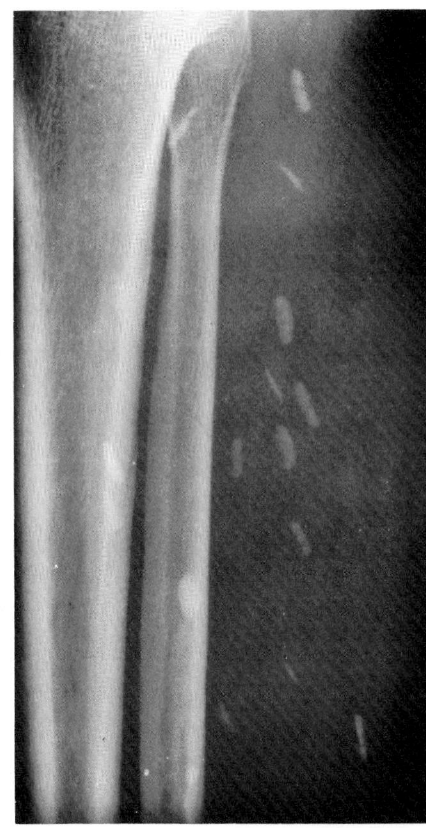

B

**Figure 18.49. Cysticercosis.** (A) Frontal and (B) lateral views of the calf region demonstrate multiple linear and oval calcifications in the soft tissues.

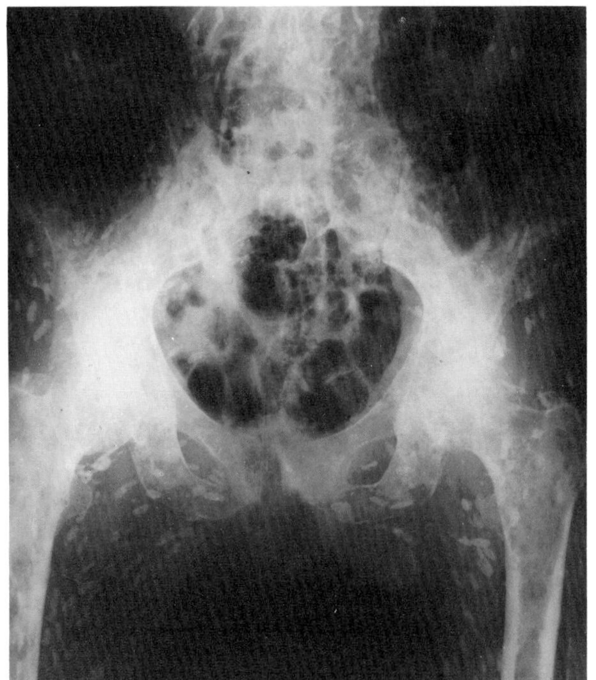

**Figure 18.50. Cysticercosis.** A radiograph of the abdomen and pelvis shows multiple calcified cysticerci in the muscles of the thighs, abdomen, and gluteal regions. (From Keats,[56] with permission from the publisher.)

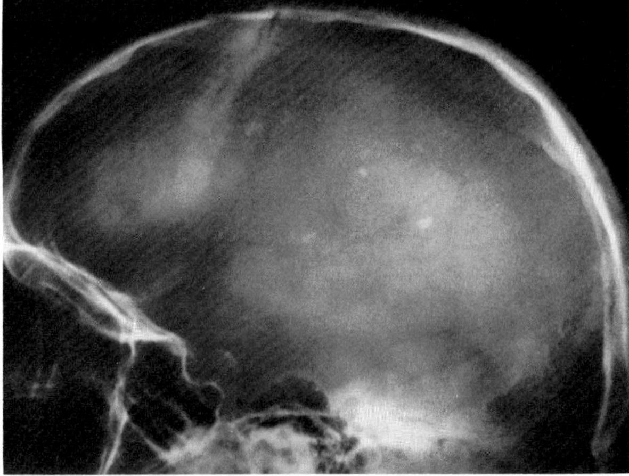

**Figure 18.51. Cysticercosis.** Small blebs of intracranial calcification in a young Guatemalan immigrant with headaches and seizures.

of the surrounding muscle bundle, unlike the random distribution of soft tissue calcifications in the Ehlers-Danlos syndrome.

Central nervous system involvement commonly occurs and can produce epilepsy, mental disturbance, loss of vision, and even a fulminating disease that resembles acute encephalitis. Intracranial calcification is less com-

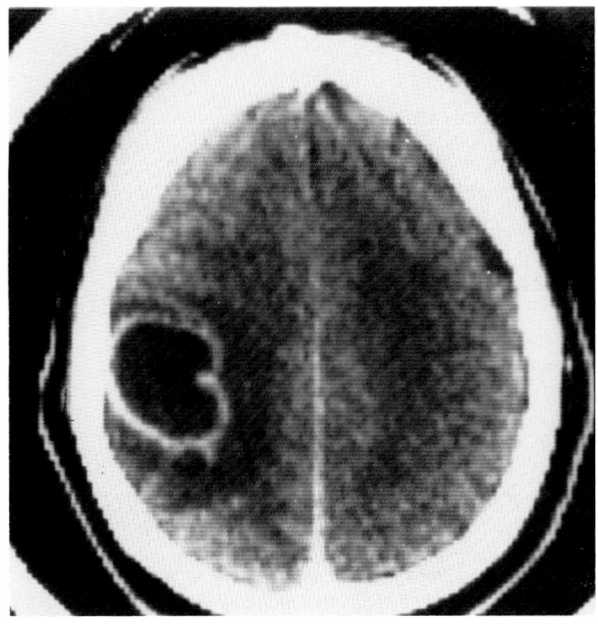

A

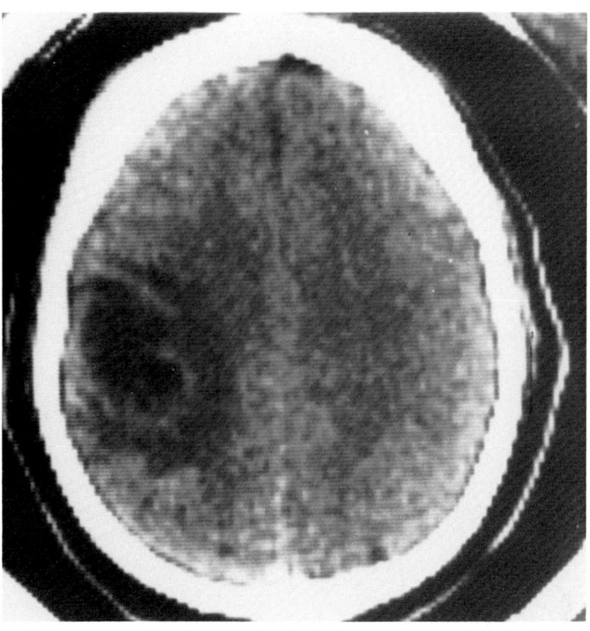

B

**Figure 18.52. Cysticercosis.** (A) A precontrast CT scan shows a primarily low density area in the right frontoparietal region. The ring of increased density around the lesion is vaguely evident initially but becomes readily apparent after contrast enhancement (B). (From Zee, Segall, Miller, et al,[55] with permission from the publisher.)

mon than that in the muscles and soft tissues. It often appears as a tiny central calcification (representing the scolex) surrounded by an area of radiolucency and rimmed by calcium deposition in the overlying cyst capsule. Large cysts, which can appear as cerebral tumors, and cysts within the ventricles, which can cause hydrocephalus, are best seen on CT.[18,54–56]

## Echinococciasis

Echinococciasis is a tissue infection of humans caused by the larval stage of a small tapeworm, most commonly *Echinococcus granulosus*. Dogs, sheep, cattle, and camels are the major intermediate hosts. The ova that are ingested by humans penetrate the intestinal mucosa and enter the portal circulation. Most are filtered out by the liver or lung, but some escape into the general circulation to involve kidney, bones, brain, and other tissues. The larvae that are not destroyed develop into hydatid cysts, which have a double wall composed of thick outer membrane (exocyst) and thin inner wall of germinal cells (endocyst). The cysts may grow slowly over a period of years (mimicking a tumor), rupture into adjacent tissues, or become calcified.

Hydatid cysts are the most frequent cause of hepatic calcification in endemic areas. Patients infected with the common *Echinococcus granulosus* typically have complete oval or circular calcification at the periphery of the mother cyst. Within the mother cyst, there may be multiple daughter cysts with arclike calcifications. Hydatid

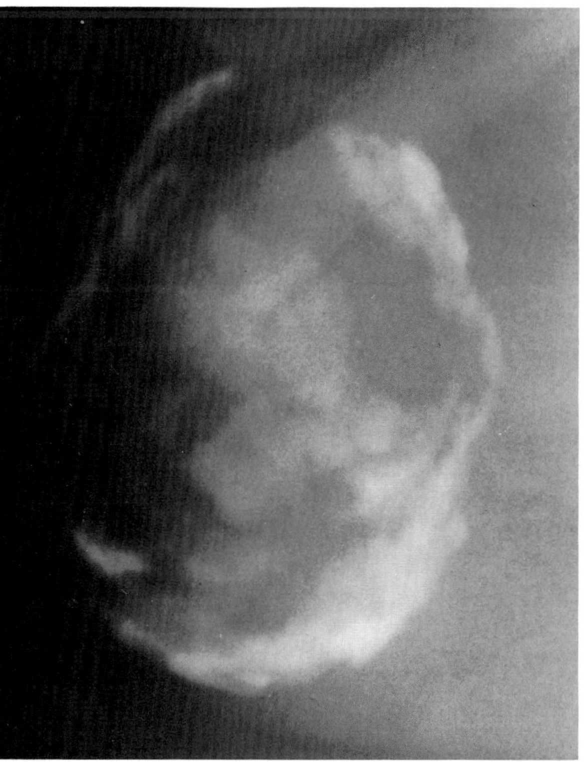

**Figure 18.53. Hydatid liver cyst (*Echinococcus granulosus*).** There is complete calcification at the periphery of the mother cyst. Within the mother cyst are several small arclike calcifications representing daughter cysts. (From Eisenberg,[69] with permission from the publisher.)

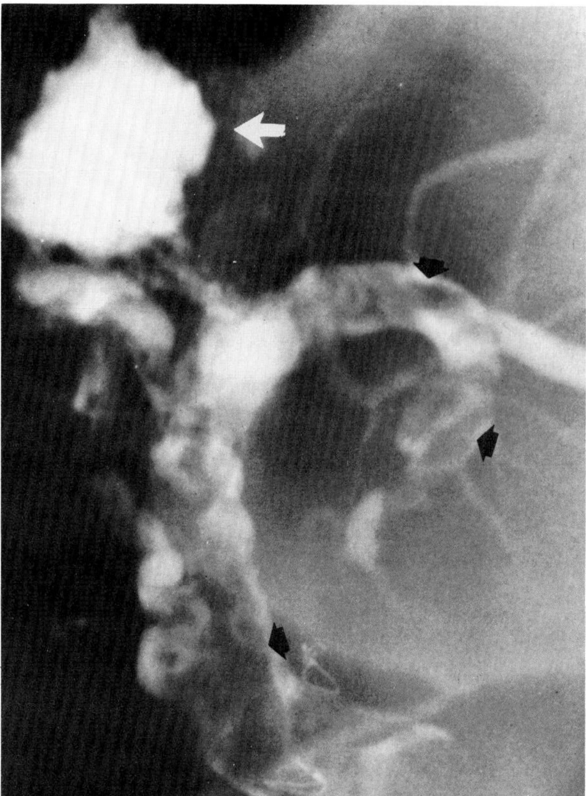

**Figure 18.54. Hydatid disease of the liver and biliary tree.** A cholangiogram shows multiple cysts presenting as filling defects (black arrows) in the bile ducts. Note the contrast material filling a large communicating cystic cavity (white arrow) in the liver parenchyma.

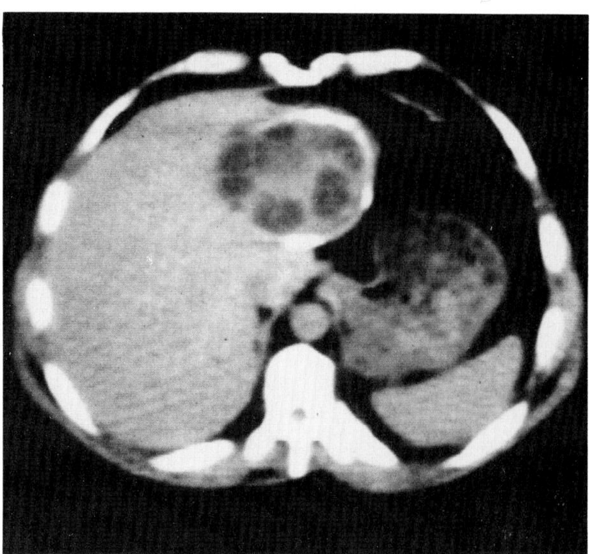

**Figure 18.55. Echinococcal cyst of the liver.** A CT scan shows a well-defined cyst in the left lobe of the liver; the cyst contains multiple small daughter cysts and is rimmed by peripheral calcification. (From Scherer, Weinzierl, Sturm, et al,[68] with permission from the publisher.)

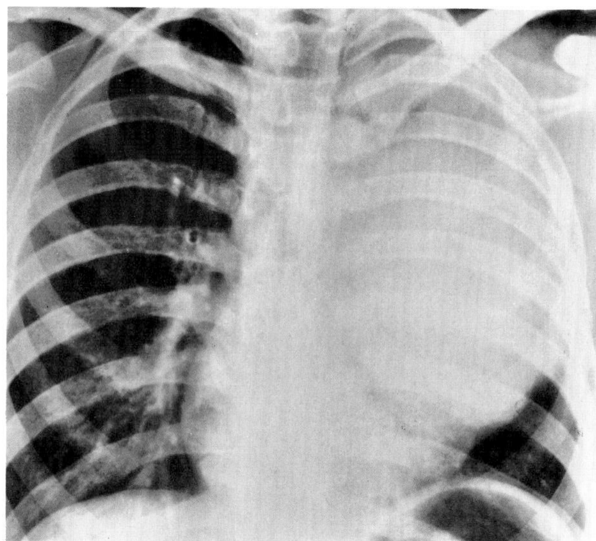

**Figure 18.56. Pulmonary echinococcal cyst.** The huge mass fills most of the left hemithorax.

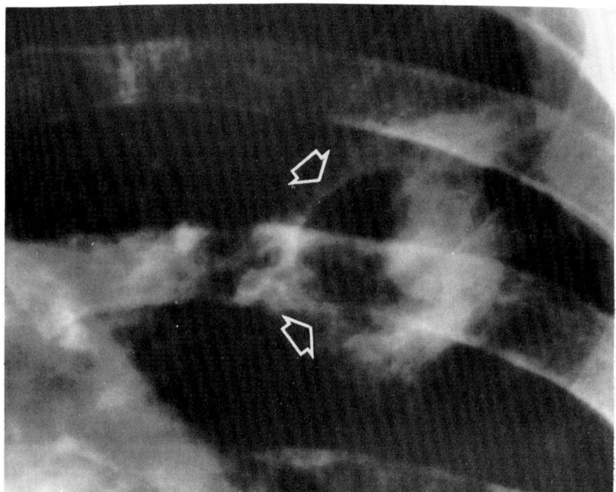

**Figure 18.57. Meniscus sign in pulmonary echinococcal cyst.** A crescent of air (arrows) is seen about the periphery of the cyst.

cyst calcification generally develops 5 to 10 years after the liver has been infected and can be present in either active or inactive cysts. Extensive dense calcifications suggest quiescence of the parasitic process; segmental calcification (nonhomogeneous, striped, trabeculated) suggests cystic activity and may be an indication for surgery.

Hydatid cysts of the spleen also often calcify. These cysts are frequently multiple and tend to have thicker and coarser rims of peripheral calcification than simple splenic cysts.

Echinococcal disease frequently causes diffuse hepatomegaly that may be associated with an enlarging liver mass, portal hypertension, splenomegaly, jaundice,

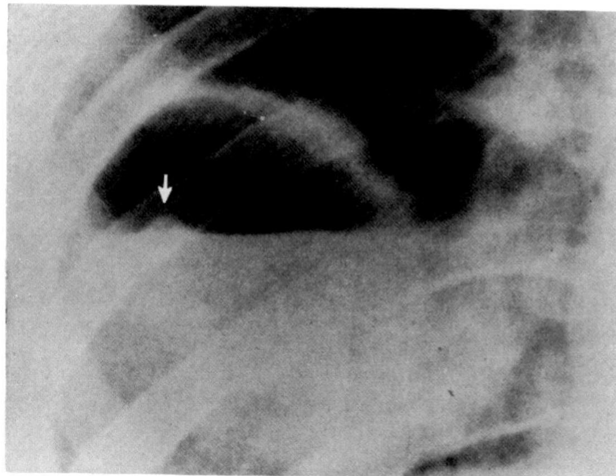

**Figure 18.58. Waterlily sign in pulmonary echinococcal cyst.** The endocyst membranes (arrow) are floating on the surface of fluid within a ruptured echinococcal cyst. (From Kegel and Fatemi,[61] with permission from the publisher.)

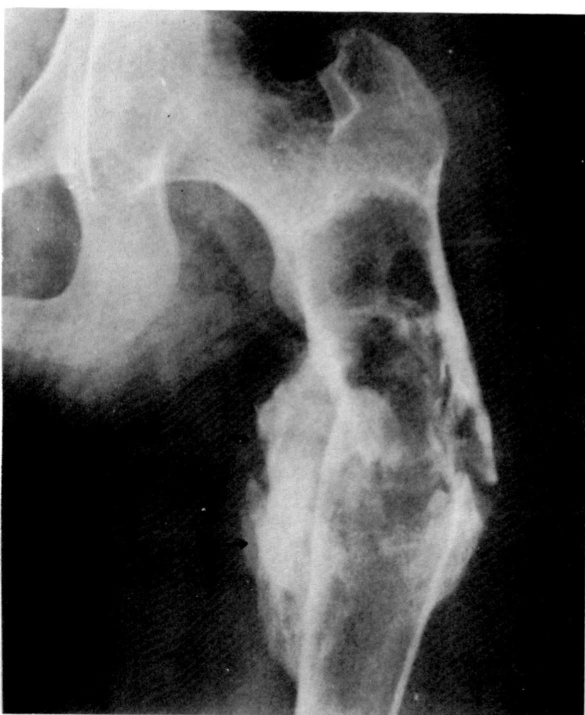

**Figure 18.59. Skeletal echinococciasis.** A destructive lesion in the proximal femoral shaft led to the development of a pathologic fracture. Extensive callus (arrows) indicates healing about the fracture site. (From Cockshott and Middlemiss,[72] with permission from the publisher.)

and ascites, producing a pattern often indistinguishable from that of carcinoma of the liver. CT and ultrasound are of value in demonstrating uncalcified or partially calcified hydatid cysts. On angiograms, vessels are stretched around the avascular cysts, and there is usually a virtually diagnostic halo of contrast material (representing the exocyst) surrounding the lesion.

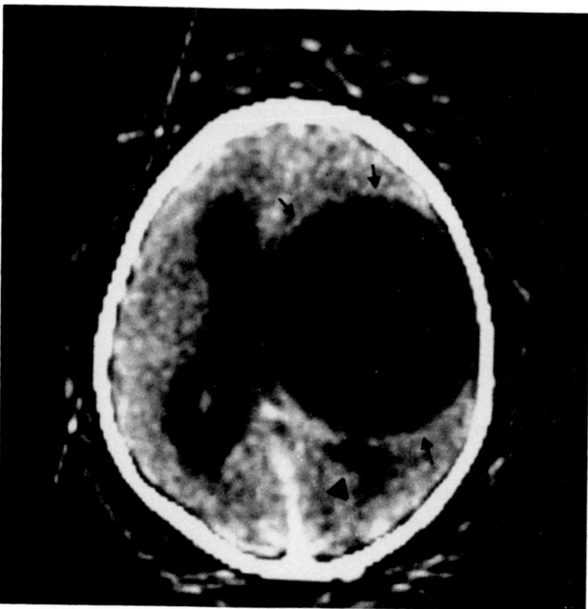

**Figure 18.60. Echinococcal cyst of the brain.** A CT scan demonstrates a huge right supratentorial echinococcal cyst (arrows). The right ventricle is partially visible posterior and medial to the cyst (arrowhead). The left ventricle is enlarged. (From Abbassioun, Rahmat, Ameli, et al,[64] with permission from the publisher.)

Large parent cysts in the liver can communicate with the biliary tree. The periodic shedding of daughter cysts into the bile duct causes recurrent episodes of biliary colic and can produce round or irregular filling defects in the bile duct or cyst cavity. Daughter cysts can be trapped in the region of the ampulla and cause common duct obstruction.

Pulmonary echinococcal cysts appear as sharply circumscribed, spherical or oval masses surrounded by normal lung tissue. The cysts are often multiple and may be up to 10 cm in diameter. Some are of bizarre, irregular shape, probably because of impingement on relatively rigid structures (e.g., bronchovascular bundles) as the cyst grows. Rupture between the exocyst and the pericyst (the external capsule contributed by the host tissue) permits the entry of air between these layers and produces a characteristic thin crescent of air around the periphery of the cyst (meniscus sign). Direct communication with a bronchus allows the cyst contents to escape and leads to the development of an air-fluid level. Residual cyst membranes floating on the fluid within the cyst produce the classic waterlily sign (sign of the camalote). Peripherally located cysts may rupture into the pleural cavity and result in a hydropneumothorax. Unlike hydatid cysts in other areas, pulmonary cysts calcify extremely rarely, presumably because they grow so rapidly and rupture so frequently that calcification has no chance to develop.

On plain abdominal radiographs, renal hydatid cysts may appear as single or multiple calcified masses. An uncalcified and intact hydatid cyst causes distortion of

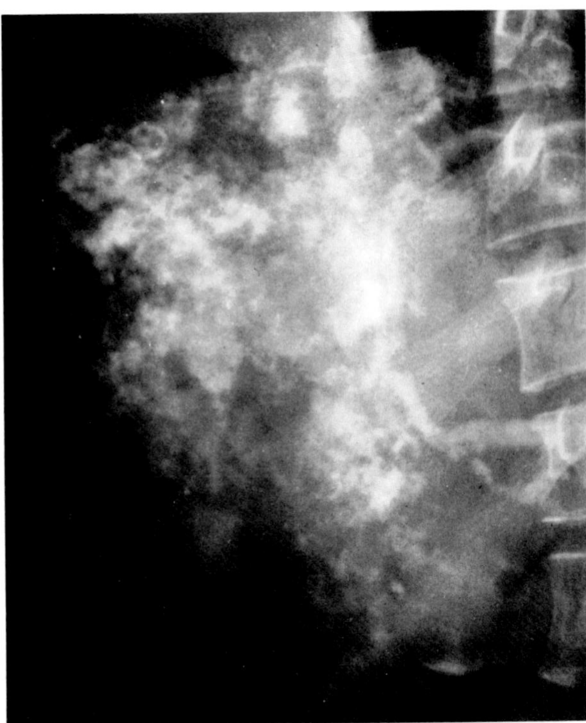

**Figure 18.61. Alveolar hydatid disease (*Echinococcus multilocularis*).** Multiple small radiolucencies are surrounded by rings of calcification, which, in turn, lie within large areas of amorphous calcification. (From Thompson, Chisholm, and Tank,[57] with permission from the publisher.)

the renal collecting system and a mass effect indistinguishable from other space-occupying lesions in the kidney. The rupture of a cyst into the renal collecting system can cause hematuria and renal colic. Urographic contrast material can diffuse among a packed mass of daughter cysts and produce a typical mottled appearance, or can completely fill a renal cavity and simulate a renal abscess, tuberculosis, or papillary necrosis.

Bone involvement occurs in about 1 percent of the patients with hydatid disease and most commonly affects the vertebral bodies, the pelvis, and the sacrum. The infiltration of daughter cysts into the bone causes round or irregular areas of lytic destruction that are often multiloculated and resemble a bunch of grapes. Periosteal new bone formation and reactive sclerosis are infrequent, except in response to secondary infection. Ill-defined multiple loculated areas of destruction less frequently develop at the distal ends of long tubular bones.

Intracerebral hydatid cysts cause increased intracranial pressure and simulate other avascular space-occupying lesions. Plain skull radiographs occasionally demonstrate calcification of the cysts.

A rarer and more malignant form of hydatid disease is the alveolar type, due to *Echinococcus multilocularis*. The natural intermediate hosts of this organism are small rodents, and, unlike the generally indolent *Echinococcus granulosus* disease, alveolar hydatid disease can be a fulminant, even fatal, condition. The striking radiographic feature of alveolar hydatid disease is liver calcification, which typically appears as multiple small radiolucencies, measuring 2 to 4 mm in diameter, surrounded by rings of calcification, which, in turn, lie within larger areas of amorphous calcifications up to 10 to 12 cm in diameter.

The combination of hepatomegaly along with portal hypertension, splenomegaly, jaundice, and ascites may produce a pattern indistinguishable from that of carcinoma of the liver.[57–68,72]

## REFERENCES


1. Wilson ES: Pleuropulmonary amebiasis. *AJR* 111:518–524, 1971.
2. Cardoso JM, Kimura K, Stoopen M, et al: Radiology of invasive amebiasis of the colon. *AJR* 128:935–941, 1977.
3. Rogers LF, Ralls PW, Boswell WD, et al: Amebiasis: Unusual radiographic manifestations. *AJR* 135:1253–1257, 1980.
4. Kolawole PM, Lewis EA: Radiologic observations on intestinal amebiasis. *AJR* 122:257–265, 1974.
5. Godard JE, Hansen RA: Interstitial pulmonary edema in acute malaria. *Radiology* 101:523–524, 1971.
6. Heineman HS; The clinical syndrome of malaria in the United States. *Arch Intern Med* 129:607–616, 1972.
7. Reeder MM, Simao C: Chagas' myocardiopathy. *Semin Roentgenol* 4:374–382, 1969.
8. Reeder MM, Hamilton LC: Radiologic diagnosis of tropical diseases of the gastrointestinal tract. *Radiol Clin North Am* 7:57–81, 1969.
9. Mussbichler H: Radiologic study of intracranial calcifications in congenital toxoplasmosis. *Acta Radiol [Diagn]* 7:369–379, 1968.
10. Tucker AS: Intracranial calcification in infants (toxoplasmosis and cytomegalic inclusion disease). *AJR* 86:458–461, 1961.
11. Theologides A, Kennedy BJ: Clinical manifestations of toxoplasmosis in the adult. *Arch Intern Med* 117:536–540, 1966.
12. Enzmann BR, Brant-Zawadzki M, Britt RH: Computed tomography of central nervous system infections in immunosuppressed patients. *Am J Neuroradiol* 1:239–243, 1980.
13. Feinberg SB, Lester RG, Burke B: The roentgen findings in *Pneumocystis carinii* pneumonia. *Radiology* 76:594–599, 1961.
14. Forrest JV: Radiographic findings in *Pneumocystis carinii* pneumonia. *Radiology* 103:539–544, 1972.
15. Burke BA, Good RA: *Pneumocystis carinii* infection. *Medicine* 52:23–51, 1973.
16. Fisher CH, Oh KS, Bayless TM, et al: Current perspectives on giardiasis. *AJR* 125:207–217, 1975.
17. Marshak RH, Ruoff M, Lindner AE: Roentgen manifestations of giardiasis. *AJR* 104:557–560, 1968.
18. Reeder MM: Tropical diseases of the soft tissues. *Semin Roentgenol* 8:47–71, 1973.
19. Akisada M, Tani S: Filarial chyluria in Japan. *Radiology* 90:311–317, 1968.
20. Herlinger H: Pulmonary changes in tropical eosinophilia. *Br J Radiol* 36:889–901, 1963.
21. Khoo FY, Danaraj TJ: The roentgenographic appearance of eosinophilic lung (tropical eosinophilia). *AJR* 83:251–259, 1960.
22. Williams I: Calcification in loiasis. *J Fac Radiol* 6:142–144, 1955.
23. Beskin CA, Colvin SH, Beaver PC: Pulmonary dirofilariasis. *JAMA* 198:655–656, 1966.
24. Nevel OT, Masry NA, Castell DO, et al: Schistosomal disease of the colon: A reversible form of polyposis. *Gastroenterology* 67:939–943, 1974.

25. Phillips JS, Cockrill H, Jorge E, et al: Radiographic evaluation of patients with schistosomiasis. *Radiology* 114:31–37, 1975.
26. James WB: Urological manifestations in *Schistosoma hematobium* infestations. *Br J Radiol* 36:40–45, 1963.
27. Al-Ghorab MM: Radiological manifestation of genitourinary bilharziasis. *Clin Radiol* 19:100–111, 1968.
28. Farid Z: Chronic pulmonary schistosomiasis. *Am Rev Tuberc* 79:119–123, 1959.
29. El-Badawi AA: Bilharzial polyps of the urinary bladder. *Br J Urol* 38:24–35, 1966.
30. Darlak JJ, Moskowitz M, Kattan KR: Calcifications in the liver. *Radiol Clin North Am* 18:209–219, 1980.
31. Reeder MM, Hamilton LC: Tropical diseases of the colon. *Semin Roentgenol* 3:62–80, 1968.
32. Fisher RM, Cremin BJ: Rectal bleeding due to *Trichuris trichiuria*. *Br J Radiol* 43:214–215, 1970.
33. Phills JA, Harrold AJ, Whiteman GV, et al: Pulmonary infiltrates, asthma and eosinophilia due to *Ascaris* infestation in man. *N Engl J Med* 286:965–970, 1972.
34. Weissberg DL, Berk RN: Ascariasis of the gastrointestinal tract. *Gastrointest Radiol. 3:415–418, 1978.*
35. Ellman BA, Wynne JM, Freeman A: Intestinal ascariasis: New plain-film features. *AJR* 135:37–42, 1980.
36. Okumura M, Nakashima Y, Curti P, et al: Acute intestinal obstruction by ascariasis: Analysis of 455 cases. *Rev Inst Med Trop Sao Paulo* 16:292–300, 1974.
37. Kuipers FC: Eosinophilic phlegmonous inflammation of the alimentary canal caused by a parasite from the herring. *Pathol Microbiol* 27:925–932, 1964.
38. Richman RH, Lewicki AN: Right ileal-colitis secondary to anisakiasis. *AJR* 119:329–331, 1973.
39. Chuttani HK, Puri SK, Misra RC: Small intestine in hookworm disease. *Gastroenterology* 53:381–388, 1967.
40. Gilles HM, Watson-Williams EJ, Ball PA: Hookworm infection and anaemia. *Q J Med* 33:1–24, 1964.
41. Sheehy TW, Meroney WH, Cox RS, et al: Hookworm disease and malabsorption. *Gastroenterology* 42:148–156, 1962.
42. Kalman EH: Creeping eruption with transient pulmonary infiltration. *Radiology* 62:222–226, 1954.
43. Drasin GF, Moss JP, Cheng SH: *Strongyloides sterocoralis* colitis: Findings in four cases. *Radiology* 126:619–621, 1978.
44. Berkman YM, Rabinowitz J: Gastrointestinal manifestations of strongyloidiasis. *AJR* 115:306–311, 1972.
45. Purtilo DT, Meyers WM, Connor DH: Fatal strongyloidiasis in immunosuppressed patients. *Am J Med* 56:488–493, 1974.
46. Kim SK, Walker AE: Cerebral paragonimiasis. *Acta Psychiatr Neurol Scand* 36:153–157, 1961.
47. Oh SJ: Roentgen findings in cerebral paragonimiasis. *Radiology* 90:292–299, 1968.
48. Suwanik R, Harinsuta C: Pulmonary paragonimiasis. *AJR* 81:236–244, 1959.
49. Ogakwu M, Nwokolo C: Radiological findings in pulmonary paragonimiasis as seen in Nigeria. *Br J Radiol* 46:699–705, 1973.
50. Ameres JP, Levine MP, DeBasi HP: Acalculous clonorchiasis obstructing the common bile duct. *Am Surg* 42:170–172, 1976.
51. Belgraier AH: Common bile duct obstruction due to *Fasciola hepatica*. *NY State J Med* 76:936–937, 1976.
52. Monroe LS, Norton RA: Roentgenographic signs produced by *Taenia saginata*. *Am J Dig Dis* 7:519–522, 1962.
53. Hamilton JB: *Taenia saginata*. *Radiology* 47:64–65, 1946.
54. Carbajal JR, Palacios E, Azar-Kiah B, et al: Radiology of cysticercosis of the central nervous system including computed tomography. *Radiology* 125:127–131, 1977.
55. Zee CS, Segall HD, Miller C, et al: Unusual neuroradiological features of intracranial cysticercosis. *Radiology* 137:397–407, 1980.
56. Keats TE: Cysticercosis: Roentgen manifestations. *Mo Med* 58:457–459, 1961.
57. Thompson WM, Chisholm DP, Tank R: Plain-film roentgenographic findings in alveolar hydatid cyst: *Echinococcus multilocularis*. *AJR* 116:345–348, 1972.
58. Harris JD: Rupture of hydatid cysts of the liver into the biliary tracts. *Br J Surg* 52:210–214, 1965.
59. Balikian JB, Mudarris FF: Hydatid disease of the lungs. *AJR* 122:692–707, 1975.
60. McPhail JL, Arora TS: Intrathoracic hydatid disease. *Dis Chest* 52:772–781, 1967.
61. Kegel RFC, Fatemi A: The ruptured pulmonary hydatid cyst. *Radiology* 76:60–64, 1961.
62. Shawket TN, Al-Waidh M: Hydatid cysts of the kidney. *Br J Urol* 46:371–376, 1974.
63. Kirkland K: Urological aspects of hydatid disease. *Br J Urol* 38:241–254, 1966.
64. Abbassioun K, Rahmat H, Ameli NO, et al: CT in hydatid cyst of the brain. *J Neurosurg* 49:408–411, 1978.
65. Duran H, Fernandez L, Castresana F, et al: Osseous hydatidosis. *J Bone Joint Surg* 60A:685–690, 1978.
66. Gharbi HA, Hassine W, Brauner MW, et al: Ultrasound examination of the hydatid liver. *Radiology* 139:459–463, 1981.
67. Gonzalez LR, Marcos J, Illanas M, et al: Radiologic aspects of hepatic echinococcus. *Radiology* 130:21–27, 1979.
68. Scherer U, Weinzierl M, Sturm R, et al: Computed tomography in hydatid disease of the liver. *J Comput Assist Tomogr* 2:612–617, 1978.
69. Eisenberg RL: *Gastrointestinal Radiology: A Pattern Approach*. Philadelphia, JB Lippincott, 1983.
70. Ralls PW, Meyers HI, Lapin SA, et al: Gray-scale ultrasonography of hepatic amoebic abscesses. *Radiology* 132:125–132, 1979.
71. Kelly WM, Brant-Zawadzki M: Acquired immunodeficiency syndrome: Neuroradiologic findings. *Radiology* 149:485–492, 1983.
72. Cockshott WP, Middlemiss H: *Clinical Radiology in the Tropics*. Edinburgh, Churchill Livingstone, 1979.
73. Nakata H, Takeda K, Nakayama T: Radiological diagnosis of acute gastric anisakiasis. *Radiology* 135:49–53, 1980.
74. Manzano C, Thomas MA, Valenzuela C: Trichuriasis. *Pediatr Radiol* 8:76–78, 1979.

# PART THREE
# DISORDERS OF THE HEART

## Chapter Nineteen
# Diagnostic Modalities

## CHEST RADIOGRAPHY

Plain chest radiography is a useful, relatively inexpensive, noninvasive screening procedure for patients with suspected cardiac disease. Chest radiographs are of value in detecting enlargement of specific chambers of the heart, assessing the pulmonary vascular pattern, and determining the size and configuration of the great vessels. In addition to routine frontal and lateral projections, oblique views or four views of the chest with the esophagus filled with barium are usually obtained to visualize specific portions of the heart. Fluoroscopy is also often of value in detecting calcification within the cardiac valves, coronary arteries, pericardium, or myocardium, as well as in analyzing the size and pulsations of the cardiac chambers and great vessels.

### Cardiac Silhouette

The enlargement of a cardiac chamber may reflect either dilatation or muscular hypertrophy. Dilatation of a cardiac chamber is the result of an increased volume of blood within it (left-to-right shunt, valvular insufficiency) and is usually readily detectable on chest radiographs by a change in cardiac size and contour. Muscular hypertrophy, which develops when blood flow from a cardiac chamber is blocked (atresia, valvular or postvalvular stenosis), often results in thickening of the ventricular wall at the expense of the chamber cavity and usually produces only slight cardiac enlargement or alteration of the cardiac silhouette.

The right atrium is the most difficult chamber to assess. On posteroanterior (PA) and left anterior oblique (LAO) views, right atrial enlargement may cause bulging or an increased prominence in the curvature of the right cardiac border.

Enlargement of the right ventricle is best seen on lateral views, on which the ventricle fills in the lower retrosternal space. The enlarged ventricle also causes lateral and upward displacement of the radiographic cardiac apex on PA views and an anterior convexity just below the origin of the pulmonary artery on right anterior oblique (RAO) views.

Because the esophagus is in direct contact with the left atrium, any enlargement of this chamber produces a discrete posterior indentation on the barium-filled esophagus that is best seen on lateral or RAO projections. On lateral views, posterior displacement of the left main bronchus is another manifestation of left atrial enlargement. On PA projections, enlargement of the left atrium can cause a bulge in the region of the left atrial appendage, a widening of the carina, and an elevation of the left main bronchus. More pronounced left atrial enlargement may produce the characteristic "double-contour" configuration because of the projection of the enlarged left atrium through the normal right atrial silhouette.

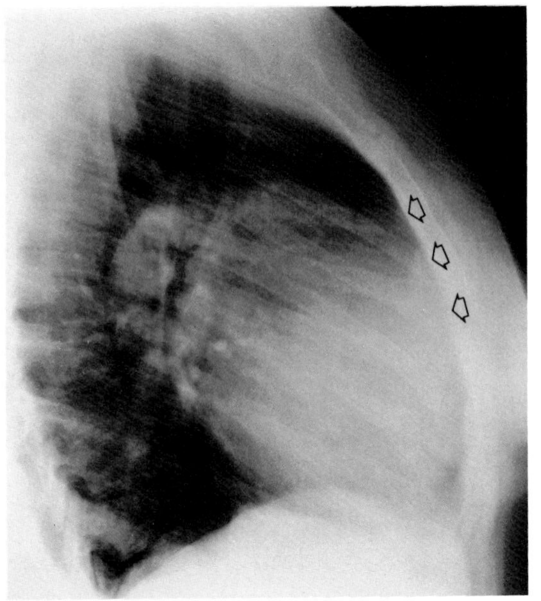

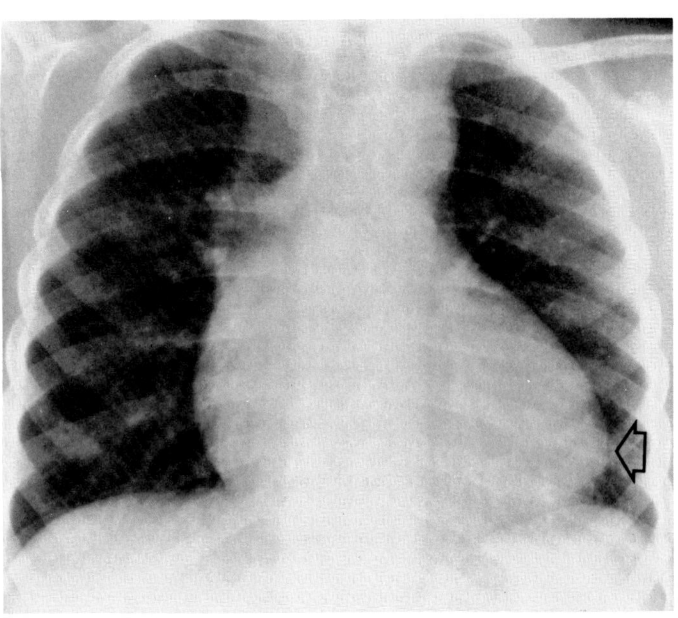

A B

**Figure 19.1. Right ventricular enlargement.** (A) A lateral view in a patient with primary pulmonary hypertension shows the enlarged right ventricle filling in most of the retrosternal space (arrows). (B) A frontal view of another patient with tetralogy of Fallot shows right ventricular enlargement as a lateral and upward displacement of the radiographic cardiac apex (arrow).

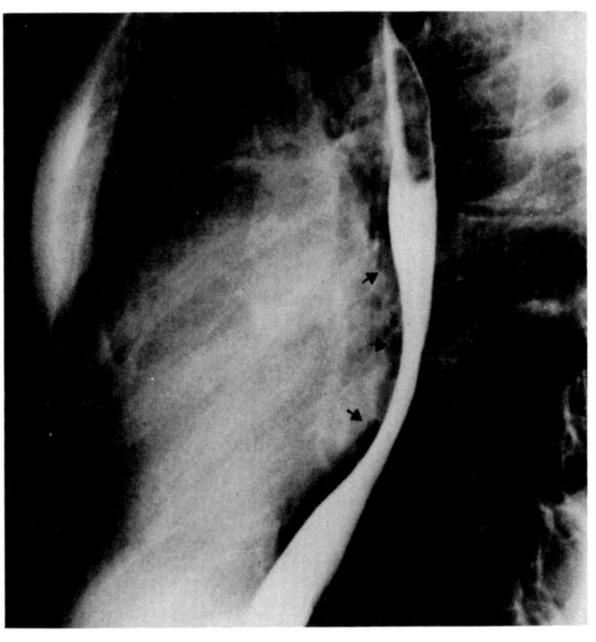

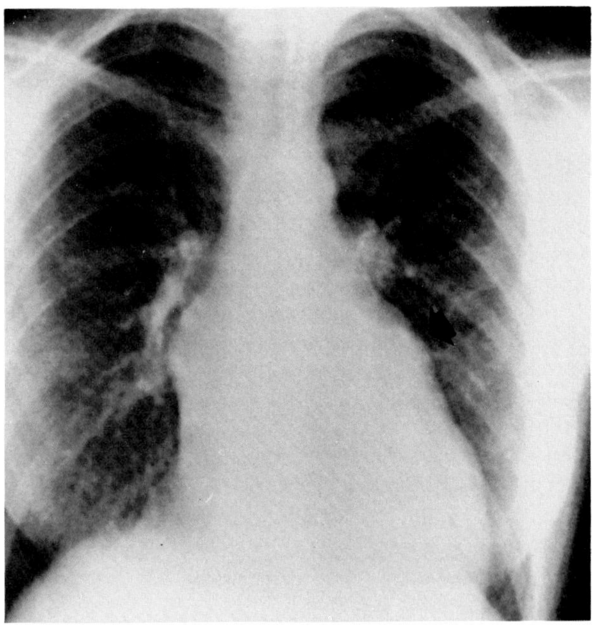

A B

**Figure 19.2. Left atrial enlargement.** (A) On a lateral view, the enlargement of the chamber produces a discrete posterior indentation on the barium-filled esophagus (arrows). (B) On a frontal view, the enlarged left atrium produces a bulge (arrow) in the region of the left atrial appendage. In this patient, underlying mitral stenosis has led to an increase in pulmonary venous pressure and interstitial edema, which produce fuzziness of lower lobe vessels and a redistribution of blood flow to distended upper lobe veins.

Left ventricular enlargement produces downward displacement of the cardiac apex on PA views and a rounded bulge of the lower half of the posterior cardiac silhouette on LAO and lateral views.

Although many measurements have been suggested for the precise evaluation of cardiac size, the only widely used one is the cardiothoracic ratio. In normal persons, the maximal diameter of the cardiac silhouette divided by the maximal internal thoracic diameter is less than 0.50. However, the cardiothoracic ratio should be used only as a guideline; precise evaluation of the size of individual cardiac chambers requires echocardiography.

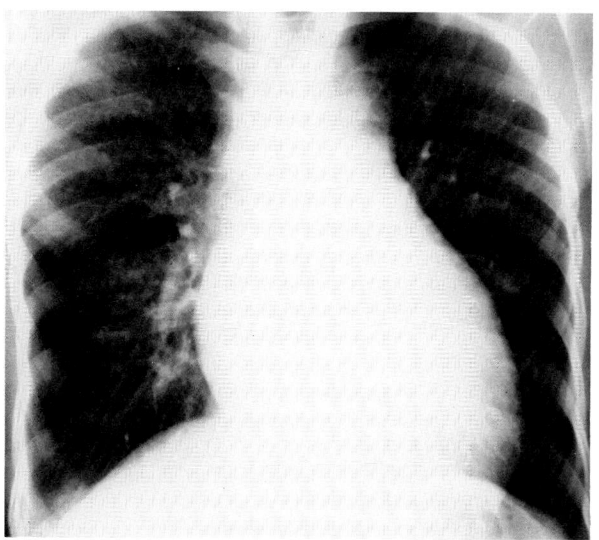
A

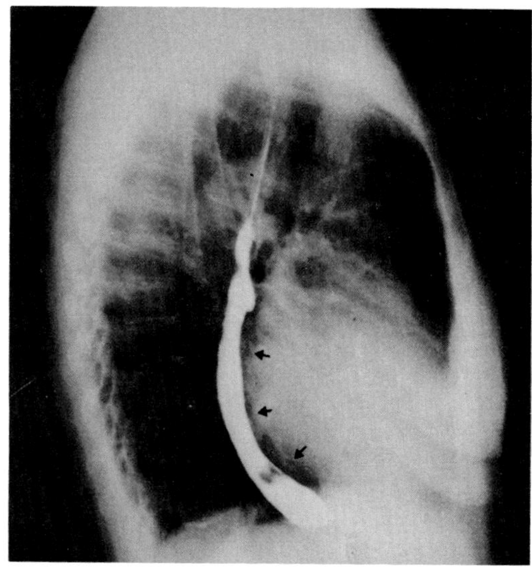
B

**Figure 19.3. Left ventricular enlargement.** (A) A frontal view shows downward displacement of the cardiac apex. (B) On a lateral view, the bulging of the lower half of the posterior cardiac silhouette causes a broad indentation on the barium-filled esophagus (arrows).

## Pulmonary Vasculature

The size of the pulmonary arteries and veins is proportional to the flow within them. Increased pulmonary flow, as in left-to-right shunts or high-output states, causes diffuse dilatation and tortuosity of the pulmonary arteries and veins throughout the lungs. In contrast, pulmonary arterial hypertension causes dilatation of the main pulmonary artery and its central branches but rapid tapering or "pruning" of peripheral arteries. An increase in pulmonary venous pressure causes interstitial edema and fuzziness of lower lobe vessels and a redistribution of blood flow to distended upper lobe veins. In severe cases, alveolar pulmonary edema and pleural effusions may develop. Decreased pulmonary vascularity reflects diminished pulmonary blood flow, which may be seen with right outflow tract obstruction or pulmonary thromboembolic disease. Diminished pulmonary blood flow causes focal or generalized hyperlucency of the lungs along with thin, stringy blood vessels.

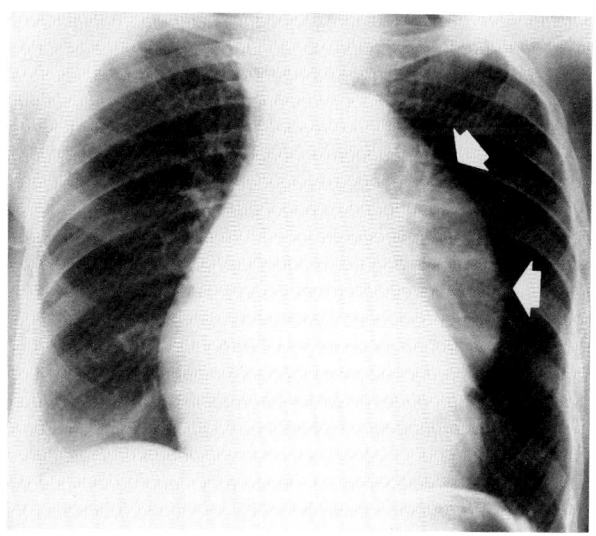

**Figure 19.4. Enlargement of the descending aorta.** The marked elongation and tortuosity of the descending aorta (arrows) is due to severe atherosclerotic disease.

## Great Vessels

Enlargement of the ascending aorta suggests poststenotic dilatation due to valvular aortic stenosis, increased aortic blood flow due to aortic valve insufficiency or great vessel left-to-right shunts (patent ductus arteriosus, persistent truncus arteriosus), or increased pressure within the aorta (hypertension). Prominence of the descending aorta may reflect the elongation and tortuosity of atherosclerotic disease. Enlargement of the aortic knob may be caused by prestenotic dilatation in a patient with coarctation of the aorta. A small aorta, best detected in the region of the aortic knob, indicates decreased aortic blood flow (certain left-to-right shunts) or aortic hypoplasia (hypoplastic left heart syndrome, supravalvular aortic stenosis).

The main pulmonary artery is dilated (convex) in patients with left-to-right shunts or intracardiac mixing lesions (increased pulmonary blood flow), pulmonary valvular stenosis (poststenotic dilatation), and pulmonary valve insufficiency. A small (concave) pulmonary artery is seen in patients with right outflow tract obstruction (tetralogy of Fallot, hypoplastic right heart syndrome, Ebstein's anomaly) and in patients with conditions in which the position of the artery is abnormal (transposition of the great vessels, persistent truncus arteriosus).[1]

## ECHOCARDIOGRAPHY

Echocardiography utilizes short pulses of ultrasound at a frequency of about 2 to 5 mHz to image the heart and great vessels. In M-mode or time-motion echocardiography, ultrasound waves are transmitted and received along a single line to produce an image of high temporal resolution. Newer echocardiographic techniques provide a cross-sectional or two-dimensional view that results in a tomographic image of high spatial resolution. When multiple images are obtained each second, the motion of the heart can be viewed in real time.

Echocardiography is a major noninvasive modality for demonstrating stenotic and regurgitant lesions in valvular heart disease; valvular vegetations in infective endocarditis; the size, wall thickness, and function of the left ventricle; and pericardial effusion.[2,3]

## RADIONUCLIDE IMAGING OF THE HEART

Radionuclide imaging of the heart can be used to assess ventricular function (radionuclide ventriculography), to identify and quantify intracardiac shunts, to study acute myocardial infarction, and to assess myocardial perfusion.

Radionuclide ventriculography can delineate the chambers of the heart and the great vessels. In the first-pass method, a computer records the initial transit of an intravenously injected bolus of isotope through the right heart chambers, lungs, and left heart chambers. In the equilibrium, or gated, technique, changes in intraventrical isotope activity that occur during many cardiac cycles are measured, averaged, and related to the timing reference of the QRS complex of the electrocardiogram. These techniques permit the calculation of right and left ventricular ejection fractions, ventricular volumes, and ejection and filling rates, and provide measurements that correlate closely with those obtained with more invasive catheterization methods.

In patients with left-to-right shunts, a modification of first-pass radionuclide ventriculography permits the determination of the ratio of pulmonary to systemic flow and thus the size of the shunt. The results closely correlate with those of cardiac catheterization. The techniques of myocardial perfusion imaging and acute infarct scintigraphy are discussed in Chapter 23 in the sections "Ischemic Heart Disease" and "Acute Myocardial Infarction," respectively.[4,5]

## CARDIAC CATHETERIZATION AND CORONARY ANGIOGRAPHY

The major invasive imaging modalities for diseases of the heart are cardiac catheterization and coronary angiography. In these procedures, high-speed x-ray motion pictures (cineangiography) are obtained following the selective injection of contrast material into the coronary arteries or cardiac chambers.

In many cardiac disorders, cardiac catheterization and coronary angiography are required to provide precise information about the anatomy and dynamic physiology of the heart. In patients with acquired valvular heart disease, cardiac catheterization can demonstrate the site and degree of stenotic and regurgitant lesions. Angiography is usually necessary to characterize the primary

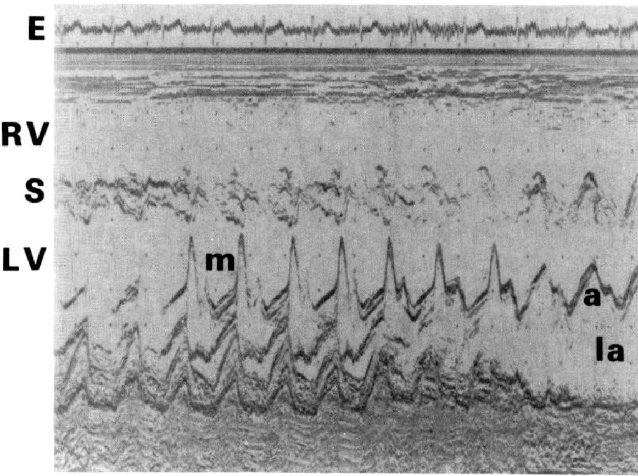


**Figure 19.5. Normal echocardiogram.** E = electrocardiogram; RV = right ventricle; S = interventricular septum; LV = left ventricle; m = mitral valve; a = aortic valve; la = left atrium.

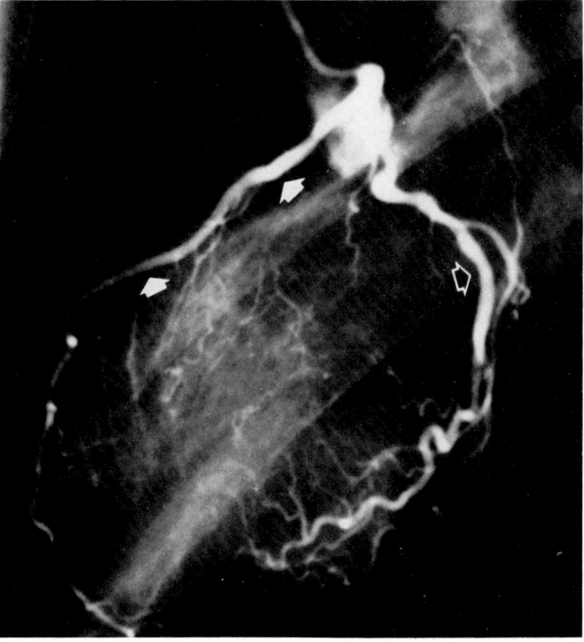

**Figure 19.6. Normal left coronary arteriogram in the lateral projection.** The closed arrows point to the left anterior descending artery; the open arrow points to the circumflex branch. (Courtesy of Martin J. Lipton, M.D.)

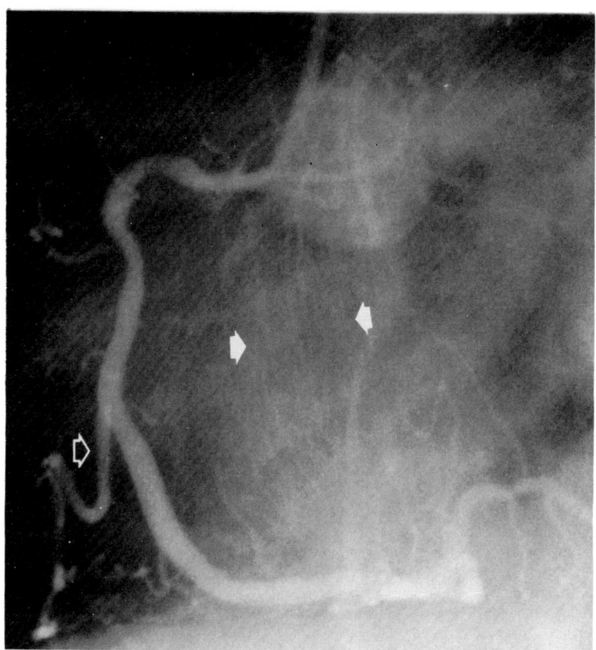

**Figure 19.7. Normal selective right coronary arteriogram.** The open arrow points to the acute marginal branch; the closed arrows point to septal branches serving as collateral vessels to fill the left anterior descending artery, which was totally occluded in this patient. (Courtesy of Martin J. Lipton, M.D.)

defect in patients with congenital heart disease, as well as to determine whether associated lesions are present. Patients with chest pain of undetermined cause may require angiographic visualization of the coronary arteries to determine the presence and extent of atherosclerotic coronary disease or coronary artery spasm. Cardiac catheterization studies may be of value in assessing the postoperative condition following such procedures as the insertion of prosthetic valves, coronary artery bypass grafting, and the correction of congenital defects. In patients with evidence of pulmonary hypertension or suspected myocardial or pericardial disease, cardiac catheterization may be required to exclude a lesion that could be treated surgically.

The techniques and interpretation of cardiac catheterization and coronary angiography are beyond the scope of this text.[6,7]

## REFERENCES


1. Swischuk LE: *Plain Film Interpretation in Congenital Heart Disease*. Baltimore, Williams & Wilkins, 1979.
2. Feigenbaum H: *Echocardiography*. Philadelphia, Lea & Febiger, 1981.
3. Popp RL, Rubenson DS, Tucker CR, et al: Echocardiography: M-mode and two-dimensional methods. *Ann Intern Med* 93:844–856, 1980.
4. Berger HJ, Zaret BL: Nuclear cardiology. *N Engl J Med* 305:799–807, 855–865, 1981.
5. Mason DT: *Principles of Noninvasive Cardiac Imaging: Echocardiography and Nuclear Cardiology*. New York, Le Jacq, 1980.
6. Ross J, Peterson KL: Cardiac catheterization and angiography, in Petersdorf RG, Adams RD, Braunwald E, et al (eds): *Harrison's Principles of Internal Medicine*. New York, McGraw-Hill, 1983, 1335–1343.
7. Wynne J, O'Rourke RA, Braunwald E: Noninvasive methods of cardiac examination. Roentgenography, phonocardiography, echocardiography, and radionuclide techniques, in Petersdorf RG, Adams RD, Braunwald E, et al (eds): *Harrison's Principles of Internal Medicine*. New York, McGraw-Hill, 1983, pp 1330–1335.

**Chapter Twenty**

# Functional Abnormalities of the Heart

## CONGESTIVE HEART FAILURE

Left-sided heart failure produces a classic radiographic appearance of cardiac enlargement, redistribution of pulmonary venous blood flow (enlarged superior pulmonary veins and decreased caliber of the veins draining the lower lung), interstitial and alveolar edema, and pleural effusions. (See the sections "Pulmonary Edema" in this chapter and "Pleural Effusion" in Chapter 35.) In acute left ventricular failure secondary to coronary thrombosis, however, there may be severe pulmonary congestion and edema with very little cardiac enlargement.

In right-sided heart failure, there is dilatation of the right ventricle and right atrium. The transmission of increased pressure may cause dilatation of the superior vena cava and widening of the right superior mediastinum. The enlargement of a congested liver may elevate the right hemidiaphragm.

Because right ventricular failure usually is secondary to failure of the left ventricle, the overall radiographic pattern tends to reflect evidence of both disorders.[1]

## PULMONARY EDEMA

Pulmonary edema refers to an abnormal accumulation of fluid in the extravascular pulmonary tissues. The most common cause of pulmonary edema is an elevation of the pulmonary venous pressure. This is most often due to left-sided heart failure, but may also be caused by pulmonary venous obstruction (mitral valve disease, left atrial tumor) or lymphatic blockade (fibrotic, inflammatory, or metastatic disease involving the mediastinal lymph nodes). Other causes of pulmonary edema include uremia, narcotic overdose, exposure to noxious fumes, excessive oxygen, high altitudes, fat embolism, adult respiratory distress syndrome, and various neurologic abnormalities.

Transudation of fluid into the interstitial spaces of the lungs is the earliest stage of pulmonary edema. However, in patients with congestive heart failure or pulmonary venous hypertension, increased pulmonary venous pressure first appears as a redistribution of blood flow from the lower to the upper lung zones. This phenomenon, probably due to reflex venous spasm, causes prominent enlargement of the superior pulmonary veins and decreased caliber of the veins draining the inferior portions of the lung. Edema fluid in the interstitial space causes a loss of the normal sharp definition of pulmonary vascular markings. Accentuation of the perihilar vascular markings produces a perihilar haze. Fluid in the interlobular septa produces characteristic thin horizontal lines of increased density at the axillary margins of the lung margin inferiorly (Kerley B lines). The accumulation of fluid in the subpleural space causes thickening of the interlobar fissures. This is best noted adjacent to the minor fissure on the right but may also be observed along the major fissures on lateral projections.

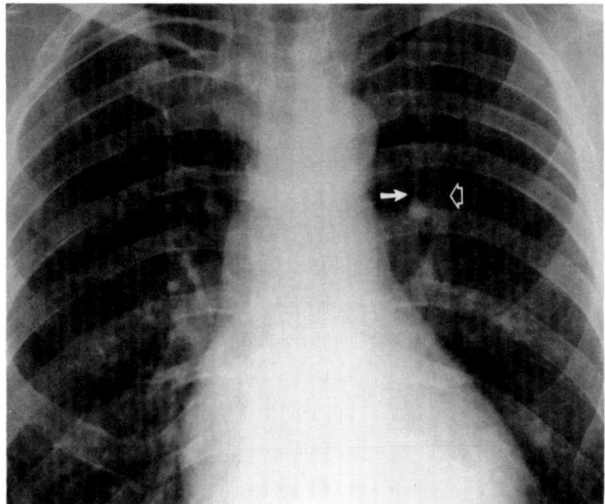

**Figure 20.1. Redistribution of blood flow in pulmonary edema.** Striking dilatation of the pulmonary veins draining the upper lobes. On the left, the pulmonary vein (open arrow) is larger than the more medial, corresponding artery (closed arrow), the reverse of the normal situation.

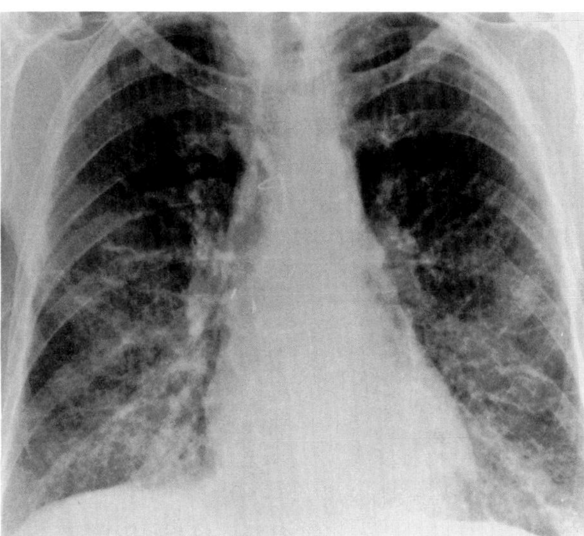

**Figure 20.2. Pulmonary edema.** Edema fluid in the interstitial space causes a loss of the normal sharp definition of pulmonary vascular markings and a perihilar haze. At the bases, note the thin horizontal lines of increased density (Kerley B lines) that represent fluid in the interlobular septa.

A further increase in pulmonary venous pressure leads to the development of alveolar and/or pleural transudates. Alveolar edema appears as irregular, poorly defined patchy densities scattered throughout the lungs. The classic radiographic finding of alveolar pulmonary edema is the butterfly (or bat's wing) pattern, a diffuse, bilaterally symmetric, fan-shaped infiltration that is most prominent in the central portion of the lungs and fades toward the periphery. The apex, base, and lateral portions of each lung are relatively spared. Occasionally, alveolar pulmonary edema may be unilateral.

Pleural effusion associated with pulmonary edema usually occurs on the right side. If bilateral, the effusion tends to be more marked on the right. There is often an associated thickening of the interlobar fissures.

Following the adequate treatment of pulmonary edema, the interstitial, alveolar, and pleural abnormalities may disappear within several hours. Loculated pleural fluid within a fissure (especially the minor fissure) may resorb more slowly and appear as a sharply defined, elliptical or circular density that simulates a solid parenchymal mass (phantom tumor).

Most patients with pulmonary edema due to congestive failure or other heart disease have evidence of cardiomegaly. One major exception is the patient with left ventricular failure due to coronary thrombosis, who may have severe pulmonary congestion with little cardiac enlargement. In noncardiogenic causes of pulmonary edema, the heart often remains normal in size.[1–6,12]

## CARDIAC ARRHYTHMIAS

The major role of chest radiography in the patient with an arrhythmia is to aid in detecting the underlying cardiac anomaly. Left atrial enlargement or characteristic mitral valve calcification suggests mitral valve disease. A generalized increase in the pulmonary vascularity may reflect a high-output state, as in patients with hyperthyroidism. Calcification of a coronary artery indicates atherosclerotic change; in a patient with heart block, cal-

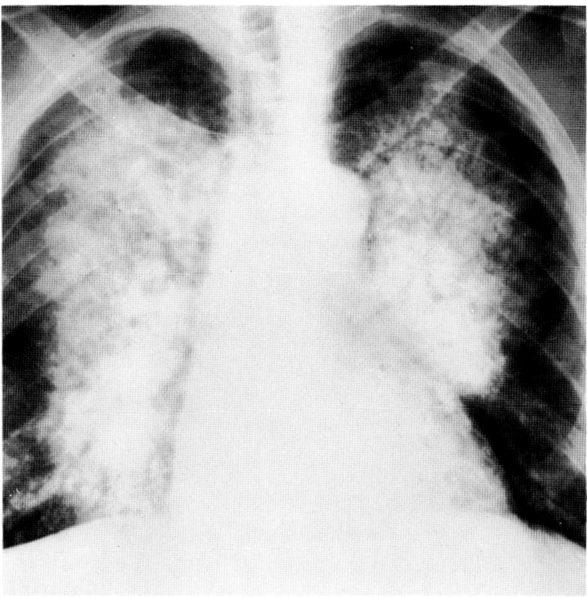

**Figure 20.3. Pulmonary edema.** Diffuse, bilaterally symmetric infiltration of the central portion of the lungs along with relative sparing of the periphery produces the butterfly, or bat's wing, pattern. The margins of the edematous lung are sharply defined. The consolidation is fairly homogeneous and is associated with a well-defined air bronchogram on both sides. (From Fraser and Pare,[2] with permission from the publisher.)

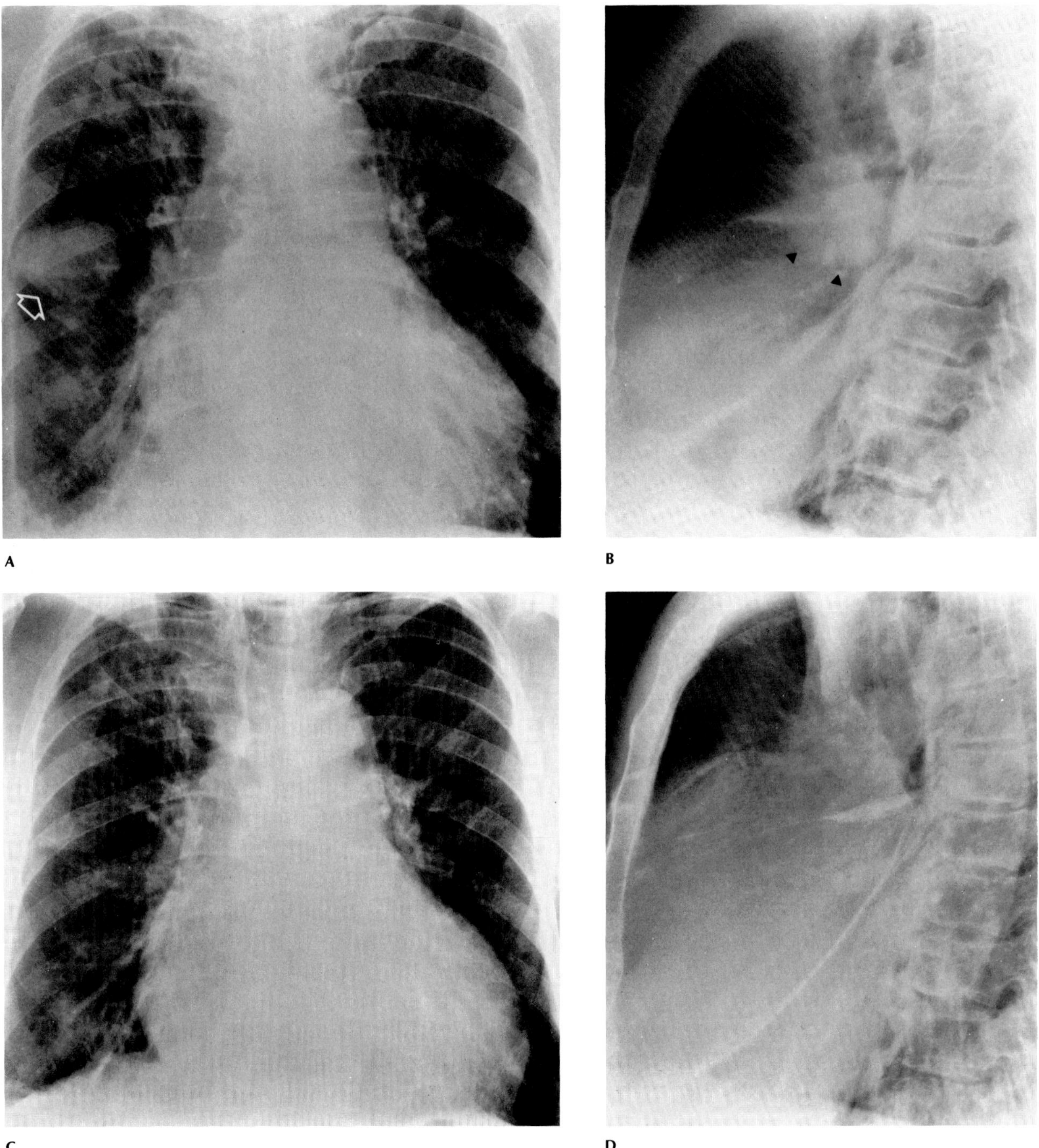


**Figure 20.4. Phantom tumor of pulmonary edema.** (A) Frontal and (B) lateral views of the chest demonstrate a sharply defined elliptical density (arrows) in the right midlung resembling a solid parenchymal mass. Note the associated right pleural effusion and the signs of pulmonary vascular congestion. (C) Frontal and (D) lateral views of the chest following an improvement in the patient's cardiac status show a rapid regression of the fluid density, clearly indicating the true nature of the phantom tumor. (From Eisenberg,[11] with permission from the publisher.)

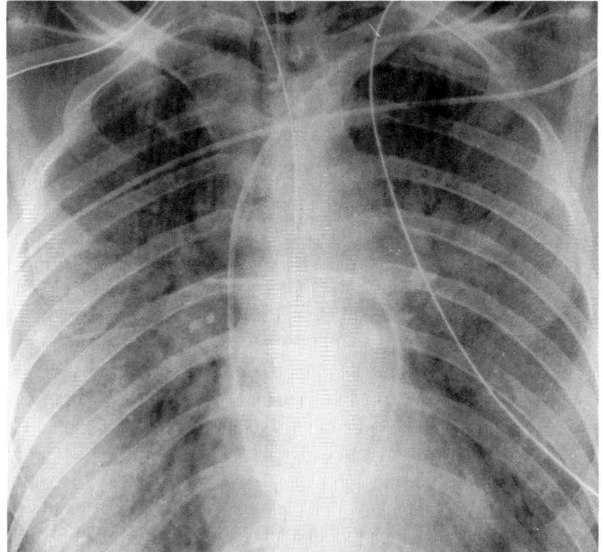

**Figure 20.5. Salicylate-induced pulmonary edema.** A supine chest radiograph of a 36-year-old man with a serum salicylate level of 63 mg/dl shows a bilateral alveolar pattern indicative of pulmonary edema. Unlike the appearance of pulmonary edema due to congestive heart failure, the cardiac silhouette is normal in this patient, who had no evidence of cardiac disease. (From Walters, Woodring, and Stelling, et al,[12] with permission from the publisher.)

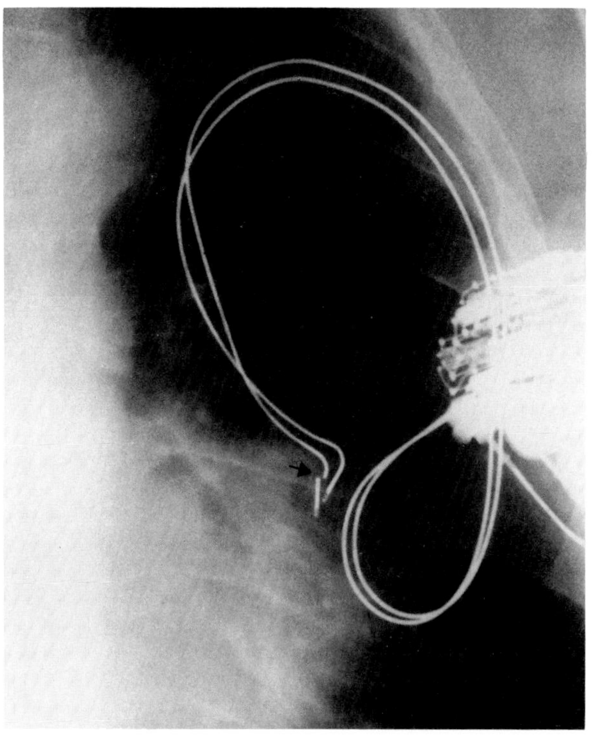

**Figure 20.6. Fracture of a cardiac pacemaker wire** (arrow).

cification in the upper portion of the interventricular septum may indicate regional infarction.

Radiographic evaluation is also of value in assessing the permanent cardiac pacemakers used in patients with cardiac arrhythmias. Frontal and lateral chest radiographs can demonstrate the position of opaque transvenous electrodes, which should extend to the apex of the right ventricle. Fractures of cardiac pacemaker wires can also be detected. When epicardial electrodes are employed, these fractures tend to occur at fixation points in the myocardium; when transvenous leads are used, they tend to occur at points of flexion over the margins of the clavicle or ribs. Some subtle fractures may only be demonstrated on oblique views. Chest radiographs can also assess the condition of pacemaker batteries. Fully charged mercury batteries demonstrate a uniform lucent ring of electrolytes that is sharply demarcated from the surrounding opaque rings of metallic material. As the battery fails, the lucent ring becomes thinner, and its inner border becomes irregular because of metallic deposits.[7–10]

## REFERENCES


1. Juhl JH: *Essentials of Roentgen Interpretation*. Philadelphia, JB Lippincott, 1981.
2. Fraser RG, Pare JAP:*Diagnosis of Diseases of the Chest*. Philadelphia, WB Saunders, 1979.
3. Azimi F, Wolson AH, Dalinka MK, et al: Unilateral pulmonary edema: Differential diagnosis. *Australas Radiol* 19:20–26, 1975.
4. Staub NC: Pathogenesis of pulmonary edema: State of the art review. *Am Rev Respir Dis* 109:358–367, 1974.
5. Heitzman ER, Ziter FM: Acute interstitial pulmonary edema. *AJR* 98:291–297, 1966.
6. Rigler LG, Surprenant EL: Pulmonary edema. *Semin Roentgenol* 2:33–46, 1967.
7. Rosenbaum HD: Roentgen demonstration of broken cardiac pacemaker wires. *Radiology* 84:933–936, 1965.
8. Lillehei CW, Cruz AB, Johnsrude I, et al: New method of assessing the state of charge of implanted cardiac pacemaker batteries. *Am J Cardiol* 16:717–721, 1965.
9. Walter WH: Radiographic identificatrion of commonly used pulse generators. *JAMA* 215:1974–1979, 1971.
10. Sorkin RP, Schuurmann BJ, Simon AB: Radiology aspects of permanent cardiac pacemakers. *Radiology* 119:281–286, 1976.
11. Eisenberg RL: *Atlas of Signs in Radiology*. Philadelphia, JB Lippincott, 1981.
12. Walters JS, Woodring JH, Stelling CB, et al: Salicylate-induced pulmonary edema. *Radiology* 146:289–293, 1983.

## Chapter Twenty-One
# Congenital Heart Disease

## COMMUNICATIONS BETWEEN THE SYSTEMIC AND PULMONARY CIRCULATIONS WITHOUT CYANOSIS (LEFT-TO-RIGHT SHUNTS)

The accurate diagnosis of cardiovascular malformations is essential since many affected babies can now be cured by aggressive medical and surgical management. In many cases, a precise diagnosis can be suggested by evaluating the pulmonary vascularity, the heart size, the position and size of the aorta, the size of the main pulmonary artery segments, and whether the patient is cyanotic (Table 21.1). Confirmation of the precise congenital defect requires angiocardiography.

### ATRIAL SEPTAL DEFECT

Atrial septal defect, the most common congenital cardiac lesion, permits free communication between the two atria. Because the left atrial pressure is usually higher than the pressure in the right atrium, the resulting shunt is from left to right. The magnitude of the shunt depends on the size of the defect, the relative compliance of the ventricles, and the differences in atrial pressure. The left-to-right shunt causes increased pulmonary blood flow and overloading of the right ventricle. This produces a radiographic appearance of enlargement of the right ventricle, the right atrium, and the pulmonary outflow tract, along with a diffuse increase in pulmonary vascularity. Unlike patients with ventricular septal defect and patent ductus arteriosus, patients with uncomplicated atrial septal defect show no evidence of left atrial or left ventricular enlargement. The shunting of blood away from the left side of the heart into the pulmonary circulation causes decreased flow through the aorta. Thus the aorta tends to be smaller than normal, unlike the enlarged aorta seen in patent ductus arteriosus.

Like any left-to-right shunt, an atrial septal defect can be complicated by the development of pulmonary hypertension (Eisenmenger physiology). This is caused by increased vascular resistance within the pulmonary arteries related to chronically increased flow through the pulmonary circulation. Pulmonary hypertension appears as an increased fullness of the central pulmonary arteries with abrupt narrowing and pruning of peripheral vessels, which makes the periphery of the lung appear more lucent. The elevation of pulmonary arterial pressure tends to balance or even reverse the interatrial shunt and thus ease the volume overloading of the right side of the heart;

**Table 21.1. Classification of Congenital Heart Disease**

| | Pulmonary Vascularity | | |
|---|---|---|---|
| | Increased | Decreased | Normal |
| Cyanotic | Increased blood flow<br>Persistent truncus arteriosus<br>Transposition of great arteries<br>Single ventricle<br>Complete endocardial cushion defect<br>Pulmonary venous obstruction<br>Total anomalous pulmonary venous return<br>Malformations obstructing pulmonary venous flow<br>Pulmonary hypertension (Eisenmenger physiology)<br>Atrial septal defect<br>Ventricular septal defect<br>Patent ductus arteriosus | Right-to-Left shunt<br>Tetralogy of Fallot<br>Trilogy of Fallot<br>Tricuspid atresia<br>Ebstein's anomaly<br>Pulmonary stenosis and transposition<br>Pulmonary stenosis and atrial septal defect | |
| Acyanotic | Increased blood flow (right-to-left shunt)<br>Atrial septal defect<br>Ventricular septal defect<br>Patent ductus arteriosus<br>Aorticopulmonary septal defect<br>Ruptured aortic sinus aneurysm<br>Coronary arteriovenous fistula<br>Partial anomalous pulmonary venous return<br>Pulmonary venous congestion<br>Heart failure | | Aortic stenosis<br>Coarctation of the aorta<br>Pulmonary stenosis<br>Endocardial fibroelastosis<br>Anomalous pulmonary origin of coronary artery |

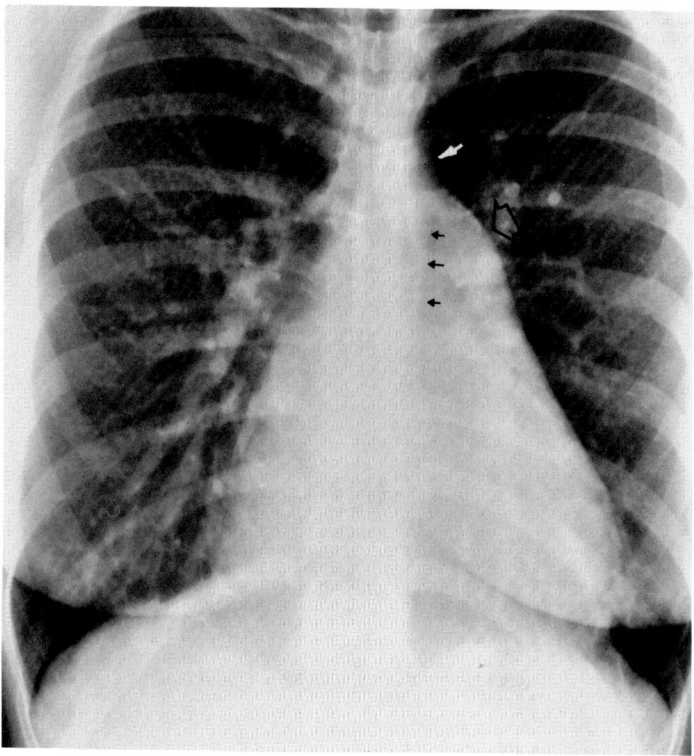

A

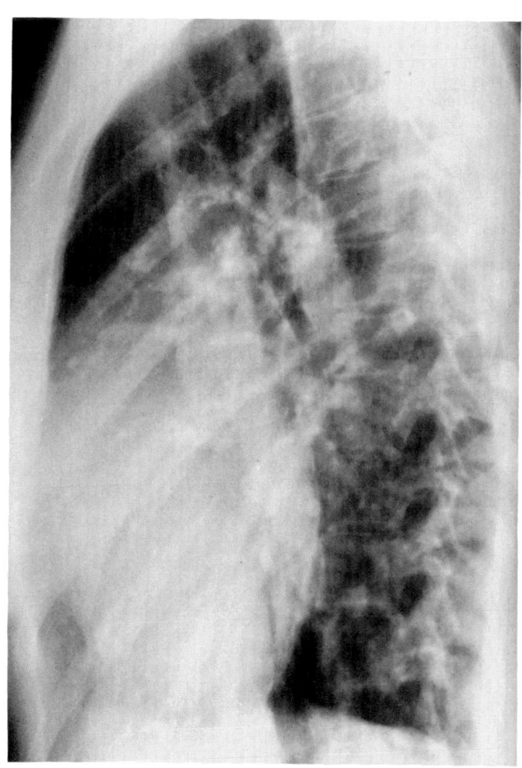

B

**Figure 21.1. Atrial septal defect.** (A) Frontal and (B) lateral views of the chest demonstrate cardiomegaly along with an increase in pulmonary vascularity reflecting the left-to-right shunt. Filling in of the restrosternal space indicates enlargement of the right ventricle. The small aortic knob (white arrow) and descending thoracic aorta (small black arrows) are dwarfed by the enlarged pulmonary outflow tract (large black arrow).

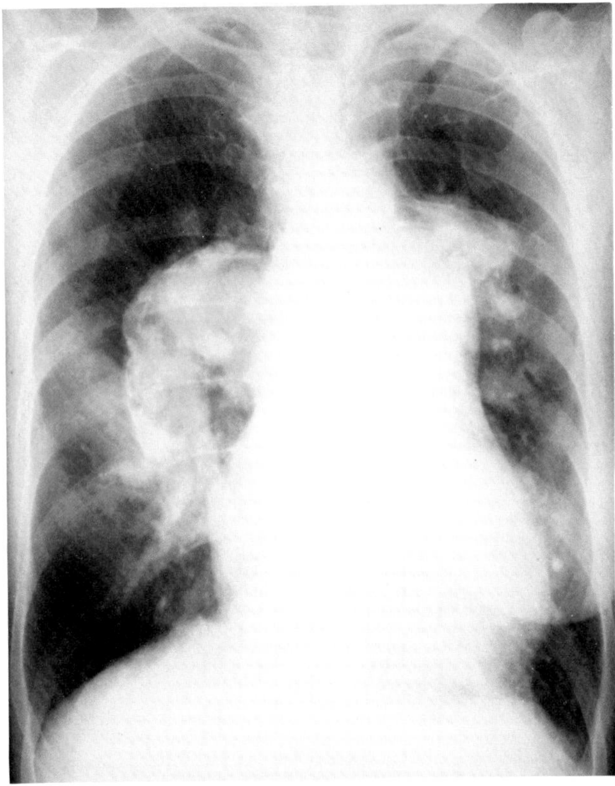

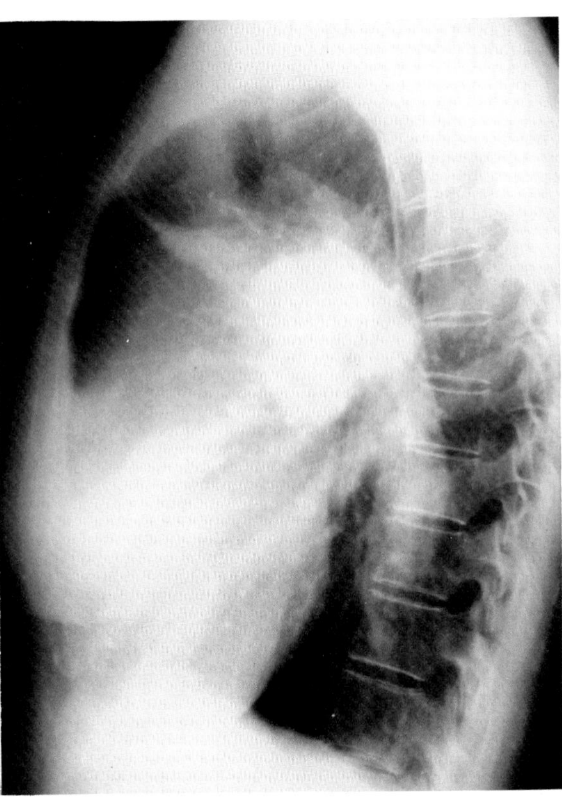

A B

**Figure 21.2. Eisenmenger physiology in atrial septal defect.** (A) Frontal and (B) lateral films demonstrate slight but definite cardiomegaly and a great increase in the size of the pulmonary trunk. The right and left pulmonary artery branches are huge, but the peripheral pulmonary vasculature is relatively sparse. Long-standing pulmonary hypertension has produced degenerative intimal changes in the pulmonary arteries, and the arteries have become densely calcified. (From Cooley and Schreiber,[83] with permission from the publisher.)

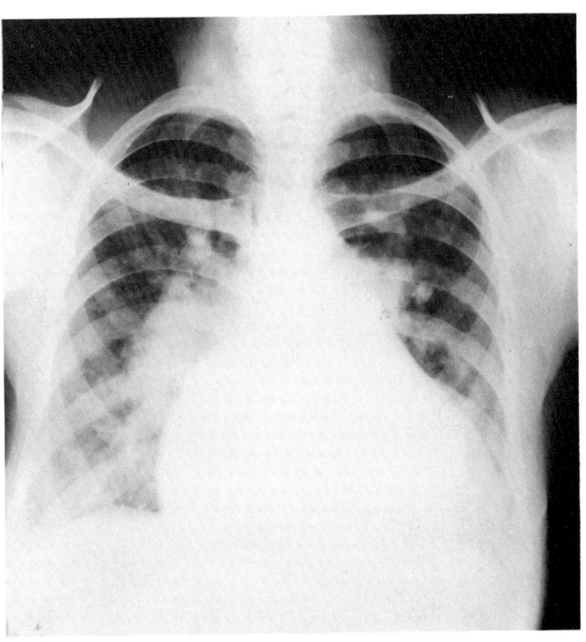

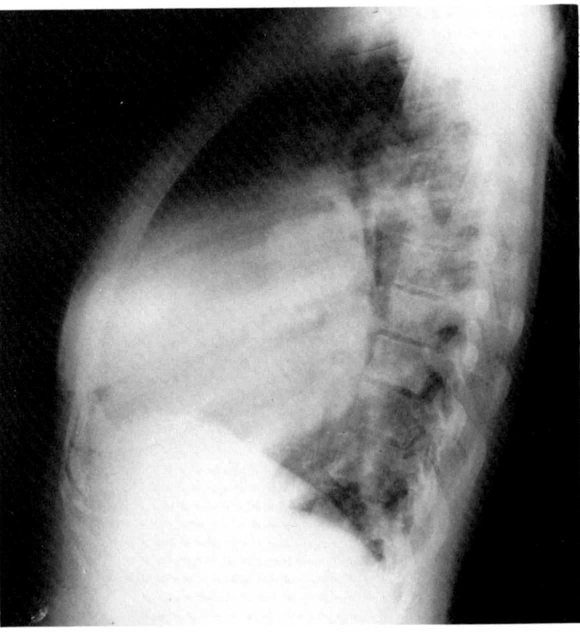

A B

**Figure 21.3. Lutembacher's syndrome.** (A) Frontal and (B) lateral views demonstrate marked cardiomegaly and a striking enlargement of the pulmonary outflow tract and main pulmonary arteries in a patient with atrial septal defect combined with mitral stenosis. (From Cooley and Schreiber,[83] with permission from the publisher.)

this results in a more apparent right ventricular hypertrophy but a decrease in the overall cardiac size.

Atrial septal defect can occur in association with many other cardiac anomalies. When the defect is combined with mitral stenosis (Lutembacher's syndrome), there is a much greater increase in the work load of the right ventricle than when there are uncomplicated atrial septal defects of similar size. This combination results in extreme enlargement of the pulmonary outflow tract and main pulmonary arteries, which tends to dominate the radiographic appearance.

The definitive diagnosis of atrial septal defect requires angiocardiography, which can demonstrate the size and location of the septal defect. The defect is best seen with the patient in the left anterior oblique position, which brings the septum into profile.[1–5]

## ENDOCARDIAL CUSHION DEFECT

In this congenital anomaly, most often seen in children with Down's syndrome, a low atrial septal defect (ostium primum type) is associated with a high ventricular septal defect. Associated abnormalities in the mitral and the tricuspid valves can lead to a common atrioventricular canal. There is extensive bidirectional shunting of blood involving all the cardiac chambers. However, since the basic shunts are from left to right, the pulmonary vascularity is increased. Enlargement of all the cardiac chambers gives the heart a nonspecific globular configuration, bulging nearly equally on both sides of the spine.

Angiography demonstrates virtually immediate opacification of all four cardiac chambers along with simultaneous opacification of the aorta and pulmonary artery. A cleft mitral valve, seen in all patients with endocardial cushion defect, produces a characteristic gooseneck deformity of the subaortic outflow segment of the left ventricle.[6–8]

## VENTRICULAR SEPTAL DEFECT

Ventricular septal defect is a common congenital cardiac anomaly that may be found in isolated form or in association with other abnormalities. Because the left ventricular pressure is usually higher than the pressure in the right ventricle, the resulting shunt is from left to right. The magnitude of the shunt depends on the size of the defect and the differences in ventricular pressure. Large

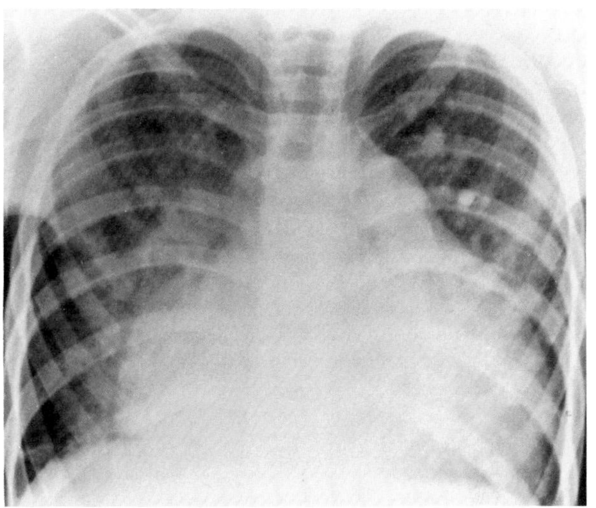

**Figure 21.4. Endocardial cushion defect.** There is globular enlargement of the heart with increased pulmonary vascularity.

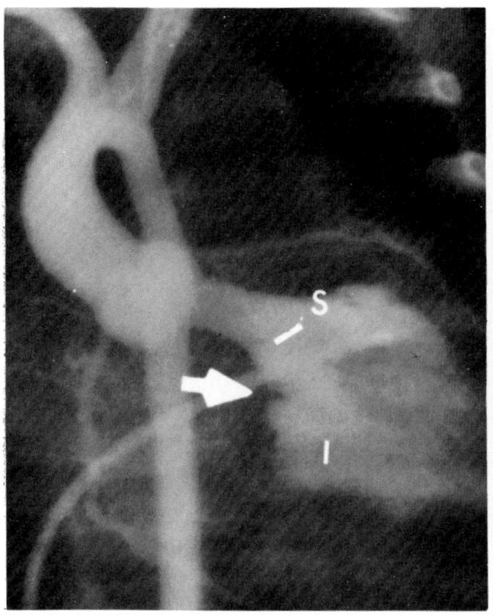



A

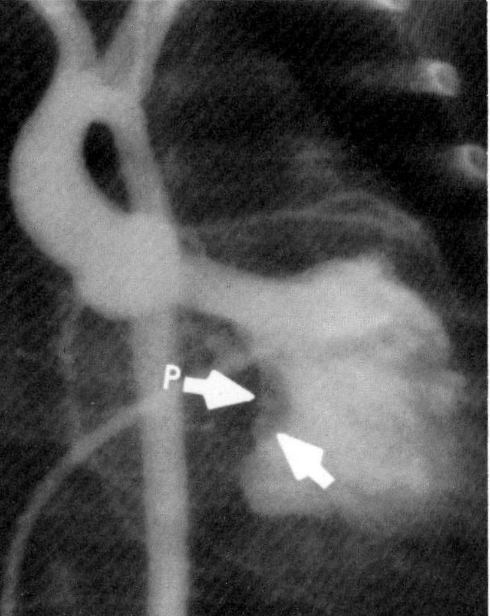



B

**Figure 21.5. Complete atrioventricular canal.** (A) A left ventricular angiogram (frontal view) in early systole shows the cleft (arrow) between the superior (S) and inferior (I) segments of the anterior mitral leaflet located along the right contour of the ventricle. There is no evidence of mitral insufficiency or an interventricular shunt. (B) In diastole, the ventricular outflow tract is narrowed and lies in a more horizontal position than normal. The right border of the ventricle can be followed directly into the scooped-out margin (arrow) of the interventricular septum. The attachment of the posterior mitral leaflet (P) is also visible because of a thin layer of contrast material trapped between the leaflet and the posterior ventricular wall. (From Baron,[3] with permission from the publisher.)

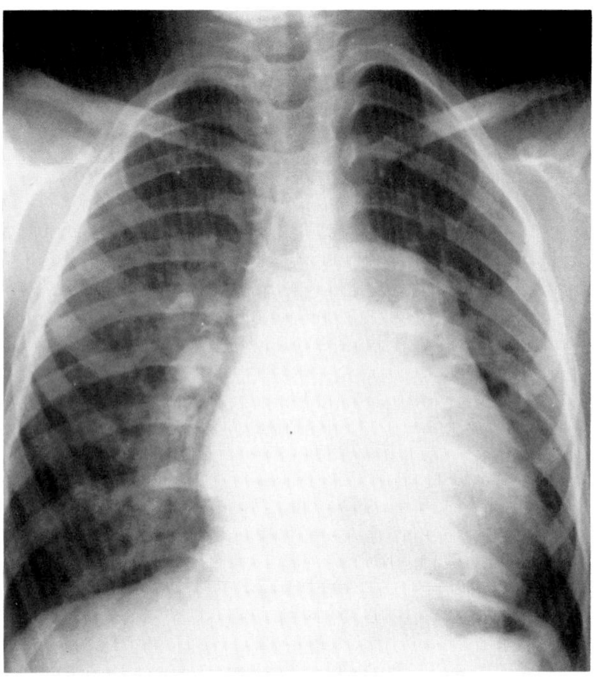

Figure 21.6. **Ventricular septal defect.** A plain chest radiograph shows the heart to be enlarged and somewhat triangular. The pulmonary trunk is very large and overshadows the normal-sized aorta, which seems small by comparison. The pulmonary artery branches in the hilum and in the periphery of the lung are enlarged, and the pulmonary vascular volume is increased. The biventricular and the left atrial enlargement were verified by electrocardiography and esophogram. (From Cooley and Schreiber,[83] with permission from the publisher.)

shunts produce dramatic radiographic abnormalities; small defects may produce no changes detectable on chest radiographs. The left-to-right shunt causes increased pulmonary blood flow and, consequently, increased pulmonary venous return. This leads to diastolic overloading and enlargment of the left atrium and left ventricle. Because shunting occurs primarily in systole and any blood directed to the right ventricle immediately goes into the pulmonary artery, there is no overloading of the right ventricle; thus no right ventricular enlargement is seen. With larger defects, however, the right ventricle becomes subject to the same stresses as the left ventricle and enlargement of both of these chambers becomes apparent radiographically. At times, an extremely large defect produces a functional "single ventricle," in which there is effectively no pressure difference between the right and left ventricles. This condition appears radiographically as marked nonspecific globular enlargement of the heart with engorgement of the pulmonary vascularity. As with atrial septal defect, the shunting of blood away from the left side of the heart into the pulmonary circulation causes decreased aortic flow, and thus the aorta appears small or normal in size.

Like any left-to-right shunt, ventricular septal defect can be complicated by the development of pulmonary hypertension (Eisenmenger physiology). Pulmonary hypertension appears radiographically as an increased fullness of the central pulmonary arteries with abrupt narrowing and pruning of peripheral vessels, which makes the periphery of the lung appear more lucent. The elevation of pulmonary arterial pressure tends to balance or even reverse the interventricular shunt, thus easing the volume overloading of the right side of the heart.

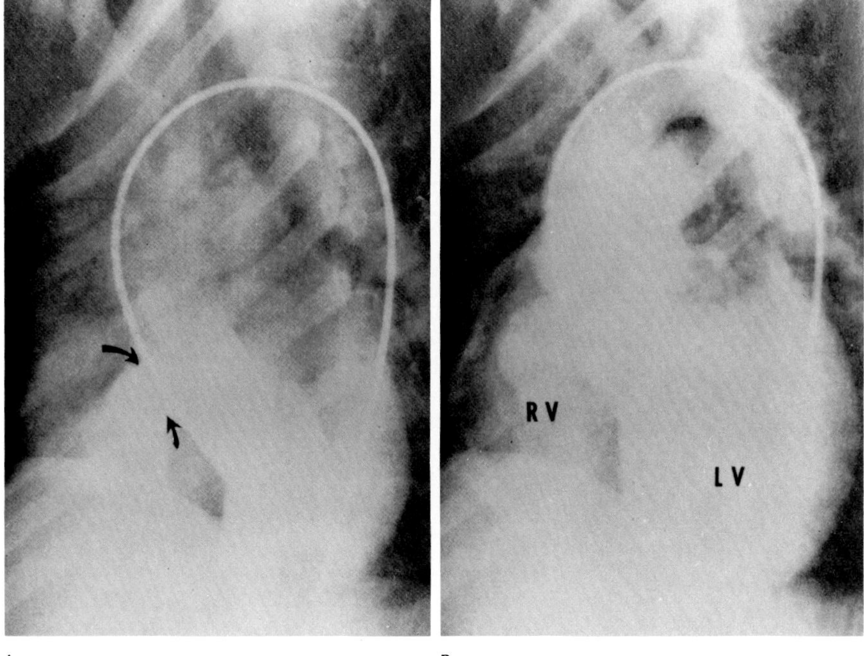


**Figure 21.7. Ventricular septal defect.** Two films from a left ventriculogram made at half-second intervals demonstrate opacification of the left ventricle (LV) with rapid filling of the right ventricle (RV) through a high interventricular septal defect (arrows). (From Dotter,[87] with permission from the publisher.)

This may produce a pattern of almost pure right ventricular hypertrophy, with the left atrial and left ventricular size returning to normal.

The definitive diagnosis of ventricular septal defect requires angiocardiography to demonstrate the size and precise location of the septal defect.[1,4,9,83,87]

## PATENT DUCTUS ARTERIOSUS

The ductus arteriosus is a vessel that extends from the bifurcation of the pulmonary artery to join the aorta just distal to the left subclavian artery. It serves to shunt blood from the pulmonary artery into the systemic circulation during intrauterine life. Persistence of the ductus arteriosus, which normally closes soon after birth, results in a left-to-right shunt. The flow of blood from the higher-pressure aorta to the lower-pressure pulmonary artery causes increased pulmonary blood flow, and an excess volume of blood is returned to the left atrium and left ventricle. Radiographically, there is enlargement of the left atrium, the left ventricle, and the central pulmonary arteries, along with a diffuse increase in pulmonary vascularity. The increased blood flow through the aorta proximal to the shunt produces a prominent aortic knob, in contrast to the small or normal-sized aorta seen in atrial and ventricular septal defect. In addition, the aortic

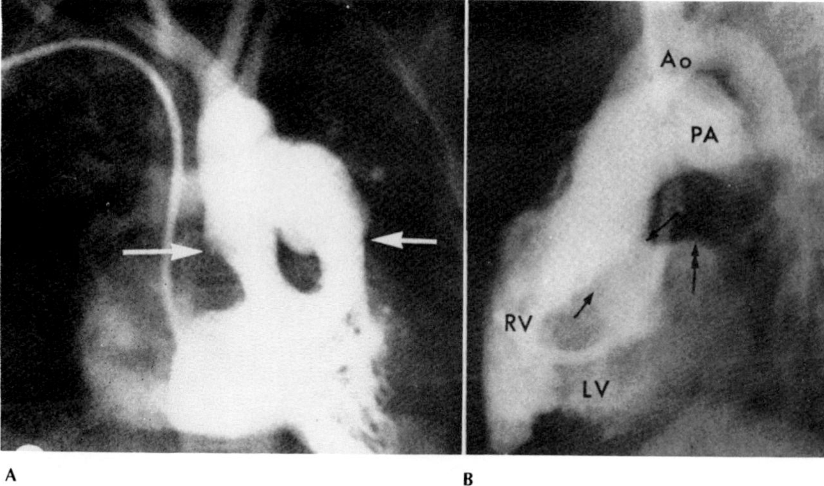


**Figure 21.8. Single ventricle.** A right ventriculogram in a patient whose great arteries both originate from the right ventricle and are side by side. (A) A frontal view shows muscular tracts leading from the right ventricle to both great arteries, the valves of which (arrows) are at the same horizontal level. (B) A lateral view shows the anteriorly situated right ventricle (RV) communicating with the left ventricle (LV) via a ventricular septal defect (single arrows). The anterior mitral leaflet is not continuous with either semilunar valve. The double arrow indicates the base of the anterior mitral leaflet. PA = pulmonary artery; Ao = aorta. (From Hallermann, Kincaid, Ritter, et al,[59] with permission from the publisher.)

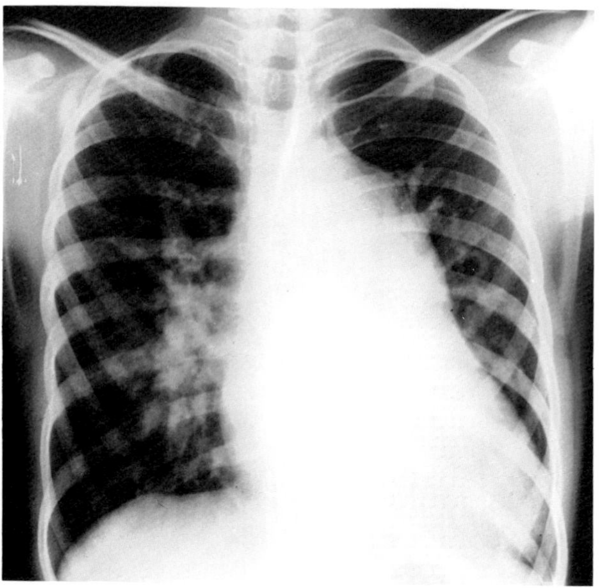


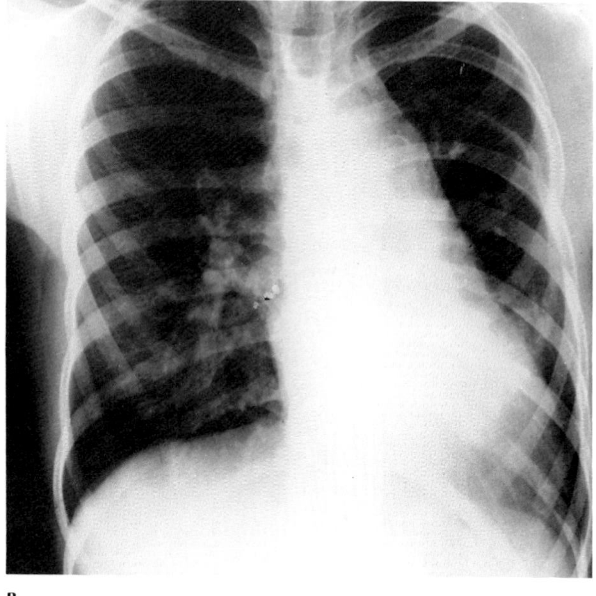


**Figure 21.9. Patent ductus arteriosus.** (A) A preoperative frontal chest film demonstrates cardiomegaly with enlargement of the left atrium, the left ventricle, and the central pulmonary arteries. There is a diffuse increase in pulmonary vascularity. (B) One year following surgery, there has been a decrease in the size and a change in the contour of the heart. Note the marked decrease in pulmonary engorgement. (From Cooley and Schreiber,[83] with permission from the publisher.)

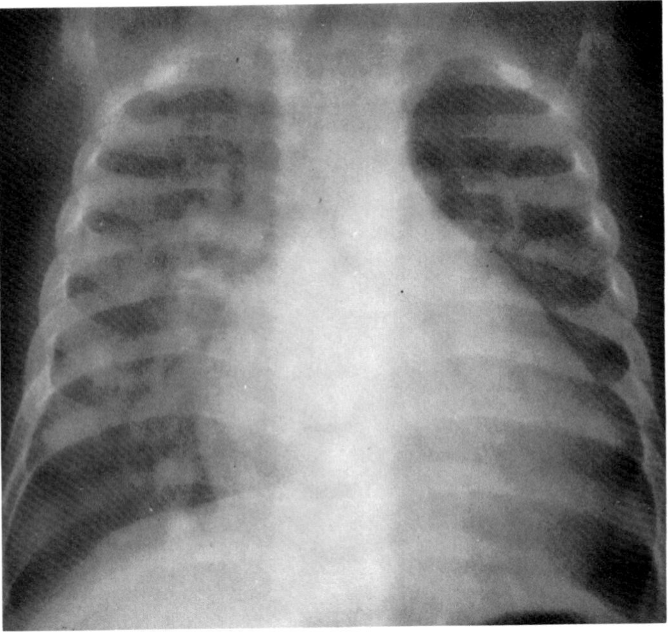


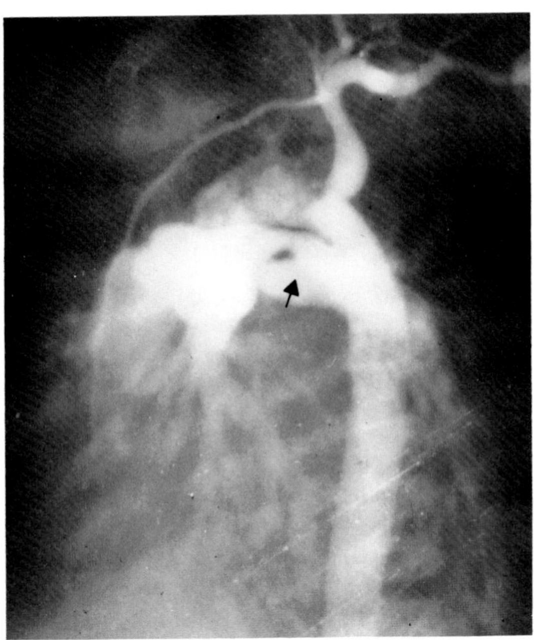


**Figure 21.10. Patent ductus arteriosus.** (A) A plain chest radiograph demonstrates cardiomegaly with increased pulmonary vascularity. (B) An aortogram shows persistent patency of the ductus arteriosus. (From Cooley and Schreiber,[83] with permission from the publisher.)

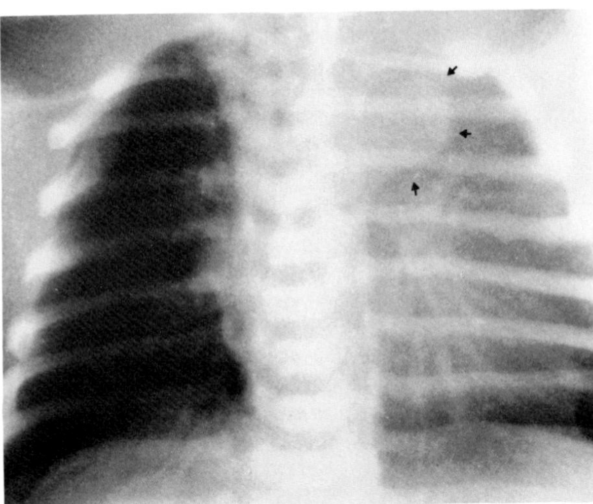

**Figure 21.11. Patent ductus arteriosus.** A convex bulge (arrows) on the left side of the superior mediastinum represents dilatation of the aortic end of the ductus ("ductus bump").

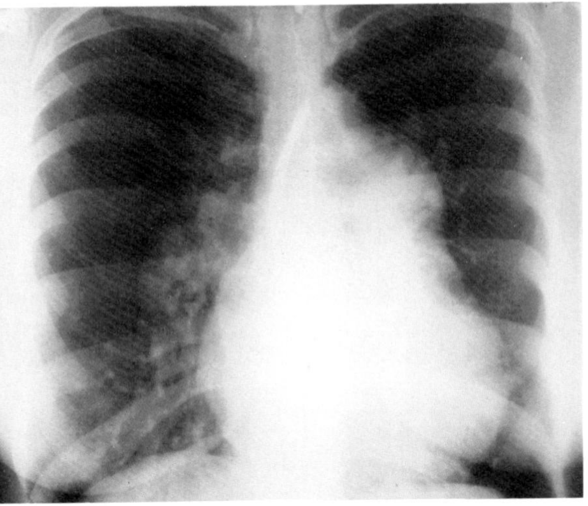

**Figure 21.12. Eisenmenger physiology in patent ductus arteriosus.** There is an increased fullness of the central pulmonary arteries with an abrupt narrowing and a paucity of peripheral vessels.

end of the ductus (infundibulum) is often dilated and produces a convex bulge on the left border of the aorta just below the knob.

Like any left-to-right shunt, patent ductus arteriosus can be complicated by the development of pulmonary hypertension (Eisenmenger physiology), which causes an increased fullness of the central pulmonary arteries with abrupt narrowing and pruning of peripheral vessels. The elevation of pulmonary arterial pressure causes right ventricular hypertrophy and enlargement, which may dominate the radiographic appearance. As the pulmonary blood flow and the pulmonary venous return decrease, the left atrial and left ventricular dilatation subside.

Retrograde arteriography can outline the patent ductus arteriosus and demonstrate the shunting of blood from the aorta into the pulmonary circulation.[1,4,10,83]

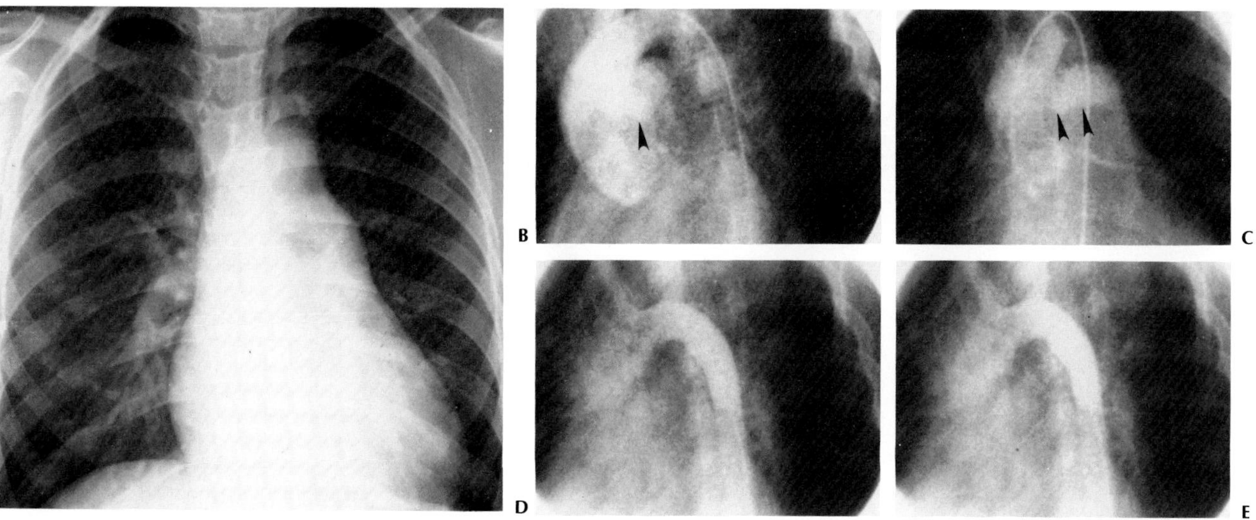

**Figure 21.13. Aortopulmonary septal defect.** (A) A plain film of the chest demonstrates enlargement of the left ventricle, a low position of the apex, and an increase in pulmonary vascularity. (B, C) The injection of contrast material into the ascending aorta shows rapid shunting into the pulmonary arteries (arrows). (D, E) The injection of contrast material into the descending aorta does not show a shunt, thus confirming that the shunt is in the ascending aorta. (From Cooley and Schreiber,[83] with permission from the publisher.)

## AORTOPULMONARY SEPTAL DEFECT (AORTOPULMONARY WINDOW)

Aortopulmonary septal defect is an uncommon anomaly in which a communication between the pulmonary artery and the aorta (just above their valves) is caused by a failure of complete septation of the primitive truncus arteriosus. Aortopulmonary septal defect may be difficult to distinguish from patent ductus arteriosus because both appear radiographically as enlargement of the left ventricle, the left atrium, and the central pulmonary arteries, and a diffuse increase in pulmonary vascularity. However, because in aortopulmonary septal defect the shunt arises in the ascending aorta proximal to the knob, the aortic knob is usually less prominent than in patent ductus arteriosus.[1,11,12,83]

## AORTIC SINUS ANEURYSM AND FISTULA

The aortic sinuses (sinuses of Valsalva) are three dilatations in the root of the aorta just above the aortic valve. An aneurysm of the sinus of Valsalva is a rare congenital anomaly that usually involves the sinus above the right cusp of the aortic valve. A large aneurysm may produce a local bulge in the right anterolateral cardiac contour; calcium may be apparent in the aneurysm wall. A definitive diagnosis is made by opacification of the aneurysm during aortography.

Rupture of the aneurysm, usually into the right ventricle but occasionally into the right atrium, causes a sudden large left-to-right shunt that produces the acute onset of chest pain, shortness of breath, and a cardiac murmur. Radiographically, there is a rapid increase in the heart size along with right ventricular enlargement, dilatation of the main pulmonary arterial trunk, and a diffuse increase in pulmonary vascularity. At aortography, the injected contrast material can be demonstrated opacifying the right heart chambers.[1,12,13]

## CORONARY ARTERIOVENOUS FISTULA

A communication between a coronary artery and a cardiac chamber or a pulmonary artery is an unusual congenital anomaly. The right coronary artery is most often involved and communicates with, in order of frequency, the right ventricle, the right atrium, the coronary sinus, or the pulmonary artery. This communication between the coronary artery and the pulmonary vessels or the right side of the heart produces a left-to-right shunt.

Coronary arteriovenous fistula causes enlargement of the main pulmonary artery and a diffuse increase in pulmonary vascularity. A fistula into the right ventricle results in enlargement of this chamber; fistulization into the right atrium causes enlargement of both the right atrium and the right ventricle.

A definitive diagnosis is made by coronary arteriography, which demonstrates the dilated and tortuous coronary artery and its abnormal communication with the heart or pulmonary artery.[14–17]

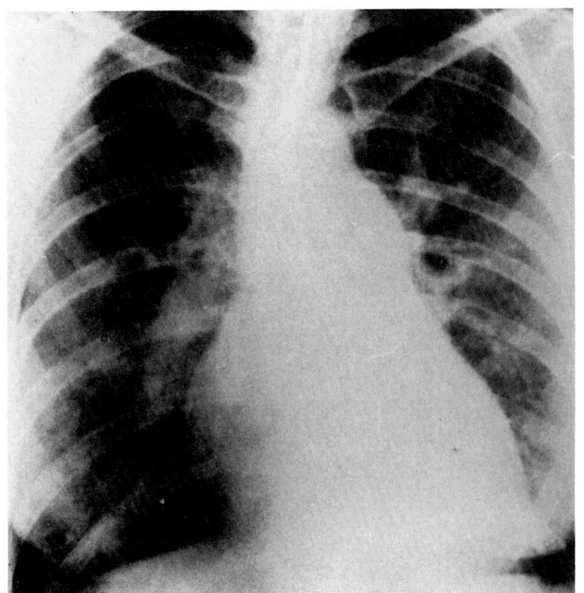
A

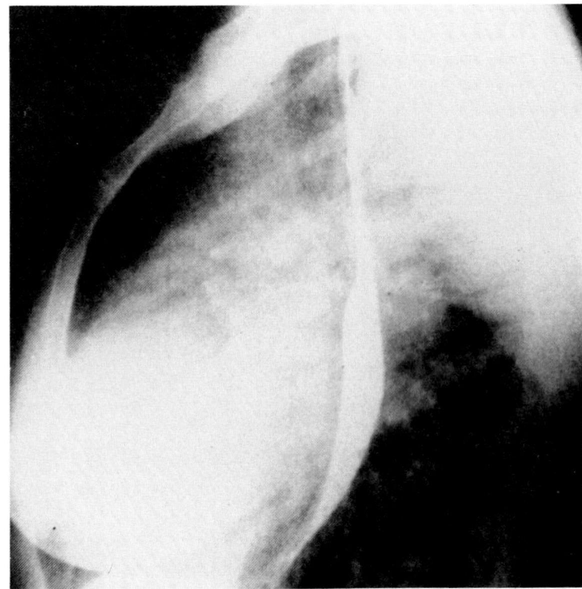
B

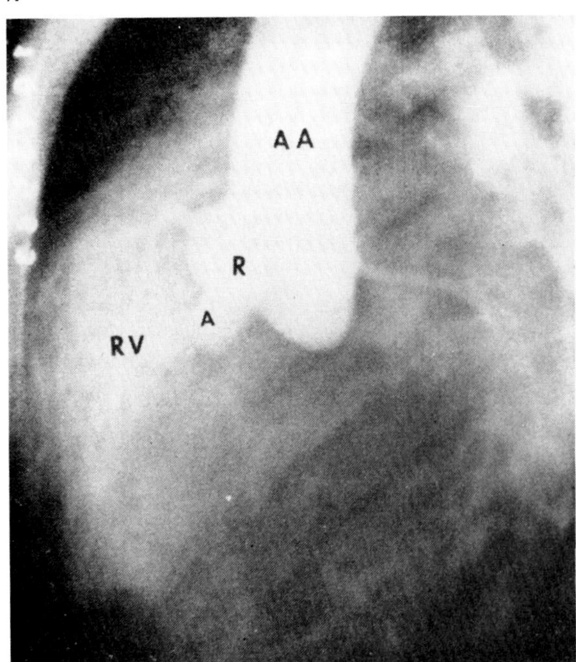

C

**Figure 21.14. Ruptured right aortic sinus aneurysm.** (A) Frontal and (B) lateral chest radiographs demonstrate cardiomegaly and increased pulmonary vascularity. (C) A lateral projection from a selective thoracic aortogram shows an aneurysm (A) of the right aortic sinus (R) projecting into the outflow tract of the right ventricle (RV). The contrast medium has opacified the right ventricle through the aneurysm. (From Elliott,[12] with permission from the publisher.)

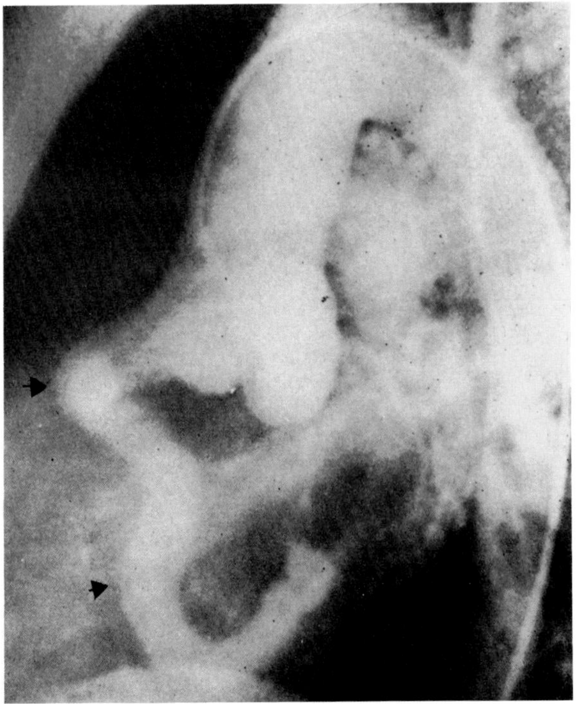

**Figure 21.15. Coronary arteriovenous fistula.** A lateral view from an angiocardiogram shows a huge right coronary artery (arrow) draining into the right ventricle. (From Steinberg and Holswade,[16] with permission from the publisher.)

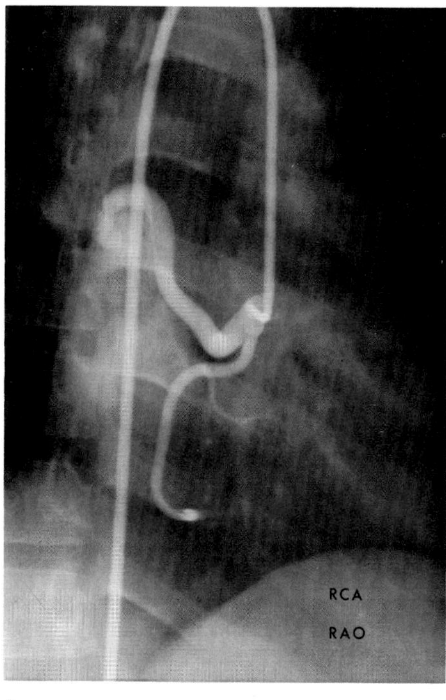


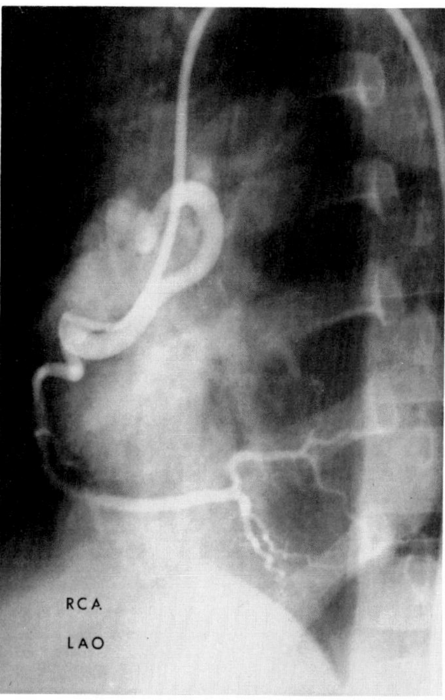

A B

**Figure 21.16. Coronary arteriovenous fistula.** (A) Right anterior oblique and (B) left anterior oblique views from a right coronary arteriogram demonstrate marked enlargement of the sinoatrial branch that forms a direct fistulous communication between the right coronary artery and the superior vena cava. Note the opacification of the cavity of the right atrium. (From Morettin,[17] with permission from the publisher.)

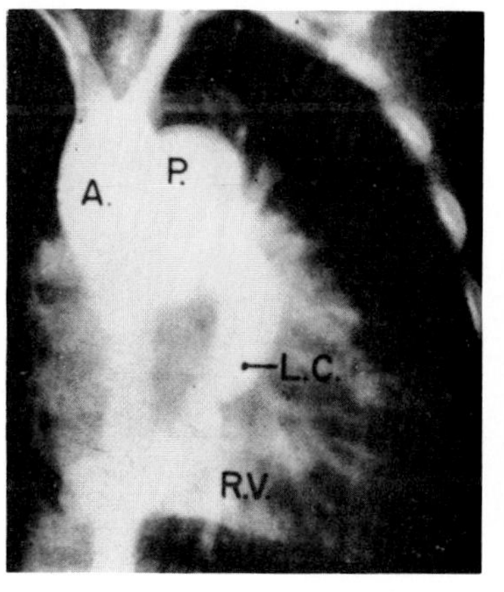


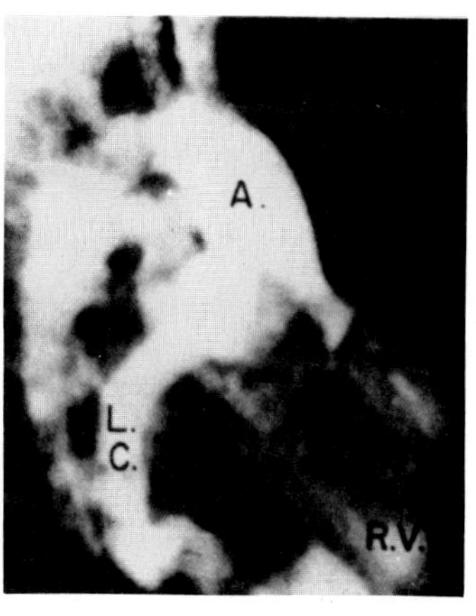


A B

**Figure 21.17. Coronary arteriovenous fistula.** (A) Frontal and (B) lateral projections from a thoracic aortogram show opacification of a dilated, tortuous, and saccular left circumflex (LC) coronary artery that communicates with the right ventricle (RV). There is subsequent opacification of the pulmonary trunk (P). (From Elliott,[12] with permission from the publisher.)

## ANOMALOUS PULMONARY ORIGIN OF THE CORONARY ARTERY

In this rare anomaly, the blood supply to the myocardium is affected because the left coronary artery arises from the pulmonary artery. In infancy, before collateral circulation can be established, this condition causes ischemia of the left side of the heart with impaired contractility leading to left ventricular enlargement and increasing pulmonary vascular congestion. The development of collateral circulation from the normal right coronary artery to the anomalous left coronary artery and then on to the pulmonary artery produces a large left-to-right shunt with a corresponding increase in the pulmonary vascularity.

Angiocardiography demonstrates that only the right coronary artery fills from the aorta. Retrograde flow via intercoronary anastomoses fills the left coronary artery and subsequently opacifies the pulmonary artery.[1,18,19]

## PERSISTENT TRUNCUS ARTERIOSUS

In persistent truncus arteriosus, the failure of the common truncus arteriosus to divide normally into the aorta and pulmonary artery results in a single large arterial trunk that receives the outflow of blood from both ventricles. This single vessel usually overrides the ventricular septum, is almost always associated with a high ventricular septal defect, and supplies the systemic, pulmonary, and coronary circulations. The pulmonary arteries usually arise in one of several patterns from the truncus. In type IV truncus arteriosus, the pulmonary

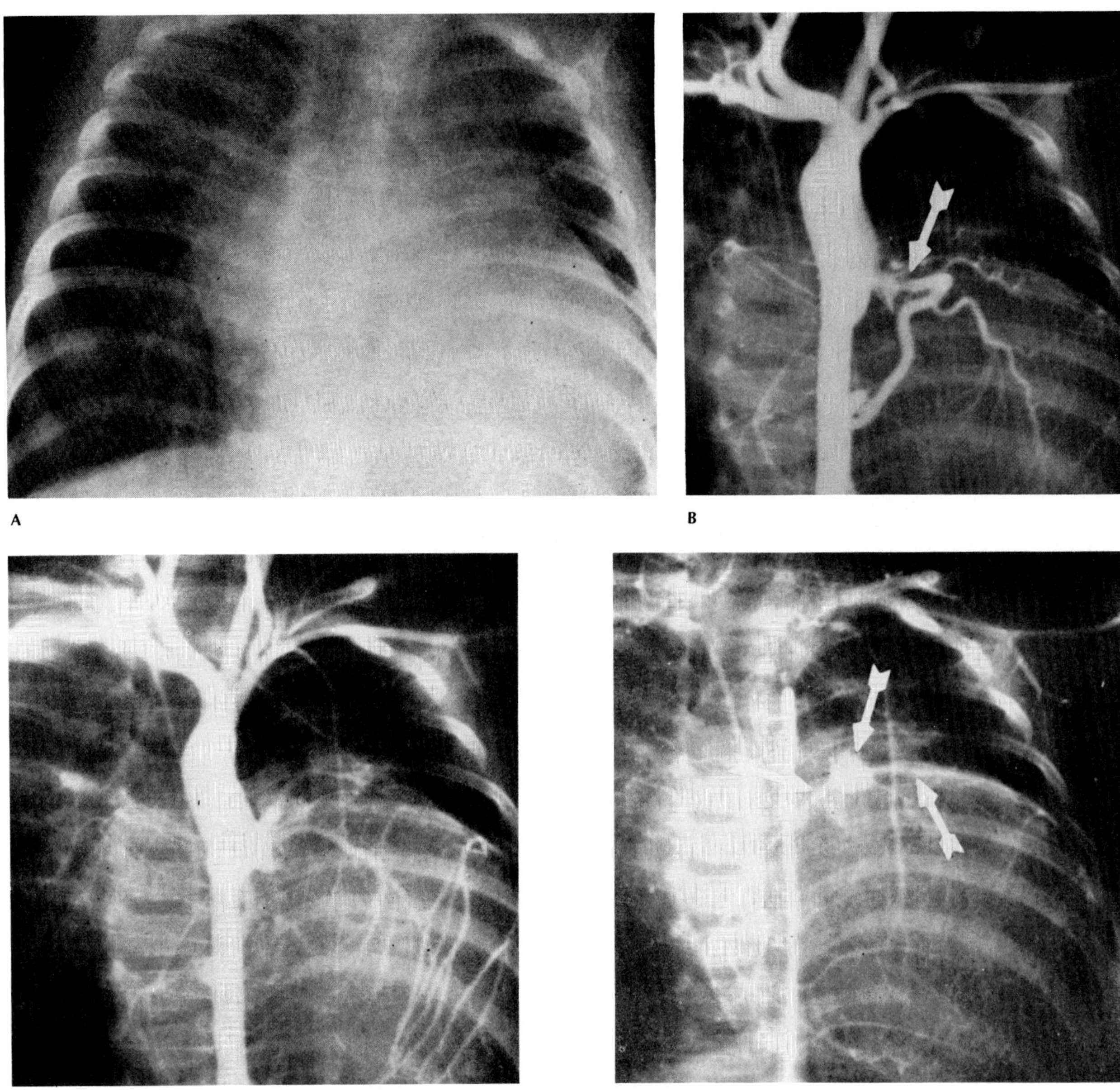


**Figure 21.18. Anomalous origin of the left coronary artery from the pulmonary artery.** (A) A frontal radiograph shows massive cardiomegaly. There is partial atelectasis of the right upper lobe. (B) A frontal view from an aortogram shows the enlarged right coronary artery (arrow). (C) Later in the series, the anastomotic channels between the right and left coronary arteries were visualized. (D) There is retrograde filling of the pulmonary artery (upper arrow); the lower arrows point to anterior descending and circumflex branches of the left coronary artery. (From Stein, Hagstrom, Ehlers, et al,[19] with permission from the publisher.)

arteries are absent and the pulmonary circulation is supplied via bronchial arteries or other collateral vessels arising from the descending aorta.

The truncus arteriosus is usually large and may appear as a bulge in the region of the ascending arch. In most forms of the anomaly, the pulmonary outflow tract appears concave, because its origin from the truncus is more medial than its normal origin from the right ventricle. A right-sided aortic arch occurs in about 30 percent of the patients. There is often a high positioning of the transverse arch of the aorta and elevation of the left pulmonary artery.

Much of the blood flow through the truncus is preferentially directed into the lower-pressure pulmonary arterial circulation; this produces a left-to-right shunt with an increase in the pulmonary vascularity. Indeed per-

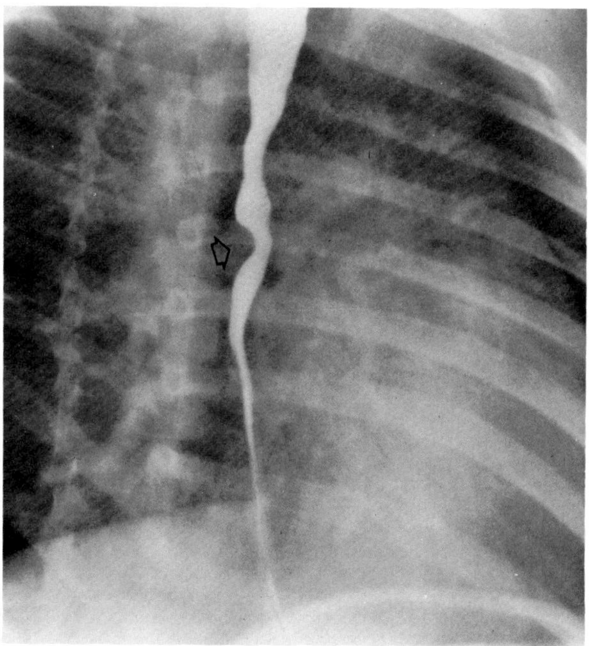

A

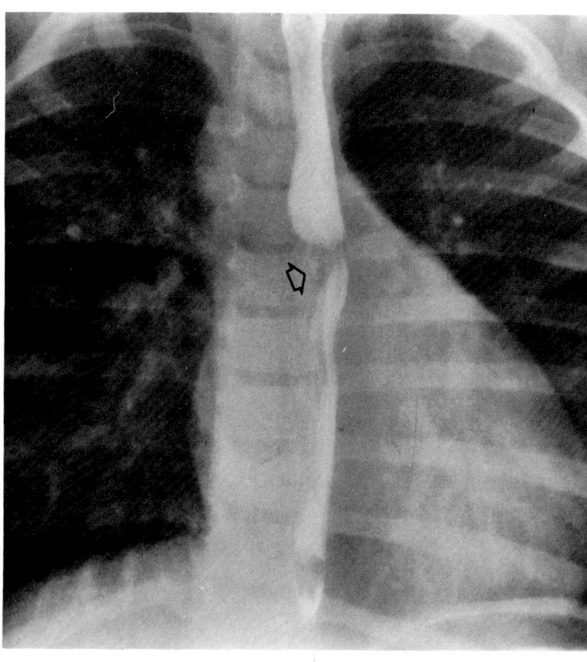

B

**Figure 21.19. Persistent truncus arteriosus.** (A) A plain oblique chest radiograph demonstrates a characteristic indentation (arrow) on the posterior wall of the esophagus; the indentation is somewhat lower than that usually seen with an aberrant left subclavian artery. (B) A frontal view demonstrates a right-sided impression on the esophagus (arrow). Note the right aortic arch.

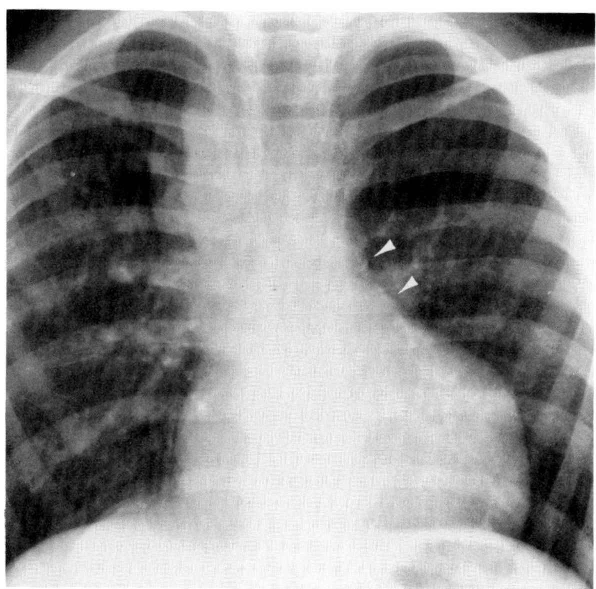

**Figure 21.20. Persistent truncus arteriosus.** A frontal chest film shows the characteristic concave appearance of the pulmonary outflow tract (arrowheads). Note the right-sided aortic arch.

sistent truncus arteriosus is the only condition in which there is a combination of increased pulmonary vascularity and a concave pulmonary artery. Because it is working against systemic pressures, the right ventricle hypertrophies and may become markedly enlarged, resulting in elevation of the cardiac apex simulating the tetralogy of Fallot. Right ventricular prominence in truncus arteriosus is usually more striking than in the tetralogy of Fallot; in addition, the pulmonary vascularity is increased in the former but decreased in the latter. Increased pulmonary venous flow to the left side of the heart causes both the left atrium and the left ventricle to dilate and enlarge. If pulmonary hypertension develops, blood flow through the pulmonary circulation diminishes and the size of the left-to-right shunt decreases.

In the type IV anomaly, in which the vessels supplying the lungs do not arise directly from the truncus, the pulmonary vascularity is strikingly diminished. The bronchial arteries supplying the lungs often appear as small vessels extending outward from the hila in a fine brushlike pattern. Decreased pulmonary vascularity can also develop in patients with hypoplasia of one or both of the pulmonary arteries arising directly from the truncus. Bronchial collaterals can become quite large and can produce discrete indentations on the posterior wall of the esophagus at a level somewhat lower than the impression seen with an aberrant left subclavian artery.

Angiocardiography with a catheter in the right ventricle demonstrates a common trunk and rapid filling of both the pulmonary arteries and the aorta. This technique also demonstrates the precise site of origin of the pulmonary arteries or the presence of bronchial collaterals supplying the pulmonary circulation.[1,20–22,83]

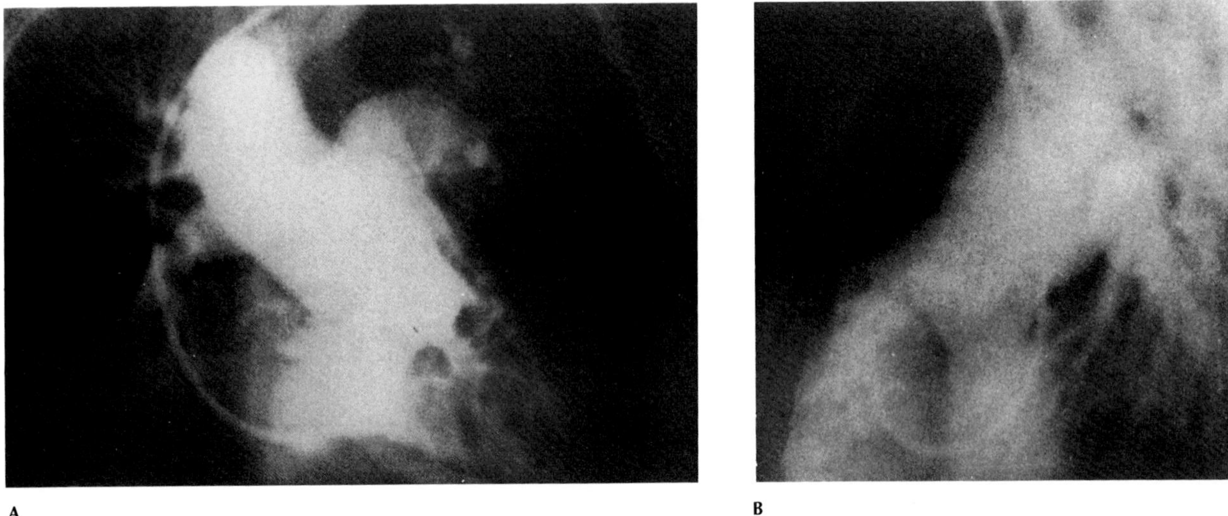


**Figure 21.21. Persistent truncus arteriosus.** (A) Frontal and (B) left lateral angiocardiograms obtained after the injection of contrast material into the right ventricle demonstrate the large truncus arteriosus. Both ventricles are enlarged, and a ventricular septal defect is present immediately below the truncus. A single pulmonary trunk arises from the left lateral aspect of the common arterial trunk. The left pulmonary artery is higher than the right. The aortic arch is on the right, and the brachiocephalic vessels are mirror images of the normal ones. (From Hallermann, Kincaid, Tsakiris, et al,[20] with permission from the publisher.)

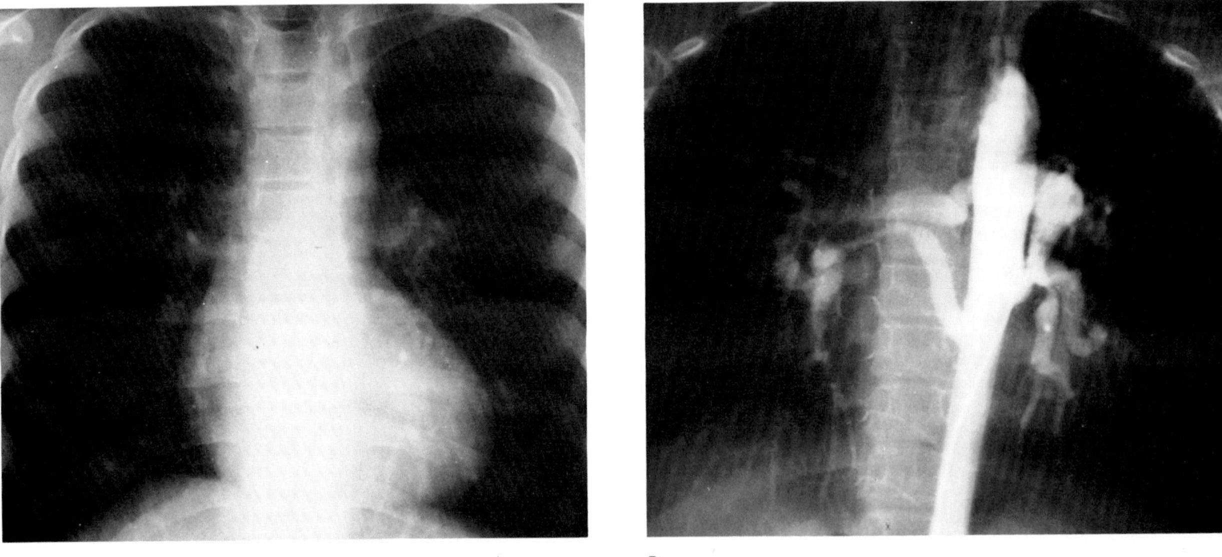


**Figure 21.22. Persistent truncus arteriosus (type IV).** (A) A plain chest radiograph shows the pulmonary vascularity to be strikingly diminished. (B) Angiography shows that most of the supply to both lungs originates from two large arteries arising from the descending aorta. (From Cooley and Schreiber,[83] with permission from the publisher.)

## Pseudotruncus

Pseudotruncus arteriosus refers to a condition in which a single vessel arises from the heart but is accompanied by a remnant of atretic pulmonary artery. The single vessel draining blood from the heart is the aorta, not a persistent truncus arteriosus; in addition, although the pulmonary artery is atretic, it still arises from the right ventricle and not the truncus. Pseudotruncus is essentially the same as the tetralogy of Fallot with pulmonary atresia (see the section "Tetralogy of Fallot" in this chapter).[1,23]

# OBSTRUCTING VALVULAR AND VASCULAR LESIONS —WITH OR WITHOUT ASSOCIATED RIGHT-TO-LEFT SHUNTS—

## PULMONARY STENOSIS WITH INTACT VENTRICULAR SEPTUM

Congenital pulmonary stenosis is a common anomaly that may be found in isolated form or in combination with other abnormalities. The stenosis is most commonly at the level of the pulmonary valve; supravalvular or subvalvular (infundibular) stenosis can also occur.

Patients with mild degrees of pulmonary stenosis may demonstrate minimal or no abnormality on chest radiographs. More severe stenosis causes systolic overloading and hypertrophy of the right ventricle. This initially produces no change in the heart size, but may eventually cause some lateral displacement of the cardiac apex and soft tissue fullness in the retrosternal space. The right atrium is sometimes enlarged. Pulmonary valvular stenosis is characterized by poststenotic dilatation of the pulmonary artery, often associated with dilatation of the left main pulmonary artery. This poststenotic dilatation of the pulmonary artery is not seen in patients with supravalvular or subvalvular (infundibular) pulmonary stenosis. The pulmonary vascularity is usually normal in all types of pulmonary stenosis; there occasionally may be increased pulmonary blood flow to the left lung, possibly related to the preferential streaming of blood through the stenotic pulmonary valve into the left main pulmonary artery.

In adolescents and young adults, especially women, it is essential to differentiate the poststenotic dilatation of the pulmonary artery caused by pulmonary valvular stenosis from idiopathic dilatation of the pulmonary artery, which is a normal radiographic appearance at this age.

Angiocardiography with the injection of contrast material into the right ventricle demonstrates the site and severity of pulmonary stenosis and can document any coexistent cardiac malformations.[1,24,25,83]

## TETRALOGY OF FALLOT

Tetralogy of Fallot is the most common cause of cyanotic congenital heart disease beyond the immediate neonatal period. The four components of this malformation are (1) high ventricular septal defect, (2) obstruction to right ventricular outflow (usually infundibular pulmonary stenosis), (3) overriding of the aortic orifice above the ventricular defect, and (4) right ventricular hypertrophy. Pulmonary stenosis causes an elevation of pressure in the right ventricle and hypertrophy of that chamber. The ventricular septal defect and the overriding of the aorta produce right-to-left shunting of unoxygenated venous blood from the right ventricle into the systemic circulation.

The severity of symptoms is related to the degree of the pulmonary stenosis; patients with minimal pulmonary stenosis ("pink," or balanced, tetralogy) are gen-

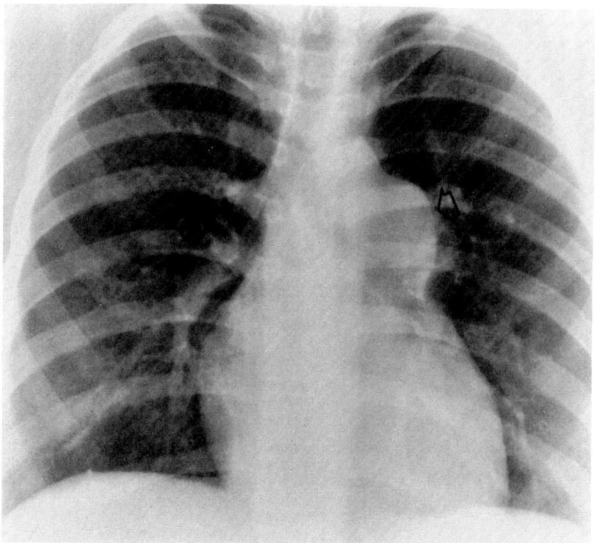

**Figure 21.23. Valvular pulmonary stenosis.** Severe poststenotic dilatation of the pulmonary outflow tract (arrow). The heart size and pulmonary vascularity remain within normal limits.

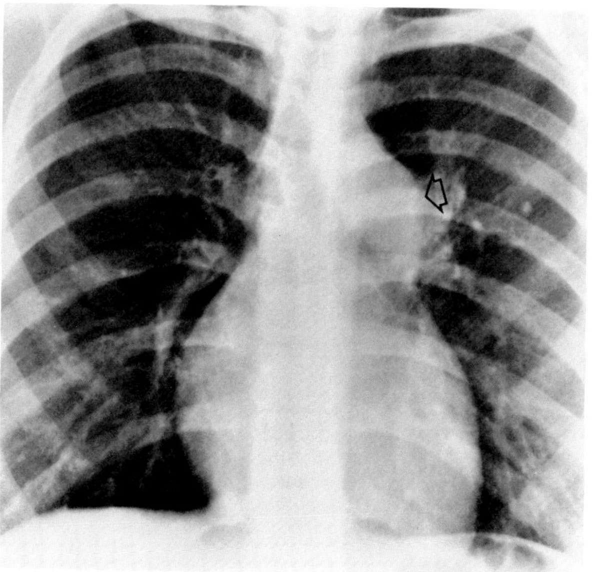

**Figure 21.24. Idiopathic dilatation of the pulmonary artery.** A plain chest radiograph in a normal young woman demonstrates a prominent pulmonary artery (arrow), which simulates the poststenotic dilatation associated with pulmonary valvular stenosis.

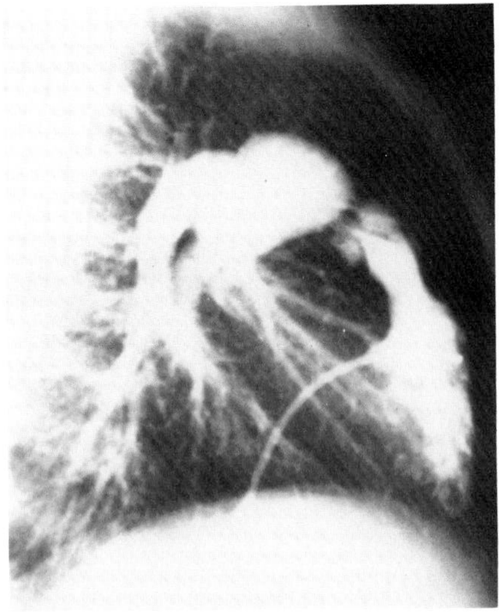


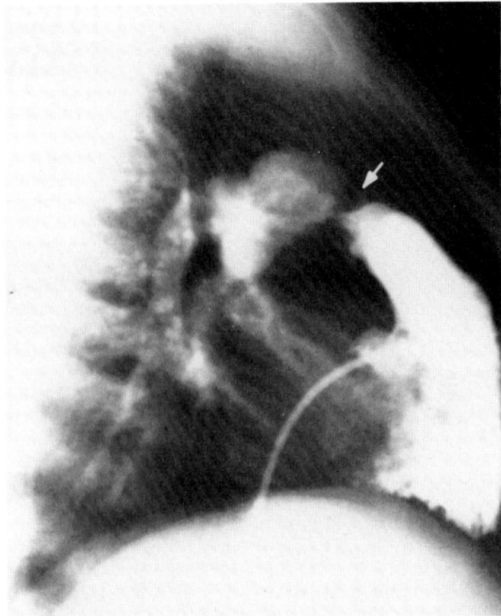


**Figure 21.25. Supravalvular pulmonary stenosis.** Contrast injection into the right ventricle in (A) systole and (B) diastole demonstrates a slightly thickened, incompletely opened, and probably dysplastic pulmonary valve. About 2 cm distal to the valve is an area of pronounced supravalvular stenosis (arrow); just distal to this is an area of poststenotic dilatation that involves only the pulmonary trunk. The right and left pulmonary arteries are not dilated. Supravalvular pulmonary stenosis may exist as an isolated lesion or be combined with other forms of congenital heart disease, in this case with mild to moderate valvular pulmonary stenosis. (From Cooley and Schreiber,[83] with permission from the publisher.)

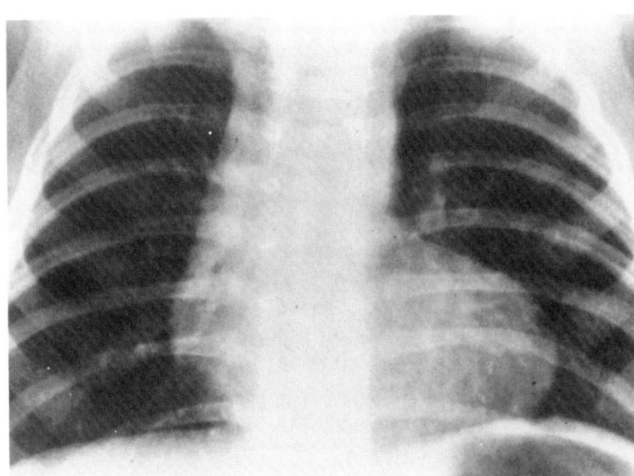

**Figure 21.26. Tetralogy of Fallot.** A plain chest radiograph demonstrates lateral displacement and upward tilting of a prominent left cardiac apex, producing the characteristic *coeur en sabot* appearance.

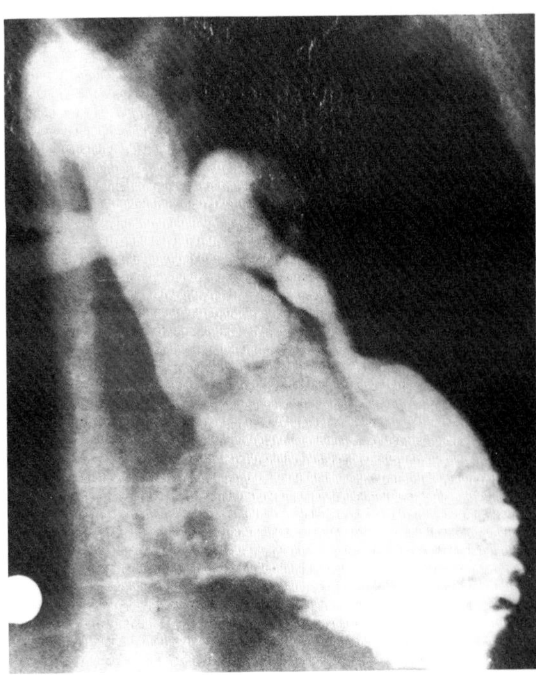


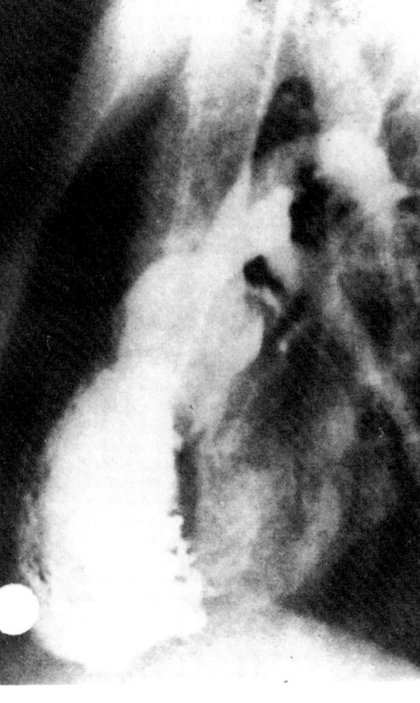


**Figure 21.27. Tetralogy of Fallot.** (A) Frontal and (B) lateral angiocardiograms demonstrate moderate stenosis of the ostium infundibulum. The right ventricular output resistance is severely elevated because of the long, narrow infundibular chamber. At surgery, the pulmonary valve was found to be severely stenotic. (From Kirklin,[28] with permission from the publisher.)

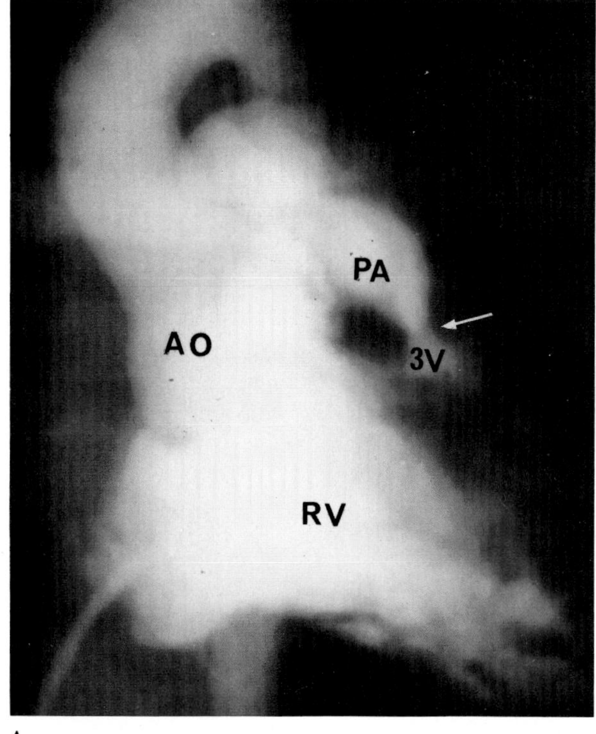


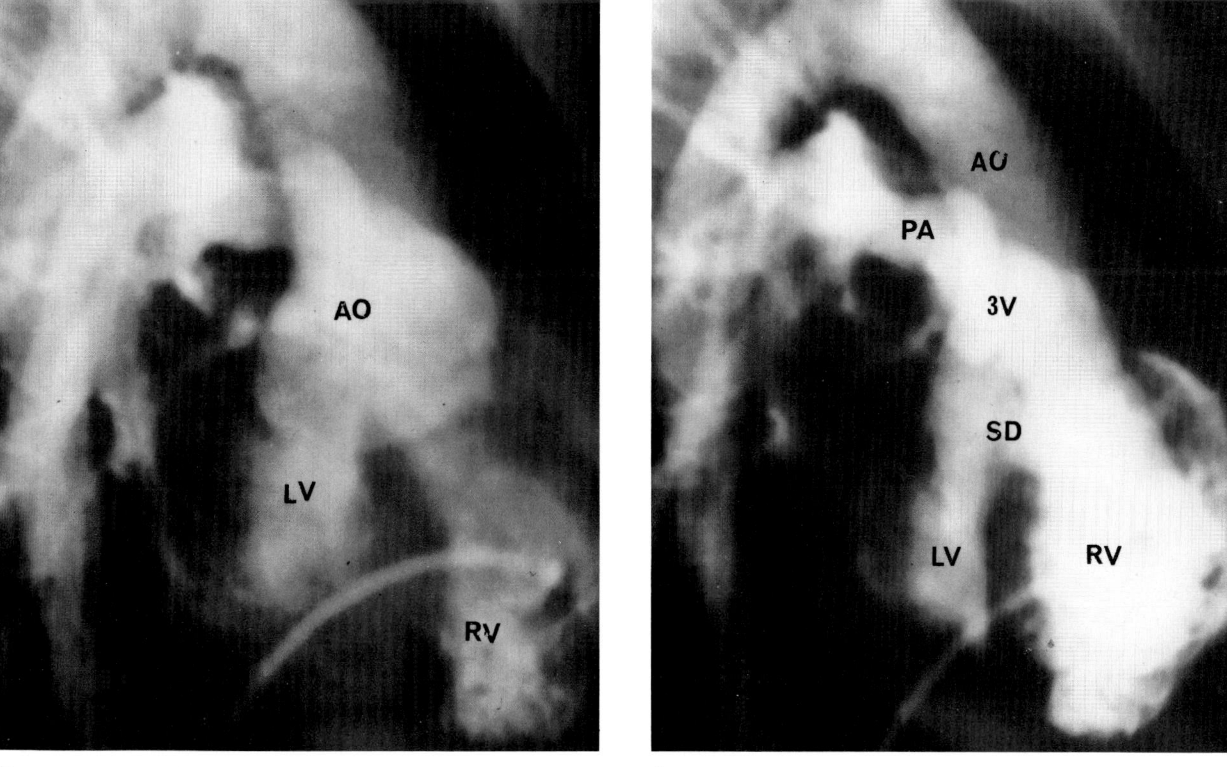


**Figure 21.28. Tetralogy of Fallot.** (A) Frontal and (B, C) lateral projections from a selective right ventricular angiogram demonstrate the components of the tetralogy of Fallot with infundibular and mild valvular pulmonary stenosis. The aorta is much larger than the pulmonary artery and straddles the ventricular septum. AO = aorta; PA = pulmonary artery; SD = septal defect; VS = ventricular septum; RV = right ventricle; 3V = infundibular chamber. The arrow indicates the position of the pulmonary valve. (From Cooley and Schrieber,[83] with permission from the publisher.)

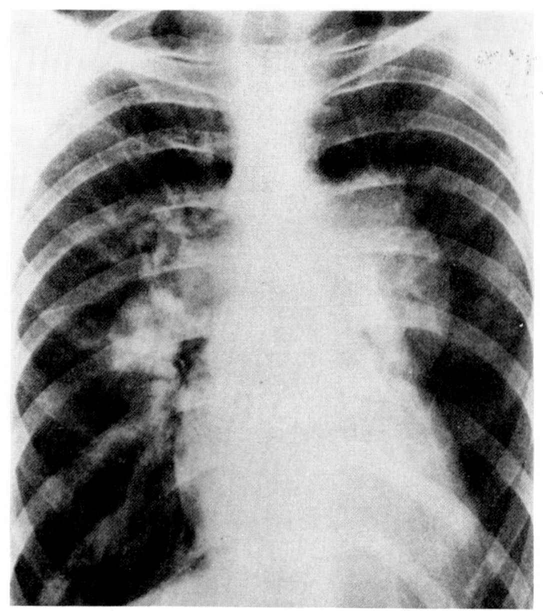

**Figure 21.29. Tetralogy of Fallot.** Pulmonary hypertension following a Blalock-Taussig shunt. A frontal chest radiograph demonstrates severe pulmonary hypertension with massive enlargement of the left and right main pulmonary arteries. (From Marr, Giargiana, and White,[26] with permission from the publisher.)

erally asymptomatic and have only subtle radiographic abnormalities. Because the pulmonary stenosis is usually infundibular, the pulmonary artery is flat or concave; poststenotic dilatation of the pulmonary artery develops only in the infrequent patient with valvular stenosis. The pulmonary vascularity almost always is diffusely decreased. Associated hypoplasia or total absence of the left pulmonary artery may rarely produce unequal pulmonary vascularity. With severe pulmonary stenosis, collateral blood flow through the lungs via dilated bronchial arteries can cause a brushlike reticular appearance.

Enlargement of the right ventricle causes upward and lateral displacement of the cardiac apex and a rounded configuration of the lower left margin of the heart. This results in the classic *coeur en sabot* appearance resembling the curved-toe portion of a wooden shoe. The overall heart size, however, is usually normal.

A right-sided aorta is present in about one-fourth of the patients with the tetralogy of Fallot. The size of the aorta depends on the degree of right outflow tract obstruction. Severe pulmonary stenosis, which essentially forces the aorta to drain both ventricles, causes a pronounced bulging of the ascending aorta and prominence of the aortic knob.

Angiocardiography demonsrates the infundibular pulmonary stenosis and the simultaneous filling of the aorta and pulmonary vessels from the right ventricle. It can also demonstrate associated cardiovascular anomalies, especially stenoses of the pulmonary artery or its branches.

The establishment of a subclavian-pulmonary artery anastomosis (Blalock-Taussig shunt) is one type of surgical correction of the tetralogy of Fallot. This may produce pulmonary hypertension and postoperative enlargement of the central pulmonary arteries. Rib notching similar to that occurring in coarctation of the aorta may develop on the side of the shunt.[1,26–29,83]

## TRILOGY OF FALLOT

The trilogy of Fallot refers to the combination of pulmonary valvular stenosis with an intact ventricular septum and an interatrial shunt (patent foramen ovale or true atrial septal defect). The trilogy of Fallot differs from the tetralogy of Fallot in that the ventricular septum is intact in the former; thus shunting can occur only at the atrial level. Because of increased pressure on the right side of the heart due to the pulmonary stenosis, the interatrial shunt is right to left and the patient is cyanotic.

Radiographically, the trilogy of Fallot presents the unusual combination of poststenotic dilatation of the pulmonary artery and diminished pulmonary vascularity. The heart often remains normal in size, though there is usually some evidence of right ventricular hypertrophy. The associated development of tricuspid insufficiency may cause marked dilatation of the right atrium and right ventricle.

Angiocardiography with the injection of contrast material into the vena cava or right atrium demonstrates the atrial defect with rapid filling of the left side of the heart and almost simultaneous opacification of the pulmonary artery and aorta.[1]

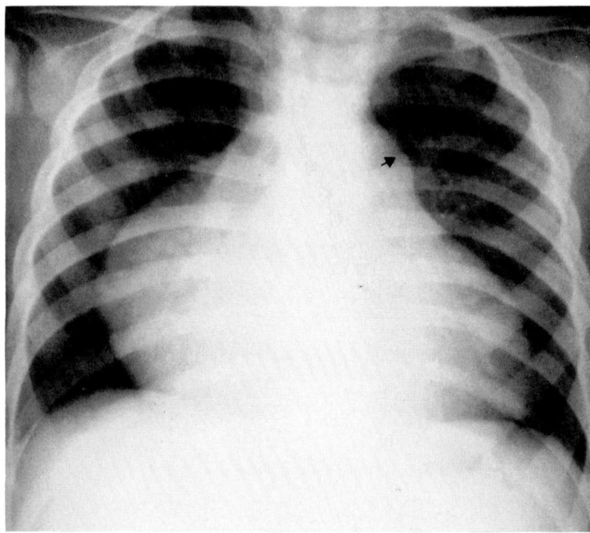

**Figure 21.30. Trilogy of Fallot.** There is enormous right atrial and moderate right ventricular enlargement. The pulmonary artery (arrow) shows marked poststenotic dilatation, while the pulmonary vascularity is decreased. (From Swischuk,[1] with permission from the publisher.)

## EBSTEIN'S ANOMALY

Ebstein's anomaly consists of downward displacement of an incompetent tricuspid valve into the right ventricle so that the upper portion of the right ventricle is effectively incorporated into the right atrium. Because the atrialized portion of the ventricle has a very thin wall and feeble muscular contractions, the large right atrium cannot empty properly. Functional obstruction to right atrial emptying produces increased right atrial pressures and a right-to-left atrial shunt, usually through a patent foramen ovale.

The clinical and radiographic findings depend on the size of the atrial septal defect, the degree of right ventricular atrialization, and the extent of right heart hypoplasia. In patients with a minimal anomaly, the cardiac size and the pulmonary vascularity may be essentially normal. In severe cases, there is marked nonspecific cardiomegaly. Enlargement of the right atrium causes a bulging of the right border of the heart and displaces the right ventricular outflow tract upward and outward, producing a characteristic squared or boxed appearance of the heart. There is a narrow vascular pedicle and a small aortic arch. Decreased right ventricular output associated with the right-to-left shunt produces a flat or concave pulmonary artery segment and a pattern of decreased pulmonary vascularity.

Angiocardiography demonstrates slow filling and emptying of the markedly enlarged right atrium and flow through the interatrial shunt. The selective injection of contrast material into the right ventricle can identify the level of the tricuspid valve and indicate whether tricuspid insufficiency is present.[1,30–32,83]

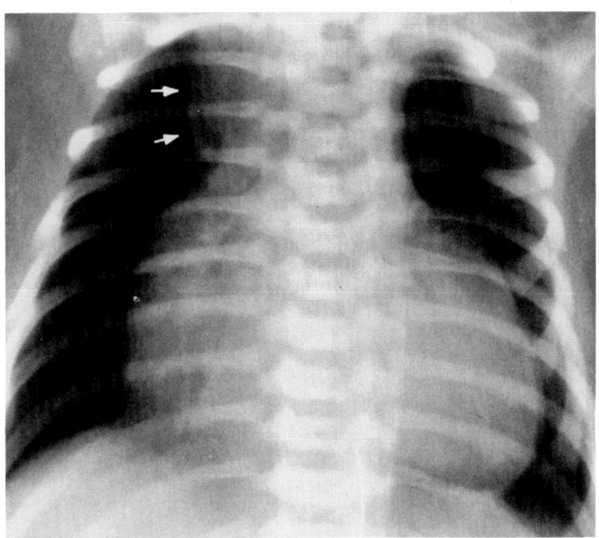

**Figure 21.31. Ebstein's anomaly.** A plain chest radiograph demonstrates enlargement of the right atrium. This enlargement causes upward and outward bulging of the right heart border and a characteristic squared appearance of the heart. Widening of the right side of the superior portion of the mediastinum (arrows) reflects marked dilatation of the superior vena cava due to right ventricular failure.

## TRICUSPID ATRESIA

Atresia of the tricuspid valve is always associated with a right-to-left shunt at the atrial level (patent foramen ovale or true atrial septal defect). A ventricular septal defect or a patent ductus arteriosus is also usually present. The right ventricle, right ventricular outflow tract, and pulmonary artery are hypoplastic. Pulmonary stenosis and/or transposition of the great vessels may be associated anomalies.

The size of the atrial shunt greatly affects the radiographic appearance. A small atrial septal defect causes a marked elevation of right atrial pressures and a striking enlargement of this chamber; in contrast, large atrial shunts result in only minimal right atrial enlargement. Although the left ventricle rather than the right ventricle is enlarged in tricuspid atresia, the left border of the heart is rounded with apparent elevation of the apex. This appearance, which simulates right ventricular enlargement, is due to abnormal rotation of the heart by the markedly enlarged right atrium. The hypoplastic right ventricle may result in flattening of the right border of the heart. Although the left atrium is often enlarged, it is infrequently identified radiographically since this chamber is so malpositioned that it does not rest directly against the esophagus. Most patients have some degree of pulmonary stenosis and thus decreased pulmonary vascularity.

Tricuspid atresia without pulmonary stenosis, usually associated with transposition of the great vessels, appears radiographically as marked cardiomegaly and increased pulmonary vascularity. With transposition, the aorta is small, rather than of normal size or slightly enlarged as in tricuspid atresia without transposition. In patients with transposition, both the right atrium and the left atrium lie to the left of the aorta, and the right atrial appendage lies directly above the left. This produces a characteristic bulge high on the left cardiac border that is associated with flatness of the right border of the heart.

Angiocardiography demonstrates an enlarged right atrium, an atretic tricuspid valve, and rapid opacification of the left side of the heart via a right-to-left shunt. The hypoplastic right ventricle appears as a triangular radiolucent space (right ventricular window) along the inferior margin of the heart.[1,33–35,83]

## COARCTATION OF THE AORTA

Coarctation of the aorta is a congenital anomaly in which there is an area of constriction in the aorta. In the more common "adult" type, the aortic narrowing occurs at or

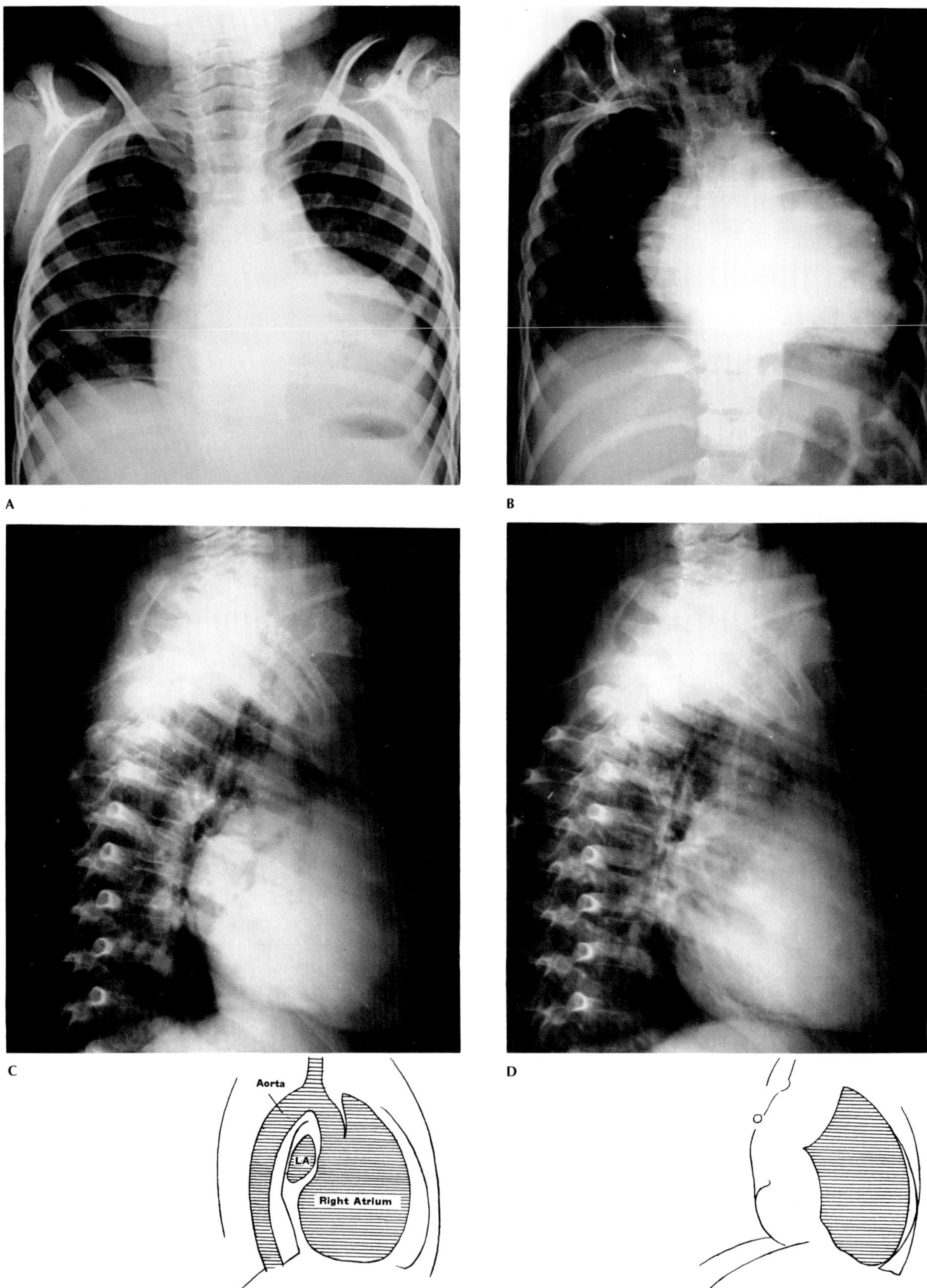


**Figure 21.32. Ebstein's anomaly.** (A) A plain chest radiograph demonstrates considerable enlargement of the heart with an increase in the radiolucency of the lungs. At fluoroscopy, the pulsations of the cardiac borders were diminished. (B) A frontal view from an angiocardiogram demonstrates opacification of almost the entire heart. This appearance was unchanged for several seconds. (C) A lateral view from an angiocardiogram 4 seconds after the injection of contrast material shows a large dilated chamber consisting of the right atrium and that portion of the right ventricle which is associated with the deformity of the tricuspid valve. The left atrium (LA) filled through a patent foramen ovale. (D) Two seconds later, contrast material continues to stagnate in this large chamber. (From Cooley and Schreiber,[83] with permission from the publisher.)

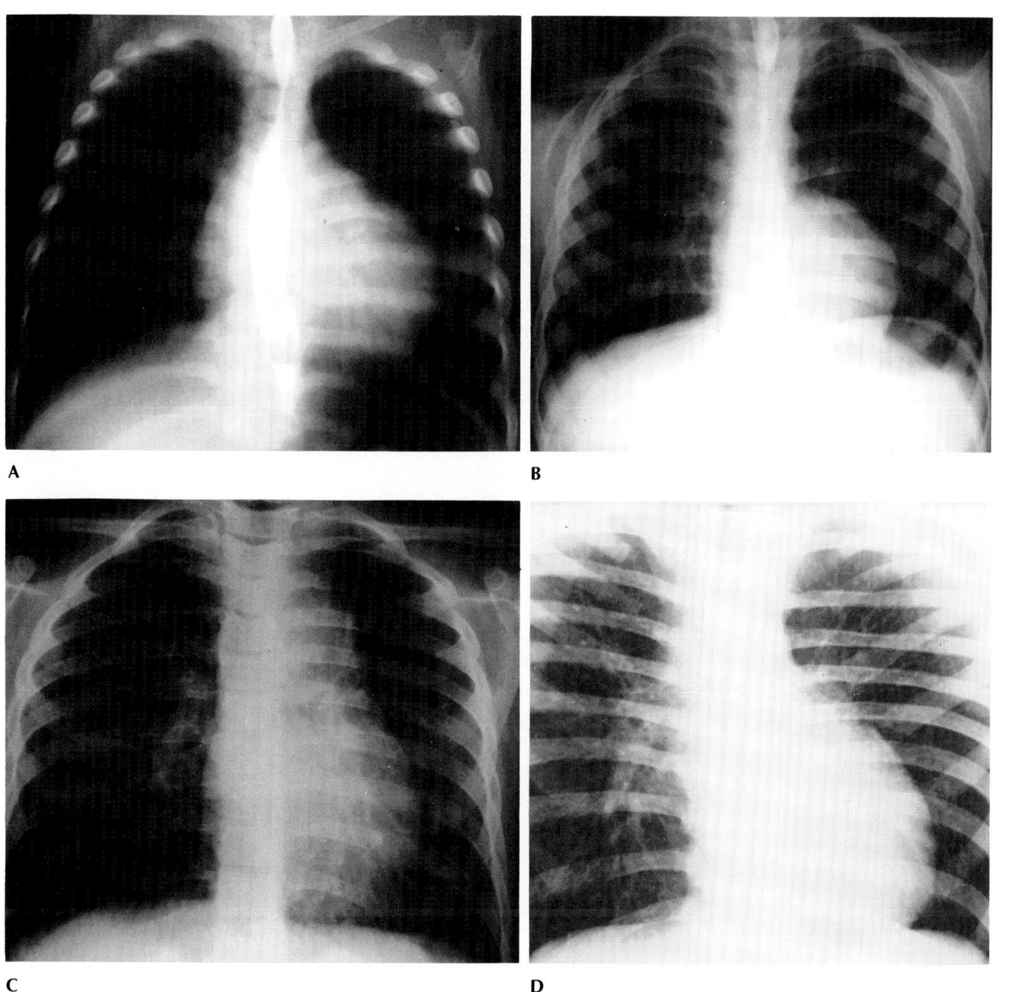

**Figure 21.33. Variable contour of the heart in four different examples of tricuspid atresia.** (A) The left border is somewhat elongated and rounded in the most commonly encountered contour. The lungs are hypovascular. In a smaller number of cases, the left midborder is definitely concave (B). (C) In this case, corrected transposition of the great vessels is associated with tricuspid atresia. (D) The right contour of the heart is flattened; this represents only a small minority of cases. (From Cooley and Schreiber,[83] with permission from the publisher.)

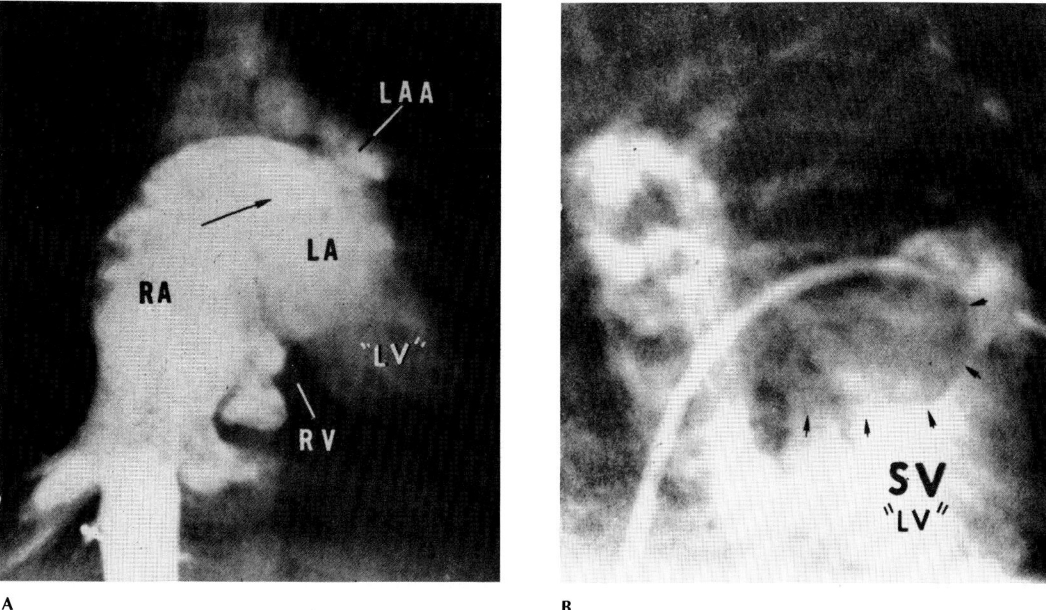


**Figure 21.34. Angiocardiograms in two cases of tricuspid atresia.** (A) Selective right atrial injection shows the typical sequence of filling from right atrium to left atrium to left ventricle. The nonopacified area below and to the left of the right atrium is termed the *right ventricular window* (RV). (B) A selective left ventriculogram (SV-"LV") shows the enlarged anterior leaflet of the mitral valve as a radiolucent filling defect. The valve and annulus (arrows) have an eccentric position. (From Elliott, Van Mierop, Gleason, et al,[14] with permission from the publisher.)

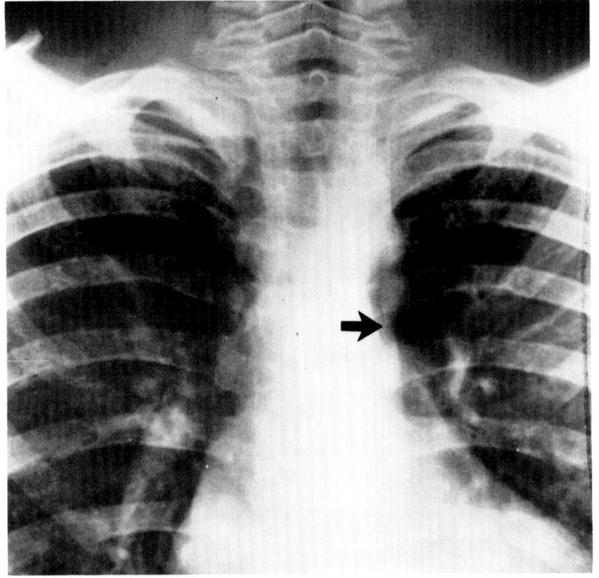

A

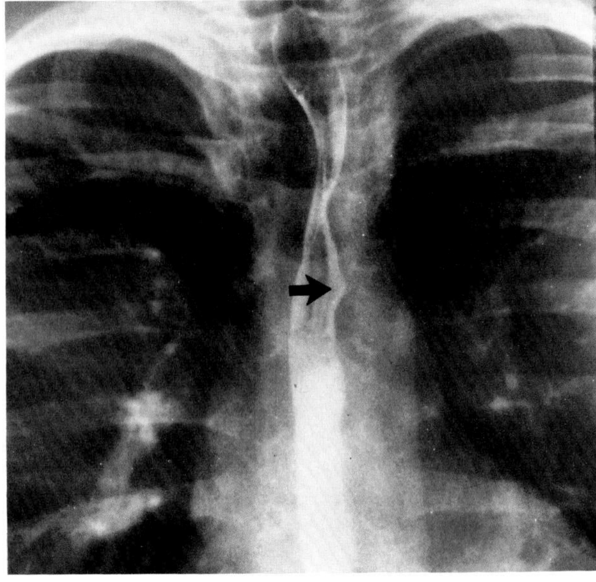

B

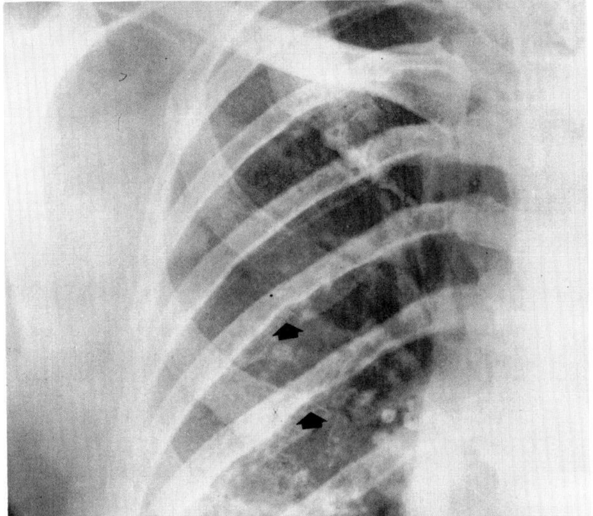

C

**Figure 21.35. Coarctation of the aorta.** (A) A plain chest radiograph demonstrates the figure-3 sign (the arrow points to the center of the 3). The upper bulge represents prestenotic dilatation; the lower bulge represents poststenotic dilatation. (B) An esophogram demonstrates the reverse figure-3 sign (the arrow points to the center of the reverse figure-3). (C) Rib notching. There is notching of the posterior fourth through eighth ribs (the arrows point to examples). (From Swischuk,[1] with permission from the publisher.)

just distal to the level of the ductus arteriosus; in the much less frequent "infantile" variety, a long segment of narrowing lies proximal to the ductus. In the latter type of coarctation, a right-to-left shunt (always a patent ductus arteriosus; often a ventricular septal defect) is present so that blood can be delivered from the pulmonary artery to the descending aorta and systemic circulation. Coarctation is often associated with other forms of congenital heart disease; there is a relatively high incidence in women with Turner's syndrome.

The narrowing of the aorta causes systolic overloading and hypertrophy of the left ventricle. This produces a "well-rounded" appearance of the left cardiac border simulating aortic stenosis. The relative obstruction of aortic blood flow leads to the progressive development of collateral circulation. This is often seen radiographically as rib notching (usually involving the posterior fourth to eighth ribs) due to pressure erosion by dilated and pulsating intercostal collateral vessels. The first and second ribs do not become notched, since their intercostals do not serve as collaterals. Because rib notching rarely appears before the sixth or seventh year, this sign is of no value in assessing the possibility of coarctation of the aorta in children. The dilatation of mammary artery collaterals can produce retrosternal notching.

Coarctation of the aorta often causes two bulges in the region of the aortic knob that produce a characteristic figure-3 sign on plain chest radiographs and a reverse figure-3 or figure-E impression on the barium-filled esophagus. The more cephalad bulge represents dila-

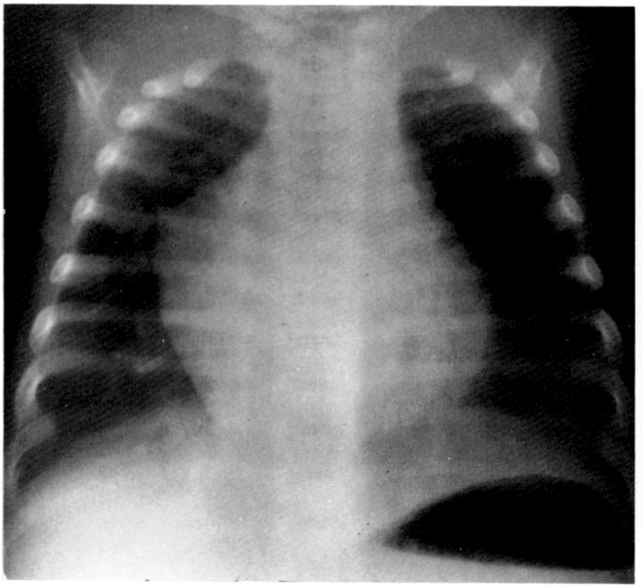
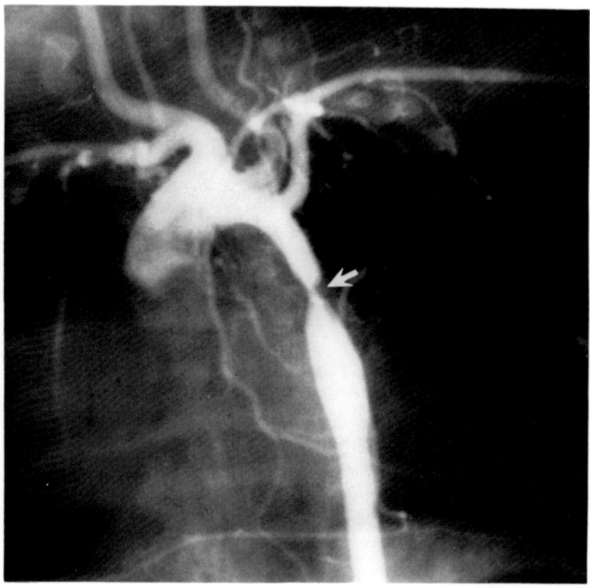

A B

**Figure 21.36. Coarctation of the aorta in infancy.** (A) A frontal chest radiograph of a 2-month-old boy with symptoms of heart failure shows enlargement of the heart with a convex right atrial contour. (B) An aortogram demonstrates a tight stenosis of the aorta (arrow) with no evidence of collateral circulation. (From Cooley and Schreiber,[83] with permission from the publisher.)

tation of the proximal aorta and the base of the left subclavian artery (prestenotic dilatation); the lower bulge reflects poststenotic aortic dilatation.

In the infantile type of coarctation, increased pulmonary blood flow and right ventricular systolic overloading are added to the pattern of left ventricular hypertrophy. Right ventricular enlargement, prominence of the pulmonary artery, and increased pulmonary vascularity are usually seen. The aorta is normal or small, and there is no radiographic evidence of rib notching or the figure-3 or figure-E signs.

Aortography can accurately localize the site of obstruction, determine the length of the coarctation, and identify any associated cardiac malformations.[1,36–38,83]

## Pseudocoarctation of the Aorta

Pseudocoarctation of the aorta is an uncommon condition in which a tortuous, dilated, and kinked aortic arch may produce radiographic signs suggestive of true coarctation of the aorta. On frontal chest radiographs, two bulges in the region of the aortic knob can produce a well-demarcated figure-3 sign. These bulges, which represent the dilated portions of the aorta just proximal and distal to the kink, may also indent the barium-filled esophagus. The upper bulge is usually somewhat higher than the normal aortic knob and can simulate a left superior mediastinal tumor. The indentation of the aorta at the site of kinking can often be identified on left anterior oblique and lateral views. Because there is no obstruction or hemodynamic abnormality, there are no intercostal collateral vessels or rib notching.

Aortography demonstrates that the aorta is extremely kinked without an area of true coarctation or a pressure gradient.[39–41]

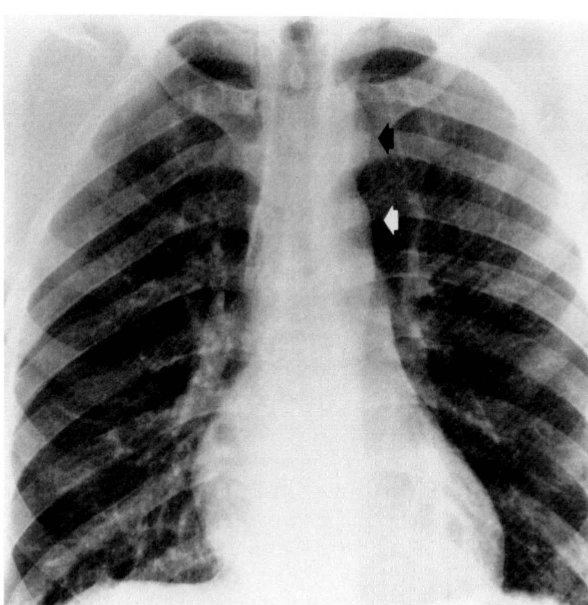

**Figure 21.37. Pseudocoarctation of the aorta.** A frontal view of the chest demonstrates two bulges (arrows) producing a well-demarcated figure-3 sign in the region of the aortic knob. The upper bulge (black arrow) is higher than the normal aortic knob and simulates a mediastinal mass. Because there is no hemodynamic abnormality, the heart is normal in size and there is no rib notching.

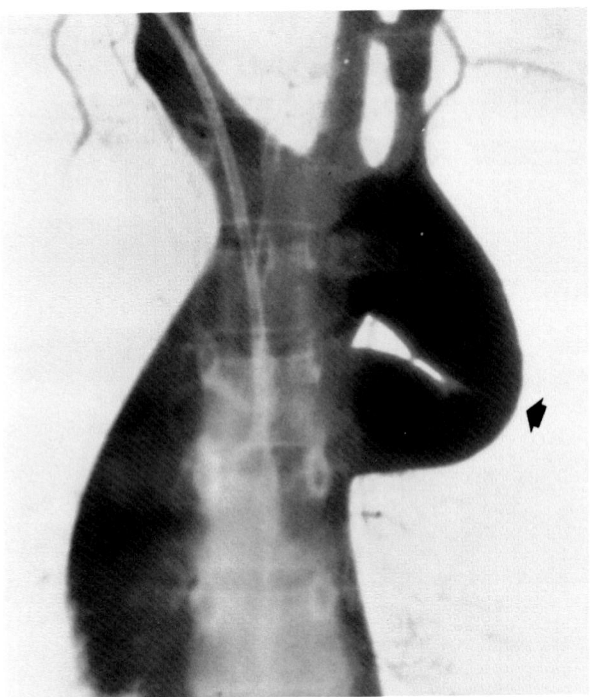

**Figure 21.38. Pseudocoarctation of the aorta.** An aortogram demonstrates extreme kinking of the descending aorta (arrow) without an area of true coarctation.

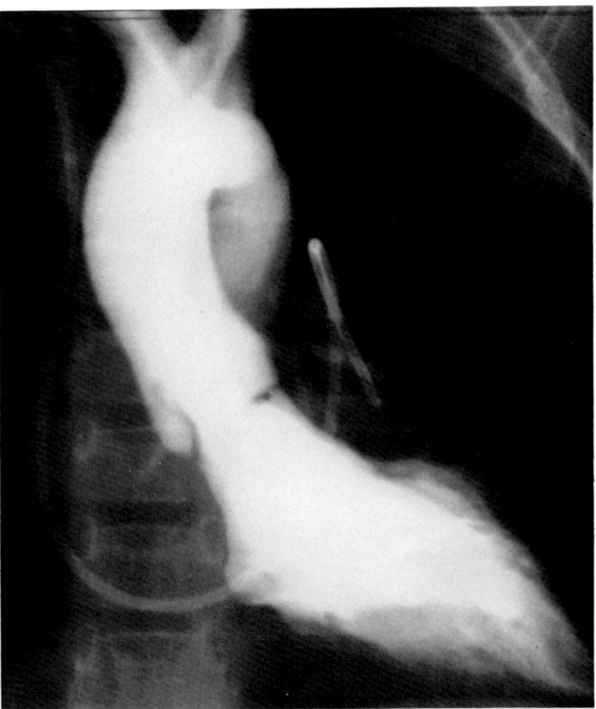

**Figure 21.39. Valvular aortic stenosis.** A frontal projection from an angiocardiogram during systole shows irregular thickening of aortic valve leaflets and relative rigidity of the left coronary cusp. (From Takekaw, Kincaid, Titus, et al,[42] with permission from the publisher.)

## CONGENITAL AORTIC STENOSIS

Congenital aortic stenosis can be divided into valvular, subvalvular, and supravalvular types. Valvular stenosis is caused by cusp deformity and is often associated with a bicuspid valve. Subvalvular aortic stenosis may be the result of a thin membranous diaphragm or a fibrous ring encircling the left ventricular outflow tract just beneath the base of the aortic valve or, more commonly, an idiopathic hypertrophy of the left ventricular outflow tract and interventricular septum that obstructs blood flow in systole. In supravalvular aortic stenosis, the localized or diffuse narrowing of the ascending aorta originates just above the level of the coronary arteries at the superior margin of the sinuses of Valsalva. In this type of aortic stenosis, the coronary arteries are subjected to the elevated pressures of the left ventricle and are often dilated and tortuous.

Regardless of the precise lesion, congenital aortic stenosis causes systolic overloading and hypertrophy of the left ventricle. This produces an increased convexity or prominence of the left border of the heart. The overall heart size is often normal or only slightly enlarged; substantial cardiomegaly reflects left ventricular failure and dilatation. The pulmonary vascularity is normal.

Valvular aortic stenosis often causes poststenotic dilatation of the ascending aorta, which produces bulging or widening of the right superior mediastinal silhouette on frontal views. This finding is infrequent and less prominent in subvalvular and supravalvular stenosis. Calcification of a stenotic aortic valve increases with aging and usually indicates a high-grade narrowing.

Angiocardiography can demonstrate the precise site and severity of congenital aortic stenosis.[1,42–47]

## HYPOPLASTIC LEFT-SIDED HEART SYNDROME

The hypoplastic left-sided heart syndrome consists of several conditions in which underdevelopment of the left side of the heart is related to an obstructive lesion (stenosis or atresia of the mitral valve, aortic valve, or aortic arch). A severe impairment of the blood flow from the left side of the heart causes marked pulmonary venous hypertension; thus this syndrome is one of the causes of congestive heart failure in the first week of life. Interstitial edema may produce a diffuse reticular pattern in the lungs. In addition to increased pulmonary vascularity, there is progressive cardiomegaly with the heart having a somewhat globular or oval-shaped configuration due to a combination of right atrial and right ventricular enlargement.[1,48–51,83]

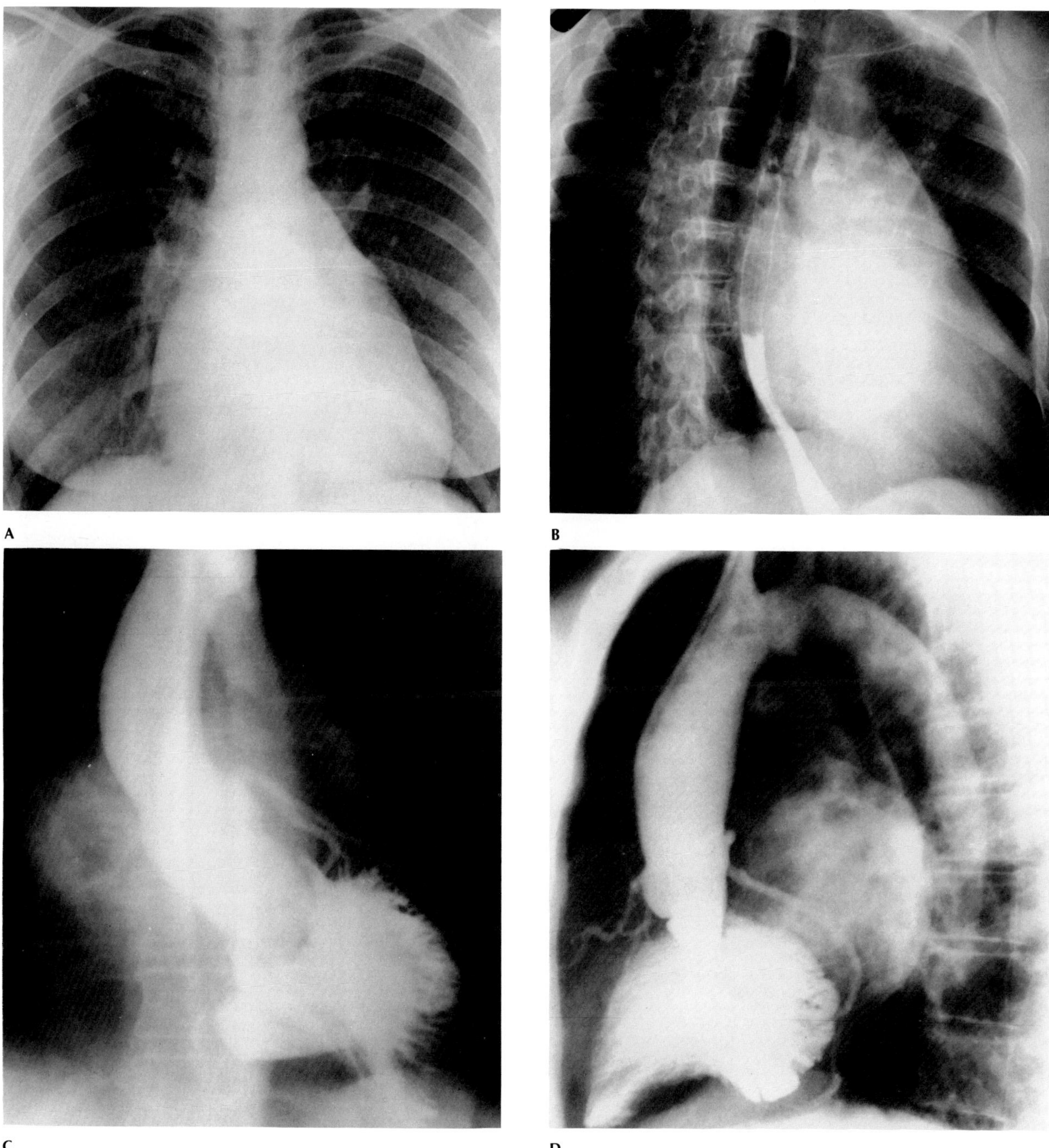


**Figure 21.40. Subaortic stenosis.** (A) A plain chest radiograph shows moderate enlargement of the left ventricle and left atrium. (B) An esophogram in the right anterior oblique projection shows moderate enlargement of the left atrium. (C) Frontal and (D) lateral views from an angiocardiogram demonstrate a muscular ridge protruding from the upper portion of the ventricular septum (arrow); the ridge is about 2 cm below the aortic valve and encroaches on the outflow tract of the left ventricle. The left ventricular outflow tract in diastole has a typical inverted-cone shape (B). There is moderate poststenotic dilatation of the ascending aorta and marked hypertrophy of the left ventricular wall and papillary muscles. Mitral insufficiency also is evident. (From Takekaw, Kincaid, Titus, et al,[42] with permission from the publisher.)

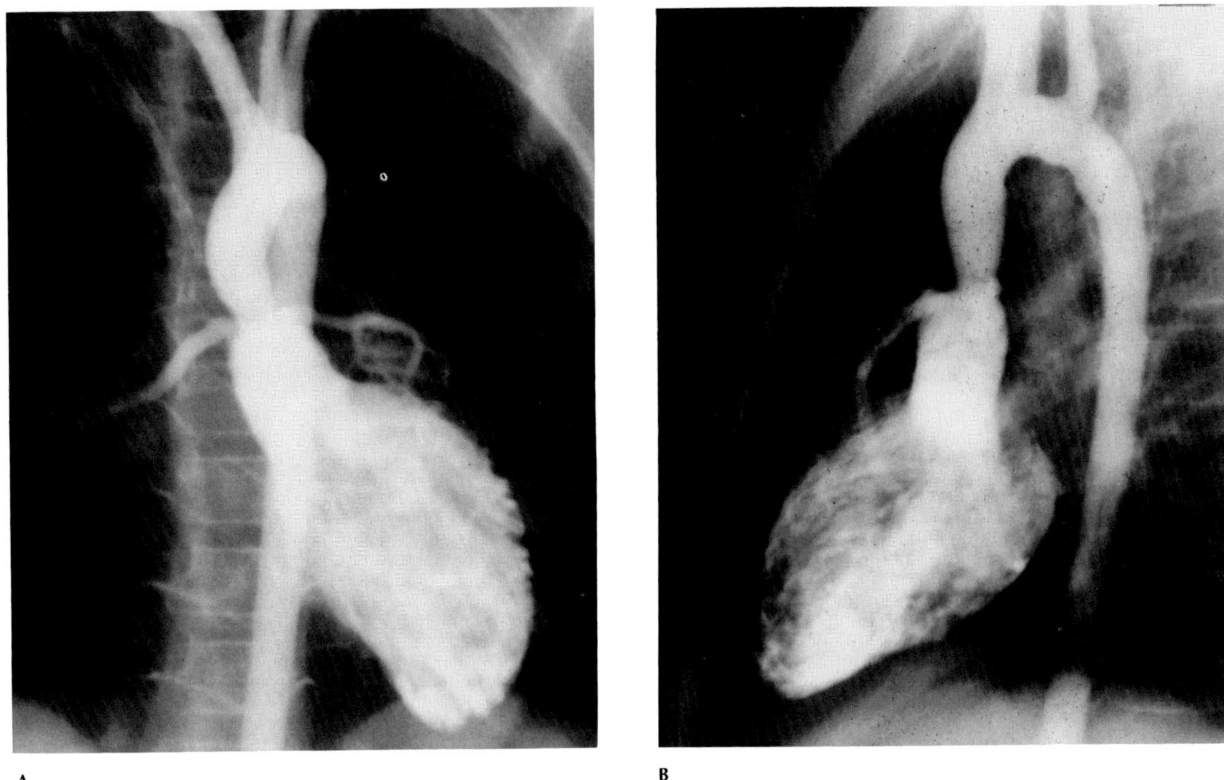

A B

**Figure 21.41. Supravalvular aortic stenosis.** (A) Frontal and (B) lateral views from an angiocardiogram during diastole demonstrate a narrowed segment located just above the coronary ostia. (From Takekaw, Kincaid, Titus, et al,[42] with permission from the publisher.)

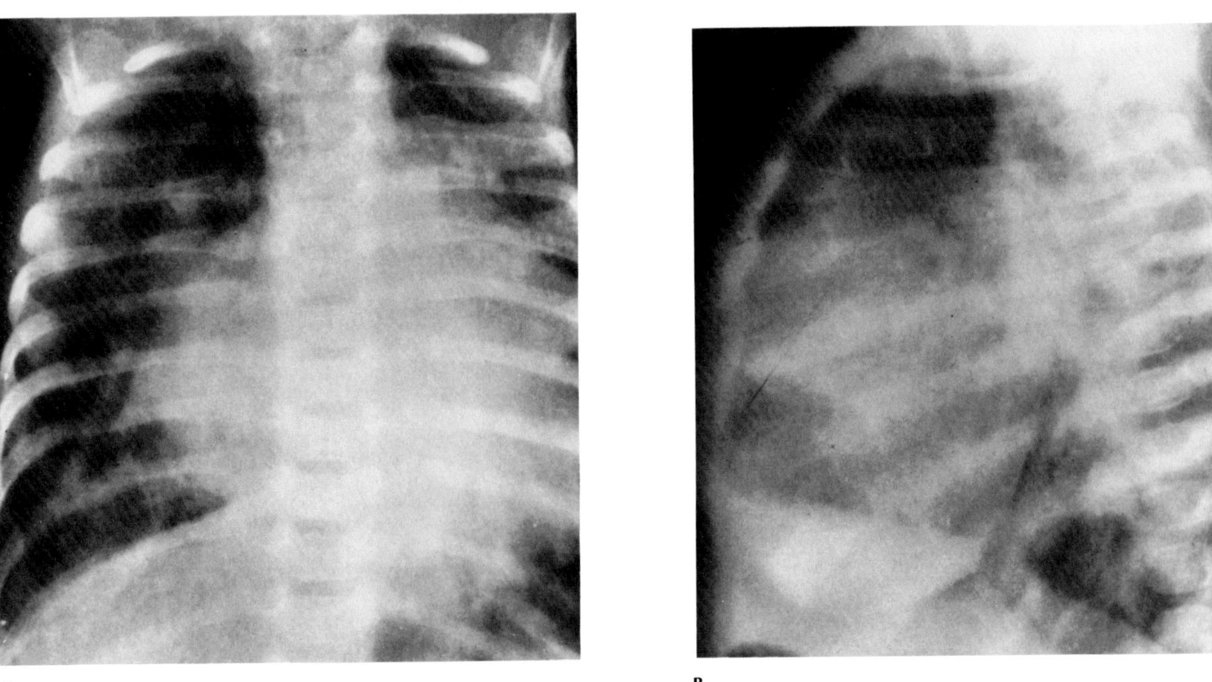

A B

**Figure 21.42. Hypoplastic left-sided heart.** (A) Frontal and (B) lateral views of the chest demonstrate gross cardiomegaly and severe pulmonary venous congestion. Note the enlargement of the right atrium and right ventricle.

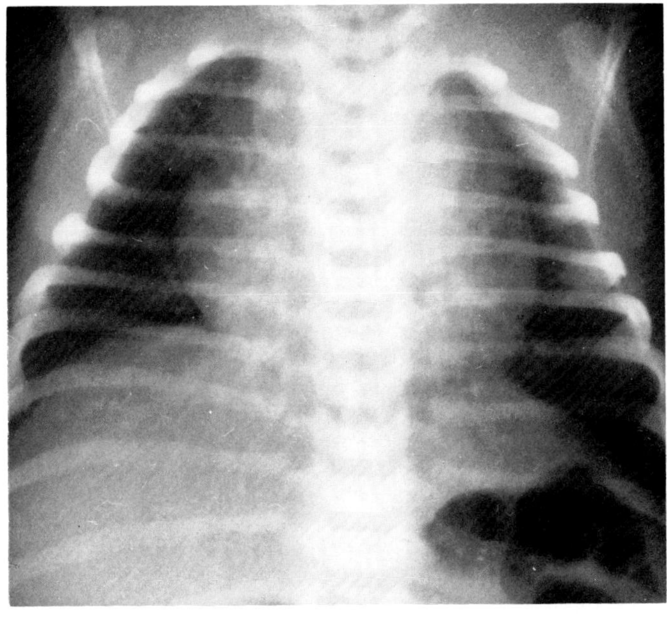
A

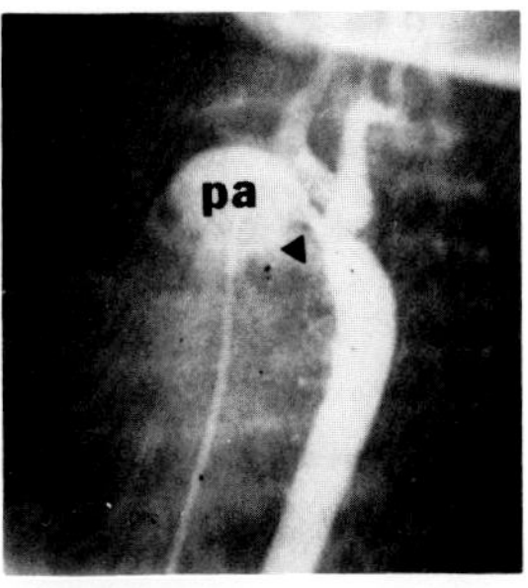

B

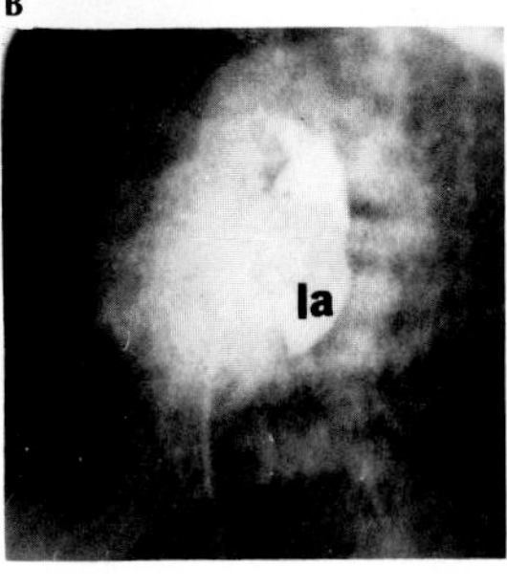

C

**Figure 21.43. Hypoplastic left-sided heart.** (A) A plain chest radiograph demonstrates globular cardiomegaly in a 2-day-old infant with mitral atresia, a small high ventricular septal deflect, and hypoplasia of the aortic valve and ascending aorta. (B) An angiocardiogram with the injection of contrast material into a dilated pulmonary artery (pa) shows a narrow ductus arteriosus (arrowhead) and a small aortic arch and brachiocephalic vessels. (C) Following the injection of contrast material into the left atrium (la), there is rapid passage of contrast material into the right side of the heart and the pulmonary artery. (From Cooley and Schreiber,[83] with permission from the publisher.)

## MALFORMATIONS OBSTRUCTING PULMONARY VENOUS FLOW

### Mitral Stenosis

Congenital mitral stenosis is an uncommon anomaly. The radiographic findings are similar to those of acquired mitral stenosis and include enlargement of the left atrium and pulmonary venous congestion. Secondary pulmonary hypertension causes enlargement of the pulmonary artery and hypertrophy of the right ventricle.

### Cor Triatriatum

Cor triatriatum is a rare anomaly in which an incomplete fibromuscular diaphragm divides the left atrium into a posterosuperior chamber, into which the pulmonary veins drain, and an anteroinferior chamber, which communicates with the mitral valve and the atrial appendage. The diaphragm contains one or more openings, the size and number of which determine the degree of obstruction to pulmonary venous return. The radiographic appearance is similar to that of congenital mitral stenosis, though left atrial enlargement is usually absent. Pulmonary arteriography demonstrates the double-chambered left atrium and the obstructing membrane.

### Congenital Pulmonary Vein Stenosis

In this malformation, localized narrowing at or near the junctions of the pulmonary veins and the left atrium, or hypoplasia of individual pulmonary veins, produces the typical radiographic appearance of pulmonary venous congestion.[52–55]

# ABNORMALITIES IN THE ORIGINS OF THE GREAT ARTERIES AND VEINS: THE TRANSPOSITIONS

## TRANSPOSITION OF THE GREAT ARTERIES

Failure of normal spiraling of the primitive truncus arteriosus leads to a spectrum of abnormalities of positioning of the great arteries. The aorta always arises from the right ventricle to the right of and anterior to the pulmonary artery. The pulmonary artery, however, can arise in three ways: solely from the left ventricle (complete transposition of the great arteries), from both the right and the left ventricles (Taussig-Bing anomaly), or from the right ventricle with the aorta (double-outlet right ventricle).

Angiocardiography can demonstrate the abnormal positions of the great vessels and the site of origin of the pulmonary artery.

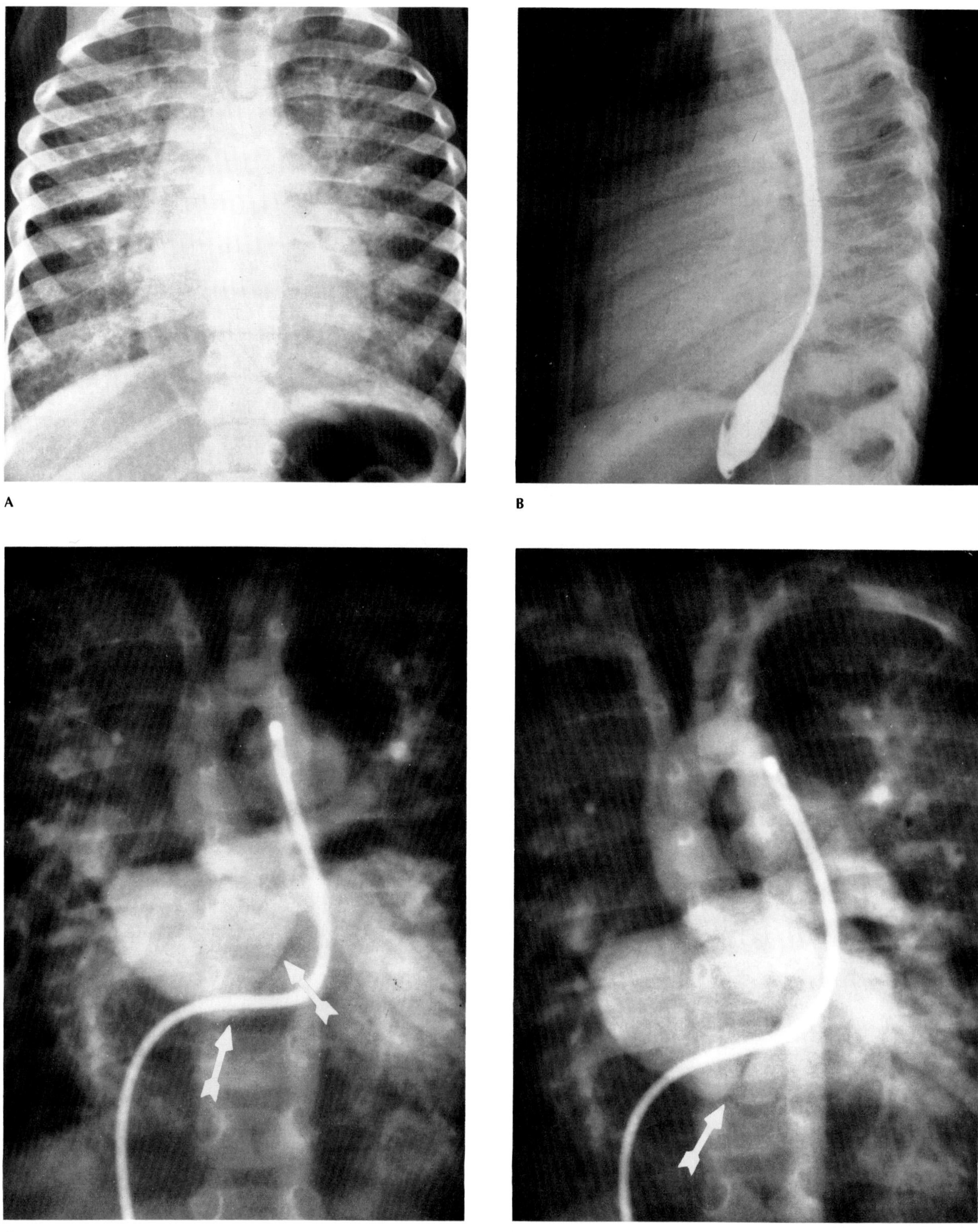

**Figure 21.44. Cor triatriatum.** (A) Frontal chest radiograph shows pulmonary edema, small pleural effusions, and cardiomegaly with enlargement of the right heart and central pulmonary arteries. Widening of the subcarinal angle suggests some left atrial enlargement. (B) A lateral esophogram demonstrates posterior displacement due to considerable left atrial enlargement. (C) An angiocardiogram in ventricular diastole shows opacification of the pulmonary veins and the left side of the heart. Note the enlarged dorsal left atrial compartment overlying the spine and extending to the right, separated by a curved membrane (arrows) from the smaller left atrial compartment containing the mitral valve. (D) A film taken in late ventricular systole shortly after (C) shows an increase in size and apparent opacification of the more ventral compartment and straightening of the membrane (arrow) after closure of the mitral valve. (From Ellis, Griffiths, Jesse, et al,[54] with permission from the publisher.)

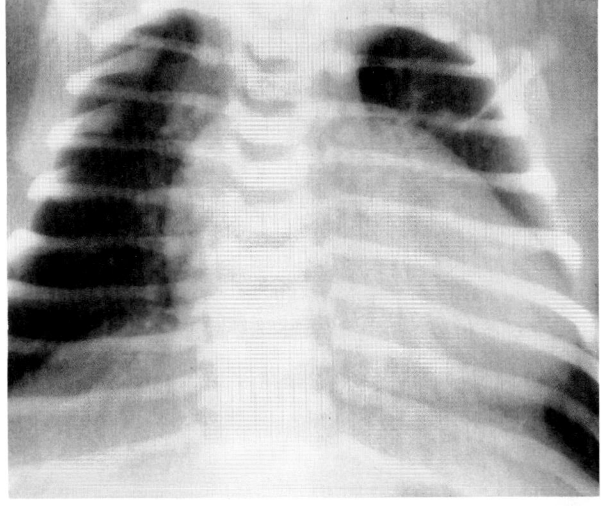

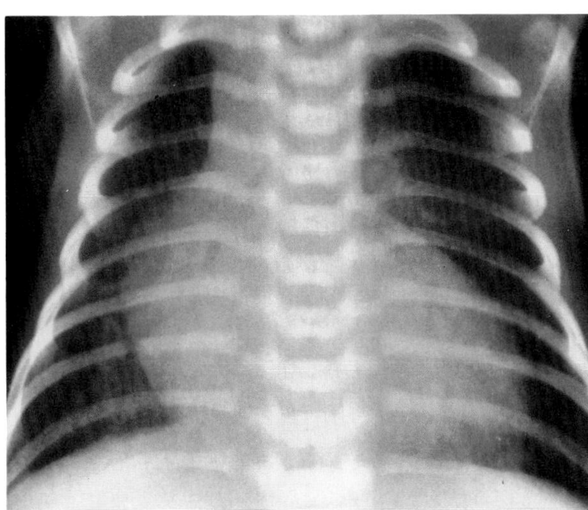

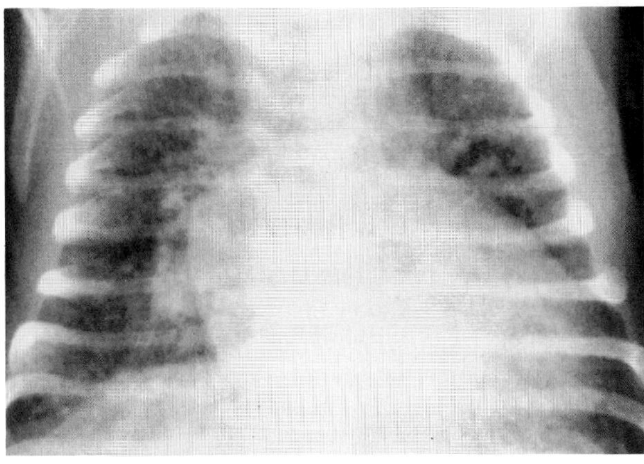

**Figure 21.45. Complete transposition.** In three different patients, biventricular enlargement produces the typical oval or egg-shaped configuration of the heart with varying degrees of pulmonary vascular congestion. Note the narrowing of the vascular pedicle due to the superimposition of the abnormally positioned aorta and pulmonary artery.

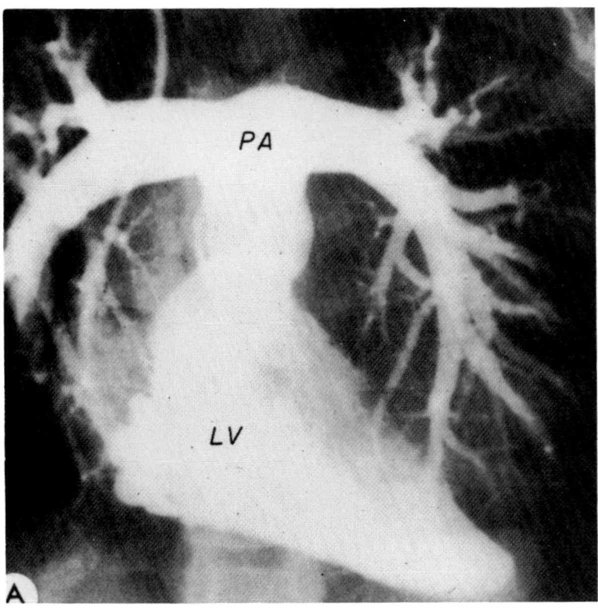


A

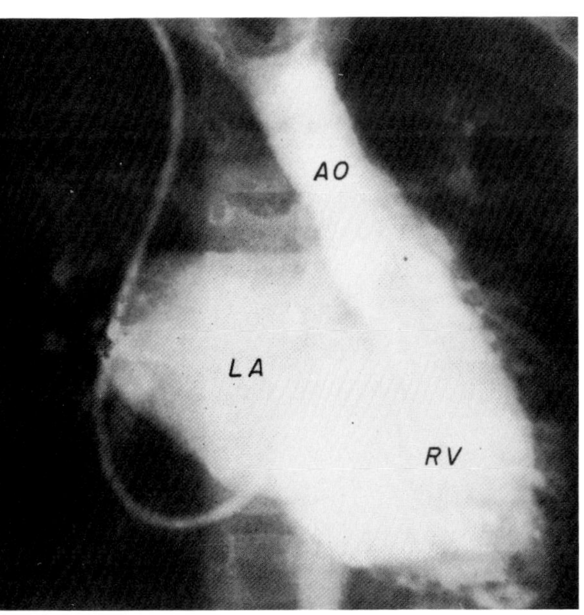


B

**Figure 21.46. Complete transposition.** (A) The injection of contrast material into the morphologic left ventricle (LV), which is located to the right, demonstrates prompt filling of the pulmonary artery (PA). Atrioventricular valve insufficiency permits reflux into the associated atrium. (B) The aorta (AO) originates from the morphologic right ventricle (RV). During systole, contrast material flows from the right ventricle through the abnormal left atrioventricular valve to the dilated left atrium (LA). (From Barcia, Kincaid, Davis, et al,[84] with permission from the publisher.)

## Complete Transposition of the Great Arteries

In this anomaly, the normal relation of the aorta and the pulmonary artery are reversed: the aorta arises anteriorly from the right ventricle and the pulmonary artery originates posteriorly from the left ventricle. Because this results in two separate circulations, intracardiac or extracardiac shunts (atrial or ventricular septal defects; patent ductus arteriosus) must be present for survival. Shunts through atrial and ventricular septal defects are bidirectional and permit the mixing of oxygenated and unoxygenated blood. In contrast, shunts through the foramen ovale and ductus arteriosus are predominantly right to left and thus of less value. Although the shunts are essential for initial survival, they eventually result in preferential blood flow into the lower-pressure pulmonary circulation. This leads to pulmonary vascular congestion, increased pulmonary venous return, diastolic overloading of the heart, and early congestive failure. Concomitant pulmonary stenosis, which occurs in up to 20 percent of the patients, acts as a protective device in decreasing pulmonary blood flow and consequently diminishing diastolic overloading of the heart.

During the first few days of life, the heart is usually normal or nearly normal in size and the pulmonary vascularity is within normal limits. However, pulmonary blood flow soon increases, and venous congestion adds to the radiographic appearance of increased pulmonary vascularity. The enlargement of both ventricles produces a typical oval or egg-shaped configuration of the heart. In patients with associated pulmonary stenosis, the pulmonary vasculature is decreased and the heart is more normal in size.

A characteristic finding is narrowing of the vascular pedicle on frontal projections. This is due to the superimposition of the abnormally positioned aorta and pulmonary artery combined with the absence of normal thymic tissue because of stress atrophy. On lateral projections, the anterior position of the aorta with respect to the pulmonary artery causes widening of the vascular pedicle.[1,56,57,84]

## Taussig-Bing Anomaly

In this condition, the aorta arises from the right ventricle while the pulmonary artery overrides the ventricular septum and receives blood from both ventricles. The pulmonary artery lies to the left of and slightly posterior to the aorta. A high ventricular septal defect is also present.

The radiographic findings of generalized cardiomegaly, increased pulmonary vascularity, and pulmonary venous congestion are indistinguishable from those seen in complete transposition of the great arteries. In the Taussig-Bing anomaly, however, there is the more frequent development of pulmonary hypertension, which diminishes the heart size and results in a smaller increase in the pulmonary vascularity.[1]

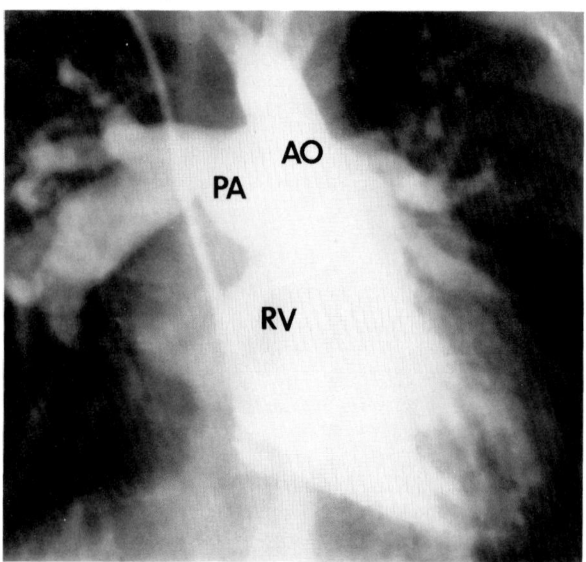


A

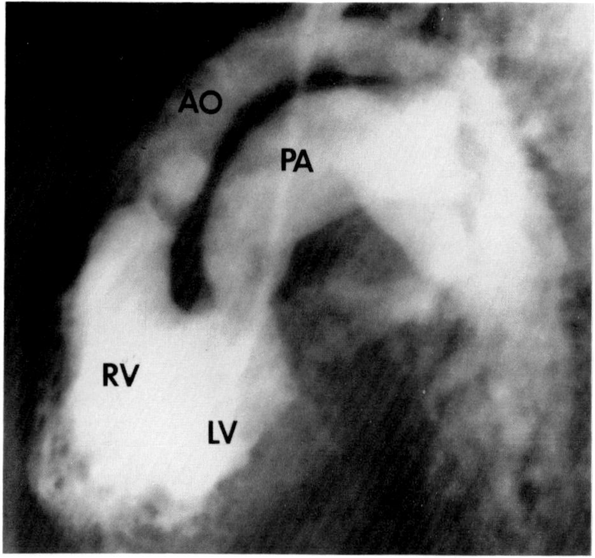


B

**Figure 21.47. Complete transposition.** (A) Frontal and (B) lateral views from an angiocardiogram demonstrate contrast material in the right ventricle (RV), which is situated anteriorly and to the right. It communicates through a large ventricular septal defect with the left ventricle (LV), which is located posteriorly and to the left. The transposed aorta (AO) originates from the right ventricular infundibulum directly in front of the pulmonary artery (PA), which arises from the left ventricle. The aortic valve is located higher and slightly to the right relative to the pulmonary valve. (From Barcia, Kincaid, Davis, et al,[84] with permission from the publisher.)

## Double-Outlet Right Ventricle

In this rare anomaly, both the aorta and the pulmonary artery arise from the right ventricle. A left-to-right ventricular septal defect permits oxygenated blood from the left ventricle to pass to the right ventricle and then on to the systemic circulation.

As with other types of transposition, there is generalized cardiomegaly and increased pulmonary vascularity. In double-outlet right ventricle, however, the aorta and the pulmonary artery tend to have a more side-to-side configuration, and thus the cardiac waist is wider on frontal views than in other types of transposition.[1,58,59]

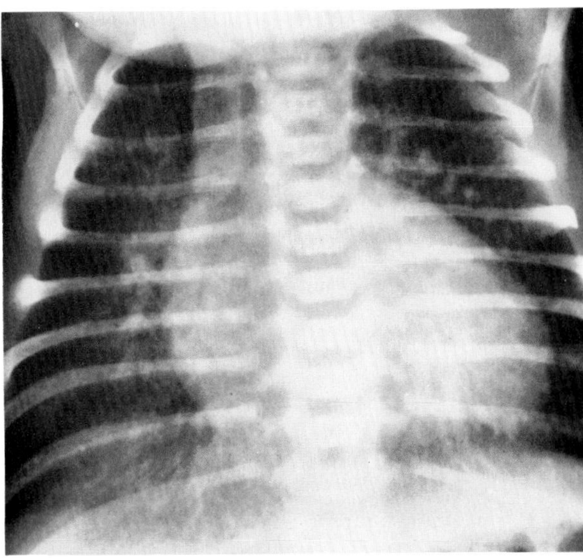

**Figure 21.48. Taussig-Bing anomaly.** A frontal view of the chest shows engorged pulmonary vasculature, oval cardiomegaly, and a laterally pointing apex.

## Corrected Transposition

In this anomaly, there is both transposition of the origins of the aorta and pulmonary artery (aortic root anterior to and to the left of the pulmonary artery) and inversion of the ventricles and their accompanying atrioventricular valves. Thus blood from the systemic circulation returns to the right atrium and then continues on through the functioning right ventricle (anatomic left ventricle) to the lungs via the posteriorly placed pulmonary artery. Oxygenated blood returns from the lungs into the left atrium

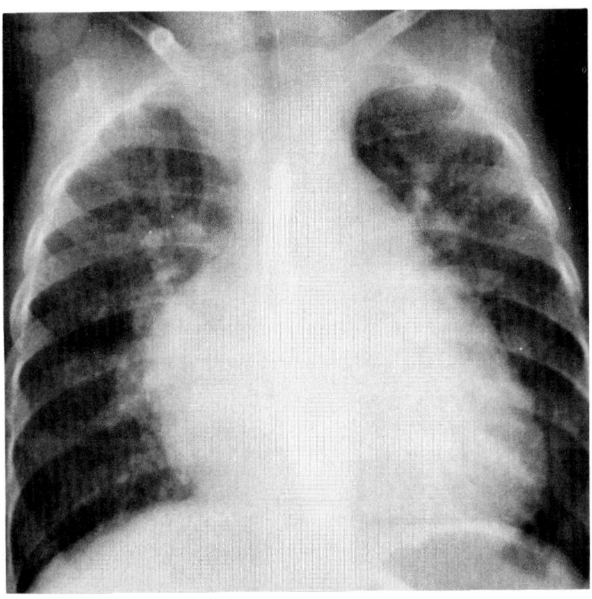

**Figure 21.49. Double-outlet right ventricle.** Generalized cardiomegaly with increased pulmonary vascularity. Because the aorta and the pulmonary artery have a more side-to-side configuration, the cardiac waist is relatively wider than in other types of transpositions.

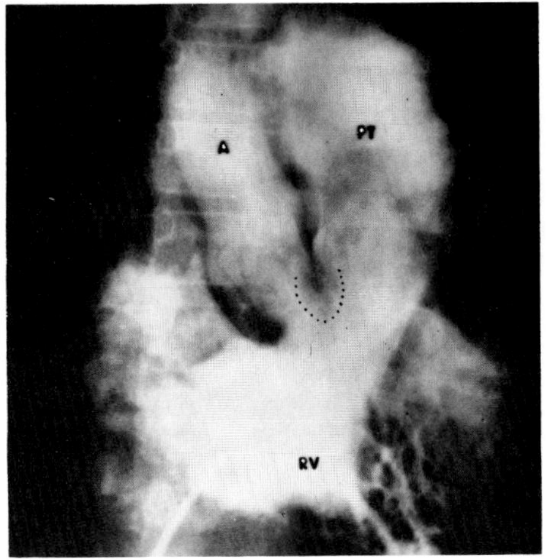


A

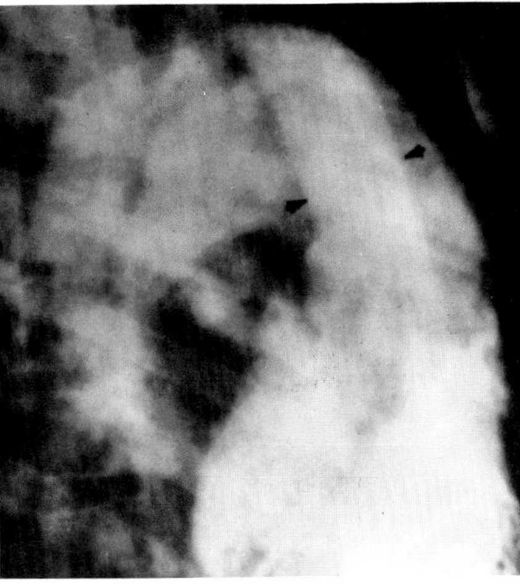

B

**Figure 21.50. Double-outlet right ventricle.** (A) A frontal view from a selective right ventriculogram shows simultaneous and equal opacification of both great vessels from the right ventricle (RV). The ventricular septal defect lies immediately beneath the crista supraventricularis (dotted line). (B) A lateral view shows the aorta (arrows) superimposed over the posterior two-thirds of the pulmonary trunk. A = aorta; PT = pulmonary trunk. (From Carey and Edwards,[58] with permission from the publisher.)

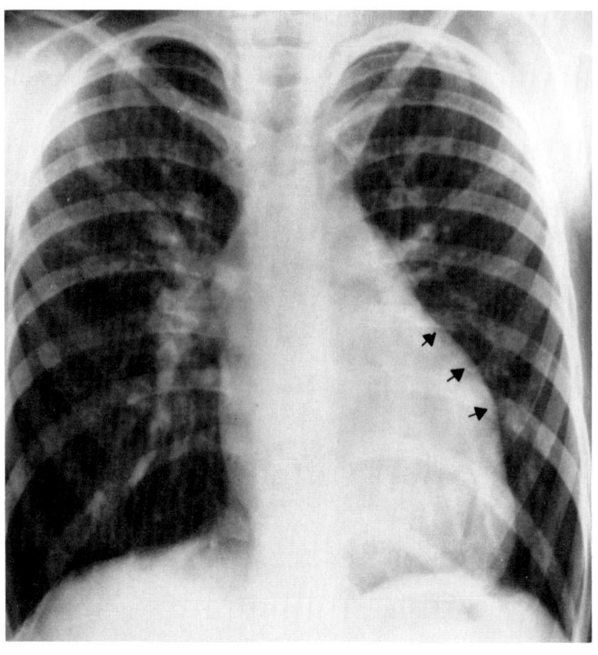

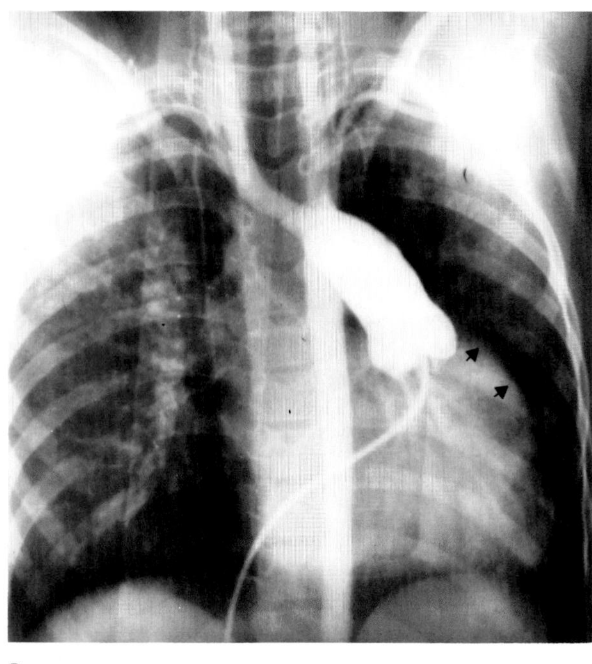

A B

**Figure 21.51. Corrected transposition with ventricular septal defect.** (A) There is fullness of the upper left border of the heart (arrows). Because of the left-to-right ventricular shunt, the pulmonary vasculature is engorged. (B) A film from an angiocardiogram demonstrates the inverted aorta and right ventricular outflow tract (arrows). (From Swischuk,[1] with permission from the publisher.)

and functioning left ventricle (anatomic right ventricle) and then flows into the systemic circulation through the anteriorly placed aorta. In patients with uncomplicated corrected transposition, there is no functional circulatory abnormality and the patient is usually acyanotic and often asymptomatic. However, most patients are symptomatic because of associated cardiovascular anomalies, especially ventricular septal defect or pulmonary stenosis.

A characteristic radiographic finding in corrected transposition is a smooth bulging of the upper left border of the heart due to the displaced ascending aorta and right ventricular outflow tract. This single convexity replaces the normal double bulge of the aortic knob and pulmonary artery segment. Although the pulmonary artery does not form a part of the left heart border, in its lower and more medial position it may produce an unusual indentation on the barium-filled esophagus below the normal aortic impression. Because there is no shunt in uncomplicated corrected transposition, the pulmonary vascularity is normal and the heart is not enlarged. However, if there is an associated ventricular septal defect, the resulting left-to-right shunt produces cardiomegaly and increased pulmonary vascularity.

Angiocardiography can demonstrate the low, medial origin of the pulmonary artery and the high, lateral, and anterior position of the aorta, as well as any associated cardiac anomalies.[1,60–62]

## ANOMALIES OF VENOUS RETURN

### Total Anomalous Pulmonary Venous Return

In total anomalous pulmonary venous return (TAPVR), all the pulmonary veins connect either to the right atrium directly or to the systemic veins or their tributaries. Because all pulmonary venous blood returns to the right atrium, a right-to-left shunt through an interatrial communication (atrial septal defect or patent foramen ovale) is necessary for survival; without it no blood would reach the left side of the heart or the systemic circulation.

TAPVR has been divided into four types on the basis of the site of insertion of the anomalous pulmonary veins, which almost always unite to form a single vessel posterior to the heart before entering a cardiac chamber or systemic vein. In the type I anomaly (50 percent), the anomalous pulmonary vein enters either the persistent left superior vena cava, the left innominate vein, the right superior vena cava, or the azygos vein. In the type II anomaly (30 percent), the anomalous pulmonary vein enters directly into the right atrium or, more commonly, into the coronary sinus. In the type III anomaly, the abnormal vein travels with the esophagus through the diaphragm and inserts into a systemic vein or, more frequently, into the portal vein. Type IV is a mixture of the first three types. Selective pulmonary arteriography can demonstrate the precise pattern of drainage of the anomalous pulmonary veins.

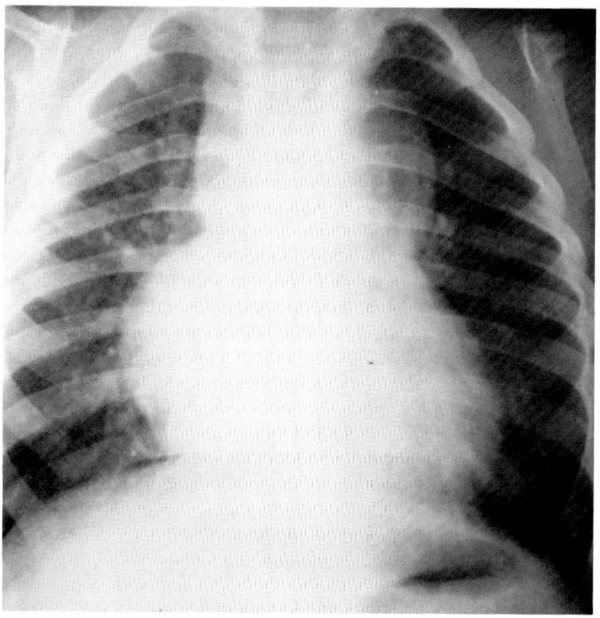
A

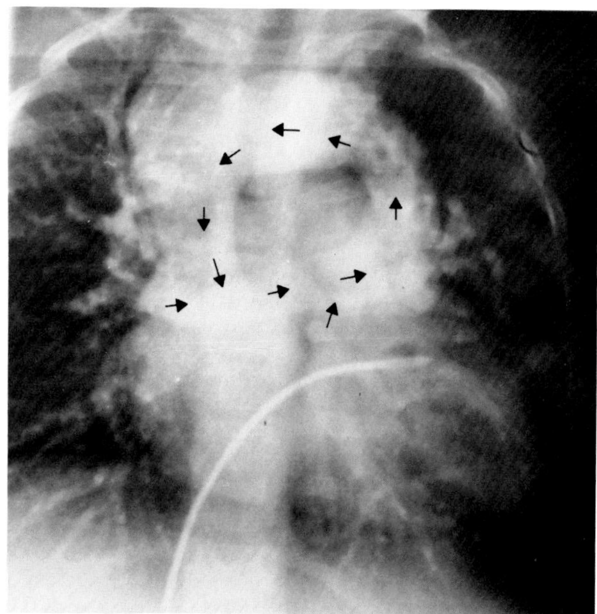
B

**Figure 21.52. Total anomalous pulmonary venous return (type I).** (A) A frontal chest radiograph demonstrates a snowman, or figure-8, heart with right atrial and right ventricular enlargement. The widening of the superior mediastinum is due to the large, anomalous inverted-U-shaped vein. Pulmonary vascularity is greatly increased. The large pulmonary artery is hidden in the superior mediastinal silhouette. (B) An angiocardiogram demonstrates that all the pulmonary veins drain into the inverted-U-shaped vessel that eventually empties into the superior vena cava (arrows). The widening of the mediastinum produced by this vessel causes the snowman heart. (From Swischuk,[1] with permission from the publisher.)

In types I and II, the large functional left-to-right shunt causes enlargement of the right atrium and right ventricle, dilatation of the pulmonary artery, and increased pulmonary vascularity. Emptying of the anomalous pulmonary vein into a persistent left superior vena cava produces the characteristic "snowman," or figure-8, configuration of the cardiovascular silhouette. The anomalous pulmonary vein causes a marked widening of the superior mediastinum (upper part of the snowman). The bottom part of the snowman is caused by the combination of right atrial and right ventricular enlargement. Angiocardiography demonstrates that the pulmonary veins drain into an inverted-U-shaped vessel, which eventually empties into the superior vena cava.

In the type III anomaly, the anomalous pulmonary vein is almost obstructed as it travels through the diaphragm. This seriously reduces pulmonary blood flow and markedly diminishes left-to-right shunting, resulting in a normal-sized heart and a stringy and reticular appearance of the pulmonary arteries. Obstruction of the anomalous pulmonary vein causes pulmonary venous congestion and interstitial edema. As the anomalous vein descends through the diaphragm, it may produce an anterior indentation on the lower portion of the adjacent barium-filled esophagus, slightly below the expected site of a left atrial impression. The combination of a normal-

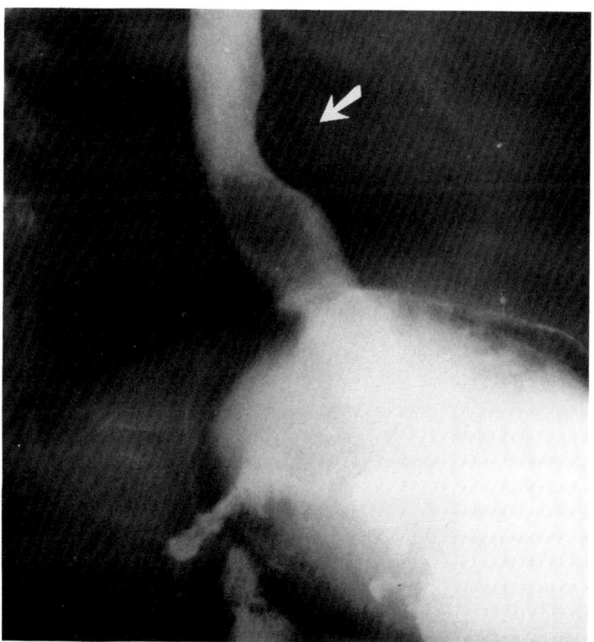

**Figure 21.53. Total anomalous pulmonary venous return (type III).** There is an anterior indentation on the lower border of the barium-filled esophagus (arrow) slightly below the expected site of the left atrium. (From Eisenberg,[88] with permission from the publisher.)

sized heart, congestive failure, and cyanosis in an infant with no heart murmurs is highly suggestive of TAPVR below the diaphragm.[1,63–66]

### Partial Anomalous Pulmonary Venous Return

In this condition, one or more of the pulmonary veins is connected to the right atrium or its tributaries. This produces a left-to-right shunt that may be asymptomatic or produce clinical and radiographic findings that are virtually indistinguishable from those of atrial septal defect (with which partial anomalous pulmonary venous return is often associated). The degree of pulmonary artery enlargement and increased pulmonary vascularity is related to the size of the shunt.

Anomalous venous drainage of the right lung by a large vein emptying into the inferior vena cava is occasionally associated with hypoplasia of the right lung. This combination is termed the "scimitar syndrome" because of the characteristic radiographic appearance of the anomalous venous channel as a crescent-like band of density, which curves like a Turkish sword through the right lower lung toward the midline in the direction of the inferior vena cava. The associated hypoplasia of the right lung may cause a small right hilum and displacement of the heart into the right hemithorax.[67]

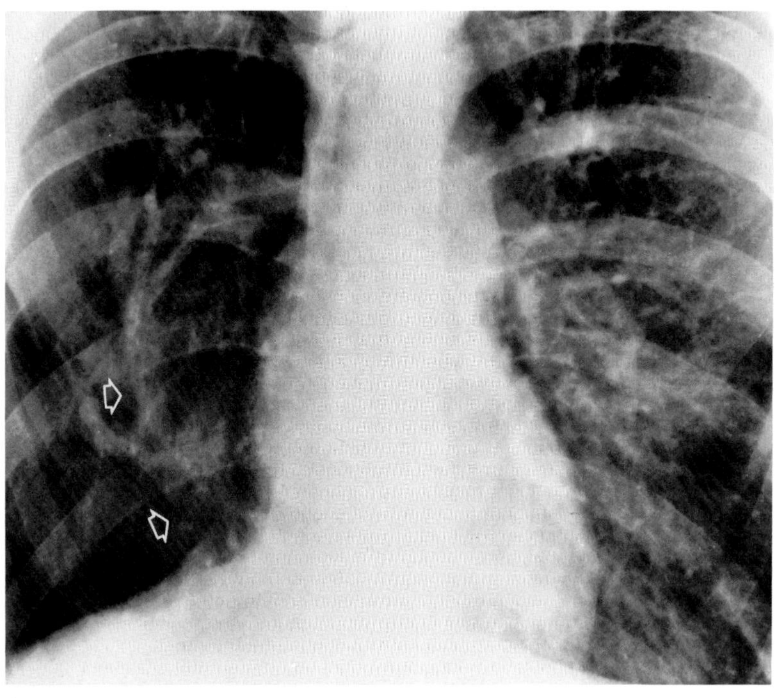
A

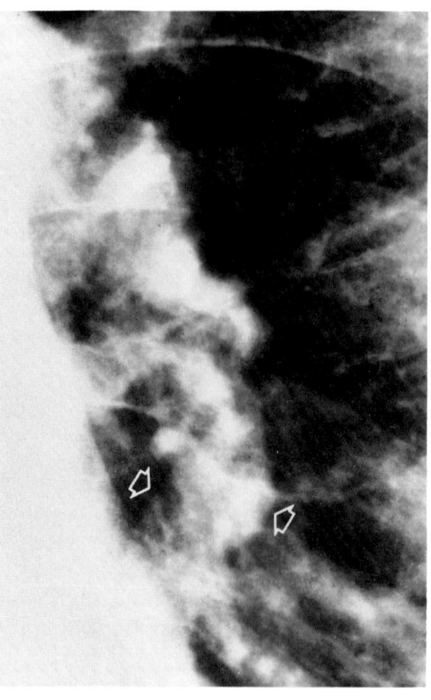
B

**Figure 21.54. Partial anomalous pulmonary venous return.** Two examples (A, B) of curvilinear venous pathways (arrows) resembling a Turkish scimitar. (From Eisenberg,[88] with permission from the publisher.)

# MALPOSITIONS OF THE HEART; ANOMALIES OF THE AORTA AND ITS BRANCHES; MISCELLANEOUS CONDITIONS

## MALPOSITIONS OF THE HEART

In evaluating cardiac malpositions, it is important to assess the position of both the apex of the heart and the abdominal viscera (the latter best indicated by the gas-filled gastric fundus). In dextrocardia, the cardiac apex is located in the right side of the chest. Mesocardia refers to a midline position of the cardiac apex; levocardia is the term for the normal position of the cardiac apex on the left side. Situs solitus indicates that the abdominal viscera are normally situated, with the gastric fundus on the left. Situs inversus refers to an abnormal right-sided location of the gastric fundus.

The most common malposition of the heart is mirror-image dextrocardia, in which the apex points to the right and there is complete inversion of the cardiac chambers.

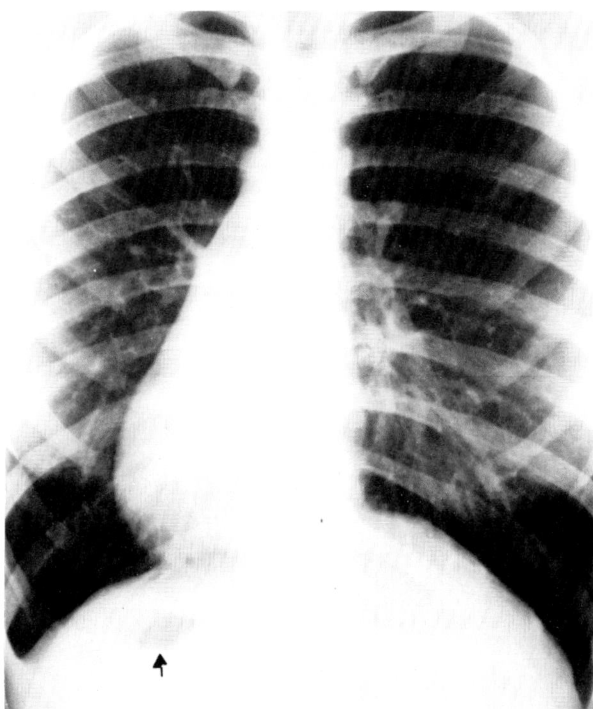

**Figure 21.55. Mirror-image dextrocardia with situs inversus.** The apex of the heart points to the right and the gastric bubble is on the right (arrow). (From Swischuk,[1] with permission from the publisher.)

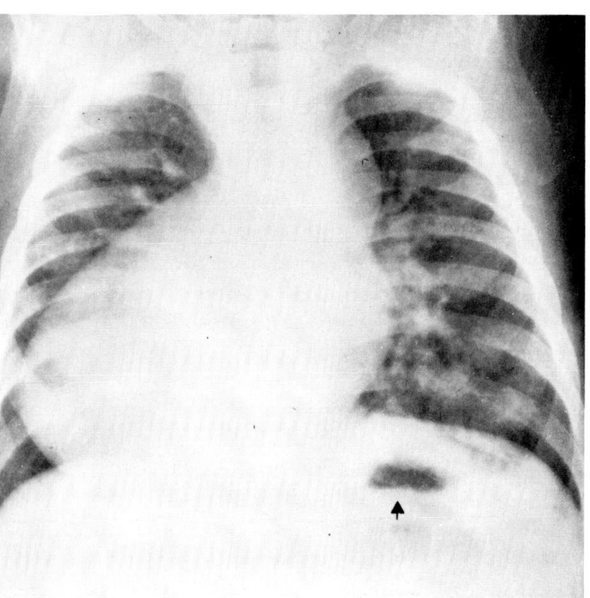

**Figure 21.56. Dextroversion (situs solitus).** The cardiac apex points to the right while the gastric bubble (arrow) is on the left. The bizarre cardiac configuration in this cyanotic patient reflects underlying congenital heart disease. (From Swischuk,[1] with permission from the publisher.)

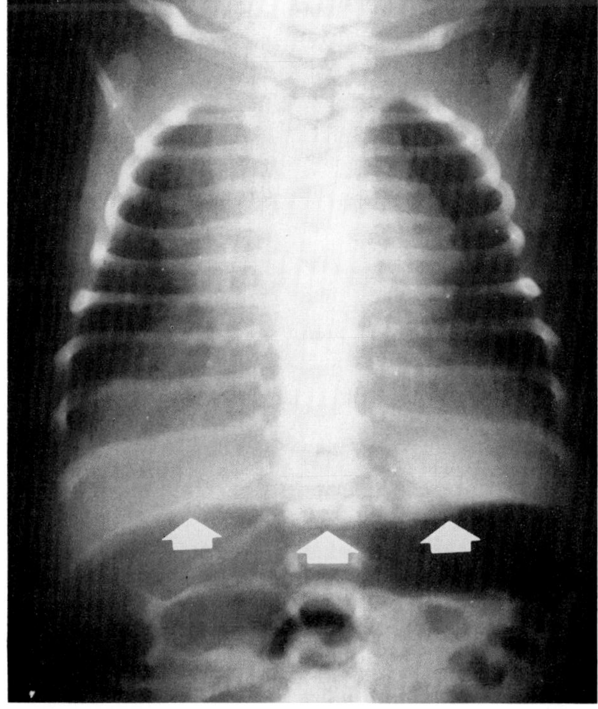

A

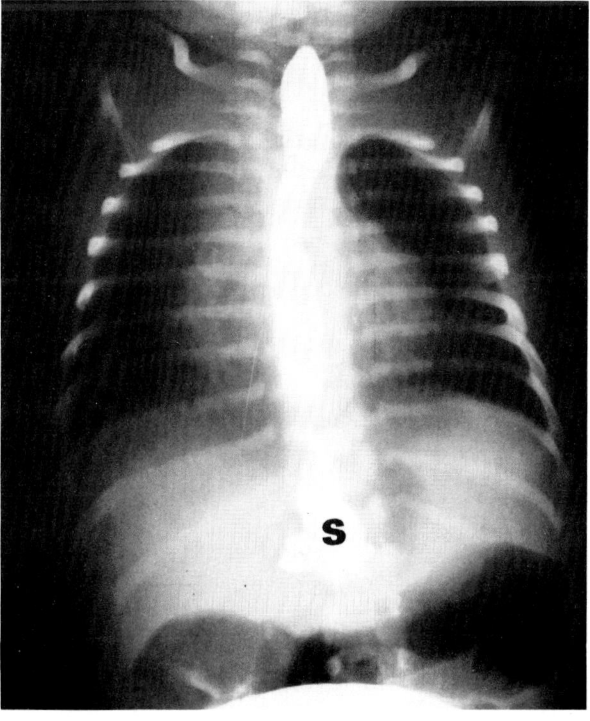


B

**Figure 21.57. Polysplenia.** (A) A soft tissue density (arrows) representing the liver extends from abdominal wall to abdominal wall. The heart is enlarged and there is slight vascular congestion. (B) An esophogram shows a small (microgastria) stomach (S). (From Swischuk,[1] with permission from the publisher.)

Situs inversus is invariably present and there is usually no significant congenital heart disease.

Dextroversion refers to a right-sided position of the cardiac apex without situs inversus. This type of cardiac malposition is commonly associated with cyanotic congenital heart disease, especially some form of transposition of the great arteries.

Radiographically, mirror-image dextrocardia appears as an exact reversal of the normal pattern. In dextroversion, however, the cardiac silhouette is usually abnormal and varies according to the specific underlying congenital cardiac anomaly.

In mesocardia, the heart lies halfway between the normal levoposition and the abnormal dextroposition. The heart has a somewhat globular configuration with the apex pointing anteriorly and a greater-than-normal portion of the cardiac silhouette lying to the right of the spine.

A normal position of the heart (levocardia) associated with situs inversus is an extremely rare anomaly that is almost always associated with congenital heart disease.

It is important to distinguish intrinsic cardiac malposition from malposition due to extracardiac causes, such as agenesis or hypoplasia of the pulmonary artery or lung, inflammatory retraction of the heart or mediastinum, large pleural effusions, or fibrothorax.

In the asplenia or polysplenia syndromes, the visceral situs is indeterminate and there is a high incidence of complex congenital cardiac anomalies that sometimes produces a bizarre configuration of the heart. Obstruction of the right outflow tract commonly results in severely decreased pulmonary vascularity. A symmetric midline liver or a small midline stomach can often be identified. Absence of the spleen or the presence of multiple accessory spleens can be demonstrated on radionuclide studies.[68–73]

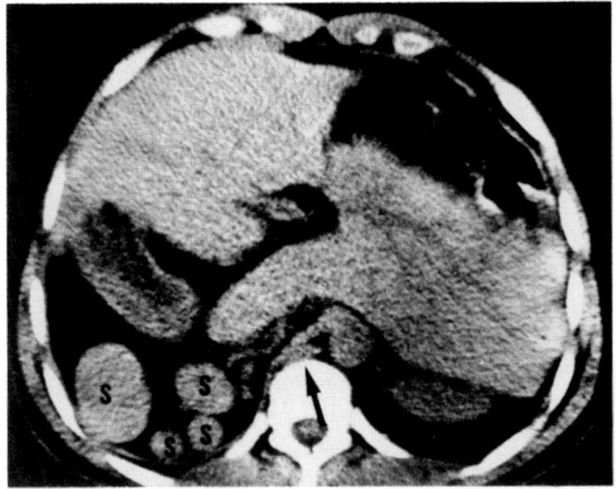


A

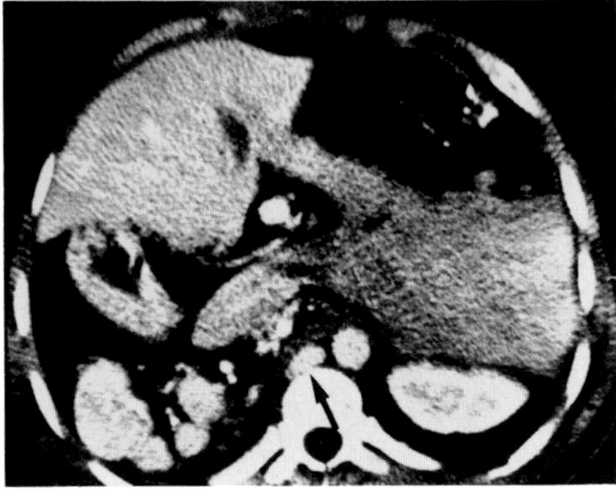

B

**Figure 21.58. Polysplenia.** (A) A precontrast CT scan demonstrates abdominal situs inversus. The liver is located centrally and the splenic silhouette is absent. Several round masses (S) are present in the right hypochondrium. Note the oval density (arrow) behind the right diaphragmatic crus. (B) On the postcontrast scan, the multiple nodules in the right hypochondrium show enhancement consistent with splenic tissue. Enhancement of the right retrocrural density (arrow) identifies it as a vascular structure (hemiazygos vein). The normal inferior vena cava in the caudate lobe of the liver is absent. (From De Maeyer, Wilms, and Baert,[86] with permission from the publisher.)

## ANOMALIES OF THE AORTIC ARCH AND ITS BRANCHES

### Right Aortic Arch

A right-sided position of the aortic arch is the most common aortic anomaly. This condition is easily detected on plain chest radiographs by the absence of the characteristic aortic knob on the left and the presence of a slightly higher bulge on the right. The trachea is deviated to the left, and the barium-filled esophagus is indented on the right.

When the aortic arch is right-sided, the descending aorta can run on either the right or the left. If it descends on the right, the brachiocephalic vessels can originate in one of three ways. With the mirror-image pattern, no vessels cross the mediastinum posterior to the esophagus, and, consequently, there is no esophageal indentation on lateral projections. This anomaly is frequently

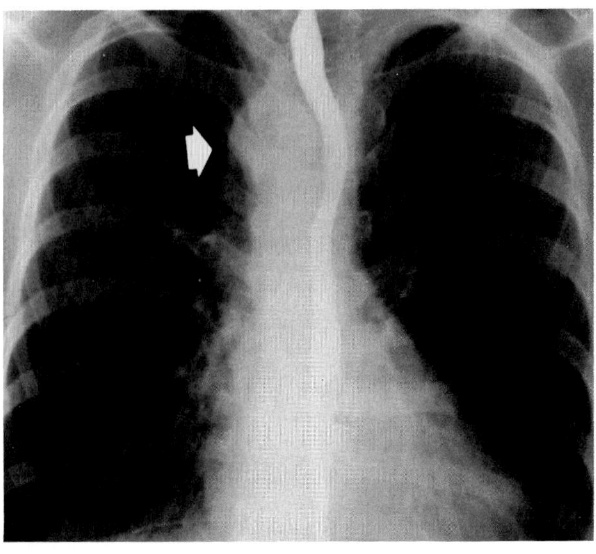

**Figure 21.59. Right aortic arch (arrow).**

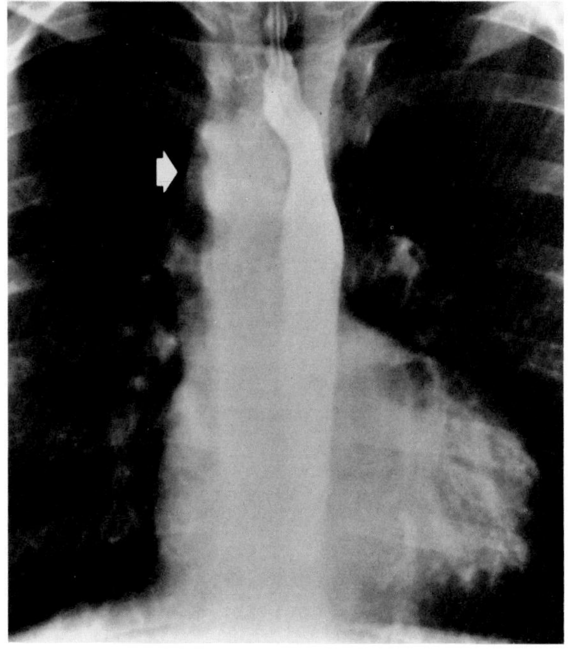


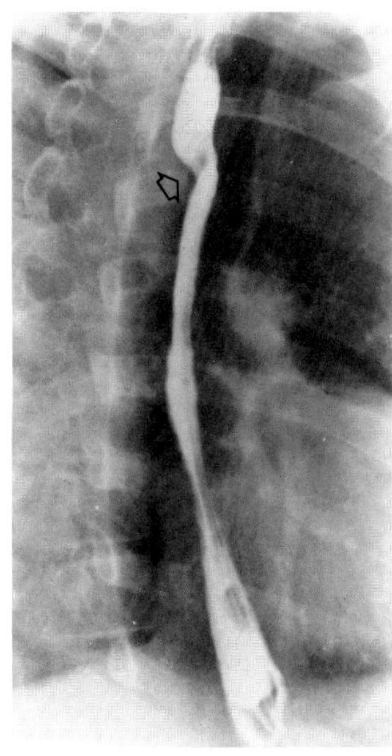


**Figure 21.60. Right aortic arch with aberrant left subclavian artery.** (A) A frontal view from an esophogram demonstrates the right aortic arch (arrow). (B) An oblique posterior impression on the esophagus (arrow) represents the left subclavian artery as it courses to reach the left upper extremity. (From Eisenberg,[85] with permission from the publisher.)

associated with congenital heart disease, primarily the tetralogy of Fallot.

The other two anomalies associated with a right-sided arch and right descending aorta differ with respect to the origin of the left subclavian artery. In the more common type, the left subclavian artery arises as the most distal branch of the aorta (the reverse of the aberrant right subclavian artery, which originates from a left aortic arch). In order to reach the left upper extremity, the left subclavian artery must course across the mediastinum posterior to the esophagus, producing a characteristic oblique posterior indentation on the esophagus. Almost all patients with this anomaly have no associated congenital heart disease.

In the second type of anomaly, the left subclavian artery is atretic at its base and totally isolated from the aorta (isolated left subclavian artery). In this rare condition, the left subclavian artery receives blood from the left pulmonary artery, or from the aorta in a circuitous fashion through retrograde flow by way of the ipsilateral vertebral artery (congenital subclavian steal syndrome). The tenuous blood supply in patients with this condition often results in decreased pulses and ischemia of the left upper extremity.

A right aortic arch with a left descending aorta is an uncommon anomaly. Because the aortic knob is on the right and the aorta descends on the left, the transverse portion must cross the mediastinum. This usually occurs posterior to the esophagus, resulting in a prominent posterior esophageal indentation that tends to be more transverse and much larger than that seen with an aberrant left subclavian artery.[74,75]

## Cervical Aortic Arch

In this congenital anomaly, the ascending aorta extends higher than usual so that the aortic arch is in the neck. This results in a pulsatile mass above the clavicle that may simulate an aneurysm of the subclavian, the carotid, or the innominate artery. No coexistent intracardiac congenital heart disease has been described in the few cases reported. A posterior esophageal impression is caused by the distal arch or proximal descending aorta as it courses in a retroesophageal position.[76]

## Double Aortic Arch

In most patients with a double aortic arch, the aorta ascends on the right, branches, and finally reunites on the left. The two limbs of the aorta completely encircle the trachea and esophagus, forming a ring. The anterior portion of the arch is usually smaller than the posterior part. When the aorta descends on the left (in about 75 percent of the cases), the posterior arch is higher than the anterior arch; the reverse pattern is seen if the aorta descends on the right.

On plain chest radiographs, the two aortic limbs can appear as bulges on either side of the superior mediastinum; the right one is usually larger and higher than the left. On an esophogram, a double aortic arch produces a characteristic reverse S-shaped indentation on the esophagus. The upper curve of the S is produced by the larger, posterior arch; the lower curve is related to the smaller, anterior arch. Infrequently, a patient with a dou-

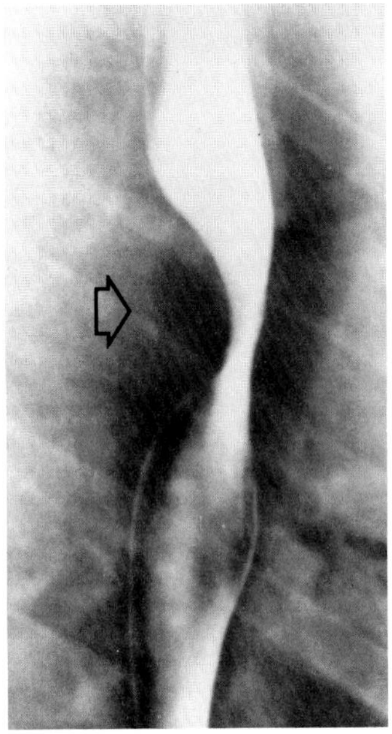

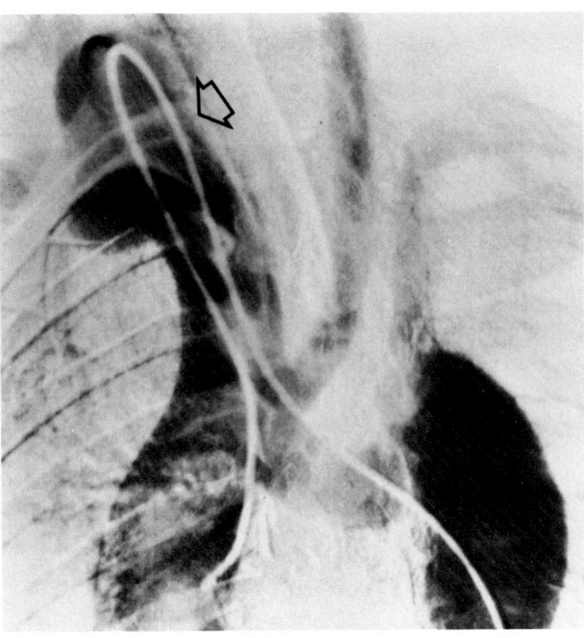

A B

**Figure 21.61. Cervical aortic arch.** (A) A posterior esophageal impression (arrow) is caused by the retroesophageal course of either the distal arch or the proximal descending aorta. (B) A subtraction film from an aortogram demonstrates the aortic arch extending into the neck (arrow). (From Eisenberg,[85] with permission from the publisher.)

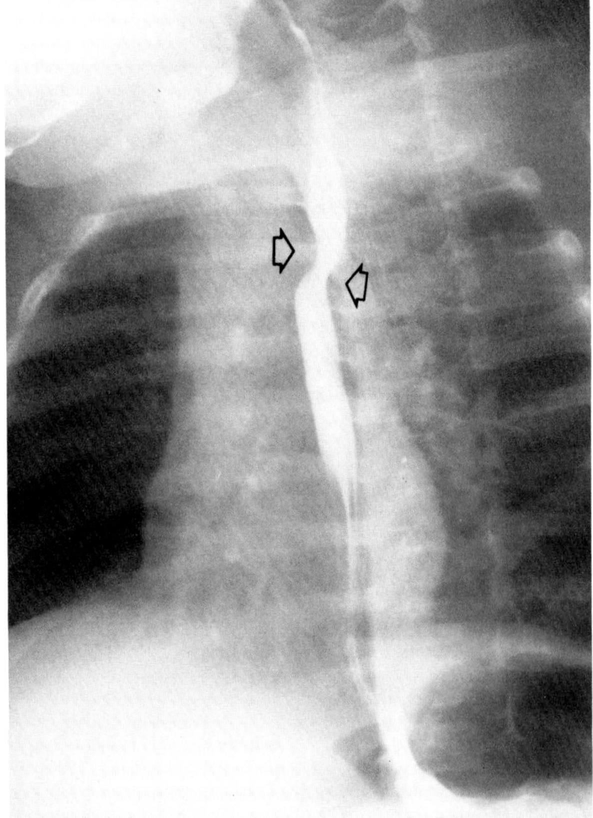

**Figure 21.62. Double aortic arch.** There is a characteristic reverse S-shaped indentation on the esophagus (arrows). As usual, the right (posterior) arch is higher and larger than the left (anterior) arch. (From Swischuk,[1] with permission from the publisher.)

ble aortic arch can have anterior and posterior esophageal indentations directly across from each other rather than in an S-shaped configuration.[1,77,78]

### Aberrant Right Subclavian Artery

An aberrant right subclavian artery arises as the last major vessel of the aortic arch, just distal to the left subclavian artery. In order to reach the right upper extremity, the aberrant right subclavian artery must course across the mediastinum behind the esophagus, and this produces a posterior esophageal indentation. On frontal views, the esophageal impression runs obliquely upward and to the right. This appearance of an aberrant right subclavian artery is so characteristic that no further radiographic investigation is required.[1,78,79]

## STRAIGHT BACK SYNDROME

The developmental loss of the normal kyphotic curvature of the thoracic spine and the consequent narrowing of the anteroposterior diameter of the chest can produce clinical signs that mimic organic heart disease. Systolic ejection murmurs and electrocardiographic findings may suggest a variety of congenital cardiac abnormalities.

In about half of the patients with the straight back syndrome, the radiographic appearance of the heart is normal. The compression of the heart between the sternum and the inwardly displaced thoracic spine can produce a "pancake" configuration simulating cardio-

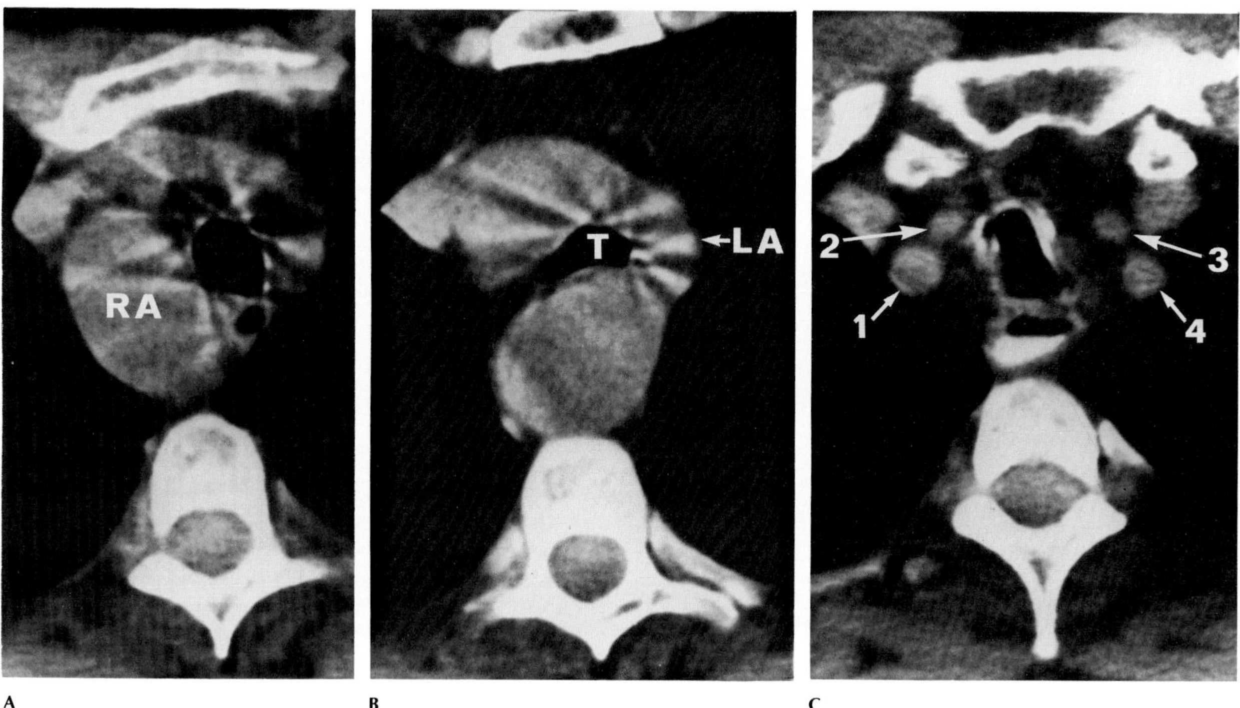


**Figure 21.63. CT of double aortic arch.** (A) An initial section shows the major right side (RA) of a double aortic arch. (B) A sequential caudal section shows anteroposterior compression of the trachea (T), and the smaller left aortic arch (LA). (C) A subsequent scan shows the four major arteries grouped around the trachea. 1 = right subclavian; 2 = right common carotid; 3 = left common carotid; 4 = left subclavian. (From McLoughlin, Weisbrod, Wise, et al,[78] with permission from the publisher.)

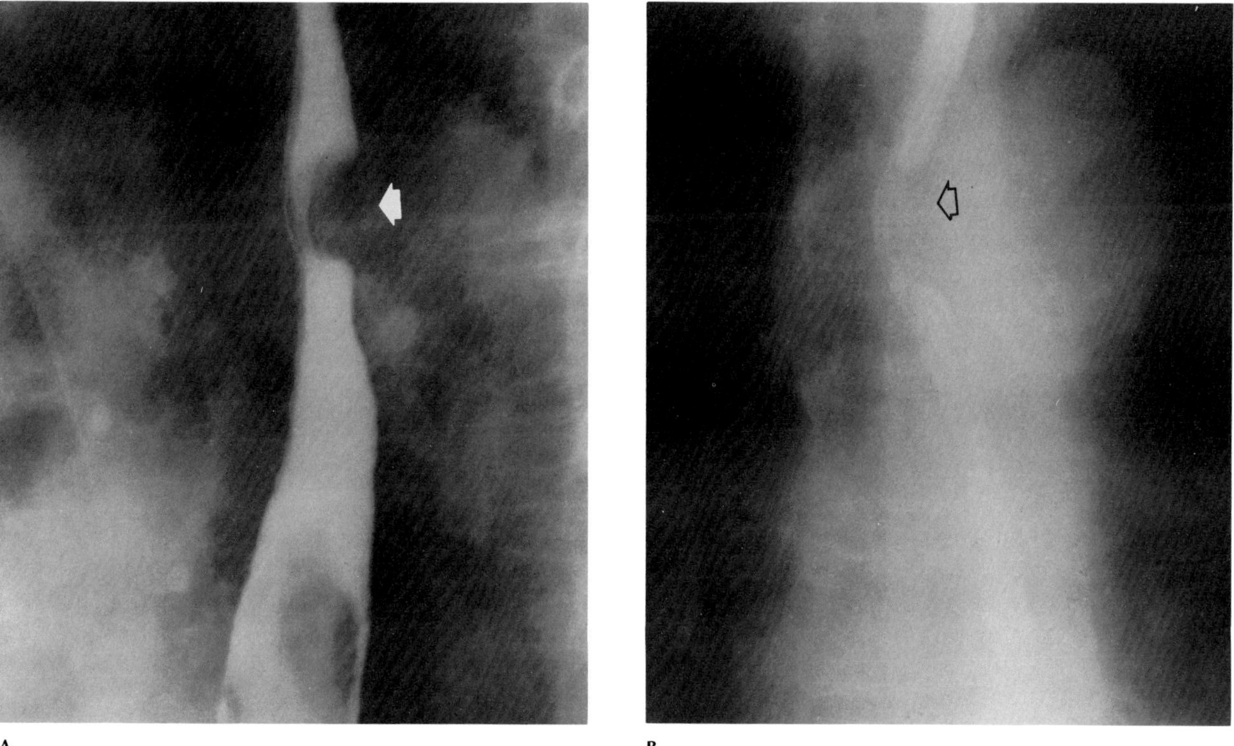


**Figure 21.64. Aberrant right subclavian artery.** (A) A lateral view from an esophogram demonstrates a posterior esophageal impression (arrow). (B) On a frontal view, the esophageal impression runs obliquely upward and to the right (arrow). (From Eisenberg,[85] with permission from the publisher.)

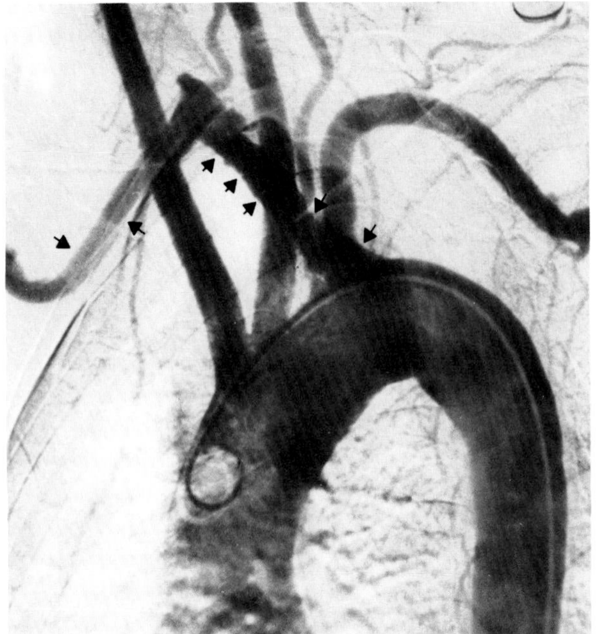

Figure 21.65. **Aberrant right subclavian artery.** A subtraction film from an arteriogram shows the aberrant vessel (arrows) arising distal to the left subclavian artery.

megaly on frontal views. Other patients have a leftward shift of the heart that produces prominence of the pulmonary artery segment.

The recognition of the straight back syndrome may prevent unnecessary cardiac catheterization and angiographic studies, all of which will be normal in patients with this condition.[80]

## ABERRANT LEFT PULMONARY ARTERY (PULMONARY SLING)

An aberrant left pulmonary artery arises from the right pulmonary artery and must cross the mediastinum to reach the left lung. As it courses between the trachea and the esophagus, the anomalous vessel produces a characteristic impression on the posterior aspect of the trachea just above the carina and a corresponding indentation on the anterior wall of the barium-filled esophagus. The aberrant left pulmonary artery may compress the tracheobronchial tree and cause respiratory distress with stridor and cyanosis.[81,82]

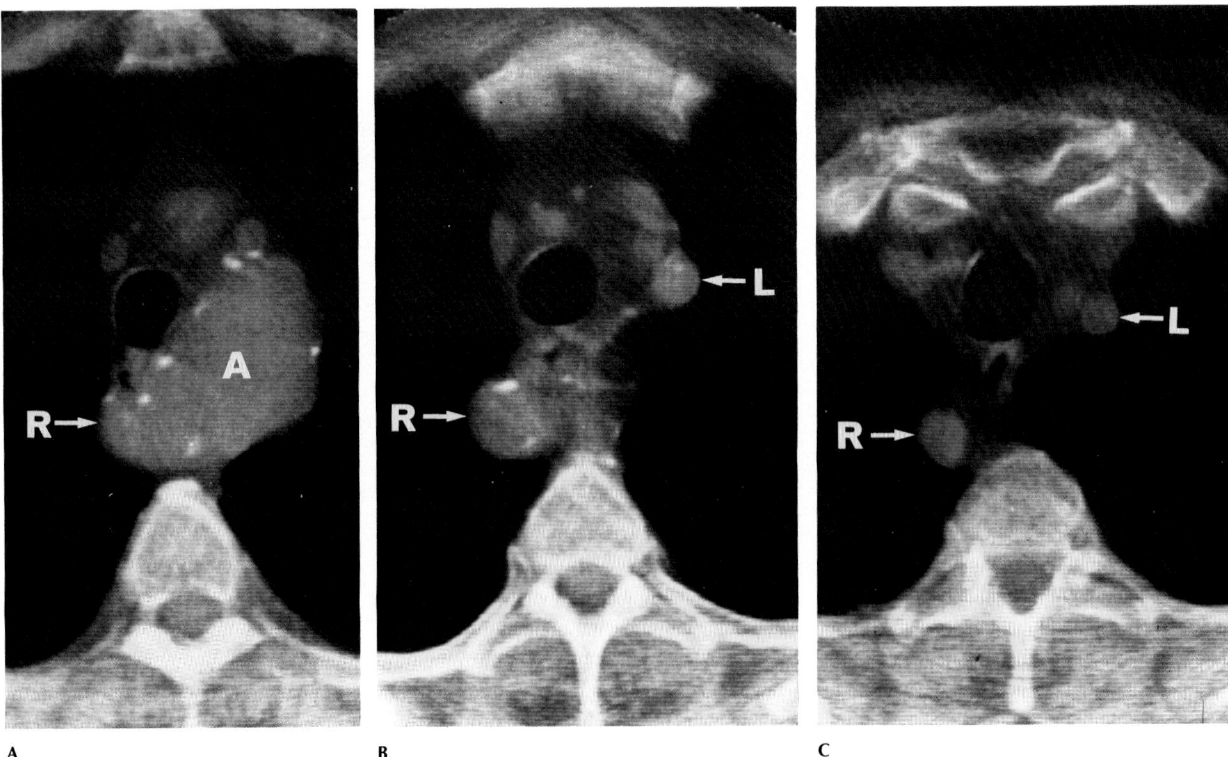


**Figure 21.66. CT of aberrant right subclavian artery.** (A) The aberrant right subclavian artery (R) arises behind the esophagus. Note the aneurysmal, partly calcified left arch (A) posterolateral to the trachea. (B, C) Sequential cranial cross sections show the course of the partly calcified, aberrant right subclavian artery (R) in the superior mediastinum. L = left subclavian artery. (From McLoughlin, Weisbrod, Wise, et al,[78] with permission from the publisher.)

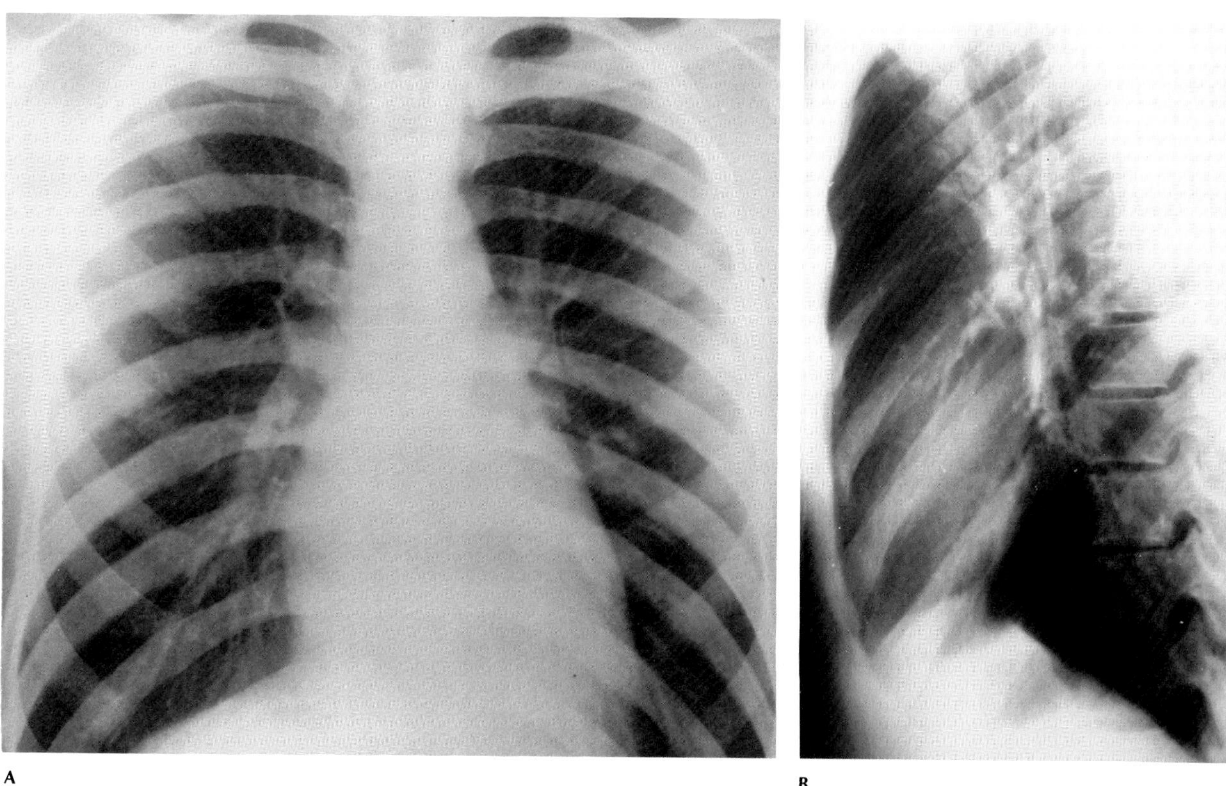


**Figure 21.67. Straight back syndrome.** (A) Frontal and (B) lateral chest radiographs demonstrate loss of the normal thoracic kyphosis and a slight pancake configuration of the heart. (Twigg, de Leon, Perloff, et al,[80] with permission from the publisher.)

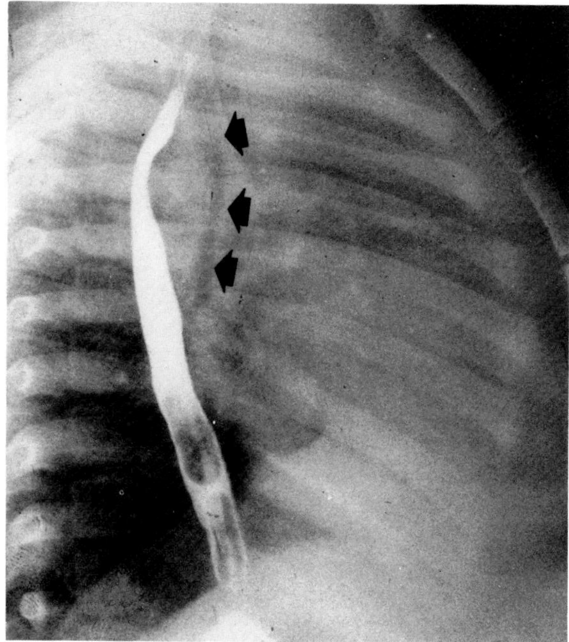

**Figure 21.68. Aberrant left pulmonary artery.** A lateral esophogram demonstrates characteristic indentation of the anterior wall of the esophagus. Note the posterior impression and anterior displacement of the trachea (arrows) by the aberrant left pulmonary artery. (From Jue, Raghib, Amplatz, et al,[81] with permission from the publisher.)

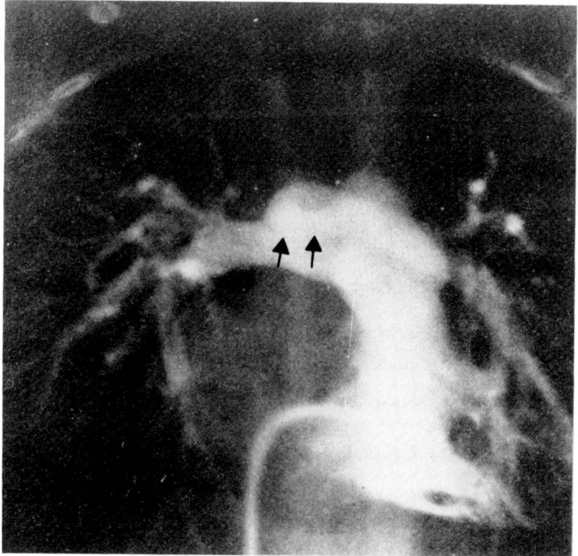

**Figure 21.69. Aberrant left pulmonary artery.** A right ventricular angiogram shows the abnormal left pulmonary artery (arrows) arising from the upper border of the main pulmonary artery on the right of the midline. The aberrant artery then crosses the midline to reach the left hilum. (From Philip, Sumerling, Fleming, et al,[82] with permission from the publisher.)

**REFERENCES**


1. Swischuk LE: *Plain Film Interpretation in Congenital Heart Disease*, Baltimore, William & Wilkins, 1979.
2. Spitz HB: The roentgenology of atrial septal defect in the adult. *Semin Roentgenol* 1:67–86, 1966.
3. Baron MG: Endocardial cushion defects. *Radiol Clin North Am* 4:43–52, 1968.
4. Spitz HB: Eisenmenger's syndrome. *Semin Roentgenol* 3:373–376, 1968.
5. Krovetz LJ: Hemodynamics of left-to-right shunts. *Radiol Clin North Am* 4:319–326, 1968.
6. Fellows K, Henschel WG, Keck E, et al: Left ventricular angiocardiography in endocardiac cushion defects: Emphasis on the lateral projection. *Ann Radiol* 15:223–230, 1972.
7. Baron MG: Endocardial cushion defects. *Radiol Clin North Am* 6:343–360, 1968.
8. Rubenstein BM, Young D, Pinals D, et al: The roentgen spectrum in persistent common atrioventricular canal. *Radiology* 86:860–864, 1966.
9. Rosenbaum HD, Leiber A, Hanson BJ, et al: Roentgen findings in ventricular septal defect. *Semin Roentgenol* 1:47–66, 1966.
10. Klatte EC, Burko H: The roentgen diagnosis of patent ductus arteriosus. *Semin Roentgenol* 1:87–101, 1966.
11. Neufeld HN, Lester RG, Adams P, et al: Aortopulmonary septal defect. *Am J Cardiol* 9:12–25, 1962.
12. Elliott LP: Other forms of left-to-right shunt. *Semin Roentgenol* 1:120–136, 1966.
13. Keene RJ, Steiner RE, Olsen EJG, et al: Aortic root aneurysm: Radiographic and pathologic features. *Clin Radiol* 22:330–340, 1971.
14. Morgan JR, Forker AD, O'Sullivan MJ, et al: Coronary arterial fistulas. Seven cases with unusual features. *Am J Cardiol* 30:432–436, 1973.
15. Neufeld HN, Lester RG, Adams P: Congenital communication of a coronary artery with a cardiac chamber or the pulmonary trunk (coronary artery fistula). *Circulation* 24:171–179, 1961.
16. Steinberg I, Holswade GR: Coronary arteriovenous fistula. *AJR* 116:82–90, 1972.
17. Morettin LB: Coronary arteriography: Uncommon observations. *Radiol Clin North Am* 14:189–208, 1976.
18. Lundquist C, Amplatz K: Anomalous origin of the left coronary artery from the pulmonary artery. *AJR* 95:611–620, 1965.
19. Stein HL, Hagstrom JWC, Ehlers KH, et al: Anomalous origin of the left coronary artery from the pulmonary artery: Angiographic and pathologic study of a case. *AJR* 93:320–330, 1965.
20. Hallermann FJ, Kincaid OW, Tsakiris AGL, et al: Persistent truncus arteriosus: A radiographic and angiographic study. *AJR* 107:827–834, 1969.
21. Dalinka MK, Rubinstein BM, Lopez F: The roentgen findings in truncus arteriosus. *J Can Assoc Radiol* 21:85–90, 1970.
22. Collett RW, Edwards JE: Persistent truncus arteriosus: A classification according to anatomic types. *Surg Clin North Am* 29:245–270, 1949.
23. Lafargue RT, Vogel JHK, Pryor R, et al: Pseudotruncus arteriosus. *Am J Cardiol* 19:239–242, 1967.
24. Chen JTT, Robinson AE, Goodrich JK: Uneven distribution of pulmonary blood flow between left and right lungs in isolated valvular pulmonary stenosis. *AJR* 107:343–350, 1969.
25. Singleton EB, Leachman RD, Rosenberg HS: Congenital abnormalities of the pulmonary arteries. *AJR* 91:487–499, 1964.
26. Marr K, Giargiana FA, White RI: Radiographic diagnosis of pulmonary hypertension following Blalock-Taussig shunts in patients with tetralogy of Fallot. *AJR* 122:125–132, 1974.
27. Daves ML: Roentgenology of tetralogy of Fallot. *Semin Roentgenol* 3:337–390, 1968.
28. Kirklin JW: The tetralogy of Fallot. *AJR* 102:253–266, 1968.
29. Wilson WJ, Amplatz K: Unequal vascularity in tetralogy of Fallot. *AJR* 100:318–321, 1967.
30. Amplatz K, Lester RG, Schiebler GL, et al: The roentgenologic features of Ebstein's anomaly of the tricuspid valve. *AJR* 81:788–794, 1959.
31. Deutsch V, Wexler L, Blieden LC, et al: Ebstein's anomaly of tricuspid valve: Critical review of roentgenologic features and additional angiographic signs. *AJR* 125:395–411, 1975.
32. Elliott LP, Hartmann AF: The right ventricular infundibulum in Ebstein's anomaly of the tricuspid valve. *Radiology* 89:694–700, 1967.
33. Kieffer SA, Carey LS: Tricuspid atresia with normal root: Roentgen-anatomic correlation. *Radiology* 80:605–615, 1963.
34. Elliott LP, Van Mierop LHS, Gleason B, et al: The roentgenology of tricuspid atresia. *Semin Roentgenol* 3:339–409, 1968.
35. Guller B, Kincaid OW, Ritter DG, et al: Angiocardiographic findings in tricuspid atresia: Correlation with hemodynamic and morphologic features. *Radiology* 93:531–540, 1969.
36. Robinson AE, Capp MP, Chen JTT, et al: Left-sided obstructive diseases of the heart and great vessels. *Semin Roentgenol* 3:410–419, 1968.
37. Sloan RD, Cooley RN: Coarctation of the aorta: Roentgenologic aspect of one hundred and twenty-five surgically confirmed cases. *Radiology* 61:701–721, 1953.
38. Figley MM: Accessory roentgen signs of coarctation of the aorta. *Radiology* 62:671–687, 1954.
39. Griffin JF: Congenital kinking of the aorta (pseudocoarctation). *N Engl J Med* 271:726–728, 1964.
40. Steinberg I: Anomalies (pseudocoarctation) of the arch of the aorta. *AJR* 88:73–92, 1962.
41. Hoeffel JC, Henry M, Mentre B, et al: Pseudocoarctation or congenital kinking of the aorta: Radiologic considerations. *Am Heart J* 89:428–436, 1975.
42. Takekawa SB, Kincaid OW, Titus JL, et al: Congenital aortic stenosis. *AJR* 98:800–821, 1966.
43. Freimanis AK, Wooley CS, Mechstroth CU, et al: Roentgenographic aspects of congenital left ventricular outflow tract obstruction. *AJR* 95:573–591, 1965.
44. Baltaxe HA, Moller JH, Amplatz K: Membranous subaortic stenosis and its associated malformations. *Radiology* 95:287–291, 1970.
45. Batson GA, Urquhart W, Sideris DA: Radiological features in aortic stenosis. *Clin Radiol* 23:140–143, 1972.
46. Freundlich IM, McMurray JT, Lehman JS: Idiopathic hypertrophic subaortic stenosis. *AJR* 100:284–289, 1967.
47. Kupic EA, Abrams HL: Supravalvular aortic stenosis. *AJR* 98:822–835, 1966.
48. Noonan JA, Nadas AS: The hypoplastic left heart syndrome: An analysis of 101 cases. *Pediatr Clin North Am* 5:1029–1056, 1958.
49. Gerald B: Combined aortic and mitral obstruction in early infancy. The hypoplastic left heart syndrome. *Radiology* 88:1100–1104, 1967.
50. Kallfelz HC, Hauke H: Radiological and angiocardiographic findings in infants with congenital obstructive mitral and aortic lesions. *Ann Radiol* 12:313–320, 1969.
51. Folger GM, Saied A: A new roentgenographic sign of hypoplastic left heart. *Chest* 64:298–302, 1973.
52. McLoughlin MF: Cor triatriatum sinister: The role of radiology in the diagnosis of this rare but curable anomaly. *Clin Radiol* 21:287–296, 1970.
53. Mortensson W: Radiologic diagnosis of cor triatriatum in infants. *Pediatr Radiol* 1:92–95, 1973.
54. Ellis K, Griffiths SP, Jesse MJ, et al: Cor triatriatum: Angiographic demonstration of the obstructing left atrial membrane. *AJR* 92:669–675, 1964.
55. Hilbish TF, Cooley RN: Congenital mitral stenosis. Roentgen study of its manifestations. *AJR* 76:743–757, 1956.
56. Carey LS, Elliott LP: Complete transposition of the great vessels: Roentgenographic findings. *AJR* 91:529–532, 1964.
57. Guerin R, Soto B, Karp RB, et al: Transposition of the great arteries: Determination of the position of the great arteries in conventional chest roentgenograms. *AJR* 110:747–756, 1970.

58. Carey LS, Edwards JE: Roentgenographic features in case with origin of both great vessels from the right ventricle without pulmonary stenosis. *AJR* 93:269–297, 1965.
59. Hallermann FJ, Kincaid WW, Ritter DG, et al: Angiocardiographic and anatomic findings in origin of both great arteries from the right ventricle. *AJR* 109:51–66, 1970.
60. Ellis K, Morgan BC, Blumenthal S, et al: Congenitally corrected transposition of the great vessels. *Radiology* 79:35–49, 1962.
61. Du Bois R, Dupuis C, Remy J: Corrected transposition of the great vessels. Some aspects of standard radiography. *Ann Radiol* 9:119–134, 1966.
62. Carey LS, Ruttenberg HD: Roentgenographic features of congenital corrected transposition of the great vessels: A comparative study of 33 cases with a roentgenographic classification based on the associated malformations and hemodynamic states. *AJR* 92:623–651, 1964.
63. Darling RC, Rothney WB, Craig JM: Total pulmonary venous drainage into the right side of the heart: A report of seventeen autopsy cases not associated with other major cardiovascular anomalies. *Lab Invest* 6:44–65, 1957.
64. Lester RG, Mauck HP, Grubb WL: Anomalous pulmonary venous return to the right side of the heart. *Semin Roentgenol* 1:102–119, 1966.
65. Levin B, White H: Total anomalous pulmonary venous drainage into the portal system. *Radiology* 76:894–901, 1961.
66. Eisen S, Elliott LP: A plain film sign of total anomalous pulmonary venous connection below the diaphragm. *AJR* 102:372–379, 1968.
67. Roehm TU, Jue KL, Amplatz K: Radiographic features of the scimitar syndrome. *Radiology* 86:856–859, 1966.
68. Ellis K, Fleming RJ, Griffiths SP, et al: New concepts in dextrocardia. *AJR* 97:295–313, 1966.
69. Freedom RM, Fellows KE: Radiographic visceral pattern in the asplenia syndrome. *Radiology* 107:387–391, 1973.
70. Randall PA, Moller JH, Amplatz K: The spleen and congenital heart disease. *AJR* 119:551–559, 1973.
71. Steinberg I, Ayres SM: Roentgen features of dextrorotation of the heart: A report of 39 cases. *AJR* 91:340–363, 1964.
72. Van Praagh R, van Praagh S, Vlad P, et al: Diagnosis of anatomic types of congenital dextrocardia. *Am J Cardiol* 15:234–247, 1955.
73. Vaughan TJ, Hawkins IF, Elliott LP: Diagnosis of polysplenia syndrome. *Radiology* 101:511–518, 1971.
74. Stewart JR, Kincaid OW, Titus JL: Right arotic arch: Plain film diagnosis and significance. *AJR* 97:377–389, 1966.
75. Shuford WH, Sybers RG, Edwards FK: The three types of right aortic arch. *AJR* 109:67–74, 1970.
76. Shuford WH, Sybers RG, Milledge RD, et al: The cervical aortic arch. *AJR* 116:519–527, 1972.
77. Gross RE, Neuhauser EBD: Compression of trachea or esophagus by vascular anomalies. *Pediatrics* 7:69–83, 1951.
78. McLoughlin MJ, Weisbrod G, Wise DJ, et al: Computed tomography in congenital anomalies of the aortic arch and great vessels. *Radiology* 138:399–403, 1981.
79. Klinkhamer AC: Aberrant right subclavian artery: Clinical and roentgenologic aspects. *AJR* 97:438–446, 1966.
80. Twigg HL, de Leon AC, Perloff JK, et al: The straight back syndrome: Radiographic manifestations. *Radiology* 88:274–277, 1967.
81. Jue KL, Raghib G, Amplatz K, et al: Anomalous origin of the left pulmonary artery from the right pulmonary artery. *AJR* 95:598–610, 1965.
82. Philip T, Sumerling MD, Fleming J, et al: Aberrant left pulmonary artery. *Clin Radiol* 23:153–159, 1972.
83. Cooley RN, Schreiber MH: *Radiology of the Heart and Great Vessels*. Baltimore, Williams & Wilkins, 1978.
84. Barcia A, Kincaid OW, Davis GD, et al: Transposition of the great arteries. *AJR* 100:249–261, 1967.
85. Eisenberg RL: *Gastrointestinal Radiology: A Pattern Approach*. Philadelphia, JB Lippincott, 1983.
86. DeMaeyer P, Wilms G, Baert AL: Polysplenia. *J Comput Assist Tomogr* 5:104–105, 1981.
87. Dotter CT: Left ventricular and systemic arterial catheterization. *AJR* 83:969–984, 1960.
88. Eisenberg RL: *Atlas of Signs in Radiology*. Philadelphia, JB Lippincott, 1984.

## Chapter Twenty-Two
# Acquired Valvular Heart Disease

## RHEUMATIC FEVER

The major radiographic abnormality in patients with rheumatic fever is nonspecific symmetric enlargement of the heart. This may be due to myocarditis, mitral valve dysfunction (caused by acute valvulitis or acute left ventricular dilatation), or, less frequently, pericardial effusion. Fluoroscopy demonstrates a diminished amplitude of cardiac pulsations. Congestive heart failure complicating acute rheumatic fever can cause a pattern of pulmonary edema that in the past has been described as "rheumatic pneumonia."

Although rheumatic fever may be associated with symptoms of inflammation involving multiple joints, there usually are no radiographic changes other than periarticular swelling. In rare cases, deformities of the hands (and occasionally the feet) may develop following the resolution of a severe attack of rheumatic fever (Jaccoud's arthritis). In this condition, muscular atrophy, ulnar deviation, and flexion of the metacarpophalangeal or distal joints are due to periarticular fascial and tendon fibrosis rather than synovitis. Unlike most types of inflammatory arthritis, rheumatic fever does not cause joint space narrowing or bone erosion.[1–3]

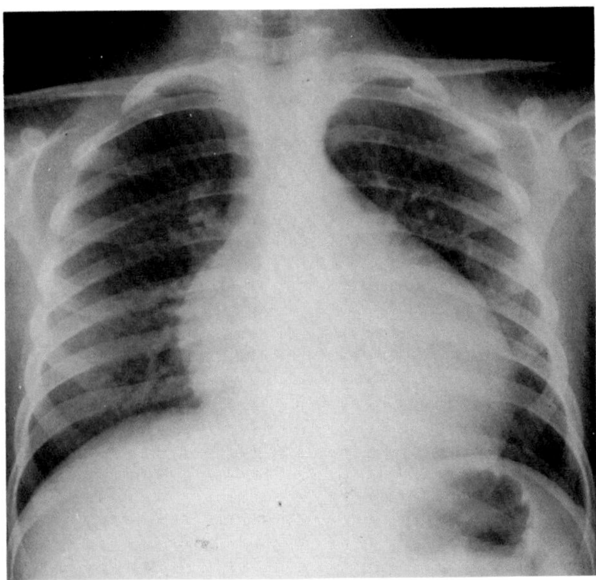

**Figure 22.1. Acute rheumatic fever in a child.** There is generalized cardiac enlargement with a small pericardial effusion. The lungs are clear. (From Cooley and Schreiber,[1] with permission from the publisher.)

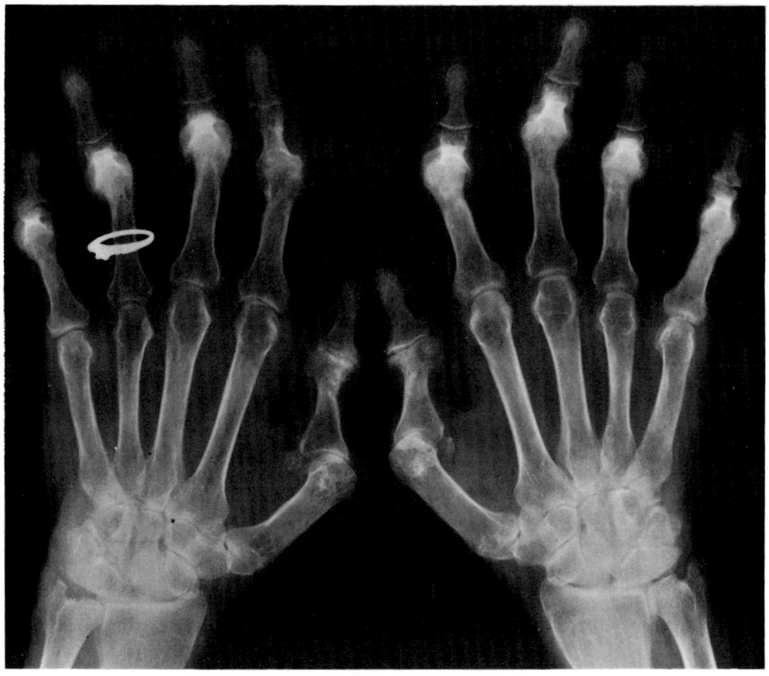
A

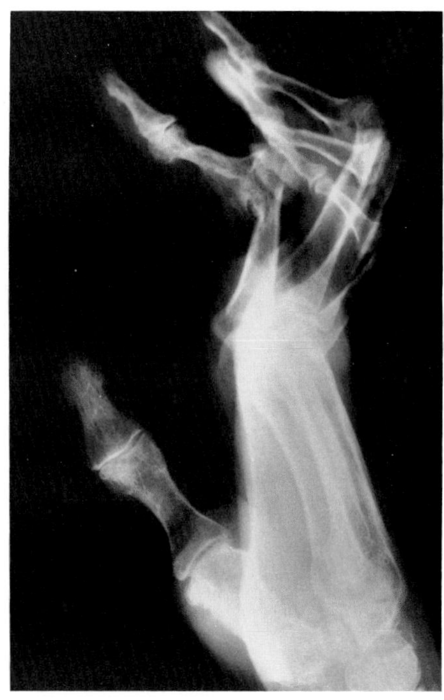
B

**Figure 22.2. Jaccoud's arthritis.** (A) Frontal views of the hands and wrists and (B) a lateral view of the right hand demonstrate mild ulnar deviation with pronounced flexion of the proximal interphalangeal joints. There is ulnar deviation at the first metacarpophalangeal joints. Note that there is no evidence of joint space narrowing or bone erosion.

## MITRAL STENOSIS

Stenosis of the mitral valve, the most common rheumatic valvular lesion, results from diffuse thickening of the valve by fibrous tissue and/or calcific deposits. The obstruction of blood flow from the left atrium into the left ventricle during diastole causes increased pressure in the left atrium that is transmitted to the pulmonary veins and eventually extends to the pulmonary arteries and the right side of the heart.

The radiographic changes in mitral stenosis depend on the degree of obstruction of the mitral valve. The most common finding is enlargement of the left atrium. This causes straightening or convexity of the upper left border of the heart, representing enlargement of the atrial appendage. The enlarged left atrium produces a characteristic anterior impression and posterior displacement of the barium-filled esophagus that begins about 2 cm below the carina and is best seen on lateral and right anterior oblique projections. Other radiographic signs of left atrial enlargement include posterior displacement of the left main bronchus, widening of the carina, and a characteristic "double contour" configuration due to the projection of the enlarged left atrium through the normal right atrial silhouette.

Increased pulmonary venous pressure causes a redistribution of blood flow to the upper lobes and an indistinct quality of the lower lobe vessels due to interstitial edema. A further increase in pulmonary venous pressure causes distension of interlobular septa and lymphatics with edema fluid. This produces Kerley B lines, which are fine, dense horizontal lines in the lower lungs that extend at right angles to the pleural surface. The development of pulmonary arterial hypertension causes enlargement of the pulmonary outflow tract and central pulmonary arteries, narrowing of the peripheral vessels, and enlargement of the right ventricle. The left ventricle is normal in size, and the aortic knob is small and inconspicuous because of decreased left ventricular output. Overall cardiac enlargement is minimal in patients with mild or moderate mitral stenosis; gross cardiac enlargement reflects the development of ventricular dilatation and failure.

Calcification of the mitral valve or left atrial wall, best demonstrated by fluoroscopy, can develop in patients with long-standing severe mitral stenosis. In patients who have had multiple episodes of hemoptysis, deposits of hemosiderin in the interstitial tissues may appear radiographically as fine granular or miliary shadows throughout the lungs. Multiple calcific nodules that progress to ossification uncommonly develop in the lower portions of the lungs in areas of chronic interstitial edema.

Echocardiography is the most sensitive and most specific noninvasive method for diagnosing mitral stenosis. M-mode tracings show calcification and thickening of the mitral valve as multilayered echoes or a thickening of the echo pattern. A reduction in the EF slope reflects the failure of the anterior leaflet of the mitral valve to

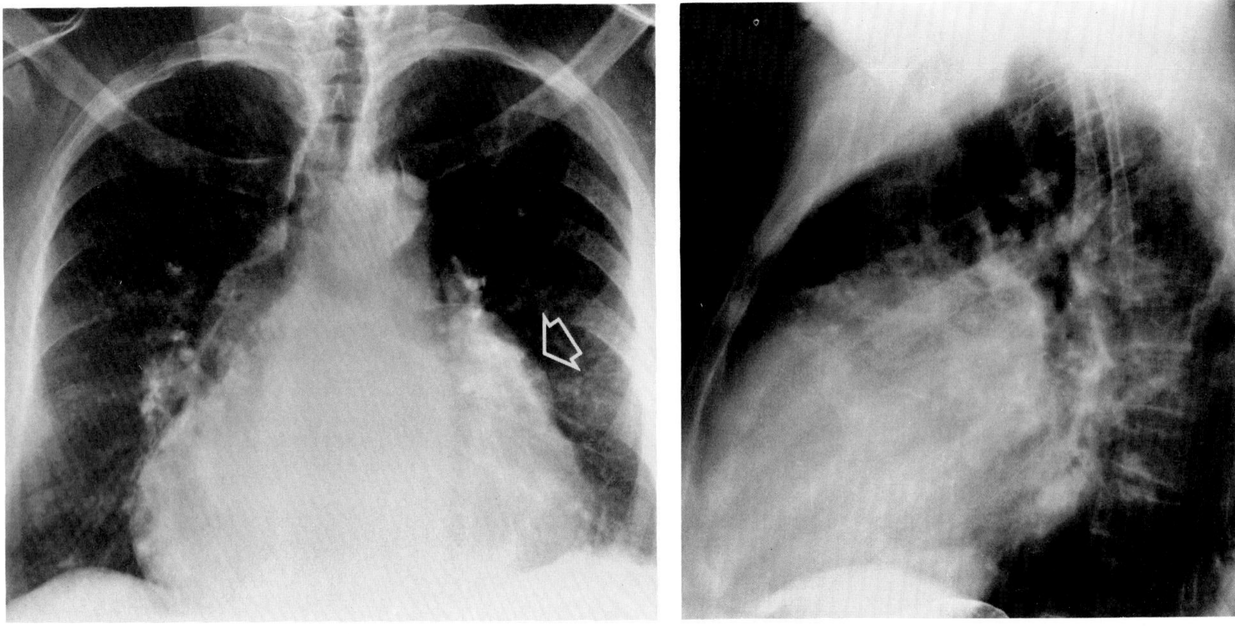

A B

**Figure 22.3. Mitral stenosis.** (A) Frontal and (B) lateral views of the chest demonstrate cardiomegaly with enlargement of the left atrium and right ventricle. Left atrial enlargement produces a convexity of the upper left border of the heart (arrow); right ventricular enlargement causes obliteration of the retrosternal air space. There also is some redistribution of pulmonary blood flow to the upper lobes.

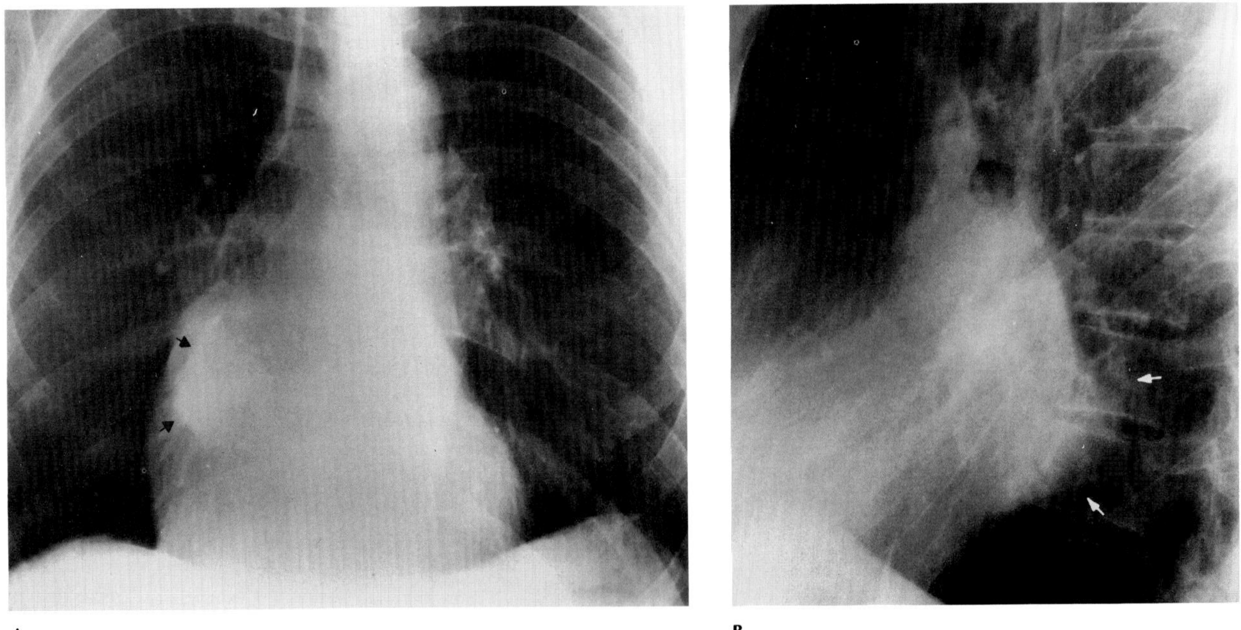

A B

**Figure 22.4. Mitral stenosis.** (A) A frontal chest radiograph demonstrates a double contour (arrows) representing the increased density of an enlarged left atrium. (B) A lateral view confirms the left atrial enlargement (arrows) in this patient with rheumatic heart disease.

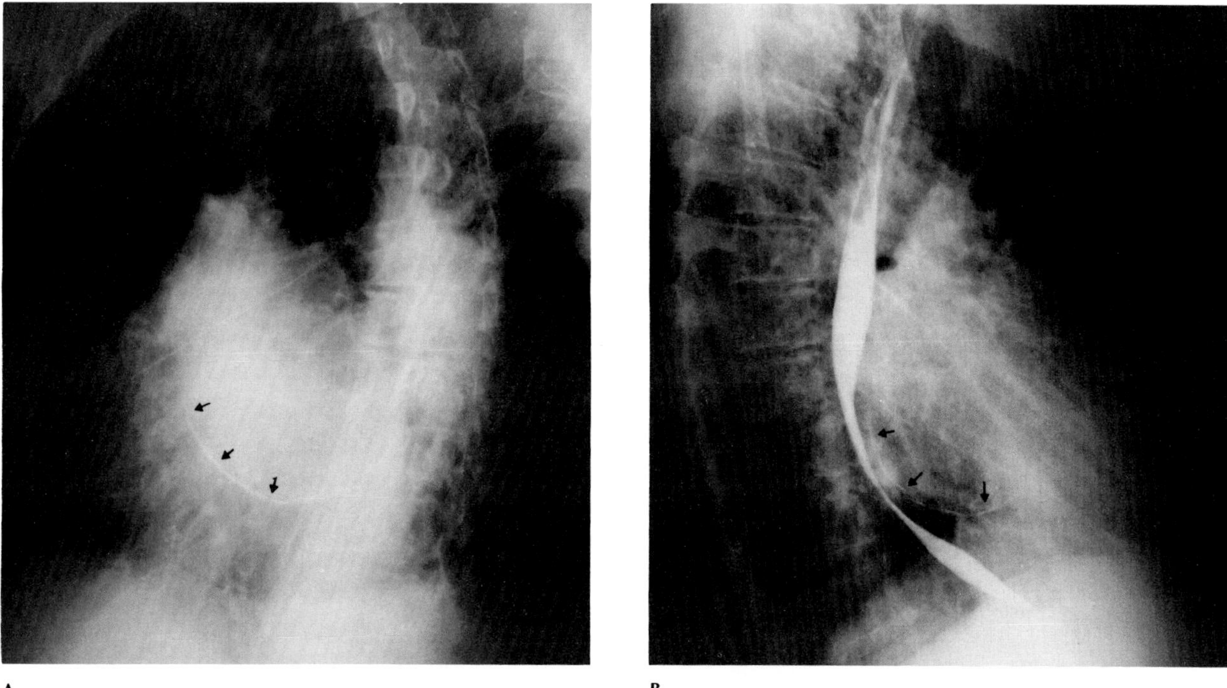


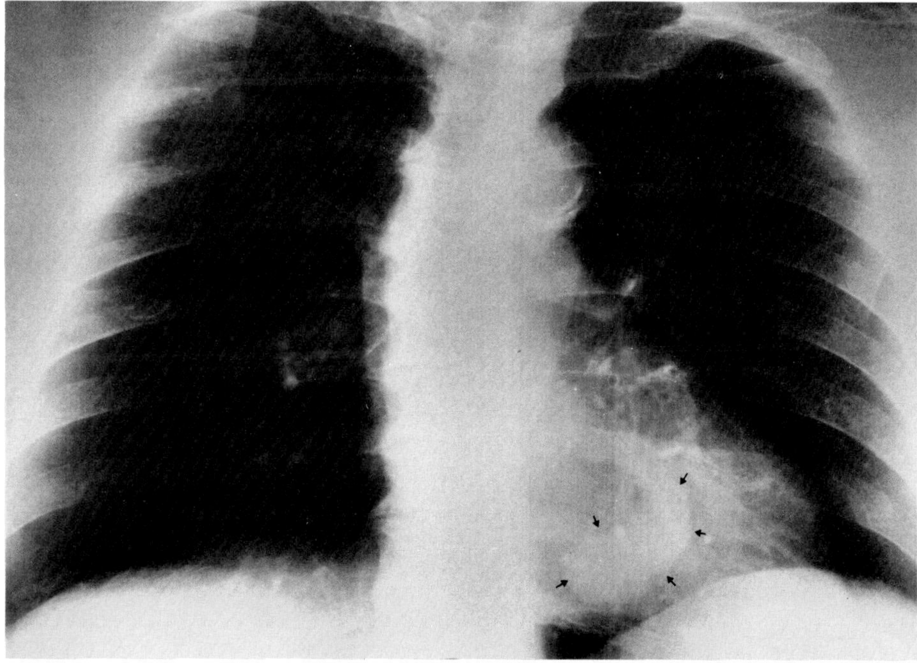


**Figure 22.5. Mitral stenosis.** (A) An overpenetrated film in the left anterior oblique projection shows left atrial calcification (arrows). (B) A lateral view with barium in the esophagus shows enlargement of the left atrium and calcification of the wall of this chamber (arrows). (C) An overpenetrated frontal film shows dense calcification of the mitral annulus (arrows). (From Vickers, Kincaid, Ellis, et al,[5] with permission from the publisher.)

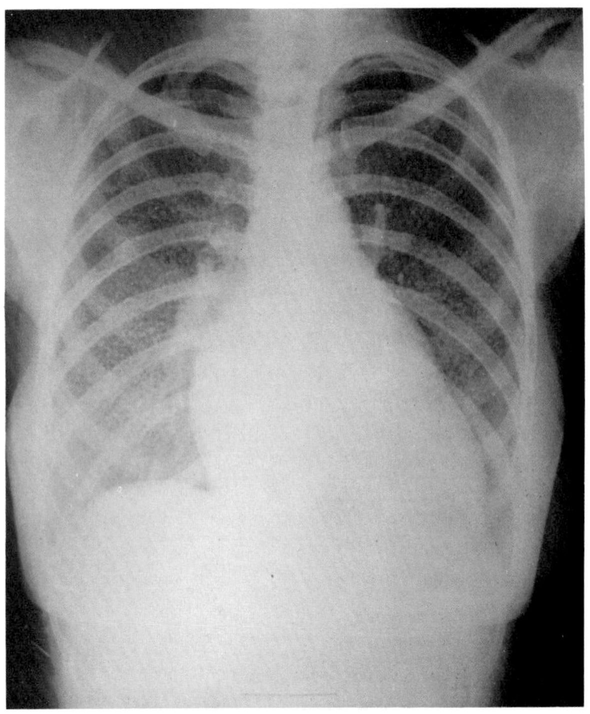
A

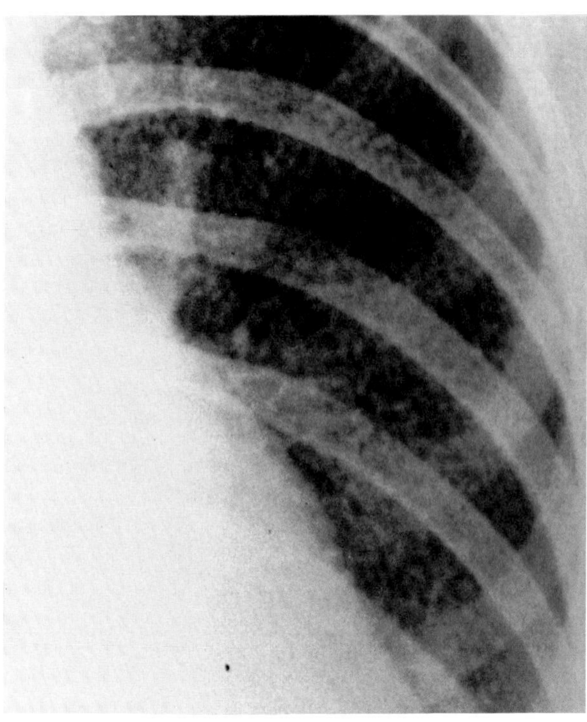
B

**Figure 22.6. Pulmonary hemosiderosis in long-standing mitral valve disease.** (A) A plain chest radiograph and (B) an enlargement of the left upper lung demonstrate miliary nodules of hemosiderin deposits throughout the lungs.

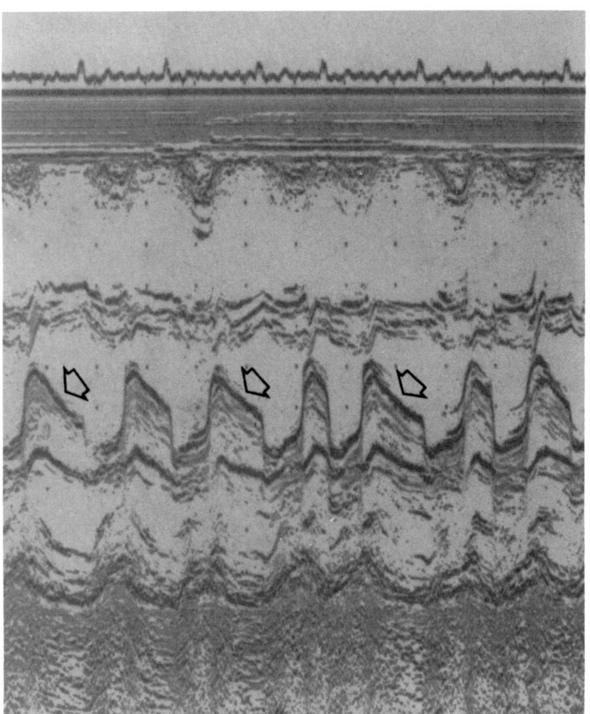

**Figure 22.7. Mitral stenosis.** An echocardiogram demonstrates thickening of the mitral valve with a decreased slope (arrows). Note that the posterior mitral valve leaflet moves anteriorly during diastole instead of posteriorly, a finding diagnostic of mitral stenosis.

float back to midposition in mid-diastole. In most patients, the posterior mitral valve leaflet moves anteriorly during diastole instead of posteriorly, a finding diagnostic of mitral stenosis. Two-dimensional imaging permits the determination of the mitral orifice area and the demonstration of left atrial enlargement.[1,4–6]

## MITRAL REGURGITATION (INSUFFICIENCY)

Although most often due to rheumatic heart disease, mitral regurgitation may also be caused by the rupture of chordae tendineae, by papillary muscle dysfunction, or by severe left ventricular dilatation that distorts the mitral annulus.

In pure mitral insufficiency, the regurgitation of blood during ventricular systole causes overfilling and dilatation of the left atrium. In most cases, the left atrium is considerably larger in mitral regurgitation than in mitral stenosis; occasionally, an enormous left atrium can form both the right and left borders of the heart on frontal projections. An increased volume of blood flowing from the dilated left atrium to the left ventricle in diastole increases the left ventricular work load and leads to dilatation and hypertrophy of this chamber. This causes downward displacement of the cardiac apex and rounding of the lower left border of the heart. The aortic knob

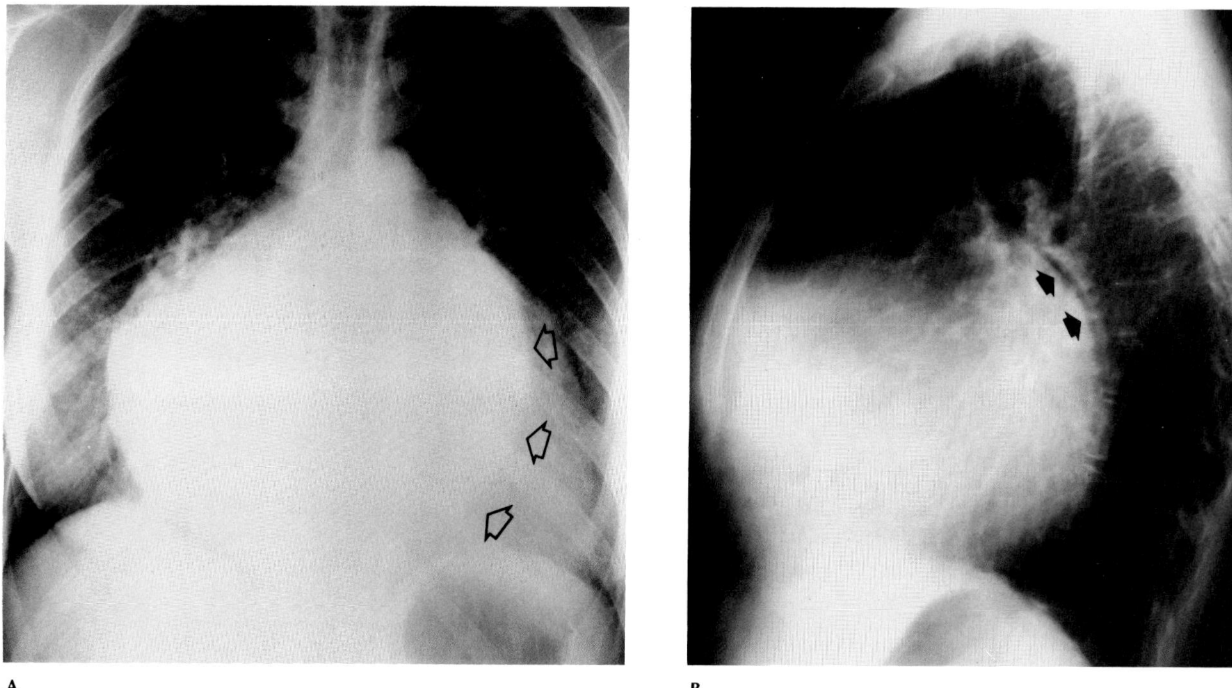

**Figure 22.8. Mitral regurgitation.** (A) Frontal and (B) lateral views of the chest demonstrate gross cardiomegaly with enlargement of the left ventricle and left atrium. Note the striking double-contour configuration (open arrows) and elevation of the left main bronchus (closed arrows), characteristic signs of left atrial enlargement. The aortic knob is normal in size, and there is no evidence of pulmonary venous congestion.

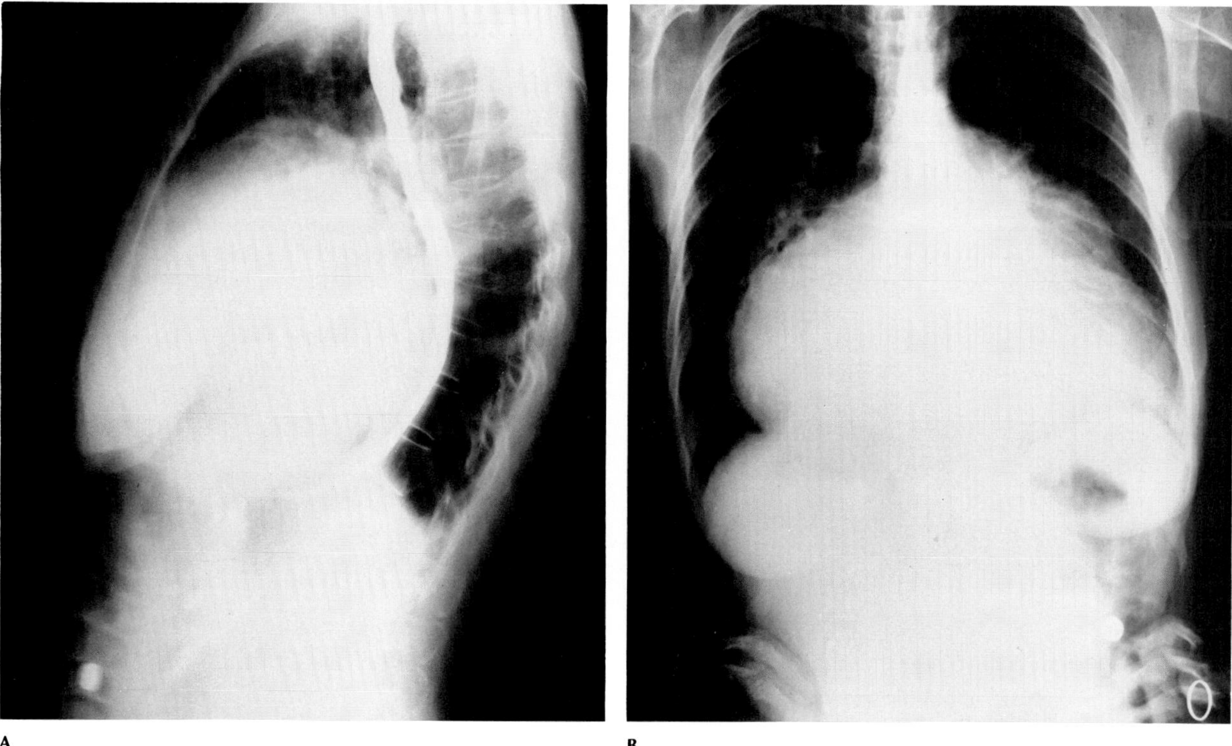

**Figure 22.9. Mitral regurgitation.** (A) Lateral and (B) frontal views of the chest show marked dilatation of the left atrium, which forms almost the entire right border of the heart. (From Cooley and Schreiber,[1] with permission from the publisher.)

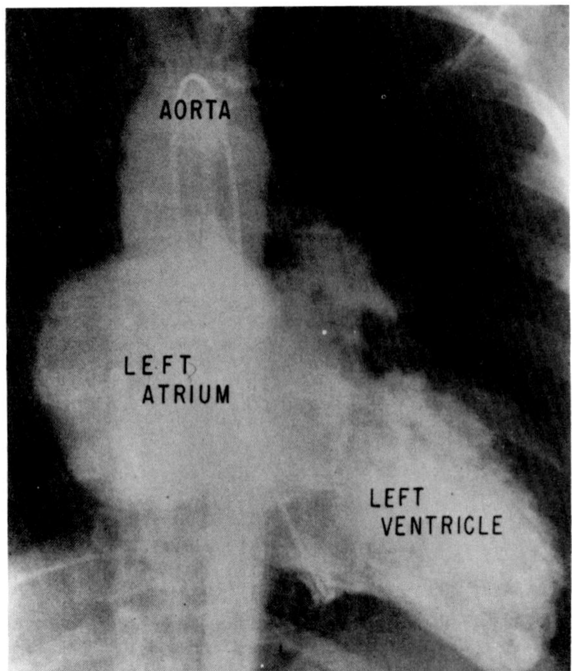


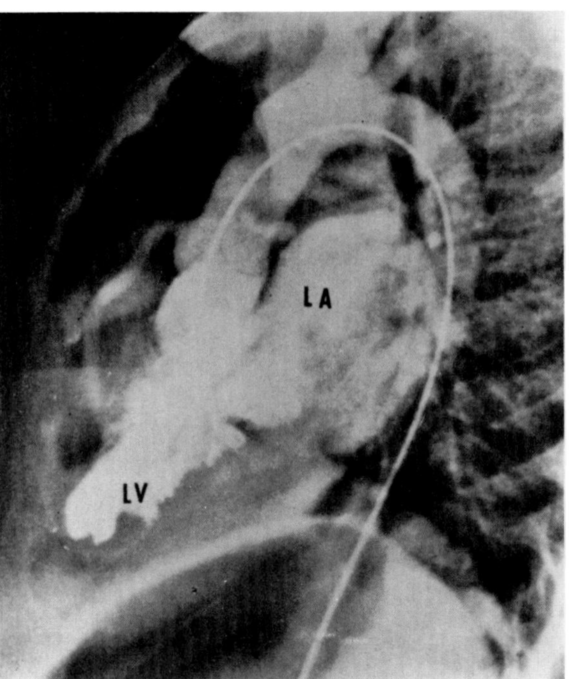


**Figure 22.10. Mitral regurgitation.** (A) An angiocardiogram with injection of contrast material into the left ventricle shows an enlarged left ventricle, a normal thoracic aorta, and a dilated left atrium, which fills through mitral regurgitation. (B) A lateral view from an angiocardiogram with injection into the left ventricle shows contrast material regurgitating through the mitral orifice into the dilated left atrium. The left ventricular wall is thickened, but the aortic valve cusps are normal. There is possible enlargement of the right coronary artery. (From Dotter,[15] with permission from the publisher.)

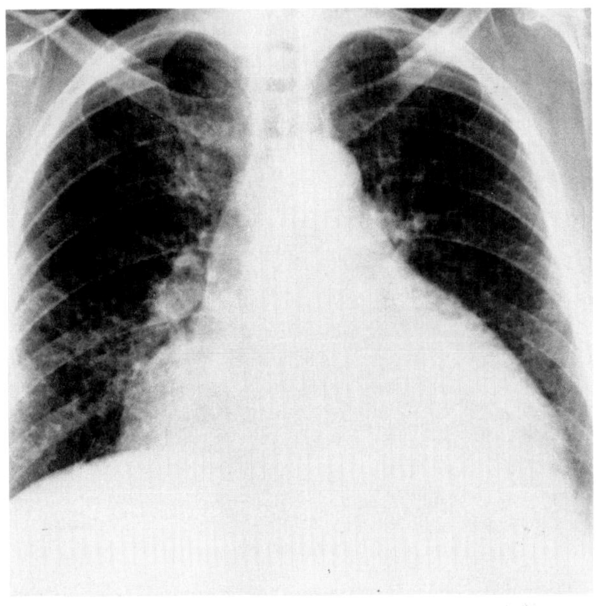

**Figure 22.11. Combined mitral regurgitation and mitral stenosis.** A frontal chest film shows marked cardiomegaly with left atrial and left ventricular prominence. Note the pulmonary venous congestion (redistribution of blood flow) that may develop when mitral regurgitation is combined with mitral stenosis.

tends to be normal in size. Because the mean pulmonary venous pressure is lower in mitral regurgitation than in mitral stenosis, evidence of pulmonary venous congestion is less frequent and less prominent.

Mitral regurgitation is often associated with mitral stenosis, and the combination may produce a confusing radiographic appearance. In general, an enormous left atrium and marked cardiomegaly (including left ventricular enlargement) suggest mitral regurgitation, as does a relatively normal appearance of the pulmonary vasculature. Calcification of the mitral valve is rare in pure mitral regurgitation; moderate or marked calcification is often found in patients with some degree of insufficiency associated with stenosis.[7,8]

## AORTIC STENOSIS

Aortic stenosis may be due to rheumatic heart disease, a congenital valvular deformity (especially a bicuspid valve), or a degenerative process of aging (idiopathic calcific stenosis). Aortic valve disease due to rheumatic heart disease is rarely an isolated process and is most commonly associated with a significant lesion of the mitral valve.

The obstruction to left ventricular outflow in aortic stenosis increases the work load of the left ventricle. The resulting concentric left ventricular hypertrophy without dilatation produces only some rounding of the cardiac apex on frontal chest radiographs and slight backward displacement on lateral views. The overall size of the heart remains within normal limits until left ventricular failure develops. Failure leads to left ventricular dilatation, which causes elongation of the left cardiac border with the apex moving downward and to the left. This eventually may be accompanied by left atrial enlargement, pulmonary venous congestion, and prominence of the right ventricle and pulmonary artery.

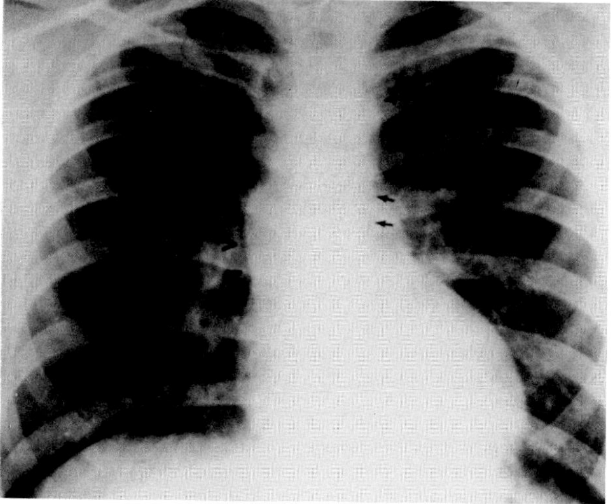

**Figure 22.12. Aortic stenosis.** A frontal chest film shows prominence of the left ventricle with poststenotic dilatation of the ascending aorta (arrowheads). The aortic knob and the descending aorta (arrows) are normal. (Courtesy of Marvin Belasco, M.D.)

Significant aortic stenosis is usually associated with poststenotic dilatation of the ascending aorta. This produces lateral bulging of the right upper margin of the heart. The aortic knob is normal or small.

Aortic valve calcification, best demonstrated on fluoroscopic examination, is a common finding and indicates that the aortic stenosis is severe.

Echocardiography reveals a thickening of the left ventricular wall in patients with aortic stenosis. Calcification of the valve produces multiple bright, thick echoes from within the aortic root, without the normal boxlike separation of the aortic cusps during diastole. Echocardiography is of particular value for identifying valvular abnormalities such as mitral stenosis and aortic regurgitation, both of which sometimes accompany aortic stenosis, and for differentiating valvular from obstructive hypertrophic cardiomyopathy.[9–10]

## AORTIC REGURGITATION (INSUFFICIENCY)

Although most commonly caused by rheumatic heart disease, aortic regurgitation may be due to syphilis, in-

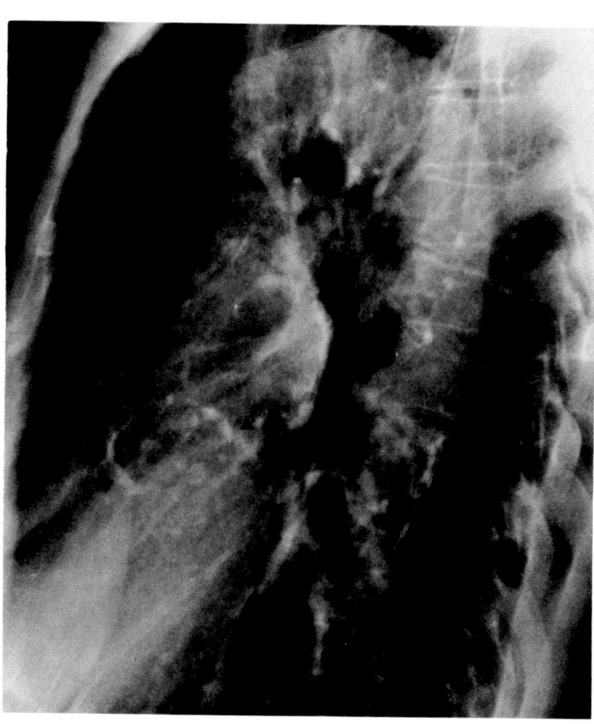

A

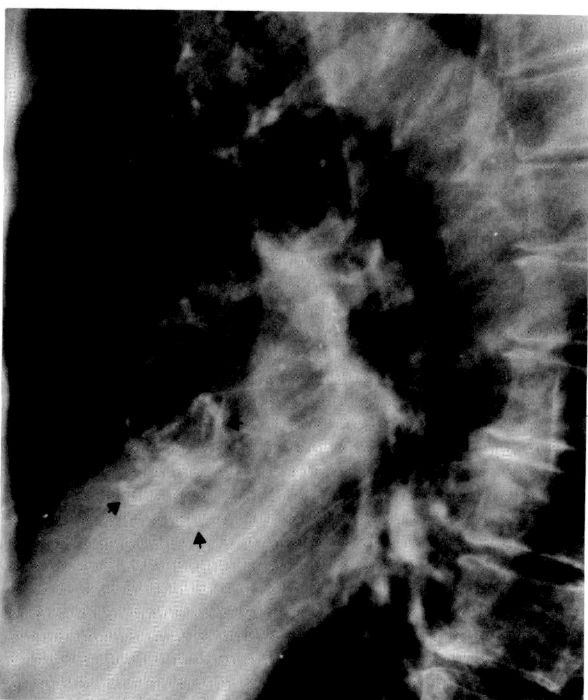

B

**Figure 22.13. Aortic stenosis.** Calcification in (A) the aortic annulus (arrows) and (B) the three leaflets of the aortic valve (arrows).

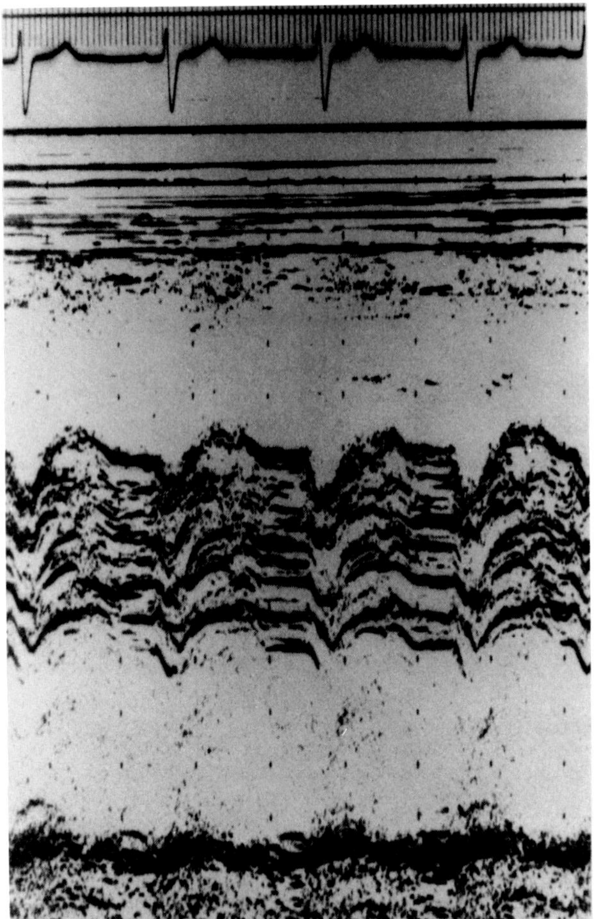

**Figure 22.14. Aortic stenosis.** An echocardiogram shows multiple continuous echoes within the aortic root during both diastole and systole, indicating thickened, degenerated aortic valve cusps. The numerous and dense echoes suggest calcification of the cusps. (From Newell, Higgins, and Kelley,[16] with permission from the publisher.)

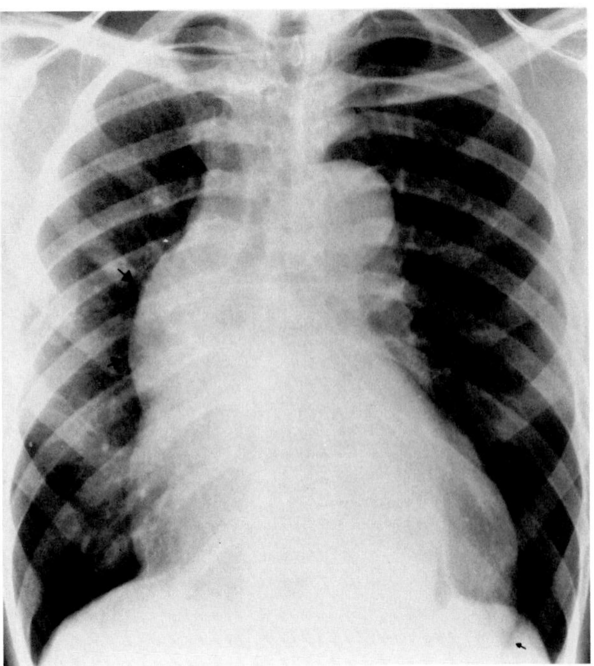

**Figure 22.15. Aortic regurgitation.** A frontal chest radiograph shows left ventricular enlargement with downward and lateral displacement of the cardiac apex. Note that the cardiac shadow exaorta is markedly dilated (arrows), suggesting some underlying aortic stenosis.

fective endocarditis, dissecting aneurysm, or Marfan's syndrome. Reflux of blood from the aorta during diastole causes volume overloading of the left ventricle and both dilatation and hypertrophy of that chamber. This causes downward, lateral, and posterior displacement of the cardiac apex; frequently, the cardiac shadow extends below the dome of the left hemidiaphragm on routine frontal projections. In the absence of congestive heart failure, the pulmonary vascularity remains normal. As the left ventricle fails, signs of pulmonary venous congestion develop along with left atrial enlargement due to relative mitral insufficiency.

The ascending aorta and the aortic knob are often moderately dilated in patients with aortic regurgitation. Marked dilatation, especially of the ascending aorta, suggests combined aortic stenosis and aortic insufficiency.[1]

## TRICUSPID VALVE DISEASE

Acquired tricuspid valve disease most commonly is the result of rheumatic heart disease. Rarely an isolated lesion, rheumatic tricuspid disease is almost always associated with mitral or aortic valve disease, most often mitral stenosis. Tricuspid regurgitation is usually functional and secondary to marked dilatation of the right ventricle in patients with right-sided heart failure. Rare causes of isolated tricuspid valve disease include carcinoid syndrome and endomyocardial fibrosis.

Tricuspid stenosis and tricuspid regurgitation produce a similar radiographic appearance, though enlargement of the right ventricle may be seen in the latter. The radiographic hallmark of tricuspid valve disease is enlargement of the right atrium due to the pooling of an excessive volume of blood. Right atrial enlargement, which can be extreme, causes bulging or increased prominence of the curvature of the right border of the heart. The superior vena cava is often dilated. The coexistence of mitral or aortic valve disease results in a complicated radiographic appearance. For example, combined tricuspid and mitral stenosis causes less pulmonary venous congestion than is seen in pure mitral valve disease.[11]

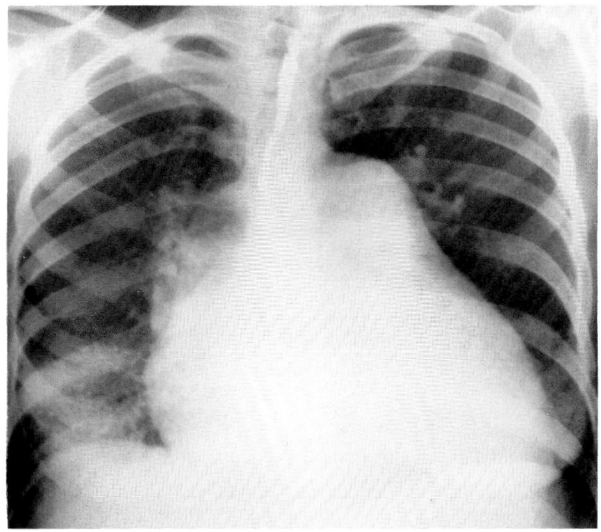
A

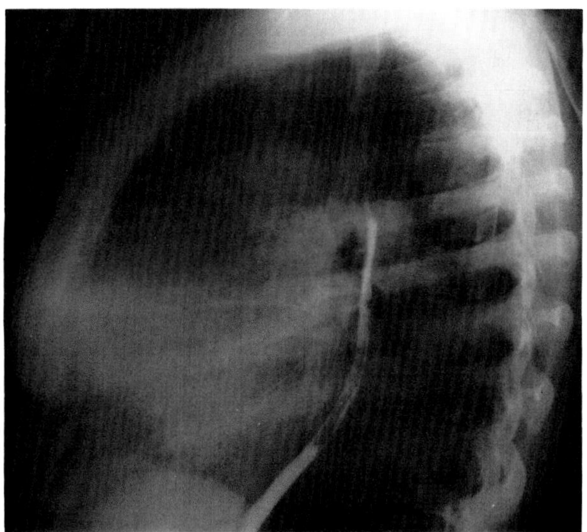
B

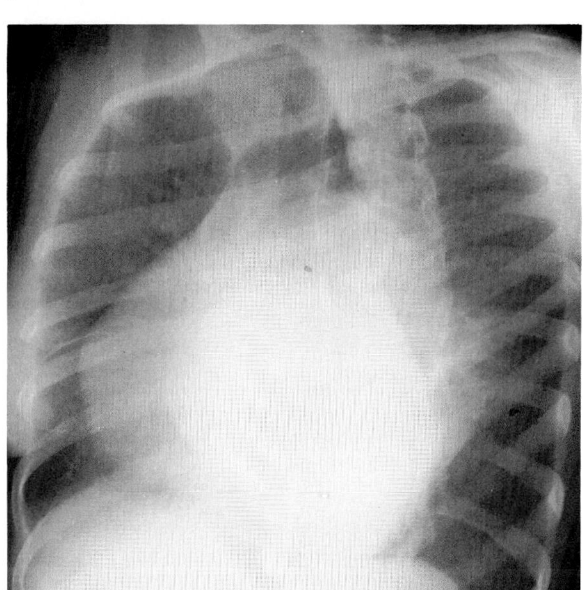
C

**Figure 22.16. Tricuspid insufficiency.** (A) Frontal, (B) lateral, and (C) left anterior oblique projections. The striking right atrial enlargement is best seen in Figures A and C. The lateral view (B) shows enlargement of the left atrium from concomitant mitral valve stenosis. The upper lobe veins, particularly on the right, are dilated. (From Cooley and Schreiber,[1] with permission from the publisher.)

## PULMONARY VALVE DISEASE

Pulmonary stenosis is usually a congenital anomaly; the radiographic features have been previously described (see the section "Pulmonary Stenosis with Intact Ventricular Septum" in Chapter 21). Pulmonary insufficiency is rare and usually results from infective endocarditis or dilatation of the pulmonary valve ring secondary to severe pulmonary hypertension. Pulmonary insufficiency may produce enlargement of the right ventricle and pulmonary artery; often no radiographic abnormality is noted.[1]

## INFECTIVE ENDOCARDITIS

Plain radiography is of little value in patients with infective endocarditis. The cardiac silhouette may be normal or demonstrate evidence of previous valvular heart disease or congestive heart failure. Abdominal radiographs often show substantial splenomegaly, which may suggest subacute infectious endocarditis in a febrile patient with appropriate cardiac murmurs.

Echocardiography is the only noninvasive procedure that can detect the valvular vegetations that are the hallmark of infective endocarditis. On the echocardiogram, these vegetations appear as masses of shaggy echoes producing an irregular thickening of the affected valves. The abnormal thickening of the valves may simulate aortic fibrosis and calcification, with or without stenosis; however, in endocarditis the cusps appear to have unrestricted motion, rather than being fixed as in intrinsic aortic valve disease.

Plain chest radiographs can demonstrate complications of infectious endocarditis. Septic pulmonary emboli from right-sided infective endocarditis can appear as small, scattered areas of consolidation mimicking pneumonia. These infiltrates often excavate, forming typical thin-walled cavities simulating pneumatoceles. Excretory urography, computed tomography, or arteri-

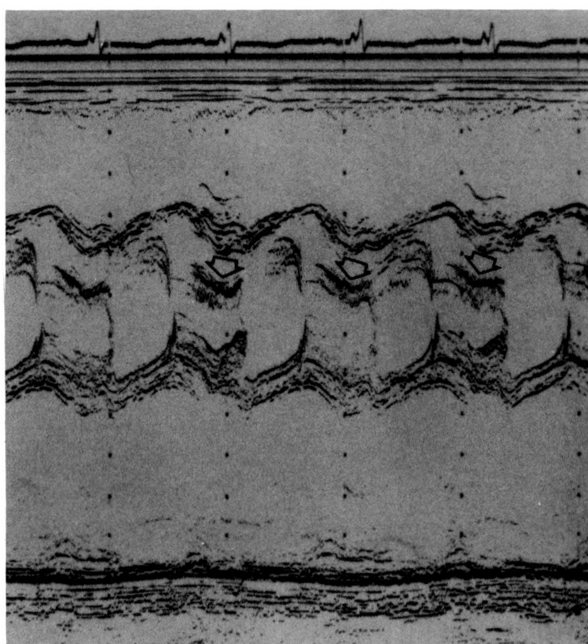

**Figure 22.17. Infective endocarditis.** An echocardiogram demonstrates vegetations as masses of shaggy echoes producing an irregular thickening of the aortic valve (arrows).

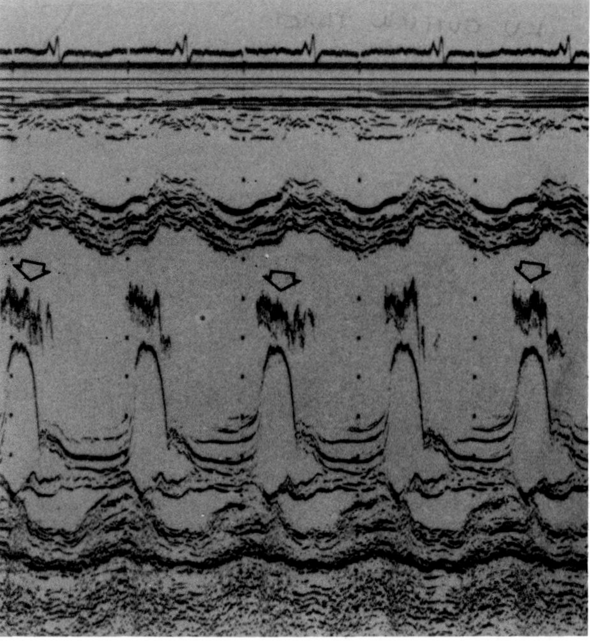

**Figure 22.18. Infective endocarditis.** In this patient, vegetations on the aortic valve (arrows) are projected into the left ventricular outflow tract just above the mitral valve.

ography can demonstrate renal infarctions from septic emboli involving the aortic valve. Acute perforation or rupture of a cusp of the aortic valve causes the typical radiographic appearance of aortic regurgitation and congestive heart failure. Rupture into a chamber on the right side of the heart produces the characteristic appearance of a left-to-right shunt.[12–14]

## REFERENCES


1. Cooley RN, Schreiber MH: *Radiology of the Heart and Great Vessels.* Baltimore, Williams & Wilkins, 1978.
2. Twigg HL, Smith BF: Jaccoud's arthritis. *Radiology* 80:417–421, 1963.
3. Murphy WA, Staple TW: Jaccoud's arthropathy reviewed. *AJR* 118:300–304, 1973.
4. Simon G: The value of radiology in critical mitral stenosis—An amendment. *Clin Radiol* 23:145–151, 1972.
5. Vickers SCW, Kincaid OW, Ellis FH, et al: Left atrial calcification. *Radiology* 72:569–575, 1959.
6. McAfee JG, Biondetti P: Roentgenologic follow-up on 150 consecutive mitral commissurotomy patients. *AJR* 78:213–225, 1957.
7. Steiner RE, Jacobson G, Dinsmore R, et al: Mitral regurgitation. *Clin Radiol* 14:113–125, 1963.
8. De Sanctis RW, Dean DC, Bland EF: Extreme left atrial enlargement: Some characteristic features. *Circulation* 29:14–23,1964.
9. Batson GA, Urquhart W, Sideris DA: Radiologic features in aortic stenosis. *Clin Radiol* 23:140–148, 1972.
10. Lehman JS, Florence H, Schimert AP, et al: Acquired aortic valvular stenosis. *Radiology* 81:24–37, 1963.
11. Tilloson PM, Steinberg I: Roentgen featues of rheumatic tricuspid stenosis. *AJR* 87:948–961, 1962.
12. Jaffe RB, Koschmann EB: Septic pulmonary emboli. *Radiology* 96:527–532, 1970.
13. Dillon JC, Feigenbaum H, Konecke LL, et al: Echocardiographic manifestations of valvular vegetations. *Am Heart J* 86:698–704, 1973.
14. Roy P, Tajik AJ, Guiliani ER, et al: Spectrum of echocardiographic findings in bacterial endocarditis. *Circulation* 53:474–482, 1976.
15. Dotter CT: Left ventricular and systemic arterial catheterization. *AJR* 83:969–984, 1960.
16. Newell JD, Higgins CB, Kelley MJ: Radiographic-echocardiographic approach to acquired heart disease. *Radiol Clin North Am* 18:387–402, 1980.

## Chapter Twenty-Three
# Other Diseases of the Heart

### ISCHEMIC HEART DISEASE

Ischemic heart disease is caused by oxygen deprivation of the myocardium resulting from reduced perfusion via the coronary arteries. In most patients, ischemic heart disease is the result of atherosclerotic lesions in the coronary circulation. The degree of luminal narrowing determines whether an atherosclerotic lesion causes significant and clinically evident ischemia.

Plain chest radiographs are usually normal or nonspecific in most patients with ischemic heart disease. Cardiac enlargement, especially involving the left ventricle, is a nonspecific finding that usually reflects the presence of a large quantity of infarcted myocardium. Calcification of a coronary artery, though infrequently visualized on routine chest radiographs, strongly suggests the presence of hemodynamically significant coronary artery disease. Coronary artery calcification may appear as punctate, patchy, or linear densities. Calcification primarily involves the circumflex and anterior descending branches of the left coronary artery and is most commonly seen along the left margin of the heart below the pulmonary artery segment.

Cardiac fluoroscopy is far more sensitive than plain chest radiography in demonstrating coronary artery calcification. There is controversy about the prognostic significance of fluoroscopically identified coronary artery calcification in patients with ischemic heart disease. In patients under the age of 50, coronary artery calcification is a strong predictor of major narrowing in women and a moderate predictor of significant stenosis in men. The identification of calcification in older patients is not as useful, since calcification is observed in many elderly patients who do not manifest significant coronary obstructive disease.

Radionuclide thallium perfusion scanning is a physiologic study that primarily assesses the regional blood flow to the myocardium. Focal decreases in thallium uptake that are observed immediately after exercise but no longer identified on delayed scans usually indicate transient ischemia associated with significant coronary artery stenosis or spasm. Postexercise focal defects that are unchanged on delayed scans more frequently reflect scar formation. A normal thallium exercise scan makes a diagnosis of myocardial ischemia unlikely, though in about 10 percent of patients with significant obstructive disease the presence of sufficient collateral vessels can prevent the radionuclide demonstration of regional ischemia.

Radionuclide ventriculography with multiple-gated blood pool images of the left ventricle after the injection of pertechnetate-labeled red blood cells can provide a measure of the ventricular volume and ejection fraction at rest and during exercise. This modality can demon-

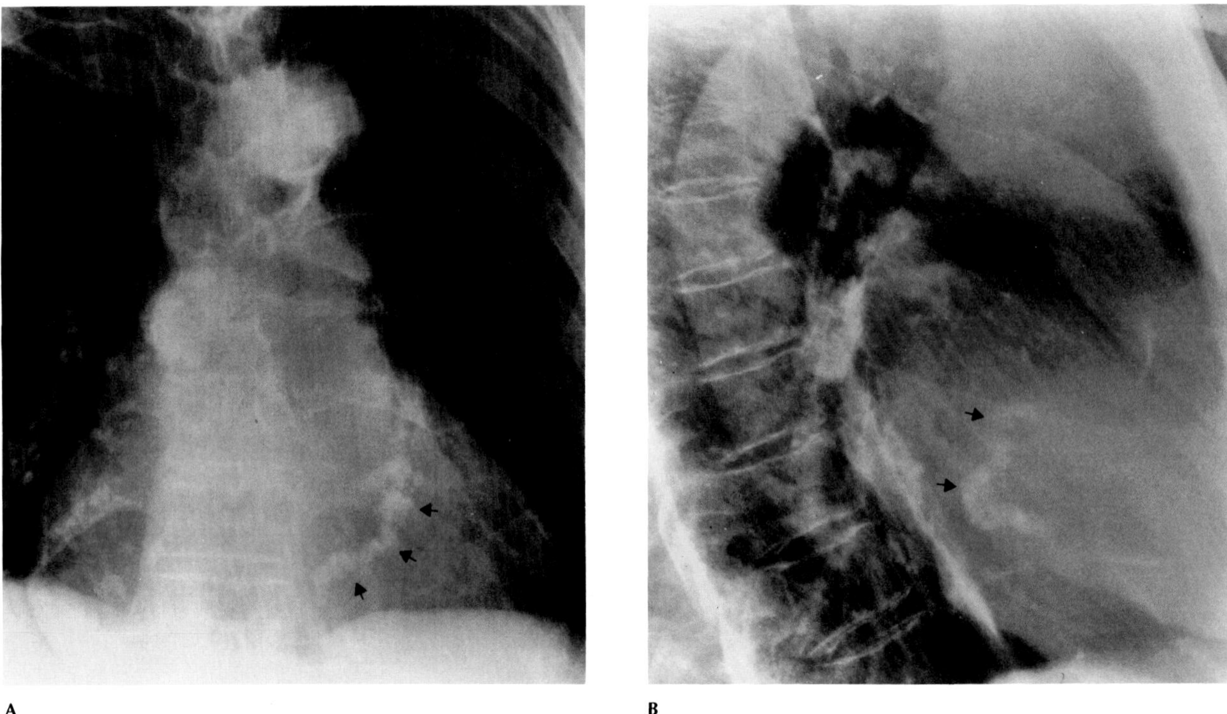


**Figure 23.1. Ischemic heart disease.** (A) Frontal and (B) lateral views of the chest demonstrate cardiomegaly with typical linear calcification in a coronary artery (arrows).

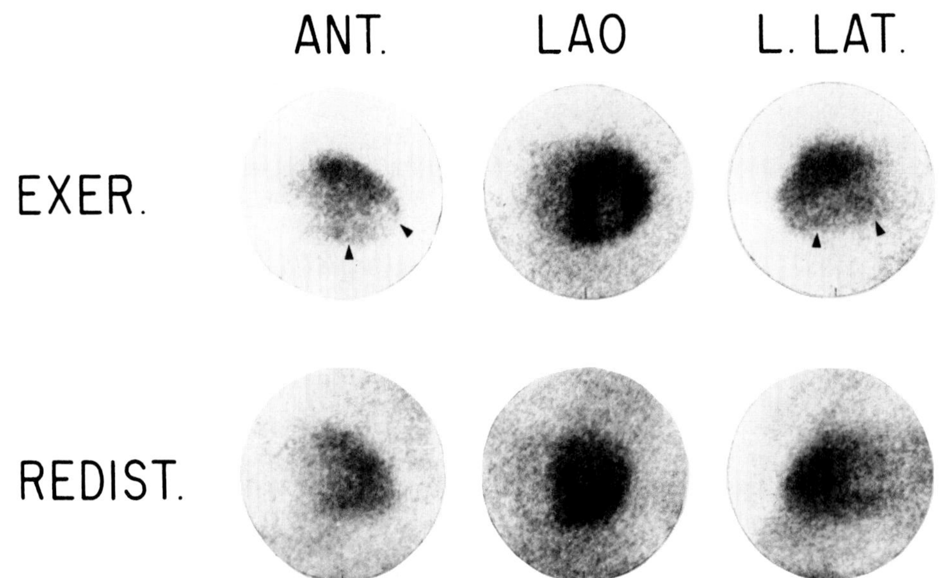


**Figure 23.2. Radionuclide thallium perfusion scanning of ischemic heart disease.** This 56-year-old man with right coronary and circumflex artery disease showed a focal diminished uptake of thallium 201 (arrowheads) immediately following exercise (EXER), which was terminated by fatigue without angina at 8 minutes. The patient reached a heart rate of 160. Note the almost complete redistribution in the inferior wall at 4 hours (bottom row), a characteristic finding in stress-induced ischemia. ANT = anterior; LAO = left anterior oblique; L LAT = left lateral. (From Ashburn and Tubau,[1] with permission from the publisher.)

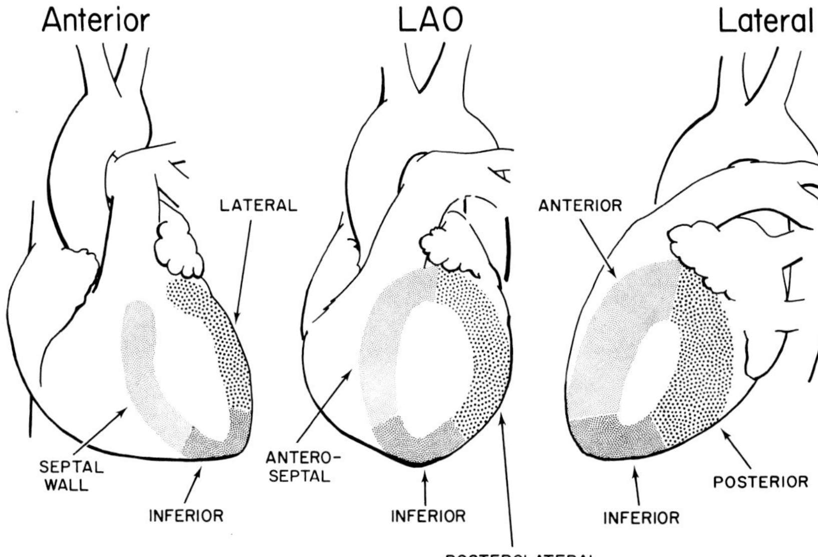


**Figure 23.3. Radionuclide thallium perfusion scanning of ischemic heart disease.** A line drawing of anterior, left anterior oblique (LAO), and left lateral (LLAT) views shows the portion of the left ventricular myocardium seen in each position with thallium scanning. The myocardium perpendicular to the scintillation camera face is seen best because of the "end-on" effect with increased depth of radioactivity. (From Parkey,[7] with permission from the publisher.)

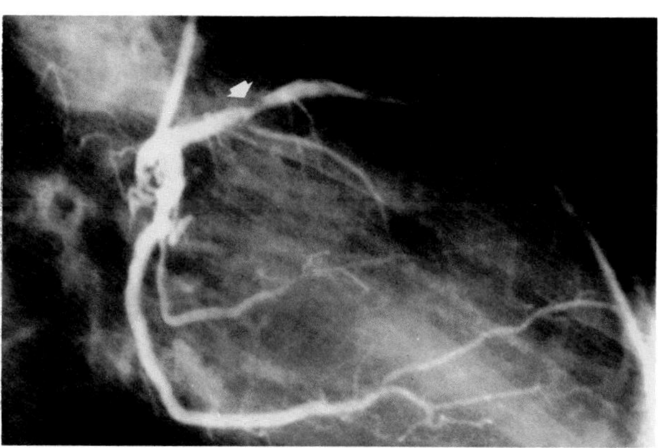

**Figure 23.4. Coronary arteriogram of ischemic heart disease.** Selective left coronary arteriogram in the right anterior oblique projection demonstrates a 75 percent stenosis (arrow) of the left anterior descending artery. (Courtesy of Martin J. Lipton, M.D.)

strate disorders of global left ventricular function and of localized wall motion. Patients with ischemic heart disease show no increase and often a decrease in the ejection fraction during exercise; normal persons increase the ejection fraction by more than 5 percent. Although there are a large number of false-positive results with the ejection fraction technique, one study reported that the combination of exercise thallium imaging and the ejection fraction technique detected all the patients with significant coronary artery disease.

Like all forms of exercise testing, radionuclide thallium scanning and the measurement of the ejection fraction require close physician monitoring.

Coronary arteriography is generally considered the definitive test for determining the presence and assessing the severity of coronary artery disease. About 30 percent of significant stenoses involve a single vessel, most commonly the anterior descending artery. Two vessels are involved in 30 percent, and significant stenosis of the three main vessels can be demonstrated in the remaining 40 percent. Almost all patients with complete occlusion of one coronary artery demonstrate stenotic lesions in one or both of the other two major vessels. Coronary arteriography also demonstrates congenital vascular anomalies (malformation, aneurysm, fistula) and can exclude coronary artery disease in those patients who have angina due to aortic valvular disease.[1–7]

## Aortocoronary Bypass Graft

Aortocoronary bypass graft, usually using sections of saphenous vein, is an increasingly popular procedure in patients with ischemic heart disease. Arteriography has been the procedure of choice in demonstrating the patency and functional efficiency of aortocoronary bypass grafts. Patent functioning grafts demonstrate prompt clearing of contrast material and good filling of the grafted artery. Stenotic or malfunctioning grafts demonstrate areas of narrowing; filling defects; slow flow and delayed washout of contrast material; or even retrograde filling of the graft if there has been complete proximal occlusion.

New, less invasive techniques can also permit the accurate evaluation of graft patency. High-speed computerized transmission tomography following the intravenous injection of contrast material can evaluate graft patency and also determine the quality of the graft. Graft patency can also be demonstrated by digital angiography following intravenous contrast injection.[8–11]

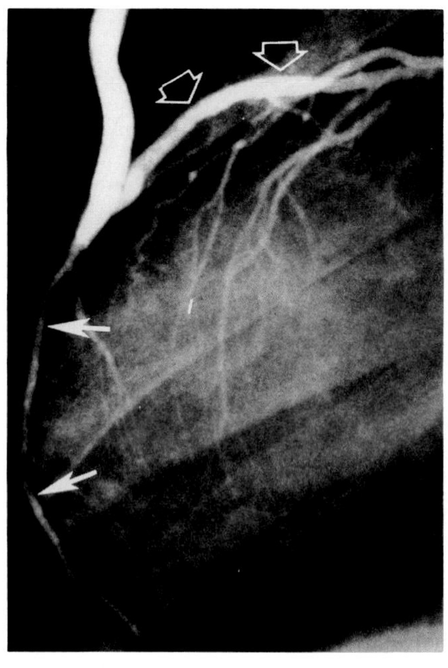

A

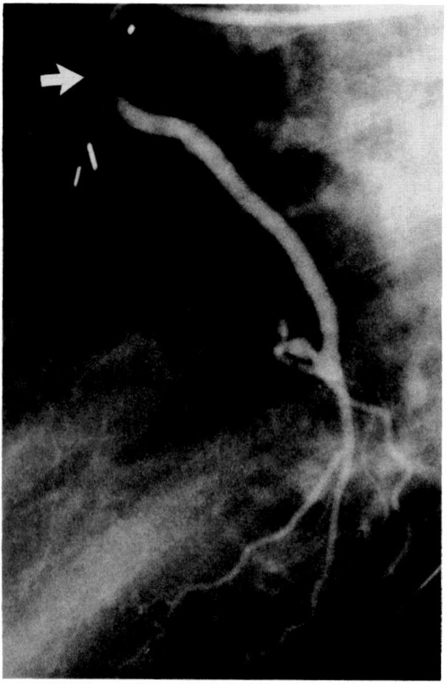

B

**Figure 23.5. Angiography of a normal aortocoronary bypass graft.** (A) With the patient in a lateral position, selective injection of contrast material into a graft to the left anterior descending coronary artery demonstrates a widely patent graft that is anastomosed to the midportion of the left anterior descending coronary artery. Both antegrade distal filling (closed arrows) and retrograde filling of the proximal left anterior descending coronary artery (open arrows) and its diagonal and septal branches are evident. (B) The selective injection of contrast material into a graft to the circumflex marginal coronary system (lateral projection) demonstrates a proximal course anterior to the pulmonary artery (arrow), and a posterior course to the distal insertion. There is good antegrade distal filling (open arrow) and minimal retrograde proximal filling (arrowhead). (From Guthaner and Wexler,[8] with permission from the publisher.)

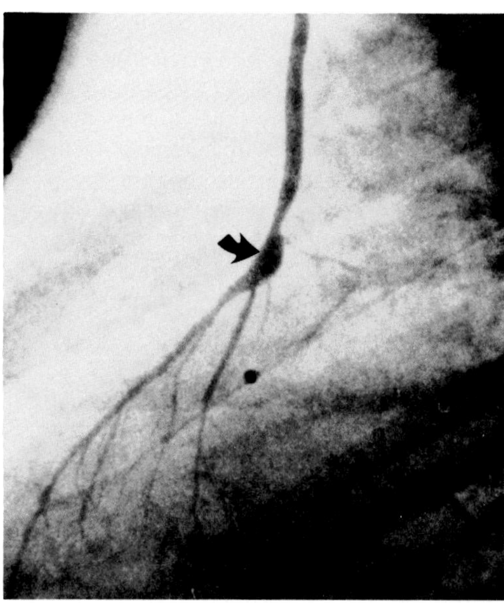

A

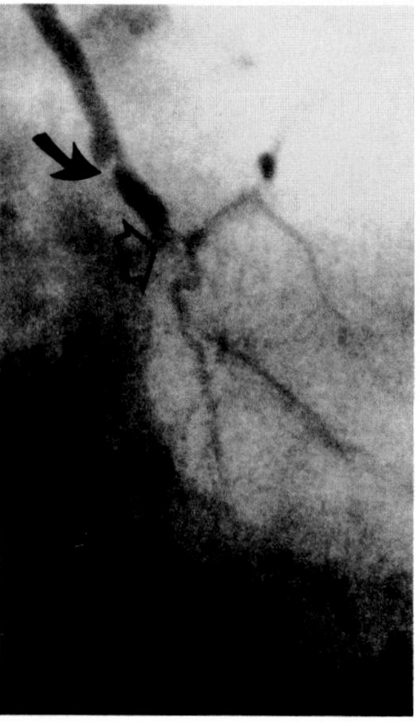

B

**Figure 23.6. Angiography of aortocoronary bypass graft dysfunction.** (A) The selective injection of contrast material into a graft to the left anterior descending coronary artery (lateral projection) demonstrates a narrowing of the distal graft and a pseudoaneurysm at the distal anastomotic site (arrow). There is good filling of the distal left anterior descending coronary artery, but poor filling is evident proximally. (B) At late follow-up, a localized segment of irregularity and stenosis (arrow) attributable to intimal hyperplasia and atherosclerosis is visualized. Occlusion at the distal anastomosis (open arrow) has occurred, and there is filling of only a few diagonal branches. The aneurysm no longer fills. (From Guthaner and Wexler,[8] with permission from the publisher.)

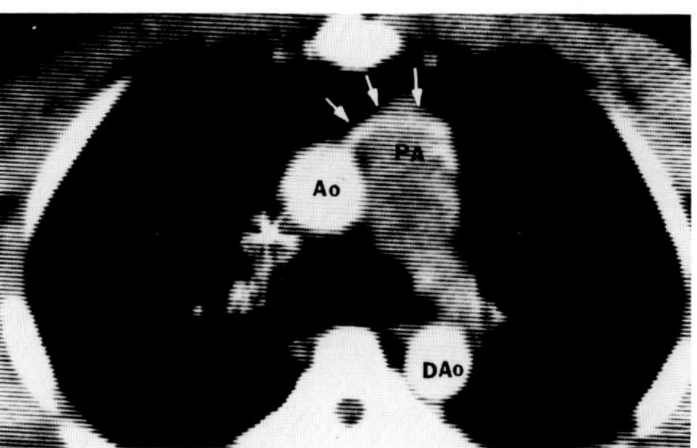


**Figure 23.7. Aortocoronary bypass graft.** A CT scan was performed through the level of the main pulmonary artery (PA) following the intravenous administration of a bolus of contrast material. During opacification of the ascending aorta (Ao) and descending aorta (DAo), the graft to the left anterior descending coronary artery (arrows) is seen arising from the ascending aorta and coursing anterior to the pulmonary artery. Of incidental note is a streak artifact from the central venous line in the superior vena cava. (From Guthaner and Wexler,[8] with permission from the publisher.)

## ACUTE MYOCARDIAL INFARCTION

Plain chest radiographs are entirely normal in many, if not most, patients sustaining an initial myocardial infarction. The heart size is often difficult to gauge on initial films since they are usually anteroposterior or portable radiographs that have been obtained with the patient in a supine or semiupright position. Shallow respiration during the first few days after acute myocardial infarction often results in the development of linear streaks of atelectasis at the lung bases. Evidence of pulmonary venous congestion can be detected within the initial 24 hours after an acute myocardial infarction in up to one-half of the patients; the presence and severity of this finding has been shown to have a close correlation with early and late mortality following acute myocardial infarction. The abrupt development of pulmonary edema suggests a complication of acute myocardial infarction, such as rupture of the interventricular septum or a papillary muscle.

Radionuclide scanning using $^{99m}$Tc-pyrophosphate or other compounds that are sequestered by acutely infarcted myocardium (infarct avid myocardial scintigraphy) is a new noninvasive technique that aids in the detection, localization, and classification of myocardial necrosis. The focal cardiac uptake of isotope ("hot spot") is a highly accurate indicator of transmural acute myocardial infarction. A diffuse pattern of cardiac uptake, however, has a considerably lower specificity and predictive value. Almost 15 percent of patients without acute myocardial infarction demonstrate the diffuse pattern; this pattern is frequently the only abnormality displayed on scans in patients with subendocardial myocardial infarction.

Infarct avid myocardial scintigraphy is of value in patients with suspected acute infarction when other clinical and laboratory evidence has been nondiagnostic. It is especially helpful in patients who present more than 48 hours after the onset of symptoms, when the very sensitive myocardial serum enzyme levels have usually returned to normal. Because increased isotope uptake persists several months after infarction, a positive infarct avid myocardial scan in a patient with a prior myocardial infarction may not always be diagnostic of an acute myocardial infarction. In addition, almost half of the patients with unstable angina without definite acute infarction also have positive scans.

Myocardial infarction causes a regional decrease in coronary blood flow that appears on radionuclide thallium scans as an area of diminished isotope activity within the outline of the left ventricle. This technique gives positive results in almost all patients with acute myocardial infarctions who are scanned within 6 hours after the onset of symptoms. Thus thallium scanning can be used effectively in selecting those patients with persistent precordial pain who should be admitted to the coronary care unit. A normal thallium scan (a uniform uptake of isotope on all views) obtained within the early hours following the suspected event can provide reasonable assurance that an acute myocardial infarction is unlikely to have occurred. Unfortunately, thallium scanning may be unable to differentiate acute infarction from unstable angina with transient (reversible) ischemia or scar formation secondary to a previous acute infarction.

Areas of myocardial ischemia appear on computed tomography (CT) as regions of decreased attenuation value because of the increased water content due to intramyocardial cellular edema. The injection of intravenous contrast material has a biphasic effect on the infarcted area. During the first 1 to 2 minutes after injection (perfusion phase), the attenuation value of normal myocardium increases, while the nonperfused infarct is minimally enhanced or nonenhanced. On the other hand, scans obtained 5 to 10 minutes after contrast injection demonstrate a relatively increased accumulation of contrast material in the infarcted area (hot spot). Because of its excellent spatial resolution, CT may be more precise than infarct avid myocardial scintigraphy in defining the

A

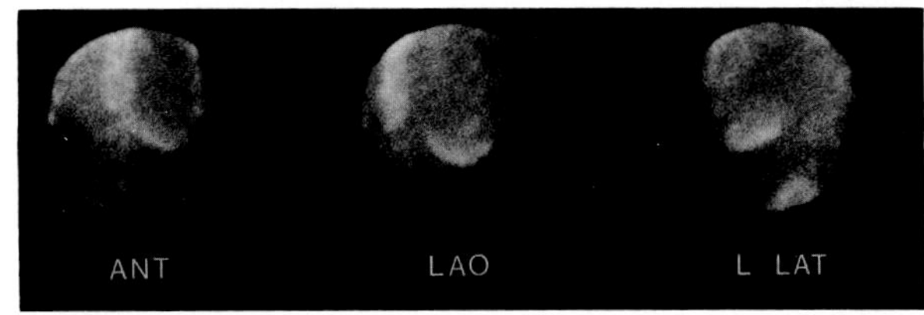


B

C

**Figure 23.8. Infarct avid myocardial scintigraphy in myocardial infarction.** (A) Anterior inferior myocardial infarction. (B) Inferior posterior myocardial infarction. (C) A diffuse pattern of cardiac uptake in a patient with acute subendocardial myocardial infarction. ANT = anterior; LAO = left anterior oblique; L LAT = left lateral. (From Holman and Wynne,[12] with permission from the publisher.)

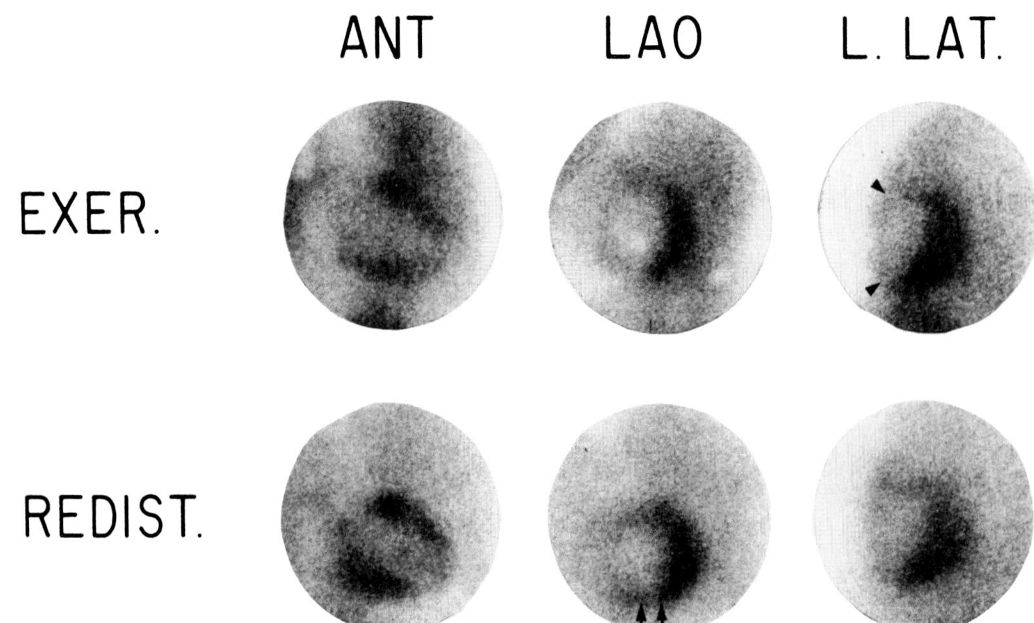


**Figure 23.9. Radionuclide thallium perfusion scan in myocardial infarction.** A 36-year-old man with a history of several prior acute myocardial infarctions that were localized electrocardiographically to the anteroseptal region was exercised to a heart rate of only 130. At this point exercise was discontinued because of dyspnea and angina. On the images obtained immediately after exercise (EXER), extensive regions of reduced uptake of thallium 201 can be seen on all three projections. This reduced uptake is most marked in the anterior wall (arrowheads), as seen on the left lateral (L LAT) view. Virtually no redistribution (REDIST) into this region occurred at 4 hours (bottom row), a finding suggestive of extensive scar formation. Note the small region at the apex (double arrows) in which the redistribution of thallium did occur. This suggests the presence of transiently ischemic yet viable myocardium in this area. The pulmonary accumulation of isotope following exercise (top left) is a finding often associated with chronic left ventricular failure. Submaximal exercise stress is suggested by the accumulation of isotope in the liver (beneath the heart). ANT = anterior; LAO = left anterior oblique. (From Ashburn and Tubau,[1] with permission from the publisher.)

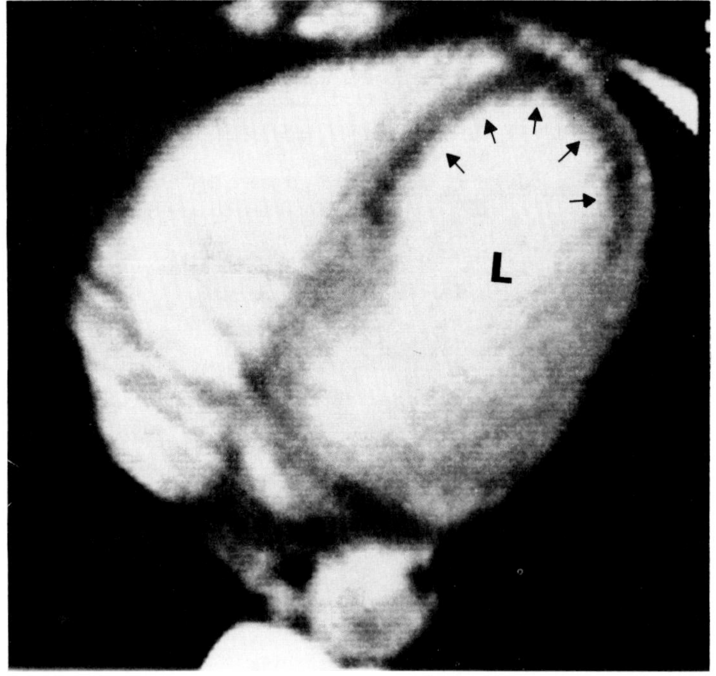


**Figure 23.10. CT scan in myocardial infarction.** A CT scan through the level of the cardiac ventricles in a patient with an acute (3-day-old) myocardial infarction shows both ventricular cavities as contrast-enhanced regions. The left ventricle (L) is surrounded by less dense myocardium than is the right ventricle. The large infarcted region is represented by a low-density zone (arrows). It is anterior in location and extends into the septum that lies between the two ventricular chambers. (From Lipton and Boyd,[27] with permission from the publisher.)

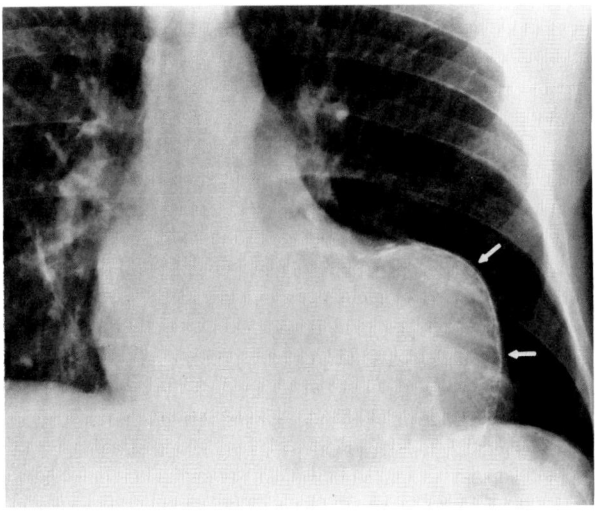

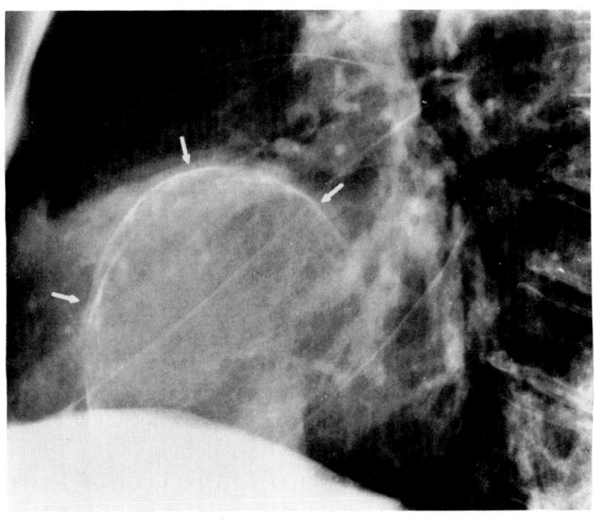

A B

**Figure 23.11. Ventricular aneurysm.** (A) Frontal and (B) lateral views of the chest demonstrate a bulging and curvilinear peripheral calcification (arrows) along the lower left cardiac border near the apex. Note the relatively anterior position of the aneurysm on the lateral view.

site, transmural extent, and size of acute myocardial infarctions. In addition, CT can detect myocardial necrosis as early as 2 hours after the onset of symptoms. CT can not only directly image the infarct itself, but can also identify an associated left ventricular aneurysm and mural thrombi.[3,6,12–14]

## Ventricular Aneurysm

Ventricular aneurysm is most commonly a complication of myocardial infarction; trauma and inflammation are infrequent underlying conditions. Weakening of the myocardial wall permits the development of a local bulge at the site of infarct. On frontal projections, a ventricular aneurysm most often produces a focal bulging or diffuse prominence along the lower left border of the heart near the apex; the aneurysm is located anteriorly on lateral views. Curvilinear calcification in the wall of the aneurysm is an infrequent but important finding. At fluoroscopy, a ventricular aneurysm demonstrates paradoxical or extremely limited pulsation.

Angiocardiography can demonstrate the thin-walled ventricular aneurysm, its paradoxical bulging during systole, and the size, shape, and amount of thrombus within it. It can also differentiate this entity from cardiac tumors, adjacent mediastinal masses, pericardial cysts, or localized pericardial effusions.

False aneurysms of the left ventricle usually result from the pericardial containment of localized rupture of an infarcted area of the ventricular wall. In contrast to true ventricular aneurysms, which have an anterolateral position, most false aneurysms arise from the posterior border of the heart. On frontal views, the false aneurysm may appear as a left retrocardiac double density. False aneurysms tend to be much larger than true aneurysms and often appear to increase in size on serial radiographs. Angiocardiography can opacify the false aneurysm and often demonstrates the narrow ostium connecting it with the left ventricle.[15–18]

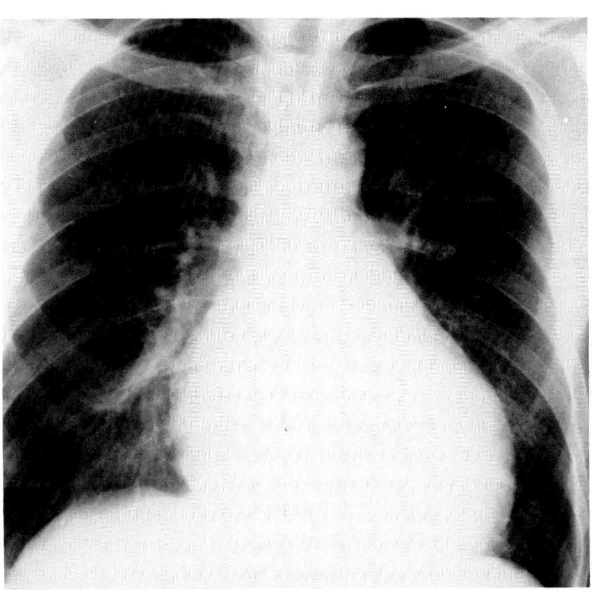

**Figure 23.12. Dressler's syndrome.** A chest film obtained 3 weeks after an acute myocardial infarction demonstrates a large pericardial effusion appearing as a rapid increase in heart size in comparison with an essentially normal-sized heart 1 week previously.

## Postmyocardial Infarction Syndrome (Dressler's Syndrome)

The postmyocardial infarction syndrome, which probably represents an autoimmune pericarditis, pleuritis, and pneumonitis, is characterized by fever and pleuropericardial chest pain that begin 1 to 6 weeks after acute myocardial infarction. Pericardial effusion, pleural effusion, and basilar pneumonia, occurring alone or in combination, can be demonstrated in most patients. Pericardial effusion appears as a rapid increase in heart size. About half of the pleural effusions are bilateral, though they are usually greater on the left side. Ill-defined patches of air space consolidation, often with linear streaks of atelectasis, may be bilateral or only involve the left base. The findings on chest radiography are often of value in distinguishing the postmyocardial infarction syndrome from pulmonary thromboembolism or recurrent myocardial infarction.[19,20]

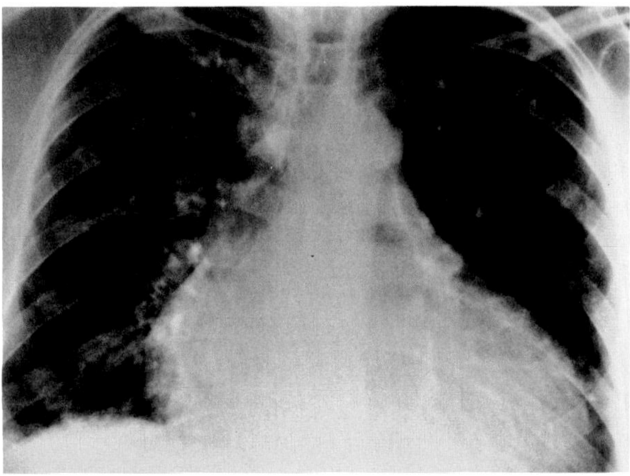

**Figure 23.13. Alcoholic cardiomyopathy.** There is generalized cardiac enlargement that involves all chambers but has a left ventricular predominance. Pulmonary vascular congestion and a right pleural effusion are also evident. The radiographic appearance simulates that of pericardial effusion.

## CARDIOMYOPATHY

The cardiomyopathies are diseases that involve the myocardium primarily, rather than as a result of hypertension or of congenital, valvular, coronary arterial, or pericardial abnormalities. The cardiomyopathies have been divided pathophysiologically into three main groups. Dilated (congestive) cardiomyopathy refers to impairment of systolic pump function, which leads to cardiac enlargement and often produces symptoms of congestive heart failure. This group includes cardiomyopathy due to such diverse causes as alcoholism, drugs, infection, neuromuscular disease, hyperthyroidism, myxedema, anemia, and beriberi; it also includes the rare cardiomyopathy that develops in the postpartum period. Restrictive cardiomyopathy is due to fibrosis, hypertrophy, or infiltration of the myocardium that makes the ventricular walls excessively rigid and impedes ventricular filling. Examples of this group include endomyocardial fibrosis, eosinophilic endomyocardial disease, amyloidosis, glycogen storage disease, hemochromatosis, sarcoidosis, and metastases to the heart. Hypertrophic (obstructive) cardiomyopathy is characterized by left ventricular hypertrophy, as in patients with idiopathic hypertrophic subaortic stenosis.

The radiographic manifestations of cardiomyopathy are similar regardless of the underlying cause. There is usually generalized cardiac enlargement involving both sides of the heart, often with left ventricular predominance. The degree of cardiac enlargement varies considerably depending on the severity of the illness, its chronicity, and the extent of myocardial damage. The development of left ventricular failure leads to pulmonary venous congestion. In many instances, the diffuse

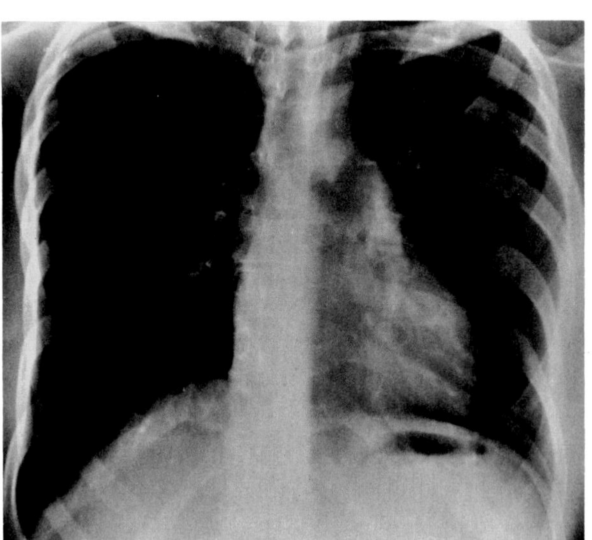

A

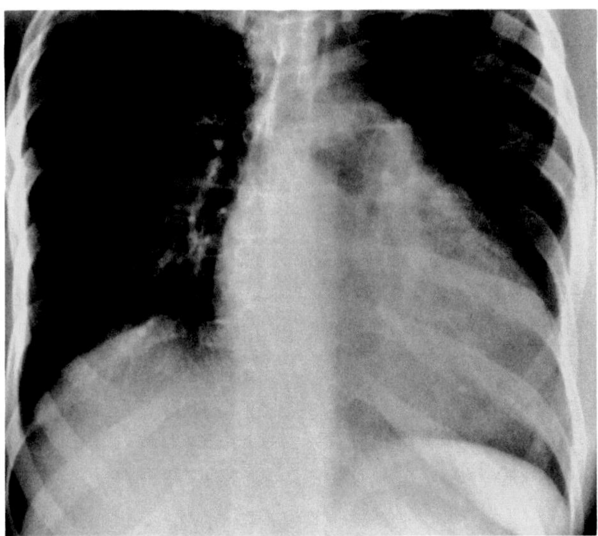

B

**Figure 23.14. Drug-induced cardiomyopathy.** (A) An initial chest radiograph in a 25-year-old woman with carcinoma of the breast demonstrates a right mastectomy but a normal-sized heart. (B) After several months of doxorubicin hydrochloride (Adriamycin) therapy, there is dramatic enlargement of the heart with striking left ventricular prominence.

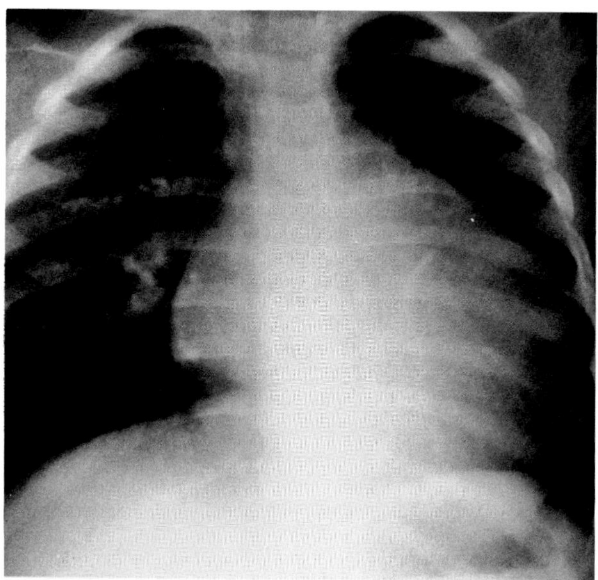

**Figure 23.15. Glycogen storage disease.** Generalized cardiac enlargement with a globular configuration.

cardiac enlargement simulates the appearance in pericardial effusion, and echocardiography or angiography is required for differentiation.

Fluoroscopy in patients with dilated (congestive) and restrictive cardiomyopathy shows dramatic reductions in cardiac pulsations because of the decreased contractility of the heart. In contrast, in patients with hypertrophic cardiomyopathy, pronounced systolic contraction of the ventricles causes excessive cardiac pulsations.[21]

## TUMORS OF THE HEART

Primary tumors of the heart are rare. About three-quarters are histologically benign; malignant primary tumors are almost all sarcomas.

Myxoma is the most common primary cardiac tumor, accounting for up to one-half of all cases. Almost all arise in the atrium, particularly the left. Because the tumor is usually pedunculated, it often causes intermittent obstruction or traumatic injury to the mitral or the tricuspid valve. Therefore, the most common clinical presentation is symptoms of mitral or tricuspid stenosis or regurgitation. Fragmentation of the tumor may produce showers of systemic or pulmonary emboli.

The radiographic findings of cardiac myxoma often resemble those of mitral or tricuspid valve disease, depending upon the site of origin of the tumor. Left atrial myxoma is suggested whenever there is radiographic evidence of mitral stenosis without a history of rheumatic fever, an inconstant and changing murmur, or embolic phenomena occurring without atrial fibrillation. The heart size and the pulmonary vasculature usually remain normal until valvular dysfunction develops. Atrial myxomas calcify in about 10 percent of the cases. The calcification is best seen at fluoroscopy, where it may present the pathognomonic appearance of a calcified mass prolapsing into the ventricle during diastole.

Echocardiography can accurately demonstrate the site of attachment and size of a cardiac myxoma. Angiocardiography shows the tumor as a well-defined filling defect in the opacified atrium. However, catheterization of the chamber from which the tumor originates runs the risk of dislodging the tumor emboli, and therefore this procedure is often not performed if adequate information is available from noninvasive procedures.

Primary sarcomas of the heart often project outward from the cardiac surface and result in bizarre configurations of the cardiac silhouette. Invasion of the pericardial space can produce a huge pericardial effusion and generalized cardiomegaly. Obstruction or filling of cardiac chambers by tumor can present a varied array of radiographic manifestations.

Metastatic tumors of the heart are far more common than primary neoplasms. The most frequent sources of cardiac metastases are bronchogenic carcinoma and carcinoma of the breast, followed by malignant melanoma, lymphoma, and leukemia. Cardiac metastases infrequently cause clinical manifestations and are rarely the

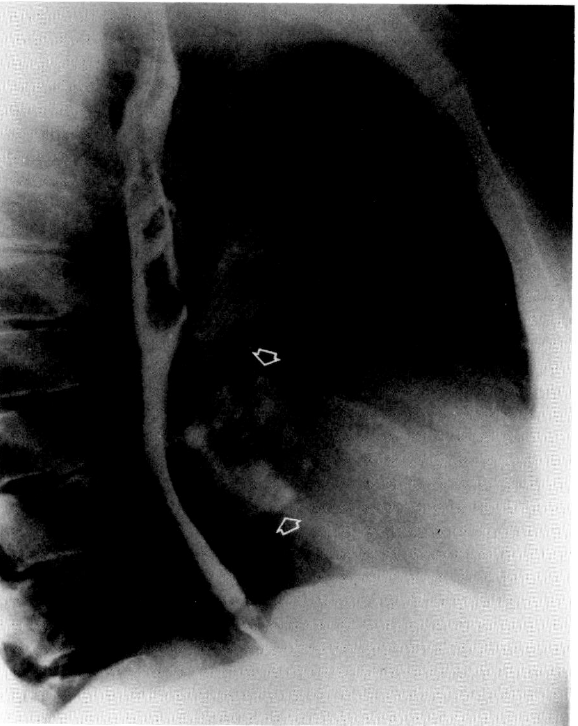

**Figure 23.16. Left atrial myxoma.** The arrows point to calcification within the tumor. The myxoma causes dysfunction of the mitral valve with resulting left atrial enlargement that causes an impression on the barium-filled esophagus. (From Eisenberg,[28] with permission from the publisher.)

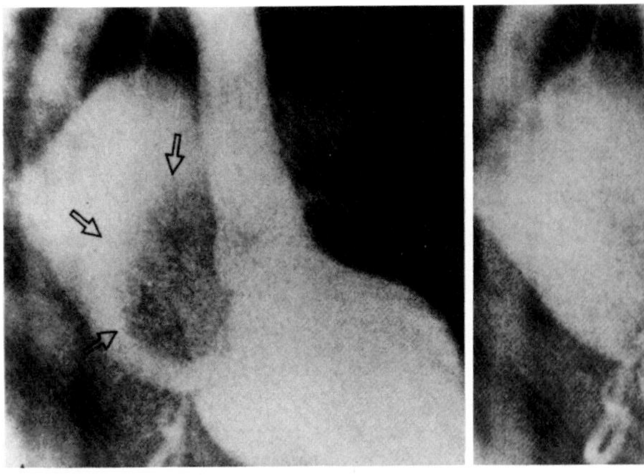
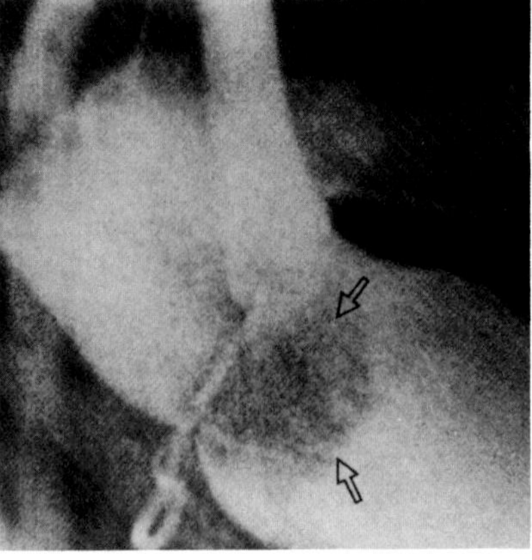

A B

**Figure 23.17. Left atrial myxoma.** (A) A frame from a cineangiogram obtained with the patient in the right anterior oblique projection shows a large filling defect (arrows) in the opacified left atrium. The edges of the filling defect are sharp, and it is close to the mitral valve. (B) On a later frame, the tumor mass (arrows) has prolapsed through the mitral valve into the left ventricle, which is maximally distended in diastole. (From Abrams, Adams, and Grant,[22] with permission from the publisher.)

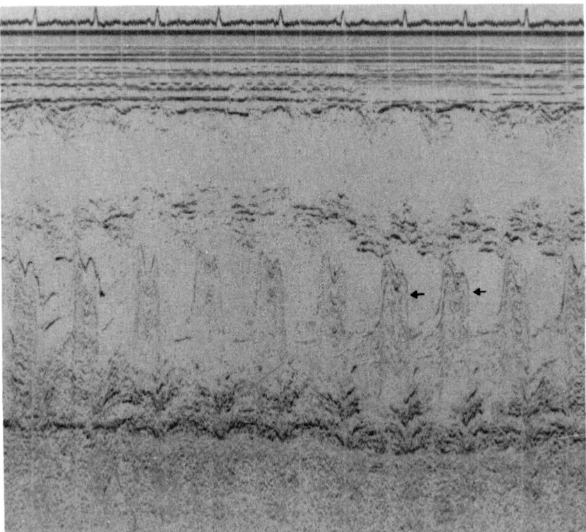

**Figure 23.18. Left atrial myxoma.** Echocardiography demonstrates the tumor as increased echoes (arrows) prolapsing through the mitral valve during diastole.

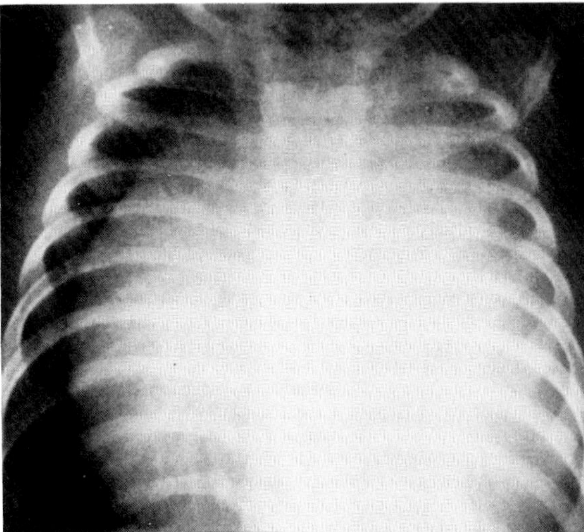

**Figure 23.19. Rhabdomyosarcoma of the pericardium.** A frontal chest radiograph in a 3-month-old infant with a history of vomiting and coughing after feeding shows a large mass in the left hemithorax contiguous with the cardiac shadow and producing a marked increase in overall cardiac size. At surgery, the tumor was found to arise from the right ventricular pericardium and to be firmly attached to it. (From Abrams, Adams, and Grant,[22] with permission from the publisher.)

cause of death. Chest radiographs usually demonstrate diffuse cardiac enlargement secondary to a malignant pericardial effusion. An irregular cardiac mass developing in a patient with known malignant disease is virtually diagnostic of cardiac metastases. Echocardiography can demonstrate a malignant pericardial effusion and even visualize larger metastases. Discrete metastatic deposits and bizarre chamber deformity can be shown by angiocardiography.[22]

## ENDOCARDIAL FIBROELASTOSIS

Endocardial fibroelastosis, characterized by diffuse thickening of the left ventricular endocardium with collagen and elastic tissue, is one of the most common causes of cardiac failure during the first year of life. Radiographically, the heart is usually strikingly enlarged and globular, often with left-sided prominence. In addition to left ventricular hypertrophy and dilatation, dramatic left atrial enlargement can develop from the often-associated mitral regurgitation. The aortic knob is small because of decreased left ventricular output. The pulmonary vasculature remains normal until congestive heart failure supervenes.

Angiocardiography demonstrates a dilated and poorly contracting left ventricle with a thickened wall. Injection of contrast material into the left ventricle usually shows considerable mitral regurgitation that outlines an enlarged left atrium.[23–25]

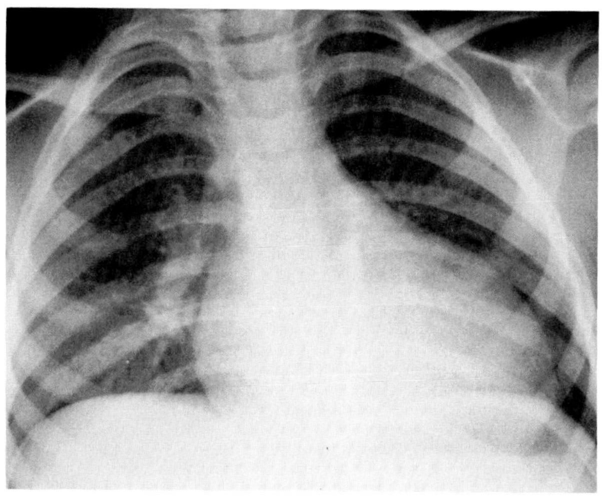

A

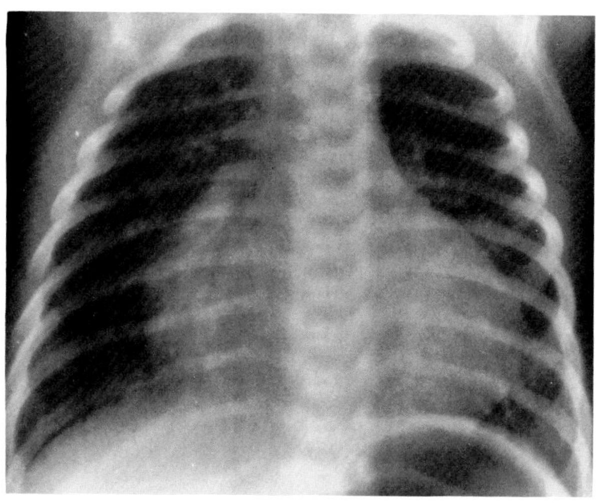

B

**Figure 23.20. Endocardial fibroelastosis.** (A) A frontal chest radiograph demonstrates cardiomegaly with left-sided prominence, a small aortic knob, and mild pulmonary vascular engorgement. (B) In this patient, there is a more globular configuration of the heart with engorgement of pulmonary veins due to congestive heart failure.

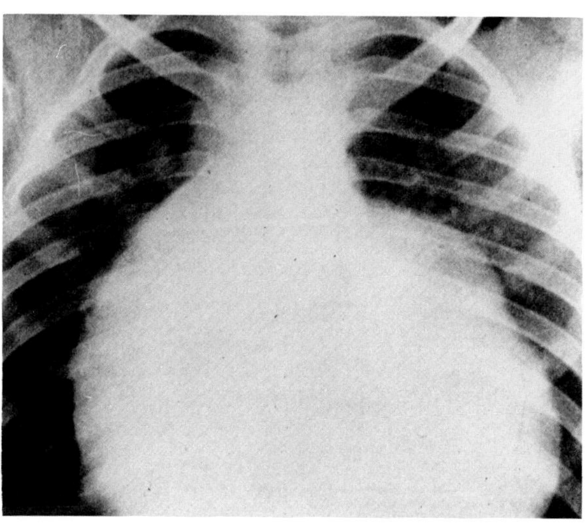

A

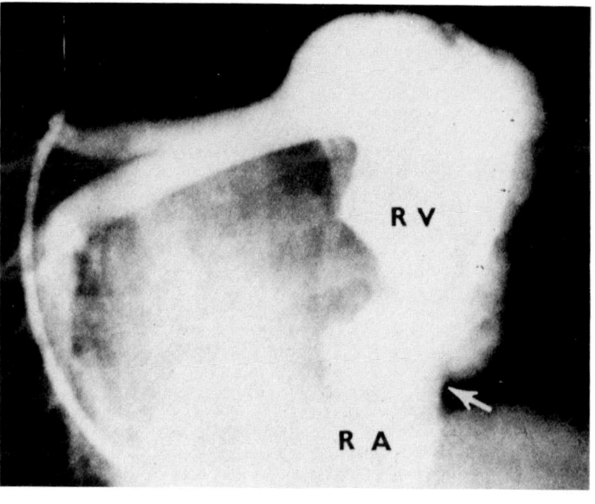


B

**Figure 23.21. Right-sided endomyocardial fibrosis.** (A) A plain chest radiograph and (B) an angiocardiogram demonstrate huge cardiomegaly with a greatly dilated right atrium but no pericardial fluid. There is dilatation of the right ventricular outflow tract. The arrow in B indicates the position of the tricuspid valve. Cicatricial obliteration of the apex of the right ventricle is present. (From Cockshott,[26] with permission from the publisher.)

## ENDOMYOCARDIAL FIBROSIS

This progressive tropical disease of unknown etiology is characterized by fibrous endocardial lesions in the inflow portion of the right or left ventricle, or both. The fibrous tissue thickens and contracts, reducing the size of the affected ventricular cavity. Involvement of the papillary muscles results in atrioventricular incompetence.

In right-sided disease, constriction of the right ventricle and incompetence of the tricuspid valve cause cardiomegaly with right ventricular prominence, a bulging pulmonary artery, and decreased pulmonary vascularity due to low cardiac output. Linear calcification in the right ventricular wall is a characteristic, though infrequent, finding.

In left-sided disease, pulmonary hypertension and mitral regurgitation are the major hemodynamic disturbances. The heart is moderately enlarged and has a pattern simulating mitral valve disease. Fibrous obliteration of the left ventricle leads to the development of pulmonary vascular congestion. Associated pulmonary arterial hypertension causes enlargement of central pulmonary arteries with rapid tapering distally. Pericardial and pleural effusions commonly occur.[26]

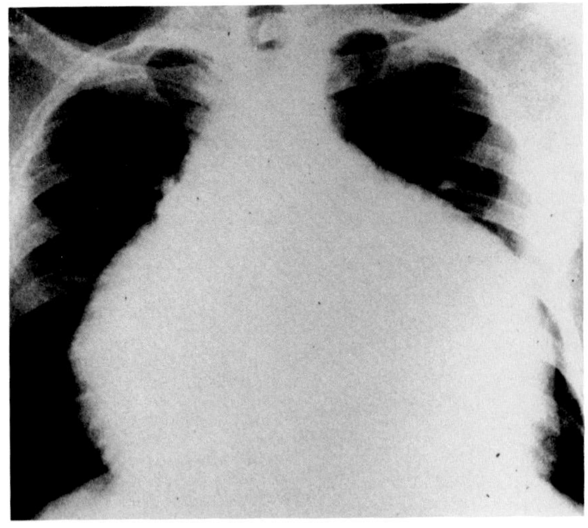
A

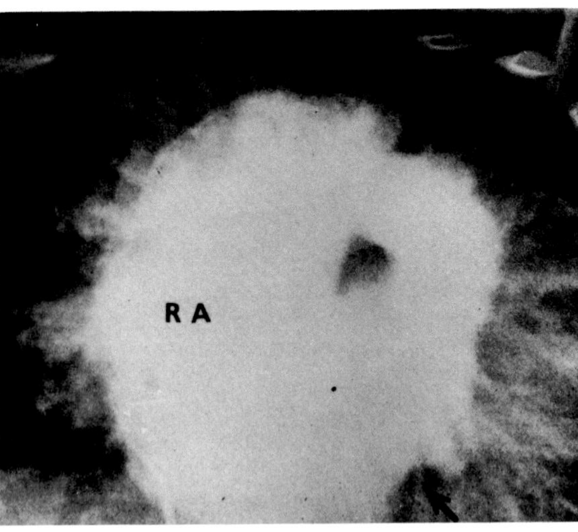

B

**Figure 23.22. Right-sided endomyocardial fibrosis.** (A) A plain chest radiograph and (B) an angiocardiogram demonstrate a huge pericardial effusion and a large right atrium. The arrow in B indicates an area of obliterated right ventricular apex. (From Cockshott,[26] with permission from the publisher.)

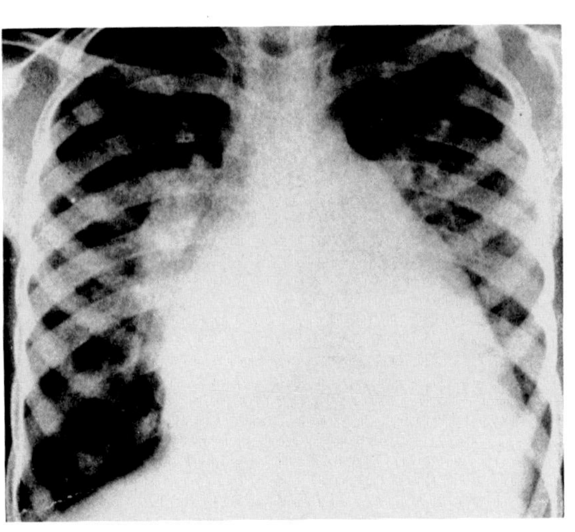
A

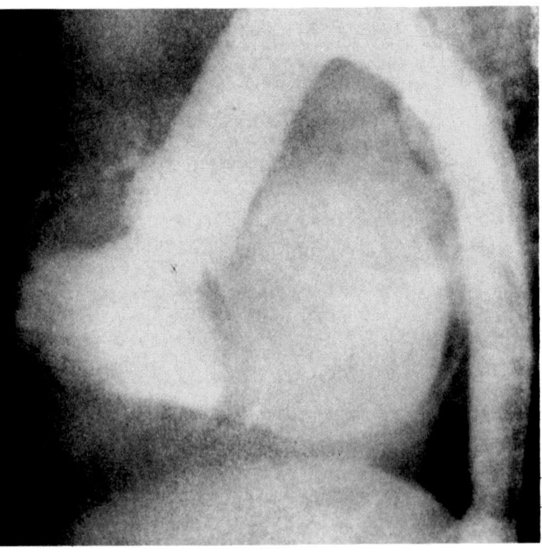
B

**Figure 23.23. Left-sided endomyocardial fibrosis.** (A) A plain chest radiograph and (B) an angiocardiogram demonstrate involvement of papillary muscle, mitral insufficiency, and severe pulmonary hypertension. The left ventricle is small and deformed, and the left atrium is large. (From Cockshott,[26] with permission from the publisher.)

## REFERENCES


1. Ashburn WL, Tubau J: Myocardial perfusion imaging in ischemic heart disease. *Radiol Clin North Am* 18:467–486, 1980.
2. Rifkin R, Parisi AF, Folland E: Coronary calcification in the diagnosis of coronary artery disease. *Am J Cardiol* 44:141–147, 1979.
3. Ritchie JL, Zaret BL, Strauss HW, et al: Myocardial imaging with thallium 201: A multicenter study in patients with angina pectoris or acute myocardial infarction. *Am J Cardiol* 42:345–350, 1978.
4. Caldwell JH, Hamilton GW, Sorenson SG, et al: The detection of coronary artery disease with radionuclide techniques: A comparison of rest-exercise thallium imaging and ejection fraction response. *Circulation* 61:610–619, 1980.
5. Bartel AG, Chen JT, Peter RH, et al: The significance of coronary calcification detected by fluoroscopy. A report of 360 patients. *Circulation* 49:1247–1253, 1974.
6. Higgins CB, Lipton MJ: Chest pain, in Eisenberg RL, Amberg JR (eds): *Critical Diagnostic Pathways in Radiology: An Algorithmic Approach*. Philadelphia, JB Lippincott, 1981.
7. Parkey RW: Imaging of acute myocardial infarction. *J Nucl Med* 17:771–775, 1973.
8. Guthaner DF, Wexler L: New aspects of coronary angiography. *Radiol Clin North Am* 18:501–514, 1980.
9. Lipton MJ, Higgins CB: Evaluation of ischemic heart disease

by computerized transmission tomography. *Radiol Clin North Am* 18:557–576, 1980.

10. Baltaxe HA, Levin DC: Angiographic demonstration of complications related to the saphenous aortocoronary bypass procedure. *AJR* 119:484–492, 1973.
11. Rosch J, Judkins MP, Green GS, et al: Aortocoronary venous bypass grafts: Angiographic study of 84 cases: *Radiology* 102:567–573, 1972.
12. Holman BL, Wynne J: Infarct avid (hot spot) myocardial scintigraphy. *Radiol Clin North Am* 18:487–499, 1980.
13. Tudor J, Maurer BJ, Wray R, et al: Lung shadows after acute myocardial infarction. *Clin Radiol* 24:365–369, 1973.
14. Grande P, Christiansen C, Pedersen A, et al: Optimal diagnosis in acute myocardial infarction: A cost effectiveness study. *Circulation* 61:723–728, 1980.
15. Steinberg I: Angiocardiographic findings in ventricular aneurysm due to arteriosclerotic myocardial infarction. *AJR* 97:321–337, 1966.
16. Higgins CB, Lipton MJ: Radiography of acute myocardial infarction. *Radiol Clin North Am* 18:359–368, 1980.
17. Kittredge RD, Gamboa B, Kemp HJ: Radiographic visualization of left ventricular aneurysms on lateral chest films. *AJR* 126:1140–1146, 1976.
18. Higgins CB, Lipton MJ, Johnson AD, et al: False aneurysms of the left ventricle. *Radiology* 127:21–27, 1978.
19. Dressler W: The post-myocardial infarction syndrome. *Arch Intern Med* 103:28–42, 1959.
20. Levin EJ, Bryk D: Dressler's syndrome. *Radiology* 87:731–736, 1966.
21. Steiner RE: The roentgen features of the cardiomyopathies. *Semin Roentgenol* 4:311–327, 1969.
22. Abrams HL, Adams DF, Grant HA: The radiology of tumors of the heart. *Radiol Clin North Am* 9:299–326, 1971.
23. Swischuk LE: *Plain Film Interpretation in Congenital Heart Diseases,* Baltimore, William & Wilkins, 1979.
24. Amplatz K: Endocardial fibroelastosis. *Semin Roentgenol* 4:360–366, 1969.
25. Johnsrude IS, Carey LS: Roentgenographic manifestations of endocardial fibroelastosis. *AJR* 94:109–121, 1965.
26. Cockshott WP: Cardiomyopathy in the tropics: Endomyocardial fibrosis. *Semin Roentgenol* 4:367–373, 1969.
27. Lipton MJ, Boyd DP: Cardiac computed tomography. *Proc IEEE* 71:301, 1983.
28. Eisenberg RL: *Gastrointestinal Radiology: A Pattern Approach.* Philadelphia, JB Lippincott, 1983.

## Chapter Twenty-Four
# Pericardial Disease

## PERICARDIAL EFFUSION

Echocardiography is the most effective imaging technique for demonstrating pericardial effusions and has largely replaced other methods. In patients with small effusions, pericardial fluid appears as a relatively echo-free space between the posterior pericardium and the posterior left ventricular epicardium. Larger effusions also produce an echo-free space between the anterior right ventricle and the parietal pericardium just beneath the anterior chest wall. If necessary, the diagnosis can be confirmed by angiocardiography, which demonstrates an excessive distance (>5 mm) between the opacified right atrium and the outer border of the cardiac silhouette.

Computed tomography (CT) can demonstrate pericardial effusions as small as 50 ml of fluid, a sensitivity similar to that of echocardiography. The CT density of effusions varies depending on the protein, cellular, and fat content of the fluid. In the supine patient, a small effusion tends to accumulate posteriorly. Larger effusions form a layer of fluid that surrounds the heart, separating the cardiac chambers from mediastinal fat, parietal pleura, mediastinal organs, or the lungs.

On plain chest radiographs, a pericardial effusion causes enlargement of the cardiac silhouette. At least 200 ml of fluid must be present before a pericardial

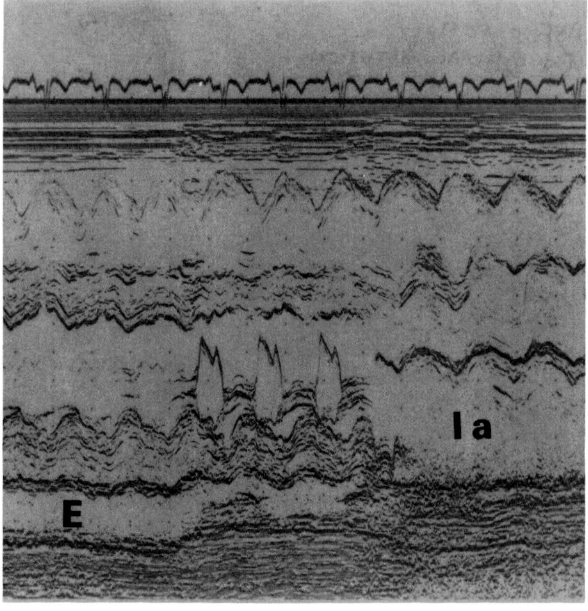


**Figure 24.1. Pericardial effusion.** An echocardiogram demonstrates a large posterior effusion (E) that causes separation of the pericardium from the epicardium. Note that the effusion stops at the region of the left atrium (la), in contrast to a pleural effusion that parallels the entire cardiac silhouette.

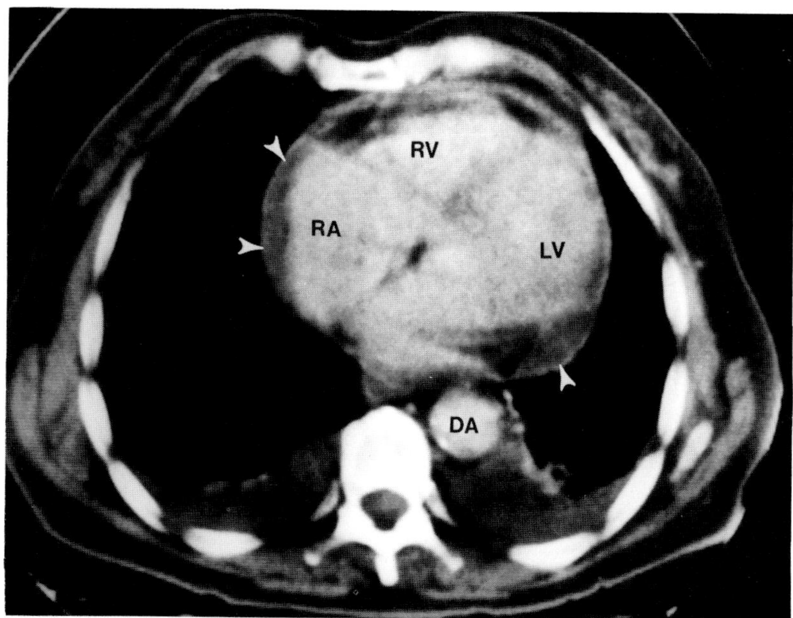


**Figure 24.2. Pericardial effusion.** A CT scan following the injection of intravenous contrast material shows the pericardial effusion as a low-density area (arrowheads) that is clearly demarcated from the contrast-enhanced blood in the intracardiac chambers and descending aorta. Note the bilateral pleural effusions posteriorly. RA = right atrium; RV = right ventricle; LV = left ventricle; DA = descending aorta. (From Sagel,[5] with permission from the publisher.)

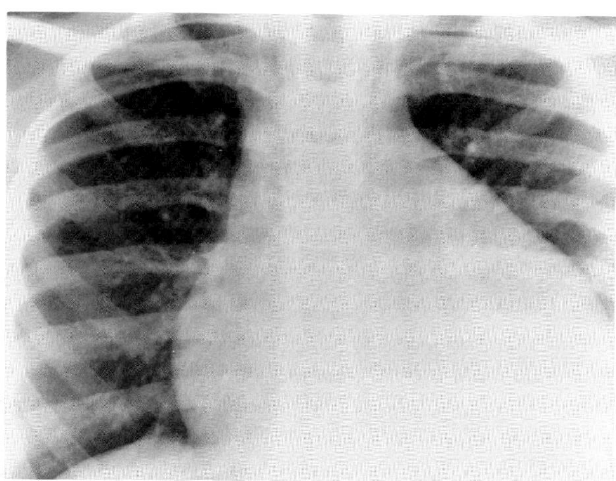

**Figure 24.3. Pericardial effusion.** There is generalized enlargement of the cardiac silhouette with an unusually acute right cardiophrenic angle. Although there is a small right pleural effusion, the pulmonary vascularity is essentially normal, unlike the venous congestion and pulmonary edema that might be expected if this appearance represented true cardiomegaly without pericardial effusion.

effusion can be detected by this technique. With small or moderate effusions, cardiac enlargement is generalized and the cardiophrenic angle typically appears more acute than normal. As the amount of fluid increases, there is obliteration of normal cardiac markings and the cardiophrenic angle. At fluoroscopy, cardiac pulsations are diminished or almost absent. Rapid enlargement of the cardiac silhouette, especially in the absence of pulmonary vascular engorgement suggestive of congestive heart failure, is highly suggestive of pericardial effusion.

The epicardial fat pad sign is a good indication of pericardial effusion (or thickening) on routine lateral chest radiographs. Normally, the anterior mediastinal and subepicardial fat can be visualized on either side of a pencil-thin line representing the normal pericardium. Separation of the subepicardial fat stripe from the anterior mediastinal fat by a soft tissue density of more than 2 mm is virtually pathognomonic of pericardial effusion or thickening. A positive demonstration of the epicardial fat pad sign is significant; the absence of the sign is of no diagnostic value.

Angiography, intravenous carbon dioxide injection, and a pericardial tap with air injection have been used in the past to demonstrate pericardial effusion; however, echocardiography has effectively replaced these invasive techniques.[1–5]

## CARDIAC TAMPONADE

The accumulation of fluid in the pericardial space can cause severe obstruction to the inflow of blood to the ventricles (cardiac tamponade); this leads to diminished cardiac output, systemic venous congestion, hypotension, and circulatory shock. In acute tamponade, which most often results from bleeding into the pericardial space following surgery or trauma or occurs in patients with tuberculosis or tumor, the rapid development of pericardial effusion may produce severe alterations of cardiac function with minimal changes in the radiographic cardiac silhouette. In contrast, chronic or subacute pericardial effusions permit the stretching of the pericardium to adapt to the increased volume of fluid. This may occur in renal failure during dialysis, pericarditis of tuberculous, viral, or postirradiation etiology, or metastatic peri-

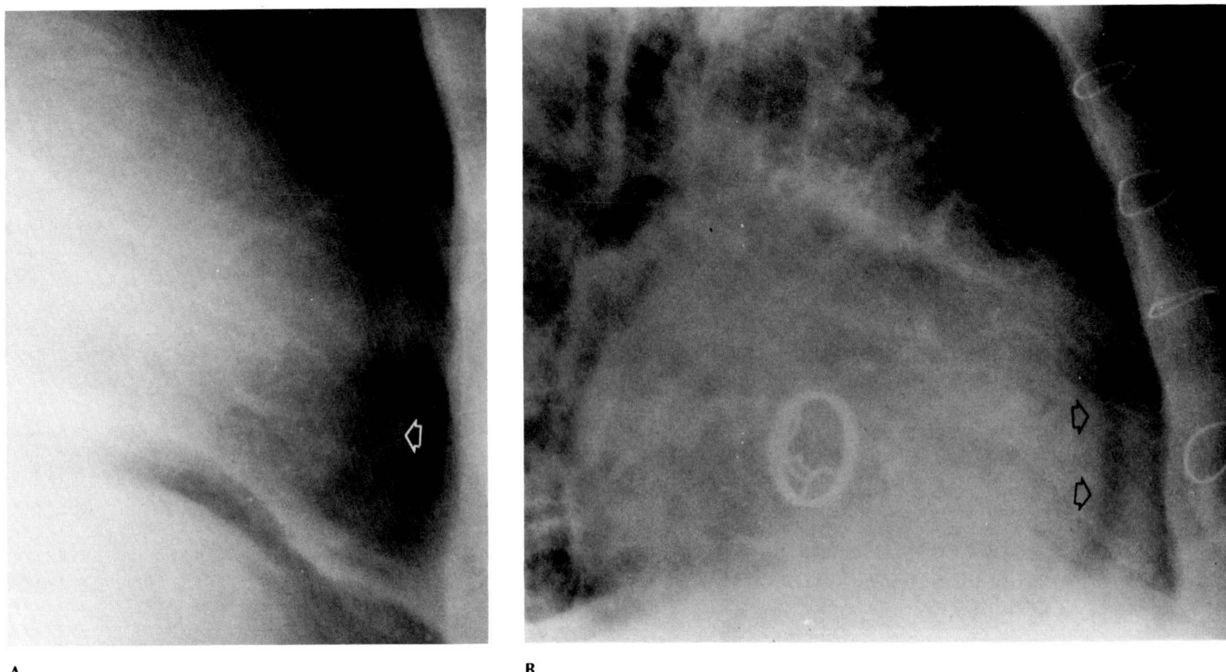


**Figure 24.4. Epicardial fat pad sign in pericardial effusion.** (A) In a normal person, a thin, relatively dense line (arrow) representing the normal pericardium lies between the anterior mediastinal and subepicardial fat. (B) A lateral chest radiograph demonstrates a wide soft tissue density separating the subepicardial fat stripe (arrows) from the anterior mediastinal fat; this is a virtually pathognomonic sign of pericardial effusion or thickening. (From Eisenberg,[13] with permission from the publisher.)

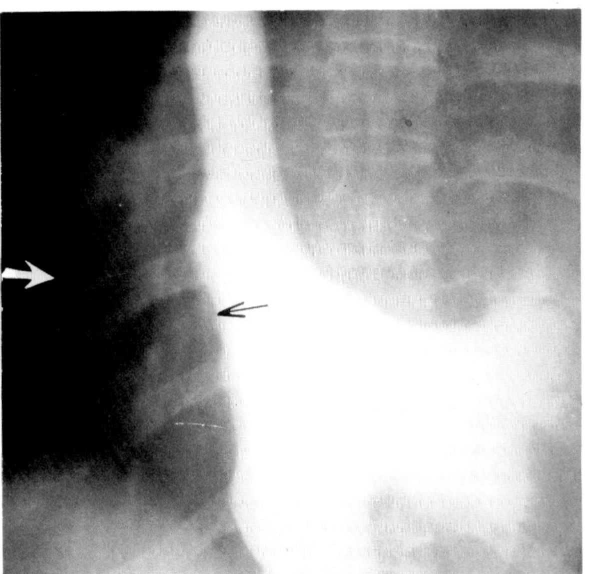

**Figure 24.5. Cardiac tamponade from tuberculous pericarditis.** An angiocardiogram shows marked separation of the right atrium (black arrow) from the right border of the heart (white arrow) by a large pericardial collection. The lateral border of the right atrium has been compressed into a concave shape. There was prolonged opacification of the superior and the inferior vena cava. (From Spitz and Holmes,[6] with permission from the publisher.)

cardial disease. Large effusions, which may be massive without causing tamponade, produce diffuse enlargement of the cardiac silhouette with relatively minimal changes in the pulmonary vasculature. The development of obstruction to systemic venous return to the right atrium causes unusual prominence of the superior vena cava and a decrease in pulmonary vascularity.

Echocardiography in patients with cardiac tamponade demonstrates an exaggerated motion of the anterior wall of the right ventricle, significant changes in the right ventricular diameter with respiration, and anterior as well as posterior accumulation of fluid in the pericardium.[6,7,16]

## CHRONIC CONSTRICTIVE PERICARDITIS

In chronic constrictive pericarditis, the pericardial cavity is obliterated by scarred granulation tissue resulting from the healing of an acute fibrinous or serofibrinous pericarditis. Although in the past chronic constrictive pericarditis has usually been of tuberculous origin, the condition now more commonly follows pyogenic or viral infection, trauma, irradiation, neoplastic disease, connective tissue disorders, or uremia treated by chronic dialysis.

Adherence and contraction of the visceral and parietal pericardium prevent the ventricles from filling normally,

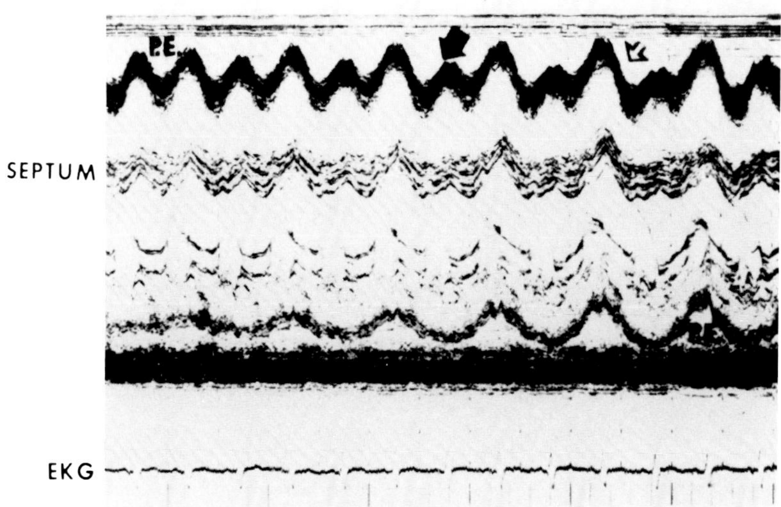


**Figure 24.6. Cardiac tamponade.** An echocardiogram shows a large pericardial effusion located both posteriorly and anteriorly. There is hyperdynamic movement of the anterior wall of the right ventricle. Findings characteristic of cardiac tamponade are variations in the dimensions of the ventricles with respiratory phases, and a mid-diastolic notch on the anterior wall of the right ventricle. (From Salcedo,[16] with permission from the publisher.)

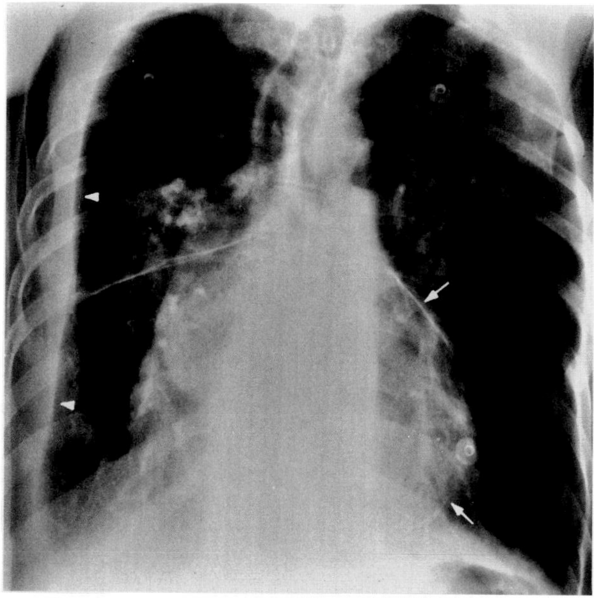

**Figure 24.7. Chronic constrictive pericarditis.** A lateral decubitus view of the chest demonstrates moderate enlargement of the cardiac silhouette and a large right pleural effusion (arrowheads). Note the calcified plaque (arrows) within the pericardium.

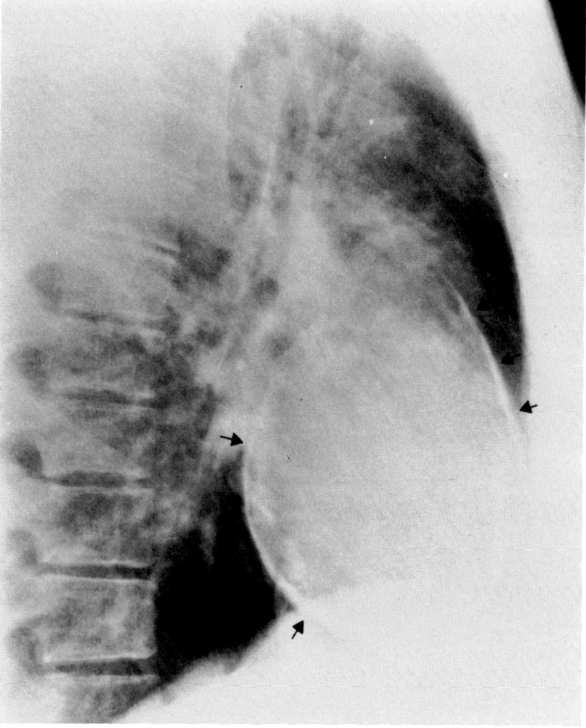

**Figure 24.8. Chronic constrictive pericarditis.** Dense calcification in the pericardium (arrows) completely surrounds a normal-sized heart.

resulting in decreased stroke volume and increased venous pressure. On plain chest radiographs, the heart size is often normal or small; mild or moderate enlargement of the cardiac silhouette may occur. Localized pericardial thickening or adhesions between the pleura and the pericardium may cause irregularity of the cardiac border. Pulmonary venous congestion, pleural effusion, and prominence of the superior vena cava are commonly seen.

Calcified plaques in the thickened pericardium are present in about half of the patients with constrictive pericarditis. Though the heaviest deposits of calcium are located anteriorly, posterior calcification and calcification of the pericardium adjacent to the diaphragm can often be seen. At times the heart appears to be encased in a calcific shell. The presence of typical calcification in a patient with clinical signs of constrictive pericarditis is virtually pathognomonic.

Fluoroscopy demonstrates decreased cardiac pulsations and may permit the detection of cardiac calcifications that are not noticed on plain chest radiographs.

The diagnosis of constrictive pericarditis can be confirmed by angiocardiography, which demonstrates thickening of the combined pericardium and atrial wall along with flattening and rigidity of the lateral border of the right atrium. Venous hypertension and a prolonged circulation time may simulate congestive heart failure; however, in constrictive pericarditis the cardiac chambers are normal in size and not dilated.

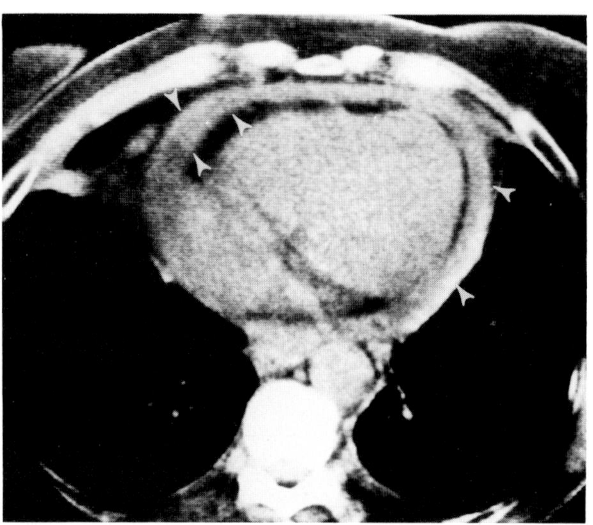

**Figure 24.9. Pericardial thickening in chronic constrictive pericarditis.** A CT scan performed during an infusion of contrast material shows enhancement of the soft-tissue-density pericardium (arrowheads), which is up to 6 mm thick. (From Sagel,[5] with permission from the publisher.)

CT is highly sensitive in detecting pericardial thickening. By accurately differentiating a normal from a thickened pericardium, CT offers great promise in solving the difficult clinical problem of separating constrictive pericarditis (thickened pericardium) from restrictive cardiomyopathy (normal pericardium), two conditions that are virtually indistinguishable on the basis of clinical signs and symptoms and hemodynamic findings.[3,5,8,9]

## PERICARDIAL CYST

Pericardial cyst is a fluid-filled mass that typically appears as a round or lobulated, sharply demarcated density in the right cardiophrenic angle that touches both the anterior chest wall and the anterior portion of the right hemidiaphragm. Less frequently, a pericardial cyst may arise in the left cardiophrenic angle or as a mediastinal mass not touching the diaphragm. Unlike diverticula, true pericardial cysts do not communicate with the pericardial cavity. Although most pericardial cysts are found in asymptomatic persons, the pressure of large cysts on adjacent organs can produce symptoms.

Radiographically, a pericardial cyst must be differentiated from a large pericardial fat pad, a diaphragmatic hernia (especially through the foramen of Morgagni), an aneurysm of the heart and great vessels, and cardiac,

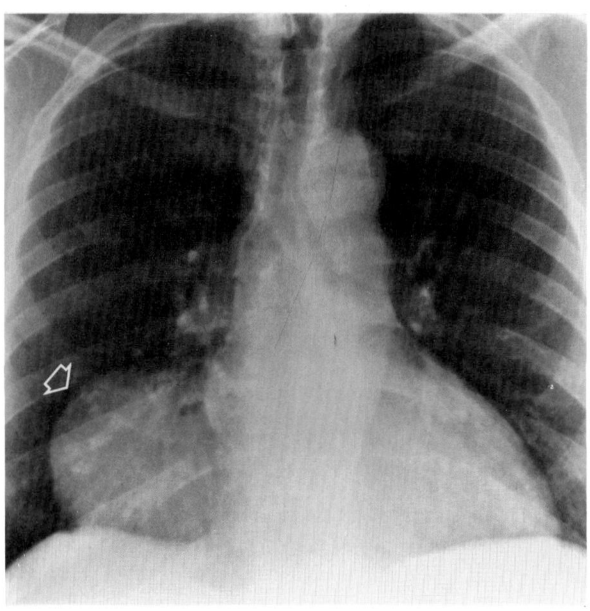

A

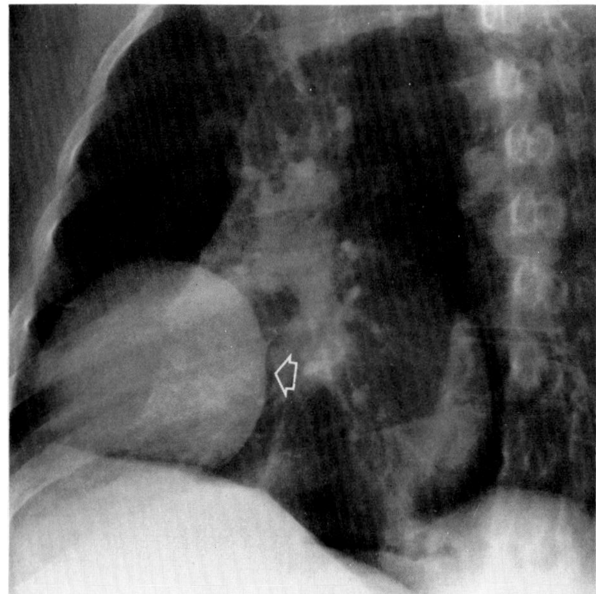

B

**Figure 24.10. Pericardial cyst.** (A) Frontal and (B) oblique views demonstrate a smooth mass (arrows) in the right cardiophrenic angle. (From Eisenberg,[15] with permission from the publisher.)

pericardial, or mediastinal cysts and tumors. Ultrasound can demonstrate the cystic and unilocular nature of the mass. Contrast-enhanced CT can exclude the possibility of a solid mass or aneurysm as well as differentiate the cystic mass from fatty ones, such as pericardial fat pads and lipomas.[10,11]

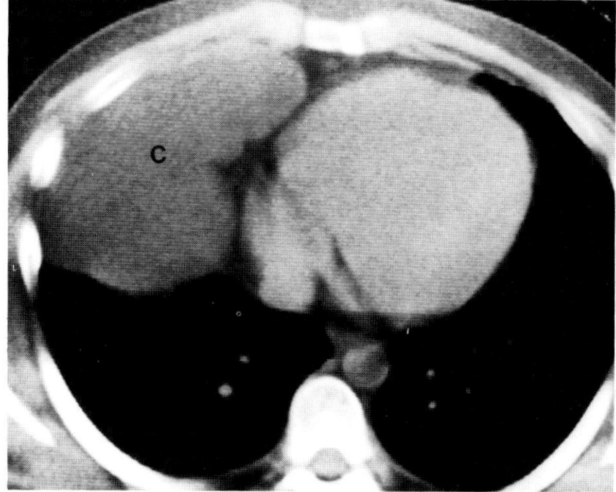


**Figure 24.11. Pericardial cyst.** A CT scan shows a typical large, homogeneous, near-water-density mass (C) in the right cardiophrenic angle. (From Sagel and Aronberg,[11] with permission from the publisher.)

## CONGENITAL ABSENCE OF THE PERICARDIUM

Absence of the pericardium is an infrequent lesion that almost always occurs on the left side. Complete absence of the left pericardium causes the heart to be displaced to the left with concomitant clockwise rotation, so that the left border is formed by the left ventricle. The pulmonary artery is long and more sharply defined than usual because the interposed lung produces a lucent zone between the pulmonary artery and the aorta.

Partial absence of the left pericardium most commonly occurs in the region of the left atrium; herniation of the left atrial appendage through the defect typically causes a bulge in the left border of the heart below the level of the main pulmonary artery.

A partial right-sided pericardial defect permits the protrusion of a part of the right atrium, which causes an abnormal bulge along the upper right border of the heart in the vicinity of the superior vena cava.[12]

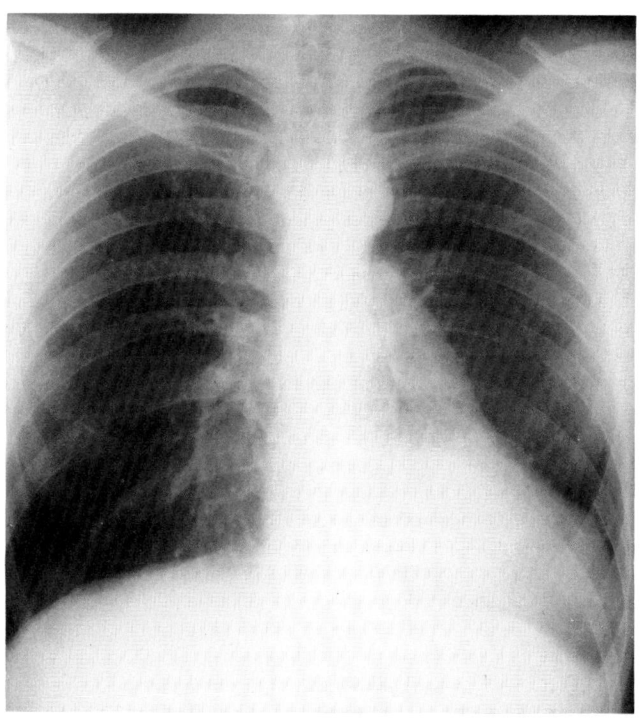

A

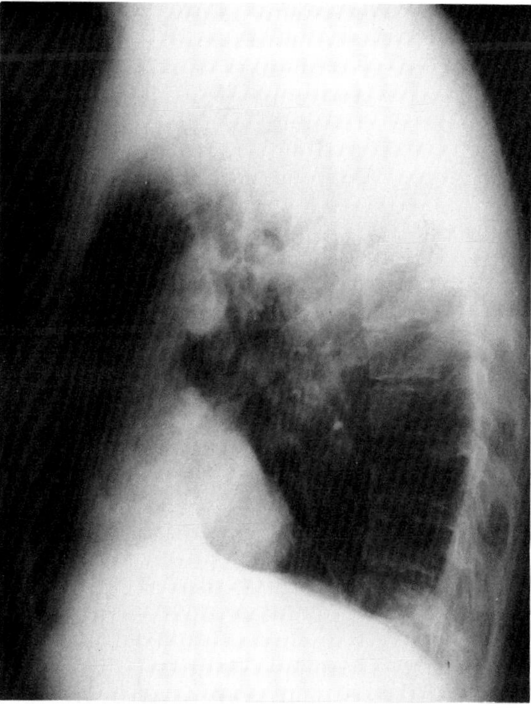

B

**Figure 24.12. Complete congenital absence of the pericardium.** (A) Frontal and (B) lateral views of the chest show prominence of the pulmonary trunk with an increased length of its arc. The pulmonary artery is unusually well defined on both frontal and lateral films. The heart appears to be unrestrained and to sag toward the left, its shape conforming to that of the left hemidiaphragm. The lingula is partly and irregularly consolidated. At surgery, the heart was shifted toward the left and covered a part of the left hemidiaphragm. Adhesions between the lingula and the heart had produced a distortion of the left upper lobe bronchus and atelectasis of the lingula. (From Cooley and Schreiber,[14] with permission from the publisher.)

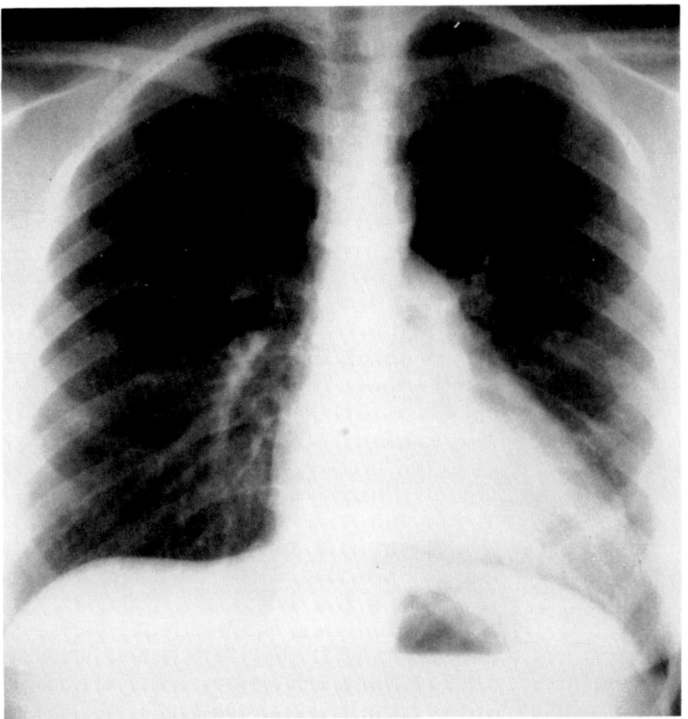
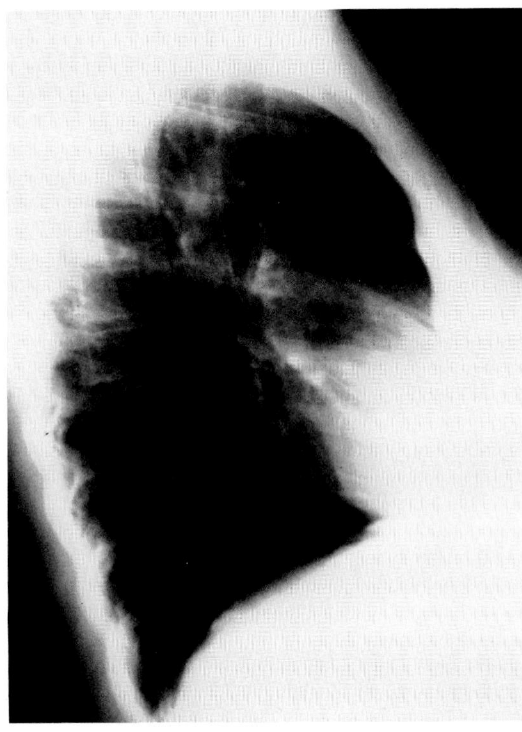

A B

**Figure 24.13. Partial absence of the left side of the pericardium.** (A) Frontal and (B) lateral views of the chest in an asymptomatic woman show sharp definition of the cardiac contour in the upper portion of the left side and over the posterior surface. Fluoroscopy showed exaggerated pulsations of the pulmonary artery segment. (From Cooley and Schreiber,[14] with permission from the publisher.)

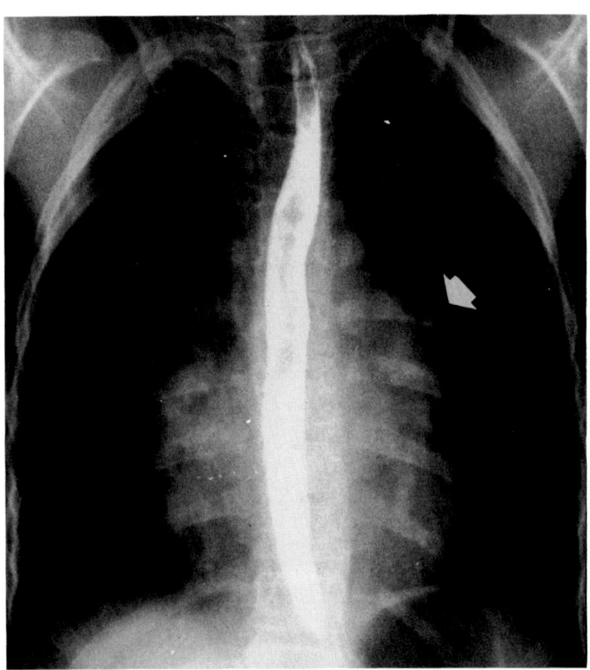

**Figure 24.14. Partial absence of the upper left portion of the pericardium.** The localized bulging of the upper left portion of the cardiac silhouette (arrow) represents herniation of the left atrial appendage. (From Cooley and Schreiber,[14] with permission from the publisher.)

## REFERENCES


1. Lane EJ, Carsky EW: Epicardial fat: Lateral plain film analysis in normals and in pericardial effusion. *Radiology* 91:1–5, 1968.
2. Feigenbaum H: Echocardiographic diagnosis of pericardial effusion. *Am J Cardiol* 26:475–479, 1970.
3. Lipton MJ, Herfkens RJ, Gamsu G: Computed tomography of the heart and pericardium, in Moss AA, Gamsu G, Genant HK (eds): *Computed Tomography of the Body*. Philadelphia, WB Saunders, 1983.
4. Wong BYS, Kyo RL, McArthur R: Diagnosis of pericardial effusion by CT. *Chest* 81:177–181, 1982.
5. Sagel SS: Lung, pleura, pericardium, and chest wall, in Lee JKT, Sagel SS, Stanley RJ (eds): *Computed Body Tomography*. New York, Raven Press, 1983, pp 99–130.
6. Spitz HB, Holmes JC: Right atrial contour in cardiac tamponade. *Radiology* 103:69–74, 1972.
7. Stein L, Shubin H, Weil NH: Recognition and management of pericardial tamponade. *JAMA* 225:503–505, 1973.
8. Steinberg I, Hagstrom JWC: Angiocardiography in diagnosis of effusive-restrictive pericarditis. *AJR* 102:305–319, 1968.
9. Cornell SH, Rossi NP: Roentgenographic findings in constrictive pericarditis. Analysis of 21 cases. *AJR* 102:301–304, 1968.
10. Feigin DS, Fenoglio JJ, McAllister HA, et al: Pericardial cysts: A radiologic-pathologic correlation and review. *Radiology* 125:15–20, 1977.
11. Sagel SS, Aronberg DJ: Thoracic anatomy and mediastinum, in Lee JKT, Sagel SS, Stanley RD (eds): *Computed Body Tomography*, New York, Raven Press, 1983, pp 55–98.

12. Glover LB, Barcia A, Reeves PJ: Congenital absence of the pericardium. A review of the literature with demonstration of a previously unreported fluoroscopic finding. *AJR* 106:542–549, 1969.
13. Eisenberg RL: *Atlas of Signs in Radiology*. Philadelphia, JB Lippincott, 1984.
14. Cooley RN, Schreiber MH: *Radiology of the Heart and Great Vessels*. Baltimore, Williams & Wilkins, 1978.
15. Eisenberg RL: *Gastrointestinal Radiology: A Pattern Approach*. Phildelphia, JB Lippincott, 1983.
16. Salcedo EE: *Atlas of Echocardiography*. Philadelphia, WB Saunders, 1978.

# PART FOUR
# DISORDERS OF THE VASCULAR SYSTEM

**Chapter Twenty-Five**

# Hypertensive Vascular Disease

## RENOVASCULAR HYPERTENSION

The vast majority of patients with elevated blood pressure have essential, or idiopathic, hypertension. Nevertheless, about 6 percent have hypertension that is curable by surgery. Although some patients in this group have adrenal abnormalities (Cushing's syndrome, primary aldosteronism, pheochromocytoma), other endocrine dysfunction, or coarctation of the aorta, the vast majority have renal parenchymal or vascular disease as the underlying cause of hypertension.

The standard screening test for renovascular hypertension has been the rapid-sequence excretory urogram, in which radiographs are obtained at each of the first 5 minutes following the rapid injection of contrast material. Features suggesting renal ischemia include (1) unilateral delayed appearance and excretion of contrast material, (2) difference in kidney size greater than 1.5 cm, (3) irregular contour of the renal silhouette, suggesting segmental infarction or atrophy, (4) indentations on the ureter or renal pelvis due to dilated, tortuous ureteral arterial collaterals, and (5) hyperconcentration of contrast material in the collecting system of the smaller kidney on delayed films. Although up to 25 percent of patients with renovascular hypertension have a normal rapid-sequence excretory urogram, this modality is also of value in detecting other causes of hypertension, such as tumor, pyelonephritis, polycystic kidneys, or renal infarction.

Arteriography is the most accurate screening examination for detecting renovascular lesions. The most common cause of renal artery obstruction is arteriosclerotic narrowing, which most commonly occurs in the proximal portion of the vessel close to its origin from the aorta.

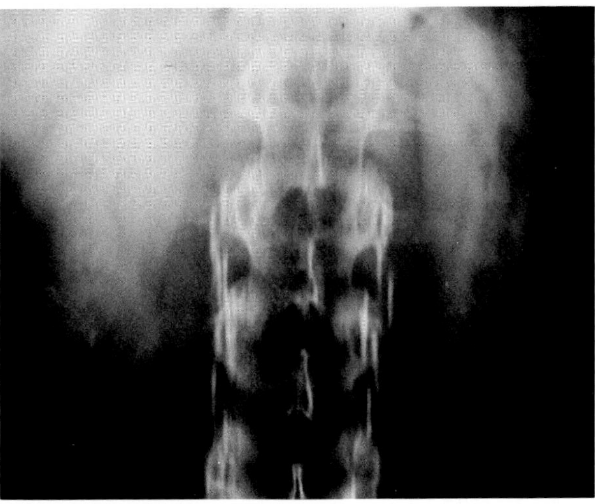

**Figure 25.1. Renovascular hypertension.** A film from a rapid-sequence urogram obtained 3 minutes after the injection of contrast material shows no calyceal opacification on the left in a patient with left renal artery stenosis. (From Burko, Smith, Kirchner, et al,[1] with permission from the publisher.)

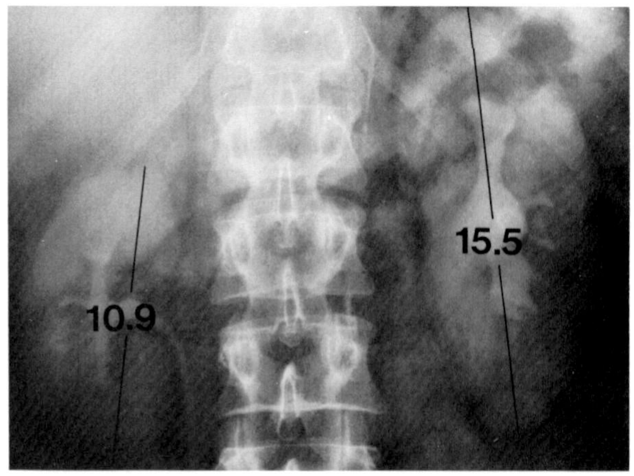


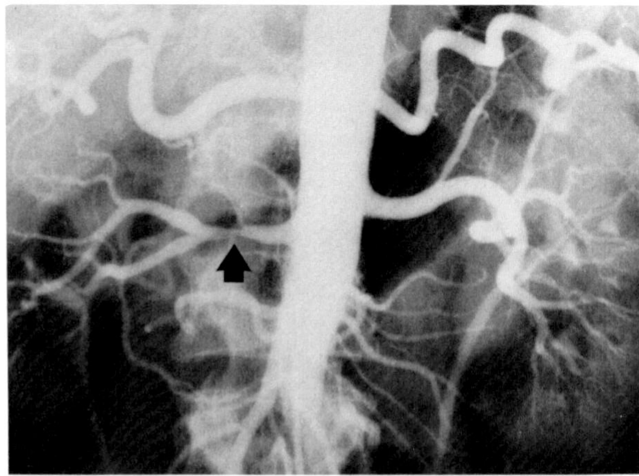

A B

**Figure 25.2. Renovascular hypertension.** Diminished size of the right kidney (A) due to renal artery stenosis (B). (From Burko, Smith, Kirchner, et al,[1] with permission from the publisher.)

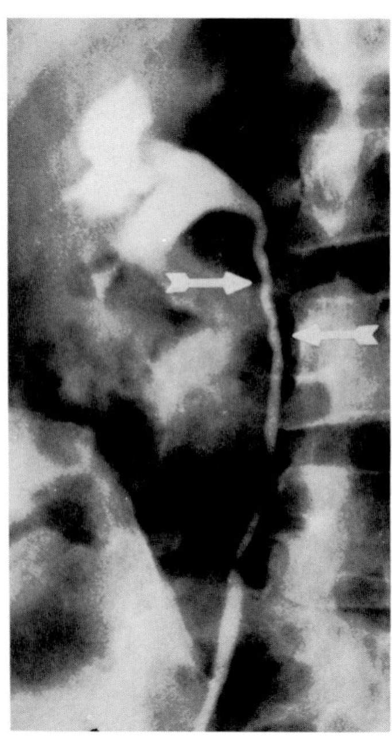

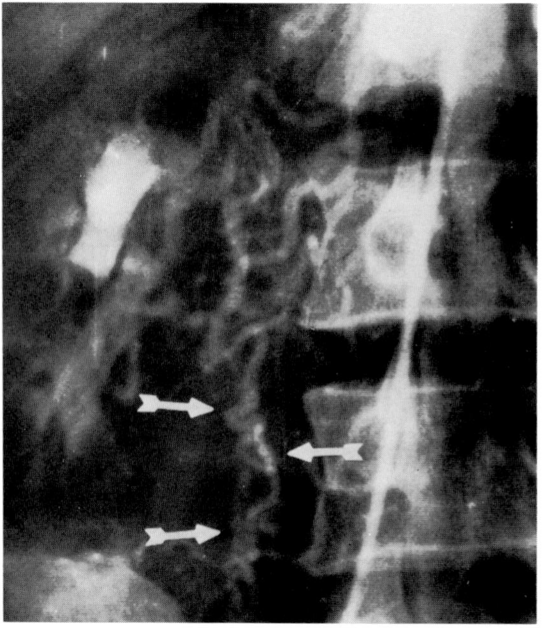

A B

**Figure 25.3. Renovascular hypertension.** (A) A right retrograde urogram demonstrates notching of the proximal ureter (arrows). (B) An aortogram shows that the ureteral defects were caused by extrinsic pressure from enlarged collateral arteries (arrows) surrounding the ureter in this patient with chronic renal artery stenosis. (From Witten, Myers, and Utz,[7] with permission from the publisher.)

Poststenotic dilatation is common. Bilateral renal artery stenoses are noted in up to one-third of the patients. Oblique projections, which demonstrate the vessel origins in profile, are often required to demonstrate renal artery stenosis.

The other major cause of renovascular hypertension is fibromuscular dysplasia. This disease is most frequent in young adult women and is often bilateral. The most common radiographic appearance of fibromuscular dysplasia is the string-of-beads pattern, in which there are alternating areas of narrowing and dilatation (representing microaneurysms). Smooth, concentric stenoses less frequently occur. Even the most minor contour deformity in a lesion of fibromuscular dysplasia may be of significance, since at surgery it is often seen to represent a pinhole diaphragm.

The mere presence of a renovascular lesion does not mean that it is the cause of hypertension; indeed, many normotensive patients have severe renal artery disease. Therefore, bilateral renal vein catheterization for the measurement of plasma renin activity is used to assess the functional significance of any stenotic lesion. The demonstration of a renal vein renin ratio of 1.5:1 or greater (abnormal side to normal side) indicates, with an accuracy rate of about 85 percent, the functional significance of a lesion. When arteriography demonstrates

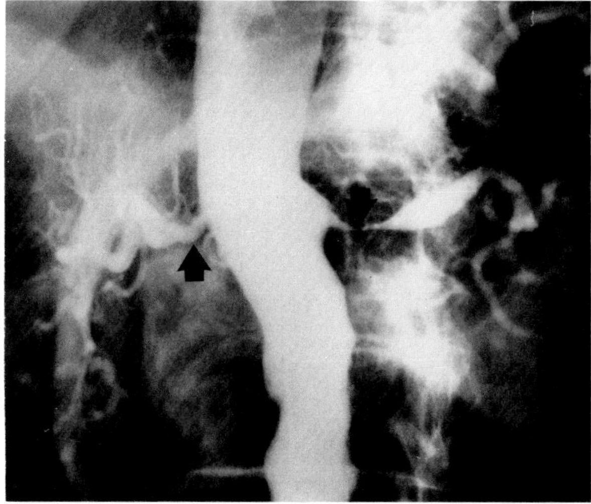

**Figure 25.4. Renovascular hypertension.** Typical bilateral arteriosclerotic renal artery stenoses (arrows). (From Burko, Smith, Kirchner, et al,[1] with permission from the publisher.)

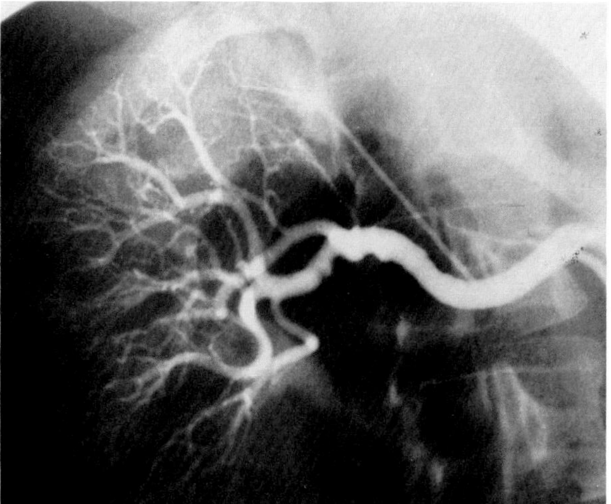

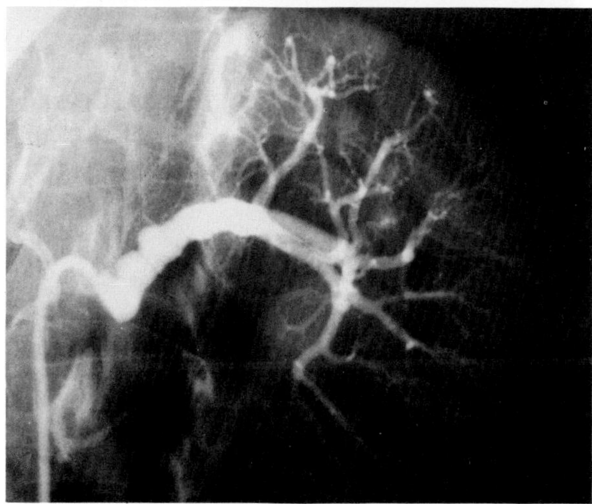

**Figure 25.5. Renovascular hypertension.** Selective injections of contrast material into both renal arteries reveal the typical string-of-beads pattern of fibromuscular disease bilaterally. (From Burko, Smith, Kirchner, et al,[1] with permission from the publisher.)

obstructing lesions in the branches of a renal artery, an attempt should be made to obtain blood samples from the main branches of the renal vein in order to identify a localized intrarenal arterial lesion that may be responsible for the hypertension.

Surgery has been the traditional treatment for a patient with arteriographically demonstrated renal artery stenosis and confirmatory renal vein renin studies. Recently, percutaneous transluminal angioplasty has been shown to be effective in dilating renal artery stenoses. Although insufficient time has elapsed to compare the long-term results of renal transluminal angioplasty with corrective surgical procedures, initial studies demonstrate the improvement or cure of hypertension in about 80 percent of the patients treated with angioplasty.[1–6]

## OTHER RADIOGRAPHIC MANIFESTATIONS OF HYPERTENSION

The increased work load of the left ventricle due to chronic hypertension initially causes concentric hypertrophy, which produces little if any change in the radiographic appearance of the cardiac silhouette. Eventually, the continued strain leads to dilatation and enlargement of the left ventricle along with downward displacement of the cardiac apex, which often projects below the left hemidiaphragm. Aortic tortuosity with prominence of the ascending portion commonly occurs.

Plain radiographs may be of value in demonstrating the cause of hypertension of nonrenovascular origin. When widening of the superior mediastinum due to increased fat deposition is associated with osteopenia and compression changes in the dorsal vertebrae, Cushing's syndrome is suggested. Inordinate dilatation of the ascending aorta, widening of the left superior mediastinum, the figure-E or figure-3 sign, and rib notching are consistent with coarctation of the aorta. Pheochromocytoma may produce a paravertebral mass; hyperparathyroidism from chronic renal disease may cause typical erosions of the distal clavicles.

Ultrasound and computed tomography can demonstrate adrenal or renal masses causing hypertension (see the sections "Cushing's Syndrome," "Aldosteronism," and "Pheochromocytoma" in Chapter 4).[1]

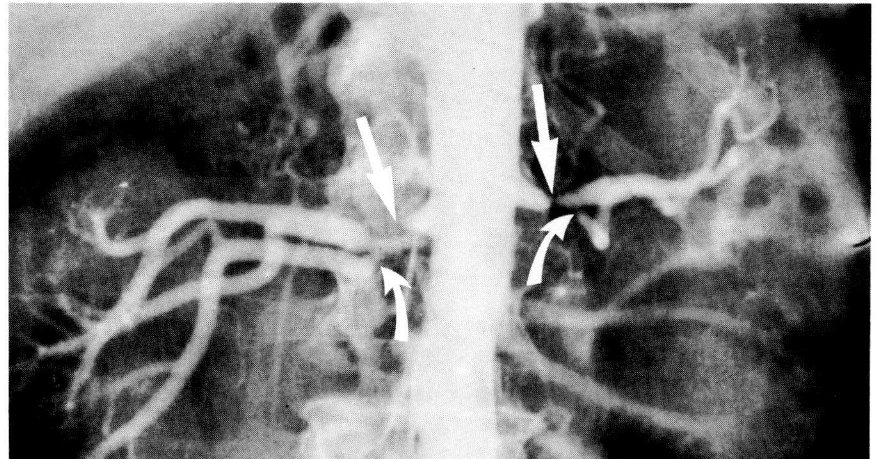

A

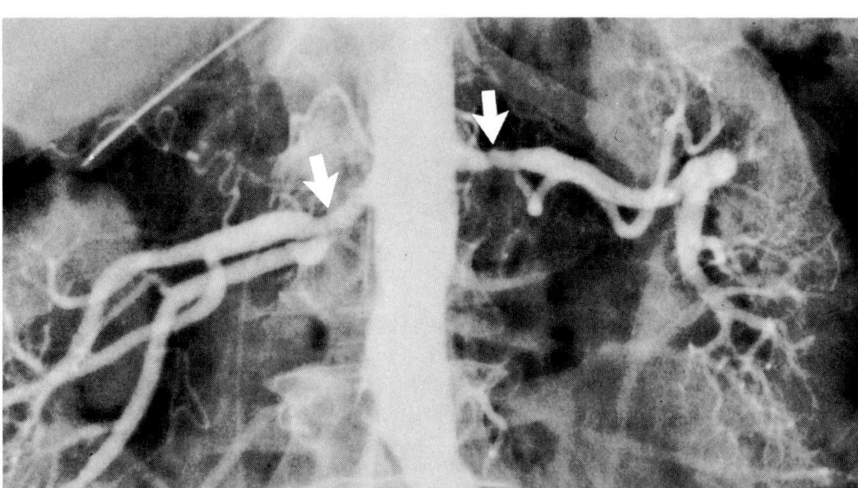

B

**Figure 25.6. Percutaneous transluminal angioplasty for renovascular hypertension.** (A) An abdominal aortogram demonstrates severe bilateral stenoses of the main renal arteries (straight arrows) and stenoses at the origins of early bifurcations (curved arrows). (B) An aortogram following angioplasty of both renal arteries (performed during one sitting) shows irregularities in the areas of the previous arteriosclerotic narrowing (arrows) but an improved residual lumen in both main renal arteries. Stenoses are still present at the origins of the early bifurcating branches of both main renal arteries. Following the procedure, the patient's previous severe hypertension was controllable to normal levels with only minimal dosages of diuretic medication. (From Waltman,[6] with permission from the publisher.)

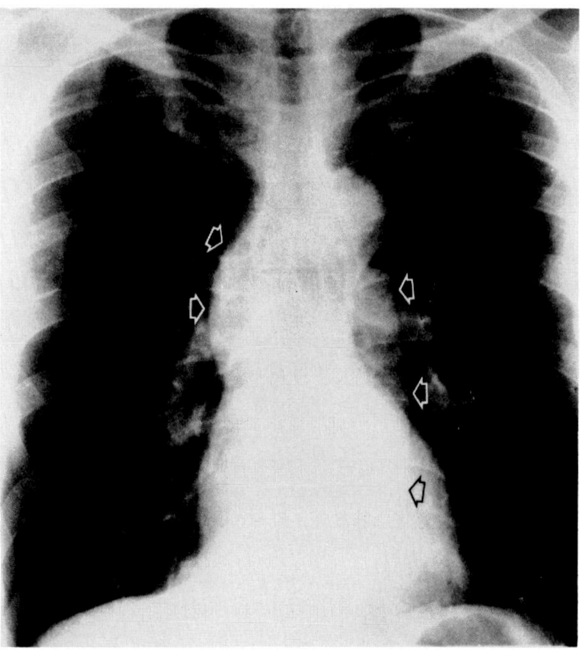

A

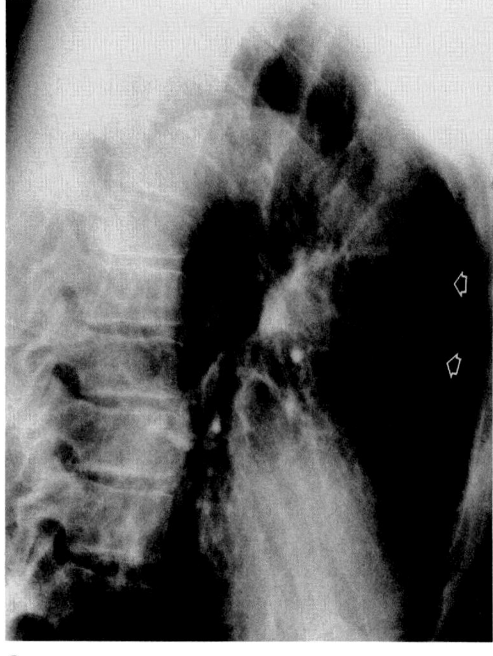

B

**Figure 25.7. Hypertension.** (A) Frontal and (B) lateral views of the chest demonstrate characteristic tortuosity of the aorta (arrows), especially the ascending portion. Because the elevated blood pressure has caused left ventricular hypertrophy without dilatation, the radiographic appearance of the cardiac silhouette remains normal.

## REFERENCES


1. Burko H, Smith CW, Kirchner SG, et al: Hypertension, in Eisenberg RL, Amberg JR (eds): *Critical Diagnostic Pathways in Radiology: An Algorithmic Approach.* Philadelphia, JB Lippincott, 1981, pp 199–211.
2. Kincaid OW, Davis GD, Hallermann FJ: Fibromuscular dysplasia of the renal arteries. Arteriographic features, classification and observations on natural history of the disease. *AJR* 104:271–282, 1968.
3. Bookstein JJ, Abrams HL, Buenger RE, et al: Radiologic aspects of renovascular hypertension. *JAMA* 220:1225–1230, 221:368–374, 1972.
4. Judson WE, Helmar OM: Diagnostic and prognostic values of renin activity in renovenous plasma in renovascular hypertension. *Hypertension* 13:79–89, 1965.
5. Martin EC, Mattern RF, Baer L, et al: Renal angioplasty for hypertension: Predictive factors for long-term success. *AJR* 137:921–924, 1981.
6. Waltman AC: Percutaneous transluminal angioplasty for renal artery stenosis, in Athanasoulis CA, Pfister RC, Greene RE, et al (eds): *Interventional Radiology.* Philadelphia, WB Saunders, 1982.
7. Witten DM, Myers GH, Utz DC: *Emmett's Clinical Urography.* Philadelphia, WB Saunders, 1977.

## Chapter Twenty-Six
# Diseases of the Aorta

### ANEURYSMS OF THE AORTA

A true aortic aneurysm is a circumscribed area of localized widening that involves all three layers of the vessel wall. The destruction of elastic fibers in the media permits the remaining fibrous tissue to stretch, resulting in an increased diameter of the aorta. Most true aortic aneurysms are the result of arteriosclerosis, though cystic medial necrosis, trauma, and syphilis and other infections may be the underlying cause. In false aneurysms, most of which are due to trauma, disruption of the intimal and medial segments of the wall permits expansion of the aorta so that the wall of the aneurysm consists of only adventitia and/or perivascular clot.

#### Aneurysms of the Abdominal Aorta

Three-fourths of all aortic aneurysms occur in the abdominal aorta, just below the renal arteries; almost all are caused by arteriosclerosis. Plain abdominal radiographs can demonstrate curvilinear calcification in the wall of the aneurysm. This may outline a large soft tissue mass or, if both the medial and lateral walls of the aorta are calcified, it may define the enlarged dimensions of the vessel. Lateral projections may demonstrate anterior bulging of the calcified wall of the aneurysm and the virtually pathognomonic, though rare, finding of vertebral body erosion.

Ultrasound is the modality of choice for demonstrating an abdominal aortic aneurysm. The sonographic definition of aneurysmal dilatation of the abdominal aorta is an enlargement of the structure to a diameter greater than 3 cm. Ultrasound can demonstrate intraluminal clot, and the noninvasive nature of ultrasound permits the serial estimation of aneurysm size. Computed tomography (CT) can also demonstrate the location and extent of an abdominal aortic aneurysm, as well as providing some evidence about the risk of rupture.

The major value of arteriography in patients with an abdominal aortic aneurysm is as a presurgical road map, to define the extent of the lesion and whether the renal arteries or other major branches are involved. Because the aneurysm may be filled with clot, aortography often underestimates the extent of aneurysmal dilatation.

#### Aneurysms of the Descending Aorta

Aneurysms of the descending aorta most frequently arise just distal to the origin of the left subclavian artery; almost all are due to arteriosclerosis. Radiographically, they appear as sharply marginated, fusiform or saccular masses that bulge into the left lung and cannot be separated

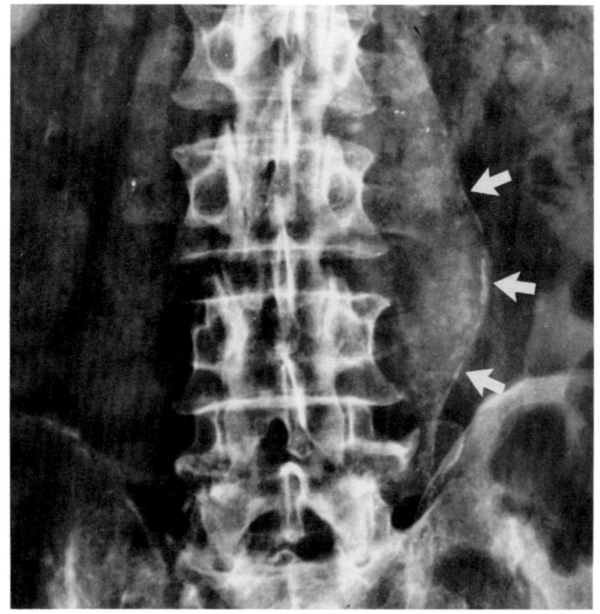

A

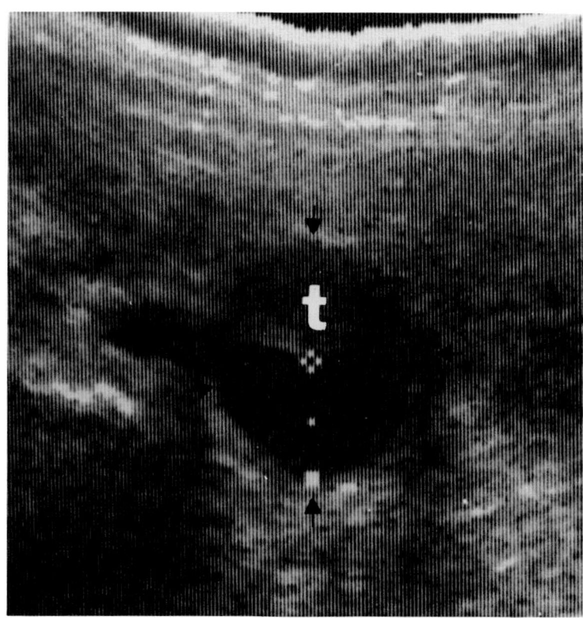


B

C

**Figure 26.1. Abdominal aortic aneurysm.** (A) A plain film of the abdomen demonstrates calcification in the wall of an aneurysm (arrows). (B) Ultrasound shows a large abdominal aortic aneurysm (arrows) with an echogenic thrombus (t) filling the anterior half of the sac. (C) A CT scan of an abdominal aortic aneurysm shows a low-density thrombus (t) surrounding the blood-filled lumen (L). The wall of the aneurysm contains high-density calcification (arrow).

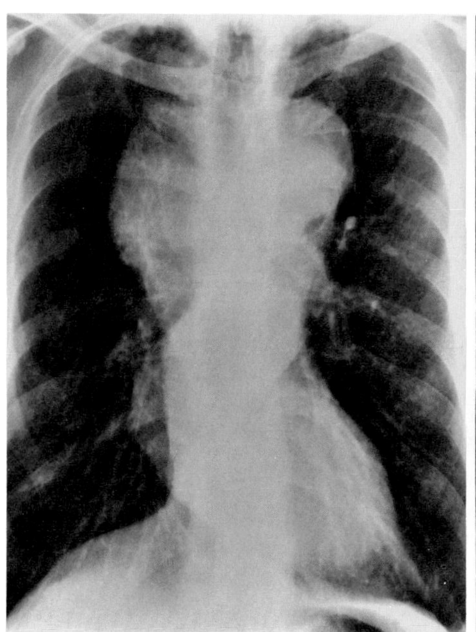

A

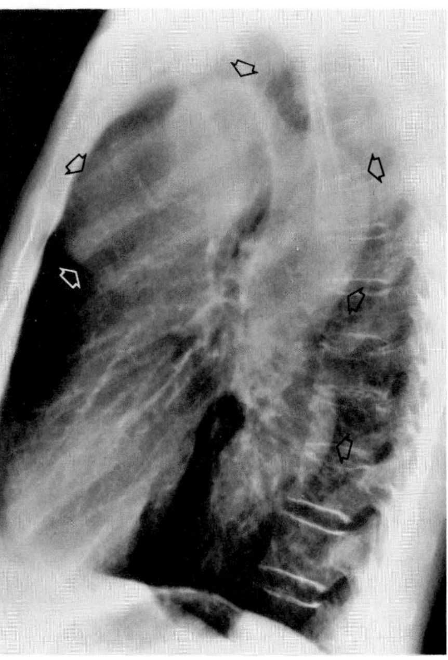

B

**Figure 26.2. Aneurysm of the thoracic aorta.** (A) Frontal and (B) lateral views of the chest demonstrate marked dilatation of both the ascending and the descending portions of the thoracic aorta (arrows).

from the aorta on any projection. Their walls often contain curvilinear calcification. Large masses displace or impress the adjacent trachea and esophagus; continuous pulsatile pressure can rarely cause erosions of the anterior aspects of the vertebral bodies. Rupture or leakage causes the border of the aneurysm to become indistinct.

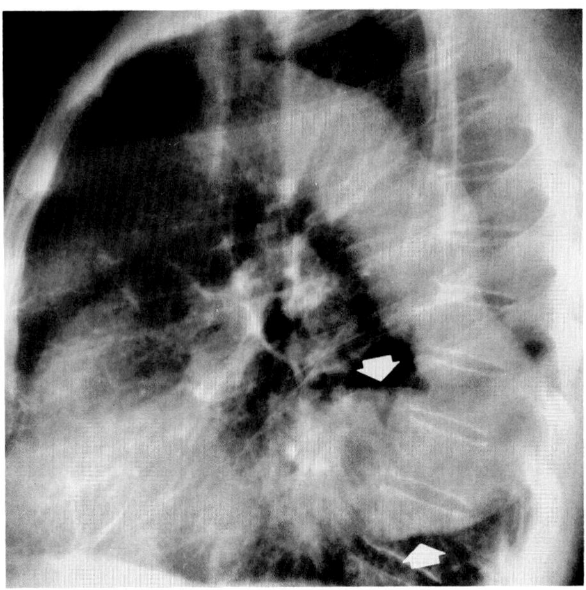

**Figure 26.3. Aneurysm of the descending thoracic aorta.** A lateral view of the chest demonstrates aneurysmal dilatation of the lower thoracic aorta (arrows). Note the marked tortuosity of the remainder of the descending thoracic aorta.

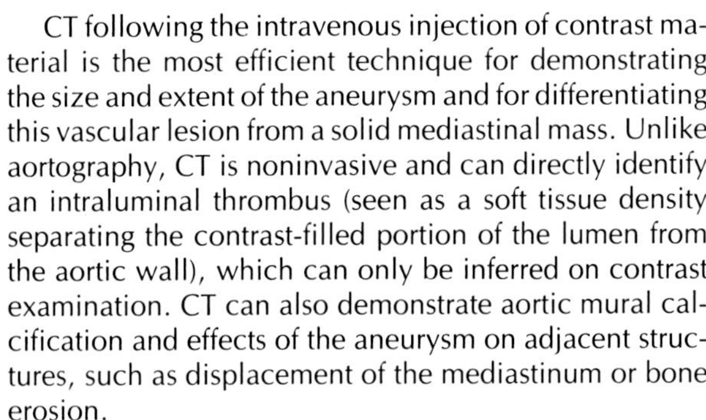

CT following the intravenous injection of contrast material is the most efficient technique for demonstrating the size and extent of the aneurysm and for differentiating this vascular lesion from a solid mediastinal mass. Unlike aortography, CT is noninvasive and can directly identify an intraluminal thrombus (seen as a soft tissue density separating the contrast-filled portion of the lumen from the aortic wall), which can only be inferred on contrast examination. CT can also demonstrate aortic mural calcification and effects of the aneurysm on adjacent structures, such as displacement of the mediastinum or bone erosion.

## Aneurysms of the Ascending Aorta

Though in the past considered pathognomonic of syphilitic aortitis, aneurysms of the ascending aorta are now most frequently due to cystic medial necrosis, which may occur in association with Marfan's syndrome or as a response to hypertension or to mere aging of the aorta; some cases are of unknown cause. Plain frontal chest radiographs may demonstrate increased prominence or a discrete mass involving the right side of the mediastinum. There is filling in of the retrosternal space on lateral views. Pulsatile pressure erosion may involve the adjacent sternum or anterior ribs. Mural calcification may indicate the size of the aneurysmal dilatation.

Aneurysms of the ascending aorta can be demonstrated by echocardiography, CT, or aortography. The latter can also confirm aortic regurgitation, which sometimes accompanies an aneurysm of the ascending aorta and leads to left ventricular failure.[1–5]

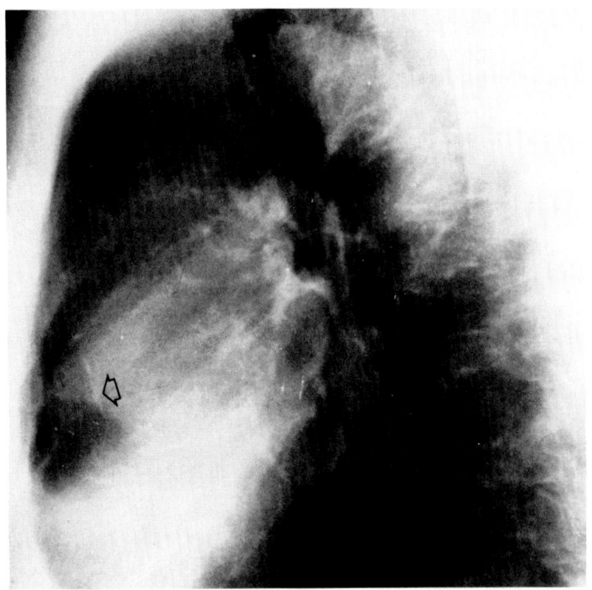

A

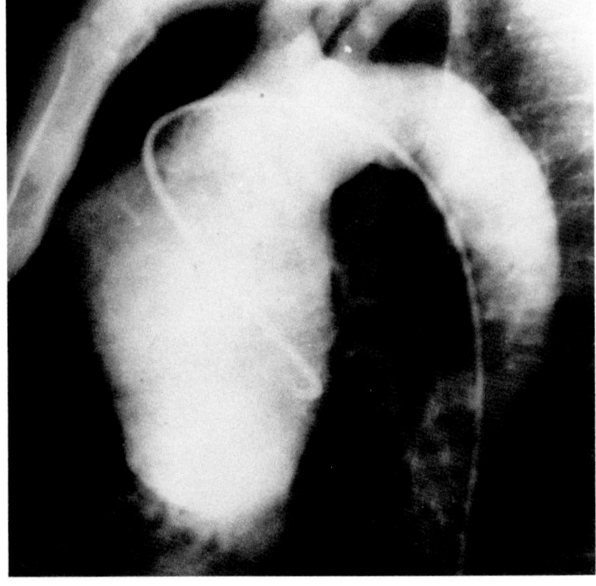

B

**Figure 26.4. Aneurysm of the ascending aorta.** (A) A lateral chest radiograph and (B) an aortogram demonstrate a large saccular aneurysm of the ascending aorta. Note the calcification in the wall of the aneurysm (arrow). (From Ovenfors and Godwin,[1] with permission from the publisher.)

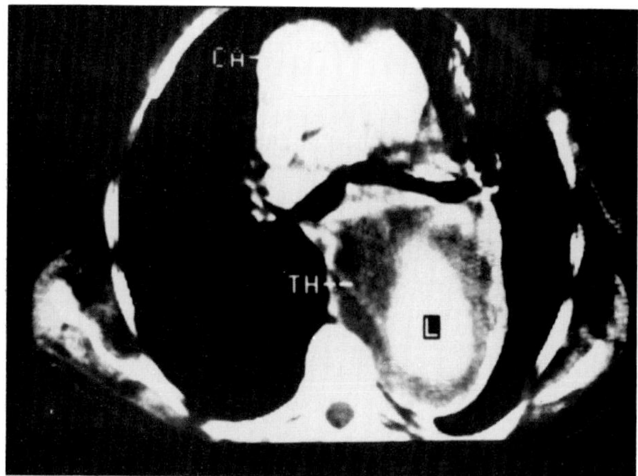


**Figure 26.5. Aneurysms of the ascending and descending thoracic aorta.** Following intravenous contrast injection, a CT scan obtained at a level just below the carina demonstrates a markedly dilated ascending (OA) and descending (L) aorta. Note the large mural thrombus (TH) surrounding the lumen (L) of the descending aorta. (From Ovenfors and Godwin,[1] with permission from the publisher.)

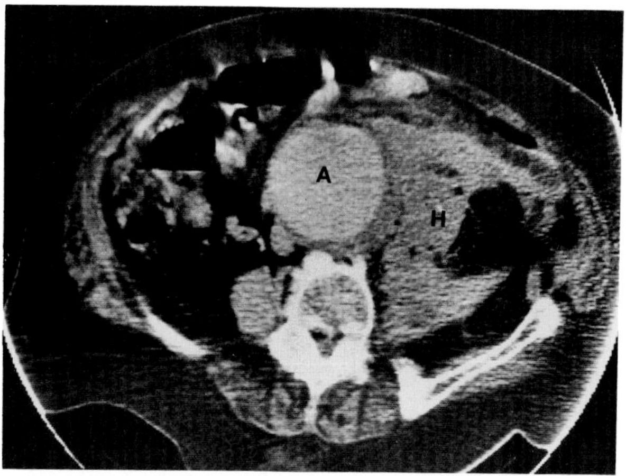


**Figure 26.6. Leaking abdominal aortic aneurysm.** A CT scan shows a large abdominal aortic aneurysm (A) that has bled and given rise to an extensive retroperitonal hematoma (H).

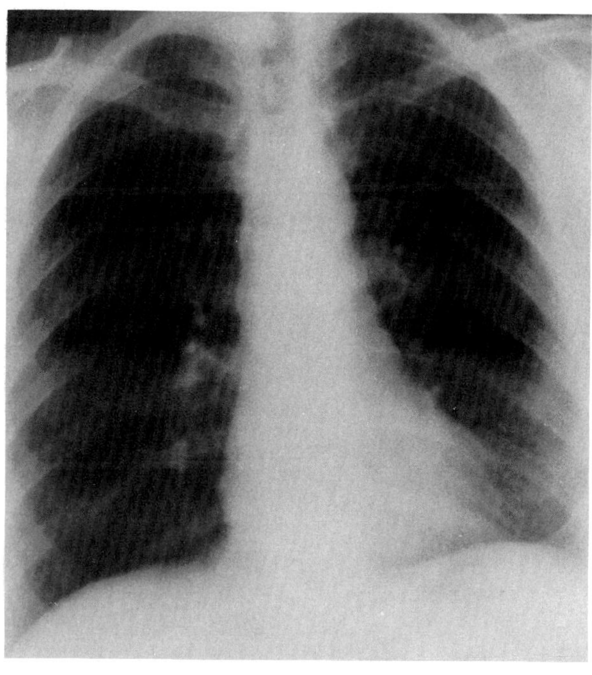

A

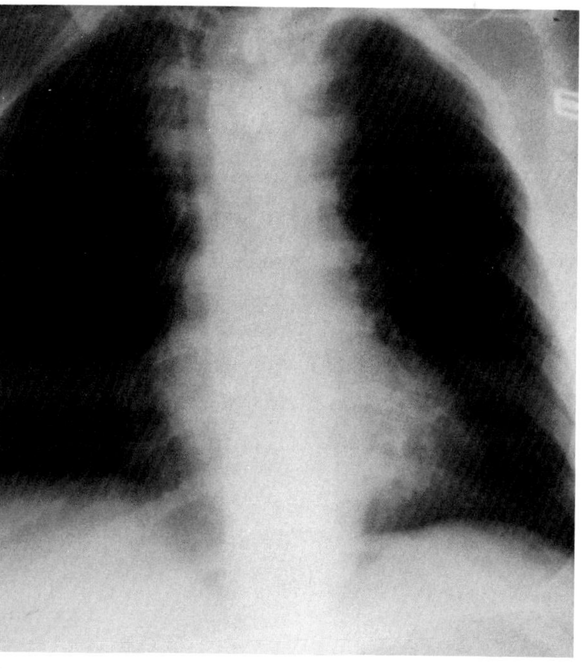

B

**Figure 26.7. Aortic rupture after deceleration trauma (traffic accident).** (A) A frontal chest radiograph taken immediately following trauma demonstrates widening of the upper mediastinum and loss of a discrete aortic knob shadow when compared with a chest film taken 3 months later (B). There is also displacement of the trachea to the right. (From Ovenfors and Godwin,[1] with permission from the publisher.)

## TRAUMATIC ANEURYSM OF THE AORTA

Traumatic rupture of the aorta is a potentially fatal complication of closed chest trauma (rapid deceleration, blast, compression). In almost all cases, the aortic tear occurs just distal to the left subclavian artery at the site of the ductus arteriosus. The intima and media are torn; only the intact adventitia holds the vessel together. At times, the adventitia is also partially torn, resulting in a false aneurysm.

On plain chest radiographs, hemorrhage into the mediastinum causes widening of the mediastinal silhouette

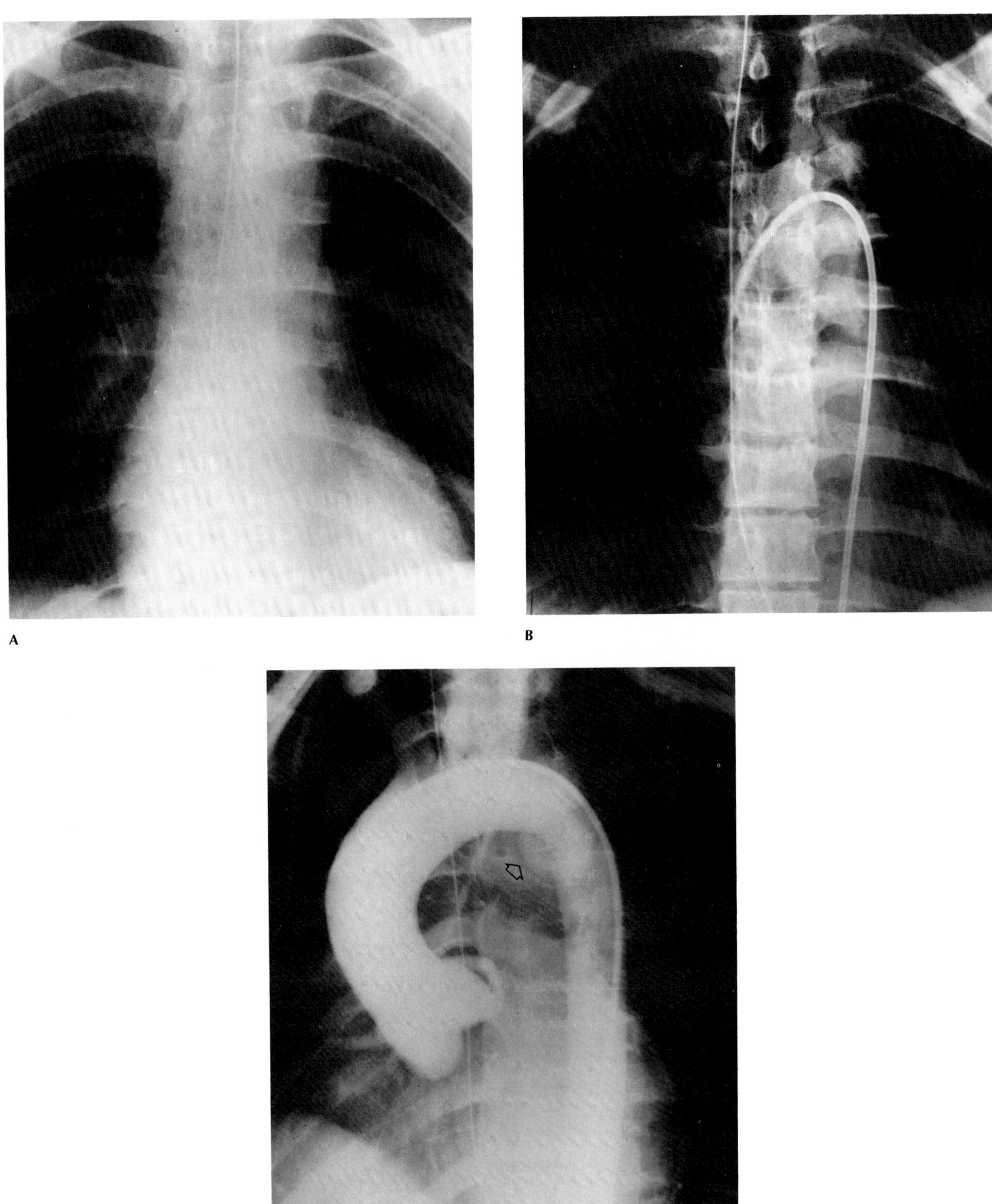

**Figure 26.8. Esophageal tube displacement sign in traumatic aortic aneurysm.** (A) An admission chest radiograph demonstrates widening of the mediastinum. The nasogastric tube is in the midline. (B) A chest film obtained 8 hours later demonstrates displacement of the nasogastric tube to the right. (C) Arteriography demonstrates a pseudoaneurysm (arrow) at the isthmus of the aorta. (From Tisnado, Tsai, Als, et al,[9] with permission from the publisher.)

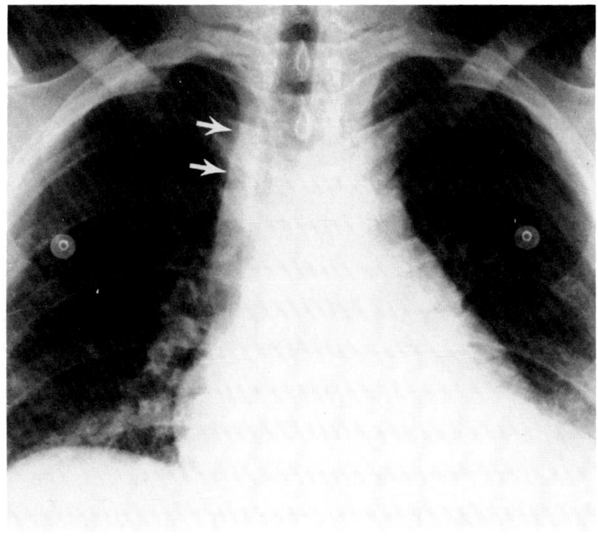

A

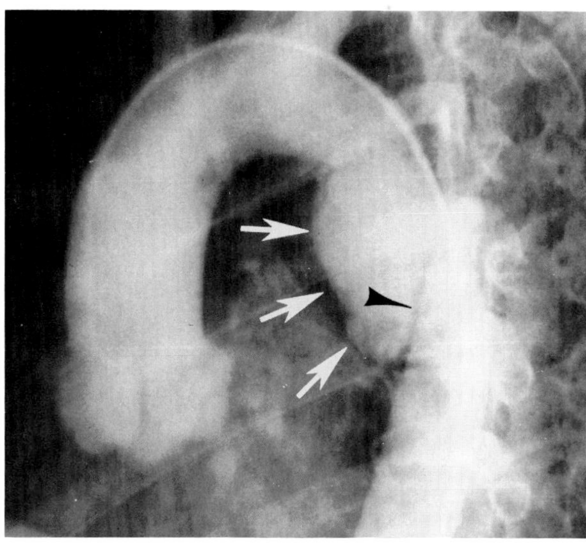

B

**Figure 26.9. Widened right paratracheal stripe sign in aortic rupture.** (A) A supine chest radiograph in a patient with blunt chest trauma shows thickening of the right paratracheal stripe (arrows), which measures 1 cm in width. (B) An aortogram demonstrates a pseudoaneurysm at the level of the aortic isthmus (arrows). The arrowhead indicates an intimal flap. (From Woodring, Pulmano, and Stevens,[10] with permission from the publisher.)

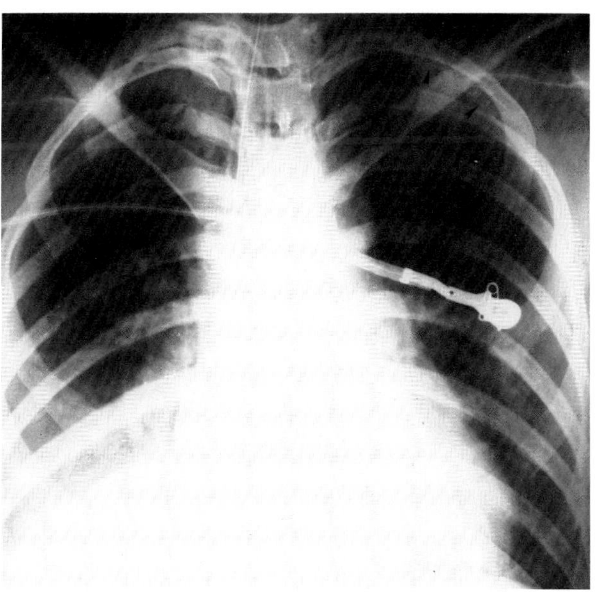

A

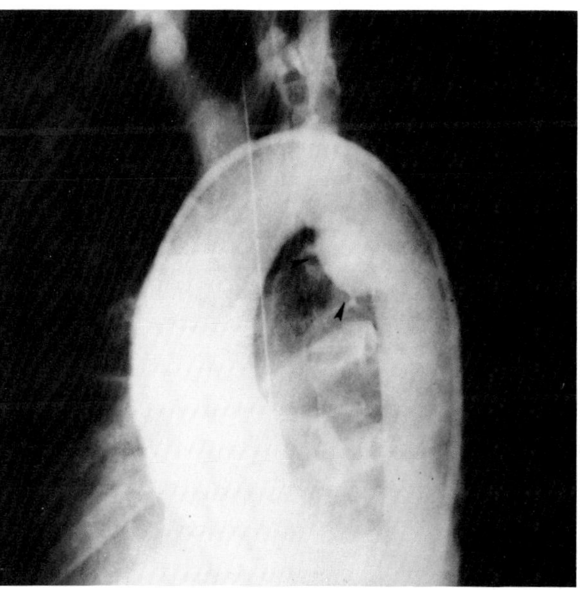

B

**Figure 26.10. Apical pleural cap sign in aortic rupture.** (A) Arrows point to a collection of fluid over the left apex in a patient with a traumatic aortic aneurysm. Note the widening of the superior mediastinum and shift of the endotracheal and esophageal tubes to the right. There are fractures of the right upper ribs and left scapula. (B) An aortogram demonstrates the traumatic aortic aneurysm (arrows). (From Swischuk,[22] with permission from the publisher.)

and loss of a discrete aortic knob shadow. Associated rib or sternal fractures may be apparent. However, because nonspecific mediastinal widening is a frequent finding, especially on the supine, anteroposterior portable radiographs taken after trauma, other plain-film signs are of diagnostic importance.

The displacement of an opaque nasogastric tube to the right, indicative of esophageal displacement, has been reported as the most reliable plain-film sign for the diagnosis of acute traumatic rupture of the thoracic aorta. Blunt trauma to the chest and back produces a hematoma in the anterior and posterior mediastinum, respectively.

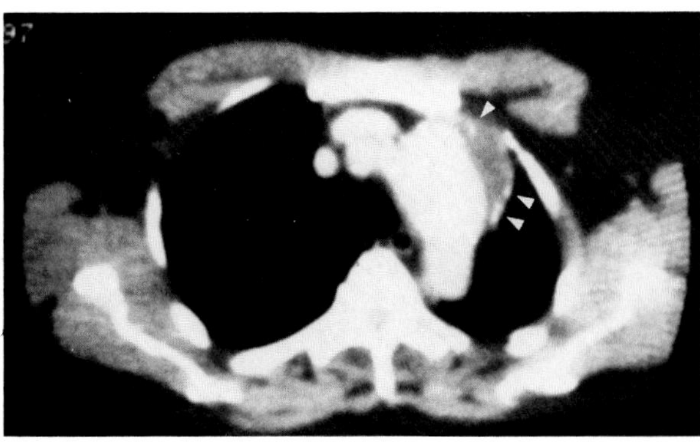

A

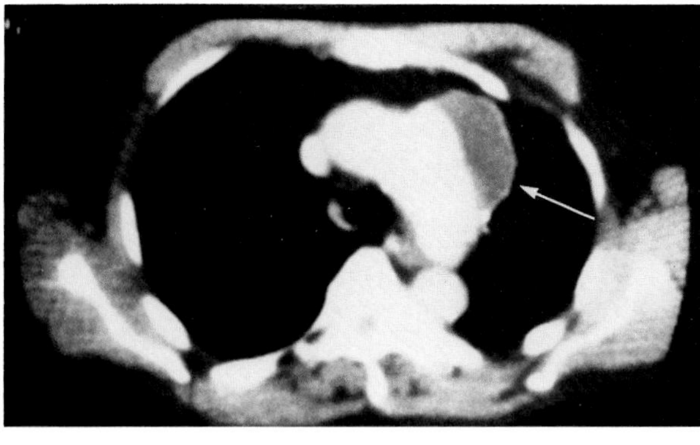

B

**Figure 26.11. Chronic traumatic aortic aneurysm.** CT scans through (A) and slightly below (B) the aortic arch after the intravenous injection of contrast material demonstrate calcification in the wall of the aneurysm (arrowheads) and a large filling defect consisting of thrombus (arrow). (From Ovenfors and Godwin,[1] with permission from the publisher.)

Although it may cause widening of the mediastinum visible on frontal chest radiographs, the hematoma is not strategically positioned to shift the relation between the aorta and the esophagus or trachea, both of which are located in the middle mediastinum close to the aortic arch and isthmus. In contrast, a hematoma in the middle mediastinum caused by a traumatic aortic aneurysm does displace the esophagus and trachea. Therefore, the demonstration of esophageal tube displacement is considered an indication for emergency aortography.

In patients with blunt trauma to the chest, widening of the right paratracheal stripe is frequently associated with major arterial injury. In contrast, patients with stripes of normal width have no underlying aortic rupture. Widening of the right paratracheal stripe is a nonspecific finding that may also be associated with lymphadenopathy, mediastinal inflammation, pleural thickening or effusion, and tracheal disease.

In patients with trauma to the chest, the appearance of a collection of blood over the apex of the left lung is an indication of a traumatic aortic aneurysm. This apical pleural cap sign results from the leakage of blood from a ruptured aortic arch into the pleural space over the left lung. This subtle, but characteristic, sign is an indication for emergency aortography.

Emergency aortography is the definitive examination for the evaluation of possible laceration or rupture of the aorta. A laceration or rupture typically appears as a saccular false aneurysm with an irregular wall, often combined with sharply defined linear defects produced by infolding of the intima and media.

Occasionally, a small aortic rupture with intact adventitia may be undetected at the time of the initial injury. This results in a chronic traumatic aneurysm that may cause a slow, progressive enlargement in the region of the aortic knob detectable on serial radiographs. Calcification may develop in the wall of the aneurysm. Aortography is required to confirm the presence of a chronic saccular post-traumatic aneurysm.[1,6–10,16,22,24]

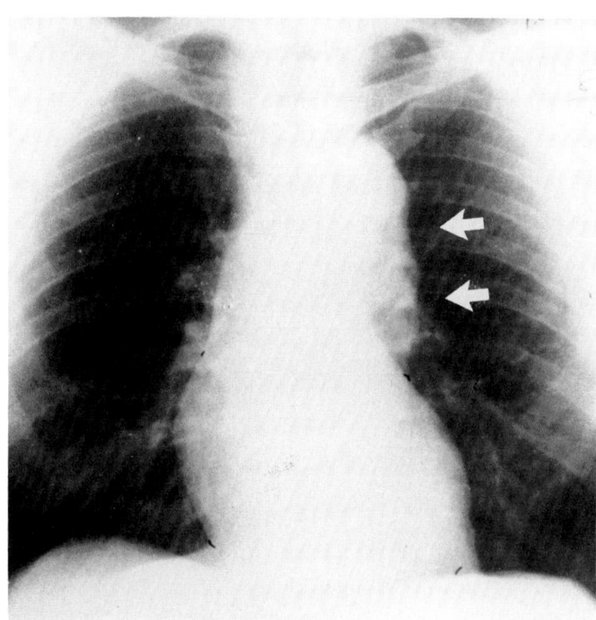

**Figure 26.12. Aortic dissection.** A plain chest radiograph shows only diffuse widening of the descending aorta with an irregular, wavy left border. In this patient the dissection involved both the thoracic and the abdominal aorta. (From Ovenfors and Godwin,[1] with permission from the publisher.)

## DISSECTION OF THE AORTA

Dissection of the aorta is a potentially life-threatening condition in which intimal disruption permits blood to enter the wall of the aorta and separate its layers. Most aortic dissections occur in patients with arterial hypertension. Some are secondary to trauma, while others are due to a congenital defect, such as Marfan's syndrome.

The majority of aortic dissections begin as a tear in the intima immediately above the aortic valve. In two-thirds of these (type I) the dissection continues into the

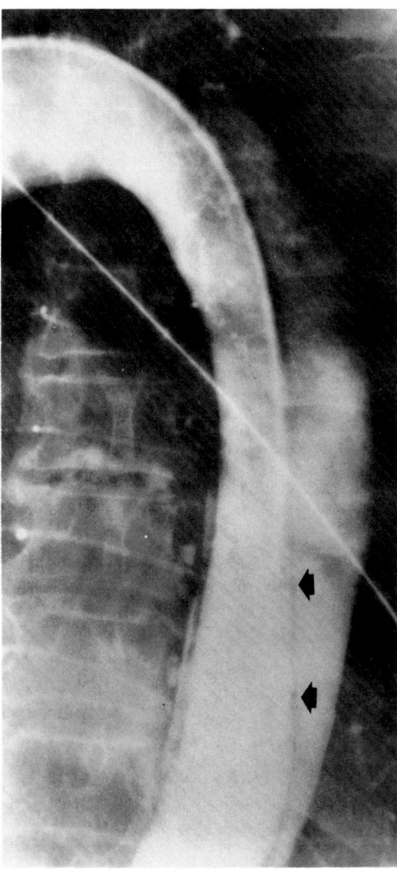

**Figure 26.13. Aortic dissection.** An aortogram demonstrates a thin radiolucent intimal flap (arrows) separating the true and false aortic channels. (From Eisenberg,[23] with permission from the publisher.)

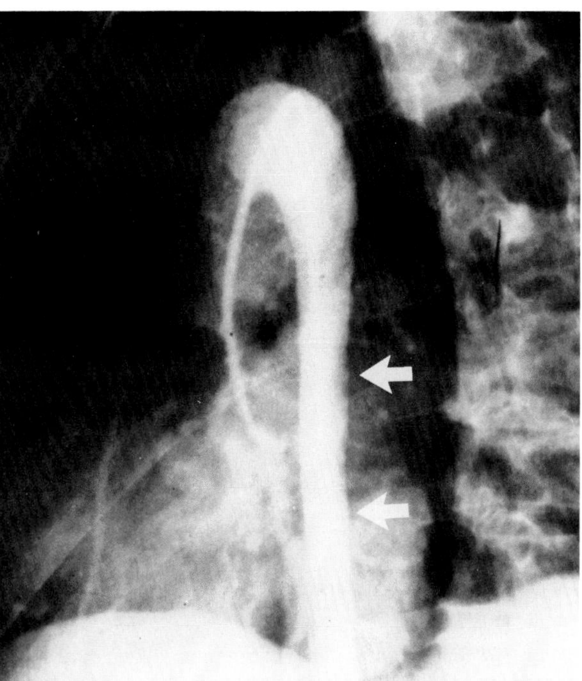

**Figure 26.14. Aortic dissection.** An aortogram shows narrowing of the true lumen (arrow) with nonfilling of the false lumen, which was filled with clot. (From Ovenfors and Godwin,[1] with permission from the publisher.)

descending aorta, often extending as far as the level of the iliac bifurcation. In the remainder (type II) the dissection is limited to the ascending aorta and stops at the origin of the brachiocephalic vessels. In type III disease the dissection begins in the thoracic aorta distal to the subclavian artery and extends for a variable distance proximal and distal to the original site.

On plain chest radiographs, aortic dissection causes progressive widening of the aortic shadow, which may have an irregular or wavy outer border. This widening of the mediastinum, however, may be indistinguishable from an aortic aneurysm or a soft tissue mass. In such cases, the appearance of calcification within the lesion often provides the major clue to the proper diagnosis. Since calcifications are located within the intima, the thickness of the aortic wall can be estimated by observing the distance between the calcification and the outer border of the aortic shadow. In the case of aneurysms, the intima closely follows the contour of the aneurysmal sac, and the distance between the calcification and the outer border of the aorta is only a few millimeters. In dissections, however, the accumulation of blood within the media displaces the intima inward, thus separating the calcium deposits from the aortic contour.

An abrupt change in the configuration of the aorta suggests the presence of dissection. If the ascending aorta is involved, the aortic valve is frequently distorted, resulting in valvular insufficiency and left ventricular dilatation. Dissections often rupture outward through the adventitia and may result in massive hemothorax (usually on the left side); rupture into the pericardium causes acute tamponade and sudden death. At times the dissection may produce a mediastinal hematoma tnat obscures the underlying aortic rupture.

Although the chest film frequently suggests the presence of a dissection of the thoracic aorta, the appearance is rarely pathognomonic. A definite diagnosis of dissection must be made on the basis of CT or aortography.

In the patient with a suspected acute dissection of the thoracic aorta, aortography is the procedure of choice. The diagnosis of aortic dissection depends on the demonstration of two channels separated by a thin radiolucent intimal flap. The true lumen generally runs on the inner curvature of the ascending and descending aorta with the false lumen spiraling around it. The false channel may be filled with clot and be impossible to opacify with contrast material. In this case, the diagnosis can be made by demonstrating narrowing and compression of the true channel. Thickening of the aortic wall (>5 mm) may suggest dissection, though a similar appearance can

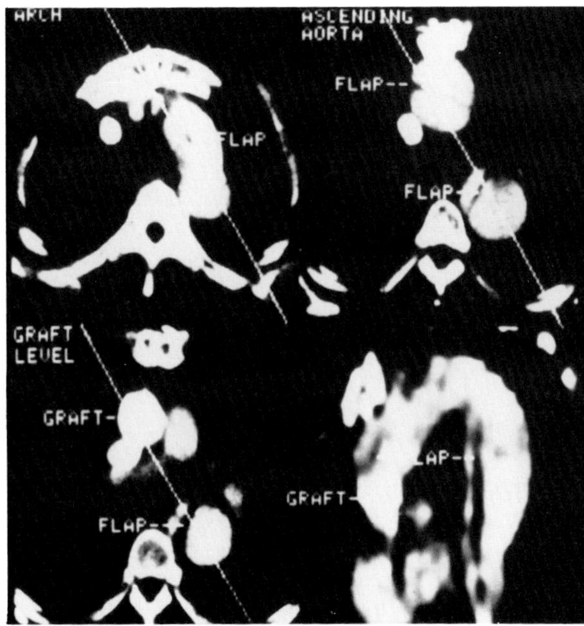


**Figure 26.15. Recurrent aortic dissection.** CT scans through the arch (upper left), ascending and descending aorta (upper right), and at the level of a graft inserted for a previous dissection (lower left) show the typical intimal flap. The lower right picture is a reformatted image of the aorta in an oblique plane that is indicated by the dotted lines on the previous scans. (From Ovenfors and Godwin,[1] with permission from the publisher.)

be due to mural thrombosis, aortitis, neoplasm, or abundant periaortic fat.

When both the true and the false lumen fill with contrast material, it is important to demonstrate the distal re-entry point where the false lumen joins the true lumen. Extension of the dissection to the aortic valve may produce aortic regurgitation with reflux of contrast material into the left ventricle. If the dissection extends below the diaphragm, arteriography can provide presurgical information about which major vessel branches from the aorta are blocked and which remain patent.

In the patient in whom nonacute aortic dissection is considered in the differential diagnosis of chest pain, CT offers a low-risk alternative diagnostic method to aortography. CT can determine the location and extent of a suspected dissection. By demonstrating whether the dissection involves the ascending aorta or if there is rupture into the pericardium or pleura, CT can aid in determining whether an immediate operation is required. The major limitation of CT in the patient with suspected acute dissection is that only a single level can be evaluated on a given slice, as opposed to the rapid series of images covering a wide field that can be achieved with aortography.

CT can demonstrate aortic dissection as a double channel with an intimal flap. With dynamic scanning and the rapid injection of a bolus of contrast material, differential filling of the true and false channels can be observed as with aortography. If partial filling of the aorta with a thrombus or hematoma is demonstrated, inward displacement of intimal calcification can differentiate dissection from a simple aortic aneurysm. Other findings consistent with dissection include an irregular caliber of the aorta and compression of the true lumen.[1,11–15]

## ARTERIOSCLEROTIC OCCLUSIVE DISEASE OF THE AORTA

Arteriosclerosis is the most common disease of the aorta. Although arteriosclerosis is generally considered a degenerative condition of old age, intimal thickening, plaque formation, and vascular narrowing develop in a substantial number of younger patients, especially those with diabetes mellitus, hypertension, and familial disorders of lipid metabolism. Arteriosclerotic occlusive disease is most frequent in the abdominal aorta (below the renal arteries), where it involves the terminal part of the aorta and extends for a variable distance into the iliac and femoral arteries. The classic symptom is claudication, which is present in the buttocks and thighs or in the calves. Severe occlusive disease associated with poor collateral circulation causes marked ischemia that may lead to pain at rest or tissue necrosis and gangrene.

Although plain chest and abdominal radiographs may suggest the presence of aortic arteriosclerosis, aortography is required to demonstrate the severity and extent of the disease. The aorta appears diffusely narrowed and has irregular vascular contours. Nicks, filling defects, and calcifications may be seen along the vessel walls, indicating intimal atheromas, mural thrombi, and de-

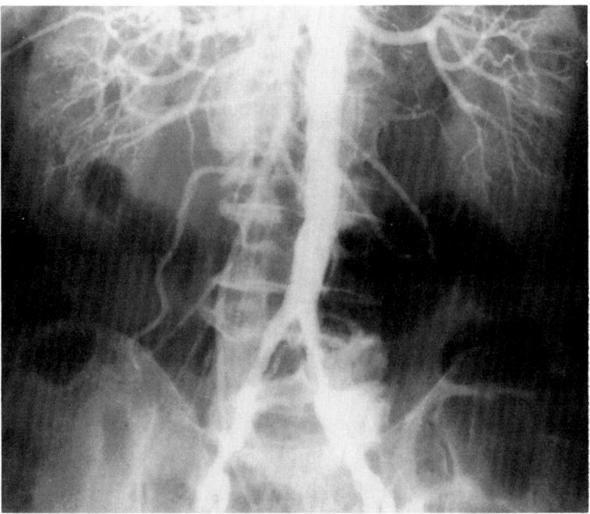

**Figure 26.16. Arteriosclerosis of the aorta.** An arteriogram demonstrates irregular narrowing of the infrarenal aorta and iliac arteries.

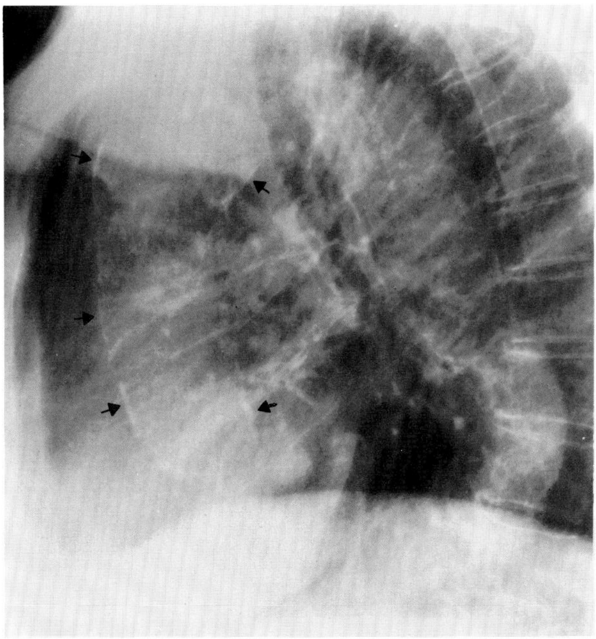

**Figure 26.17. Arteriosclerosis of the aorta.** A lateral view of the chest demonstrates calcification of the anterior and posterior walls of the ascending aorta (arrows). The descending thoracic aorta is markedly tortuous.

generative calcification, respectively. Dysfunction of the elastic fibers of the aorta leads to elongation, tortuosity, and dilatation of the aorta, occasionally with saccular dilatations or frank aneurysms along the vessel's course. Progressive thrombosis or acute embolization can cause complete aortic obstruction.

In the thoracic region, plain chest radiographs demonstrate elongation and tortuosity of the aorta as increased prominence or bulging of the mediastinal contours. The aortic knob often appears somewhat enlarged. Linear plaques of calcification, most commonly seen in the aortic knob and transverse arch, indicate the presence of arteriosclerotic disease. In severe cases, the entire aorta may be outlined by extensive calcification in its wall. Calcification is also commonly seen in the abdominal aorta, where it may be best appreciated on lateral projections.

Elongation and tortuosity of the descending aorta produce a broadened and convex left paraspinal shadow. Oblique and lateral views may be required to demonstrate that this represents a tortuous aorta rather than a mass lesion. If necessary, aortography or contrast-enhanced CT can confirm that the mass represents a dilated, tortuous aorta. A tortuous descending thoracic aorta may also cause a typical sickle-like deformity on the adjacent barium-filled esophagus, displacing this organ anteriorly and to the left.[17]

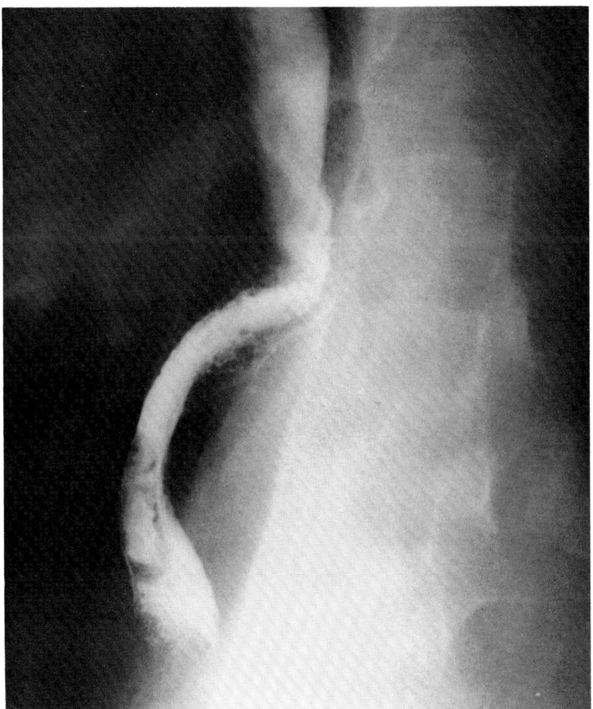

A

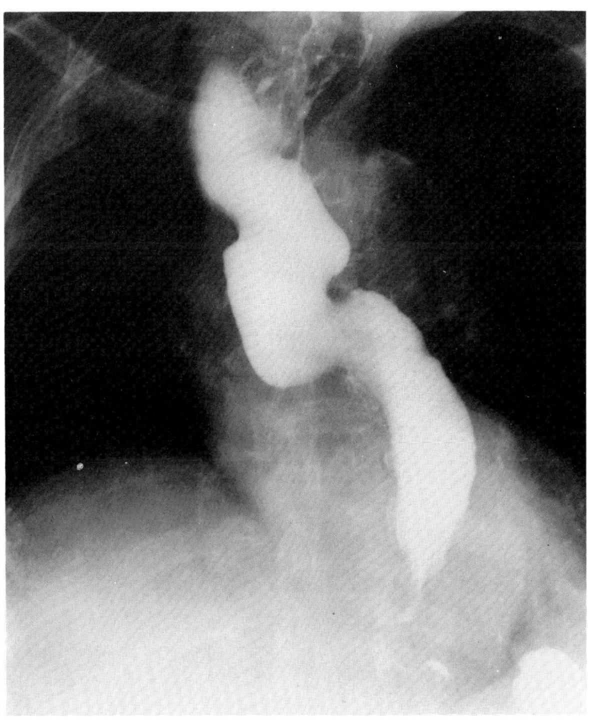

B

**Figure 26.18. Arteriosclerosis of the aorta.** Tortuosity of the descending thoracic aorta produces characteristic displacement of the esophagus anteriorly (A) and to the left (B). The retraction of the upper esophagus to the right, which was caused by chronic inflammatory disease, simulates an extrinsic mass arising from the opposite side. (From Eisenberg,[23] with permission from the publisher.)

## SYPHILITIC AORTITIS

Syphilitic aortitis predominantly involves the ascending aorta because of the predilection of the infecting treponemas for its rich lymphatic supply. Radiographically, this produces dilatation and prominence of the ascending aorta, which may even extend more laterally than the right border of the heart. Thin, curvilinear streaks of calcification frequently develop. These have been contrasted with the thicker, more irregular plaques that typically occur in arteriosclerotic disease. Although calcification of the ascending aorta has been considered virtually pathognomonic of syphilitic aortitis, this disease is now so rare that almost all ascending aorta calcification is due to arteriosclerotic disease.

Syphilitic aortitis can lead to inflammation of the aortic valvular ring, which may result in clinical and radiographic evidence of aortic regurgitation. About one-third of the patients with syphilitic aortitis develop a narrowing of the coronary ostia, which may lead to symptoms of ischemic heart disease. Saccular thoracic aneurysms may also develop.[18]

## TAKAYASU'S DISEASE

Takayasu's disease ("pulseless" disease) is a nonspecific obstructive arteritis, primarily affecting young women, in which granulation tissue destroys the media of large vessels and the resultant mural scarring causes luminal narrowing and occlusion. Though the disease has a particular predilection for the aortic arch, it can involve any portion of the aorta or its major branches.

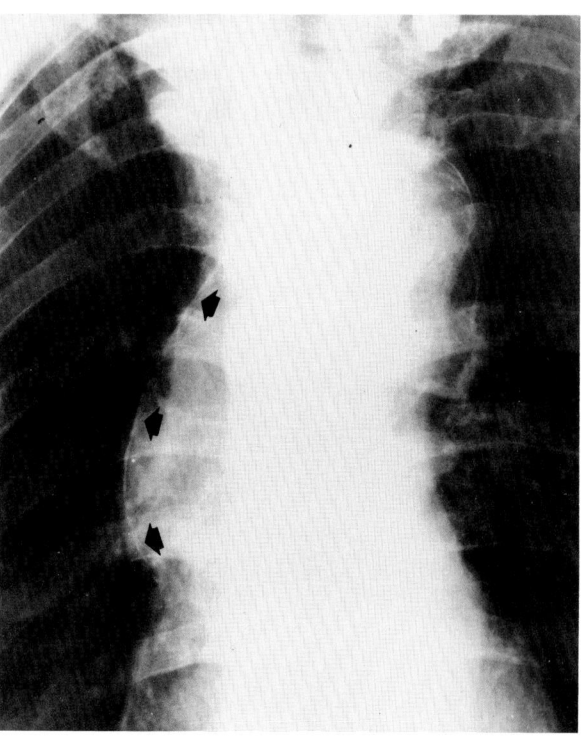

**Figure 26.19. Syphilitic aortitis.** There is aneurysmal dilatation of the ascending aorta with extensive linear calcification of the wall (arrows). Some calcification is also seen in the distal aortic arch.

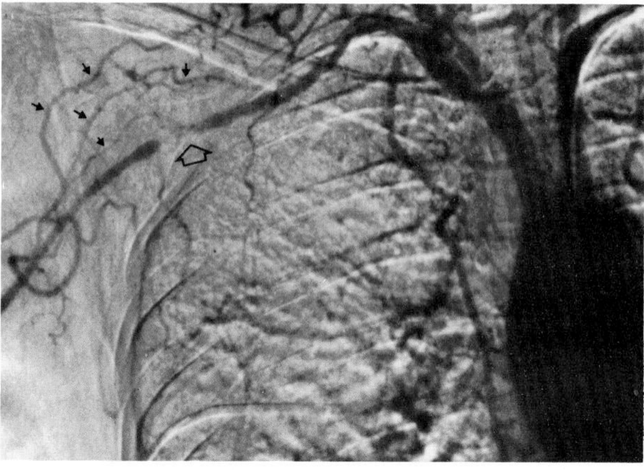

**Figure 26.20. Takayasu's disease.** An arch aortogram shows almost complete occlusion of the right subclavian artery at the level of the clavicle (open arrow) with reconstitution by numerous collaterals (closed arrows).

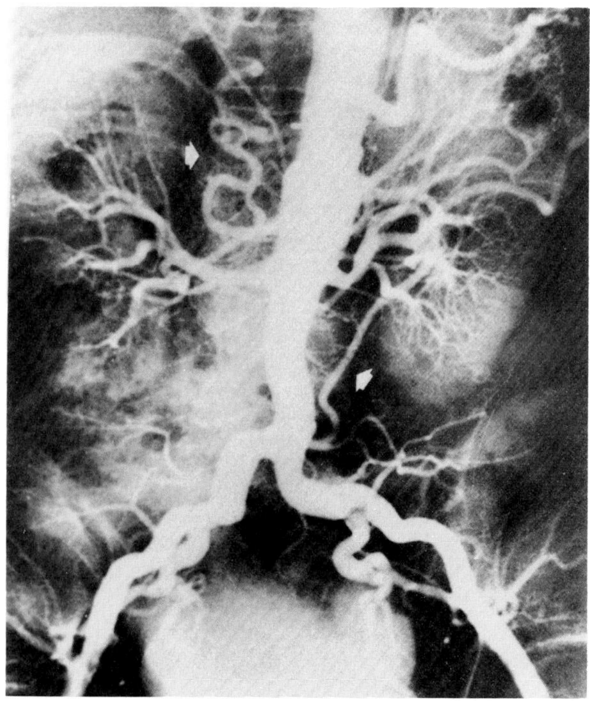

**Figure 26.21. Takayasu's disease.** An aortogram demonstrates marked irregularity of the abdominal aorta. The hepatic artery is occluded and is revascularized through the pancreaticoduodenal arcade (top arrow). The superior mesenteric artery is occluded, and a large left colic artery (bottom arrow) ascends to supply the superior mesenteric territory. Marked tortuosity of the iliac arteries suggests previous arteritis. (From Lande and Rossi,[21] with permission from the publisher.)

Plain chest radiographs demonstrate widening of the aorta and irregularity of its contours, findings that are best appreciated on serial films. Linear calcifications frequently occur. Aortic insufficiency develops in up to 20 percent of the patients because of dilatation of the aortic root.

In the later occlusive phase, aortography demonstrates segmental narrowing or occlusion of the aorta or one or more of its major branches. Unlike arteriosclerosis, Takayasu's disease causes arterial narrowing that is often smooth and tapering.

The combination of aortic dilatation and occlusive disease of one or more major aortic branches in a young woman with fever and constitutional symptoms suggests Takayasu's disease as the underlying cause.[19–21]

## REFERENCES


1. Ovenfors CO, Godwin JD: Aortic aneurysms and dissections, in Eisenberg RL, Amberg JR (eds): *Critical Diagnostic Pathways in Radiology: An Algorithmic Approach*. Phildelphia, JB Lippincott, 1981, pp. 63–75.
2. Symbas PN, Baldwin BJ, Silverman ME, et al: Marfan's syndrome with aneurysm of the ascending aorta and aortic regurgitation. *Am J Cardiol* 25:483–489, 1970.
3. Higgins CB, Silverman NR, Harris RD, et al: Localized aneurysms of the descending thoracic aorta. *Clin Radiol* 26:475–482, 1975.
4. Gramiak R, Shah PM: Echocardiography of the aortic root. *Invest Radiol* 3:356–366, 1968.
5. Beachley MC, Ranniger K: Radiographic findings in aneurysms of the aorta. *CRC Crit Rev Clin Radiol Nucl Med* 7:291–338, 1976.
6. Pinet F, Michaud P, Amiel M, et al: Aortography of traumatic aneurysms of the thoracic aorta: A study of 10 cases. *Ann Radiol* 16:11–19, 1973.
7. Sutorious DJ, Schreiber JT, Helmsworth JA: Traumatic disruption of the thoracic aorta. *J Trauma* 13:583–590, 1973.
8. Simeone JF, Minagi H, Putman CE: Traumatic disruption of the thoracic aorta: Significance of the left apical extrapleural cap. *Radiology* 117:265–268, 1975.
9. Tisnado J, Tsai FY, Als A, et al: A new radiographic sign of acute traumatic rupture of the thoracic aorta: Displacement of the nasogastric tube to the right. *Radiology* 125:603–608, 1977.
10. Woodring JH, Pulmano CM, Stevens RK: The right paratracheal stripe in blunt chest trauma. *Radiology* 143:605–608, 1982.
11. Hayashi K, Meaney TF, Zelch JV, et al: Aortographic analysis of aorta dissection. *AJR* 122:769–782, 1974.
12. Wyman SM: Dissecting aneurysm of the thoracic aorta: Its roentgen recognition. *AJR* 78:247–255, 1957.
13. Beachley MC, Ranniger K, Roth FJ: Roentgenologic evaluation of dissecting aneurysms of the aorta. *AJR* 121:617–625, 1974.
14. Barcia TC, Livoni JP: Indications for angiography in blunt thoracic trauma. *Radiology* 147:15–19, 1983.
15. Gundry SR, Williams S, Burney RE, et al: Indications for aortography after blunt chest trauma. *Invest Radiol* 18:230–237, 1983.
16. Sefczek DM, Sefczek RJ, Deeb ZL: Radiographic signs of acute traumatic rupture of the thoracic aorta. *AJR* 141:1259–1262, 1983.
17. Lipchik EO, Rogoff SM: The abnormal abdominal aorta: Arteriosclerosis and other diseases, in Abrams HL (ed): *Abrams Angiography: Vascular and Interventional Radiology*. Boston, Little, Brown, 1983, pp 1057–1078.
18. Kampmeier RH: The late manifestations of syphilis: Skeletal, visceral, and cardiovascular. *Med Clin North Am* 48:667–697, 1964.
19. Berkman YM, Lande A: Chest roentgenography as a window to the diagnosis of Takayasu's arteritis. *AJR* 125:842–846, 1975.
20. Deutsch W, Wexler L, Deutsch H: Takayasu's arteritis. *AJR* 122:13–28, 1974.
21. Lande A, Rossi P: The value of total aortography in the diagnosis of Takayasu's arteritis. *Radiology* 114:287–297, 1975.
22. Swischuk LE: *Emergency Radiology of the Acutely Ill or Injured Child*. Baltimore, Williams & Wilkins, 1979.
23. Eisenberg RL: *Gastrointestinal Radiology: A Pattern Approach*. Philadelphia, JB Lippincott, 1983.
24. Woodring JH, Loh FK, Kryscio RJ: Mediastinal hemorrhage: An evaluation of radiographic manifestations. *Radiology* 151:15–21, 1984.

## Chapter Twenty-Seven
# Vascular Diseases of the Extremities

### ACUTE ARTERIAL OCCLUSION

Acute arterial occlusion most commonly is the result of embolization. Emboli in the peripheral arteries usually arise from the heart; atrial fibrillation, recurrent cardiac arrhythmias, myocardial infarction with a mural thrombus, and infective endocarditis are the most common predisposing conditions. In the lower extremities, emboli may also arise from ulcerated plaques and from aneurysms of the aorta and the femoral and popliteal arteries. Emboli most frequently occlude arteries in the leg, especially the femoral or popliteal vessels. Although emboli may obstruct a normal vessel, they more frequently obstruct an already-compromised arteriosclerotic artery.

The success of therapy for acute embolic occlusion depends on the rapid recognition of the clinical problem and the location of the obstruction. Arteriography is the procedure of choice to confirm the clinical diagnosis and demonstrate the extent of occlusion, degree of collateral circulation, and condition of the distal runoff vessels. Embolic occlusion typically appears as an abrupt termination of the contrast column along with a proximal curved margin reflecting the nonopaque embolus protruding into the contrast-filled lumen. In acute occlusion there is usually little, if any, evidence of collateral circulation.

Although emergency embolectomy has been the traditional treatment for acute arterial occlusion, the intra-arterial infusion of streptokinase via a catheter placed immediately proximal to the occlusion is becoming a more common therapeutic alternative, especially in the patient who is a poor surgical risk.[1,2]

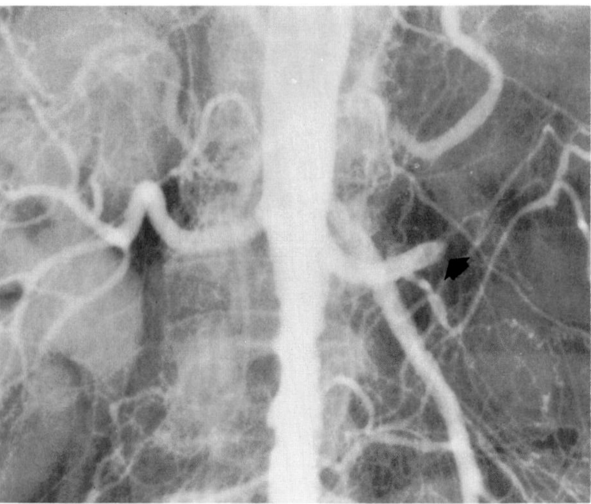

**Figure 27.1. Acute embolic occlusion of the left renal artery.** Abrupt termination of the contrast column in the left renal artery (arrow). Note the irregular contour of the infrarenal aorta, which represents arteriosclerotic disease. The hydronephrosis of the right kidney was due to obstruction of the right ureter by carcinoma of the endometrium.

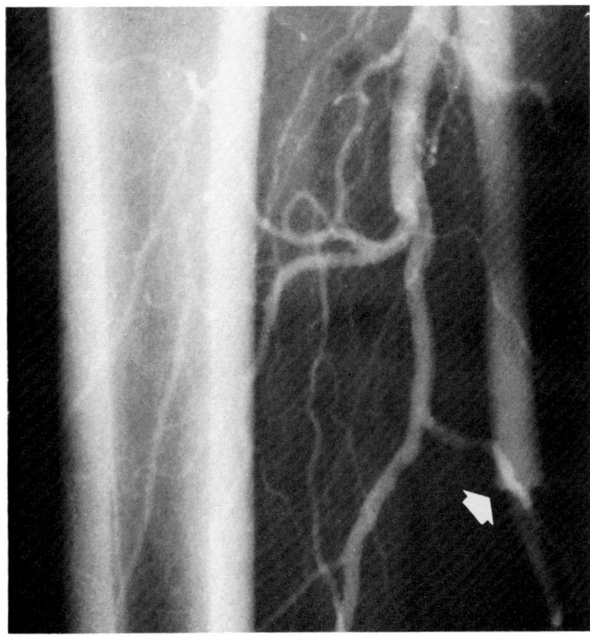

**Figure 27.2. Acute embolic occlusion of the right superficial femoral artery.** Abrupt termination of the contrast column (arrow) with minimal collateral circulation.

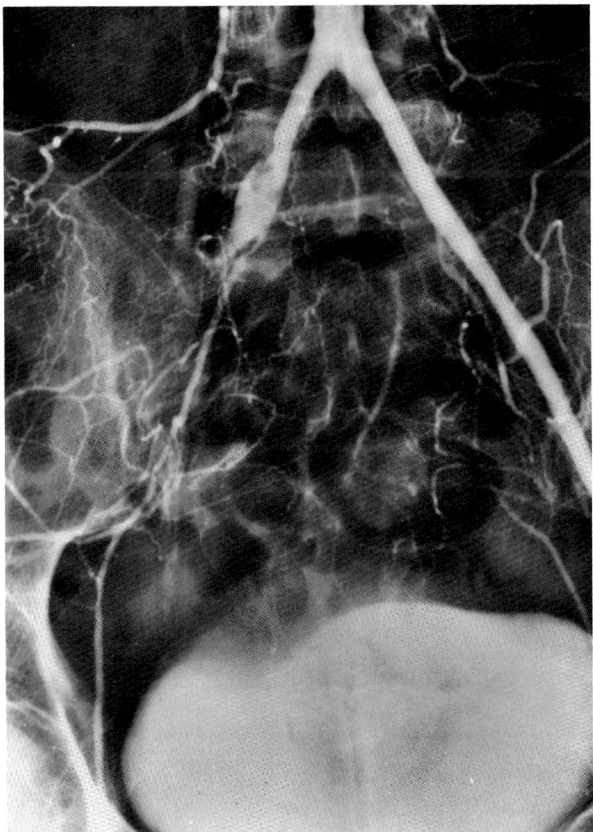

A

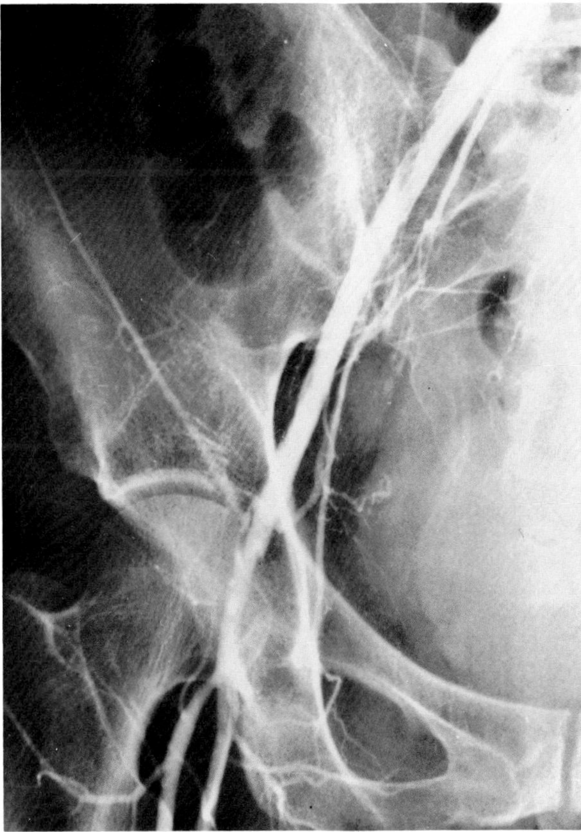

B

**Figure 27.3. Low-dose streptokinase therapy for acute arterial occlusion.** (A) Following angioplasty of the right external iliac artery, there was a loss of the right common femoral pulse. Repeat arteriography from a left femoral approach shows complete occlusion of the external iliac artery at its origin. (B) After 8 hours of streptokinase administration (5000 units per hour), an arteriogram demonstrates lysis of the thrombus and re-establishment of the lumen. One year after angioplasty, the patient had a normal femoral pulse and a patent artery. (From Katzen and van Breda,[1] with permission from the publisher.)

## ARTERIOSCLEROSIS OF THE EXTREMITIES

The major cause of vascular disease of the extremities is arteriosclerosis, in which intimal plaques lead to progressive narrowing and often complete occlusion of large and medium-sized arteries. In the abdomen, the disease primarily involves the aorta and the common iliac arteries, often sparing the external iliac vessels. Distal to the inguinal ligament, arteriosclerotic narrowing most commonly affects the superficial femoral artery, especially in the region of the adductor canal. In such cases, the profunda femoris, which anastomoses freely with popliteal artery branches around the knee, provides the major collateral network between the pelvis and the lower leg. If the popliteal artery is occluded, collateral circulation to reconstitute the distal runoff arteries develops from small, unnamed branches arising proximally from the superficial femoral or profunda femoris arteries or their branches.

Arteriosclerotic plaques often calcify and appear as irregularly distributed densities along the course of an artery. The calcifications are often elongated or somewhat triangular; their presence bears little relation to the severity of the vascular disease. Small vessel calcification, especially in the hands and feet, is often seen in patients with accelerated arteriosclerosis, especially those with diabetes mellitus.

Peripheral vascular calcification may also develop as a result of Mönckeberg's sclerosis, in which calcium deposition follows the degeneration of smooth muscle cells in the media of medium-sized muscular arteries. This condition is common in the elderly and in patients on long-term corticosteroid therapy; it is especially severe and accelerated in persons with diabetes mellitus. The calcification in Mönckeberg's sclerosis appears as closely spaced, fine concentric rings. It is generally diffuse, involving long segments of a vessel or multiple vessels. The process most often involves the femoral, popliteal, and tibial arteries, though moderate-sized vessels in the hands and feet may also be affected, especially in diabetes. The medial changes alone do not narrow the lumen and thus have little effect on the circulation; however, in the lower extremities, medial sclerosis is often associated with arteriosclerosis, which leads to arterial occlusion.

Doppler ultrasound is an effective noninvasive technique for screening patients with clinically suspected peripheral arteriosclerotic disease. Definitive diagnosis requires arteriographic demonstration of the peripheral vascular tree. Evidence of arteriosclerosis includes diffuse vascular narrowing, irregularity of the lumen, filling defects, and patchy alterations of density in the contrast column due to impinging plaques. In patients with severe stenosis or obstruction, arteriography demonstrates the degree and source of collateral circulation as well as the

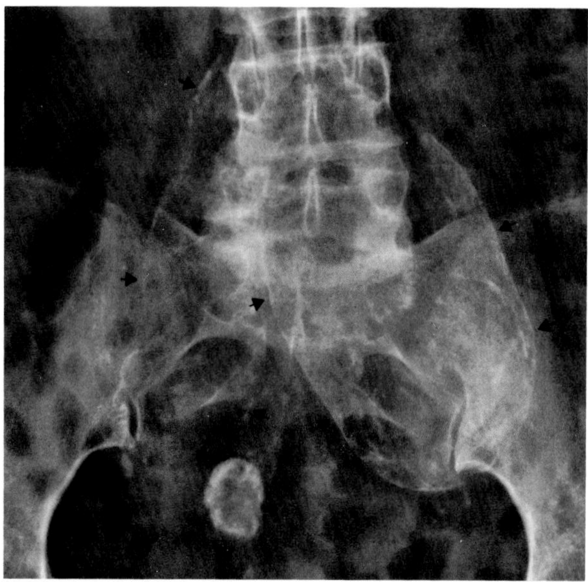

**Figure 27.4. Arteriosclerosis of the extremities.** There are calcified plaques in the walls of aneurysms of the lower abdominal aorta and both common iliac arteries (arrows). (From Eisenberg,[6] with permission from the publisher.)

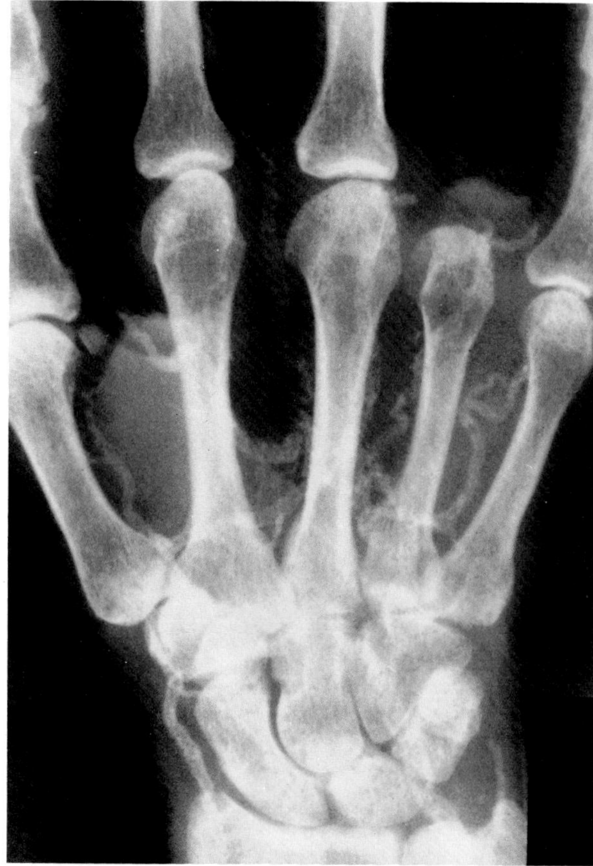

**Figure 27.5. Mönckeberg's sclerosis.** A frontal view of the hand and wrist in a diabetic demonstrates typical calcification of the media in moderate-sized vessels. Note the prior surgical resection of the phalanges of the fourth digit.

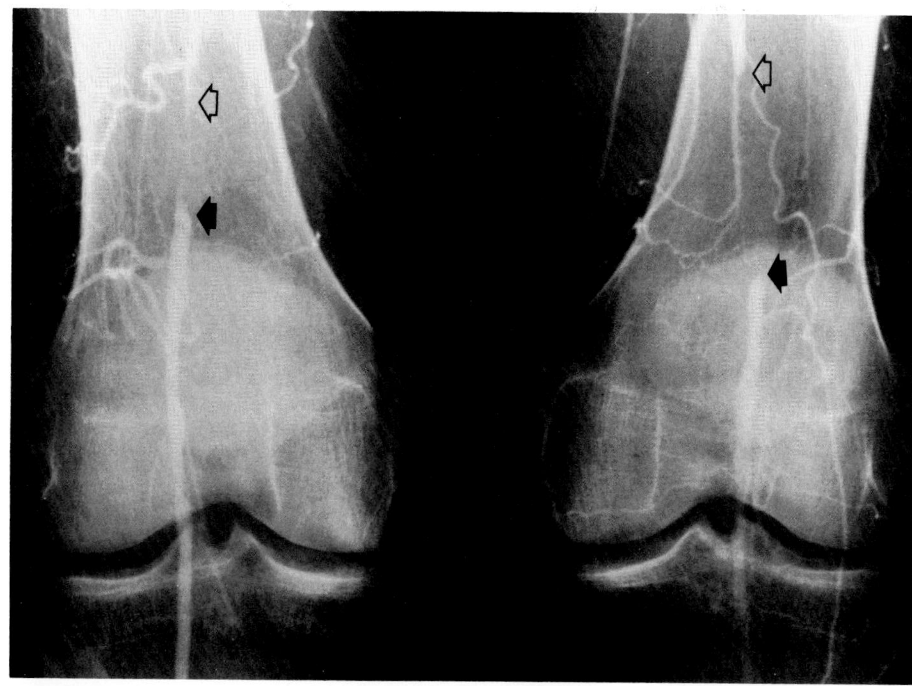

**Figure 27.6. Bilateral arteriosclerotic occlusion of the superficial femoral arteries.** An arteriogram demonstrates bilateral occlusion of the distal superficial femoral arteries (closed arrows) with reconstitution by collateral vessels (open arrows).

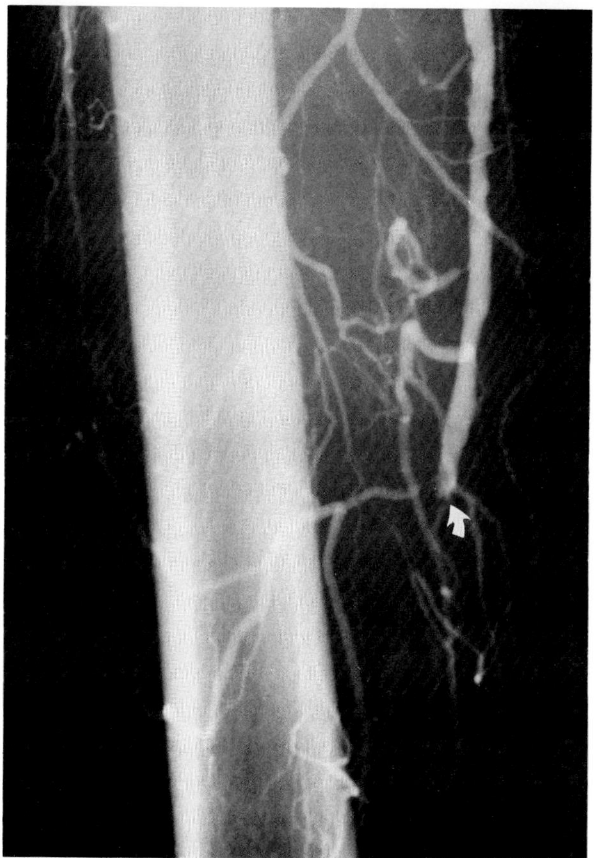

A

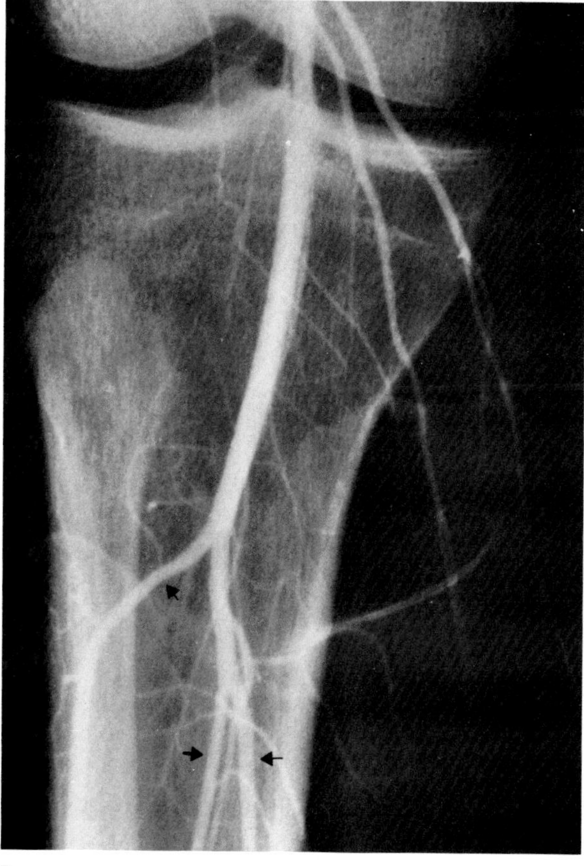

B

**Figure 27.7. Arteriosclerotic occlusion with excellent runoff.** (A) An arteriogram demonstrates occlusion of the superficial femoral artery (arrow) in the midthigh. (B) A subsequent film shows reconstitution of the popliteal artery with normal-appearing runoff vessels (arrows).

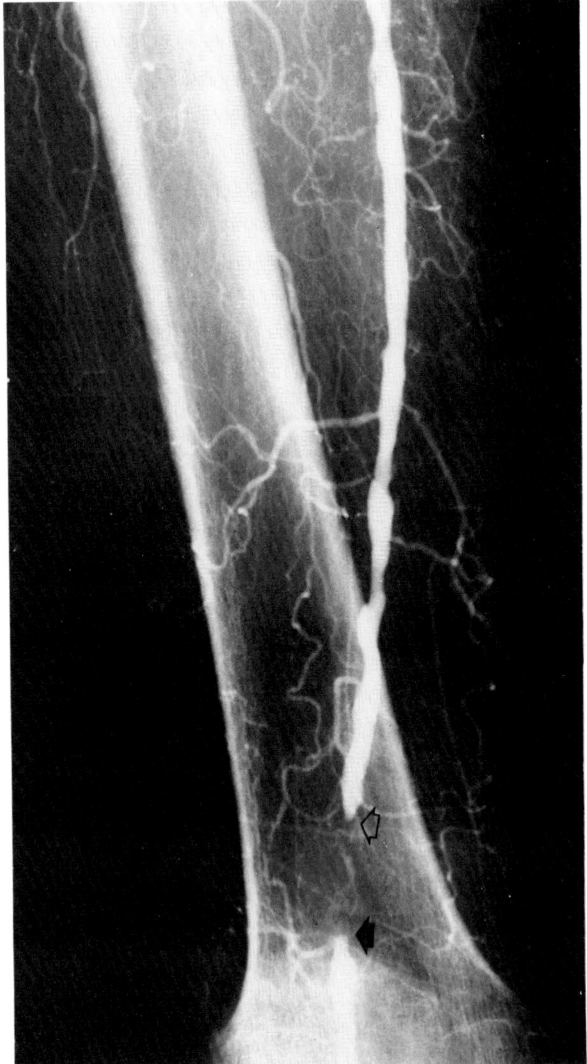

A

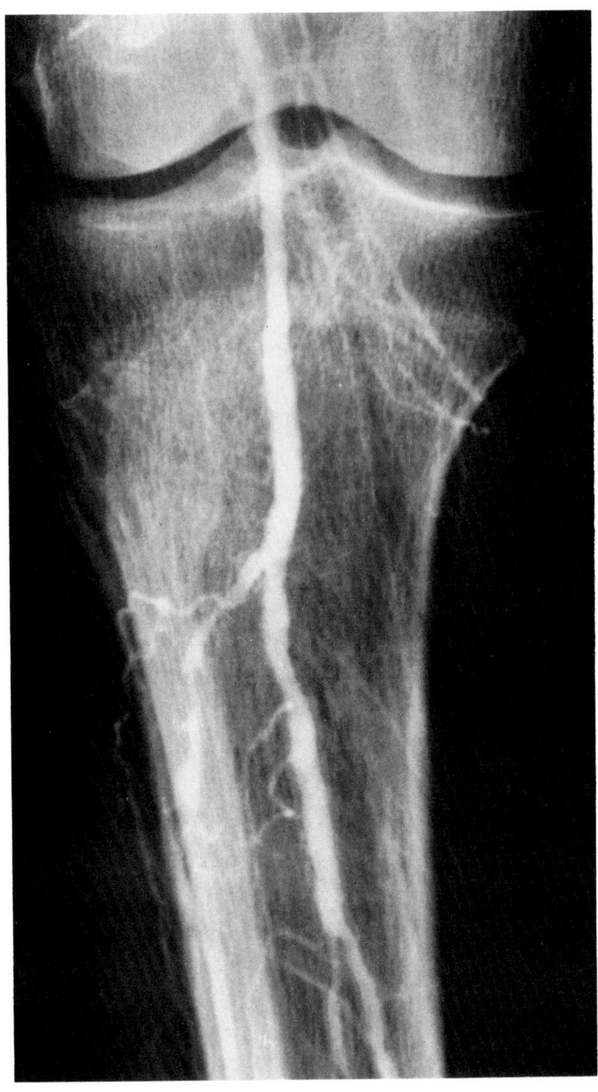

B

**Figure 27.8. Arteriosclerotic occlusion with poor runoff in a diabetic.** (A) An arteriogram demonstrates marked irregularity of the superficial femoral artery with distal occlusion (open arrows) and reconstitution of the vessel as it approaches the knee (closed arrows). (B) A subsequent film shows poor runoff with virtual occlusion of two of the trifurcation vessels and substantial narrowing of branches of the third.

status of the distal runoff vessels. In general, the major factor determining the probability of successful surgical revascularization for lower extremity occlusive vascular disease is the condition of the inflow arteries (aorta and iliac vessels) and the distal runoff arteries (the terminal branches of the popliteal artery below the knee). Combined inflow and outflow disease is a poor prognostic sign.

Concomitant diabetes mellitus tends to increase the severity of peripheral arteriosclerosis. Diabetics have a higher incidence of lesions, especially in the runoff arteries, and a more severe degree of obstruction than nondiabetics. About two-thirds of the combined lesions of the inflow arteries and distal runoff vessels occur in diabetic patients.

Percutaneous transluminal angioplasty using a balloon catheter is now a recognized procedure for the alleviation of symptoms of peripheral ischemia in patients with arteriosclerosis. Initial studies indicate vessel patency rates similar to those following reconstructive surgery. Percutaneous transluminal angioplasty is of special value in providing limb salvage and prolonged relief of ischemia in patients who would not benefit from surgery.[2–5]

## TERMINAL AORTIC OCCLUSION (LERICHE SYNDROME)

The Leriche syndrome, which refers to the combination of hip, thigh, and buttock claudication, along with impotence in the male, indicates occlusion of the terminal aorta below the level of the renal arteries. Aortography

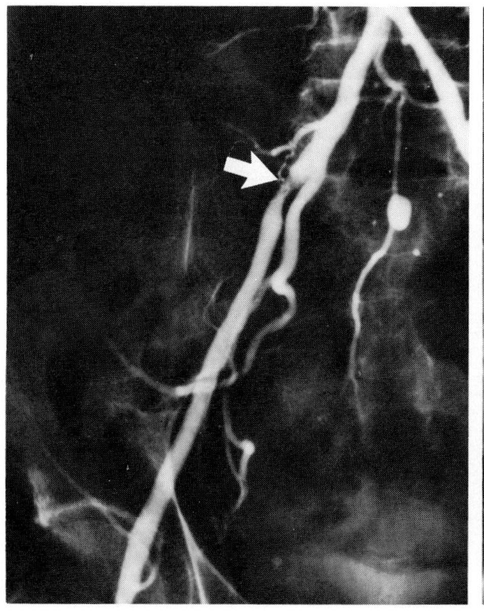

A

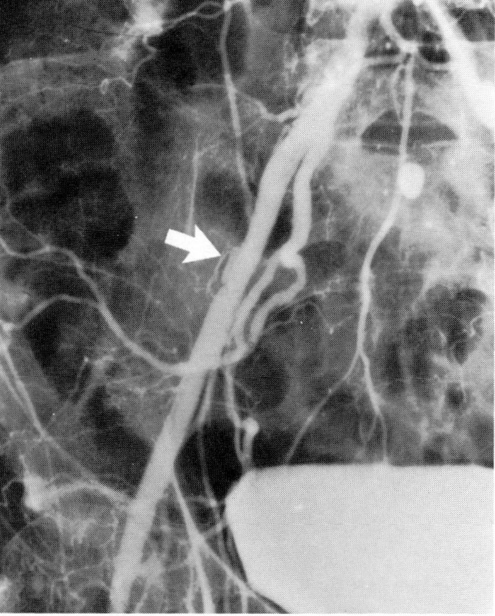

B

**Figure 27.9. Percutaneous transluminal angioplasty of a hemodynamically significant right external iliac artery stenosis.** (A) An initial film in a patient with claudication demonstrates narrowing of the proximal right external iliac artery (arrow). (B) Following angioplasty, there is relief of the stenosis (arrow), and hemodynamic improvement. (From Waltman,[5] with permission from the publisher.)

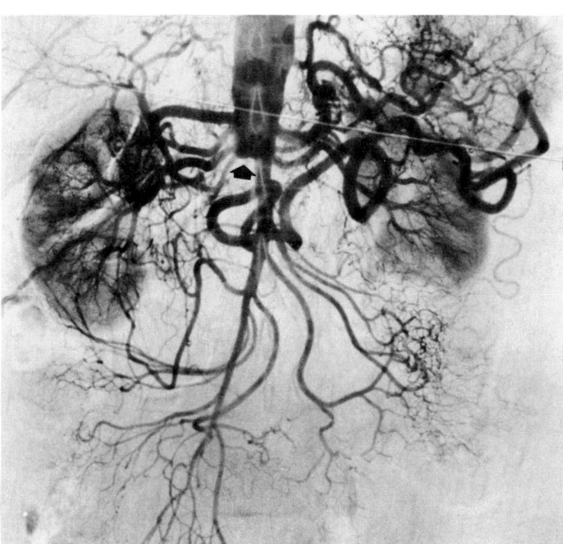

A

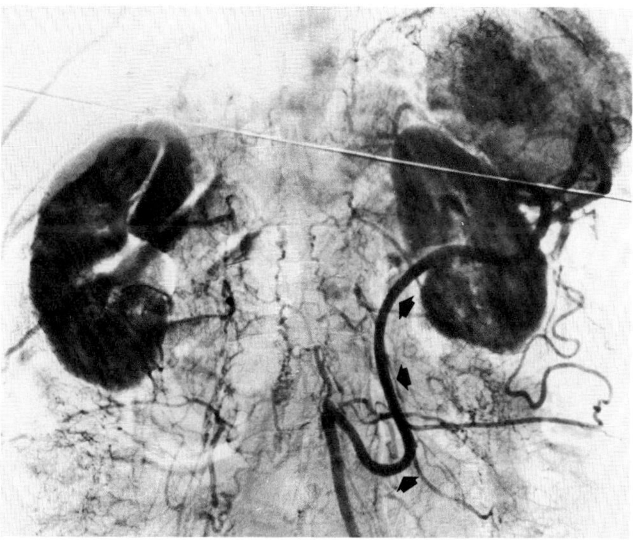

B

**Figure 27.10. Leriche syndrome.** (A) A subtraction film from an aortogram demonstrates occlusion of the distal aorta (arrows) just below the level of the renal arteries. (B) A film obtained 4 seconds later shows a large marginal artery of Drummond (arrows) serving as a major collateral vessel.

demonstrates obstruction of the distal aorta and extensive collateral circulation (especially via the mesenteric, lumbar, and circumflex arteries).[2]

## FIBROMUSCULAR DYSPLASIA

Fibromuscular dysplasia is a nonatheromatous angiopathy of unknown cause that involves various small and medium-sized arteries, especially the renal and extracranial cephalic vessels. At arteriography, fibromuscular dysplasia typically produces multiple arterial dilatations separated by irregularly spaced concentric stenoses (string-of-beads sign). This appearance must be differentiated from circular spastic arterial contractions due to vascular irritation, in which the constrictions are more regular and evenly spaced and occur without the dilatation of intervening segments.[7,8]

## THROMBOANGIITIS OBLITERANS (BUERGER'S DISEASE)

Thromboangiitis obliterans is a nonatheromatous disorder consisting of an arteritis and phlebitis that typically

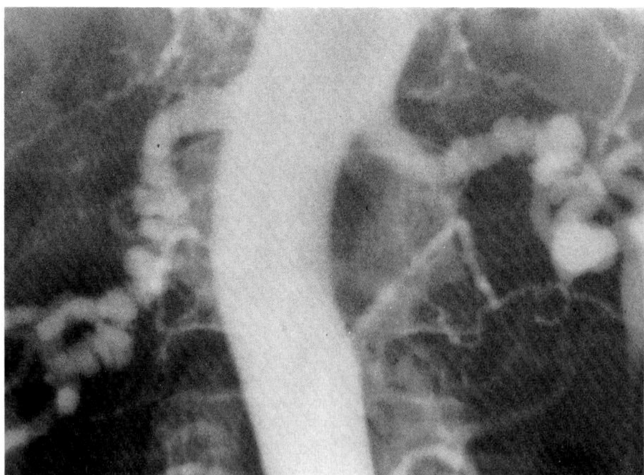

**Figure 27.11. Fibromuscular dysplasia.** The string-of-beads sign involving both renal arteries.

occur in young men and may lead to nonhealing ulcers and gangrene. Unlike arteriosclerosis, thromboangiitis obliterans produces pain at rest rather than intermittent claudication and tends to begin in the smaller arteries of the hands and feet rather than in the medium or large arteries. The exact pathogenesis is obscure, though there appears to be a definite relation with tobacco smoking or chewing.

Arteriography in a patient with thromboangiitis obliterans usually demonstrates one or more stenoses or occlusions of small arteries; vessels proximal to the site of narrowing or obstruction generally have smooth contours and appear normal. Reconstitution distal to the obstruction is either lacking or extremely poor, with visualization primarily of multiple tortuous collateral vessels that tend to run parallel to the main occluded artery. Arterial wall calcification is uncommon.[8,9]

## ARTERIOVENOUS FISTULAS

An arteriovenous fistula, which may be congenital or acquired, is an abnormal "short circuit" between an adjacent artery and vein. Acquired fistulas almost always involve penetrating trauma from gunshot or stab wounds, shrapnel particles, or fracture fragments. They usually occur in the extremities, where adjacent arteries and veins are relatively close to the surface. A palpable thrill, continuous murmur, or venous dilatation are suggestive findings. Large arteriovenous shunts may increase venous return and lead to high-output cardiac failure with cardiomegaly and increased pulmonary vascularity. Chronic fistulas may cause varicose veins, phleboliths, and severe limb edema.

In acquired arteriovenous fistulas, arteriography typically demonstrates one or several markedly dilated feeding branches with rapid filling of an adjacent vein, which is also dilated because of the increased blood flow and the direct exposure to arterial pressure. The arterial circulation distal to the fistulous connection is often poorly opacified, since much of the contrast material passes directly into the fistulized vein. Extravasation of blood at the time of initial arterial injury can lead to the formation of a false aneurysm.

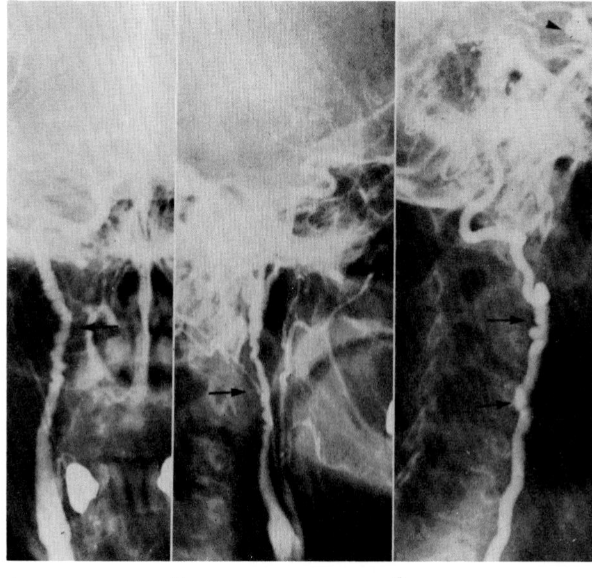


**Figure 27.12. Fibromuscular dysplasia of the cervical arteries.** (A) A frontal view from a right carotid arteriogram, (B) a lateral view from a left carotid arteriogram, and (C) an oblique view from a left vertebral arteriogram demonstrate multiple stenoses with small aneurysms (arrows). An associated berry aneurysm of the basilar artery is present in C (arrowhead). (From Grollman, Lecky, and Rosch,[8] with permission from the publisher.)

Congenital arteriovenous fistulas can produce bizarre patterns of dilated feeding arteries and draining veins. Arteriography is important to define the site and extent of arterial supply, to assess the feasibility of surgical extirpation, and to help establish the adequacy of surgical resection.[2,10–13]

## THORACIC OUTLET SYNDROME

Thoracic outlet syndrome refers to muscular or skeletal compression of the neurovascular bundle as it courses to the upper extremity. The scalenus anticus muscle or a cervical rib is often implicated as an underlying cause. However, cervical ribs occur in less than half of the patients with thoracic outlet syndrome and are often present in asymptomatic individuals.

Arteriographic demonstration (in a neutral position) of localized compression or torsion of the subclavian artery with poststenotic dilatation distal to one of the classic points of compression (the interscalene tunnel, the costoclavicular space, or beneath the tendon of the pectoralis minor muscle) is strongly suggestive of a he-

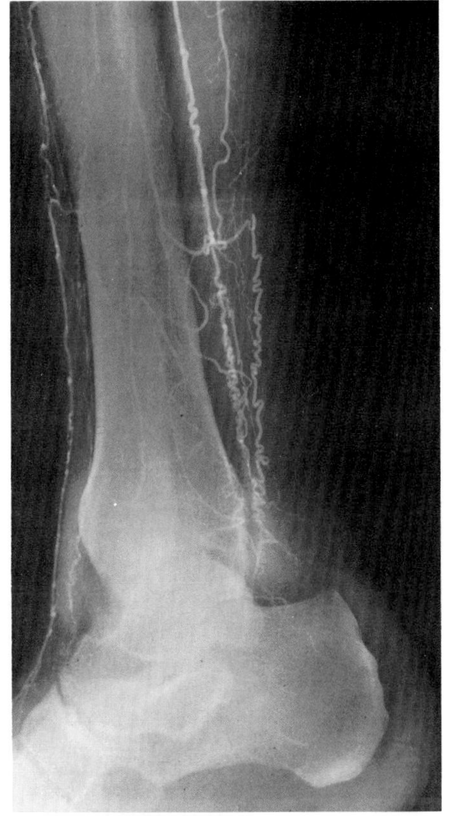

A

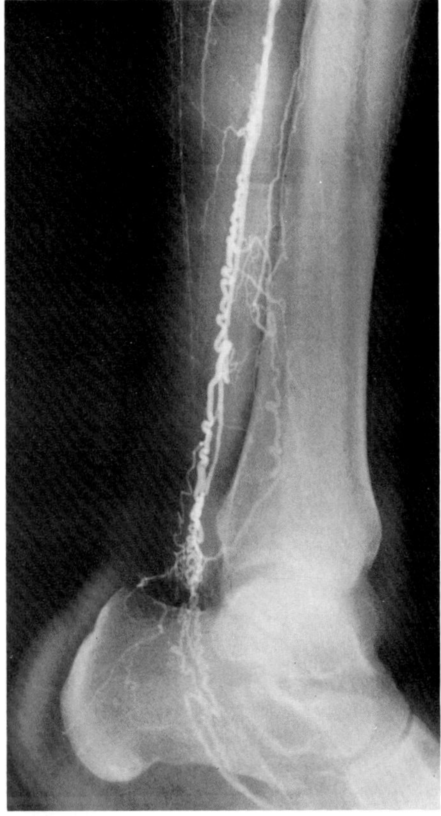

B

**Figure 27.13. Thromboangiitis obliterans.** Views of the right leg (A) and left leg (B) 7 seconds after the injection of contrast material in both common femoral arteries demonstrate long collateral channels showing a great number of fine, serrated loops. The ultimate direction of these collateral vessels is parallel to occluded arteries and close to them, mainly in the posterior tibial region. Other films demonstrated occlusion of the main arteries of the legs. (From Roy,[9] with permission from the publisher.)

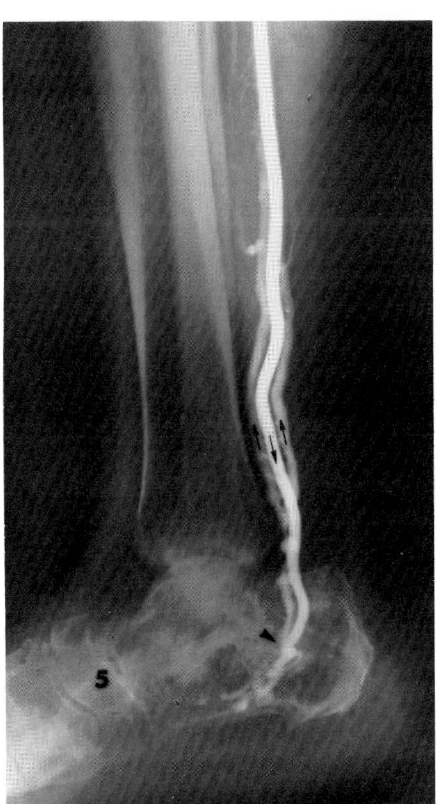


A

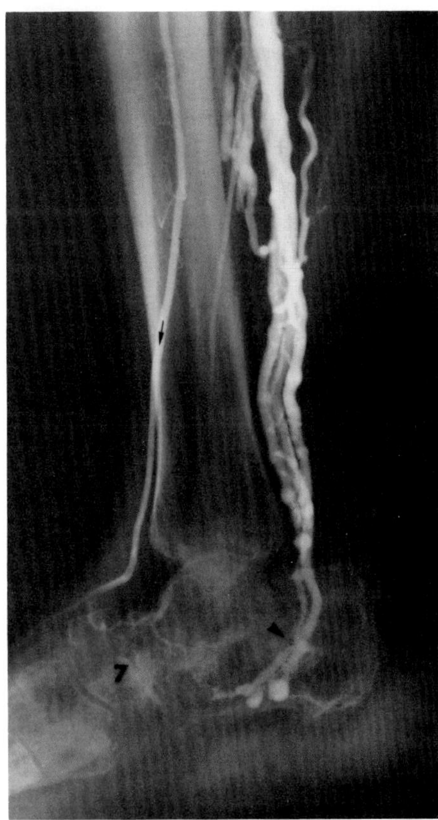


B

**Figure 27.14. Traumatic arteriovenous fistula of the plantar artery.** Films obtained 5 seconds (A) and 7 seconds (B) after the retrograde injection of contrast material in the right common femoral artery demonstrate rapid, markedly increased flow in the enlarged posterior tibial artery and almost instantaneous venous return. The arrowhead points to the site of the arteriovenous fistula. Descending arrows indicate arterial flow; ascending arrows point out venous return. (From Roy,[9] with permission from the publisher.)

modynamically significant obstructive lesion. Additional studies obtained with the patient in the erect position during hyperabduction maneuvers, which cause contraction of the scalenus and pectoralis minor muscles, may demonstrate marked narrowing or even complete obliteration of the subclavian artery that was not detectable on routine supine arteriographic views. Arteriography is also important in identifying patients with multiple lesions of the subclavian artery in the thoracic outlet region, persons who otherwise would have poor surgical results attributable to the failure to recognize and surgically correct a second complicating lesion.[14,33]

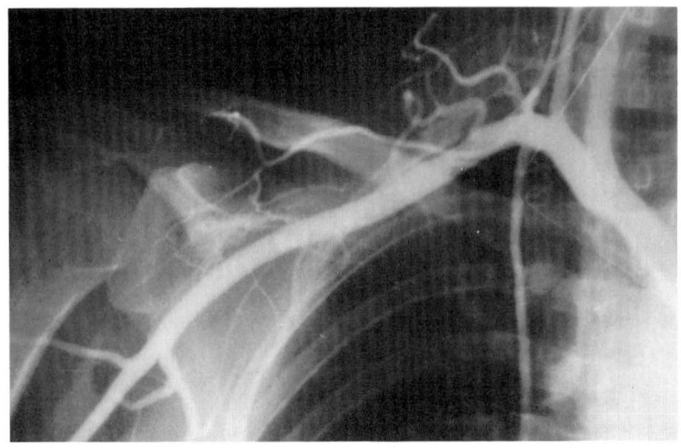

A

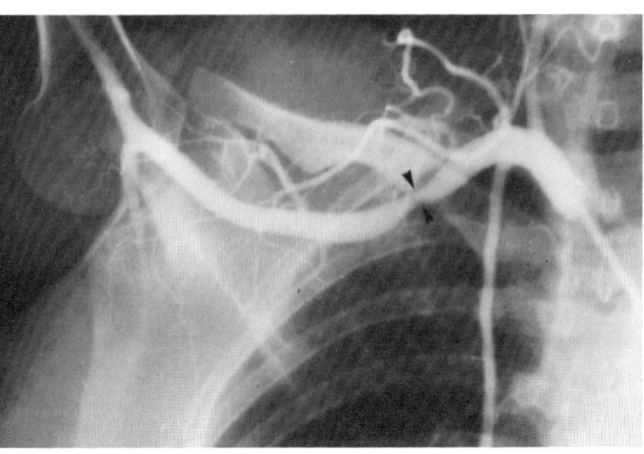

B

**Figure 27.15. Thoracic outlet syndrome.** (A) An arteriogram obtained with the right arm at rest shows the subclavian artery to be normal in caliber. (B) A film obtained with the right arm elevated in hyperabduction demonstrates significant stenosis of the subclavian artery (arrows) due to extrinsic compression at the point where the clavicle and the first rib cross each other. (From Roy,[9] with permission from the publisher.)

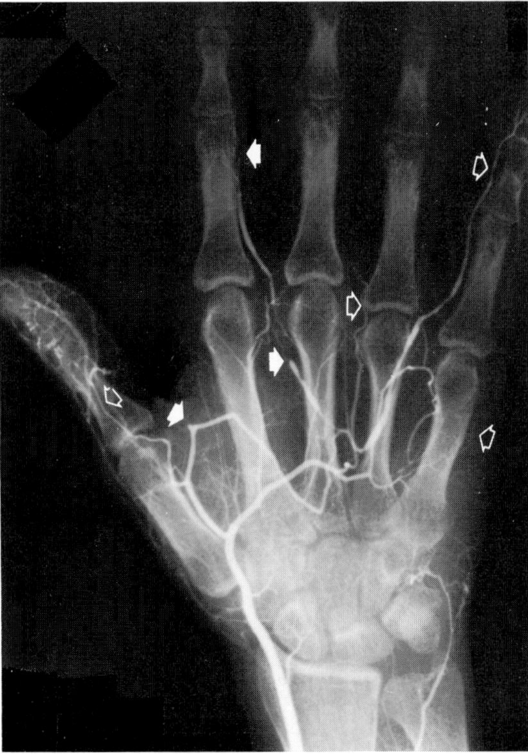

A

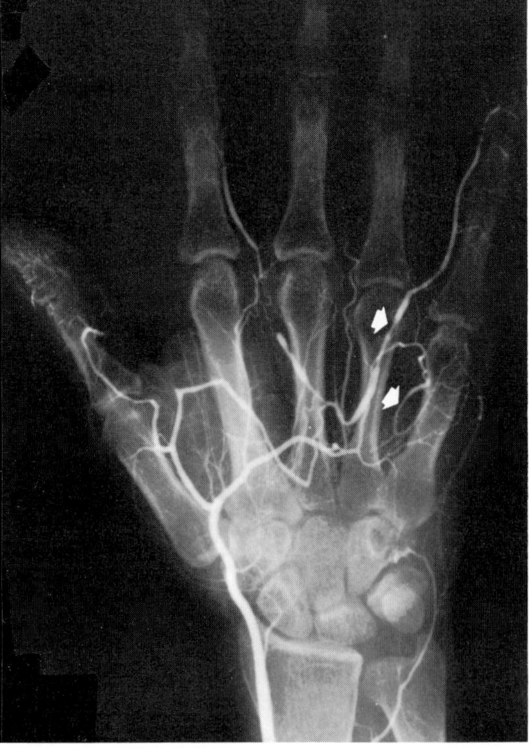

B

**Figure 27.16. Raynaud's phenomenon.** (A) An arteriogram of the hand in a 29-year-old woman shows occlusion of several common and proper digital arteries (closed arrows) and spasm involving multiple vessels (open arrows). The ulnar artery is occluded at the wrist. (B) Following the intra-arterial injection of reserpine, there is a reversal of the spasm in several of the vessels. The anatomically narrowed segments are now more obvious (arrows). (From Kadir and Athanasoulis,[15] with permission from the publisher.)

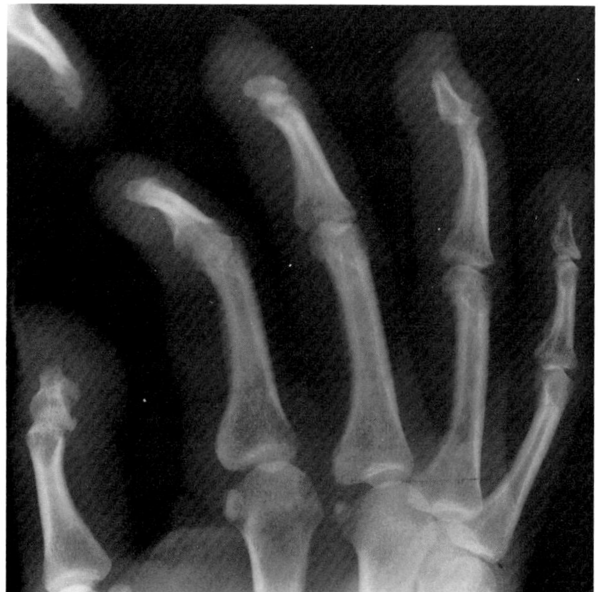

**Figure 27.17. Raynaud's phenomenon.** There are severe trophic changes involving the distal phalanges with resorption of the terminal tufts.

## RAYNAUD'S PHENOMENON

Raynaud's phenomenon refers to a cold sensitivity that occurs almost exclusively in women and produces symptoms of peripheral arterial spasm, especially in the upper limbs. It may be either an isolated finding or a symptom of a more severe underlying condition, such as scleroderma or another connective tissue disorder.

Arteriography may demonstrate multiple sites of stenosis or occlusion that primarily involve the digital arteries of the fingers. Prolonged spasm leads to the development of trophic changes. These include soft tissue atrophy in the fingertips and trophic osteolysis and resorption of the terminal tufts, which causes the distal phalanges to appear shortened and tapered. Discrete calcium deposits may develop in the fingertips, especially in patients with scleroderma.[8,9,15]

## ERGOTISM

Ergotism is a condition due to overdose of or sensitivity to ergot-derivative drugs, which are used primarily for the treatment of migraine headaches or postpartum hemorrhage. The vasoconstrictor nature of ergot drugs can cause neurologic symptoms of convulsions or coma and

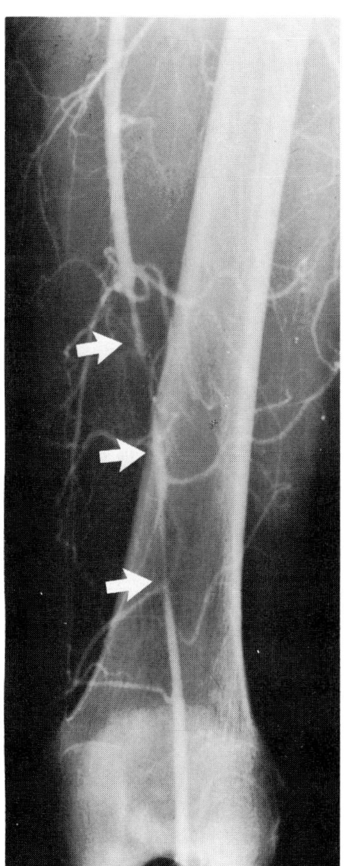

A

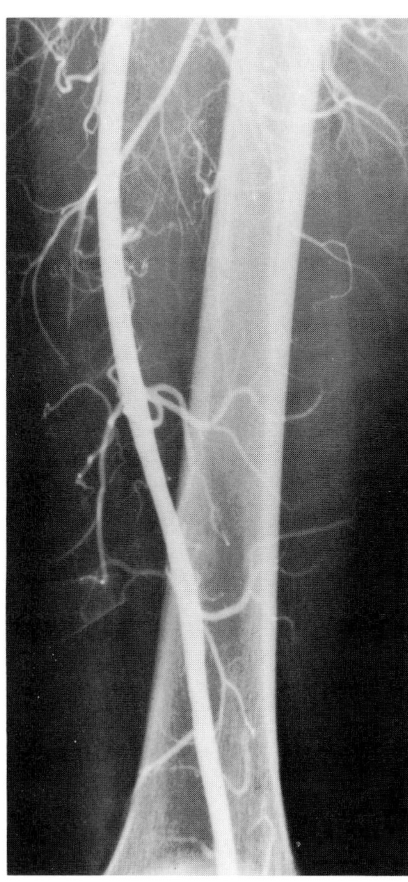

B

**Figure 27.18. Ergotism.** (A) A femoral arteriogram in a 45-year-old woman receiving ergotamine for migraine attacks demonstrates occlusion of the superficial femoral artery (arrows). (B) Two weeks after the drug was discontinued, the vessel lumen is restored. (From Kadir and Athanasoulis,[15] with permission from the publisher.)

peripheral ischemic changes that most often affect the lower extremities and may even lead to gangrene of the digits.

Arteriography demonstrates bilateral and symmetric arterial spasm with tapering or threadlike narrowing and even total occlusion. The femoral, popliteal, and trifurcation arteries are most often affected. In patients who have taken ergot drugs chronically, extensive collateral vessels may develop. The stasis of blood distal to areas of spasm, or a direct toxic effect of the ergot drugs on the vascular endothelium, may lead to thrombus formation and the presence of small filling defects within the vascular lumen. The peripheral arterial changes are completely reversible once the ergot drug is discontinued or following the continuous intra-arterial and intravenous infusion of sodium nitroprusside.[15–17]

## VARICOSE VEINS

Varicose veins are dilated, elongated, and tortuous vessels with incompetent valves, which most commonly involve the saphenous system. They may be limited to the superficial veins (primary) or be secondary to obstruction and valvular incompetence of the deep venous system (secondary). The resulting venous stasis may lead to the development of phleboliths, calcified thrombi within a vein that appear as rounded densities and often contain lucent centers. Phleboliths are also commonly seen in pelvic veins and in the dilated venous spaces of a cavernous hemangioma. Chronic venous stasis may also lead to periosteal new bone formation along the tibial and fibular shafts as well as the development of plaquelike calcification in the chronically congested subcutaneous tissues.

Although the diagnosis of varicose veins is primarily a clinical observation, venography is of value in demonstrating the competency of perforating veins and the patency of the deep venous system, especially if surgical intervention is considered. Following the application of a tourniquet to occlude superficial flow, the peripheral injection of contrast material opacifies the deep venous system. Filling of the superficial veins indicates the presence of incompetent perforating veins above the level of the tourniquet.[32]

## ACUTE VENOUS THROMBOSIS

Deep vein thrombosis, which primarily involves the lower extremity, is a major source of potentially fatal pulmonary embolism. Precipitating factors in the development of venous thrombosis include trauma, bacterial infection, prolonged bed rest, oral contraceptives, and disseminated intravascular coagulation. At times, deep vein thrombosis may be the earliest symptom of an unsuspected malignancy of the pancreas, the lung, or the gastrointestinal system.

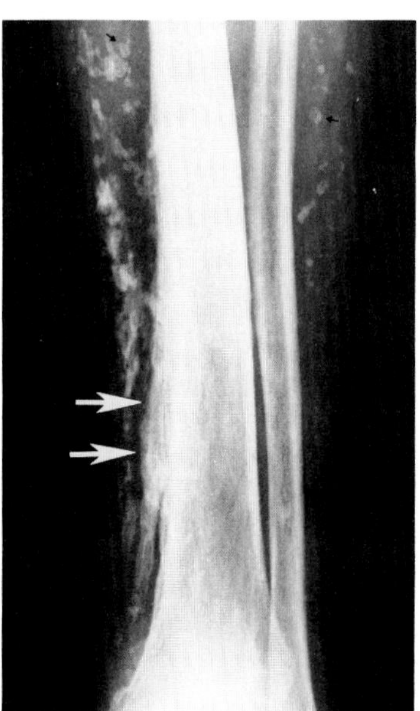

**Figure 27.19. Varicose veins.** A plain film of the lower leg shows multiple round and oval calcifications in the soft tissues. These phleboliths represent calcified thrombi, some of which have characteristic lucent centers (black arrows). Note the extensive periosteal new bone formation along the medial aspect of the tibial shaft (white arrows) caused by long-standing vascular stasis.

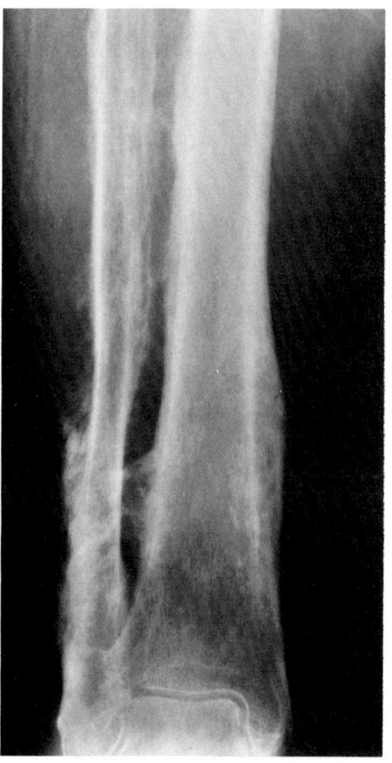

**Figure 27.20. Varicose veins.** Chronic venous stasis has led to striking periosteal new bone formation cloaking the tibia and fibula.

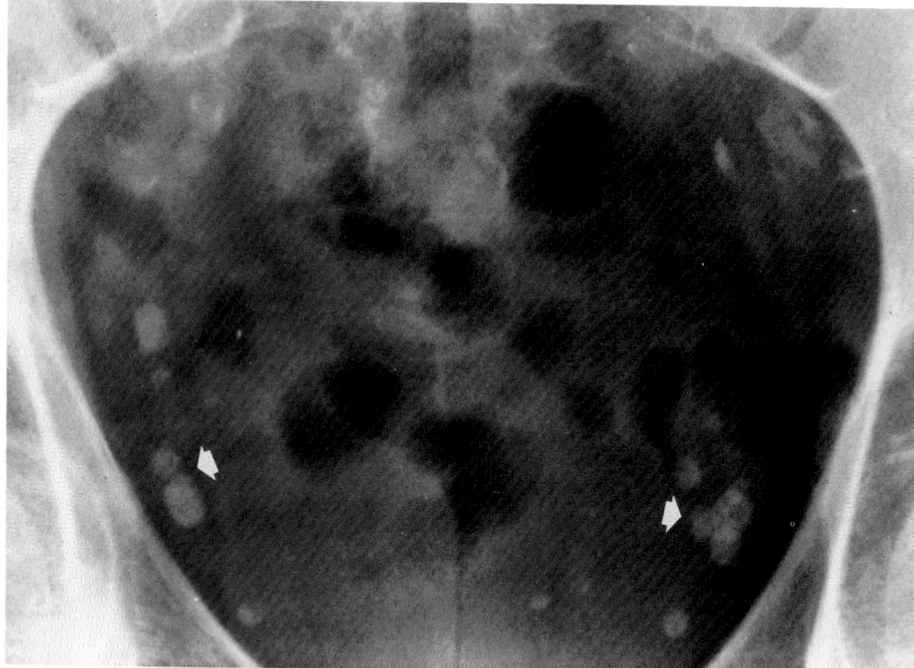
A

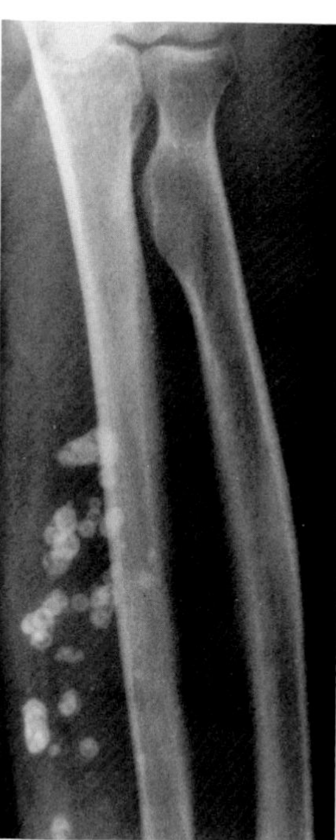
B

**Figure 27.21. Other examples of phleboliths.** (A) Pelvic veins (arrows). (B) Cavernous hemangioma of the arm.

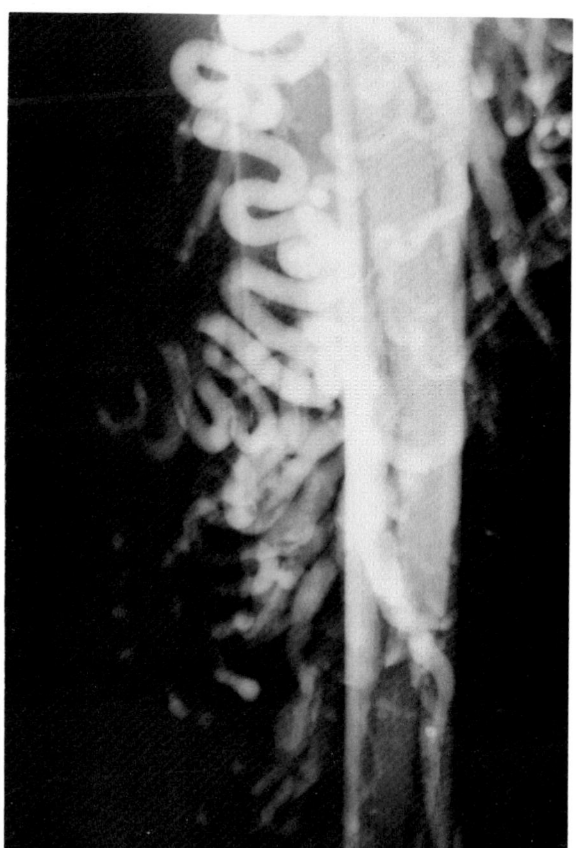

**Figure 27.22. Varicose veins.** A lower extremity venogram shows multiple tortuous, dilated venous structures.

A precise diagnosis of deep vein thrombosis requires contrast venography, which can demonstrate the major venous channels and their tributaries from the foot to the inferior vena cava. The identification of a constant filling defect, representing the actual thrombosis, is conclusive evidence of deep vein thrombosis. Venographic findings that are highly suggestive of, though not conclusive for, the diagnosis of deep vein thrombosis include the abrupt ending of the opaque column in a vein, the nonfilling of one or more veins that are normally opacified, and extensive collateral venous circulation.

Because venography is an invasive technique, other modalities have been developed for detecting deep vein thrombosis. Doppler ultrasound demonstrates changes in the velocity of venous blood flow. It is of special value in showing thrombotic occlusion of major venous pathways in the popliteal and femoral regions, though it is less reliable in the evaluation of calf and muscular veins. An abnormal Doppler ultrasound examination is not specific for thrombosis, since similar changes in venous blood flow may be caused by congestive heart failure, extensive leg edema, local soft tissue masses, and limited arterial inflow into the extremity.

The radioactive fibrinogen uptake test is based on the incorporation of $^{125}$I-fibrinogen into actively forming thrombus, where it can be detected and localized by external scanning. Because the radionuclide must be given before the initiation of thrombosis, the test cannot detect pre-existent thrombi and is thus primarily limited to prospective studies. This radionuclide study has very high sensitivity in the calf but substantially lower sensitivity above the midthigh level because of background

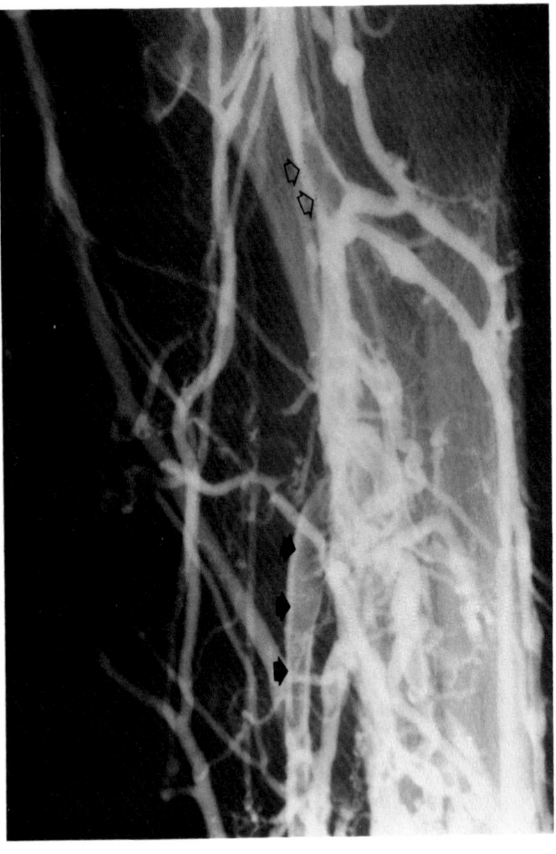

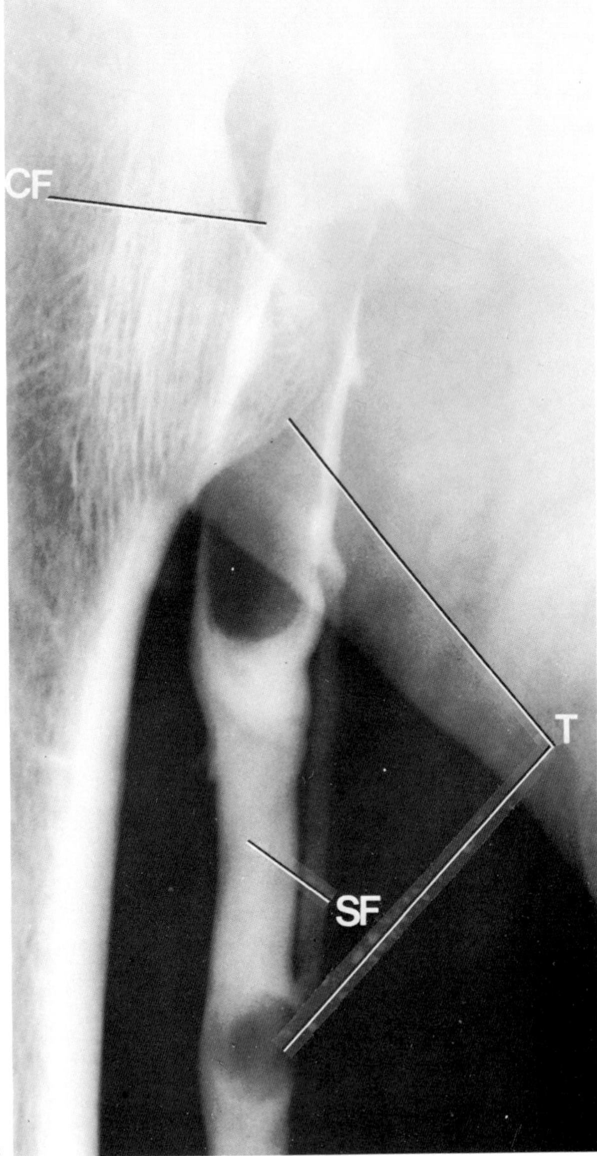


A

B

**Figure 27.23. Acute venous thrombosis.** (A) A venogram shows lucent filling defects (arrows) in several deep veins of the lower leg. (B) In another patient, a venogram shows lucent thrombi (T) in the common femoral vein (CF) and superficial femoral vein (SF). The more proximal, larger thrombus appears to have only a small area of attachment to the vein wall. (B, from Rabinov and Paulin,[32] with permission from the publisher.)

activity. Another approach to the nuclear medicine detection of deep vein thrombosis is combining radionuclide venography with simultaneous lung scanning, which both demonstrates the status of the major deep veins and provides a baseline perfusion lung scan.

A negative test result with the newer noninvasive techniques excludes deep vein thrombosis in large proximal venous channels with a high degree of accuracy. False-negative examinations result from the occasional failure of even large thrombi to cause sufficient obstruction of flow to be detected.[18–24,32]

## VENA CAVA OBSTRUCTION

Obstruction of the inferior vena cava most commonly involves the middle and lower portions and is most frequently caused by retroperitoneal malignancy (especially carcinoma of the kidney with extension of tumor from the renal vein into the inferior vena cava) and inflammatory processes (abscess, retroperitoneal fibrosis). Intrinsic obstruction of the inferior vena cava is usually secondary to long-standing extrinsic pressure on the cava with resultant stasis and thrombosis.

Contrast inferior vena cavography has traditionally been the major technique for demonstrating the site of obstruction and for showing tumor masses extending into the lumen. Cavographic findings of extrinsic obstruction include displacement or angulation, local narrowing, distortion of the contour, blurring of the margins, changes in the density of the opacified circulation, and actual invasion of the vena cava with a resultant thrombus or occlusion.

Ultrasound was the first noninvasive method for dem-

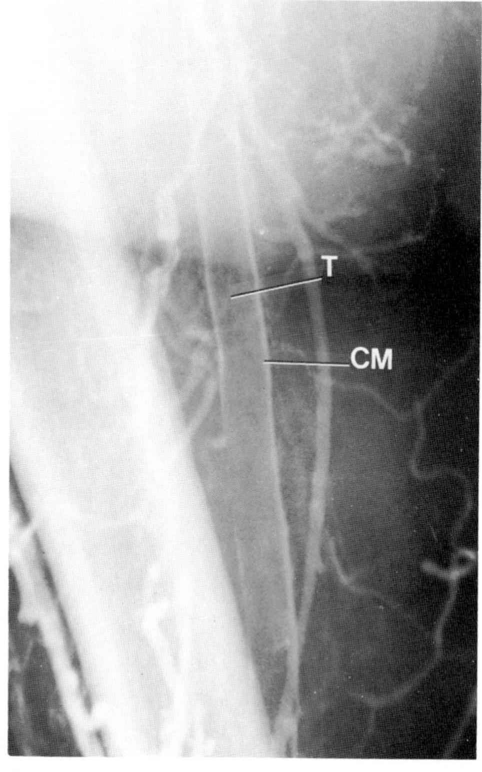


A

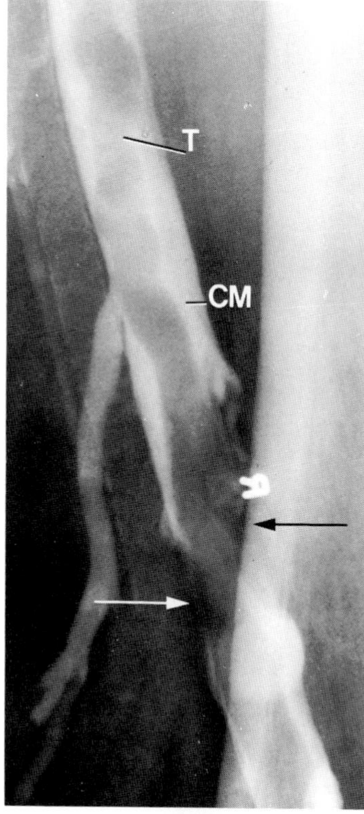


B

**Figure 27.24. Morphologic changes in the appearance of venous thrombosis according to thrombus age.** (A) A fresh thrombus (T) virtually fills the lumen of the vein. A thin layer of contrast material (CM) almost completely separates the thrombus from the vein wall. (B) An aging thrombus (T) shows the early stages of retraction. The thrombus has shrunk and is separated from the vein wall except for attachments at several locations (arrows). The layer of contrast material (CM) between the thrombus and the vein wall is much wider. The margins of the thrombus have an undulating contour. (From Rabinov and Paulin,[12] with permission from the publisher.)

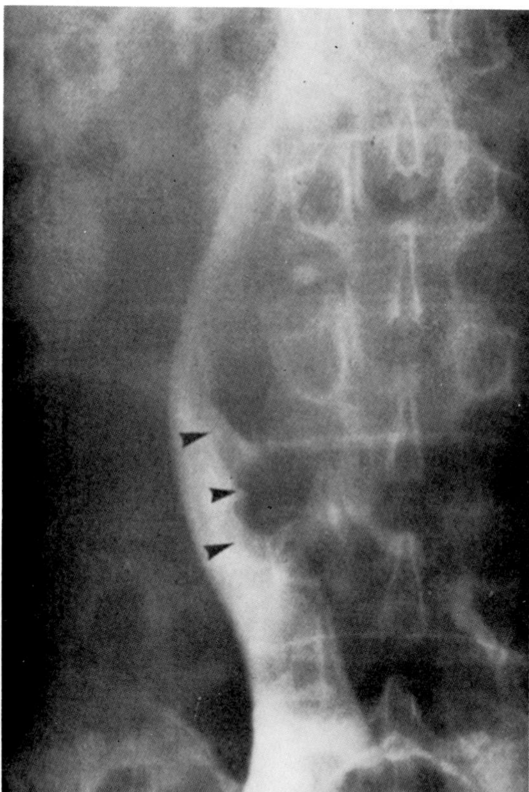

A

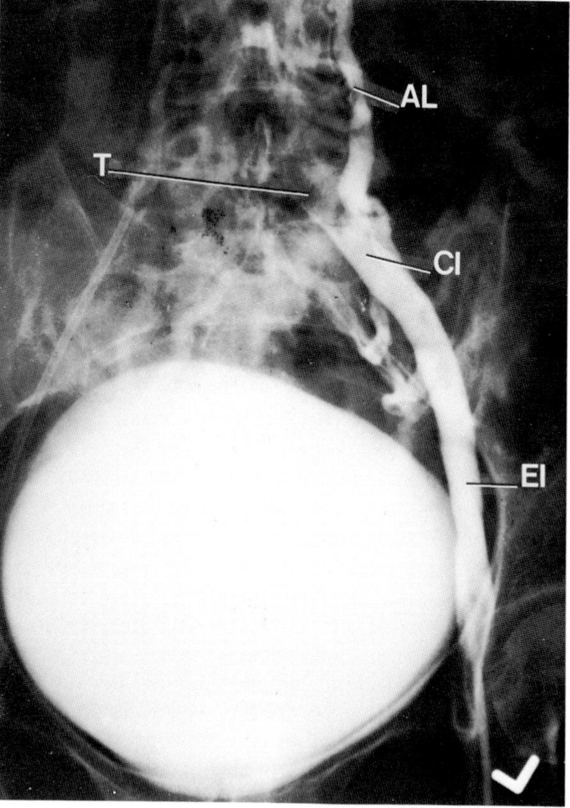


B

**Figure 27.25. Venograms demonstrating obstruction of the inferior vena cava.** (A) Leiomyoma (arrows) producing an irregular bilobed deformity with marked narrowing. (B) Thrombotic occlusion of the upper part of the common iliac vein (CI) with collateral flow through the ascending lumbar vein (AL). The patient presented with chronic swelling in the calf and thigh. EI = external iliac veins; T = thrombus. (From Rabinov and Paulin,[12] with permission from the publisher.)

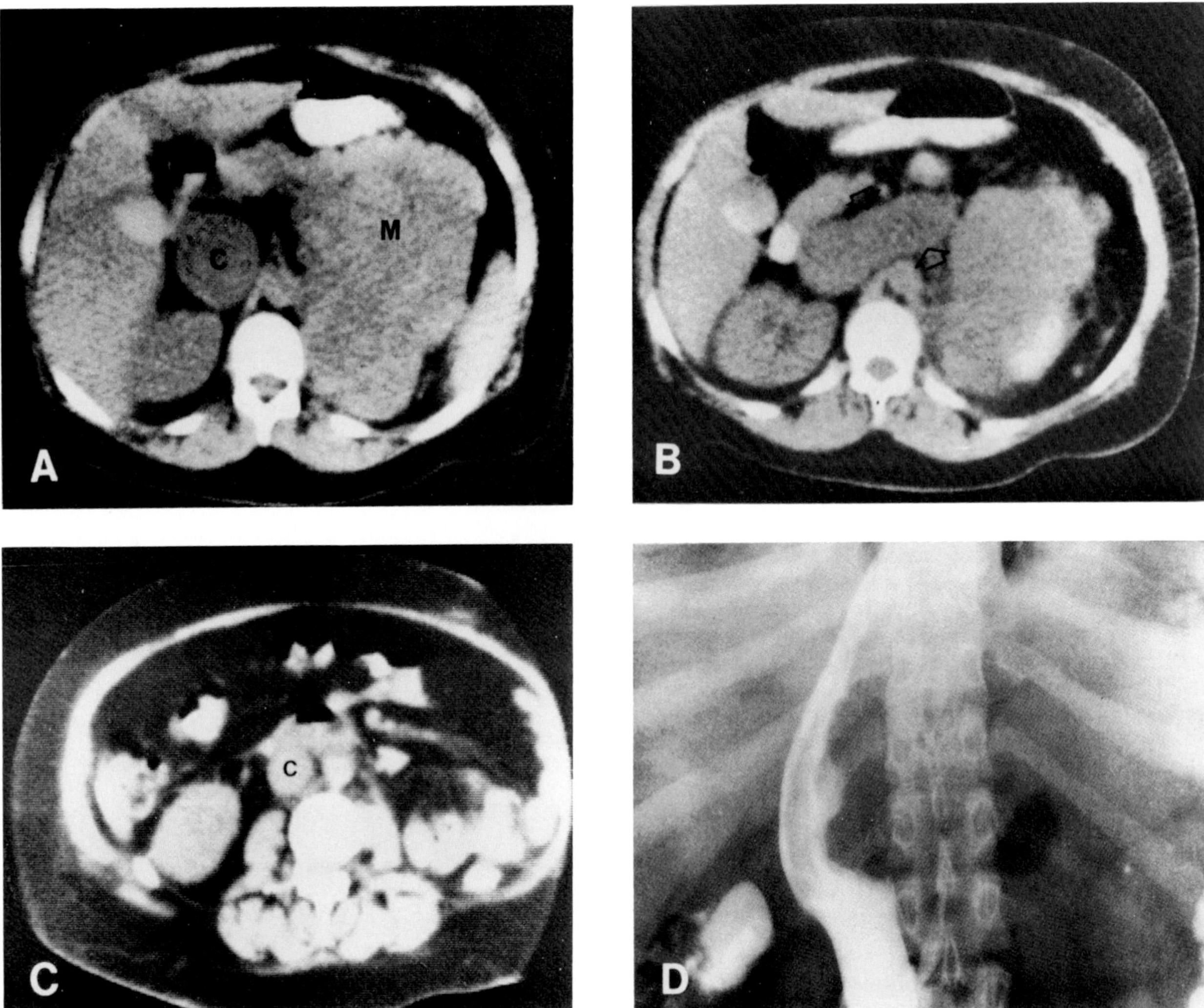


**Figure 27.26. Vena cava obstruction.** (A) A CT scan through the upper abdomen shows a mass (M) inseparable from the anterior aspect of the upper pole of the left kidney. The inferior vena cava (C) is markedly dilated. (B) A scan 2 cm caudal from A shows a markedly enlarged left renal vein (arrows) entering the dilated inferior vena cava. (C) A scan 4 cm caudal from B shows the normal diameter of the inferior vena cava (C). (D) An inferior vena cavogram shows displacement and possible invasion of the vena cava by a mass. The cavogram fails to demonstrate the enlarged diameter of the vena cava because contrast material does not outline the left wall. (From Marks, Korobkin, Callen, et al,[27] with permission from the publisher.)

onstrating both the position and intrinsic abnormalities of the inferior vena cava. More recently, computed tomography (CT) with dynamic scanning following the intravenous administration of contrast material has been shown to produce studies of the inferior vena cava comparable in all respects to the more invasive standard cavography. If the lumen of the cava is only partially occluded, the thrombus appears as a low-density filling defect surrounded by the contrast-containing blood. An obstructing tumor thrombus can cause massive enlargement of the vena cava with an abrupt decrease in caliber distally.

The extensive collateral venous circulation that develops following inferior vena cava obstruction may produce pressure effects on the urinary and gastrointestinal tracts that cause a tear-shaped bladder, medial displacement and notching of the ureters, elevation of the sigmoid colon, and extrinsic narrowing of the rectum.[25–28]

Obstruction of the superior vena cava is discussed in the section "Superior Vena Caval Syndrome" in Chapter 35.

## LYMPHEDEMA

An abnormal accumulation of lymph in the extremities may be due to a primary developmental anomaly or, more commonly, may be secondary to inflammatory,

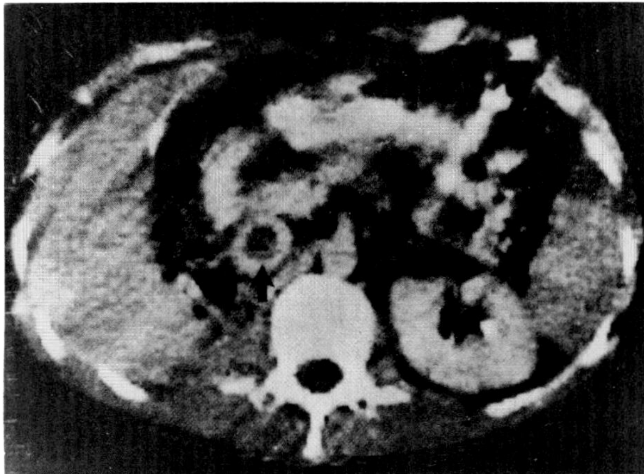

**Figure 27.27. Tumor thrombus in the inferior vena cava.** Although the diameter of the inferior vena cava is within normal limits, the CT scan shows an intraluminal tumor thrombus appearing as a low-density filling defect (arrow) surrounded by contrast material. (From Marks, Korobkin, Callen, et al,[27] with permission from the publisher.)

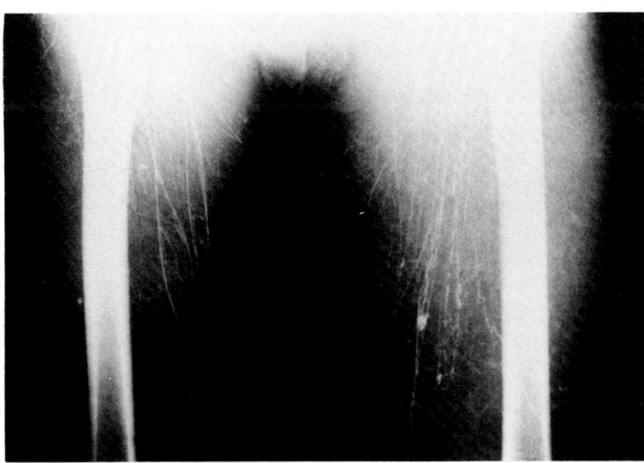

**Figure 27.28. Primary hyperplastic lymphedema.** A bipedal lymphogram demonstrates an increased number of lymphatic channels in both legs. (From Gamba, Silverman, Ling, et al,[31] with permission from the publisher.)

traumatic, parasitic, or neoplastic obstruction of the lymphatic system.

In primary lymphedema (Milroy's disease, lymphedema precox), lymphography may demonstrate either hyperplasia or hypoplasia of the lymph vessels. In hyperplasia, the dilatation and tortuosity of the lymphatics may be segmental or involve large portions of the lymph system. There is often associated dermal backflow, in which valvular insufficiency permits contrast filling of multiple small, superficial lymphatics. A decreased number and size of the lymph vessels is seen in hypoplastic conditions; at times the lymphatic channels are virtually absent. CT shows a coarse, nonenhancing reticular pattern in an enlarged subcutaneous compartment. This mo-

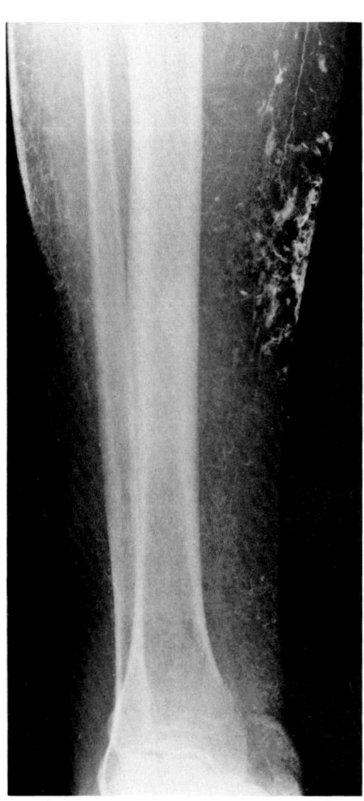

**Figure 27.29. Primary hypoplastic lymphedema.** A lymphogram demonstrates a decreased number and size of lymph vessels in the lower leg.

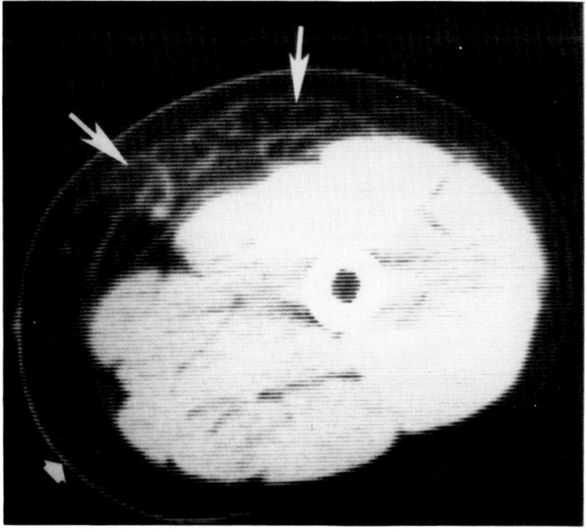

**Figure 27.30. Primary lymphedema.** A CT scan demonstrates multiple reticular soft tissue structures (arrows) representing dilated lymphatics in the enlarged subcutaneous compartment of the left thigh. Mild skin thickening also is present (arrowhead). (From Gamba, Silverman, Ling, et al,[31] with permission from the publisher.)

dality can also exclude secondary lymphedema due to an obstructing mass by demonstrating a normal retroperitoneum and pelvis.

Secondary lymphedema may result from the obstruction of lymph nodes or channels (lymphangitis, filariasis, malignant disease), trauma (direct injury, burns, surgery, radiation therapy), or venous obstruction (in which the lymphatic system is unable to adequately drain the increased amount of fluid within the soft tissues). At lymphography, the lymph vessels peripheral to the site of occlusion are distended and tortuous, and there is often backflow into dermal and interstitial lymphatics.[29–31]

**REFERENCES**


1. Katzen BT, van Breda A: Low dose streptokinase in the treatment of arterial occlusions. *AJR* 136:1171–1178, 1981.
2. Bron KM: Femoral arteriography, in Abrams HL (ed): *Abrams Angiography: Vascular and Interventional Radiology*. Boston, Little, Brown, 1983, pp 1835–1876.
3. Tamura S, Sniderman KW, Beinart C, et al: Percutaneous transluminal angioplasty of the popliteal artery and its branches. *Radiology* 143:645–648, 1982.
4. Freiman DB, Spence R, Gatenby R, et al: Transluminal angioplasty of the iliac and femoral arteries: Follow-up results without anticoagulation. *Radiology* 141:347–350, 1981.
5. Waltman AC: Percutaneous transluminal angioplasty of the iliac and deep femoral arteries, in Athanasoulis CA, Pfister RC, Greene RE, et al (eds): *Interventional Radiology*. Philadelphia, WB Saunders, 1982.
6. Eisenberg RL: *Gastrointestinal Radiology: A Pattern Approach*. Philadelphia, JB Lippincott, 1983.
7. Osborn AG, Anderson RE: Angiographic spectrum of cephalocervical fibromuscular dysplasia. *Stroke* 8:617–626, 1977.
8. Grollman JH, Lecky JW, Rosch J: Miscellaneous diseases of arteries, or, all arterial lesions aren't fat. *Semin Roentgenol* 5:306–321, 1970.
9. Roy P: Peripheral angiography in ischemic arterial disease of the limbs. *Radiol Clin North Am* 5:467–496, 1967.
10. Hewitt RL, Smith AD, Drapanas T: Acute traumatic arteriovenous fistulas. *J Trauma* 13:901–906, 1973.
11. Steinberg I, Tillotson PM, Halpern M: Roentgenography of systemic (congenital and traumatic) arteriovenous fistulas. *AJR* 89:343–357, 1963.
12. Thomas ML, Andress MR: Angiography and venous dysplasia. *AJR* 113:722–731, 1971.
13. Bliznak J, Staple TW: Radiology of angiodysplasias of the limb. *Radiology* 110:35–44, 1974.
14. Lang EK: Arteriographic diagnosis of the thoracic outlet syndrome. *Radiology* 84:296–303, 1965.
15. Kadir S, Athanasoulis CA: Peripheral vasospastic disorders: Management with intra-arterial infusion of vasodilatory drugs, in Athanasoulis CA, Pfister RC, Greene RE (eds): *Interventional Radiology*. Philadelphia, WB Saunders, 1982.
16. Bagby RJ, Cooper RD: Angiography in ergotism. *AJR* 116:179–186, 1972.
17. O'Dell CW, Davis GB, Johnson AD, et al: Sodium nitroprusside in the treatment of ergotism. *Radiology* 124:73–74, 1977.
18. Dean RH: Radionuclide venography and simultaneous lung scanning: Evaluation of clinical application, in Bernstein EF (ed): *Noninvasive Diagnostic Techniques in Vascular Diseases*. St. Louis, CV Mosby, 1978.
19. Barnes RW: Doppler ultrasonic diagnosis of venous disease, in Bernstein EF (ed): *Noninvasive Diagnostic Techniques in Vascular Diseases*. St. Louis, CV Mosby, 1978.
20. O'Donnell JA, Lipp J, Hobson RW: New methods of testing for deep venous thrombosis. *Am Surg* 44:121–132, 1978.
21. Pollak EW: Choice of tests for diagnosis of venous thrombosis. *Vasc Surg* 11:219–224, 1977.
22. Kakkar VV: Fibrinogen uptake test for detection of deep vein thrombosis—A review of current practice. *Semin Nucl Med* 7:229–244, 1977.
23. Holden RW, Klatte EC, Park HM, et al: Efficacy of noninvasive modalities for diagnosis of thrombophlebitis. *Radiology* 141:63–66, 1981.
24. Vine HS, Hillman B, Hassel SJ: Deep venous thrombosis: Predictive value of signs and symptoms. *AJR* 136:167–171, 1981.
25. Amoe HE, Lewis RE: Urographic and barium enema appearance in inferior vena caval obstruction. *Radiology* 108:307–308, 1973.
26. Crummy AB, Hipona FA: The aortic impression in inferior vena cavography. *Clin Radiol* 15:130–134, 1964.
27. Marks WM, Korobkin M, Callen PW, et al: CT diagnosis of tumor thrombosis of the renal vein and inferior vena cava. *AJR* 131:843–846, 1978.
28. Steele JR, Sones PJ, Heffner LT: The detection of inferior vena caval thrombosis with computed tomography. *Radiology* 128:385–386, 1978.
29. Fuchs WA: Lymphangiopathies, in Abrams HL (ed): *Abrams Angiography: Vascular and Interventive Radiology*. Boston, Little, Brown, 1983, pp 2023–2040.
30. Kinmonth JB: *The Lymphatics: Diseases, Lymphography and Surgery*. London, Arnold, 1972.
31. Gamba JL, Silverman PM, Ling D, et al: Primary lower extremity lymphedema: CT diagnosis. *Radiology* 149:218–219, 1983.
32. Rabinov K, Paulin S: Venography of the lower extremities, in Abrams HL (ed): *Abrams Angiography: Vascular and Interventive Radiology*. Boston, Little, Brown, 1983, pp 1887–1922.
33. Lang EK: Arteriography of the thoracic outlet, in Abrams HL (ed): *Abrams Angiography: Vascular and Interventional Radiology*. Boston, Little, Brown, 1983, pp 1001–1018.

# PART FIVE

# DISORDERS OF THE RESPIRATORY SYSTEM

## Chapter Twenty-Eight

# Lung Disease Caused by Immunologic and Environmental Injury

### ASTHMA

Asthma is a disease in which widespread narrowing of the airways reflects an increased responsiveness of the tracheobronchial tree to a multiplicity of stimuli. The disease is characterized clinically by wheezing, prolongation of the expiratory phase of respiration, dyspnea, and cough.

Early in the course of the disease, chest radiographs obtained between acute episodes demonstrate no abnormalities. During an acute asthmatic attack, hyperinflation causes increased lucency of the lungs, flattening of the hemidiaphragms, and an increase in the retrosternal air space. In contrast to emphysema, the pulmonary vascular markings in asthma are of normal caliber. In patients with chronic asthma, especially those with a history of repeated episodes of infection, thickening of bronchial walls can produce parallel or slightly tapered line shadows outside the boundary of the pulmonary hila ("tramlines"). Inspissated mucous plugs, which may appear as elongated opacities with undulating borders, can cause transient parenchymal densities due to segmental atelectasis. Severe coughing and straining during an acute attack can rarely lead to pneumothorax, interstitial emphysema, and pneumomediastinum.

The major importance of chest radiography in patients with asthma is to exclude other causes of diffuse wheezing throughout the chest (emphysema, bronchiectasis, obstruction of the trachea or major bronchi) and to detect superimposed pneumonia.

Chronic asthma complicated by recurrent pulmonary infections can lead to pulmonary fibrosis and emphysema. Bronchial obstruction caused by mucoid impaction can permit the accumulation of secretions in the obstructed segment, leading to repeated infections and bronchiectasis.[1,2]

### DRUG-INDUCED HYPERSENSITIVITY LUNG DISEASE

Iatrogenic pulmonary disease may be an immunologic reaction to a variety of drugs. The symptoms of drug-induced pulmonary disease are usually nonspecific and consist of acute or insidious cough and dyspnea, with or without fever. Although there is some overlap, several basic radiographic patterns may be produced.

Diffuse, acute alveolar infiltrates that are usually bilateral and resemble pulmonary edema may represent a reaction to antibiotics such as penicillin, sulfonamides, and chlorpropamide. Alveolar infiltrates are often patchy and tend to have a peripheral distribution with perihilar

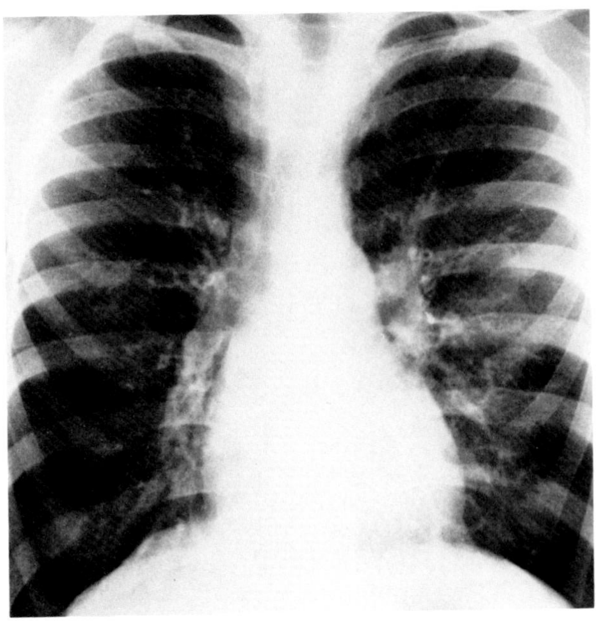

A

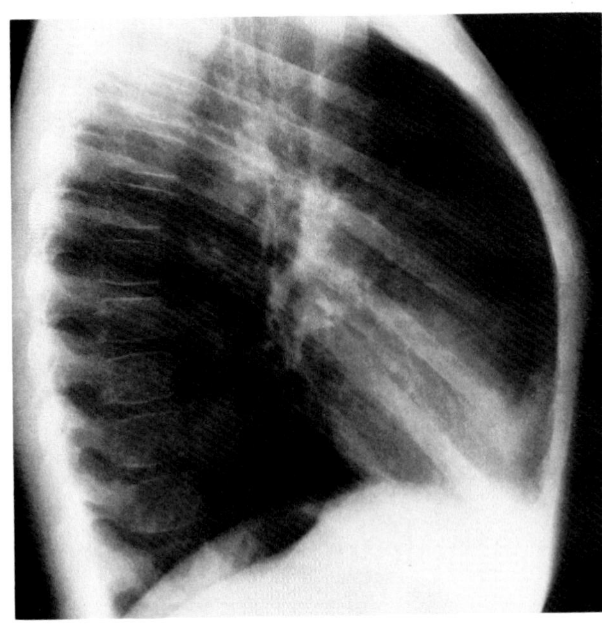

B

**Figure 28.1. Asthma.** (A) Frontal and (B) lateral views of the chest demonstrate hyperexpansion of the lungs with depression of the hemidiaphragms, increased anteroposterior diameter of the chest and retrosternal air space, and prominence of the interstitial structures. The heart and pulmonary vascularity are normal.

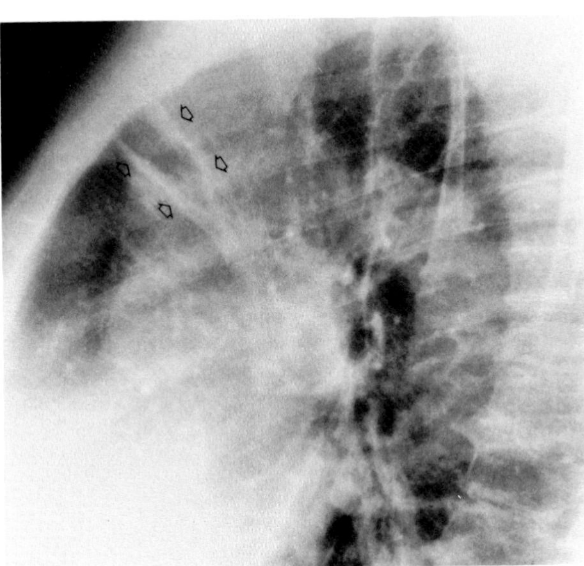

**Figure 28.2. Asthma with mucoid impaction.** Inspissation of tenacious mucoid sputum within several subsegmental bronchi in the upper lobe produces two elongated opacities that join together in a V (arrow), the apex of which is toward the hilum.

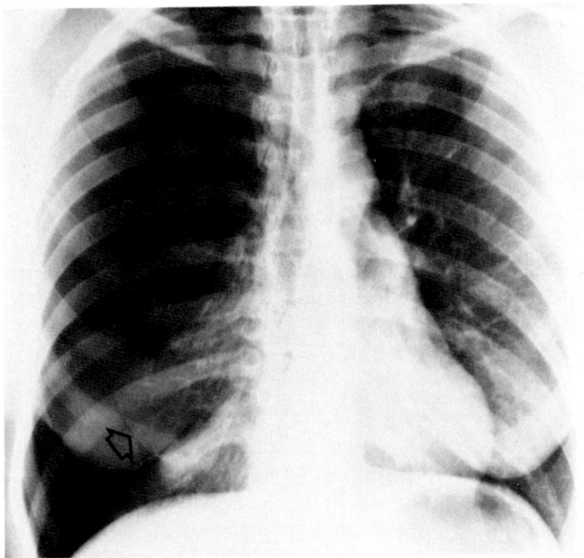

**Figure 28.3. Spontaneous pneumothorax in asthma.** Severe coughing and straining during an acute attack led to the development of a large pneumothorax with complete collapse of the right lung (arrow).

sparing, similar to the pattern in idiopathic eosinophilic pneumonia. This latter appearance is characteristic of the pulmonary disease induced by nitrofurantoin, which also may cause a more chronic interstitial infiltrate.

An acute, diffuse interstitial pattern resembling interstitial edema that rapidly progresses to patchy alveolar consolidation can be a complication of such drugs as methotrexate, chlorothiazide (Diuril), and phenytoin (Dilantin). Diffuse interstitial pulmonary disease may be caused by busulfan, bleomycin, cyclophosphamide (Cytoxan), and methysergide (Sansert). Nitrofurantoin and methotrexate can also cause a chronic interstitial pattern.

Procainamide (Pronestyl), hydralazine, and isoniazid can cause a pleuropulmonary reaction (pleural effusion,

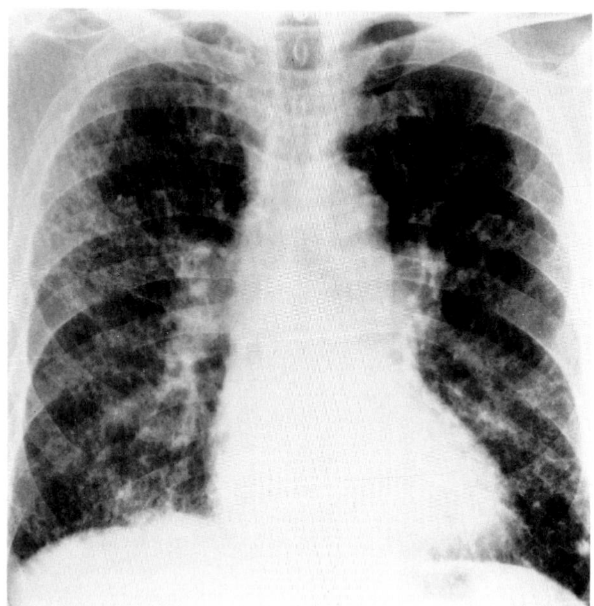

**Figure 28.4. Asthma.** Recurrent pulmonary infections have led to the development of diffuse pulmonary fibrosis.

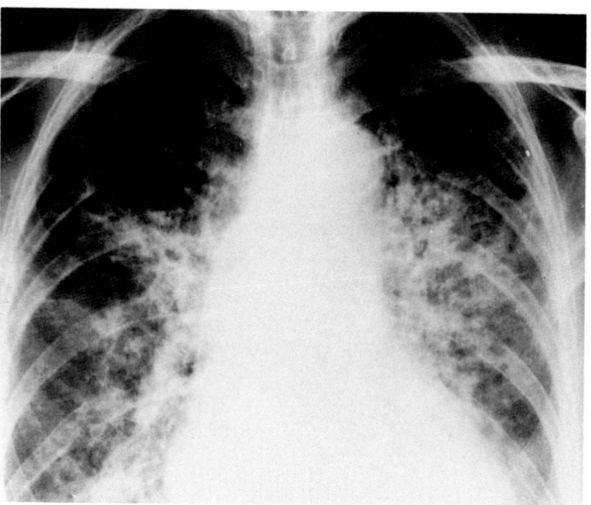

**Figure 28.6. Nitrofurantoin-induced lung disease.** A mixed alveolar and interstitial pattern in an elderly woman who presented with progressive cough and dyspnea following the long-term use of nitrofurantoin for recurrent urinary tract infections. (From Brettner, Heitzman, and Woodin,[5] with permission from the publisher.)

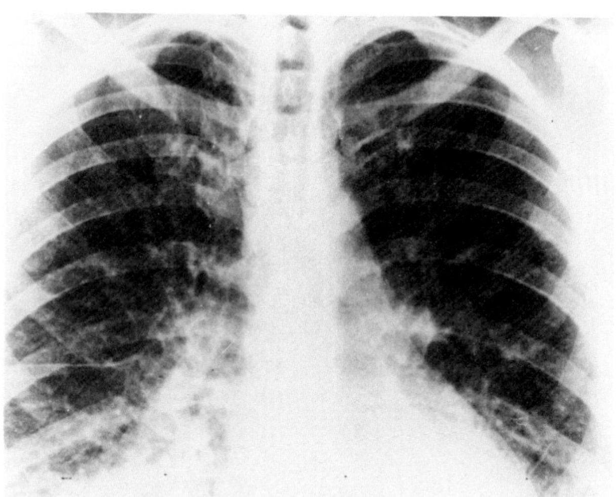

**Figure 28.5. Nitrofurantoin-induced lung disease.** A chest radiograph obtained during an acute episode shows a diffuse interstitial pattern with bilateral pleural effusions. The patient presented with recurrent episodes of undiagnosed pulmonary edema, manifested clinically by dyspnea and left-sided chest pain. (From Brettner, Heitzman, and Woodin,[5] with permission from the publisher.)

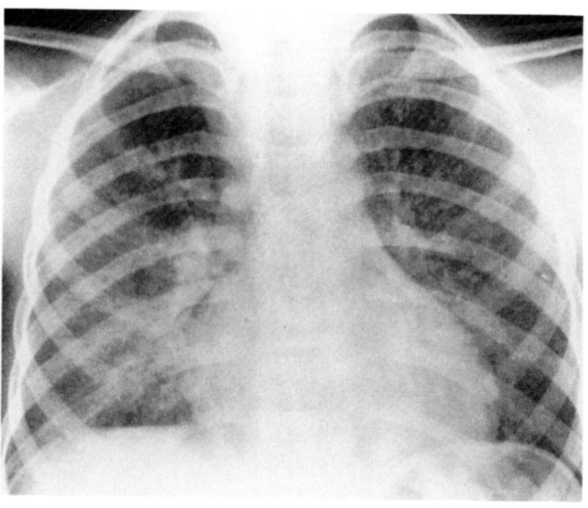

**Figure 28.7. Methotrexate-induced lung disease.** A diffuse interstitial pattern with patches of alveolar consolidation in a child treated for myelogenous leukemia. After the cessation of methotrexate therapy, there was rapid clinical and radiographic improvement. (From Brettner, Heitzman, and Woodin,[5] with permission from the publisher.)

basilar infiltrates, pericardial effusion) that is indistinguishable from the pattern in systemic lupus erythematosus.

Bilateral hilar adenopathy may be a manifestation of hypersensitivity to anticonvulsants, especially phenylhydantoin.

In susceptible persons, aspirin and other nonsteroidal anti-inflammatory agents can cause an acute asthmatic episode.

Regardless of the radiographic pattern, the discontinuation of the offending drug usually leads to clinical recovery and relatively prompt disappearance of the pulmonary changes. In severe cases, steroid therapy may be of value.[3–5,39]

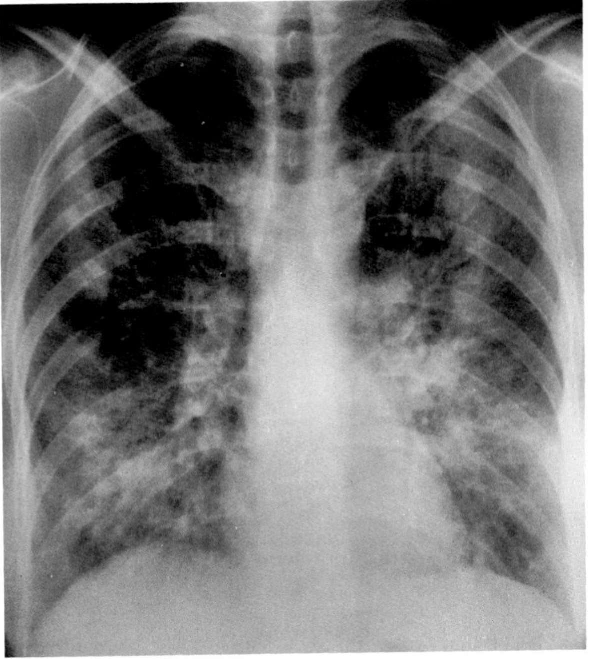

**Figure 28.8. Methotrexate-induced lung disease.** There are diffuse, bilateral, patchy densities that were changeable and fleeting during the course of illness. The radiographic findings cleared completely after steroid therapy. (From Sostman, Putman, and Gamsu,[39] with permission from the publisher.)

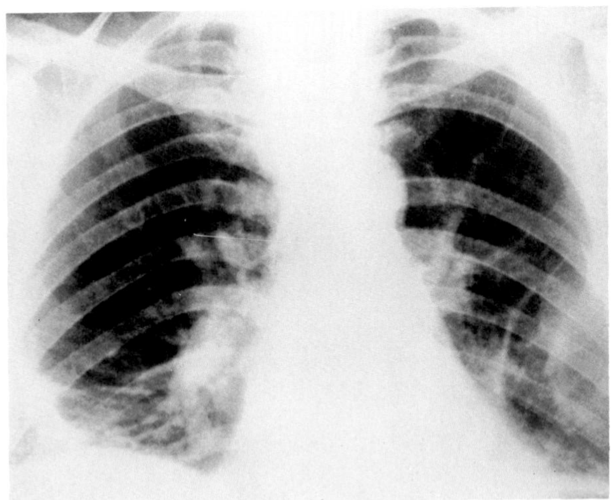

**Figure 28.9. Procainamide-induced lung disease.** Linear, basilar, and parenchymal shadows accompanied by a pleural effusion in a middle-aged man who presented clinically with lupus erythematosus after procainamide therapy. The radiographic and clinical findings cleared slowly after the cessation of drug therapy. (From Brettner, Heitzman, and Woodin,[5] with permission from the publisher.)

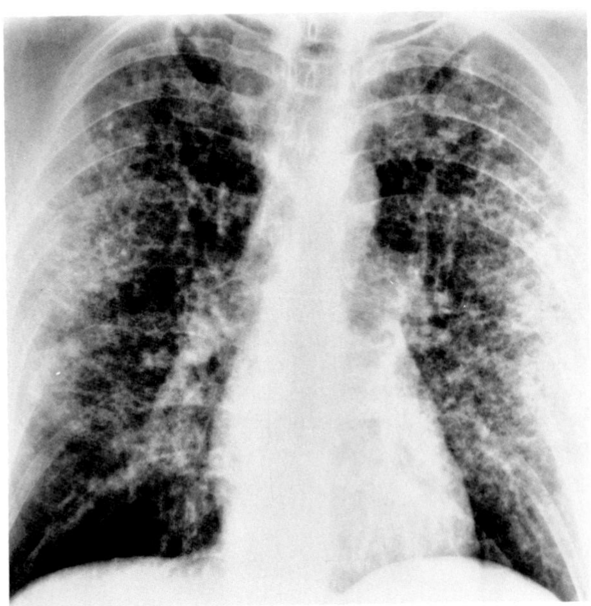

**Figure 28.10. Busulfan-induced lung disease.** Diffuse interstitial pulmonary disease in an elderly woman with myeloma treated with busulfan.

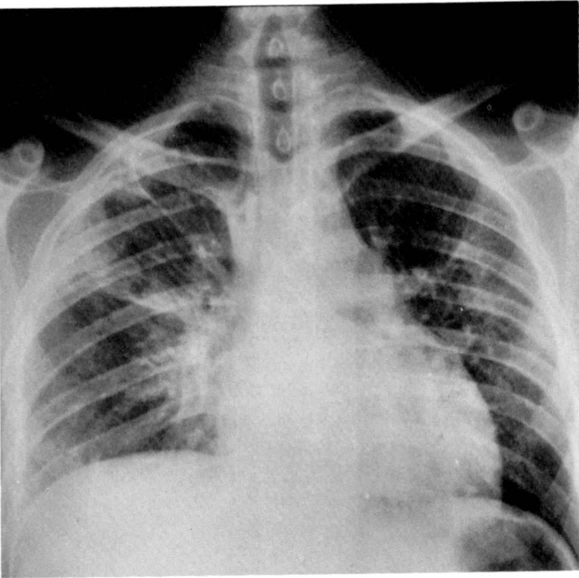

**Figure 28.11. Bleomycin-induced lung disease.** Bilateral linear densities with scattered patches of peripheral infiltrate. (From Sostman, Putman, and Gamsu,[39] with permission from the publisher.)

## PNEUMOCONIOSIS

Occupational exposure to inorganic dusts can cause severe pulmonary disease and a spectrum of radiographic findings. The reaction of the lungs to inhaled dust depends on multiple factors, such as the chemical nature of the dust; the size, concentration, distribution, and clearance of dust particles; the duration of exposure; and individual susceptibility.

The majority of inorganic dusts, if inhaled in sufficient quantity over a prolonged time, cause pulmonary fibrosis. Even though some dusts (tin, iron, barium) are "inert" and not fibrogenic, prolonged exposure still can produce some degree of functional impairment. Most cases of pneumoconiosis develop only after many years of dust exposure; occasionally, severe progressive lung disease may develop after a relatively brief exposure.

In patients with pneumoconiosis, the International Labor Organization has developed an extensive standardized scheme to classify chest radiographs according to the nature and size of the pulmonary opacities and the extent of parenchymal involvement. Nevertheless, it is important to remember that classifications based only on chest radiographs may overestimate or underestimate the functional impact of pneumoconiosis. The degree of involvement on the chest radiograph may be quite extensive in a patient with minimal impairment of pulmonary function. Conversely, chest radiographs may show only mild pulmonary changes in patients with severe pulmonary disease.

Because the linear and nodular fibrosis of most of the pneumoconioses is nonspecific, a history of occupational exposure to a specific dust is required to make the proper diagnosis.[1,6]

### Asbestosis

Asbestosis is a diffuse interstitial fibrosing disease of the lungs that is directly related to the intensity and duration of exposure to asbestos fibers. Once the pulmonary lesion is established, it progresses even though the exposure is not continued. The radiographic hallmark of asbestosis is involvement of the pleura, which implies only exposure and not necessarily pulmonary impairment. The pleural thickening is generally in the form of linear plaques, which are most often along the lower chest wall and diaphragm. Local areas of thin pleural thickening may often be overlooked or ascribed to companion rib shadows. Because pleural plaquing most commonly develops along the posterolateral or anterolateral portion of the thorax, oblique views should be standard in the radiographic investigation of a patient suspected of having asbestosis.

The calcification of pleural plaques, especially in the form of thin, curvilinear densities conforming to the upper surfaces of the diaphragm bilaterally, is virtually pathognomonic of asbestosis, though it is seen only in about 20 percent of the patients with the disease. Pleural calcification generally does not develop until at least 20 years after the first exposure to asbestos.

In the lungs, round or irregular opacities produce a reticulonodular pattern in patients with asbestosis. Initially, a fine reticulation predominantly affects the lower lung and is associated with a ground-glass appearance, which probably is the result of combined pleural thickening and early interstitial pneumonitis or fibrosis. More extensive parenchymal and pleural changes may obscure the heart border (and diaphragm), producing the so-called shaggy heart. Large conglomerate opacities

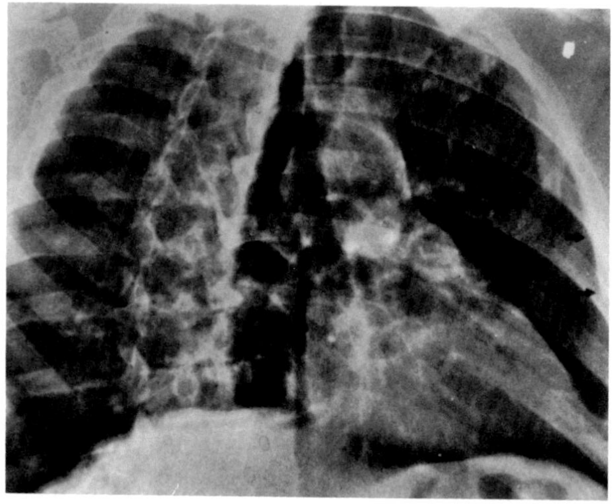

A

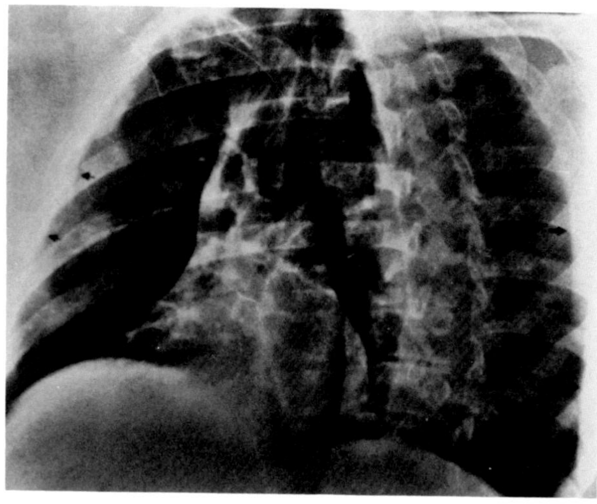

B

**Figure 28.12. Pleural plaques in asbestosis.** (A) A right oblique view shows en face pleural plaques projected in profile, posteriorly on the right and anteriorly on the left (arrows). (B) On the left oblique view, the converse is noted. The costophrenic angles are spared on all projections. (From Sargent, Jacobson, and Gordonson,[10] with permission from the publisher.)

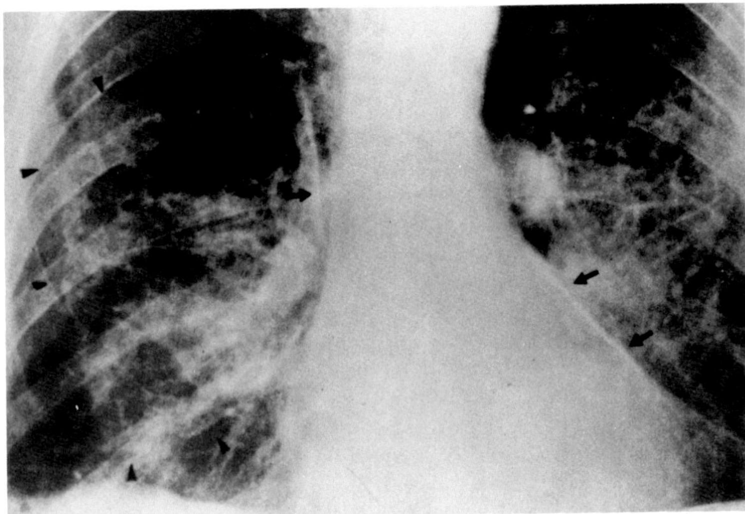

A

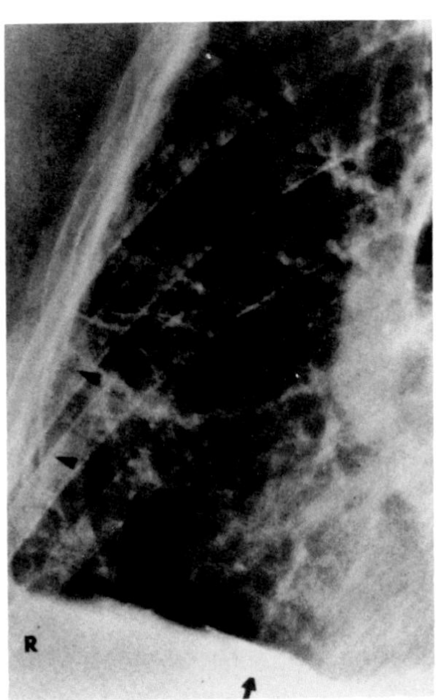

B

**Figure 28.13. Pleural plaques in asbestosis.** (A) A frontal view shows en face calcifications on the right (arrowheads); linear calcifications in profile in the mediastinal reflection of the pleura on the right and in the pericardium on the left (transverse arrows); and linear calcification in the left diaphragmatic pleura (vertical arrow). (B) A left oblique film shows linear pleural calcification in profile in the area of the central tendon of the right hemidiaphragm (arrow). The en face plaques of A now appear in profile as extensive linear calcifications (arrowheads) adjacent to anterior ribs. (From Sargent, Jacobson, and Gordonson,[10] with permission from the publisher.)

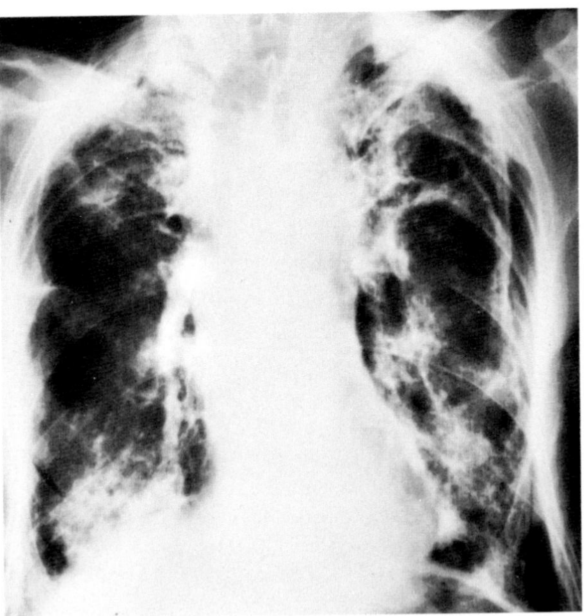

A

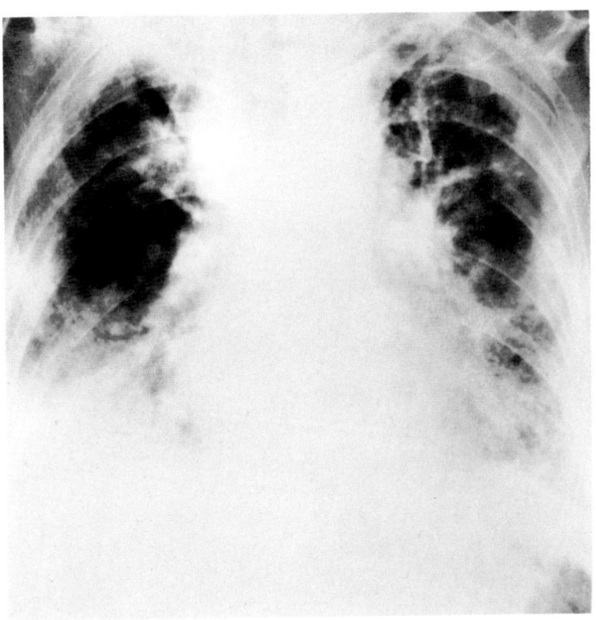

B

**Figure 28.14. Asbestosis.** (A) A frontal radiograph demonstrates severe disorganization of lung architecture with generalized coarse reticulation, which has become confluent in the right base and obliterates the right hemidiaphragm. There is marked pleural thickening, particularly in the apical and axillary regions. A spontaneous pneumothorax is present on the left. (B) A frontal radiograph obtained 4 months later demonstrates marked deterioration in the appearance of the chest and obliteration of the heart borders and diaphragm. A further loss of lung volume is also noted. (From Fraser and Pare,[1] with permission from the publisher.)

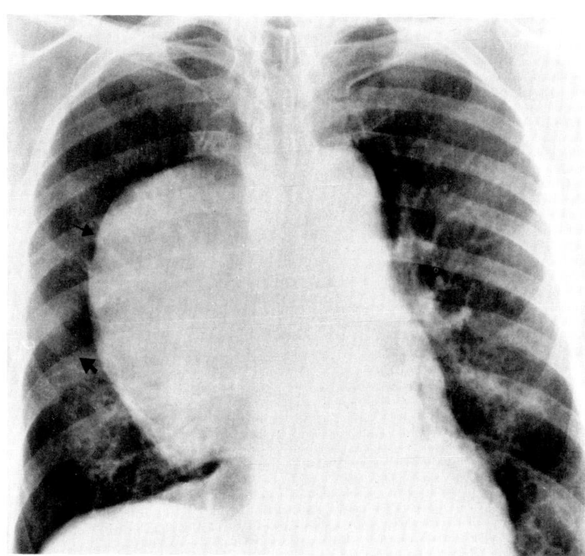

**Figure 28.15. Mesothelioma.** There is a huge, homogeneous soft tissue mass (arrows) arising from the mediastinal pleura and projecting into the right hemithorax. The patient had only mild underlying interstitial fibrosis and no pleural plaquing.

measuring 1 cm or more in diameter may develop in patients with extensive interstitial fibrosis. These opacities may be well-defined or ill-defined, are often multiple and nonsegmental, and predominantly involve the lower lung, in contrast to the upper lobe predominance of the conglomerate opacities in silicosis.

About 80 percent of pleural or peritoneal mesotheliomas develop in patients with asbestos exposure. The risk of developing this type of tumor peaks 30 to 35 years after the initial exposure. Bronchogenic carcinoma is also unusually common in patients with asbestosis, especially those who are cigarette smokers.[1,7–10]

## Silicosis

The inhalation of high concentrations of silicon dioxide is a cause of an occupational lung disease that primarily affects workers engaged in mining, foundry work, and sandblasting. Although acute silicosis can develop within 10 months of exposure in workers exposed to sandblasting in confined spaces, 15 to 20 years of long-term, relatively less intense exposure is required to produce radiographic changes.

The earliest radiographic change in silicosis is a fine interstitial reticular pattern that appears to accentuate the bronchovascular markings. Small transverse lines of density (Kerley B lines) are often seen above the costophrenic angles. The classic radiographic pattern consists of multiple nodular shadows scattered throughout the lungs. The shadows tend to spare the apices and bases, which may appear hyperlucent from associated emphysema. The nodules tend to be fairly well circumscribed and of uniform density. Calcification of the nodules may occur.

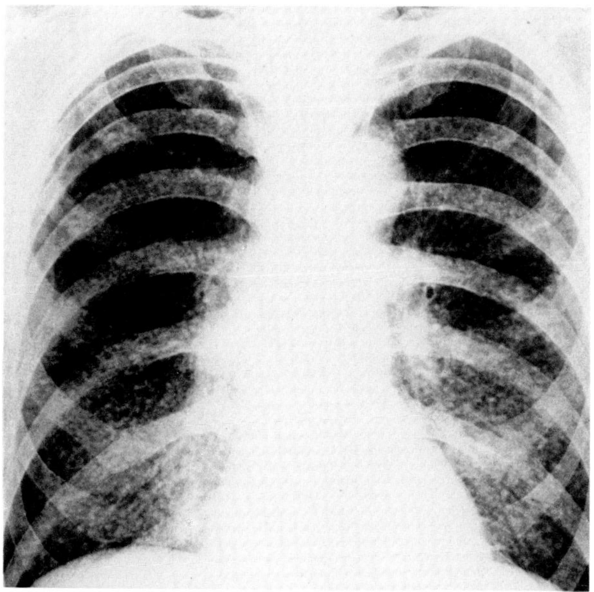

**Figure 28.16. Silicosis.** Multiple miliary nodular shadows scattered throughout both lungs.

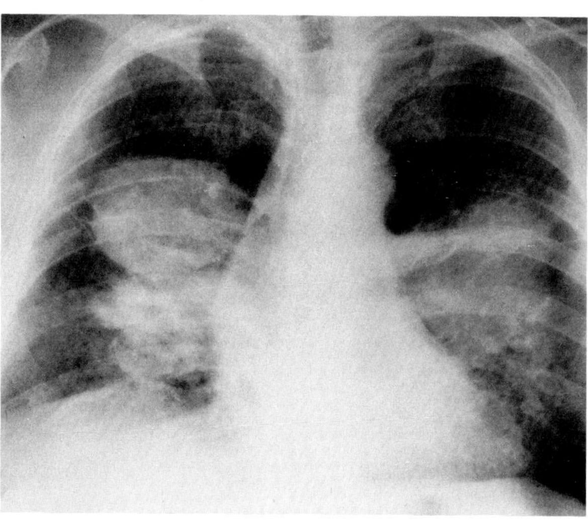

**Figure 28.17. Progressive massive fibrosis in silicosis.** Large, irregular nodules in both perihilar regions.

Unlike asbestosis, in silicosis there is infrequently any evidence of pleural abnormality.

As the pulmonary nodules increase in size, they tend to coalesce and form nonsegmental conglomerates of irregular masses in excess of 1 cm in diameter (progressive massive fibrosis). These masses are usually bilateral and relatively symmetric and are almost always restricted to the upper half of the lungs. The conglomerate fibrotic lesions may cavitate as a result of either central ischemic necrosis or tuberculous caseation. They commonly develop in the midzone or periphery of the lung and tend to migrate later toward the hilum, leaving overinflated

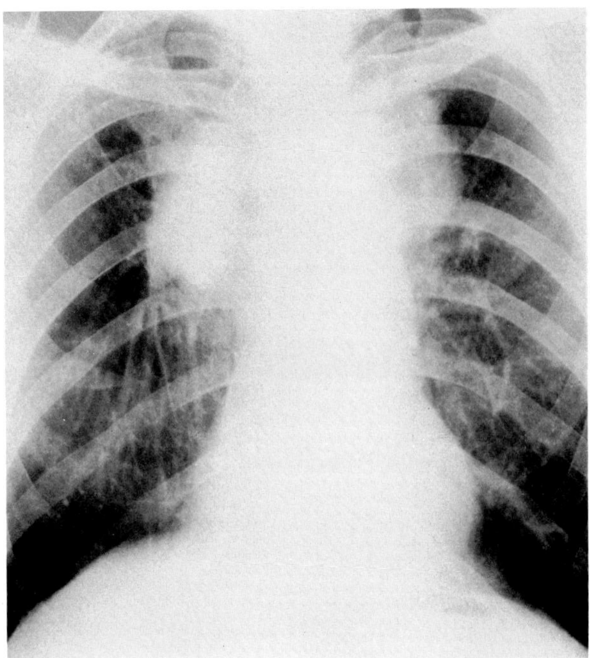

**Figure 28.18. Progressive massive fibrosis in silicosis.** Nonsegmental areas of homogeneous density in both upper lobes.

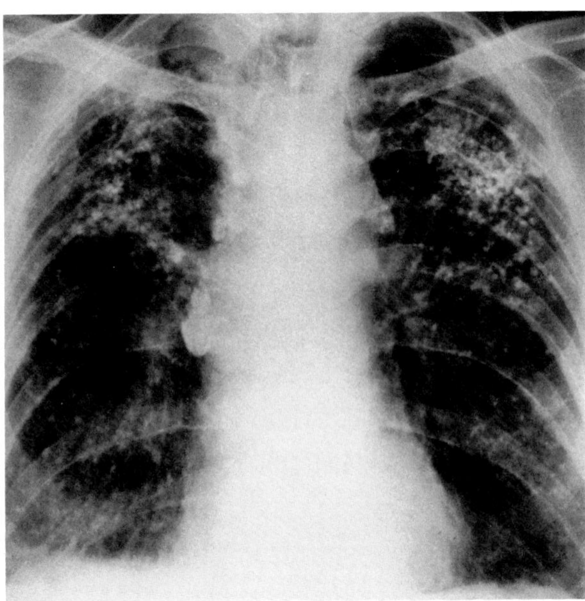

**Figure 28.19. Silicosis.** Bilateral calcific densities that tend to conglomerate in the upper lobes with upward retraction of the hila.

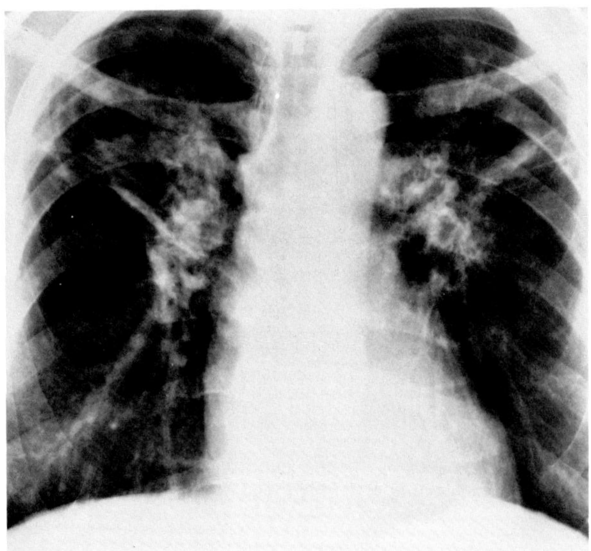

A

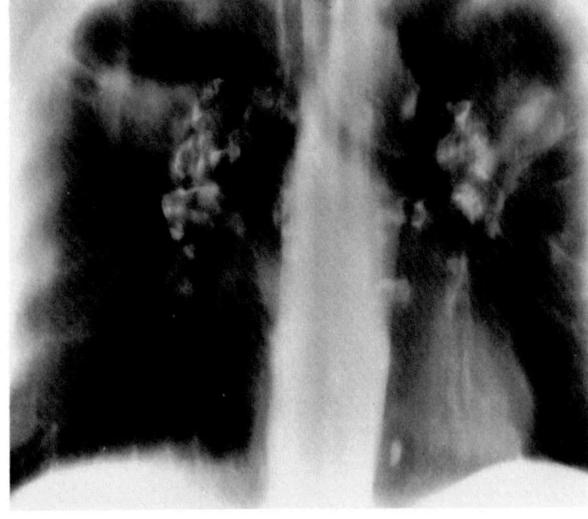

B

**Figure 28.20. Eggshell calcification in silicosis.** (A) A plain film and (B) a tomogram of the chest demonstrate characteristic eggshell lymph node calcification associated with bilateral perihilar masses.

emphysematous lung tissue between the consolidation and the pleural surface. A single, large homogeneous mass in the perihilar area of one lung may closely simulate bronchogenic carcinoma. The more extensive the progressive massive fibrosis, the less the apparent nodularity in the remainder of the lungs. Severe generalized emphysema and localized blebs can also be seen.

Hilar lymph node enlargement is a common radiographic finding that may occur at any stage of silicosis. The deposition of calcium salts in the periphery of enlarged lymph nodes produces the characteristic eggshell appearance, which is virtually pathognomonic of silicosis, though occasionally seen in sarcoidosis. Fibrosis can eventually cause hilar nodes to decrease in size.

Silicosis predisposes to the development of tuberculous infection. This complication may be suspected if

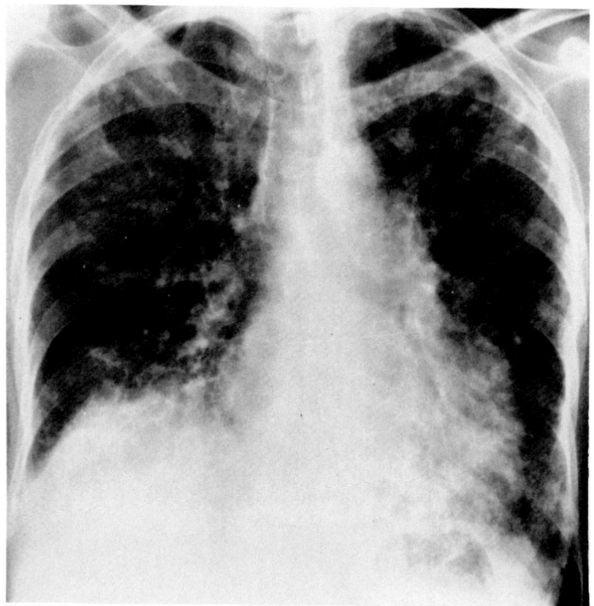

**Figure 28.21. Talcosis.** A frontal chest film shows a diffuse reticulonodular pattern throughout both lungs. Confluent areas of pulmonary infiltrate obscure the borders of the heart and hemidiaphragm.

cavitation develops in the conglomerate fibrotic densities, if the disease is markedly asymmetric, or if there is extensive pleural involvement.[11–14]

## Talcosis

Severe pulmonary disease can develop in persons with long-term exposure to high concentrations of talc (magnesium silicate) dust in mining and milling operations. The radiographic hallmark of talcosis is pleural plaque formation, which may be massive, is often bizarre in shape, and may extend over much of the surface of both lungs. Pericardial involvement can also occur. Findings in the lung parenchyma are similar to those in asbestosis and present a radiographic pattern of general haziness, nodulation, and reticulation, with relative sparing of the apices and costophrenic sulci. The confluence of nodules can produce large pulmonary opacities.[15–17]

## Coal Worker's Pneumoconiosis

Coal miners, especially those working with anthracite (hard coal), are susceptible to developing pneumoconiosis by inhaling high concentrations of coal dust. Initially, multiple small, irregular opacities produce a reticular pattern similar to that of silicosis. However, the nodules tend to be somewhat less well defined than those of silicosis and are of a granular density, unlike the homogeneous density of silicosis nodules. With advanced disease, a pattern of progressive massive fibrosis can develop. This appears as one or more masses of fibrous tissue with smooth, well-defined lateral borders that tend to parallel the rib cage and are projected several centimeters from it. Cavitation within the large nodules can be caused by ischemic necrosis or superimposed tuberculosis. As with the conglomerate shadows of silicosis,

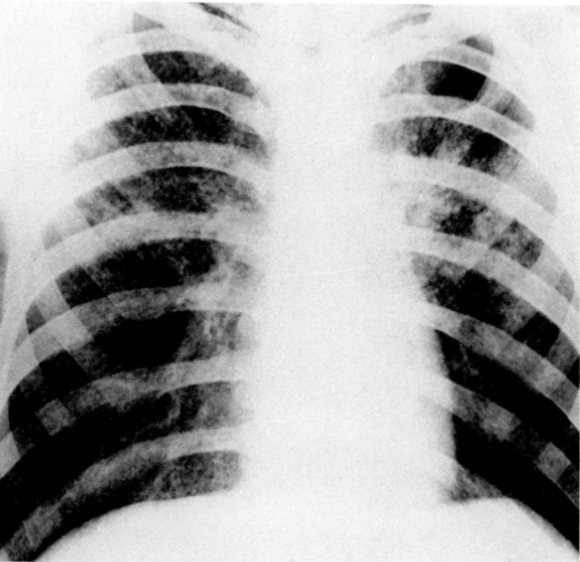

**Figure 28.22. Coal worker's pneumoconiosis.** Multiple small, irregular opacifications throughout both lungs.

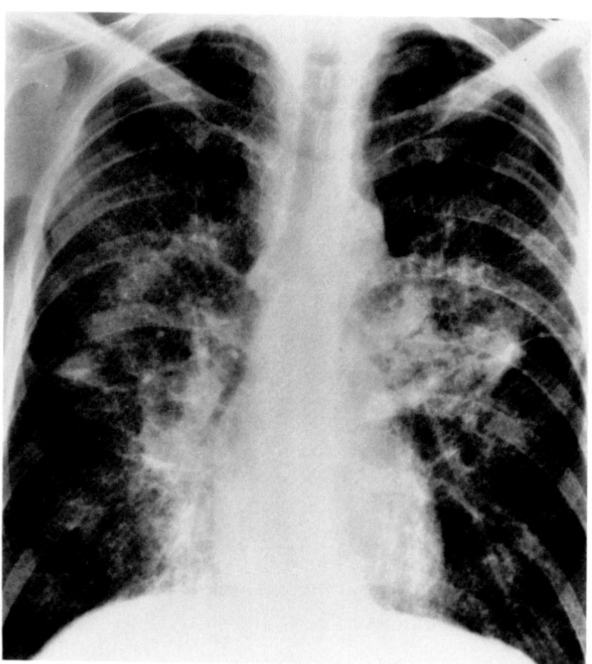

**Figure 28.23. Coal worker's pneumoconiosis.** Ill-defined masses of fibrous tissue in the perihilar region extend to the right base.

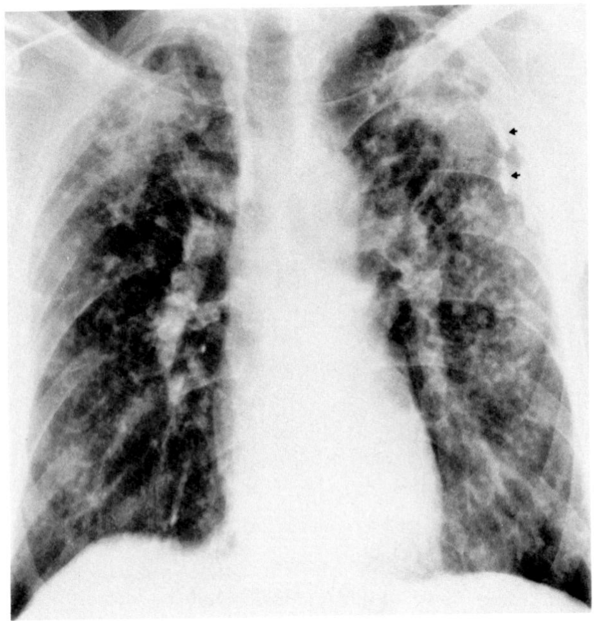

**Figure 28.24. Coal worker's pneumoconiosis.** Bilateral fibrous masses in the apices with upward retraction of the hila. There is pleural calcification (arrows) in the left apex.

a large fibrous mass in coal worker's pneumoconiosis gradually migrates toward the hilum, leaving a zone of overinflated emphysematous lung between it and the chest wall. A single large homogeneous mass in the perihilar area of one lung may simulate bronchogenic carcinoma; the detection of the underlying background of pneumoconiosis is essential for the proper diagnosis.[18,19,40]

## Caplan's Syndrome

Caplan's syndrome refers to the combination of rheumatoid arthritis and the characteristic fibrotic conglomerate lesions of progressive massive fibrosis. The syndrome was initially described in coal miners but subsequently has been found in patients with a number of multiple rounded, peripheral nodules that range from 0.5 to 5 cm in diameter. Most often seen in patients who have subcutaneous rheumatoid nodules, the pulmonary nodules in Caplan's syndrome tend to be more regular in contour and more peripherally located than the typical masses of progressive massive fibrosis. The nodules may cavitate and are often superimposed on a reticulonodular background. The pulmonary nodules may appear at intervals and often precede the exacerbation of arthritic symptoms.[20,21]

## Berylliosis

The inhalation of beryllium dust can cause both acute and chronic disease, especially in refinery workers.

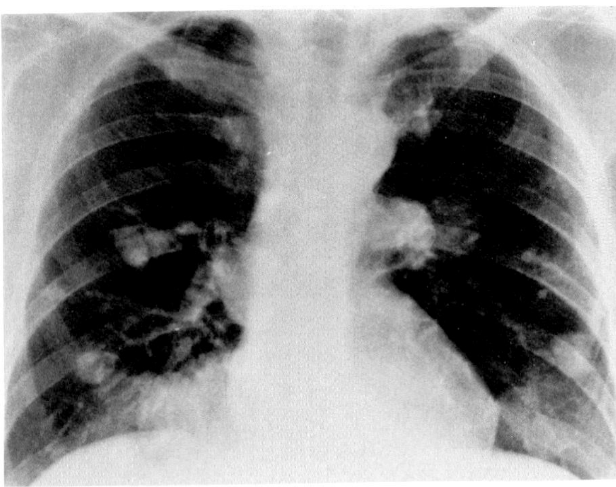

**Figure 28.25. Caplan's syndrome.** Multiple well-circumscribed, rounded nodules of varying size in a patient with subcutaneous rheumatoid nodules.

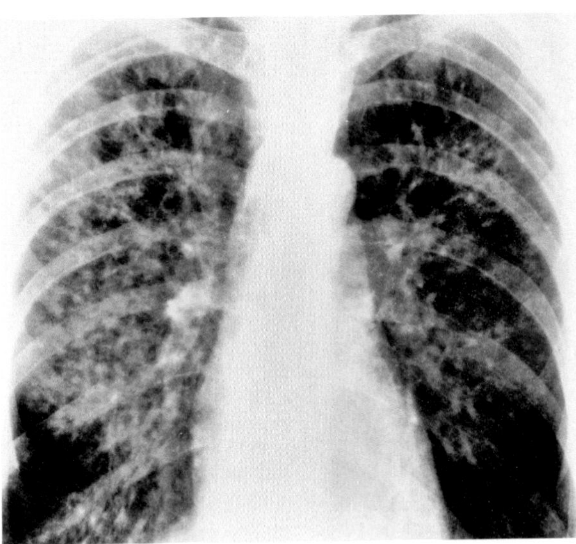

**Figure 28.26. Chronic berylliosis.** A frontal radiograph demonstrates a diffuse reticulonodular pattern throughout both lungs. There is relative sparing of the apices and bases.

Overwhelming exposure can produce fulminating acute berylliosis, characterized by acute pulmonary edema and hemorrhage that may be rapidly fatal. The subacute disease consists of a chemical pneumonitis that appears radiographically several weeks after the onset of symptoms. Diffuse, symmetric, bilateral haziness may develop into irregular patchy densities scattered widely throughout both lungs. Complete radiographic clearing may take several months.

Chronic berylliosis is a widespread systemic disease due to prolonged exposure that produces lesions in the lungs, lymph nodes, liver, spleen, kidney, and heart. The radiographic pattern may vary from diffuse, finely granular haziness with a tendency to spare the apices and bases to ill-defined nodules of moderate size that are scattered diffusely throughout both lungs. The concomitant enlargement of hilar shadows, primarily due to the engorgement of pulmonary vessels secondary to pulmonary hypertension, may produce an appearance simulating that of reticulonodular sarcoidosis. With advanced disease, coarse interstitial fibrosis may be associated with decreased lung volume and areas of emphysema, especially in the upper lobes. Spontaneous pneumothorax occurs in about 10 percent of the patients.[22–25]

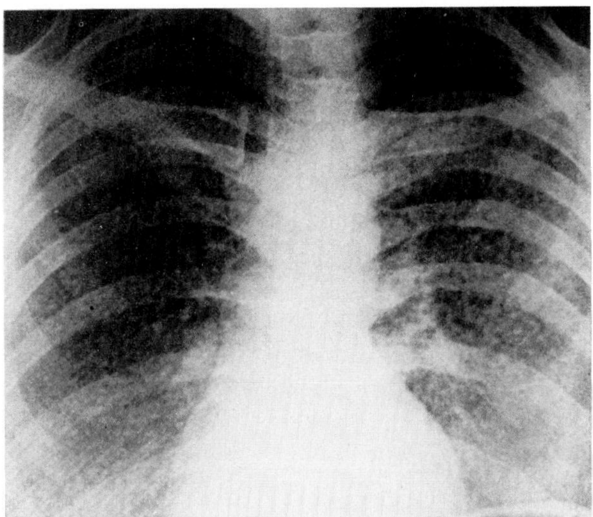

**Figure 28.27. Siderosis.** Multiple soft tissue nodules throughout the lungs represent an accumulation of iron oxide in the lymphatics and interstitial tissues. (From Sander,[26] with permission from the publisher.)

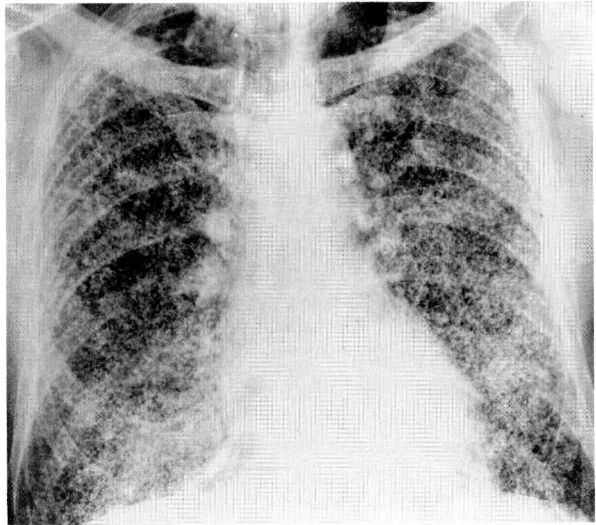

**Figure 28.28. Stannosis.** Multiple tiny shadows of high density distributed evenly throughout the lungs of a tin smelter.

## Pneumoconiosis Due to Other Inorganic Dusts

Particles of some inhaled inorganic dusts may lodge in the lung and produce nodular densities that are sharply defined and uniformly distributed throughout both lungs. Because these dusts incite no fibrosis or other reaction, the radiographic findings develop in patients with no clinical evidence of pulmonary disease.

Siderosis is due to the accumulation of iron oxide in the lymphatics and interstitial tissue of the lungs. This disease occurs most commonly in iron miners, arc welders, and silver polishers. The resulting nodules are of rather low density compared with those of silicosis. Unlike most pneumoconioses, in siderosis the radiographic abnormality may disappear partly or completely when the patient is removed from dust exposure.

Tin smelters may develop stannosis, which produces a radiographic pattern of multiple tiny shadows of high density distributed evenly throughout the lungs. Baritosis is due to the deposition of barium sulfate in the lungs; it is most common in workers in barium mines. Because of the high radiopacity of barium, the discrete shadows in the lungs are extremely dense. The lesions typically regress after the patient is removed from the dust-filled environment. The inhalation of cerium dioxide and antimony salts can also cause multiple nodules throughout the lungs.[26–28]

## Pneumoconiosis Due to Organic Dusts

The inhalation of organic dusts may lead to the development of a hypersensitivity pneumonitis. Many of these diseases are named for the specific setting in which the

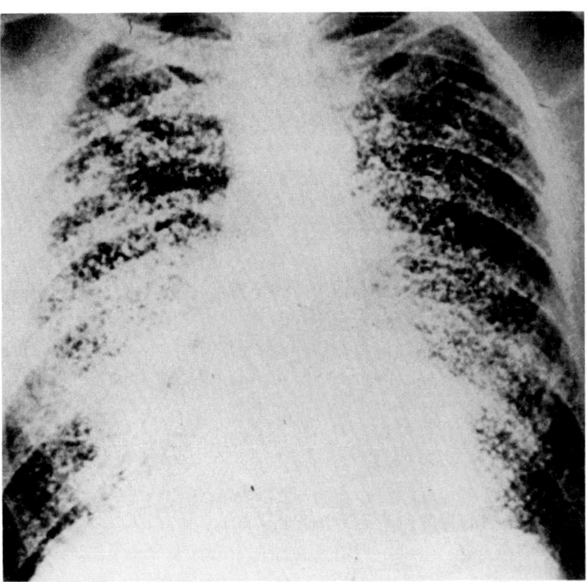

**Figure 28.29. Baritosis.** Multiple dense nodules throughout the lungs in an asymptomatic person who had a heavy occupational exposure to fine barium dust. (From Sander,[26] with permission from the publisher.)

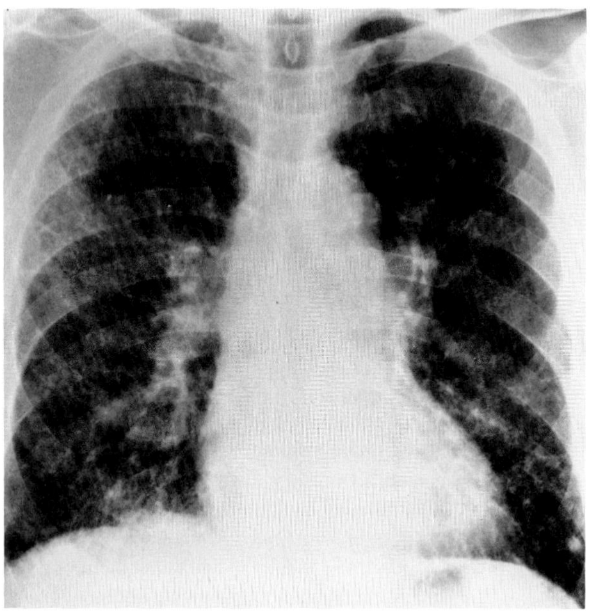

**Figure 28.30. Byssinosis.** Prolonged exposure has resulted in irreversible pulmonary insufficiency and radiographic evidence of interstitial fibrosis.

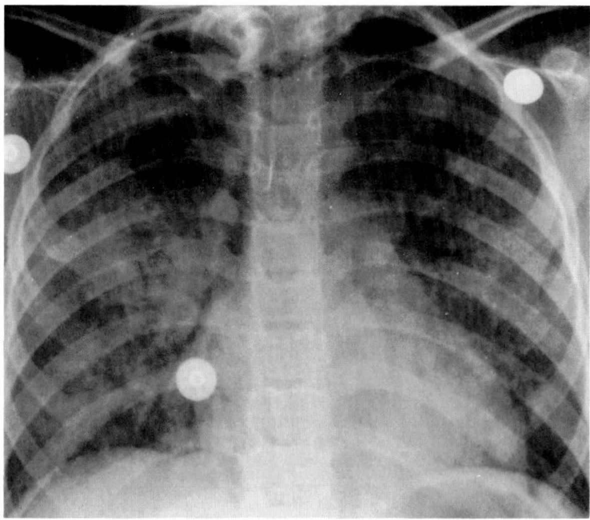

**Figure 28.31. Farmer's lung.** There are multiple confluent and nodular densities throughout both lungs, with relative sparing of the apices and bases.

disease is found. Because most present a nonspecific radiographic pattern, information about occupational and other environmental exposures must be sought when these conditions are suspected.

Byssinosis is a pulmonary disease that develops primarily in cotton-mill workers exposed to large amounts of cotton dust. Symptoms of dyspnea typically develop on the first day of the work week ("Monday chest tightness") and gradually decrease as the week progresses. During the reversible stages of the disease, the radiographic findings are minimal and consist of nonspecific accentuation of perihilar markings. Prolonged exposure can result in irreversible pulmonary insufficiency with radiographic evidence of interstitial fibrosis and emphysema.

Bagassosis is a hypersensitivity pneumonitis that develops primarily in sugarcane workers who inhale bagasse, the fibrous material that remains after the sugar-containing juice has been extracted from sugarcane. Radiographically, prominent peribronchial markings around the hila may be associated with a fine granularity that is diffusely distributed throughout the lungs.

Farmer's lung is a hypersensitivity pneumonitis resulting from exposure to moldy hay, grain, or silage. In the acute phase, there is a pattern of discrete, though poorly defined, small nodules that produce a diffuse, fine granular density that tends to spare the apices and bases. Extensive involvement produces alveolar consolidation (allergic alveolitis). If a patient continues to be exposed to the mold, permanent changes of interstitial fibrosis and pulmonary emphysema may develop.

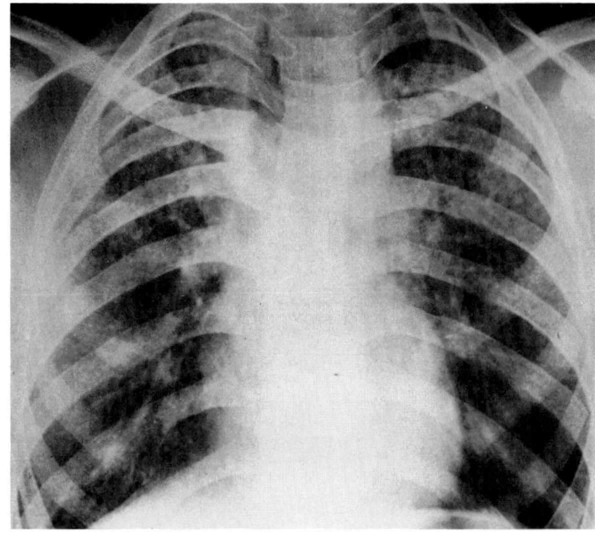

**Figure 28.32. Pigeon-breeder's disease.** A diffuse reticulonodular infiltrate predominantly involves the perihilar and upper lobe regions.

Other diverse diseases associated with the inhalation of organic dusts include pigeon-breeder's lung, maple bark disease (sawmill or paper-mill workers), and mushroom-worker's lung.

Hypersensitivity pneumonitis should be strongly suggested by the following sequence of radiograph changes: the diffuse involvement of both lungs by a persistent reticular or nodular pattern, the superimposition of an acinar pattern with acute exacerbations, and the resolution of the latter pattern within a few days of the initial acute clinical presentation.[1,29–31]

## Inhalation of Toxic Gases

Gases and liquids in a finely dispersed state can produce acute and sometimes chronic damage of the air passages and pulmonary parenchyma. Soluble gases (sulfur dioxide, ammonia, chlorine) tend to cause irritation of the mucous membranes of the eyes and upper respiratory tracts; less soluble gases (nitrogen dioxide, ozone, phosgene, and high concentrations of oxygen) penetrate deeply into the lungs and produce alveolocapillary damage. The concentration of the gas and the duration of the exposure are the predominant factors determining the clinical presentation and radiographic findings.

Intense acute exposure to noxious gases usually causes the development of perihilar infiltrates and pulmonary edema. Less intense exposure or less irritating gases produce patchy pneumonitis, often with interstitial emphysema. If the patient survives the acute exposure, the radiographic abnormalities tend to clear as the patient recovers. Chronic exposure or severe pulmonary damage due to the initial exposure can lead to severe interstitial fibrosis and emphysema.

Smoke inhalation in fire fighters and fire victims is the most common cause of chemical pneumonitis. Carbon monoxide poisoning may be the result of an industrial accident or a suicide attempt. The ingestion or inhalation of hydrocarbons (kerosene, gasoline, furniture polish, turpentine) is the leading cause of poisoning in children. The ingested hydrocarbon is absorbed through the gastrointestinal tract and is carried by the bloodstream to the lungs, where it adds to the pulmonary injury. In addition to producing perihilar infiltrates and pulmonary edema, hydrocarbon inhalation in children can lead to the formation of pneumatoceles simulating those in staphylococcal pneumonia.

The decomposition of fresh silage produces large quantities of the gaseous oxides of nitrogen that can cause a chemical pneumonitis (silo-filler's disease). In this condition, the initial clinical and radiographic changes develop and clear rapidly. A recurrence of the disease usually develops several weeks after the remission of symptoms. This is characterized by fever, progressive dyspnea, cyanosis, and cough. Radiographs obtained during this phase demonstrate widely scattered, small nodular densities resembling the lesions of acute miliary tuberculosis. Confluence of the nodules can produce patchy infiltrates. Progressive interstitial fibrosis and emphysema can develop.

The herbicide paraquat can cause an acute chemical pneumonitis, especially in patients who ingest marijuana that has been sprayed with this substance. The ingestion of large doses produces acute pulmonary edema and hemorrhage and may be rapidly fatal.

A prolonged exposure to mechanical ventilation with high concentrations of oxygen can lead to the progressive deterioration of pulmonary function. This complication most often affects infants treated with oxygen for hyaline membrane disease. After 4 to 10 days of therapy, there is nearly complete opacification of the lungs with a clearly defined air-bronchogram pattern. This appearance is soon replaced by widespread multiple, small, round radiolucencies resembling bullae, which give the lungs a spongelike appearance. Large emphysematous areas develop in infants who survive longer than 1 month and who require continuous oxygen therapy. In adults, an initial exudative reaction produces a diffuse pulmonary edema pattern. In the later proliferative phase, linear and nodular densities are seen, most commonly in the bases.[32–35]

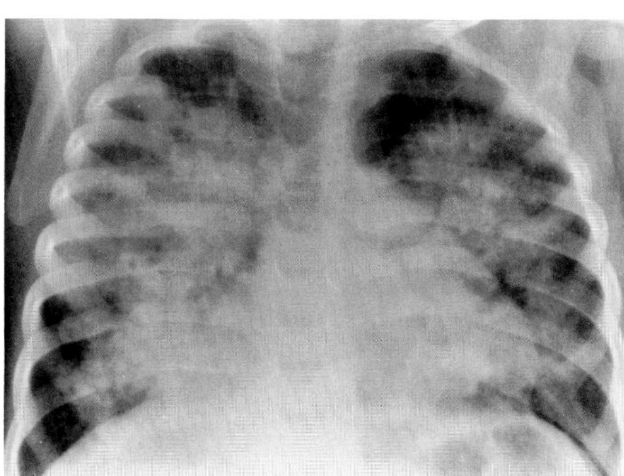

**Figure 28.33. Hydrocarbon poisoning.** There is a diffuse pulmonary edema pattern, with the alveolar consolidation most prominent in the central portions of the lung.

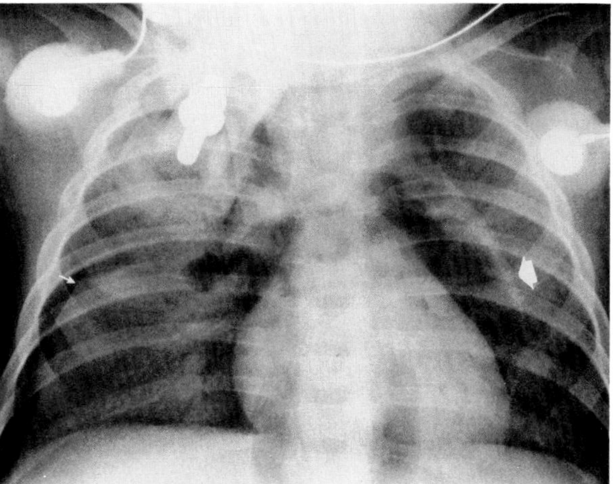

**Figure 28.34. Hydrocarbon poisoning.** Following intubation and positive-pressure ventilation, this child developed a pneumothorax (small arrow) and pneumomediastinum (large arrow). Note that the stiffness of the lungs has prevented substantial collapse.

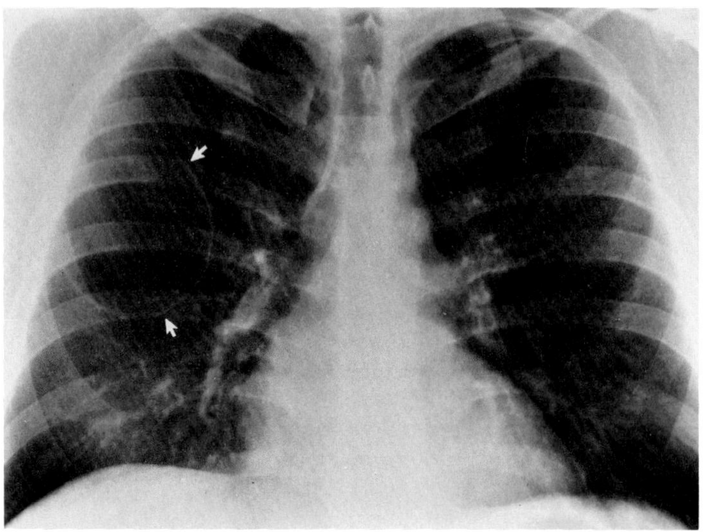

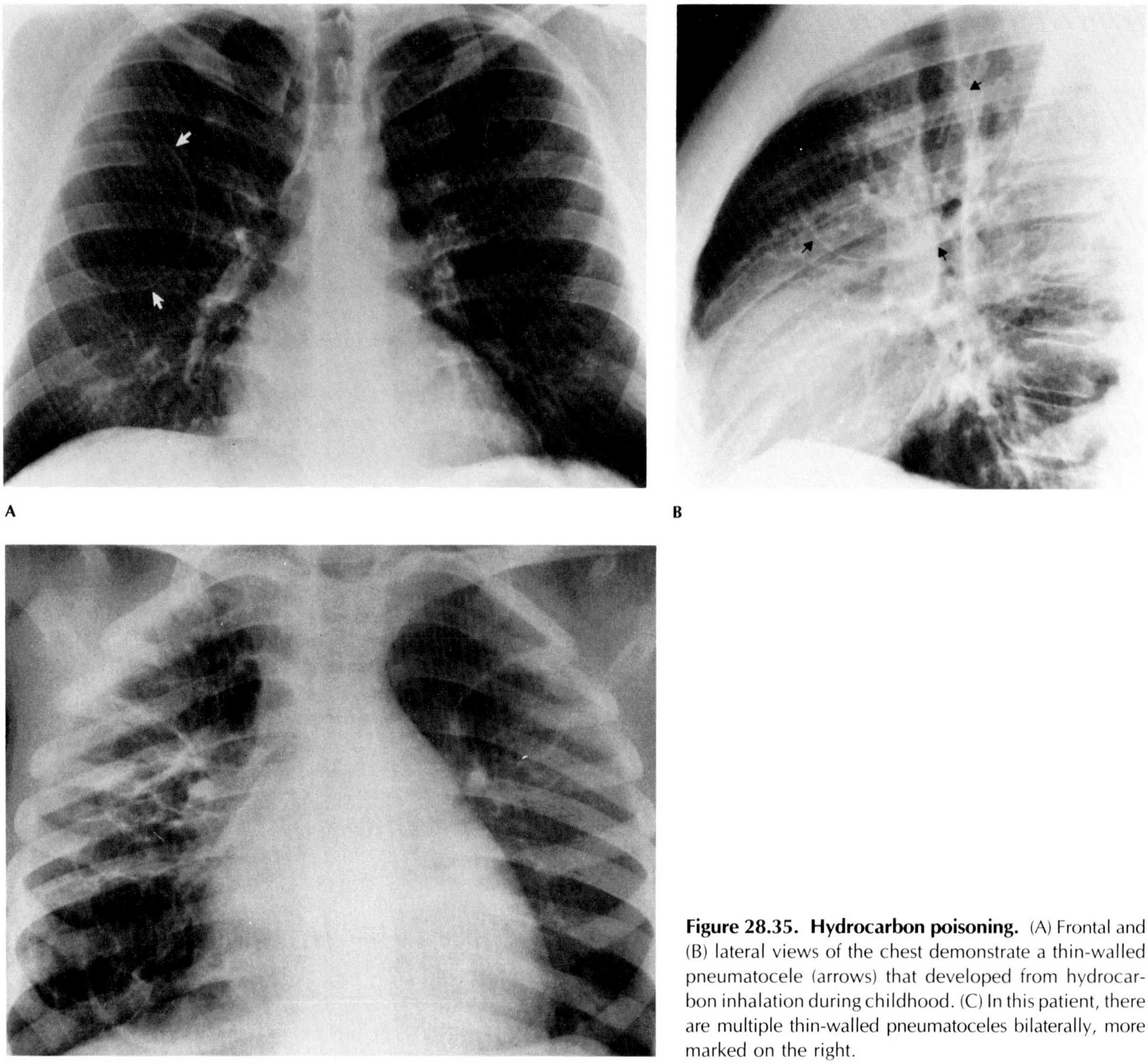

A

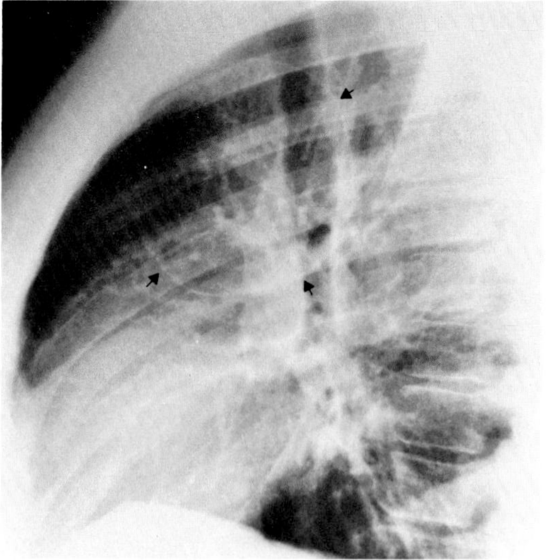

B

C

**Figure 28.35. Hydrocarbon poisoning.** (A) Frontal and (B) lateral views of the chest demonstrate a thin-walled pneumatocele (arrows) that developed from hydrocarbon inhalation during childhood. (C) In this patient, there are multiple thin-walled pneumatoceles bilaterally, more marked on the right.

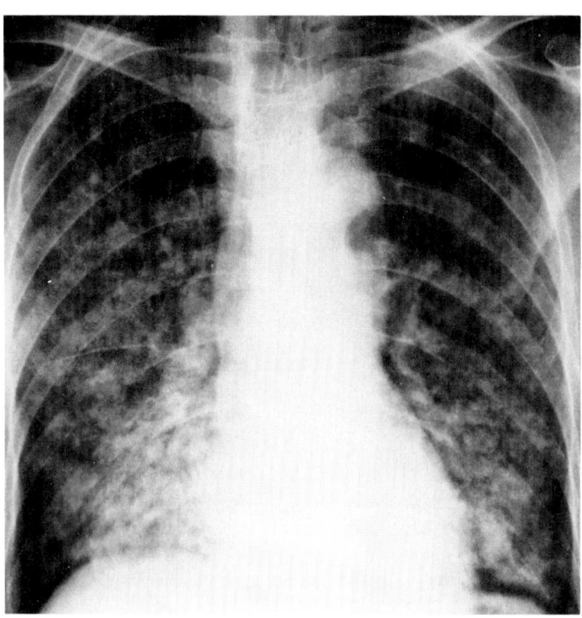

**Figure 28.36. Silo-filler's disease.** A chemical pneumonitis produces small nodular densities widely scattered throughout both lungs. Confluence of the nodules at the bases results in patchy infiltrates.

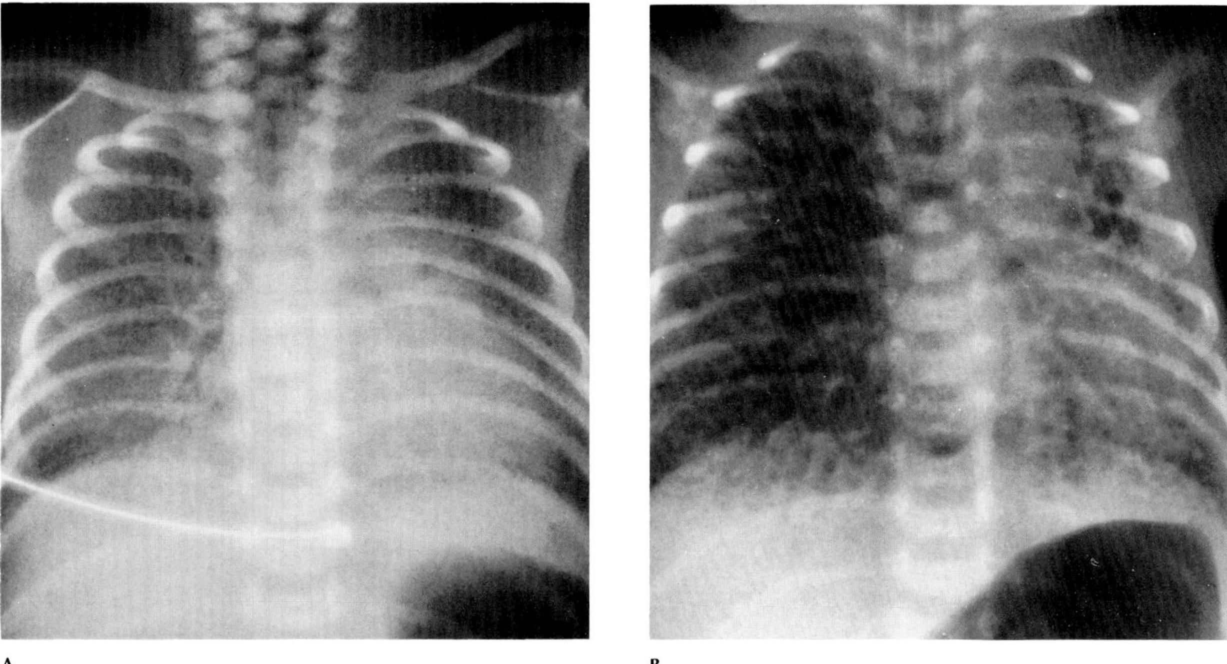

A B

**Figure 28.37. Oxygen toxicity.** (A) A chest radiograph in an infant demonstrates a typical granular parenchymal pattern with air bronchograms due to hyaline membrane disease. (B) Following intensive oxygen therapy, multiple small round lucencies resembling bullae have developed, and this gives the lungs a spongelike appearance.

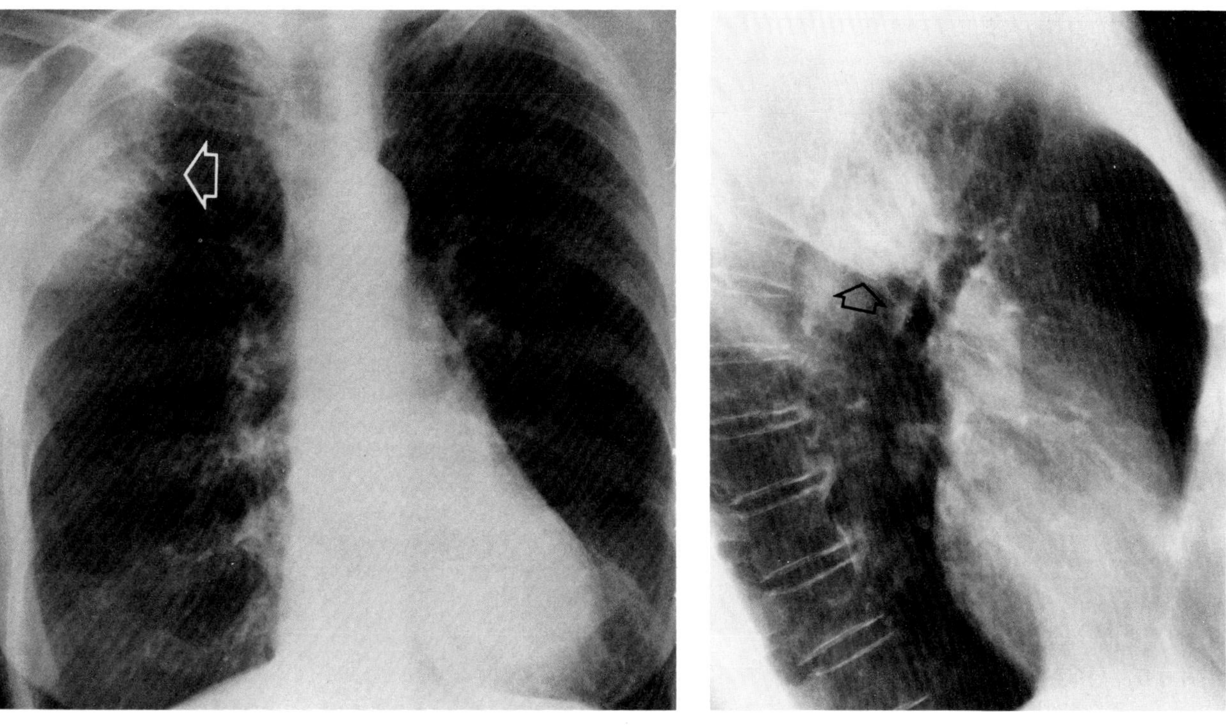

A B

**Figure 28.38. Lipoid pneumonia.** (A) Frontal and (B) lateral views of the chest demonstrate an air space consolidation in the posterior segment of the right upper lobe (arrows). Note the prominence of interstitial reticular markings leading from the right hilum to the infiltrate.

## Lipoid Pneumonia

Lipoid pneumonia is a pulmonary disease caused by the aspiration of various vegetable, animal, or mineral oils into the lungs. The condition most frequently develops in children, who may aspirate animal oil in milk and cod liver oil preparations. In adults, the most common cause of lipoid pneumonia is mineral oil taken for the treatment of constipation. Less frequent causes include the chronic use of oil in nose drops, and bronchography.

The aspiration of irritating oils initially produces a granular pattern of small, scattered alveolar densities, predominantly in the perihilar and lower lobe areas. As the oil is taken from the alveolar spaces by macrophages that pass into the interstitial space, a fine reticular pattern is produced. The lipoid distension of lymphatic vessels in the interlobular septa is manifest as Kerley B lines at the lateral aspects of the lung bases just above the costophrenic sulci. Reactive fibrosis can cause a loss of volume of the affected segments. Lipoid pneumonia can also appear as a granulomatous-lipoid mass, which varies considerably in size (up to 10 cm in diameter) and may have either a poorly defined or a sharply circumscribed margin. An oil granuloma can simulate bronchogenic carcinoma.

Because lipoid pneumonia produces fewer symptoms than comparable involvement by an infectious process, most patients are asymptomatic, and the abnormality is usually discovered on a screening chest radiograph.[36–38]

### REFERENCES


1. Fraser RG, Pare JAP: *Diagnosis of Diseases of the Chest,* Philadelphia, WB Saunders, 1979.
2. Tsai SH, Jenne JW: Mucoid impaction of the bronchi. *AJR* 96:953–961, 1966.
3. Ansell G: Radiological manifestations of drug-induced disease. *Clin Radiol* 20;133–148, 1969.
4. Horowitz AL, Friedman M, Smith J, et al: The pulmonary changes of bleomycin toxicity. *Radiology* 106:65–68, 1973.
5. Brettner A, Heitzman RE, Woodin WG: Pulmonary complications of drug therapy. *Radiology* 96:31–38, 1970.
6. *Guidelines for the Use of International Labour Office Classification of Radiographs of Pneumoconiosis.* Occupational Safety and Health Sciences 22 (Revised 1980). Geneva, ILO, 1980.
7. Fletcher DE, Edge JR: The early radiologic changes in pulmonary and pleural asbestosis. *Clin Radiol* 21:355–365, 1970.
8. Borow M, Couston A, Livornene L, et al: Mesothelioma following exposure to asbestos. A review of 72 cases. *Chest* 64:641–646, 1973.
9. Anton HC: Multiple pleural plaques. *Br J Radiol* 41: 341–348,1968.

9a. Freundlich IM, Greening RR: Asbestosis and associated medical problems. *Radiology* 89:224–229, 1967.

10. Sargent EN, Jacobson G, Gordonson JS: Pleural plaques: A signpost of abestosis dust inhalation. *Semin Roentgenol* 12:287–297, 1977.
11. Michel RD, Morris JF: Acute silicosis. *Arch Intern Med* 113:850–855, 1964.
12. Pendergrass EP: Silicosis and a few of the other pneumoconioses: Observations on certain aspects of the problem with emphasis on the role of the radiologist. *AJR* 80:1–41, 1958.
13. Greening RR, Heslep JH: The roentgenology of silicosis. *Semin Roentgenol* 2:265–275, 1967.
14. Grayson CE, Blumenfeld H: "Egg shell" calcifications in silicosis. *Radiology* 53:216–226, 1949.


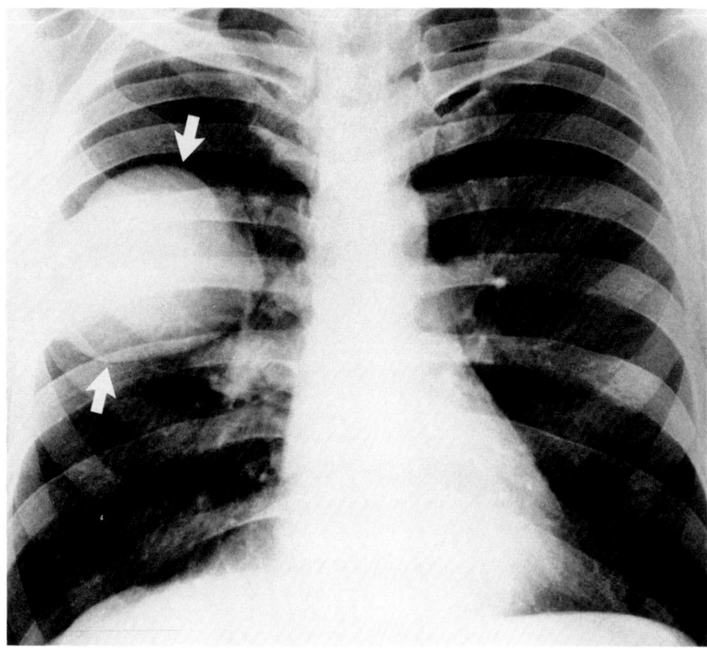

A

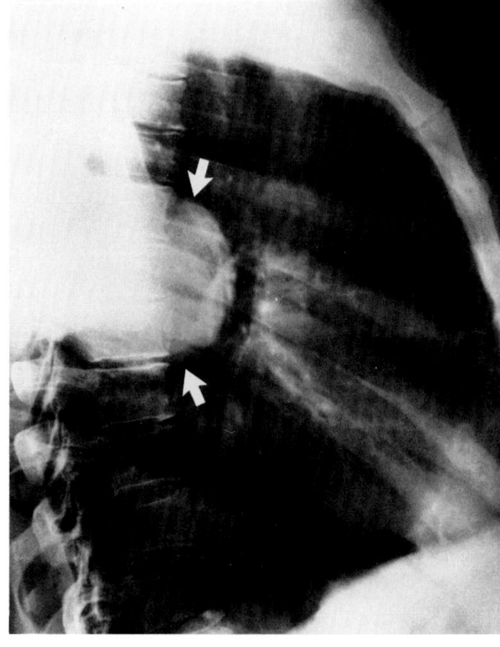

B

**Figure 28.39. Lipoid pneumonia.** (A) Frontal and (B) lateral views of the chest demonstrate a sharply demarcated granulomatous-lipoid mass (arrows) simulating a neoplastic process.

15. Siegal W, Smith AR, Greenburg L: The dust inhaled in tremolite talc mining, including roentgenologic findings in talc workers. *AJR* 49:11–18, 1943.
16. Kleinfeld M, Messite J, Tabershaw IR: Talc pneumoconiosis. *AMA Arch Industr Health* 12:66–72, 1955.
17. Porro FW, Patton JR, Hobbs AA: Pneumoconiosis in the talc industry. *AJR* 47:507–514, 1942.
18. Williams JL, Moller GA: Solitary mass in the lungs of coal miners. *AJR* 117:765–770, 1973.
19. Morgan WKC: Respiratory disease in coal miners. *JAMA* 231:1347–1348, 1975.
20. Morgan WKC, Lapp NL: Respiratory disease in coal miners. State of the art. *Am Rev Respir Dis* 113:531–559, 1976.
21. Caplan A: Certain unusual radiological appearances in the chest of coal miners suffering from rheumatoid arthritis. *Thorax* 8:29–34, 1953.
22. Gary JE, Schatzki R: Radiological abnormalities in chronic pulmonary disease due to beryllium. *AMA Arch Industr Health* 19:118–120, 1959.
23. Tebrock HE: Beryllium poisoning (berylliosis). X-ray manifestations and advances in treatment. *Am J Surg* 90:120–121, 1955.
24. Weber AL, Stoeckle JD, Hardy HL: Roentgenologic patterns in long-standing beryllium disease. *AJR* 93:879–890, 1965.
25. Sander OA: Berylliosis. *Semin Roentgenol* 2:306–311, 1967.
26. Sander OA: The nonfibrogenic (benign) pneumoconioses. *Semin Roentgenol* 2:312–319, 1967.
27. Robertson AJ, Whitaker PH: Radiological changes in pneumoconiosis due to tin oxide. *J Fac Radiol* 6:224–228, 1955.
28. Pendergrass EP, Greening RR: Baritosis. Report of a case. *AMA Arch Industr Health* 7:44–48, 1953.
29. Frank RG: Farmer's lung: A form of pneumoconiosis due to organic dusts. *AJR* 79:189–215, 1957.
30. Bristol LJ: Pneumoconioses caused by asbestos and by other siliceous and nonsiliceous dusts. *Semin Roentgenol* 2:283–292, 1967.
31. Hargreave FE, Pepys J, Holford-Strevens V: Bagassosis. *Lancet* 1:619–620, 1968.
32. Northway WH, Rosan RC, Porter DY: Pulmonary disease following respiratory therapy of hyaline membrane disease. Bronchopulmonary dysplasia. *N Engl J Med* 276:357–368, 1967.
33. Cornelius EA, Betlach EH: Silo-filler's disease. *Radiology* 74:232–238, 1960.
34. Thurlbeck WM, Thurlbeck SM: Pulmonary effects of paraquat poisoning. *Chest* (Suppl) 69:276–280, 1976.
35. Bonte FJ, Reynolds J: Hydrocarbon pneumonitis. *Radiology* 71:391–397, 1958.
36. Siddons AHM: Oil granuloma of the lung. Report of three cases. *Br Med J* 1:305–309, 1958.
37. Generux GP: Lipoids in the lungs: Radiologic-pathologic correlation. *J Can Assoc Radiol* 21:2–9, 1970.
38. Brody JS, Levin B: Interlobular septa thickening in lipoid pneumonia. *AJR* 88:1061–1069, 1962.
39. Sostman HD, Putman CE, Gamsu G: Review: Diagnosis of chemotherapy lung. *AJR* 136:33–41, 1981.
40. Davies D: Disability and coal worker's pneumoconiosis. *Br Med J* 2:652–659, 1974.
41. Sone S, Higashira T, Kotake T, et al: Pulmonary manifestations in acute carbon monoxide poisoning. *AJR* 120:865–871, 1974.

# Chapter Twenty-Nine
# Pulmonary Infections

## PNEUMONIA

Acute pneumonia can be caused by a variety of organisms, many of which are discussed in separate sections elsewhere in the book (Part Two). Regardless of etiology, pneumonias tend to conform to one of three basic radiographic patterns.

Alveolar or air space pneumonia, exemplified by pneumococcal pneumonia, is produced by an organism that causes an inflammatory exudate that spreads from one alveolus to the next via communicating channels (pores of Kohn, canals of Lambert). Since segmental boundaries do not impede the passage of air or fluid via these pores, the exudate of acute alveolar pneumonia can spread throughout the lung periphery and produce a nonsegmental distribution. Because the infection is propagated by centrifugal spread, the advancing margin of consolidation is frequently quite smooth and sharply circumscribed.

Consolidation of the lung parenchyma with little or no involvement of the conducting airways produces the characteristic air-bronchogram sign. The sharp contrast between air within the bronchial tree and the surrounding airless parenchyma permits the normally invisible bronchial air column to become radiographically visible. The appearance of an air bronchogram requires the presence of air within the bronchial tree, implying that the bronchus is not completely occluded at its origin. The presence of an air bronchogram excludes a pleural or mediastinal lesion since there are no bronchi in these regions.

Because air in the alveoli is replaced by an equal or almost equal quantity of fluid or tissue, and since the airways leading to the affected portions of the lung remain patent, there is no evidence of volume loss in alveolar pneumonia.

Bronchopneumonia, typified by staphylococcal infection, originates in the airways and spreads to peribronchial alveoli. Because interalveolar spread in the peripheral air spaces is minimal, the consolidation tends to maintain a distribution corresponding to the involved pulmonary segment. Inflammation causing bronchial or bronchiolar obstruction causes atelectasis; air bronchograms are absent.

Interstitial pneumonia is most commonly produced by viral and mycoplasmal infections. In this type of pneumonia, the inflammatory process predominantly involves the alveolar septa and interstitial supporting structures of the lung, producing a linear or reticular pattern. When seen on end, the thickened interstitium may appear as multiple small nodular densities.

Extensive inflammation can cause a mixed pattern of alveolar, bronchial, and interstitial pneumonia.[1]

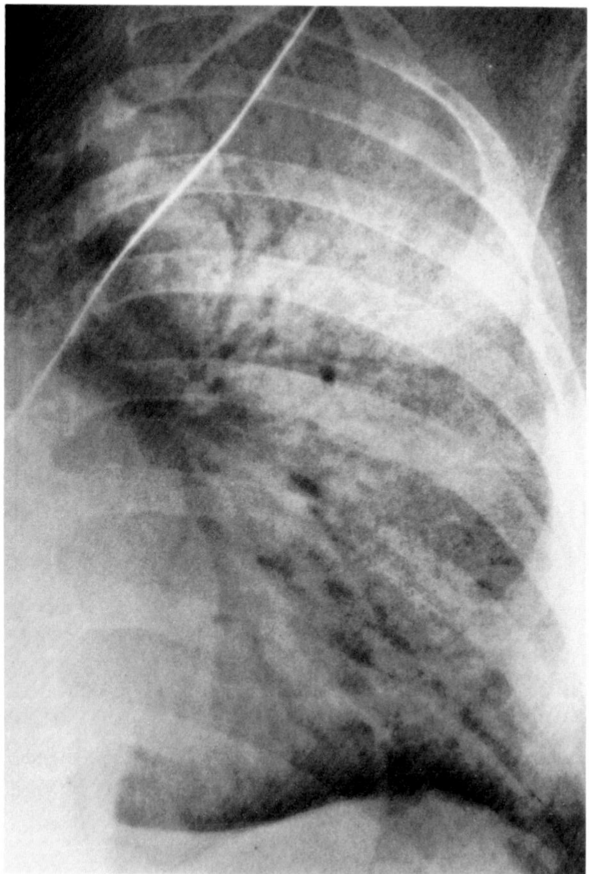

**Figure 29.1. Air-bronchogram sign.** A frontal chest radiograph demonstrates air within the intrapulmonary bronchi in a patient with diffuse alveolar pneumonia of the left lung.

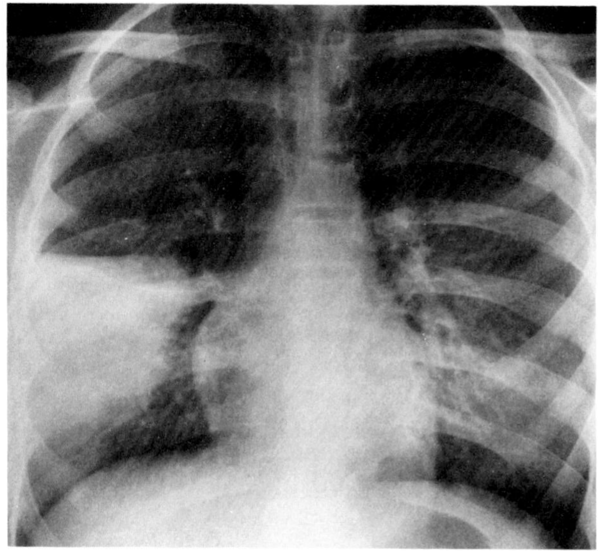

**Figure 29.2. Staphylococcal bronchopneumonia.** The air space consolidation is confined to the lateral segment of the right middle lobe. Note the absence of air bronchograms.

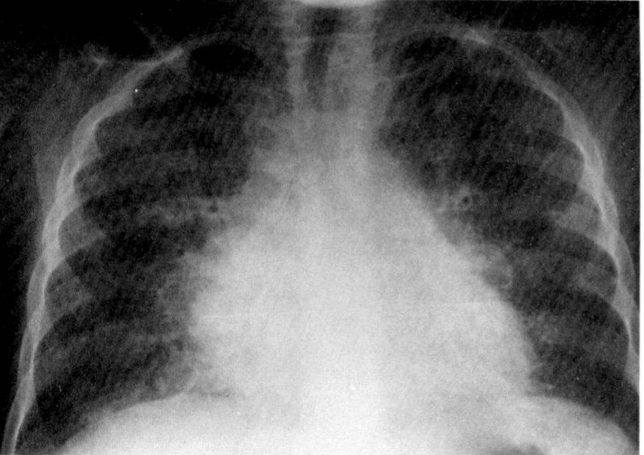

**Figure 29.3. Viral pneumonia.** The diffuse inflammatory process predominantly involves the alveolar septa and interstitial supporting structures, producing a reticular pattern. At times, the thickened interstitium seen on end appears as multiple small nodular densities.

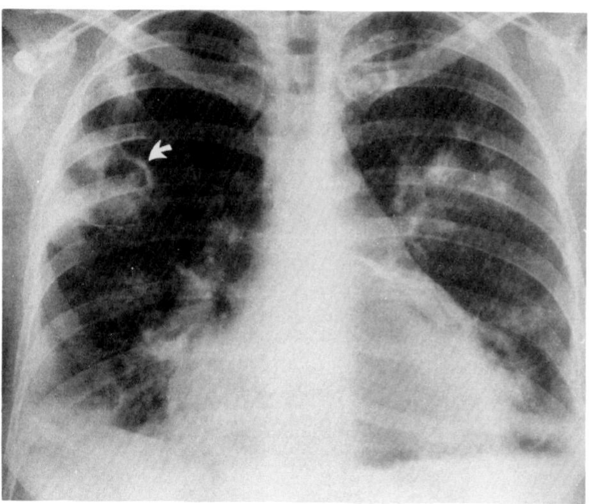

**Figure 29.4. Aspiration pneumonia.** Patches of alveolar consolidation in the upper and lower lobes developed bilaterally a few days after the aspiration of vomitus. The multiple abscesses (arrow points to the largest) represent secondary infection of the infiltrate. There are bilateral pleural effusions.

## Aspiration Pneumonia

The aspiration of esophageal contents can occur in patients with esophageal obstruction (tumor, stricture, achalasia), diverticula (Zenker's), or neuromuscular disturbances in swallowing. The anatomic distribution of pulmonary changes is affected by gravity. Thus the posterior segments of the upper and lower lobes are most commonly affected, especially in debilitated or bedridden patients. The typical radiographic appearance is that of air space consolidation with some degree of atelectasis, often involving several segments. The pneumonia tends to resolve slowly and may lead to abscess for-

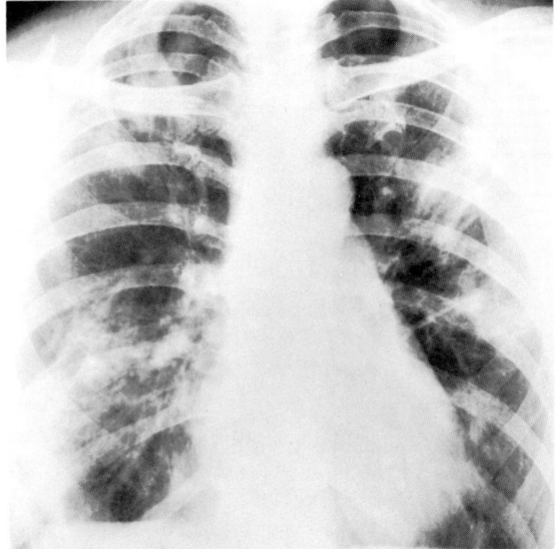

A

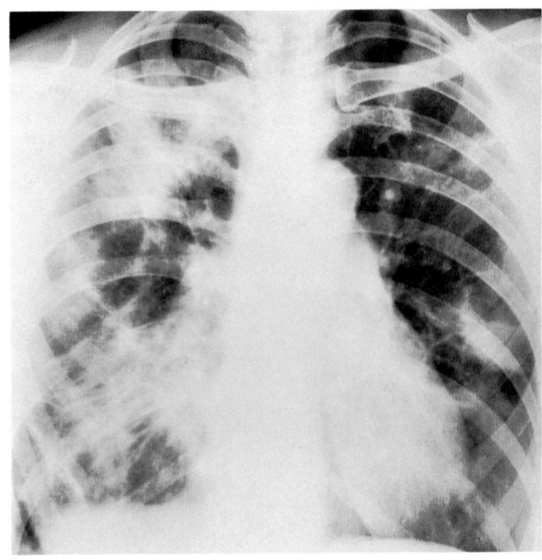

B

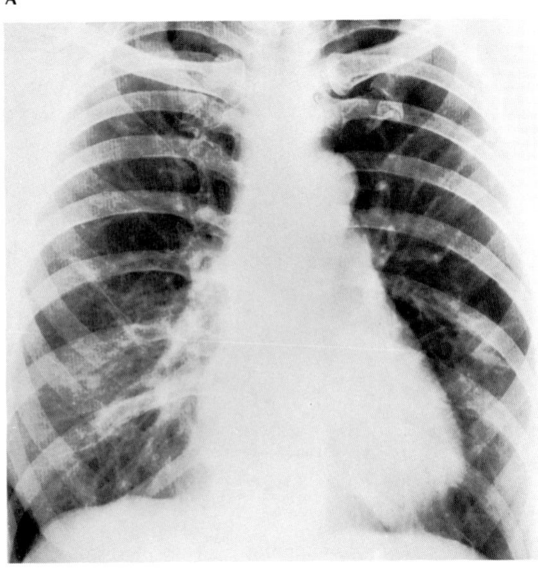

C

**Figure 29.5. Idiopathic eosinophilic pneumonia.** (A) An initial frontal chest radiograph shows numerous bilateral areas of consolidation that have no precise segmental distribution. Note particularly the broad shadow of increased density along the lower axillary zone of the right lung. At this time, the patient's total white blood cell count was 11,000 per cubic millimeter with 1700 (15 percent) eosinophils. (B) One week later, the anatomic distribution of the consolidations has changed considerably, being more extensive in the right upper and lower lobes and less extensive in the left upper lobe. At this time, the total white blood cell count had increased to 14,000 per cubic millimeter with 20 percent eosinophils. (C) One week later, following adrenocorticotropic hormone (ACTH) therapy, the white blood cell count had returned to normal levels, the eosinophilia had disappeared, and the radiographic abnormalities had completely resolved. (From Fraser and Pare,[1] with permission from the publisher.)

mation. Chronic aspiration may result in an irregular accentuation of linear markings representing residual fibrosis.

The aspiration of liquid gastric contents (general anesthesia, tracheostomy, coma, trauma) may produce multiple alveolar densities that are distributed widely and diffusely throughout both lungs and present a radiographic appearance of widespread pneumonia or massive pulmonary edema. Early diagnosis and the prompt institution of corticosteroid and antibiotic therapy are essential to improve the otherwise grave prognosis.[1]

### Idiopathic Eosinophilic Pneumonia (Loeffler's Syndrome)

Idiopathic eosinophilic pneumonia refers to a radiographic pattern of transient, rapidly changing, nonsegmental areas of parenchymal consolidation that is associated with blood eosinophilia. A similar appearance can develop secondary to parasites (filariasis, ascariasis, cutaneous larva migrans), drug therapy (nitrofurantoin), and fungal infections (hypersensitivity bronchopulmonary aspergillosis). When caused by an identifiable extrinsic agent, the disease usually is acute and responds promptly to the removal of the offending organism or drug. When no obvious cause is detectable, the pulmonary consolidation and eosinophilia tend to be more prolonged and persistent, though there usually is a dramatic response to steroids.

Serial radiographs demonstrate the transitory and migratory nature of the patchy parenchymal consolidations. The infiltrates are often located in the periphery of the lung, running more or less parallel to the lateral chest wall and simulating a pleural process.[1,2]

## LUNG ABSCESS

A lung abscess is a necrotic area of pulmonary parenchyma containing purulent material. Lung abscesses may be complications of bacterial pneumonia, bronchial obstruction, aspiration (food or vomitus), or a foreign body. Less frequently, abscesses arise from the hematogenous spread of organisms to the lungs either in a patient with diffuse bacteremia or as a result of septic emboli.

A lung abscess may result from the stasis of secretions and pneumonia distal to a bronchial obstruction caused by carcinoma or a foreign body. It also can be a complication of pulmonary infarction. A lung abscess is often the result of the aspiration of foreign material (food or vomitus). Thus it is generally located in those segments that are most dependent at the time of the aspiration: the posterior segments of the upper lobes or the superior segments of the lower lobes, especially on the right, when the patient is supine, or the basilar segments of the lower lobes when the patient is upright.

The earliest radiographic finding of lung abscess is a spherical density that characteristically has a dense center with a hazy, poorly defined periphery. If there is communication with the bronchial tree, the fluid contents of the cavity are partly replaced by air, producing a typical air-fluid level within the abscess. A cavitary lung abscess usually has a thickened wall with a shaggy, irregular inner margin.

Computed tomography (CT) may be of value in differentiating a lung abscess from a pyopneumothorax, both of which may appear on plain radiographs as a "cavitary" lesion with an air-fluid level adjacent to the chest wall. This differentiation is critical since a pyopneumothorax is managed by drainage through a thoracostomy tube while a lung abscess is appropriately treated with antibiotic therapy and postural drainage. A lung abscess typically is round, has an irregular thick wall, and lacks a discrete boundary between the lesion and the lung parenchyma; a pyopneumothorax tends to have a regular smooth shape with a sharply defined border between the lesion and the lung. In pyopneumothorax, scanning the patient in different positions (prone, decubitus) usually demonstrates a change in the configuration of the cavity and unequal fluid levels that closely approximate the chest wall. In lung abscess, the air-fluid level remains of equal length in all dimensions.[1,3–5,16]

## BRONCHIECTASIS

Bronchiectasis refers to the permanent abnormal dilatation of one or more large bronchi due to the destruction of the elastic and muscular components of the bronchial wall. The origin of the destructive process is nearly always a bacterial infection, which may be a primary event (suppurative necrotizing pneumonia) or secondary to local or systemic abnormalities that impair defense mechanisms and promote bacterial growth. Conditions predisposing to the development of bronchiectasis include

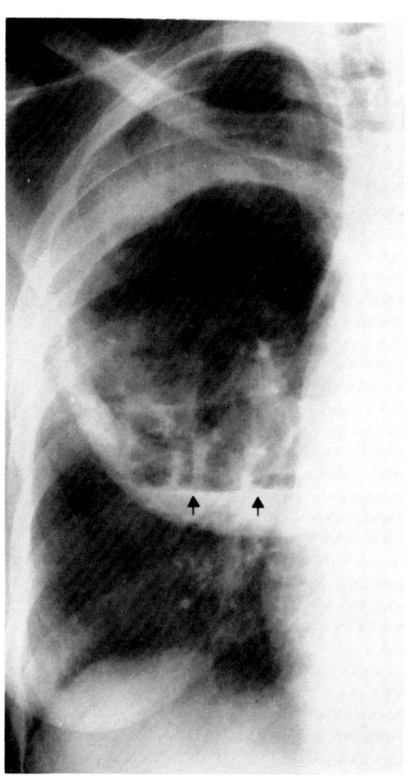
A

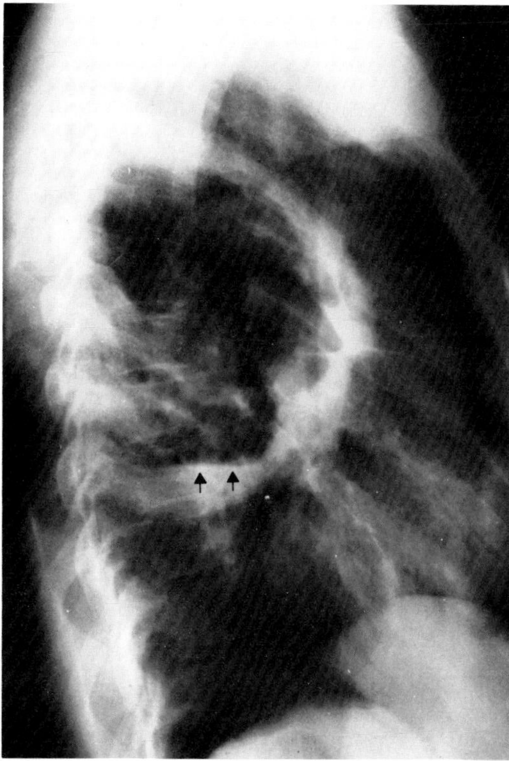
B

**Figure 29.6. Lung abscess.** (A) Frontal and (B) lateral views of the chest demonstrate a huge abscess in the superior segment of the right lower lobe. The cavitary lung abscess has a thickened wall with a shaggy, irregular inner margin. An air-fluid level (arrows) is seen at the base. (From Landay, Christensen, Bynum, et al,[3] with permission from the publisher.)

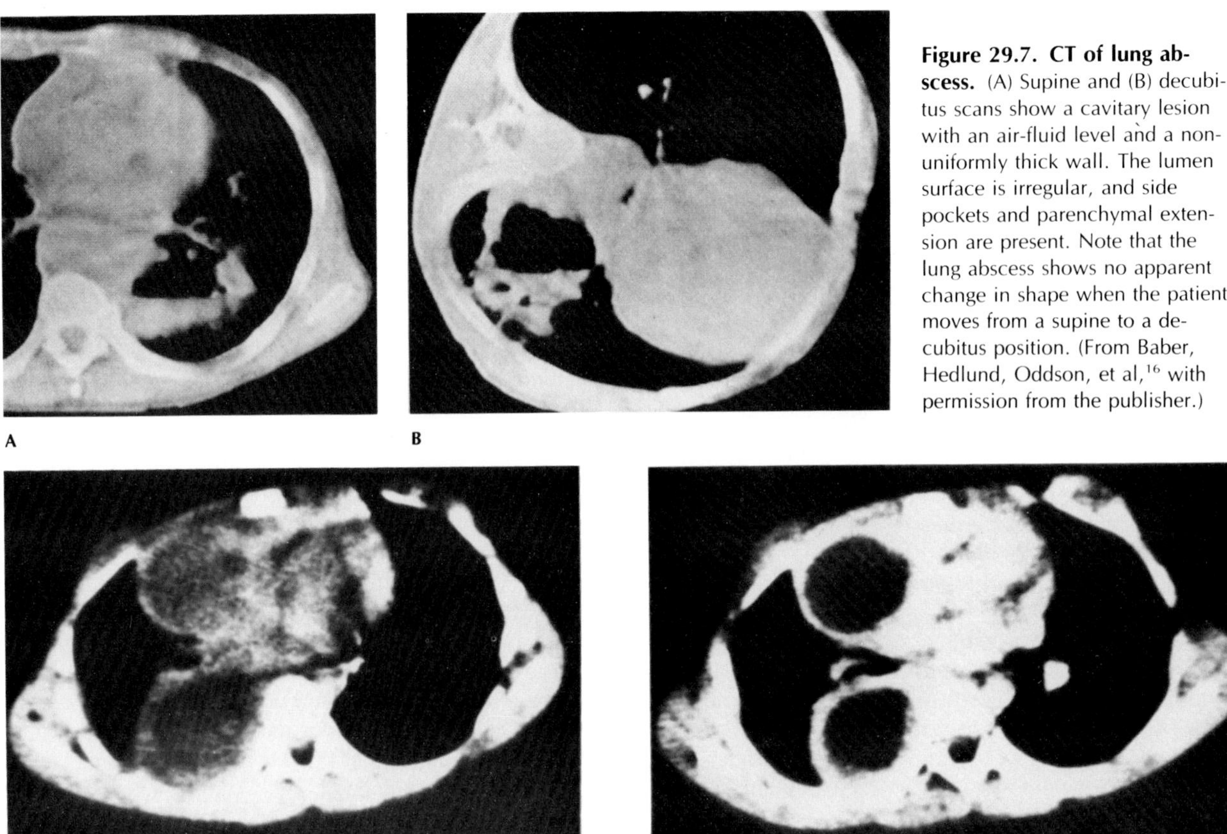

**Figure 29.7. CT of lung abscess.** (A) Supine and (B) decubitus scans show a cavitary lesion with an air-fluid level and a nonuniformly thick wall. The lumen surface is irregular, and side pockets and parenchymal extension are present. Note that the lung abscess shows no apparent change in shape when the patient moves from a supine to a decubitus position. (From Baber, Hedlund, Oddson, et al,[16] with permission from the publisher.)

**Figure 29.8. CT of staphylococcal empyema.** (A) A nonenhanced scan shows two large pleural-based masses and one small lateral mass. The margin between these masses and the adjacent lung is sharply defined and smooth. (B) With contrast enhancement, a narrow band of uniform width is seen surrounding a less dense core. The margin between the core and wall appears to be smooth and regular. (From Baber, Hedlund, Oddson, et al,[16] with permission from the publisher.)

immunodeficiency diseases; cystic fibrosis; and the Kartagener's syndrome, in which bronchiectasis is associated with dextrocardia and sinusitis. Since the advent of antibiotic therapy, the incidence of bronchiectasis has substantially decreased.

The patient with bronchiectasis typically presents with a chronic productive cough, often associated with recurrent episodes of acute pneumonia and hemoptysis. The disease usually involves the basal segments of the lower lobes and is bilateral in about half of the cases. Plain chest radiographs often demonstrate findings highly suggestive of bronchiectasis, though it is important to remember that a normal chest film does not exclude the disease. Coarseness and a loss of definition of interstitial markings in specific segmental areas of the lungs relate to peribronchial fibrosis and retained secretions. An associated loss of volume, due to fibrosis and atelectasis, causes crowding of the interstitial markings. Thickening of bronchial walls, peribronchial fibrosis, and alveolar collapse can produce parallel or slightly tapered line shadows outside the boundary of the pulmonary hila ("tramlines"). In more advanced disease, oval or circular cystic spaces can develop. These cystic dilatations can be up to 2 cm in diameter and often contain air-fluid levels. In very severe disease, coarse interstitial fibrosis surrounding local emphysematous spaces can produce a honeycomb pattern. The uninvolved sections of the lung may demonstrate signs of compensatory overinflation.

Although plain radiographs may strongly suggest the diagnosis of bronchiectasis, bronchography is necessary to unequivocally establish the diagnosis. Three basic bronchographic patterns may be produced. Cylindrical (fusiform) bronchiectasis refers to mild to moderate widening of bronchi that retain a relatively regular outline and tend to end squarely and abruptly. In saccular (cystic) bronchiectasis, there is marked bronchial dilatation that increases progressively toward the periphery, where contrast material pools within the ballooned outline of large cystic spaces. Varicose bronchiectasis is an intermediate form in which the bronchi are irregular and have a beaded appearance simulating varicose veins.

"Reversible" bronchiectasis refers to the bronchial dilatation that often is a manifestation of acute pneumonia. The bronchial dilatation results from retained secretions and atelectasis that almost always accompany a resolving pneumonia. The bronchial tree may not completely return to normal until up to 6 months following the resolution of the pneumonia.[1,6–8]

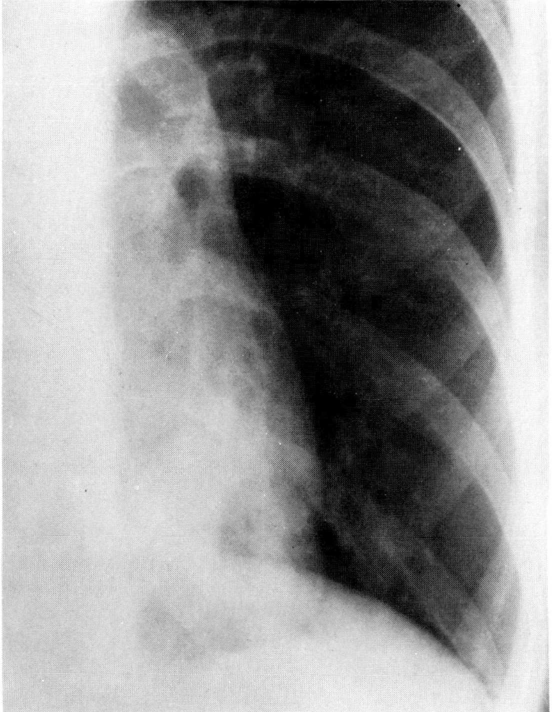

A

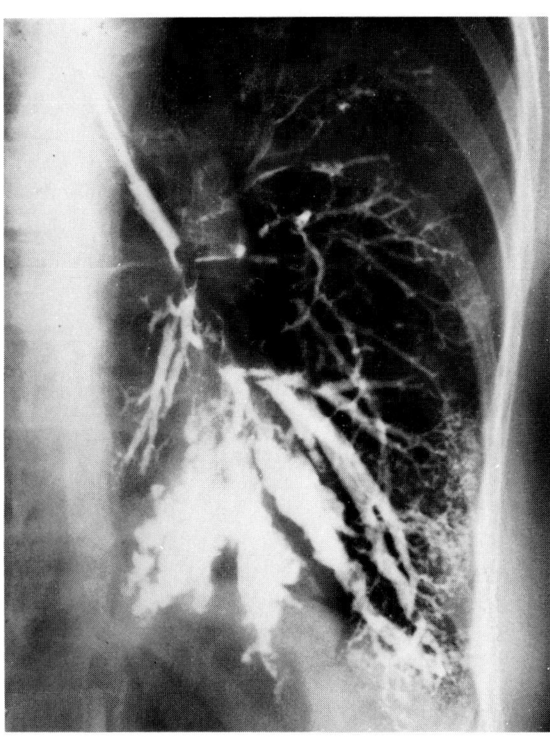

B

**Figure 29.9. Bronchiectasis.** (A) A frontal radiograph of the lower half of the left lung shows several broad, slightly divergent line shadows situated in the bronchovascular distribution of the left lower lobe; the shadows represent retained mucus and pus within bronchiectatic segments. (B) A bronchogram shows severe varicose bronchiectasis of several of the basal bronchi of the left lower lobe, producing a "gloved-fingers" appearance. (From Fraser and Pare,[1] with permission from the publisher.)

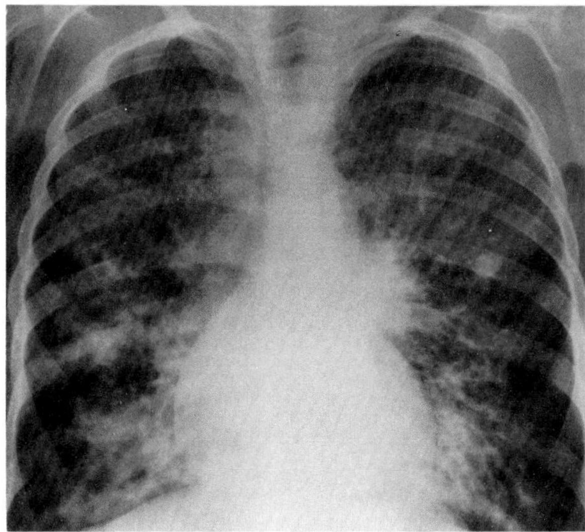

**Figure 29.10. Bronchiectasis.** There is crowding of thickened interstitial markings at the bases, due to fibrosis and atelectasis, in a patient with bronchiectasis secondary to cystic fibrosis. A lesser degree of thickening and crowding of interstitial markings can be seen in the upper lung zones.

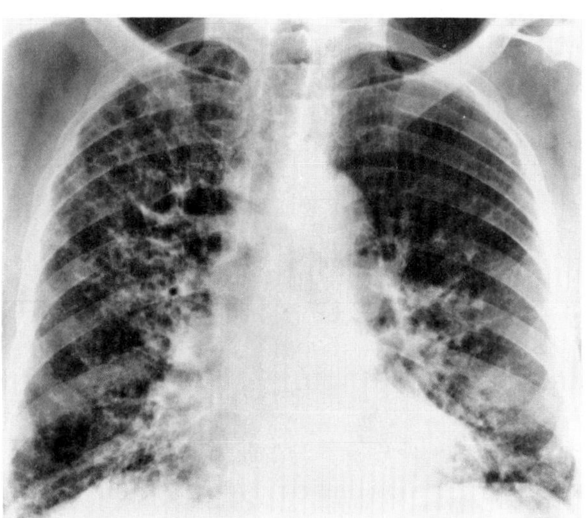

**Figure 29.11. Chronic bronchiectasis.** There is severe coarsening of interstitial markings involving the bases and right upper lobe. Oval and circular cystic spaces, producing somewhat of a honeycomb pattern, are best seen in the right upper lobe.

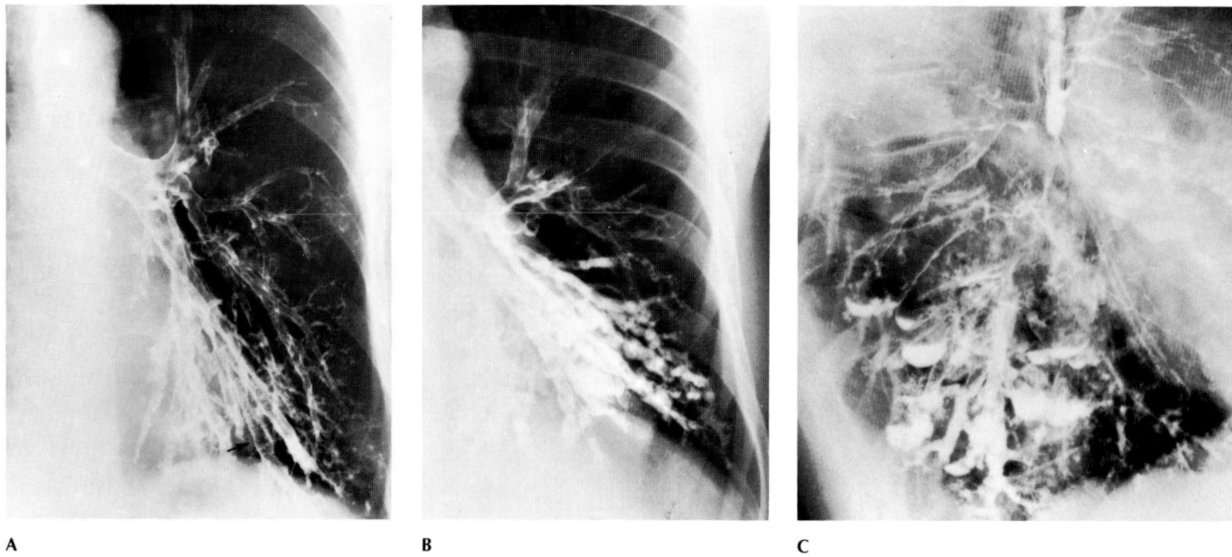


**Figure 29.12. Bronchiectasis.** (A) Cylindrical (fusiform) bronchiectasis. Uniform dilatation of all basal bronchi of the left lower lobe. All bronchiectatic segments end abruptly, and there is little or no peripheral filling. The remainder of the bronchial tree is normal. (B) Varicose bronchiectasis. There is extensive dilatation of all basal bronchi of the left lower lobe. The dilatation is not uniform but is characterized by numerous local constrictions that give the bronchi a configuration resembling varicose veins. There is a notable absence of peripheral filling. (C) Cystic (saccular) bronchiectasis. Numerous cystic spaces containing contrast material in the left lower lung. There are contrast-fluid levels in many areas. (From Fraser and Pare,[1] with permission from the publisher.)

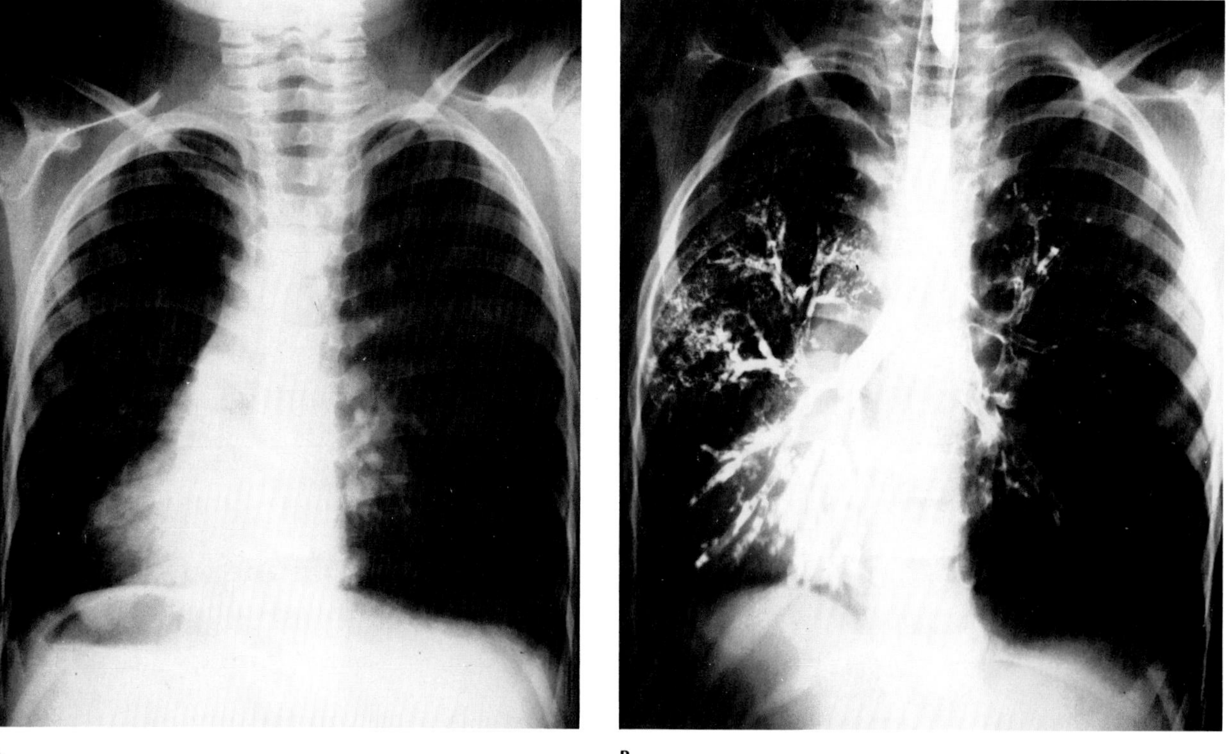


**Figure 29.13. Kartagener's syndrome.** (A) A frontal chest radiograph shows mirror-image dextrocardia with situs inversus. (B) A film from a bronchogram demonstrates bronchiectasis. The combination of situs inversus, bronchiectasis, and sinusitis is known as Kartagener's triad. (From Cooley and Schreiber,[17] with permission from the publisher.)

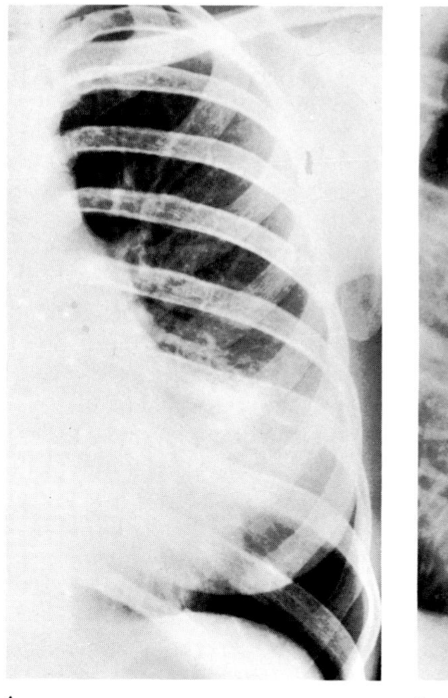
A

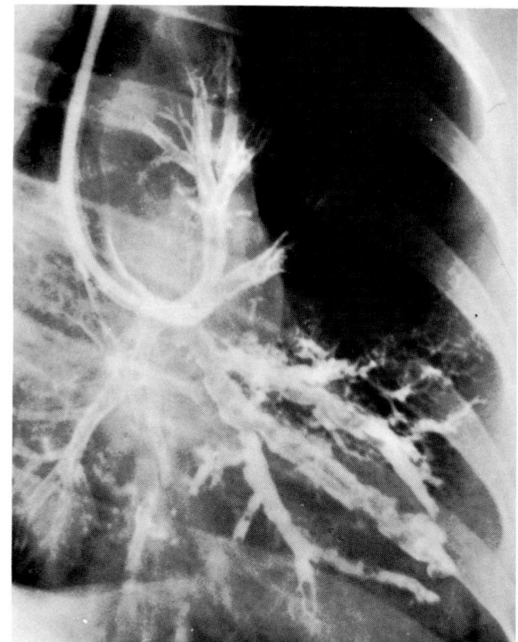
B

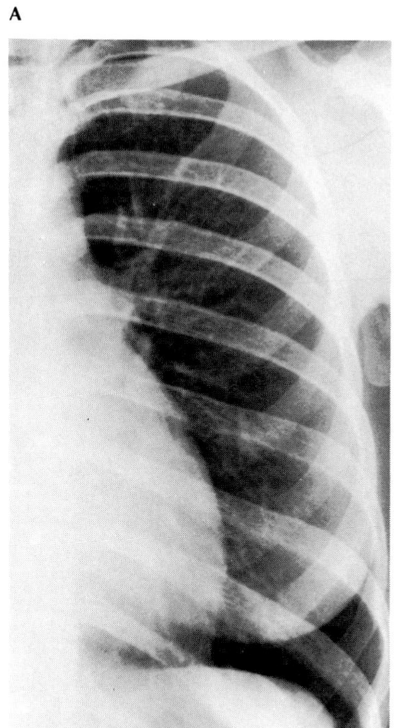
C

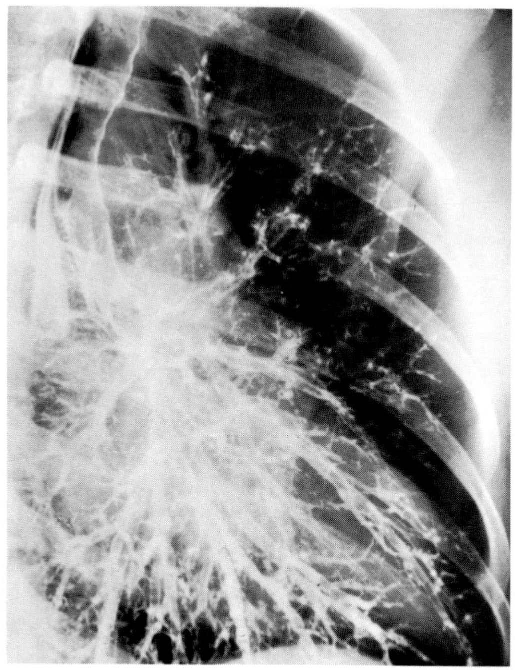
D

**Figure 29.14. Reversible bronchiectasis.** (A) An initial film of a 26-year-old woman shows acute pneumonia of the lingula. During each day of her convalescence, she coughed up a cup of greenish, purulent sputum. Because of this symptom, she was referred for bronchography 3 weeks after the onset of the acute pneumonia. The bronchogram (B) reveals moderate cylindrical and varicose bronchiectasis of both segments of the lingula. Following supportive therapy, her symptoms gradually diminished. (C) A frontal view of the left hemithorax 1 year after the acute pneumonia reveals a normal appearance of the left lung. (D) Bronchography performed at this time shows a normal bronchial tree. The lingular segments show none of the deformity observed a year previously. (From Fraser and Pare,[1] with permission from the publisher.)

## BRONCHOLITHIASIS

Broncholithiasis is usually caused by a calcified bronchopulmonary lymph node that erodes into a bronchus. Therefore, it is generally a late complication of one of the three common granulomatous infections: histoplasmosis, tuberculosis, and coccidioidomycosis. Rare causes of intraluminal broncholiths include calcification of aspirated food or tissue that was retained in the airway a long time, and fragmentation of a calcified bronchial cartilage because of necrosis of the bronchial wall and bronchiectasis.

Erosion of a contiguous calcified granuloma through the bronchial wall is often accompanied by paroxysms of coughing and intermittent hemoptysis. The lodging of a broncholith within a bronchus can cause obstruction with distal atelectasis and infection. A pathognomonic finding is the coughing and expectoration of chalky sediment, sandy particles, or stones.

The diagnosis of broncholithiasis should be suspected

in any patient with recurrent cough and hemoptysis whose chest radiograph shows multiple calcifications in the lung and/or mediastinal lymph nodes. After the patient has expectorated one or more stones, serial radiographs may demonstrate some of the calcified particles to have disappeared. Conventional tomography may be of value in demonstrating the precise relation of the calcification to a bronchus.[9,10]

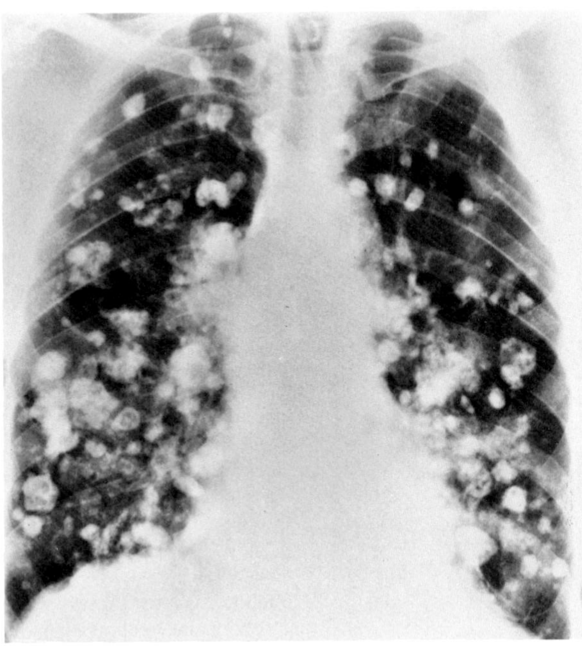

**Figure 29.15. Diffuse broncholithiasis.** Innumerable calcified masses throughout the lungs.

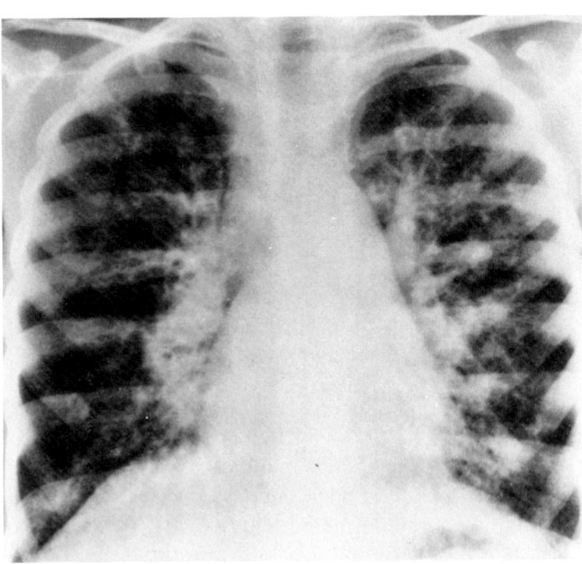

**Figure 29.16. Cystic fibrosis.** Diffuse peribronchial thickening appears as a perihilar infiltrate associated with hyperexpansion and flattening of the hemidiaphragms.

## CYSTIC FIBROSIS (MUCOVISCIDOSIS)

Cystic fibrosis is a hereditary disorder characterized by abnormally thick secretions from exocrine mucous glands; pancreatic insufficiency; and an increased concentration of sodium and chloride in the sweat. Because exocrine glands that secrete mucus are most plentiful in the pancreas and tracheobronchial tree, the radiographic manifestations of cystic fibrosis most commonly involve the respiratory and gastrointestinal systems.

Viscid mucus in the bronchial tree causes a radiographic pattern typical of extensive obstruction of medium-sized and small airways. The inflammatory thickening of bronchial walls produces accentuation of the linear interstitial markings. Air trapping and collateral air drift cause focal areas of overinflation distal to the obstructed segmental bronchi. Recurrent pulmonary infection is common and generally perihilar in distribution, leading to irregular, stringy prominence of markings extending outward from the hila on both sides. Repeated infections may result in bronchiectasis; multiple small bronchiolar abscesses may develop. The prominence of perihilar markings and the involvement of the lung parenchyma contiguous to the heart may cause the nonspecific shaggy heart appearance. In advanced cystic fibrosis, there is a combination of coarse interstitial markings, patchy nodular densities, and marked hyperexpansion of the lungs. Extensive pulmonary infection leading to respiratory failure is the major cause of death.

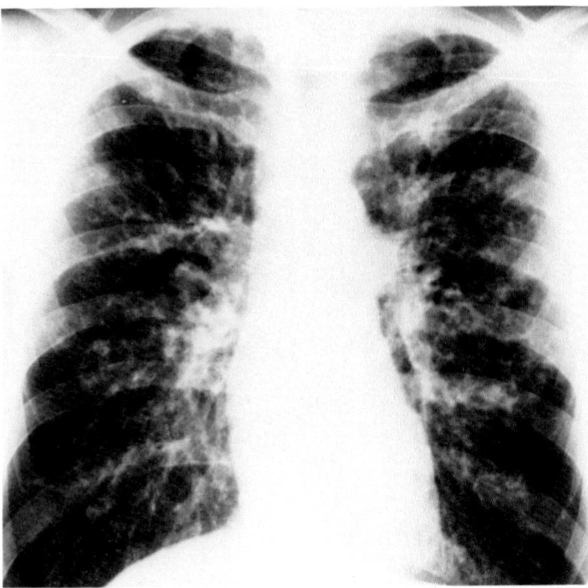

**Figure 29.17. Cystic fibrosis.** There is diffuse, coarse thickening of pulmonary interstitial markings with cyst formation in the periphery. Enlargement of the central pulmonary vessels with relatively rapid tapering is consistent with the patient's pulmonary hypertension. Although the patient is in clinical right-sided heart failure, the heart is small because of the pronounced hyperinflation of the lungs and low position of the hemidiaphragms.

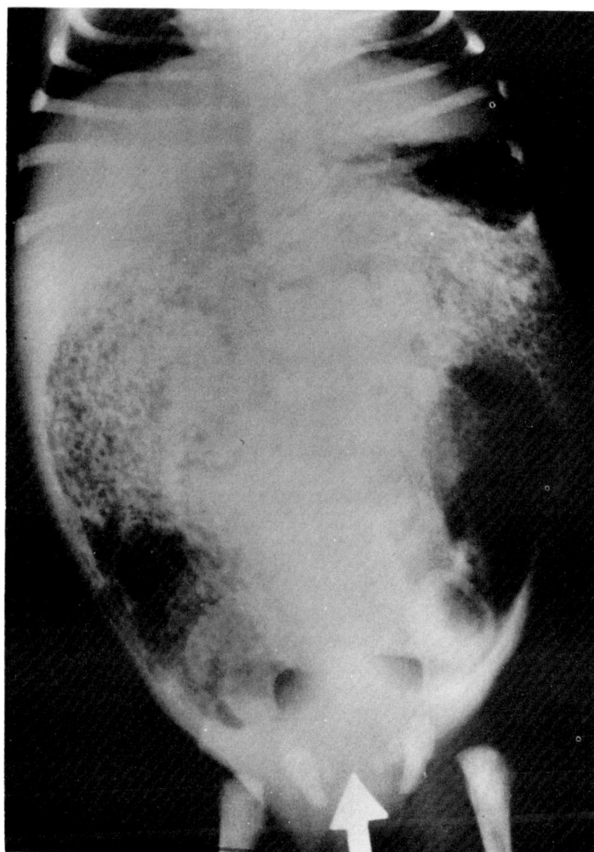

**Figure 29.18. Meconium ileus.** A plain abdominal radiograph of an infant demonstrates massive distention of the small bowel and a profound soap-bubble effect of air mixed with meconium. (From Swischuk,[18] with permission from the publisher.)

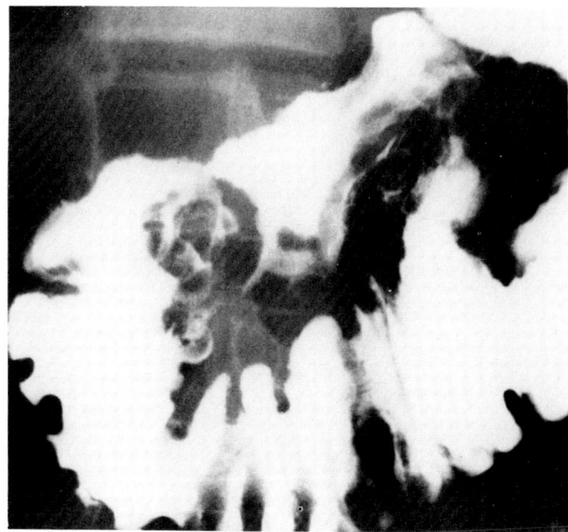

**Figure 29.19. Cystic fibrosis.** There is coarse thickening of the folds in the duodenum. (From Eisenberg,[19] with permission from the publisher.)

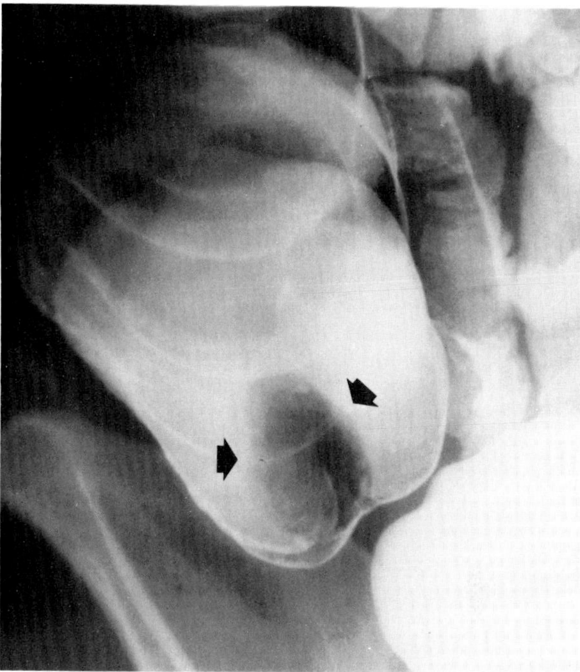

**Figure 29.20. Cystic fibrosis.** An adherent fecalith presents as a tumor-like mass (arrows) in the cecum. (From Eisenberg,[19] with permission from the publisher.)

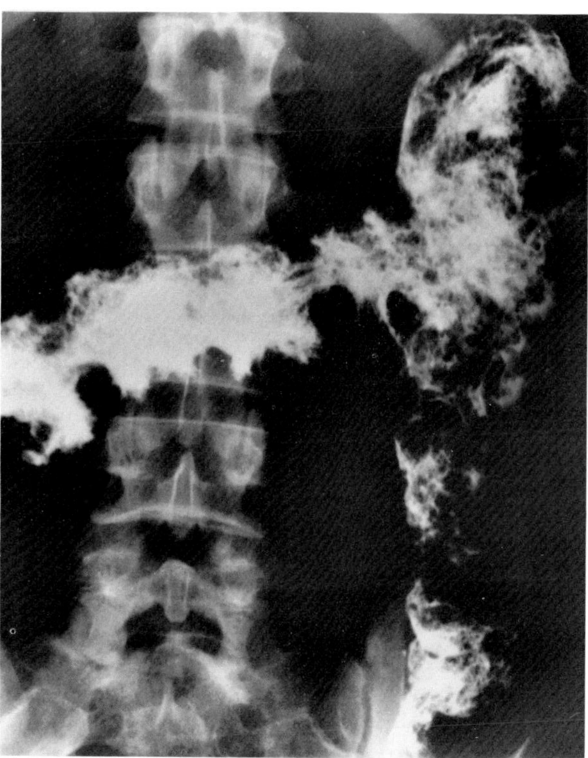

**Figure 29.21. Cystic fibrosis.** Residual secretions produce multiple, poorly defined filling defects in the colon simulating diffuse polyposis.

Cystic fibrosis is the most common cause of meconium ileus, in which a distal small bowel obstruction develops in infants when the thick and sticky meconium cannot be readily propelled through the bowel. The unusually viscid intestinal contents in infants with meconium ileus appear radiographically as a bubbly or frothy pattern superimposed on numerous dilated loops of small bowel. On barium enema examination, the colon is seen to have a very small caliber (microcolon), since it has not been used during fetal life. The escape of meconium into the peritoneal cavity incites a chemical peritonitis, which can lead to the development of multiple small calcific deposits scattered widely throughout the abdomen.

In older children or adults, inspissated intestinal contents in the distal ileum can cause small bowel obstruction, the "meconium ileus equivalent" syndrome that is virtually pathognomonic of cystic fibrosis. The sticky fecal material in patients with cystic fibrosis can form a persistent tumor-like mass in the colon, particularly in the cecum. On barium enema examination, this adherent fecalith can produce a filling defect that often persists on repeat studies and can simulate a colonic neoplasm. An adherent fecalith can occasionally be the leading point of an intussusception. Because of adherent collections of viscid mucus, adequate cleansing is rarely achieved before barium enema examination. The residual thick secretions can cause multiple, poorly defined filling defects that give the colonic mucosa a hyperplastic appearance simulating diffuse polyposis.

A thickened, coarse fold pattern in the duodenum is commonly demonstrated in patients with cystic fibrosis. Associated findings include nodular indentations along the duodenal wall; smudging or poor definition of the mucosal fold pattern; and redundancy, distortion, and kinking of the duodenal contour. These changes are usually confined to the first and second portions of the duodenum, though the thickened fold pattern occasionally extends into the proximal jejunum. Though the cause of duodenal fold thickening in cystic fibrosis is obscure, it is postulated that the lack of pancreatic bicarbonate precludes adequate buffering of the normal amounts of gastric acid, which causes mucosal irritation and muscular contractions that produce the thickened mucosal folds.

A small, contracted, poorly functioning gallbladder can be seen in up to 50 percent of the patients with cystic fibrosis of the pancreas. These microgallbladders have marginal irregularities and multiple weblike trabeculations and are filled with thick, tenacious, colorless bile and mucus. Gallstones are frequently seen in these patients. Inspissation of viscous bile can obstruct the biliary tree and cause cirrhosis of the liver with radiographic manifestations of portal hypertension such as splenomegaly and esophageal varices. Calcification of the pancreas can occur.[11–15,19,20]

## REFERENCES


1. Fraser RG, Pare JAP: *Diagnosis of Diseases of the Chest.* Philadelphia, WB Saunders, 1979.
2. Citro LA, Gordon ME, Miller WT: Eosinophilic lung disease. *AJR* 117:787–792, 1973.
3. Landay MJ, Christensen EE, Bynum LJ, et al: Anaerobic pleural and pulmonary infections. *AJR* 134:233–239, 1980.
4. Sagel SS: Lung, pleura, pericardium, and chest wall, in Lee JKT, Sagel SS, Stanley RJ (eds): *Computed Body Tomography.* New York, Raven Press, 1983.
5. Juhl H: *Essentials of Roentgen Interpretation.* Philadelphia, JB Lippincott, 1981.
6. Reid L: Reduction in bronchial subdivisions in bronchiectasis. *Thorax* 5:233–241, 1950.
7. Gudbjerg CE: Roentgenologic diagnosis of bronchiectasis. An analysis of 112 cases. *Acta Radiol* 43:209–226, 1955.
8. Nelson SW, Christoforidis A: Reversible bronchiectasis. *Radiology* 71:75–79, 1958.
9. Baum, GL, Bernstein IL, Schwarz J: Broncholithiasis produced by histoplasmosis. *Am Rev Tuberc* 77:162–167, 1968.
10. Weed LA, Anderson HA: Etiology of broncholithiasis. *Dis Chest* 37:270–277, 1960.
11. L'Heureux PR, Isenberg JN, Sharp HL, et al: Gallbladder disease in cystic fibrosis. *AJR* 138:953–956, 1977.
12. Taussig LM, Saldino RM, di Sant'Agnese PA: Radiographic abnormalities of the duodenum and small bowel in cystic fibrosis of the pancreas (mucoviscidosis). *Radiology* 106: 369–376, 1973.
13. Berk RN, Lee FA: The late gastrointestinal manifestations of cystic fibrosis of the pancreas. *Radiology* 106:377–381, 1973.
14. Tucker AS, Matthews LW, Doershuk CF: Roentgen diagnosis of complications of cystic fibrosis. *AJR* 89:1048–1059, 1963.
15. Polgar G, Denton R: Cystic fibrosis in adults. Studies of pulmonary function and some physical properties of bronchial mucus. *Am Rev Respir Dis* 85:319–327, 1962.
16. Baber CE, Hedlund LW, Putman CE, et al: Differentiating empyemas and peripheral pulmonary abscesses: Value of computed tomography. *Radiology* 135:755–758, 1980.
17. Cooley RN, Schreiber MH: *Radiology of the Heart and Great Vessels.* Baltimore, Williams & Wilkins, 1978.
18. Swischuk LE: *Radiology of the Newborn and Young Infant.* Baltimore, Williams & Wilkins, 1980.
19. Eisenberg RL: *Gastrointestinal Radiology: A Pattern Approach.* Philadelphia, JB Lippincott, 1983.
20. Phelan MS, Fine DR, Zentler-Munro PL, et al: Radiographic abnormalities of the duodenum in cystic fibrosis. *Clin Radiol* 34:573–577, 1983.

## Chapter Thirty
# Diffuse Lung Disease

## CHRONIC OBSTRUCTIVE LUNG DISEASE

Chronic obstructive lung disease is defined as a condition in which there is chronic obstruction to airflow due to chronic bronchitis and/or emphysema. Chronic bronchitis is characterized by excessive tracheobronchial mucus production leading to the obstruction of small airways. Emphysema refers to the distention of the air spaces distal to the terminal bronchiole caused by the destruction of alveolar walls and the obstruction of small airways. Predisposing factors to chronic obstructive lung disease include cigarette smoking, infection, air pollution, and occupational exposure to noxious agents.

### Chronic Bronchitis

The clinical criterion of chronic bronchitis is a chronic cough with expectoration that occurs on most days during at least 3 consecutive months for more than 2 consecutive years. Although abnormalities in the lungs on chest films may suggest the disease, chronic bronchitis cannot be unequivocally diagnosed radiographically. In addition, 40 to 50 percent of patients with chronic bronchial disease demonstrate no changes on chest radiographs.

The most common radiographic manifestation of chronic bronchitis is a generalized increase in bronchovascular markings, especially in the lower lungs ("dirty chest"). Thickening of bronchial walls and peribronchial inflammation can produce parallel or slightly tapered tubular line shadows ("tramlines") outside the boundary

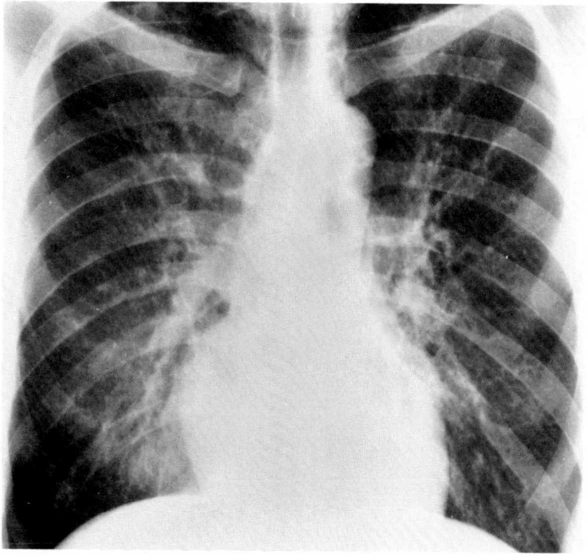

**Figure 30.1. Chronic bronchitis.** A generalized increase in bronchovascular markings produces the "dirty chest" appearance. There is hyperexpansion of the lungs and flattening of the hemidiaphragms.

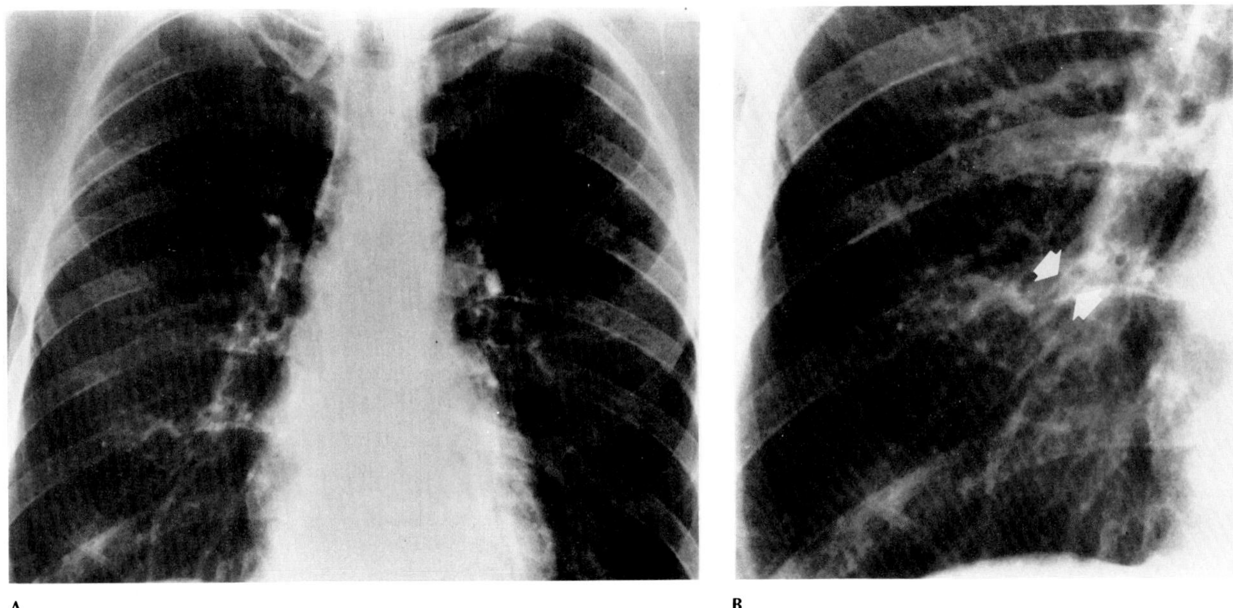


**Figure 30.2. Chronic bronchitis.** (A) Full and (B) coned frontal views of the right lower lung demonstrate parallel line shadows (tramlines) outside the boundary of the pulmonary hilum.

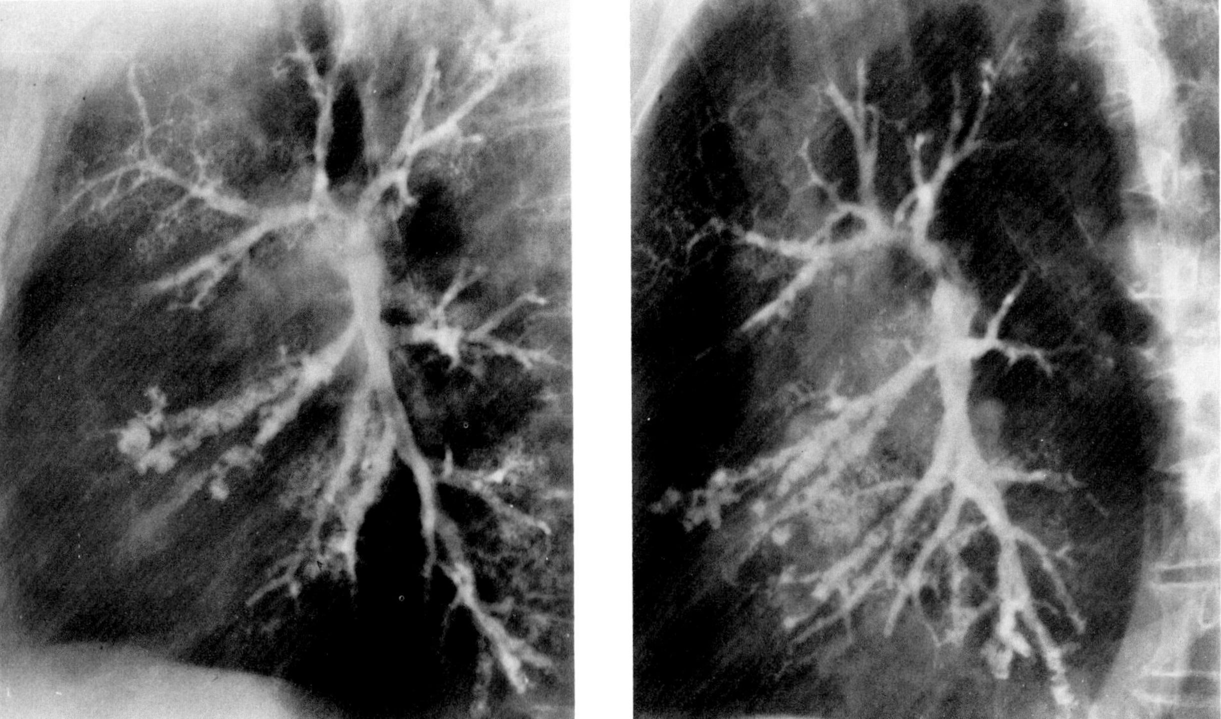

**Figure 30.3. Chronic bronchitis.** Two views from a bronchogram in a patient with severe disease demonstrate areas of saccular dilatation with filling of only the large bronchi and first- and second-generation branches; this produces a leafless-tree appearance. Note the irregularity, or beading, of the bronchial lumina and the small, irregular outpouchings representing dilated ducts of mucous glands.

of the pulmonary hila. However, the mere prominence of basilar markings does not necessarily indicate chronic bronchitis, since there is a wide variation in the normal appearance. Thickening of bronchial shadows can often best be identified when the structures are visualized end-on in the perihilar region. Although overinflation of the lungs is often considered suggestive of chronic bronchitis, this finding should more properly be ascribed to emphysema.

Although certain bronchographic signs are virtually diagnostic of chronic bronchitis, it is doubtful whether bronchography is indicated in patients with clinically suspected chronic bronchitis, except to exclude potentially curable bronchiectasis. Bronchographic evidence of chronic bronchitis includes the following: small, irregular outpouchings, arising from the bronchial lumen, that represent dilated ducts of mucous glands; abrupt termination of smaller bronchi or bronchioles with squared or truncated endings; irregularity, or "beading," of the bronchial lumen; and bulbous cystlike expansion of terminal bronchi (bronchiolectasis).

## Pulmonary Emphysema

There appear to be two basic mechanisms in the pathogenesis of emphysema: airway obstruction and "elastolysis," acting independently or together. The primary site of airway obstruction in emphysema is in the bronchioles, where mucus plugging or bronchiolar destruction of chemical or infectious origin produces increased airway resistance. Collateral air drift permits the ventilation of lung parenchyma served by the obstructed bronchioles. Air trapping leads to alveolar distension and eventually to the rupture of the alveolar septa. Defects in collagen or elastic tissue in older persons probably contribute to the development of emphysema. In young patients, chronic obstructive lung disease is found in a variety of hereditary disorders of connective tissue (osteogenesis imperfecta, Marfan's syndrome, generalized elastolysis). A deficiency of the antiproteolytic enzyme $\alpha_1$ antitrypsin is a hereditary disorder in which the uninhibited destruction of elastic and connective tissue produces severe emphysema that affects young patients and has a striking lower lobe predominance.

The radiographic signs of emphysema are related to pulmonary overinflation, alterations in the pulmonary vasculature (oligemia), and bulla formation.

The hallmark of pulmonary overinflation is flattening of the domes of the diaphragm. Although a low position of the diaphragm is often considered an indication of overexpansion, depression of the diaphragm to the level of the eleventh rib posteriorly is commonly seen at full inspiration in normal persons. A superiorly concave configuration of the diaphragm in adults is virtually diagnostic of emphysema. A similar pattern in children can result from severe acute overinflation of the lungs as in bronchiolitis. Increased size and lucency of the retrosternal air space, the distance between the posterior sternum and the anterior wall of the ascending aorta, is a

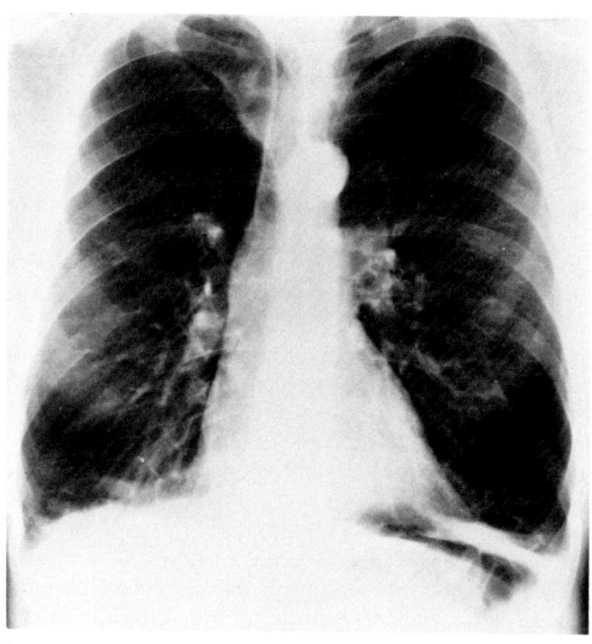

A

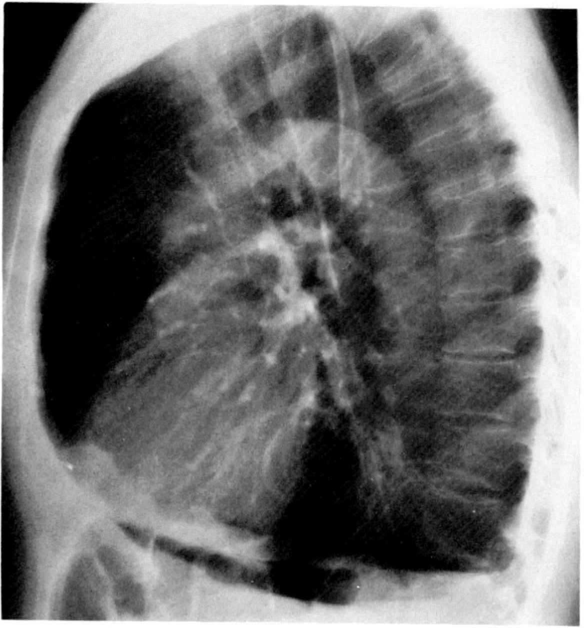

B

**Figure 30.4. Pulmonary emphysema.** (A) Frontal and (B) lateral views of the chest demonstrate severe overinflation of the lungs along with flattening and even a superiorly concave configuration of the hemidiaphragm. There also is an increased size and lucency of the retrosternal air space, an increase in the anteroposterior diameter of the chest, and a reduction in the number and caliber of peripheral pulmonary arteries. The patient was a nonsmoking 30-year-old with $\alpha_1$ antitrypsin deficiency.

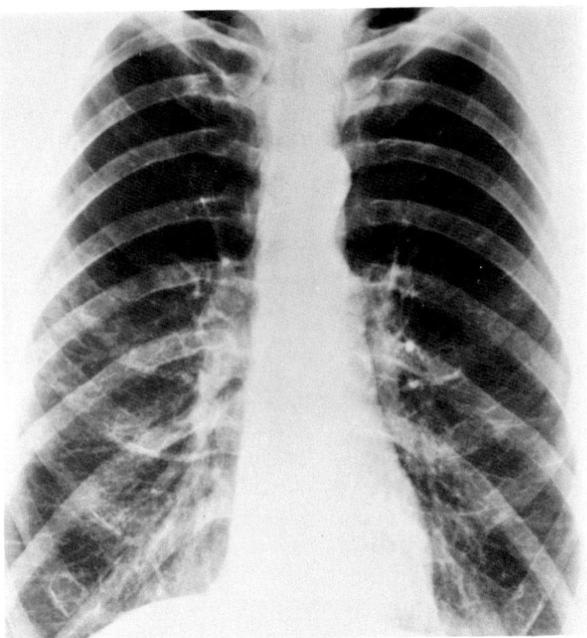

**Figure 30.5. Pulmonary emphysema.** An example of the increased-markings pattern with apical blebs. Note the narrow transverse diameter and vertical orientation of the heart, due to severe pulmonary overinflation.

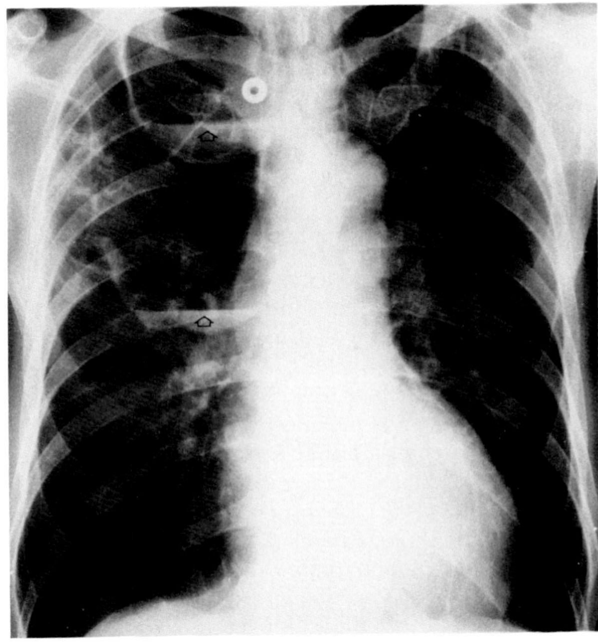

**Figure 30.6. Pulmonary emphysema.** Large bullae in the right upper lung. The presence of air-fluid levels (arrows) within these cystic spaces indicates superimposed infection.

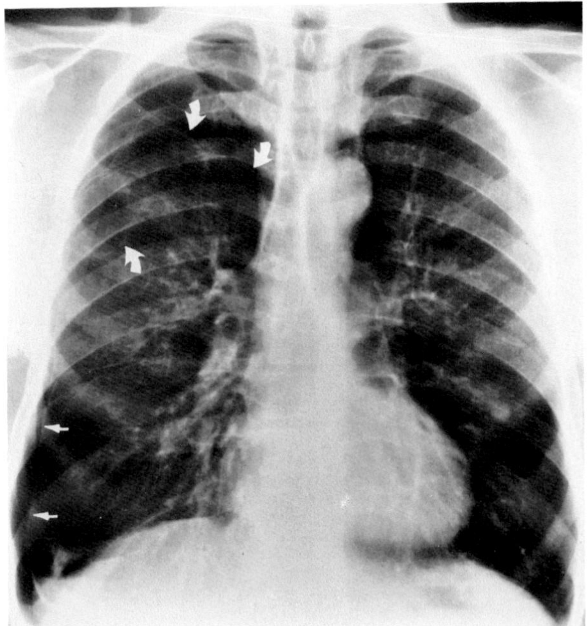

**Figure 30.7. Pulmonary emphysema.** A small right pneumothorax (straight arrows), due to the rupture of a bulla. The curved arrows point to the walls of three of the multiple bullae in the right upper lung.

sign of pulmonary overinflation. In addition, the separation of the aorta from the sternum extends more inferiorly than in a normal person. Other findings that have been described as signs of emphysema include an increase in the anteroposterior diameter of the chest, hyperlucency of the lungs, anterior bowing of the sternum, accentuation of the thoracic kyphosis, and horizontally inclined, widely spaced ribs. Serration of the diaphragm may develop because of muscle fiber hypertrophy or thickening of visceral pleura attached to septa between the basal bullae. Air trapping may be detected fluoroscopically as a restriction of the diaphragmatic excursion to one interspace or less (normally it occurs over two or three interspaces).

A reduction in the number and caliber of the peripheral arteries, best seen on whole-lung conventional tomography, is the major vascular change in patients with emphysema. The vascular diminution is often asymmetric and usually coincides with areas of increased lucency of the lung. The diversion of blood from more severely affected areas causes dilatation of vessels in those segments least involved. Peripheral vascular deficiency is often localized to one or more areas of both lungs, with the vasculature in the remaining portions being normal or increased in caliber. With the development of pulmonary hypertension, prominence of the main and central pulmonary arteries is seen; this further accentuates the appearance of rapid tapering of peripheral vessels. The heart typically has a narrowed trans-

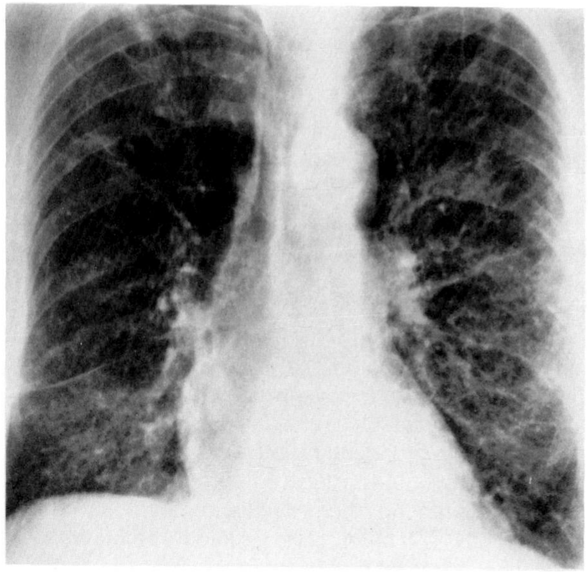
A

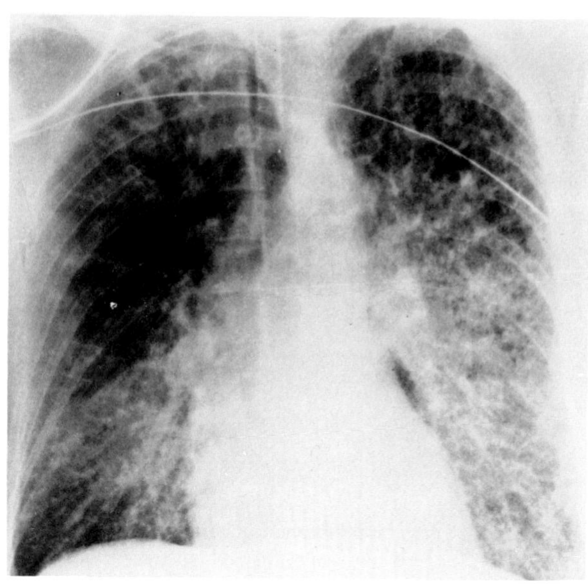
B

**Figure 30.8. Congestive heart failure in pulmonary emphysema.** (A) An initial chest radiograph demonstrates a paucity of vascular markings in the right mid and upper zones along with increased interstitial markings elsewhere. (B) With the onset of congestive heart failure, there is patchy interstitial and alveolar edema that does not affect the segments in which the vascularity previously had been severely diminished.

verse diameter and vertical orientation. Development of cor pulmonale leads to evidence of right ventricular enlargement.

Bullae are local, air-containing cystic spaces within the lung, ranging from 1 to 2 cm in diameter up to an entire hemithorax. They may be single or multiple, and their walls are usually of hairline thickness, so it may be difficult to distinguish them from involved parenchyma. The large, radiolucent, air-filled sacs are found predominantly at the apices or at the bases. They are essentially devoid of lung markings, though they may contain fine fibrous septa. Bullae may become so large that they cause respiratory insufficiency by compression of the remaining, relatively normal lung; a single huge lesion may even mimic a pneumothorax.

A less common radiographic appearance of emphysema is the increased-markings pattern. Instead of being attenuated and diminished in caliber, the vascular markings in this condition are more prominent than normal and tend to be irregular in contour and indistinct in definition. This "dirty chest" appearance suggests severe chronic bronchitis. The increased-markings type of emphysema is almost invariably associated with signs of pulmonary arterial hypertension leading to right ventricular enlargement.

At bronchography, the general overinflation in emphysema results in typical splaying of the bronchial radicals and widening of their angles of bifurcation. Contrast material fills small rounded structures, probably representing enlarged emphysematous spaces, at the termination of peripheral bronchioles. Bronchioles often end abruptly with a squared, rounded, or tapered configuration. Several blind bronchioles may appear to arise from a single point, producing a characteristic "spider" deformity.

The development of left-sided heart failure in patients with emphysema can produce confusing patterns of pulmonary edema. Interstitial edema can cause the diminution or disappearance of signs of overinflation, especially in the lower lung. Rather than the typical diffuse alveolar pattern of pulmonary edema, in patients with emphysema there is patchy involvement that does not affect those segments in which the vascularity has been severely diminished.[1–8]

## CONGENITAL (INFANTILE) LOBAR EMPHYSEMA

Congenital lobar emphysema is characterized by severe respiratory distress due to the overinflation of a pulmonary lobe, usually either an upper lobe or the right middle lobe. The disorder reflects bronchial obstruction that can be due to pressure from an abnormal vessel, the malformation of bronchial cartilage, redundant bronchial mucosa, or an unrecognized infection. Although congenital lobar emphysema is often considered a surgical emergency, the clinical and radiographic manifestations may regress spontaneously without the need for thoracotomy and lobar resection.

Congenital lobar emphysema classically appears on chest radiographs as marked radiolucency and increased

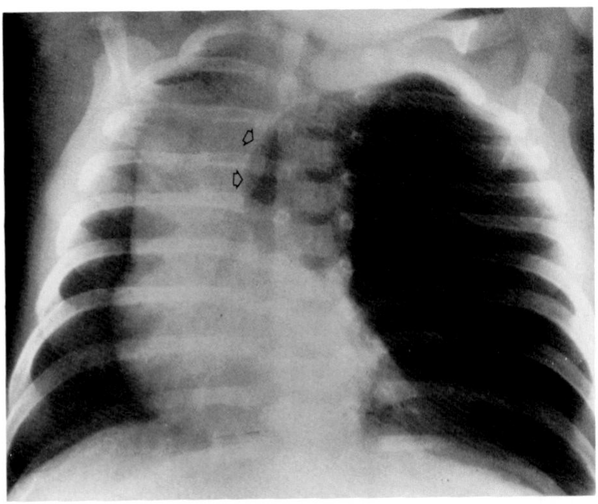

**Figure 30.9. Congenital lobar emphysema.** Severe overdistension of the left upper lobe causes marked radiolucency of the left hemithorax along with depression of the ipsilateral hemidiaphragm and displacement of the mediastinum into the right hemithorax. The hyperinflated left upper lobe has herniated into the right side of the chest (arrows). Note the small and widely separated bronchovascular markings in the lucent left lung.

volume of the affected lobe. There is depression of the ipsilateral hemidiaphragm and displacement of the mediastinum into the contralateral hemithorax. The overdistended lung segments may herniate across to the normal side. Compression atelectasis of other lobes of the ipsilateral lung may occur. Air trapping causes further displacement of the heart and mediastinum toward the normal (contralateral) side during expiration and their partial return toward the midline during inspiration. Vascular markings in the affected lobe are widely separated and small. The identification of vascular markings is important because it permits differentiation from congenital air cysts, loculated pneumothorax, and postpneumonic pneumatocele. Infrequently, the involved lobe is opaque, probably due to an accumulation of intrapulmonary fluid distal to the bronchial obstruction. This fluid usually drains spontaneously, leaving the overexpanded lobe radiolucent.[9,10]

## UNILATERAL HYPERLUCENT LUNG (SWYER-JAMES OR MACLEOD'S SYNDROME)

Unilateral hyperlucent lung is a condition that is usually detected in childhood and is probably related to repeated infections that result in obliterative bronchitis and bronchiolitis. The peripheral parenchyma is ventilated by collateral air drift; chronic expiratory obstruction leads to air trapping and overdistension.

The radiographic manifestations of unilateral hyperlucent lung are virtually pathognomonic. Chest films obtained at maximal inspiration demonstrate a striking difference in the radiolucency of the two lungs or, in the segmental form, between the affected lobe and the uninvolved segments of lung. The increased lucency is not caused by a relative increase in air, but rather is due to decreased perfusion with the peripheral pulmonary markings being severely narrowed and attenuated. The volume of the affected lobe or lung is either comparable or reduced, but not enlarged. Because of airway obstruction and air trapping, the involved portion of lung does not expand or contract normally on respiration. Therefore, the mediastinum shifts away from the affected side on expiration and toward the affected side on inspiration. The motion of the associated hemidiaphragm and rib cage is also decreased.

Bronchography demonstrates the irregular dilatation and abrupt squared or tapered termination of involved segmental bronchi. The peripheral bronchiolar radicals infrequently fill with contrast material, and there is no evidence of obstruction of the larger bronchi. At arteriography, the pulmonary artery to the affected lung is small and there is greatly reduced circulation.[1,11,12]

## DIFFUSE INFILTRATIVE DISEASES

Many pathologic conditions produce diffuse pulmonary involvement affecting all lobes of both lungs. These disorders produce several recognizable radiographic patterns that may be of value in making the correct diagnosis.

The *acinar*, or *alveolar*, pattern refers to air space consolidations that appear as small densities with ill-defined, irregular margins. In many patients there are areas of confluence of acinar shadows separated by normal air-containing parenchyma; this produces a pattern of numerous fluffy, poorly defined densities throughout the lungs. Confluent alveolar densities may involve the central and midportions of the lungs more severely than the peripheral regions, and this results in a *butterfly* or *batwing* pattern. A pathognomonic sign of alveolar consolidation is the air bronchogram, indicating the presence of a patent small bronchus coursing through the area of parenchymal consolidation. Major causes of a diffuse acinar pattern include pulmonary edema and hemorrhage, pneumonia (often of unusual cause), hyaline membrane disease, the bronchogenic spread of tuberculosis or fungal disease, alveolar proteinosis and microlithiasis, alveolar cell carcinoma, lymphoma, and the alveolar form of sarcoidosis.

*Interstitial* lung disease refers to cellular (inflammatory or neoplastic) or noncellular (edema, collagen, or other substances) infiltration into the supporting structures of the lung (bronchial walls, blood vessels, lymphatics). Interstitial infiltrates may appear as fine reticular or linear densities that appear to accompany the bronchovascular markings; indeed, minimal interstitial changes may be mistaken for an accentuation of normal pulmonary mark-

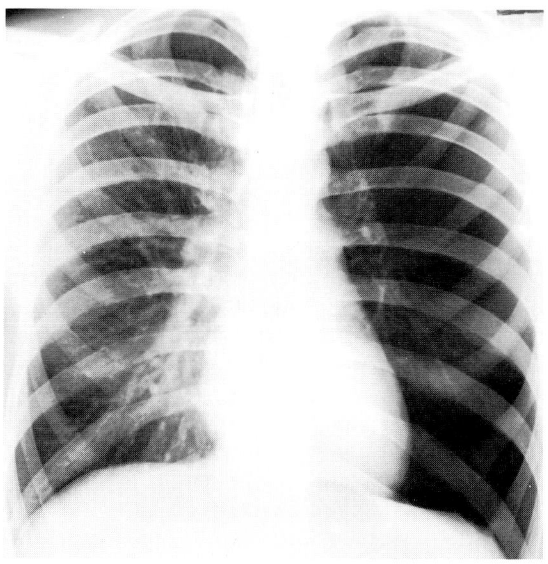

A

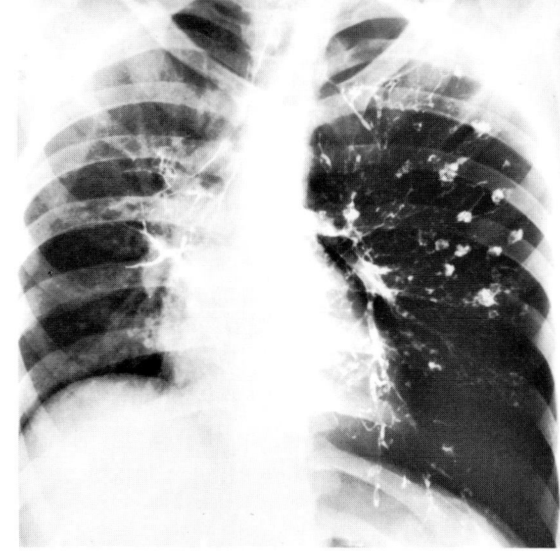

B

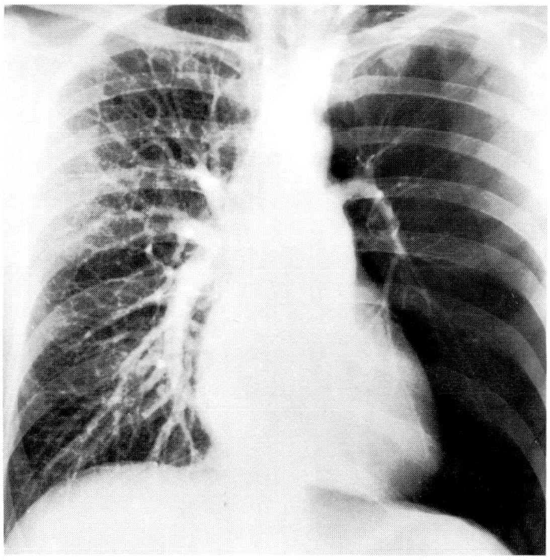

C

**Figure 30.10. Unilateral hyperlucent lung.** (A) A frontal radiograph exposed at total lung capacity reveals a marked discrepancy in the radiolucency of the two lungs, the left showing severe oligemia. The left hilar shadow is diminutive. The left lung appears to be of approximately normal volume compared with the right. (B) A frontal radiograph at residual volume following bronchography demonstrates severe air trapping in the left lung and little change in volume from total lung capacity. Since the deflation of the right lung is normal, the mediastinum has swung sharply to the right. The bronchial tree shows less bronchiectasis than usual with this condition, although there is filling of numerous large spaces typical of emphysema. (C) A pulmonary arteriogram shows the discrepancy in blood flow to the two lungs. The left pulmonary artery is present, although diminutive, differentiating this appearance from congenital absence of the left pulmonary artery. (From Fraser and Pare,[1] with permission from the publisher.)

ings. In nodular interstitial disease (e.g., disseminated hematogenous infections such as miliary tuberculosis), the radiographic pattern consists of innumerable discrete nodular densities that range from barely detectable to up to 5 mm in diameter. Unlike the irregular, ill-defined acinar (alveolar) densities, the individual interstitial nodule is round or slightly oval and has clearly defined borders. A mixture of nodular deposits and diffuse linear thickening throughout the interstitial space produces the *reticulonodular* pattern.

An interstitial disease that involves the interlobular septa (e.g., pulmonary edema, lymphangitic metastases) produces Kerley lines, three types of linear shadows that are related to lymphatic dilatation. The most frequent of these are the B lines, short (1 to 3 cm), thin (1 to 2 mm), faint linear shadows that are arranged in a horizontal stepladder pattern and are seen best in the lateral portion of the lung bases extending to the pleura.

A sign of severe interstitial fibrosis is the *honeycomb lung*, which refers to a very coarse reticular pattern in which the air spaces between the thickened linear interstitial opacities measure 5 mm or more in diameter.

To aid in differential diagnosis, some authors have described five basic radiographic patterns of diffuse interstitial pulmonary disease. The *connective tissue* pattern consists of reticular, nodular, and reticulonodular densities that may progress to a classic honeycomb pattern. It is seen most often in patients with sarcoidosis, pneumoconiosis, connective tissue disease, histiocytosis X, and interstitial fibrosis. The addition of septal (Kerley) lines to the reticulonodular pattern of interstitial disease produces the *lymphatic* pattern, which is relatively specific for pulmonary edema and lymphangitic metastases. The *bronchial* pattern, seen in bronchiectasis, chronic bronchitis, and cystic fibrosis, refers to a mixture of parallel lines, ring shadows, and coarse reticulation. The *miliary*, or *deposition*, pattern is composed of miliary mottling, small nodulations, and reticular changes and is seen most commonly in occupational diseases, the hematogenous spread of tuberculosis or metastases, and pulmonary hemosiderosis. A fifth, *vascular* pattern refers to engorgement of the normal arterial and venous markings in such conditions as pulmonary edema, embolic disease, and the arteritis of connective tissue disorders.[13]

## Histiocytosis X

Histiocytosis X refers to three disorders of unknown etiology (Letterer-Siwe disease, Hand-Schüller-Christian disease, eosinophilic granuloma) in which there is proliferation of, and infiltration by, histiocytes into various tissues. These conditions may produce diffuse and bilaterally symmetric pulmonary disease that, unlike many other diffuse diseases of the lungs, does not predominantly affect the bases and tends to be more extensive in the upper zones. In the granulomatous, or active, stage of the disease, small individual lesions produce a nodular pattern. In later stages the disease becomes more reticulonodular and eventually produces a very coarse reticular pattern, especially in the upper lung zones, that often results in a typical cystic appearance of the honeycomb lung. Spontaneous pneumothorax often occurs, due to the rupture of a subpleural emphysematous bleb, and may be the first indication of pulmonary disease.[14–16]

## Idiopathic Pulmonary Hemosiderosis and Goodpasture's Syndrome

Idiopathic pulmonary hemosiderosis is an uncommon disease that generally affects infants and children and is rare in persons more than 20 years of age. Bleeding into the lung interstitium causes iron deficiency anemia and hemoptysis and the typical finding of hemosiderin-filled macrophages in the sputum. The condition consists of repeated episodes of pulmonary hemorrhage with intervening periods of remission.

The radiographic appearance primarily depends on the number of hemorrhagic episodes that have occurred. Initially, there is patchy air space consolidation that is usually diffusely scattered throughout the lungs but may be most prominent in the perihilar region, where it may simulate pulmonary edema. Hilar adenopathy occasionally develops. Within a few days after the acute episode, blood in the alveolar spaces is transported by macrophages to the interstitial space and lymphatics. Radiographically, this is associated with the disappearance of the fluffy alveolar consolidation and its replacement by a reticular pattern with a distribution identical to that of the air space disease. The reticular pattern gradually diminishes, and the chest radiograph usually returns to normal within 2 weeks following the initial acute episode. With repeated episodes, an increased amount of hemosiderin deposited in the interstitial tissue incites progressive interstitial fibrosis. Thus there is only partial clearing after each exacerbation, and a fine reticular pattern persists on serial chest radiographs. The prognosis of idiopathic pulmonary hemosiderosis is poor; death usually is the result of an acute episode of bleeding, though it may be related to respiratory insufficiency due to extensive interstitial fibrosis.

An identical radiographic appearance may develop in Goodpasture's syndrome, in which pulmonary hemorrhage is associated with glomerulonephritis. Unlike idiopathic pulmonary hemosiderosis, which primarily occurs in children, Goodpasture's syndrome usually oc-

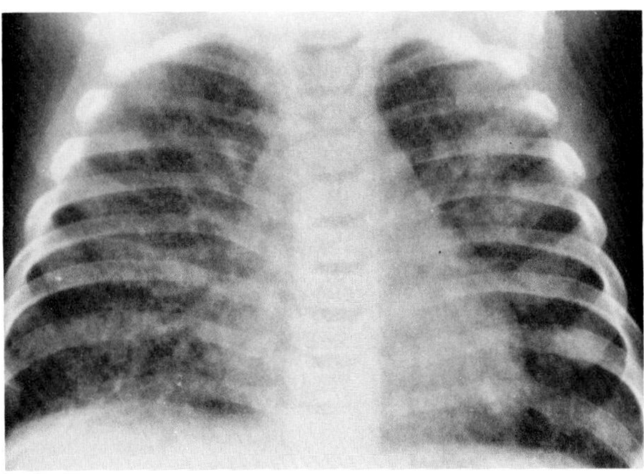

**Figure 30.11. Letterer-Siwe disease.** Diffuse and bilaterally symmetric infiltrates have a predominantly nodular pattern. Some areas of confluence are seen.

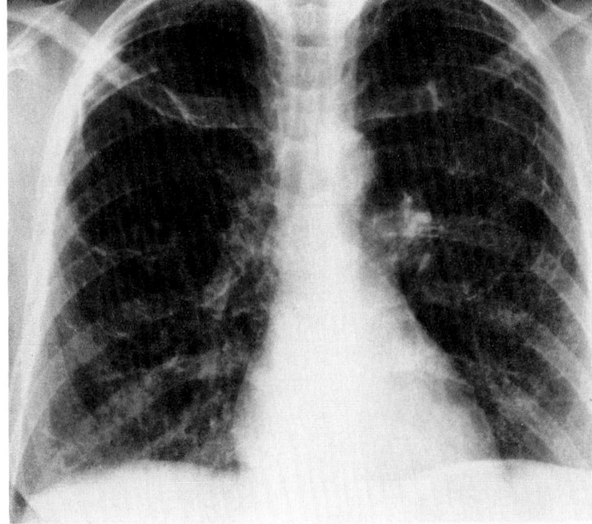

**Figure 30.12. Histiocytosis.** A coarse reticular pattern with pronounced bleb formation in the upper zones.

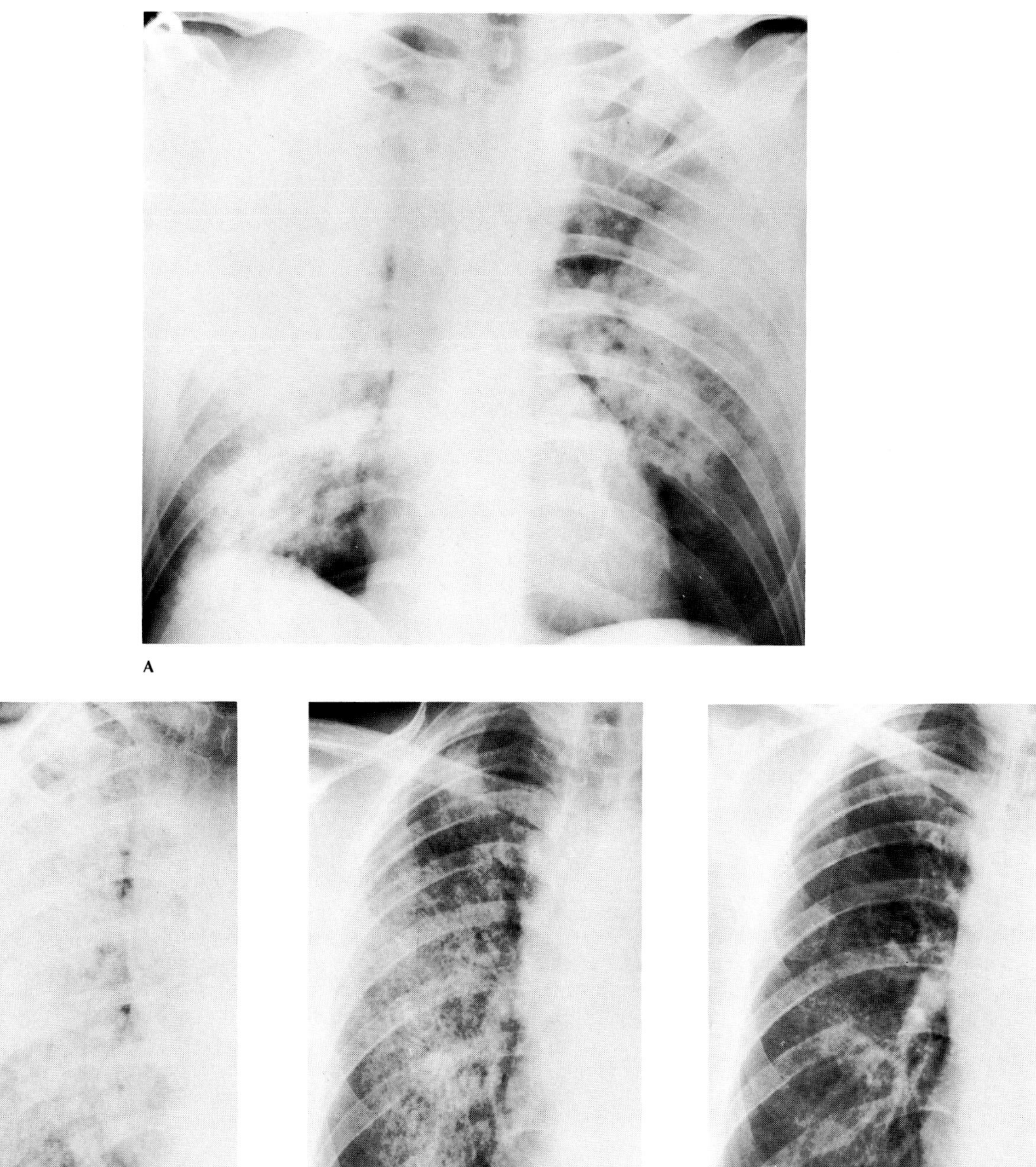

A

B C D

**Figure 30.13. Goodpasture's syndrome.** (A) A frontal chest film in a patient with massive pulmonary hemorrhage demonstrates extensive bilateral pulmonary consolidation, which is confluent in most areas. The distribution and homogeneity of the disease, and the presence of well-defined air bronchograms, indicate air space consolidation from either edema or hemorrhage. In view of the normal heart size, the latter was considered the more likely possibility. (B) Three days later, the pattern is somewhat granular. (C) Ten days after the initial episode, the pattern has become distinctly reticular. (D) Six days later, only a fine reticular pattern remains in an anatomic distribution identical to that of the original involvement. One month later, a fresh episode of massive pulmonary hemorrhage resulted in death from respiratory insufficiency; at necropsy, massive pulmonary hemorrhage, subacute glomerulonephritis, and necrotizing arteritis and thrombosis of small splenic arteries were found; the final diagnosis was Goodpasture's syndrome. (From Fraser and Pare,[1] with permission from the publisher.)

curs in young adult men. Goodpasture's syndrome is most likely an autoimmune disease, and circulating antibodies against glomerular and alveolar basement membrane are usually found in patients with this condition.[1,17–19]

## Pulmonary Alveolar Proteinosis

Pulmonary alveolar proteinosis is a rare disease of unknown etiology characterized by the progressive deposition of a granular lipoprotein material within the air spaces of the lung. This results in a typical radiographic pattern of bilateral and symmetric air space consolidation that often appears to extend outward from the hilum in a butterfly or batwing pattern. Although this appearance may closely simulate pulmonary edema, the normal heart size and the absence of other signs of pulmonary venous hypertension permit the differentiation between these two disorders. A confluence of the alveolar densities often produces irregular, poorly defined patchy shadows scattered throughout both lungs.

In contrast to the relatively uniform and extensive radiographic changes in the lungs, the clinical manifestations of pulmonary alveolar proteinosis are varied and often mild. The disease may improve either spontaneously or following bronchial lavage or repeated intermittent positive-pressure breathing. Pulmonary alveolar proteinosis is fatal in about one-third of the cases; death is caused by respiratory failure or superimposed infection (primarily due to oppportunistic organisms, especially *Nocardia*).[1,20,21]

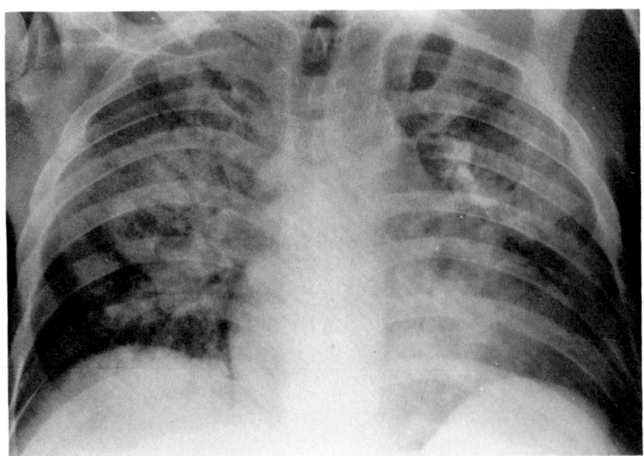

**Figure 30.14. Pulmonary alveolar proteinosis.** Diffuse, bilateral air space consolidation predominantly involves the central portions of the lung and simulates pulmonary edema. The patient was asymptomatic, and serial radiographs over several months showed little change.

## Pulmonary Alveolar Microlithiasis

Pulmonary alveolar microlithiasis is a rare disorder in which innumerable tiny calcified particles are distributed diffusely throughout the alveoli of both lungs. There is a striking lack of association between the radiographic and clinical findings in this condition. Most patients are asymptomatic, and the disease is only discovered on a screening chest radiograph; dyspnea on exertion, cough, and respiratory insufficiency may develop.

Pulmonary alveolar microlithiasis presents the pathognomonic radiographic appearance of fine, sandlike micronodules diffusely scattered throughout both lungs. Unlike the miliary nodular pattern seen in tuberculosis, pneumoconiosis, and alveolar sarcoidosis, in pulmonary alveolar microlithiasis the nodules are especially dense and sharply defined. The overall density of the lungs is greatest in the lower zone, due to the increased thickness of the lung in this region rather than selectively greater involvement. The microliths may cause a basilar density so pronounced that it completely obliterates the cardiac margins (vanishing heart phenomenon). A classic finding is the appearance of the water density of the pleura as a black line instead of a white one (black pleura sign), because the pleura lies between the bony rib cage and the extremely dense calcified pulmonary infiltrate.[22–24]

## Diffuse Interstitial Pulmonary Fibrosis

A broad spectrum of pathologic processes can produce the identical pattern of *diffuse interstitial pulmonary fibrosis*. These disorders are often lumped together under the term *usual organizing interstitial pneumonia* (UIP), which describes the predominance of fibrotic reaction with a minimal cellular infiltrate that occurs in all of them. The precise cause of UIP can be identified in about half of the patients and may be due to such conditions as infections, irradiation, oxygen inhalation, prolonged circulatory failure, and connective tissue disorders. The other patients, in whom a specific diagnosis cannot be made, are generally described as having diffuse idiopathic pulmonary fibrosis. Although the term *Hamman-Rich disease* is sometimes used interchangeably with *diffuse idiopathic pulmonary fibrosis,* the former term should more properly be applied only in patients with a specific pattern of acute, rapidly progressive, diffuse interstitial pneumonitis with fibrosis.

The earliest radiographic abnormalities in diffuse interstitial pulmonary fibrosis consist of fine linear, reticular streaks that are predominantly found in the lung bases. With increasing fibrosis, the reticular or reticulonodular pattern becomes more coarse and may progress to the end-stage appearance of the classic honeycomb lung. Serial radiographs over several years often demonstrate a progressive loss of lung volume with an increasing elevation of the diaphragm.[13]

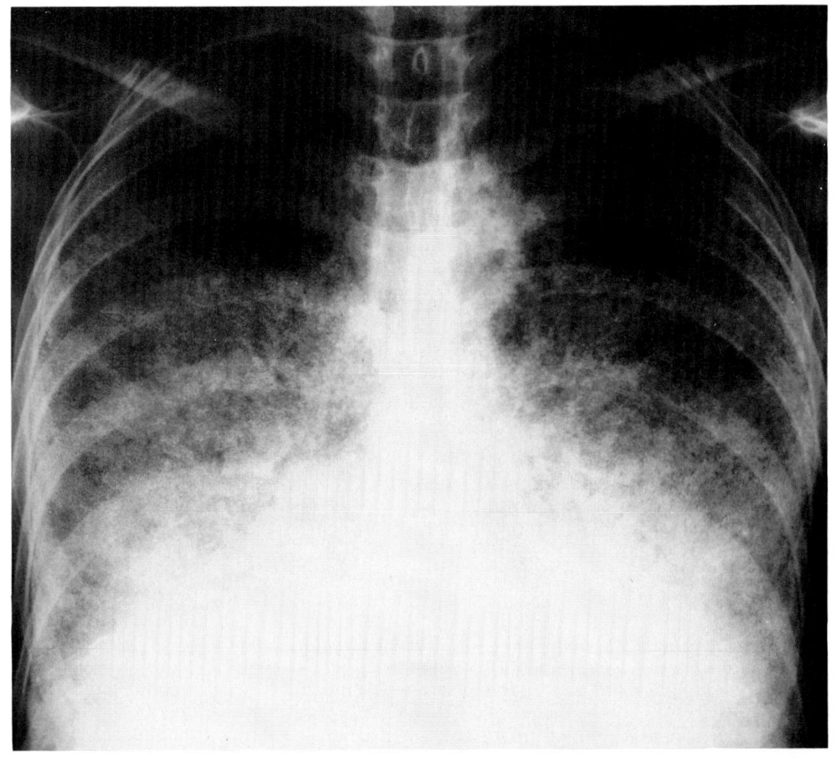

A

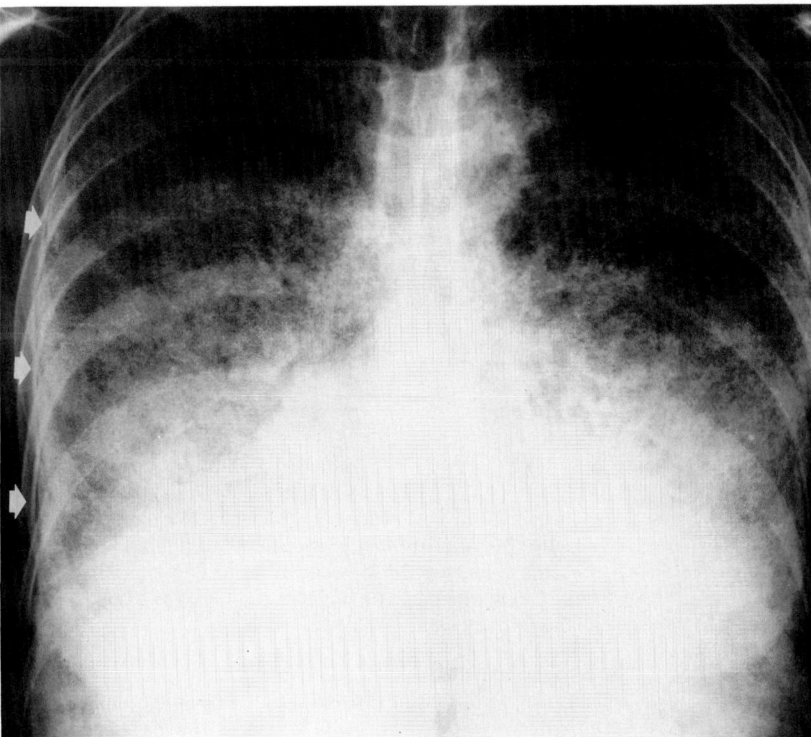

B

**Figure 30.15. Pulmonary alveolar microlithiasis.** (A) A full chest radiograph and (B) a penetrated coned view of the right upper lung demonstrate the nearly uniform distribution of typical fine, sandlike mottling in this patient with alveolar microlithiasis. The tangential shadow of the pleura is displayed along the lateral wall of the chest as a dark, lucent strip (arrows). The strip appears this way because of its relative radiotransparency (water density) when seen between the ribs and the markedly radiopaque lungs that are characteristic of this disease. (From Theros, Reeder, and Eckert,[18] with permission from the publisher.)

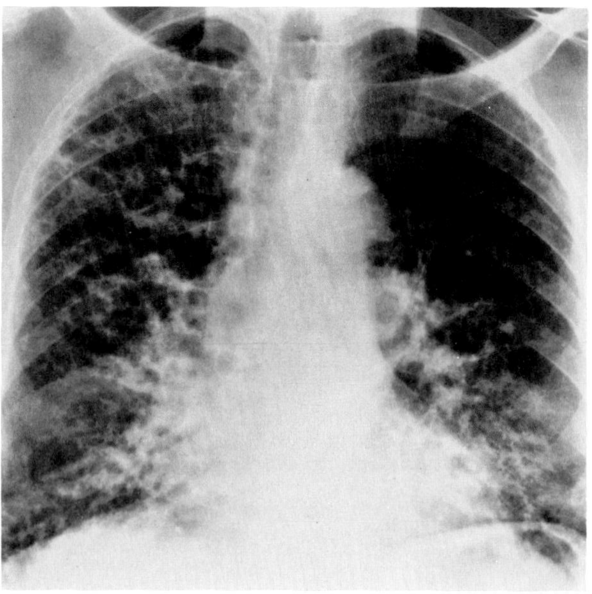


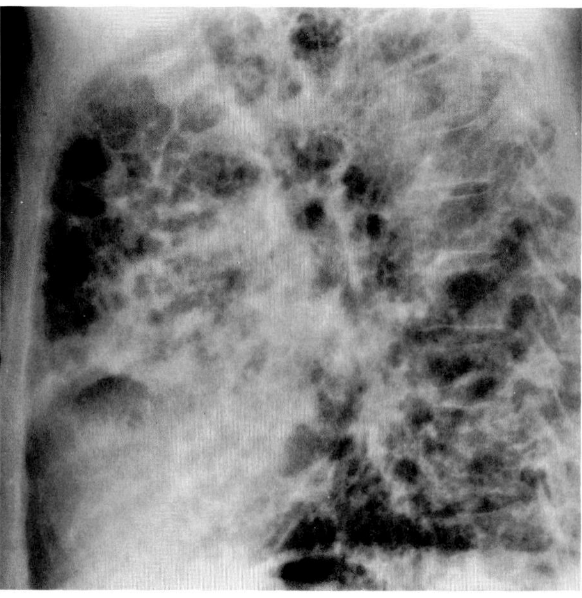


**Figure 30.16. Diffuse interstitial pulmonary fibrosis.** (A) Frontal and (B) lateral views of the chest demonstrate a coarse reticular pattern indicating pronounced fibrosis. Intervening small areas of lucency produce the appearance of a honeycomb lung, especially in the right upper lobe.

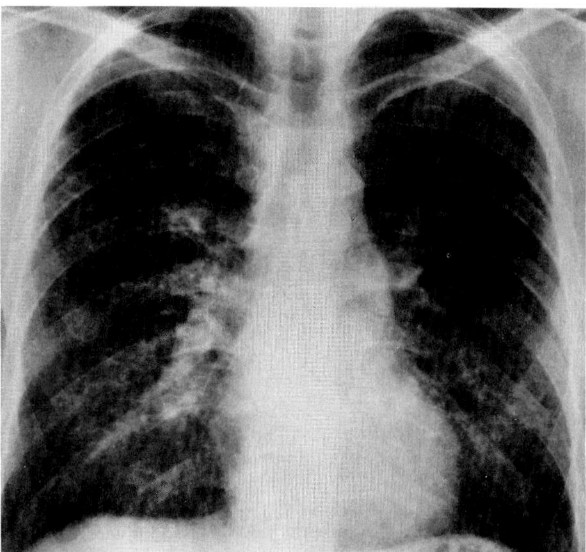

**Figure 30.17. Diffuse interstitial pulmonary fibrosis.** The bilateral pattern of fine, linear, reticular streaks predominantly involves the lower portions of the lungs.

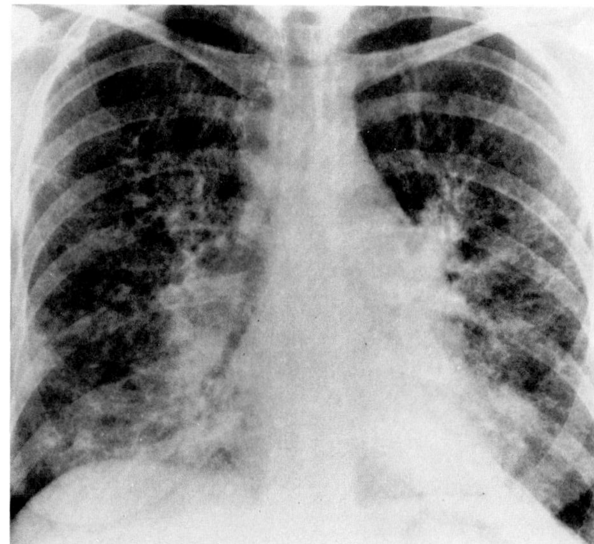

**Figure 30.18. Desquamative interstitial pneumonia.** There is a diffuse reticulonodular pattern, indicating interstitial disease, combined with bibasilar air space consolidation that obscures the borders of the heart.

## Desquamative Interstitial Pneumonia

Desquamative interstitial pneumonia (DIP) is a disease of unknown etiology in which there is striking hyperplasia of alveolar lining cells and an accumulation of macrophages in the distal air spaces of the lungs. Because the alveolar disease is often followed by a progressive interstitial fibrosis, some consider that DIP most likely represents an early stage in the development of the more commonly observed UIP.

The classic radiographic appearance of DIP is a ground-glass opacification of the lungs that is bilaterally symmetric and predominantly involves the basal regions. A suggestive finding is the presence of triangular opacifications extending into both lung bases from the hila and consisting of combined reticulonodular and alveolar patterns (indicating combined interstitial and air space disease). The progressive loss of lung volume may lead to elevation of the diaphragm and cause segmental atelectasis. As the disorder progresses to UIP, diffuse fibrosis and a honeycomb lung pattern may develop.

In many patients, the minimal radiographic abnormalities correlate poorly with the prominent clinical symptoms (dyspnea, weight loss, persistent cough) and significant abnormalities of pulmonary function. As with most diffuse infiltrative diseases of the lung, a definitive diagnosis can only be made by lung biopsy.[25–27]

## REFERENCES

1. Fraser RG, Pare JAP: *Diagnosis of Diseases of the Chest.* Philadelphia, WB Saunders, 1979.
2. Simon G, Galbraith HJB: Radiology of chronic bronchitis. *Lancet* 265:850–852, 1953.
3. Bates DV, Gordon CA, Paul GI, et al: Chronic bronchitis: Report on the third and fourth stages of the coordinated study of chronic bronchitis in the Department of Veterans Affairs, Canada. *Med Serv J Can* 22:5–22, 1966.
4. Eriksson S: Pulmonary emphysema and alpha$_1$-antitrypsin deficiency. *Acta Med Scand* 175:197–205, 1964.
5. Simon G, Pride MB, Jones NL, et al: Relation between abnormalities in the chest radiographs and changes in pulmonary function in chronic bronchitis and emphysema. *Thorax* 28: 15–23, 1973.
6. Barden RP: The interpretation of some radiologic signs of abnormal pulmonary function. *Radiology* 59:41–47, 1952.
7. Heitzmann ER, Markarian B, Solomon J: Chronic obstructive pulmonary disease. *Radiol Clin North Am* 11:49–61, 1973.
8. Simon G: Chronic bronchitis and emphysema: A symposium. III. Pathological findings and radiological changes in chronic bronchitis and emphysema. *Br J Radiol* 32:292–305, 1959.
9. Fagan CJ, Swischuk LE: The opaque lung in lobar emphysema. *AJR* 114:300–304, 1972.
10. Broadhair GD: Nonoperative management of lobar emphysema, long-term follow-up. *Radiology* 102:125–127, 1972.
11. Swyer PR, James GCW: A case of unilateral pulmonary emphysema. *Thorax* 8:133–136, 1953.
12. MacLeod WM: Abnormal transradiancy of one lung. *Thorax* 9:117–120, 1954.
13. Johnson TH, Gajaraj A, Feist JH: Patterns of pulmonary interstitial disease. *AJR* 109:516–521, 1970.
14. Dunmore LA, El-Khoury SA: Eosinophilic granuloma of the lungs. A report of three cases in Negro patients. *Am Rev Respir Dis* 90:789–791, 1964.
15. Lewis JG: Eosinophilic granuloma and its variants with special reference to lung involvement. A report of twelve patients. *Q J Med* 33:337–359, 1964.
16. Nadeau PJ, Ellis FH, Harrison EG, et al: Primary pulmonary histiocytosis X. *Dis Chest* 37:325–339, 1960.
17. Soergel KH, Sommers SC: Idiopathic pulmonary hemosiderosis and related syndromes. *Am J Med* 32:499–511, 1962.
18. Theros EG, Reeder MM, Eckert JF: An exercise in radiologic-pathologic correlation. *Radiology* 90:784–791, 1968.
19. Sybers RG, Sybers JL, Dickie HA, et al: Roentgenographic aspects of hemorrhagic pulmonary-renal disease (Goodpasture's syndrome). *AJR* 94:674–680, 1965.
20. Rosen SH, Castleman B, Liebow AA: Pulmonary alveolar proteinosis. *N Engl J Med* 258:1123–1142, 1958.
21. Mendenhall E, Solu S, Easom HF: Pulmonary alveolar proteinosis. *Am Rev Respir Dis* 84:876–880, 1961.
22. Balikian JP, Fuleihan FJD, Nucho CN: Pulmonary alveolar microlithiasis: Report of five cases with special reference to roentgen manifestations. *AJR* 103:509–518, 1968.
23. Felson B: The roentgen diagnosis of disseminated pulmonary alveolar diseases. *Semin Roentgenol* 2:3–16, 1967.
24. Sosman MC, Dodd GD, Jones WD, et al: Familial occurrence of pulmonary alveolar microlithiasis. *AJR* 77:947–1012, 1957.
25. Liebow AA, Steer A, Billingsley J: Desquamative interstitial pneumonia. *Am J Med* 39:369–404, 1965.
26. Crystal RG, Fulmer JD, Roberts WC, et al: Idiopathic pulmonary fibrosis. *Ann Intern Med* 85:769–788, 1976.
27. Lemire P, Bettez P, Gelinas M, et al: Patterns of desquamative interstitial pneumonia (DIP) and diffuse interstitial pulmonary fibrosis (DIPF). *AJR* 115:479–487, 1972.

## Chapter Thirty-One
# Vascular Disease of the Lung

## PULMONARY HYPERTENSION

Elevated pressure in the pulmonary arterial circulation may be related to primary vascular disease, pleuropulmonary disease, or cardiac disorders. The radiographic appearance varies widely depending on the underlying etiology.

A major cause of pulmonary arterial hypertension of vascular origin is multiple pulmonary emboli occurring over many years. The emboli cause a reduction of the pulmonary vascular bed, thus increasing the resistance in the pulmonary circulation and leading to pulmonary hypertension. In addition to thrombi, other materials that can embolize to the lungs and cause pulmonary hypertension include metastatic neoplasms, schistosomal parasites, and material injected intravenously by drug addicts. Pulmonary hypertension can also be caused by an obliterative arteritis associated with Takayasu's disease or the connective tissues disorders.

The classic radiographic appearance of pulmonary arterial hypertension is enlargement of the hilar pulmonary arteries with abrupt narrowing of distal vessels, resulting in a striking discrepancy between the central and peripheral vasculature. Rapid tapering, or pruning, of the distal vessels leads to an increased lucency of the lung periphery. Diffuse pulmonary oligemia of vascular origin can be differentiated from that seen in emphysema

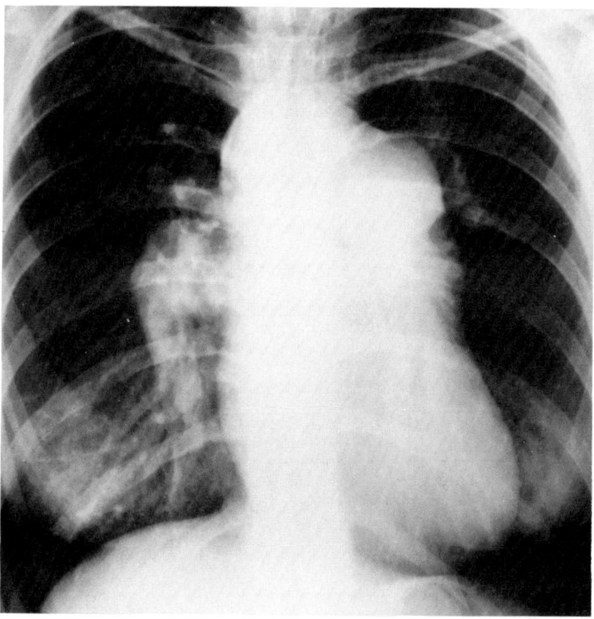

**Figure 31.1. Pulmonary hypertension associated with a left-to-right shunt.** A frontal chest film in a patient with atrial septal defect and Eisenmenger physiology demonstrates a huge pulmonary outflow tract and central pulmonary arteries with abrupt tapering and sparse peripheral vasculature. The heart is normal in size. (From Spitz,[3] with permission from the publisher.)

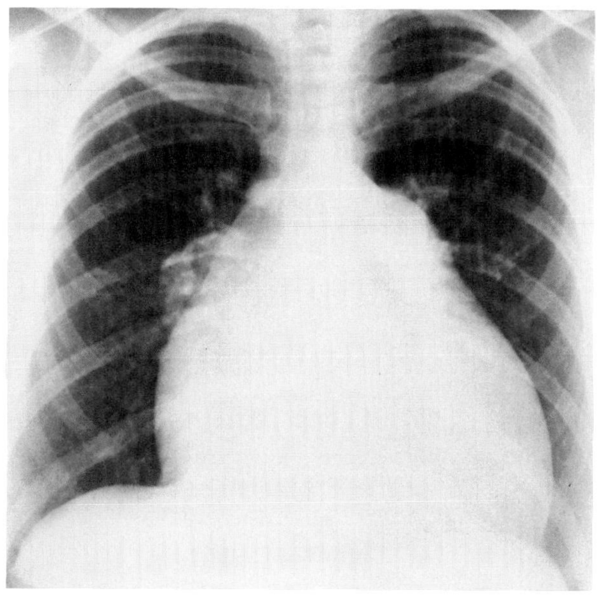
A

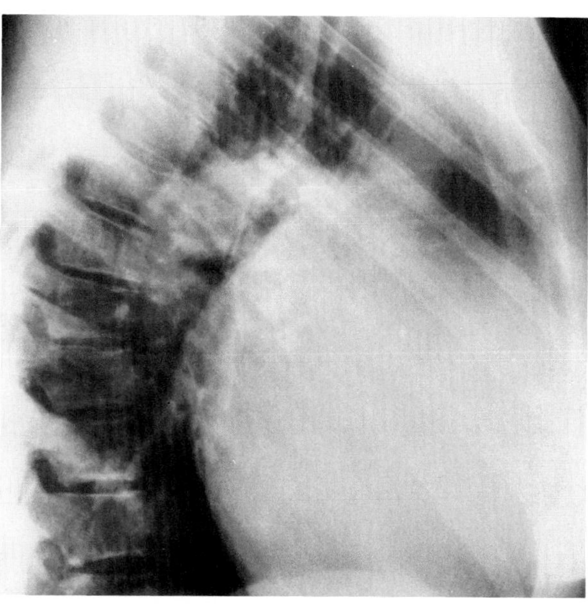
B

**Figure 31.2. Primary pulmonary hypertension.** (A) Frontal and (B) lateral views of the chest show marked globular cardiomegaly with prominence of the pulmonary trunk and central pulmonary arteries. The peripheral pulmonary vascularity is strikingly reduced. Right ventricular enlargement has caused obliteration of the retrosternal air space on the lateral view.

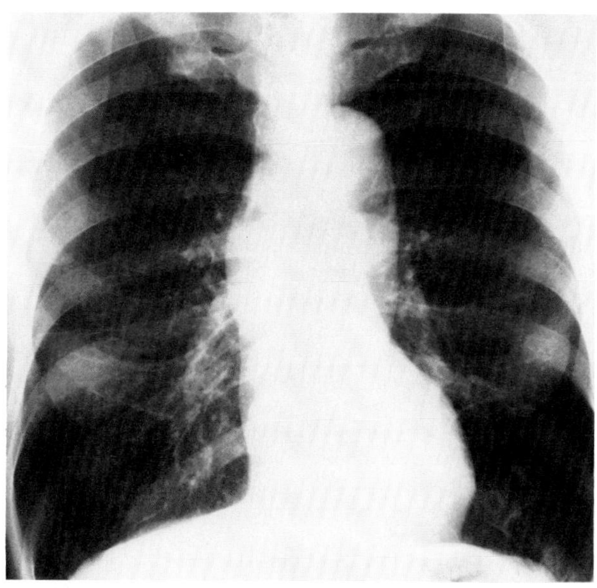

**Figure 31.3. Pulmonary hypertension associated with emphysema.** A severe bilateral decrease in the peripheral vascularity has led to the development of increased pulmonary arterial pressure.

because of the absence of the overinflation seen in patients with chronic obstructive pulmonary disease.

Chronic pulmonary hypertension eventually leads to right ventricular hypertrophy. Radiographically, enlargement of the right ventricle causes obliteration of the retrosternal air space on lateral chest films.

Increased flow through the pulmonary vessels accompanies congenital heart defects with left-to-right shunts (e.g., atrial septal defect, ventricular septal defect, patent ductus arteriosus). The typical radiographic appearance is an increase in the caliber of all the pulmonary arteries throughout the lungs. Because the degree of enlargement of the main and hilar pulmonary arteries is proportional to that of the intrapulmonary vessels, the arteries taper gradually and proportionately distally as long as peripheral resistance remains normal. Peripheral vascular markings that normally are invisible can often be detected. The increased flow eventually leads to an elevation in vasomotor tone that causes increased pulmonary resistance, pulmonary hypertension, and even reversal of flow through the shunt (Eisenmenger physiology).

Primary pulmonary hypertension is an unusual cause of elevated pulmonary arterial presssure that typically develops in women between the ages of 20 and 40.

Emphysema and chronic bronchitis can cause pulmonary hypertension because of hypoxemia and the destruction of the microvasculature of the lung. Diffuse alveolar or interstitial disease can cause increased resistance in the pulmonary circulation by limiting the distensibility of the pulmonary vascular tree. Underventilation of normal lungs (alveolar hypoventilation syndromes) causes hypoxia and acidosis that may lead to vasospasm and the development of pulmonary hypertension. Decreased pulmonary ventilation may be due to obesity, upper airway obstruction (chronic enlarged adenoids), severe kyphoscoliosis, continuous depression of the respiratory center by drugs, or chronic exposure to high altitudes.

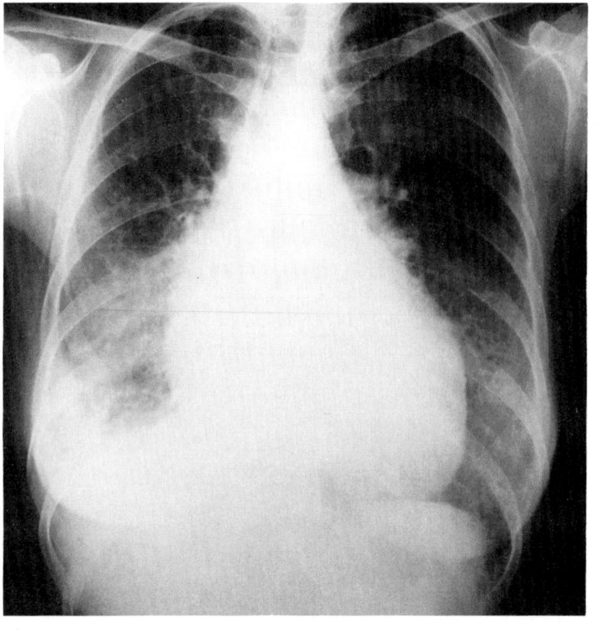

A

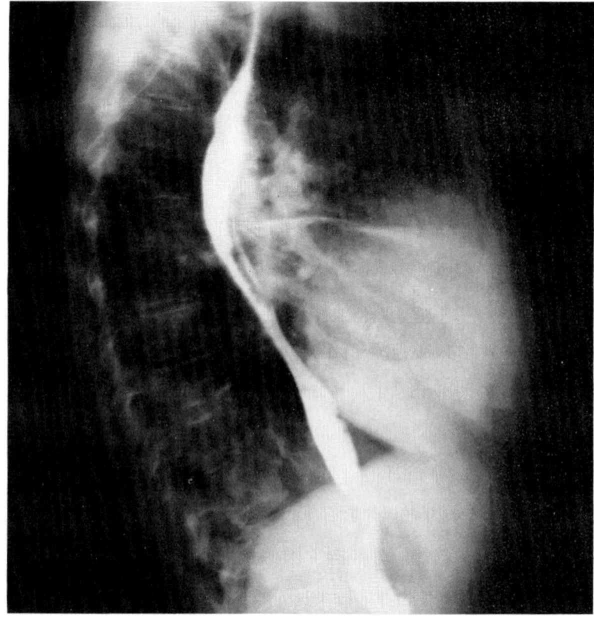

B

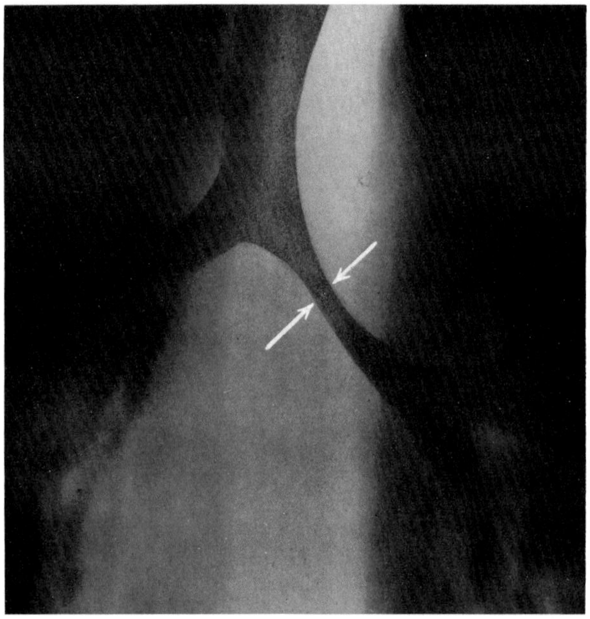

C

**Figure 31.4. Pulmonary hypertension associated with compression of pulmonary veins.** (A) Frontal and (B) lateral views of the chest show marked hypertrophy and dilatation of the right ventricle. Although the radiographic appearance of lung stasis resembles that seen in long-standing mitral valvular disease or other obstruction of the left side of the heart, at necropsy the left atrium was not dilated and the posterior displacement of the esophagus was due to enlarged main pulmonary arteries. (C) A tomogram of the region of the tracheal bifurcation shows narrowing (retouched) of the left main stem bronchus due to surrounding chronic mediastinal inflammation and fibrosis, which was the underlying cause for compression of the pulmonary veins. (From Cooley and Schreiber,[27] with permission from the publisher.)

Prolonged increased pulmonary venous pressure due to mitral stenosis or chronic congestive failure can also lead to narrowing of the pulmonary arterioles and secondary pulmonary arterial hypertension. In these patients, the development of pulmonary arterial hypertension serves to decrease the pressure in the pulmonary venous circulation, thus decreasing the degree of venous engorgement and often causing the disappearance of septal (Kerley) lines.[1–3]

## COR PULMONALE

Cor pulmonale refers to enlargement of the right ventricle secondary to pulmonary dysfunction. It is invariably associated with pulmonary hypertension, which usually reflects an increase in pulmonary vascular resistance to blood flow through small muscular arteries and arterioles. Obliteration, obstruction, or reduction of the pulmonary vascular tree may result from primary pulmonary

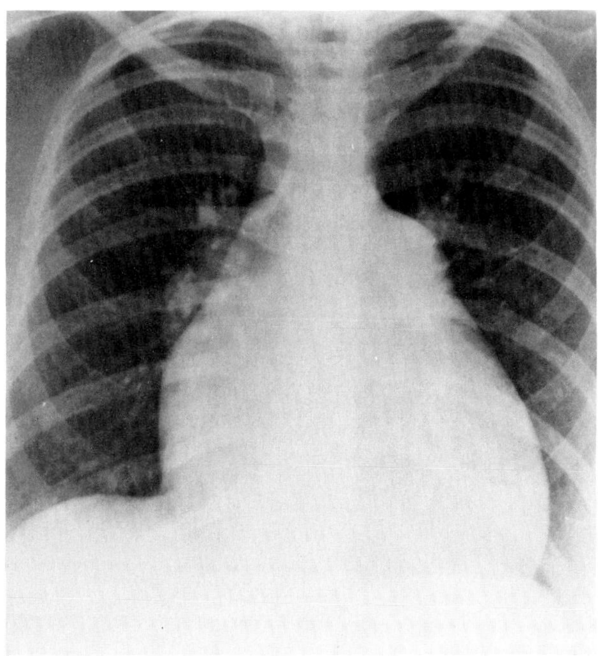
A

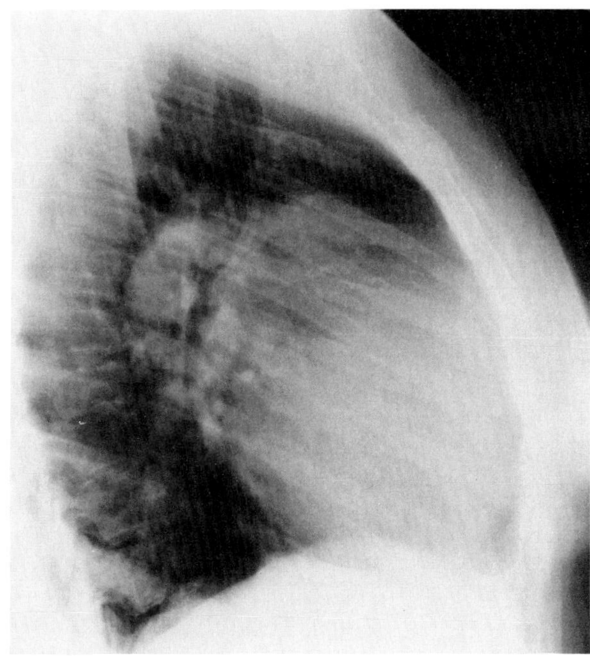
B

**Figure 31.5. Cor pulmonale.** In a patient with primary pulmonary hypertension, enlargement of the right ventricle causes cardiomegaly with lateral displacement of the cardiac apex on a frontal view (A) and filling in of the retrosternal space on a lateral view (B). Note the prominence of the pulmonary outflow tract and the markedly dilated central pulmonary vessels.

hypertension, recurrent pulmonary emboli, chronic left-to-right shunts, diffuse interstitial or emphysematous lung disease, or pneumoconiosis. Hypoxia and acidosis, from extrapulmonary causes such as kyphoscoliosis, neuromuscular disorders, extreme obesity, hypercapnia, or alveolar hypoventilation, cause a vasomotor increase in pulmonary vascular resistance. An increase in cardiac output or blood viscosity can also cause overloading of the right ventricle.

Concentric hypertrophy of the right ventricle in cor pulmonale may produce no apparent increase in cardiac size. Dilatation of the right ventricle causes cardiomegaly with lateral and upward displacement of the cardiac apex on frontal views and filling in of the retrosternal space on lateral views. The left atrium and left ventricle are normal in appearance unless there is underlying cardiac disease affecting these chambers. The development of right-sided heart failure results in increased cardiac enlargement and may produce widening of the superior mediastinum because of dilatation of the superior vena cava, elevation of the right hemidiaphragm from hepatic congestion, and pleural effusion.

The major value of plain chest radiographs in patients with cor pulmonale is to demonstrate signs of pulmonary arterial hypertension and to detect evidence of a possibly correctable underlying pulmonary disorder. Pulmonary arterial hypertension causes an increased fullness of the central pulmonary arteries with abrupt narrowing and pruning of peripheral vessels. Disorders of the lung that predispose to pulmonary arterial hypertension and cor pulmonale and that can be identified on chest radiographs include emphysema, pneumoconiosis, and diffuse fibrotic interstitial disease of any etiology.[1]

## PULMONARY THROMBOEMBOLISM

Pulmonary thromboembolism is a potentially fatal condition that is by far the most common pathologic process involving the lungs of hospitalized patients. In about 80 percent of the patients with this disorder, the condition does not cause symptoms and is thus unrecognized. Even when symptomatic, pulmonary thromboembolism is frequently misdiagnosed. More than 95 percent of pulmonary emboli arise from thrombi in the deep venous system of the lower extremities. The remainder come from thrombi occurring in the right side of the heart or in brachial or cervical veins. Most embolic occlusions occur in the lower lobes because of the preferential blood flow to these regions.

The physiologic consequences of thromboembolic occlusion of the pulmonary arteries depend on the size of the embolic mass and the general state of the pulmonary circulation. In young persons with good cardiovascular function and adequate collateral circulation, the occlusion of a large central vessel may be associated with minimal, if any, functional impairment. In contrast,

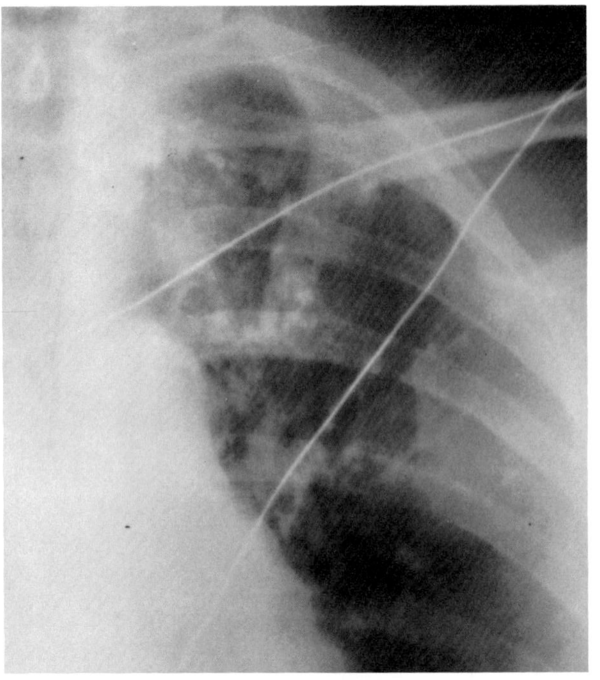

A

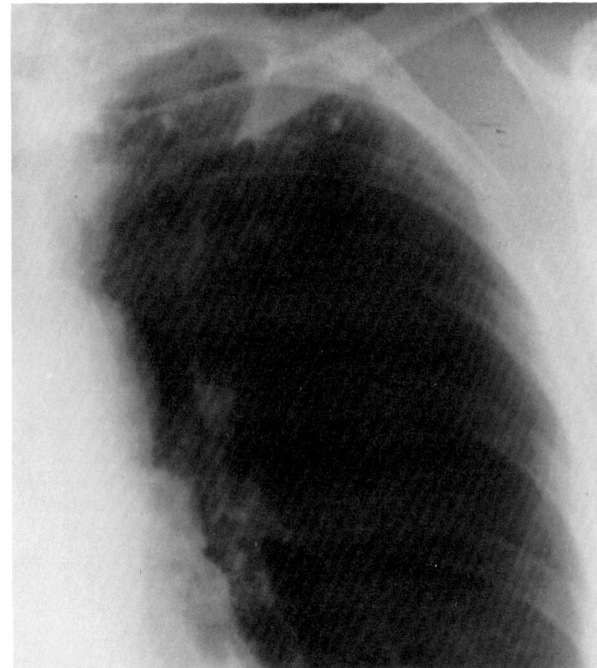

B

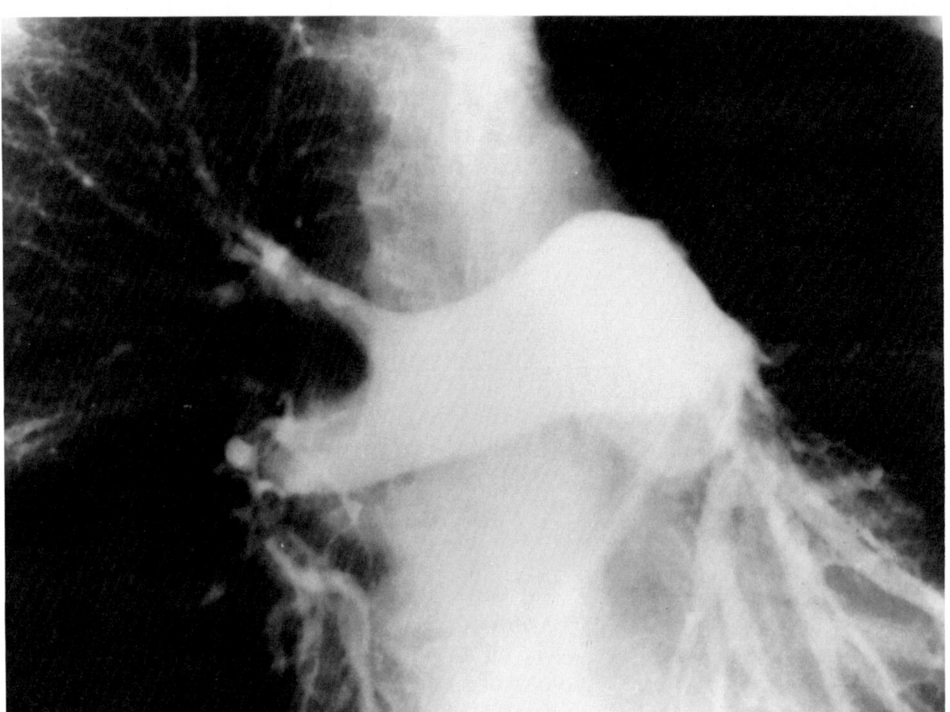

C

**Figure 31.6. Westermark's sign of pulmonary thromboembolism.** (A) A baseline chest radiograph demonstrates normal vascularity in the left upper lobe. (B) Striking hyperlucency of the left upper lobe coincides with the onset of the patient's symptoms. (C) An arteriogram performed the same day as B demonstrates an occluding clot in the left upper lobe and multiple emboli in the right lung. (From Julien,[4] with permission from the publisher.)

in patients with cardiovascular disease or severe debilitating illnesses, pulmonary vascular occlusion often leads to infarction. Arterial occlusion can also cause consolidation of distal lung parenchyma that reflects hemorrhage and edema without actual lung necrosis.

Most patients with thromboembolism without infarction have a normal chest radiograph. Nevertheless, some subtle, yet distinctive, abnormalities on plain radiographs can strongly suggest the diagnosis. A large-vessel pulmonary thromboembolism causes a focal reduction in blood volume without a substantial change in air or tissue volume; this leads to focal pulmonary oligemia and relative lucency of the involved portion of lung (Westermark's sign). This finding is best seen when oli-

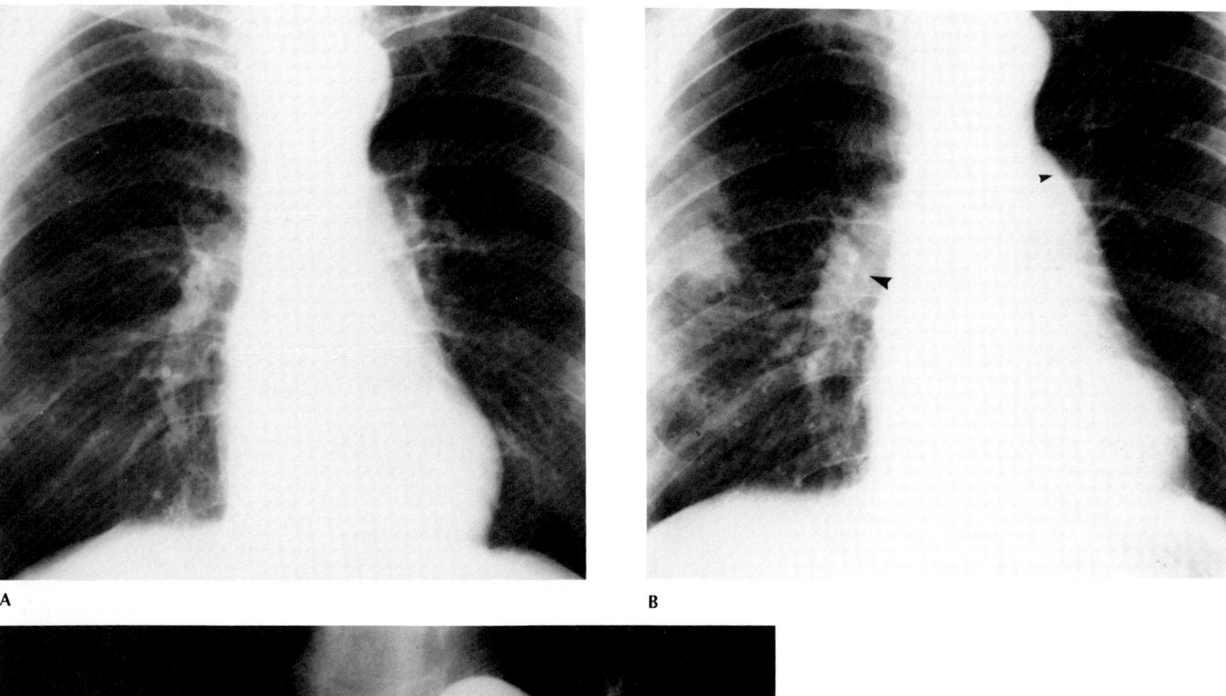


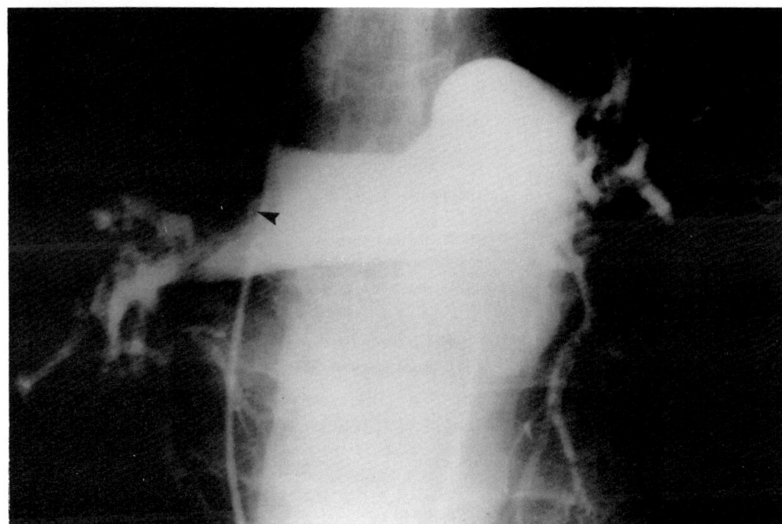


**Figure 31.7. Fleischner's sign of pulmonary thromboembolism.** (A) A baseline chest radiograph demonstrates normal-sized pulmonary arteries. (B) Enlargement of the main pulmonary artery (small arrow) and right pulmonary artery (large arrow) coincides with the onset of the patient's symptoms. (C) An arteriogram demonstrates multiple bilateral pulmonary emboli and a large right saddle embolus (arrow). (From Julien,[4] with permission from the publisher.)

gemia due to deprivation of pulmonary arterial supply to a major part or all of the lung is contrasted with increased pulmonary arterial blood flow to the unoccluded lung. Widespread small-vessel thromboembolism can cause diffuse peripheral oligemia. Another sign of pulmonary thromboembolism is enlargement of the ipsilateral main pulmonary artery due to pulmonary hypertension or distension of the vessel by bulk thrombus (Fleischner's "plump hilus" sign). This sign is of most value when serial radiographs demonstrate progressive enlargement of the affected vessel. In addition to an increase in size, there is usually abrupt distal tapering of the occluded vessel, often with sudden termination ("knuckle" or "sausage" sign).

Thromboembolism with infarction appears radiographically as an area of consolidated lung parenchyma, regardless of whether there is simple hemorrhage and edema or frank tissue necrosis. A highly characteristic, though uncommon, appearance of pulmonary infarction is a pleural-based, wedge-shaped density that has a rounded apex (Hampton's hump) and is most commonly seen at the base of the lung, often in the costophrenic sulcus. In many instances, an infarction produces a nonspecific parenchymal density that simulates an acute pneumonia. It is often said that infarction invariably extends to a visceral pleural surface, though this is of little diagnostic value since most pneumonias have a similar appearance. The pattern of resolution of the consolidation is of value in distinguishing among acute inflammatory processes, pulmonary hemorrhage and edema, and frank necrosis. Pulmonary infarctions tend to gradually reduce in size while retaining the same general

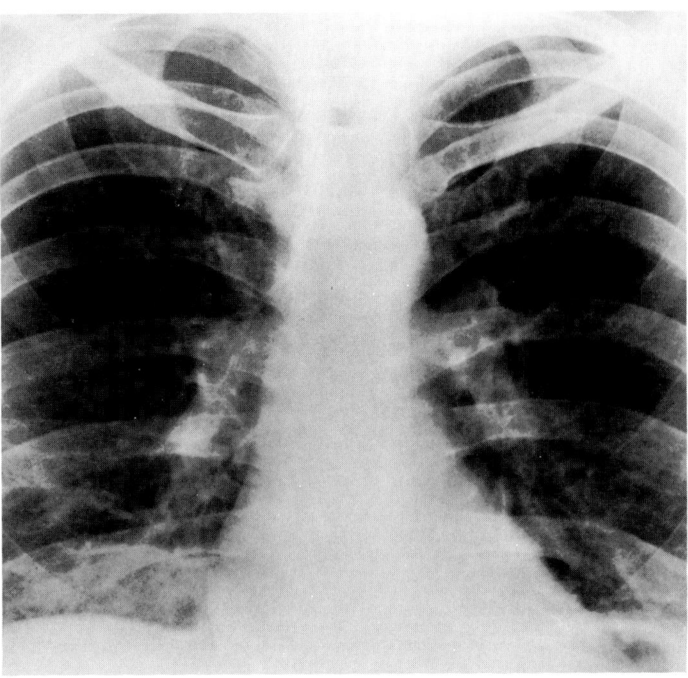
A

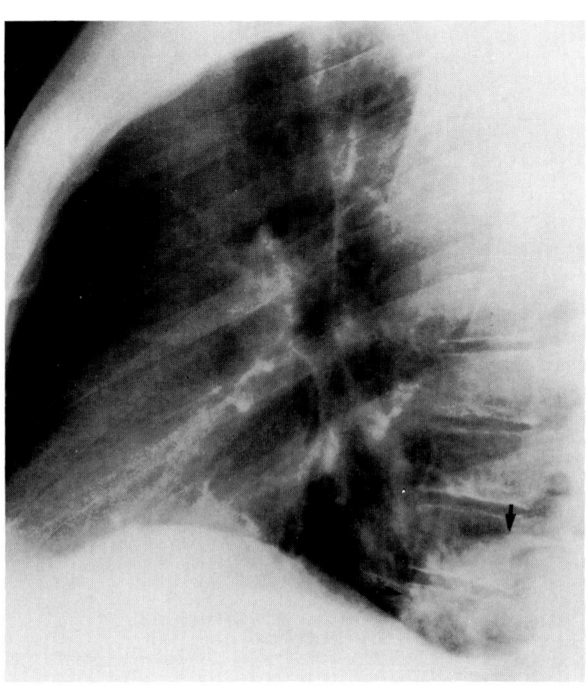
B

**Figure 31.8. Hampton's hump in pulmonary thromboembolism.** (A) Frontal and (B) lateral views of the chest demonstrate a fairly well circumscribed shadow of homogeneous density occupying the posterior basal segment of the right lower lobe. On the lateral projection, the shadow has the shape of a truncated cone that is convex toward the hilum (arrows). A small effusion can be identified on the lateral projection. This combination of abnormalities is highly suggestive of pulmonary infarction; the history and biochemical findings were compatible with the diagnosis. (From Fraser and Pare,[1] with permission from the publisher.)

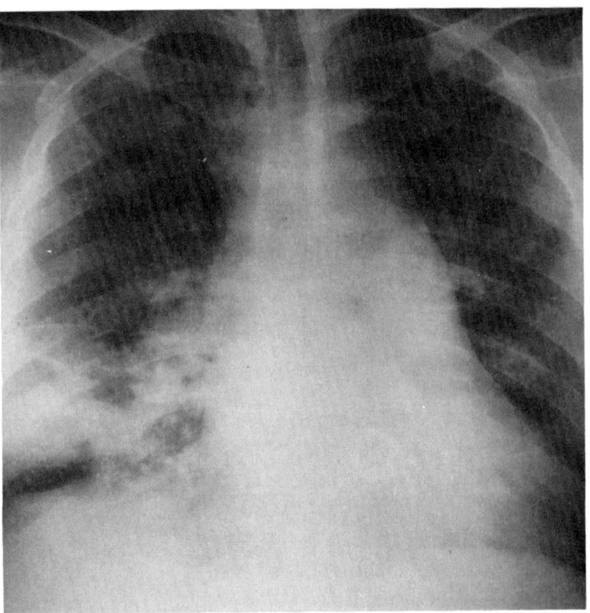
A

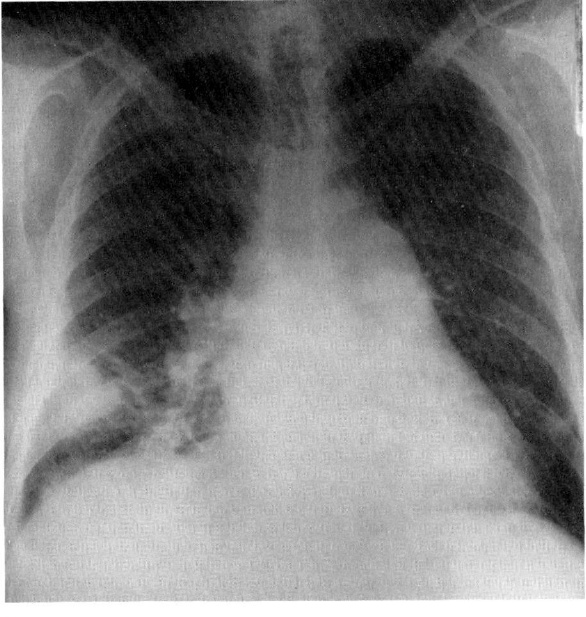
B

**Figure 31.9. Resolution of pulmonary infarction.** (A) A frontal chest film made 3 days after open-heart surgery demonstrates a very irregular density at the right base. The major differential diagnosis was between pneumonia and pulmonary embolization with infarction. (B) On a film obtained 5 days later, the dense lesion is seen to have reduced in size yet to have retained the same general configuration seen on the initial view. The diagnosis of pulmonary embolism was confirmed on a radionuclide lung scan. (From Woesner, Sanders, and White,[28] with permission from the publisher.)

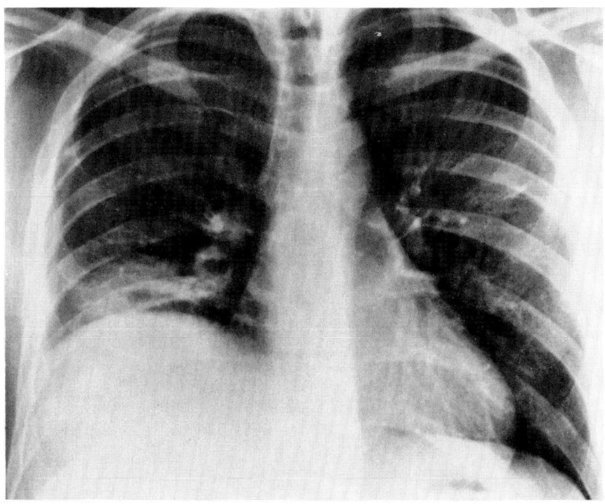


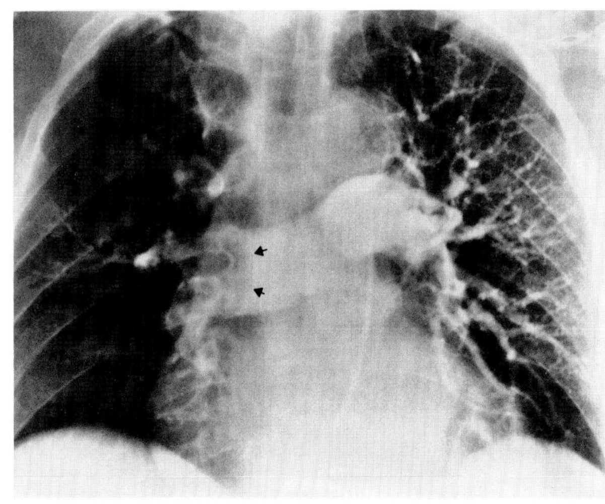


**Figure 31.10. Pulmonary thromboembolism.** (A) A plain chest radiograph demonstrates atelectasis at the right base, which is associated with elevation of the right hemidiaphragm that represents a large subpulmonic pleural effusion. (B) A pulmonary arteriogram shows virtually complete obstruction (arrows) of the right pulmonary artery.

configuration seen on initial views. There is resorption of the perimeters of the infarct with preservation of the pleural base. In contrast, the resolution of pneumonia tends to be patchy and is characterized by a fading of the radiographic density throughout the entire involved area. Parenchymal hemorrhage and edema generally clear within 4 to 7 days; the resolution of necrotic lung tissue usually requires 3 weeks or more.

Line shadows are a frequent radiographic appearance of pulmonary embolism and infarction. These usually basilar densities may reflect platelike atelectasis, parenchymal scarring, or thrombosed arteries and veins. Pleural effusion is a common manifestation of thromboembolic disease and almost always indicates some degree of infarction, even if no parenchymal consolidation can be detected.

Because the chest radiograph is usually either normal or nonspecific, the radionuclide lung scan is generally considered the most effective screening test for significant pulmonary thromboembolic disease. The perfusion lung scan is very sensitive in detecting thromboembolic disease; in one large study, there were no instances in which the lung scan was completely normal when an arteriogram showed pulmonary emboli. The area of lung distal to an embolus appears as a perfusion defect on the radionuclide scan. Unfortunately, false-positive scans may be recorded in portions of the lung that are poorly perfused because of impaired ventilation, even if no obstructing vascular lesion is present. Therefore, the concomitant use of ventilation lung scans increases the diagnostic accuracy. The ventilation scan is usually relatively normal in patients with pulmonary thromboembolism, while it generally demonstrates defects corresponding to the zones of underperfusion in patients with chronic pulmonary disease. Nevertheless, although combined ventilation and perfusion radionuclide scanning may offer excellent evidence of the presence of pulmonary thromboembolism, many cases are equivocal and require pulmonary arteriography.

Pulmonary arteriography is the definitive technique for evaluating the patient with suspected thromboembolic disease. However, it is associated with a small, though definite, risk of morbidity and mortality. The unequivocal arteriographic diagnosis of pulmonary thromboembolism requires the demonstration of an abrupt occlusion (cutoff) of a pulmonary artery or a persistent intraluminal filling defect within it. Secondary signs of thromboembolic disease include tortuous, abruptly tapering peripheral vessels with a paucity of branching vessels (pruning); areas of oligemia or avascularity; and a prolongation of the arterial phase that is usually accompanied by slow filling and emptying of the pulmonary veins. It must be emphasized that secondary signs can also be found in nonembolic conditions that produce defects on the radionuclide perfusion scan.[1,4–13,28]

## NONTHROMBOTIC EMBOLISM

### Fat Embolism

Almost all pulmonary fat emboli are the result of trauma, especially leg fractures, in which marrow fat enters torn peripheral veins and is trapped by the pulmonary circulation. The patient is typically asymptomatic for the first 1 to 2 days, after which sudden cardiopulmonary and neurologic deterioration occurs. The characteristic appearance on chest radiographs consists of diffuse air space consolidation due to alveolar hemorrhage and

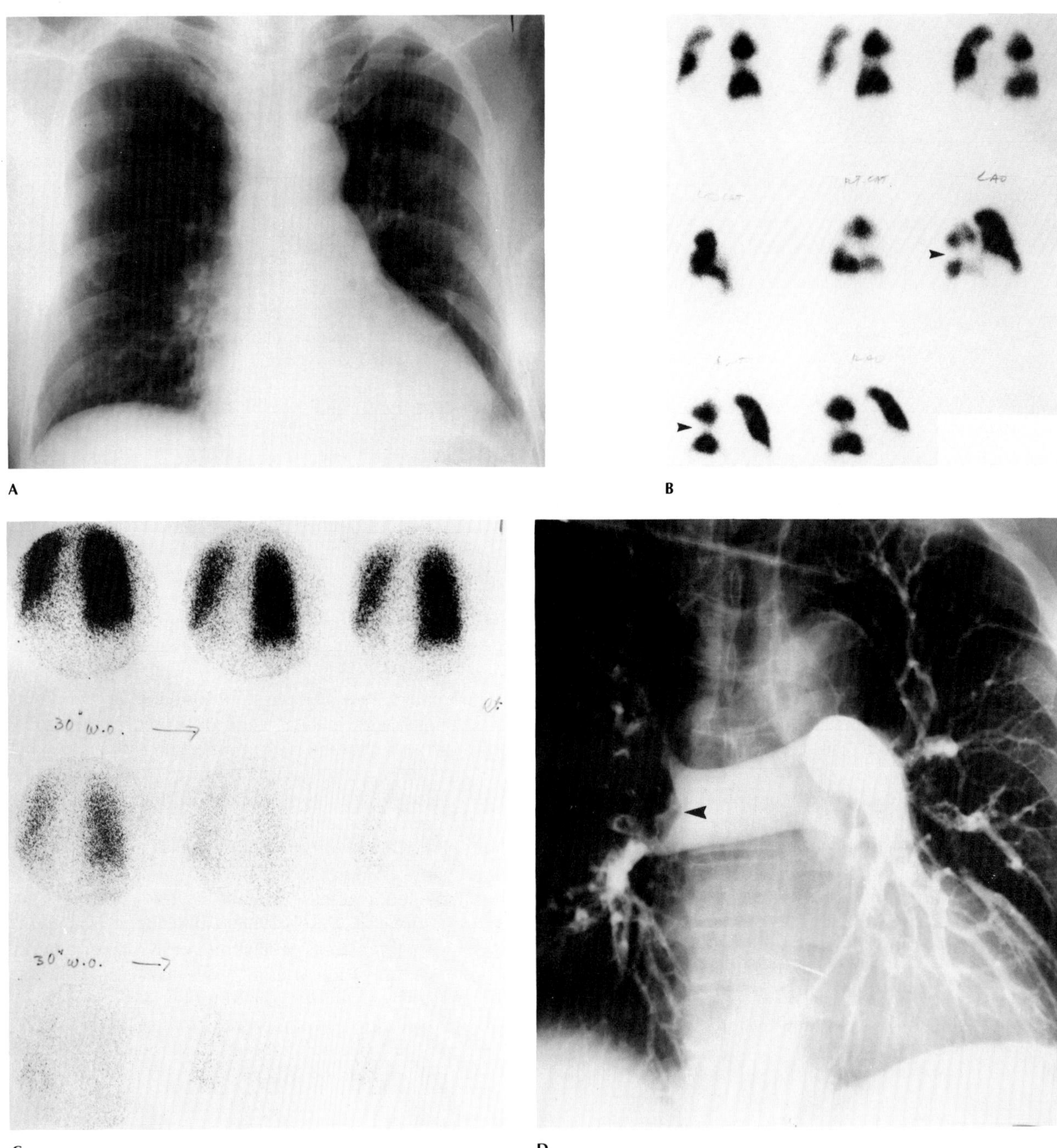


**Figure 31.11. Pulmonary thromboembolism.** (A) A chest radiograph shows a small left effusion but is otherwise normal. (B) A perfusion lung scan demonstrates segmental perfusion defects and a lobar defect in the right lung (arrow). (C) Normal xenon scan on the right. (D) An arteriogram demonstrates multiple bilateral emboli and a large saddle embolus at the bifurcation of the right pulmonary artery (arrow). (From Julien,[4] with permission from the publisher.)

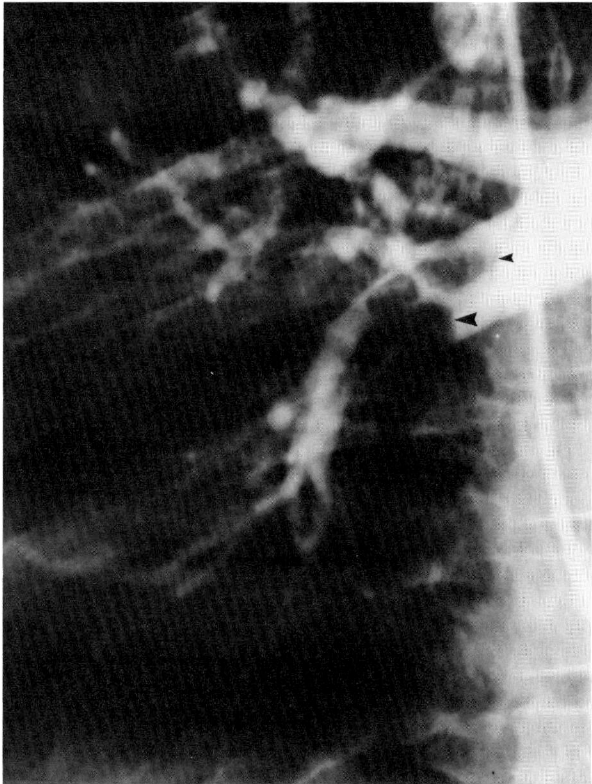

**Figure 31.12. Pulmonary thromboembolism.** A pulmonary arteriogram demonstrates an intraluminal filling defect (small arrow) and vascular cutoff (large arrow). (From Julien,[4] with permission from the publisher.)

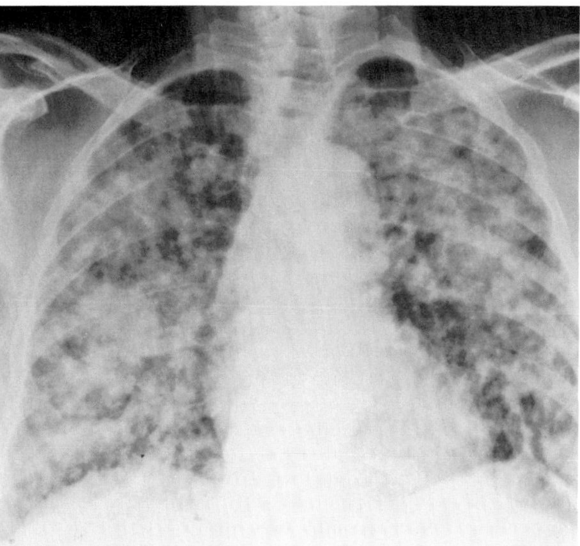

**Figure 31.13. Fat embolism.** A frontal chest radiograph 3 days after leg fracture demonstrates diffuse bilateral air space consolidation due to alveolar hemorrhage and edema. Unlike cardiogenic pulmonary edema, the distribution in this patient is predominantly peripheral rather than central, and the heart is not enlarged.

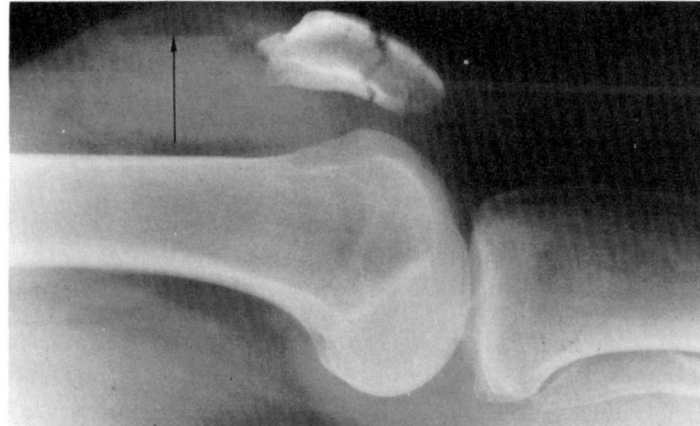

**Figure 31.14. Fat-blood interface associated with a patellar fracture.** A recumbent radiograph of the knee obtained using a horizontal beam demonstrates a fat-blood interface (arrow) within a large suprapatellar effusion. Marrow fat that enters torn peripheral vessels can be trapped by the pulmonary circulation and can lead to the development of diffuse alveolar consolidation. (From Wenzel,[14] with permission from the publisher.)

edema. Unlike cardiogenic pulmonary edema, the distribution in fat embolism is predominantly peripheral rather than central and involves the basal regions to a greater degree. In addition, there is no evidence of cardiac enlargement and there are no signs of pulmonary venous hypertension or interstitial edema. The typical latent period and slow resolution (7 to 10 days) serve to differentiate traumatic fat embolism from traumatic lung contusion, which invariably appears immediately after injury and clears rapidly (within 1 day).[14–17]

## Amniotic Fluid Embolism

The entrance of amniotic fluid containing particulate matter into the maternal circulation during spontaneous delivery or cesarean section can cause sudden and massive obstruction of the pulmonary vascular bed, leading to shock and, often, death. Because the condition is often rapidly fatal, radiographs are infrequently obtained; most of the rare nonfatal cases are incorrectly diagnosed. The diffuse air space consolidation seen on chest radiographs is virtually indistinguishable from the appearance caused by other forms of acute pulmonary edema.[18–20]

## Septic Embolism

Septic emboli primarily arise from either the heart (bacterial endocarditis of the tricuspid valve; ventricular septal defect) or the peripheral veins (septic thrombophlebitis). Many patients have a history of intravenous drug abuse. Septic emboli are almost always multiple and appear radiographically as ill-defined, round or wedge-shaped densities in the periphery of the lungs. They often present a migratory pattern, first appearing in one area and then in another as the older lesions resolve. Cavitation frequently develops.[21]

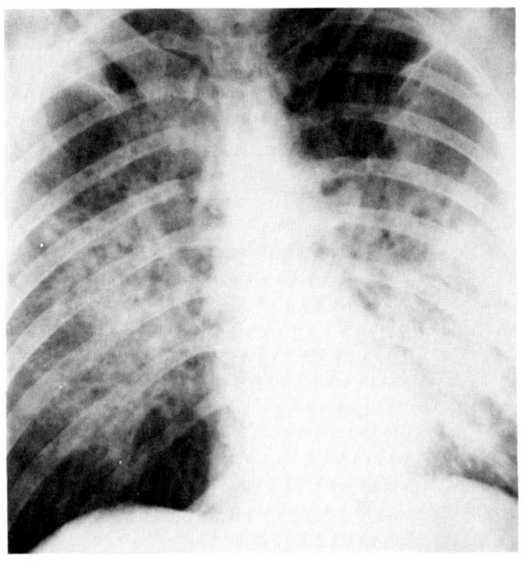

A

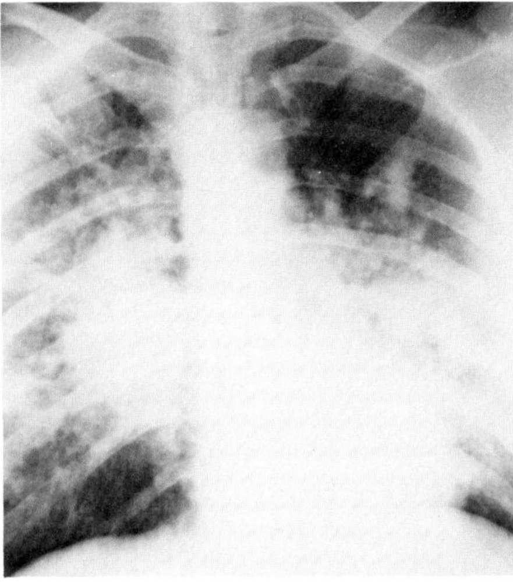

B

**Figure 31.15. Amniotic fluid embolism.** (A) An initial film, 6 hours after the onset of acute symptoms, shows heavy bilateral perihilar infiltrates. (B) Twelve hours later, the infiltrates have become more confluent in the perihilar zones. (From Arnold, Gardner, and Goodman,[18] with permission from the publisher.)

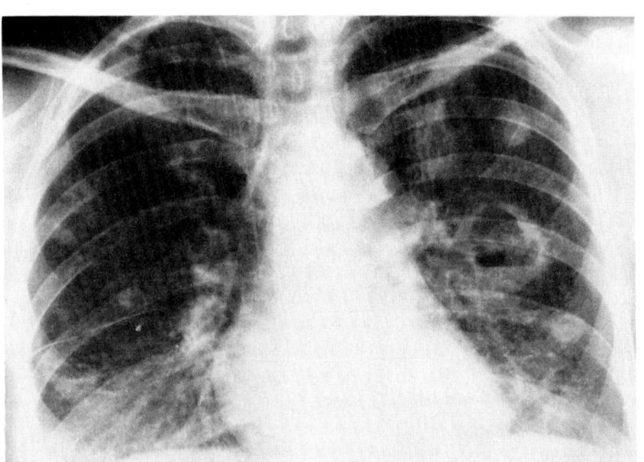

**Figure 31.16. Septic pulmonary emboli.** Round lesions, many with cavitation, are seen throughout the lungs in this intravenous drug abuser with staphylococcal tricuspid endocarditis.

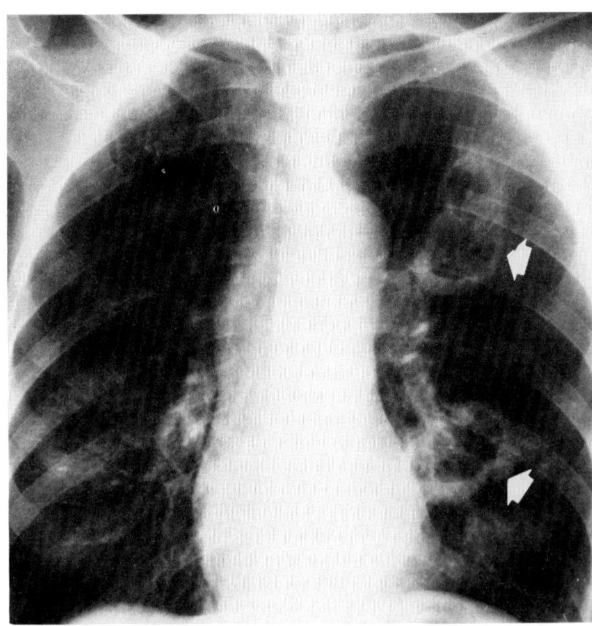

**Figure 31.17. Septic pulmonary emboli.** Large cavitary lesions are seen in the left lung (arrows) in this intravenous drug abuser with septic thrombophlebitis.

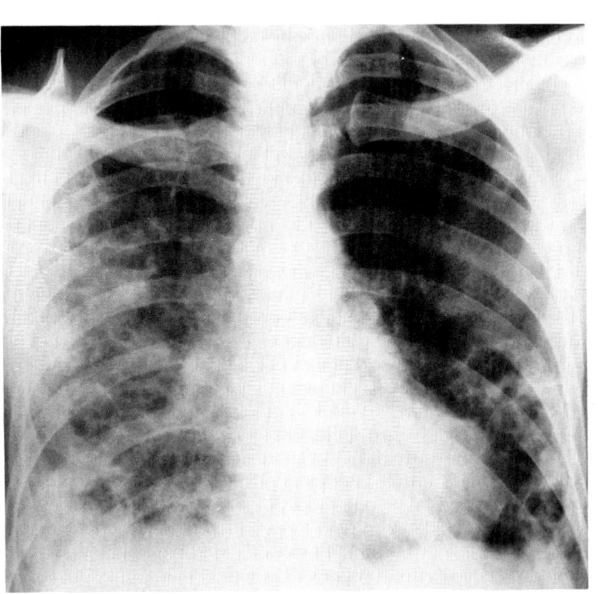

**Figure 31.18. Septic pulmonary emboli.** Thin-walled cavitary lesions are noted in both lungs in this intravenous drug abuser with staphylococcal tricuspid endocarditis. (From Jaffe and Koschmann,[21] with permission from the publisher.)

## PULMONARY ARTERIOVENOUS FISTULA

Pulmonary arteriovenous fistulas are abnormal vascular communications from a pulmonary artery to a pulmonary vein. About one-third of the patients with pulmonary arteriovenous fistulas have multiple lesions in the lungs; up to two-thirds of the patients with these malformations in the lungs have similar arteriovenous communications elsewhere (hereditary hemorrhagic telangiectasia).

A pulmonary arteriovenous fistula typically appears as a round or oval, lobulated, soft tissue mass that is most commonly situated in the lower lobes. A pathognomonic finding is the identification of the feeding and draining vessels; this may be difficult on plain radiographs and often requires tomography. The feeding artery appears to relate to the hilum, while the draining vein extends toward the left atrium. The vascular nature of the lesion may be seen at fluoroscopy by the demonstration of active pulsations in the mass itself as well

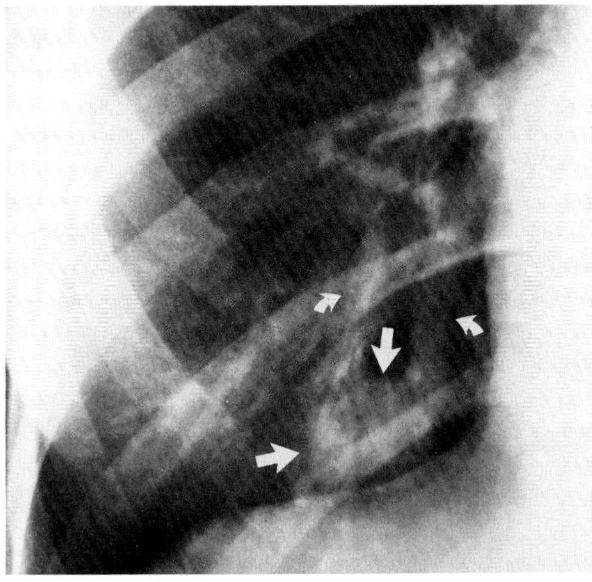

A

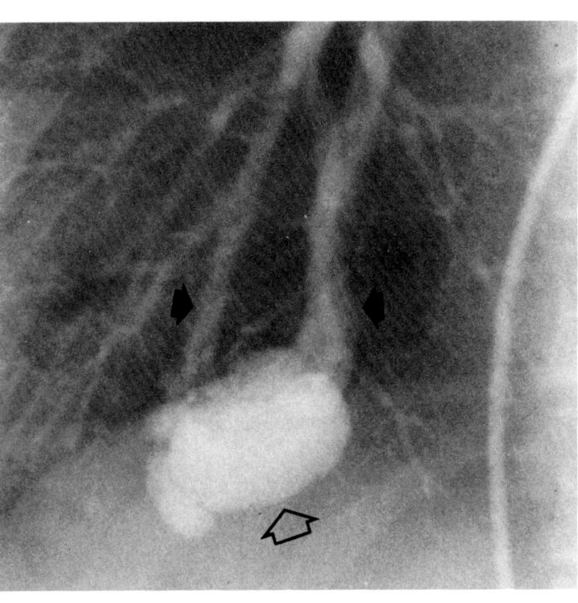

B

**Figure 31.19. Pulmonary arteriovenous fistula.** (A) A coned view of the right lung shows a round soft tissue mass (straight arrows) at the left base. Feeding and draining vessels (curved arrows) extend to the lesion. (B) An arteriogram clearly shows the feeding artery and draining vein (closed arrows) associated with the arteriovenous malformation.

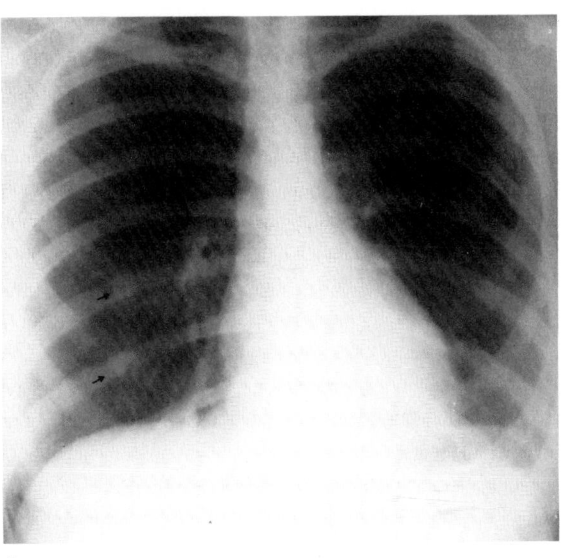

A

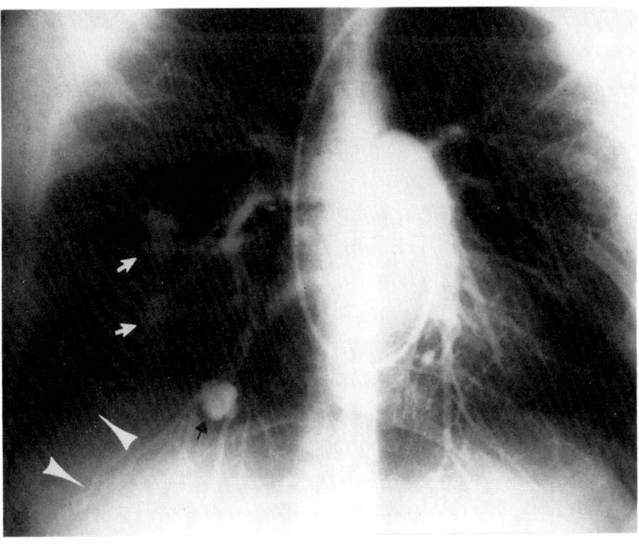

B

**Figure 31.20. Multiple pulmonary arteriovenous fistulas.** (A) A frontal chest film demonstrates two nodular densities in the lower portion of the right lung. (B) A pulmonary arteriogram shows these and other fistulas in the right lung (arrows). Arrowheads point to vascular loops in which a sizable artery merges imperceptibly with a sizable vein. (From Cooley and Schreiber,[27] with permission from the publisher.)

as a decrease and increase in its size during the Valsalva and Müller maneuvers, respectively. Pulmonary arteriography is required before surgical removal, not only to confirm the diagnosis but also to detect smaller, unsuspected vascular malformations that cannot be identified on routine radiographs.[1,22–27]

## REFERENCES


1. Fraser RG, Pare JAP: *Diagnosis of Diseases of the Chest*. Philadelphia, WB Saunders, 1979.
2. Walcott G, Burchell HB, Brown AL: Primary pulmonary hypertension. *Am J Med* 49:70–79, 1970.
3. Spitz HB: Eisenmenger's syndrome. *Semin Roentgenol* 3:373–376, 1968.
4. Julien P: Pulmonary embolism, in Eisenberg RL, Amberg JR (eds): *Critical Diagnostic Pathways in Radiology: An Algorithmic Approach*. Philadelphia, JB Lippincott, 1981.
5. Freiman DG, Suyemoto J, Wessler S: Frequency of pulmonary thromboembolism in man. *N Engl J Med* 272:1278–1280, 1965.
6. Fleischner FG: Pulmonary embolism. *Clin Radiol* 13:169–181, 1962.
7. Heitzman ER, Markarian B, Dailey ET: Pulmonary thromboembolic disease. *Radiology* 103:529–537, 1972.
8. Stein GN, Chen JT, Goldstein F, et al: The importance of chest roentgenography in the diagnosis of pulmonary embolism. *AJR* 81:255–263, 1959.
9. Sagel SS, Greenspan RH: Nonuniform pulmonary arterial perfusion: Pulmonary embolism? *Radiology* 99:541–548, 1971.
10. Biello DR, Matter AG, McKnight RC, et al: Ventilation perfusion studies in suspected pulmonary embolism. *AJR* 133:1033–1037, 1979.
11. Dalen JE, Brooks HL, Johnson LW, et al: Pulmonary angiography in acute pulmonary embolism: Indications, techniques and results in 367 patients. *Am Heart J* 81:175–185, 1971.
12. Moses DC, Silver TM, Bookstein JJ: The complementary roles of chest radiology, lung scanning, and selective pulmonary angiography in the diagnosis of pulmonary embolism. *Radiology* 109:179–182, 1974.
13. The Urokinase–Pulmonary Embolism Trial: A national cooperative study. *Circulation* (Suppl II) 47:1–108, 1973.
14. Wenzel WW: The FBI sign. *Rocky Mt Med J* 69:71–72, 1972.
15. Berrigan TJ, Carsky EW, Heitzman ER: Fat embolism. Roentgenographic pathologic correlation in three cases. *AJR* 96:967–971, 1966.
16. Maruyama Y, Little JB: Roentgen manifestations of traumatic pulmonary fat embolism. *Radiology* 79:945–952, 1962.
17. Feldman F, Ellis K, Green WM: The fat embolism syndrome. *Radiology* 114:535–542, 1975.
18. Arnold HR, Gardner JE, Goodman PH: Amniotic pulmonary embolism. *Radiology* 77:629–634, 1961.
19. Cornell SH: Amniotic pulmonary embolism. *AJR* 89:1084–1086, 1963.
20. Steiner PE, Lushbaugh CC: Maternal pulmonary embolism by amniotic fluid as a cause of obstetric shock and unexpected deaths in obstetrics, *JAMA* 117:1245–1248, 1941.
21. Jaffe RB, Koschmann EB: Septic pulmonary emboli. *Radiology* 96:527–532, 1970.
22. Revill D, Matts SGF: Pulmonary arteriovenous aneurysm in hereditary telangiectasia. *Br J Tuberc* 52:222–224, 1959.
23. Steinberg I, Finby N: Roentgen manifestations of pulmonary arteriovenous fistula: Diagnosis and treatment of four new cases. *AJR* 78:234–238, 1957.
24. Stork WJ: Pulmonary arteriovenous fistulas. *AJR* 74:441–448, 1955.
25. Sagel SS, Greenspan RH: Minute pulmonary arteriovenous fistulas demonstrated by magnification pulmonary angiography. *Radiology* 97:529–534, 1970.
26. Hoffman R, Rabens R: Evolving pulmonary nodules: Multiple pulmonary arteriovenous fistulas. *AJR* 120:861–867, 1974.
27. Cooley RN, Schreiber MH: *Radiology of the Heart and Great Vessels*. Baltimore, Williams & Wilkins, 1978.
28. Woesner ME, Sanders I, White GW: The melting sign in resolving transient pulmonary infarction. *AJR* 111:782–790, 1971.

## Chapter Thirty-Two

# Diseases of the Upper Respiratory Tract and Airways

## DISEASES OF THE UPPER RESPIRATORY TRACT

### Epiglottitis

Acute infections of the epiglottis, most commonly caused by *Hemophilus influenzae* in children, cause thickening of epiglottic tissue and aryepiglottic folds and almost complete obliteration of the valleculae and pyriform sinuses. On lateral views of the neck, a rounded thickening of the epiglottic shadow gives it the configuration and approximate size of an adult's thumb, in contrast to the normal, narrow epiglottic shadow resembling an adult's little finger. The prompt recognition of acute epiglottitis is imperative, since the condition may result in abrupt complete airway obstruction.

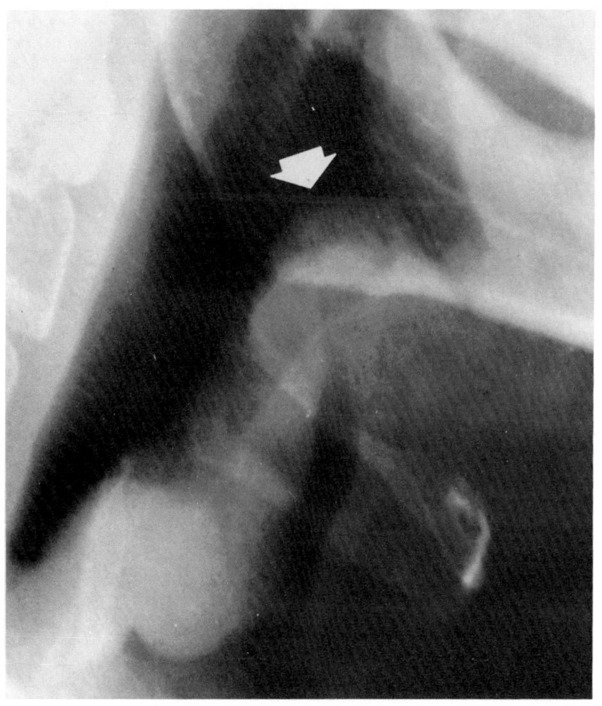

**Figure 32.1. Epiglottitis.** A lateral radiograph of the neck demonstrates a wide, rounded configuration of the inflamed epiglottis (arrow). (From Podgore and Bass,[2] with permission from the publisher.)

### Croup (Acute Laryngotracheobronchitis)

Croup is primarily a viral infection of young children that produces inflammatory obstructive swelling localized to the subglottic portion of the trachea. On frontal radiographs of the lower neck, there is a characteristic smooth, fusiform, tapered narrowing of the subglottic airway (gothic-arch sign), unlike the broad shouldering seen normally.

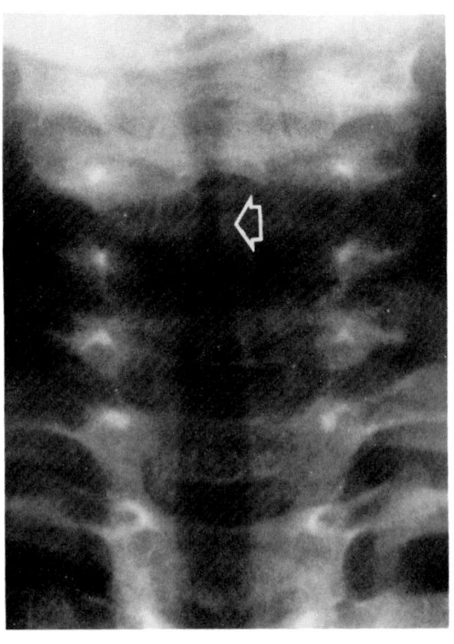
A

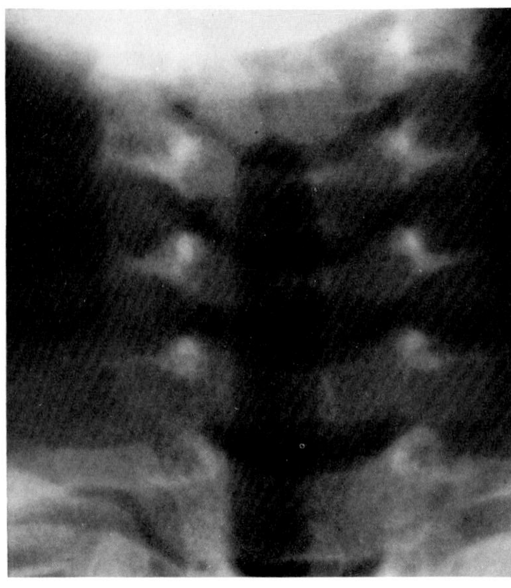
B

**Figure 32.2. Croup.** (A) Smooth, tapered narrowing of the subglottic portion of the trachea (gothic-arch sign). (B) A normal trachea with broad shouldering in the subglottic region.

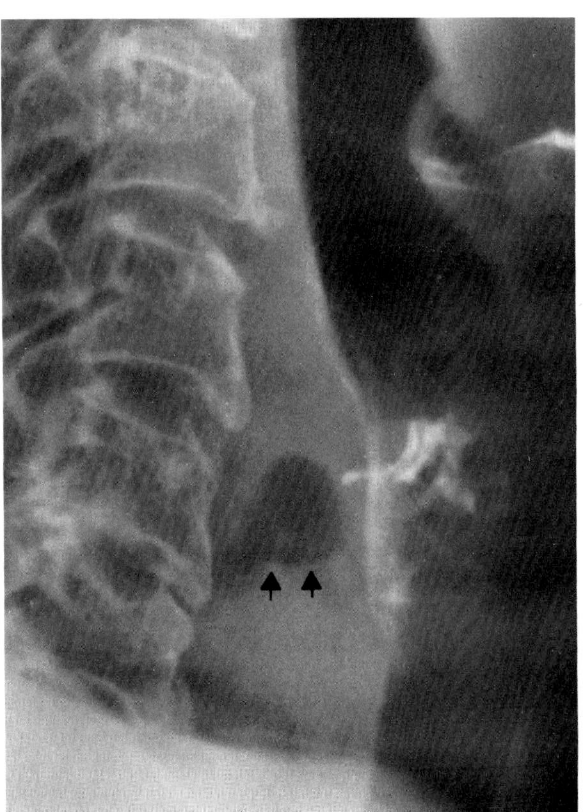

**Figure 32.3. Retropharyngeal abscess.** A collection of air in the region of the hypopharynx and proximal cervical esophagus (arrows) due to perforation of the pyriform sinus by a foreign body.

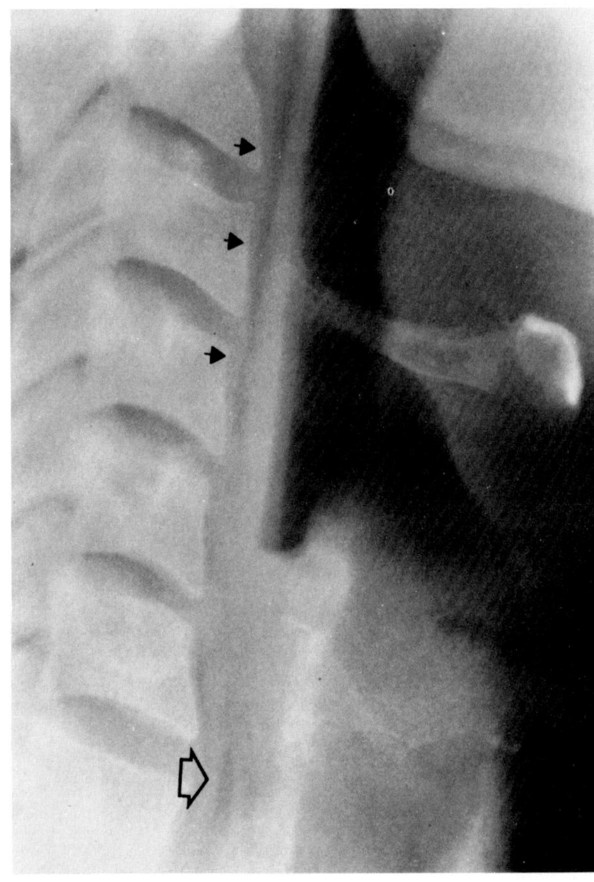

**Figure 32.4. Retropharyngeal abscess.** In the retropharyngeal region, air (large arrow) has dissected upward (small arrows) in the prevertebral space. The patient complained of fullness in the neck 1 day after a chicken bone stuck in his throat.

### Retropharyngeal Abscess

Retropharyngeal abscess is most common in children, in whom it is usually a complication of acute bacterial pharyngitis. In adults, retropharyngeal abscess may result from an injury to the posterior pharyngeal wall by a sharp object or may be secondary to middle ear or parotid infection. The characteristic radiographic appearance is a rounded thickening of the prevertebral tissues that causes forward displacement, compression, and even complete obliteration of the esophagus, larynx, and trachea. An air-fluid level may sometimes be seen within the mass. In infants and young children, it is essential that the patient's neck be extended when the lateral film is obtained, since the upper airway is so flexible that a pseudomass or pseudostenosis may be produced if the neck is flexed or incompletely extended.[1–3]

## DISEASES OF THE SINUSES

### Sinusitis

Viral infection of the upper respiratory tract may lead to obstruction of drainage of the paranasal sinuses and the development of localized pain, tenderness, and fever. Radiographically, acute or chronic sinusitis causes mucosal thickening that appears as a soft tissue density lining the walls of the involved sinuses. The maxillary antra are most commonly affected and are best visualized on the Water's projection. Although occasionally seen in chronic sinusitis, an air-fluid level in the sinus is usually considered a manifestation of acute inflammatory disease. The destruction of the bony sinus wall is an ominous sign indicating secondary osteomyelitis. Virulent fungal infections of the sinus (mucormycosis, actinomycosis) often cause diffuse bone destruction simulating malignant disease.

An unusual form of chronic sinusitis is associated with bronchiectasis and situs inversus in Kartagener's syndrome.

### Tumors

The most common benign tumor of the paranasal sinuses is osteoma, a dense mass of solid cortical bone that usually arises in the frontal or ethmoid region and may be associated with multiple polypoid lesions of the intestine (Gardner's syndrome). Carcinoma and other malignant tumors produce soft tissue masses associated with diffuse bone destruction without reactive sclerosis.[3]

## DISORDERS OF THE TRACHEA

### Tracheal Stenosis

Persistent tracheal narrowing of varying length and severity may result from external trauma or be secondary to intubation or tracheostomy. Plain chest radiographs or conventional mediastinal tomograms demonstrate tracheal narrowing that shows little or no change during the various phases of respiration.[4–6]

### Tracheobronchomegaly

Tracheobronchomegaly refers to dilatation of the tracheobronchial tree due to a weakness of muscular and

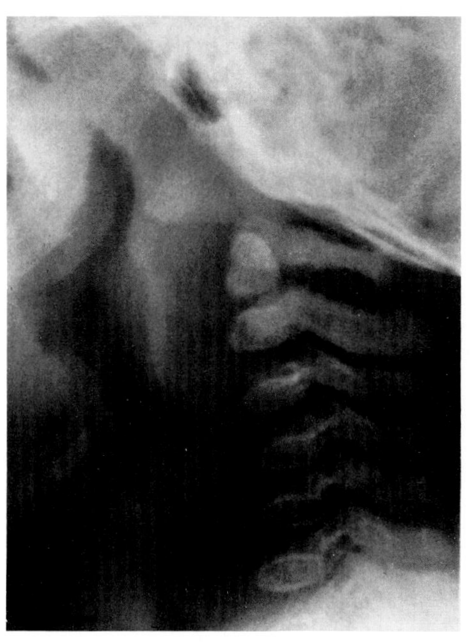

A

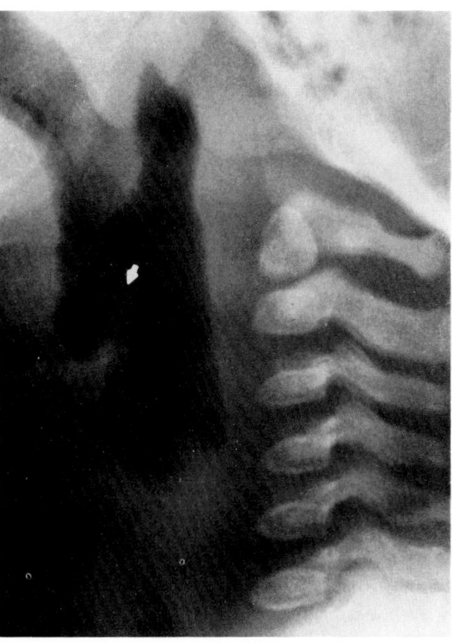

B

**Figure 32.5. Pseudoretropharyngeal abscess.**
(A) A diffuse soft tissue prominence in the retropharyngeal region in a young child. The epiglottis cannot be visualized on the film. (B) A repeat study with the head slightly more extended demonstrates a normal precervical space and epiglottis (arrow).

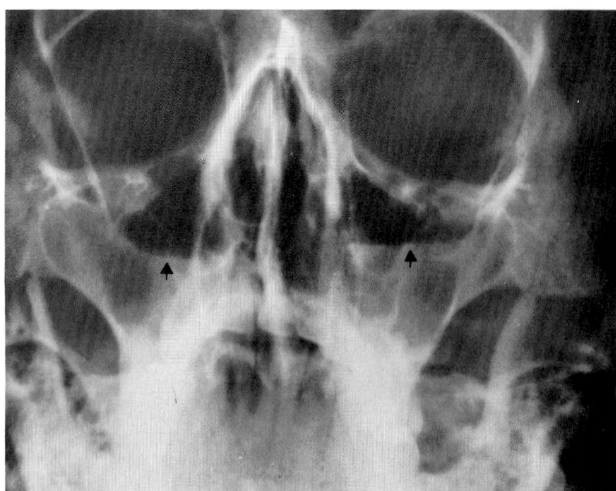

**Figure 32.6. Acute sinusitis.** There is mucosal thickening involving most of the paranasal sinuses and air-fluid levels (arrows) in both maxillary antra.

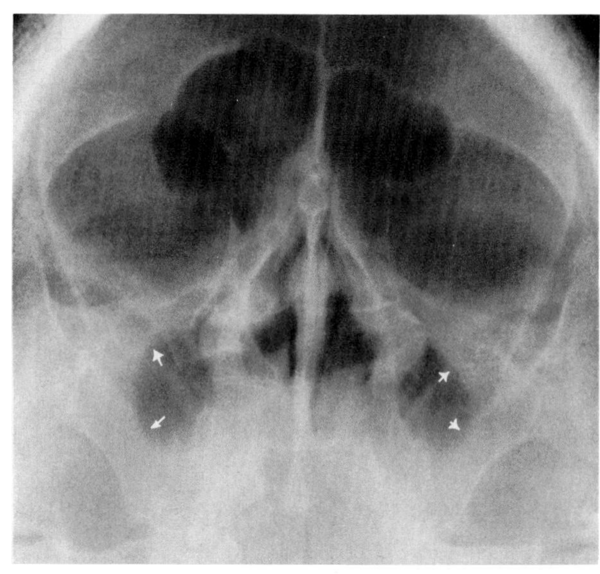

**Figure 32.7. Chronic sinusitis.** Mucosal thickening appears as a soft tissue density (arrows) lining the walls of the maxillary antra.

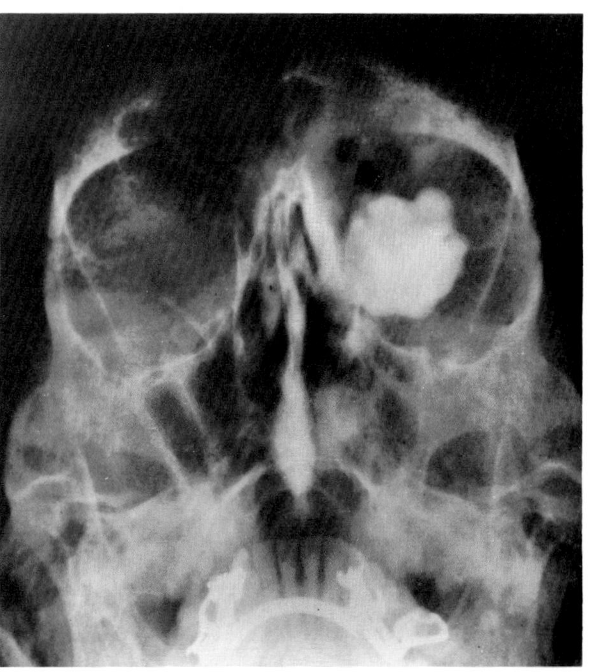

A

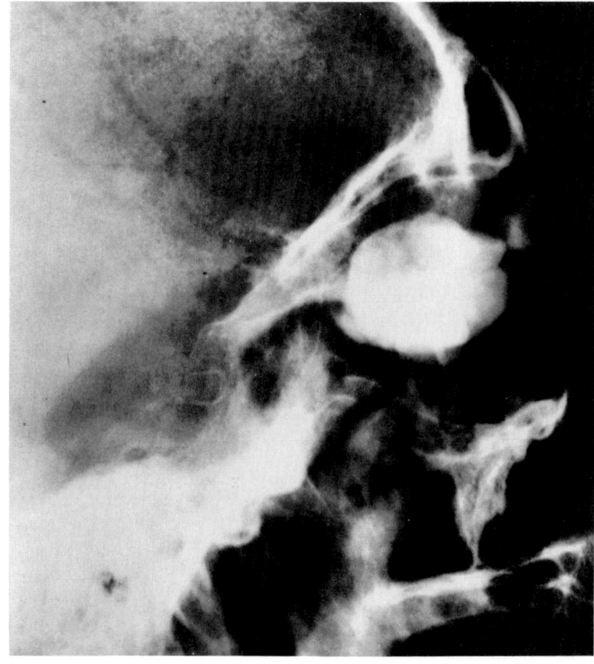

B

**Figure 32.8. Osteoma of the ethmoid sinus.** (A) Frontal and (B) lateral views of the skull demonstrate a large, extremely dense osteoma filling and expanding the left ethmoid sinus.

elastic tissue. The increased compliance of the tracheal and bronchial walls causes easy collapsibility during forced expiration and leads to symptoms that are usually indistinguishable from those caused by chronic bronchitis or bronchiectasis. Plain chest radiographs, especially those obtained in the lateral projection, usually demonstrate an increased caliber of the trachea and major bronchi. The air columns tend to have an irregular corrugated appearance due to the protrusion of redundant mucosa between the cartilagenous rings ("tracheal diverticulosis"). Inefficiency of the cough mechanism permits mucus retention that leads to recurrent infection and radiographic evidence of chronic inflammatory disease.[5,7–9]

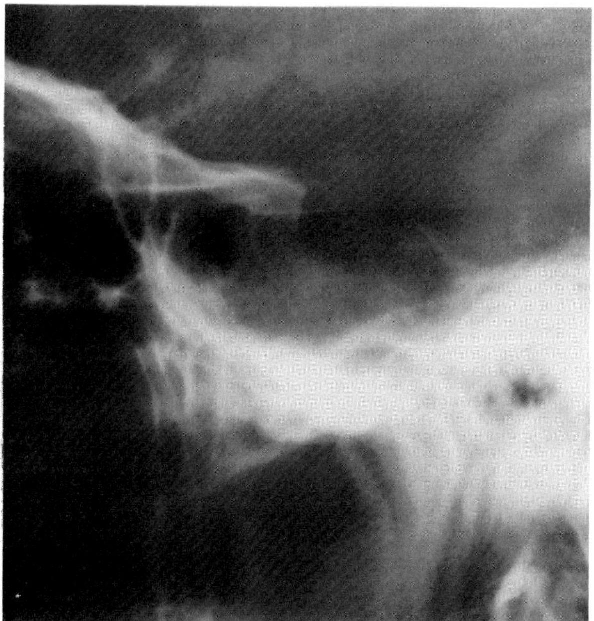

**Figure 32.9. Squamous cell carcinoma of the sphenoid sinus.** There is a large soft tissue mass with complete destruction of the floor and posterior part of the sella turcica.

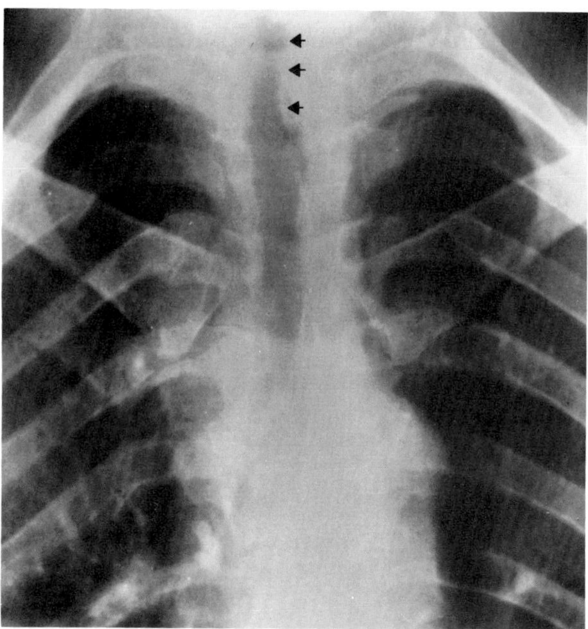

**Figure 32.10. Tracheal stenosis.** A plain chest radiograph demonstrates narrowing of the trachea (arrows) following prolonged intubation.

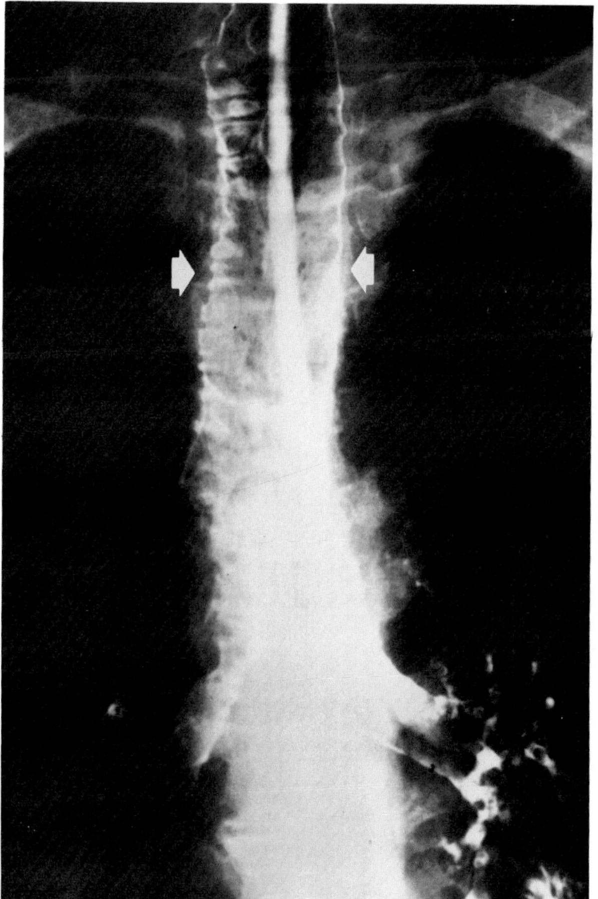

**Figure 32.11. Tracheobronchomegaly.** Contrast examination shows diffuse widening of the trachea (arrows), with outpouchings of the tracheal wall between the cartilaginous rings. A bronchographic catheter is present. (From Choplin, Wehunt, and Theros,[5] with permission from the publisher.)

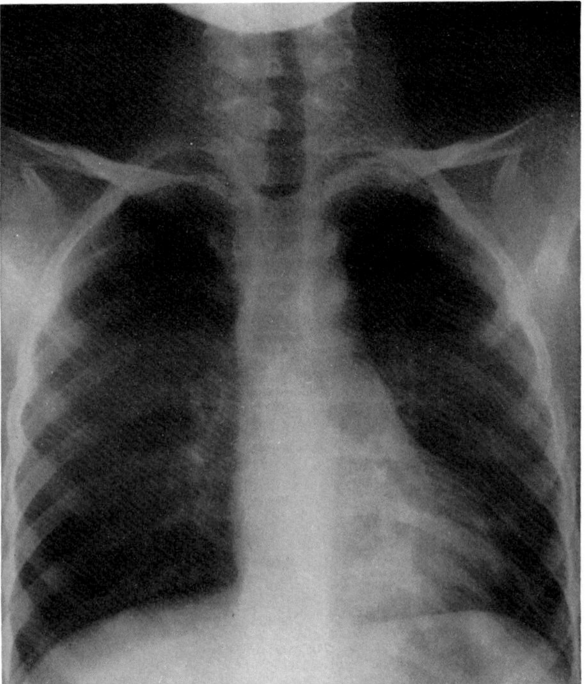

A

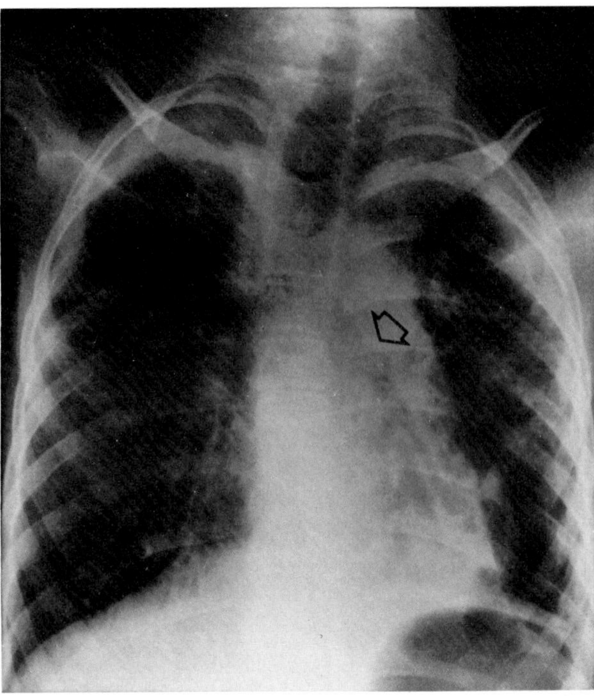

B

**Figure 32.12. Progressive tracheobronchomegaly.** (A) An initial film at 10 years of age shows moderate enlargement of the trachea. (B) On a film taken 1½ years later, the thoracic tracheal diameter and the main bronchi are enlarged. This is best seen in the left main stem bronchus (arrow). (From Lallemand, Chagnon, Buriot, et al,[10] with permission from the publisher.)

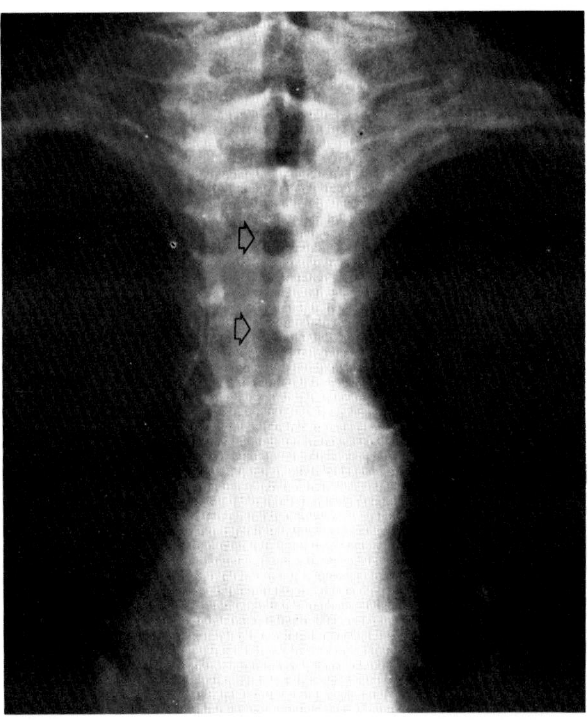

**Figure 32.13. Relapsing polychondritis.** There is narrowing of the trachea from the subglottic region to its bifurcation (arrows) in this patient with long-standing disease. (From Choplin, Wehunt, and Theros,[5] with permission from the publisher.)

## Relapsing Polychondritis

Relapsing polychondritis is an unusual inflammatory and degenerative process that affects cartilage throughout the body, including the tracheobronchial tree, joints, ear lobes, and nose. Involvement of the trachea and major bronchi causes severe irregular narrowing of the air column that is unchanging on forced inspiration or expiration. Erosive changes in the skeleton, especially the hands and feet, spine, and sacroiliac joints, closely simulate rheumatoid arthritis.[5,10,11]

## Saber Sheath Trachea

The "saber sheath" trachea refers to a sign of chronic obstructive pulmonary disease in which the coronal diameter of the intrathoracic trachea is half or less that of the sagittal diameter. The lateral walls of the trachea are usually thickened, and there is often evidence of ossification of the cartilaginous rings. The shape of the trachea abruptly changes to a normal, rounded configuration at the thoracic outlet.[12]

## Tracheal Tumors

Carcinoma of the trachea is a rare tumor that causes symptoms of chronic upper airway obstruction. Well-

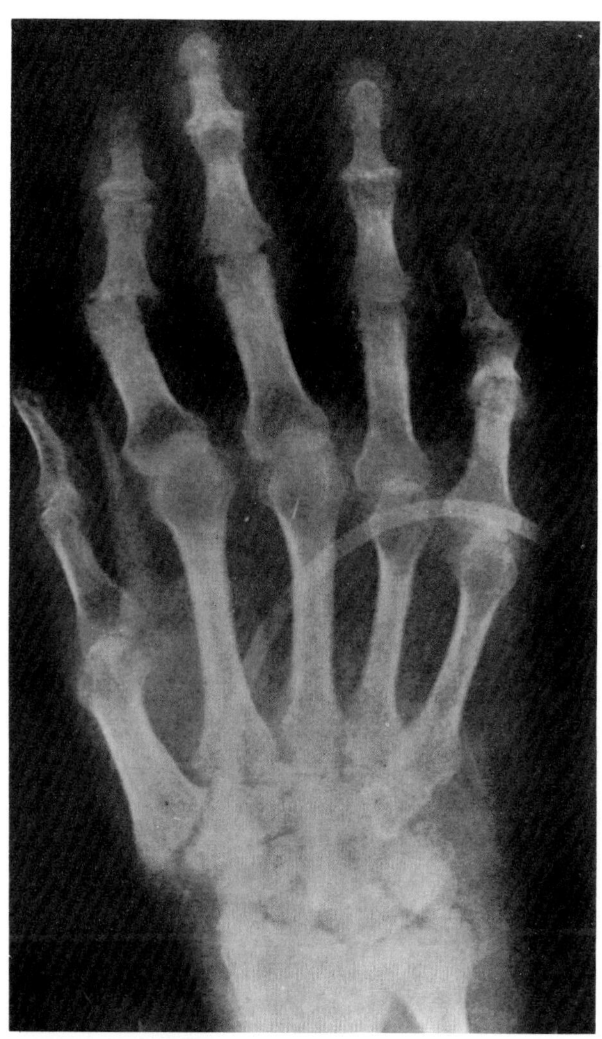

**Figure 32.14. Relapsing polychondritis.** A radiograph of the hand demonstrates soft tissue swelling over the ulnar styloid process, swelling of the fingers, and severe erosive changes at the interphalangeal joint cartilages. The terminal tufts of the fingers are spared, and bone mineralization is good. Destruction of portions of the joint surfaces in the wrist and in the metacarpophalangeal joints is also noted. (From Johnson, Mital, Rodnan, et al,[11] with permission from the publisher.)

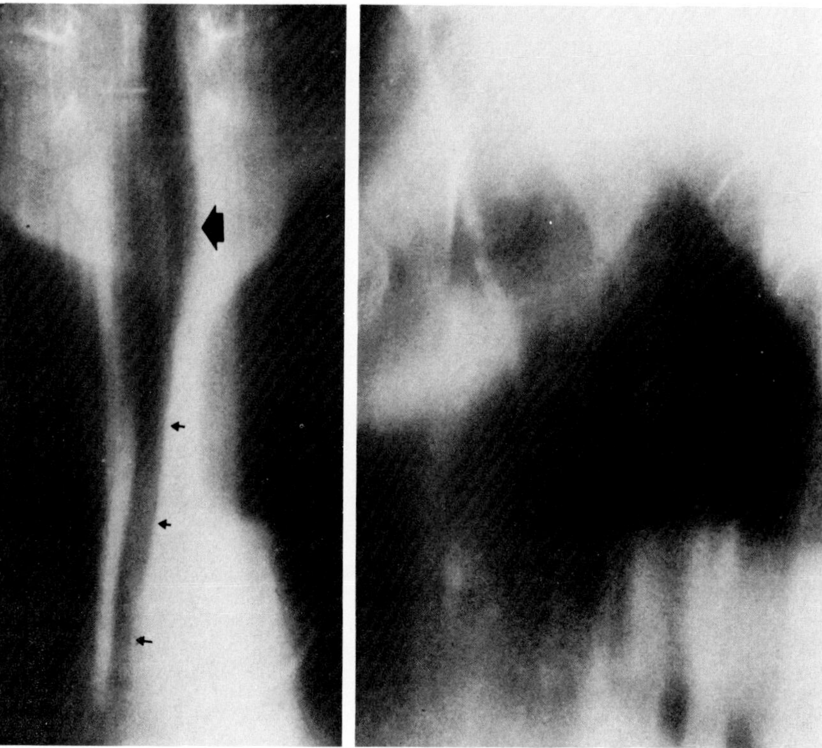



**Figure 32.15. Saber sheath trachea.** (A) Frontal and (B) lateral tomographic sections in a patient with chronic obstructive pulmonary disease demonstrate severe coronal narrowing of the intrathoracic trachea (small arrows) with an abrupt change to a more rounded cross-sectional shape at the thoracic outlet (large arrow). Calcific densities are present in the tracheal rings. (From Greene and Lechner,[12] with permission from the publisher.)

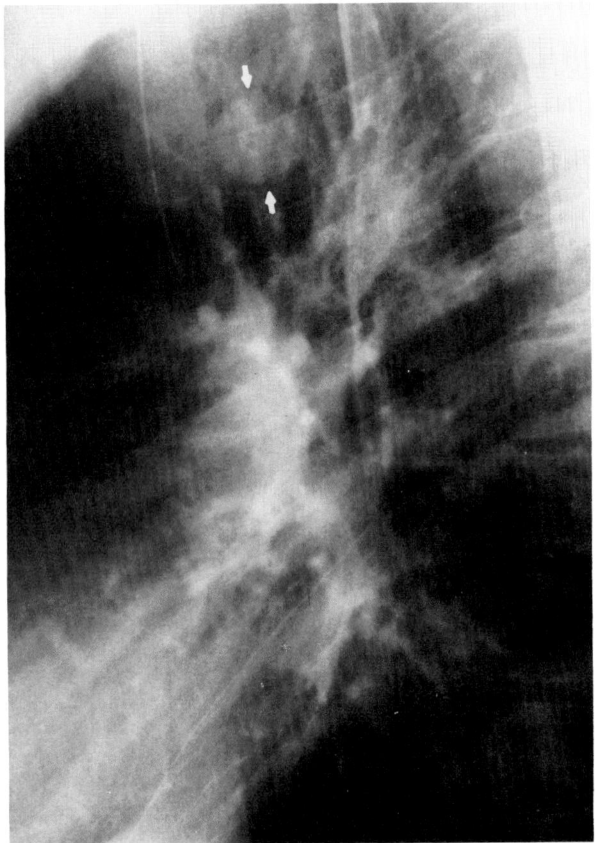
A

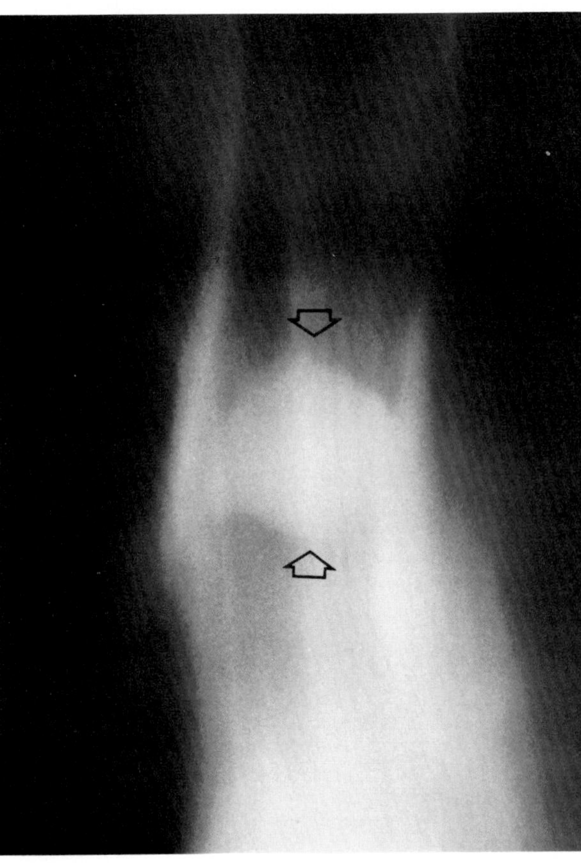
B

**Figure 32.16. Carcinoma of the trachea.** (A) A lateral view of the chest shows an ill-defined soft tissue density (arrows) within the tracheal air column. (B) Tomography more clearly shows the mass.

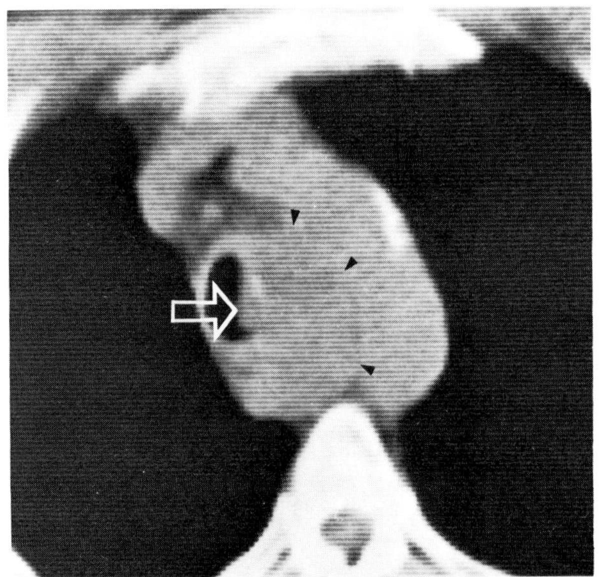

**Figure 32.17. Carcinoma of the trachea.** A CT scan shows the intratracheal mass (open arrow) with mediastinal extension (arrowheads) of the tumor. (From Gamsu and Webb,[14] with permission from the publisher.)

penetrated plain films of the chest may demonstrate tracheal narrowing or a polypoid mass projecting into the radiolucent tracheal air column. Mediastinal tomography is often of value in more clearly showing the lesion. In patients with primary or secondary neoplasms involving the trachea, computed tomography is accurate in defining the intraluminal presence of tumor, the degree of airway compression, and the extratracheal extension of tumor.[6,13,14,15]

## REFERENCES


1. Dunbar JS: Epiglottitis and croup. *J Can Assoc Radiol* 12:86–95, 1961.
2. Podgore JK, Bass JW: The "thumb sign" and "little finger sign" in acute epiglottitis. *J Pediatr* 88:154–155, 1976.
3. Juhl JH: *Essentials of Roentgen Interpretation*. Philadelphia, JB Lippincott, 1981.
4. James AE, McMillan AS, Eaton SB, et al: Roentgenology of tracheal stenosis resulting from cuffed tracheostomy tubes. *AJR* 109:455–466, 1970.
5. Choplin RH, Wehunt WD, Theros EG: Diffuse lesions of the trachea. *Semin Roentgenol* 18:38–50, 1983.
6. Gamsu G: Computed tomography of the trachea and central bronchi, in Moss AA, Gamsu G, Genant HK (eds): *Computed Tomography of the Body*. Philadelphia, WB Saunders, 1983.

7. Campbell AH, Young IF: Tracheobronchial collapse, a variant of obstructive respiratory disease. *Br J Dis Chest* 57:174–181, 1963.
8. Bateson EM, Woo-Ming M: Tracheobronchomegaly. *Clin Radiol* 24:354–358, 1973.
9. Lallemand D, Chagnon S, Buriot D, et al: Tracheomegaly and immune deficiency syndromes in childhood. *Ann Radiol* 24:67–72, 1981.
10. Dolan DL, Lemmon GB, Teitelbaum SL: Relapsing polychondritis. An analytical literature review and studies on pathogenesis. *Am J Med* 41:285–299, 1966.
11. Johnson TH, Mital N, Rodnan GP, et al: Relapsing polychondritis. *Radiology* 106:313–315, 1973.
12. Greene R, Lechner GL: "Saber-sheath" trachea. A clinical and functional study of marked coronal narrowing of intrathoracic trachea. *Radiology* 15:265–268, 1975.
13. Janower ML, Grillo HC, MacMillan AS, et al: The radiological appearance of carcinoma of the trachea. *Radiology* 96:39–43, 1970.
14. Gamsu G, Webb WR: Computed tomography of the trachea: Normal and abnormal. *AJR* 139:321–326, 1982.
15. Felson B: Neoplasms of the trachea and main stem bronchi. *Semin Roentgenol* 18:23–37, 1983.

## Chapter Thirty-Three

# Tumors of the Lung

### BRONCHOGENIC CARCINOMA

Bronchogenic carcinoma produces a broad spectrum of radiographic abnormalities that depend upon the site of the tumor and its relation to the bronchial tree. The tumor may appear as a discrete mass or be undetectable and identified only by virtue of secondary postobstructive changes caused by tumor within or compressing a bronchus.

The solitary pulmonary nodule within the lung parenchyma presents a diagnostic dilemma between malignancy and a benign process (tumor or granuloma). Although malignant tumors generally have ill-defined, irregular, or fuzzy borders in contrast to the sharp margins of benign lesions, many exceptions to this rule are reported. Benign lesions tend to be smaller (less than 2 cm in diameter) than malignant ones. The presence of central target or popcorn calcification is almost certain proof of an inflammatory cause. A comparison with previous films is essential, since a nodule that is unchanged in size on radiographs made 2 or more years previously can be assumed to be benign. The growth rate of a solitary pulmonary nodule (doubling time) has been used to determine the likelihood of its being malignant. A pulmonary nodule that doubles in volume in less than

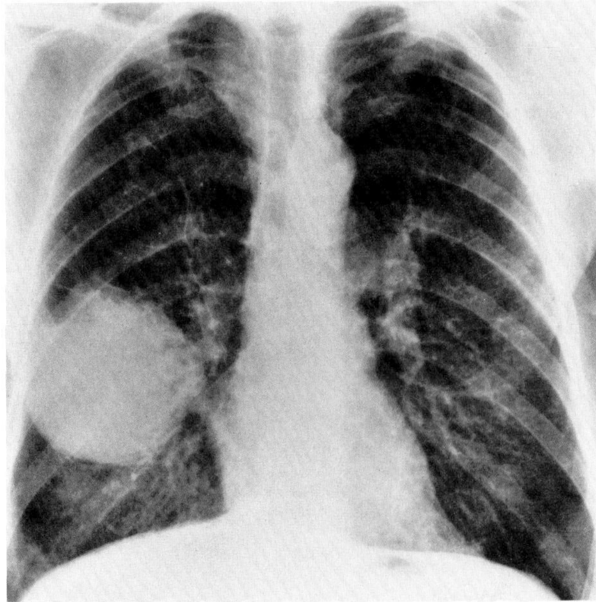

**Figure 33.1. Bronchogenic carcinoma.** There is a large soft tissue mass in the right lower lobe. The borders are relatively irregular and there is no evidence of calcification. Previous films showed only emphysematous changes with bullae in the upper lobes.

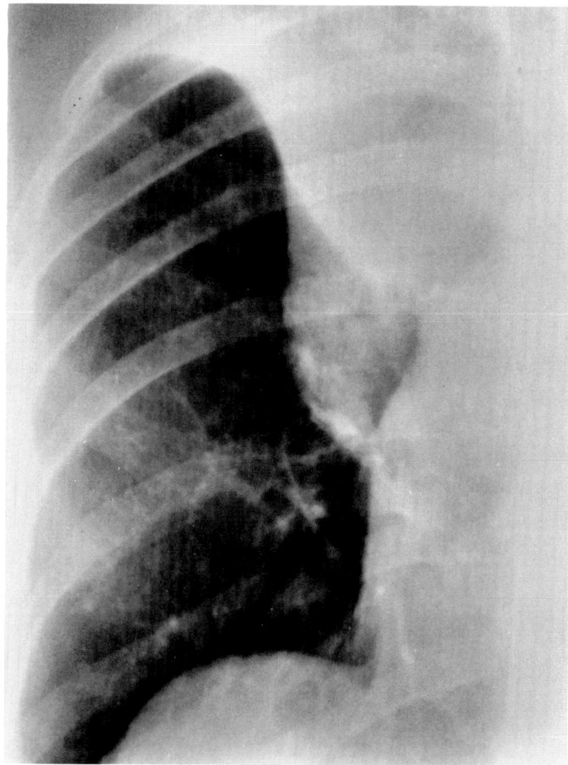

**Figure 33.2. Bronchogenic carcinoma.** A frontal chest radiograph demonstrates a typical reverse S-shaped curve (S sign of Golden) representing collapse of the right upper lobe associated with malignant bronchial obstruction.

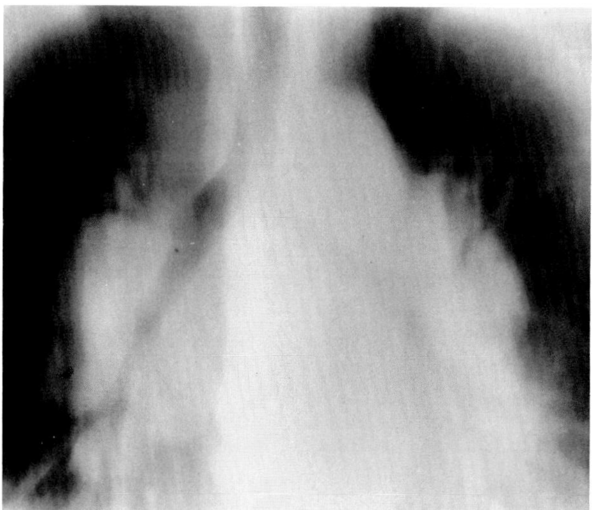

**Figure 33.3. Bronchogenic carcinoma.** Tomography demonstrates bilateral bulky hilar adenopathy typical of oat cell carcinoma.

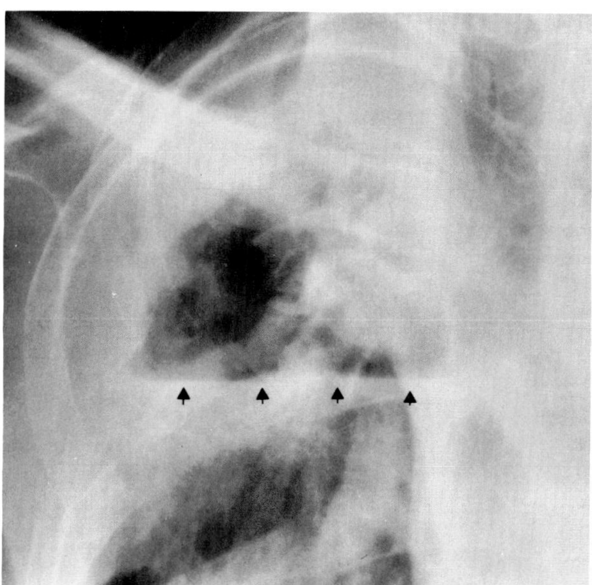

**Figure 33.4. Bronchogenic carcinoma.** A large, cavitary, right upper lobe mass with an air-fluid level (arrows) and associated rib destruction.

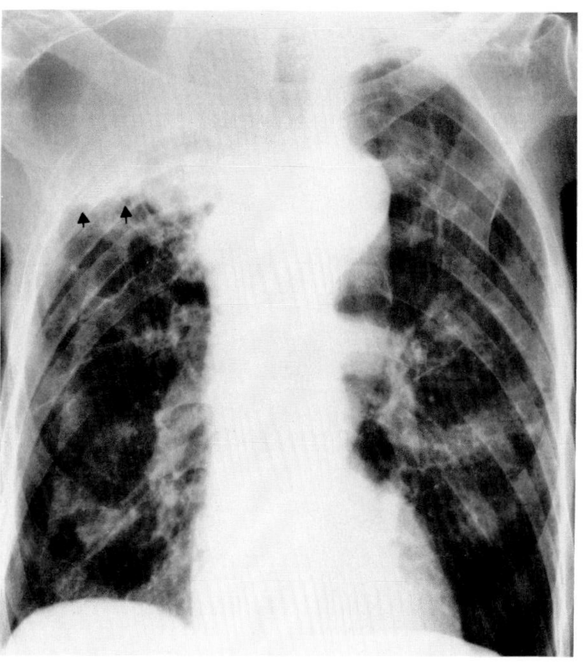

**Figure 33.5. Pancoast's tumor.** Increased density in the right apex (arrows) is associated with some destruction of the underlying ribs. Although this appearance may simulate benign apical pleural thickening, the marked asymmetry and irregularity of the right apical mass should suggest the diagnosis of bronchogenic carcinoma.

1 month or more than 18 months is usually benign. However, the overlapping of growth rates of benign and malignant lesions, particularly among rapidly growing nodules, makes the use of doubling time unreliable as an absolute indicator of malignancy. In the patient over 35 to 40 years of age, a solitary pulmonary nodule should be resected unless it can be unequivocally demonstrated to be benign.

Airway obstruction by bronchogenic carcinoma may cause segmental atelectasis and often leads to postob-

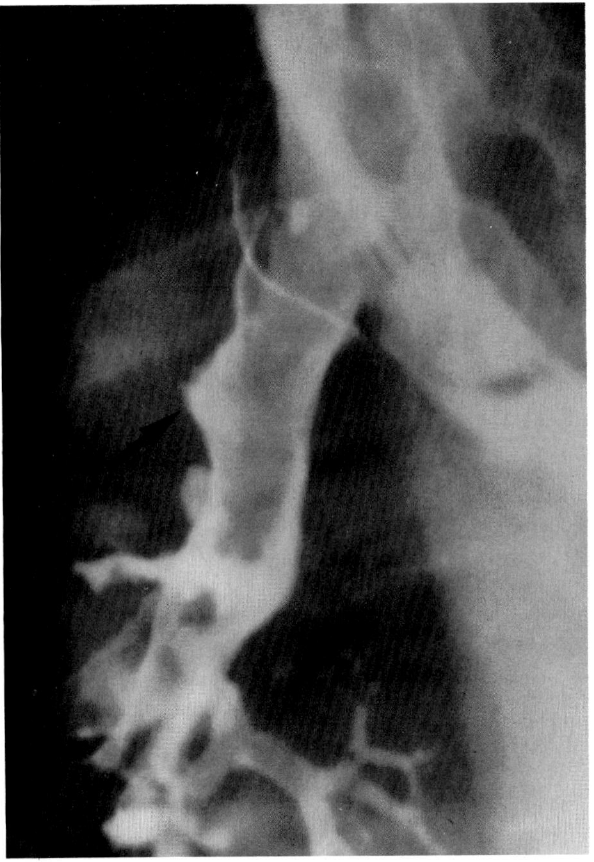

**Figure 33.6. Bronchogenic carcinoma.** A bronchogram in a man with an ill-defined right perihilar mass shows abrupt termination (arrow) of the right upper lobe bronchus at its origin. (From Eisenberg,[27] with permission from the publisher.)

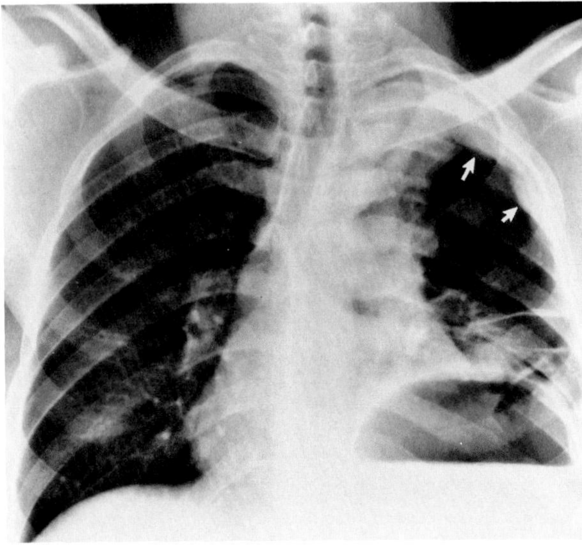

**Figure 33.7. Pleural metastases (arrows) from bronchogenic carcinoma.** There is elevation of the left hemidiaphragm due to phrenic nerve involvement, and postobstructive atelectatic change secondary to the left perihilar lesion.

structive pneumonia that develops distal to the obstructed bronchus. An important sign differentiating postobstructive pneumonia from simple inflammatory disease is the absence of an air bronchogram in the former, since this sign can only be detected if there is a patent airway leading to the area of consolidation. A characteristic radiographic sign of tumor-induced atelectasis of the right upper lobe is the S sign of Golden, in which the upper, laterally concave segment of the S is formed by the elevated minor fissure and the lower medial convexity is produced by the tumor mass, which has caused bronchial narrowing and is responsible for the collapse. Other signs of airway obstruction due to bronchogenic carcinoma include incomplete resolution of a pneumonia, and focal emphysema due to a check-valve obstruction caused by the endobronchial lesion.

Unilateral enlargement of the hilum, best appreciated on serial chest radiographs, may be the earliest sign of bronchogenic carcinoma. The enlarged hilum represents either a primary carcinoma arising in the major hilar bronchus or metastases to enlarged pulmonary lymph nodes from a small primary lesion elsewhere in the lung. Tomography may be of value in assessing whether a hilar mass represents tumor or enlarged vascular structures, as well as in searching for focal bronchial narrowing produced by the tumor. Bulky hilar adenopathy, often bilateral, is characteristic of oat cell carcinoma. Generalized enlargement of mediastinal lymph nodes producing an undulating or lobular contour of a widened mediastinum is usually due to spread from an undifferentiated carcinoma.

Cavitation is common in bronchogenic carcinoma. It most often involves upper lung lesions and represents central necrosis of the neoplasm. The cavities usually resemble acute lung abscesses and have thick walls with an irregular, often nodular, inner surface.

Bronchogenic carcinoma arising in the apex of the lung (superior pulmonary sulcus, or Pancoast's, tumor) may be difficult to diagnose because the increased density simulates the apical pleural thickening often seen in older patients. Therefore, irregular or markedly asymmetric "pleural thickening" in the apex should elicit a high degree of suspicion. The destruction of adjacent ribs or vertebrae, Horner's syndrome, or pain down the arm should suggest an underlying malignant process in the apex of the lung.

Local spread and distant metastases from bronchogenic carcinoma can produce multiple radiographic abnormalities. Direct lymphatic spread can cause enlargement of hilar or mediastinal lymph nodes. Pleural effusion, sometimes large enough to obscure the underlying primary tumor, may reflect hilar lymphatic obstruction or direct extension or metastases to the pleura. Hematogenous spread to ribs or lymphatic metastasis to the visceral pleura may produce an extrapleural mass, with or without rib destruction. Involvement of the phrenic nerve can cause paralysis and elevation of the ipsilateral hemidiaphragm. On the right, extension of tumor in the su-

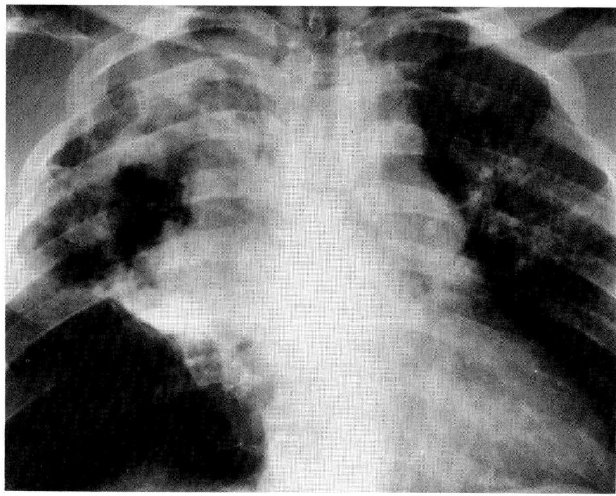
A

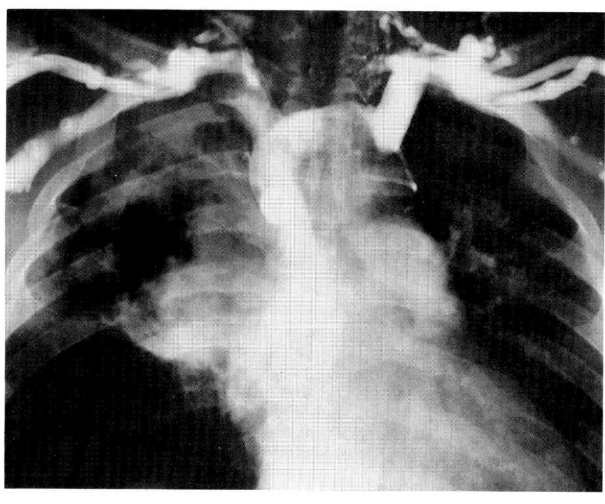
B

**Figure 33.8. Obstruction of the superior vena cava by a right upper lobe tumor.** (A) A frontal view of the chest shows a bulky, irregular mass filling much of the right upper lobe. (B) Bilateral upper extremity venograms show almost complete occlusion of the superior vena cava by the large malignant neoplasm.

prahilar region can cause obstruction of the superior vena cava. Metastases to bone are frequent and predominantly osteolytic. Osteoblastic metastases suggest oat cell carcinoma or adenocarcinoma, since they are very rare with squamous cell lesions. Hematogenous metastases to other portions of the lung occasionally occur, causing scattered masses of varying size that are usually smaller than the primary lesion. Metastases to the bowel produce single or multiple filling defects that often have central ulcers and a bull's-eye appearance.

Extrathoracic, nonmetastatic manifestations of bronchogenic carcinoma include hypertrophic pulmonary osteoarthropathy (see the section "Hypertrophic Osteoarthropathy" in Chapter 64), which often regresses after resection of the primary lung tumor; endocrine syndromes caused by ectopic hormone production by the tumor; migratory thrombophlebitis; and neuromuscular abnormalities.[1–8]

## BRONCHIOLAR (ALVEOLAR CELL) CARCINOMA

Unlike typical squamous cell bronchogenic carcinoma, bronchiolar carcinoma has no predilection for either sex and no apparent relation to cigarette smoking. The tumor most frequently appears as a well-circumscribed, peripheral solitary nodule, which may vary in size from 1 cm or less in diameter to a huge mass involving most of a lobe. The margins of the tumor are usually rather well circumscribed, though the mass may be poorly defined and simulate an area of focal pneumonia. The mass often contains an air bronchogram, a finding that is never associated with solitary nodules caused by bronchogenic carcinoma or a granuloma. The presence of a "pleural

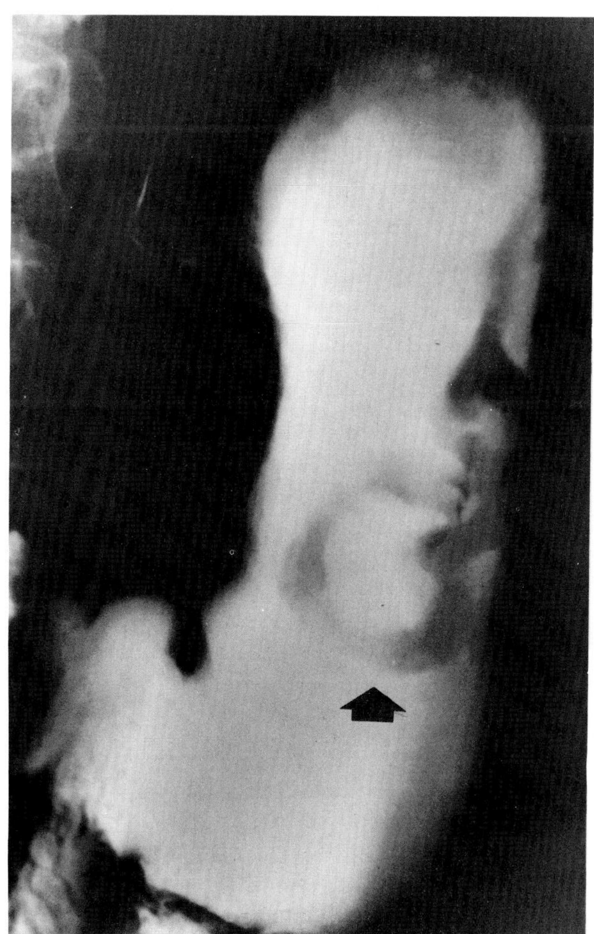

**Figure 33.9. Metastatic bronchogenic carcinoma.** There is a large, ulcerated filling defect in the stomach (arrow).

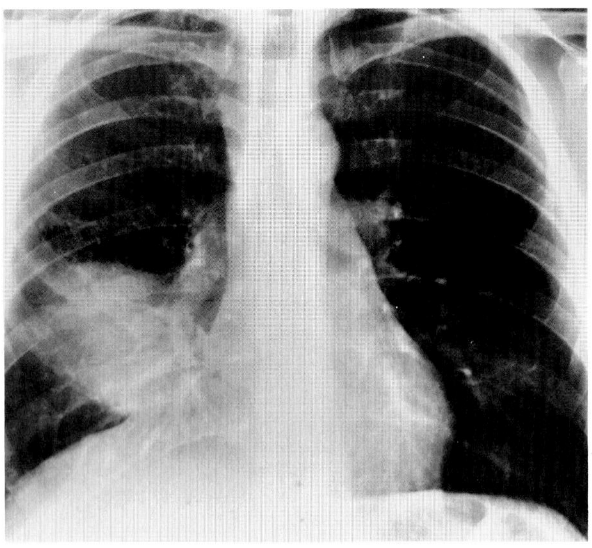

**Figure 33.10. Alveolar cell carcinoma.** A large, well-circumscribed tumor mass represents lobar consolidation.

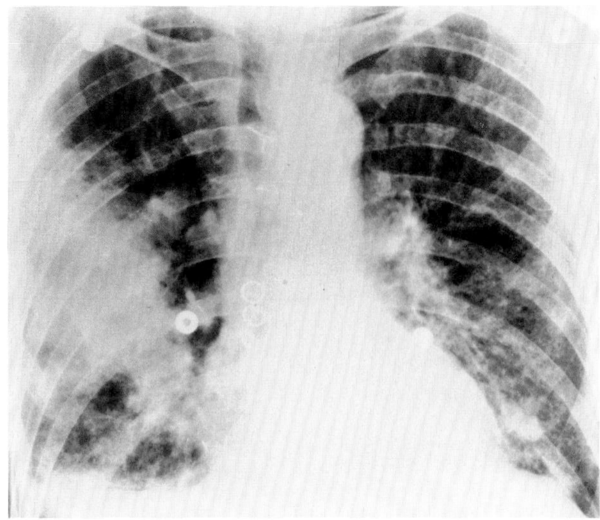

**Figure 33.11. Alveolar cell carcinoma.** The right-sided mass is patchy and ill-defined and simulates an area of focal pneumonia. The patient had previously undergone a coronary artery bypass procedure.

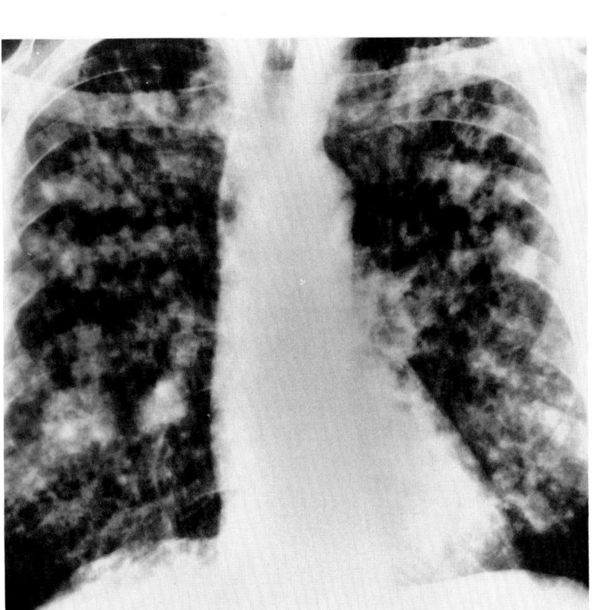

**Figure 33.12. Alveolar cell carcinoma.** Multiple poorly defined nodules are scattered throughout both lungs.

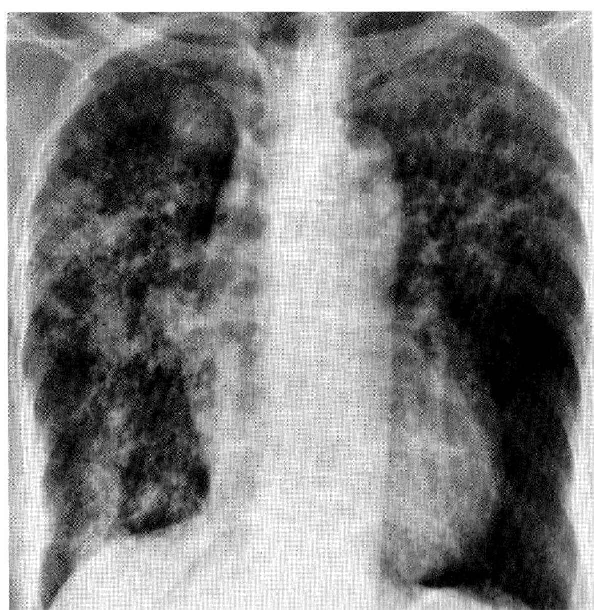

**Figure 33.13. Alveolar cell carcinoma.** The pronounced reticulonodular infiltrate that diffusely involves both lungs represents bronchogenic spread.

tail," a thin, untapered linear density arising from the lateral margin of a peripheral mass and extending to the pleural surface, was originally described as pathognomonic of bronchiolar carcinoma. Although an identical appearance may occur in granulomatous disease, when associated with an air bronchogram a pleural tail is highly suggestive of bronchiolar carcinoma.

The less common diffuse form of bronchiolar carcinoma presents a radiographic pattern varying from poorly defined nodules scattered throughout both lungs to irregular pulmonary infiltrates, often with air bronchograms. There is usually extensive involvement of one lung; bronchogenic spread commonly occurs and may lead to diffuse bilateral involvement.[9–11]

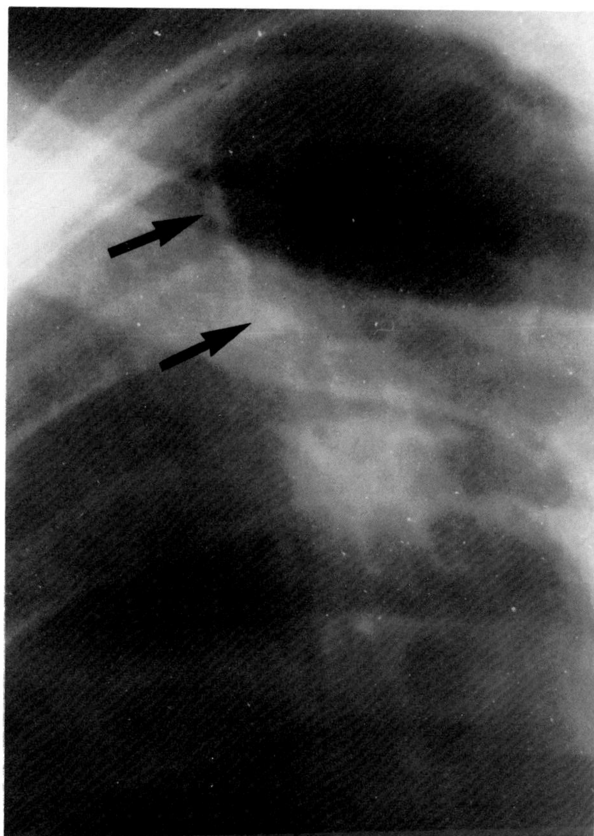

**Figure 33.14. Alveolar cell carcinoma.** A well-defined pleural tail (arrows) arises from a spiculated mass. (From Webb,[1] with permission from the publisher.)

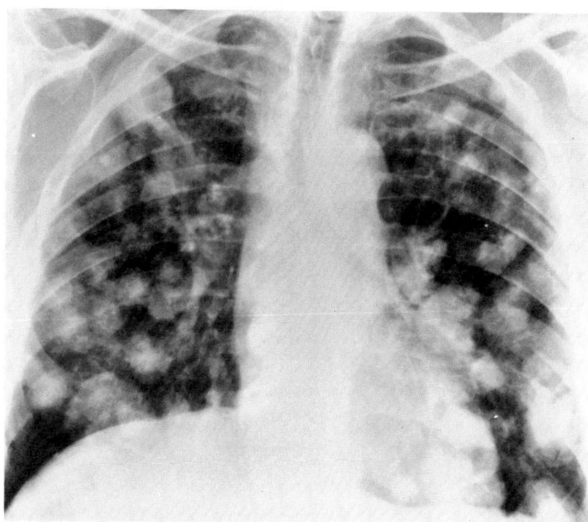

**Figure 33.15. Hematogenous metastases.** Multiple well-circumscribed nodules are scattered diffusely throughout both lungs.

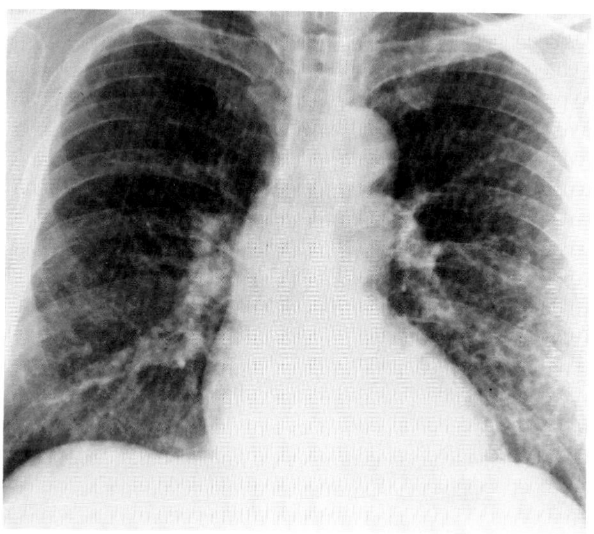

**Figure 33.16. Metastatic thyroid carcinoma.** Multiple fine miliary nodules are scattered throughout both lungs.

## METASTASES TO THE LUNG

Up to one-third of patients with cancer develop pulmonary metastases; in about one-half of these patients, the only demonstrable metastases are confined to the lungs. Pulmonary metastases may develop from hematogenous or lymphatic spread, most commonly from musculoskeletal sarcomas, myeloma, and carcinomas of the breast, urogenital tract, thyroid, and colon. Carcinomas of the breast, esophagus, or stomach may directly extend to involve the lungs; primary lung lesions may metastasize by endobronchial spread.

Hematogenous metastases typically appear radiographically as multiple, relatively well circumscribed, round or oval nodules throughout the lungs. The pattern may vary from fine miliary nodules produced by highly vascular tumors (renal or thyroid carcinoma; sarcoma of bone; trophoblastic disease) to huge, well-defined masses (cannonball lesions) caused by metastatic sarcomas. Carcinoma of the thyroid typically causes a snowstorm of metastatic deposits yet remains unchanged for a prolonged period because of a very low grade of malignancy.

Solitary metastases, which occur in about 25 percent of the cases, may be indistinguishable from primary bronchogenic carcinomas or benign granulomas. They usually have smooth or slightly lobulated margins and predominantly occur in the lower lobes. It is important to remember that the identification or prior resection of a primary neoplasm elsewhere does not necessarily indicate that a solitary mass in the lung is metastatic. Computed tomography (CT) may permit the detection of additional pulmonary masses that cannot be seen on standard chest radiographs, though CT may be unable to differentiate small metastatic nodules from noncalcified be-

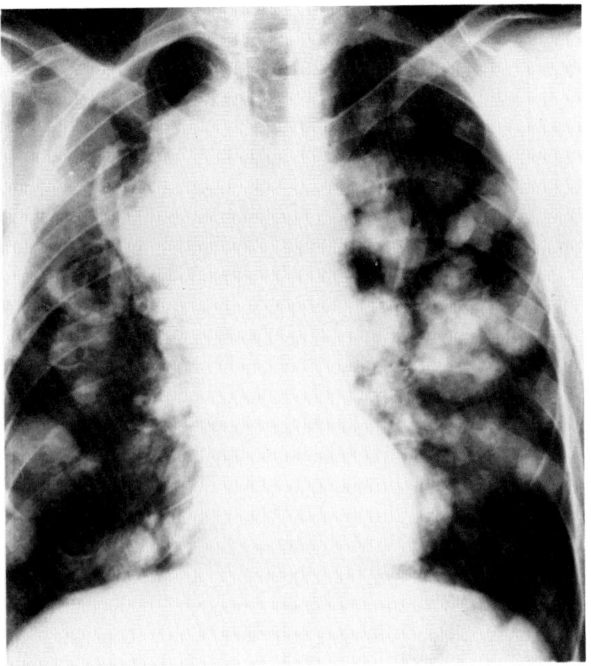

**Figure 33.17. Metastatic sarcoma.** Multiple large nodules, some of which have cavitated. A large soft tissue mass in the upper right mediastinum represents nodal metastases that obstructed the right upper lobe bronchus and led to bullous changes in the right apex.

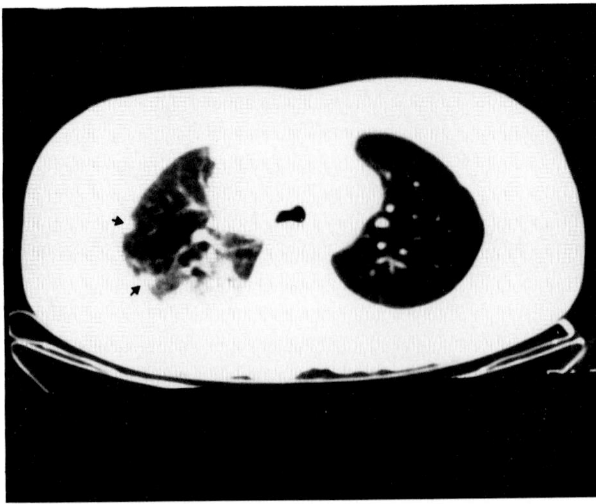

**Figure 33.18. Subpleural pulmonary metastases.** A CT scan shows two small metastatic nodules (arrows) that could not be detected on plain chest radiographs or whole-lung tomograms.

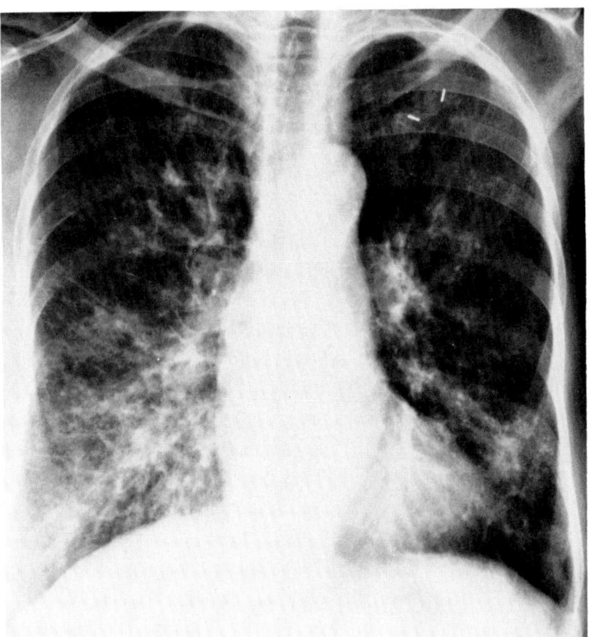

A

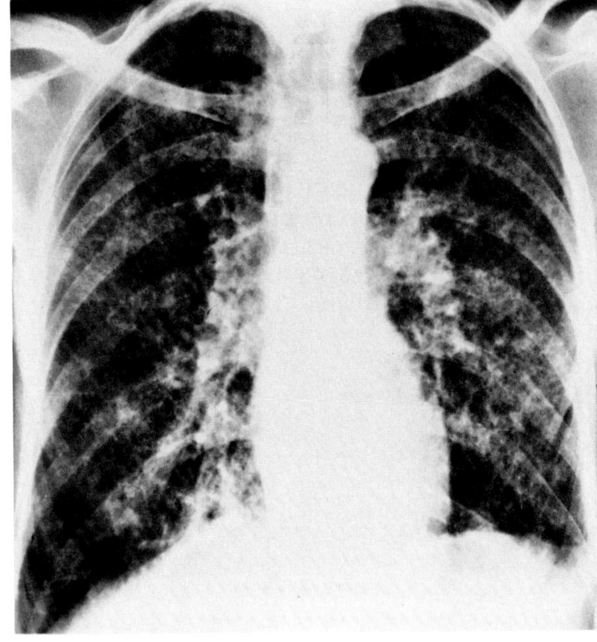

B

**Figure 33.19. Lymphangitic metastatic spread.** (A) Coarsened bronchovascular markings of irregular contour and poor definition primarily involve the right lower lung. Note the septal (Kerley) lines and the left mastectomy in this patient with carcinoma of the breast. (B) In this patient with metastatic carcinoma of the stomach, a superimposed nodular component representing hematogenous deposits produces a coarse reticulonodular pattern.

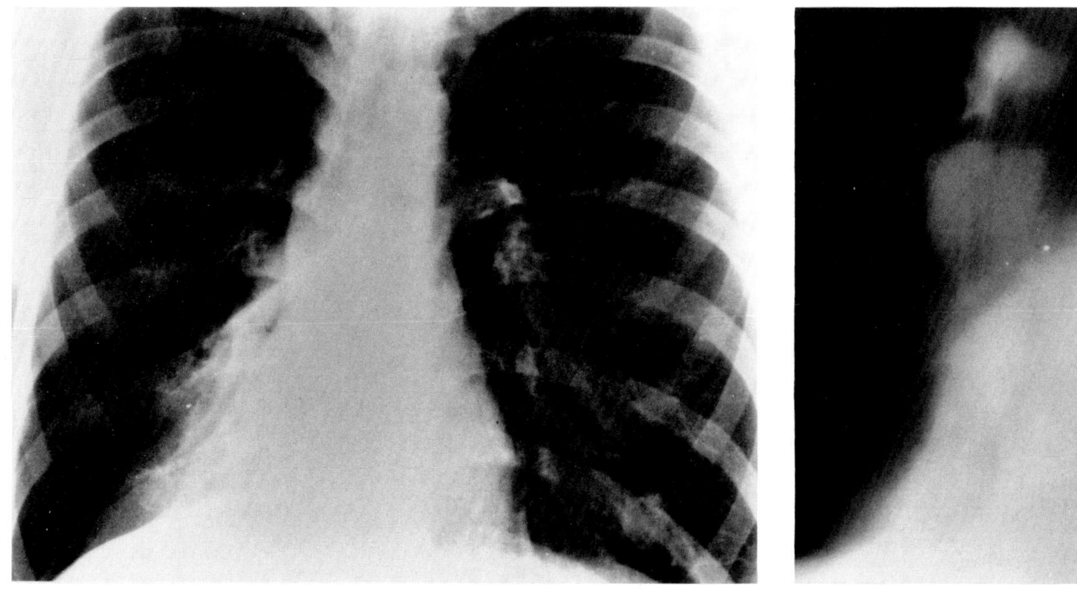

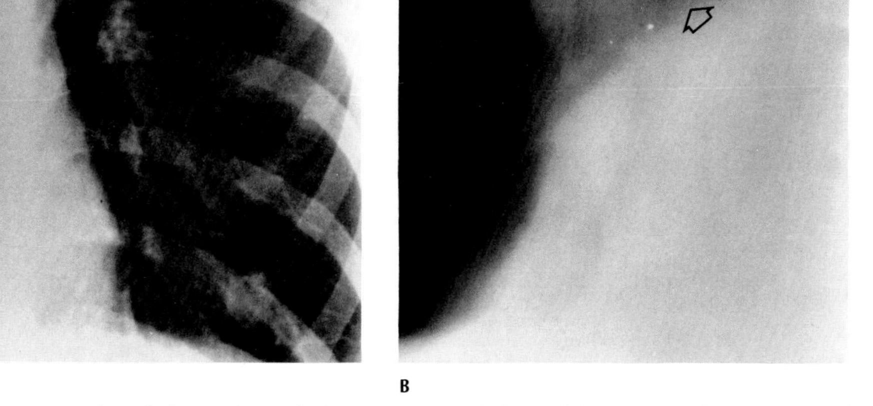

A B

**Figure 33.20. Central bronchial adenoma.** (A) A frontal chest radiograph demonstrates a right lower lobe density with obscuration of the right hemidiaphragm and relative preservation of the right border of the heart, consistent with right lower lobe collapse. (B) Tomography shows an ill-defined mass (arrow) causing a high-grade obstruction of the right lower lobe bronchus.

nign healed granulomas. With certain primary tumors (osteogenic sarcoma, choriocarcinoma), the growth rate of a malignant pulmonary nodule may be a valuable criterion, since these tumors tend to have doubling times of less than 1 month.

Lymphangitic metastatic spread throughout the lungs most commonly is a complication of carcinoma of the breast, stomach, thyroid, pancreas, larynx, cervix, and prostate. The radiographic appearance consists of coarsened bronchovascular markings of irregular contour and poor definition that are most prominent in the lower lobes and simulate interstitial pulmonary edema. A superimposed nodular component representing hematogenous deposits may produce a coarse reticulonodular pattern. Hilar lymph node enlargement and pleural effusion may also be present.[12–16]

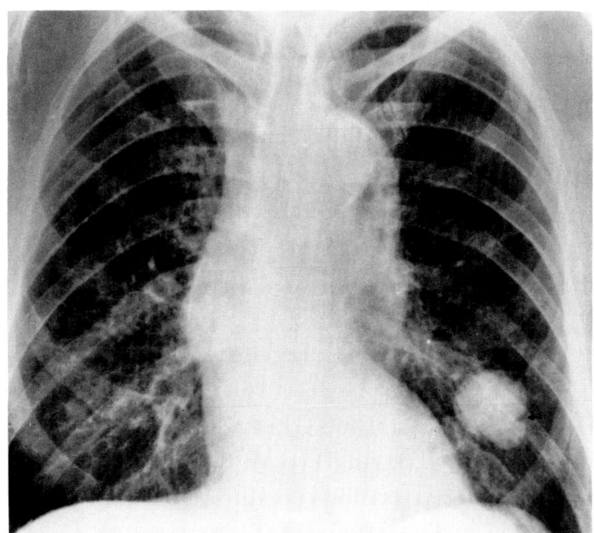

**Figure 33.21. Peripheral bronchial adenoma.** There is a nonspecific solitary pulmonary nodule at the left base. Note the notched indentation of the lateral wall (arrow) of the mass. Although this "Rigler notch" sign was initially described as pathognomonic of malignancy, an identical appearance is commonly seen in benign processes.

## BENIGN TUMORS OF THE LUNG

### Bronchial Adenoma

Bronchial adenomas are neoplasms of low-grade malignancy that constitute about 1 percent of all pulmonary neoplasms. Histologically, about 90 percent of bronchial adenomas are carcinoids; the remainder are cylindromas that tend to be more central in location and more invasive than the carcinoid type. Bronchial adenomas are at least as common in women as in men and are found in a younger age group than bronchogenic carcinoma. Hemoptysis and recurrent pneumonia are the most common symptoms; bony metastases, though rare, tend to be osteoblastic and to arise in patients with the carcinoid syndrome.

Because about 80 percent of bronchial adenomas are located centrally in major or segmental bronchi, bronchial obstruction with peripheral atelectasis and postobstructive pneumonitis is the most common radiographic

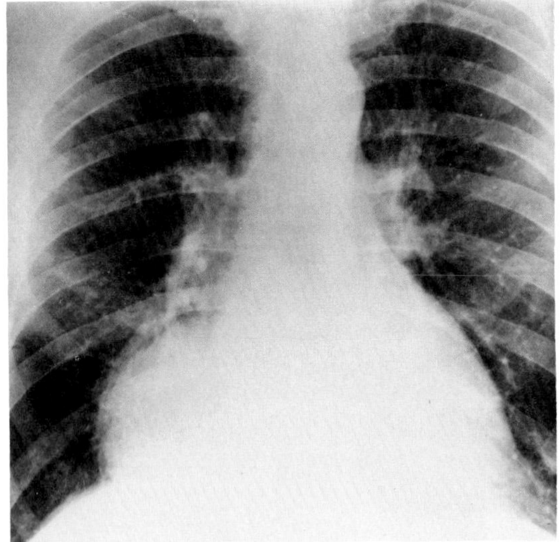
A

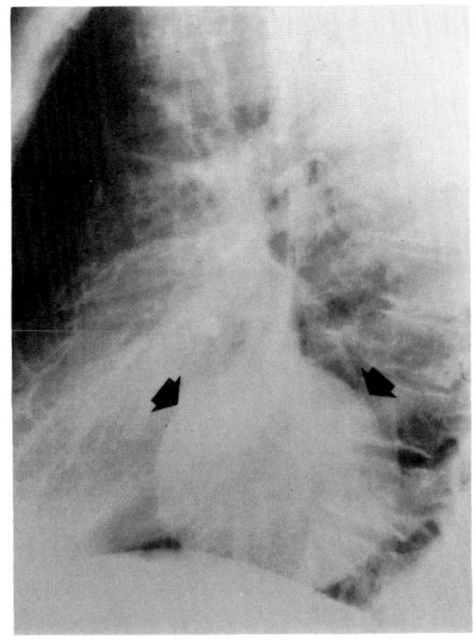
B

**Figure 33.22. Hamartoma.** (A) A frontal view of the chest shows a large mass (arrow) in the right cardiophrenic angle; the mass mimics a pericardial cyst or herniation through the foramen of Morgagni, both of which tend to occur at this site. (B) A lateral view shows the mass to be posterior (arrows), effectively excluding the other diagnostic possibilities. The mass is indistinguishable from other benign or malignant processes within the lung.

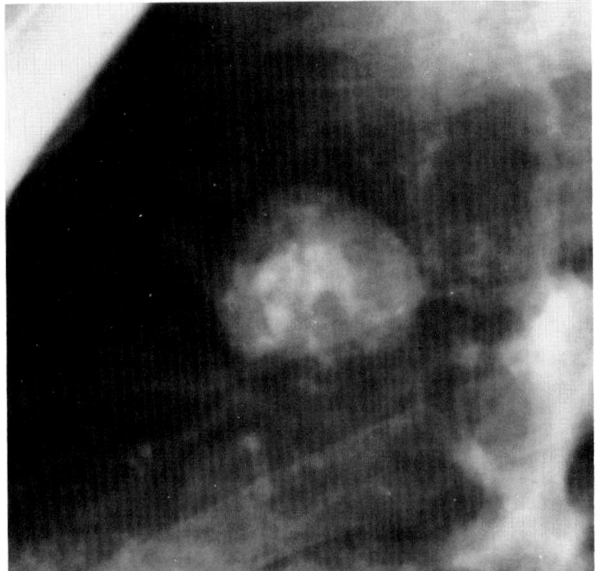

**Figure 33.23. Hamartoma.** The well-circumscribed solitary nodule contains characteristic irregular scattered calcifications (popcorn pattern).

finding. This characteristically produces a homogeneous increase in density corresponding exactly to a lobe or one or more segments, usually with a substantial loss of volume. If large enough, a central bronchial adenoma that has caused peripheral atelectasis and pneumonia may be identifiable as a discrete, lobulated, soft tissue mass. If the tumor is too small to obstruct the lumen, the chest radiograph appears normal. Tomography may demonstrate the rounded tumor mass within an air-filled bronchus. Peripheral bronchial adenomas do not cause bronchial obstruction and appear as nonspecific solitary pulmonary nodules.[17–19]

## Hamartoma

Unlike bronchial adenomas, hamartomas are completely benign, and almost all are located in the periphery of the lung. They typically appear as small, well-circumscribed solitary nodules that may demonstrate slow and, rarely, rapid growth on serial films. Although fewer than one-third of these tumors are calcified, the characteristic popcorn pattern of irregular, scattered calcifications is virtually pathognomonic. Tomography may be of value in showing these calcifications as well as in demonstrating the unusual small peripheral lucencies due to islands of fat that are reported to be characteristic of hamartomas. Fewer than 10 percent of hamartomas arise within the bronchi, where they may produce obstructive changes simulating those of bronchial adenoma.

Unless the characteristic pattern of calcification can be demonstrated, a hamartoma cannot be differentiated from a primary or metastatic malignancy, and thus surgical excision is required.[20–23]

## Bronchial Papilloma

Papillomas are the most common pharyngeal tumors in children but are rare in adults. They may be single or multiple and, when multiple, may extend down the tracheobronchial tree into the lungs. Papillomas typically obstruct airways and produce a radiographic appearance of recurrent pneumonia, atelectasis, and fibrosis simulating a chronic inflammatory disease. Tracheal tomography may demonstrate multiple filling defects in the air column (see Figs. 33.24 and 33.25).

A solitary intrabronchial papilloma is a rare lesion that is radiographically undetectable until bronchial obstruction leads to focal pneumonia or atelectasis. Bronchography may show a nodular intrabronchial defect indistinguishable from adenoma or carcinoma.[24–26]

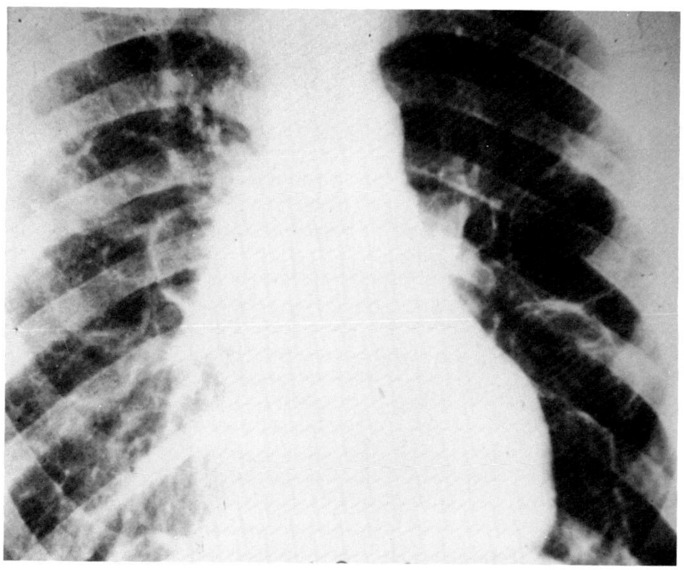
A

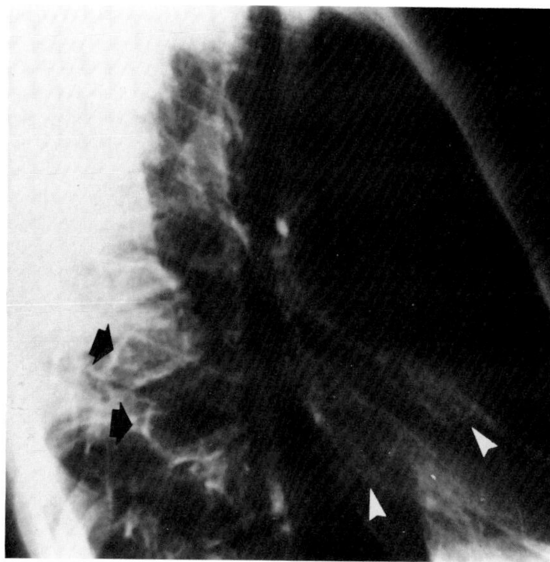
B

**Figure 33.24. Papillomatosis of the lower respiratory tract.** (A) Frontal and (B) lateral chest radiographs demonstrate multiple nodules, some of which are cavitating (arrows). These multiple thin-walled empty cysts are diagnostic of papillomatosis in a youngster. (From Felson,[28] with permission from the publisher.)

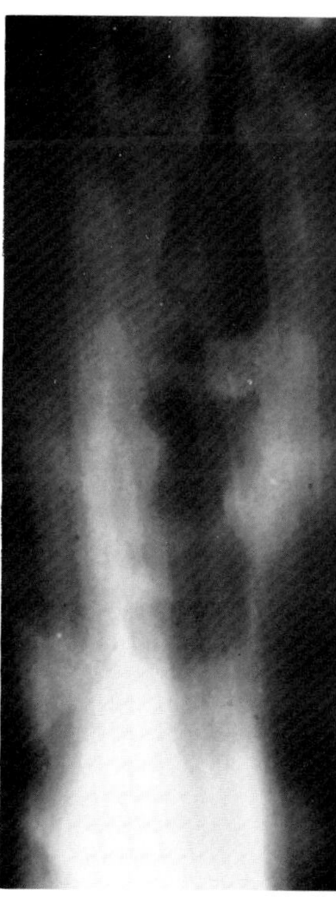

**Figure 33.25. Papillomatosis of the trachea.** A frontal tomogram shows innumerable nodules that are 3 to 10 mm in diameter and involve the entire trachea. (From Felson,[28] with permission from the publisher.)

## REFERENCES


1. Webb WR: The solitary pulmonary nodule, in Eisenberg RL, Amberg JR (eds): *Critical Diagnostic Pathways in Radiology: An Algorithmic Approach*. Philadelphia, JB Lippincott, 1981.
2. Fraser RG, Pare JAP: *Diagnosis of Diseases of the Chest*. Philadelphia, WB Saunders, 1979.
3. Emerson GL, Emerson MS, Sherwood CE: The natural history of carcinoma of the lung. *J Thorac Cardiovasc Surg* 37:291–304, 1959.
4. Cohen S, Sossain MS: Primary carcinoma of the lung. A review of 417 histologically proved cases. *Dis Chest* 49:67–74, 1966.
5. Nathan MH, Collins VP, Adams RA: Differentiation of benign and malignant pulmonary nodules by growth rate. *Radiology* 79:221–231, 1962.
6. O'Keefe ME, Good CA, McDonald JR: Calcification in solitary nodules of the lung. *AJR* 77:1023–1033, 1957.
7. Napoli LD, Hansen HH, Muggia FM, et al: The incidence of osseous involvement in lung cancer with special reference to the development of osteoblastic changes. *Radiology* 108:17–21, 1973.
8. Higgins GA, Shields FW, Keehn RJ: The solitary pulmonary nodule. *Arch Surg* 110:570–575, 1975.
9. Ludington LG, Verska JJ, Howard T, et al: Bronchiolar carcinoma (alveolar cell), another great imitator; A review of 41 cases. *Chest* 61:622–628, 1972.
10. Webb WR: The pleural tail sign. *Radiology* 127:309–313, 1978.
11. Kittridge RD, Sherman RS: Roentgen findings in terminal bronchiolar carcinoma. *AJR* 87:875–883, 1962.
12. Aronberg DJ, Sagel SS: Pulmonary metastases, in Eisenberg RL, Amberg JR (eds): *Critical Diagnostic Pathways in Radiology: An Algorithmic Approach*. Philadelphia, JB Lippincott, 1981.
13. Mintzer RA, Malave SR, Neiman HL, et al: Computed vs conventional tomography in evaluation of primary and secondary pulmonary neoplasms. *Radiology* 132:653–659, 1979.
14. Adkins PC, Wesselhoeft CW, Newman W, et al: Thoracotomy on the patient with previous malignancy: Metastasis or new primary? *J Thorac Cardiovasc Surg* 56:351–361, 1968.
15. McGee AR, Warren R: Carcinoma metastatic from the thyroid to the lungs. A twenty-four year radiographic follow-up. *Radiology* 87:516–517, 1966.
16. Janower ML, Blennerhassett JB: Lymphangitic spread of metastatic cancer to the lung: A radiologic-pathologic classification. *Radiology* 101:267–273, 1971.
17. Bower G: Bronchial adenoma. A review of twenty-eight cases. *Am Rev Respir Dis* 92:558–563, 1965.

18. Giusta PE, Stassa G: The multiple presentations of bronchial adenomas. *Radiology* 93:1013–1019, 1969.
19. Good CA, Harrington SW: Asymptomatic bronchial adenoma. *Proc Mayo Clin* 28:577–586, 1953.
20. Bateson EM, Abbott EK: Mixed tumors of the lung, or hamartochondromas. A review of the radiological appearances of cases published in the literature and a report of fifteen new cases. *Clin Radiol* 11:232–241, 1960.
21. Sagel SS, Ablow RC: Hamartoma: On occasion a rapidly growing tumor of the lung. *Radiology* 91:971–972, 1968.
22. Metys R: Roentgen symptomatology of pulmonary chondrohamartomas. *Fortschr Roentgenstr* 106:90–94, 1967.
23. Bateson EM: An analysis of 155 solitary lung lesions illustrating the differential diagnosis of mixed tumours of the lung. *Clin Radiol* 16:51–65, 1965.
24. Greenfield H, Herman PG: Papillomatosis of the trachea and bronchi. *AJR* 89:45–49, 1963.
25. Kaufman G, Klopstock R: Papillomatosis of the respiratory tract. *Am Rev Respir Dis* 88:839–846, 1963.
26. Singer DB, Greenberg FD, Harrison GM: Papillomatosis of the lung. *Am Rev Respir Dis* 94:777–780, 1966.
27. Eisenberg RL: *Atlas of Signs in Radiology*. Philadelphia, JB Lippincott, 1984.
28. Felson B: Neoplasms of the trachea and main stem bronchi. *Semin Roentgenol* 18:23–27, 1983.

## Chapter Thirty-Four
# Miscellaneous Diseases of the Respiratory System

## ADULT RESPIRATORY DISTRESS SYNDROME

*Adult respiratory distress syndrome* is a term used to describe a clinical picture of severe, unexpected, and life-threatening acute respiratory distress developing in patients with no major underlying lung disease and occurring after a widely diverse group of clinical insults. A complete breakdown in the integrity of the lung, which results in a marked increase in the permeability of the microvascular bed and alveolar epithelium, leads to the abrupt onset of marked dyspnea, increased respiratory effort, and severe hypoxemia, all of which are associated with radiographic evidence of widespread air space consolidation. Conditions that may lead to the adult respiratory distress syndrome include diffuse pulmonary infections, the aspiration or inhalation of toxins and irritants, nonthoracic trauma with hypotension (shock lung), septicemia, drug overdose, and postcardiopulmonary bypass.

The earliest abnormalities on chest radiographs characteristically develop up to 12 hours after the clinical onset of respiratory failure and consist of patchy, ill-defined areas of alveolar consolidation scattered throughout both lungs. Unlike cardiogenic pulmonary edema, in the adult respiratory distress syndrome the heart is usually of normal size, and there is no evidence of redistribution of blood flow to the upper zones. During the second day, the patchy zones of consolidation become more extensive and coalescent; pleural effusion is uncommon, and its presence should suggest a complicating acute pneumonia or pulmonary infarction. Near the end of the first week, the radiographic appearance quickly improves; the previously homogeneous consol-

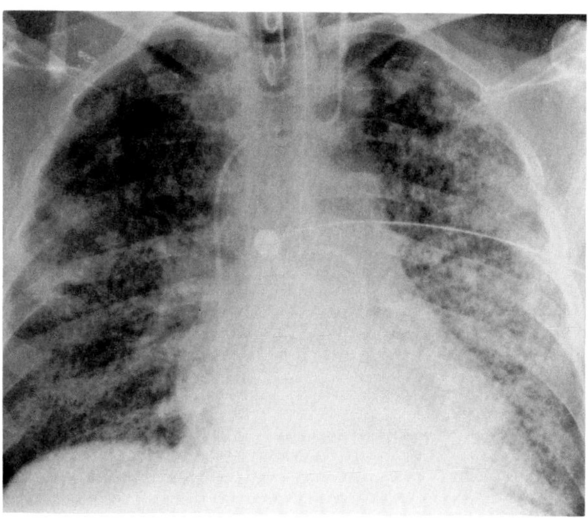

**Figure 34.1. Adult respiratory distress syndrome.** There are ill-defined areas of alveolar consolidation with some coalescence scattered throughout both lungs.

idation becomes inhomogeneous, suggesting a reduction in the amount of alveolar edema. Focal areas of consolidation representing acute pneumonia may appear. Continuous positive-pressure ventilation may cause diffuse interstitial emphysema, which may lead to pneumomediastinum and pneumothorax. After 1 week, diffuse interstitial and patchy air space fibrosis produce a coarse reticular pattern. If the patient recovers completely, the radiographic abnormalities may clear completely and there may be only a mild decrease in pulmonary function.[1–4]

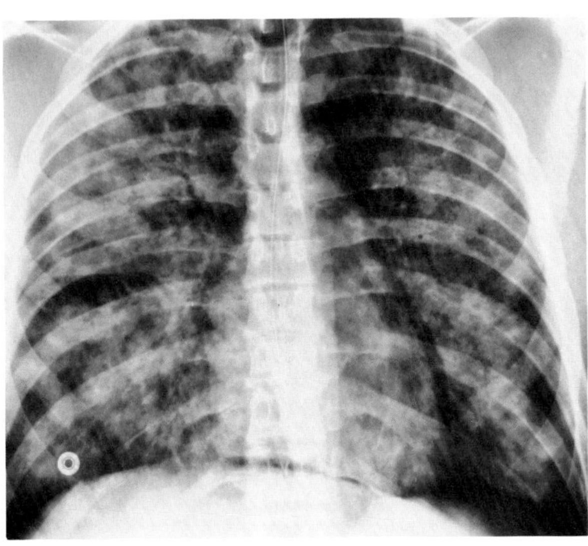

**Figure 34.2. Adult respiratory distress syndrome.** Continuous positive-pressure ventilation has caused diffuse interstitial emphysema, pneumothorax, and pneumoperitoneum to be superimposed upon a pattern of diffuse alveolar densities.

## ATELECTASIS

Atelectasis refers to a condition in which there is diminished air within the lung associated with reduced lung volume. Atelectasis is most commonly the result of bronchial obstruction, which may be due to a neoplasm, a foreign body, or inspissated mucous plugs. Nonobstructive atelectasis may be the result of lung compression by pneumothorax, pleural fluid, a tumor, or large bullae. Atelectasis may also be related to inactivation of the pulmonary surfactant system (adult respiratory distress syndrome, hyaline membrane disease, radiation pneumonitis) or may result from local or general pulmonary fibrosis.

There are many radiographic signs of atelectasis. The most common is a local increase in density, caused by airless lung, that may vary from thin platelike streaks to the collapse of an entire lung. The indirect signs of atelectasis reflect compensatory processes. These include elevation of the ipsilateral hemidiaphragm; displacement of the heart, mediastinum, and hilum toward the atelectatic segment; compensatory overinflation of the remainder of the ipsilateral lung; and narrowing of the interspaces on the affected side. An important direct sign of atelectasis is the displacement of interlobar fissures, which shift and become bowed, conforming to the contour of the collapsed segment.

The increased density of an affected portion of the lung may be enhanced by edema, blood, and/or retained secretions. Mediastinal shift may be associated with herniation of the opposite lung across the midline into the involved hemithorax.

Lobar atelectasis produces characteristic radiographic patterns. With collapse of the right upper lobe, the minor fissure and the upper half of the major fissure approxi-

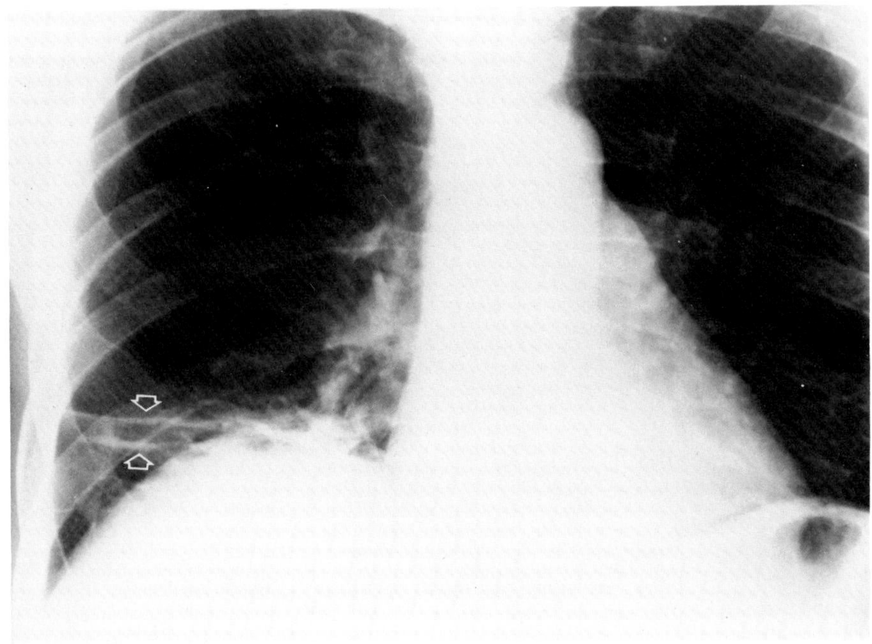

**Figure 34.3. Platelike atelectasis.** There are two horizontal linear streaks of density (arrows) in the left lower lung. Note the concomitant elevation of the ipsilateral hemidiaphragm.

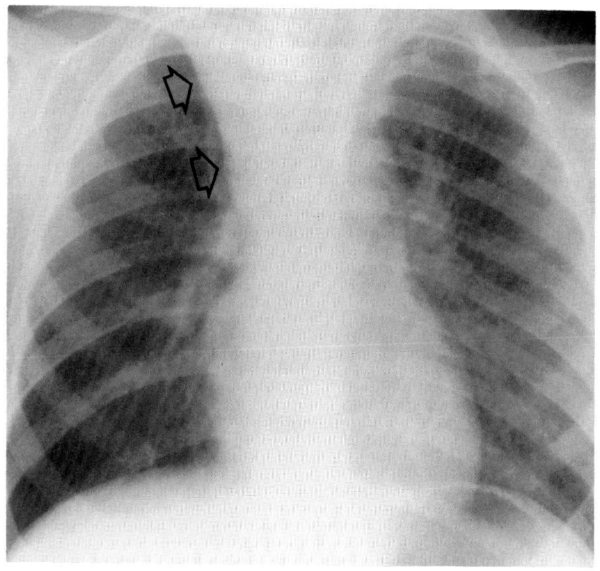


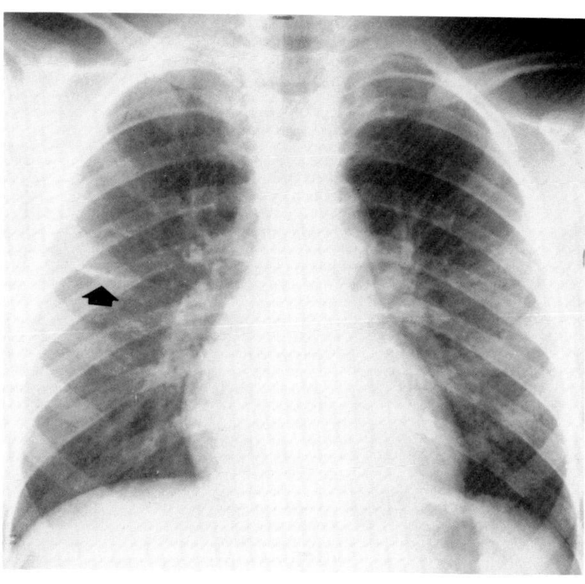


**Figure 34.4. Right upper lobe collapse.** (A) An initial chest radiograph demonstrates the collapsed right upper lobe, which appears as a homogeneous soft tissue mass (arrows) in the right apex along the upper mediastinum. (B) Following relief of the atelectasis, the soft tissue mass has disappeared and the minor fissure (arrow) has reappeared.

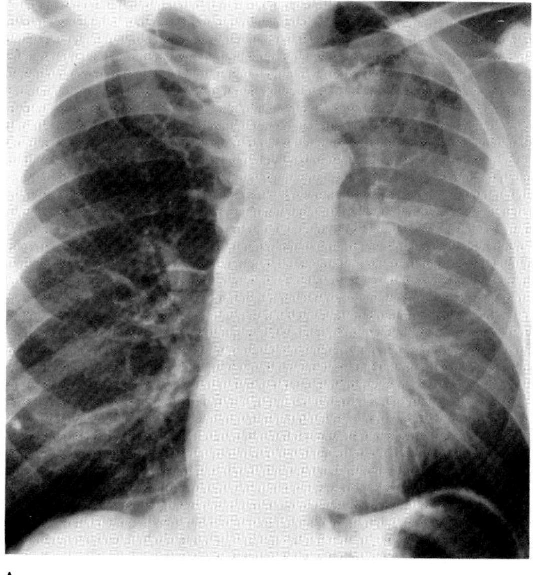


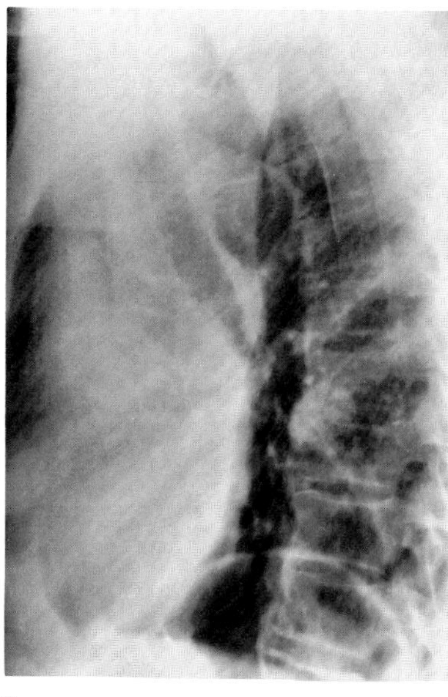


**Figure 34.5. Left upper lobe collapse.** (A) A frontal chest radiograph demonstrates a generalized increase in the density of the left hemithorax with no obliteration of the aortic knob or proximal descending aorta. The visualized vascular markings reflect lower lobe vessels. (B) A lateral view confirms the anterior position of the collapsed left upper lobe.

mate by shifting upward, and the collapsed lobe shrinks to occupy the apex and upper mediastinum. Left upper lobe collapse also produces an anterior and medial density, though it is more difficult to recognize on the frontal projection because of the absence of a minor fissure on that side. Although atelectasis of the right middle lobe causes only minimal silhouetting of the right border of the heart on frontal views, it is easy to demonstrate on lateral projections by the approximation of the minor fissure and the lower half of the major fissure, which results in a narrow band of increased density overlying the heart. The lower lobes collapse medially and posteriorly. On frontal views the collapsed lobe appears as a triangular band of increased density extending downward and posteriorly from the hilum. On the left, a radiograph of sufficient penetration is required to demonstrate the triangular shadow of the collapsed lower lobe behind the shadow of the heart.

An important iatrogenic cause of atelectasis is the improper placement of an endotracheal tube below the

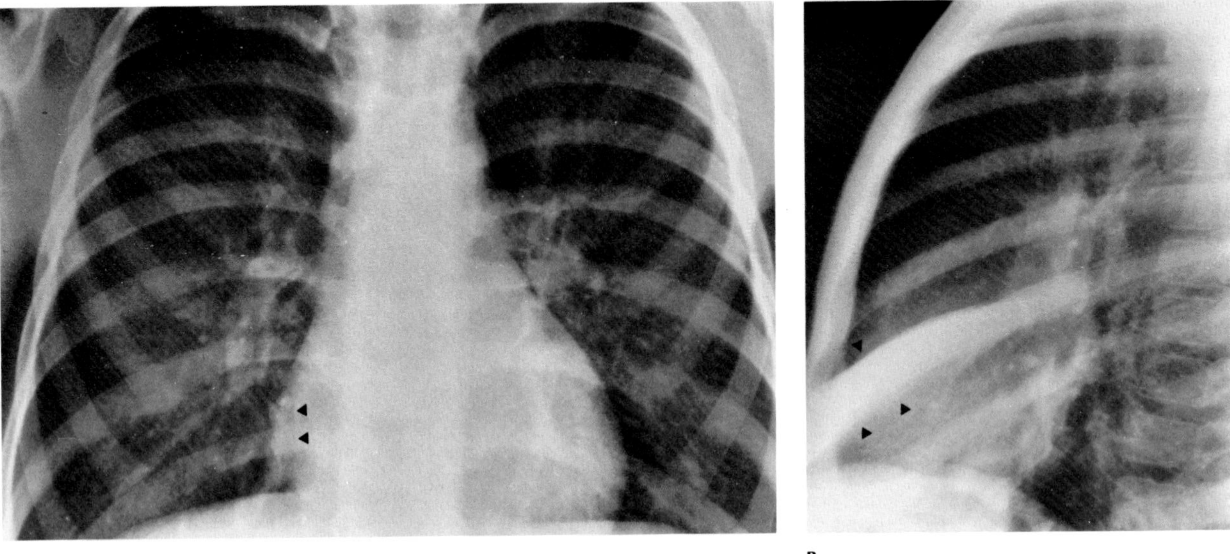


**Figure 34.6. Right middle lobe collapse.** (A) A frontal chest radiograph demonstrates minimal obliteration of the lower part of the right border of the heart (arrows). (B) A lateral view demonstrates collapse of the right middle lobe (arrows).

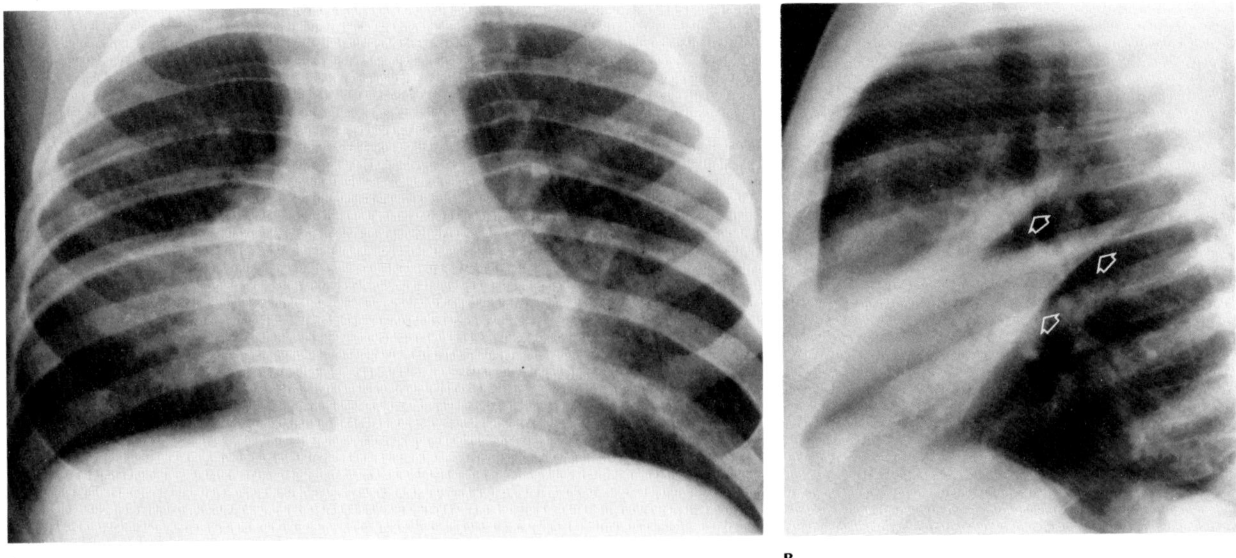


**Figure 34.7. Right middle lobe and lingula collapse.** (A) A frontal chest radiograph demonstrates obliteration of both the right and the left borders of the heart. (B) A lateral view demonstrates collapse of both the right middle lobe and the lingula (arrows).

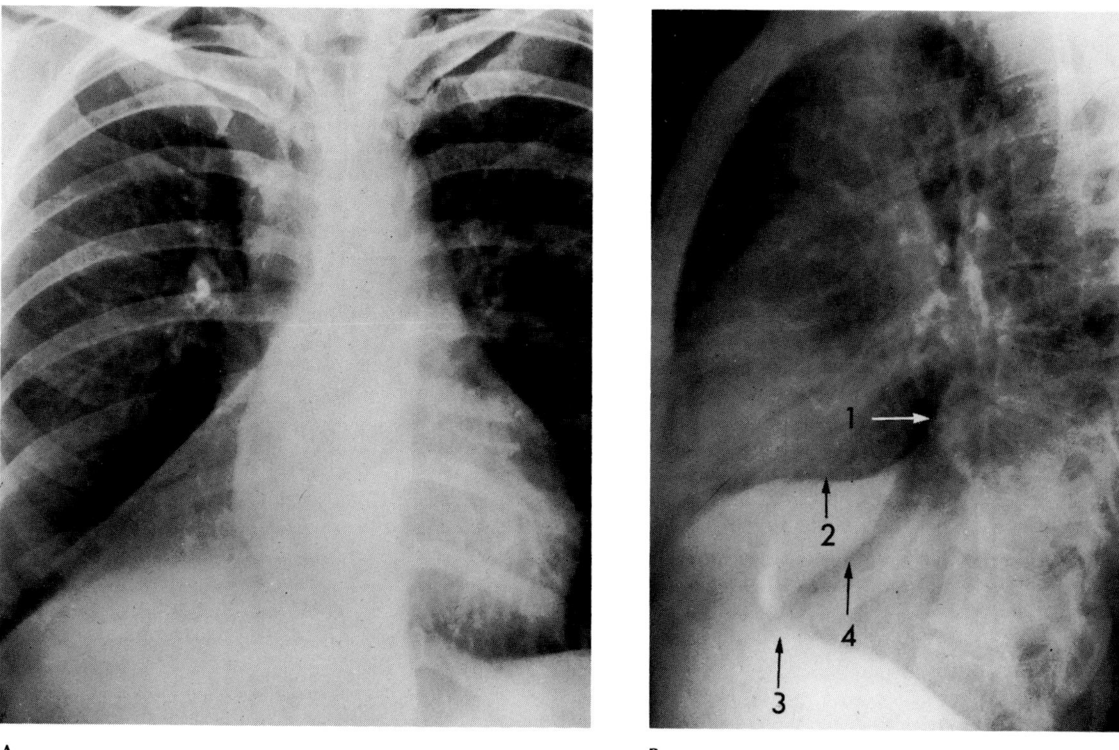


**Figure 34.8. Right lower lobe collapse.** (A) A frontal chest radiograph demonstrates a right lower lung density with preservation of the right border of the heart. The right hemidiaphragm is obscured. (B) A lateral view confirms the presence of right lower lobe collapse (due to bronchogenic carcinoma) with posterior displacement of the major fissure (1). The elevated right hemidiaphragm (2) is obliterated posteriorly by the airless right lower lobe, and the anterior third of the left hemidiaphragm (3) is obscured by the bottom of the heart. The overlapping shadows of the back of the heart (4), which lies in the left thorax, and the right hemidiaphragm simulate interlobar effusion. (From Felson,[28] with permission from the publisher.)

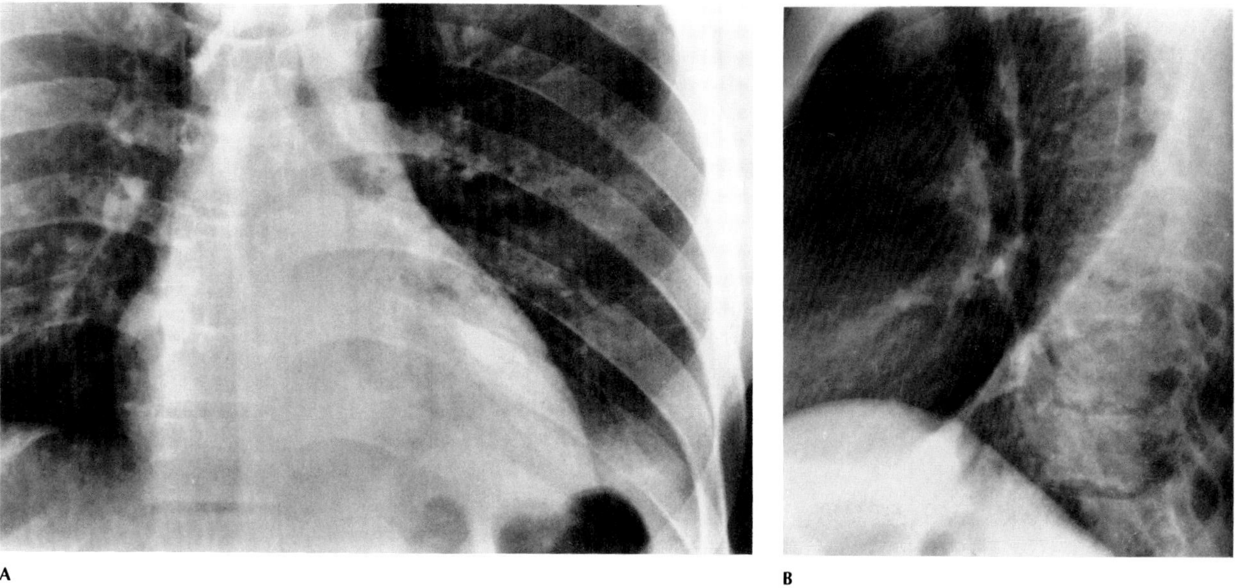


**Figure 34.9. Left lower lobe collapse.** (A) A frontal chest radiograph demonstrates obliteration of the descending thoracic aorta as well as obscuration of much of the left hemidiaphragm. (B) A lateral view confirms the posterior position of the collapsed left lower lobe.

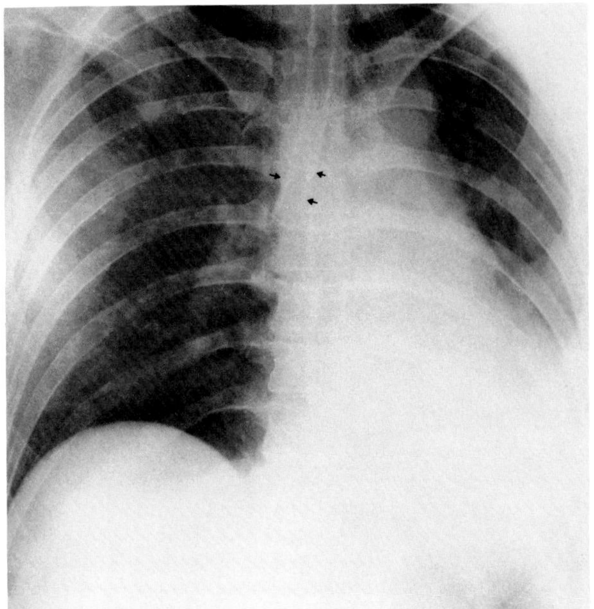

**Figure 34.10. Iatrogenic atelectasis.** The collapse of the left lung, especially the left lower lobe, is due to an endotracheal tube (arrows) in the right main stem bronchus that effectively blocks the passage of air into the left bronchial tree.

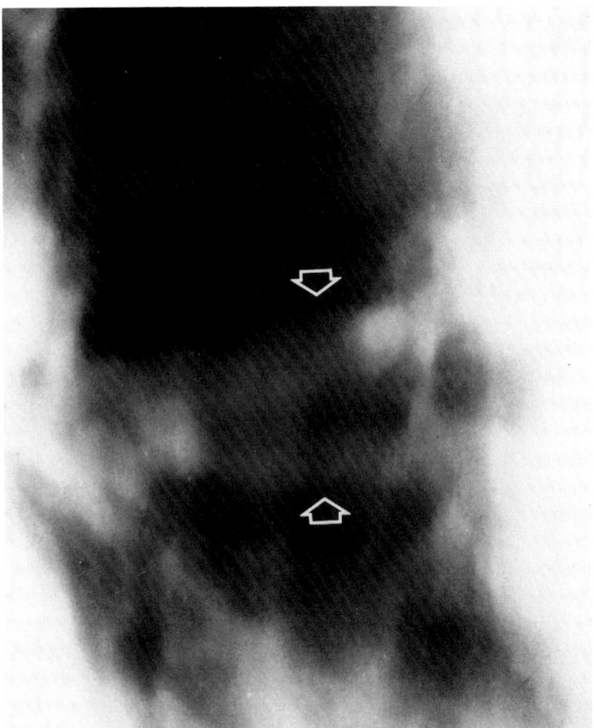

**Figure 34.11. Mucoid impaction.** Tomography of the lung demonstrates two mucoid impactions (arrows) in the lingula that form a V, the apex of which is toward the hilum. (From Tsai and Jenne,[6] with permission from the publisher.)

level of the carina. Because of geometric factors, the endotracheal tube tends to enter the right main stem bronchus, effectively blocking the left bronchial tree and causing the collapse of part or all of the left lung.[1,5,28]

## MUCOID IMPACTION

Mucoid impaction refers to the inspissation of tenacious mucoid sputum within one or more segmental or subsegmental bronchi. Predominantly involving the upper lobes, the condition is often associated with asthma and bronchitis and, occasionally, with cystic fibrosis and allergic pulmonary aspergillosis.

On chest radiographs or conventional tomography, a mucoid impaction typically appears as an elongated opacity with undulating borders. A pair of impactions frequently forms a V, the apex of which is toward the hilum (V sign). The involvement of numerous adjacent bronchi produces the cluster-of-grapes sign.[6]

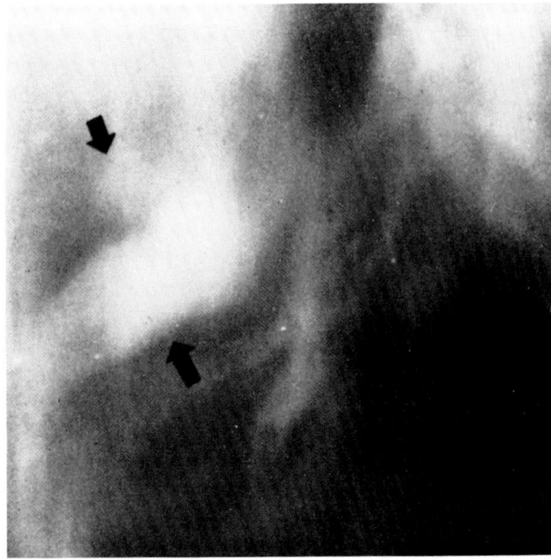

**Figure 34.12. Mucoid impaction.** Tomography of the lung demonstrates two mucoid impactions (arrows) in the superior segment of the right lower lobe. The scalloped borders suggest a cluster-of-grapes pattern. (From Tsai and Jenne,[6] with permission from the publisher.)

## INTRABRONCHIAL FOREIGN BODIES

The aspiration of solid foreign bodies into the tracheobronchial tree occurs almost exclusively in young children. Although some foreign bodies are radiopaque and easily detected on plain chest radiographs, most aspi-

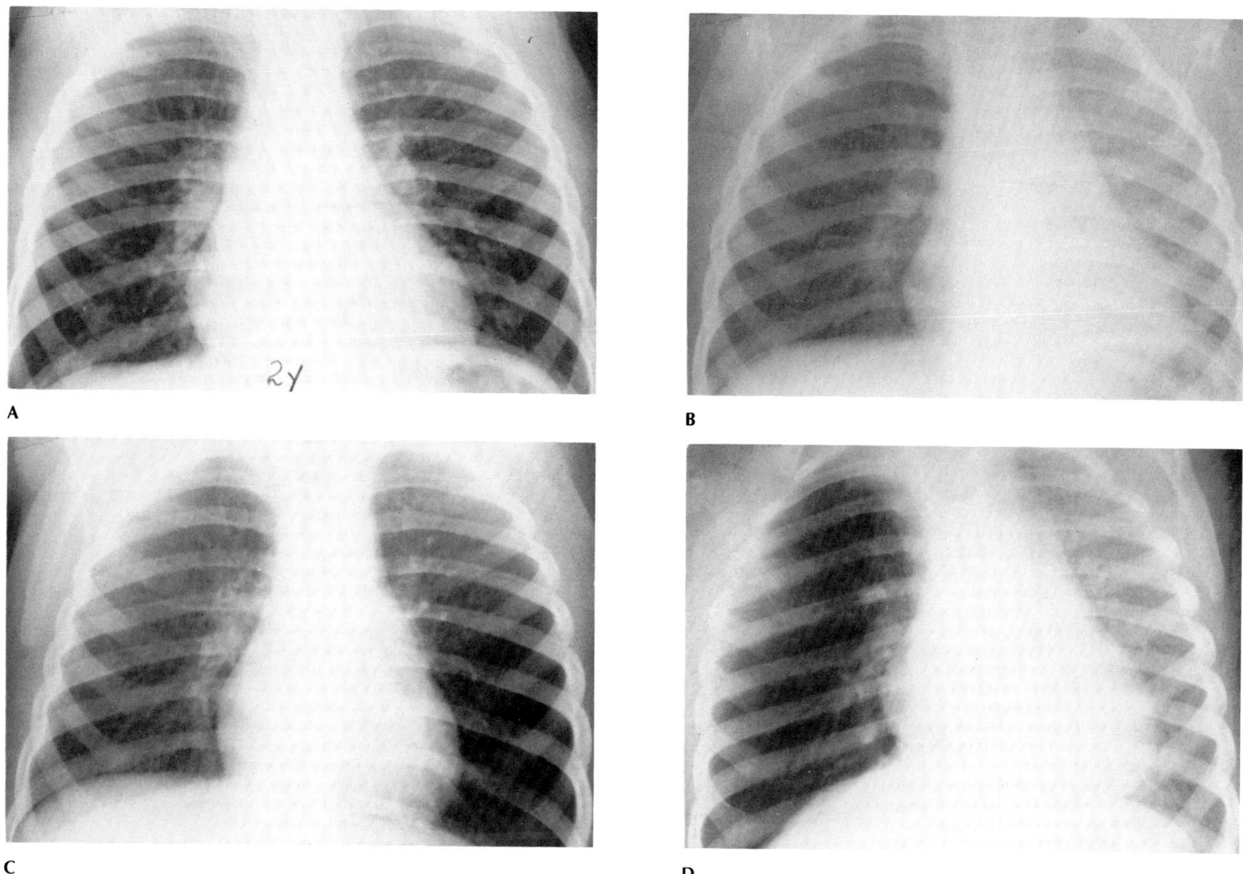


**Figure 34.13. A 2-year-old boy with a peanut in the right main bronchus.** During inspiration (A), the lungs are well-aerated. Air trapping in the right lung is seen during expiration (B) and with the right side down (C). The normal left lung is underaerated when that side is down (D). (From Capitanio and Kirkpatrick,[8] with permission from the publisher.)

rated foreign bodies are nonopaque and can only be diagnosed by observing secondary signs in the lungs caused by partial or complete bronchial obstruction. The lower lobes are almost always involved, the right more often than the left.

The complete obstruction of a major bronchus leads to resorption of trapped air, alveolar collapse, and atelectasis of the involved segment or lobe. Extensive volume loss causes a shift of the heart and mediastinal structures toward the affected side along with elevation of the ipsilateral hemidiaphragm and narrowing of the intercostal spaces. There may even be herniation of the normal, contralateral lung across the mediastinum.

Partial bronchial obstruction may produce air trapping as a result of a check-valve phenomenon. Air freely passes the partial obstruction during inspiration, but remains trapped distally as the bronchus contracts normally during expiration. Hyperaeration of the affected lobe causes a shift of the heart and mediastinum toward the normal, contralateral side, a finding that is accentuated during forced expiration because the hyperaerated segment does not contract. This classic appearance of partial bronchial obstruction is dramatically demonstrated at fluoroscopy as the mediastinum shifts away from the affected side during deep expiration and returns toward the midline on full inspiration. Lateral decubitus views may also be of value in demonstrating air trapping. In normal persons, the inflation of the dependent lung tends to be less than that of the upper lung. However, if there is air trapping in the dependent lung, the affected lobe or segment tends to remain hyperlucent. Radionuclide lung scanning may also aid in detecting endobronchial foreign bodies by demonstrating striking perfusion defects in the involved segment or lobe.

A malpositioned endotracheal tube can act as an intrabronchial foreign body. The tube tends to extend down the right main stem bronchus, causing hyperlucency of the right lung and obstructive atelectasis of the left lung.[7–10]

## BRONCHOPULMONARY SEQUESTRATION

Bronchopulmonary sequestration is a congenital pulmonary malformation in which a portion of pulmonary tissue is detached from the remainder of the normal lung

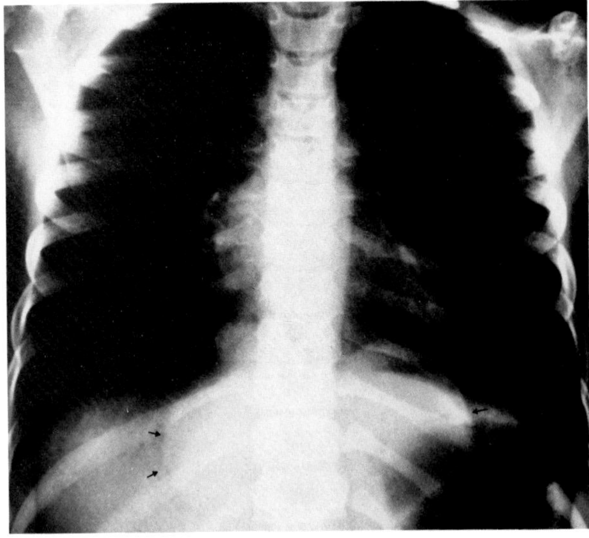

A

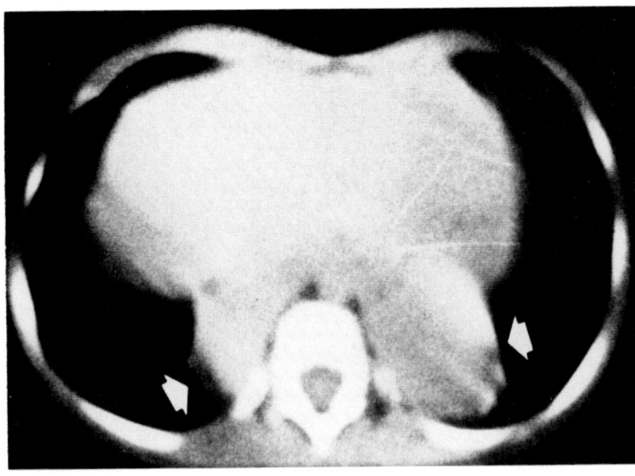

B

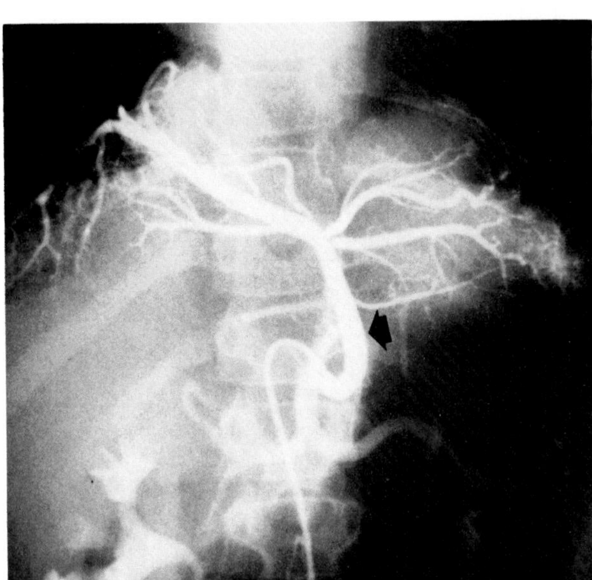

C

**Figure 34.14. Bilateral pulmonary sequestration.** (A) A frontal view of the chest shows bilateral oval, slightly lobulated paravertebral masses (arrows) in the juxtadiaphragmatic region. (B) A CT scan demonstrates the bilateral paravertebral masses (arrows), which are isodense with the diaphragm. (C) A selective angiogram of a large anomalous artery arising from the celiac trunk shows several branches supplying the bilateral paravertebral masses. The venous drainage was via the pulmonary veins. (From Wimbish, Agha, and Brady,[26] with permission from the publisher.)

and receives its blood supply from a systemic artery. In the more common intralobar sequestration, the nonfunctioning portion of lung lies contiguous to normal lung parenchyma and within the same visceral pleural envelope. An extralobar sequestration is enclosed within its own pleural membrane and is usually within or below the diaphragm, most commonly on the left. Both types of sequestration receive their blood supply from branches of the aorta rather than from the pulmonary artery. An intralobar sequestration drains into the pulmonary veins, while an extralobar sequestration drains via the systemic venous sytem (inferior vena cava, azygos or hemiazygos system, or portal venous system).

A simple intralobar sequestration is asymptomatic and appears as a round, oval, or triangular sharply circumscribed soft tissue mass situated in the posterior portion of the lower lobe (usually the left) contiguous to the diaphragm. If infection has resulted in communication with the airways or contiguous lung tissue, the sequestration appears as an air-containing cystic mass, often very large or multiple; the mass may contain an air-fluid level and simulate an abscess, a bronchiectatic cavity, or an infected cyst.

Extralobar sequestration appears radiographically as a homogeneous intrathoracic or intra-abdominal soft tissue mass. Because the extralobar sequestered segment is surrounded by its own pleural sac, there is much less chance of its becoming infected and containing air.

The definitive diagnosis of bronchopulmonary sequestration requires arteriographic demonstration of the anomalous feeding vessel arising from the aorta.[1,11–13,26]

## LYMPHANGIOMYOMATOSIS

Pulmonary lymphangiomyomatosis is a rare disorder of young or middle-aged women that is part of a general syndrome characterized by the excessive accumulation of muscle in relation to extrapulmonary lymphatics. The

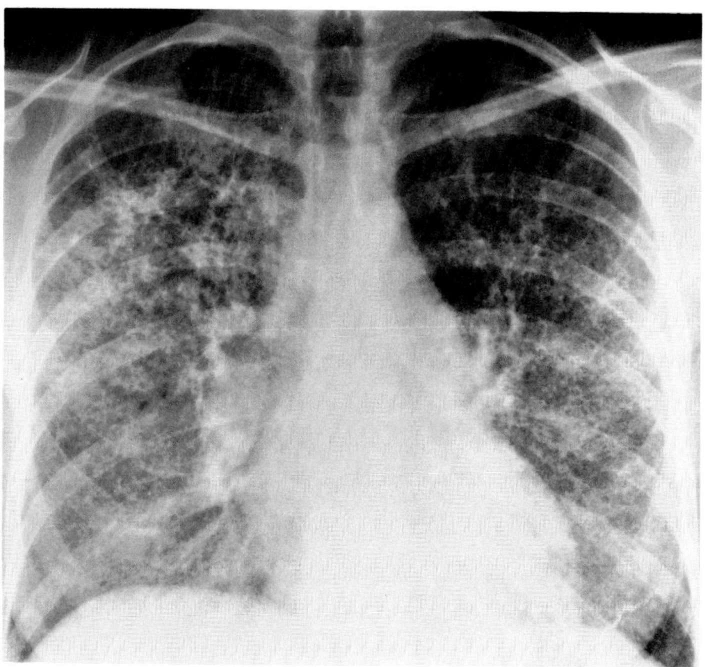

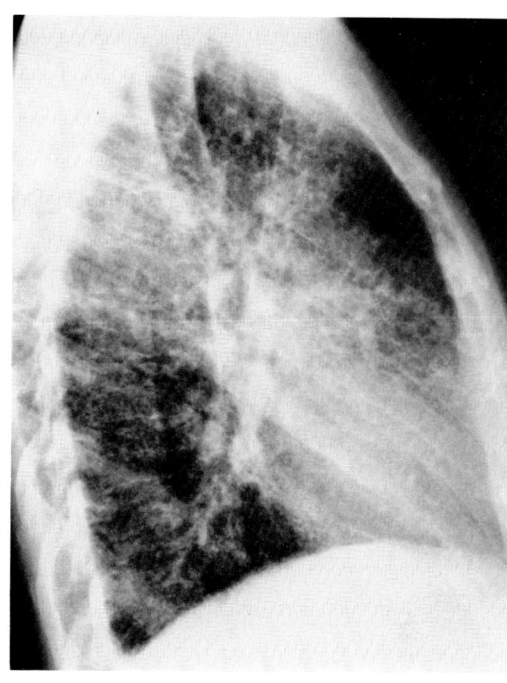

A B

**Figure 34.15. Lymphangiomyomatosis.** (A) Frontal and (B) lateral views of the chest demonstrate a diffuse reticulonodular interstitial pattern throughout both lungs. The honeycomb quality is best appreciated on the lateral view.

hallmark of lymphangiomyomatosis is repeated episodes of chylous pleural effusion due to mediastinal lymphangiomyomas that cause a fistulous connection between the thoracic duct and the pleura. Diffuse lymphangiomyomatous involvement of the smooth muscles of bronchioles, pulmonary vessels, and lymphatics produces a coarse reticulonodular interstitial pattern in the lungs. This can lead to honeycombing and progressive respiratory insufficiency. Spontaneous pneumothorax is a common complication and may be the presenting symptom.[14,15]

## TRAUMA TO THE LUNGS

Pulmonary contusion, the most common pulmonary complication of blunt chest trauma, refers to the exudation of edema and blood into both the air spaces and the interstitium of the lung. Radiographically, pulmonary contusion produces a pattern that varies from irregular patchy areas of air space consolidation to an extensive homogeneous density involving almost an entire lung. In the absence of an appropriate clinical history of trauma or evidence of rib fractures, pulmonary contusion may be indistinguishable from pneumonia. Resolution typically occurs rapidly, with complete clearing within 2 weeks.

Pulmonary laceration is primarily associated with penetrating injuries and is seen less commonly in blunt trauma. Pulmonary laceration can result in the development of one or more cystic spaces within the lung; these spaces may remain air-filled (traumatic air cyst) or, more commonly, may fill partly or completely with blood (pulmonary hematoma). Traumatic air cysts appear as oval or spherical cystic spaces located subpleurally in peripheral areas of the lung. Bleeding into a cyst, which

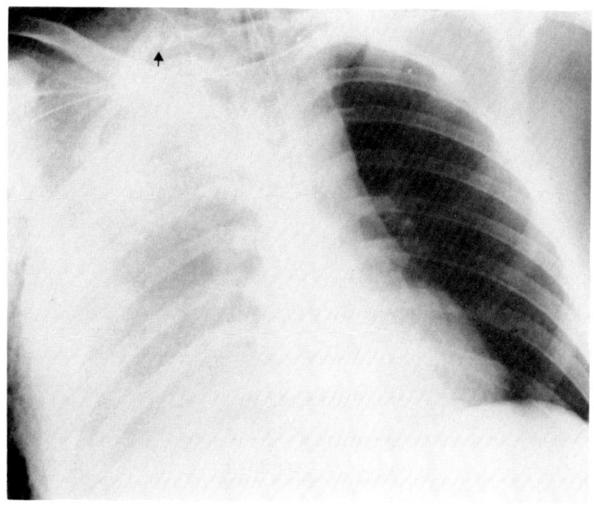

**Figure 34.16. Pulmonary contusion.** Following blunt trauma, there is almost complete homogeneous opacification of the right chest with minimal air bronchograms centrally. There is a fracture of the right first rib (arrow), and subcutaneous emphysema in the soft tissues of the right hemithorax.

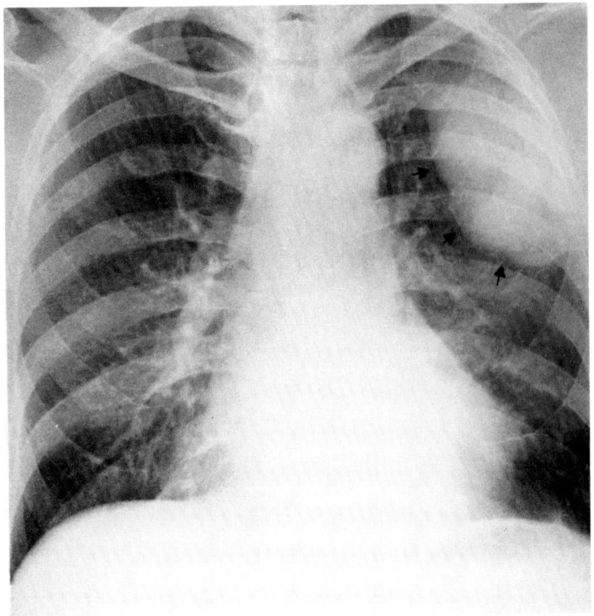

**Figure 34.17. Pulmonary hematoma.** There is a large extrapleural density (arrows) over the left upper lobe.

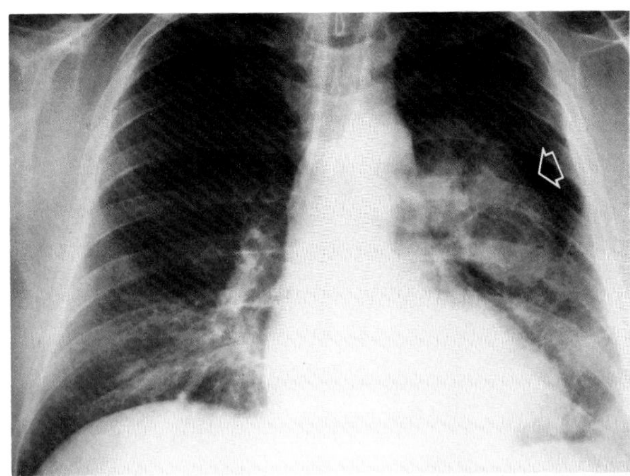

**Figure 34.18. Pulmonary hematoma.** Following a stab wound, a homogeneous kidney-shaped opacity (arrow) developed in the superior segment of the left lower lobe. There is blunting of the left costophrenic angle.

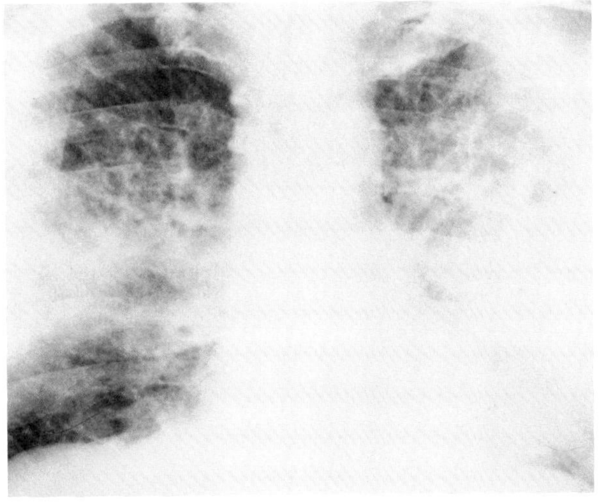

A

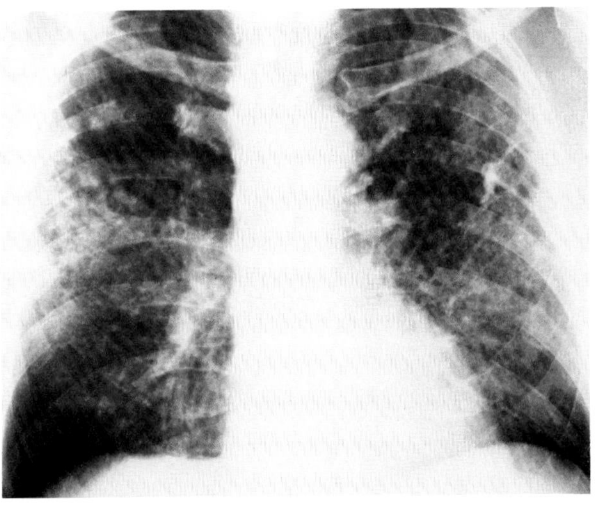

B

**Figure 34.19. Pulmonary hemorrhage.** (A) Diffuse bilateral air space consolidation developed in a patient on high-dose anticoagulant therapy. (B) With resolution of the hematoma, a reticular pattern developed in the same distribution as the alveolar infiltrate.

results from the rupture of capillaries, produces a homogeneous, well-circumscribed mass of water density. Traumatic lung cysts and hematomas generally decrease progressively in size, though they may persist for several months.

Computed tomography (CT) can aid in defining the extent of injury. CT demonstrates focal and diffuse patchy parenchymal densities corresponding to the changes on plain radiographs.[16–19]

## PULMONARY AGENESIS AND HYPOPLASIA

This developmental anomaly may vary from the complete absence of a lung with no trace of bronchial or vascular supply or of parenchymal tissue (agenesis) to hypoplasia of the bronchus with a rudimentary, undeveloped lung. Radiographically, there is total or almost total absence of aerated lung in one hemithorax. The markedly reduced volume leads to approximation of the

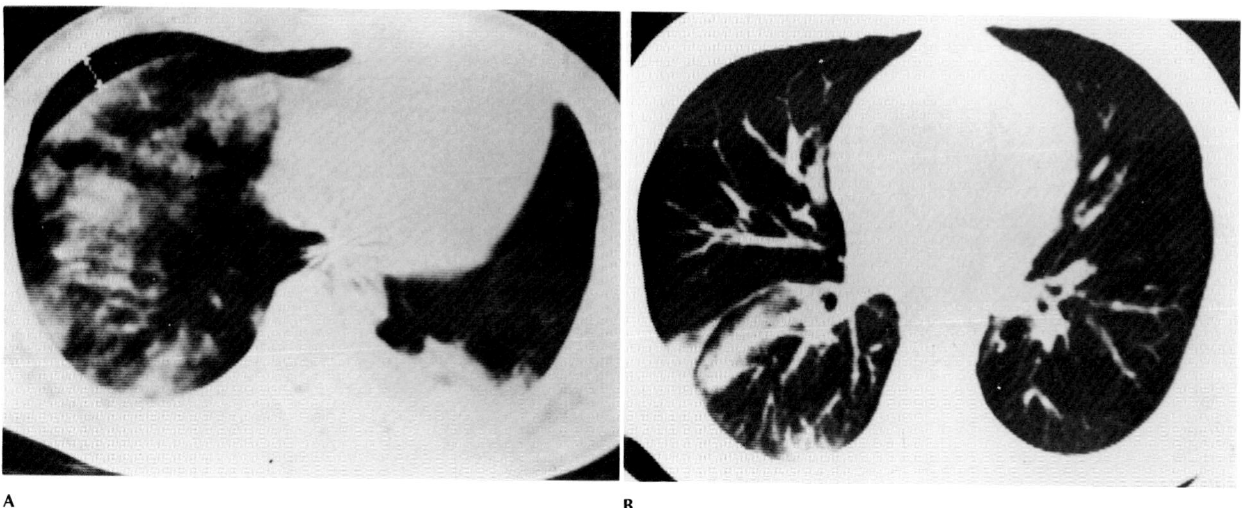


**Figure 34.20. CT in lung trauma.** (A) Diffuse pulmonary contusion appears as multiple patchy parenchymal densities within the right lung, associated with a small right pneumothorax (dotted line). Focal contusion is seen posteriorly in the left lung. (B) Pulmonary laceration secondary to a gunshot wound. A CT scan performed 2 days after the trauma indicates a cylindrical pulmonary laceration in the right lower lobe. Blood is present within the major fissure. (From Toombs, Sandler, and Lester,[19] with permission from the publisher.)

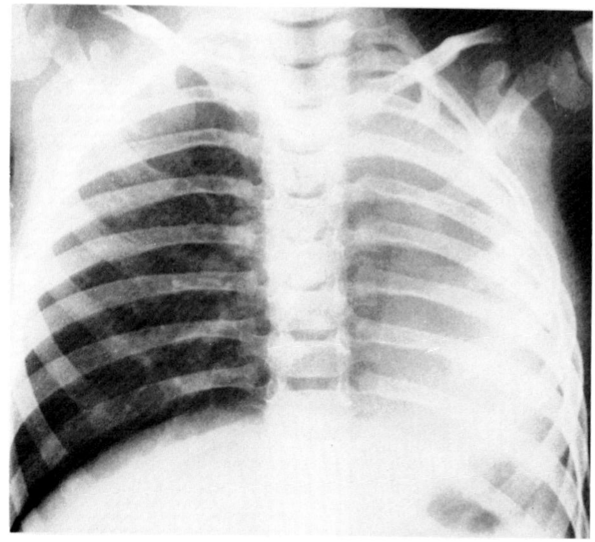

**Figure 34.21. Agenesis of the left lung in a child.** There is virtually total absence of aerated lung in the left hemithorax. The right lung is markedly overinflated and has herniated across the midline. The entire mediastinum lies within the left hemithorax. The chest wall is asymmetric, and the ribs are somewhat more closely together on the left.

ribs, elevation of the ipsilateral hemidiaphragm, and a shift of the mediastinum. The contralateral, normal lung shows compensatory hyperinflation and often herniates across the mediastinum to the affected side.

Tomography, bronchography, and angiography may be required to establish the degree of underdevelopment of the lung or to differentiate agenesis from other conditions that may closely mimic it radiographically (total atelectasis from any cause; severe bronchiectasis with collapse; advanced fibrothorax).

More than half of the patients with agenesis of the lung have other congenital anomalies, most frequently congenital heart disease, anomalies of the great vessels, congenital diaphragmatic hernia, and skeletal anomalies.

Although many patients with pulmonary agenesis are totally asymptomatic and live normal lives, the anomaly appears to predispose to respiratory infection.[20–23,27]

## MIDDLE LOBE SYNDROME

The term *middle lobe syndrome* refers to chronic atelectasis of the right middle lobe that predisposes to recurrent pneumonia and irreversible fibrosis. It may reflect bronchial obstruction caused by a neoplasm, by inflammatory stenosis, or by extrinsic pressure from an enlarged hilar node, usually of tuberculous origin.

On frontal chest radiographs, the collapsed middle lobe appears as an area of increased density that obliterates a portion of the right border of the heart but may be difficult to detect. It usually can be easily outlined on the lateral film as a wedge-shaped or triangular area of density that is sharply bounded above and below by normally aerated lung. The apex of the triangle is at the hilum and the base at the anterior inferior thoracic wall. Occasionally, the collapsed middle lobe can only be seen on a lordotic view.[24,25]

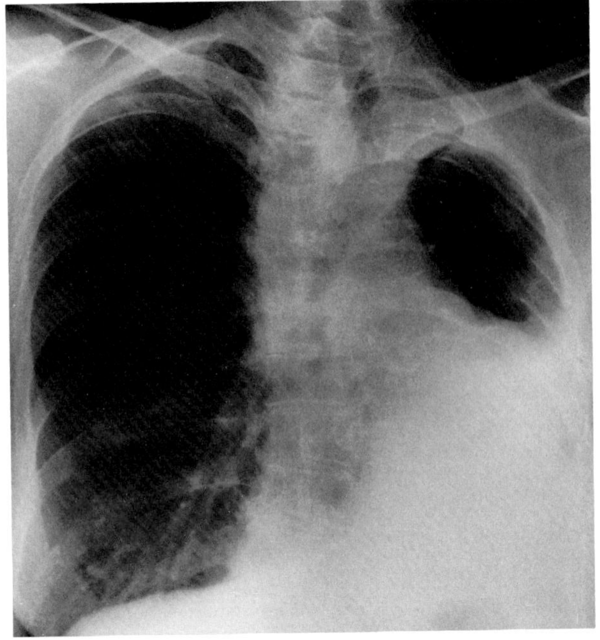

A

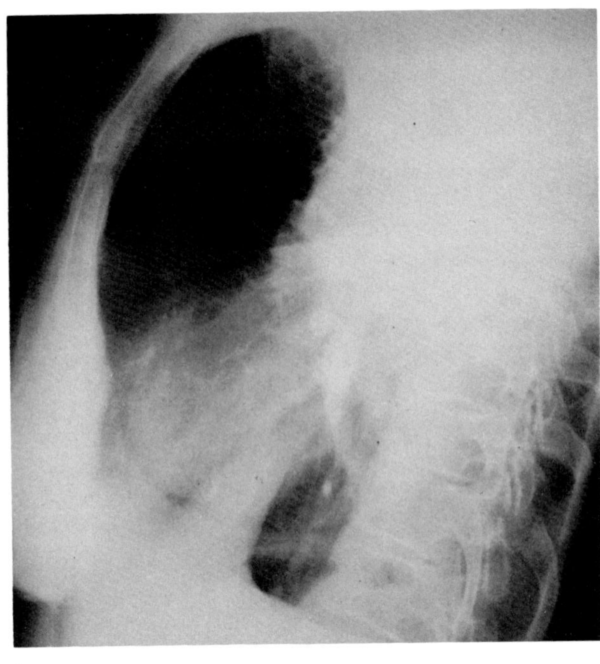

B

**Figure 34.22. Agenesis of the left lung in an adult.** (A) A frontal chest radiograph shows uniform opacity in the lower left hemithorax with crowding of the left upper ribs. The trachea is deviated to the left, and the left leaf of the diaphragm is obscured. The mass to the right of the spine is the ascending aorta. (B) A left lateral view shows anterior herniation of the right lung. (From Shenoy, Culver, and Pirson,[27] with permission from the publisher.)

## REFERENCES


1. Fraser RG, Pare JAP: *Diagnosis of Diseases of the Chest*. Philadelphia, WB Saunders, 1979.
2. Petty TL, Ashbough BG: The adult respiratory distress syndrome. Clinical features, factors influencing prognosis and principles of management. *Chest* 60:233–239, 1971.
3. Joffe N: The adult respiratory distress syndrome. *AJR* 122:719–723, 1974.
4. Ostendorf P, Birzle H, Vogel W, et al: Pulmonary radiographic abnormalities in shock. Roentgen-clinical-pathological correlation. *Radiology* 115:257–263, 1975.
5. Robbins LL, Hale CH: The roentgen appearance of lobar and segmental collapse of the lung. *Radiology* 44:543, 1945; 45:23; 120; 260; 347, 1945.
6. Tsai SH, Jenne JW: Mucoid impaction of the bronchi. *AJR* 96:953–961, 1966.
7. Brown BS, Ma H, Dunbar JS, et al: Foreign bodies in the tracheobronchial tree in childhood. *J Can Assoc Radiol* 14:158–171, 1963.
8. Capitanio MA, Kirkpatrick JA: The lateral decubitus film: An aid in determining air-trapping in children. *Radiology* 103:460–461, 1972.
9. Davis CM: Inhaled foreign bodies in children. An analysis of forty cases. *Arch Dis Child* 41:402–406, 1966.
10. Rudavsky AZ, Leonidas JC, Abramson AL: Lung scanning for the detection of endobronchial foreign bodies in infants and children: Clinical and experimental studies. *Radiology* 108:629–633, 1973.
11. Ranniger K, Valvassori GE: Angiographic diagnosis of intralobar pulmonary sequestration. *AJR* 92:540–546, 1964.
12. Sutton D, Samuel RH: Thoracic aortography in intra-lobar lung sequestration. *Clin Radiol* 14:317–321, 1963.
13. Felson B: The many faces of pulmonary sequestration. *Semin Roentgenol* 7:3–16, 1972.
14. Corrin B, Liebow AA, Friedman PJ: Pulmonary lymphangiomyomatosis. *Am J Pathol* 79:348–367, 1975.
15. Silverstein EF, Ellis K, Wolff M, et al: Pulmonary lymphangiomyomatosis. *AJR* 120:832–850, 1974.
16. Williams JR, Stembridge VA: Pulmonary contusion secondary to nonpenetrating chest trauma. *AJR* 91:284–290, 1964.
17. Wiot JF: The radiologic manifestations of blunt chest trauma. *JAMA* 231:500–503, 1975.
18. Stevens E, Templeton AW: Traumatic nonpenetrating lung contusion. *Radiology* 85:247–252, 1965.
19. Toombs BD, Sandler CM, Lester RG: Computed tomography of chest trauma. *Radiology* 140:733–738, 1981.
20. Boyden EA: Developmental anomalies of the lungs. *Am J Surg* 89:79–89, 1955.
21. Soulen RL, Cohen RV: Plain film recognition of pulmonary agenesis in the adult. *Chest* 60:185–187, 1971.
22. Steinberg I, Stein HL: Angiocardiography in diagnosis of agenesis of a lung. *AJR* 96:991–1006, 1966.
23. Felson B: Pulmonary agenesis and related anomalies. *Semin Roentgenol* 7:17–30, 1972.
24. Albo RJ, Grimes OF: The middle lobe syndrome. *Dis Chest* 50:509–518, 1966.
25. Juhl JH: *Essentials of Roentgen Interpretation*. Philadelphia, JB Lippincott, 1981.
26. Wimbish KJ, Agha FP, Brady TM: Bilateral pulmonary sequestration: CT appearance. *AJR* 140:689–690, 1983.
27. Shenoy SS, Culver GJ, Pirson HS: Agenesis of the lung in an adult. *AJR* 133:755–757, 1979.
28. Felson B: *Chest Roentgenology*. Philadelphia, WB Saunders, 1973.

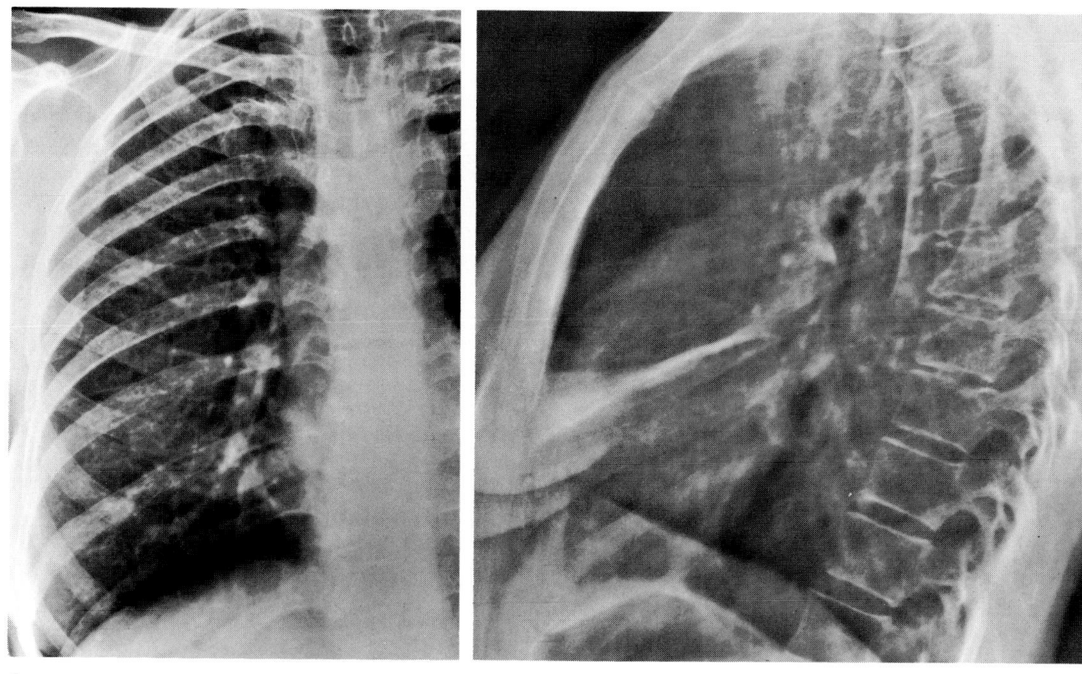


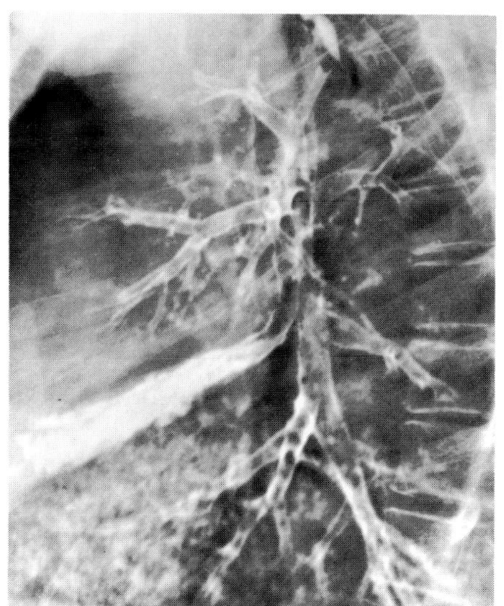


**Figure 34.23. Chronic middle lobe syndrome.** (A) A frontal view of the right lung shows obliteration of the right border of the heart, which suggests airlessness of the right middle lobe. (B) On a lateral projection, a thin triangular shadow represents the airless right middle lobe between the approximated minor and major fissures. (C) A right bronchogram shows severe crowding of the bronchiectatic middle lobe segmental bronchi. Note that the middle lobe bronchus is widely patent. (From Fraser and Pare,[1] with permission from the publisher.)

## Chapter Thirty-Five

# Diseases of the Pleura, Mediastinum, and Diaphragm

## DISEASES OF THE PLEURA

### Pleural Effusion

Pleural effusion is a nonspecific finding that may be due to the increased permeability of the parietal pleura secondary to inflammatory or neoplastic disease or may reflect an alteration in hydrostatic and oncotic forces, as in congestive heart failure. The most common infectious cause is tuberculosis, which often appears as an isolated unilateral pleural effusion without radiographically demonstrable parenchymal disease. Effusion with a pulmonary infiltrate suggests a pulmonary infarct or pneumonia. Neoplastic causes of pleural effusion include bronchogenic carcinoma, metastatic carcinoma (most commonly from the breast), lymphoma, and primary pleural mesothelioma. Rheumatoid arthritis may cause pleural effusions, especially in males with subcutaneous nodules, as can other connective tissue diseases such as systemic lupus erythematosus and Wegener's granulomatosus.

Congestive heart failure is a major cause of pleural effusion. Although the effusion is frequently bilateral, congestive heart failure is the most common cause of a unilateral pleural effusion that, for unknown reasons, is almost invariably on the right side. Indeed, in a patient with cardiac failure who has a unilateral left-sided effusion, some other cause, such as pulmonary embolism, should be sought.

A variety of abdominal diseases may produce sympathetic effusions. These include recent abdominal surgery; ascites; subphrenic abscess; and pancreatitis, in which a left-sided pleural effusion is common. Women with nonmetastatic pelvic tumors may develop Meigs's syndrome, a combination of ascites and pleural effusion that is usually right-sided. A radiographic picture identical to pleural effusion can be caused by a hemothorax following direct trauma to the chest, or by a chylothorax due to leakage of thoracic duct lymph into the pleural space. A chylothorax may develop secondary to trauma to the thoracic duct or to obstruction of the duct by a malignant process or mediastinal fibrosis.

Blunting of the normally sharp angle between the diaphragm and the rib cage along with an upward concave border of the fluid level (meniscus) is the earliest radiographic finding in pleural effusion. Because the costophrenic sulci are deeper posteriorly than laterally, small pleural effusions collect in the posterior costophrenic sulcus and are seen earliest on the routine lateral view. As much as 300 to 400 ml of pleural fluid may be present and still not produce blunting of the lateral costophrenic angles on erect frontal views of the chest. Larger amounts of pleural fluid produce a homogeneous density that extends higher laterally than medially (because of differ-

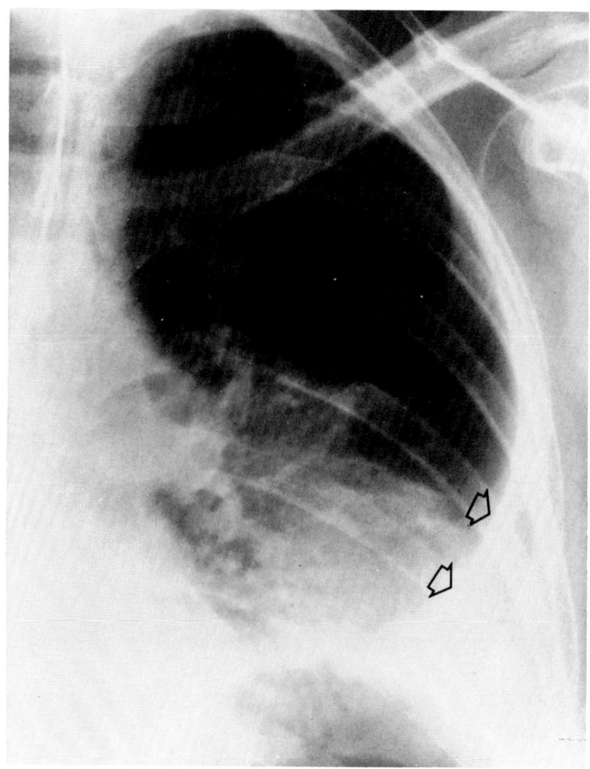

**Figure 35.1. Pleural effusion.** There is blunting of the normally sharp angle between the diaphragm and the rib cage (arrows) along with a characteristic upward concave border (meniscus) of the fluid level.

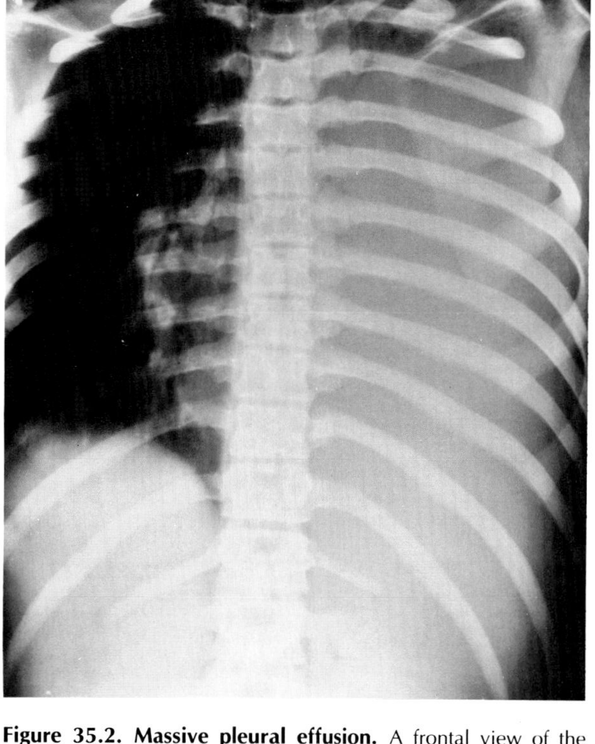

**Figure 35.2. Massive pleural effusion.** A frontal view of the chest in a patient with severe coccidioidomycosis demonstrates complete homogeneous opacification of the left hemithorax. The massive pleural effusion must be associated with virtually complete collapse of the left lung, since there is no contralateral shift of the mediastinal structures.

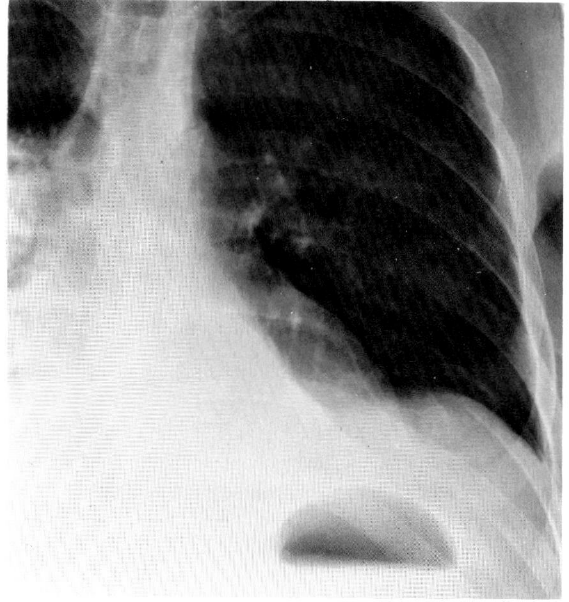

A

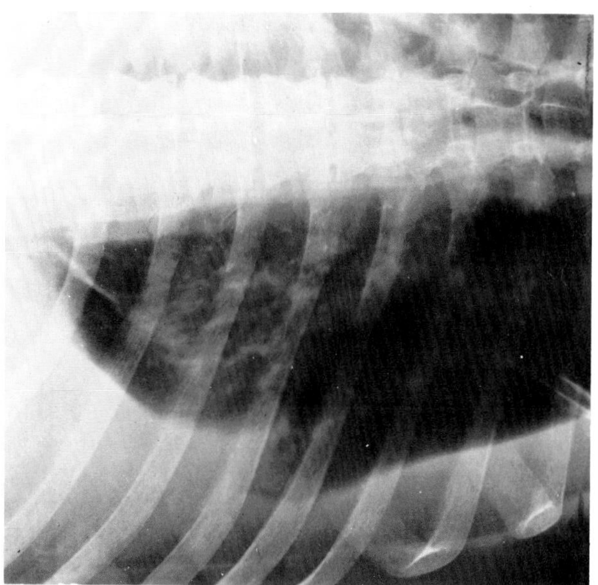

B

**Figure 35.3. Pleural effusion.** (A) A frontal view of the left chest demonstrates a large distance between the gastric air bubble and the top of the false left hemidiaphragm in this patient with a large subpulmonic effusion. Note the retrocardiac paraspinal density that simulates a left lower lobe infiltrate or atelectasis. (B) A left lateral decubitus view shows that the retrocardiac density represents a large amount of free fluid. The step-off of the fluid along the left chest wall is related to the difference in the elastic recoil of the upper and the lower lobes. (From Vix,[49] with permission from the publisher.)

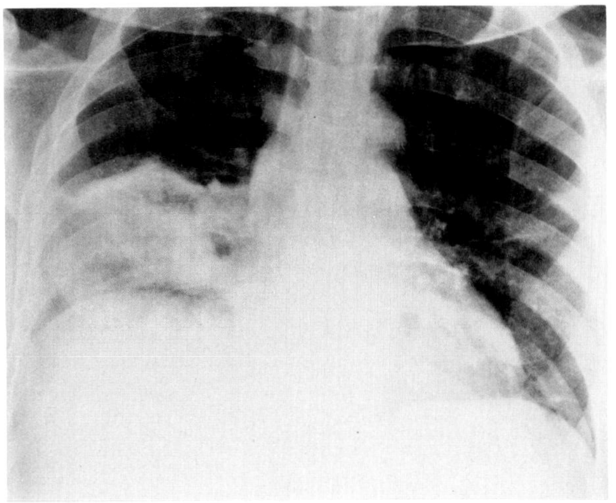

A

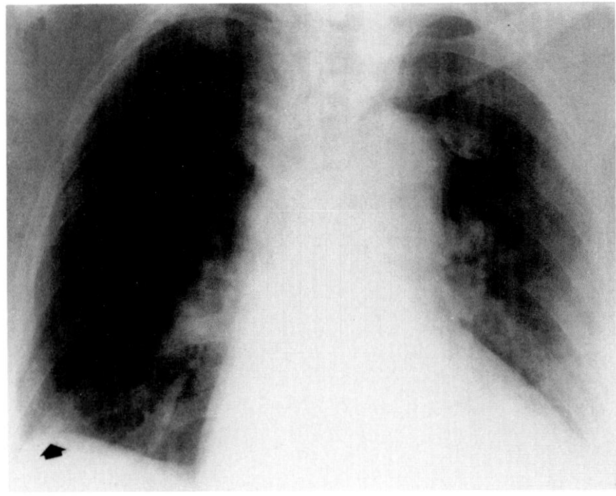

B

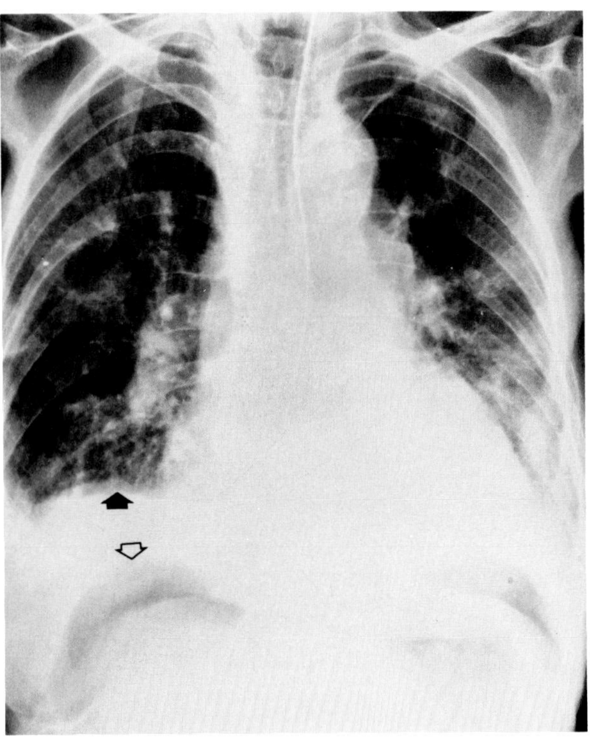

C

**Figure 35.4. Subpulmonic pleural effusion.** (A) In this patient with an extensive right lower lobe pneumonia, the apparent elevation of the right hemidiaphragm represents pleural fluid that has collected below the inferior surface of the lung. (B) An unusually lateral position of the apex of the false hemidiaphragm (pseudodiaphragmatic contour) in a patient with subpulmonic effusion. (C) In this patient with coexisting pneumoperitoneum, the right subpulmonic effusion extends from the tip of the subdiaphragmatic gas collection (open arrow) to the base of the right lung (closed arrow).

ences in the elastic recoil of the lung) and may obscure the diaphragm and adjacent borders of the heart. If the superior pleural line is horizontal and clearly defined (air-fluid level), it indicates that some air is present along with the fluid in the pleural space. Massive effusions may compress contiguous lung, as in a pneumothorax, or act as a space-occupying process displacing the heart and mediastinum to the contralateral side.

A small pleural effusion suspected on conventional films can be easily confirmed on films obtained in the lateral decubitus position (horizontal beam with the affected side down). As little as 5 ml of pleural fluid can be seen layering out as a linear density of fluid along the dependent chest wall. If the fluid is loculated, however, it will not layer out in the decubitus position and may be indistinguishable from pleural thickening, which also may blunt the costophrenic sulcus. If pleural fluid is suspected but not confirmed on a decubitus view, ultrasound is often of value in differentiating a loculated effusion from pleural thickening.

Pleural effusions may be missed on frontal radiographs taken with the patient in the supine position since the fluid layers out posteriorly (most dependent part of the thorax), and the lateral costophrenic sulci may remain sharp. In such patients, pleural effusion produces a generalized hazy shadow over the entire

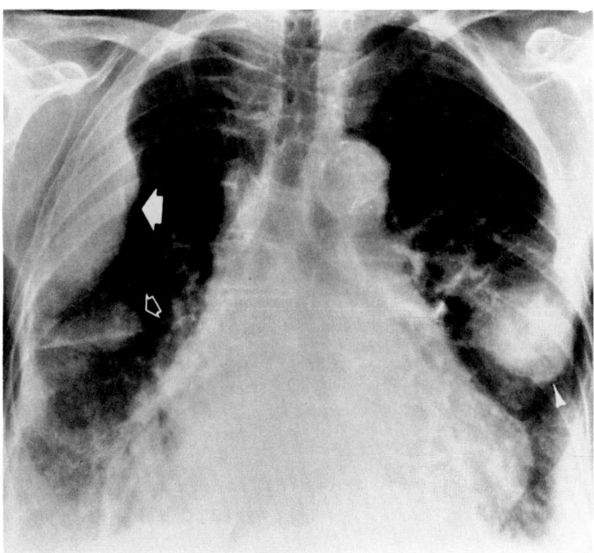
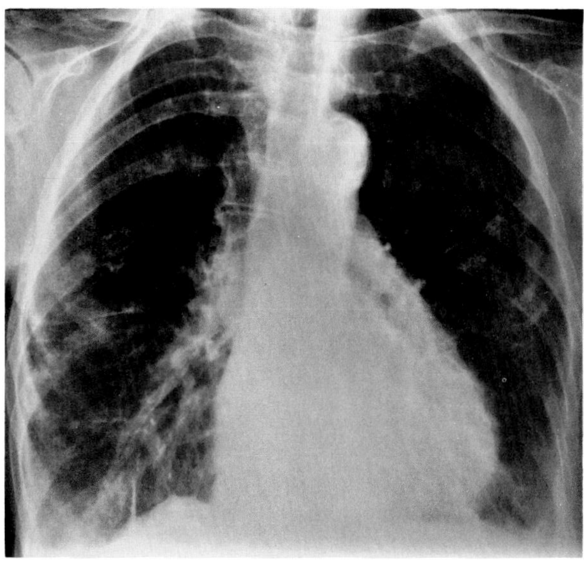

A B

**Figure 35.5. Phantom tumors.** (A) During an episode of congestive heart failure, a frontal chest radiograph demonstrates marked cardiomegaly with bilateral pleural effusion. Note the fluid collections along the lateral chest wall (closed arrow), in the minor fissure (open arrow), and in the left major fissure (arrowhead). (B) With improvement in the patient's cardiac status, the phantom tumors have disappeared. Bilateral small pleural effusions persist.

hemithorax that does not obliterate the normal lung markings and may be difficult to detect when unilateral and impossible to detect when bilateral.

Pleural fluid may collect below the inferior surface of the lung (subpulmonic effusion) and give the radiographic appearance of an elevated hemidiaphragm. In such patients, the presence of a subpulmonic effusion can be suggested by the blunting of the posterior costophrenic angle on lateral views, by an unusually lateral position of the apex of the false hemidiaphragm (pseudodiaphragmatic contour), and by an increased distance between the gastric air bubble and the apparent upper border of the false left hemidiaphragm. The presence of pleural fluid can be easily confirmed on lateral decubitus views.

In patients with congestive heart failure, a loculated effusion can develop in an interlobar fissure and produce a round or oval density resembling a solitary pulmonary nodule. As the patient's cardiac status improves, subsequent studies demonstrate decreased size or complete resolution of these phantom tumors.[1–5,66]

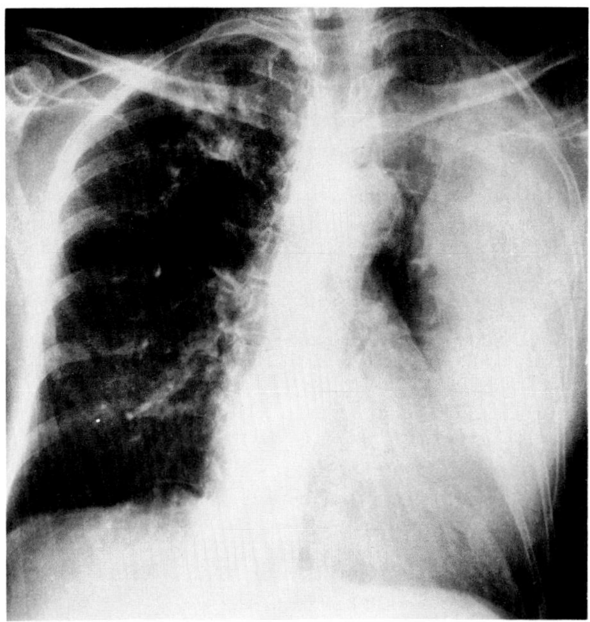

**Figure 35.6. Empyema.** A large soft tissue mass fills much of the left hemithorax.

## Empyema

Empyema (pyothorax) refers to the presence of infected liquid or frank pus in the pleural space. Usually the result of the spread of a contiguous infection (e.g., bacterial pneumonia, subdiaphragmatic abscess, lung abscess, esophageal perforation), empyemas may also follow thoracic surgery, trauma, or instrumentation of the pleural space. Radiographically, empyemas are initially indistinguishable from pleural effusion. A subacute or chronic empyema becomes loculated and appears as a discrete mass density that may vary in size from a large lesion filling much of the hemithorax to a small mass along the chest wall or in an interlobar fissure. Air within a free or loculated empyema causes an air-fluid level and indicates communication with a bronchus or the skin sur-

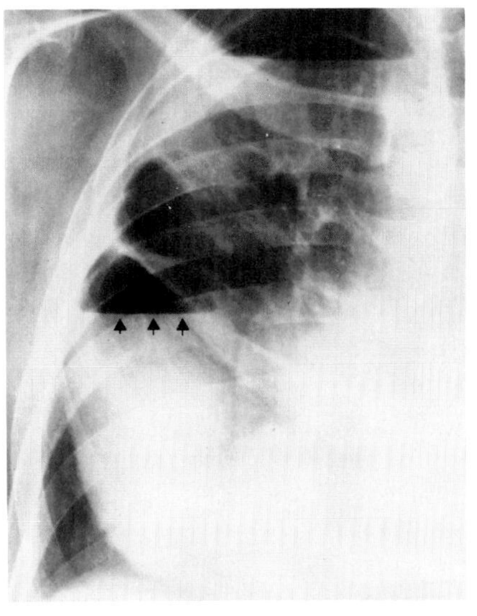

A

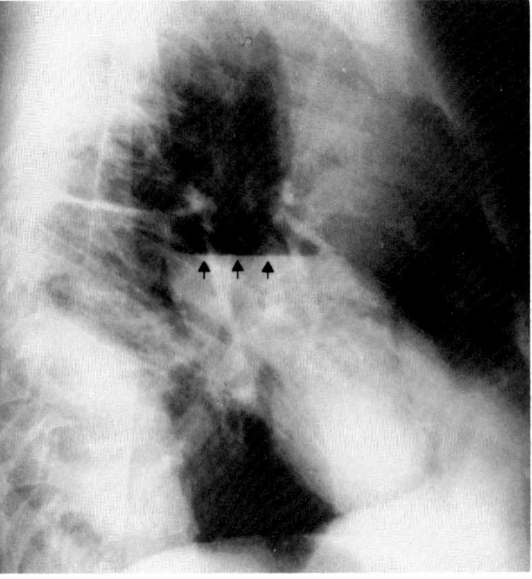

B

**Figure 35.7. Multilocular empyema and bronchopleural fistula.** (A) Frontal and (B) lateral views demonstrate a large multilocular empyema in a patient with an anaerobic lung infection. The air-fluid level within the empyema indicates communication with a bronchus. (From Landay, Christensen, Bynum, et al,[50] with permission from the publisher.)

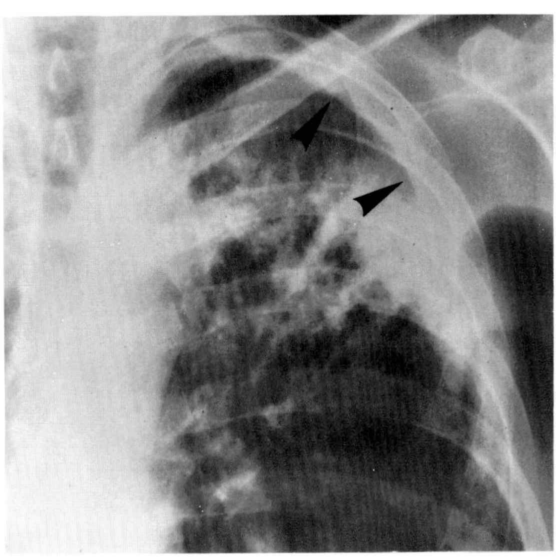

A

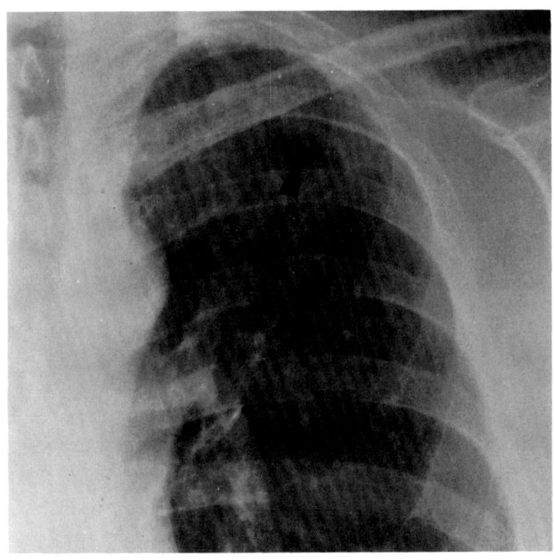

B

**Figure 35.8. Pleural thickening.** (A) There is upper lobe infiltration with overlying pleural thickening (arrows), characteristic of histoplasmosis. Neither the pleural nor the parenchymal abnormalities were present 16 days earlier. (B) Two years later the veil-like density with an edge (arrow) that does not correspond to an anatomic structure is characteristic of pleural thickening. (From Vix,[49] with permission from the publisher.)

face (bronchopleural fistula, post-thoracentesis). Aspiration of an empyema may be performed under fluoroscopic guidance or, if it is a loculated mass adjacent to the chest wall, via ultrasound.

At times, an intrathoracic, fluid-containing cavitary lesion adjacent to the chest wall may be consistent with either a lung abscess or empyema and thus pose a difficult diagnostic problem. On computed tomography (CT), abscesses have an irregular shape and a relatively thick wall that is not uniformly wide and does not have a discrete boundary between the lesion and the lung parenchyma (see Figure 29.7). In contrast, empyemas have a regularly shaped lumen, a smooth inner surface, and a sharply defined border between the lesion and the lung (see Figure 29.8).[6–8,50]

## Pleural Thickening

Thickening and fibrosis of the pleura is most commonly the result of a remote pleural inflammation or effusion. It appears radiographically as a soft tissue density interposed between the inner aspect of the ribs and the adjacent air-containing lungs. Pleural thickening may vary

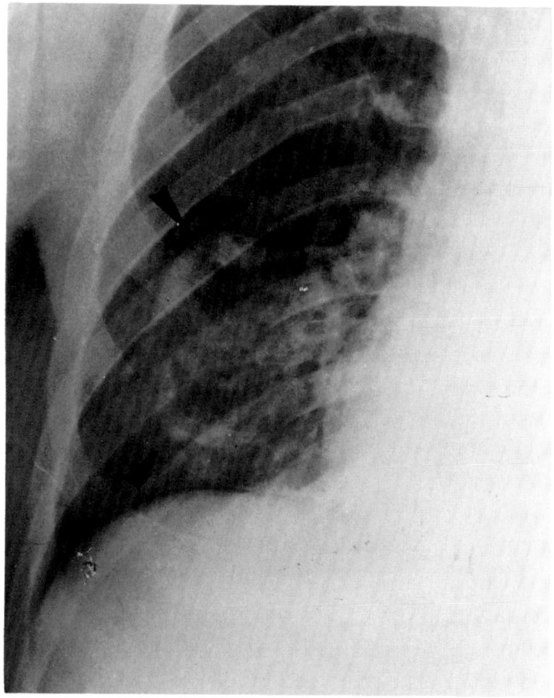


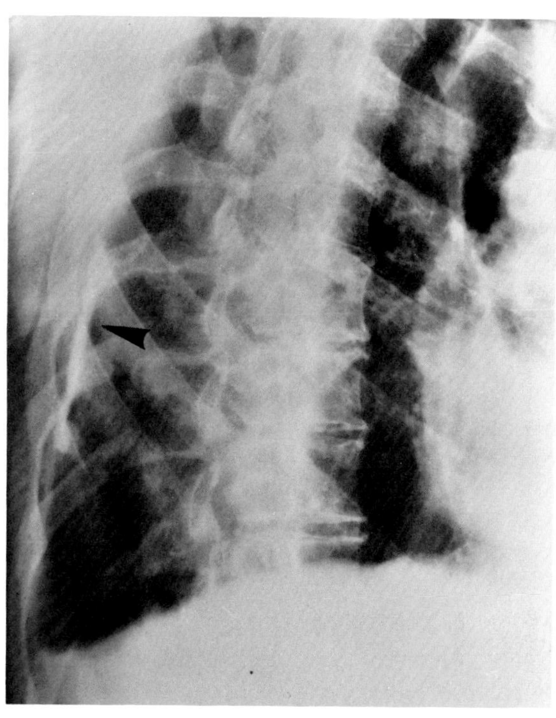


**Figure 35.9. Calcified thickened pleura.** (A) The density in the lower lung (arrows) has a well-defined irregular border closely resembling a cavity. (B) An oblique view, however, shows a linear density (arrow) paralleling but separated from the chest wall. This profile appearance is pathognomonic of calcification in a thickened pleura. (From Vix,[49] with permission from the publisher.)

from a thin linear band to extensive irregular masses along the chest wall. Pleural thickening often causes blunting of the costophrenic angle, which tends to be more sharply angulated than the meniscus-shaped configuration of pleural effusion. However, pleural thickening and effusion often cannot be differentiated on routine radiographs, and a lateral decubitus view may be required to demonstrate that the pleural thickening remains unchanged with alterations in patient position. Extensive bilateral pleural thickening suggests asbestosis, especially if associated with linear calcific plaques in the diaphragmatic pleura.

Irregular symmetric thickening of the apical pleura is a common finding on frontal chest radiographs. Although this apical cap is often ascribed to tuberculosis, the appearance probably represents a fibrotic reaction to nonspecific pulmonary inflammation. Asymmetry of apical pleural thickening or a changed appearance on sequential films may be an early manifestation of Pancoast's tumor.[9,49]

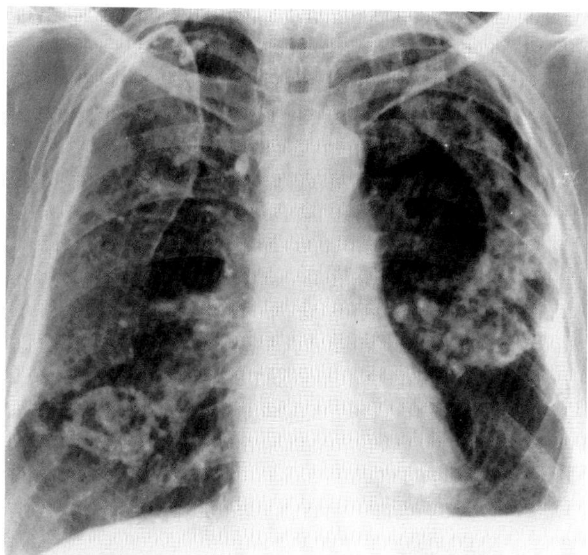

**Figure 35.10. Pleural calcification in tuberculosis.** Bilateral broad continuous sheets of calcification overlie much of the lung surface.

## Pleural Calcification

Pleural calcification may appear as single or multiple discrete plaques overlying the chest wall or as a broad continuous sheet overlying the entire lung surface. Unilateral pleural calcification is usually secondary to old inflammatory disease (tuberculosis, empyema) or to traumatic hemorrhage into the pleural space. Bilateral pleural calcification usually represents asbestosis, especially if associated with calcified plaques in the parietal pleura of the diaphragm.[5]

## Pneumothorax

Pneumothorax refers to the presence of air in the pleural cavity that results in a partial or complete collapse of the lung. In adults, pneumothorax is associated with a broad spectrum of underlying pulmonary diseases including diffuse interstitial fibrosis, bullous emphysema, and respiratory distress syndrome. In neonates, pneumothorax is a frequent complication of hyaline membrane disease that requires prolonged assisted ventilation. Pneumothorax may be post-traumatic or iatrogenic (thoracentesis or an attempted insertion of a central venous catheter into the subclavian vein), with air entering the pleural space following a penetrating injury to the chest that tears the pleura or damages the lungs. In women, recurrent catamenial pneumothorax, usually right-sided, can occur during menstrual flow (possibly related to intrathoracic endometriosis). Spontaneous pneumothorax refers to air in the pleural cavity that occurs in an otherwise healthy person. This condition develops most commonly in men between 20 and 40 years of age, is frequently recurrent, and probably results from the leakage of air into the pleural space due to the rupture of small, often radiographically undetectable, blebs on the surface of the visceral pleura.

A pneumothorax appears radiographically as a hyperlucent area in which all pulmonary markings are absent. The radiographic hallmark of pneumothorax is the demonstration of the visceral pleural line, which is outlined centrally by air within the lung and peripherally by air within the pleural space. A large pneumothorax can cause the collapse of an entire lung. Chest radiographs for pneumothorax should be taken with the patient in the upright position; in the supine position, the upward movement of air with approximation of the visceral and parietal pleurae laterally may obscure its presence. To identify small pneumothoraxes, a radiograph should be obtained with the lung in full expiration. This maneuver causes the lung to decrease in volume and become relatively denser, while the volume of air in the pleural space remains constant, thereby providing a smaller visceral pleural surface in contact with the pneumothorax, which is then easier to define. Another technique for demonstrating small pneumothoraxes is a film obtained with a horizontal x-ray beam while the patient is in the lateral decubitus position. In this position, air rising to the highest point in the hemithorax is more clearly visible over the lateral chest wall than on erect views, in which a small amount of air in the apical region may be obscured by overlying bony densities. If a pneumothorax is associated with the tearing of adhesions in the pleural space, a hemopneumothorax with an air-fluid level may develop.

Small pneumothoraxes usually reabsorb spontaneously; larger pneumothoraxes require thoracostomy tube drainage. Failure of the lung to completely re-expand following chest tube suction may be related to extensive adhesions from chronic disease; the occlusion of a major bronchus; or a bronchopleural fistula, a large continuing air leak through a major communication between the pleural space and lung that is usually associated with an inflammatory pulmonary process.

Tension pneumothorax is a medical emergency in which air continues to enter the pleural space but cannot

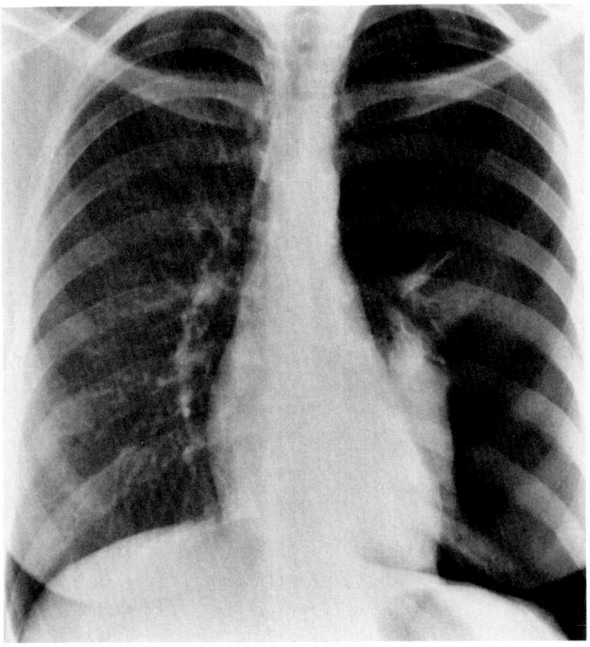

**Figure 35.11. Spontaneous pneumothorax.** Complete collapse of the left lung that developed in an otherwise healthy young woman.

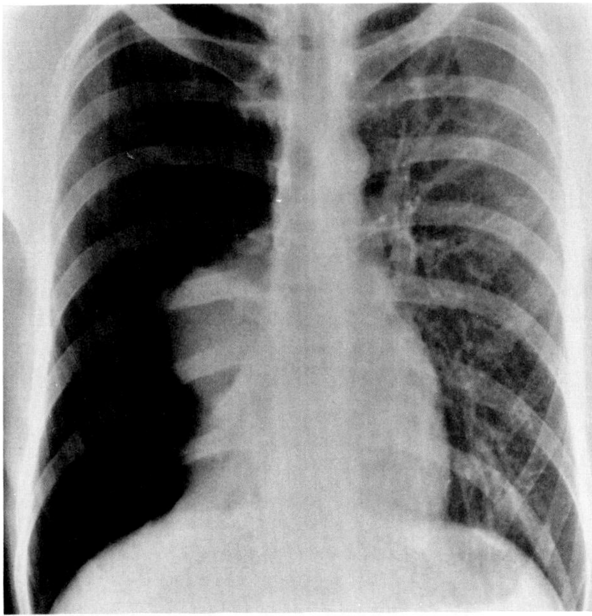

**Figure 35.12. Pneumothorax.** Complete collapse of the right lung that developed following an unsuccessful attempt to introduce a central venous pressure line through the right subclavian vein.

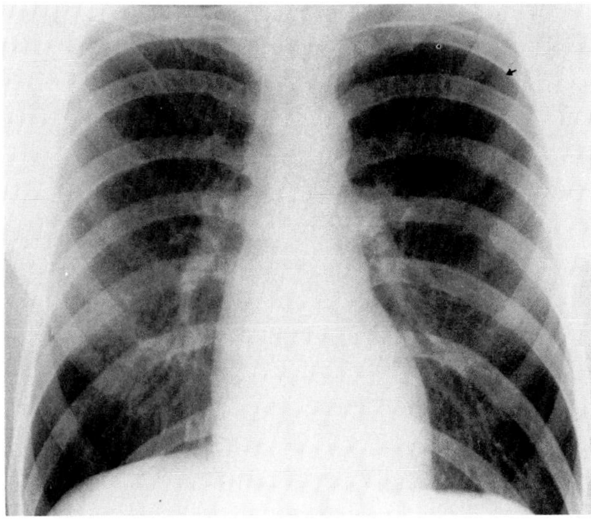
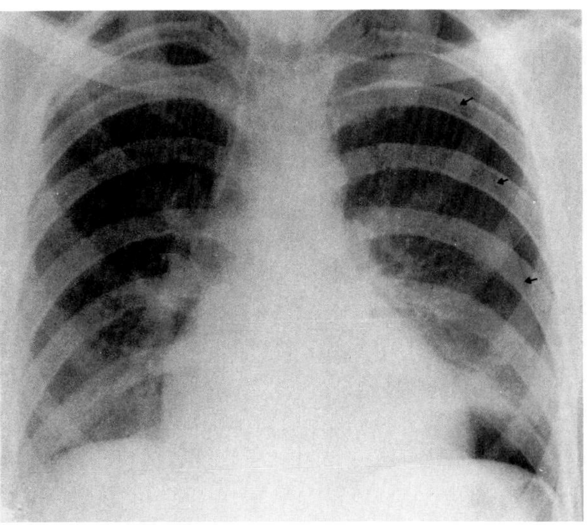

A B

**Figure 35.13. Pneumothorax.** (A) On a routine frontal chest film, there is a faint rim of pleura (arrow) at the left apex separated from the thoracic wall by an area containing air but no pulmonary vasculature. (B) On an expiratory film, the left pneumothorax (arrows) is clearly seen.

leave it. The accumulation of air within the pleural space causes complete collapse of the lung on that side, provided there are no adhesions to prevent it. As the tension pneumothorax enlarges, there is ipsilateral depression of the diaphragm. The heart and mediastinal structures are shifted toward the unaffected side, severely compromising cardiac output because of the positive intrathoracic pressure that decreases venous return to the heart. If a tension pneumothorax is not treated promptly, the resulting circulatory collapse may be fatal.[5,10,11]

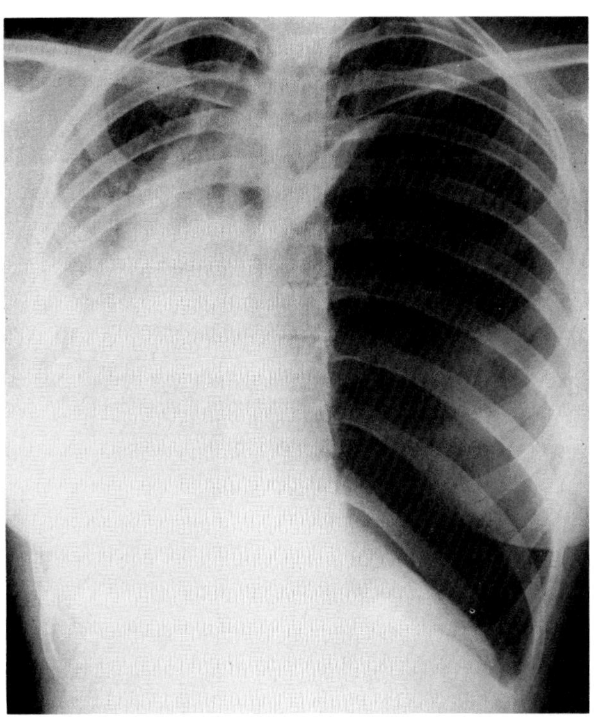

**Figure 35.14. Tension pneumothorax.** The left hemithorax is completely radiolucent with a total absence of vascular markings. There is a dramatic shift of the heart and mediastinum to the right. The left hemidiaphragm is markedly depressed, and there is spreading of the left ribs.

## Pleural Tumors

The most important primary neoplasm of the pleura is mesothelioma, which can exist in two widely divergent forms. The benign, localized mesothelioma is a solitary, sharply circumscribed mass of homogeneous density that arises on a pleural surface and only occasionally causes a pleural effusion. Often asymptomatic and curable by surgical resection, the tumor is frequently associated with hypertrophic osteoarthropathy, which, if seen in the presence of a large local intrathoracic mass, strongly suggests the possibility of benign pleural mesothelioma.

Diffuse pleural mesothelioma is a highly malignant tumor with a well-documented relation to asbestos exposure. Radiographically, malignant mesothelioma appears as an irregular scalloped or nodular density of the pleural space at the periphery of the lung or within an interlobar fissure. The tumor extends rapidly and may surround the entire lung or spread to the pericardium or opposite pleural surface. Malignant mesothelioma is frequently associated with a large pleural effusion that may obscure the underlying neoplasm. Even with massive effusions, a substantial shift of the mediastinum toward the contralateral side infrequently occurs because of the fixation of the mediastinum by tumor.

Metastatic pleural tumors occur much more frequently than primary malignant neoplasms. The most common primary tumors that metastasize to the pleura

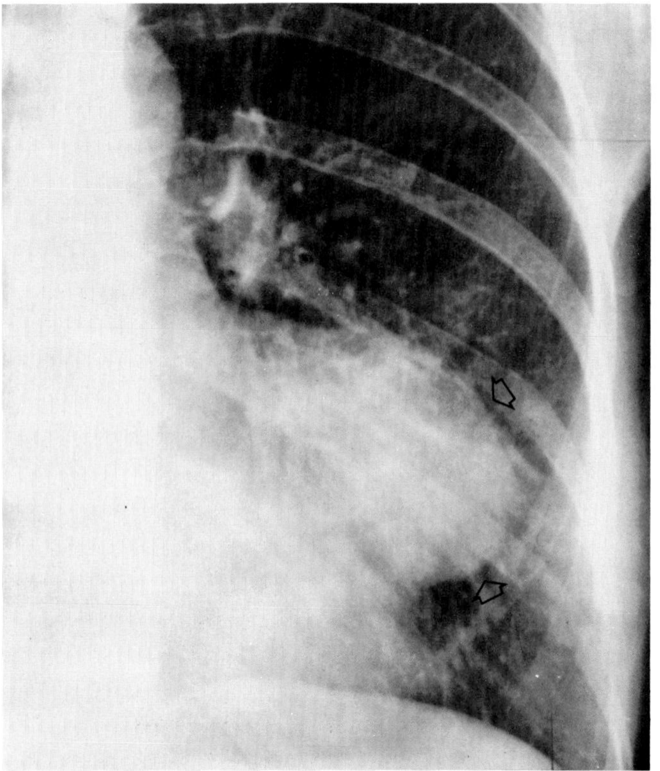

A

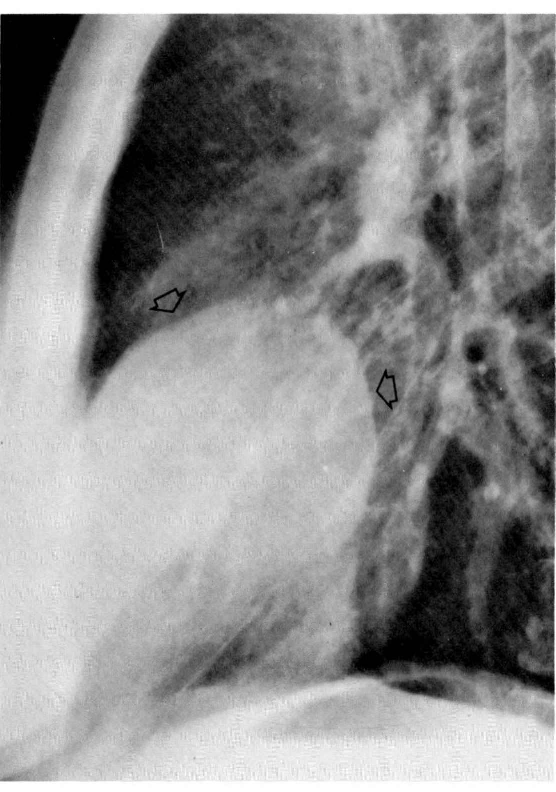

B

**Figure 35.15. Localized fibrous mesothelioma.** (A) Frontal and (B) lateral views of the chest demonstrate a large paracardiac mass (arrows). At surgery the lesion was found to be attached to the lingula by a narrow pedicle. (From Ellis and Wolff,[15] with permission from the publisher.)

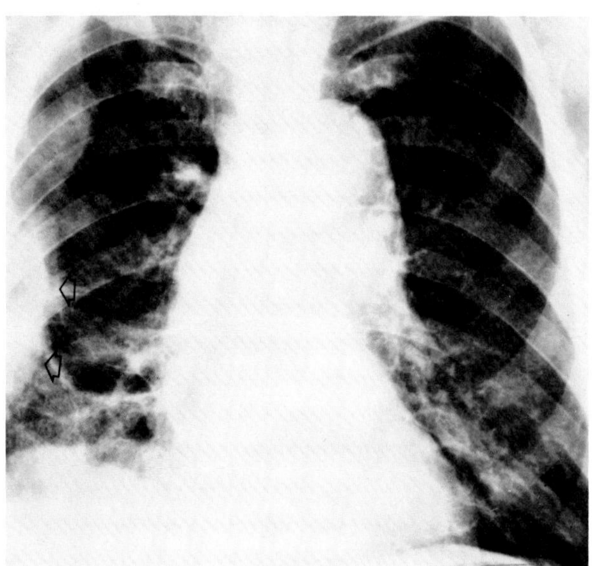

**Figure 35.16. Diffuse pleural mesothelioma.** Multiple masses thicken the right pleura (arrows) in this 80-year-old man with chronic asbestos exposure. (From Ellis and Wolff,[15] with permission from the publisher.)

are neoplasms of the bronchus, breast, ovary, and gastrointestinal tract. Because the actual metastatic deposits are often very small and radiographically undetectable, a large pleural effusion may be the only indication of pleural metastases on conventional chest examinations. CT is a more sensitive modality for discovering small pleural metastases that would otherwise remain undetectable on standard chest radiographs.[12–16]

## DISEASES OF THE MEDIASTINUM

### Mediastinal Masses

Because various types of mediastinal masses tend to occur predominantly in specific locations, the mediastinum is often divided into anterior, middle, and posterior compartments. The anterior compartment extends from the sternum back to the trachea and anterior border of the heart. The middle mediastinal compartment contains the heart and great vessels, the central tracheobronchial tree and lymph nodes, and the phrenic nerves. The posterior compartment is composed of the space behind the pericardium.

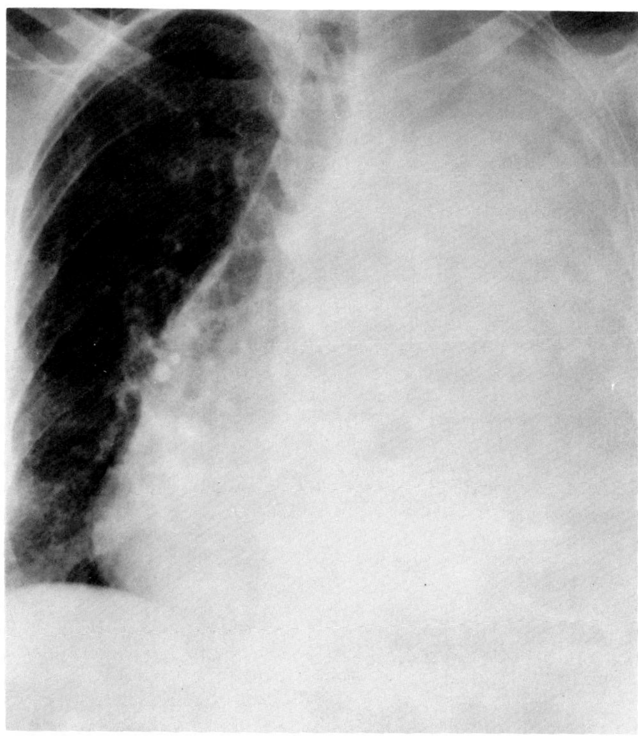

A

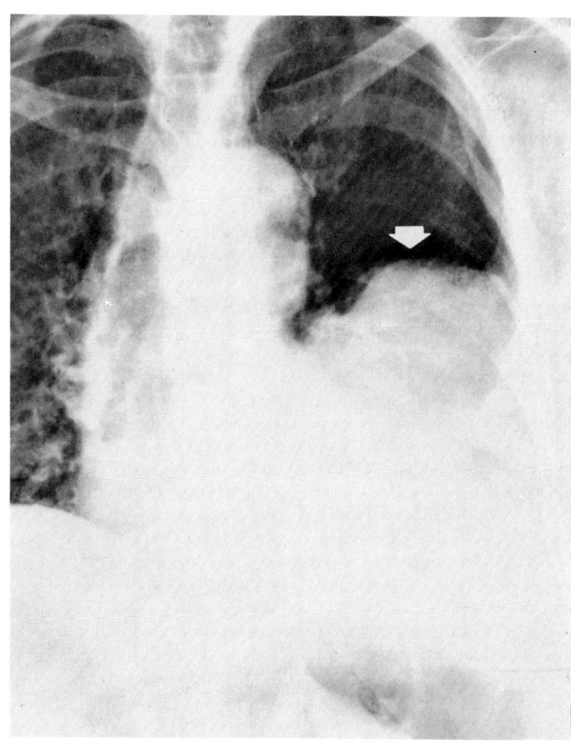

B

**Figure 35.17. Diffuse pleural mesothelioma.** (A) An initial film demonstrates a massive left pleural effusion. (B) Following thoracentesis, the top of a large mass (arrow) is seen. (From Ellis and Wolff,[15] with permission from the publisher.)

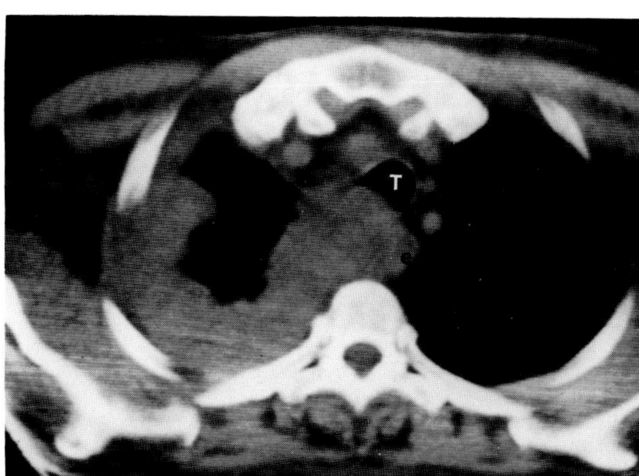


**Figure 35.18. CT of pleural mesothelioma.** An extensive lobulated neoplasm encompassing the right lung is seen extending into the mediastinum and compressing the trachea (T) and esophagus (e). (From Sagel,[16] with permission from the publisher.)

Major lesions of the anterior mediastinum include thymomas, teratodermoids, thyroid masses, lipomas, and lymphoma. The middle mediastinum is involved with lymph node enlargement (lymphoma, metastatic carcinoma, granulomatous processes), bronchogenic cysts, vascular anomalies, and various masses situated in the anterior cardiophrenic angle (pleuropericardial cysts, foramen of Morgagni hernias). The posterior mediastinum is the site of neurogenic tumors, neurenteric cysts, aneurysms of the descending aorta, and extramedullary hematopoiesis.

The configuration of a mediastinal mass depends to a large degree on its consistency. Cystic masses are often compressed between blood vessels and the tracheobronchial tree, producing a multiloculated appearance. In contrast, solid masses tend to compress and displace adjacent structures.

About one-third of the patients with mediastinal masses are asymptomatic, and the lesion is detected on a routine chest radiograph. Chest pain, cough, dyspnea, and symptoms due to compression or invasion of structures in the mediastinum (dysphagia, hoarseness due to recurrent laryngeal nerve involvement, superior vena caval obstruction) are highly suggestive of malignancy. In addition to plain chest radiographs, conventional tomography and contrast studies of the esophagus may be of value in defining the anatomic location and borders of the mass. CT of the chest following the intravenous injection of contrast may aid in distinguishing between vascular and nonvascular lesions as well as in determining whether a lesion is cystic and thus most probably benign.[17–22]

## Teratodermoid Tumor

Teratodermoids are solid or cystic anterior mediastinal masses that most frequently arise at the root of the great vessels. Most cystic lesions are benign and appear as round or oval masses that have smooth contours. The 20 percent of teratodermoids that are malignant are usually more lobulated and less clearly defined than their benign counterparts. Calcification is often present in the wall of a teratodermoid tumor, though this is of little diagnostic value since similar peripheral calcification may also occur in thymomas. Rarely, a precise diagnosis of a dermoid cyst can be made if the mass contains bones or teeth or sufficient fatty material to appear lucent on chest radiographs. A cystic tumor can infrequently become infected and perforate into surrounding structures.

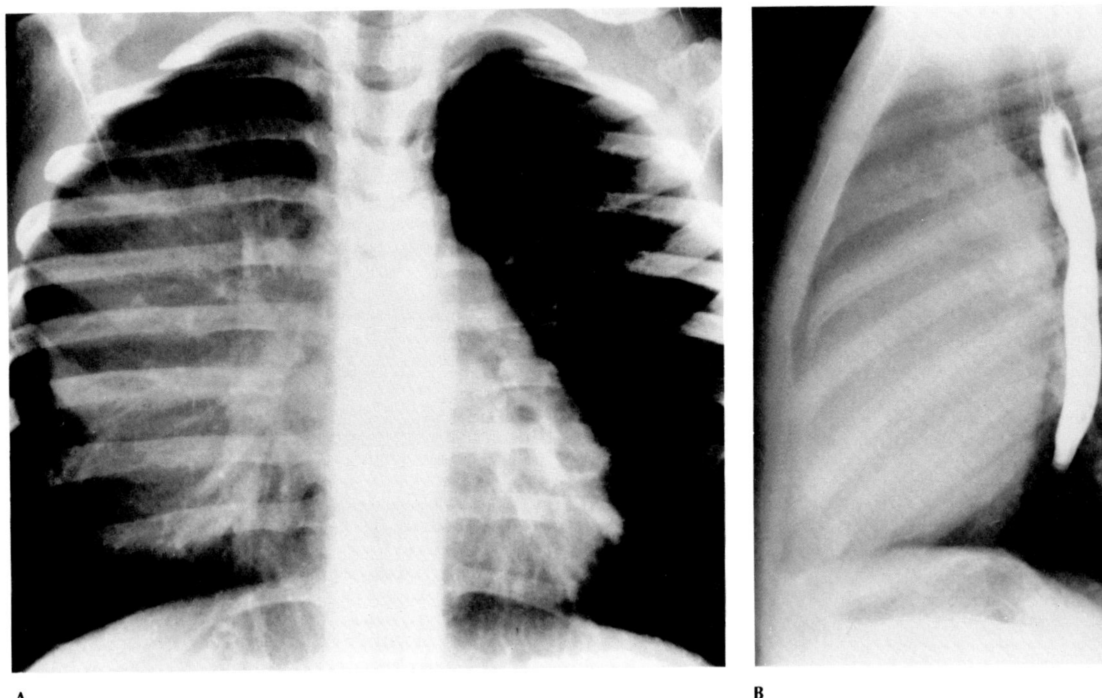


**Figure 35.19. Teratodermoid tumor.** (A) Frontal and (B) lateral views of the chest demonstrate a huge smooth anterior mediastinal mass that is confluent with the right and superior borders of the heart.

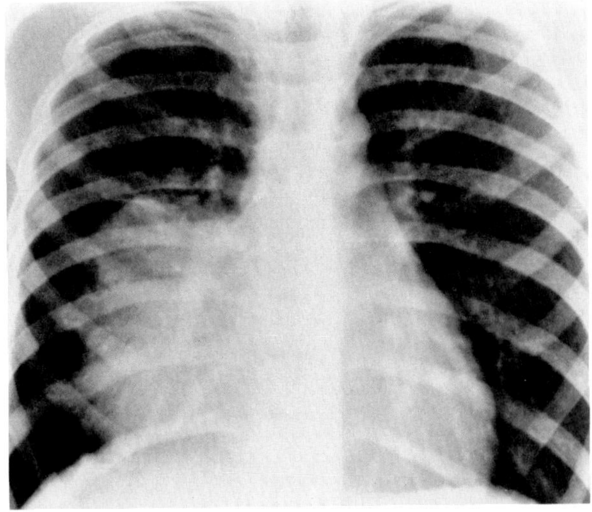

**Figure 35.20. Teratodermoid tumor.** A frontal chest radiograph demonstrates a large lobulated mass arising in the typical location near the base of the heart and projecting to the right.

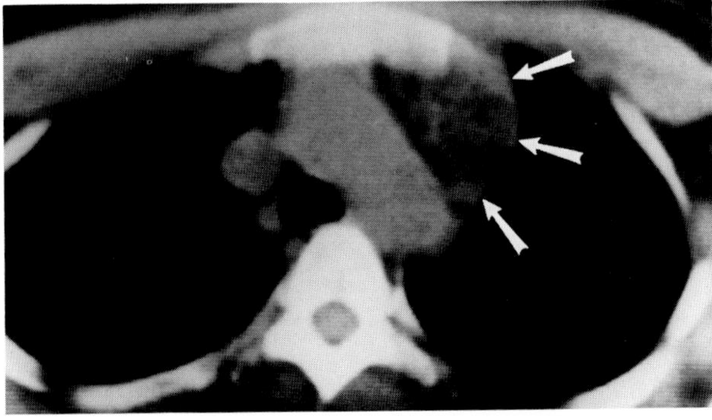

**Figure 35.21. Teratodermoid tumor.** A CT scan demonstrates an inhomogeneous anterior mediastinal mass (arrows) that lies lateral to the aortic arch and contains fat, near-water, and some tissue densities. (From Sagel and Aronberg,[26] with permission from the publisher.)

On CT, a teratodermoid tumor should be suspected when fat, and perhaps calcification, is seen intermixed within an anterior mediastinal mass.[23–26]

## Thymoma

Thymomas are anterior mediastinal tumors that are usually located near the junction of the heart and great vessels. They generally appear as well-circumscribed, smooth or lobulated masses that may displace the heart and great vessels posteriorly. Midline tumors may be impossible to detect on frontal projections, though they are readily seen on oblique or lateral views. Thymomas often protrude to one side of the mediastinum; bilateral extension of a tumor suggests malignancy, as does a rapid increase in size. Mottled calcification within the lesion may occur; peripheral calcification suggests a benign thymic cyst. Although a rapid increase in size and the presence of symptoms are highly suggestive of malignancy, the differentiation between benign and malignant tumors can rarely be made.

Thymolipomas are benign neoplasms arising from fatty tissue within the thymus. Because these tumors are usually asymptomatic, they often grow very large and tend to extend inferiorly to the anterior aspect of the diaphragm, thus leaving the superior mediastinal space clear. CT can demonstrate the fat content of this benign lesion.

About one-fourth of thymomas are malignant, though they rarely metastasize. Myasthenia gravis occurs in almost half of the patients with thymoma; however, only about 20 percent of patients with myasthenia gravis have a thymic tumor. Because many of these thymomas cannot be detected on plain chest radiographs, CT is recommended as a screening procedure in patients with myasthenia gravis. This modality is sensitive in diagnosing thymomas in patients older than 40 years; in younger patients, however, CT often cannot differentiate

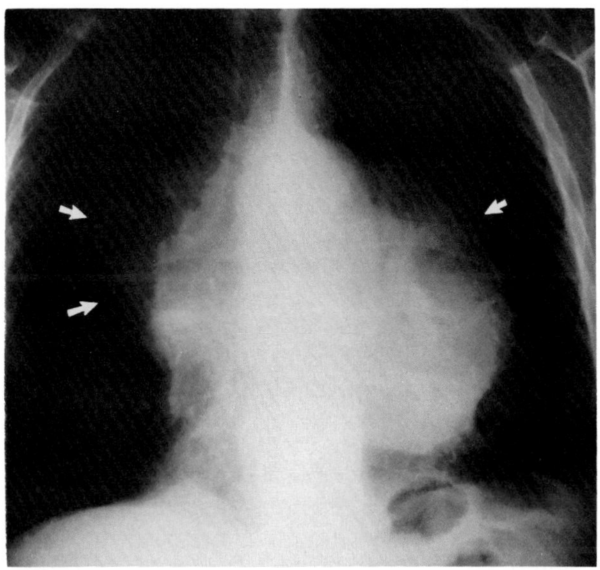

A

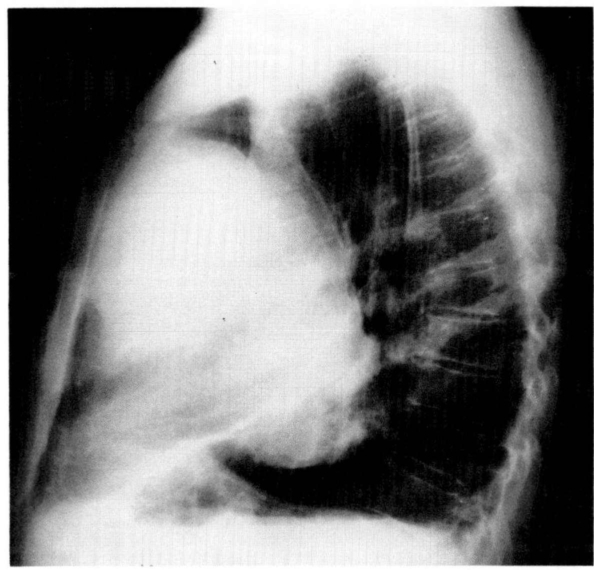

B

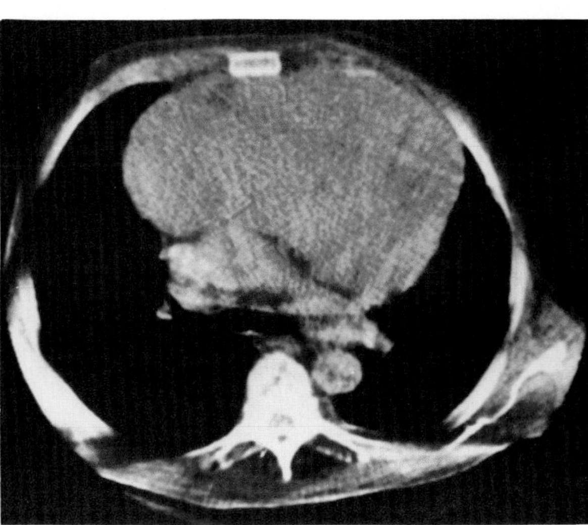

C

**Figure 35.22. Thymoma.** (A) A frontal view of the chest shows a large, bilateral, lobulated mass (arrows) extending to both sides of the mediastinum. (B) A lateral view shows filling of the anterior cardiac space by a mass and posterior displacement of the left side of the heart. (C) A CT scan shows the enormous size of the soft tissue mass and the posterior displacement of other mediastinal structures. No difference in density can be seen between the mass and the heart behind it. (From Feigin,[69] with permission from the publisher.)

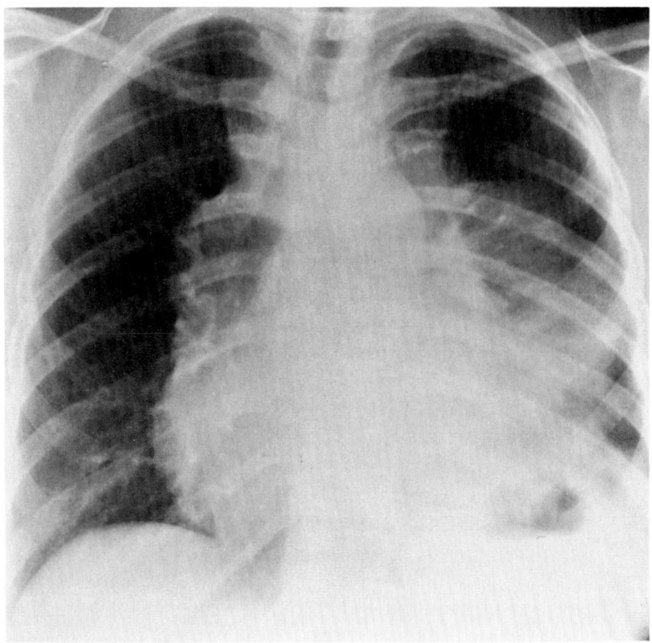

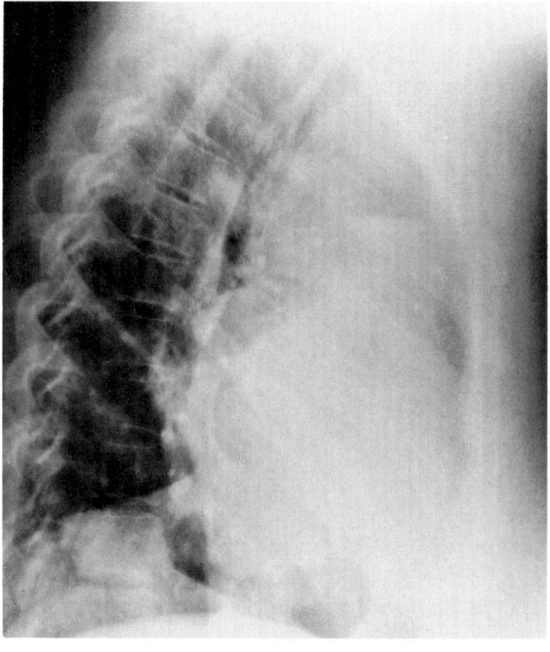

A B

**Figure 35.23. Thymoma.** (A) Frontal and (B) lateral views of the chest demonstrate a huge, lobulated anterior mediastinal mass that obscures the cardiac margins and completely fills the retrosternal air space.

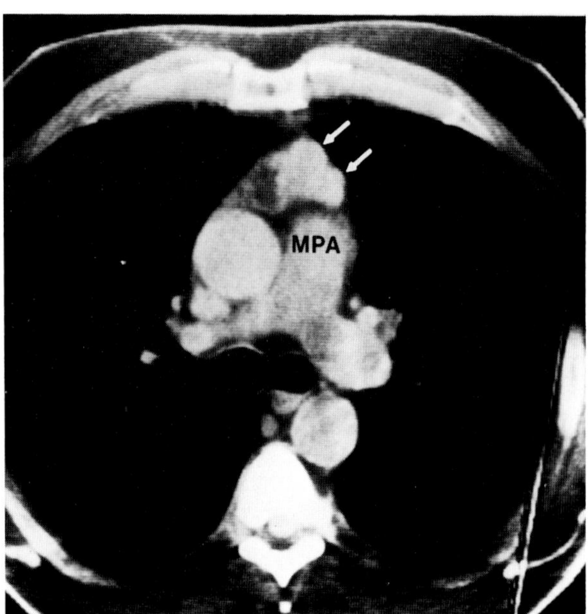


**Figure 35.24. Thymoma.** A CT scan demonstrates a slightly lobulated mass (arrows) anterior to the main pulmonary artery (MPA) in a patient with myasthenia gravis. (From Sagel and Aronberg,[26] with permission from the publisher.)

thymomas from normal thymus or thymic hyperplasia. CT is also of value in suggesting or excluding a diagnosis of thymoma; in differentiating thymoma from thymic hyperplasia, and benign cysts from solid tumors; and in defining the extent of a thymic neoplasm.[27–30,69]

## Substernal Thyroid

Thyroid masses (intrathoracic goiters) may arise from a lower pole or isthmus of the thyroid and extend to the anterior mediastinum in front of the trachea. They appear radiographically as sharply defined, smooth or lobulated masses that displace the trachea laterally and posteriorly and frequently compress it. Amorphous calcification may be seen within the lesion. Intrathoracic thyroid tissue occasionally may produce a mass in the posterior mediastinum behind the trachea, almost always on the right side. In this location, the mass displaces the trachea anteriorly and the esophagus posteriorly.

Most patients with intrathoracic goiters are asymptomatic, and the lesion is discovered on a screening chest radiograph. Symptoms usually relate to tracheal compression, or to hoarseness due to involvement of the recurrent laryngeal nerve.

The uptake of radioactive iodine may permit the definitive diagnosis of an intrathoracic goiter on radionuclide scanning. However, since these lesions are seldom functioning, a negative scan does not exclude the possibility of intrathoracic thyroid tissue.

On serial scans, CT demonstrates the substernal thyroid contiguous with the cervical portion of the gland rather than as a separate ectopic lesion. Because of the increased iodine content, retrosternal thyroid tissue tends to appear denser than surrounding mediastinal structures or the thoracic muscles and demonstrates prolonged enhancement following the intravenous administration of contrast material. A substernal thyroid is often nonhomogeneous and contains discrete, nonenhancing, low-density areas.[21,31–33]

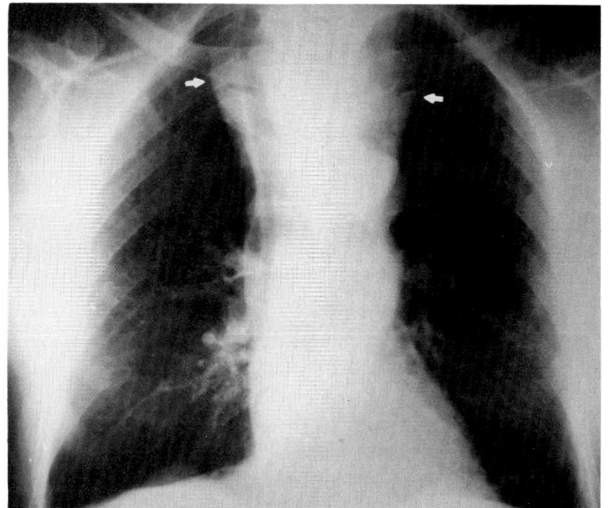
A

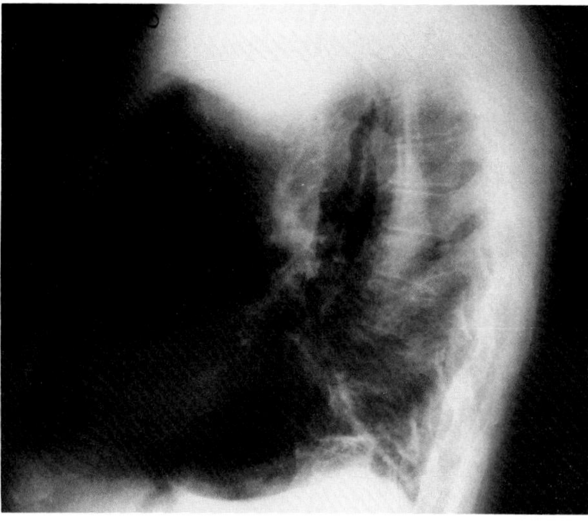
B

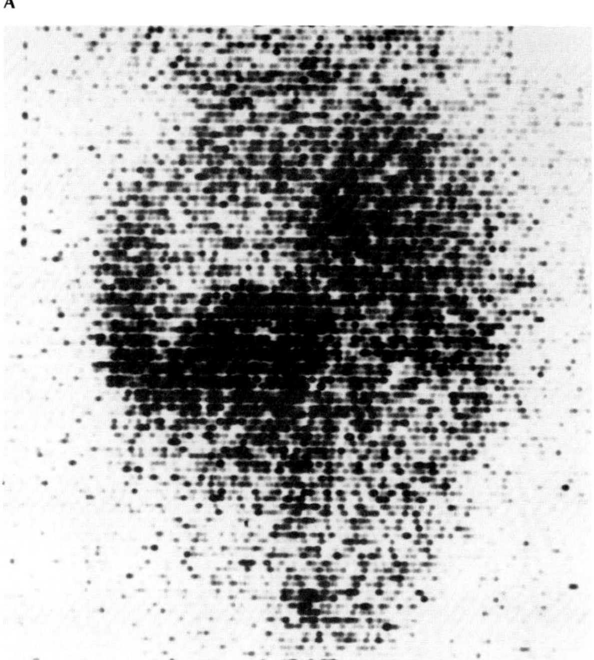
C

**Figure 35.25. Substernal thyroid.** (A) A frontal view of the chest shows marked widening of the superior mediastinum to both sides (arrows) and severe deviation of the trachea to the right. (B) A lateral view shows posterior displacement of the trachea by the large superior mediastinal mass. (C) An iodine 131 scan shows markedly increased uptake of the radionuclide in the area of the mass seen on the radiographs. (Courtesy of the Armed Forces Institute of Pathology.)

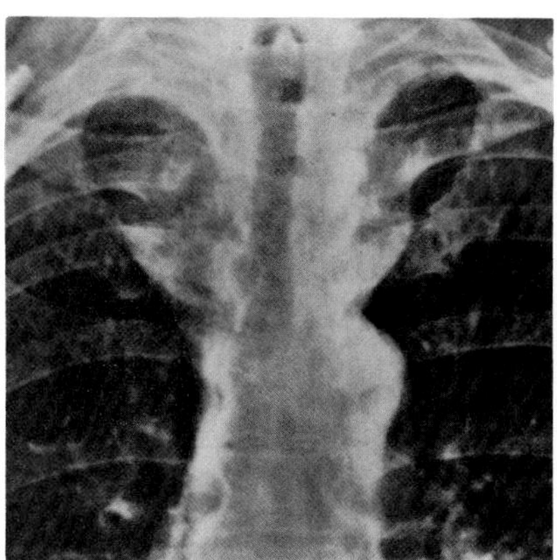
A

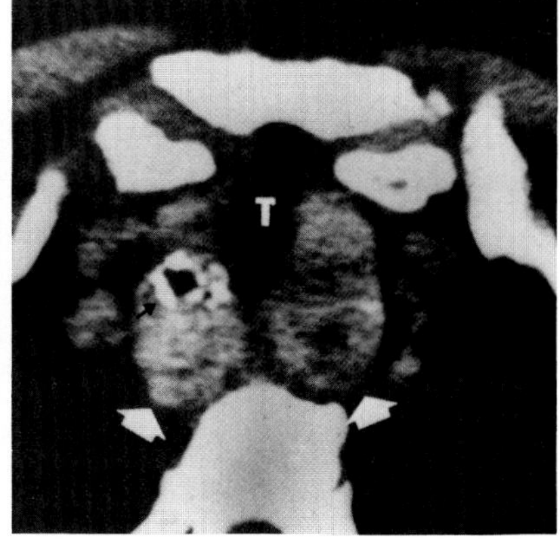

B

**Figure 35.26. Substernal thyroid.** (A) A frontal chest film shows symmetric widening of the superior mediastinum. There is slight compression without displacement of the trachea, an appearance consistent with bilateral extension of nontoxic goiter. (B) A postcontrast CT scan at the level of the manubrium shows the goiter (arrows) extending on both sides of the trachea (T) and narrowing it. Although the goiter generally has higher attenuation than the anterior thoracic musculature, focal low-density areas are present bilaterally. On the right, a ring of calcification surrounds one of these low-density areas, representing a cyst with a calcified wall. The great vessels are lateral to the goiter on both sides and form the outer margins of the mediastinum. (From Bashist, Ellis, and Gold,[32] with permission from the publisher.)

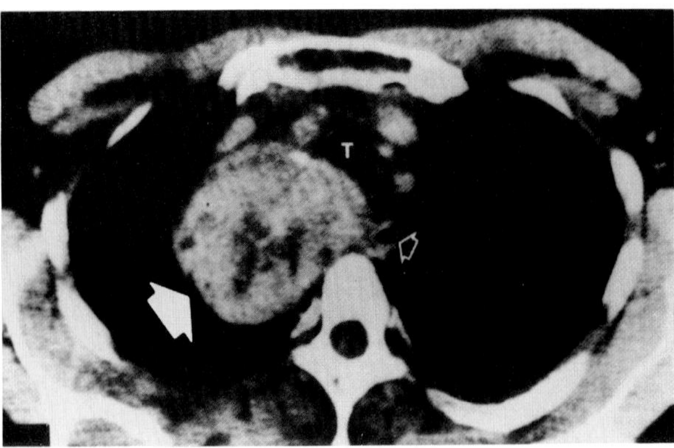


**Figure 35.27. Substernal thyroid.** A postcontrast CT scan at the level of the manubrium shows a large goiter (closed arrows) pushing the trachea (T) anteriorly and separating the trachea from the esophagus (open arrow). The goiter has sharply defined borders, low-density areas within it, and central and peripheral calcification. (From Bashist, Ellis, and Gold,[32] with permission from the publisher.)

## Enlargement of Mediastinal Lymph Nodes

Lymph node enlargement is a common cause of mediastinal masses. Bilateral, though usually asymmetric, mediastinal lymphadenopathy is seen on the initial chest radiographs of more than one-third of the patients with lymphoma. Unlike sarcoidosis, another common cause of hilar and paratracheal adenopathy, lymphoma also frequently involves the anterior mediastinal and retrosternal nodes. Mediastinal and hilar lymphadenopathy is also the most common radiographic appearance of leukemia within the thorax, occurring in about 25 percent of patients with this condition. Metastatic cancer, especially from the lung, upper gastrointestinal tract, prostate, or kidneys, is another major cause of mediastinal lymphadenopathy. Other causes of mediastinal and hilar lymph node enlargement include granulomatous disease (tuberculosis, histoplasmosis), infectious mononucleosis, and lymph node hyperplasia.

CT is the most valuable radiographic technique for demonstrating the enlargement of mediastinal lymph nodes. It must be remembered, however, that CT cannot distinguish malignant from benign causes of mediastinal lymph node enlargement; in addition, a neoplasm that replaces the normal nodal architecture but does not cause enlargement will not be detected on CT scanning. Nevertheless, CT permits the detection of lymph node enlargement at an earlier stage than is possible with any other radiographic method; by precisely depicting which group of lymph nodes are involved, this modality can assess whether a tissue diagnosis can best be obtained by transcervical mediastinoscopy or parasternal mediastinotomy.[5,24,26,34,35,68]

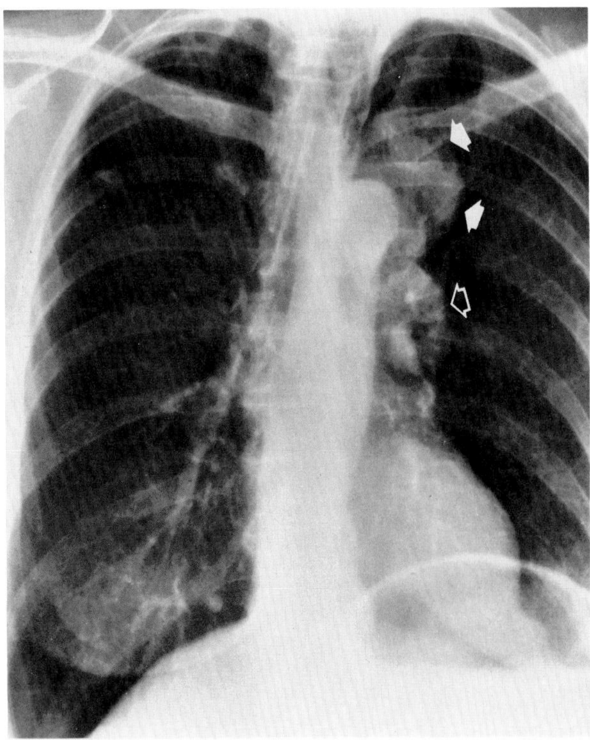

A

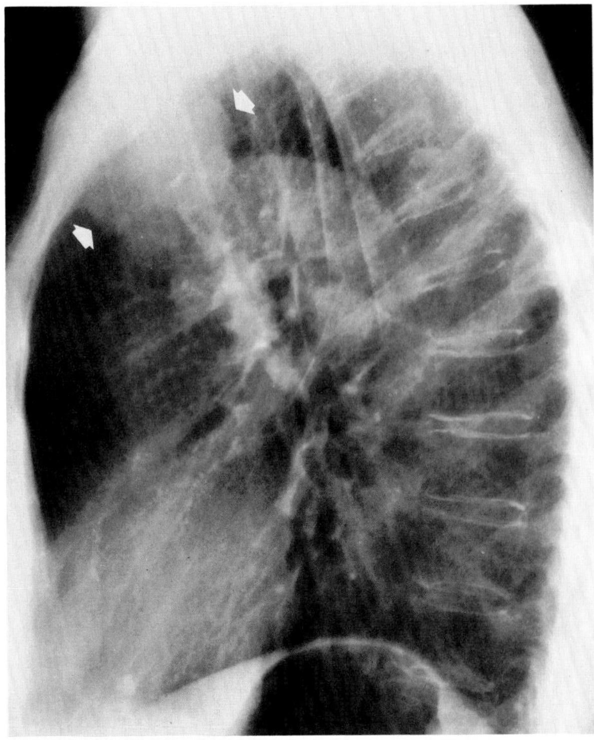

B

**Figure 35.28. Lymphadenopathy due to oat cell carcinoma of the lung.** (A) Frontal and (B) lateral views of the chest demonstrate enlargement of anterior mediastinal lymph nodes (closed arrows) in addition to left hilar adenopathy (open arrow).

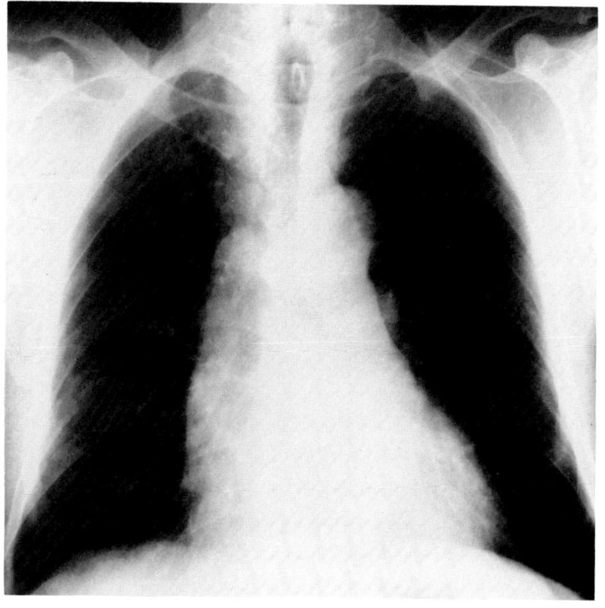
A

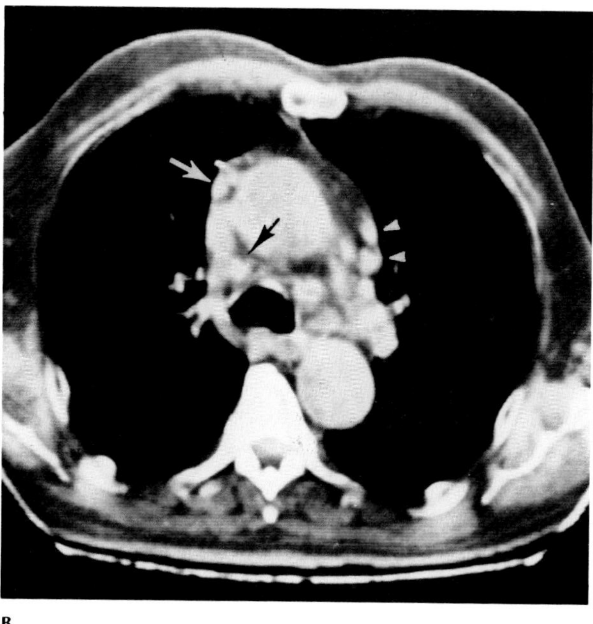
B

**Figure 35.29. Mediastinal lymphoma.** (A) A frontal chest radiograph of a 78-year-old man with fatigue and weight loss reveals mediastinal widening, thought most likely to represent only tortuous vessels. (B) A CT scan demonstrates enlarged lymph nodes in the anterior mediastinum (white arrow), in the paratracheal region accessible to mediastinoscopy (black arrow), and in the aortopulmonary window (white arrowheads). Mediastinoscopy disclosed histiocytic lymphoma. (From Baron, Levitt, Sagel, et al,[68] with permission from the publisher.)

## Vascular Causes of Mediastinal Masses

Aneurysms of the great vessels can appear radiographically as mediastinal masses. Lesions of the ascending arch usually extend anteriorly and to the right. An aneurysm of the transverse arch causes a middle mediastinal mass that characteristically obliterates the aortic window. Aneurysms of the descending aorta may produce posterior mediastinal masses that tend to project to the left. At times, the descending thoracic aorta may be so elongated that it assumes a horizontal position just above the diaphragm, curving sharply to the right and anteriorly and simulating a posterior mediastinal mass. Traumatic aneurysms of the thoracic aorta typically involve the posterior portion of the descending arch, just beyond the origin of the left subclavian artery. They tend to project to the left and often become calcified.

Aneurysms of the innominate artery appear as smooth, well-defined masses in the right superior paramediastinal area, extending upward from the aortic arch. An identical radiographic appearance may be due to buckling of the innominate artery. This common condition, occurring in almost 20 percent of the patients with hypertension or arteriosclerosis, is due to dilatation and elongation of the thoracic aorta that moves the arch cephalad and forces the superiorly fixed innominate artery to buckle to the right.

Other congenital anomalies of the aorta that can appear as mediastinal masses include aortic vascular ring, which produces a "double" aortic knob and a vessel posterior to the esophagus; pseudocoarctation of the aorta, in which elongation and buckling of the aorta produce a left paramediastinal "mass"; cervical aortic arch, in which the arch extends into the soft tissues of the neck before turning downward on itself to become the descending aorta; and the three types of right aortic arch.

Dilatation of the azygos and hemiazygos venous systems can produce the radiographic appearance of a mediastinal mass. Enlargement of these veins may be due to such conditions as portal vein obstruction, anomalous pulmonary venous drainage, acquired or congenital occlusion of the inferior or superior vena cava, and elevated central venous pressure of any cause. Dilatation of the azygos vein is seen as a round or oval density in the right tracheobronchial angle that measures more than 10 mm in diameter on standard erect films. The differentiation of a dilated azygos vein from an enlarged azygos lymph node or mediastinal tumor can be made by demonstrating the definite difference in the diameter of the shadow on radiographs exposed with the patient in supine and erect positions. If uncertainty persists, venography can provide a precise diagnosis.

Dilatation of the posterior portions of the azygos or hemiazygos veins may cause widening and irregularity of the paraspinal line on the right and left sides, respectively.

Dilatation of the superior vena cava gives the distinctive radiographic appearance of a smooth, well-defined widening of the right side of the mediastinum. This mediastinal widening changes considerably with the phase

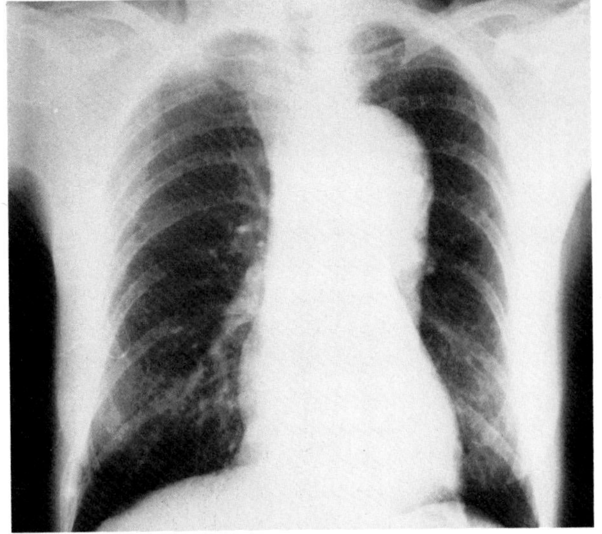

A

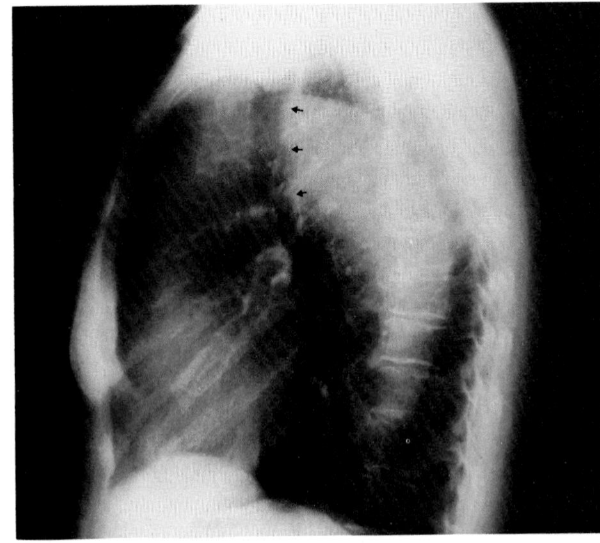

B

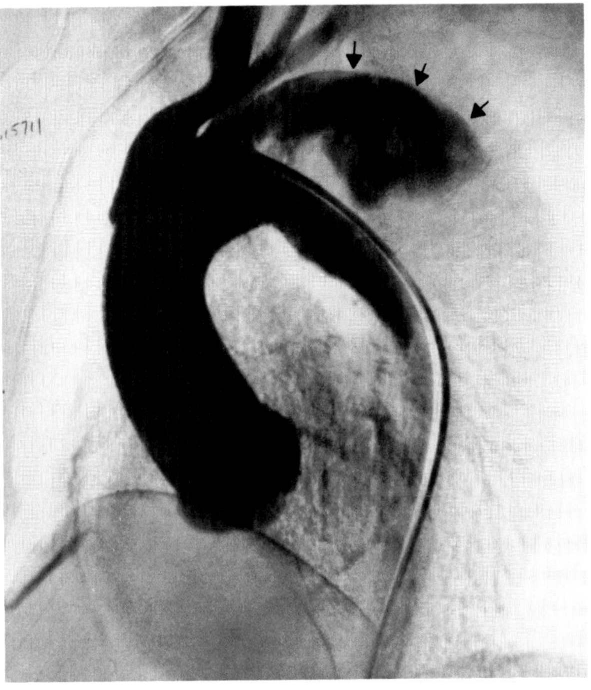

C

**Figure 35.30. Dissecting aortic aneurysm.** (A) A frontal view of the chest in a 54-year-old woman who had experienced sharp, sudden chest pain shows massive dilatation of the aortic knob and descending thoracic aorta. (B) A lateral view confirms the aortic enlargement and shows some anterior displacement of the trachea (arrows). (C) An oblique view of an aortogram shows contrast material entering a large dissection (arrows) that originates near the origin of the left subclavian artery. (From Feigin,[69] with permission from the publisher.)

of respiration and with alterations in intrathoracic pressure (Valsalva and Müller maneuvers). Dilatation of the superior vena cava is usually the result of increased central venous pressure due to congestive heart failure; cardiac tamponade due to pericardial effusion; or constrictive pericarditis. Absence of the right superior vena cava is a rare anomaly that causes the venous drainage from the upper part of the body to flow through a left-sided superior vena cava, thus producing a straight, left-sided mediastinal density that overlies the aortic arch and proximal descending aorta.

CT is of value in demonstrating vascular anomalies or dilatation as the cause for mediastinal widening. Curvilinear calcification can often be seen at the periphery of arteriosclerotic aneurysms. Unlike a solid mediastinal lesion, a vascular structure enhances to the same extent as adjacent normal vessels following the injection of a bolus of intravenous contrast material.[5,36–43]

## Lipoma and Lipomatosis

Lipomas are benign tumors that most commonly occur in the anterior mediastinal compartment. Though they appear as nonspecific masses on plain radiographs, CT almost always can demonstrate the fatty nature of the lesions and thus provide a definitive diagnosis.

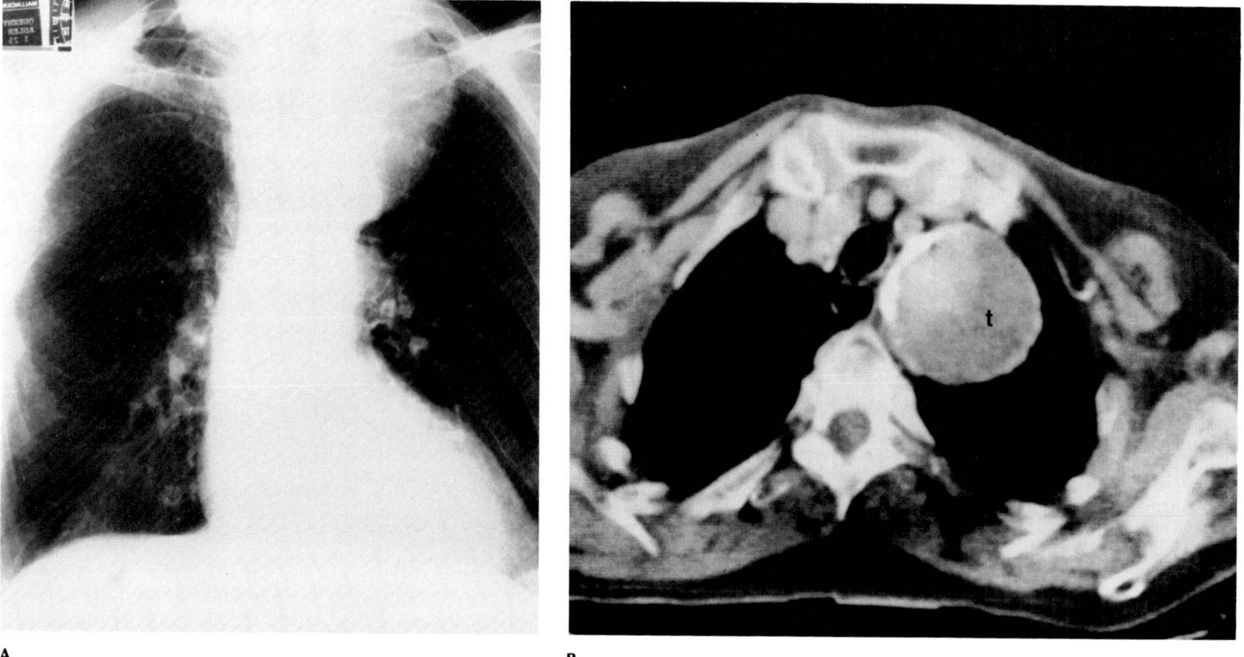

A B

**Figure 35.31. Aneurysm of the left subclavian artery.** (A) A frontal view of the chest in an 82-year-old woman without chest symptoms demonstrates left superior mediastinal widening. (B) A CT scan following the injection of contrast material shows that the mediastinal widening is caused by a left subclavian artery aneurysm that is partially filled with thrombus (t). (From Baron, Levitt, Sagel, et al,[68] with permission from the publisher.)

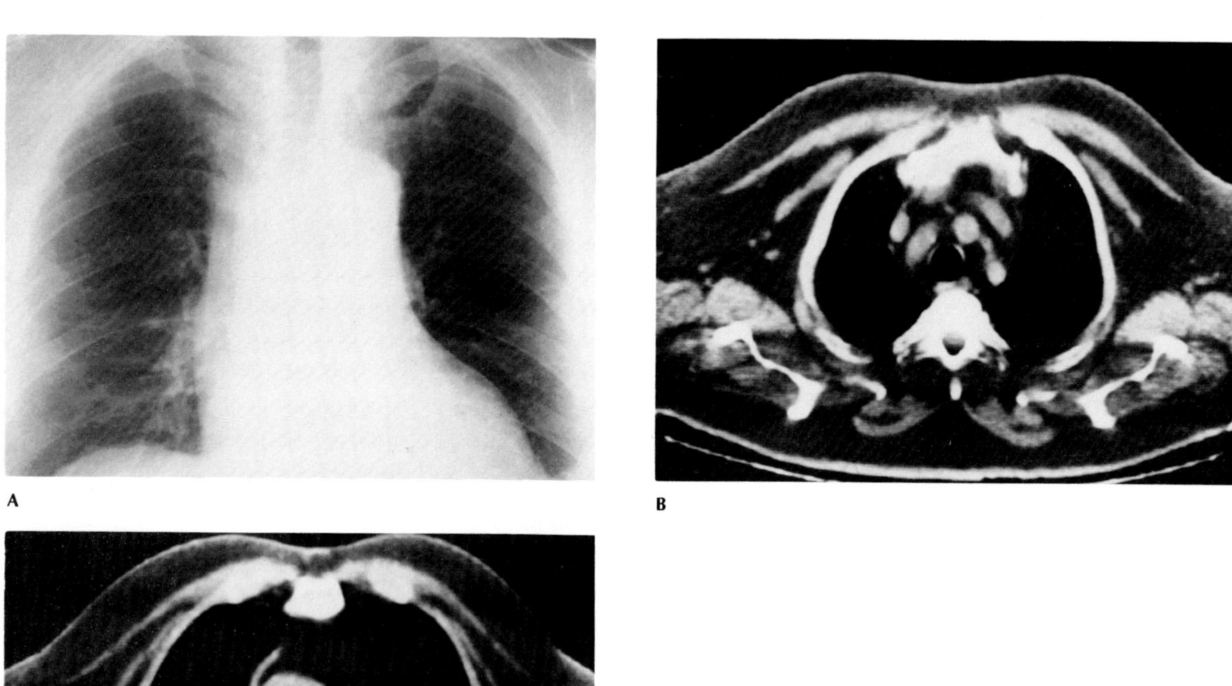

A B C

**Figure 35.32. Mediastinal lipomatosis.** (A) A frontal chest radiograph shows widening of the mediastinum. (B,C) CT scans through the upper mediastinum demonstrate marked tortuosity of the great vessels, along with abundant mediastinal fat, accounting for the widening. No discrete aneurysm or mass is present. (From Baron, Levitt, Sagel, et al,[68] with permission from the publisher.)

An unusual accumulation of fat within the mediastinum may develop in patients with Cushing's syndrome or in those receiving long-term corticosteroid therapy. This mediastinal lipomatosis appears radiographically as smooth and bilaterally symmetric mediastinal widening extending from the thoracic inlet to the hila.[25,29,44]

## Neurogenic Tumors

Neurogenic tumors are found almost exclusively in the posterior mediastinum near the paravertebral gutter. Most of these neoplasms arise from peripheral nerves (neurofibromas, schwannomas) or sympathetic ganglia (ganglioneuromas). Other neurogenic tumors in the posterior mediastinum include neuroblastomas, pheochromocytomas, and paragangliomas (chemodectomas).

All neurogenic tumors present the similar radiographic appearance of a sharply circumscribed, round or oval mass that is usually smooth but may be lobulated. Both benign and malignant neurogenic tumors may cause erosions of the vertebral bodies and pedicles, widening of the intervertebral foramina, and erosion and spreading of the ribs. Some slight differences may aid in differentiating among the neurogenic tumors, though precise

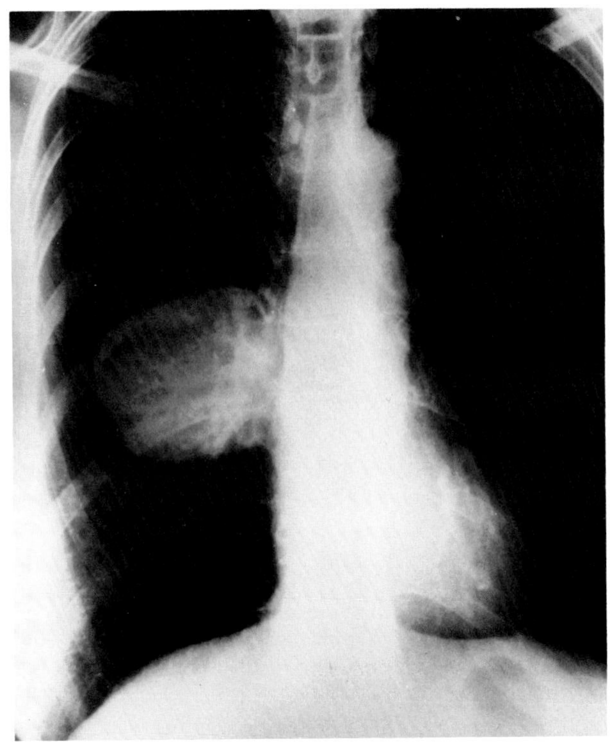

A

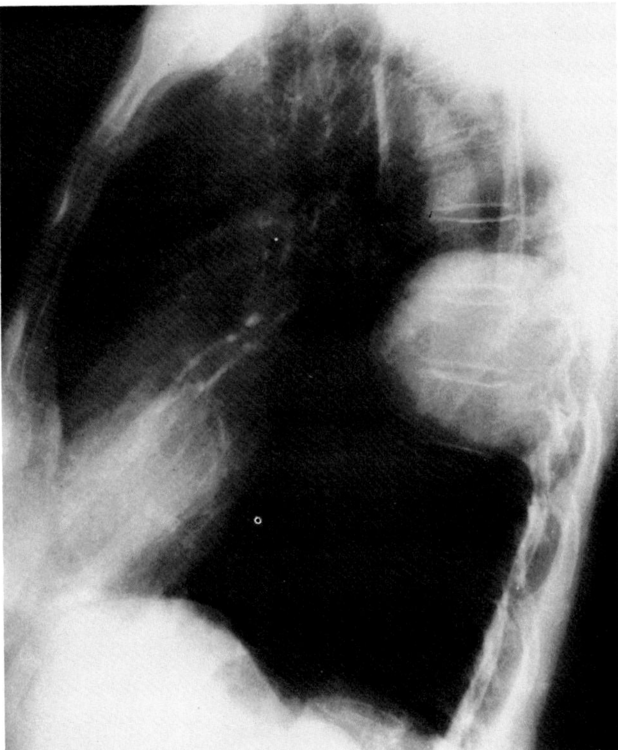

B

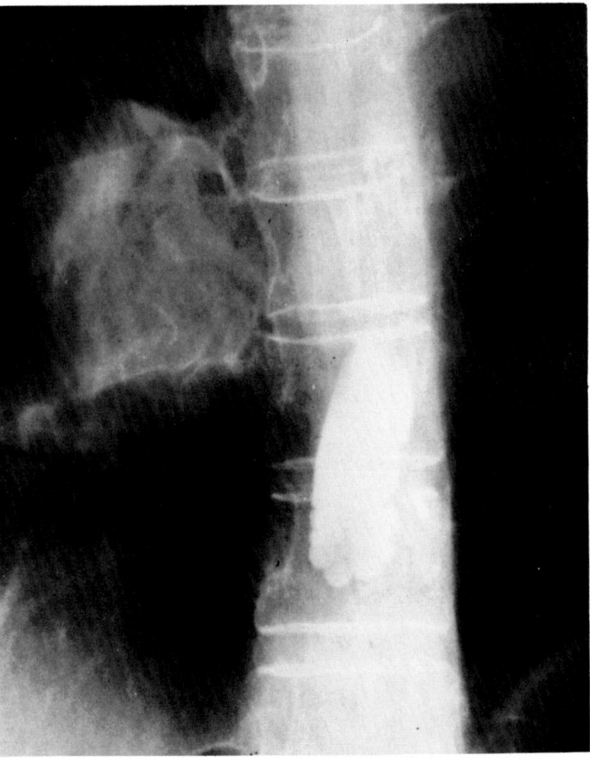

C

**Figure 35.33. Neurogenic tumor.** (A) Frontal and (B) lateral views demonstrate a large right posterior mediastinal mass. (C) A myelogram with the patient in a head-down position shows the extradural mass displacing the subarachnoid space toward the left and obstructing it. Note the erosion of the pedicle of the vertebra adjacent to the center of the mass. (Courtesy of the Armed Forces Institute of Pathology.)

radiographic distinction is usually not possible. Neurofibromas and schwannomas tend to be round masses with a narrow mediastinal base, whereas ganglioneuromas are often more elongated with a broad mediastinal base. Neurofibromas originating in a nerve root may produce a mass that is part inside and part outside the spinal canal. Myelography can demonstrate the intraspinal component of this dumbbell- or hourglass-shaped tumor.[20,45]

## Bronchogenic Cyst

Bronchogenic cysts are congenital lesions that are lined with ciliated columnar epithelium and result from abnormal budding or branching of the tracheobronchial tree during its embryologic development. Mediastinal bronchial cysts usually appear radiographically as sharply defined masses of homogeneous density that most commonly arise just inferior to the carina and often protrude slightly to the right, thus overlapping the right hilar shadow. Most are oval or round, though the shape may vary with the phase of respiration. Bronchial cysts in adults often become huge without causing symptoms; in children, however, symptomatic compression of the tracheobronchial tree is not uncommon.

Bronchial cysts arising in the lung parenchyma appear as nonspecific solitary pulmonary nodules that are initially detected on screening chest radiographs.

Bronchogenic cysts do not communicate with the tracheobronchial tree unless they become infected. This is a frequent occurrence with cysts arising in the pulmonary parenchyma, but is rare with mediastinal cysts. A communicating cyst contains air, with or without fluid.

Most bronchogenic cysts have a homogeneous near-water attenuation value on CT and do not enhance after intravenous contrast administration. At times, bronchogenic cysts may be filled with thick viscid secretions, and this results in a higher attenuation value simulating a solid neoplasm.[46–48]

## Gastroenteric Cyst (Duplication)

Gastroenteric cysts are lined by gastric or intestinal epithelium and represent a failure of complete cannulation of the originally solid esophagus to produce a hollow tube. Although most are asymptomatic, a gastroenteric cyst may become infected and form an abscess. A cyst containing acid-secreting cells may be complicated by ulceration, perforation, and hemorrhage.

Gastroenteric cysts appear as large, round or oval, posterior mediastinal densities that usually extend to one side of the midline. Although most have no connection with the gastrointestinal tract, communication

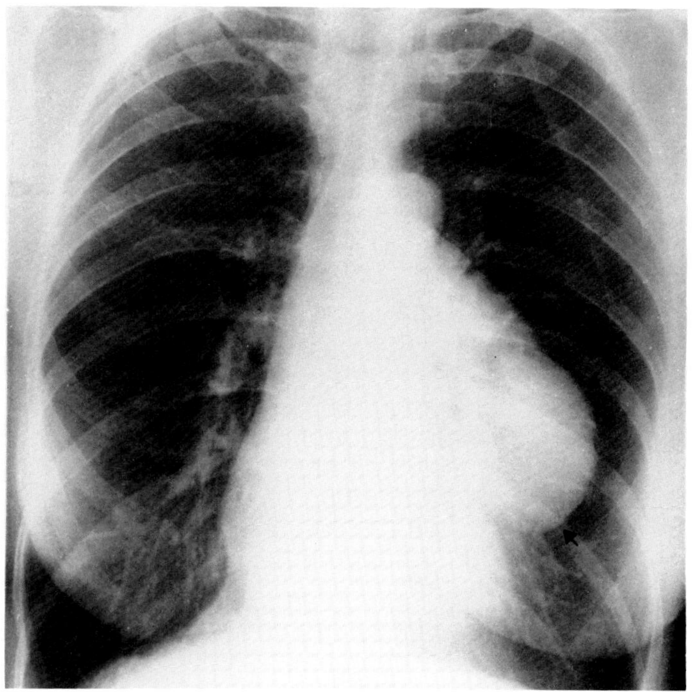

A

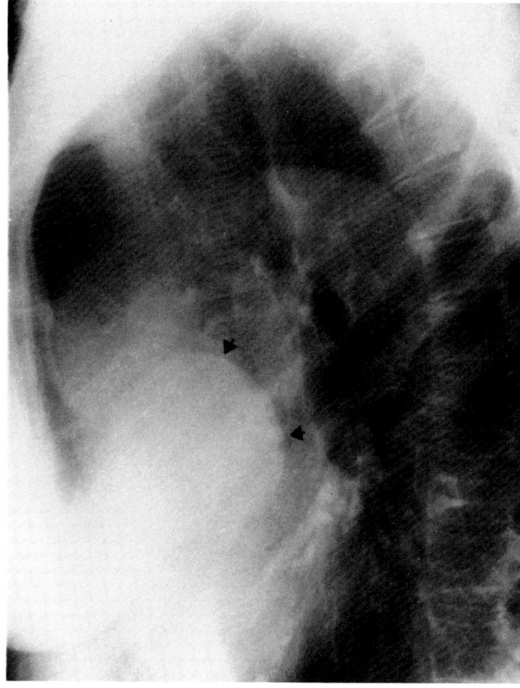

B

**Figure 35.34. Bronchogenic cyst.** (A) Frontal and (B) lateral views of the chest demonstrate a large, smooth-walled, spherical middle mediastinal mass (arrows) projecting into the lung and left hilum.

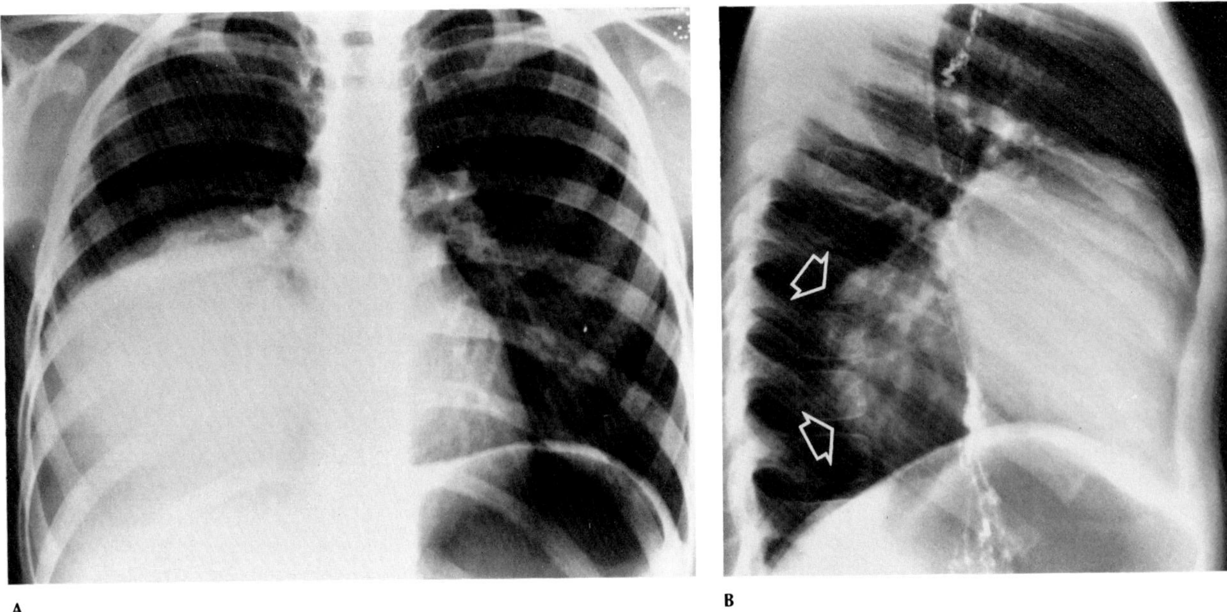

**Figure 35.35. Bronchogenic cyst.** (A) Frontal and (B) lateral views of the chest demonstrate a huge middle mediastinal mass (arrows) protruding to the right and filling much of the lower half of the right hemithorax. The patient was asymptomatic.

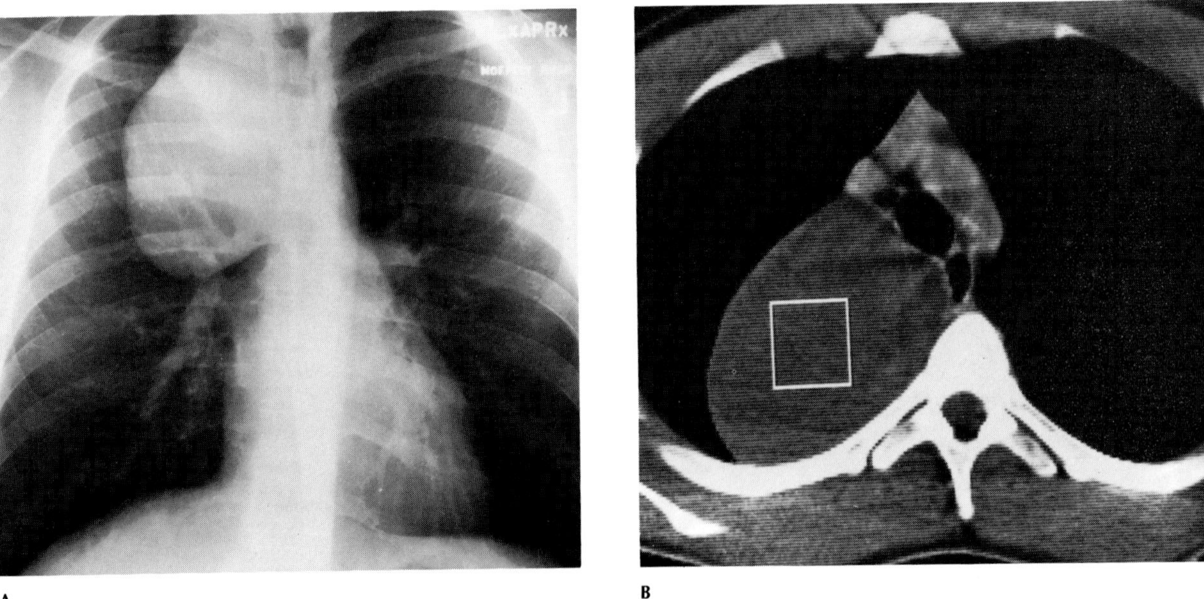

**Figure 35.36. Bronchogenic cyst.** (A) A frontal chest radiograph shows a large right upper mediastinal mass in a young man with an incidental upper respiratory infection. (B) A CT scan shows the mass extending from the right of the trachea to the posterior chest wall. The cyst has a uniform appearance and near-water density; it extended vertically from the lower pole of the thyroid gland to the carina. (From Gamsu,[48] with permission from the publisher.)

with the alimentary canal may allow the passage of air into the cyst. On an esophagram, a gastroenteric cyst may produce an eccentric compression on the barium-filled esophagus simulating an intramural or mediastinal mass. CT demonstrates a gastroenteric cyst as a well-circumscribed, homogeneous mass without contrast enhancement.[21,24,51]

## Neurenteric Cyst

A neurenteric cyst is a sharply defined, round or oval, homogeneous mass that results from the incomplete separation of endoderm from the notochordal plate during early embryonic life. It is connected by a stalk to the meninges and commonly to a portion of the gastrointestinal tract. Congenital defects of the thoracic spine are often associated.[21,52]

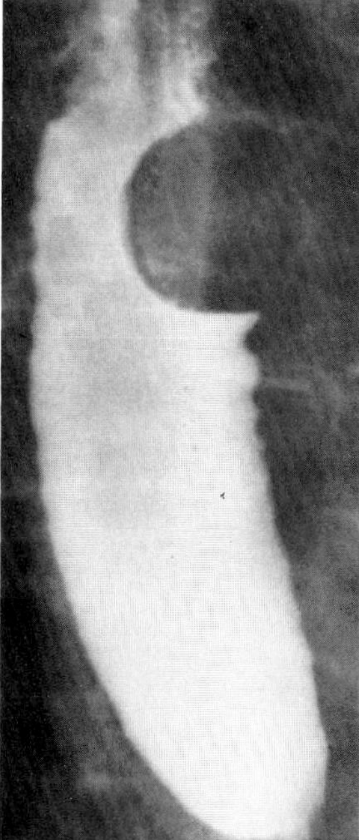

**Figure 35.37. Gastroenteric cyst.** An eccentric compression on the barium-filled esophagus simulates an intramural mass. (From Eisenberg,[51] with permission from the publisher.)

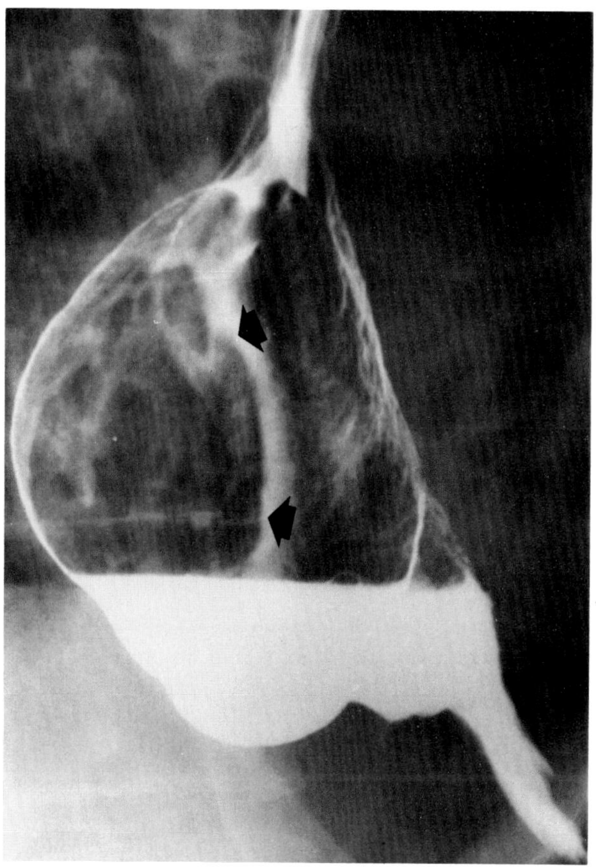

**Figure 35.38. Gastroenteric cyst with postsurgical communication with the esophageal lumen.** Arrows point to the duplication cyst. (From Eisenberg,[51] with permission from the publisher.)

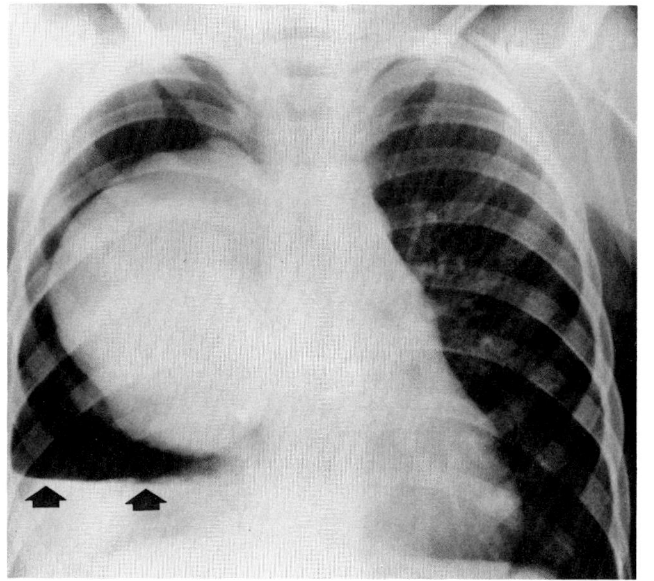

A

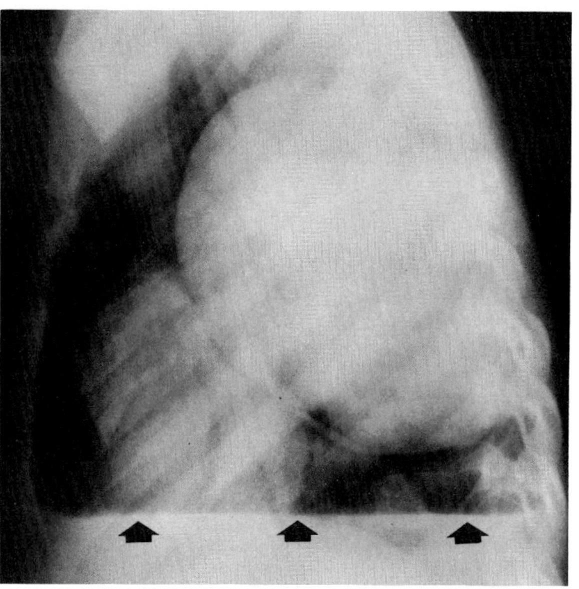

B

**Figure 35.39. Neurenteric cyst.** (A) Frontal and (B) lateral views of the chest demonstrate a large, oval, homogeneous mass in the posterior mediastinum. Note the right hydropneumothorax (arrows) with long air-fluid level that developed as a complication of diagnostic needle biopsy.

## Superior Vena Caval Syndrome

Obstruction of the superior vena cava is almost invariably due to malignant disease; about 75 percent of the cases are secondary to bronchogenic carcinoma, and the remainder are due to lymphoma. Infrequent causes of the syndrome include fibrosing mediastinitis, retrosternal thyroid, and an aortic aneurysm.

The injection of contrast material into a vein of the upper extremity can demonstrate the site of superior vena cava occlusion as well as the collateral circulation through the azygos and hemiazygos venous systems. Radionuclide studies can also demonstrate the extensive collateral circulation resulting from superior vena cava obstruction. Collateral flow of venous blood from the head and neck to the heart via the periesophageal plexus frequently produces "downhill" varices of the upper esophagus (see Figure 51.15). Concomitant enlargement of intercostal vein collaterals occasionally causes rib notching.[53,54]

## Pneumomediastinum (Mediastinal Emphysema)

Air within the mediastinal space may appear spontaneously or be the result of chest trauma, perforation of the esophagus or tracheobronchial tree, or the spread of air along fascial planes from the neck, peritoneal cavity, or retroperitoneal space. Spontaneous pneumomediastinum usually results from a sudden rise in intra-alveolar pressure (e.g., severe coughing, vomiting, or straining) that causes alveolar rupture and the dissection of air along blood vessels in the interstitial space to the hilum and mediastinum. Air may also extend peripherally and rupture into the pleural space, causing a pneumothorax that often is associated with pneumomediastinum (especially in neonates).

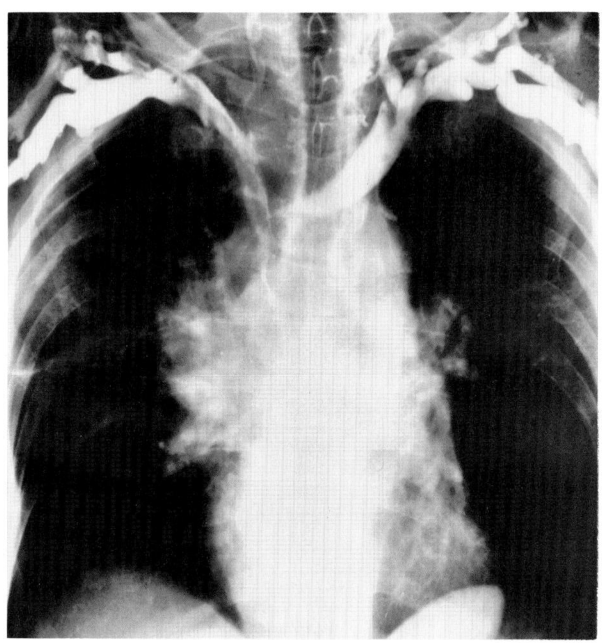

**Figure 35.40. Superior vena cava obstruction due to bronchogenic carcinoma.** Bilateral upper extremity venograms show virtual occlusion of the superior vena cava by a large malignant neoplasm in the right hilar and perihilar region.

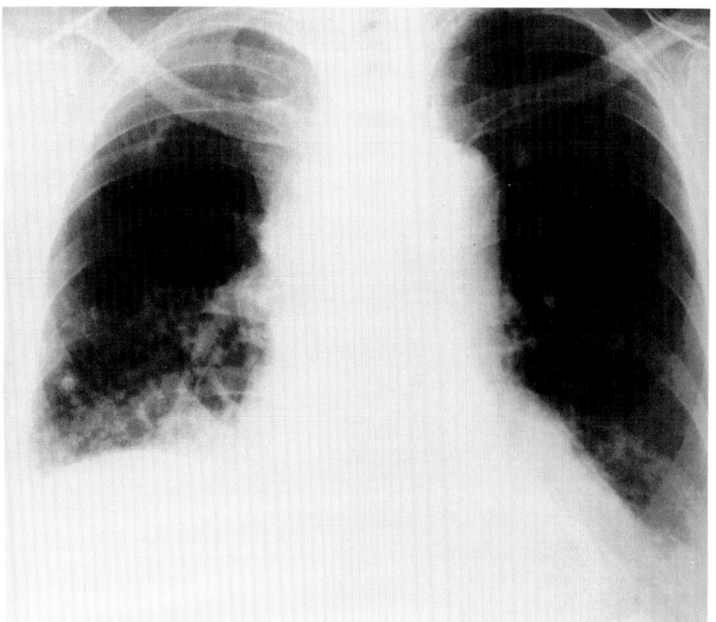

A

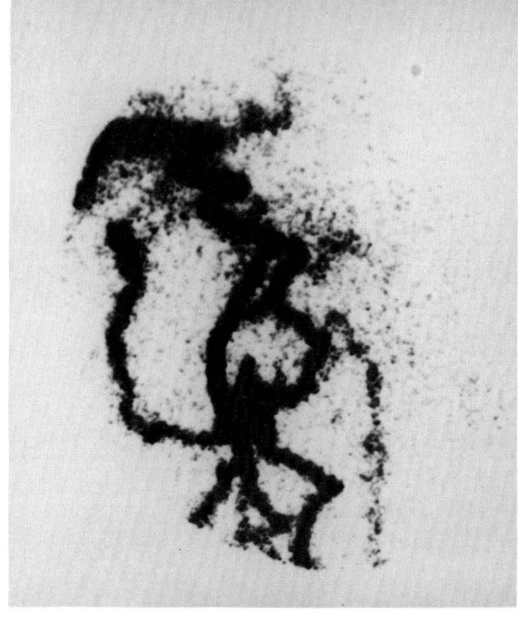

B

**Figure 35.41. Superior vena cava obstruction due to bronchogenic carcinoma.** (A) A plain film of the chest shows widening of the upper mediastinum. (B) A radionuclide scan shows extensive venous collaterals bypassing the superior vena cava obstruction.

On frontal chest radiographs, air causes lateral displacement of the mediastinal pleura, which appears as a long linear opacity that is parallel to the heart border but separated from it by gas. On lateral projections, gas typically collects retrosternally, extending in streaks downward and anterior to the heart. Chest radiographs may also demonstrate air outlining the pulmonary arterial trunk and aorta, as well as dissecting into the soft tissues of the neck. Another sign of pneumomediastinum is the interposition of gas between the heart and the diaphragm, permitting visualization of the normally silhouetted central portion of the diaphragm in continuity with the lateral portions (continuous-diaphragm sign).

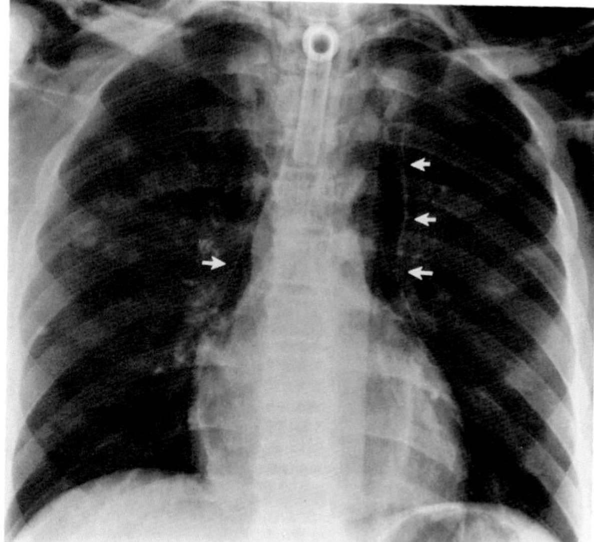

**Figure 35.42. Pneumomediastinum.** The mediastinal pleura is displaced laterally and appears as a long linear opacity (arrows) parallel to the heart border but separated from it by gas.

In infants, mediastinal air causes elevation of the thymus. Loculated air confined to one side produces an appearance similar to that of a windblown spinnaker sail; bilateral mediastinal air elevates both thymic lobes to produce an angel-wings configuration. Another sign of pneumomediastinum in infants is a collection of free mediastinal air between the parietal pleura and the diaphragm of either hemithorax (extrapleural air sign). The collection is limited above by a sharp pleural stripe, is located posterior to the domes of the diaphragm, and, unlike a pneumothorax, does not shift with a change in body position.[55–58]

## Acute Mediastinitis

Acute infections of the mediastinum are rare and usually result from esophageal perforation, either from erosion from a primary carcinoma or impacted foreign body, as

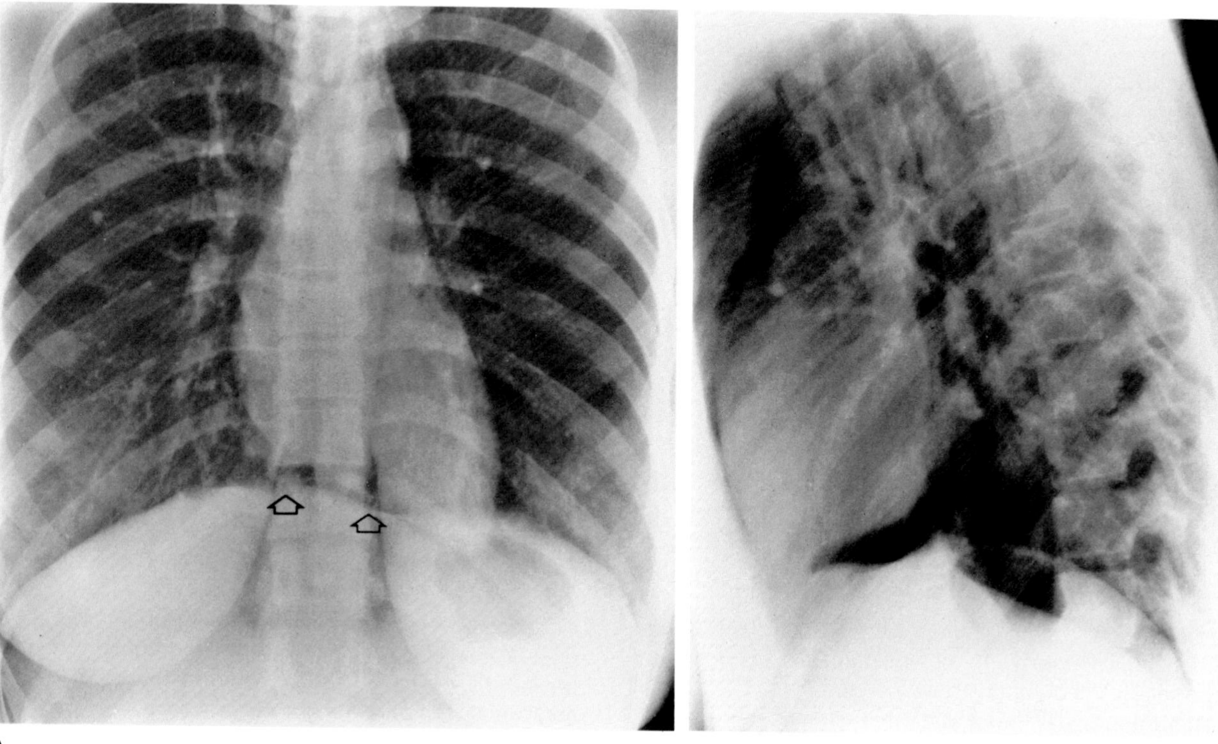


**Figure 35.43. Pneumomediastinum.** (A) A frontal chest radiograph demonstrates the interposition of gas between the heart and the diaphragm that permits visualization of the central portion of the diaphragm in continuity with the lateral portions (arrows). (B) The lateral view confirms the diagnosis of pneumomediastinum. (From Levin,[55] with permission from the publisher.)

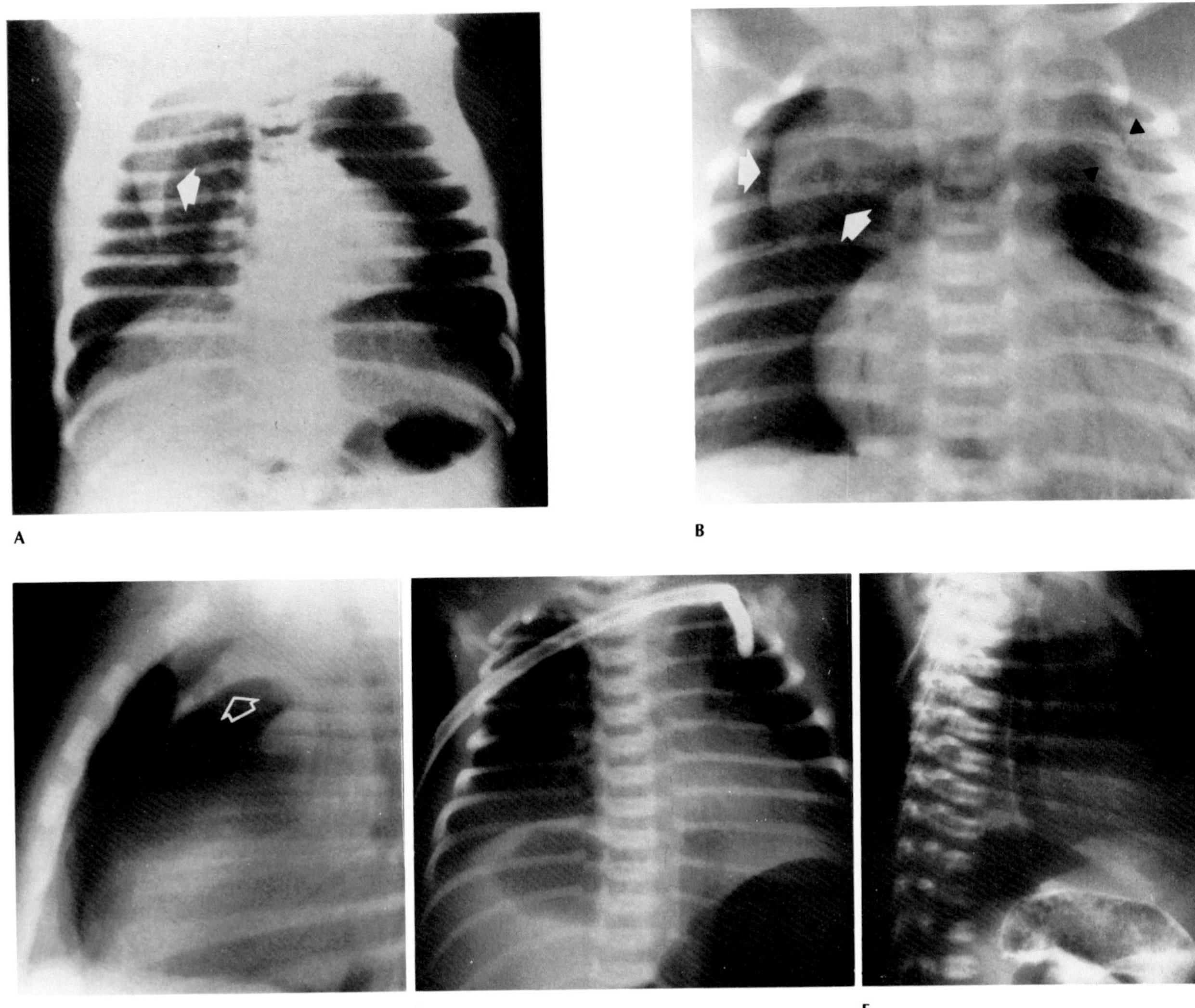


**Figure 35.44. Pneumomediastinum in infants.** (A) A crescentic spinnaker sail shadow extending out over the right lung (arrow) represents the displaced right lobe of the thymus. (B) Elevation of both thymic lobes by mediastinal air (arrows) produces the angel-wings sign. (C) A lateral projection shows the anterior mediastinal air lifting the thymus off the pericardium and great vessels (arrows). (D) A frontal chest film demonstrates a prominent extrapleural air sign on the right that appears to lie below the right hemidiaphragm. (E) A lateral film shows that the air collection is limited above by a well-defined pleural stripe. Barium defines the stomach. (A, from Moseley,[57] with permission from the publisher; B and C, from Eisenberg,[70] with permission from the publisher; D and E, from Lillard and Allen,[56] with permission from the publisher.)

a complication of instrumentation, or spontaneously (usually after vomiting). Direct extension of infection from contiguous soft tissues (lymph nodes, pleura, pericardium, retropharyngeal space) is a very rare cause.

Acute mediastinitis appears radiographically as widening of the mediastinum that is usually most pronounced superiorly. If the infection is secondary to esophageal perforation, radiolucent air may be visible within the mediastinum and often extends into the soft tissues of the neck. Formation of a chronic abscess may produce a large, tumor-like mediastinal mass, often containing an air-fluid level.

## Chronic Mediastinitis

Chronic mediastinitis may reflect either predominantly granulomatous or fibrotic involvement. Granulomatous mediastinitis is usually caused by chronic inflammation, most likely due to histoplasmosis or tuberculosis. The etiology of sclerosing mediastinitis (idiopathic mediastinal fibrosis) is usually unknown, though some cases are associated with a similar fibrotic process elsewhere (idiopathic retroperitoneal fibrosis). Both disorders cause nonspecific widening of the upper half of the mediastinum. A common complication is obstruction of the

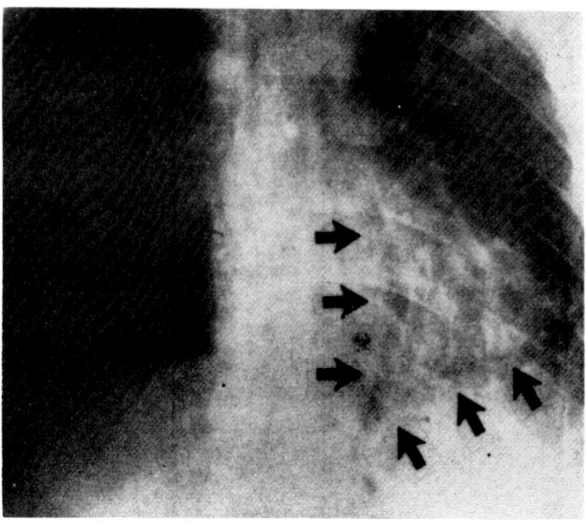

A

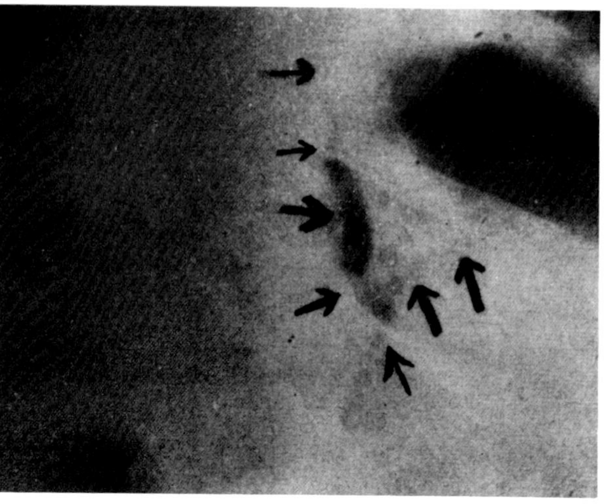

B

**Figure 35.45. Acute mediastinitis due to spontaneous rupture of the esophagus.** Two views demonstrate linear lucent shadows (arrows) which represent localized mediastinal emphysema and correspond to the fascial planes of the mediastinal and diphragmatic pleurae in the region of the lower esophagus. (From Naclerio,[62] with permission from the publisher.)

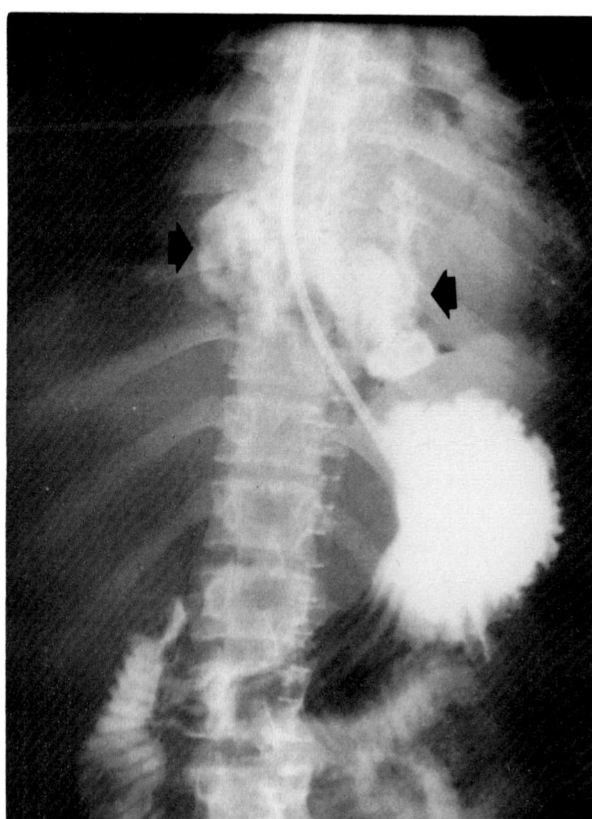

**Figure 35.46. Acute mediastinitis due to Boerhaave syndrome.** An esophogram demonstrates extravasation of contrast material (arrows) into the lower mediastinum through a transmural perforation. (From Eisenberg,[51] with permission from the publisher.)

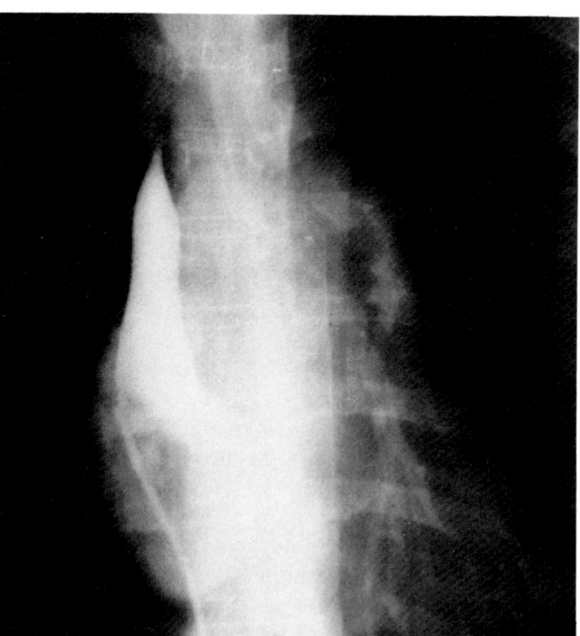

**Figure 35.47. Chronic sclerosing mediastinitis.** A venogram shows smooth tapering of the lower portion of the superior vena cava. This 38-year-old woman had varicosities over her upper abdomen and lower chest. (From Feigin, Eggleston, and Siegelman,[63] with permission from the publisher.)

superior vena cava; this may lead to the development of dilated collateral mediastinal veins, which produce a somewhat lobulated paratracheal mass projecting more to the right than to the left. Compression of the esophagus and tracheobronchial tree occurs much less commonly.[29,59–63]

## DISEASES OF THE DIAPHRAGM

The diaphragm is the major muscle of respiration that separates the thoracic and abdominal cavities. Radiographically, the height of the diaphragm varies considerably with the phase of respiration. On full inspiration, the diaphragm is usually at about the level of the tenth posterior intercostal space. On expiration, it may appear two or three intercostal spaces higher. The average range of diaphragmatic motion with respiration is 3 to 6 cm, though in patients with emphysema this may be substantially reduced. The level of the diaphragm falls as a patient moves from a supine to an upright position. In an erect patient, the dome of the diaphragm tends to be about half an interspace higher on the right than on the left. However, in about 10 percent of patients, the hemidiaphragms are at the same height, or the left is higher than the right.

### Diaphragmatic Paralysis

Elevation of one or both leaves of the diaphragm can be caused by paralysis resulting from any process that interferes with the normal function of the phrenic nerve. This may be due to inadvertent surgical transection of the phrenic nerve, involvement of the nerve by primary bronchogenic carcinoma or metastatic malignancy in the mediastinum, or a variety of intrinsic neurologic diseases. The radiographic hallmark of diaphragmatic paralysis is paradoxical movement of the diaphragm that is best demonstrated at fluoroscopy by having the patient sniff. This rapid but shallow inspiration causes a quick downward thrust of a normal leaf of the diaphragm, while a paralyzed hemidiaphragm tends to ascend with inspiration because of the increased intra-abdominal pressure. During expiration, the normal hemidiaphragm rises and the paralyzed one descends. Although small amounts of paradoxical excursion of the diaphragm on sniffing are not infrequent in normal persons, a marked degree of paradoxical motion is a valuable aid in differentiating paralysis of the diaphragm from limited diaphragmatic motion secondary to intrathoracic or intra-abdominal disease.[49,50,64]

### Eventration of the Diaphragm

Eventration of the diaphragm is a rare congenital abnormality in which one hemidiaphragm (very rarely both) is hypoplastic, consisting of a thin membranous sheet attached peripherally to normal muscle at points of origin from the rib cage. This weakened structure permits the upward movement of abdominal contents into the thoracic cage and leads to the radiographic appearance of a localized bulging or generalized elevation of the diaphragm. The condition is usually asymptomatic and is most common on the left. At fluoroscopy, movement of the affected hemidiaphragm may be normal or diminished; paradoxical diaphragmatic motion is occasionally demonstrated, though much less frequently than in patients with diaphragmatic paralysis. The cardiomediastinal structures may be displaced toward the contralateral side. Eventration must be distinguished from diaphragmatic hernias through which abdominal contents are displaced into the chest. Oral administration of barium should permit the differentiation between the normal contours of the bowel below a diaphragmatic eventration and the crowding of these structures and narrowing of their afferent and efferent limbs when trapped in a hernia sac.[49,51,65]

### Other Causes of Elevation of the Diaphragm

Diffuse elevation of one or both leaves of the diaphragm can be caused by ascites, obesity, pregnancy, or any other process in which the intra-abdominal volume is increased. Intra-abdominal inflammatory diseases, such as subphrenic abscess, can lead to elevation of a hemi-

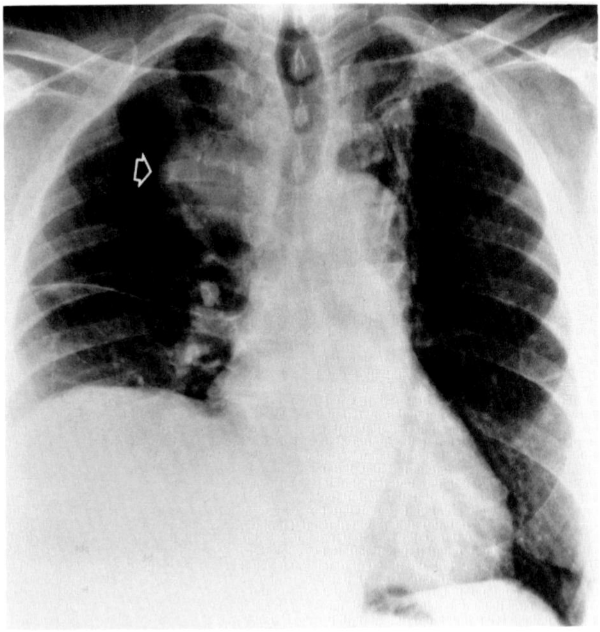

**Figure 35.48. Diaphragmatic paralysis.** Paralysis of the right hemidiaphragm due to involvement of the phrenic nerve by primary bronchogenic carcinoma (arrow). (From Eisenberg,[51] with permission from the publisher.)

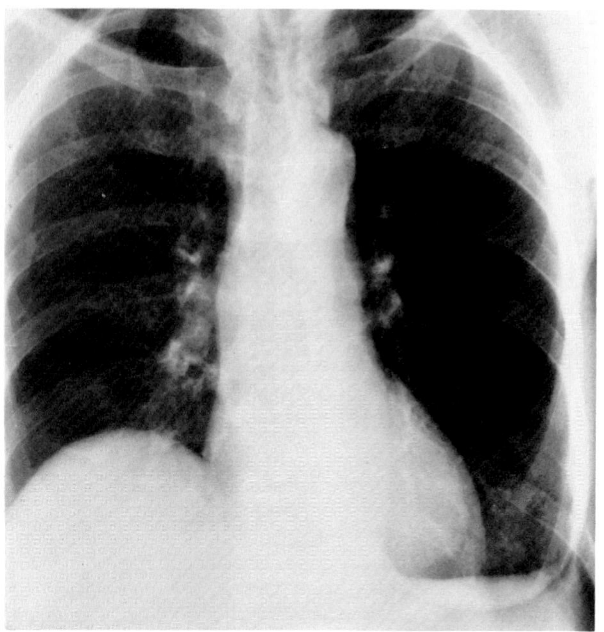
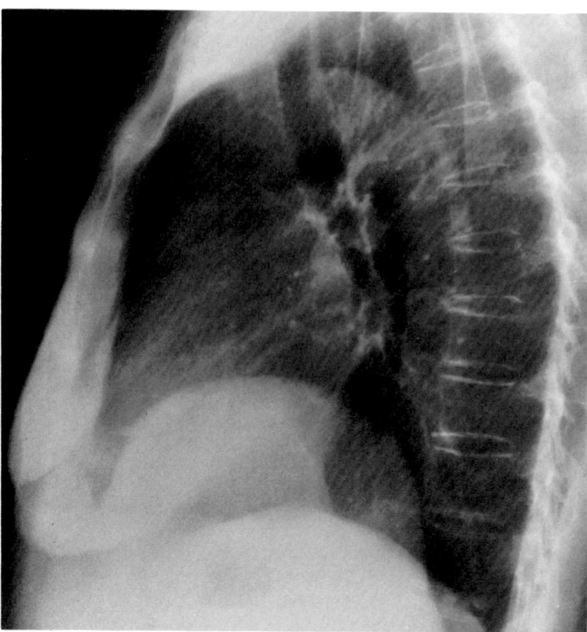

A B

**Figure 35.49. Eventration of the right hemidiaphragm.** (A) Frontal and (B) lateral projections.

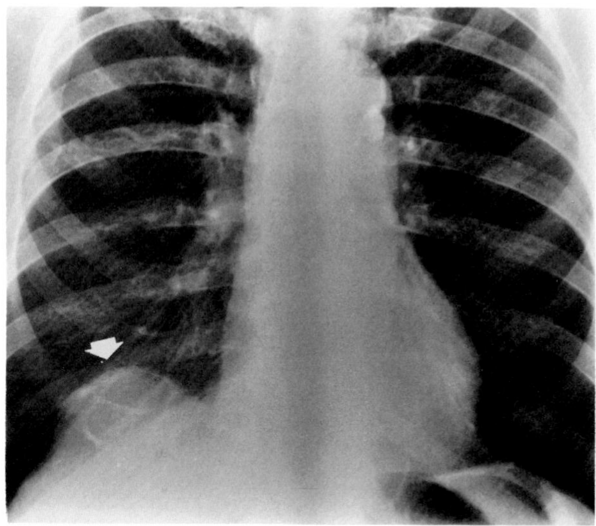

**Figure 35.50. Partial eventration of the right hemidiaphragm (arrow).**

diaphragm with severe limitation of diaphragmatic motion. Cystic or tumor masses arising in the upper quadrants can cause localized or generalized bulging of the diaphragm. Acute intrathoracic processes (chest-wall injury, atelectasis, pulmonary embolism) can produce diaphragmatic elevation due to splinting of the diaphragm. An apparent elevation of the hemidiaphragm may be due to subpulmonic pleural effusion, which can be correctly diagnosed on a chest radiograph performed with the patient in the lateral decubitus position.[51]

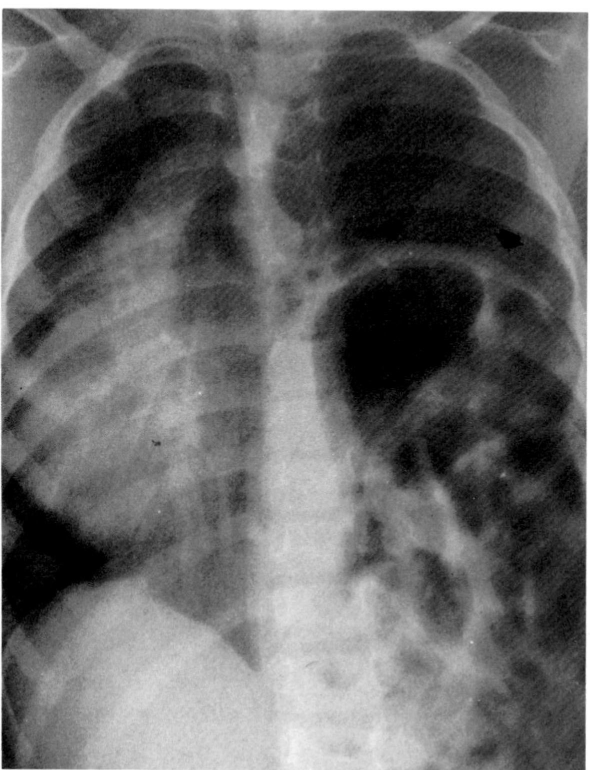

**Figure 35.51. Total eventration of the left hemidiaphragm of a child.** The cardiomediastinal structures are displaced toward the right. The dome of the left hemidiaphragm is at the level of the sixth posterior rib (arrows); however, the bowel contents remain below the level of the hemidiaphragm. (From Eisenberg,[51] with permission from the publisher.)

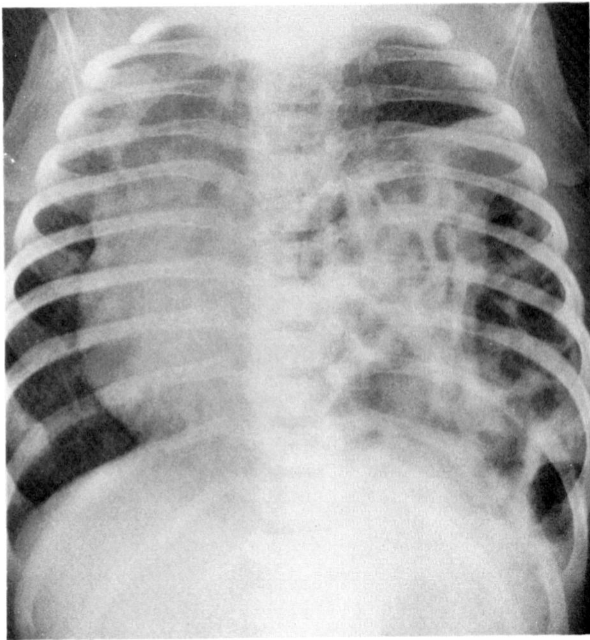

**Figure 35.52. Congenital diaphragmatic hernia.** A plain chest radiograph demonstrates multiple radiolucencies in the left chest due to gas-filled loops of bowel. Unlike the total eventration in Figure 35.51, the margin of the left hemidiaphragm cannot be identified. (From Eisenberg,[51] with permission from the publisher.)

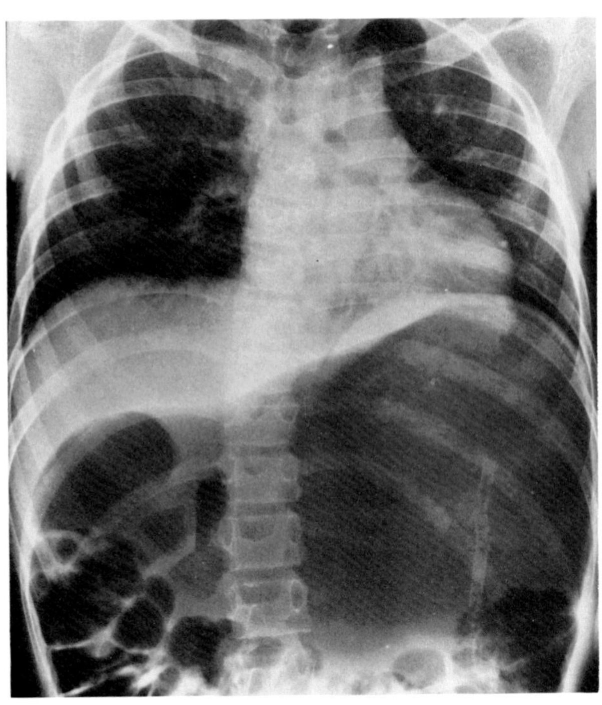

**Figure 35.53. Acute gastric dilatation.** There is diffuse elevation of both leaves of the diaphragm. (From Eisenberg,[51] with permission from the publisher.)

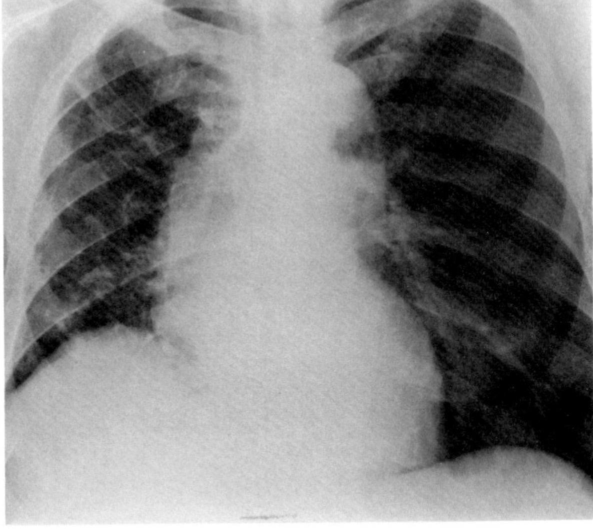

**Figure 35.54. Huge syphilitic gumma of the liver.** There is elevation of the right hemidiaphragm. (From Eisenberg,[51] with permission from the publisher.)

## REFERENCES


1. Fleischner FG: Atypical arrangement of free pleural effusion. *Radiol Clin North Am* 1:347–354, 1963.
2. Petersen JA: Recognition of infrapulmonary pleural effusion. *Radiology* 74:34–41, 1960.
3. Davis S, Gardner F, Ovist G: The shape of a pleural effusion. *Br Med J* 1:436–444, 1963.
4. Collins JD, Burwell D, Furmanski S, et al: Minimal detectable pleural effusions: A roentgen pathology model. *Radiology* 105:51–53, 1972.
5. Fraser RG, Pare JAP: *Diagnosis of Diseases of the Chest*. Philadelphia, WB Saunders, 1979.
6. Sandweiss DA, Hanson JC, Gosink BB, et al: Ultrasound in diagnosis, localization, and treatment of loculated pleural empyema. *Ann Intern Med* 82:50–53, 1975.
7. Doust BD, Baum JK, Maklad NF, et al: Ultrasonic evaluation of pleural opacities. *Radiology* 114:135–140, 1975.
8. Baber CE, Hedlund LW, Oddson TA, et al: Differentiating empyemas and peripheral pulmonary abscesses: The value of computed tomography. *Radiology* 135:755–758, 1980.
9. Renner RR, Markarian B, Pernice NJ, et al: The apical cap. *Radiology* 110:569–573, 1974.
10. Inouye WY, Berggren RB, Johnson J: Spontaneous pneumothorax: Treatment and mortality. *Dis Chest* 51:67–73, 1967.
11. Sherin RPN, Hepper NGG, Payne WS: Recurrent spontaneous pneumothorax concurrent with menses. *Mayo Clin Proc* 49:98–101, 1974.
12. Alexander E, Clark RA, Colley DP, et al: CT of malignant pleural mesothelioma. *AJR* 137:287–291, 1981.
13. Freundlich IM, Greening RR: Asbestosis and associated medical problems. *Radiology* 89:224–229, 1967.
14. Solomon A: Radiological features of diffuse mesothelioma. *Environ Res* 3:330–334, 1970.
15. Ellis K, Wolff M: Mesotheliomas and secondary tumors of the pleura. *Semin Roentgenol* 12:303–311, 1977.
16. Sagel SS: Lung, pleura, pericardium, and chest wall, in Lee JKT, Sagel SS, Stanley RJ (eds): *Computed Body Tomography*. New York, Raven Press, 1983, pp 99–130.
17. Jost RG, Sagel FS, Stanley RJ, et al: Computed tomography of the thorax. *Radiology* 126:125–136, 1978.

18. Pugatch RD, Faling LJ, Robbins AH, et al: CT diagnosis of benign mediastinal abnormalities. *AJR* 134:685–694, 1980.
19. Boyd DP, Midell AI: Mediastinal cysts and tumors. An analysis of 86 cases. *Surg Clin North Am* 48:493–502, 1968.
20. Wychulis AR, Payne WS, Clagett OT, et al: Surgical treatment of mediastinal tumors. A forty-year experience. *J Thorac Cardiovasc Surg* 62:379–392, 1972.
21. Leigh TF: Mass lesions of the mediastinum. *Radiol Clin North Am* 1:377–390, 1963.
22. Benjamin SP, McCormack LJ, Effler DB, et al: Critical review: "Primary tumours of the mediastinum." *Chest* 62:297–303, 1972.
23. Templeton AW: Malignant mediastinal teratoma with bone metastases. A case report. *Radiology* 76:245–247, 1961.
24. Lyons HA, Calvy GL, Sammons BP: The diagnosis and classification of mediastinal masses. A study of 782 cases. *Ann Intern Med* 51:897–932, 1959.
25. Zylak CJ, Pallie W, Jackson R: Correlated anatomy and computed tomography: A module on the mediastinum. *RadioGraphics* 2:555–592, 1982.
26. Sagel SS, Aronberg BJ: Thoracic anatomy and mediastinum, in Lee JKT, Sagel SS, Stanley RJ (eds): *Computed Body Tomography*. New York, Raven Press, 1983, pp 55–98.
27. Baron RL, Lee JKT, Sagel SS, et al: Computed tomography of the abnormal thymus. *Radiology* 142:127–134, 1982.
28. Fon GT, Bein ME, Mancuso AA, et al: Computed tomography of the anterior mediastinum in myasthenia gravis. *Radiology* 142:135–141, 1982.
29. Leigh TF, Weens HS: Roentgen aspects of mediastinal lesions. *Semin Roentgenol* 4:59–69, 1969.
30. Rosenthal T, Hertz M, Samra Y, et al: Thymoma: Clinical and additional radiological signs. *Chest* 65:428–430, 1974.
31. Rietz KA, Werner B: Intrathoracic goiter. *Acta Chir Scand* 119:379–388, 1960.
32. Bashist B, Ellis K, Gold RP: Computed tomography of intrathoracic goiters. *AJR* 140:455–460, 1983.
33. Glazer GM, Axel L, Moss AA: CT diagnosis of mediastinal thyroid. *AJR* 138:495–498, 1982.
34. Castleman B (ed): Case records of the Massachusetts General Hospital, Case 40011. *N Engl J Med* 250:26–30, 1954.
35. Bogsch A: Roentgen observations in infectious mononucleosis. *Fortschr Roentgenstr* 82:785–789, 1955.
36. Felson B: *Chest Roentgenology*. Philadelphia, WB Saunders, 1973.
37. Schneider HJ, Felson B: Buckling of the innominate artery simulating aneurysm and tumor. *AJR* 85:1106–1110, 1961.
38. Cheng TO: Pseudocoarctation of the aorta. An important consideration in the differential diagnosis of the superior mediastinal mass. *Am J Med* 49:551–555, 1970.
39. Fleming JS, Gibson RV: Absent right superior vena cava as an isolated anomaly. *Br J Radiol* 37:696–697, 1964.
40. Moncada R, Shannon M, Miller R, et al: The cervical aortic arch. *AJR* 125:591–601, 1975.
41. Shuford WH, Sybers RG, Edwards FK: The three types of right aortic arch. *AJR* 109:67–74, 1970.
42. McLoughlin MJ, Weisbrod G, Wise DJ, et al: Computed tomography in congenital anomalies of the aortic arch and great vessels. *Radiology* 138:399–403, 1981.
43. Baron RL, Gutierrez FR, Sagel SS, et al: CT of anomalies of the mediastinal vessels. *AJR* 137:571–576, 1981.
44. Price JE, Rigler LG: Widening of the mediastinum resulting from fat accumulation. *Radiology* 96:497–500, 1970.
45. Reed JC, Hallet KK, Feigin BS: Neural tumors of the thorax: Subject review from the AFIP. *Radiology* 126:9–17, 1978.
46. Rogers LF, Osmer JC: Bronchogenic cyst: A review of forty-six cases. *AJR* 91:273–283, 1964.
47. Marvasti MA, Mitchell GE, Burke WA, et al: Misleading density of mediastinal cysts on computerized tomography. *Ann Thorac Surg* 31:167–170, 1981.
48. Gamsu G: Computed tomography of the mediastinum, in Moss AA, Gamsu G, Genant HK (eds): *Computed Tomography of the Body*. Philadelphia, WB Saunders, 1983.
49. Vix VA: Roentgen manifestations of pleural disease. *Semin Roentgenol* 12:277–286, 1977.
50. Landay MJ, Christensen EE, Bynum LJ, et al: Anaerobic plural and pulmonary infections. *AJR* 134:233–241, 1980.
51. Eisenberg RL: *Gastrointestinal Radiology: A Pattern Approach*. Philadelphia, JB Lippincott, 1983.
52. Neuhauser EBD, Harris GBC, Berrett A: Roentgenographic features of neurenteric cyst. *AJR* 79:235–240, 1958.
53. Felson B, Lessure AP: "Downhill" varices of the esophagus. *Dis Chest* 46:740–746, 1964.
54. Mikkelsen WJ: Varices of the upper esophagus in superior vena caval obstruction. *Radiology* 81:945–948, 1963.
55. Levin B: The continuous diaphragm sign. A newly-recognized sign of pneumomediastinum. *Clin Radiol* 24:337–338, 1973.
56. Lillard RL, Allen RP: The extrapleural air sign in pneumomediastinum. *Radiology* 85:1093–1098, 1965.
57. Moseley JE: Loculated pneumomediastinum in the newborn: Thymic "spinnaker sail" sign. *Radiology* 75:788–790, 1960.
58. Felson B: The mediastinum. *Semin Roentgenol* 4:41–56, 1969.
59. Schowengerdt CG, Suyemoto R, Main FB: Granulomatous and fibrous mediastinitis—A review and analysis of 180 cases. *J Thorac Cardiovasc Surg* 57:365–379, 1969.
60. Goodwin RA, Nickell JA, Des Prez RM: Mediastinal fibrosis complicating healed primary histoplasmosis and tuberculosis. *Medicine* 51:227–246, 1972.
61. Salmon HW: Combined mediastinal and retroperitoneal fibrosis. *Thorax* 23:158–164, 1968.
62. Naclerio NA: The "V" sign in the diagnosis of spontaneous rupture of the esophagus (an early roentgen clue). *Am J Surg* 93:291–298, 1957.
63. Feigin DS, Eggleston JC, Siegelman SS: The multiple roentgen manifestations of sclerosing mediastinitis. *Johns Hopkins Med J* 144:1–8, 1979.
64. Alexander C: Diaphragm movements and the diagnosis of diaphragmatic paralysis. *Clin Radiol* 17:79–83, 1966.
65. Thomas T: Nonparalytic eventration of the diaphragm. *J Thorac Cardiovasc Surg* 55:586–593, 1968.
66. Campbell JA: The diaphragm in roentgenology of the chest. *Radiol Clin North Am* 1:395–410, 1963.
67. Riley EA: Idiopathic diaphragmatic paralysis. *Am J Med* 32:404–415, 1962.
68. Baron RL, Levitt RG, Sagel SS, et al: Computed tomography in the evaluation of mediastinal widening. *Radiology* 138:107–114, 1981.
69. Feigin DS: Mediastinal masses, in Eisenberg RL, Amberg JR (eds): *Critical Diagnostic Pathways in Radiology: An Algorithmic Approach*. Philadelphia, JB Lippincott, 1981.
70. Eisenberg RL: *Atlas of Signs in Radiology*. Philadelphia, JB Lippincott, 1984.

# PART SIX

# DISEASES OF THE KIDNEYS AND URINARY TRACT

## Chapter Thirty-Six
# Renal Failure

## ACUTE RENAL FAILURE

Acute renal failure refers to a rapid deterioration in kidney function that is sufficient to result in the accumulation of nitrogenous wastes in the body. In prerenal failure (due to hypovolemia, cardiac failure, or bilateral vascular obstruction), impaired renal perfusion causes an acute reduction of the glomerular filtration rate and decreased endogenous urinary creatinine clearance. Obstruction of urine outflow from both kidneys, most commonly due to prostatic disease or functional (organic or drug-induced neuropathy) obstruction of the bladder neck, produces postrenal failure. Acute renal failure may also be the result of specific renal diseases, such as glomerulonephritis, interstitial nephritis, bilateral acute pyelonephritis, and malignant hypertension. Other causes of acute renal failure include nephrotoxic agents (aminoglycoside antibiotics, radiographic contrast material, anesthetic agents, heavy metals, organic solvents), intravascular hemolysis, and large amounts of myoglobin in the circulation due to trauma, muscle ischemia, or increased muscular oxygen consumption.

Because it is independent of renal function, ultrasound is especially useful in the evaluation of patients with acute renal failure. In addition to demonstrating dilatation of the ureters and pelves due to postobstructive hydronephrosis, ultrasound can assess renal size and the presence of focal kidney lesions or diffuse renal cystic disease. In the patient with prerenal azotemia, ultrasound can aid in distinguishing hypovolemia from right-sided heart failure; in the latter there is dilatation of the

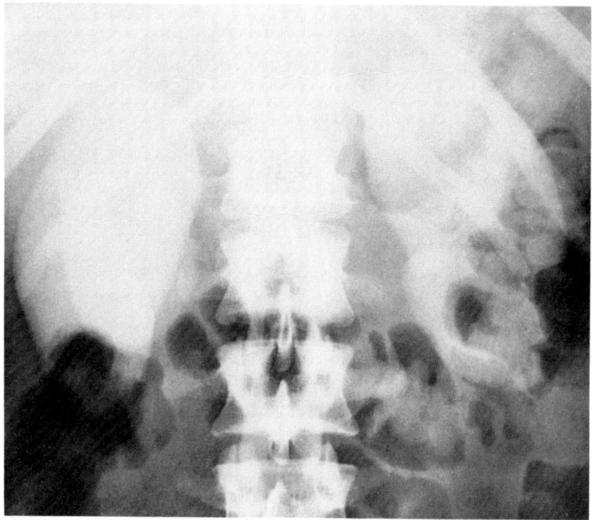

**Figure 36.1. Acute renal failure secondary to myoglobinemia following trauma.** A film from an excretory urogram 20 minutes after the injection of contrast material shows bilateral persistent nephrograms with no calyceal filling.

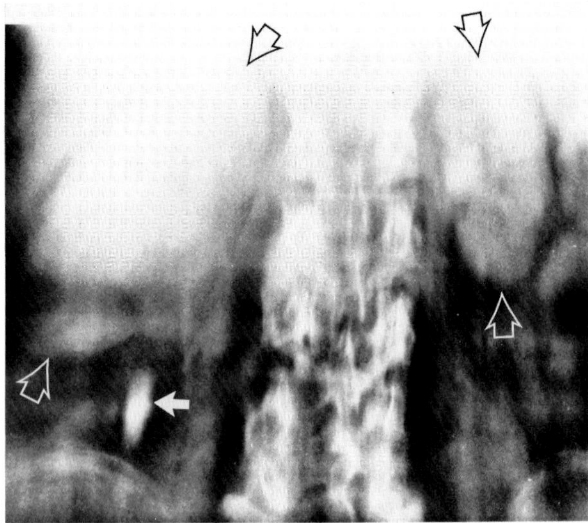

**Figure 36.2. Acute renal failure due to bilateral disease of different etiologies.** The patient has a marked renal artery stenosis on the left, which accounts for the small left kidney. She developed acute renal failure when a calculus (solid arrow) completely obstructed the right ureter. (From Swenson,[2] with permission from the publisher.)

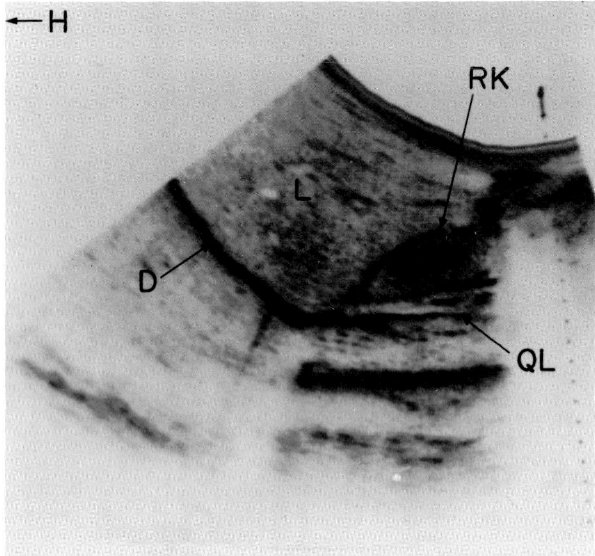


**Figure 36.3. Renal failure due to chronic glomerulonephritis.** A parasagittal sonogram demonstrates a tiny right kidney (RK) with marked thinning of the renal parenchyma. The echogenicity of the renal tissue greatly exceeds that of the adjacent liver (L). The medullary pyramids are no longer distinguishable. Scans of the left kidney show similar findings. Although ultrasound cannot access renal function, it is obvious that this patient's kidneys are severely and irreversably damaged. H = head; D = diaphragm; QL = quadratus lumborum muscle. (From Swenson,[2] with permission from the publisher.)

inferior vena cava and hepatic veins that does not occur in patients with low circulating blood volume. Ultrasound can also be used to localize the kidney for percutaneous renal biopsy.

Plain film tomography can often demonstrate renal size and contours. Bilaterally enlarged, smooth kidneys suggest acute renal parenchymal dysfunction; small kidneys usually indicate chronic, pre-existing renal disease. Plain film tomograms can also demonstrate bilateral renal calcification, which may suggest either secondary hyperparathyroidism due to chronic renal disease, or bilateral calculi that have obstructed both ureters to produce postrenal failure.

Excretory urography in the patient with acute renal failure demonstrates bilateral renal enlargement with a delayed but prolonged nephrogram; vicarious excretion of contrast material by the liver occasionally results in opacification of the gallbladder. When ultrasound is equivocal, excretory urography can exclude hydronephrosis and postrenal failure. In most instances, excretory urography is unnecessary in the patient with acute renal failure, especially in view of the controversy about the possible harmful effects of intravenous contrast material on the kidneys of such patients.

Both venous digital subtraction angiography and radionuclide scanning following the intravenous injection of iodo-orthohippurate can be employed to exclude acute bilateral renal artery embolization as a cause of prerenal failure. The intra-arterial injection of contrast material is generally contraindicated in patients with acute renal failure.[1–4]

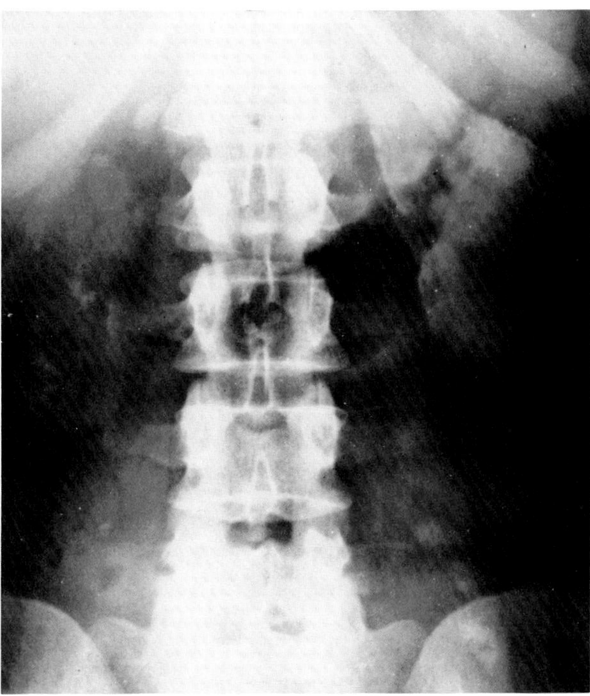

**Figure 36.4. Bilateral acute renal cortical necrosis.** There is punctate calcification within the right kidney and a peripheral rim of calcification surrounding the left kidney.

## BILATERAL ACUTE RENAL CORTICAL NECROSIS

Acute vascular compromise of the kidneys may cause bilateral cortical necrosis (sparing the medulla) and acute renal failure. This rare condition may be associated with severe burns, multiple fractures, internal hemorrhage, transfusions of incompatible blood, and complications of pregnancy, especially abruptio placentae.

Excretory urography demonstrates diffuse renal enlargement with poor function; the pelves and calyces appear normal on retrograde studies. Selective renal arteriography shows prolongation of transit time, nonopacification of intralobular arteries, and distortion of the cortical nephrogram.

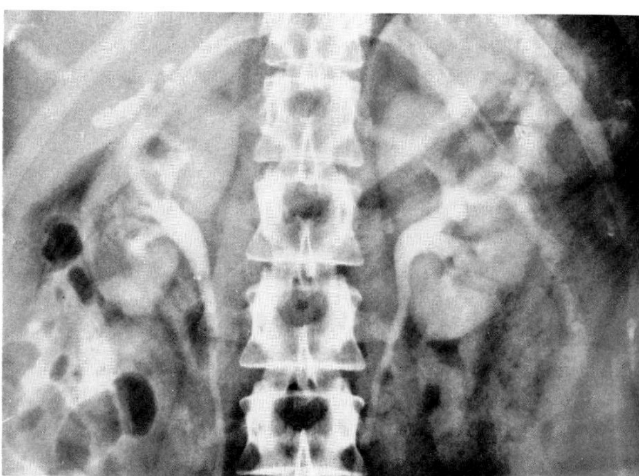

**Figure 36.5. Chronic renal failure due to chronic glomerulonephritis.** An excretory urogram demonstrates bilateral small kidneys with smooth contours and normal calyces. (From Swenson,[2] with permission from the publisher.)

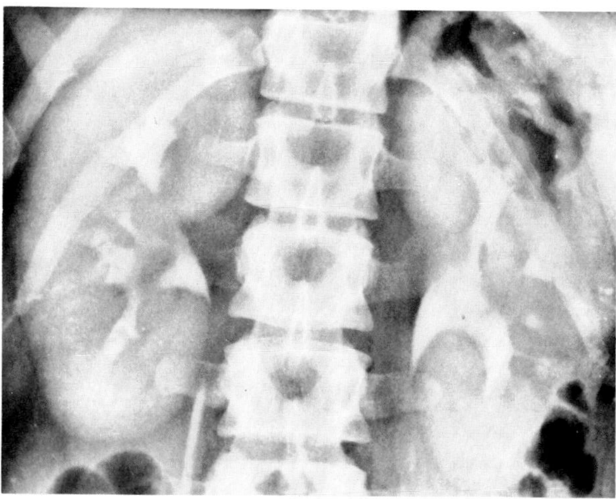

**Figure 36.6. Chronic renal failure due to nephrotic syndrome.** An excretory urogram demonstrates bilateral large, smooth kidneys with normal calyces. (From Swenson,[2] with permission from the publisher.)

The classic radiographic feature of acute cortical necrosis in surviving patients is the development of calcification confined to the renal cortex. The typical punctate or linear (tramline) calcification often occurs within a month of the onset of the disease.[1,5–8]

## CHRONIC RENAL FAILURE

Like acute renal failure, chronic kidney dysfunction may reflect prerenal, postrenal, or intrinsic kidney disease. Therefore, underlying causes of chronic renal failure include bilateral renal artery stenosis, bilateral obstructive uropathy, and intrinsic renal disorders such as chronic glomerulonephritis, pyelonephritis, interstitial nephritis, and familial cystic diseases.

Because it is independent of renal function, ultrasound is often the initial procedure in the evaluation of patients with chronic renal failure. Ultrasound is of special value in diagnosing treatable diseases such as hydronephrosis, and intrarenal or perirenal infections. This modality can also assess renal size and the presence of focal kidney lesions or diffuse renal cystic disease, as well as localize the kidneys for percutaneous renal biopsy.

Even in patients with chronic renal failure and uremia, high-dose excretory urography with tomography may produce sufficient opacification to be of diagnostic value. An initial plain abdominal radiograph may demonstrate bilateral renal calcification (nephrocalcinosis) or obstructing ureteral calculi. Small kidneys with smooth contours suggest chronic glomerulonephritis, nephrosclerosis, or bilateral renal artery stenosis. Small kidneys

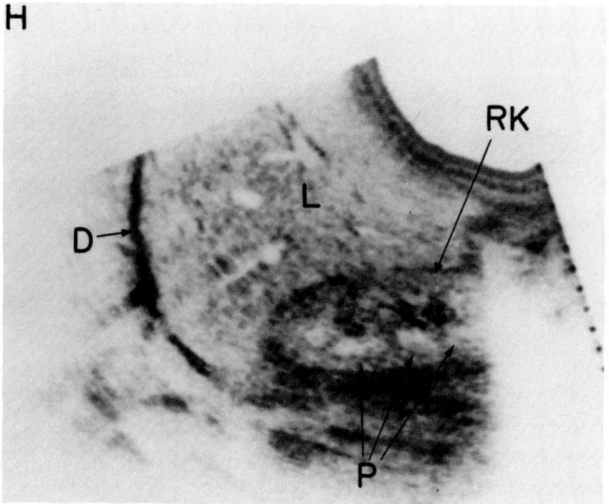


**Figure 36.7. Chronic renal failure.** A parasagittal sonogram of the right kidney (RK) shows that the renal cortical tissue has increased in echogenicity to such an extent that its echogenicity is now greater than that of the hepatic parenchyma (L). The renal medullary pyramids (P) are clearly visible in the right kidney. H = head; D = diaphragm. (From Swenson,[2] with permission from the publisher.)

with irregular contours, thin cortices, and typical patchy calyceal clubbing are consistent with chronic pyelonephritis. Large kidneys suggest obstructive uropathy, infiltrative processes (lymphoma, myeloma, amyloidosis), renal vein thrombosis, polycystic kidney disease, or superimposed acute renal failure. Details of the urographic findings in the above conditions are found in sections describing the specific conditions. It must be stressed that in patients whose renal failure is due to certain causes (advanced diabetic nephropathy, multiple myeloma, amyloidosis, hyperuricosuria), intravenous urography is potentially dangerous and contraindicated.

Renal arteriography is primarily of value in patients with chronic renal failure due to polyarteritis nodosa, bilateral renal artery stenosis or occlusion, and medullary cystic disease. (The virtually pathognomonic angiographic findings in these conditions are described in the relevant sections.) With other causes of chronic renal failure, the nonspecific arteriographic findings (decreased size of the main renal and the intrarenal arteries, prolonged arterial opacification, decreased or absent vascularity in the periphery) tend to be of little additional diagnostic value.[9–12]

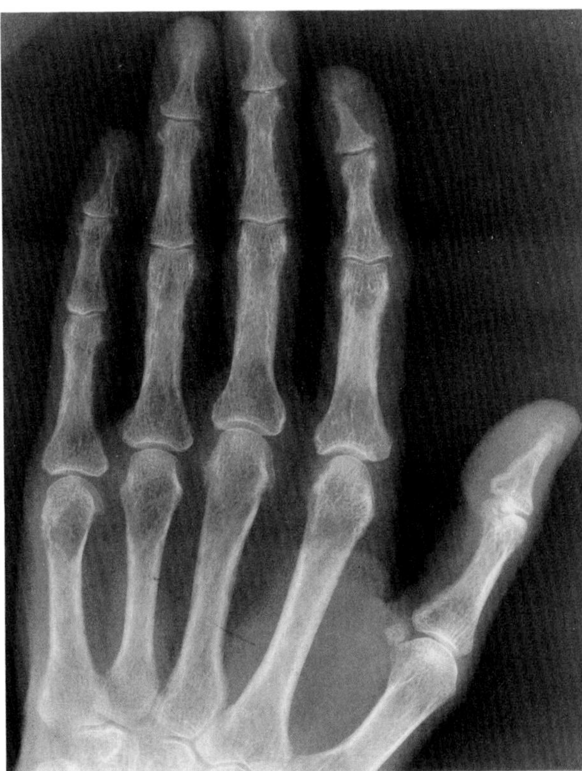

**Figure 36.8. Renal osteodystrophy.** Diffuse subperiosteal bone resorption secondary to hyperparathyroidism primarily involves the radial aspects of the proximal and middle phalanges. Erosive changes affect the tufts of several distal phalanges.

## EFFECTS OF CHRONIC RENAL FAILURE ON OTHER BODY SYSTEMS

### Renal Osteodystrophy

Secondary hyperparathyroidism develops in almost all patients with chronic renal failure, especially since survival times can now be prolonged by dialysis. The typical radiographic changes (refer to the section "Hyperparathyroidism" in Chapter 59) are best seen in the hands and include subperiosteal resorption of the phalanges (especially along the radial aspects), erosion of the terminal tufts, and a pattern of joint erosions that may superficially resemble rheumatoid arthritis. Magnification techniques often demonstrate cortical bone erosion with widening of the haversian canals that makes the cortical bone look like cancellous bone. The development of

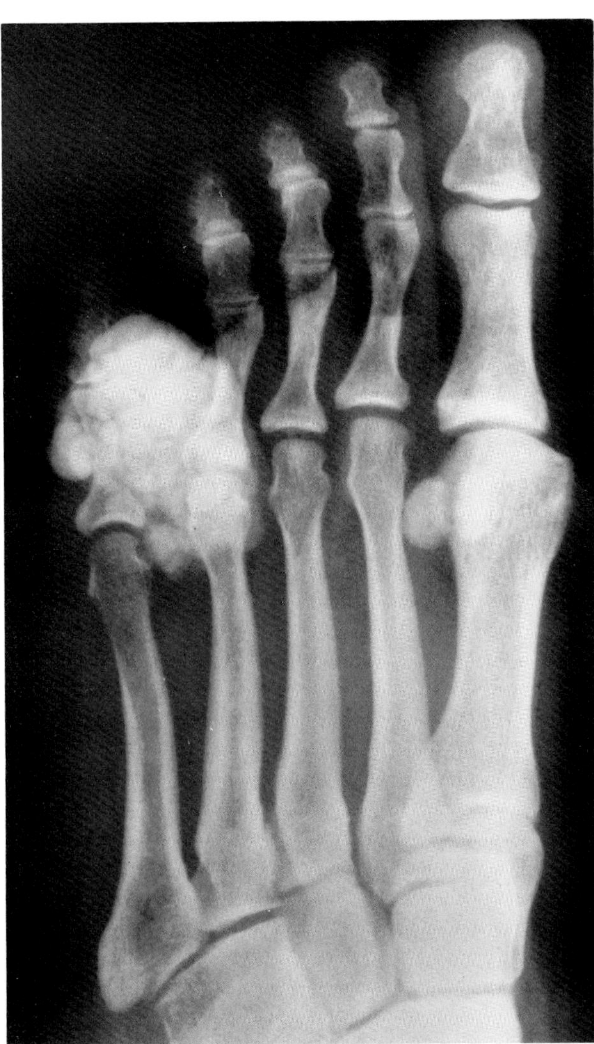

**Figure 36.9. Renal osteodystrophy.** A dense mass of tumoral calcification in the soft tissues of the foot is caused by hyperparathyroidism secondary to chronic renal failure.

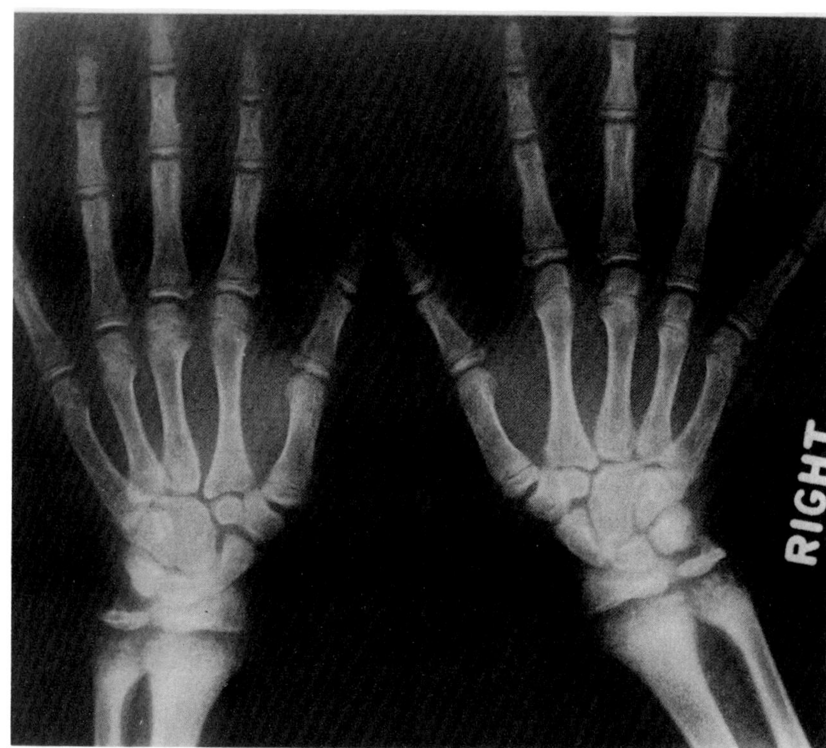


**Figure 36.10. Renal rickets.** Fraying and cupping of radial and ulnar metaphyses in a 14-year-old boy with florid rickets secondary to long-standing pyelonephritis and chronic renal failure. (From Shapiro,[28] with permission from the publisher.)

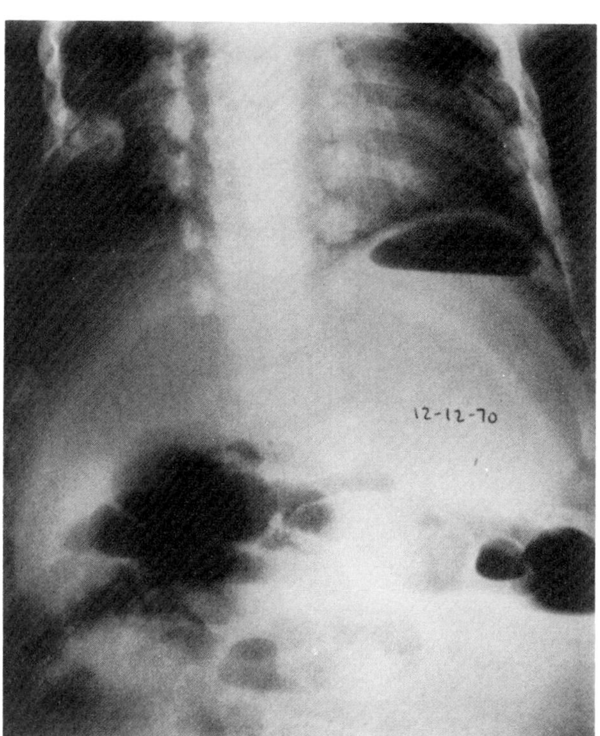


**Figure 36.11. Renal rickets.** Prominent rachitic rosary in a 2-year-old boy with rickets secondary to renal failure resulting from chronic pyelonephritis in a solitary hypoplastic kidney. (From Shapiro,[28] with permission from the publisher.)

thick bands of increased density (osteosclerosis) adjacent to the superior and inferior margins of vertebral bodies can produce the characteristic "rugger-jersey" spine. Other signs of secondary hyperparathyroidism that may develop in patients with chronic renal failure include vascular and periarticular soft tissue calcifications, brown tumors, and "salt-and-pepper" demineralization of the skull.

In children with chronic renal failure, an irregular overgrowth of noncalcified osteoid causes a pattern of frayed, irregular, and cuffed metaphyses, widening of the epiphyseal line, and bone demineralization, findings that are identical to those in ordinary rickets, and are thus termed *renal rickets*.[13–15,28]

## Chest

Pulmonary edema is a common complication in patients with chronic renal failure. Though most often related to congestive heart failure, pulmonary edema may complicate uremia even in the absence of volume overload. This "uremic lung" appears radiographically as characteristic perihilar alveolar densities producing a butterfly or batwing appearance. In uremic pulmonary edema not due to congestive heart failure, the cardiac silhouette is often normal and there is generally no redistribution of pulmonary blood flow.

Enlargement of the cardiac silhouette in patients with chronic renal failure may result from congestive heart failure or pericardial effusion. Echocardiography is the

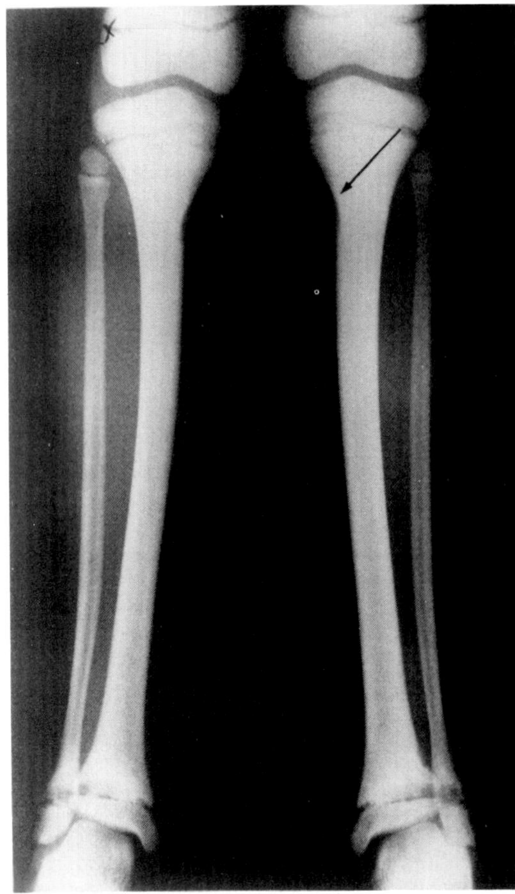

**Figure 36.12. Renal rickets.** Sclerosis of the long bones in an 11-year-old boy with chronic glomerulonephritis, rickets, and secondary hyperparathyroidism. In addition to the increased skeletal density, note the widened zone of provisional calcification at the ankles and the subperiosteal resorption along the medial margins of the upper tibial shafts (arrow).

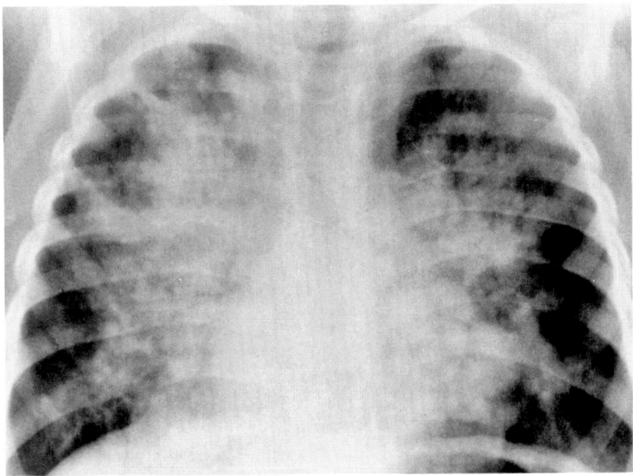

**Figure 36.13. Chronic renal failure.** A frontal chest film demonstrates typical perihilar alveolar densities producing the butterfly pattern of uremic lung. Unlike in pulmonary edema due to congestive heart failure, in chronic renal failure the cardiac silhouette is normal in size.

imaging modality of choice for demonstrating pericardial effusion as the cause of apparent cardiomegaly.

Pleural effusion can be the result of uncomplicated uremia, though it more commonly results from associated congestive heart failure or a metabolic or infectious complication. Patients with chronic renal failure, especially those on dialysis or those who have had a transplant, are prone to develop pulmonary infections due to opportunistic organisms that may produce a bizarre array of radiographic patterns.[16]

## Gastrointestinal Tract

Thickening and irregularity of folds in the duodenal bulb and the second portion of the duodenum may be seen in patients with uremia, especially those undergoing chronic dialysis. This appearance simulates pancreatitis, a disease that frequently complicates prolonged uremia and may be responsible for producing the radiographic pattern. A greater-than-normal incidence of peptic ulcer disease is also reported in patients with chronic renal failure.[17]

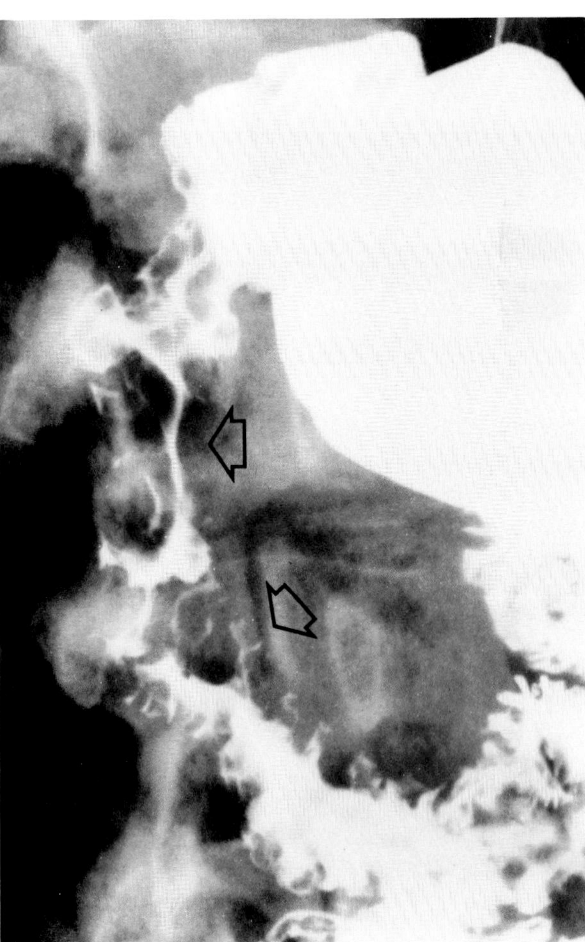

**Figure 36.14. Chronic renal failure.** Irregular thickening of folds in the duodenal sweep simulates pancreatitis.

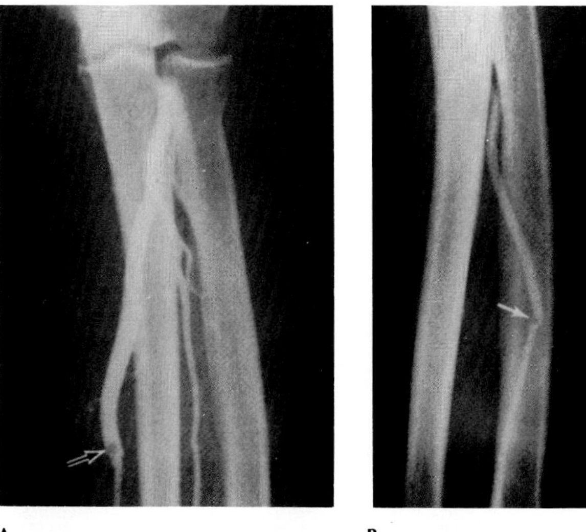

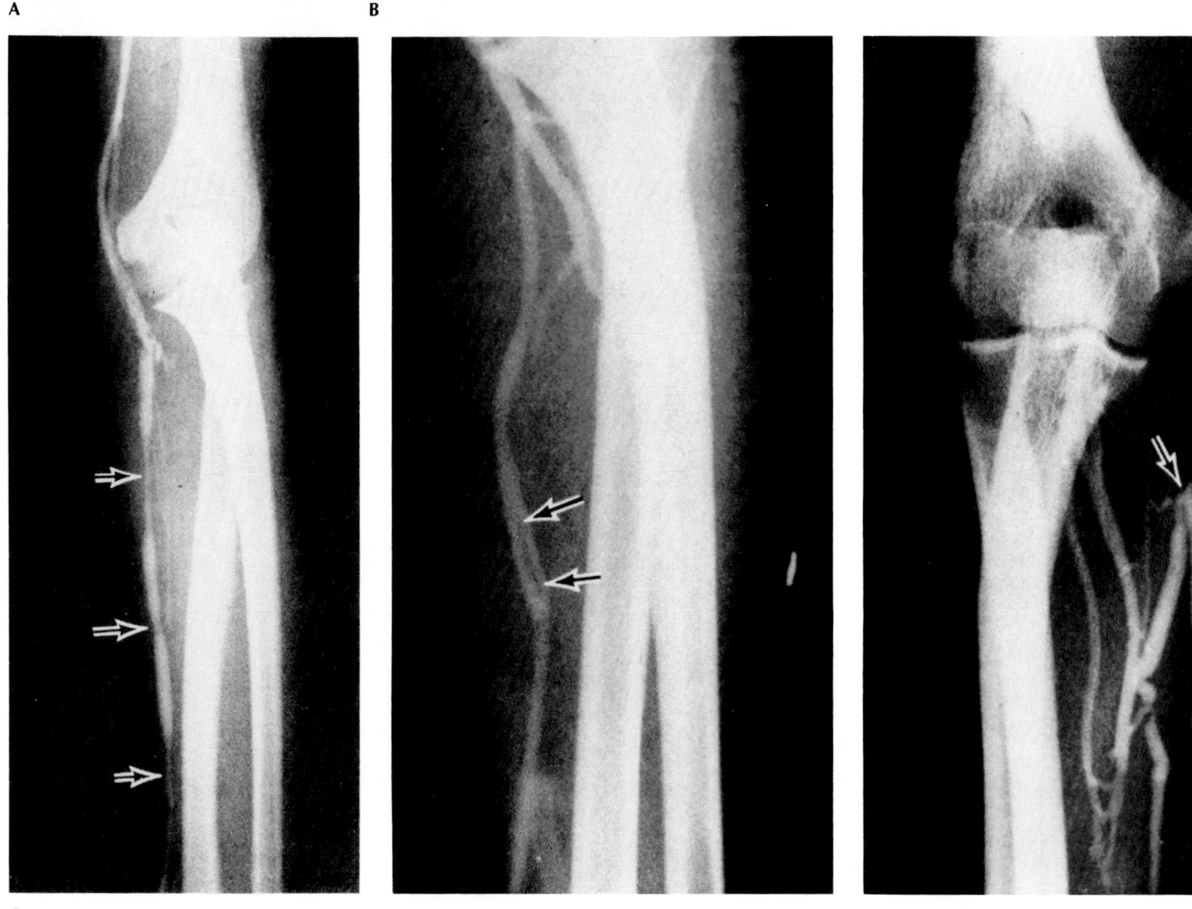

**Figure 36.15. Arteriographic demonstration of hemodialysis shunt complications.** (A) The rounded lucency of a thrombus (arrow) is seen just above the catheter. (B) A thrombus (arrow) markedly narrows the catheter-arterial junction. (C) Three areas of venous narrowing (arrows), probably due to thrombophlebitis, are greatly impeding peripheral flow. (D) A long thin thrombus (arrows) in the vein above the catheter. (E) An occluding venous thrombus (arrow) about 3 inches above the cannula site causes the contrast material to use collateral veins and thus diminishes the total flow. (From Schwartz and Teplick,[18] with permission from the publisher.)

## DIALYSIS AND TRANSPLANTATION

### Radiographic Sequelae of Hemodialysis

Hemodialysis requires the creation of an arteriovenous fistula or conduit that is best evaluated by angiography. Complications that may require the revision of a shunt include arterial or venous thrombosis, arteritis, thrombophlebitis, pseudoaneurysm formation, and leakage around the shunt. Graft stenoses can be successfully treated by percutaneous balloon dilatation.

The need to use heparin during the hemodialysis procedure may lead to such complications as subdural hematoma and retroperitoneal, gastrointestinal, pericardial, and pleural hemorrhage.

Other complications in hemodialysis patients are related to underlying uremia and secondary hyperparathyroidism and include peptic ulcer, renal osteodystrophy, and periarticular and arterial calcifications. Uremic pleural and pericardial effusions and congestive failure often respond dramatically to the institution of effective hemodialysis.[9,18–20]

## Renal Transplantation

Radionuclide scanning and ultrasound are the major imaging modalities for evaluating complications of renal transplantation. Baseline evaluations are generally performed within 24 hours following surgery, since serial

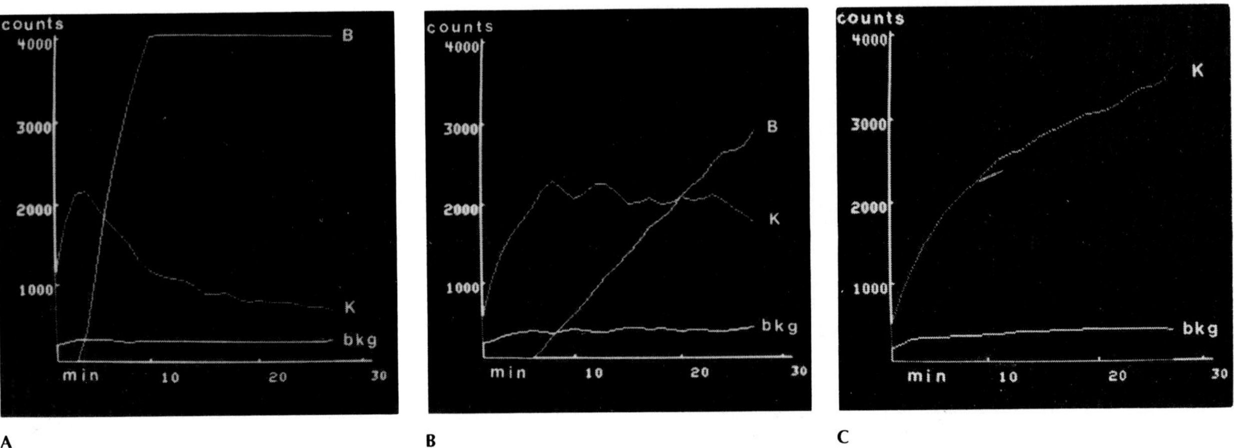


**Figure 36.16. Radionuclide scanning in the evaluation of renal transplantation.** (A) A normal $^{131}$I Hippuran scan shows rapid uptake of isotope by the kidney (K). As the isotope activity within the kidney decreases, there is a rapid increase in isotope activity in the bladder (B). (B) Nine days after transplantation, there is a marked reduction in the excretion of isotope, reflecting acute rejection. (C) In another patient, a scan obtained 3 days after transplantation shows no evidence of excretion, reflecting acute tubular necrosis. bkg = background activity. (From Baum, Vincent, Lyons, et al,[29] with permission from the publisher.)

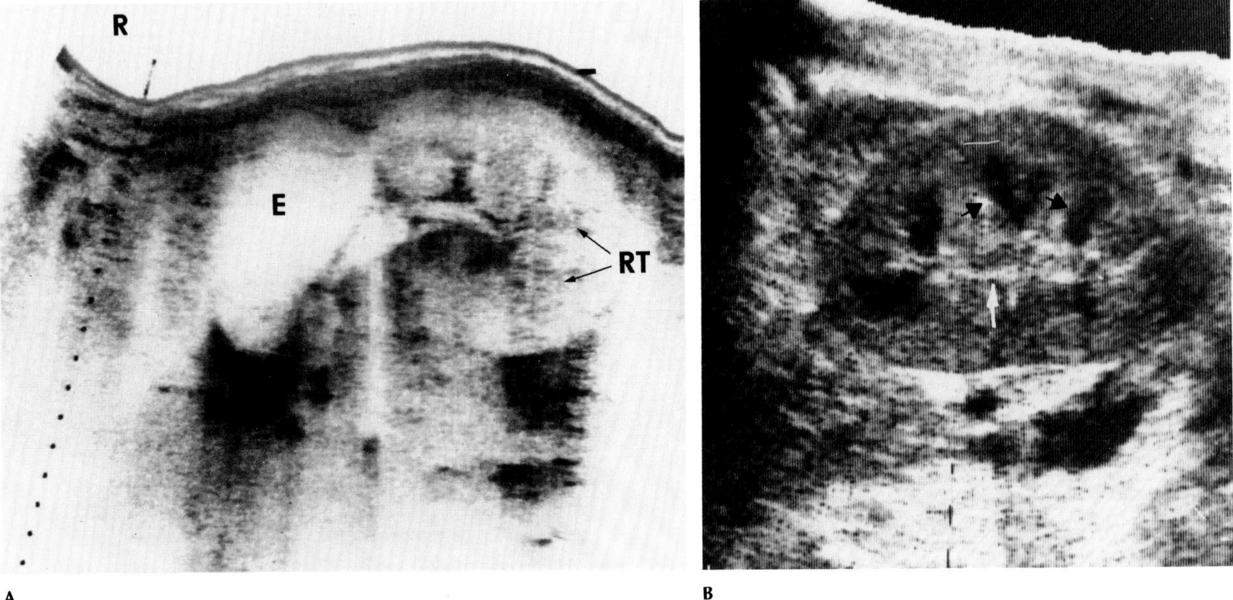


**Figure 36.17. Renal transplant rejection.** (A) A transverse sonogram shows that the renal transplant (RT) has become huge and has lost its corticomedullary definition. The renal vascular pedicle is compressed as it enters the renal hilum. An effusion (E), sometimes seen with acute transplant rejection, is noted medial to the kidney. (B) A sagittal supine sonogram shows enlargement of the renal transplant with increased sonolucency of the medullary pyramids and thinning of the central echogenic hilar structures (white arrow). (A, from Swenson,[2] with permission from the publisher; B, from Eisenberg, Krebs, Ratcliff, et al,[30] with permission from the publisher.)

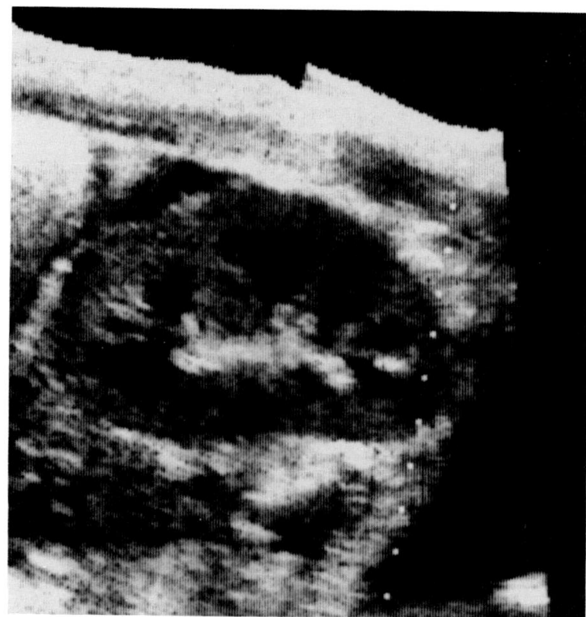

**Figure 36.18. Renal transplant rejection.** A sagittal sonogram shows focal areas of increased echogenicity in the markedly dilated medullary pyramids of an enlarged renal allograft.

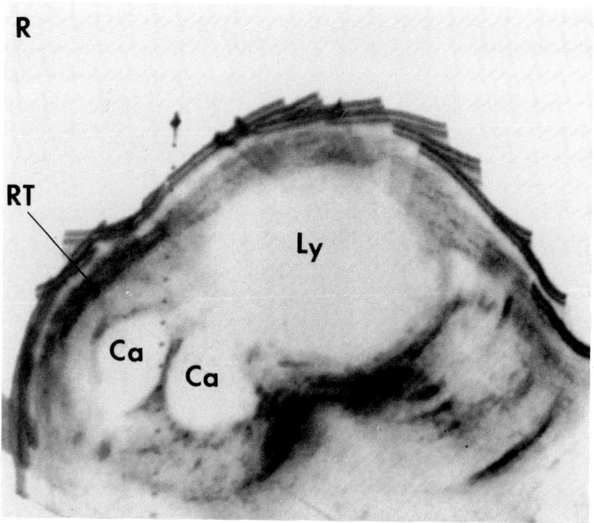


**Figure 36.19. Renal transplant rejection.** A transverse sonogram of the right iliac fossa shows a large lymphocele (Ly) obstructing the transplanted kidney (RT) and causing gross dilatation of the calyces (Ca) and slight thinning of the overlying renal parenchyma. A small amount of ascites is seen adjacent to the lymphocele. (From Swenson,[2] with permission from the publisher.)

studies are often more valuable than individual examinations. This is especially important in attempting to distinguish between rejection and acute tubular necrosis, the two major post-transplant complications.

Radionuclide studies demonstrate renal perfusion (the distribution and integrity of the vascular supply using $^{99m}$Tc DTPA), clearance (the extraction of $^{131}$I Hippuran from the blood), tubular transit time, and bladder activity (to assess possible obstruction). In acute tubular necrosis, there is usually relative preservation of perfusion compared with clearance; both are usually affected to a similar degree in rejection and other causes of decreased renal function. With severe rejection, perfusion and clearance are absent. The transplanted kidney appears as decreased or absent activity relative to background, and no radionuclide reaches the bladder.

Because it is independent of renal function, ultrasound may permit the evaluation of kidneys that cannot be adequately visualized by radionuclide studies. Sonographic findings suggesting transplant rejection include enlargement of the pyramids (containing focal areas of decreased echogenicity), a hyperechogenic cortex, distortion of the renal outline due to localized areas of swelling, and poor definition of renal sinus echoes (normally highly echogenic). At present, the major role of ultrasound in renal transplantation is the detection of transplant hydronephrosis and peritransplant fluid collections. Fluid collections seen following a renal transplant include lymphoceles, abscesses, hematomas, and uriniferous pseudocysts. Although ultrasound is sensitive for the detection of fluid, it is not specific for the fluid type; thus each of these fluid collections may present a similar appearance.

The major role of arteriography or digital subtraction angiography in the failing renal transplant is to detect transplant renal artery stenosis after a radionuclide study has suggested a vascular obstruction. Percutaneous transluminal balloon dilatation of a transplant artery stenosis can be performed to control hypertension and preserve renal function. Noninvasive radionuclide and ultrasound techniques have effectively supplanted arteriography in assessing transplant rejection.

If metallic clips are attached to the upper and lower poles of the transplanted kidney at surgery, serial plain abdominal radiographs may be used to demonstrate changes in the size of the transplanted kidney. During the first year, the measured distance normally increases up to 15 percent because of normal compensatory hypertrophy. Larger increases indicate the presence of some abnormality requiring further investigation. Displacement of the clips suggests the development of a large retroperitoneal fluid collection pushing the transplanted kidney.

Because abnormalities of the transplanted kidney are commonly associated with a major decrease in renal function, excretory urography generally shows minimal or no opacification of the urinary tract and is thus of limited value. At times, excretory urography can be useful in demonstrating the site of urinary obstruction or extravasation.

The overall incidence of de novo malignancies occuring in organ transplant recipients is 6 percent, 100

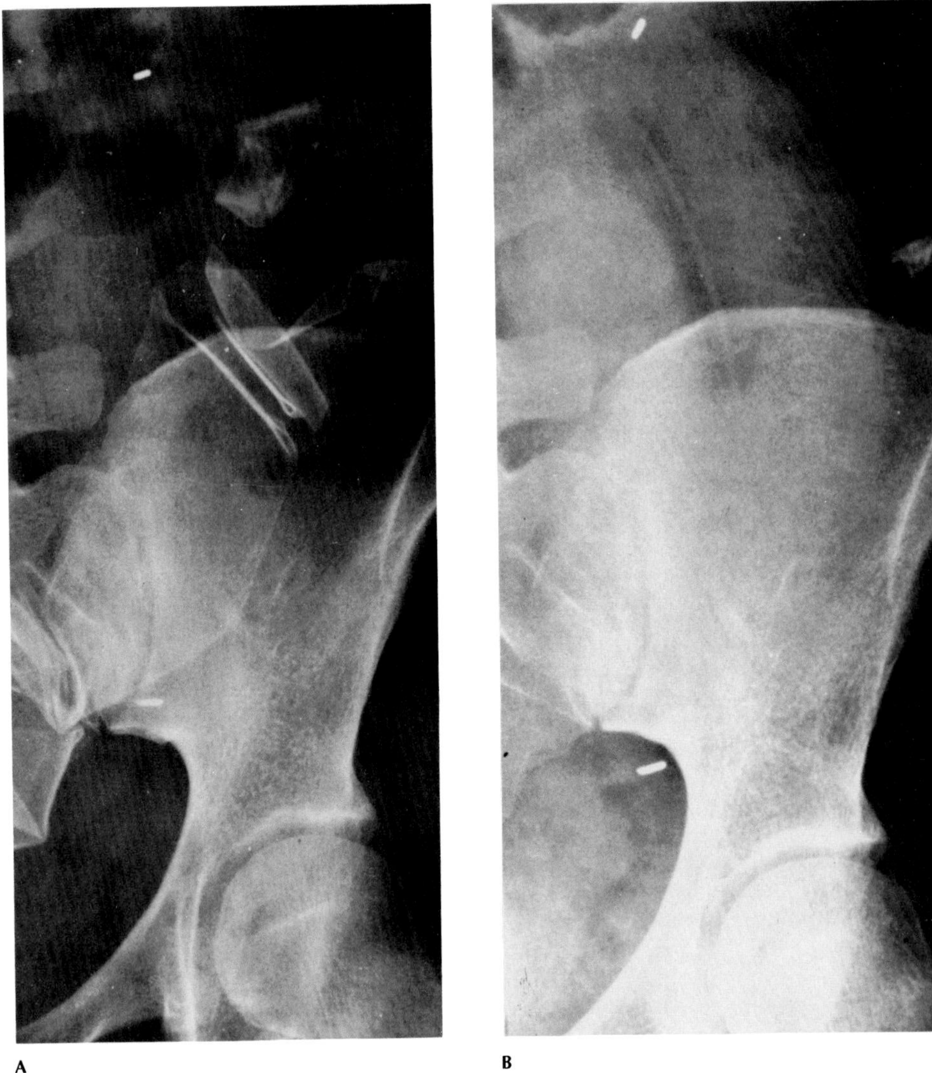

**Figure 36.20. Value of attaching silver clips to the poles of the transplanted kidney.** (A) An intraoperative radiograph shows silver clips marking the upper and lower poles of the transplanted kidney. Opaque drainage tubes are also visible. (B) One day later, the transplanted kidney appears significantly larger because its vein has thrombosed. (From Swenson,[2] with permission from the publisher.)

times that for the age-matched general population. Malignant lymphomas make up 20 percent of the post-transplantation malignancies, the second most common after carcinoma of the skin and lips. Unlike lymphomas in nontransplant patients, lymphomas in transplant recipients have a high incidence of central nervous system involvement; in more than 80 percent of these patients, the brain is the only site of lymphomatous disease. The most common CT appearance of cerebral lymphoma is one or more masses that contain a central lucency and show peripheral enhancement after the administration of intravenous contrast material. Extracranial lymphomas developing in the gastrointestinal tract, liver, lymph nodes, and chest tend to have rapid growth patterns and often are atypical in their radiographic appearance compared with lymphomas that occur in the general population.[2,21–27]

## REFERENCES


1. Davidson AJ: *Radiologic Diagnosis of Renal Parenchymal Disease*. Philadelphia, WB Saunders, 1977.
2. Swenson RS: Renal failure, renal dialysis, and renal transplantation, in Friedland GW, Filly R, Goris ML, Gross D, Kempson RL, Korobkin M, Thurber BD, and Walter J (eds): *Uroradiology: An Integrated Approach*. New York, Churchill Livingstone, 1983.

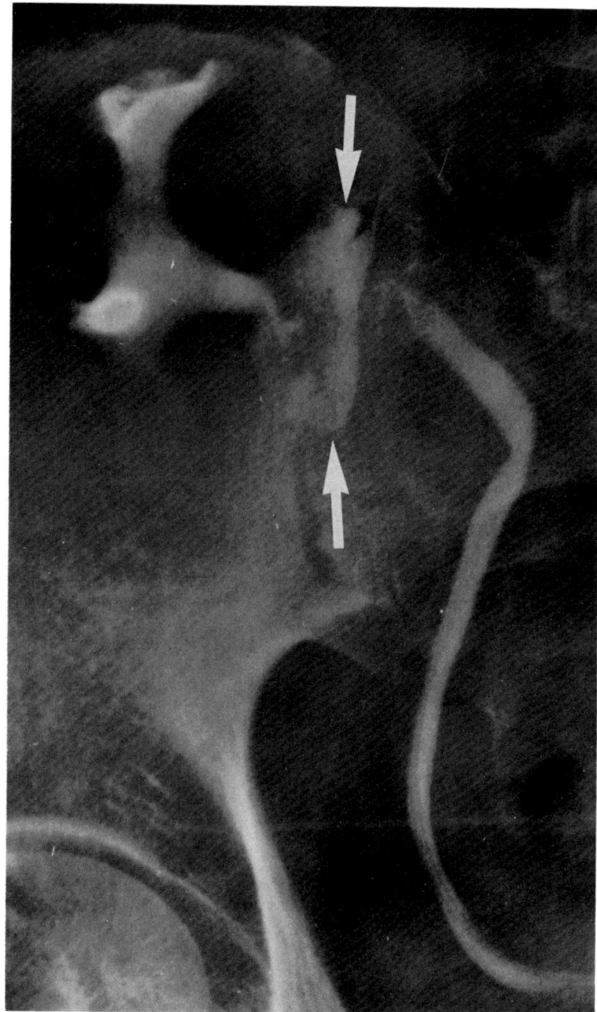

**Figure 36.21. Urine extravasation following renal transplantation.** An excretory urogram shows extravasation from the site of the ureteroureteral anastomosis (arrows). (From Swenson,[2] with permission from the publisher.)

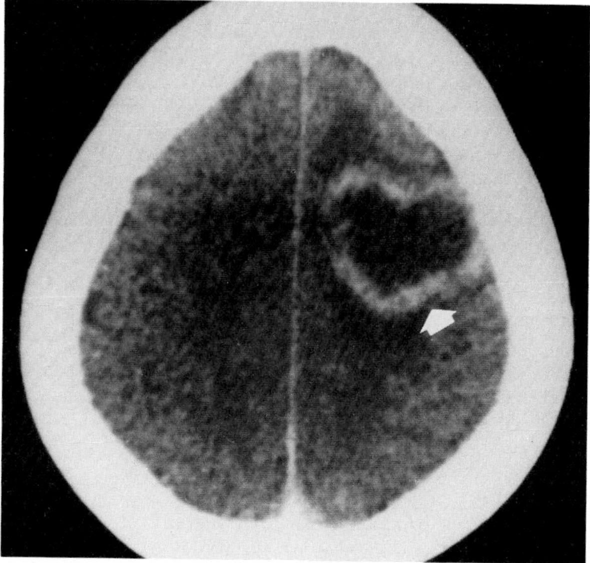

**Figure 36.22. Lymphoma developing after renal transplantation.** A CT scan following the administration of intravenous contrast material shows a heart-shaped, peripherally enhancing, centrally lucent lesion (arrow) situated in the right frontoparietal region. There is moderate surrounding edema. (From Tubman, Frick, and Hanto,[27] with permission from the publisher.)

3. Older RA, Korobkin M, Cleeve BM, et al: Contrast-induced renal failure: Persistent nephrogram as clue to early detection. *AJR* 134:399–401, 1980.
4. Shea TE, Pfister RC: Opacification of the gallbladder by urographic contrast material. *AJR* 107:763–768, 1969.
5. Palmer FJ: Renal cortical calcification. *Clin Radiol* 21:175–177, 1970.
6. Kleinknecht D, Grunfeld JP, Gomez PC, et al: Diagnostic procedures and long-term prognosis in bilateral renal cortical necrosis. *Kidney Int* 4:390–400, 1973.
7. Whelan JG, Ling JT, Davis LA: Antimortem roentgen manifestations of bilateral renal cortical necrosis. *Radiology* 89:682–689, 1967.
8. McAllister WH, Nedelman SH: The roentgen manifestations of bilateral renal cortical necrosis. *AJR* 86:129–135, 1961.
9. Freidland: in Friedland GW, Filly R, Goris ML, Gross D, Kempson RL, Korobkin M, Thurber BD, and Walter J (ed): *Uroradiology: An Integrated Approach*. New York, Churchill Livingstone, 1983.
10. Halpern M: Angiography in chronic renal disease and renal failure. *Radiol Clin North Am* 10:467–494, 1972.
11. Bosniak MA, Schweizer RD: Urographic findings in patients with renal failure. *Radiol Clin North Am* 10:433–445, 1972.
12. Davidson AJ, Talner LB: Lack of specificity of renal angiography in the diagnosis of renal parenchymal disease: A point of view. *Invest Radiol* 8:90–95, 1973.
13. Carr D, Davidson JK, McMillan M, et al: Renal osteodystrophy: An underdiagnosed condition. *Clin Radiol* 31:55–59, 1980.
14. Sundaram M, Joyce PF, Shields JB, et al: Terminal phalangeal tufts: Earliest site of renal osteodystrophy findings in hemodialysis patients. *AJR* 133:25–29, 1979.
15. Sundaram M, Wolverson MK, Heiberg E, et al: Erosive azotemic osteodystrophy. *AJR* 136:363–367, 1981.
16. Schwartz EE, Onesti G: the cardiopulmonary manifestations of uremia and renal transplantation. *Radiol Clin North Am* 10:569–581, 1972.
17. Wiener SN, Vertes V, Shapira H: The upper gastrointestinal tract in patients undergoing chronic dialysis. *Radiology* 92:110–114, 1969.
18. Schwartz C, Teplick JG: Radiologic consideration in maintenance dialysis. *Radiol Clin North Am* 10:511–519, 1972.
19. Leonard A, Shapira FL: Subdural hematoma in regularly hemodialyzed patients. *Ann Intern Med* 82:650–658, 1975.
20. Mirahmandi KS, Coburn JW, Bluestone R: Calcific periarthritis and hemodialysis. *JAMA* 223:548–549, 1973.
21. Ney C, Friedenberg RM: *Radiographic Atlas of the Genitourinary System*. Philadelphia, JB Lippincott, 1981.
22. Stables DP, Klingensmith WC, Johnson ML: Renal transplantation, in Rosenfield AT, Glickman MG, Hodson J (eds): *Diagnostic Imaging of Renal Disease*. New York, Appleton-Century-Crofts, 1979.
23. Kuhlmann U, Vetter W, Furrer J, et al: Renovascular hypertension: Treatment by percutaneous transluminal dilatation. *Ann Intern Med* 92:1–6, 1980.
24. Silver TM, Campbell D, Wicks JD, et al: Peritransplant fluid collections: Ultrasound evaluation and clinical significance. *Radiology* 138:141–151, 1981.

25. Singh A, Cohen WN: Renal allograft rejection: Sonography and scintigraphy. *AJR* 135:73–77, 1980.
26. Burgener FA, Schabel SI: The radiographic size of renal transplants. *Radiology* 117:547–550, 1975.
27. Tubman DE, Frick MP, Hanto DW: Lymphoma after organ transplantation: Radiologic manifestations in the central nervous system, thorax, and abdomen. *Radiology* 149:625–631, 1984.
28. Shapiro R: Radiologic aspects of renal osteodystrophy. *Radiol Clin North Am* 10:557–569, 1972.
29. Baum S, Vincent NR, Lyons KP, et al: *Atlas of Nuclear Medicine Imaging* New York, Appleton-Century-Crofts, 1981.
30. Eisenberg RL, Krebs CA, Ratcliff SL, et al: Renal ultrasound: Test your interpretation. *RadioGraphics* 2:153–178, 1982.

## Chapter Thirty-Seven
# Glomerular and Tubular Diseases

## GLOMERULAR DISEASES

### Glomerulonephritis

Alterations of the structural and functional integrity of the glomerular capillary circulation, associated with hematuria, proteinuria, reduced glomerular filtration rate, and hypertension, can be the result of a broad spectrum of disorders. Glomerulonephritis can reflect primary glomerular disease or infections (especially due to group A beta-hemolytic streptococci), or can be a facet of multisystem diseases (systemic lupus erythematosus, Goodpasture's syndrome, Henoch-Schönlein purpura, systemic necrotizing vasculitis, amyloidosis, diabetic nephropathy, neoplastic disease).

The excretory urographic findings in glomerulonephritis depend on the duration and severity of the disease process and on the level of renal function. In patients with acute glomerulonephritis, the kidneys may be normal or diffusely increased in size with smooth contours and normal calyces. A loss of renal substance in chronic glomerulonephritis produces bilateral small kidneys. The renal outline remains smooth, and the collecting system is normal, unlike the irregular contours and blunted calyces seen in chronic pyelonephritis. The density of the nephrogram and the amount of contrast material in the pelvocalyceal system vary with the severity of the disease. An infrequent finding is diffuse fine granular calcifications in the renal parenchyma, a pattern which is similar to that in bilateral cortical necrosis and almost pathognomonic of chronic glomerulonephritis when associated with bilateral small kidneys.

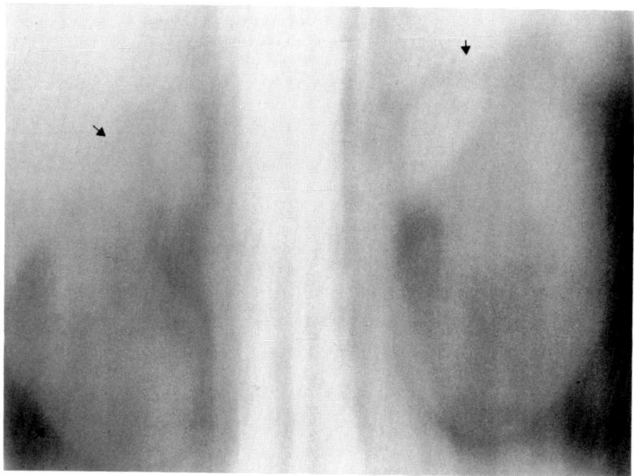

**Figure 37.1. Rapidly progressive acute glomerulonephritis.** A nephrotomogram demonstrates bilateral large kidneys with smooth contours (arrows). The nephrographic density, although faint, was maximal at this time. The calyces were never visualized. (From Davidson,[1] with permission from the publisher.)

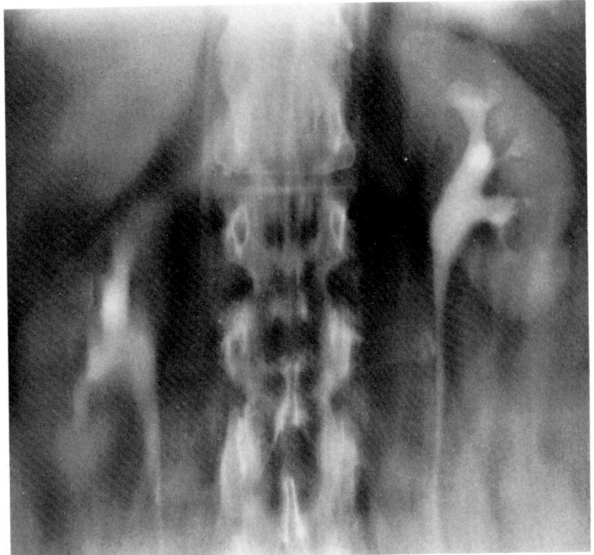

**Figure 37.2. Chronic glomerulonephritis.** A nephrotomogram shows bilateral small, smooth kidneys. The uniform reduction in parenchymal thickness is particularly apparent in the right kidney. Note that the pelvocalyceal system is well-opacified and without the irregular contours and blunted calyces seen in chronic pyelonephritis.

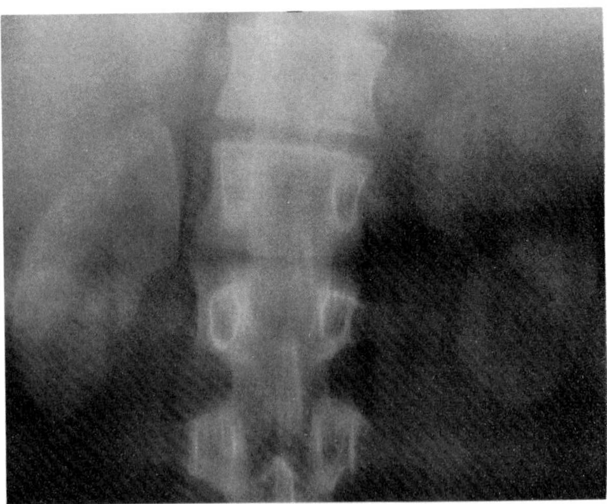

**Figure 37.3. Chronic glomerulonephritis.** A plain film tomogram shows bilateral small, smooth kidneys with diffuse fine calcification in the renal parenchyma, an appearance similar to that in bilateral cortical necrosis.

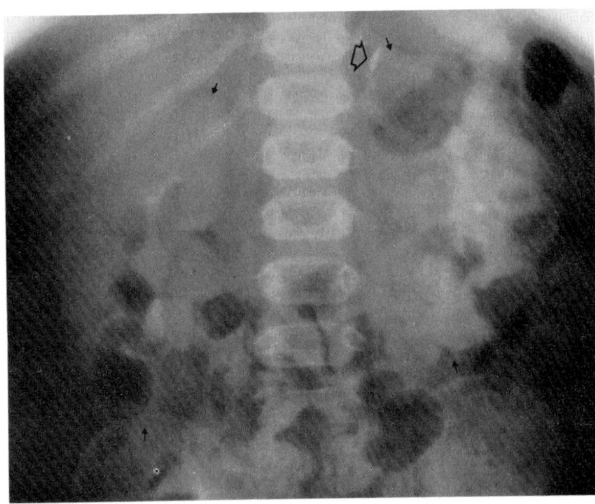

A

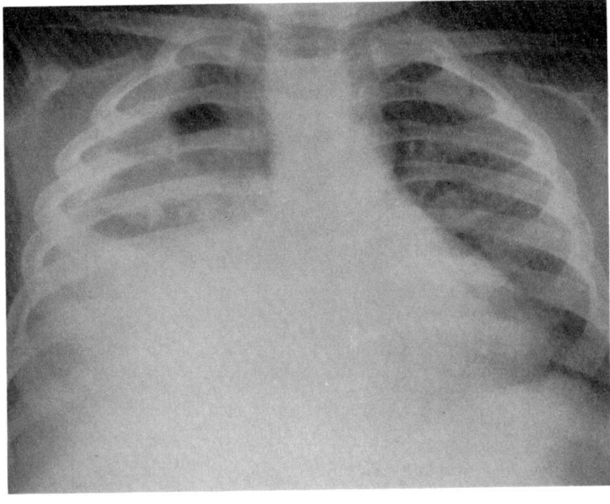

B

**Figure 37.4. Nephrotic syndrome.** (A) An excretory urogram demonstrates striking enlargement of both kidneys. (Arrows point to the tips of the upper and lower poles of the kidneys.) There is moderate opacification of the collecting systems. Of incidental note is calcification in the left adrenal gland (open arrow). (B) A chest radiograph shows diffuse cardiomegaly with a large right pleural effusion, which is situated both along the lateral chest wall and in a subpulmonic location.

Arteriography in patients with chronic glomerulonephritis demonstrates a decreased caliber of the major renal and intrarenal arteries along with a loss of peripheral branches, a lack of normal tapering, and prolonged arterial opacification. However, arteriography is generally of little diagnostic value since an identical nonspecific pattern can be seen in other forms of bilateral diffuse parenchymal renal disease.[1–4]

## Nephrotic Syndrome

The nephrotic syndrome, characterized by albuminuria, hypoalbuminemia, hyperlipidemia, and edema, is the common end point of a variety of disease processes that damage the glomerular capillary wall and permit the excessive glomerular leakage of plasma proteins into the urine. Causes of the nephrotic syndrome include glo-

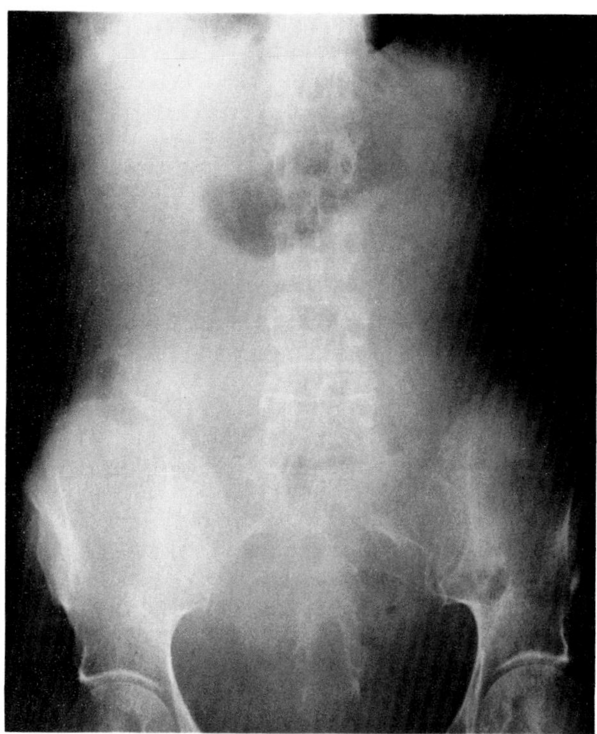

**Figure 37.5. Nephrotic syndrome.** Severe ascites obscures the visceral margins and causes a ground-glass appearance of the abdomen.

merulonephritis, immunologic disorders, toxic injuries, metabolic abnormalities, biochemical defects, and vascular disorders. Biopsy of the kidney generally is required to determine the underlying renal disorder.

The excretory urogram is normal in many patients with nephrotic syndrome. Renal edema may cause kidney enlargement with associated stretching and thinning of the collecting system. The nephrogram and the density of contrast material in the pelvocalyceal system usually remain normal. Progression of the underlying disease may lead to the development of renal failure, resulting in small kidneys of smooth contour with poor opacification of the collecting systems.

Pleural effusion, often subpulmonic, and ascites can often be demonstrated in patients with the nephrotic syndrome. Renal vein thrombosis, either unilateral or bilateral, is a serious complication of nephrotic syndrome, rather than an underlying cause as previously considered.[5,6]

### Alport's Syndrome

Alport's syndrome is a hereditary nephritis that frequently coexists with sensorineural deafness and, occasionally, with ocular abnormalities. The renal disease generally appears at an early age, usually as recurrent episodes of hematuria. Males are more frequently and severely affected than females and tend to suffer slowly progressive renal insufficiency that terminates in end-stage kidney failure at an early age. In females the disease usually is not progressive, and death due to chronic nephritis is uncommon.

The urographic findings in Alport's syndrome are indistinguishable from those in other forms of glomerulonephritis and consist of small, smooth kidneys with impaired excretion of contrast material. The nonspecific arteriographic changes are similar to those in other types of chronic glomerulonephritis.[7,8]

## TUBULOINTERSTITIAL DISEASES OF THE KIDNEY

### Hypercalcemic Nephropathy

Chronic hypercalcemia, which may occur in primary hyperparathyroidism, sarcoidosis, multiple myeloma, vitamin D intoxication, or metastatic bone disease, can cause tubulointerstitial damage and progressive renal insufficiency. Plain abdominal radiographs demonstrate calcification within the renal parenchyma (nephrocalcinosis) as well as nephrolithiasis, which is due to the hypercalcuria that often accompanies hypercalcemia (see the sections "Nephrolithiasis" and "Nephrocalcinosis" in Chapter 40).

### Vesicoureteral Reflux

Reflux of urine from the bladder into the ureters and even into the renal pelves may occur during voiding or with an elevation of pressure in the bladder. The precise relation between vesicoureteral reflux and urinary tract infection is controversial and unclear. However, it is evident that reflux combined with urinary tract infection can produce severe renal scarring, and dilated and tortuous ureters that may require reimplantation.

Voiding cystourethrography is the examination most commonly used to detect the retrograde filling of one or both ureters in patients with vesicoureteral reflux. An alternative approach is the radionuclide voiding cystogram, which has a lower radiation dose and can quantify the degree and extent of vesicoureteral reflux.

In addition to caliectasis, renal scarring, and ureteral dilatation, excretory urography in children with vesicoureteral reflux and infection may demonstrate an unusual striated appearance in the renal pelvis and ureter.

In the child with recurrent urinary tract infections, an excretory urogram is currently recommended as the initial imaging procedure. If the upper urinary tracts are normal, no further radiographic studies are warranted. However, an abnormal upper urinary tract should be further evaluated by contrast or radionuclide cystography to determine the grade of reflux and the subsequent therapy.[9–12]

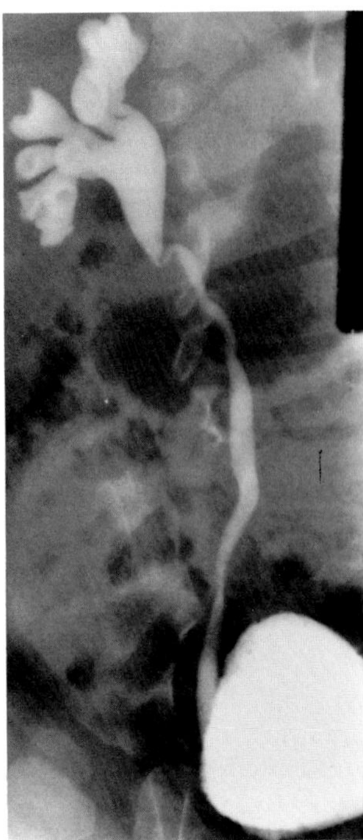

**Figure 37.6. Vesicoureteral reflux.** A cystogram demonstrates retrograde filling of the right ureter and pelvocalyceal system from the bladder.

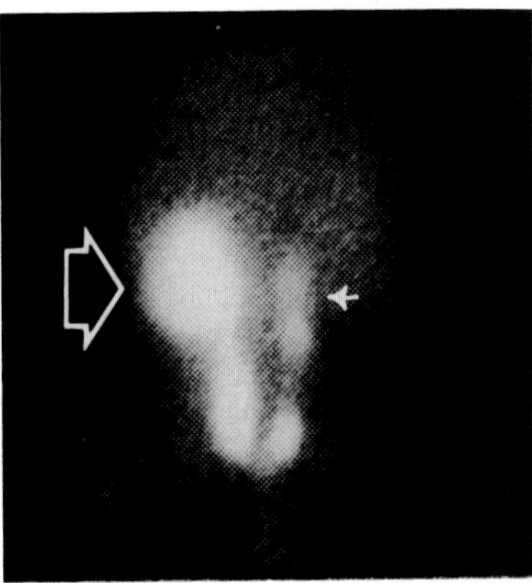

**Figure 37.7. Vesicoureteral reflux.** A posterior view of a radionuclide voiding cystogram shows reflux from the bladder to the ureter (small arrow) on the right and to the ureter and renal pelvis (large arrow) on the left. (From Baum, Vincent, Lyons, et al,[31] with permission from the publisher.)

## Radiation Nephritis

Inclusion of the kidney within the radiation field can lead to diffuse renal ischemia and vasculitis that become apparent after a latent period of 6 to 12 months following radiation therapy. Some decrease in renal function is often noted, though the renal collection system appears normal on retrograde studies. Hypertension, which may reach malignant proportions, is a not-infrequent serious complication. Chronic radiation nephritis is characterized by progressive ischemic atrophy and decreased function involving the affected kidney.[13–15]

## HEREDITARY TUBULAR DISORDERS

### Adult Polycystic Disease

The adult form of polycystic kidney disease is an inherited disorder in which multiple cysts of varying size cause lobulated enlargement of the kidneys and progressive renal impairment that is presumably due to cystic compression of nephrons that causes localized intrarenal obstruction. One-third of the patients with adult polycystic kidney disease have associated cysts of the liver, which do not interfere with hepatic function. About 10 percent have one or more saccular (berry) aneurysms of cerebral arteries, which may rupture and produce a fatal subarachnoid hemorrhage.

Many patients with adult polycystic disease are hypertensive, a condition that may cause further deterioration of renal function as well as increase the likelihood that a cerebral aneurysm will rupture. Because patients tend to be asymptomatic during the first 3 decades of life, early diagnosis is made either by chance or as the result of a specific search prompted by a positive family history.

Excretory urography demonstrates enlarged kidneys with a multilobulated contour. The pelvic and infundibular structures are elongated, effaced, and often displaced around larger cysts to produce a crescentic outline. The nephrogram typically has a distinctive mottled or Swiss-cheese pattern due to the presence of innumerable lucent cysts of varying size throughout the kidneys. Plaques of calcification occasionally occur in cyst walls.

The enlarged kidneys in some patients with polycystic kidney disease have smooth outlines and normal-appearing collecting structures. This pattern is seen primarily in young, asymptomatic patients in whom the cysts are sufficiently numerous to enlarge the kidney but not yet big enough to distort the renal contour or collecting structures. In these patients, high-dose urography with nephrotomography is necessary to demonstrate the characteristic radiolucencies in the nephrogram.

Ultrasound in patients with adult polycystic kidney disease demonstrates grossly enlarged kidneys containing multiple cysts that vary markedly in size and are randomly distributed throughout the kidney. The concomitant demonstration of hepatic cysts further strengthens the diagnosis. Ultrasound is also of value in screening family members of a patient known to have this hereditary disorder. In patients with bilateral kidney enlargement and poor renal function, ultrasound permits the differentiation of polycystic kidney disease from multiple solid masses.

Arteriography in patients with adult polycystic disease demonstrates stretching, elongation, and displacement of intrarenal arterial branches by the multiple cysts. During the nephrogram phase, the multiple cysts appear as lucent regions of varying size randomly distributed through the kidney. In patients with associated liver cysts, hepatic

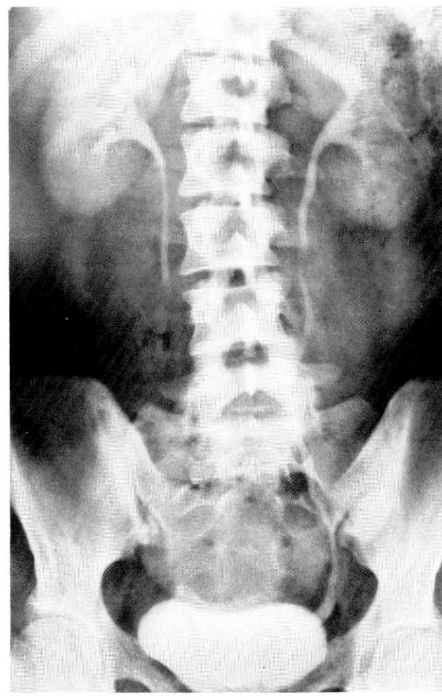


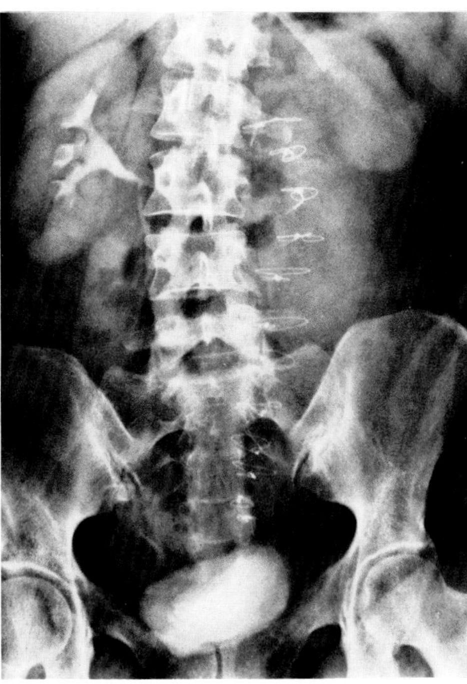


**Figure 37.8. Radiation nephritis.** (A) An excretory urogram performed before radiation therapy shows a normal urinary tract. The soft tissue mass above the bladder represents a myomatous uterus. Several years later the patient developed abdominal lymphoma and received radiation therapy to a wide port that included the spleen. (B) An excretory urogram 5 years after radiation therapy shows that the left kidney has shrunk markedly. Delayed films showed no contrast material in the collecting system. Although a large left paraspinous mass deviates the axis of the left kidney, ultrasound revealed no obstruction; the intensity of the nephrogram excludes long-standing obstructive atrophy. (From Gross,[15] with permission from the publisher.)

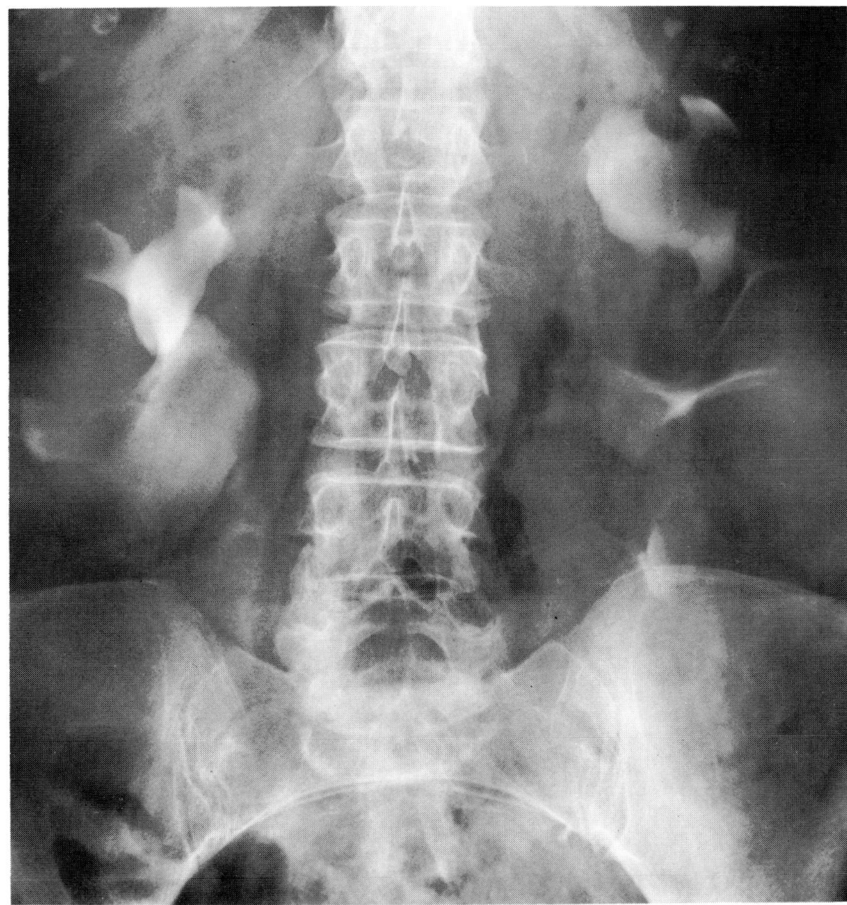

**Figure 37.9. Adult polycystic disease.** An excretory urogram shows marked multifocal enlargement of both kidneys, focal displacement of the collecting structures, and normal opacification. (From Davidson,[1] with permission from the publisher.)

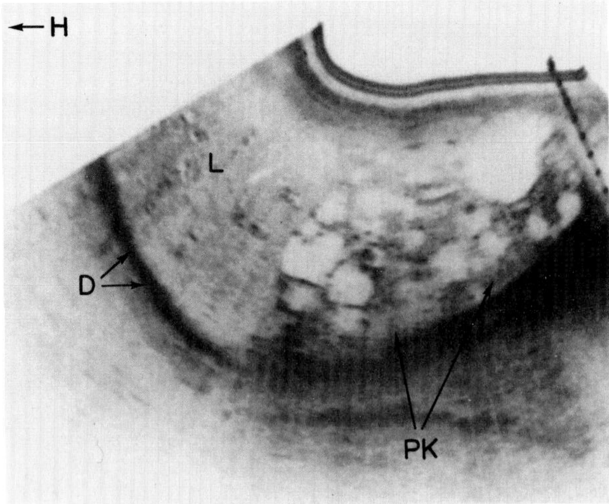


**Figure 37.10. Adult polycystic disease.** A parasagittal sonogram of the right kidney (RK) in a patient with advanced polycystic renal disease shows a random distribution of multiple cysts that vary dramatically in size. The normal reniform contour is maintained. L = liver; D = diaphragm; H = head. (From Friedland, Filly, Goris, et al,[9] with permission from the publisher.)

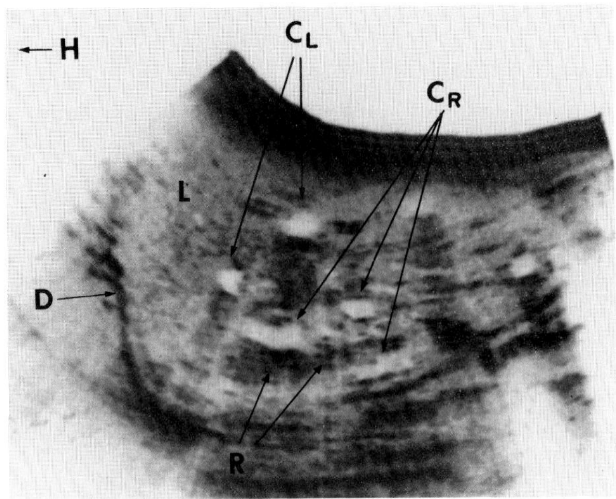


**Figure 37.11. Polycystic kidney and liver disease.** A parasagittal sonogram in a young, asymptomatic member of a family afflicted with polycystic disease. There are multiple cysts ($C_R$, $C_L$) in the right kidney (R) and liver (L). The finding of cysts in the kidney and the liver is characteristic of polycystic disease. D = diaphragm; H = head. (From Friedland, Filly, Goris, et al,[9] with permission from the publisher.)

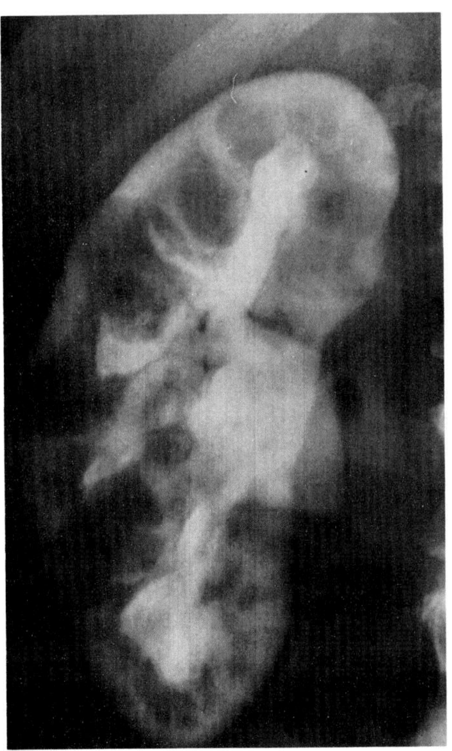

A

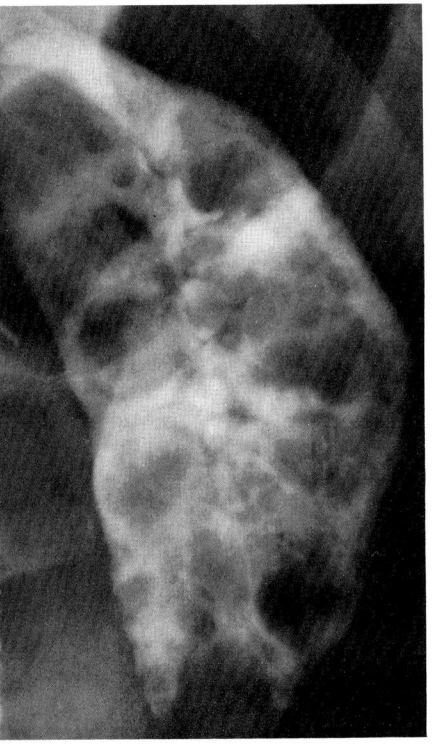

B

**Figure 37.12. Adult polycystic disease.** Nephrogram phases from selective arteriograms of the right (A) and left (B) kidneys demonstrate innumerable cysts throughout both kidneys. The cysts range from pinhead size to 2 cm in diameter and tend to be larger on the right. Of incidental note is opacification of the right adrenal gland. (From Bosniak and Ambos,[17] with permission from the publisher.)

arteriography demonstrates multiple hypovascular masses displacing the intrahepatic vessels. Arteriography is also of value in excluding tumor neovascularity in any patient with adult polycystic kidney disease who is suspected of harboring a coexistent renal carcinoma.[1,9,16–21]

## Polycystic Renal Disease in Infants and Children

Infantile polycystic disease is a rare, usually fatal, autosomal recessive disorder that manifests itself at birth by diffusely enlarged kidneys, renal failure, and mal-

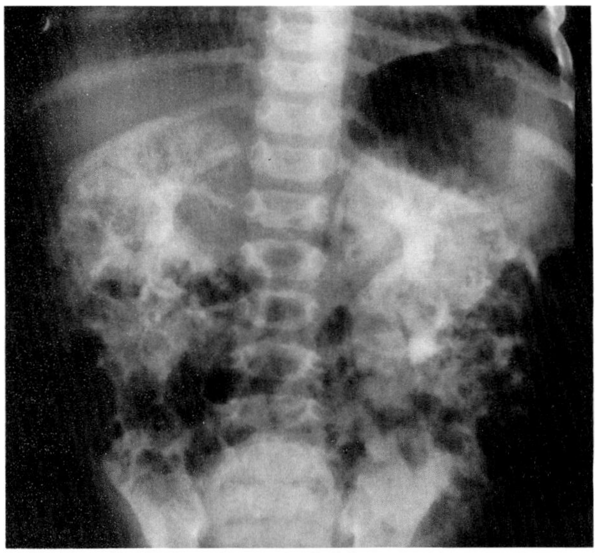


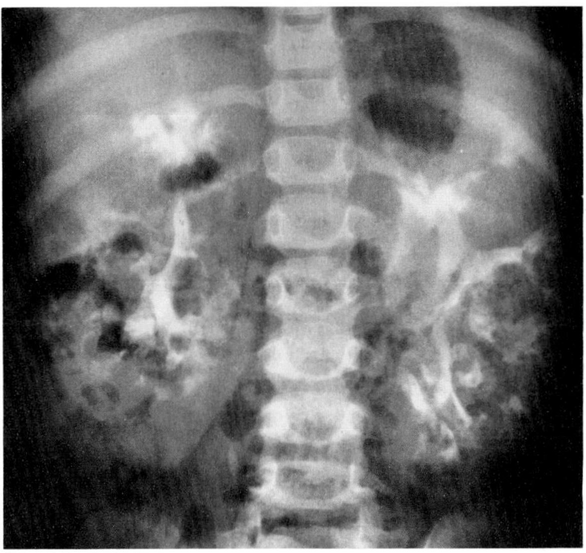


**Figure 37.13. Infantile polycystic kidney disease.** (A) An excretory urogram in this 1-year-old asymptomatic boy with large, palpable abdominal masses demonstrates renal enlargement with characteristic streaky densities leading to the calyceal tips. There is only minimal distortion of the calyces. (B) In this 3-year-old boy with hematuria, an excretory urogram shows greatly enlarged kidneys with distortion of the calyces, infundibular structures, and pelves. Note the streaky densities associated with the renal medullary areas, indicating dilatation of the tubules typical of this condition. (From Bosniak and Ambos,[17] with permission from the publisher.)

development of intrahepatic bile ducts. In the childhood form, the renal abnormality is usually milder but is associated with severe congenital hepatic fibrosis and portal hypertension.

In infantile polycystic disease, plain films demonstrate the kidneys as large soft tissue masses occupying much of the abdomen and displacing the stomach and bowel. The cortical margins are smooth, unlike the irregular renal contours in adult polycystic disease that are due to the protrusion of innumerable cysts from the kidney surface. When renal function is sufficient, excretory urography results in a striking nephrogram in which a streaky pattern of alternating dense and lucent bands reflects contrast material puddling in elongated cystic spaces that radiate perpendicular to the cortical surface. At ultrasound, there is distortion of the intraparenchymal architecture, though the individual cysts are too small to be visualized.

In the childhood form, the kidneys are only moderately enlarged, and the calyces are widened and blunted without the splayed appearance seen in the adult type of polycystic disease. Opacification of ectatic collecting ducts in the renal pyramids may produce a brush border appearance similar to that seen in mild forms of medullary sponge kidney.[5,9,17,22,23]

## Medullary Sponge Kidney

Medullary sponge kidney (renal tubular ectasia) is characterized by cystic dilatation of the distal collecting tubules in the renal pyramids. The ectatic changes may be limited to a single pyramid but are usually more extensive and bilateral, though not necessarily symmetric. Although renal function is preserved, tubular stasis predisposes to calculus formation and pyelonephritis. Medullary sponge kidney is generally asymptomatic, except when medullary calculi become dislodged and produce renal colic or hematuria.

Plain abdominal radiographs often demonstrate multiple small, smoothly rounded calculi that occur in clusters or in a fanlike arrangement in the papillary tip of one or more renal pyramids. Excretory urography demonstrates the ectatic tubules as fine linear striations of contrast material producing a brush border pattern. With increasing dilatation, the tubules become more cystic and simulate a cluster of grapes; there is also broadening, increased cupping, and distortion of the calyces.[1,5,24,25]

## Medullary Cystic Disease

Medullary cystic disease is a rare inherited cause of renal failure in which there are numerous cysts of varying size in the medullary and corticomedullary areas of the kidney. The condition is characterized by the insidious development of anemia, polyuria, an inability to concentrate urine, salt wasting, and progressive uremia. The kidneys are normal- or small-sized and have smooth contours. Decreased renal function usually leads to the impaired excretion of contrast material at excretory urography. In most cases, the cysts are too small to distort the pelvocalyceal system; because the cysts do not involve the outer cortex, the renal contour remains intact.

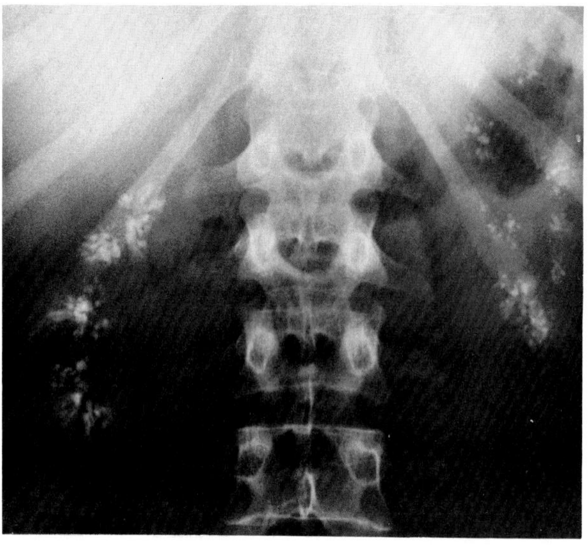
A

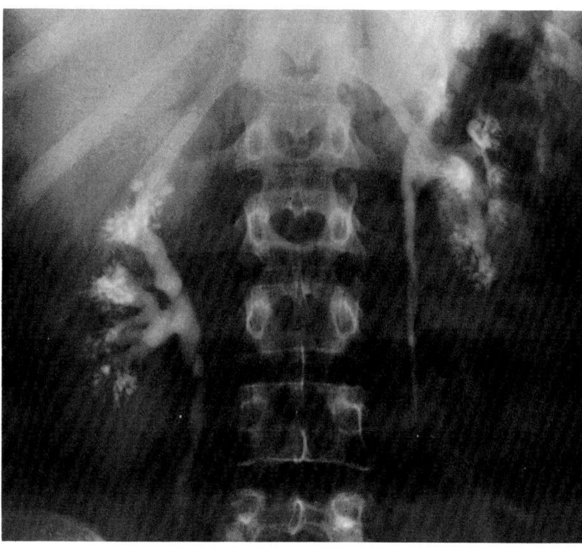
B

**Figure 37.14. Medullary sponge kidney.** (A) A plain abdominal radiograph demonstrates multiple small, smoothly rounded calculi occurring in clusters and a fanlike arrangement in the papillary tips of multiple renal pyramids. (B) Excretory urography confirms the location of the calculi.

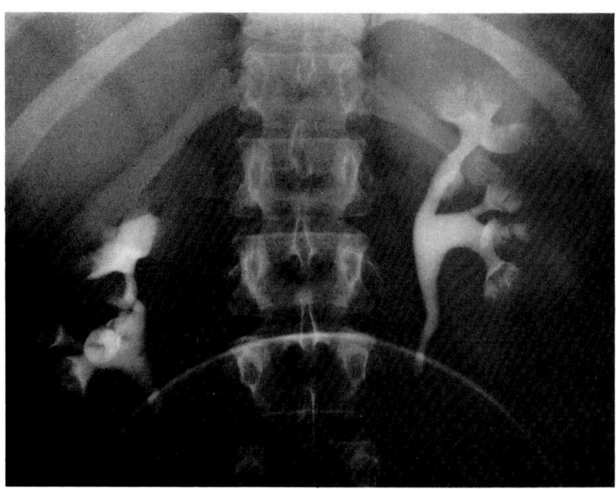

**Figure 37.15. Medullary sponge kidney.** On an excretory urogram, the ectatic tubules appear as fine linear striations of contrast producing the characteristic brush border pattern.

However, in some patients the medullary cysts are large enough to be demonstrated as sharply defined lucent defects on nephrotomograms.

Because of its more intense stain, the arteriographic nephrogram may be superior to the urographic nephrogram for visualizing the medullary cysts. The cortex appears uniformly thin, and the corticomedullary junction is poorly defined.[1,26,32]

## Congenital Multicystic Kidney Disease

Multicystic kidney disease is a nonhereditary congenital dysplasia, usually unilateral and asymptomatic, in which the kidney is composed almost entirely of large thin-walled cysts with only a little solid renal tissue. The disorder is the most frequent cause of an abdominal mass in the newborn and must be differentiated from hydronephrosis or a neoplasm.

Excretory urography demonstrates the complete functional failure of the involved kidney. However, the walls of the individual cysts are vascularized and may become slightly opaque during the total body opacification phase of urography. This permits the visualization of the cysts as round lucent areas separated from each other by slightly opacified septa (cluster-of-grapes sign). Retrograde pyelography demonstrates an atretic ureter that often has a blind proximal end. Arteriography typically shows an absent or severely atretic renal artery on the involved side. Ultrasound is of value in demonstrating the disorganized pattern of cysts and the lack of renal parenchyma and reniform contour in multicystic kidney disease, in contrast to the precise organization of symmetrically positioned fluid-filled spaces in hydronephrosis due to congenital ureteropelvic junction obstruction.

Because congenital multicystic kidney disease is usually asymptomatic, the unilateral, multilobulated, nonfunctioning mass in the area normally occupied by the kidney may not be detected until adulthood. Plain abdominal radiographs may demonstrate thin curvilinear calcification outlining the cyst walls. At excretory urography, the contralateral kidney is usually enlarged due to compensatory hypertrophy.[1,9,27–30]

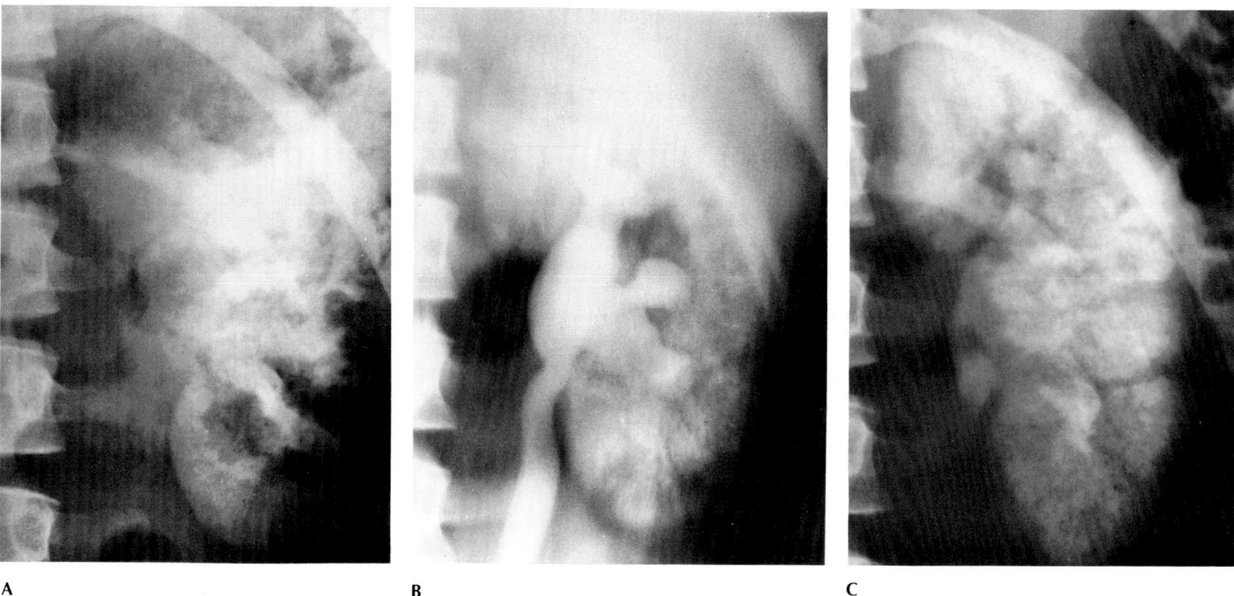


**Figure 37.16. Medullary cystic disease.** (A) A 4-minute radiograph from an excretory urogram shows a normal-sized kidney with a smooth margin, delayed contrast excretion, and a poor but homogeneous nephrogram over the entire kidney. (B) A tomogram taken at 10 minutes shows opacification of the collecting system with mild blunting of the calyces. The nephrogram is composed of numerous streaky contrast collections radiating from the calyces to the periphery. (C) A radiograph taken at 120 minutes shows a higher-density nephrogram confined to the medulla. This is probably caused by contrast accumulation in dilated tubules. The cortex and the columns of Bertin are recognizable as radiolucent areas. (From Burgener and Spataro,[32] with permission from the publisher.)

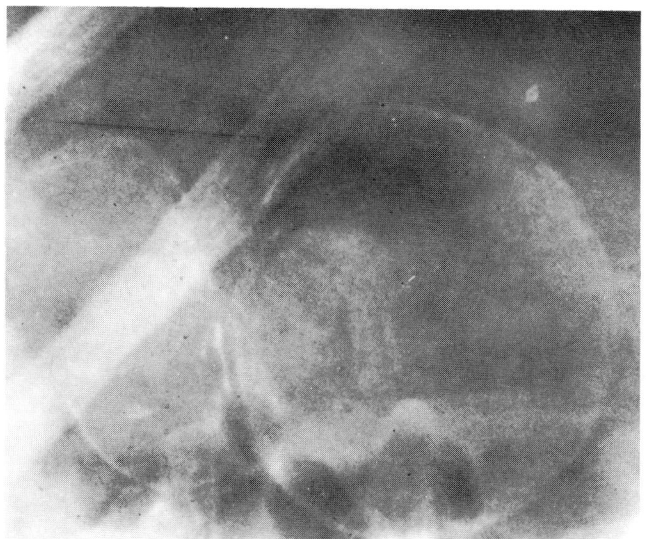

**Figure 37.17. Congenital multicystic kidney.** A plain film of the abdomen demonstrates multiple thin curvilinear calcifications outlining the cysts. (From Kyaw,[29] with permission from the publisher.)

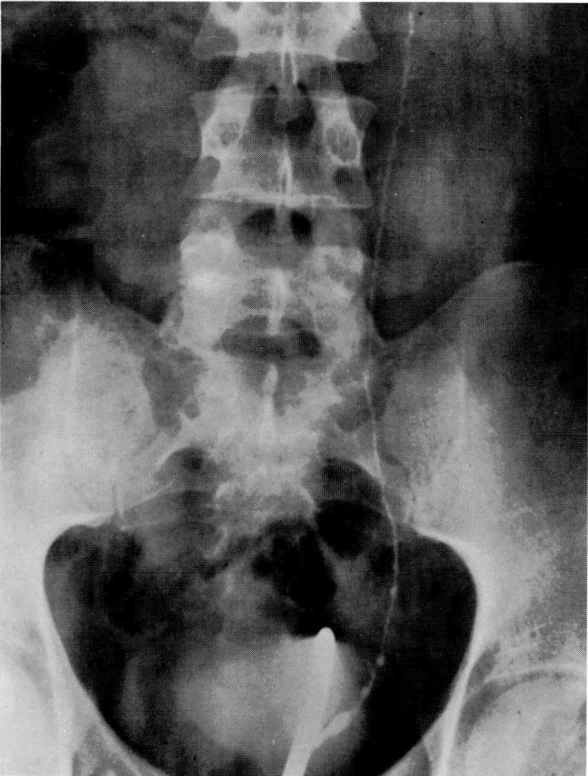

**Figure 37.18. Congenital multicystic kidney.** A retrograde pyelogram demonstrates atresia of the proximal ureter. (From Kyaw,[29] with permission from the publisher.)

## REFERENCES


1. Davidson AJ: *Radiologic Diagnosis of Renal Parenchymal Disease*, Philadelphia, WB Saunders, 1977.
2. Mena E, Bookstein JJ, Gikas PW: Angiographic diagnosis of renal parenchymal disease. Chronic glomerulonephritis, chronic pyelonephritis, and arteriolonephrosclerosis. *Radiology* 108:523–532, 1973.
3. Boyd RM, Warren L, Garrow DG: Renal size in various nephropathies. *AJR* 119:723–726, 1973.
4. Esposito WJ: Specific nephrocalcinosis of chronic glomerulonephritis. *AJR* 101:688–691, 1967.
5. Witten DM, Myers GH, Utz BC: *Clinical Urography*, Philadelphia, WB Saunders, 1977.
6. Robinson T, Rabinowitz JG: The nephrotic syndrome. *Radiol Clin North Am* 10:495–511, 1972.
7. Chuang VP, Reuter SR: Angiographic features of Alport's syndrome. Hereditary nephritis. *AJR* 121:539–543, 1974.
8. Perkoff GT: The hereditary renal diseases. *N Engl J Med* 277:79–85, 1967.
9. Friedland GW, Filly R, Goris ML, Gross D, Kempson RL, Korobkin M, Thurber BD, and Walter J (eds): *Uroradiology: An Integrated Approach*. New York, Churchill Livingstone, 1983.
10. Colodny AH, Lebowitz RL: A plea for grading vesicoureteral reflux. *Urology* 4:357–358, 1974.
11. Conway JJ, King LR, Bellman AB, et al: Detection of vesicoureteral reflux with radionuclide cystography. *AJR* 115:720–727, 1972.
12. Friedland GW, Forsberg L: Striation of the renal pelvis in children. *Clin Radiol* 23:58–60, 1972.
13. Aron BS, Schlesinger A: Complications of radiation therapy: The genitourinary tract. *Semin Roentgenol* 9:65–74, 1974.
14. Madrazo A, Schwarz G, Churg J: Radiation nephritis: A review. *J Urol* 114:822–827, 1975.
15. Gross D: The effect of complications of surgical, radiation, and medical therapy on the urinary tract, in Friedland GW, Filly R, Goris ML, Gross D, Kempson R, Korobkin M, Thurber BD, and Walter J (eds): *Uroradiology: An Integrated Approach*. New York, Churchill Livingstone, 1983.
16. Halpern M, Dalrymple G, Young J: The nephrogram in polycystic disease: An important radiographic sign. *J Urol* 103:21–23, 1970.
17. Bosniak MA, Ambos MA: Polycystic kidney disease. *Semin Roentgenol* 10:133–143, 1975.
18. Lalli AF, Poirier VC: Urographic analysis of the development of polycystic kidney disease. *AJR* 119:705–711, 1973.
19. Wolf B, Rosenfield AT, Taylor KJW, et al: Pre-symptomatic diagnosis of adult onset of polycystic kidney disease by ultrasonography. *Clin Genet* 14:1–7, 1978.
20. Lufkin EG, Alfrey AC, Trucksess ME, et al: Polycystic kidney disease: Earlier diagnosis using ultrasound. *Urology* 4:5–9, 1974.
21. Cornell SH: Angiography in polycystic disease of the kidneys. *J Urol* 38:505–510, 1970.
22. Gwinn JL, Landing BH: Cystic diseases of the kidneys in infants and children. *Radiol Clin North Am* 6:191–204, 1968.
23. Potter EL: *Normal and Abnormal Development of the Kidney*. Chicago, Year Book Medical Publishers, 1972.
24. Palubinskas AJ: Medullary sponge kidney. *Radiology* 76:911–918, 1961.
25. Lalli AF: Medullary sponge kidney disease. *Radiology* 92:92–96, 1969.
26. Mena E, Bookstein JJ, McDonald FD, et al: Angiographic findings in renal medullary cystic disease. *Radiology* 110:277–281, 1974.
27. Griscom NT: The roentgenology of neonatal abdominal masses. *AJR* 93:447–463, 1965.
28. Lachman RS, Lindstrom RR, Hirose FM: The "septation sign" in multicystic dysplastic kidney. *Pediatr Radiol* 3:117–119, 1975.
29. Kyaw M: The radiological diagnosis of congenital multicystic kidney. "Radiological triad." *Clin Radiol* 25:45–62, 1974.
30. Becker JA, Robinson T: Congenital multicystic kidney in the adult. *J Can Assoc Radiol* 21:165–169, 1970.
31. Baum S, Vincent NR, Lyons KP, et al: *Atlas of Nuclear Medicine Imaging*. New York, Appleton-Century-Crofts, 1981.
32. Burgener FA, Spataro RF: Early medullary cystic disease. *Radiology* 130:321–322, 1979.

## Chapter Thirty-Eight

# Urinary Tract Infection, Pyelonephritis, and Related Conditions

## PYELONEPHRITIS

### Acute Pyelonephritis

In most patients with acute pyelonephritis, the excretory urogram is normal. Urographic abnormalities seen in acute pyelonephritis include generalized enlargement of the kidney on the symptomatic side, delayed calyceal opacification, and decreased density of the contrast material. Focal polar swelling and calyceal compression may reflect localized abscess formation. A characteristic finding is linear striations in the renal pelvis, which probably represent acute mucosal edema.

Renal ultrasound in patients with acute pyelonephritis demonstrates a normal or enlarged kidney with preservation of intraparenchymal anatomy. This modality is especially valuable for detecting acute focal bacterial nephritis (the most common form of acute pyelonephritis encountered in children with reflux nephropathy and infection), which appears as an area of decreased echogenicity simulating a tumor or an intrarenal abscess. Ultrasound also can exclude renal or perirenal abscess and can sometimes detect totally unexpected conditions such as pyonephrosis (acutely infected hydronephrotic collecting system).

Contrast-enhanced computed tomography (CT) in patients with acute pyelonephritis may demonstrate a striated

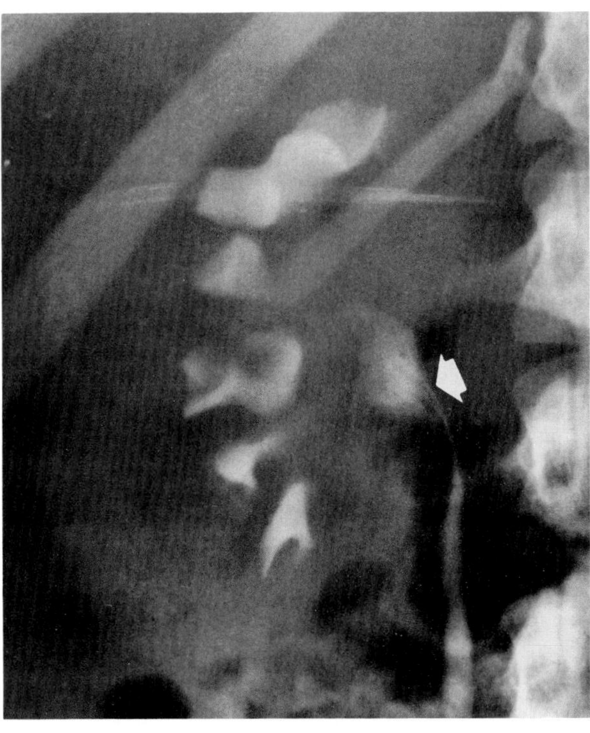

**Figure 38.1. Acute pyelonephritis.** Linear striations (arrow) in the renal pelvis represent acute mucosal edema.

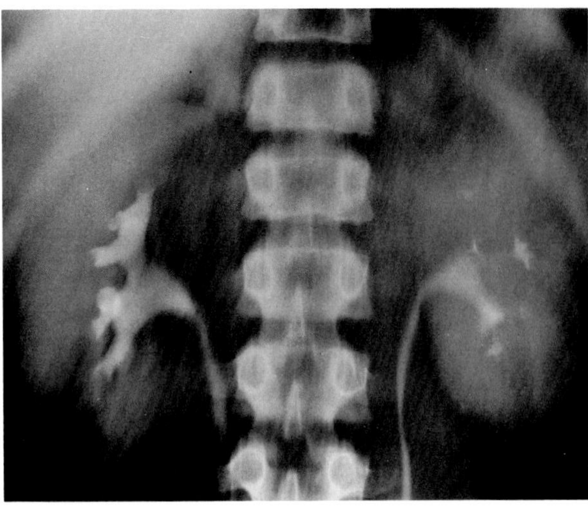

**Figure 38.2. Acute focal bacterial nephritis.** A nephrotomogram shows a relatively lucent area in the upper pole of the left kidney. The calyces draining the upper pole fill poorly. (From Friedland, Filly, Goris, et al,[3] with permission from the publisher.)

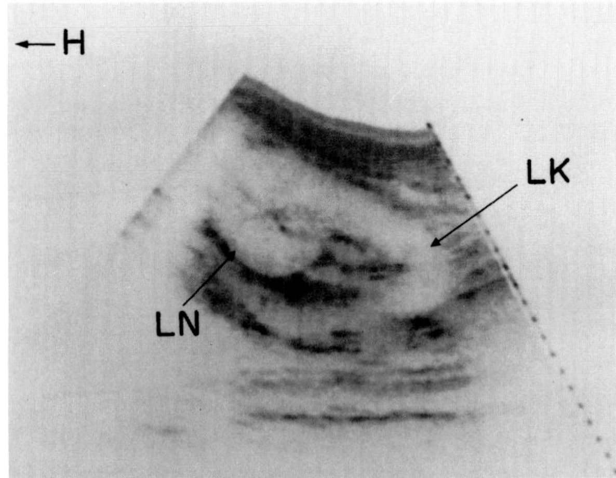


**Figure 38.3. Acute focal bacterial nephritis.** A prone parasagittal sonogram of the left kidney (LK) demonstrates acute focal bacterial nephritis (LN) as focal prominence of the renal parenchyma with poor definition of medullary pyramids in the upper pole. (From Friedland, Filly, Goris, et al,[3] with permission from the publisher.)

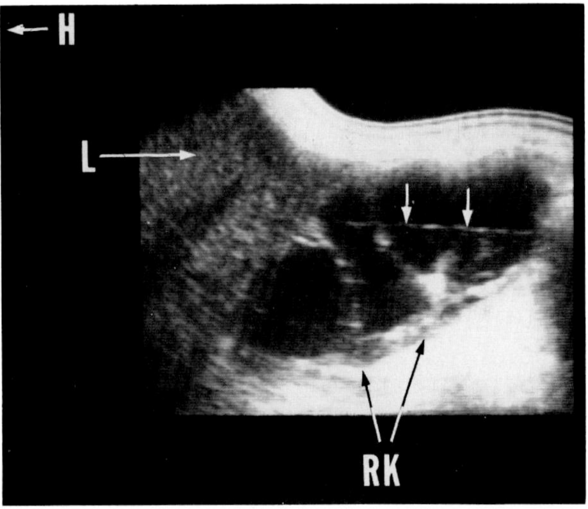


**Figure 38.4. Pyonephrosis.** A parasagittal sonogram of the right kidney (RK) demonstrates marked hydronephrosis and a characteristic fluid level (arrows). This fluid level indicates sediment within the renal collecting system and is a typical finding of pyonephrosis. L = liver; H = head. (From Friedland, Filly, Goris, et al,[3] with permission from the publisher.)

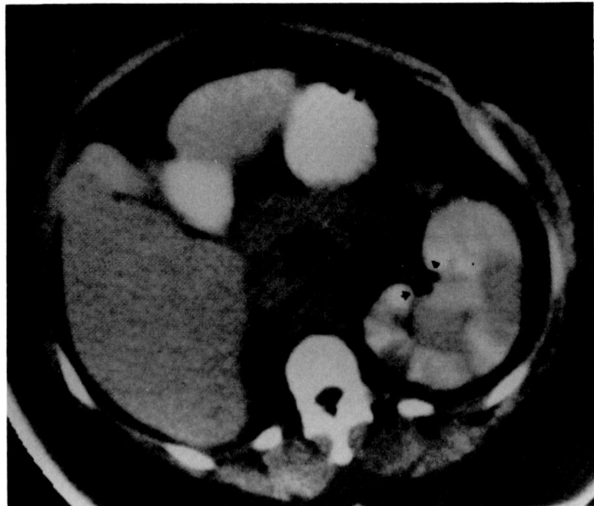

**Figure 38.5. Acute pyelonephritis.** A postcontrast CT scan demonstrates striated renal parenchyma and air in the left renal collecting system (arrows). (From Hoffman, Mindelzun, and Anderson,[8] with permission from the publisher.)

appearance of radially oriented zones of increased density within the affected kidney. Acute focal bacterial nephritis may appear as single or multiple, poorly marginated mass lesions.[1-8]

## Chronic Pyelonephritis

Patchy calyceal clubbing with overlying parenchymal scarring is the urographic hallmark of chronic pyelonephritis. Initially, there is blunting of the calyces, which then become rounded or clubbed. Fibrotic scarring causes a cortical depression overlying the dilated calyx. Progressive cortical atrophy and thinning may be so extensive that the tip of the blunted calyx appears to lie directly beneath the renal capsule. The urographic findings may be unilateral or bilateral and are often most pronounced at the poles. If calyceal changes are minimal, the overlying cortical depressions may simulate lobar infarctions or normal kidney lobulations. However, in chronic pye-

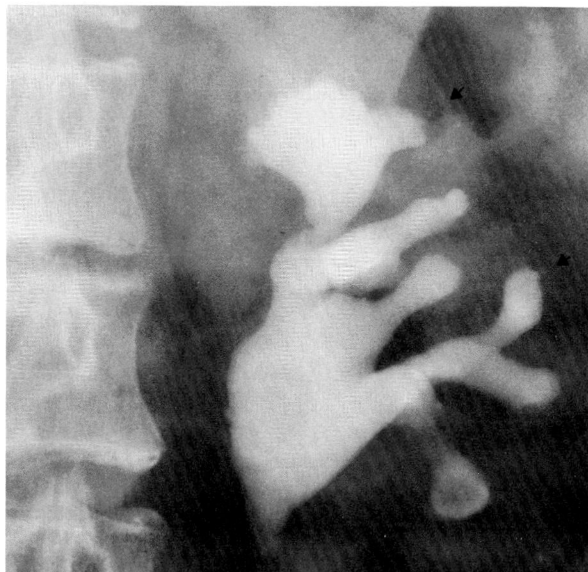

**Figure 38.6. Chronic pyelonephritis.** There is diffuse rounded clubbing of multiple calyces with atrophy and thinning of the overlying renal parenchyma (arrows indicate the outer margin of the kidney).

lonephritis the cortical depression lies directly over a calyx, rather than between calyces as in lobar infarctions or congenital lobulation. Chronic pyelonephritis may progress to end-stage renal disease with small, usually irregular, poorly functioning kidneys.

In children, chronic pyelonephritis inhibits the growth of all or a portion of the affected kidney. In the absence of an appropriate history, this appearance may be mistaken for a congenital anomaly.[1-4,9]

## Emphysematous Pyelonephritis

Emphysematous pyelonephritis is a severe form of acute parenchymal and perirenal infection with gas-forming bacteria that virtually only occurs in diabetics and causes an acute necrosis of the entire kidney. The presence of radiolucent gas shadows within and around the kidney is pathognomonic of emphysematous pyelonephritis, a surgical emergency that is lethal if treated medically.[10,11]

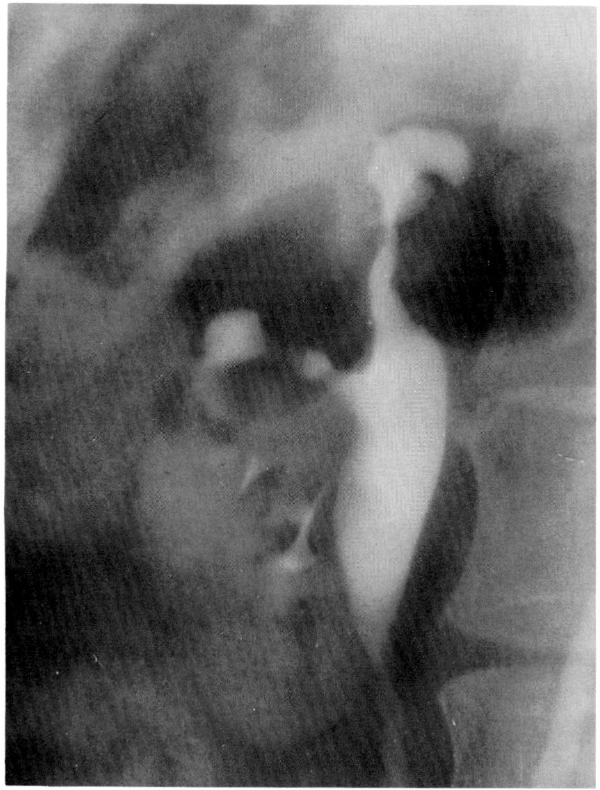

A

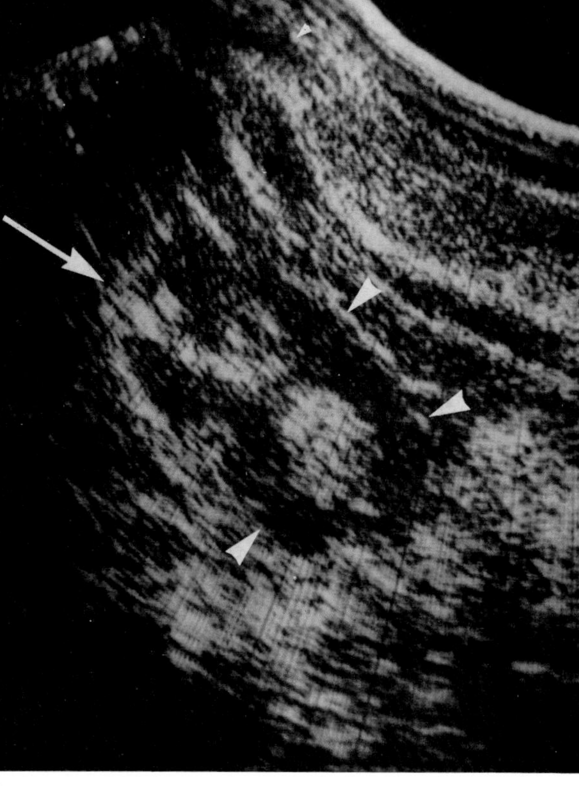

B

**Figure 38.7. Chronic pyelonephritis.** (A) An excretory urogram demonstrates focal areas of clubbing with a loss of renal parenchyma in the upper pole of the right kidney. (B) Ultrasound examination of the right kidney (arrowheads) with the patient prone shows a focal loss of renal parenchyma and extension of the calyces peripherally from the renal sinus to the renal margin. Note the associated focal area of increased echogenicity due to fibrosis (arrow) in the upper pole. (From Kay, Rosenfield, Taylor, et al,[9] with permission from the publisher.)

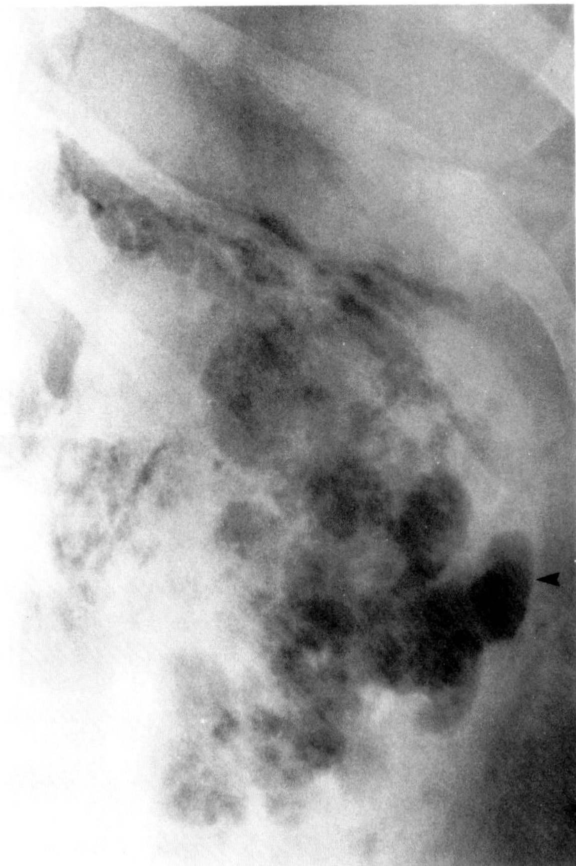

**Figure 38.8. Emphysematous pyelonephritis.** A plain film of the left upper quadrant of the abdomen demonstrates extensive perinephric and parenchymal gas. Except for four or five pockets of gas in the overlying splenic flexure of the colon, one of which is indicated by the arrow, the remainder of the gas is pathologic and indicates extensive tissue necrosis. (From Friedland, Filly, Goris, et al,[3] with permission from the publisher.)

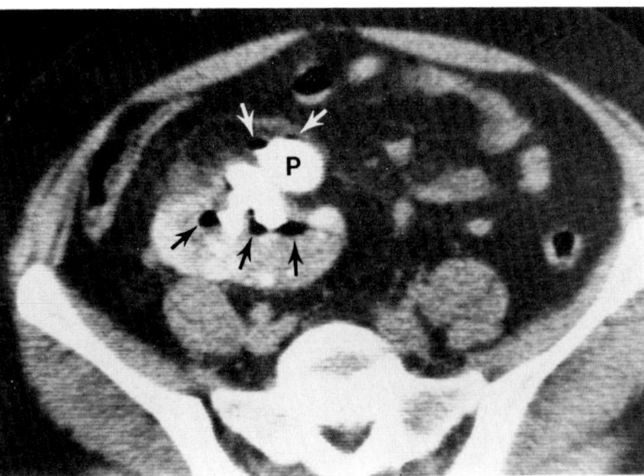


**Figure 38.9. Emphysematous pyelonephritis.** A CT scan following the intravenous administration of contrast material shows inhomogeneous opacification of a pelvic kidney. Both contrast material and gas (arrows) are present in the collecting structures and renal pelvis (P). (From Friedland, Filly, Goris, et al,[3] with permission from the publisher.)

### Xanthogranulomatous Pyelonephritis

Xanthogranulomatous pyelonephritis refers to the unusual nodular replacement of renal parenchyma by large lipid-filled macrophages (foam cells) that may develop in chronically infected kidneys. The term *xanthogranulomatous* is derived from the yellow color of these lipid-filled masses, which may appear as multiple small nodules coalescing to form several large masses, as a single large granulomatous mass, or as diffuse replacement of the renal parenchyma. An obstructing calculus and a long history of urinary tract infection with gram-negative bacteria are almost always present.

The classic radiographic appearance of xanthogranulomatous pyelonephritis is an enlarged, nonfunctioning kidney that is associated with an obstructing radiopaque calculus at the ureteropelvic junction. Retrograde pyelography demonstrates pelvocaliectasis caused by the obstructing stone or, less frequently, an inflammatory stricture or tumor. In the tumefactive form, single or multiple irregular inflammatory masses distort the opacified collecting system or renal margins in a pattern simulating renal abscess or neoplastic disease.

Arteriography is nonspecific in patients with xanthogranulomatous pyelonephritis. Stretching of interlobar and arcuate vessels may reflect underlying hydronephrosis or inflammatory masses. The demonstration of fine, wispy neovascularity, especially in the region of the renal pelvis, may suggest an underlying transitional-cell malignant neoplasm.

Ultrasound shows the involved kidney to be markedly enlarged, with a large central echogenic area and an increased parenchymal anechoic pattern. CT may be of value in showing the fatty consistency of the xanthogranulomatous mass.[12–16]

## PAPILLARY NECROSIS

Papillary necrosis is characterized by ischemic coagulative necrosis involving a varying amount of the medullary papillae and pyramids. It is most often seen in patients with diabetes, pyelonephritis, urinary tract infection or obstruction, sickle cell disease, or phenacetin abuse. Radiographic signs arise when a cleavage plane develops in a zone of ischemia and communicates with the calyx. Though often overlooked, on excretory urography the fissures appear as faint streaks of contrast material oriented parallel to the long axis of the papilla. They usually extend from the fornix, though they can also arise in the papillary tips.

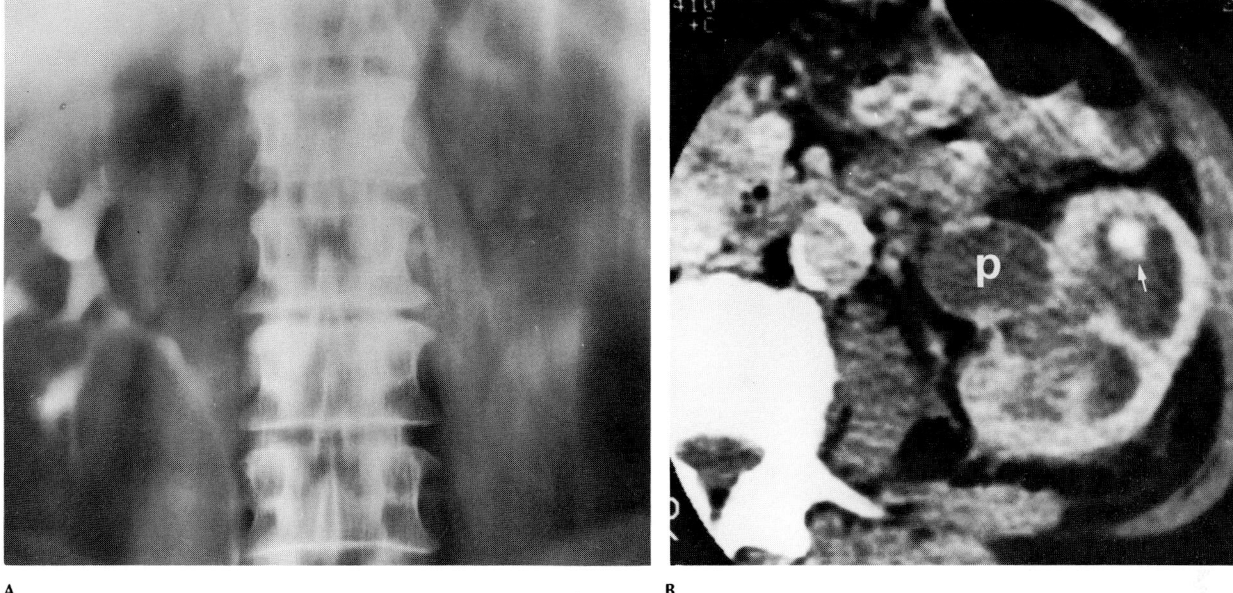


**Figure 38.10. Xanthogranulomatous pyelonephritis.** (A) An excretory urogram demonstrates nonfunction of the left kidney. (B) A CT scan shows prolonged opacification of the left renal cortex. The renal pelvis (p) is filled with pus, as are the intrarenal collecting structures. The high-density focus (arrow) is a renal calculus. (From Moss,[16] with permission from the publisher.)

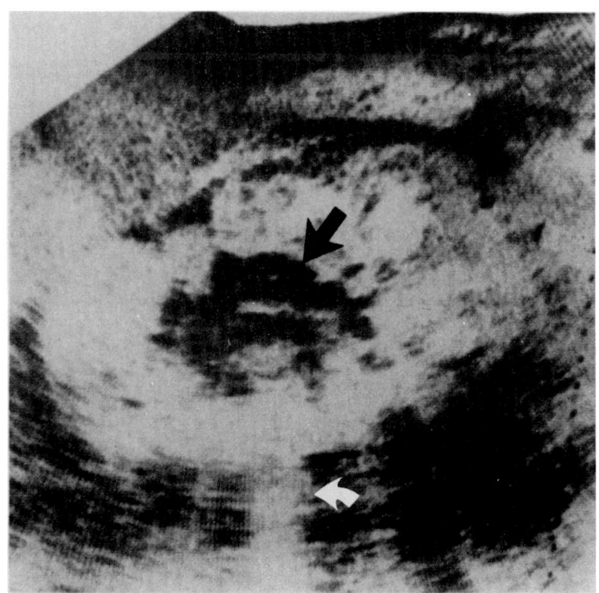

**Figure 38.11. Xanthogranulomatous pyelonephritis.** A supine longitudinal sonogram shows a diffusely enlarged right kidney with a central echogenic focus (black arrow), acoustic shadowing (white arrow), and scattered loculated cystic areas containing purulent material. (From Van Kirk, Go, and Wedel,[15] with permission from the publisher.)

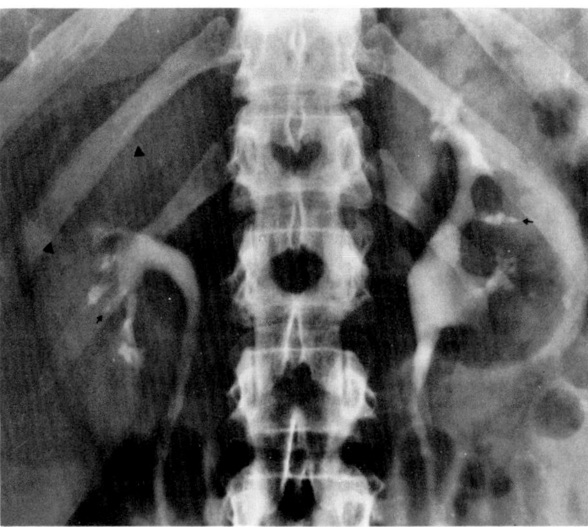

**Figure 38.12. Papillary necrosis.** There are striking central cavitations within several papillae (arrows). An incidental finding is a large upper pole cyst in the right kidney (arrowheads).

Papillary necrosis is best detected when there is cavitation of the central portion of the papilla or complete sloughing of the papillary tip. Cavitation may be central or eccentric; its long axis parallels that of the papilla. The shape of the cavity varies from long and thin to short and bulbous; the margins may be sharp or irregular. When a piece of medullary tissue has been completely separated from the rest of the renal parenchyma, a ring of contrast material is seen surrounding the triangular lucent filling defect that represents the sloughed necrotic tissue. The calyx that remains has a round, saccular, or club-shaped configuration. The sloughed papilla may stay in place and become calcified (characteristically ring-shaped with a lucent center), or it may pass down

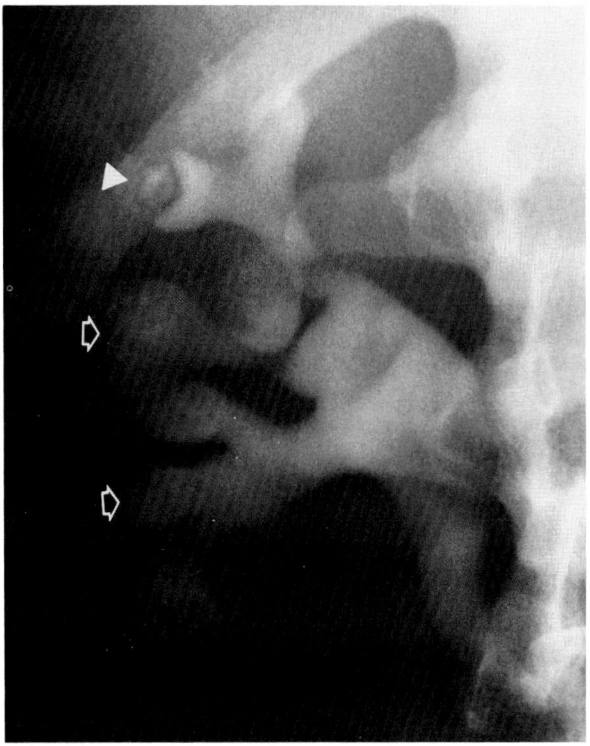

**Figure 38.13. Papillary necrosis.** Total sloughing of many of the papillae has resulted in smooth, rounded calyceal cavities (open arrows). Some central cavitation persists (arrowhead).

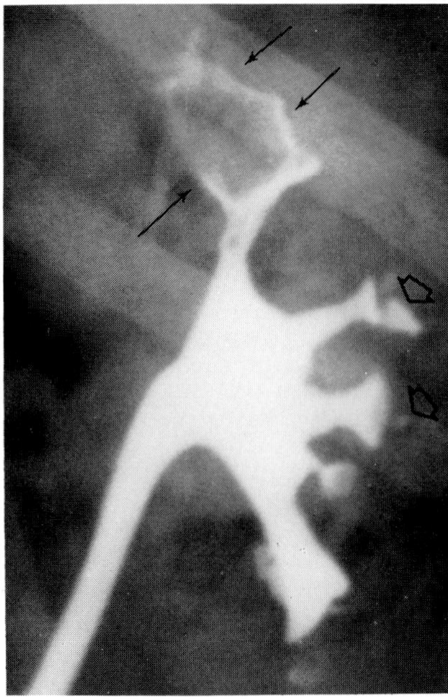

**Figure 38.14. Papillary necrosis.** A ring of contrast (long arrows) surrounds a triangular lucent filling defect, which represents an almost complete papilla that has been separated from the rest of the renal parenchyma. The short arrows point to less severe extension of contrast material from the calyces into the papilla. (From Ney and Friedenberg,[21] with permission from the publisher.)

the ureter, where it may simulate a stone and even cause obstruction.

About half of the patients with papillary necrosis have shrunken kidneys and uniform narrowing of the renal parenchyma. The outer margin is often wavy because of renal scarring located between the calyces.[1,3,17–21]

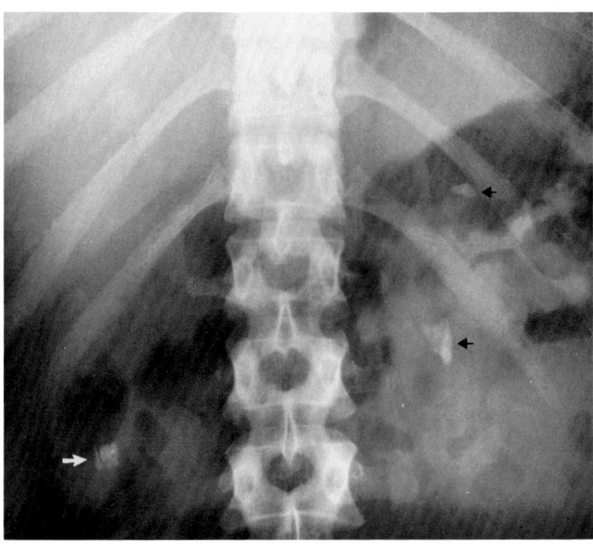

**Figure 38.15. Papillary necrosis.** Typical ring-shaped calcifications (arrows) are associated with sloughing of the entire papillary tip. (From Davidson,[1] with permission from the publisher.)

## REFERENCES


1. Davidson AJ: *Radiologic Diagnosis of Renal Parenchymal Disease*, Philadelphia, WB Saunders, 1977.
2. Witten DM, Myers GH, Utz BC: *Clinical Urography*, Philadelphia, WB Saunders, 1977.
3. Friedland GW, Filly R, Goris ML, Gross D, Kempson RL, Korobkin M, Thurber BD, and Walter J (eds): *Uroradiology: An Integrated Approach*. New York, Churchill Livingstone, 1983.
4. Hodson CJ: Radiology in pyelonephritis. *Curr Probl Radiol* 2:1–32, 1972.
5. Winberg J, Claesson J, Jacobsson B, et al: Renal growth after acute pyelonephritis in childhood: An epidemiological approach, in Hodson CJ, Smith TK (eds): *Reflux Nephropathy*. New York, Masson, 1979.
6. Silver TM, Kass EJ, Thornbury JR, et al: The radiological spectrum of acute pyelonephritis in adults and adolescents. *Radiology* 118:65–71, 1976.
7. Gold RP, McClennan BL, Rottenberg RR: CT appearance of acute inflammatory disease of the renal interstitium. *AJR* 141:343–349, 1983.
8. Hoffman EP, Mindelzun RE, Anderson RU: Computed tomography in acute pyelonephritis associated with diabetes. *Radiology* 134:691–695, 1980.
9. Kay CJ, Rosenfield AT, Taylor KJW, et al: Ultrasonic characteristics of chronic atrophic pyelonephritis. *AJR* 132:47–53, 1979.

10. Dunn SR, DeWolf WC, Gonzales R: Emphysematous pyelonephritis: Report of three cases treated by nephrectomy. *J Urol* 114:343–347, 1975.
11. Freiha FS, Messing EM, Gross DM: Emphysematous pyelonephritis. *J. Contin Ed Urol* 18:9–12, 1979.
12. Palubinskas AJ: Xanthogranulomatous pyelonephritis. *Semin Roentgenol* 6:331–337, 1971.
13. Levin DC, Gordon D, Kinkhabwala M, et al: Reticular neovascularity in malignant and inflammatory renal masses. *Radiology* 120:61–68, 1976.
14. Beachley MC, Ranninger K, Roth FJ: Xanthogranulomatous pyelonephritis. *AJR* 121:500–507, 1974.
15. Van Kirk OC, Go RT, Wedel VJ: Sonographic features of xanthogranulomatous pyelonephritis. *AJR* 134:1035–1039, 1980.
16. Moss AA: Computed tomography of the kidneys, in Moss AA, Gamsu G, Genant HK (eds): *Computed Tomography of the Body*. Philadelphia, WB Saunders, 1983.
17. Lalli AS: Renal papillary necrosis. *AJR* 114:741–745, 1972.
18. Hare WSC, Poynter JD: The radiology of renal papillary necrosis as seen in analgesic nephropathy. *Clin Radiol* 25:423–443, 1974.
19. Lindvall N: The radiologic changes of renal papillary necrosis. *Kidney Int* 13:93–104, 1978.
20. Eckert DE, Jonutis AJ, Davidson AJ: The incidence and manifestations of urographic papillary abnormalities in patients with S hemoglobinopathies. *Radiology* 113:59–63, 1974.
21. Ney C, Friedenberg RM: *Radiographic Atlas of the Genitourinary System*. Philadelphia, JB Lippincott, 1981.

## Chapter Thirty-Nine
# Vascular Injury to the Kidney

### ACUTE RENAL ARTERY OCCLUSION

Acute renal artery occlusion is most commonly caused by an embolism from the heart in patients with mitral stenosis and atrial fibrillation, infective endocarditis, or mural thrombi overlying a myocardial infarct. Blunt trauma to the abdomen or back is a less frequent cause. Acute renal artery occlusion results in coagulation necrosis in the region supplied by the obstructed artery; if a major intrarenal arterial branch is involved, the size of the resulting wedge-shaped infarct varies with the level of the occlusion.

Excretory urography in the patient with occlusion of the main renal artery characteristically demonstrates a nonfunctioning kidney on the affected side. A peripheral rim of opacified cortex may be observed during the nephrogram phase, probably reflecting viable renal cortex perfused by perforating collateral vessels from the renal capsule. Occlusion of a branch of the renal artery may produce segmental infarction, which leads to either complete nonvisualization of the kidney or a local failure of calyceal filling with a triangular nephrographic defect whose base is in the subcapsular region. Retrograde studies demonstrate a normal collecting system which, in combination with a nonfunctioning kidney, is almost pathognomonic of complete arterial occlusion. Although a similar pattern can develop with severe radiation nephritis, the clinical history should permit differentiation between these two entities.

Arteriography is the definitive procedure for demonstrating partial or complete occlusion of the main renal artery or its major branches. Complete occlusion of a segmental artery produces a paucity of vessels and a diminished nephrogram in the involved area. The embolus itself may be identified as an intra-arterial filling defect. A midstream aortogram is necessary before selective renal arteriography to prevent a nephrogram defect due to an accessory renal artery from being mistakenly attributed to segmental arterial occlusion. It is also important to study both renal arteries, since embolization frequently involves the renal vasculature bilaterally.

Following total renal infarction, the kidney usually remains nonfunctioning and decreases in size. In segmental infarction, the kidney tends to regain function, though there is residual scarring that produces an irregular renal contour with cortical depressions situated between calyces rather than directly over the calyces as in chronic pyelonephritis.

Computed tomography (CT) has recently been used for the prompt and noninvasive diagnosis of renal infarction. Infarcted tissue characteristically appears as an area of low attenuation, which is surrounded by a higher-attenuation subcapsular rim on contrast-enhanced scans. Although a similar subcapsular rim may be seen in renal

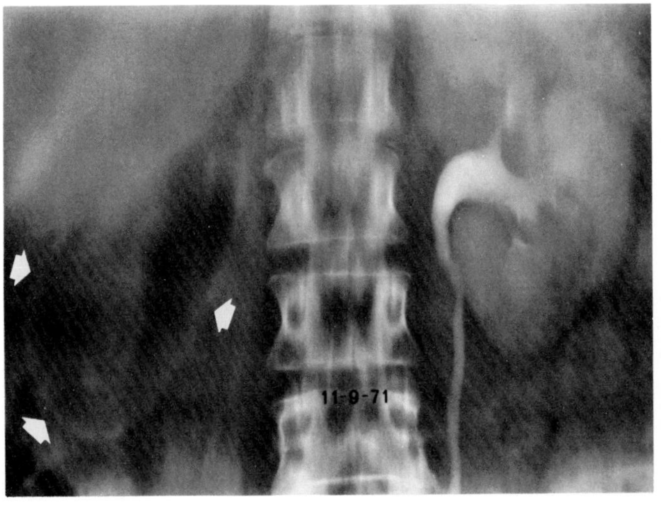


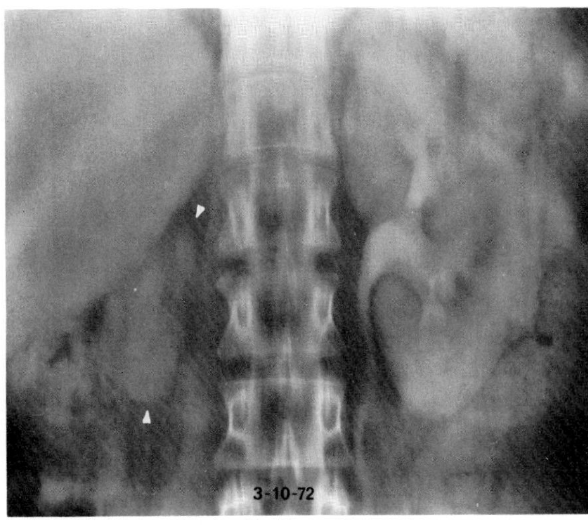


A B

**Figure 39.1. Acute renal artery occlusion.** (A) An initial nephrotomogram demonstrates a thin cortical rim surrounding the right kidney (arrows), reflecting viable renal cortex perfused by perforating collateral vessels from the renal capsule. (B) Four months later a repeat nephrotomogram shows a marked decrease in the size of the atrophic right kidney (arrows). (From Paul and Stephenson,[2] with permission from the publisher.)

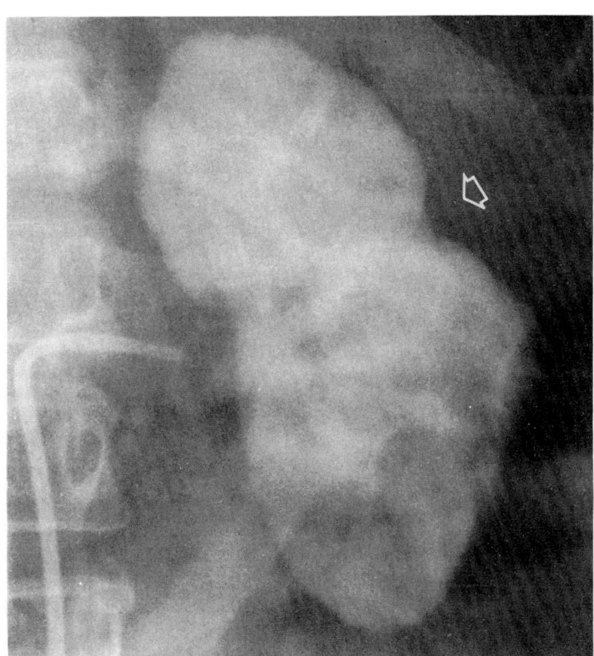

**Figure 39.2. Segmental renal infarction.** A film from the nephrogram phase of a selective arteriogram demonstrates a typical peripheral triangular defect with its base in the subcapsular region (arrow).

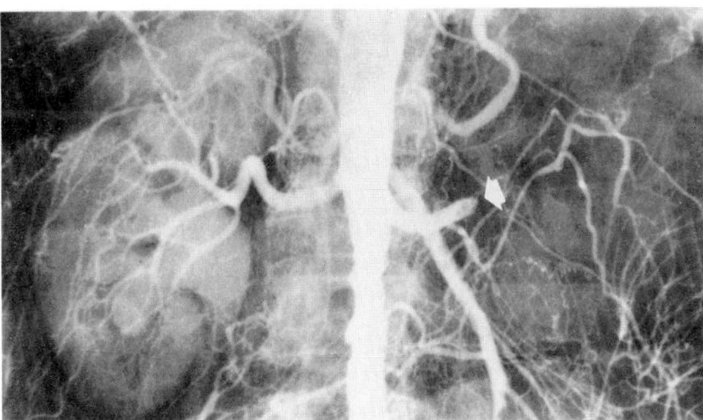

**Figure 39.3. Acute renal artery aneurysm.** There is abrupt termination of the contrast column in the left renal artery (arrow). The irregular contour of the infrarenal aorta represents arteriosclerotic disease. Hydronephrosis of the right kidney was due to obstruction of the right ureter by carcinoma of the endometrium.

vein thrombosis and acute tubular necrosis, the kidney is generally enlarged with renal vein thrombosis while normal or small with renal infarction, and the bilaterality of acute tubular necrosis differs from the predominantly unilateral distribution of renal infarction.[1-7]

## RENAL ARTERY STENOSIS

This is discussed in the section "Renovascular Hypertension" in Chapter 25.

## ARTERIOLAR NEPHROSCLEROSIS

Arteriolar nephrosclerosis refers to the intimal thickening of medium- and small-sized arteries and arterioles in the kidney that presumably is responsible for most of the loss of renal function that accompanies normal aging. In younger persons, arteriolar nephrosclerosis is primarily related to a proliferative endarteritis due to acceler-

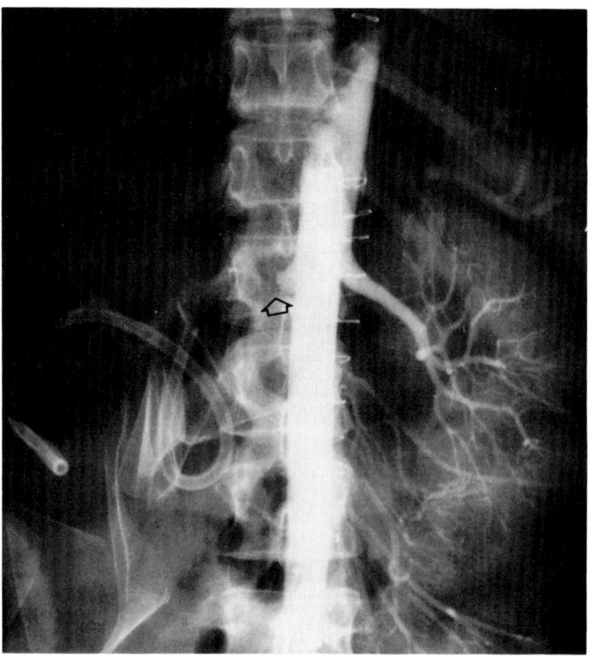

A

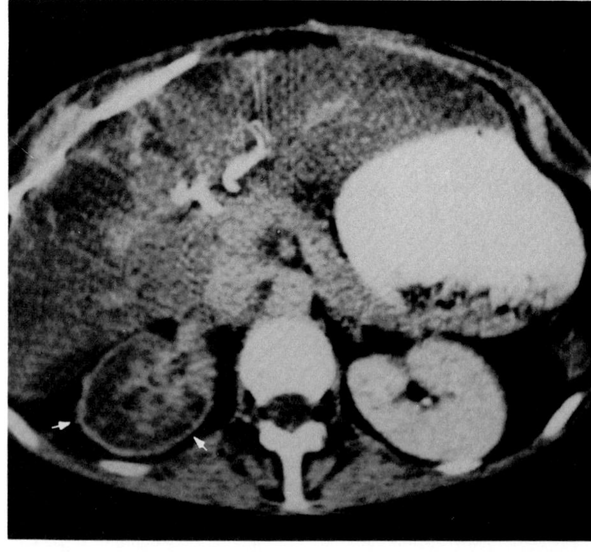

B

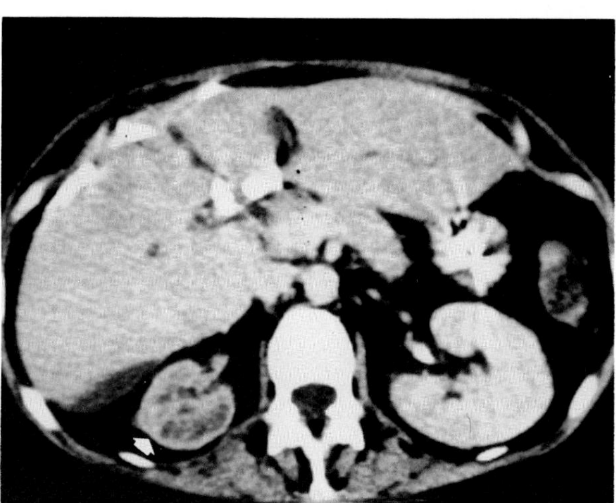

C

**Figure 39.4. Traumatic renal artery occlusion.** (A) An aortogram demonstrates avulsion of the right renal artery (arrow) and an absent right nephrogram. (B) A CT scan 3 days later, after the administration of both oral and intravenous contrast material, demonstrates the infarcted tissue as an area of low attenuation surrounded by a higher-attenuation subcapsular rim (arrows). (C) One month later a postcontrast scan shows a small, nonfunctioning right kidney with a persistent cortical rim (arrow). Note the hypertrophied left kidney. (From Glazer, Francis, Brady, et al,[6] with permission from the publisher.)

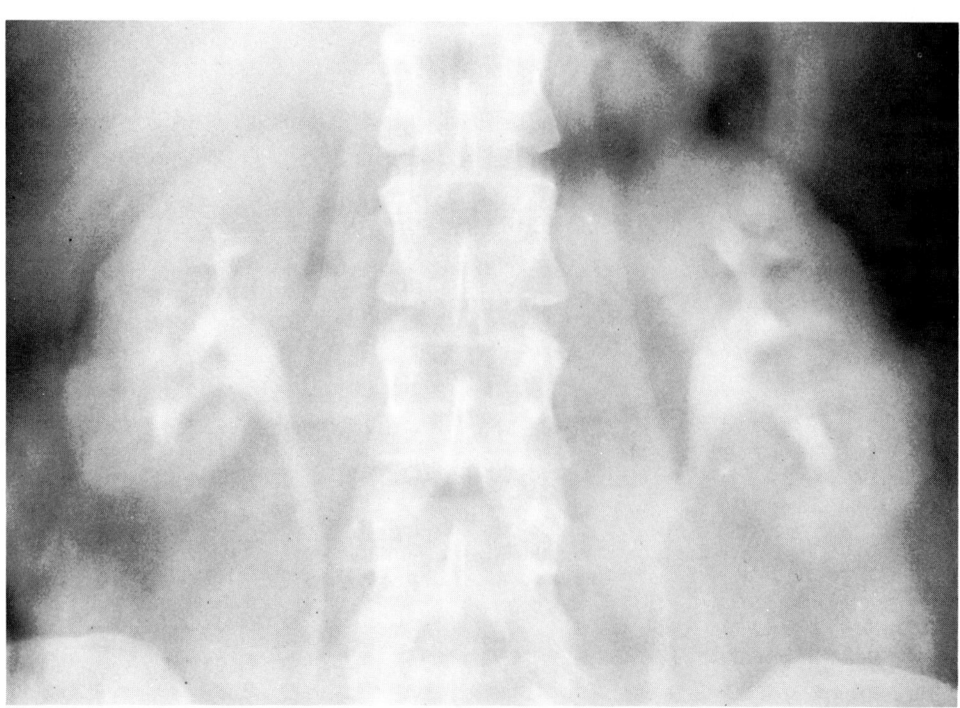

**Figure 39.5. Benign nephrosclerosis.** An excretory urogram shows bilaterally small kidneys with several shallow infarct scars. The pelvocalyceal systems and renal opacification are normal. (From Davidson,[14] with permission from the publisher.)

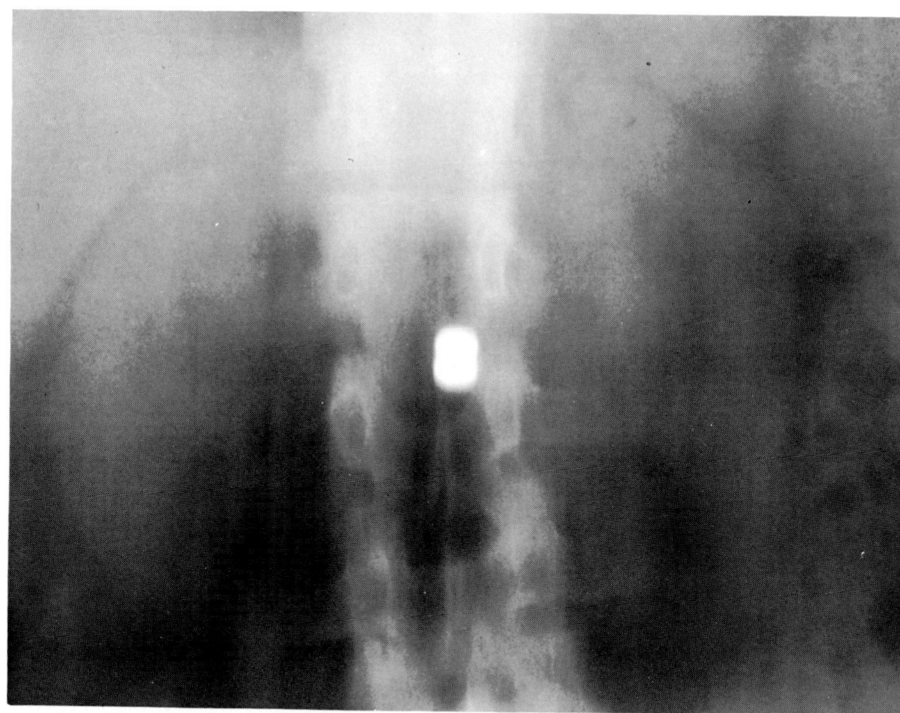

**Figure 39.6. Malignant nephrosclerosis.** A nephrotomogram obtained 5 minutes after the injection of contrast material shows minimum opacification of small, smooth kidneys. (From Davidson,[14] with permission from the publisher.)

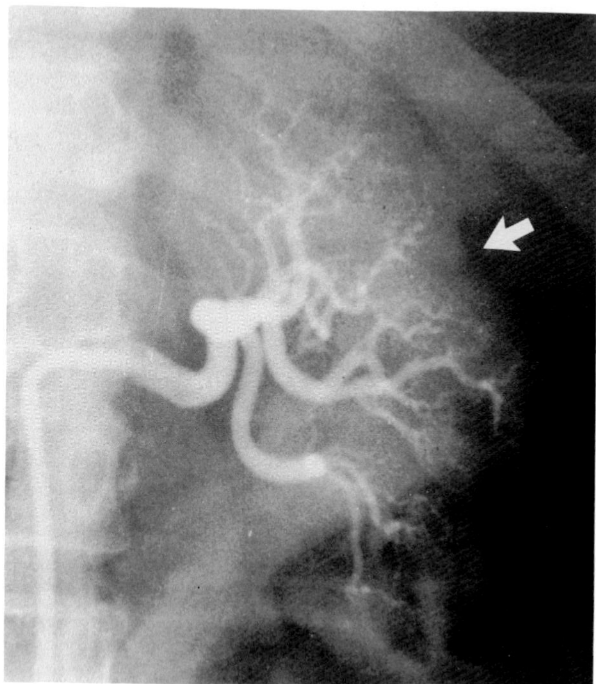

**Figure 39.7. Arteriolar nephrosclerosis.** A selective left renal arteriogram shows the characteristic tortuosity and rapid tapering of interlobar arteries and their branches. The irregular renal outline (arrow) reflects an infarct scar.

ated or malignant hypertension. The severity of nephrosclerosis is directly proportional to the degree of hypertension, and the condition may lead to the development of uremia with bilateral impairment of renal function.

In "benign" hypertension with moderate changes, the kidneys are symmetrically reduced in size, and their contours remain smooth except for the occasional development of random, shallow infarct scars. Kidney opacification and the pelvocalyceal structures are normal. Renal arteriography may initially be normal. As nephrosclerosis progresses, the interlobar arteries and their branches become tortuous (corkscrew appearance). The cortex becomes thinned and develops a finely granular pattern due to multiple small areas of cortical scarring.

In progressive malignant nephrosclerosis, the kidneys become further decreased in size. Reduced function produces a diminished nephrogram on excretory urograms. The main renal arteries become smaller in proportion to the diminished size of the kidneys. The interlobar arteries are tortuous and markedly reduced in size, and there is pruning of their branches. The intrarenal circulation time is prolonged, and there is no evidence of cortical opacification. Overall, the arteriographic appearance in malignant nephrosclerosis is often indistinguishable from that seen in chronic glomerulonephritis.[8–10]

## RENAL VEIN THROMBOSIS

Renal vein thrombosis occurs most frequently in children who are severely dehydrated. In the adult, thrombosis is most often a complication of another renal disease (chronic glomerulonephritis, amyloidosis, pyelonephritis), trauma, the extension of a thrombus from the inferior vena cava, and direct invasion or extrinsic pressure secondary to renal tumors. It is an uncommon cause of the nephrotic syndrome.

Renal vein thrombosis may be unilateral or bilateral. The clinical and radiographic findings are greatly influenced by the rapidity with which venous occlusion occurs. Sudden total occlusion causes striking kidney enlargement with minimal or no opacification on excretory urograms. If unresolved, this acute venous occlusion leads to the urographic appearance of a small, atrophic, nonfunctioning kidney.

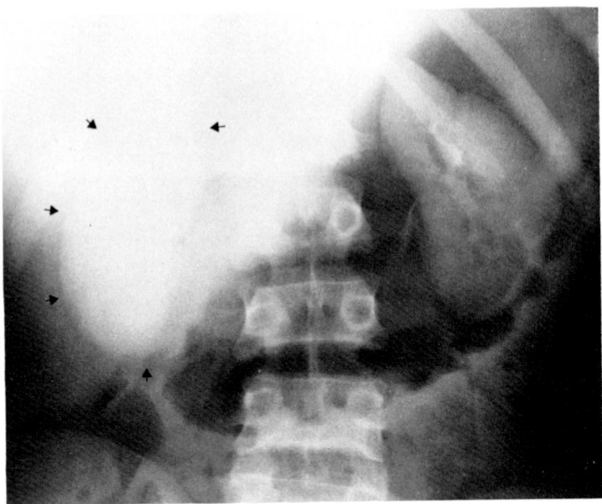

**Figure 39.8. Renal vein thrombosis.** A film of the right kidney taken 5 minutes after contrast material injection demonstrates a dense nephrogram (hyperconcentration) and the absence of calyceal filling (diminished excretion).

When venous occlusion is partial or accompanied by adequate collateral formation, the kidney is large and smooth but with some degree of contrast excretion. There is stretching and thinning of the collecting system due to surrounding interstitial edema. Enlargement of collateral pathways for renal venous outflow (gonadal and ureteric veins) produces characteristic notching of the upper ureter.

The confirmation of renal vein thrombosis requires venographic demonstration of vessel occlusion or a localized filling defect. Inferior vena cavography with a catheter placed well below the renal veins is often performed initially to exclude caval thrombosis with proximal extension before the renal veins are catheterized directly. Opacification of the renal veins may be enhanced if renal blood flow is temporarily decreased by the injection of epinephrine into the renal artery immediately before the venous study.

During the acute congestive phase of renal vein thrombosis, arteriography demonstrates stretching and narrowing of the intrarenal arteries. There is prolongation of the arterial phase and diminished or absent venous opacification.[8,11–16,20]

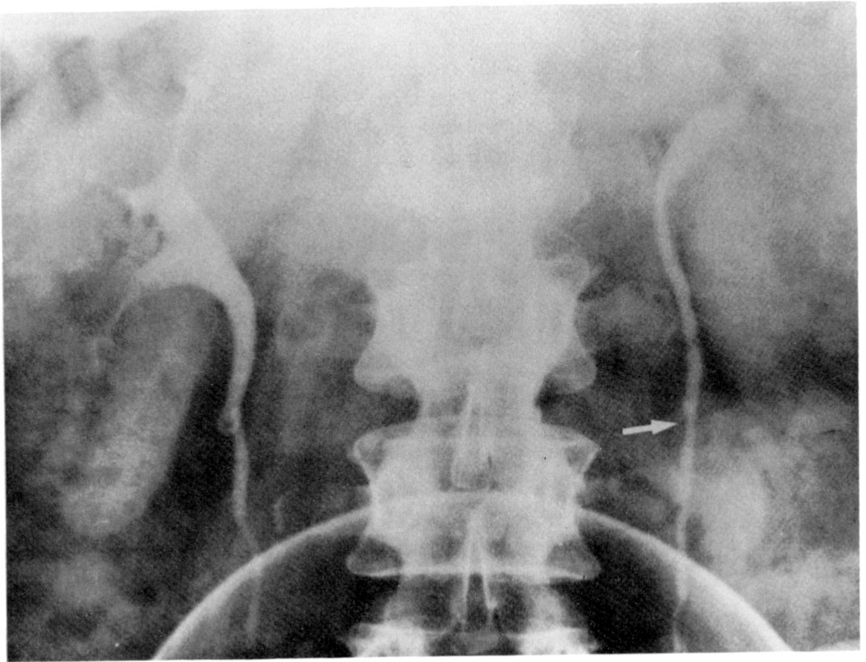

A

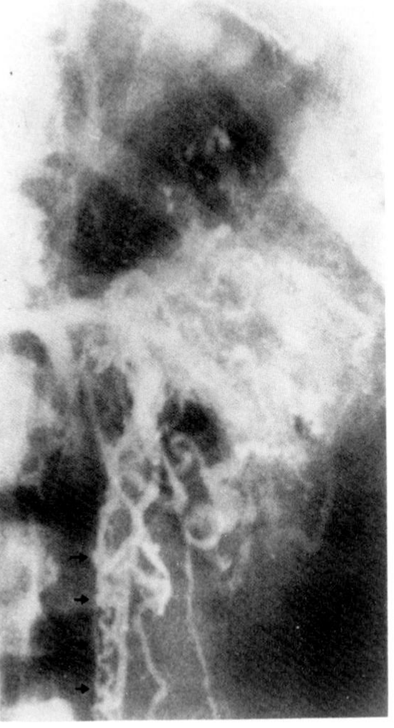

B

**Figure 39.9. Renal vein thrombosis.** (A) An excretory urogram shows characteristic notching (arrow) of the upper ureter. There is enlargement of the left kidney with poor calyceal function due to compression from parenchymal engorgement. (B) In another patient with renal vein thrombosis and ureteral notching, a venogram demonstrates exuberant periureteral collaterals (arrows). (B, from Bradley, Jacobs, Trew, et al,[16] with permission from the publisher.)

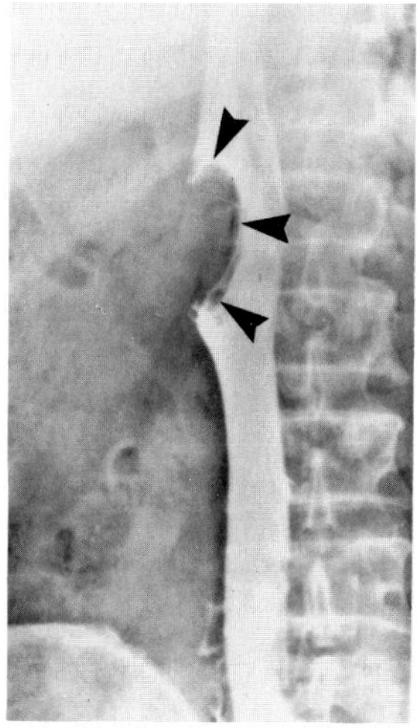

**Figure 39.10. Renal vein thrombosis.** An inferior vena cavogram shows a large mass protruding into the lumen of the inferior vena cava (arrowheads). The filling defect represents a thrombus, which extended into the inferior vena cava from the right renal vein. (From Friedland, Filly, Goris, et al,[19] with permission from the publisher.)

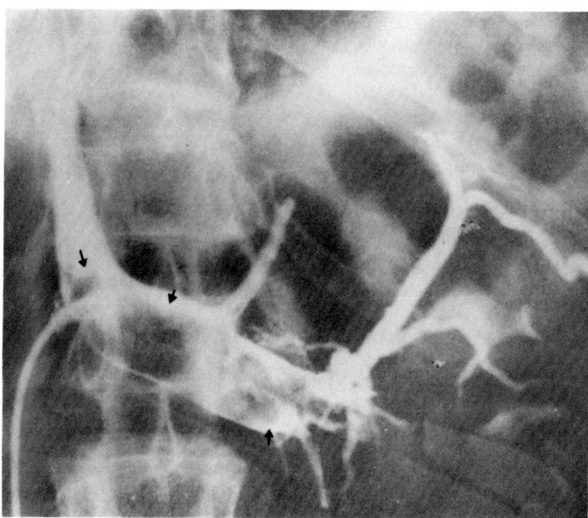

**Figure 39.11. Renal vein thrombosis.** A renal venogram demonstrates a large filling defect in the left renal vein that extends into the vena cava. (From Cohn, Lee, Hopper, et al,[20] with permission from the publisher.)

## POLYARTERITIS NODOSA

Polyarteritis nodosa is a connective tissue disorder in which a necrotizing inflammation involves all layers of the walls of small arteries, arterioles, and veins. More than half of the patients with polyarteritis nodosa have renal disease; hypertension and hematuria are common. Arteriography typically demonstrates multiple aneurysms of the segmental, interlobar, arcuate, and interlobular arteries. The spontaneous rupture of an intrarenal aneurysm may cause a renal infarct or perinephric or parenchymal hematoma. The occlusion of small renal vessels may lead to renal infarction and severe hypertension. Collateral vessels developing around an arterial occlusion may be so prominent as to suggest a vascular neoplasm.[17,18]

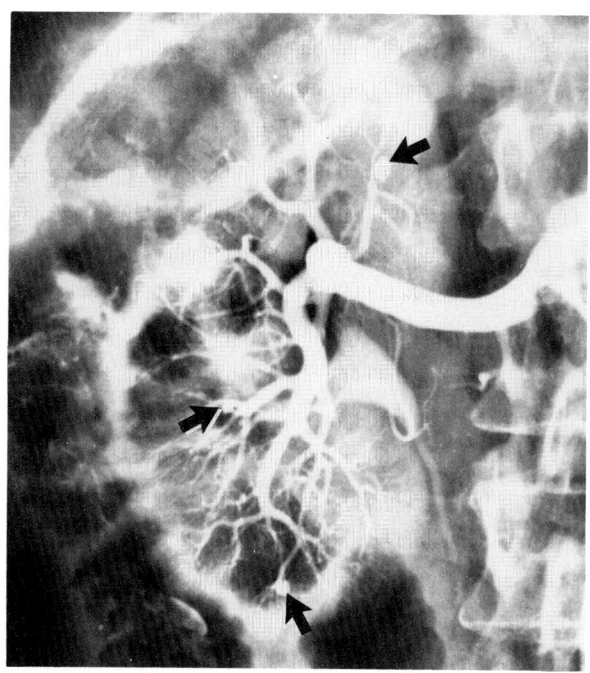

**Figure 39.12. Polyarteritis nodosa.** A selective right renal arteriogram demonstrates microaneurysms (arrows) arising from the arcuate arteries. (From Friedland, Filly, Goris, et al,[19] with permission from the publisher.)

### REFERENCES


1. Witten DM, Myers GH, Utz BC: *Clinical Urography*. Philadelphia, WB Saunders, 1977.
2. Paul GJ, Stephenson TF: The cortical rim sign in renal infarction. *Radiology* 122:338, 1977.
3. Janower ML, Weber AL: Radiological evaluation of acute renal infarction. *AJR* 95:309–317, 1965.
4. Scatliff JH, Cuttino JT, Winfield HG, et al: Angiographic evaluation of renal infarction. *AJR* 108:674–690, 1970.
5. Siegelman SS, Caplan LH: Acute segmental renal artery embolism: A distinctive urographic and arteriographic complex. *Radiology* 88:509–512, 1967.
6. Glazer GM, Francis IR, Brady TM, et al: Computed tomography of renal infarction. *AJR* 140:721–727, 1983.
7. Hann L, Pfister RC: Renal subcapsular rim sign. New etiologies and pathogenesis. *AJR* 138:51–54, 1982.
8. Ney C, Friedenberg RM: *Radiographic Atlas of the Genitourinary System*, Philadelphia, JB Lippincott, 1981.
9. Gill WM, Pudvan WR: The arteriographic diagnosis of renal parenchymal diseases. *Radiology* 96:81–84, 1970.
10. Mena E, Bookstein JJ, Gikas BW: Angiographic diagnosis of renal parenchymal disease—chronic glomerulonephritis, chronic pyelonephritis, and arteriolonephrosclerosis. *Radiology* 108:523–532, 1973.
11. Chait A, Stoane L, Moskowitz H, et al: Renal vein thrombosis. *Radiology* 90:886–896, 1968.
12. Mulhern CB, Arger PH, Miller WT, et al: The specificity of renal vein thrombosis. *AJR* 125:291–299, 1975.

13. Olin TB, Reuter SR: A pharmacoangiographic method for improving nephrophlebography. *Radiology* 85:1036–1042, 1965.
14. Davidson AJ: *Radiologic Diagnosis of Renal Parenchymal Disease,* Philadelphia, WB Saunders, 1977.
15. Scanlon GT: The radiographic changes in renal vein thrombosis. *Radiology* 80:208–211, 1963.
16. Bradley WG, Jacobs RP, Trew PA, et al: Renal vein thrombosis: Occurrence in membranous glomerulonephropathy and lupus nephritis. *Radiology* 139:571–576, 1981.
17. Fleming RJ, Stern LZ: Multiple intraparenchymal aneurysms and polyarteritis nodosa. *Radiology* 84:100–103, 1965.
18. Peterson C, Willerson JT, Doppman JL, et al: Polyarteritis nodosa with bilateral renal artery aneurysm and perirenal haematomas: Angiographic and nephrotomographic features. *Br J Radiol* 43:62–71, 1970.
19. Friedland GW, Filly R, Goris ML, Gross D, Kempson RL, Korobkin M, Thurber ED, and Walter J (eds): *Uroradiology: An Integrated Approach.* New York, Churchill Livingstone, 1983.
20. Cohn LH, Lee J, Hopper J, et al: The treatment of bilateral renal vein thrombosis and nephrotic syndrome. *Surgery* 64:387–396, 1968.

## Chapter Forty
# Nephrolithiasis and Urinary Tract Obstruction

## NEPHROLITHIASIS

Nephrolithiasis refers to the development of stones within the pelvocalyceal system of the kidney, in contrast to nephrocalcinosis, in which calcium deposits occur within the renal parenchyma. The etiology of renal stones is varied and often reflects an underlying metabolic abnormality. Urinary stasis and infection are also important factors in promoting calculus formation.

More than 80 percent of symptomatic renal calculi are radiopaque and detectable on plain abdominal radiographs. Stones composed of calcium phosphate and calcium oxalate usually have uniform dense radiopacity. They may develop secondary to hyperparathyroidism, renal tubular acidosis, hyperoxaluria, or any cause of increased calcium excretion in the urine. In at least 20 percent of the patients, there is no obvious cause for the development of calcium stones.

Completely radiolucent calculi contain no calcium and are composed of pure uric acid or urates, xanthine, or matrix concretions that are a combination of mucoprotein and mucopolysaccharide. Struvite (magnesium ammonium phosphate) stones are moderately radiopaque but have a variable internal density. Found mainly in women, these common and potentially dangerous calculi form in the presence of chronic urinary tract infection with *Proteus* organisms. Cystine calculi, though

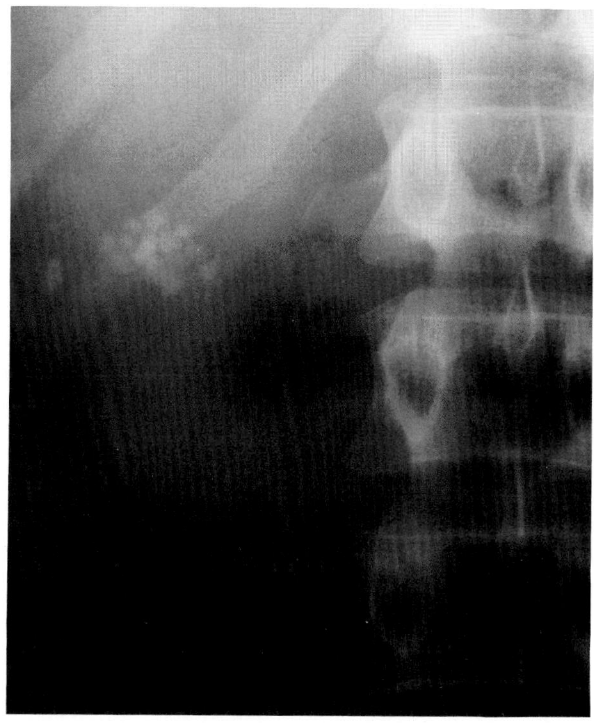

**Figure 40.1. Nephrolithiasis.** Multiple radiopaque calculi in the renal pelvis.

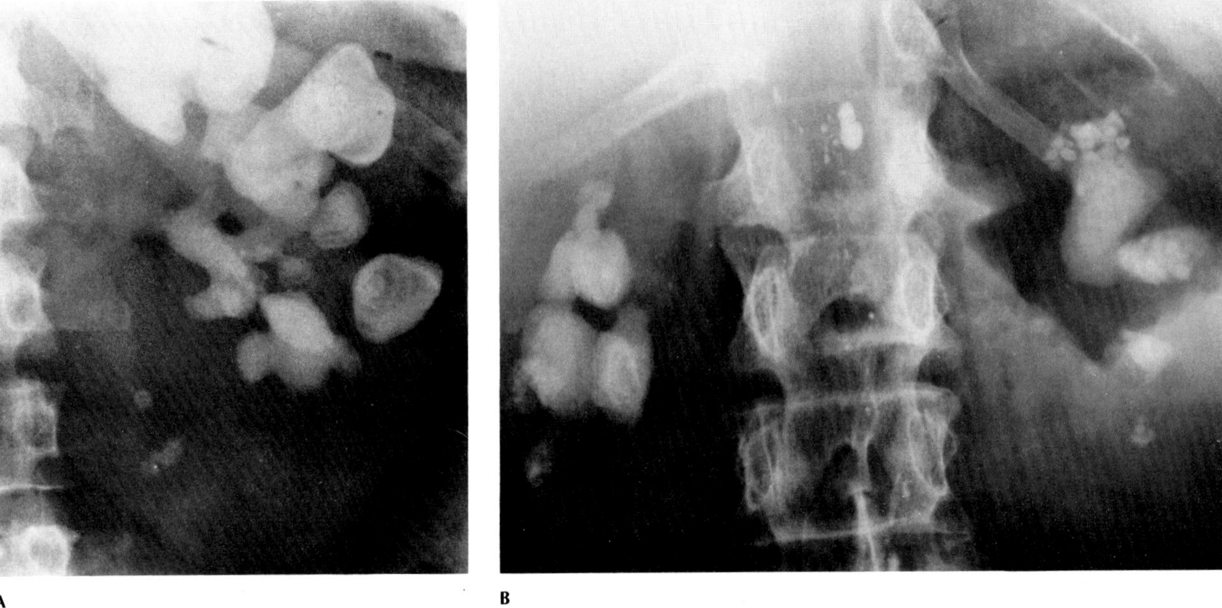

A B

**Figure 40.2. Staghorn calculi.** (A) Unilateral and (B) bilateral calculi filling the renal pelves.

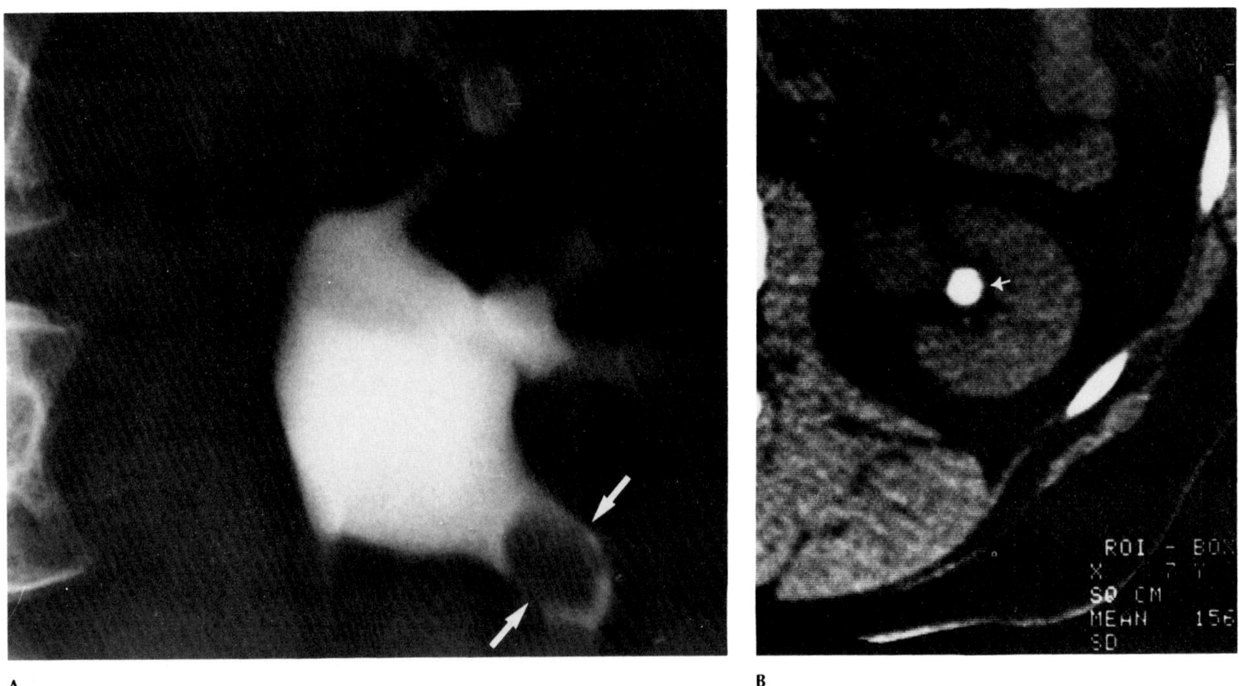


A B

**Figure 40.3. Calculus in the renal pelvis.** (A) An excretory urogram shows a large calyceal filling defect (arrows). (B) A CT scan without contrast enhancement shows a calyceal mass of high attenuation (312 H) representing a uric acid stone. (From Parienty, Ducellier, Pradel, et al,[4] with permission from the publisher.)

often considered nonopaque, are usually moderately opaque and present a frosted, or ground-glass, appearance. Renal calculi can be laminated as a result of the deposition of alternate layers of densely radiopaque material (calcium phosphate, calcium oxalate) and material of relatively low radiodensity (magnesium ammonium phosphate, urate).

Excretory urography may be performed to accurately localize suspected calculi within the pelvocalyceal system or to detect otherwise invisible nonopaque stones, which appear as filling defects in the contrast-filled collecting system. In patients with renal colic due to an obstructing calculus in the ureter, excretory urography may demonstrate the point of cutoff as well as dilatation

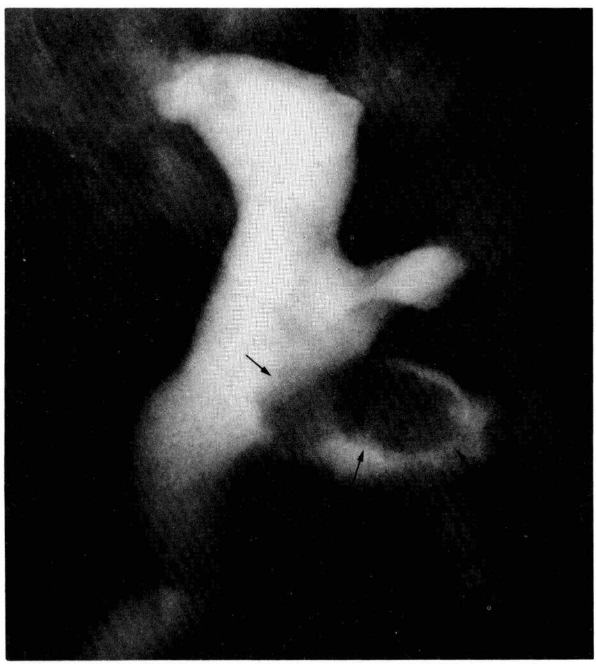


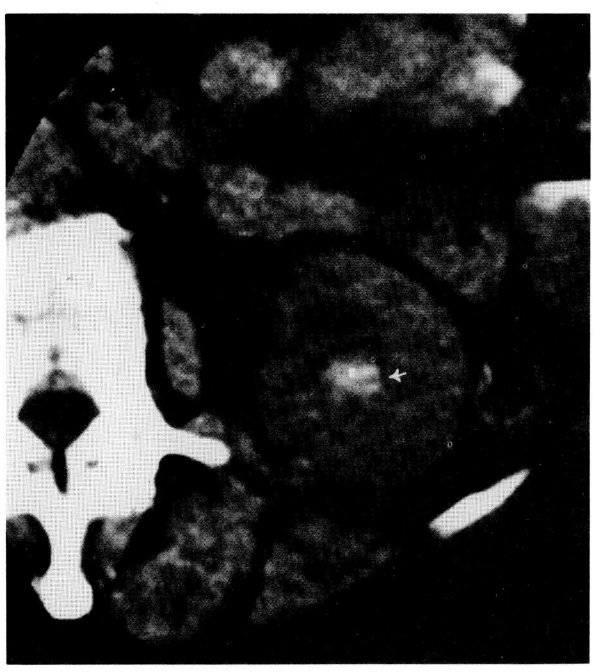


**Figure 40.4. Blood clot.** (A) An excretory urogram shows a filling defect with a smooth contour (arrows). (B) A CT scan shows a round, sharply marginated, dense mass (arrow) situated in a papilla and impinging on the nonopacified renal pelvis. The CT number of the blood clot is 62 H, a density higher than that of transitional cell carcinoma but lower than that of a nonopaque stone. (From Parienty, Ducellier, Pradel, et al,[4] with permission from the publisher.)

of the proximal ureter and pelvocalyceal system. With acute obstruction, the intrapelvic pressure may increase to such an extent that there is little or no glomerular filtration, resulting in a delayed but prolonged nephrogram and a lack of calyceal filling on the affected side. If necessary, retrograde pyelography may be performed to confirm the diagnosis of an obstructing stone (see the section "Urinary Tract Obstruction" in this chapter).

On computed tomography (CT), stones in the renal collecting system, regardless of their composition, tend to have high attenuation values. This can permit their distinction from other causes of pelvocalyceal filling defects, such as blood clots and transitional-cell tumors, that appear as lower density masses on CT scans.

In addition to the clinical history and laboratory examinations, additional radiographic studies may provide a clue to the underlying cause of renal calculi. Radiographs of the pelvis, lumbar spine, and hands may provide evidence of hyperparathyroidism, Cushing's disease, myeloma, osteolytic metastases, Paget's disease, or gout; chest films may reveal the characteristic lymphadenopathy or parenchymal changes of sarcoidosis.[1-4]

## NEPHROCALCINOSIS

Nephrocalcinosis refers to radiographically detectable diffuse calcium deposition within the renal parenchyma, chiefly within the medullary pyramids. The calcification varies from a few scattered punctate densities to very dense and extensive calcifications throughout both kidneys.

Nephrocalcinosis occurs in about 25 percent of patients with primary hyperparathyroidism. The excessive secretion of parathyroid hormone increases osteoclast activity, resulting in deossification of the skeleton and hypercalcemia. Bone destruction in patients with metastatic carcinoma leads to a release of excessive amounts of calcium from osseous structures. Patients with primary carcinomas, especially of the lung or kidney, may develop a paraneoplastic syndrome with hypercalcemia and nephrocalcinosis; the syndrome appears to be related to the tumor's inappropriate secretion of specific humoral factors. Deossification of the skeleton and subsequent nephrocalcinosis can also occur in patients with severe osteoporosis (due to immobilization, menopause, senility) or Cushing's disease, and in persons receiving steroid therapy.

Increased intestinal absorption of calcium can lead to nephrocalcinosis. In patients with sarcoidosis, an increased intestinal sensitivity to vitamin D results in an excessive absorption of dietary calcium. A similar mechanism occurs in patients with hypervitaminosis D; large amounts of vitamin D also promote the dissolution of calcium salts from bone. Patients with the milk-alkali syndrome have a long history of excessive calcium ingestion, usually in the form of milk and antacids containing calcium carbonate. The large tubular load of calcium

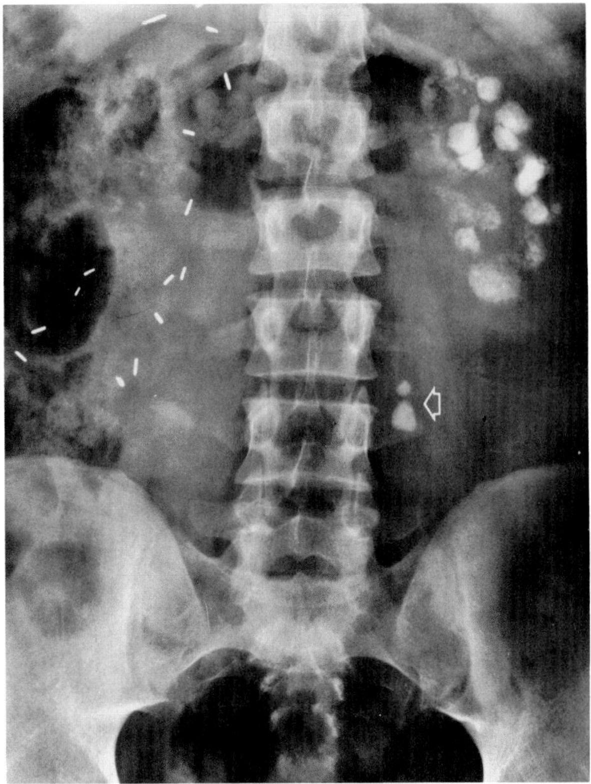

A

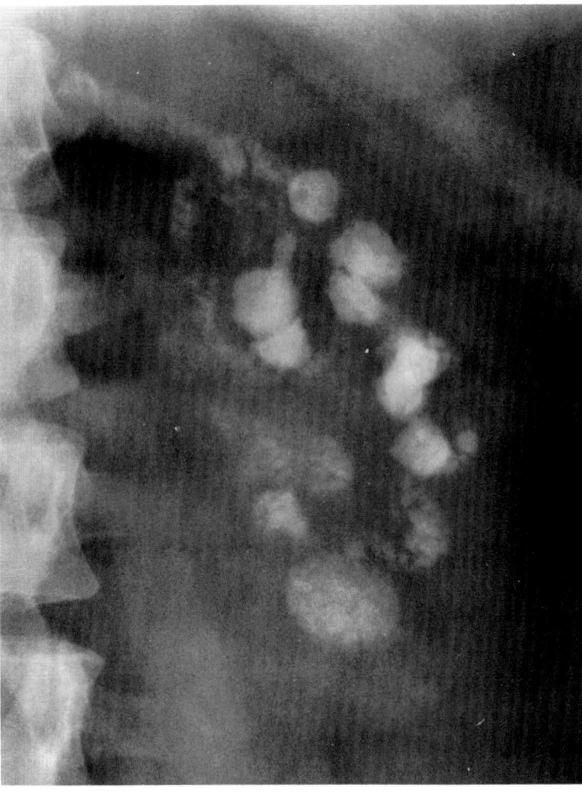

B

**Figure 40.5. Renal tubular acidosis causing nephrocalcinosis.** (A) An abdominal radiograph demonstrates diffuse calcification in the medullary pyramids of the left kidney. In addition, two stones (one of which is causing an obstruction) are seen in the midportion of the left ureter (arrow). The patient had previously undergone a right nephrectomy. (B) A close-up view of the left kidney demonstrates the intrarenal calcification.

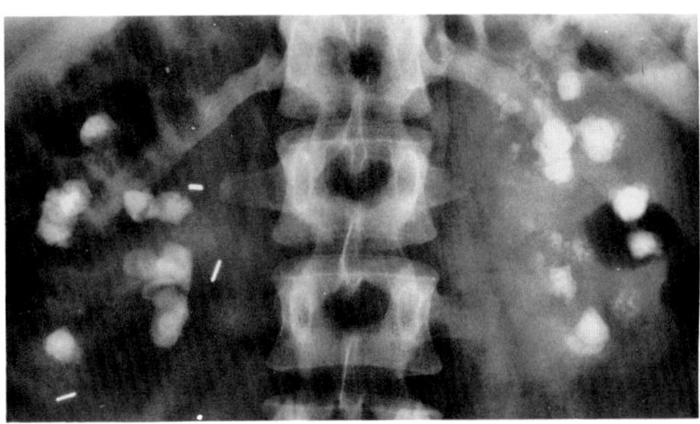

**Figure 40.6. Milk-alkali syndrome causing nephrocalcinosis.**

and phosphate in the presence of alkaline urine and interstitial fluid causes nephrocalcinosis.

Renal tubular acidosis is a disorder in which the kidney is unable to excrete an acid urine (below pH 5.4) because the distal nephron cannot secrete hydrogen against a concentration gradient. This disorder typically produces very dense and extensive parenchymal calcification, often associated with staghorn calculi. Patients with renal tubular acidosis frequently suffer from osteomalacia; children with the disorder may demonstrate the so-called renal rickets.

Calcification within cystic dilatations of the distal collecting ducts is a manifestation of medullary sponge kidney (see the section "Medullary Sponge Kidney" in Chapter 37). The calculi are usually small and round, tending to cluster around the apices of the pyramids. Hyperoxaluria produces nephrocalcinosis by the interstitial deposition of calcium oxalate (see the section "Hyperoxaluria" in Chapter 3). The primary form is a rare inherited metabolic disease; secondary hyperoxaluria occurs in association with intestinal diseases (especially Crohn's disease), in which the increased absorption of dietary oxalate is related to the inflammatory process. Nephrocalcinosis is a common finding in patients with renal papillary necrosis (see the section "Papillary Necrosis" in Chapter 38). The necrotic papilla can remain in situ and become calcified or become detached and serve as a nidus for calculus formation.

When nephrocalcinosis is detected on plain abdominal radiographs, a chest radiograph and a skeletal survey are indicated. The chest radiograph may demonstrate sarcoidosis or metastases; the skeletal survey may reveal a bone lesion or evidence of hyperparathyroidism that has caused hypercalcemia.[5–7]

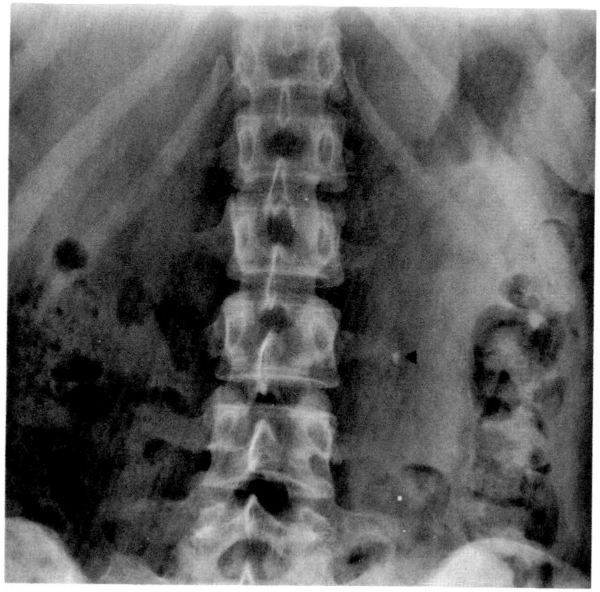
A

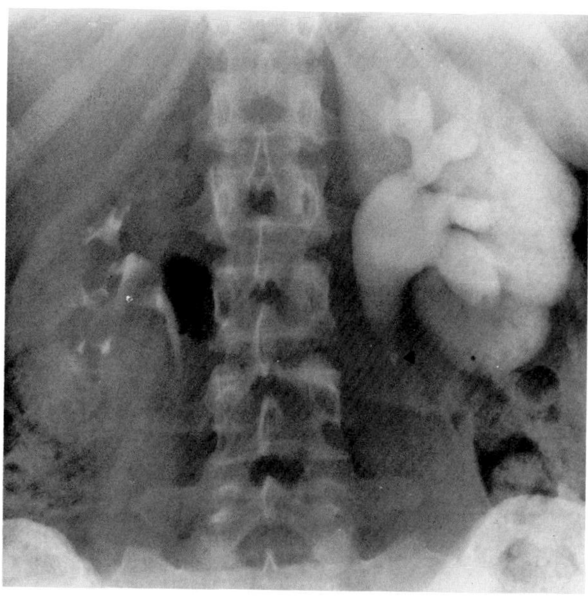
B

**Figure 40.7. Obstructing ureteral calculus.** (A) A plain abdominal radiograph demonstrates a calcification (arrow) overlying the left transverse process of L3. (B) An excretory urogram demonstrates a prolonged nephrogram and marked dilatation of the collecting system and pelvis proximal to the obstructing stone (arrow).

## OTHER CALCULI IN THE GENITOURINARY TRACT

### Ureteral Calculi

Ureteral calculi are extremely common, and their detection is clinically important. They are usually small, irregular, and poorly calcified and are therefore easily missed on abdominal radiographs that are not of good quality. Calculi most commonly lodge in the lower portion of the ureter, especially at the ureterovesical junction and at the pelvic brim. They are often oval, with their long axes paralleling the course of the ureter. Ureteral calculi must be differentiated from the far more common phleboliths, which are spherical and are located in the lateral portion of the pelvis below a line joining the ischial spines. In contrast, ureteral calculi are situated medially above the interspinous line.

Nonopaque ureteral calculi may be demonstrated as lucent filling defects on excretory urograms. If renal function is insufficient, retrograde pyelography is required to demonstrate calculi in the ureter.[18]

### Bladder Calculi

Stone formation in the bladder is primarily a disorder of elderly men with obstruction or infection of the lower urinary tract. Frequently associated lesions include bladder outlet obstruction, urethral strictures, neurogenic bladder, bladder diverticula, and cystoceles. At times, upper urinary tract stones migrate down the ureter and are retained in the bladder.

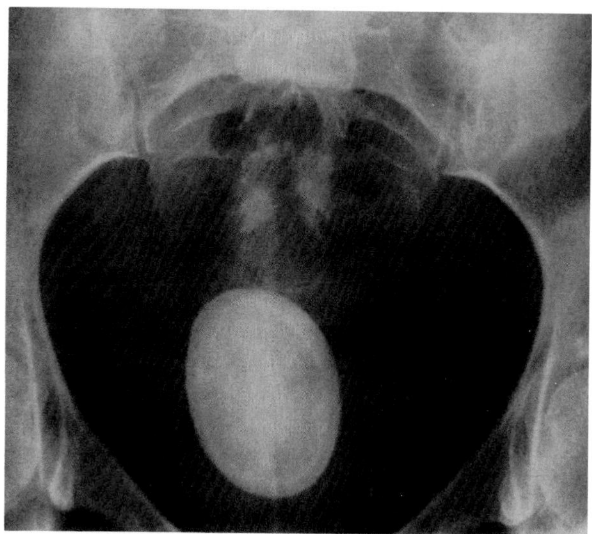

**Figure 40.8. Single huge, laminated, calcified bladder stone.**

Bladder calculi can be single or multiple. They vary in size from tiny concretions, each the size of a grain of sand, to an enormous single calculus occupying the entire bladder lumen. When located in a bladder diverticulum, calculi are occasionally identified in an unusual position close to the lateral pelvic wall.

Most bladder calculi are circular and oval in outline; however, almost any shape can be encountered. They can be amorphous, laminated, or even spiculated. One unusual type with a characteristic radiographic appearance is the hard burr, or jackstone, variety, which gets

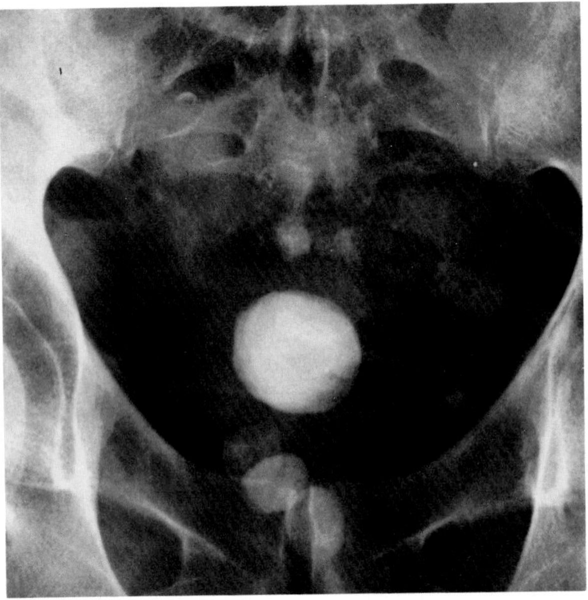

**Figure 40.9. Multiple bladder stones of varying sizes.**

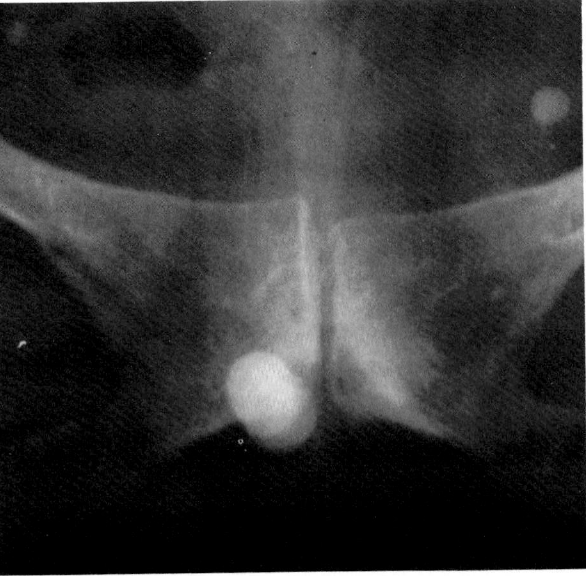

**Figure 40.10. Calcified calculi within a urethral diverticulum.** The stones are in a characteristic location in the subpubic angle of the pelvis close to the midline.

its name from the many irregular prongs that project from its surface. Dumbbell-shaped stones, with one end lodged in a diverticulum and the other projecting into the bladder, are not uncommon.[18]

### Urethral Calculi

Urethral calculi are easily recognized because of their unique location in the subpubic angle of the pelvis at or close to the midline. In males, they occur in the prostatic or bulbous urethra, usually proximal to an obstruction. In females, urethral calculi are almost always associated with diverticula and infection.

### Urachal Calculus

A solitary urachal calculus can appear as an oval or dumbbell-shaped opacity that lies at or close to the midline of the upper pelvis and is superimposed on the sacrum. On lateral projections, a urachal calculus is readily distinguished by its extreme anterior position; on cystograms, the superior portion of the bladder is pear-shaped and points upward toward the stone.

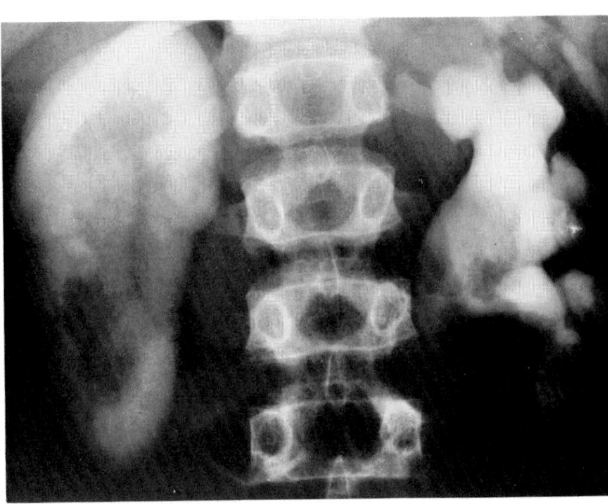

**Figure 40.11. Acute urinary tract obstruction.** There is a prolonged and intensified obstructive nephrogram of the right kidney. On the left, there is marked dilatation of the pelvocalyceal system but no persistent nephrogram. This reflects an intermittent chronic obstruction on that side. (From Eisenberg,[19] with permission from the publisher.)

## URINARY TRACT OBSTRUCTION (OBSTRUCTIVE NEPHROPATHY)

Urinary tract obstruction produces anatomic and functional changes that vary with the rapidity of onset, the degree of occlusion, and the distance between the kidney and the obstructing lesion. Obstruction to urine flow can result from intrinsic or extrinsic mechanical blockade as well as from functional defects not associated with fixed occlusion of the urinary drainage system. Normal points of narrowing, such as the ureteropelvic and ureterovesical junctions, the bladder neck, and the urethral meatus, are common sites of obstruction. Blockage above the level of the bladder causes unilateral dilatation of the ureter (hydroureter) and renal pelvocalyceal sys-

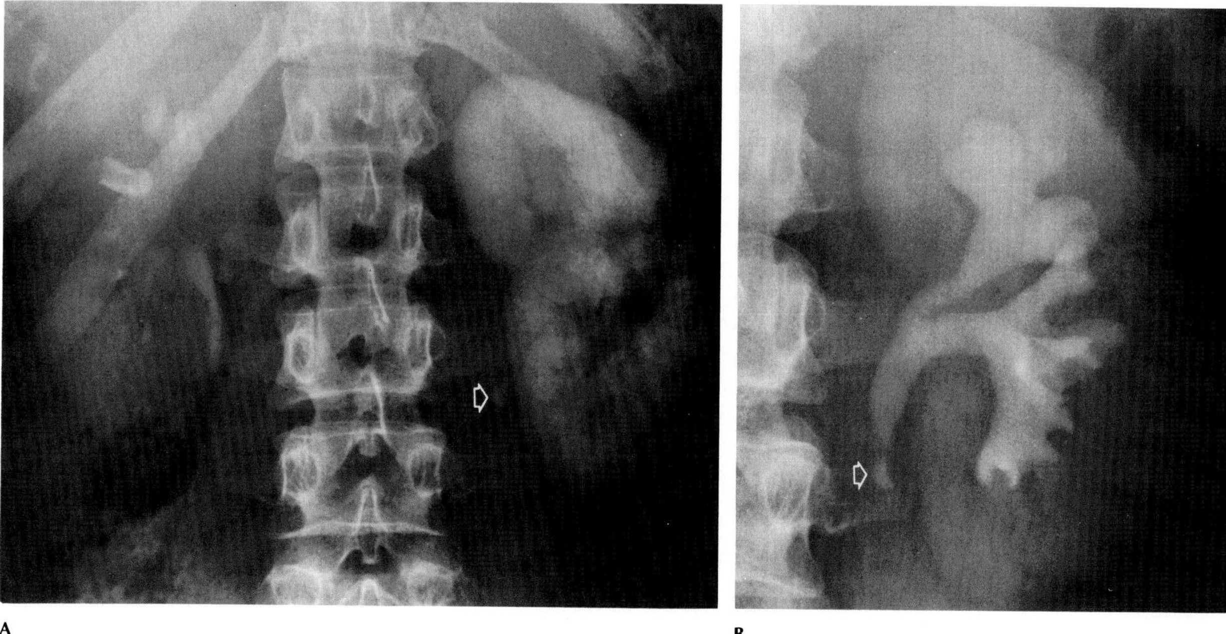

**Figure 40.12. Acute urinary tract obstruction.** (A) An excretory urogram demonstrates a prolonged nephrogram on the left with fine cortical striations (alternating radiolucent and radiopaque lines) and no calyceal filling. An arrow points to the obstructing stone in the proximal left ureter. (B) On a delayed film obtained 4 hours after the injection of contrast material, there is marked dilatation of the pelvocalyceal system and proximal ureter to the level of the obstructing stone (arrow).

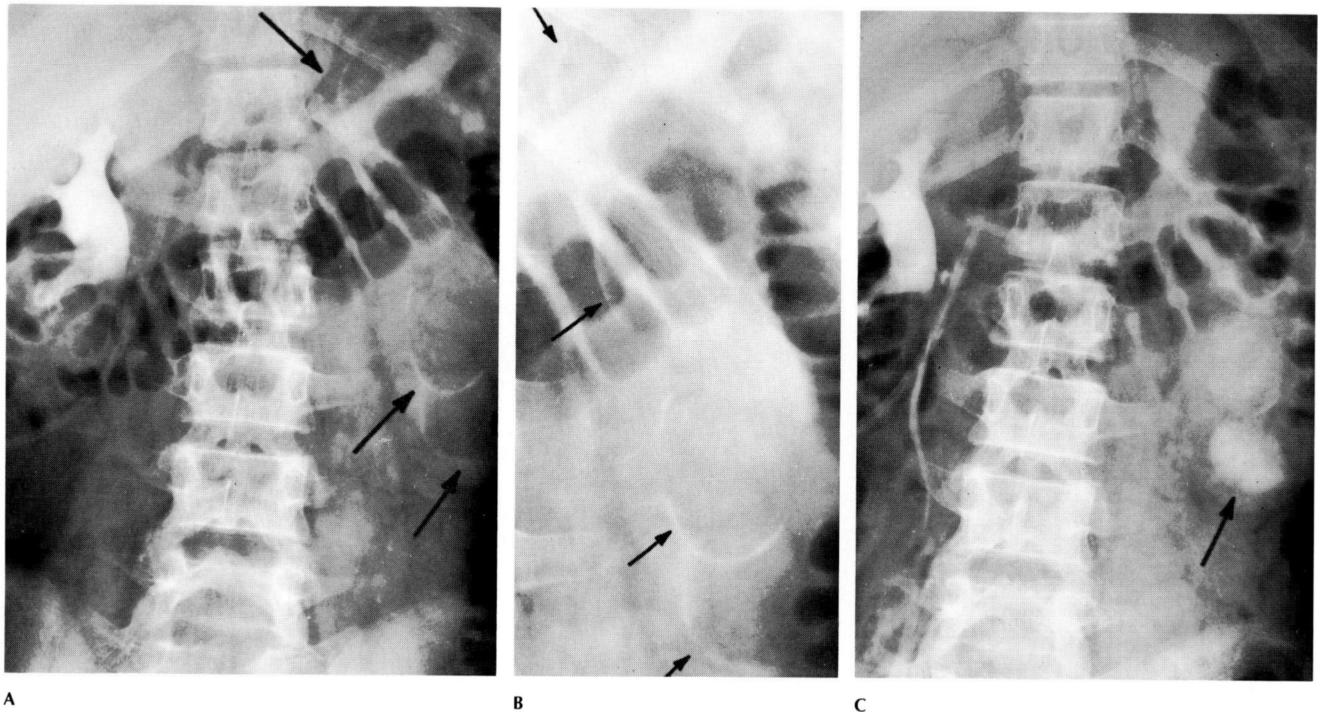

**Figure 40.13. Crescent sign of chronic urinary tract obstruction.** (A) An excretory urogram demonstrates good opacification of the right renal pelvis and calyces. The pelvis and calyces of the left kidney are not opacified, but crescentic collections of contrast material (arrows) are seen in the renal parenchyma overlying dilated calyces. (B) Enlargement of part of the film shown in A. The crescent sign of hydronephrosis is exaggerated (arrows). (C) A delayed film obtained 1 hour later than B shows some contrast material collecting in the dilated calyces (arrow). The crescent sign has almost completely disappeared. (From Witten, Myers, and Utz,[18] with permission from the publisher.)

tem (hydronephrosis); if the lesion is at or below the level of the bladder, as in prostatic hypertrophy or tumor, bilateral involvement is the rule.

In children, congenital malformations (ureteropelvic junction narrowing, ureterocele, retrocaval ureter, posterior urethral valve) are the most common forms of mechanical obstruction. In adults, pelvic tumors, calculi, and urethral strictures are major causes. Obstructive uropathy may also result from extrinsic neoplasms (carcinoma of the cervix or colon, retroperitoneal lymphoma) or inflammatory disorders (Crohn's disease, retroperitoneal fibrosis). Functional defects causing an impairment of urine flow include neurogenic bladder and vesicoureteral reflux.

In acute urinary tract obstruction, diminished filtration of urographic contrast material results in delayed parenchymal opacification compared with the nonobstructed kidney. The nephrogram eventually becomes more dense than normal because of a decreased rate of fluid flow through the tubules, which results in enhanced water reabsorption by the nephrons and greater concentration of the contrast material. There is delayed and decreased pelvocalyceal filling because of dilatation and elevated pressure in the collecting system. The radiographic study may have to be prolonged up to 48 hours after the administration of contrast material to determine the precise site of obstruction.

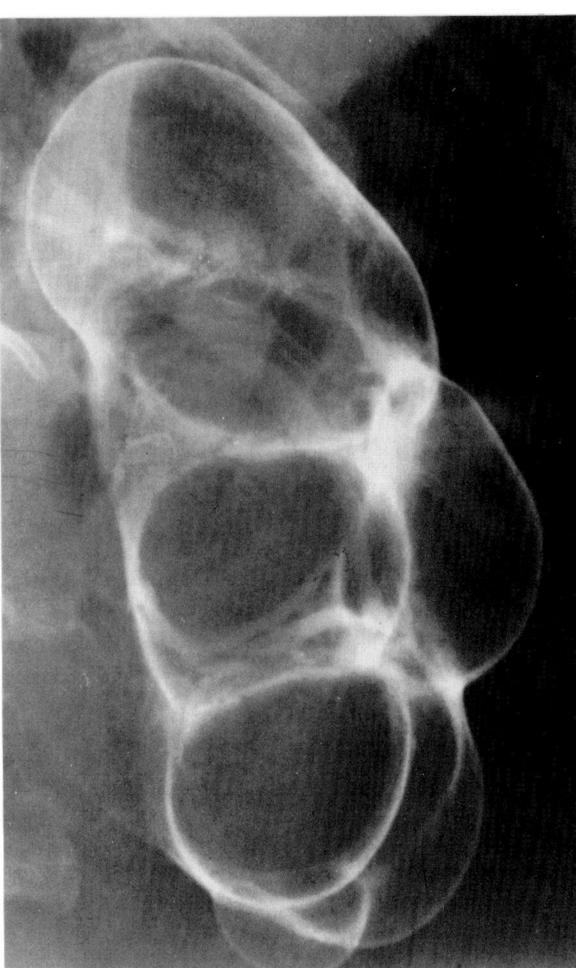

**Figure 40.14. Rim sign of chronic urinary tract obstruction.** A film from a selective arteriogram in a patient with a ureteropelvic junction obstruction demonstrates a nephrogram of remaining renal parenchyma rimming the dilated calyces. Unlike the crescent sign, the rim sign can be present even if the kidney has no residual excretory function. (From Elkin,[10] with permission from the publisher.)

In the patient with acute urinary tract obstruction, the kidney is generally enlarged and the calyces are moderately dilated. An unusual finding is alternating radiolucent and radiopaque striations in the nephrogram, presumably representing contrast material filling dilated tubules in the medullary rays. Another uncommon urographic finding in acute unilateral obstruction (usually due to ureteral stone) is opacification of the gallbladder 8 to 24 hours after the injection of contrast material. This "vicarious excretion" is related to increased hepatic excretion of contrast material due to prolonged plasma disappearance time.

As obstruction becomes more chronic, the predominant urographic finding is a markedly dilated pelvocalyceal system and ureter proximal to the obstruction. Initially, obstructive back pressure causes flattening of the normal calyceal concavity and blunting of the sharp peripheral angle produced by the papilla as it juts into the calyx. Prolonged increased pressure causes progressive papillary atrophy, leading to calyceal clubbing, in which the concavity produced by the papilla is reversed. Gradual enlargement of the calyces and renal pelvis with progressive destruction of renal parenchyma may continue until the kidney becomes a nonfunctioning hydronephrotic sac in which normal anatomy is obliterated. With severely decreased renal function, early nephrogram films may demonstrate thin curvilinear collections of contrast material in the compressed renal parenchyma overlying nonopacified, dilated calyces (crescent sign). This reflects the accumulation of contrast material in collecting tubules that have been flattened and displaced by the hydronephrotic calyces so that they lie parallel to the renal convexity and near its surface. Delayed films may eventually demonstrate opacification of the dilated collecting system itself as contrast material passes downward from the tubules into the large reservoir of unopacified urine.

If there is a low obstruction near the ureterovesical junction, the ureter may appear markedly dilated and tortuous. Obstruction at or below the level of the bladder produces thickening, trabeculation, and diverticula of the bladder wall.

Whenever possible, the site of obstruction should be demonstrated. Although high-dose excretory urography with delayed films may accomplish this purpose, retrograde or antegrade pyelography is often required. In the latter, a catheter or needle is placed percutaneously into the dilated collecting system under ultrasonic or

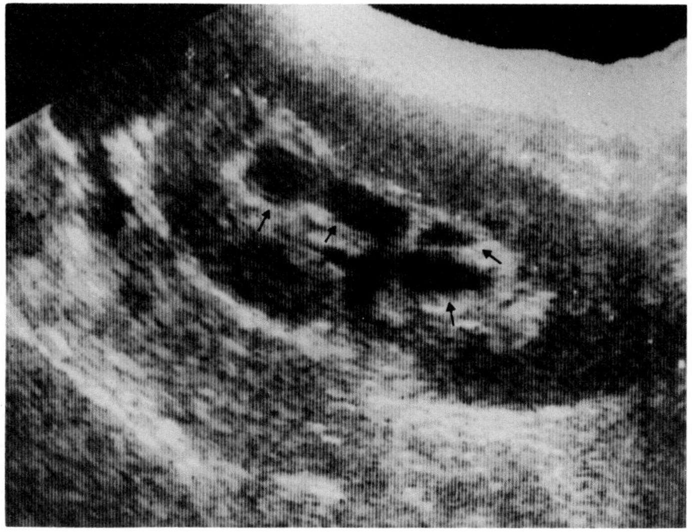
A

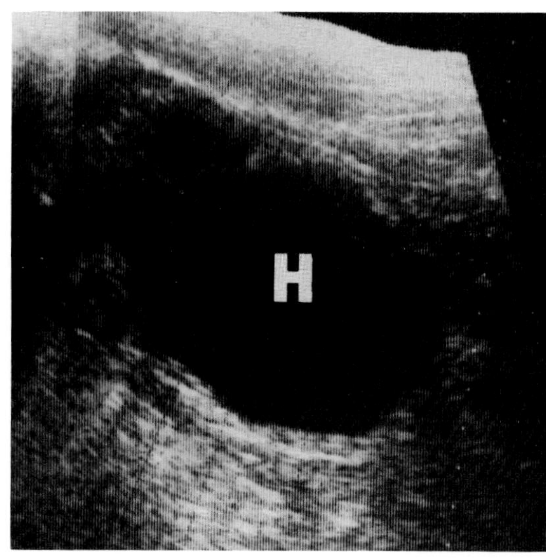

B

**Figure 40.15. Ultrasound of hydronephrosis.** (A) In a patient with moderate disease, the dilated calyces and pelvis appear as echo-free sacs (arrows) separated by septa of compressed tissue and vessels. (B) In a patient with severe hydronephrosis, the intervening septa have disappeared, leaving a large fluid-filled sac (H) with no evidence of internal structure and no normal parenchyma apparent at its margins.

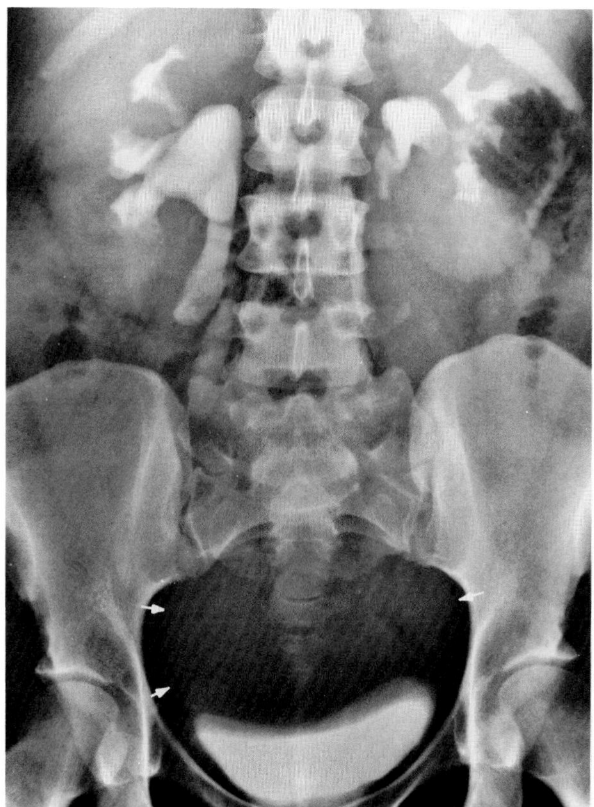

**Figure 40.16. Hydronephrosis of pregnancy.** An excretory urogram performed 3 days postpartum demonstrates bilateral large kidneys with dilatation of the ureters and pelvocalyceal systems, especially on the right. The large pelvic mass (arrows) indenting the superior surface of the bladder represents the uterus, which still is causing extrinsic pressure on the ureters.

fluoroscopic guidance, and contrast material is then introduced. This approach has the added advantage of providing immediate and certain decompression of a unilateral obstructing lesion.

Ultrasound is of particular value in detecting hydronephrosis in patients with such severe urinary tract obstruction and renal dysfunction that there is no opacification of the kidneys and collecting systems on excretory urograms. When hydronephrosis has reached the nonfunctioning stage, the renal outline is enlarged and the central medullary echoes are distorted. The dilated calyces and pelvis become large, hydronephrotic, echo-free sacs separated by septa of compressed tissue and vessels. These fluid-filled sacs vary in size and are usually fewer than those in a multicystic or polycystic kidney. The septa separating the echo-free regions in hydronephrosis are fewer, thicker, and longer than the septa between cysts in a multicystic kidney. With increased duration and severity of hydronephrosis, the intervening septa may disappear, leaving a large fluid-filled sac with no evidence of internal structure and no normal parenchyma apparent at its margins.

Arteriography in hydronephrosis demonstrates attenuated, sparse, displaced arteries, a prolonged circulation time, and a characteristic nephrogram of thin rims of opacified remaining renal parenchyma surrounding the dilated lucent calyces. However, arteriography is rarely indicated since ultrasound and urography usually permit the correct diagnosis and percutaneous antegrade pyelography is the preferred approach in difficult cases.

In children, hydronephrosis is a common cause of an abdominal mass. Ultrasound is the procedure of choice in differentiating hydronephrosis from renal malformations or tumors of the kidney or adrenal.

A physiologic form of hydronephrosis often develops during pregnancy. The enlarging uterine mass and physiologic hormonal changes cause extrinsic pressure on the ureter, leading to progressive dilatation of the proximal collecting systems. Though often bilateral, the dilatation usually is more prominent and develops earlier on the right side. Following delivery, the urinary tract returns to normal within several weeks. In some women, however, persistent dilatation of the ovarian vein can compress the ureter and result in prolonged postpartum hydronephrosis.[8–17]

## REFERENCES


1. McAfee JG, Donner MW: Differential diagnosis of calcifications encountered in abdominal radiographs. *Am J Med Sci* 243:609–650, 1962.
2. Boyce WH: Radiology in the diagnosis and surgery of renal calculi. *Radiol Clin North Am* 3:89–102, 1965.
3. Finlayson B: Renal lithiasis in review. *Urol Clin North Am* 1:181–212, 1974.
4. Parienty RA, Ducellier R, Pradel J-M, et al: Diagnostic value of CT numbers in pelvocalyceal filling defects. *Radiology* 145:743–747, 1982.
5. Courey WB, Pfister RC: The radiographic findings in renal tubular acidosis. *Radiology* 105:497–503, 1972.
6. Eisenberg RL: *Gastrointestinal Radiology: A Pattern Approach,* Philadelphia, JB Lippincott, 1983.
7. Creel L: Radiological aspects of nephrocalcinosis. *Clin Radiol* 13:218–230, 1962.
8. Davidson AJ: *Radiologic Diagnosis of Renal Parenchymal Disease,* Philadelphia, WB Saunders, 1977.
9. Paul GJ, Stephenson TF: The cortical rim sign in renal infarction. *Radiology* 122:338, 1977.
10. Elkin M: Obstructive uropathy and uremia. *Radiol Clin North Am* 10:447–458, 1972.
11. Bigongiari LR, Patel SK. Appelman H, et al: Medullary rays. Visualization during excretory urography. *AJR* 125:795–803, 1975.
12. Sokoloff J, Talner LB: The heterotopic excretion of sodium iothalamate. *Br J Radiol* 46:571–577, 1973.
13. LeVine M, Allen A, Stein JL, et al: The crescent sign. *Radiology* 81:971–973, 1963.
14. Pfister RC, Yoder IC, Newhouse JH: Percutaneous uroradiologic procedures. *Semin Roentgenol* 16:135–151, 1981.
15. Dure-Smith P: Pregnancy dilatation of the urinary tract. *Radiology* 96:545–550, 1970.
16. Ney C, Friedenberg RM: *Radiographic Atlas of the Genitourinary System,* Philadelphia, JB Lippincott, 1981.
17. Sanders RC, Bearman S: B-scan ultrasound in the diagnosis of hydronephrosis. *Radiology* 108:375–379, 1973.
18. Witten DM, Myers GH, Utz DC: *Emmett's Clinical Urography.* Philadelphia, WB Saunders, 1977.
19. Eisenberg RL: *Atlas of Signs in Radiology,* Philadelphia, JB Lippincott, 1984.

## Chapter Forty-One
# Tumors of the Urinary Tract

## BENIGN RENAL TUMORS

### Adenoma

Renal adenomas are benign cortical tumors that are usually small and asymptomatic and discovered incidentally. If an adenoma reaches substantial size, excretory urography may demonstrate distortion of the pelvocalyceal system without evidence of calyceal destruction. On the nephrogram phase, the tumor appears as a smooth, relatively lucent mass that may be indistinguishable from a cyst. The not-infrequent association of a thick capsule can suggest a malignant lesion. Ultrasound demonstrates a renal adenoma as a solid mass indistinguishable from other neoplastic lesions.[1,2]

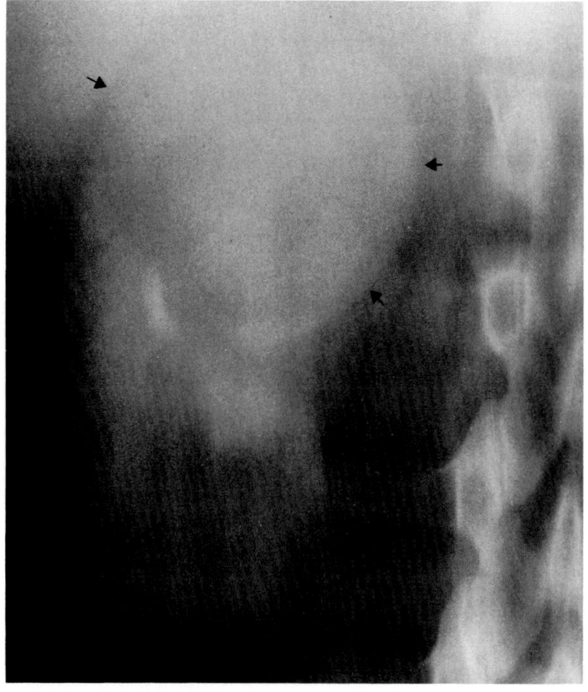

**Figure 41.1. Renal adenoma.** A nephrotomogram demonstrates the tumor as a smooth, relatively lucent mass (arrows) that is indistinguishable from a cyst.

### Hamartoma (Angiomyolipoma)

Most hamartomas occur as isolated, unilateral kidney lesions in otherwise normal persons. However, these benign tumors also develop in a large percentage of patients with tuberous sclerosis, in whom involvement is usually multifocal and bilateral. On plain abdominal radiographs, the tumor may present as a well-defined area of mottled lucency representing the large amount of fat within the lesion. At excretory urography, the single

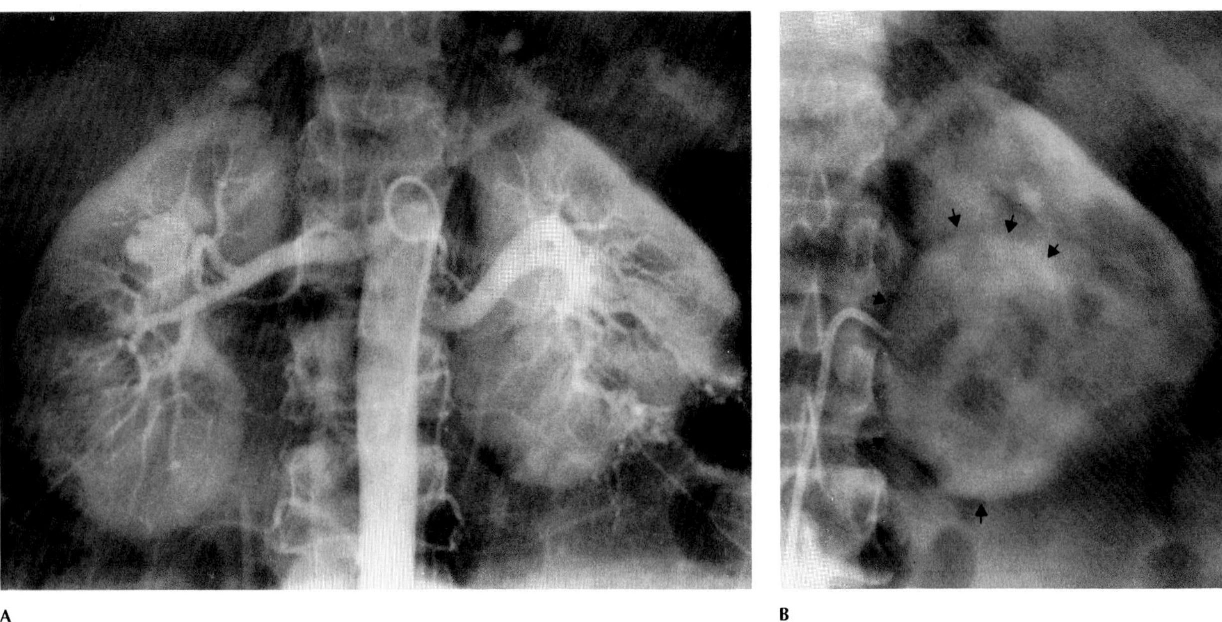

A B

**Figure 41.2. Renal adenoma.** (A) Arterial and (B) nephrogram phases of arteriography demonstrate a mass (arrows) that is isodense with normal renal parenchyma and shows no evidence of abnormal neovascularity.

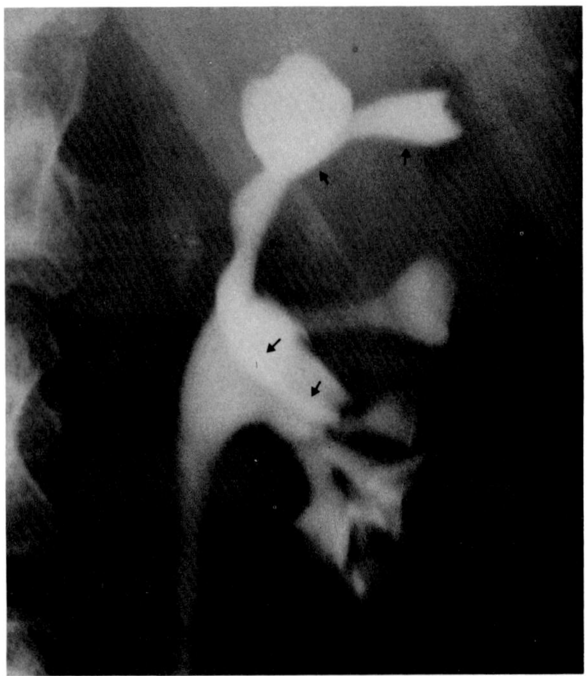

**Figure 41.3. Renal hamartoma.** An excretory urogram demonstrates a large left renal mass distorting and displacing the pelvocalyceal system (arrows).

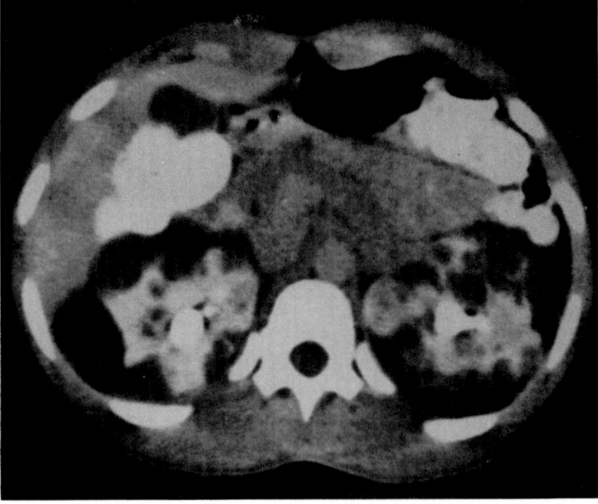

**Figure 41.4. Multiple renal hamartomas in tuberous sclerosis.** A CT scan demonstrates innumerable low-attenuation masses within both kidneys.

or multiple masses enlarge the kidney and distort and displace the pelvis and calyces. In patients with tuberous sclerosis, the multiple, bilateral mass lesions may simulate adult polycystic kidney disease. Ultrasound can differentiate the intensely echogenic masses of angiomyolipomas from the echo-free cysts of polycystic disease. The characteristic high fat content of an angiomyolipoma can be well demonstrated by computed tomography.

Arteriography demonstrates angiomyolipomas as hypervascular lesions that are typically associated with a circumferential arrangement of feeding vessels, multiple aneurysmal dilatations of medium-sized arteries, pools of contrast material, and a sunburst, or whorled, nephrographic pattern. However, hypernephromas may have an identical angiographic appearance. Arteriovenous shunting with early venous filling, often seen in hypernephromas, is unusual with angiomyolipomas.[3–5]

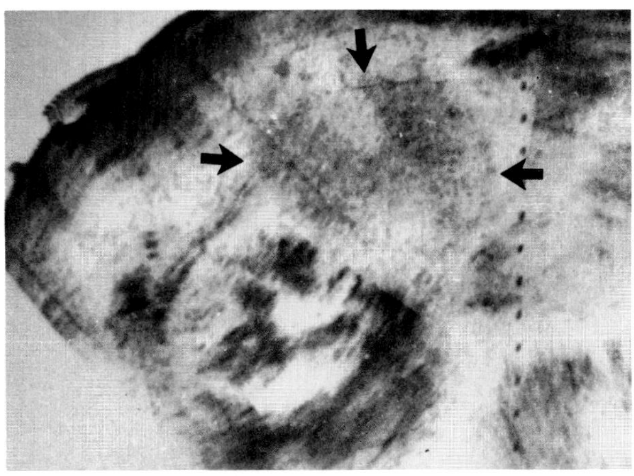

A

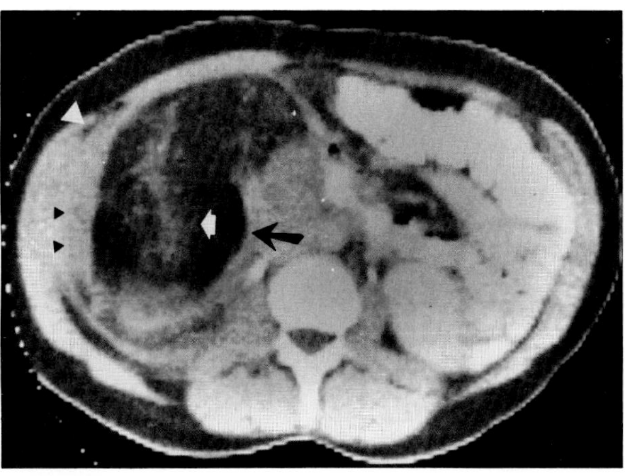

B

**Figure 41.5. Renal hamartoma.** (A) A transverse supine sonogram demonstrates a generally echogenic mass (arrows) arising from the kidney. Less echogenic foci within the mass represent hemorrhage within the fatty tumor. (B) A CT scan 5 cm caudally shows a fatty mass (long arrow) intermixed with (short arrow) and surrounded by (arrowheads) areas of tissue density, representing intratumoral and perinephric hemorrhage, respectively. (From Hartman, Goldman, Friedman, et al,[5] with permission from the publisher.)

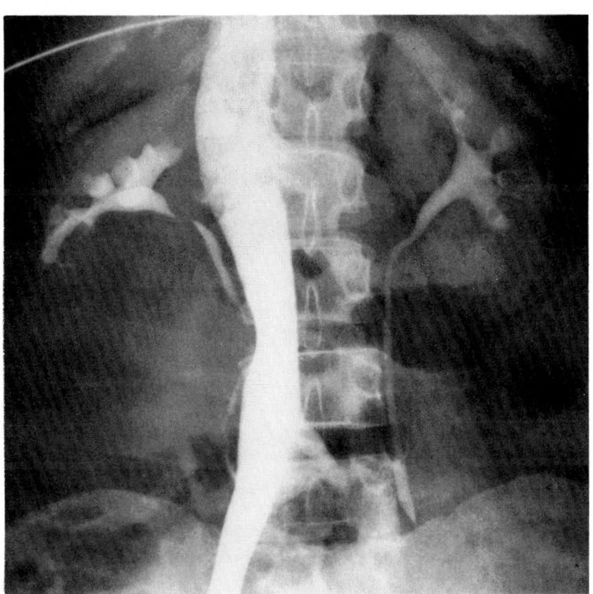

A

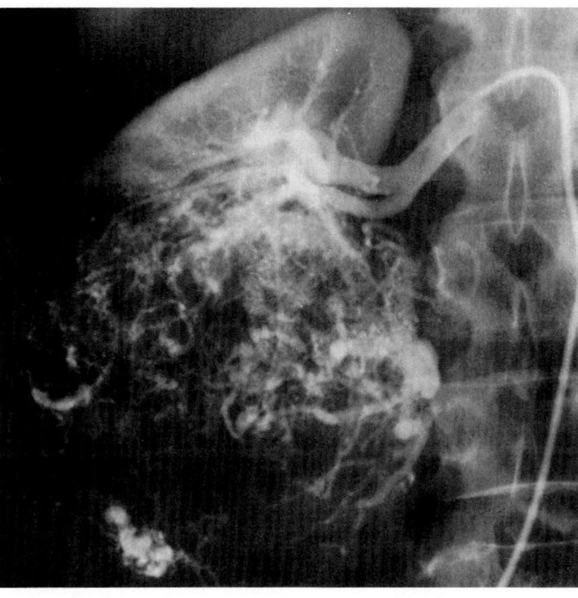

B

**Figure 41.6. Renal hamartoma.** (A) A combined excretory urogram and inferior vena cavogram shows a large mass in the lower pole of the right kidney with displacement but no invasion of the pelvocalyceal system and inferior vena cava. (B) Arteriography demonstrates the mass to be hypervascular. The overall radiographic appearance is indistinguishable from renal cell carcinoma.

### Reninoma

Reninomas are uncommon benign tumors of the juxtaglomerular cells. These small tumors are associated with increased renin secretion and hypertension, both of which usually disappear with resection of the tumor. Reninomas appear as solid tumors on ultrasound and are hypovascular on arteriography. Because the tumor tends to be isodense with normal renal parenchyma on computed tomography (CT), it often may only be seen after contrast enhancement.[6]

## RENAL CYST

Simple renal cysts are the most common unifocal masses of the kidney. They are fluid-filled and usually unilocular, though septa sometimes divide the cyst into chambers that may or may not communicate with each other. Cysts vary in size and may occur at single or multiple sites in one or both kidneys. Thin curvilinear calcifications can be demonstrated in the wall of about 3 percent of simple renal cysts. However, this peripheral curvilinear calcification is not pathognomonic of a be-

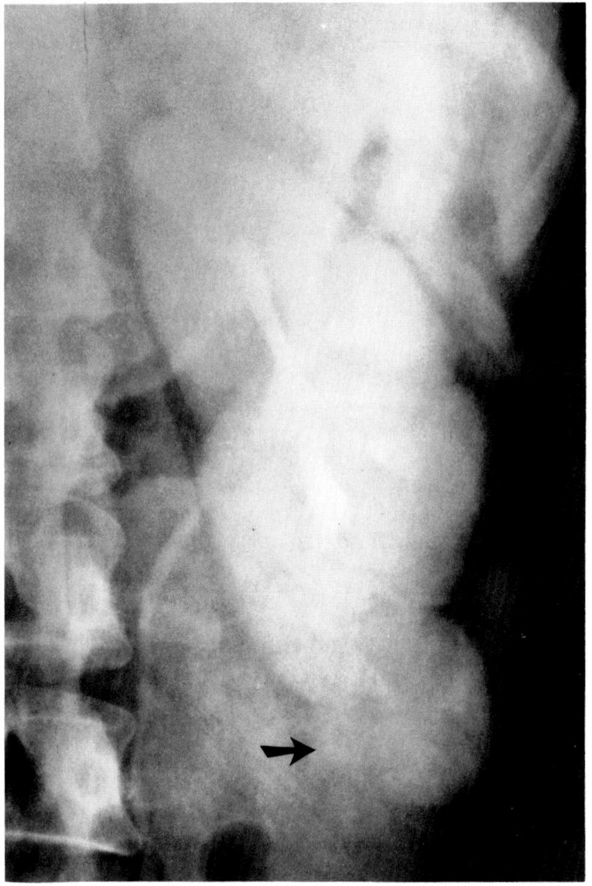

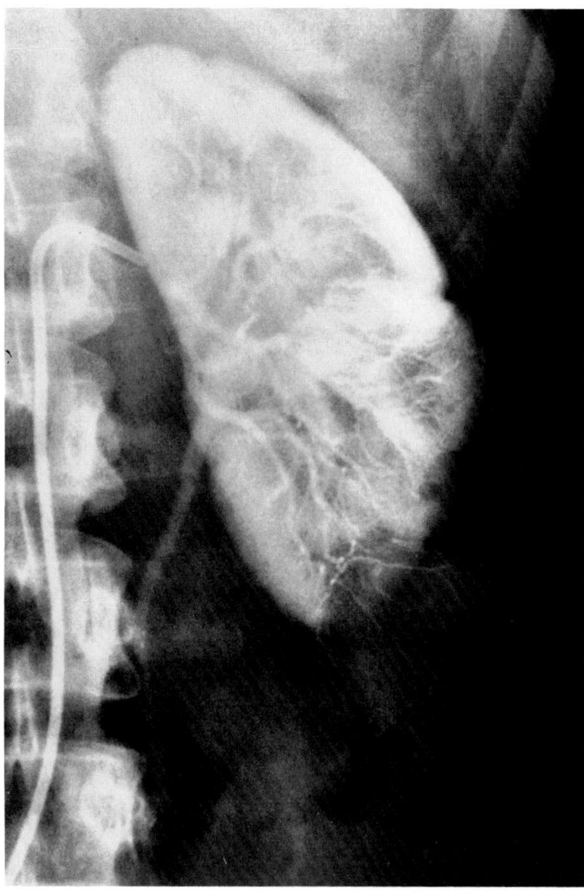

A B

**Figure 41.7. Reninoma.** (A) An excretory urogram reveals a 3.5-cm mass arising from the lower pole of the left kidney (arrow). (B) A selective left renal arteriogram confirms the tumor and demonstrates its hypovascular nature. (From Dunnick, Hartman, Ford, et al,[6] with permission from the publisher.)

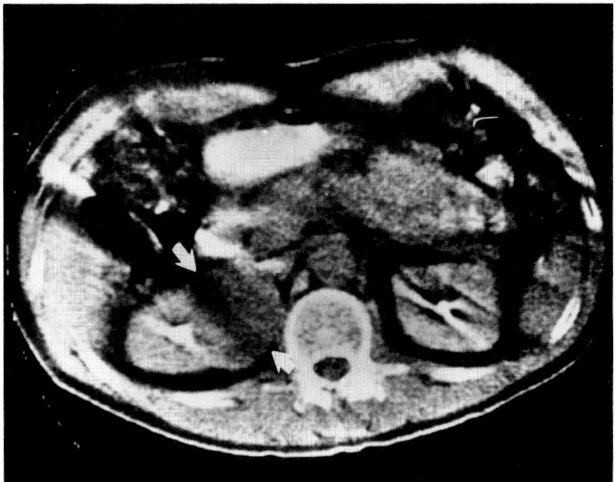

**Figure 41.8. Reninoma.** A CT examination in another patient demonstrates the tumor as a discrete mass (arrow) that does not enhance as much as the normal kidney following contrast administration. (From Dunnick, Hartman, Ford, et al,[6] with permission from the publisher.)

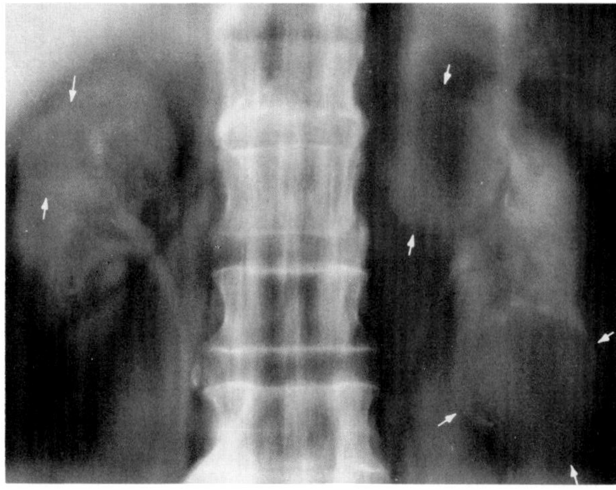

**Figure 41.9. Multiple renal cysts.** A nephrotomogram demonstrates bilateral renal cysts (arrows).

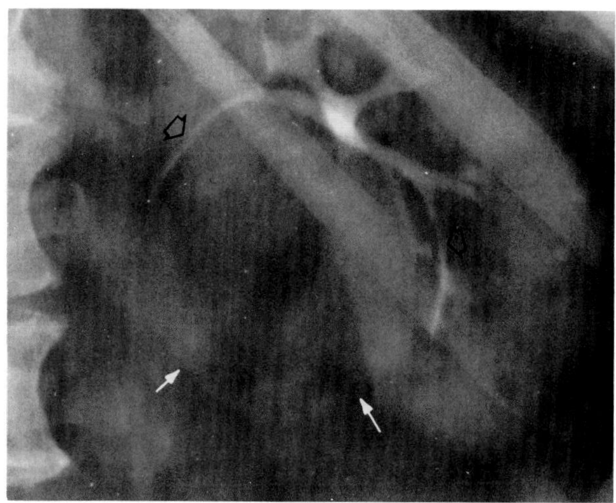

**Figure 41.10. Renal cyst.** An excretory urogram demonstrates a large left renal cyst causing smooth displacement of the attenuated lower pole calyces (open arrows). The solid arrows indicate the inferior extent of the cyst.

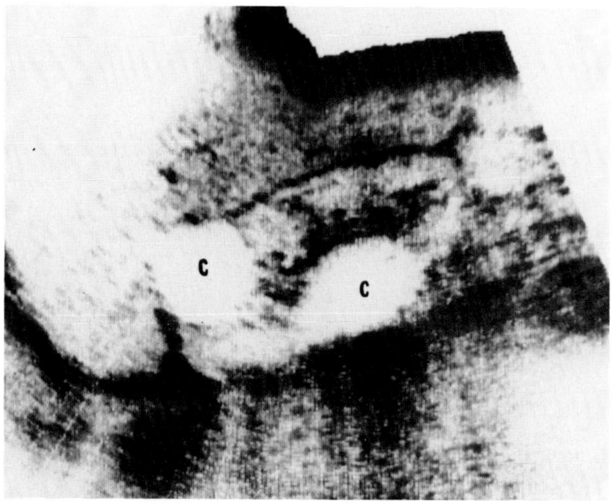


**Figure 41.11. Renal cyst.** Ultrasound demonstrates two renal cysts as anechoic fluid-filled masses (C) with strongly enhanced posterior walls. (From Eisenberg,[34] with permission from the publisher.)

nign process, since in one series 20 percent of the kidney masses with this pattern proved to be malignant lesions.

As a simple renal cyst slowly increases in size, its protruding portion elevates the adjacent edges of the cortex. This cortical margin appears on nephrotomography or arteriography as a very thin, smooth radiopaque rim about the bulging lucent cyst (beak sign). Although the beak sign is generally considered to be characteristic of benign renal cysts, it is merely a reflection of the slow expansion of the mass and thus may occasionally be seen in slow-growing solid lesions, including carcinoma. Thickening of the rim about a lucent mass suggests bleeding into a cyst, cyst infection, or a malignant lesion. Renal cysts cause focal displacement of adjacent portions of the pelvocalyceal system. The displaced, attenuated collecting structures remain smooth, unlike the shagginess and obliteration that often occurs when focal displacement is due to a malignant neoplasm.

Ultrasound is the modality of choice in distinguishing fluid-filled simple cysts from solid mass lesions. Fluid-filled masses classically appear as anechoic structures with strongly enhanced posterior walls, in contrast to solid or complex lesions that appear as echo-filled masses without posterior wall enhancement. CT is also highly accurate in detecting and characterizing simple renal cysts. On unenhanced scans, the cyst has a uniform attenuation value near that of water. Following the injection of contrast material, a simple cyst becomes more apparent as the contrast material is concentrated by the surrounding normal parenchyma. The cyst itself shows no change in attenuation value, unlike a solid renal neoplasm, which always shows a small but definite increase in density.

Because cystic necrosis can occur within an otherwise solid tumor and since a tumor can arise from the wall

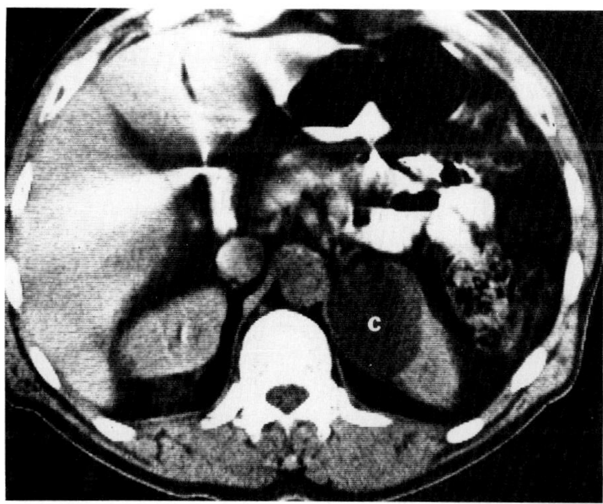


**Figure 41.12. Renal cyst.** A CT scan shows a benign renal cyst as a nonenhancing left renal mass (C) with a sharply marginated border and thin wall. (From Eisenberg,[34] with permission from the publisher.)

of a cyst, many symptomatic sonolucent masses are evaluated by percutaneous needle aspiration under ultrasound guidance. Fluid from a renal cyst can be clearly differentiated from that obtained from an abscess or renal tumor in terms of its color, bacteria, cells, fat and protein content, and activity of lactic dehydrogenase. The introduction of contrast material or air after the cyst fluid has been removed demonstrates the inner wall of the cyst and further decreases the possibility of missing a malignant neoplasm. Minor irregularities of the walls of simple cysts are not uncommon. However, a markedly irregular

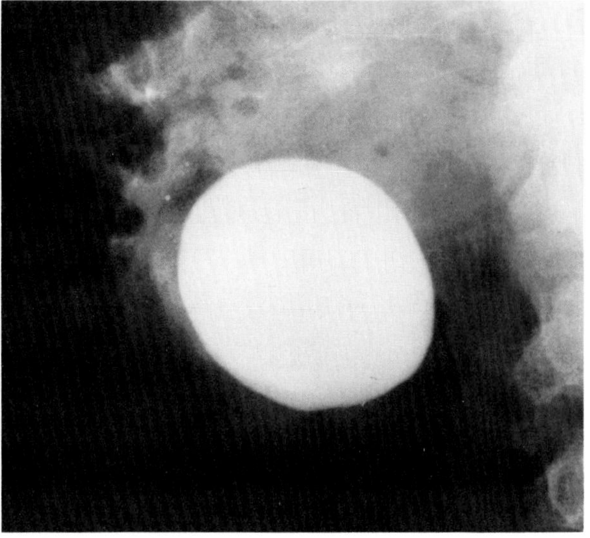

A

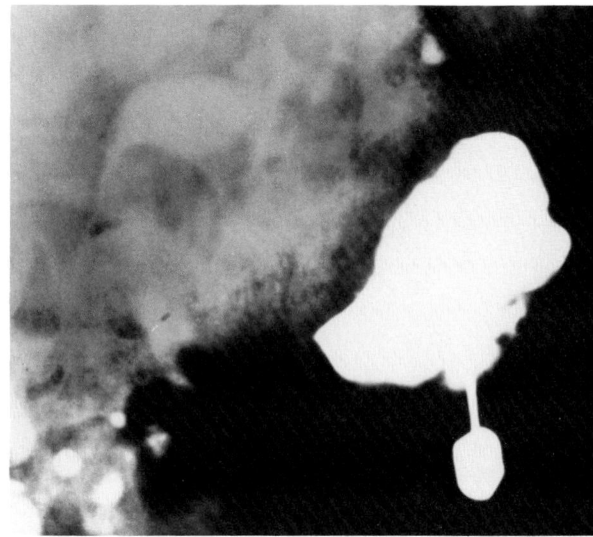

B

**Figure 41.13. Renal cyst puncture.** (A) The instillation of contrast material shows the smooth inner wall characteristic of a benign cyst. (B) In another patient, the introduction of contrast material reveals the markedly irregular inner border of a necrotic hypernephroma. (From Eisenberg,[34] with permission from the publisher.)

inner border may be the only sign of a tumor in that location.

Because of the accuracy of CT and ultrasound, cyst puncture is rarely necessary if these modalities provide unequivocal evidence of a simple cyst. However, because abscesses, cystic or necrotic tumors, and inflammatory or hemorrhagic cysts can mimic simple cysts, cyst puncture should be performed if there is an atypical appearance or a strong clinical suspicion of abscess, or if the patient has hematuria or hypertension.

Ultrasound and CT have virtually precluded the need for routine arteriography of renal cysts. The nephrogram phase of arteriography can demonstrate the beak and thin rim signs better than urography and clearly indicates the avascular nature of a simple cyst. The presence of neovascularity, abnormal stains within the mass, early venous opacification, or pericapsular draining veins essentially eliminates the possibility of a simple cyst.[3,7–13]

## RENAL PSEUDOTUMOR

Congenital renal pseudotumors (columns of Bertin) are masses of normal cortical tissue that are located within the renal parenchyma. They most commonly develop at the junction of the middle and upper thirds of a duplex kidney. The combination of a focal "mass" apparently arising from the surface of the kidney with compression and splaying of the pelvocalyceal system may produce an appearance simulating a true neoplasm. On the nephrogram phase of arteriography, the pseudotumor appears as a masslike nodule with a dense blush. It usually is surrounded on its medial aspect by a radiolucent halo of peri-infundibular fat. The correct diagnosis can be made by radionuclide scanning, which reveals normal or even increased uptake of isotope in the pseudotumor in contrast to the decreased uptake in other renal masses.[7,14]

## MALIGNANT RENAL TUMORS

### Renal Cell Carcinoma (Hypernephroma)

Renal cell carcinoma is the most common renal neoplasm, predominantly occurring in patients over the age of 40. About 10 percent of hypernephromas contain calcification that is usually located in reactive fibrous zones about areas of tumor necrosis. In the differentiation of solid tumors from fluid-filled benign cysts, the location of calcium within the mass is more important than the pattern of calcification. Of all masses containing calcium in a nonperipheral location, almost 90 percent are malignant. Although peripheral curvilinear calcification is much more suggestive of a benign cyst, hypernephromas can have a calcified fibrous pseudocapsule that presents an identical radiographic appearance.

Hypernephromas typically produce urographic evidence of localized bulging or generalized renal enlargement. The tumor initially causes elongation of adjacent calyces; progressive enlargement and infiltration lead to distortion, narrowing, or obliteration of part or all of the collecting system. Large tumors may partially obstruct the pelvis or upper ureter and cause proximal dilatation.

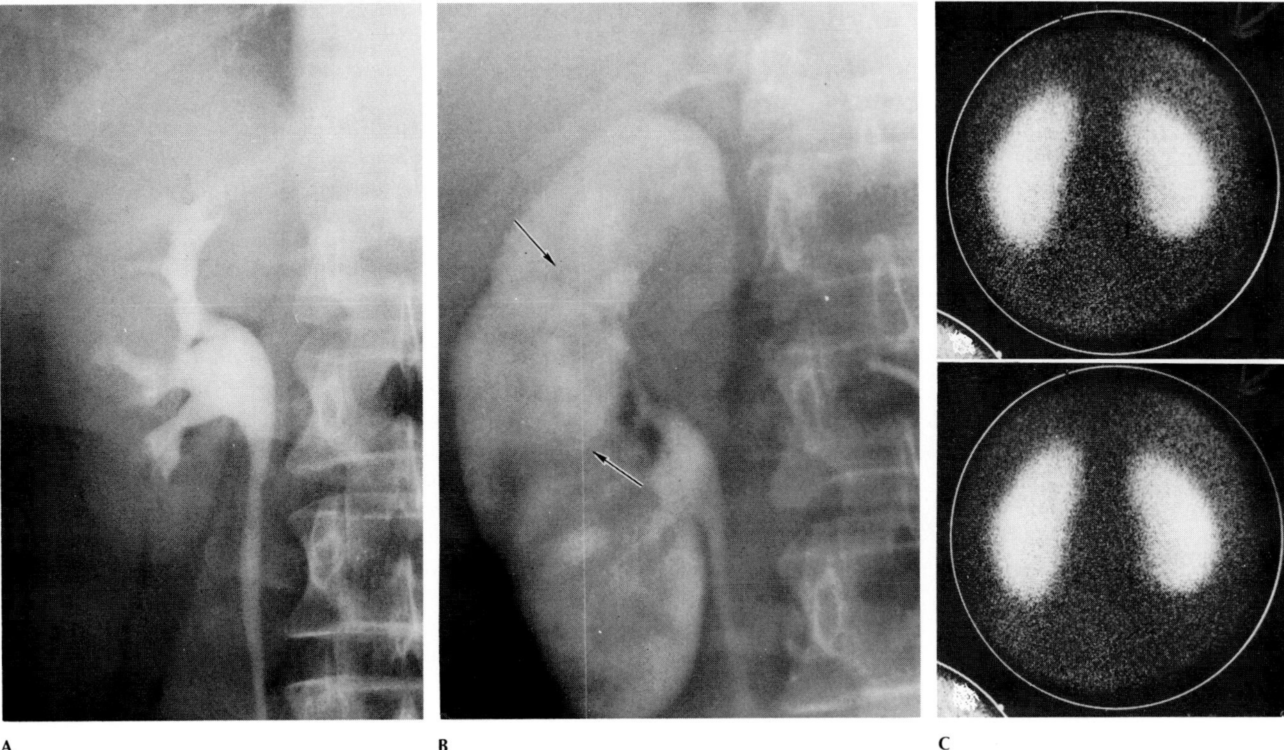


**Figure 41.14. Congenital renal pseudotumor.** (A) An excretory urogram shows displacement of the upper calyceal system of the right kidney. (B) A film from the nephrogram phase of a selective right renal arteriogram shows the large cortical column (arrows), which appears denser than the medullary substance. (C) A renal scan shows normal functioning parenchyma. (From Ney and Friedenberg,[14] with permission from the publisher.)

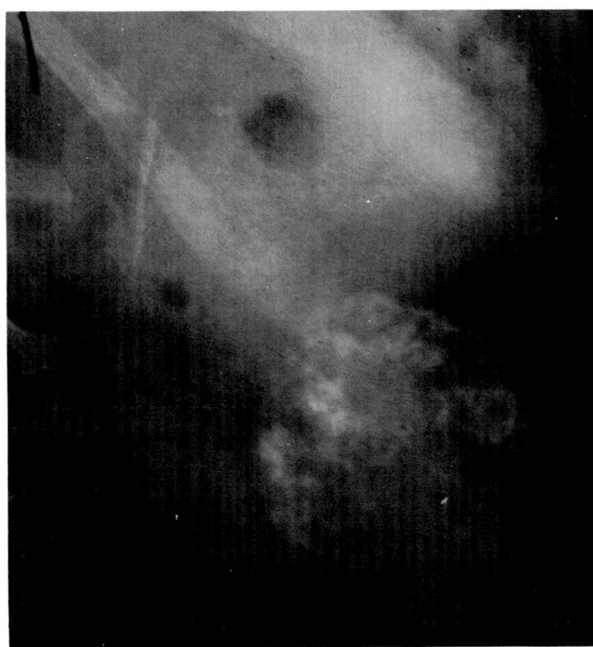

**Figure 41.15. Calcification in a renal cell carcinoma.** Mottled or punctate calcification that appears to be within a mass, especially in the absence of peripheral calcification, is highly indicative of a malignant lesion. (From Daniel, Hartman, and Witten,[10] with permission from the publisher.)

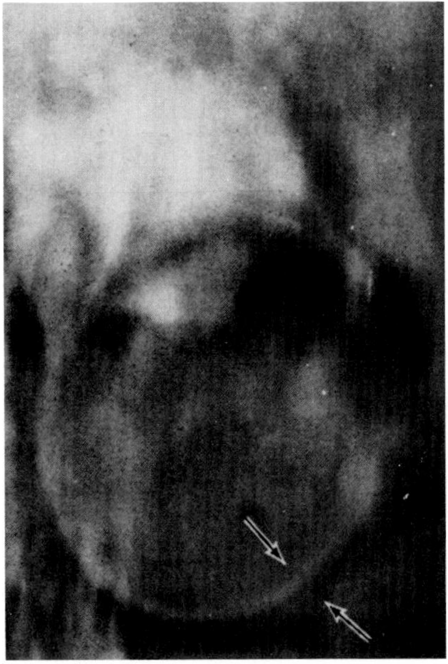

**Figure 41.16. Renal cell carcinoma.** A nephrotomogram demonstrates a lucent, well-demarcated renal mass with a thick wall (arrows). (From Bosniak and Faegenburg,[35] with permission from the publisher.)

Complete loss of function on excretory urography usually indicates tumor invasion of the renal vein.

On nephrotomography, a hypernephroma generally appears as a mass with indistinct outlines and a density similar to that of normal parenchyma, unlike the classic radiolucent mass with sharp margins and a thin wall that represents a benign cyst. Necrotic neoplasms can also appear cystic, though they are usually surrounded by thick, irregular walls.

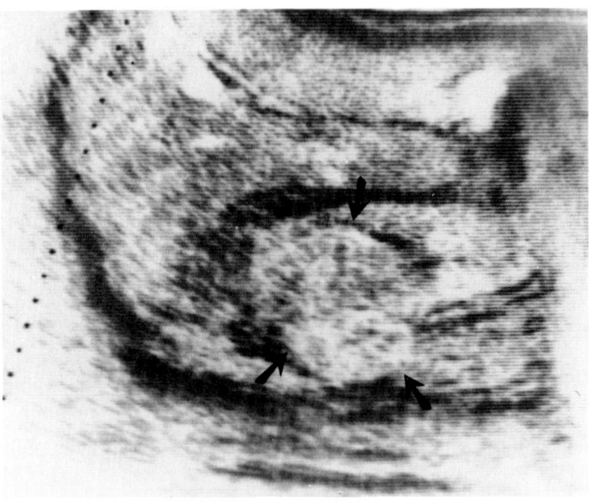

**Figure 41.17. Renal cell carcinoma.** Ultrasound demonstrates an echo-filled solid mass (arrows) without evidence of posterior enhancement.

On ultrasound, a renal cell carcinoma typically appears as a solid mass with numerous internal echoes and no evidence of acoustic enhancement. Unlike renal cysts, hypernephromas often demonstrate an irregular or poorly visualized interface with the remaining normal renal parenchyma. However, since cystic necrosis can occur within an otherwise solid tumor and, conversely, a tumor can arise from the wall of a cyst, sonolucent masses are often evaluated by percutaneous needle aspiration (under ultrasonic guidance) followed by an analysis of the cyst fluid and the introduction of contrast material or air (see the section "Renal Cyst" in this chapter). Ultrasound can also demonstrate the extension of renal tumor into the renal vein and inferior vena cava as well as metastases to the retroperitoneal lymph nodes.

CT is of great value in both the diagnosis and the staging of hypernephroma. On unenhanced scans, a solid neoplasm is often inhomogeneous and has an attenuation value near that of normal renal parenchyma, unlike a simple cyst, which has a uniform attenuation value near that of water. Following the injection of contrast material, a simple cyst shows no change in attenuation value, while a solid renal neoplasm demonstrates a small but definite increase in density that is probably due primarily to vascular perfusion. However, this increased density is much less than that of the surrounding normal parenchyma, which also tends to concentrate the contrast material, and thus renal neoplasms become more apparent on contrast-enhanced scans. As with conventional nephrotomography, the CT demonstration of a thickened or indefinable wall enclosing an otherwise cystic mass projecting beyond the renal contours makes

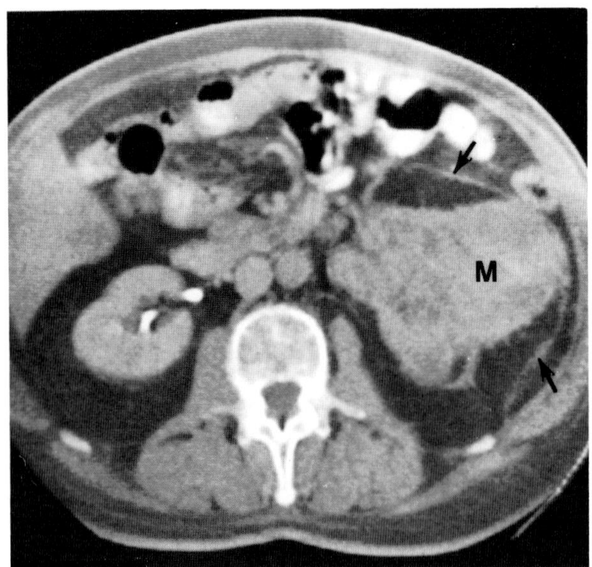


A

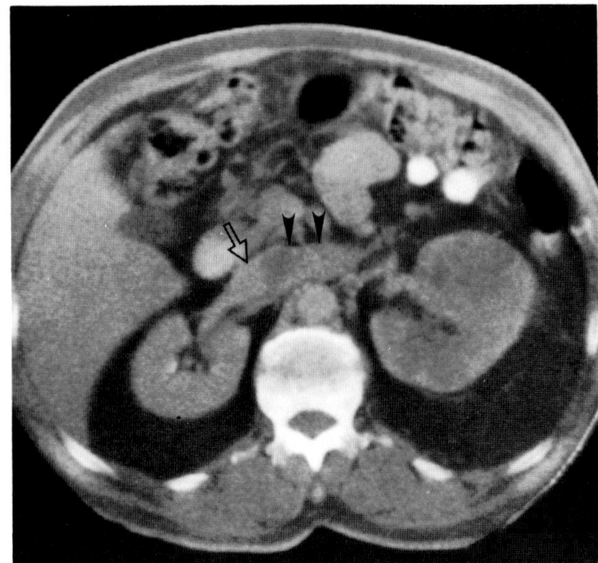

B

**Figure 41.18. Renal cell carcinoma with left renal vein invasion.** (A) A CT scan shows a large carcinoma (M) of the left kidney with thickening of Gerota's fascia (arrows). (B) A postcontrast scan 2 cm cephalad shows a dilated left renal vein filled with tumor thrombus (arrowheads). The thrombus extends to the inferior vena cava (open arrow). (From McClennan and Lee,[24] with permission from the publisher.)

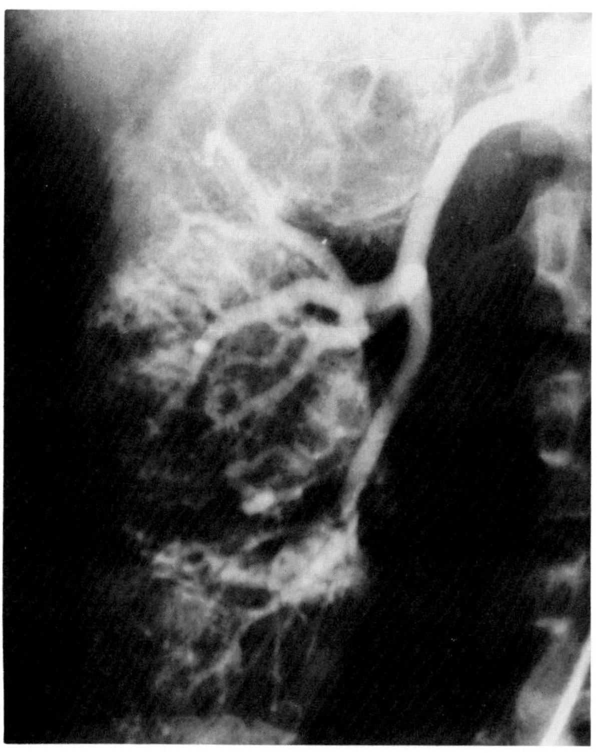

A

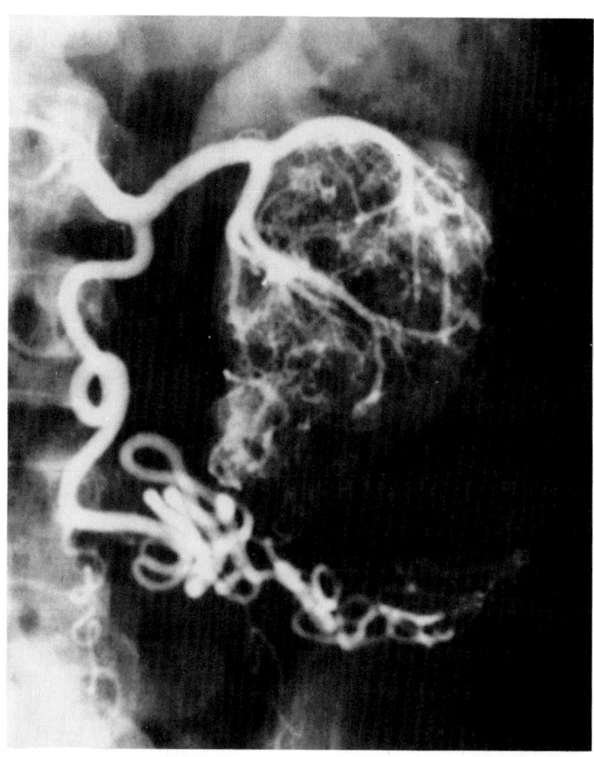

B

**Figure 41.19. Bilateral renal cell carcinomas.** (A) Right and (B) left renal arteriograms demonstrate large hypervascular masses in both kidneys.

the diagnosis of a simple benign cyst less likely and suggests an infected cyst, an abscess, or a neoplasm.

CT is an accurate method for detecting the local and regional spread of hypernephroma. It can usually distinguish between neoplasms confined to the renal capsule and those that have extended beyond it. Spread of the neoplasm to Gerota's fascia, the diaphragmatic crura, and the psoas, paraspinal, and abdominal wall muscles can be well shown. CT is also the most accurate method for detecting enlargement of para-aortic, paracaval, and retrocrural lymph nodes as well as spread to the ipsilateral renal vein and inferior vena cava.

Most renal cell carcinomas are hypervascular and demonstrate arteriographic evidence of neovascularity (fine tortuous tumor vessels of irregular size that are chaotically distributed throughout the mass). Microaneurysms, small arteriovenous communications, and lakes of pooled contrast material are often seen. Multiple arteriovenous communications may cause early opacification of the main renal vein. Tumor necrosis appears as areas of radiolucency within the mass.

Some hypernephromas are hypovascular and appear on arteriography as radiolucent masses with a relatively sharp interface between the tumor and adjacent normally stained renal tissue. Occasionally, an avascular tumor may simulate a benign cyst. Although these avascular or hypovascular tumors may show no detectable pathologic vessels, the nephrogram usually demonstrates the lucent

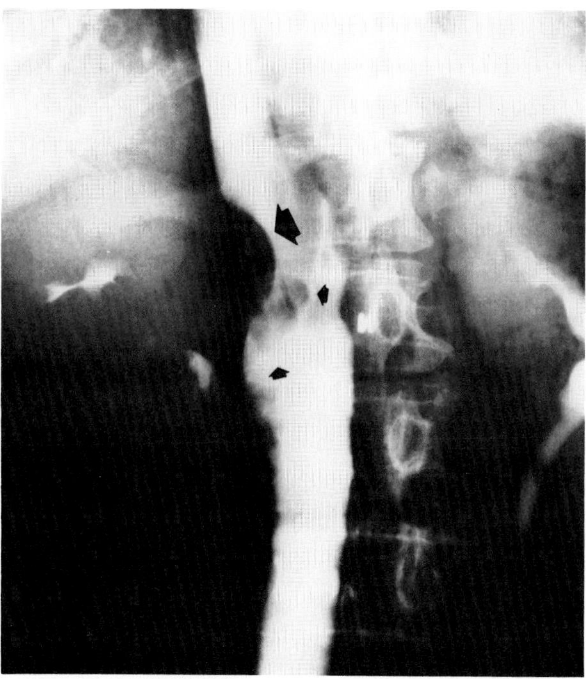

**Figure 41.20. Renal cell carcinoma.** An inferior vena cavogram demonstrates the extension of tumor thrombus (small arrows) from the right renal vein. Metastatic involvement of lymph nodes produced the large extrinsic impression on the inferior vena cava (large arrow). (From Eisenberg,[34] with permission from the publisher.)

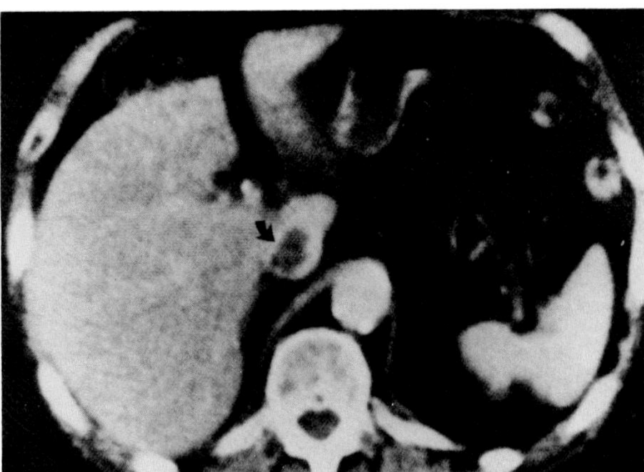

**Figure 41.21. Renal cell carcinoma.** A CT scan shows intraluminal tumor thrombus as a low-density filling defect (arrow) within the inferior vena cava. (From Marks, Korobkin, Callen, et al,[36] with permission from the publisher.)

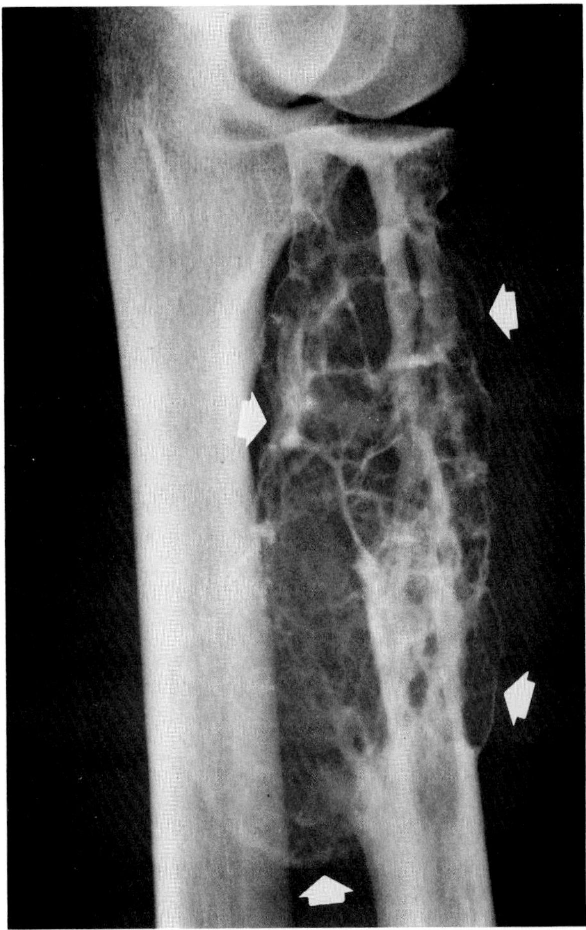

**Figure 41.22. Renal cell carcinoma metastatic to bone.** A typical expansile bubbly lesion (arrows) in the proximal shaft of the radius.

mass to have an irregular margin, often with a thickened wall.

Local renal artery infusion of epinephrine a few seconds before the injection of contrast material may be used to further evaluate patients with nonvascular or minimally vascular tumors. It can also be employed in patients in whom it is difficult to categorize a lesion in the peripelvic region because of overlying arteries. Epinephrine causes constriction of normal arteries, although it has little effect on the neovascularity associated with tumor vessels (without musculature in their walls). As the normal vessels constrict, the neoplastic vessels are accentuated and produce a "tumor blush."

Another pharmacoangiographic technique is to inject a vasodilator (tolazoline, papaverine) before contrast injection. The vasodilator may permit small abnormal feeding vessels (invisible on conventional arteriography) to enlarge sufficiently to become visible. This technique is of special value in differentiating minimally vascular neoplasms from completely avascular simple renal cysts.

In addition to establishing the diagnosis of hypernephroma, arteriography has often been used to demonstrate metastases, to detect tumors in the contralateral kidney (which occur in 1 to 2 percent of cases), and to illustrate the vascular supply to the kidney. CT has effectively replaced arteriography in demonstrating metastases to the liver or abdominal lymph nodes, extension into the inferior vena cava, and the presence of bilateral lesions. There is controversy among surgeons about the need for at least a midstream aortogram to ascertain the number of renal arteries to the affected kidney and to demonstrate any vessels that are distorted by a large mass.

Embolic occlusion of the renal artery to reduce the blood supply to large hypervascular tumors is an interventive technique that can be used in the treatment of renal cell carcinoma. This procedure reduces the bulk of the mass and lessens operative blood loss, thus making the surgical approach easier.[3,7,10,12,15–24,35,36]

## Nephroblastoma (Wilms's Tumor)

Wilms's tumor is the most common abdominal neoplasm of infancy and childhood. The lesion arises from embryonic renal tissue, may be bilateral, and tends to become very large and appear as a palpable mass.

Wilms's tumor must be differentiated from neuroblastoma, a tumor of adrenal medullary origin that is the second most common malignancy in children. Peripheral cystic calcification occurs in about 10 percent of Wilms's tumors, in contrast to the fine, granular or stippled calcification seen in about half of the cases of neuroblastoma. At excretory urography, the intrarenal Wilms's tumor causes marked distortion and displacement of the pelvocalyceal system. The major effect of the extrarenal neuroblastoma is to displace the entire kidney downward and laterally; because the kidney itself is usually not invaded, there is no distortion of the pelvocalyceal system.

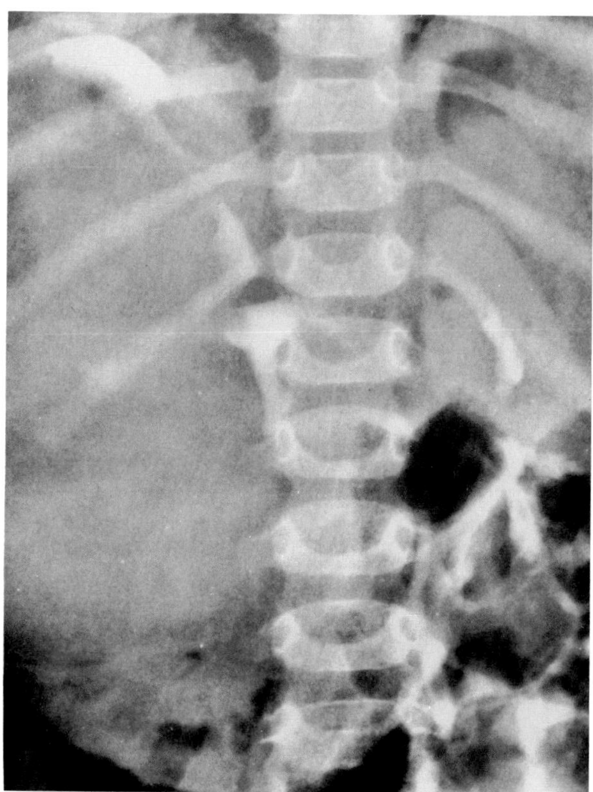

**Figure 41.23. Wilms's tumor.** An excretory urogram demonstrates a huge mass in the right kidney that distorts and displaces the pelvocalyceal sytem. (From Friedland, Filly, Goris, et al,[7] with permission from the publisher.)

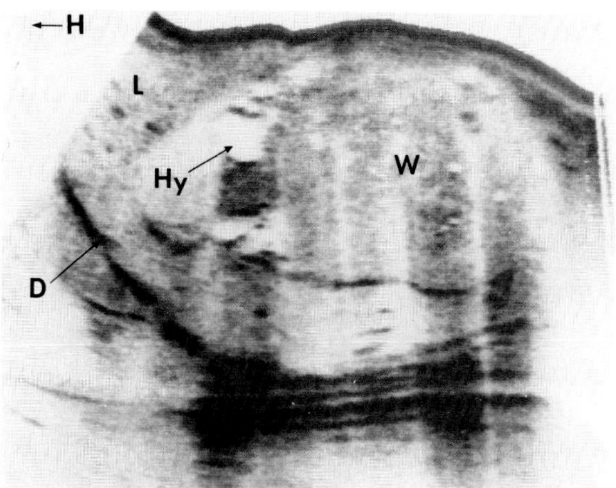


**Figure 41.24. Wilms's tumor.** A parasagittal supine sonogram demonstrates a huge Wilms's tumor (W) involving the lower pole of the right kidney and resulting in hydronephrosis in the upper collecting system (Hy). Wilms's tumors tend to have a moderately low internal echogenicity and, as in this patient, often contain multiple tiny cystic spaces. The large mass dramatically displaces the liver (L). H = head; D = diaphragm. (From Friedland, Filly, Goris, et al,[7] with permission from the publisher.)

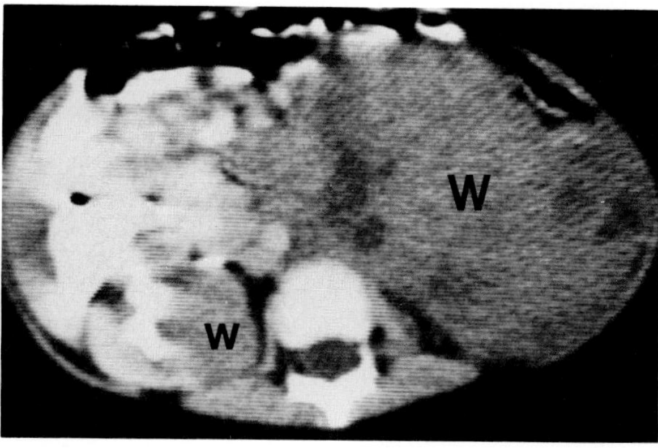


**Figure 41.25. Bilateral Wilms's tumors.** A CT scan shows a huge left renal mass (W) that crosses the midline. A small separate mass (W) also exists in the right kidney. (From Friedland, Filly, Goris, et al,[7] with permission from the publisher.)

Ultrasound is of value in distinguishing Wilms's tumors from hydronephrosis, another major cause of a palpable renal mass in a child. Wilms's tumors typically have a solid appearance with gross distortion of the renal morphology, unlike the precise organization of symmetrically positioned fluid-filled spaces in hydronephrosis. Ultrasound can also demonstrate the intrarenal location of Wilms's tumor, in contrast to the extrarenal origin of a neuroblastoma. Although it entails the use of ionizing radiation, CT can show the full extent of the tumor (including invasion of the inferior vena cava) as well as detecting a recurrence of the neoplasm following surgical removal.[7,25,26]

## TUMORS OF THE URINARY COLLECTING SYSTEM

Tumors of the renal pelvis, ureters, and urinary bladder arise from transitional epithelia. These urothelial cancers are the most important cause of hematuria in patients over the age of 35 years and often occur (and recur) at multiple sites in the urinary tract in an affected patient.

### Bladder

The wall of the urinary bladder is the most common site of involvement of transitional-cell cancer. Plain radiographs may demonstrate punctate, coarse, or linear calcifications that are usually encrusted on the surface of the tumor but occasionally lie within it. On excretory urograms, urothelial bladder cancer appears as one or more polypoid defects arising from the bladder wall or as focal bladder wall thickening. However, urography can detect only about 60 percent of bladder carcinomas because most are small when first symptomatic and are

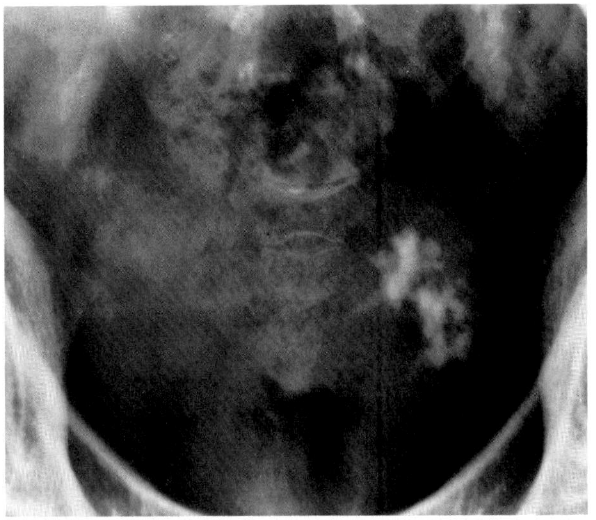

**Figure 41.26. Calcified transitional-cell carcinoma of the bladder.** Coarse tumor calcification was associated with an intravesical mass on excretory urography. (From Eisenberg,[37] with permission from the publisher.)

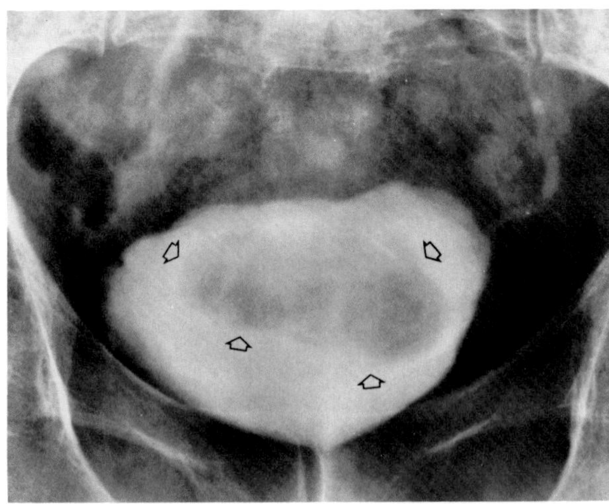

**Figure 41.27. Transitional-cell carcinoma of the bladder.** A large, irregular filling defect (arrows) in the bladder.

located on the trigone, where they can be difficult to visualize. Therefore, all patients with lower urinary tract hematuria should undergo cystoscopy.

CT with full distension of the bladder demonstrates a urothelial tumor as a mass projecting into the bladder lumen or as focal thickening of the bladder wall. This modality is the method of choice for preoperative staging since it can determine the presence and degree of extravesical distension and pelvic side wall involvement of a bladder neoplasm. It can also demonstrate enlarged pelvic or para-aortic lymph nodes and thus indicate a site for percutaneous aspiration biopsy.[7,27–29]

## Renal Pelvis

Transitional-cell cancers of the renal pelvis appear as smooth or irregular polypoid filling defects on excretory urograms. Although these nonopaque filling defects in the pelvocalyceal system must be considered cancer until proved otherwise, a number of other conditions (stones, blood clots, sloughed papillae, fungus balls, pyeloureteritis cystica) can produce an identical apperance. The presence of a stippled pattern of contrast material within the filling defect suggests a transitional-cell tumor, since it reflects trapping of small amounts of contrast material within the interstices of papillary tumor fronds. All patients suspected of having a carcinoma involving the upper collecting system require cystoscopy to exclude a coincident bladder cancer.

At arteriography, the tumor is relatively hypovascular with no pooling of contrast material or arteriovenous shunting. Fine neovascularity may become evident after the administration of a vasodilator. In preliminary studies, CT has had a high degree of accuracy in detecting and characterizing filling defects of the renal pelvis and ureter.[30–32]

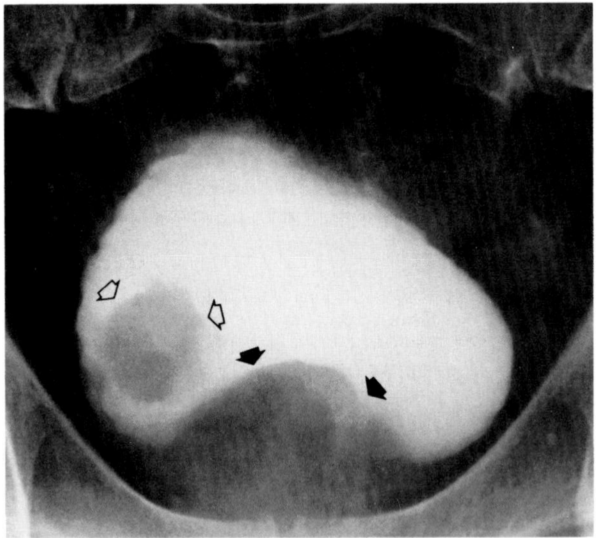

**Figure 41.28. Transitional-cell carcinoma of the bladder.** An irregular filling defect on the right side of the bladder (open arrows). Note the large filling defect (closed arrows) at the base of the bladder representing benign prostatic hypertrophy.

## Ureter

Transitional-cell carcinomas of the ureter are rare. Most appear as smooth or mulberry-shaped filling defects. There is often localized dilatation of the ureter below the level of an expanding intraluminal tumor, in contrast to ure-

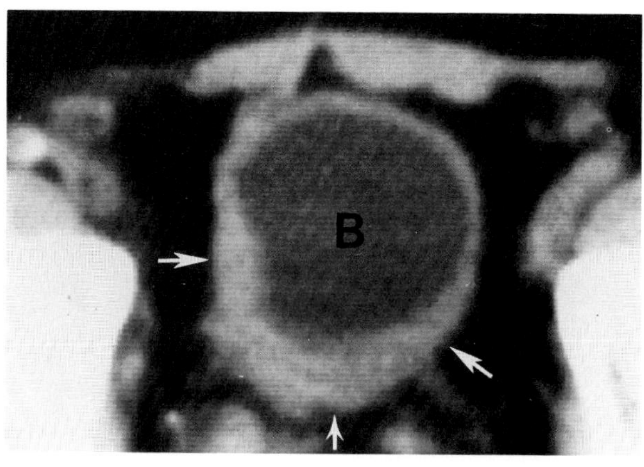


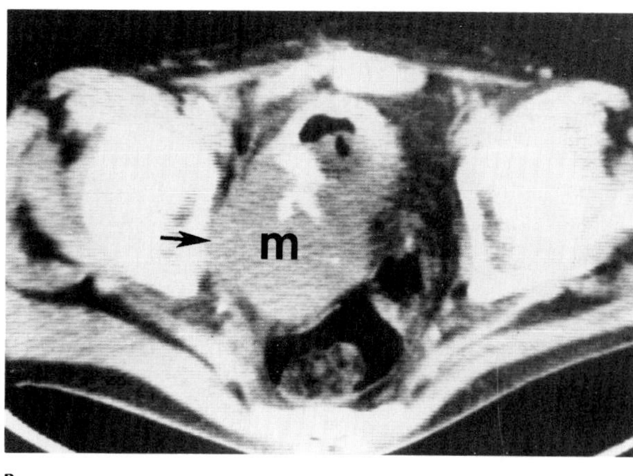


A B

**Figure 41.29. CT of bladder carcinoma.** (A) Focal thickening of the posterior wall (arrows). (B) In this patient, an extensive bladder cancer (m) has extended to involve the side wall of the pelvis (arrow). (From Friedland, Filly, Goris, et al,[7] with permission from the publisher.)

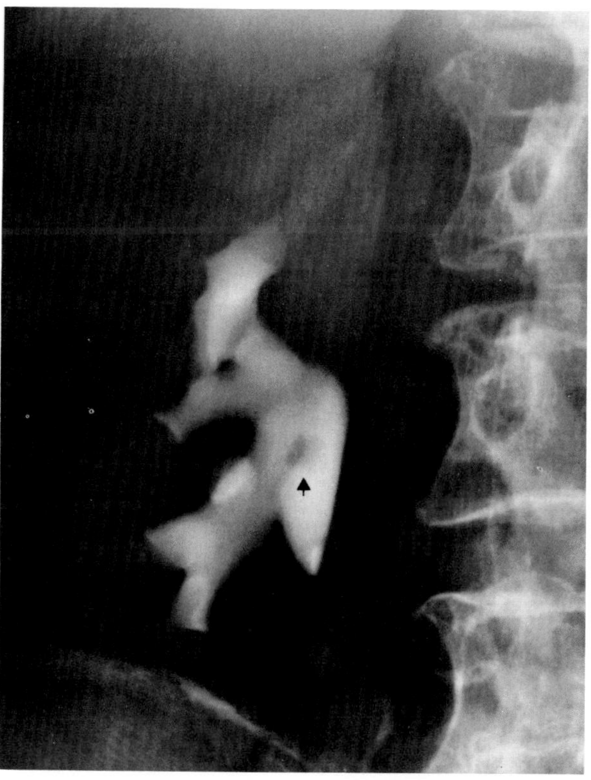

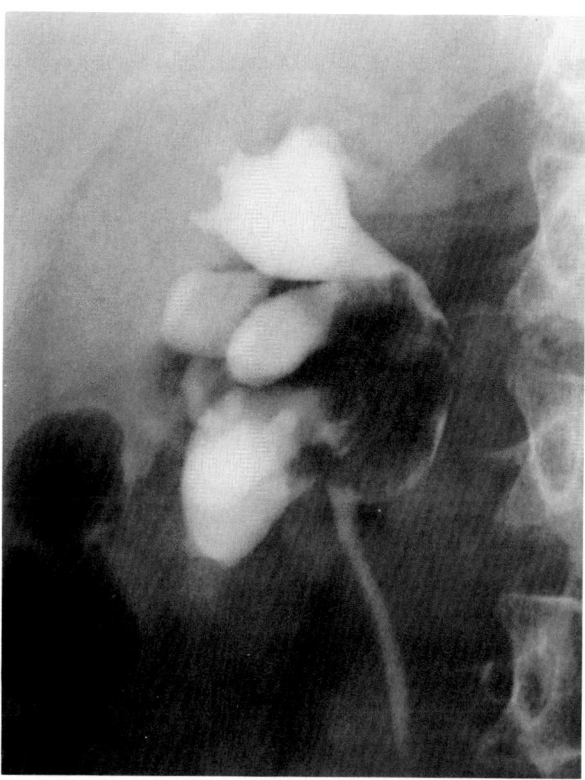

A B

**Figure 41.30. Transitional-cell carcinomas of the renal pelvis in two patients.** (A) A small filling defect (arrow) in the renal pelvis simulates a blood clot, stone, fungus ball, or sloughed papilla. (B) A huge mass fills virtually all of the renal pelvis.

teral collapse distal to an obstructing stone. Retrograde pyelography may demonstrate a characteristic meniscus appearance to the superior border of the contrast material, producing a wineglass sign outlining the lower margin of the tumor.

Urothelial carcinomas can also appear as ureteral strictures. These are usually irregular in shape with overhanging margins, though they can also appear completely smooth and concentric with tapering edges. Carcinomatous strictures may simulate ureteral narrowing due to inflammatory processes, calculi, or extrinsic compression.[14,33]

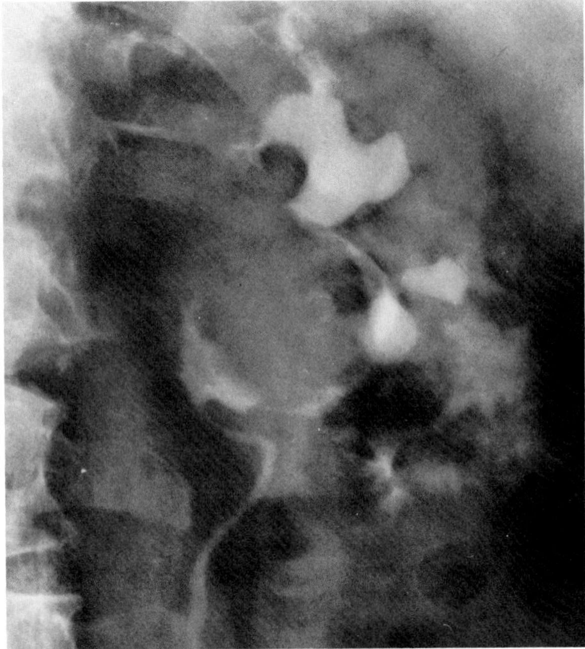

**Figure 41.31. Transitional-cell carcinoma of the renal pelvis.** A large mass causes distortion of the pelvocalyceal system and deviation of the ureter.

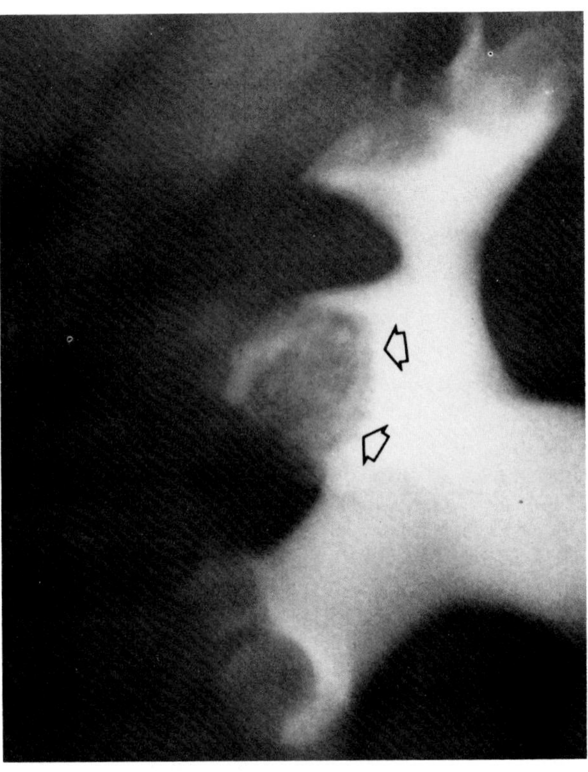

**Figure 41.32. Transitional-cell carcinoma of the renal pelvis.** A small filling defect occupies an interpolar calyx (arrow). Although the defect might at first glance be mistaken for a large but otherwise normal papilla, the numerous small contrast stipples as well as the suggestively irregular border make its neoplastic nature evident. (From McLean, Pollack, and Banner,[29] with permission from the publisher.)

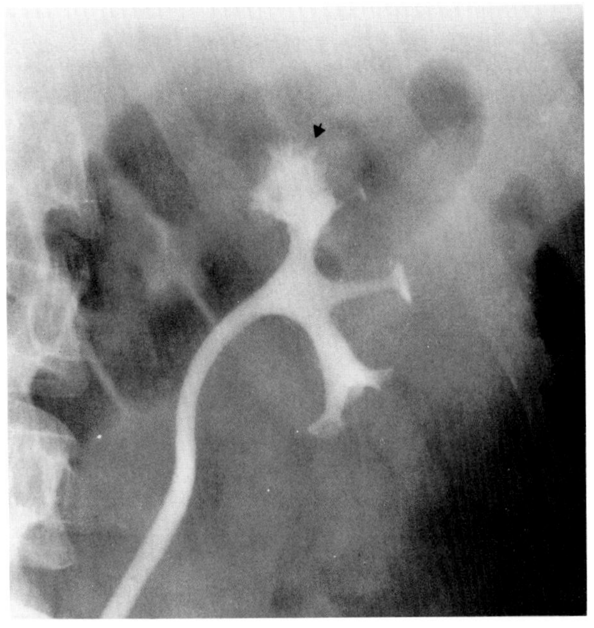

A

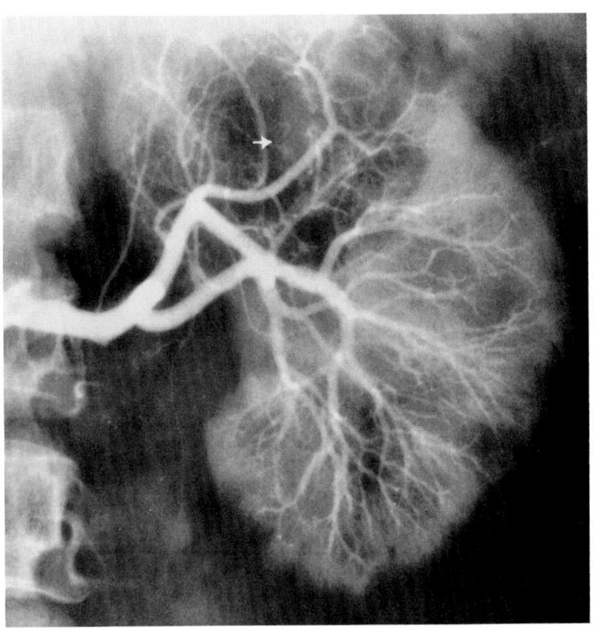

B

**Figure 41.33. Transitional-cell carcinoma of the renal pelvis.** (A) Excretory urogram demonstrates striking irregularity of the upper pole calyx (arrow) and a surrounding mass. (B) At arteriography the tumor is relatively hypovascular but with some fine neovascularity (arrow).

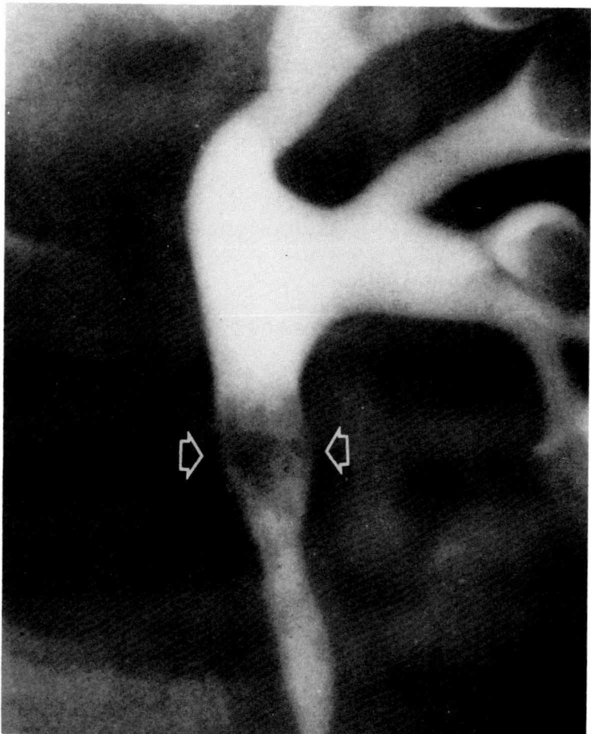

**Figure 41.34. Transitional-cell carcinoma of the ureter.** The presence of stippling throughout this proximal ureteral filling defect (arrows) and the suggestive papillary contour of its proximal and distal margins allowed the correct diagnosis of transitional-cell carcinoma to be made preoperatively. (From McLean, Pollack, and Banner,[31] with permission from the publisher.)

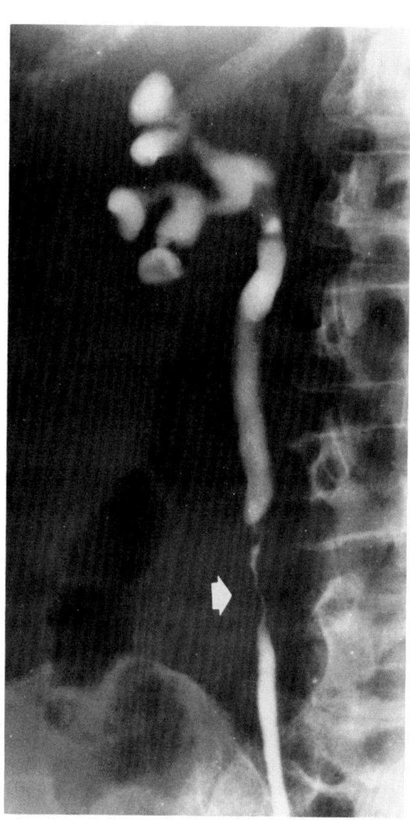

**Figure 41.35. Transitional-cell carcinoma of the ureter.** There is an irregular stricture (arrow) with proximal ureteral and pelvocalyceal dilatation.

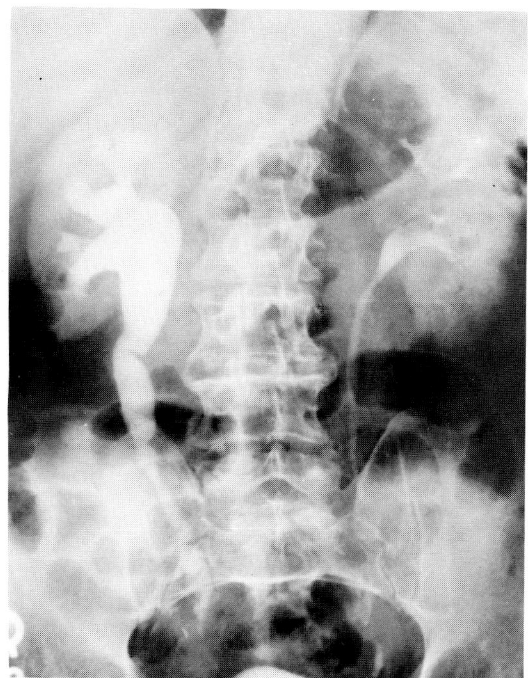

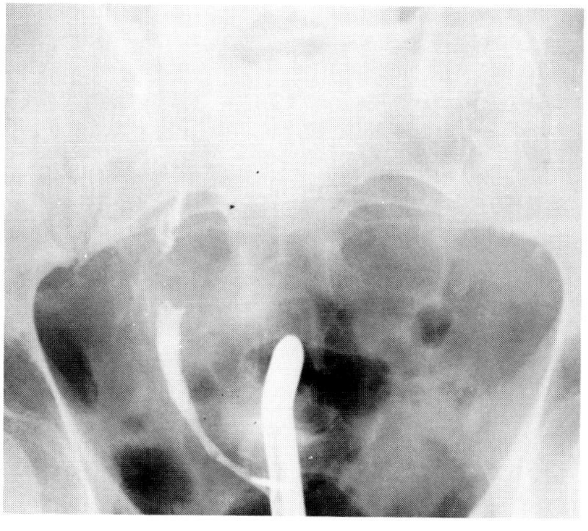

B

**Figure 41.36. Transitional-cell carcinoma of the ureter.** (A) An excretory urogram shows right pyelocaliectasis and proximal ureterectasis from a distal ureteral obstruction. The obstructing tumor is not well defined. (B) A retrograde urogram clearly shows the characteristic cupped appearance of the obstructing ureteral neoplasm. (From Ney and Friedenberg,[14] with permission from the publisher.)

## REFERENCES


1. Rabinowitz JG, Wolf BS, Goldman RH: Roentgen features of renal adenomas. *Radiology* 84:263–269, 1965.
2. Antoine JE, Kopperman M: Renal adenoma: Retrospect and prospect. *AJR* 119:727–730, 1973.
3. Davidson AJ: *Radiologic Diagnosis of Renal Parenchymal Disease*, Philadelphia, WB Saunders, 1977.
4. Becker JA, Kinkhabwala M, Pollack H, et al: Angiomyolipoma (hamartoma) of the kidney. An angiographic review. *Acta Radiol (Diagn)* 14:561–568, 1973.
5. Hartman DS, Goldman SM, Friedman AC, et al: Angiomyolipoma: Ultrasonic-pathologic correlation. *Radiology* 139: 451–458, 1981.
6. Dunnick NR, Hartman DS, Ford KK, et al: The radiology of juxtaglomerular tumors. *Radiology* 147:321–326, 1983.
7. Friedland GW, Filly R, Goris ML, Gross D, Kempson RL, Korobkin M, Thurber BD, and Walter J (eds): *Uroradiology: An Integrated Approach*. New York, Churchill Livingstone, 1983.
8. McClennan BL, Stanley RJ, Melson GL, et al: CT of the renal cyst: Is cyst aspiration necessary? *AJR* 133:671–675, 1979.
9. Silverman JF, Kilhenny C: Tumor in the wall of a simple renal cyst. Report of a case. *Radiology* 93:95–98, 1969.
10. Daniel WW, Hartman GW, Witten DM, et al: Calcified renal masses. A review of 10 years experience at the Mayo Clinic. *Radiology* 103:503–508, 1972.
11. Goldberg BB, Pollack HM: Ultrasonically-guided renal cyst aspiration. *J Urol* 109:5–7, 1973.
12. Korobkin M, Kressel HY, Moss AA, et al: Computed tomographic angiography of the body. *Radiology* 126:807–811, 1978.
13. Lloyd LK, Witten DM, Bueschen AJ, et al: Enhanced detection of asymptomatic renal masses with routine tomography during excretory urography. *Urology* 11:523–528, 1978.
14. Ney C, Friedenberg RM: *Radiographic Atlas of the Genitourinary System*, Philadelphia, JB Lippincott, 1981.
15. Love L, Churchill R, Reynes C, et al: Computed tomography staging of renal carcinoma. *Urol Radiol* 1:3–11, 1979.
16. Levine E, Lee KR, Weigel J: Preoperative determination of abdominal extent of renal-cell carcinoma by computed tomography. *Radiology* 132:395–398, 1979.
17. Zeman RK, Cronan JJ, Viscomi GN, et al: Coordinated imaging in the detection and characterization of renal masses. *Crit Rev Clin Radiol* 15:273–288, 1981.
18. Bosniak MA, Ambos MM, Madayag MA, et al: Epinephrine-enhanced renal angiography in renal mass lesions: Is it worth performing? *AJR* 129:647–652, 1977.
19. Chuang VP, Fried AM: High-dose renal pharmacoangiography in the assessment of hypovascular renal neoplasms. *AJR* 131:807–811, 1978.
20. Ekelund L: Pharmacoangiography of the kidney: An overview. *Urol Radiol* 2:9–18, 1980.
21. Goldstein HM, Green B, Weaver RM: Ultrasonic detection of renal tumor extension into the inferior vena cava. *AJR* 130:1083–1085, 1978.
22. Watson RC, Fleming RJ, Evans JA: Arteriography in the diagnosis of renal carcinoma. *Radiology* 91:888–897, 1968.
23. Ben-Menachem Y, Crigler CM, Corriere JN: Elective transcatheter renal artery occlusion prior to nephrectomy. *J Urol* 114:355–359, 1975.
24. McClennan BL, Lee JKT: Kidney, in Lee JKT, Sagel SS, Stanley RJ (eds): *Computed Body Tomography*. New York, Raven Press, 1983, pp 341–378.
25. Kuhn JP, Berger PE: Computed tomographic imaging of abdominal abnormalities in infancy and childhood. *Pediatr Ann* 9:200–209, 1980.
26. Kaufman RA, Holt JF, Heidelberger KP: Calcification in primary and metastatic Wilms's tumor. *AJR* 130:783–785, 1978.
27. Hodson NH, Husband JE, MacDonald JF: The role of computed tomography in the staging of bladder cancer. *Clin Radiol* 30:389–395, 1979.
28. O'Cleireachain F, Awad SA, Prentice RSA: Gross calcification in bladder tumor. *Urology* 3:642–643, 1974.
29. McCarthy JP, Gavrell GJ, Leblanc GA: Transitional cell carcinoma of the bladder in patients under 30 years of age. *Urology* 13:487–489, 1979.
30. Goldman SM, Meng CH, White RI, et al: Transitional cell tumors of the kidney: How diagnostic is the angiogram? *AJR* 129:99–105, 1977.
31. McLean GK, Pollack HM, Banner MP: The "stipple sign"—Urographic harbinger of transitional cell neoplasms. *Urol Radiol* 1:77–83, 1979.
32. Parienty RA, Ducellier R, Pradel J-M, et al: Diagnostic value of CT numbers in pelvocalyceal filling defects. *Radiology* 145:743–747, 1982.
33. Burger R, Spjut HJ: Primary ureteral carcinoma. *Urology* 4:40–47, 1974.
34. Eisenberg RL: Renal Masses. In Eisenberg RL, Amberg JR (eds). *Critical Diagnostic Pathways in Radiology: An Algorithmic Approach*. Philadelphia, JB Lippincott, 1981.
35. Bosniak MA, Faegenburg D: The thick-wall sign: An important finding in nephrotomography. *Radiology* 84:692–698, 1965.
36. Marks WM, Korobkin M, Callen PW, et al: CT diagnosis of tumor thrombosis of the renal vein and inferior vena cava. *AJR* 131:843–846, 1978.
37. Eisenberg RL: *Atlas of Signs in Radiology*. Philadelphia, JB Lippincott, 1984.

## Chapter Forty-Two
# Renal Anomalies and Miscellaneous Renal Disorders

## ANOMALIES OF THE KIDNEY AND URINARY TRACT

Congenital anomalies of the urinary tract are common, and most are of little clinical significance. Nevertheless, some anomalies are potentially dangerous, and since they are usually amenable to surgery, prompt radiographic diagnosis is essential.

### Anomalies of the Kidney

#### ANOMALIES OF NUMBER

Unilateral renal agenesis (solitary kidney) is a rare anomaly that may be associated with a variety of other congenital malformations. Before the diagnosis can be made, it is essential to exclude a nonfunctioning, diseased kidney or a prior nephrectomy. Ultrasound or computed tomography (CT) can demonstrate the absence of renal tissue. If the left kidney is congenitally absent, the anatomic splenic flexure of the colon occupies the renal fossa and thus lies in a more medial and posterior position than normal. This can be suggested by nephrotomographic demonstration of sharply outlined gas or fecal material in the plane of the renal fossa and confirmed by barium enema examination. At cystoscopy, the ureteral orifice and one-half of the trigone are almost always absent. The solitary kidney tends to be larger than expected, reflecting compensatory hypertrophy.

Supernumerary kidney is also a rare anomaly. The third kidney is usually small and rudimentary and possesses a separate pelvis, ureter, and blood supply.[1]

#### ANOMALIES OF SIZE AND FORM

A small, hypoplastic kidney often appears as a miniature replica of a normal kidney, with good function and a normal relation between the amount of parenchyma and the size of the collecting system. Less often, the kidney has a bizarre appearance with distorted calyces extending to the renal border. Renal hypoplasia must be differentiated from an acquired atrophic kidney, which is small and contracted because of vascular or inflammatory disease that has reduced the volume of renal parenchyma. At times, the differentiation between these two entities may require arteriographic demonstration of the aortic orifice of the renal artery, which is small in the hypoplastic kidney but of normal size in the atrophic kidney.

Compensatory hypertrophy is an acquired condition that develops when one kidney is forced to perform the function normally carried out by two kidneys. This phenomenon may follow unilateral renal agenesis, atrophy, or nephrectomy. The ability of the kidney to undergo

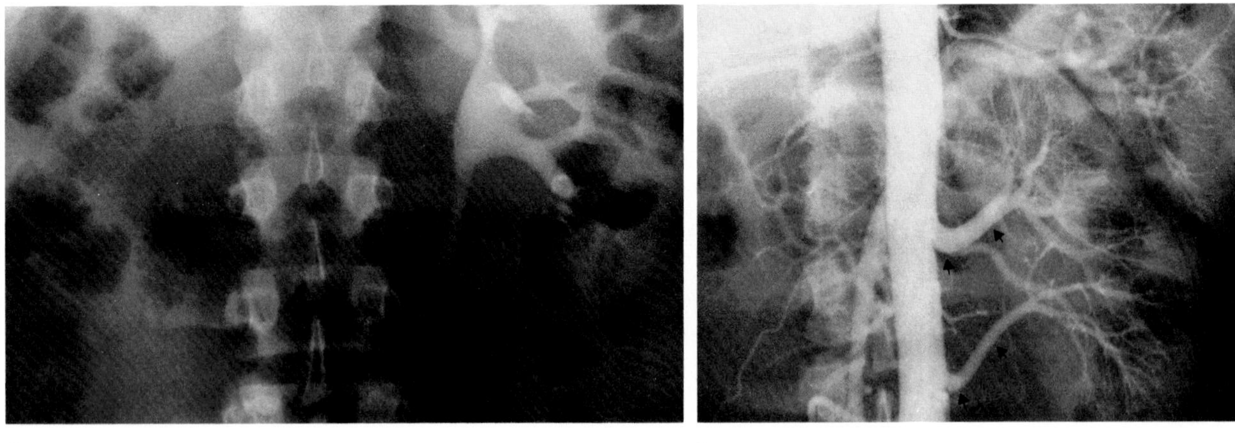

A B

**Figure 42.1. Solitary kidney.** (A) A film from an excretory urogram demonstrates a normal left kidney with no evidence of right renal tissue. (B) An aortogram shows two renal arteries to the left kidney (arrows) and no evidence of a right renal artery, thus confirming the diagnosis of unilateral renal agenesis.

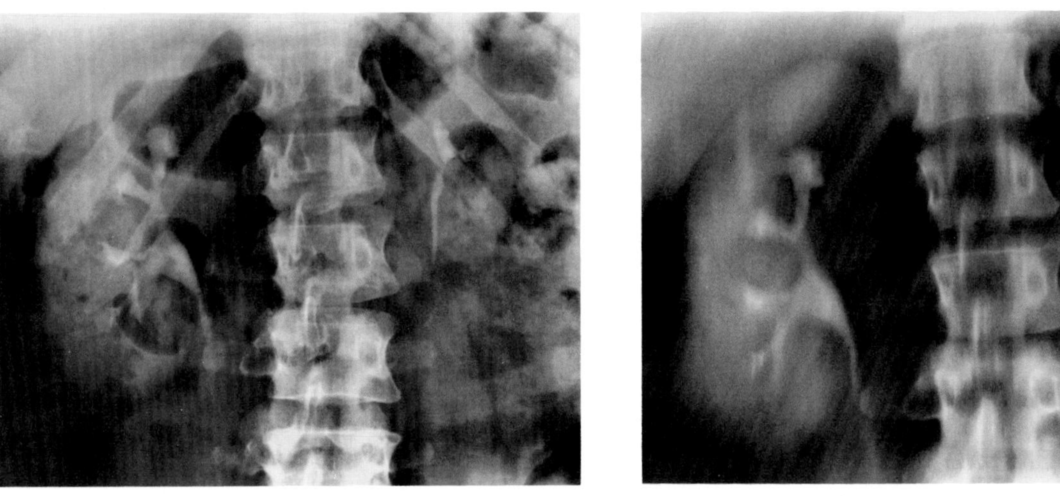

A B

**Figure 42.2. Hypoplastic kidney.** (A) A plain film from an excretory urogram and (B) a tomographic view show a small left kidney that appears as a miniature replica of a normal kidney. The tiny left kidney has good function and a normal relation between the amount of parenchyma and the size of the collecting system. Note the compensatory hypertrophy of the right kidney.

compensatory hypertrophy is greatest in children and diminishes in adulthood.[2]

#### ANOMALIES OF ROTATION

Malrotation of one or both kidneys may produce a bizarre appearance of the renal parenchyma, calyces, and pelvis that suggests a pathologic condition when in reality the kidney is otherwise entirely normal. When the renal pelvis is situated in an anterior or lateral position, the upper part of the ureter often appears displaced laterally, suggesting an extrinsic mass. The elongated pelvis of the malrotated kidney may mimic obstructive dilatation.[3]

#### FUSION ANOMALIES

Horseshoe kidney is the most common type of fusion anomaly. In this condition, both kidneys are malrotated and their lower poles are joined by a band of normal renal parenchyma (isthmus) or connective tissue. The characteristic urographic features include a vertical or reversed longitudinal axis of the kidneys (upper poles tilted away from the spine), a demonstration on the nephrogram of the parenchymal isthmus (if present) connecting the lower poles, and a projection of the lower calyces medial to the upper ureters on the frontal projection. The pelves are often large and flabby and may simulate obstruction. Conversely, true ureteropelvic junction obstruction may develop because of the unusual course of the ureter, which arises high in the renal pelvis, passes over the isthmus, and may kink at the ureteropelvic junction.

Complete fusion of the kidneys is a rare anomaly that produces a single irregular mass that has no resemblance to a renal structure. The resulting bizarre appearance has been given such varied names as disk, cake, lump, and doughnut kidney. The fused kidney usually fails to ascend during development and therefore appears as a

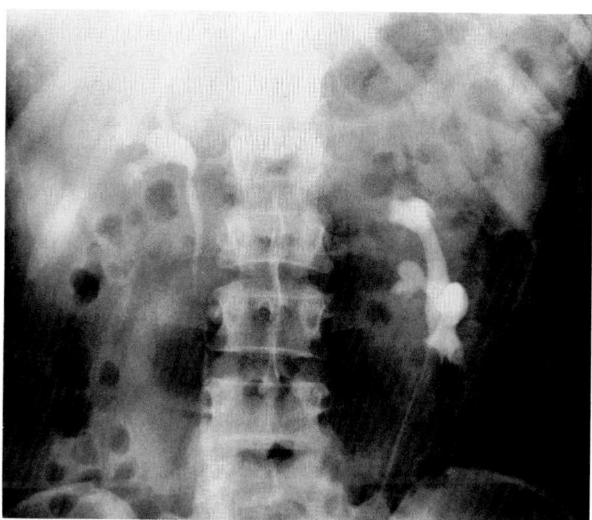

**Figure 42.3. Malrotation of the left kidney.** Note the apparent lateral displacement of the upper ureter and the elongation of the pelvis.

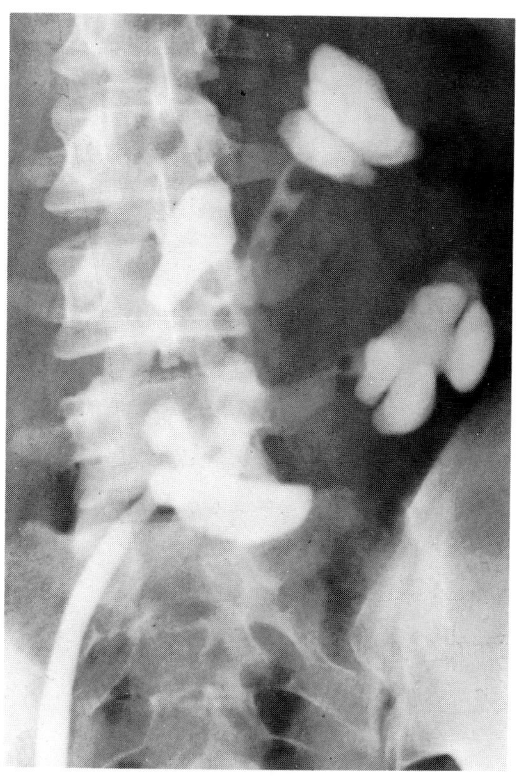

**Figure 42.4. Disk kidney.** At cystoscopy, only one ureteral orifice was found in the bladder. The injection of contrast material into this orifice demonstrates a centrally placed, irregularly defined, single renal mass with four calyceal systems extending out from a small, central renal pelvis. Severe hydronephrosis and pyelonephritis were present. (From Ney and Friedenberg,[4] with permission from the publisher.)

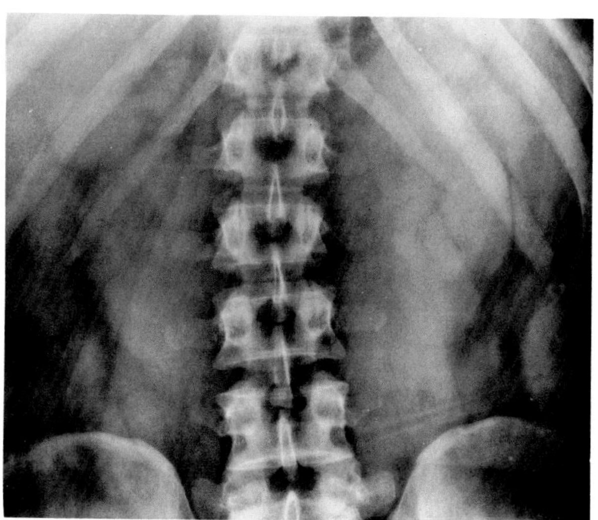

A

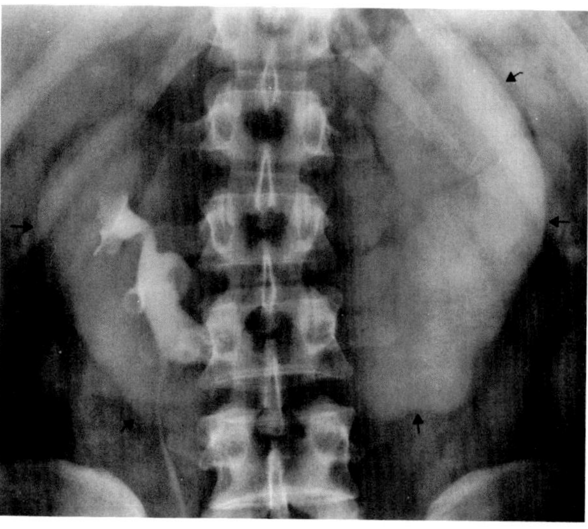

B

**Figure 42.5. Horseshoe kidney.** (A) A plain abdominal radiograph and (B) an excretory urogram demonstrate the outline of a horseshoe kidney (arrows). On the left there is a prolonged nephrogram and a delay in calyceal filling, both caused by an obstructing stone at the ureteropelvic junction on that side.

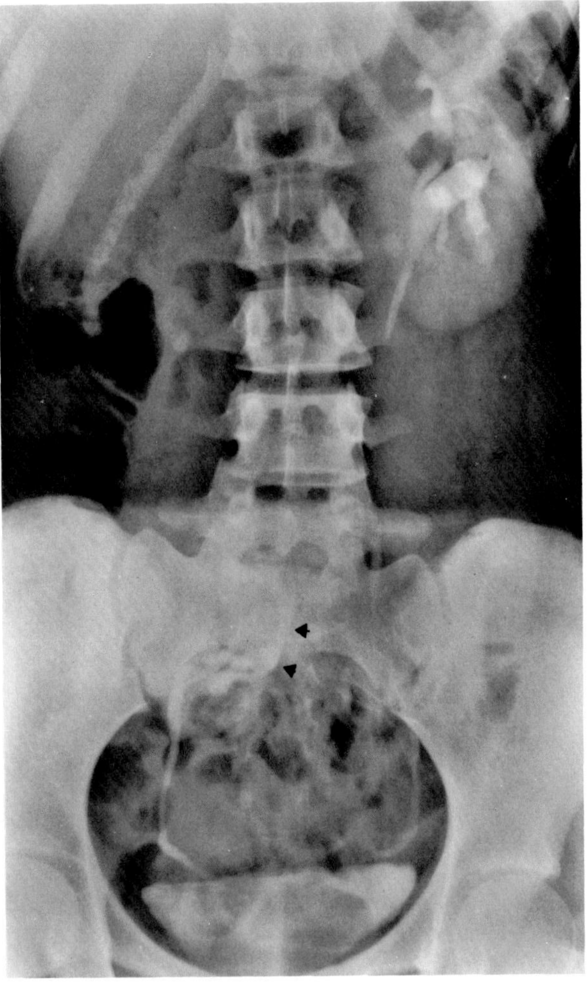

**Figure 42.6. Unilateral pelvic kidney.** An excretory urogram shows a normally positioned and functioning left kidney. The collecting system of the right pelvic kidney (arrows) is difficult to identify and might be missed if the right ureter were not seen on this full view of the abdomen.

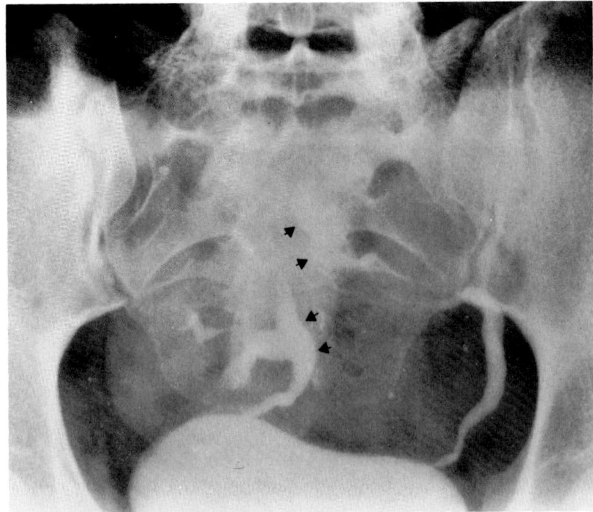

**Figure 42.7. Bilateral pelvic kidneys.** Arrows point to the collecting systems.

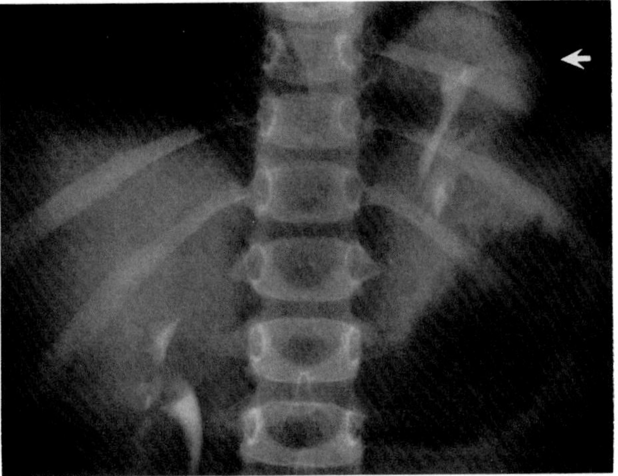

**Figure 42.8. Intrathoracic kidney (arrow).**

single mass of tissue within the bony pelvis. Poor emptying of the two pelvocalyceal systems, caused by the abdominal position of the ureteropelvic junction and ureteral deviation, often results in infection and calculus formation.[3,4]

#### ANOMALIES OF POSITION

Abnormally positioned kidneys (renal ectopia) may be found in various locations, from the true pelvis (pelvic kidney) to above the diaphragm (intrathoracic kidney). Whenever only one kidney is seen on excretory urography, a full view of the abdomen is essential to search for an ectopic kidney. Although the ectopic kidney usually functions, the nephrogram and the pelvocalyceal system may be obscured by overlying bone and fecal contents. Crossed ectopia refers to a condition in which an ectopic kidney lies on the same side as the normal kidney and is most commonly fused with it (unilateral fused kidney).[4]

### Anomalies of the Renal Pelvis and Ureter

#### DUPLICATION

Duplication (duplex kidney) is a common anomaly that may vary from a simple bifid pelvis to a completely double pelvis, ureter, and ureterovesical orifice. If unilateral, the duplex kidney is usually longer than the contralateral, normal one. The upper pole pelvis tends to be much smaller than that of the lower pole and often consists of only a single minor calyx. The ureter draining the upper renal segment enters the bladder below the ureter draining the lower renal segment. The orifice of the lower ureter may be ectopic, most frequently emptying into the posterior urethra.

Complete duplication can be complicated by obstruction or by vesicoureteral reflux with infection. Vesicoureteral reflux and infection more commonly involve the ureter draining the lower renal segment; obstruction more frequently affects the upper pole, where it can cause a hydronephrotic mass that displaces and compresses the lower calyces.[5–6]

**Figure 42.9. Complete duplication of the renal pelvis and ureter on the right.** Calyceal dilatation and a paucity of parenchyma in the lower pole reflect cortical atrophy. The left kidney had been removed surgically and the relative enlargement of the renal parenchyma in the upper pole reflects compensatory hypertrophy.

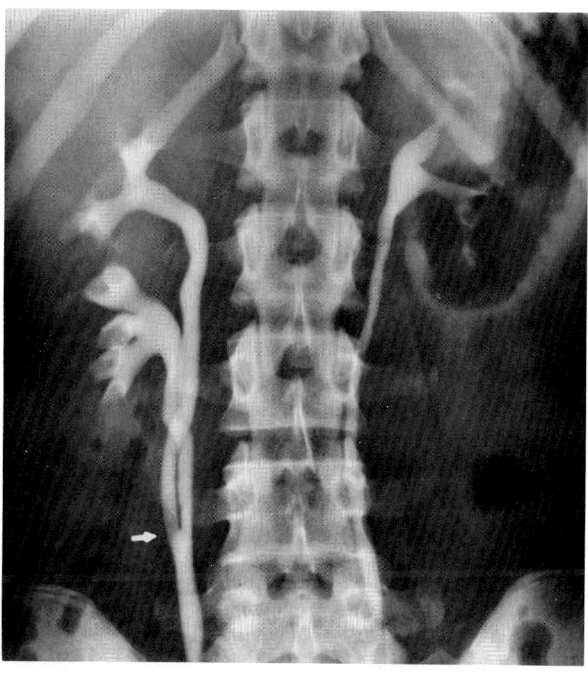

**Figure 42.10. Complete duplication of the renal pelvis and upper ureter on the right.** The two ureters fuse into a single trunk at the L4 level (arrow).

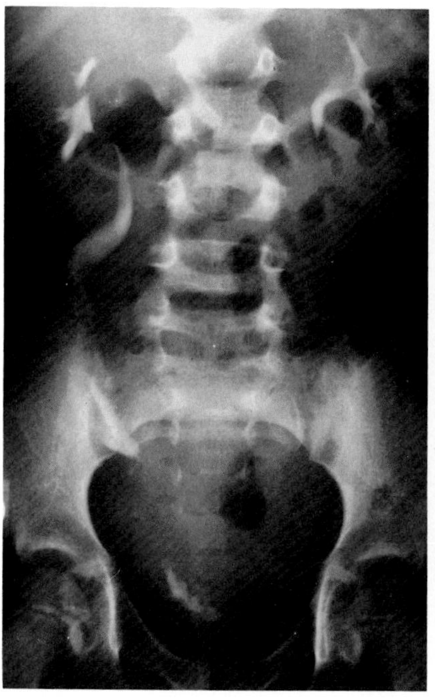

A

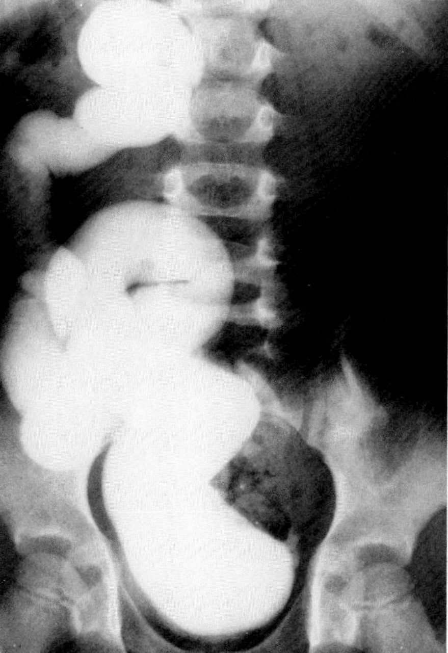

B

**Figure 42.11. Complete duplication with obstruction.** (A) An excretory urogram demonstrates dilatation and lateral displacement of the right ureter. Ureteral duplication was not suspected in this patient. (B) A delayed film following a voiding cystogram demonstrates contrast material filling a markedly dilated and tortuous ureter to the upper segment. This ureter, which was not seen on the excretory urogram, has caused lateral displacement of the ureter to the lower segment. (From Amar,[5] with permission from the publisher.)

### RETROCAVAL URETER

This developmental defect of the vena cava, rather than the ureter itself, occurs almost exclusively on the right side. The ureter loops around the vena cava by passing behind it, then bending forward between the vena cava and the aorta before finally coursing to the right again in front of the vena cava. Excretory urography demonstrates an abrupt medial swing of the ureter, which usually lies over or medial to the vertebral pedicles at the L4–5 level. Compression of the ureter between the inferior vena cava and posterior abdominal wall often causes narrowing or obstruction of the ureter with progressive hydronephrosis.[1–4]

### URETEROCELE

A ureterocele is a cystic dilatation of the distal intravesical segment of the ureter with protrusion into the bladder. In the simple (adult) type, the opening of the ureter

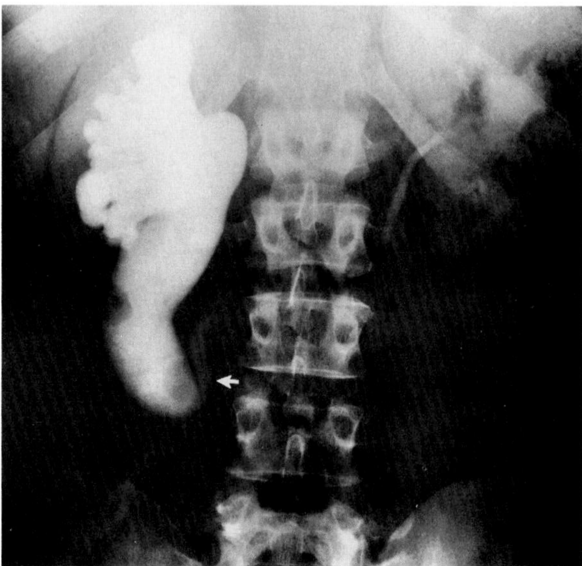

**Figure 42.12. Retrocaval ureter.** An excretory urogram shows an abrupt medial swing of the ureter at the L4 level. Compression of the ureter (arrow) between the inferior vena cava and the posterior abdominal wall has caused ureteral obstruction and hydronephrosis.

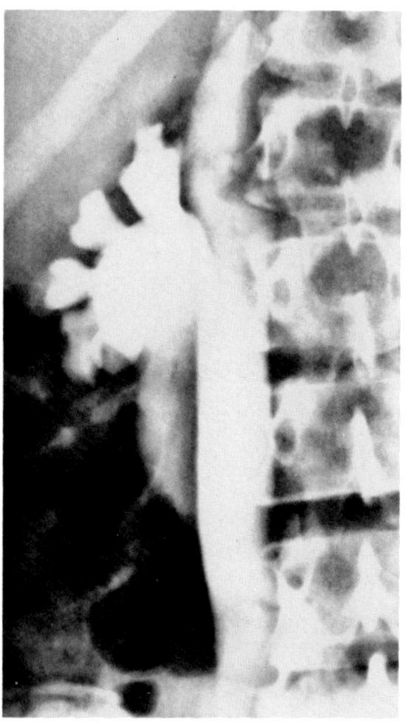

**Figure 42.13. Retrocaval ureter.** A combined excretory urogram and inferior vena cavogram shows the right ureter looping around the inferior vena cava. (From Friedland,[1] with permission from the publisher.)

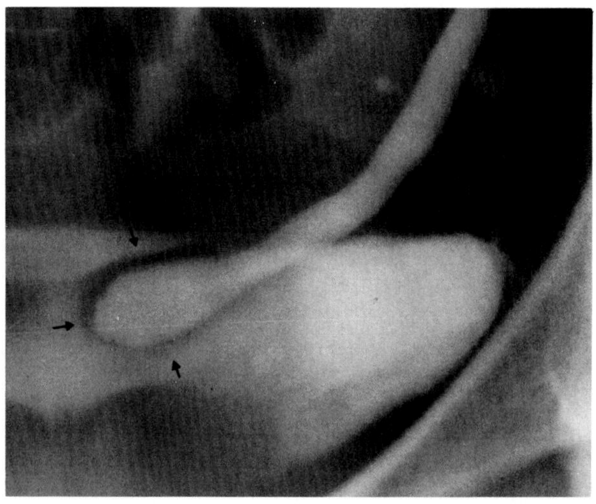

A

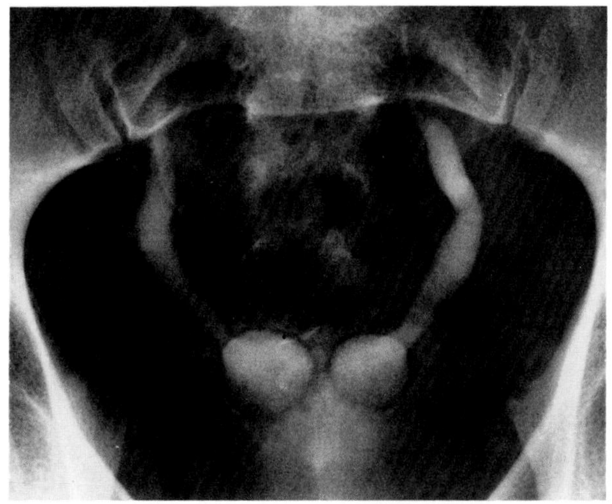

B

**Figure 42.14. Simple ureterocele.** (A) An excretory urogram demonstrates an oval density of opacified urine in the dilated distal segment of the ureter. Note the thin radiolucent halo (cobra head sign). (B) Bilateral simple ureteroceles.

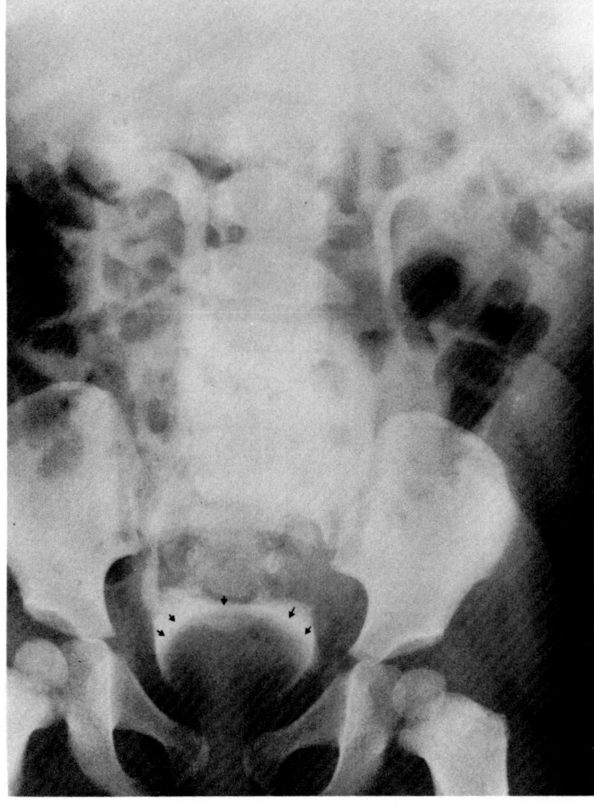

A

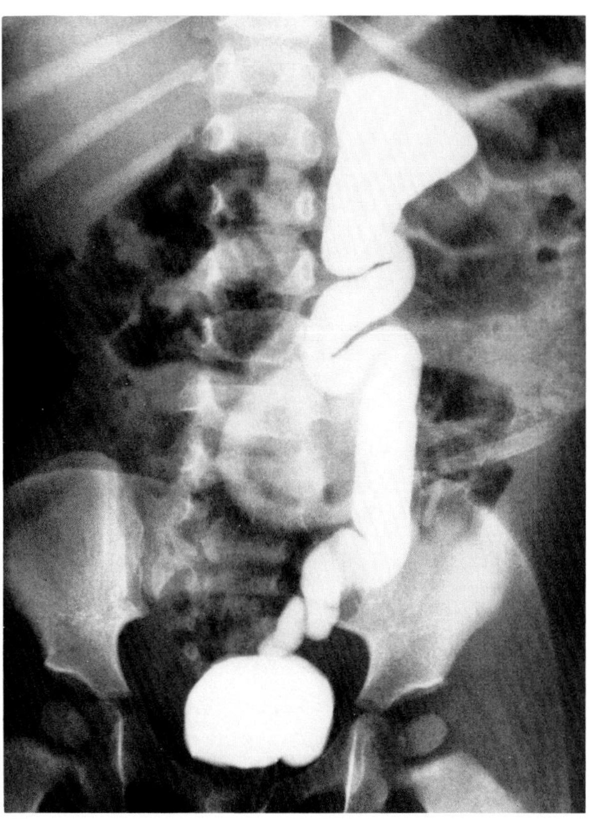

B

**Figure 42.15. Ectopic ureterocele.** (A) An excretory urogram demonstrates a large lucency (arrows) filling much of the bladder. There is slight downward and lateral displacement of the visualized collecting system on the left. (B) A cystogram shows contrast material refluxing to fill the markedly dilated collecting system draining the upper pole of the left kidney. Note the severe dilatation and tortuosity of the left ureter.

is situated at or near its normal position in the bladder, usually with stenosis of the ureteral orifice and with varying degrees of dilatation of the proximal ureter. The stenosis leads to prolapse of the distal ureter into the bladder and dilatation of the lumen of the prolapsed segment. The appearance on excretory urography depends on whether opaque medium fills the ureterocele. If it is filled, the lesion appears as a round or oval density surrounded by a thin radiolucent halo representing the wall of the prolapsed ureter and the mucosa of the bladder (cobra head sign). When the ureterocele is not filled with contrast material, it appears as a radiolucent mass within the opacified bladder in the region of the ureteral orifice.

Ectopic ureteroceles are found almost exclusively in infants and children; the vast majority are associated with ureteral duplication. On excretory urography, an ectopic ureterocele typically appears as a large, eccentric filling defect encroaching on the floor of the bladder. The ureterocele arises from the ureter draining the upper segment of the duplicated collecting sytem. A mass effect, representing hydronephrosis, often involves the upper pole of the kidney and causes downward and lateral displacement of the lower portion of the collecting system. Contralateral hydronephrosis may develop if the ectopic ureterocele extends across the trigone and obstructs the contralateral ureteral orifice.[1–4]

## Anomalies of the Bladder and Urethra

### EXSTROPHY OF THE BLADDER

In this congenital anomaly the anterior wall of the bladder and the lower anterior abdominal wall are absent. Plain radiographs demonstrate the pathognomonic appearance of wide separation of the symphysis pubis.[7]

### BLADDER EARS

Bladder ears are lateral protrusions of the urinary bladder in infants that represent transitory extraperitoneal herniation. Usually bilateral, bladder ears occur in about 10 percent of all infants under the age of 1 year, disappear spontaneously, and are considered a normal variant.[24]

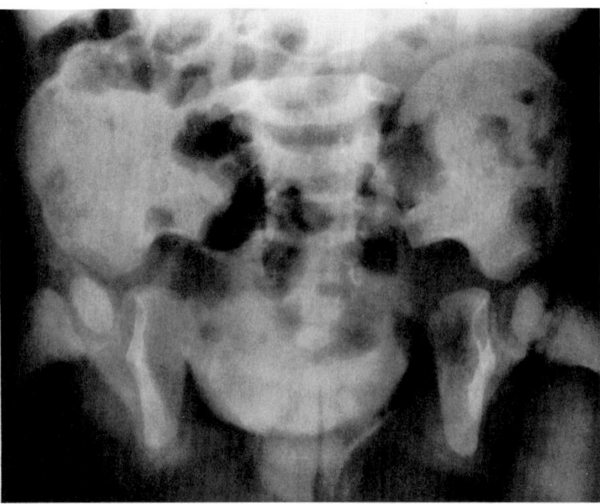

**Figure 42.16. Exstrophy of the bladder.** A plain radiograph demonstrates wide separation of the symphysis pubis.

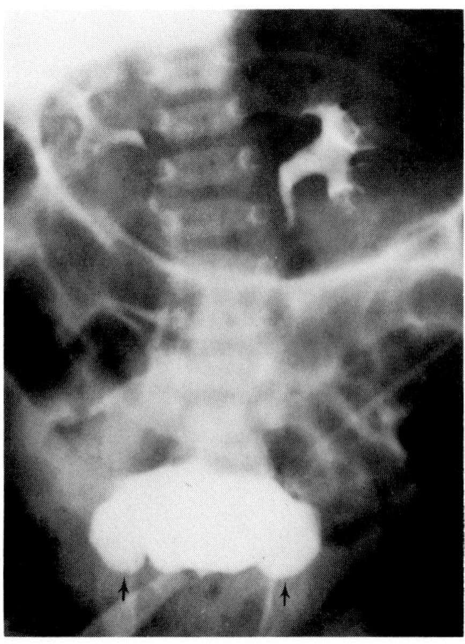

**Figure 42.17. Bladder ears.** The lateral walls of the bladder are displaced downward (arrows) and project into the inguinal ring. This is a normal occurrence in infants under the age of 1 year. (From Rabinowitz,[24] with permission from the publisher.)

#### PRUNE-BELLY SYNDROME

In this rare condition that occurs almost exclusively in males, congenital enlargement of the bladder along with hydronephrosis and hydroureter is associated with absence or hypoplasia of the abdominal muscles. Bulging of the flanks, due to a lack of support by the abdominal muscles, is a characteristic finding on plain radiographs. The term *prune belly* refers to the wrinkled appearance of the skin due to absence of the abdominal muscles.[1–4]

#### PATENT URACHUS AND URACHAL CYST

The urachus is a canal connecting the bladder with the allantoic duct in fetal life. It permanently closes soon after birth and persists as a thin fibrous cord running between the bladder and umbilicus. A completely patent urachus can be demonstrated by the injection of contrast material either at the umbilical opening or into the bladder. The internal type of persistent urachus appears on retrograde cystography as a contrast-filled, cystlike blind pouch that extends from the upper margin of the bladder in the direction of the umbilicus. Urachal cysts, which develop when the midportion of the urachus is patent while the cephalic and caudal ends are closed, often go undetected until infection occurs or until a mass develops in an uninfected cyst. A urachal cyst can appear on ultrasound as an echo-free mass compressing the dome of the bladder and lying between it and the umbilicus.[8–10]

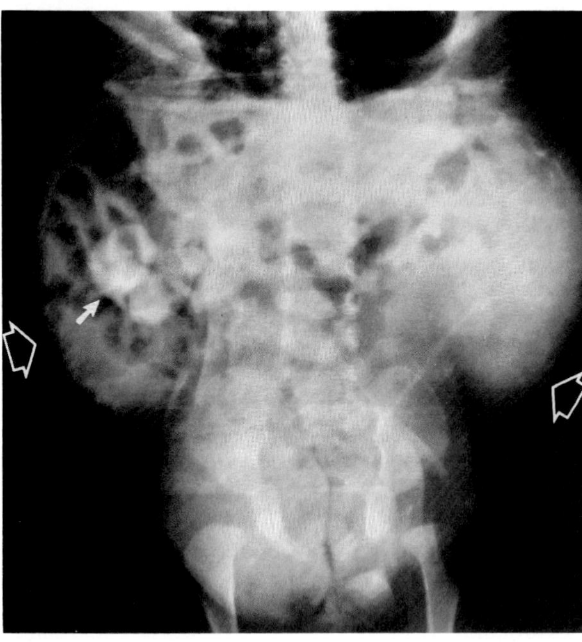

**Figure 42.18. Prune-belly syndrome.** Pronounced bulging of the flanks (open arrows) is due to a lack of support by the hypoplastic abdominal muscles. The patient had multiple genitourinary anomalies including hydronephrosis of the right collecting system (arrow).

#### POSTERIOR URETHRAL VALVES

Posterior urethral valves are thin transverse membranes, found almost exclusively in males, that cause outlet obstruction and may lead to severe hydronephrosis and renal damage. Voiding cystourethrography with the patient in the steep oblique or lateral position demonstrates elongation and dilatation of the posterior urethra and a characteristic thin, sail-like lucent defect representing the bulging valve. Although usually normal in width, the bladder neck often appears narrowed because of the disparity in size between it and the urethra that bulges posteriorly beneath it.[11]

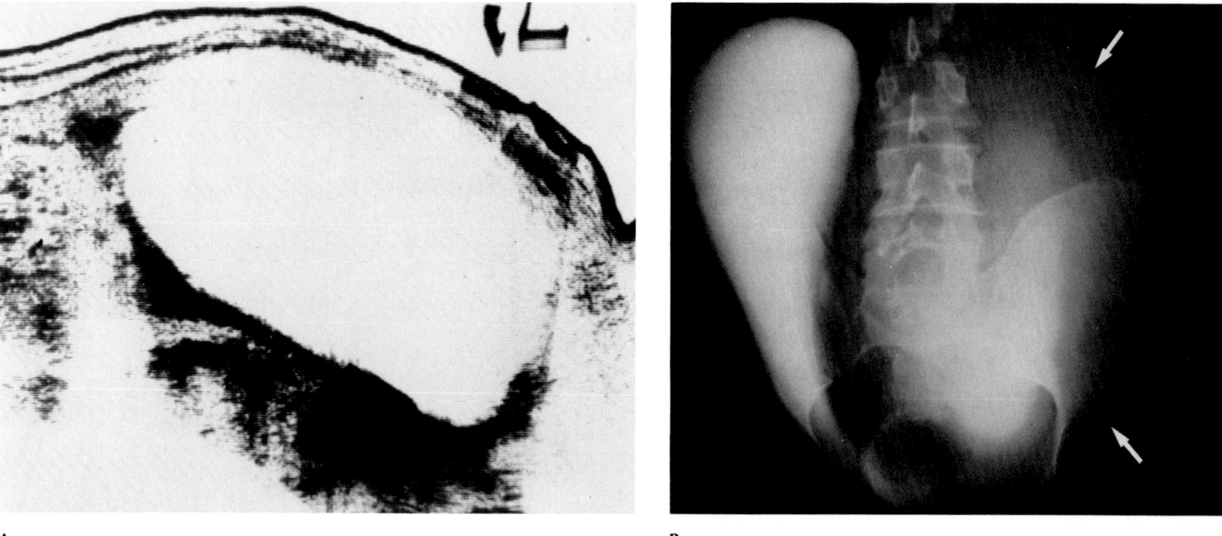

A B

**Figure 42.19. Urachal cyst.** (A) A longitudinal sonogram taken after catheterization of the bladder shows a large anterior cystic mass that is separate from the bladder. (B) A cystogram shows the compressed and displaced bladder. A catheter has been placed into the urachal cyst, fluid removed, and contrast material injected to show the size of the cyst (arrows) and its relation to the bladder. (From Spataro, Davis, McLachlan, et al,[10] with permission from the publisher.)

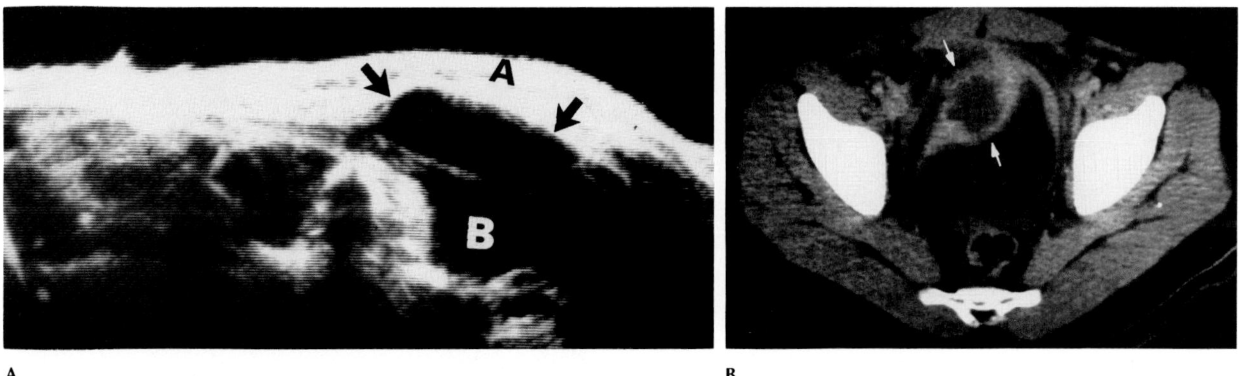


A B

**Figure 42.20. Infected urachal cyst.** (A) A longitudinal sonogram shows the apposition of an infected, abscess-forming urachal cyst (arrows) to the anterior wall of the bladder (B). A = anterior abdominal wall. (B) In another patient, a contrast-enhanced CT scan shows an irregular, thick-walled cystic structure involving the intramural portion of the dome of the bladder. (From Spataro, Davis, McLachlan, et al,[10] with permission from the publisher.)

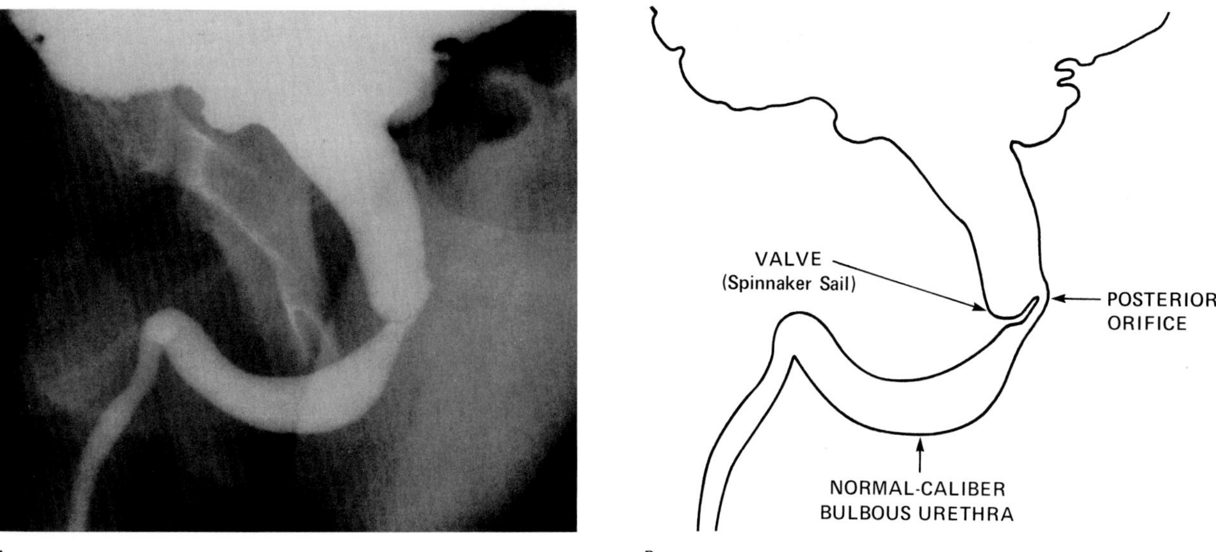


A B

**Figure 42.21. Posterior urethral valves.** (A) A voiding cystourethrogram and (B) a diagram show the spinnaker-sail shape and posterior orifice of the valve. Note that the caliber of the bulbous urethra is normal. (From Friedland, Fair, Govan, et al,[11] with permission from the publisher.)

## RENAL SINUS LIPOMATOSIS

Small to moderate amounts of loose fatty and fibrous tissue are always present in the renal sinus surrounding the pelvis and calyces in the renal hilum. Renal sinus lipomatosis refers to the accumulation of excessive amounts of fat in the renal sinus. Often a normal variant, renal sinus lipomatosis may develop in obese persons or in a patient whose kidney has lost renal parenchyma as a result of ischemia, infarction, or infection.

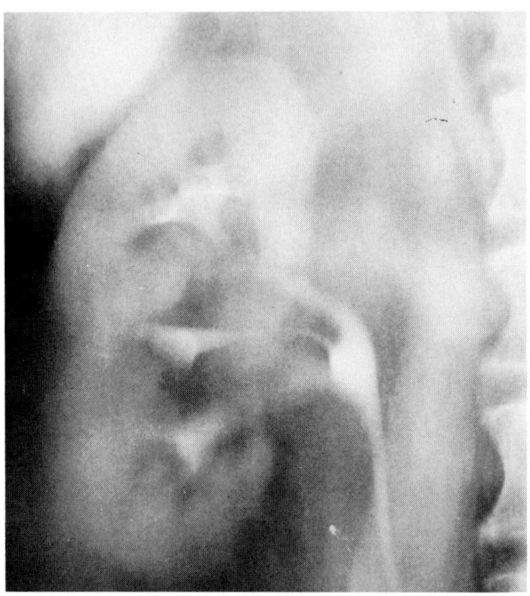

**Figure 42.22. Renal sinus lipomatosis.** A nephrotomogram shows increased radiolucency (fat) around the renal sinuses and calyces. The excessive fatty deposition causes stretching and elongation of the pelvocalyceal system. (From Ney and Friedenberg,[4] with permission from the publisher.)

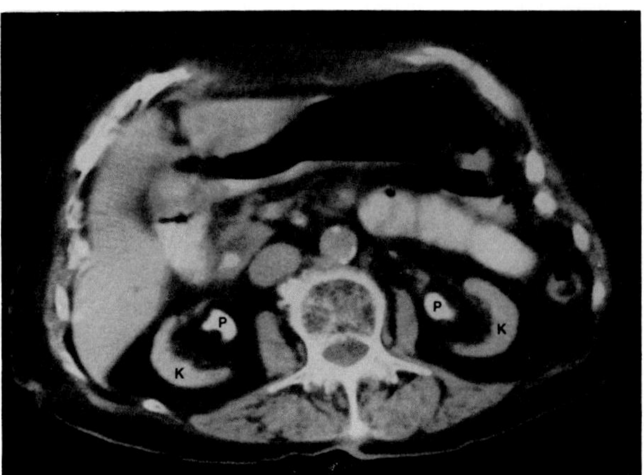


**Figure 42.23. Renal sinus lipomatosis.** A CT scan through both kidneys shows a large amount of fat density separating the contrast-filled renal pelves (P) from the renal parenchyma (K).

Although renal sinus lipomatosis is asymptomatic and without clinical significance, the excessive fatty deposition may cause pelvocalyceal deformities that simulate solitary or multiple peripelvic masses. The nephrotomographic demonstration of the characteristic lucencies adjacent to the collecting system can usually exclude a renal tumor.[2,12]

## PYELOURETERITIS CYSTICA

Pyeloureteritis cystica refers to the presence of multiple inflammatory cysts—varying in size from 1 mm to large cysts that may partially obstruct the lumen—that project from the urothelial lining of the ureter and, less frequently, the renal pelvis and bladder. Although the etiology is not known, there is a definite association with chronic urinary tract infection. The urographic findings are virtually pathognomonic and consist of multiple small,

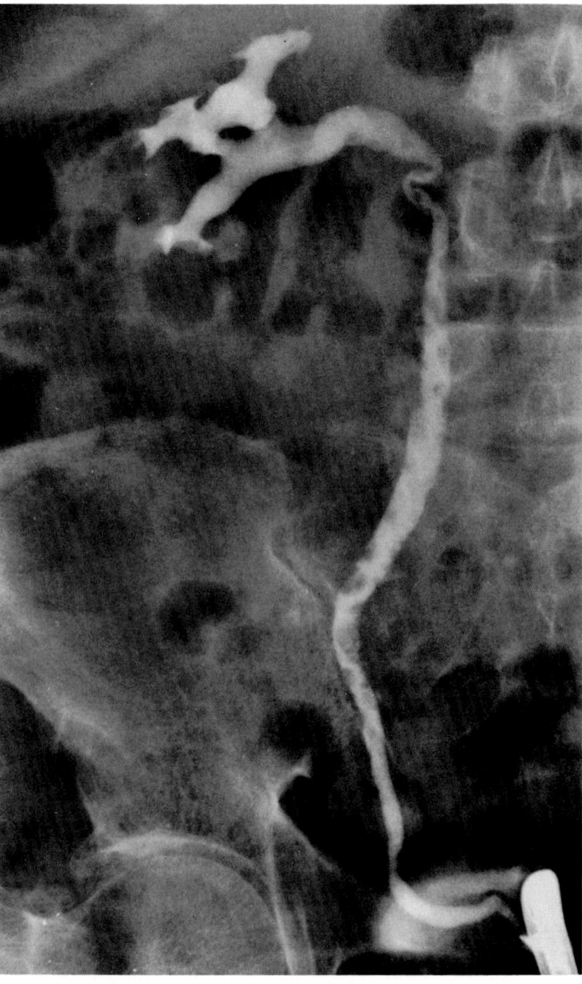

**Figure 42.24. Pyeloureteritis cystica.** Multiple small, smooth lucent filling defects within the contrast-filled ureter and pelvis in a patient with chronic urinary tract infection.

smooth, rounded lucent filling defects within the contrast-filled ureter or pelvis. When the cysts are viewed in profile, a characteristic scalloping is seen. The radiographic appearance of pyeloureteritis cystica must be differentiated from that of air bubbles, which do not have a constant position, tend to coalesce into large defects with a change in the patient's position, and do not produce scalloping. A single cyst or a few scattered cysts in the renal pelvis or ureter may simulate a transitional-cell tumor; because pyeloureteritis cystica may persist and remain unchanged over many years, comparison with previous excretory urograms is extremely helpful in differentiating these benign cysts from malignant lesions.[2]

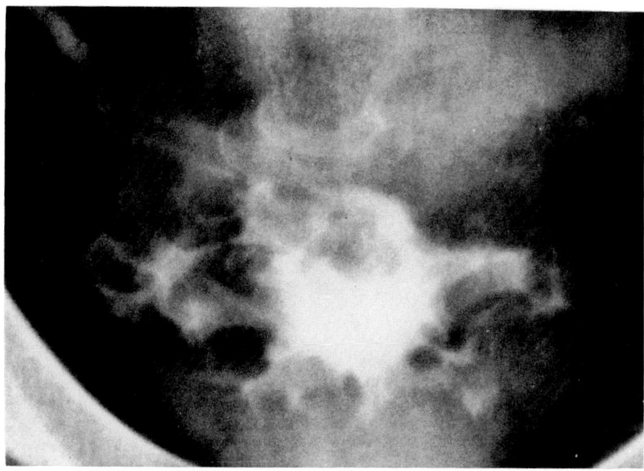

**Figure 42.25. Malakoplakia of the bladder.** A postvoiding excretory urogram shows multiple smooth, nodular filling defects. (From Elliott, Moloney, and Clement,[13] with permission from the publisher.)

## MALAKOPLAKIA

Malakoplakia is an uncommon chronic inflammatory disease that most often involves the bladder but can also occur elsewhere in the urinary tract. The condition predominantly affects women, most of whom have a history of recurrent or chronic urinary tract infection. It usually presents with dysuria and hematuria. Lesions in the bladder produce multiple smooth, nodular filling defects that suggest a neoplastic process or severe cystitis with greatly edematous mucosa. Involvement of the ureter may cause general dilatation with multiple filling defects or a scalloped appearance simulating ureteritis cystica; occasionally, extensive disease can obstruct the ureter and simulate a primary tumor or stricture.[2,13,14]

## CYSTITIS

Although acute inflammation of the urinary bladder generally does not produce changes detectable on excretory urography, chronic cystitis causes a decrease in bladder size that is often associated with serration of the bladder wall. In candidal cystitis, fungus balls may produce lucent filling defects in the opacified bladder. Similar lucent filling defects in the bladder may reflect blood clots in patients with the hemorrhagic cystitis that may complicate cyclophosphamide (Cytoxan) administration in the treatment of leukemia and lymphoma. A dramatic radiographic appearance is produced by emphysematous cystitis, an inflammatory disease of the bladder that most often occurs in diabetic patients and is caused by gas-forming bacteria. Characteristic plain-film findings

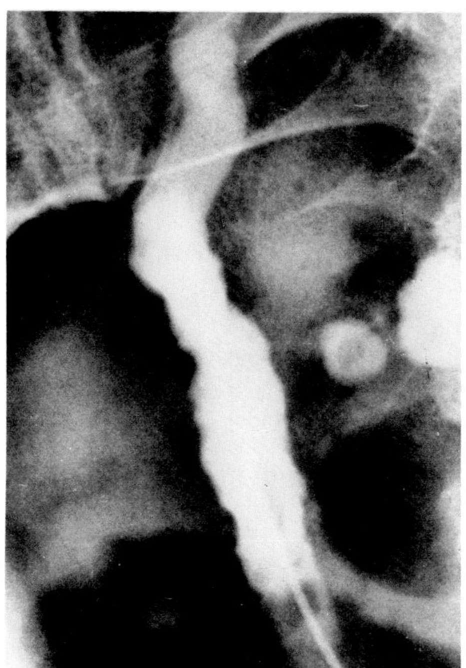

A

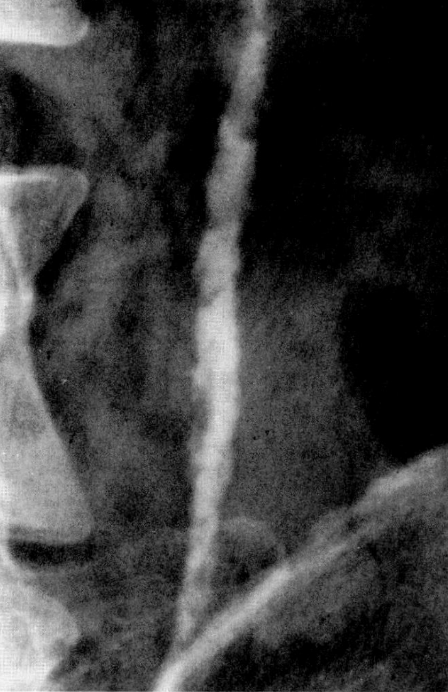

B

**Figure 42.26. Malakoplakia of the ureter.** (A) A magnification view of the distal right ureter shows multiple filling defects simulating ureteritis cystica. (B) A retrograde pyelogram of the left ureter shows feathery irregularity simulating papillary transitional-cell carcinoma. (From Elliott, Moloney, and Clement,[13] with permission from the publisher.)

are a ring of lucent gas outlining all or a part of the bladder wall, and the presence of gas within the bladder lumen.[15,16,27]

## NEUROGENIC BLADDER

Disease or injury involving the spinal cord or the peripheral nerves supplying the bladder results in changes in bladder function that may produce either incontinence or the retention of urine. Upper motor neuron lesions tend to produce a small, spastic, trabeculated bladder with a pointed dome (pine tree bladder). Although this sign is usually considered pathognomonic of spastic neurogenic bladder, an identical appearance may be seen in patients who have no neurologic disease but only simple bladder outlet infection, often with superimposed urinary tract obstruction. A large, atonic bladder with little or no trabeculation suggests diabetes, tabes dorsalis, or syringomyelia, though it also may be of psychogenic origin in a patient with no neurologic disease.[3,4,17,18]

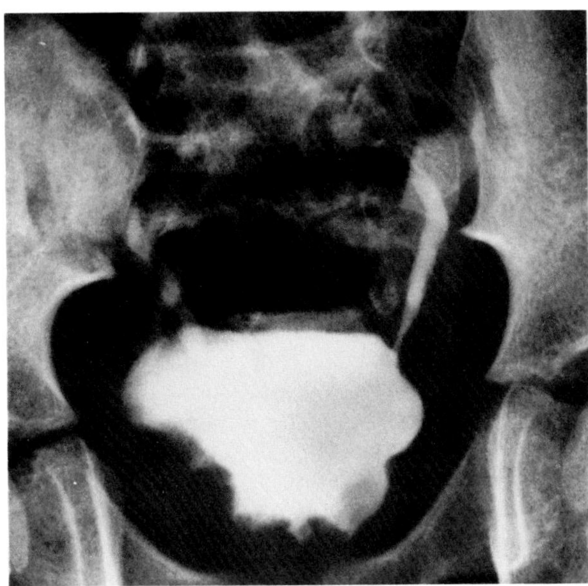

**Figure 42.27. Cystitis.** An excretory urogram in a 5-year-old boy with hematuria shows irregular, lobulated filling defects (representing intense mucosal edema) at the base of the bladder.

## RETROPERITONEAL FIBROSIS

In this disorder, a fibrosing inflammatory process envelops the retroperitoneal structures but usually does not invade them. The disease may extend from the kidneys down to the pelvic brim and can cause partial or complete obstruction of the ureters and vascular and lymphatic structures. Although the etiology is unknown, many cases have been associated with drug ingestion, principally methysergide for migraine, but also ergot derivatives, phenacetin, and methyldopa. Fibrosis in the retroperitoneum may coexist with similar fibrotic processes in other sites (fibrosing mediastinitis, sclerosing cholangitis, retro-orbital pseudotumor, retractile mesenteritis, and Riedel's thyroiditis).

The classic urographic appearance consists of dilatation of the proximal urinary tract with narrowing and frequently medial deviation of both ureters between L4 and S2. Ureteral narrowing is smooth and conical and may involve a long segment. CT and ultrasound can demonstrate the localized fibrous mass and show a de-

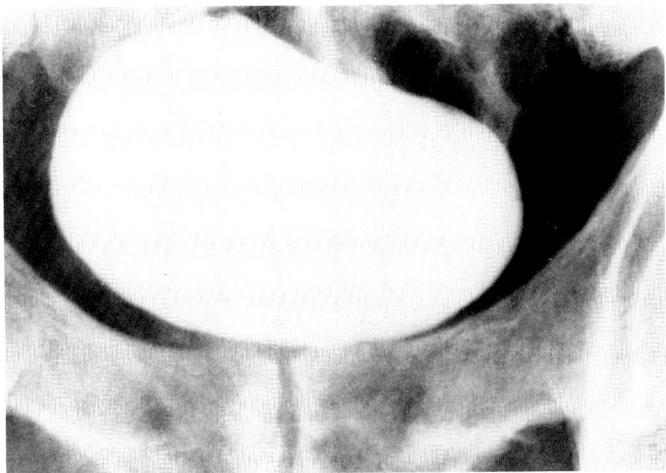

A

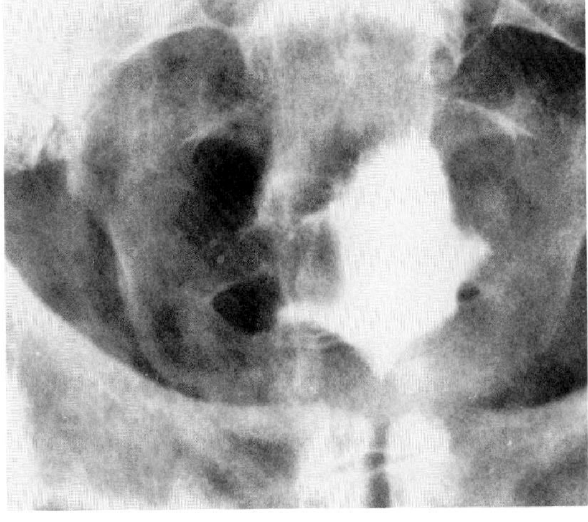

B

**Figure 42.28. Cyclophosphamide (Cytoxan) cystitis.** (A) The bladder appears normal prior to treatment. (B) Six months following repeated cycles of cyclophosphamide therapy, the bladder volume is markedly contracted. The bladder wall appears ulcerated and edematous. The patient experienced severe irritative symptoms along with urge incontinence and repeated episodes of gross hematuria. (From Gross,[25] with permission from the publisher.)

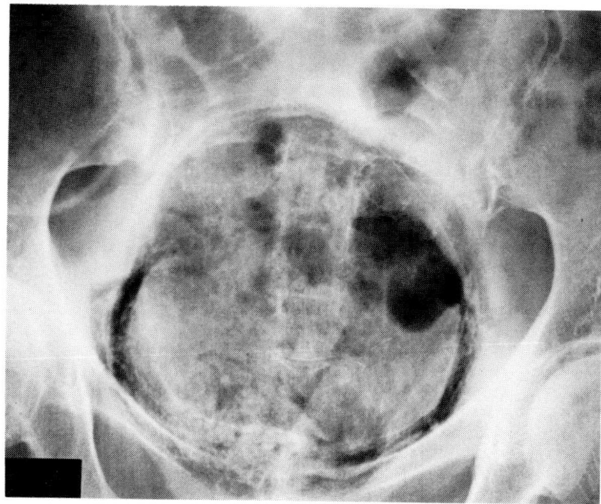

**Figure 42.29. Emphysematous cystitis.** A plain film of the pelvis shows radiolucent gas in the wall of the bladder. (From Ney and Friedenberg,[4] with permission from the publisher.)

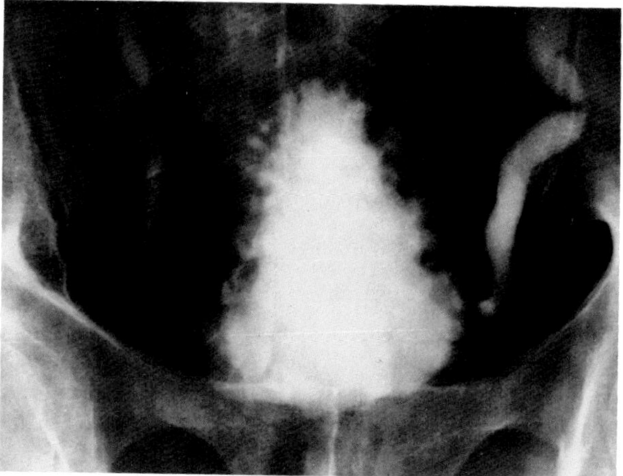

**Figure 42.30. Neurogenic bladder.** A small, spastic, trabeculated bladder with a pointed dome (pine tree bladder). (From Elkin,[17] with permission from the publisher.)

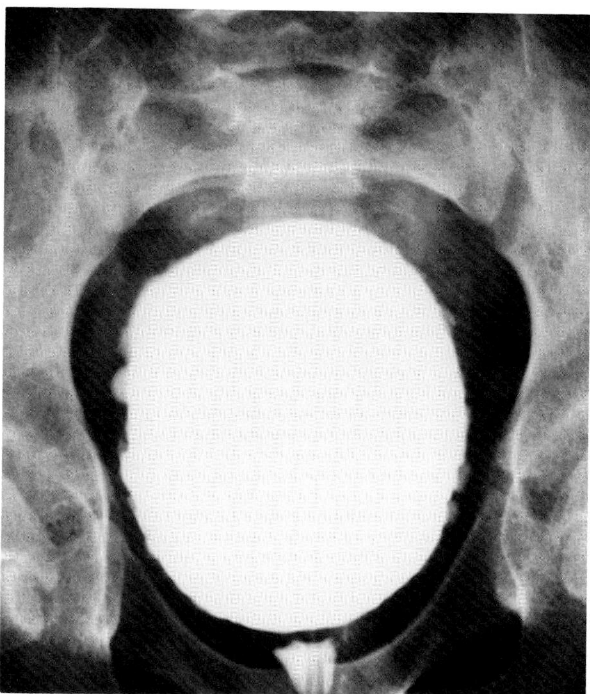

**Figure 42.31. Neurogenic bladder.** A large, atonic bladder with minimal cellule formation in a child with traumatic paralysis of the lower extremities.

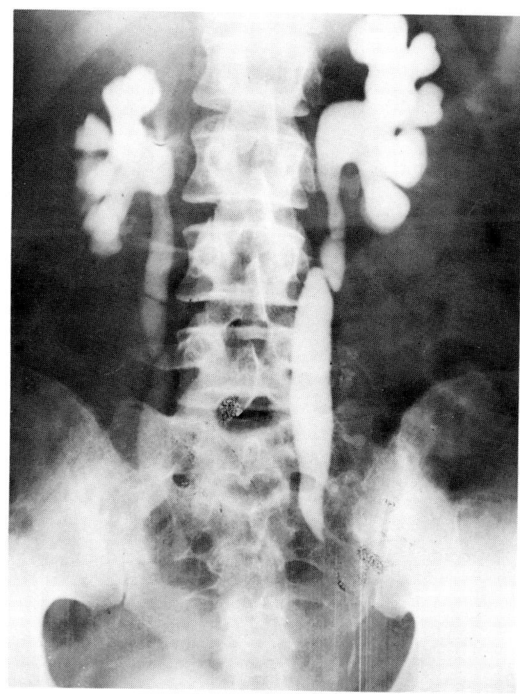

**Figure 42.32. Retroperitoneal fibrosis.** An excretory urogram demonstrates marked bilateral hydronephrosis with bilateral ureterectasis above the level of the sacral promontory. Below this point both ureters, where visualized, appear to be normal in caliber. No definite ureteral deviation is seen. An excretory urogram performed 1 year previously was entirely within normal limits. (From Ney and Friedenberg,[4] with permission from the publisher.)

crease in size spontaneously or in response to steroid therapy. Sonographic findings consist of an extensive retroperitoneal, extrarenal, anechoic, well-marginated, and irregularly contoured mass. CT shows the paraspinal mass to be isodense with the surrounding muscles. Inferior vena cavography may reveal narrowing and medial displacement of the vena cava at the site of involvement.[4,19,20]

## PELVIC LIPOMATOSIS

Pelvic lipomatosis is a benign condition in which there is an increased depositon of normal, mature adipose tissue in the extraperitoneal pelvic soft tissues around

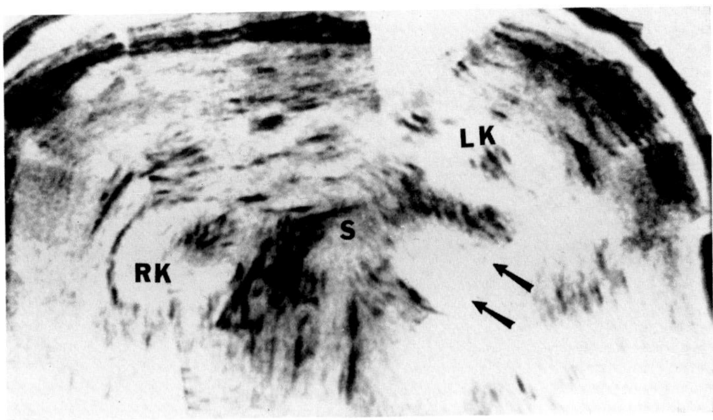


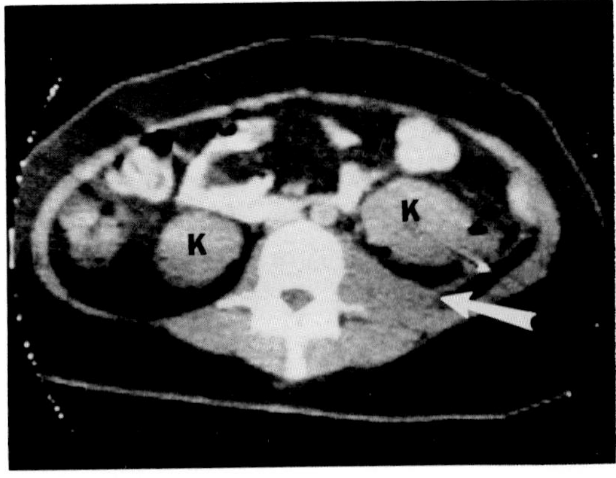


A B

**Figure 42.33. Retroperitoneal fibrosis.** (A) A transverse sonogram of the abdomen shows anterior displacement of the left kidney (LK) by a plaquelike anechoic mass (arrows) adjacent to the spine (s). RK = right kidney. (B) A CT scan demonstrates a similar finding (arrow). The smooth, well-delineated anterior margin of the mass has the same density as the kidneys (K) and the muscles of the back. (From Fagan, Larrieu, Amparo,[20] with permission from the publisher.)

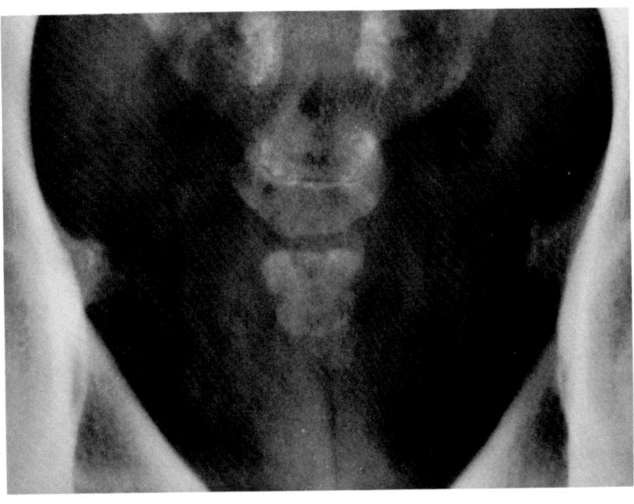

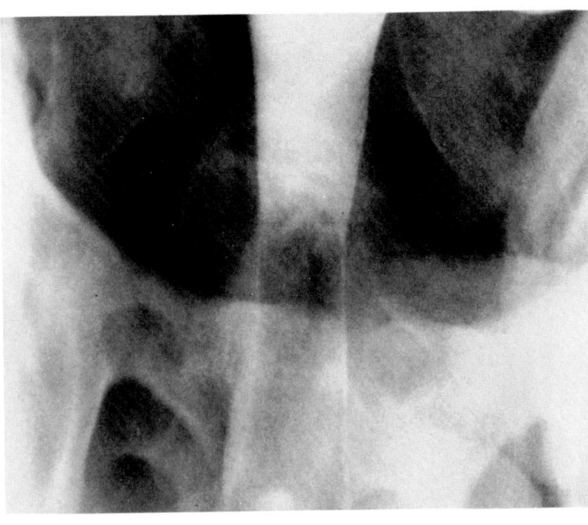

A B

**Figure 42.34. Pelvic lipomatosis.** (A) A plain abdominal radiograph demonstrates increased radiolucency in the pelvis caused by the excessive deposition of fat. (B) Smooth narrowing of the rectum and sigmoid is caused by the extrinsic fatty mass. (From Eisenberg,[26] with permission from the publisher.)

the urinary bladder, rectum, and prostate. Almost all reported caes have been in men. The clinical symptoms are nonspecific and include urinary tract infection, increased frequency of urination, constipation, and low back and abdominal pain.

Plain films of the abdomen reveal an increased radiolucency in the pelvis caused by the excessive deposition of fat. On excretory urography, the contrast-filled bladder is elevated, elongated, and compressed into a teardrop- or gourd-shaped configuration. Pelvic fat accumulating in the peripelvic and periureteric areas elevates and compresses the ureters and may cause hydronephrosis and urinary stasis.

Barium enema examination demonstrates vertical elongation of the sigmoid colon with narrowing of the rectum and sigmoid by the extrinsic fatty mass. The lumen of the bowel is distensible, and the colonic mucosa remains intact. On lateral views, there is widening of the retrorectal space.

CT can confirm the diagnosis of pelvic lipomatosis by demonstrating that the excessive pelvic soft tissues compressing the bladder and rectum have the same attenuation as normal subcutaneous fat.[21–23]

## REFERENCES

1. Friedland GW: Congenital anomalies of the urinary tract, in Friedland GW, Filly R, Goris ML, Gross D, Kempson RL, Korobkin M, Thurber BD, and Walter J (eds): *Uroradiology: An Integrated Approach*. New York, Churchill Livingstone, 1983.

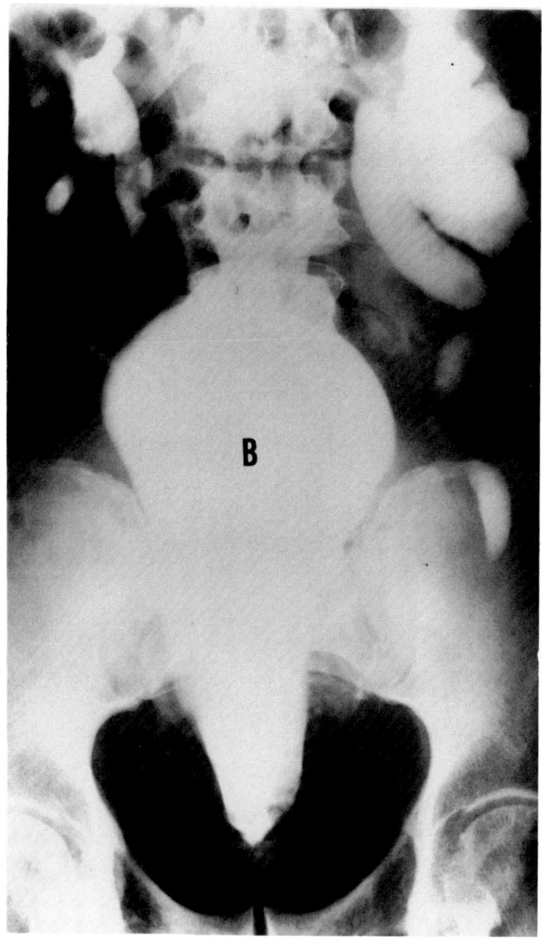


A

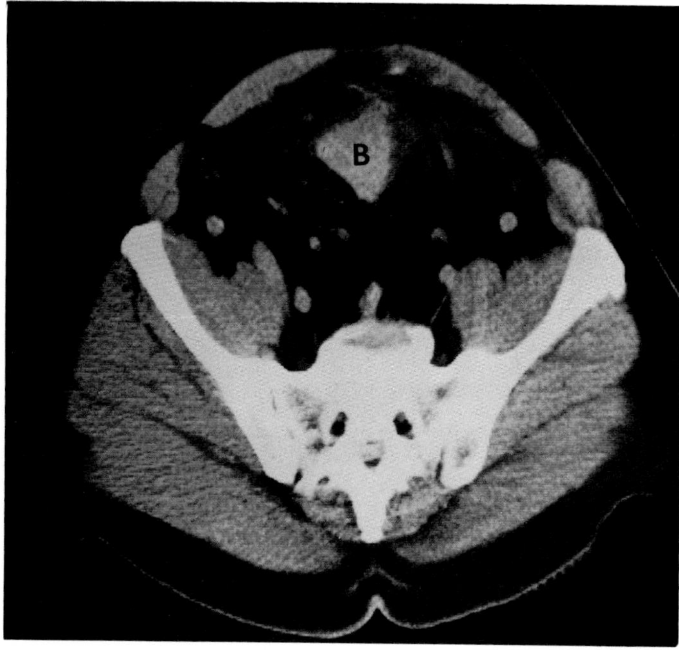


B

**Figure 42.35. Pelvic lipomatosis.** (A) An excretory urogram demonstrates bilateral hydronephrosis, displacement of the left ureter, and an abnormal pear-shaped configuration of the urinary bladder (B). There could be multiple causes for this appearance. (B) A CT scan reveals that the compressed bladder (B) is surrounded by low-density fat, thus confirming the diagnosis of pelvic lipomatosis. (From Laing,[27] with permission from the publisher.)


2. Witten DM, Myers GH, Utz BC: *Clinical Urography*. Philadelphia, WB Saunders, 1977.
3. Juhl JH: *Essentials of Roentgen Interpretation*, Philadelphia, JB Lippincott, 1981.
4. Ney C, Friedenberg RM: *Radiographic Atlas of the Genitourinary System*, Philadelphia, JB Lippincott, 1981.
5. Amar AD: Lateral ureteral displacement: Sign of nonvisualized duplication. *J Urol* 105:638–641, 1971.
6. Lebowitz RL, Griscom NT: Neonatal hydronephrosis: 146 cases. *Radiol Clin North Am* 15:49–59, 1977.
7. White P, Lebowitz RL: Exstrophy of the bladder. *Radiol Clin North Am* 15:93–106, 1977.
8. Morin ME, Tan A, Baker DA, et al: Urachal cyst in the adult: Ultrasound diagnosis. *AJR* 132:831–832, 1979.
9. Ney C, Friedenberg RM: Radiographic findings in anomalies of the urachus. *J Urol* 99:288–291, 1968.
10. Spataro RF, Davis RS, McLachlan MSF, et al: Urachal abnormalities in the adult. *Radiology* 149:659–663, 1983.
11. Friedland GW, Fair WR, Govan DE, et al: Posterior urethral valves. *Clin Radiol* 27:367–373, 1977.
12. Faegenburg D, Bosniak M, Evans JA: Renal sinus lipomatosis: Its demonstration by nephrotomography. *Radiology* 83: 987–997, 1964.
13. Elliott GB, Moloney PJ, Clement JG: Malacoplakia of the urinary tract. *AJR* 116:830–837, 1972.
14. Schneiderman C, Simon MA: Malakoplakia of the urinary tract. *J Urol* 100:694–698, 1968.
15. Bailey H: Cystitis emphysematosa. *AJR* 86:850–862, 1961.
16. Texter JH, Koontz WW, McWilliams NB: Hemorrhagic cystitis as a complication of the management of pediatric neoplasms. *Urol Surg* 29:47–48, 1979.
17. Elkin M: *Radiology of the Urinary System*. Boston, Little, Brown, 1980.
18. Kendall AR, Karafin L: Classification of neurogenic bladder disease. *Urol Clin North Am* 1:37–44, 1974.
19. Arger PH, Stolz JL, Miller WT: Retroperitoneal fibrosis: An analysis of the clinical spectrum and roentgenographic signs. *AJR* 119:810–821, 1973.
20. Fagan CJ, Larrieu AJ, Amparo EG: Retroperitoneal fibrosis: Ultrasound and CT features. *AJR* 133:239–243, 1979.
21. Friedman AC, Hartman DS, Sherman J, et al: Computed tomography of abdominal fatty masses. *Radiology* 139:415–429, 1981.
22. Moss AA, Clark RE, Goldberg HI, et al: Pelvic lipomatosis: A roentgenographic diagnosis. *AJR* 115:411–419, 1972.
23. Crane DB, Smith MJV: Pelvic lipomatosis. *J Urol* 118:547–550, 1977.
24. Rabinowitz JG: *Pediatric Radiology*. Philadelphia, JB Lippincott, 1978.
25. Gross D: The effect of complications of surgery, radiation, and medical therapy on the urinary tract, in Friedland GW, Filly R, Goris ML, Gross D, Kempson RL, Korobkin M, Thurber BD, and Walter J (eds): *Uroradiology: An Integrated Approach*. New York, Churchill Livingstone, 1983.
26. Eisenberg RL: *Gastrointestinal Radiology: A Pattern Approach*. Philadelphia, JB Lippincott, 1983.
27. Laing FC: Pelvic masses, in Eisenberg RL, Amberg JR (eds): *Critical Diagnostic Pathways in Radiology: An Algorithmic Approach*. Philadelphia, JB Lippincott, 1981.

# PART SEVEN
# DISORDERS OF THE ALIMENTARY TRACT

## Chapter Forty-Three

# Diseases of the Esophagus

## MOTILITY DISORDERS

### Abnormalities in Swallowing

Abnormalities in swallowing may be caused by motor incoordination of the pharynx and upper esophageal sphincter due to primary striated muscle disease (myasthenia gravis, myotonic distrophy, polymyositis) or to diseases of the central and peripheral nervous systems (bulbar poliomyelitis, syringomyelia, multiple sclerosis). Radiographic evaluation of the pharyngoesophageal region requires cineradiography or videotape studies because of the rapidity with which the contrast material passes through this area. Decreased motility or a loss of the proper sequence of movement is best demonstrated with the patient in the recumbent position, since in the erect position gravity can simulate the effect of peristalsis and hide subtle abnormalities. Prolonged retention of a substantial amount of barium in the valleculae and pyriform sinuses after swallowing (vallecular sign) is often considered the radiographic counterpart of clinical dysphagia.[1]

### Cricopharyngeal Dysfunction

Cricopharyngeal achalasia is the failure of pharyngeal peristalsis to coordinate with relaxation of the upper esophageal sphincter. This condition, which occurs without mechanical obstruction or esophageal stenosis, is the result of any lesion that interferes with the complex neuromuscular activity in this region. Incomplete relax-

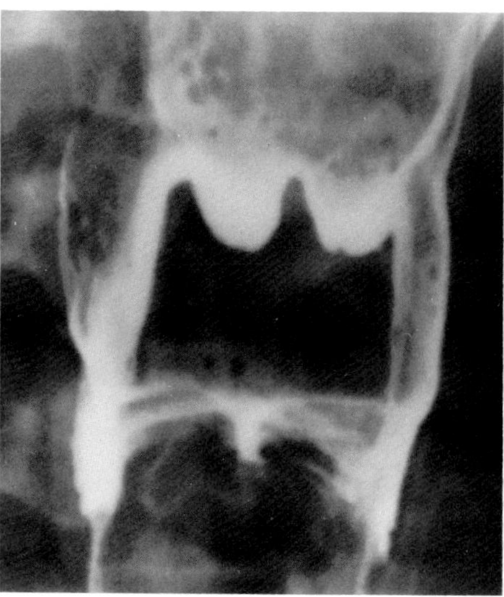

**Figure 43.1. Vallecular sign.** Prolonged retention of barium in the valleculae and pyriform sinuses after swallowing in a patient with clinical dysphagia. (From Eisenberg,[60] with permission from the publisher.)

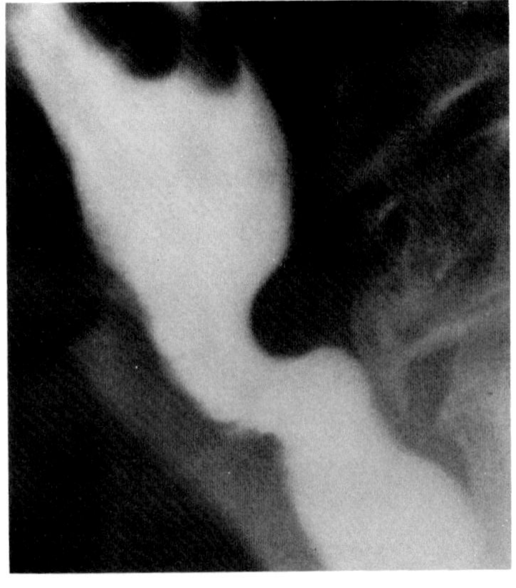

A

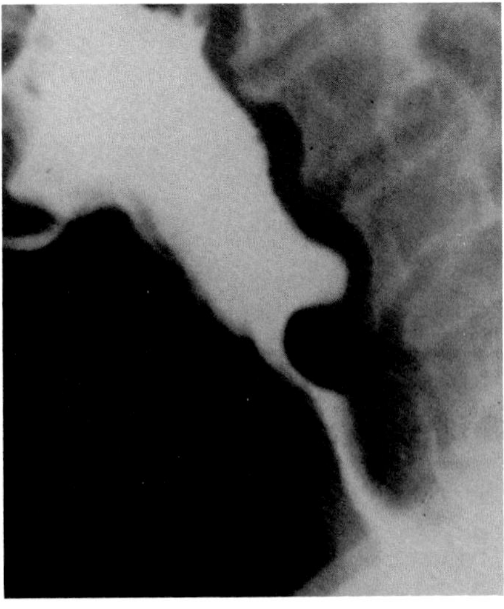

B

**Figure 43.2. Cricopharyngeal achalasia.** (A, B) Two examples of posterior impressions on the esophagus at the level of the pharyngoesophageal junction. (From Eisenberg,[55] with permission from the publisher.)

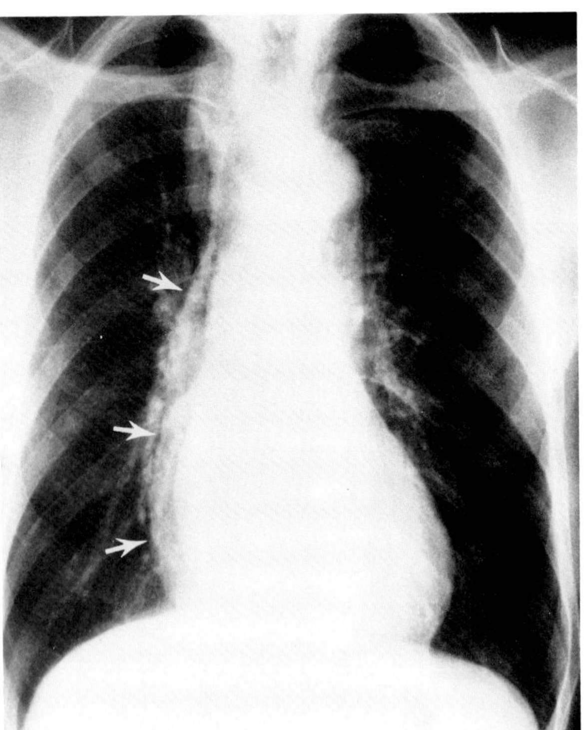

A

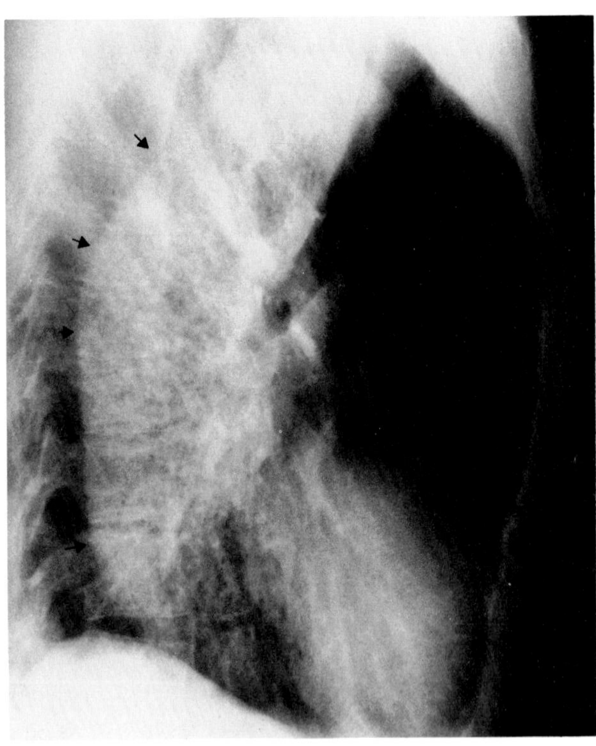

B

**Figure 43.3. Achalasia.** (A) A frontal chest radiograph demonstrates the margin of the dilated, tortuous esophagus (arrows) parallel to the right border of the heart. (B) A lateral chest film shows a mixture of fluid and air density within the dilated esophagus (arrows).

ation of the cricopharyngeus muscle produces the characteristic radiographic appearance of a hemispherical or horizontal, shelflike posterior protrusion into the barium-filled pharyngoesophageal junction at approximately the C5–6 level. The posterior location of this defect distinguishes it from an esophageal web, which is usually situated anteriorly. Although the presence of this cricopharyngeus impression indicates the existence of some physiologic abnormality, lesser degrees of it may not be associated with clinical symptoms. More significant neuromuscular abnormalities can result in dysphagia by acting as obstructions to the passage of the bolus. In severe disease, swallowing can result in an overflow of ingested material into the larynx and trachea, and in pulmonary complications of aspiration.

Cricopharyngeal achalasia has been suggested as an important factor in the development of posterior pharyngeal (Zenker's) diverticula.[2]

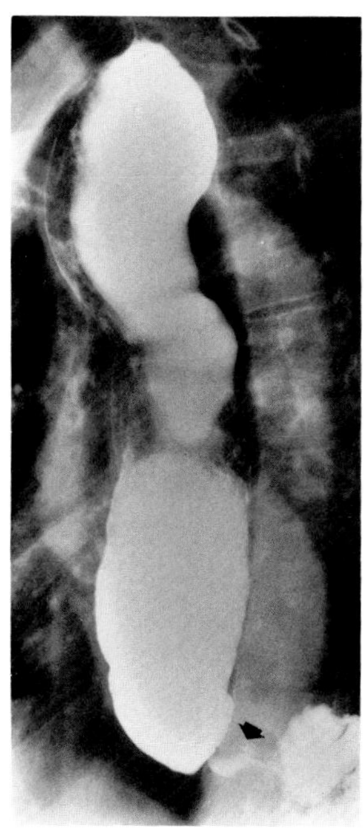

**Figure 43.4. Achalasia.** Failure of relaxation of the region of the distal esophageal sphincter (arrow) with severe proximal dilatation. (From Eisenberg,[55] with permission from the publisher.)

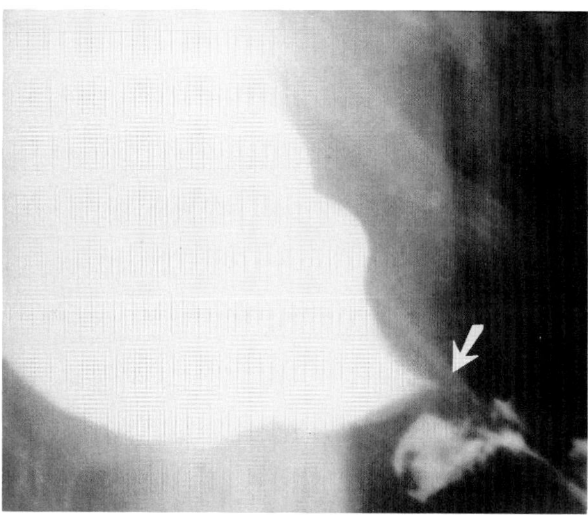

**Figure 43.5. Achalasia.** Characteristic narrowing of the distal esophageal segment (beak sign) (arrow).

## Achalasia

Achalasia is a disorder of motility in which marked dilatation, elongation, and tortuosity of the esophagus are caused by incomplete relaxation of the lower esophageal sphincter combined with a failure of normal peristalsis in the smooth muscle portion of the esophagus. Although the precise pathogenesis of achalasia is not known, the most accepted explanation is a defect in the cholinergic innervation of the esophagus related to a paucity or absence of ganglion cells in the myenteric plexuses (Auerbach) of the distal esophageal wall. This cause of achalasia is supported by the Mecholyl test, in which the injection of the parasympathomimetic drug causes denervation hypersensitivity (Cannon's law) and results in violent segmental contractions of the esophagus.

Plain chest radiographs are frequently sufficient for the diagnosis of achalasia. The dilated, tortuous esophagus can appear as a widened mediastinum, often with an air-fluid level, that is usually on the right side adjacent to the cardiac shadow. Associated aspiration frequently leads to chronic interstitial pulmonary disease or intermittent episodes of acute pneumonia. The air bubble of the gastric fundus on upright films is small or totally absent.

Following the ingestion of barium, there is disordered motility with weak, nonpropulsive, dysrhythmic peristaltic waves that are ineffective in propelling the bolus into the stomach. The hallmark of achalasia is a gradually tapered, smooth, conical narrowing of the distal esophageal segment that extends some 1 to 3 cm in length (beak appearance). Sequential radiographs, especially with the patient in the erect position, demonstrate small spurts of barium entering the stomach through the narrowed distal segment.

Central and peripheral neuropathy of various causes can result in a radiographic pattern identical to achalasia, as can destruction of the myenteric plexuses due to infestation by the protozoan *Trypanosoma cruzi* in Chagas' disease.

Malignant lesions can produce an achalasia pattern by several mechanisms. Metastases to the midbrain or vagal nuclei, or direct extension of the tumor to involve the vagus nerve, can result in a failure of relaxation of the lower esophageal sphincter. A similar pattern can be produced by a carcinoma of the distal esophagus or a malignant lesion in the gastric cardia that invades the esophagus and destroys ganglion cells in the myenteric plexus. It is essential to distinguish between nonmalignant causes of the achalasia pattern and carcinoma of the distal esophagus or gastric cardia. Patients with classic benign achalasia tend to be younger and have symptoms of shorter duration. Radiographically, in achalasia there tends to be deranged peristalsis, persistence of the normal mucosal pattern, and a gradual zone of transition, in contrast to normal peristalsis, mucosal destruction or nodularity, and a sharply defined, more rapid transition zone between normal esophagus and neoplasm.

Achalasia may be treated by surgical disruption of the muscular fibers about the lower esophageal sphincter (Heller myotomy) or by balloon dilatation under fluoroscopic control. In the latter procedure, forceful dilatation of a pneumatic bougie, positioned with its midpoint at the narrowest level of the gastroesophageal junction, causes the expansion of this segment to the desired degree of dilatation. The most serious, albeit uncommon, complication of balloon dilatation of the lower esophageal sphincter is esophageal rupture, which may not be radiographically detectable on an esopho-

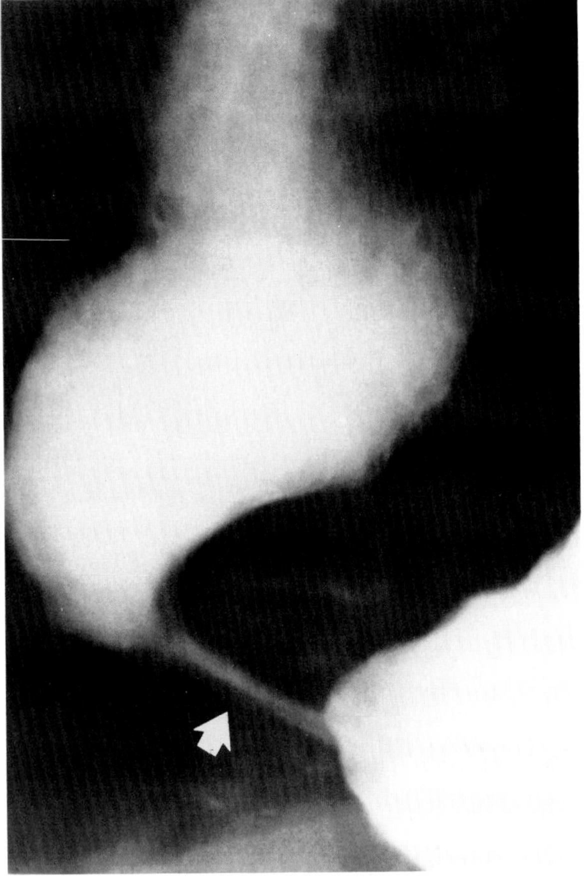

**Figure 43.6. Achalasia.** A small spurt of barium (arrow) enters the stomach through the narrowed distal segment (jet effect). (From Eisenberg,[55] with permission from the publisher.)

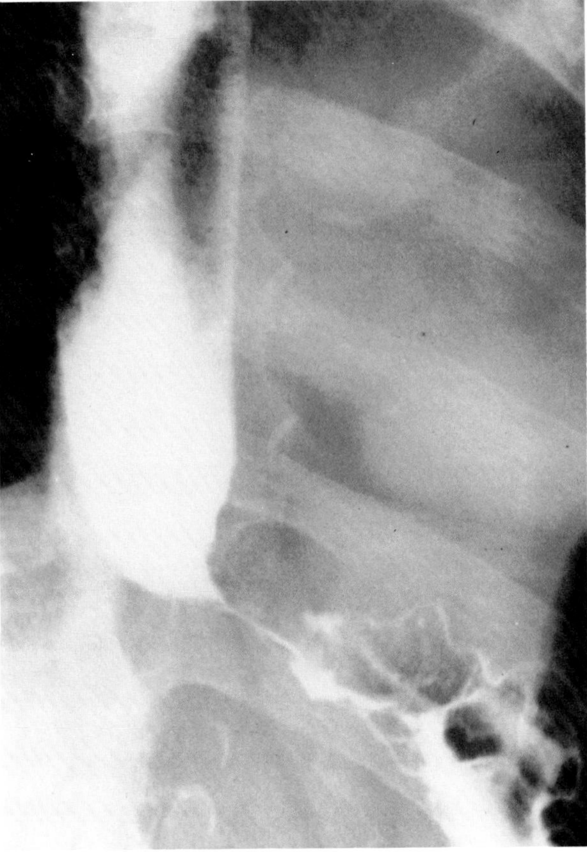

**Figure 43.7. Achalasia pattern.** In this patient, esophageal dilatation with distal narrowing is caused by the proximal extension of carcinoma of the fundus of the stomach. (From Eisenberg,[55] with permission from the publisher.)

gram obtained immediately after dilatation. Persistent or increasing symptoms in a patient who has undergone this procedure should suggest the possibility of delayed esophageal perforation and the need for a repeat radiographic evaluation.[3–10]

## Scleroderma (Progressive Systemic Sclerosis)

Scleroderma frequently involves the esophagus, sometimes even before the characteristic skin changes become evident. Degeneration and atrophy of smooth muscle in the lower half to two-thirds of the esophagus with subsequent replacement of esophageal musculature by fibrosis lead to atony of the esophagus and a failure of peristaltic activity. Incompetence of the lower esophageal sphincter permits the reflux of acid-pepsin gastric secretions into the distal esophagus. In almost half of the patients, this reflux leads to peptic esophagitis and stricture formation that cause heartburn and severe dysphagia. Because the upper third of the esophagus is composed primarily of striated muscle, which is infrequently affected by scleroderma, a barium swallow demonstrates a normal stripping wave that clears the upper esophagus but stops at about the level of the aortic arch. Distal to that point there is minimal or absent peristaltic activity and, with the patient in a recumbent position, barium will remain for a long time in the dilated, atonic esophagus. Multiple radiographs obtained several minutes apart can be effectively superimposed on each other. In contrast to achalasia, when the patient with scleroderma is placed in the upright position the barium flows rapidly through the widely patent region of the lower esophageal sphincter.[3–6]

## Diffuse Esophageal Spasm

A variety of diverse disorders may produce abnormal motility in the lower two-thirds of the esophagus. When mild, there may be few or no symptoms; severe forms of diffuse esophagus spasm produce retrosternal pain and dysphagia.

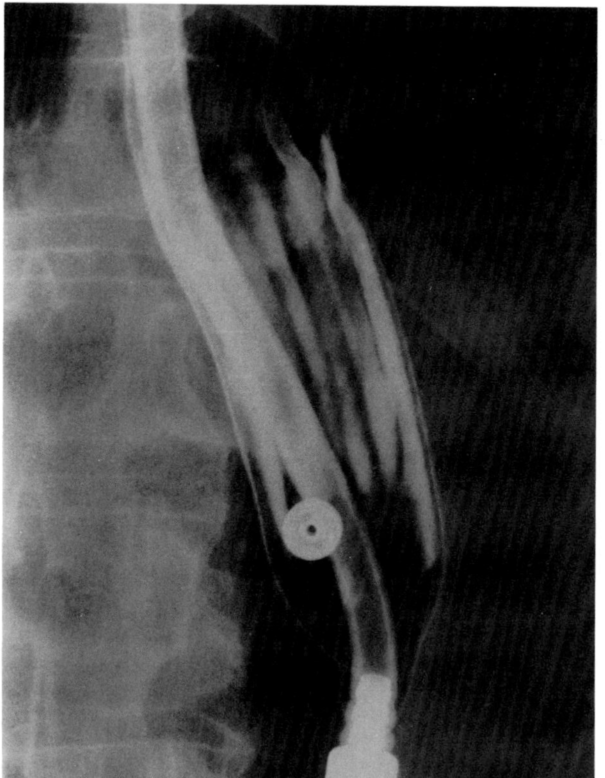

**Figure 43.8. Achalasia.** A pneumatic bougie for balloon dilatation of the esophagus. (From Eisenberg,[55] with permission from the publisher.)

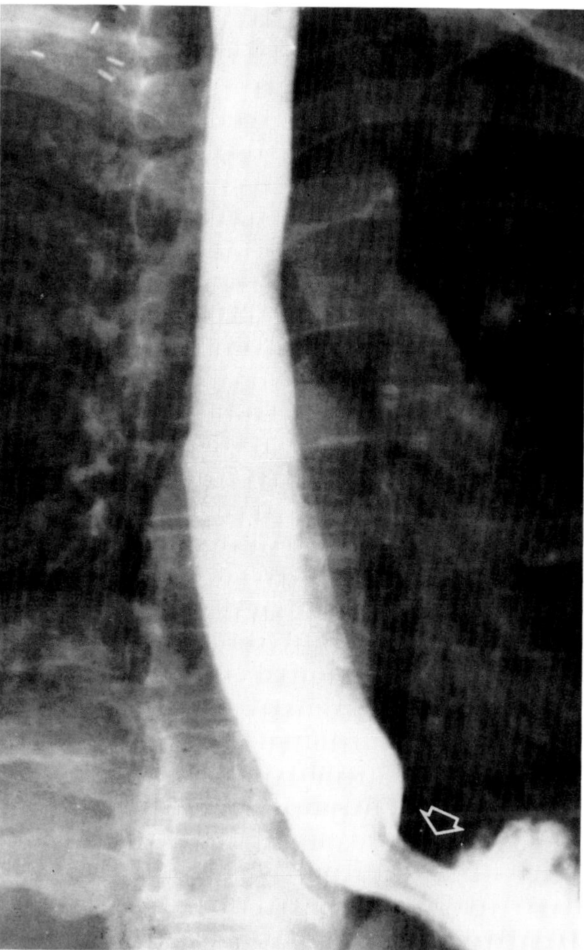

**Figure 43.9. Scleroderma.** The esophagus is dilated and atonic, and the esophagogastric junction is patulous (arrow). The surgical clips are from a cervical sympathectomy for Raynaud's phenomenon. (From Eisenberg,[55] with permission from the publisher.)

The mildest abnormal motility disorder is tertiary contractions. These are multiple, irregular, ringlike contractions that are nonpropulsive and appear and disappear rapidly, following each other from top to bottom with such speed that they appear to occur simultaneously. Radiographically, tertiary contractions are seen as asymmetric indentations of unequal width and depth along the esophagus, with pointed, rounded, or truncated projections between them. Tertiary contractions are most commonly seen in persons with presbyesophagus, a motility disturbance associated with aging. A similar pattern may be seen in patients with esophageal inflammation (esophagitis, corrosive agents, infections, radiation injury), early stages of achalasia, obstruction of the cardia, and neuromuscular abnormalities, especially diabetes mellitus.

In the severe form of diffuse esophageal spasm, tertiary contractions of abnormally high amplitude can obliterate the lumen and cause compartmentalization of the barium column. These segmental, nonpropulsive contractions can be accompanied by severe pain and cause the barium to be displaced both proximally and distally from the site of spasm, producing a bizarre contour of the lower esophagus with a corkscrew pattern of transient sacculations or pseudodiverticula (rosary bead esophagus).[11]

## INFLAMMATORY DISORDERS

### Reflux Esophagitis

Reflux esophagitis develops when the lower esophageal sphincter fails to act as an effective barrier to the entry of gastric acid contents into the distal esophagus. Although there is a higher-than-normal likelihood of gastroesophageal reflux in patients with sliding hiatal hernias, reflux esophagitis can be endoscopically demonstrated in only about one-quarter of patients with them. Conversely, esophagitis is often encountered in patients in whom no hiatal hernia can be demonstrated.

A number of radiographic approaches have been suggested for the demonstration of gastroesophageal reflux. One procedure is to increase intra-abdominal pressure by straight-leg raising or manual pressure on the abdomen, with or without a Valsalva maneuver. Turning the patient from prone to supine or vice versa may dem-

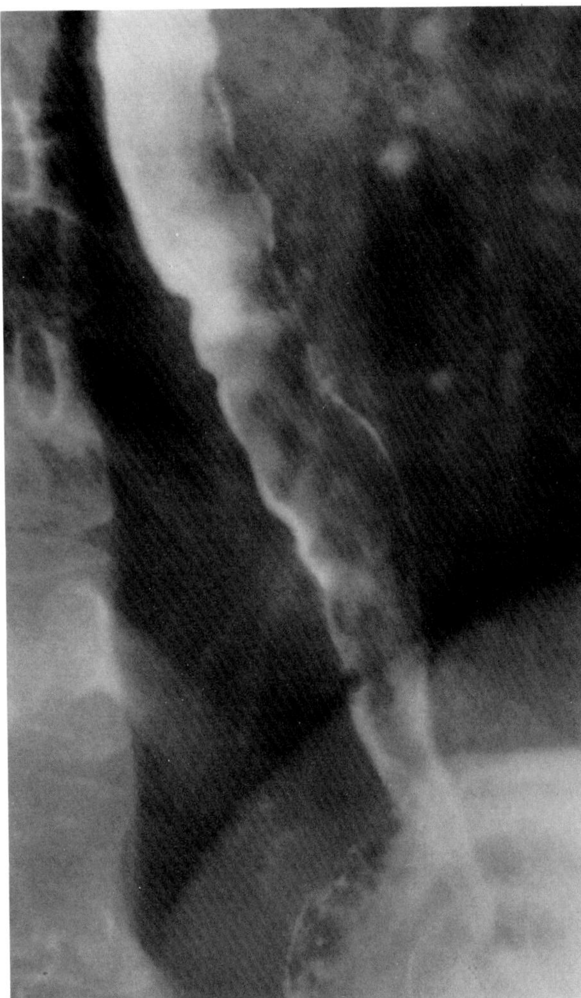

**Figure 43.10. Tertiary contractions.**

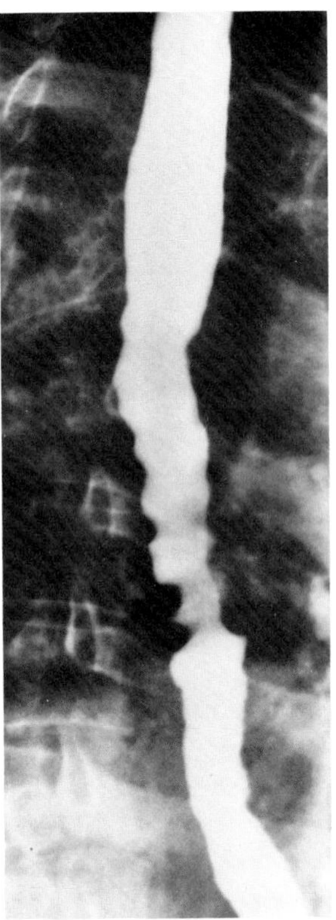

**Figure 43.11. Tertiary contractions.** Disordered esophageal motility secondary to radiation therapy for carcinoma of the lung. The disordered contractions begin at the upper margin of the treatment port. (From Rogers and Goldstein,[56] with permission from the publisher.)

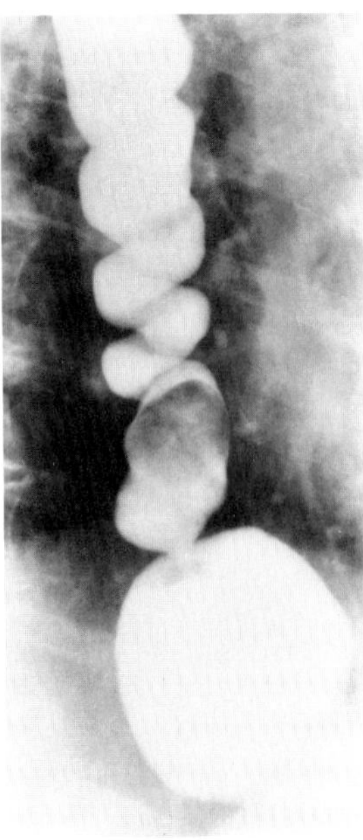

**Figure 43.12. Diffuse esophageal spasm.** High-amplitude contractions irregularly narrow the lumen of the esophagus.

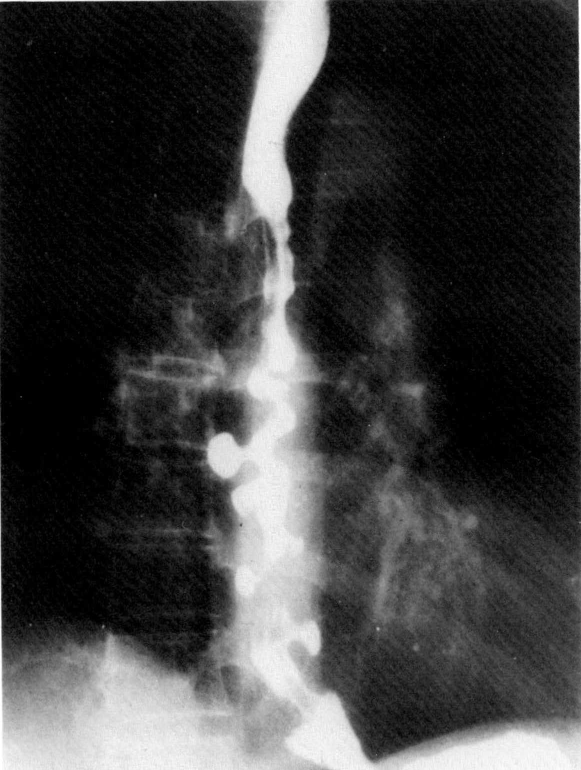

**Figure 43.13. Diffuse esophageal spasm.** Pseudodiverticula are in a corkscrew pattern. (From Eisenberg,[55] with permission from the publisher.)

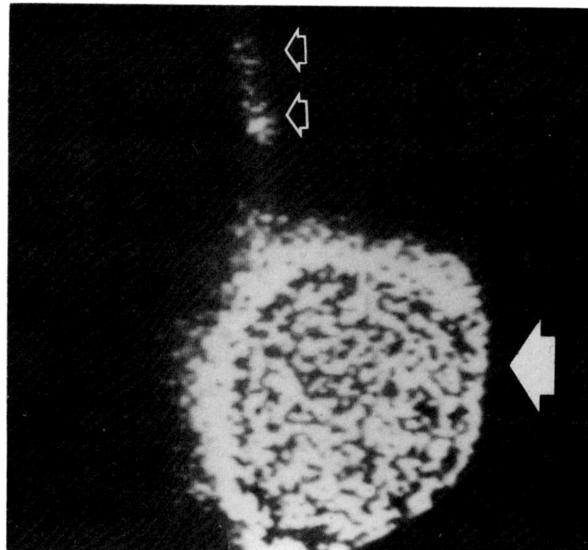

**Figure 43.14. Radionuclide demonstration of gastroesophageal reflux.** Note the reflux of radionuclide into the esophagus (small, open arrows) from the stomach (large, solid arrow) following the oral administration of $^{99m}$Tc DTPA. (From Eisenberg,[55] with permission from the publisher.)

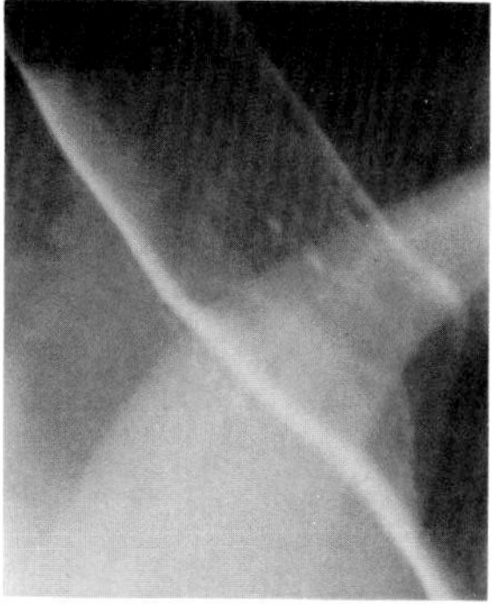

**Figure 43.15. Reflux esophagitis.** A double-contrast study demonstrates superficial ulcerations or erosions as streaks or dots of contrast material superimposed on the flat mucosa of the distal esophagus. (From Eisenberg,[55] with permission from the publisher.)

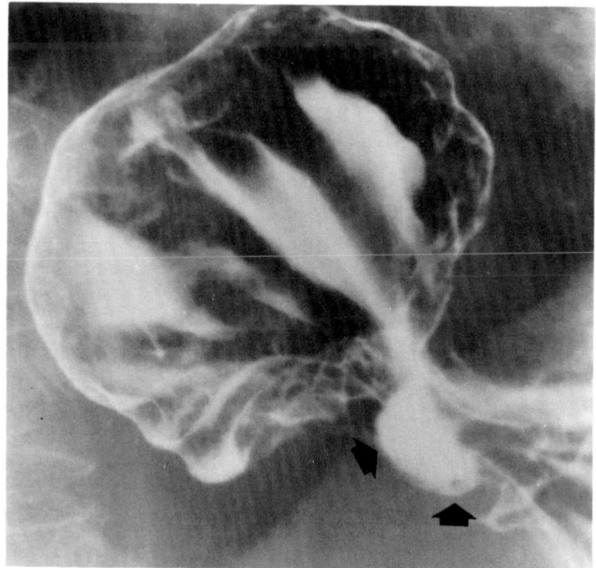

**Figure 43.16. Reflux esophagitis.** A penetrating ulcer (arrows) in a large hiatal hernia sac. (From Eisenberg,[55] with permission from the publisher.)

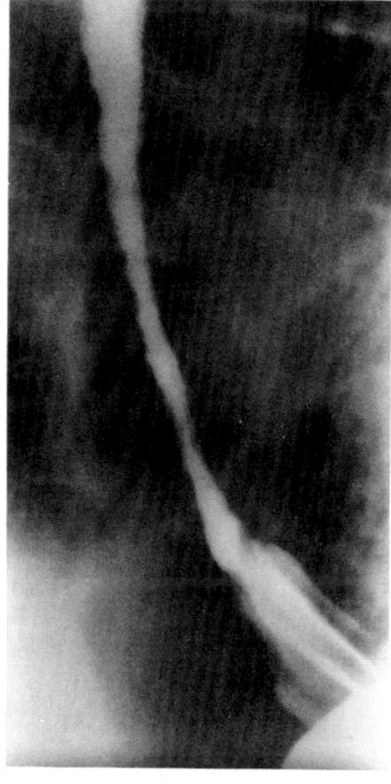

**Figure 43.17. Reflux esophagitis.** A long esophageal stricture with an associated hiatal hernia. (From Eisenberg,[55] with permission from the publisher.)

onstrate reflux of barium from the stomach into the esophagus. A new radionuclide technique for demonstrating and measuring gastroesophageal reflux is to scan the lower esophagus and stomach after the oral administration of $^{99m}$Tc DTPA. If no spontaneous reflux of radionuclide from the stomach into the distal esophagus is observed, an abdominal binder is used to raise intragastric pressure. It must be remembered, however, that the failure to demonstrate reflux radiographically does not exclude the possibility that a patient's esophagitis is related to reflux. As long as typical radiographic findings of reflux esophagitis are noted, there is little reason to persist in strenuous efforts to actually demonstrate retrograde flow of barium from the stomach into the esophagus.

The earliest radiographic findings in reflux esophagitis are detectable on double-contrast studies. They consist of superficial ulcerations or erosions that appear as streaks or dots of barium superimposed on the flat mucosa of the distal esophagus. In single-contrast studies of patients with esophagitis, the outer borders of the barium-filled esophagus are not sharply seen, but rather have a hazy, serrated appearance with shallow, irregular protrusions that are indicative of erosions of varying length and depth. Widening and coarsening of edematous longitudinal folds can simulate filling defects. In addition to diffuse erosion, reflux esophagitis can result in large, discrete penetrating ulcers in the distal esophagus or in a hiatal hernia sac. These ulcers have a radiographic appearance similar to benign gastric ulcers due to chronic peptic disease and not infrequently penetrate through the wall of the esophagus, causing massive bleeding or perforation.

Fibrotic healing of diffuse reflux esophagitis or a localized penetrating ulcer may cause narrowing of the distal esophagus. Strictures secondary to reflux esophagitis tend to be smooth and tapering with no demonstrable mucosal pattern.[4,12]

## Barrett's Esophagus

Barrett's esophagus is a condition related to severe reflux esophagitis in which the normal stratified squamous lining of the lower esophagus is destroyed and replaced by columnar epithelium similar to that of the stomach. Peptic ulceration in Barrett's esophagus occurs anywhere along the columnar epithelium, particularly at the squamocolumnar junction, which can even be as high as the aortic arch. Unlike the shallow ulceration that reflux esophagitis usually causes in the squamous epithelium, the Barrett's ulcer tends to be deep and penetrating and identical to peptic gastric ulceration. Although a sliding hiatal hernia with gastroesophageal reflux is commonly demonstrated, the Barrett's ulcer is usually separated from the hiatal hernia by a variable length of normal-appearing esophagus, in contrast to reflux esophagitis, in which the distal esophagus is abnormal down to the level of the hernia.

Fibrous healing of ulceration in a Barrett's esophagus often leads to a smooth, tapered stricture. In addition to postinflammatory stricture, Barrett's esophagus has an unusually high propensity for developing malignancy in the columnar-lined portion. In one series, adenocarcinoma, which is usually very rare in the esophagus, developed in 10 percent of the patients with Barrett's esophagus.

Radionuclide examination with intravenous pertechnetate can be used to demonstrate a Barrett's esophagus. Because this radionuclide is actively taken up by the gastric type of mucosa, continuous concentration of the isotope in the distal esophagus to a level that corresponds approximately to that of the ulcer or stricture is indicative of a Barrett's esophagus and may obviate the need for mucosal biopsy.[13–15]

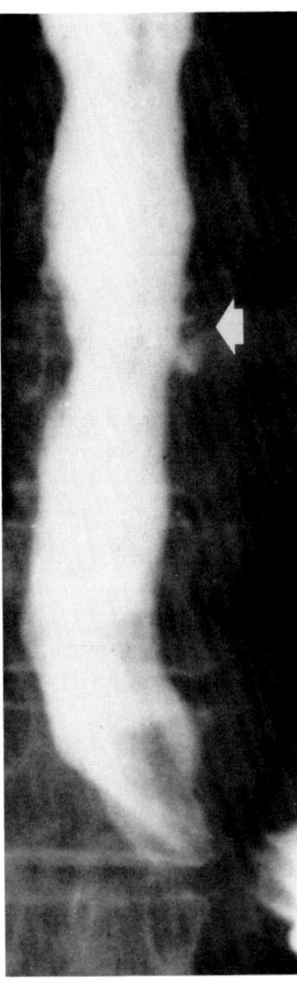

**Figure 43.18. Barrett's esophagus.** Ulcerations (arrow) have developed at a distance from the esophagogastric junction. (From Eisenberg,[55] with permission from the publisher.)

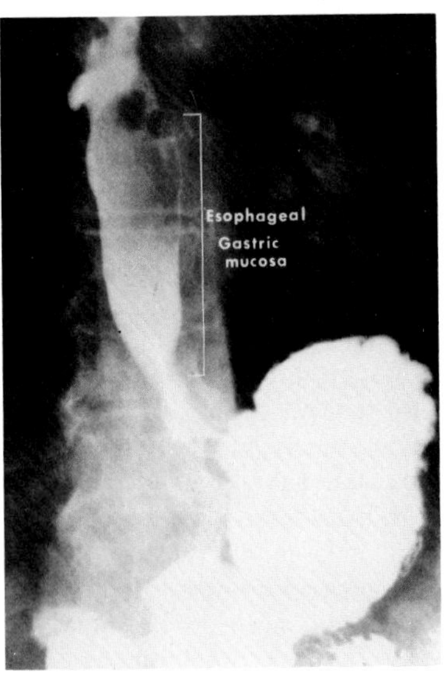


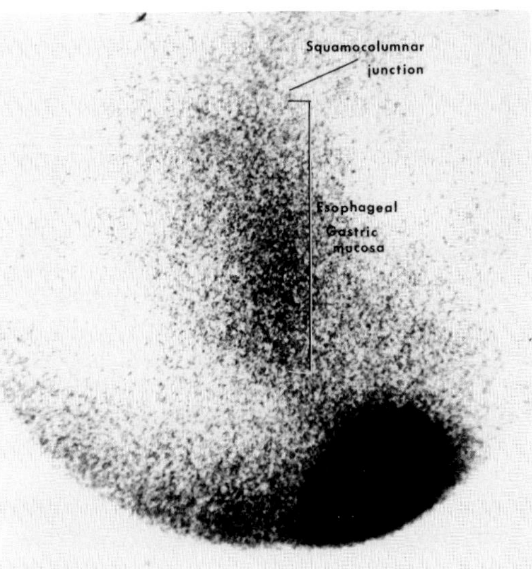


**Figure 43.19. Barrett's esophagus.** (A) Esophogram. (B) A $^{99m}$Tc pertechnetate scintigram. The uptake of isotope extends above the esophagogastric junction to a level comparable to that of the esophageal stricture and ulcer on the radiograph. (From Berquist, Nolan, Stephens, et al,[13] with permission from the publisher.)

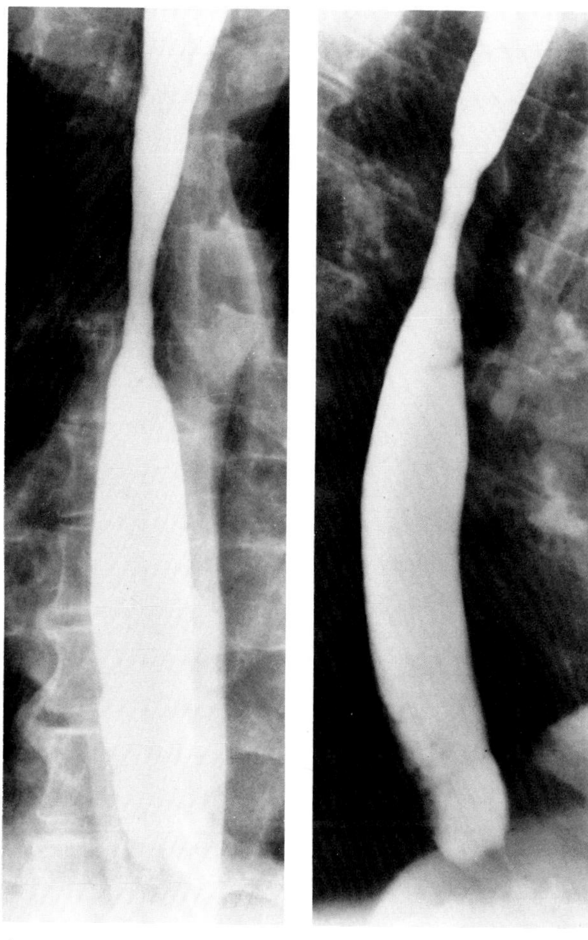


**Figure 43.20. Barrett's esophagus.** (A) Frontal and (B) lateral projections demonstrate a smooth stricture in the upper thoracic esophagus. (From Eisenberg,[55] with permission from the publisher.)

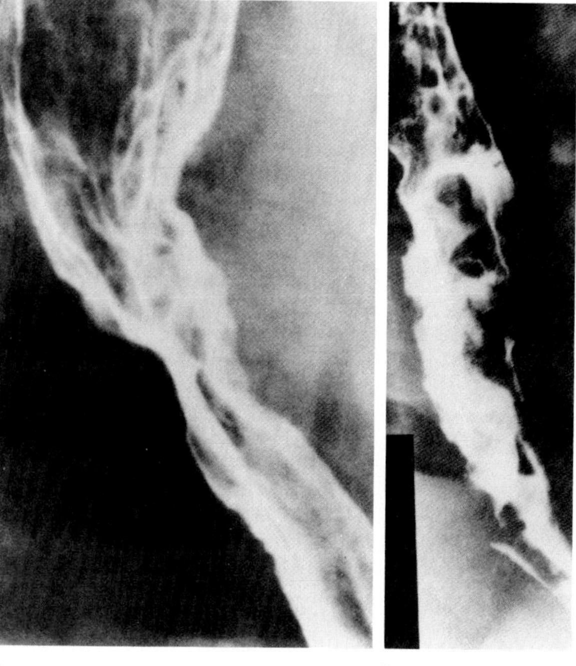


**Figure 43.21. Adenocarcinoma in Barrett's esophagus.** (A) An irregular, infiltrating stricture. (B) Thickened, tortuous esophageal folds (varicoid appearance) due to the submucosal spread of tumor. (From Levine, Caroline, Thompson, et al,[57] with permission from the publisher.)

## Acute Esophagitis

Although the reflux of gastric acid contents is the most common cause of acute esophagitis, infectious and granulomatous disorders, physical injury (caustic agents, radiation injury), and medication may produce a similar inflammatory response. The esophagus is often dilated with a loss of effective peristalsis. Nonpropulsive peristaltic waves, ranging from mild tertiary contractions to severe segmental spasms, are an early finding.

In candidal and herpetic esophagitis, the classic radiographic appearance is a shaggy marginal contour caused by deep ulcerations and sloughing of the mucosa. The intervening esophageal contour is nodular with a thickened mucosa and plaquelike filling defects producing an irregular cobblestone pattern.

The ingestion of alkaline corrosive agents produces acute inflammatory changes in the esophagus. Superficial penetration of the toxic agent results in only minimal ulceration. Deeper penetration into the submucosal and muscular layers causes sloughing of destroyed tissue and deep ulceration.

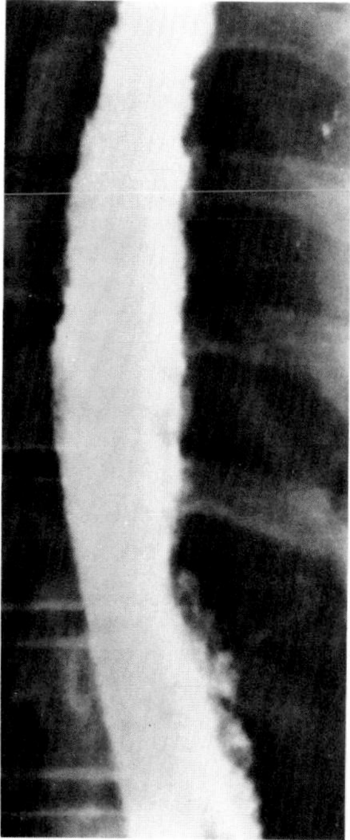

**Figure 43.22. Candidal esophagitis.** The shaggy marginal contour is caused by deep ulcerations and sloughing of the mucosa. (From Eisenberg,[55] with permission from the publisher.)

Doses of radiation of greater than 4500 rad frequently lead to severe esophagitis. The radiographic appearance is indistinguishable from that of candidal esophagitis, a far more common condition in patients undergoing chemotherapy and radiation therapy for malignant disease.

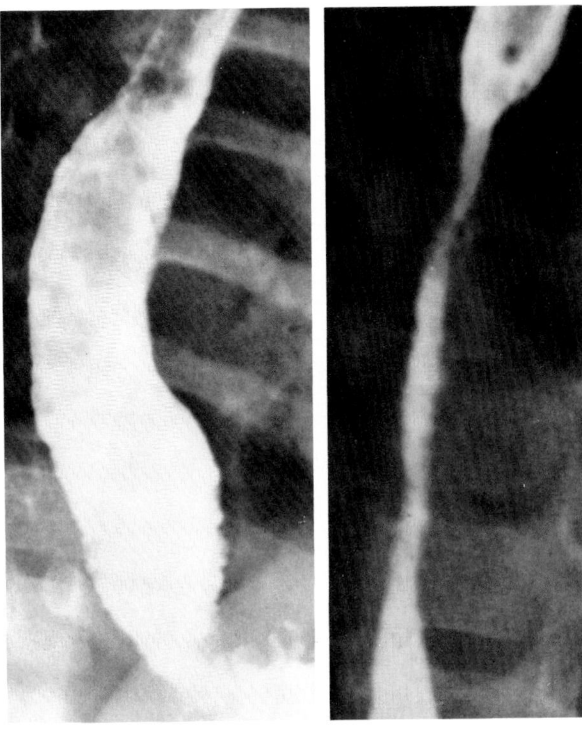


**Figure 43.23. Caustic esophagitis.** (A) A dilated, boggy esophagus with ulceration 8 days after the ingestion of a corrosive agent. (B) Stricture formation is evident on an esophogram performed 3 months after the caustic injury. (From Eisenberg,[55] with permission from the publisher.)

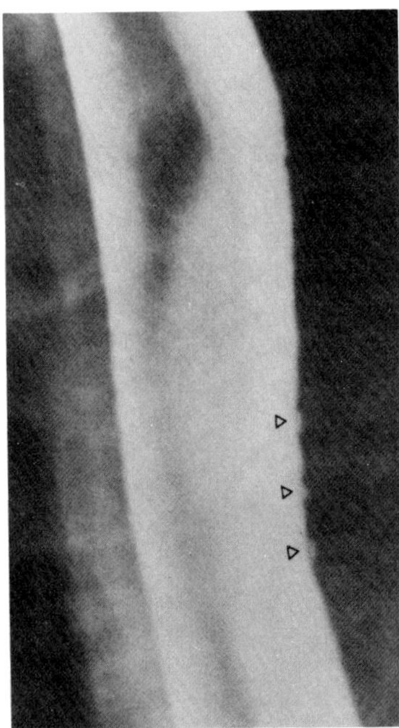

**Figure 43.24. Quinidine-induced esophagitis.** Multiple small ulcers (arrows) are seen projecting along the left lateral wall of the esophagus. (From Creteur, Laufer, Kressel, et al,[58] with permission from the publisher.)

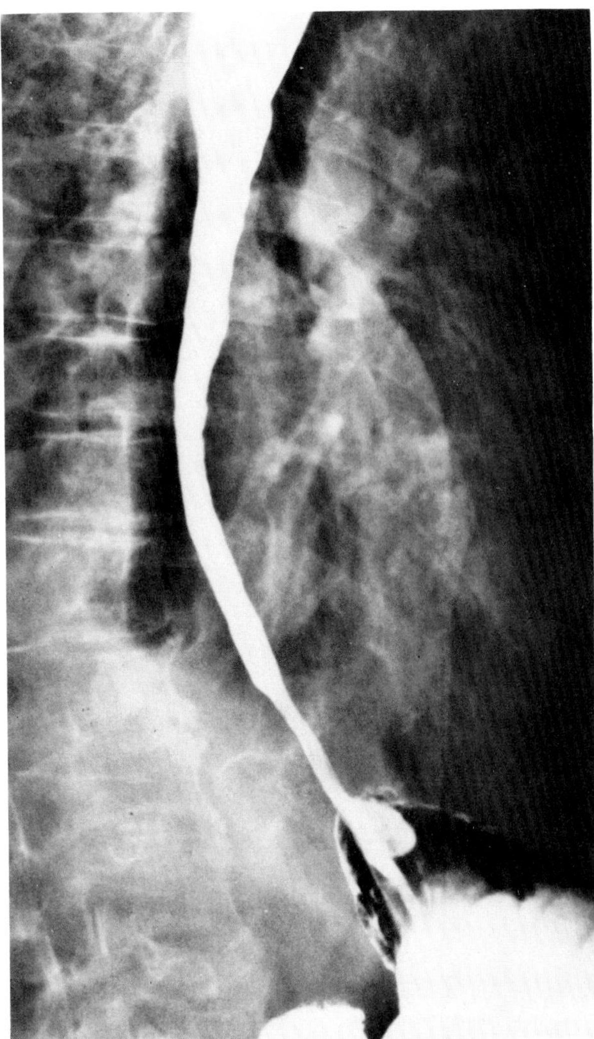

**Figure 43.25. Caustic stricture.** An extensive area of narrowing, involving almost the entire thoracic esophagus, due to lye ingestion. (From Eisenberg,[55] with permission from the publisher.)

Drug-induced esophagitis may occur in patients who have delayed esophageal transit time, which permits prolonged mucosal contact with the ingested substance. The most common drug causing esophageal ulceration is potassium chloride in tablet form. Other medications implicated as causes of esophagitis include tablets of tetracycline, emepronium bromide, quinidine, and ascorbic acid, all of which are weak caustic agents that are innocuous when they pass rapidly through the esophagus.

Healing of the intense mucosal and intramural inflammation of acute esophagitis may lead to marked fibrosis and stricture formation. These benign strictures tend to be long lesions with tapered margins and relatively smooth mucosal surfaces, in contrast to the irregular narrowing, mucosal destruction, and overhanging margins that are generally associated with malignant processes.[16–21]

## TUMORS

### Benign Intramural Tumors of the Esophagus

Leiomyomas are the most common benign tumors of the esophagus. Most are asymptomatic and discovered during radiographic examinations for nonesophageal complaints. In the few patients with symptoms, dysphagia and substernal pain are the major complaints. Leiomyomas are most frequently found in the lower third of the esophagus. The vast majority are located intramurally and have little, if any, tendency to undergo malignant transformation. When viewed in profile, an esophageal leiomyoma appears radiographically as a characteristic smooth, crescent-shaped defect on one side of the lumen with stretched, but intact, linear mucosal folds. In contrast to the gradual and smooth curve of an extrinsic lesion, the defect in the barium column caused by an intramural leiomyoma is generally abrupt and sharply angled. There is no evidence of infiltration or undercutting at the tumor margin. Other sharply defined intramural defects with smooth margins include lipomas, fibromas, neurofibromas, and duplication cysts.[22,23]

### Carcinoma of the Esophagus

Progressive dysphagia in a person over 40 years of age must be assumed to be due to cancer until proved otherwise. Because the symptoms of esophageal carcinoma

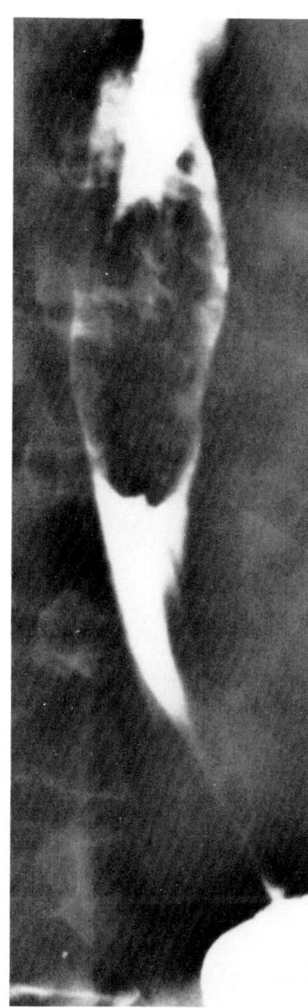

**Figure 43.27. Fibrovascular polyp.** A bulky, sausage-shaped mass with a mildly lobulated surface. (From Eisenberg,[55] with permission from the publisher.)

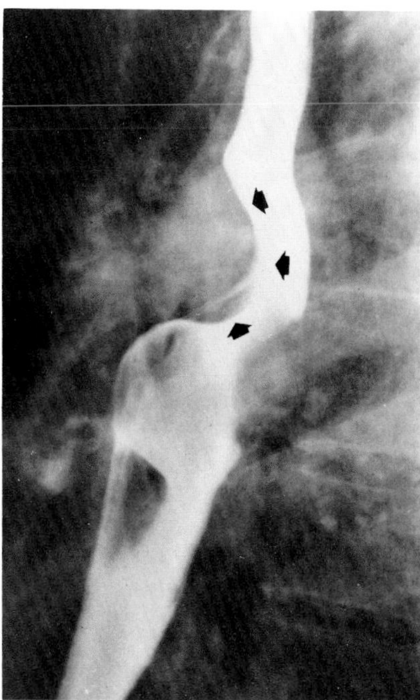

**Figure 43.26. Leiomyoma.** A smooth, rounded intramural defect in the barium column (arrows). (From Eisenberg,[55] with permission from the publisher.)

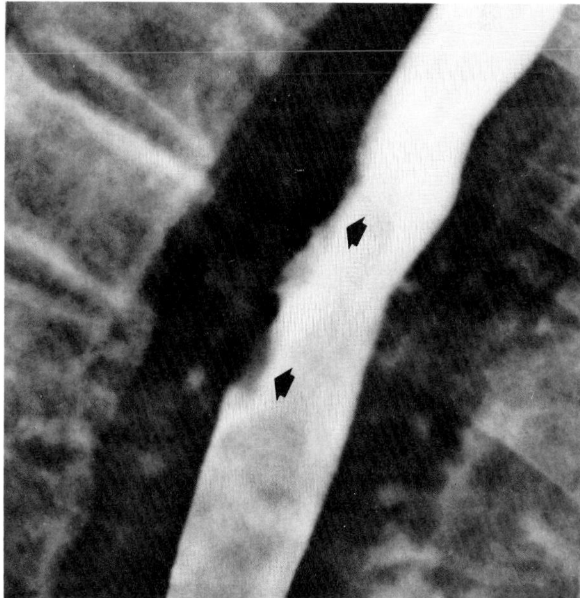

**Figure 43.28. Early squamous cell carcinoma of the esophagus.** A flat, plaquelike lesion (arrows) involves the posterior wall of the esophagus. (From Eisenberg,[55] with permission from the publisher.)

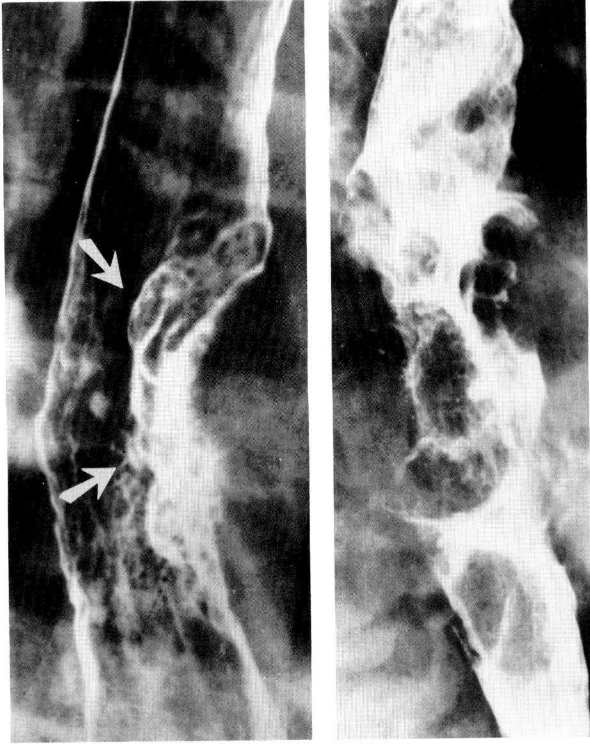


**Figure 43.29. Squamous cell carcinoma of the esophagus.** (A) A localized polypoid mass with ulceration (arrows). (B) A bulky, irregular filling defect with destruction of mucosal folds. (From Eisenberg,[55] with permission from the publisher.)

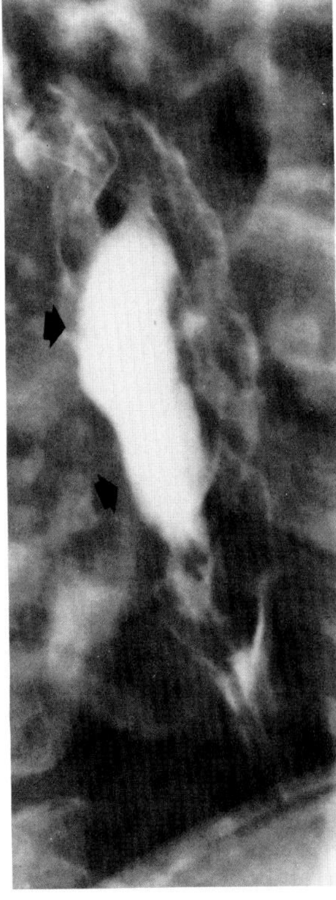

**Figure 43.30. Primary ulcerative carcinoma of the esophagus.** A characteristic meniscoid ulceration (arrows) is surrounded by a tumor mass. (From Eisenberg,[55] with permission from the publisher.)

tend to appear late in the course of the disease, and since the lack of limiting serosa commonly permits direct extension of the tumor by the time of initial diagnosis, carcinoma of the esophagus has a dismal prognosis. Most carcinomas of the esophagus are of the squamous cell type. About half arise in the middle third; of the remainder, slightly more occur in the lower than in the upper third. The incidence of carcinoma of the esophagus is far higher in men than in women. There is a strong correlation between excessive alcohol intake, smoking, and esophageal carcinoma. The incidence of esophageal carcinoma is also increased in patients with severe corrosive esophagitis, Barrett's esophagus, and untreated achalasia.

The earliest radiographic appearance of infiltrating carcinoma of the esophagus is a flat, plaquelike lesion, occasionally with central ulceration, that involves one wall of the esophagus. At this stage there may be minimal reduction in the caliber of the lumen, and the lesion is seen to best advantage on double-contrast views of the distended esophagus. Unless the patient is carefully examined in various projections, this earliest form of esophageal carcinoma can be missed. As the infiltrating cancer progresses, luminal irregularities indicating mucosal destruction are noted. Advanced lesions encircle the lumen completely, causing annular constrictions with overhanging margins and, often, some degree of obstruction. The lumen through the stenotic area is irregular, and mucosal folds are absent or severely ulcerated. Carcinoma of the esophagus can appear as a localized polypoid mass, often with deep ulceration and a fungating appearance. In the relatively uncommon primary ulcerative esophageal carcinoma, ulceration of virtually all of an eccentric, flat mass produces a meniscoid ap-

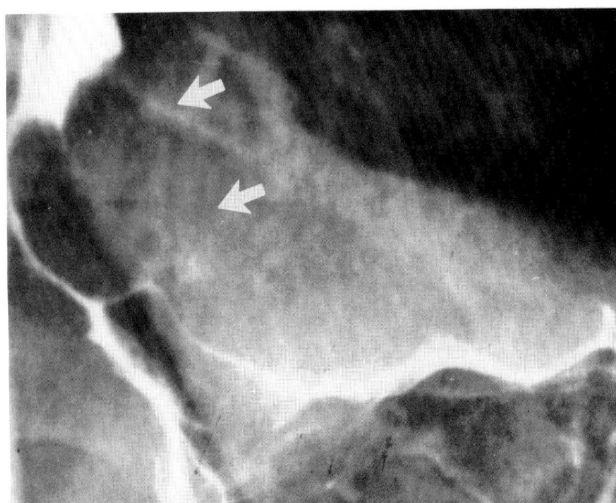

**Figure 43.31. Adenocarcinoma of the fundus involving the esophagus.** An irregular tumor of the superior aspect of the fundus extends proximally as a large mass (arrows), almost obstructing the distal esophagus. (From Eisenberg,[55] with permission from the publisher.)

pearance analogous to the Carman sign seen with gastric malignancy.

More than 10 percent of adenocarcinomas of the stomach arising near the cardia invade the lower esophagus at an early stage and can cause symptoms of esophageal obstruction. This process can produce an irregularly narrowed, sometimes ulcerated lesion simulating carcinoma of the distal esophagus. Careful examination of the cardia is necessary to demonstrate the gastric origin of the tumor.

Luminal obstruction due to carcinoma causes proximal dilatation of the esophagus and may result in aspiration pneumonia. Extension of the tumor to contiguous mediastinal structures may lead to fistula formation, especially between the esophagus and the respiratory tract.

Computed tomography (CT) has become a major staging method in patients with esophageal carcinoma. It can provide information on tumor size, extension, and resectibility that was previously only available at thoracotomy. The major CT findings of carcinoma of the esophagus are a soft tissue mass and focal wall thickening with an eccentric or irregular lumen. Evidence of the spread of tumor includes the obliteration of the fat planes between the esophagus and adjacent structures (left atrium, aorta), the formation of a sinus tract or fistula to the tracheobronchial tree, and evidence of metastatic disease, which primarily appears as enlargement of mediastinal, retrocrural, left gastric, or celiac lymph nodes, or low-density masses in the liver. The major problem in using CT to demonstrate tumor invasion into adjacent structures is the lack of adequate mediastinal and abdominal fat planes around the esophagus that may be encountered in these often severely cachectic patients.[15,23–29]

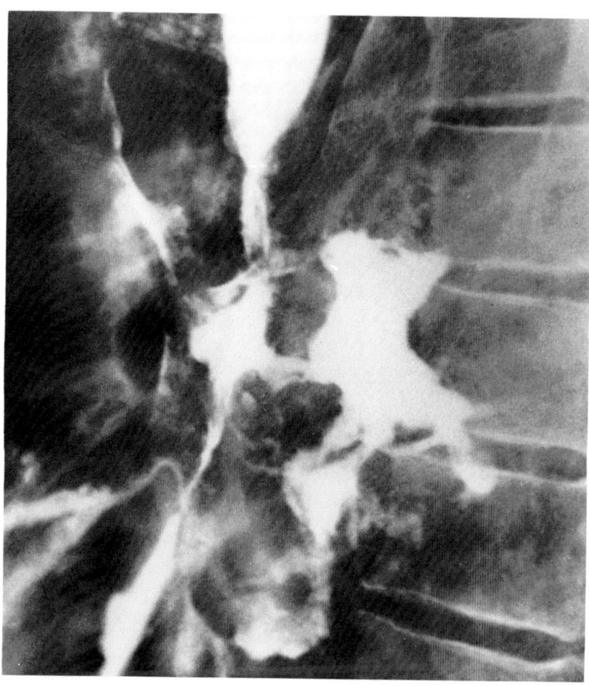

**Figure 43.32. Esophagorespiratory fistula secondary to squamous cell carcinoma of the esophagus.** (From Eisenberg,[55] with permission from the publisher.)

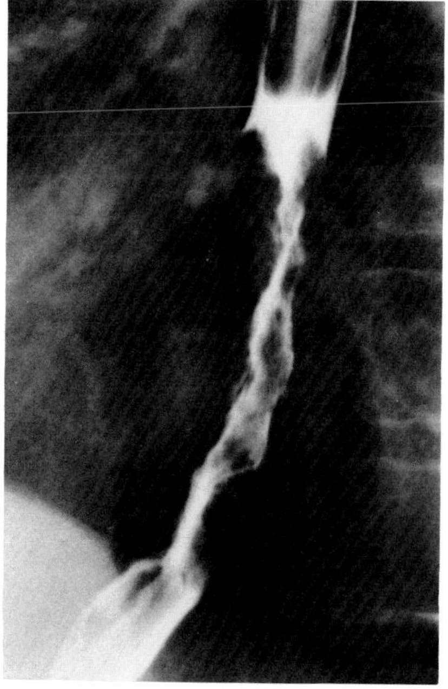

A

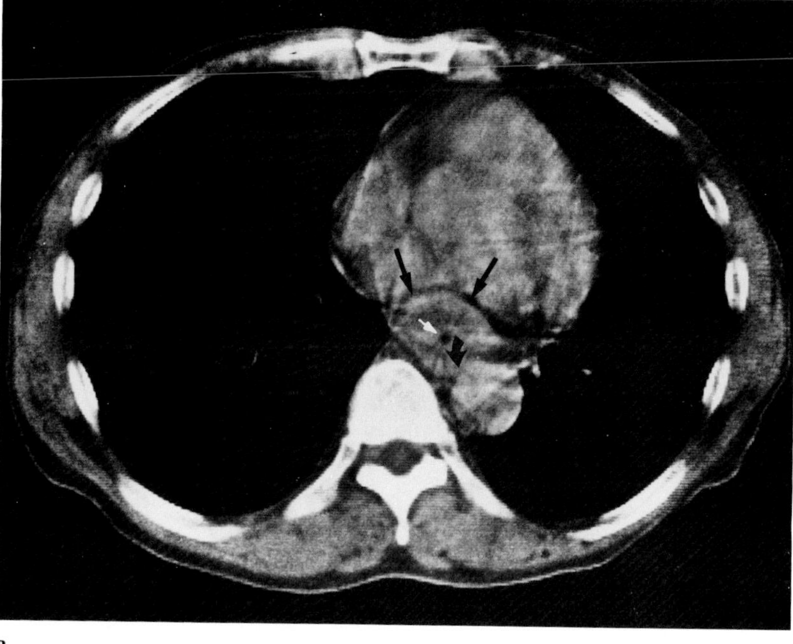

B

**Figure 43.33. CT staging of esophageal carcinoma.** (A) An esophogram demonstrates an infiltrating lesion causing irregular narrowing of the distal esophagus. (B) A CT scan shows circumferential narrowing of the lumen (white arrow) by the bulky carcinoma (straight black arrows). Obliteration of the fat plane adjacent to the aorta (curved arrow) indicates mediastinal invasion.

## OTHER ESOPHAGEAL DISORDERS

### Esophageal Diverticula

Esophageal diverticula are of two types: traction, containing all esophageal layers; and pulsion, composed of mucosa and submucosa herniating through the muscularis.

A Zenker's diverticulum arises in the upper esophagus, its neck lying in the midline of the posterior wall at the pharyngoesophageal junction (approximately the C5–6 level). The development of a Zenker's diverticulum is apparently related to premature contraction or other motor incoordination of the cricopharyngeus muscle; this produces increased intraluminal pressure and a pulsion diverticulum at a point of anatomic weakness in the posterior pharyngoesophageal wall. Although usually asymptomatic, the diverticulum may fill with food and secretions and enlarge to such an extent that it compresses the esophagus at the level of the thoracic inlet and produces esophageal obstruction. On plain radiographs of the neck, a Zenker's diverticulum may appear as a widening of the retrotracheal soft tissue space, often with an air-fluid level. Following the administration of barium, the diverticulum appears as a saccular outpouching protruding from the esophageal lumen and connected to it by a relatively narrow neck. Because the diverticulum arises from the posterior wall of the esophagus, it is best visualized on lateral projections. As the diverticulum enlarges, the sac extends downward and posteriorly, displacing and narrowing the adjacent esophageal lumen.

Diverticula of the thoracic portion of the esophagus usually arise opposite the bifurcation of the trachea in the region of the hilum of the lung. These are almost invariably traction diverticula that develop in response to the pull of fibrous adhesions following infection of mediastinal lymph nodes. Usually asymptomatic, thoracic diverticula appear radiographically as funnel- or

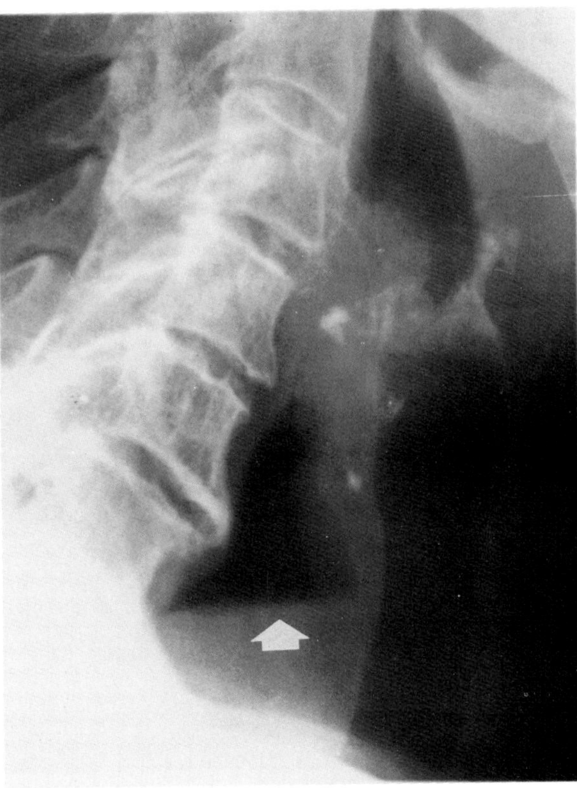

**Figure 43.34. Zenker's diverticulum.** An air-fluid level within the diverticulum (arrow) is seen in the retrotracheal soft tissue space. (From Eisenberg,[55] with permission from the publisher.)

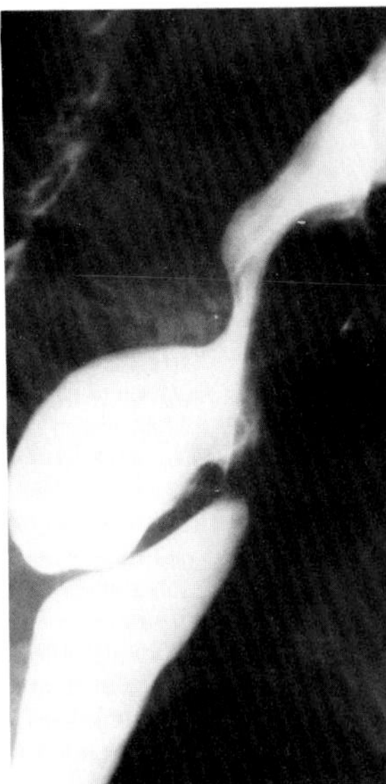

**Figure 43.35. Zenker's diverticulum.** An oblique view demonstrates a large Zenker's diverticulum almost occluding the esophageal lumen. (From Eisenberg,[55] with permission from the publisher.)

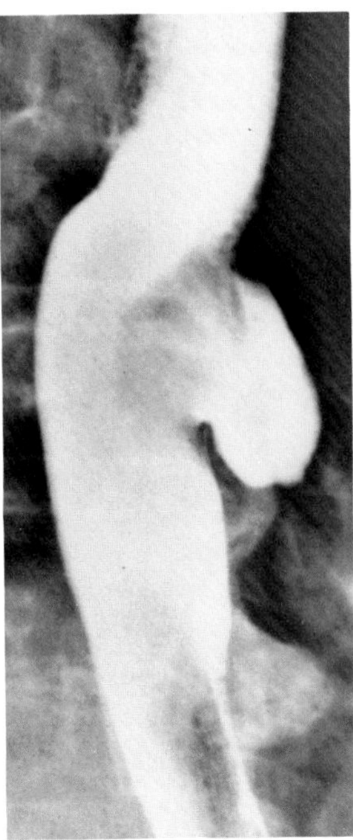

**Figure 43.36. Traction diverticulum of the midthoracic esophagus.** (From Eisenberg,[55] with permission from the publisher.)

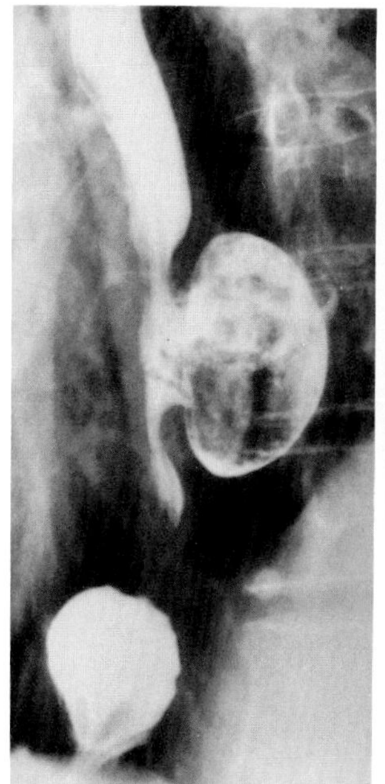

**Figure 43.37. Epiphrenic diverticulum.** (From Eisenberg,[55] with permission from the publisher.)

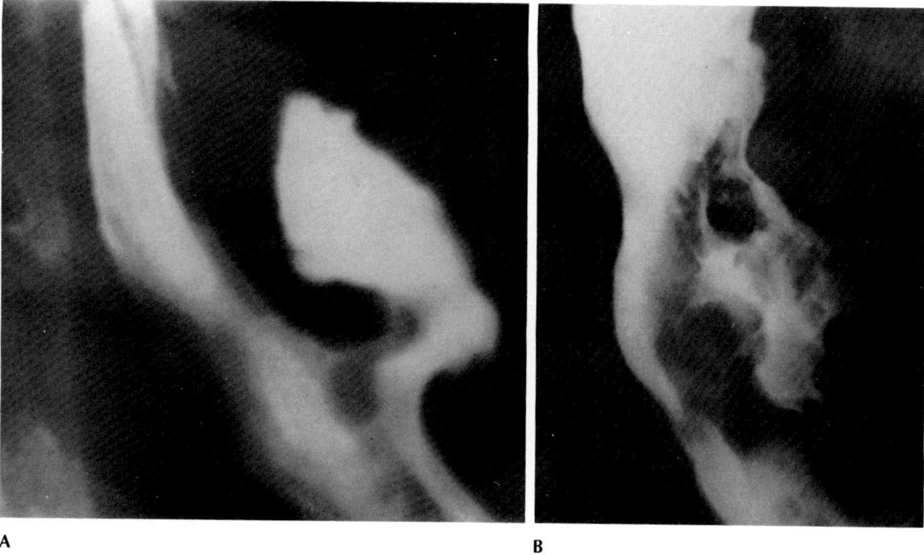

**Figure 43.38. Carcinoma arising in an epiphrenic diverticulum.** (A) An initial esophogram demonstrates a mild irregularity of the diverticulum that was not appreciated at the time of the examination. (B) Six months later, the large, ulcerating carcinoma of the esophagus is obvious. (From Eisenberg,[55] with permission from the publisher.)

cone-shaped, barium-filled sacs extending horizontally or slightly inferiorly from the midesophagus.

Epiphrenic diverticula are of the pulsion type and occur in the distal 10 cm of the esophagus. These diverticula probably result from increased intraluminal pressure in the lower esophageal segment due to incoordination of esophageal peristalsis and sphincter relaxation. Usually asymptomatic, epiphrenic diverticula can cause dysphagia if large enough to compress the adjacent esophagus.

Pulsion diverticula in the cervical or distal esophagus are usually smooth in contour. Irregularity of the diverticulum in either of these regions should suggest the possibility of infection or malignancy.

Intramural esophageal pseudodiverticulosis can simulate diverticular involvement of the esophagus. In this extremely rare disorder, numerous small (1 to 3 mm), flask-shaped outpouchings represent dilated ducts coming from submucosal glands. There is frequently a smooth stricture in the upper esophagus. The radiographic appearance of multiple ulcer-like projections has been likened to a chain of beads and to the Rokitansky-Aschoff sinuses in adenomyomatosis of the gallbladder.[30–33]

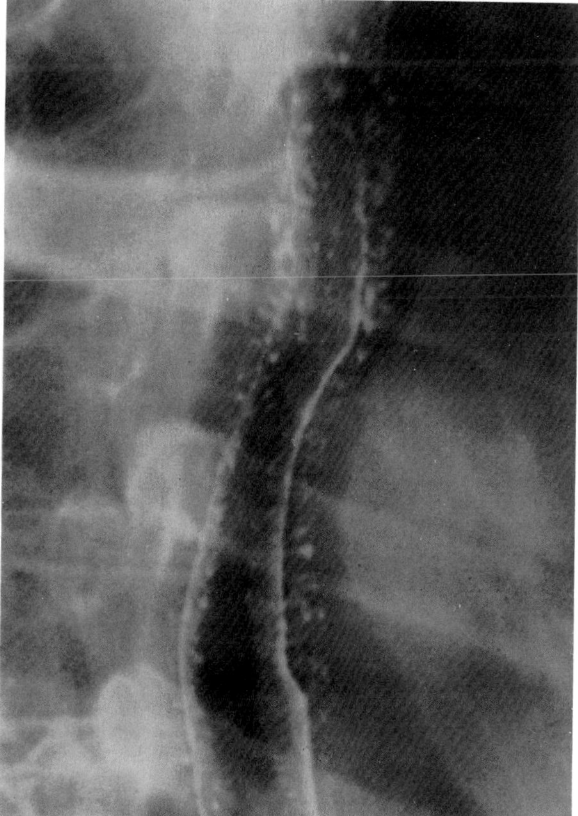

**Figure 43.39. Intramural esophageal pseudodiverticulosis.** The numerous diverticular outpouchings represent dilated ducts coming from submucosal glands in the wall of the esophagus. (From Eisenberg,[55] with permission from the publisher.)

## Esophageal Web

Congenital esophageal webs are smooth, thin, delicate membranes covered with normal-appearing mucosa. Most congenital esophageal webs are situated in the cervical esophagus within 2 cm of the pharyngoesophageal junc-

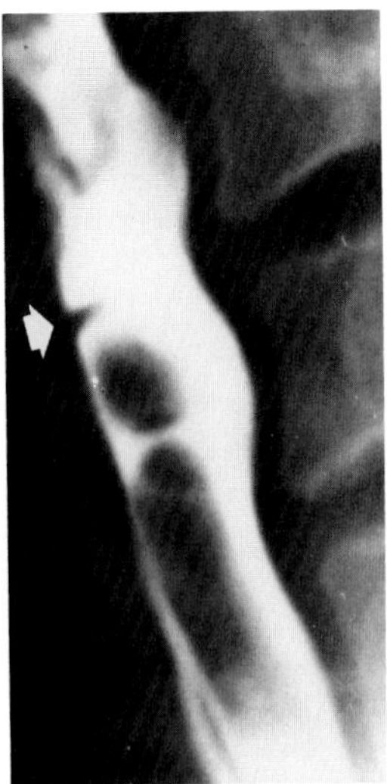

**Figure 43.40. Esophageal web.** A smooth, thin membrane covered with normal-appearing mucosa protrudes into the esophageal lumen (arrow). (From Eisenberg,[55] with permission from the publisher.)

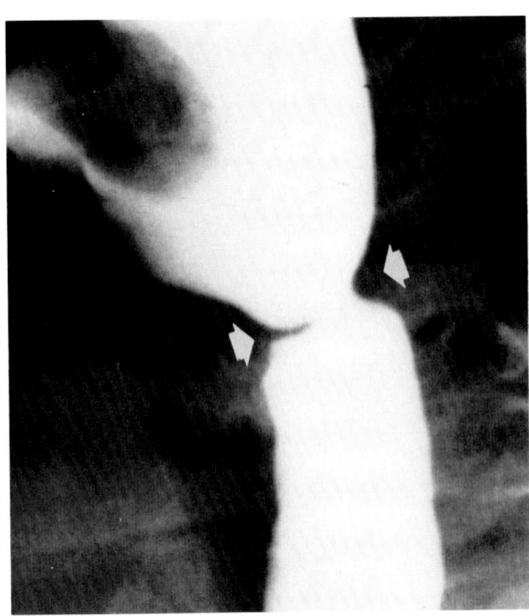

**Figure 43.41. Circumferential esophageal web (arrows).** (From Eisenberg,[55] with permission from the publisher.)

tion. The typical transverse web arises from the anterior wall of the esophagus (never from the posterior wall) and is best seen on lateral projections. It forms a right angle with the wall of the esophagus and protrudes into the esophageal lumen. A web may occasionally be circumferential and appear as a symmetric, annular radiolucent band that concentrically narrows the barium-filled esophagus. Esophageal webs can usually be seen only when the barium-filled esophagus is maximally distended. Because rapid peristalsis permits cervical esophageal webs to be visible for only a fraction of a second, cinefluorography is often required for their radiographic demonstration.

An esophageal web is usually an incidental finding of no clinical importance. Many diseases have been associated with esophageal webs, though in most cases the association appears to be coincidental. A causal relationship seems likely in certain skin diseases, such as epidermolysis bullosa and benign mucosal pemphigoid, in which healing of mucous membrane inflammation leads to scarring. A controversial entity is the Plummer-Vinson syndrome (sideropenic dysphagia), in which esophageal webs have been reported in middle-aged women with dysphagia and iron deficiency anemia. A predisposition to developing pharyngeal carcinoma has also been reported in this condition. However, because of the relative infrequency with which many of the component signs and symptoms have occurred together, many consider that an esophageal web in this "syndrome" merely represents a coincidental scar or adhesion from one of the variety of disease processes with which esophageal webs have occasionally been associated.[34–37]

### Lower Esophageal (Schatzki) Ring

The lower esophageal (Schatzki) ring is a smooth, concentric narrowing several centimeters above the diaphragm that marks the junction between the esophageal and gastric mucosae. The ring is constant, unchanging in position, and only visible when the esophagus above and below it is fully distended with barium. Although usually asymptomatic, lower esophageal rings with a narrowed luminal opening (less than 12 mm in diameter) are often associated with long-standing symptoms of intermittent dysphagia, particularly when large chunks of solid material are ingested.[38]

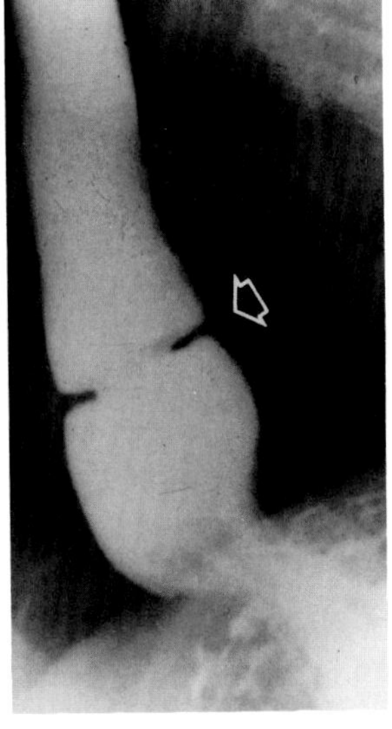

**Figure 43.42. Lower esophageal (Schatzki) ring.** Smooth, concentric narrowing of the distal esophagus (arrow) marks the junction between the esophageal and gastric mucosae. (From Eisenberg,[55] with permission from the publisher.)

### Hiatal Hernia

Hiatal hernia is the most frequent abnormality detected on upper gastrointestinal examination. Its broad radiographic spectrum ranges from large esophagogastric hernias, in which much of the stomach lies within the thoracic cavity and there is a predisposition to volvulus, to small hernias that emerge above the diaphragm only under certain circumstances (related to changes in intra-abdominal or intrathoracic pressure) and easily slide back into the abdomen through the hiatus. The symptoms associated with hiatal hernia, as well as its complications (esophagitis, esophageal ulcer, esophageal stenosis), are related to the presence of gastroesophageal reflux rather than the hiatal hernia itself. Most hiatal hernias are asymptomatic and clinically of no importance.

The intrathoracic location of a hiatal hernia is often difficult to demonstrate, since it may be impossible radiographically to precisely locate the exact levels of the diaphragmatic opening and the junction between the esophagus and stomach. An exception is the lower esophageal ring (Schatzki ring), which indicates that the transition zone between the esophageal and gastric mucosae is above the diaphragm and thus denotes the existence of a hiatal hernia. A large hiatal hernia may appear as a soft tissue mass in the posterior mediastinum, often containing a prominent air-fluid level. The presence of obvious gastric folds above the diaphragm, folds that are coarse and tortuous, unlike the straight, fine parallel folds of the distal esophagus, is a sign of hiatal hernia. Small hiatal hernias must be differentiated from the phrenic ampulla. A bolus of swallowed barium usually hesitates briefly at the lower end of the esophagus before filling a somewhat flared portion of the esophagus (vestibule) just proximal to the stomach. This site of hesitation represents the uppermost portion of an area of

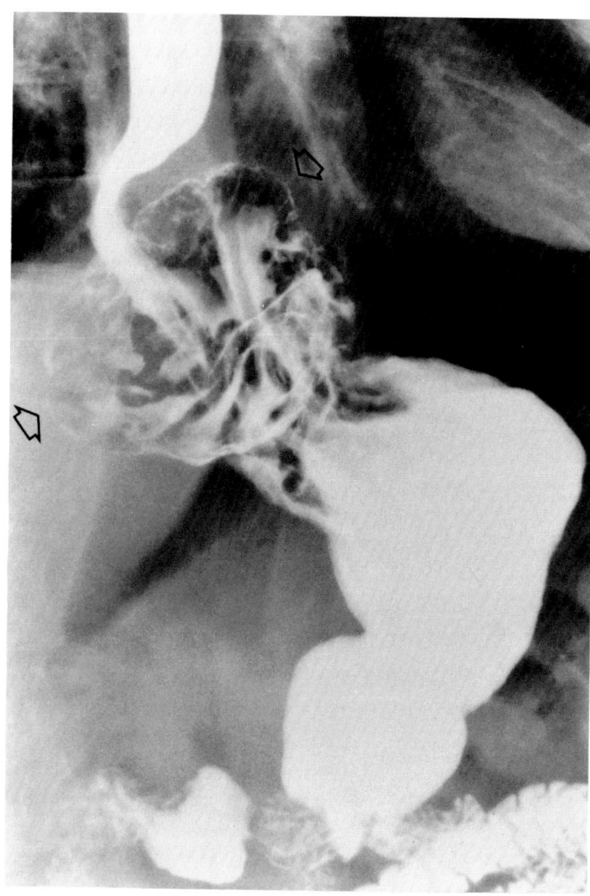

**Figure 43.43. Large hiatal hernia (arrows).**

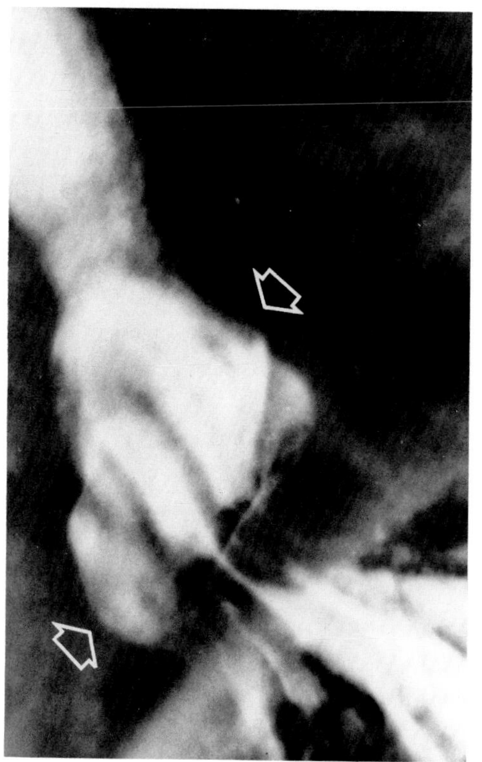

A

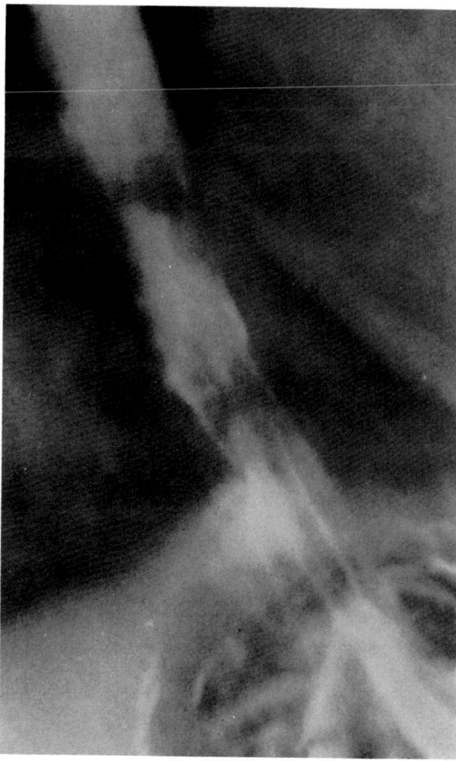

B

**Figure 43.44. Sliding hiatal hernia.** (A) A moderate hiatal hernia is visible above the left hemidiaphragm (arrows). (B) Reduction of the hiatal hernia. The marginal irregularity and thickened folds in the esophagus suggest esophagitis. (From Eisenberg,[55] with permission from the publisher.)

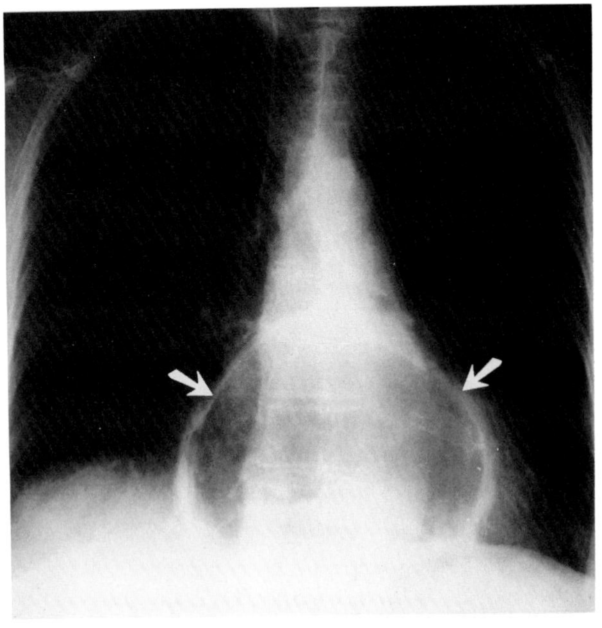


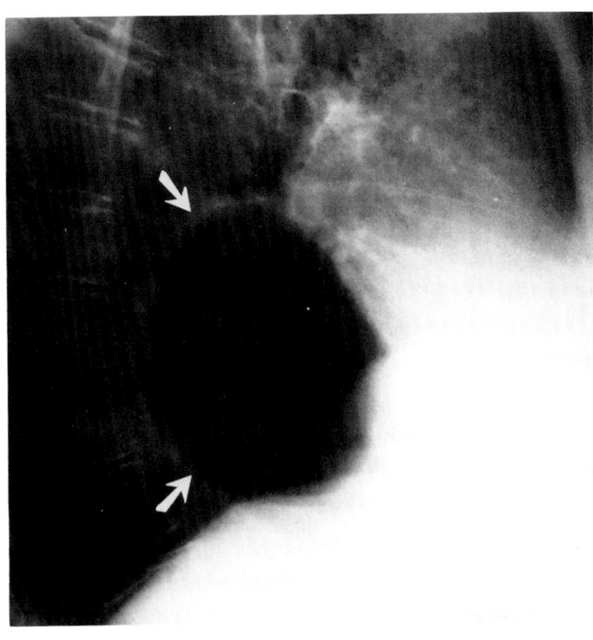


**Figure 43.45. Hiatal hernia.** (A) Frontal and (B) lateral views of the chest demonstrate a huge, air-filled hiatal hernia that appears as a mediastinal mass (arrows). (From Eisenberg,[55] with permission from the publisher.)

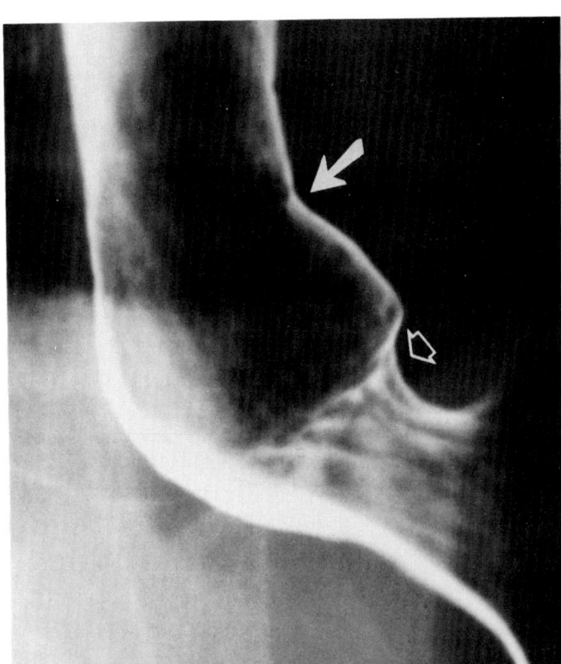

**Figure 43.46. A ring.** An indentation (solid arrow) separates the tubular esophagus above from the dilated vestibule below. The open arrow marks the level of the junction between the esophageal and gastric mucosae. (From Eisenberg,[55] with permission from the publisher.)

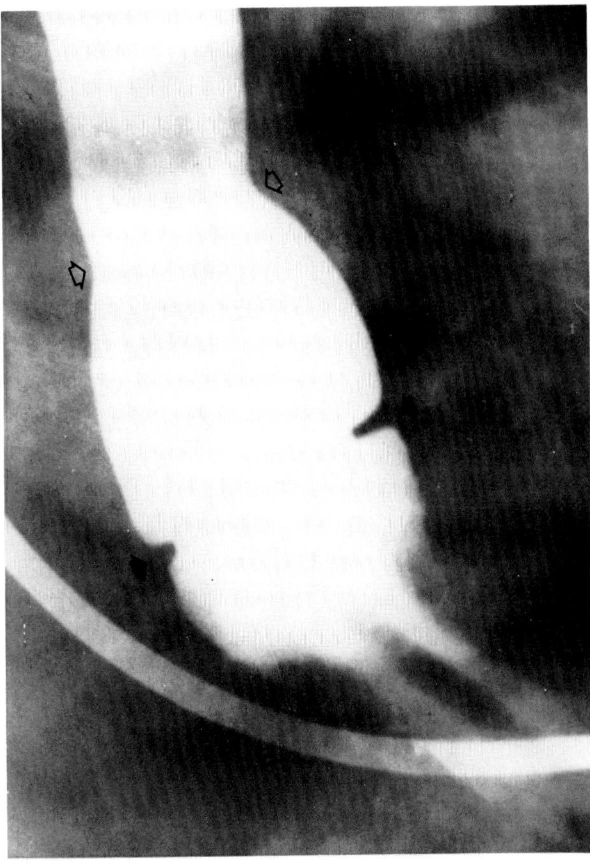

**Figure 43.47. B ring.** The presence of a second indentation (solid arrows) below the A ring (open arrows) implies the existence of a hiatal hernia. (From Eisenberg,[55] with permission from the publisher.)

high resting pressure (A ring) that appears radiographically as an indentation separating the tubular esophagus above from the slightly dilated vestibule (phrenic ampulla) below. A second indentation above the diaphragm (B ring) is usually considered to represent the site of transition between the vestibule and the gastric cardia, implying the existence of a hiatal hernia.[39–41]

## Paraesophageal Hernia

A paraesophageal hernia is progressive herniation of the stomach anterior to the esophagus, usually through a widened esophageal hiatus but occasionally through a separate adjacent gap in the diaphragm. Unlike the situation with a hiatal hernia, the terminal esophagus in a patient with a paraesophageal hernia remains in its normal position, and the esophagogastric junction is situated below the diaphragm. As herniation progresses, increasing amounts of the greater curvature roll into the hernia sac, inverting the stomach so that the greater curvature lies uppermost (rolling hernia). At times, spleen, omentum, transverse colon, or small bowel may accompany the herniated stomach into the thorax. Plain radiographs of the chest may demonstrate a paraesophageal hernia as a round soft tissue density posterior to the heart, occasionally containing an air-fluid level. Unlike sliding hiatal hernias, paraesophageal hernias are associated with normal functioning of the gastroesophageal junction, and therefore reflux esophagitis rarely occurs.[42,43]

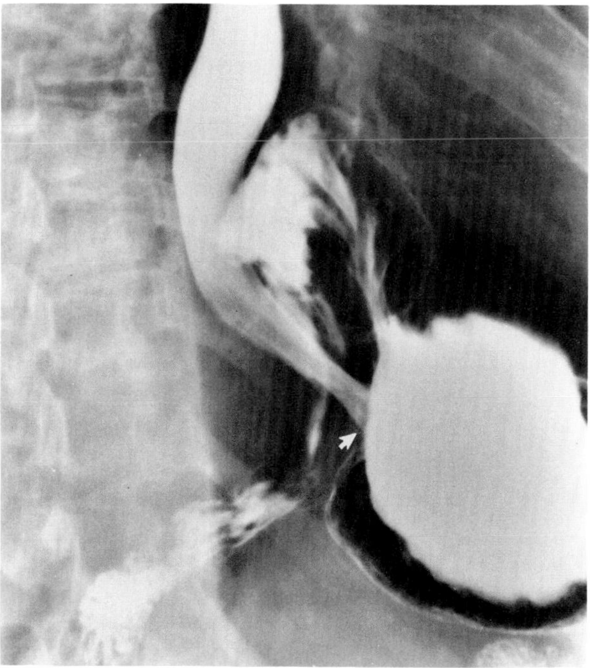

**Figure 43.48. Paraesophageal hernia.** The esophagogastric junction (arrow) remains below the level of the left hemidiaphragm. There is an associated gastric volvulus with a portion of the stomach located within the hernia sac.

## Other Diaphragmatic Hernias

Diaphragmatic hernias through orifices other than the esophageal hiatus are due to congenital abnormalities in the formation of the diaphragm or to trauma. Many small diaphragmatic hernias contain only omentum. Larger lesions usually include parts of the stomach, transverse colon, and greater omentum. They infrequently contain small bowel, cecum, liver, or pancreas. The symptoms caused by large diaphragmatic hernias depend on the size of the hernia, the viscera that have herniated, and whether the herniated structures are fixed or capable of sliding back and forth between the abdomen and the thorax. Because the stomach is usually situated within large diaphragmatic hernias, some degree of postprandial distress is common. Symptoms of intermittent partial bowel obstruction are frequent and, if there is interference with the blood supply of the portion of intestine that is within the hernia sac, strangulation and gangrenous necrosis may result.

Herniation through the anteromedial foramina of Morgagni can occur on either side of the attachment of the diaphragm to the sternum. Most common on the right, herniation through the foramen of Morgagni typically appears radiographically as a large, smoothly marginated soft tissue mass in the right cardiophrenic angle. On lateral views, the anterior position of the hernia is evident. Herniation through the posterolateral foramina of Bochdalek more commonly occurs on the left and appears as a mass in the posterior portion of the thorax. When a loop of gas-filled bowel is seen within the soft tissue mass, the diagnosis of a diaphragmatic hernia is easy to make. However, if the hernia contains only omentum and no gas-filled bowel, on plain chest radiographs it may be impossible to distinguish from a mediastinal mass.

Traumatic diaphragmatic hernias most commonly follow direct laceration by a knife, bullet, or other penetrating object, but can also occur as a result of a marked increase in the intra-abdominal pressure. They occur much more frequently on the left and appear radiographically as herniated bowel contents above the suspected level of the diaphragm. Because the herniated viscera can parallel the diaphragm on both frontal and lateral projections, the radiographic appearance can simulate eventration or diaphragmatic paralysis. The administration of barium by mouth or by rectum may be required to demonstrate the relation of the gastrointenstinal tract to the diaphragm. A major differential point is the constriction of the afferent and efferent loops of bowel as they traverse a laceration in the diaphragm, in contrast to the wide separation of loops that is typically seen with eventration or paralysis of the diaphragm.[44–47]

## Perforation of the Esophagus

Perforation of the esophagus may be a complication of esophagitis, peptic ulcer, neoplasm, external trauma, or

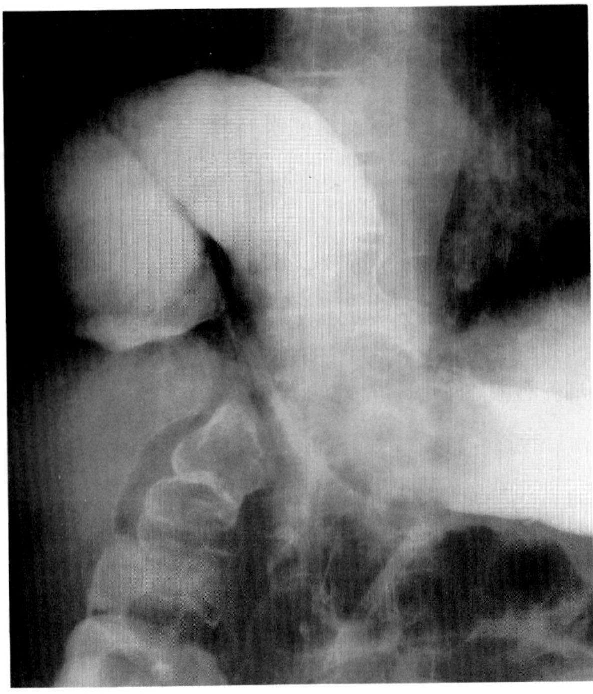
A

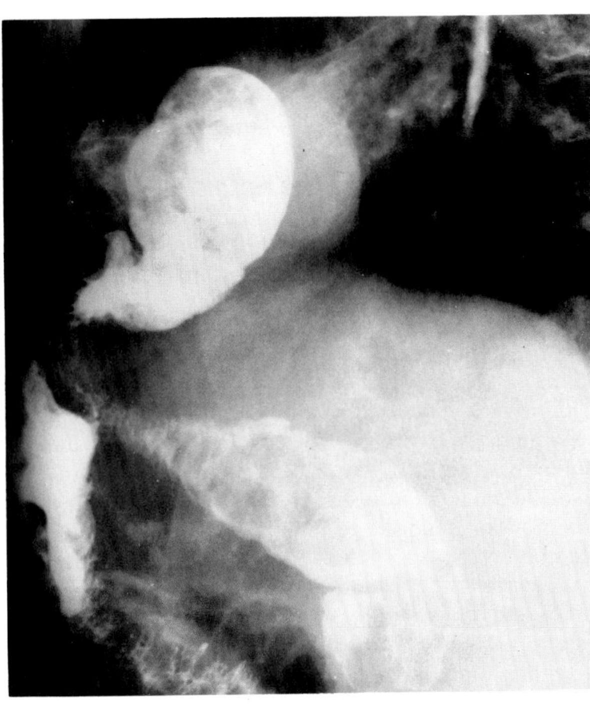
B

**Figure 43.49. Morgagni hernia.** (A) Frontal and (B) lateral views demonstrate barium-filled bowel within a hernia sac that lies anteriorly and to the right.

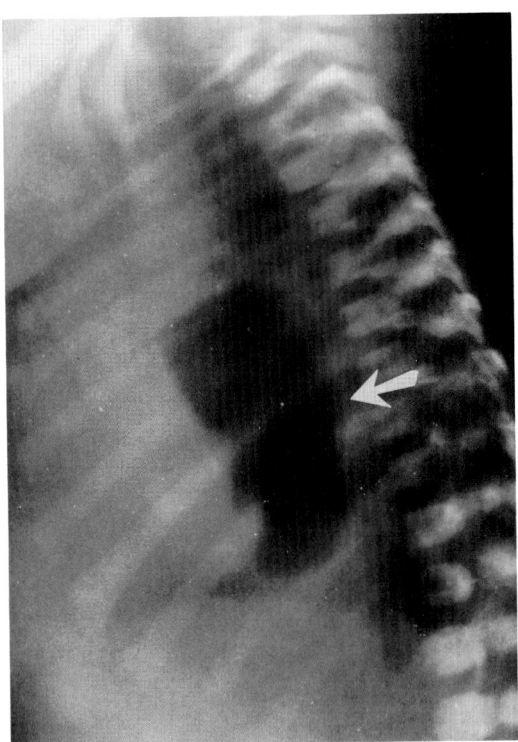

**Figure 43.50. Herniation through the foramen of Bochdalek.** A gas-filled loop of bowel (arrow) is visible posteriorly in the thoracic cavity. (From Eisenberg,[55] with permission from the publisher.)

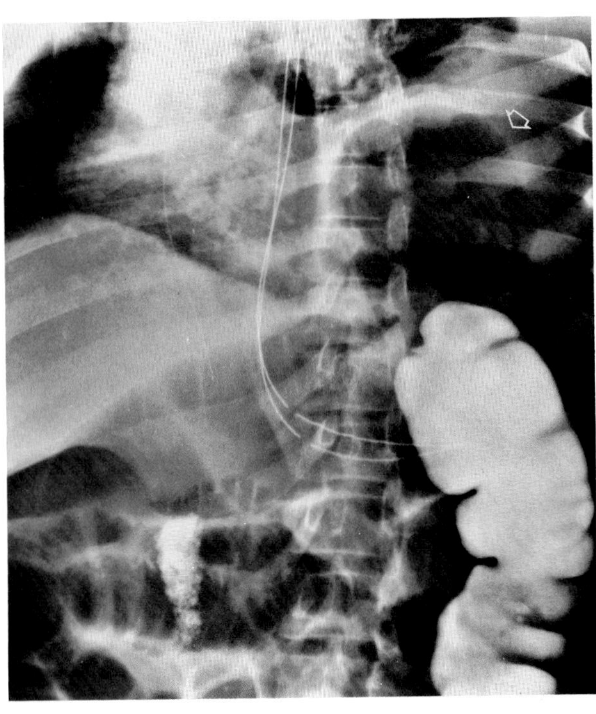

**Figure 43.51. Post-traumatic diaphragmatic hernia.** There is herniation of a portion of the splenic flexure (arrow) with obstruction to the retrograde flow of barium. (From Eisenberg,[55] with permission from the publisher.)

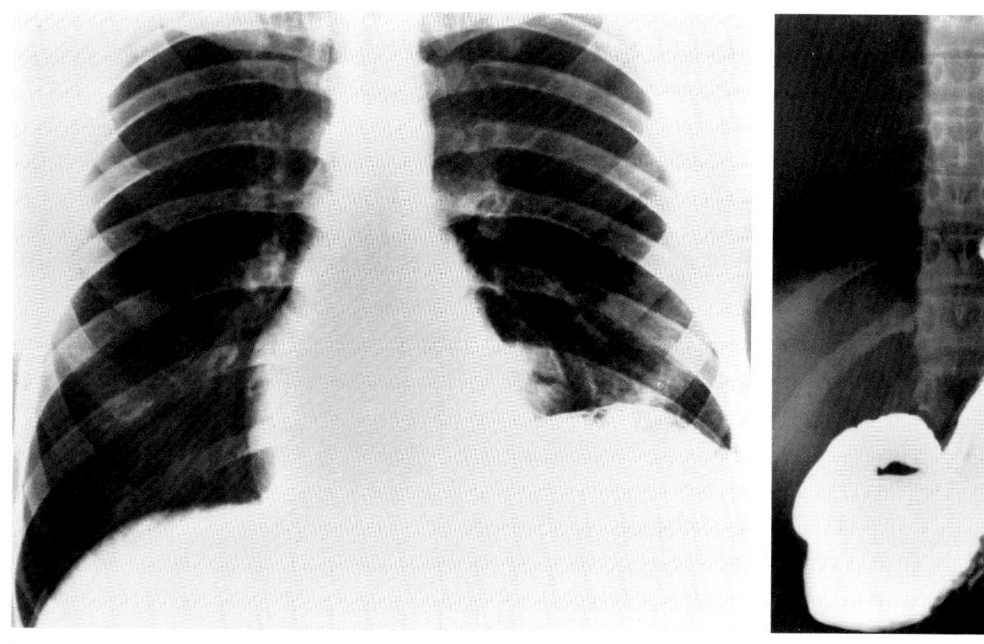

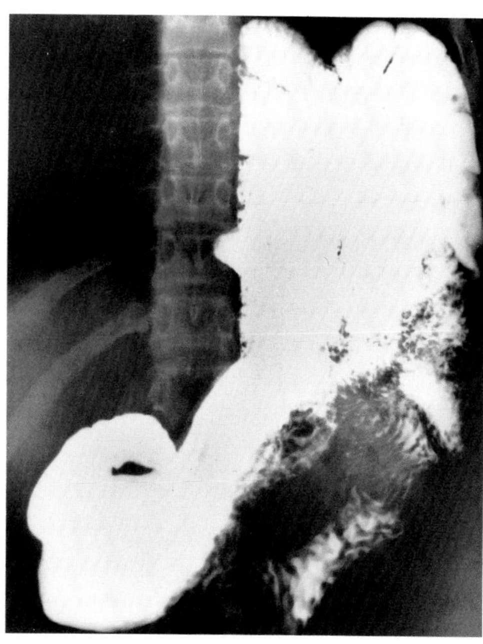

A B

**Figure 43.52. Post-traumatic diaphragmatic hernia.** (A) On a frontal projection the radiographic appearance simulates eventration or diaphragmatic paralysis. (B) The administration of barium clearly demonstrates herniation of bowel contents into the chest. (From Eisenberg,[55] with permission from the publisher.)

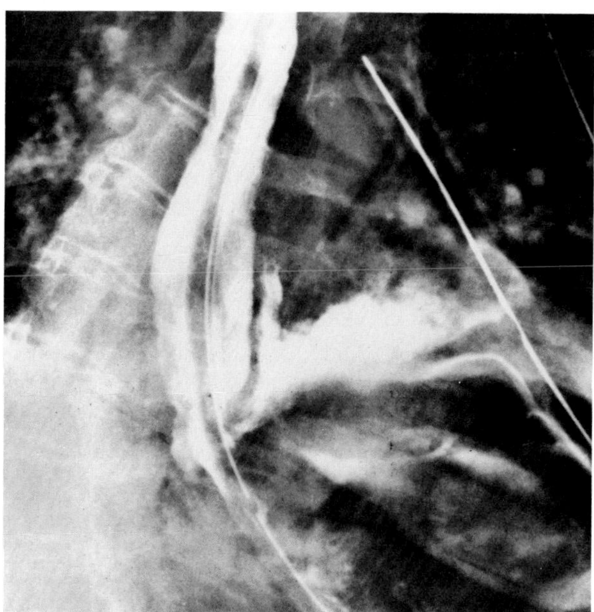

**Figure 43.53. Boerhaave syndrome.** The esophageal rupture occurred as a complication of severe vomiting in an elderly alcoholic. (From Eisenberg,[55] with permission from the publisher.)

instrumentation. At times, perforation of a previously healthy esophagus can result from severe vomiting or coughing, often from dietary or alcoholic indiscretion. Complete rupture of the wall of the esophagus (Boerhaave syndrome) may cause the sudden development of severe epigastric pain simulating myocardial infarction, pancreatitis, or a ruptured abdominal viscus. In the Mallory-Weiss syndrome, an increase in intraluminal and intramural pressure associated with vomiting following an alcoholic bout causes superficial mucosal lacerations or fissures near the esophagogastric junction that produce severe and occasionally exsanguinating hemorrhage.

Transmural esophageal perforation can lead to free air in the mediastinum or periesophageal soft tissues. The administration of radiopaque contrast material may demonstrate extravasation through a transmural perforation or an intramural dissecting channel separated by an intervening lucent line from the normal esophageal lumen (double-barrel esophagus). Fistulous communication between the esophagus and the tracheobronchial tree may produce an esophagorespiratory fistula.[48–51]

## Foreign Bodies

A wide spectrum of foreign bodies can become impacted in the esophagus, usually in the cervical esophagus at or just above the level of the thoracic inlet. Most metallic objects, such as pins, coins, and small toys, are very radiopaque and are easily visualized on radiographs or during fluoroscopy. Objects made of aluminum and some light alloys may be impossible to detect radiographically because the density of these metals is almost equal to that of soft tissue. It is essential that any suspected foreign body be evaluated on two views to be certain that the object projected over the esophagus truly lies within it.

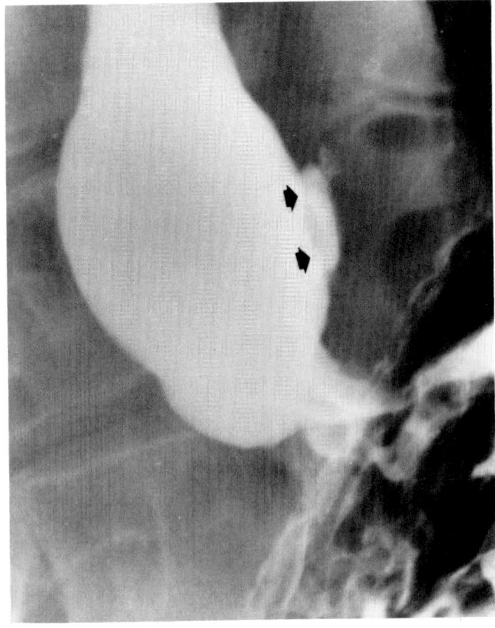

A

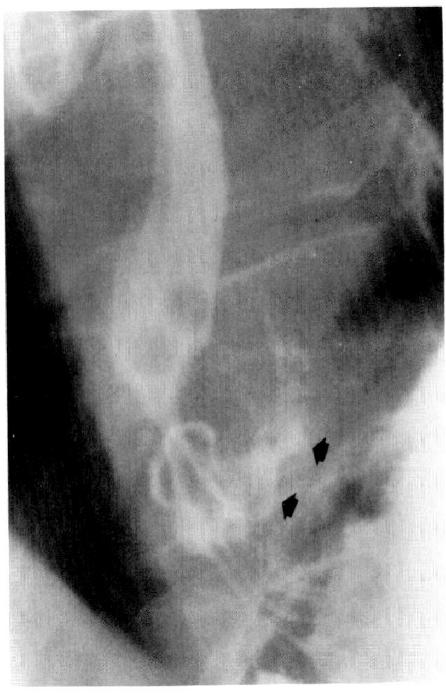

B

**Figure 43.54. Mallory-Weiss tears.** (A) A tear confined to the esophageal mucosa (arrows). (B) Penetration of barium into the wall of the esophagus. (From Eisenberg,[55] with permission from the publisher.)

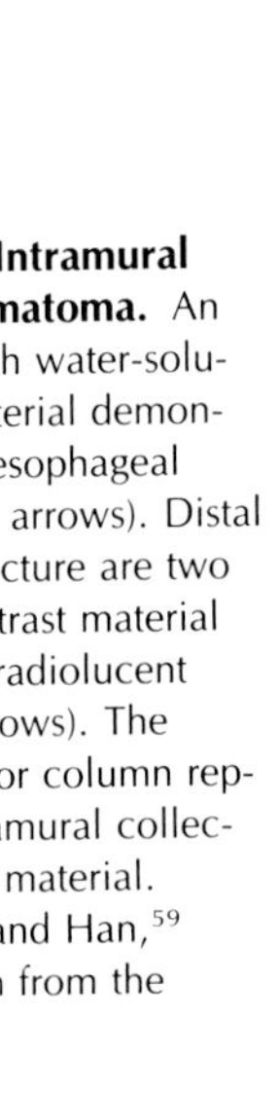

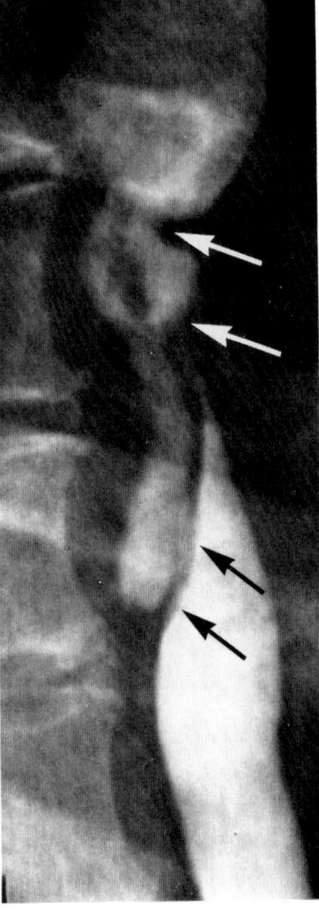

**Figure 43.55. Intramural esophageal hematoma.** An esophogram with water-soluble contrast material demonstrates two midesophageal strictures (white arrows). Distal to the lower stricture are two columns of contrast material separated by a radiolucent stripe (black arrows). The smaller, posterior column represents the intramural collection of contrast material. (From Bradley and Han,[59] with permission from the publisher.)

**Figure 43.56. Fishbone impacted in the lower cervical esophagus.** (From Eisenberg,[55] with permission from the publisher.)

Nonopaque foreign bodies in the esophagus, especially pieces of poorly chewed meat, can be demonstrated only after the ingestion of barium. Such foreign bodies usually become impacted in the distal esophagus just above the level of the diaphragm and are often associated with a distal stricture. These intraluminal filling defects usually have an irregular surface and may resemble a completely obstructing carcinoma. Impactions may also be due to strictures in the cervical portion of the esophagus.[52]

## Extrinsic Disease Involving the Esophagus

Compression and displacement of the cervical esophagus may be due to enlargement of the thyroid, parathyroid, or lymph nodes, as well as to abscesses or hematomas in the periesophageal soft tissues and to neoplastic or inflammatory disease of the spine. Anterior marginal osteophytes of the cervical spine can produce smooth, regular indentations on the posterior wall of the cervical esophagus. Although usually asymp-

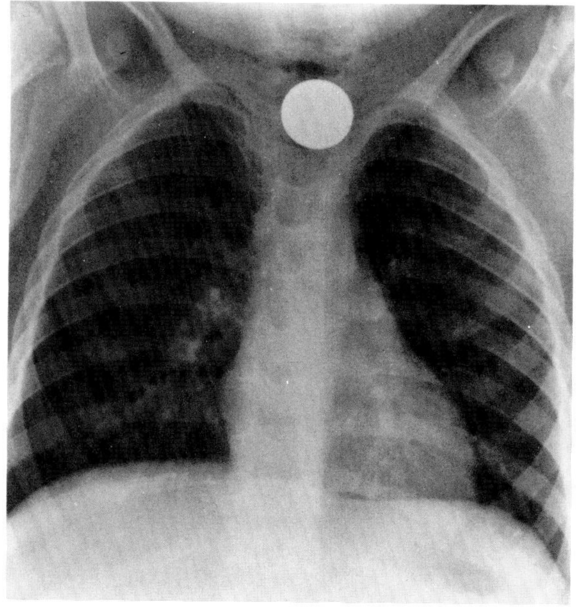

A

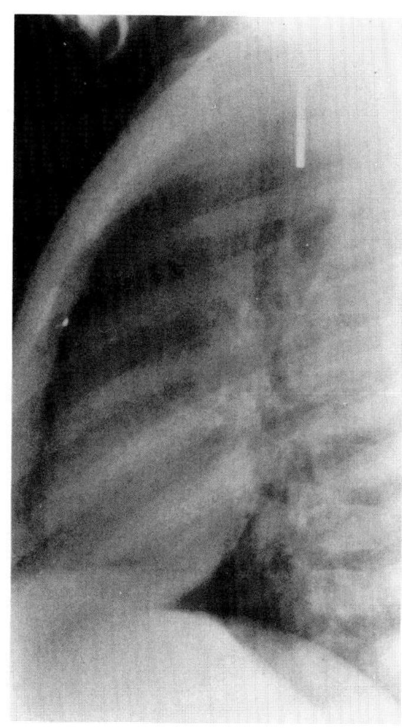

B

**Figure 43.57. Metallic coin impacted in the esophagus.** (A) Frontal and (B) lateral projections. (From Eisenberg,[55] with permission from the publisher.)

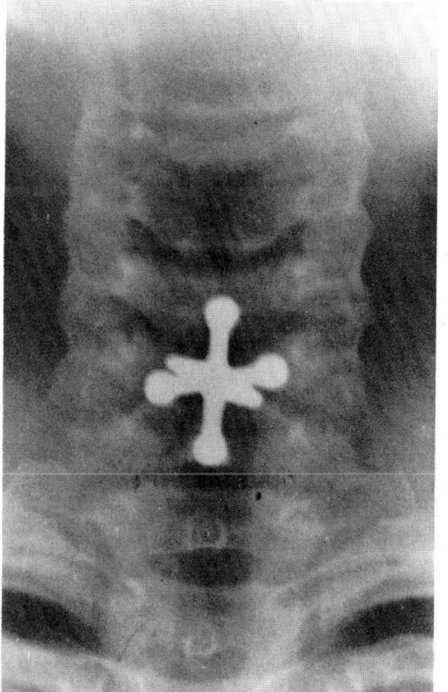

A

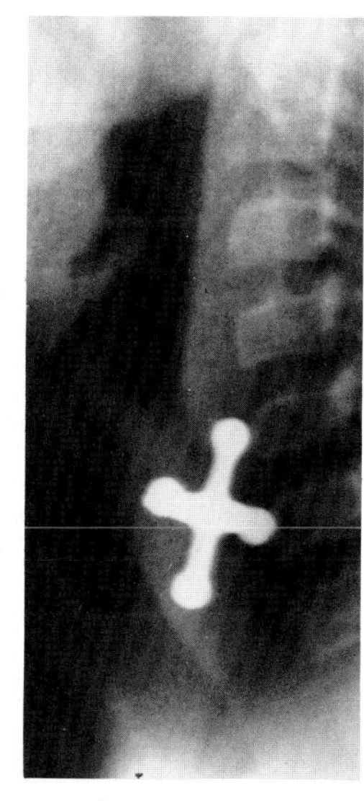

B

**Figure 43.58. Metallic jack impacted in the esophagus.** (A) Frontal and (B) lateral projections. (From Eisenberg,[55] with permission from the publisher.)

**Figure 43.59. Cherry pit impacted in the cervical esophagus proximal to a caustic stricture.** (From Eisenberg,[55] with permission from the publisher.)

tomatic, these osteophytes may impinge on the cervical esophagus and produce pharyngoesophageal dysfunction and dysphagia. Profuse osteophytosis from vertebral margins in patients with diffuse idiopathic skeletal hyperostosis (DISH, or Forrestier's disease) is especially likely to interfere with pharyngeal function.

Two structures normally indent the anterior and lateral aspects of the thoracic esophagus. The more cephalad normal impression, which is due to the transverse arch of the aorta (aortic knob), is more prominent as the aorta becomes increasingly dilated and tortuous with age. The more caudal impression is caused by the left main stem bronchus.

Cardiomegaly, pericardial disease, and vascular abnormalities such as aortic aneurysm and aberrant right subclavian artery may displace and compress the lumen of the thoracic esophagus. Indeed, displacement of the esophagus is often a valuable sign of left atrial or left ventricular enlargement. Any mass lesion adjacent to the esophagus that arises within the mediastinum, lungs,

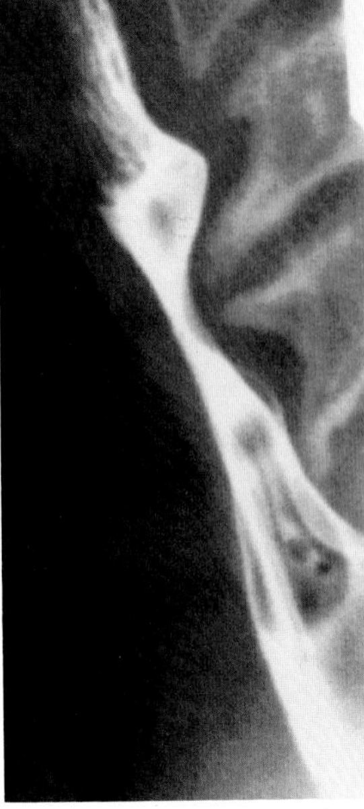

**Figure 43.60. Anterior marginal osteophytes of the cervical spine compressing the esophagus.** A smooth, regular indentation is seen on the posterior wall at the level of an intervertebral disk space. (From Eisenberg,[55] with permission from the publisher.)

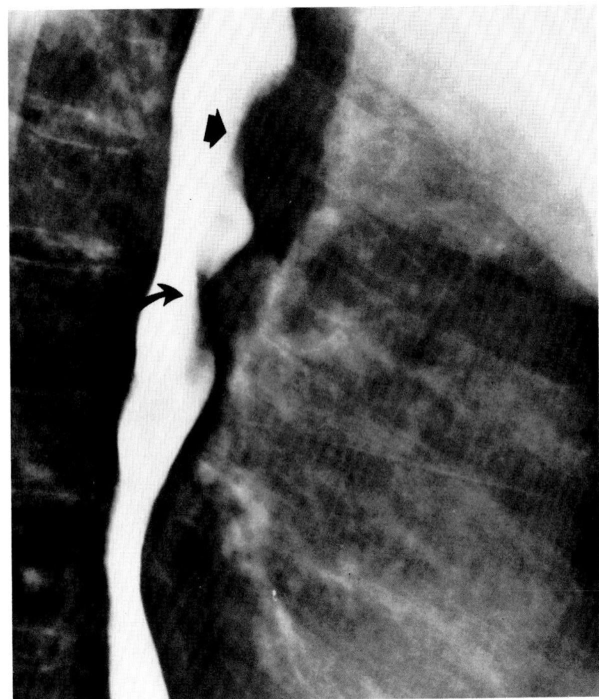

**Figure 43.61. Normal esophageal impressions.** The short arrow points to the aorta; the long arrow points to the left main stem bronchus. (From Eisenberg,[55] with permission from the publisher.)

trachea, or lymph nodes can also impress the barium-filled esophagus. Depending on the size and position of the mass, there can be a focal or broad impression on the esophagus and displacement of the esophagus in any direction.[53,54]

## REFERENCES


1. Arendt J, Wolf A: The vallecular sign: Its diagnosis and clinical significance. *AJR* 57:435–445, 1947.
2. Seaman WB: Functional disorders of the pharyngo-esophageal junction. *Radiol Clin North Am* 7:113–119, 1969.
3. Dodds WJ: Current concepts of esophageal motor function: Clinical implications for radiology. *AJR* 128:549–561, 1977.
4. Donner MW, Saba GP, Martinez CR: Diffuse disease of the esophagus: A practical approach. *Semin Roentgenol* 16:198–213, 1981.
5. Margulis AR, Koehler RE: Radiologic diagnosis of disordered esophageal motility: A unified approach. *Radiol Clin North Am* 14:429–439, 1976.
6. Seaman WB: Pathophysiology of the esophagus. *Semin Roentgenol* 16:214–227, 1981.
7. Shulze KS, Goresky CA, Jabbari M, et al: Esophageal achalasia associated with gastric carcinoma. *Can Med Assoc J* 1:857–864, 1975.
8. Simeone J, Burrell M, Toffler R: Esophageal aperistalsis secondary to metastatic invasion of the myenteric plexus. *AJR* 127:862–864, 1976.
9. Stewart ET, Miller WN, Hogan WJ, et al: Desirability of roentgen esophageal examination immediately after pneumatic dilatation for achalasia. *Radiology* 130:589–591, 1979.
10. Zegel HG, Kressel HY, Levine GM, et al: Delayed esophageal perforation after pneumatic dilatation for the treatment of achalasia. *Gastrointest Radiol* 4:219–221, 1979.
11. Bennett JR, Hendrix TR: Diffuse esophageal spasm: A disorder with more than one cause. *Curr Clin Concepts* 59:273–279, 1970.
12. Crummy AB: The water test in the evaluation of gastroesophageal reflux: Its correlation with pyrosis. *Radiology* 78:501–504, 1966.


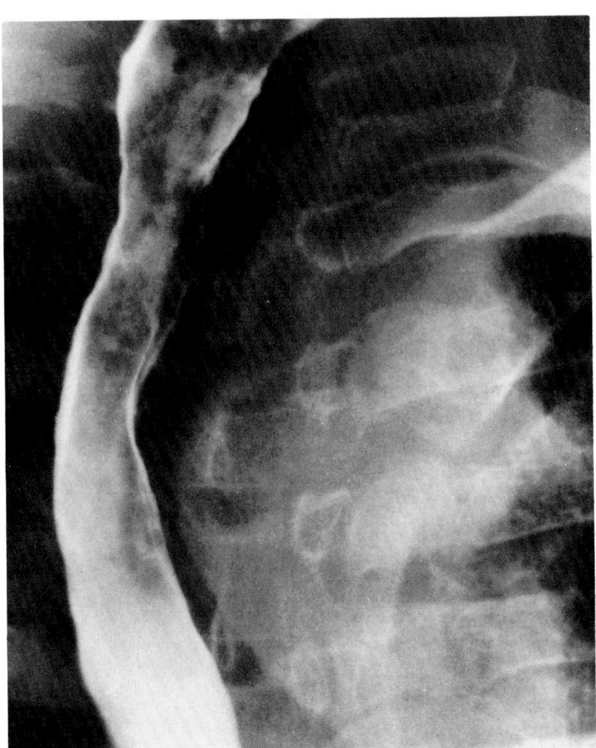

**Figure 43.62. Squamous cell carcinoma of the lung impressing the upper thoracic esophagus.** (From Eisenberg,[55] with permission from the publisher.)

13. Berquist HT, Nolan NG, Stephens DH, et al: Radioisotope scintigraphy in the diagnosis of Barrett's esophagus. *AJR* 123:401–411, 1975.
14. Robbins AH, Hermos JA, Schimmel EM, et al: The columnar-lined esophagus: An analysis of 26 cases. *Radiology* 123:1–7, 1977.
15. Cho KJ, Hunter TB, Whitehouse WM: The columnar epithelial-lined esophagus and its association with adenocarcinoma of the esophagus. *Radiology* 115:563–568, 1975.
16. Boal DKB, Newburger TE, Teel RL: Esophagitis induced by combined radiation and Adriamycin. *AJR* 132:567–570, 1979.
17. Franken EA: Caustic damage of the gastrointestinal tract: Roentgen features. *AJR* 118:77–85, 1973.
18. Lewicki AM, Moore JP: Esophageal moniliasis: A review of common and less frequent characteristics. *AJR* 125:218–225, 1975.
19. Meyers C, Durkin MG, Love L: Radiographic findings in herpetic esophagitis. *Radiology* 119:21–22, 1976.
20. Muhletaler CA, Gerlock AJ, de Soto L, et al: Acid corrosive esophagitis: Radiographic findings. *AJR* 134:1137–1140, 1980.
21. Teplick JG, Teplick SK, Ominsky SH, et al: Esophagitis caused by oral medication. *Radiology* 134:23–25, 1980.
22. Jang GC, Clouse ME, Fleischner FG: Fibrovascular polyp: A benign intraluminal tumor of the esophagus. *Radiology* 92:1196–1200, 1969.
23. Goldstein HM, Zornoza J, Hopens T: Intrinsic diseases of the adult esophagus: Benign and malignant tumors. *Semin Roentgenol* 16:183–197, 1981.
24. Koehler RE, Moss AA, Margulis AR: Early radiographic manifestations of carcinoma of the esophagus. *Radiology* 119:1–5, 1976.
25. Lansing PB, Ferrante WA, Ochsner JL: Carcinoma of the esophagus at the site of lye stricture. *Am J Surg* 118:108–111, 1973.
26. Gloyna RE, Zornoza J, Goldstein HM: Primary ulcerative carcinoma of the esophagus. *AJR* 129:599–600, 1977.
27. Thompson WM, Halvorsen RA, Foster WL, et al: Computed tomography for staging esophageal and gastroesophageal cancer. *AJR* 141:951–958, 1983.
28. Moss AA, Schnyder P, Thoeni RF, et al: Esophageal carcinoma: Pretherapy staging by computed tomography. *AJR* 136: 1051–1056, 1981.
29. Picus D, Balfe DM, Koehler RE, et al: Computed tomography in the staging of esophageal carcinoma. *Radiology* 146: 433–438, 1983.
30. Beauchamp JM, Nice CM, Belanger MA, et al: Esophageal intramural pseudodiverticulosis. *Radiology* 113:273–276, 1974.
31. Bruggeman LL, Seaman WB: Epiphrenic diverticula: An analysis of 80 cases. *AJR* 119:266–276, 1973.
32. Royd RM, Bogoch A, Greig JH, et al: Esophageal intramural pseudodiverticulum. *Radiology* 113:267–270, 1974.
33. Wychulis RA. Gunnlaugsson HG, Clagett OT: Carcinoma occuring in pharyngo-esophageal diverticulum. *Surgery* 66:976–979, 1969.
34. Clements JL, Cox GW, Torres WE, et al: Cervical esophageal webs: A roentgen anatomic correlation. *AJR* 121:221–231, 1974.
35. Han SY, Mishas AA: Circumferential web of the upper esophagus. *Gastrointest Radiol* 3:7–9, 1978.
36. Nosher JL, Campbell WL, Seaman WB: The clinical significance of cervical esophageal and hypopharyngeal webs. *Radiology* 117:45–47, 1975.
37. Hutton CF: Plummer-Vinson syndrome. *Br J Radiol* 29:81–85, 1956.
38. Schatzki R, Gary JE: The lower esophageal ring. *AJR* 75:246–261, 1956.
39. Stein GN, Finkelstein A: Hiatal hernia: Roentgen incidence and diagnosis. *Am J Dig Dis* 5:77–87, 1960.
40. Vestby G, Aaksus T: Incidence of sliding hiatus hernia. *Invest Radiol* 1:379–385, 1966.
41. Wolf BS, Brahms SA, Khilnani MT: The incidence of hiatal hernia in routine barium meal examinations. *J Mount Sinai Hosp NY* 26:598–600, 1959.
42. Hill LD: Incarcerated paraesophageal hernia: A surgical emergency. *Am J Surg* 126:286–291, 1973.
43. Hoyt T, Kyaw MM: Acquired paraesophageal and disparaesophageal hernias: Complications of hiatal hernia repair. *AJR* 121:248–251, 1974.
44. Ahrand T, Thompson B: Hernia of the foramen of Bochdalek in the adult. *Am J Surg* 122:612–615, 1971.
45. Hood R: Traumatic diaphragmatic hernia. *Ann Thorac Surg* 12:311–324, 1971.
46. Rennel CL: Foramen of Morgagni hernia with volvulus of the stomach. *AJR* 117:248–250, 1973.
47. Bartley O, Wickbom I: Roentgenologic diagnosis of rupture of the diaphragm. *Acta Radiol [Diagn]* 53:33–41, 1960.
48. Butler ML: Radiologic diagnosis of Mallory-Weiss syndrome. *Br J Radiol* 46:553–554, 1973.
49. Lowman RM, Goldman, R, Stern H: The roentgen aspects of intramural dissections of the esophagus: The mucosal stripe sign. *Radiology* 93:1329–1331, 1969.
50. Thompson NW, Ernst CB, Fry WJ: The spectrum of emetogenic injury to the esophagus and stomach. *Am J Surg* 113:13–26, 1967.
51. Maglinte DDT, Edwards MC: Spontaneous closure of esophageal tear in Boerhaave's syndrome. *Gastrointest Radiol* 4:223–225, 1979.
52. Smith PC, Swischuk LE, Fagan CJ: An elusive and often unsuspected cause of stridor or pneumonia (the esophageal foreign body). *AJR* 122:80–89, 1974.
53. Resnick D, Shaul SR, Robins JM: Diffuse idiopathic skeletal hyperostosis (DISH): Forrestier's disease with extraspinal manifestations. *Radiology* 115:513–524, 1975.
54. Swischuk LE: *Plain Film Interpretation in Congenital Heart Disease*, Baltimore, William & Wilkins, 1979.
55. Eisenberg RL: *Gastrointestinal Radiology: A Pattern Approach*. Philadelphia, JB Lippincott, 1983.
56. Rogers LF, Goldstein HM: Roentgen manifestations of radiation injury to the gastrointestinal tract. *Gastrointest Radiol* 2:281–291, 1977.
57. Levine MS, Caroline D, Thompson JJ, et al: Adenocarcinoma of the esophagus: Relationship to Barrett mucosa. *Radiology* 150:305–309, 1984.
58. Creteur V, Laufer I, Kressel HY, et al: Drug-induced esophagitis detected by double-contrast radiography. *Radiology* 147: 365–369, 1983.
59. Bradley JL, Han SY: Intramural hematoma (incomplete perforation) of the esophagus associated with esophageal dilatation. *Radiology* 130:59–62, 1979.
60. Eisenberg RL: *Atlas of Signs in Radiology*. Philadelphia, JB Lippincott, 1984.

## Chapter Forty-Four
# Peptic Ulcer Disease

### DUODENAL ULCER

Duodenal ulcer is the most common manifestation of peptic disease. More than 95 percent of duodenal ulcers occur in the first portion of the duodenum (duodenal bulb). An unequivocal diagnosis of active duodenal ulcer requires the demonstration of the ulcer crater, which in about 90 percent of cases can be seen using double-contrast techniques. When seen in profile, the ulcer crater is a small collection of barium projecting from the lumen. When seen en face, the ulcer niche appears as a rounded or linear collection of contrast material surrounded by lucent folds that often radiate toward the crater as they do with gastric ulcers. Secondary signs of duodenal ulcer disease are thickening of mucosal folds and deformity of the duodenal bulb. Acute ulcers incite muscular spasm leading to marginal deformity of the duodenal bulb that may be inconstant and vary during the examination. With chronic ulceration, fibrosis and scarring cause a fixed deformity that persists even though the ulcer heals. The degree of deformity is not directly related to the size of the ulcer; small ulcers may produce large deformities, and huge ulcers may produce little alteration in bulb contour. Symmetric narrowing of the duodenal bulb in its midportion, associated with dilatation of the inferior and superior recesses at the base of the bulb (pseudo-diverticula), may produce the typical cloverleaf deform-

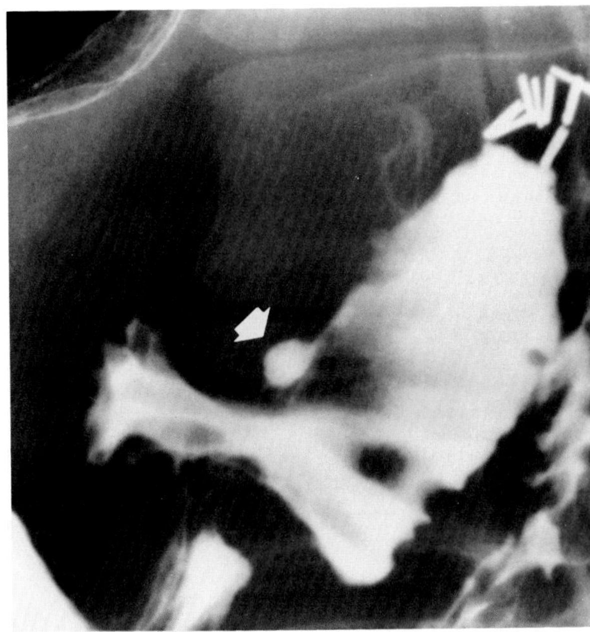

**Figure 44.1. Duodenal ulcer.** The ulcer niche appears as a rounded collection of barium (arrow) surrounded by thickened edematous folds. There is moderate deformity of the duodenal bulb.

ity of chronic duodenal ulcer disease. An apparent eccentric position of the pyloric canal may be due to muscular spasm associated with an ulcer at the base of the bulb.

It must be emphasized that although secondary signs such as thickened folds, spasm, and deformity of the duodenal bulb are indications of peptic disease, the demonstration of the crater itself is necessary for the diagnosis of an active duodenal ulcer. However, once the diagnosis of duodenal ulcer disease is clearly established radiographically or by endoscopy, there is usually no reason to repeat the upper gastrointestinal examination. The patient should be treated symptomatically, and the physician should not depend on radiographic confirmation of healing or recurrence of an active duodenal ulcer.

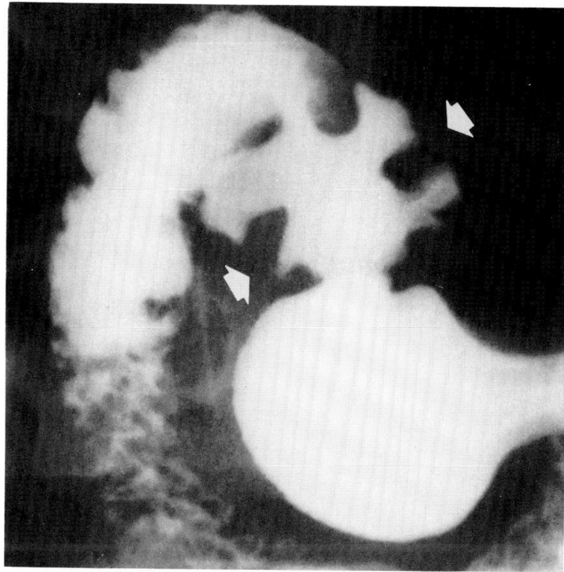

**Figure 44.2. Duodenal ulcer disease.** A typical cloverleaf deformity (arrows). (From Eisenberg,[23] with permission from the publisher.)

Although the vast majority of duodenal ulcers are small (less than 1 cm in diameter) and involve only a small portion of the duodenal bulb, some duodenal ulcers are extremely large (3 to 6 cm) and completely replace the bulb. Unlike the normal duodenal bulb or a duodenal diverticulum, these giant duodenal ulcers are rigid-walled cavities that lack a normal mucosal pattern and remain constant in size and shape throughout the gastrointestinal examination.[1,2]

## POSTBULBAR ULCER

Ulceration in the postbulbar region represents only about 5 percent of duodenal ulcers secondary to benign peptic disease. Although postbulbar ulcerations are often difficult to detect radiographically, their identification is important because they so frequently are the cause of obstruction, pancreatitis, gastrointestinal bleeding, and atypical abdominal pain. Hyperactive peristalsis, mucosal edema, and poor barium coating can obscure the ulcer niche. Severe spasm of the duodenum in the area of ulceration can narrow and deform the lumen and prevent barium from filling the ulcer crater.

The classic radiographic appearance of a benign postbulbar ulcer is a shallow, flattened niche on the medial aspect (rarely the lateral) of the upper second portion of

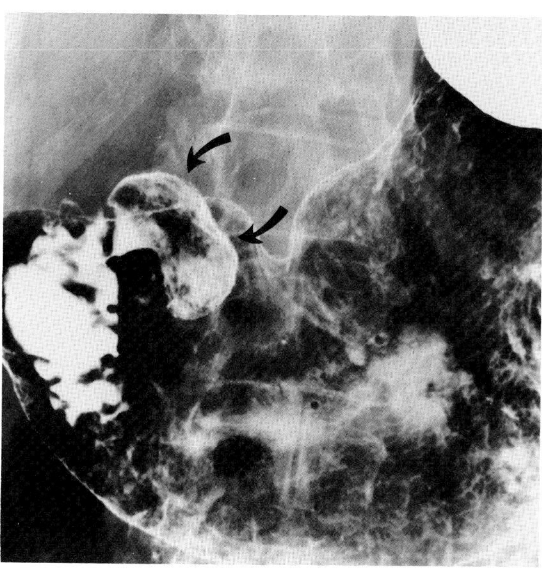

A

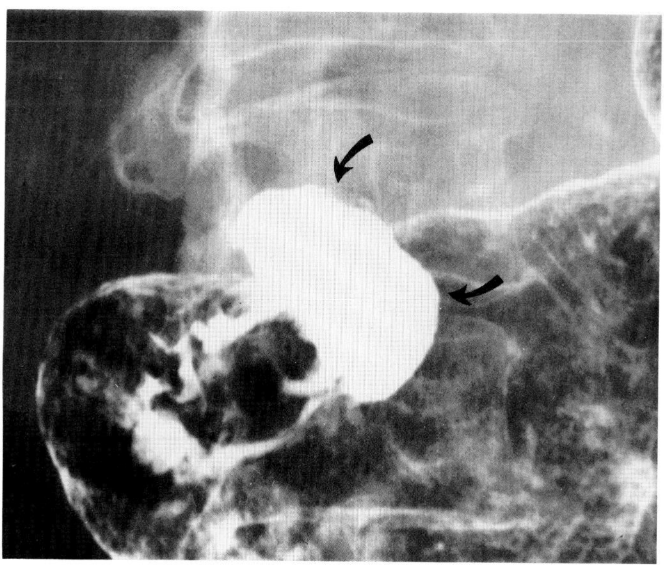

B

**Figure 44.3. Giant duodenal ulcer.** There is little change in the appearance of the rigid-walled cavity (arrows) in (A) air-contrast and (B) barium-filled views. (From Eisenberg, Margulis, Moss, et al,[1] with permission from the publisher.)

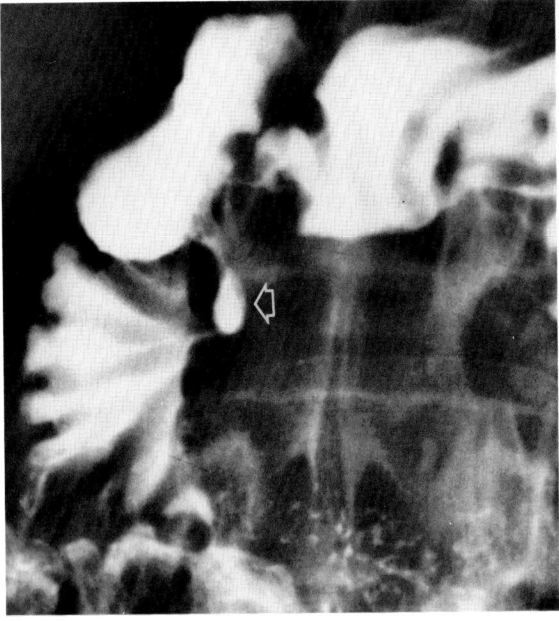
A

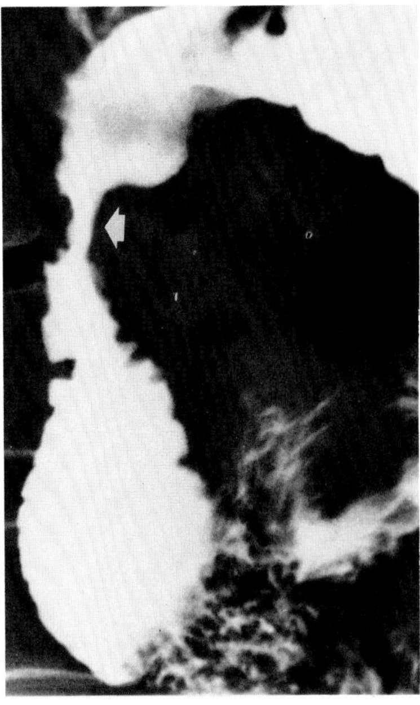
B

**Figure 44.4. Postbulbar ulcer and ring stricture.** (A) A postbulbar ulcer (arrow) with duodenal folds radiating to the edge of the crater. (B) Circumferential narrowing (arrow) of the lumen represents fibrotic healing in the region of the ulcer. (From Eisenberg,[23] with permission from the publisher.)

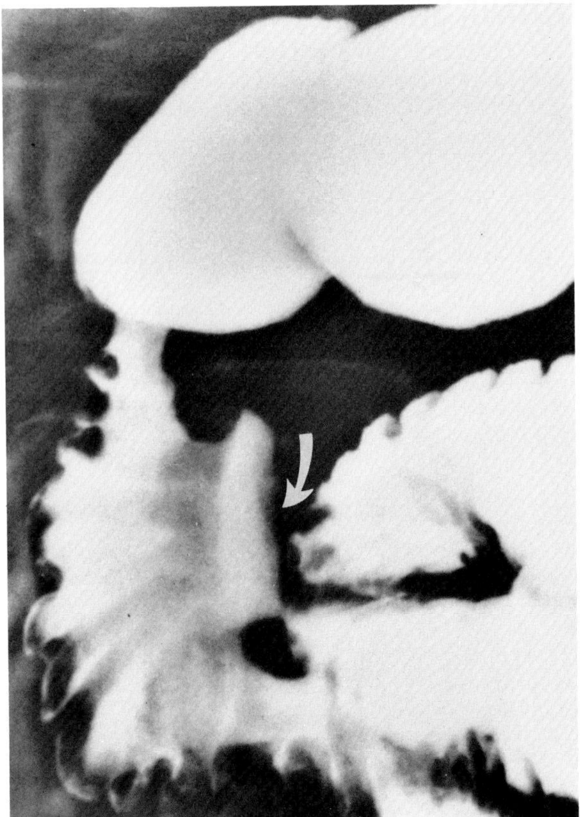

**Figure 44.5. Postbulbar ulcer.** A large ulcer (arrow) on the medial wall of the second portion of the duodenum has penetrated into the head of the pancreas. (From Eisenberg,[23] with permission from the publisher.)

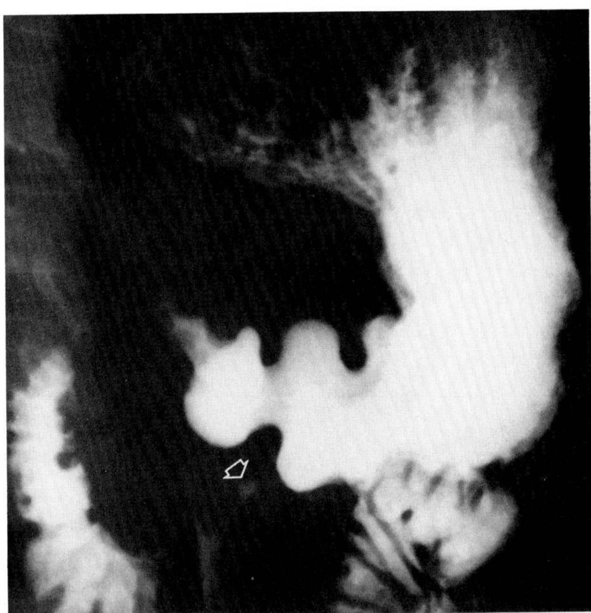

**Figure 44.6. Postbulbar ulcer in the Zollinger-Ellison syndrome.** An ulcer (arrow) is seen in the fourth portion of the duodenum. Note the thickened gastric and duodenal folds. (From Eisenberg,[23] with permission from the publisher.)

the duodenum or just past the apex of the duodenal bulb. Intense spasm often produces an incisura, an indentation defect on the opposite duodenal margin at the same level as the ulcer crater. This causes eccentric narrowing of the lumen that may persist if there is chronic ulceration or fibrotic scarring during the healing phase of the postbulbar ulcer.

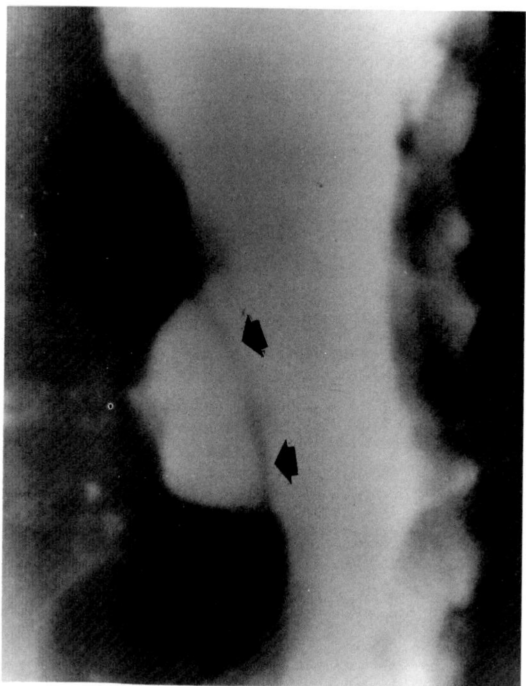

**Figure 44.7. Benign gastric ulcer.** The ulcer clearly projects beyond the gastric lumen on this profile view. The Hampton line appears as a thin, sharply demarcated, lucent line with parallel straight margins (arrows) at the base of the ulcer crater. (From Eisenberg,[23] with permission from the publisher.)

Peptic ulcers arising distal to the mid-descending duodenum are rare. Single or multiple ulcers in this location strongly suggest the Zollinger-Ellison syndrome.[3–5]

## GASTRIC ULCER

The detection of gastric ulcers and the decision about whether these represent benign or malignant processes are major parts of the upper gastrointestinal examination. Up to 95 percent of gastric ulcers can be revealed using double-contrast techniques. Certain technical factors preclude the demonstration of a small percentage of gastric ulcers. The ulcer can be shallow or filled with residual blood, mucus, food, or necrotic tissue that prevents barium from filling it. Similarly, the margins of an ulcer can be so edematous that barium cannot enter it; a small ulcer can be obscured by large rugal folds. In contrast, a false-positive ulcer-like pattern can be caused by barium trapped between gastric folds.

The classic sign of a benign gastric ulcer on profile views is penetration–the clear projection of the ulcer outside the normal barium-filled gastric lumen because the ulcer represents an excavation in the wall of the stomach. Three other signs of benignancy on profile views are the Hampton line, the ulcer collar, and the ulcer mound, all of which are related to undermining of the mucosa (relatively resistant to peptic digestion) that appears to overhang the more rapidly destroyed submucosa. The amount of mucosal edema due to inflammatory exudate determines which of these three signs is

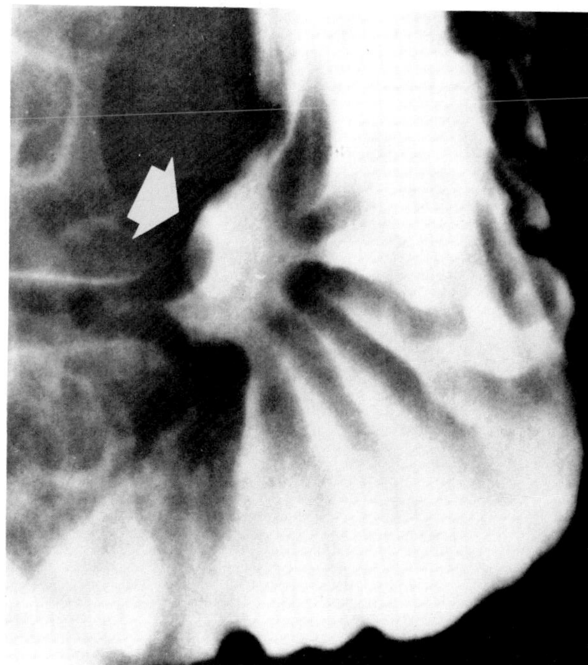

**Figure 44.8. Benign gastric ulcer.** Thickened gastric folds radiate toward the crater (arrow). (From Eisenberg,[23] with permission from the publisher.)

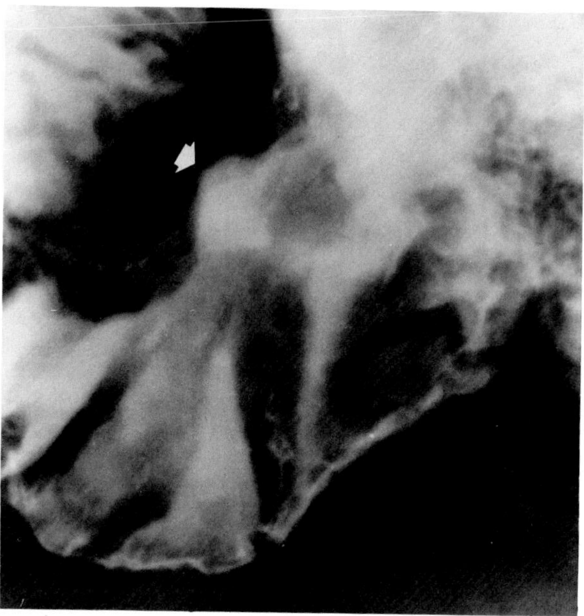

**Figure 44.9. Malignant gastric ulcer.** Thick folds radiate to an irregular mound of tissue around the ulcer (arrow). (From Eisenberg,[23] with permission from the publisher.)

produced. When visualized en face, the ulcer appears as a persistent collection of barium surrounded by a halo of edema. A hallmark of benign gastric ulcer is radiation of mucosal folds to the edge of the crater. However, since radiating folds can be identified in both malignant and benign ulcers, the character of the folds must be carefully assessed. If the folds are smooth and slender and appear to extend into the edge of the crater, the ulcer is most likely benign. In contrast, irregular folds that merge into a mound of polypoid tissue around the crater suggest malignancy. Although the size, shape, number, and location of gastric ulcers were formerly suggested as criteria for distinguishing between benign and malignant lesions, these findings are of little practical value. One exception is ulcers in the gastric fundus above the level of the cardia, essentially all of which are malignant. A lesser curvature ulcer may cause intense muscular spasm and an indentation on the opposite wall (incisura). The incisura may be mistaken for a gastric mass unless the underlying lesser curvature ulcer is demonstrated.

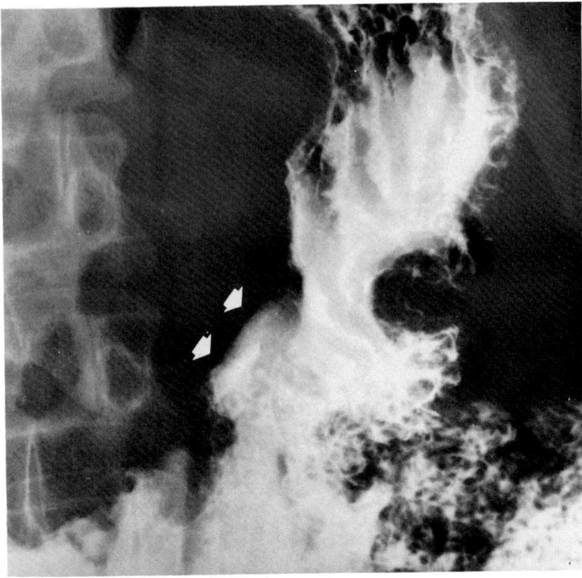

**Figure 44.10. Benign gastric ulcer.** A large incisura simulates a filling defect on the greater curvature. The incisura is incited by a long ulcer (arrows) on the lesser curvature. (From Eisenberg,[23] with permission from the publisher.)

At times, it may be difficult to decide whether a persistent collection of barium in the stomach or elsewhere in the gastrointestinal tract represents an acute ulceration or a nonulcerating deformity. If the barium collection has an elliptical configuration, the orientation of the long axis of the ellipse can be an indicator of the nature of the pathologic process (ellipse sign). If the long axis is parralel to the lumen, the collection represents an acute ulceration. If the long axis is perpendicular to the lumen, the collection represents a deformity without acute ulceration.

An abrupt transition between the normal mucosa and the abnormal tissue surrounding a gastric ulcer is characteristic of a malignant lesion, in contrast to the diffuse and almost imperceptible transition between the normal gastric mucosa and the mound of edema surrounding a benign ulcer. Neoplastic tissue surrounding a malignant ulcer is usually nodular, unlike the smooth contour of the edematous mound around a benign ulcer. A malig-

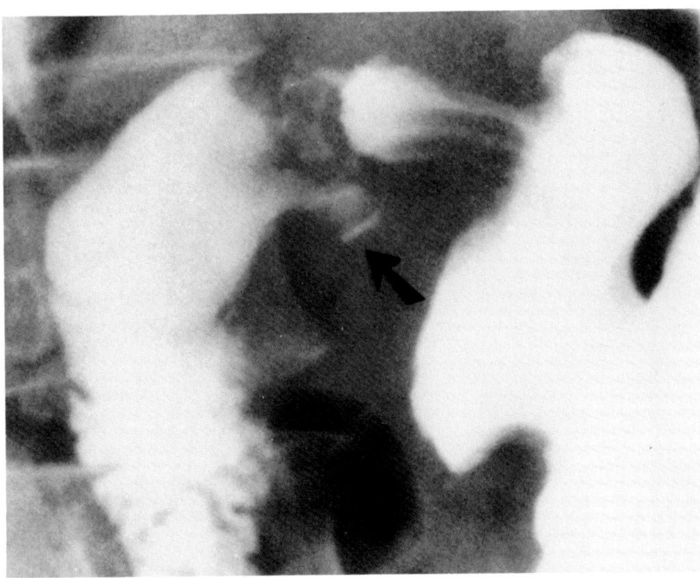

A

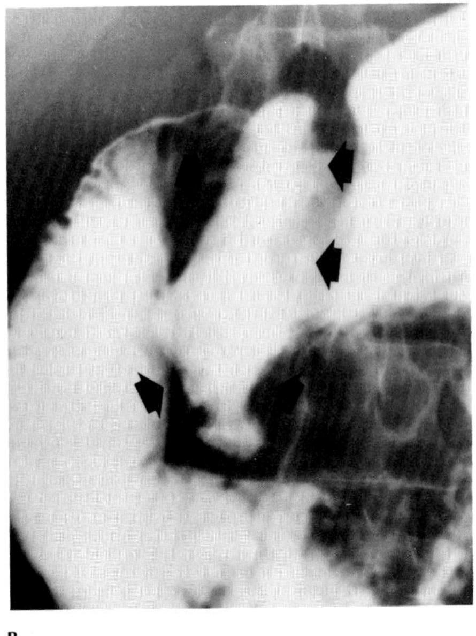

B

**Figure 44.11. Ellipse sign.** (A) A benign gastric ulcer appears as a persistent barium collection (arrow) running parallel to the lumen. (B) The long axis of the bizarre barium collection (arrows) is perpendicular to the lumen in this postulcer deformity. (From Eisenberg and Hedgcock,[6] with permission from the publisher.)

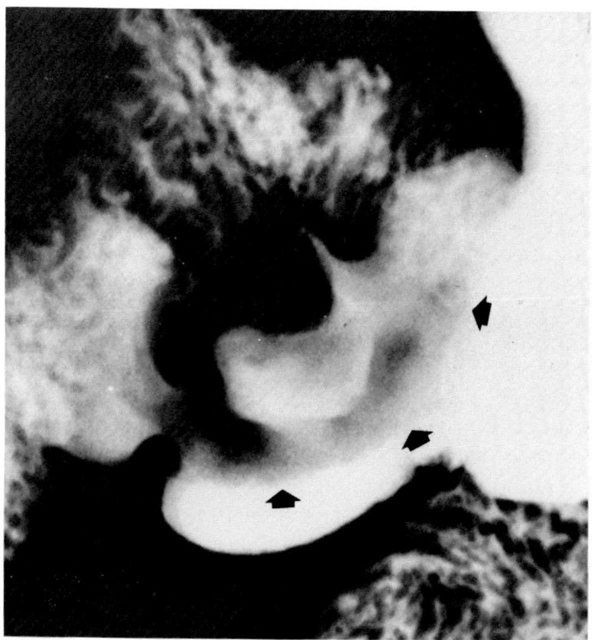

**Figure 44.12. Carman's meniscus sign in malignant gastric ulcer.** The huge ulcer has a semicircular configuration with its inner margin convex toward the lumen. This is surrounded by the radiolucent shadow of an elevated ridge of neoplastic tissue (arrows). (From Eisenberg,[23] with permission from the publisher.)

nant ulcer does not penetrate beyond the normal gastric lumen but remains within it, because the ulcer merely represents a necrotic area within an intramural or intraluminal mass. Carman's meniscus sign is diagnostic of a specific type of ulcerated neoplasm. When examined in profile with compression, this ulcer is seen to have a semicircular (meniscoid) configuration that is convex toward the lumen and surrounded by the lucent shadow of an elevated ridge of neoplastic tissue.

The vast majority of gastric ulcers (more than 95 percent) are benign and heal completely with medical therapy. Most benign ulcers diminish to one-half or less their original size within 3 weeks and show complete healing within 6 weeks. Complete healing does not necessarily mean that the stomach returns to an absolutely normal radiographic appearance; bizarre deformities can result because of fibrotic retraction and stiffening of the wall of the stomach. It is essential to remember that many malignant ulcers show significant healing, though there is almost never complete disappearance of the ulcer crater.

The role of endoscopy in evaluating patients with gastric ulcers is controversial. At present, endoscopy is indicated whenever the radiographic findings are not typical of a benign ulcer, if healing of the ulcer does not progress at the expected rate, or if the mucosa surrounding a healed ulcer crater has a nodular surface or any other feature suggestive of an underlying early gastric cancer.[6–9]

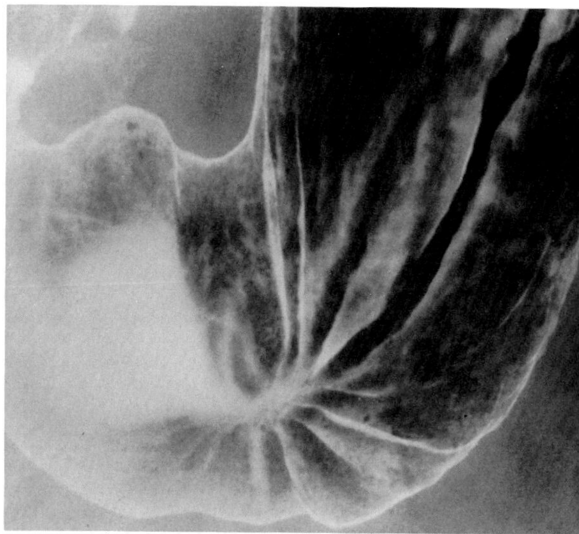

**Figure 44.13. Gastric ulcer scar.** Endoscopy demonstrated complete healing of the benign gastric ulcer. (From Eisenberg,[23] with permission from the publisher.)

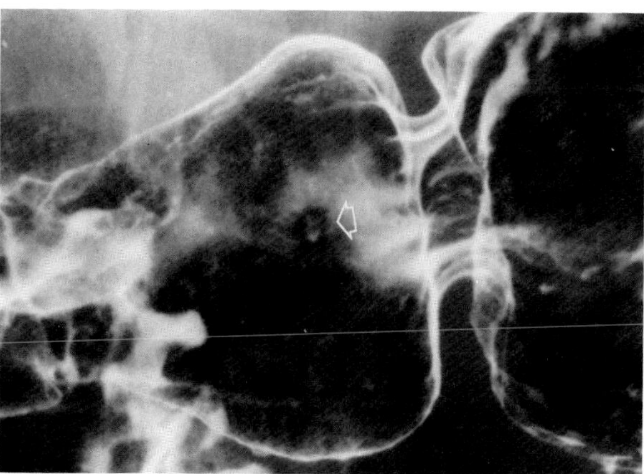

**Figure 44.14. Superficial gastric erosions.** Tiny flecks of barium, representing erosions (arrow), are surrounded by radiolucent halos, representing mounds of edematous mucosa. (From Eisenberg,[23] with permission from the publisher.)

## SUPERFICIAL GASTRIC EROSIONS

Superficial gastric erosions are defects in the epithelium of the stomach that do not penetrate beyond the muscularis mucosae. Because they are very small and shallow, superficial gastric erosions have rarely been demonstrated on conventional upper gastrointestinal examinations. With the increasing use of double-contrast techniques, however, more than half of the superficial gastric erosions noted endoscopically can be demonstrated radiographically. The classic radiographic appearance of a superficial gastric erosion is a tiny fleck of barium, which represents the erosion, surrounded by a radiolucent halo, which represents a mound of edematous mucosa. Possible factors implicated in the production of superficial gastric erosions include alcohol, anti-inflammatory agents (e.g., aspirin, steroids), analgesics, Crohn's disease, and candidiasis.[10–12]

## COMPLICATIONS OF PEPTIC ULCER DISEASE

### Hemorrhage

Peptic ulcer disease is the most common cause of acute upper gastrointestinal bleeding. The proper diagnostic approach depends on the severity and activity of the hemorrhage. Patients with massive hemorrhage, whose vital signs cannot be maintained, require emergency surgery for both diagnostic and therapeutic purposes. In the patient with acute upper gastrointestinal hemorrhage who has relatively stable vital signs, endoscopy, not a barium study, is the initial diagnostic procedure of choice. The 95 percent accuracy of endoscopy is substantially higher than the accuracy of barium studies, even using double-contrast techniques. Barium examinations may be technically difficult to perform in the patient with acute hemorrhage since the presence of fresh and clotted blood severely affects the quality of mucosal coating. Unlike barium studies, endoscopy allows the physician not only to visualize the abnormality but also to say with some certainty whether or not a specific lesion is the cause of the acute bleeding episode. This is important because a significant number of patients with acute upper gastrointestinal hemorrhage are found to have more than one lesion. If emergency endoscopy is not available, a double-contrast upper gastrointestinal series should be performed following vigorous lavage through a large-bore tube to remove clots that could simulate a gastric mass or obscure an acute ulcer crater.

In a patient with active bleeding at a rate of at least 0.5 to 1 ml per minute, selective celiac and superior mesenteric arteriography can often demonstrate extravasation of contrast material from the vascular tree and thus localize the site of bleeding. In addition to peptic ulcer disease, virtually all other causes of acute upper gastrointestinal hemorrhage (gastritis, varices, arteriovenous malformations, Mallory-Weiss tears) can be diagnosed and treated via the arteriographic catheter. The specific cause of the hemorrhage generally determines whether vasoconstrictive agents or embolization will be the most effective therapy. In patients with massive hemorrhage, arteriography and transcatheter therapeutic measures may allow time for vascular volume replacement and a more stable patient before surgical intervention.

Radionuclide scanning with $^{99m}$Tc-labeled sulfur colloid or red blood cells can accurately localize gastrointestinal bleeding with minimum discomfort and risk. The addition of rapid sequential images (radionuclide angiogram) to the routine static scans increases the rate of lesion detection. Although more sensitive, radionuclide scanning is less specific than arteriography and offers no therapeutic value.[13–15]

### Perforation

Perforation is an occasional complication of peptic disease, especially involving ulcers in the pyloroduodenal region. Free perforation of a peptic ulcer is the most frequent cause of pneumoperitoneum with peritonitis. With the patient in an erect position, as little as 1 ml of air can be demonstrated as a sickle-shaped lucency beneath the domes of the diaphragm. Free intraperitoneal gas is easiest to recognize on the right side between the diaphragm and the homogeneous density of the liver. On the left, the normal gas and fluid shadows present

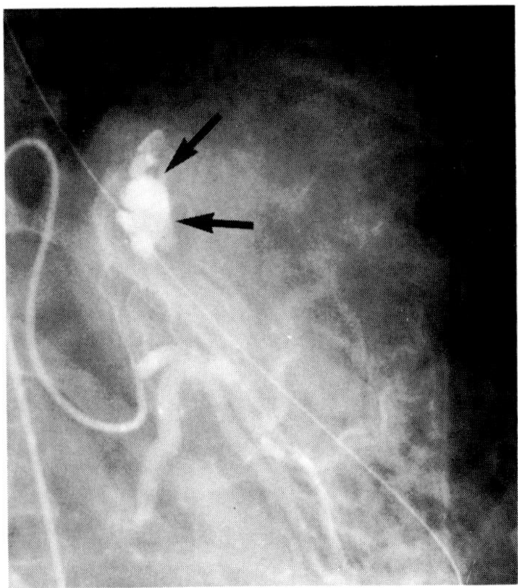

A

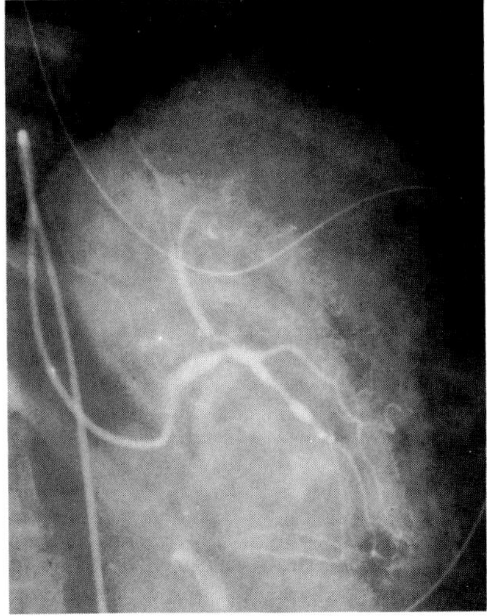

B

**Figure 44.15. Upper gastrointestinal hemorrhage controlled by intra-arterial vasopressin infusion.** (A) A late film from a left gastric arteriogram shows contrast extravasation (arrows) in the gastric fundus. (B) Vasopressin infused into the left gastric artery constricts its branches and stops the bleeding. (From Athanasoulis,[26] with permission from the publisher.)

in the fundus of the stomach can be confusing. Free air is shown to best advantage if the patient remains in an upright position for 10 minutes before a radiograph is obtained. If the patient is too ill to sit or stand, a lateral decubitus view (preferably with the patient on his or her left side) can be used. In this position, free gas moves to the right and collects between the lateral margin of the liver and the abdominal wall. On supine views of the abdomen, free intraperitoneal air accumulates between intestinal loops and is much more difficult to demonstrate. However, a large quantity of air can be diagnosed indirectly because it permits visualization of the outer margins of the intestinal wall. The distinct demonstration of the inner and outer contours of the bowel wall is often the only sign of pneumoperitoneum in patients in such poor condition that they cannot be turned on their side or be examined upright.

It is important to remember that in about 30 percent of perforated peptic ulcers, no free intraperitoneal gas can be identified. Therefore, the failure to demonstrate a pneumoperitoneum is of no value in excluding the possibility of a perforated ulcer. In patients with suspected perforation and no pneumoperitoneum, extensive extravasation from the upper gastrointestinal tract can often be demonstrated following the oral administration of a small amount of water-soluble contrast material.

Ulcers of the posterior wall of the duodenum tend to penetrate into the pancreas and frequently result in increased levels of serum amylase. Less commonly, duodenal ulcers may penetrate into the liver, biliary tract, colon, or lesser sac, as well as dissecting along retroperitoneal fascial planes. Penetrating ulcers involving the pancreas can cause radiographic changes of pancreatitis such as serrations and spiculations of the mucosa of the descending duodenum, effacement of folds, and a mass effect.

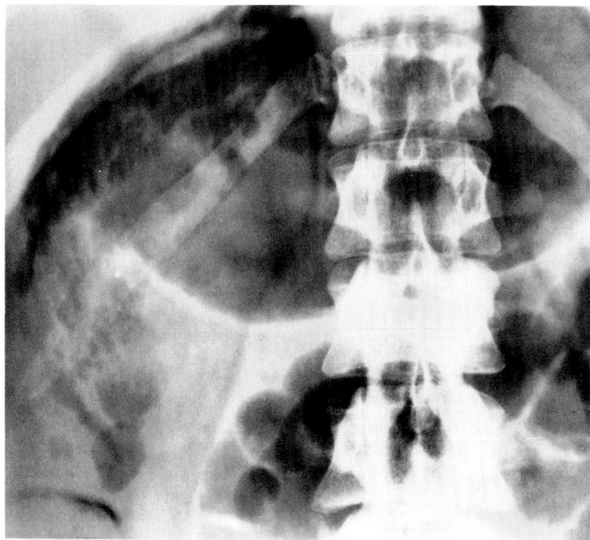

**Figure 44.16. Perforated duodenal ulcer.** Gas in the retroperitoneum outlines the kidney and the undersurface of the liver. (From Eisenberg,[23] with permission from the publisher.)

Perforation of a peptic ulcer may become walled off and form an abscess cavity. A persistent connection to the lumen may result in an air-fluid level on upright abdominal radiographs and may permit barium to fill the abscess and sinus tract during a contrast examination. If the communication with the lumen has closed, the abscess may appear as a soft tissue density producing a mass effect on the stomach or duodenum.[13,16]

## Gastric Outlet Obstruction

Peptic ulcer disease is by far the most common cause of gastric outlet obstruction in adults. The obstructing lesion in peptic ulcer disease is usually in the duodenum, occasionally in the pyloric channel or prepyloric gastric antrum, and rarely in the body of the stomach. Narrowing of the lumen due to peptic ulcer disease can result in spasm, acute inflammation and edema, muscular hypertrophy, or contraction of scar tissue. Most patients with peptic ulcer disease causing pyloric obstruction have a long history of ulcer symptoms and characteristically present with vomiting.

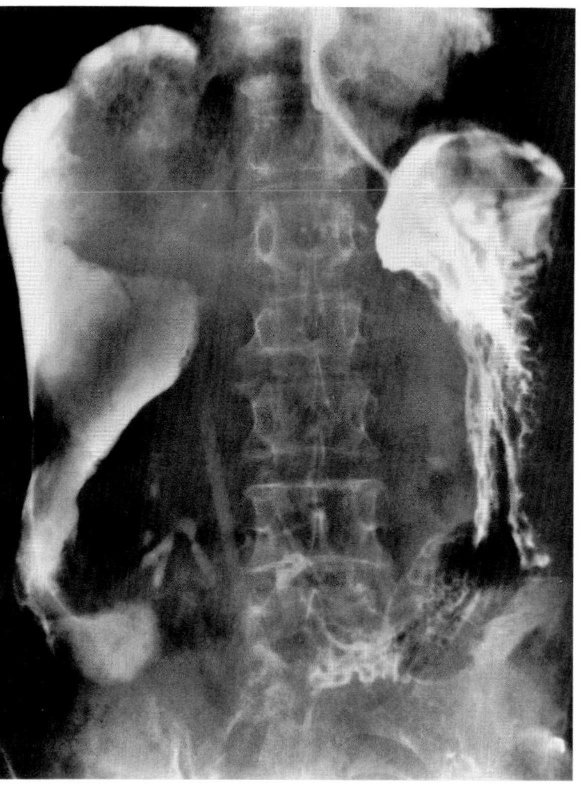

**Figure 44.17. Perforated duodenal ulcer.** There is extensive extravasation from the upper gastrointestinal tract following the oral administration of contrast material. (From Eisenberg,[23] with permission from the publisher.)

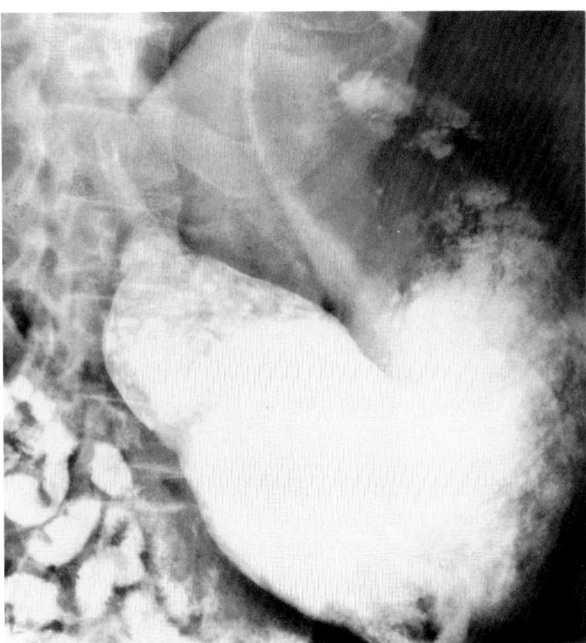

**Figure 44.18. Gastric outlet obstruction caused by peptic ulcer disease.** (From Eisenberg,[23] with permission from the publisher.)

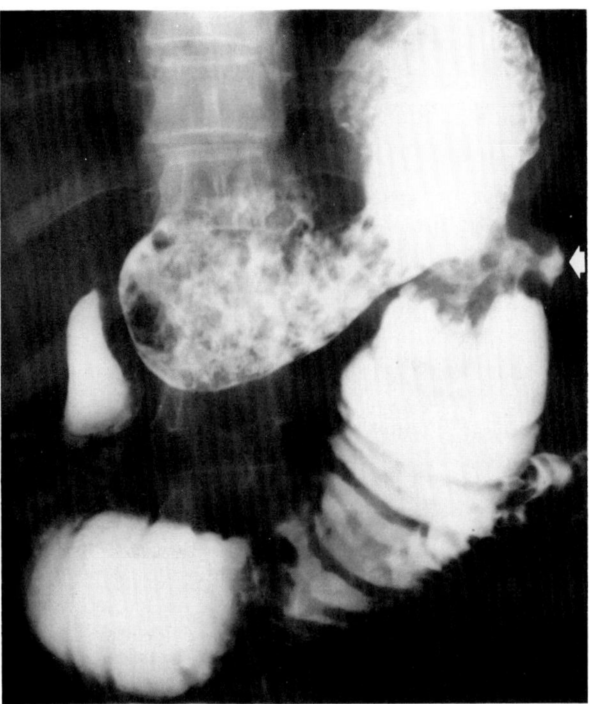

**Figure 44.19. Recurrent (marginal) ulceration following gastrojejunostomy.** The ulcer (arrow) appears in the jejunum within the first few centimeters of the anastomosis. (From Eisenberg,[23] with permission from the publisher.)

Plain abdominal radiographs often demonstrate the shadowy outline of a distended stomach. On barium examination, a mottled density of nonopaque material represents excessive overnight gastric residual. There is a marked delay in gastric emptying, and barium is often retained in the stomach for 24 hours or longer. The critical differential diagnosis is between benign peptic disease and a malignant cause of gastric outlet obstruction. Whenever possible, delayed films should be obtained to look for barium in the duodenal bulb. A distorted, scarred bulb with pseudodiverticula formation suggests peptic ulcer disease as the most likely cause. Conversely, a radiographically normal duodenal bulb increases the likelihood of underlying malignant disease. In many patients the distinction between a benign or a malignant cause of gastric outlet obstruction cannot be made on the basis of barium studies, and endoscopy or surgical exploration is required.[13]

## COMPLICATIONS AND SYNDROMES AFTER PEPTIC ULCER SURGERY

### Recurrent Ulceration

Recurrent (marginal) ulceration develops in about 5 percent of patients following surgery for peptic ulcer disease. Recurrence is much more common following vagotomy and pyloroplasty (6 to 8 percent) than after vagotomy and antrectomy (fewer than 2 percent). The marginal ulcer is really not a recurrent one, but rather represents a new ulceration due to the residual action of acid and pepsin on the sensitive intestinal mucosa. Thus it is usually situated in the jejunum within the first few centimeters of the anastomosis. Because marginal ulcers are rarely found on the gastric side of the anastomosis, the development of postoperative ulceration at this site should suggest the possibility of gastric stump malignancy. In most cases, the site of the original ulcer for which surgery was performed is the duodenum; marginal ulcers are much less frequent following surgery for gastric ulcers.

Because up to half of marginal ulcers are too superficial or shallow to be detected radiographically, endoscopy is often required for diagnosis. Overlapping jejunal mucosal folds or surgical deformity can hide an ulcer; conversely, barium trapped between converging gastric or jejunal folds about the distorted anastomotic site can simulate an ulcer niche.

Marginal ulcers may appear radiographically as small, circumscribed collections of barium or huge, contrast-filled masses that mimic postoperative pseudodiverticula. Secondary signs of marginal ulceration include edema of the duodenal or jejunal folds at the anastomotic site, flattening and rigidity of the jejunum adjacent to the ulcer, and wide separation of jejunal and gastric segments.[23]

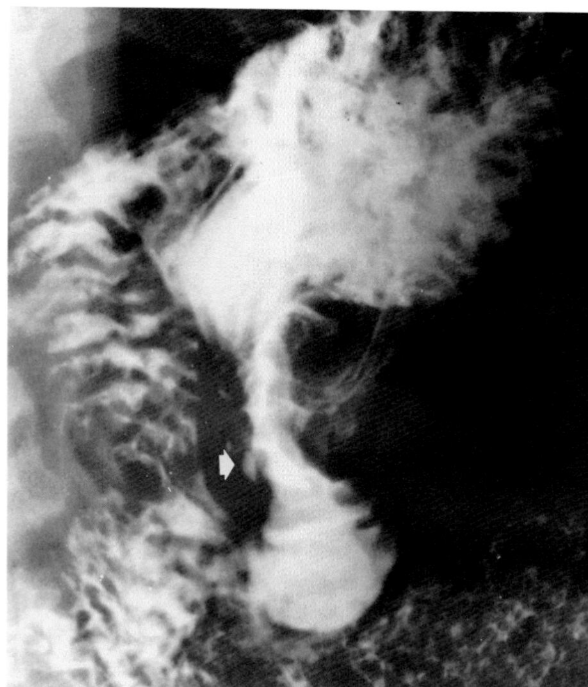

**Figure 44.20. Recurrent (marginal) ulceration (arrow) following Billroth II anastomosis.** Marked edema of jejunal folds at the anastomotic site suggests recurrent ulcer disease. There is also narrowing of the stoma with relative separation of the gastric and jejunal segments. (From Eisenberg,[23] with permission from the publisher.)

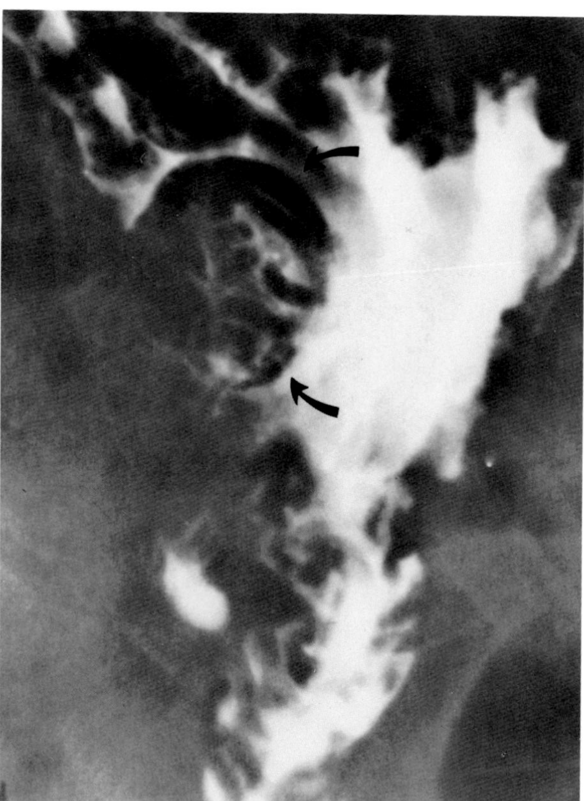

**Figure 44.21. Afferent loop syndrome.** Retrograde jejunogastric intussusception (afferent loop) produces a large, sharply defined filling defect (arrows). (From Eisenberg,[23] with permission from the publisher.)

## Afferent Loop Syndrome

Distension and poor drainage of the afferent intestinal loop of a Billroth II anastomosis may cause abdominal bloating and pain following eating, often associated with nausea and vomiting. Because of complete or partial obstruction, it may be difficult to demonstrate the afferent loop on the upper gastrointestinal examination. Complete obstruction of the afferent loop at the anastomosis may be caused by retrograde jejunogastric intussusception, which appears radiographically as a clearly defined spherical or ovoid intraluminal filling defect in the gastric remnant. If the obstruction is proximal to the anastomosis, barium may fill a short segment of the afferent loop to the point of obstruction. If there is preferential filling of the afferent loop and no organic obstruction, barium examination demonstrates a dilated afferent loop that fills rapidly and empties slowly and incompletely.[18]

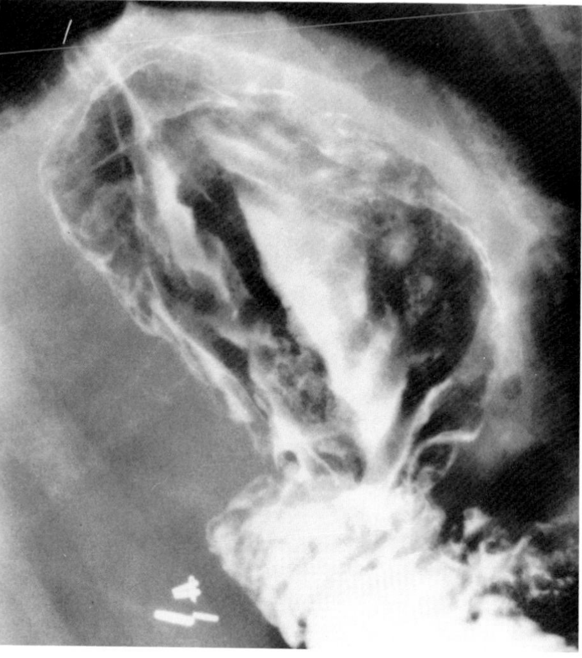

**Figure 44.22. Bile reflux gastritis.** There is thickening of rugal folds in the gastric remnant.

## Bile Reflux Gastritis

Highly alkaline digestive secretions are normally prevented from entering the stomach by an intact pylorus. When the pyloric mechanism is destroyed or circumvented by partial gastric resection, free reflux can produce

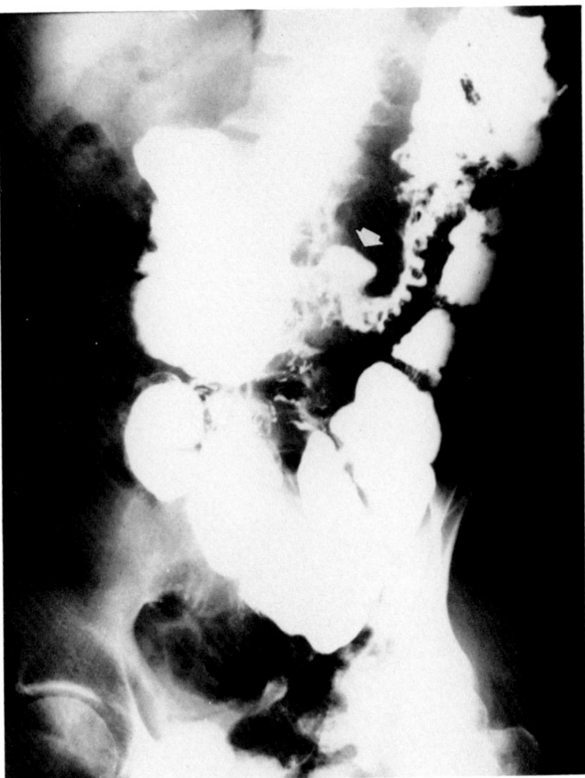

**Figure 44.23. Gastrojejunocolic fistula.** A large anastomotic ulcer (arrow) is visible near the site of previous gastroenterostomy. Partial filling of the colon is seen on the upper gastrointestinal series. (From Thoeny, Hodgson, and Scudamore,[17] with permission from the publisher.)

severe gastritis and ulceration. Bile reflux gastritis appears radiographically as thickened folds in the gastric remnant. The most severe changes tend to occur near the anastomosis. Ulcerations due to bile reflux gastritis occur on the gastric side of the remnant, unlike true marginal ulcerations that occur on the jejunal side of the anastomosis.[19]

### Gastrojejunocolic Fistula

A fistulous communication between the stomach, jejunum, and colon (gastrojejunocolic fistula) or directly between the stomach and colon is a grave complication of marginal ulceration after gastric surgery for peptic ulcer disease. Most patients with this condition (there is a heavy predominance in men) have diarrhea and weight loss; pain, vomiting, and bleeding are common. These fistulas are usually demonstrated during a barium enema examination in which contrast material is observed to extend directly from the transverse colon into the stomach. These postsurgical fistulas are associated with a high mortality rate, especially if recognized late.[17]

### Carcinoma after Partial Gastrectomy

The development of recurrent symptoms after 10 years or more of relatively good health following partial gastrectomy for ulcer disease should suggest the possibility of gastric stump carcinoma. The incidence of carcinoma in the gastric remnant is two to six times higher than in the intact stomach. Gastric stump carcinoma may be difficult to demonstrate on barium examination. The tumor usually appears as an irregular polypoid mass at the anastomotic margin or in the gastric remnant. A decrease in the size of the gastric remnant can occur secondary to uniform infiltration by carcinoma. This sign of malignancy requires comparison with previous studies and is an excellent reason for obtaining a baseline upper gastrointestinal examination several months after partial gastrectomy.

Gastric stump malignancy can also appear as a marginal ulceration near the anastomosis. Patients with this type of carcinoma tend to have a long symptom-free period averaging in excess of 20 years, in contrast to patients with benign mucosal ulcerations, which usually occur within 2 years of surgery. Since benign marginal ulcerations usually occur on the jejunal side of the anastomosis, any ulcer on the gastric side should be considered malignant until proved otherwise.[20]

## ZOLLINGER-ELLISON SYNDROME (GASTRINOMA)

The Zollinger-Ellison syndrome is caused by a non-beta islet cell tumor of the pancreas that continually secretes gastrin. The persistently high blood level of this hormone results in a strong stimulus to the gastric parietal cells that causes voluminous gastric hypersecretion and hyperacidity and a clinical picture of severe, often intractable, peptic ulcer disease. Although most ulcers in patients with the Zollinger-Ellison syndrome occur in the stomach or duodenal bulb, up to 25 percent may be in an atypical location in more distal portions of the duodenum and in the jejunum. The unusual location of the peptic ulcers seen distal to the duodenal bulb is due to the excessively large volume of highly acid gastric fluid that bathes the duodenum and proximal jejunum and overwhelms the alkaline biliary and pancreatic secretions. The resultant chemical enteritis produces severe ulceration and inflammatory fold thickening in the vulnerable distal duodenum and proximal jejunum, which are not normally exposed to such an acid environment. Because the normal succus entericus in the more distal portions of the small bowel dilutes the acid gastric contents and raises the pH, the mucosal pattern of the distal jejunum and the ileum is usually normal. Giant duodenal ulcers, though not pathognomonic, should suggest the possibility of the Zollinger-Ellison syndrome.

In the stomach, the Zollinger-Ellison syndrome is associated with diffuse thickening of gastric folds and ex-

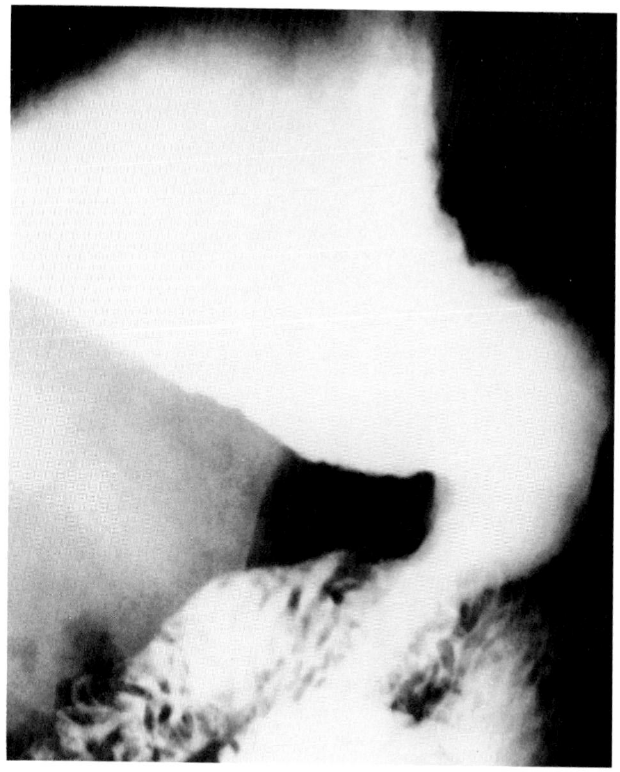
A

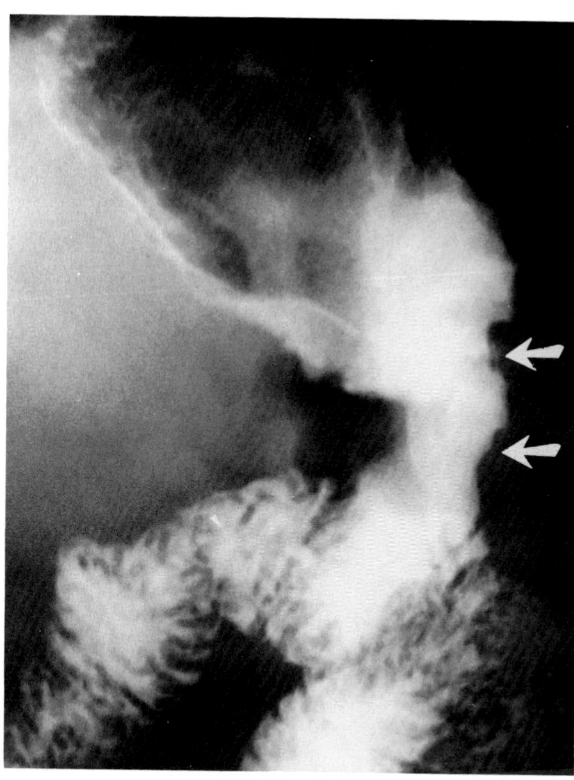
B

**Figure 44.24. Gastric stump carcinoma.** (A) A normal gastric remnant and Billroth II anastomosis following surgery for peptic disease. (B) Irregular narrowing of the perianastomotic region (arrows) several years later represents a gastric stump carcinoma. (From Eisenberg,[23] with permission from the publisher.)

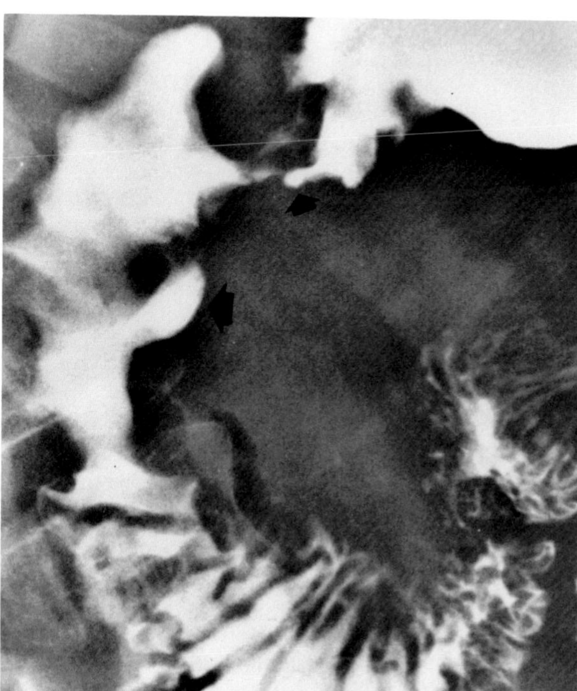

**Figure 44.25. Zollinger-Ellison syndrome.** There is diffuse thickening of folds in the proximal duodenal sweep with bulbar and postbulbar ulceration (arrows). (From Eisenberg,[23] with permission from the publisher.)

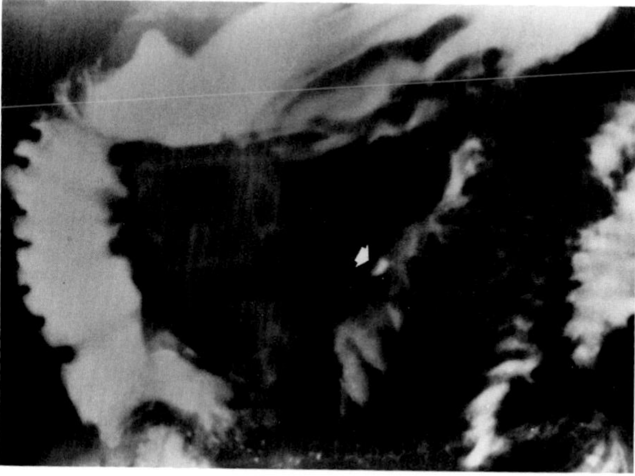

**Figure 44.26. Zollinger-Ellison syndrome.** An ulcer (arrow) is seen in the fourth portion of the duodenum. Note the thickened gastric and duodenal folds. (From Eisenberg,[23] with permission from the publisher.)

cessive fasting gastric secretions without any demonstrable gastric outlet obstruction. The duodenal ulcers, though radiographically indistinguishable from those secondary to benign peptic disease, usually fail to respond to traditional medical and surgical therapy. The large amount of fluid entering the small intestine from the hypersecreting stomach leads to small bowel dila-

**Figure 44.27. Zollinger-Ellison syndrome.** There is prominent thickening of gastric and duodenal folds, and dilatation and fold thickening in the small bowel. Excessive secretions in the small bowel cause the barium to have a granular, indistinct quality. (From Eisenberg,[23] with permission from the publisher.)

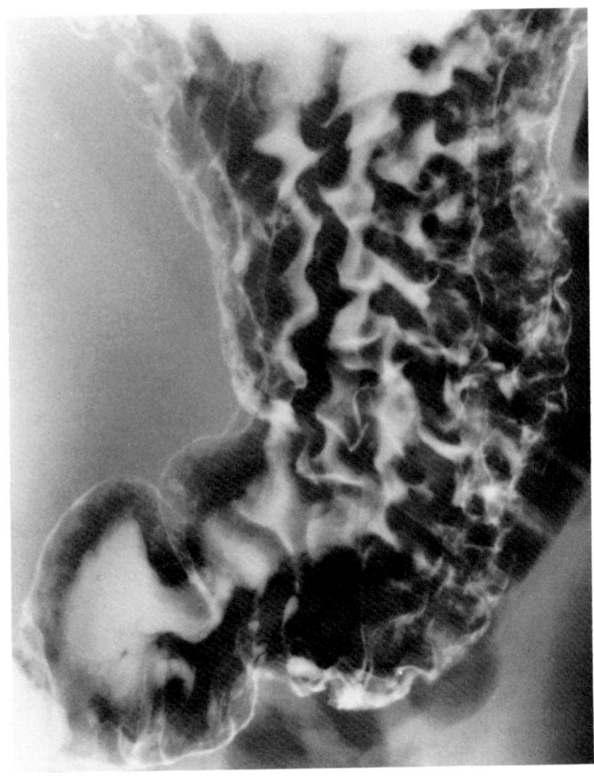

**Figure 44.28. Zollinger-Ellison syndrome.** There is diffuse thickening of gastric folds.

tation, thickening of duodenal and jejunal folds, and a considerable dilution of the barium, which appears to have a gray, watery appearance.

About 90 percent of gastrinomas are found within the pancreas. The remainder lie in ectopic locations such as the stomach, duodenum, or splenic hilum. Fewer than half of gastrinomas can be identified by CT or pancreatic arteriography. Although most gastrinomas are very small, about 50 percent are malignant. Metastases (usually to regional lymph nodes or the liver) continue to secrete gastrin and stimulate the gastric parietal cell mass even after the primary tumor of the pancreas has been removed or a total pancreatectomy has been performed. Partial gastric resection or vagotomy with pyloroplasty is almost invariably followed by prompt and often fulminant ulcer recurrence. Therefore, the surgical procedure of choice is the removal of the entire target organ (i.e., the parietal cell mass of the stomach) by total gastrectomy.

The Zollinger-Ellison syndrome frequently coexists with multiple endocrine adenomatosis. In addition to the pancreas, the most commonly involved endocrine glands are the adrenals, parathyroids, and pituitary.[4,21,22]

## STRESS AND DRUG-ASSOCIATED ULCERS

A variety of acute ulcerative lesions of the stomach and duodenum are clinically distinct from, though radiographically identical to, those arising from chronic peptic ulcer disease. Stress ulcers can develop rapidly in patients with shock, burns, sepsis, and severe trauma. They most commonly present with gastrointestinal hemorrhage, which may be substantial. Although gastric acid appears to be involved in the production of these acute stress ulcers, there is usually no evidence of acid hypersecretion. Often multiple, stress ulcers are frequently too superficial to be detected on barium examination and may only be identified by upper gastrointestinal endoscopy. Celiac or selective left gastric arteriography may be of value in localizing a site of hemorrhage, and intra-arterial infusion of vasopressin may decrease or stop the bleeding.

Anti-inflammatory agents such as aspirin, steroids, phenylbutazone, and indomethacin have been associated with an increased incidence of ulcer disease and upper gastrointestinal hemorrhage. Therefore, the use of these substances should be limited in patients with peptic ulcer disease. Radiographically, a superficial gastric erosion due to anti-inflammatory agents usually appears as a tiny fleck of barium surrounded by a radiolucent halo of edematous mucosa.[8,9,24]

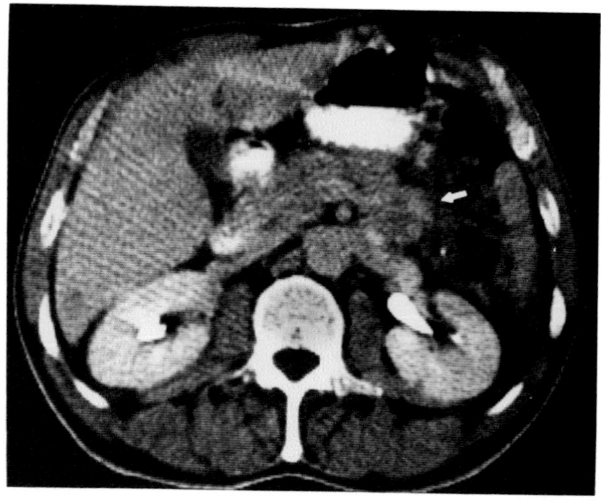

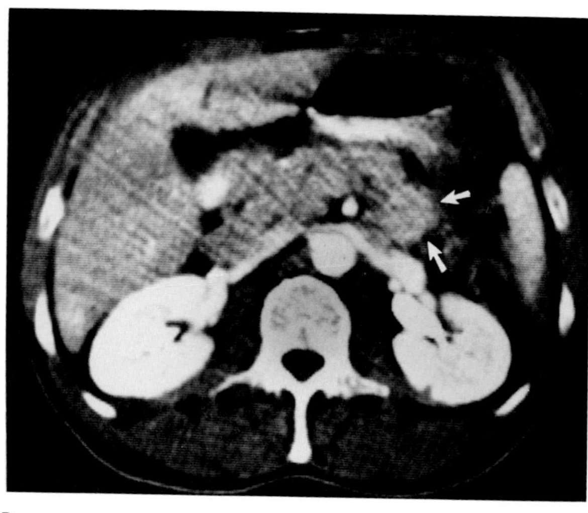

A B

**Figure 44.29. Gastrinoma in Zollinger-Ellison syndrome.** (A) A routine CT scan following the administration of contrast material demonstrates a suspicious lobulation (arrow) in the body of the pancreas. (B) A repeat examination following the infusion of a bolus of contrast material reveals a vascular blush (arrows) characteristic of islet cell tumors. At operation, a 3-cm gastrinoma was removed. (From Stark, Moss, Goldberg, et al,[21] with permission from the publisher.)

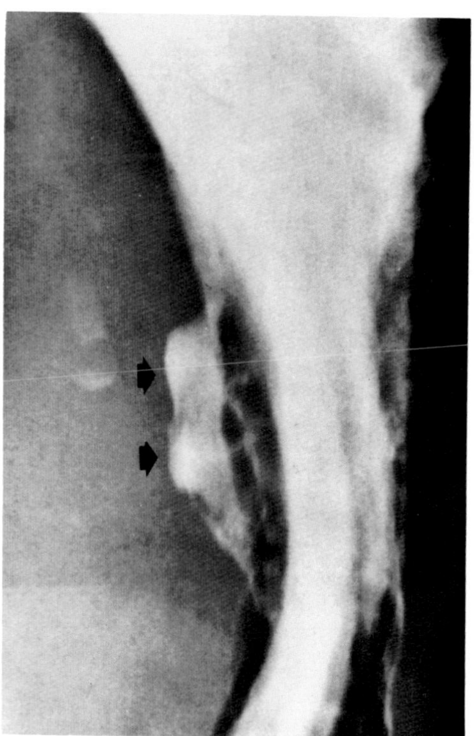

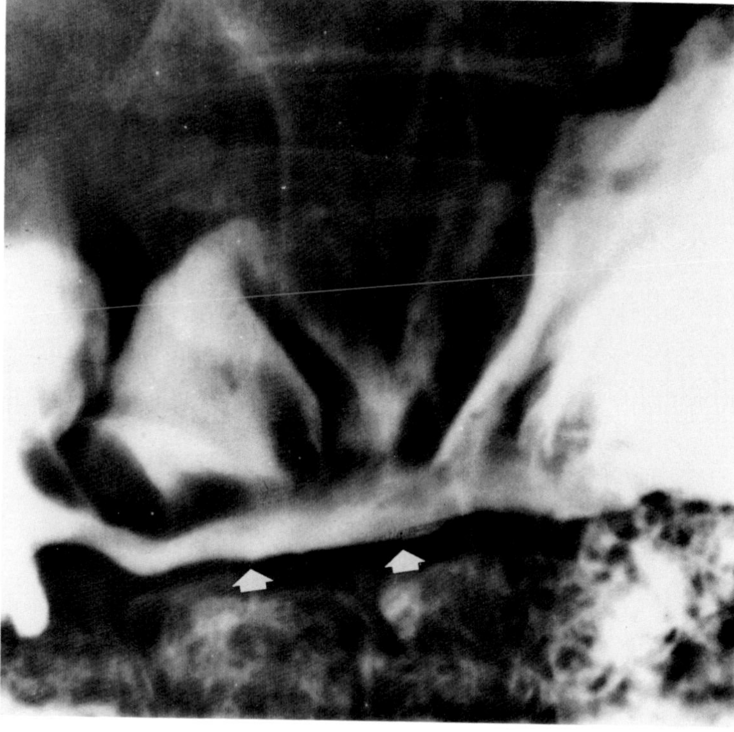

A B

**Figure 44.30. Benign stress ulcers.** Two examples of long ulcers (arrows) in (A) the body and (B) the antrum of the stomach. (From Eisenberg,[23] with permission from the publisher.)

## BRUNNER'S GLAND HYPERPLASIA

Brunner's glands are elaborately branched acinar glands containing both mucus and serous secretory cells. They are arranged in 0.5- to 1-mm lobules and fill much of the submucosal space of the duodenal bulb and proximal half of the second portion of the duodenum. The alkaline secretions from Brunner's glands are rich in mucus and bicarbonate and protect the sensitive duodenal mucosa from erosion by stomach acid.

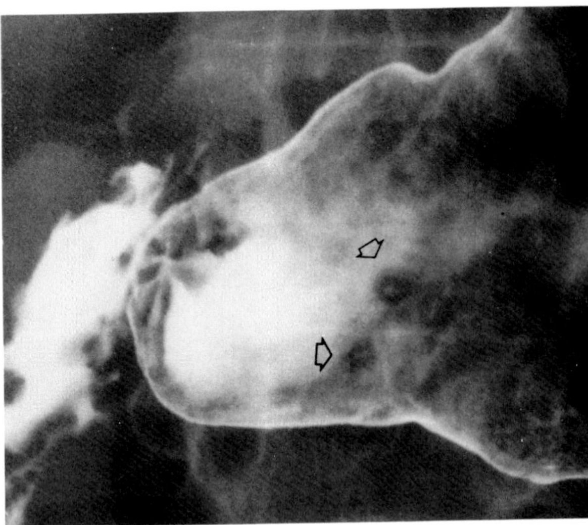

**Figure 44.31. Drug-associated ulcers.** Superficial gastric erosions (arrows) in a patient taking anti-inflammatory medication.

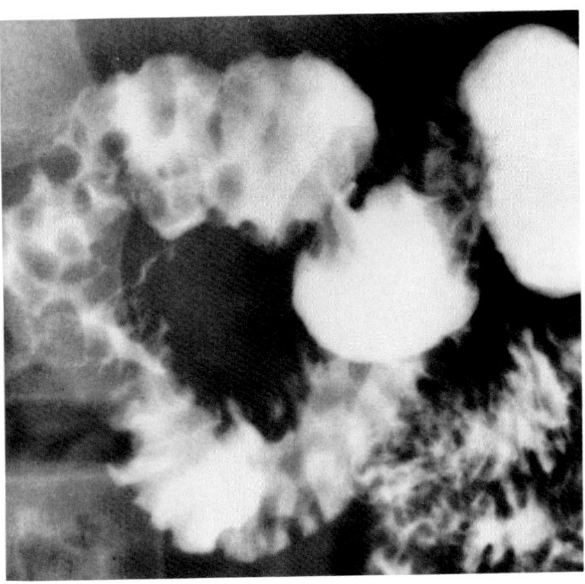

**Figure 44.32. Brunner's gland hyperplasia.** Multiple nodules produce a cobblestone appearance involving the duodenal bulb and proximal sweep. (From Eisenberg,[23] with permission from the publisher.)

Hyperplasia of Brunner's glands probably represents a response of the duodenal mucosa to peptic ulcer disease. The condition typically appears radiographically as generalized nodular thickening of folds, which is usually limited to the first portion of the duodenum. The so-called Brunner's gland adenoma, which appears as a discrete filling defect in the bulb or second portion of the duodenum, represents a localized area of hypertrophy and hyperplasia rather than a true neoplasm.[25]

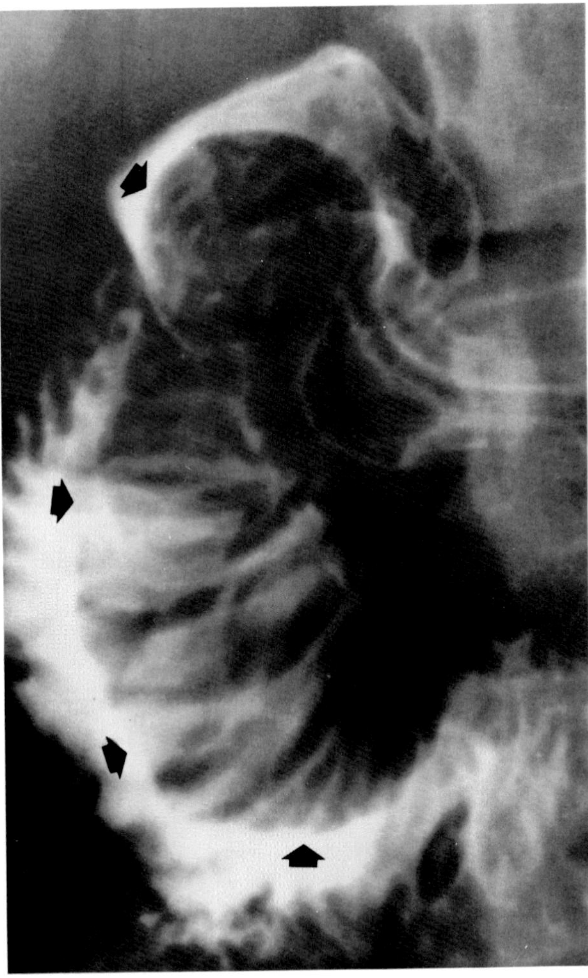

**Figure 44.33. Brunner's gland adenoma.** A large filling defect (arrows) involves the duodenal bulb and sweep. This proved histologically to represent a large area of hypertrophy and hyperplasia of Brunner's glands, rather than a true neoplasm. (From Eisenberg,[23] with permission from the publisher.)

## REFERENCES


1. Eisenberg RL, Margulis AR, Moss AA: Giant duodenal ulcers. *Gastrointest Radiol* 1:347–353, 1978.
2. Op den Orth JO: Duodenum, in Margulis AR, Burhenne HJ (eds): *Alimentary Tract Radiology*, St. Louis, CV Mosby, 1983, pp 800–831.
3. Bilbao MK, Frische LH, Rosch J, et al: Postbulbar duodenal ulcer and ring-stricture: Cause and effect. *Radiology* 100:27–35, 1971.
4. Zborlaske FS, Amberg JR: Detection of the Zollinger-Ellison syndrome: The radiologist's responsibility. *AJR* 104:529–543, 1968.
5. Kaufman SA, Levene G: Post-bulbar duodenal ulcer. *Radiology* 69:848–852, 1957.
6. Eisenberg RL, Hedgcock MW: The ellipse sign: An aid in the diagnosis of acute ulcers. *J Can Assoc Radiol* 30:26–29, 1979.
7. Gelfand DW, Ott DJ: Gastric ulcer scars. *Radiology* 140:37–43, 1981.
8. Nelson SW: The discovery of gastric ulcers and the differential

diagnosis between benignancy and malignancy. *Radiol Clin North Am* 7:5–25, 1969.
9. Wolf BS: Observations on roentgen features of benign and malignant gastric ulcers. *Semin Roentgenol* 6:140–150, 1971.
10. Laufer I, Costopoulous L: Early lesions of Crohn's disease. *AJR* 130:307–311, 1978.
11. Cronan J, Burrell M, Trepeta R: Aphthoid ulcerations in gastric candidiasis. *Radiology* 134:607–611, 1980.
12. Poplack W, Paul RE, Goldsmith M, et al: Demonstration of erosive gastritis by the double-contrast technique. *Radiology* 117:519–521, 1975.
13. Walker CO: Complications of peptic ulcer disease and indications for surgery, in Sleisinger MH, Fordtran JS (eds): *Gastrointestinal Disease.* Philadelphia, WB Saunders, 1978.
14. Thoeni RF, Cello JP: A critical look at the accuracy of endoscopy and double-contrast radiography of the upper gastrointestinal tract in patients with substantial upper GI hemorrhage. *Radiology* 135:305–308, 1980.
15. Berger RB, Zemon RK, Gottshalk A: The technetium-99m-sulfur colloid angiogram in suspected gastrointestinal bleeding. *Radiology* 147:555–588, 1983.
16. Jacobson G, Berne CJ, Meyers HI, et al: Examination of patients with suspected ulcer using water soluble contrast medium. *AJR* 86:37–49, 1961.
17. Thoeny RH, Hodgson JR, Scudamore HH: The roentgenologic diagnosis of gastrocolic and gastrojejunocolic fistulas. *AJR* 83:876–881, 1960.
18. Jay BS, Burrell M: Iatrogenic problems following gastric surgery. *Gastrointest Radiol* 2:239–257, 1977.
19. Herrington JL, Sawyers JL, Whitehead WA: Surgical management of reflux gastritis. *Ann Surg* 180:526–537, 1974.
20. Burrell M, Touloukian JS, Curtis AM: Roentgen manifestations of carcinoma in the gastric remnant. *Gastrointest Radiol* 5:331–341, 1980.
21. Stark DD, Moss AA, Goldberg HI, et al: CT of pancreatic islet cell tumors. *Radiology* 150:491–494, 1984.
22. Christoforidis AJ, Nelson SW: Radiological manifestations of ulcerogenic tumors of the pancreas. *JAMA* 198:511–516, 1966.
23. Eisenberg RL: *Gastrointestinal Radiology: A Pattern Approach.* Philadelphia, JB Lippincott, 1983.
24. Laufer I, Hamilton J, Mullens JE: Demonstration of superficial gastric erosions by double-contrast radiography. *Gastroenterology* 68:387–391, 1975.
25. Weinberg PE, Levin B: Hyperplasia of Brunner's glands. *Radiology* 84:259–262, 1965.
26. Athanasoulis CA: Upper gastrointestinal bleeding of arteriocapillary origin, in Athanasoulis CA, Pfister RC, Greene RE et al (eds): *Interventive Radiology.* Philadelphia, WB Saunders, 1982.

# Chapter Forty-Five
# Diseases of the Stomach

## CANCER OF THE STOMACH

### Carcinoma

Carcinoma of the stomach has a dismal prognosis because symptoms are rarely noted until the disease is far advanced. The incidence of gastric cancer varies widely throughout the world. It is very high in Japan, Chile, and parts of Eastern Europe, while low in the United States, where for an unknown reason the incidence of the disease has been decreasing. Several conditions appear to predispose persons to the development of carcinoma of the stomach. Atrophic gastric mucosa, as in pernicious anemia, especially when associated with intestinal metaplasia, appears to be a predisposing factor. Adenomatous gastric polyps more than 2 cm in diameter frequently contain carcinoma, although it is unclear whether these uncommon large polyps are malignant from the outset or whether they were originally benign. There is also an increased risk of gastric cancer 10 to 20 years following a Billroth II partial gastrectomy for peptic ulcer disease.

Gastric carcinoma can present a broad spectrum of radiographic appearances. Tumor infiltration of the gastric wall may stimulate a desmoplastic response, which produces diffuse thickening, narrowing, and fixation of the stomach wall (linitis plastica pattern). The involved stomach is contracted into a tubular structure without normal pliability. This scirrhous process usually begins near the pylorus and progresses slowly upward, the fundus being the area least involved. Gastric carcinoma can

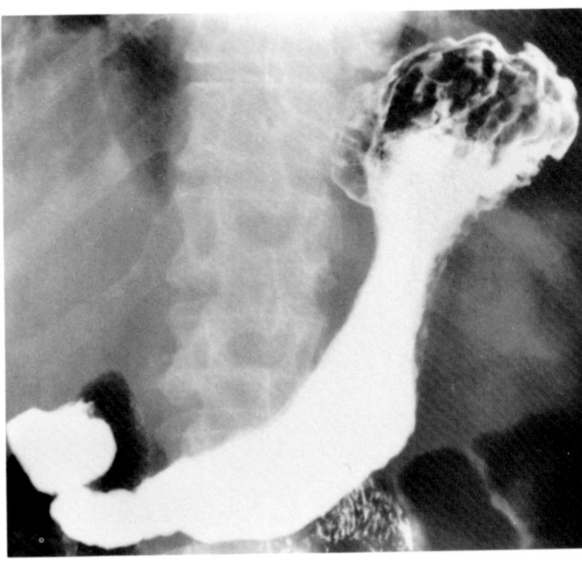

**Figure 45.1. Linitis plastica pattern in carcinoma of the stomach.** Diffuse narrowing and fixation of the stomach wall. (From Eisenberg,[46] with permission from the publisher.)

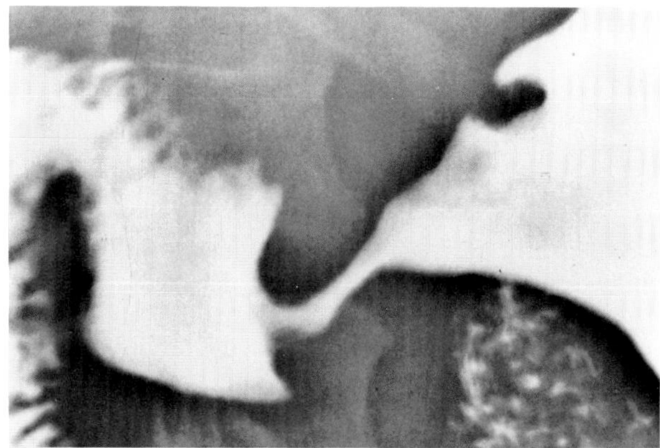

**Figure 45.2. Carcinoma of the stomach.** A localized stricture in the gastric antrum. (From Eisenberg,[46] with permission from the publisher.)

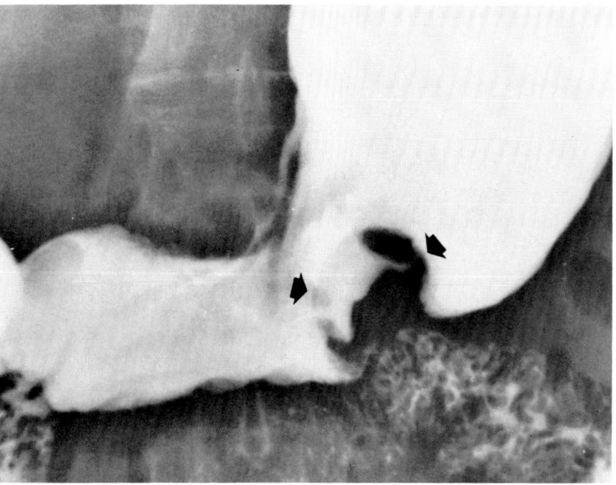

**Figure 45.3. Carcinoma of the stomach.** A huge ulcer is evident in a fungating polypoid mass (arrows). (From Eisenberg,[46] with permission from the publisher.)

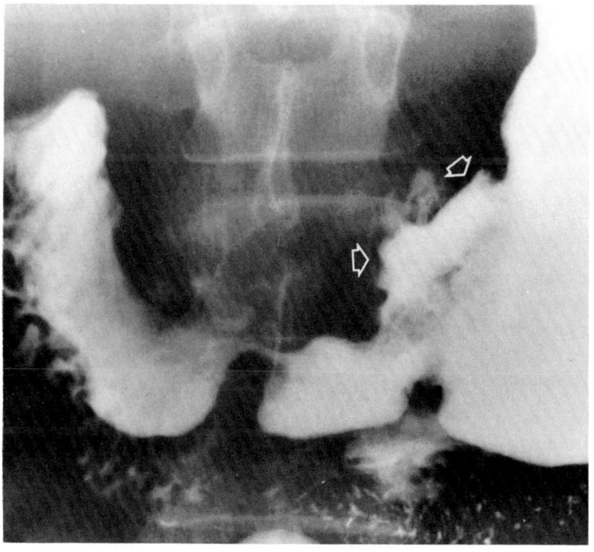

**Figure 45.4. Carcinoma of the stomach.** There is an abrupt transition between the normal mucosa and the abnormal tissue surrounding the irregular malignant gastric ulcer (arrows). (From Eisenberg,[46] with permission from the publisher.)

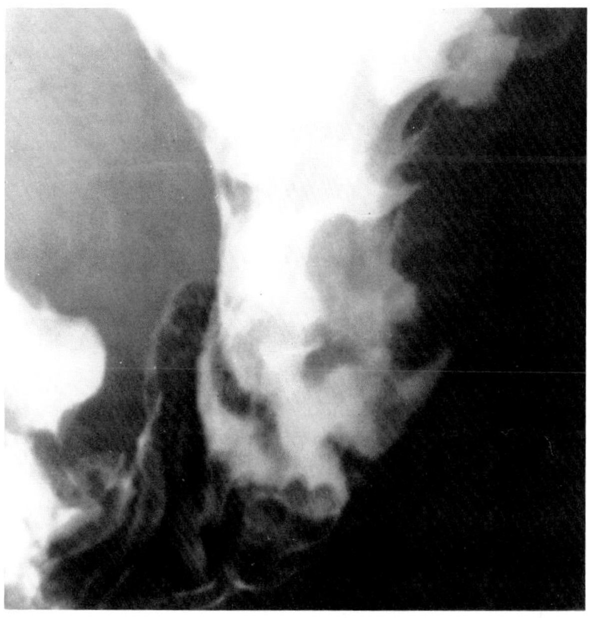

**Figure 45.5. Carcinoma of the stomach.** The pattern of enlarged, tortuous, and coarse rugal folds simulates lymphoma. (From Eisenberg,[46] with permission from the publisher.)

also cause segmental narrowing of the stomach. At an early stage, this may appear as a plaquelike infiltrative lesion along one curvature that progresses to form a constricting lesion similar to that produced by annular carcinoma of the colon.

Polypoid masses larger than 2 cm, particularly sessile ones, are often malignant, though many are benign. Irregularity and ulceration suggest malignancy; a stalk, pliability of the wall of the stomach, normal-appearing gastric folds extending to the tumor, and unimpaired peristalsis are signs of benignancy. Mottled, granular calcific deposits in association with a gastric mass suggest a mucinous adenocarcinoma of the stomach.

Ulceration can develop in any gastric carcinoma. The radiographic appearance of malignant ulceration runs the gamut from shallow erosions in relatively superficial mucosal lesions to huge excavations within fungating polypoid masses. Signs of malignant ulcer include a lack of penetration beyond the normal gastric lumen; an abrupt transition between the nodular surrounding neoplastic tissue and the normal mucosa; and adjacent infiltration, rigidity, and mucosal destruction. Although many malignant ulcers show significant healing in response to therapy, there is almost never complete dis-

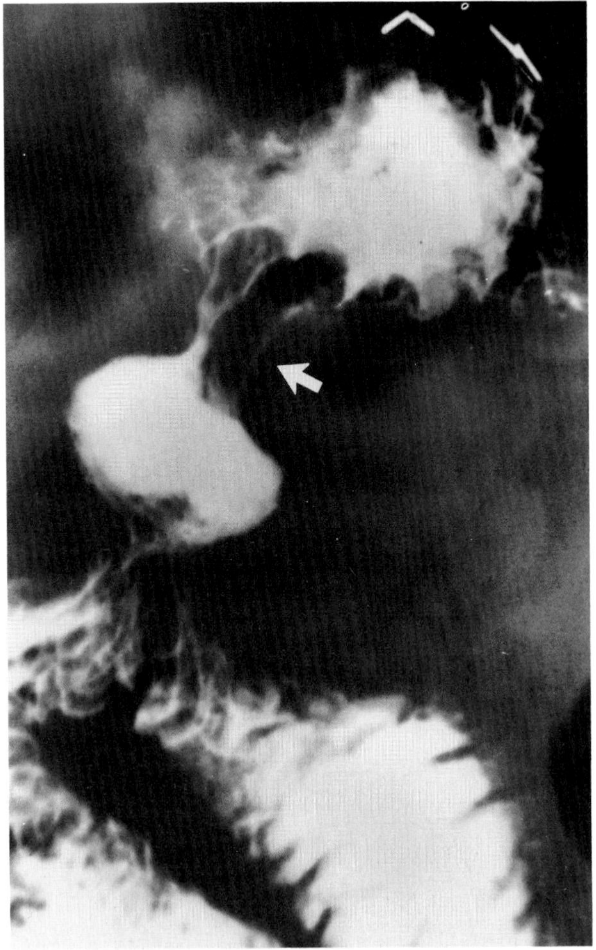

**Figure 45.6. Gastric stump carcinoma.** Tumor infiltration causes narrowing of the lumen (arrow). (From Eisenberg,[46] with permission from the publisher.)

appearance of the ulcer crater. Endoscopy is indicated if there are radiographic findings suspicious of malignancy or if the ulcer does not heal at the expected rate.

Gastric carcinoma infrequently presents a radiographic pattern of enlarged, tortuous, and coarse gastric folds simulating lymphoma. Unlike most cases of diffuse infiltrating adenocarcinoma, this form of the disease shows preservation of a relatively normal gastric volume, pliability, and peristaltic activity.

In the fundus, carcinoma frequently extends proximally to involve the distal esophagus, producing a radiographic appearance that mimics achalasia.

Following surgery, the recurrence of a gastric carcinoma can cause a defect in the gastric remnant, infiltration of the wall with straightening and loss of normal distensibility, or mucosal destruction with superficial ulceration. The major sign of recurrence at the anastomosis is symmetric or eccentric narrowing with local mucosal effacement.

Computed tomography (CT) is of major value in the staging and treatment planning of gastric carcinoma as well as in assessing the response to therapy and in detecting recurrence. Carcinoma of the stomach may appear as concentric or focal thickening of the gastric wall or as an intraluminal mass. Obliteration of the perigastric fat planes is a reliable indicator of the extragastric spread of tumor. CT can demonstrate direct tumor extension to intra-abdominal organs and distant metastases to the liver, ovary, adrenals, kidney, and peritoneum.[1–6]

## Lymphoma

As elsewhere in the bowel, gastric lymphoma is a great imitator of both benign and malignant disease. One manifestation of lymphoma of the stomach is a large, bulky

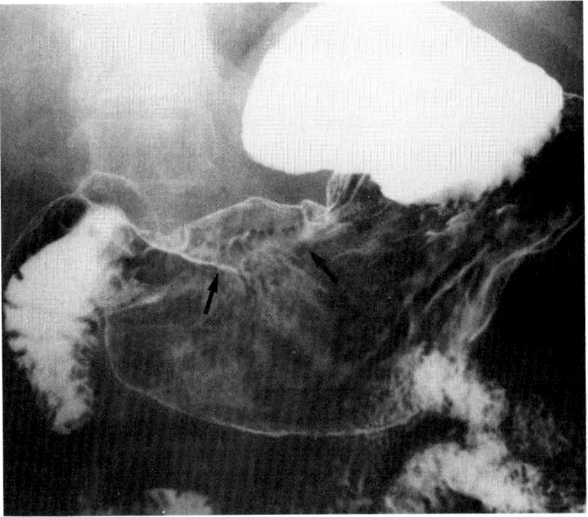

A

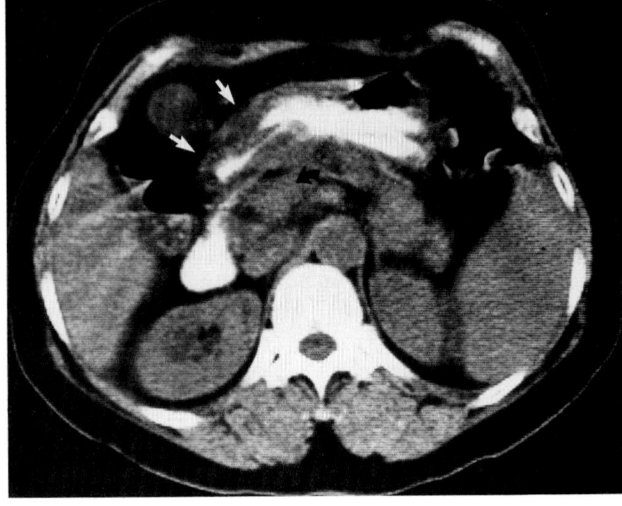

B

**Figure 45.7. CT staging of gastric carcinoma.** (A) A double-contrast study demonstrates a large lesser curvature mass (arrows). (B) A CT scan shows narrowing of the antrum by the gastric carcinoma (white arrows) and adjacent nodal metastases (curved arrow).

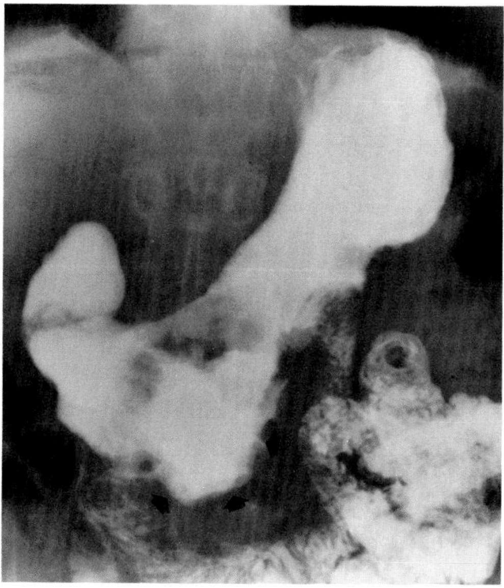

**Figure 45.8. Lymphoma of the stomach.** A huge, irregular ulcer (arrows) is visible in a neoplastic gastric mass. (From Eisenberg,[46] with permission from the publisher.)

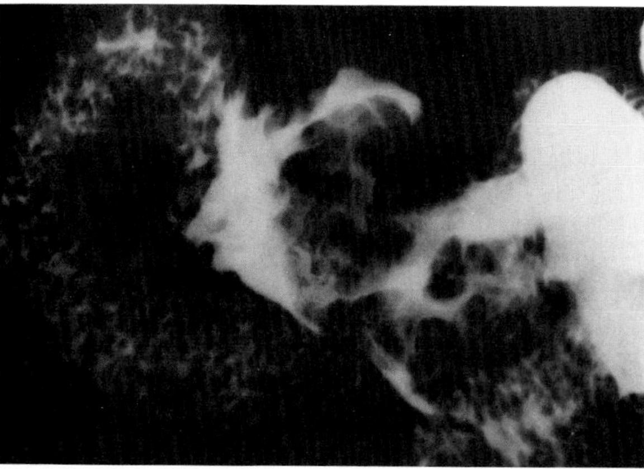

**Figure 45.9. Lymphoma of the stomach.** Multiple polypoid filling defects with generalized thickening of folds involve the antrum and duodenal bulb. (From Eisenberg,[46] with permission from the publisher.)

polypoid lesion, usually irregular and ulcerated, that can be difficult to differentiate from gastric carcinoma. These polyps can be combined with thickened folds (infiltrative form of lymphoma) or separated by a normal-appearing mucosal pattern, unlike the atrophic mucosal background that is seen with multiple carcinomatous polyps in patients with pernicious anemia. A multiplicity of ulcerated masses suggests lymphoma, as does relative flexibility of the gastric wall.

Thickening, distortion, and nodularity of gastric rugal folds simulating Ménétrier's disease is another pattern of lymphoma of the stomach. If the enlarged rugal folds predominantly involve the distal portion of the stomach and the lesser curvature, or if there is some loss of pliability of the gastric wall, lymphoma is more likely. However, if the process stops at the incisura and spares the lesser curvature, if there is no ulceration or true rigidity, or if excess mucus can be demonstrated, Ménétrier's disease is the probable diagnosis.

Invasion of the gastric wall by an infiltrative type of lymphoma can cause a severe desmoplastic reaction and a radiographic pattern that mimics the linitis plastica appearance of scirrhous carcinoma. Unlike the rigidity and fixation of scirrhous carcinoma, residual peristalsis and flexibility of the stomach wall are often preserved in lymphoma.

Enlargement of the spleen and an extrinsic impression on the somach by retrogastric and other regional lymph nodes suggest lymphoma as the underlying disorder.

On CT, gastric lymphoma tends to produce bulky masses and a lobulated inner contour of the gastric wall representing thickened gastric rugae. However, gastric lymphoma may also produce smooth, concentric wall thickening or a focal mass simulating adenocarcinoma of the stomach. The demonstration of other signs of lymphoma (splenomegaly, diffuse retroperitoneal and mesenteric lymphadenopathy), when present, suggests the correct histologic diagnosis.[7–12]

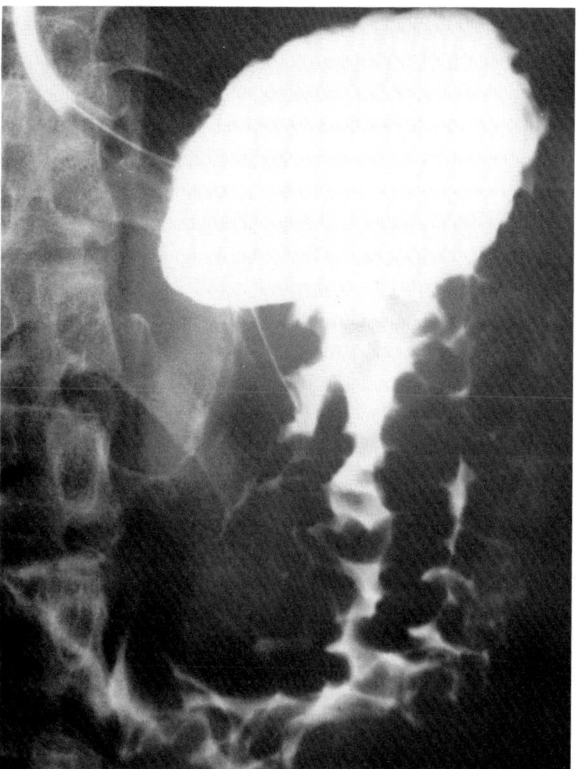

**Figure 45.10. Lymphoma of the stomach.** There is diffuse thickening, distortion, and nodularity of gastric folds. (From Eisenberg,[46] with permission from the publisher.)

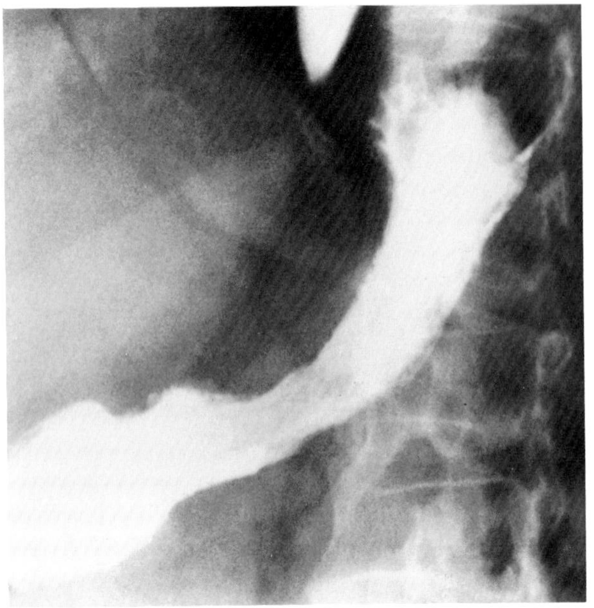

Figure 45.11. Lymphoma of the stomach. A severe desmoplastic reaction produces the linitis plastica pattern mimicking scirrhous carcinoma. (From Eisenberg,[46] with permission from the publisher.)

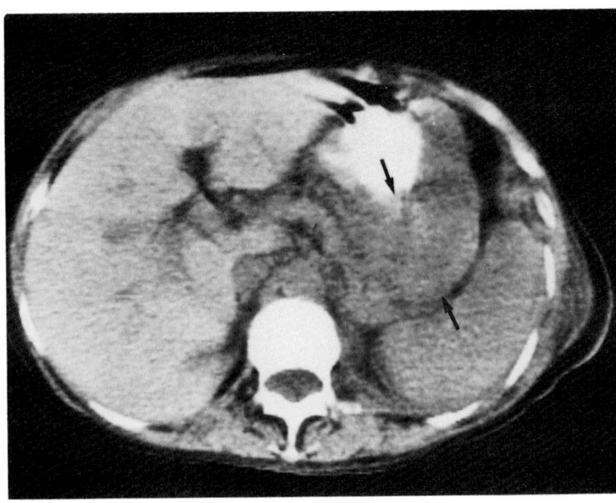

Figure 45.12. Lymphoma of the stomach. CT demonstrates diffuse thickening of the gastric wall (arrows).

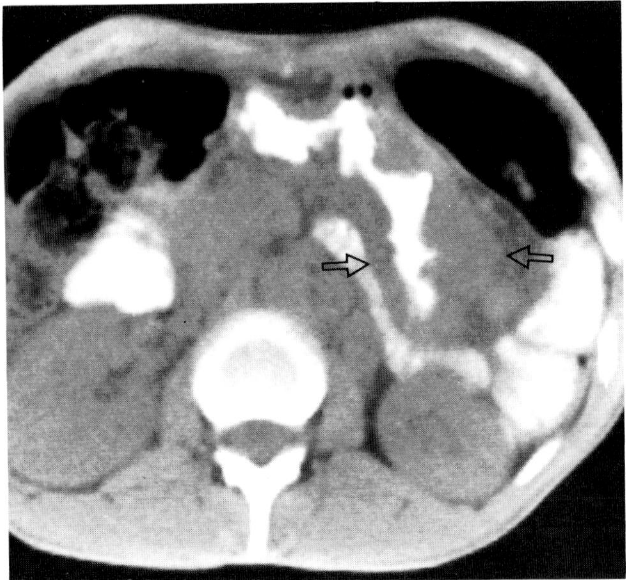

A

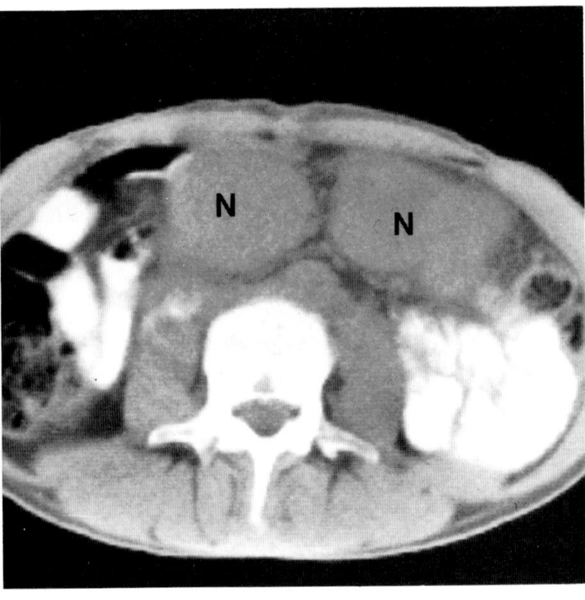


B

Figure 45.13. Lymphoma of the stomach. (A) A CT scan shows a circumferential intramural mass (arrows) causing gross distortion of the contrast-filled gastric lumen. (B) The presence of large mesenteric (N) and periaortic nodes suggests the correct histologic diagnosis. (From Mauro and Koehler,[10] with permission from the publisher.)

## Leiomyosarcoma

Gastric leiomyosarcomas are large, bulky tumors most often found in the body of the stomach. Although originally arising in an intramural location, leiomyosarcomas often appear as intraluminal, occasionally pedunculated masses. They frequently undergo extensive central necrosis, causing ulceration and gastrointestinal bleeding. Extensive spread into surrounding tissues is common (as are metastases to the liver, omentum, and retroperito-

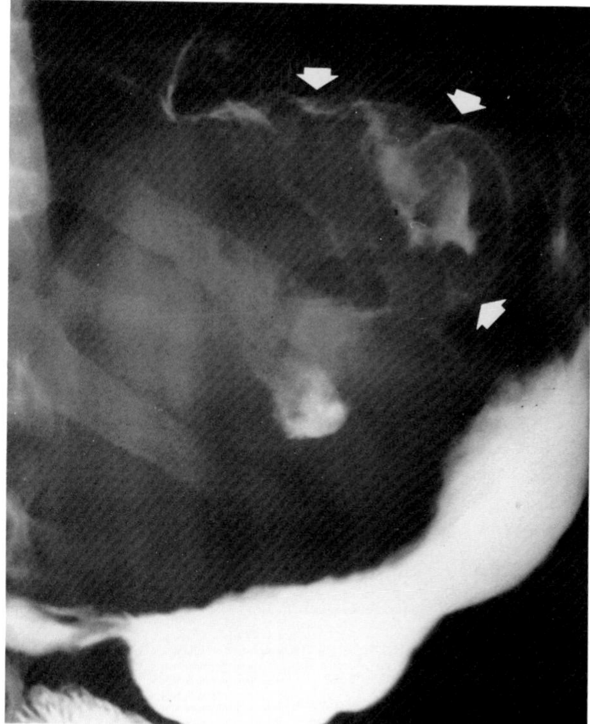

Figure 45.14. **Leiomyosarcoma of the stomach.** The large fundal mass (arrows) shows marked exophytic extension and ulceration. (From Eisenberg,[46] with permission from the publisher.)

neum), and the resulting large exogastric component may suggest an extrinsic lesion. It is frequently impossible to radiographically differentiate leiomyosarcoma from benign leiomyoma, though the presence of a large exogastric mass suggests the presence of malignancy.

CT may demonstrate either the primary intragastric lesion or the large extraluminal component of a leiomyosarcoma. Characteristic findings in tumors of this histologic type are small foci of calcification and well-defined, low-density areas within the mass representing either areas of necrosis and liquefaction or a cystic component to the tumor. Unlike gastric adenocarcinoma or lymphoma, gastric leiomyosarcoma commonly metastasizes to the liver and lung, while spread to regional lymph nodes is unusual.[4,13]

## BENIGN TUMORS OF THE STOMACH

### Epithelial Polyps

With the increased use of double-contrast techniques, gastric polyps are being detected more frequently. In one series, polyps were demonstrated in 1.6 percent of gastrointestinal examinations. Although most gastric polyps are asymptomatic and discovered as incidental findings, some can bleed and produce hematemesis or melena. The vast majority of epithelial polyps of the stomach can be divided into two groups: hyperplastic (regenerative)

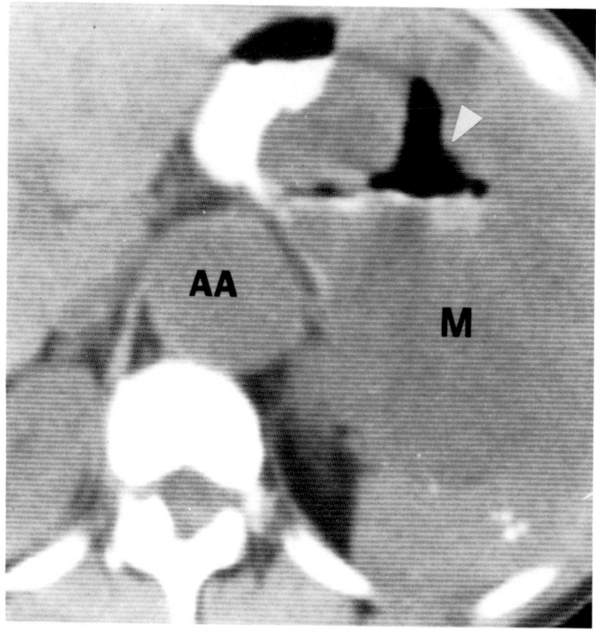


A

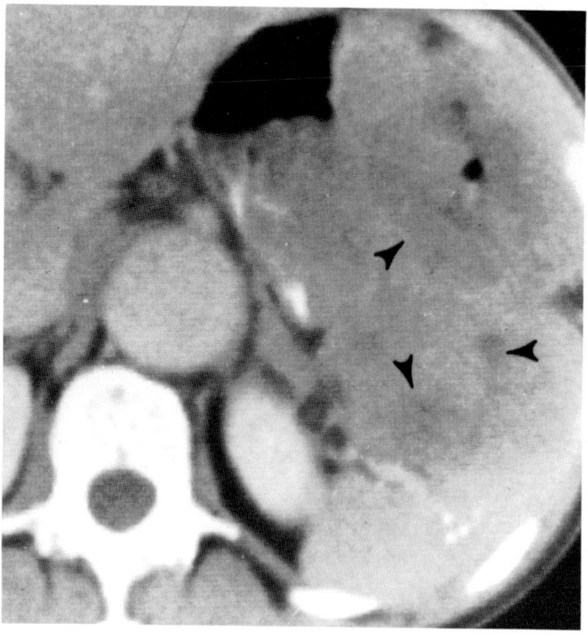

B

Figure 45.15. **Leiomyosarcoma of the stomach.** (A) A CT scan shows a large mass (M) that is primarily extragastric. There is contrast material in the distorted gastric lumen as well as in an excavation (arrowhead) within the tumor. An abdominal aortic aneurysm (AA) is also present. (B) Following the administration of intravenous contrast material, low-density areas (arrowheads) characteristic of leiomyosarcoma are evident within the mass. (From Mauro and Koehler,[10] with permission from the publisher.)

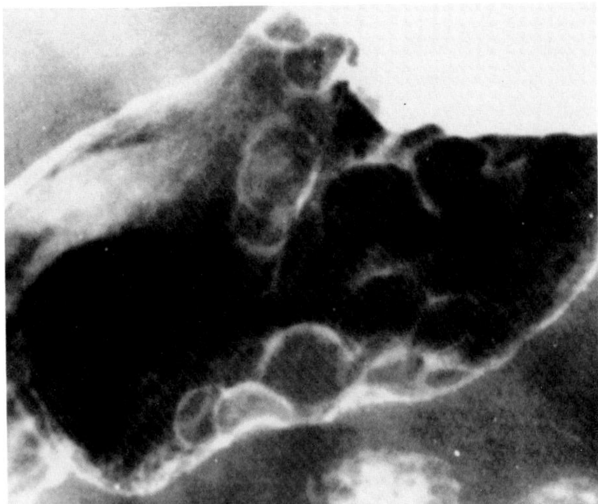

**Figure 45.16. Hyperplastic polyps.** There are multiple smooth filling defects of similar size. (From Eisenberg,[46] with permission from the publisher.)

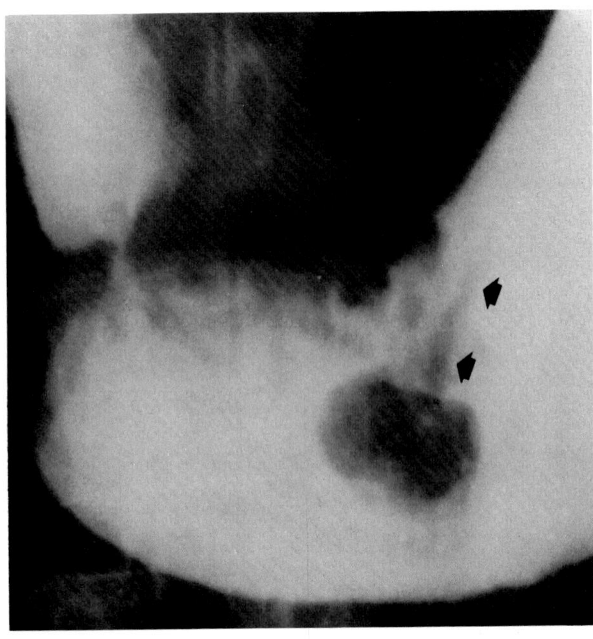

**Figure 45.17. Adenomatous polyp.** A long, thin pedicle (arrows) extends from the head of the polyp to the stomach wall. (From Eisenberg,[46] with permission from the publisher.)

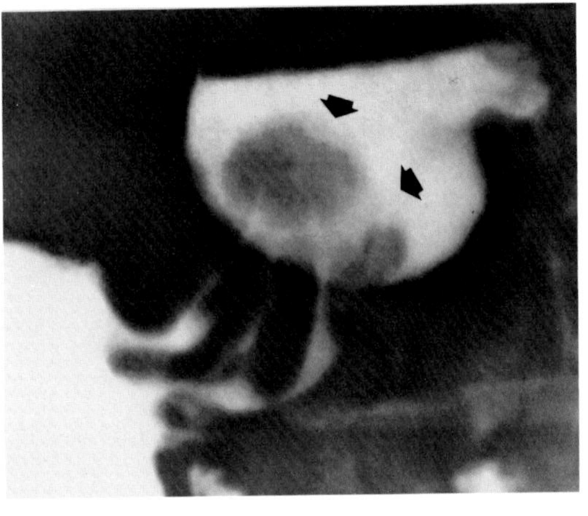

A

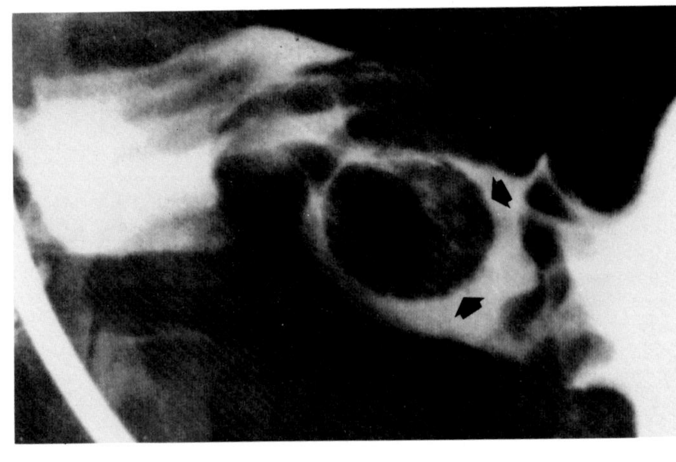

B

**Figure 45.18. Prolapsing antral polyp.** (A) A prolapsed polyp appears as a solitary filling defect (arrows) in the base of the duodenal bulb. (B) With reduction of the prolapse, the true origin of the polyp within the antrum becomes evident (arrows). (From Eisenberg,[46] with permission from the publisher.)

polyps and adenomas. Although there is some overlap, these two types of polyps can usually be distinguished by their radiographic appearance.

Hyperplastic polyps are the most common causes of discrete filling defects in the stomach. Not true neoplasms, these polyps are the result of excessive regenerative hyperplasia in an area of chronic gastritis. They are typically asymptomatic, small (average, 1 cm), often multiple, and randomly distributed throughout the stomach. The sharply defined filling defects have smooth contours and no evidence of contrast material within them. Although malignant transformation virtually never occurs, hyperplastic polyps can be associated with an independent, coexisting carcinoma elsewhere in the stomach.

Adenomas are true neoplasms that are composed of dysplastic glands and are capable of continued growth. They have a definite tendency toward malignant transformation, and the reported incidence of this complication increases with the size of the polyp (average, about 40 percent). Most adenomas are large (greater than 2 cm), sessile or pedunculated lesions with a lobulated

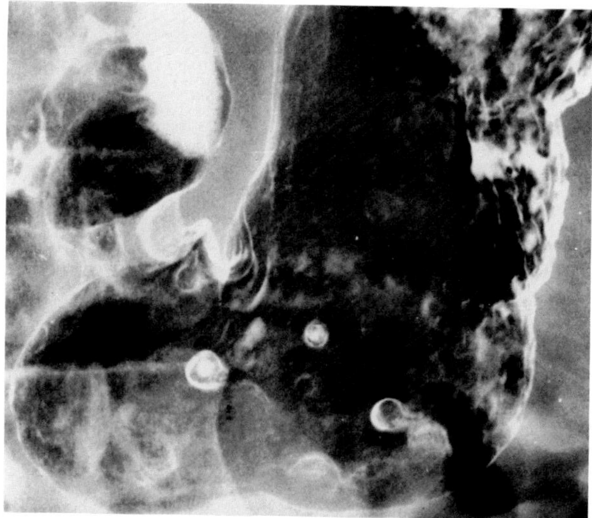

**Figure 45.19. Peutz-Jeghers syndrome.** Multiple hamartomas of the stomach are seen in this patient with small bowel polyps and mucocutaneous pigmentation. (From Eisenberg,[46] with permission from the publisher.)

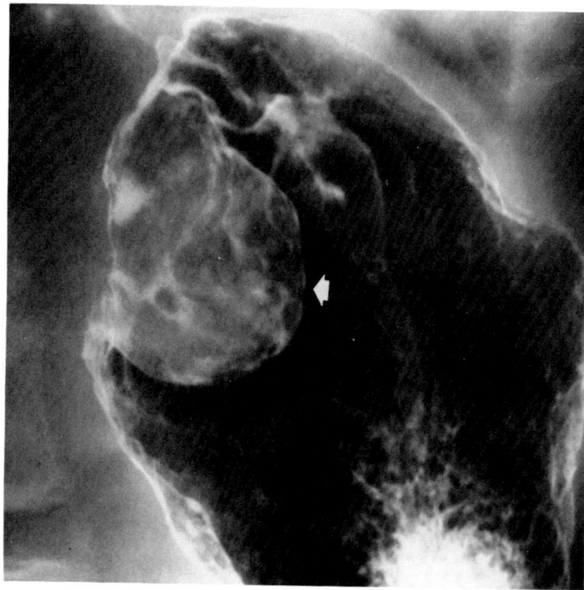

**Figure 45.20. Leiomyoma of the fundus (arrow).** (From Eisenberg,[46] with permission from the publisher.)

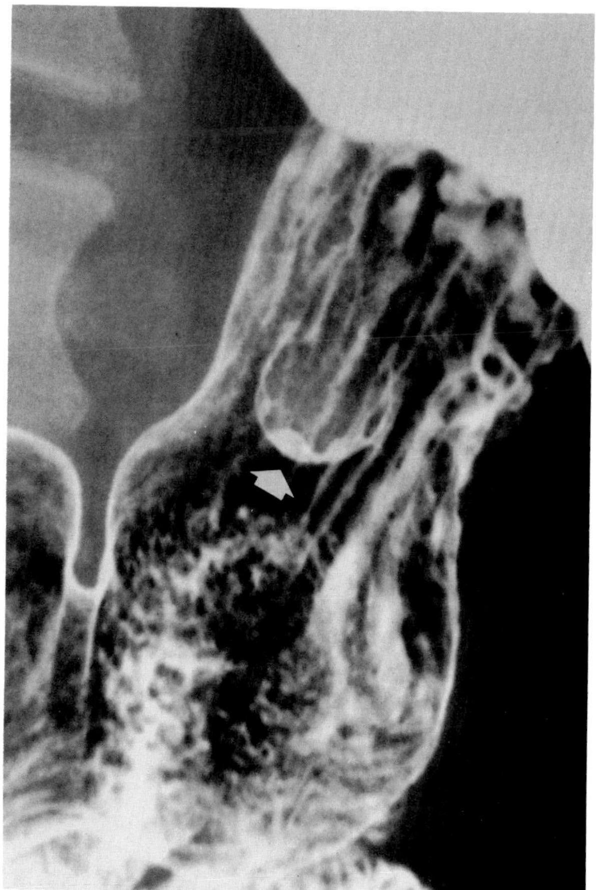

**Figure 45.21. Lipoma of the stomach.** There is a smooth polypoid mass in the body of the stomach (arrow). (From Eisenberg,[46] with permission from the publisher.)

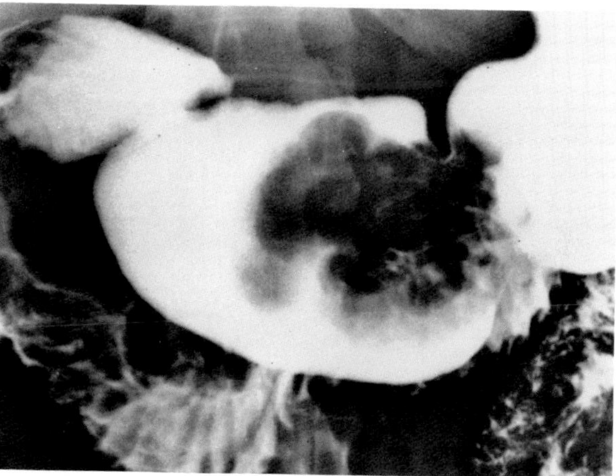

**Figure 45.22. Leiomyoblastoma of the stomach.** The antral mass is huge and irregular. (From Eisenberg,[46] with permission from the publisher.)

surface. Contrast material entering deep fissures and furrows in the polyp tends to produce a papillary or villous appearance. Adenomatous polyps are usually single and are most commonly found in the antrum. A large, pedunculated antral polyp can prolapse into the duodenum and cause obstruction.

Both hyperplastic and adenomatous polyps tend to develop in patients with chronic atrophic gastritis, a condition known to be associated with a high incidence of carcinoma. Thus even though a gastric polyp is proved to be benign, the entire stomach must be carefully examined for the possibility of a coexisting carcinoma.

There is a higher-than-normal incidence of adenomatous and hyperplastic gastric polyps in patients with familial polyposis of the colon or the Cronkhite-Canada syndrome. Multiple hamartomatous gastric polyps, with essentially no malignant potential, may be seen in patients with Peutz-Jeghers syndrome, Cowden's disease, or juvenile polyposis.[14–17]

### Leiomyoma and Rare Benign Tumors

Spindle-cell tumors constitute the overwhelming majority of benign submucosal gastric neoplasms. These lesions vary in size from tiny nodules, often discovered incidentally at laparotomy or autopsy, to bulky tumors with large intraluminal components that can be associated with hemorrhage, obstruction, or perforation. Exogastric extension of these tumors can mimic extrinsic compression of the stomach by normal or enlarged liver, spleen, pancreas, or kidney. It may be extremely difficult to distinguish radiographically between benign spindle-cell tumors and their malignant counterparts. Although large, markedly irregular filling defects with prominent ulcerations suggest malignancy, a radiographically benign tumor can be histologically malignant.

Leiomyoma is the most common spindle-cell tumor of the stomach. Usually single rather than multiple, leiomyomas may appear as small, rounded filling defects simulating sessile polyps or as large lesions predominantly located in intraluminal, intramural, or extramural locations. Because of their tendency toward central necrosis and ulceration, bleeding is common as the tumor grows.

Lipomas, fibromas, hemangiomas, leiomyoblastomas (granular cell tumors), and neurogenic tumors, all of which can be radiographically indistinguishable from leiomyomas, are far less common submucosal gastric neoplasms.[18,19]

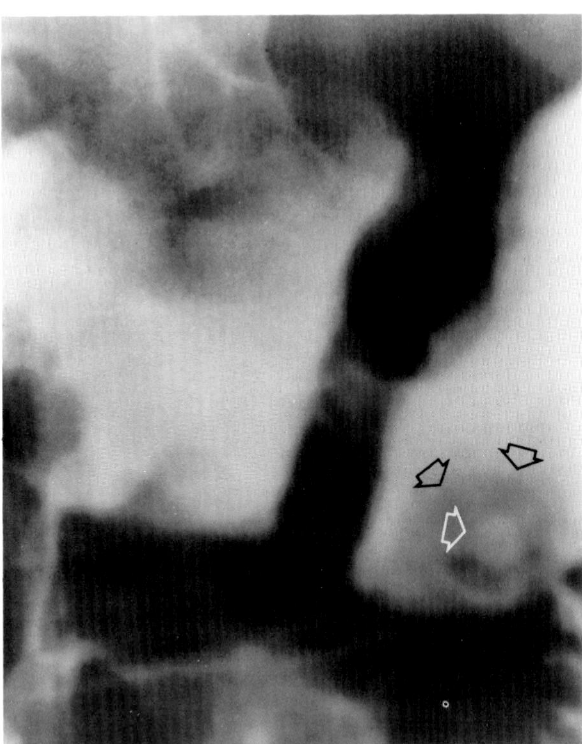

**Figure 45.23. Heterotopic (aberrant) pancreas.** Umbilication of a rudimentary pancreatic duct, rather than necrotic ulceration, causes central opacification (white arrow) within the soft tissue mass (black arrows) in the distal antrum, producing a bull's-eye pattern. (From Eisenberg,[46] with permission from the publisher.)

## PSEUDOTUMORS

### Heterotopic (Aberrant) Pancreas

Aberrant pancreatic tissue (heterotopic pancreas) is a non-neoplastic embryologic anomaly that can be found in many areas of the gastrointestinal tract but is most common on the distal greater curvature of the gastric antrum within 3 to 6 cm of the pylorus. Heterotopic pancreas is usually asymptomatic, though some patients may complain of vague abdominal pain, nausea, and occasional vomiting; bleeding can occur if the overlying mucosa becomes ulcerated. Radiographically, heterotopic pancreas appears as a smooth submucosal mass, rarely more than 2 cm in diameter. A characteristic radiographic sign is a central collection of barium (dimple) within the mass, which simulates an ulcer but actually represents filling of miniature ductlike structures present in the nodule of aberrant pancreatic tissue.[20]

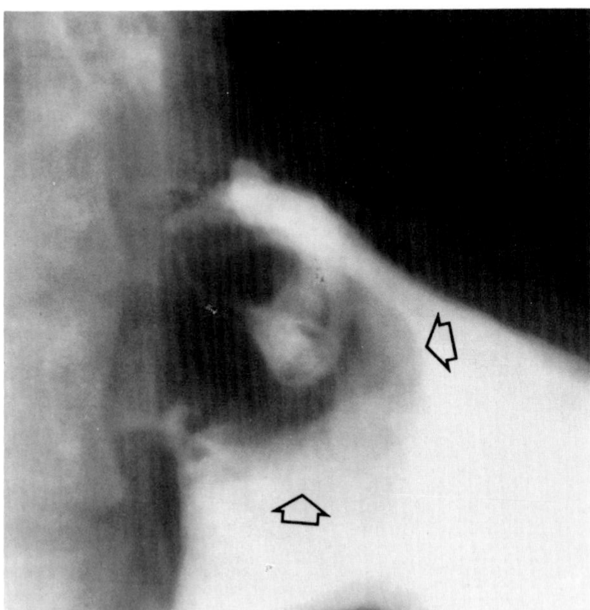

**Figure 45.24. Fundal mass caused by Nissen fundoplication.** The distal esophagus appears to pass through the center of the concave mass (arrows). (From Eisenberg,[46] with permission from the publisher.)

## Fundoplication Deformity (Nissen Repair)

Nissen fundoplication can produce a pseudotumor in the gastric fundus. This method of hiatal hernia repair involves wrapping the gastric fundus around the lower esophagus to create an intra-abdominal esophageal segment with a natural valve mechanism at the esophagogastric junction. The surgical procedure characteristically results in a prominent filling defect at the esophagogastric junction that is generally smoothly marginated and symmetric on both sides of the distal esophagus. The demonstration of a preserved lumen and mucosal pattern of the distal esophagus as it appears to pass through the center of the concave mass (pseudotumor) excludes the possibility of a neoplastic process.[21–23]

## Gastric Varices

Gastric varices are usually associated with esophageal varices and represent dilated peripheral branches of the left gastric and short gastric veins due to portal hypertension. The presence of gastric varices without esophageal varices is a very specific sign of isolated splenic vein occlusion, most commonly secondary to pancreatitis or pancreatic carcinoma. Gastric varices most commonly appear in the fundus as thickened, tortuous mucosal folds or multiple smooth, lobulated filling defects projecting between curvilinear, crescentic collections of barium. Unlike neoplastic processes, gastric varices demonstrate considerable alteration in size and shape when the patient changes position and phase of respiration.

Although unusual, gastric varices may also occur in the antrum and body of the stomach.[24,25]

## Prolapsed Antral Mucosa

Redundant mucosa of the gastric antrum can prolapse through the pylorus under the influence of active peri-

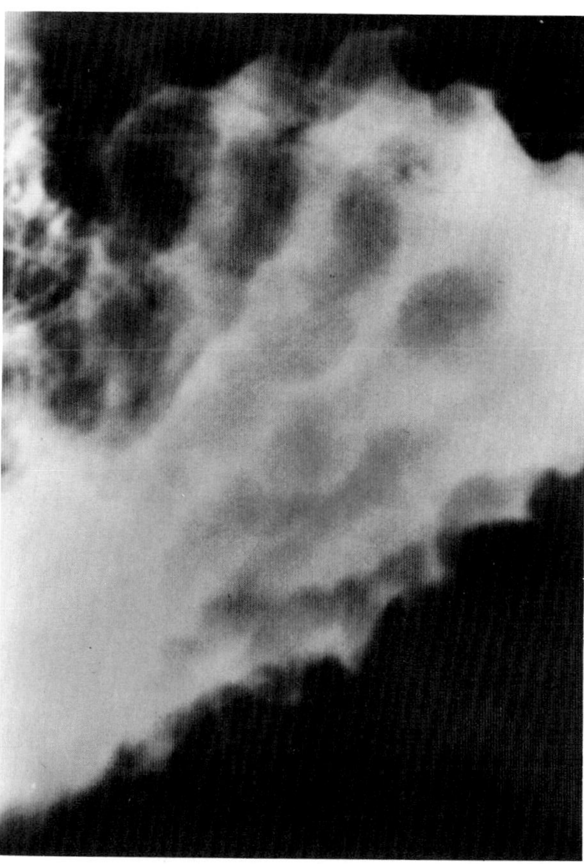

**Figure 45.25. Fundal gastric varices.** Multiple smooth, lobulated filling defects represent the dilated venous structures. (From Eisenberg,[46] with permission from the publisher.)

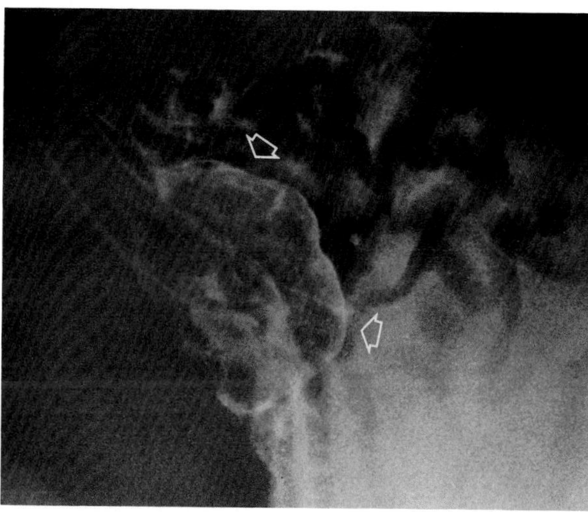

**Figure 45.26. Gastric varix.** A single fundal mass (arrows) in the region of the esophagogastric junction simulates a neoplastic process. (From Eisenberg,[46] with permission from the publisher.)

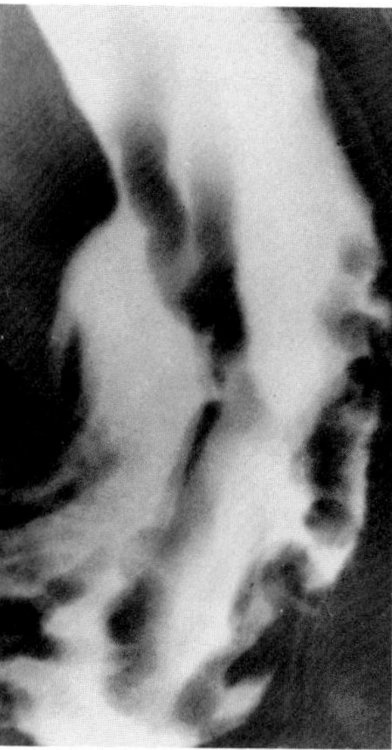

**Figure 45.27. Nonfundal gastric varices.** Varices appear as lobulated filling defects or giant folds in the body of the stomach, particularly along the greater curvature. There was no evidence of fundal varices or splenomegaly. (From Sos, Meyers, and Baltaxe,[25] with permission from the publisher.)

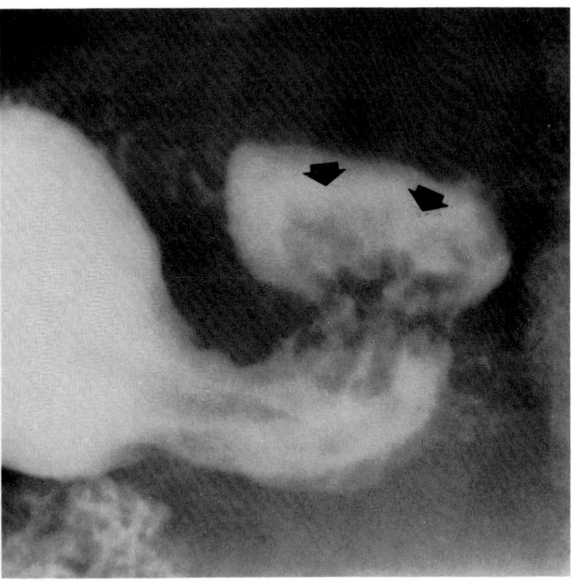

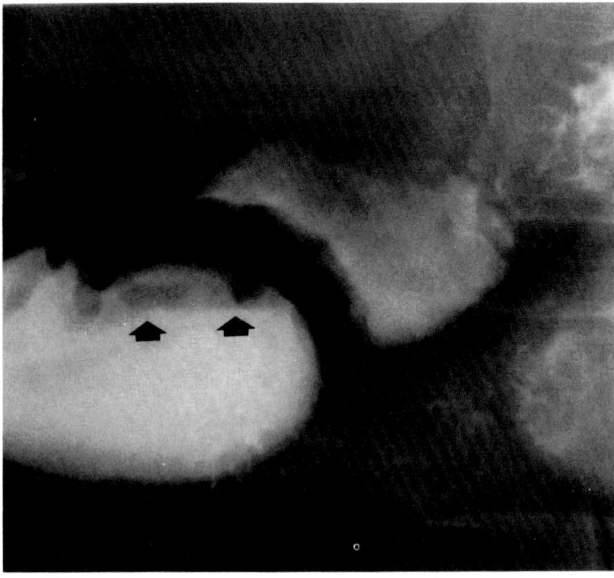

A B

**Figure 45.28. Gastric antral folds prolapsed through the pylorus.** (A) The appearance of a mass (arrows) in the duodenal bulb. (B) With reduction of the prolapse, the mass in the base of the bulb disappears and the redundant antral folds become evident (arrows). (From Eisenberg,[46] with permission from the publisher.)

stalsis, resulting in single or lobulated filling defects at the base of the duodenal bulb. Mucosal folds in the prepyloric area of the stomach can usually be traced through the pylorus to the base of the bulb, where they become continuous with the characteristic mushroom-, umbrella-, or cauliflower-shaped prolapsed mass. Under fluoroscopy, mucosal prolapse can be detected as a gastric peristaltic wave passes through the antrum. As the wave relaxes, the mucosal folds tend to return into the antrum, and the defect in the base of the bulb diminishes or completely disappears. Some degree of mucosal prolapse is frequently observed during gastrointestinal examinations. Antral prolapse is generally considered to be asymptomatic, though associated ulceration and bleeding have occasionally been reported.

## GASTRITIS

### Acute Erosive Gastritis

Acute erosive gastritis (multiple gastric erosions) can be caused by such agents as alcohol and anti-inflammatory drugs (aspirin, steroids, phenylbutazone, indomethacin). Although the erosions in this condition are too superficial to be detected on standard upper gastrointestinal series, they may be demonstrated by double-contrast techniques. A superficial gastric erosion typically appears as a tiny fleck of barium surrounded by a radiolucent halo, which represents a mound of edematous mucosa. The resultant target lesions are usually multiple, though a solitary erosion is occasionally demonstrated.[46]

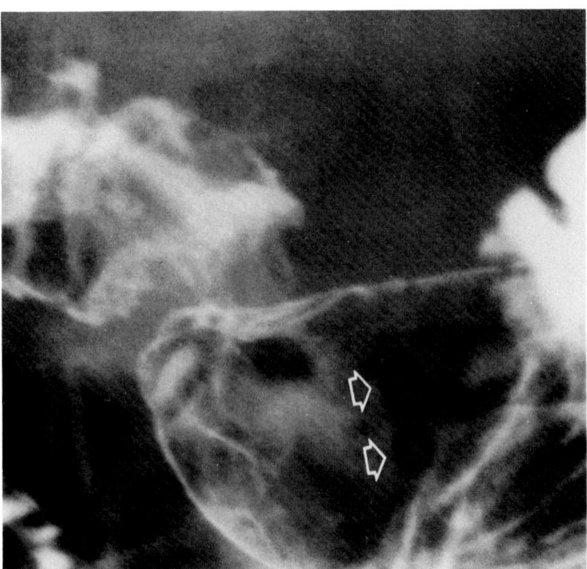

**Figure 45.29. Acute erosive gastritis.** Superficial gastric erosions appear as tiny flecks of barium surrounded by radiolucent halos of edematous mucosa.

### Chronic Atrophic Gastritis

In chronic atrophic gastritis, severe mucosal atrophy may cause thinning and a relative absence of mucosal folds, with the fundus or entire stomach having a bald appearance. This is a nonspecific radiographic pattern that can be related to such factors as age, malnutrition, med-

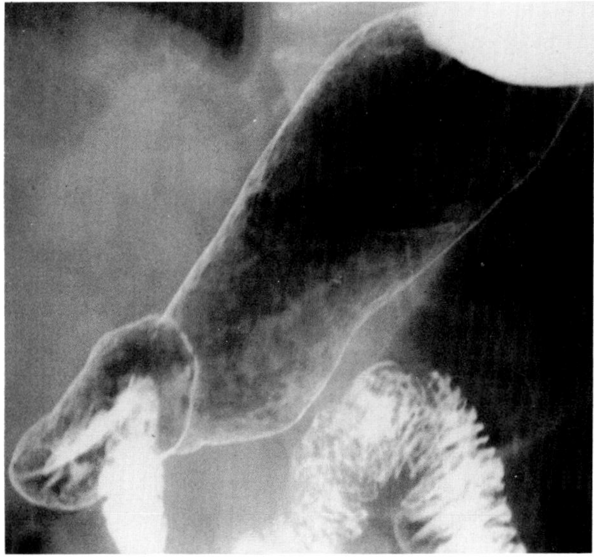

**Figure 45.30. Chronic atrophic gastritis.** There is a relative absence of folds in this patient with a long drinking history. (From Eisenberg,[46] with permission from the publisher.)

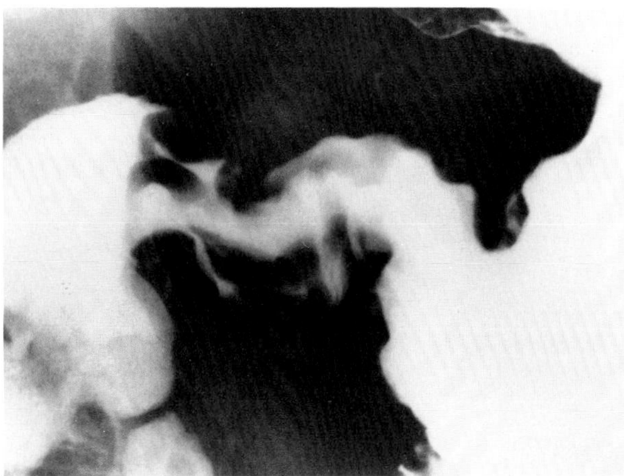

**Figure 45.31. Antral gastritis.** Thickening of gastric rugal folds is confined to the antrum. (From Eisenberg,[46] with permission from the publisher.)

ication, and complications of alcoholism; it also occurs in patients with pernicious anemia. In the elderly, an atrophic stomach may be hypotonic and show limited distensibility, simulating scirrhous carcinoma.[46]

## Antral Gastritis

Thickening of mucosal folds localized to the antrum is a controversial entity that most likely reflects one end of the spectrum of peptic ulcer disease. Isolated antral gastritis appears without fold thickening or acute ulceration in the duodenal bulb. The term *antral gastritis* is actually a misnomer, since in most cases there is gastroscopic evidence of disease elsewhere in the stomach that is not radiographically detectable. In addition to fold thickening, antral gastritis can refer to transient antral spasm and a lack of normal distension that may simulate neoplastic disease. The radiographic findings of antral gastritis may persist even when the patient is clinically well but tend to increase when the typically ulcer-like symptoms occur.[47]

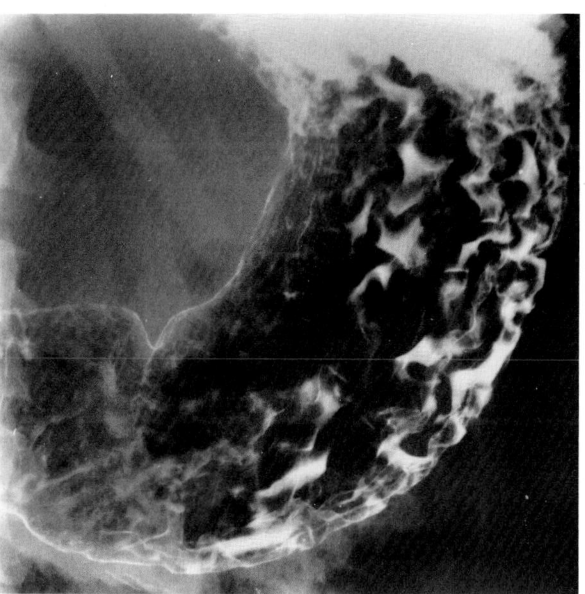

**Figure 45.32. Gastric hyperplasia.** Diffuse thickening of gastric rugal folds. (From Eisenberg,[46] with permission from the publisher.)

## Gastric Hyperplasia (Hypertrophic Gastritis)

Although often termed *hypertrophic gastritis*, this condition is characterized by thickening of the mucosa that reflects an increased number (hyperplasia) rather than enlargement (hypertrophy) of individual mucosal epithelial cells. This controversial entity appears to be related to chronic inflammation of the gastric mucosa that may represent the upper end of the spectrum of increased fundal gland mass seen in patients with peptic ulcer disease. Gastric hyperplasia appears radiographically as thickening of gastric folds, especially in the fundus and proximal half of the stomach. Although this pattern may simulate neoplastic disease, the thickened folds of gastric hyperplasia can usually be obliterated by compression, in contrast to folds infiltrated by neoplasm, which cannot be obliterated by compression or by gastric distension.[47]

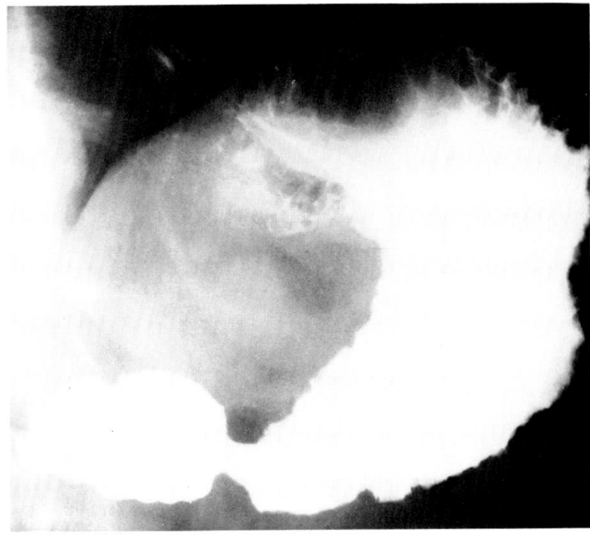


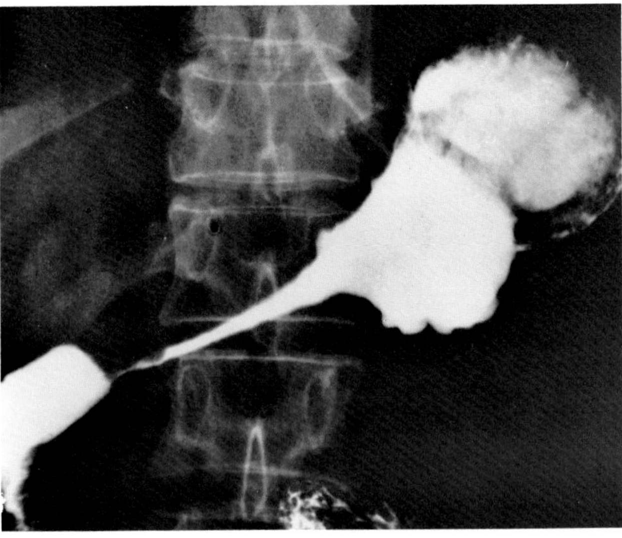


**Figure 45.33. Corrosive gastritis.** (A) Diffuse ulceration involving the body and antrum of the stomach is due to the ingestion of hydrochloric acid. (B) Several weeks later, there is a severe corrosive stricture of the antrum. (From Eisenberg,[46] with permission from the publisher.)

## Bile Reflux Gastritis

This is discussed in the section "Bile Reflux Gastritis" in Chapter 44.

## Corrosive Gastritis

The ingestion of corrosive agents results in a severe form of acute gastritis characterized by intense mucosal edema and inflammation. Radiographically, thickened gastric folds are associated with mucosal ulcerations, atony, and rigidity. A fixed, open pylorus is usually seen, probably due to extensive damage to the muscular layer. The presence of gas in the wall of the stomach after the ingestion of corrosive agents is an ominous sign; free gastric perforation may occur.

Corrosive gastritis predominantly involves the antrum, though the entire stomach may be involved. Unlike their effects on the esophagus, ingested acids generally produce more severe gastric damage than ingested alkalies.

The acute inflammatory reaction of corrosive gastritis heals by fibrosis and scarring, which results in stricturing of the antrum within several weeks of the initial injury. In patients who have rigidity and narrowing of the stomach without a history of corrosive ingestion, the clinical symptoms of weight loss and early satiety, combined with the radiographic pattern of narrowing of the stomach (linitis plastica), can be impossible to distinguish from gastric malignancy.[26–28]

## Phlegmonous Gastritis

Phlegmonous gastritis is an extremely rare condition in which bacterial invasion causes thickening of the wall of the stomach associated with a discolored mucosa and an edematous submucosa. Most cases of phlegmonous gastritis are due to alpha-hemolytic streptococci, though pneumococci, staphylococci, *Escherichia coli*, or gas-forming bacteria can also be the causative organisms. Phlegmonous gastritis usually produces gastric narrowing (linitis plastica) due to diffuse thickening of the stomach wall, a pattern often indistinguishable from infiltrating carcinoma. Radiographic differentiation between

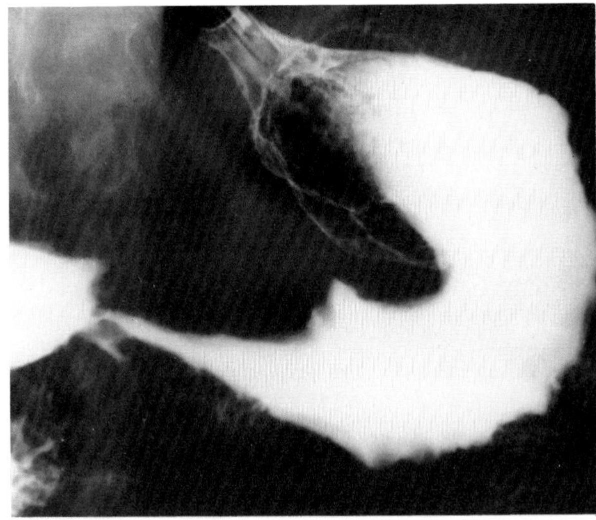

**Figure 45.34. Phlegmonous gastritis.** There is irregular narrowing of the antrum and distal body of the stomach with effacement of mucosal folds along the lesser curvature and marked thickening of folds along the greater curvature. (From Turner and Beachley,[28] with permission from the publisher.)

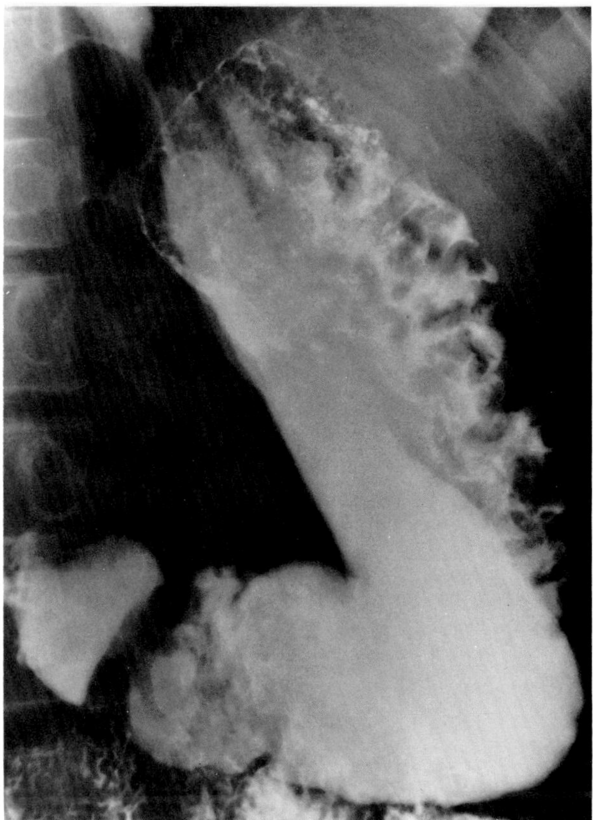

**Figure 45.35. Ménétrier's disease.** Fold thickening involves the greater curvature of the fundus and body and spares the lesser curvature and antrum. (From Eisenberg,[46] with permission from the publisher.)

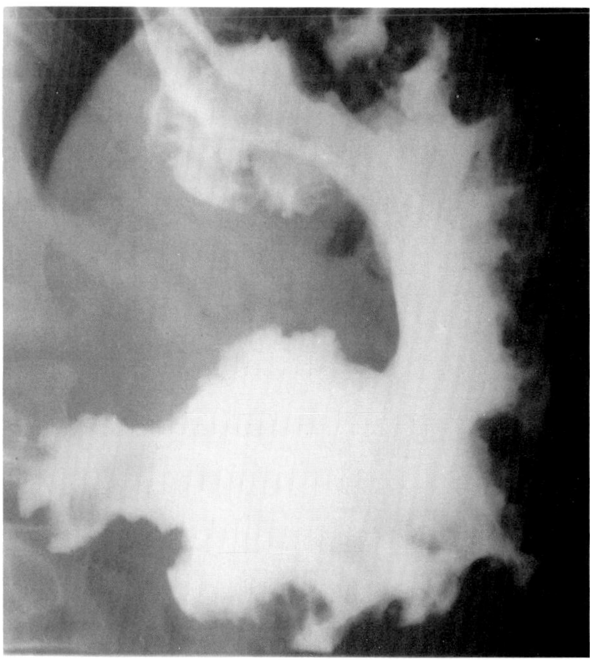

**Figure 45.36. Ménétrier's disease.** Generalized rugal fold thickening involves the entire stomach.

these two entities is possible only if bubbles of gas can be demonstrated in the wall of the stomach. This signifies the development of emphysematous gastritis, an extremely lethal form of bacterial invasion of the stomach wall.[29]

## Ménétrier's Disease

Ménétrier's disease (giant hypertrophic gastritis) is an uncommon disorder of unknown cause that is characterized by massive enlargement of rugal folds due to hyperplasia of the gastric glands. There is usually hyposecretion of acid and excessive secretion of gastric mucus. A loss of protein into the lumen of the stomach may result in hypoproteinemia and edema. The thickened gastric rugae become contorted and folded on each other in a convolutional pattern suggestive of the gyri and sulci of the brain. Enlarged rugal folds are particularly prominent along the greater curvature. Although the disorder is classically described as a lesion of the fundus and body, involvement of the entire stomach can occur. The disease can be diffuse or localized, and the transition between normal and pathologic folds is usually abrupt.

Radiographically, affected rugal folds are thick, tortuous, and angular with no uniformity in pattern or direction. When seen on end, the folds may closely simulate polypoid filling defects. Lines of barium can be seen perpendicular to the stomach because of spicules of contrast material trapped by apposed giant rugal folds.

Ménétrier's disease must be differentiated from a malignant process, especially lymphoma. If the thickened rugal folds predominantly involve the fundus and spare the lesser curvature, if there is no ulceration or true rigidity, or if excess mucus can be demonstrated, Ménétrier's disease is the probable diagnosis. If the enlarged rugal folds predominantly involve the distal portion of the stomach and the lesser curvature, or if there is some loss of pliability of the gastric wall, lymphoma is more likely.[30–32]

# OTHER DISEASES OF THE STOMACH

## Acute Gastric Dilatation

Acute gastric dilatation refers to sudden and excessive distension of the stomach by fluid and gas, usually accompanied by vomiting, dehydration, and peripheral vascular collapse. Within minutes or hours, a normal stomach can expand into a hyperemic, cyanotic, atonic sac that fills the abdomen.

Most cases of acute gastric dilatation occur during the first several days after abdominal surgery, though the incidence of this postoperative complication has dramatically decreased with the advent of nasogastric suction, improved anesthetics, close monitoring of acid-

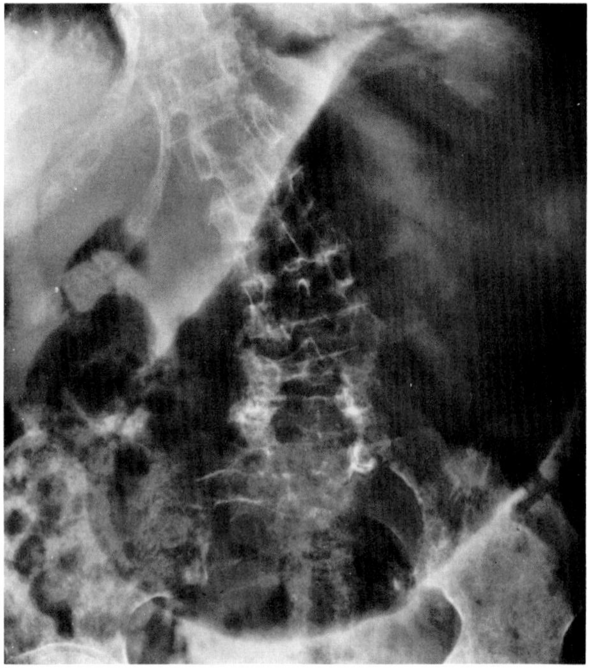

**Figure 45.37. Acute gastric dilatation.** A plain abdominal radiograph demonstrates a huge quantity of air and fluid filling a massively enlarged stomach that extends to the lower portion of the pelvis. (From Eisenberg,[46] with permission from the publisher.)

base and electrolyte balance, and meticulous care in the handling of tissues at surgery. Acute gastric dilatation can also be related to abdominal trauma, severe pain, peritonitis, electrolyte and acid-base imbalance, diabetic acidosis, the use of body casts, or large doses of anticholinergic drugs.

Plain abdominal radiographs in the patient with acute gastric dilatation demonstrate huge quantities of air and fluid filling a massively enlarged stomach that can extend even to the floor of the pelvis. The administration of barium demonstrates a large amount of solid gastric residue and can eliminate the possibility of an organic gastric outlet obstruction.[33,34]

### Adult Hypertrophic Pyloric Stenosis

The histologic, anatomic, and radiographic abnormalities in adult hypertrophic pyloric stenosis are indistinguishable from those in the infantile form. Unlike children, however, adults with hypertrophic pyloric stenosis infrequently have high-grade gastric outlet obstruction. About half of all patients with demonstrable pyloric hypertrophy have concomitant gastric ulceration. This probably reflects the development of an ulcer due to delayed gastric emptying, which interferes with the passage of semisolid food and results in increased gastrin production and consequent hyperacidity.

Elongation and narrowing of the pyloric canal are characteristic radiographic findings in adult hypertrophic

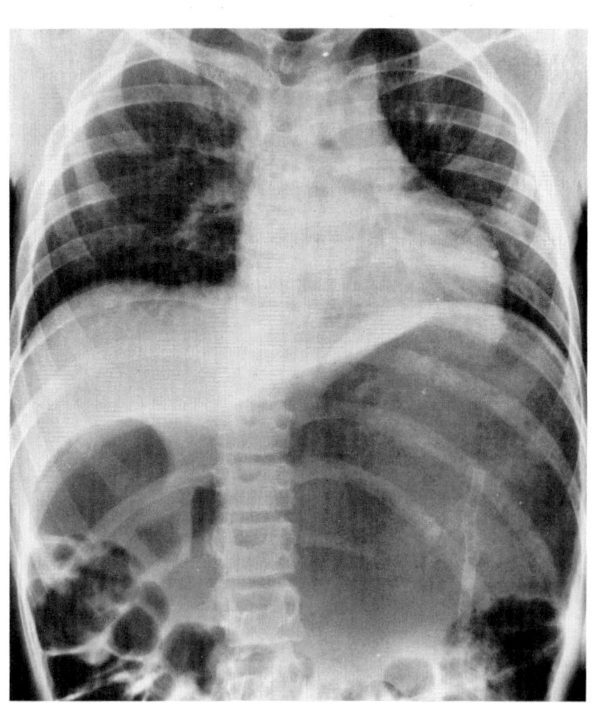

A

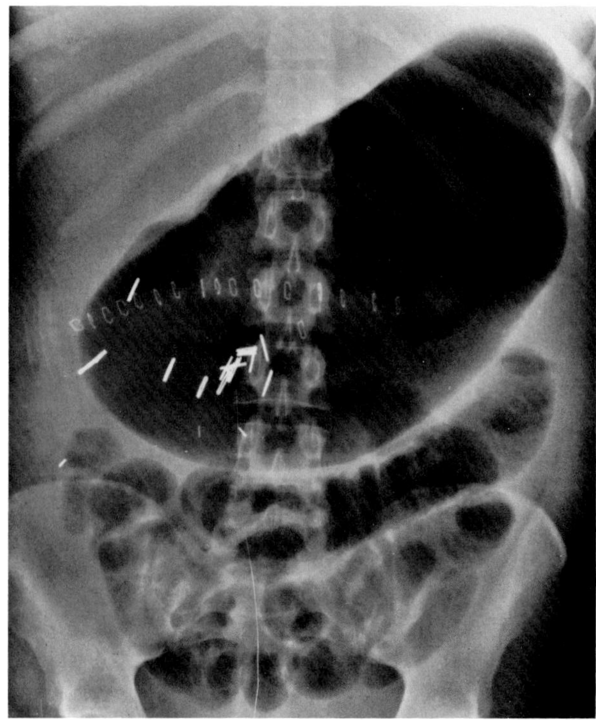

B

**Figure 45.38. Two examples of acute gastric dilatation.** (A) Post-traumatic. (B) Following abdominal surgery.

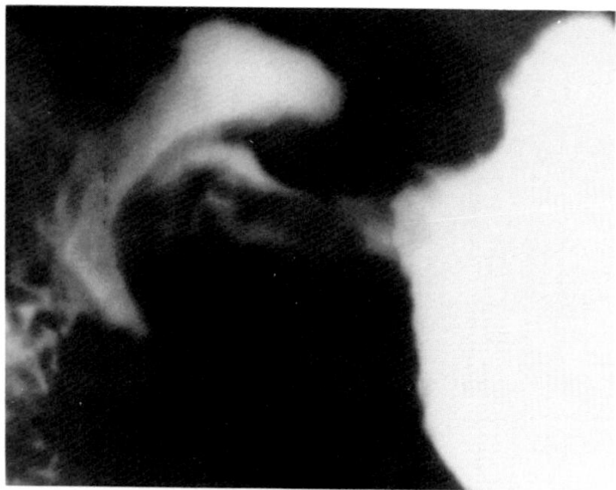

**Figure 45.39. Adult hypertrophic pyloric stenosis.** There is narrowing and elongation of the pyloric canal along with a characteristic concave, crescentic indentation at the base of the duodenal bulb. (From Eisenberg,[46] with permission from the publisher.)

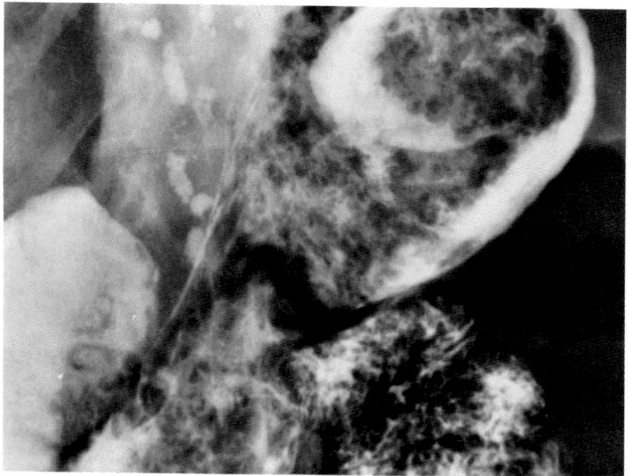

**Figure 45.40. Bezoar.** Infiltration of contrast material into the interstices of the mass results in a characteristic mottled appearance. (From Eisenberg,[46] with permission from the publisher.)

pyloric stenosis. The proximal end of the narrowed pylorus merges gradually with the contiguous stomach, resulting in a smooth, round junction without the shoulders suggestive of a malignant neoplasm. Although seen in only a minority of cases, a small triangular outpouching from the greater curvature side of the narrowed pyloric canal is pathognomonic (Twining recess). It probably represents a protrusion of the mucosa between adjacent bundles of the hypertrophied circular muscle of the pylorus and antrum. A classic sign of adult hypertrophic pyloric stenosis is a symmetric, concave, crescentic indentation of the base of the duodenal bulb. This mushroom-shaped defect is presumably due to partial invagination of the hypertrophied muscle mass into the bulb.[35]

## Mallory-Weiss Syndrome

This condition is discussed in the section "Perforation of the Esophagus" in Chapter 43.

## Bezoars and Foreign Bodies in the Stomach

A bezoar is an intragastric mass composed of accumulated ingested material. Phytobezoars, which are composed of undigested vegetable material, have classically been associated with the eating of unripe persimmons, a fruit containing substances that coagulate on contact with gastric acid to produce a sticky gelatinous material, which then traps seeds, skin, and other foodstuffs. Trichobezoars (hairballs) occur predominantly in females, especially those with schizophrenia or other mental disorder. Bezoars in the gastric remnant are a common complication following partial gastric resection with Billroth I or II anastomoses. The chief constituent of postgastrectomy bezoars is the fibrous, pithy component of

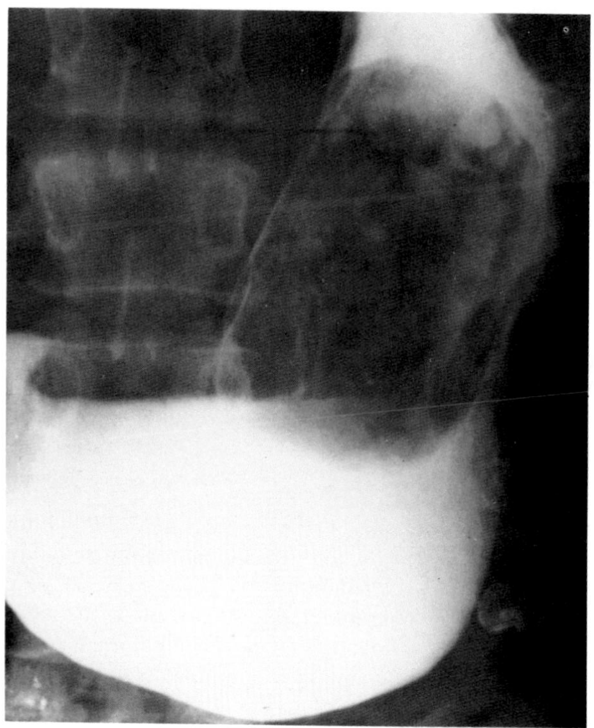

**Figure 45.41. Glue bezoar in a young model-airplane builder.** The smooth mass simulates an enormous air bubble. (From Eisenberg,[46] with permission from the publisher.)

fruits and vegetables, the consumption of which should be reduced as much as possible in postgastrectomy patients.

Symptoms of gastric bezoars result from the mechanical presence of the foreign body. They include cramplike epigastric pain and a sense of dragging, fullness, or heaviness in the upper abdomen. The incidence of associated peptic ulcers is high, especially with the more

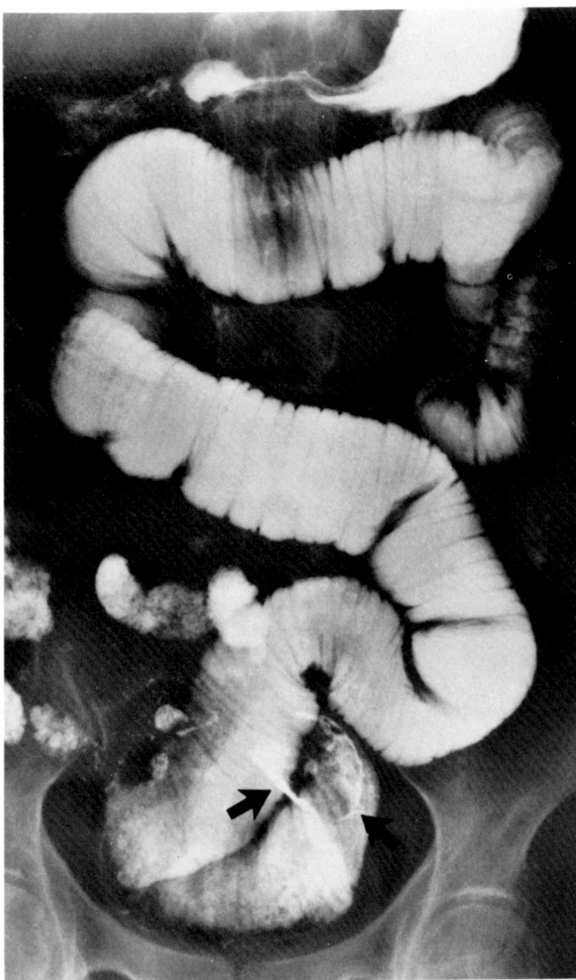

**Figure 45.42. Impacted bezoar (arrows) causing small bowel obstruction.** (From Eisenberg,[46] with permission from the publisher.)

abrasive phytobezoars. When bezoars are large, symptoms of pyloric obstruction can clinically simulate symptoms of a gastric carcinoma.

Plain abdominal radiographs often show a bezoar as a soft tissue mass floating in the stomach at the air-fluid interface. On barium studies, contrast material coating the mass and infiltrating into the interstices results in a characteristic mottled or streaked appearance. The filling defect may occasionally be completely smooth, simulating an enormous gas bubble that is freely movable within the stomach.

Small foreign bodies such as coins, marbles, or even closed safety pins usually pass through the stomach and bowel without difficulty. Elongated, sharp objects such as needles, toothpicks, or open safety pins may hold up at some point and cause obstruction, ulceration, bleeding, abscess, or peritonitis. Metallic foreign bodies appear opaque on plain abdominal radiographs. Nonmetallic foreign bodies appear as lucent filling defects within the barium-filled stomach.[36,37]

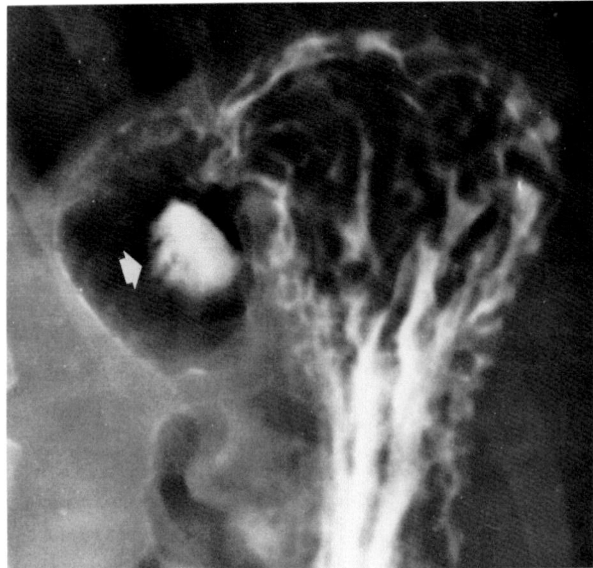

**Figure 45.43. Gastric diverticulum in the fundus.** Pooling of barium (arrow) simulates an acute ulceration. (From Eisenberg,[46] with permission from the publisher.)

## Gastric Diverticula

Gastric diverticula usually occur just above the cardia on the posterior wall near the lesser curvature. Almost always asymptomatic, these uncommon lesions have a characteristic radiographic appearance. A large gastric diverticulum that fails to fill with gas or barium may mimic a smooth-bordered submucosal mass. On repeat examination, barium can usually be demonstrated to enter the diverticulum, thereby establishing the diagnosis. At times, a collection of barium may pool in a gastric diverticulum and mimic an acute ulceration.

Antral diverticula are rare and may simulate an ulceration, though the absence of edema and spasm should permit the proper diagnosis.

## Gastric Volvulus or Torsion

Gastric volvulus is an uncommon acquired twist of the stomach upon itself that can lead to gastric outlet obstruction. It usually occurs in conjunction with a large paraesophageal hernia or eventration of the diaphragm that permits part or all of the stomach to assume an intrathoracic position. Organoaxial volvulus refers to rotation of the stomach upward along its long axis (a line connecting the cardia with the pylorus). In this condition, the antrum moves from an inferior to a superior position. In the mesenteroaxial type of gastric volvulus, the stomach rotates from right to left or left to right around the long axis of the gastrohepatic omentum (a line connecting the middle of the lesser curvature with the middle of the greater curvature).

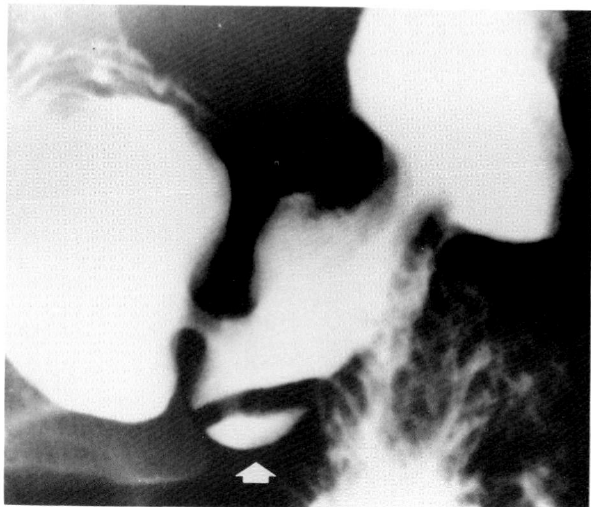

**Figure 45.44. Antral diverticulum (arrow).** (From Eisenberg,[46] with permission from the publisher.)

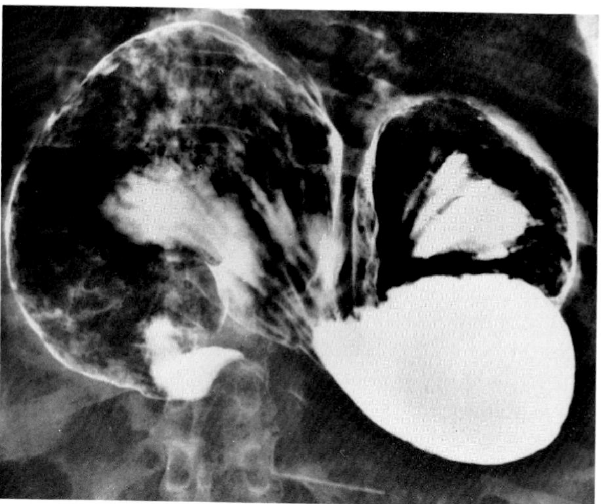

**Figure 45.45. Gastric volvulus.** The stomach is trapped in a large hiatal hernia sac. (From Eisenberg,[46] with permission from the publisher.)

Gastric volvulus can be asymptomatic if there is no outlet obstruction or vascular compromise. Acute volvulus associated with interference of the blood supply is a surgical emergency with a high mortality rate.

The radiographic signs of gastric volvulus are characteristic. They include a double air-fluid level on upright films, inversion of the stomach with the greater curvature above the level of the lesser curvature, positioning of the cardia and pylorus at the same level, and downward pointing of the pylorus and duodenum.[38]

## Pseudolymphoma

Pseudolymphoma of the stomach (gastric lymphoid hyperplasia) is a benign proliferation of lymphoid tissue that can clinically and histologically simulate malignant lymphoma. Although the etiology of this condition remains obscure, it is considered to represent a nonspecific late reaction to chronic peptic ulcer disease. Most patients with pseudolymphoma have a long history of gastrointestinal complaints, usually without a palpable abdominal mass, in contrast to the short and devastating course that is typical of gastric carcinoma and some malignant lymphomas.

A large ulcer surrounded by a mass and associated with regional or generalized enlargement of the rugal folds is characteristic of pseudolymphoma. The ulcer is usually well-defined and looks benign, though it may be poorly defined and irregular and simulate a malignant lesion. Other manifestations of pseudolymphoma of the stomach include a tumor mass, large gastric rugal folds without ulceration, and a constricting lesion of the body or antrum of the stomach.

Because in many cases malignant lymphoma cannot be excluded on frozen section or biopsy, most patients

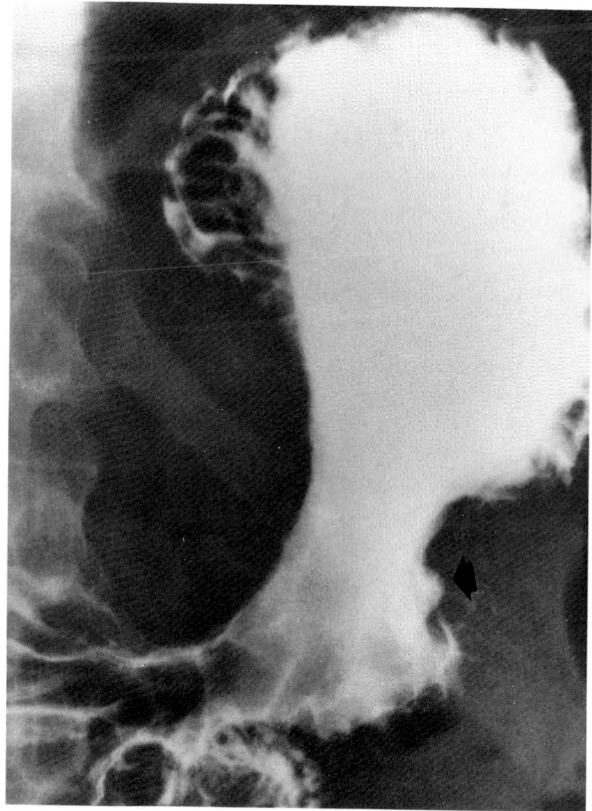

**Figure 45.46. Gastric pseudolymphoma.** A greater curvature ulcer (arrow) is surrounded by a soft tissue mass and is associated with regional enlargement of rugal folds. (From Eisenberg,[46] with permission from the publisher.)

with pseudolymphoma of the stomach require partial gastrectomy for diagnosis.[39,40]

## Eosinophilic Gastritis and Eosinophilic Granuloma

Eosinophilic gastritis causes thickening of the muscular layer of the wall of the stomach due to edema and a diffuse infiltrate of predominantly mature eosinophils. It primarily involves the antrum and produces irregular narrowing and rigidity (linitis plastica pattern) that may simulate an infiltrating carcinoma. The disease is characterized by peripheral eosinophilia and the development of gastrointestinal symptoms and signs following the ingestion of specific foods. Although eosinophilic gastritis can simulate a more aggressive process, it is essentially a benign condition that is self-limited and often completely returns to normal after steroid therapy.

Eosinophilic granuloma (inflammatory fibroid polyp) of the stomach is a benign polypoid mass containing a nonspecific inflammatory infiltrate. It is unrelated to the lesion of the same name in bone and lung. Although the polyp is infiltrated by varying numbers of eosinophils, it differs from eosinophilic gastritis in that it is not associated with any food allergy or peripheral eosinophilia. Most eosinophilic granulomas of the stomach are solitary and sessile and appear as sharply defined, smooth-bordered, round or oval filling defects. Unlike eosinophilic gastritis, the inflammatory fibroid polyp of eosinophilic granuloma does not respond to steroid therapy and may

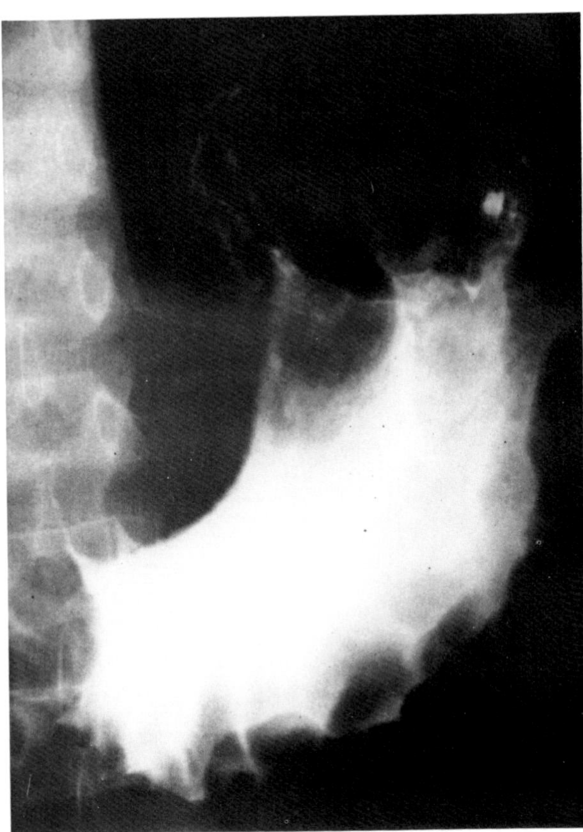

**Figure 45.47. Eosinophilic gastritis.** Marked enlargement of rugal folds is caused by a diffuse infiltration of eosinophilic leukocytes. (From Eisenberg,[46] with permission from the publisher.)

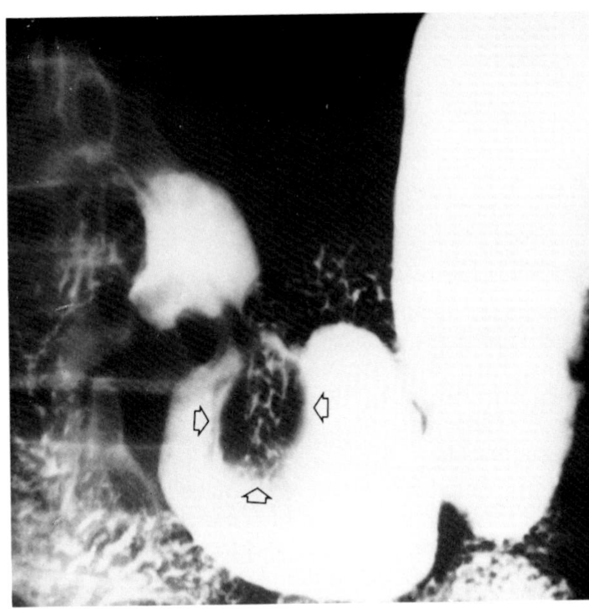

**Figure 45.48. Eosinophilic granuloma.** There is a sharply defined, smooth-bordered, oval filling defect in the gastric antrum (arrows). (From Eisenberg,[46] with permission from the publisher.)

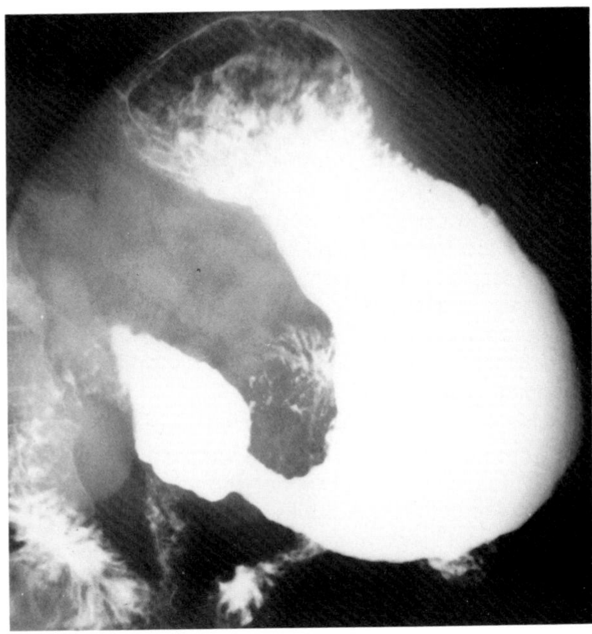

**Figure 45.49. Tuberculosis of the stomach.** Fibrotic healing produces narrowing and rigidity of the distal antrum. (From Eisenberg,[46] with permission from the publisher.)

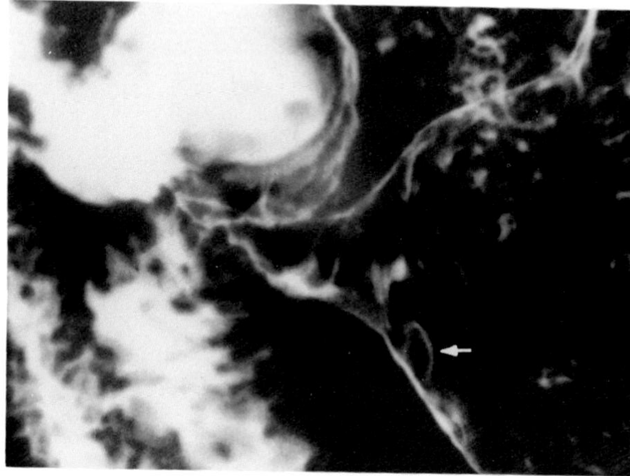

**Figure 45.50. Tertiary syphilis of the stomach.** Narrowing of the distal antrum with several gummatous polypoid masses (arrow).

require excision because of intermittent pyloric obstruction or bleeding secondary to erosion of the overlying mucosa.[41–44]

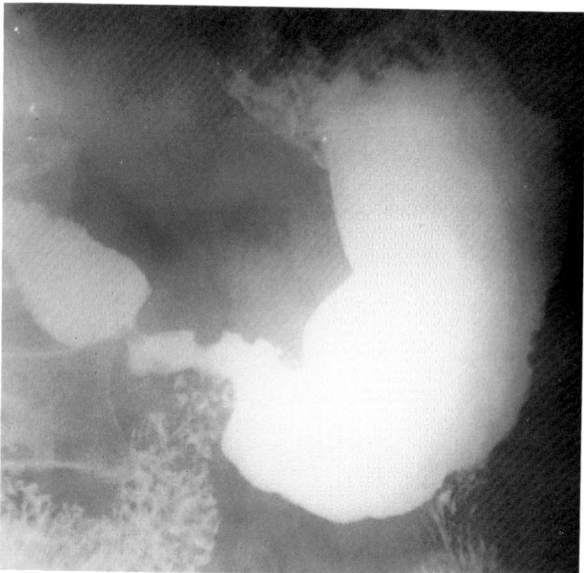

**Figure 45.51. Sarcoidosis of the stomach.** There is irregular narrowing of the distal antrum. (From Eisenberg,[46] with permission from the publisher.)

## Granulomatous Infections of the Stomach

Tuberculosis of the stomach is a rare condition that causes rigidity of the distal stomach and a linitis plastica pattern due to diffuse inflammation or fibrotic healing. Ulcerations and fistulas between the antrum and small bowel can simulate the radiographic appearance of gastric involvement by Crohn's disease.

Tertiary syphilis involving the stomach is now an exceedingly rare disease. Although discrete, nodular, gumma-like lesions can occur, diffuse involvement of the stomach is more common. Swelling and thickening of the gastric wall can result in mural rigidity and narrowing of the lumen indistinguishable from scirrhous carcinoma. Like most infiltrative granulomatous diseases of the stomach, syphilis has a predilection for involving the antrum.

About 10 percent of patients with sarcoidosis have evidence of stomach involvement on gastroscopic biopsy, though most are asymptomatic. Localized sarcoid granulomas can produce discrete mass defects. Diffuse lesions cause severe mural thickening and luminal narrowing, predominantly in the antrum. Ulcerations or erosions can lead to acute upper gastrointestinal hemorrhage.

In extremely rare instances, herpes simplex, histoplasmosis, actinomycosis, and candidiasis can also infiltrate the wall of the stomach.[45]

### REFERENCES


1. Nelson SW: The discovery of gastric ulcers and the differential diagnosis between benignancy and malignancy. *Radiol Clin North Am* 7:5–25, 1969.
2. Wolf BS: Observations on roentgen features of benign and malignant gastric ulcers. *Semin Roentgenol* 6:140–150, 1971.
3. Bachman AL, Parmer EA: Radiographic diagnosis of recurrence following resection for gastric cancer. *Radiology* 84:913–924, 1965.
4. Balfe DM, Koehler RE, Karsteadt N, et al: Computed tomography of gastric neoplasms. *Radiology* 140:431–436, 1981.
5. Moss AA, Schnyder P, Candardjis G, et al: Computed tomography of benign and malignant gastric abnormalities. *J Clin Gastroenterol* 2:401–409, 1980.
6. Lee KR, Levine E, Moffat RE, et al: Computed tomographic staging of malignant gastric neoplasms. *Radiology* 133:151–155, 1979.
7. Kline TS, Goldstein F: Malignant lymphoma of the stomach. *Cancer* 32:961–968, 1973.
8. Menuch LS: Gastric lymphoma: A radiologic diagnosis. *Gastrointest Radiol* 1:159–161, 1976.
9. Privett JTJ, Davies ER, Roylance J: The radiological features of gastric lymphoma. *Clin Radiol* 28:457–463, 1977.
10. Mauro MA, Koehler RE: Alimentary tract, in Lee JKT, Sagel SS, Stanley RJ (eds): *Computed Body Tomography*. New York, Raven Press, 1983, pp 307–340.
11. Buy JN, Moss AA: Computed tomography of gastric lymphoma. *AJR* 138:859–865, 1982.
12. Megibow AJ, Balthazar EJ, Naidich DP, et al: Computed tomography of gastrointestinal lymphoma. *AJR* 141:541–547, 1983.
13. Burgess JN, Dockerty MB, Remine RH: Sarcomatous lesions of the stomach. *Ann Surg* 173:758–766, 1971.
14. Kaye JJ, Stassa G: Mimicry and deception in the diagnosis of tumors of the gastric cardia. *AJR* 110:295–303, 1970.
15. Op den Orth JO, Dekker W: Gastric adenomas. *Radiology* 141:289–293, 1981.
16. Gordon R, Laufer I, Kressel HY: Gastric polyps found on routine double-contrast examination of the stomach. *Radiology* 134:27–30, 1980.
17. Ming SC: Malignant potential of gastric polyps. *Gastrointest Radiol* 1:121–125, 1976.
18. Kavlie H, White TT: Leiomyomas of the upper gastrointestinal tract. *Surgery* 71:842–848, 1972.

19. Faegenburg D, Farman J, Dallemand S, et al: Leiomyoblastoma of the stomach: Report of nine cases. *Radiology* 117:297–300, 1975.
20. Thoeni RF, Gedgaudas RK: Ectopic pancreas: Usual and unusual features. *Gastrointest Radiol* 5:37–42, 1980.
21. Feigen DS, James AE, Stitik FP, et al: The radiological appearance of hiatal hernia repairs. *Radiology* 110:71–77, 1974.
22. Skucas J, Mangla JC, Adams JP, et al: An evaluation of the Nissen fundoplication. *Radiology* 118:539–543, 1976.
23. Thoeni RF, Moss AA: The radiographic appearance of complications following Nissen fundoplication. *Radiology* 131:17–21, 1979.
24. Muhletaler CA, Gerlock AJ, Goncharenko V, et al: Gastric varices secondary to splenic vein occlusion: Radiographic diagnosis and clinical significance. *Radiology* 132:593–598, 1979.
25. Sos T, Meyers MA, Baltaxe HA: Nonfundic gastric varices. *Radiology* 105:579–580, 1972.
26. Muhletaler CA, Gerlock AJ, de Soto L, et al: Gastroduodenal lesions of ingested acids: Radiographic findings. *AJR* 135:1247–1252, 1980.
27. Poteshman N: Corrosive gastritis due to hydrochloric acid ingestion. *AJR* 99:182–185, 1967.
28. Martel W: Radiologic features of esophagogastritis secondary to extremely caustic agents. *Radiology* 103:31–36, 1972.
29. Turner MA, Beachley MC, Stanley D: Phlegmonous gastritis. *AJR* 133:527–528, 1979.
30. Olmsted WW, Cooper PH, Madewell JE: Involvement of the gastric antrum in Ménétrier's disease. *AJR* 126:524–529, 1976.
31. Press AJ: Practical significance of gastric rugal folds. *AJR* 125:172–183, 1975.
32. Reese BF, Hodgson JR, Dockerty MB: Giant hypertrophy of the gastric mucosa (Ménétrier's disease): A correlation of the roentgenographic, pathologic and clinical findings. *AJR* 88:619–626, 1962.
33. Joffe N: Some unusual roentgenologic findings associated with marked gastric dilatation. *AJR* 119:291–299, 1973.
34. Rimer DG: Gastric retention without mechanical obstruction. *Arch Intern Med* 117:287–299, 1966.
35. Kreel L, Ellis H: Pyloric stenosis in adults: A clinical and radiological study of 100 consecutive patients. *Gut* 6:253–261, 1965.
36. Goldstein HM, Cohen LE, Hagne RO, et al: Gastric bezoars: A frequent complication in the postoperative ulcer patient. *Radiology* 107:341–344, 1973.
37. Rogers LF, Davis EK, Harle TS: Phytobezoar formation and food boli following gastric surgery. *AJR* 119:280–290, 1973.
38. Gerson DE, Lewicki AM: Intrathoracic stomach: When does it obstruct? *Radiology* 119:257–264, 1976.
39. Chiles JT, Platz CE: The radiographic manifestations of pseudolymphoma of the stomach. *Radiology* 116:551–556, 1975.
40. Martel W, Abell MR, Allan TNK: Lymphoreticular hyperplasia of the stomach (pseudolymphoma). *AJR* 127:261–265, 1976.
41. Allman RM, Cavanagh RC, Helwig EB, et al: Inflammatory fibroid polyp. *Radiology* 127:69–73, 1978.
42. Goldberg HI, O'Kieffe D, Jenis EH, et al: Diffuse eosinophilic gastroenteritis. *AJR* 119:342–351, 1973.
43. Klein MC, Hargrove RL, Sleisinger MH, et al: Eosinophilic gastroenteritis. *Medicine* 49:299–319, 1970.
44. Wehnut WD, Olmsted WW, Neiman HL, et al: Eosinophilic gastritis. *Radiology* 120:85–89, 1976.
45. McLaughlin JS, Van Eck W, Thayer W, et al: Gastric sarcoidosis. *Ann Surg* 153:283–288, 1961.
46. Eisenberg RL: *Gastrointestinal Radiology: A Pattern Approach*. Philadelphia, JB Lippincott, 1983.
47. Juhl JH: *Essentials of Roentgen Interpretation*. Philadelphia, JB Lippincott, 1981.

## Chapter Forty-Six
# Disorders of Absorption

### INTESTINAL MALABSORPTION

The term *malabsorption disorders* refers to a multitude of conditions in which there is defective absorption of carbohydrates, proteins, and fats from the small bowel. Regardless of the cause, malabsorption results in steatorrhea—the passage of bulky, foul-smelling, high-fat-content stools. Many of the diseases causing malabsorption produce radiographic abnormalities in the small bowel, though malabsorption can exist without any detectable small bowel changes. The degree of clinical malabsorption is not closely correlated with the appearance or severity of the radiographic changes.

Clinical malabsorption is generally associated with two major radiographic patterns of small bowel abnormalities. Small bowel dilatation with normal folds is seen in such causes of malabsorption as sprue, scleroderma, and lactase deficiency. A pattern of generalized, irregular, distorted small bowel folds is seen with infiltrative causes of malabsorption such as Whipple's disease, giardiasis, lymphoma, amyloidosis, eosinophilic enteritis, and intestinal lymphangiectasia. Other disorders associated with malabsorption, such as multiple jejunal diverticula, small bowel fistulas, and Crohn's disease, can produce a broad spectrum of radiographic small bowel abnormalities.[1]

### SPRUE

The classic disease of malabsorption is sprue. This term is used to refer to any of three diseases, all of which are similar clinically: idiopathic (nontropical) sprue, tropical sprue, and celiac disease of children. Sprue typically appears radiographically as dilatation of the small bowel with normal folds. The degree of dilatation is generally related to the severity of disease and tends to be best visualized in the mid and distal jejunum. Although the appearance of long and tortuous loops sometimes superficially resembles that of mechanical obstruction, the dilated loops in sprue are flaccid and contract poorly. Segmentation and flocculation of barium have traditionally been considered indicative of sprue, as well as other malabsorption disorders. However, in recent years the use of nondispersible barium with suspending agents has all but eliminated these radiographic signs.

An excessive amount of fluid in the bowel lumen is a constant phenomenon in patients with sprue. This fluid may represent either increased movement of water into the lumen (hypersecretion) or deficient absorption of water by the deranged mucosa. The excessive fluid causes the barium in the small bowel to have a coarse, granular appearance, unlike barium in the normal intestine, which has a homogeneous quality.

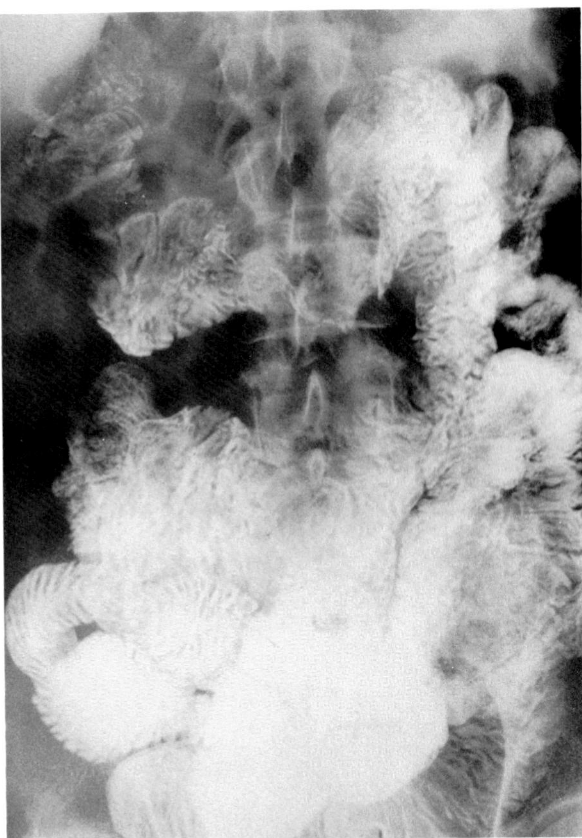

**Figure 46.1. Intestinal malabsorption.** Small bowel dilatation with normal folds in a patient with malabsorption due to sprue. (From Eisenberg,[2] with permission from the publisher.)

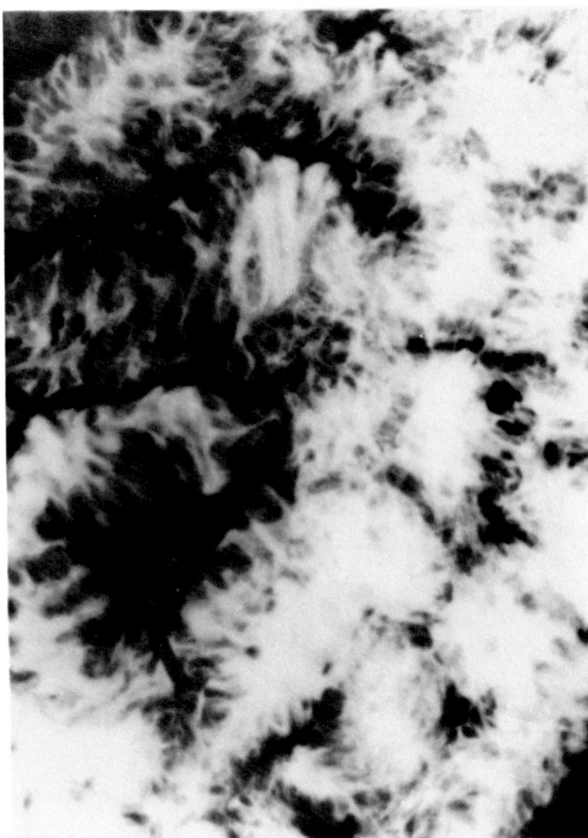

**Figure 46.2. Intestinal malabsorption.** Irregular thickening of small bowel folds in a patient with malabsorption due to amyloidosis. (From Eisenberg,[2] with permission from the publisher.)

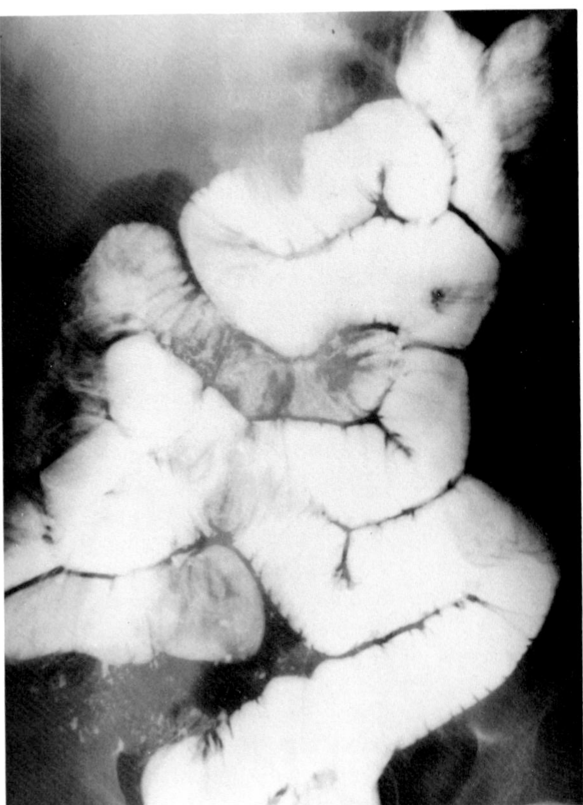

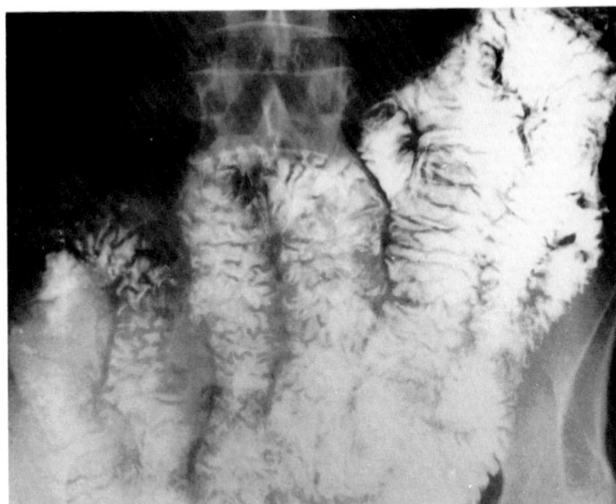

**Figure 46.4. Idiopathic (nontropical) sprue.** The barium in the dilated loops of small bowel has a coarse, granular appearance due to hypersecretion. (From Eisenberg,[2] with permission from the publisher.)

**Figure 46.3. Idiopathic (nontropical) sprue.** There is diffuse dilatation of the entire small bowel with pronounced hypersecretion. (From Eisenberg,[2] with permission from the publisher.)

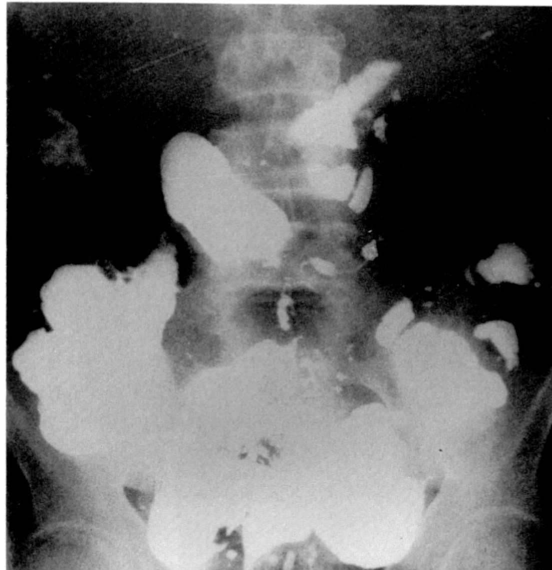

**Figure 46.5. *Moulage* sign in sprue.** Barium-filled loops of small bowel in a patient with sprue show smooth contours and unindented, molded margins. (From Marshak, Lindner, and Maklansky,[21] with permission from the publisher.)

The *moulage* sign describes a radiographic appearance of the jejunum in sprue. The French term means "moulding" or "casting" and refers to the smooth contour and unindented margins of barium-filled small bowel loops. This tubular appearance in sprue is probably due to atrophy and effacement of the jejunal mucosal folds.

Intussusception is a common finding in patients with sprue. The diagnosis is based on the typical pattern of a localized filling defect with stretched and thin valvulae conniventes overlying it (coiled-spring appearance). Because the intussusceptions in sprue are transient and nonobstructing, they are often missed on a single examination.

The diagnosis of sprue is made by jejunal biopsy, which demonstrates flattening, broadening, coalescence, and sometimes complete atrophy of intestinal villi. A characteristic finding of idiopathic sprue and celiac disease of children is dramatic clinical and histologic improvement after the patient has been placed on a diet free from gluten (the water-insoluble protein fraction of cereal grains). The radiographic and jejunal biopsy findings in tropical sprue are identical to those seen in idiopathic sprue. However, clinical improvement in tropical sprue follows folic acid or antibiotic therapy rather than gluten withdrawal.[2]

## SCLERODERMA OF THE SMALL BOWEL

Scleroderma involving the small bowel can cause the malabsorption syndrome and a radiographic pattern of dilated small bowel with normal folds. Smooth muscle atrophy and deposition of connective tissue in the sub-

**Figure 46.6. Intussusception in idiopathic (nontropical) sprue.** Note the characteristic coiled-spring appearance (arrows). (From Eisenberg,[2] with permission from the publisher.)

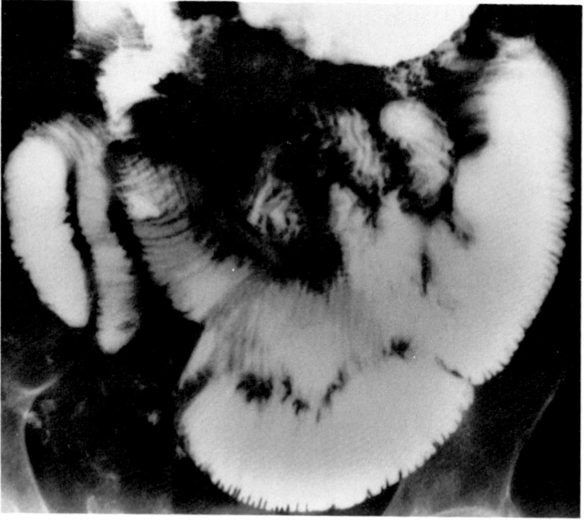

**Figure 46.7. Scleroderma.** Diffuse dilatation of small bowel loops. (From Eisenberg,[2] with permission from the publisher.)

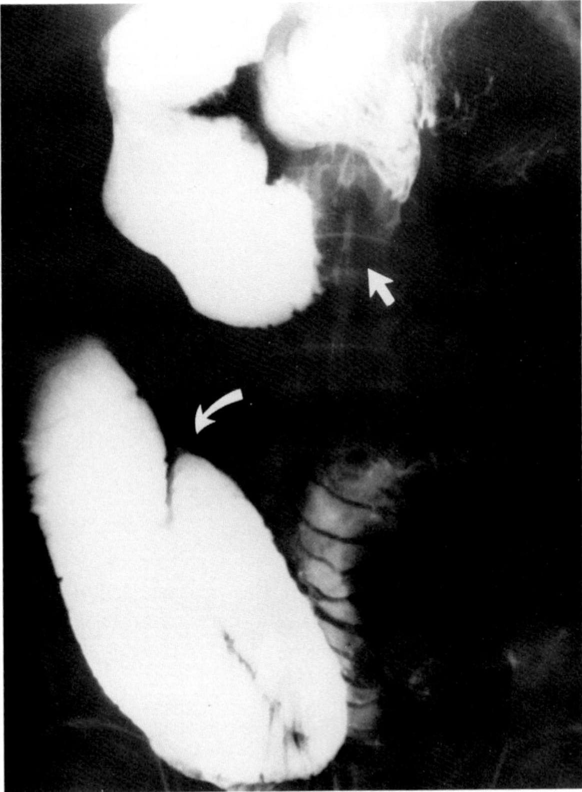

**Figure 46.8. Scleroderma.** There is diffuse dilatation of distal small bowel loops (curved arrow) in addition to severe dilatation of the duodenum proximal to the site at which the transverse portion passes between the aorta and the superior mesenteric artery (straight arrow). (From Eisenberg,[2] with permission from the publisher.)

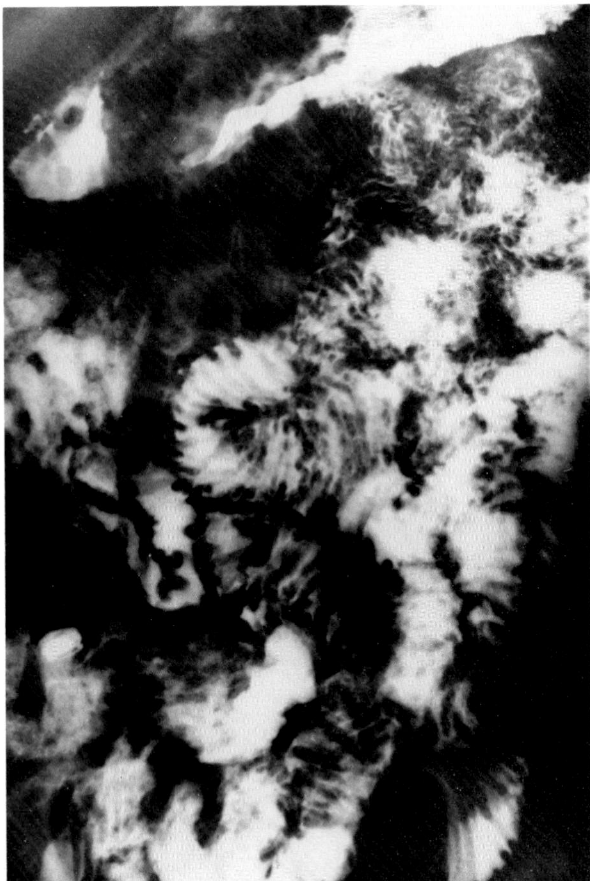

**Figure 46.9. Whipple's disease.** There is diffuse irregular thickening of small bowel folds. (From Eisenberg,[2] with permission from the publisher.)

mucosal, muscular, and serosal layers result in hypomotility of the small bowel and an extremely prolonged transit time that leads to stasis of intestinal contents. The small bowel dilatation is often most marked in the duodenum proximal to the site at which the transverse portion passes between the aorta and the superior mesenteric artery, though the entire small bowel can be diffusely involved. Pseudosacculations, large broad-necked outpouchings simulating small bowel diverticula, are characteristic of scleroderma.

Scleroderma and sprue can usually be readily distinguished from one another on the basis of radiographic findings. In contrast to the appearance in scleroderma, the small bowel in patients with sprue has relatively normal motility, is most often dilated in the mid and distal jejunum, and generally demonstrates increased secretions.[3,4]

## WHIPPLE'S DISEASE

In Whipple's disease, the lamina propria is extensively infiltrated by large macrophages filled with multiple glycoprotein granules that react positively to the periodic acid Schiff (PAS) stain. In patients with active disease, numerous small, gram-positve bacilli also invade the involved tissues and produce a distinctive, usually diagnostic, histologic pattern. Most patients with Whipple's disease present with a generalized malabsorption syndrome; diarrhea is a prominent complaint. Extraintestinal symptoms (arthritis, fever, lymphadenopathy) are common and may precede gastrointestinal complaints by many years.

The radiographic hallmark of Whipple's disease is extensive irregular thickening and distortion of small bowel folds, most frequently seen in the duodenum and proximal jejunum; in severe cases, the distal jejunum and ileum may also be abnormal. The lumen of the bowel can be normal or slightly dilated. There is often evidence of hypersecretion that can lead to hypoalbuminemia due to an excessive loss of protein into the gastrointestinal tract. Thickening of the mesentery, infiltration of the bowel wall, and enlargement of the mesenteric lymph nodes causes separation of small bowel loops in patients with Whipple's disease. If the intestinal villi swell to such an

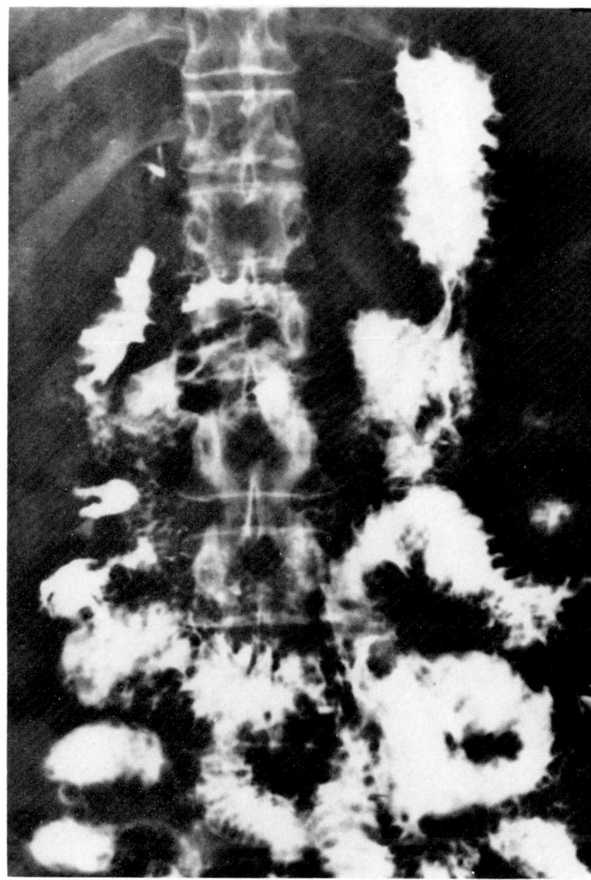

**Figure 46.10. Lymphoma.** There is diffuse irregular thickening of small bowel folds with mesenteric involvement and separation of bowel loops. (From Eisenberg,[2] with permission from the publisher.)

extent that they become bulbous structures large enough to be macroscopically visible, they may appear as innumerable, sandlike filling defects superimposed on the irregularly thickened fold pattern.

Following successful treatment of Whipple's disease with antibiotic therapy, the irregular thickening of small bowel folds becomes less apparent, and the radiographic appearance may even revert to normal.[5–7]

## INTESTINAL LYMPHOMA

Intestinal lymphoma can either originate in the small bowel (primary) or be a manifestation of a disseminated lymphomatous process that affects many organs (secondary). In either case, the disease can be localized to a single intestinal segment, be multifocal, or diffusely involve most of the small bowel. Intestinal lymphoma is most frequent in the ileum, where the greatest amount of lymphoid tissue is present. Primary diffuse intestinal lymphoma is rare in the Western world but relatively

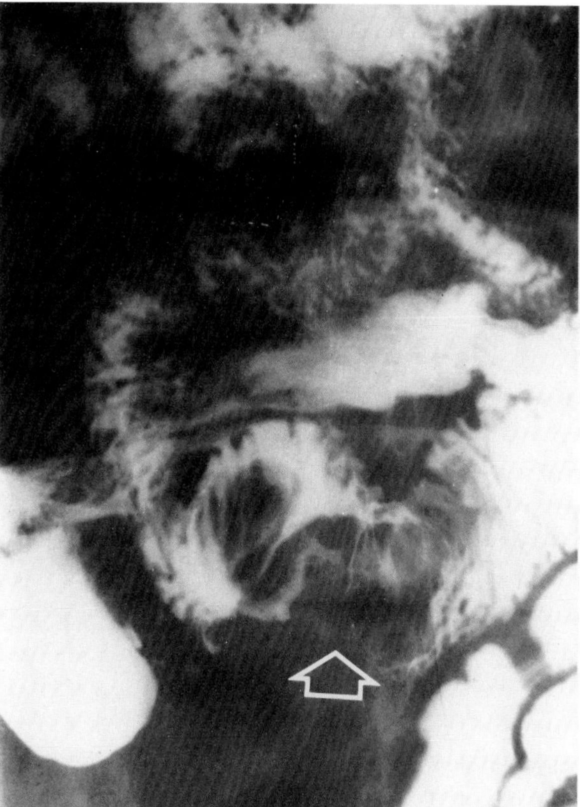

**Figure 46.11. Lymphoma.** A large, bulky, irregular lesion (arrow). (From Eisenberg,[2] with permission from the publisher.)

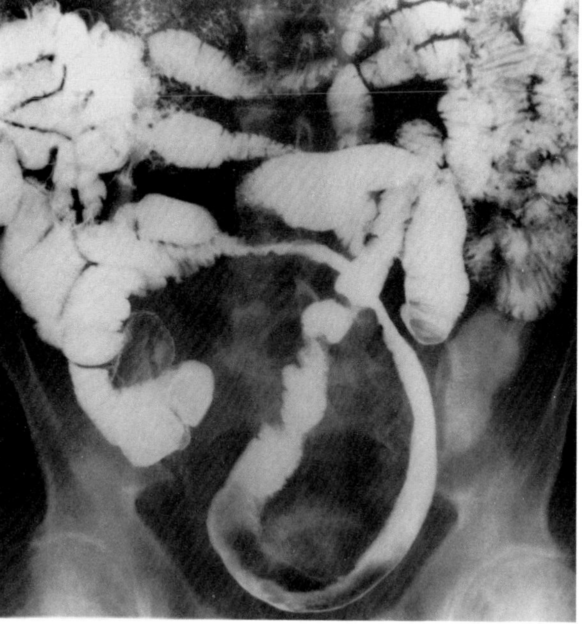

**Figure 46.12. Lymphoma.** There is smooth narrowing of a segment of ileum with obliteration of mucosal folds. The elongated bowel loop is displaced around a large suprapubic mass. (From Eisenberg,[2] with permission from the publisher.)

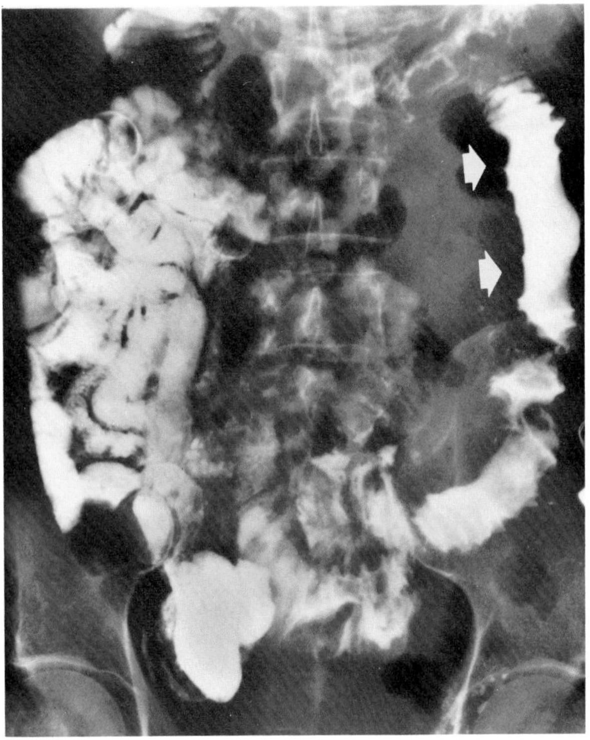

**Figure 46.13. Aneurysmal lymphoma.** Localized dilatation of a segment of small bowel (arrows) is due to sloughing of the necrotic central core of the neoplastic mass. (From Eisenberg,[2] with permission from the publisher.)

frequent in adolescents and young adults in populations native to Middle Eastern countries. Symptoms and signs of intestinal malabsorption are a prominent feature, and the radiographic appearance of small bowel dilatation with normal folds may be indistinguishable from sprue. Indeed, diffuse intestinal lymphoma may complicate long-standing sprue, and this complication should be considered whenever a patient with sprue shows a sudden refractoriness to treatment or develops fever, bowel perforation, or hemorrhage.

Lymphomatous infiltration of the bowel wall causes irregular thickening and distortion of mucosal folds. Segmental constrictions may occur. Involvement of mesenteric lymph nodes results in separation of bowel loops. Lymphoma may appear as single or multiple, discrete polypoid masses that are often large and bulky and have irregular ulcerations. If there is no substantial intramural extension, the intraluminal mass can be drawn forward by peristalsis to form a pseudopedicle and can even become the leading edge of an intussusception that obstructs the intestinal lumen. If lymphoma produces large masses in the bowel that necrose and cavitate, the central core may slough into the bowel lumen and produce aneurysmal dilatation of the bowel.

Computed tomography (CT) in the patient with intestinal lymphoma can demonstrate localized thickening of the bowel wall as well as exophytic and mesenteric tumor masses. The major value of this modality is to detect disease in other areas of the abdomen, especially enlargement of the mesenteric and retroperitoneal lymph nodes and spread of tumor to the liver, spleen, kidneys, and adrenals.[8–12]

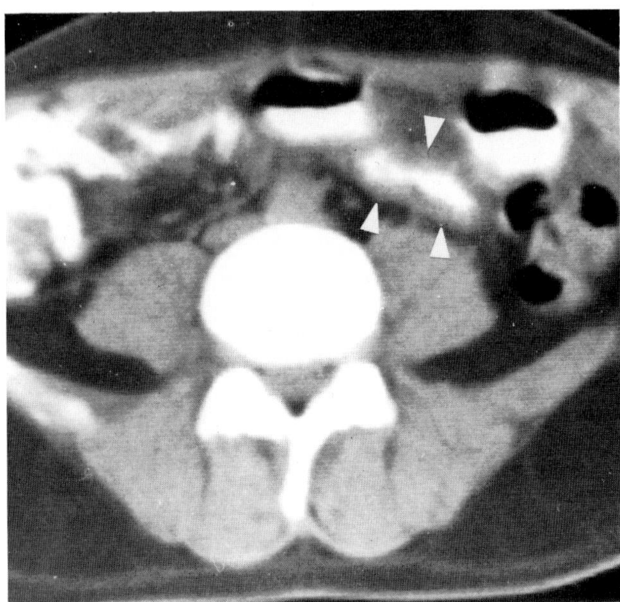

A

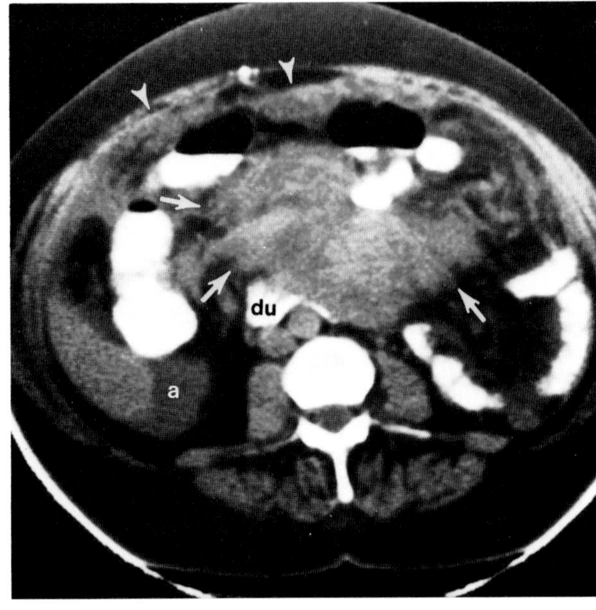


B

**Figure 46.14. CT in intestinal lymphoma.** (A) Uniform thickening of the wall (arrowheads) in the affected segment of ileum. (B) In another patient with abdominal lymphoma, there is infiltration of the mesentery (arrows) and omentum (arrowheads), compression of the third portion of the duodenum (du), and ascites (a) within the hepatorenal space. (A, from Mauro and Koehler,[12] with permission from the publisher; B, from Levitt,[11] with permission from the publisher.)

## EOSINOPHILIC ENTERITIS

Diffuse infiltration of the small bowel by eosinophilic leukocytes produces the thickened folds seen in eosinophilic enteritis. The jejunum is most prominently involved, though the entire small bowel is sometimes affected. Concomitant eosinophilic infiltration of the stomach is common. Involvement of the mucosa and lamina propria initially results in regular thickening of small bowel folds. More extensive, transmural involvement causes irregular fold thickening, angulation, and a sawtooth contour of the small bowel. These findings can be associated with rigidity of the bowel and hyperplastic mesenteric nodes that simulate Crohn's disease. However, peripheral eosinophilia and the typical history of gastrointestinal symptoms related to the ingestion of specific foods usually permit the proper diagnosis of eosinophilic enteritis.[13]

**Figure 46.15. Eosinophilic enteritis.** Irregular thickening of folds primarily involves the jejunum. No concomitant involvement of the stomach is identified. (From Eisenberg,[2] with permission from the publisher.)

## LACTASE DEFICIENCY

Lactase deficiency, the most common of the disaccharidase-deficiency syndromes, is an isolated enzyme defect in which the patient is unable to hydrolyze and absorb lactose properly. In some population groups, such as North American blacks, Mexicans, and Chinese, more than 75 percent of all adults have lactase deficiency. Even among whites in North American and northern Europe, groups that are the least affected, this enzyme defect can be demonstrated in 5 to 15 percent of adults. The diagnosis of lactase deficiency should be suggested whenever a patient experiences abdominal discomfort, cramps, and watery diarrhea within 30 minutes to several hours of ingesting milk or milk products. Although conventional small bowel examinations are normal in patients with lactase deficiency, the addition of 25 to 100 g of lactose to the barium mixture results in marked dilatation of the small bowel with dilution of barium, rapid transit time, and reproduction of symptoms.[14,15]

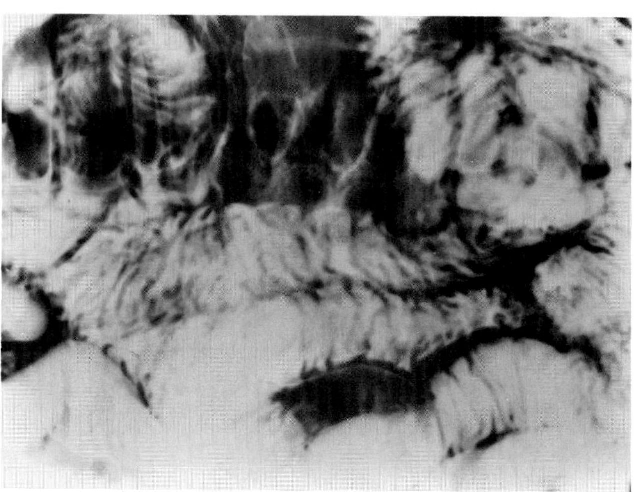

A

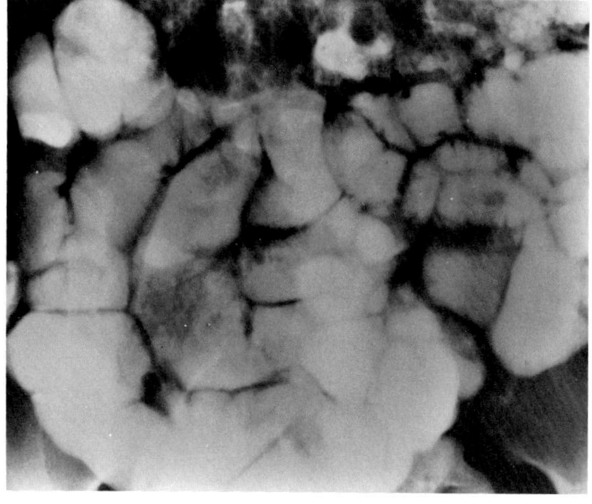

B

**Figure 46.16. Lactase deficiency.** (A) A normal conventional small bowel examination. (B) After the addition of 50 g of lactose to the barium mixture, there is marked dilatation of the small bowel with dilution of barium, rapid transit time, and reproduction of symptoms. (From Eisenberg,[2] with permission from the publisher.)

## HYPOGAMMAGLOBULINEMIA

Hypogammaglobulinemia is often associated with nodular lymphoid hyperplasia, which primarily involves the jejunum but can occur throughout the small bowel. The chronically reduced concentration of serum globulins results in an increased susceptibility to respiratory and other infections. Diarrhea and malabsorption are frequent complaints; *Giardia lamblia* infestation can be demonstrated in up to 90 percent of the patients with this condition.

The hyperplastic lymph follicles in the lamina propria of a patient with nodular lymphoid hyperplasia cause effacement of the villous pattern, making the bowel appear to be studded with innumerable tiny polypoid masses uniformly distributed through the involved segment of the intestine. The filling defects are round and regular in outline and have no recognizable ulceration. If there is no associated disease, the sandlike nodules of lymphoid hyperplasia are superimposed on a background of normal small bowel folds. When *Giardia* infection is also present, however, the underlying fold pattern is irregularly thickened and grossly distorted.

It is important to note that in children and young adults, the presence of multiple small, symmetric nodules of lymphoid hyperplasia in the terminal ileum is a normal finding.[16,17]

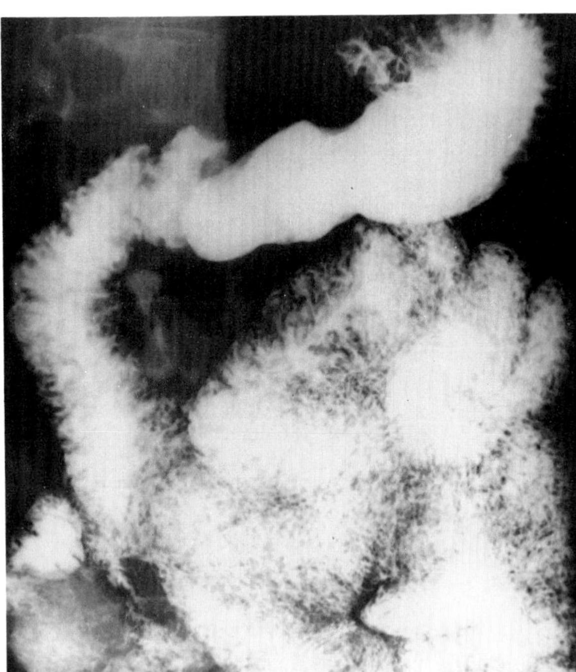

**Figure 46.17. Nodular lymphoid hyperplasia.** Innumerable filling defects are uniformly distributed throughout the involved segments of small bowel. The underlying small bowel fold pattern is normal in this patient, who had no evidence of associated disease. (From Eisenberg,[2] with permission from the publisher.)

## INTESTINAL LYMPHANGIECTASIA

Intestinal lymphangiectasia is a disorder that primarily affects children and young adults and is characterized by increased enteric loss of protein, hypoproteinemia, edema, lymphocytopenia, malabsorption, and gross dilatation of the lymphatics in the small bowel mucosa and submucosa. The primary form probably represents a congenital mechanical block to lymphatic outflow; secondary intestinal lymphangiectasia is a complication of inflammatory or neoplastic lymphadenopathy.

Intestinal lymphangiectasia causes regular thickening of small bowel mucosal folds due to a combination of intestinal edema and lymphatic dilatation. Intestinal edema in patients with lymphangiectasia can be due to either lymphatic obstruction or a severe loss of protein into the gastrointestinal tract. The dilated, telangiectatic lymphatic vessels in the small bowel cause marked enlargement of the villi, which produces diffuse nodularity and a sandlike pattern as well as radiographic evidence of hypersecretion. Diffuse intestinal edema in a patient (especially a child or young adult) with no evidence of liver, kidney, or heart disease should suggest the diagnosis of intestinal lymphangiectasia.[18,19]

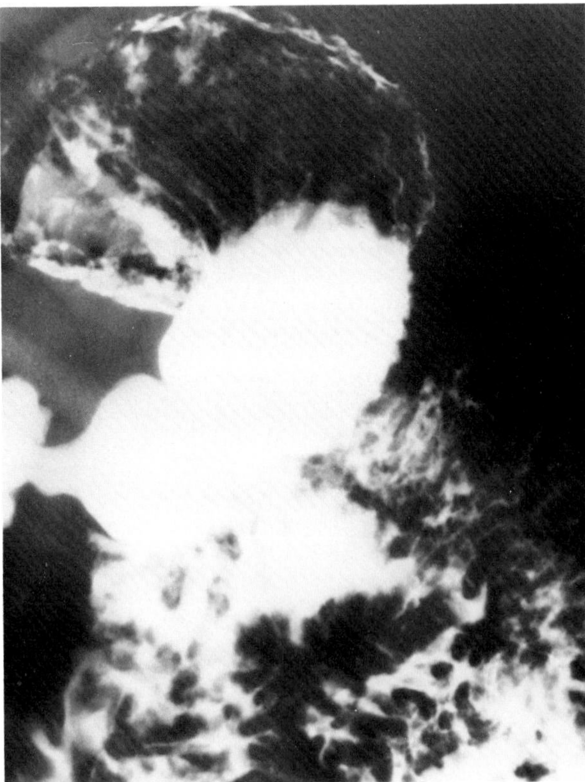

**Figure 46.18. Nodular lymphoid hyperplasia in a patient with an immune deficiency and *Giardia lamblia* infestation.** The relatively larger nodules are superimposed on an irregularly thickened and grossly distorted underlying fold pattern. (From Eisenberg,[2] with permission from the publisher.)

## ABETALIPOPROTEINEMIA

Abetalipoproteinemia is a rare, recessively inherited disease manifested clinically by malabsorption of fat, progressive neurologic deterioration, and retinitis pigmentosa. An inability to produce the apoprotein moiety of the beta lipoprotein leads to a failure to transport lipid material out of the intestinal epithelial cell. This results in fat malabsorption, with no plasma beta lipoproteins and markedly reduced serum levels of cholesterol, phospholipids, carotenoids, and vitamin A. Acanthocytosis, a thorny appearance of the red blood cells, is a characteristic finding. Jejunal biopsy reveals the pathognomonic accumulation of foamy lipid material in the cytoplasm of intestinal epithelial cells.

Abetalipoproteinemia appears radiographically as small bowel dilatation with mild to moderate thickening of mucosal folds. Although most marked in the duodenum and jejunum, fold thickening is present throughout the small bowel. The folds may be uniformly thickened, demonstrate an irregular, more disorganized appearance, or even have a nodular pattern.[20]

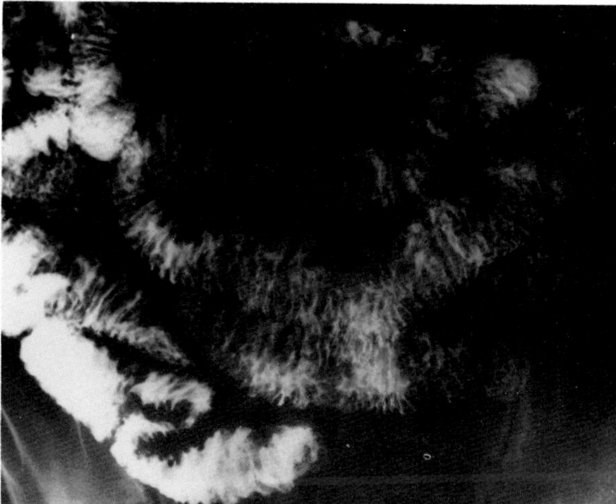

**Figure 46.19. Intestinal lymphangiectasia.** Lymphatic obstruction is due to infiltration of the bowel wall and mesentery by metastatic carcinoma. (From Eisenberg,[2] with permission from the publisher.)

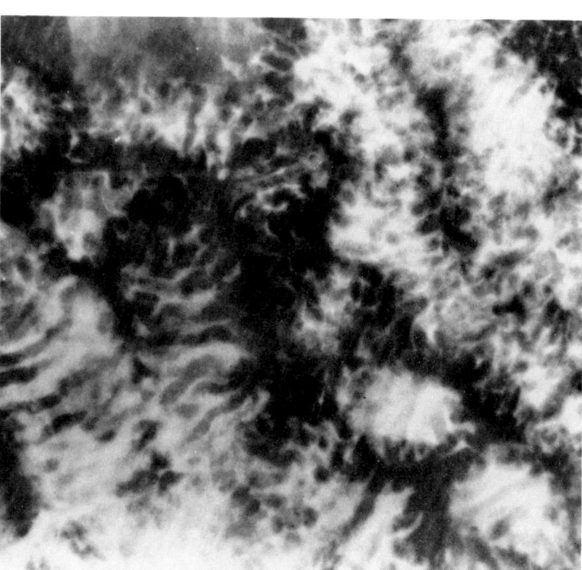

**Figure 46.20. Intestinal lymphangiectasia.** Note the diffuse nodularity and the irregular folds. (From Eisenberg,[2] with permission from the publisher.)

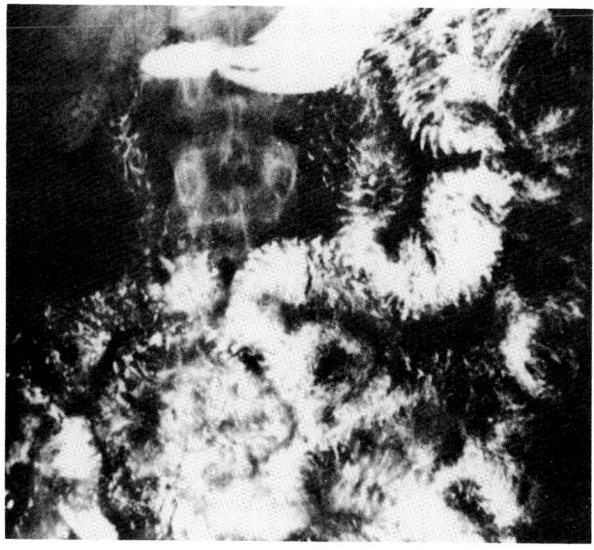

A

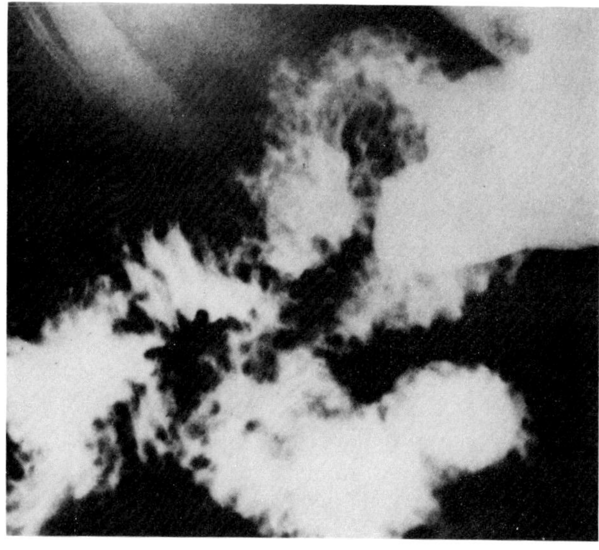

B

**Figure 46.21. Abetalipoproteinemia.** (A) A moderately disorganized fold pattern. (B) A nodular, or cobblestone, pattern of folds in the duodenum and jejunum. (From Weinstein,[20] with permission from the publisher.)

## REFERENCES


1. Isbell RG, Carlson HT, Hoffman HN: Roentgenologic pathologic correlation in malabsorption syndromes. *AJR* 107: 158–169, 1969.
2. Eisenberg, RL: *Gastrointestinal Radiology: A Pattern Approach.* Philadelphia, JB Lippincott, 1983.
3. Horowitz AL, Meyers MA: The "hide-bound" small bowel of scleroderma: Characteristic mucosal fold pattern. *AJR* 119: 332–334, 1973.
4. Poirier T, Rankin G: Gastrointestinal manifestations of progressive systemic sclerosis, based on a review of 364 cases. *Am J Gastroenterol* 58:30–44, 1972.
5. Clemett AR, Marshak RH: Whipple's disease: Roentgen features and differential diagnosis. *Radiol Clin North Am* 7:105–111, 1969.
6. Philips RL, Carlson HC: The roentgenographic and clinical findings in Whipple's disease. *AJR* 123:268–273, 1975.
7. Rice RP, Roufail W, Reeves RJ: The roentgen diagnosis of Whipple's disease (intestinal lipodystrophy) with emphasis on improvement following antibiotic therapy. *Radiology* 88: 295–301, 1967.
8. Balikian JP, Nassar NT, Shama'A MH, et al: Primary lymphomas of the small intestine including the duodenum: A roentgen analysis of 29 cases. *AJR* 107:131–141, 1969.
9. Marshak RH, Lindner AE, Maklansky DM: Lymphoreticular disorders of the gastrointestinal tract: Roentgenographic features. *Gastrointest Radiol* 4:103–120, 1979.
10. Koehler RE: Neoplasms of the small bowel, in Margulis AR, Burhenne HJ (eds): *Alimentary Tract Radiology*. St. Louis, Mosby, 1982, pp 962–980.
11. Levitt RG: Abdominal wall and peritoneal cavity, in Lee JKT, Sagel SS, Stanley RJ (eds): *Computed Body Tomography*. New York, Raven Press, 1983, pp 287–306.
12. Mauro MA, Koehler RE: Alimentary tract, in Lee JKT, Sagel SS, Stanley RJ (eds): *Computed Body Tomography*. New York, Raven Press, 1983, pp 307–340.
13. Goldberg HI, O'Kieffe D, Denis EH, et al: Diffuse eosinophilic enteritis. *AJR* 119:342–351, 1973.
14. Morrison WJ, Christopher NL, Bayless TM, et al: Low lactase levels: Evaluation of the radiologic diagnosis. *Radiology* 111:513–518, 1974.
15. Laws JW, Neale G: Radiological diagnosis of disaccharidase deficiency. *Lancet* 2:139–143, 1966.
16. Ament ME, Rubin CE: Relations of giardiasis to abnormal intestinal structure and function in gastrointestinal immunodeficiency syndromes. *Gastroenterology* 62:216–226, 1972.
17. Hodgson JR, Hoffman HN, Huivenga KA: Roentgenologic features of lymphoid hyperplasia of the small intestines associated with dysgammaglobulinemia. *Radiology* 88:883–888, 1967.
18. Shimkin PM, Waldmann TA, Krugman RL: Intestinal lymphangiectasia. *AJR* 110:827–841, 1970.
19. Olmsted WW, Madewell JE: Lymphangiectasia of the small intestine: Description and pathophysiology of the roentgenographic signs. *Gastrointest Radiol* 1:241–243, 1976.
20. Weinstein MA, Pearson KD, Agus FG: Abetalipoproteinemia. *Radiology* 108:269–273, 1973.
21. Marshak RH, Lindner AE, Maklansky D: Malabsorption, in Margulis AR, Burhenne HJ (eds): *Alimentary Tract Radiology*. St. Louis, Mosby, 1982.

## Chapter Forty-Seven
# Inflammatory Bowel Disease

## CROHN'S DISEASE (REGIONAL ENTERITIS)

Crohn's disease is a chronic inflammatory disorder of unknown cause that can occur at any age but is most common in the second and third decades. It most often involves the terminal ileum but can affect any part of the gastrointestinal tract. The granulomatous process is frequently discontinuous, with diseased segments of bowel separated by apparently healthy portions. In Crohn's disease, there is diffuse transmural inflammation with edema and infiltration of lymphocytes and plasmocytes in all layers of the intestinal wall. Ulceration is common, and intramural tracking within the submucosal and muscular layers is not infrequent.

The clinical spectrum of Crohn's disease is broad, ranging from an indolent course with unpredictable exacerbations and remissions to severe diarrhea and an acute abdomen. Extraintestinal complications (large joint migratory polyarthritis, ankylosing spondylitis, sclerosing cholangitis) occur with a higher frequency than normal in patients with Crohn's disease. In the genitourinary system, infections can result from enterovesical fistulas, hydronephrosis can develop from ureteral obstruction due to the involvement of the ureter in the granulomatous inflammatory process, and renal oxalate stones can be caused by an increased absorption of dietary oxalate and consequent hyperoxaluria. Deficient absorption of bile salts distorts the ratio of bile salts to cholesterol and predisposes to the development of cholesterol stones in the gallbladder. Almost one-third of all patients hospitalized for Crohn's disease eventually develop a small bowel obstruction. Fistula formation is seen in at least half of the patients with chronic Crohn's disease, and

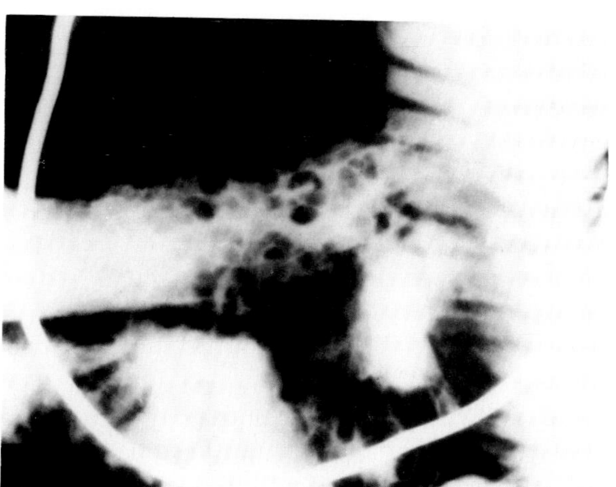

**Figure 47.1. Crohn's disease of the small bowel.** A cobblestone appearance is produced by transverse and longitudinal ulcerations separating islands of thickened mucosa and submucosa.

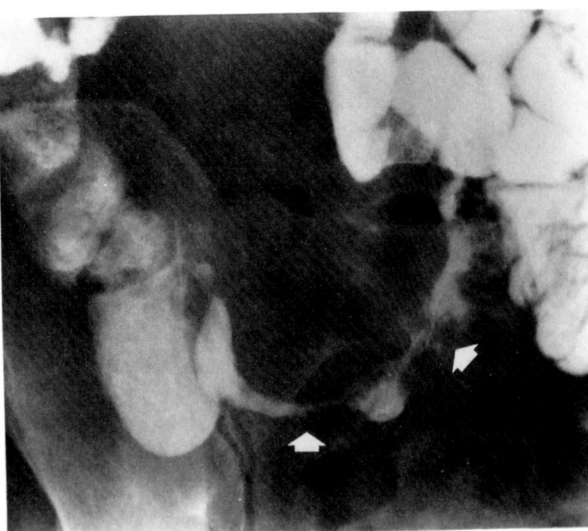

**Figure 47.2. String sign in Crohn's disease of the small bowel.** The mucosal pattern is lost in a severely narrowed, rigid segment of the terminal ileum (arrows). (From Eisenberg,[38] with permission from the publisher.)

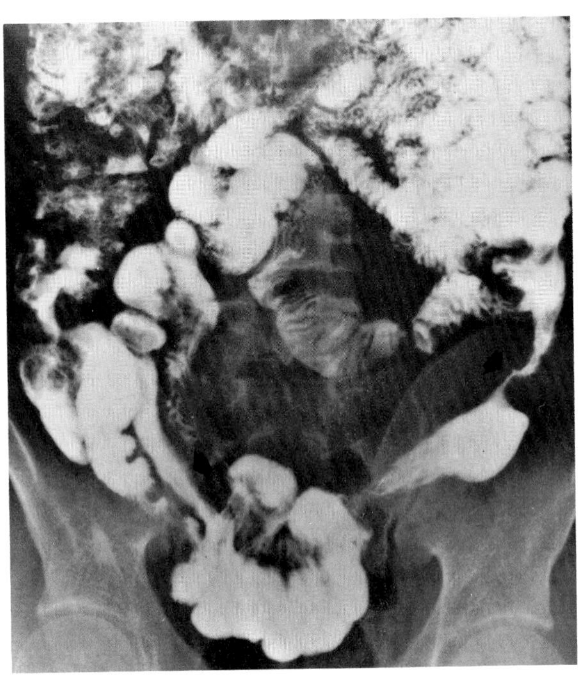

**Figure 47.3. Skip lesions in Crohn's disease of the small bowel.** The arrows point to widely separated areas of disease. (From Eisenberg,[38] with permission from the publisher.)

**Figure 47.4. Crohn's disease of the small bowel.** Marked thickening of the mesentery and mesenteric nodes produces a lobulated mass widely separating small bowel loops.

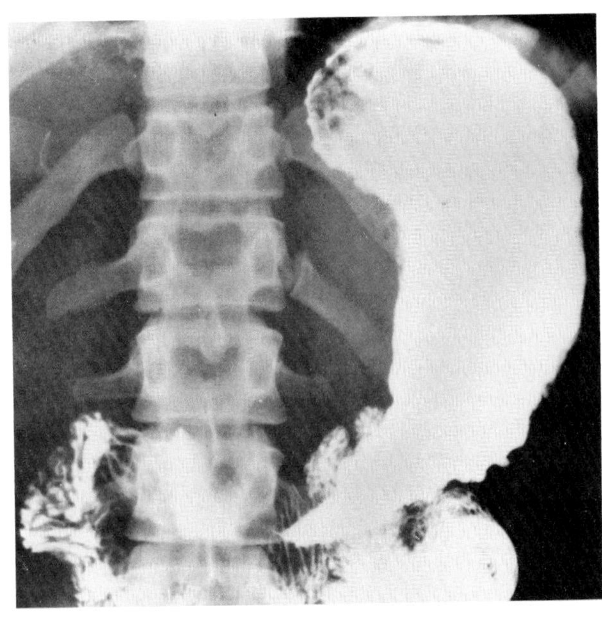

**Figure 47.5. Crohn's disease of the stomach.** Smooth, tubular narrowing of the antrum flares out into a relatively normal gastric body and fundus. (From Farman, Faegenburg, Dallemand, et al,[1] with permission from the publisher.)

chronic indurated rectal fissures and fistulas with associated perirectal abscesses occur in about one-third. After surgical resection of an involved segment of small bowel, there is a high incidence of Crohn's disease recurring adjacent to the anastomosis.

Radiographically, the earliest small bowel changes in Crohn's disease include irregular thickening and distortion of the valvulae conniventes due to submucosal inflammation and edema. Transverse and longitudinal ulcerations can separate islands of thickened mucosa and submucosa, leading to a characteristic rough cobblestone appearance. Rigid thickening of the entire bowel wall produces pipelike narrowing. Continued inflammation and fibrosis can result in a severely narrowed, rigid segment of small bowel in which the mucosal pat-

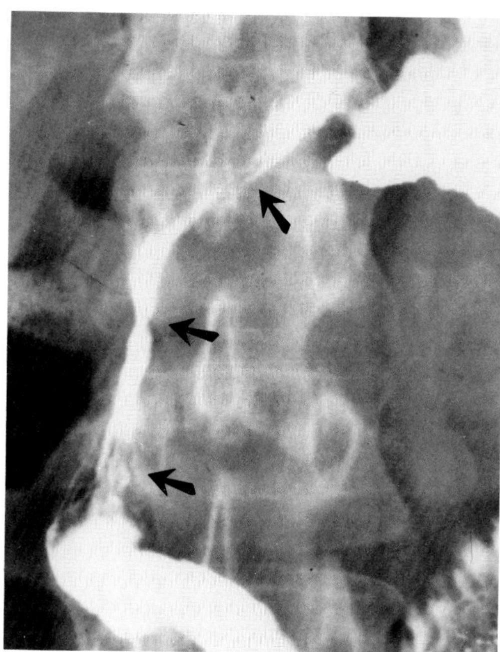

**Figure 47.6. Crohn's disease of the duodenum.** There is a long stenosis (arrows) with effacement of the normal mucosal pattern involving the first and second portions of the duodenum. (From Eisenberg,[38] with permission from the publisher.)

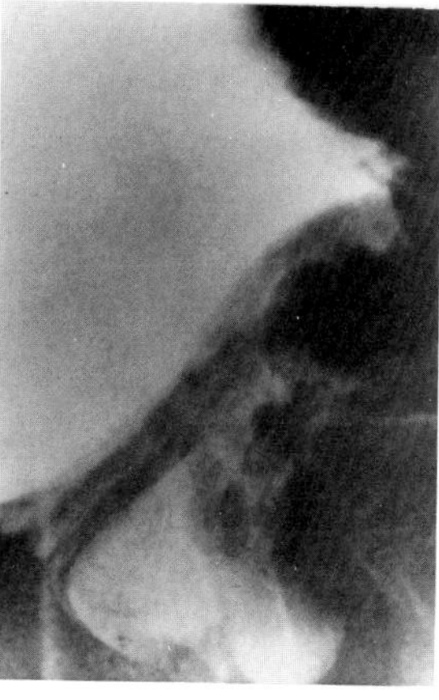

**Figure 47.7. Crohn's disease of the appendix.** A pressure film from a barium enema examination reveals a large extrinsic mass impinging upon the cecal tip and medial cecal wall. Note the normal mucosa and the distensibility of the terminal ileum. (From Threatt and Appelman,[6] with permission from the publisher.)

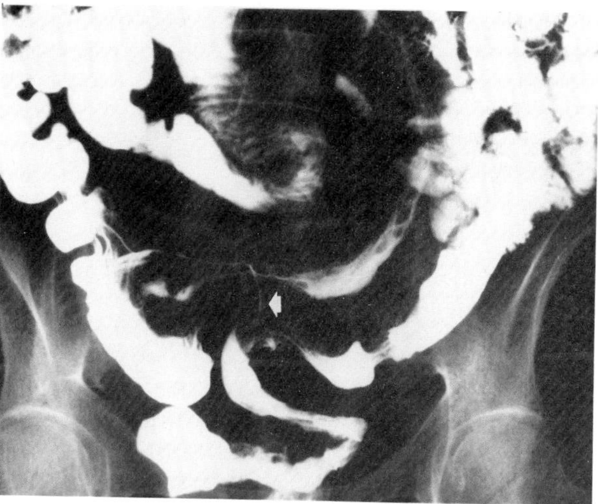

**Figure 47.8. Crohn's disease.** There is a fistula between the distal ileum and the sigmoid (arrow). (From Eisenberg,[38] with permission from the publisher.)

tern is lost (string sign). When several areas of small bowel are diseased, involved segments of varying length are often sharply separated from radiographically normal segments (skip lesions).

Loops of small bowel involved with Crohn's disease often are separated from one another due to thickening of the bowel wall and infiltration of the mesentery. Mass effects on involved loops can be produced by adjacent abscesses, thickened indurated mesentery, or enlarged and matted lymph nodes. Irregular fibrotic strictures are common and can lead to intestinal obstruction. Localized perforations and fistulas from the small bowel to other visceral organs can sometimes be demonstrated.

Crohn's disease can affect virtually any portion of the gastrointestinal tract, though there are almost always associated lesions of the small bowel. In the esophagus, Crohn's disease can rarely produce diffuse or discrete esophageal ulceration, severe luminal narrowing, and esophagorespiratory fistulas. Crohn's disease involving the stomach typically results in a smooth, tubular antrum that is poorly distensible and exhibits sluggish peristalsis. The narrowed antrum flares out into a normal gastric body and fundus, giving the appearance of a ram's horn. Because the adjacent duodenal bulb and proximal sweep are almost always involved, the radiographic appearance of diffuse narrowing (pseudo-Billroth-I pattern) can mimic that in a patient who has undergone partial gastrectomy and Billroth I anastomosis. In addition to antral narrowing, there can be cobblestoning of antral folds along with fissures and ulcerations. The ulcerations begin as superficial gastric erosions, which appear as tiny flecks of barium surrounded by a radiolucent mound of edematous mucosa. These "aphthous" gastric ulcers can progress to deeper ulcers, scarring, and stenosis. Up to two-thirds of the patients with Crohn's disease of the stomach develop gastric outlet obstruction.

In the duodenum, Crohn's disease can cause nodularity and postbulbar ulceration that can only be differentiated from benign peptic ulcer disease because of concomitant distal small bowel involvement. Thickening of mucosal folds produces a cobblestone appearance. Fibrotic healing causes fusiform and concentric narrowing of the duodenum that may lead to partial obstruction.

Rarely, Crohn's disease is limited to the appendix and shows no evidence of terminal ileal involvement. This condition usually occurs in young adults with no previous gastrointestinal problems who present with an acute onset of abdominal pain and a palpable right lower quadrant mass simulating acute appendicitis. Barium enema examination generally demonstrates a large extrinsic mass

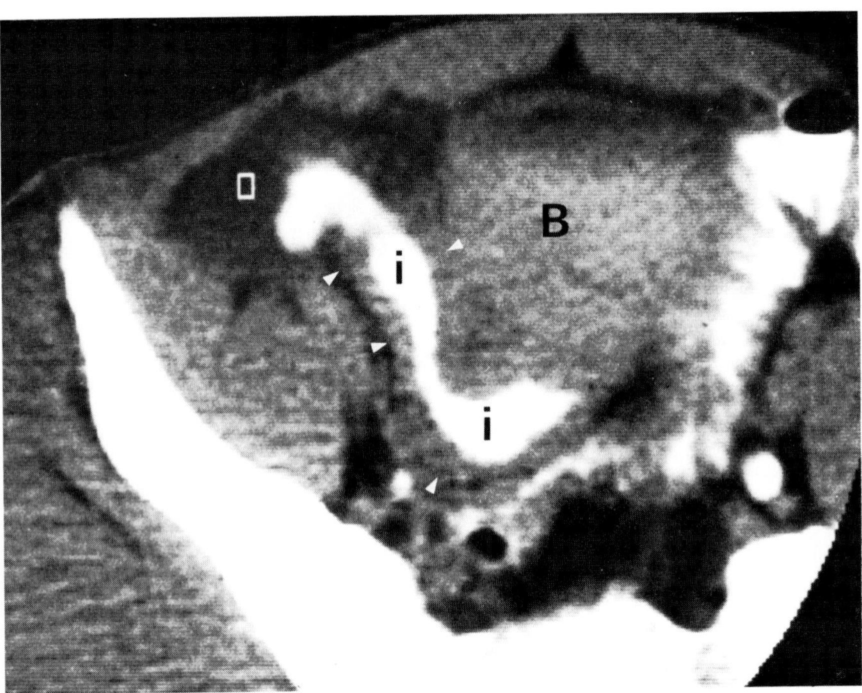


**Figure 47.9. Crohn's disease.** A CT scan shows the thickened wall (arrowheads) of a long abnormal segment of contrast-filled ileum (i). Note the mass of water density (box), which proved to be a mesenteric abscess. B = bladder. (From Moss and Thoeni,[9] with permission from the publisher.)

impinging upon the cecal tip and medial cecal wall. The terminal ileum is distensible and has a normal mucosal pattern.

Fistula formation is a hallmark of chronic Crohn's disease, found in at least half of all patients with this condition. The diffuse inflammation of the serosa and mesentery in Crohn's disease causes involved loops of bowel to be firmly matted together by fibrous peritoneal and mesenteric bands. Fistulas apparently begin as ulcerations that burrow through the bowel wall into adjacent loops of small bowel and colon. Fistulas between segments of bowel can cause severe nutritional problems if they bypass extensive areas of intestinal absorptive surface; the recirculation of intestinal contents and subsequent stasis can permit bacterial overgrowth and malabsorption. In addition to fistulas between loops of bowel, a characteristic finding in Crohn's disease is the appearance of fistulous tracts ending blindly in abscess cavities surrounded by dense inflammatory tissue. These abscess cavities are situated intraperitoneally, retroperitoneally, or deep within the mesentery and can produce palpable masses, persistent fever, or pain. Although less common than bowel-bowel fistulas, internal fistulas extending from the bowel to the bladder or vagina can occur in patients with Crohn's disease. External gastrointestinal fistulas are commonly encountered in patients with Crohn's disease. They usually extend to the perianal area and produce chronic indurated rectal fistulas with associated fissures and perirectal abscesses. (Crohn's colitis is discussed later in this section.)

Computed tomography (CT) is of value in demonstrating mural, serosal, and mesenteric abnormalities in Crohn's disease and in defining the nature of mass effects, separation, or displacement of small bowel segments seen on barium studies. CT can show thickening of the bowel wall, fibrofatty proliferation of mesenteric fat, mesenteric abscesses, inflammatory reactions of the mesentery, and mesenteric lymphadenopathy.[1–9]

## INFLAMMATORY DISEASES OF THE COLON

Ulcerative inflammation of the colon or rectum is a nonspecific response to a host of harmful agents and processes. In many cases, an ulcerating colitis can be attributed to a specific infectious disease, systemic disorder, or toxic agent. However, in a large group of patients, a precise cause cannot be determined. Most of these "nonspecific" inflammatory diseases of the colon are generally placed into one of two categories: ulcerative colitis or Crohn's disease. Although radiographic and pathologic criteria have been established for distinguishing between these two processes, there is a substantial overlap in practice. In at least 10 percent of colectomy specimens from patients with an ulcerating colitis, it is impossible to distinguish between ulcerative colitis and Crohn's disease even with careful gross inspection and multiple microscopic sections. Features of ulcerative colitis and Crohn's disease often coexist, making a precise histologic diagnosis difficult.

### Ulcerative Colitis

Ulcerative colitis is primarily a disease of young adults, the peak incidence being in persons between 20 and 40

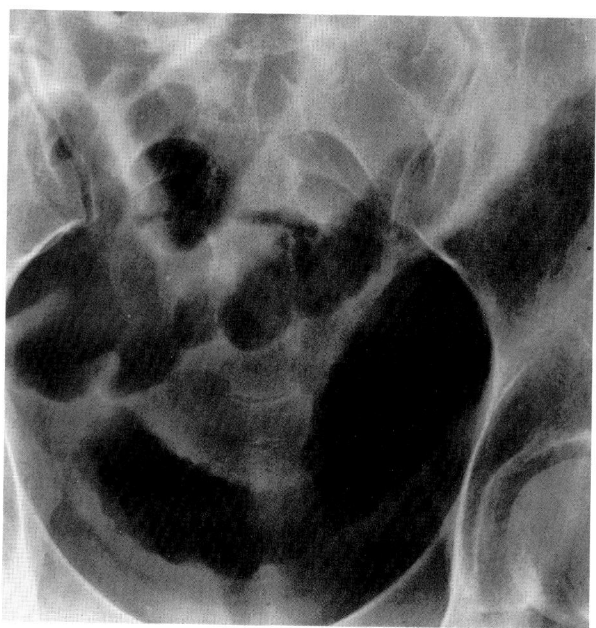

**Figure 47.10. Ulcerative colitis.** A plain abdominal radiograph demonstrates nodular protrusions of hyperplastic mucosa and the loss of haustral markings involving essentially the entire sigmoid colon. (From Eisenberg,[38] with permission from the publisher.)

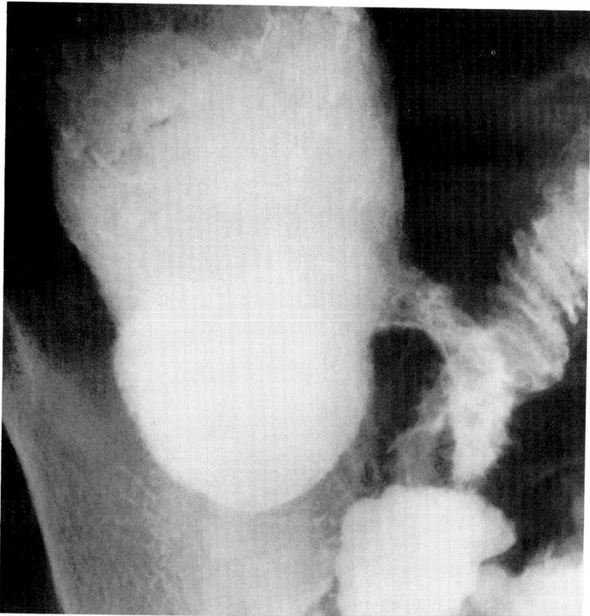

**Figure 47.11. Backwash ileitis in ulcerative colitis.** The ileocecal valve is gaping, and there are inflammatory changes in the terminal ileum. (From Eisenberg,[38] with permission from the publisher.)

years of age. Ulcerative colitis is highly variable in severity, clinical course, and ultimate prognosis. The onset of the disease, as well as subsequent exacerbations, can be insidious or abrupt. The major symptoms include bloody diarrhea, abdominal pain, fever, and weight loss. A characteristic feature of ulcerative colitis is alternating periods of remission and exacerbation. Most patients have intermittent episodes of symptoms with complete remission between attacks. In fewer than 15 percent of the patients, ulcerative colitis is an acute fulminant process. Patients with this form of the disease have severe hemorrhagic diarrhea, fever, systemic toxicity, and electrolyte depletion, as well as a far higher incidence of severe complications such as toxic megacolon and free perforation into the peritoneal cavity.

Extracolonic manifestations of ulcerative colitis are relatively common and include spondylitis, peripheral arthritis, iritis, skin disorders (erythema nodosum, pyoderma gangrenosum), and various liver abnormalities (pericholangitis, fatty infiltration, sclerosing cholangitis, cholangiocarcinoma, and chronic active hepatitis).

In the radiographic evaluation of a patient with known or suspected ulcerative colitis, plain abdominal radiographs are essential. Large nodular protrusions of hyperplastic mucosa, deep ulcers outlined by intraluminal gas, or polypoid changes along with a loss of haustral markings suggest the diagnosis. Plain abdominal radiographs can also demonstrate evidence of toxic megacolon or free intraperitoneal gas, contraindications to barium enema examinations in patients with acute colitis.

Ulcerative colitis has a strong tendency to begin in the rectosigmoid. Although by radiographic criteria alone the rectum appears normal in about 20 percent of the patients with ulcerative colitis, true rectal sparing is infrequent, and there is usually evidence of disease on sigmoidoscopy or rectal biopsy. Although ulcerative colitis may spread to involve the entire colon (pancolitis), isolated right colon disease with a normal left colon does not occur. The disease is almost always continuous, without evidence of the skip areas seen in Crohn's disease. Except for "backwash ileitis" (minimal inflammatory changes involving a short segment of terminal ileum), ulcerative colitis does not involve the small bowel, a feature distinguishing it from Crohn's disease, which may involve both the large and the small intestine.

Because of reports of complications after the use of routine purgatives and cleansing enemas, the patient with suspected or known ulcerative colitis should receive minimal preparation of the colon before a barium enema examination. If time permits, the safest preparation is several days of a clear liquid diet with gentle, small-volume enemas the night before and the morning of the examination. Strong laxatives such as castor oil are contraindicated. In patients with acute disease, it may be wisest to postpone the barium enema, because the increased intraluminal pressure during the examination may precipitate the development of toxic megacolon.

On double-contrast studies, the earliest detectable radiographic abnormality in ulcerative colitis is fine granularity of the mucosa corresponding to the hyperemia

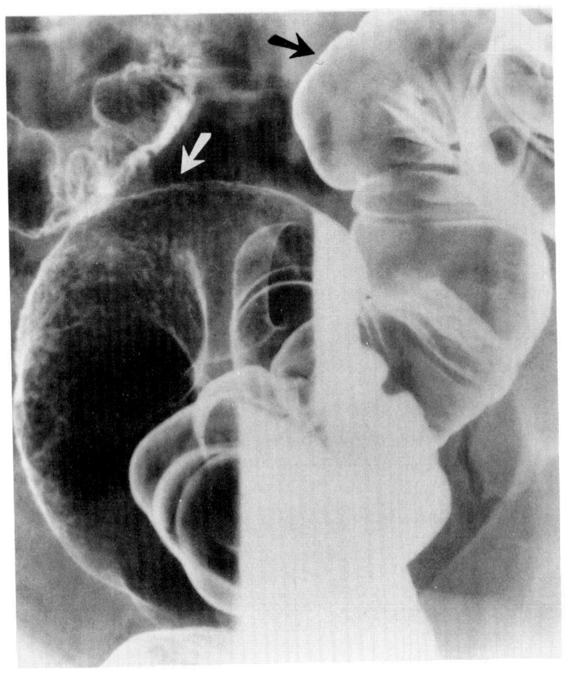
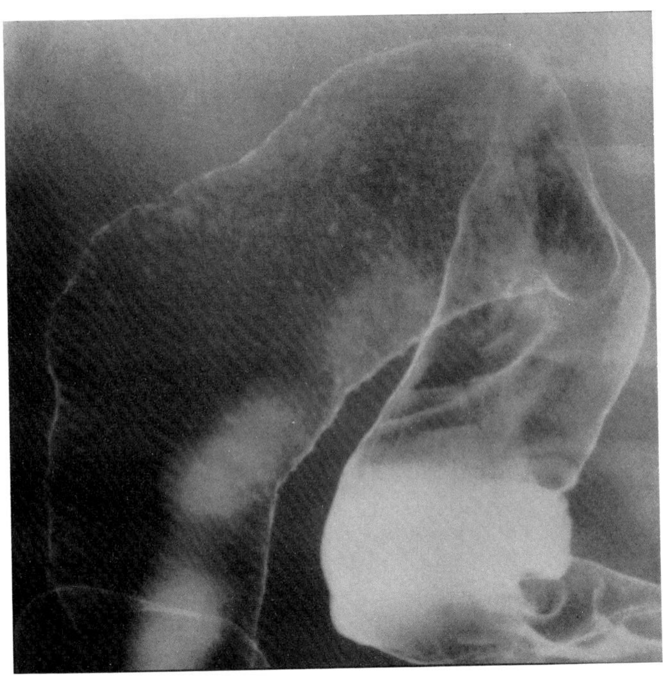

A B

**Figure 47.12. Early ulcerative colitis.** (A) The distal rectosigmoid mucosa (white arrow) is finely granular compared with the normal-appearing mucosa (black arrow) in the more proximal colon. (B) Typical stippled mucosal pattern. (From Eisenberg,[38] with permission from the publisher.)

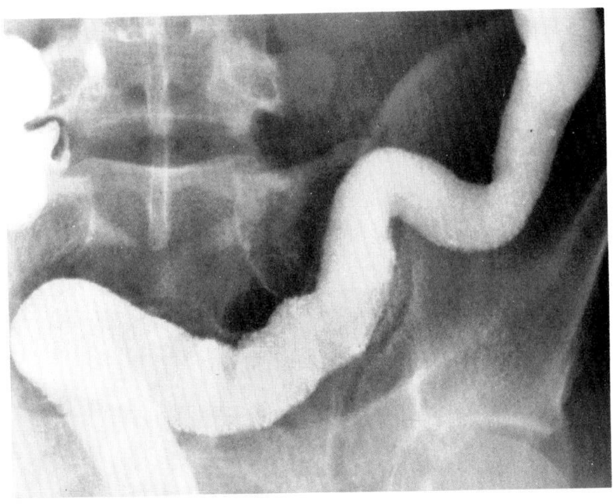

**Figure 47.13. Early ulcerative colitis.** The hazy quality of the bowel contour is due to edema, excessive mucus, and tiny ulcerations.

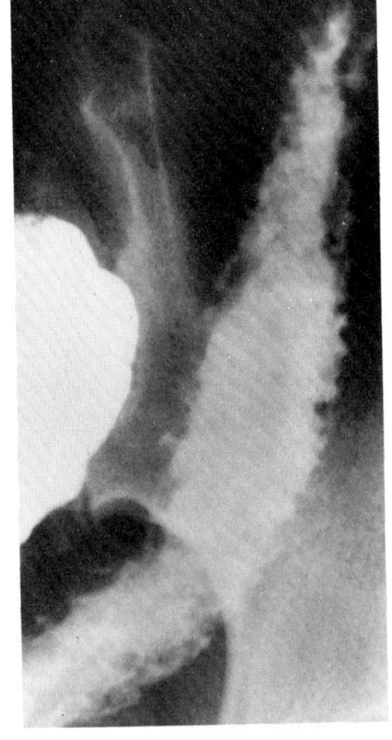

**Figure 47.14. Ulcerative colitis.** Progression of the disease results in deep ulceration (collar-button ulcers) into the mucosal layer. (From Eisenberg,[38] with permission from the publisher.)

and edema seen endoscopically. Once superficial ulcers develop, small flecks of adherent barium produce a stippled mucosal pattern. On full-column examination, an early finding in ulcerative colitis is a hazy or fuzzy quality of the bowel contour with serrated margins that is related to edema, excessive mucus, and tiny ulcerations. It is essential that these hazy, asymmetric, nonuniform ulcerations be distinguished from innominate lines, tiny spicules that mimic ulcerations but are symmetric and sharply defined and represent barium penetration into normal grooves that are present on the surface of the colonic mucosa.

On single-contrast studies, the postevacuation film is frequently of great value in the detection of the early changes of ulcerative colitis. Unlike the fine crinkled pattern of criss-crossing thin mucosal folds seen in the normal colon, the folds in ulcerative colitis become thickened, indistinct, and coarsely nodular and tend to course in a longitudinal direction. The thin coating of barium on the surface appears finely stippled because of countless tiny ulcers that cause numerous spikelike projections when seen in profile.

As the disease progresses, the ulcerations become deeper. Extension into the submucosa may produce broad-based ulcers with a collar-button appearance, a nonspecific pattern that can be seen in numerous forms of ulcerating colitis. Perirectal inflammation can cause widening of the soft tissue space between the anterior sacrum and the posterior rectum (retrorectal space).

With chronic disease, fibrosis and muscular spasm cause progressive shortening and rigidity of the colon. The haustral pattern is absent and the bowel contour is relatively smooth because of ulcer healing and subsequent re-epithelialization. Eventually, the colon may appear as a symmetric, rigid tubular structure (lead-pipe colon).[10–17]

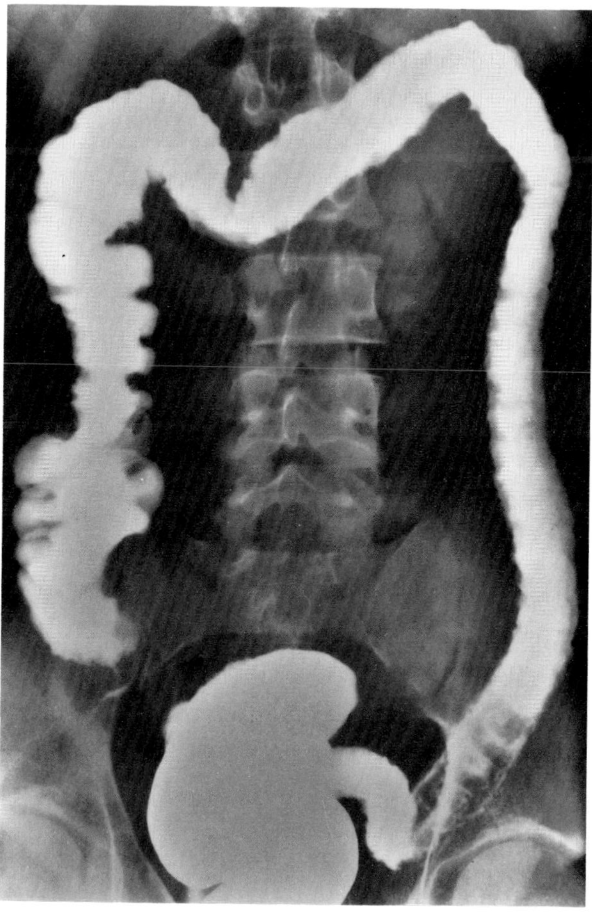

**Figure 47.15. Chronic ulcerative colitis.** Fibrosis and muscular spasm cause shortening and rigidity of the colon and a loss of haustral markings. (From Eisenberg,[38] with permission from the publisher.)

## Complications of Ulcerative Colitis

### TOXIC MEGACOLON

Toxic megacolon is a dramatic and ominous complication of ulcerative colitis. It is characterized by extreme dilatation of a segment of colon, or of an entire diseased colon, combined with systemic toxicity (abdominal pain and tenderness, tachycardia, fever, and leukocytosis). Pathologically, specimens from patients with toxic megacolon show extensive deep ulcerations and acute inflammation involving the muscular layer of the colon and often extending to the serosa. The wall of the colon is extremely thin and friable, predisposing to perforation.

In most patients with toxic megacolon, a simple plain film of the abdomen demonstrating marked distension of the colon ($>$ 5.5 cm) is diagnostic. With the patient supine, the transverse colon is the portion most prominently distended (since colonic gas rises to the highest segment). The distal descending colon and the sigmoid are less frequently dilated; distension of the rectum is uncommon. Gas within the dilated segment of colon is frequently sufficient to silhouette the mucosa and often reveals multiple broad-based, nodular, pseudopolypoid projections extending into the lumen, as well as gas-filled crevices that probably represent deep ulcers between the nodular masses.

The major complication of toxic megacolon is spontaneous perforation of the colon, which can be dramatic and sudden and can cause irreversible shock. Toxic megacolon is associated with a mortality rate of up to 30 percent, though in some hospitals this has been somewhat reduced by early diagnosis and aggressive surgical therapy. Because there is such a high danger

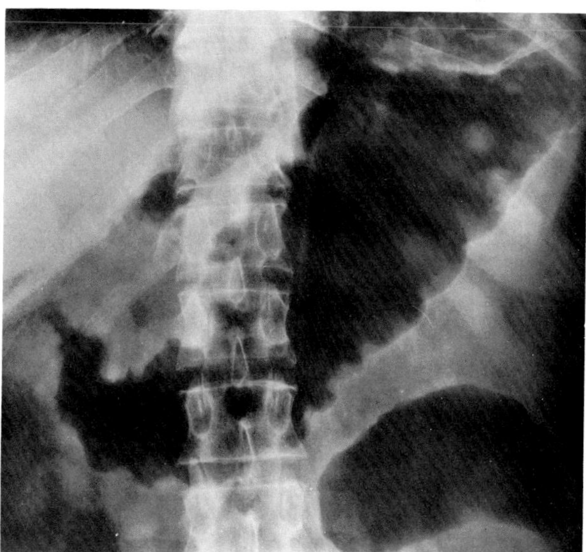

**Figure 47.16. Toxic megacolon in ulcerative colitis.** There is dilatation of the transverse colon with multiple pseudopolypoid projections extending into the lumen. (From Eisenberg,[38] with permission from the publisher.)

of spontaneous perforation, barium enema examination is absolutely contraindicated during a recognized attack of toxic megacolon.[18–20]

#### FISTULA FORMATION

Rectovaginal fistulas occur in about 2 to 3 percent of women with ulcerative colitis. These fistulas frequently do not heal after local surgical repair; colectomy with ileostomy or a temporary diverting procedure is often required. Colovesical fistulas can also develop in ulcerative colitis. Although fistulization between loops of bowel does occur in ulcerative colitis, it is much less common than in Crohn's disease.[38]

#### PSEUDOPOLYPS

Pseudopolyps are islands of hyperplastic, inflamed regenerating mucosa that remain between areas of ulceration. Inflammatory pseudopolyps in ulcerative colitis are usually small and uniform in appearance, producing a somewhat nodular pattern. On rare occasions, however, the mucosa can undergo such extreme hyperplasia that the pseudopolyps appear as large masses simulating fungating polypoid tumors. Pseudopolyps in ulcerative colitis are usually associated with radiographic evidence of an inflammatory process and a characteristic history of chronic diarrhea. In the occasional case in which the inflammatory process has healed and the only residual radiographic abnormality is a large number of pseudopolyps, it can be impossible to distinguish this appearance from familial polyposis.[38]

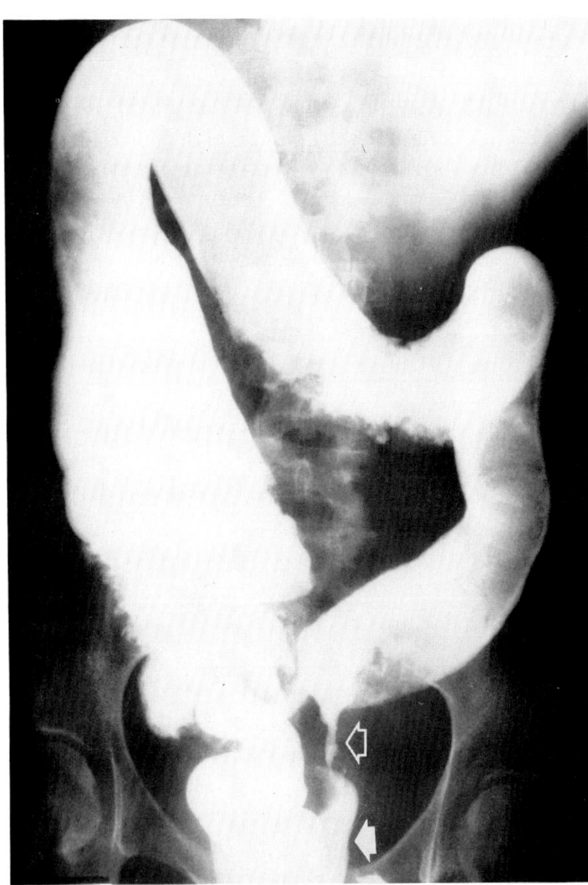

**Figure 47.17. Colovesical fistula in ulcerative colitis.** The open arrow points to the fistula; the closed arrow points to contrast material in the bladder. (From Eisenberg,[38] with permission from the publisher.)

#### COLONIC STRICTURES

Benign colonic strictures develop in up to 10 percent of patients with chronic ulcerative colitis. The most common site of benign stricture is the rectum, followed next in frequency by the transverse colon. Usually asymptomatic, the strictures tend to be short (2 to 3 cm) but can extend up to 30 cm in length. Although most commonly single, strictures in ulcerative colitis can be multiple, especially in patients with universal colonic disease.

Radiographically, a stricture due to ulcerative colitis has a typically benign appearance with a concentric lumen, smooth contours, and fusiform, pliable tapering margins. Occasionally, the stricture is somewhat eccentric and has irregular contours, simulating malignancy. Because carcinoma in patients with ulcerative colitis can have a radiographic appearance indistinguishable from that of benign stricture, colonoscopy or surgery is frequently required to make this differentiation.[21]

#### CARCINOMA OF THE COLON

Carcinoma of the colon is about 10 times more frequent in patients with ulcerative colitis than in the general population. The incidence of cancer is related to the duration of the colitis, the age of the patient at the time of onset, and the linear extent of the disease; it is not related to the severity or activity of the inflammatory

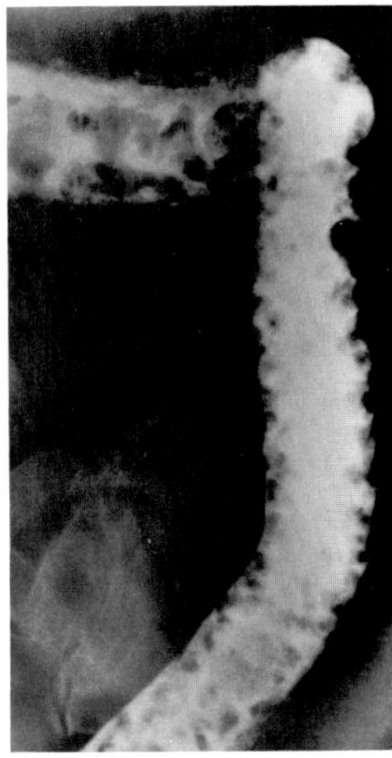

**Figure 47.18. Pseudopolyposis in acute ulcerative colitis.** Irregular filling defects are associated with radiographic evidence of a severe ulcerating process. (From Eisenberg,[38] with permission from the publisher.)

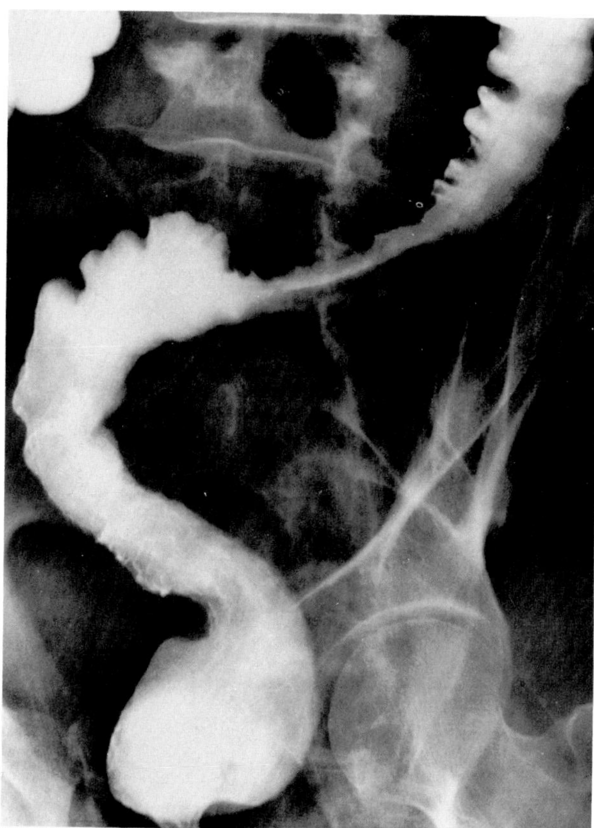

**Figure 47.19. Chronic ulcerative colitis.** A benign stricture is visible in the sigmoid colon (solid arrow). Note the ulcerative changes in the upper rectum and proximal sigmoid colon. (From Eisenberg,[38] with permission from the publisher.)

process. During the first 10 years of disease, there is only a small risk of malignancy. Thereafter, however, it is estimated that there is a 20 percent chance per decade that a patient with ulcerative colitis will develop carcinoma. The incidence of malignancy is far higher in patients who develop ulcerative colitis before the age of 25 and in those with colitis involving the entire colon. Malignant lesions in ulcerative colitis generally occur at a much younger age than in the general population and tend to be extremely virulent. Because cancer in ulcerative colitis is multicentric in up to 20 percent of the cases, atypical in its early appearance, and rapidly metastasizing, the diagnosis is often difficult to make and the prognosis is poor.

Carcinoma of the colon in patients with chronic ulcerative colitis often appears as a filiform stricture rather than having the more characteristic polypoid or apple-core appearance of a primary colonic malignancy. The tumor is typically a narrowed segment with an eccentric lumen, irregular contours, and margins that are rigid and tapered. Because it is frequently difficult to distinguish carcinoma from benign stricture in patients with ulcerative colitis, colonoscopy or surgery is often required for an unequivocal diagnosis to be made.[22]

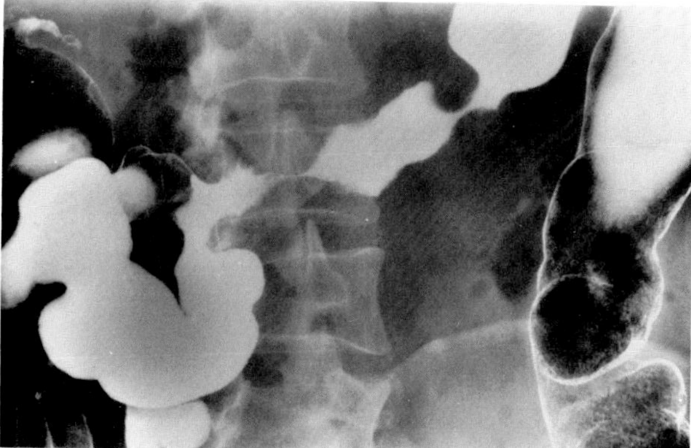

**Figure 47.20. Carcinoma of the colon developing in a patient with long-standing chronic ulcerative colitis.** A long, irregular lesion with a bizarre pattern is visible in the transverse colon. Note the pseudopolyps in the visualized portion of the descending colon.

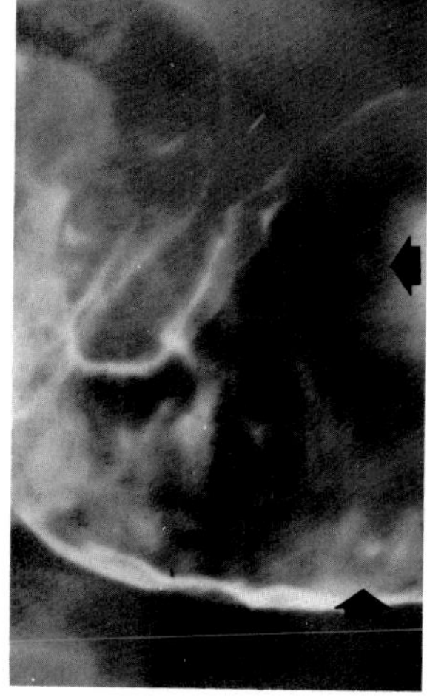

**Figure 47.21. Aphthoid ulcers of Crohn's colitis.** Punctate collections of barium are surrounded by lucent halos of edema (arrows). (From Eisenberg,[38] with permission from the publisher.)

## Crohn's Colitis

Crohn's disease of the colon is identical to the same pathologic process involving the small bowel and must be distinguished from ulcerative colitis, the other main cause of "nonspecific" inflammatory disease of the colon. The proximal portion of the colon is most frequently involved in Crohn's disease; concomitant disease of the terminal ileum is seen in up to 80 percent of the patients. Unlike ulcerative colitis, in Crohn's colitis the rectum is often spared and isolated rectal disease very rarely occurs. Crohn's disease usually has a patchy distribution, with involvement of multiple noncontiguous segments of colon (skip lesions), unlike the continuous colonic involvement in ulcerative colitis. Crohn's colitis is a transmural disease, affecting all layers of the colon, unlike ulcerative colitis, which is primarily a mucosal disease. Although granulomas can be demonstrated in only about half of the patients with Crohn's disease of the

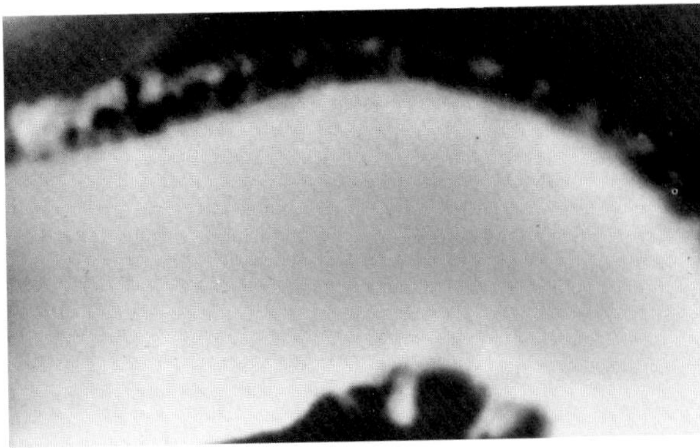

**Figure 47.22. Collar-button ulcers in Crohn's colitis.** The ulcers are distributed asymmetrically around the circumference of the bowel, in contrast to the more uniform and monotonous pattern of ulceration in ulcerative colitis. (From Eisenberg,[38] with permission from the publisher.)

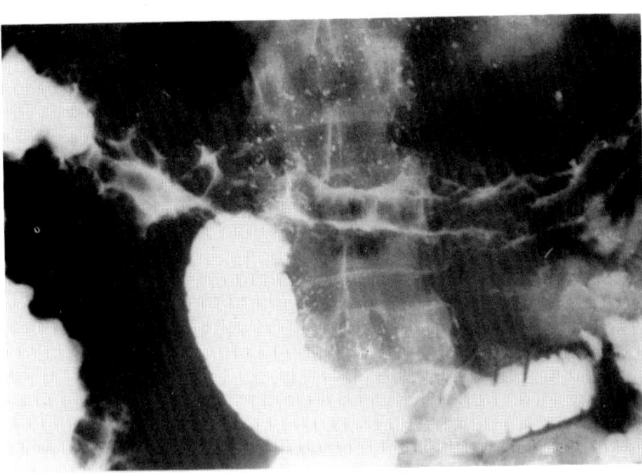

**Figure 47.23. Crohn's colitis.** Deep, linear, transverse and longitudinal ulcers. (From Eisenberg,[38] with permission from the publisher.)

colon, they are virtually a specific histopathologic feature in that they are not seen in patients with ulcerative colitis.

Perianal or perirectal abnormalities (fissures, abscesses, fistulas) occur at some point during the course of disease in half of the patients with Crohn's colitis but are rare in ulcerative colitis.

The earliest radiographic findings of Crohn's disease of the colon are seen on double-contrast examinations. Isolated, tiny, discrete erosions (aphthoid ulcers) appear as punctate collections of barium with a thin halo of edema around them. Aphthoid ulcers in Crohn's disease have a patchy distribution against a background of normal mucosa, unlike the blanket of abnormal granular mucosa seen in ulcerative colitis. These aphthoid ulcers are not specific for Crohn's disease; morphologically similar lesions can occur in other inflammatory conditions of the colon, such as amebic colitis, tuberculosis, *Yersinia* colitis, and Behçet's syndrome.

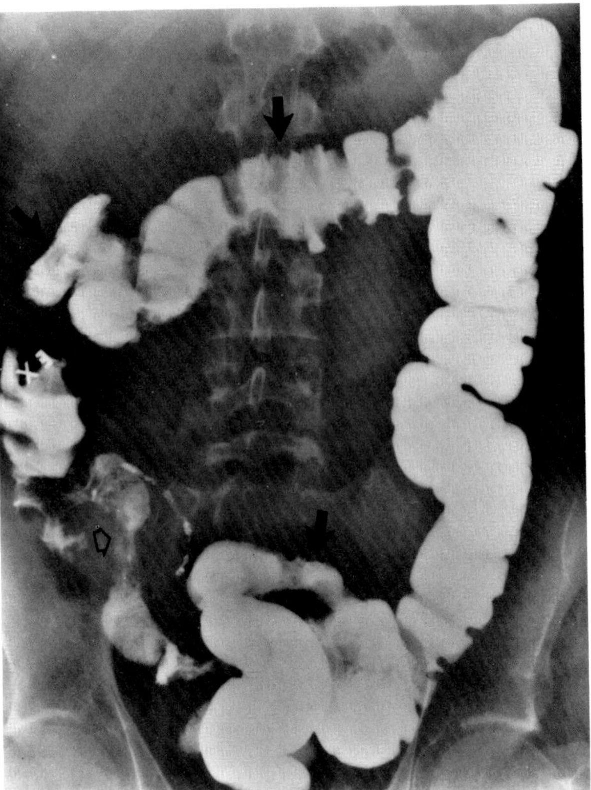

**Figure 47.24. Crohn's colitis.** Areas of involved colon (skip lesions) in the ascending, transverse, and sigmoid regions (solid arrows) are separated by normal-appearing segments. Note the inflammatory changes affecting the distal ileum (open arrow). (From Eisenberg,[38] with permission from the publisher.)

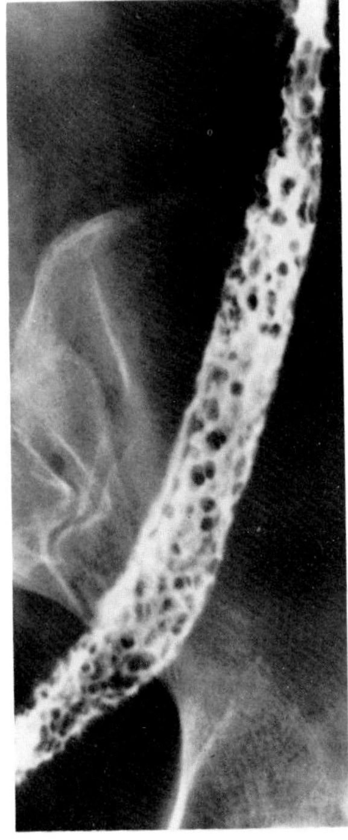

**Figure 47.25. Pseudopolyposis in Crohn's colitis.** The swollen and edematous mucosa surrounded by long, deep, linear ulcers and transverse fissures produces a coarsened nodular cobblestone appearance.

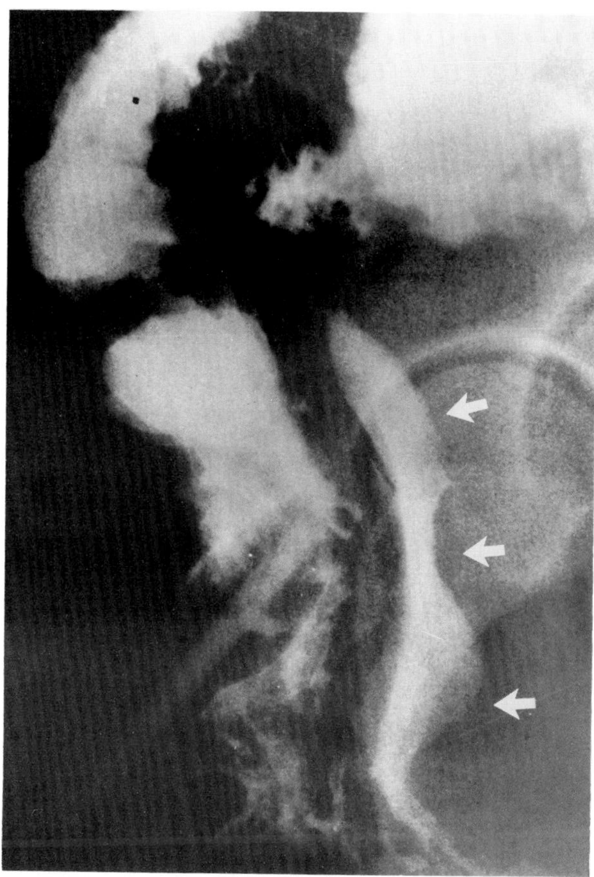

**Figure 47.26. Rectovaginal fistula in Crohn's disease.** The arrows point to contrast material in the vagina. (From Eisenberg,[38] with permission from the publisher.)

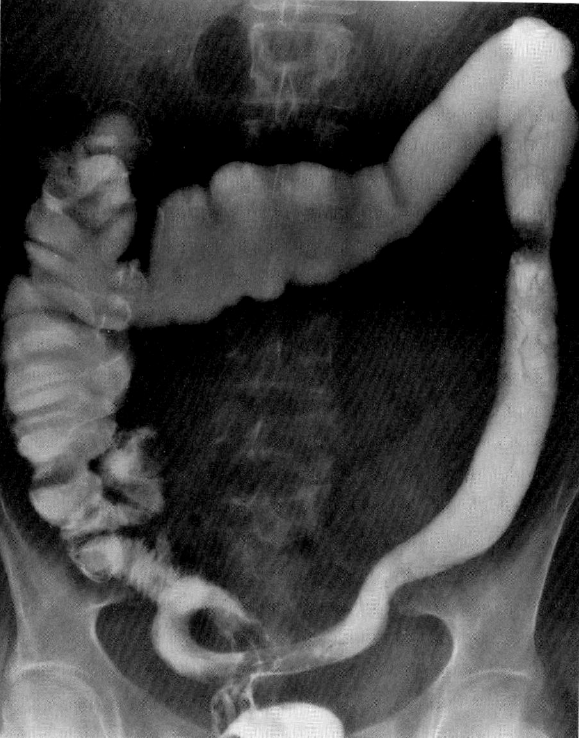

**Figure 47.27. Chronic Crohn's colitis.** Foreshortening and loss of haustrations involving the colon distal to the hepatic flexure simulate the appearance of chronic ulcerative colitis. (From Eisenberg,[38] with permission from the publisher.)

As Crohn's colitis progresses, the ulcers become deeper and more irregular, with a great variation in size, shape, and overall appearance. The distribution of ulcers around the circumference of the bowel in Crohn's disease is random and asymmetric, not uniform and monotonous as in ulcerative colitis. Deep, linear, transverse and longitudinal ulcers often separate intervening mounds of edematous but nonulcerating mucosa, thereby producing a characteristic cobblestone appearance. If the penetrating ulcers extend beyond the contour of the bowel, they can coalesce to form long tracks running parallel to the longitudinal axis of the colon. The penetration of ulcers into adjacent loops of bowel or into the bladder, vagina, or abdominal wall causes fistulas that can often be demonstrated radiographically.

Thickening of the bowel wall due to transmural inflammation and intramural fibrosis leads to narrowing of the lumen and stricture formation. Occasionally, an eccentric stricture with a suggestion of overhanging edges can be difficult to distinguish from annular carcinoma. In most instances, however, characteristic features of Crohn's disease elsewhere in the colon (deep ulcerations, pseudopolyposis, skip lesions, sinus tracts, fistulas) clearly indicate the correct diagnosis.

Patients with Crohn's colitis appear to have a higher incidence of developing colon cancer than the general population, although this association is less striking than that between colon cancer and ulcerative colitis. Carcinoma complicating Crohn's colitis is most common in the proximal portion of the colon. It usually appears radiographically as a fungating mass with typical malignant features, unlike the mildly irregular stricture characteristic of colon cancer in patients with ulcerative colitis.[11–15,17,39]

## Infectious Colitis

Diffuse or segmental colitis can be a manifestation of numerous infectious processes caused by bacterial agents (*Shigella, Salmonella, Mycobacterium tuberculosis,* gonococcus, *Yersinia, Campylobacter fetus*), protozoans (amebiasis, schistosomiasis), and viruses (lymphogranuloma venereum, herpes zoster, cytomegalovirus) and worms (strongyloidiasis). Most of these organisms (discussed in Part Two) produce a nonspecific ulcerative inflammation of the colon that can simulate ulcerative or Crohn's colitis.[23–35]

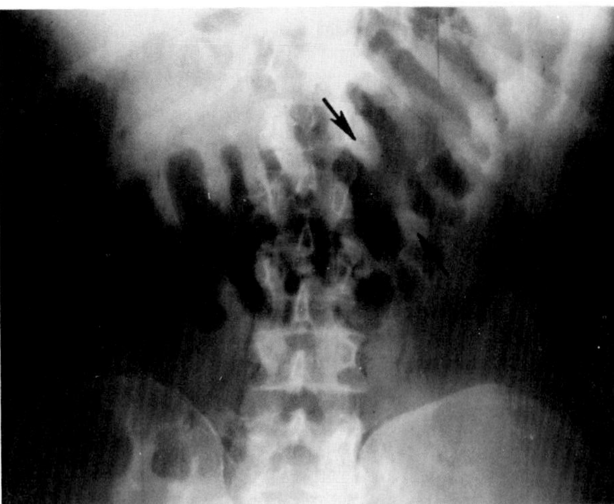

**Figure 47.28. Pseudomembranous colitis.** A plain abdominal radiograph demonstrates wide transverse bands of thickened colonic wall (arrows). (From Stanley, Melson, and Tedesco,[36] with permission from the publisher.)

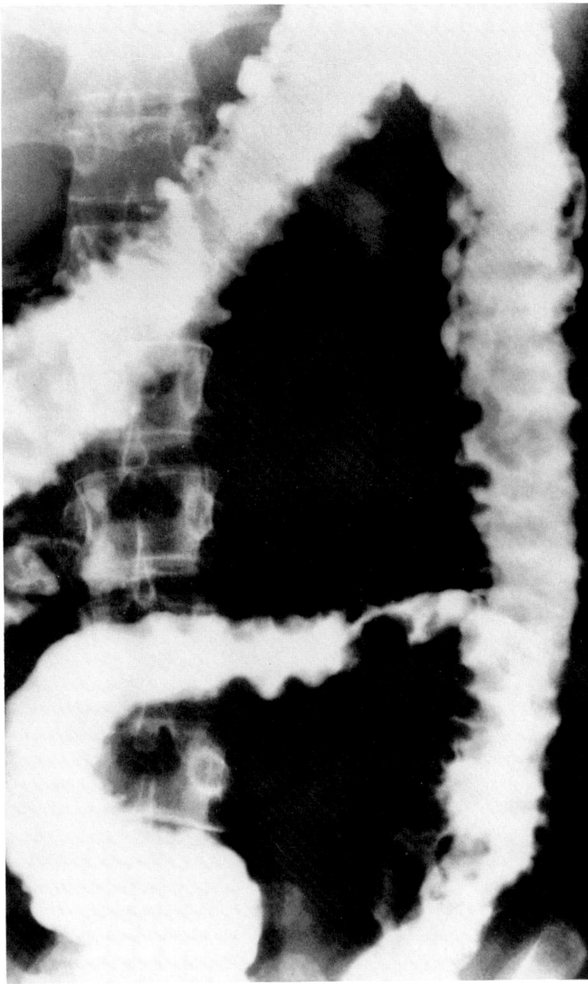

**Figure 47.29. Pseudomembranous colitis.** The barium column has a shaggy and irregular appearance because of the pseudomembrane and superficial necrosis with mucosal ulceration. (From Eisenberg,[38] with permission from the publisher.)

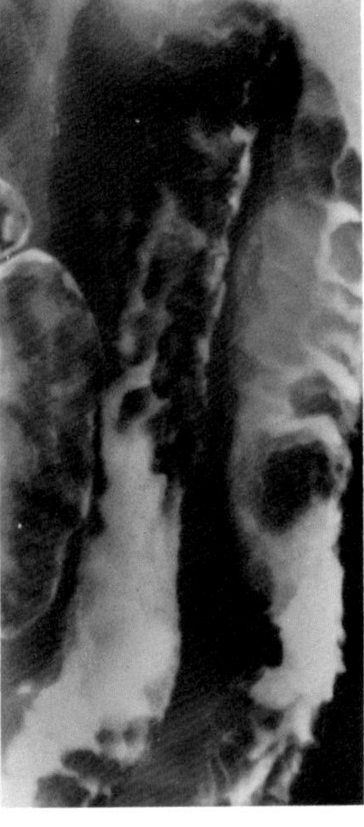

**Figure 47.30. Pseudomembranous colitis.** The pseudomembranes appear as multiple flat, raised lesions distributed circumferentially about the margin of the colon. (From Eisenberg,[38] with permission from the publisher.)

## Pseudomembranous Colitis

Pseudomembranous colitis is a potentially serious complication of antibiotic therapy which can develop following the administration of such drugs as lincomycin, clindamycin, ampicillin, and tetracycline. This condition most commonly arises after the administration of oral antibiotics, though it can also develop following intravenous therapy. The demonstration of an overgrowth of *Clostridium difficile* in the stools of a high percentage of patients with pseudomembranous colitis suggests that the condition is caused by a resistant strain of the bacterium that elaborates a cytotoxic substance which causes necrosis of the colonic epithelium.

The clinical hallmark of pseudomembranous colitis is debilitating diarrhea. Indeed, this complication should be suspected in any patient receiving antibiotics who suddenly experiences copious diarrhea and signs of abdominal cramps, tenderness, or peritonitis. Most patients recover uneventfully after the withdrawal of the offending antibiotic and the institution of adequate fluid and electrolyte replacement. If untreated, however, the condition may be fatal.

Plain abdominal radiographs in severe cases of pseudomembranous colitis can demonstrate moderate, diffuse gaseous distension of the colon. The haustral markings are edematous and distorted, producing wide transverse bands of thickened colonic wall. Barium enema examination is contraindicated in patients with severe pseudomembranous colitis. In mild cases, or as the condition subsides, a low-pressure barium enema study can be performed with caution. The barium column appears shaggy and irregular because of the characteristic pseudomembrane and superficial necrosis. Multiple flat, raised lesions may be distributed circumferentially about the margin of the colon. Mucosal ulcerations simulating those of other forms of ulcerating colitis are frequently seen. In many cases, however, this serrated outline actually represents barium interposed between the plaquelike membranes rather than true ulceration with surrounding edema.[36–37]

## REFERENCES

1. Farman J, Faegenburg D, Dallemand S, et al: Crohn's disease of the stomach: The "ram's horn" sign. *AJR* 123:242–251, 1975.
2. Nelson SW: Some interesting and unusual manifestations of Crohn's disease ("regional enteritis") of the stomach, duodenum, and small intestine. *AJR* 107:86–101, 1969.
3. Gonzalez G, Kennedy T: Crohn's disease of the stomach. *Radiology* 113:27–29, 1974.
4. Ariyama J, Weahlin L, Lindstrom CG, et al: Gastroduodenal erosions in Crohn's disease. *Gastrointest Radiol* 5:121–125, 1980.
5. Cynn WS, Chon HK, Gurejhian PA, et al: Crohn's disease of the esophagus. *AJR* 125:359–362, 1975.
6. Threatt B, Appelman H: Crohn's disease of the appendix presenting as acute appendicitis. *Radiology* 110:313–317, 1974.
7. Korelitz BI: Colonic-duodenal fistula in Crohn's disease. *Dig Dis* 22:1040–1048, 1977.
8. Goldberg HI, Gore RM, Margulis AR, et al: Computed tomography in the evaluation of Crohn's disease. *AJR* 140:277–282, 1983.
9. Moss AA, Thoeni RF: Computed tomography of the gastrointestinal tract, in Moss AA, Gamsu G, Genant HK (eds): *Computed Tomography of the Body*. Philadelphia, WB Saunders, 1983, pp 535–598.
10. Bartram CI: Radiology in the current assessment of ulcerative colitis. *Gastrointest Radiol* 1:383–392, 1977.
11. Greenstein AJ, Janowitz HD, Sachar DV: The extra-intestinal complications of Crohn's disease and ulcerative colitis: A study of 700 patients. *Medicine* 55:401–412, 1976.
12. Kelvin FM, Oddson TA, Rice RP, et al: Double-contrast barium enema in Crohn's disease and ulcerative colitis. *AJR* 131:207–213, 1978.
13. Laufer I, Mullens JE, Hamilton J: Correlation of endoscopy and double-contrast radiography in the early stages of ulcerative and granulomatous colitis. *Radiology* 118:1–5, 1976.
14. Margulis AR: Radiology of ulcerating colitis. *Radiology* 105:251–273, 1972.
15. Thoeni RF, Margulis AR: Radiology in inflammatory disease of the colon: An area of increased interest for the modern clinician. *Invest Radiol* 4:281–292, 1980.
16. Bargen JA: Chronic ulcerative colitis. Diagnostic and therapeutic problems: A lifelong study. *AJR* 99:5–17, 1967.
17. Sommers SC: Ulcerative and granulomatous colitis. *AJR* 133:817–823, 1978.
18. Binder SD, Patterson JF, Glotzer BJ: Toxic megacolon in ulcerative colitis. *Gastroenterology* 66:909–915, 1974.
19. Kirsner JB: Toxic megacolon complicating ulcerative colitis: Current perspectives. *Gastroenterology* 66:1088–1090, 1974.
20. Wruble LD, Bronstein MW: Toxic dilatation of the colon following barium enema examination during the quiescent stage of chronic ulcerative colitis. *Am J Dig Dis* 13:918–924, 1968.
21. Glouston SJ, McGovern VJ: The nature of the benign strictures in ulcerative colitis. *N Engl J Med* 281:290–295, 1969.
22. James EM, Carlson H: Chronic ulcerative colitis and colon cancer: Can radiographic appearance predict survival patterns? *AJR* 130:825–830, 1978.
23. Rogers LS, Ralls PW, Boswell WD, et al: Amebiasis: Unusual radiographic manifestations. *AJR* 135:1253–1257, 1980.
24. Owen RL, Hill JL: Rectal and pharyngeal gonorrhea in homosexual men. *JAMA* 220:1315–1318, 1972.
25. Menuck LS, Brahme F, Amberg JR, et al: Colonic changes of herpes zoster. *AJR* 127:273–276, 1976.
26. Lachman R, Soong J, Wishon G, et al: *Yersinia* colitis. *Gastrointest Radiol* 2:133–135, 1977.
27. Farman J, Rabinowitz JG, Meyers MA: Roentgenology of infectious colitis. *AJR* 119:375–381. 1973.
28. Drasin GS, Moss JP, Cheng SH: *Strongyloides stercoralis* colitis: Findings in four cases. *Radiology* 126:619–621, 1978.
29. Cho SR, Tisnado J, Liu CI, et al: Bleeding cytomegalovirus ulcer of the colon: Barium enema and angiography. *AJR* 136:1213–1215, 1981.
30. Chait A: Schistosomiasis mansoni: Roentgenologic observations in a nonendemic area. *AJR* 90:688–708, 1963.
31. Carrera GF, Young S, Lewicki AM: Intestinal tuberculosis. *Gastrointest Radiol* 1:147–155, 1976.
32. Cardoso JM, Kimura K, Stoopen M, et al: Radiology of invasive amebiasis of the colon. *AJR* 128:935–941, 1977.
33. Blaser MJ, Parsons RB, Wang WLL: Acute colitis caused by *Campylobacter fetus jejuni*. *Gastroenterology* 78:448–453, 1980.
34. Balthazar EJ, Bryk D: Segmental tuberculosis of the colon: Radiographic features in seven cases. *Gastrointest Radiol* 5:75–80, 1980.
35. Annamunthodo H, Marryatt J: Barium studies in intestinal lymphogranuloma venereum. *Br J Radiol* 34:53–57, 1961.
36. Stanley RJ, Melson GL, Tedesco FJ: The spectrum of radiographic findings in antibiotic-related pseudomembranous colitis. *Radiology* 111:519–524, 1974.
37. Stanley RJ, Melson GL, Tedesco FJ, et al: Plain-film findings in severe pseudomembranous colitis. *Radiology* 118:7–11, 1976.
38. Eisenberg RL: *Gastrointestinal Radiology: A Pattern Approach*. Philadelphia, JB Lippincott, 1983.
39. Kerber GW, Frank PH: Carcinoma of the small intestine and colon as a complication of Crohn disease. *Radiology* 150:639–645, 1984.

## Chapter Forty-Eight
# Diseases of the Small and Large Intestine

## DISORDERS OF INTESTINAL MOTILITY

### DIVERTICULOSIS OF THE SMALL BOWEL

#### Duodenal Diverticula

Diverticula of the duodenum are common lesions that most frequently arise along the medial border of the descending duodenum in the periampullary region. On barium examinations, duodenal diverticula typically have a smooth, rounded shape and generally change configuration during the course of the study. The lack of inflammatory reaction (spasm, distortion of mucosal folds) permits a duodenal diverticulum to be differentiated from a postbulbar ulcer. Inconstant filling defects representing inspissated food particles, blood clots, or gas can sometimes be identified in a duodenal diverticulum.

Although the overwhelming majority of duodenal diverticula are asymptomatic, serious complications can develop. Duodenal diverticulitis mimics numerous abdominal diseases (cholecystitis, peptic ulcer disease, pancreatitis) and may lead to hemorrhage, perforation, abscesses, and fistulas. Anomalous insertion of the common bile duct and pancreatic duct into a duodenal diverticulum may interfere with the normal emptying

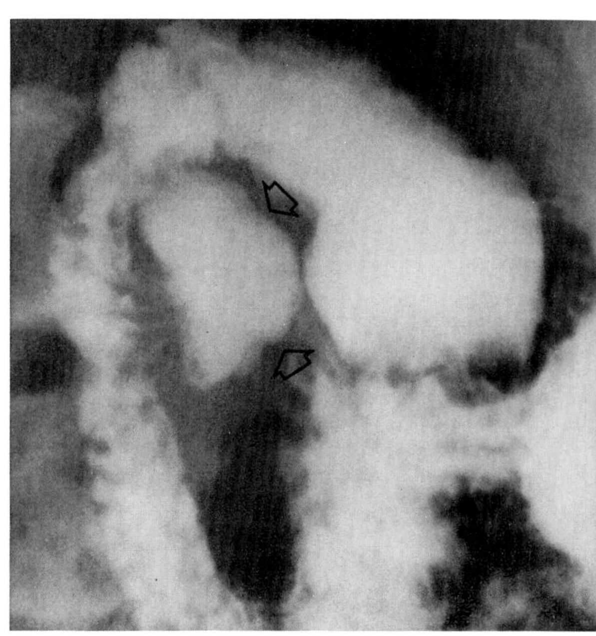

**Figure 48.1. Duodenal diverticulum (arrows) arising from the second portion of the duodenum.**

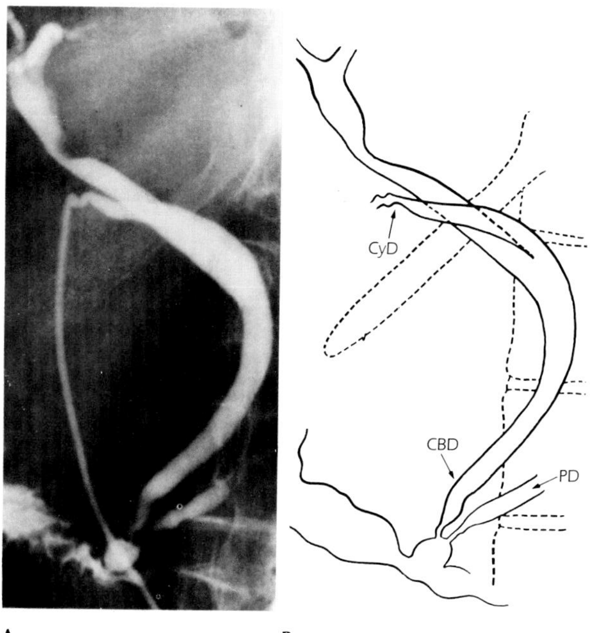


A B

**Figure 48.2. Insertion of the common bile duct and pancreatic duct into the dome of a small duodenal diverticulum.** (A) An operative cholangiogram and (B) a corresponding line drawing. (From Nelson and Burhenne,[1] with permission from the publisher.)

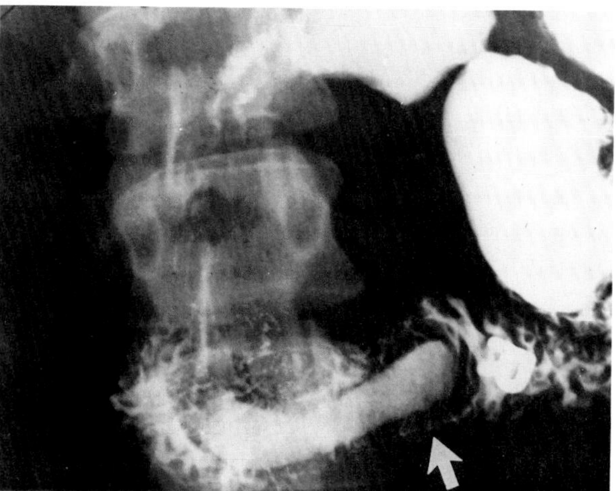

**Figure 48.3. Halo sign of intraluminal duodenal diverticulum.** The diverticulum appears as a saclike collection of contrast material surrounded by a lucent halo (arrow). (From Heilbrun and Boyden,[74] with permission from the publisher.)

mechanism of the ductal systems and predispose to obstructive biliary and pancreatic disease. The absence of an ampullary sphincter mechanism permits the spontaneous reflux of barium from the diverticulum into the common bile duct, and this can be a cause of ascending infection.[1,2]

## Intraluminal Duodenal Diverticulum

An intraluminal duodenal diverticulum is a sac of duodenal mucosa originating in the second portion of the duodenum near the papilla of Vater. The diverticulum forms from a congenital duodenal web or diaphragm because of mechanical factors such as forward pressure by food and strong peristaltic activity. When filled with barium, the intraluminal duodenal diverticulum appears as a finger-like sac separated from contrast material in the duodenal lumen by a radiolucent band representing the wall of the diverticulum (halo sign).[3,74]

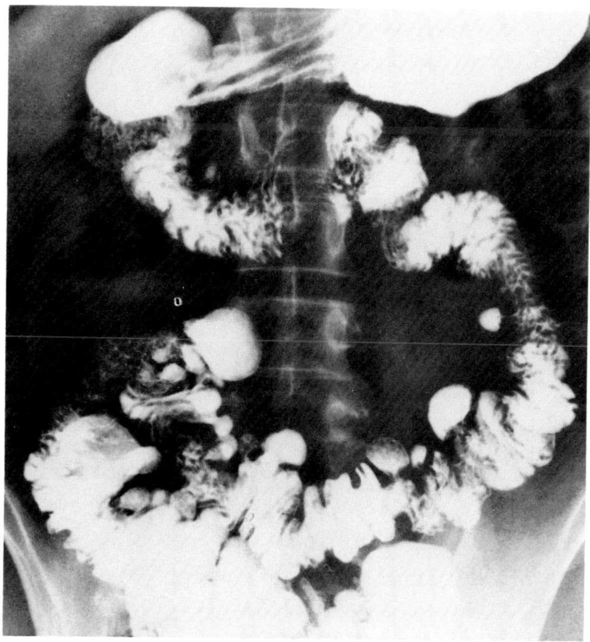

**Figure 48.4. Jejunal diverticulosis.** (From Eisenberg,[32] with permission from the publisher.)

## Multiple Jejunal Diverticula

Jejunal diverticula are thin-walled outpouchings that lack muscular components and may permit intestinal stasis and bacterial overgrowth. Multiple large jejunal diverticula can lead to malabsorption and megaloblastic anemia. Jejunal diverticulosis is one of the leading gastrointestinal causes of pneumoperitoneum in the absence of peritonitis or surgery. Other rare complications of jejunal diverticulosis include bleeding, acute inflammation, enterolith formation, and impaction of a foreign body, which can ultimately result in perforation with peritonitis.[4,5]

## Meckel's Diverticulum

Meckel's diverticulum represents a persistence of the rudimentary omphalomesenteric duct (embryonic communication between the gut and yolk sac) and is the most frequent congenital anomaly of the intestinal tract. Meckel's diverticulum usually arises within 100 cm of the ileocecal valve and opens into the antimesenteric side of the ileum, unlike other diverticula, which arise on the mesenteric side of the small bowel.

In most patients, Meckel's diverticulum remains asymptomatic throughout life. In children, the most common symptom is bleeding, which is almost invariably

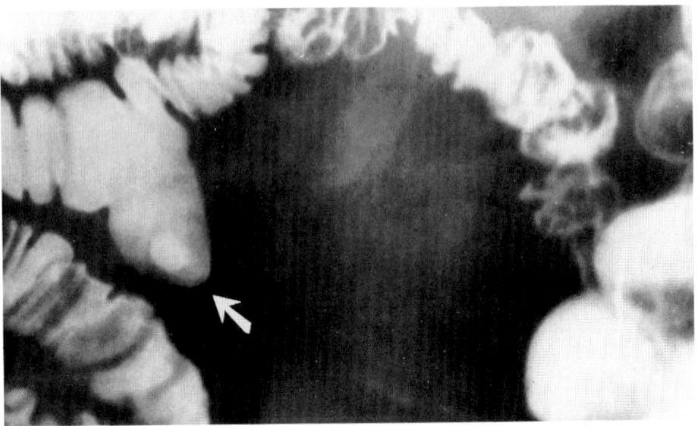

**Figure 48.5. Meckel's diverticulum (arrow).** The mouth of the diverticular sac has a width approximately equal to that of the intestinal lumen. Note the small diverticulum (area of increased density) arising from it. (From Eisenberg,[32] with permission from the publisher.)

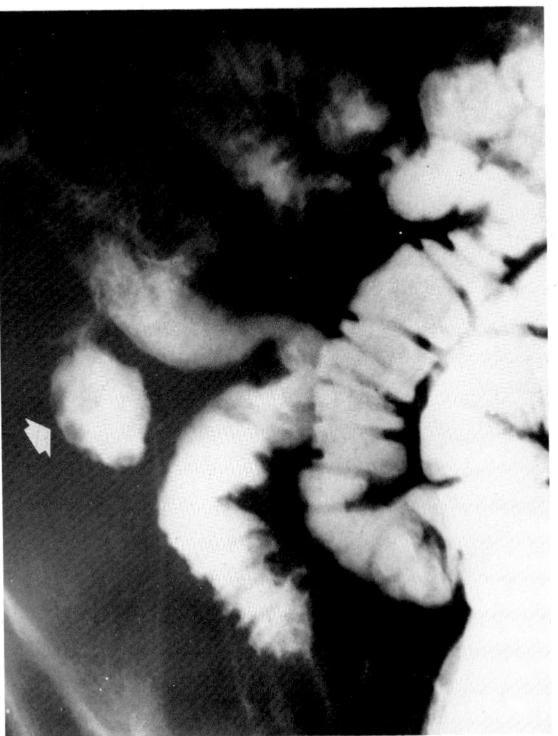

**Figure 48.6. Meckel's diverticulum.** The multiple defects (arrow) represent ectopic gastric mucosa. (From Eisenberg,[32] with permission from the publisher.)

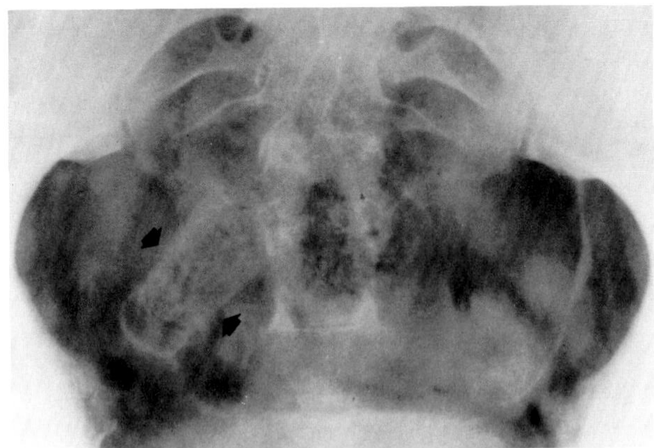

**Figure 48.7. Calcified enterolith (arrows) in a Meckel's diverticulum.** (From Eisenberg,[32] with permission from the publisher.)

the result of ulceration of ileal mucosa adjacent to heterotopic gastric mucosa in the diverticulum. Because this ectopic gastric mucosa has an affinity for pertechnetate, radionuclide imaging can be helpful in demonstrating the lesion. In young adults, inflammation in a Meckel's diverticulum can produce a clinical picture indistinguishable from that of acute appendicitis. The most common symptom in adults is intestinal obstruction, which can be due to invagination and intussusception of the diverticulum, volvulus, inflammation, or adhesions.

Despite the relative frequency of Meckel's diverticula, preoperative radiographic demonstration is unusual. If a diverticulum is not large, it can be difficult to distinguish from normal loops of small bowel unless careful compression films are obtained. A Meckel's diverticulum appears radiographically as an outpouching arising from the antimesenteric side of the distal ileum. The mouth of the sac is wide, often equal to the width of the intestinal lumen itself. Filling defects in the diverticulum strongly suggest the presence of ectopic gastric mucosa. Impaired drainage from a Meckel's diverticulum may lead to stasis and formation of an enterolith, a smooth stone with radiopaque laminated calcifications.[6–8]

## DIVERTICULAR DISEASE OF THE COLON

### Diverticulosis

Colonic diverticula are acquired herniations of mucosa and submucosa through the muscular layers of the bowel wall. The incidence of colonic diverticulosis increases with age. Rare in persons below the age of 30, diverticula can be demonstrated in up to half of persons over 60. Diverticula occur most commonly in the sigmoid colon and decrease in frequency in the proximal colon. Although most patients with diverticulosis have no symptoms, a substantial number have chronic or intermittent lower abdominal pain, frequently related to meals or emotional stress, and alternating bouts of diarrhea and constipation. Bleeding may be caused by inflammatory erosion of penetrating branches of the vasa recta at the base of the diverticulum.

Colonic diverticula appear radiographically as round or oval outpouchings of barium projecting beyond the confines of the lumen. They vary in size from barely visible dimples to saclike structures 2 cm or more in diameter. Giant sigmoid diverticula of up to 25 cm in diameter, which probably reflect slowly progressing chronic diverticular abscesses, may appear as large, well-circumscribed, lucent cystic structures in the lower abdomen.

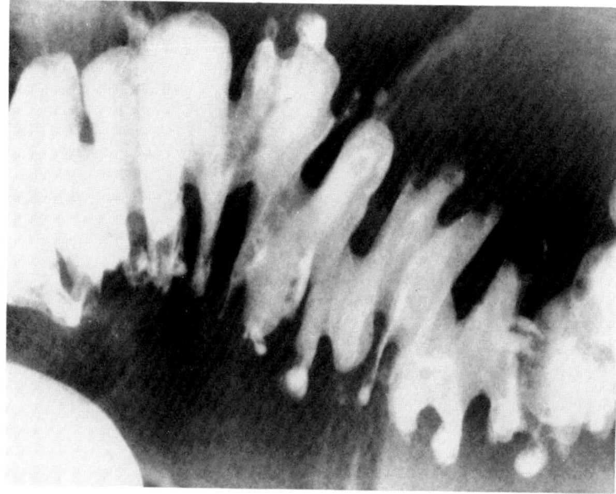

**Figure 48.8. Colonic diverticulosis.** Multiple diverticula and deep criss-crossing ridges of thickened circular muscle produce a characteristic sawtooth configuration. (From Eisenberg,[32] with permission from the publisher.)

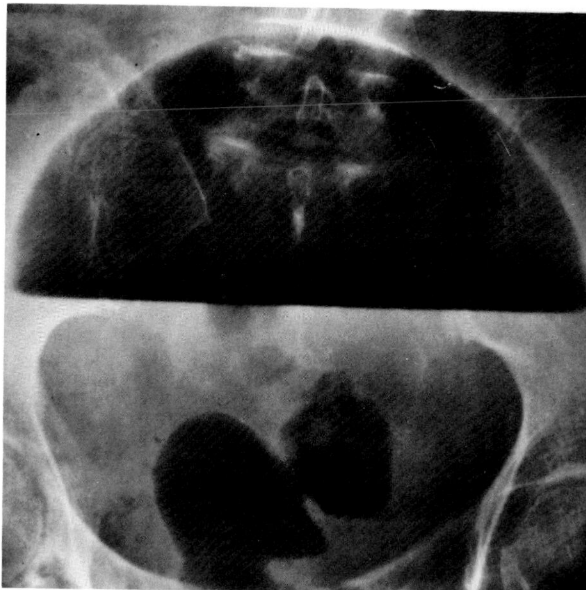

**Figure 48.9. Giant sigmoid diverticulum.** A plain abdominal radiograph demonstrates a huge, walled-off, chronic diverticular abscess with a gas-fluid level. (From Eisenberg,[32] with permission from the publisher.)

Diverticula are usually multiple and tend to occur in clusters, though a solitary diverticulum is occasionally found. With multiple diverticula, deep criss-crossing ridges of thickened circular muscle can produce a characteristic series of sacculations (sawtooth configuration).

Massive hemorrhage is sometimes due to a bleeding diverticulum, particularly in the right colon. In such patients, selective mesenteric arteriography can be both diagnostic in localizing the bleeding site and therapeutic if vasoconstrictive drugs are infused intra-arterially to control hemorrhage.[9,10]

## Diverticulitis

Diverticulitis is a complication of diverticular disease of the colon, especially in the sigmoid region, in which perforation of a diverticulum leads to the development of a peridiverticular abscess. It is estimated that up to 20 percent of patients with diverticulosis eventually develop acute diverticulitis. Inspissation of retained fecal material trapped in a diverticulum by the narrow opening of the diverticular neck causes inflammation of the mucosal lining that leads to perforation of the diverticulum. This usually results in a localized peridiverticular abscess that is walled off by fibrous adhesions. Free intraperitoneal perforation is rare. The inflammatory process may localize within the wall of the colon and produce an intramural mass, or it may dissect around the colon, causing segmental narrowing of the lumen. Subserosal

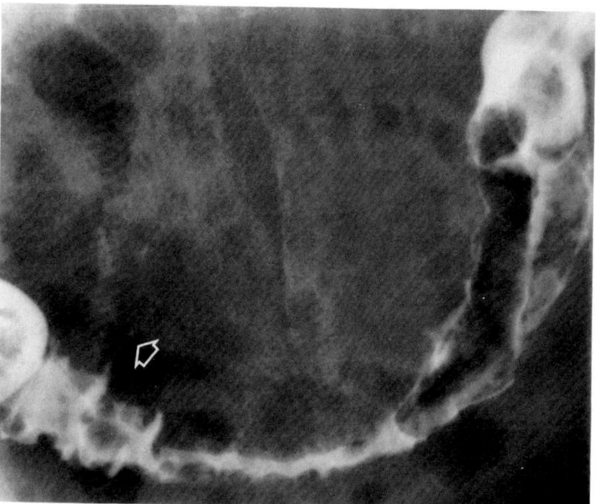

**Figure 48.10. Acute sigmoid diverticulitis.** A thin projection of contrast material (arrow) implies extravasation from the colonic lumen. Note the severe spasm of the sigmoid colon due to the intense adjacent inflammation. (From Eisenberg,[32] with permission from the publisher.)

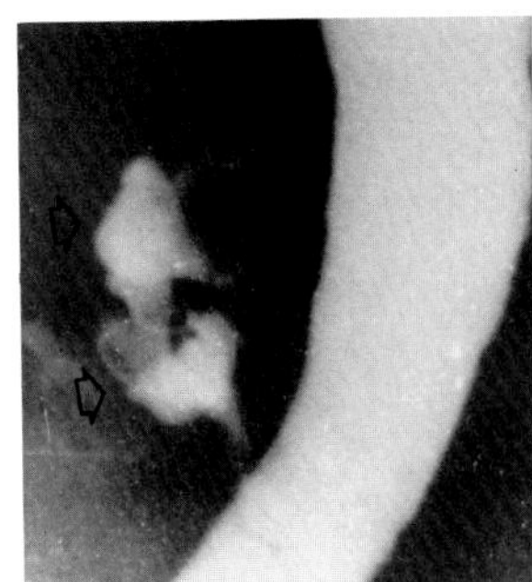

**Figure 48.11. Diverticulitis.** Contrast material fills a pericolic abscess (arrows). (From Eisenberg,[32] with permission from the publisher.)

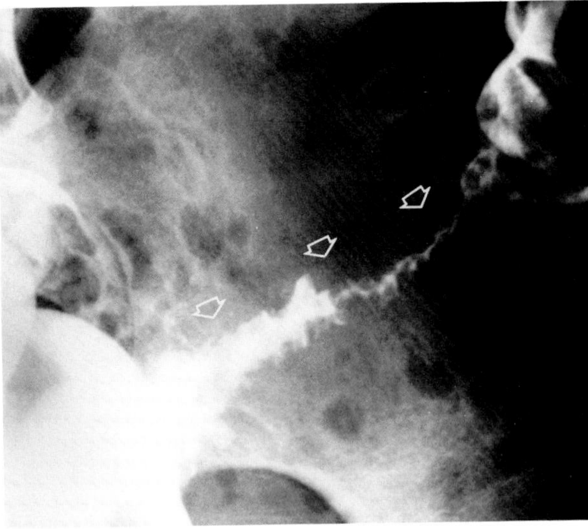

**Figure 48.12. Diverticulitis.** Severe narrowing of a long involved portion of the sigmoid colon (arrows) in a patient with no radiographically detectable diverticula. (From Eisenberg,[32] with permission from the publisher.)

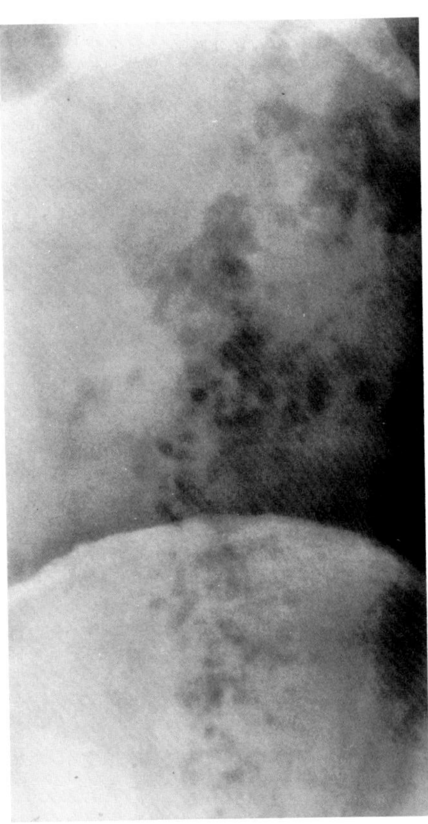

A

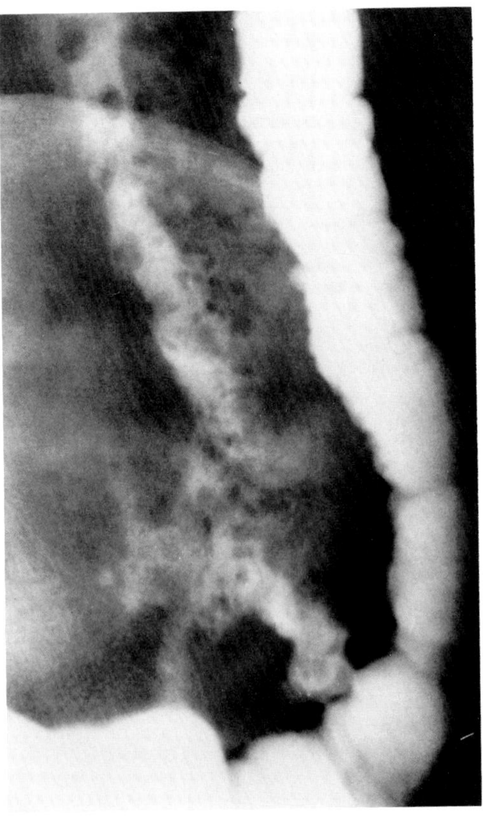

B

**Figure 48.13. Colon perforation in diverticulitis.** (A) A plain abdominal radiograph demonstrates multiple gas bubbles along the course of the descending colon. (B) A contrast study shows extravasation from the colon. (From Eisenberg,[32] with permission from the publisher.)

extension of the inflammatory process along the colon can involve adjacent diverticula, resulting in a longitudinal sinus tract. A common complication of diverticulitis is the development of fistulas to adjacent organs (bladder, vagina, ureter, small bowel, colon).

The radiographic diagnosis of diverticulitis requires direct or indirect evidence of diverticular perforation. The most specific sign is extravasation, which can appear either as a tiny projection of contrast material from the tip of a diverticulum or as obvious filling of a pericolic

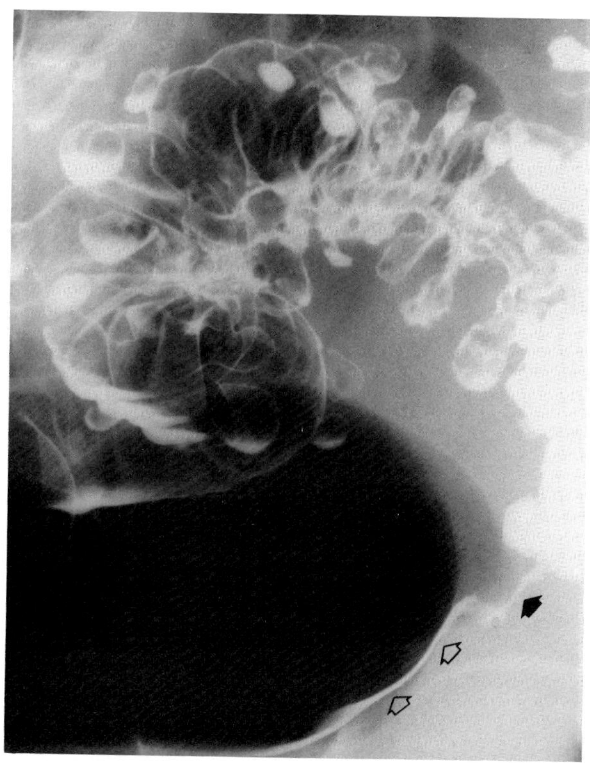

**Figure 48.14. Colovesical fistula in diverticulitis.** A barium enema examination demonstrates barium in the fistulous tract (solid arrow) between the sigmoid colon and the bladder. Barium can also be seen lining the base of the gas-filled bladder (open arrows). (From Eisenberg,[32] with permission from the publisher.)

abscess. A more common, though somewhat less specific, sign of diverticulitis is the demonstration of a pericolic soft tissue mass that is due to a localized abscess and represents a walled-off perforation. This extraluminal mass appears as a filling defect causing eccentric narrowing of the bowel lumen. The adjacent diverticula are spastic, irritable, and attenuated and frequently seem to drape over the mass. It is important to remember, however, that a peridiverticular abscess caused by diverticulitis can occur without radiographically detectable diverticula.

Severe spasm or fibrotic healing of diverticulitis can cause rigidity and progressive narrowing of the colon that simulate annular carcinoma. Although radiographic distinction from carcinoma may be impossible, findings favoring the diagnosis of diverticulitis include the involvement of a relatively long segment, a gradual transition from diseased to normal colon, a relative preservation of mucosal detail, and fistulous tracts and intramural abscesses. Significant obstruction to the flow of barium can prevent the identification of these distinguishing features. The intravenous administration of glucagon may relax spasm due to inflammation, thereby permitting the differentiation of acute diverticulitis from an annular carcinoma. Nevertheless, colonoscopy or surgery may be required for a definitive diagnosis to be made.

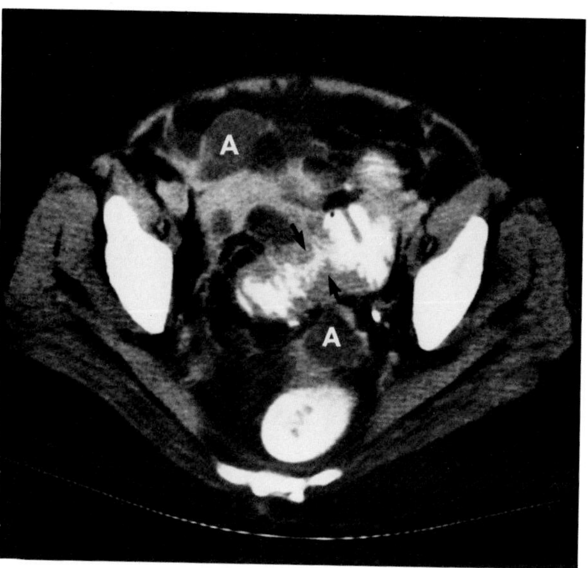


**Figure 48.15. Acute diverticulitis.** A CT scan demonstrates narrowing of the lumen of the sigmoid colon (arrows) and multiple adjacent paracolonic abscesses (A). (Courtesy of R. Brooke Jeffrey, M.D.)

Many radiologists are unwilling to perform a barium enema examination in a patient with acute diverticulitis, lest the increased pressure caused by the enema result in additional bacteria-laden luminal contents being extravasated through the perforated diverticulum. However, after several days of medical management (bowel rest, antibiotics), the inflammatory reaction has usually subsided enough to permit a safe enema examination. If perforation is suspected and an immediate diagnosis is required, a water-soluble contrast agent should be used.

Computed tomography (CT) demonstrates a pericolonic diverticular abscess as a mass with a thick wall of soft tissue density and a low-density center that may contain gas or, if in communication with the lumen, contrast material. Uncomplicated diverticulosis of the colon may appear on CT as thickening of the colonic wall due to circular muscle hypertrophy.[10,32]

## MEGACOLON

### Aganglionic Megacolon (Hirschsprung's Disease)

Aganglionic megacolon is a congenital disorder causing massive dilatation of the large bowel and prolonged retention of fecal material within the colon. Most common in males, the disorder is usually seen in infants who present with constipation, abdominal distension, and vomiting.

The diagnosis of aganglionic megacolon can often be made from a plain abdominal radiograph. Fecal matter

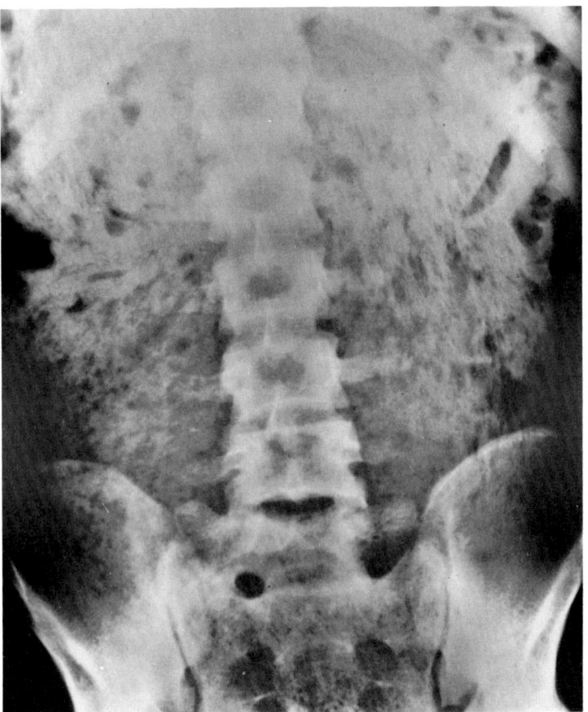

**Figure 48.16. Aganglionic megacolon in a child.** A plain abdominal radiograph demonstrates fecal material and gas within a severely dilated colon. (From Eisenberg,[32] with permission from the publisher.)

and gas within a severely dilated colon produce the typical mottled shadow of fecal impaction. On lateral views, the rectum or rectosigmoid is not distended and contains little or no gas or feces. On barium enema examination, the rectum appears to be of essentially normal character. At some point in the upper rectum or distal sigmoid, there is an abrupt transition to an area of grossly dilated bowel. It is important to remember that the narrowed, relatively normal appearing distal colon is actually the abnormal segment in which there is marked diminution or complete absence of ganglion cells in the myenteric plexus. In contrast, the severely dilated proximal colon has a normal pattern of innervation.

Rarely, aganglionosis involves the entire colon. In this instance, the colon appears normal or small (microcolon) on barium enema studies. Marked distension of the small bowel may simulate a distal small bowel obstruction.[11,12]

## Chronic Idiopathic Megacolon (Psychogenic Megacolon)

Chronic idiopathic megacolon, which has its onset later in childhood than does Hirschsprung's disease, is characterized by huge dilatation of a feces-filled rectum, often with distension of the entire colon. Unlike aganglionic megacolon, in chronic idiopathic megacolon there is no

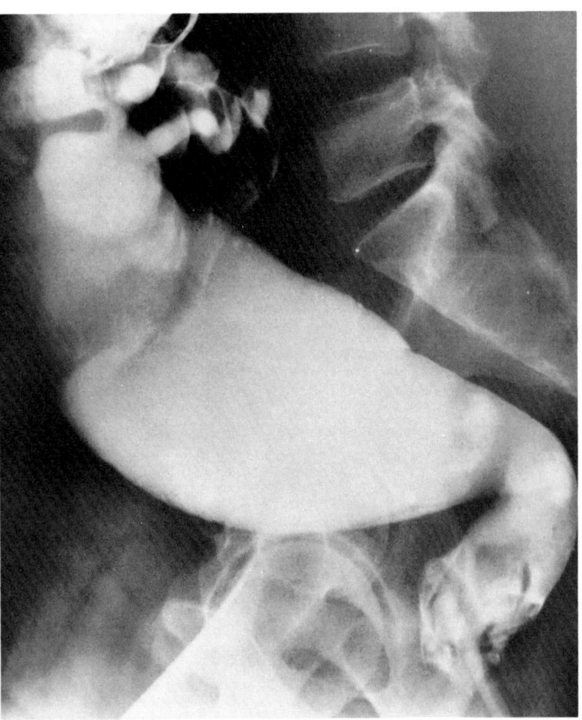

A

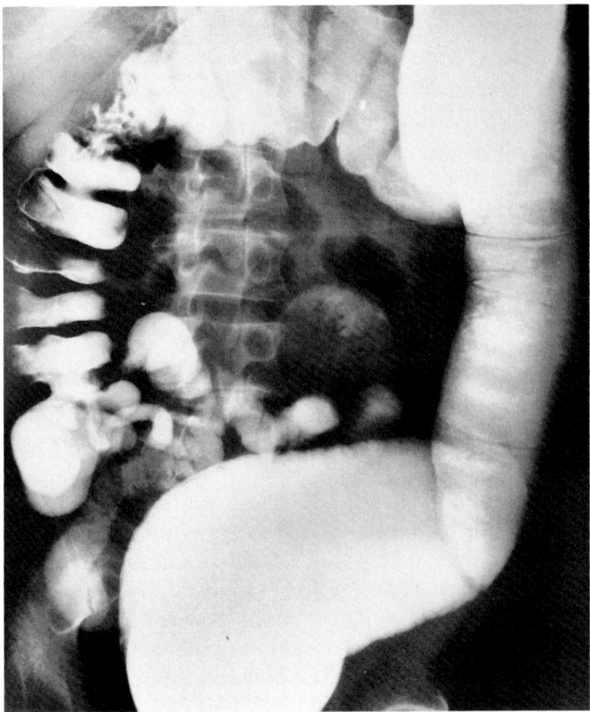

B

**Figure 48.17. Aganglionic megacolon in an adult.** (A) On a lateral view of the rectosigmoid, an abrupt transition between the normal caliber and massive dilatation of the bowel is evident. (B) A frontal view demonstrates severe dilatation of the descending and transverse portions of the colon. (From Eisenberg,[32] with permission from the publisher.)

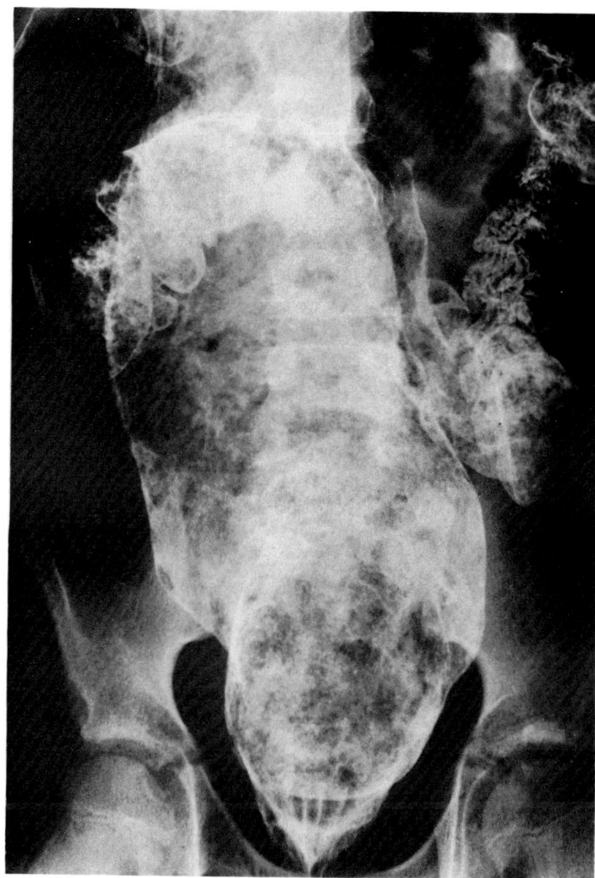

**Figure 48.18. Chronic idiopathic megacolon in a child.** There is huge dilatation of the feces-filled rectum. (From Eisenberg,[32] with permission from the publisher.)

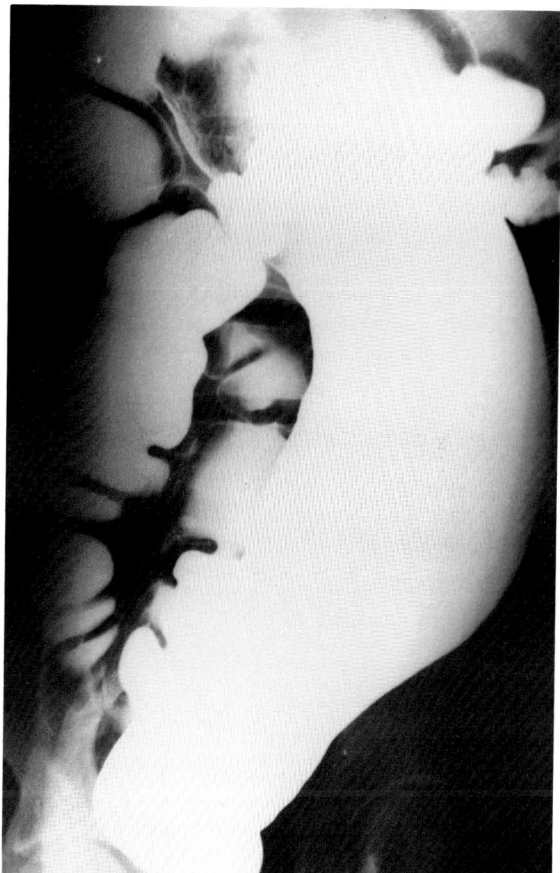

**Figure 48.19. Chagas' disease.** There is striking elongation and dilatation of the rectosigmoid colon.

narrowed segment, and rectal biopsy discloses the normal complement of ganglion cells in Auerbach's plexus.[13]

### Acquired Megacolon

In Chagas' disease, destruction of the colonic myenteric plexuses by the protozoan *Trypanosoma cruzi* causes striking elongation and dilatation, especially of the rectosigmoid and the descending colon. Acquired megacolon can also be found in adults with severe neurologic or psychologic disorders and in those with abnormal colonic motility (myxedema; infiltrative diseases such as amyloidosis and scleroderma; narcotic drugs).[13]

### Toxic Megacolon

This is discussed in the section "Toxic Megacolon" in Chapter 47.

### Intestinal Pseudo-Obstruction

This is discussed in the section "Chronic Idiopathic Intestinal Pseudo-Obstruction" in Chapter 49.

## FUNCTIONAL DISTURBANCES OF THE COLON

### Irritable Bowel Syndrome

The term *irritable bowel syndrome* refers to one of three clinical variants that have an alteration in intestinal motility as the underlying pathophysiologic abnormality. Patients with this condition may complain primarily of chronic abdominal pain and constipation (spastic colitis), chronic intermittent watery diarrhea, often without pain, or alternating bouts of constipation and diarrhea. Although there are no specific radiographic findings in the irritable bowel syndrome, patients with this condition usually undergo a barium enema examination to exclude another colonic disorder as the cause of the symptoms. When the patient is symptomatic, the barium enema may demonstrate areas of irritability and spasticity, increased segmentation, and accentuated haustration. However, similar radiographic findings may be observed in normal asymptomatic persons, especially those who have received laxatives and enemas.[14,15]

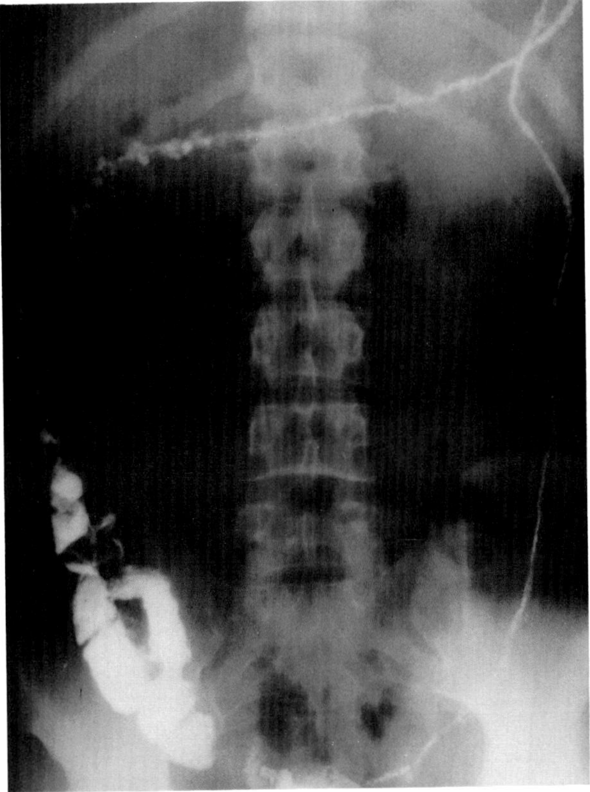

**Figure 48.20. Irritable bowel syndrome.** There is severe spastic narrowing of virtually the entire colon.

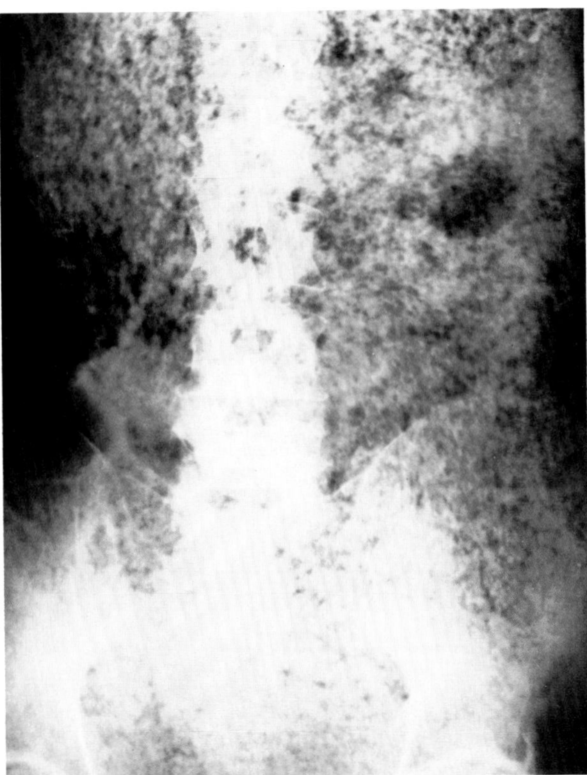

**Figure 48.21. Fecal impaction.** The characteristic mottled density of feces throughout the colon is diagnostic on this plain abdominal radiograph. (From Eisenberg,[32] with permission from the publisher.)

## Chronic Constipation and Fecal Impaction

Chronic constipation may be caused by mechanical obstruction (e.g., carcinoma, stricture) or be of functional origin, as in bedridden elderly patients or persons with improper bowel habits. Plain abdominal radiographs may demonstrate a tremendously dilated, tortuous colon and rectum filled with a large fecal residue.

The incomplete evacuation of feces over a prolonged period can result in the formation of a fecal impaction, a large, firm, immovable mass of stool in the rectum that may cause large bowel obstruction. Fecal impactions most commonly develop in elderly, debilitated, or sedentary persons. They can occur in patients who have been inactive for long periods (e.g., because of myocardial infarction, traction), in narcotic addicts, in patients on large dosages of tranquilizers, and in children with megacolon or psychogenic problems.

The symptoms of fecal impaction usually consist of vague rectal fullness and nonspecific abdominal discomfort. A common complaint is overflow diarrhea, the uncontrolled passage of small amounts of watery and semiformed stool around a large obstructing impaction. In elderly, bedridden patients, it is essential that this overflow phenomenon be recognized as secondary to fecal impaction rather than perceived as true diarrhea.

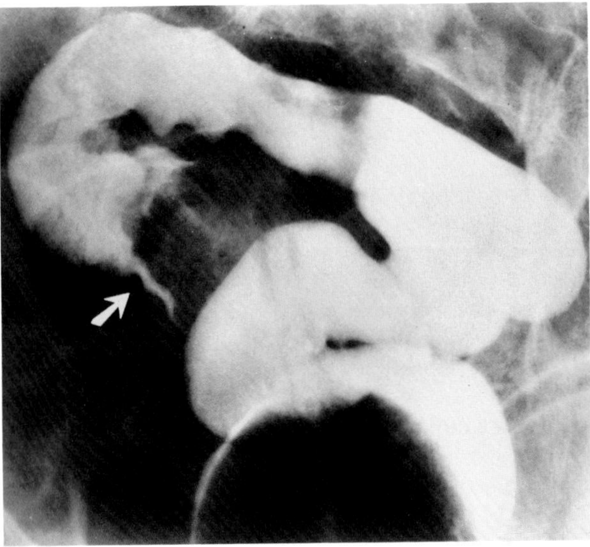

**Figure 48.22. Inspissated fecalith obstructing the sigmoid colon (arrow).** The radiographic appearance simulates a primary colonic malignancy. (From Eisenberg,[32] with permission from the publisher.)

Plain radiographs of the pelvis are usually diagnostic of fecal impaction. The fecal mass within the rectum typically appears as a mottled soft tissue density containing multiple small, irregular lucent areas that reflect pockets of gas within the mass. Contrast studies demonstrate a large, irregular intraluminal mass that can occasionally be confused with colonic malignancy. For fecal impaction to be confirmed as the cause of large bowel obstruction, an enema of water-soluble contrast material should be used instead of barium; hypertonic contrast medium draws fluid into the bowel and can aid in breaking up the fecal mass.[32]

### Cathartic Colon

Cathartic colon is due to the prolonged use of stimulant-irritant cathartics (e.g., castor oil, phenophthalein, cascara, senna, podophyllum). The typical patient with this condition is a woman of middle age who has habitually used irritant cathartics for more than 15 years. Ironically, the patient often initially denies the use of cathartics and complains only of constipation. Because prolonged stimulation of the colon by irritant laxatives results in neuromuscular incoordination and an inability of the colonic musculature to produce adequate contractile force without external stimulants, the patient with cathartic colon is often unable to have a bowel movement without laxative assistance.

The radiographic appearance of cathartic colon is similar to that of "burned-out" chronic ulcerative colitis. In contrast to ulcerative colitis, however, the absent or diminished haustral markings, bizarre contractions, and the inconstant areas of narrowing in cathartic colon primarily involve the right colon. In severe cases, the left side of the colon can also be affected, though the sigmoid and rectum usually appear normal. The mucosal pattern is linear or smooth; ulcerations are not seen. The ileocecal valve is frequently flattened and gaping, simulating the backwash ileitis seen in ulcerative colitis. Shortening of the ascending colon can be severe, but unlike the rigid, tubular bowel in chronic ulcerative colitis, the shortened segment in a cathartic colon remains remarkably distensible. Transient areas of narrowing of the bowel lumen can be seen. Although the radiographic appearance of cathartic colon may mimic chronic ulcerative colitis, the two entities can usually be separated clinically because of the history of long-term constipation and laxative use in the patient with cathartic colon in contrast to the complaint of diarrhea in most patients with inflammatory bowel disease.[16,17]

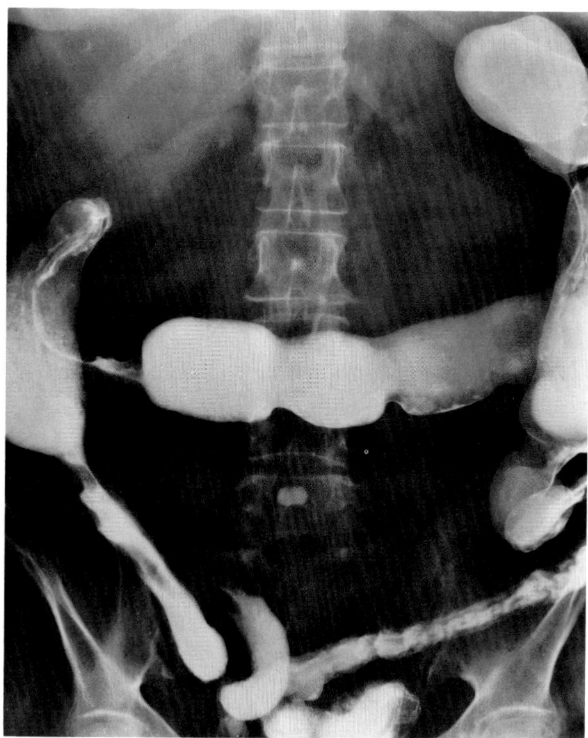

**Figure 48.23. Cathartic colon.** Bizarre contractions and irregular areas of narrowing primarily involve the right colon. (From Eisenberg,[32] with permission from the publisher.)

## VASCULAR DISORDERS OF THE INTESTINE

### GASTROINTESTINAL HEMORRHAGE

The diagnostic approach to gastrointestinal hemorrhage depends on whether the patient is suffering from acute or chronic blood loss and whether the suspected bleeding site is located in the upper or lower portions of the gastrointestinal tract.

#### Acute Upper Gastrointestinal Hemorrhage

In patients with suspected acute upper gastrointestinal hemorrhage, the prevailing medical opinion favors endoscopy as the initial diagnostic procedure of choice. The endoscopist can visualize superficial mucosal lesions, such as esophagitis, gastritis, and superficial ero-

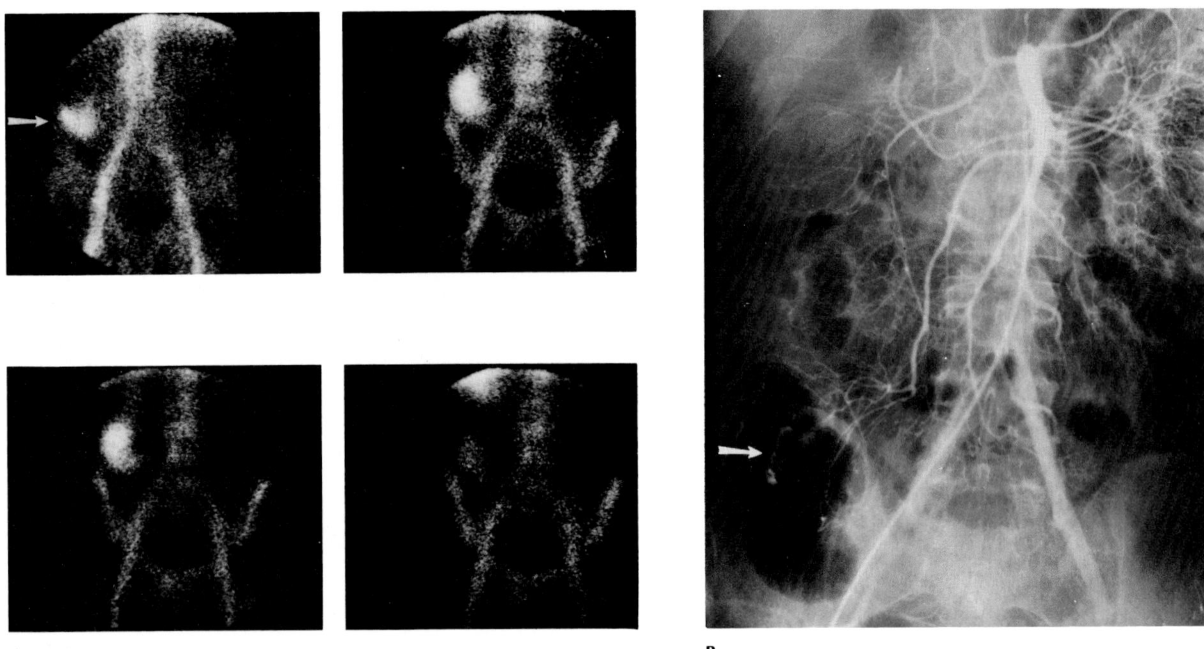

A B

**Figure 48.24. Bleeding cecal lesion due to ischemia.** (A) A radionuclide scan shows a bleeding site in the right lower quadrant (arrow). On subsequent examinations, the increased isotope activity gradually moves up toward the liver (top right); in the final scan (lower right), there is very faint activity at the original site. (B) A superior mesenteric arteriogram performed immediately after completion of the radionuclide scan shows corresponding extravasation of contrast material in the cecal region (arrow). (From Alavi,[75] with permission from the publisher.)

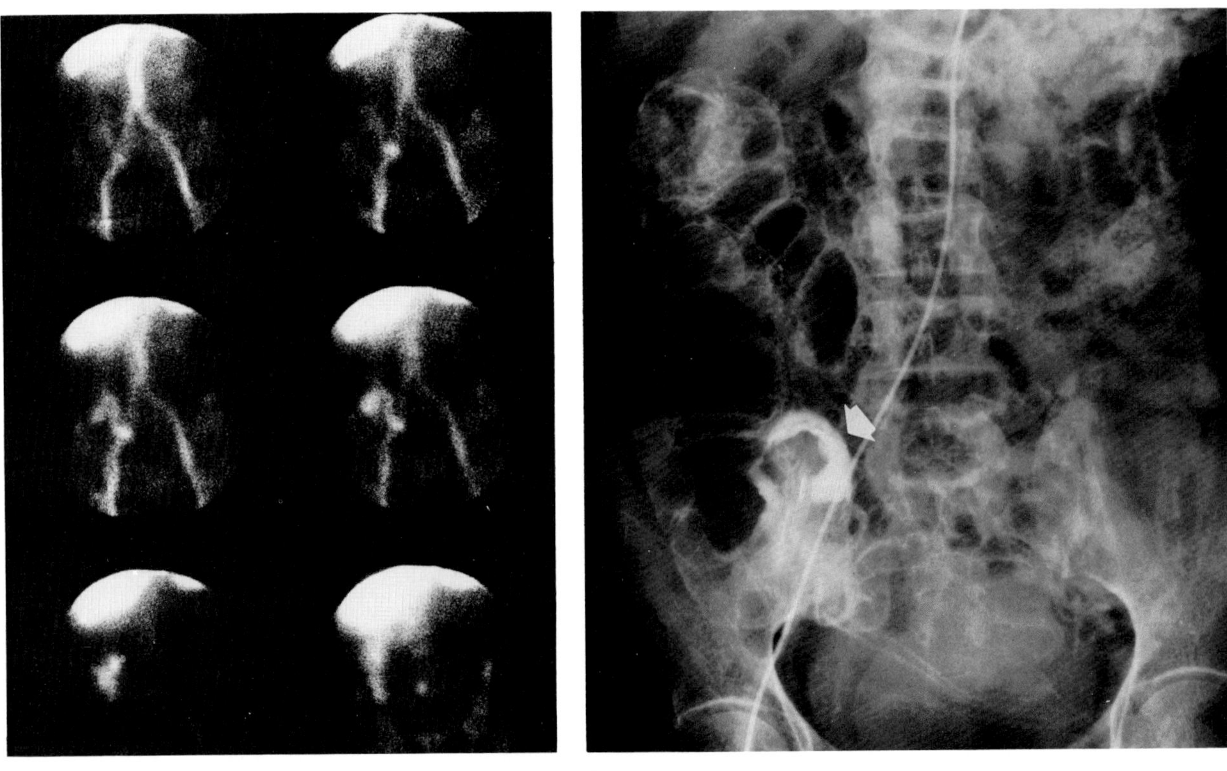

A B

**Figure 48.25. Bleeding in the terminal ileum.** (A) Shortly after the injection of $^{99m}$Tc sulfur colloid, a point of extravasation is noted in the right lower quadrant overlying the iliac vessels (left top). The activity gradually moves upward and laterally into a wider lumen (top right, middle left and right). Finally, the extravasated blood moves upward toward the liver (lower left and right). These findings were interpreted to represent a bleeding site in the terminal ileum, very close to the ileocecal valve. (B) A late phase of a superior mesenteric artery angiogram performed very shortly after the radionuclide scan shows extravasation of contrast material exactly at the site (arrow) shown in the scan. (From Alavi,[75] with permission from the publisher.)

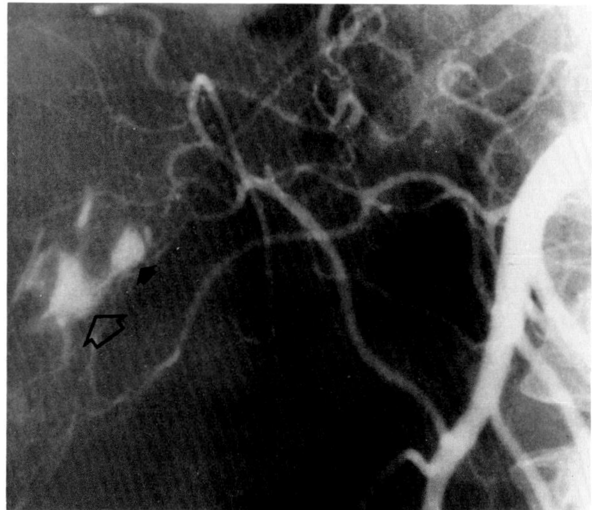

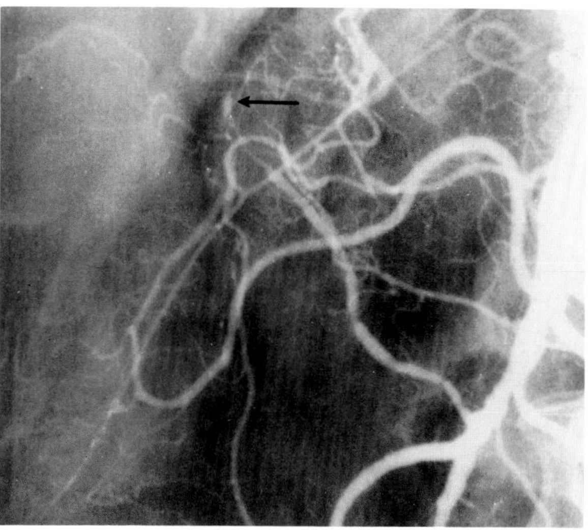

A B

**Figure 48.26. Acute bleeding from diverticulosis.** (A) A superior mesenteric arteriogram shows an acute diverticular bleed, with contrast material in the diverticulum (closed arrow) and spilling out into the bowel lumen (open arrow). (B) After vasopressin failed to control the bleeding, a Gelfoam clot soaked in contrast material (arrow) occluded the bleeding artery, permanently controlling the hemorrhage. (From Goldberger and Bookstein,[21] with permission from the publisher.)

sions, as well as varices, Mallory-Weiss tears, and penetrating ulcers. Unlike barium studies, endoscopy enables the physician not only to visualize the abnormality but also to usually say with some certainty whether or not a specific lesion is the cause of the acute bleeding episode. This is important because a substantial number of patients with acute upper gastrointestinal hemorrhage are found to have more than one lesion.

Bleeding at a rate of at least 0.5 to 1 ml per minute usually can be detected by selective arteriography. This modality should be utilized whenever upper gastrointestinal hemorrhage is too rapid for efficient endoscopy. In addition to demonstrating the specific cause of the hemorrhage, arteriography can be used therapeutically for the infusion of vasoconstrictor agents (vasopressin) or for the direct injection of embolic material into the artery perfusing the bleeding site. If there is clinical evidence that acute bleeding has stopped, however, arteriography should not be considered as the first diagnostic approach.

Radionuclide studies using technetium 99m–labeled sulfur colloid or technetium 99m-labeled red blood cells can accurately localize gastrointestinal bleeding with minimum discomfort and risk. In experimental studies, bleeding as slow as 0.05 to 0.1 ml per minute has been detected. The addition of rapid sequential scanning (radionuclide angiogram) to routine static scintigraphy for gastrointestinal bleeding permits more accurate localization of the bleeding site. Although more sensitive, radionuclide scanning is less specific than arteriography and has no therapeutic value.[18–21,75]

## Acute Lower Gastrointestinal Hemorrhage

In patients with acute lower gastrointestinal hemorrhage, most of whom are bleeding from diverticulosis, arteriography is generally considered the procedure of choice. Arteriography also is highly accurate in detecting bleeding from arteriovenous malformations, the second leading cause of massive acute lower gastrointestinal bleeding. Particularly in cases of diverticular hemorrhage, transcatheter hemostasis from vasopressor infusion or embolization has been highly successful; in many patients surgery may be avoided entirely. In those patients in whom surgery is ultimately required, the combined diagnostic and therapeutic value of arteriography permits both stabilization of the patient and resection of the colon limited to the area of the bleeding site. As in upper gastrointestinal bleeding, radionuclide studies are sensitive in determining the bleeding site, though they are of no therapeutic value.

The barium enema has no role to play in the evaluation of patients with acute massive lower gastrointestinal hemorrhage. Benign polyps and carcinomas of the colon, the major disorders detected by barium studies, rarely present as acute lower gastrointestinal bleeding. Although a barium examination may localize potential bleeding sources, it will not necessarily define the bleeding site. In addition, large amounts of barium within the colon make the results of subsequent colonoscopy or angiography extremely difficult to interpret.

In contrast to the diagnostic value of endoscopy in acute upper gastrointestinal hemorrhage, the role of co-

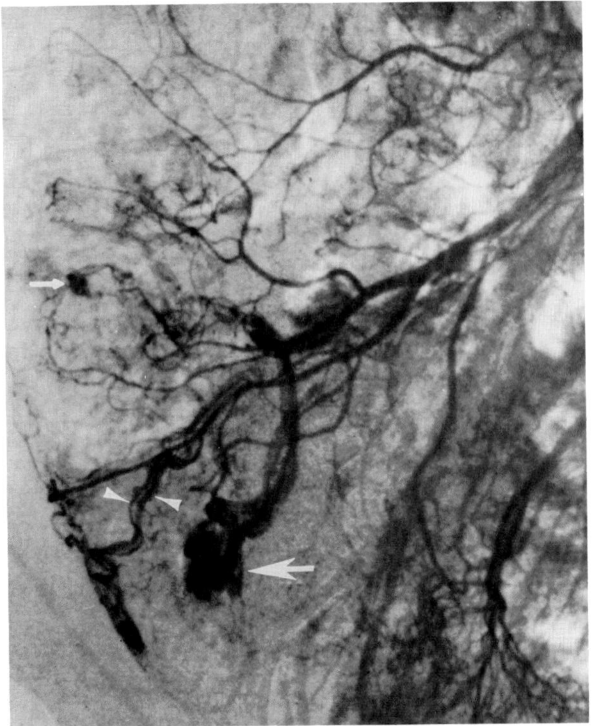

**Figure 48.27. Bleeding from an arteriovenous malformation.** A selective arteriogram demonstrates multiple arteriovenous malformations (AVMs) in the cecum and right colon. A large AVM (large white arrow), a small AVM (small white arrow), and a parallel artery and draining vein simultaneously filled (arrowheads) are all well-demonstrated. (From Goldberger,[18] with permission from the publisher.)

lonoscopy in acute lower gastrointestinal bleeding is severely limited. The presence of large amounts of blood in the colon often negates the value of direct endoscopic examination. Proctosigmoidoscopy, however, is of some value in identifying sources of bleeding in the rectum as well as in documenting the presence of fresh blood originating above this level.[18–21,75]

## Chronic Gastrointestinal Bleeding

In the patient with chronic gastrointestinal bleeding, barium studies (especially air-contrast examinations) are generally considered to be the screening procedures of choice. If both upper and lower gastrointestinal series are negative, endoscopy is indicated. Barium and endoscopic studies are clearly complementary from a diagnostic standpoint. This is especially true for the colon, in which colonoscopy has certain blind spots and complete evaluation is not universally accomplished. Tumors and inflammatory diseases of the small bowel that are not identified on conventional barium studies may be uncovered by enteroclysis. In this technique, large volumes of barium and water or air are rapidly instilled following the oral placement of a tube into the distal duodenum. If no source of bleeding has been discovered, abdominal arteriography should be performed to detect angiodysplasias that can only be demonstrated by this modality.[22–24]

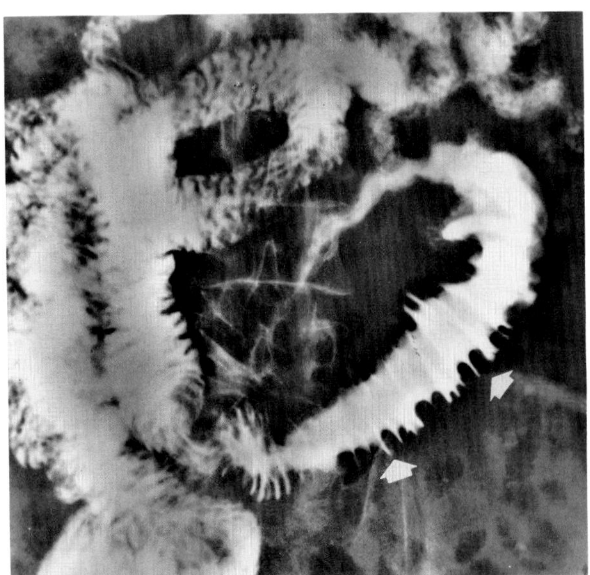

A

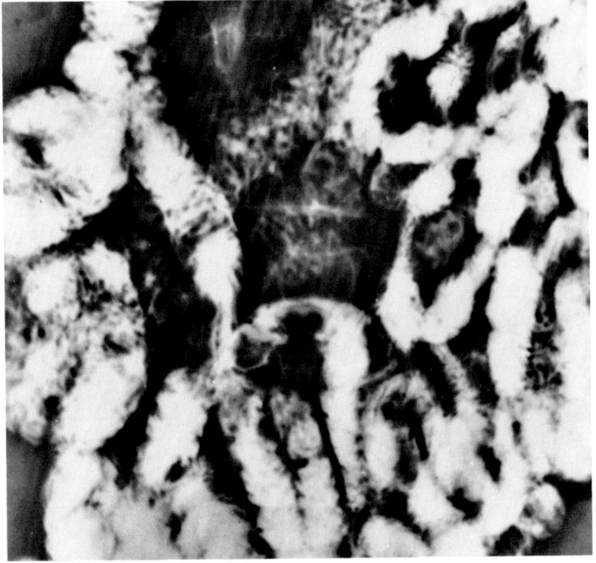

B

**Figure 48.28. Small bowel ischemia.** (A) Segmental ischemia causes a picket-fence pattern of regular thickening of small bowel folds (arrows). (B) Following conservative therapy, there is complete resolution of the ischemic process. (From Eisenberg,[32] with permission from the publisher.)

## MESENTERIC ISCHEMIA AND INFARCTION

Ischemic bowel disease is relatively common among elderly patients with cardiac failure and/or arteriosclerosis. It can also be caused by thrombosis or embolic occlusion of a major mesenteric artery or vein or its peripheral branches. In women of childbearing age, ischemic disease can be a complication of the use of hormonal contraceptive pills. Trauma, vasculitis, surgery on the abdominal aorta, and the endarteritis with vascular occlusion that may follow radiation therapy (> 5000 rad) to the small bowel can also result in ischemic bowel disease. The presenting symptom is usually colicky periumbilical pain, which may become more diffuse and continuous. Vomiting, abdominal distension, and bloody diarrhea also commonly occur.

The broad spectrum of radiographic appearances in patients with mesenteric vascular insufficiency depends on the rapidity of onset, the length of intestine involved, and the extent of collateral circulation. Rapid occlusion of the major mesenteric vessels results in massive bowel necrosis, and death from peritonitis and shock. However, if segmental rather than major vessels are involved, and if the occlusion is slow enough to allow collateral blood flow to develop, total intestinal infarction will not occur. In these cases, the relative degrees of damage and healing give rise to a variety of radiographic findings.

In patients with acute obstruction of the superior or inferior mesenteric arteries or veins, the appearance on plain abdominal radiographs may vary from an apparently normal gas pattern to a diffuse adynamic ileus involving both small and large bowel. Mucosal edema and intramural hemorrhage may cause thumbprinting—sharply defined, finger-like marginal indentations along the contours of the wall of the small bowel. One sign of acute occlusion is a relatively small amount of gas that remains in the generally straight or curvilinear lumen of a rigid, edematous segment of bowel and does not change in distribution on upright or decubitus radiographs (rigid loop sign). Nonspecific findings are a gasless abdomen in a symptomatic patient, and completely fluid-filled, gasless loops that may simulate a soft-tissue mass (pseudotumor). A loss of mucosal integrity, and/or increased intraluminal pressure in the bowel, can permit a potentially fatal bacterial invasion of the bowel wall, which produces crescentic linear gas collections in the wall of ischemic loops of bowel. Gas in the bowel wall typically disappears once blood flow to the affected segment of bowel improves. However, the concomitant finding of gas in the portal vein indicates irreversible intestinal necrosis and is a poor prognostic sign.

In patients with acute extensive bowel ischemia, arteriography often can demonstrate occlusion of a major mesenteric artery. Prompt surgery for the removal of the thrombus or embolus often permits revascularization without the need for intestinal resection.[25-29]

### Gas in the Portal Vein

Except when associated with umbilical venous catheterization in infants, and in rare instances in adults, the presence of gas in the portal venous system is of grave

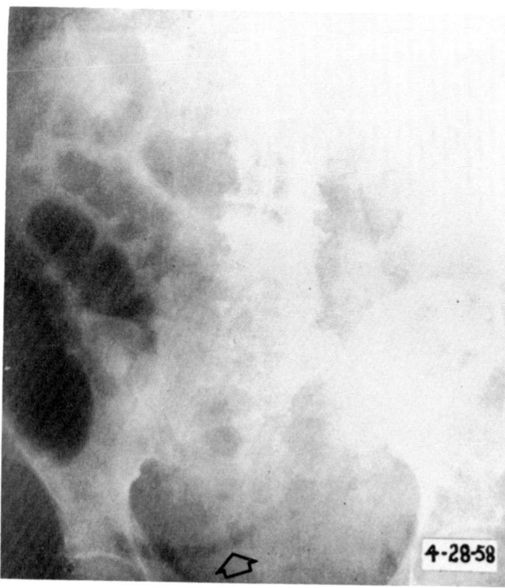


**Figure 48.29. Rigid loop sign is mesenteric venous occlusion.** A supine abdominal film demonstrates a narrow, sickle-shaped collection of gas in the right lower quadrant. (From Nelson and Eggleston,[29] with permission from the publisher.)

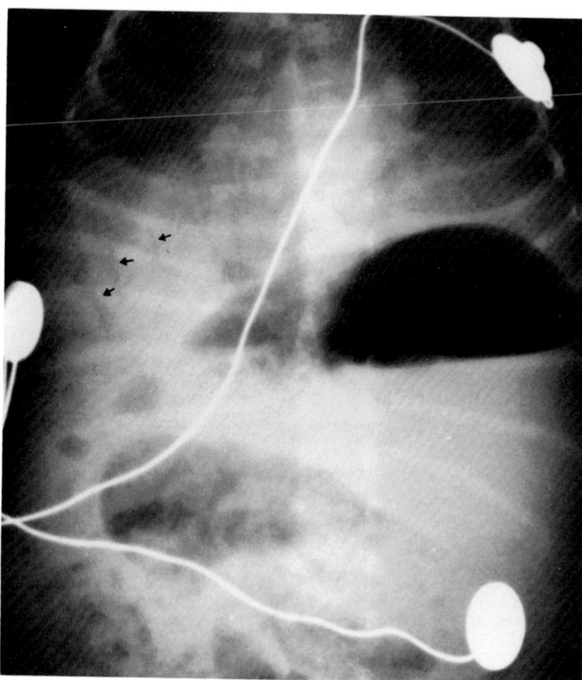

**Figure 48.30. Portal vein gas.** Characteristic radiographic appearance of tubular lucencies extending to the edge of the liver (arrows) in an infant who died from acute necrotizing enterocolitis. (From Eisenberg,[32] with permission from the publisher.)

prognostic significance and a sign of imminent death. Two major mechanisms are postulated to result in the radiographic appearance of gas in the portal veins. Mechanical intestinal obstruction or mesenteric artery occlusion may lead to a loss of intestinal mucosal integrity (necrosis), permitting intraluminal bowel gas to penetrate vessel walls and flow to the liver. A second mechanism suggests that either local intestinal necrosis is followed by infection of the bowel wall by gas-producing organisms, or that bowel necrosis is initially caused by an overwhelming enterocolitis. In either situation, gas in the bowel wall enters the portal venous system and lodges peripherally in the liver.

In children, the most common cause of gas in the portal veins is necrotizing enterocolitis, an often-fatal clinical syndrome characterized by abdominal distension, bloody vomitus and stools, and shock. In adults, most cases of gas in the portal veins are associated with mesenteric arterial occlusion and bowel infarction. In infants, a benign type of portal vein gas, which does not reflect a potentially fatal disorder, is caused by the inadvertent injection of air during umbilical venous catheterization or during drug administration through the catheter.

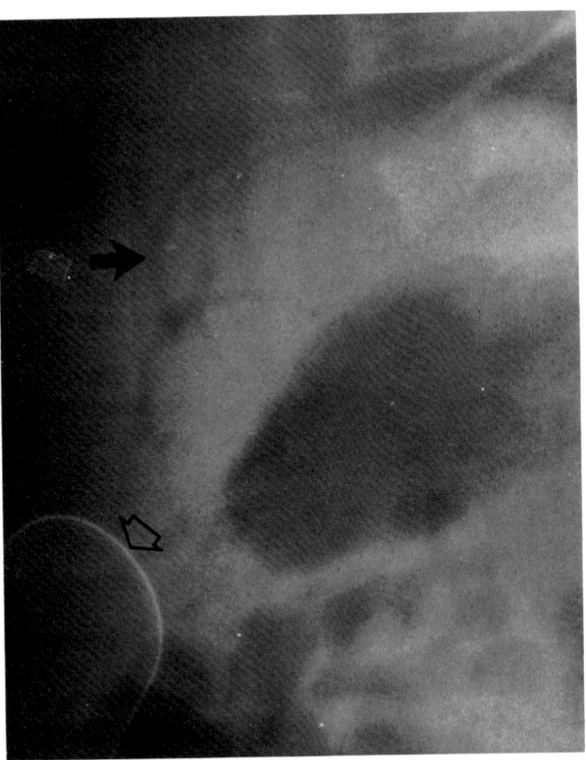

**Figure 48.31. Portal vein gas.** In this newborn male infant with mild respiratory distress, which cleared rapidly, gas in the portal vein (closed arrow) was caused by the inadvertent injection of air during umbilical catheterization and was of no clinical significance. The open arrow points to the umbilical catheter. (Swaim and Gerald,[30] with permission from the publisher.)

Gas in the portal veins must be distinguished from gas in the biliary system secondary to fistula formation between the bile duct and the gastrointestinal tract. When gas embolizes into the portal venous system, bubbles are carried into the fine peripheral radicles in the liver by the centrifugal flow of portal venous blood. This presents a characteristic radiographic pattern of radiating tubular radiolucencies branching from the porta hepatis to the edge of the liver. The visualization of gas in the outermost 2 cm of the liver is considered presumptive evidence of portal vein gas. In contrast, gas in the biliary tree is found in the larger, more centrally situated bile ducts. It is prevented from entering finer radicles by the continuous centripetal flow of the secreted bile.[30–32]

## ISCHEMIC COLITIS

Ischemic colitis is characterized by the abrupt onset of lower abdominal pain and rectal bleeding. Diarrhea is common, as is abdominal tenderness on physical examination. Most patients are over the age of 50, and many have a history of prior cardiovascular disease.

The radiographic appearance of ischemic colitis depends on the phase of the process during which the patient is examined. Because the mucosa is the layer most dependent on intact vascularity, fine superficial ulceration associated with inflammatory edema is the earliest radiographic sign of ischemic colitis. This causes the outer margin of the barium-filled colon to appear serrated, simulating ulcerative colitis. Unlike ulcerative colitis, however, in ischemic colitis the rectum is usually

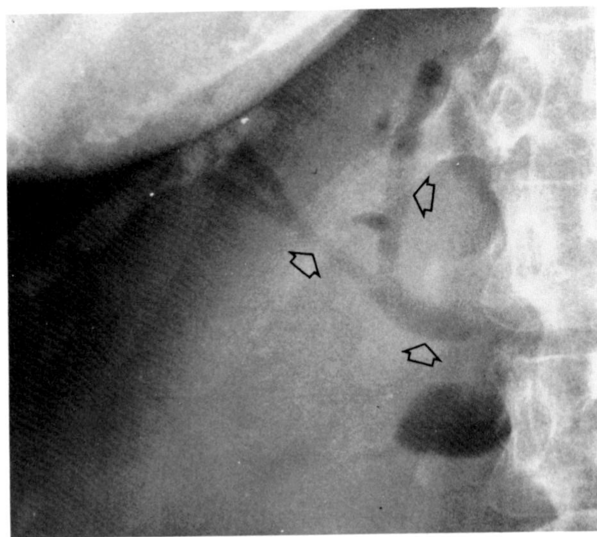

**Figure 48.32. Gas in the biliary tree.** Unlike portal venous gas, which involves the fine peripheral radicles in the liver, gas in the biliary tree is found in the larger, more centrally situated bile ducts (arrows). (From Eisenberg,[32] with permission from the publisher.)

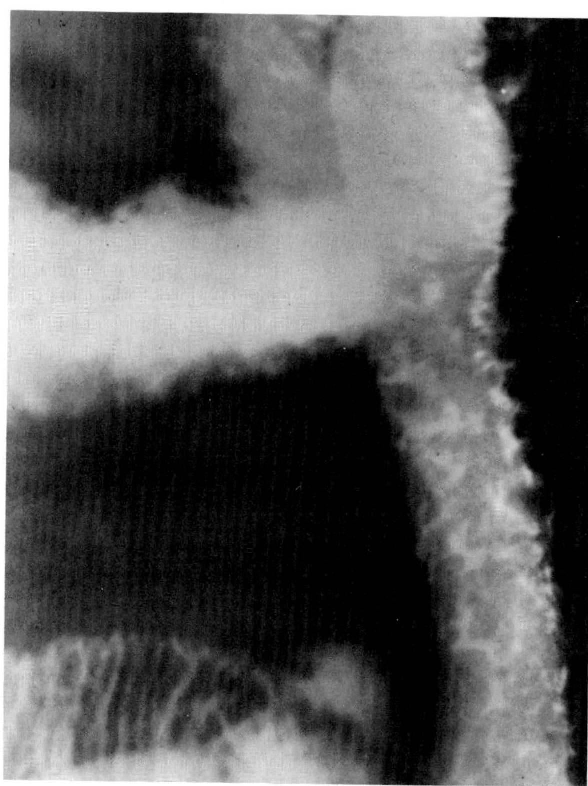

**Figure 48.33. Ischemic colitis.** Superficial ulcers and inflammatory edema produce a serrated outer margin of the barium-filled colon simulating ulcerative colitis. (From Eisenberg,[32] with permission from the publisher.)

spared. As the disease progresses, deep penetrating ulcers, pseudopolyposis, and "thumbprinting" can be demonstrated. Thumbprinting refers to sharply defined, finger-like marginal indentations along the contours of the colon wall. Although these well-circumscribed filling defects seen on plain abdominal radiographs or barium studies are generally considered to be manifestations of colon ischemia or hemorrhage, thumbprinting can occur in any inflammatory or neoplastic disease that produces polypoid masses or substantial enlargement of mucosal folds.

In most cases, the radiographic appearance of the colon returns to normal within 1 month if good collateral circulation is established. Extensive fibrosis during the healing phase can cause tubular narrowing and a smooth stricture. If blood flow is insufficient, acute bowel necrosis and perforation may result.

The differentiation of ischemic colitis from ulcerative colitis and Crohn's disease of the colon can be extremely difficult on radiographic or pathologic examination alone. In such cases, the characteristic acute episode of abdominal pain and bleeding, the rapid progression of radiographic findings, and the low rate of recurrence usually permit ischemic colitis to be readily distinguished from ulcerative colitis and Crohn's disease of the colon, which are typically more indolent, chronic, and recurring.

Although some cases of chronic infarction are caused by embolization or thrombosis of a mesenteric artery, the majority of cases do not have a demonstrable arterial occlusion.[33–35]

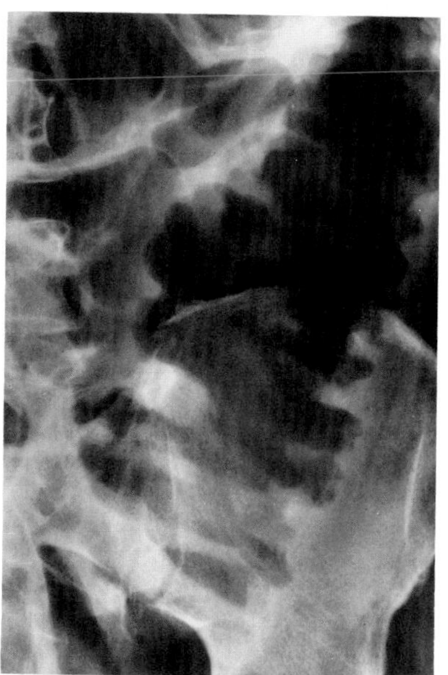

A

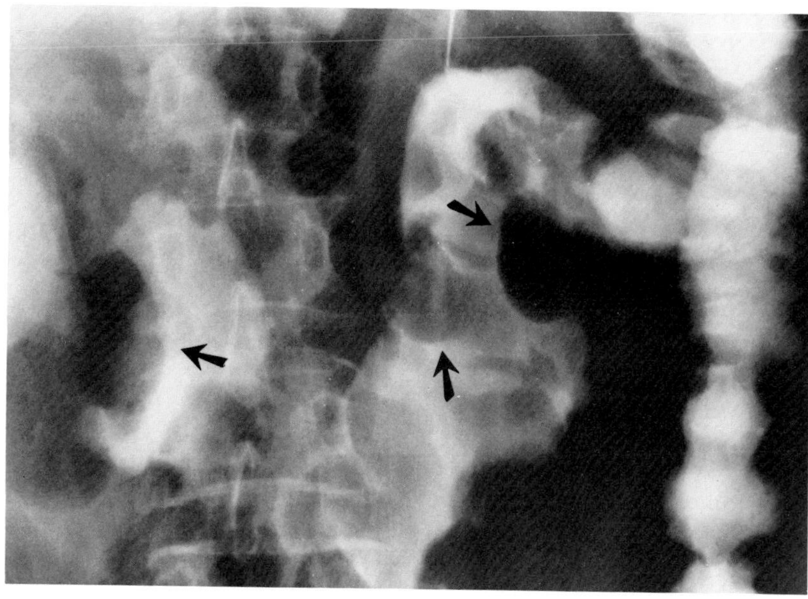

B

**Figure 48.34. Thumbprinting in ischemic colitis.** (A) A plain abdominal film demonstrates soft tissue polypoid densities that protrude into the lumen of the descending colon in this patient with acute abdominal pain and rectal bleeding. (B) A barium enema examination shows multiple filling defects (arrows) indenting the margins of the transverse and descending portions of the colon. (From Eisenberg,[32] with permission from the publisher.)

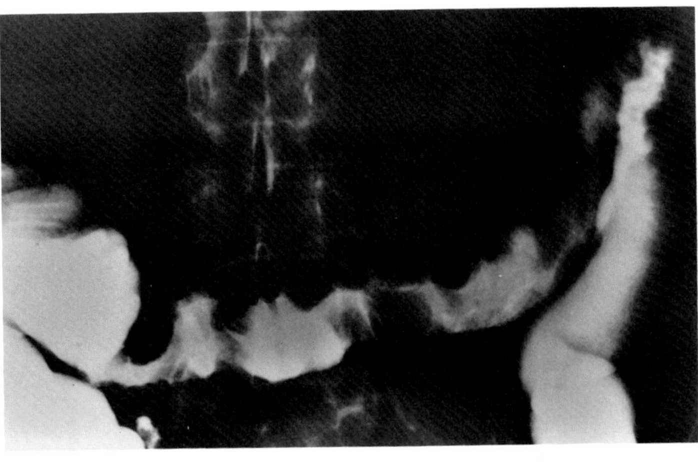

A

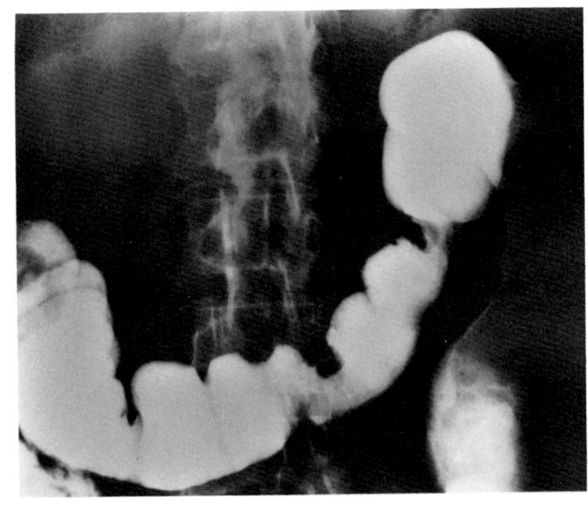

B

**Figure 48.35. Reversibility of ischemic colitis.** (A) Thumbprinting involves the transverse colon during an acute ischemic attack. (B) One week later, the ischemic changes have reversed, and the patient has clinically recovered. (From Eisenberg,[32] with permission from the publisher.)

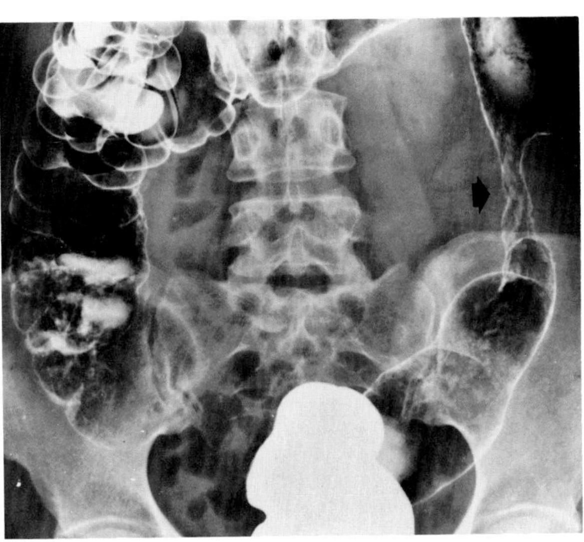

**Figure 48.36. Ischemic colitis.** A stricture in the descending colon (arrow) followed healing of the ischemic episode. (From Eisenberg, Montgomery, and Margulis,[33] with permission from the publisher.)

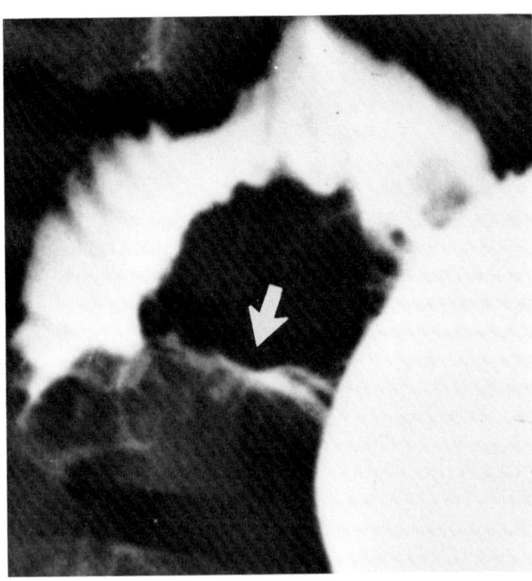

**Figure 48.37. Intramural duodenal hematoma.** This high-grade stenotic lesion (arrow) was seen in a young child who had been kicked in the abdomen by his father. (From Eisenberg,[32] with permission from the publisher.)

## INTRAMURAL HEMATOMA OF THE DUODENUM

Because the retroperitoneal portions of the duodenum are the most fixed segments of the small bowel, external blunt trauma predominantly affects this area. Hemorrhage into the bowel wall produces a tumor-like intramural mass with well-defined margins that often causes partial or even complete obstruction of the lumen. Duodenal intramural hematomas can also be secondary to anticoagulant therapy or to an abnormal bleeding diathesis.[32]

## PRIMARY NONSPECIFIC ULCERATION

### Small Intestine

Unexplained ulcers of the small intestine may be caused by ischemic necrosis of the mucosa secondary to mesenteric thrombosis or embolism, vasculitis, incarcerated hernia, or aortic surgery. In many cases, nonspecific ulceration occurs in patients receiving enteric-coated drugs known to be irritating to the mucosa (potassium chloride, thiazides).

Nonspecific ulcerations of the small intestine are often shallow and are usually difficult to demonstrate radio-

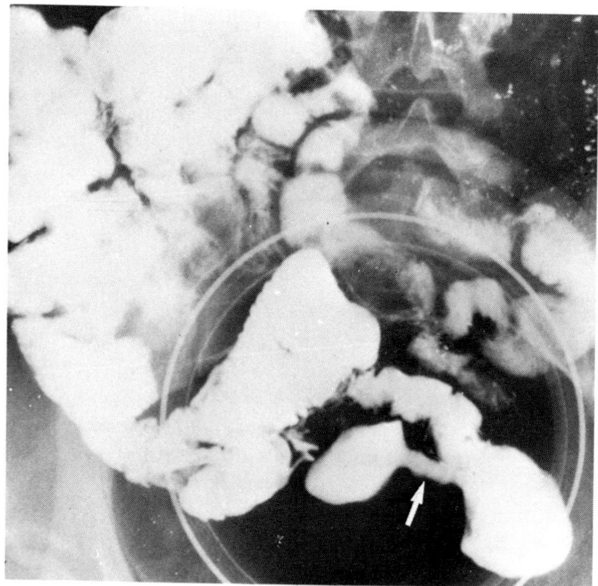

**Figure 48.38. Nonspecific ulcer of the small bowel.** There is a short, symmetric stenosis (arrow) in the proximal ileum. (From Margulis and Burhenne,[76] with permission from the publisher.)

graphically. The associated inflammatory reaction may cause irregular thickening of mucosal folds, luminal narrowing, and thickening of the mesentery that produce a pattern indistinguishable from Crohn's disease. Fibrotic healing often causes stricture formation and partial small bowel obstruction.[36–38]

### Colon

Nonspecific ulceration of the colon is a diagnosis of exclusion that is rarely made preoperatively. Although various causes have been suggested (peptic ulceration, solitary diverticulitis, drugs, mucosal trauma, infection, vascular disease), no precise cause has been identified. The clinical symptoms of nonspecific ulcers depend on their location. Ulcers in the ascending colon usually present acutely, mimicking appendicitis. Nonspecific ulcers of the transverse and descending colon have a more insidious onset and suggest carcinoma, obstruc-

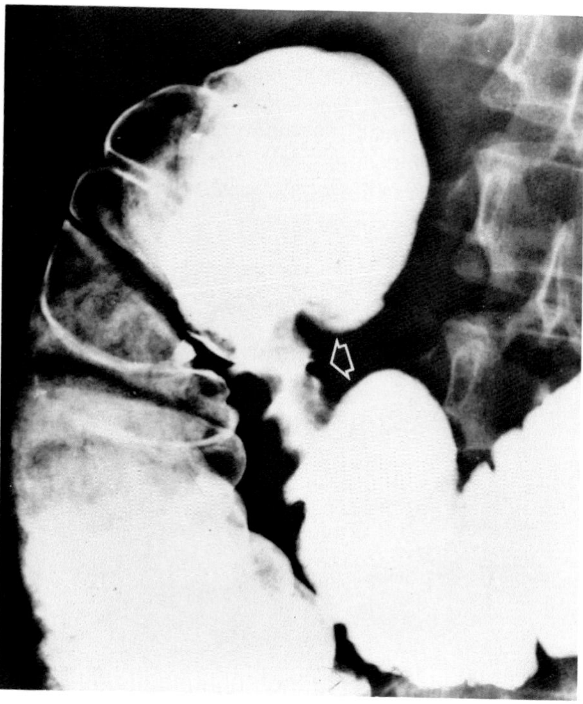

**Figure 48.39. Nonspecific ulcer of the colon.** There is an area of narrowing in the proximal transverse colon with ulceration along its inferior aspect and marginal spiculation (arrow). (From Gardiner and Bird,[39] with permission from the publisher.)

tion, or diverticulitis. Severe complications include perforation with secondary peritonitis, frank hemorrhage, and stricture.

More than half of all nonspecific colonic ulcers occur in the cecum and ascending colon in the region of the ileocecal valve. Most are single, though multiple ulcers are present in up to 20 percent of cases. They range from small, superficial ulcers to transmural lesions involving the entire circumference of the colon. In most cases, nonspecific benign ulcerations are associated with an intense inflammatory reaction and appear as large filling defects, usually without radiographically demonstrable ulceration. Fibrotic healing can cause smooth or irregular areas of narrowing that can be indistinguishable from carcinoma.[39,40]

# TUMORS OF THE SMALL INTESTINE

## BENIGN TUMORS

### Leiomyomas

Leiomyomas are the most common benign neoplasms of the small bowel. They are almost always single and, although found at all levels of the small bowel, are most frequent in the jejunum. These smooth muscle tumors arise as subserosal or submucosal lesions but can extend intraluminally and become pedunculated. Although the surface of a leiomyoma exhibits a rich blood supply, the central portion of the tumor is often virtually avascular and tends to undergo central necrosis and ulceration causing gastrointestinal hemorrhage. A leiomyoma with a large intraluminal component can be the leading point of an intussusception.

An intramural leiomyoma is seen in profile as a characteristic broad filling defect, its base wider than its projection into the lumen. Retention of barium in a superficial mucosal ulceration that communicates with a relatively deep pit in the tumor is a characteristic finding.

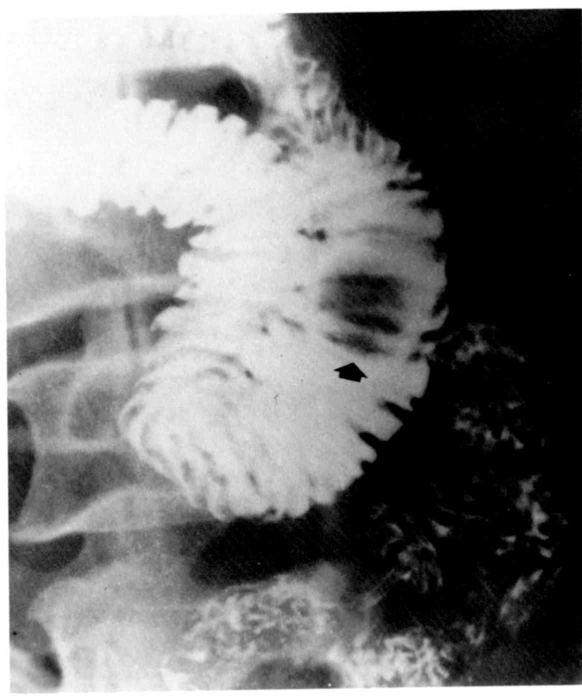

**Figure 48.40. Intramural leiomyoma of the jejunum.** The small intramural filling defect (arrow) is visible on the en face view. (From Good,[41] with permission from the publisher.)

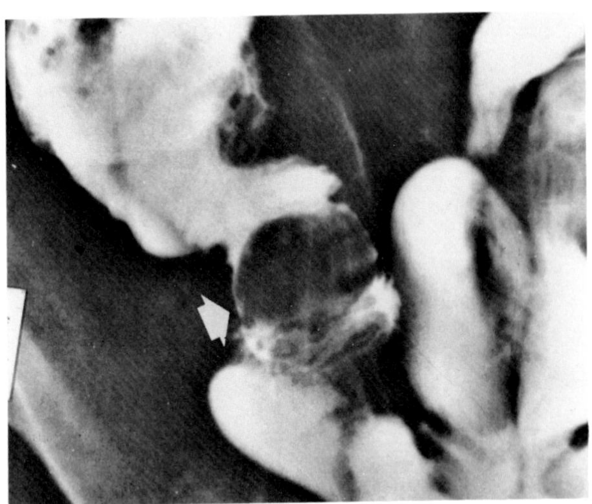

**Figure 48.41. Adenoma of the terminal ileum.** The smooth polypoid mass (arrow) fills most of the lumen. (From Eisenberg,[32] with permission from the publisher.)

Some leiomyomas project from the serosal surface and are detectable radiographically only when they are large enough to displace adjacent barium-filled loops of small bowel.[41]

## Adenomas

Polypoid adenomas are the second most common benign small bowel neoplasms. They can be found throughout the small bowel but are most frequent in the ileum. Most adenomas are single, well-circumscribed polyps, though multiple lesions do occur. These polyps are usually intraluminal and pedunculated and are therefore prone to act as the leading point of an intussusception.

Brunner's gland adenomas are not truly neoplastic but represent localized areas of hypertrophy or hyperplasia of submucosal duodenal glands. Brunner's glands are elaborately branched acinar glands containing both mucous and serous secretory cells. They are most common in the duodenal bulb and proximal half of the second portion of the duodenum, though they can also be found in the distal antrum and distal duodenum. The alkaline secretions from Brunner's glands are rich in mucus and bicarbonate, which protect the sensitive duodenal mucosa from erosion by stomach acid. Brunner's gland hyperplasia probably represents a response of the duodenal mucosa to peptic ulcer disease. The much more common diffuse form of Brunner's gland hyperplasia appears as generalized nodular thickening of folds, usually limited to the first portion of the duodenum.[41]

## Lipomas

Lipomas are the third most frequent benign tumors of the small bowel. Most are found in the distal ileum and ileocecal valve area, though they can occur anywhere in the small bowel. Lipomas arise in the submucosa but tend to protrude into the lumen and, if pedunculated, can be associated with intussusception. On barium studies, lipomas characteristically appear as intraluminal filling defects with a smooth surface and a broad base of attachment indicating their intramural origin. The fatty consistency of these tumors permits them to be easily deformed by palpation.[41]

## Angiomas

Angiomas can be divided into hemangiomas (true tumors composed of endothelium-lined, blood-containing spaces) and telangiectasias (dilatation of an existing vascular structure). Though far less common than other benign small bowel tumors, hemangiomas are clinically important because of their propensity for bleeding. Hemangiomas of the small bowel are generally multiple. Most are relatively sessile lesions that are frequently missed

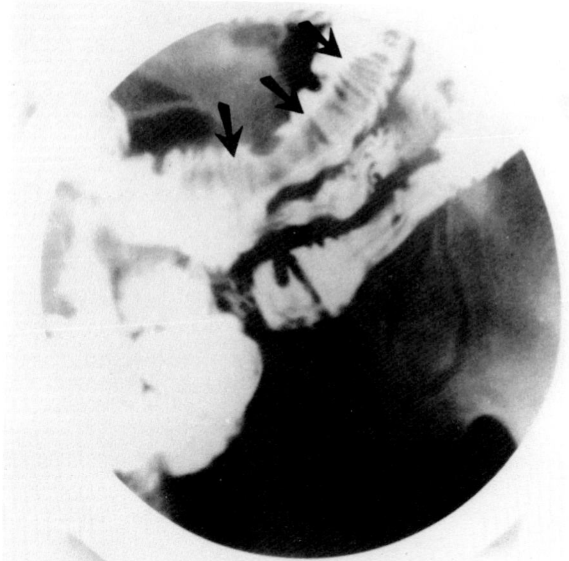

**Figure 48.42. Lipoma of the small bowel.** A long pedunculated tumor produces an intraluminal filling defect (arrows). (From Good,[41] with permission from the publisher.)

on barium studies because of their small size and easy compressibility. The uncommon combination of phleboliths and single or multiple filling defects in the small bowel is pathognomonic of hemangioma.

Telangiectasias can be associated with several abnormalities, the best known of which is the Osler-Rendu-Weber syndrome, in which there is a familial history of repeated hemorrhage from the nasopharynx and the gastrointestinal tract and multiple telangiectatic lesions involving the nasopharyngeal, buccal, and gastrointestinal mucosae.[41]

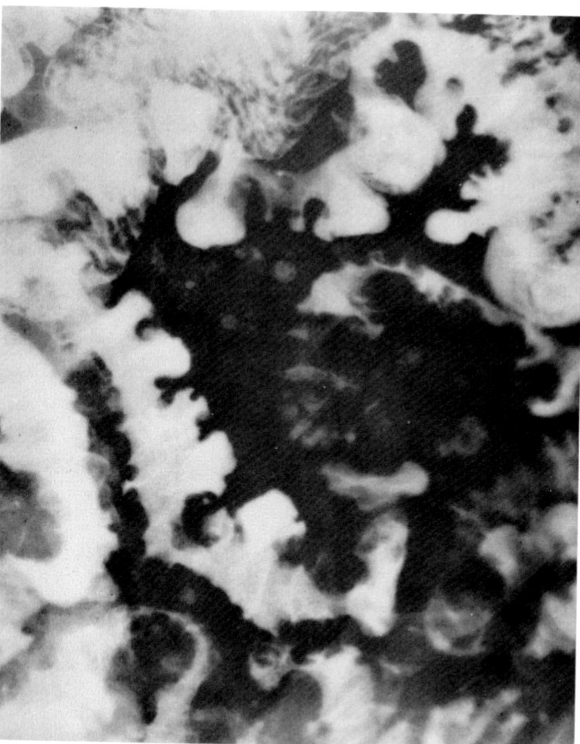

**Figure 48.43. Hemangiomatosis of the small bowel and mesentery.** Characteristic phleboliths are associated with multiple filling defects in the small bowel. (From Eisenberg,[32] with permission from the publisher.)

## MALIGNANT TUMORS

In contrast to benign tumors, which are most commonly asymptomatic, malignant neoplasms of the small bowel are frequently associated with fever, weight loss, anorexia, bleeding, and a palpable abdominal mass. Although a given type of primary tumor can be found anywhere in the small bowel, carcinomas tend to cluster around the ligament of Treitz (i.e., distal duodenum and proximal jejunum), while sarcomas occur more frequently in the ileum. Like benign tumors, a malignant lesion with a large intraluminal component can be the leading point of an intussusception.

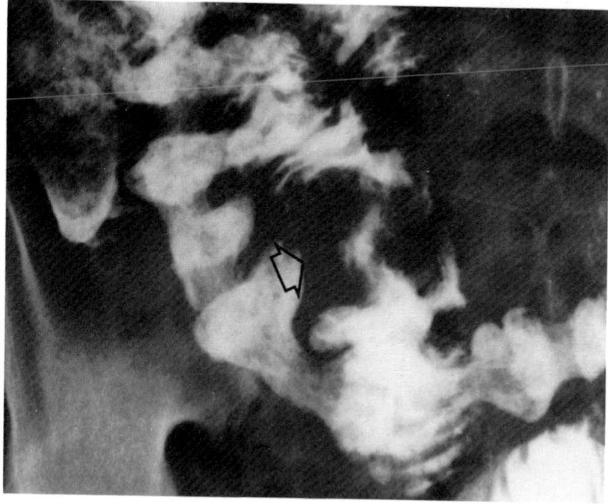

**Figure 48.44. Primary adenocarcinoma of the ileum.** An annular constricting lesion (arrow). (From Eisenberg,[32] with permission from the publisher.)

### Adenocarcinomas

Adenocarcinomas are the most common malignant tumors of the small bowel. They tend to be aggressively invasive and extend rapidly around the circumference of the bowel, inciting a fibrotic reaction and luminal narrowing that soon cause obstruction. Adenocarcinomas of the small bowel occasionally appear as broad-based intraluminal masses; extremely rarely they appear as pedunculated polyps.[32,42]

## Leiomyosarcomas

Leiomyosarcomas are most often large, bulky, irregular lesions usually more than 5 cm in diameter. Like benign leiomyomas, leiomyosarcomas have a tendency for central necrosis and ulceration leading to massive gastrointestinal hemorrhage and the radiographic appearance of an umbilicated lesion. Because more than two-thirds of leiomyosarcomas primarily project into the peritoneal cavity, the major manifestation of this type of tumor is the displacement of adjacent, uninvolved, barium-filled loops of small bowel.[32]

## Lymphomas

This is discussed in the section "Intestinal Lymphoma" in Chapter 46.

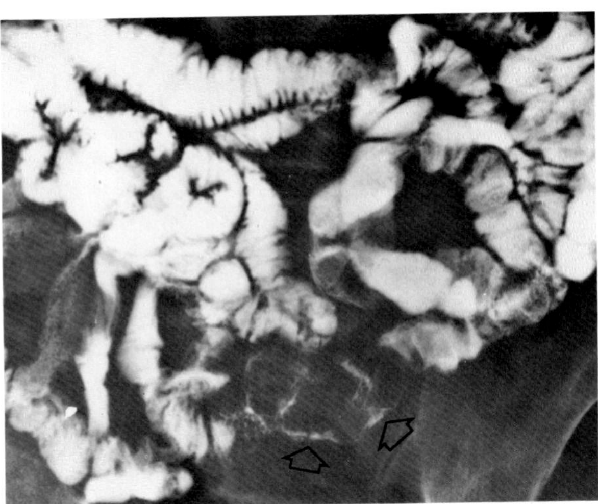

**Figure 48.45. Leiomyosarcoma of the small bowel.** Note the large, bulky, irregular lesion (arrow). (From Eisenberg,[32] with permission from the publisher.)

**Figure 48.46. Carcinoid tumor.** A polypoid filling defect (arrow) in the small bowel. (From Eisenberg,[32] with permission from the publisher.)

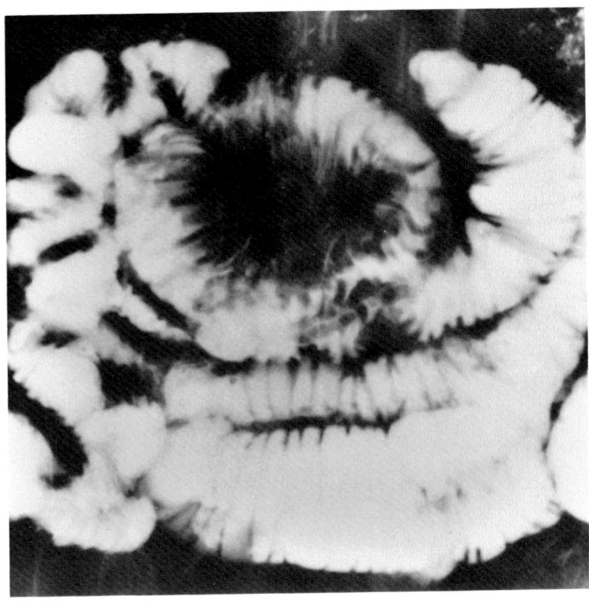

**Figure 48.47. Carcinoid tumor.** Separation of bowel loops, luminal narrowing, and fibrotic tethering of mucosal folds. (From Eisenberg,[32] with permission from the publisher.)

## Carcinoid Tumors

Carcinoid tumors are the most common primary neoplasms of the small bowel. The "rule of one-third" has been applied to small bowel carcinoids: they account for one-third of gastrointestinal carcinoid tumors, one-third of them show metastases, one-third present with a second malignancy, and about one-third are multiple. Although carcinoids of the small bowel can occur at any site, they are most frequently seen in the ileum.

Small carcinoids appear as small, sharply defined, submucosal lesions. At this stage of development, they are usually asymptomatic and are rarely detected on routine small bowel examination. Most carcinoids tend to grow extraluminally, infiltrating the bowel wall, the lymphatic channels, and eventually the regional lymph nodes and mesentery. The local release of serotonin leads to hypertrophic muscular thickening and an intense desmoplastic response, which results in the characteristic radiographic appearance of diffuse luminal narrowing and separation, fixation, and abrupt angulation of intestinal loops.[43,44]

# TUMORS OF THE COLON

## COLONIC POLYPS

Colonic polyps are small masses of tissue, with or without a stalk, that arise from the mucosa and project into the lumen of the bowel. Colonic polyps are common lesions; clinical and autopsy studies have indicated a 12.5 percent incidence of polyps in the general population.

In the past, more than 90 percent of all colonic polyps were considered to be hyperplastic (metaplastic). Hyperplastic polyps are smooth, sessile mucosal elevations less than 5 mm in size. Hyperplastic polyps arise from excessive cellular proliferation in the crypts of Lieberkühn but maintain their cellular differentiation and have no malignant potential or clinical significance. Some studies, however, indicate that the large majority of small polyps (less than 5 mm) identified on barium studies or colonoscopy are adenomatous, rather than hyperplastic, and thus have the potential for growth and malignant transformation. These findings have led to a more aggressive approach to the clinical management of polyps, and these small lesions are now removed if found incidentally during proctoscopy or colonoscopy.

Adenomatous polyps are true neoplasms that can contain malignant foci. About 80 percent of adenomatous polyps occur in the rectum and sigmoid colon. The frequency of polyps rises with increasing age, and persons with one polyp have a greater risk of having two or more polyps than a person in the general population has of having even a single polyp.

The radiographic detection of small polypoid lesions requires meticulous colon preparation and examination. Poor preparation may result in a polyp being overlooked or confused with fecal material. Diverticula seen en face rather than in profile can appear to lie within the lumen, rather than projecting beyond it, and may be difficult to distinguish from polyps. Rotation of the patient usually demonstrates that the diverticulum truly extends beyond the colonic lumen. A small adenomatous polyp can easily be hidden by a cluster of diverticula.

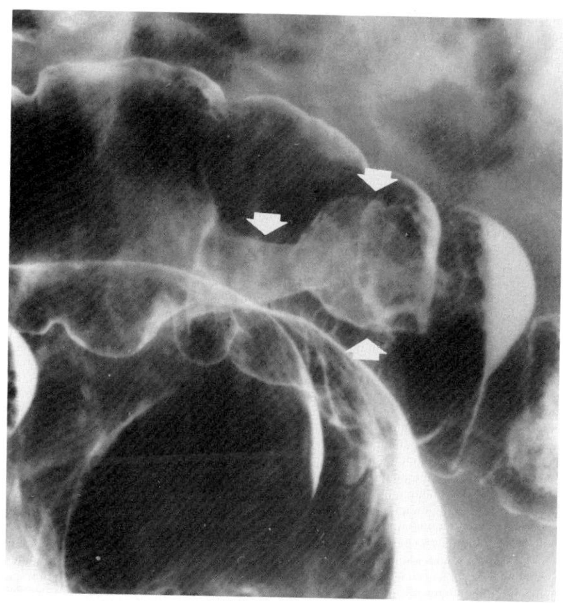

**Figure 48.49. Pedunculated adenomatous polyp (arrows).** (From Eisenberg,[32] with permission from the publisher.)

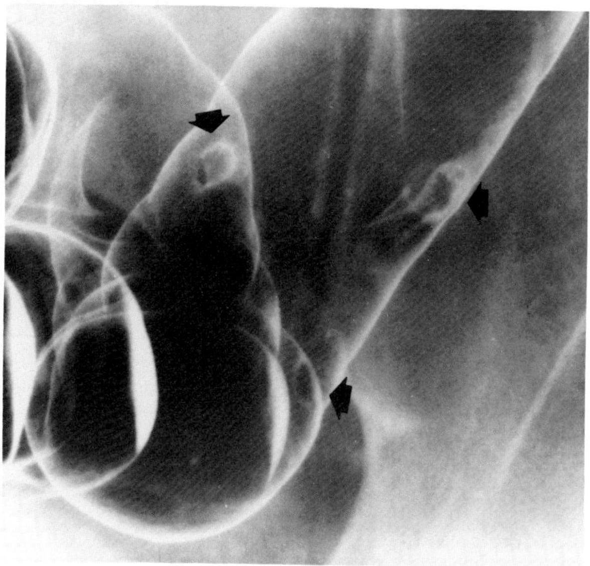

**Figure 48.50. Three benign adenomatous polyps (arrows).** (From Eisenberg,[32] with permission from the publisher.)

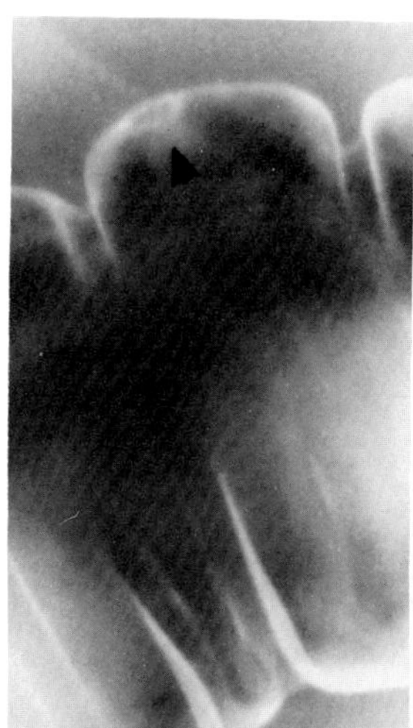

**Figure 48.48. Hyperplastic (metaplastic) colonic polyp.** A smooth, sessile mucosal elevation (arrow) 2 mm in diameter.

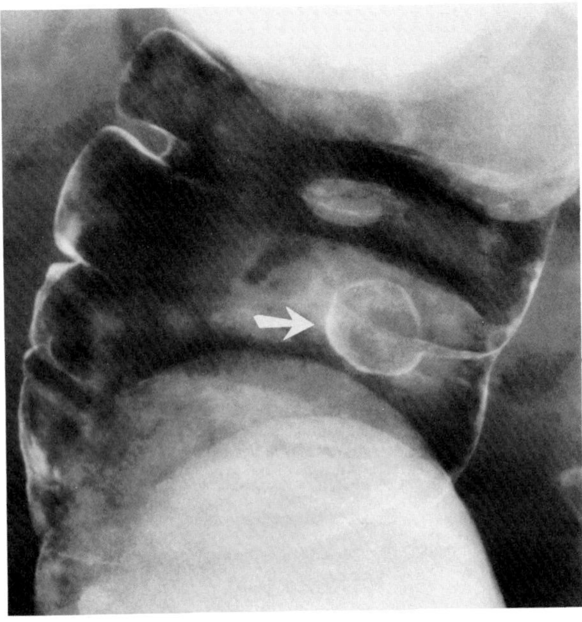

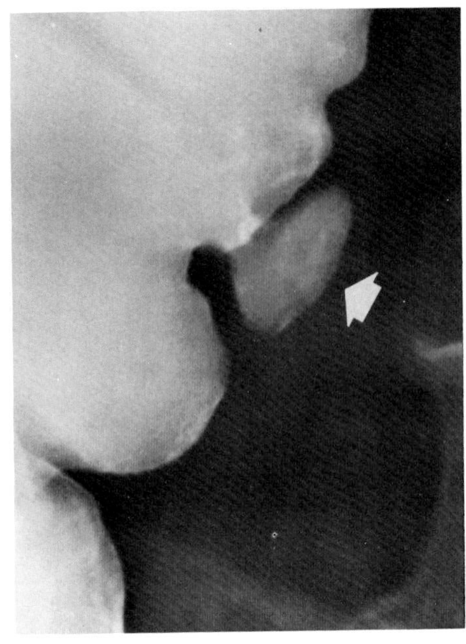

A B

**Figure 48.51. Diverticulum mimicking a colonic polyp.** (A) On the en face projection, the barium-coated diverticulum (arrow) simulates a polyp lying within the lumen. (B) A radiograph obtained after rotation of the patient demonstrates that the diverticulum fills with barium (arrow) and clearly extends beyond the colonic lumen. (From Eisenberg,[32] with permission from the publisher.)

Sessile polyps are flat lesions attached to the mucosa by a broad base. Peristaltic waves and the flow of the fecal stream may cause traction on a sessile polyp and force the underlying normal mucosa to be drawn out into a pedicle or stalk. On a profile view, the pedicle may appear as a linear lucency in the barium-filled colon or be thinly coated by barium in an air-contrast examination. When seen en face (with the central beam parallel to the long axis of the pedicle), the barium-coated pedicle is seen as a small white circle within a larger circle of barium covering the body of the polyp (target sign). The radiographic demonstration of a pedicle 2 cm or more in length is virtually pathognomonic of a benign polyp. If a malignant polyp is pedunculated, the stalk is usually short, thick, and irregular.[45,46,80,81]

## LIPOMAS

Lipomas of the colon appear radiographically as circular or ovoid, sharply defined, smooth filling defects in the barium column. A pathognomonic diagnostic feature of lipomas is their changeability in size and shape during the course of barium enema examination. Because these tumors are extremely soft, their configuration can be altered by palpation and extrinsic pressure. Thus a malleable lipoma that appears round or oval on filled films characteristically becomes elongated (sausage- or banana-shaped) on postevacuation films in which the colon is contracted.

Other spindle-cell tumors (leiomyoma, fibroma, neurofibroma) are rare in the colon. Malignant spindle-cell tumors are extremely rare. They tend to be much larger and more irregular than their benign counterparts, although differentiation between benign and malignant submucosal tumors can be extremely difficult.[47]

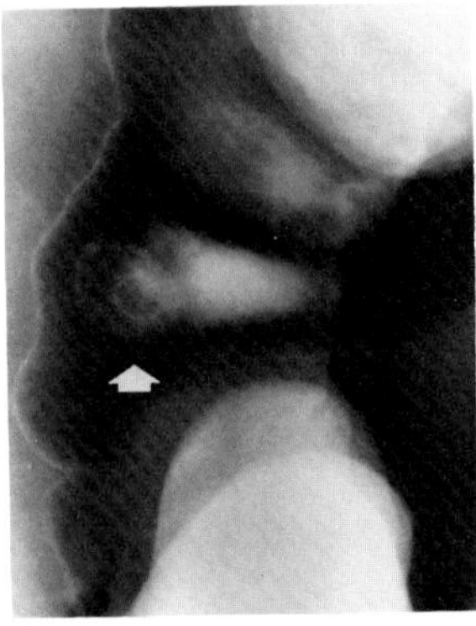

**Figure 48.52. Target sign of pedunculated colonic polyp.** (From Eisenberg,[32] with permission from the publisher.)

## VILLOUS ADENOMAS

Villous adenomas of the colon are polypoid masses consisting of innumerable villous fronds that give the surface a corrugated appearance. Most are solitary and are located in the rectosigmoid area. Villous adenomas have a far greater malignant potential than adenomatous polyps; about 40 percent of villous adenomas demonstrate infiltrating carcinoma, usually at the base. Villous adenomas also tend to be much larger than adenomatous polyps; about 75 percent exceed 2 cm in diameter,

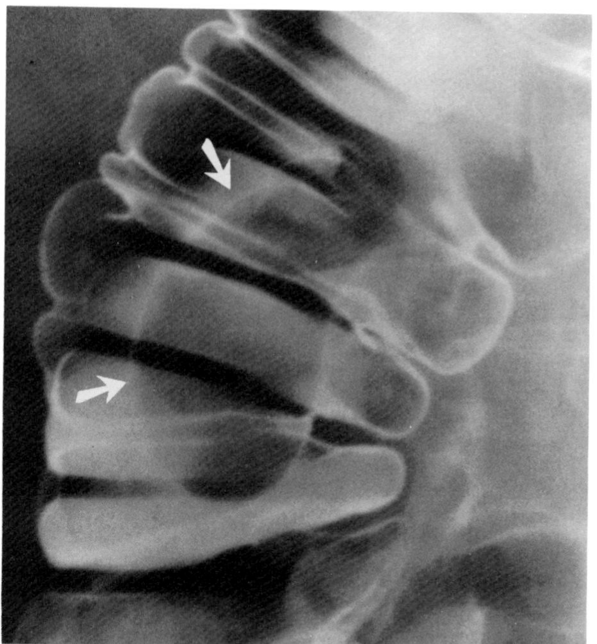

**Figure 48.53. Lipoma of the ascending colon.** The mass is extremely lucent and has smooth margins and a teardrop shape (arrows). (From Eisenberg,[32] with permission from the publisher.)

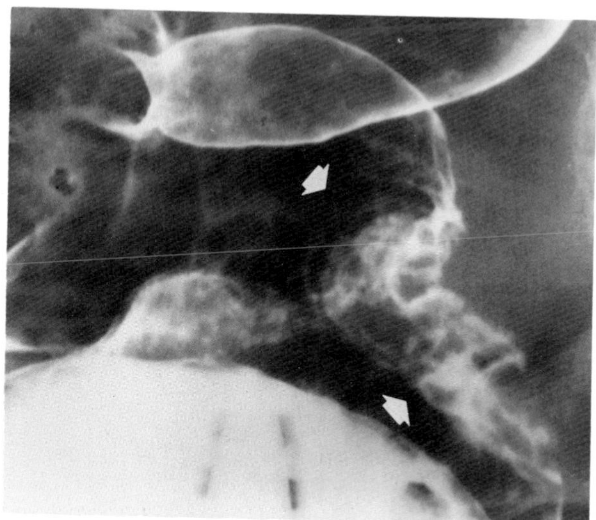

**Figure 48.54. Benign villous adenoma of the rectum (arrows).** Barium is seen filling the deep clefts between the multiple fronds. (From Eisenberg,[32] with permission from the publisher.)

whereas only 5 percent of adenomatous polyps reach this size.

Although infrequent, large amounts of mucus may be secreted into the lumen across the papillary surface of a villous adenoma. This mucous diarrhea causes severe fluid, protein, and electrolyte (especially potassium) depletion.

Radiographically, villous adenomas classically appear as bulky tumors with a spongelike pattern (bouquet of flowers) caused by barium filling of deep clefts between the multiple fronds. This radiographic feature is best demonstrated on the postevacuation view, in which barium remains within the interstices of the villous tumor. Because the lobular, irregular tumor has a soft consistency, its appearance can change on serial films and with palpation. The adjacent bowel wall is pliable and distensible, since the tumor does not incite a desmoplastic response.

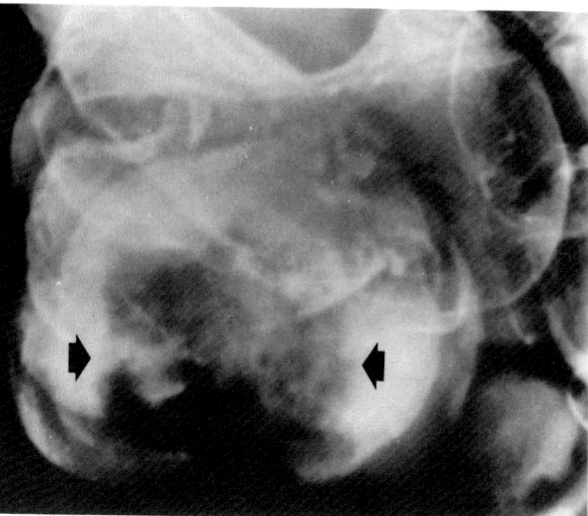

**Figure 48.55. Villous adenocarcinoma of the cecum.** A huge irregular mass (arrows) is visible. Barium fills the interstices of the frondlike tumor. (From Eisenberg,[32] with permission from the publisher.)

Although large size, ulceration, and indentation of the tumor base have been suggested as signs of malignancy, no radiographic findings in villous adenoma are sufficient to exclude malignant degeneration. Because invasive carcinoma in villous adenoma is usually found at the base of the lesion rather than on the surface, biopsies can be unreliable as a result of inadequate tissue sampling. Therefore, even a benign-appearing villous adenoma should be totally excised.[48]

## INTESTINAL POLYPOSIS SYNDROMES

The intestinal polyposis syndromes are a diverse group of conditions that differ widely in the histology of the polyps, the incidence of extracolonic polyps, extra-abdominal manifestations, and the potential for developing malignant disease. An intestinal polyposis disorder should be suspected when a polyp is demonstrated in a young person, when multiple polyps are found in any person, or when carcinoma of the colon is found in a patient under 40 years of age. If one of the hereditary forms of intestinal polyposis is diagnosed, the patient's immediate family should be studied so that a potentially fatal disease is not missed in its premalignant stage.

## Familial Polyposis

Familial polyposis is an inherited disease (autosomal dominant) characterized by multiple adenomatous polyps almost exclusively limited to the colon and rectum. The polyps are not present at birth but tend to appear in childhood and adolescence. They may cause diarrhea or rectal bleeding, though many patients with familial polyposis are asymptomatic and only discovered during routine investigation of relatives of a patient known to have the disease. There is no evidence of extra-intestinal involvement.

On barium enema examination, the polyps appear as sessile or pedunculated lesions scattered throughout the colon. Although the rectum and left colon are involved more frequently than the right, myriads of polyps often blanket the entire length of the colon. With diffuse disease, the colon can appear to be poorly prepared; however, in familial polyposis, the true adenomatous polyps remain fixed in position with palpation, unlike retained fecal material, which is usually freely movable.

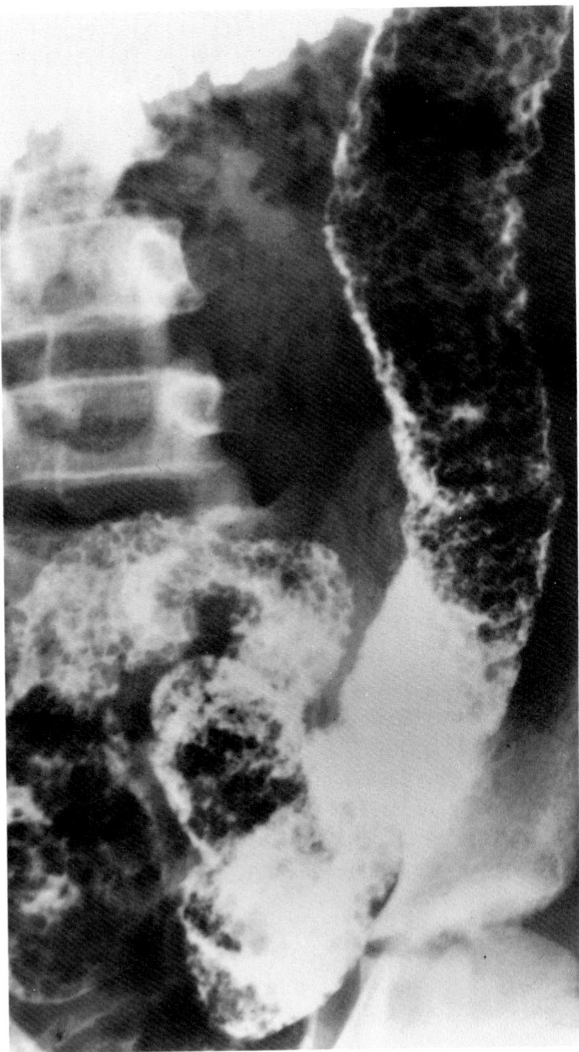

**Figure 48.56. Familial polyposis.** Innumerable adenomatous polyps blanket the descending and sigmoid colon. (From Eisenberg,[32] with permission from the publisher.)

Because patients with familial polyposis have virtually a 100 percent risk of developing carcinoma of the colon or rectum, total colectomy is usually recommended at the time of diagnosis.[49]

## Gardner's Syndrome

Gardner's syndrome is an inherited disorder (autosomal dominant) in which diffuse colonic polyposis is associated with bony abnormalities and soft tissue tumors. Osteomas are common, especially in the paranasal sinuses. Sebaceous cysts and subcutaneous fibromas, leiomyomas, and lipomas are often seen. Exostoses and cortical thickening can involve the long bones and the ribs. Dental abnormalities are frequent and include odontomas, extra teeth, unerupted teeth, and a propensity toward numerous caries.

The distribution and appearance of the adenomatous polyps in Gardner's syndrome are indistinguishable from the pattern in familial polyposis. The polyps are almost always limited to the colon and rectum; extracolonic polyps occasionally occur in the small bowel and stomach. Like patients with familial polyposis, patients with Gardner's syndrome have almost a 100 percent risk of developing carcinoma of the colon or rectum. Therefore,

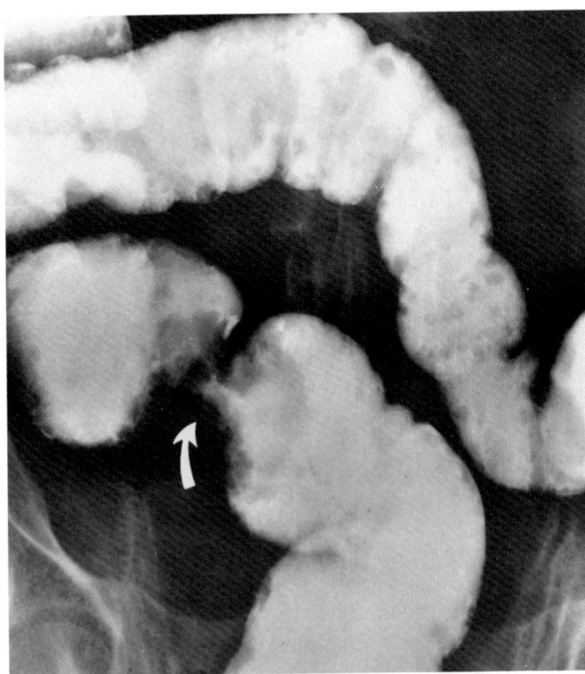

**Figure 48.57. Carcinoma of the sigmoid (arrow) developing in a patient with long-standing familial polyposis.** (From Eisenberg,[32] with permission from the publisher.)

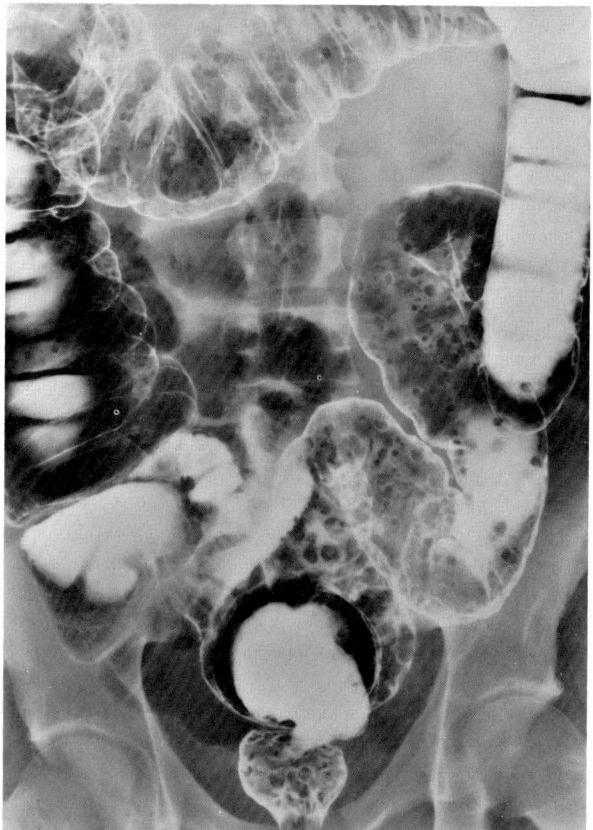

**Figure 48.58. Gardner's syndrome.** Innumerable adenomatous polyps throughout the colon present a radiographic appearance indistinguishable from that of familial polyposis. (From Eisenberg,[32] with permission from the publisher.)

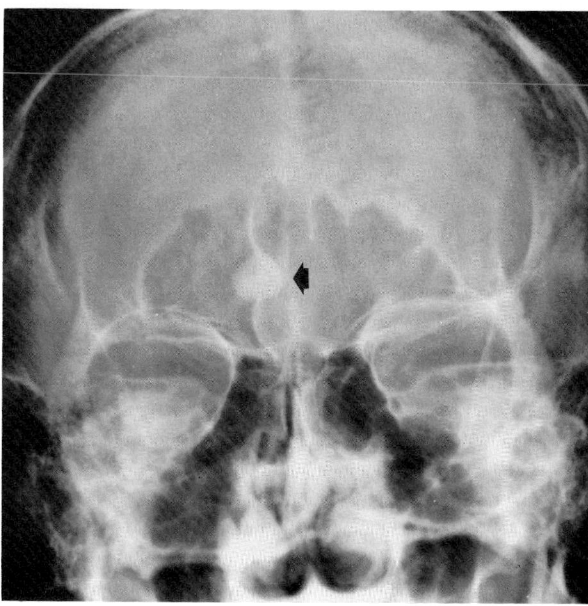

**Figure 48.59. Gardner's syndrome.** An osteoma (arrow) is present in the frontal sinus. (From Eisenberg,[32] with permission from the publisher.)

a total colectomy is recommended. In addition, patients with Gardner's syndrome appear to have a predilection toward small bowel malignancies, particularly in the pancreaticoduodenal region.[49,50]

## Peutz-Jeghers Syndrome

Peutz-Jeghers syndrome is an inherited disorder (autosomal dominant) in which multiple gastrointestinal polyposis is associated with mucocutaneous pigmentation. The syndrome usually develops during childhood or adolescence. The excessive melanin deposits characteristic of Peutz-Jeghers syndrome are flat and small and occur predominantly on the lips and buccal mucosa. They can also be seen on the face, abdomen, genitalia, hands, and feet. The hamartomatous polyps in Peutz-Jeghers syndrome are primarily found in the small bowel but can occur in the stomach and colon. The polyps are benign and apparently do not undergo malignant transformation. However, about 3 percent of the patients with the syndrome develop adenocarcinomas of the intestinal tract, and 5 percent of the women with the disease have ovarian cysts or tumors.[49,51]

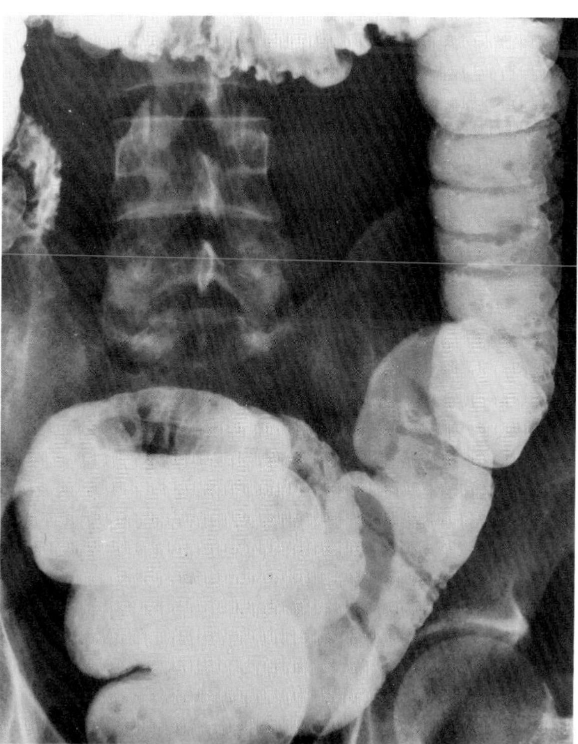

**Figure 48.60. Peutz-Jeghers syndrome.** Multiple colonic hamartomas are evident in this patient, who also demonstrated abnormal mucocutaneous pigmentation. (From Eisenberg,[32] with permission from the publisher.)

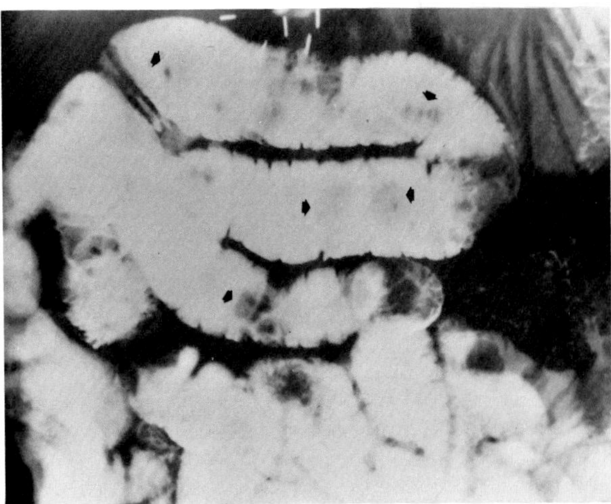

**Figure 48.61. Peutz-Jeghers syndrome.** Multiple small bowel hamartomas are present in a patient with abnormal mucocutaneous pigmentation. The arrows point to a few of the many filling defects in the barium column. (From Eisenberg,[32] with permission from the publisher.)

## Juvenile Polyposis

Juvenile polyposis is a disorder in which multiple hamartomatous or inflammatory polyps are found in the colon or in both the large and the small bowel. The polyps are benign with no malignant potential, usually develop in children, and tend to autoamputate or regress. Therefore, surgical removal of juvenile polyps is indicated only if there are significant or repeated episodes of rectal bleeding or intussusception.[52]

## Cronkhite-Canada Syndrome

The Cronkhite-Canada syndrome is a rare disorder in which multiple hamartomatous or inflammatory juvenile polyps are associated with hyperpigmentation, alopecia, and atrophy and subsequent loss of fingernails and toenails. Diffuse polyposis usually affects the stomach and colon; small bowel polyps are less frequent. The Cronkhite-Canada syndrome appears much later in life than other intestinal polyposis syndromes (average age of onset, over 50) and may be accompanied by malabsorption and severe diarrhea which result in substantial electrolyte and protein loss. Although the polyps are benign, the disease is often relentlessly progressive and may be fatal within 1 year of diagnosis.[49,53]

## Turcot Syndrome

Turcot syndrome (glioma-polyposis syndrome) is the association of multiple colonic adenomatous polyps with malignant tumors of the central nervous system. Patients with this extremely rare syndrome usually present in the second decade of life with neurologic complaints caused by a brain tumor, or with diarrhea due to colonic polyposis. Although an increased incidence of colorectal carcinoma has been reported in patients with Turcot syndrome, the precise malignant potential of the colonic polyps is unknown since most patients with this syndrome have died of central nervous system tumors at a very young age.[49]

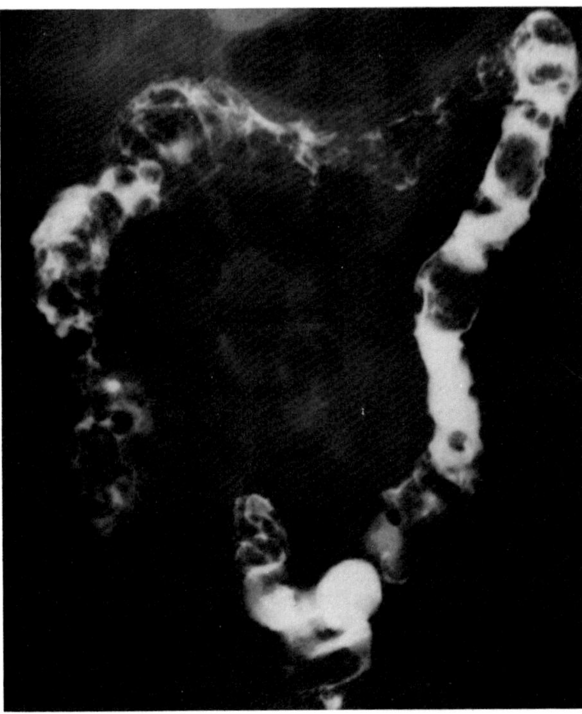

**Figure 48.62. Juvenile polyposis.** A postevacuation radiograph demonstrates many filling defects of various sizes throughout the colon. (From Schwartz and McCauley,[53] with permission from the publisher.)

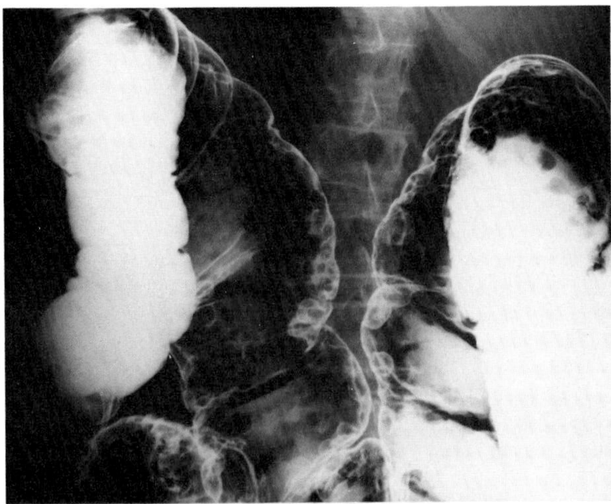

**Figure 48.63. Cronkhite-Canada syndrome.** Multiple polypoid lesions simulate familial polyposis. (From Dodds,[49] with permission from the publisher.)

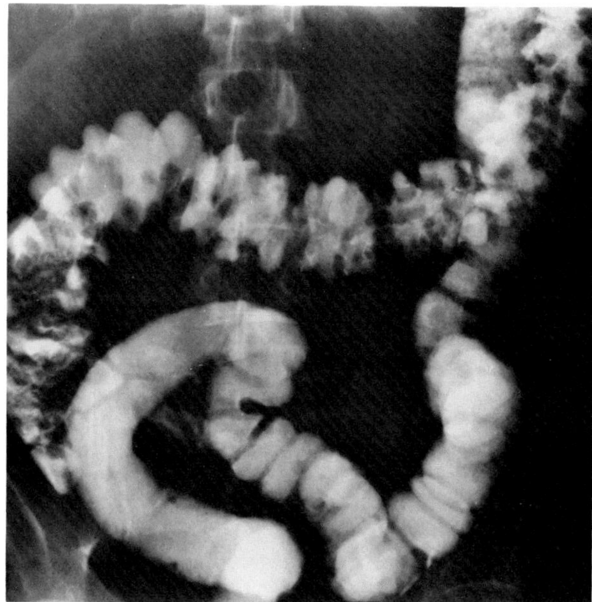

**Figure 48.64. Turcot syndrome.** Multiple colonic adenomatous polyps were found in this young patient, who died of a malignant glioma. (From Eisenberg,[32] with permission from the publisher.)

## CANCER OF THE COLON

Cancer of the colon kills more persons than any malignant tumor other than cancer of the lung in men and breast cancer in women. Even with some improvement in lesion detection and in medical and surgical therapy, the 5 year survival rate has changed little in the last several decades and remains about 40 percent. Because cancer of the colon is curable if discovered early in the course, delay in diagnosis is the most significant factor in the poor prognosis. Adenocarcinoma of the colon and rectum is primarily a disease of older persons, with a peak incidence in the 50- to 70-year range. About 60 to 75 percent of colon carcinomas occur in the rectum and sigmoid; the cecum is the second most common site of malignancy.

Because there is considerable evidence to suggest that many, if not most, carcinomas of the colon arise in preexisting villous or adenomatous polyps, the early diagnosis of colonic cancer is basically an exercise in polyp detection. Although the double-contrast barium enema examination appears to have greater sensitivity in detecting polyps, reports of lesions missed on this study but identified with single-contrast techniques indicate that the two procedures have complementary value. Malignant polyps tend to be sessile lesions with an irregular or lobulated surface, unlike benign adenomatous polyps that are usually smooth and often pedunculated. Other radiographic criteria suggesting that a polyp is malignant include large size (especially if greater than 2 cm in diameter), retraction or indentation (puckering) of the colon wall seen on profile view at the site of origin of a

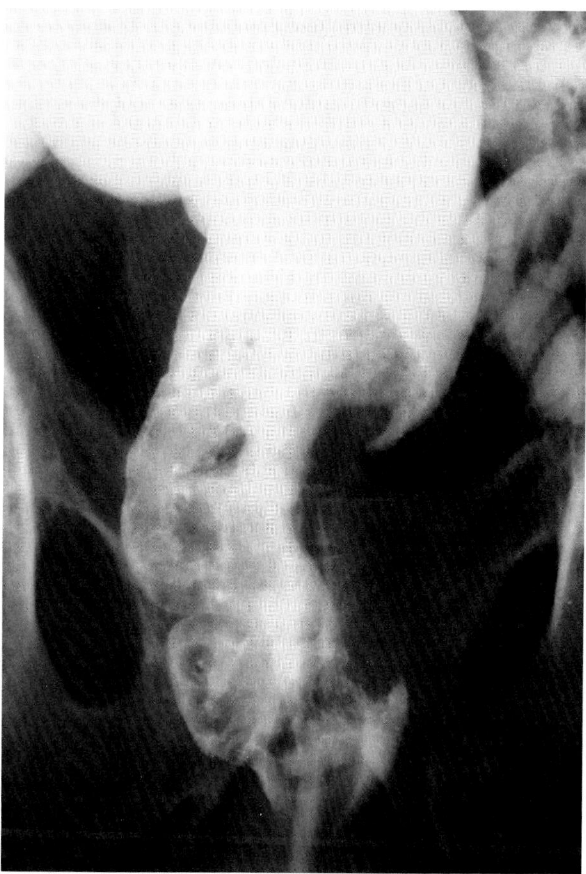

**Figure 48.65. Carcinoma of the rectum.** This bulky lesion could be felt on rectal examination. (From Eisenberg,[32] with permission from the publisher.)

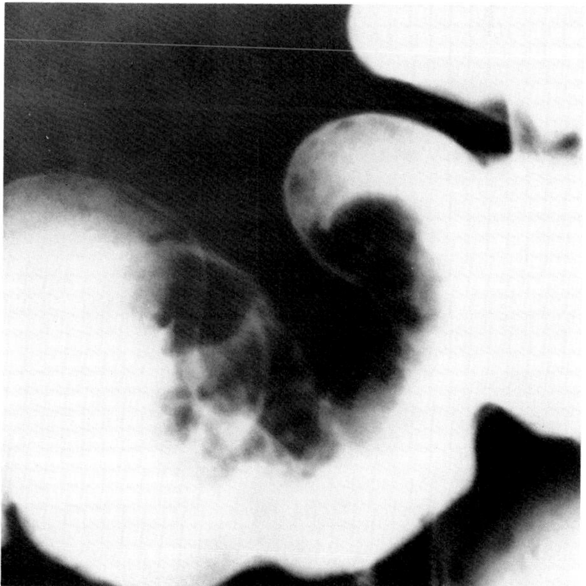

**Figure 48.66. Malignant sessile colonic polyp.** Retraction or indentation (puckering) of the colon wall is seen on a profile view. (From Eisenberg,[32] with permission from the publisher.)

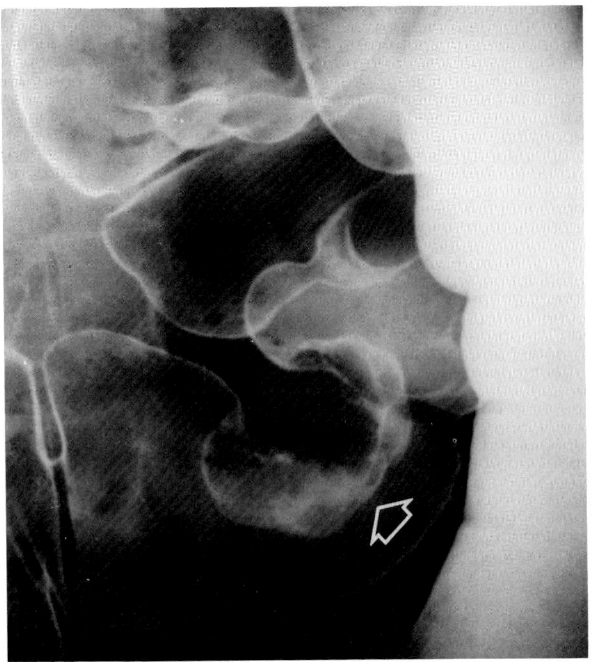

**Figure 48.67. Saddle cancer of the colon.** The tumor (arrow) appears to sit on the upper margin of the distal transverse colon like a saddle on a horse. (From Eisenberg,[32] with permission from the publisher.)

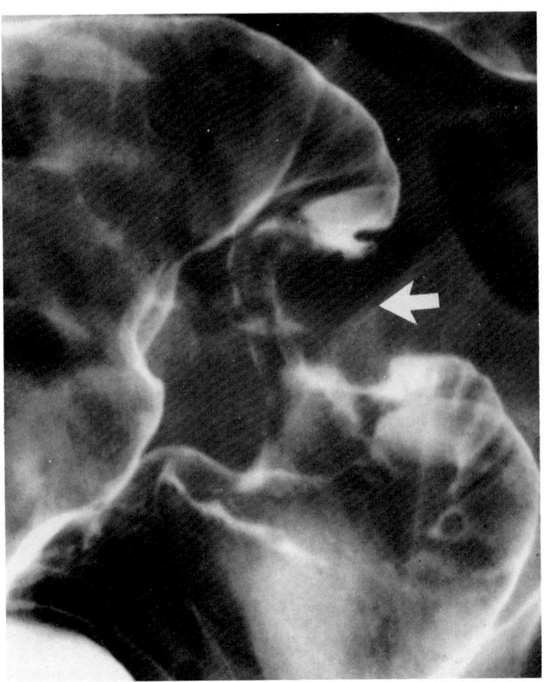

**Figure 48.68. Annular carcinoma of the colon.** The lesion is relatively short and has sharply defined proximal and distal margins (arrow).

sessile polyp, and evidence of interval growth of a polyp on sequential examinations.

Annular carcinoma (apple-core, napkin-ring) is one of the most typical forms of primary colonic malignancy. Annular carcinomas appear to arise from flat plaques of tumor (saddle lesions) that involve only a portion of the circumference of the colon wall. Unless there is meticulous care in searching for an area of minimal straightening or slight contour defects, the small and subtle, but lethal, saddle carcinomas can easily be overlooked. As the tumor grows, it characteristically infiltrates the bowel wall rather than forming a bulky intraluminal mass. This produces a classic bilateral contour defect with ulcerated mucosa, eccentric and irregular lumen, and overhanging margins. Progressive constriction of the bowel can cause complete colonic obstruction, most commonly in the sigmoid region.

In scirrhous carcinoma, a rare variant of annular carcinoma of the colon, an intense desmoplastic reaction infiltrates the bowel wall with dense fibrous tissue. As the tumor grows, it spreads circumferentially and longitudinally, producing a long segment of narrowed bowel and often obstructive symptoms.

Ulceration is common in carcinoma of the colon. It can vary from an excavation within a large fungating mass to mucosal destruction within an annular apple-core tumor.

A patient with carcinoma of the colon has a 1 percent risk of having multiple synchronous colon cancers. Therefore, it is still essential to carefully examine the rest of the colon once an obviously malignant lesion has been detected. In addition, such a patient has a 3 percent risk of developing additional metachronous cancers at a later date.

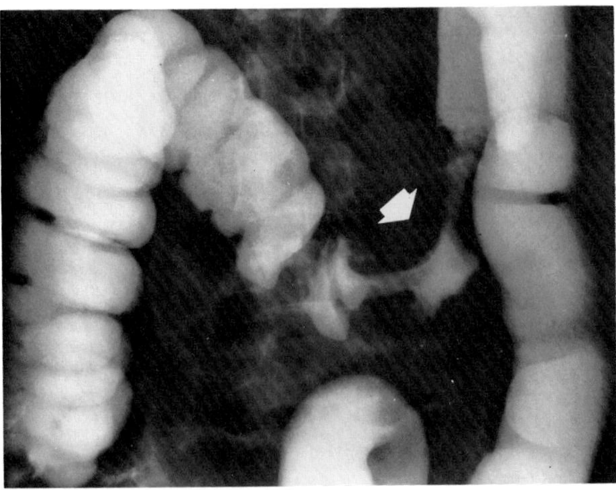

**Figure 48.69. Scirrhous carcinoma of the colon.** The long, circumferentially narrowed area (arrow) simulates segmental colonic encasement due to metastatic disease. (From Eisenberg,[32] with permission from the publisher.)

CT is a major modality for staging carcinoma of the colon and in assessing tumor recurrence. In the rectosigmoid region, adenocarcinoma causes asymmetric or circumferential thickening of the bowel wall with narrowing and deformity of the lumen. CT can demonstrate

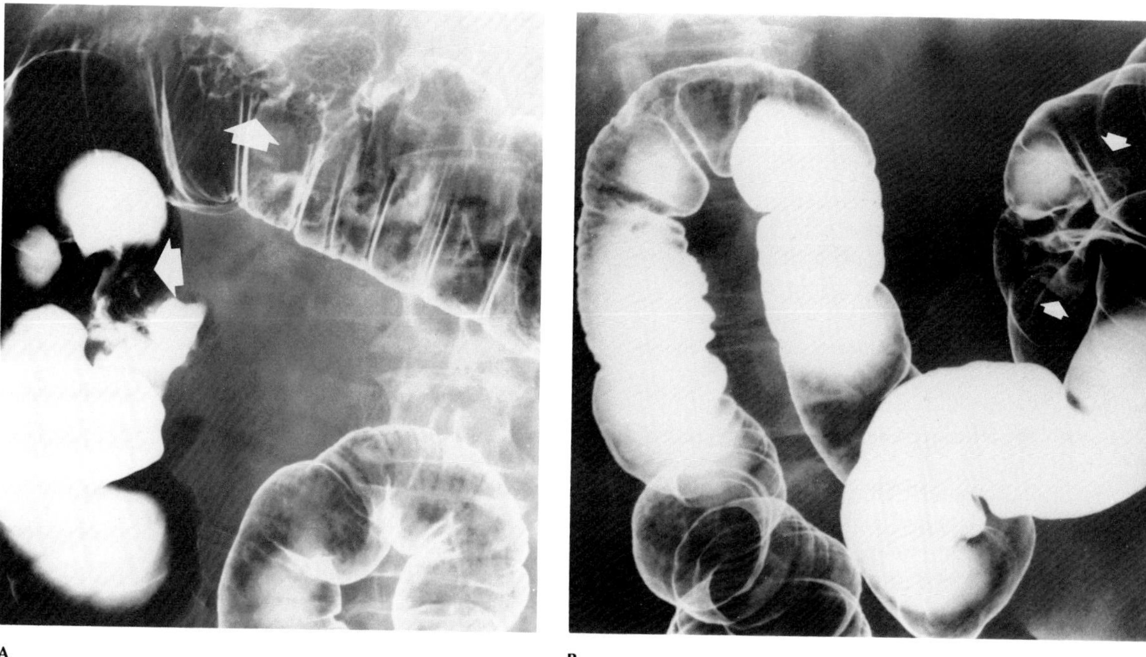

Figure 48.70. Multiple synchronous carcinomas of the colon. (A) Carcinomas of the ascending and transverse portions of the colon (arrows). (B) Carcinoma (arrows) within a tortuous descending colon. (From Eisenberg,[32] with permission from the publisher.)

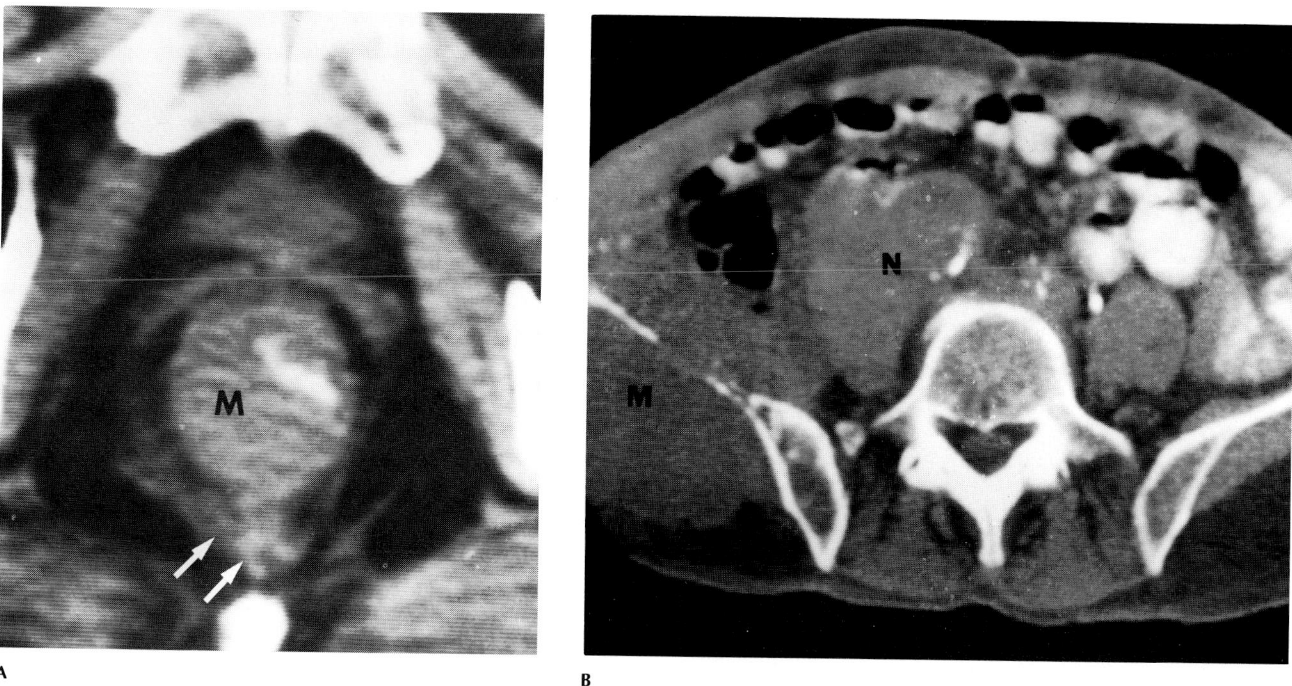


Figure 48.71. CT of recurrent colonic carcinoma. (A) A soft tissue mass extends beyond the bowel wall to invade the mesentery (arrows). (B) After a right colectomy, there is a large mass (M) destroying the right ilium and marked retroperitoneal lymph node enlargement (N). (A, from Moss and Thoeni,[77] with permission from the publisher; B, from Mauro and Koehler,[78] with permission from the publisher.)

local extension of tumor to the pelvic musculature, bladder, prostate, seminal vesicles, and ovaries by showing the obliteration of fat planes between the colon and these adjacent structures. Lymphadenopathy, metastases to the adrenals or liver, and masses in the abdominal wall or mesentery can also be detected by this technique. Following surgical resection, CT can demonstrate tumor recurrence as focal bowel wall thickening that may be accompanied by extension of the mass into adjacent muscles, organs, bone, or the pelvic sidewalls.[46,54–62]

## METASTASES TO THE COLON

Metastases to the colon can arise from direct invasion, intraperitoneal seeding, or hematogenous or lymphangitic spread. Direct invasion of the colon from a contiguous primary tumor indicates a locally aggressive lesion that has broken through fascial planes. In men, the most common primary tumor is advanced carcinoma of the prostate gland, which spreads posteriorly across the rectogenital septum. The most frequent presentation of prostatic carcinoma metastatic to the colon is a long, asymmetric, annular stricture that often has irregular scalloped margins caused by intramural tumor nodules or by edema infiltrating the bowel wall. A large, smooth, concave pressure defect on the anterior aspect of the rectosigmoid due to metastases occasionally may be severe enough to obstruct the colon. Invasion of the anterior rectal wall produces a fungating, ulcerated mass that closely simulates primary rectal carcinoma. In women, direct invasion from a noncontiguous primary tumor is usually related to a pelvic mass arising in the ovary or uterus. Invasion of the bowel wall produces a mass effect that is often of great length and does not demonstrate overhanging margins. An associated desmoplastic reaction causes angulation and tethering of mucosal folds and can even lead to the development of an annular stricture.

Carcinomas of the stomach and pancreas are noncontiguous primary tumors that can spread to the colon along mesenteric reflections. Primary carcinoma of the stomach (usually scirrhous) extends down the gastrocolic ligament to involve the transverse colon along its superior haustral border. Pancreatic carcinomas spread downward through the transverse mesocolon to predominantly involve the inferior aspect of the transverse colon. Both lesions cause fixation and nodularity of the transverse colon; progressive involvement can lead to circumferential colonic narrowing.

Primary abdominal malignancies can extend into the peritoneal cavity and shed tumor cells into ascitic fluid. The sites of serosal bowel metastases caused by this intraperitoneal seeding reflect the predictable course of malignant ascites that is determined by mesenteric reflections, peritoneal recesses, and the forces of gravity and negative intra-abdominal pressure. In more than half of the cases, intraperitoneal metastases grow in the region of the pouch of Douglas (the lower extension of the peritoneal reflection between the rectosigmoid and the urinary bladder at the level of the lower-second to upper-fourth sacral segments), where they produce a nodular mass or the characteristic pattern of fixed transverse parallel folds. Spread along the distal small bowel mesentery in the right lower quadrant can cause a smooth or lobulated extrinsic mass that indents the medial and inferior borders of the cecum below the level of the ileocecal valve. Involvement of the sigmoid mesocolon is usually localized to the superior border and typically incites an intense desmoplastic reaction leading to tethering or retraction of folds. When seen en face on double-contrast

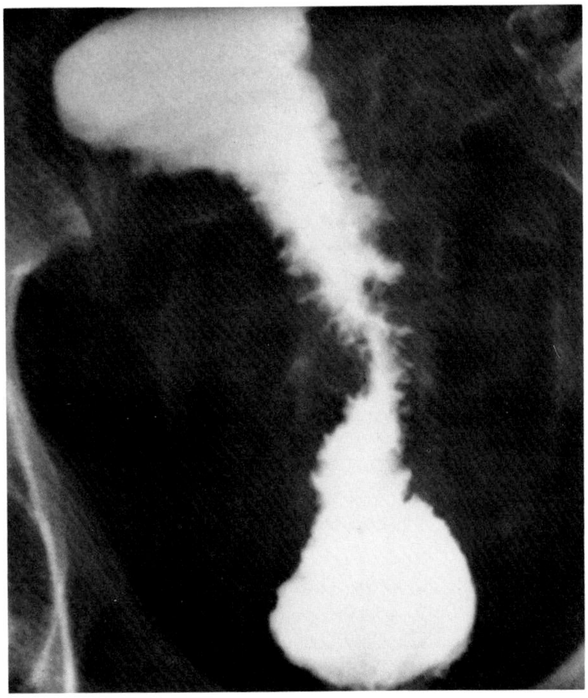

**Figure 48.72. Carcinoma of the prostate involving the rectum.** Circumferential involvement causes diffuse rectal narrowing and ulceration. (From Eisenberg,[32] with permission from the publisher.)

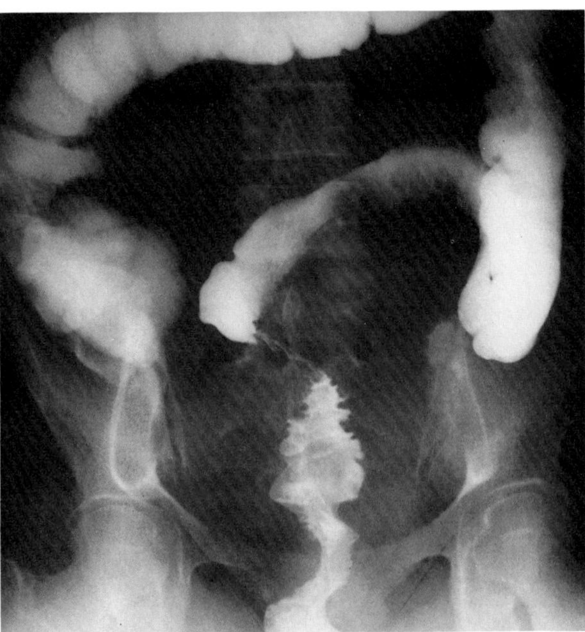

**Figure 48.73. Metastatic cystadenocarcinoma of the ovary.** In addition to the mass effect of the tumor, an associated desmoplastic reaction causes tethering of mucosal folds and an annular stricture. (From Eisenberg,[32] with permission from the publisher.)

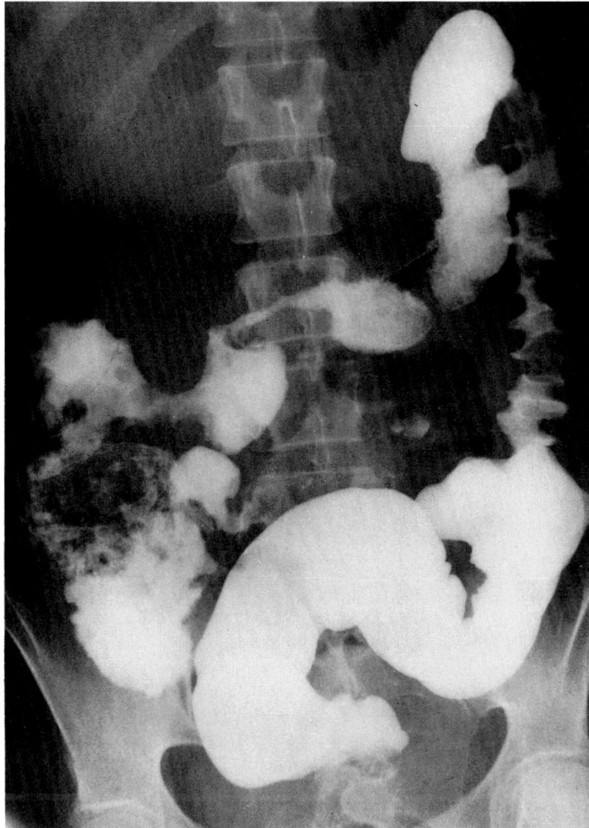

**Figure 48.74. Carcinoma of the stomach metastatic to the colon.** Diffuse, irregular narrowing involves the ascending, transverse, and descending portions of the colon; ulcerations and mucosal edema simulate Crohn's disease. (From Eisenberg,[32] with permission from the publisher.)

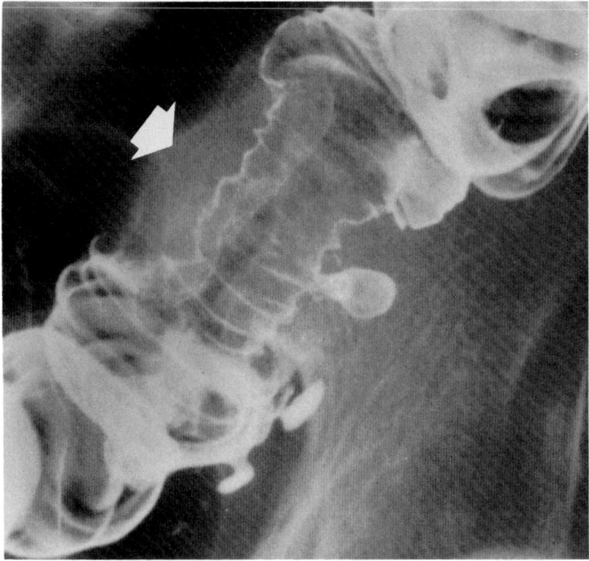

**Figure 48.76. Intraperitoneal seeding of undifferentiated carcinoma involving the sigmoid mesocolon.** There is a mass effect and tethering localized to the superior border of the sigmoid colon (arrow). (From Eisenberg,[32] with permission from the publisher.)

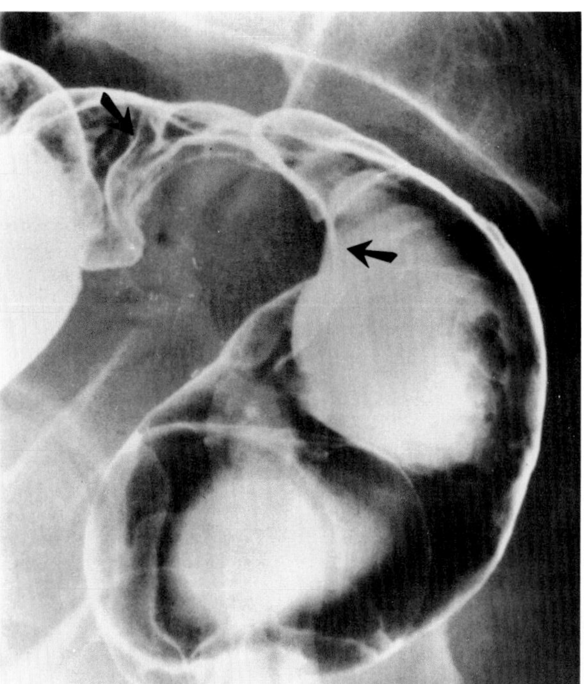

**Figure 48.75. Intraperitoneal metastases from carcinoma of the pancreas.** The nodular mass in the region of the pouch of Douglas (arrows) was clinically palpable (Blumer's shelf). (From Eisenberg,[32] with permission from the publisher.)

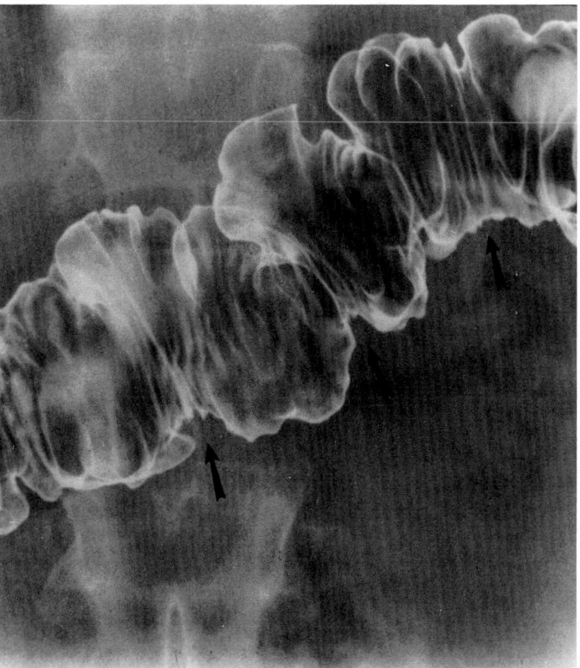

**Figure 48.77. Striped colon sign of metastatic serosal implants.** A double-contrast barium enema study demonstrates numerous transverse folds of the transverse colon (arrows). (From Ginaldi, Lindell, and Zornoza,[65] with permission from the publisher.)

studies, this tethering appears to be projected through the colonic lumen as transverse folds that do not completely cross the lumen of the colon (striped colon).

The hematogenous spread of tumor, especially due to carcinoma of the breast, tends to cause thickening and rigidity of a long segment of colon. This pattern, which is due to densely cellular submucosal metastatic deposits, simulates a primary infiltrating scirrhous carcinoma. At times, metastatic breast carcinoma can mimic primary inflammatory colitis and produce such radio-

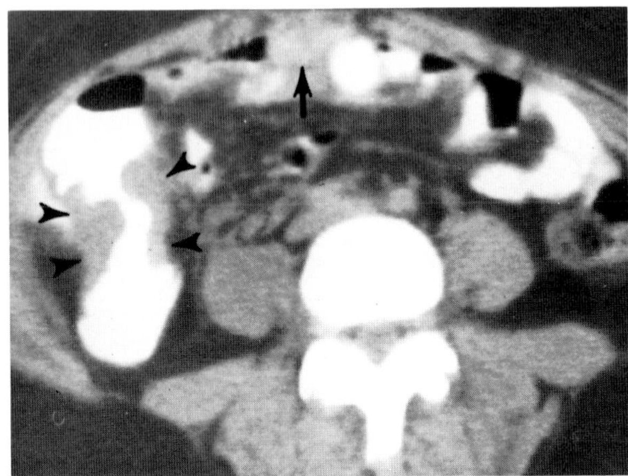

**Figure 48.78. Carcinoma of the breast metastatic to the colon.** A CT scan shows circumferential thickening (arrowheads) of the colonic wall and omental metastases (arrow). (From Mauro and Koehler,[78] with permission from the publisher.)

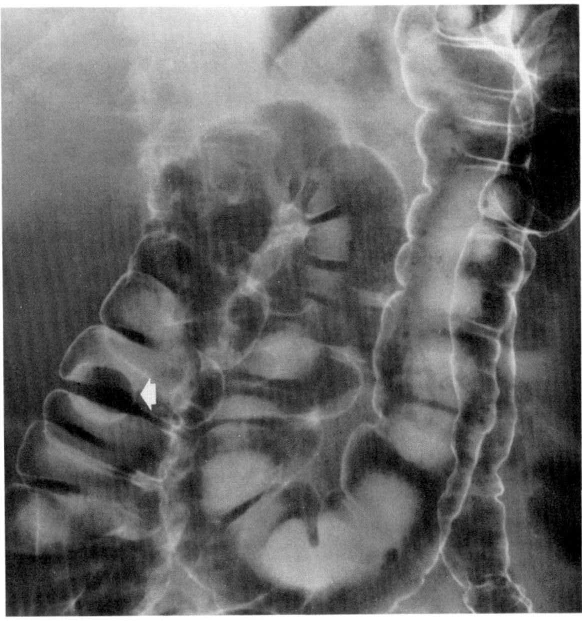

**Figure 48.79. Lymphoma of the colon.** A smooth polypoid mass (arrow). (From Eisenberg,[32] with permission from the publisher.)

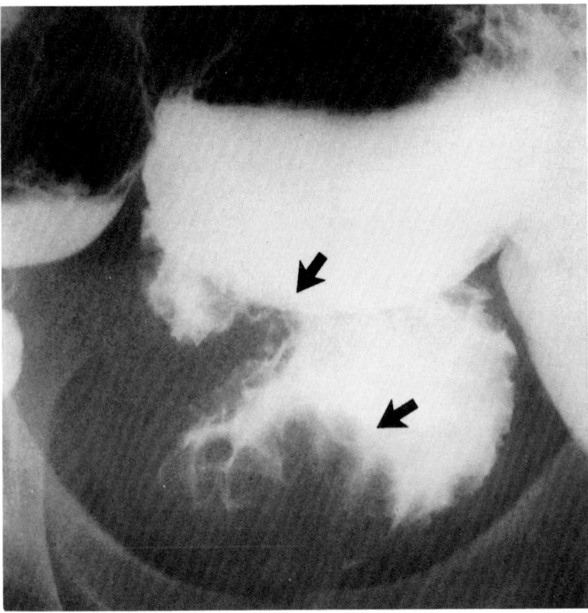

**Figure 48.80. Lymphoma of the colon.** A bulky, irregular, ulcerated mass involves much of the rectum (arrows). (From Eisenberg,[32] with permission from the publisher.)

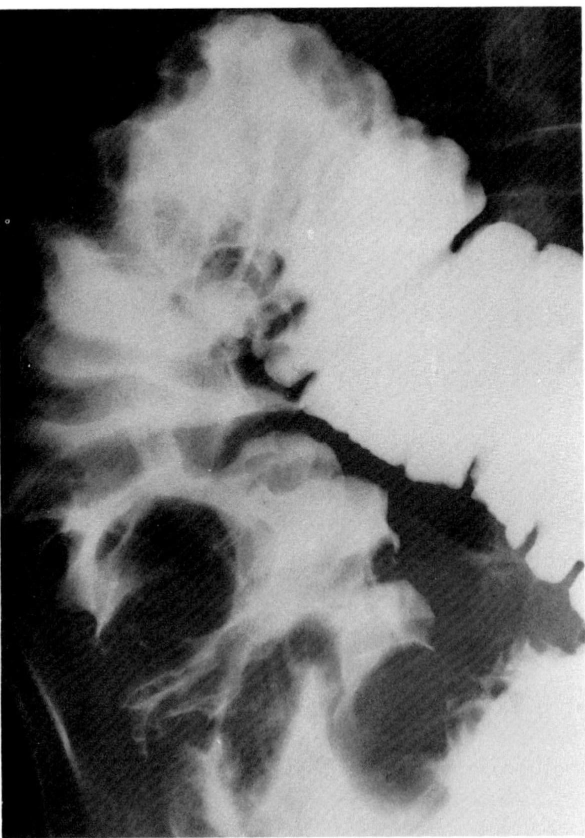

**Figure 48.81. Lymphoma of the colon.** Multiple irregular, nodular filling defects. (From Eisenberg,[32] with permission from the publisher.)

graphic findings as mucosal thickening, nodular masses, multiple and eccentric strictures, and spiculations.

CT can demonstrate mural and mesenteric masses and stretching of colonic loops due to metastases to the colon. This modality can also be of value in differentiating a primary colonic tumor from secondary spread of pancreatic or gastric carcinoma.[32,63–65]

### LYMPHOMA OF THE COLON

Although the gastrointestinal tract is the most common location of primary extranodal lymphoma, the colon is the segment of gut that is least often affected. Localized lymphoma can appear as a single, smooth or lobulated polypoid mass that is radiographically indistinguishable from polypoid carcinoma. Unlike carcinoma, localized lymphoma tends to be unusually bulky and to extend over a longer segment of the colon. Diffuse submucosal infiltration can produce multiple nodules simulating familial polyposis or irregular thumbprinting suggesting ischemic colitis. Lymphoma occasionally appears as an area of localized narrowing simulating annular carcinoma. Subserosal lymphoma may develop a large extracolonic component that displaces adjacent abdominal structures. Multiple subserosal masses may mimic mesenteric metastases.[66–68]

## MISCELLANEOUS CONDITIONS

### HEMORRHOIDS

Internal hemorrhoids can produce single or multiple rectal filling defects that simulate polyps. The proper diagnosis can easily be made by inspection, visual examination, or direct vision through the anoscope.

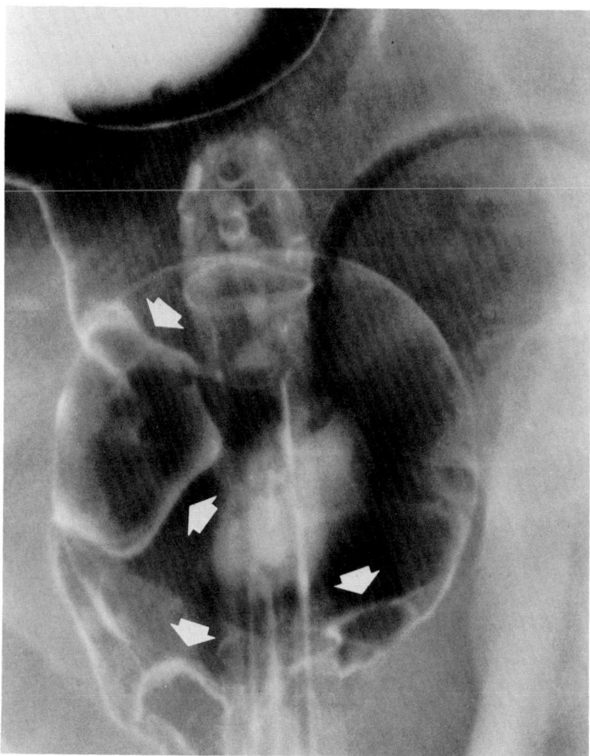

**Figure 48.82. Internal hemorrhoids.** Multiple rectal filling defects (arrows) simulate polyps. (From Eisenberg,[32] with permission from the publisher.)

### ENDOMETRIOSIS OF THE BOWEL

Endometriosis is the presence of heterotopic foci of endometrium in an extrauterine location. Endometriosis involving the bowel primarily affects those segments that are situated in the pelvis, especially the rectosigmoid colon. Because endometriosis is usually clinically apparent only when ovarian function is active, most women who are symptomatic from endometriosis are between 20 and 45 years of age. The typical gastrointestinal complaint is abdominal cramps and diarrhea during the menstrual period.

Growth of endometrial tissue in the wall of the bowel may produce an eccentric intramural filling defect sim-

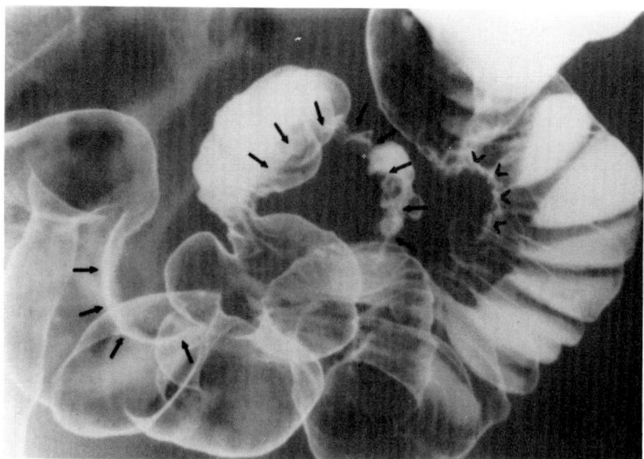

**Figure 48.83. Endometriosis.** A radiograph of the sigmoid colon demonstrates three endometrial implants (arrows and arrowheads). The most distal lesion has a smooth interface with the bowel wall, indicating no intramural invasion. The two more proximal lesions demonstrate crenulations (arrowheads) that indicate intramural or submucosal invasion. (From Gedgaudas, Kelvin, Thompson, et al,[79] with permission from the publisher.)

ulating a flat saddle cancer. In contrast to primary colonic malignancy, the underlying mucosal pattern in endometriosis usually remains intact or is pleated because of secondary fibrosis. Endometriosis can also appear as a constricting lesion simulating annular carcinoma. Radiographic findings favoring a diagnosis of endometriosis are an intact mucosa, a long lesion with tapered margins, and the absence of ulceration within the mass. Repeated shedding of endometrial tissue and blood into the peritoneal cavity can lead to the development of dense adhesive bands causing extrinsic obstruction of the bowel.[69]

## COLITIS CYSTICA PROFUNDA

In colitis cystica profunda, large epithelium-lined mucous cysts up to 2 cm in diameter form in the submucosal layer of the colon, most commonly in the rectum and sigmoid. The disorder most frequently affects young adults who present with recurrent rectal bleeding, mucous discharge, and diarrhea. Colitis cystica profunda usually appears radiographically as multiple, irregular colonic filling defects suggesting adenomatous polyps. Barium filling of the clefts between the nodular polypoid lesions can mimic ulceration; scalloping of the colon can simulate the thumbprinting of ischemic colitis. Occasionally, only a single rectal mass mimicking a sessile polyp is seen. Colitis cystica profunda is a benign condition with no malignant potential; surgery is indicated only in patients in whom there is significant bleeding or associated rectal prolapse.[70]

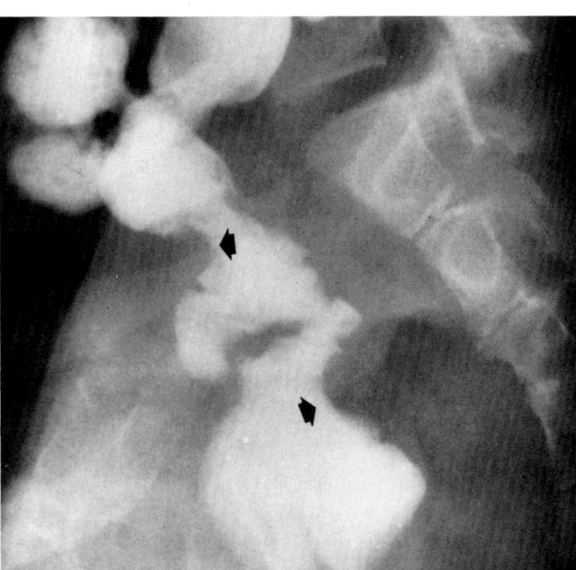

**Figure 48.84. Colitis cystica profunda.** Widening of the retrorectal space accompanies multiple intraluminal filling defects (arrows) in the rectum. (From Ledesma-Medina, Reid, and Girdany,[70] with permission from the publisher.)

## LIPOMATOSIS OF THE ILEOCECAL VALVE

Lipomatosis of the ileocecal valve is characterized by enlargement of the valve due to benign submucosal fatty infiltration. It is found predominantly in women and is uncommon in persons under 40 years of age. Many persons are asymptomatic, and prominence of the valve is usually an incidental finding at barium enema examination.

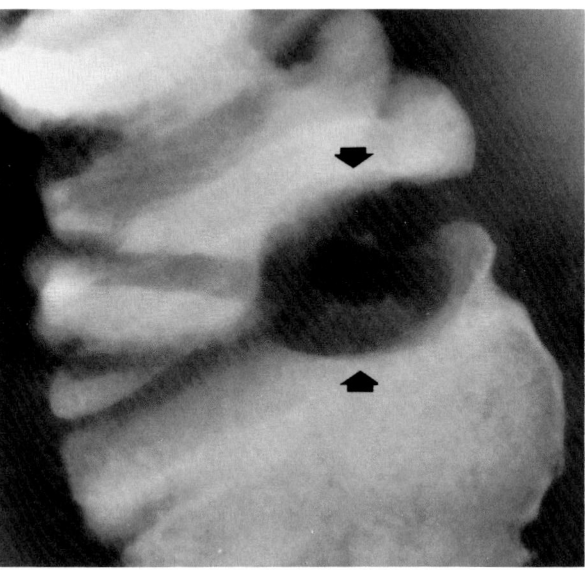

**Figure 48.85. Lipomatosis of the ileocecal valve.** Smooth, masslike enlargement (arrows). (From Eisenberg,[32] with permission from the publisher.)

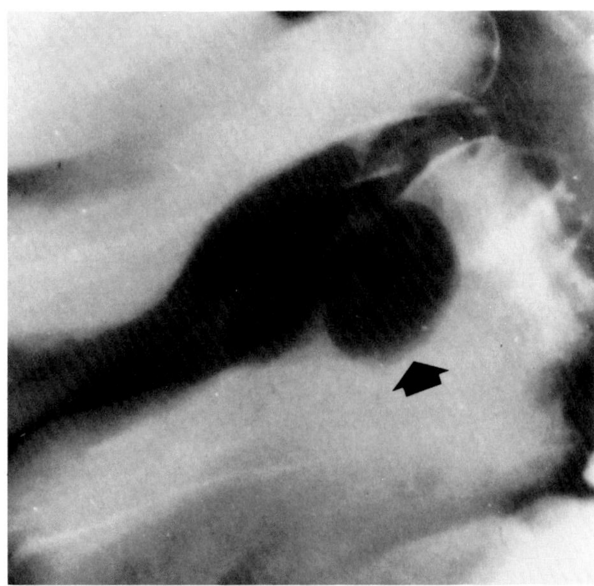

**Figure 48.86. Lipoma of the ileocecal valve.** A sharply circumscribed, smooth, rounded mass arises from the lower lip of the valve (arrow). (From Eisenberg,[32] with permission from the publisher.)

Lipomatosis of the ileocecal valve appears radiographically as a smooth, masslike enlargement that is sharply demarcated from the surrounding bowel mucosa. The valve can be lobulated and slightly irregular because of contraction of the muscularis, but the surface remains smooth and the valve is changeable in size and shape. If barium can be refluxed back into the terminal ileum, demonstration of the characteristic stellate appearance of the ileal mucosa on the en face view reveals that the process is benign.

A sharply circumscribed, smooth rounded mass arising from the ileocecal valve usually represents a lipoma. In contrast to lipomatous infiltration, lipomas of the ileocecal valve have a true capsule and are confined to only one portion of the valve. Like tumors of this cell type elsewhere in the bowel, lipomas have a soft consistency that may permit a changing appearance on serial films. Lipomas of the ileocecal valve are rarely of clinical significance unless they become so large as to cause substantial bleeding or episodes of intussusception.[71–73]

## REFERENCES


1. Nelson JH, Burhenne HJ: Anomalous biliary and pancreatic duct insertion into duodenal diverticula. *Radiology* 120:49–52, 1976.
2. Millard JR, Ziter FMH, Slover WP: Giant duodenal diverticula. *AJR* 121:334–337, 1974.
3. Loudan JCH, Norton GI: Intraluminal duodenal diverticulum. *AJR* 90:756–760, 1963.
4. Jiustra PE, Killoran PJ, Root JA, et al: Jejunal diverticulitis. *Radiology* 125:609–611, 1977.
5. Dunn V, Nelson JA: Jejunal diverticulosis and chronic pneumoperitoneum. *Gastrointest Radiol* 4:165–168, 1979.
6. Dalinka MK, Wunder JS: Meckel's diverticulum and its complications with emphasis on roentgenologic demonstration. *Radiology* 106:295–298, 1973.
7. White AF, Oh KS, Weber AL, et al: Radiologic manifestations of Meckel's diverticulum. *AJR* 118:86–94, 1973.
8. Maglinte DDT, Elmore MF, Isenberg M, et al: Meckel diverticulum: Radiologic demonstration by enteroclysis. *AJR* 134:925–932, 1980.
9. Rosenberg RF, Naidich JB: Plain-film recognition of giant colonic diverticulum. *Am J Gastroenterol* 76:59–69, 1981.
10. Almy TP, Howell DA: Diverticular disease of colon. *N Engl J Med* 302:324–331, 1980.
11. Berdon WE, Baker DH: The roentgenographic diagnosis of Hirschsprung's disease in infancy. *AJR* 93:432–446, 1965.
12. Hope JW, Borns PS, Burg PK: Radiologic manifestation of Hirschsprung's disease in infancy. *AJR* 95:217–229, 1965.
13. Bryk D, Soong KY: Colonic ileus and its differential roentgen diagnosis. *AJR* 101:329–337, 1967.
14. Ferguson A, Sircus W, Eastwood MA: Frequency of "functional" gastrointestinal disorders. *Lancet* 2:613–614, 1977.
15. Lumsden K, Charudhary MA, Truelove SC: The irritable colon syndrome. *Clin Radiol* 14:54–61, 1963.
16. Urso FP, Urso MJ, Lee CH: The cathartic colon: Pathological findings and radiological/pathological correlation. *Radiology* 116:557–559, 1975.
17. Kim SK, Gerle RD, Rozanski R: Cathartic colitis. *AJR* 131:1079–1081, 1978.
18. Goldberger LE: Gastrointestinal hemorrhage, in Eisenberg RL, Amberg JR (eds): *Critical Diagnostic Pathways in Radiology: An Algorithmic Approach*. Philadelphia, JB Lippincott, 1981, pp 114–126.
19. Berger RB, Zeman RK, Gottschalk A: The technetium-99m-sulfur colloid angiogram in suspected gastrointestinal bleeding. *Radiology* 147:555–558, 1983.
20. Johnsrude IS, Jackson DC: The role of the radiologist in acute gastrointestinal bleeding. *Gastrointest Radiol* 3:357–368, 1978.
21. Goldberger LE, Bookstein JJ: Transcatheter embolization for treatment of diverticular hemorrhage. *Radiology* 122:613–617, 1977.
22. Miller KD, Tutton RH, Bell KA, et al: Angiodysplasia of the colon. *Radiology* 132:309–313, 1979.
23. Miller RE, Sellink JL: Enteroclysis: The small bowel enema. How to succeed and how to fail. *Gastrointest Radiol* 4:269–283, 1979.
24. Laufer I, Smith NC, Mullen JE: The radiological demonstration of colorectal polyps undetected by endoscopy. *Gastroenterology* 70:167–170, 1976.
25. Wittenberg J, Athanasoulis CA, Shapiro JH, et al: A radiological approach to the patient with acute, extensive bowel ischemia. *Radiology* 106:13–24, 1973.
26. Ghahremani GG, Meyers MA, Farman J, et al: Ischemic disease of the small bowel and colon associated with oral contraceptives. *Gastrointest Radiol* 2:221–228, 1977.
27. Marshak RH, Lindner AE, Maclansky D: Ischemia of the small intestine. *Am J Gastroenterol* 66:309–400, 1976.
28. Schwartz S, Boley S, Schultz L, et al: A survey of vascular diseases of the small intestine. *Semin Roentgenol* 1:178–218, 1966.
29. Nelson SW, Eggleston W: Findings on plain roentgenograms of the abdomen associated with mesenteric vascular occlusion with a possible new sign of mesenteric venous thrombosis. *AJR* 83:886–894, 1960.
30. Swaim TJ, Gerald B: Hepatic portal venous gas in infants without subsequent death. *Radiology* 94:343–345, 1970.
31. Sisk PB: Gas in the portal venous system. *Radiology* 77:103–107, 1961.
32. Eisenberg RL: *Gastrointestinal Radiology: A Pattern Approach*, Philadelphia, JB Lippincott, 1983.
33. Eisenberg RL, Montgomery CK, Margulis AR: Colitis in the elderly: Ischemic colitis mimicking ulcerative and granulomatous colitis. *AJR* 133:1113–1118, 1979.
34. Williams LF, Bosniak MA, Wittenberg J, et al: Ischemic colitis. *Am J Surg* 117:254–264, 1969.
35. Schwartz S, Boley S, Lash J, et al: Roentgenologic aspects of reversible vascular occlusion of the colon and its relationship to ulcerative colitis. *Radiology* 80:625–635, 1963.
36. Davies DR, Brightmore T: Idiopathic and drug-induced ulceration of the small intestine. *Br J Surg* 57:134–138, 1970.
37. Baker DR, Schrader WH, Hitchcock CR: Small-bowel ulceration apparently associated with thiazide and potassium therapy. *JAMA* 190:586–588, 1964.
38. Pringot J, Goncette L, Ponette E, et al: Nonstenotic ulcers of the small bowel. *RadioGraphics* 4:357–375, 1984.
39. Gardiner GA, Bird CR: Nonspecific ulcers of the colon resembling annular carcinoma. *Radiology* 137:331–334, 1980.
40. Brodey PA, Hill RP, Baron S: Benign ulceration of the cecum. *Radiology* 122:323–327, 1977.
41. Good CA: Tumors of the small intestine. *AJR* 89:685–705, 1963.
42. Bruno MS, Sein HD: Primary malignant and benign tumors of the duodenum. *Arch Intern Med* 125:670–679, 1970.
43. Balthazar EJ: Carcinoid tumors of the alimentary tract: Radiographic diagnosis. *Gastrointest Radiol* 3:47–56, 1978.
44. Bancks NH, Goldstein HM, Dodd GD: The roentgenologic spectrum of small intestinal carcinoid tumors. *AJR* 123:274–280, 1975.
45. Thoeni RF, Menuck L: Comparison of barium enema and colonoscopy in the detection of small colonic polyps. *Radiology* 124:631–635, 1977.
46. Ott DJ, Gelfand DW: Colorectal tumors. Pathology and detection. *AJR* 131:691–695, 1978.
47. Wolf BS: Lipoma of the colon. *JAMA* 235:2225–2226, 1976.

48. Delamarre J, Descombs P, Marti R, et al: Villous tumors of the colon and rectum: Double-contrast study of 47 cases. *Gastrointest Radiol* 5:69–73, 1980.
49. Dodds WJ: Clinical and roentgen features of intestinal polyposis syndromes. *Gastrointest Radiol* 1:127–142, 1976.
50. Dolan KD, Seibert J, Seibert RW: Gardner's syndrome. *AJR* 119:359–364, 1973.
51. Godard JE, Dodds WJ, Phillips JC, et al: Peutz-Jeghers syndrome: Clinical and roentgenographic features. *AJR* 113:313–324, 1971.
52. Schwartz AM, McCauley RGK: Juvenile gastrointestinal polyposis. *Radiology* 121:441–444, 1976.
53. Koehler TR, Kyaw MM, Fenlon JW: Diffuse gastrointestinal polyposis with ectodermal changes. Cronkhite-Canada syndrome. *Radiology* 103:589–594, 1972.
54. Maglinte DDT, Keller KJ, Miller RE, et al: Colon and rectal carcinoma: Spacial distribution and detection. *Radiology* 147:669–672, 1983.
55. Dreyfuss JR, Benacerraf B: Saddle cancers of the colon and their progression to annular carcinomas. *Radiology* 129: 289–293, 1978.
56. Lane N, Fenoglio CM: Observations on the adenoma as precursor to ordinary large bowel carcinoma. *Gastrointest Radiol* 1:111–119, 1976.
57. Muto T, Bussey HJR, Morson BC: The evolution of cancer of the colon and rectum. *Cancer* 36:2251–2270, 1975.
58. Kagan AR, Steckel RJ: Colon polyposis and cancer. *AJR* 131:1065–1067, 1978.
59. McCartney WH, Hoffer PB: The value of carcinoembryonic antigen (CEA) as an adjunct to the radiological colon examination in the diagnosis of malignancy. *Radiology* 110:325–328, 1974.
60. Thoeni RF, Moss AA, Schnyder P, et al: Detection and staging of primary rectal and rectosigmoid cancer by computed tomography. *Radiology* 141:135–138, 1981.
61. Moss AA, Thoeni RF, Schnyder P, et al: Value of computed tomography in the detection and staging of recurrent rectal carcinomas. *J Comput Assist Tomogr* 5:870–874, 1981.
62. Zaunbauer W, Haertel M, Fuchs WA: Computed tomography in carcinoma of the rectum. *Gastrointest Radiol* 6:79–84, 1981.
63. Meyers MA: *Dynamic Radiology of the Abdomen: Normal and Pathologic Anatomy*, New York, Springer-Verlag, 1976.
64. Meyers MA, Oliphant M, Teixidor H: Metastatic carcinoma simulating inflammatory colitis. *AJR* 123:74–83, 1975.
65. Ginaldi S, Lindell MM, Zornoza J: The striped colon: A new radiographic observation in metastatic serosal implants. *AJR* 134:453–455, 1980.
66. O'Connell BJ, Thompson AJ: Lymphoma of the colon. The spectrum of radiologic changes. *Gastrointest Radiol* 2:377–385, 1978.
67. Loehr WJ, Mujahed Z, Zahn FD, et al: Primary lymphoma of the gastrointestinal tract: A review of 100 cases. *Ann Surg* 170:232–238, 1969.
68. Pochaczevsky R, Sherman RS: Diffuse lymphomatous disease of the colon: Its roentgen appearance. *AJR* 87:670–683, 1962.
69. Spjut HJ, Perkins DE: Endometriosis of the sigmoid colon and rectum. *AJR* 82:1070–1075, 1959.
70. Ledesma-Medina J, Reid BS, Girdany BR: Colitis cystica profunda. *AJR* 131:529–530, 1978.
71. Short WS, Smith BD, Hoy RJ: Roentgenologic evaluation of the prominent or unusual ileocecal valve. *Med Radiogr Photogr* 52:2–26, 1976.
72. Berk RN, Davis GB, Cholhaffy EB: Lipomatosis of the ileocecal valve. *AJR* 119:323–328, 1973.
73. Boquist L, Bergdahl L, Andersson A: Lipomatosis of the ileocecal valve. *Cancer* 29:136–140, 1972.
74. Heilbrun N, Boyden EA: Intraluminal duodenal diverticula. *Radiology* 82:887–894, 1964.
75. Alavi A: Detection of gastrointestinal bleeding with $^{99m}$Tc-sulfur colloid. *Semin Nucl Med* 12:126–134, 1982.
76. Margulis AR, Burhenne HJ: *Alimentary Tract Radiology*. St. Louis, CV Mosby, 1984.
77. Moss AA, Thoeni RF: Computed tomography of the gastrointestinal tract, in Moss AA, Gamsu G, Genant HK (eds): *Computed Tomography of the Body*. Philadelphia, WB Saunders, 1983, pp 535–598.
78. Mauro MA, Koehler RE: Alimentary tract, in Lee JKT, Sagel SS, Stanley RJ (eds): *Computed Body Tomography*. New York, Raven Press, 1983.
79. Gedgaudas RK, Kelvin FM, Thompson WM, et al: Value of the preoperative barium-enema examination in the assessment of pelvic masses. *Radiology* 146:609–616, 1983.
80. Feczko PJ, Bernstein MA, Halpert RD, et al: Small colonic polyps: A reappraisal of their significance. *Radiology* 152: 301–303, 1984.
81. Granqvist S, Gabrielson N, Sundelin B: Diminutive colonic polyps—clinical significance and management. *Endoscopy* 11:36–42, 1979.

## Chapter Forty-Nine
# Intestinal Obstruction

### SMALL BOWEL OBSTRUCTION

Fibrous adhesions caused by previous surgery or peritonitis account for almost 75 percent of all small bowel obstructions. External hernias (inguinal, femoral, umbilical, incisional) are the second most frequent cause. Other general causes of mechanical small bowel obstruction include luminal occlusion (gallstone, intussusception) and intrinsic lesions of the bowel wall (neoplastic or inflammatory strictures, vascular insufficiency).

Distended loops of small bowel containing gas and fluid can usually be recognized within 3 to 5 hours of the onset of complete obstruction. Almost all gas proximal to a small bowel obstruction represents swallowed air. On upright or lateral decubitus views, the interface between gas and fluid forms a straight horizontal margin. Gas-fluid levels are occasionally present normally, but more than two gas-fluid levels in small bowel distal to the duodenum is generally considered to be abnormal. Although the presence of gas-fluid levels at different heights in the same loop has traditionally been considered excellent evidence for mechanical obstruction, an identical pattern can also be demonstrated in some patients with adynamic ileus.

As time passes, the small bowel may become so distended as to be almost indistinguishable from colon. To make the critical differentiation between small and large bowel obstruction, it is essential to determine which loops of bowel contain abnormally large amounts of air. Small bowel loops generally occupy the more central portion of the abdomen, whereas colonic loops are positioned laterally around the periphery of the abdomen or inferiorly in the pelvis. Gas within the lumen of the small bowel outlines the thin valvulae conniventes, which completely encircle the bowel. In contrast, colonic haustral markings are thicker and farther apart and occupy only a portion of the transverse diameter of the bowel.

The site of obstruction can usually be predicted with considerable accuracy if the number and position of dilated bowel loops are analyzed. The presence of a few dilated loops of small bowel located high in the abdomen (in the center or slightly to the left) indicates an obstruction in the distal duodenum or jejunum. The involvement of more small bowel loops suggests a lower obstruction. As additional loops are affected, they appear to be placed one above the other upward and to the left, producing a characteristic stepladder appearance. The point of obstruction is always distal to the lowest loop of dilated bowel.

In patients with complete mechanical small bowel obstruction, little or no gas is found in the colon. This is a valuable point in the differentiation between mechanical obstruction and adynamic ileus, in which gas is seen within distended loops throughout the bowel.

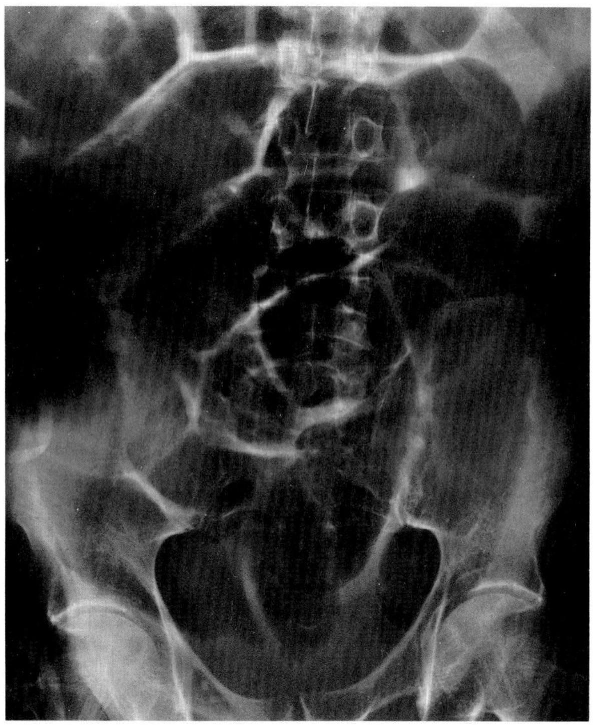

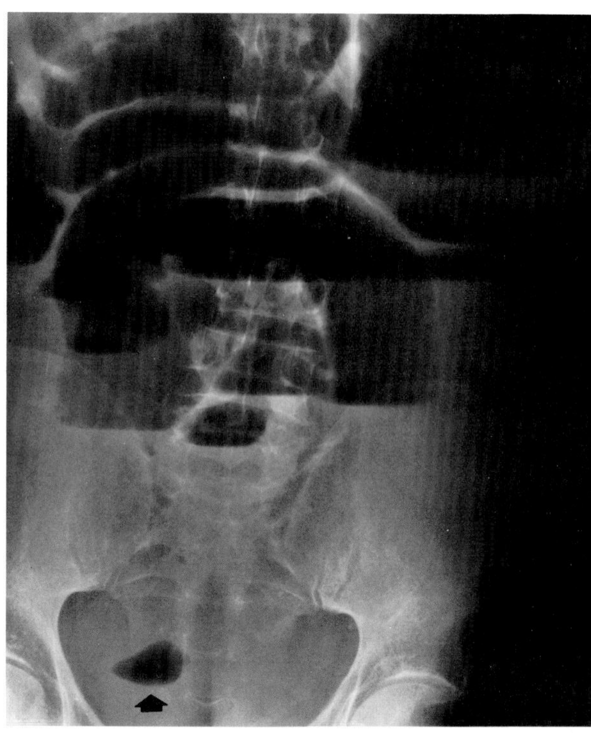

A B

**Figure 49.1. Small bowel obstruction.** (A) Supine and (B) upright views demonstrate large amounts of gas in dilated loops of small bowel but only a single, small collection of gas (arrow) in the colon. (From Eisenberg,[13] with permission from the publisher.)

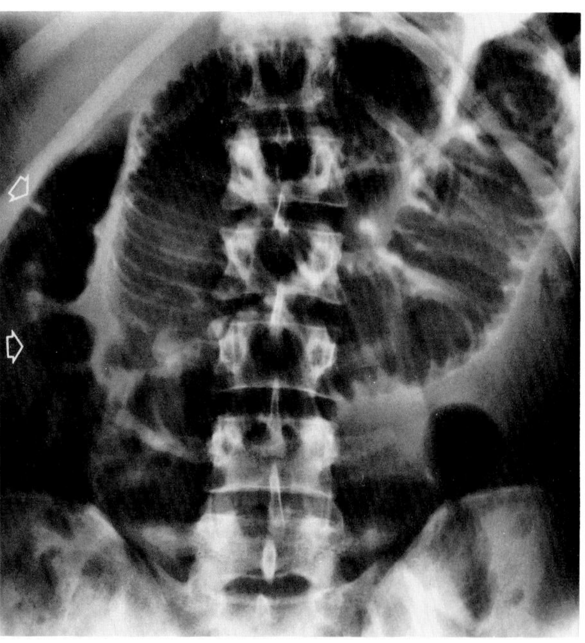

**Figure 49.2. Small bowel obstruction.** The dilated loops of small bowel occupy the central portion of the abdomen, and the non-dilated cecum and ascending colon are positioned laterally around the periphery of the abdomen (arrows). Gas within the lumen of the bowel outlines the valvulae conniventes, which completely encircle the bowel. (From Eisenberg,[13] with permission from the publisher.)

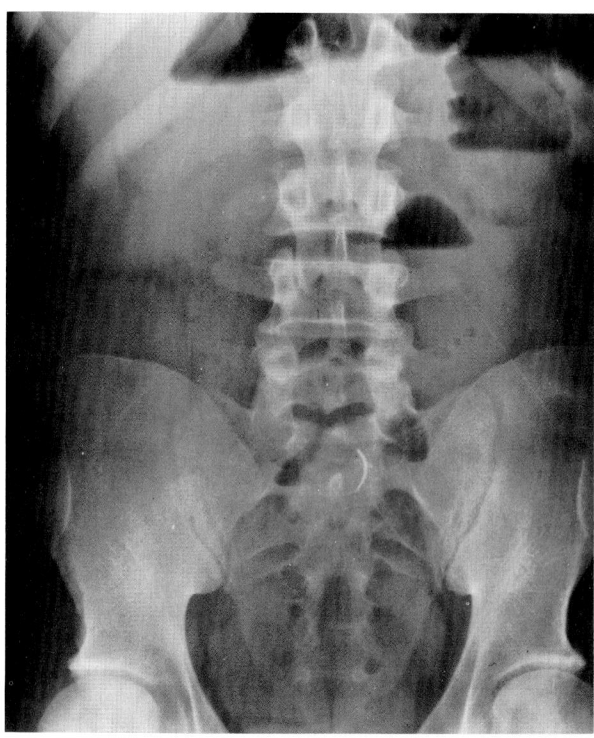

**Figure 49.3. Jejunal obstruction.** The few dilated loops of small bowel are located in the left upper abdomen. (From Eisenberg,[13] with permission from the publisher.)

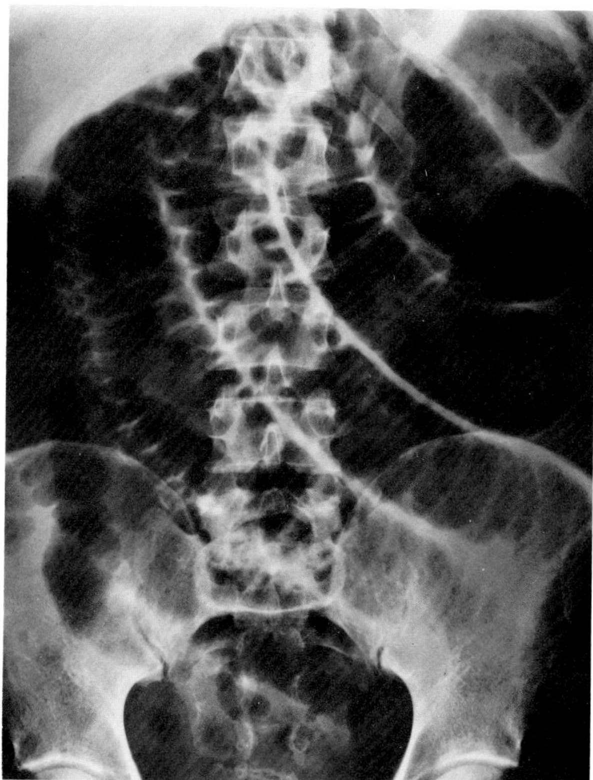

**Figure 49.4. Low small bowel obstruction.** The dilated loops of gas-filled bowel appear to be placed one above the other, upward and to the left, producing a characteristic stepladder appearance. (From Eisenberg,[13] with permission from the publisher.)

**Figure 49.5. Small bowel obstruction.** The antegrade administration of barium demonstrates the precise site of small bowel obstruction. A radiolucent gallstone (arrow) is causing the distal ileal obstruction. (From Eisenberg,[13] with permission from the publisher.)

Although a small amount of gas or fecal accumulations may be present at an early stage of a small bowel obstruction, the detection of a large amount of gas in the colon effectively eliminates this diagnosis.

The bowel proximal to an obstruction can contain no gas but be completely filled with fluid. This may produce a confusing picture of a normal-appearing abdomen or a large soft tissue abdominal mass.

Plain abdominal radiographs are occasionally not sufficient for a distinction to be made between small and large bowel obstruction. In these instances, a carefully performed barium enema examination will document or eliminate the possibility of large bowel obstruction. If it is necessary to determine the precise site of small bowel obstruction, barium can be administered in either a retrograde or an antegrade manner. Oral barium (*not* water-soluble agents) is the most effective contrast material for demonstrating the site of small bowel obstruction. The large amount of fluid proximal to a small bowel obstruction prevents any trapped barium from hardening or increasing the degree of obstruction. The density of barium permits excellent visualization far into the intestine, unlike water-soluble agents, which are lost to sight because of dilution and absorption. It must be emphasized that if plain radiographs clearly demonstrate a mechanical small bowel obstruction, any contrast examination is superfluous.

Strangulation of bowel due to interference with the blood supply is a serious complication of small bowel obstruction. In a closed-loop obstruction (volvulus, incarcerated hernia), both the afferent and efferent limbs of a loop of bowel become obstructed. The involved segments are usually filled with fluid and appear radiographically as a tumor-like soft tissue mass (pseudotumor). A closed loop is a clinically dangerous form of obstruction, since the continuing outpouring of fluid into the enclosed space can raise the intraluminal pressure and rapidly lead to occlusion of the blood supply to that segment of bowel. Because venous pressure is normally lower than arterial pressure, blockage of venous outflow from the strangulated segment occurs before obstruction of the mesenteric arterial supply. Ischemia can rapidly cause necrosis of the bowel with sepsis, peritonitis, and a potentially fatal outcome.[1–4]

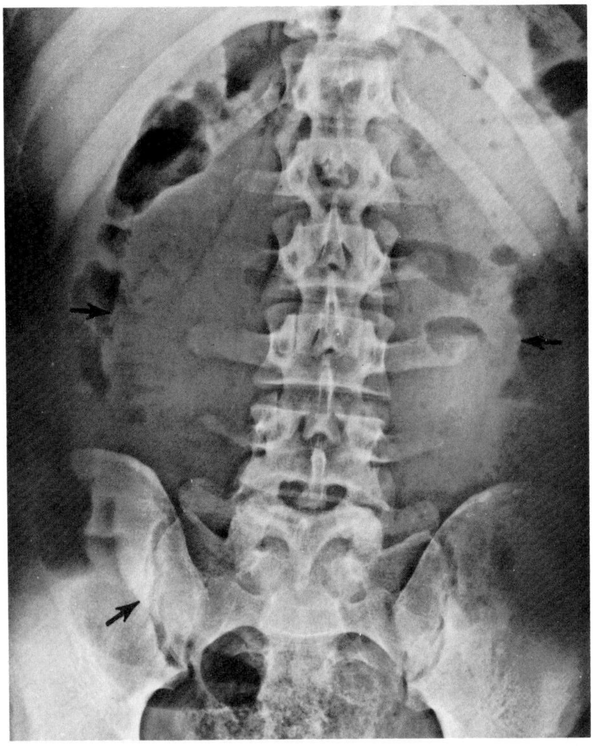

A

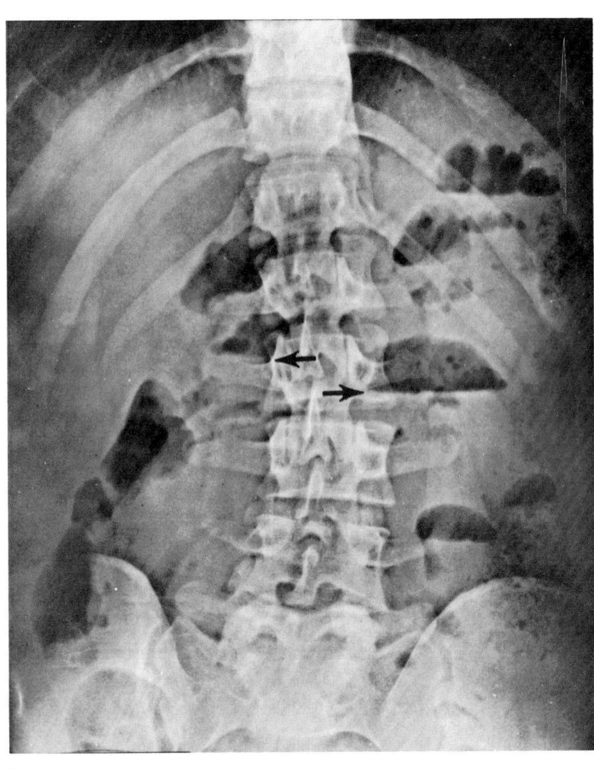

B

**Figure 49.6. Pseudotumor of small bowel obstruction.** (A) A supine film shows fluid-filled loops as a tumor-like density in the mid-abdomen with a polycyclic outline indenting adjacent gas-containing loops (arrows). (B) An upright film shows fluid levels in the pseudotumor (arrows). (From Bryk,[3] with permission from the publisher.)

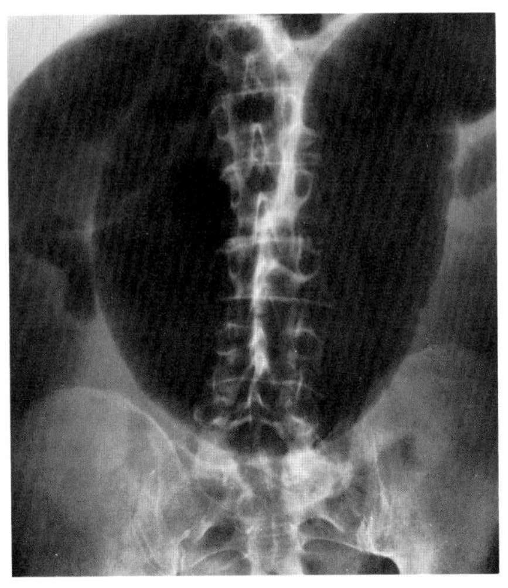

A

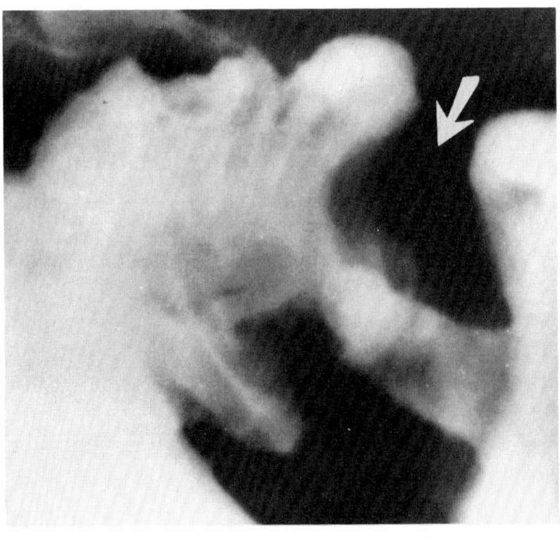

B

**Figure 49.7. Large bowel obstruction caused by annular carcinoma of the sigmoid.** (A) A plain abdominal radiograph demonstrates pronounced dilatation of the gas-filled transverse and ascending colon. (B) A barium enema demonstrates a typical apple-core lesion (arrow) producing the colonic obstruction. (From Eisenberg,[13] with permission from the publisher.)

## LARGE BOWEL OBSTRUCTION

About 70 percent of large bowel obstructions are secondary to primary colonic carcinoma. Diverticulitis and volvulus account for most other cases. Colonic obstructions tend to be more subacute than small bowel obstructions; the symptoms develop more slowly, and fewer fluid and electrolyte disturbances are produced.

The radiographic appearance of colonic obstruction depends on the competency of the ileocecal valve. If the ileocecal valve is competent, obstruction causes a large, dilated colon with a markedly distended, thin-walled cecum and little small bowel gas. The colon distal to the obstruction is usually collapsed and free of gas. If the ileocecal valve is incompetent, there is distension of gas-filled loops of both colon and small bowel that may simulate adynamic ileus.

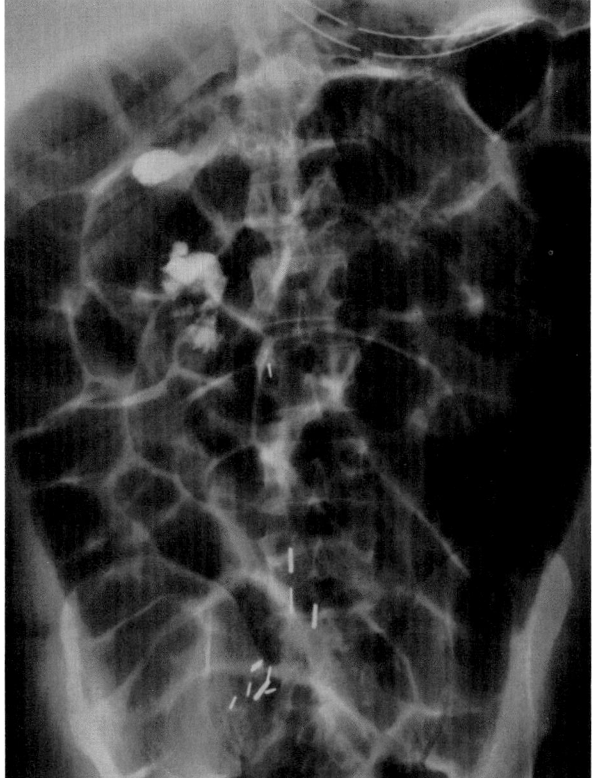

**Figure 49.8. Large bowel obstruction due to torsion of the splenic flexure as it enters a traumatic diaphragmatic hernia.** Because of the incompetent ileocecal valve, there is diffuse dilatation of gas-filled loops of both the colon and the small bowel; this produces a radiographic pattern that suggests adynamic ileus. (From Eisenberg,[13] with permission from the publisher.)

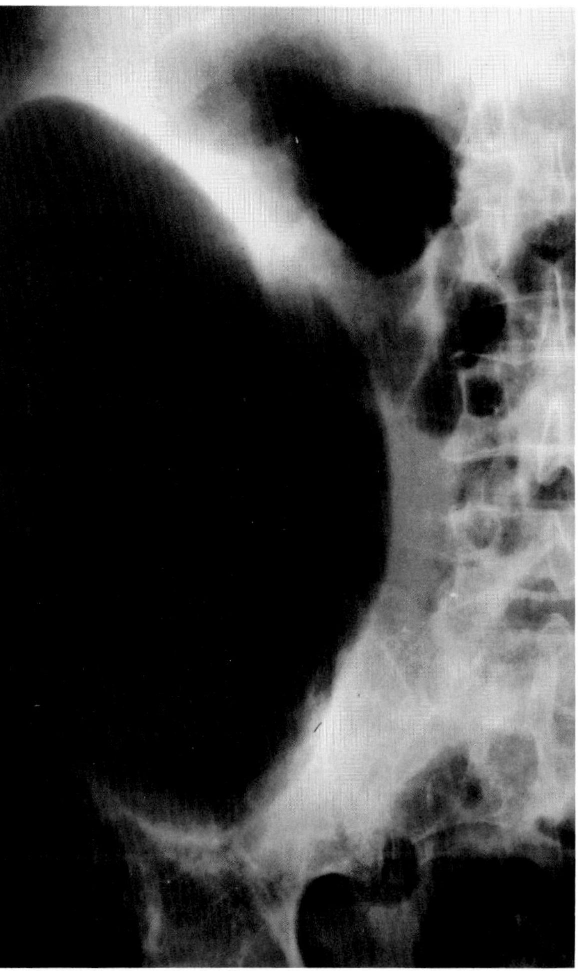

**Figure 49.9. Large bowel obstruction.** Huge dilatation of the cecum (13 cm in diameter) without perforation. (From Eisenberg,[13] with permission from the publisher.)

The major danger in colonic obstruction is perforation. If the ileocecal valve is competent, the colon behaves like a closed loop, and the increased pressure due to the obstruction cannot be dissipated. Because the cecum is spherical and has a large diameter, it is the most likely site for perforation. In acute colonic obstruction, perforation is very likely if the cecum distends to more than 10 cm; in intermittent or chronic obstruction, however, the cecal wall can become hypertrophied, and the diameter of the cecum can greatly exceed 10 cm without perforation. In the patient with suspected large bowel obstruction, a low-pressure barium enema can be safely performed and will demonstrate the site and often the cause of the obstruction.[5]

## INTUSSUSCEPTION

Intussusception is a major cause of bowel obstruction in children; it is much less common in adults. Intussusception is the invagination of one part of the intestinal tract into another due to peristalsis, which forces the proximal segment of bowel to move distally within the ensheathing outer portion. Once such a lead point has been established, it gradually progresses forward and causes increased obstruction. This can compromise the vascular supply and produce ischemic necrosis of the intussuscepted bowel.

In children, intussusception is most common in the region of the ileocecal valve. The clinical onset tends to be abrupt, with severe abdominal pain, blood in the stools ("currant jelly"), and often a palpable right-sided mass. If the diagnosis is made early and therapy instituted promptly, the mortality rate of intussusception in children is less than 1 percent. However, if treatment is delayed more than 48 hours after the onset of symptoms, the mortality rate increases dramatically. In adults, intussusception is often chronic or subacute and is characterized by irregular recurrent episodes of colicky pain, nausea, and vomiting. A specific cause of intussusception often cannot be detected in children. In adults, however, the leading edge is frequently a pedunculated polypoid tumor or an inflammatory mass.

Radiographically, an intussusception produces the classic coiled-spring appearance of barium trapped be-

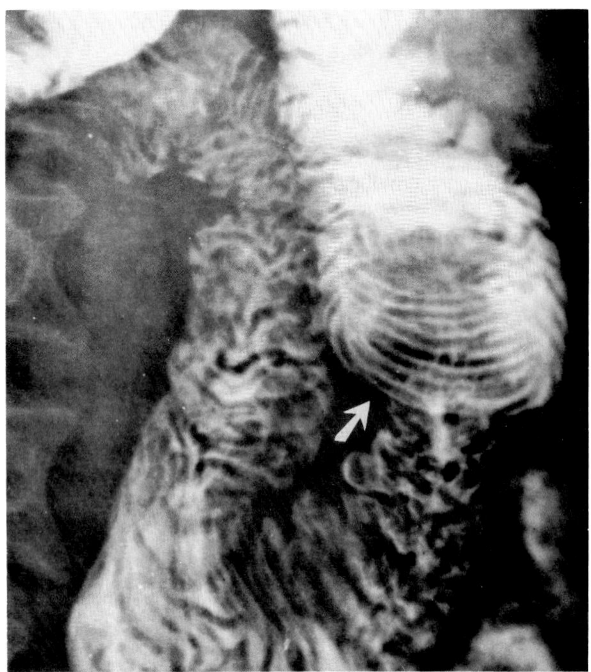

**Figure 49.10. Jejunojejunal intussusception causing small bowel obstruction.** Note the characteristic coiled-spring appearance (arrow). (From Eisenberg,[13] with permission from the publisher.)

tween the intussusceptum and the surrounding portions of bowel. Reduction of a colocolic intussusception can sometimes be accomplished by barium enema examination, though great care must be exercised to prevent excessive intraluminal pressure and consequent colonic perforation. If a colonic intussusception is reduced in an adult, a repeat barium enema examination is necessary to determine whether an underlying polyp or tumor is present.

On ultrasound, intussusception can produce a target-like appearance (multiple concentric ring sign), because edematous bowel wall within the intestinal invagination presents a series of interfaces that give rise to a central dense echo core with a surrounding sonolucent zone. The double areas of sonodensity and sonolucency correspond to contents in the lumen and the edematous bowel wall of the intussuscipiens, the interface between the intussuscipiens and the intussusceptum, and edema in the wall of the intussusceptum.

Computed tomography (CT) often can demonstrate an intussusception by identifying the individual layers of bowel wall, contrast material, and mesenteric fat. If mesenteric fat is seen, three individual layers of bowel wall can be identified; in portions of the intussusception where mesenteric fat is not present, only two layers are visible. Because of the asymmetric location of the invaginated mesenteric fat, the lumen of the intussusceptum often is eccentrically positioned.[1,5–9]

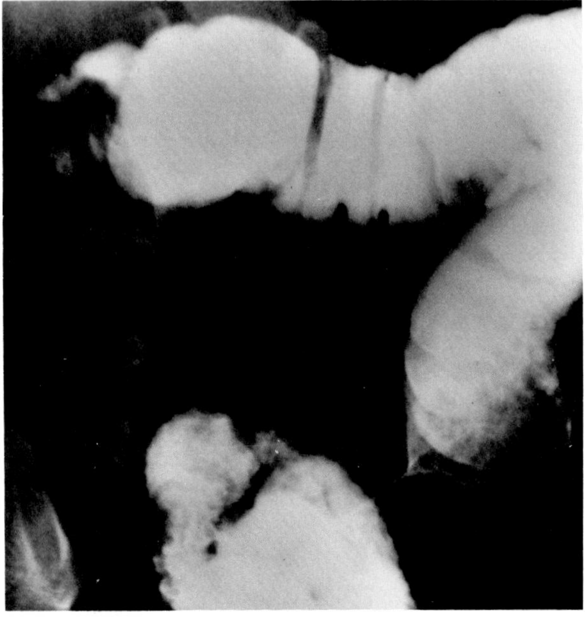

A

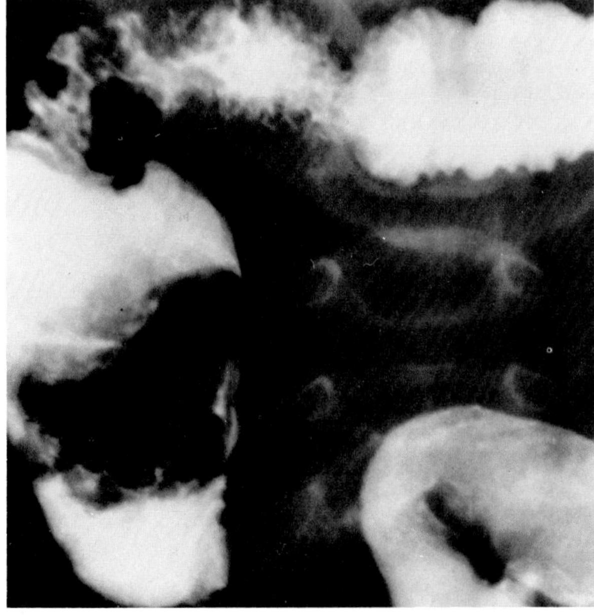

B

**Figure 49.11. Ileocolic intussusception in a child.** (A) Complete obstruction in the region of the hepatic flexure. (B) A gentle barium enema, which has succeeded in reducing the intussusception, reveals a multilobulated mass in the region of the ileocecal valve. (From Eisenberg,[13] with permission from the publisher.)

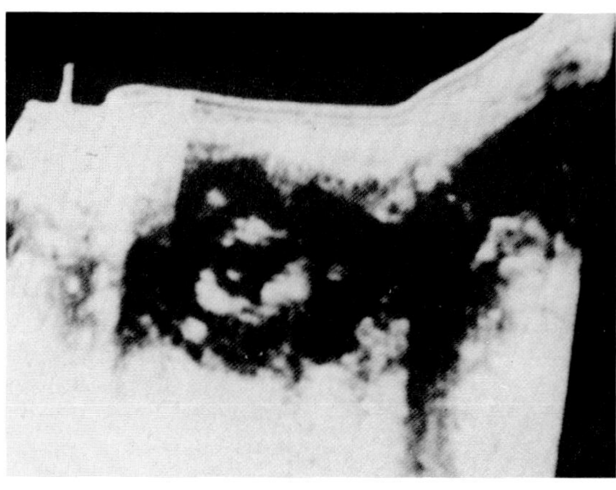

A

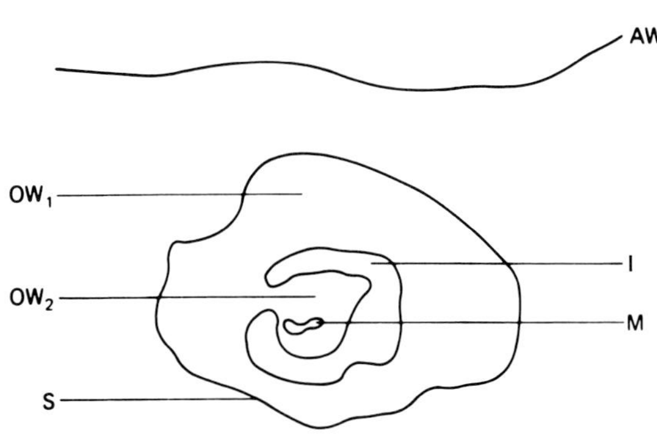


B

**Figure 49.12. Ileocolic intussusception.** (A) A supine longitudinal sonogram and (B) a diagram demonstrate the multiple concentric ring shadows of an intussusception. AW = anterior abdominal wall; $OW_1$ = edematous wall of intussusceptum; I = interface of intussuscipiens and intussusceptum; $OW_2$ = edematous wall of intussuscipiens; M = mucosa of intussuscipiens; S = serosal coat of large bowel. (From Holt and Samuel,[6] with permission from the publisher.)

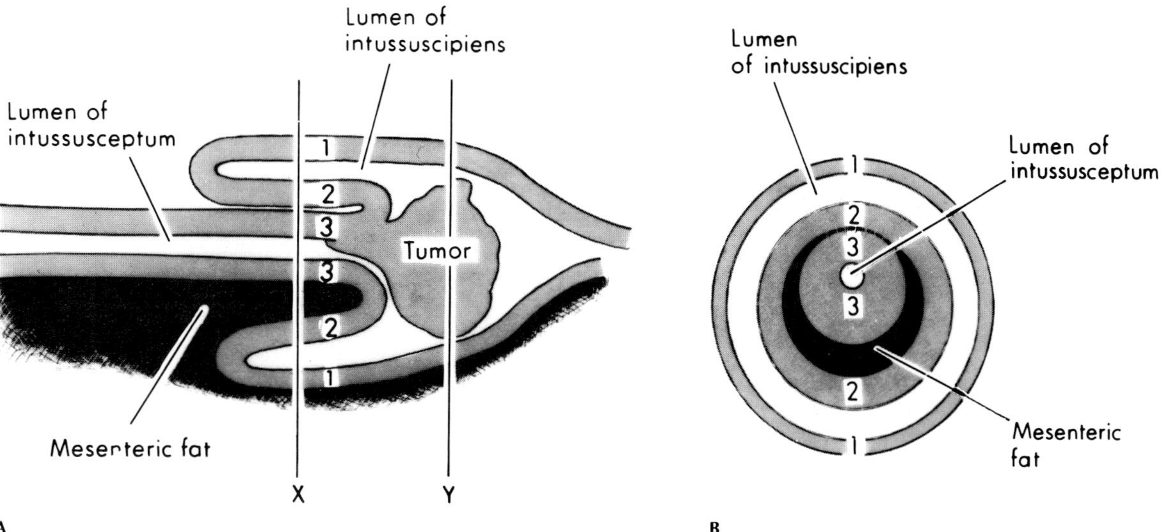


A

B

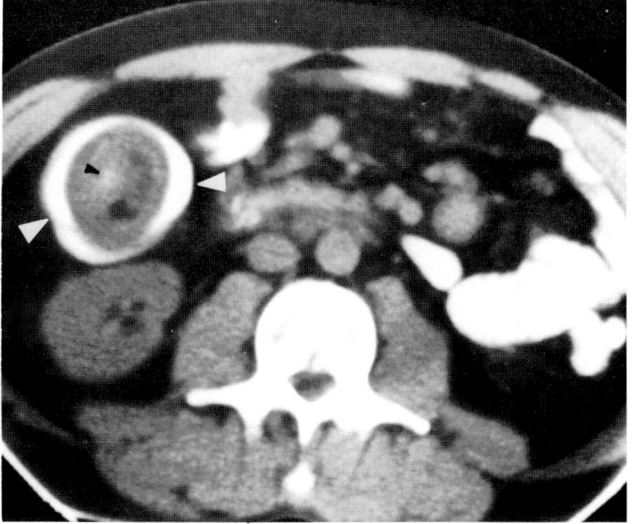

C

**Figure 49.13. CT of intussusception.** (A) A diagram of a longitudinal section through an intussusception. 1 = wall of intussuscipiens; 2, 3 = folded layers of bowel wall that constitute the intussusceptum. Note how invaginated mesenteric fat is attached to one side of the intussusceptum and separates the wall of the intussusceptum into two individual layers. (B) A diagram of the cross section of an intussusceptum at level X in A. Note the eccentric lumen of the intussusceptum. Where mesenteric fat is present, three distinct layers (1, 2, 3) can be seen. Only two layers are visible where there is no mesenteric fat, because layers 2 and 3 cannot be distinguished. (C) In this patient with an ileocolic intussusception, a cross-sectional view shows contrast material in the ascending colon (white arrowheads) surrounding the intussuscepted ileum and associated mesenteric fat. A tiny amount of contrast material is seen in the ileal lumen (black arrowhead). Lymphoma of the ileum formed the leading mass. (From Mauro and Koehler,[9] with permission from the publisher.)

## VOLVULUS OF THE COLON

Because torsion of the bowel usually requires a long, movable mesentery, volvulus of the large bowel most frequently involves the cecum and sigmoid colon. The transverse colon, which has a short mesentery, is rarely affected by volvulus.

### Cecal Volvulus

The ascending colon and the cecum may have a long mesentery as a fault of rotation and fixation during the development of the gut. This situation predisposes to volvulus, with the cecum twisting on its long axis. It should be stressed, however, that only a few patients with a hypermobile cecum ever develop cecal volvulus.

In cecal volvulus, the distended cecum tends to be displaced upward and to the left, though it can be found anywhere within the abdomen. A pathognomonic sign of cecal volvulus is a kidney-shaped mass (representing the twisted cecum) with the torqued and thickened mesentery mimicking the renal pelvis. A barium enema examination is usually required for definite confirmation of the diagnosis. This study demonstrates obstruction of the contrast column at the level of the stenosis, with the tapered edge of the column pointing toward the site of torsion.[5,10]

### Sigmoid Volvulus

A long, redundant loop of sigmoid colon can undergo a twist on its mesenteric axis and form a closed-loop obstruction. In sigmoid volvulus, the greatly inflated sigmoid loop appears as an inverted-U-shaped shadow that rises out of the pelvis in a vertical or oblique direction and can even reach the level of the diaphragm. The affected loop appears devoid of haustral markings and has a sausage or balloon shape. On supine radiographs, there are often three dense, curved lines running downward and converging toward the point of stenosis. These lines appear to end in a small tumor-like density that corresponds to the twisted mesenteric loop. The central and most constant line is a dense midline crease produced by the two walls of the torqued loop lying pressed together. The other two lines, less frequently seen, are made up of the outer margins of the closed loop joined with the medial walls of the cecum on the right and the descending colon on the left. When a barium enema examination is performed on a patient with sigmoid volvulus, the flow of contrast material ceases at the obstruction and the rectum becomes distended. The lumen tapers toward the site of stenosis, and a pathognomonic bird's beak appearance is produced.[5,10,11]

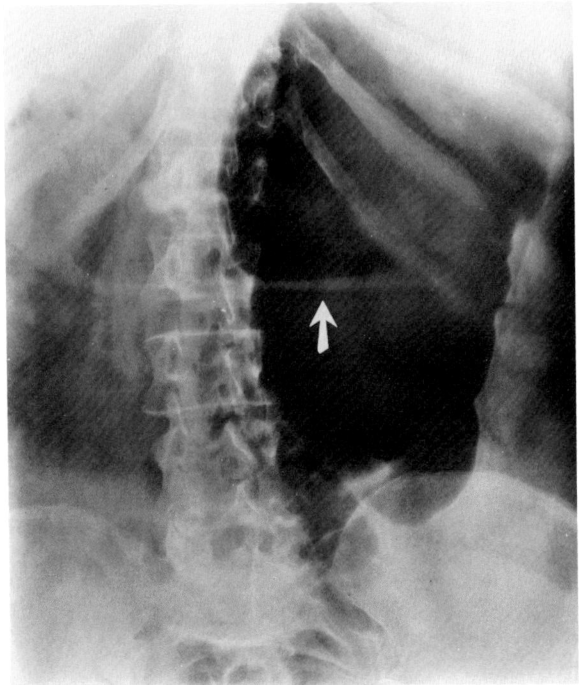

A

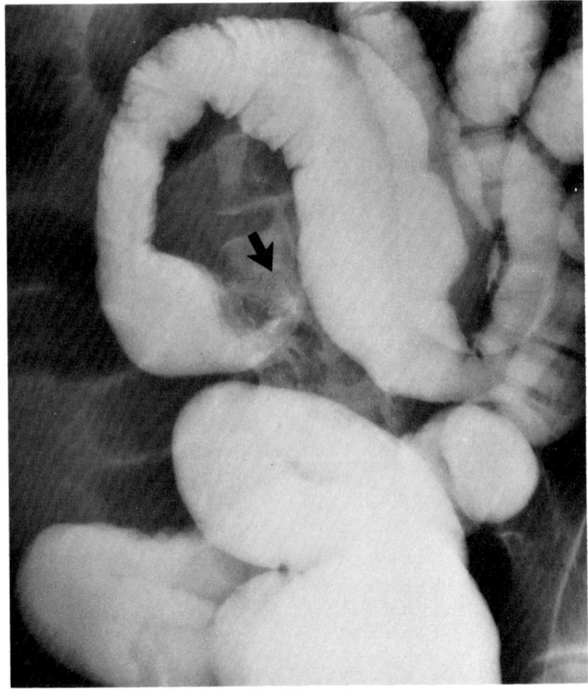

B

**Figure 49.14. Cecal volvulus.** (A) The dilated, gas-filled cecum appears as a kidney-shaped mass, with the torqued and thickened mesentery (arrow) mimicking the renal pelvis. (B) A barium enema examination demonstrates obstruction of the contrast column at the level of the stenosis, with the tapered edge of the column pointing toward the site of torsion (arrow). (From Eisenberg,[13] with permission from the publisher.)

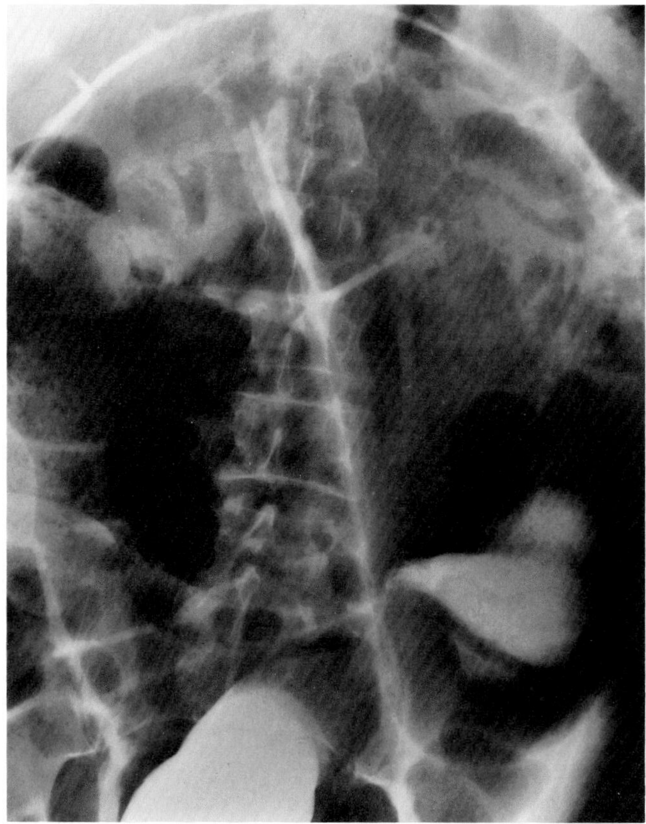
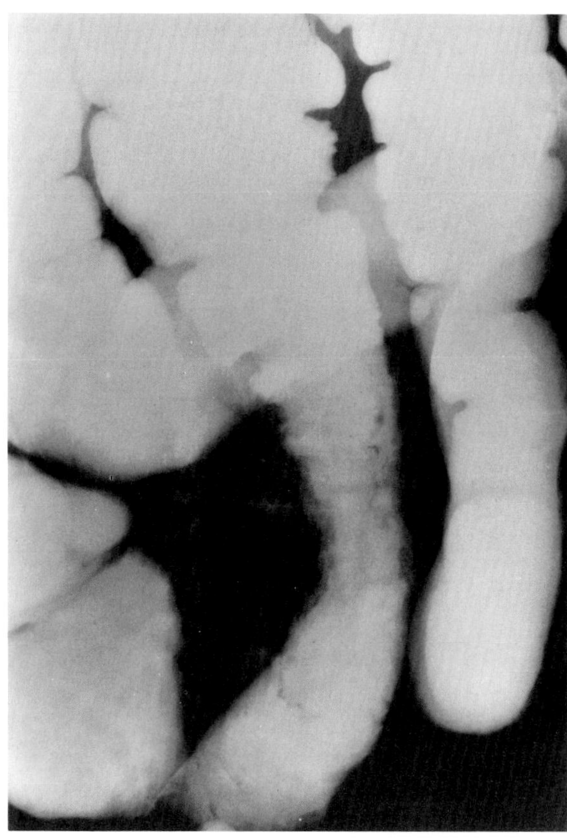

A B

**Figure 49.15. Sigmoid volvulus.** (A) The massively dilated loop of sigmoid appears as an inverted-U-shaped shadow rising out of the pelvis. (B) A barium enema examination following reduction of the volvulus demonstrates the severely ectatic sigmoid colon. (From Eisenberg,[13] with permission from the publisher.)

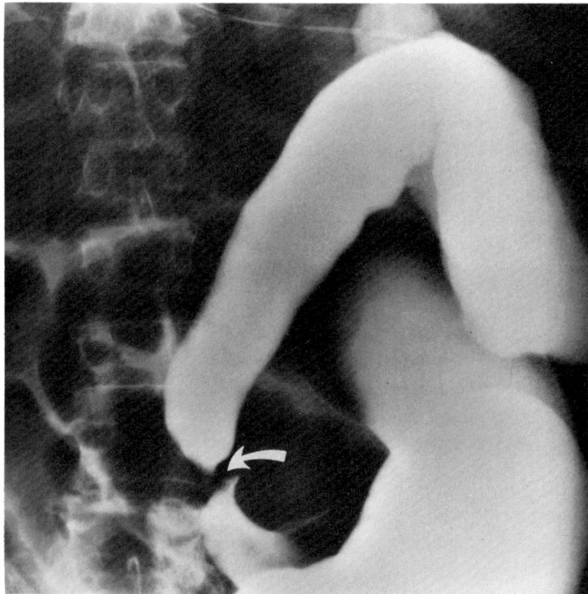

**Figure 49.16. Sigmoid volvulus.** A barium enema examination demonstrates luminal tapering at the site of stenosis (arrow), producing the characteristic bird's beak configuration. (From Eisenberg,[13] with permission from the publisher.)

## ADYNAMIC ILEUS

Adynamic ileus is a common disorder of intestinal motor activity in which fluid and gas do not progress normally through a nonobstructed small and large bowel. A variety of neural, humoral, and metabolic factors can precipitate reflexes that inhibit intestinal motility. Adynamic ileus occurs to some extent in almost every patient who undergoes abdominal surgery. Other causes of adynamic ileus include peritonitis, medication (atropine-like effect, electrolyte and metabolic disorders), and trauma. The clinical appearance of patients with adynamic ileus varies from minimal symptoms to generalized abdominal distension with a marked decrease in the frequency and intensity of bowel sounds.

The radiographic hallmark of adynamic ileus is the retention of large amounts of gas and fluid in dilated small and large bowel. Unlike the appearance in mechanical small bowel obstruction, the entire small and large bowel in adynamic ileus appears almost uniformly dilated, with no demonstrable point of obstruction.[12,13]

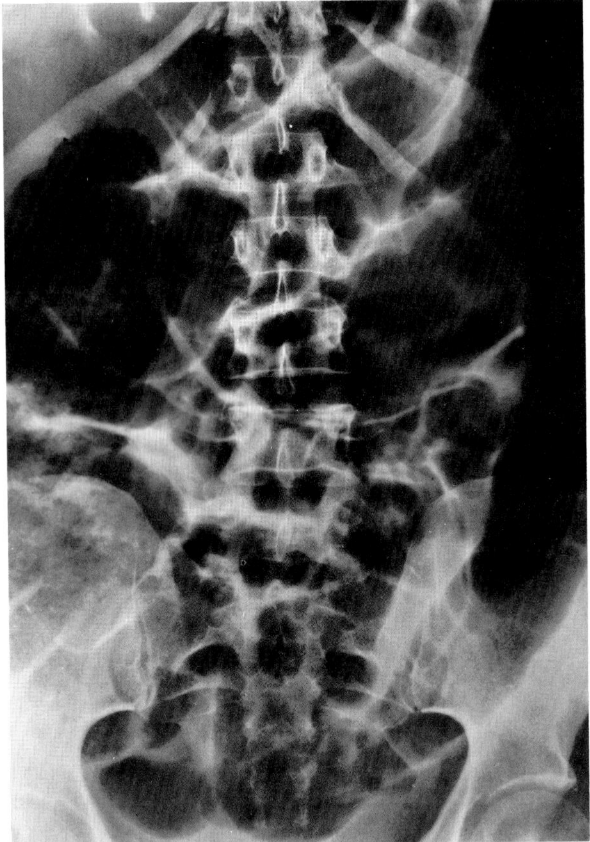

**Figure 49.17. Adynamic ileus.** Large amounts of gas and fluid are retained in loops of dilated small and large bowel. The entire small and large bowel appears almost uniformly dilated, with no demonstrable point of obstruction.

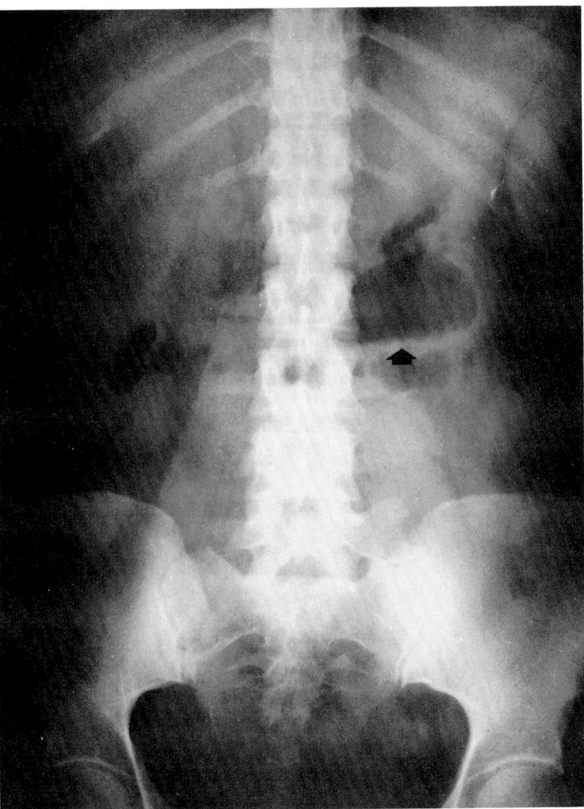

**Figure 49.18. Localized ileus (arrow) in a patient with acute pancreatitis.** (From Eisenberg,[13] with permission from the publisher.)

## VARIANTS OF ADYNAMIC ILEUS

### Localized Ileus

An isolated distended loop of small or large bowel reflecting localized adynamic ileus (sentinel loop) is often associated with an adjacent acute inflammatory process. The portion of the bowel involved can offer a clue to the underlying disease. Localized segments of the jejunum or transverse colon are frequently dilated in patients with acute pancreatitis. Similarly, the hepatic flexure of the colon can be distended in acute cholecystitis, the terminal ileum can be dilated in acute appendicitis, the descending colon can be distended in acute diverticulitis, and dilated loops can be seen along the course of the ureter in acute ureteral colic. Unfortunately, isolated segments of distended small bowel are commonly seen in patients with abdominal pain, and the "sentinel loop" may be found guarding the wrong area.[13]

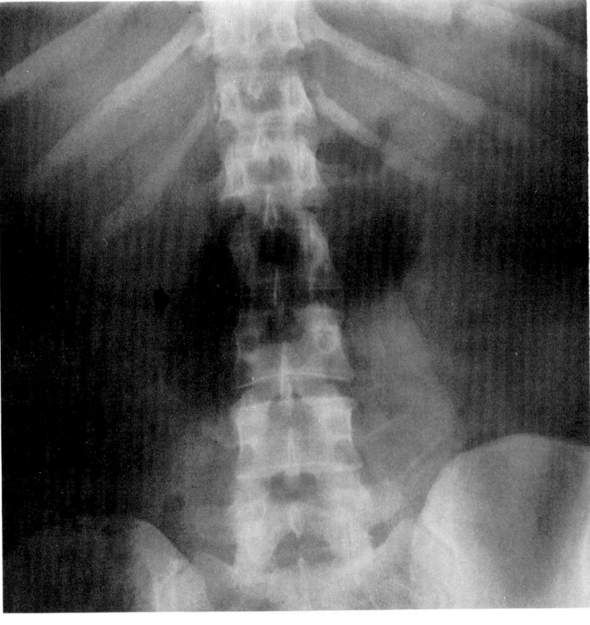

**Figure 49.19. Localized ileus in a patient with acute ureteral colic.** The arrow points to the impacted ureteral stone. (From Eisenberg,[13] with permission from the publisher.)

### Colonic Ileus

Colonic ileus refers to selective or disproportionate gaseous distension of the large bowel without an organic obstruction. Massive distension of the cecum, which is often horizontally oriented, characteristically dominates the radiographic appearance. Although the pathogenesis of colonic ileus is not known, it is probably related to an imbalance between the sympathetic and parasympathetic innervation to the large bowel. Colonic ileus usually accompanies or follows an acute abdominal inflammatory process or abdominal surgery. The clinical presentation and the findings on plain abdominal radiographs simulate those of mechanical obstruction of the colon. A barium enema examination is usually necessary to exclude an obstructing lesion.[14,15]

## CHRONIC IDIOPATHIC INTESTINAL PSEUDO-OBSTRUCTION

Chronic idiopathic intestinal pseudo-obstruction is a rare condition characterized by repeated bouts of signs and symptoms of mechanical obstruction without a demonstrable organic lesion. Although the etiology is unclear, chronic idiopathic intestinal pseudo-obstruction may be related to an intrinsic smooth muscle lesion or to an abnormality of the intramural nerve plexuses. Radiographically, there is pronounced distension of the bowel (especially the small intestine) mimicking intestinal obstruction. Recognition of the true nature of this nonobstructive condition is essential to prevent the patient from undergoing an unnecessary and unrevealing laparotomy.[16–18,23]

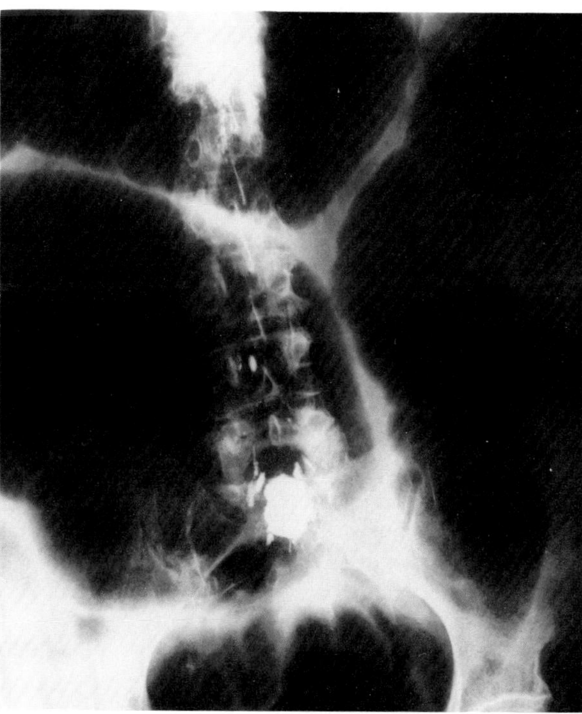

**Figure 49.20. Colonic ileus in a patient with severe diabetes and hypokalemia.** (From Eisenberg,[13] with permission from the publisher.)

## DUODENAL DILATATION (SUPERIOR MESENTERIC ARTERY SYNDROME)

The transverse portion of the duodenum lies in a fixed position in a retroperitoneal location. It is situated in a closed compartment bounded anteriorly by the root of the mesentery (containing the superior mesenteric artery, vein, and nerve) and posteriorly by the aorta and lumbar spine. Any factor that compresses or fills the compartment can cause narrowing of the transverse duodenum and subsequent proximal duodenal dilatation and stasis (superior mesenteric artery syndrome). Even in normal persons, there is often a transient delay of barium passage

**Figure 49.21. Chronic idiopathic intestinal pseudo-obstruction.** Diffuse small bowel dilatation with normal folds simulates mechanical obstruction. (From Maldanado, Gregg, Green, et al,[18] with permission from the publisher.)

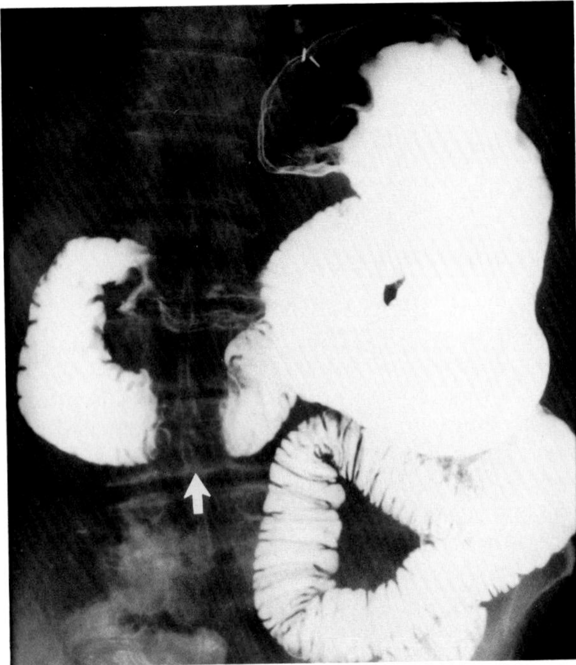

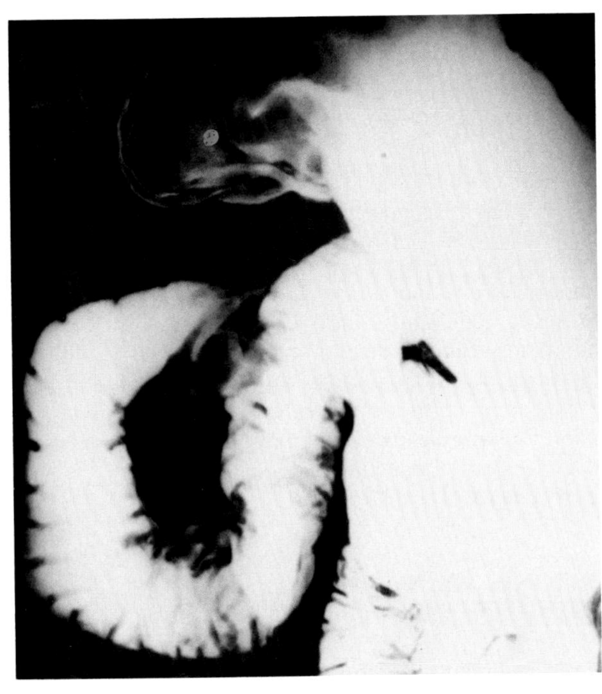

A B

**Figure 49.22. Normal patient with transient proximal duodenal dilatation.** (A) A frontal projection shows apparent obstruction of the third portion of the duodenum (arrow), suggesting the superior mesenteric artery syndrome. (B) A right anterior oblique view obtained slightly later shows the duodenal sweep to be entirely normal, without any evidence of organic obstruction. (From Eisenberg,[13] with permission from the publisher.)

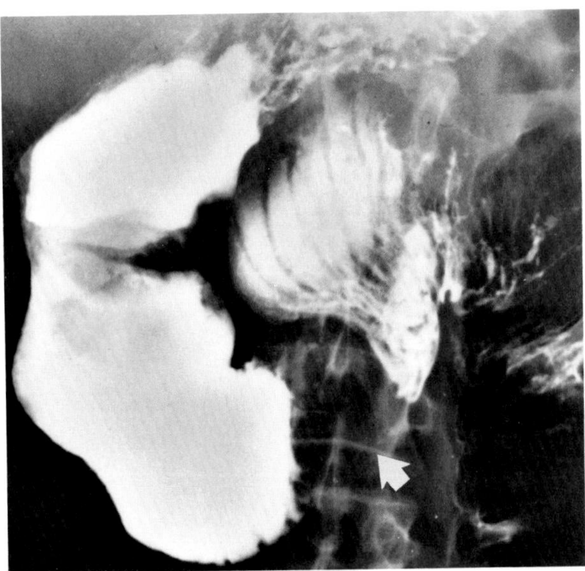

**Figure 49.23. Scleroderma causing superior mesenteric artery syndrome.** There is severe atony and dilatation of the duodenum proximal to the aorticomesenteric angle (arrow). (From Eisenberg,[13] with permission from the publisher.)

and mild, inconstant duodenal dilatation proximal to the point at which the transverse duodenum crosses the spine.

Conditions that tend to close the nutcracker-like jaws of the aorticomesenteric angle (substantial loss of weight by asthenic persons, prolonged bed rest or immobilization in the supine position) result in some degree of compression of the transverse portion of the duodenum. In patients with diseases causing reduced duodenal peristaltic activity (scleroderma and other connective tissue disorders), the combination of lumbar spine, aorta, and superior mesenteric artery can constitute enough of a barrier to cause significant obstruction of the transverse duodenum. Inflammatory thickening of the bowel wall or mesenteric root (pancreatitis, Crohn's disease) or metastases to the mesentery or mesenteric nodes also can lead to relative duodenal obstruction.

Regardless of the underlying pathologic mechanism, the radiographic appearance is almost identical in all patients with the superior mesenteric artery syndrome. Pronounced dilatation of the first and second portions of the duodenum is associated with a vertical, linear extrinsic pressure defect in the transverse portion of the duodenum overlying the spine. The duodenal mucosal folds are intact but compressed. To differentiate the superior mesenteric artery syndrome from an organic obstruction, the patient should be turned to a prone, left decubitus, or knee-chest position. With the mesenteric traction drag on the transverse duodenum thus decreased (through widening of the aorticomesenteric angle), barium can usually be seen to promptly pass the "obstruction," confirming the diagnosis of superior mesenteric artery syndrome. Because the aorticomesenteric angle is not reduced in patients with thickening of the bowel wall or root of the mesentery, postural change provides much less relief of duodenal compression in these conditions.[19–22]

## REFERENCES

1. Schwartz FS: The differential diagnosis of intestinal obstruction. *Semin Roentgenol* 8:323–338, 1973.
2. Levin B: Mechanical small bowel obstruction. *Semin Roentgenol* 8:281–297, 1973.
3. Bryk D: Strangulating obstruction of the bowel. A re-evaluation of radiographic criteria. *AJR* 130:835–843, 1978.
4. Williams JL: Obstruction of the small intestine. *Radiol Clin North Am* 2:21–31, 1964.
5. Love L: Large bowel obstruction. *Semin Roentgenol* 8:299–322, 1973.
6. Holt S, Samuel E: Multiple concentric ring sign in the ultrasonographic diagnosis of intussusception. *Gastrointest Radiol* 3:307–309, 1978.
7. Weissberg DL, Scheible W, Leopold GR: Ultrasonographic appearance of adult intussusception. *Radiology* 124:791–792, 1977.
8. Parienty RA, Lepreux JF, Gruson B: Sonographic and CT features of ileocolic intussusception. *AJR* 136:608–610, 1981.
9. Mauro MA, Koehler RE: Alimentary tract, in Lee JKT, Sagel SS, Stanley RJ (eds): *Computed Body Tomography*. New York, Raven Press, 1983, pp 307–340.
10. Kerry RL, Lee F, Ransom HK: Roentgenologic examination in the diagnosis and treatment of colon volvulus. *AJR* 113:343–348, 1971.
11. Siroospour D, Berardi RS: Volvulus of the sigmoid colon: A ten-year study. *Dis Colon Rectum* 19:535–541, 1976.
12. Seaman WB: Motor dysfunction of the gastrointestinal tract. *AJR* 116:235–244, 1972.
13. Eisenberg RL: *Gastrointestinal Radiology: A Pattern Approach*, Philadelphia, JB Lippincott, 1983.
14. Bryk D, Soong KY: Colonic ileus and its differential roentgen diagnosis. *AJR* 101:329–337, 1967.
15. Meyers MA: Colonic ileus. *Gastrointest Radiol* 2:37–40, 1977.
16. Moss AA, Goldberg HI, Brotman M: Idiopathic intestinal pseudo-obstruction. *AJR* 115:312–317, 1972.
17. Schuffler MD, Rohrmann CA, Templeton FE: The radiographic manifestations of idiopathic intestinal pseudo-obstruction. *AJR* 127:729–736, 1976.
18. Maldonado JE, Gregg JA, Green PA, et al: Chronic idiopathic intestinal pseudo-obstruction. *Am J Med* 49:203–212, 1970.
19. Berk RN, Coulson EB: The body cast syndrome. *Radiology* 94:303–305, 1970.
20. Gondos B: Duodenal compression defect and the "superior mesenteric artery syndrome." *Radiology* 123:575–580, 1977.
21. Fischer HW: The big duodenum. *AJR* 83:861–875, 1960.
22. Simon M, Lerner MA: Duodenal compression by the mesenteric root in acute pancreatitis and in inflammatory conditions of the bowel. *Radiology* 79:75–81, 1962.
23. Teixidor HS, Heneghan MA: Idiopathic intestinal pseudo-obstruction in a family. *Gastrointest Radiol* 3:91–95, 1978.

## Chapter Fifty
# Diseases of the Appendix, Peritoneum, and Mesentery

### APPENDICITIS

Acute appendicitis is the most common inflammatory disease of the right lower quadrant. Occlusion of the neck of the appendix by a fecalith or by postinflammatory scarring creates a closed-loop obstruction within the organ. Because of inadequate drainage, fluid accumulates in the obstructed portion and serves as a breeding ground for bacteria. High intraluminal pressure causes distension and thinning of the appendix distal to the obstruction and may lead to gangrene and perforation. If the process evolves slowly, adjacent organs (terminal ileum, cecum, omentum) may wall off the appendiceal area so that a localized abscess develops; rapid progression of vascular impairment may cause free perforation and generalized peritonitis.

The clinical symptoms of acute appendicitis are usually so characteristic that there is no need for routine radiographs to make the correct diagnosis. The presence of severe right lower quadrant pain, rebound tenderness, low-grade fever, and slight leukocytosis, especially in young adults, is presumed to be evidence of appendicitis. However, in some patients, especially the elderly, the clinical findings may be obscure or minimal. In these instances, radiographic examination may be necessary for prompt diagnosis and surgical intervention before perforation occurs.

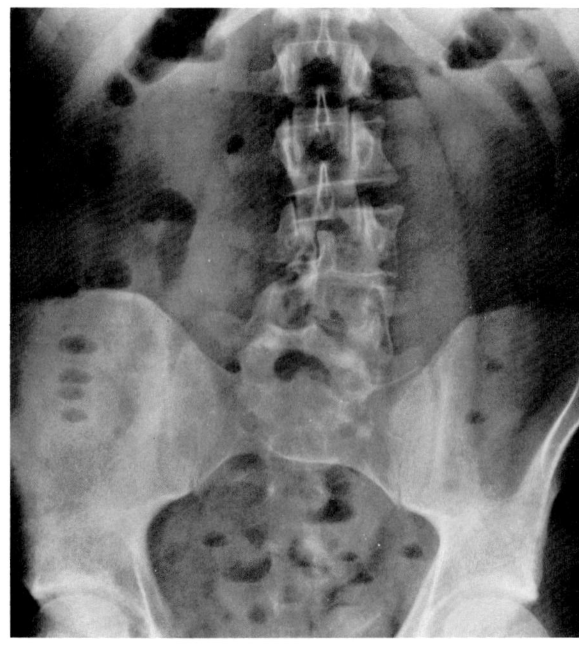

**Figure 50.1. Appendicitis.** A plain abdominal radiograph demonstrates air-fluid levels in the right lower quadrant with obliteration of the right psoas margin and scoliosis of the lumbar spine convex to the left.

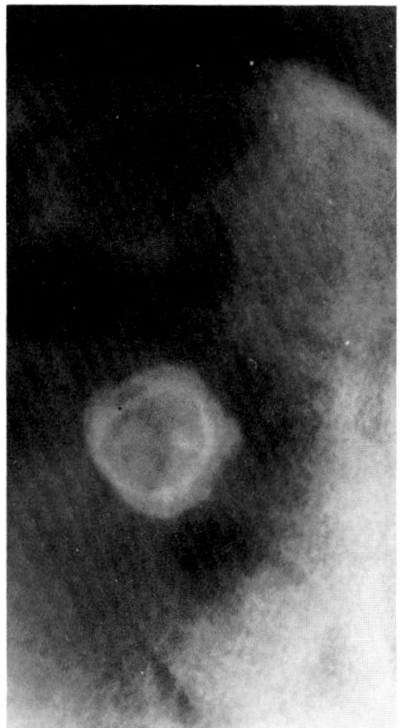

**Figure 50.2. Appendicolith.** Laminated right lower quadrant calcification in a patient with symptoms of acute appendicitis.

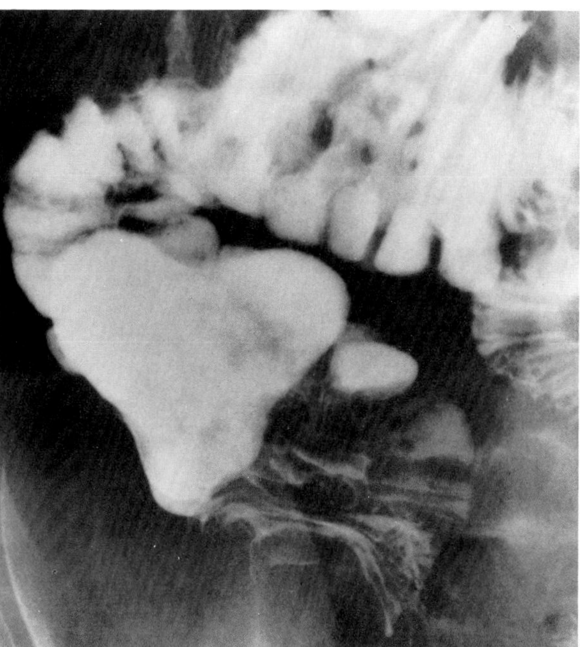

**Figure 50.3. Periappendiceal abscess.** There is fixation and a mass effect at the base of the cecum with no filling of the appendix. (From Eisenberg,[31] with permission from the publisher.)

Plain abdominal radiographs may demonstrate a distended loop of small bowel in the appendiceal area (sentinel loop), often with an air-fluid level, as well as obliteration of the right psoas margin and scoliosis of the lumbar spine convex to the left. Free peritoneal gas is rarely seen. A round or oval, laminated calcified fecalith in the appendix (appendicolith) is found in up to 15 percent of cases of acute appendicitis. Surgical experience suggests that the presence of an appendicolith in combination with symptoms of acute appendicitis usually implies that the appendix is gangrenous and likely to perforate. Most appendicoliths are located in the right lower quadrant. Depending on the length and position of the appendix, however, an appendicolith can also be seen in the pelvis or in the right upper quadrant (retrocecal appendix), where it can simulate a gallstone. An appendicolith located near the midline can mimic a ureteral stone; this is of great significance, since an inflamed appendix in this region can cause hematuria and suggest renal colic rather than appendicitis.

Because of the danger of perforation, barium enema examination is usually avoided in acute appendicitis. If it is performed, an irregular impression of the base of the cecum (due to inflammatory edema), in association with failure of barium to enter the appendix, has been considered virtually pathognomonic of acute appendicitis or appendiceal abscess. Nevertheless, in the absence of other signs, failure of barium to fill the appendix is not a reliable sign of appendicitis, since the appendix does not fill in about 20 percent of normal patients. Partial filling of the appendix with distortion in its shape or caliber strongly suggests acute appendicitis, especially if there is a cecal impression. In contrast, a patent appendiceal lumen effectively excludes the diagnosis of acute appendicitis, especially when barium extends to fill the rounded appendiceal tip.

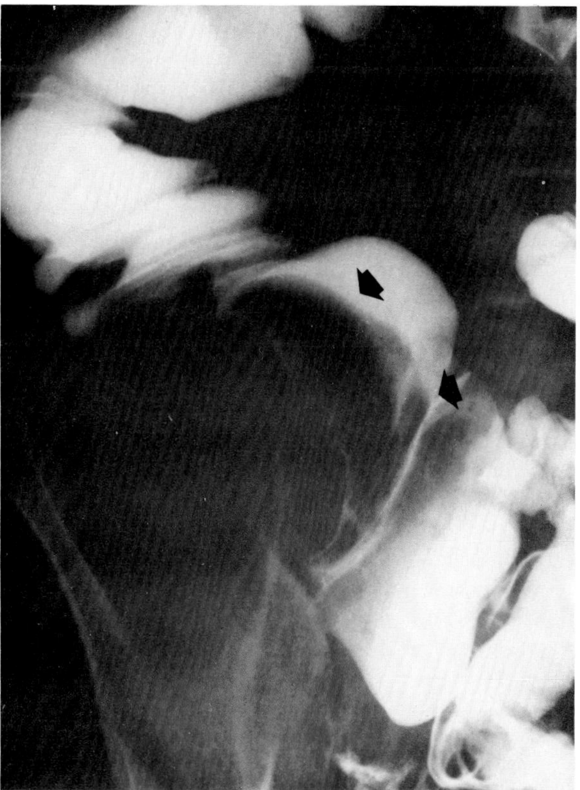

**Figure 50.4. Periappendiceal abscess.** This large extrinsic mass involving the lateral aspect of the ascending colon (arrows) was seen in a patient with a ruptured retrocecal appendix. (From Eisenberg,[31] with permission from the publisher.)

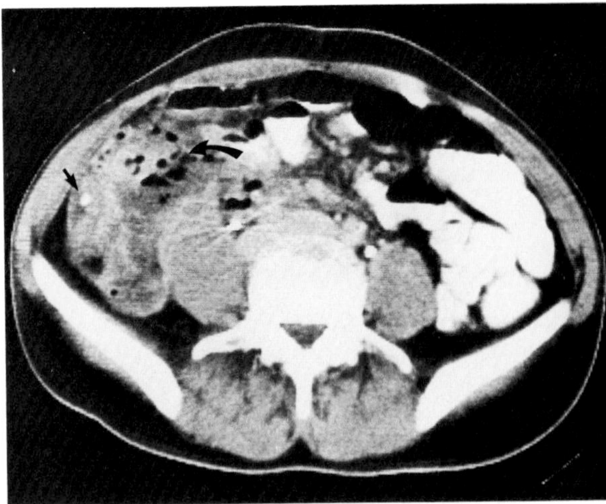

**Figure 50.5. Ruptured appendix with periappendiceal abscess.** A CT scan demonstrates a large soft tissue mass with ectopic gas (curved arrow) and a high-density appendicolith (arrow).

An appendiceal abscess usually causes an irregular eccentric defect at the base of the cecum, most commonly on the medial aspect. In a patient with a retrocecal appendix, the mass involves the colon more proximally, often on its lateral aspect. In rare instances, barium enters the abscess cavity itself, implying that the appendiceal lumen has remained partially patent. Plain abdominal radiographs occasionally demonstrate an appendiceal abscess as a soft tissue mass with a mottled gas pattern.

Computed tomography (CT) can demonstrate an appendiceal abscess, which most frequently is located in the right lower quadrant but may occur in the cul-de-sac or right upper abdomen. The appendiceal abscess appears as an oval or round mass of soft tissue density that may contain gas and occasionally a densely calcified appendicolith.[1–4]

## PERITONITIS

In acute peritonitis the motor activity of the bowel is decreased, and the intestinal lumen becomes distended with gas and fluid. This appears radiographically as adynamic ileus, uniform dilatation of the entire small and large bowel with no demonstrable point of obstruction. Edema of the walls of the small bowel leads to an increased distance between intestinal loops. Peritoneal inflammation and edema may cause obliteration of the properitoneal flank stripe. Peritonitis due to a perforated viscus may be associated with free intraperitoneal gas.

Subsiding generalized peritonitis or a more localized intra-abdominal disease process or injury may result in the formation of localized collections of pus. These intraperitoneal abscesses appear as soft tissue masses displacing and separating small bowel loops. A critical radiographic sign of an intraperitoneal abscess is the presence of extraluminal bowel gas, which can appear as discrete round lucencies, multiple small lucencies ("soap bubbles"), or linear radiolucent shadows that follow fascial plains. Localized ileus (sentinel loop) is often seen adjacent to an intraperitoneal abscess, but this is a nonspecific finding.

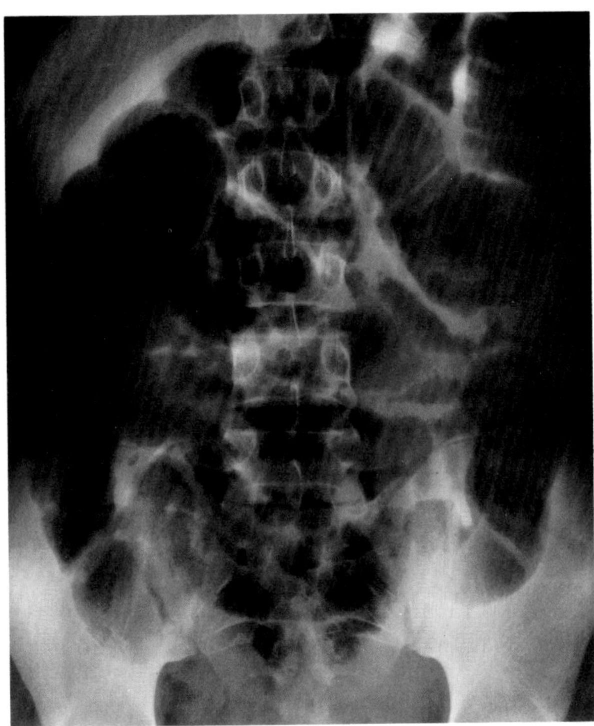

**Figure 50.6. Peritonitis.** Adynamic ileus with dilatation of gas-filled small and large bowel.

## PSEUDOMYXOMA PERITONEI

Rupture of a mucocele of the appendix or a mucinous ovarian cyst can lead to the development of pseudomyxoma peritonei, a condition characterized by epithelial implants on the peritoneal surface and massive accumulation of gelatinous ascites. The radiographic demonstration of a sudden decrease in size of the mucocele may indicate that rupture has occured. Curvilinear calcifications may develop on the periphery of the jelly-like masses secondary to pseudomyxoma peritonei. There may also be widespread abdominal calcifications that are annular in appearance and tend to be most numerous in the pelvis.

Ultrasound and CT findings suggesting the diagnosis of pseudomyxoma peritonei are peritoneal scalloping of the liver margin and ascitic septation. Scalloping refers to extrinsic pressure on the border of the liver by adjacent peritoneal implants without liver parenchymal metastases. Septations refer to the margins of mucinous nodules of low-attenuation material in what otherwise would appear to be ascites.[5–8]

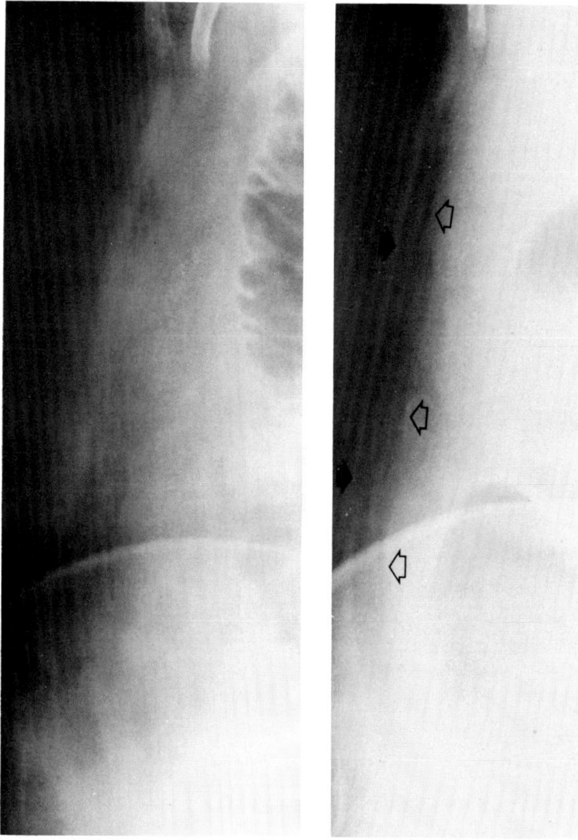

A B

**Figure 50.7. Distortion of properitoneal flank stripe in peritonitis.** (A) Irregular density of the retroperitoneal flank stripe in a patient with peritonitis. In some portions, the lateral margin of the flank stripe is faintly visible. The medial margin is completely obscured. (B) Normal radiographic appearance of the same patient's flank stripe (arrows) 2 years earlier. (From Harris and Harris,[32] with permission from the publisher.)

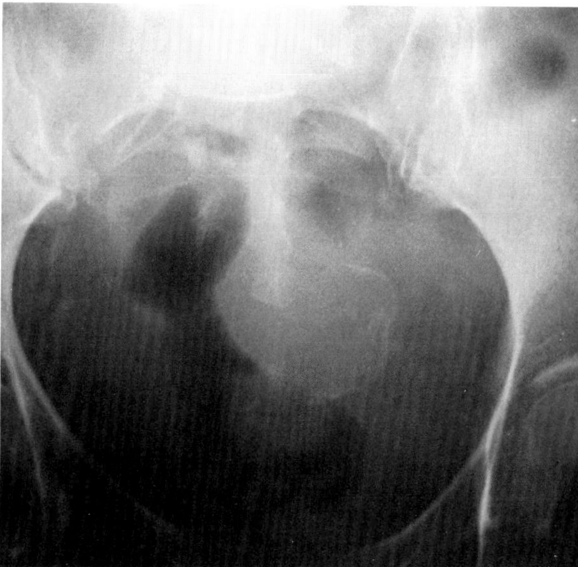

**Figure 50.8. Pseudomyxoma peritonei.** Curvilinear calcifications developed at the periphery of the jelly-like masses. (From Eisenberg,[31] with permission from the publisher.)

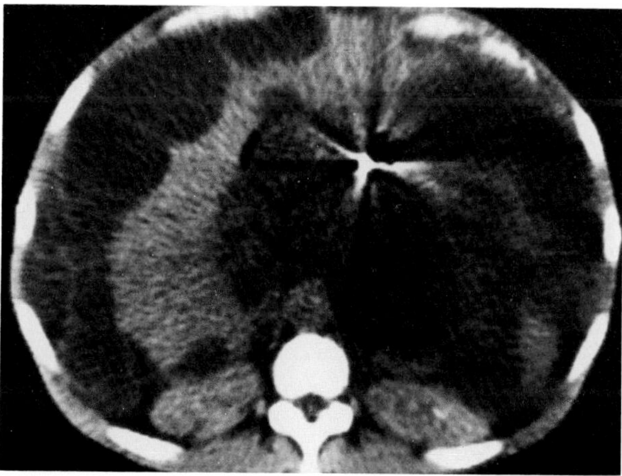

**Figure 50.9. Pseudomyxoma peritonei.** A CT scan through the inferior aspect of the right hepatic lobe demonstrates septations (best seen on the lateral aspect of the liver) and scalloping of the liver by contiguous tumor. (From Seshul and Coulam,[8] with permission from the publisher.)

## MALIGNANT DISEASE OF THE PERITONEUM

Intraperitoneal seeding of metastases occurs when tumor cells float freely in ascitic fluid and implant themselves on peritoneal surfaces. Metastatic tumors to the peritoneum commonly occur in the terminal stages of cancer (mostly adenocarcinoma) of the intraperitoneal organs. Neoplasms of the ovary, stomach, and colon are especially prone to widespread seeding of the peritoneal surfaces. Major areas of intraperitoneal seeding include the pouch of Douglas at the rectosigmoid junction; the right lower quadrant at the lower end of the small bowel mesentery; the left lower quadrant along the superior border of the sigmoid mesocolon and colon; and the right pericolic gutter lateral to the cecum and ascending colon. Metastases to the peritoneum usually produce large volumes of ascites, and the diagnosis of intraperitoneal carcinomatosis can often be made by cytologic examination of aspirated ascitic fluid. In addition to the separation of intestinal loops by ascites, peritoneal carcinomatosis can cause mesenteric masses, nodular impressions, or angulated segments of small bowel. Secondary neoplastic involvement of the small bowel may produce stretching and fixation of mucosal folds transverse to the longitudinal axis of the bowel lumen.

Metastatic disease to the peritoneum can produce generalized calcification on plain abdominal radiographs. The granular or sandlike psammomatous calcification of ovarian cystadenocarcinoma can be confined

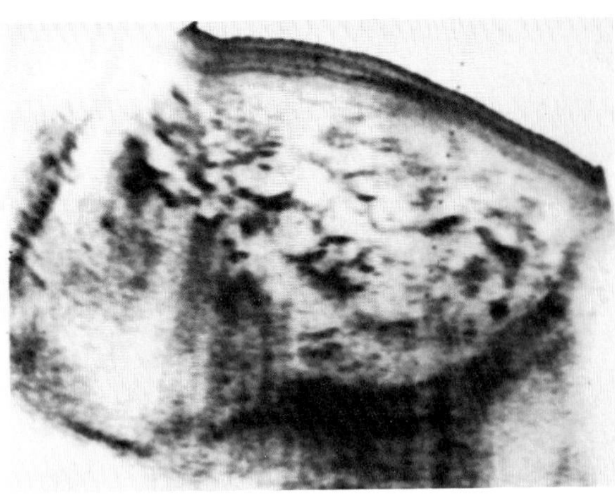

A

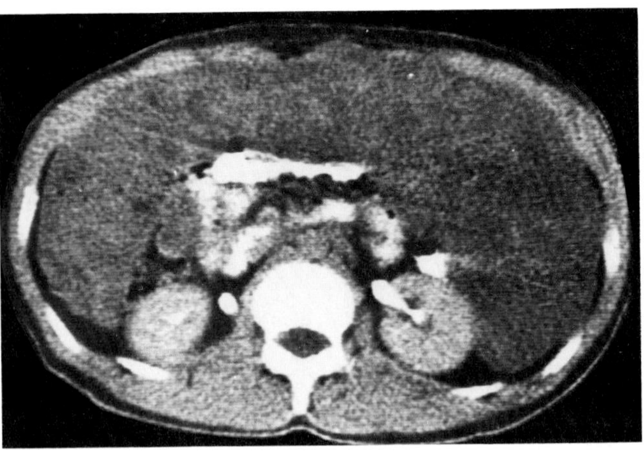

B

**Figure 50.10. Pseudomyxoma peritonei.** (A) A sagittal sonogram demonstrates multiple septations throughout the peritoneal cavity. (B) A CT scan shows that the pseudomyxoma peritonei extends from one colonic gutter to the other. Poorly defined septations are seen throughout the abdomen, along with abnormal posterior displacement of contrast-filled bowel. (From Seshul and Coulam,[8] with permission from the publisher.)

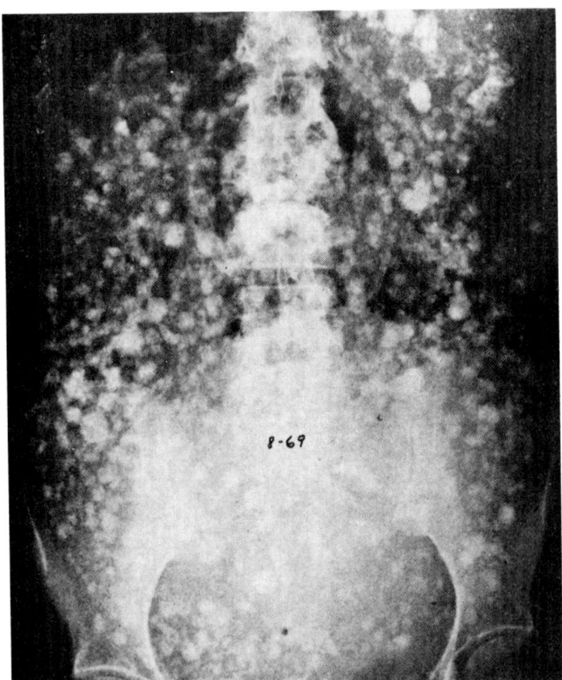


**Figure 50.11. Malignant disease of the peritoneum.** Bizarre masses of calcification not conforming to any organ are seen in this patient with an undifferentiated abdominal malignancy. (From Dalinka, Lally, Azimi, et al,[9] with permission from the publisher.)

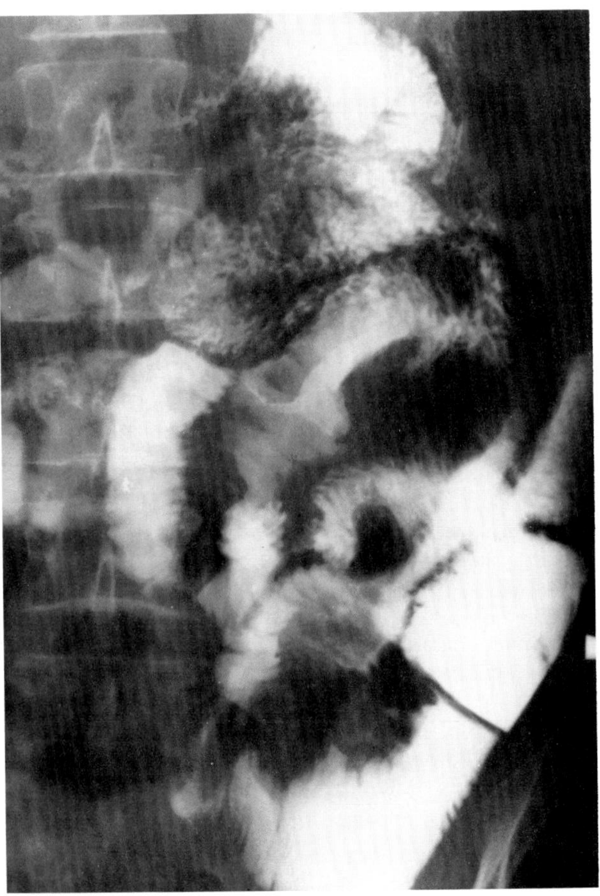

**Figure 50.12. Malignant disease of the peritoneum.** Multiple nodular masses cause separation of small bowel loops in this patient with leiomyosarcoma of the ileum metastatic to the small bowel and mesentery. (From Eisenberg,[31] with permission from the publisher.)

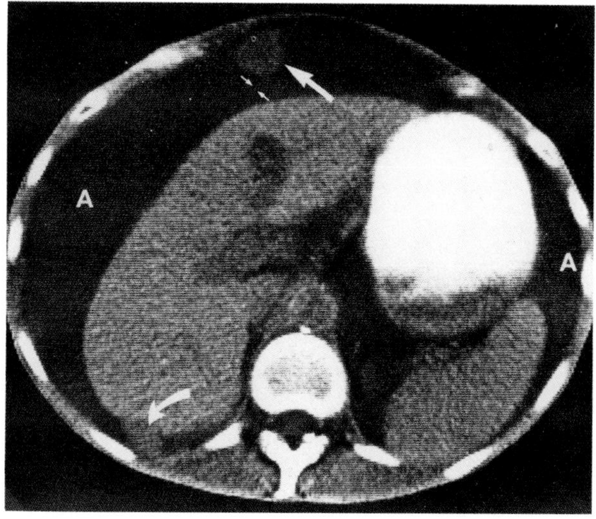


A

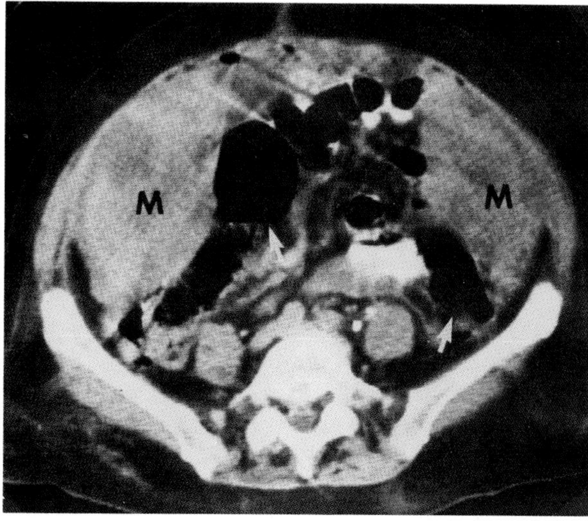


B

**Figure 50.13. CT appearance of peritoneal implants.** (A) A nodular implant (large straight arrow) due to ovarian carcinoma is located adjacent to the falciform ligament (small arrows). A second nodular implant is seen along the lateral peritoneal surface of the liver (curved arrow). There is massive ascites (A) in the peritoneal cavity. (B) Peritoneal metastases from endometrial carcinoma produce sheetlike peritoneal masses (M) that displace the bowel (arrows) centrally. (From Jeffrey,[18] with permission from the publisher.)

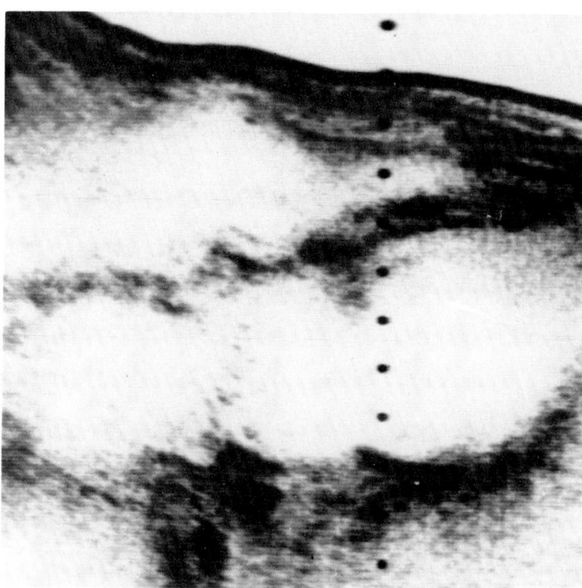

**Figure 50.14. Malignant disease of the peritoneum.** A parasagittal sonogram demonstrates a homogeneous, sonolucent mass around a central area of linear echogenicity representing the mesentery (sandwich sign) in a patient with lymphoma. (From Mueller, Ferrucci, Harbin, et al,[14] with permission from the publisher.)

to the primary tumor or diffusely involve metastases throughout the abdomen. Bizarre masses of calcification that do not conform to any organ have been described in undifferentiated abdominal malignancies. Patients with this condition have large soft tissue masses of multiple linear or nodular calcific densities that can coalesce to form distinctive conglomerate masses.

Mesenteric and omental metastases can be demonstrated on ultrasound and CT. On ultrasound, they appear as nodular, sheetlike, or irregular solid masses with variable echogenicity. Small peritoneal nodules can be demonstrated in patients with a large amount of ascites that permits the peritoneal lining to be clearly seen within the echo-free ascitic fluid. Because the quality of CT scans is unaffected by excessive bowel gas or mesenteric fat, this modality is the procedure of choice in the initial radiographic examination of patients with known or suspected mesenteric or peritoneal metastases. Direct spread of gastrointestinal or genital tumors along peritoneal surfaces appears as infiltration of surrounding fat by soft tissue density and encasement of bowel loops within the metastatic deposit. Diffuse neoplastic infiltration of the greater omentum produces a soft tissue mass separating the colon or small bowel from the anterior abdominal wall and obliteration of normal fat planes (omental cake). Intraperitoneal seeding of neoplasm results in soft tissue masses associated with ascites. Hematogenous spread of tumor causes thickening of mesenteric leaves or focal bowel wall thickening, which is occasionally associated with recognizable ulceration simulating the bull's-eye lesions seen on barium studies. Malignant involvement of mesenteric nodes produces single or multiple soft tissue masses. A characteristic pattern of lymphomatous involvement of the mesentery is a lobulated, confluent mass infiltrating the mesenteric leaves and encasing the superior mesenteric artery and veins. This produces the ultrasonic "sandwich" sign, a lobulated, sonolucent mass surrounding a centrally positioned, linear echogenic area.

Primary neoplasms of the peritoneum (mesotheliomas) are extremely rare. They are usually seen in middle-aged or elderly persons, predominantly men. Like

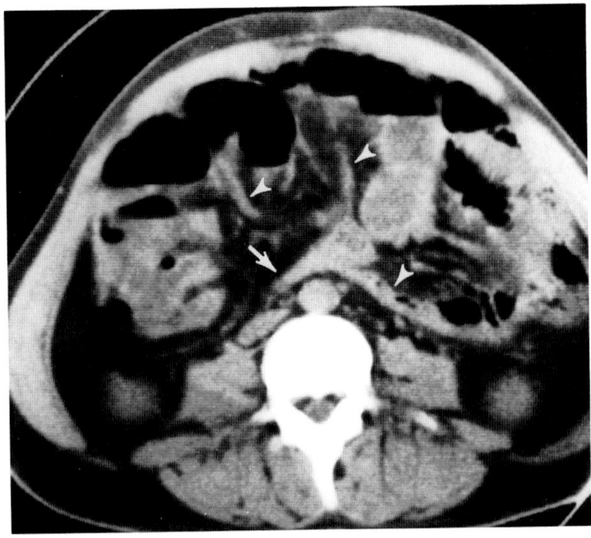
A

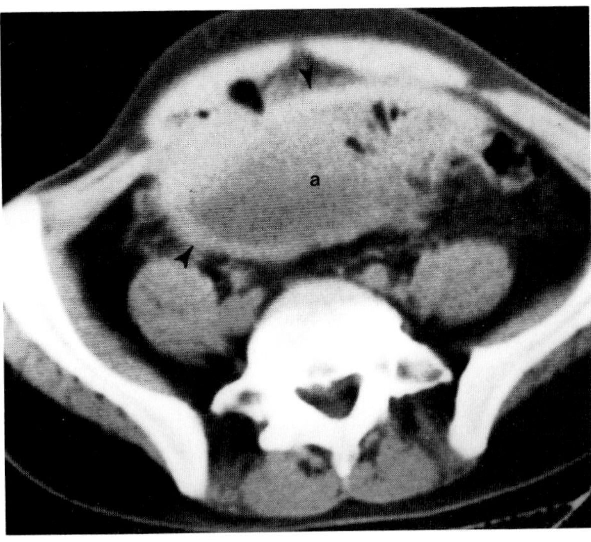

B

**Figure 50.15. Primary peritoneal mesothelioma.** (A) A CT scan through the lower abdomen shows thickening of mesenteric leaves (arrowheads) including the mesenteric root (arrow). (B) A scan through the upper pelvis shows loculated ascites (a) surrounded by the mesothelioma (arrowheads) infiltrating the mesentery. (From Levitt,[33] with permission from the publisher.)

tumors of the same cell type involving the pleura, peritoneal mesotheliomas appear to be closely related to exposure to asbestos. In addition to having a large bulk, peritoneal mesotheliomas are associated with severe ascites, which may produce wide separation of small bowel loops or become loculated and appear as a mass.

CT in patients with primary peritoneal mesothelioma demonstrates focal plaquelike masses and thickening of the mesenteric leaves reflecting infiltration of the tumor along the peritoneal surfaces. Ultrasound- or CT-guided aspiration biopsy may permit a precise tissue diagnosis and obviate the need for surgical intervention to distinguish peritoneal mesothelioma from diffuse peritoneal carcinomatosis, which often has a similar radiographic appearance.[9–18]

## PNEUMATOSIS INTESTINALIS

Gas in the bowel wall (pneumatosis intestinalis) can exist as an isolated entity or in conjunction with a broad spectrum of diseases of the gastrointestinal tract or respiratory system. In primary pneumatosis intestinalis (about 15 percent of cases), no respiratory or other gastrointestinal abnormality is present. Primary pneumatosis usually occurs in adults and mainly involves the colon. Secondary pneumatosis intestinalis (about 85 percent of cases) more commonly involves the small bowel and is associated with a wide variety of pre-existing disorders. In the primary form, gas collections usually appear cystic; in the secondary type, a linear distribution of gas is generally seen.

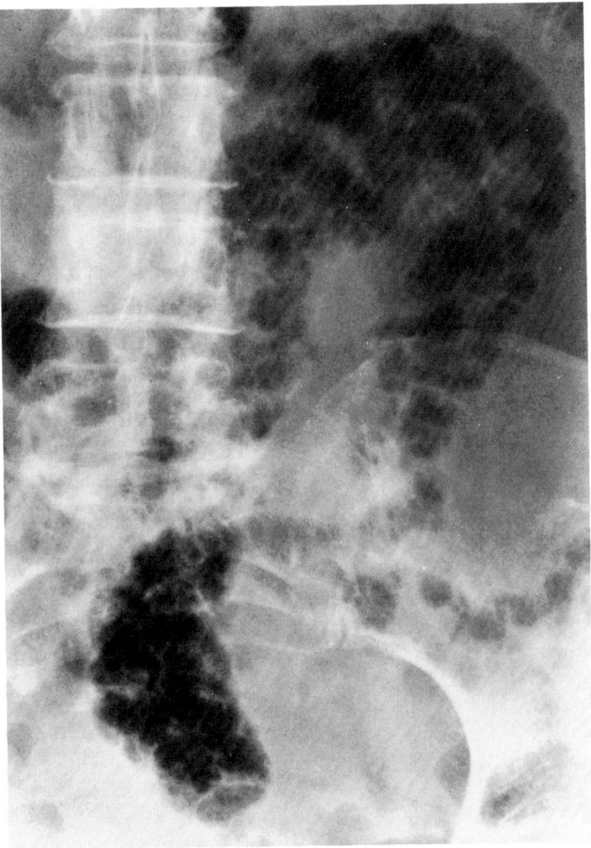

**Figure 50.16. Primary pneumatosis intestinalis in an asymptomatic man.** Radiolucent clusters of gas-filled cysts are seen along the contours of the bowel. (From Eisenberg,[31] with permission from the publisher.)

### Primary Pneumatosis Intestinalis

Primary pneumatosis intestinalis is a relative rare, benign condition characterized pathologically by multiple thin-walled, noncommunicating, gas-filled cysts in the subserosal or submucosal layer of the bowel. The overlying mucosa is entirely normal, as is the muscularis. The appearance of radiolucent clusters of cysts along the contours of the bowel is diagnostic of primary pneumatosis intestinalis. On barium examinations, the filling defects are seen to lie between the lumen (outlined by contrast material) and the water density of the outer wall of the bowel. The radiographic pattern of pneumatosis can simulate more severe gastrointestinal conditions. Small cysts may be confused with tiny polyps. Larger cysts can produce scalloped defects simulating inflammatory pseudopolyps or the thumbprinting seen with intramural hemorrhage. At times, the cysts of pneumatosis intestinalis concentrically compress the lumen, causing gas shadows that extend on either side of the bowel contour and surround a thin, irregular stream of barium, thus mimicking the appearance of an annular carcinoma. To differentiate pneumatosis intestinalis from these other conditions, it is important to note the striking lucency of the gas-filled cysts in contrast to the soft tissue density of an intraluminal or intramural lesion. Other distinguishing factors are the compressibility of the cysts on palpation and the frequent occurrence of asymptomatic pneumoperitoneum. Primary pneumatosis intestinalis usually requires no treatment and resolves spontaneously.[19]

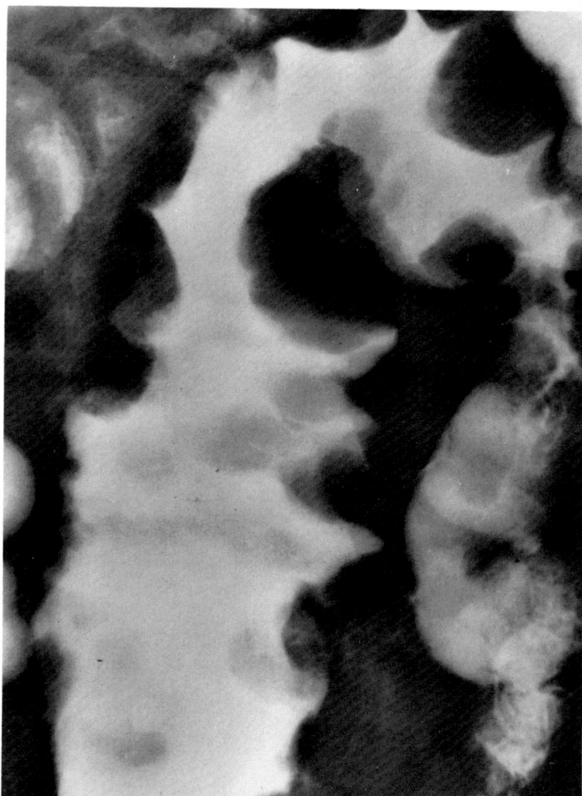

**Figure 50.17. Primary pneumatosis intestinalis in an asymptomatic elderly man.** The large gas-filled cysts produce scalloped defects in the colon simulating inflammatory pseudopolyps or the thumbprinting seen with intramural hemorrhage. (From Eisenberg,[31] with permission from the publisher.)

### Secondary Pneumatosis Intestinalis

Secondary pneumatosis intestinalis most commonly reflects gastrointestinal disease with bowel necrosis. In infants, pneumatosis intestinalis suggests an underlying necrotizing enterocolitis, which primarily occurs in premature or debilitated infants, most commonly affects the ileum and right colon, and has a very low survival rate. This condition is characterized by a frothy or bubbly appearance of gas in the wall of diseased bowel loops. The appearance often resembles that of fecal material in the right colon. However, it must be remembered that although this feces-like appearance is perfectly normal in adults, it is always abnormal in premature infants. The gas in the wall of the colon in necrotizing enterocolitis is probably related to mucosal necrosis and the subsequent passage of intraluminal gas into the bowel wall. This may be complicated by intraluminal gas-forming organisms that also penetrate the diseased mucosa to reach the inner layers of the intestinal wall. Dissection of air into the intrahepatic branches of the portal vein is an ominous sign.

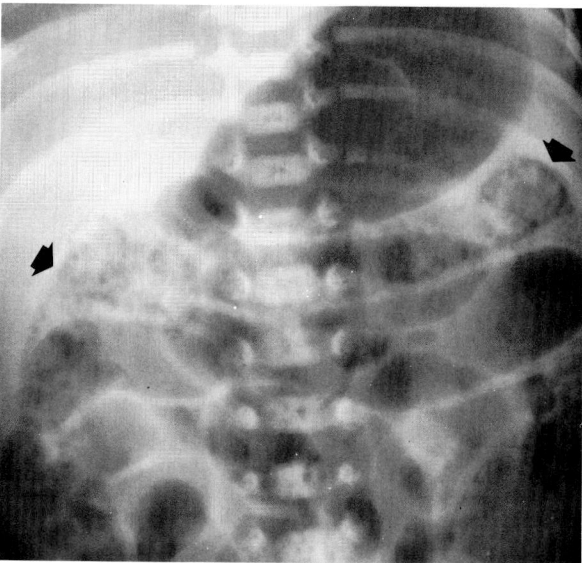

**Figure 50.18. Secondary pneumatosis intestinalis in a premature infant with underlying necrotizing enterocolitis.** The bubbly appearance of gas in the wall of diseased colon resembles fecal material (arrows); although this appearance is normal in adults, it is always abnormal in premature infants. (From Eisenberg,[31] with permission from the publisher.)

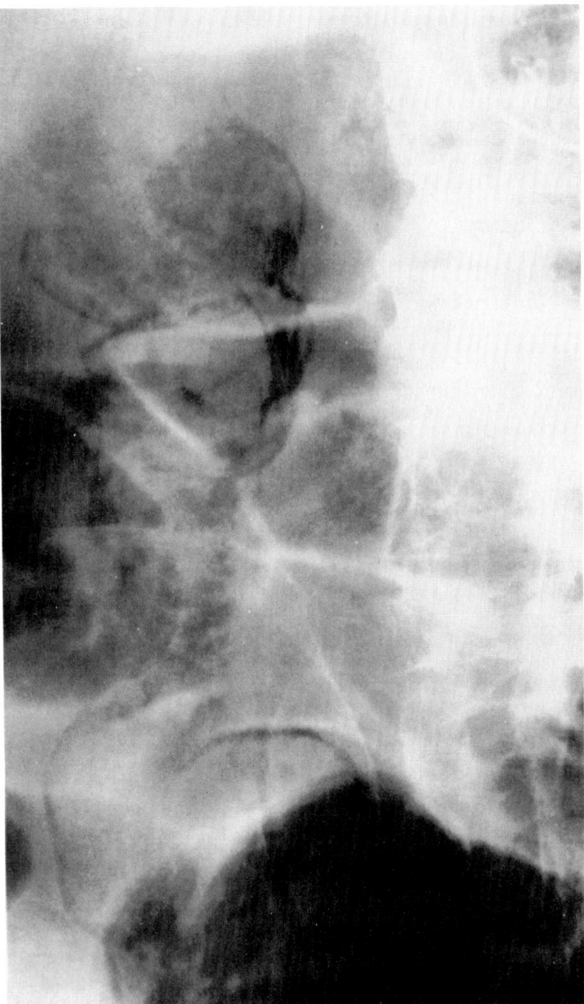

**Figure 50.19. Pneumatosis intestinalis secondary to mesenteric arterial thrombosis.** Crescentic linear collections of gas are seen in the walls of ischemic bowel loops.

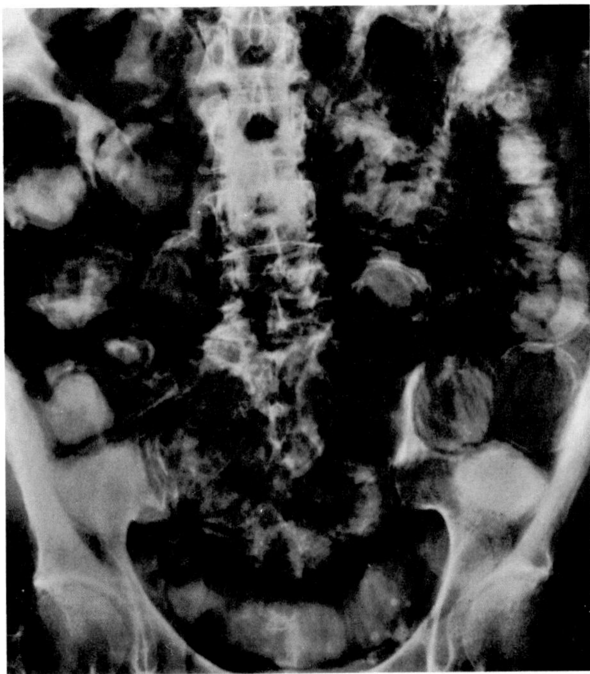

**Figure 50.20. Pneumatosis intestinalis secondary to primary amyloidosis of the small bowel.** Linear collections of gas are seen to parallel essentially the entire course of the small bowel. (From Eisenberg,[31] with permission from the publisher.)

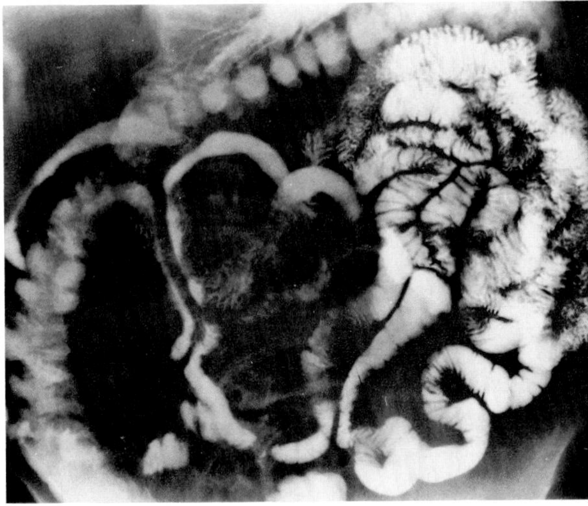

**Figure 50.21. Retractile mesenteritis.** The marked separation of small bowel loops remained constant on successive studies. (From Clemett and Tracht,[26] with permission from the publisher.)

In adults, secondary pneumatosis intestinalis suggests bowel necrosis due to mesenteric arterial or venous thrombosis. Secondary pneumatosis intestinalis may be related to mucosal ischemia, as in strangulating obstructions, or to mucosal destruction by infectious organisms or powerful corrosive agents.

Secondary pneumatosis intestinalis may also develop in the absence of necrosis of the bowel wall. Any gastrointestinal tract lesion that results in mucosal ulceration or intestinal obstruction (obstructive peptic ulcer disease, severe pyloric stenosis, scleroderma, inflammatory bowel disease) may be associated with gas in the bowel wall. Gas in the bowel wall is an uncommon complication of gastrointestinal endoscopy. Severe obstructive pulmonary disease can also be associated with the development of pneumatosis intestinalis. Partial bronchial obstruction and coughing presumably cause alveolar rupture, with subsequent dissection of gas along peribronchial and perivascular tissue planes into the mediastinum. Gas can then pass through the various hiatuses in the diaphragm to reach the retroperitoneal area, from which it dissects between the leaves of the mesentery to eventually reach subserosal and submucosal locations in the bowel wall.[19–24]

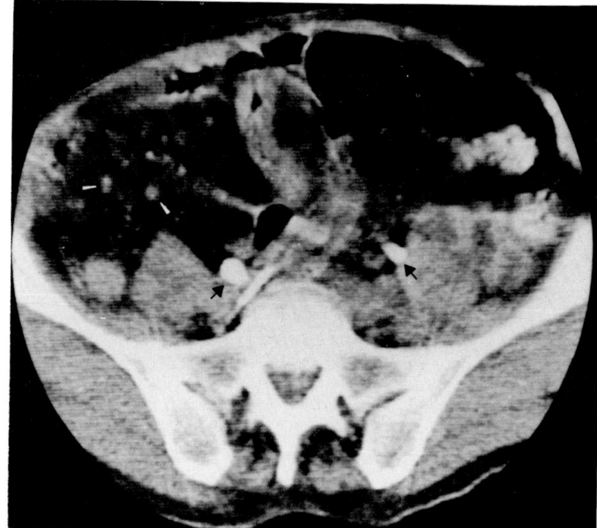

**Figure 50.22. Retractile mesenteritis.** CT scan shows a poorly defined but fairly localized fat-density mass within the right lower quadrant. Multiple small areas of increased density within the mass (white arrowheads) represent enhanced neurovascular bundles due to the administration of contrast material during the performance of the CT scan. Both ureters are well-opacified (black arrows). (From Seigel, Kuhns, Borlaza, et al,[28] with permission from the publisher.)

## RETRACTILE MESENTERITIS

Retractile mesenteritis is a slowly progressive mesenteric inflammatory process characterized by fibrofatty thickening and sclerosis of the mesentery. Three major pathologic features are usually present to some extent: fibrosis, inflammation, and fatty infiltration. When fibrosis is the dominant feature, the disease is known as retractile mesenteritis. When fatty infiltration is the most prominent feature, the condition is called lipomatosis or isolated lipodystrophy of the mesentery. Mesenteric panniculitis is the term used whenever chronic inflammation is the major pathologic feature. For all practical purposes, these three different terms describe the same process, or perhaps different stages of a single disease.

The small bowel mesentery is the usual site of origin of retractile mesenteritis, though the sigmoid mesentery can also be affected. The radiographic appearance is that of a diffuse mesenteric mass that separates and displaces small bowel loops. When prominent fibrosis causes adhesions and retractions, the bowel tends to be drawn into a central mass with kinking, angulation, and conglomeration of adherent loops.

CT demonstrates retractile mesenteritis as a localized fat-density mass containing areas of increased density.[25–28]

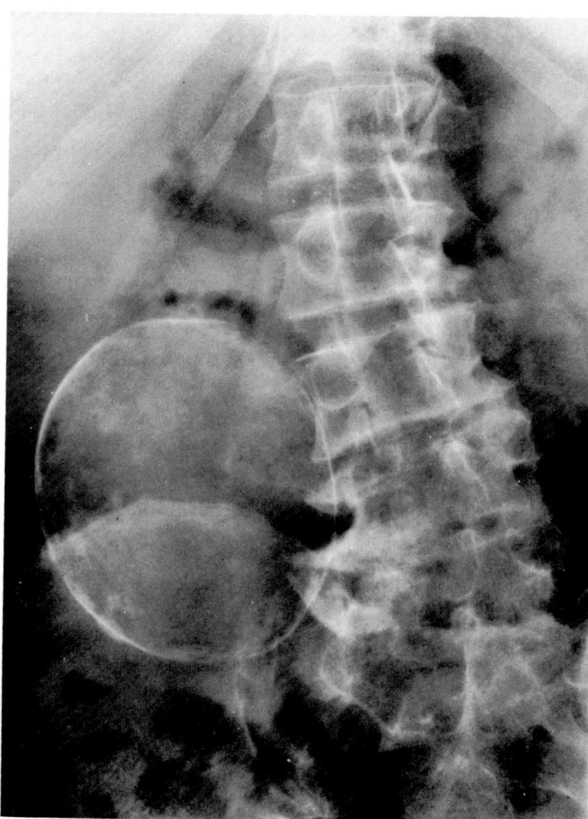

**Figure 50.23. Calcified mesenteric cyst.** (From Eisenberg,[31] with permission from the publisher.)

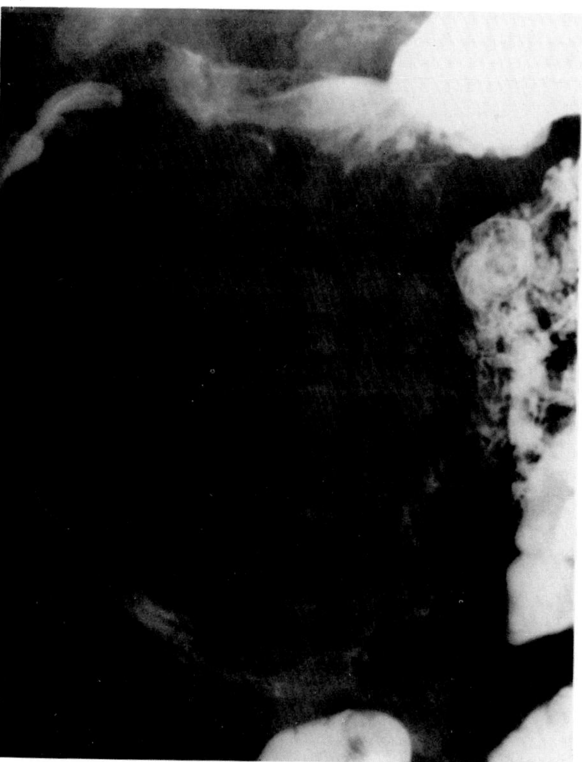

**Figure 50.24. Cystic lymphangioma of the mesentery.** There is dramatic widening of the duodenal sweep and scattered clumps of calcification within the lesion. (From Eisenberg,[31] with permission from the publisher.)

## MESENTERIC CYSTS

Cystic lymphangiomas are the most common type of mesenteric cysts. These benign, unilocular or multilocular cystic structures contain serous or chylous fluid. Cystic lymphangiomas can be the result of congenital or developmental misplacement and obliteration of draining lymphatics or be secondary to acquired lymphatic obstruction (e.g., trauma). As with lymphangiectasia, protein-losing enteropathy and hypoproteinemic edema may occur. Radiographically, mesenteric cysts appear as round, sharply outlined, soft tissue masses that displace adjacent viscera, especially the small bowel. Calcification of the wall may occur. On ultrasound, a mesenteric cyst appears as a well-outlined, sonolucent, transonic abdominal mass.[29–30]

### REFERENCES


1. Beneventano TC, Schein CJ, Jacobson HG: The roentgen aspects of some appendiceal abnormalities. *AJR* 96:344–360, 1966.
2. Figiel LS, Figiel SJ: Barium examination of cecum in appendicitis. *Acta Radiol* 57:469–480, 1962.
3. Schey WL: Use of barium in the diagnosis of appendicitis in children. *AJR* 118:95–103, 1973.
4. Soter CS: The contribution of the radiologist to the diagnosis of acute appendicitis. *Semin Roentgenol* 8:375–388, 1973.
5. Ghosh BC, Huvos AG, Whiteley HW: Pseudomyxoma peritonei. *Dis Colon Rectum* 15:420–425, 1972.
6. Gibbs NM: Mucinous cystadenocarcinoma of vermiform appendix with particular reference to mucocele and pseudomyxoma peritonei. *J Clin Pathol* 26:413–421, 1973.
7. Parsons J, Gray JF, Thorbjarmarson B: Pseudomyxoma peritonei. *Arch Surg* 101:545–549, 1970.
8. Seshul MB, Coulam CM: Pseudomyxoma peritonei: Computed tomography and sonography. *AJR* 136:803–806, 1981.
9. Dalinka MK, Lally JF, Azimi F, et al: Calcification in undifferentiated abdominal malignancies. *Clin Radiol* 26:115–119, 1975.
10. Teplick JG, Haskin ME, Alavi A: Calcified intraperitoneal metastases from ovarian carcinoma. *AJR* 127:1003–1006, 1976.
11. Moncada R, Cooper RA, Garces M, et al: Calcified metastases from malignant ovarian neoplasm: A review of the literature. *Radiology* 113:31–35, 1974.
12. Banner MP, Gohel VK: Peritoneal mesothelioma. *Radiology* 129:637–640, 1978.
13. Meyers MA: Metastatic seeding along the small bowel mesentery: Roentgen features. *AJR* 123:67–73, 1975.
14. Mueller PR, Ferrucci JT, Harbin WP, et al: Appearance of lymphomatous involvement of the mesentery by ultrasonography and body computed tomography: The "sandwich sign." *Radiology* 134:467–473, 1980.
15. Yeh HC: Ultrasonography of peritoneal tumors. *Radiology* 133:419–424, 1979.
16. Reuter K, Raptopoulous V, Reale F, et al: Diagnosis of peritoneal mesothelioma: Computed tomography, sonography, and fine-needle aspiration biopsy. *AJR* 140:1189–1194, 1983.
17. Levitt RG, Sagel FS, Stanley RJ: Detection of neoplastic involvement of the mesentery and omentum by computed tomography. *AJR* 131:835–838, 1978.
18. Jeffrey RB: Computed tomography of the peritoneal cavity and mesentery, in Moss AA, Gamsu G, Genant HK (eds): *Computed Tomography of the Body*. Philadelphia, WB Saunders, 1983, pp 955–986.
19. Marshak RH, Lindner AE, Maklansky D: Pneumatosis cystoides coli. *Gastrointest Radiol* 2:85–89, 1977.
20. Meyers MA, Ghahremani GG, Clements JL, et al: Pneumatosis intestinalis. *Gastrointest Radiol* 2:91–105, 1977.
21. Mueller CS, Morehead R, Alter A, et al: Pneumatosis intestinalis in collagen disorders. *AJR* 114:300–305, 1972.
22. Olmsted WW, Madewell JE: Pneumatosis cystoides intestinalis: A pathophysiologic explanation of the roentgenographic signs. *Gastrointest Radiol* 1:177–181, 1976.
23. Robinson AE, Grossman H, Brumely GW: Pneumatosis intestinalis in the neonate. *AJR* 120:333–341, 1974.
24. Seaman WB, Fleming RJ, Baker DH: Pneumatosis intestinalis of the small bowel. *Semin Roentgenol* 1:234–242, 1966.
25. Aach RD, Kahn LI, Frech RS: Obstruction of the small intestine due to retractile mesenteritis. *Gastroenterology* 54:594–598, 1968.
26. Clemett AR, Tracht DG: The roentgen diagnosis of retractile mesenteritis. *AJR* 107:787, 1969.
27. Tedeschi CG, Botta GC: Retractile mesenteritis. *N Engl J Med* 266:1035–1040, 1962.
28. Seigel RS, Kuhns LR, Borlaza GS, et al: Computed tomography and angiography in ileal carcinoid tumor and retractile mesenteritis. *Radiology* 134:437–440, 1980.
29. Leonidas JC, Kopel FB, Danese CA: Mesenteric cyst associated with protein loss in the gastrointestinal tract. *AJR* 112:150–154, 1971.
30. Mittelstaedt C: Ultrasonic diagnosis of omental cysts. *Radiology* 117:673–676, 1975.
31. Eisenberg RL: *Gastrointestinal Radiology: A Pattern Approach*. Philadelphia, JB Lippincott, 1983.
32. Harris JH, Harris WH: *The Radiology of Emergency Medicine*. Baltimore, Williams & Wilkins, 1981.
33. Levitt RG: Abdominal wall and peritoneal cavity, in Lee JKT, Sagel SS, Stanley RJ (eds): *Computed Body Tomography*. New York, Raven Press, 1983, pp 287–306.

# PART EIGHT
# DISORDERS OF THE HEPATOBILIARY SYSTEM AND PANCREAS

## Chapter Fifty-One
# Cirrhosis of the Liver: Major Sequelae

### FATTY INFILTRATION OF THE LIVER

Fatty infiltration of the liver is the result of excessive deposition of triglycerides, which occurs in cirrhosis and other hepatic disorders. Computed tomography (CT) is the most sensitive modality for demonstrating fatty infiltration. In normal individuals, the mean liver CT number is never lower than that of the spleen; in fatty infiltration, the liver has a much lower attenuation value than the spleen. A common finding is the appearance of the portal veins as high-density structures surrounded by a background of low density caused by hepatic fat, the opposite of the normal pattern of portal veins as low-density channels on noncontrast scans. In fatty infiltration of the liver due to cirrhosis, there often is prominence of the caudate lobe associated with shrinkage of the right lobe, which results from both actual caudate lobe enlargement and greater fibrotic scarring and shrinkage of the right hepatic lobe.[1-3]

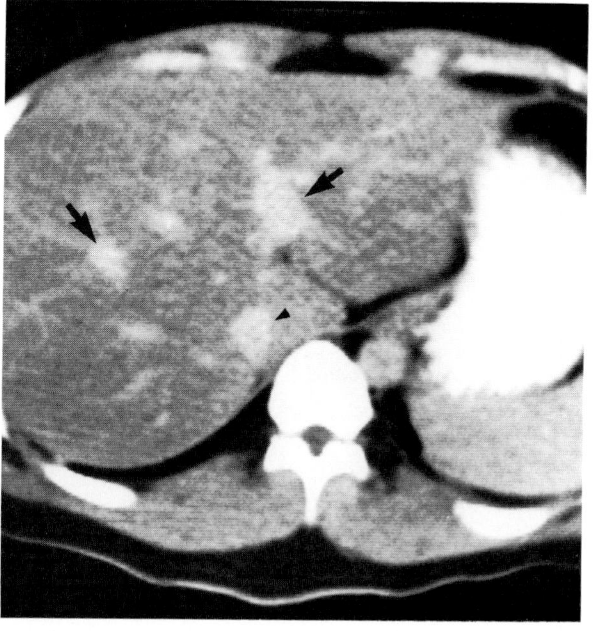

**Figure 51.1. Diffuse fatty infiltration of the liver resulting from congenital hypertriglyceridemia.** Precontrast CT scan demonstrates that the liver has a lower attenuation value than the spleen. The inferior vena cava (arrowhead) and portal vein (arrows) stand out as high-density structures. (From Moss,[3] with permission from the publisher.)

### HEPATIC VEIN OCCLUSION (BUDD-CHIARI SYNDROME)

Occlusion of the major hepatic veins (Budd-Chiari syndrome) or their small intrahepatic branches may lead to portal hypertension with hepatomegaly, ascites, and abdominal pain. Hepatic venous occlusion may be due to primary veno-occlusive disease with thrombosis or be secondary to obstruction of the veins by tumor, intrahepatic disease, polycythemia, or hypercoagulopathy states.

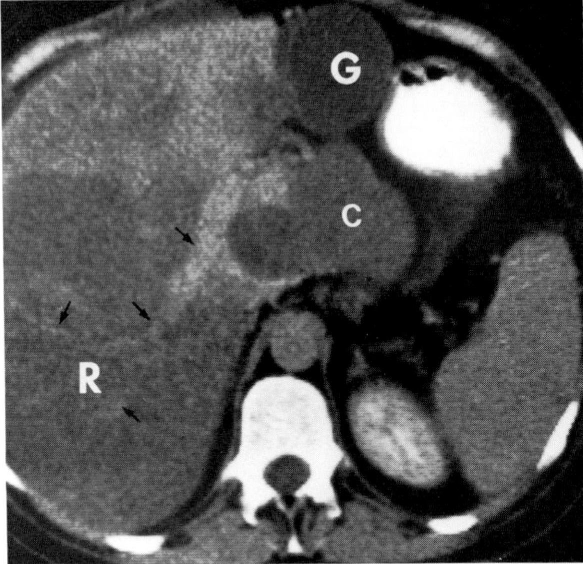


**Figure 51.2. Patchy fatty infiltration of the liver resulting from cirrhosis.** The right (R) and caudate (c) lobes of the liver are replaced by fat to a degree that makes the density almost equal to that of the gallbladder (G). The medial segment of the left hepatic lobe has a higher CT density but contains foci of low attenuation. The spleen is large and the caudate lobe is prominent. The portal vein (arrow) courses normally through the center of the right hepatic lobe, distinguishing fatty infiltration from a low-density tumor. (From Moss,[3] with permission from the publisher.)

Direct hepatic venography following passage of a catheter through the right atrium demonstrates complete obstruction of a major hepatic vein as a spider web vascular pattern radiating from the catheter tip with no opacification of the normal vein and its branches. Incomplete obstruction leads to the development of a coarse network of smaller intrahepatic veins that converge at the junction between the hepatic vein and the vena cava. The precise site of obstruction can be demonstrated if the catheter crosses the narrowed segment and opacifies the proximal obstructed portion of the vein.

CT following the administration of intravenous contrast material can demonstrate a patchy area of increased attenuation emanating from the retrohepatic part of the inferior vena cava in a fanlike distribution that reflects extensive collateral circulation within the liver. Ultrasound shows abnormal intrahepatic vascular structures that do not connect with the portal or hepatic venous systems or bile ducts; no normal hepatic veins are seen.[4–6]

## PORTAL VEIN OCCLUSION

Intrahepatic or extrahepatic occlusion of the portal vein most commonly is due to thrombosis secondary to preexisting portal hypertension. Inflammatory and neoplastic processes and trauma also can cause portal vein occlusion.

The venous phase of selective splenic or superior mesenteric arteriography demonstrates nonfilling or obstruction of the portal vein, often with opacification of collateral venous channels to systemic veins. CT shows portal vein thrombosis as a rounded filling defect within the portal vein that does not enhance after intravenous contrast administration. Calcified clot in the portal vein occasionally may be seen on plain radiographs as a linear opaque density crossing the vertebral column.[7,8]

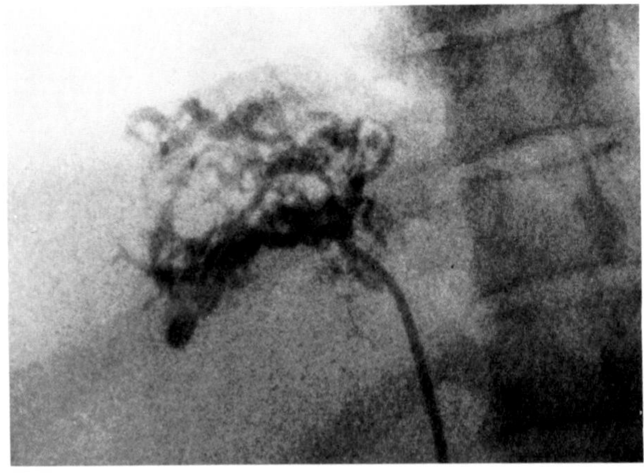

A

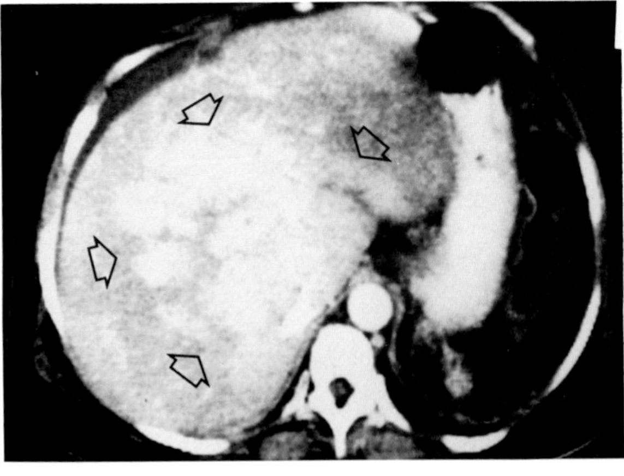

B

**Figure 51.3. Hepatic vein occlusion.** (A) A hepatic venogram shows obstruction of the major hepatic veins with a typical pattern of collateral vessels. (B) A CT scan following intravenous injection of contrast material shows a patchy, fan-shaped pattern of enhancement (arrows) characteristic of hepatic vein obstruction. The precontrast scan revealed a homogeneous, normal-appearing liver. (From Harter, Gross, Hilaire, et al,[4] with permission from the publisher.)

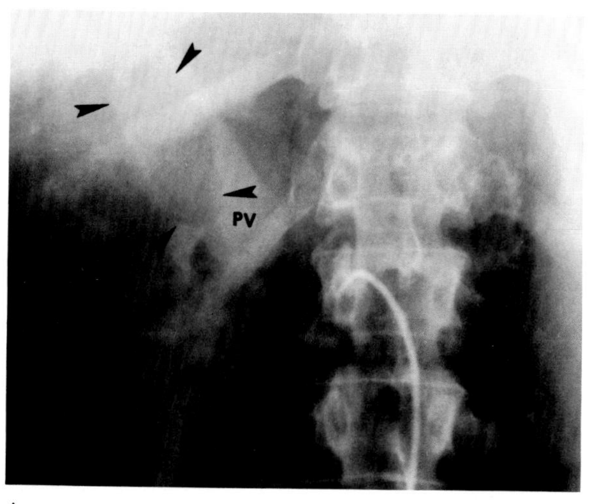


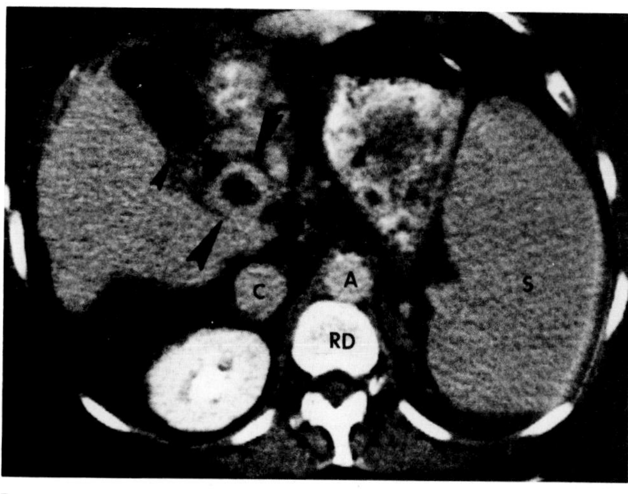


**Figure 51.4. Portal vein occlusion.** (A) Indirect splenoportography at 15 seconds after injection of contrast material into the superior mesenteric artery demonstrates obliteration of the splenic vein (SV) and a thrombus (arrows) within the opacified portal vein (PV). (B) A CT scan following contrast material injection shows the portal vein thrombus as a low-density region surrounded by a peripheral, ring-enhanced portal vein (large arrows). A gallstone is incidentally noted (small arrow). A = aorta; c = inferior vena cava. (From Vujic, Rogers, and LeVeen,[7] with permission from the publisher.)

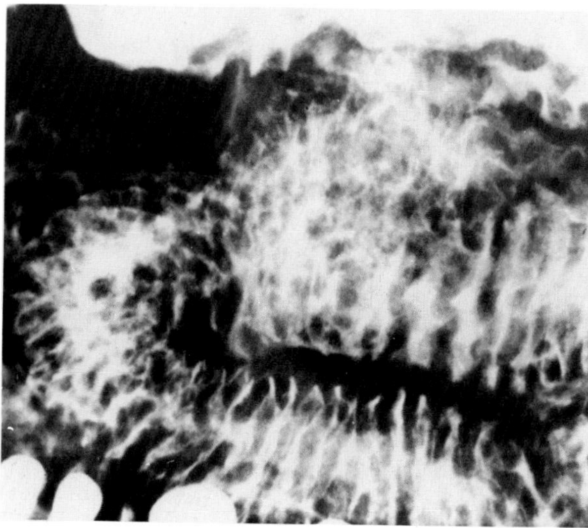

**Figure 51.5. Hypoproteinemia.** Regular thickening of small bowel folds in a patient with alcoholic cirrhosis. (From Eisenberg,[18] with permission from the publisher.)

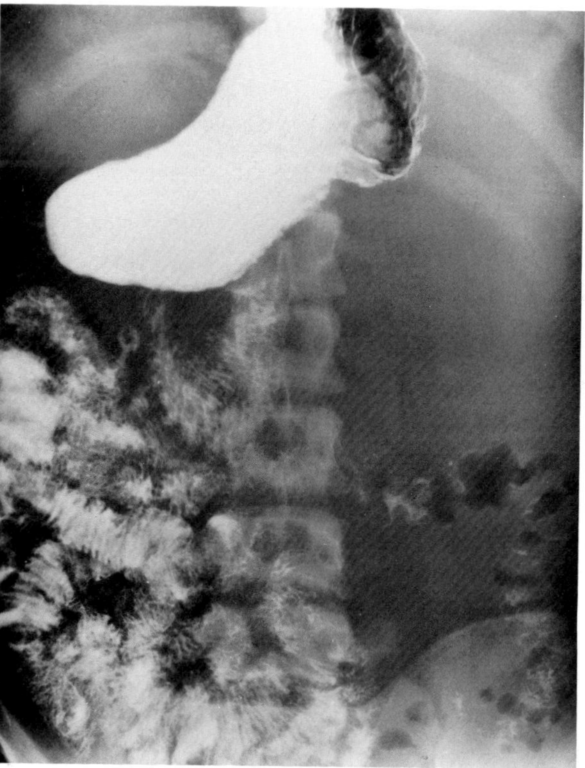

**Figure 51.6. Hypoproteinemia.** In this patient with alcoholic cirrhosis, regular thickening of small bowel folds is associated with marked splenomegaly due to portal hypertension. (From Eisenberg,[18] with permission from the publisher.)

## HYPOPROTEINEMIA

Hypoproteinemia most commonly results from alcoholic cirrhosis. When the serum albumin level is 2 g/dl or lower, cell-free infiltrate accumulates in the bowel wall and causes regular, uniform thickening of small bowel folds. Other causes of hypoproteinemia producing a similar radiographic pattern include kidney disease (nephrotic syndrome) and protein loss from the gastrointestinal tract (due to such protein-losing enteropathies as Ménétrier's disease, Crohn's disease, Whipple's disease, lymphoma, carcinoma, ulcerative colitis, and intestinal lymphangiectasia).[9,10]

## ASCITES

The accumulation of ascitic fluid in the peritoneal cavity can be caused by abnormalities in venous pressure, plasma colloid osmotic pressure, hepatic lymph formation, splanchnic lymphatic drainage, renal sodium and water excretion, or subperitoneal capillary permeability. In almost 75 percent of patients with ascites, the underlying disease is hepatic cirrhosis, in which there is an elevated portal venous pressure and decreased serum albumin level. Extrahepatic portal venous obstruction can also produce ascites, though it rarely does so in the absence of liver disease or hypoalbuminemia. The permeability of subperitoneal capillaries is increased in a broad spectrum of inflammatory and neoplastic diseases. Large amounts of intraperitoneal fluid are seen in patients with peritonitis secondary to infectious processes such as bacterial infection, tuberculosis, typhoid fever, and various fungal and parasitic infestations. Altered capillary permeability is probably responsible for the development of ascites in patients with peritoneal carcinomatosis, as well as in those with myxedema, ovarian disease, and allergic vasculitis. Primary or metastatic diseases of the lymphatic system can obstruct lymphatic vessels or the thoracic duct and produce ascites by blocking the normal splanchnic drainage.

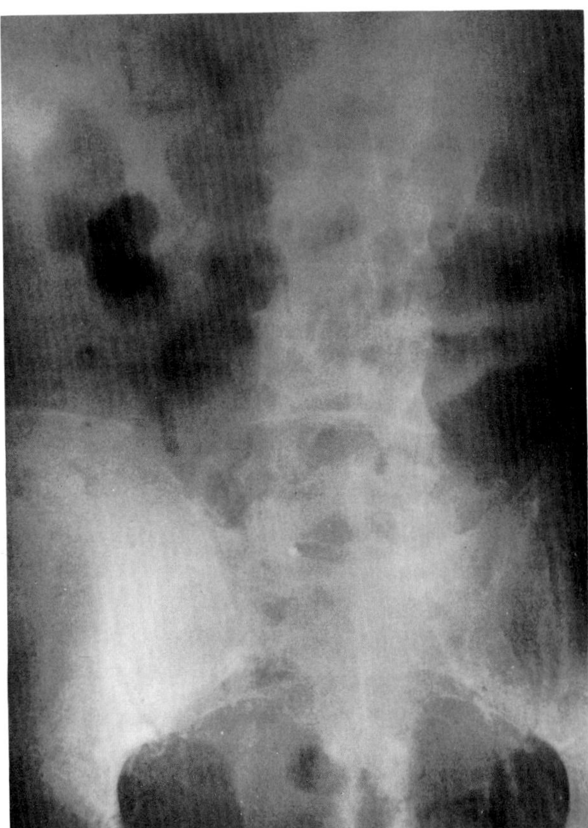

**Figure 51.7. Ascites.** General abdominal haziness (ground-glass appearance).

Large amounts of ascitic fluid are easily detectable on plain abdominal radiographs as a general abdominal haziness (ground-glass appearance). There may be elevation of the diaphragm due to the increased volume of abdominal contents. With the patient in a supine position, the peritoneal fluid tends to gravitate to dependent portions of the pelvis and accumulate within the pelvic peritoneal reflections, thus filling the recesses on both sides of the bladder and producing a symmetric density resembling dog's ears. Smaller amounts of fluid (800 to 1000 ml) may widen the flank stripe and obliterate the right lateral inferior margin of the liver (hepatic angle). On barium examination, the presence of ascitic fluid causes separation of small bowel loops.

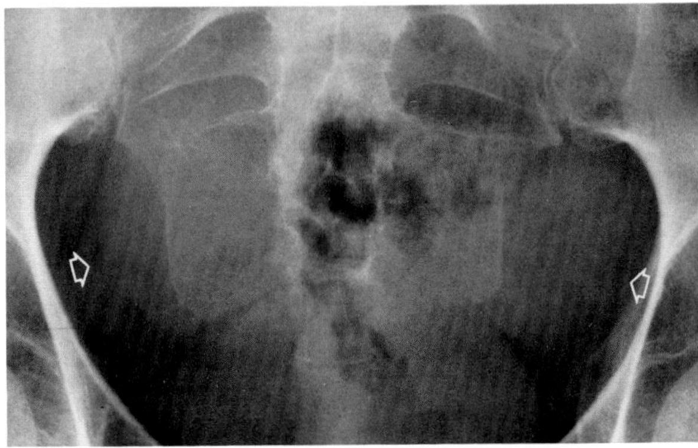

**Figure 51.8. Ascites.** A supine abdominal radiograph demonstrates the accumulation of a large amount of ascitic fluid within the pelvic peritoneal reflections (arrows).

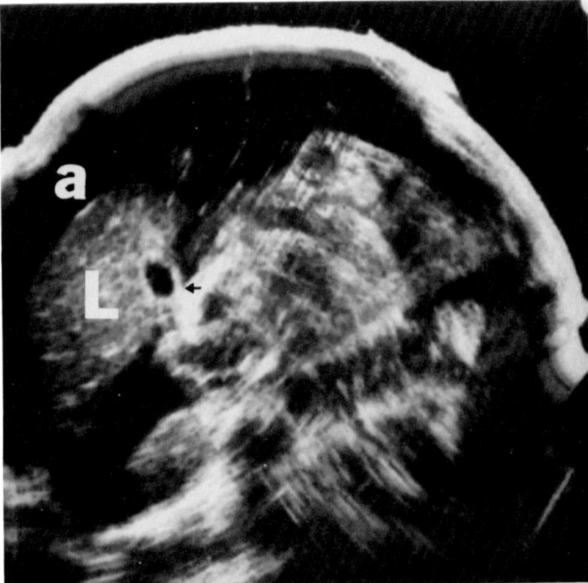


**Figure 51.9. Ascites.** A large amount of sonolucent ascitic fluid (a) separates the liver (L) and other soft tissue structures from the anterior abdominal wall. Note the relative thickness of the gallbladder wall (arrow).

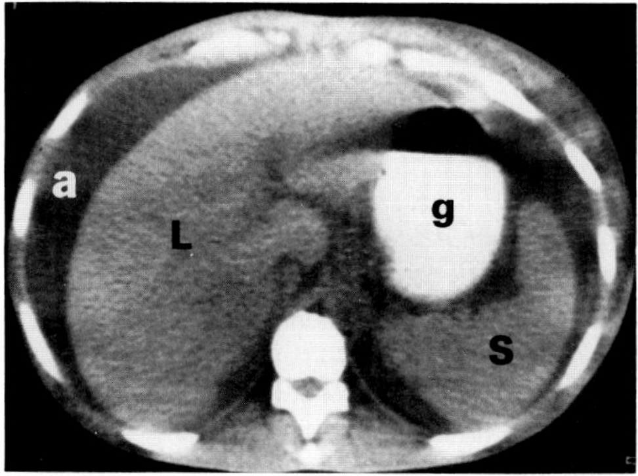


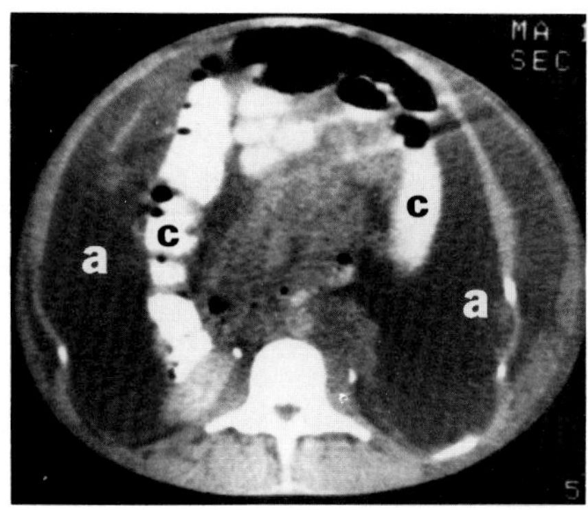


**Figure 51.10. Ascites.** (A) A CT scan through the upper abdomen shows low density ascitic fluid (a) lateral to the liver (L) and spleen (S) and separating these structures from the abdominal wall. g = barium in the stomach. (B) A CT scan through the lower abdomen shows a huge amount of low-density ascitic fluid (a), with medial displacement of the ascending and descending colon (c).

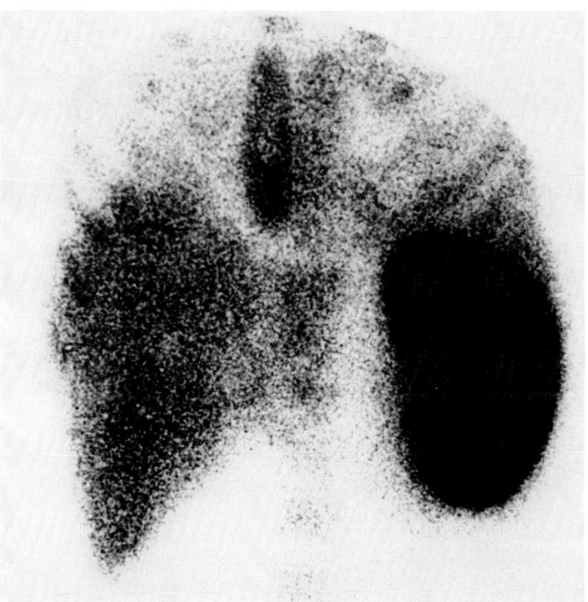

**Figure 51.11. Congestive splenomegaly.** A radionuclide liver-spleen scan shows hepatosplenomegaly with markedly increased uptake in the spleen, vertebrae, and ribs and decreased flow to the liver; the pattern is consistent with cirrhosis.

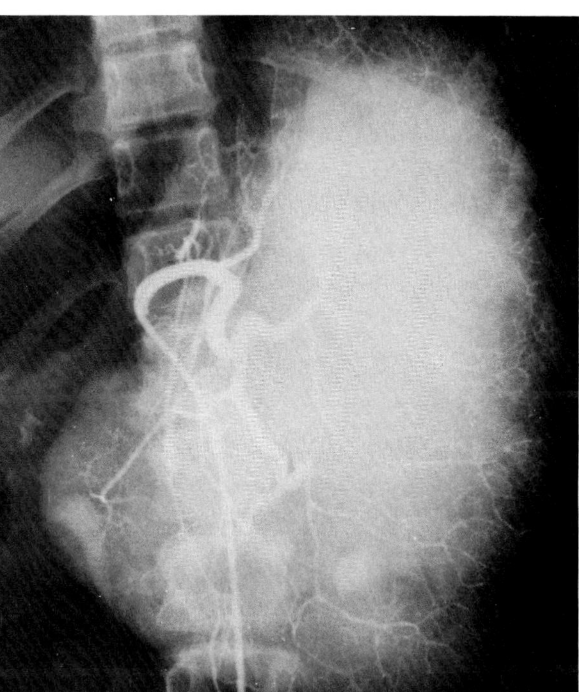

**Figure 51.12. Congestive splenomegaly.** A film from a splenic arteriogram in a patient with portal vein obstruction demonstrates a huge spleen with prolonged arterial and capillary phases.

Ultrasound demonstrates ascites as mobile, echo-free fluid regions shaped by adjacent structures. The smallest volumes of fluid in the supine patient appear first around the inferior tip of the right lobe of the liver, the superior right flank, and in the cul-de-sac of the pelvis. Fluid then collects in the paracolic gutters, as well as lateral and anterior to the liver. The distribution of fluid is not determined solely by gravity but is influenced by volume, the boundaries of peritoneal compartments, the fluid density, and intraperitoneal pressures, as well as the origin of the fluid.

CT shows ascites as an extravisceral collection of fluid with an attenuation value less than that of adjacent soft tissue organs. This modality is of special value in patients with noncirrhotic ascites, in whom it may detect the underlying lesion producing the excessive amounts of intra-abdominal fluid.[11–14]

## CONGESTIVE SPLENOMEGALY

Passive congestion, fibrosis, and siderosis produce enlargement of the spleen in most patients with significant and long-standing portal hypertension due to cirrhosis. Nevertheless, the absence of splenomegaly does not exclude portal hypertension, since splenic size correlates poorly with the level of portal pressure. In addition to evidence of splenomegaly on plain abdominal radiographs, radionuclide liver-spleen scans show significant uptake of radioactivity in the spleen and bone marrow and relatively decreased uptake in the liver, reflecting the shunting of blood away from the liver that develops in cirrhosis.

## ESOPHAGEAL VARICES

Esophageal varices are dilated veins in the subepithelial connective tissue that are most commonly the result of portal hypertension. Increased pressure in the portal venous system is usually secondary to cirrhosis of the liver. Other causes of portal hypertension include obstruction of the portal or splenic veins by carcinoma of the pancreas, pancreatitis, or inflammatory diseases of the retroperitoneum, and high-viscosity low-flow states (e.g., polycythemia), which predispose to intravascular thrombosis. In patients with portal hypertension, much of the portal blood cannot flow along its normal pathway through the liver to the inferior vena cava and then on to the heart. Instead, it must go by a circuitous collateral route through the coronary vein, across the esophagogastric hiatus, and into the periesophageal plexus before reaching the azygos or hemiazygos systems, superior vena cava, and right atrium. The periesophageal plexus communicates with veins in the submucosa of the esophagus and gastric cardia. Increased blood flow through these veins causes the development of esophageal (and gastric) varices.

Bleeding, the major complication of esophageal varices, is nearly as likely to occur in small varices as in large and more extensive lesions. Immediate death from exsanguination occurs in about 10 to 15 percent of patients with variceal bleeding. The diagnosis of esophageal varices in patients with cirrhotic liver disease implies significant portal venous hypertension and is an ominous sign. Up to 90 percent of the deaths from liver disease in patients with cirrhosis occur within 2 years of the diagnosis of varices.

The radiographic demonstration of esophageal varices requires precise technique. Multiple radiographs must be taken during the resting stage following a swallow of barium. Complete filling of the esophagus with barium may obscure varices; powerful contractions of the esophagus may squeeze blood out of the varices and make them impossible to detect. The characteristic radiographic appearance of esophageal varices is serpiginous thickening of folds, which appear as round or oval filling defects resembling the beads of a rosary. Initially, there is only mild thickening of folds and irregularity of the esophageal outline. Distension with barium hides these thickened folds and causes an irregularly notched (worm-eaten) appearance of the esophageal border. Once the typical tortuous, ribbon-like defects are visible, the radiographic diagnosis of esophageal varices is usually easy to make. In patients with severe portal hypertension, varices can be demonstrated throughout the entire thoracic esophagus.

Esophageal varices are infrequently demonstrated in the absence of portal hypertension. They have been observed in patients with noncirrhotic liver disease such as metastatic carcinoma, carcinoma of the liver, and congestive failure. "Downhill" varices are produced when venous blood from the head and neck cannot reach the heart because of an obstruction of the superior vena cava due to tumors or inflammatory disease in the medias-

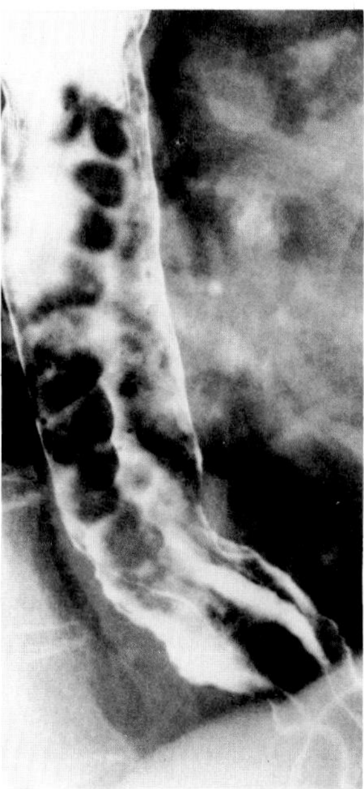

**Figure 51.13. Esophageal varices.** A line of round filling defects represents dilated veins in the subepithelial connective tissue.

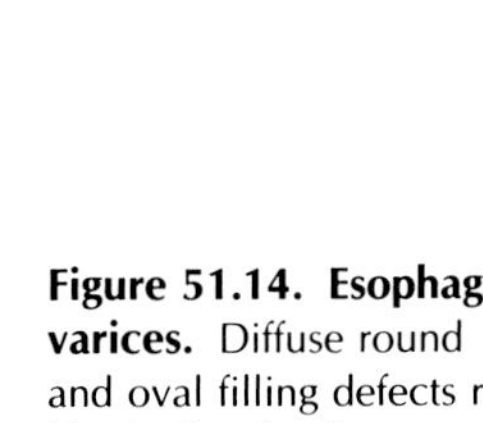

**Figure 51.14. Esophageal varices.** Diffuse round and oval filling defects resemble the beads of a rosary.

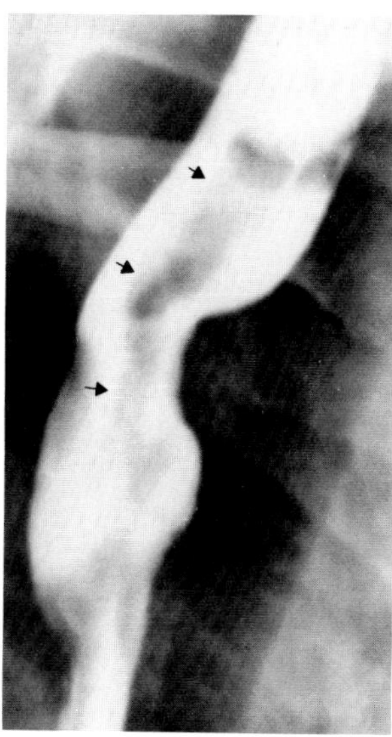

**Figure 51.15. "Downhill" varices.** Varices are confined to the upper esophagus (arrows) in a patient with carcinomatous obstruction of the superior vena cava.

tinum. In this situation, blood flows "downhill" through the azygos-hemiazygos system, the periesophageal plexus, and the coronary veins before eventually entering the portal vein, through which it flows to the inferior vena cava and the right atrium.[15-17]

## REFERENCES


1. Piekarski J, Goldberg HI, Royal SA, et al: Difference between liver and spleen CT numbers in the normal adult: Usefulness in predicting the presence of diffuse liver disease. *Radiology* 137:727–735, 1980.
2. Harbin WP, Robert NJ, Ferrucci JT: Diagnosis of cirrhosis based on regional changes in hepatic morphology: A radiological and pathological analysis. *Radiology* 135:273–281, 1980.
3. Moss AA: Computed tomography of the hepatobiliary system, in Moss AA, Gamsu G., Genant HK (eds): *Computed Tomography of the Body*. Philadelphia, WB Saunders, 1983.
4. Harter LP, Gross BH, St. Hilaire J, et al: CT and sonographic appearance of hepatic vein obstruction. *AJR* 139:176–178, 1982.
5. Deutsch V, Rosenthal T, Adar R, et al: Budd-Chiari syndrome: The study of angiographic findings and remarks on etiology. *AJR* 116:430–439, 1972.
6. Kreel L, Freston JW, Clain D: Vascular radiology in Budd-Chiari syndrome. *Br J Radiol* 40:755–759, 1967.
7. Vujic I, Rogers CI, LeVeen HH: Computed tomographic detection of portal vein thrombosis. *Radiology* 135:697–698, 1980.
8. Dotter RF: Extrahepatic portal obstruction in childhood and its angiographic diagnosis. *AJR* 112:143–150, 1971.
9. Marshak RH, Khilnani MT, Eliasoph J, et al: Intestinal edema. *AJR* 101:379–387, 1967.
10. Marshak RH, Wolf BS, Cohen N, et al: Protein-losing disorders of the gastrointestinal tract: Roentgen features. *Radiology* 77:893–906, 1961.
11. Keeffe EJ, Gagliardi RA, Pfister RC: The roentgenographic evaluation of ascites. *AJR* 101:388–396, 1967.
12. Margulies M, Stoane L: Hepatic angle in roentgen evaluation of peritoneal fluid. *Radiology* 88:51–56, 1967.
13. Gefter WB, Arger PH, Edell SL: Sonographic patterns of ascites. *Semin Ultrasound* 2:226–232, 1981.
14. Jolles H, Coulam CM: CT of ascites. *AJR* 135:315–322, 1980.
15. Cockerill EM, Miller RE, Chernish SM, et al: Optimal visualization of esophageal varices. *AJR* 126:512–523, 1976.
16. Nelson SW: The roentgenologic diagnosis of esophageal varices. *AJR* 77:599–611, 1957.
17. Felson B, Lessure AP: "Downhill" varices of the esophagus. *Dis Chest* 46:740–746, 1964.
18. Eisenberg, RL: *Gastrointestinal Radiology: A Pattern Approach*. Philadelphia, JB Lippincott, 1983.

## Chapter Fifty-Two
# Tumors of the Liver

## HEPATOCELLULAR CARCINOMA

In the United States, primary liver cell carcinoma most commonly occurs in patients with underlying diffuse hepatocellular disease, especially alcoholic or postnecrotic cirrhosis. The clinical presentation varies from mild right upper quadrant discomfort and weight loss to hemorrhagic shock from massive intraperitoneal bleeding, which reflects rupture of the tumor into the peritoneal cavity. Invasion of the biliary tree may produce obstructive jaundice. Although primary liver carcinoma is relatively infrequent in North and South America and Europe, the incidence of hepatocellular carcinoma is dramatically higher in Africa and Asia, where this tumor may account for up to one-third of all types of malignancy.

Computed tomography (CT) is the modality of choice in diagnosing hepatocellular carcinoma. The tumor appears as a large mass, with an attenuation value close to normal parenchyma, that tends to alter the contour of the liver by projecting beyond its outer margin. Following the bolus administration of intravenous contrast material, there usually is dense, diffuse, and nonuniform enhancement of the tumor. Unlike metastases, hepatocellular carcinoma tends to be solitary or produce a small number of lesions. Demonstration of one or a few large focal lesions in association with a pattern of generalized cirrhosis strongly suggests the possibility of hepatocellular carcinoma complicating underlying diffuse liver disease.

In the past, selective celiac or hepatic angiography was the major modality for diagnosing hepatocellular carcinoma. The tumor is usually highly vascular and produces a large conglomerate mass or a diffuse nodular pattern. Characteristic features of hepatocellular carcinoma are its tendency to invade the hepatic and portal venous systems and a high incidence of arteriovenous shunting.

Ultrasound can be used to detect hepatocellular carcinoma, as well as to identify tumor thrombus involving the inferior vena cava and portal and hepatic veins. However, since many patients with hepatocellular carcinoma have isoechoic satellite lesions, ultrasound may be of limited value in determining the extent of parenchymal involvement. Hepatic scintigraphy with $^{99m}$Tc sulfur colloid typically demonstrates hepatocellular carcinoma as a photopenic area, which has strong uptake of isotope on gallium scans. However, the degree of spatial resolution afforded by hepatic scintigraphy often is insufficient for the precise detection of small lesions.

Although the overall prognosis of hepatocellular carcinoma remains bleak, radiographic efforts should be directed toward determining whether the patient can successfully undergo hepatic resection. CT with dynamic bolus scanning is the best technique for the precise de-

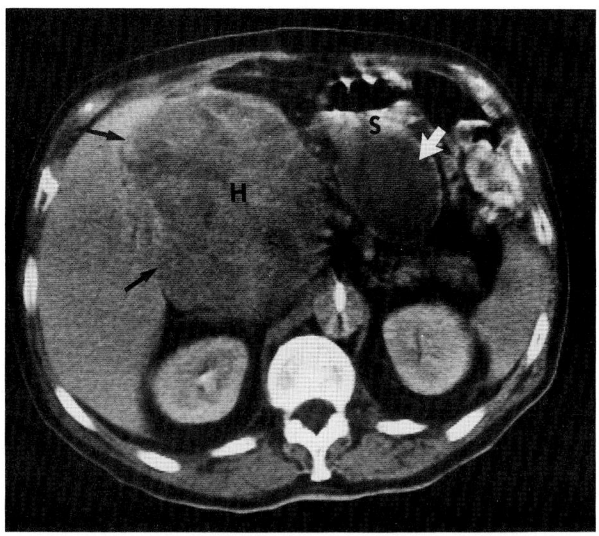


A

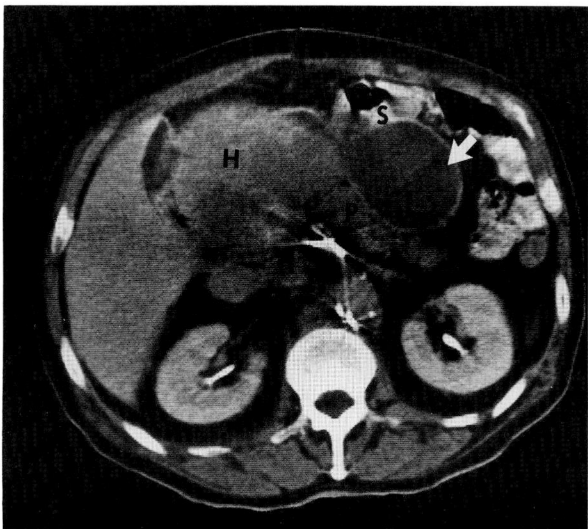


B

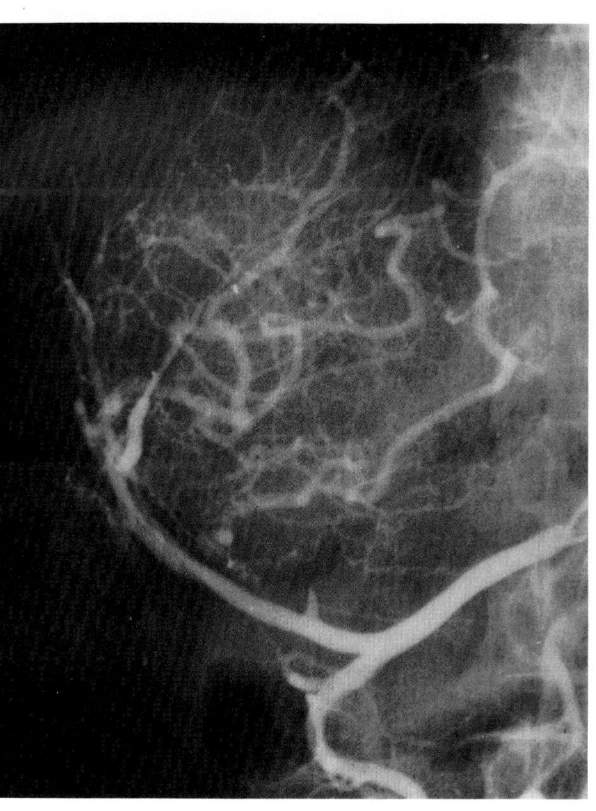

C

**Figure 52.1. Hepatocellular carcinoma (H).** (A) A CT scan shows a huge mass with an attenuation value slightly less than that of normal liver. The black arrows point to the hepatoma–normal liver interface. Of incidental note is a pancreatic pseudocyst (white arrow) in the lesser sac between the stomach (S) and pancreas. (B) On a slightly lower scan there is absence of the fat plane surrounding the head of the pancreas (P), indicating invasion of the pancreas by the tumor. (C) A hepatic arteriogram demonstrates diffuse neovascularity within the mass.

termination of the extent of parenchymal involvement and the lobar distribution of tumor. This modality also can determine extrahepatic spread to adjacent organs and lymph nodes, as well as tumor thrombus within the portal or hepatic veins. If the tumor appears resectable on CT, arteriography is usually indicated to demonstrate the precise surgical anatomy and to detect small, staining tumor nodules that cannot be identified with noninvasive imaging techniques.[1–7]

## BENIGN TUMORS OF THE LIVER

### Hemangioma

Cavernous hemangiomas are the most common benign tumors of the liver. Most are single, small, and asymptomatic and are found incidentally at surgery or autopsy or during unrelated radiographic procedures. Large symptomatic hemangiomas may present as palpable

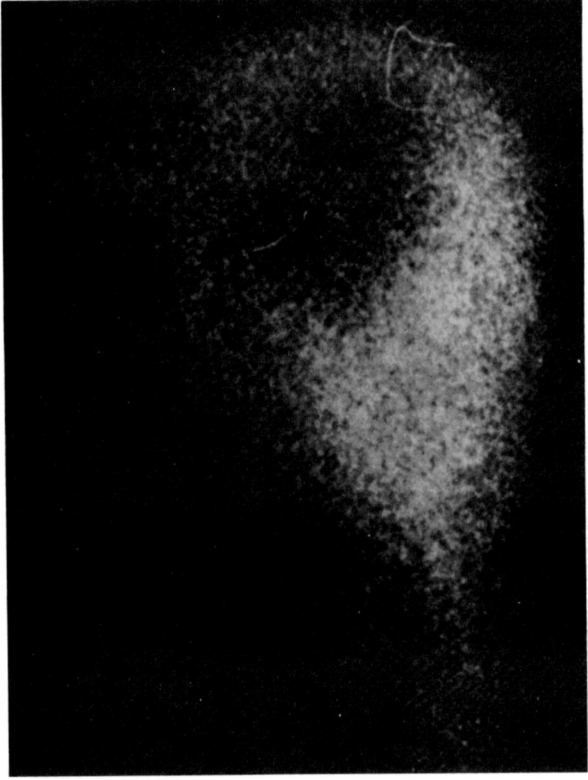
A

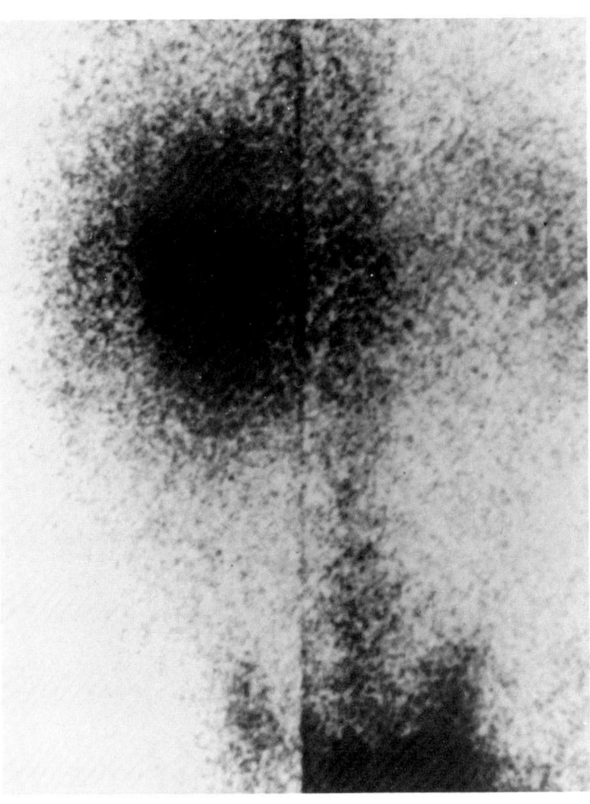
B

**Figure 52.2. Radionuclide scanning for hepatocellular carcinoma.** (A) A $^{99m}$Tc sulfur colloid scan (posterior view) demonstrates the tumor as a large defect in the liver. (B) A $^{67}$Ga scan shows marked uptake of the radionuclide in the distribution of the colloid scan abnormality. (From Scheible,[25] with permission from the publisher.)

masses; spontaneous rupture infrequently causes massive intraperitoneal hemorrhage.

Hepatic arteriography is the most reliable method for diagnosing cavernous hemangiomas. The intensely hypervascular tumor is fed by normal arterial branches and contains ringlike clusters of small vascular spaces in which contrast material pools during the arterial phase and lasts well into the venous phase. The portal vein is normal, and there is no evidence of arteriovenous shunting.

Ultrasound demonstrates small hemangiomas as discrete echogenic masses within the liver. Larger tumors have a variable hypo- or hyperechoic sonographic appearance that is indistinguishable from other hepatic neoplasms. Noncontrast CT scans show a cavernous hemangioma as a well-circumscribed, low-density area. Following bolus injection of intravenous contrast material, the periphery of the lesion becomes hyperdense, and serial scans show a centripetally advancing border of enhancement as the central area of low density becomes progressively smaller. On $^{99m}$Tc radionuclide scans, a hemangioma appears as a nonspecific area of decreased isotope uptake. Dynamic blood pool imaging may reveal the increased tumor uptake, though both false-negative and false-positive diagnoses may result.[8–12]

## Focal Nodular Hyperplasia

Focal nodular hyperplasia is a controversial entity that appears to represent an uncommon benign tumor of the liver composed of normal hepatocytes and Kupffer cells. A characteristic morphologic feature of the lesion is a central stellate, fibrous scar with peripherally radiating septa that divide the mass into lobules. Focal nodular hyperplasia often is subcapsular in location or pedunculated along the inferior margin of the liver; it infrequently is situated deep within the hepatic parenchyma.

At arteriography, focal nodular hyperplasia appears as solitary or multiple hypervascular masses that typically have a central feeding vessel emanating from the region of the central scar and a stellate pattern of increased vascularity. In some patients, the vascular supply may arise peripherally and enter directly into the mass without a spoked-wheel effect.

The appearance of focal nodular hyperplasia on CT and ultrasound is often nonspecific. The lesion usually appears as a low-density mass on noncontrast scans and often demonstrates significant enhancement following the intravenous administration of contrast material. Occasionally, the central scar may be large enough to ap-

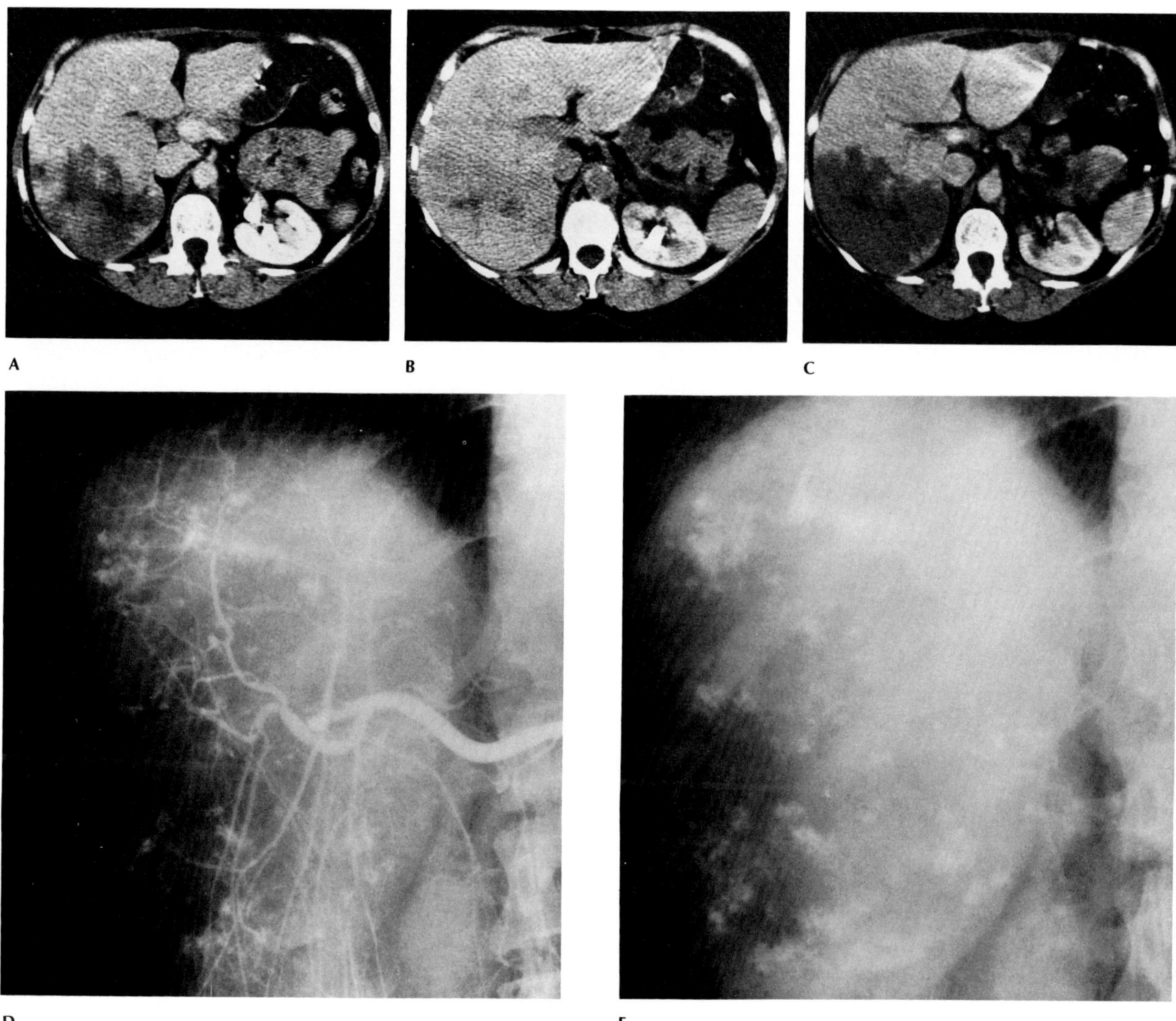


**Figure 52.3. Cavernous hemangioma.** (A) An initial CT scan following bolus injection of contrast material demonstrates a large low-density lesion in the posterior segment of the right lobe of the liver. (B,C) Delayed CT scans show progressive enhancement of the lesion until it becomes nearly isodense with normal hepatic parenchyma. (D) The arterial phase of a hepatic arteriogram shows characteristic pooling of contrast material within sinusoidal spaces, which lasts through the venous phase (E).

pear as a relatively low-density stellate area within a generally enhancing mass. On ultrasound, focal nodular hyperplasia may appear as hypo- or hyperechoic regions or areas of mixed echogenicity.

The presence of Kupffer cells within the mass of focal nodular hyperplasia permits a variable degree of isotope uptake on radionuclide scans, ranging from a focus of decreased uptake to an area of increased activity compared with the normal surrounding parenchyma. The normal uptake of $^{99m}$Tc virtually excludes other hepatic neoplasms that do not contain Kupffer cells.[12–16]

## Adenoma

Hepatic adenomas are benign lesions that generally are solitary and composed entirely of hepatocytes without Kupffer cells. The tumor has a strong hormonal association and most often occurs in women taking oral contraceptives. When seen in men, the tumor may be associated with hormonal therapy of carcinoma of the prostate. Spontaneous hemorrhage, sometimes of life-threatening proportions, is relatively common.

The CT and ultrasound features of hepatic adenomas are nonspecific. On CT scans, the tumor appears as a low-density or almost isodense mass that demonstrates a variable degree of enhancement. Recent hemorrhage appears as a central area of increased attenuation, while a remote bleeding episode may produce a central area of low attenuation reflecting an evolving hematoma or central cellular necrosis. On ultrasound, hepatic adenomas tend to be echogenic but indistinguishable from other benign and malignant conditions.

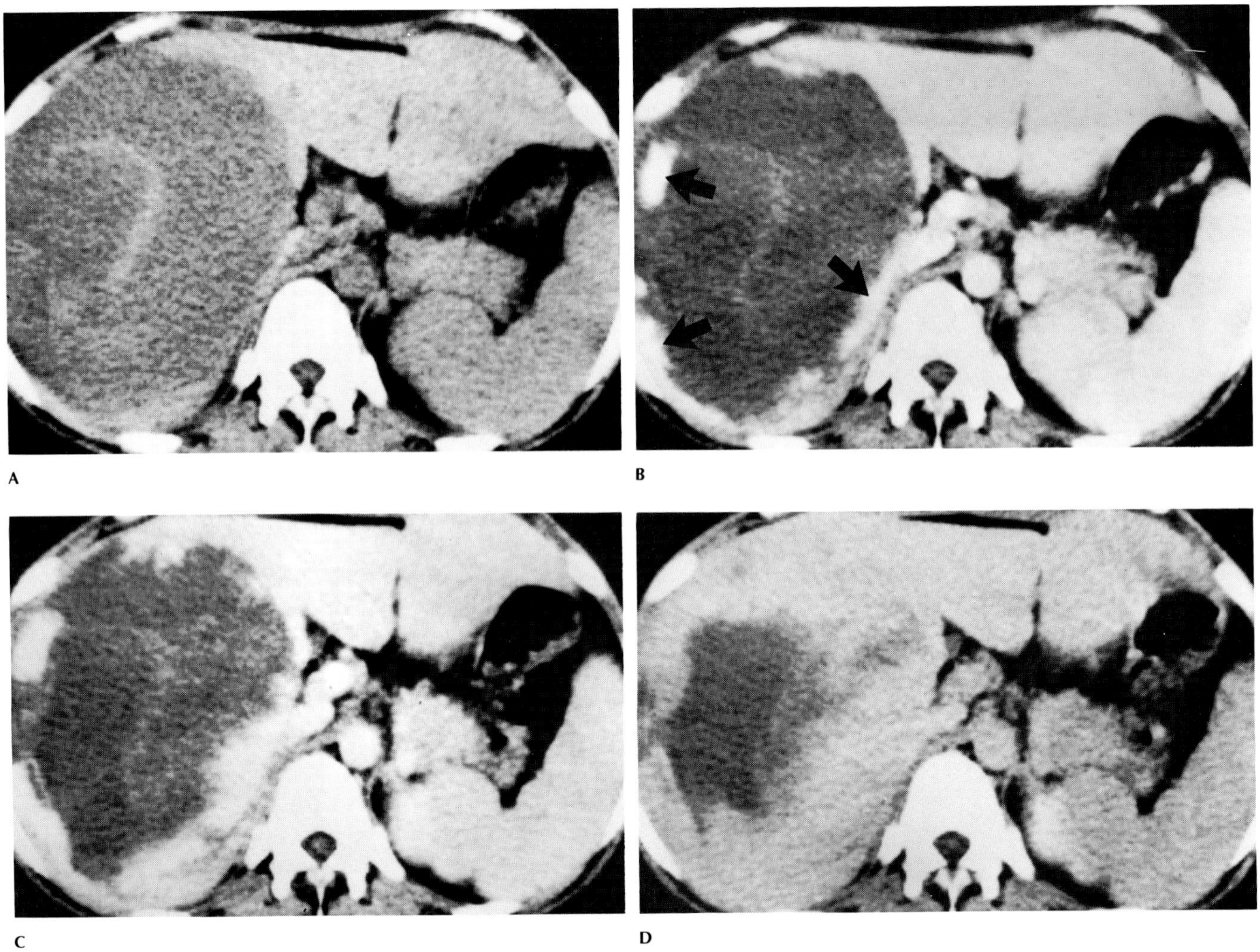

**Figure 52.4. Cavernous hemangioma.** (A) An initial noncontrast CT scan demonstrates a large low-density mass filling the right lobe of the liver. (B) A CT scan obtained 30 seconds after a bolus injection of contrast material demonstrates typical intense peripheral enhancement (arrows). (C) A CT scan 1 minute after contrast injection shows the gradual spread of contrast material toward the center of the hemangioma. (D) A CT scan obtained 4 minutes later reveals continued progressive contrast enhancement of the center of the hemangioma. The peripheral part of the hemangioma is almost isodense with the normal liver, making the hemangioma appear smaller than on the initial scan. (From Moss,[26] with permission from the publisher.)

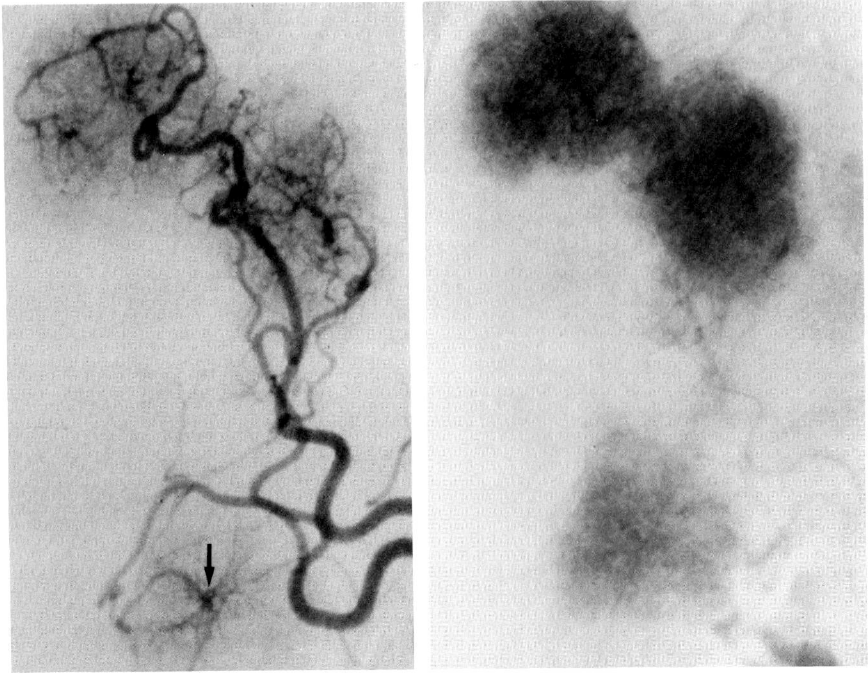

**Figure 52.5. Focal nodular hyperplasia.** (A) A film from the arterial phase of a hepatic arteriogram demonstrates several hypervascular masses. Note the spoked-wheel appearance (arrow) of the lower mass. (B) There is dense staining of the lesion during the capillary phase of the arteriogram.

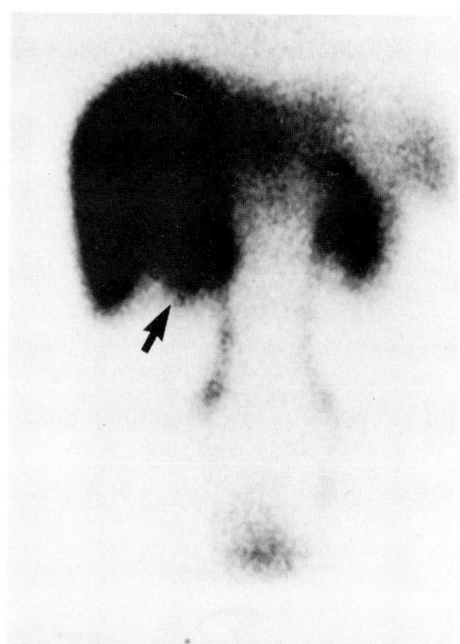

A

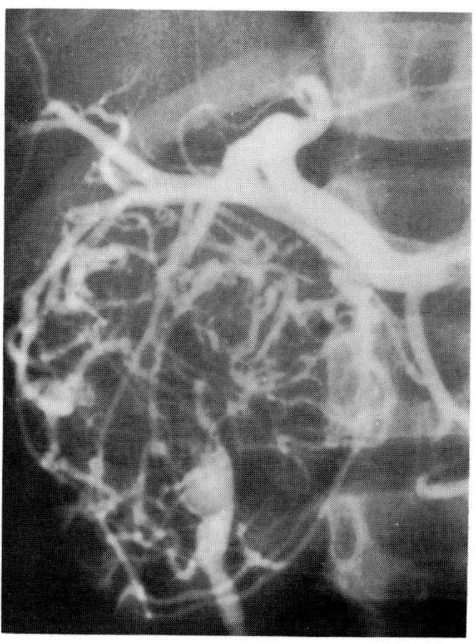

B

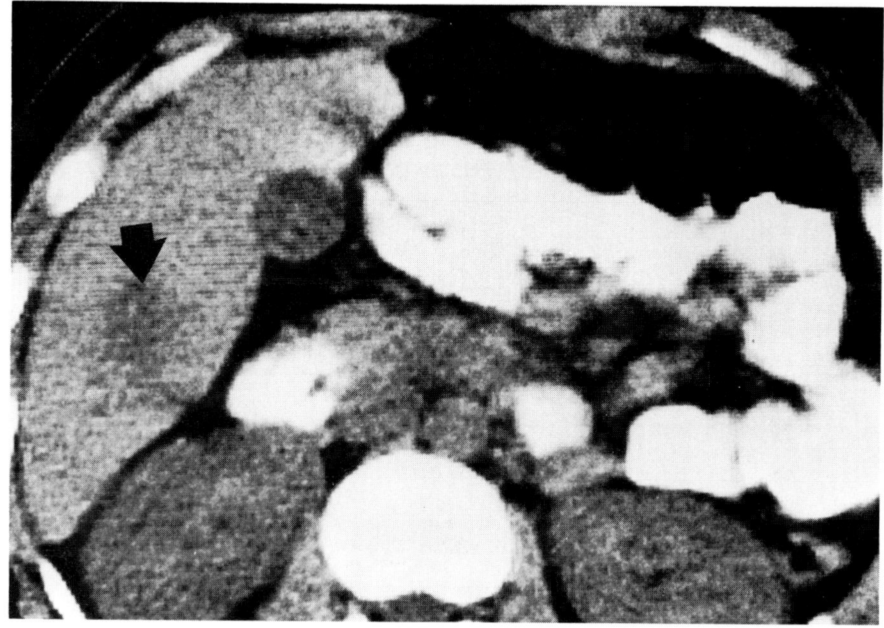

C

**Figure 52.6. Focal nodular hyperplasia.** (A) A radionuclide sulfur colloid scan demonstrates a mass lesion (arrow) that arises in the right lobe of the liver and concentrates the isotope normally. (B) An arteriogram shows a hypervascular mass with vessels entering from the periphery. (C) A noncontrast CT scan reveals the zone of focal nodular hyperplasia as an area of low attenuation (arrow), which is indistinguishable from a primary or secondary hepatic neoplasm. The lesion became isodense following the administration of contrast material. (From Moss,[26] with permission from the publisher.)

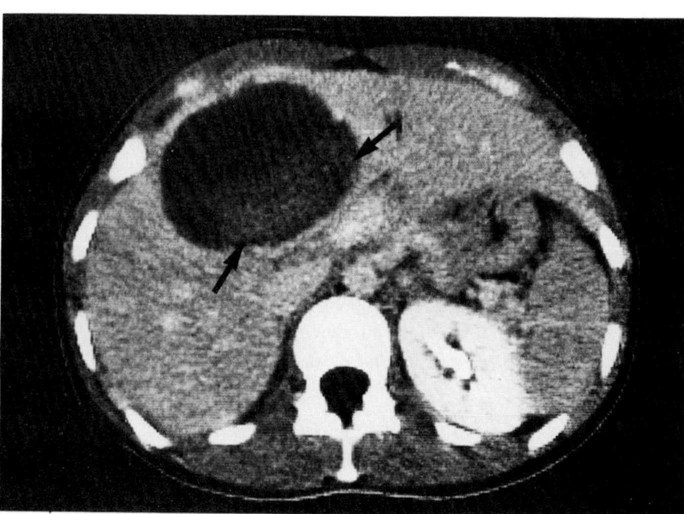

**Figure 52.7. Hepatic adenoma.** A CT scan demonstrates a large low-density mass within the liver. Note the area of higher density (arrows), which represents a blood clot, along the posterior aspect of the lesion.

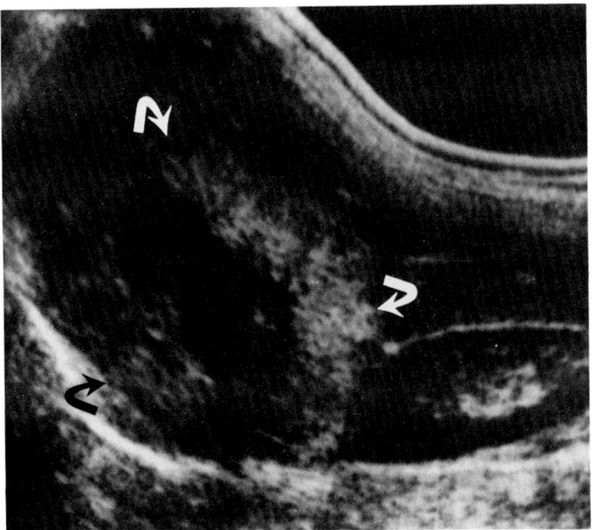

**Figure 52.8. Hepatic adenoma.** A longitudinal sonogram shows a large dense lesion with a central fluid area in the right lobe of the liver (arrows). (From Scheible,[25] with permission from the publisher.)

Because hepatic adenomas do not contain Kupffer cells, the lesions invariably appear as photopenic defects on radionuclide hepatic scans. On arteriograms, the tumor is about equally likely to appear as a hypervascular or a hypovascular mass.[12–16]

## Hemangioendothelioma

Hemangioendotheliomas are the most common hepatic lesions producing symptoms during infancy. The vast majority of these tumors present before 6 months of age. Although they are generally considered benign, there have been rare reports of distant metastases. Extensive arteriovenous shunting within the lesion may lead to high-output congestive heart failure.

Plain abdominal radiographs may demonstrate a soft tissue mass and speckled calcifications. Although sonography is useful in localizing the mass to the liver, the tumor has a nonspecific pattern and may appear as a hypoechoic, complex, or hyperechoic lesion. Dynamic CT scanning using a bolus injection of intravenous contrast material typically shows focal areas of low atten-

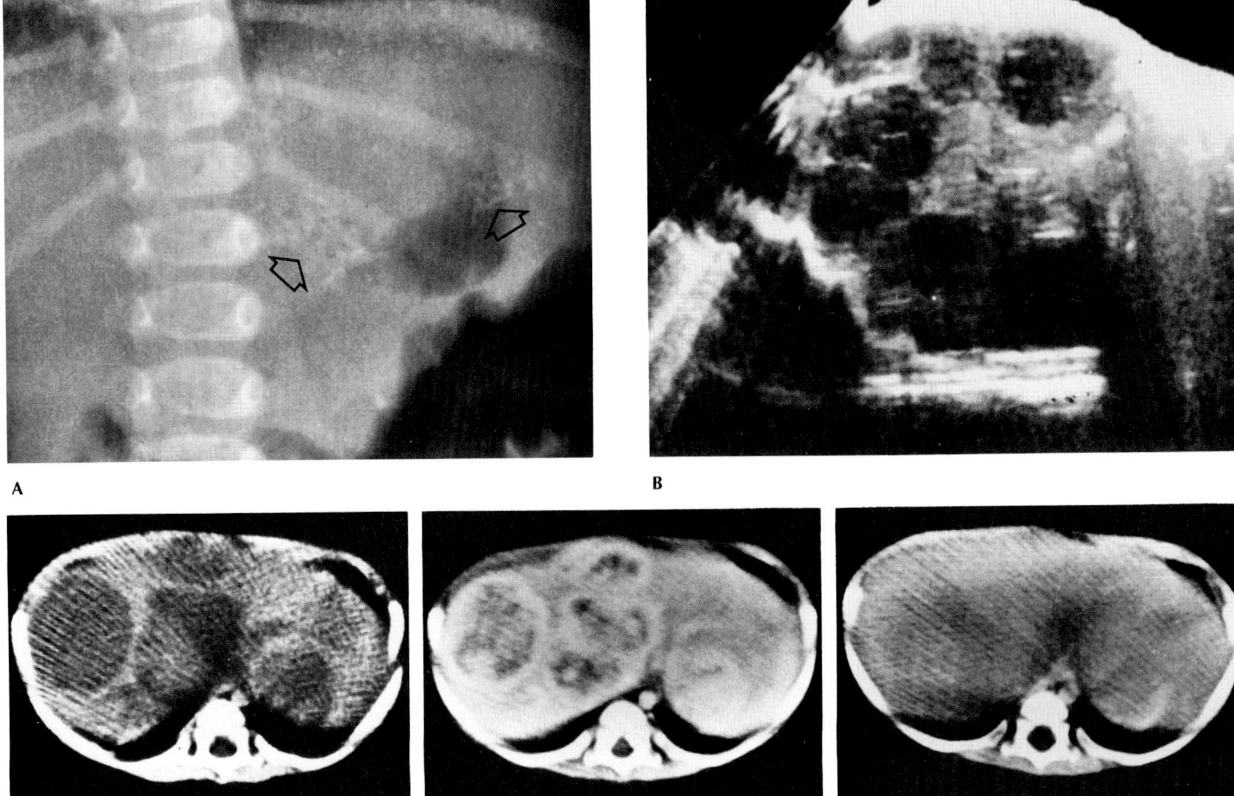


**Figure 52.9. Infantile hemangioendothelioma of the liver.** (A) A plain film of the abdomen demonstrates a fine, speckled pattern of calcification (arrows) in the left lobe of the liver. (B) A sagittal sonogram shows multiple discrete, hypoechoic solid masses. (C) A precontrast CT scan shows multiple well-demarcated masses of lower density than the surrounding liver. (D) An immediate CT scan following the bolus injection of contrast material shows early edge enhancement of all lesions. (E) A delayed postcontrast CT scan shows the tumor nearly isodense with the surrounding liver. (From Dachman, Lichtenstein, Friedman, et al,[12] with permission from the publisher.)

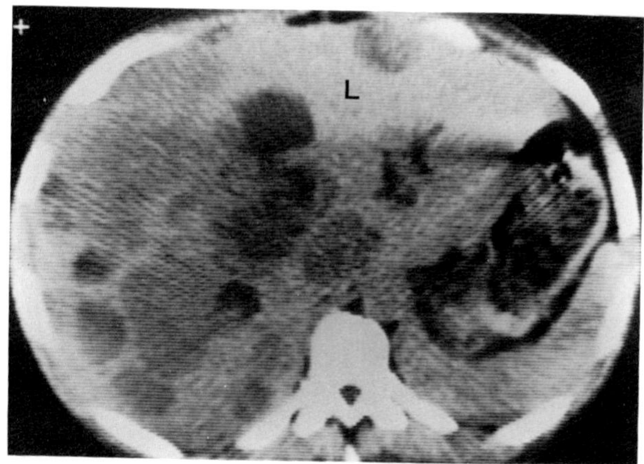


**Figure 52.10. Hepatic metastases.** A CT scan in a patient with testicular carcinoma demonstrates multiple metastases of variable size and shape within the enlarged liver. Each of the lesions is of lower attenuation than normal liver parenchyma, as seen in the anterior part of the left lobe (L). (From Scheible,[25] with permission from the publisher.)

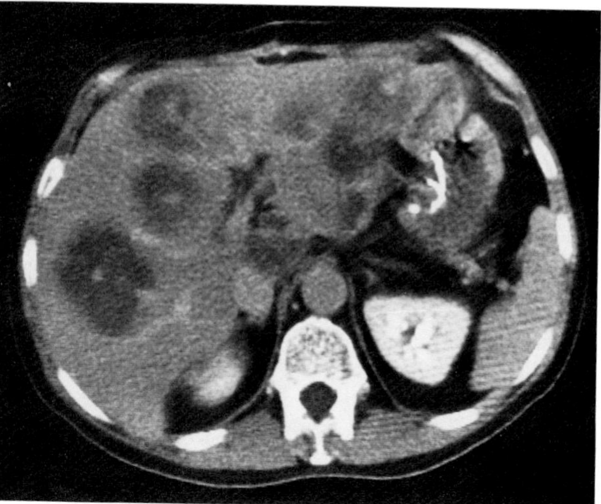

**Figure 52.11. Hepatic metastases.** A CT scan demonstrates multiple low-density metastases with high-density centers.

uation which, on delayed scans, may become isodense with normal liver and thus mimic the appearance of cavernous hemangioma in adults. Arteriography shows pooling of contrast material in vascular lakes, combined with evidence of arteriovenous shunting and enlarged tortuous feeding vessels.

Because spontaneous involution usually starts within 6 to 8 months, vigorous medical treatment of congestive heart failure is an alternative to surgical therapy. If a hemangioendothelioma is considered likely and surgery is not performed, the lesion may be followed by ultrasound and should regress gradually.

In adults, hemangioendotheliomas are rare malignant vascular tumors that may develop in association with chronic exposure to vinyl chloride or the administration of Thorotrast.[17]

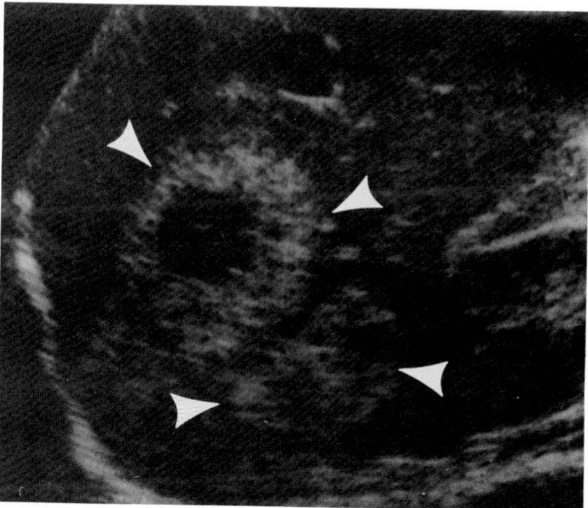

**Figure 52.12. Hepatic metastases.** Ultrasound demonstrates two large lesions (arrowheads) adjacent to one another in the posterior right lobe. The masses are more echogenic than normal hepatic parenchyma. The larger lesion has undergone central necrosis, resulting in a fluid-appearing center. (From Scheible,[25] with permission from the publisher.)

## HEPATIC METASTASES

Metastatic neoplasms are by far the most common malignant tumors involving the hepatic parenchyma. Although some types of metastases (especially mucinous carcinoma of the colon or rectum) may produce diffuse, finely granular calcifications seen on plain radiographs, the diagnosis of hepatic metastases usually requires CT, ultrasound, or radionuclide studies.

CT is probably the most sensitive technique for detecting hepatic metastases. In addition, CT-guided needle aspiration of suspected metastatic deposits is a useful adjunct for cytologic confirmation. Most metastases are relatively well marginated and appear less dense than normal liver parenchyma. Metastases rarely may have an attenuation value higher than liver parenchyma, either because of diffuse calcification or recent hemorrhage or because of fatty infiltration of the surrounding hepatic tissue. Though frequently detectable on noncontrast scans, most metastatic lesions are best seen as areas of increased density adjacent to normally enhancing hepatic parenchyma following the administration of intravenous contrast material. Cystic metastases (sarcomas, melanoma, ovarian and colon carcinoma) may closely simulate benign cysts, though they often have somewhat shaggy and irregular walls. Amorphous punctate deposits of calcification within an area of diminished

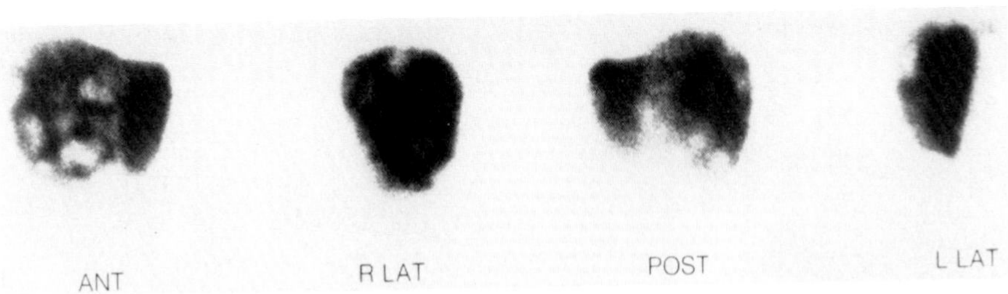


**Figure 52.13. Hepatic metastases.** A radionuclide scan demonstrates multiple photopenic defects within the liver. The spleen is markedly enlarged and displaced anteriorly by the prominent left lobe of the liver. (From Baum, Vincent, and Lyons,[27] with permission from the publisher.)

density may be seen in metastases from mucin-producing tumors (carcinomas of the gastrointestinal tract).

Ultrasound and radionuclide scans can demonstrate hepatic metastases, though these modalities are slightly less sensitive than CT. CT has the additional advantage of being able to detect extrahepatic metastases, such as metastases to abdominal lymph nodes.

Although arteriography can demonstrate metastases as multiple areas of tumor staining and neovascularity with stretching and distortion of intrahepatic vessels, this technique has been supplanted by less invasive procedures.[18–24]

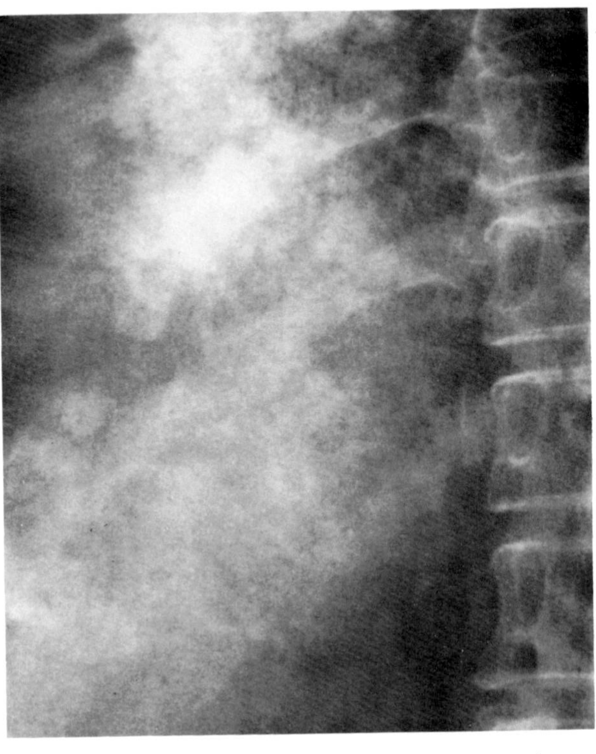

**Figure 52.14. Hepatic metastases.** A diffuse, finely granular pattern of calcification within the liver is due to metastases from colloid carcinoma of the colon. (From Eisenberg,[28] with permission from the publisher.)

**REFERENCES**


1. Itai Y, Araki T, Furui S, et al: Differential diagnosis of hepatic masses on computed tomography, with particular reference to hepatocellular carcinoma. *J Comput Assist Tomogr* 5: 834–842, 1981.
2. Kunstlinger F, Federle MP, Moss AA, et al: Computed tomography of hepatocellular carcinoma. *AJR* 134:431–437, 1980.
3. Freeny PC: Portal vein tumor thrombus: Demonstration by computed tomographic arteriography. *J Comput Assist Tomogr* 4:263–264, 1980.
4. Cottone M, Marceno MP, Maringhini A, et al: Ultrasound in the diagnosis of hepatocellular carcinoma associated with cirrhosis. *Radiology* 147:517–519, 1983.
5. Kamin PD, Bernardino ME, Green B: Ultrasound manifestations of hepatocellular carcinoma. *Radiology* 131:459–461, 1979.
6. Inamoto K, Sugiki K, Yamasaki H, et al: Computed tomography and angiography of hepatocellular carcinoma. *J Comput Assist Tomogr* 4:832–839, 1980.
7. Itai Y, Nishikawa J, Tasaka A: Computed tomography in the evaluation of hepatocellular carcinoma. *Radiology* 131: 165–170, 1979.
8. Barnett PH, Zerhouni EA, White RI, et al: Computed tomography in the diagnosis of cavernous hemangioma of the liver. *AJR* 134:439–447, 1980.
9. Johnson CM, Sheedy PH, Stanson AW, et al: Computed tomography and angiography of cavernous hemangiomas of the liver. *Radiology* 138:115–121, 1981.
10. Itai Y, Ohtomo K, Araki T, et al: Computed tomography and sonography of cavernous hemangioma of the liver. *AJR* 141:315–320, 1983.
11. Freeny PC, Vimont TR, Barnett DC: Cavernous hemangioma of the liver: Ultrasonography, arteriography, and computed tomography. *Radiology* 132:143–148, 1979.
12. Stanley RJ, Sagel SS, Levitt RG: Computed tomography of the liver. *Radiol Clin North Am* 15:331–348, 1977.
13. Piers DA, Houthoff HJ, Crom RAF: Hot spot liver scan in focal nodular hyperplasia. *AJR* 135:1289–1292, 1980.
14. Biersack HJ, Thelen M, Torres JF, et al: Focal nodular hyperplasia of the liver as established by $^{99m}$Tc sulfur colloid and HIDA scintigraphy. *Radiology* 137:187–190, 1980.
15. Rogers JV, Mack LA, Freeny PC, et al: Hepatic focal nodular hyperplasia: Angiography, CT, sonography, scintigraphy. *AJR* 137:983–990, 1981.
16. Engel MA, Marks DS, Sandler MA, et al: Differentiation of focal intrahepatic lesions with $^{99m}$Tc-red blood cell imaging. *Radiology* 146:777–782, 1983.
17. Dachman AH, Lichtenstein JE, Friedman AC, et al: Infantile hemangioendothelioma of the liver. *AJR* 140:1091–1096, 1983.
18. Bernardino ME: Computed tomography of calcified liver metastases. *J Comput Assist Tomogr* 3:32–35, 1979.
19. Federle MP, Filly RA, Moss AA: Cystic hepatic neoplasms. *AJR* 136:345–348, 1981.
20. Snow JH, Goldstein HM, Wallace S: Comparison of scintigraphy, sonography and computed tomography in the evaluation of hepatic neoplasms. *AJR* 132:915–922, 1979.
21. Knopf DR, Torres WE, Fajman WJ, et al: Liver lesions: Comparative accuracy of scintigraphy and computed tomography. *AJR* 138:623–628, 1982.

22. Smith TJ, Kemeny MM, Sugerbaker PH, et al: A prospective study of hepatic imaging in the detection of metastatic disease. *Ann Surg* 195:486–493, 1982.
23. Alderson PO, Adams DF, McNeil BJ, et al: Computed tomography, ultrasound and scintigraphy of the liver in patients with colon or breast carcinoma: A prospective comparison. *Radiology* 149:225–231, 1983.
24. Berland LL, Lawson TL, Foley WD, et al: Comparison of pre- and post-contrast CT in hepatic masses. *AJR* 138:853–859, 1982.
25. Scheible FW: Hepatic masses, in Eisenberg RL, Amberg JR (eds): *Critical Diagnostic Pathways in Radiology: An Algorithmic Approach.* Philadelphia, J.B. Lippincott, 1981, pp 151–163.
26. Moss AA: Computed tomography of the hepatobiliary system, in Moss AA, Gamsu G, Genant HK: *Computed Tomography of the Body*. Philadelphia, W.B. Saunders, 1983, pp 599–698.
27. Baum S, Vincent NR, Lyons KP, et al: *Atlas of Nuclear Medicine Imaging*. New York, Appleton-Century-Crofts, 1981.
28. Eisenberg RL: *Gastrointestinal Radiology: A Pattern Approach*. Philadelphia, J.B. Lippincott, 1983.

## Chapter Fifty-Three

# Diseases of the Gallbladder

## RADIOGRAPHIC APPROACH TO THE JAUNDICED PATIENT

Ultrasound is usually the initial procedure of choice in differentiating biliary obstruction from hepatocellular disease as the cause of jaundice. This modality will be positive in almost 95 percent of the patients with dilated bile ducts. False-negative results occasionally occur in patients with sclerosing cholangitis, in which the ducts do not dilate, or in partial obstruction of the biliary tree by small common duct stones. In addition, dilated ducts filled with thick bile may be sonographically indistinguishable from adjacent liver parenchyma. Nevertheless, the ultrasound demonstration of a nondilated biliary tree effectively excludes extrahepatic obstruction.

Computed tomography (CT) is as sensitive as ultrasound in demonstrating a dilated biliary system in patients with obstructive jaundice. However, ultrasound is usually the initial imaging procedure, since it is substantially less expensive and involves no ionizing radiation. CT is much more accurate than ultrasound in determining the level and cause of biliary obstruction. In patients with malignant biliary obstruction, CT is of considerable value in detecting intra-abdominal nodal metastases as well as in the preoperative assessment of the extent of local invasion. CT is more reliable than sonography for detecting stones in the common bile duct, though small cholesterol stones of a density similar to bile may escape detection.

If ultrasound and CT fail to precisely define the exact site and cause of biliary obstruction, direct cholangiography is indicated. This is most often accomplished by

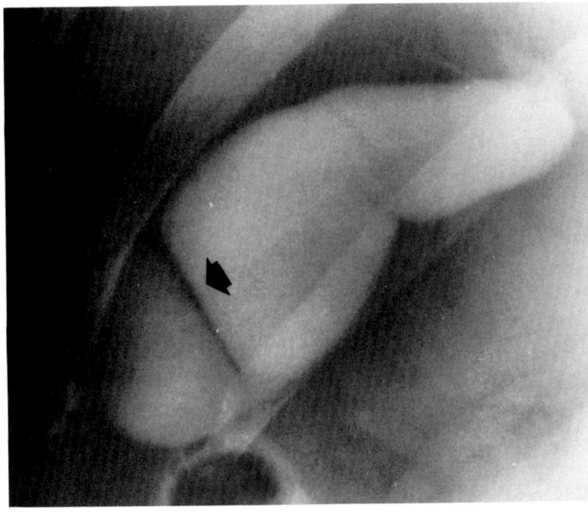

**Figure 53.1. Phrygian cap.** An incomplete septum (arrow) extends across the fundus of the gallbladder, partially separating it from the body. (From Eisenberg,[2] with permission from the publisher.)

the percutaneous transhepatic approach with a thin needle; endoscopic retrograde cholangiopancreatography (if available) can also be employed, especially when abnormal bleeding parameters make the percutaneous approach too risky.[1]

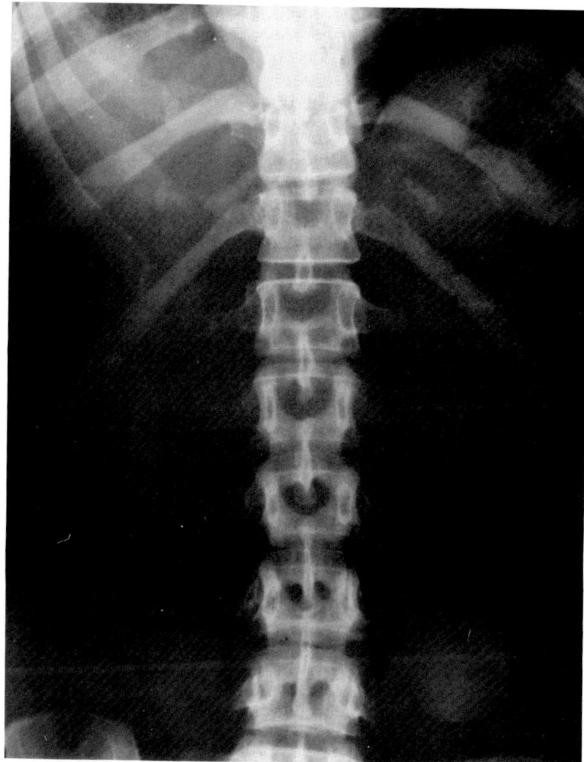

**Figure 53.2. Left-sided gallbladder.** A functioning gallbladder is situated in an anomalous position in the left lower quadrant (arrow).

## CONGENITAL ANOMALIES OF THE GALLBLADDER

Abnormalities in number, size, and shape of the gallbladder not infrequently occur. Infants and children may demonstrate hypoplasia of the gallbladder, which appears as little more than a small, rudimentary pouch at the end of the cystic duct. Congenital multiseptate gallbladder is a hyperplastic structure with multiple intercommunicating septa dividing the lumen of the gallbladder. It appears on oral cholecystography as a small gallbladder with a honeycomb pattern. Stasis of bile in a multiseptate gallbladder predisposes to infection and gallstone formation. The Phrygian cap is a developmental anomaly in which an incomplete septum extends across the fundus of the gallbladder, partially separating it from the body. Although of no clinical significance, this congenital deformity must be differentiated from localized adenomyomatosis. Anomalies of position include left-sided gallbladder, intrahepatic gallbladder, and the "floating" gallbladder, which may predispose to acute torsion, volvulus, or herniation of the gallbladder.[2]

## GALLSTONES

Gallstones can develop whenever bile contains insufficient bile salts and lecithin in proportion to cholesterol to maintain the cholesterol in solution. This situation can result from a decrease in the amount of bile salts present (because of decreased reabsorption in the terminal ileum secondary to inflammatory disease or surgical resection) or can be caused by increased hepatic synthesis of cholesterol. Because cholesterol is not radiopaque, most gallstones are radiolucent and visible only on contrast

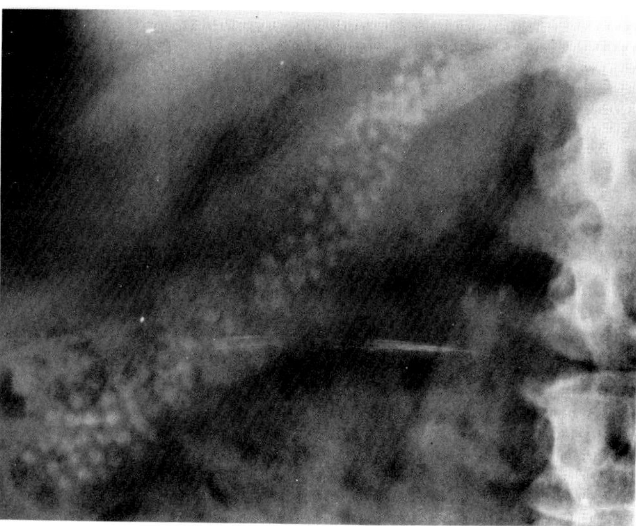

A

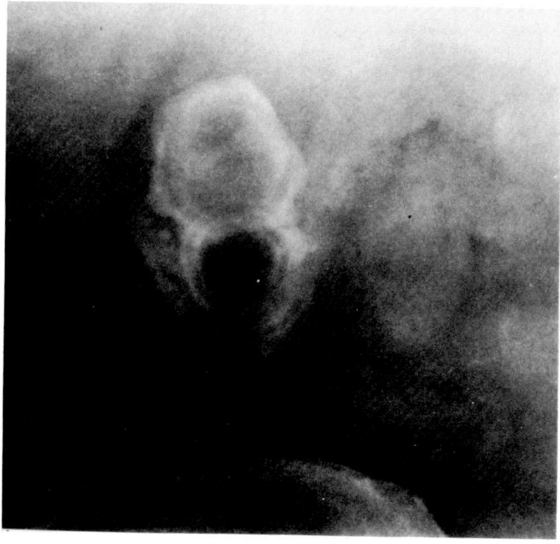

B

**Figure 53.3. Radiopaque gallstones.** (A) Multiple small radiopaque stones in a large gallbladder. (B) Two gallstones with laminated calcification.

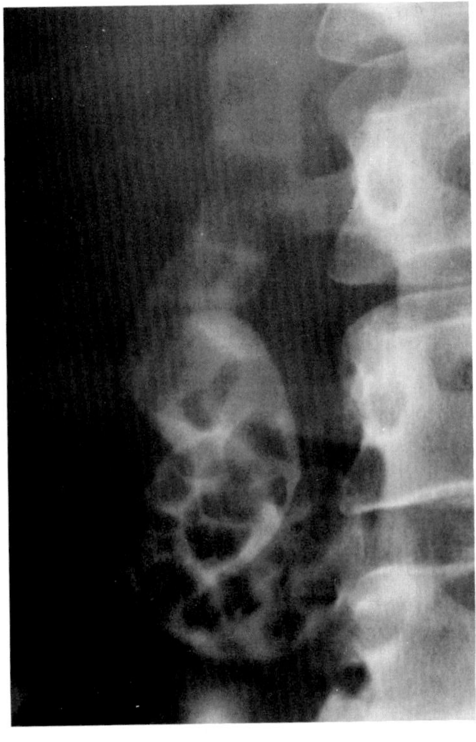
A

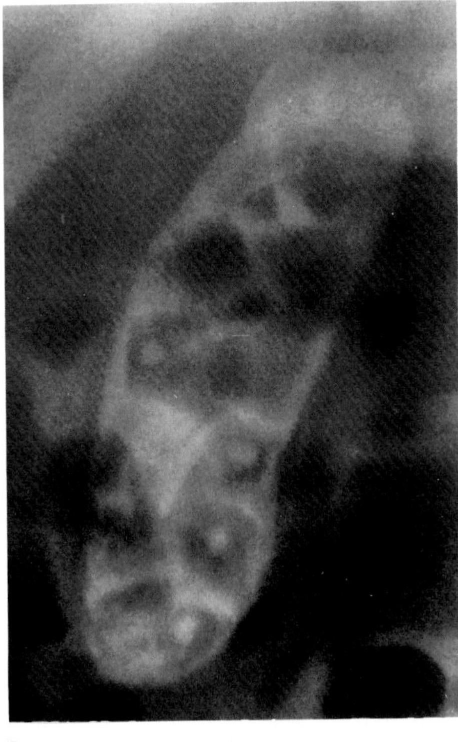
B

**Figure 53.4. Radiolucent gallstones.** (A) Multiple lucent gallstones. (B) Multiple lucent gallstones, many of which contain a central nidus of calcification.

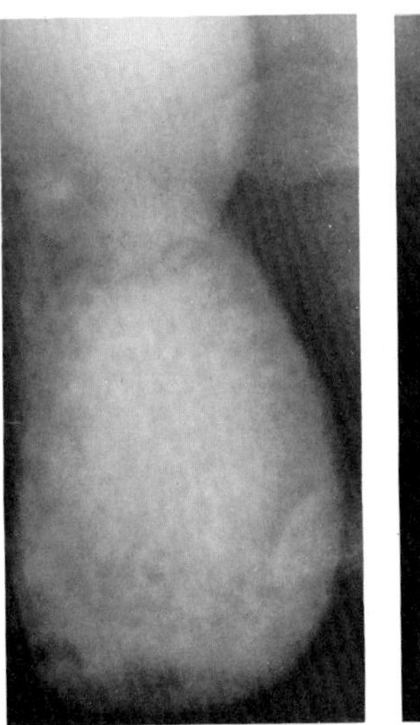
A

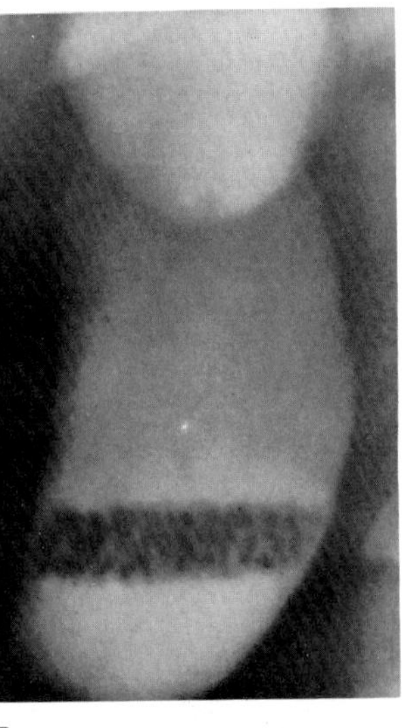
B

**Figure 53.5. Layering of gallstones.** (A) With the patient supine, the stones are poorly defined and have a gravel-like consistency. (B) On an erect film taken with a horizontal beam, the innumerable gallstones layer out and are easily seen.

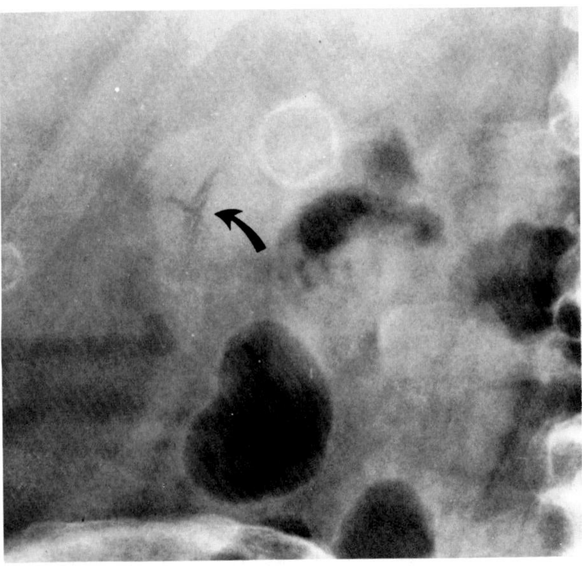

**Figure 53.6. Fissuring in a gallstone.** "Mercedes-Benz" sign (arrow). Note the adjacent gallstone with a radiopaque rim.

examinations. In up to 20 percent of the patients, however, gallstones are composed of calcium bilirubinate or are of mixed composition and contain sufficient calcium to be detectable on plain abdominal radiographs. Gallstones can have a central nidus of calcification, be laminated (alternating opaque and lucent rings), or have calcification around the periphery. Occasionally, a nonopaque stone may contain gas-filled fissures that produce the "Mercedes-Benz" sign, a characteristic triradiate pattern similar to the German automobile trademark.

Oral cholecystography (OCG) has been the traditional technique for the diagnosis of gallstones, though it is rapidly being replaced by ultrasound (see below). Gallstones appear as freely movable filling defects in the

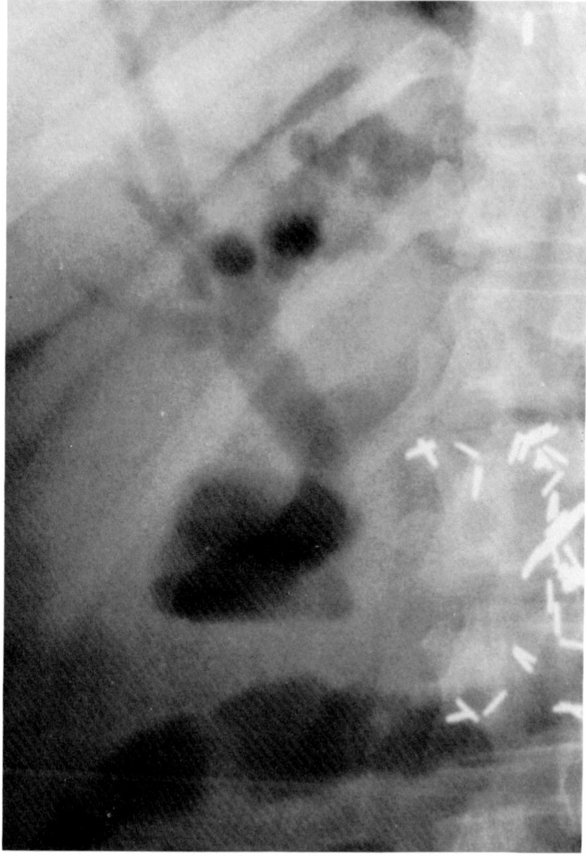


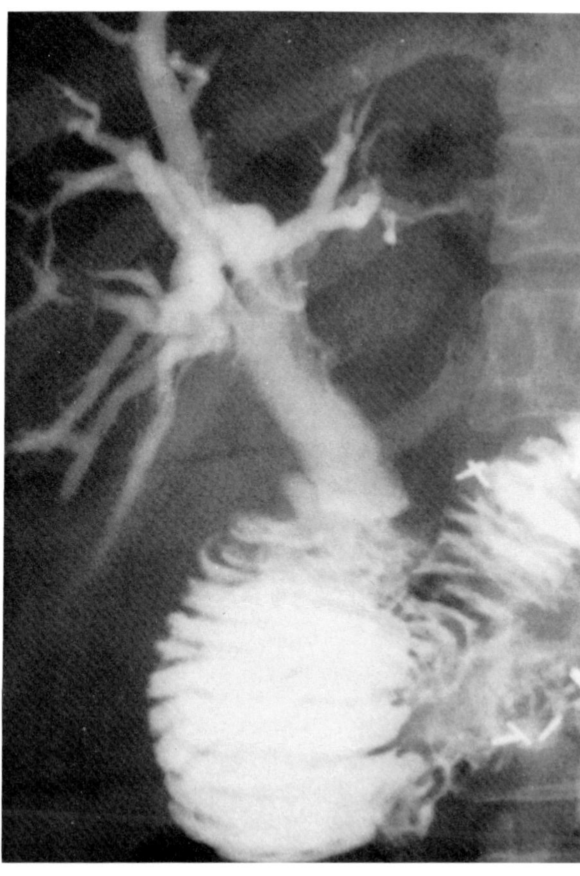


**Figure 53.11. Fistula formation.** (A) Gas in the biliary tree and (B) pancreaticoduodenal reflux of barium are seen in a patient who had undergone a surgical procedure to relieve biliary obstruction. (From Eisenberg,[2] with permission from the publisher.)

are a cause of nonvisualization of the gallbladder, since they permit bile to flow directly and continuously into the intestine and result in nonfilling of the gallbladder.

A walled-off perforation of the gallbladder complicating acute cholecystitis produces a pericholecystic abscess. This appears on ultrasound as a hypoechoic collection of fluid in the gallbladder fossa. On barium studies, a pericholecystic abscess may cause an irregular mass impression on the lateral aspect of duodenal sweep, which is often associated with such inflammatory changes as fold thickening and spiculation.[9–11]

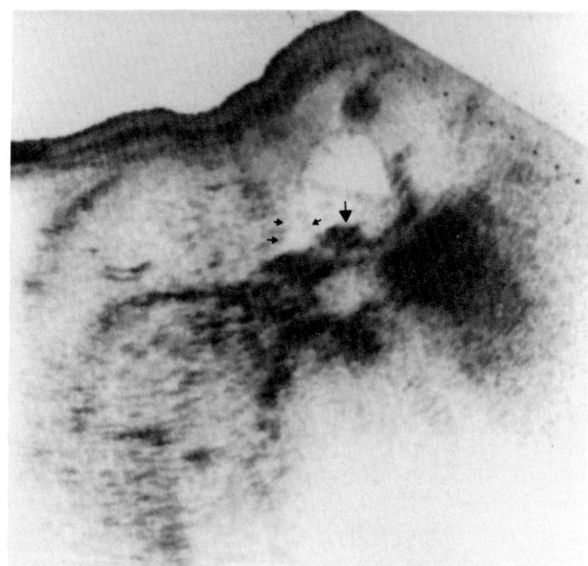

**Figure 53.12. Acute cholecystitis with infected fluid surrounding the gallbladder.** A transverse sonogram shows sludge and stones (large arrow) within the gallbladder. In addition, there is a pericholecystic fluid collection (small arrows), which contained a purulent exudate. (Courtesy of Gretchen A.W. Gooding, M.D.)

## Gallstone Ileus

The combination of mechanical small bowel obstruction, gas or barium in the biliary tree, and an opaque calculus or lucent filling defect in the ileum or jejunum is virtually pathognomonic of gallstone ileus. In this condition, which primarily occurs in elderly women, a large gallstone enters the small bowel by way of a fistula from the gallbladder or common bile duct to the duodenum. As the stone temporarily lodges at various levels in the small bowel, intermittent symptoms of abdominal cramps,

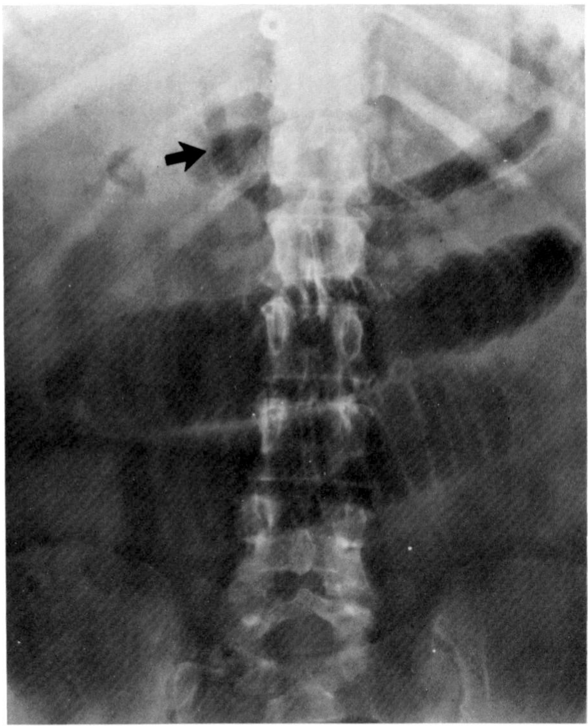

**Figure 53.13. Gallstone ileus.** A plain abdominal radiograph demonstrates dilated, gas-filled loops of obstructed small bowel combined with gas in the biliary tree (arrow). (From Eisenberg,[2] with permission from the publisher.)

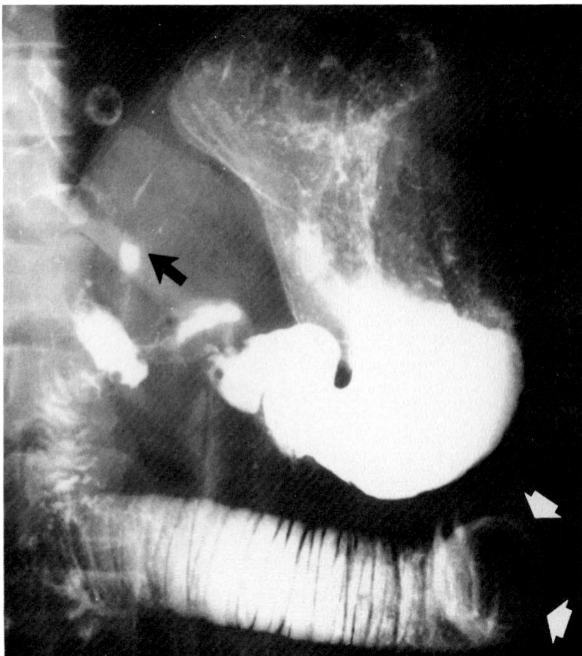

**Figure 53.14. Gallstone ileus.** An upper gastrointestinal series demonstrates the obstructing stone (white arrows) and barium in the biliary tree (black arrows).

nausea, and vomiting can simulate a recurrent partial obstruction. Complete obstruction develops when the gallstone finally reaches a portion of bowel too narrow to allow further progression. This usually occurs in the ileum near the ileocecal valve, which is the narrowest segment of the small bowel, provided that the more proximal small bowel is of normal caliber.[12]

## Limey (Milk of Calcium) Bile

Milk of calcium bile is a condition in which the gallbladder becomes filled with an accumulation of bile that is rendered radiopaque because of a high concentration of calcium carbonate. On plain abdominal radiographs there is a diffuse, hazy opacification of bile with a layering effect. The disorder is secondary to chronic cholecystitis and is accompanied by thickening of the gallbladder wall and obstruction of the cystic duct. Because the increased density of bile makes the entire gallbladder opaque and simulates the appearance of a normal gallbladder filled with contrast material, it is necessary that a preliminary abdominal film be obtained before an oral cholecystographic contrast agent is administered in order to diagnose this disorder.[2]

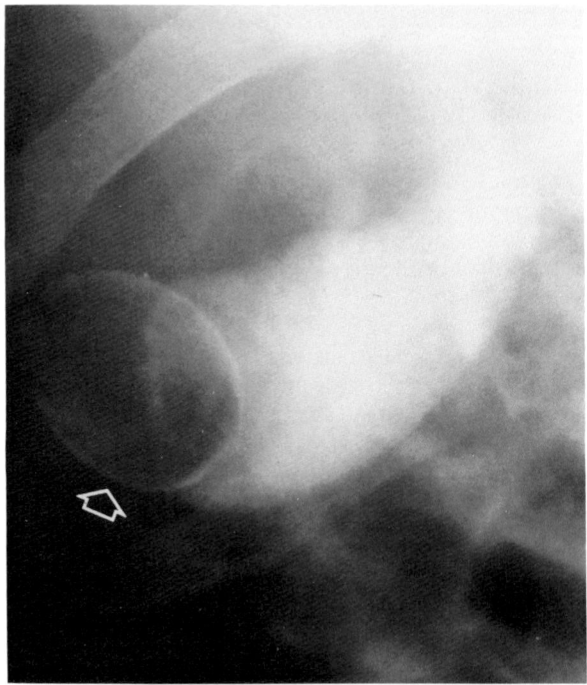

**Figure 53.15. Milk of calcium bile.** Although this film simulates an oral cholecystogram, the patient had not received any cholecystographic contrast agent. Note that the gallbladder also contains a large lucent stone (arrow). (From Eisenberg,[2] with permission from the publisher.)

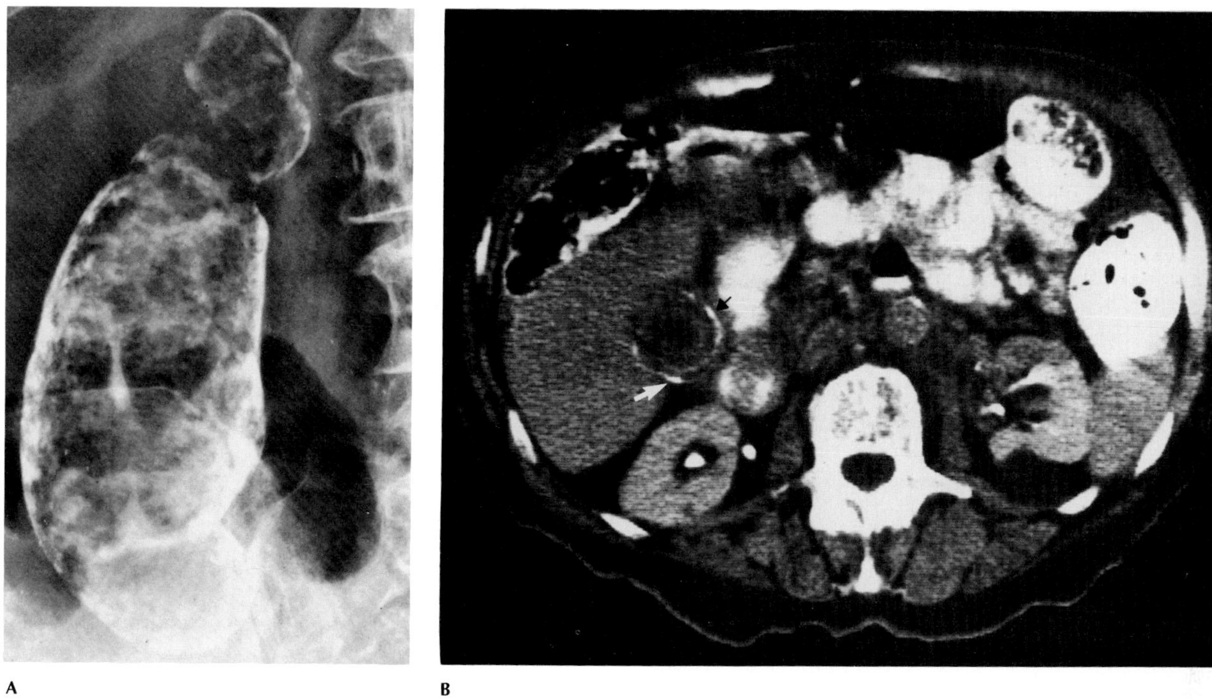

A B

**Figure 53.16. Porcelain gallbladder.** (A) A plain radiograph demonstrates extensive mural calcification around the perimeter of the gallbladder. (B) A CT scan in another patient shows calcification within the gallbladder wall (arrows).

## Porcelain Gallbladder

*Porcelain gallbladder* refers to extensive mural calcification around the perimeter of the gallbladder, which forms an oval density that corresponds to the size and shape of the organ. The term reflects the blue discoloration and brittle consistency of the gallbladder wall. The calcification in a porcelain gallbladder can appear as a broad continuous band in the muscular layers or be multiple and punctate and occur in the glandular spaces of the mucosa. The detection of extensive calcification in the wall of the gallbladder should suggest the possibility of carcinoma. Although a porcelain gallbladder is uncommon in cases of carcinoma of the gallbladder, there is a striking incidence of carcinoma in porcelain gallbladders (up to 60 percent of the cases). Therefore, even if they are asymptomatic, patients with porcelain gallbladders are usually subjected to prophylactic cholecystectomy.[2,8]

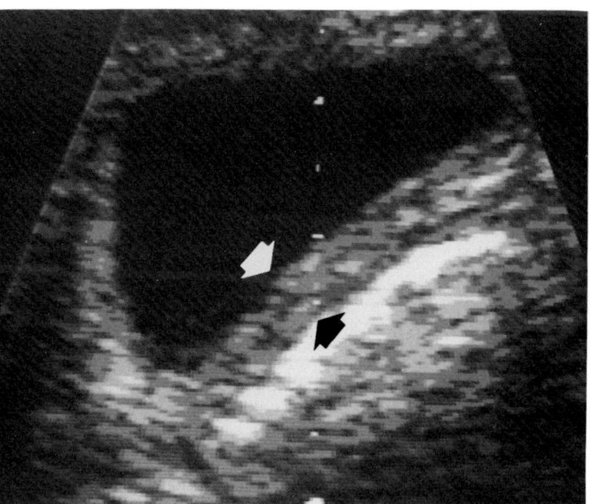

**Figure 53.17. Acalculous cholecystitis.** Ultrasound demonstrates an enlarged gallbladder with a thickened, edematous wall (arrows). There is no evidence of gallstones or posterior acoustic shadowing.

# ACALCULOUS CHOLECYSTITIS

In up to 10 percent of patients with acute cholecystitis, calculi obstructing the cystic duct are not found at surgery, and no underlying explanation for acalculous inflammation is found. This condition is most commonly seen after significant trauma or burns or in patients receiving prolonged hyperalimentation. Ultrasound demonstrates a large, tense, static gallbladder, often with progressive edema of the gallbladder wall and the presence of sludge without discrete stones. Radionuclide scans may show cystic duct obstruction due to associated edema.

## EMPHYSEMATOUS CHOLECYSTITIS

Emphysematous cholecystitis is a rare condition in which the growth of gas-forming organisms (*Escherichia coli, Clostridium welchii* or *perfringens*) in the gallbladder is facilitated by stasis and ischemia caused by cystic duct obstruction (most often by stones). Emphysematous cholecystitis occurs most frequently in elderly men and in patients with poorly controlled diabetes mellitus. Plain abdominal radiographs demonstrate gas in the gallbladder lumen that dissects into the wall or pericholecystic tissues to produce the pathognomonic appearance of a rim of translucent bubbles or streaks of gas outside of and roughly parallel to the gallbladder lumen. Ultrasound may suggest emphysematous cholecystitis if there is marked irregular thickening of the gallbladder wall with multiple echoes representing sloughed mucosa or intraluminal debris.[13]

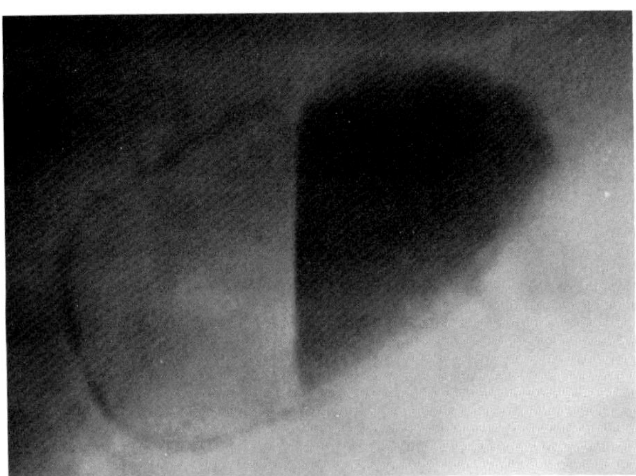

**Figure 53.18. Emphysematous cholecystitis.** There is gas within both the lumen and the wall of the gallbladder. (From Eisenberg,[2] with permission from the publisher.)

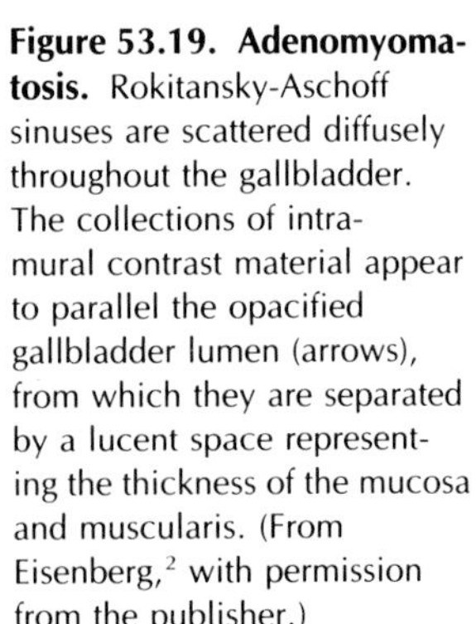

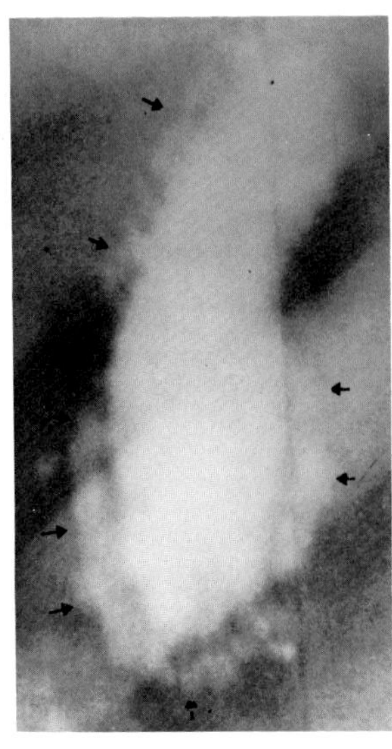

**Figure 53.19. Adenomyomatosis.** Rokitansky-Aschoff sinuses are scattered diffusely throughout the gallbladder. The collections of intramural contrast material appear to parallel the opacified gallbladder lumen (arrows), from which they are separated by a lucent space representing the thickness of the mucosa and muscularis. (From Eisenberg,[2] with permission from the publisher.)

## HYPERPLASTIC CHOLECYSTOSES

The hyperplastic cholecystoses are noninflammatory disorders of the gallbladder that consist of benign proliferation of normal tissue elements. Radiographically, they are also associated with functional abnormalities of the gallbladder, such as hyperconcentration, hypercontractility, and hyperexcretion.

### Adenomyomatosis

Adenomyomatosis is a proliferation of surface epithelium with glandlike formations and outpouchings of the mucosa into or through the thickened muscular layer. The sinuses (Rokitansky-Aschoff) associated with this form of intramural diverticulosis can be limited to a single segment or be scattered diffusely throughout the gallbladder.

Radiographically, the Rokitansky-Aschoff sinuses appear as single or multiple oval collections of contrast material projected just outside the lumen of the gallbladder. These opaque dots range in diameter from pinpoint size to 10 mm. When multiple and viewed tangentially, they resemble a string of beads closely applied to the circumference of the opacified gallbladder lumen. The clear line separating the opaque sacs from the gallbladder cavity represents the thickness of the mucosa and muscularis. Annular thickening can result in a pattern of multiple septal folds. Gallstones are often present

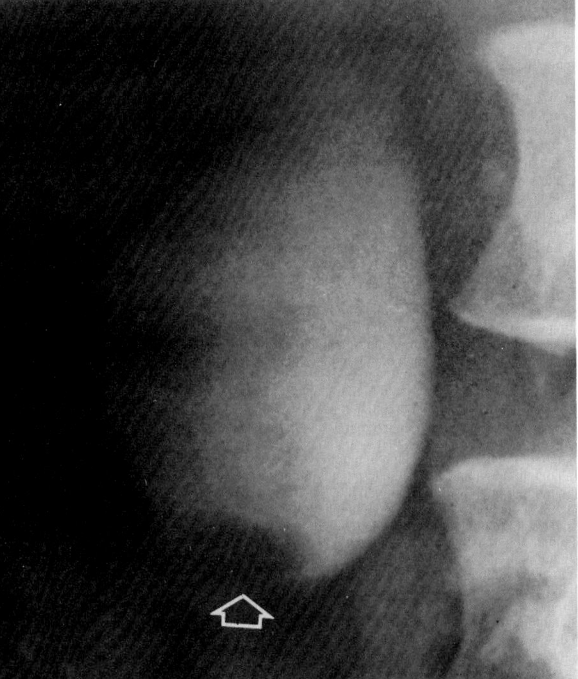

**Figure 53.20. Solitary adenomyoma.** A broad mass (arrow) is evident at the tip of the fundus of the gallbladder. (From Eisenberg,[2] with permission from the publisher.)

in patients with adenomyomatosis; they can be seen in a diverticular sinus or in the fundal portion of a septated gallbladder.

An adenomyoma is a single filling defect in the gallbladder that reflects a localized form of adenomyomatosis rather than a true neoplasm. It is almost invariably situated at the tip of the fundus and appears as an intramural mass projecting into the gallbladder, often with an opaque central speck of contrast material representing umbilication of the mound.[14,15]

## Cholesterolosis

Cholesterolosis (strawberry gallbladder) is characterized by abnormal deposits of cholesterol esters in fat-laden macrophages in the lamina propria of the gallbladder wall. This fatty material causes coarse yellow, speckled masses on the surface of a reddened, hyperemic gallbladder mucosa, an appearance resembling strawberry seeds. Cholesterolosis can produce single or multiple small polypoid filling defects in the opacified gallbladder. These lesions can occur in any portion of the gallbladder and have no malignant potential.

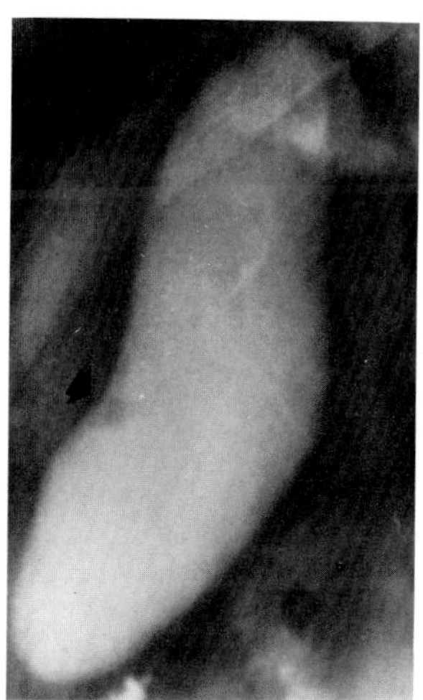

**Figure 53.21. Cholesterol polyp (arrow).** (From Eisenberg,[2] with permission from the publisher.)

The filling defects in cholesterolosis (as in adenomyomatosis) are best seen on radiographs made after partial emptying of the gallbladder. Compression or a fatty meal can also be employed to demonstrate the lesions to better advantage. The filling defects are fixed in position with respect to the gallbladder wall, in contrast to gallstones, which move freely in response to gravity and with changes in patient position.[14,15]

## CARCINOMA OF THE GALLBLADDER

Primary carcinoma of the gallbladder usually develops in elderly patients with a history of gallstones. In most instances, direct hepatic invasion or regional lymph node metastases precedes the diagnosis and leads to a dismal prognosis. The clinical presentation is varied, ranging from vague right upper quadrant pain mimicking biliary colic to progressive jaundice.

Both ultrasound and CT may be of value in demonstrating carcinoma of the gallbladder and the extent of tumor spread. Ultrasound may demonstrate gallbladder carcinoma as focal thickening of the gallbladder wall with adjacent hypoechoic masses. Direct extension to the liver appears as a low-density mass on contrast-enhanced CT scans.

Carcinoma of the gallbladder usually causes nonvisualization of the organ on oral cholecystography. This is due to obstruction of the cystic duct by an invasive tumor or to the presence of associated chronic cholecystitis or cholelithiasis, which almost invariably accompanies gallbladder cancer. On rare occasions, noninvasive carcinoma appears as a solitary fixed polyp or an irregular mural filling defect in a well-opacified gallbladder.

On barium upper gastrointestinal series, carcinoma of the gallbladder can cause a large mass impression that indents or invades the duodenal bulb and sweep, usually displacing them inferiorly and medially. If ulceration into

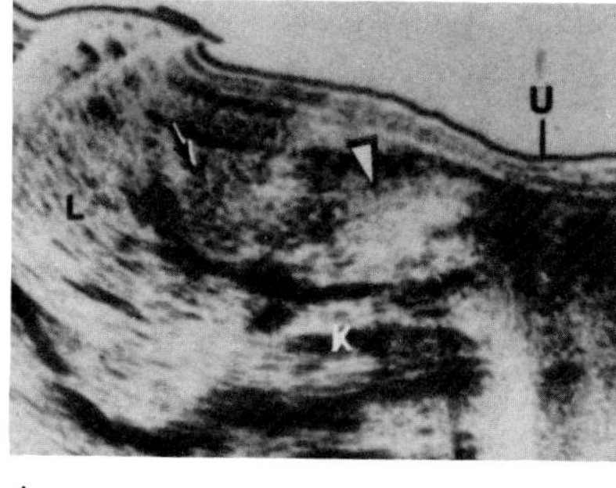


A

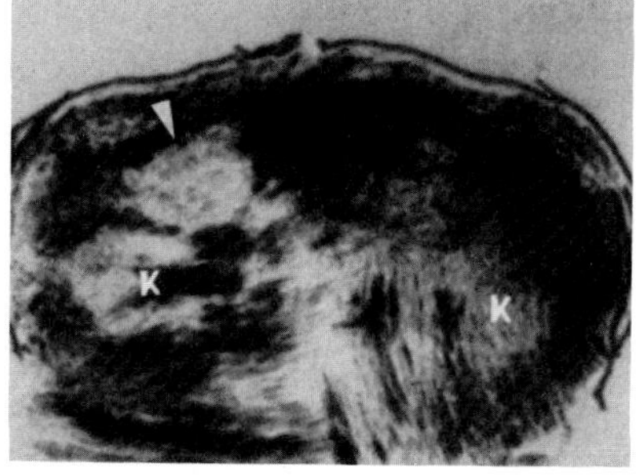


B

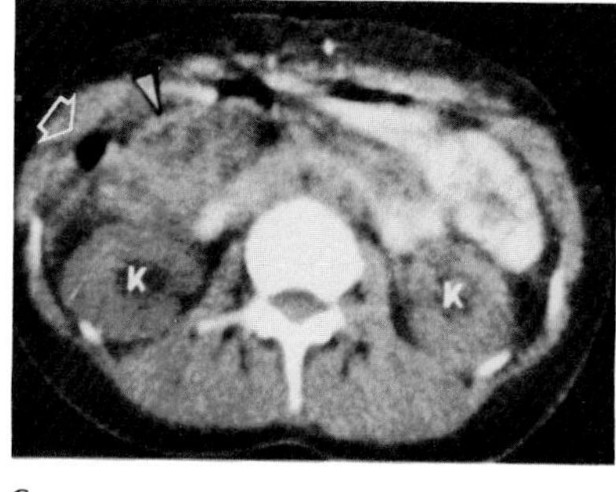


C

**Figure 53.22. Carcinoma of the gallbladder.** (A) A longitudinal sonogram 4 cm to the right of the midline and (B) a transverse scan 11 cm above the umbilicus (U) demonstrate a carcinoma of the gallbladder (white arrowhead) with a highly echogenic liver lesion (arrow). (C) On CT, the metastatic lesion (arrow) is more lucent than the gallbladder tumor (white arrowhead). K = kidney; L = liver. (From Yeh,[17] with permission from the publisher.)

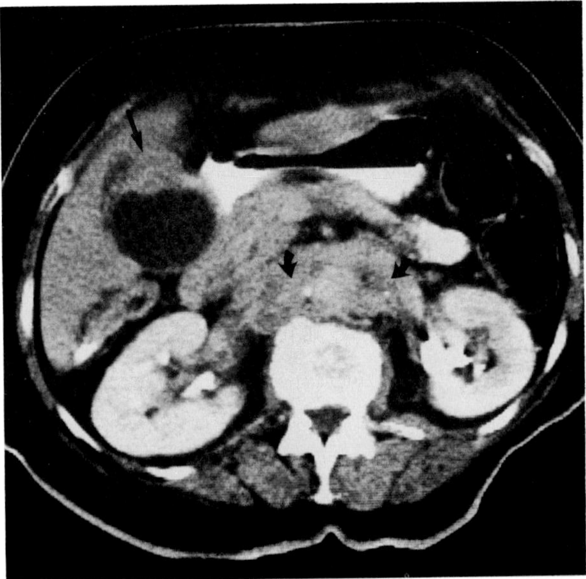

**Figure 53.23. Carcinoma of the gallbladder.** A CT scan demonstrates a soft tissue mass along the anterior wall of the gallbladder (arrow). Note the para-aortic nodal metastases (curved arrows).

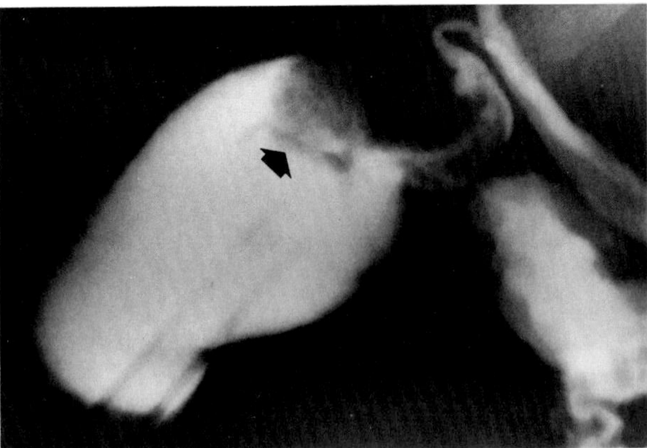

**Figure 53.24. Carcinoma of the gallbladder.** There is an irregular mural mass (arrow), with tumor growth extending into the cystic duct. (From Eisenberg,[2] with permission from the publisher.)

the duodenum occurs, gas can be seen in the biliary ductal system.

In patients with extensive and nonresectable disease, percutaneous biliary drainage may be an alternative form of palliation for severe pruritus and malaise.[8,16–18]

## REFERENCES


1. Janes JO, Nelson JA: Jaundice, in Eisenberg RL, Amberg JR (eds): *Critical Diagnostic Pathways in Radiology: An Algorithmic Approach*. Philadelphia, JB Lippincott, 1981.
2. Eisenberg RL: *Gastrointestinal Radiology: A Pattern Approach*. Philadelphia, JB Lippincott, 1983.
3. Cooperberg PL, Burhenne HJ: Real-time ultrasonography: Diagnostic technique of choice in calculous gallbladder disease. *N Engl J Med* 300:1277–1278, 1980.
4. Krook PM, Allen FH, Bush WH, et al: Comparison of real-time cholecystosonography and oral cholecystography. *Radiology* 135:145–153, 1980.
5. Mujahed Z: Factors interfering with the opacification of a normal gallbladder. *Gastrointest Radiol* 1:183–185, 1976.
6. Berk RN, Ferrucci JT, Leopold GR: *Radiology of the Gallbladder and Bile Ducts*. Philadelphia, WB Saunders, 1983.
7. Cheng TH, Davis MA, Seltzer SE, et al: Evaluation of hepatobiliary imaging by radionuclide scintigraphy, ultrasonography, and contrast cholangiography. *Radiology* 133:761–767, 1979.
8. Berk RN, Armbuster TG, Saltzstein SL: Carcinoma in the porcelain gallbladder. *Radiology* 106:29–31, 1973.
9. Mujahed Z: Nonopacification of the gallbladder and bile ducts: A previously unreported cause. *Radiology* 112:297–298, 1974.
10. Evans RC, Kaude JV, Steinberg W: Spontaneous disappearance of large stones from the common bile duct. *Gastrointest Radiol* 5:47–48, 1980.
11. Haff RC, Wise L, Ballinger WF: Biliary-enteric fistulas. *Surg Gynecol Obstet* 133:84–88, 1971.
12. Eisenman JI, Finck EJ, O'Louhlin BJ: Gallstone ileus: A review of the roentgenographic findings and report of a new roentgen sign. *AJR* 101:361–366, 1967.
13. Grainger K: Acute emphysematous cholecystitis. *Clin Radiol* 12:66–69, 1961.
14. Jutras JA: Hyperplastic cholecystoses. *AJR* 83:795–827, 1960.
15. Jutras JA, Levesque HP: Adenomyoma and adenomyomatosis of the gallbladder: Radiologic and pathologic correlations. *Radiol Clin North Am* 4:483–500, 1966.
16. Cimmino CV: Carcinoma in a well-functioning gallbladder. *Radiology* 71:563–564, 1958.
17. Yeh HC: Ultrasonography and computed tomography of carcinoma of the gallbladder. *Radiology* 133:167–173, 1979.
18. Itai Y, Araki T, Yoshikaw K, et al: Computed tomography of gallbladder carcinoma. *Radiology* 137:713–718, 1980.
19. Harned RK: Gallbladder disease, in Eisenberg RL, Amberg JR (eds): *Critical Diagnostic Pathways in Radiology: An Algorithmic Approach*. Philadelphia, JB Lippincott, 1981, pp 127–140.

## Chapter Fifty-Four
# Diseases of the Bile Ducts

## CONGENITAL ANOMALIES OF THE BILE DUCTS

### Biliary Atresia

Biliary atresia is the most common cause of persistent neonatal jaundice. Children with this anomaly generally have an unfavorable prognosis. The entire extrahepatic biliary ductal system is usually atretic, though the common hepatic duct or common bile duct may be individually involved. The diagnosis is confirmed by surgical exploration with operative cholangiography, which demonstrates dilatation of the biliary tree proximal to the point of atresia.

### Choledochal Cyst

A choledochal cyst is a cystic or fusiform dilatation of the common bile duct and adjacent portions of the common hepatic and cystic ducts that is typically associated with localized constriction of the distal common bile duct. Concomitant dilatation of intrahepatic bile ducts has recently been recognized with increasing frequency. Although usually considered to be a congenital, devel-

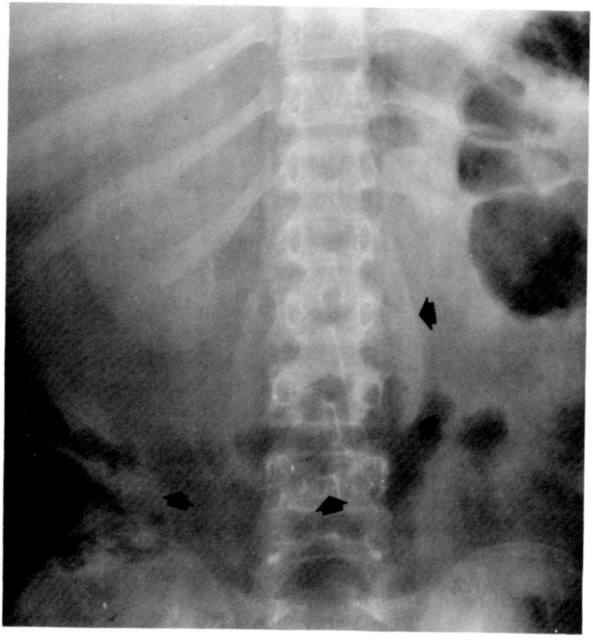

**Figure 54.1. Choledochal cyst.** A huge soft tissue mass in the right upper quadrant (arrows) displaces gas-filled loops of bowel. (From Eisenberg,[20] with permission from the publisher.)

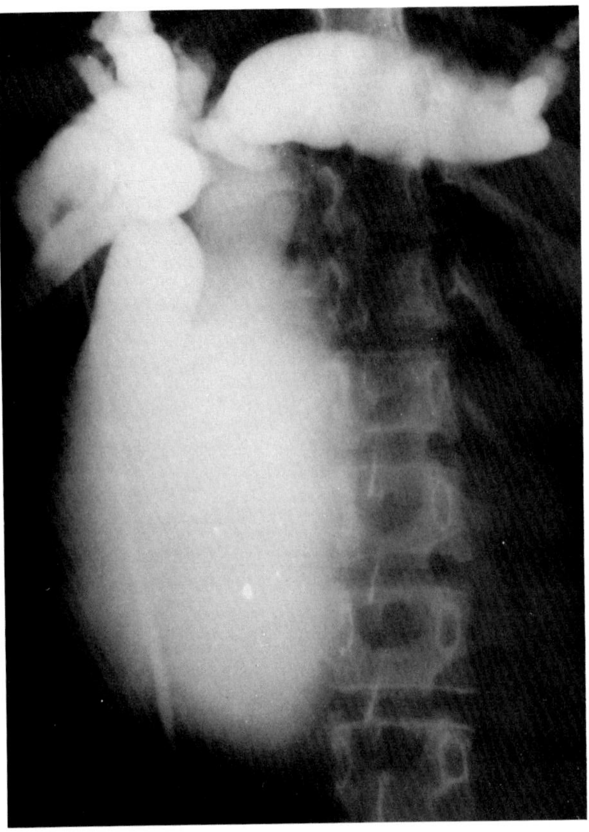

**Figure 54.2. Choledochal cyst.** Cholangiographic contrast material fills the huge fusiform dilatation of the common bile duct as well as the markedly dilated intrahepatic ducts. (From Eisenberg,[20] with permission from the publisher.)

opmental abnormality, many choledochal cysts are probably acquired lesions caused by regurgitation of pancreatic secretions into the distal common bile ducts, leading to cholangitis, gradual stricture formation, and ductal dilatation over a long period. Clinically, choledochal cysts are classically described as presenting with a triad of abdominal pain, right upper quadrant mass, and jaundice.

A soft tissue mass representing the markedly dilated bile duct is often seen on plain abdominal radiographs. An upper gastrointestinal examination can demonstrate generalized widening of the duodenal sweep or displacement of the duodenum anteriorly, inferiorly, and to the left. Oral cholecystography is usually unsuccessful. Contrast material administered via percutaneous transhepatic cholangiography can fill the dilated biliary system.

Ultrasound, computed tomography (CT), and radionuclide scanning with iminodiacetic acid derivatives are all noninvasive techniques that can establish the diagnosis of choledochal cyst.[1–4]

### Congenital Biliary Ectasia

Congenital dilatation of the intrahepatic bile ducts occurs in Caroli's disease and congenital hepatic fibrosis. In Caroli's disease, the patient usually experiences crampy abdominal pain and fever secondary to a marked stasis-induced predisposition to biliary calculus disease, cholangitis, and liver abscesses. About 80 percent of the patients with Caroli's disease have associated medullary sponge kidney. Congenital hepatic fibrosis is a far more severe disorder, which is complicated by massive periportal fibrosis that causes death at an early age as a result of liver failure and portal hypertension.

In both Caroli's disease and congenital hepatic fibrosis, cholangiography demonstrates large or small cystic spaces communicating with the intrahepatic bile ducts, producing a characteristic lollipop-tree appearance of the biliary system. On ultrasound, the dilated ducts may mimic multiple small cysts and are often associated with stones.

CT demonstrates multiple low-density branching tubular structures reflecting dilated bile ducts communicating with focal areas of increased ectasia.[5–9]

## CHOLEDOCHOLITHIASIS

Biliary calculi usually arise in the gallbladder and reach the bile ducts either by passage through the cystic duct or by fistulous erosion through the wall of the gallbladder. Calculi rarely originate in either the extrahepatic or the intrahepatic bile ducts, except in patients with congenital or acquired cystic dilatation of the bile ducts or strictures due to biliary obstruction.

Stones in the extrahepatic bile ducts tend to move freely and change location with alterations in patient position. However, a calculus can become impacted in the distal common duct and cause obstruction. Ultrasound is much less effective in the detection of impacted stones than gallbladder calculi, primarily because of gas in the duodenum that overlies the distal common duct. On percutaneous cholangiography, an impacted stone can usually be diagnosed with confidence because of the typical appearance of a smooth, sharply defined intraluminal filling defect with an upward convex contour. If the intrahepatic bile ducts are not dilated on ultrasound, endoscopic retrograde cholangiopancreatography (ERCP) is the modality of choice in demonstrating impacted bile duct stones.

An impacted stone in the distal common bile duct may incite an inflammatory reaction in the duodenal papilla. The resulting edematous papilla has a characteristic smooth, semilunar configuration with a diameter several times that of the impacted stone. Similar edematous swelling of the papilla can be secondary to acute pancreatitis or duodenal ulcer disease.

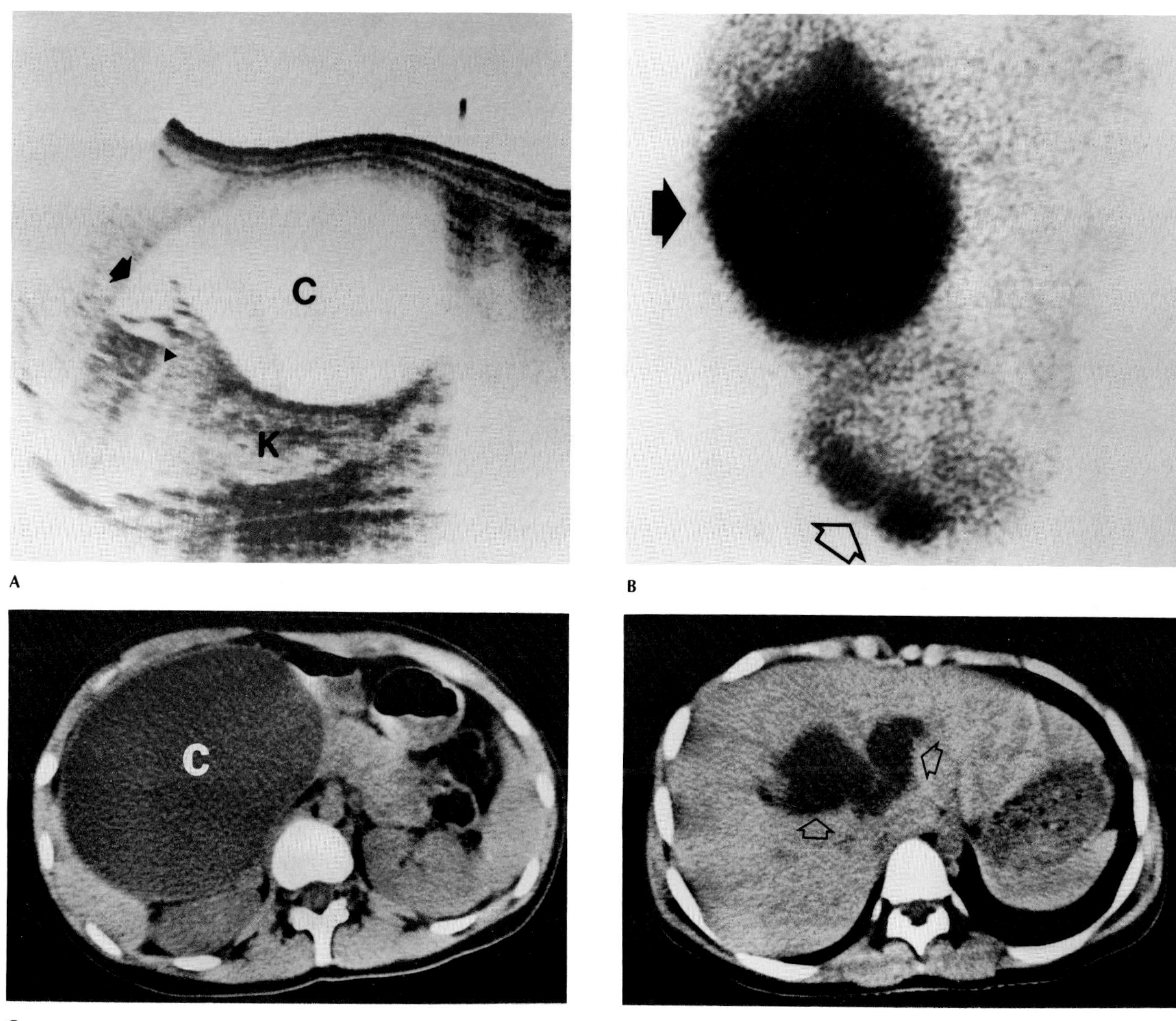


**Figure 54.3. Choledochal cyst.** (A) A longitudinal sonogram 4 cm to the right of the midline shows a large choledochal cyst (C) located in the region of the hilum of the liver and extending down to the area of the pancreatic head. Note the dilated proximal common hepatic duct (arrow) branching off from the cystic mass, with dilated central intrahepatic bile ducts (arrowhead). (B) A delayed film from a hepatic scintigram shows a large amount of isotope uptake within the choledochal cyst (arrow). There is also isotope activity within the intestines (open arrow). (C) A CT scan shows a 15-cm choledochal cyst (C). (D) A more cephalad scan shows cystic dilatation of the right and left main hepatic ducts (arrows). The peripheral intrahepatic ducts are not dilated. K = kidney. (A and B, from Han, Babcock, and Gelfand;[4] C and D, from Araki, Itai, and Tasaka,[7] with permission from the publishers.)

Common duct stones developing in patients who have had their gallbladders removed usually represent retained calculi that were not identified at the time of cholecystectomy. After cholecystectomy, stones can develop secondary to stasis in a large cystic duct remnant.

If a T tube is still in place, residual common bile duct stones may be removed by fluoroscopically guided percutaneous basket techniques. Later, ERCP is of special value since it not only can demonstrate the stones but can also facilitate their removal by endoscopic papillotomy.[10–12]

## SURGICAL OR TRAUMATIC BILIARY STRICTURES

The overwhelming majority of benign strictures of the common bile duct are related to previous biliary tract surgery. They can be caused by severing, clamping, or excessive probing of the common bile duct during the operative procedure. In many cases, the operation is completed without the surgeon being aware that an accident to a major bile duct has occurred. Postoperatively, stricture of the bile duct can be secondary to a suture,

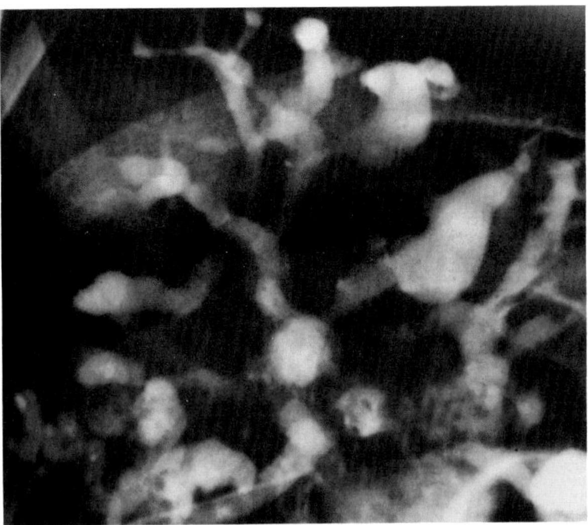

**Figure 54.4. Caroli's disease.** There is segmental saccular dilatation of the intrahepatic bile ducts throughout the liver. (From Eisenberg,[20] with permission from the publisher.)

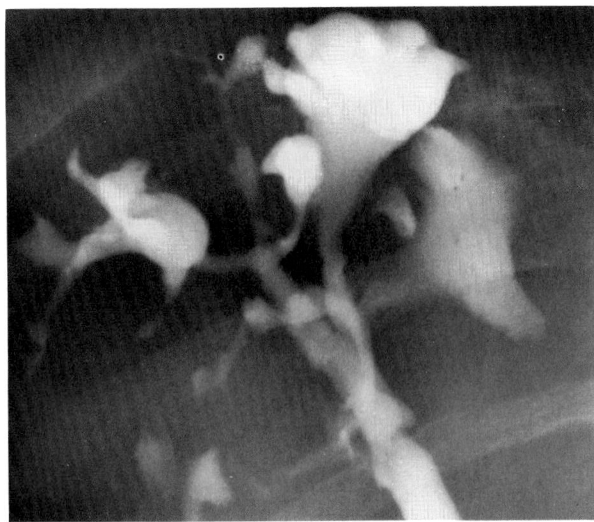

**Figure 54.5. Congenital hepatic fibrosis.** An operative cholangiogram demonstrates the "lollipop-tree" appearance of the biliary system. (From Unite, Maitem, Bagnasco, et al,[5] with permission from the publisher.)

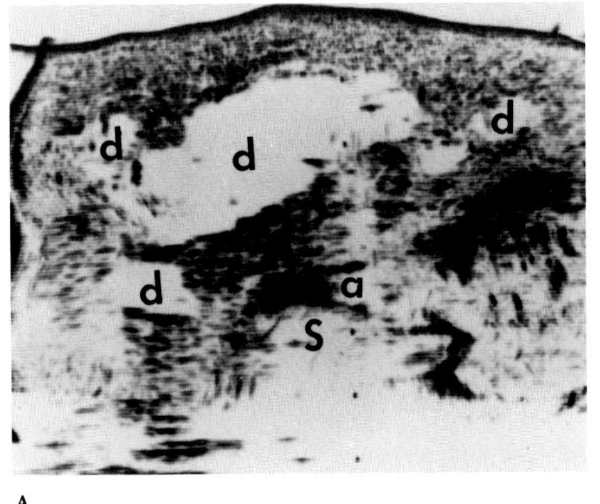


A

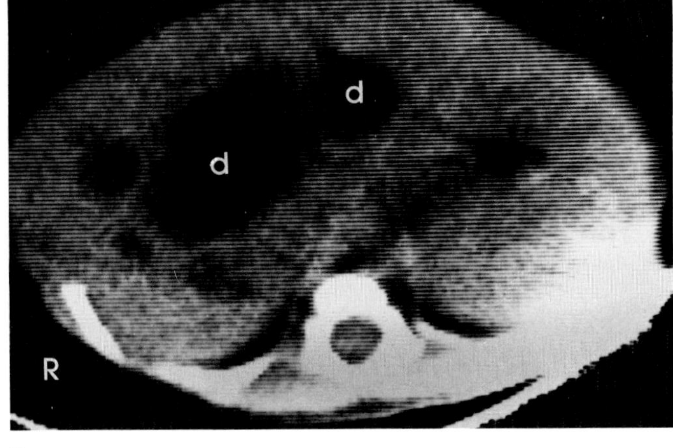


B

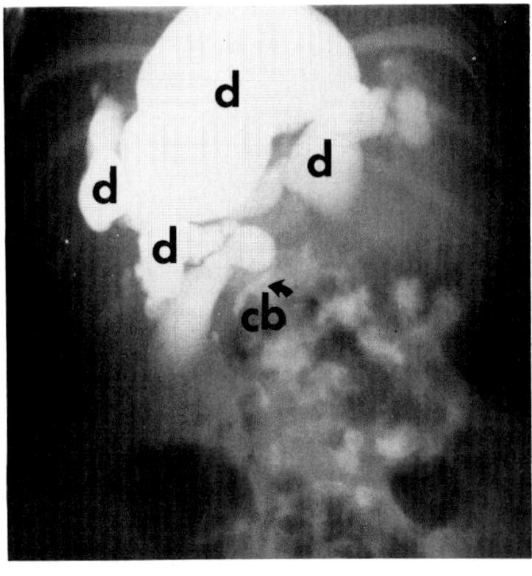


C

**Figure 54.6. Caroli's disease.** (A) A transverse supine sonogram demonstrates multiple dilated bile ducts (d) as sonolucent spaces within the liver. S = spine; a = aorta. (B) A CT scan at the same level shows the dilated ducts as low-density areas (d) that appear larger in the central part of the liver. (C) A frontal view of a transhepatic cholangiogram in a projection corresponding to A shows cystic dilatation of the distal intrahepatic ducts (d) with a normal-sized common bile duct (cb). (From Mittelstaedt, Volberg, and Fischer,[8] with permission from the publisher.)

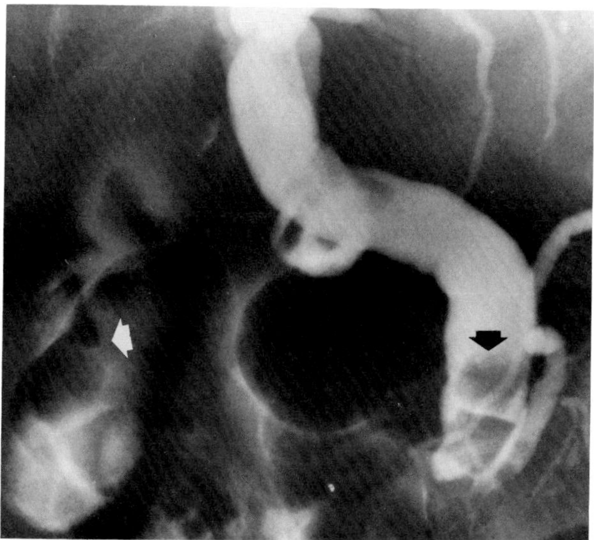

**Figure 54.7. Choledocholithiasis.** Calculi are seen within both the common bile duct (black arrow) and the gallbladder (white arrow). (From Eisenberg,[20] with permission from the publisher.)

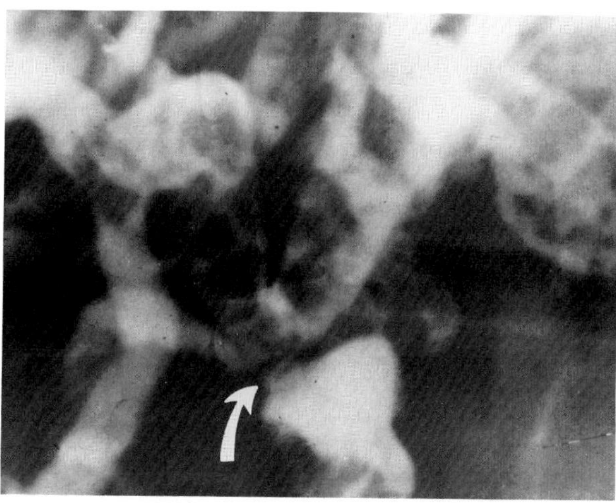

**Figure 54.8. Choledocholithiasis.** There are multiple hepatic duct stones in a patient who developed a stricture at the junction of the left and right hepatic ducts (arrow). This junction was the site of an anastomosis with the jejunum for a previous distal bile duct stricture. (From Eisenberg,[20] with permission from the publisher.)

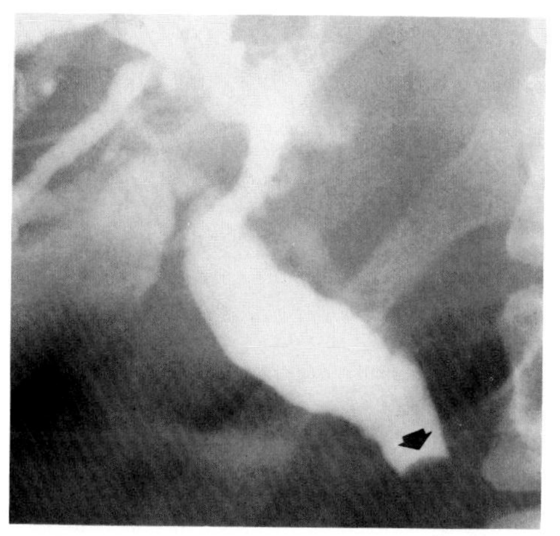

A

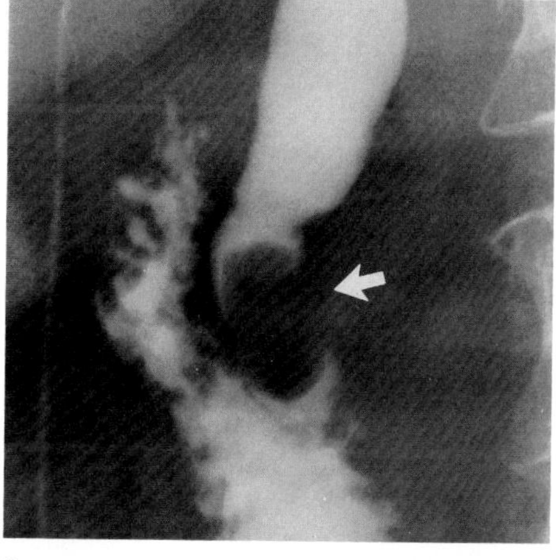

B

**Figure 54.9. Impacted ampullary stones (arrows).** (A) Characteristic smooth, concave intraluminal filling defect. (B) Unusual peanut-shaped configuration. (From Eisenberg,[20] with permission from the publisher.)

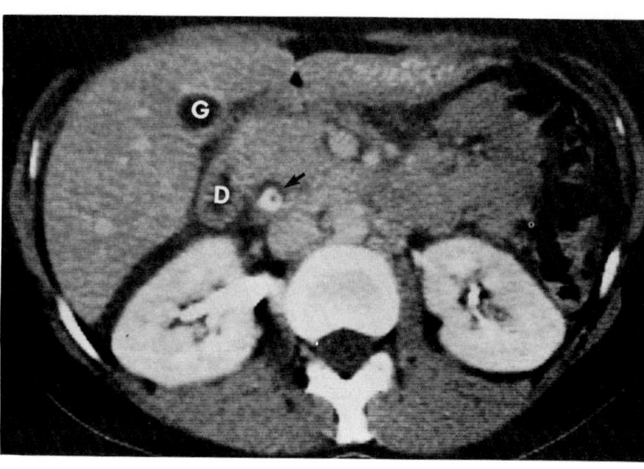


A

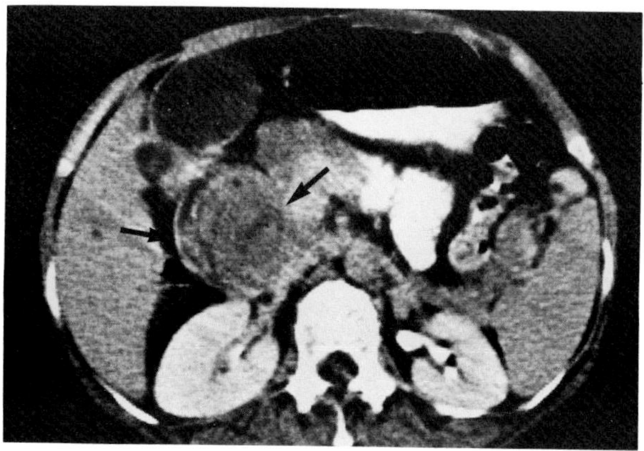

B

**Figure 54.10. CT of common duct stones.** (A) Calcified calculus (arrow). G = gallbladder; D = duodenum. (B) Large, impacted pigment stone (arrows), with marked dilatation of the common bile duct. (From Jeffrey, Federle, Laing, et al,[21] with permission from the publisher.)

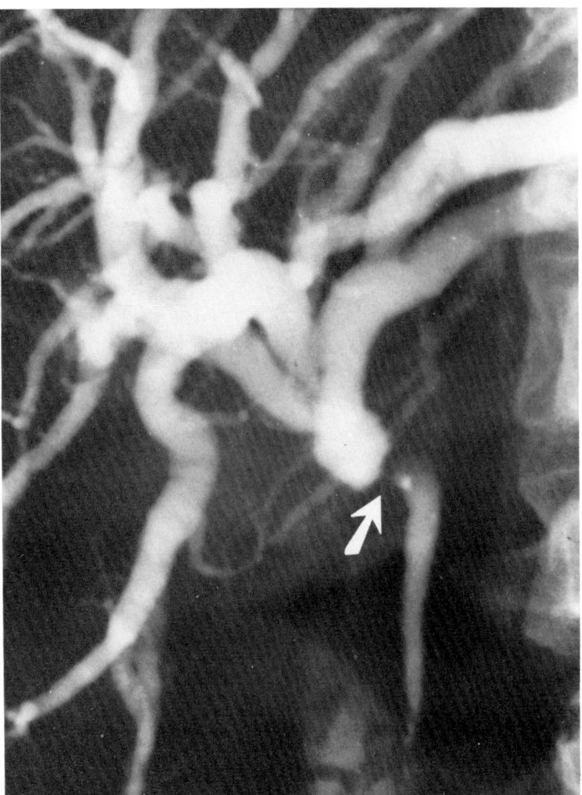

**Figure 54.11. Benign stricture of the common bile duct (arrow).** The stricture was related to previous biliary tract surgery. (From Eisenberg,[20] with permission from the publisher.)

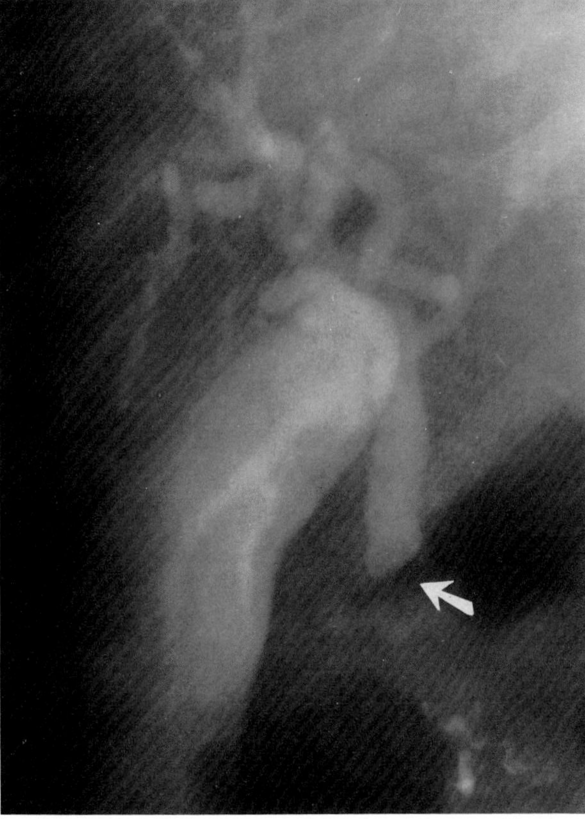

**Figure 54.12. Benign stricture of the common bile duct (arrow).** This stricture followed blunt abdominal trauma.

biliary leakage, or prolonged T-tube placement. Infrequently, biliary stricture can result from blunt abdominal trauma with torsion injury to the common bile duct.

Radiographically, the postoperative stricture is generally smooth and concentric, with the obstructed end appearing funnel-shaped or convex distally, in contrast to the concave margin produced by an obstructing calculus. Unlike malignant lesions, benign strictures tend to involve long segments of the bile duct without total obstruction; there is usually a gradual transition to normal segments of the duct.[12,13]

## SCLEROSING CHOLANGITIS

Cholangitis is usually secondary to long-standing partial obstruction of the common bile duct, which can be due to biliary calculi, parasitic infestation, malignancy, or prior surgery. Primary sclerosing cholangitis is a rare disease in which diffuse thickening and stenosis of the bile ducts develop without calculi in the gallbladder or common duct (unless clearly coincidental), a history of previous operative trauma, or evidence of malignant disease. Many cases of primary sclerosing cholangitis occur in patients with inflammatory bowel disease (both Crohn's disease and chronic ulcerative colitis), though the precise incidence and cause of this relationship are unclear. A higher than normal incidence of primary sclerosing cholangitis has also been described in patients with retroperitoneal fibrosis, mediastinal fibrosis, Riedel's thyroiditis, and retro-orbital tumors.

The diffuse periductal fibrosis in patients with sclerosing cholangitis leads to biliary strictures that are usually multiple and of variable length. Beading of the duct occurs between the narrowed segments; the degree of dilatation varies. The extrahepatic ducts are almost always involved, often with progressive involvement of the intrahepatic ducts. The smaller radicles are obliterated, resulting in a pruned-tree appearance. Because intrahepatic ductal involvement may make percutaneous transhepatic cholangiography difficult or impossible, endoscopic retrograde cholangiopancreatography is often the radiographic technique of choice.

It is sometimes very difficult, both radiographically and histologically, to distinguish between sclerosing cholangitis and diffuse sclerosing carcinoma of the bile duct, especially when the periductal glands are distorted by fibrosis and inflammation. Consequently, patients with the diagnosis of sclerosing cholangitis should be fol-

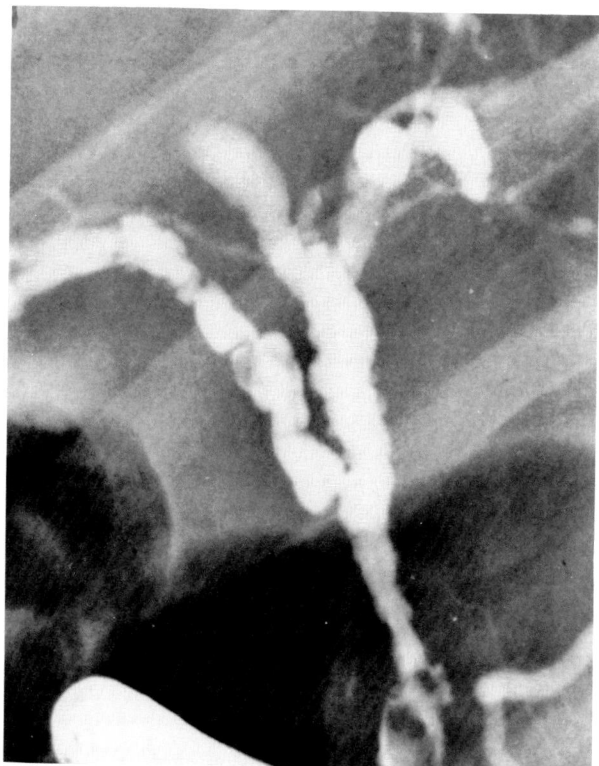

**Figure 54.13. Primary sclerosing cholangitis.** The patient suffered from chronic ulcerative colitis. (From Eisenberg,[20] with permission from the publisher.)

lowed regularly to exclude a misdiagnosed common duct carcinoma. The presence of an associated fibrosing disease (fibrosing mediastinitis or mesenteritis, retroperitoneal fibrosis) supports the diagnosis of sclerosing cholangitis.[12–16]

## TUMORS OF THE BILE DUCTS

### Malignant Tumors

Primary carcinomas of the bile ducts (cholangiocarcinoma) are almost invariably adenocarcinomas that usually present with obstructive jaundice. Carcinoma can occur at any site along the bile ducts. The most common locations are in the retroduodenal or supraduodenal segments of the common bile duct and in the common hepatic duct at the carina. Because of their infiltrative nature, most bile duct carcinomas are far advanced at the time of diagnosis, with regional lymph node metastases and extension along the bile ducts. Tumors arising at the junction of the right and left hepatic ducts (Klatskin tumors) behave as distinct clinical entities; they tend to grow slowly and to metastasize late.

Cholangiocarcinoma most commonly presents radiographically as a short, well-demarcated, segmental constriction. The tumor usually begins as a plaquelike lesion of the wall that infiltrates and spreads along the duct in both directions. An extensive desmoplastic response tends to produce diffuse narrowing of the duct. An obstructing

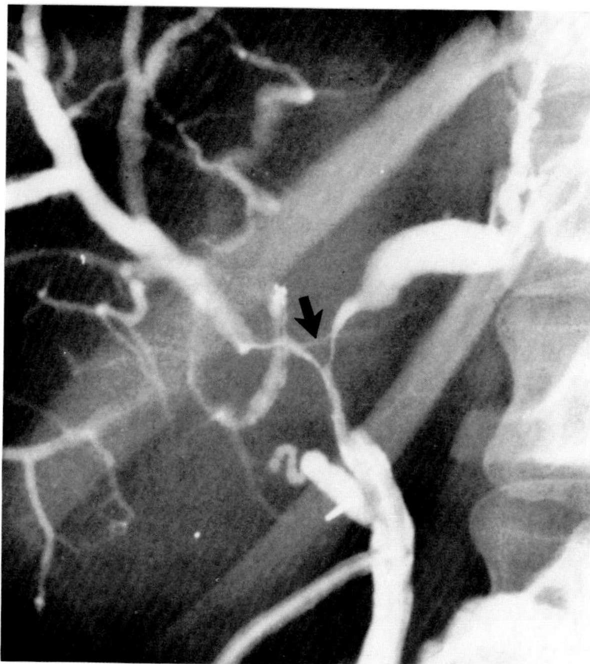

**Figure 54.14. Klatskin tumor.** A sclerosing cholangiocarcinoma (arrow) arises at the junction of the right and left hepatic ducts. (From Eisenberg,[20] with permission from the publisher.)

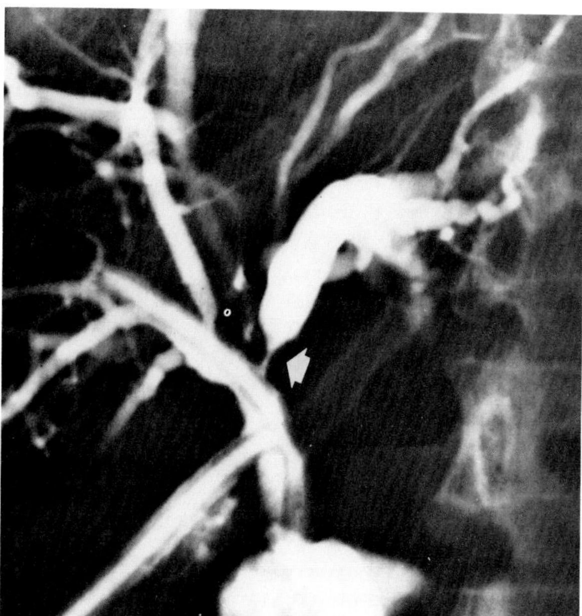

**Figure 54.15. Cholangiocarcinoma.** There is a short, well-demarcated, segmental constriction (arrow) at the origin of the left hepatic duct. (From Eisenberg,[20] with permission from the publisher.)

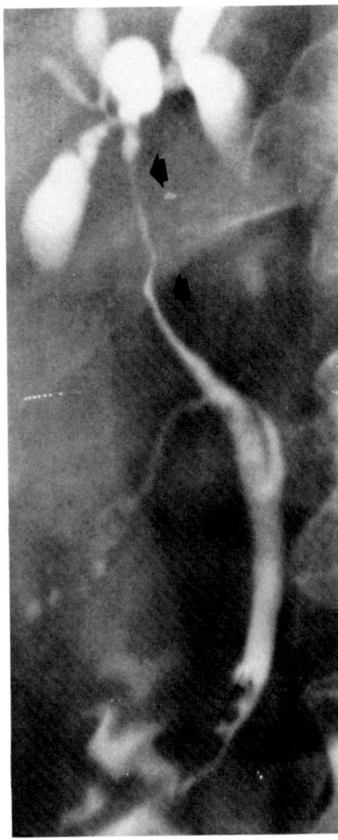

**Figure 54.16. Cholangiocarcinoma.** There is severe narrowing of a long segment of the common hepatic duct. (From Eisenberg,[20] with permission from the publisher.)

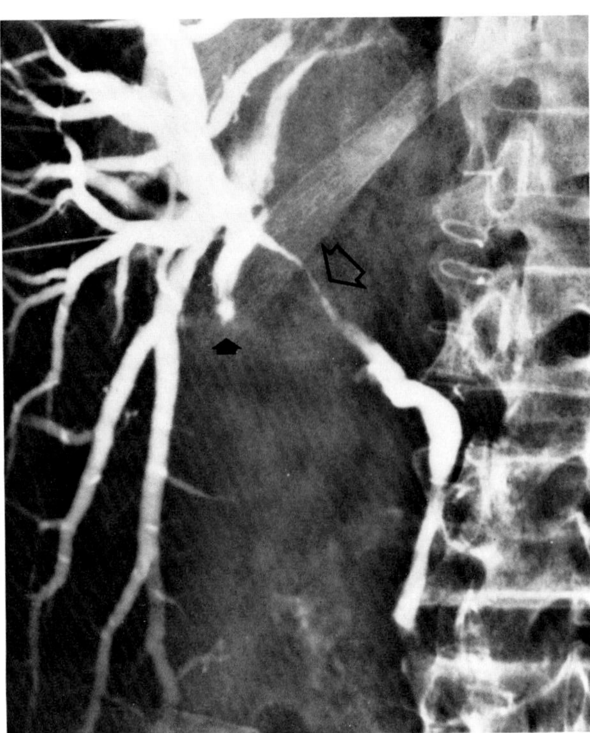

**Figure 54.17. Multicentric cholangiocarcinoma.** In addition to severe narrowing of the common hepatic duct (open arrow), there is obstruction of a branch of the right hepatic duct (solid arrow) and poor filling of the left hepatic radicles.

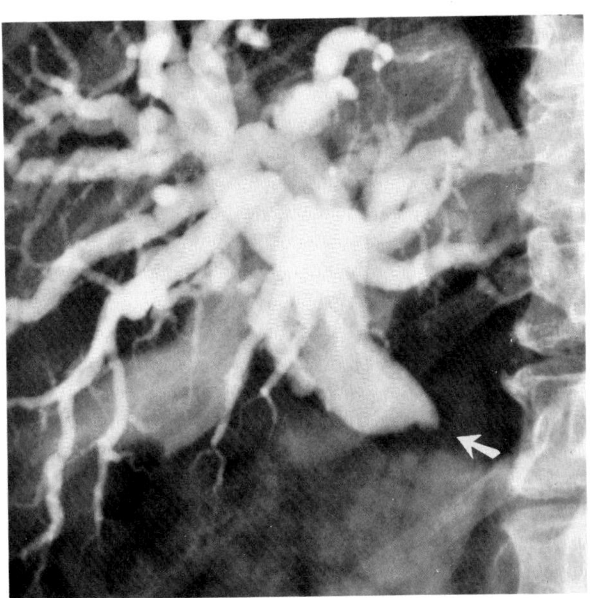

**Figure 54.18. Extrinsic obstruction of the common bile duct (arrow).** The obstruction was due to nodal metastases from carcinoma of the colon.

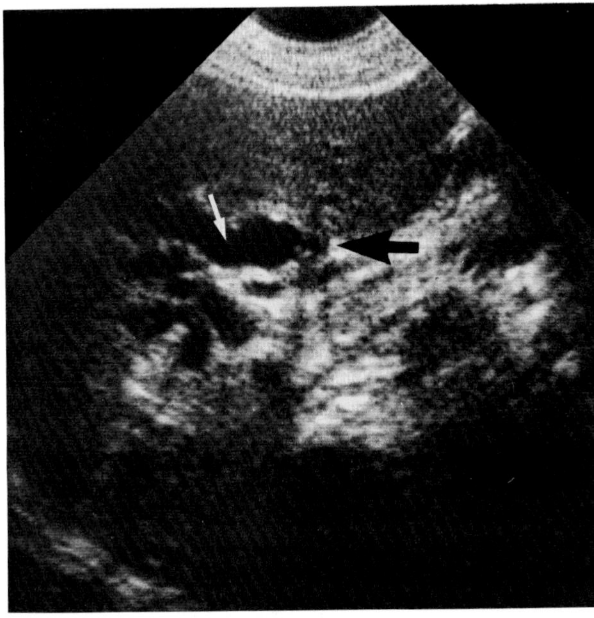

**Figure 54.19. Cholangiocarcinoma.** A transverse sonogram through the liver demonstrates the tumor (black arrow) obstructing the common hepatic duct (white arrow).

tumor most often causes abrupt occlusion of the common bile duct with proximal dilatation. Although there is usually little intraluminal extension of tumor, cholangiocarcinoma occasionally appears as a discrete, bulky polypoid tumor producing a filling defect in the opacified ductal system. Rarely, cholangiocarcinoma is multicentric and, because of the fibrosing nature of the disease, can produce an appearance of stricture-like narrowing with proximal dilatation that is indistinguishable from sclerosing cholangitis.

Narrowing of the main hepatic or common bile ducts can also be produced by adjacent primary malignancies or metastases. Carcinoma of the head of the pancreas can encircle and asymmetrically narrow or obstruct the common bile duct. Similarly, primary carcinoma of the duodenum or gallbladder can extend to involve the bile ducts. Metastases to lymph nodes in the porta hepatis or along the medial margin of the descending duodenum can cause extrinsic obstruction of bile ducts. These metastases are usually secondary to primary malignancies of the gastrointestinal tract but can also represent spread from carcinoma of the lung or breast. The diffuse desmoplastic reaction evoked by metastases can simulate the appearance of primary cholangiocarcinoma.

Although direct invasion of periductal structures and the porta hepatis at the time of clinical presentation usually precludes surgical resection of a cholangiocarcinoma, successful percutaneous or surgically placed biliary drainage catheters can often prolong patient survival and improve the quality of life.[17–19]

**Figure 54.20. Adenomatous polyp.** Intravenous cholangiography demonstrates a smooth filling defect (arrow) in the common bile duct. (From Eisenberg,[20] with permission from the publisher.)

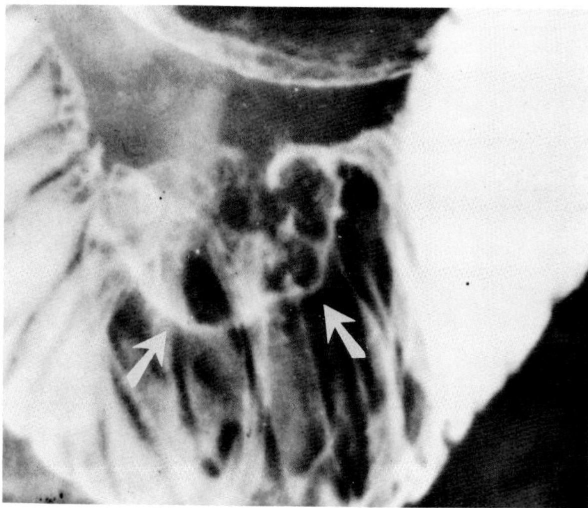

**Figure 54.21. Villous adenocarcinoma of the ampulla.** An upper gastrointestinal series demonstrates irregular enlargement of the papilla (arrows). (From Eisenberg,[20] with permission from the publisher.)

## Benign Tumors

Benign neoplasms of the bile ducts are extremely rare. Most are papillomas or adenomas that arise in the distal portion of the biliary tree. Adenomas are usually single,

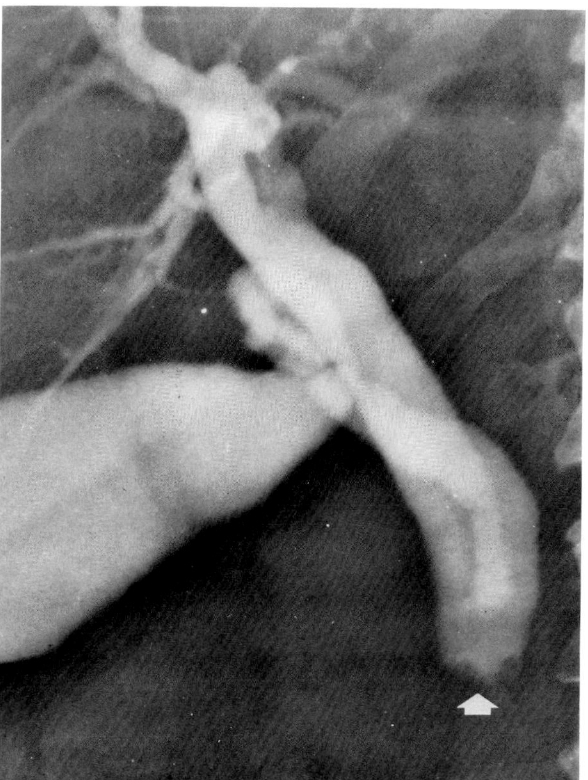

**Figure 54.22. Villous adenoma of the ampulla of Vater.** A transhepatic cholangiogram shows a lobulated obstruction of the distal common bile duct (arrow). (From Eisenberg,[20] with permission from the publisher.)

whereas papillomas are not uncommonly of multifocal origin. Benign tumors of the bile ducts usually appear radiographically as small polypoid filling defects associated with some degree of obstruction. Because their margins tend to be smoothly rounded, these benign tumors often closely resemble biliary stones.[20]

## CARCINOMA OF THE PAPILLA AND AMPULLA OF VATER

Papillary or ampullary carcinoma may appear on upper gastrointestinal studies as enlargement of the papilla, similar to that caused by an impacted common duct stone, acute pancreatitis, or duodenal ulcer disease. Unlike these nonmalignant processes, carcinoma produces a papillary surface that is often irregular and may demonstrate local erosion and infiltration. Papillary carcinoma occasionally has a smooth surface and appears identical to a benign edematous process.

Obstruction of the bile duct at its distal end suggests the possibility of carcinoma of the ampulla of Vater. These small neoplasms can appear as irregular or polypoid masses or merely cause distal common bile duct obstruction without a demonstrable tumor mass.[10]

### REFERENCES


1. Babbitt DP, Starshak RJ, Sty JR: Choledochal cyst: Pathogenesis, diagnosis, and surgical implications. *Applied Radiol* Nov–Dec, 1981.
2. Rabinowitz JG, Kinkhabwala MN, Rose JS: Rim sign in choledochal cyst: Additional diagnostic feature. *J Can Assoc Radiol* 24:226–230, 1973.
3. Araki T, Itai Y, Tasaka A: Computed tomography of choledochal cyst. *AJR* 135:729–734, 1980.
4. Han BK, Babcock DS, Gelfand MH: Choledochal cyst with bile duct dilatation: Sonography and $^{99m}$Tc IDA cholescintigraphy. *AJR* 136:1075–1079, 1981.
5. Unite I, Maitem A, Bagnasco FM, et al: Congenital hepatic fibrosis associated with renal tubular ectasia. *Radiology* 109:565–570, 1973.
6. Lucaya J, Gomez JL, Molino C, et al: Congenital dilatation of the intrahepatic bile ducts (Caroli's disease). *Radiology* 127:746, 1978.
7. Caroli J: Diseases of the intrahepatic biliary tree. *Clin Gastroenterol* 2:147–161, 1973.
8. Mittelstaedt CA, Volberg FM, Fischer GJ, et al: Caroli's disease: Sonographic findings. *AJR* 134:585–587, 1980.
9. Kaiser JA, Mall JC, Salmen BJ, et al: Diagnosis of Caroli's disease by computed tomography. *Radiology* 132:661–664, 1979.
10. Berk RN, Ferrucci J, Leopold GR: *Radiology of the Gallbladder and Bile Ducts*. Philadelphia, WB Saunders, 1983.
11. Way LW: Retained common duct stones. *Surg Clin North Am* 53:1169–1190, 1973.
12. Menuck L, Amberg J: The bile ducts. *Radiol Clin North Am* 14:499–523, 1976.
13. Larsen CR, Scholz FJ, Wise RE: Diseases of the biliary ducts. *Semin Roentgenol* 11:259–267, 1976.
14. Geisse G, Melson GL, Tedesco FH, et al: Stenosing lesions of the biliary tree. Evaluation with endoscopic retrograde cholangiopancreatography and percutaneous transhepatic cholangiography. *AJR* 123:378–385, 1975.
15. Krieger J, Seaman WB, Porter MR: The roentgenologic appearance of sclerosing cholangitis. *Radiology* 94:369–374, 1970.
16. Rohrmann CA, Ansel HJ, Freeny PC, et al: Cholangiographic abnormalities in patients with inflammatory bowel disease. *Radiology* 127:635–641, 1978.
17. Van Sonnenberg E, Ferrucci JT: Bile duct obstruction in hepatocellular carcinoma (hepatoma): Clinical and cholangiographic characteristics. *Radiology* 130:7–13, 1979.
18. Klatskin G: Adenocarcinoma of the hepatic duct at its bifurcation within the porta hepatis: An unusual tumor with distinctive clinical and pathological features. *Am J Med* 38:241–256, 1965.
19. Legge BA, Carlson HC: Cholangiographic appearance of primary carcinoma of the bile ducts. *Radiology* 102:259–266, 1972.
20. Eisenberg RL: *Gastrointestinal Radiology: A Pattern Approach*. Philadelphia, JB Lippincott, 1983.
21. Jeffrey RB, Federle MP, Laing F, et al: Computed tomography of choledocholithiasis. *AJR* 140:1179–1185, 1983.

## Chapter Fifty-Five
# Diseases of the Pancreas

### RADIOGRAPHIC APPROACH TO DISEASES OF THE PANCREAS

As a noninvasive screening procedure, ultrasound is often the initial imaging modality in patients with suspected pancreatic disease. Ultrasound can outline the contour of the gland, and the internal echographic appearance can suggest underlying edema, inflammation, and calcification. Although there is overlap between benign and malignant conditions, ultrasound can successfully identify about 90 percent of patients with an abnormal pancreas. The major limitation of ultrasound is the frequent occurrence of excessive overlying intestinal gas, especially in patients with acute pancreatitis, which can severely limit the technical quality of the examination. A recent barium contrast examination can also interfere with the performance of ultrasound studies.

High-resolution computed tomography (CT) is the most effective imaging modality for detecting a pancreatic lesion, assessing its extent, and defining its cause. Oral water-soluble contrast agents are often used to opacify the stomach and duodenum and thus distinguish these organs from the pancreas. Although CT can be difficult to interpret in thin patients with little mesenteric or retroperitoneal fat, large amounts of intestinal gas present no problem, a major advantage of this modality over ultrasound.

In patients with pancreatitis, plain abdominal radiographs can demonstrate calcifications or a mass in the region of the pancreas, focal dilatation of adjacent bowel loops due to adynamic ileus, and, rarely, a mottled appearance due to saponification of the peritoneal fat by pancreatic enzymes. However, results of plain radiographs rarely influence the treatment of patients with suspected pancreatitis and are of no value in assessing the patient with suspected pancreatic neoplasm.

The value of barium studies in patients with pancreatic disease is severely limited, because they detect changes only when the pathologic process displaces or infiltrates the bowel. Large pancreatic lesions may not be seen on barium studies, even using hypotonic techniques. Barium studies are far more sensitive for detecting lesions in the head of the pancreas and the region of the ampulla of Vater than for processes involving the body or tail of the gland. Even when an abnormality is detected, barium studies rarely permit distinction between benign and malignant disease. The high sensitivity and specificity of the newer imaging modalities, such as ultrasound and CT, have essentially eliminated the role of barium studies in the evaluation of inflammatory or malignant diseases of the pancreas.

The inadequacies of barium studies led to the extensive use of selective catheterization of the celiac and superior mesenteric arteries, especially with subselective and magnification techniques, in the evaluation of patients with suspected pancreatic disease. However, the less expensive and safer modalities of ultrasound and CT have effectively supplanted arteriography in these patients. Although arteriography has been of value in staging pancreatic malignancy in terms of surgical resectability, this role is rapidly being taken over by CT.

Endoscopic retrograde cholangiopancreatography (ERCP) is primarily an endoscopic procedure, generally not directly performed by radiologists, in which contrast administration under fluoroscopy is used to document the appearance of the pancreatic duct or biliary tree. Because of the endoscopic skill required to be routinely successful, ERCP is much less readily available than other diagnostic modalities for evaluating the pancreas. ERCP can establish the diagnosis of pancreatic carcinoma with great accuracy, especially when combined with aspiration cytology via the cannula. However, with ERCP it may be difficult to separate benign from malignant strictures in patients with chronic pancreatitis. The modality is of little value in staging disease, since it does not provide information about invasion of local structures or distant metastases to the liver. ERCP is usually contraindicated in patients with acute pancreatitis.

Fine-needle aspiration cytology of the pancreas under ultrasound or CT guidance is now an acceptable and rapid method for establishing the definitive diagnosis of a pancreatic neoplasm. Because it is common to find a small primary malignancy with much surrounding inflammatory disease, tissue sampling errors may occur; therefore, a negative cytology does not exclude malignancy. The limiting factor of fine-needle aspiration is that it requires a good cytologist to make an accurate histologic diagnosis. Although seeding of tumor along the needle tract is a theoretically possible complication, fine-needle aspiration cytology has been essentially free of any complications.[1]

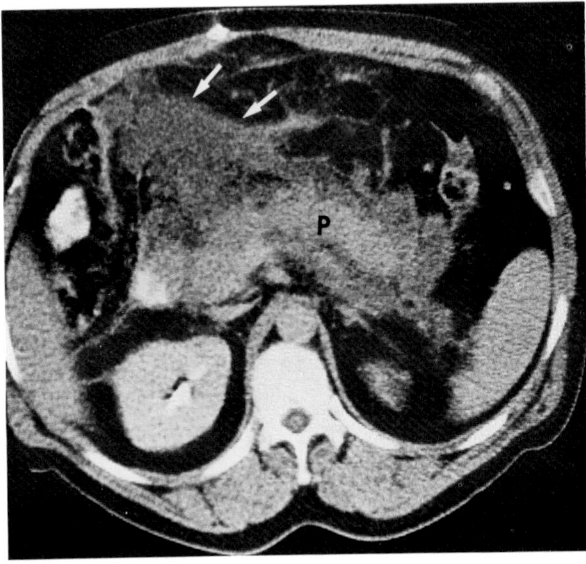


**Figure 55.1. Acute pancreatitis.** A CT scan demonstrates diffuse enlargement of the pancreas (P) with obliteration of peripancreatic fat planes by the inflammatory process. Note the extension of the inflammatory reaction into the transverse mesocolon (arrows). (From Jeffrey, Federle, Laing,[24] with permission from the publisher.)

## ACUTE PANCREATITIS

CT and ultrasound are the imaging modalities that can most precisely define the degree of pancreatic inflammation and the pathways of its spread throughout the abdomen. They also are of great clinical importance in

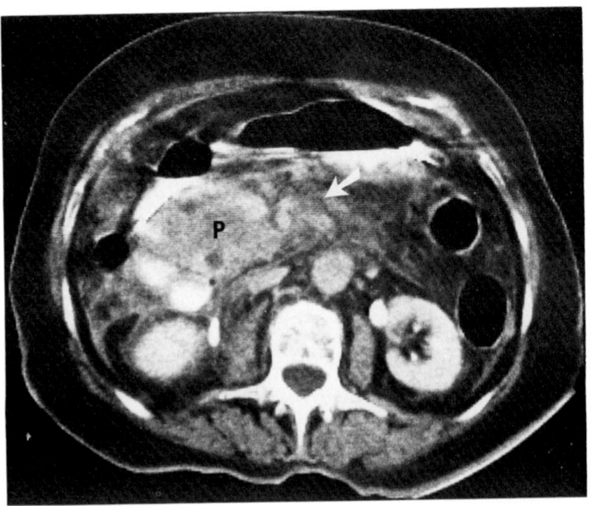


A

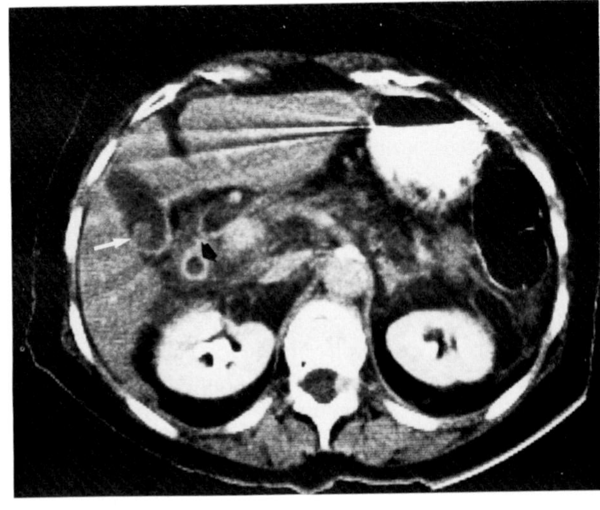

B

**Figure 55.2. Acute gallstone pancreatitis.** There is enlargement of the head of the pancreas (P) with inflammatory reaction surrounding peripancreatic fat planes. A stone (white arrow) is seen in the gallbladder, and the common bile duct is enlarged (black arrow).

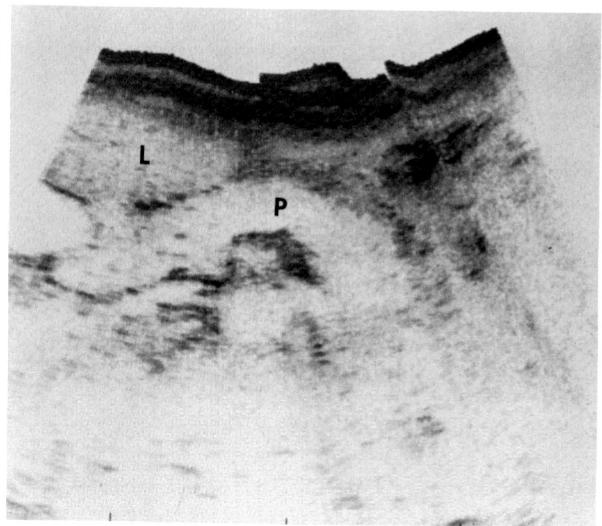


**Figure 55.3. Acute pancreatitis.** A transverse sonogram demonstrates diffuse enlargement of the gland with retention of its normal shape. Note the relative sonolucency of the pancreas (P) when compared to the echogenicity of the adjacent liver (L).

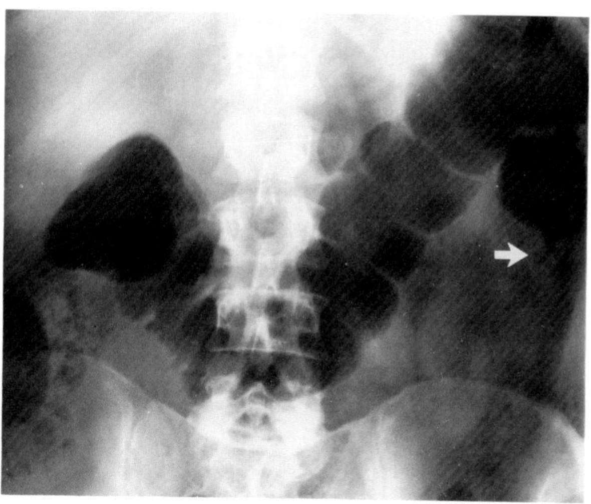

**Figure 55.4. Colon cutoff sign of acute pancreatitis.** The colon gas column is abruptly cut off just distal to the splenic flexure (arrow).

the early diagnosis of complications of acute pancreatitis, such as abscess, hemorrhage, and pseudocyst formation.

CT in acute pancreatitis demonstrates diffuse or focal enlargement of the gland. In the normal patient, the margins of the pancreas are sharply delineated by surrounding peripancreatic fat. Extrapancreatic spread of inflammation and edema (especially into the anterior perirenal space, lesser peritoneal sac, and transverse mesocolon) obscures the peripancreatic soft tissues and often thickens the surrounding fascial planes.

Acute pancreatitis may alter both the size and the parenchymal echogenicity of the gland on ultrasound. Although the pancreas usually enlarges symmetrically and retains its initial shape, nonspecific enlargement of the pancreatic head or tail can simulate focal pancreatic carcinoma. The accompanying interstitial inflammatory edema causes the pancreas to appear relatively sonolucent when compared to the adjacent liver. In the presence of hemorrhage or necrosis, however, fat, clotted blood, or peripancreatic debris may produce areas of increased echogenicity. Ultrasound is an excellent method for demonstrating cholelithiasis, an important cause of acute pancreatitis; however, it is less precise than CT in defining fascial compartments and pathways of extraperitoneal spread of inflammation. One limitation of ultrasound in patients with acute pancreatitis is the frequent occurrence of adynamic ileus with excessive intestinal gas, which may prevent adequate visualization of the gland.

Plain abdominal radiographs are often normal in the patient with acute pancreatitis; even when abnormal, the findings are usually nonspecific and consistent with any intra-abdominal inflammatory disease. The most fre-

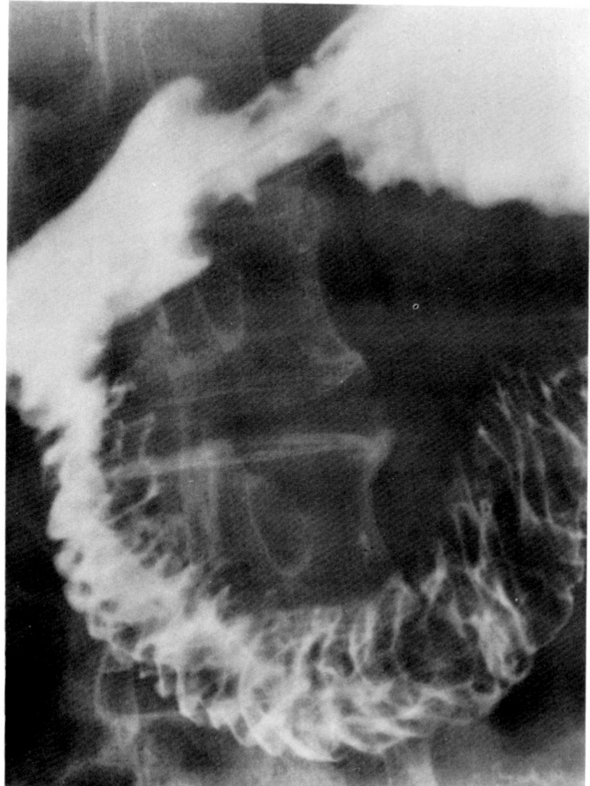

**Figure 55.5. Acute pancreatitis.** Barium upper gastrointestinal series demonstrates thickening of duodenal folds with spiculations. There is widening of the duodenal sweep with pressure effects on the medial border. (From Eisenberg,[6] with permission from the publisher.)

quent abnormalities include a localized adynamic ileus, usually involving the jejunum ("sentinel loop"); generalized ileus with diffuse gas-fluid levels; isolated distension of the duodenal sweep; and localized distension of the transverse colon to the level of the splenic flexure (colon cutoff sign). Pancreatic calcifications indicate that the patient has chronic pancreatitis and may suggest an exacerbation of the inflammatory disease.

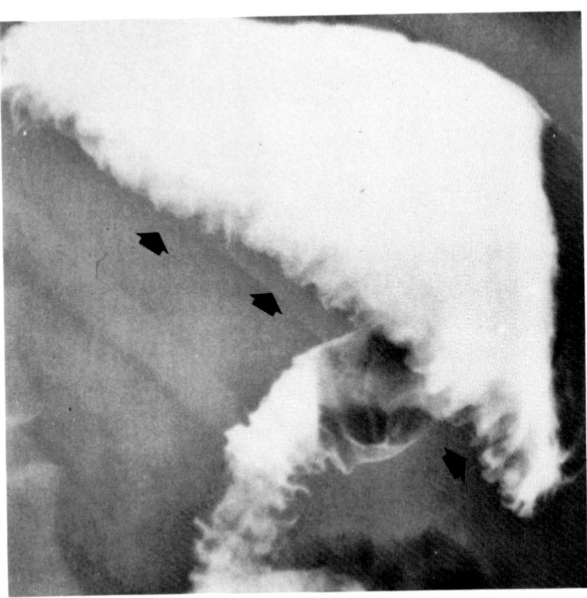

**Figure 55.6. Acute pancreatitis.** Prominence of rugal folds on the posterior wall of the stomach (arrows) is associated with a large retrogastric mass. (From Eisenberg,[6] with permission from the publisher.)

Plain chest radiographs may demonstrate elevation of the left hemidiaphragm, linear streaks of atelectasis at the bases, and pleural effusions, nonspecific findings that can be seen in patients with upper abdominal inflammatory disease from any cause.

Barium upper gastrointestinal series often demonstrate changes in the duodenal sweep in patients with acute pancreatitis. These include thickened folds in the duodenum, widening of the sweep with pressure effects on the medial border, and swelling of the papilla (Poppel's sign). Prominent rugal folds may be seen on the posterior and greater curvature aspects of the stomach. Spread of pancreatic enzymes may cause spasm, thickened folds, and obstruction in the transverse colon or ileocecal region. However, the availability of ultrasound and CT has essentially eliminated the usefulness of barium examinations in the diagnosis of acute pancreatitis.[2–6]

## COMPLICATIONS OF ACUTE PANCREATITIS

### Pancreatic Pseudocysts

Pancreatic pseudocysts are loculated fluid collections arising from inflammation, necrosis, or hemorrhage associated with acute pancreatitis or trauma. On ultrasound, a pseudocyst typically appears as an echo-free cystic structure with a sharp posterior wall. Hemorrhage into the pseudocyst produces a complex fluid collection

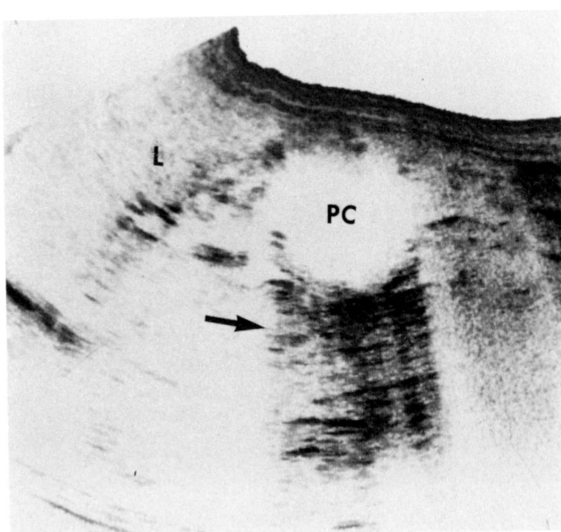


**Figure 55.7. Pancreatic pseudocyst.** A longitudinal sonogram of the right upper quadrant demonstrates an irregularly marginated pseudocyst (PC) with acoustic shadowing (arrow). L = liver.

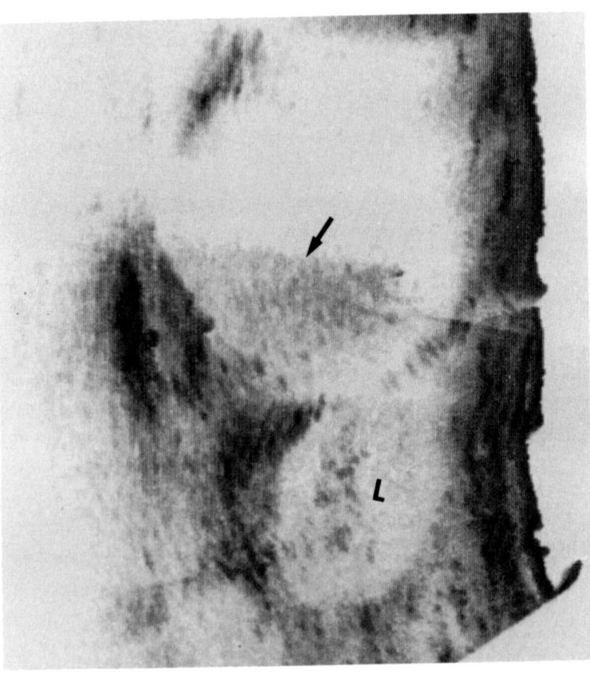


**Figure 55.8. Pancreatic pseudocyst.** An erect sonogram demonstrates a fluid-debris level (arrow) within the pseudocyst. L = left kidney.

containing septations or echogenic areas. CT demonstrates pseudocysts as sharply marginated, fluid-filled collections that are often best delineated following the administration of intravenous contrast material. Because of its ability to image the entire body, CT may demonstrate pseudocysts that have dissected superiorly into the mediastinum or to other ectopic locations, such as the lumbar or inguinal regions or within the liver, spleen, or kidney.

Large pseudocysts are visible on plain radiographs of the abdomen when they displace the gas-filled stomach and bowel. Similarly, pseudocysts in the head of the pancreas can cause pressure defects and widening of the duodenal sweep, while those arising from the body or tail of the pancreas can displace and deform the stomach, proximal jejunum, or colon. However, since ultrasound is highly accurate in diagnosing pancreatic pseudocysts, it has completely replaced plain abdominal radiographs and barium studies, which display only indirect signs.

Pseudocysts may undergo spontaneous resolution or persist as chronic collections that may require surgical

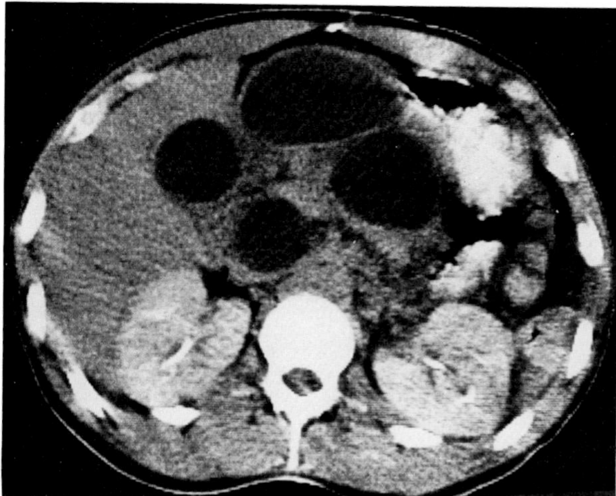

**Figure 55.9. Multiple pancreatic pseudocysts.** A CT scan following the intravenous administration of contrast material demonstrates four sharply marginated, fluid-filled collections.

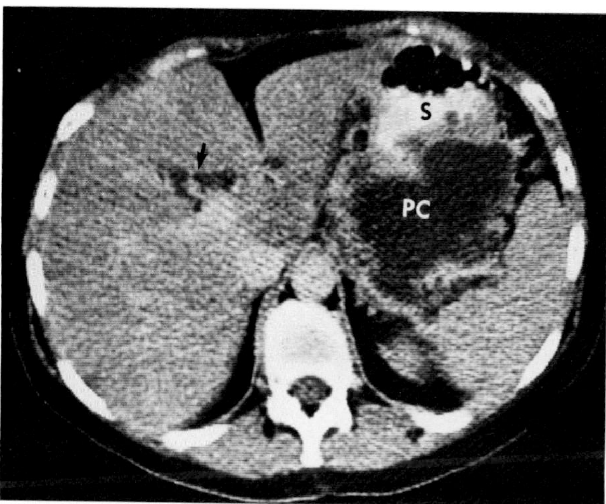


**Figure 55.10. Ectopic pancreatic pseudocyst.** A CT scan shows the pseudocyst (PC) in the superior recess of the lesser sac posterior to the stomach (S). Note the dilated intrahepatic bile ducts (arrow).

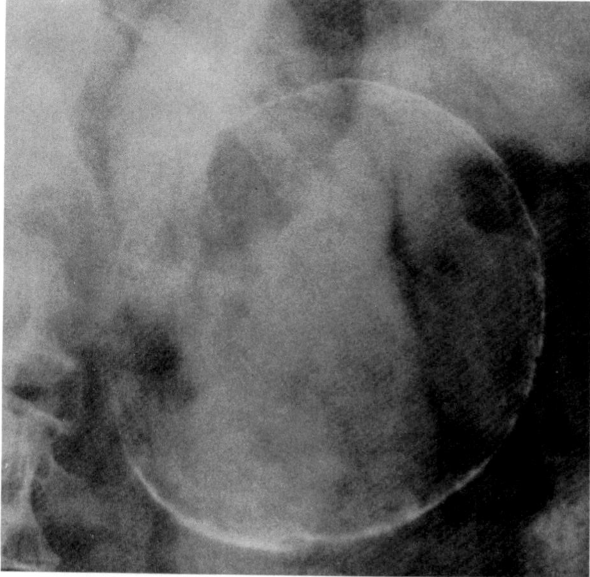

**Figure 55.11. Calcified pancreatic pseudocyst.** A plain abdominal radiograph demonstrates a rim of calcification outlining the wall of the huge pseudocyst. (From Eisenberg,[6] with permission from the publisher.)

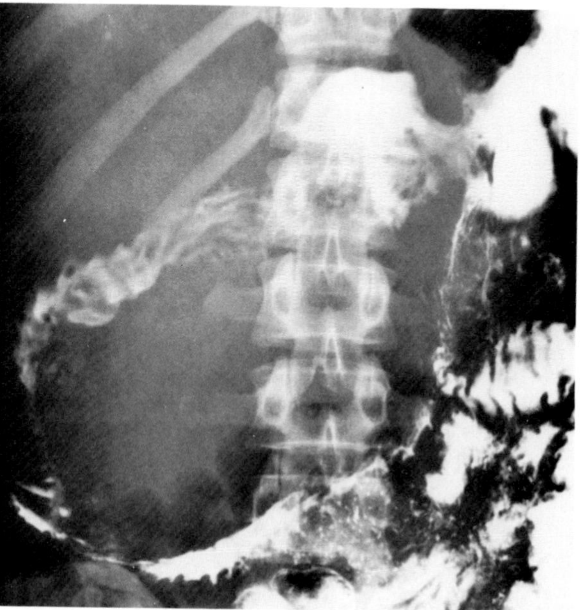

**Figure 55.12. Pancreatic pseudocyst.** There is huge enlargement of the duodenal sweep. (From Eisenberg,[6] with permission from the publisher.)

intervention. Because ultrasound is relatively inexpensive and without ionizing radiation, serial sonograms are usually used to monitor the progression of a pancreatic pseudocyst.[2,5,7–9]

### Pancreatic Abscess

Pancreatic abscess is a serious and often fatal complication in severe cases of acute pancreatitis. Early CT scanning in patients with suspected pancreatic abscess may significantly lower the morbidity and mortality by permitting prompt surgical intervention. The presence of gas within the pancreatic bed strongly suggests a pancreatic abscess, though this finding may also be demonstrated in patients with a pancreatic pseudocyst that erodes into the gastrointestinal tract without abscess formation. Because pancreatic abscesses often have a nonspecific CT appearance, diagnostic needle aspiration under CT guidance is an extremely useful adjunct for early diagnosis.

On plain films of the abdomen, pancreatic abscess may produce mottled areas of increased and decreased density in the region of the pancreatic bed.[10,11]

### Intra-abdominal Hemorrhage

Severe hemorrhage is an infrequent, though potentially lethal, complication of necrotizing pancreatitis. The proteolytic enzymes liberated during an attack of acute pancreatitis may cause erosion of the adventitia of adjacent vessels and produce gastrointestinal hemorrhage or pseudoaneurysms. Selective visceral arteriography may demonstrate the bleeding site; transcatheter embolization may be successful in controlling bleeding and may obviate surgical intervention. Hemorrhage within a pancreatic pseudocyst may appear as increased echogenicity on ultrasound or as an increase in attenuation values on CT.[6]

## CHRONIC PANCREATITIS

Pancreatic calcifications are an essentially pathognomonic finding in chronic pancreatitis, developing in about one-third of patients with this disease. The small, irregular calcifications are seen most frequently in the head of the pancreas and can extend upward and to the left to involve the body and tail of the organ. Although usually confined to the parenchyma, pancreatic calcifications can also be found within the ductal system.

Alteration of the intrinsic echo pattern is a major feature of the ultrasound diagnosis of chronic pancreatitis. This reflects generalized or local increased tissue reflectivity that is particularly obvious in calcific pancreatitis but can also be due to fibrosis in noncalcific disease. The pancreas often is atrophic in chronic pancreatitis; in subacute disease or with acute recurrence, however,

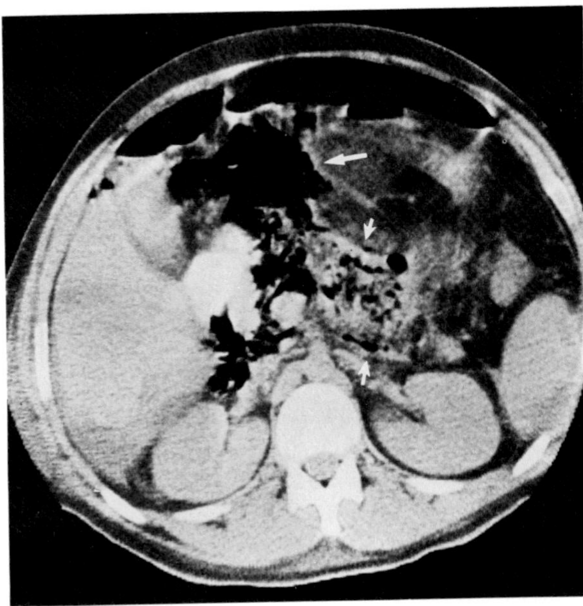

**Figure 55.13. Pancreatic abscess.** A CT scan demonstrates a gas-containing abscess (small arrows) in the pancreatic bed, with marked anterior extension (large arrow) of the inflammatory process.

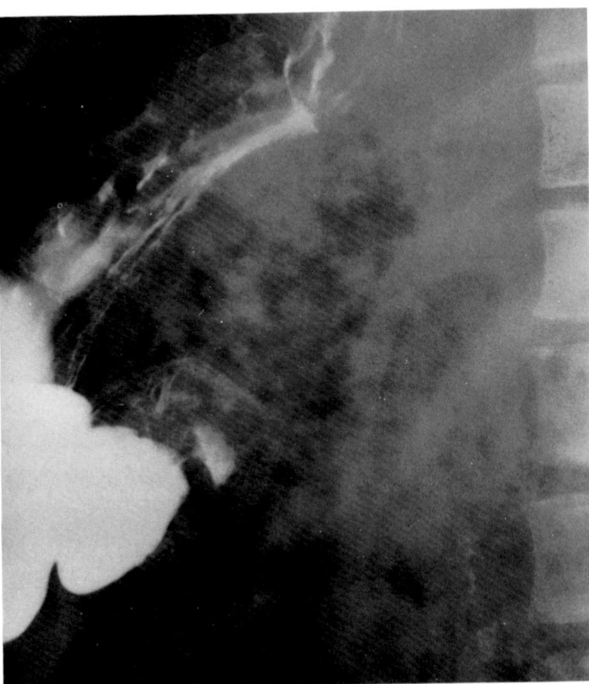

**Figure 55.14. Pancreatic abscess.** The characteristic mottled pattern of speckled radiolucencies, with normal fat intermingled with areas of water density, involves much of the retroperitoneal space.

the pancreas can be significantly enlarged and the echo pattern can become increasingly sonolucent. Dilatation of the pancreatic duct due to gland atrophy and obstruction can be seen, though a similar pattern can be produced by ductal obstruction in pancreatic cancer. CT can also demonstrate ductal dilatation, calcification, and atrophy of the gland in patients with chronic pancreatitis. How-

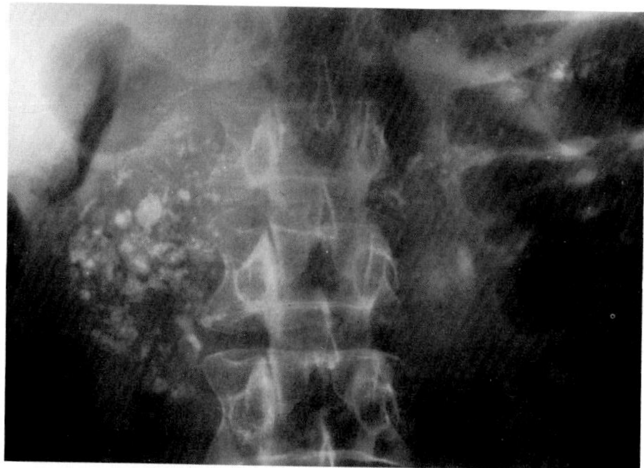

**Figure 55.15. Chronic pancreatitis.** Diffuse pancreatic calcifications in chronic alcoholic pancreatitis.

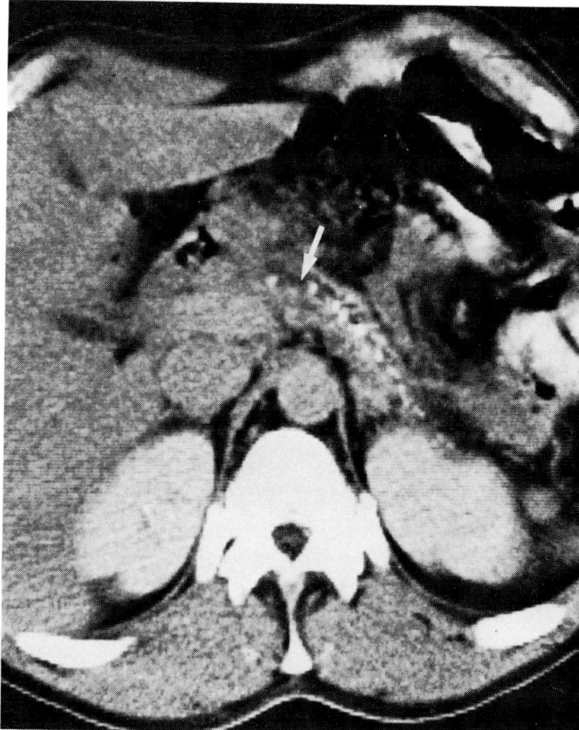

**Figure 55.17. Chronic pancreatitis.** A CT scan shows pancreatic atrophy along with multiple intraductal calculi and dilatation of the pancreatic duct (arrow). The calcifications were not seen on plain abdominal radiographs. (From Federle and Goldberg,[14] with permission from the publisher.)

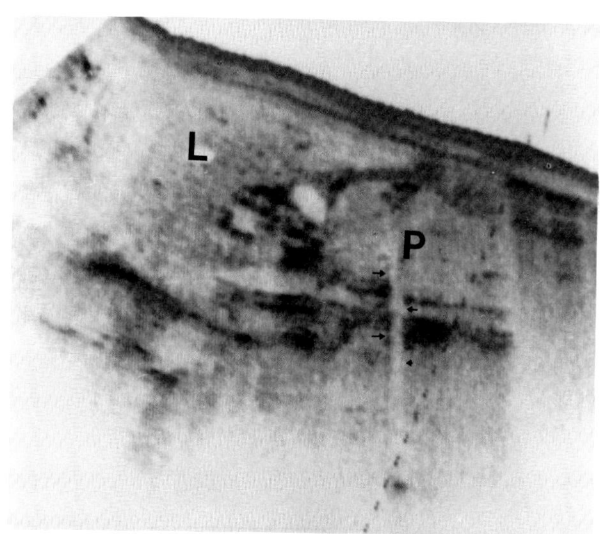


**Figure 55.16. Chronic pancreatitis.** A transverse sonogram shows a large pancreas (P) containing calcification which produces acoustic shadowing (arrows). The intrahepatic bile ducts are not enlarged. L = liver. (Courtesy of Gretchen A. W. Gooding, M.D.)

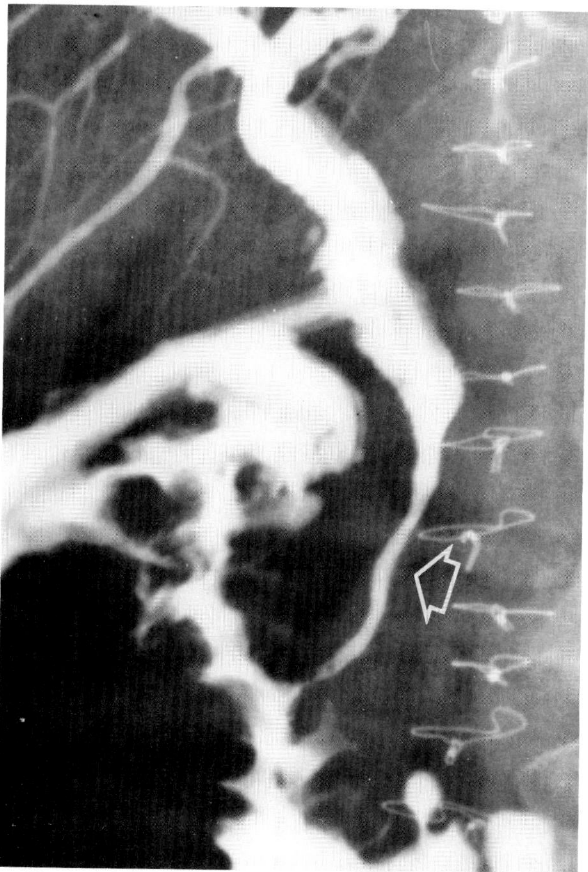

**Figure 55.18. Chronic pancreatitis.** A cholangiogram demonstrates smooth narrowing of the intrahepatic portion of the common bile duct (arrow). Note the associated irregular thickening of folds in the adjacent second portion of the duodenum. (From Eisenberg,[6] with permission from the publisher.)

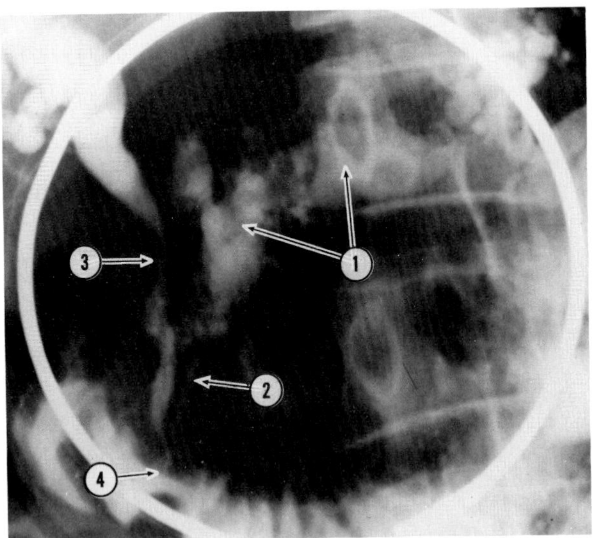


**Figure 55.19. Chronic pancreatitis.** An ERCP shows ectasia and irregularity of the main pancreatic duct (1) and its lateral branches. The main pancreatic duct in the head of the pancreas is narrowed (2). The common duct is also stenotic as it passes through the head of the pancreas (3). Note that the location of the papilla of Vater (4) must be clearly demonstrated radiographically in order to evaluate the length of the stricture. Cellular material for cytologic study can be easily obtained from the stricture of the main pancreatic duct at the time of the ERCP. (From Stewart,[1] with permission from the publisher.)

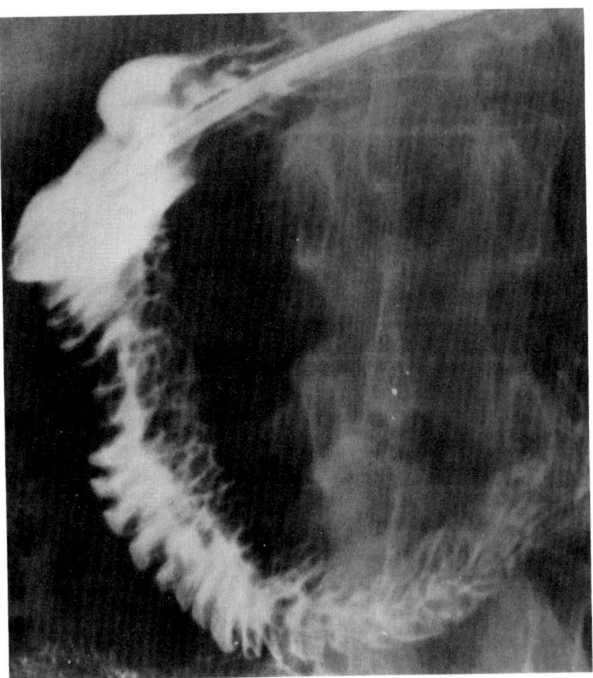

**Figure 55.20. Chronic pancreatitis.** Upper gastrointestinal series shows mucosal spiculation and a mass impression on the inner border of the descending duodenum from the markedly enlarged head of the pancreas. (From Eisenberg,[6] with permission from the publisher.)

ever, since similar information can be obtained less expensively and without ionizing irradiation by ultrasound, CT is usually reserved for patients with chronic pancreatitis in whom technical factors make ultrasound suboptimal.

Narrowing of the distal common bile duct, often with smooth tapering, can be demonstrated on transhepatic cholangiography in patients with obstructive jaundice due to chronic pancreatitis. ERCP can show dilatation, irregularity, tortuosity, or diffuse narrowing of the main pancreatic duct and lateral branches, as well as ductal calcifications, communicating pseudocysts, or total occlusion of the duct by a distal calculus. The ERCP findings may be a major factor in determining the operative approach in patients with chronic pancreatitis who have intractable pain.

Enlargement of the pancreatic head can cause widening and pressure changes on the inner aspect of the duodenal sweep on barium studies. This produces narrowing of the lumen or a double-contour effect. Associated spiculation can make this appearance indistinguishable from pancreatic carcinoma. Severe pancreatic insufficiency can lead to the development of a nonspecific malabsorption pattern resembling sprue.[2,12–14]

## CYSTIC FIBROSIS (MUCOVISCIDOSIS)

The increased viscidity of mucus excreted by the exocrine glands in this hereditary disorder causes a variety of radiographic changes that primarily affect the respiratory and gastrointestinal systems. Pulmonary involvement develops in infancy and childhood, with focal areas of atelectasis and recurrent pulmonary infections due to bronchiolar obstruction by viscid mucus. Generalized irregular thickening of linear markings throughout the lungs and almost invariable hyperinflation produce a typical appearance of severe chronic lung disease that often leads to cor pulmonale.

Meconium ileus, obstruction of the small bowel by thick mucus, affects about 10 percent of newborns with cystic fibrosis. Bowel perforation with subsequent peritonitis not infrequently occurs. The extraluminal meconium can become calcified in an amorphous pattern. In older children or adults, a viscid fecal bolus in the terminal ileum or cecum can cause small bowel obstruction (meconium ileus equivalent syndrome) or appear as a tumor-like defect on barium studies. Because adequate cleansing prior to barium enema examination is rarely achieved in patients with cystic fibrosis, the residual thick secretions may produce multiple, poorly defined filling defects that give the colonic mucosa a hyperplastic appearance simulating polyposis. Less common gastrointestinal tract abnormalities in patients with cystic fibrosis include thickening and nodular indentations of mucosal folds in the duodenum and proximal jejunum and a malabsorption pattern due to pancreatic enzyme deficiency.[15,16]

## CANCER OF THE PANCREAS

Because it is noninvasive and relatively inexpensive, ultrasound is often the initial screening procedure in a patient suspected of pancreatic carcinoma. Ultrasound can demonstrate most tumors over 2 cm in diameter that lie in the head of the pancreas; lesions in the body and tail of the pancreas are more difficult to detect by this modality. Pancreatic carcinoma typically causes the gland to have an irregular contour and a semisolid pattern of intrinsic echoes. Associated signs can include compression of the vena cava and superior mesenteric vein, dilatation of the biliary tract, and demonstration of hepatic metastases.

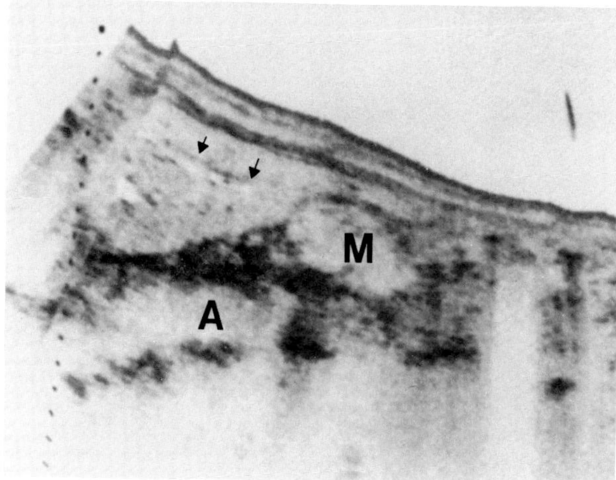


**Figure 55.21. Carcinoma of the pancreas.** A longitudinal sonogram demonstrates an irregular mass (M) with a semisolid pattern of intrinsic echoes. There is associated dilatation of the intrahepatic bile ducts (arrows). A = aorta. (Courtesy of Gretchen A. W. Gooding, M.D.)

CT is the most effective modality for detecting a pancreatic cancer in any portion of the gland and for defining its extent. CT can demonstrate the mass of the tumor as well as ductal dilatation and invasion of neighboring structures. Following the administration of intravenous contrast material, the relatively avascular tumor appears as an area of decreased attenuation when compared with the normal pancreas.

CT is a valuable noninvasive tool for the staging of pancreatic carcinoma and may prevent needless surgery in patients with nonresectable lesions. This technique may permit detection of hepatic metastases or involvement of regional vessels and adjacent retroperitoneal lymph nodes. The subtle finding of encasement of the superior mesenteric, splenic, or hepatic arteries strongly suggests unresectable carcinoma with extrapancreatic extension.

Cytologic examination of tissue obtained by percutaneous fine-needle aspiration under ultrasound or CT guidance can often provide the precise histologic diagnosis of a neoplastic mass, thus obviating surgical intervention.

Carcinoma of the head of the pancreas causing obstructive jaundice can produce narrowing of the distal common bile duct on transhepatic cholangiography. Although their appearances may overlap, pancreatic carcinoma tends to cause irregularity or abrupt cutoff of the

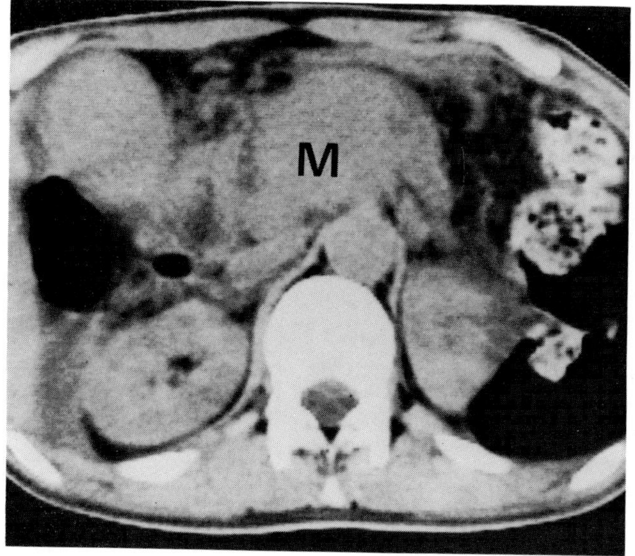


A

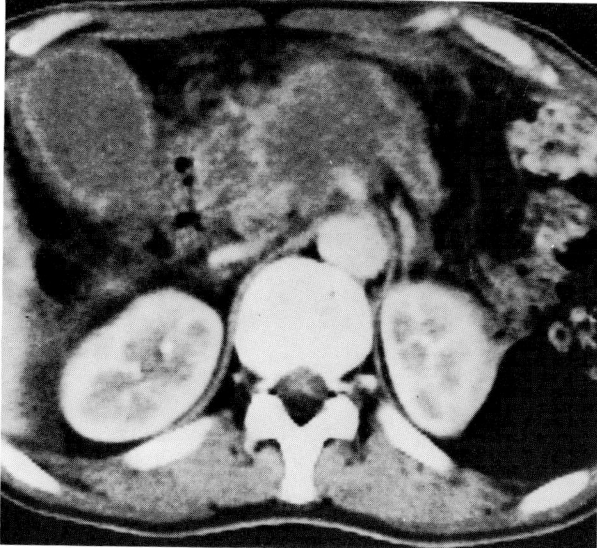

B

**Figure 55.22. Carcinoma of the pancreas.** (A) A noncontrast CT scan demonstrates a homogeneous mass (M) in the body of the pancreas. (B) After the administration of an intravenous bolus of contrast material, a CT scan at the time of maximum aortic contrast shows enhancement of the surrounding vascular structures and normal pancreatic parenchyma while the pancreatic carcinoma remains unchanged and thus appears as a low-density mass. (From Federle and Goldberg,[14] with permission from the publisher.)

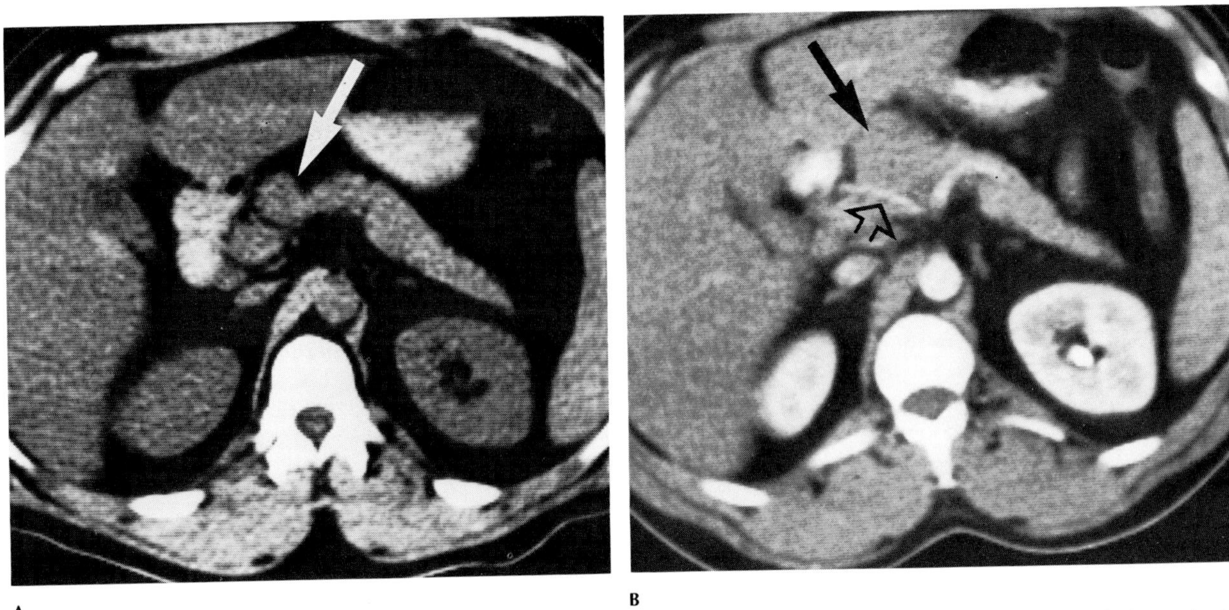


**Figure 55.23. Rapid growth and arterial encasement of pancreatic carcinoma.** (A) A CT scan demonstrates a focal change in the shape of the ventral contour of the pancreas at the junction of the body and head (arrow). No enlargement of the pancreatic tissue is present. This was initially interpreted as representing an anatomic variant. (B) Three months later, a repeat CT scan shows a focal tumor mass (closed arrow) in the location of the focal contour abnormality seen in A. A dynamic CT scan after an intravenous bolus injection of contrast material demonstrates the splenic and hepatic arteries at the base of the tumor. Note that the hepatic artery (open arrow) has an irregular contour. Arteriography showed encasement by this unresectable tumor. (From Federle and Goldberg,[14] with permission from the publisher.)

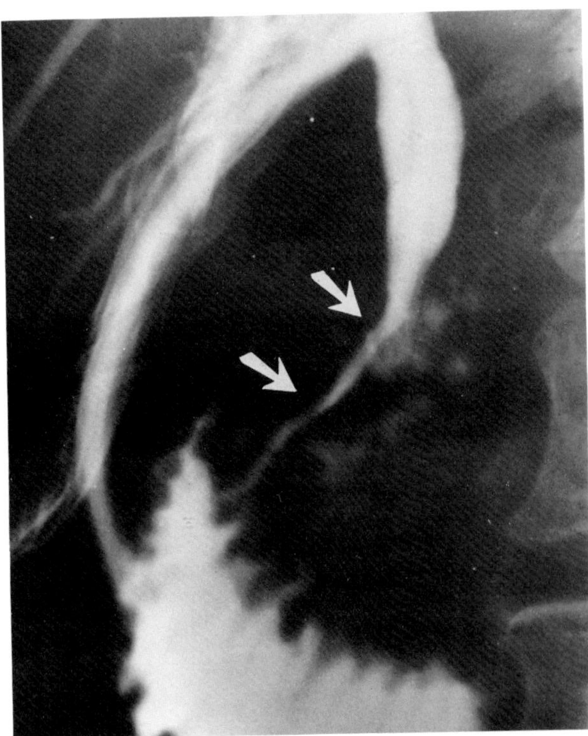

**Figure 55.24. Carcinoma of the head of the pancreas.** A cholangiogram demonstrates irregular narrowing of the common bile duct (arrows). The calcifications reflect underlying chronic pancreatitis. (From Eisenberg,[6] with permission from the publisher.)

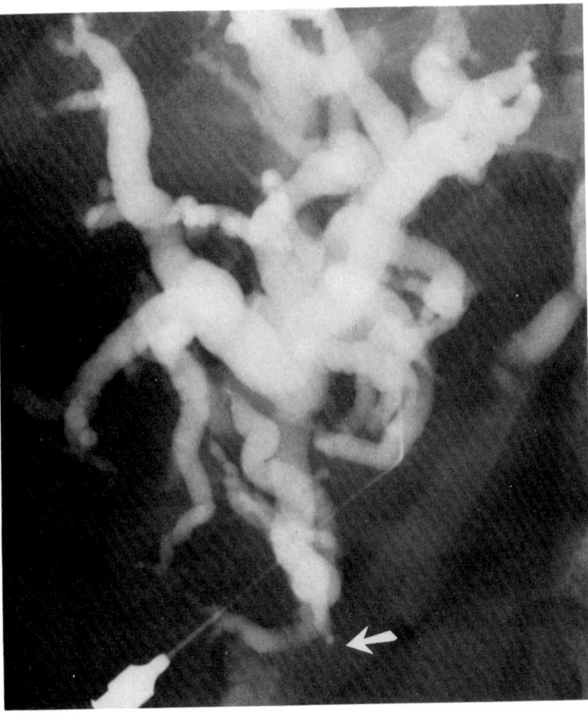

**Figure 55.25. Carcinoma of the pancreas.** A cholangiogram demonstrates abrupt obstruction of the bile duct (arrow) with dilatation of the proximal biliary tree. (From Eisenberg,[6] with permission from the publisher.)

common bile duct, while chronic pancreatitis shows a tapering and often undulating narrowing of the distal common bile duct. Percutaneous biliary drainage may represent an alternative to surgical intervention in the palliative treatment of pancreatic carcinoma causing obstruction of the bile ducts. Although the transhepatic insertion of biliary drainage tubes does not alter the dismal prognosis, it can reduce patient morbidity and the need for hospitalization.

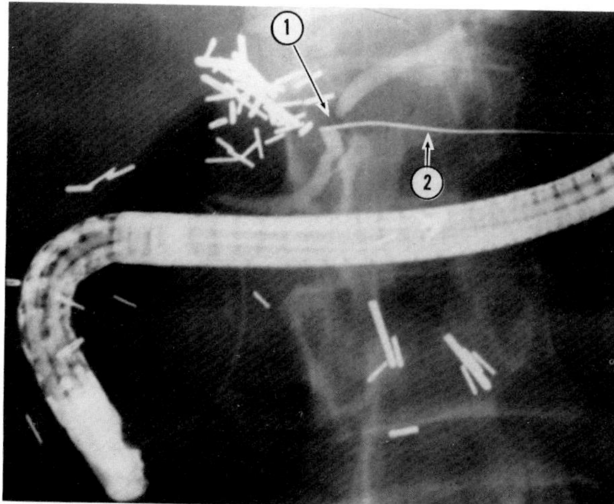


**Figure 55.26. Carcinoma of the pancreas.** An ERCP shows sharp amputation of the main pancreatic duct (1) at the level of the takeoff of the duct of Santorini (2). Complete obstruction at this point, associated with abdominal pain, is characteristic of carcinoma. (From Stewart,[1] with permission from the publisher.)

ERCP characteristically shows stenosis or obstruction of the pancreatic and/or common bile duct in patients with pancreatic cancer. However, the differentiation between carcinoma and chronic pancreatitis by ERCP can be difficult if both diseases are present.

Barium upper gastrointestinal studies demonstrate distortion of the mucosal pattern and configuration of the duodenal sweep in about half of the patients with carcinoma of the head of the pancreas. Early or small lesions, however, rarely produce detectable radiographic abnormalities; tumors of the body or tail of the pancreas must be quite large to be visible on barium examinations. The mass of the tumor causes focal or generalized indentation on the inner surface of the sweep with effacement and fixation of mucosal folds. Extension of the tumor can impress the gastric antrum (antral pad sign) and narrow the antrum or duodenal sweep. Although Frostberg's reverse figure-3 sign (impressions on the duodenal sweep above and below the papilla of Vater) was originally considered pathognomonic of pancreatic malignancy, an identical appearance can be seen in patients with inflammatory disease of the pancreas and postbulbar ulcers.

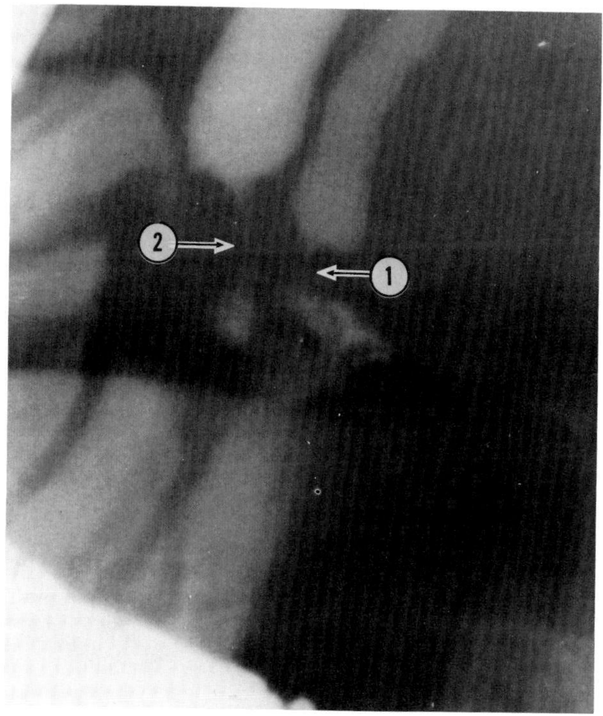


A

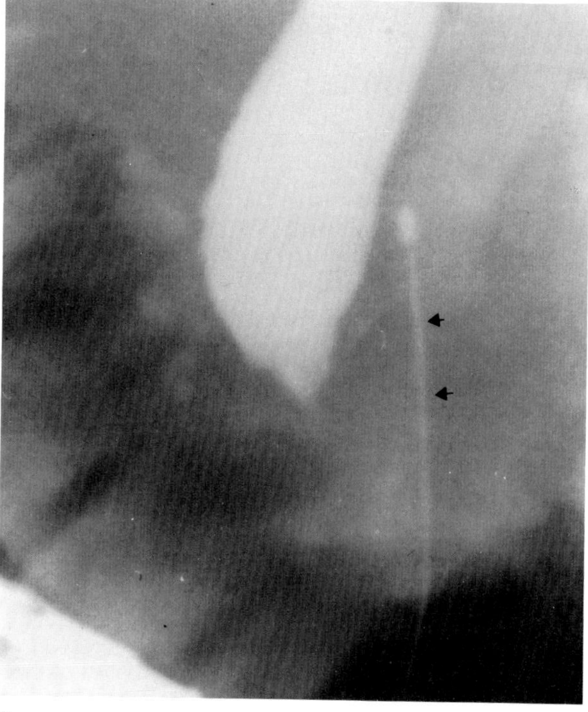

B

**Figure 55.27. Carcinoma of the pancreas.** (A) An ERCP shows a short stricture of the main pancreatic duct (1) and the common duct (2). (B) A magnified film demonstrates the cytology brush (arrows) inserted into the pancreatic duct at the time of the ERCP. (From Stewart,[1] with permission from the publisher.)

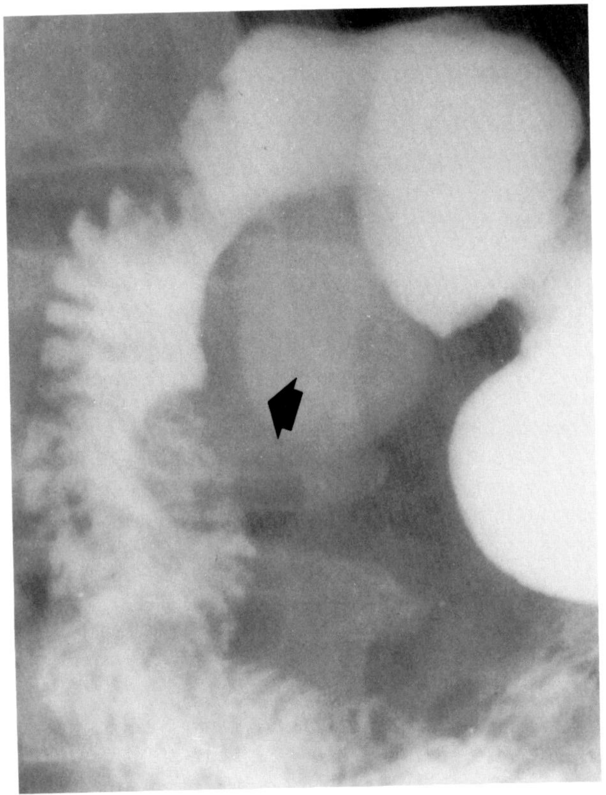

A

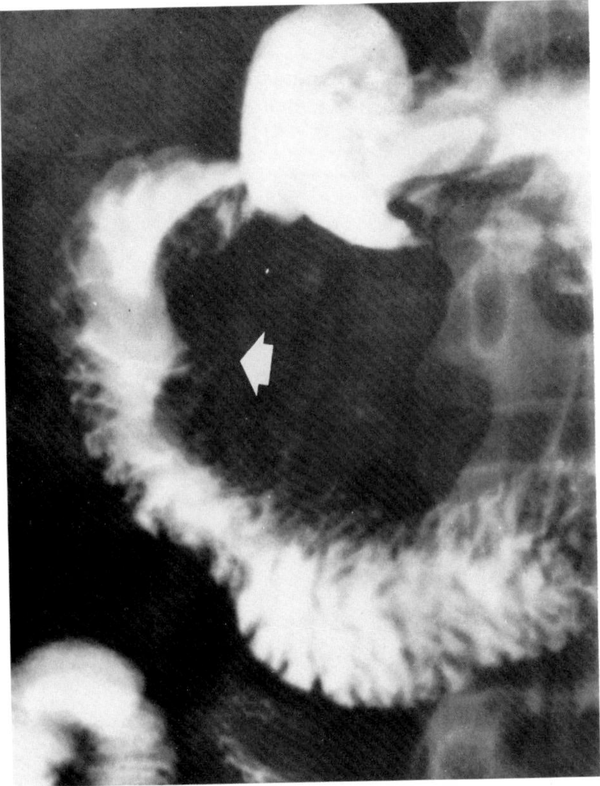

B

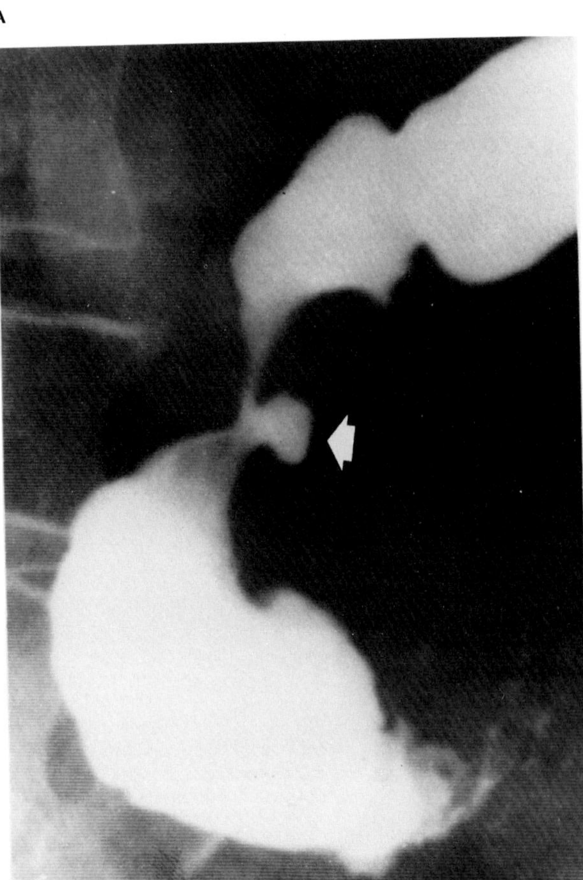

C

**Figure 55.28. Frostberg's reverse figure-3 sign (arrows) in pancreatic disease.** (A) Carcinoma of the head of the pancreas. (B) Severe acute pancreatitis. (C) Large postbulbar ulcer. (From Eisenberg,[6] with permission from the publisher.)

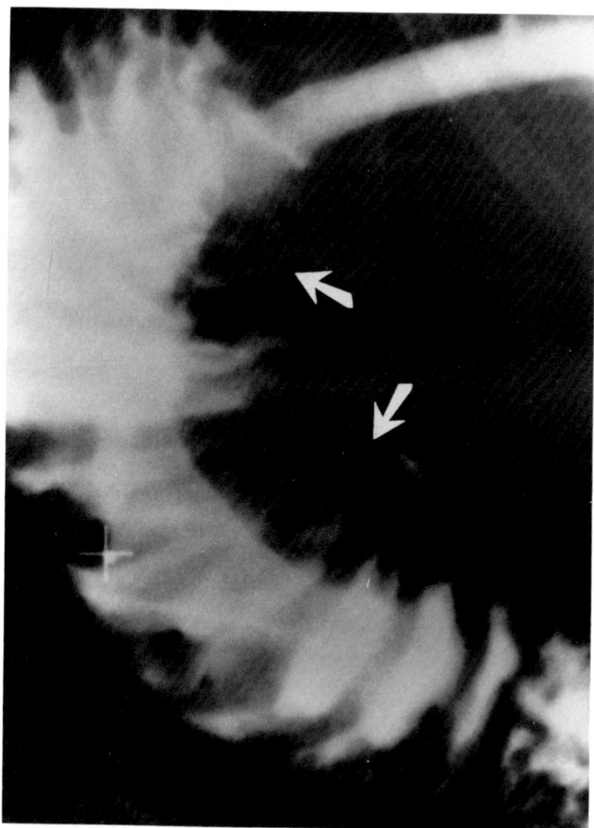

**Figure 55.29. Carcinoma of the pancreas.** Upper gastrointestinal series demonstrates spiculation of duodenal folds (arrows) with a mass effect on the medial wall of the descending duodenum. (From Eisenberg,[6] with permission from the publisher.)

Prior to the advent of ultrasound and CT, pancreatic arteriography was frequently employed to detect carcinoma of the pancreas. Its use is now limited to detecting endocrine diseases of the pancreas, which are often related to highly vascular tumors, and staging pancreatic malignancy in terms of surgical resectability. Arteriographic signs of unresectability include hepatic metastases; invasion or occlusion of portal or superior mesenteric veins; and encasement of celiac, superior mesenteric, gastric, or intestinal arteries.[6,13,14,17–22]

## ANNULAR PANCREAS

Annular pancreas is an anomalous ring of pancreatic tissue encircling the duodenal lumen that can cause intestinal obstruction in the neonate or the adult. Infants with symptomatic annular pancreas have the characteristic radiographic pattern of the double-bubble sign. Unlike duodenal atresia, which has a similar appearance, annular pancreas almost always results in incomplete obstruction, so that a small but recognizable amount of gas can be demonstrated within the bowel distal to the

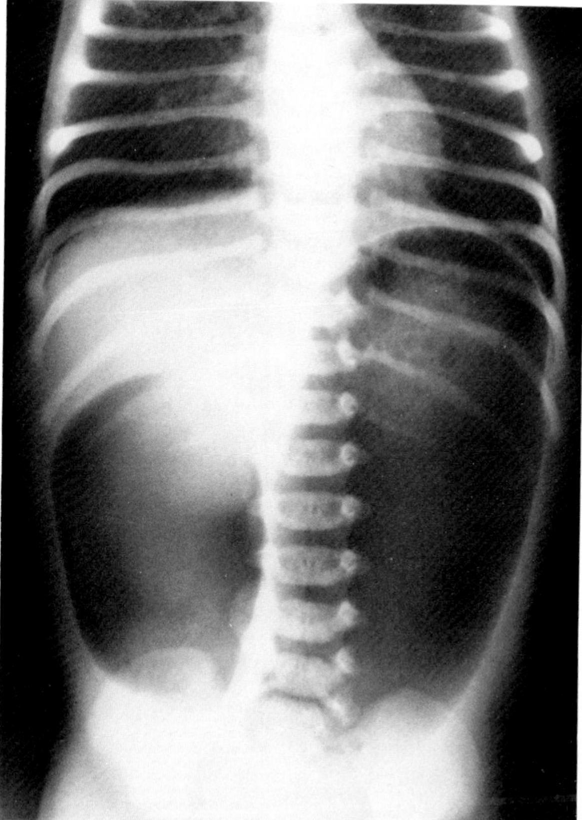

**Figure 55.30. Double-bubble sign in annular pancreas.** An abdominal film shows huge dilatation of the stomach (left bubble) and duodenal bulb (right bubble). (From Eisenberg,[6] with permission from the publisher.)

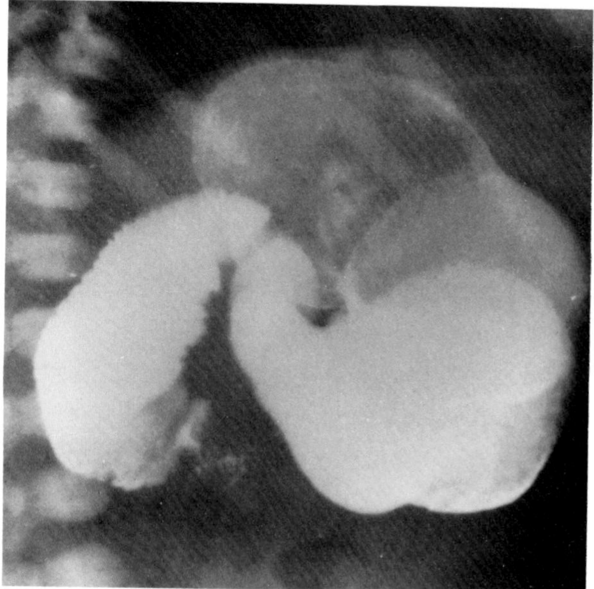

**Figure 55.31. Annular pancreas.** The presence of small amounts of gas distal to the obstruction indicates that the stenosis is incomplete. (From Eisenberg,[6] with permission from the publisher.)

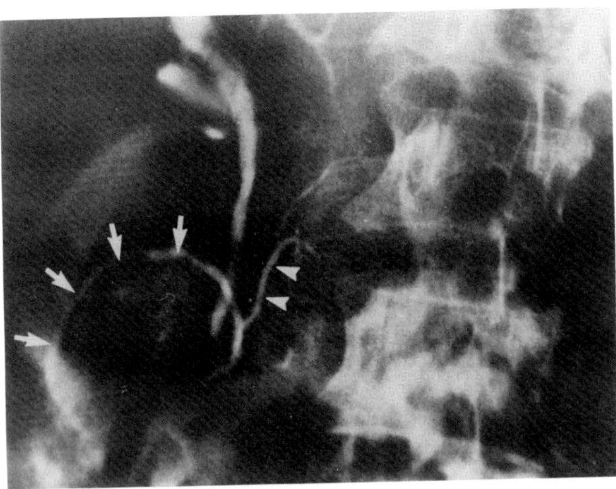

**Figure 55.32. Annular pancreas.** An ERCP shows contrast material in the duct of the annular pancreas (arrows), in the duct of "ventral pancreas" (arrowheads), and in the common bile duct. (From Glazer and Margulis,[23] with permission from the publisher.)

level of the high-grade duodenal stenosis. On barium studies, an annular pancreas can produce either a single extrinsic impression on the lateral aspect of the second portion of the duodenum or symmetric masses with effacement but no destruction of the mucosa and dilatation of the duodenum proximal to the annular band. If the diagnosis is in doubt, ERCP can demonstrate the precise ductal anatomy.[23]

## PANCREATIC ENDOCRINE TUMORS

These tumors are discussed in the section "Hormone-Secreting Tumors of the Pancreas" in Chapter 4.

### REFERENCES


1. Stewart ET: Pancreatic disease, in Eisenberg RL, Amberg JR (eds): *Critical Diagnostic Pathways in Radiology: An Algorithmic Approach*. Philadelphia, JB Lippincott, 1981.
2. Sarti DA, King W: The ultrasonic findings in inflammatory pancreatic diseases. *Semin Ultrasound* 1:178–191,1980.
3. Brascho DJ, Reynolds TN, Zanca P: The radiographic "colon cut-off sign" in acute pancreatitis. *Radiology* 79:763–768, 1962.
4. Poppel MH: The manifestations of relapsing pancreatitis. *Radiology* 62:514–521, 1954.
5. Silverstein W, Isikoff MB, Hill MC, et al: Diagnostic imaging of acute pancreatitis: Prospective study using CT and sonography. *AJR* 157:497–502, 1981.
6. Eisenberg RL: *Gastrointestinal Radiology: A Pattern Approach*, Philadelphia, JB Lippincott, 1983.
7. Siegelmann SS, Copeland BE, Saba GP, et al: CT of fluid collections associated with pancreatitis. *AJR* 134:1121–1132, 1980.
8. Kolmannskog F, Kolbenstvedt A, Aakhus T: Computed tomography in inflammatory mass lesions following acute pancreatitis. *J Comput Assist Tomogr* 5:169–172, 1981.
9. Shockman AT, Marasco JA: Pseudocysts of the pancreas. *AJR* 101:628–638, 1967.
10. Berenson JE, Spitz HB, Felson B: The abdominal fat necrosis sign. *Radiology* 100:567–571, 1971.
11. Federle MP, Jeffrey RB, Crass RA, et al: Computed tomography of pancreatic abscesses. *AJR* 136:879–882, 1981.
12. Ferrucci JT, Wittenberg J, Black EB, et al: Computed body tomography in chronic pancreatitis. *Radiology* 130:175–182, 1979.
13. Buonocore E: Transhepatic percutaneous cholangiography. *Radiol Clin North Am* 14:527–542, 1976.
14. Federle MP, Goldberg HI: Computed tomography of the pancreas, in Moss AA, Gamsu G, Genant HK (eds): *Computed Tomography of the Body*. Philadelphia, WB Saunders, 1983, pp 699–762.
15. Leonidas J, Berdon WE, Baker DH, et al: Meconium ileus and its complications. *AJR* 108:598–609, 1970.
16. Taussig LM, Saldino RM, di Sant'Agnese PA: Radiographic abnormalities of the duodenum and small bowel in cystic fibrosis of the pancreas (mucoviscidosis). *Radiology* 106:369–373.
17. Eaton SB, Benedict KT, Ferrucci JT, et al: Hypotonic duodenography. *Radiol Clin North Am* 8:125–137, 1970.
18. Goldstein HM, Neiman HL, Bookstein JJ: Angiographic evaluation of pancreatic disease. *Radiology* 112:275–282, 1974.
19. Sheedy PF, Stephens DH, Hattery RR, et al: Computed tomography in the evaluation of patients with suspected carcinoma of the pancreas. *Radiology* 124:731–737, 1977.
20. Lee JKT, Stanley RJ, Melson GL, et al: Pancreatic imaging by ultrasound and computed tomography. *Radiol Clin North Am* 16:105–117, 1979.
21. Weyman PJ, Stanley RJ, Levitt RG: Computed tomography in the evaluation of the pancreas. *Semin Roentgenol* 16:301–311, 1981.
22. Mitty AJ, Efemidis SC, Yeh HC: Impact of fine-needle biopsy on management of patients with carcinoma of the pancreas. *AJR* 137:1119–1121, 1981.
23. Glazer GM, Margulis AR: Annular pancreas: Etiology and diagnosis using endoscopic retrograde cholangiopancreatography. *Radiology* 133:303–306, 1979.
24. Jeffrey RB, Federle MD, Laing FC: Computed tomography of mesenteric involvement in fulminant pancreatitis. *Radiology* 147:185–192, 1983.

# PART NINE

# DISORDERS OF THE HEMATOPOIETIC SYSTEM

## Chapter Fifty-Six
# Red Blood Cell Disorders

## IRON-DEFICIENCY ANEMIA

Skeletal abnormalities are occasionally seen in children with severe and chronic iron-deficiency anemia. An increased erythropoietic response causes marrow hyperplasia that results in widening of the diploic space of the skull and thinning of the outer table. The long bones, especially the hands, may demonstrate osteoporosis with thinned cortices and coarsened trabeculae.

Although iron-deficiency anemia per se produces no radiographic bone changes in adults, chronic anemia can lead to an increase in cardiac output, cardiomegaly, and high-output heart failure.[1–3]

## MEGALOBLASTIC ANEMIA

The most common cause of vitamin $B_{12}$ deficiency is pernicious anemia, due to inadequate intrinsic factor secretion related to atrophy of the gastric mucosa. Gastric atrophy presents radiographically as a tubular stomach with a bald appearance that reflects a decrease or absence of the usually prominent rugal folds. It must be emphasized, however, that radiographic findings of atrophic gastritis are often seen in older persons with no evidence of pernicious anemia.

A deficiency of folic acid (and vitamin $B_{12}$) may also be related to intestinal malabsorption. The administration of barium may demonstrate a spruelike pattern of small bowel dilatation with normal folds, an infiltrative

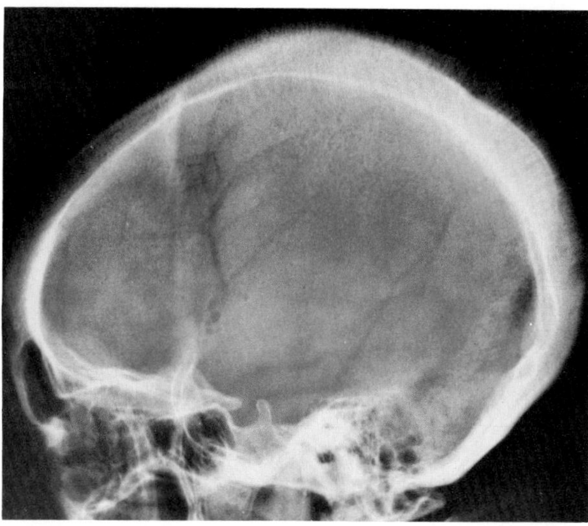

**Figure 56.1. Iron-deficiency anemia.** A lateral view of the skull in a child demonstrates widening of the diploic space with thinning of the outer table. Radial striations of trabeculae between the tables of the skull produce a hair-on-end appearance.

pattern of generalized irregular, distorted small bowel folds, or multiple jejunal diverticula.[4]

## HEREDITARY SPHEROCYTOSIS

Because the anemia of hereditary spherocytosis is usually mild, bone changes are infrequent. With severe anemia, marrow hyperplasia can cause radiographic abnormalities that resemble those seen in Cooley's anemia, except that they are of lesser degree and are usually limited to the skull (widening of the diploe, displacement and thinning of the outer table, hair-on-end appearance).

Extramedullary erythropoiesis may produce paravertebral masses visible on chest radiographs. Excessive red blood cell destruction may lead to the formation of bilirubin stones that appear as lucent filling defects in the opacified gallbladder.[1]

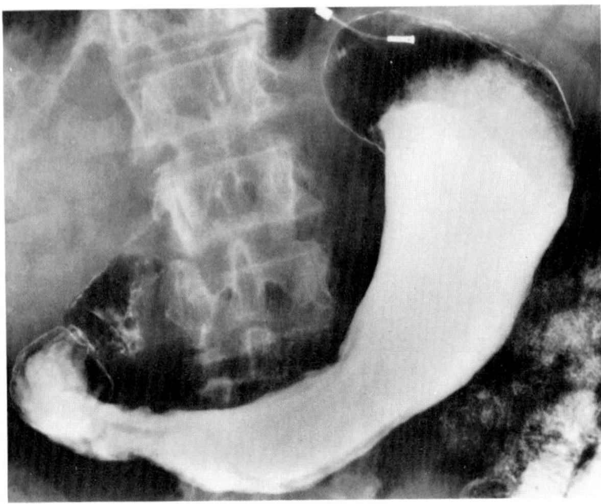

**Figure 56.2. Megaloblastic anemia.** Chronic atrophic gastritis presents as a tubular stomach with a striking decrease in the usually prominent rugal folds.

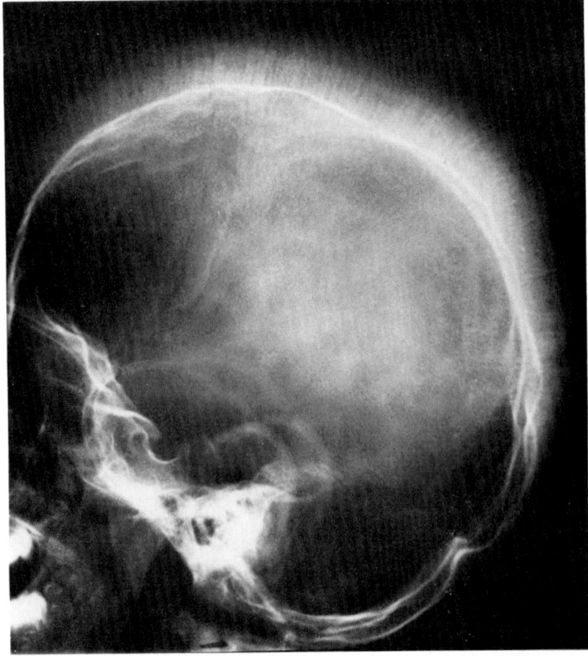

**Figure 56.3. Hereditary spherocytosis.** A lateral view of the skull in a child demonstrates thinning of the outer table with vertical striations producing a striking hair-on-end appearance.

## SICKLE CELL ANEMIA

The radiographic abnormalities in sickle cell disease are a consequence of chronic anemia with secondary bone marrow hyperplasia and of focal ischemia and infarction in multiple tissues due to sludging of the abnormally shaped red blood cells. The findings are similar to those seen in other hemolytic anemias, especially thalassemia, though in sickle cell disease the radiographic changes are less frequent and tend to be less severe.

The major radiographic changes in sickle cell anemia are found in the skeletal system. Marrow hyperplasia causes generalized osteoporosis with resorption of secondary trabeculae and coarsening and accentuation of primary trabeculae. In the spine, expansile pressure of the adjacent intervertebral disks produces characteristic biconcave indentations on both the superior and the inferior margins of the softened vertebral bodies ("fish vertebrae"). Another typical appearance is the development of localized steplike central depressions of multiple vertebral end plates. This is probably caused by circulatory stasis and ischemia, which retard growth in the central portion of the vertebral cartilaginous growth plate, while the periphery of the growth plate, which has a different blood supply, continues to grow at a more normal rate.

In the long bones, marrow hyperplasia causes widening of the medullary spaces with thinning of the cortices and coarsening of the trabecular pattern. Bone infarcts commonly occur. In infants and children, these

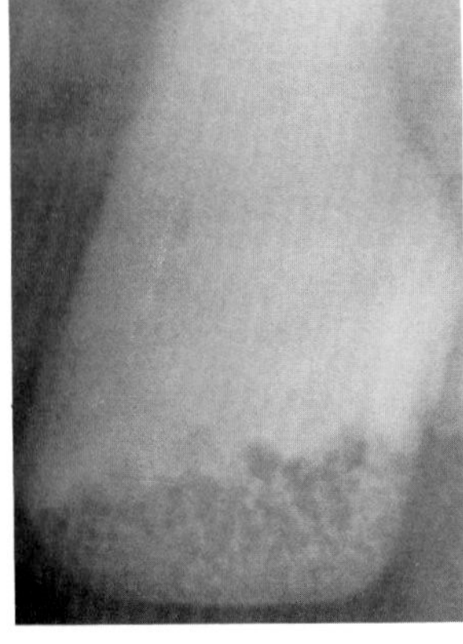

**Figure 56.4. Hereditary spherocytosis.** Multiple lucent filling defects in the opacified gallbladder represent innumerable bilirubin stones resulting from excessive red blood cell destruction.

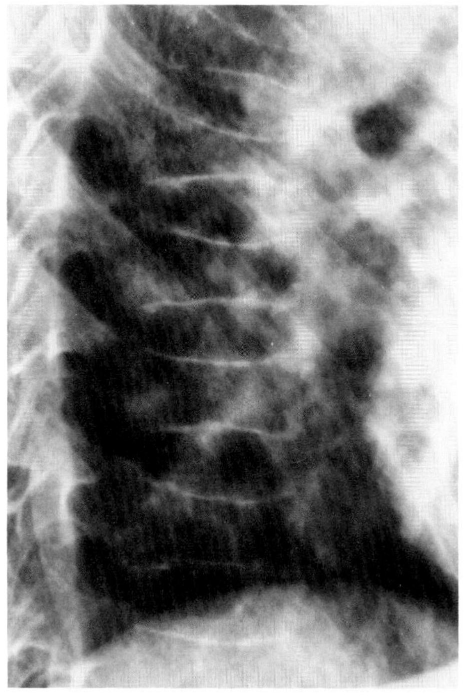

A

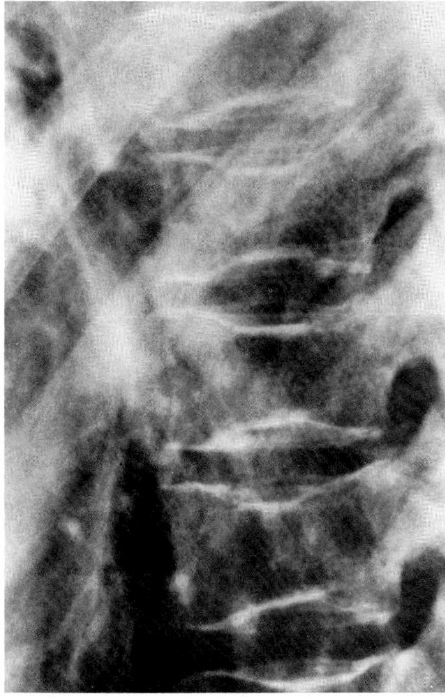

B

**Figure 56.5. Spine changes in sickle cell anemia.** (A) Biconcave indentations on both the superior and inferior margins of the soft vertebral bodies produce the characteristic "fish vertebrae." (B) Localized steplike central depressions of multiple vertebral end plates.

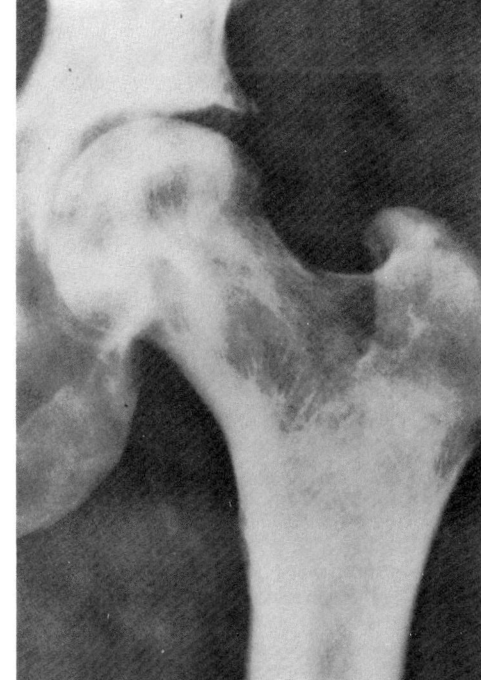

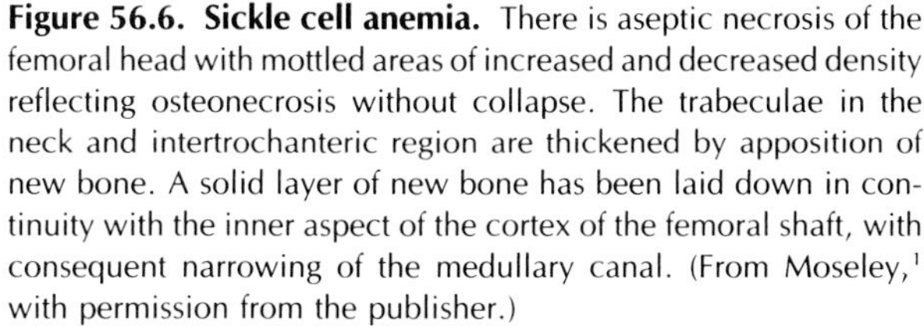

**Figure 56.6. Sickle cell anemia.** There is aseptic necrosis of the femoral head with mottled areas of increased and decreased density reflecting osteonecrosis without collapse. The trabeculae in the neck and intertrochanteric region are thickened by apposition of new bone. A solid layer of new bone has been laid down in continuity with the inner aspect of the cortex of the femoral shaft, with consequent narrowing of the medullary canal. (From Moseley,[1] with permission from the publisher.)

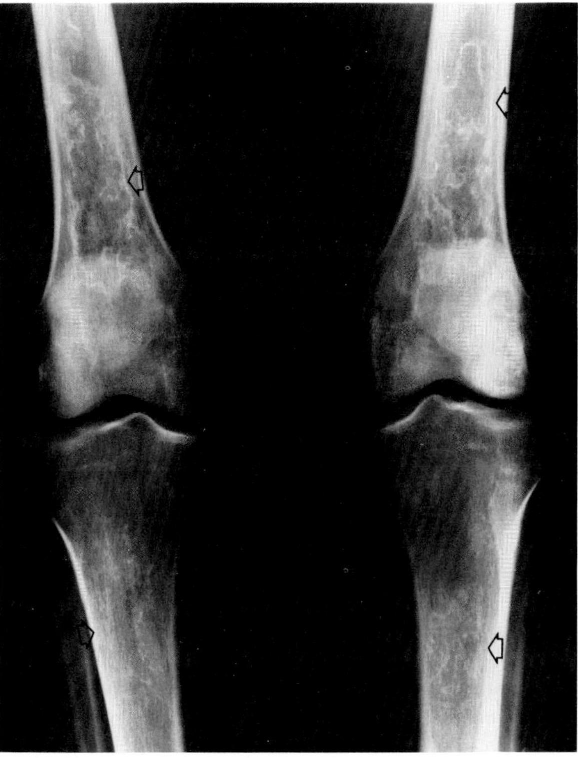

**Figure 56.7. Sickle cell anemia.** Multiple bone infarcts in the femurs and tibias (arrows) are surrounded by peripheral rims of calcification.

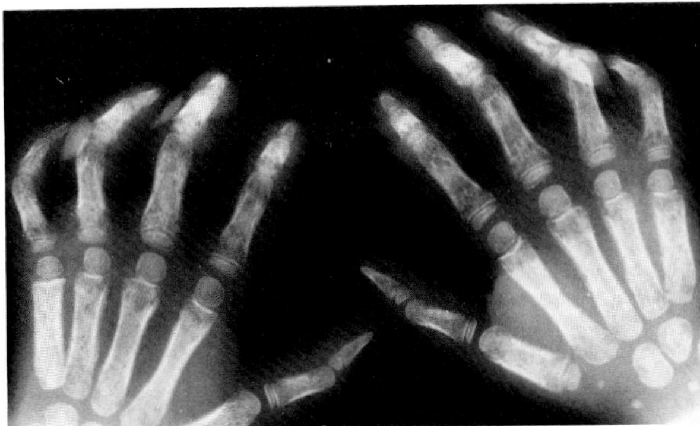

**Figure 56.8. Hand-foot syndrome in sickle cell anemia.** Diffuse destruction of the shaft of multiple phalanges and metacarpals is due to infarction. There are reactive bone changes with sclerosis and periosteal thickening.

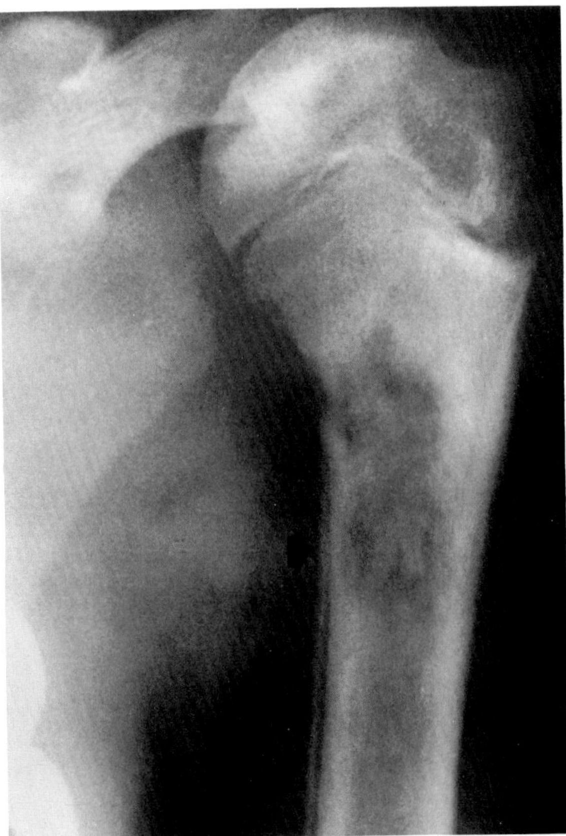

**Figure 56.10. Acute osteomyelitis in sickle cell anemia.** There is diffuse lytic destruction of the proximal humerus with extensive periosteal reaction (arrow).

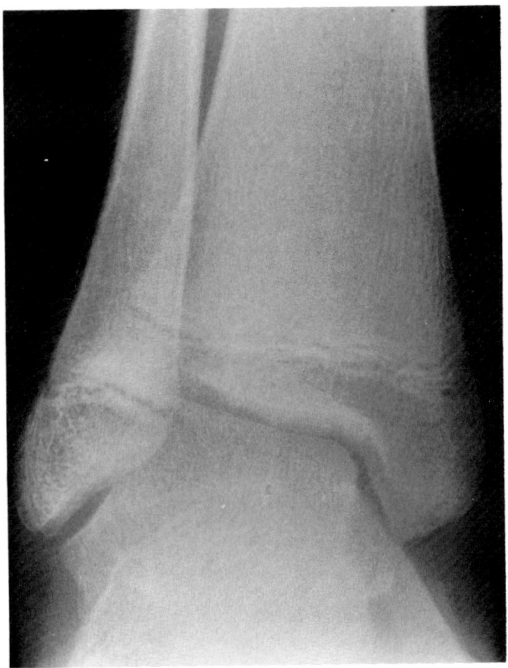

**Figure 56.9. Tibiotalar slant in sickle cell anemia.**

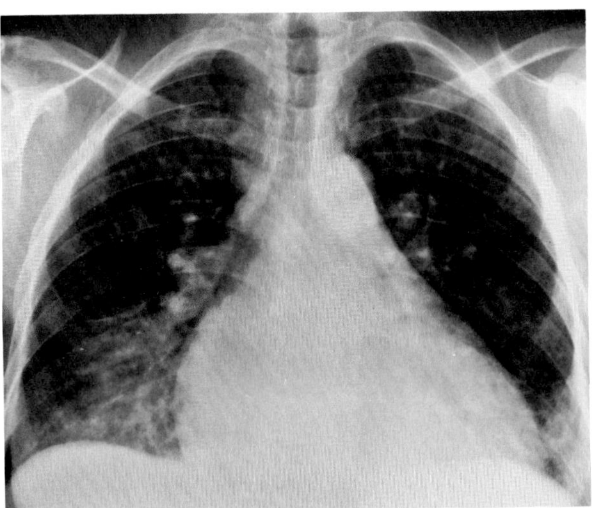

**Figure 56.11. Sickle cell anemia.** Marked cardiomegaly with a generalized increase in pulmonary vascular markings is seen in a patient with high-output failure.

most frequently involve the small bones of the hands and feet, producing an irregular area of bone destruction with overlying periosteal calcification that may be indistinguishable from osteomyelitis. In older children and adults, bone infarction may initially appear as an ill-defined lucent area which then becomes irregularly calcified. It may be associated with endosteal cortical thickening that encroaches on the marrow cavity and may even be separated from the original cortex by a thin lucent line to produce a bone-within-a-bone appearance. Ischemic necrosis of bone ends simulating Perthes' disease is common and primarily involves the femoral or humeral heads. Bone infarctions affecting the epiphyseal cartilaginous plates often lead to growth disturbances of bone with shortening and epiphyseal deformity. In the leg, this may

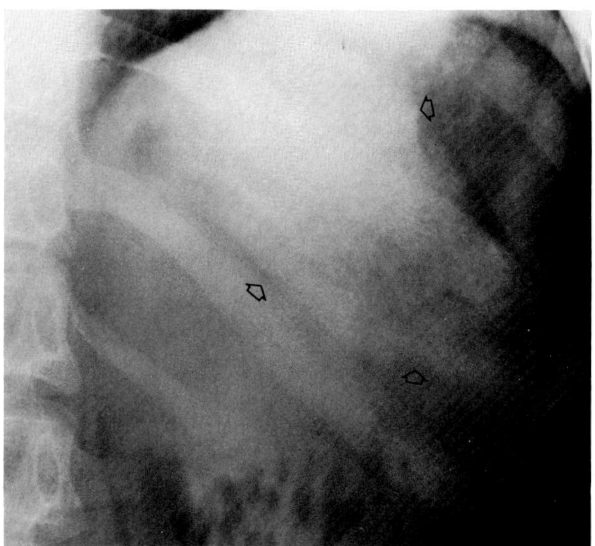

**Figure 56.12. Sickle cell anemia.** There is a generalized increase in splenic density (arrows) due to calcium deposits secondary to splenic infarctions.

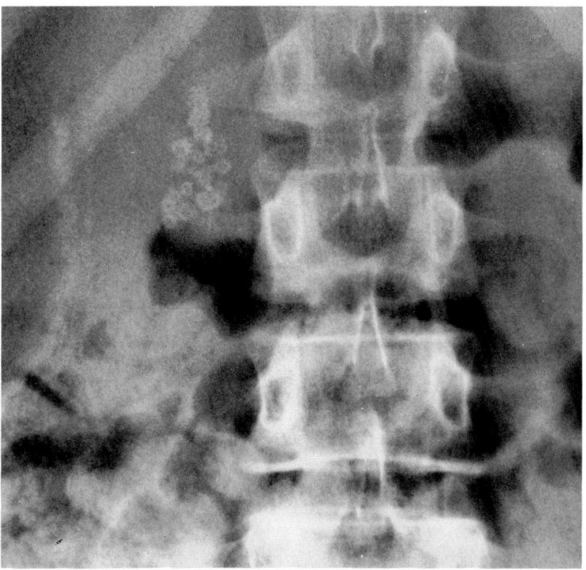

**Figure 56.13. Sickle cell anemia.** Multiple calcified gallstones have formed as a result of repeated episodes of hemolysis.

lead to an angled deformity of the ankle mortise (tibiotalar slant), in which premature closure of the lateral tibial epiphysis results in a relatively high position of the lateral aspect of these two bones.

Radiographic changes in the skull in sickle cell disease are much less prominent than those in thalassemia. Generalized osteoporosis and widening of the diploe with thinning of the outer table are the most common findings. Vertical hair-on-end striations are infrequent.

Acute osteomyelitis, often caused by *Salmonella* infection, is a common complication in sickle cell disease. The resulting lytic destruction and periosteal reaction may be extensive, often involving the entire shaft and multiple bones. Radiographically, it may be impossible to distinguish between osteomyelitis and bone infarction without infection. Pathologic fractures are common.

The most common extraskeletal abnormality in sickle cell disease is cardiomegaly due to severe anemia and increased cardiac output. The heart has a globular configuration reflecting enlargement of all chambers. Increased pulmonary blood flow produces engorgement of the pulmonary vessels, giving a plethoric appearance to the lungs. Pulmonary infarction, pulmonary edema and congestive failure, and pneumonia are frequent complications.

Renal abnormalities can be demonstrated by excretory urography in about two-thirds of the patients with sickle cell disease. The most common finding is calyceal clubbing caused by pericalyceal fibrosis, secondary to vascular stasis rather than to chronic pyelonephritis. In sickle cell disease, unlike chronic pyelonephritis, the cortical tissue tends to be preserved and the renal outline is usually regular. Enlargement of one or both kidneys is often seen. A serious complication of sickle cell disease is renal papillary necrosis, which is probably related to vessel obstruction within the papillae and may produce sinuses or cavity formation within one or more papillae. Although extensive medullary and papillary changes may be present, they usually do not severely impair renal function. Renal failure in sickle cell disease is limited to patients with extensive glomerular damage, the cause of which is unclear.

A urographic appearance reported to be characteristic of sickle cell trait is widening and deepening of the calyces, due to impressions produced by prominent papillae, without blunting or widening of the fornices.

The spleen may be enlarged early in the course of sickle cell disease. However, repeated episodes of infarction and fibrosis lead to progressive shrinkage and even to autosplenectomy. Rounded punctate calcium deposits secondary to splenic infarctions may produce a generalized increase in splenic density. There is an increased incidence of gallstone formation in patients with sickle cell disease, presumably due to repeated episodes of hemolysis. During sickle cell crisis, plain abdominal radiographs may demonstrate adynamic ileus with dilatation of large and small bowel.[1,5–10]

## THALASSEMIA

Thalassemia is a hereditary disorder of hemoglobin synthesis that causes a severe anemia. It occurs predominantly in persons living about the Mediterranean, especially those of Italian, Greek, or Sicilian descent. Thalassemia major (Cooley's anemia) is a severe anemia of infants and children that produces more dramatic ra-

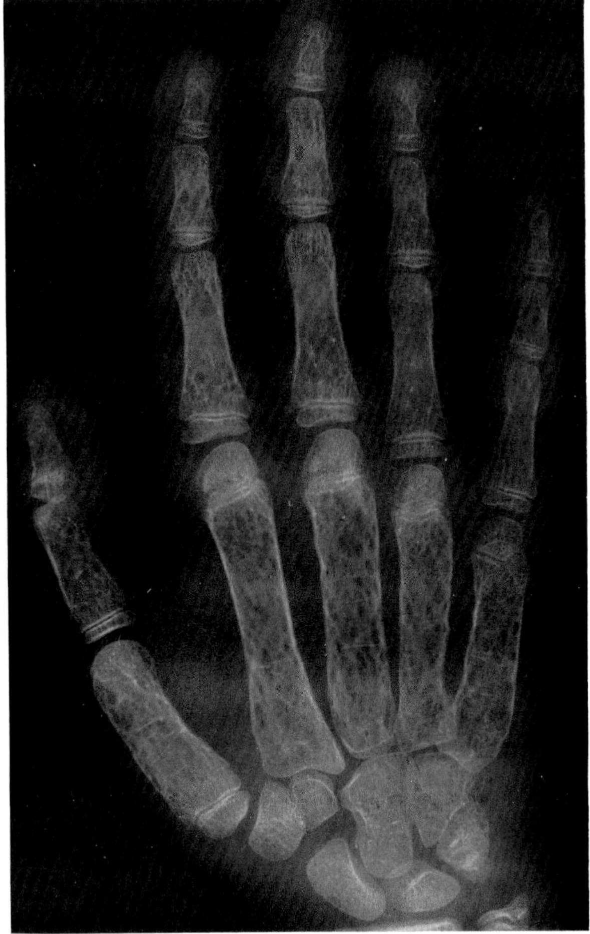
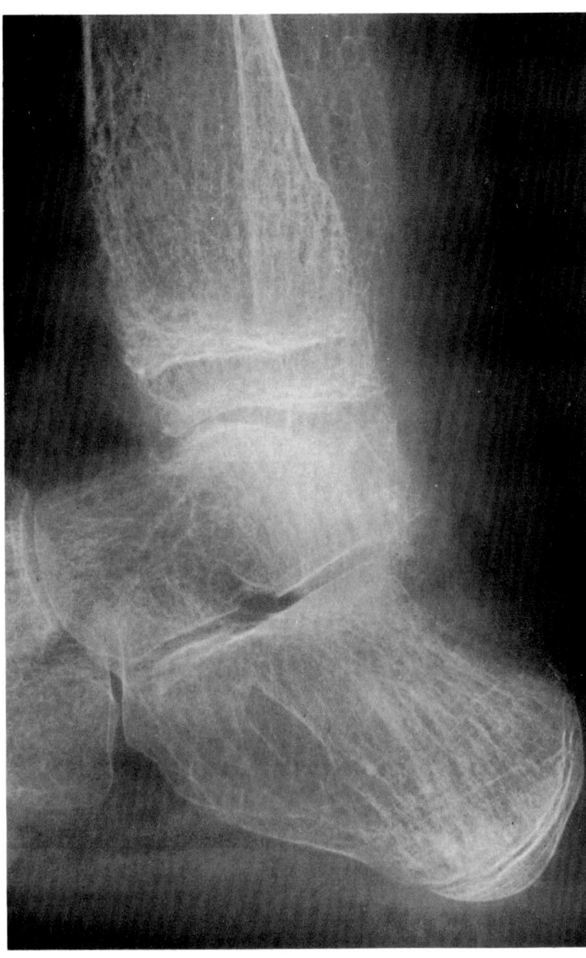

A B

**Figure 56.14. Thalassemia.** (A) A frontal view of the hand and (B) a lateral view of the ankle demonstrate pronounced widening of the medullary spaces with thinning of the cortices. Note the absence of normal modeling due to the pressure of the expanding marrow space. Localized radiolucencies simulating multiple osteolytic lesions represent tumorous collections of hyperplastic marrow.

diographic findings than any other of the childhood anemias. Thalassemia minor and the other hemoglobulinopathies cause similar but less marked radiographic abnormalities.

Extensive marrow hyperplasia, the result of ineffective erythropoiesis and rapid destruction of newly formed red blood cells, causes pronounced widening of the medullary spaces and thinning of the cortices. As the fine secondary trabeculae are resorbed, new bone is laid down on the surviving trabeculae, thickening them and producing a coarsened pattern. Normal modeling of long bones does not occur, because the expanding marrow flattens or even bulges the normally concave surfaces of the shafts. Focal collections of hyperplastic marrow cause localized radiolucencies that have the appearance of multiple osteolytic lesions.

In the growing skeleton there is severe retardation of bone maturation, with resulting skeletal dwarfism and infantilism. Premature fusion of epiphyses in long bones is common and often eccentric, causing bone deformity by tilting the epiphysis toward the site of fusion. This most commonly occurs at the upper end of the humerus, where medial fusion of the epiphysis produces a varus deformity of the humerus and shortening of the upper arm.

In the skull there is widening of the diploic space and thinning or complete obliteration of the outer table. When the hyperplastic marrow perforates or destroys the outer table, it proliferates under the invisible periosteum, and new bone spicules are laid down perpendicular to the inner table. This produces the characteristic hair-on-end appearance of vertical striations in a radial pattern. Marrow hyperplasia in the facial bones causes lack of pneumatization of the paranasal and mastoid sinuses that is usually not seen in any other anemia. Only the ethmoid sinuses are spared, because of the lack of marrow activity in the ethmoidal bone. Marrow overgrowth in the maxilla causes lateral displacement of the orbits and forward

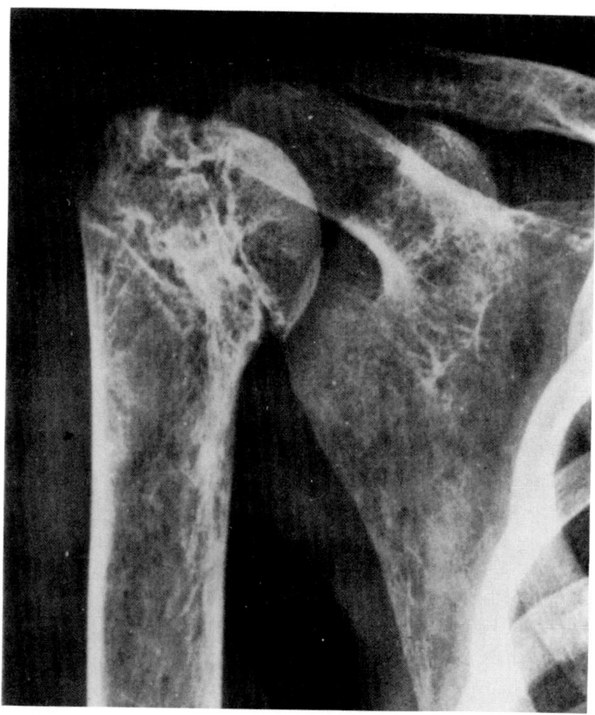

**Figure 56.15. Thalassemia.** In addition to typical findings of marrow hyperplasia, premature fusion of the medial aspect of the proximal humeral epiphysis produces a varus deformity with shortening of the upper arm. (From Moseley,[1] with permission from the publisher.)

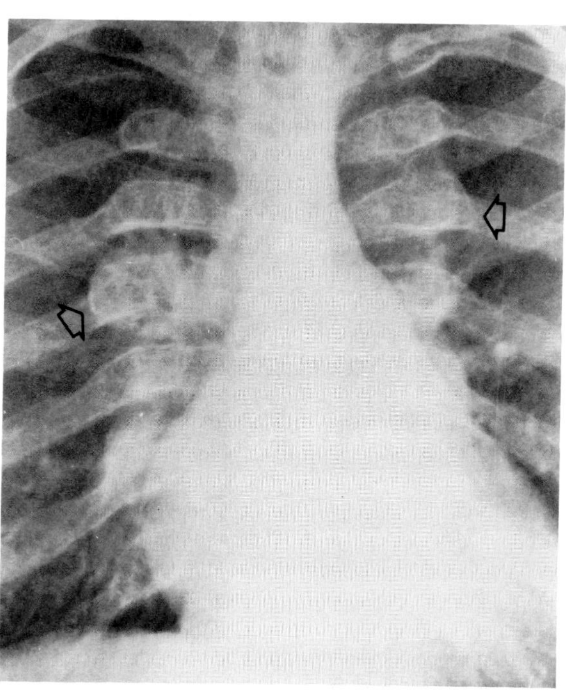

**Figure 56.16. Thalassemia.** There is characteristic bulbous enlargement of the posterior rib ends. (From Moseley,[1] with permission from the publisher.)

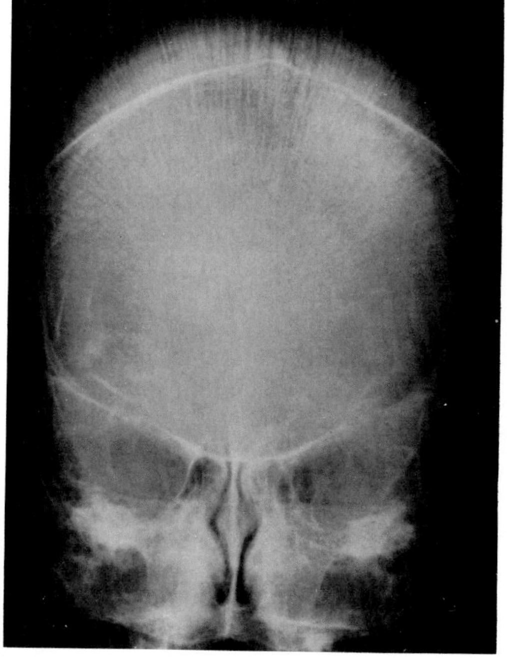

A

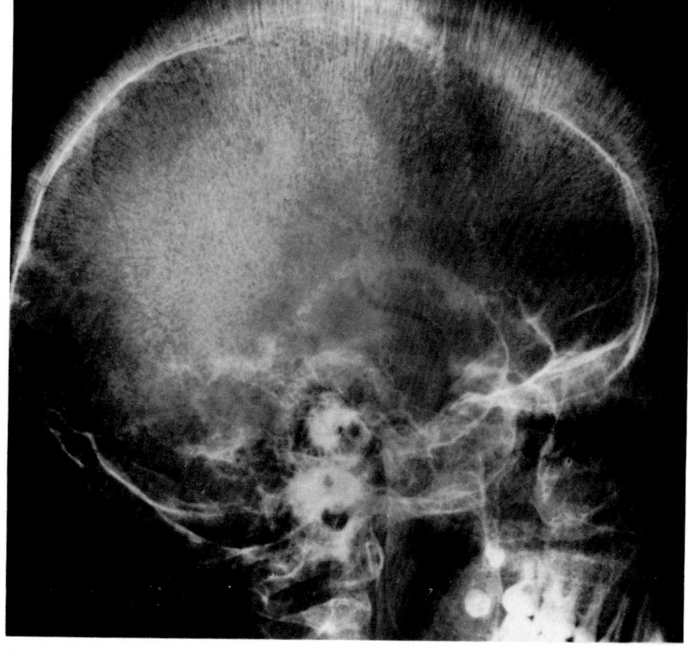

B

**Figure 56.17. Thalassemia.** (A) Frontal and (B) lateral views of the skull demonstrate marked widening of the diploic space with destruction of the outer table and coarse radial spiculation producing the characteristic hair-on-end appearance. Note the normal appearance of the calvarium inferior to the internal occipital protuberance, an area which is normally deficient in marrow and not involved in the congenital anemias. The visualized paranasal sinuses are poorly pneumatized.

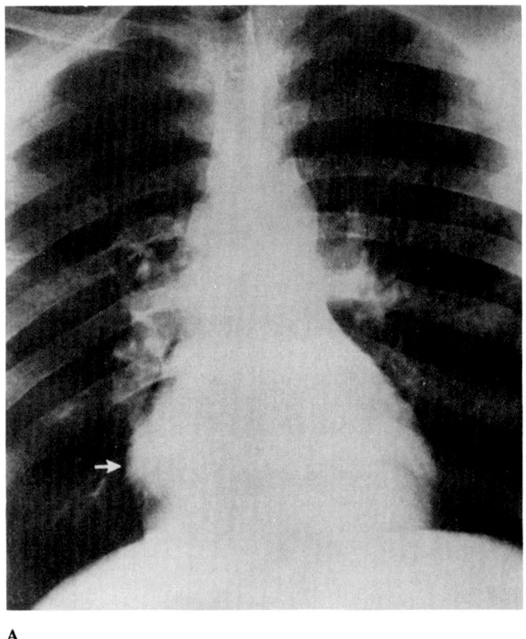
A

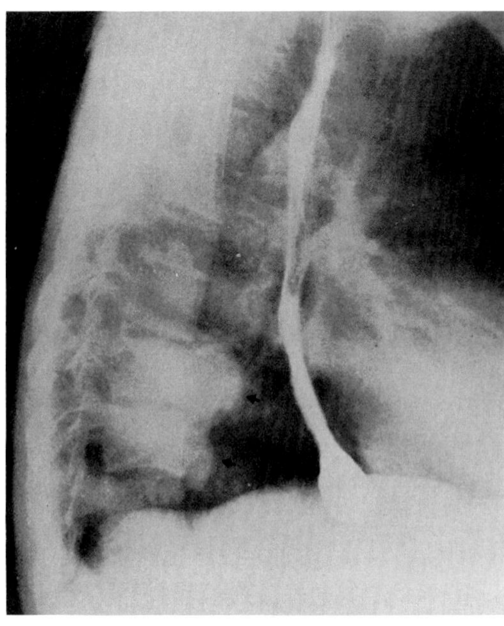
B

**Figure 56.18. Extramedullary hematopoiesis in thalassemia.** (A) Frontal and (B) lateral views of the chest demonstrate lobulated posterior mediastinal masses of hematopoietic tissue (arrows) bilaterally in the lower thoracic region. (From Leigh,[15] with permission from the publisher.)

displacement of the upper central incisors, producing malocclusion and overbite and the "rodent facies" typical of thalassemia.

The vertebrae are osteoporotic and have a coarsened trabecular pattern. Bulbous enlargement of posterior rib ends has been reported to be virtually pathognomonic of thalassemia.

Extramedullary hematopoiesis is a compensatory mechanism of the reticuloendothelial system (liver, spleen, lymph nodes) in patients with prolonged erthyrocyte deficiency due to the destruction of red blood cells or the inability of normal blood-forming organs to produce them. Paravertebral collections of hematopoietic tissue may appear on chest radiographs as single or multiple, smooth or lobulated, posterior mediastinal masses that are usually located at the lower thoracic levels.

Excessive destruction of red blood cells can lead to such radiographically detectable complications as hemochromatosis, hyperuricemia, and bilirubin gallstones.[1,11,12]

## CONGENITAL APLASTIC ANEMIA (FANCONI ANEMIA)

Fanconi anemia is an inherited disorder in which there is a constellation of transport defects in the proximal tubule of the kidney. Clinically, there is a severe aplastic anemia with pancytopenia that is often associated with brown skin pigmentation and multiple skeletal and renal anomalies. The disease is slowly progressive and usually fatal within 5 years after the onset of anemia.

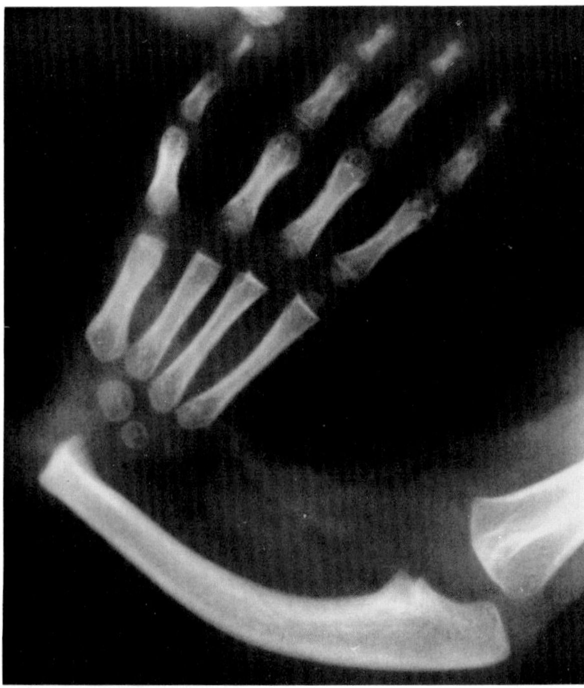

**Figure 56.19. Congenital aplastic anemia (Fanconi anemia).** There is aplasia of the radius with absence of the thumb.

The most characteristic skeletal abnormalities in Fanconi anemia are absence or hypoplasia of the thumb and anomalies of the radial side of the wrist. Hypoplasia or absence of the radius and hypoplasia of one or more bones of the hand are also common findings. Dwarfism, retarded bone age, and microcephaly are often associated. Renal anomalies are present in about 25 percent of cases and include aplasia, ectopia, and horseshoe kidney.[1,13,14]

## MYELOPHTHISIC ANEMIA

Infiltration of the bone marrow with tumor cells, or encroachment on marrow cavities due to cortical thickening, can result in severe anemia and pancytopenia. Tumors may arise from cells that are indigenous to the bone marrow (leukemia, lymphoma, myeloma), or the marrow may be invaded by extensive metastases to bone (carcinomas of the breast, prostate, lung, thyroid). Less common causes of marrow replacement include lipid storage disorders (Gaucher's disease), osteopetrosis, and myelofibrosis. In addition to destructive or osteoblastic skeletal changes, there may be splenomegaly due to extramedullary hematopoiesis.

### REFERENCES

1. Moseley JE: Skeletal changes in the anemias. *Semin Roentgenol* 9:169–184, 1974.
2. Powell JW, Weens HS, Wenger NK: The skull roentgenogram in iron-deficiency anemia and in secondary polycythemia. *AJR* 95:143–147, 1965.
3. Agarwal KN, Dhar N, Bhardwaj OP: Roentgenologic changes in iron-deficiency anemia. *AJR* 110:635–637, 1970.
4. Laws JW, Pitman RG: Radiologic features of primary pernicious anemia. *Br J Radiol* 33:229–237, 1969.
5. Sebes JI, Diggs LW: Radiographic changes of the skull in sickle cell anemia. *AJR* 132:373–377, 1979.
6. Reynolds J: Radiologic manifestations of sickle cell hemoglobulinopathy. *JAMA* 238:247–250, 1977.
7. Shaub MS, Rosen R, Boswell W, et al: Tibiotalar slant: A new observation in sickle cell anemia. *Radiology* 117: 551–552, 1975.
8. Schwartz AM, Homer MJ, McCauley RGK: "Step-off" vertebral body: Gaucher's disease versus sickle cell hemoglobulinopathy. *AJR* 132:81–85, 1979.
9. McCall IW, Moule N, Desai P, et al: Urographic findings in homozygous sickle cell disease. *Radiology* 126:99–104, 1978.
10. Ney C, Friedenberg RM: *Radiographic Atlas of the Genitourinary System*. Philadelphia, JB Lippincott, 1981.
11. Korsten J, Grossman H, Winchester PH, et al: Extramedullary hematopoiesis in patients with thalassemia anemia. *Radiology* 95:257–264, 1970.
12. Caffey J: Cooley's anemia: A review of the roentgenographic findings in the skeleton. *AJR* 78:381–391, 1957.
13. Juhl JH, Wesenberg RL, Gwinn JL: Roentgen findings in Fanconi's anemia. *Radiology* 89:646–653, 1967.
14. Minagi H, Steinbach HL: Roentgen appearance of anomalies associated with hypoplastic anemias of childhood. *AJR* 97:100–109, 1966.
15. Leigh TF: Mass lesions of the mediastinum. *Radiol Clin North Am* 1:377–393, 1963.

# Chapter Fifty-Seven
# Clotting Disorders

## IDIOPATHIC THROMBOCYTOPENIC PURPURA

Acute idiopathic thrombocytopenic purpura typically presents with the sudden onset of severe purpura 1 to 2 weeks after a sore throat or upper respiratory infection in an otherwise healthy child. In most patients, the disorder is self-limited and clears spontaneously within a few weeks. Unlike the acute form, chronic idiopathic thrombocytopenic purpura occurs primarily in young women and has an insidious onset with a relatively long history of easy bruising and menorrhagia. Because most patients with this condition have a circulating platelet autoantibody that develops without underlying disease or significant exposure to drugs, chronic idiopathic thrombocytopenic purpura is generally considered to be an autoimmune disorder.

The radiographic changes caused by either acute or chronic idiopathic thrombocytopenic purpura primarily involve the gastrointestinal tract. Hemorrhage into the small bowel produces characteristic uniform, regular thickening of mucosal folds in the affected intestinal segment. Splenomegaly is commonly present; splenectomy is often required to remove this important site of platelet destruction and major source of synthesis of platelet antibodies.[1,2]



## HENOCH-SCHÖNLEIN SYNDROME

The Henoch-Schönlein syndrome is an acute vasculitis characterized by purpura, nephritis, abdominal pain, and joint pain. The disease occurs most commonly in the first two decades, frequently develops several weeks after a streptococcal infection, and tends to be self-limited. Like other causes of vasculitis, the Henoch-Schönlein syndrome produces a radiographic pattern of regular thickening of small bowel folds simulating ischemic disease. Extensive edema and hemorrhage can lead to scalloping, thumbprinting, and separation of bowel loops. The Henoch-Schönlein syndrome can also affect the colon, causing circumferential submucosal lesions with marked luminal narrowing simulating carcinoma.

Clinical recovery, usually within 1 to 4 weeks, is accompanied by a return of the small bowel to a normal appearance.[3]

## HEREDITARY HEMORRHAGIC TELANGIECTASIA (OSLER-RENDU-WEBER DISEASE)

Hereditary hemorrhagic telangiectasia is a congenital disorder in which there is a familial history of repeated hemorrhage from the nasopharynx and gastrointestinal tract and multiple telangiectatic lesions involving the nasopharyngeal, buccal, and gastrointestinal mucosae. The most common radiographic manifestation is pulmonary arteriovenous malformations, which occur in about 15 percent of the cases. Indeed, about half of all patients with arteriovenous malformations in the lungs can be shown to have this hereditary condition. Pulmonary arteriovenous malformations appear as round or oval, homogeneous masses associated with enlarged vessels communicating with the hilum. Identification of the feeding and draining vessels, which is essential to the diagnosis, may be difficult on plain radiographs and often requires tomography. Confirmation of a pulmonary arteriovenous malformation requires pulmonary arteri-

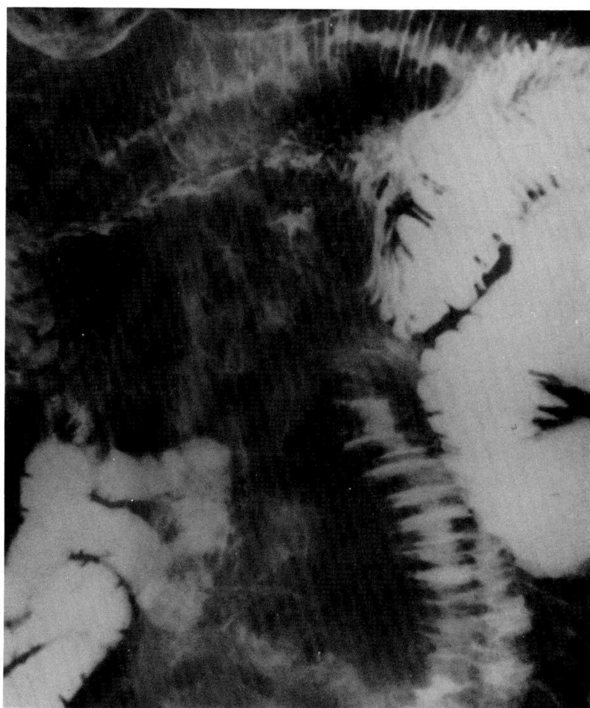

**Figure 57.1. Chronic idiopathic thrombocytopenic purpura.** Hemorrhage into the wall of the small bowel causes regular thickening of mucosal folds. (From Eisenberg,[9] with permission from the publisher.)

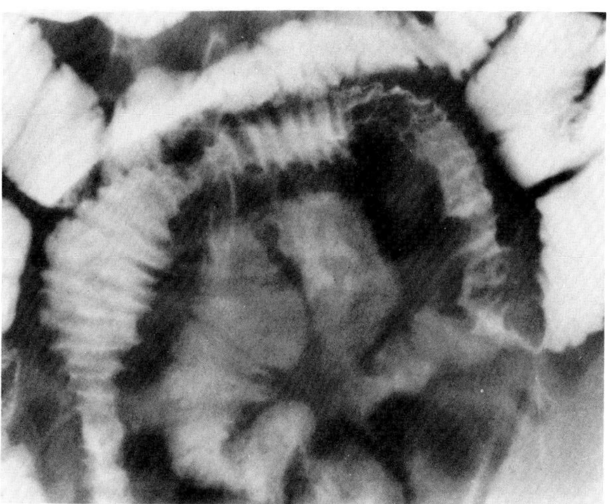

**Figure 57.2. Henoch-Schönlein syndrome.** Hemorrhage and edema within the intestinal wall causes regular thickening of small bowel folds. (From Eisenberg,[9] with permission from the publisher.)

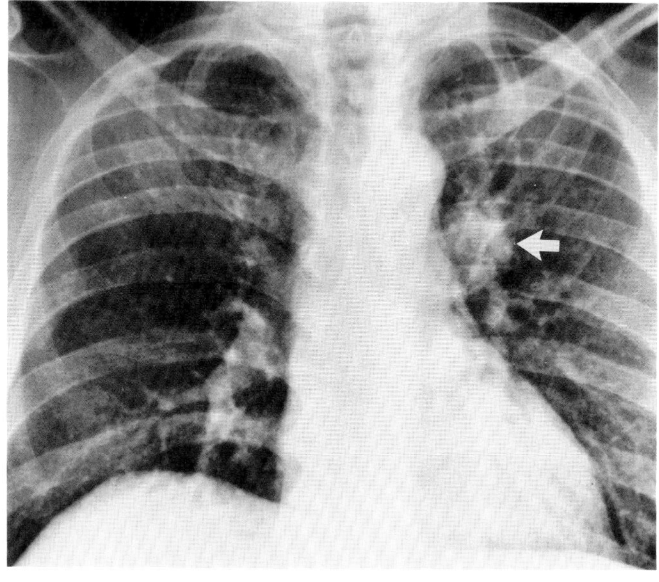

A

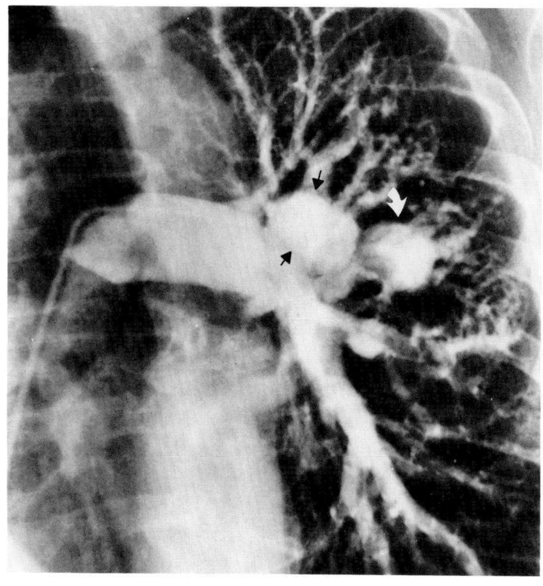

B

**Figure 57.3. Pulmonary arteriovenous malformations in hereditary hemorrhagic telangiectasia.** (A) A plain chest radiograph demonstrates an ill-defined soft tissue mass (arrow) in the left suprahilar region. (B) A pulmonary arteriogram demonstrates two vascular masses (arrows) representing pulmonary arteriovenous malformations.

ography, which often demonstrates other smaller, unsuspected vascular lesions.

Telangiectases of the bowel are rarely detected on routine barium studies and must be demonstrated by selective arteriography.[4]

## HEMOPHILIA

Hemophilia is an inherited (sex-linked recessive) anomaly of blood coagulation that appears clinically only in males. Patients with this disease have a decreased or absent serum concentration of antihemophilic globulin and suffer a lifelong tendency to spontaneous hemorrhage.

The major radiographic changes in hemophilia are complications of recurrent bleeding into the joints, which most commonly involves the knees, elbows, and ankles. Initially, the hemorrhage produces a generalized, nonspecific soft tissue prominence of the distended joint. Deposition of iron pigment may produce areas of cloudy increased density in the periarticular soft tissues. Although complete resorption of intra-articular blood may leave no residual change, subsequent episodes of bleeding result in synovial villous hypertrophy. In chronic disease, the hyperplastic synovium causes cartilage destruction and joint space narrowing and often leads to the development of multiple subchondral cysts of varying size in the immediate juxta-articular bone. Destruction of articular cartilage and continued use of the damaged joint lead to subchondral sclerosis and collapse, extensive and bizarre osteophytic spur formation, and often marked soft tissue calcification.

Intraosseous hemorrhage can lead to the development of lucent cysts in the bone distal to an affected joint. The lesions are round and osteolytic, with a thin sclerotic margin. Extensive intraosseous hemorrhage may expand and destroy a large area of bone (pseudotumor of hemophilia). A large adjacent soft tissue hemorrhage in association with bone destruction, especially in the pelvis, pubis, or femur, may suggest a sarcoma. Subperiosteal hemorrhage leads to periosteal reaction along the shaft of a long bone, with a cuff or triangle of new bone simulating a malignant tumor.

Repeated joint hemorrhages lead to increased blood flow in the region of the epiphysis and growth plate. These structures may ossify prematurely, become abnormally large, or fuse prematurely with the metaphysis. Hyperemia combined with the atrophy of bone and muscle that may follow an episode of hemarthrosis results in severe osteoporosis accentuating the trabeculae, which assume a radial distribution parallel to the long axis of the bone. Common signs suggestive, though not pathognomonic, of hemophilia include widening and deepening of the intercondylar notch of the femur and "squaring" of the inferior border of the patella. Asymmetric growth of the distal tibial epiphysis may result in "slant-

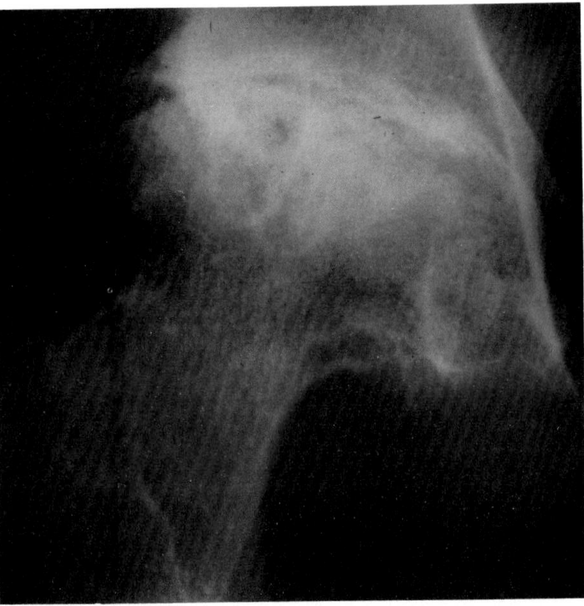

**Figure 57.4. Premature degenerative changes in the hip in hemophilia.** Numerous subarticular cysts, narrowed joint space, subchondral sclerosis, collapse, and osteophyte production in a 26-year-old man. (From Stoker and Murray,[6] with permission from the publisher.)

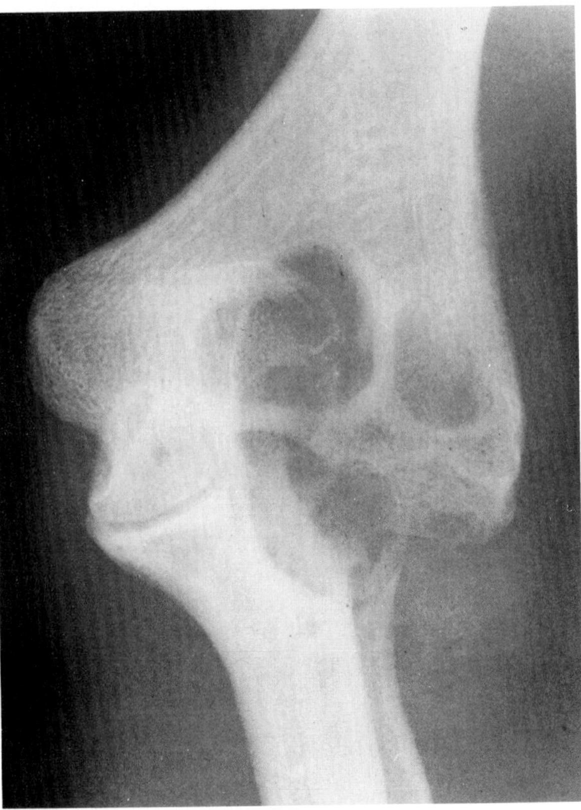

**Figure 57.5. Hemophilia of the elbow.** Large subchondral cysts.

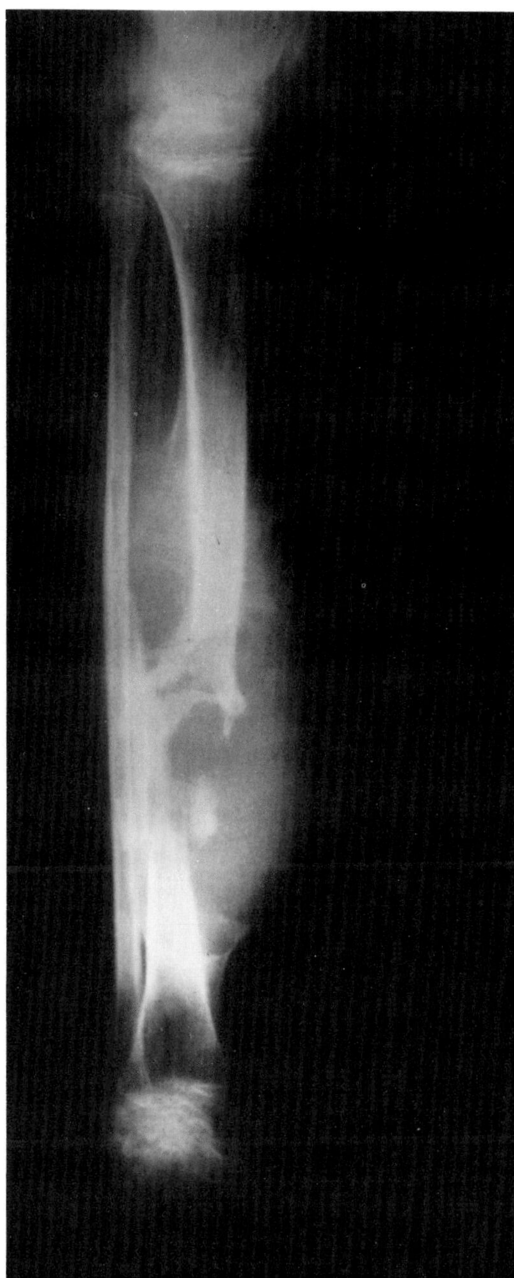

**Figure 57.6. Pseudotumor of hemophilia.** A destructive, expansile lesion of the lower tibial shaft simulates a large aneurysmal bone cyst. (From Stoker and Murray,[6] with permission from the publisher.)

ing" of the tibiotalar joint simulating the appearance in sickle cell disease. Hemarthrosis can cause occlusion of epiphyseal vessels and result in avascular necrosis. This most commonly involves the femoral and radial heads, both of which have a totally intracapsular epiphysis and are therefore especially vulnerable to deprivation of their vascular supply from compression by a tense joint effusion.

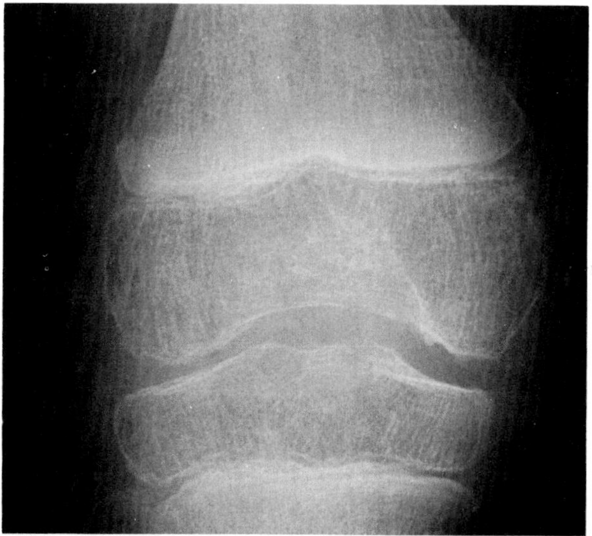

**Figure 57.7. Hemophilia of the knee.** There is demineralization and coarse trabeculation with overgrowth of the distal femoral and proximal tibial epiphyses. The intercondylar notch is moderately widened.

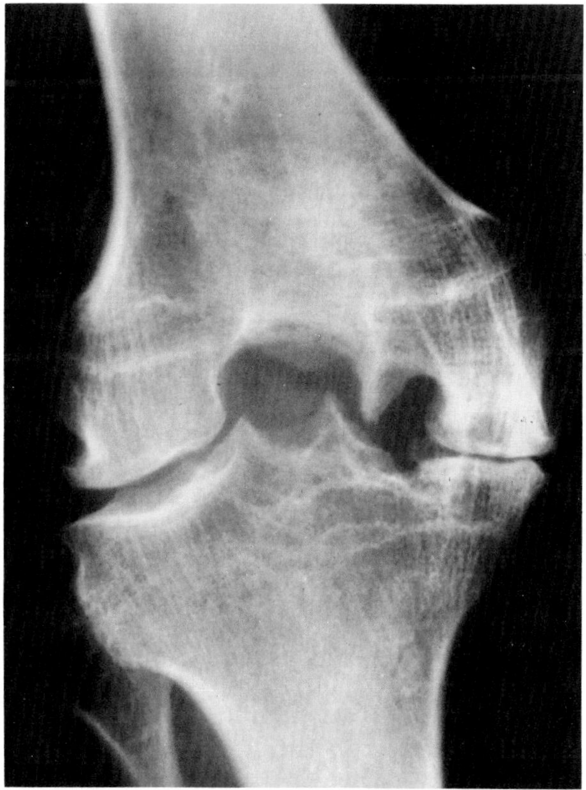

**Figure 57.8. Hemophilia of the knee.** The intercondylar notch is markedly widened with coarsened trabeculae, narrowing of the joint space, and hypertrophic spurring.

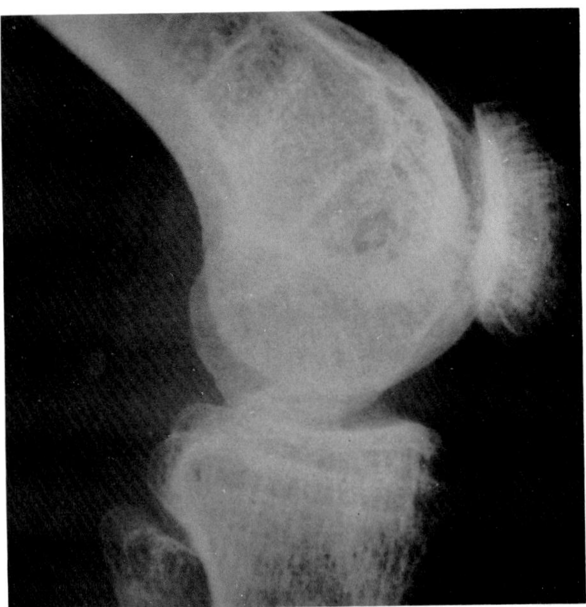

**Figure 57.9. Hemophilia.** Squaring of the inferior pole of the patella with associated growth arrest lines. (From Stoker and Murray,[6] with permission from the publisher.)

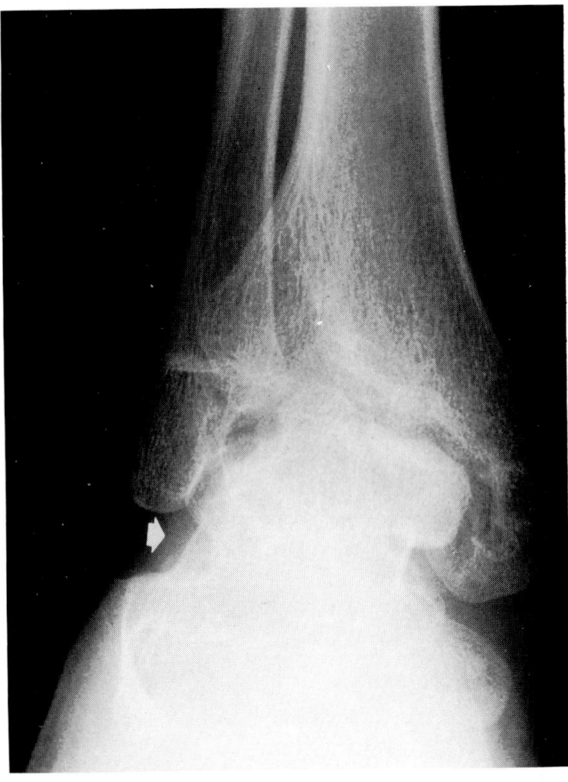

**Figure 57.10. Hemophilia of the ankle.** Tibiotalar angulation (arrow). (From Forrester, Brown, and Nesson,[10] with permission from the publisher.)

Submucosal bleeding into the wall of the gastrointestinal tract may develop in patients with hemophilia. This most commonly involves the small bowel and produces a short or long segment with regular thickening of folds. In the colon, bleeding may produce the thumbprinting pattern of sharply defined, finger-like marginal indentations along the contours of the colon wall.

Abnormalities of the urinary tract in hemophilia include intrarenal and perirenal hematomas, calyceal defects resembling papillary necrosis, and global renal enlargement. Obstructive uropathy may be caused by ureteral blood clots or by incomplete reabsorption and subsequent fibrosis of a retroperitoneal hematoma.[5–8]

## REFERENCES


1. Khilnani MT, Marshak RH, Eliasoph J, et al: Intramural intestinal hemorrhage. *AJR* 92:1061–1071, 1964.
2. Wiot JF: Intramural small intestinal hemorrhage—a differential diagnosis. *Semin Roentgenol* 1:219–233, 1966.
3. MacPherson RI: The radiologic manifestations of Henoch-Schönlein purpura. *J Can Assoc Radiol* 25:275–281, 1974.
4. Chandler D: Pulmonary and cerebral arteriovenous fistula with Osler's disease. *Arch Intern Med* 116: 277–282, 1965.
5. Dodds WJ, Spitzer RM, Friedland GW: Gastrointestinal roentgenographic manifestations of hemophilia. *AJR* 110: 413–416, 1970.
6. Stoker DJ, Murray RO: Skeletal changes in hemophilia and other bleeding disorders. *Semin Roentgenol* 9:185–193, 1974.
7. Edeiken J: *Roentgen Diagnosis of Diseases of Bones*. Baltimore, Williams & Wilkins, 1981.
8. Beck P, Evans KT: Renal abnormalities in patients with hemophilia and Christmas disease. *Clin Radiol* 23:349–354, 1972.
9. Eisenberg RL: *Gastrointestinal Radiology: A Pattern Approach*, Philadelphia, JB Lippincott, 1983.
10. Forrester DM, Brown JC, Nesson JW: *The Radiology of Bone Disease*. Philadelphia, WB Saunders, 1978.

## Chapter Fifty-Eight

# Other Hematologic Disorders

## POLYCYTHEMIA

### Primary Polycythemia (Polycythemia Vera)

Polycythemia vera is a hematologic disorder characterized by hyperplasia of the bone marrow that results in increased production of erythrocytes, granulocytes, and platelets. The disease is slowly progressive and produces symptoms associated with increased blood volume and viscosity. The hemoglobin concentration and hematocrit are markedly elevated. Cerebrovascular and peripheral vascular insufficiency are common, and many patients give a history of some thrombotic or hemorrhagic event during the course of their disease. There is an increased incidence of peptic ulcer disease, and the excessive cellular proliferation often results in hyperuricemia with secondary gout and the formation of urate stones. The spleen is often massively enlarged and may present as a left upper quadrant mass.

Increased blood volume in polycythemia vera can lead to prominence of the pulmonary vascular shadows, usually without the cardiomegaly associated with the increased pulmonary vascularity in patients with congenital heart disease. Intravascular thrombosis may cause pulmonary infarctions that appear as focal consolidations or bands of fibrosis.

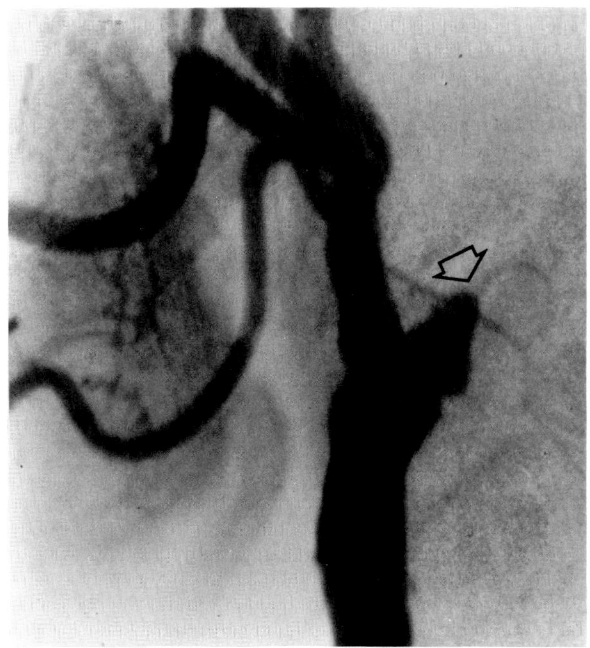

**Figure 58.1. Primary polycythemia.** A subtraction film from a carotid arteriogram in a 25-year-old man with episodes of transient cerebral ischemia demonstrates complete occlusion of the internal carotid artery (arrow) just distal to its origin.

In up to 20 percent of the cases, polycythemia vera terminates as myelofibrosis with anemia and marked splenomegaly. Radiographically, this appears as an irregular or generalized increase in bone density.[1]

### Secondary Polycythemia

Secondary polycythemia may be the result of chronic anoxia in patients with severe chronic pulmonary disease or congenital cyanotic heart disease, or may develop in persons living at high altitudes. An elevated hemoglobin concentration may also be caused by certain neoplasms (renal cell carcinoma, hepatoma, cerebellar hemangioblastoma) that result in an increased production of erythropoietin.

Because secondary polycythemia is a compensatory phenomenon, the pulmonary vasculature is normal in appearance and there is no evidence of the disease on chest radiographs. In children with severe secondary polycythemia due to cyanotic heart disease, the skull may show thickened tables and a hair-on-end appearance similar to the findings in thalassemia.[1,2]

## AGNOGENIC MYELOID METAPLASIA (MYELOFIBROSIS)

Agnogenic myeloid metaplasia (myelofibrosis) is a hematologic disorder in which gradual replacement of the marrow by fibrosis produces a varying degree of anemia and a leukemoid blood picture. Extramedullary hematopoiesis causes massive splenomegaly and often hepatomegaly and may form tumor-like masses in the posterior mediastinum that can be seen on chest radiographs. Although the disease is most commonly idiopathic, a large percentage of patients have antecedent polycythemia vera. Myelofibrosis also has been reported in association with metastatic carcinoma, chemical poisoning, chronic infection, leukemia, and histiocytosis X.

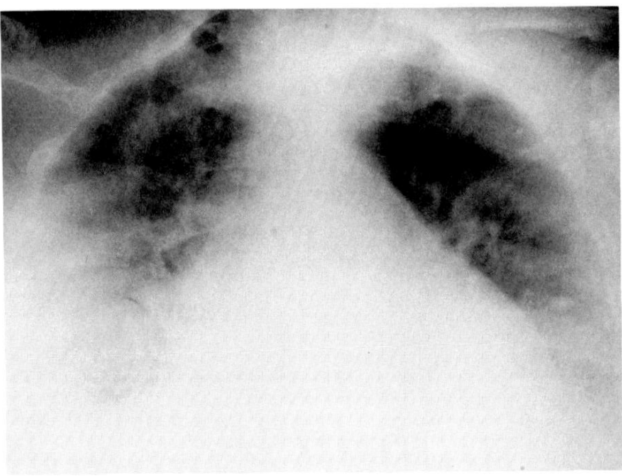

**Figure 58.2. Secondary polycythemia.** Severe hypoventilation due to profound obesity causes engorgement of pulmonary vessels. Although cardiomegaly is uncommon in polycythemia, in this case it reflects the pronounced elevation of the diaphragm due to huge abdominal girth as well as some underlying cardiac decompensation.

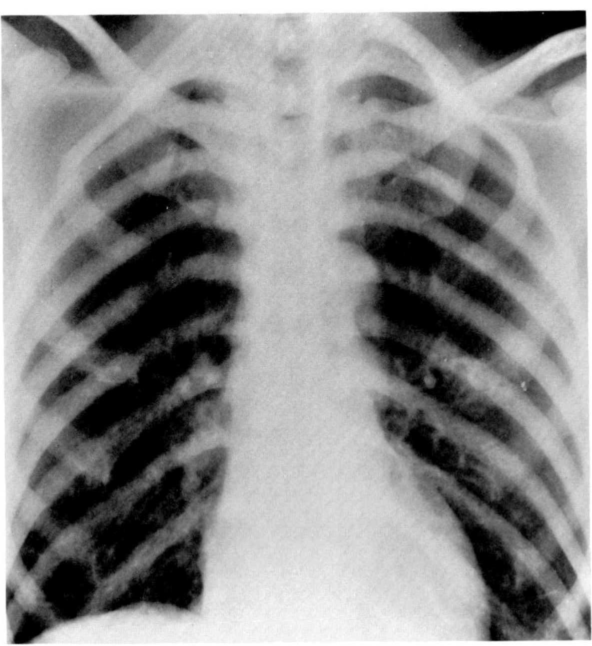

**Figure 58.3. Agnogenic myeloid metaplasia (myelofibrosis).** Diffuse uniform sclerosis of the bones of the thorax produces a jail bars appearance.

About half the patients with myelofibrosis have radiographic evidence of osteosclerosis. This primarily affects the spine, ribs, and pelvis but also can involve the long bones and skull. The typical radiographic pattern is a widespread, diffuse increase in bone density (ground-glass appearance). In the ribs, uniform obliteration of fine trabecular margins results in striking sclerosis that simulates jail bars crossing the thorax. Patchy osteosclerosis in long bones may produce a mottled moth-eaten appearance with the uninvolved bone appearing relatively osteolytic, a pattern that can suggest a destructive malignant process. In the vertebrae and long bones, endosteal cortical thickening can encroach on the medullary cavities and eventually obliterate them. In the skull, myelofibrosis may present as a generalized increase in bone density; scattered small, rounded, lucent lesions throughout the calvarium; or a mixed pattern of lucency and sclerosis.[3–5]

### REFERENCES

1. Pitman RG, Steiner RE, Szur L: Radiologic appearance of the chest in polycythemia. *Clin Radiol* 12:276–285, 1961.
2. Powell JW, Weens HS, Wenger NK: Skull roentgenogram in iron-deficiency anemia and in secondary polycythemia. *AJR* 95:143–147, 1965.
3. Jacobson HG, Fateh H, Shapira JH, et al: Agnogenic myeloid metaplasia. *Radiology* 72:716–725, 1959.
4. Meszaros WT, Sisson M: Myelofibrosis. *Radiology* 77: 958–967, 1961.
5. Pettigrew JD, Ward HP: Correlation of radiologic, histologic and clinical findings in agnogenic myeloid metaplasia. *Radiology* 93:541–548, 1969.

# PART TEN

# DISORDERS OF BONE AND BONE MINERALIZATION

## Chapter Fifty-Nine
# Disorders of Bone Mineralization

## DISORDERS OF THE PARATHYROID GLANDS

### Hyperparathyroidism

Excessive secretion of parathyroid hormone leads to a generalized disorder of calcium, phosphate, and bone metabolism that results in elevated serum levels of calcium and phosphate. Primary hyperparathyroidism may be caused by a discrete adenoma or carcinoma or by generalized hyperplasia of all glands. Other causes include nonparathyroid tumors that secrete a parathormone-like substance and the familial multiple endocrine neoplasia syndrome. Secondary hyperparathyroidism occurs more frequently than the primary form and is most often due to chronic renal failure.

The radiographic findings of primary and secondary hyperparathyroidism are similar, except that in the secondary form brown tumors (focal areas of bone destruction) are rare and osteosclerosis is more common. The earliest change is subperiosteal bone resorption, which particularly involves the radial margins of the middle phalanges at the proximal metaphyseal-diaphyseal junction, the distal clavicles, and the medial aspect of the upper third of the tibia. Loss of normal cortical definition is followed by an irregular lacy resorption, with the endosteal margin initially remaining intact. Other sites of subperiosteal bone resorption in hyperparathyroidism include the upper margins of the ribs, the pubis, ischia, calcaneus, and proximal humeri. Erosions of the terminal tufts of the fingers and loss of the lamina dura of the teeth are nonspecific findings that also occur in this condition.

Generalized loss of bone density may produce a ground-glass appearance. Brown tumors may become large and expansile and even simulate a malignant process. Pathologic fractures may lead to bizarre deformities. Irregular demineralization of the calvarium produces the characteristic salt-and-pepper skull.

A generalized increase in bone density (osteosclerosis) may develop in patients with hyperparathyroidism, especially when secondary to renal failure. Thick bands of increased density adjacent to the superior and inferior margins of vertebral bodies produce the characteristic "rugger jersey" spine.

Soft tissue calcification is common, especially in secondary hyperparathyroidism. Calcific deposits may develop in vessels, articular cartilages, menisci, joint capsules, and periarticular tissues. Hypercalcemia and hypocalcinuria may result in nephrocalcinosis and urinary tract calculi. An increased incidence of pancreatic calculi and pancreatitis, peptic ulcer, and gallstones also has been reported in patients with hyperparathyroidism.

Preoperative localization of a functioning parathyroid adenoma can be a difficult radiographic problem. Plain

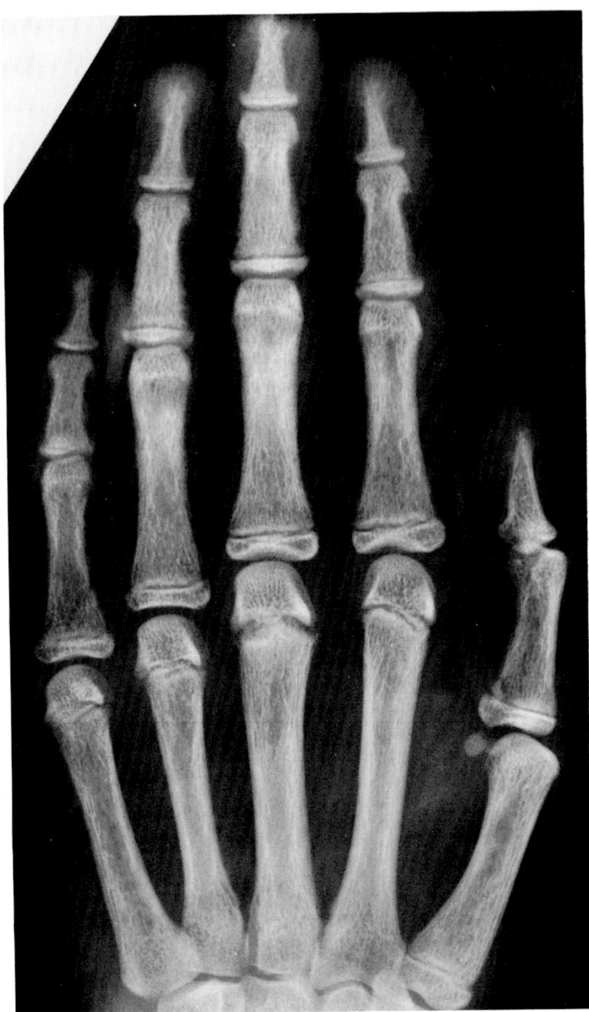

**Figure 59.1. Hyperparathyroidism.** Subperiosteal bone resorption predominantly involves the radial margins of the middle phalanges of the second, third, and fourth digits.

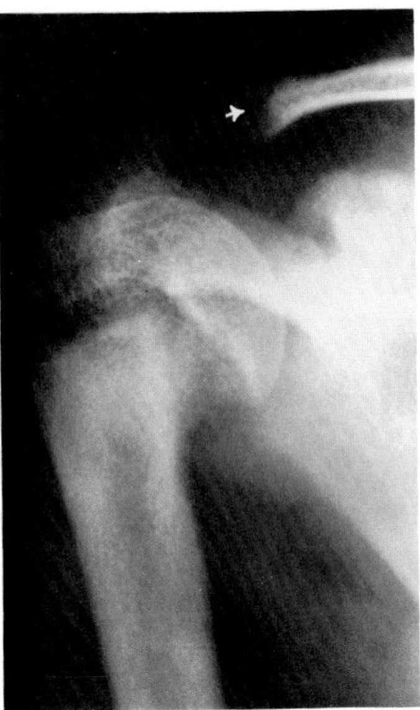

**Figure 59.2. Hyperparathyroidism.** Metaphyseal subperiosteal resorption beneath the proximal humeral head has led to a pathologic fracture with slippage of the humeral head. (An arrow points to the characteristic erosion of the distal clavicle.)

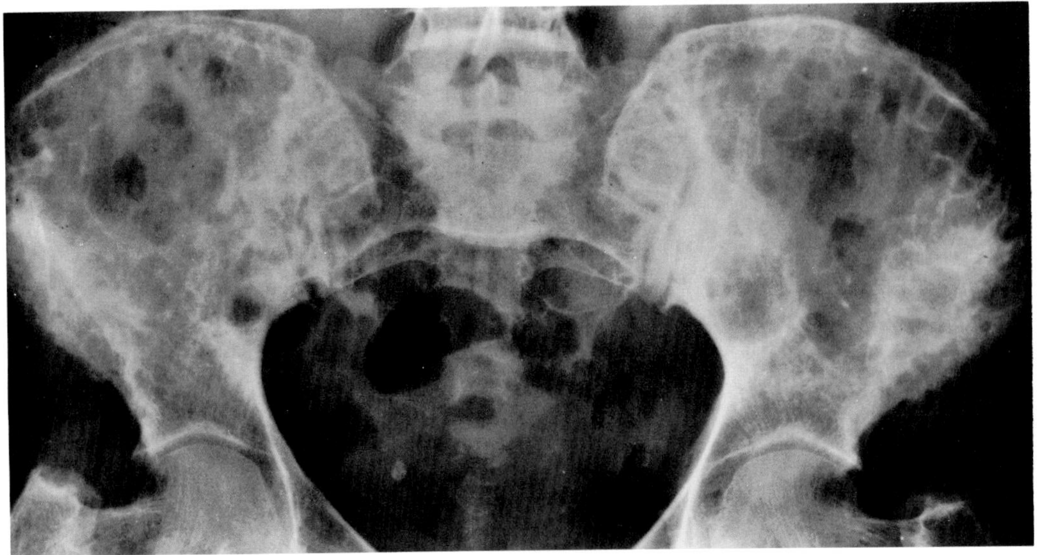

A

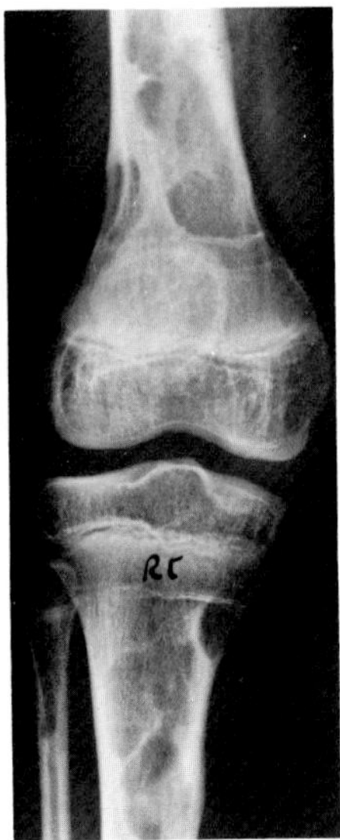

B

**Figure 59.3. Hyperparathyroidism.** Multiple brown tumors are seen in (A) the pelvis and (B) about the knee.

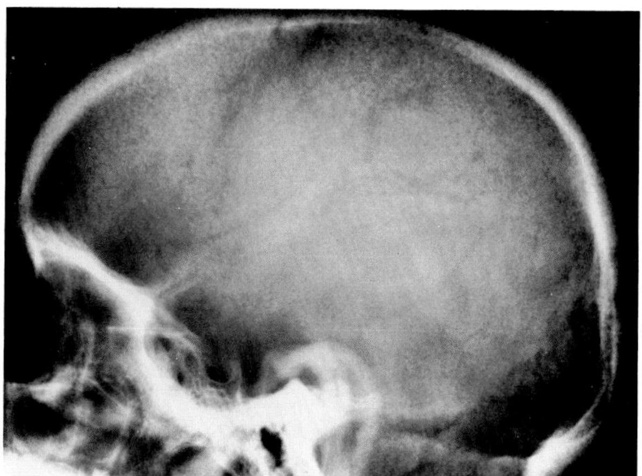

**Figure 59.4. Hyperparathyroidism.** Irregular demineralization of the calvarium produces the characteristic salt-and-pepper skull.

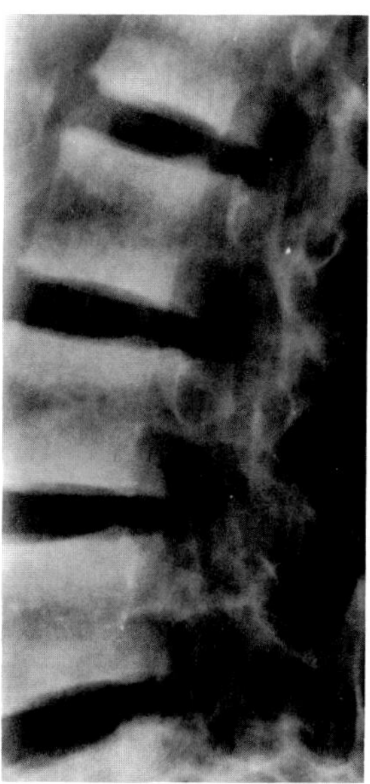

**Figure 59.5. Hyperparathyroidism.** A lateral view of the lumbar spine demonstrates osteosclerosis of the superior and inferior margins of the vertebral bodies (rugger jersey spine).

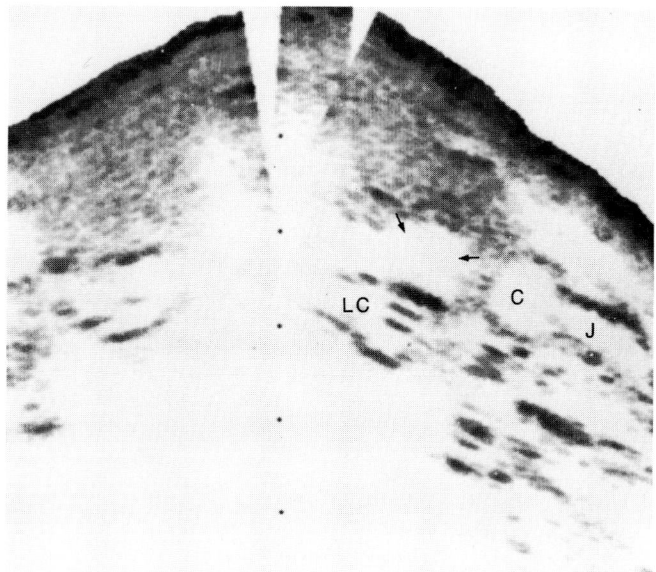


A

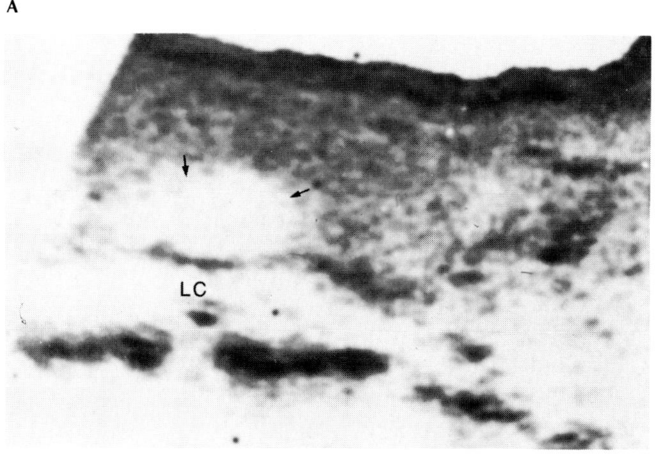


B

**Figure 59.6. Parathyroid adenoma.** (A) Transverse and (B) longitudinal sonograms through the thyroid and parathyroid regions demonstrate an adenoma of the left lower parathyroid (arrows), which is situated medial to the carotid artery (C) and jugular vein (J) and anterior to the longus colli muscle (LC) (From Mancuso,[10] with permission from the publisher.)

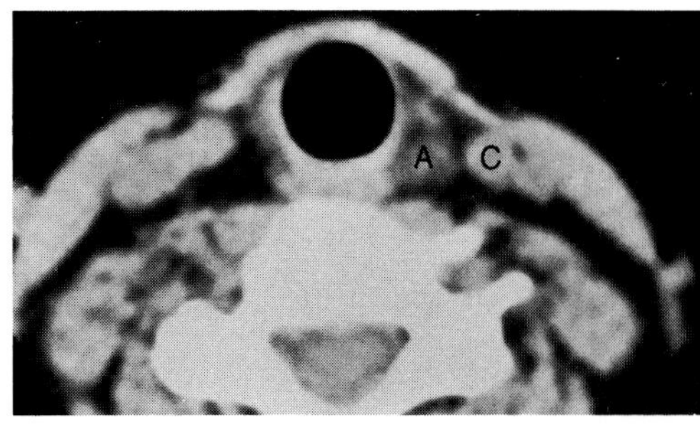


A

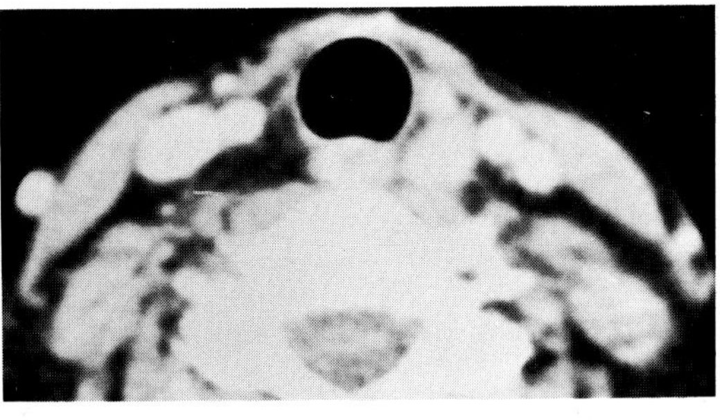

B

**Figure 59.7. Parathyroid adenoma.** (A) An initial CT scan at the level of the cricoid cartilage shows the parathyroid adenoma (A) medial to the carotid artery (C). (B) Following bolus injection of contrast material, there is marked enhancement of the adenoma. (From Mancuso,[10] with permission from the publisher.)

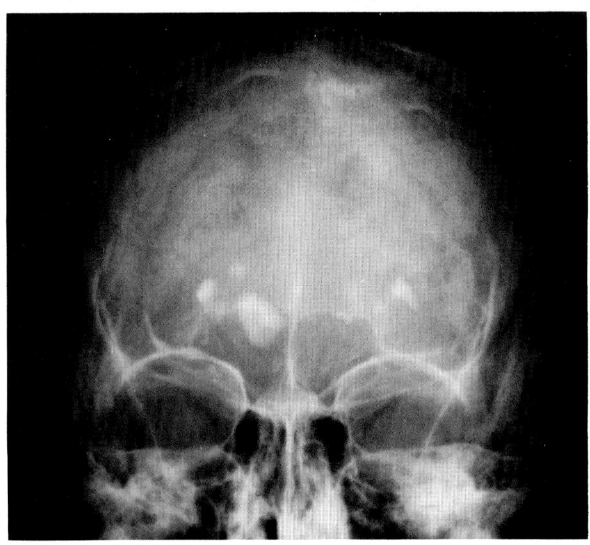

A

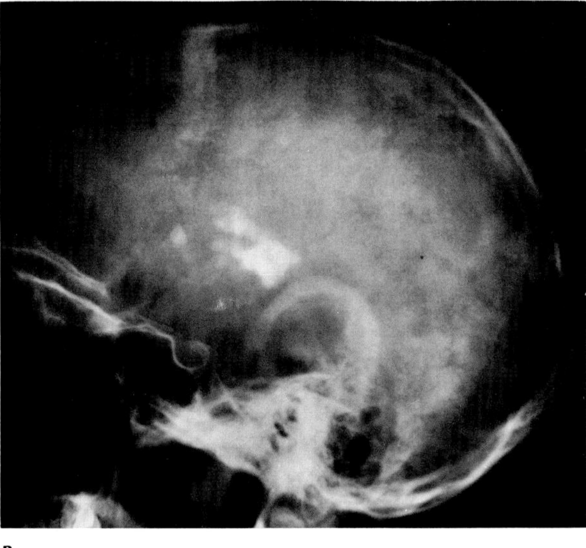

B

**Figure 59.8. Hypoparathyroidism.** (A) Frontal and (B) lateral views of the skull demonstrate calcification in the basal ganglia.

films and barium studies are of little value unless the tumor is large. Combined arteriography and venography associated with the mapping of venous sampling is probably the most accurate localizing technique. Ultrasound can demonstrate parathyroid glands more than 5 mm in size, though this modality may be unable to distinguish a dorsal thyroid lesion from a parathyroid mass. Computed tomography (CT) is especially valuable for demonstrating an aberrantly located parathyroid adenoma or hyperplastic gland in the superior mediastinum in a patient in whom neck exploration failed to demonstrate an adenoma causing hyperparathyroidism.[1–10]

## Hypoparathyroidism

Hypoparathyroidism usually results from injury or accidental removal of the glands during thyroidectomy and is not associated with any significant radiographic abnormalities. In the less common idiopathic, or primary, type of hypoparathyroidism, the major radiographic finding is cerebral calcification, especially involving the basal ganglia, dentate nuclei of the cerebellum, and choroid plexus. A pattern of increased density may develop in the long bones, usually localized to the metaphyseal area.[11–13]

## Pseudohypoparathyroidism and Pseudopseudohypoparathyroidism

Pseudohypoparathyroidism is a hereditary disorder in which there is failure of normal end organ response to normal levels of circulating parathyroid hormone. Most patients are obese and of short stature, with a round facies, corneal or lenticular opacities, brachydactyly, and mental retardation. The most common radiographic abnormality is shortening of the tubular bones of the hands and feet (especially the fourth and fifth metacarpals) and calcific or bony deposits in the skin or subcutaneous tissues. Defective metaphyseal ossification may produce a rickets-like appearance and lead to slipping of the fem-

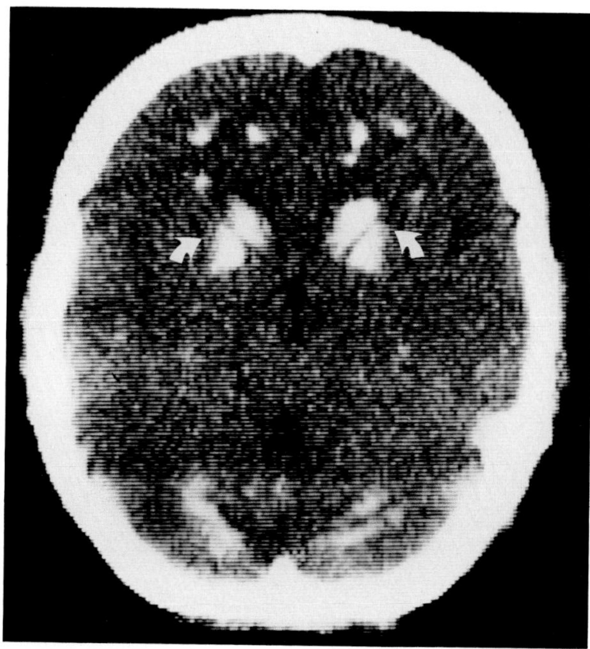

**Figure 59.9. Hypoparathyroidism.** A CT scan shows calcification in the basal ganglia (arrows). (From Cohen, Duchesneau, and Weinstein,[13] with permission from the publisher.)

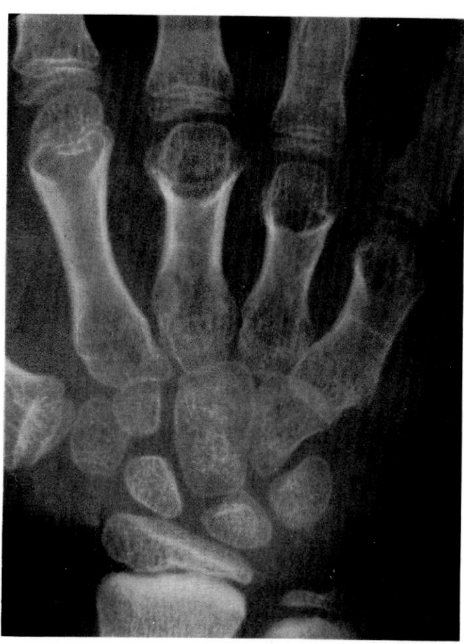

**Figure 59.10. Pseudohypoparathyroidism.** There is shortening of the third, fourth, and fifth metacarpals. Note the widening of these bones, especially in the metaphyseal regions.

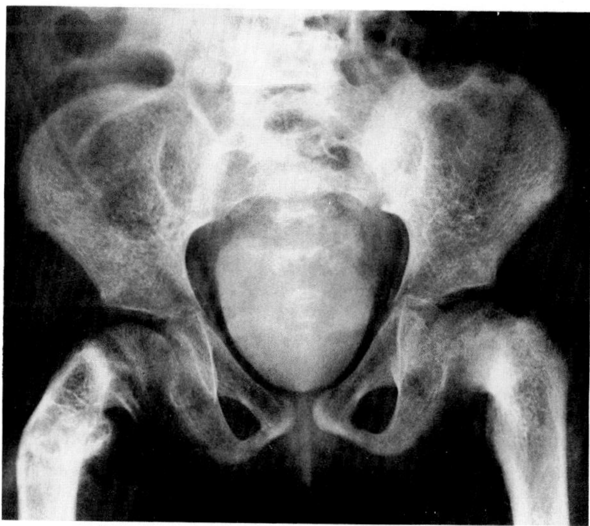

**Figure 59.11. Pseudohypoparathyroidism.** There is subperiosteal resorption of the iliac wings and metaphyses of the proximal femurs with generalized osteomalacia. Slippage of the femoral capital epiphyses has led to bilateral varus deformities of the hip.

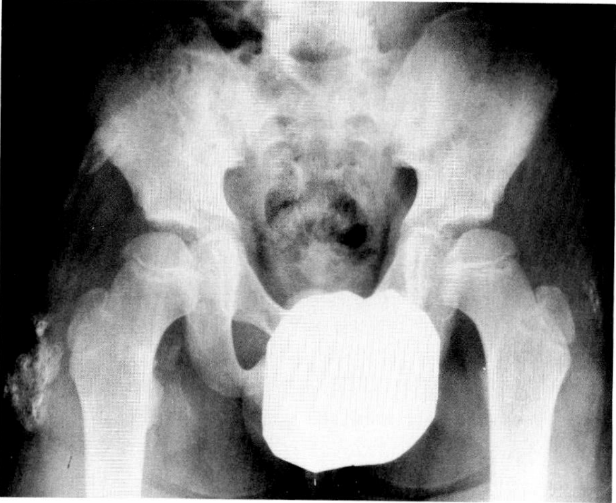

**Figure 59.12. Pseudohypoparathyroidism.** There are soft tissue calcifications lying within muscle bundles about both hip joints.

oral capital epiphyses and varus deformities about the hip. As in idiopathic hypoparathyroidism, calcification is often found in the brain, especially the basal ganglia.

Pseudopseudohypoparathyroidism refers to the presence of similar skeletal anomalies in other members of a patient's family in the absence of biochemical disturbances.[14]

## METABOLIC BONE DISEASE

### Osteoporosis

Osteoporosis is a generalized or localized deficiency of bone matrix in which the mass of bone per unit volume is decreased in amount but normal in composition. Bone is a living, constantly changing tissue, and normally there is a balance between the amount of old bone being removed and the amount of new bone replacing it. Os-

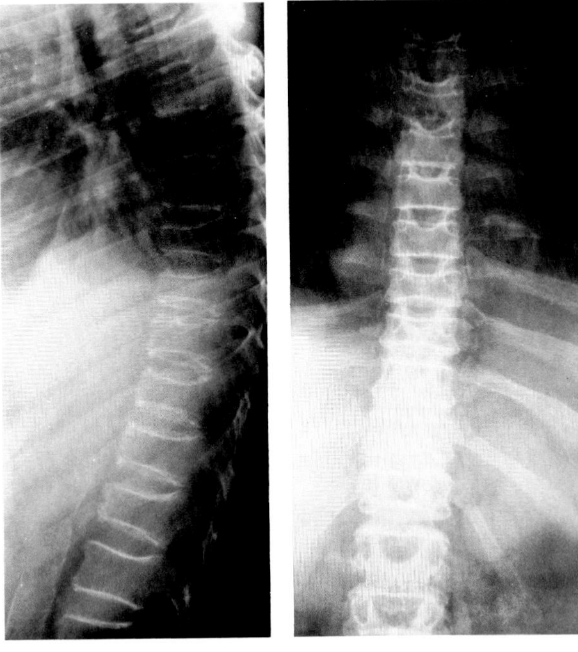

A B

**Figure 59.13. Severe osteoporosis.** (A) Lateral and (B) frontal views of the thoracolumbar spine show striking demineralization and compression of multiple vertebral bodies in a 14½-year-old girl treated with steroids for 5 years for chronic glomerulonephritis. The height age of the girl was only 9 years at this time. (From Dorst, Scott, and Hall,[25] with permission from the publisher.)

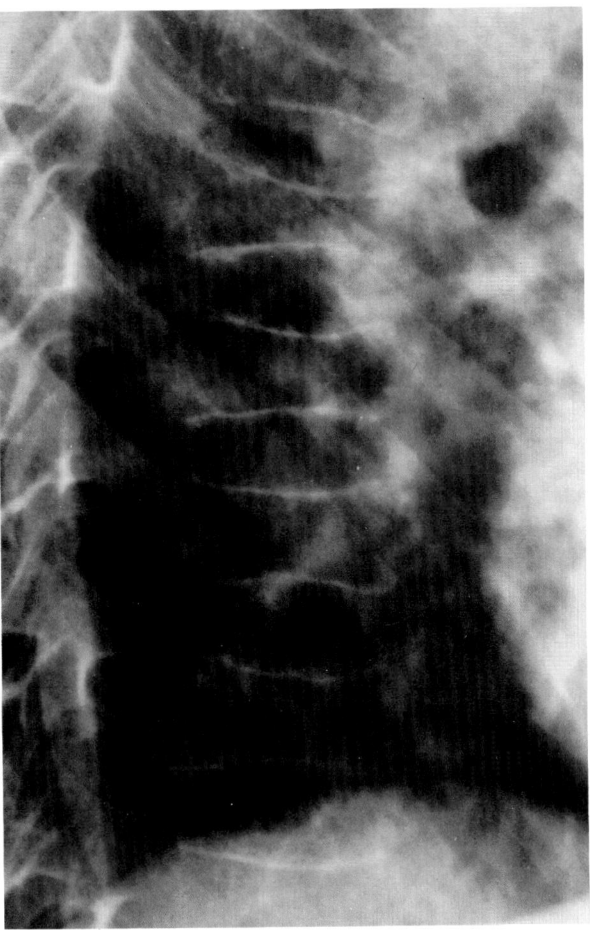

**Figure 59.14. Severe osteoporosis.** A lateral view of the thoracic spine in a patient on high-dose steroid therapy for dermatomyositis demonstrates thinning of cortical margins and biconcave deformities of vertebral bodies (fish vertebrae). (From Eisenberg,[40] with permission from the publisher.)

teoporosis is usually due to accelerated resorption of bone, though decreased bone formation may lead to osteoporosis in such entities as Cushing's syndrome, prolonged steroid administration, and disuse or immobilization osteoporosis. Loss of mineral salts causes osteoporotic bone to become more lucent than normal. This may be difficult to detect, since about 30 percent of the bone density must be lost before it can be demonstrated on routine radiographs.

The major causes of generalized osteoporosis are senility and postmenopausal hormonal changes. As a person ages, the bones lose density and become more brittle, fracturing more easily and healing more slowly. Many elderly persons also are less active and have poor diets that are deficient in protein. In postmenopausal osteoporosis there is a deficiency in the gonadal hormonal levels and decreased osteoblast activity. Other causes of generalized osteoporosis include Cushing's syndrome, steroid therapy, other endocrine dysfunctions, malnutrition and deficiency diseases, congenital disorders (osteogenesis imperfecta, homocystinuria), neoplastic disorders (multiple myeloma, metastases), and hemoglobinopathies.

Regardless of the cause, the radiographic appearance is similar in all conditions producing osteoporosis. The most striking change is cortical thinning with irregularity and resorption of the endosteal surfaces. In the long bones or the midshafts of the metacarpals and metatarsals, osteoporosis is evident if the sum of the two cortices is less than one-half the diameter of the bone. There is a decrease in the number and thickness of trabeculae in spongy bone. However, the remaining trabeculae lying along lines of stress (weight-bearing, muscle contraction) are accentuated and have an increased density and width. Although the bone density is generally decreased, the normally mineralized but thin cortex appears to be relatively dense, in contrast to the deossified spongy bone.

The radiographic changes of osteoporosis primarily involve the spine and pelvis. As the bone density of a vertebral body decreases, the cortex appears as a thin line that is relatively dense and prominent, producing the typical picture-frame pattern. Because of the severe loss of bone density, anterior wedging or compression fractures of one or more vertebral bodies may result, most commonly in the middle and lower thoracic and upper lumbar areas. The intervertebral disks may expand

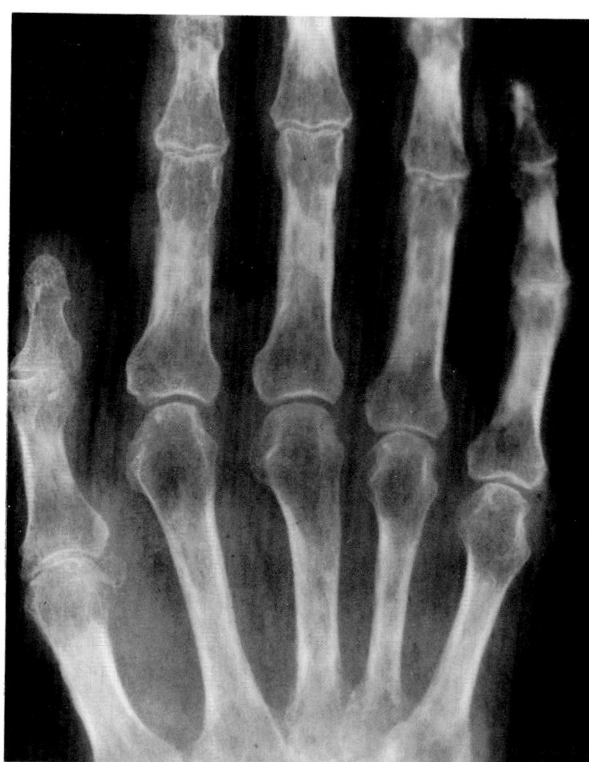

**Figure 59.15. Sudeck's atrophy.** Patchy osteoporosis predominantly affects the periarticular regions of the hand. Note the associated soft tissue swelling.

into the weakened vertebral bodies and produce characteristic concave contours of the superior and inferior disk surfaces ("fish" contour). In the skull, the calvarium may show spotty loss of density, and there is commonly deossification of the floor of the sella turcica and dorsum sellae.

Localized osteoporosis may result from disuse, neurovascular disturbances (Sudeck's atrophy), local joint disease, or inflammation. To maintain osteoblastic activity at normal levels, bones must be subjected to a normal amount of stress and muscular activity. Within a few weeks after the fracture of a bone, localized osteoporosis becomes detectable, especially distal to the site of injury. The loss of bone matrix is primarily due to inactivity, though hyperemia, impaired venous flow, and neural changes may also play some role. The cortical margin of an involved bone becomes thin but, unlike the bone destruction due to disease, never completely disappears. Sudeck's atrophy (reflex sympathetic dystrophy) is the rapid development of painful osteoporosis following rather trivial injury. Probably of neurovascular origin, Sudeck's atrophy most often involves the hands and feet with a mottled, irregular osteoporosis that primarily affects the periarticular region. The juxta-articular cortex may become extremely thin but remains intact, unlike in an arthritic process. Unusual types of localized osteoporosis include regional migratory osteoporosis (transitory osteoporosis), in which osteoporosis follows the development of severe pain about a major joint such as the hip, knee, or ankle in middle-aged or elderly adults, and transitory demineralization of the femoral head, a disorder characterized by a painful hip and progressive osteoporosis of the femoral head.[15–18]

## Osteomalacia

Osteomalacia and rickets include a variety of skeletal disorders in which there is defective mineralization of osteoid (bone matrix). The term *osteomalacia* refers to insufficient mineralization of the adult skeleton in which the epiphyseal growth plates are closed. In childhood rickets, the effect on bone already formed is the same as in the adult. However, in rickets there are also specific changes in areas of actively growing bone (epiphyses and metaphyses).

There is normally a balance between osteoid formation and mineralization. Osteomalacia is thus the result of either excessive osteoid formation or, more frequently, insufficient mineralization. Proper calcification of osteoid requires that adequate amounts of calcium and phosphorus be available at the mineralization sites. Failure of calcium and phosphorus deposition in bone matrix in osteomalacia may be due to an inadequate intake or failure of absorption of calcium, phosphorus, or vitamin D, which is necessary for intestinal absorption of calcium and phosphorus and may have a direct effect on bone. At times, the level of vitamin D is sufficient but the material is not utilized because of resistance to the action of the vitamin at end organs, such as the kidneys. Other non-nutritional causes of osteomalacia include chronic renal failure and certain renal diseases in which there is proximal tubular insufficiency without glomerular involvement.

Regardless of the cause, osteomalacia appears radiographically as a loss of bone density due to the presence of nonmineralized osteoid. Although the cortex is thinned, it may stand out more prominently than normal because of the uniform deossification of spongy bone. In contrast to osteoporosis, the cortical borders in osteomalacia are often indistinct. Fine-detail radiography of the hands in patients with osteomalacia often demonstrates intracortical striations caused by local resorption or lack of mineralization, a finding not seen in osteoporosis.

In some patients the radiographic changes in osteomalacia are indistinguishable from osteoporosis. Two features suggesting osteomalacia as the cause of bone demineralization are the presence of pseudofractures and bone deformities. Pseudofractures (Looser's lines) are ribbon-like lucent bands arising perpendicular to the cortex and extending transversely to partially or completely cross the bone. They are composed of zones of uncalcified osteoid and fibrous tissue that probably represent healing reactions to infractions. Pseudofractures are often bilateral and symmetric and are particularly common

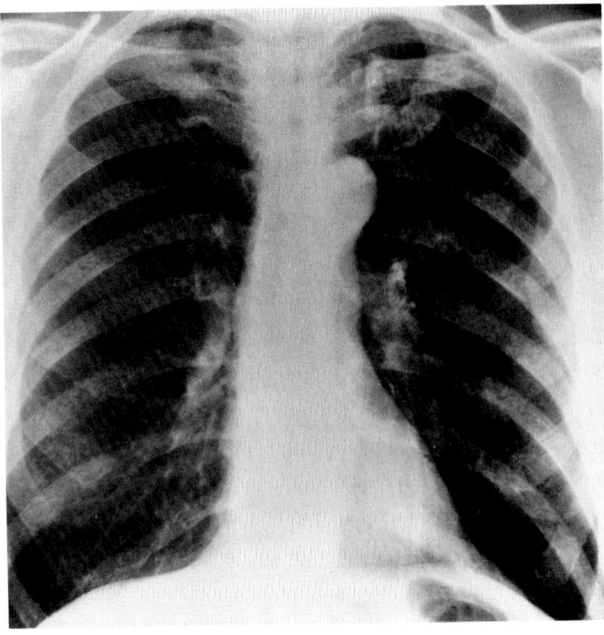

A

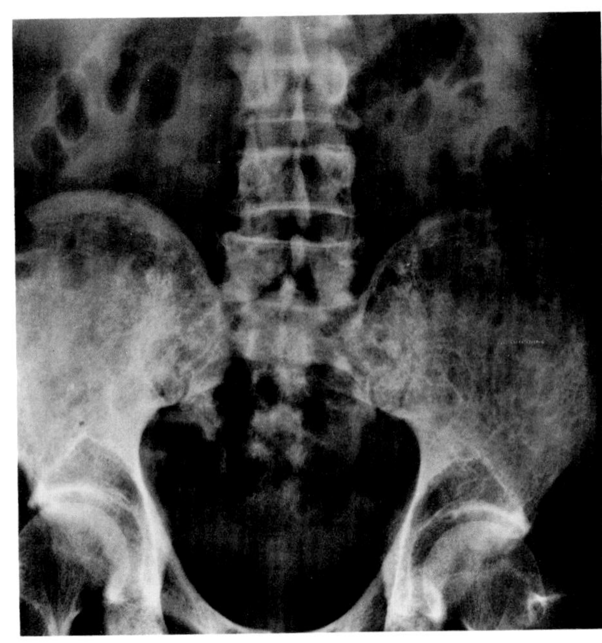

B

**Figure 59.16. Osteomalacia.** Views of (A) the chest and (B) the abdomen show generalized bony demineralization with striking prominence of residual trabeculae (especially in the ribs).

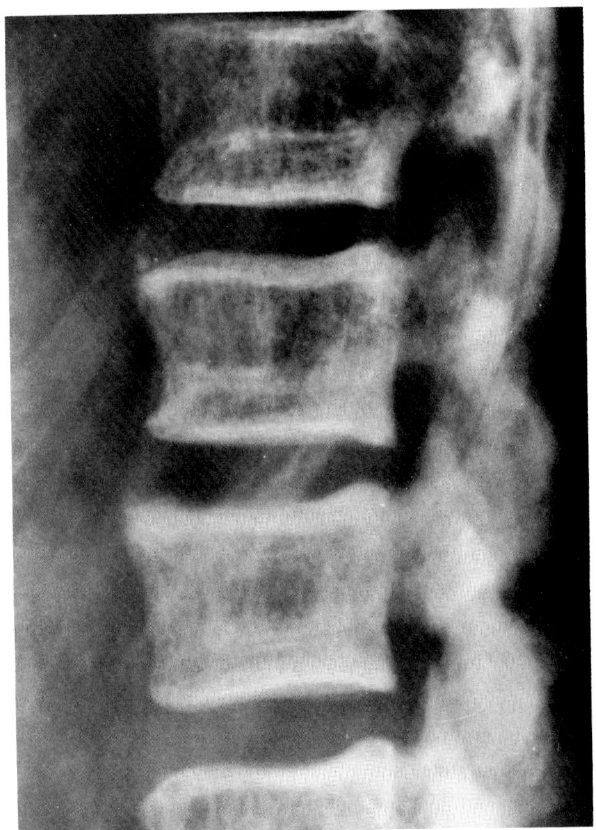

**Figure 59.17. Osteomalacia.** In this patient with a renal tubular disorder, there is striking thickening of the cortices of the vertebral bodies with increased trabeculation of spongy bone.

on the inner margin of the femoral neck and shaft, axillary border of the scapula, pubic and ischial rami, ribs, metatarsals, and bones of the forearm. Although pseudofractures are often considered pathognomonic of osteomalacia, an identical radiographic appearance may be seen in such disorders as Paget's disease, fibrous dysplasia, and osteogenesis imperfecta. If the osteomalacic process continues, pseudofractures may progress to complete fractures. Bones that are softened by osteomalacia may bend or give way as a result of weight-bearing. Bowing deformities primarily involve the pelvis, vertebral column, thorax, and proximal extremities. In the pelvis there may be inward bending of the sidewalls with deepening of the acetabular cavities (protrusio acetabuli). Vertebral bodies may show biconcave or compression deformities indistinguishable from osteoporosis. Downward molding of the skull over the upper cervical vertebrae may flatten the basal angle of the skull (platybasia).

In some patients with osteomalacia due to renal tubular disorders, hyperostosis rather than deossification may be noted. In these patients there may be a striking thickening of the cortices and increased trabeculation of spongy bone. Nevertheless, the bony architecture is abnormal and prone to fracture with relatively minimal trauma.[19–21]

## Rickets

Rickets is a systemic disease of infancy and childhood that is the equivalent of osteomalacia in the mature skel-

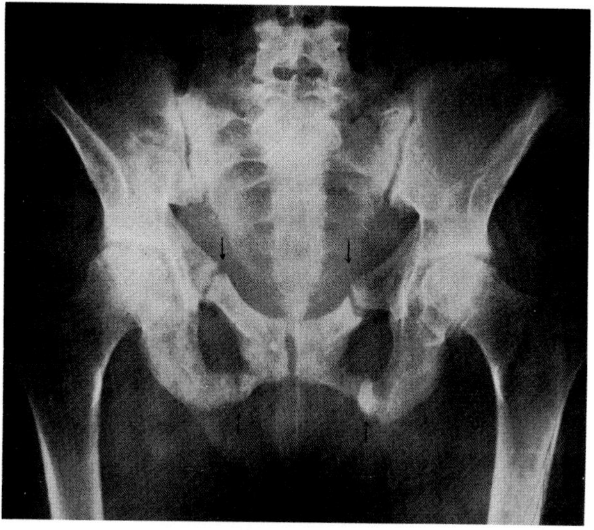
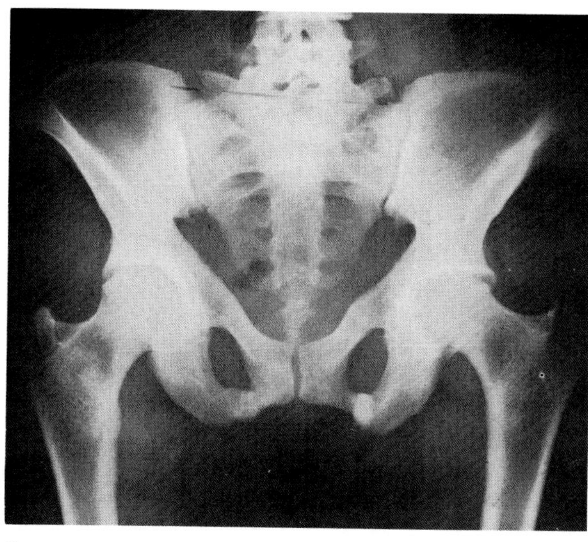

A B

**Figure 59.18. Pseudofractures in osteomalacia.** (A) Bilateral symmetric fractures that developed first as pseudofractures. (B) Healing of fractures following therapy. (From Edeiken,[22] with permission from the publisher.)

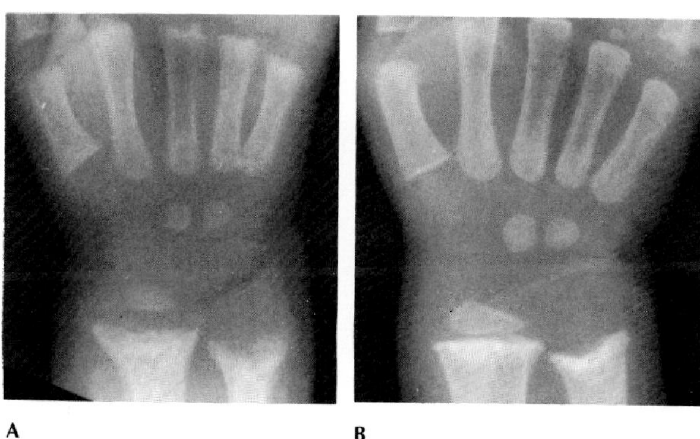

A B

**Figure 59.19. Rickets.** (A) An initial film of the wrist shows characteristic cupping and fraying of the metaphyseal ends of the radius and ulna, with disappearance of the normally sharp metaphyseal lines. (B) Following therapy with vitamin D, there is remineralization of the metaphyseal ends of the radius and ulna, sharp metaphyseal lines, and a return of the epiphyseal centers to normal density and sharpness of outline.

eton. In this condition, calcification of growing skeletal elements is defective because of a deficiency of vitamin D in the diet or a lack of exposure to ultraviolet radiation, which converts the sterols in the skin into vitamin D. Rickets is most common in premature infants and usually develops between the ages of 6 months and 1 year.

The early radiographic changes in rickets are best seen in the fastest growing portions of bones, such as the sternal ends of the ribs, proximal ends of the tibia and humerus, and distal ends of the radius and ulna. Failure of calcium deposition within the zone of provisional calcification of the epiphyseal cartilage plate results in an overgrowth of noncalcified osteoid tissue that appears radiographically as a characteristic increased distance between the ossified portion of the epiphysis and the end of the shaft. In response to the pull of muscular and ligamentous attachments, the metaphyseal ends of the bone become cupped and frayed and the normally sharp metaphyseal lines disappear. Lack of calcification leads to delayed appearance of the epiphyseal ossification centers. The ossification centers in rickets tend to have blurred margins, in contrast to scurvy, in which they are sharply outlined. There is general demineralization of the skeleton. The bones have a coarse texture because of loss of fine trabeculae and accentuation of those trabeculae that remain. Because of poor mineralization, bowing of weight-bearing bones (especially the tibia) develops once the infant begins to stand or walk. Although pseudofractures seldom occur, greenstick fractures are common. Softening of the skull bones (craniotabes) is an early finding that may be followed by frontal and parietal bossing, giving the head a boxlike appearance. The cranial sutures are widened, and fontanelle closure is often delayed. In severe disease, thin stripes may develop along the outer cortical margins of long bones. Although they resemble inflammatory periosteal calcification, these shadows actually represent zones of poorly calcified osteoid laid down by the periosteum. Excessive osteoid tissue in the sternal ends of the ribs produces a characteristic beading (rachitic rosary).

Healing of rickets begins with mineralization of the zone of provisional calcification, which widens as healing progresses. The epiphyseal centers gradually regain their normal density and sharpness of outline. Remineralization of subperiosteal osteoid appears as solid or laminated periosteal new bone formation. Although complete healing usually occurs, remineralization of the skeleton is a slow process, and residual deformities may persist.[4,22]

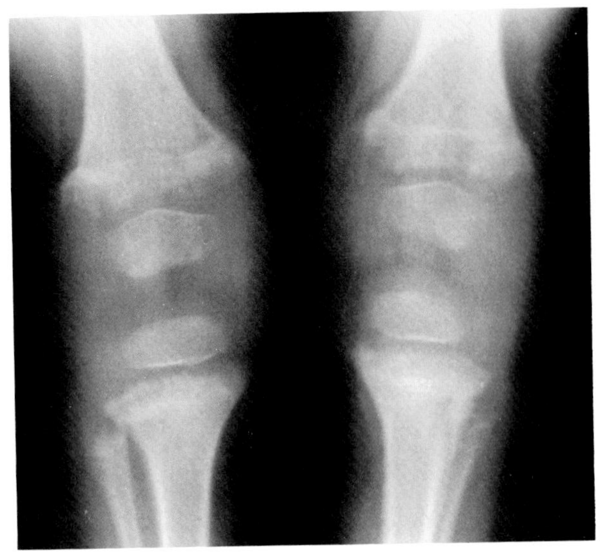
A

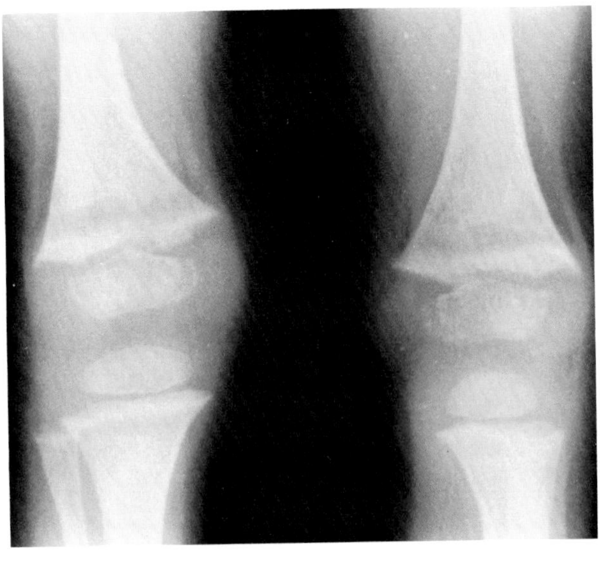
B

**Figure 59.20. Rickets.** (A) An initial film shows severe metaphyseal changes involving the distal femurs and proximal tibias and fibulas. Note the pronounced demineralization of the epiphyseal ossification centers. (B) Following vitamin D therapy, there is remineralization of the metaphyses and an almost normal appearance of the epiphyseal ossification centers.

### Hypophosphatasia

Hypophosphatasia is an inherited metabolic disorder in which a low level of alkaline phosphatase leads to defective mineralization of bone. The disorder may become manifest at various stages of life, with the most severe symptoms appearing at the earliest age of onset. Hypophosphatasia discovered in utero or in the first few days of life is generally a fatal condition in which the calvarium and many bones of the skeleton are uncalcified. When hypophosphatasia becomes apparent during infancy and childhood, the radiographic appearance closely resembles rickets. The skull contains large unossified areas simulating severe widening of the sutures. If the infant survives and calvarial recalcification occurs, there may be premature closure of the sutures. Adults with this condition are dwarfed with deformed, usually bowed, legs. They have osteomalacia that may be complicated by osteosclerosis, overgrowth of bone (especially at tendon insertions), and ossification of tendons and ligaments. A mild form of hypophosphatasia may develop in adults, in whom there is increased bone fragility, decreased stature, and bone deformities. The sternal ends of ribs may be enlarged, suggesting rickets, and the long bones may show lack of modeling with an Erlenmeyer flask deformity.[23–25]

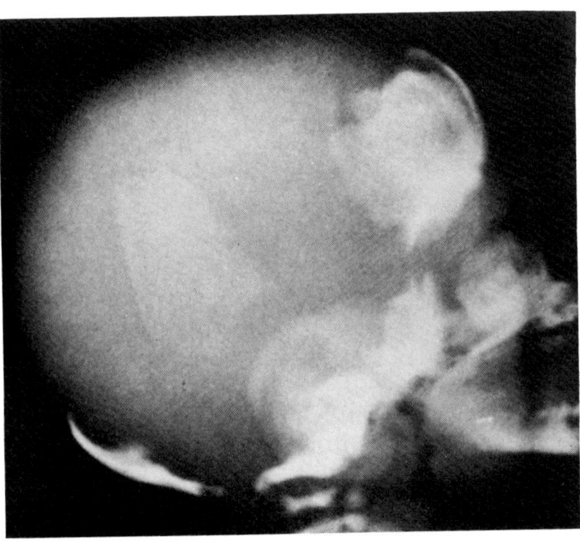

**Figure 59.21. Hypophosphatasia.** A lateral view of the skull of an infant shows large areas of uncalcified osteoid in the membranous bones in the regions adjoining the sutures and, to a lesser extent, at the base. (From James and Moule,[24] with permission from the publisher.)

## PAGET'S DISEASE

Paget's disease (osteitis deformans) is one of the most common chronic diseases of the skeleton. Destruction of bone followed by a reparative process results in weakened, deformed, and thickened bony structures. The disease is seen most commonly during middle life, affects men twice as often as women, and has been reported to occur in about 3 percent of all persons over 40 years of age. Although the destructive phase often predominates initially, there is most frequently a combination of destruction and repair; in the pelvis and weight-bearing bones of the lower extremities, the reparative

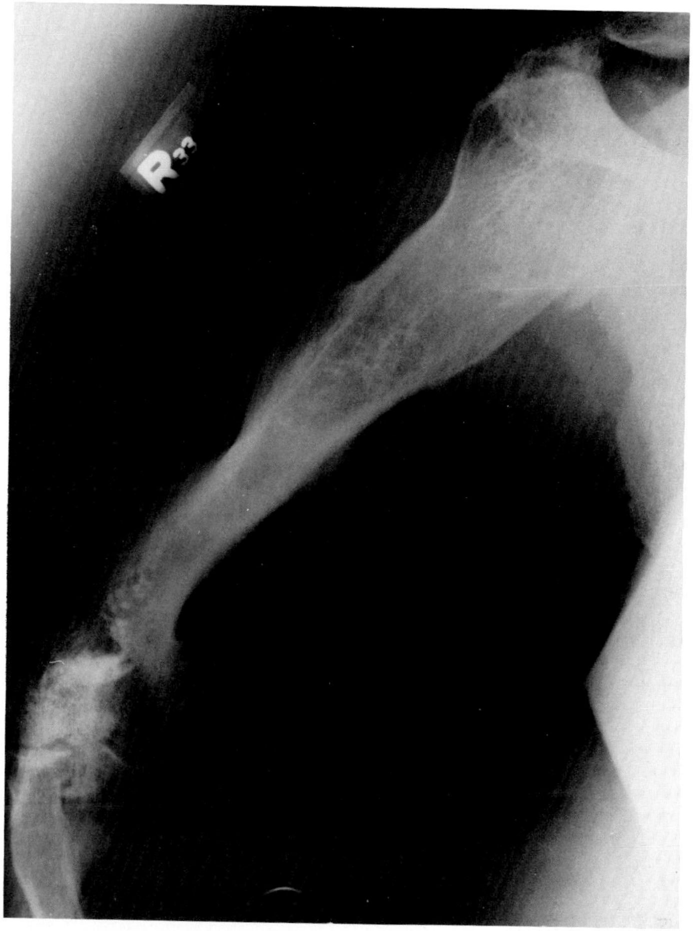

A

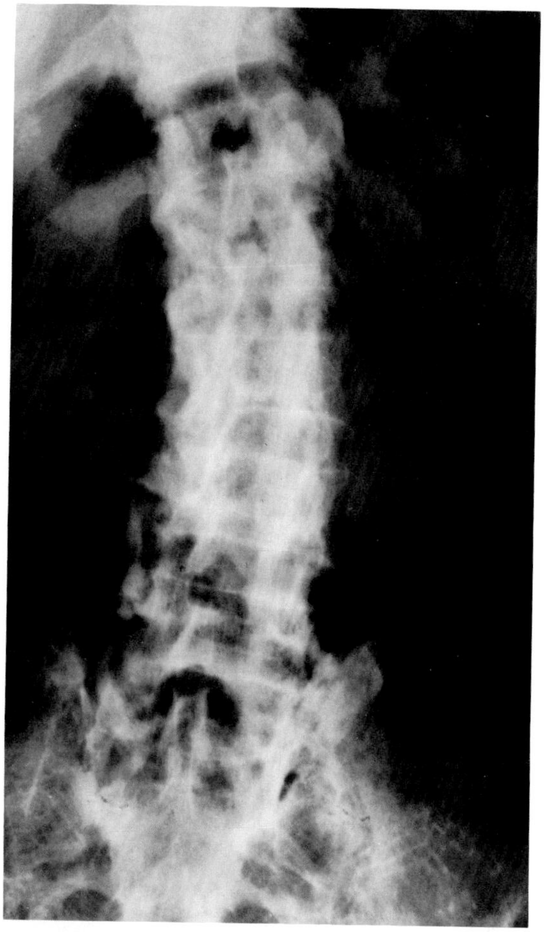

B

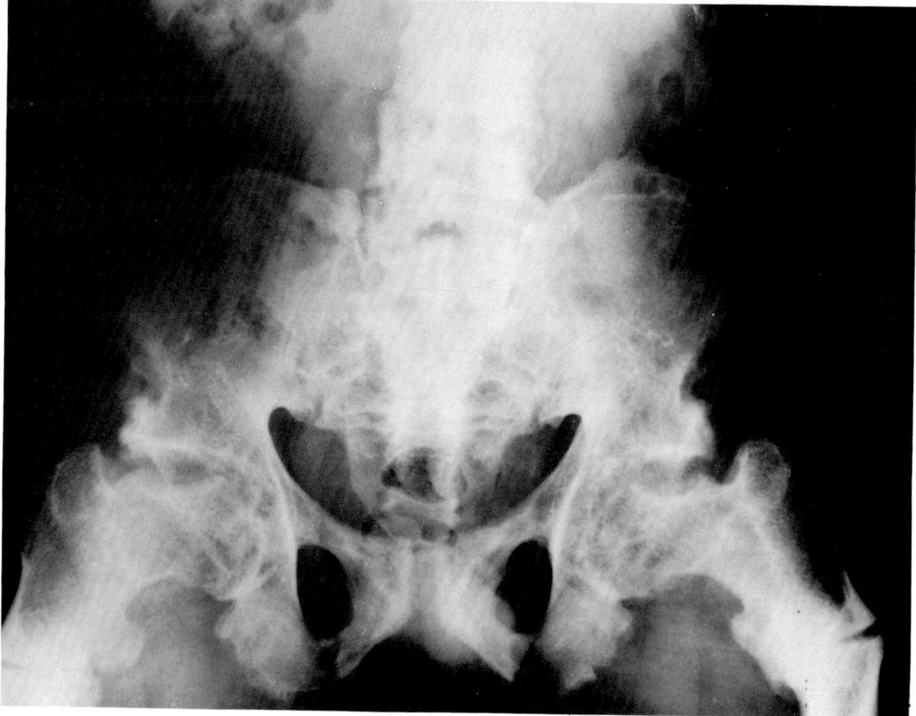

C

**Figure 59.22. Hypophosphatasia.** Severe manifestations in a 43-year-old man, 4 feet 9 inches tall. (A) There is osteomalacia of the arm with ossification of the deltoid and other muscle insertions. (B) Ossified ligaments have produced a poker spine. (C) There is ossification of tendon attachments to the pelvis and healing subtrochanteric fractures. (From Dorst, Scott, and Hall,[25] with permission from the publisher.)

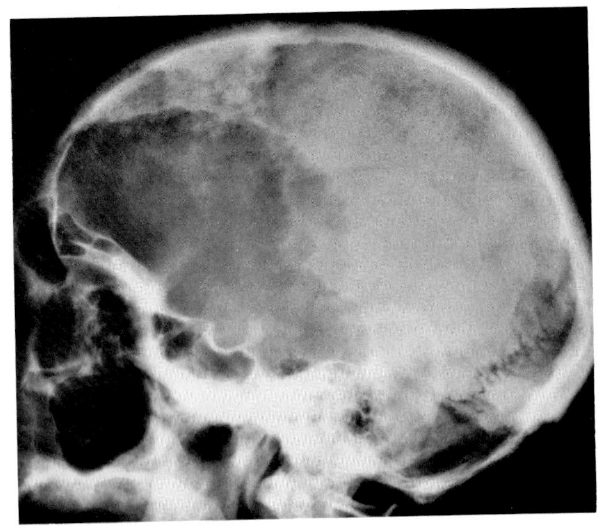

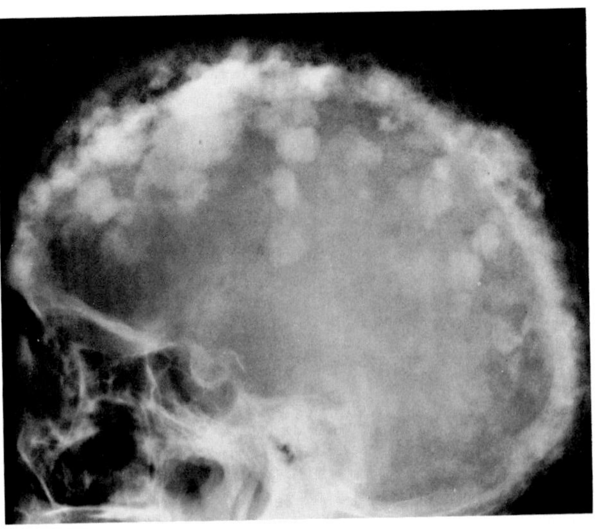

A B

**Figure 59.23. Paget's disease of the skull.** (A) A radiograph obtained during the destructive phase of the disease demonstrates an area of sharply demarcated radiolucency (osteoporosis circumscripta) that primarily involves the frontal portion of the calvarium. (B) During the reparative phase in another patient, irregular areas of sclerosis produce the characteristic cotton wool appearance of the skull. (From Eisenberg,[40] with permission from the publisher.)

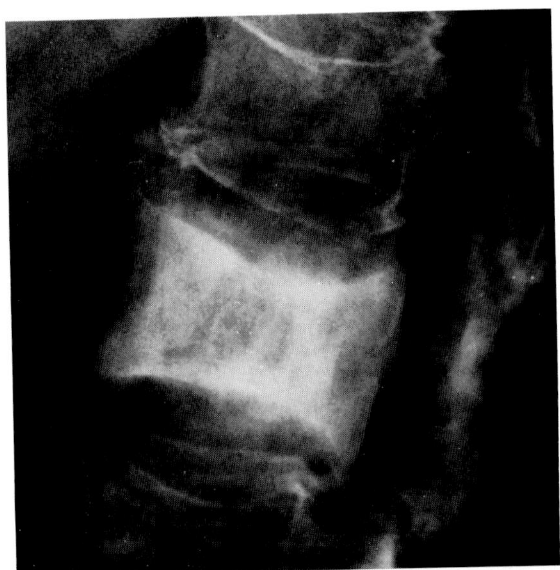

**Figure 59.24. Ivory vertebra in a patient with Paget's disease.** Note the enlargement of the vertebral body and cortical thickening. (From Edeiken,[22] with permission from the publisher.)

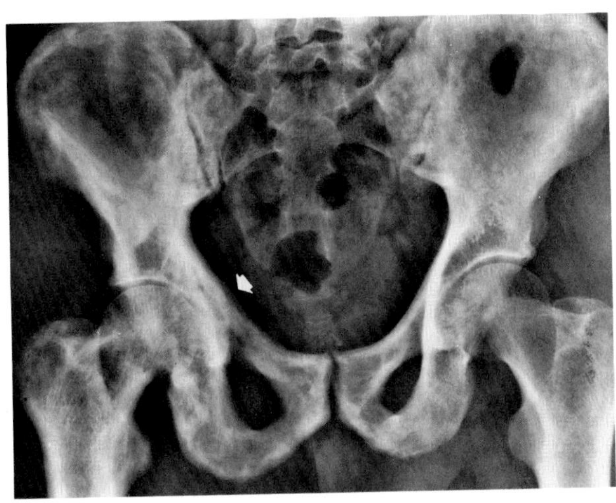

**Figure 59.25. Paget's disease of the pelvis.** There is diffuse sclerosis with cortical thickening involving the right femur and both iliac bones. Note the characteristic thickening and coarsening of the iliopectineal line (arrow) on the involved right side.

process may begin early and be the prominent feature. Often involving multiple bones, Paget's disease particularly involves the pelvis, femurs, skull, tibias, vertebrae, clavicles, and ribs.

In the skull, Paget's disease begins as an area of sharply demarcated radiolucency (osteoporosis circumscripta) that represents the destructive phase of the disease and primarily involves the outer table, sparing the inner table. Deossification begins in the frontal or occipital area and spreads slowly to encompass the major portion of the calvarium. During the reparative process, the development of irregular islands of sclerosis in the inner table is followed by thickening of the diploe and later of the outer table, resulting in a mottled, cotton wool appearance.

In the spine, Paget's disease characteristically causes enlargement of the vertebral body. Increased trabeculation, which is most prominent at the periphery of the

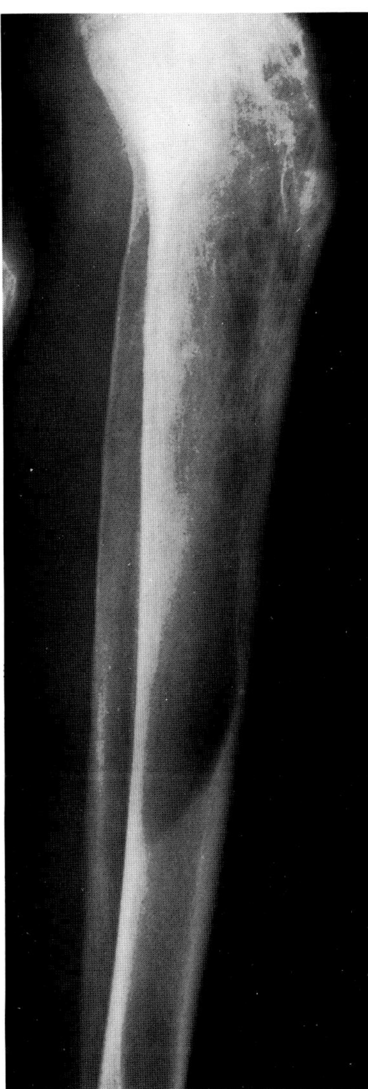

**Figure 59.26. Blade of grass sign in Paget's disease.** A sharply demarcated radiolucency with an angular configuration extends from the proximal tibia and involves two-thirds of the shaft. There is also accentuation of the trabecular pattern at the proximal aspect of the tibia. (From Greenfield,[4] with permission from the publisher.)

bone, produces a rim of thickened cortex and a picture-frame appearance. Dense sclerosis of one or more vertebral bodies (ivory vertebrae) may present a pattern simulating osteoblastic metastases or Hodgkin's disease, though in Paget's disease the vertebrae are also expanded in size.

The pelvis is the most common and often the initial site of Paget's disease. A distinctive early sign is coarsening of the trabeculae along the iliac margins that produces thickening of the pelvic brim. A combination of destructive and reparative changes is usually seen. Although diffuse sclerosis may simulate osteoblastic metastases, the characteristic cortical thickening and coarse trabeculation should suggest Paget's disease.

In the long bones, the destructive phase almost invariably begins at one end of the bone and extends along the shaft for a variable distance before ending in a typical sharply demarcated, V-shaped configuration (blade of grass appearance). In the reparative stage the bone is enlarged, with an irregularly widened cortex and coarse, thickened trabeculae. Although dense, the bones are soft and deformities are common, especially the shepherd's crook deformity of the upper femur and anterior bowing of the tibia. Transverse fractures may also occur, especially on the convex side of bowed long bones.

Paget's disease may lead to severe clinical complications. The downward thrust of the heavy head upon the softened bone of the spine may cause basilar invagination of the skull, compression of the brain stem, and numerous cranial nerve deficits. Expansion and distortion of softened vertebral bodies, sometimes with pathologic fractures, may compress the spinal cord and produce nerve root deficits. Multiple microscopic arteriovenous malformations in pagetoid bone may result in high-output cardiac failure. The most serious complication of Paget's disease is sarcomatous degeneration, which fortunately occurs in less than 1 percent of the patients with this condition. Most of the lesions are osteosarcomas, though fibrosarcomas and chondrosarcomas may occur.[4,22,26]

## HYPEROSTOSIS

An increase in bone mass (hyperostosis) appears radiographically as an increase in bone density. Hyperostosis may be due to excessive formation of new bone or decreased resorption of bone already formed. The increased bone mass may be spotty, as in osteopoikilosis, or may be uniformly distributed throughout the skeleton, as in diffuse osteopetrosis. It is unclear why certain disorders tend to evoke an osteoblastic reaction while others do not. New bone formation may occur along the surface of the trabeculae in spongy bone, along the endosteal surface of the cortex, or in the region of the periosteum. This new bone formation may be elicited by a primary or malignant process or may represent a reparative and reactive response.

### Osteopetrosis

Osteopetrosis (marble bone disease of Albers-Schönberg) is a rare hereditary bone dysplasia in which failure of the resorptive mechanism of calcified cartilage interferes with its normal replacement by mature bone. This results in a symmetric, generalized increase in bone density. Osteopetrosis varies in severity and age of clinical presentation from a fulminant, often fatal condition involving the entire skeleton at birth or in utero to an essentially asymptomatic form that is an incidental ra-

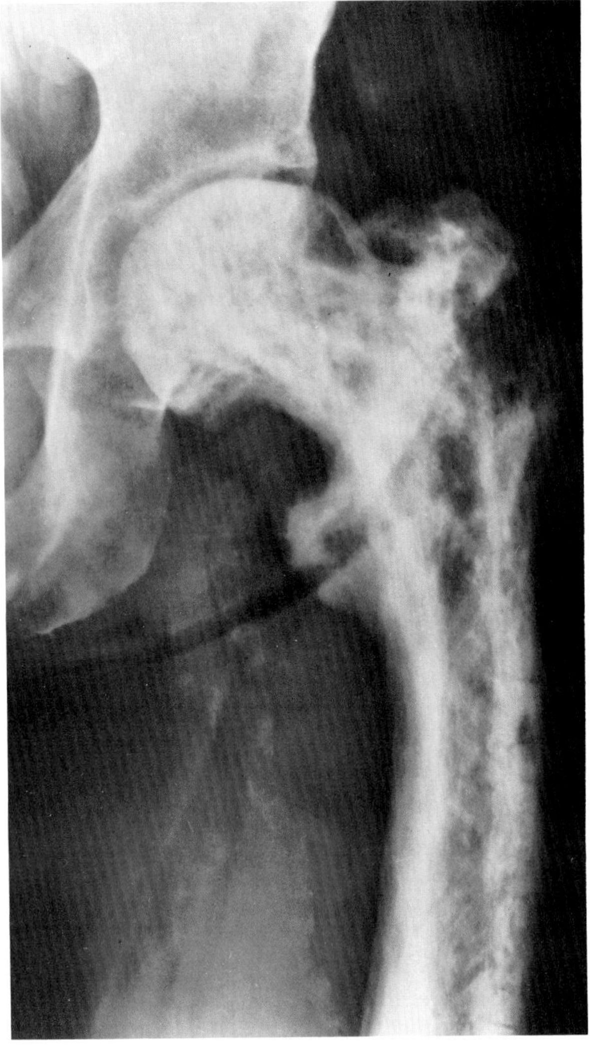

A

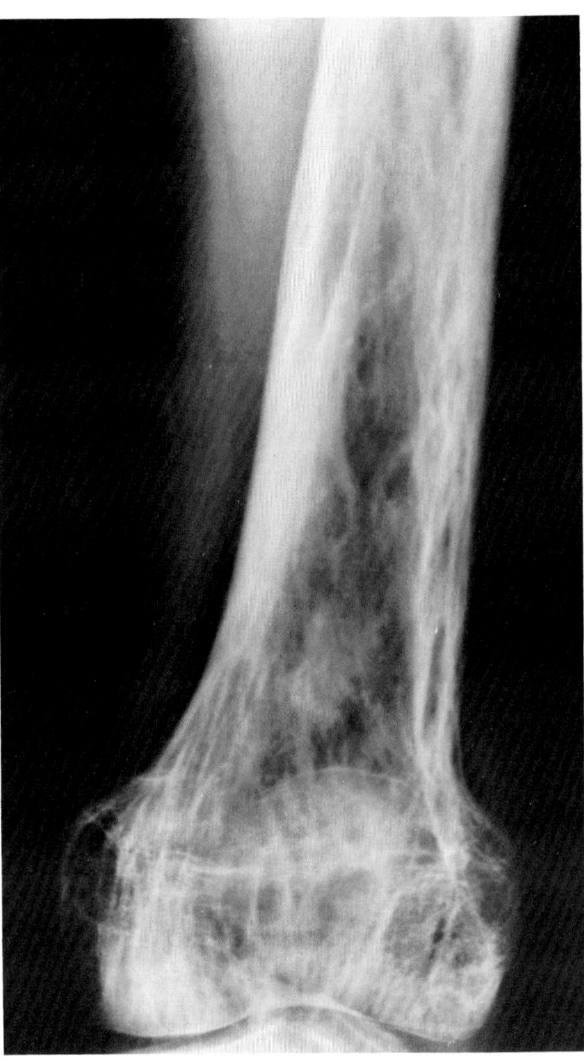

B

**Figure 59.27. Paget's disease.** Characteristic cortical thickening, destruction of fine trabeculae, and accentuation of secondary trabeculae are seen in (A) the proximal femur and (B) the distal femur.

diographic finding. The dense sclerosis of osteopetrosis obliterates the individual bony components of cortex, medullary cavity, and trabecular pattern. Lack of modeling causes widening of the metaphyseal ends. Alternating dense and lucent transverse lines in the metaphyses often occur in the long bones and vertebrae, probably reflecting the intermittent nature of the pathologic process. In the vertebrae this may produce a miniature bone inset within each vertebral body (bone-within-a-bone appearance) or increased density at the end plates ("sandwich vertebrae"). Although radiographically dense, the involved bones are brittle and fractures are a common complication, even with trivial trauma. Encroachment of bone on the marrow cavities causes a myelophthisic anemia that may lead to extramedullary hematopoiesis and enlargement of the liver, spleen, and lymph nodes. Sclerotic changes at the base of the skull may encroach on the cranial foramina and produce cranial nerve palsies, blindness, and deafness. The paranasal sinuses and mastoids may fail to develop or may become obliterated.[27,28]

## Pyknodysostosis

Pyknodysostosis is a rare hereditary dysplasia in which patients have short stature and diffusely dense, sclerotic bones. Often confused with osteopetrosis, pyknodysostosis is usually a more benign condition (clinically not associated with hepatosplenomegaly, anemia, or cranial nerve involvement) with several typical radiographic features. Although the long bones show a generalized

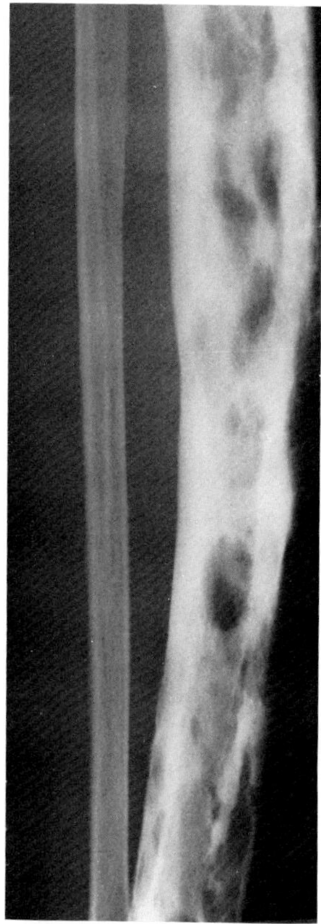

**Figure 59.28. Sarcomatous degeneration of Paget's disease.** Extensive malignant destruction is superimposed on pre-existing Paget's disease of the tibia.

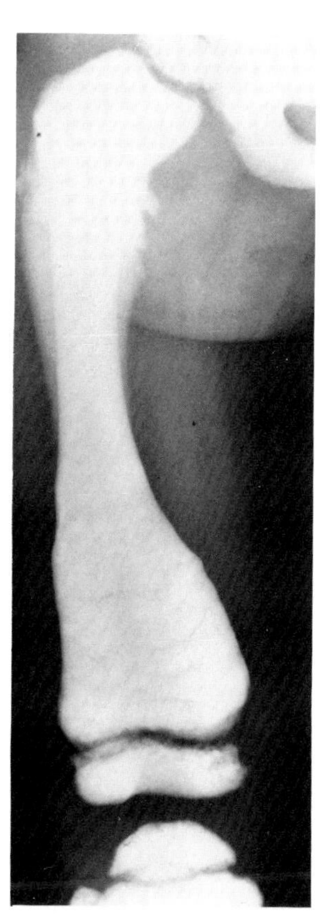

**Figure 59.29. Osteopetrosis of the femur.**

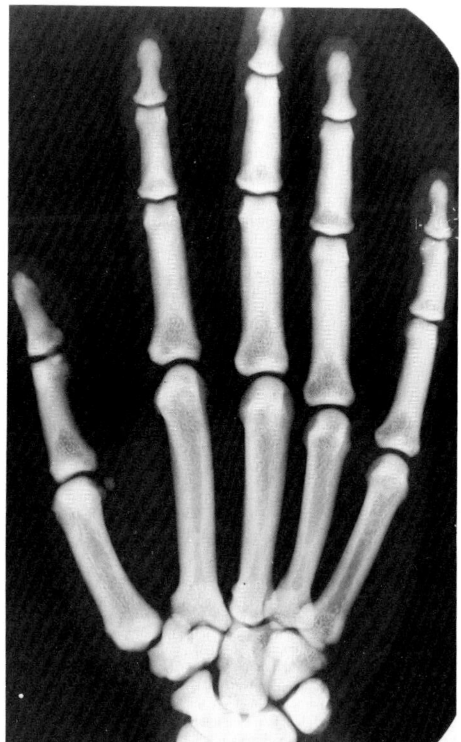

A

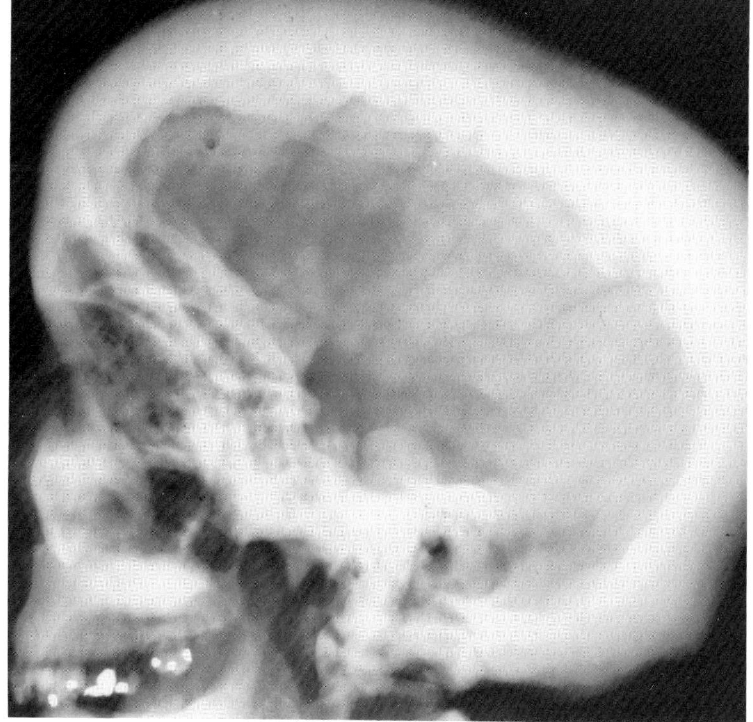

B

**Figure 59.30. Osteopetrosis of (A) the hand and (B) the skull.**

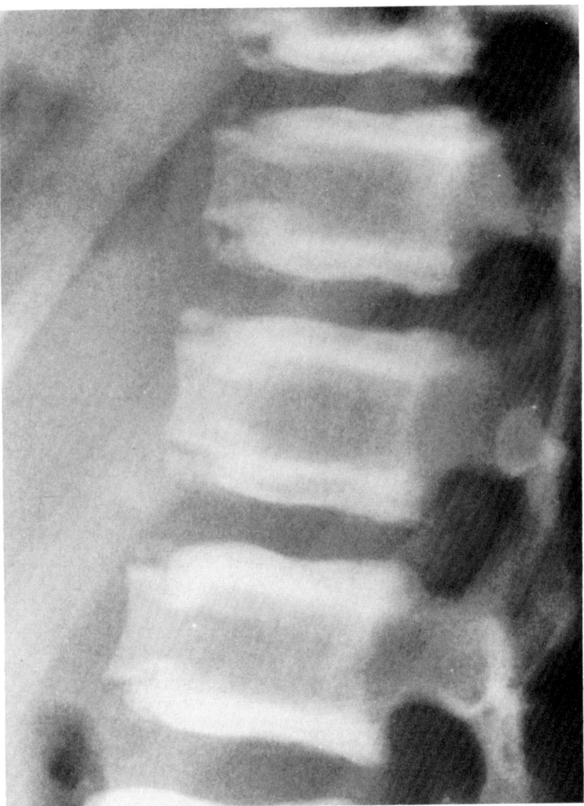

**Figure 59.31. Osteopetrosis of the vertebrae.** Increased density limited to the end plates produces a sandwich appearance.

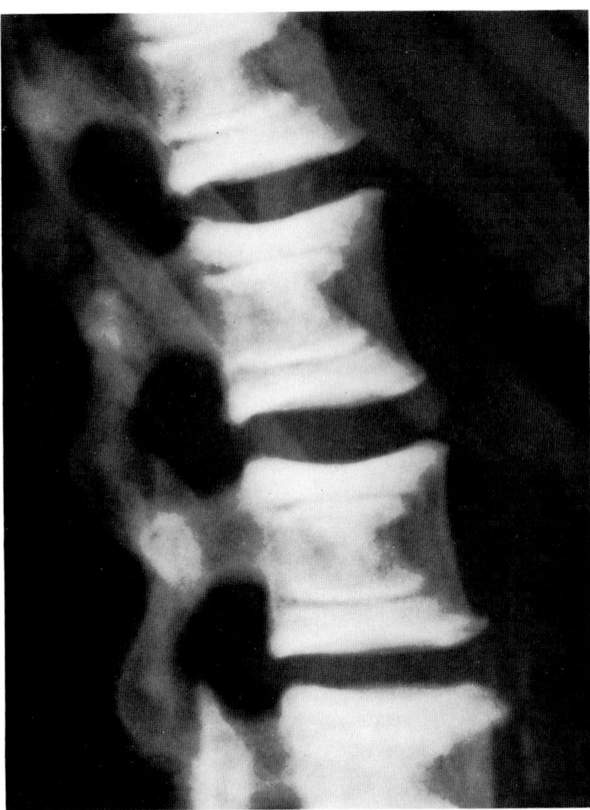

**Figure 59.32. Osteopetrosis of the vertebrae.** A miniature inset is seen in each lumbar spine vertebral body, giving a bone-within-a-bone appearance. Sclerosis at the end plates may also be seen. (From Greenfield,[4] with permission from the publisher.)

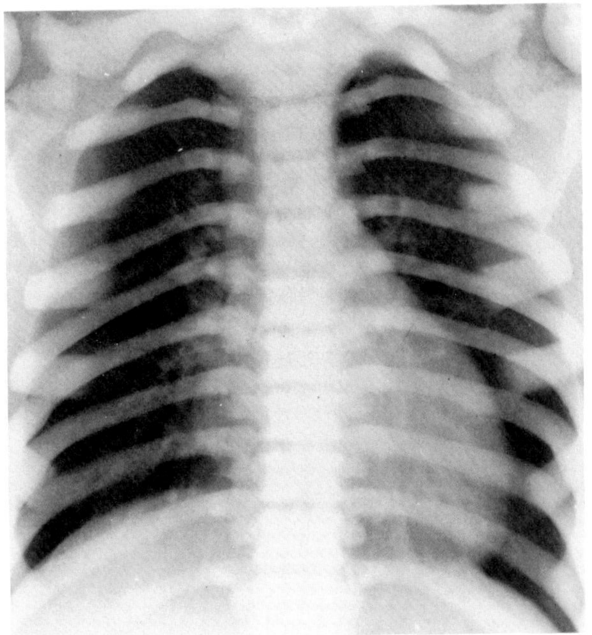

**Figure 59.33. Osteopetrosis.** A frontal chest film demonstrates generalized increased density of the ribs, dorsal spine, and clavicles. (From Eisenberg,[40] with permission from the publisher.)

increase in density with cortical thickening, the medullary cavity is preserved and there is no metaphyseal widening. Hypoplasia of the distal phalanges and absence of the terminal tufts cause the hands to be short and stubby. A characteristic finding is mandibular hypoplasia with loss of the normal mandibular angle and craniofacial disproportion. In the skull, failure of closure of the cranial sutures and fontanelles, numerous Wormian bones, and occasional dysplasia of the clavicles may simulate cleidocranial dysostosis. The sinuses and mastoids are often not pneumatized, and the cranial and facial bones are often sclerotic and thickened, though without the cranial nerve deficits seen in osteopetrosis.[29,30]

## Generalized Cortical Hyperostosis (Van Buchem's Disease)

Generalized cortical hyperostosis is a rare dysplasia characterized by symmetric sclerosis of the skull, mandible, clavicles, ribs, and diaphyses of long bones. The diaphyseal sclerosis is accompanied by thickening of the endosteal surface of the cortex, which causes widening of the cortex but does not increase the diameter of the bone. Although often asymptomatic, calvarial involvement may cause neural compression leading to optic atrophy, facial paralysis, and deafness.[31,32]

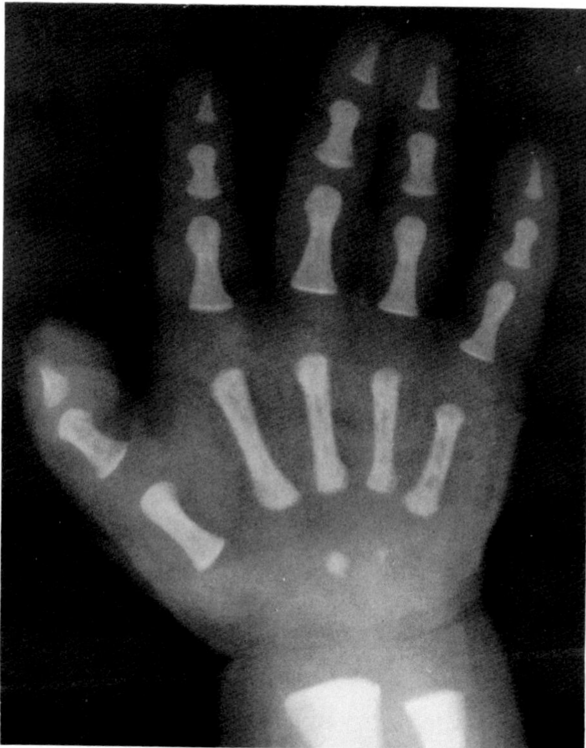

**Figure 59.34. Pyknodysostosis.** There is a generalized increase in density with cortical thickening of the bones of the hand. The distal phalanges are hypoplastic, and the terminal tufts are absent.

## Hereditary Hyperphosphatasia

Hereditary hyperphosphatasia is a rare disease of children that is characterized by bowing deformities and thickening of the long bones associated with an elevated level of serum alkaline phosphatase. A rapid turnover of lamellar bone without the laying down of compact cortical bone may cause the cortices to be denser and thinner than normal or to be markedly less dense with loose trabeculae rather than cortical bone. Thickening of the calvarium with patchy sclerosis may simulate the cotton wool appearance of Paget's disease.[33,34]

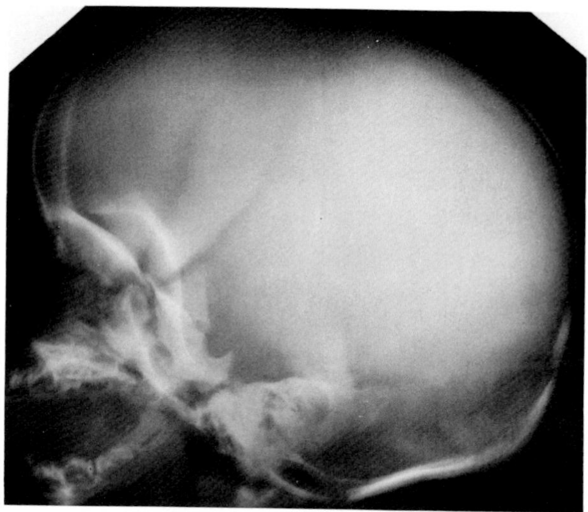

**Figure 59.35. Pyknodysostosis.** A lateral view of the skull shows mandibular hypoplasia, loss of the normal mandibular angle, and craniofacial disproportion.

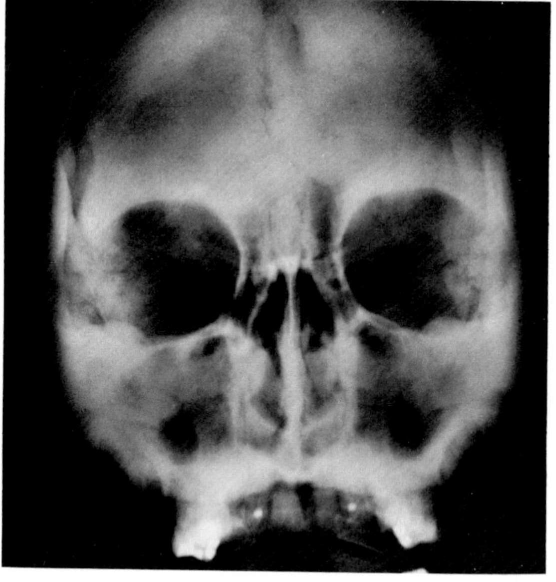

A

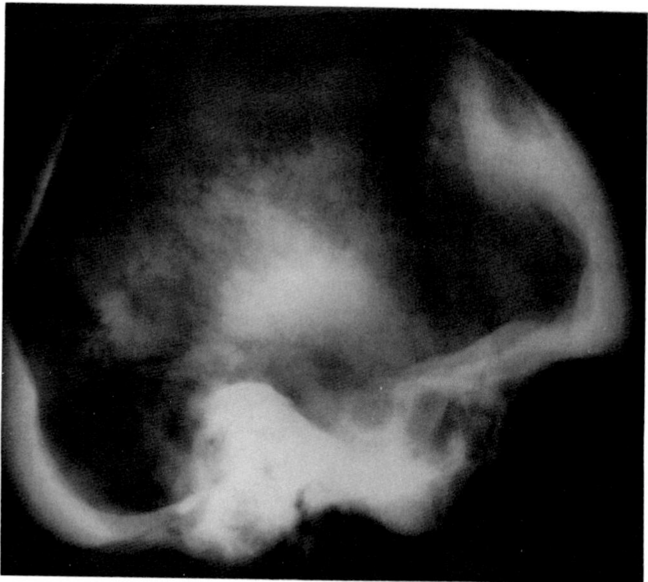

B

**Figure 59.36. Generalized cortical hyperostosis.** Dense symmetric sclerosis of the skull is seen on (A) frontal and (B) lateral views.

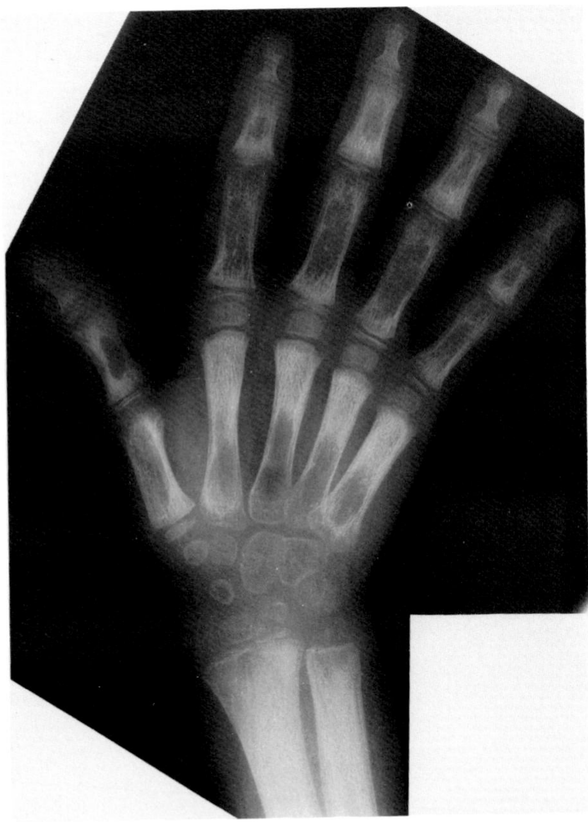

**Figure 59.37. Hereditary hyperphosphatasia.** The cortices are thinned and there is diffuse deossification. Some increase in density is seen in the metacarpals and middle phalanges.

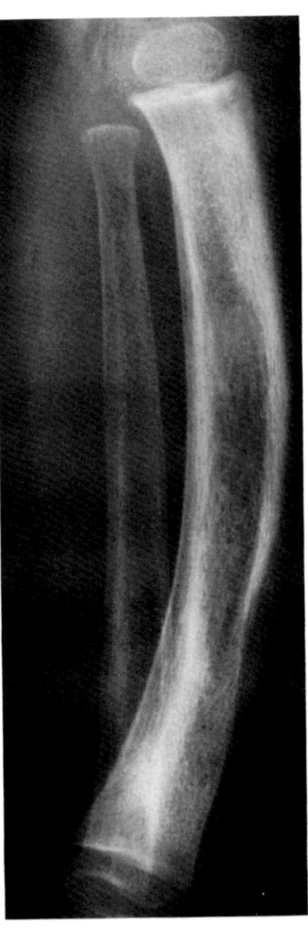

**Figure 59.38. Hereditary hyperphosphatasia.** The bones are deossified, widened, and bowed.

## Progressive Diaphyseal Dysplasia (Camurati-Engelmann Disease)

Progressive diaphyseal dysplasia is a rare disorder in which symmetric cortical thickening in the midshafts of long bones is associated with a neuromuscular dystrophy that causes a peculiar wide-based, waddling gait. Both endosteal and periosteal cortical thickening cause fusiform enlargement of the long bones. This process tends to progress and eventually involves most of the diaphyses, sparing the epiphyses and metaphyses. Encroachment on the medullary canal may cause anemia and secondary hepatosplenomegaly; amorphous increased density at the base of the skull may lead to impingement on the cranial nerves.[35,36]

## Melorheostosis

Melorheostosis is a rare disorder that appears as an irregular sclerotic thickening of the cortex, which is usually confined to one side of a single bone or multiple bones of one extremity. The condition usually occurs in childhood, with a presenting symptom of severe pain that is sometimes associated with limitation of motion, contractures, or fusion of an adjacent joint. The sclerotic process involves the inner cortex and projects into the medullary cavity. It begins at the proximal end of the bone and extends distally. The resemblance of the appearance to the flowing (Gk. *rheos*) of wax down a burning candle is responsible for the name. Involvement of the small bones of the hand and wrist, as well as the ipsilateral scapula and hemipelvis, may produce multiple small sclerotic islands of dense bone that simulate osteopoikilosis.[37]

## Osteopoikilosis

Osteopoikilosis is an asymptomatic hereditary condition characterized by the presence of multiple small, well-circumscribed, round or oval areas of increased density in cancellous bone. The sclerotic foci vary in size from 2 mm to 2 cm and produce a typical speckled appearance that primarily involves the small bones of the hands and feet, the pelvis, and the epiphyses and metaphyses of long bones.[38]

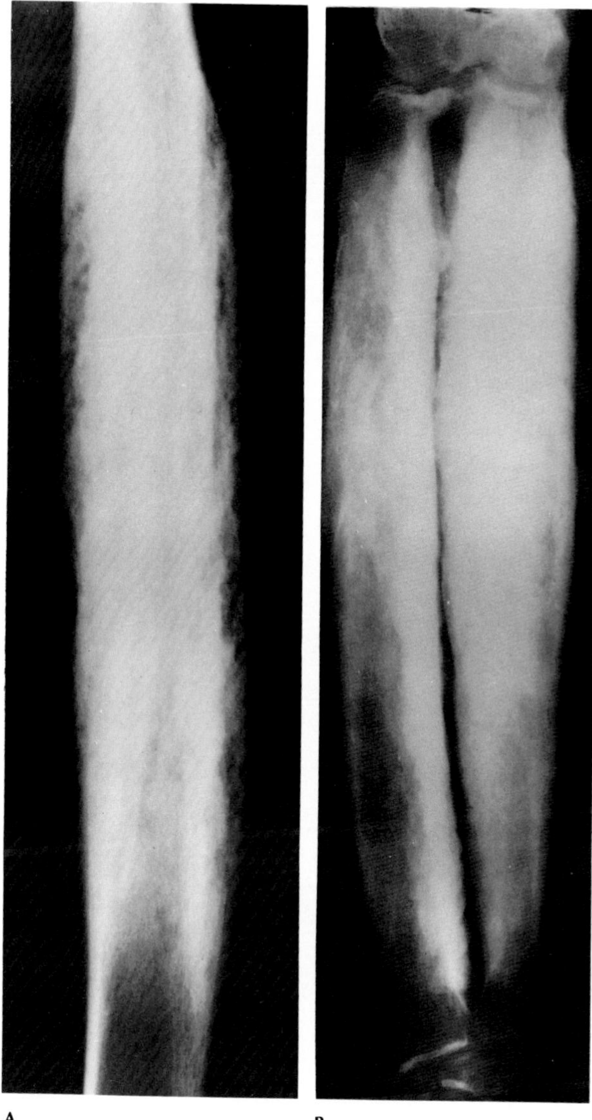

A B

**Figure 59.39. Progressive diaphyseal dysplasia.** Dense endosteal and periosteal cortical thickening causes fusiform enlargement of the midshafts of (A) the femur and (B) the radius and ulna.

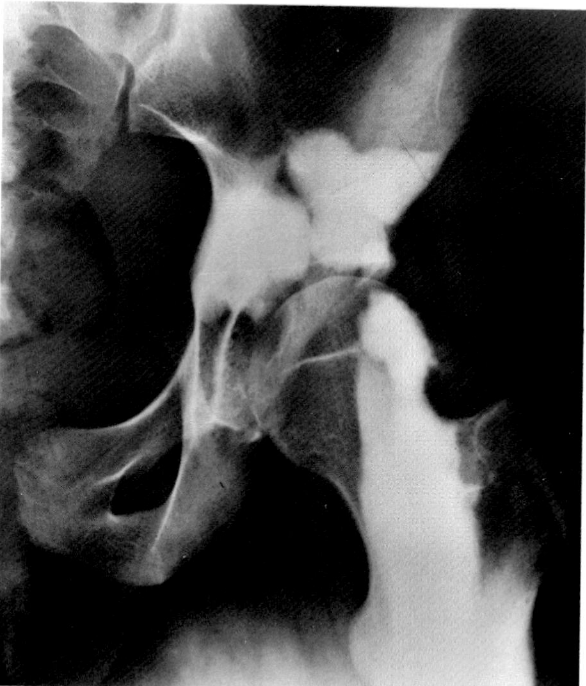

**Figure 59.40. Melorheostosis.** Dense cortical sclerosis involves the proximal femur and lower portion of the ilium.

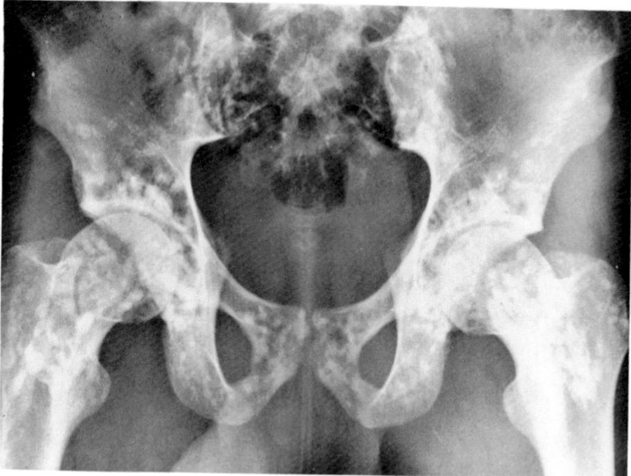

**Figure 59.41. Osteopoikilosis.** There are innumerable small, well-circumscribed areas of increased density throughout the pelvis and proximal femurs.

## Osteopathia Striata

Osteopathia striata is a rare, asymptomatic bone disorder in which an error in internal bone modeling produces characteristic dense longitudinal striations in tubular bones. Involvement of the ilium produces linear densities that radiate out from the acetabulum and fan out to the iliac crest in a sunburst pattern.[39]

## Hyperostosis Frontalis Interna

Hyperostosis frontalis interna refers to a bilateral, symmetric bony overgrowth that thickens the inner table of the frontal bone. The abnormality is almost always found in women, especially those over the age of 35, and is generally considered to be of no clinical significance. Because the irregular thickening surrounds the venous sinuses but does not obliterate them, on frontal views of the skull the superior sagittal sinus and the veins draining into it stand out as prominent radiolucent zones surrounded by dense hyperostosis.

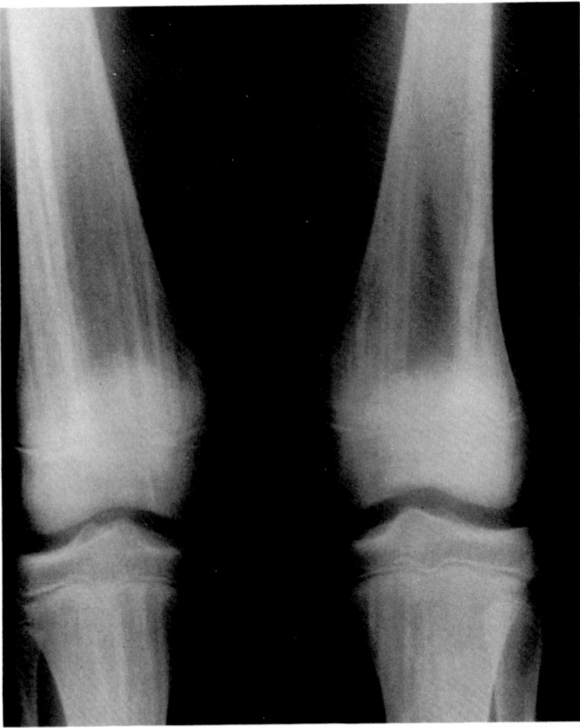

**Figure 59.42. Osteopathia striata.** There are dense longitudinal striations in the distal femur and proximal tibia.

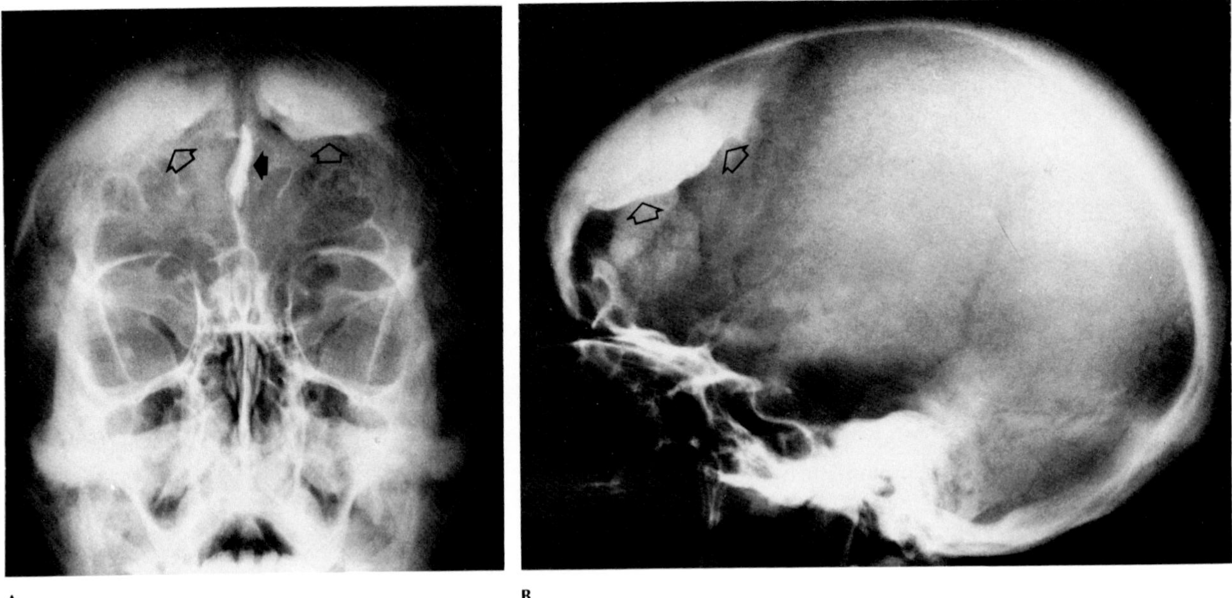

A B

**Figure 59.43. Hyperostosis frontalis interna.** (A) Frontal and (B) lateral views of the skull demonstrate bilateral symmetric osseous overgrowth that thickens the inner table of the frontal bone (open arrows). Of incidental interest is calcification in the falx (closed arrow).

## REFERENCES

1. Gleason DC, Potchen EJ: The diagnosis of hyperparathyroidism. *Radiol Clin North Am* 5:277–287, 1967.
2. Steinbach HL, Gordon GS, Eisenberg E, et al: Primary hyperparathyroidism: Correlation of roentgen, clinical and pathologic features. *AJR* 86:329–343, 1961.
3. Kuntz CH, Goldsmith RE: Selective arteriography of parathyroid adenomas. *Radiology* 102:21–28, 1972.
4. Greenfield GB: *Radiology of Bone Diseases*. Philadelphia, JB Lippincott, 1980.
5. Sample WF, Mitchell SP, Bledsoe RC: Parathyroid ultrasonography. *Radiology* 127:485–490, 1978.
6. Krudy AG, Doppman JL, Brennan MF, et al: The detection of mediastinal parathyroid glands by computed tomography, selective arteriography, and venous sampling. *Radiology* 140:739–744,1981.
7. Eisenberg H, Pallota J, Sherwood LM: Selective arteriography, venography and venous hormone assay in diagnosis and localization of parathyroid lesions. *Am J Med* 56:810–820, 1974.
8. Genant HK, Heck LL, Lanzl LH, et al: Primary hyperparathyroidism: A comprehensive study of clinical, biochemical and radiographic manifestations. *Radiology* 109:513–524, 1973.
9. Eugenidis N, Olah AJ, Hass HG: Osteosclerosis in hyperparathyroidism. *Radiology* 105:265–275, 1972.
10. Mancuso AA: Computed tomography of the neck, in Moss AA, Gamsu G, Genant HK (eds): *Computed Tomography of the Body*. Philadelphia, WB Saunders, 1983.
11. Taybi H, Keele D: Hypoparathyroidism: A review of the literature and report of two cases in sisters. *AJR* 88:432–442, 1962.
12. Bennett JC, Maffly RH, Steinbach HL: Significance of bilateral basal ganglia calcification. *Radiology* 72:368–378, 1959.
13. Cohen CR, Duchesneau PM, Weinstein MA: Calcification of the basal ganglia as visualized by computed tomography. *Radiology* 134:97–99, 1980.
14. Steinbach HL, Young DA: The roentgen appearance of pseudohypoparathyroidism (PH) and pseudo-pseudohypoparathyroidism (PPH). *AJR* 97:49–66, 1966.
15. Steinbach HL: The roentgen appearance of osteoporosis. *Radiol Clin North Am* 2:191–207, 1964.
16. Rosen RA: Transitory demineralization of the femoral head. *Radiology* 94:509–512, 1970.
17. Duncan H, Frame B, Frost HM, et al: Migratory osteolysis of the lower extremities. *Ann Intern Med* 66:1165–1173, 1967.
18. Reynolds WA, Karo JJ: Radiologic diagnosis of metabolic bone disease. *Orthop Clin North Am* 3:521–543, 1972.
19. Meema HE, Meema S: Improved roentgenologic diagnosis of osteomalacia by microscopy of hand bones. *AJR* 125:925–935, 1975.
20. Chalmers J: Osteomalacia: Review of 93 cases. *J Roy Coll Surg Edinb* 13:255–275, 1968.
21. Steinbach, HL, Noetzl M: Roentgen appearance of the skeleton in osteomalacia and rickets. *AJR* 91:955–972, 1964.
22. Edeiken J: *Roentgen Diagnosis of Diseases of Bone*. Baltimore, William & Wilkins, 1981.
23. Curranino G, Neuhauser EBD, Reyersbach GC, et al: Hypophosphatasia. *AJR* 78:392–419, 1957.
24. James W, Moule B: Hypophosphatasia. *Clin Radiol* 17:368–376, 1966.
25. Dorst JP, Scott CI, Hall JG: The radiologic assessment of short stature. *Radiol Clin North Am* 10:393–414, 1972.
26. Greditzer HG, McLeod RA, Unni KK, et al: Bone sarcomas in Paget's disease. *Radiology* 146:327–333,1983.
27. Pincus JB, Gittleman IF, Kramer B: Juvenile osteopetrosis. *Am J Dis Child* 73:458–472, 1947.
28. Yu JS, Oates RK, Walsh KH, et al: Osteopetrosis. *Arch Dis Child* 46:257–263, 1971.
29. Muthukrishnan N, Shetty MUK: Pycnodysostosis. Report of a case. *AJR* 114:247–252,1972.
30. Elmore SM: Pycnodysostosis: A review. *J Bone Joint Surg* 49A:153–162, 1967.
31. Owen RH: Van Buchem's disease. *Brit J Radiol* 49:126–132, 1976.
32. Eastman JR, Bixler D: Generalized cortical hyperostosis (van Buchem disease): Nosologic consideration. *Radiology* 125: 297–304, 1977.
33. Eyring EJ, Eisenberg E: Congenital hyperphosphatasia. *J Bone Joint Surg* 50A: 1099–1117, 1968.
34. Iancu TC, Almagor G, Friedman E, et al: Chronic familial hyperphosphatemia. *Radiology* 129:669–676, 1978.
35. Hundley JD, Wilson FC: Progressive diaphyseal dysplasia. Review of the literature and report of seven cases in one family. *J Bone Joint Surg* 55A:461–474, 1973.
36. Mottrom ME, Hill NA: Diaphyseal dysplasia. *AJR* 95:162–167, 1965.
37. Campbell CJ, Papademetriou R, Bonfiglio M: Melorheostosis. *J Bone Joint Surg* 50A:1281–1304, 1968.
38. Green AE, Ellswood WH, Collins JR: Melorheostosis and osteopoikilosis. *AJR* 87:1096–1111, 1962.
39. Carlson DH: Osteopathia striata revisited. *J Canad Assoc Radiol* 28:190–192, 1977.
40. Eisenberg RL: *Atlas of Signs in Radiology*. Philadelphia, JB Lippincott, 1984.

## Chapter Sixty
# Neoplasms of Bone

## INTRODUCTION TO BONE TUMORS

Many factors enter into the radiographic evaluation of bone tumors and the critical distinction between benign and malignant lesions. Important considerations include the age and sex of the patient, the specific bone or bones and number of bones involved, the site of the tumor within the bone (epiphysis, metaphysis, diaphysis), the site of origin (medullary canal, cortex, periosteum), the internal architecture of the lesion and sharpness of its margins, the type of periosteal reaction, the extent of soft tissue involvement, and the presence and type of calcification or ossification within the tumor mass.

Malignant lesions are generally said to be poorly defined, gradually fading at the edges and blending with the host bone, while benign tumors tend to be sharply delineated. However, these characteristics depend on the rate of growth and aggressiveness of the tumor, not necessarily on its benign or malignant potential. Rapidly growing tumors are poorly delineated, while slower growing ones permit bone reaction, clear definition from surrounding bones, and, often, sclerotic borders.

Defects in the bone cortex suggest a malignant lesion. However, in benign tumors such as aneurysmal bone cysts, the cortex may be so thin as to be radiographically invisible.

Although periosteal reaction is more commonly seen with malignant bone tumors, it may accompany benign lesions as well. Almost any type of periosteal reaction may sometimes be seen with any type of bone tumor. An exception is the solid type of periosteal reaction that almost always is limited to benign disease.

In distinguishing among malignant tumors, the age of the patient is an important criterion. During the first year of life, almost all malignant bone tumors are neuroblastomas. During the second decade, the major malignant tumors are osteosarcoma and Ewing's tumor, which can be clearly differentiated radiographically and clinically. After the age of 40, the most common bone tumors are metastases and multiple myeloma. Chondrosarcoma and osteogenic sarcoma (complicating Paget's disease) are the major primary bone malignancies in this age group.

## BENIGN BONE TUMORS

### Osteochondroma (Exostosis)

Osteochondroma is a benign projection of bone with a cartilaginous cap that probably represents a local dysplasia of cartilage at the epiphyseal growth plate. The

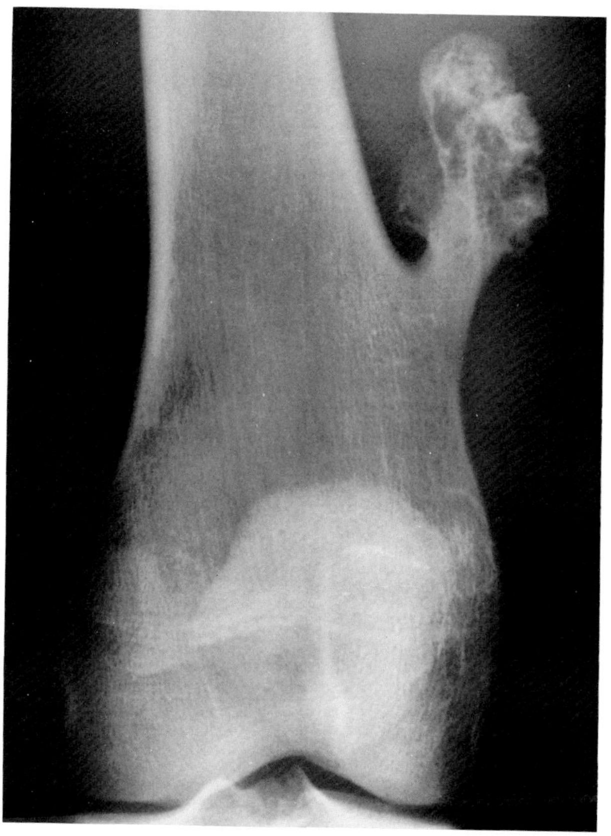

**Figure 60.1. Osteochondroma of the distal femur.** The long axis of the tumor is parallel to that of the femur and pointed away from the knee joint.

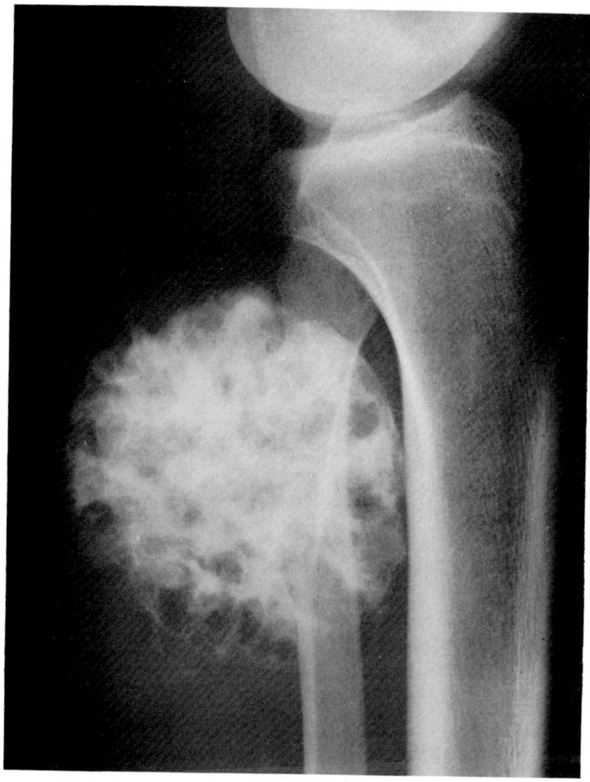

**Figure 60.2. Large osteochondroma of the proximal fibula.** The calcified and ossified mass is so extensive that the origin from the host bone is obscured.

lesion arises in childhood or the teens and continues to grow until fusion of the closest epiphyseal line. Osteochondromas are asymptomatic unless they become large enough to cause pressure symptoms or interfere with joint function. The rapid growth of a stable osteochondroma or the development of localized pain suggests malignant degeneration to a chondrosarcoma.

Osteochondromas develop in the metaphyseal region of long bones, most frequently in the femur, tibia, and humerus. The base of the osteochondroma initially grows outward at a right angle to the host bone. As the lesion grows, the pull of neighboring muscles and tendons tends to orient the tumor parallel to the long axis of the bone and pointed away from the nearest joint. The radiographic hallmark of an osteochondroma is the blending of its cortex with that of normal bone, so that there is continuity of contour between the two. Because only part of the cartilagenous cap may be calcified, the tumor is often much larger than it appears radiographically. In flat bones, an osteochondroma appears as a relatively localized area of amorphous, spotty calcification, which is often cauliflower-like in appearance. Thus pelvic osteochondromas may simulate calcified uterine fibroids, while those in the thoracic inlet may mimic thyroid calcification.[1,2]

## Enchondroma (Chondroma)

Enchondroma is a common benign cartilaginous tumor that is most frequently found in children and young adults. It arises within the medullary canal, usually near the epiphysis. The tumors are slow-growing and often multiple. Enchondromas primarily involve the small bones of the hands and feet; less frequently they are found in the long tubular bones and the ribs. As the well-demarcated tumor grows, it expands bone locally and causes thinning and endosteal scalloping of the cortex that often leads to pathologic fractures with minimal trauma. A characteristic finding of enchondroma is calcifications within the lucent matrix. These may vary from minimal stippling to large, amorphous areas of increased density.

Enchondromas are usually asymptomatic and are discovered either incidentally or when a pathologic fracture occurs. The development of severe pain or the radiographic growth of the lesion with loss of marginal definition, cortical disruption, and local periosteal reaction suggests malignant degeneration. The malignant potential of an enchondroma increases the closer the tumor is to the axial skeleton; it is extremely rare with lesions of the hands and feet.[3]

### Multiple Enchondromatosis (Ollier's Disease)

Multiple enchondromatosis is a bone dysplasia affecting the growth plate in which an excess of hypertrophic cartilage is not resorbed and ossified in a normal fashion. This causes proliferation of rounded masses or columns of decreased-density cartilage within the metaphyses and diaphyses of one or more tubular bones. The involvement is usually unilateral, and the affected bones are invariably shortened and often deformed. In long bones, the columns of radiolucent cartilage may be separated by bony septa, producing a striated appearance. In the hands and feet, the lesions are globular and expansile, often with stippled or mottled calcification. Malignant transformation of an enchondroma into a chondrosarcoma is infrequent and almost never occurs with lesions in the hands or feet.

The combination of enchondromatosis and multiple cavernous hemangiomas of the soft tissues is termed *Maffucci's syndrome*. Radiographic demonstration of calcified thrombi (phleboliths) is pathognomonic of the vascular lesions.[4,5]

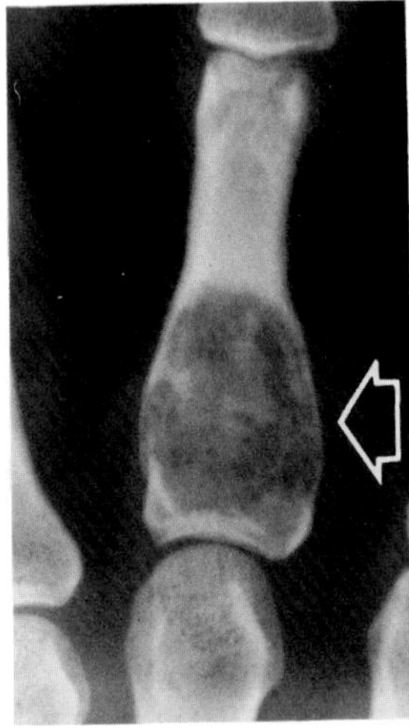

**Figure 60.3. Enchondroma.** The well-demarcated tumor expands the bone and thins the cortex.

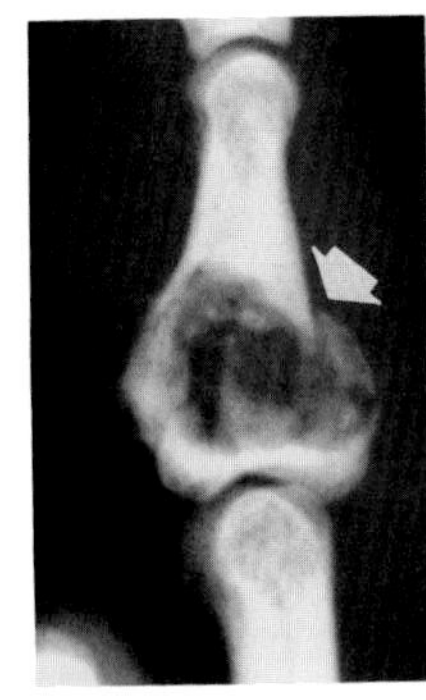

**Figure 60.4. Enchondroma with pathologic fracture (arrow).**

### Giant Cell Tumor

Giant cell tumor is an uncommon lytic lesion seen in the end of a long bone of a young adult after epiphyseal closure. Usually asymptomatic, a giant cell tumor may be associated with intermittent dull pain and a palpable tender mass. The tumor begins as an eccentric lucent lesion in the metaphysis that may extend to the immediate subarticular cortex of the bone but does not involve the joint. As the tumor expands toward the shaft, it produces a bubbly appearance that is well-demarcated from normal bone, though there is no sclerotic shell or border. There is usually no associated periosteal new bone formation unless a pathologic fracture has occurred. The classic location of a giant cell tumor is the

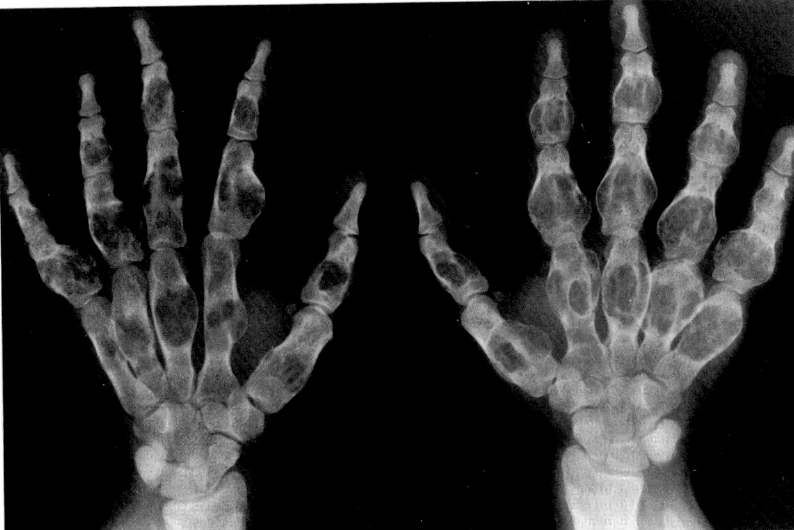

A

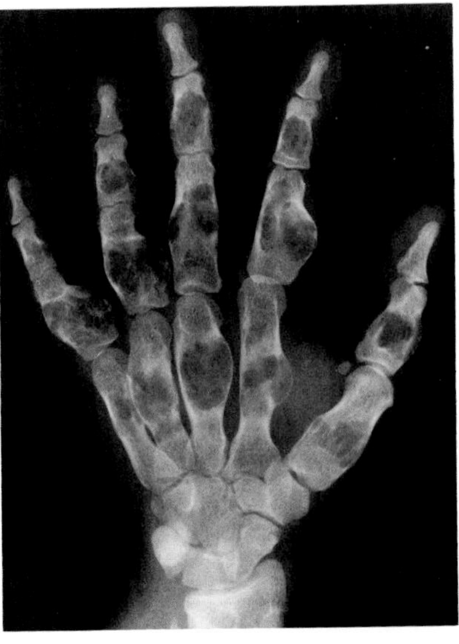

B

**Figure 60.5. Multiple enchondromatosis.** (A) Views of both hands demonstrate multiple globular and expansile lucent filling defects involving virtually all the metacarpals and the proximal and distal phalanges. (B) A coned view of the left hand.

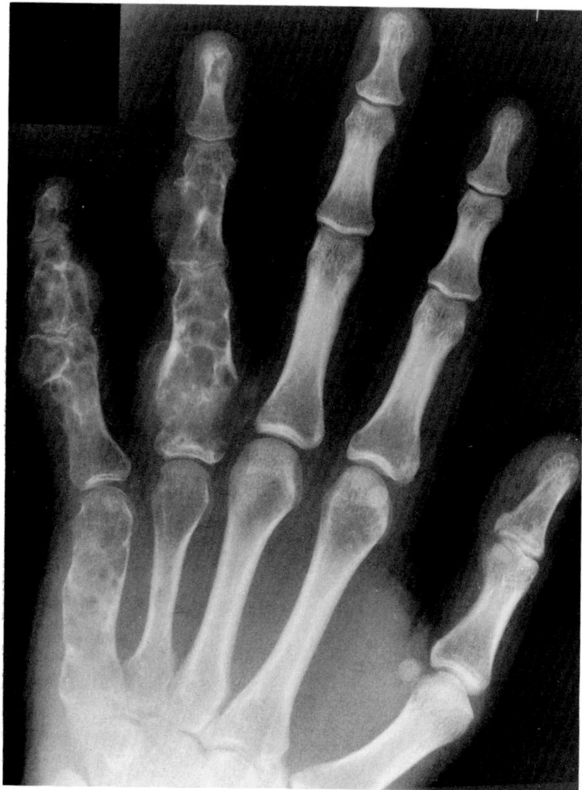

**Figure 60.6. Malignant transformation in multiple enchondromatosis.** Note the destruction of cortical margins.

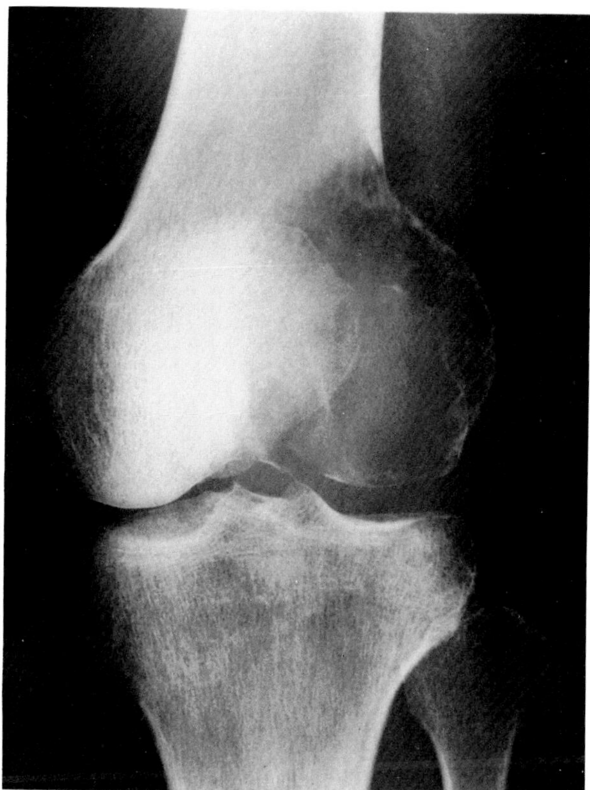

**Figure 60.8. Giant cell tumor of the distal femur.** The typical eccentric lucent lesion in the metaphysis extends to the immediate subarticular cortex.

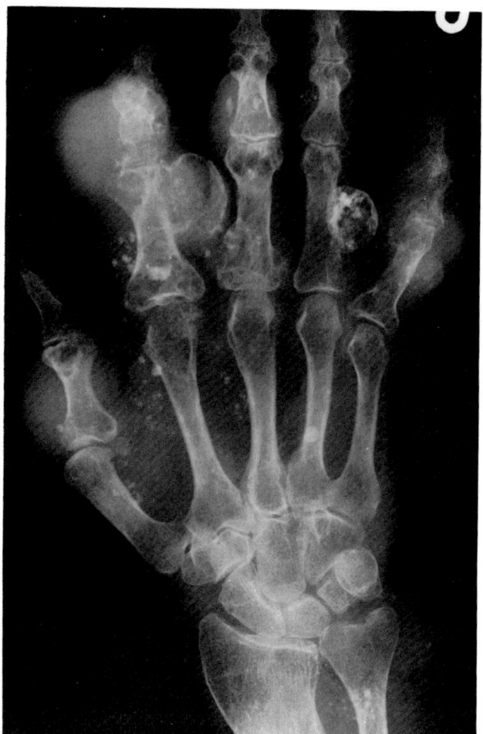

A

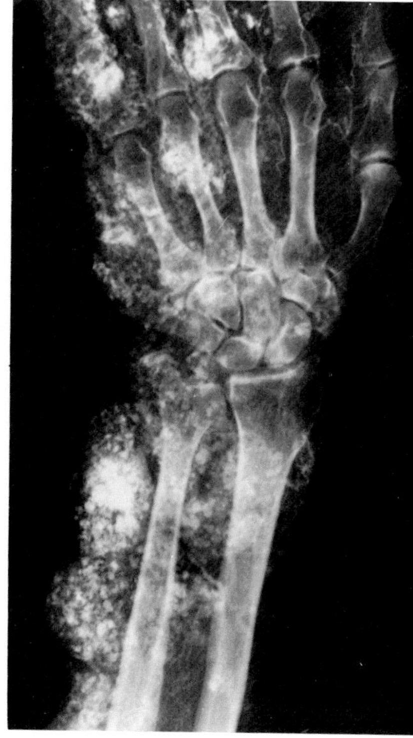

B

**Figure 60.7. Maffucci's syndrome.** (A) A plain radiograph demonstrates multiple soft tissue masses and calcified thrombi in association with expansile bony lesions. (B) A late film from an arteriogram shows contrast material filling numerous cavernous hemangiomas of the soft tissues.

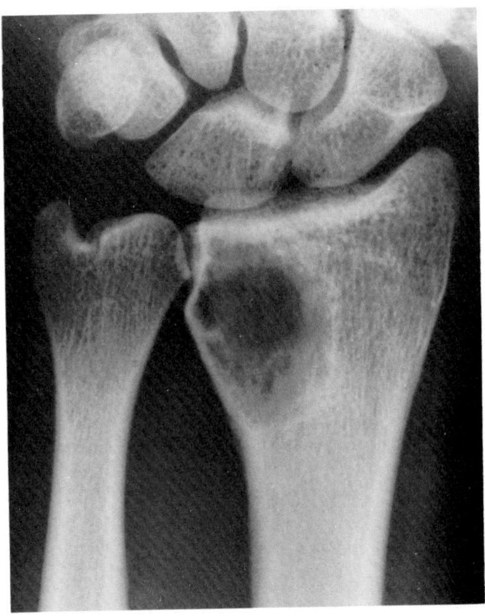

**Figure 60.9. Giant cell tumor of the distal radius.**

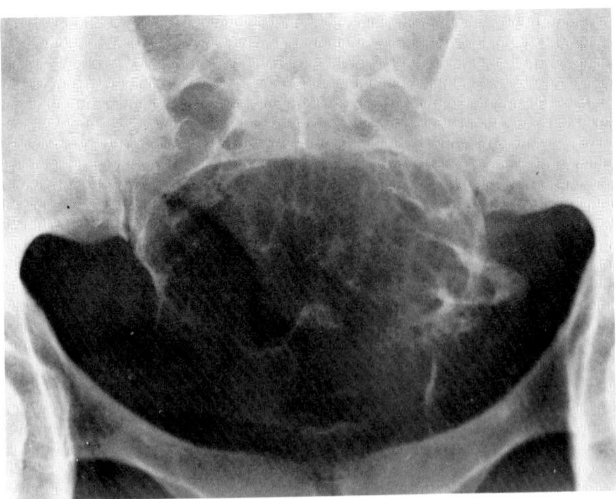

**Figure 60.10. Giant cell tumor of the sacrum.** Huge expansile lesion.

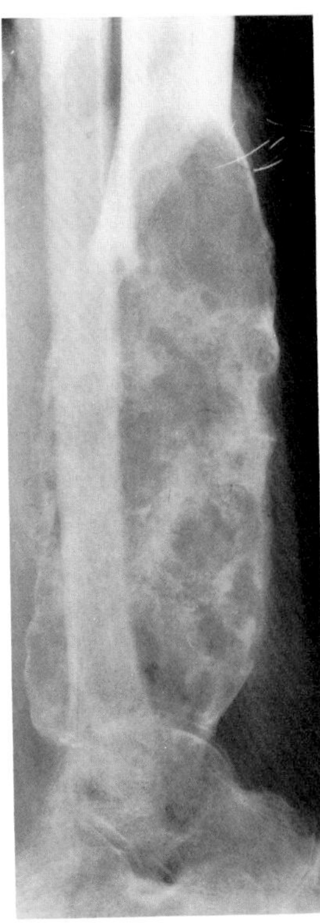

**Figure 60.11. Malignant giant cell tumor of the distal tibia.** Huge lesion with cortical destruction.

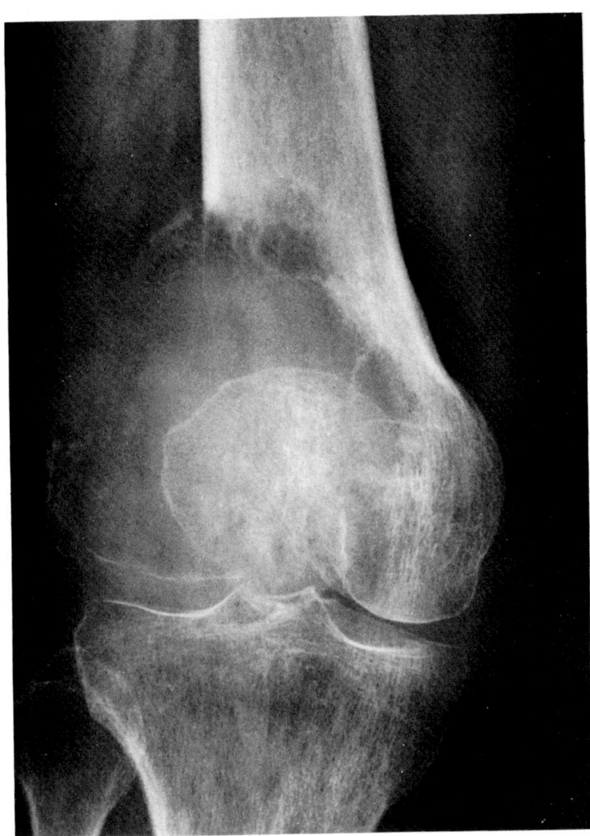

**Figure 60.12. Malignant giant cell tumor of the distal femur.** The tumor has caused cortical destruction, extends outside the host bone, and has an ill-defined margin.

distal femur or proximal tibia. Other common sites are the distal ends of the radius and ulna. The pelvis, sacrum, and vertebral bodies are occasionally involved.

Although usually benign, up to 20 percent of giant cell tumors may be malignant. Extension of a tumor through the cortex with an associated soft tissue mass, which can be well-demonstrated by computed tomography (CT), is often considered highly suggestive of malignancy. However, there is much overlap in the appearance of benign and malignant lesions; radio-

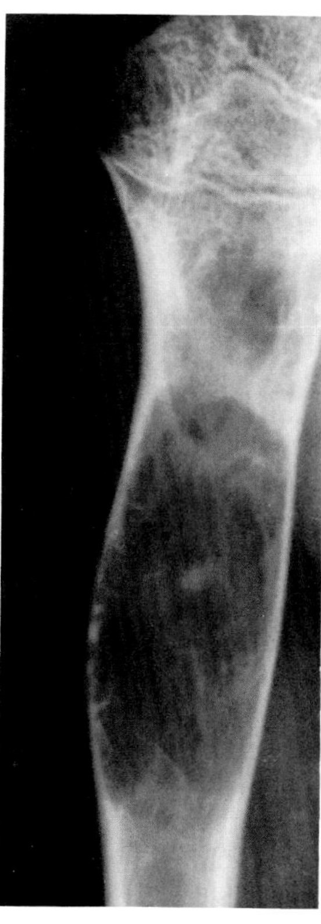

Figure 60.13. **Bone cyst in the proximal humerus.** The cyst has an oval configuration, with its long axis parallel to that of the host bone. Note the thin septa that produce a multiloculated appearance.

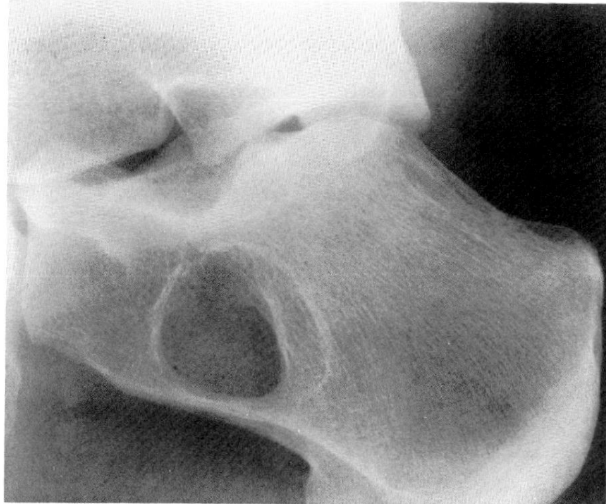

**Figure 60.14. Bone cyst in the calcaneus.**

graphically innocuous-looking tumors may grow rapidly and metastasize, while malignant-looking lesions may never recur. In many cases it is impossible to determine whether a giant cell tumor is benign or malignant on either histologic or radiographic grounds, and only the future behavior of the lesion will show whether it is malignant.[6,7]

## Simple (Unicameral) Bone Cyst

A solitary, or unicameral, bone cyst is a true fluid-filled cyst with a wall of fibrous tissue. Although not a true neoplasm, a simple bone cyst may resemble one radiographically and clinically and thus must be considered in the differential diagnosis of bone tumors. Bone cysts have been divided into two groups: active cysts, which are located immediately adjacent to the epiphyseal plate and retain growth potential; and latent cysts, which are displaced away from the growth plate by normal bone and remain as a static bone defect. Most common in males, solitary bone cysts are asymptomatic and are discovered either incidentally or following pathologic fracture. Bone cysts arise in children and adolescents in the metaphysis of a bone that is still in the active phase of growth. They are most common in the proximal humerus and femur, unlike many benign bone tumors that have a predilection for the knee region. As growth of the bone progresses, a simple bone cyst may appear to have migrated a considerable distance down the shaft, though in reality it is the epiphysis that has actually migrated away from the cyst.

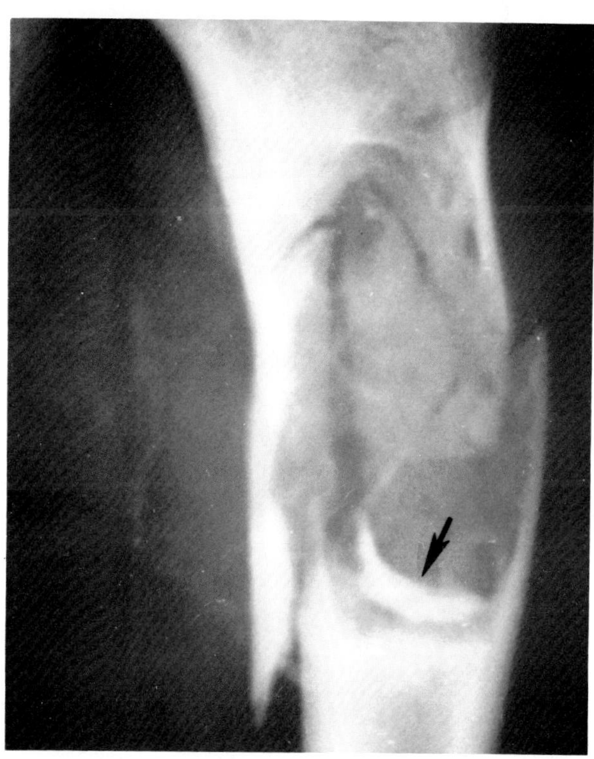

**Figure 60.15. Subtrochanteric bone cyst with pathologic fracture.** A cortical bone fragment (arrow) lies free within the cyst space. (From Reynolds,[10] with permission from the publisher.)

Radiographically, a solitary bone cyst appears as an expansile lucent lesion that is sharply demarcated from adjacent normal bone and may have a thin rim of sclerosis around it. Thin septa, representing scalloping of underlying cortex rather than true bony bridging, may produce a multiloculated appearance. A simple bone cyst has an oval configuration, with its long axis parallel to that of the host bone. It tends to be wider at the metaphyseal end and often interferes with normal modeling of the bone, though this deformity decreases as the cyst assumes a progressively more diaphyseal position with skeletal growth.

Pathologic fractures in bone cysts not infrequently occur. Because the fluid contents of the cyst offer no resistance, fragments of cortical bone are free to fall down to the dependent portion of the cyst, thus permitting differentiation of a bone cyst from radiographically similar solid processes that have a firm tissue consistency.[8-10]

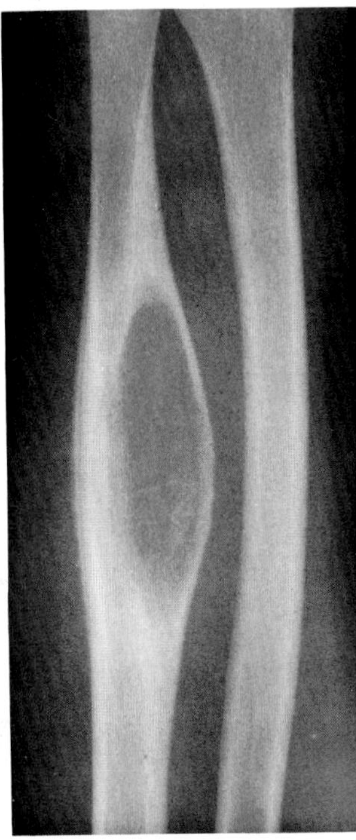

**Figure 60.16. Aneurysmal bone cyst of the radius.** An expansile, eccentric, cystlike lesion causes ballooning of the cortex and periosteal response.

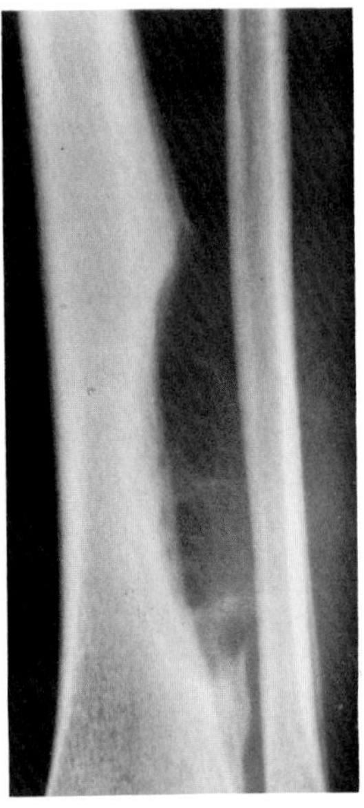

**Figure 60.17. Aneurysmal bone cyst of the tibia.** An expansile, eccentric, cystlike lesion with multiple fine internal septa. Because the severely thinned cortex is difficult to detect, the tumor resembles a malignant process. (From Eisenberg,[54] with permission from the publisher.)

## Aneurysmal Bone Cyst

An aneurysmal bone cyst, rather than being a true neoplasm or cyst, consists of numerous blood-filled arteriovenous communications. Aneurysmal bone cysts occur most frequently in children and young adults and present with mild pain of several months' duration, swelling, and restriction of movement. They most often arise in the metaphyseal area of long bones or the posterior elements of the vertebrae. In long bones, an aneurysmal bone cyst is an expansile, eccentric, cystlike lesion that causes marked ballooning of the thinned cortex. Light trabeculation and septation within the lesion may produce a multiloculated appearance; periosteal new bone may develop. An aneurysmal bone cyst may extend beyond the axis of the host bone and form a visible soft tissue mass. The cortex may be so thin that it is invisible on plain radiographs, and thus this benign lesion may be mistaken for a malignant bone tumor. In the spine, an aneurysmal bone cyst appears as an expanding, trabeculated, lucent lesion that may extend to involve several adjacent vertebrae.[11,12]

## Bone Island

Bone islands are benign solitary areas of dense compact bone that occur most commonly in the pelvis and upper femurs but may be seen in every bone except the skull. The lesions are sharply demarcated, and the margins often display thorny radiations giving a brush border. Although almost half the bone islands enlarge over a period of years and many show activity on radionuclide scans, bone islands are asymptomatic and completely benign and must be distinguished from osteoblastic metastases.[13-15]

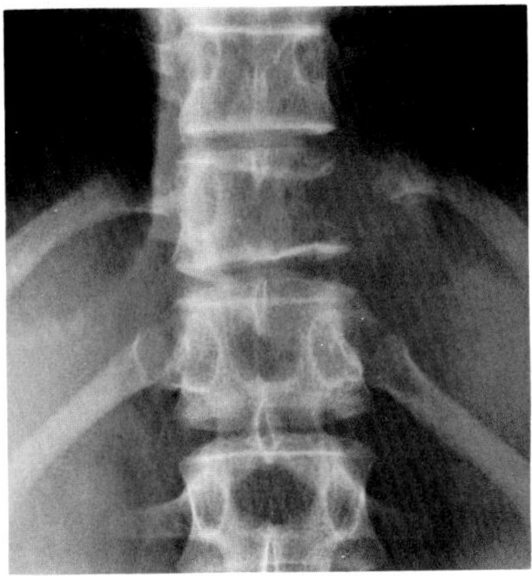

A

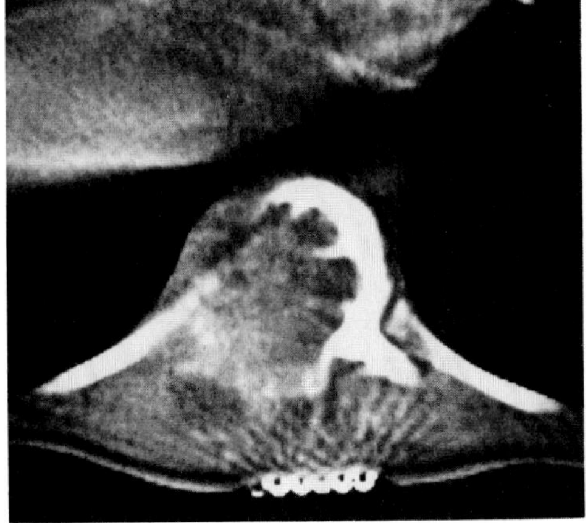

B

**Figure 60.18. Aneurysmal bone cyst of a thoracic vertebral body.** (A) There is destruction of the body and posterior elements. No peripheral shell of bone can be recognized. (B) A CT scan shows irregular destruction suggesting a malignant process. (From Beabout, McLeod, and Dahlin,[55] with permission from the publisher.)

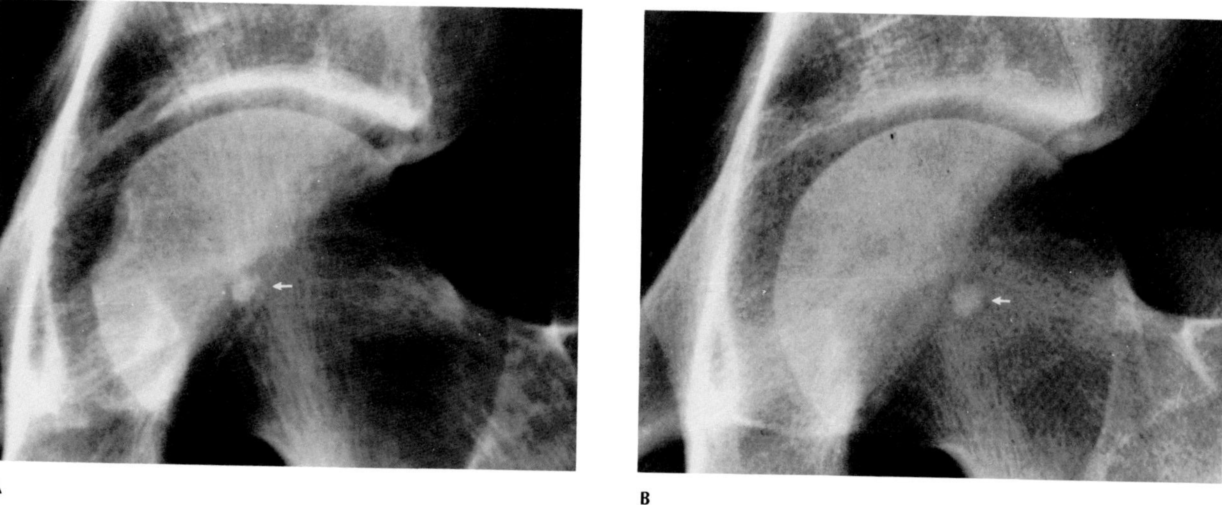

Figure 60.19. **Bone island.** (A) A solitary area of dense compact bone (arrow) in the head of the femur. (B) Twenty years later, the bone island has enlarged in size.

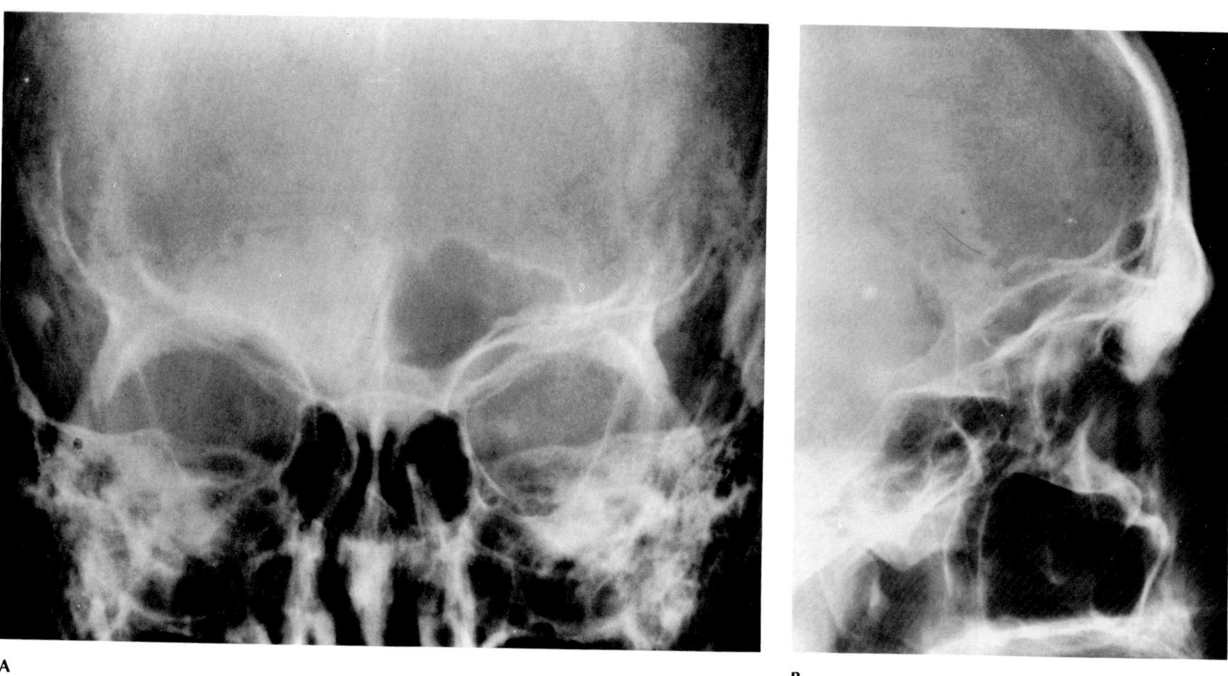

Figure 60.20. **Frontal sinus osteoma.** (A) Frontal and (B) lateral views of the skull demonstrate a large, extremely dense osteoma filling the right frontal sinus.

## Osteoma

Osteomas are flat, benign tumors of membranous bone that most often arise in the outer table of the skull, the paranasal sinuses (especially frontal and ethmoid), and the mandible. Radiographically, osteomas appear as well-circumscribed, extremely dense round lesions that are rarely larger than 2 cm in diameter. Sinus osteomas may infrequently become so large as to expand the walls of the sinus or, by obstructing the orifice, lead to mucus retention and sinusitis. Osteomas are associated with multiple colonic polyps in Gardner's syndrome.[16]

## Osteoid Osteoma

Osteoid osteoma is a benign bone tumor that usually develops in the second and third decades and is most frequent in men. The classic clinical symptom is local pain that is worse at night and is dramatically relieved

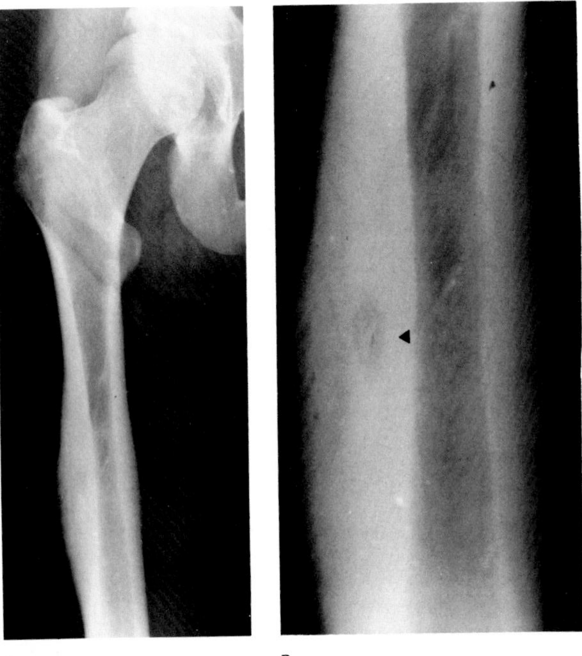


**Figure 60.21. Osteoid osteoma.** (A) Full and (B) coned views of the midshaft of the femur demonstrate a dense sclerotic zone of cortical thickening laterally, which contains a small, oval, lucent nidus (arrow).

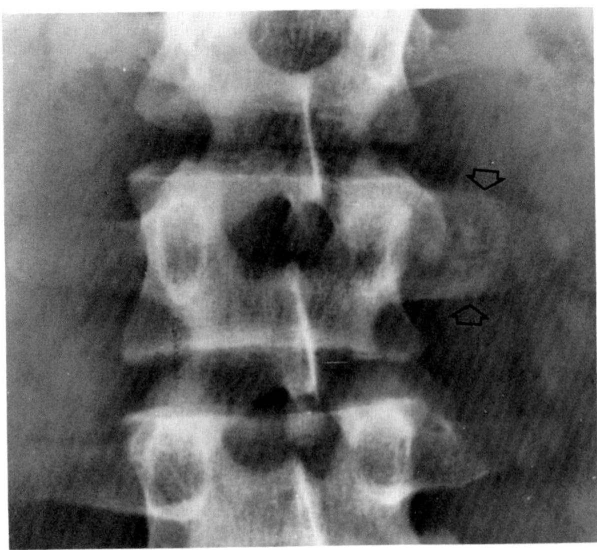

**Figure 60.22. Osteoblastoma of the lumbar spine.** A well-circumscribed, expansile lesion (arrows) involves the left transverse process of a midlumbar vertebra.

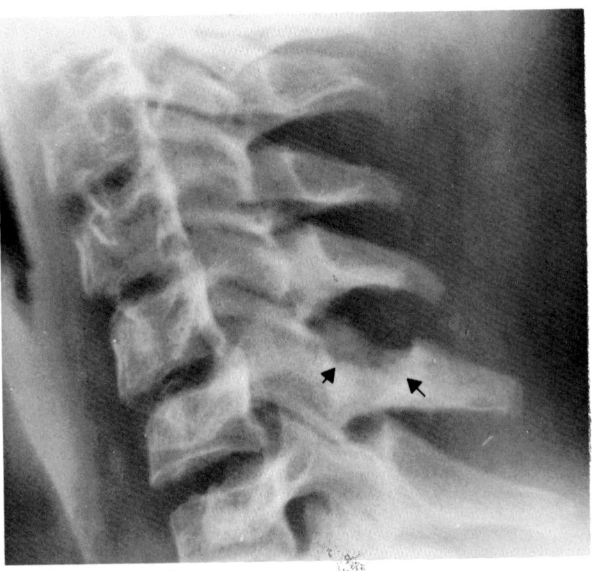

**Figure 60.23. Osteoblastoma of the cervical spine.** A sharply defined, erosive lesion (arrows) involves the superior margin of a lower cervical spinous process.

by aspirin. Occasionally, bone pain may precede any radiographic abnormality.

Most osteoid osteomas occur in the femur and tibia, though they may be seen in any of the tubular bones, the pelvis, or the vertebrae. The tumor typically presents as a small, round or oval lucent nidus, less that 1 cm in diameter, that is surrounded by a large, dense sclerotic zone of cortical thickening. Because this dense reaction may obscure the nidus on conventional radiographs, overpenetrated films or tomography (conventional or computed) may be necessary for its demonstration. The nidus may be slightly or markedly calcified. Although most frequently located within the cortex, the nidus may be in an intramedullary or subperiosteal position and be difficult to detect.

Surgical excision of the nidus is essential for cure; it is not necessary to remove the reactive calcification, even though it may form the major part of the lesion.[17–19]

## Osteoblastoma

Osteoblastoma is a rare bone neoplasm that most frequently occurs in the second decade and produces a dull aching pain, tenderness, and soft tissue swelling. About half of osteoblastomas involve the vertebral column, most frequently the neural arches and spinous processes. Tumors in the posterior elements of the spine tend to grow rapidly and expand the host bone, readily breaking through the cortex and producing a sharply defined soft tissue component that is often circumscribed by a thin calcific shell. In long bones or the small bones of the hands and feet, osteoblastomas are eccentrically located in the metaphysis or shaft and appear as well-circumscribed expansile lesions that may break through the cortex to form a soft tissue component surrounded by a fine calcific margin. Although predominantly lytic, osteoblastomas may have some internal calcification, and their aggressive appearance often simulates a malignant lesion.[20,21]

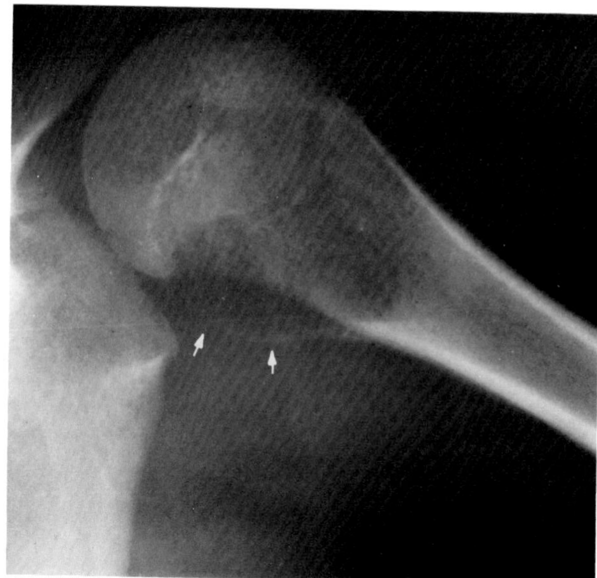

**Figure 60.24. Osteoblastoma.** An expansile, eccentric mass in the proximal humerus causes thinning of the cortex (arrows).

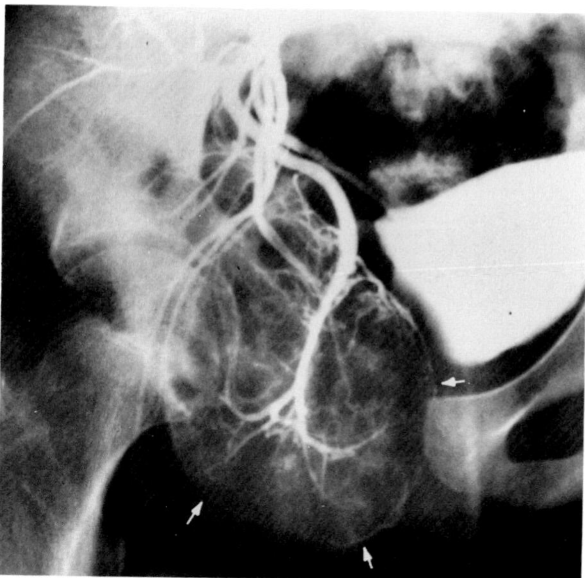

**Figure 60.25. Osteoblastoma.** A film from an arteriogram demonstrates a huge hypovascular, expansile lesion (arrows) originating in the pubic ramus.

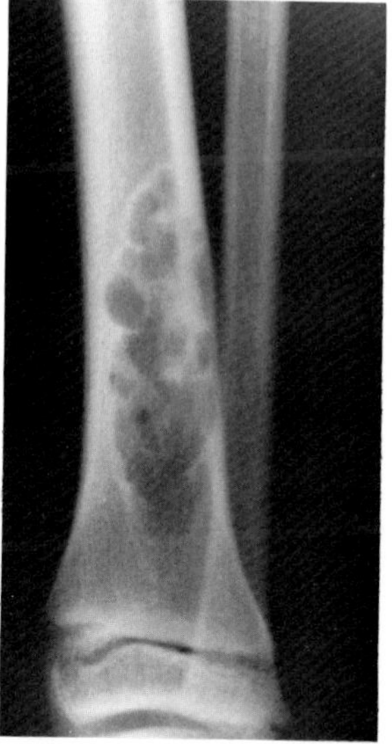

**Figure 60.26. Fibrous cortical defect.** Multilocular, eccentric lucency in the distal tibia. Note the thin, scalloped rim of sclerosis.

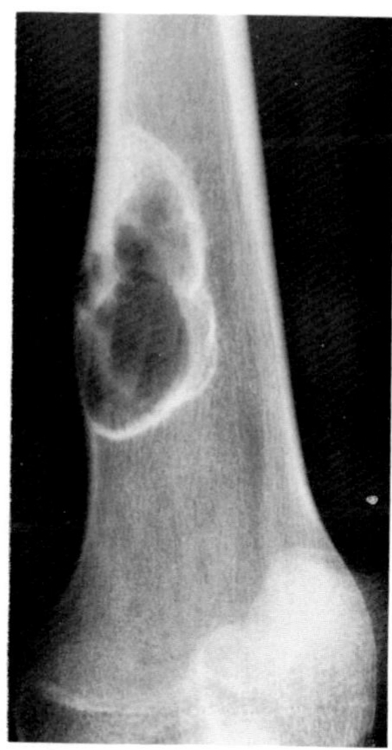

**Figure 60.27. Nonossifying fibroma.** Multilocular, eccentric lucency with a sclerotic rim in the distal femur.

## Fibrous Cortical Defect and Nonossifying Fibroma

Fibrous cortical defect is not a true neoplasm, but rather a benign and asymptomatic small focus of cellular fibrous tissue that causes an osteolytic lesion in the metaphyseal cortex of a long bone, most frequently the distal end of the femur. One or more fibrous cortical defects are estimated to develop in up to 40 percent of all healthy children. Most fibrous cortical defects regress spontaneously and disappear by the time of epiphyseal closure. If the lesion persists and continues to grow, it may eventually encroach on the medullary cavity and be termed a *nonossifying fibroma*.

Radiographically, a benign fibrous cortical defect appears as a small, often multilocular, eccentric lucency that causes cortical thinning and expansion and is sharply demarcated by a thin, scalloped rim of sclerosis. There may be several lesions in the same bone or involvement of multiple bones. Initially round, the defect soon becomes oval with its long axis parallel to that of the host bone.

A nonossifying fibroma results from continued proliferative activity of a fibrous cortical defect and is seen in older children and young adults. Except for its larger size, the radiographic appearance closely resembles a benign fibrous cortical defect and is so distinctive that no biopsy is required.[22,23]

## Chondromyxoid Fibroma

Chondromyxoid fibroma is an uncommon benign bone tumor that originates from cartilage-forming connective tissue and tends to occur in young adults. The lesion is often painful and most frequently involves the tibia, though other long and flat bones may be affected. Chondromyxoid fibroma appears radiographically as a round or ovoid, eccentric lucent lesion that arises in the metaphysis of a long bone. Although the tumor may extend into the epiphysis, it usually does not abut on the epiphyseal plate. In long bones, the axis of the tumor is

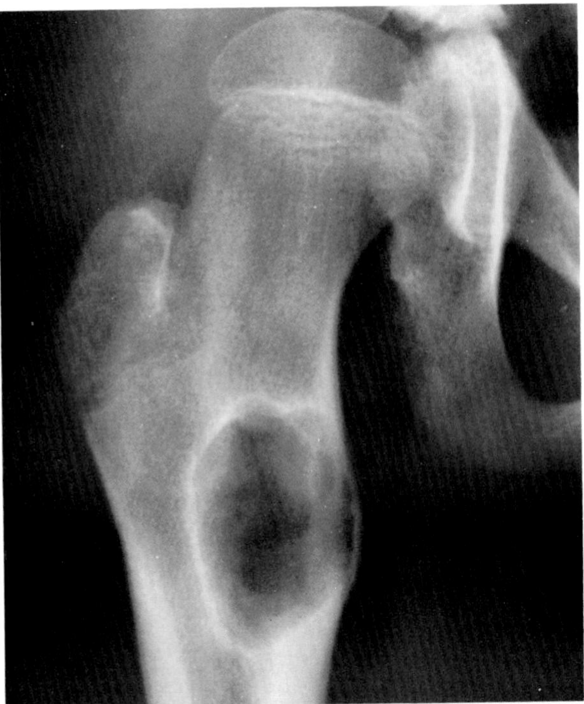

**Figure 60.28. Chondromyxoid fibroma.** Ovoid, eccentric metaphyseal lucency with thinning of the overlying cortex and a sclerotic inner margin.

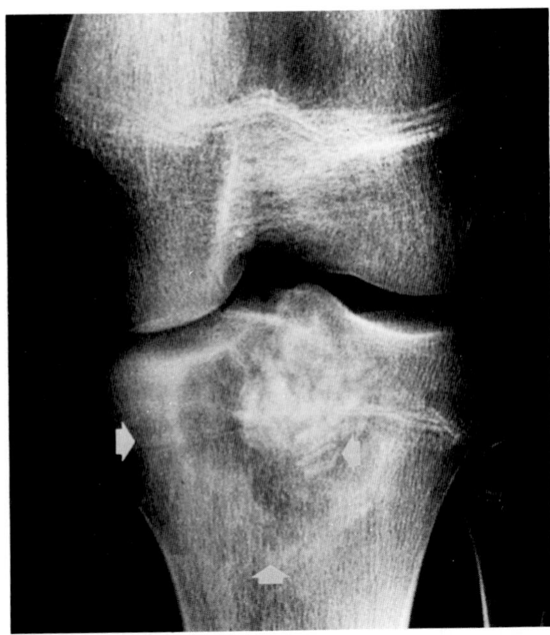

**Figure 60.29. Chondroblastoma.** There is an osteolytic lesion containing calcification (arrows) within the epiphysis. Note the open epiphyseal line. (From Edeiken,[56] with permission from the publisher.)

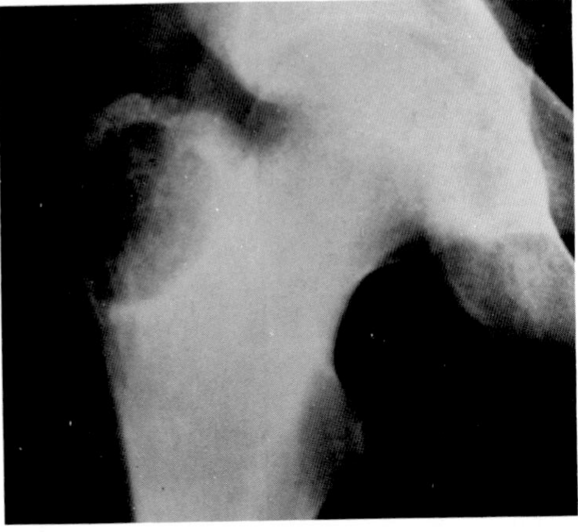

**Figure 60.30. Chondroblastoma.** Typical osteolytic area in the greater trochanter of the proximal femur. In the original films, punctate calcifications could be seen which helped make the diagnosis. (From Edeiken,[56] with permission from the publisher.)

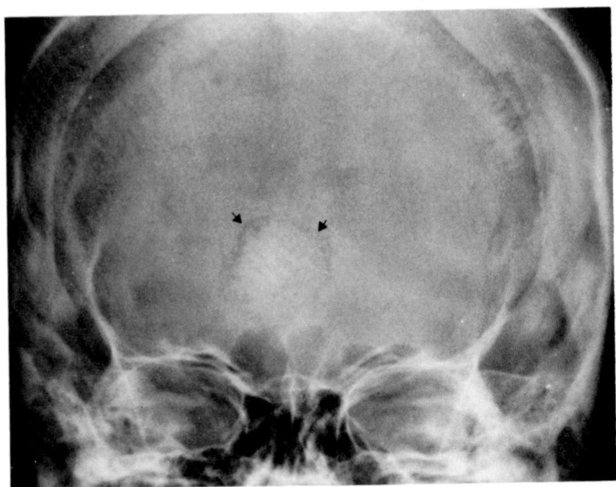

**Figure 60.31. Hemangioma of the skull (arrows).** Osseous spicules radiate from the center to produce a characteristic sunburst pattern.

oriented along the long axis of the host bone. The overlying cortex is usually bulging and thinned; the inner border is generally thick and sclerotic, often with a scalloped margin. Stippled calcification, common in chondroblastoma and other cartilaginous lesions, is much less frequently seen in chondromyxoid fibroma.[24,25]

## Chondroblastoma

Chondroblastoma is a rare benign cartilaginous tumor of the epiphysis that occurs in children and young adults before the cessation of enchondral bone growth. Seen most frequently in males, the tumor usually presents with dull pain that is referred to a joint and may be associated with local swelling and limitation of motion. Chondroblastomas are usually found in the epiphyseal centers of the femur, tibia, or humerus, frequently involving the greater trochanter of the femur and greater tuberosity of the humerus. Radiographically, chondroblastoma appears as an eccentrically situated, round or oval lucency in the epiphysis. The tumor often has a thin sclerotic rim and may contain flocculent calcification that produces a mottled, or fluffy cotton wool, appearance.[26–28]

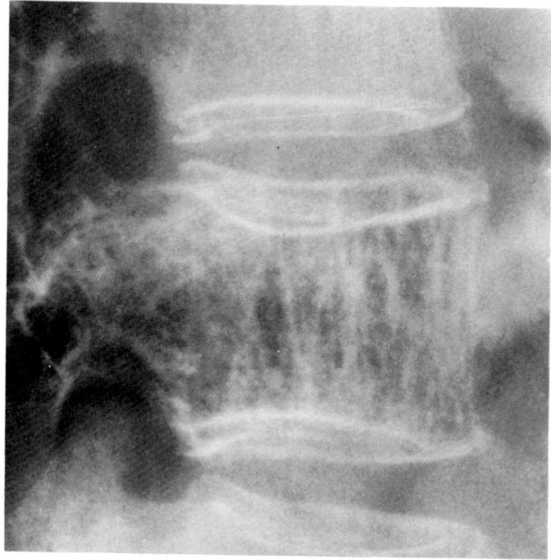

**Figure 60.32. Hemangioma of a vertebral body.** Multiple coarse, linear striations run vertically within the demineralized vertebral body.

## Hemangioma

Hemangiomas are benign, slow-growing tumors that produce characteristic findings when they involve the skull and vertebral body. In the skull, the lesion is round and lucent and often contains osseous spicules that radiate from the center to produce a typical sunburst pattern. In the vertebrae, a hemangioma produces multiple coarse linear striations running vertically within a demineralized vertebral body, an appearance that is virtually diagnostic of this tumor. Hemangiomas of other bones are rare and present nonspecific patterns that may resemble a variety of benign or malignant lesions.[29]

# MALIGNANT BONE TUMORS

## Osteogenic Sarcoma (Osteosarcoma)

Osteogenic sarcoma is the second most common primary neoplasm of bone (after multiple myeloma). Most osteogenic sarcomas arise in persons between 10 and 25 years of age. A smaller peak incidence is seen in older age groups, in whom the tumor is superimposed on a pre-existing bone disorder, particularly Paget's dis-

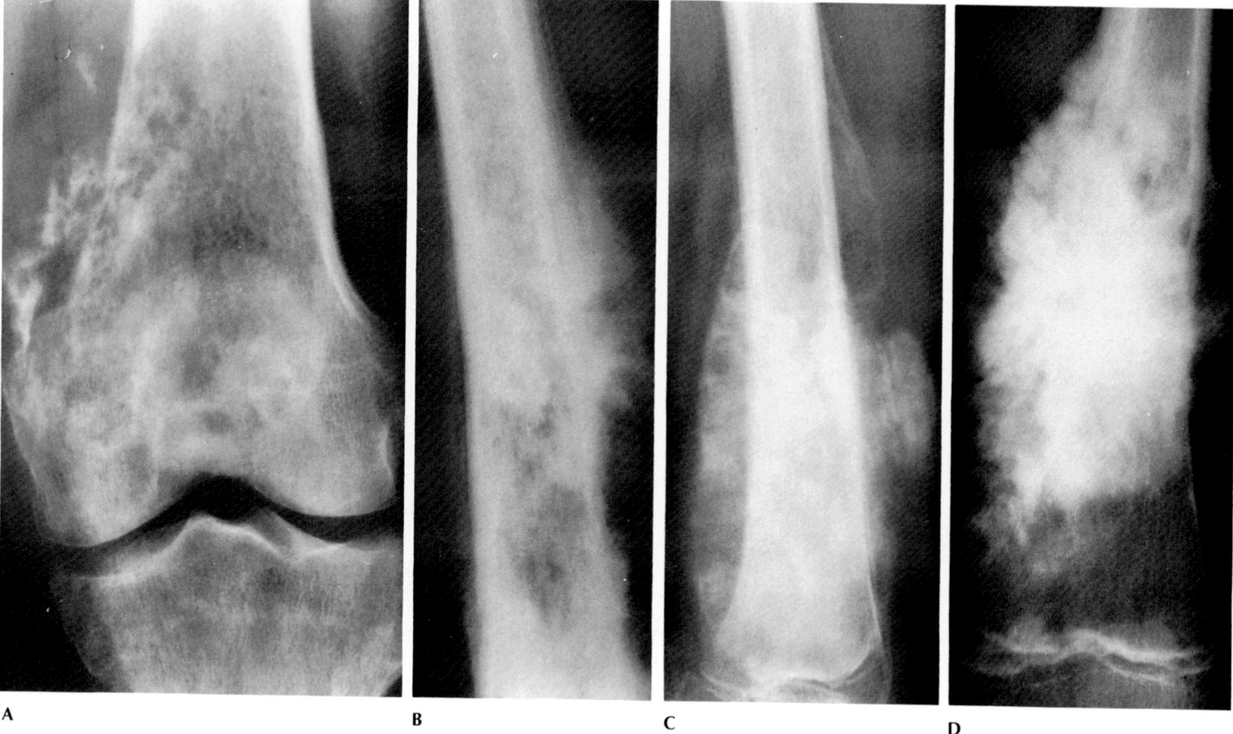


**Figure 60.33. Osteogenic sarcoma.** Four examples of osteogenic sarcoma of the femur illustrate the broad spectrum of radiographic bone changes. There are varying amounts of ragged bone destruction and exuberant, irregular periosteal response.

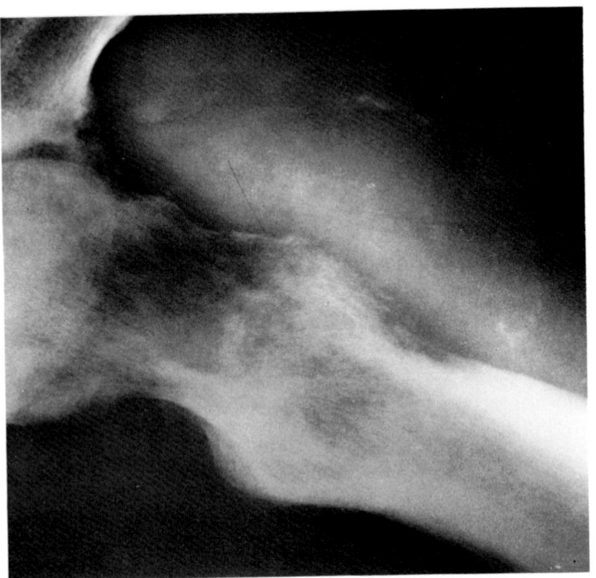

**Figure 60.34. Osteogenic sarcoma in Paget's disease.** Cortical destruction is associated with a soft tissue mass and calcification. The coarsened trabeculae and thickened cortex elsewhere in the femur are evidence of pre-existing Paget's disease.

ease. Males are affected about twice as often as females. The usual presenting complaints are local pain and swelling, sometimes followed by fever, weight loss, and secondary anemia. Pulmonary metastases develop early, and a plain chest radiograph should be obtained to exclude this unfavorable prognostic sign. If no metastases to the lung are detected by this modality, computed (not conventional) tomography should be performed.

Most osteogenic sarcomas arise in the metaphyseal portion of long bones, primarily at the distal end of the femur, the proximal end of the tibia, and the proximal end of the humerus. They may, however, arise in the diaphysis or in the soft tissues and involve any bone. In older individuals, the tumor is usually found in a flat bone.

Radiographically, osteogenic sarcomas may produce a broad spectrum of changes in bone. These vary from a purely lytic, destructive process to a sclerosing form containing extremely dense new bone. In the most frequent mixed form there is a combination of bone destruction and bone production, either of which may predominate. Bone destruction is typically ragged and uneven, and the tumor extends into the soft tissues early in its development. Periosteal reaction in osteogenic sarcoma is highly irregular and interrupted, in contrast to the more homogeneous or solid appearance accompanying benign lesions. In the classic sunburst pattern, horizontal bony spicules extend in radiating fashion into the soft tissue mass. Although elevation of the periosteum at the periphery of a lesion with subsequent new bone formation (Codman's triangle) has often been considered pathognomonic of a malignant bone tumor such as osteogenic sarcoma, it may be found in benign conditions as well.

In summary, the typical radiographic appearance of osteogenic sarcoma is that of a mixed destructive and sclerotic lesion in the metaphysis of a long bone in a young person, with associated soft tissue mass, irregular periosteal reaction, and reactive new bone formation.

As with other primary malignant tumors of bone, CT is of value in the patient with osteosarcoma to precisely define the location of the tumor and its intramedullary and extraosseous extent.

Osteosarcomas can arise in the extraskeletal soft tissues, usually in the deep soft tissues of an extremity. Unlike the tumor in bone, the extraskeletal lesion is more common in adults. It typically appears as a soft tissue mass containing either bone or cloudlike opacities representing calcification of osteoid matrix, an appearance similar to the tumor matrix mineralization seen in skeletal osteosarcoma.[30-35]

## Parosteal Sarcoma

Parosteal sarcoma is a rare malignant tumor that arises from the periosteum at the periphery of a bone. Unlike osteogenic sarcoma, parosteal sarcoma most frequently occurs in patients over 30 years of age and has a more favorable prognosis. The major clinical complaint is a slowly growing mass; pain is neither an early nor a constant symptom of this lesion. Parosteal sarcomas originate around the metaphyseal ends of long bones, primarily at either end of the femur or tibia and the proximal humerus. The lesion typically appears as a broad-based, densely ossified mass that extends outward from the host bone. The periphery of the lesion is usually lobulated, sharply demarcated from surrounding soft tissues, and somewhat less dense than the base. In early stages there is a characteristic radiolucent line that separates the dense mass of tumor bone from the cortex, except at the central stalk where the tumor is attached to the host bone. This virtually diagnostic appearance may disappear as the tumor grows. Although parosteal sarcomas tend to encircle the host bone, actual tumor invasion and destruction occur in only about 10 percent of the patients. Because it may be very difficult to differentiate parosteal sarcoma from myositis ossificans unless there is evidence of bone destruction, CT is occasionally necessary to demonstrate cortical invasion.[36,37]

## Chondrosarcoma

Chondrosarcoma is a malignant tumor of cartilaginous origin that may originate de novo or within a pre-existing cartilaginous lesion, such as an osteochondroma or an enchondroma. The tumor is about half as common as osteogenic sarcoma, develops at a later age (half the patients are more than 40 years old), grows more slowly, and metastasizes later. About 10 percent of the patients with multiple hereditary exostoses will develop chon-

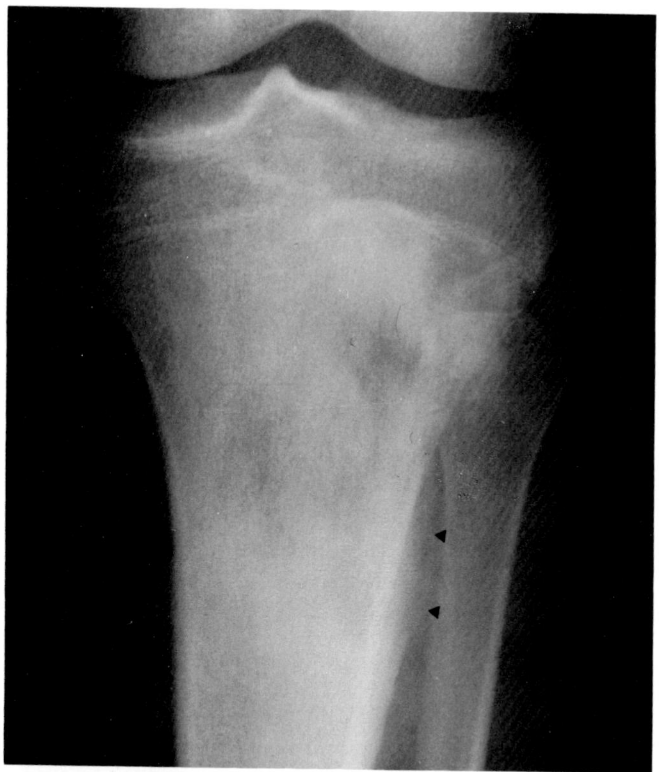

A

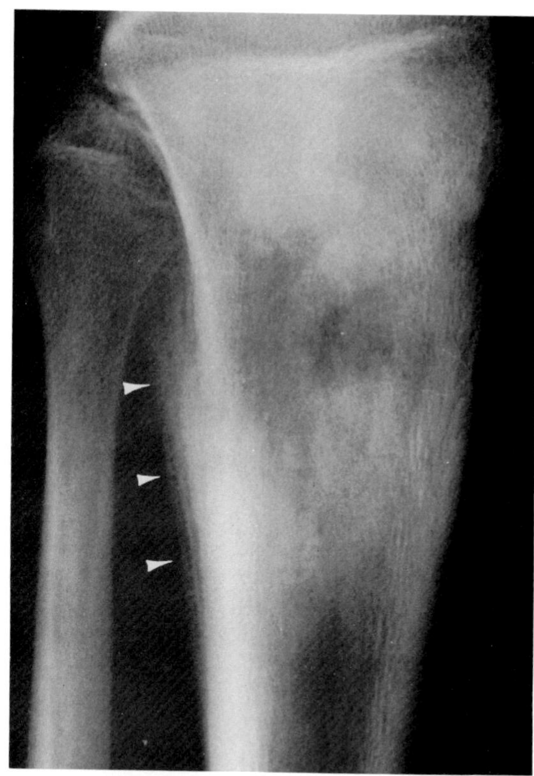

B

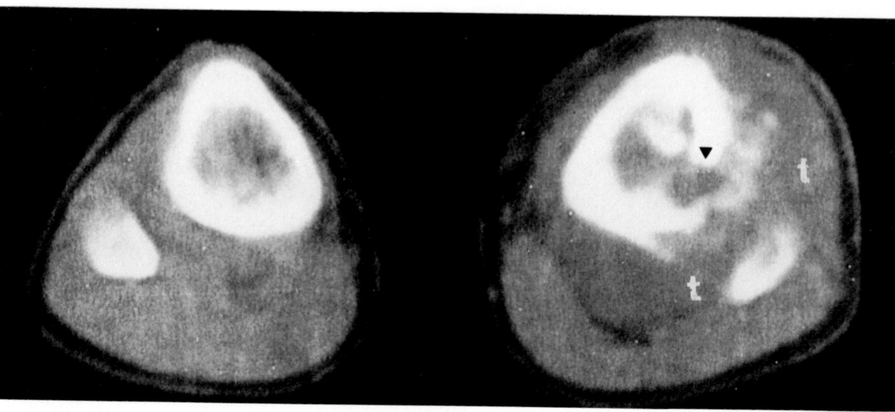


C

**Figure 60.35. Osteogenic sarcoma.** (A) Frontal and (B) lateral views demonstrate a mixed lytic and blastic lesion involving the metadiaphyseal portion of the proximal tibia, with associated lamellated periosteal bone formation (arrowheads). (C) A CT scan localizes the cortical destruction to the posterolateral portion of the tibia (arrowhead) and shows diffuse soft tissue thickening (t). (From Destouet, Gilula, and Murphy,[34] with permission from the publisher.)

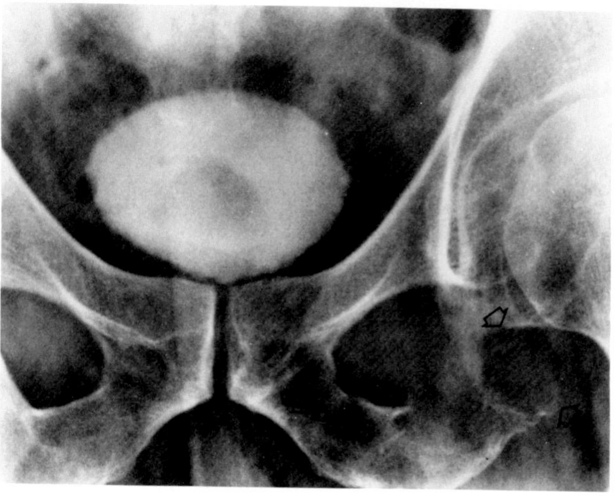

A

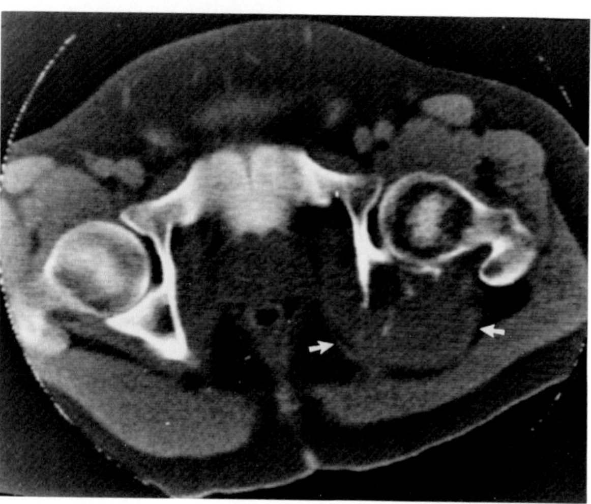

B

**Figure 60.36. Osteogenic sarcoma.** (A) A plain radiograph demonstrates destruction of the left ischium (arrows) without evidence of an associated soft tissue mass. (B) A CT scan through the level of the femoral heads confirms the destruction of the left ischium. In addition, the scan demonstrates a large soft tissue mass (arrowheads) in an area covered by the gluteus maximus muscle and separated from the rectum. The mass was not clinically palpable. (From de Santos, Bernardino, and Murray,[33] with permission from the publisher.)

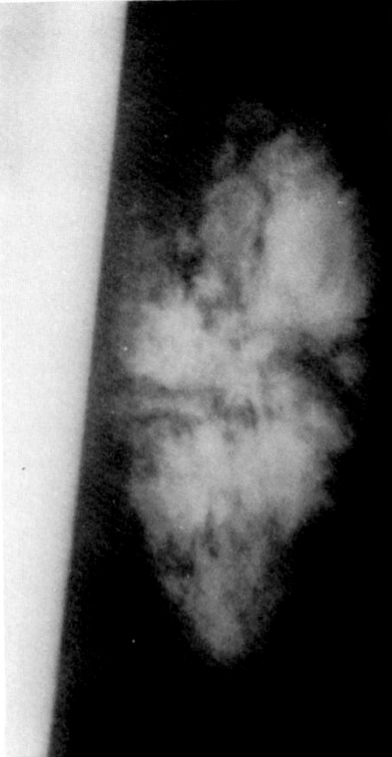

**Figure 60.37. Extraskeletal osteosarcoma of the posterolateral thigh.** (From Cavanagh,[35] with permission from the publisher.)

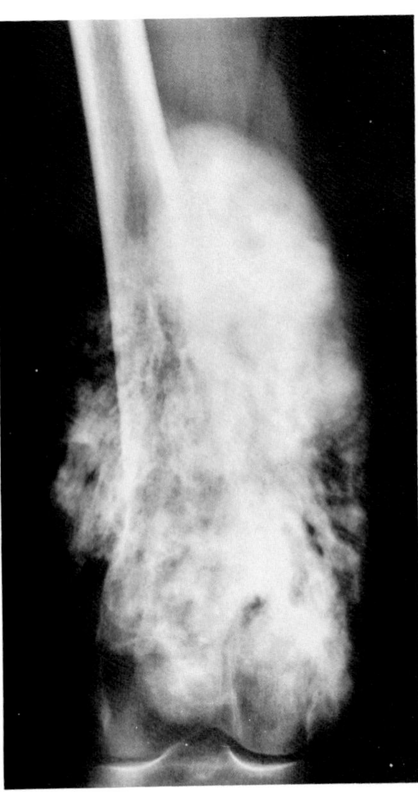

A

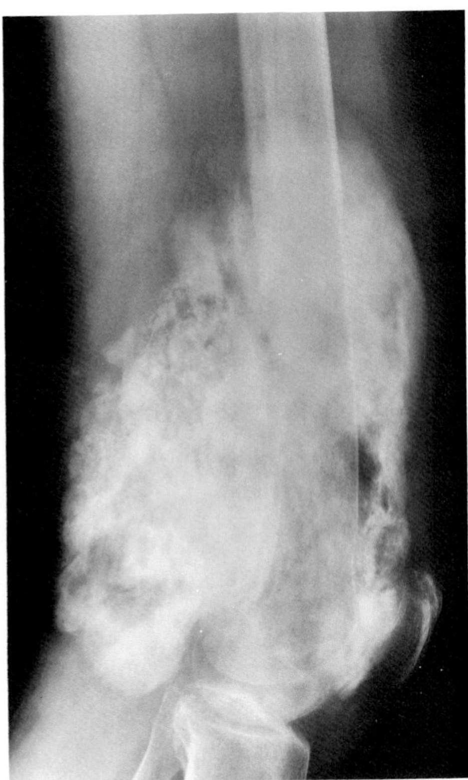

B

**Figure 60.38. Parosteal sarcoma.** (A) Frontal and (B) lateral views of the leg demonstrate a broad-based, densely ossified mass extending outward from the distal femur. The characteristic radiolucent line separating the dense mass of tumor bone from the cortex is not seen in this huge lesion.

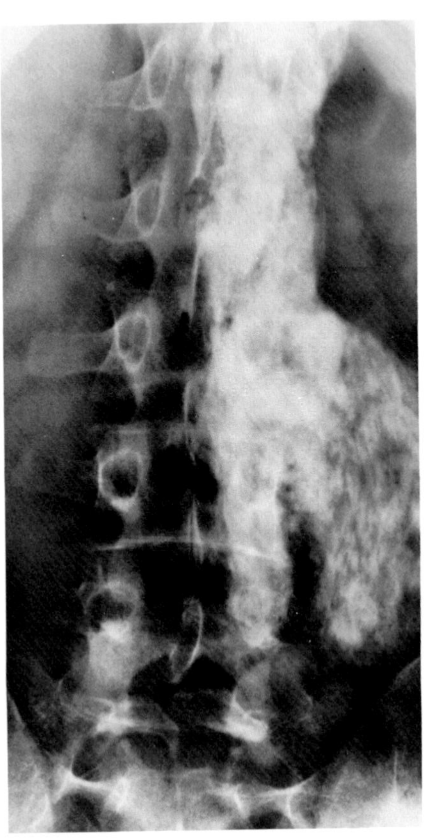

A

B

**Figure 60.39. Parosteal sarcoma.** (A) Frontal and (B) lateral views of the lumbar spine demonstrate a large, broad-based, densely ossified mass arising from the region of the posterior elements of the vertebral bodies.

drosarcoma; malignant degeneration is rare with solitary osteochondroma. Enchondromas in the trunk or in the proximal ends of a femur or humerus are especially prone to develop chondrosarcoma. The more distally the enchondroma is located, the lower its potential for malignant degeneration. Chondrosarcomas can be divided into two types, depending on their location in the bone. The more common central chondrosarcoma originates within a bone, usually a tubular bone (especially either end of the femur or the proximal end of the humerus) or the bones of the trunk. Peripheral chondrosarcoma arises outside the confines of the cortex and most commonly involves the pelvis, shoulder girdle, and ribs.

Central chondrosarcomas usually present with dull aching pain, often of long duration, with occasional noninflammatory local swelling and disability of a contiguous joint. Central chondrosarcomas have two distinct radiographic appearances. The first is a localized lucent area of osteolytic destruction in the metaphyseal end of a bone. When the rate of tumor growth exceeds that of bone repair, the margins of the lesion become irregular and ill-defined and the tumor extends to cause cortical destruction and invasion of soft tissues. The second type of central chondrosarcoma, which may be simply a later phase of the first, benign-appearing type, is an aggressive, poorly defined osteolytic lesion that blends impercep-

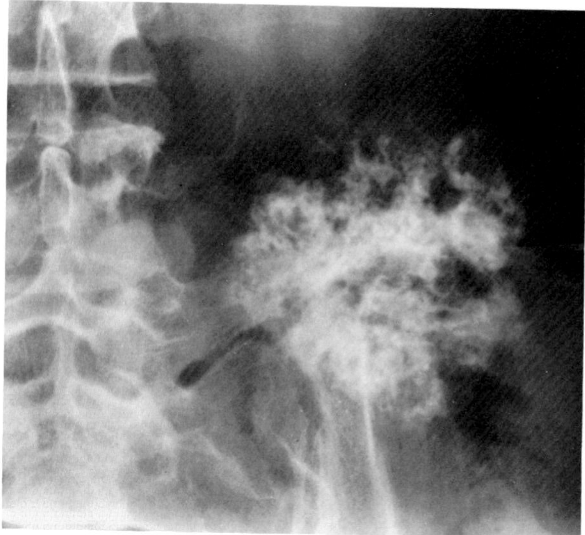

**Figure 60.40. Chondrosarcoma of the ilium.** Note the extensive calcification.

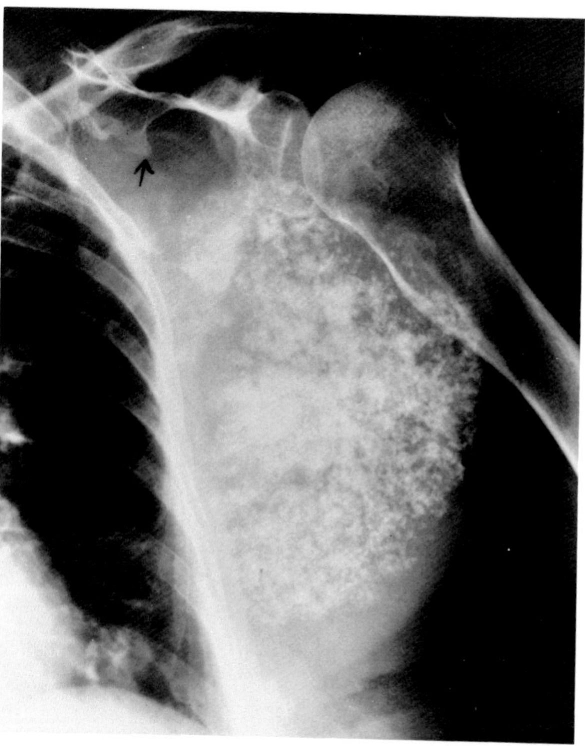

**Figure 60.41. Chondrosarcoma of the axilla.** There is extensive flocculent calcification of the cartilaginous matrix. The arrow points to a small osteochondroma in this patient with multiple hereditary exostoses.

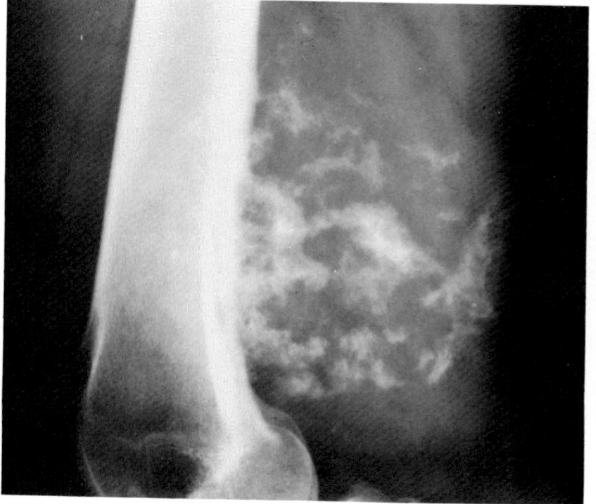

A

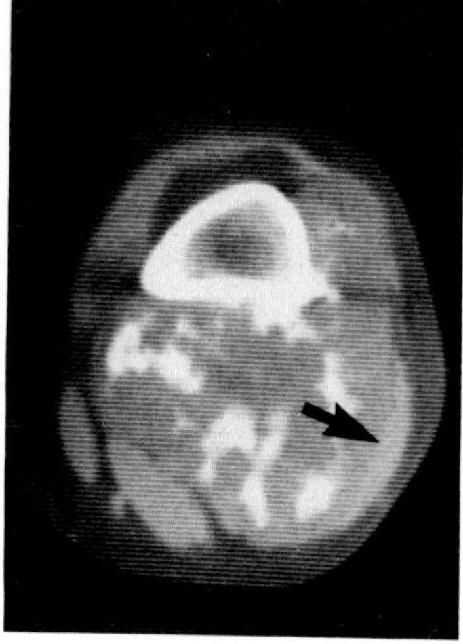

B

**Figure 60.42. Chondrosarcoma.** (A) A large exostotic chondrosarcoma contains prominent dense calcifications. (B) A CT scan shows eccentric lobular growth of the tumor with irregular distribution of calcification within the soft tissue mass. The biceps femoris muscle is compressed but not invaded (arrow). (From Rosenthal, Schiller, and Mankin,[41] with permission from the publisher.)

tibly with normal bone and can expand to replace the entire medullary cavity. The cartilaginous tissue within a chondrosarcoma can be easily recognized by the amorphous, punctate, flocculent or "snow flake" calcifications that are seen in about two-thirds of central chondrosarcomas.

Peripheral chondrosarcomas tend to occur in somewhat younger patients and present as slowly growing painless masses, most often in a patient with multiple hereditary exostoses. This tumor usually appears as a large and bulky soft tissue mass that may invade and extensively destroy the underlying bone. The mass contains characteristic spotty or flocculent calcification, often with a rather dense, radiopaque center and calcific streaks radiating toward the periphery.

Because a benign-appearing cartilaginous tumor may actually represent an early phase of chondrosarcoma, any enchondroma or osteochondroma associated with pain must be considered suggestive of malignancy, even if it has a classic benign radiographic appearance. Any growth or change in pattern of a previously static cartilaginous lesion should also suggest the development of chondrosarcoma.

CT can be of value in the preoperative differentiation between osteochondroma and chondrosarcoma. An osteochondroma appears as a bony mass with a sharply defined periphery, a more lucent but organized center with the cortex and the medullary cavity continuous with the bone from which it arises, and a thin cartilage cap. A chondrosarcoma appears as a predominantly soft tissue mass with an inhomogeneously mineralized center. There may be destruction of adjacent bone or abnormalities of adjacent soft tissues.[38-41]

## Fibrosarcoma

Fibrosarcoma is a rare primary malignant tumor of fibroblastic tissue that has not formed either osteoid or chondroid matrix. Occuring at any age, fibrosarcomas most often involve tubular bones in younger patients and flat bones in older ones. Common presenting symptoms are local pain and swelling. Fibrosarcomas tend to grow slowly and have a somewhat better prognosis than osteogenic sarcomas. Unlike most primary bone tumors, fibrosarcomas tend to metastasize to lymph nodes, and lymphography is often required before surgical intervention.

Fibrosarcomas may originate centrally or, less often, periosteally. Central tumors usually appear as irregular destructive lesions arising in the medullary cavity of the metaphysis of a long bone. They may cause thinning, expansion, and erosion of the cortex and periosteal proliferation. The periosteal type primarily arises in the pelvis, femoral neck, scapula, or rib and is associated with a large soft tissue mass without calcification or neoplastic new bone formation. As the tumor develops, there may be massive invasion of the cortex and extension into the medullary canal.[42,43]

## Ewing's Sarcoma

Ewing's sarcoma is a primary malignant tumor arising in the bone marrow. A tumor of children and young adults, Ewing's sarcoma has a peak incidence in the mid teens and is rare over the age of 30. The major clinical complaint is local pain, often of several months' duration,

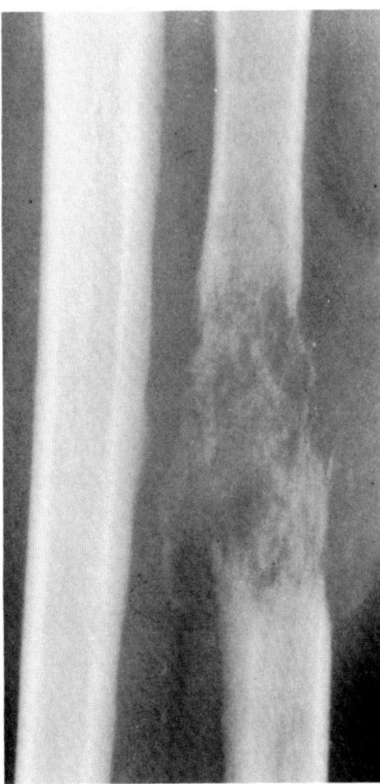

**Figure 60.43. Fibrosarcoma.** There is an irregular destructive lesion of the shaft of the radius.

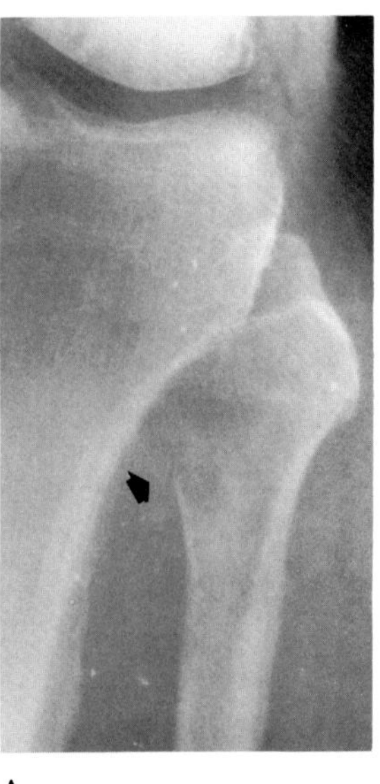

A

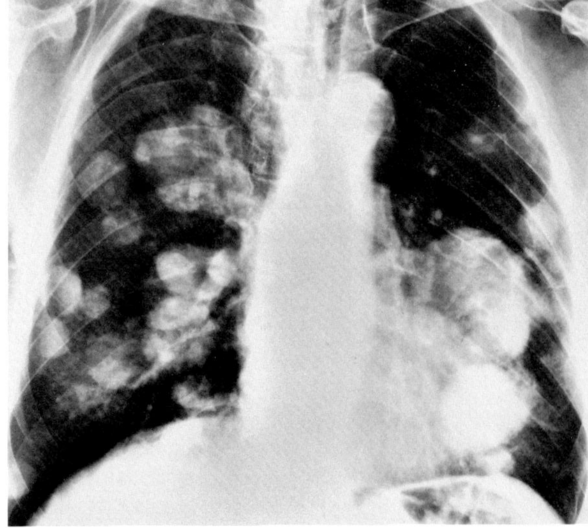

B

**Figure 60.44. Fibrosarcoma with pulmonary metastases.** (A) A small area of bone destruction in the proximal fibula (arrow) was associated with a large painful swelling in the calf. At the time of surgery, the chest radiograph was normal. (B) Five months later, extensive metastases involve both lungs.

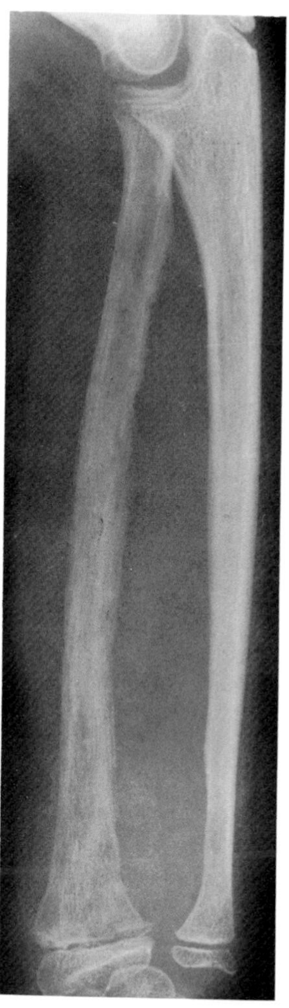

**Figure 60.45. Ewing's sarcoma.** Diffuse permeative destruction involves virtually the entire radius.

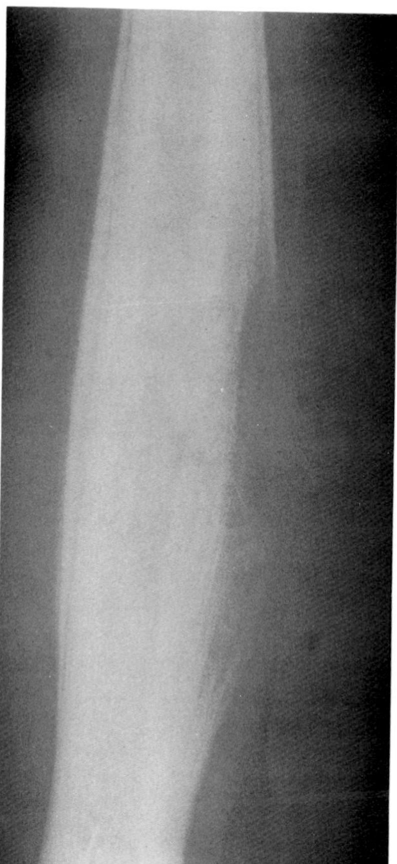

**Figure 60.46. Ewing's sarcoma.** Note the lamellated periosteal reaction on one side of the bone and the thin periosteal elevation (Codman's triangle) on the other.

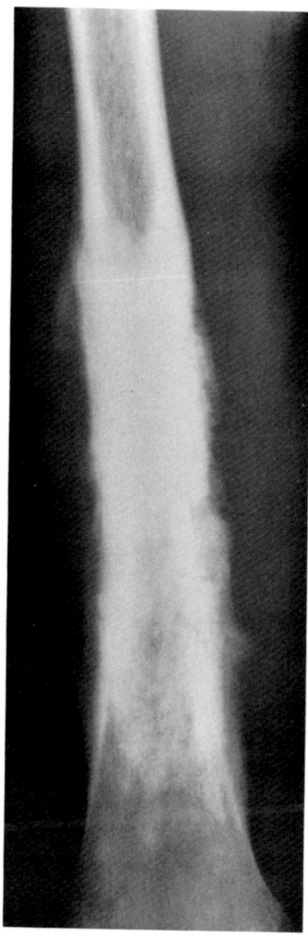

**Figure 60.47. Ewing's sarcoma.** A sunburst pattern of horizontal spicules of dense bone simulates osteogenic sarcoma.

that persistently increases in severity and may be associated with a tender soft tissue mass. Patients with this tumor characteristically have malaise and appear sick, often with fever and leukocytosis suggesting osteomyelitis. Ewing's sarcoma most commonly involves the long bones of the extremities, especially the femur and tibia. It may also arise in the pelvis, ribs, scapula, and bones of the foot (especially the calcaneus). Ewing's sarcoma tends to metastasize early to the lung and other bones.

Though seen in a minority of cases, the classic appearance of Ewing's sarcoma is an ill-defined permeative area of bone destruction that involves a large central portion of the shaft of a long bone and is associated with a fusiform lamellated periosteal reaction parallel to the shaft. However, the tumor may be metaphyseal in origin and produce either a purely lytic lesion or a mass of increased density suggesting osteogenic sarcoma. Because Ewing's sarcoma originates in the medullary cavity and tumor cells spread through the haversian system to extend to subperiosteal areas, cortical destruction is not an early finding. Tumor within the subperiosteal area elicits a periosteal reaction that may be lamellated ("onionskin"), a thin periosteal elevation (Codman's triangle), or a sunburst pattern with horizontal spicules of bone extending into a soft tissue mass. With vertebral involvement, the pattern may range from uniform sclerosis to almost completely lytic destruction. Vertebral body collapse and a paraspinal soft tissue mass are common findings.[44–47]

## Reticulum Cell Sarcoma

Reticulum cell sarcoma is a primary malignant lesion of bone that is histologically similar to Ewing's sarcoma but tends to occur in older persons (average age about 40 years). Males are affected about twice as frequently as females. Local pain and soft tissue swelling, usually of

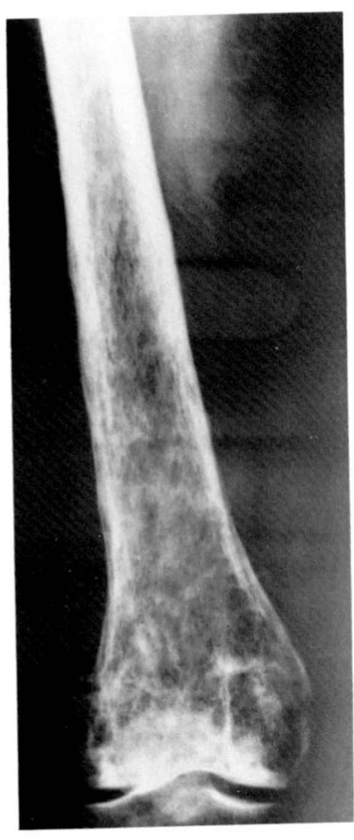

**Figure 60.48. Reticulum cell sarcoma.** Diffuse permeative destruction with mild periosteal response involves the distal half of the femur.

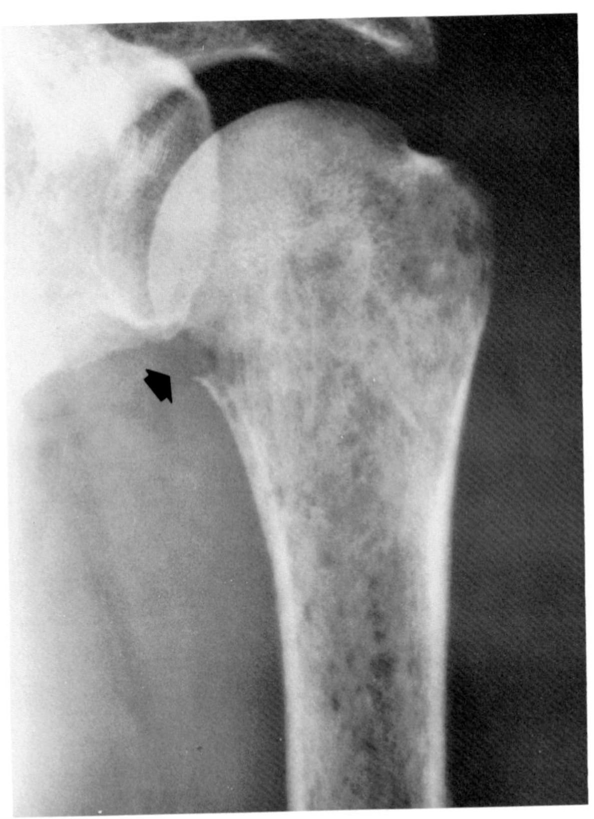

**Figure 60.49. Reticulum cell sarcoma.** A moth-eaten pattern of bone destruction in the proximal humerus is associated with a pathologic fracture (arrow).

long duration, is the major clinical presentation. Reticulum cell sarcoma, unlike Ewing's and other bone sarcomas, rarely causes systemic symptoms, and the patient displays a general well-being even when the local disease is extensive. Because symptoms are so mild and the lesion is of such long duration, pathologic fractures are far more common than with other malignant bone tumors and may at times be the initial clinical complaint. The tumor tends to metastasize late to the lymph nodes and the lungs and only rarely spreads to other bones.

Most reticulum cell sarcomas involve a long tubular bone, especially near the knee. The pelvis, scapula, ribs, and vertebrae may also be affected. Reticulum cell sarcoma characteristically produces a moth-eaten pattern of permeative bone destruction that arises in the medullary cavity and then extends to invade the cortex. The tumor incites an amorphous or lamellated periosteal reaction, which is usually not as striking as that of Ewing's sarcoma. Large soft tissue masses are commonly associated with reticulum cell sarcoma, especially with tumors of the pelvis and vertebrae, but these usually contain no neoplastic calcification or new bone formation.[48–50]

## Metastatic Malignancy to Bone

Metastases are the most common malignant bone tumors, spreading via the bloodstream or lymphatics or by direct extension. The most common primary tumors are carcinomas of the breast, lung, prostate, kidney, and thyroid. Favorite sites of metastatic spread are bones containing red marrow, such as the spine, pelvis, ribs, skull, and the upper ends of the humerus and the femur. Metastases distal to the knees and elbows are infrequent but do occur, especially with bronchogenic tumors.

The detection of skeletal metastases is critical in the management of patients with known or suspected neoplastic disease, both at the time of initial staging and during the period of continuing follow-up care. The presence of metastases may exclude some patients from the radical "curative" therapy that is offered to those without disseminated disease, thus sparing them fruitless, high-morbidity procedures for a nonremediable condition. The best screening examination for the detection of asymptomatic skeletal metastases is the radionuclide bone scan, which is unquestionably more sensitive than the radiographic "skeletal survey." Because almost half the mineral content of a bone must be lost before the loss is detectable on plain radiographs, the skeletal survey should be abandoned as a general screening examination for the detection of asymptomatic skeletal metastases. The only false-negative bone scans occur with aggressively osteolytic lesions, especially in patients with multiple myeloma. The role of plain radiographs in screening for metastases is to further evaluate focal abnormalities detected on radionuclide scanning, since a variety of lesions (e.g., infections, benign tumors, fibrous

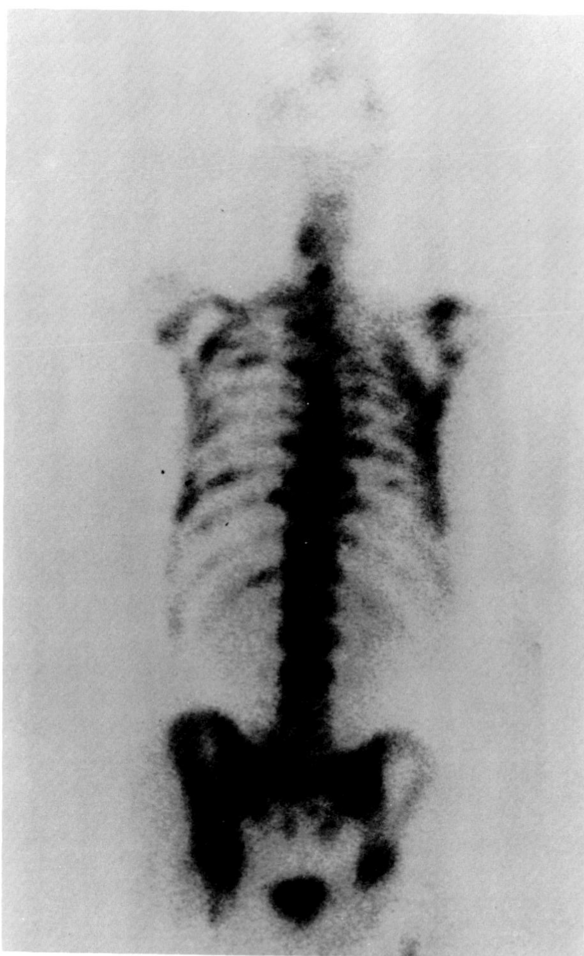

**Figure 60.50. Metastases to bone.** A radionuclide bone scan demonstrates multiple focal areas of increased labeling in the axial skeleton, representing metastases from prostate carcinoma. (From Mall,[53] with permission from the publisher.)

dysplasia, bone islands) can also give positive bone scans. Because the presence of multiple focal abnormalities is typical of metastatic disease, it is only necessary to radiographically examine a single lesion or area to confirm the diagnosis. In addition, because of the occasional false-negative bone scan, it is imperative that all *symptomatic* sites in patients with neoplastic disease be examined by plain radiographs unless the radionuclide bone scan unequivocally demonstrates diffuse metastatic disease.

Metastatic disease may produce a broad spectrum of radiographic appearances. Osteolytic metastases cause destruction without accompanying bone proliferation. The tumor cells in osteoblastic metastases incite a proliferation of reactive bone. In the mixed type of metastasis there is a combination of destruction and sclerosis, with destruction usually predominating.

Osteolytic metastases may be solitary or multiple. They develop from tumor embolic deposits in the medullary canal and eventually extend to destroy cortical bone.

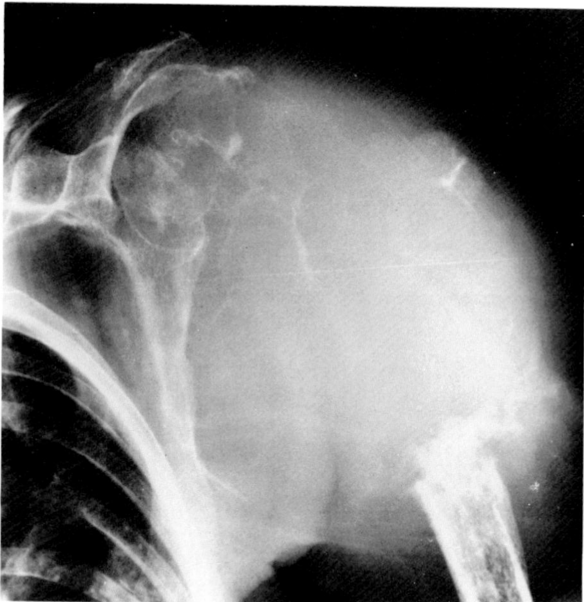

**Figure 60.51. Metastases to bone.** Osteolytic ("blowout") metastasis to the humerus from carcinoma of the kidney.

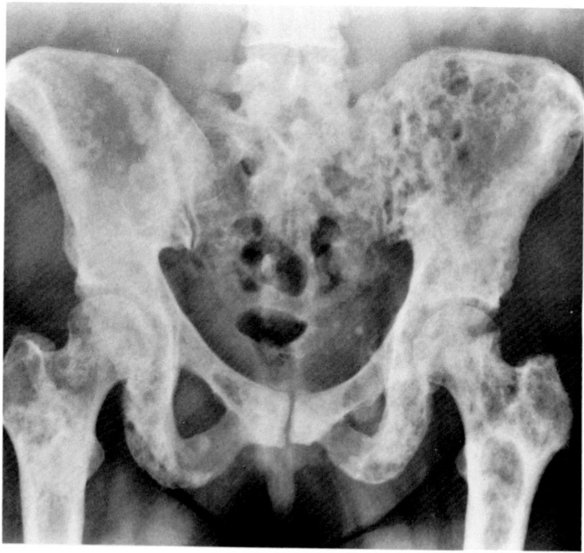

**Figure 60.52. Metastases to bone.** Diffuse lytic and blastic metastases to the pelvis from carcinoma of the breast.

The margins of the lucent lesions are irregular and poorly defined, rarely sharp and smooth. The most common primary lesions causing osteolytic metastases are carcinomas of the breast, kidney, and thyroid. Metastases from carcinomas of the kidney and thyroid typically produce a single large metastatic focus that may appear as an expansile trabeculated lesion ("blowout" metastasis). Metastases from breast carcinoma are most often multiple when first detected.

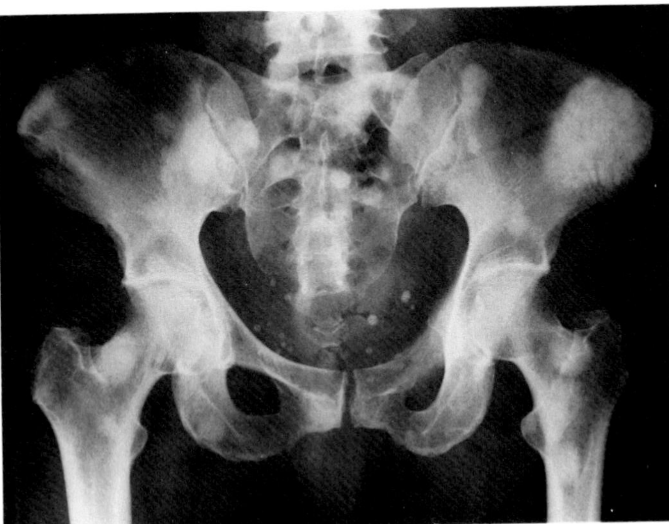

**Figure 60.53. Metastases to bone.** Diffuse osteoblastic metastases to the pelvis and proximal femurs from carcinoma of the urinary bladder.

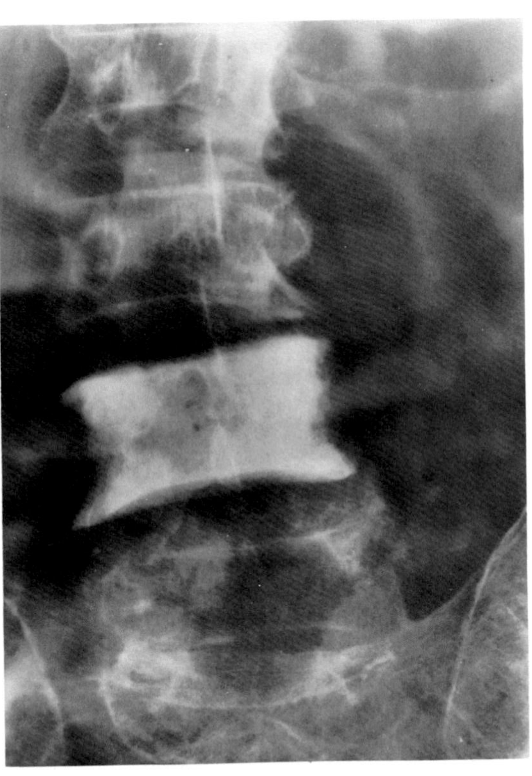

**Figure 60.54. Metastases to bone.** Ivory vertebra in a patient with osteoblastic metastases from carcinoma of the prostate. (From Eisenberg,[54] with permission from the publisher.)

Osteolytic metastases in the spine tend to involve not only the vertebral bodies but also the pedicles and neural arches. Destruction of one or more pedicles may be the earliest sign of metastatic disease and may aid in differentiating this process from multiple myeloma, in which the pedicles are much less often involved. Because cartilage is resistant to invasion by metastases, preservation of the intervertebral disk space may help to distinguish metastases from an inflammatory process. Pathologic collapse of vertebral bodies frequently occurs in advanced disease.

Osteoblastic metastases are generally considered to be evidence of slow growth in a neoplasm that has allowed time for reactive bone proliferation. In males, osteoblastic metastases are usually secondary to carcinoma of the prostate gland; carcinoma of the breast is the most common primary site of osteoblastic metastases in females. Sclerotic metastases may also be due to carcinomas of the gastrointestinal tract, lung, and urinary bladder. Osteoblastic metastases initially appear as ill-defined areas of increased density that may progress to complete loss of normal bony architecture. They may vary from small, isolated round foci of sclerotic density to a diffuse sclerosis involving most or all of a bone. In the spine, this may produce the characteristic uniform density of an "ivory" vertebral body. Ivory vertebrae may also be seen in Hodgkin's disease, in which there may be associated erosion or scalloping of the anterior margin of the vertebral body due to lymphadenopathy, and in Paget's disease, in which the vertebral body is not only sclerotic but also markedly enlarged.

The combination of destruction and sclerosis in the mixed type of metastasis causes the affected bone to have a mottled appearance, with intermixed areas of lucency and increased density.

Primary or metastatic soft tissue malignant tumors may spread to adjacent bones by direct extension. These lesions initially cause cortical erosion and destruction that may be followed by lytic changes in the medullary cavity.

CT can be employed to evaluate specific symptomatic and clinically suspicious areas after conventional modalities have failed to demonstrate or adequately characterize a lesion. Destruction of trabecular bone or replacement of marrow fat by neoplastic tissue results in an alteration of the attenuation value (CT number), which permits detection of small lesions too subtle to be recognized with other imaging modalities. In patients with known neoplastic disease who have a nonspecific focal abnormality on the radionuclide scan, demonstration of a characteristic lesion on CT scans may permit the institution of palliative radiotherapy even without a proven tissue diagnosis.[51–53]

## REFERENCES

1. Anderson RL, Popowitz L, Li JKH: An unusual sarcoma arising in a solitary osteochondroma. *J Bone Joint Surg* 51A:1199–1204, 1969.
2. Levy WM, Aegerter EE, Kirkpatrick JA: The nature of cartilaginous tumors. *Radiol Clin North Am* 2:327–336, 1964.

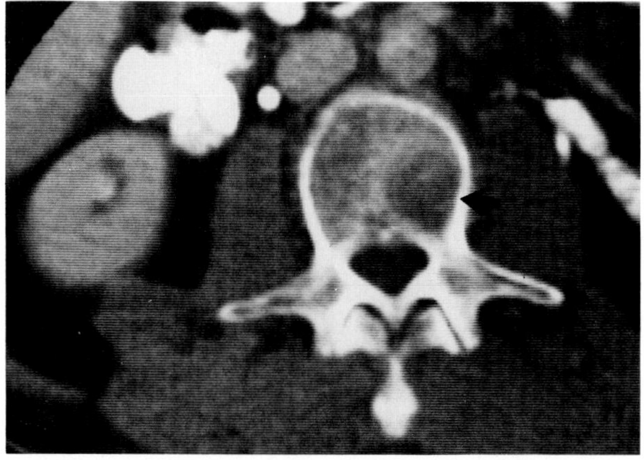

A

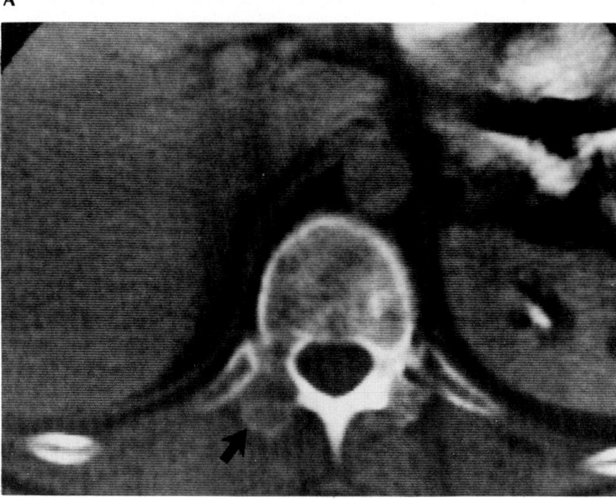

B

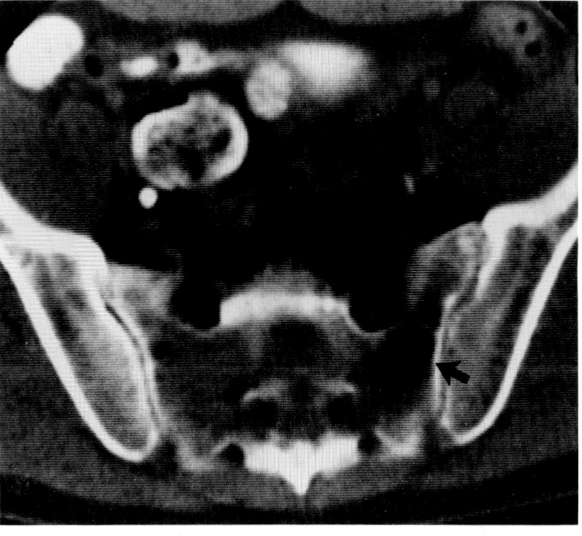

C

**Figure 60.55. CT demonstration of metastases to bone.** Examples of destructive lesions in several vertebrae and the sacrum (arrows), typical of metastatic disease, are seen in a patient in whom lumbar spine radiographs were considered equivocal. (From Mall,[53] with permission from the publisher.)


3. Hamlin JA, Adler L, Greenbaum EI: Central enchondroma: A precursor to chondrosarcoma. *J Can Assoc Radiol* 22:206–209, 1971.
4. Mainzer F, Minagi H, Steinbach HL: The variable manifestations of multiple enchondromatosis. *Radiology* 99:377–388, 1971.
5. Lewis J, Ketcham AS: Maffucci's syndrome: Functional and neoplastic significance. *J Bone Joint Surg* 55A:1465–1479, 1973.
6. McInerney DP, Middlemiss JH: Giant-cell tumor of bone. *Skel Radiol* 2:195–204, 1978.
7. Jacobs P: The diagnosis of osteoclastoma (giant-cell tumor): A radiological and pathological correlation. *Brit J Radiol* 45:121–136, 1972
8. Cohen J: Etiology of simple bone cyst. *J Bone Joint Surg* 52A:1493–1497, 1970.
9. Lodwick GS: Juvenile unicameral bone cyst. *AJR* 80:495–504, 1958.
10. Reynolds J: The "fallen fragment sign" in the diagnosis of unicameral bone cyst. *Radiology* 92:949–953, 1969.
11. Dahlin DC, Besse BE, Pugh DG, et al: Aneurysmal bone cysts. *Radiology* 64:56–65, 1955.
12. Bonakdarpour A, Levy WM, Aegerter E: Primary and secondary aneurysmal bone cyst. *Radiology* 126:75–83, 1978.
13. Blank N, Lieber A: The significance of growing bone islands. *Radiology* 85:508–511, 1965.
14. Onitsuka H: Roentgenologic aspects of bone islands. *Radiology* 123:607–612, 1977.
15. Resnick D, Nemcek AA, Haghighi P: Spinal enostoses (bone islands). *Radiology* 147:373–376, 1983.
16. Dolan KD, Seibert J, Seibert RW: Gardner's syndrome. *AJR* 119:359–364, 1975.
17. Swee RG, McLeod RA, Beabout JW: Osteoid osteoma. *Radiology* 130:117–123, 1979.
18. Freiberger RH, Loitman BS, Helpern M, et al: Osteoid osteoma. *AJR* 82:194–205, 1959.
19. Smith FW, Gilday DL: Scintigraphic appearances of osteoid osteoma. *Radiology* 137:191–195, 1980.
20. McLeod RA, Dahlin DC, Beabout JW: The spectrum of osteoblastoma. *AJR* 126:321–335, 1976.
21. Lichtenstein L, Sawyer WR: Benign osteoblastoma. *J Bone Joint Surg* 46A: 755–765, 1964.
22. Prentice AID: Variations on the fibrous cortical defect. *Clin Radiol* 25:531–533, 1974.
23. Purcel WM, Mulcahy F: Nonosteogenic fibroma of bone. *Clin Radiol* 11:51–58, 1960.
24. Feldman F, Hecht HL, Johnston AD: Chondromyxoid fibroma of bone. *Radiology* 94:249–260, 1970.
25. Turcotte B, Pugh DG, Dahlin DC: The roentgenologic aspects of chondromyxoid fibroma of bone. *AJR* 87:1085–1095, 1962.
26. McLeod RA, Beabout JW: The roentgenographic features of chondroblastoma. *AJR* 118:464–471, 1973.
27. Nolan DJ, Middlemiss H: Chondroblastoma of bone. *Clin Rad* 26:343–350, 1975.
28. Hudson TM, Hawkins IF: Radiological evaluation of chondroblastoma. *Radiology* 139:1–10, 1981.
29. Sherman RS, Wilner D: Roentgen diagnosis of hemangioma of bone. *AJR* 86:1146–1159, 1961.
30. Lee ES: Osteosarcoma: A reconnaissance. *Clin Radiol* 26:5–25, 1975.

31. de Santos LA, Rosengren JE, Wooten WB, et al: Osteogenic sarcoma after the age of 50: A radiographic evaluation. *AJR* 131:481–484, 1978.
32. Dahlin DC, Coventry MB: Osteogenic sarcoma. *J Bone Joint Surg* 49A:101–110, 1967.
33. de Santos LA, Bernardino ME Murray JA: Computer tomography in the evaluation of osteosarcoma. *AJR* 132:535–540, 1979.
34. Destouet JM, Gilula LA, Murphy WA: Computed tomography of long-bone osteosarcoma. *Radiology* 133:439–445, 1979.
35. Cavanagh RC: Tumors of the soft tissues of the extremities. *Semin Roentgenol* 8:83–89, 1973.
36. Edeiken J, Farrel C, Ackerman LV, et al: Parosteal sarcoma. *AJR* 111:579–583, 1971.
37. Smith J, Ahuja SC, Huros AG, et al: Parosteal (juxtacortical) osteogenic sarcoma, *J Can Assoc Radiol* 29:167–174, 1978.
38. Kenney PJ, Gilula LA, Murphy WA: The use of computed tomography to distinguish osteochondroma and chondrosarcoma. *Radiology* 139:129–137, 1981.
39. Reiter FB, Ackerman LV, Staple TW: Central chondrosarcoma of the appendicular skeleton. *Radiology* 105:525–530, 1972.
40. Henderson ED, Dahlin DC: Chondrosarcoma of bone. *J Bone Joint Surg* 45A:1450–1458, 1963.
41. Rosenthal DI, Schiller AL, Mankin HJ: Chondrosarcoma: Correlation of radiological and histological grade. *Radiology* 150:21–26, 1984.
42. Larson SE, Lorentzon R, Roquist L: Fibrosarcoma of bone. *J Bone Joint Surg* 58B:412–417, 1976.
43. Eyre-Brook AL, Price CHG: Fibrosarcoma of bone. *J Bone Joint Surg* 51B:20–37, 1969.
44. Sherman RS, Soong KY: Ewing's sarcoma: Its roentgen classification and diagnosis. *Radiology* 66:529–539, 1956.
45. Dahlin DC, Coventry MB, Scanlon PW: Ewing's sarcoma. *J Bone Joint Surg* 43A:185–192, 1961.
46. Vohra VG: Roentgen manifestations in Ewing's sarcoma. *Cancer* 20:727–735, 1967.
47. Ridings GR: Ewing's tumor. *Radiol Clin North Am* 2:315–325, 1964.
48. Ivins JC, Dahlin DC: Reticulum cell sarcoma of bone. *J Bone Joint Surg* 35A:835–842, 1953.
49. Sherman RS, Snyder RE: The roentgen appearance of primary reticulum cell sarcoma of bone. *AJR* 58:291–306, 1947.
50. Dolan PA: Reticulum cell sarcoma of bone. *AJR* 87:121–127, 1962.
51. Krishnamurthy GT, Tubis M, Hiss J, et al: Distribution pattern of metastatic bone disease. *JAMA* 237:2504–2506, 1977.
52. Mulvey RB: Peripheral bone metastases. *AJR* 91:155–160, 1964.
53. Mall J: Skeletal metastases, in Eisenberg RL, Amberg JR (eds): *Critical Diagnostic Pathways in Radiology: An Algorithmic Approach*. Philadelphia, JB Lippincott, 1981, pp. 367–380.
54. Eisenberg, RL: *Atlas of Signs in Radiology*. Philadelphia, JB Lippincott, 1984.
55. Beabout JW, McLeod RA, Dahlin DC: Benign tumors. *Semin Roentgenol* 14:33–43, 1979.
56. Edeiken J: *Roentgen Diagnosis of Disease of Bone*. Baltimore, Williams & Wilkins, 1981.

## Chapter Sixty-One
# Dysplasias and Chondrodystrophies

### FIBROUS DYSPLASIA

Fibrous dysplasia is a disorder that usually begins during childhood and is characterized by the proliferation of fibrous tissue within the medullary cavity. The disease may be confined to a single bone (monostotic) or the bones of one extremity, or it may be widely distributed throughout the skeleton (polyostotic). Patients with polyostotic fibrous dysplasia frequently have a unilateral predominance, often with localized pigmentations (café au lait spots) that tend to be on the same side as the bone lesions and have an irregular outline ("coast-of-Maine") in contrast to the smoothly marginated ("coast-of-California") pigmented macules seen in neurofibromatosis. About one-third of females with the polyostotic form of fibrous dysplasia also demonstrate precocious puberty (Albright's syndrome); sexual precocity is very rare in males.

The monostotic form of fibrous dysplasia primarily involves the long bones (especially the femur and tibia), ribs, and facial bones. Fibrous replacement of the medullary cavity typically produces a well-defined radiolucent area, which may vary from completely radiolucent to a homogeneous ground-glass density, depending on the amount of fibrous or osseous tissue deposited in the medullary cavity. Irregular bands of sclerosis may cross the cystlike lesion, giving it a multilocular appearance. The bone is often locally expanded, and the cortex may be eroded from within, predisposing to pathologic fractures. In severe and long-standing disease, affected bones may be bowed or deformed (e.g., the characteristic shepherd's crook deformity of the femur).

Skull lesions of fibrous dysplasia may be either lucent or sclerotic. Involvement of the base of the skull produces

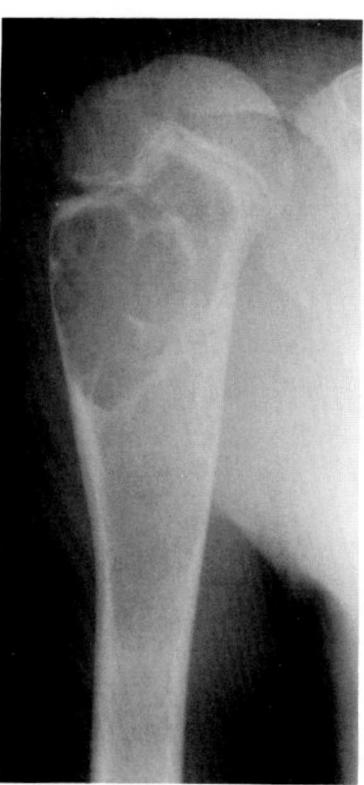

**Figure 61.1. Fibrous dysplasia of the humerus.** The typical well-defined, expansile area of radiolucency contains bands of sclerosis that produce a multilocular appearance.

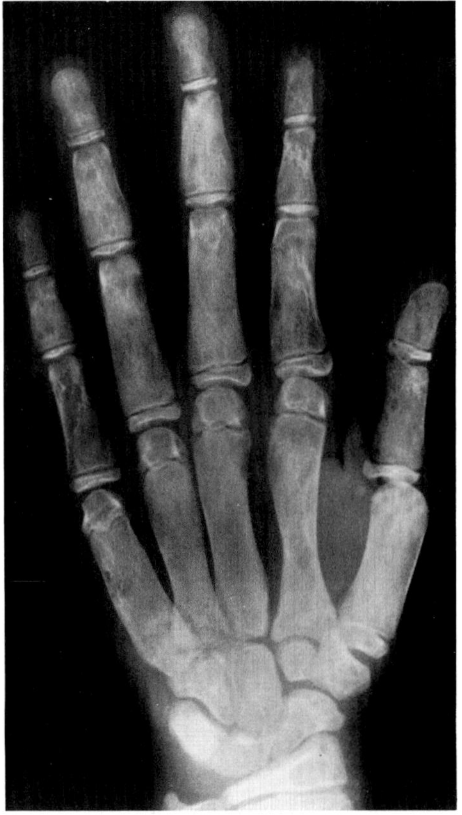


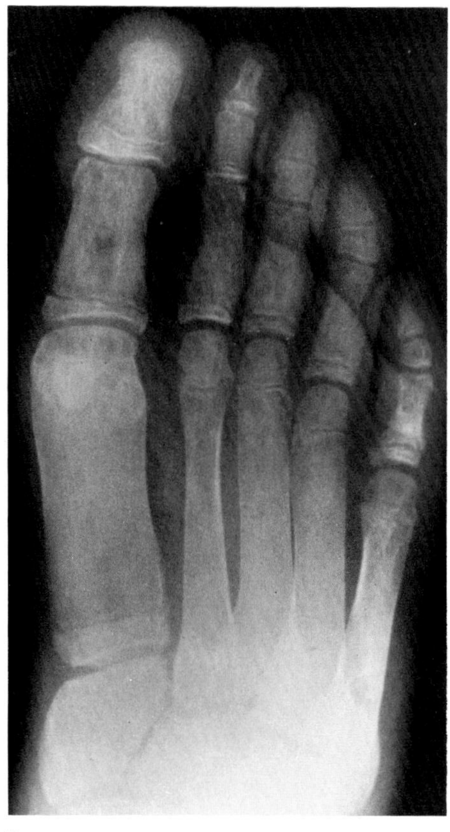


**Figure 61.2. Polyostotic fibrous dysplasia.** Films of (A) the hand and (B) the foot show a smudgy, ground-glass appearance of the medullary cavities with failure of normal modeling.

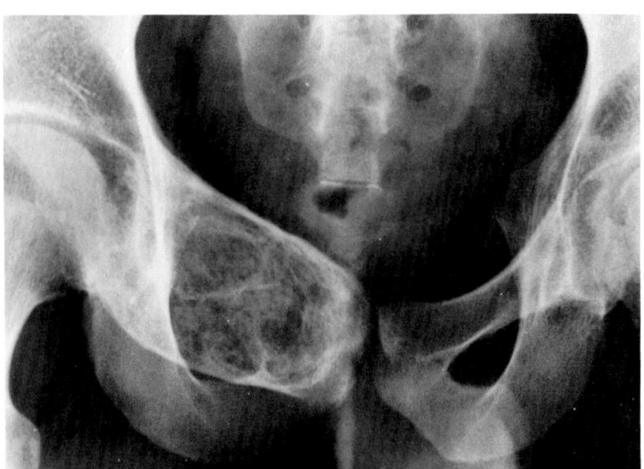

**Figure 61.3. Fibrous dysplasia of the ischium.** An expansile lesion of the superior ramus contains irregular bands of sclerosis that give it a multilocular appearance.

a sclerotic ground-glass appearance. Disease of the facial bones causes marked sclerosis and thickening, often with obliteration of the sinuses and orbits, that creates a leonine appearance (leontiasis ossea). In the calvarium, there may be multiple irregular areas of lucency with expansion of the outer table of the skull and only minimal involvement of the inner table.

Fibrous dysplasia of a rib characteristically produces an expansile lesion with a ground-glass or soap-bubble appearance; indeed, the most common cause of an expansile focal rib lesion is fibrous dysplasia.

Although reported, malignant degeneration in association with fibrous dysplasia is extremely rare.[1–3, 30]

## EPIPHYSEAL DYSPLASIAS

### Spondyloepiphyseal Dysplasia

Spondyloepiphyseal dysplasia is a rare hereditary type of dwarfism in which growth of both the extremities and the vertebrae is affected. In the congenital form, there is a general delay in ossification of the skeleton. Radiographic changes include flattening of the vertebral bodies (platyspondyly), horizontal flattening of the roof of the acetabulum with small acetabular and iliac angles, and flattening of the femoral capital epiphyses with a severe varus deformity of the femoral neck. In the noncongenital type (spondyloepiphyseal dysplasia tarda), which is transmitted as an X-linked recessive and affects only males, the failure of normal growth does not become evident until later in childhood. The disorder may not be recognized until adulthood, when the patient experiences precocious degenerative joint disease. Radiographs typ-

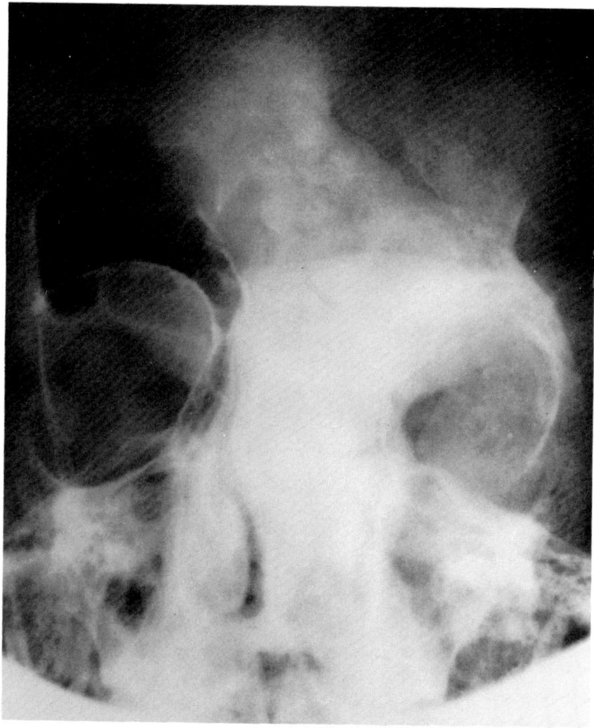

**Figure 61.4. Fibrous dysplasia of the skull and facial bones.** Widespread involvement of the left orbital region extends to adjacent bones. (From Tchang,[30] with permission from the publisher.)

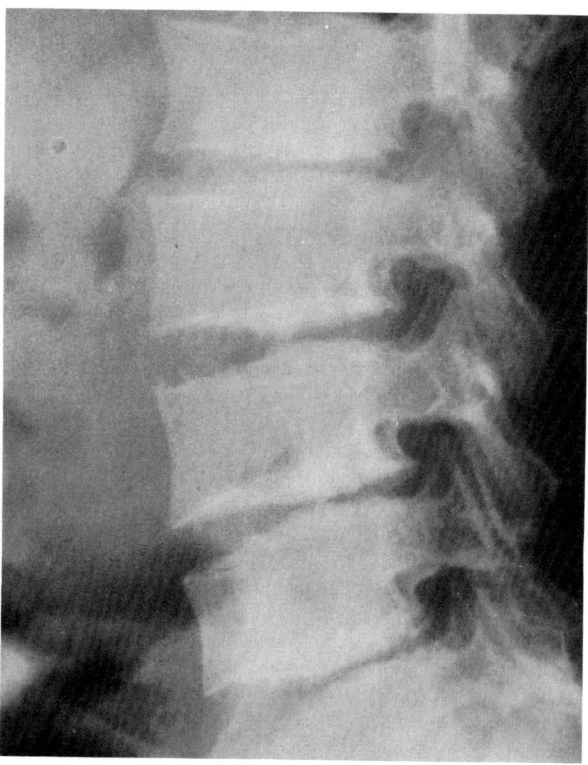

**Figure 61.6. Spondyloepiphyseal dysplasia involving the spine.** There is generalized flattening of the vertebral bodies (platyspondyly).

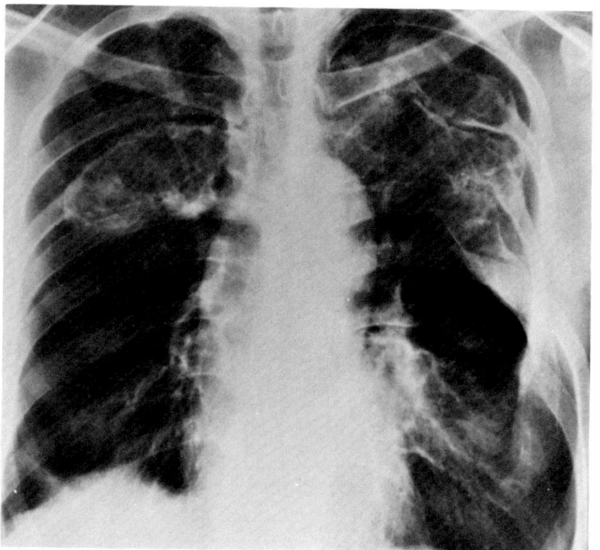

**Figure 61.5. Fibrous dysplasia of multiple ribs.** The involved ribs are expanded and have a ground-glass appearance.

ically demonstrate a small pelvis with bilateral hip osteoarthritis. The lumbar vertebral bodies are flattened, and there is often a distinctive hump-shaped mound of bone on their superior and inferior surfaces.[4–6]

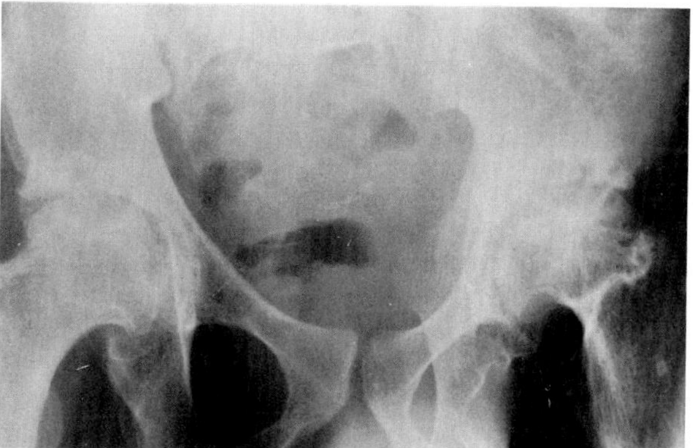

**Figure 61.7. Spondyloepiphyseal dysplasia.** Degenerative changes in the hips, especially severe on the left, are seen in a 27-year-old.

## Multiple Epiphyseal Dysplasia

Multiple epiphyseal dysplasia is characterized by bilateral, symmetric epiphyseal irregularity and hypoplasia without sclerosis. The most frequently involved sites are the hips, knees, and ankles. Deficient growth of the lateral aspect of the distal tibial epiphysis commonly produces a tibiotalar slant. The small bones of the hands and feet are short and broad, though their relative lengths

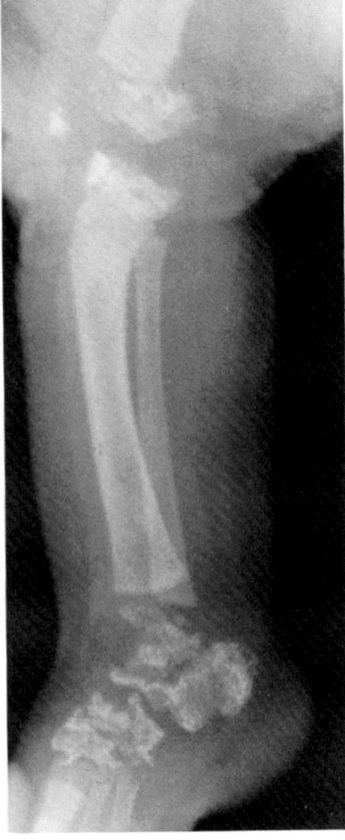

**Figure 61.8. Multiple epiphyseal dysplasia.** There is hypoplasia and irregularity of multiple epiphyseal centers.

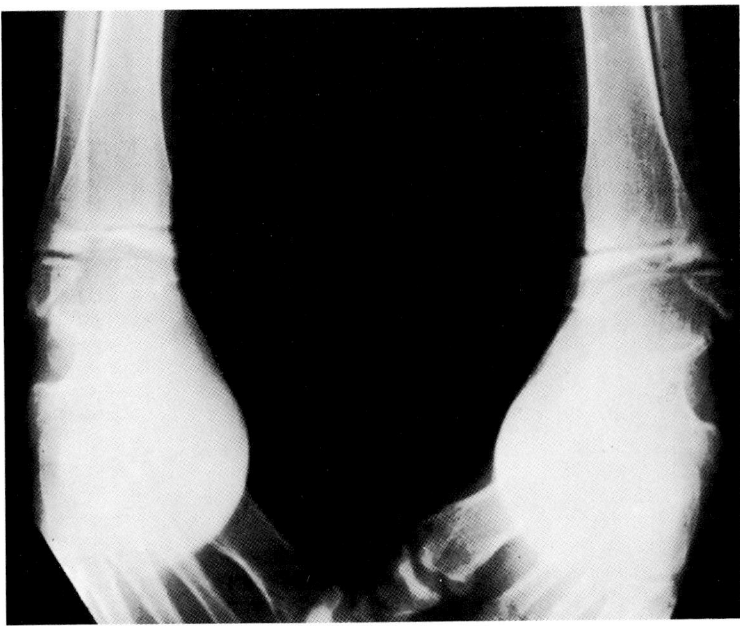

**Figure 61.9. Multiple epiphyseal dysplasia with tibiotalar angulation.** There is bilateral and symmetric irregularity of the distal tibial epiphyses, with deficiency of the ossification centers at the lateral aspects. (From Greenfield,[31] with permission from the publisher.)

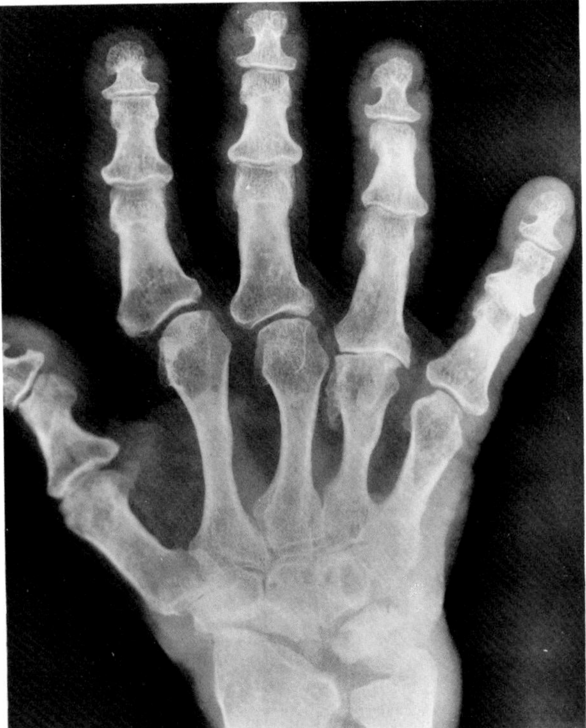

**Figure 61.10. Multiple epiphyseal dysplasia.** The small bones of the hand are short and broad, though their relative lengths are maintained.

are maintained. In children, there is often mild dwarfism with pain, stiffness, and flexion deformities of the hips and knees; in adults, the epiphyseal deformities may result in early and severe degenerative arthritis.[7,8]

## Dysplasia Epiphysialis Hemimelica

Dysplasia epiphysialis hemimelica is a rare disorder in which an eccentric cartilaginous overgrowth is limited to one-half of an epiphysis. The most frequent sites of involvement are the talus, the distal femur, and the distal tibia, though the small bones of the hands and feet may also be affected. The eccentric cartilaginous overgrowth may present as an exostosis or may cause deformity and functional impairment of an involved joint.[9,10]

## Chondrodystrophia Calcificans Congenita (Stippled Epiphyses)

Chondrodystrophia calcificans congenita is a condition characterized by multiple small punctate calcifications of varying size that occur in the respiratory cartilages and in the epiphyses before the normal time for appearance of ossification centers. In severe cases, the calcifications extend beyond the joints and into adjacent soft tissues. The most commonly involved sites are the hips, knees, shoulders, and wrists; affected bones may

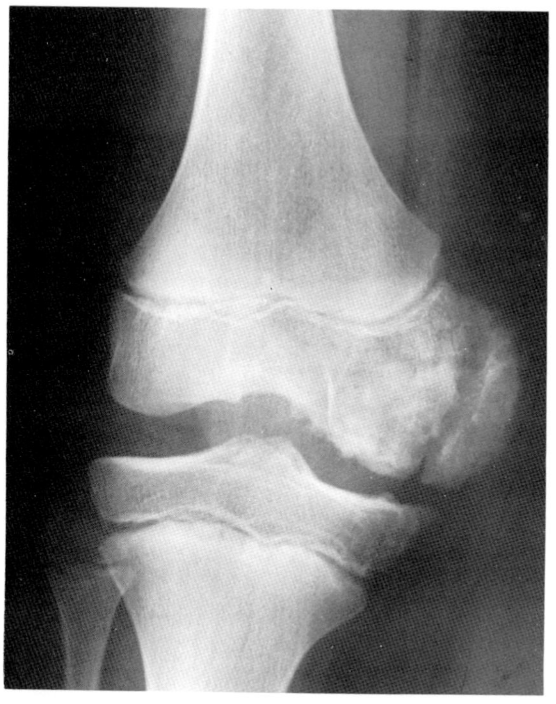

A

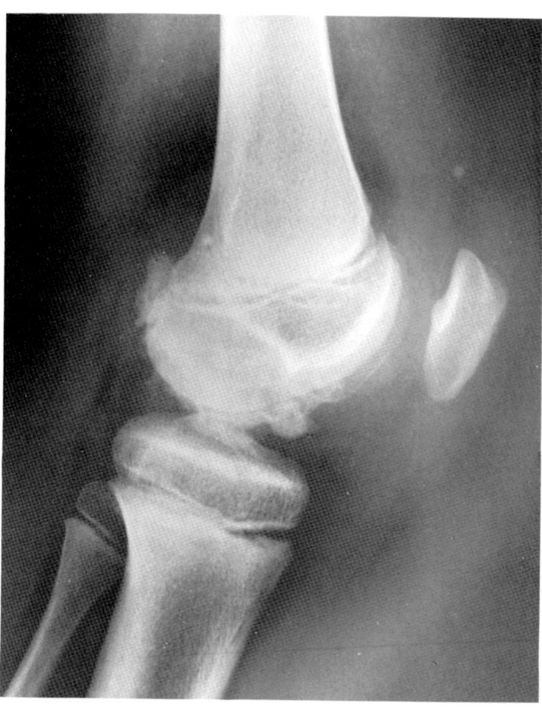

B

**Figure 61.11. Dysplasia epiphysialis hemimelica.** (A) Frontal and (B) lateral views of the knee show asymmetric enlargement of the medial half of the distal femoral epiphysis with irregularity of the medial aspect of the tibial plateau. (From Edeiken,[32] with permission from the publisher.)

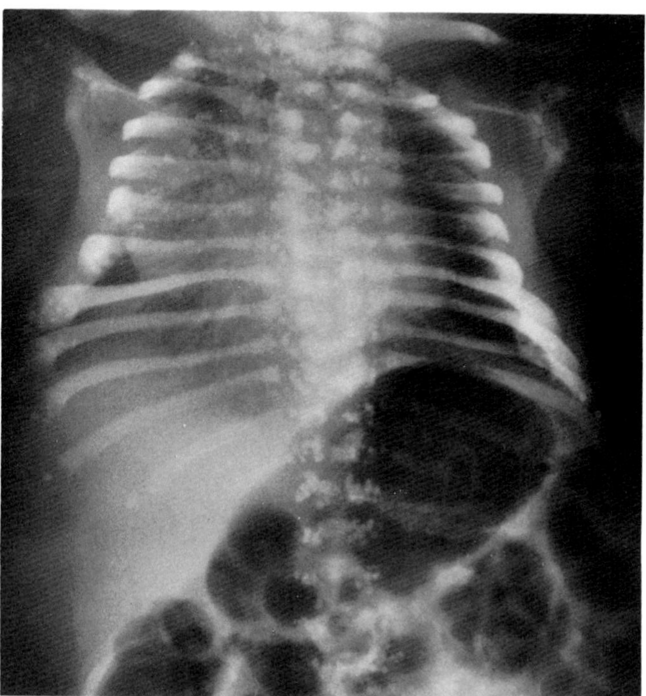

A

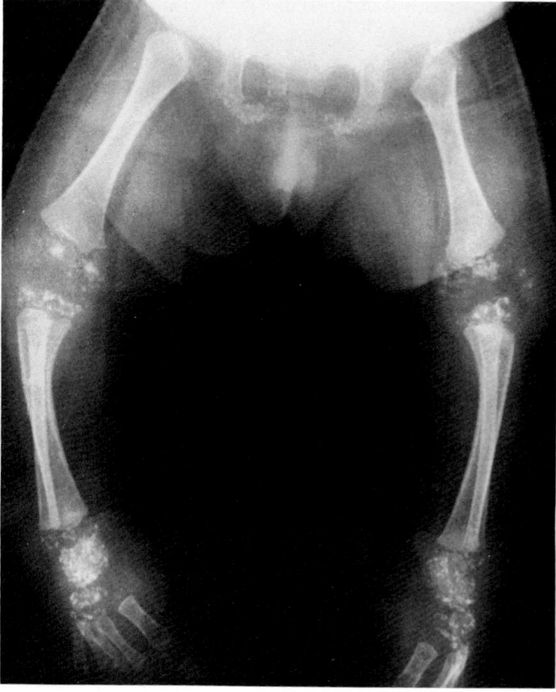

B

**Figure 61.12. Chondrodystrophia calcificans congenita.** Multiple small punctate calcifications of varying size involve virtually all the epiphyses on views of (A) the chest and upper abdomen and (B) the lower extremities.

be shortened, or the process may regress and leave no deformity. The densities may disappear by the age of 3 years or may gradually increase in size and coalesce to form a normal-appearing single ossification center.[11,12]

## PHYSEAL (GROWTH PLATE) DYSPLASIAS

### Achondroplasia

This hereditary bone dysplasia, the most common form of dwarfism, is the result of diminished proliferation of cartilage in the growth plate (decreased enchondral bone formation). Because membranous bone formation is not affected, the individual has short limbs that contrast with the nearly normal length of the trunk. Other characteristic physical features include a large head with frontal bulging, saddlenose, a prognathous jaw, and prominent buttocks that give the false impression of lumbar lordosis.

The radiographic hallmark of achondroplasia is symmetric shortening of all long bones. The proximal segments of extremities are usually more severely involved than the distal portions, so that the humerus and femur are disproportionately shorter than the radius and tibia (rhizomelia). Continued appositional growth at the metaphyses causes the ends of the long bones to be splayed. However, appositional growth at bone ends is often deficient in the center of the metaphyseal region, producing a V-shaped notch surrounding and partially burying the epiphyseal center ("ball and socket" epiphysis). The bones of the hand are short and thickened, and the fingers tend to be of the same length. Separation of the ring and middle fingers results in a "trident hand."

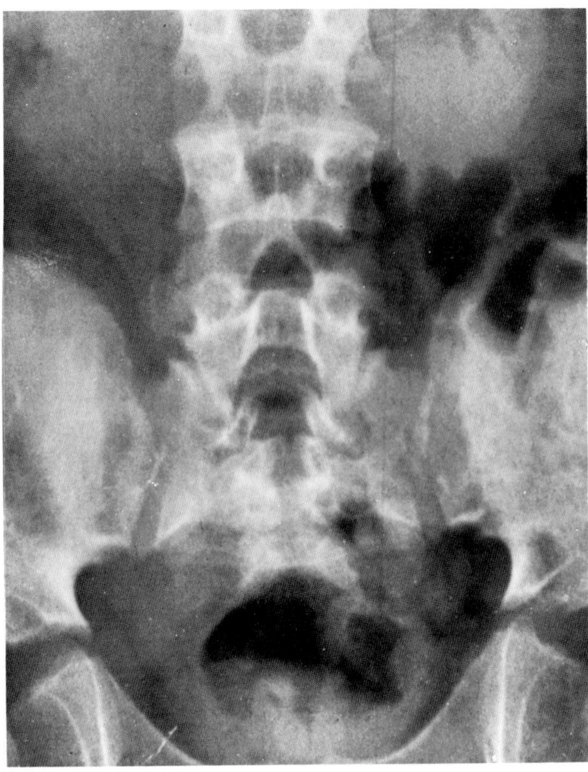

**Figure 61.14. Achondroplasia.** A frontal view of the lumbar spine shows progressive narrowing of the interpedicular spaces from above downward, the opposite of normal. Note the short, broad pelvis with small sacrosciatic notch. (From Edeiken,[32] with permission from the publisher.)

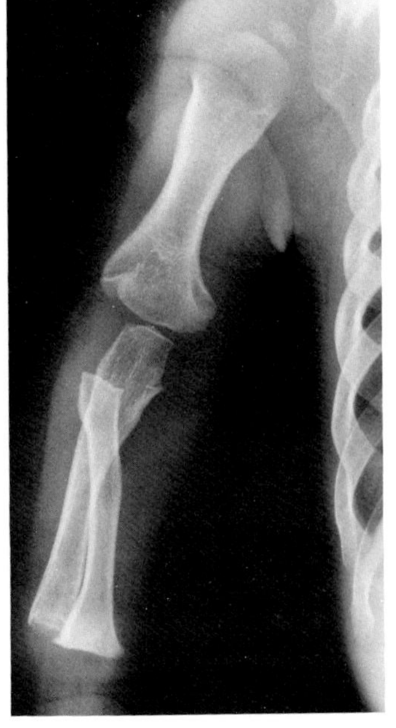

**Figure 61.13. Achondroplasia.** A frontal view of the right arm shows the humerus to be disproportionately shorter than the radius and ulna (rhizomelia).

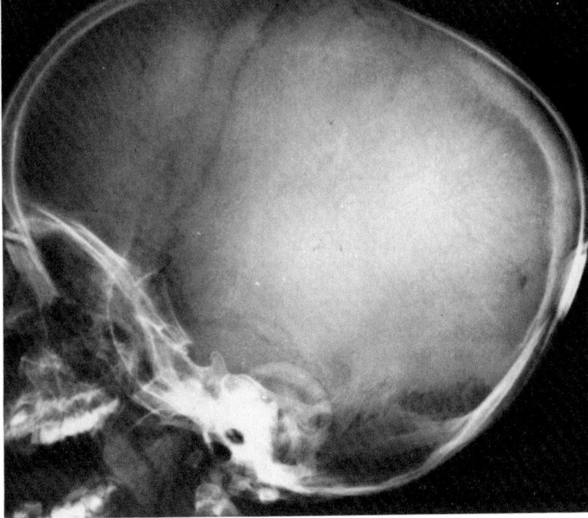

**Figure 61.15 Achondroplasia.** Enlargement of the skull with frontal bulging is the result of communicating hydrocephalus.

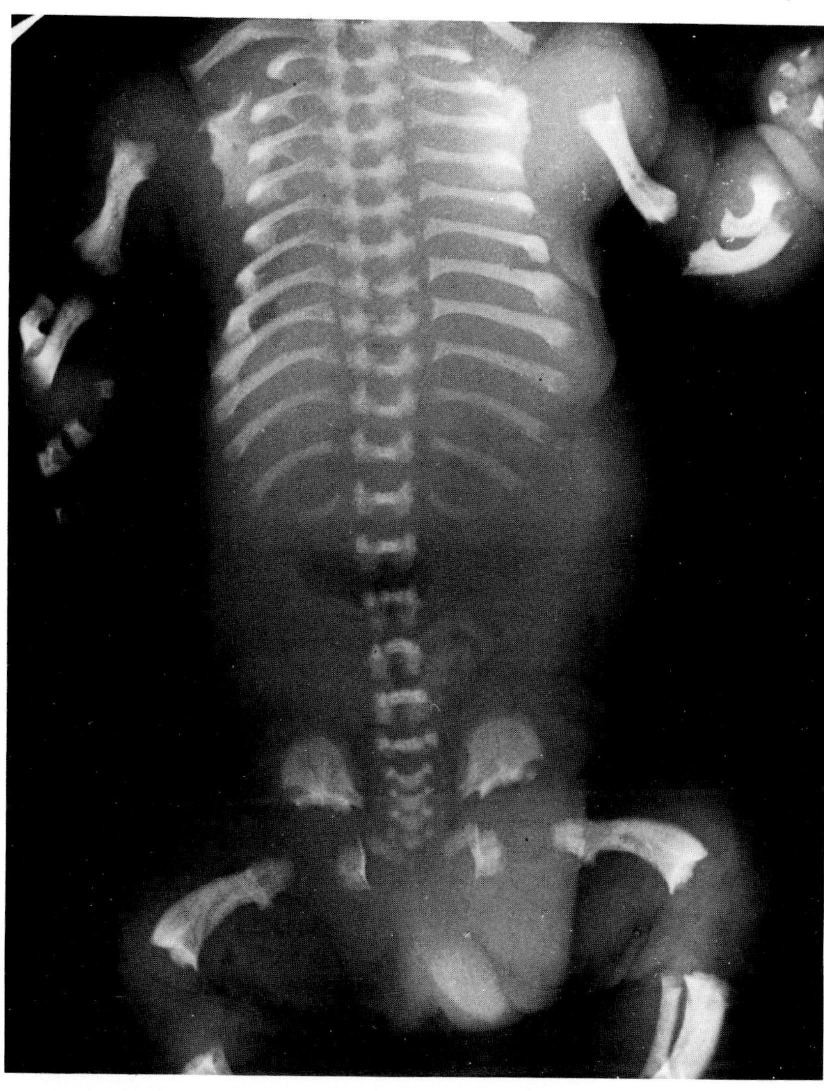

**Figure 61.16. Thanatophoric dwarfism.** There is symmetric shortening and bowing of long bones with deformity of the thoracic cage and flattening of vertebral bodies. The caudal portions of the pelvic bones are deficient. (From Edeiken,[32] with permission from the publisher.)

In the lumbar spine, the interpedicular distances narrow progressively from above downward, the opposite of normal. Narrowing of the spinal canal in both dimensions may lead to neurologic signs of cord compression and nerve root encroachment, especially in individuals with pronounced kyphoscoliosis. Anterior wedging of lumbar vertebral bodies and scalloping of their posterior margins may also be seen. The pelvis is typically short and broad, the ilia short and square, and the acetabular angles decreased.

The skull is enlarged in achondroplasia, usually because of communicating hydrocephalus due to deficient brain space in the posterior fossa caused by early closure of the intersphenoid and sphenoid-occipital synchondroses. Upward brain growth causes frontal bulging; the relatively larger size of the normally growing mandible gives the appearance of prognathism.

Thanatophoric dwarfism, which represents either a severe type of achondroplasia or a separate entity, is a cause of stillbirth or neonatal death. Typical abnormalities in this condition include a narrow thorax, flattening of vertebral bodies, and a trilobed (cloverleaf) skull deformity caused by calvarial bulging in response to premature sutural fusion.[13–16]

## Chondroectodermal Dysplasia (Ellis–Van Creveld Syndrome)

Chondroectodermal dysplasia is a rare inherited syndrome that clinically consists of shortening of tubular bones, polydactyly, defects of hair, teeth, and nails, and congenital cardiac anomalies (especially atrial septal and ventricular septal defects). Shortening of long bones is typically more severe in the distal segments of the extremities (radius, ulna, tibia, fibula). This is in contrast to achondroplasia, in which the symmetric shortening of long bones is more pronounced in the proximal segments. Fusion of carpal bones, especially the hamate and capitate, is common.[17]

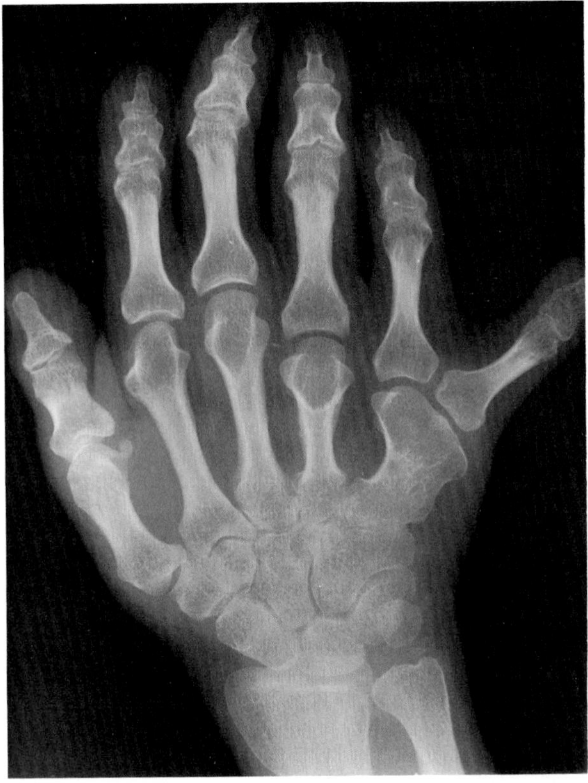

**Figure 61.17. Chondroectodermal dysplasia.** There are six digits (polydactyly) with shortening of the fourth and fifth metacarpals. The fifth metacarpal is fused with the supernumerary metacarpal.

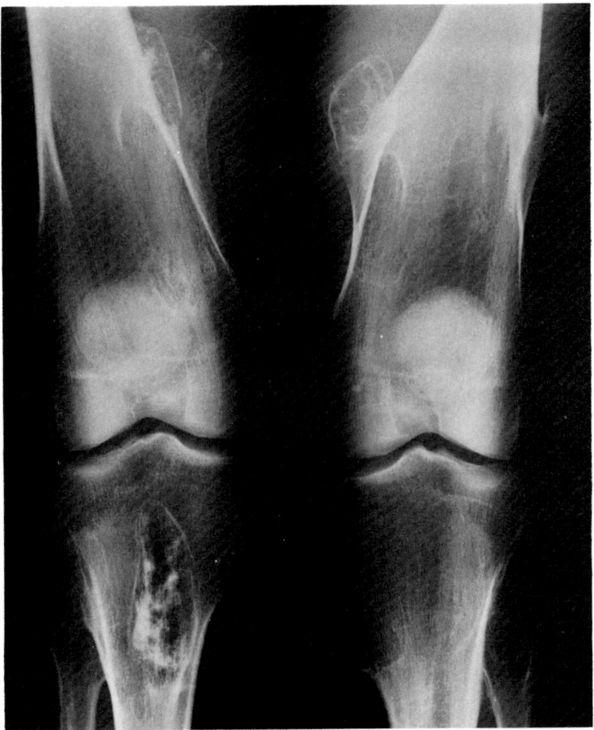

**Figure 61.18. Multiple exostoses.** There is bilateral involvement of the distal femurs and proximal tibias.

## METAPHYSEAL DYSPLASIAS

### Multiple Exostoses (Diaphyseal Aclasis)

Multiple exostoses represent an inherited bone dysplasia in which multiple osteochondromas arise from the ends of the shafts of any bones preformed in cartilage. The individual exostosis is a broad-based bony overgrowth, consisting of a cortical shell surrounding a core of cancellous bone, that arises from the metaphyseal cortex close to the epiphyseal line. The apex of the exostosis points away from the nearest joint, and the lesion is covered by a cartilaginous cap that may contain small punctate calcifications. Exostoses are most common at the sites of greatest growth, such as the knee, shoulder, elbow, and wrist. They are often asymptomatic, producing symptoms only if they interfere with the function of a joint or tendon or cause compression of a nerve. A frequent associated finding is deformity of the forearm due to shortening and bowing of the ulna.

Exostoses begin to grow in childhood and stop enlarging when the nearest epiphyseal center fuses. Sudden, painful enlargement of an exostosis long after growth should have ceased suggests sarcomatous degeneration.[18,19]

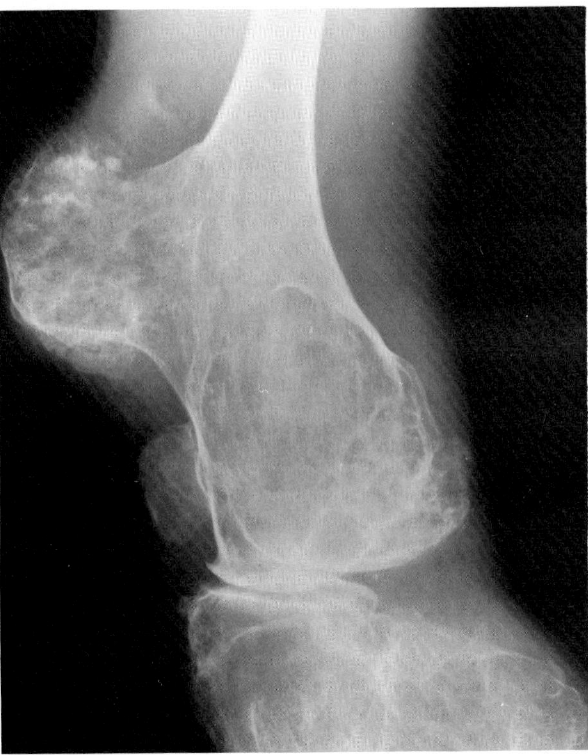

**Figure 61.19. Multiple exostoses about the knee.** This tangential view demonstrates why the exostosis of the distal femur presented as a hard mass on palpation.

## Craniometaphyseal Dysplasia

Craniometaphyseal dysplasia is a rare hereditary disorder in which failure of normal tubulation of bone is combined with hypertelorism, a broad flat nose, and defective dentition. There initially is metaphyseal lucency and diaphyseal sclerosis, which progresses to more normal mineralization and lack of modeling. In addition to the Erlenmeyer flask deformity, there may be sclerosis of the base of the skull and calvarium, lack of aeration of the paranasal sinuses and mastoids, and thickening and sclerosis of the mandible.[20,21]

**Figure 61.20. Multiple exostoses with sarcomatous degeneration.** A chondrosarcoma arising from one of the many exostoses in this patient appears as a large soft tissue mass with amorphous calcification.

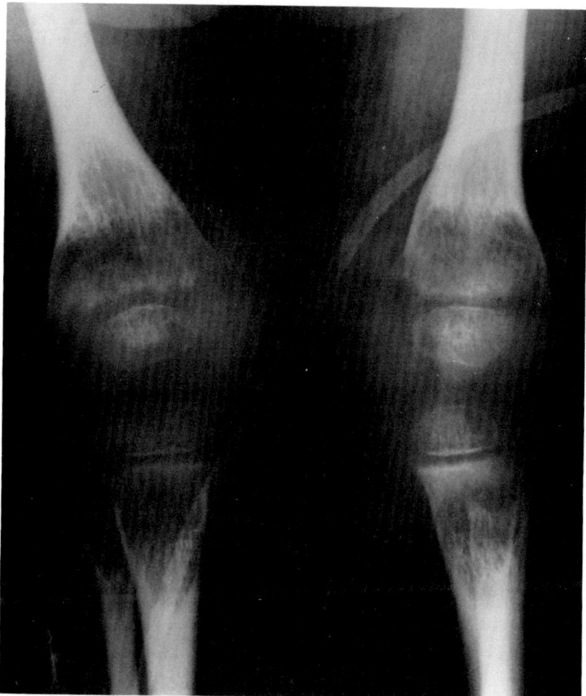

**Figure 61.21. Craniometaphyseal dysplasia.** There is severe metaphyseal lucency with diaphyseal sclerosis about the knees.

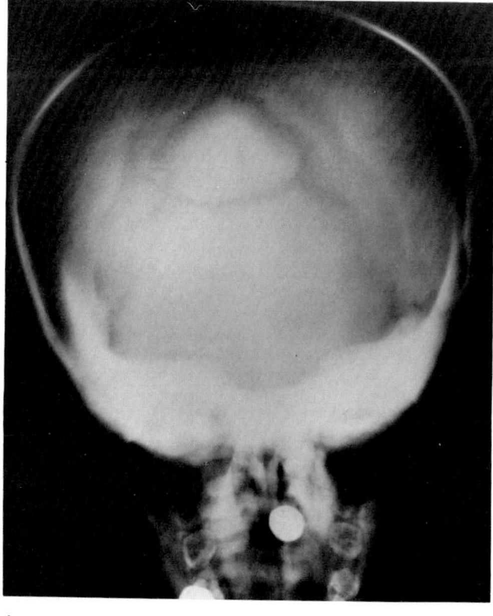

A

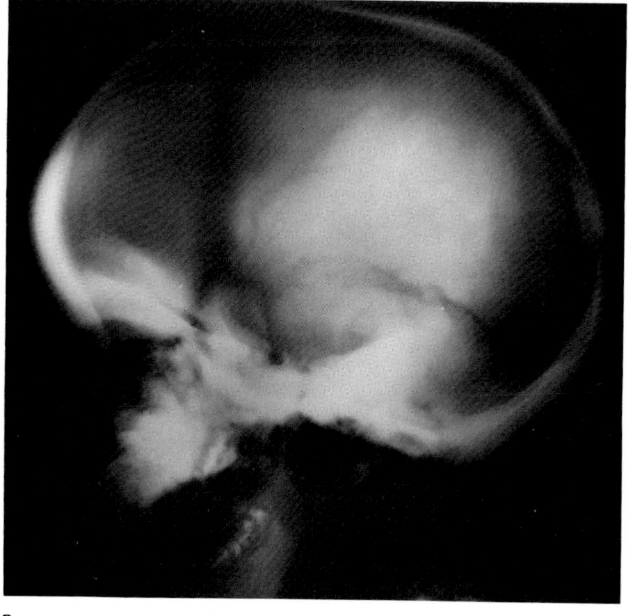

B

**Figure 61.22. Craniometaphyseal dysplasia.** (A) Towne's and (B) lateral views of the skull demonstrate dense sclerosis of the base of the skull and calvarium and lack of aeration of the paranasal sinuses. An unusual dense sutural bone is also noted.

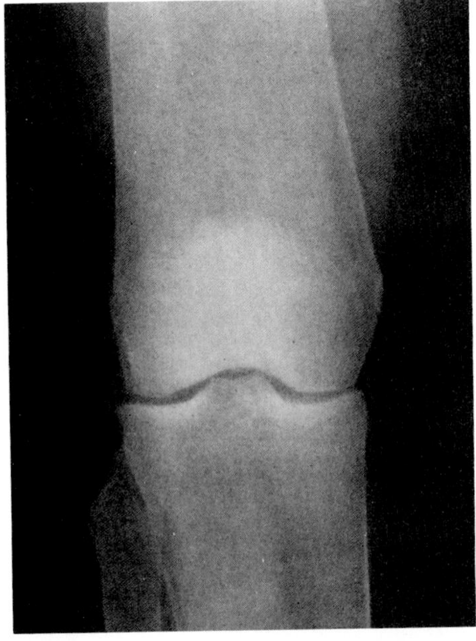


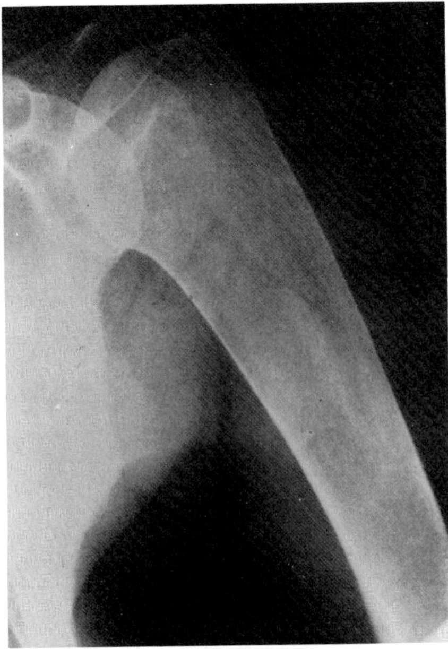


**Figure 61.23. Familial metaphyseal dysplasia.** Frontal views of (A) the knee and (B) the proximal humerus show defective modeling leading to extreme widening of the metaphyseal areas of the visualized long bones. The cortices are markedly thinned in the metaphyseal areas. (From Hermel, Gershon-Cohen, and Jones,[23] with permission from the publisher.)

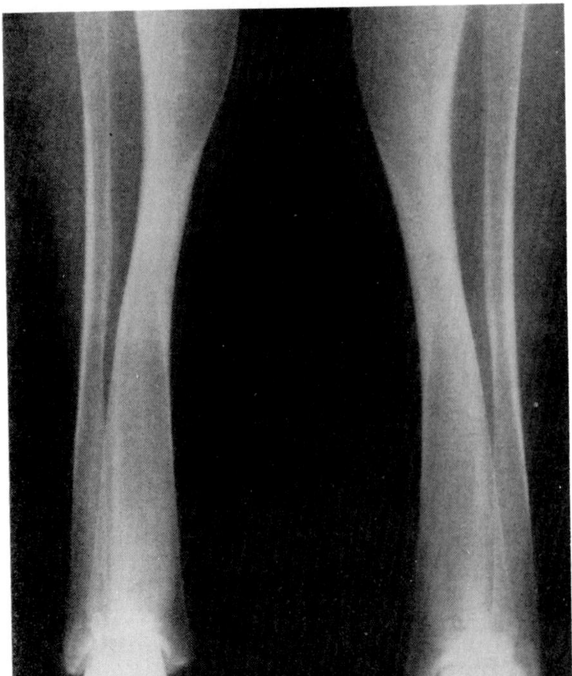

**Figure 61.24. Familial metaphyseal dysplasia.** A frontal view of both legs shows characteristic S-shaped bowing of the tibias. (From Hermel, Gershon-Cohen, and Jones,[23] with permission from the publisher.)

## Familial Metaphyseal Dysplasia (Pyle's Disease)

Familial metaphyseal dysplasia is a rare hereditary disorder characterized by symmetric paddle-shaped enlargement of the metaphyses and adjacent diaphyses of long bones (Erlenmeyer flask appearance). The thin cortices at the expanded ends of long bones make them prone to fracture.[22,23]

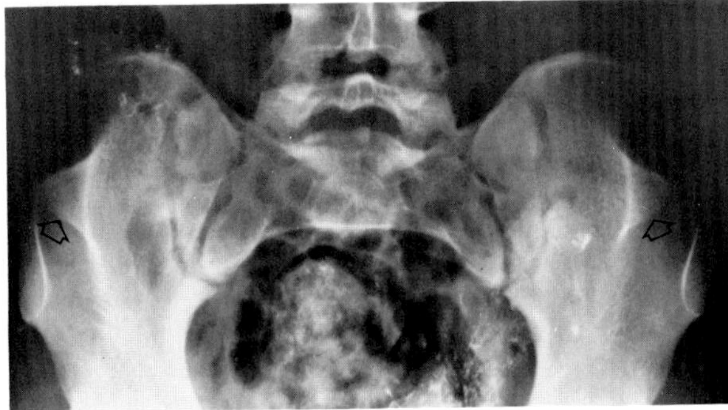

**Figure 61.25. Nail-patella syndrome.** Bilateral iliac horns (arrows).

# OTHER DYSPLASIAS

## Nail-Patella Syndrome (Fong's Disease)

The nail-patella syndrome is a hereditary condition, linked to the ABO blood groups, that is characterized by dystrophic nails, hypoplastic or absent patellas, iliac horns, elbow deformities, and renal disease. The almost pathognomonic iliac horns represent asymptomatic bilateral and symmetric cartilage-capped bony overgrowths that extend posteriorly from the iliac wings. Other skeletal abnormalities include flaring of the iliac crests and elbow deformities (increased radial carrying angle) due to hypoplasia of the radial head and capitellum.[24,25]

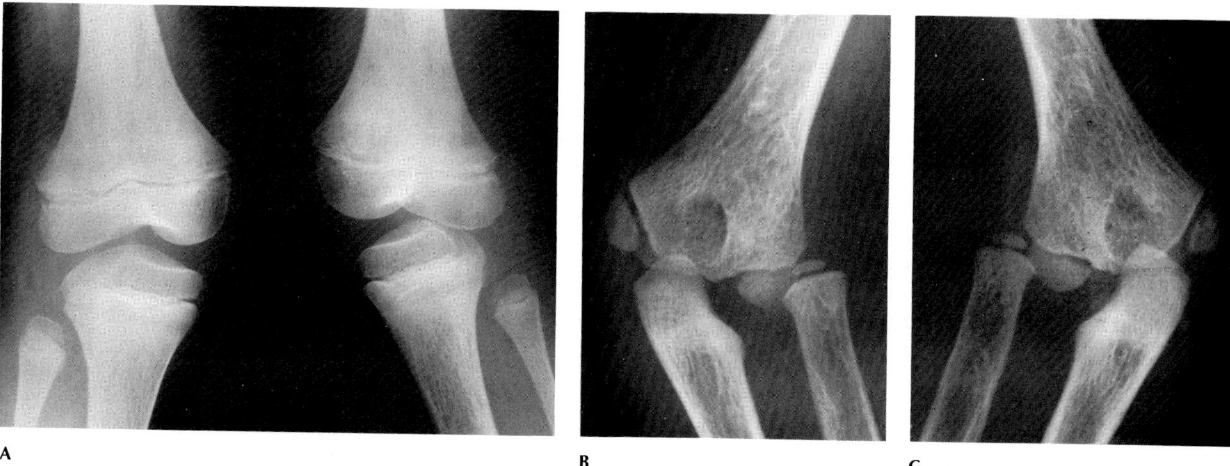

**Figure 61.26. Nail-patella syndrome.** (A) Frontal views of the knees show bilateral absence of the patellas. (B,C) Frontal views of both elbows show bilateral subluxations of the radial heads.

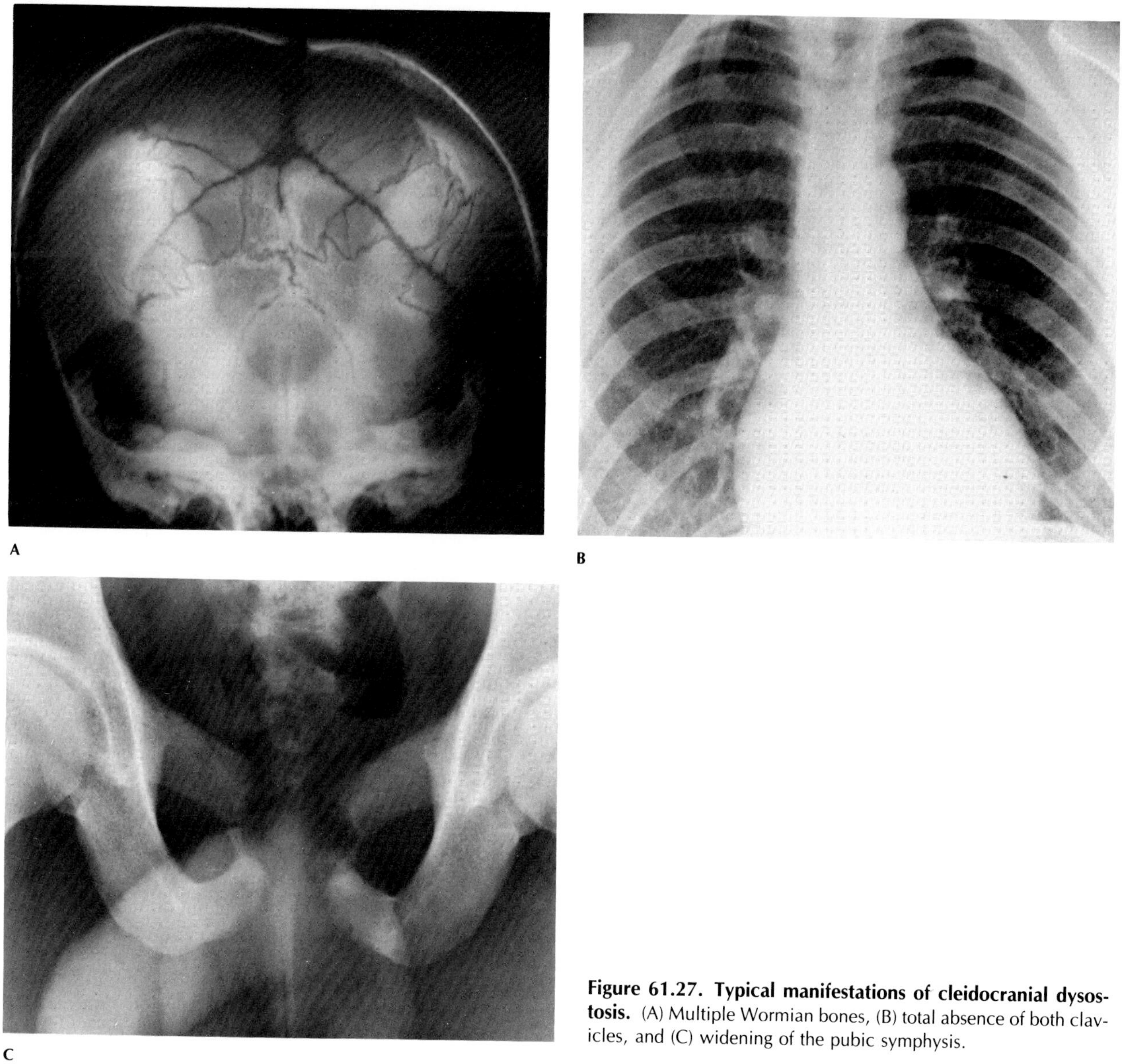

**Figure 61.27. Typical manifestations of cleidocranial dysostosis.** (A) Multiple Wormian bones, (B) total absence of both clavicles, and (C) widening of the pubic symphysis.

## Cleidocranial Dysostosis

Cleidocranial dysostosis is a congenital hereditary disorder, caused by faulty membranous bone formation, that principally involves the calvarium, clavicles, and pelvis. In the skull there is defective ossification of the calvarium, widening of sutures, persistence of the metopic suture, and the presence of multiple accessory bones along the sutures (Wormian bones). The radiographic hallmark of cleidocranial dysostosis is partial or total absence of the clavicles. Because the clavicle ossifies from three centers, any part may be absent and a variety of appearances may be produced. In the pelvis, the pubic symphysis is widened and there may be defective ossification of the ischia or pubis and hypoplasia of the iliac wings. Other associated skeletal abnormalities seen in cleidocranial dysostosis include incomplete closure of the vertebral neural arches, lateral notching of the proximal femoral epiphysis, multiple accessory ossification centers of the metacarpals and phalanges, excessive length of the second metacarpal, and pointing of the terminal tufts in the hands.[26,27]

## Osteogenesis Imperfecta

Osteogenesis imperfecta is an inherited generalized disorder of connective tissue involving bone, sclera, inner ear, teeth, skin, ligaments, tendons, and fascia. The major clinical features include blue sclerae, multiple frac-

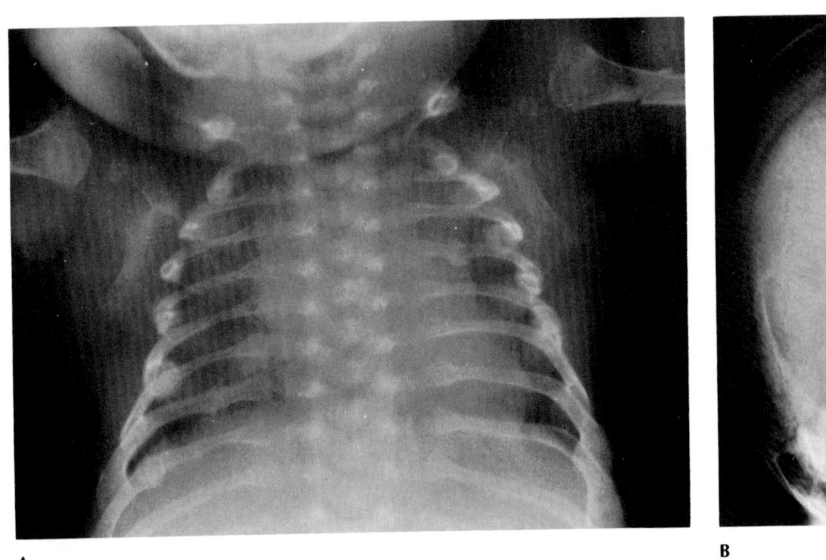

A

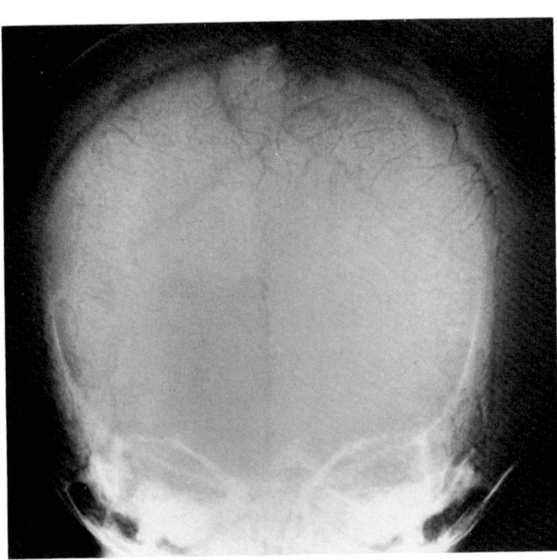

B

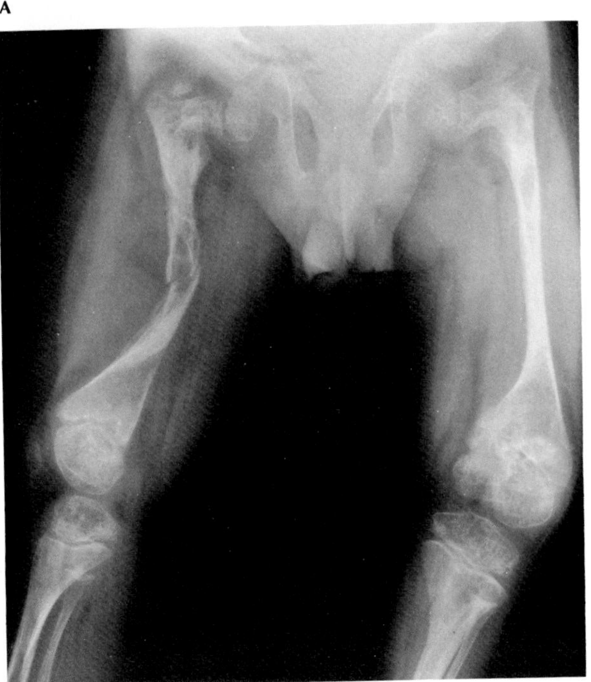

C

**Figure 61.28. Osteogenesis imperfecta.** (A) A chest film demonstrates multiple rib fractures with exuberant callus formation. There is an acute fracture of the left humerus and a healing fracture on the right. (B) A skull film shows multiple Wormian bones. (C) In the lower extremity, the bones are thin and deformed with evidence of previous fracture.

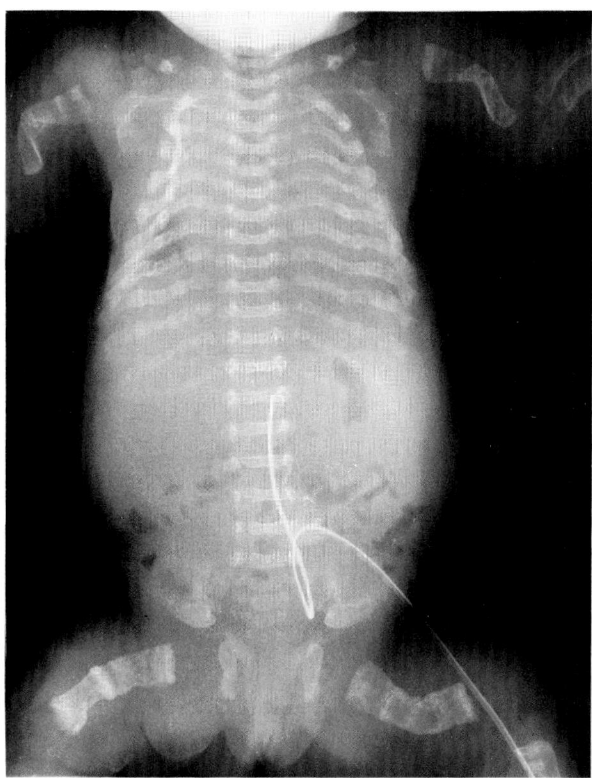

**Figure 61.29. Osteogenesis imperfecta.** Fractures of multiple ribs and long bones in an infant.

tures, hypermobility of joints, poor teeth, deafness, and cardiovascular disorders such as mitral valve prolapse or aortic regurgitation. The most severe form of this condition is osteogenesis imperfecta congenita, which develops in utero and is seen at birth as bowing and deformity of the extremities due to multiple fractures. Infants with this condition have a paper-thin skull, and death in utero or soon after birth is usually caused by intracranial hemorrhage. In the less severe form (osteogenesis imperfecta tarda), the disorder is first noted during childhood or young adulthood because of the unusual tendency for fractures, loose-jointedness, and the presence of blue sclerae, which is caused by intraocular pigment that shows through the thin, translucent outer membranes of the eyes.

The major radiographic bone changes in osteogenesis imperfecta are repeated fractures due to the severe osteoporosis and the thin, defective cortices. The fractures often heal with exuberant callus formation (often so extensive as to simulate a malignant tumor) and may cause bizarre deformities. If an infant with the congenital type of osteogenesis imperfecta survives, ossification of the skull progresses slowly, leaving wide sutures and multiple juxtasutural accessory bones (Wormian bones) that produce a mosaic appearance.[28,29]

## REFERENCES


1. Henry A: Monostotic fibrous dysplasia. *J Bone Joint Surg* 51B:300–306,1969.
2. Leeds NL, Seaman WB: Fibrous dysplasia of the skull and its differential diagnosis. *Radiology* 78:570–582, 1962.
3. Riddell DM: Malignant change in fibrous dysplasia. *J Bone Joint Surg* 46B:251–255, 1964.
4. Ford N, Silverman FC, Kozlowski K: Spondyloepiphyseal dysplasia. *AJR* 86:462–472, 1961.
5. Spranger J, Langer LO: Spondyloepiphyseal dysplasia congenita. *Radiology* 94:313–322, 1970.
6. Langer LO: Spondyloepiphyseal dysplasia tarda. *Radiology* 82:833–839, 1964.
7. Leeds NE: Epiphyseal dysplasia multiplex. *AJR* 84:506–510, 1960.
8. Shephard E: Multiple epiphyseal dysplasia. *J Bone Joint Surg* 38B:458–467, 1956.
9. Kellekamp DB, Campbell CJ, Bonfiglio M: Dysplasia epiphysialis hemimelica. *J Bone Joint Surg* 48A:746–766, 1966.
10. Keats TE: Dysplasia epiphysialis hemimelica. *Radiology* 68:558–563, 1957.
11. Mason RC, Kozlowski K: Chondrodysplasia punctata. *Radiology* 109:145–150, 1973.
12. Brogdon BG, Grow NE: Chondrodystrophia calcificans congenita. *AJR* 80:443–448, 1958.
13. Keats TE, Riddervold HO, Michaelis LL: Thanatophoric dwarfism. *AJR* 108:473–480, 1970.
14. Caffey J: Achondroplasia of the pelvis and lumbosacral spine: Some radiographic features. *AJR* 80:458–467, 1958.
15. Silverman FN: A differential diagnosis of achondroplasia. *Radiol Clin North Am* 6:223–237, 1968.
16. Langer LO, Baumann PA, Gorlin RJ: Achondroplasia. *AJR* 100:12–19, 1967.
17. Ellis RWB, Andrew JB: Chondroectodermal dysplasia. *J Bone Joint Surg* 44B: 626–636, 1962.
18. Stark JD, Adler NN, Robinson WH: Hereditary multiple exostoses. *Radiology* 59:212–215, 1952.
19. Vinstein AL, Franken EA: Hereditary multiple exostoses. *AJR* 112:405–407, 1971.
20. Schwartz E: Craniometaphyseal dysplasia. *AJR* 84:461–466, 1960.
21. Carlson DH, Harris GBC: Craniometaphyseal dysplasia. *Radiology* 103:147–152, 1972.
22. Gorlin RJ, Kozalka MF, Spranger J: Pyle's disease. *J Bone Joint Surg* 52A: 347–354, 1970.
23. Hermel MB, Gershon-Cohen J, Jones DT: Familial metaphyseal dysplasia. *AJR* 70:413–421, 1953.
24. Darlington D, Hawkins CF: Nail-patella syndrome with iliac horns and hereditary nephropathy. *J Bone Joint Surg* 49B: 164–174, 1967.
25. Zimmerman C: Iliac horns: A pathognomonic roentgen sign of familial onycho-osteodysplasia. *AJR* 86:478–483, 1961.
26. Jarvin JL, Keats TE: Cleidocranial dysostosis. *AJR* 121:5–16, 1979.
27. Keats TE: Cleidocranial dysostosis: Some atypical roentgen manifestations. *AJR* 100:71–74, 1967.
28. King JP, Bobechko WP: Osteogenesis imperfecta. *J Bone Joint Surg* 53B:72–89, 1971.
29. Navani SV, Sarzin B: Intrauterine osteogenesis imperfecta. *Br J Radiol* 40:449–452, 1967.
30. Tchang SK: The small orbit sign in supraorbital fibrous dysplasia. *J Can Assoc Radiol* 24:65–69, 1973.
31. Greenfield GB: *Radiology of Bone Diseases*. Philadelphia, JB Lippincott, 1980.
32. Edeiken J: *Roentgenology of Diseases of Bone*. Baltimore, Williams & Wilkins, 1981.

## Chapter Sixty-Two
# Infectious and Ischemic Disorders of Bone

## OSTEOMYELITIS

Osteomyelitis is caused by a broad spectrum of infectious organisms that reach bone by hematogenous spread, by extension from a contiguous site of infection, or by direct introduction of organisms (trauma or surgery). Acute hematogenous osteomyelitis tends to involve bones with rich red marrow. In infants and children, the metaphyses of long bones, especially the femur and tibia, are most often affected; staphylococci and streptococci are the most common organisms. After the age of 8 months, the epiphyseal cartilage plate acts as a barrier against the spread of infection. In adults, acute hematogenous osteomyelitis primarily occurs in the vertebrae and rarely involves the long bones. Although the incidence and severity of osteomyelitis have decreased since the advent of antibiotics, this disease has recently become more prevalent as a complication of intravenous drug abuse. In diabetic patients and those with other types of vascular insufficiency, a soft tissue infection may spread from a skin abscess or a decubitus ulcer, usually in the foot, to cause cellulitis and eventually osteomyelitis in contiguous bones.

Because the earliest changes of osteomyelitis are usually not evident on plain radiographs until about 10 days after the onset of symptoms, radionuclide bone scanning with $^{99m}$Tc pyrophosphate is the most valuable imaging modality for the early diagnosis of osteomyelitis. Increased isotope uptake, reflecting the inflammatory process and increased blood flow, is evident within hours of the onset of symptoms. Because osteomyelitis may initially result in decreased accumulation of isotope due to ischemia and because overlying cellulitis may mimic a bone infection, sequential studies and scanning with two different radionuclides (technetium and gallium) may be required.

On plain radiographs, the earliest evidence of osteomyelitis in a long bone is a localized, deep soft tissue swelling adjacent to the metaphysis. The inflammation causes displacement or obliteration of the normal fat planes adjacent to and between the deep muscle bundles, unlike skin infections, in which the initial swelling is superficial. The initial osseous change is subtle areas of metaphyseal lucency reflecting resorption of necrotic bone. Soon, bone destruction becomes more prominent, producing a ragged, moth-eaten appearance. The more virulent the organism, the larger the area of destruction. Subperiosteal spread of inflammation elevates the periosteum and stimulates the laying down of layers of new bone parallel to the shaft. This results in a lamellar periosteal reaction that is characteristic of benign diseases, especially infection. Eventually, a large amount of new bone surrounds the cortex in a thick, irregular bony sleeve (involucrum). Disruption of cortical blood supply leads

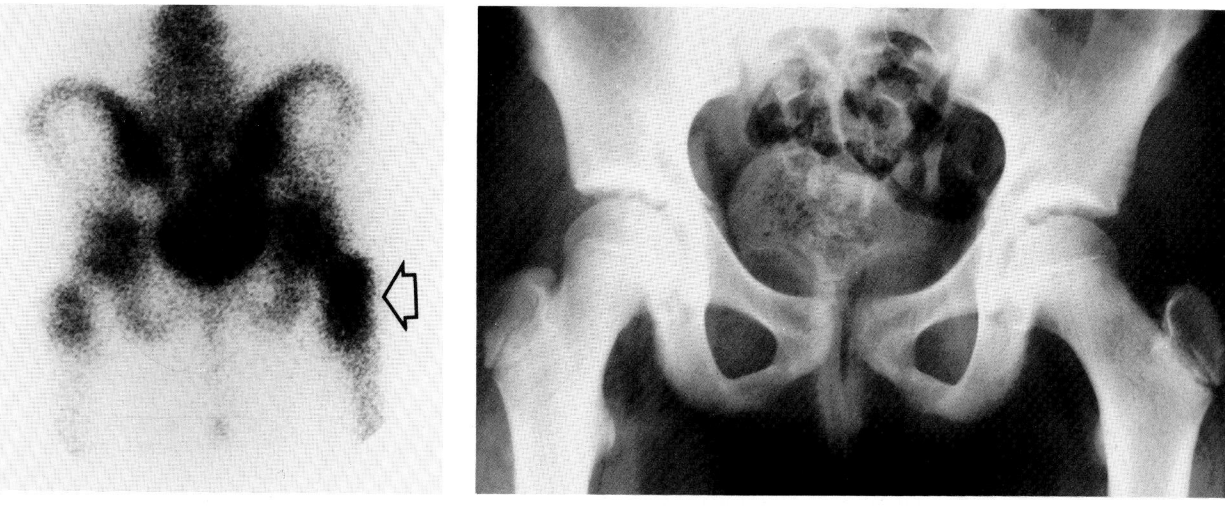

A B

**Figure 62.1. Osteomyelitis.** (A) A radionuclide bone scan (posterior view) demonstrates increased uptake of isotope in the trochanteric portion of the right femur (arrow). (B) A plain film of the pelvis and hips obtained at the same time as the radionuclide scan shows no detectable abnormality.

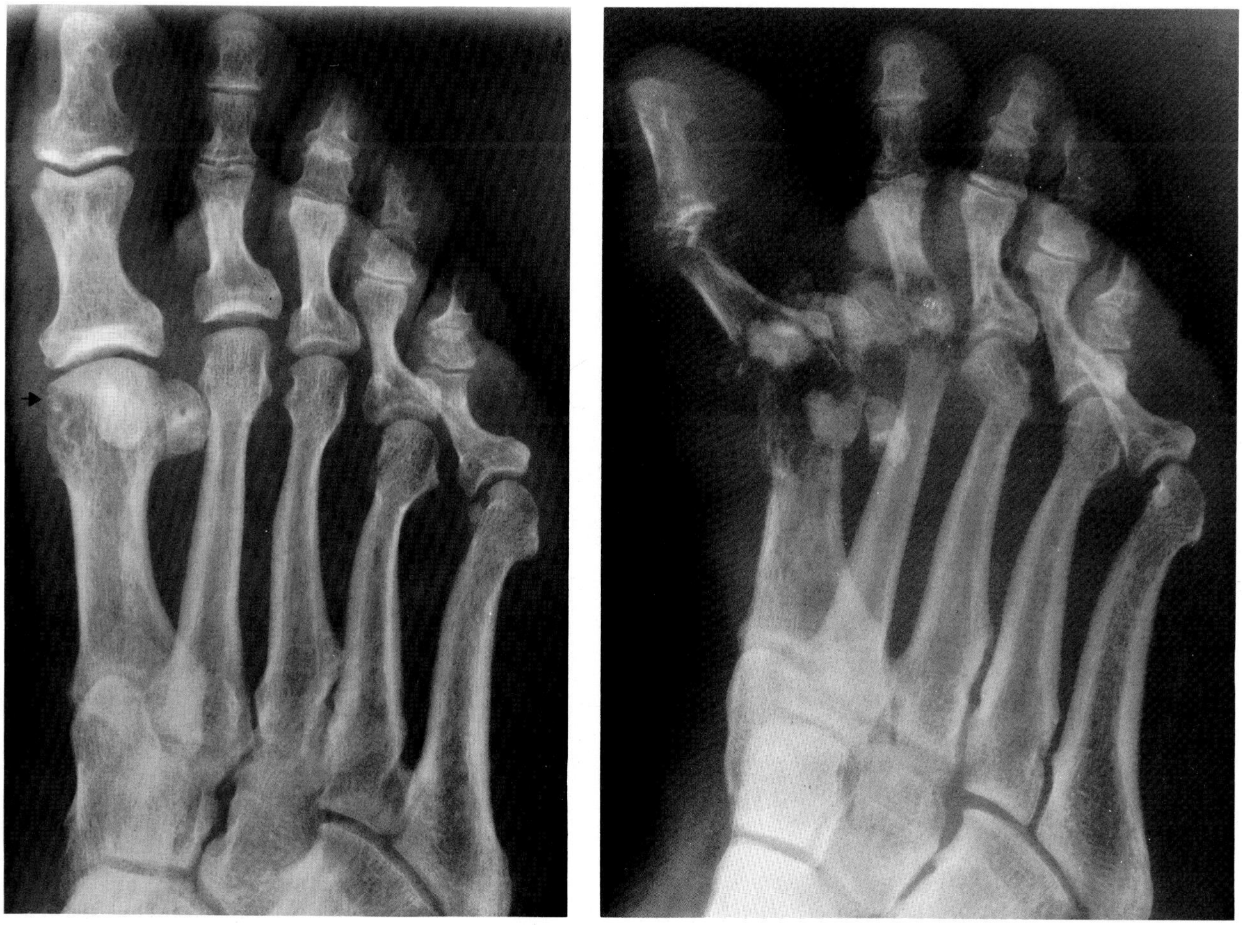

A B

**Figure 62.2. Acute osteomyelitis.** (A) An initial film of the foot in a diabetic patient with a soft tissue infection shows minimal hyperemic osteoporosis about the head of the first metatarsal with some loss of the sharp cortical outline (arrow). (B) One month later there is severe bone destruction involving not only the head of the first metatarsal but also the rest of the big toe and the second and third metatarsophalangeal joints.

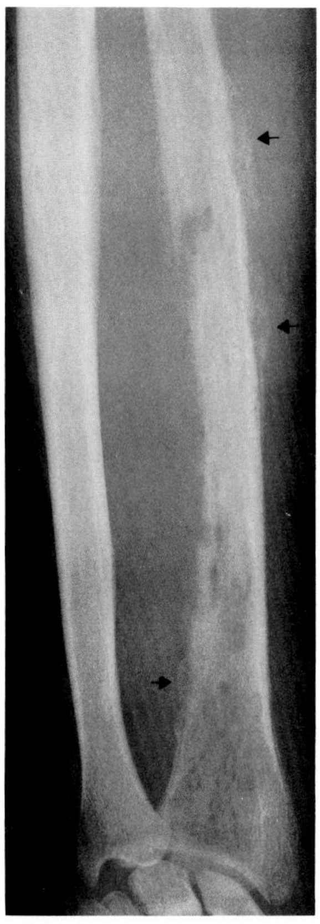

**Figure 62.3. Osteomyelitis.** A patchy pattern of bone destruction involves much of the shaft of the radius. Note the early periosteal new bone formation (arrows).

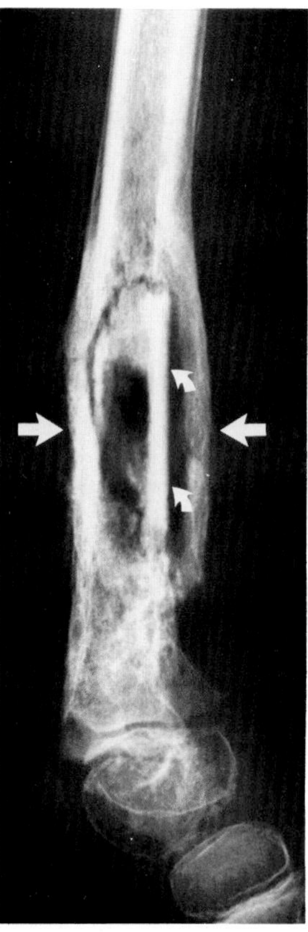

**Figure 62.4. Chronic osteomyelitis.** The involucrum (straight arrows), a thick, irregular, bony sleeve, surrounds the sequestrum (curved arrows), a residual segment of avascular dead bone.

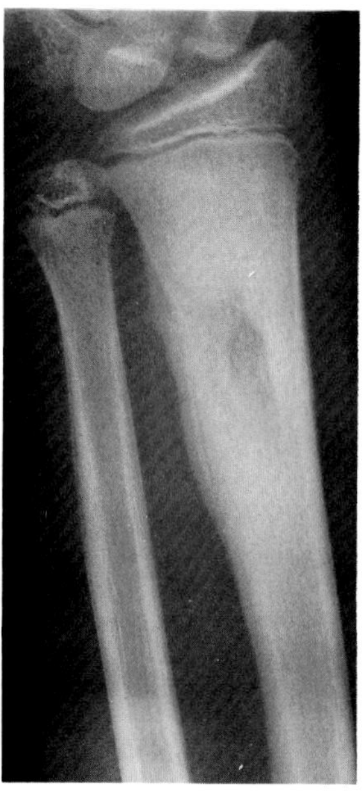

**Figure 62.5. Chronic osteomyelitis.** An ill-defined area of lucency within the distal radial shaft is almost obscured by the sclerotic periosteal new bone formation.

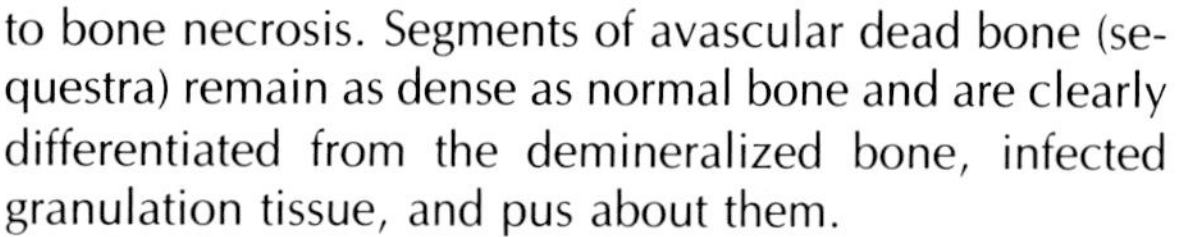

to bone necrosis. Segments of avascular dead bone (sequestra) remain as dense as normal bone and are clearly differentiated from the demineralized bone, infected granulation tissue, and pus about them.

After the acute infection has subsided, a pattern of chronic osteomyelitis develops. The bone appears thickened and sclerotic with an irregular outer margin. The cortex may became so dense that the medullary cavity is difficult to demonstrate. Reactivation of infection may appear as the recurrence of deep soft tissue swelling, periosteal calcification, or the development of lytic abscess cavities within the bone. However, plain radiographs are often inadequate to determine whether an active infection is present. Radionuclide scanning is much more sensitive and accurate in establishing a recurrence.

The earliest sign of vertebral osteomyelitis is subtle erosion of the subchondral bony plate with loss of the sharp cortical outline. This may progress to total destruction of the vertebral body associated with a paravertebral soft tissue abscess. As with osteomyelitis of the appen-

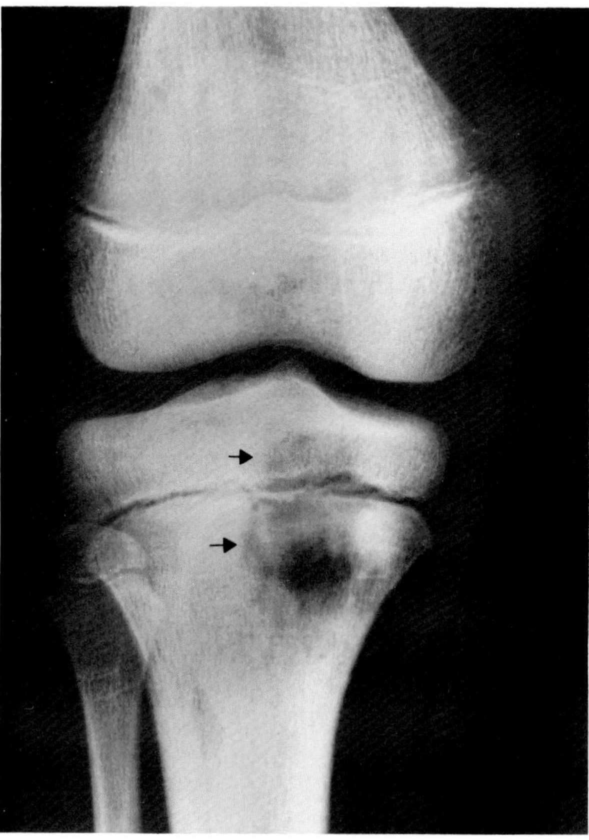

**Figure 62.6. Chronic osteomyelitis.** Lytic abscess cavities (arrows) within the bone represent reactivation of infection.

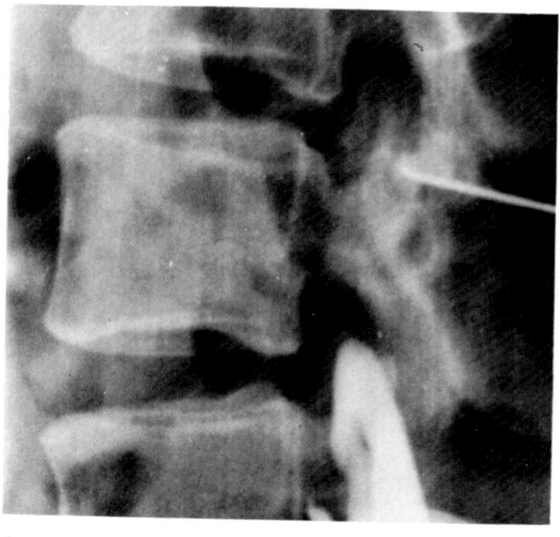

A

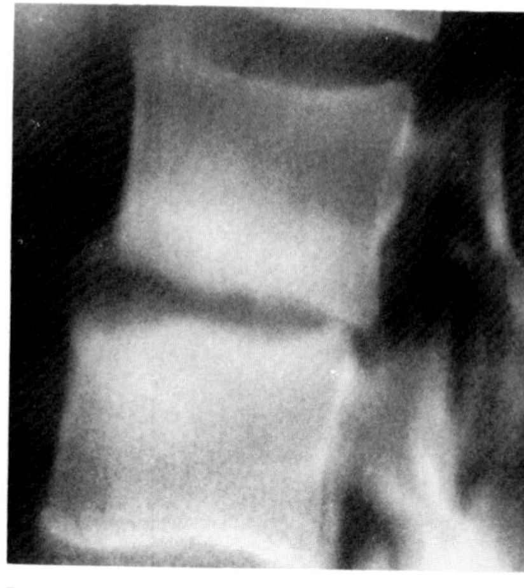

B

**Figure 62.7. Vertebral osteomyelitis resulting from a disk space infection following spinal tap.** (A) A lateral view of the lumbar spine obtained during a myelogram shows the needle in the lumbar subarachnoid space. Note the normal appearance of the adjacent vertebral bodies. (B) Two months later there is narrowing of the intervertebral disk space with irregularity of the end plates and reactive sclerosis.

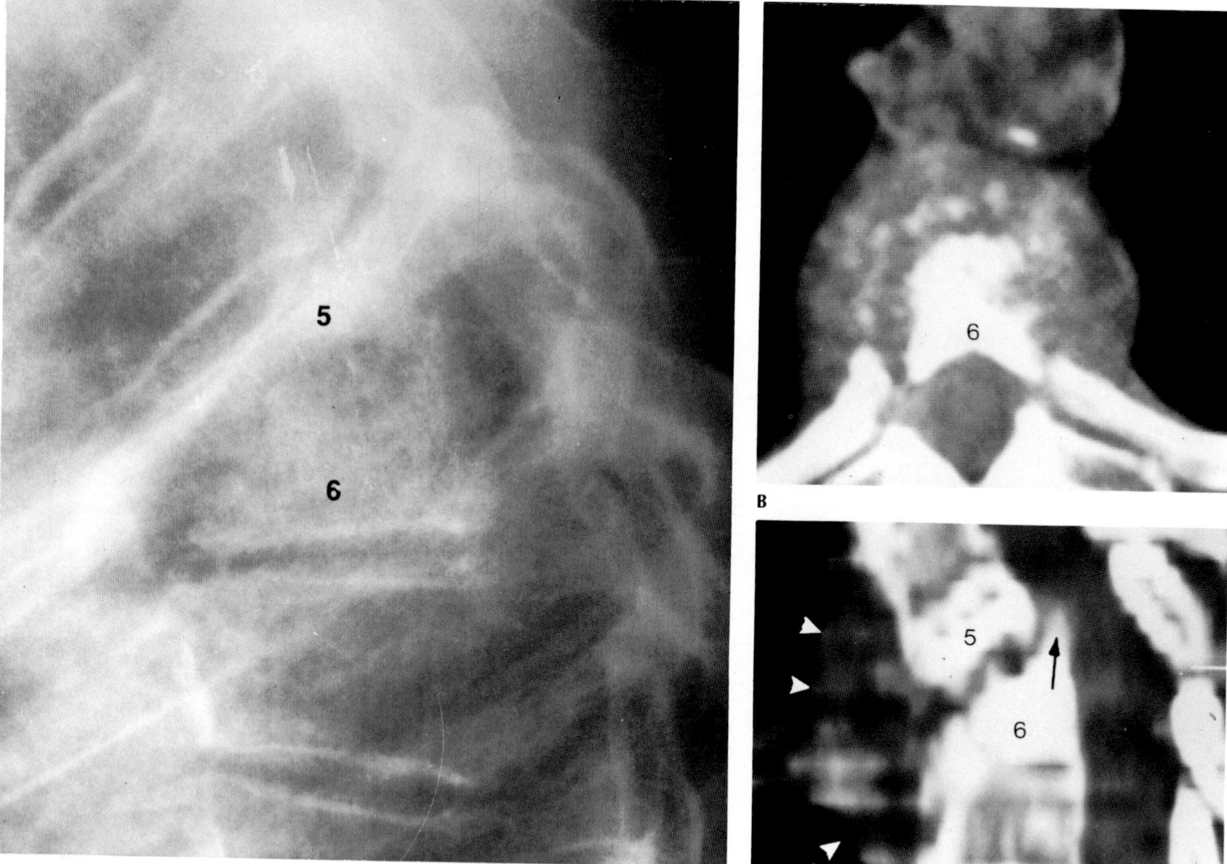


**Figure 62.8. CT of tuberculous diskitis and spondylitis.** (A) A lateral radiograph of the spine demonstrates a disk-centered destructive lesion at the T5–6 level, but does not further characterize the process. (B) A CT scan through T6 shows vertebral destruction, fragmentation, and a paraspinal mass. (C) Reformatted CT sections in a sagittal plane show that T5 is largely destroyed and that a sharp posterosuperior spur of T6 encroaches on the canal (arrow), a condition aggravated by the gibbous deformity caused by disk and bone destruction. An anterior soft tissue mass is also shown (arrowheads). (From Murphy, Gilula, Destouet, et al,[5] with permission from the publisher.)

dicular skeleton, plain radiographic changes may not develop for several weeks following infection, while radionuclide scans are positive early. Unlike neoplastic processes, osteomyelitis usually affects the intervertebral disk space and often involves adjacent vertebrae. Depending on the site of disease, anterior extension of vertebral osteomyelitis may cause retropharyngeal abscess, mediastinitis, empyema, pericarditis, subdiaphragmatic abscess, psoas muscle abscess, or peritonitis; posterior extension of inflammatory tissue can compress the spinal cord or produce meningitis if the infection penetrates the dura to enter the subarachnoid space.

In most patients, the radiographic findings, clinical history, and symptoms are sufficient to make the diagnosis of osteomyelitis. At times, however, aggressive bone destruction and bizarre periosteal reaction, especially in children, may suggest a malignant bone tumor. If there is any doubt as to the correct diagnosis, biopsy is required.

Computed tomography (CT) can be of value in the diagnosis of osteomyelitis, especially that involving the spine. This modality can precisely define the size of the surrounding soft tissue mass, its relation to nearby vital structures (aorta, spinal cord), and the presence of abscess cavities requiring surgical drainage. Serial CT scans may be used to confirm the expected decrease in size of the soft tissue mass.[1–5]

### Brodie's Abscess

Brodie's abscess refers to a chronic bone abscess of low virulence that has never had an acute stage. This painful lesion often simulates an osteoid osteoma and appears radiographically as a well-circumscribed lytic area surrounded by an irregular zone of dense sclerosis.[6]

### Garré's Sclerosing Osteomyelitis

Garré's sclerosing osteomyelitis is a rare chronic, nonsuppurative infection of bone, due to an organism of low virulence, that causes an exuberant sclerotic reaction without any bone destruction, sequestration, or periosteal response.

## ISCHEMIC NECROSIS OF BONE

Ischemic necrosis of bone results from the local loss of blood supply, which in turn can result from such varied conditions as thrombosis, vasculitis, disease of surrounding bone, or single or repeated episodes of trauma disrupting the blood supply. Among the many conditions

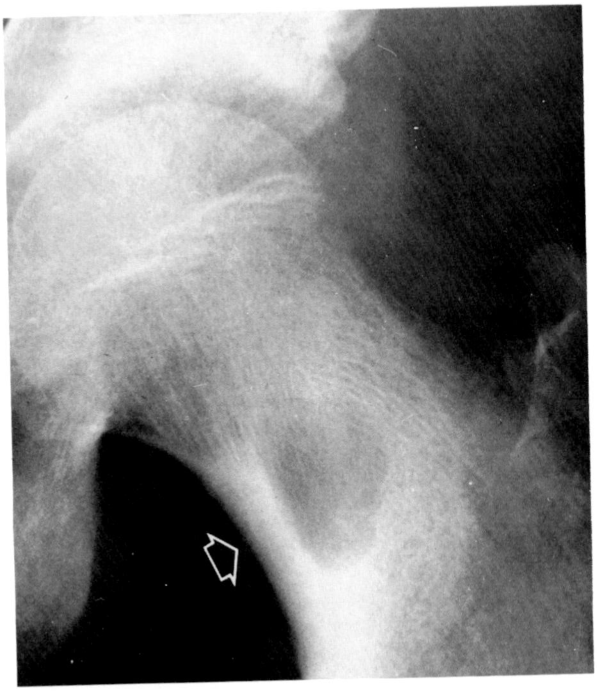

**Figure 62.9. Brodie's abscess in the femoral neck.** There is a well-defined abscess with moderate cancellous reaction and a thickly buttressed femoral neck (arrow). (From Miller, Murphy, and Gilula,[6] with permission from the publisher.)

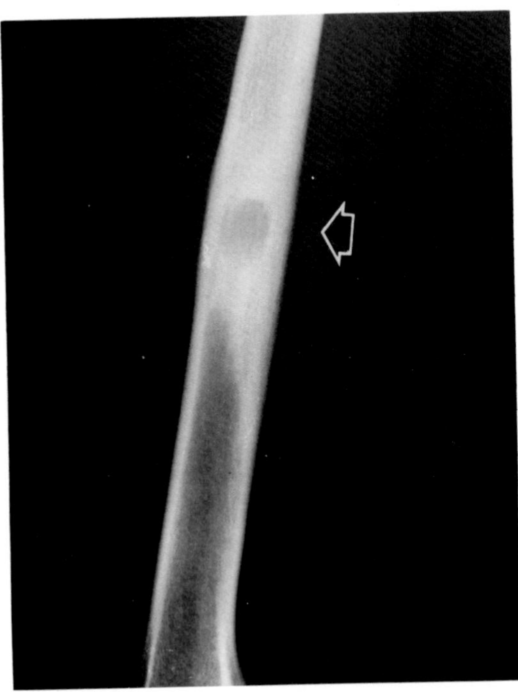

**Figure 62.10. Brodie's abscess.** A well-circumscribed lucent lesion completely fills the femoral medullary canal and is surrounded by dense endosteal sclerosis and cortical thickening (arrow). (From Miller, Murphy, and Gilula,[6] with permission from the publisher.)

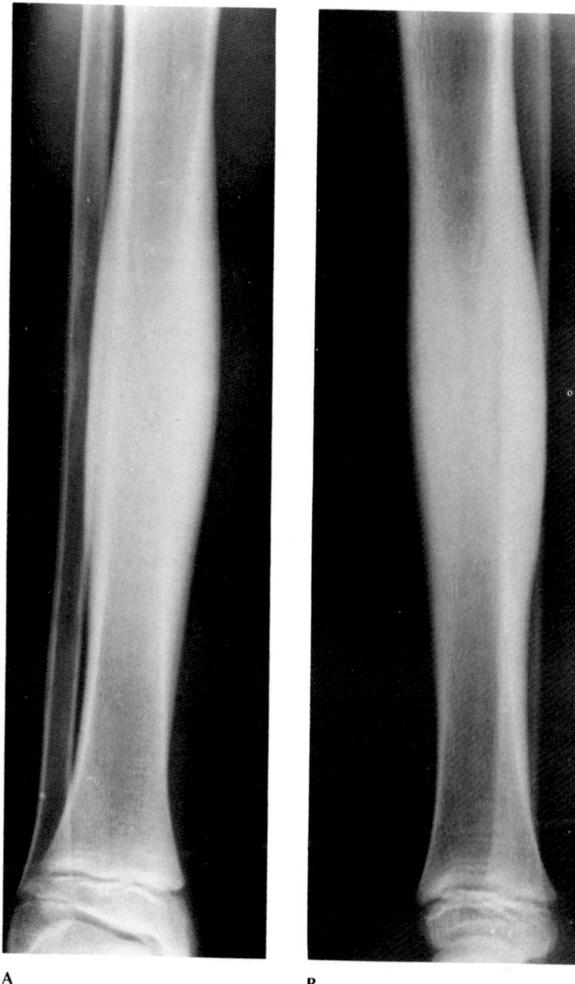


**Figure 62.11. Garré's sclerosing osteomyelitis.** (A) Frontal and (B) lateral views of the midshaft of the tibia show an exuberant sclerotic reaction without evidence of bone destruction.

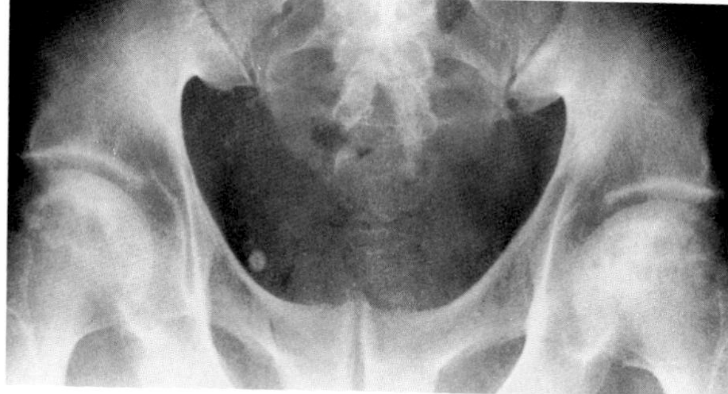

**Figure 62.12. Ischemic necrosis.** There are sclerotic changes in the femoral heads bilaterally.

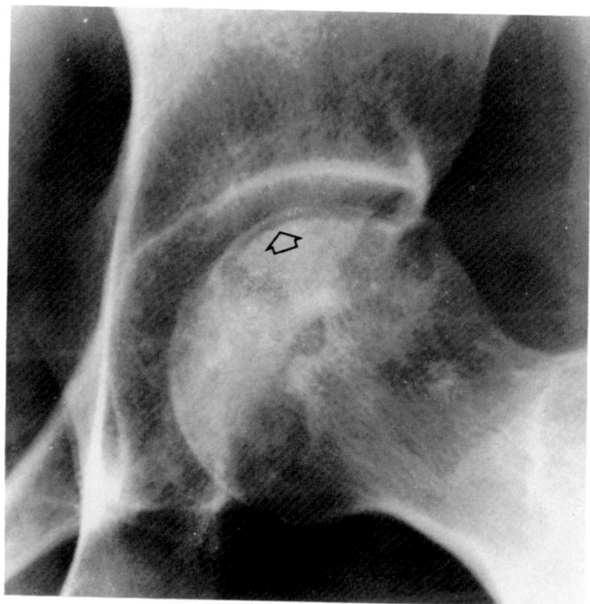

**Figure 62.13. Ischemic necrosis.** An arclike radiolucent cortical band (crescent sign) in the femoral head represents a fracture line.

associated with ischemic necrosis are acute trauma (fracture or dislocation), steroid therapy or Cushing's disease, hemolytic anemia (especially sickle cell disease), chronic alcoholism and chronic pancreatitis, Gaucher's disease, histiocytosis, radiation therapy, and caisson disease.

## Ischemic Necrosis in the Articular Ends

The femoral head is the most frequent site of ischemic necrosis involving the articular ends of bones. Initially, the ischemic bone may appear denser than adjacent viable bone. The first sign of structural failure is the development of a radiolucent subcortical band (crescent sign) representing a fracture line. This subchondral lucent line is frequently seen before the characteristic subchondral sclerosis becomes evident. As the disorder progresses, fragmentation, compression, and resorption of dead bone, along with proliferation of granulation tissue, revascularization, and production of new bone, produce a pattern of lytic and sclerotic areas with flattening of the femoral head and periosteal new bone formation. Mechanical distortion about the hip leads to uneven weight bearing and accelerated secondary osteoarthritis.[7–9]

## Infarction of the Shaft

In addition to ischemic changes in the subchondral areas of bone, infarction may involve the shaft of a long bone. In many cases, this type of bone infarction is asymptomatic and is detected only on radiographs obtained for another purpose. A mature infarct in the shaft of a bone appears as a densely calcified area in the medullary cavity. It may be sharply limited by a dense sclerotic zone or be associated with serpiginous dense streaks extend-

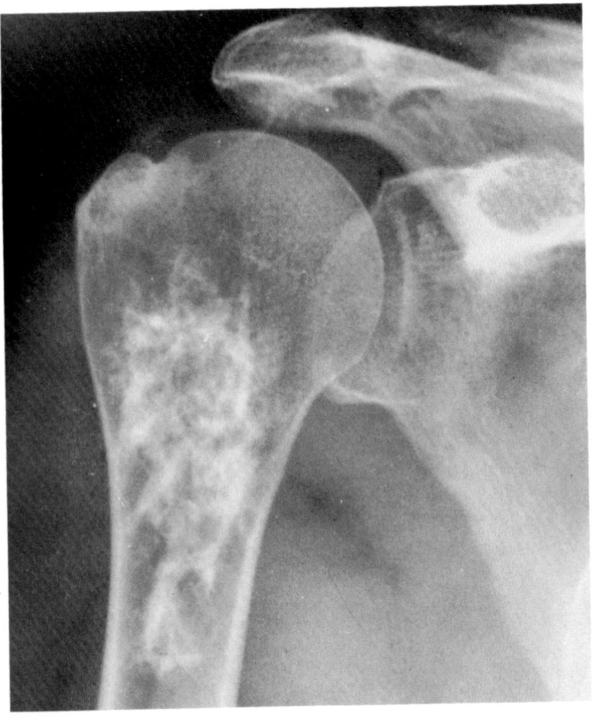

**Figure 62.14. Bone infarct.** There is a densely calcified area in the medullary cavity of the humerus with dense streaks extending from the central region.

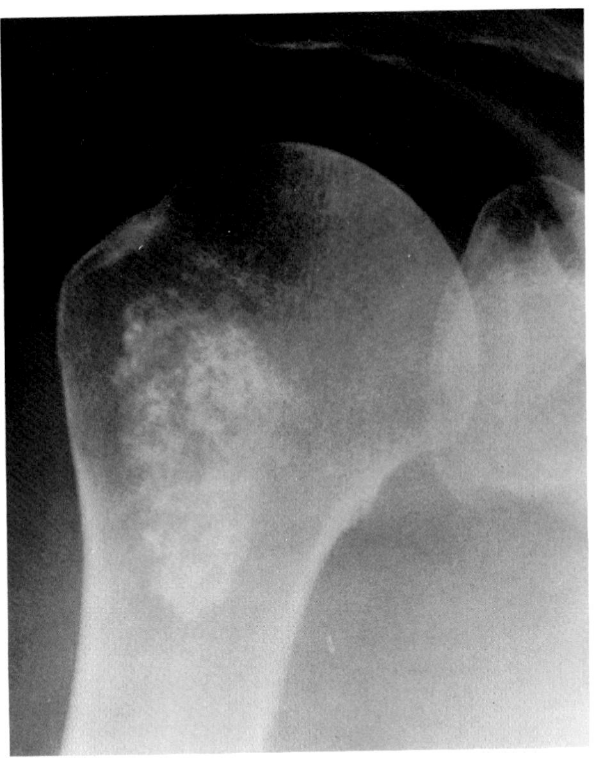

**Figure 62.15. Calcified enchondroma.** Amorphous spotty calcific densities are seen within the lesion.

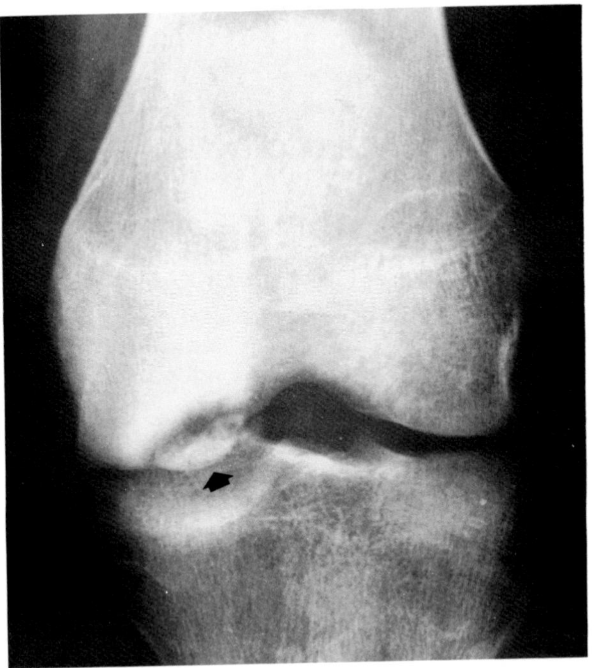

A

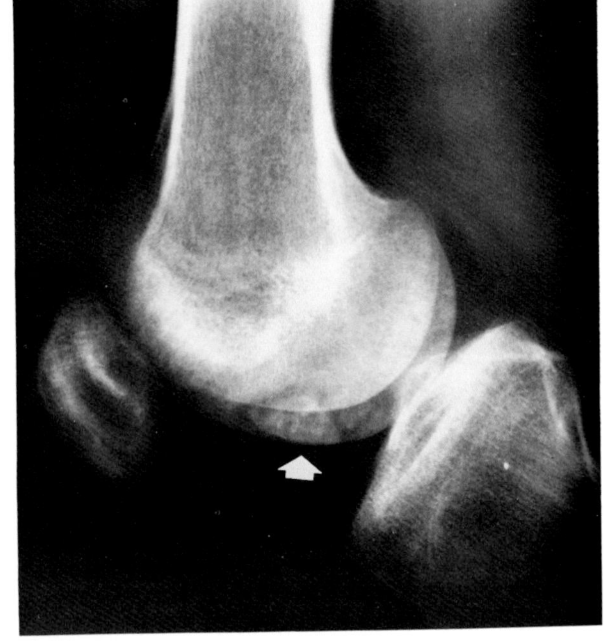

B

**Figure 62.16. Osteochondritis dissecans at the knee.** (A) Frontal and (B) lateral views of the knee demonstrate the necrotic segment (arrow) separated from the medial femoral condyle by a crescentic lucent zone.

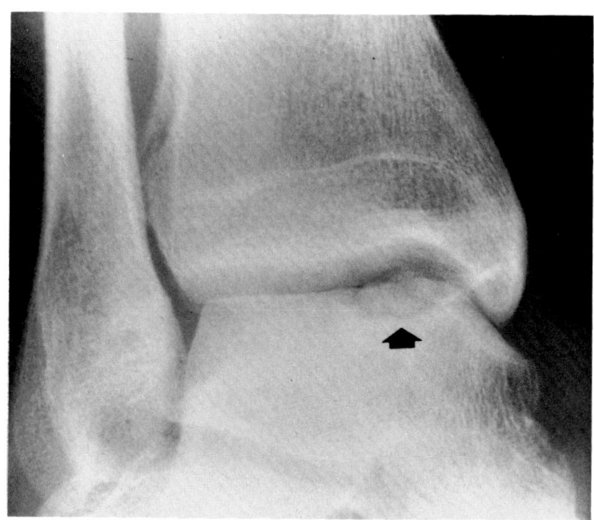

**Figure 62.17. Osteochondritis dissecans at the ankle.**

ing from the central region. Bone infarction must be differentiated from a calcified enchondroma, which contains amorphous, spotty calcific densities, is not surrounded by a sclerotic rim, and may expand the bone.[7–9]

## OSTEOCHONDRITIS DISSECANS

Osteochondritis dissecans is a localized form of ischemic necrosis that most frequently affects young males and probably is caused by trauma. It occurs primarily about the knee, usually on the lateral aspect of the medial femoral condyle. Other common locations are the ankle, femoral head, elbow, and shoulder.

In this condition, a small, round or oval necrotic segment of bone with its articular cartilage detaches and lies in a depression in the joint surface. Often denser than the surrounding bone, the necrotic segment is well-demarcated by a crescentic lucent zone. The necrotic segment may separate from the joint to form a free joint body, leaving a residual pit in the articular surface.[8]

### REFERENCES


1. Kattapuram SV, Phillips WC, Boyd R: CT in pyogenic osteomyelitis of the spine. *AJR* 140:1199–1201, 1983.
2. Capitanio MA, Kirkpatrick JA: Early roentgen observations in acute osteomyelitis. *AJR* 108:488–496, 1970.
3. Boland AL: Acute hematogenous osteomyelitis. *Orthop Clin North Am* 3:225–239, 1972.
4. Majd M: Radionuclide imaging in early detection of childhood osteomyelitis and its differentiation from cellulitis and bone infarction. *Ann Radiol* 20:9–18, 1977.
5. Murphy MA, Gilula LA, Destouet JM, et al: Musculoskeletal system, in Lee JKT, Sagel SS, Stanley RJ (eds): *Computed Body Tomography*. New York, Raven Press, 1983.
6. Miller WB, Murphy WA, Gilula LA: Brodie's abscess: Reappraisal. *Radiology* 132:15–23, 1979.
7. Edeiken J, Hodes PJ, Libscitz HI, et al: Bone ischemia. *Radiol Clin North Am* 5:515–529, 1967.
8. Greenfield GB: *Radiology of Bone Diseases*. Philadelphia, JB Lippincott, 1980.
9. Martel W, Sitterley BH: Roentgenologic manifestations of osteonecrosis. *AJR* 106:509–522, 1969.

# PART ELEVEN

# DISORDERS OF THE JOINTS, CONNECTIVE TISSUES, AND STRIATED MUSCLES

## Chapter Sixty-Three

# Rheumatoid Arthritis and Its Variants

## RHEUMATOID ARTHRITIS

Rheumatoid arthritis is a chronic systemic disease of unknown cause that primarily presents as a nonsuppurative inflammatory arthritis of peripheral joints. The average age of onset in adults is 40 years. Women are affected about three times more frequently than men. Rheumatoid arthritis usually has an insidious origin and may either run a protracted and progressive course, leading to a crippling deformity of affected joints, or undergo spontaneous remissions of variable length. The earliest sites of involvement tend to be the proximal interphalangeal joints of the hands and feet, the metacarpophalangeal joints, the intercarpal joints, the ulnar styloid process, and the distal radioulnar joints. There is usually symmetric involvement of multiple joints, and the disease often progresses proximally toward the trunk until practically every joint in the body is involved.

Rheumatoid arthritis begins as a synovial inflammatory reaction that produces a mass of thickened granulation tissue (pannus). The process begins at the periphery of the joint and may spread to cover the entire articular surface. This causes destruction of the articular cortex, followed by fibrous and even bony ankylosis.

The earliest radiographic evidence of rheumatoid arthritis is fusiform periarticular soft tissue swelling due to joint effusion and hyperplastic synovitis. Disuse and local hyperemia lead to periarticular osteoporosis that initially is confined to the portion of bone adjacent to the joint but may extend to involve the entire bone. The accumulation of joint fluid elevates the periosteum and may stimulate a thin layer of periosteal calcification that is often overlooked. Pannus and joint effusion may widen the joint space, though this is difficult to demonstrate radiographically.

Extension of the pannus from the synovial reflections onto the bone causes characteristic small foci of destruction at the edges of the joint, where articular cartilage is absent. These typical marginal erosions have poorly defined edges without a sclerotic rim and sometimes may be seen only on oblique or magnification views. Actual invasion of pannus into bone substance causes large pseudocysts, which may appear at a distance from the joint.

Destruction of articular cartilage causes narrowing of the joint space. Early joint space narrowing with preservation of the subchondral cortical margin strongly suggests rheumatoid arthritis rather than a septic arthritis, which causes extensive bony erosion. However, joint space narrowing in rheumatoid arthritis is frequently associated with extensive destruction of bone ends. This occasionally results in telescoping of the fingers and the development of an opera-glass hand (main en lorgnette deformity). The laying down of bony trabeculae across

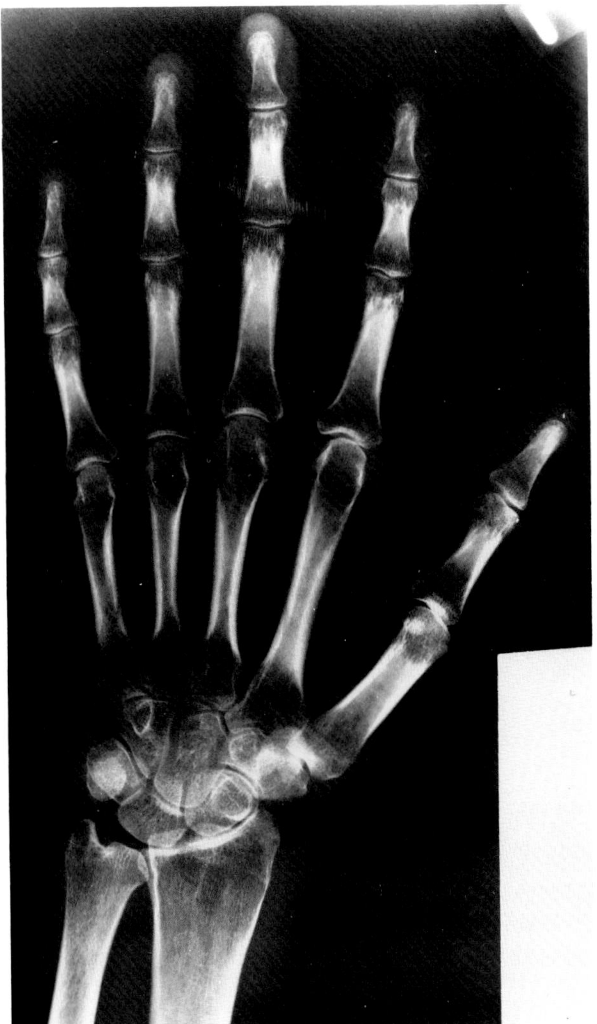

**Figure 63.1. Rheumatoid arthritis.** There is striking periarticular osteoporosis. (From Brown and Forrester,[29] with permission from the publisher.)

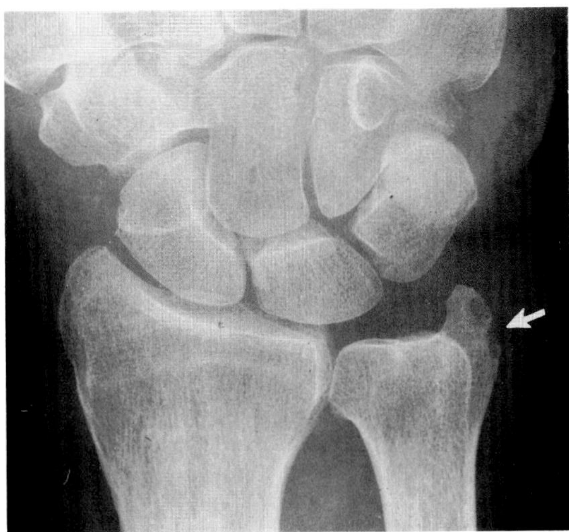

**Figure 63.2. Rheumatoid arthritis.** A frontal view of the wrist shows characteristic erosion of the ulnar styloid process (arrow) by an adjacent tenosynovitis of the extensor carpi ulnaris tendon. Note the associated soft tissue swelling.

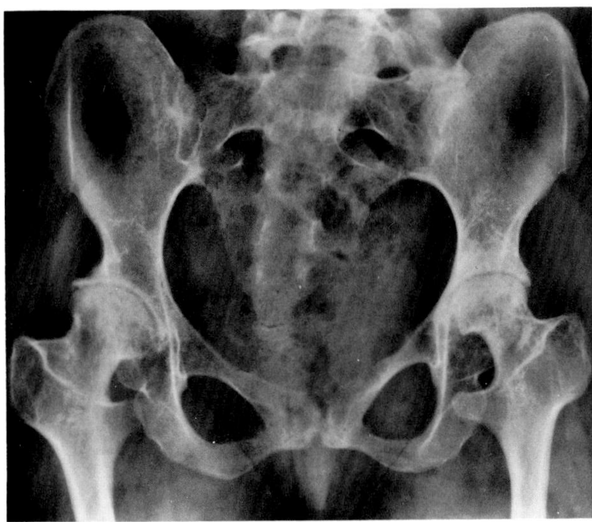

**Figure 63.3. Rheumatoid arthritis of the pelvis and hips.** There is narrowing of the hip joints bilaterally with some reactive sclerosis. Note the relative preservation of the subchondral cortical margins. In contrast to degenerative disease, in rheumatoid arthritis the joint space narrowing is symmetric and not confined to weight-bearing surfaces. Note also the obliteration of both sacroiliac joints.

a narrowed joint space may completely obliterate the joint cavity and produce solid bony ankylosis. Bony ankylosis most frequently involves the bones of the wrists and rarely affects the metacarpophalangeal joints.

A variety of contractures and subluxations may be seen in rheumatoid arthritis. In the fingers, flexion deformity of the proximal interphalangeal joint and extension deformity of the distal interphalangeal joint produces the boutonniere deformity; the reverse deformity is termed *swan neck*. Ulnar deviation of the hands is common and may mimic the findings in systemic lupus erythematosus, though in the latter there are no erosive changes. Extensions of normal bursae or rupture of joint or tendon capsules may lead to the development of giant synovial cysts. These cysts are most common about the knee, where they may dissect into the calf and produce clinical signs resembling those of deep vein thrombosis. The diagnosis of a popliteal cyst may be made by arthrography or ultrasound.

Rheumatoid arthritis of the spine primarily involves the cervical area. The most characteristic finding is atlantoaxial subluxation, an increased distance between the anterior border of the odontoid and the superior portion of the anterior arch of the atlas (normally less than 2.5 mm), which is caused by weakening of the transverse

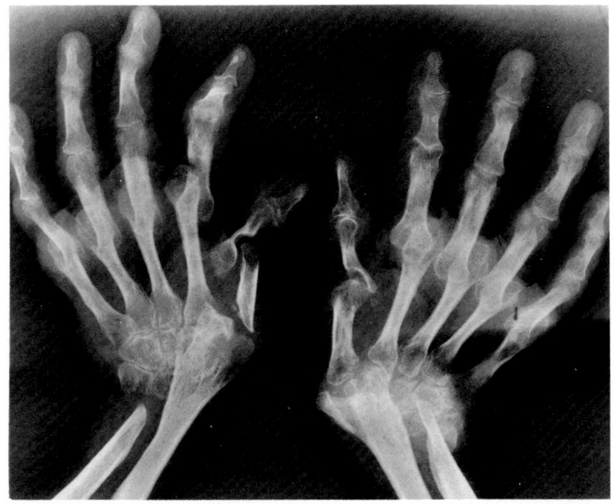

A

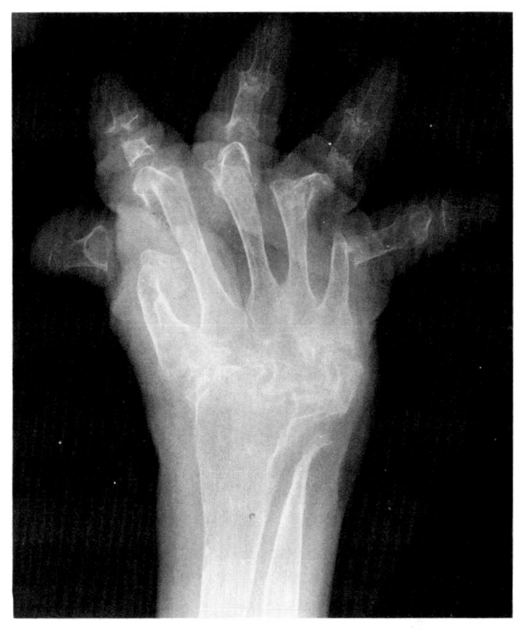

B

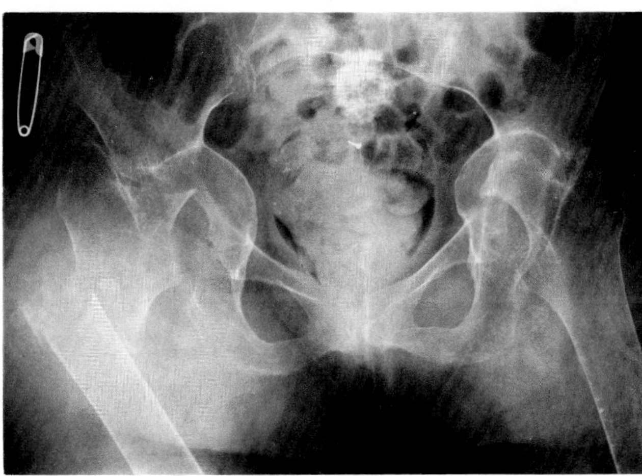

C

**Figure 63.4. Mutilating rheumatoid arthritis.** (A,B) Two examples of the opera-glass hand (main en lorgnette deformity) due to extensive destruction and telescoping of bone ends. (C) A view of the pelvis of the same patient as B shows a fracture of the proximal right femur, bilateral ankylosis of the hip joints, and severe protrusio acetabuli.

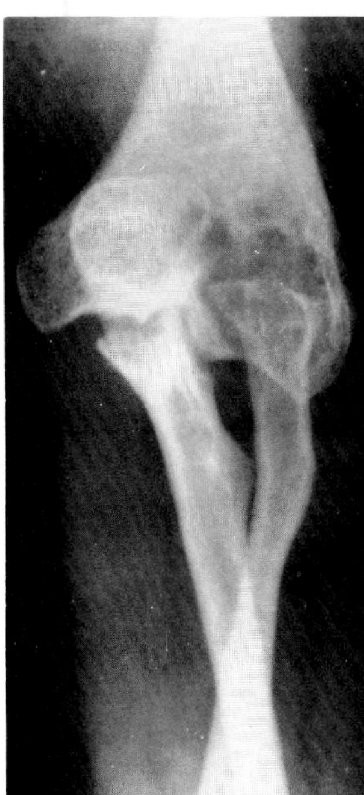

**Figure 63.5. Rheumatoid arthritis of the elbow.** There is extensive destruction and subluxation of the bony structures.

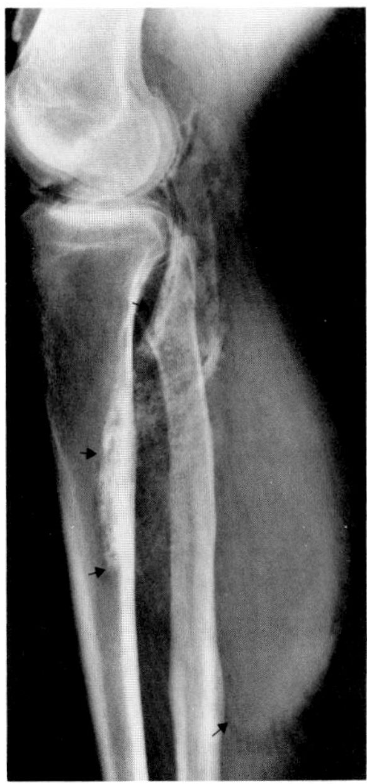

**Figure 63.6. Rheumatoid arthritis of the knee with a giant dissecting popliteal cyst.** A film from an arthrogram demonstrates a small amount of contrast material filling a huge synovial cyst (arrows) that originated in the knee and extended to involve the posterior portion of most of the lower leg.

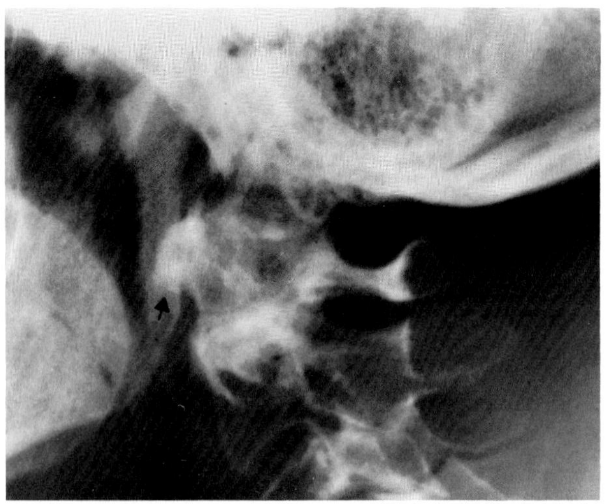

A

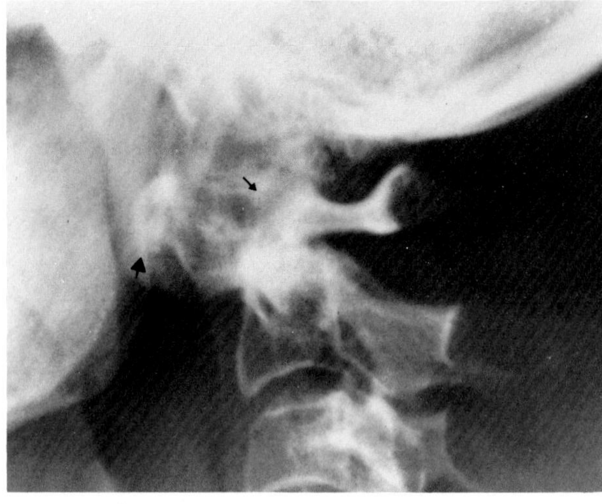

B

**Figure 63.7. Atlantoaxial subluxation in rheumatoid arthritis.** (A) A routine lateral film of the cervical spine shows a normal relation between the anterior border of the odontoid and the superior portion of the anterior arch of the atlas (arrow). (B) With flexion, there is wide separation between the anterior arch of the atlas (large arrow) and the odontoid (small arrow).

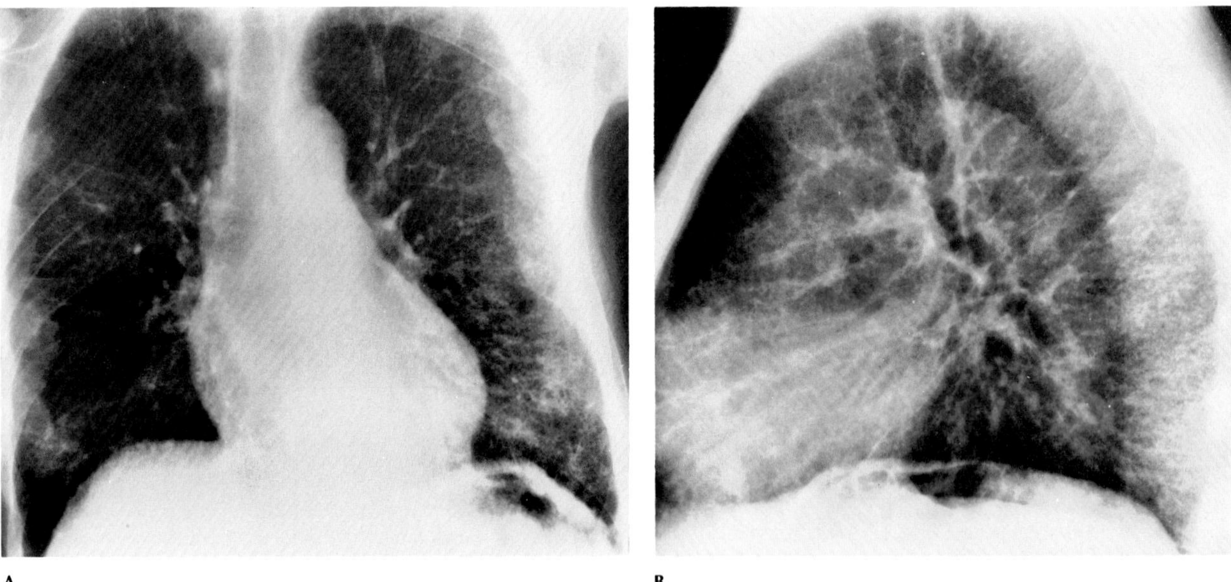

A B

**Figure 63.8. Rheumatoid lung.** (A) Frontal and (B) lateral views of the chest show diffuse thickening of the interstitial structures with prominent pleural thickening.

ligaments from synovial inflammation. Atlantoaxial subluxation may be demonstrated on routine lateral films of the cervical spine, though films taken in flexion are often necessary. The odontoid process is often eroded and the dens completely destroyed. Upward displacement of C2 may permit the dens to impinge on the upper cervical cord or medulla, producing acute neurologic symptoms requiring immediate traction or decompression. Facet joint involvement may cause subluxations at one or more vertebral levels.

Rheumatoid nodules are soft tissue masses that usually appear over the extensor surfaces on the ulnar aspect of the wrist or the olecranon but occasionally present over other bony prominences, tendons, or pressure points. These characteristic nodules are seen in no other disease and develop in about 20 percent of patients with rheumatoid arthritis.

Abnormal chest radiographs are seen in a small number of patients with rheumatoid arthritis. Pleural effusion, chronic diffuse interstitial disease, and pulmonary or

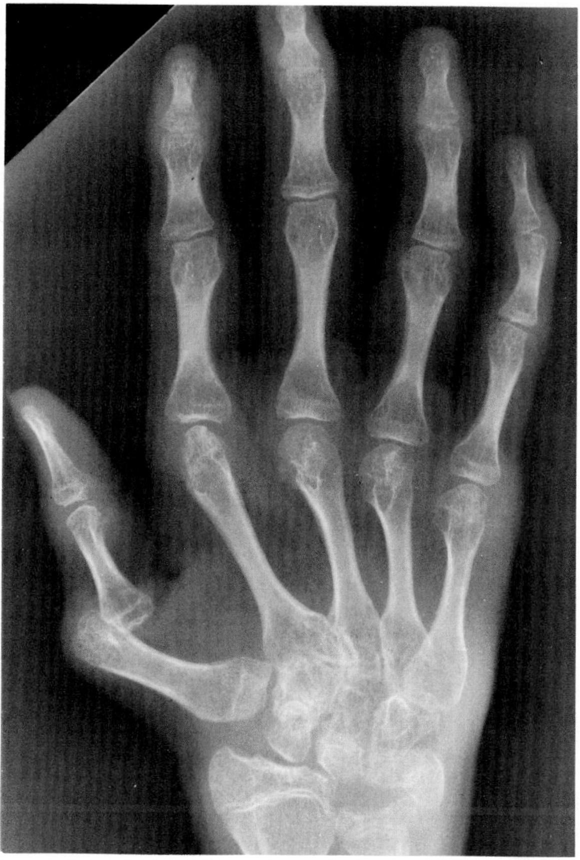
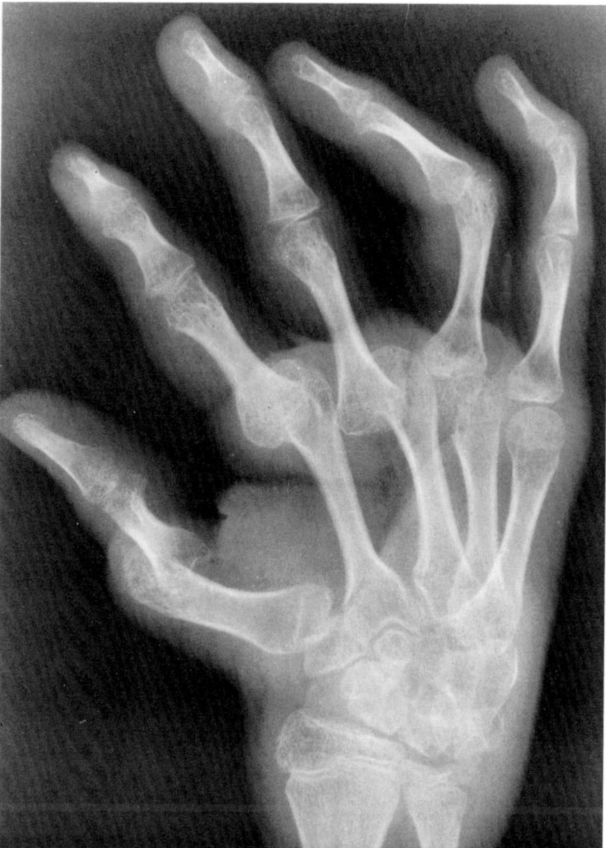

**Figure 63.9. Juvenile rheumatoid arthritis.** There is diffuse periarticular soft tissue swelling with moderate osteoporosis. Multiple subluxations are seen, especially involving the metacarpophalangeal joints.

pleural nodules may be seen in patients with rheumatoid arthritis even before the arthritic features are evident. Resorption of the distal end of the clavicle is a common finding.

Some patients with long-standing rheumatoid arthritis have associated splenomegaly and neutropenia, making them prone to infections (Felty's syndrome).[1–6]

## JUVENILE RHEUMATOID ARTHRITIS

Juvenile rheumatoid arthritis is a generalized connective tissue disease of children. Systemic manifestations are typically more severe in juvenile rheumatoid arthritis than in the adult form of the disease and may be present for several years before radiographic changes are evident. Unlike the peripheral distribution of involved joints in the adult form of rheumatoid arthritis, juvenile rheumatoid arthritis most frequently affects the areas of greatest bone growth, such as the knees, ankles, and wrists. Monarticular disease, especially in the knee, is more common than in adults. Although the radiographic findings are often similar to those of rheumatoid arthritis in adults, certain distinctive features of juvenile rheumatoid arthritis reflect the presence of the inflammatory process in an actively growing skeleton.

Periarticular soft tissue swelling and osteoporosis are early findings in juvenile rheumatoid arthritis. Extensive deossification of an affected extremity may cause bands of metaphyseal lucency mimicking the appearance in childhood leukemia. Periosteal calcification is much more common and severe than in the adult form of the disease. Joint space narrowing and articular erosion are late findings; synovial cysts infrequently occur. As in the adult form, ankylosis about the wrist and ankle is common in juvenile rheumatoid arthritis.

Juvenile rheumatoid arthritis may lead to a variety of growth disturbances. Bone growth may be initially accelerated because of local hyperemia, then delayed by early epiphyseal fusion or the administration of steroids. Overgrowth of the epiphyses of an affected joint may produce a characteristic balloon appearance, with relative constriction above and below the articular surface. Muscular atrophy accompanying overgrowth in the length of long bones may result in bowing deformities.

As in the adult form of rheumatoid arthritis, juvenile rheumatoid arthritis of the spine predominantly involves

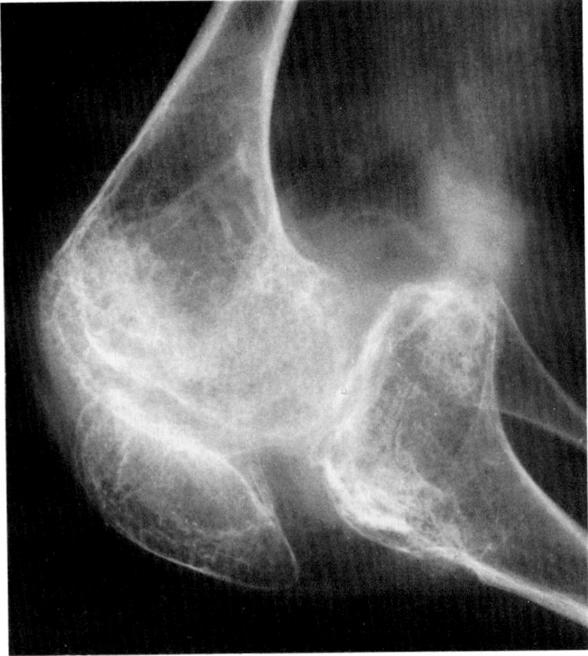

A

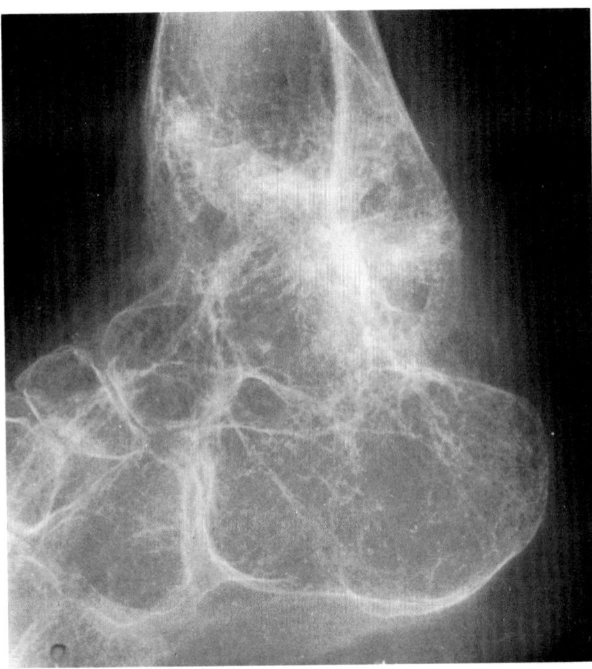

B

**Figure 63.10. Juvenile rheumatoid arthritis.** Lateral views of (A) the knee and (B) the ankle demonstrate severe demineralization of bone. There is expansion of the epiphyseal and metaphyseal areas with relative constriction of the underdeveloped diaphyseal portions. Narrowing of the tarsal joints is also seen.

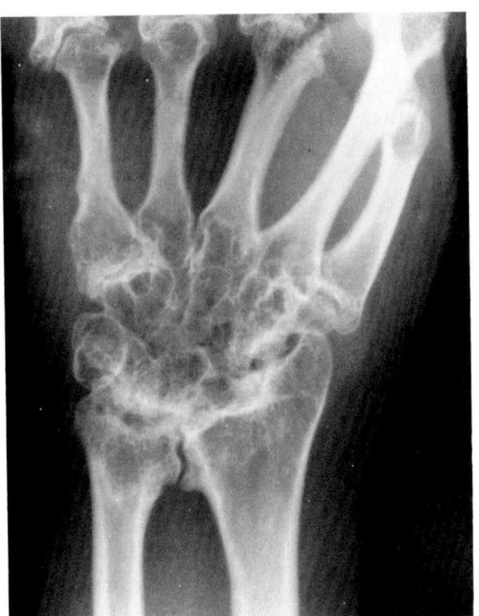

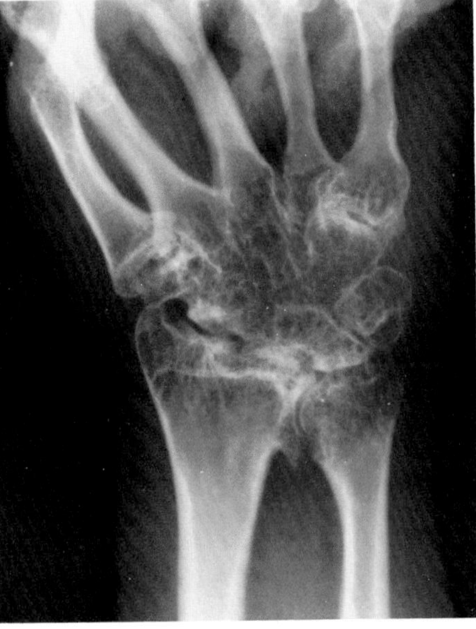

**Figure 63.11. Juvenile rheumatoid arthritis.** Views of both wrists demonstrate severe deossification of the carpal bones with joint space narrowing and even obliteration. Note the virtual ankylosis between the distal radius and ulna and the proximal carpal row.

the cervical area. Apophyseal joint ankylosis and atlantoaxial subluxation are common.

In the knees, overgrowth of epiphyses and erosion of the intercondylar notch of the femur may produce an appearance identical to that seen in hemophilia. Because the mandibular condyle is one of the most active growth centers, inflammation of the temporomandibular joints may lead to condylar erosion and micrognathia.[7–9]

## SJÖGREN'S SYNDROME

Sjögren's syndrome consists of a classic triad of dry eyes (keratoconjunctivitis sicca), dry mouth (xerostomia), and a chronic polyarthritis that occurs in half the cases and is indistinguishable from ordinary rheumatoid arthritis. Bilateral parotid gland enlargement is common, and sialography demonstrates dilatation of the ducts and alveoli

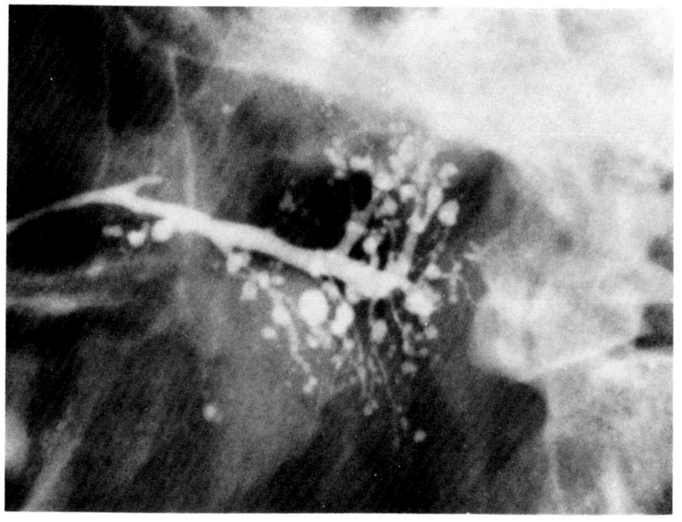
A

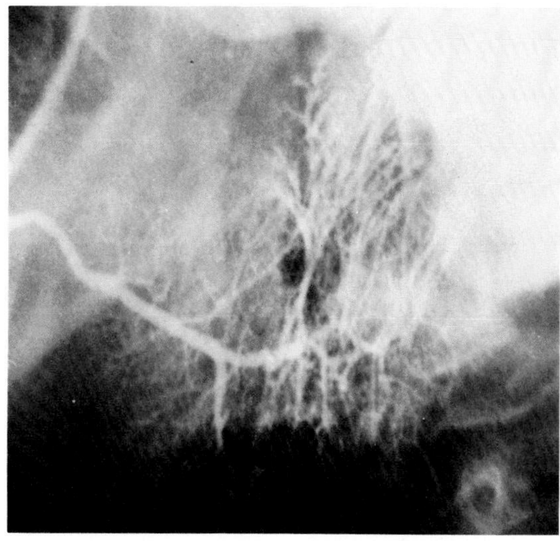
B

**Figure 63.12. Sjögren's syndrome.** (A) A sialogram demonstrates severe sialectasia with dilatation of the ducts and alveoli of the parotid gland. (B) Normal sialogram for comparison shows the fine ductal radicles.

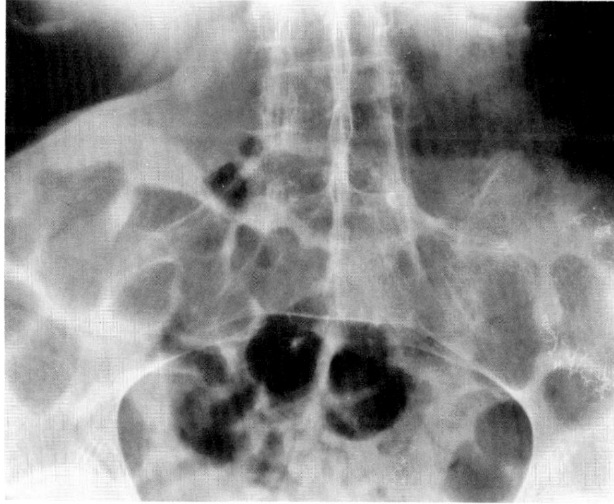

**Figure 63.13. Ankylosing spondylitis.** Bilateral symmetric obliteration of the sacroiliac joints.

of the parotid glands. Other radiographic findings in Sjögren's syndrome include interstitial fibrosis and lower lobe bronchiectasis, lipoid pneumonia (from ingestion of oils taken to combat the dry mouth), and hepatosplenomegaly. Lymphadenopathy is common in Sjögren's syndrome; malignant lymphoma occurs in up to 5 percent of the patients with this condition.[10–12]

## ANKYLOSING SPONDYLITIS (MARIE-STRÜMPELL DISEASE)

Ankylosing spondylitis is a chronic and usually progressive inflammatory disease that primarily involves the sacroiliac joints, spinal apophyseal joints, and paravertebral soft tissues. Generally considered a rheumatoid variant, ankylosing spondylitis differs from rheumatoid arthritis in that patients with the disease are generally young adult males, who very frequently have the histocompatibility antigen HLA-B27 without evidence of

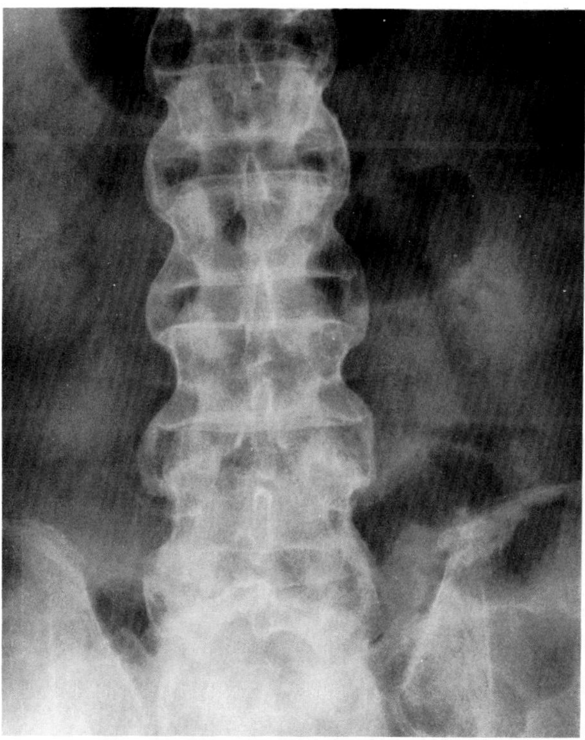

**Figure 63.14. Ankylosing spondylitis.** Extensive lateral bony bridges (syndesmophytes) connect all the lumbar vertebral bodies.

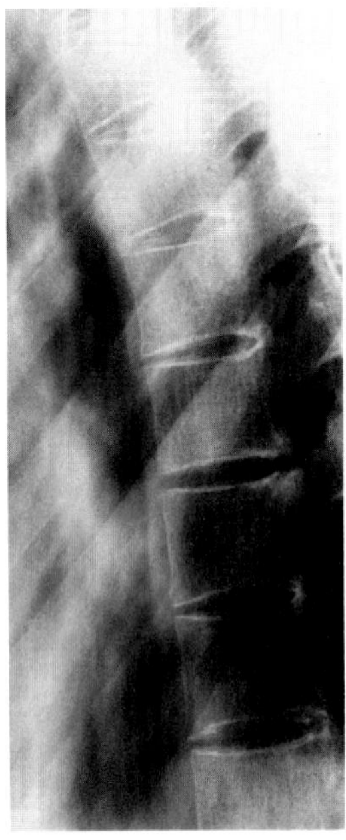

**Figure 63.15. Ankylosing spondylitis.** Squaring of the thoracic vertebral bodies.

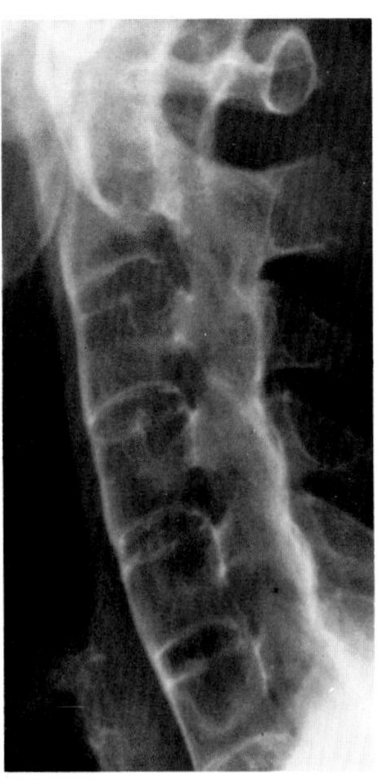

**Figure 63.16. Ankylosing spondylitis.** Extensive ossification of the longitudinal ligaments about the cervical spine.

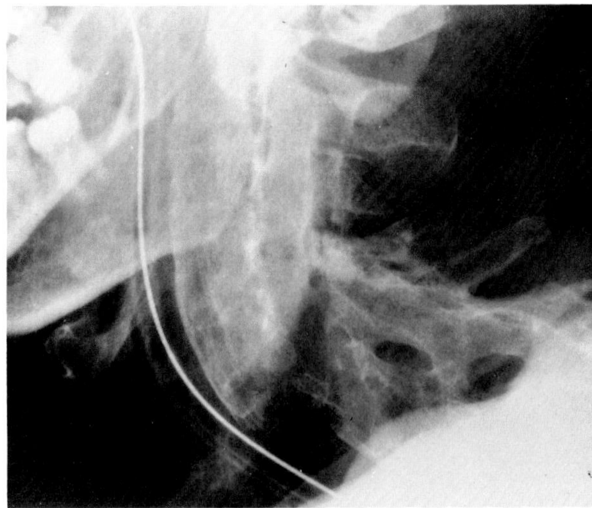

**Figure 63.17. Ankylosing spondylitis.** An oblique fracture of the midcervical spine, with anterior dislocation of the superior segment, is seen in a patient who fell while dancing and struck his head. The fracture extends through the lateral mass and lamina. Because of loss of flexibility and osteoporosis, patients with ankylosing spondylitis can suffer a fracture with relatively slight trauma.

rheumatoid factor in the serum. The earliest clinical manifestation of ankylosing spondylitis is usually intermittent low back pain and stiffness. In contrast to rheumatoid arthritis, in ankylosing spondylitis acute anterior uveitis is often seen and may precede the joint complaints; aortic insufficiency and cardiac conduction disturbances may occur.

Ankylosing spondylitis almost always begins in the sacroiliac joints, causing bilateral and usually symmetric involvement, unlike the predominantly unilateral and rarely symmetric sacroiliac joint changes that may be a late finding in rheumatoid arthritis. Early findings include blurring of the articular margins, irregular widening of the joints, and patchy sclerosis. This generally progresses to narrowing of the joint space and may lead to complete fibrous and bony ankylosis. There may be dense reactive sclerosis, or the sclerosis may become less prominent as the joint becomes obliterated.

In the lumbar spine, ankylosing spondylitis tends to initially involve the lowermost levels and progress upward to the dorsal and cervical regions. In the apophyseal joints, there is initially blurring of the articular margins followed by erosions, sclerosis, and eventual ankylosis. An erosive osteitis of the corners of the vertebral bodies produces a loss of the normal anterior concavity and a characteristic squared vertebral body. Ossification in the paravertebral tissues and longitudinal spinal ligaments combined with extensive lateral bony bridges (syndesmophytes) between vertebral bodies produces the characteristic "bamboo spine" of advanced disease. Limitation of activity leads to generalized skeletal osteoporosis and a propensity to fracture in response to stress or minor trauma. These fractures usually occur through a disk space, rather than a vertebral body, and continue through the posterior elements. At-

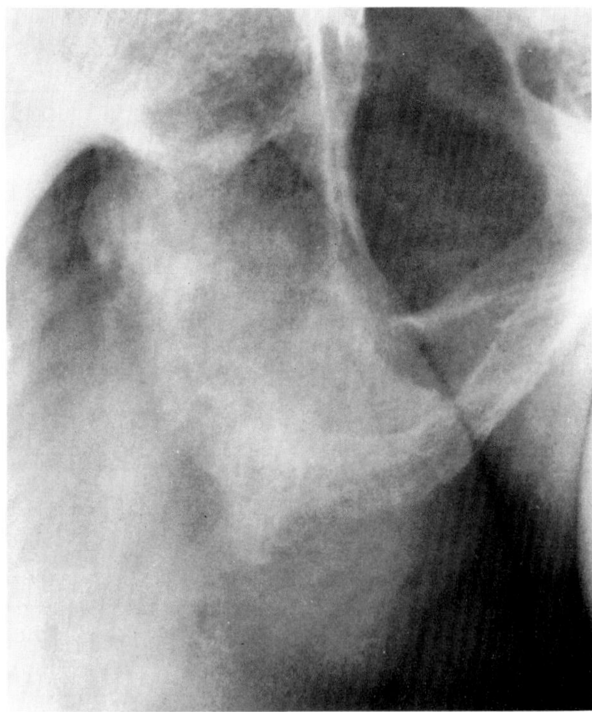

**Figure 63.18. Ankylosing spondylitis.** Irregular proliferation of new bone ("whiskering") along the inferior pubic ramus.

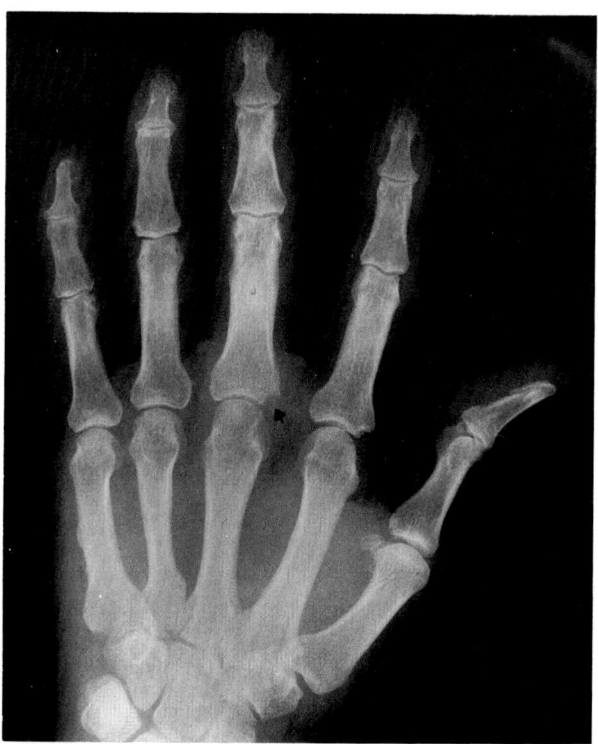

**Figure 63.19. Ankylosing spondylitis involving the hand.** Erosive changes without joint space narrowing (arrows) affect the bases of the proximal phalanges of the second and third digits.

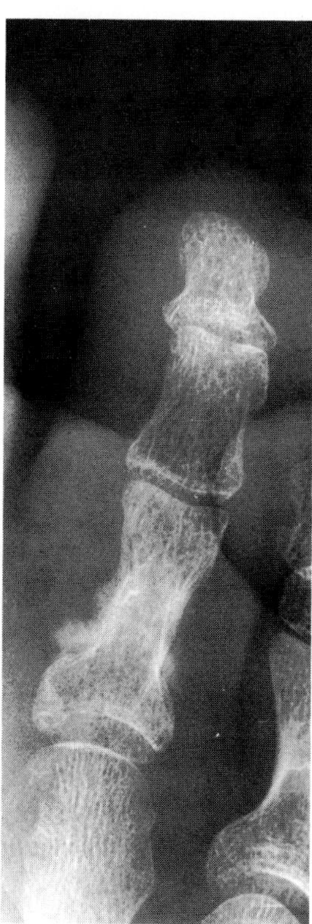

**Figure 63.20. Reiter's syndrome.** Soft tissue swelling of a toe with typical fluffy periosteal reaction about the proximal phalanx. (From Forrester, Brown, and Nesson,[1] with permission from the publisher.)

lantoaxial subluxation may develop in ankylosing spondylitis, though far less frequently than in rheumatoid arthritis.

A typical finding in ankylosing spondylitis is irregular proliferative new bone formation ("whiskering") at sites of ligamentous or muscular attachments. This fluffy periostitis most commonly involves the ischial tuberosities, iliac crests, greater trochanters of the femur, and calcaneus. The hips and shoulders are involved in up to 80 percent of the patients with ankylosing spondylitis. Although the radiographic findings are similar to those in rheumatoid arthritis, in ankylosing spondylitis there tends to be less osteoporosis and more reactive sclerosis and bony ankylosis. Peripheral joint involvement is seen in up to half the patients. The appearance in small joints resembles that in Reiter's disease or psoriatic arthritis rather than rheumatoid arthritis, in that there tends to be asymmetric involvement, less erosive change, and often a fluffy periostitis. Narrowing, erosions, and sclerosis of the temporomandibular joints are seen in about one-third of the patients with long-standing disease.

Ankylosing spondylitis is occasionally accompanied by pulmonary disease that typically appears radiographically as bilateral upper lobe fibrosis. Cavitation may occur, often associated with aspergillosis.[13–18]

## REITER'S SYNDROME

Reiter's syndrome is characterized by arthritis, urethritis, conjunctivitis, and mucocutaneous lesions. It primarily affects young adult males and appears to be a postinfectious syndrome following certain types of venereal or enteric infections. Reiter's syndrome most frequently involves the sacroiliac joints, heel, and toes. In peripheral joints, the initial change is nonspecific periarticular swelling, usually followed by joint space narrowing and marginal erosions. Although the radiographic changes often mimic rheumatoid arthritis, Reiter's syndrome tends to be asymmetric and primarily involves the feet rather than the hands. Periosteal new bone formation is common, initially appearing as a linear or laminated pattern indistinguishable from other types of arthritis. Later, a characteristic fluffy type of periostitis develops at the site of tendon insertions, especially at the attachments of the plantar fascia and Achilles tendon to the inferior and posterosuperior margins of the calcaneus. A large, irregular, fluffy spur often forms on the plantar aspect of the heel. Reiter's syndrome often causes progressive changes in the sacroiliac joints consisting of joint space narrowing, erosions, and sclerosis that may be indistinguishable from the sacroiliac involvement in ankylosing spondylitis or psoriatic arthritis. Sacroiliac involvement may be unilateral or bilateral; bilateral involvement is usually asymmetric, in contrast to the symmetric changes in ankylosing spondylitis. Reiter's syndrome tends to cause only minimal changes in the spine. The most charac-

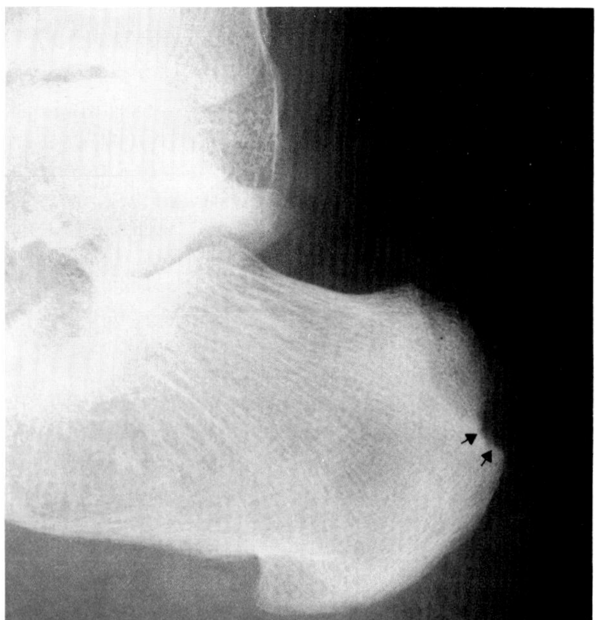

Figure 63.21. **Reiter's syndrome.** Striking bony erosion (arrows) at the site of insertion of the Achilles tendon on the posterosuperior margin of the calcaneus.

teristic features are large, nonmarginal bony spurs (syndesmophytes) that are asymmetric and bridge adjacent vertebral bodies.

In summary, Reiter's syndrome should be suggested when radiographs of a young adult male demonstrate fluffy periostitis adjacent to small joints of the foot, ankle, and calcaneus; inferior calcaneal spurs; asymmetric changes in the sacroiliac joints without significant spinal involvement; and an asymmetric peripheral arthritis that tends to involve the feet more than the hands.[19–21]

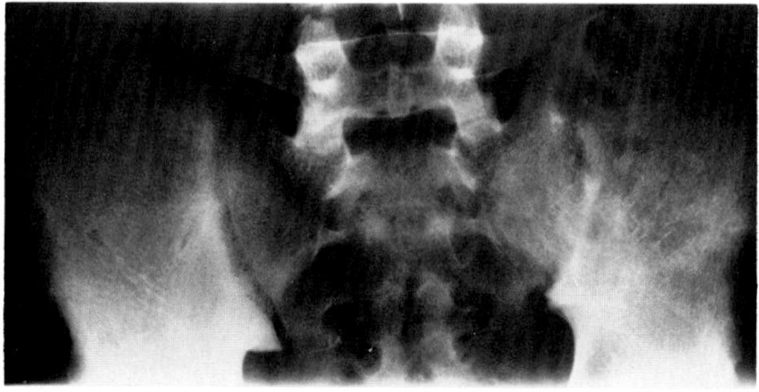

**Figure 63.22. Reiter's syndrome.** Bilateral, though asymmetric, narrowing of the sacroiliac joints with reactive sclerosis.

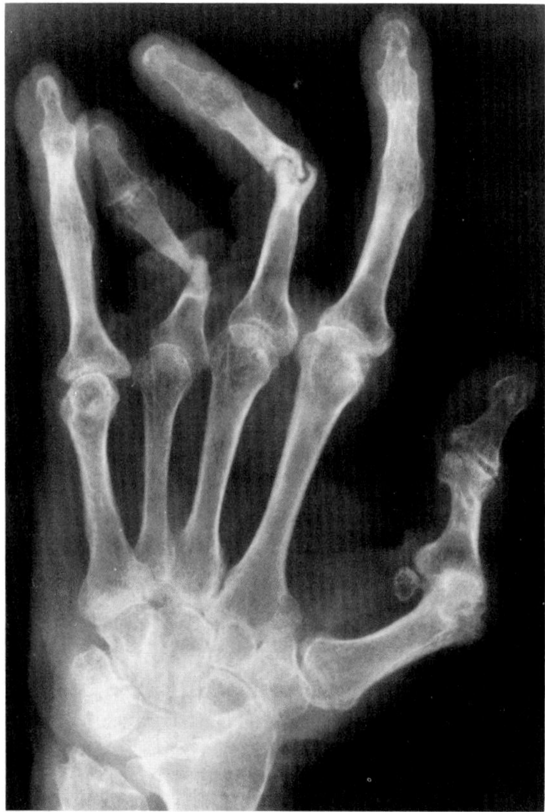

A

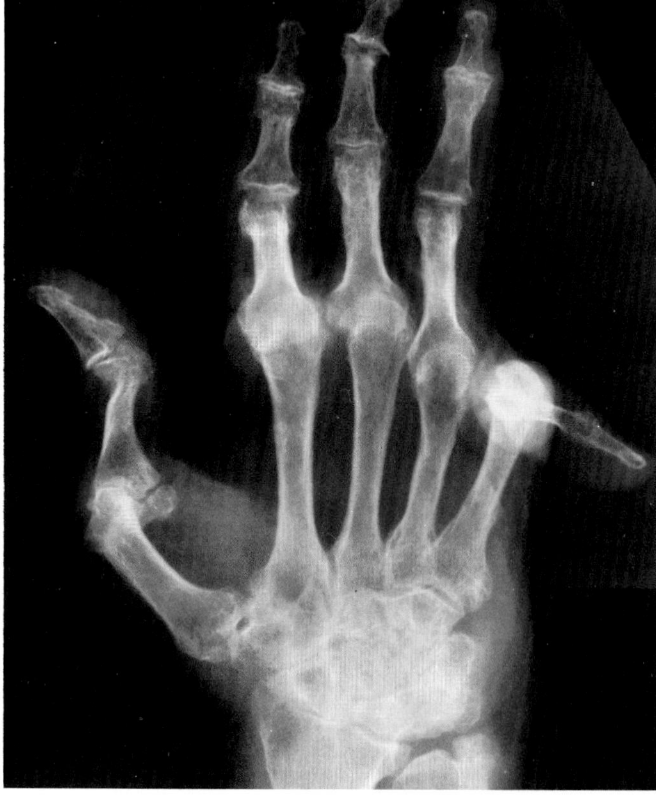

B

**Figure 63.23. Psoriatic arthritis.** Views of the left (A) and right (B) hands and wrists show a bizarre pattern of asymmetric bone destruction, subluxation, and ankylosis. Note particularly the pencil-in-cup deformity of the third proximal interphalangeal joint on the left, the dislocations about the second through fifth metacarpophalangeal joints on the right, and the bony ankylosis involving both wrists and the phalanges of the second and fifth digits on the left.

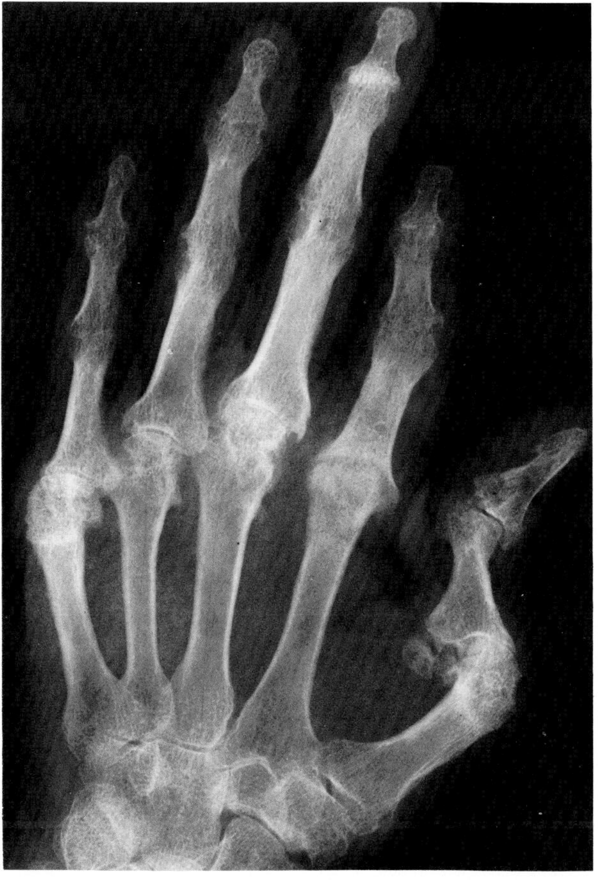


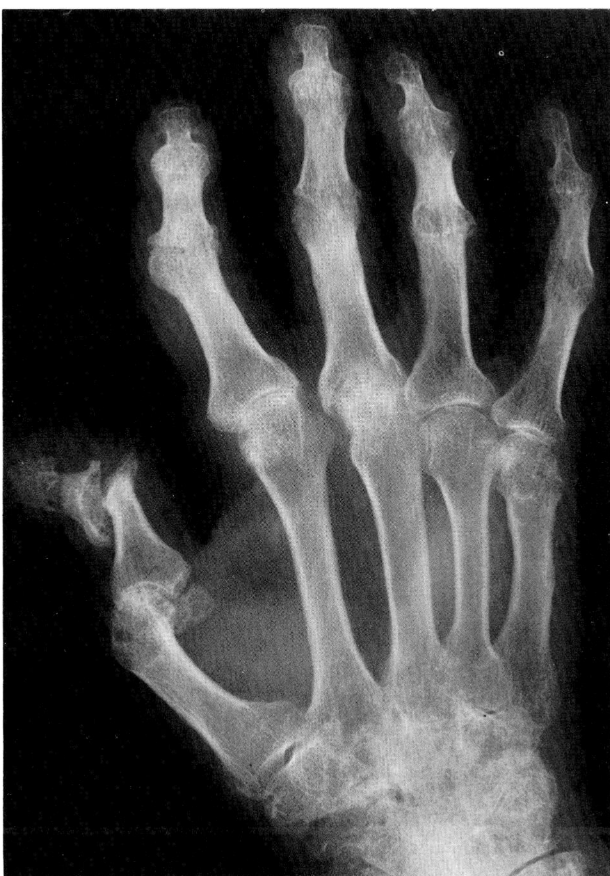


**Figure 63.24. Psoriatic arthritis.** Views of both hands and wrists (A,B) demonstrate ankylosis across many of the interphalangeal joints with scattered erosive changes involving several interphalangeal joints, most of the metacarpophalangeal joints, and the interphalangeal joint of the right thumb. Note the striking asymmetry of involvement of the carpal bones, an appearance unlike that expected in rheumatoid arthritis.

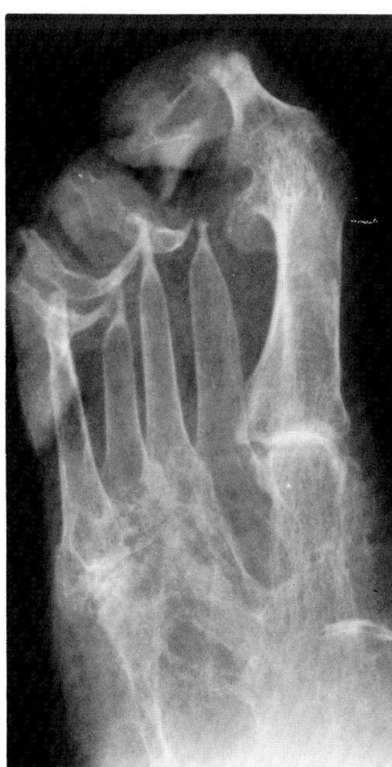

**Figure 63.25. Psoriatic arthritis.** Arthritis mutilans of the foot and ankle. There is ankylosis of almost all the tarsal joints and severe pencil-like destruction of the metatarsals and phalanges.

## PSORIATIC ARTHRITIS

Patients with psoriasis often manifest a type of arthritis that is radiographically similar to rheumatoid arthritis, with which psoriasis not uncommonly coexists. In true psoriatic arthritis, however, rheumatoid factor and rheumatoid nodules are typically not present. Like rheumatoid arthritis, psoriatic arthritis is a destructive process that produces soft tissue swelling, joint space narrowing, and periarticular erosions. Unlike rheumatoid arthritis, psoriatic arthritis predominantly involves the distal rather than the proximal interphalangeal joints of the hands and feet, produces asymmetric rather than symmetric destruction, and causes little or no periarticular osteoporosis. At times, however, both the distal and the proximal interphalangeal joints may be involved in psoriatic arthritis. Characteristic radiographic features of psoriatic arthritis include the tendency for bony ankylosis of the interphalangeal joints of the hands and feet; resorption of the terminal tufts of the distal phalanges; fluffy periosteal reaction near joints and along the shafts; calcaneal

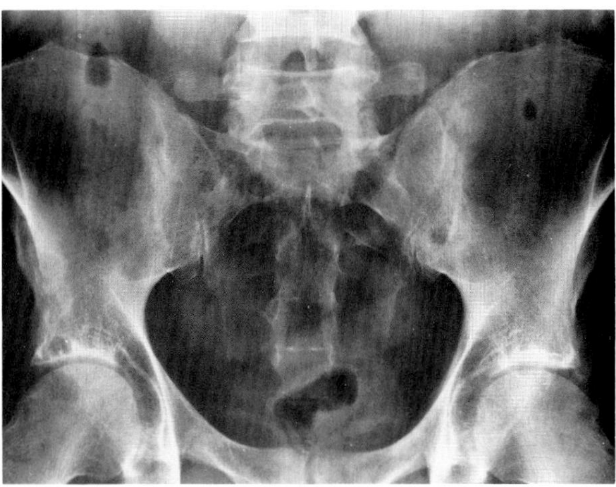

**Figure 63.26. Psoriatic arthritis.** Bilateral, though asymmetric, narrowing of the sacroiliac joints.

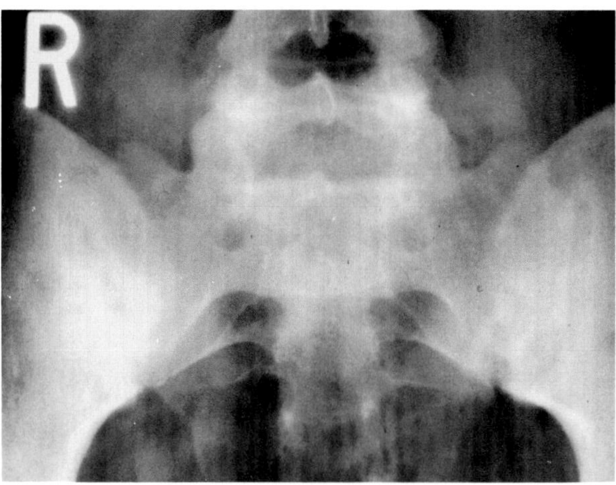


**Figure 63.27. Spondylitis in a patient with ulcerative colitis.** Symmetric involvement of the sacroiliac joints, which produces an appearance indistinguishable from that of ankylosing spondylitis.

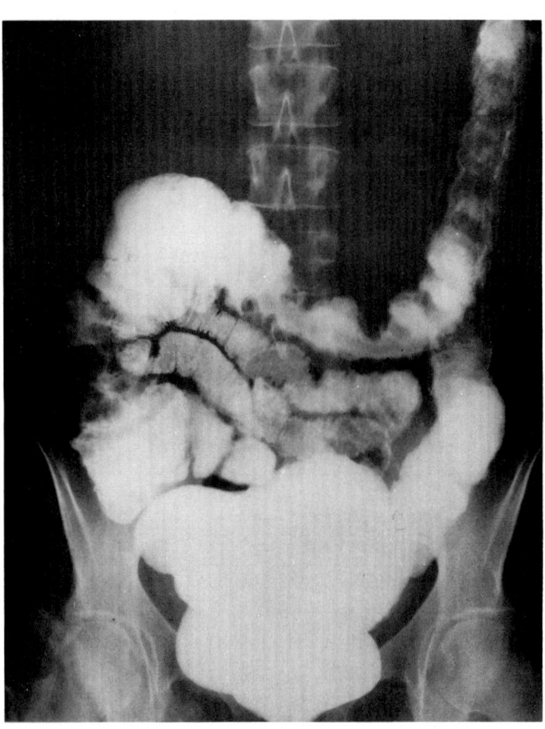

A

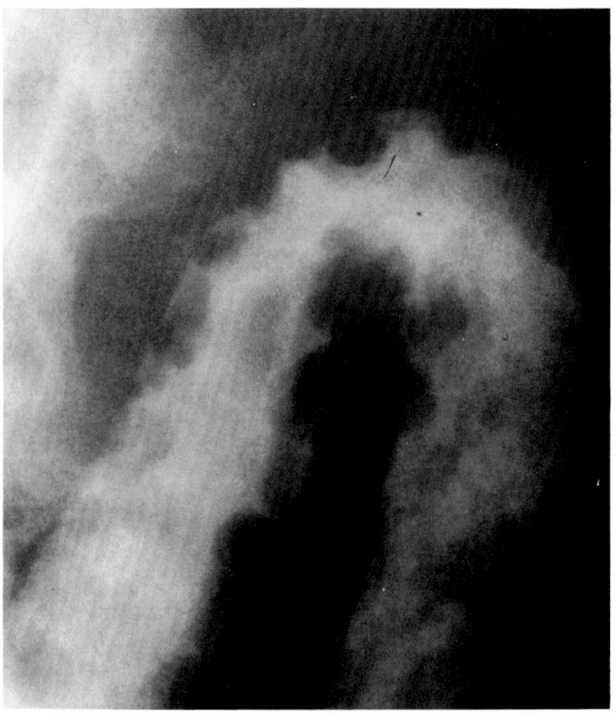

B

**Figure 63.28. Colitis in Behçet's syndrome.** (A) Barium enema examination demonstrates extensive involvement of all the large bowel except for the rectosigmoid and hepatic flexure region. The affected mucosa is nodular, ulcerated, and thickened secondary to granulomatous disease. (B) An enlarged view of the splenic flexure shows deep, circular ulcerations of uniform size and multiple nodular mucosal lesions. (From Goldstein and Crooks,[27] with permission from the publisher.)

spurs; arthritis mutilans with severe destruction and telescoping of the fingers; and a "pencil-in-cup" deformity in which destruction of the distal end of a bone produces the "pencil" that projects into a widened, cuplike erosion in the adjacent joint surface. An almost pathognomonic finding is the destruction of interphalangeal joints with abnormally wide joint spaces and sharply demarcated adjacent bony surfaces.

The sacroiliac joints in psoriatic arthritis may show unilateral or asymmetric bilateral erosive changes with joint space narrowing, reactive sclerosis, and, occasionally, bony ankylosis. Asymmetric syndesmophytes, similar to those in Reiter's disease, are seen in the thoracolumbar spine. Calcifications in the paravertebral soft tissues are common.[21–24]

## ARTHRITIS ASSOCIATED WITH INFLAMMATORY BOWEL DISEASE

Up to 25 percent of the patients with ulcerative or Crohn's colitis have some form of arthritis: spondylitis, peripheral arthritis, or coincidental rheumatoid arthritis. The spondylitis presents as symmetric involvement of the sacroiliac joints that is indistinguishable from ankylosing spondylitis. It is closely associated with the presence of the tissue histocompatibility antigen HLA-B27. In about one-third of the patients, the spondylitis antedates the onset of inflammatory bowel disease.

Peripheral arthritis associated with inflammatory bowel disease tends to be migratory and to involve large joints such as the knee and elbow. It usually follows the onset of colitis and tends to flare up during exacerbations of colonic disease. The arthritis is more frequent in patients with ulcerative colitis who have pseudopolyps and who have extensive disease that is not limited to the rectum. In Crohn's disease, peripheral arthritis is more common in patients with concomitant colitis rather than with disease limited to the small bowel. The radiographic changes are usually limited to soft tissue swelling and joint effusion. Because joint cartilage and bony apposition are generally unaffected, there usually is no residual damage. Occasionally, patients with recurrent or persistent disease may demonstrate small bony erosions and joint space narrowing.[25]

## BEHÇET'S SYNDROME

Behçet's syndrome is an uncommon multiple-system disease, most prevalent in the Middle East and Japan, that is characterized by ulcerations of the buccal and genital mucosae, ocular inflammation, and a variety of skin lesions. The underlying histopathologic lesion is a vasculitis that affects small to medium-sized arteries and veins.

Colonic involvement, which may be associated with diarrhea, abdominal pain, and bleeding, produces radiographic findings indistinguishable from other forms of inflammatory bowel disease. Diffuse mucosal thickening and ulceration may involve large segments of the colon and terminal ileum, often sparing the rectum. In some patients, multiple discrete ulcers are seen in an otherwise normal-appearing colon. Patients with colitis due to Behçet's syndrome have a high incidence of perforation and hemorrhage, both of which are life-threatening complications.

Arthritis or arthralgia develops in about half the patients with Behçet's syndrome and may precede, accompany, or follow other manifestations of the disease. The arthritis is usually asymmetric and polyarticular, most often involving large joints such as the knees, ankles, and elbows. Most patients have multiple recurrent episodes of acute inflammatory arthritis (simulating the arthritis associated with inflammatory bowel disease); however, there rarely is any radiographic evidence of residual joint damage.

Thoracic manifestations of Behçet's syndrome are rare. Thrombophlebitis may result in superior vena cava obstruction; pulmonary embolism occasionally complicates the lower extremity thrombophlebitis that not infrequently develops in this syndrome. Diffuse air space consolidation resembling pulmonary edema or extensive hemorrhage also has been reported in this condition.[26–28]

### REFERENCES


1. Forrester DM, Brown JC, Nesson JW: *The Radiology of Joint Disease*. Philadelphia, WB Saunders, 1978.
2. Martel W, Hayes JT, Duff IF: The pattern of bone erosion in the hand and wrist in rheumatoid arthritis. *Radiology* 84:204–214, 1965.
3. Cooperberg PL, Tsang I, Truelove L, et al: Gray scale ultrasound in the evaluation of rheumatoid arthritis of the knee. *Radiology* 126:759–763, 1978.
4. Resnick D: Rheumatoid arthritis of the wrist: Why the ulnar styloid? *Radiology* 112:29–35, 1974.
5. Bland JN, Van Buskirk FW, Tampas JP, et al: A study of roentgenologic criteria for rheumatoid arthritis of the cervical spine. *AJR* 95:949–954, 1965.
6. Berens DL, Lockie LM, Lin R, et al: Roentgen changes in early rheumatoid arthritis. *Radiology* 82:645–654, 1964.
7. Ansell BM, Kent PA: Radiological changes in juvenile chronic polyarthritis. *Skel Radiol* 1:129–136, 1977.
8. Schaller J, Wedgewood RJ: Juvenile rheumatoid arthritis: A review. *Pediatrics* 50:940–953, 1972.
9. Martel W, Holt JF, Cassidy JT: Roentgenologic manifestations of juvenile rheumatoid arthritis. *AJR* 88:400–423, 1962.
10. Silbiger ML, Peterson CC: Sjögren's syndrome. *AJR* 100: 554–562, 1967.
11. Lin SR, Edeiken J, Freundlich IM: The roentgenologic manifestations of Sjögren's syndrome. *Rev Interam Radiol* 3:76–81, 1968.
12. Whaley K, Blair S, Low DM, et al: Sialographic abnormalities in Sjögren's syndrome and other arthritides. *Clin Radiol* 23:474–482, 1972.
13. Braunstein EM, Martel W, Moidel R: Ankylosing spondylitis in men and women: A clinical and radiographic comparison. *Radiology* 144:91–94, 1982.
14. Resnick D: Patterns of peripheral joint disease in ankylosing spondylitis. *Radiology* 110:523–532, 1974.
15. Berens DL: Roentgen features of ankylosing spondylitis. *Clin Orthop* 74:20–33, 1971.
16. Schumacher TM, Genant HK, Kellet MJ, et al: HLA-B27 associated arthropathies. *Radiology* 126:289–297, 1978.
17. Resnick D: Temporomandibular joint involvement in ankylosing spondylitis. *Radiology* 112:587–591, 1974.
18. Hunter T, Dubo H: Spinal fractures complicating ankylosing spondylitis. *Ann Intern Med* 88:546–549, 1978.
19. Martel W, Braunstein EM, Borlaza G, et al: Radiologic features of Reiter disease. *Radiology* 132:1–10, 1979.
20. Sholkoff SD, Glickman MG, Steinbach HL: Roentgenology of Reiter's syndrome. *Radiology* 97:497–503, 1970.
21. Peterson CC, Silbiger MC: Reiter's syndrome and psoriatic

arthritis: Their roentgen spectra and some interesting similarities. *AJR* 101:860–871, 1967.

22. Killibrew K, Gold RH, Sholkoff SD: Psoriatic spondylitis. *Radiology* 108:9–16, 1973.
23. Alvila R, Pugh DC, Slocumb CH, et al: Psoriatic arthritis. A roentgenologic study. *Radiology* 75:691–701, 1960.
24. Wright V: Psoriatic arthritis: Comparative radiographic study of rheumatoid arthritis and arthritis associated with psoriasis. *Ann Rheum Dis* 20:123–132, 1961.
25. Greenstein AJ, Janowitz HD, Sachar DB: The extra-intestinal complications of Crohn's disease and ulcerative colitis: A study of 700 patients. *Medicine* 55:401–412, 1976.
26. Cadman EC, Lundberg WB, Mitchell MS: Pulmonary manifestations in Behçet's syndrome. *Arch Intern Med* 136:944–947, 1976.
27. Goldstein SJ, Crooks DJM: Colitis in Behçet's syndrome. *Radiology* 128:321–323, 1978.
28. Stanley RJ, Tedesco FJ, Melson GL, et al: The colitis of Behçet's disease: A clinical-radiographic correlation. *Radiology* 114: 603–604, 1975.
29. Brown JC, Forrester DM: Arthritis in Eisenberg RL, Amberg JR (eds): *Critical Diagnostic Pathways in Radiology: An Algorithmic Approach.* Philadelphia, JB Lippincott, 1981, pp 280–300.

# Chapter Sixty-Four
# Other Diseases of Joints and Connective Tissues

## INFECTIOUS ARTHRITIS

### Acute Bacterial Arthritis

Pyogenic organisms may gain entry into a joint by the hematogenous route, by direct extension from an adjacent focus of osteomyelitis, or from trauma to the joint (surgery, needling). Pyogenic arthritis is most often due to *Staphylococcus aureus,* though gonorrhea is the most common cause in young adults, especially women. The onset of bacterial arthritis is usually abrupt, with high fever, shaking chills, and one or a few severely tender and swollen joints.

Soft tissue swelling is the first radiographic sign of acute bacterial arthritis. In children, fluid distension of the joint capsule may cause widening of the joint space and actual subluxation, especially about the hip and shoulder. Periarticular edema displaces or obliterates adjacent tissue fat planes. Rapid destruction of articular cartilage causes joint space narrowing early in the course of disease. The earliest bone changes, which tend to appear 8 to 10 days after the onset of symptoms, are small focal erosions in the articular cortex. Because of the delay in bone changes, detection of the characteristic soft tissue abnormalities is essential for early diagnosis. Severe, untreated infections cause extensive destruction and a loss of the entire cortical outline. With healing, sclerotic bone reaction results in an irregular articular surface. If the articular cartilage has been completely destroyed, bony ankylosis usually follows.

In the spine, acute bacterial arthritis rapidly involves the intervertebral disks. There is loss of the disk spaces and destruction of the adjacent end plates, in contrast to the vertebral body involvement and preservation of disk spaces in metastatic disease.[1–3]

### Tuberculous Arthritis

Tuberculous arthritis is a chronic, indolent infection that has an insidious onset and a slowly progressive course. Usually monarticular, tuberculous arthritis primarily involves the spine, hips, and knees. Most patients have a focus of tuberculosis elsewhere in the body.

A distinctive early radiographic feature of tuberculous arthritis is extensive juxta-articular osteoporosis that precedes bone destruction, in contrast to bacterial arthritis, in which osteoporosis is a relatively late finding. Joint effusion leads to a nonspecific periarticular soft tissue swelling. Cartilage and bone destruction occur relatively late in the course of tuberculous arthritis and tend to initially involve the periphery of the joint, sparing the maximum-weight-bearing surfaces that are destroyed in

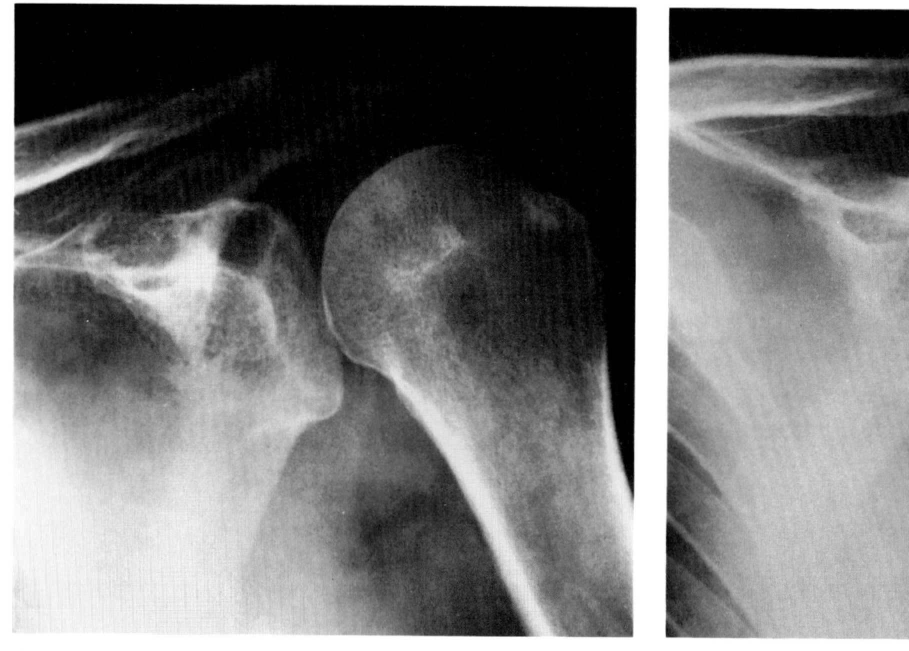
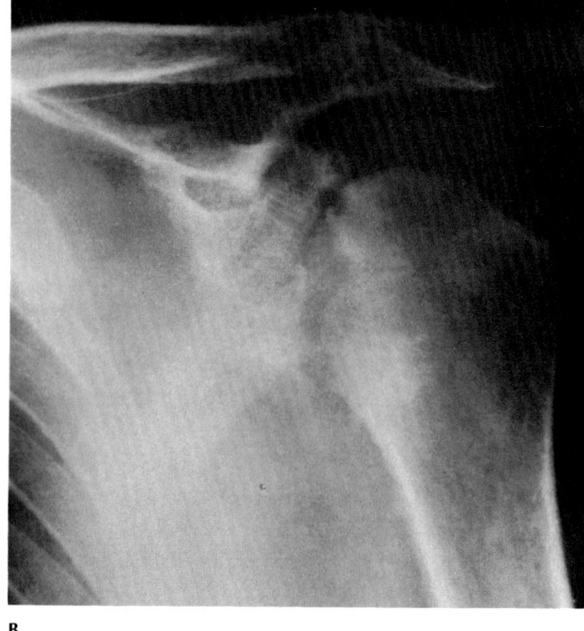

A B

**Figure 64.1. Acute staphylococcal arthritis.** (A) Several days following instrumentation of the shoulder for joint pain, there is separation of the humeral head from the glenoid fossa due to fluid within the joint space. (B) Six weeks later there is marked cartilage and bone destruction, with sclerosis on both sides of the glenohumeral joint.

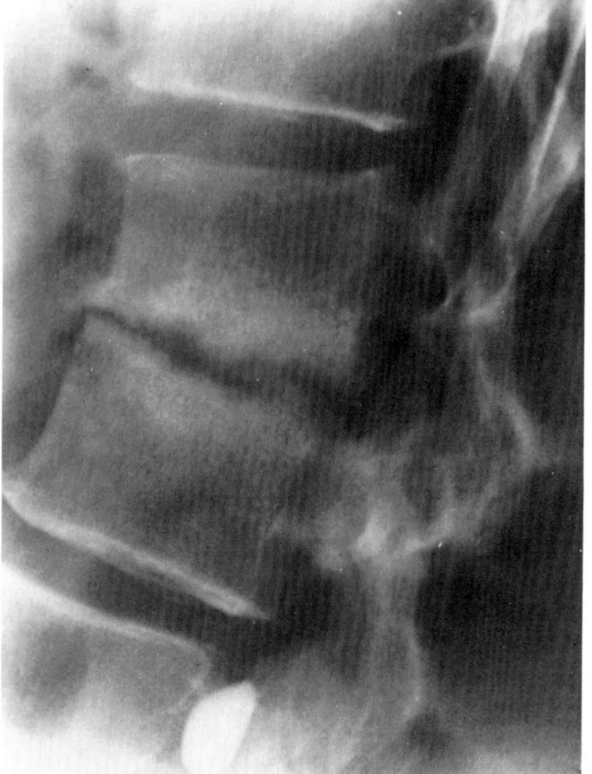

**Figure 64.2. Acute arthritis due to *Escherichia coli*.** There is disk space narrowing with irregular destruction and sclerosis of the adjacent vertebral end plates. (From Brown and Forrester,[54] with permission from the publisher.)

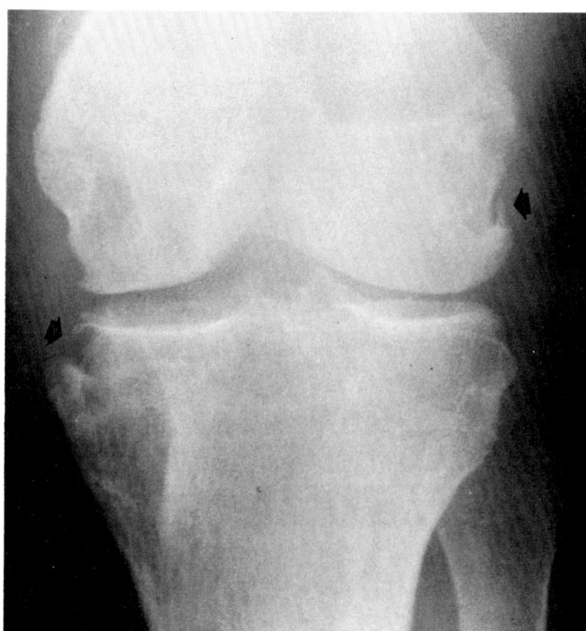

**Figure 64.3. Tuberculous arthritis of the knee.** There are destructive bone lesions (arrows) on both sides of the joint involving the medial and lateral femoral condyles and the medial aspect of the proximal tibia. Note the relative sparing of the articular cartilage and preservation of the joint space in view of the degree of bone destruction.

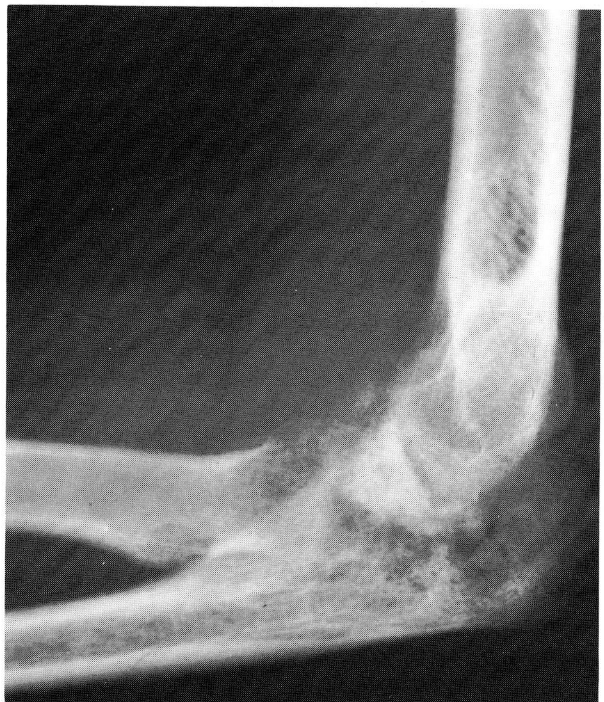

**Figure 64.4. Tuberculous arthritis of the elbow.** There is complete destruction of the joint space. Displacement of anterior and posterior fat pads and a large antecubital mass reflect marked synovial hypertrophy resulting from chronic granulomatous infection. (From Forrester, Brown, and Nesson,[6] with permission from the publisher.)

pyogenic arthritis. Therefore, joint space narrowing occurs late in tuberculous arthritis, in contrast to the early narrowing with bacterial infections. Similarly, the earliest evidence of bone destruction is usually erosion at the margins of the articular ends of bone. With progressive disease there is ragged destruction of the articular cartilage and subchondral cortex and disorganization of the joint, often with preservation of necrotic fragments of bone (sequestra) that may involve apposing surfaces ("kissing sequestra").

Tuberculous arthritis of the spine begins in the vertebral body, most commonly in the thoracic area (see the section on "Skeletal Tuberculosis" in Chapter 12). Bone destruction leads to collapse of the vertebral body and often a characteristic sharp, angular kyphosis (gibbous deformity). Extension of the infection may lead to the formation of a cold abscess that appears as a fusiform soft tissue paraspinal mass.[4]

## DEGENERATIVE JOINT DISEASE (OSTEOARTHRITIS)

Degenerative joint disease is an extremely common generalized disorder that is characterized pathologically by loss of joint cartilage and reactive new bone formation. Part of the wear and tear of the aging process, degenerative joint disease tends to predominantly affect the weight-bearing joints (spine, hip, knee, ankle) and the interphalangeal joints of the fingers. A secondary form of degenerative joint disease may develop in a joint that has been repeatedly traumatized or subjected to abnormal stresses because of orthopedic deformities, or may be a sequela of a septic or inflammatory arthritis that destroys cartilage.

The earliest radiographic findings in degenerative joint disease are narrowing of the joint space, due to thinning of the articular cartilage, and the development of small bony spurs (osteophytes) along the margins of the articular edges of the bones. In contrast to the smooth, even narrowing of the joint space in rheumatoid arthritis, the joint space narrowing in degenerative joint disease is irregular and more pronounced in that part of the joint in which weight-bearing stress is greatest and degeneration of the articular cartilage is most marked. The articular ends of the bones become increasingly dense (periarticular sclerosis). Erosion of the articular cortex may produce typical irregular, cystlike lesions with sclerotic margins in the subchondral bone near the joint. Calcific or ossified loose bodies may develop, especially at the knee. With advanced disease, relaxation of the joint capsule and other ligamentous structures may lead to subluxation. Local osteoporosis does not occur unless pain causes prolonged disuse of the joint.

In summary, the classic radiographic features of degenerative joint disease are uneven narrowing of the joint space, marginal spurring, periarticular sclerosis, and subchondral cysts.[5,6]

### Fingers

Degenerative joint disease of the hand primarily involves the distal interphalangeal joints of the fingers, though the proximal interphalangeal and first metacarpophalangeal joints may also be affected. Initially, tiny marginal spurs or small calcific flakes arise along the bases of the distal phalanges. Gradual enlargement of these spurs produces well-defined bony protuberances that appear clinically as the palpable and visible knobby thickening of Heberden's nodes. Heberden's nodes occur most frequently in women, in whom there appears to be some hereditary predisposition. Because the largest spurs form on the dorsal aspects of the articular ends of the bones, they are best demonstrated on lateral radiographs. In addition to Heberden's nodes, degenerative joint disease of the fingers produces the typical findings of joint space narrowing, articular sclerosis, and subchondral cysts.[6]

### Hip

Early changes of degenerative joint disease about the hip include increased density along the superior acetabular

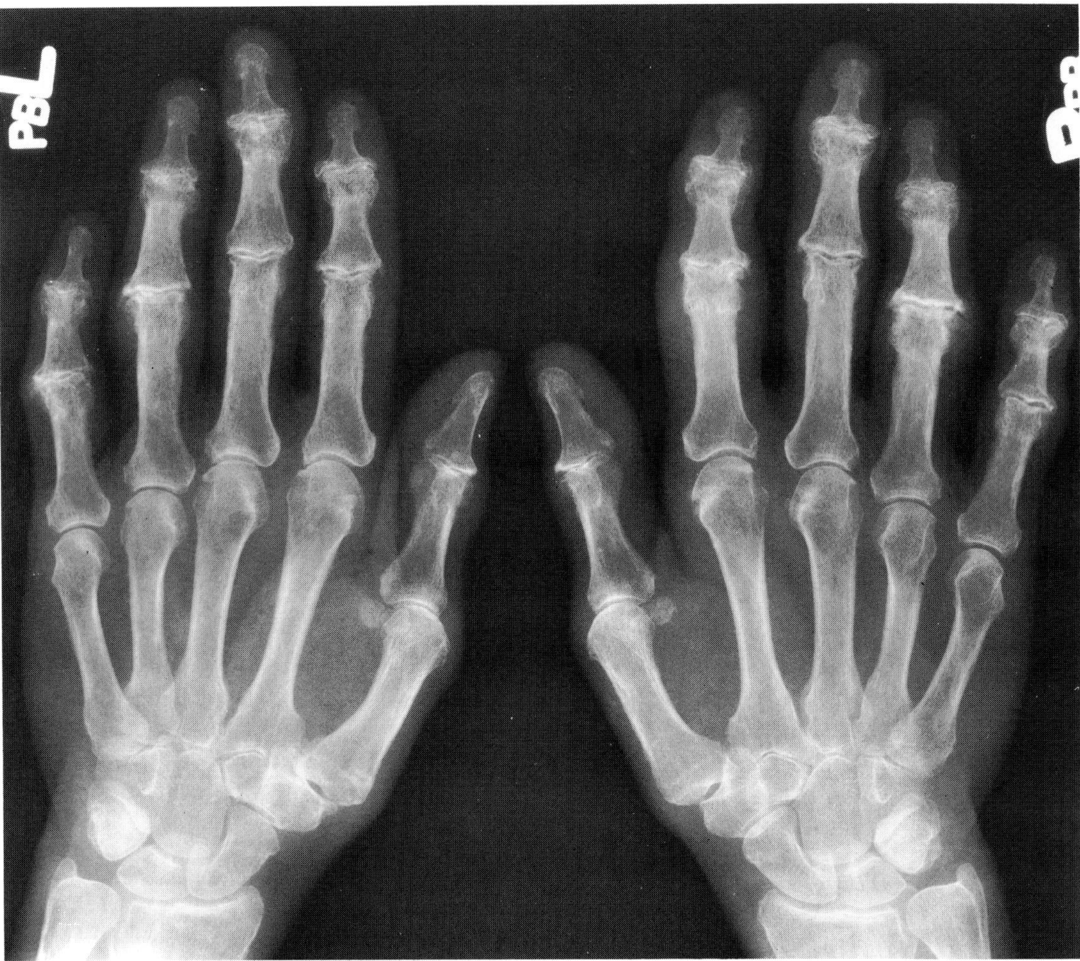


**Figure 64.5. Degenerative joint disease of the fingers.**

rim and large marginal osteophytes arising from the head of the femur. The most prominent finding is asymmetric narrowing of the joint space that predominantly involves the superior and lateral aspects of the joint, where the stress of weight-bearing is greatest. There is often the development of cystlike cavities that appear as sharply outlined rarefactions surrounded by dense sclerotic walls. Flattening of the femoral head and varus angulation may also occur.[7,8]

## Knee

The earliest change of degenerative joint disease of the knee is the development of spurring on the medial and lateral borders of the tibia and femur, the intercondylar notches, and the edges of the patella. Joint space narrowing is asymmetric and predominantly involves the medial femorotibial compartment where the weight-bearing stress is greatest. The lateral compartment is usually spared until late in the course of the disease. Calcific or bony loose bodies often form within the joint. Subchondral cysts may develop but are less frequent than in the hip. Because uneven narrowing of the joint space disturbs the weight-bearing alignment, there is often some degree of lateral subluxation of the tibia on the femur and a varus deformity, both of which may be best demonstrated on "weight-bearing" films.[9]

## Spine

The spine is composed of two distinct types of joints. The apophyseal joints and the joints of Luschka (articulations between the lateral aspects of adjacent vertebral bodies from C2 to C7) are similar to peripheral joints, and degenerative changes present as narrowing of the joint space with sclerosis and hypertrophic spur formation. There is often some degree of subluxation about the apophyseal joints, with the uppermost facets slipping forward on the ones below (spondylolisthesis). In the cervical region this slippage, combined with hyper-

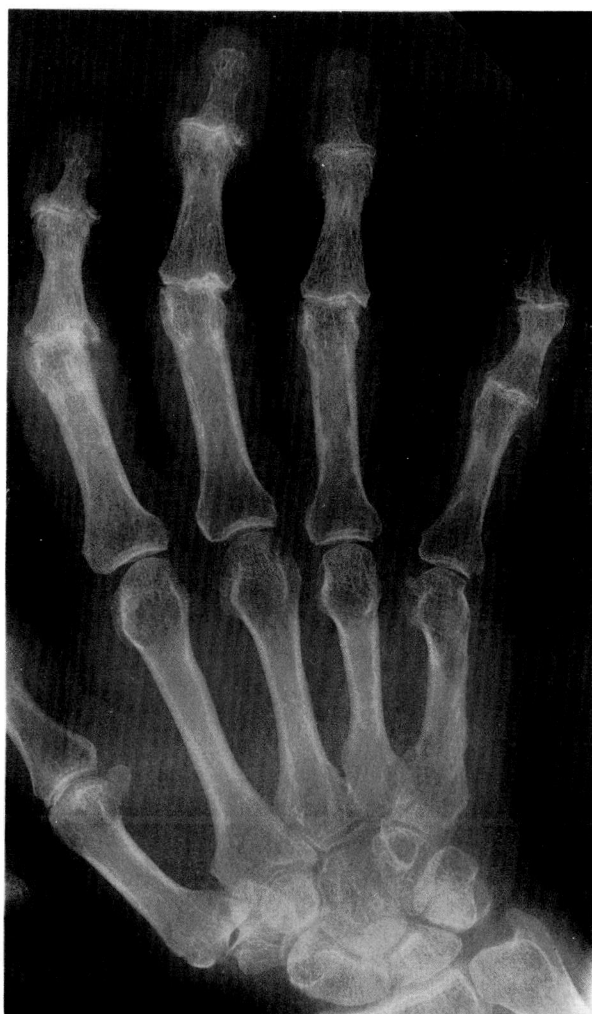

**Figure 64.6. Erosive osteoarthritis of the hand.** There is narrowing of the proximal and distal interphalangeal joints with erosions and spur formation.

trophic spurs and thinning of the adjacent intervertebral disk spaces, may cause narrowing of the corresponding neural foramina and result in clinical symptoms.

The intervertebral disk articulations are not true joints. Therefore, the presence of degenerative changes in these joints is usually termed *spondylosis*. Spondylosis is characterized by hypertrophic spur formation at the anterior, lateral, and posterior margins of the vertebral bodies. These marginal osteophytes are almost always present in patients above middle age and are particularly prone to develop in the lower cervical, lower dorsal, and lower lumbar areas. Because of the relatively narrow cervical canal, a large posterior spur may impinge upon the spinal cord and produce nerve deficits. In the lumbar spine, herniated disk disease is a far more common cause of clinical symptoms than is impingement on nerve roots by osteophytes. Disk degeneration appears as narrowing of the intervertebral spaces, which is usually associated

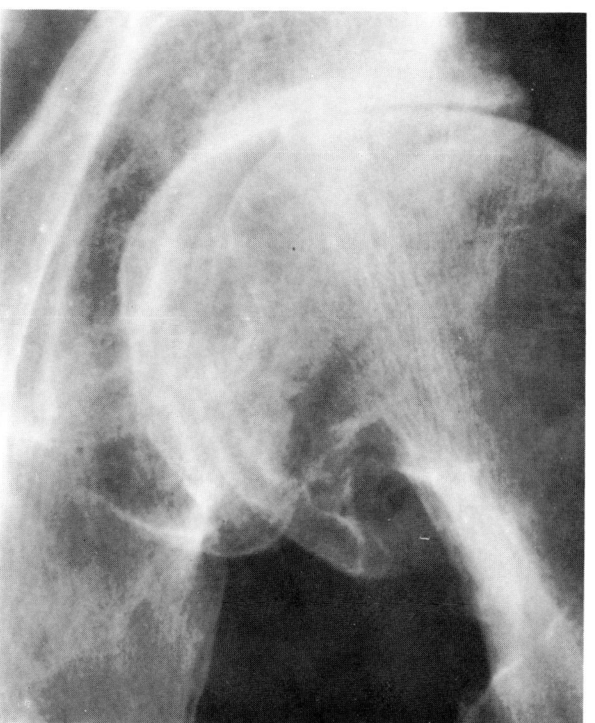

A

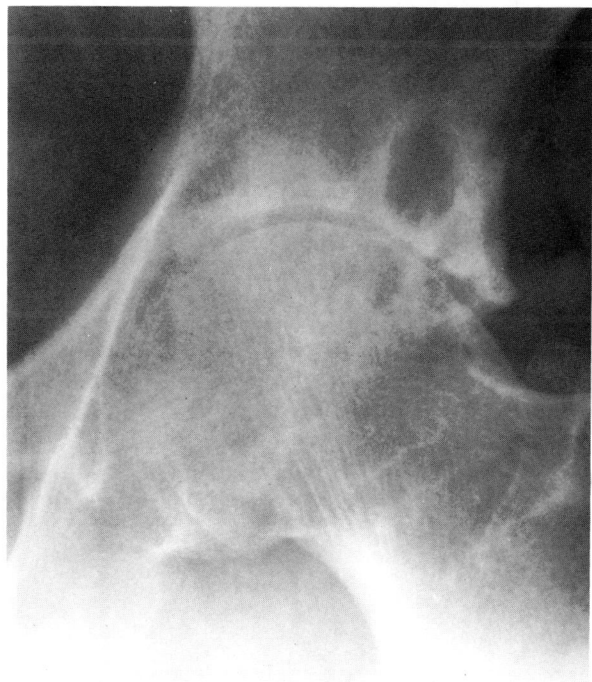

B

**Figure 64.7. Osteoarthritis of the hip.** (A) Asymmetric narrowing along the superior margin and large inferior osteophytes distinguish degenerative from inflammatory disease. (B) The subchondral cysts in the acetabulum are associated with asymmetric joint space narrowing and a small osteophyte of the superior acetabular rim. (From Forrester, Brown, and Nesson,[6] with permission from the publisher.)

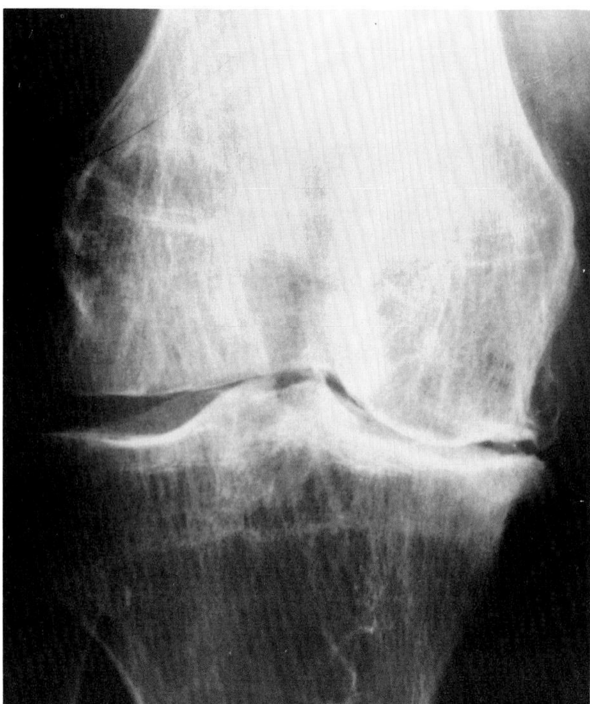

**Figure 64.8. Degenerative joint disease of the knee.** A frontal view of the knee shows hypertrophic spurring, narrowing of the medial compartment, and some subchondral sclerosis.

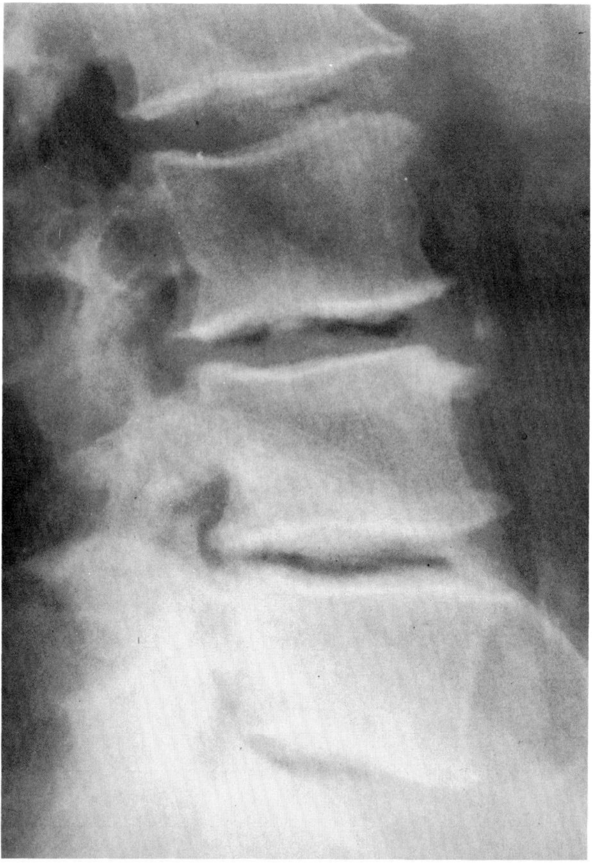

**Figure 64.9. Degenerative disk disease of the spine.** A lateral view shows hypertrophic spurring, intervertebral disk space narrowing, and reactive sclerosis. Note the linear lucent collections (vacuum phenomenon) overlying several of the intervertebral disks.

with marginal sclerosis. A characteristic finding of spondylosis, especially in the lower lumbar region, is the appearance of linear lucent collections overlying one or more intervertebral disks (vacuum phenomenon).[46,50]

## MISCELLANEOUS DISEASES

### Progressive Systemic Sclerosis (Scleroderma)

Scleroderma is a multisystem disorder characterized by fibrosis that involves the skin and a variety of internal organs, most notably the gastrointestinal tract, lungs, heart, and kidney.

Degeneration and atrophy of smooth muscle in the lower half to two-thirds of the esophagus, with subsequent replacement of the esophageal musculature by fibrosis, produce disordered esophageal motility in up to 80 percent of the patients with scleroderma, many of whom remain asymptomatic. Because the upper third of the esophagus is composed primarily of striated muscle infrequently affected by scleroderma, an esophogram demonstrates a normal stripping wave that clears the upper esophagus but stops at about the level of the aortic arch. In the early stages of scleroderma, some primary peristaltic activity and uncoordinated tertiary contractions can be observed in the lower two-thirds of the esophagus. However, these contractions are weak and infrequent and tend to disappear as the disease progresses. With the patient in the recumbent position, barium will remain for a long time in the dilated, atonic esophagus. Multiple radiographs obtained several minutes apart can be effectively superimposed on each other. In contrast to the patient with achalasia, however, when the patient with scleroderma is placed in the upright position, barium flows rapidly through the widely patent region of the lower esophageal sphincter.

Incompetence of the lower esophageal sphincter permits the reflux of acid-pepsin gastric secretions into the distal esophagus. In about 40 percent of the patients, this reflux leads to peptic esophagitis that heals with stricture formation, resulting in heartburn and severe dysphagia.

Diminished peristalsis and limited collapsibility of the esophageal walls may produce an air-filled esophagus on erect lateral chest films. Although this air esophogram sign is suggestive of scleroderma, it can also be seen in achalasia and even in normal persons. In scleroderma, the esophagus is nondistended and without an air-fluid

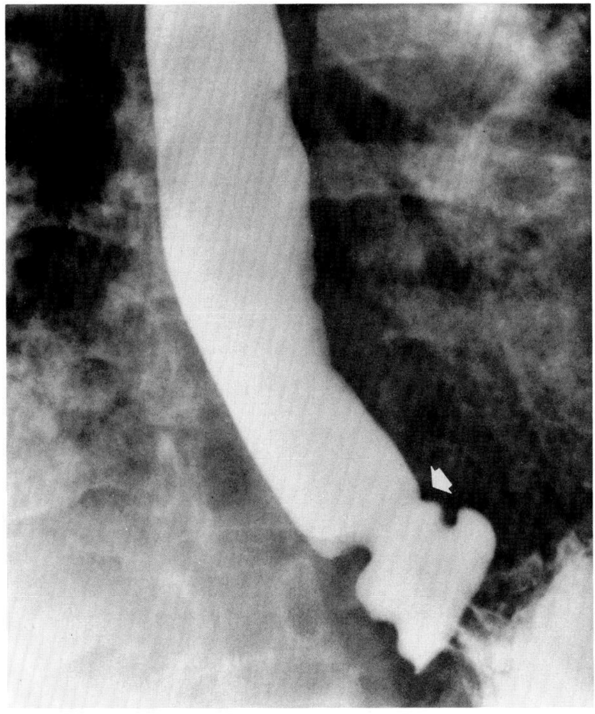

**Figure 64.10. Scleroderma.** There is dilatation of the esophagus with almost no peristalsis to the level of the moderate hiatal hernia (arrow). Severe pulmonary interstitial changes of scleroderma may be seen bilaterally in the lower lobes. (From Eisenberg,[55] with permission from the publisher.)

**Figure 64.11. Scleroderma.** There is dilatation of the small bowel with prolonged transit time and tightly packed, thin folds in a patient with malabsorption syndrome. (From Eisenberg,[55] with permission from the publisher.)

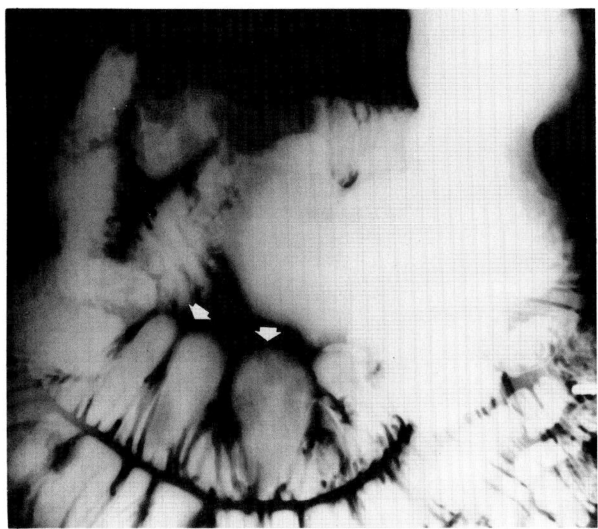

**Figure 64.12. Scleroderma.** Pseudosacculations of the small bowel (arrows) simulate jejunal diverticulosis. (From Eisenberg,[55] with permission from the publisher.)

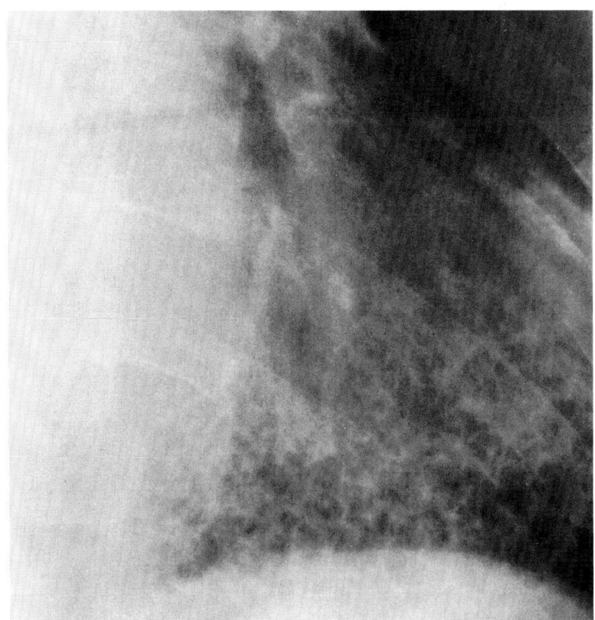

**Figure 64.13. Scleroderma.** A coned view of the left lower lung demonstrates a honeycomb pattern, with small emphysematous areas combined with fibrosis and fine nodularity.

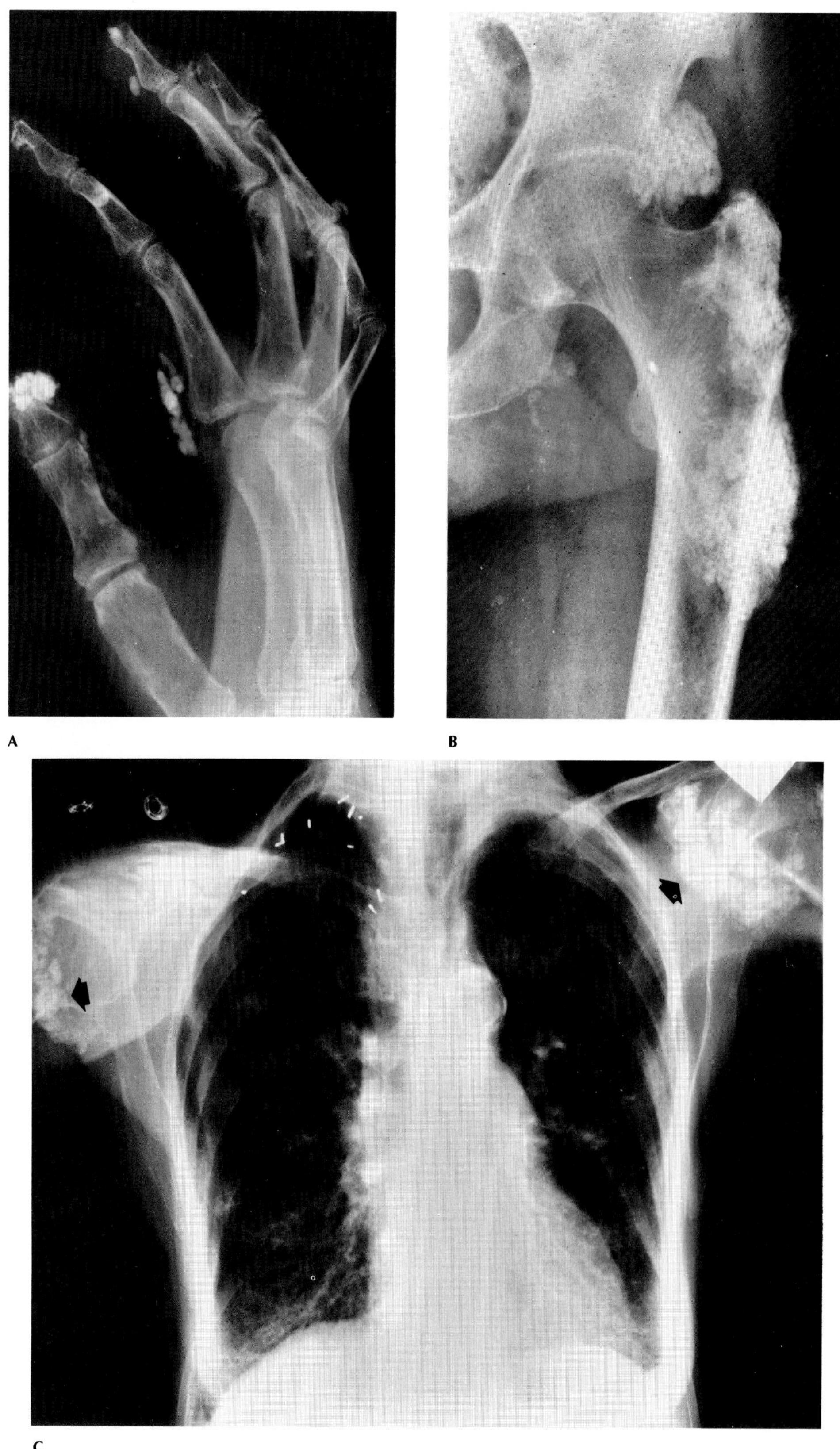


**Figure 64.14. Soft tissue calcinosis in scleroderma.** (A) Amorphous clumps of calcium in the soft tissues of the fingers. Note the trophic changes about the terminal tufts. (B) Calcifications about the hip joint and proximal femur. (C) Clumps of calcification about the shoulder joints. Note the reticulonodular interstitial pattern at both lung bases. The surgical clips overlying the right apex are from a cervical sympathectomy for the treatment of associated Raynaud's phenomenon.

level, in contrast to the prominent air-fluid level in a dilated esophagus seen in patients with achalasia or distal esophageal obstruction.

Decreased duodenal peristaltic activity in a patient with scleroderma can lead to dramatic dilatation of the duodenum proximal to the site where the transverse portion of the duodenum passes between the aorta and root of the mesentery (superior mesenteric artery syndrome) (see Figure 49.23). In the small bowel, hypomotility causes dilatation and an extremely prolonged transit time (see Figures 46.7 and 46.8). The folds are often abnormally packed together despite bowel dilatation, in sharp contrast to the typical wide longitudinal separation of valvulae seen with bowel dilatation in sprue and small bowel obstruction. Pseudosacculations—large, broad-necked outpouchings simulating small bowel diverticula—are characteristic of scleroderma and reflect the combination of smooth muscle atrophy and fibrosis accompanied by vascular occlusion. Similar pseudosacculations may arise from the antimesenteric border of the colon. Chronic intestinal distension with increased intraluminal pressure can lead to the development of pneumatosis intestinalis and even pneumoperitoneum.

In the lungs, scleroderma produces a diffuse reticular or reticulonodular pattern that tends to predominantly involve the lower lung zone. There is little or no radiographic evidence of upper-zone abnormality, at least in the early stages of the disease. Serial radiographs over several years may show progressive and uniform loss of lung volume. Small cysts up to 1 cm in diameter may be identified in the lung periphery, especially in the bases; spontaneous pneumothorax from cyst rupture has been reported. These small emphysematous areas, combined with fibrosis and fine nodularity, may produce the pattern of honeycomb lung. In scleroderma, in contrast to other connective tissue disorders, pleural involvement is uncommon, though mild pleural thickening may develop over the lung bases. Because of the disordered esophageal motility often seen in scleroderma, aspiration pneumonia is a common complication.

The most common abnormality in the musculoskeletal system is Raynaud's phenomenon, which develops in 90 percent of the patients with the skin changes of scleroderma. It is characterized by soft tissue atrophy in the fingertips with trophic osteolysis and resorption of the terminal tufts. Osteolysis of the feet, ribs, and mandibles may develop. Soft tissue calcinosis often occurs in the hands, as well as over pressure areas such as the elbows and the ischial tuberosities. Arthritic changes in the interphalangeal joints of the hands are not uncommon; in about one-fourth of the patients, the radiographic and clinical picture is indistinguishable from that of rheumatoid arthritis.

Cardiomegaly due to diffuse myocardial fibrosis may occur. Diffuse interstitial fibrosis may lead to the development of cor pulmonale.

Uniform widening of the peridontal space is reported to occur in up to 10 percent of the patients with scleroderma. Although suggestive of scleroderma, a similar appearance may be seen in other conditions, such as Paget's disease.

Although renal failure is the leading cause of death in patients with scleroderma, excretory urography is usually normal, except when arterial nephrosclerosis leads to the development of small, smooth kidneys. Arteriographic findings in scleroderma include a prolonged arterial phase, pruning of intrarenal vessels with a poor, spotty nephrogram, and lack of demarcation between the cortex and the medulla.[10–22]

## Pseudogout

Pseudogout is an inflammatory arthritis of older individuals caused by the deposition of calcium pyrophosphate dihydrate crystals in the joint. Although the cause is unknown, pseudogout probably reflects a defect in cartilage metabolism. It may present as intermittent attacks of acute joint effusion and pain or as a progressive chronic arthritis. The acute arthritis of pseudogout may be clinically indistinguishable from gout or septic arthritis; in such cases, the diagnosis is made by the identification of calcium pyrophosphate crystals in the synovial fluid.

The radiographic hallmark of pseudogout is chondrocalcinosis, the presence of calcification in hyaline cartilage or intra-articular fibrocartilage. Chondrocalcinosis

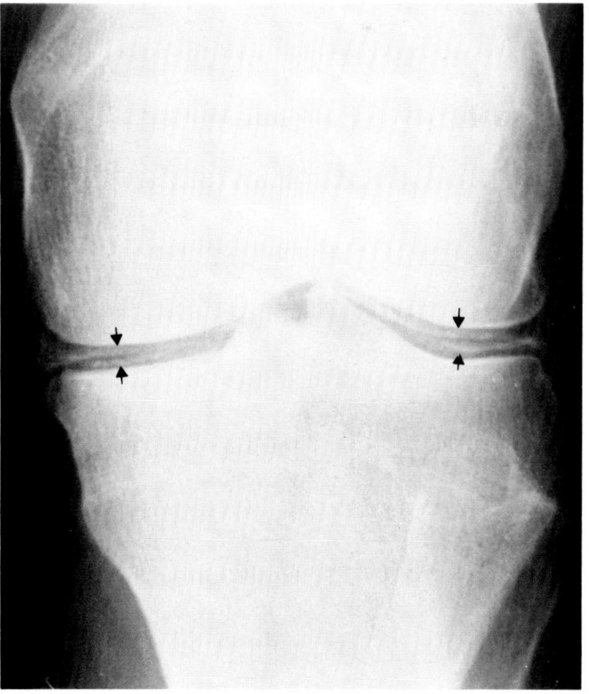

**Figure 64.15. Chondrocalcinosis in pseudogout.** Calcification of both the medial and the lateral aspects of the menisci of the knee (arrows).

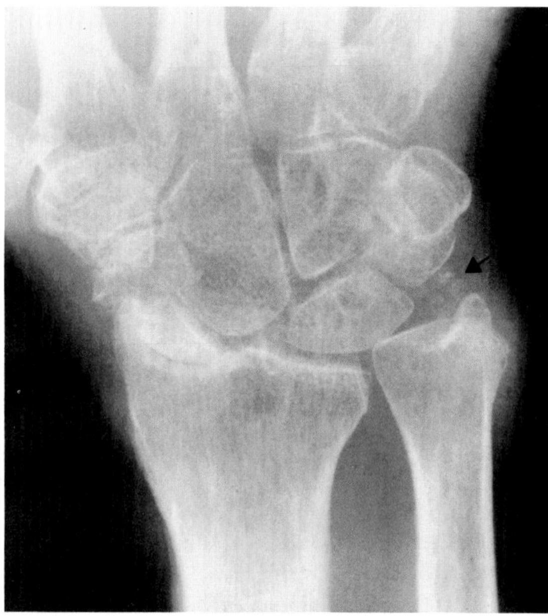

**Figure 64.16. Chondrocalcinosis in pseudogout.** Characteristic calcifications within the triangular fibrocartilage of the wrist (arrow).

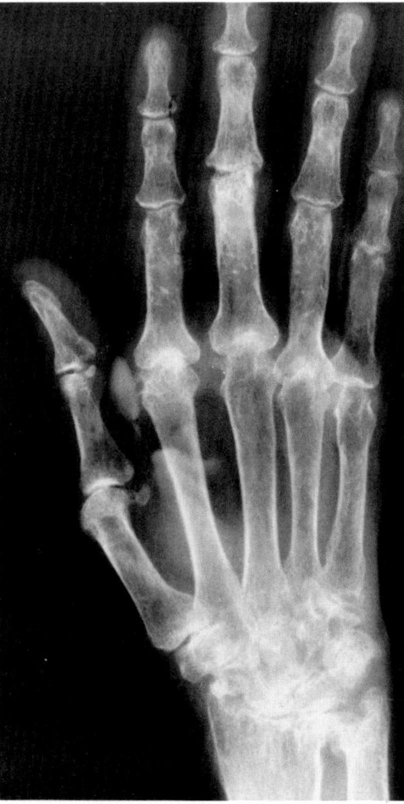

**Figure 64.17. Pseudogout arthropathy.** Severe joint space narrowing, erosive changes, and sclerosis about the wrist. Less marked changes involve the metacarpophalangeal joints and the proximal interphalangeal joint of the third digit.

can also occur in a number of systemic disorders, including diabetes mellitus, hyperparathyroidism, gout, ochronosis, Wilson's disease, hemochromatosis, and acromegaly, and in patients with degenerative joint disease. Chondrocalcinosis in pseudogout most commonly affects the knee joint. Meniscal calcification appears as dense linear deposits in the center of the knee joint. Articular cartilage calcification presents as fine linear densities parallel to the subchondral bone surfaces. Other typical sites of chondrocalcinosis in pseudogout include the articular disks of the distal radioulnar joint and the intercarpal joints of the wrists, vertical linear calcification of the symphysis pubis, and the articular cartilage and capsules of the shoulder, hip, elbow, and ankle. In the spine, there may be calcification in the annulus fibrosus of the intervertebral disk, especially in the lumbar region. In pseudogout, in contrast to ochronosis, chondrocalcinosis does not occur in the nucleus pulposus.

In addition to calcification in joint cartilage, pseudogout may produce a radiographic pattern simulating degenerative joint disease, with subchondral cyst formation, hypertrophic spurring, joint space narrowing, and subchondral sclerosis. Unlike classic degenerative joint disease, pseudogout tends to involve such joints as the radiocarpal, wrist, elbow, and shoulder, which are infrequently involved in osteoarthritis. Subchondral cyst formation occasionally may be severe and lead to bony collapse, fragmentation, and loose body formation that simulate a neuropathic joint.[23–25,60]

## Hypertrophic Osteoarthropathy

Hypertrophic osteoarthropathy is a condition characterized by the triad of periosteal new bone formation, clubbing of the fingers and toes, and arthritis. It most frequently arises in patients with primary intrathoracic neoplasms, especially bronchogenic carcinoma, and may be the first evidence of an unrecognized and asymptomatic lung tumor. Tumors of the pleura and mediastinum and chronic suppurative lung lesions (lung abscess, bronchiectasis, empyema) are also frequent causes, as are cystic fibrosis and pulmonary metastases in infants and children. Although often termed *pulmonary* hypertrophic osteoarthropathy, the syndrome may occasionally be seen in association with extrathoracic neoplasms and gastrointestinal diseases, such as biliary cirrhosis, ulcerative colitis, and Crohn's disease. The precise mechanism is unknown, though the symptoms and radiographic findings of hypertrophic osteoarthropathy tend to disappear rapidly following removal of the primary lung lesion, vagotomy, or even exploratory thoracotomy without resection. An idiopathic form of hypertrophic osteoarthropathy may rarely develop in the absence of any associated disorder.

Radiographically, the earliest change in hypertrophic osteoarthropathy is periosteal proliferation of new bone that symmetrically involves the diaphyses of tubular bones,

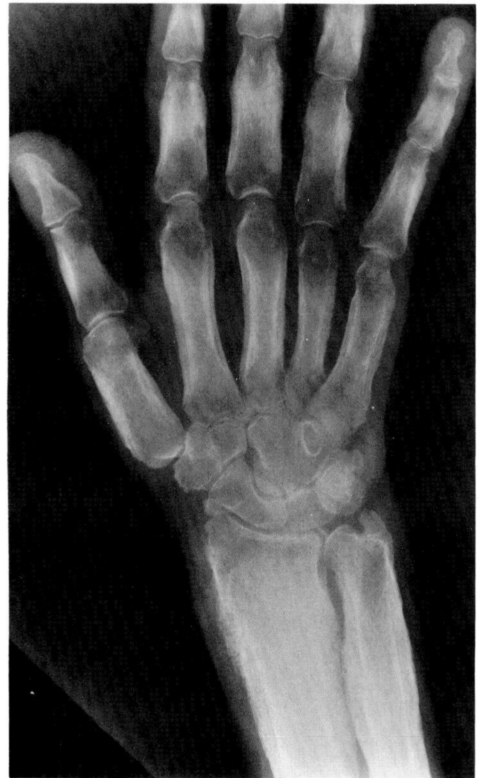
A

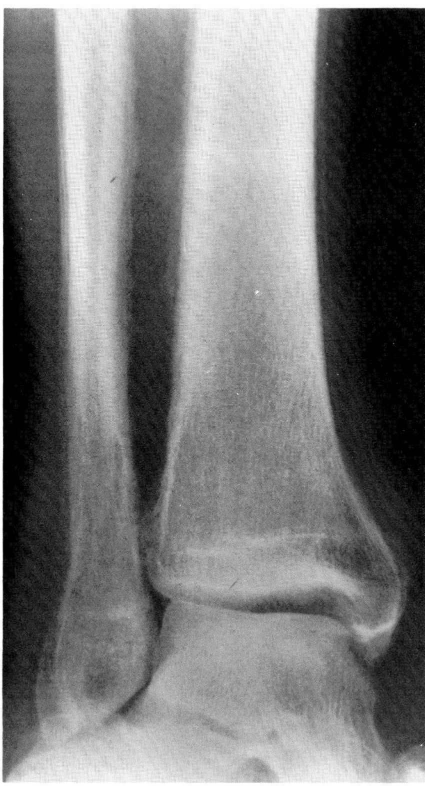
B

**Figure 64.18. Hypertrophic osteoarthropathy in two patients with bronchogenic carcinoma.** Films of the lower arm and hand (A) and the lower leg (B) demonstrate characteristic lamellated plaques of periosteal new bone. Note the irregular, undulating new bone formation affecting the distal radius and ulna. In the metacarpals, the periosteal reaction involves the diaphyses and spares the ends of these tubular bones. There is some periarticular demineralization about the metacarpophalangeal and metacarpocarpal joints but no evidence of bone erosion or cartilage destruction.

sparing the ends. The long bones of the forearm and leg are most frequently affected; the metacarpals, metatarsals, and proximal phalanges may also be involved. The periosteal new bone is initially laid down in a thin strip that is sharply demarcated from the underlying cortical bone. As it accumulates, the periosteal new bone becomes irregular, rough, and undulating and eventually fuses with the cortex. Soft tissue swelling of the distal phalanges (clubbing) is often associated with hypertrophic osteoarthropathy, yet there are no bone changes in the underlying distal phalanges. Although symptoms of arthritis are part of the clinical syndrome of hypertrophic osteoarthropathy, there is no radiographic evidence of joint erosions or cartilage destruction.[26–28]

## Pachydermoperiostosis

Pachydermoperiostosis is an inherited form of hypertrophic osteoarthropathy that is characterized by marked thickening of the skin of the extremities, face, and scalp. The disease is self-limiting, primarily involving adolescent males and progressing for several years before sta-

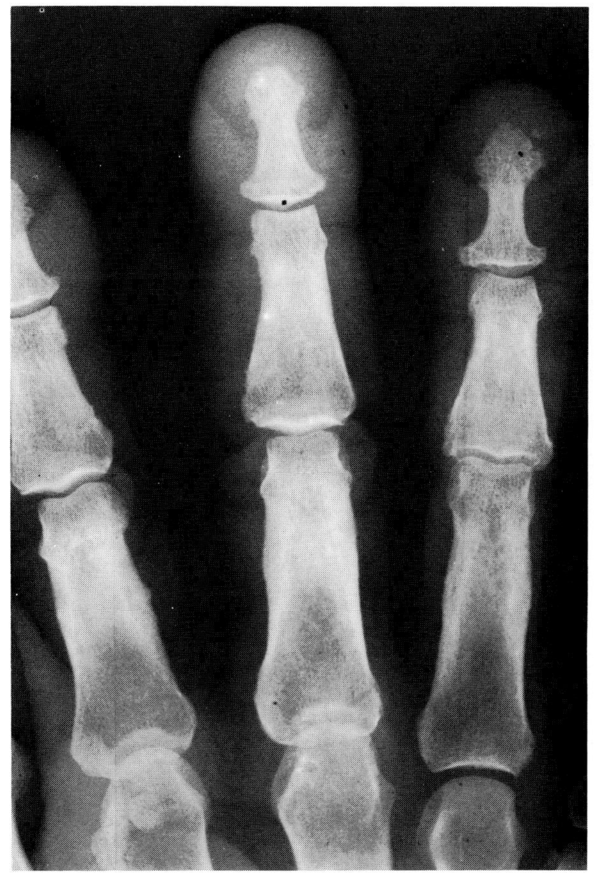

**Figure 64.19. Pachydermoperiostosis.** Elephant-like thickening of the skin causes clubbing of the distal fingers and exaggerated knuckle pads in addition to a generalized increase in the bulk of the soft tissues surrounding the phalanges. Loss of the tufts accompanies the increase in the overlying soft tissues. (From Forrester, Brown, and Nesson,[6] with permission from the publisher.)

bilizing. The radiographic features of pachydermoperiostosis are indistinguishable from other forms of hypertrophic osteoarthropathy and include irregular periosteal new bone formation and thickening of the soft tissues at the ends of the fingers (clubbing).[29]

## Neuropathic Joint Disease (Charcot Joint)

Neuropathic joint disease refers to the severe disorganization of a joint that develops in a variety of neurological disorders in which a loss of proprioception and/or deep pain sensation leads to repeated trauma on an unstable joint. Degeneration of cartilage, recurrent fractures of subchondral bone, and marked proliferation of adjacent bone lead to total disorganization of the joint. In tabes dorsalis, neuropathic disease usually involves the weight-bearing joints of the lower extremities and the lower lumbar spine. Syringomyelia is the major cause of neuropathic joint disease in the upper extremities, while diabetes and leprosy predominantly affect the hands and feet.

The earliest radiographic sign of neuropathic joint disease is soft tissue swelling that reflects persistent joint effusion. Destruction of cartilage produces narrowing of the joint space, which is followed by fractures and fragmentation of the articular surfaces. Synovial membrane calcifications develop and separate into the joint to form loose bodies. Irregular masses of new bone form at the margins of the articular surfaces. Calcific and bony debris may dissect into the soft tissues and extend about the joint and along muscle planes. Laxity of the periarticular soft tissue structures results in the severe subluxations that are characteristic of this disorder. Substantial demineralization is rare, since the absence of pain permits continued use of the joint. A neuropathic joint occasionally may demonstrate only progressive and rapid resorption of bone ends, with subluxation but no sclerosis, fragmentation, or soft tissue debris. In the hands and feet, extensive resorption may cause pencil-point narrowing of the distal ends of the metacarpals and metatarsals (arthritis mutilans).[30–32]

## Diffuse Idiopathic Skeletal Hyperostosis (Forestier's Disease)

Diffuse idiopathic skeletal hyperostosis (DISH) is characterized by laminated new bone formation and exuberant spurs involving the anterior and lateral paraspinal ligaments. Large osteophytes and continuous bony bridging may produce a corrugated appearance of the spine. The intervertebral disk spaces are preserved. DISH usually affects middle-aged and elderly men. It most frequently involves the lower thoracic spine, often extending to affect the lower cervical and entire lumbar spine. Compression of the spinal cord due to large osteophytes may lead to neurologic complications. There

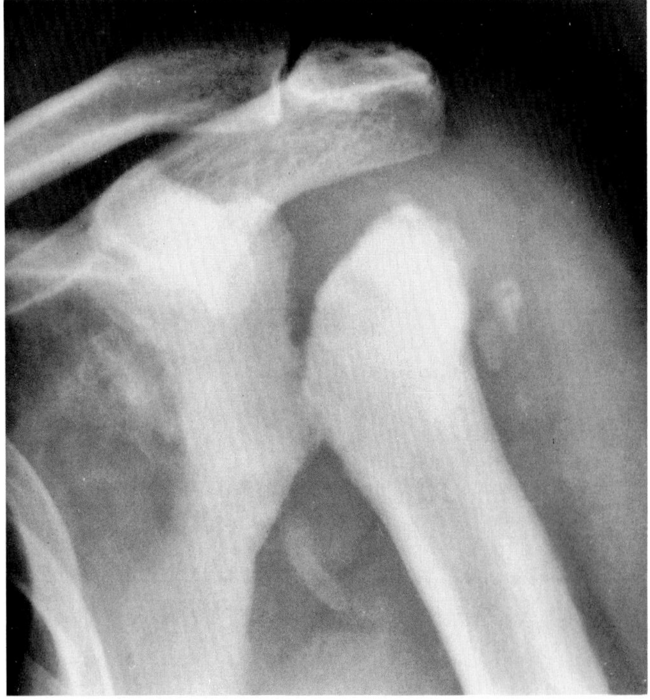

A

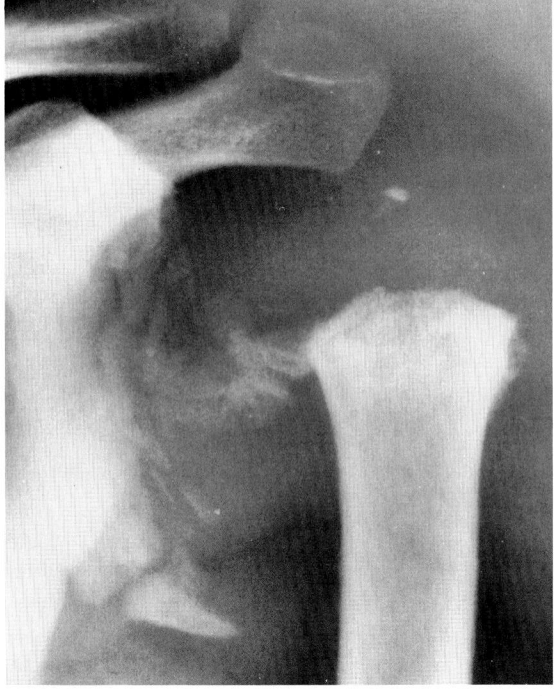

B

**Figure 64.20. Neuropathic joint disease of the shoulder.** Examples of destruction and almost complete disappearance of the humeral head with reactive sclerosis and calcific debris due to (A) syringomyelia and (B) congenital insensitivity to pain.

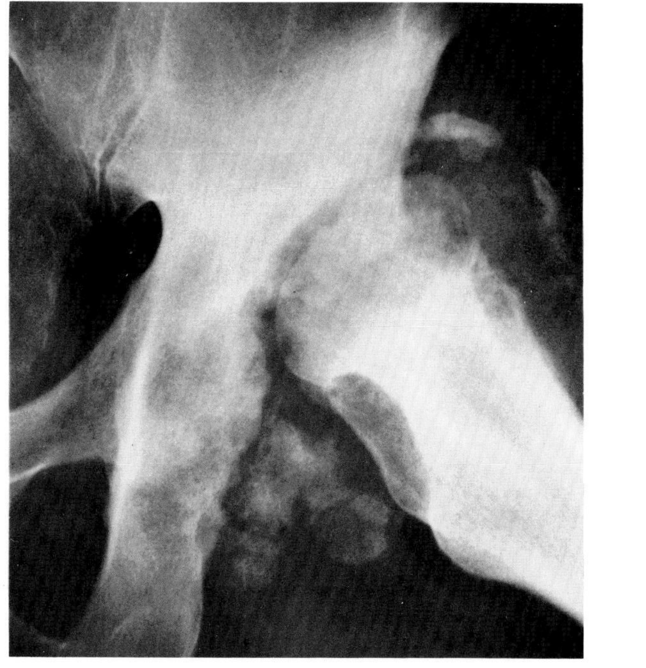
A

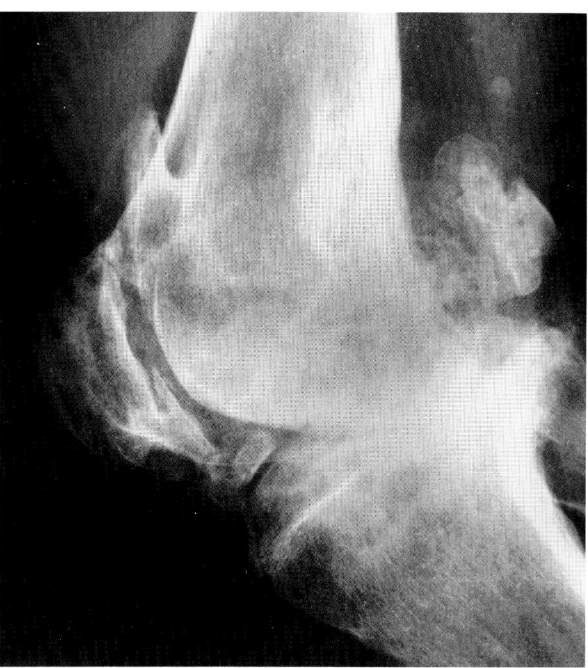
B

**Figure 64.21. Neuropathic joint disease due to tabes dorsalis.** Joint fragmentation, sclerosis, and calcific debris are seen about the hip (A) and the knee (B).

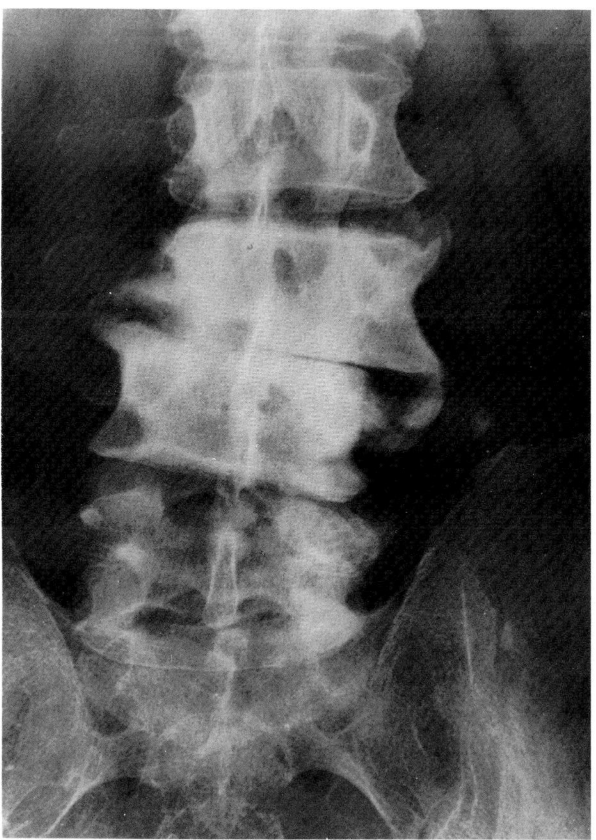
A

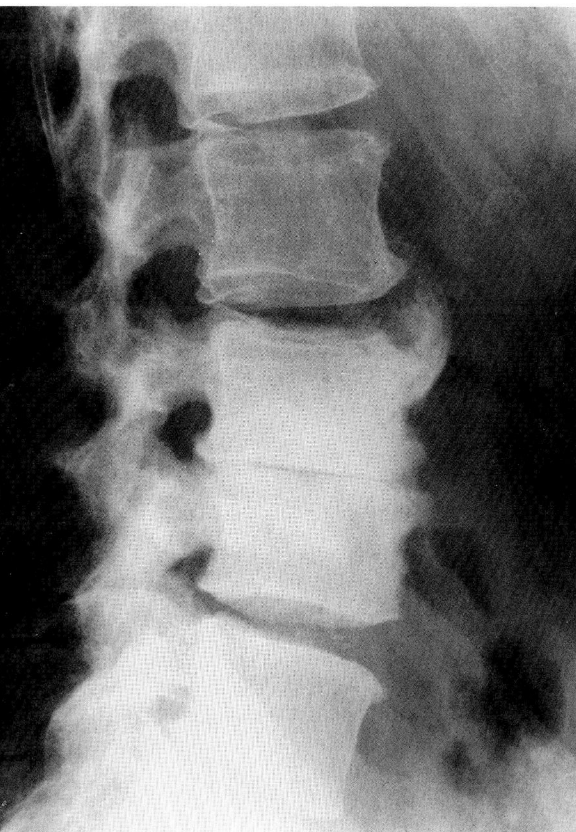
B

**Figure 64.22. Neuropathic joint disease of the spine.** (A) Frontal and (B) lateral views of the lumbosacral spine in a patient with tabes dorsalis show marked hypertrophic spurring with virtual obliteration of the L3–4 intervertebral disk space. Note the reactive sclerosis of the apposing end plates and the subluxation of the vertebral bodies seen on the frontal view.

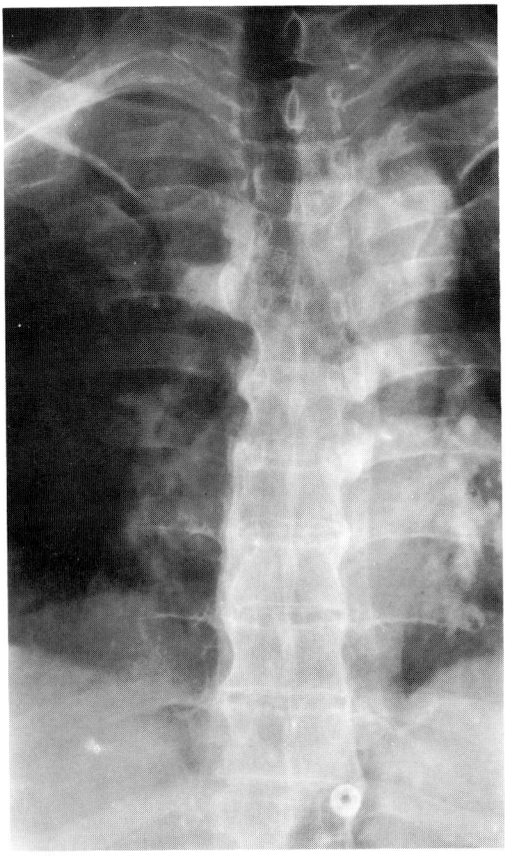

A

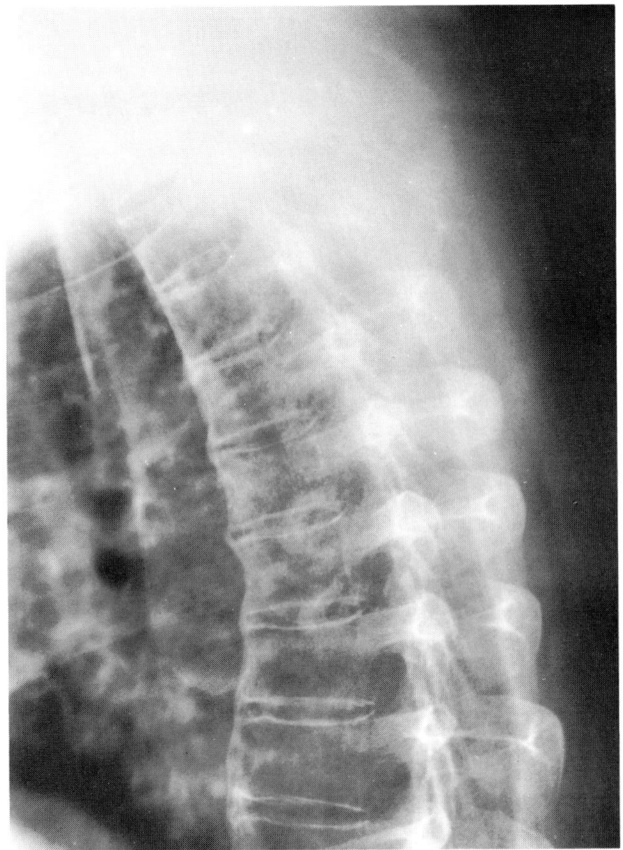

B

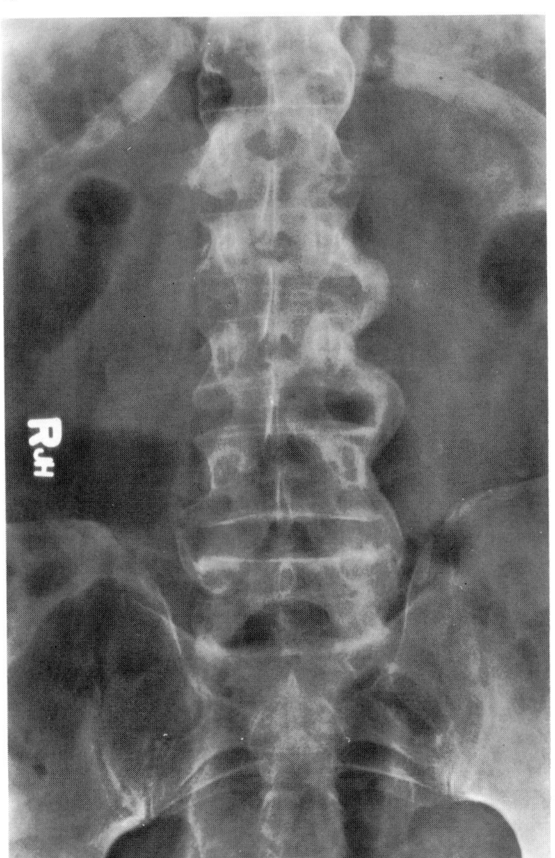

C

**Figure 64.23. Diffuse idiopathic skeletal hyperostosis.** (A) Frontal and (B) lateral views of the thoracic and upper lumbar spine demonstrate thin, vertical bony bridging producing a pattern indistinguishable from ankylosing spondylitis. (C) The normal sacroiliac and apophyseal joints exclude the possibility of ankylosing spondylitis and confirm the diagnosis of diffuse idiopathic skeletal hyperostosis. (From Forrester, Brown, and Nesson,[6] with permission from the publisher.)

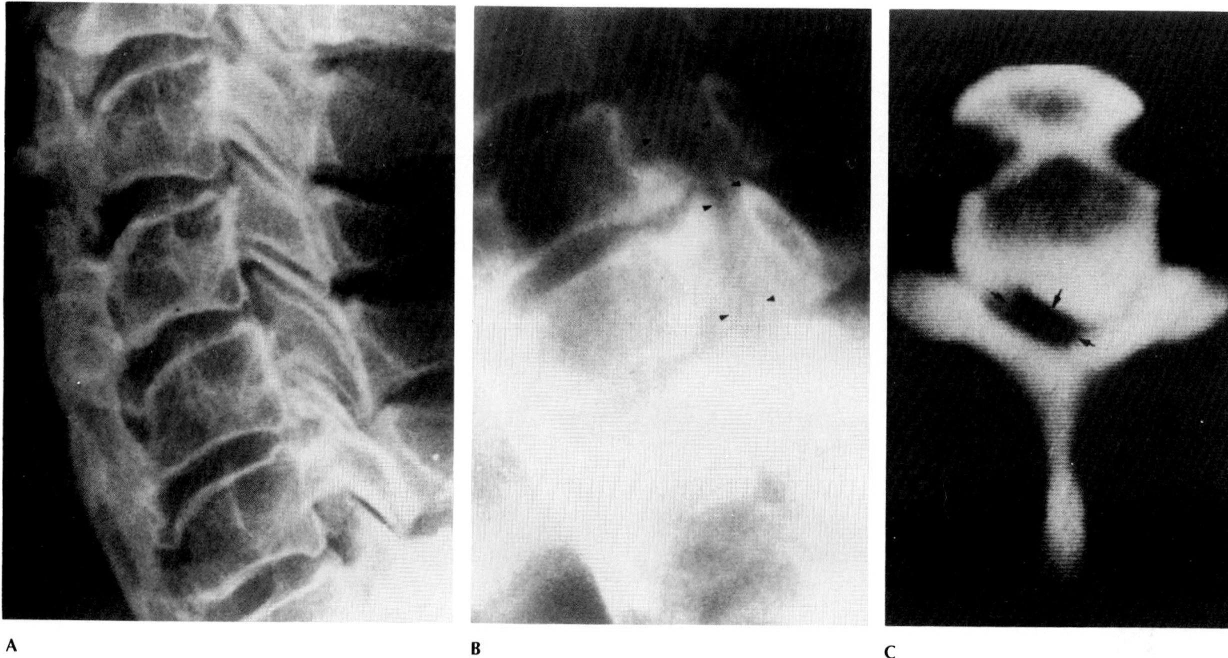

A B C

**Figure 64.24. Diffuse idiopathic skeletal hyperostosis causing spinal cord compression.** (A) A lateral radiograph of the cervical spine shows extensive ossification along the anterior vertebral bodies. The disk spaces and apophyseal joints are normal. (B) A metrizamide cervical myelogram with complex-motion tomography shows contrast material outlining the spinal cord (arrowheads), which is compressed by large posterior osteophytes. (C) A CT scan at the upper border of C6 shows metrizamide outlining the spinal cord (arrows), which is compressed and displaced to the left by posterior osteophytes. (From Alenghat, Hallet, and Kido,[35] with permission from the publisher.)

is also a tendency for hyperostoses to develop elsewhere in the body, with irregular extraspinal bone production at ligamentous attachments such as the iliac crest, ischial tuberosity, symphysis pubis, and inferior portions of the sacroiliac joints. Ossification of the iliolumbar and sacrotuberous ligaments may occur, simulating the changes seen in fluorosis. Unlike ankylosing spondylitis, DISH infrequently involves the apophyseal and sacroiliac joints and does not produce squaring of the vertebral bodies. In the hands, DISH may cause broadening and "arrowheading" of the distal phalangeal tufts, enlarged sesamoid bones, increased cortical width of tubular bones, and exostoses at the metacarpophalangeal heads.[33–35]

## Multicentric Histiocytosis (Lipoid Dermatoarthritis)

In this rare systemic disease of unknown cause, nodules containing multinucleated giant cells with PAS-positive material develop in the skin, subcutaneous tissues, synovium, and bone. A hallmark of the disease is a severe inflammation of multiple joints that progresses rapidly and leads to an incapacitating and deforming arthritis. The hands, especially the distal interphalangeal joints, are most often involved. Multiple soft tissue masses typically produce a lumpy-bumpy appearance. Resorption of the tufts of the distal phalanges frequently occurs. The severity of lipoid dermatoarthritis varies from minimal erosions of the joint surfaces to complete bony resorption. Extensive involvement may lead to arthritis mutilans or an opera-glass hand deformity. The same erosive process may also involve the knees, shoulders, and wrists; intraosseous deposits in these locations can produce large subchondral cysts.[36–38]

## Pigmented Villonodular Synovitis

Pigmented villonodular synovitis is a proliferative inflammatory lesion of unknown cause that most often occurs in young adults and presents with a history of intermittent pain and swelling of the involved joint. The disorder is usually monarticular and most commonly affects the lower extremities, especially the knee; less frequently, the ankle, hip, elbow, shoulder, or tarsal and carpal joints may be involved.

Radiographically, villonodular synovitis typically appears as a joint effusion with multiple nodular soft tissue masses that may extend beyond the joint capsule and may rarely be entirely extracapsular. Unlike synoviomas, the nodular masses of pigmented villonodular synovitis never calcify, though they may appear dense because of hemosiderin deposits within the soft tissue mass. Osseous change may result from pressure atrophy, actual invasion of bone, or both. A characteristic finding is subchondral bone erosion, particularly at the margins of

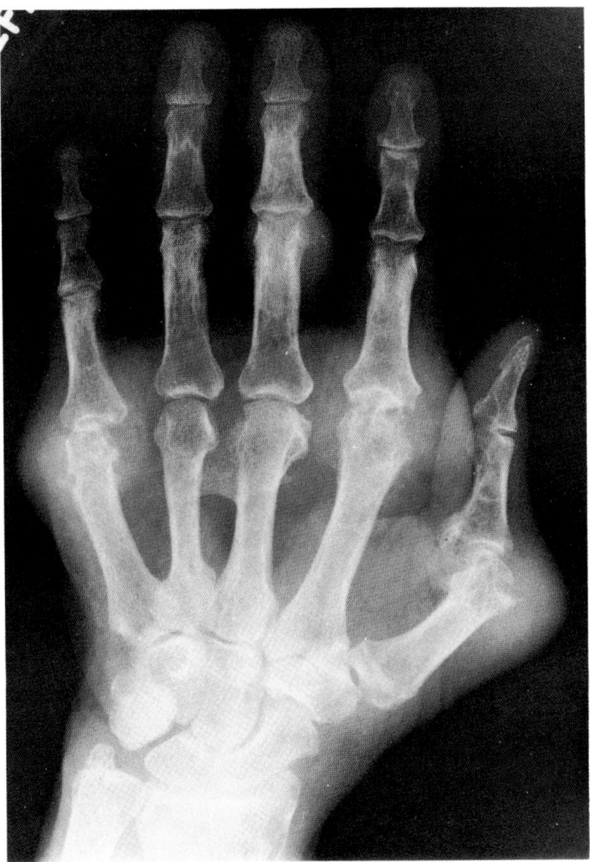

**Figure 64.25. Multicentric histocytosis.** A frontal view of the hand demonstrates multiple soft tissue masses producing a lumpy-bumpy appearance. The soft tissue deposits of multinucleated giant cells have produced erosions of juxta-articular bone. Although at this stage most of the joint spaces are spared, extensive involvement of the second metacarpophalangeal joint has given rise to total joint destruction. (From Forrester, Brown, and Nesson,[6] with permission from the publisher.)

the knees, that is often associated with small lucent defects with thin sclerotic margins affecting part or all of the adjoining ends of the bone. The cystlike defects are focal areas of deossification that are usually multiple and the result of direct invasion of adjoining bone by the synovial growth. Direct pressure erosions are common in pigmented villonodular synovitis and appear as extrinsic defects that either blend imperceptibly with surrounding normal bone or are outlined by dense margins of bone. Unlike a traumatic or infectious arthritis, in pigmented villonodular synovitis the joint space is usually preserved. Disuse osteoporosis is not a prominent feature since the disorder does not cause much disability.[39,40,56]

## Tietze's Syndrome

Tietze's syndrome consists of swelling, pain, and tenderness of the sternoclavicular and upper costochondral cartilages. A benign, self-limited condition of unknown cause, it must be distinguished from more serious causes of chest pain.

Radiographs are usually normal early in the course of the disease. At times, however, a soft tissue mass projecting from the anterior chest wall may be demonstrated on conventional tomography or on films taken tangential to the area of involvement. Hypertrophy of the involved costal cartilage combined with superimposed periosteal response and calcification may produce a striking increase in the size and density of the affected rib.[41]

## Calcific Tendinitis (Bursitis)

Tendinous calcification is a common cause of pain, limitation of motion, and disability about a joint. Calcium

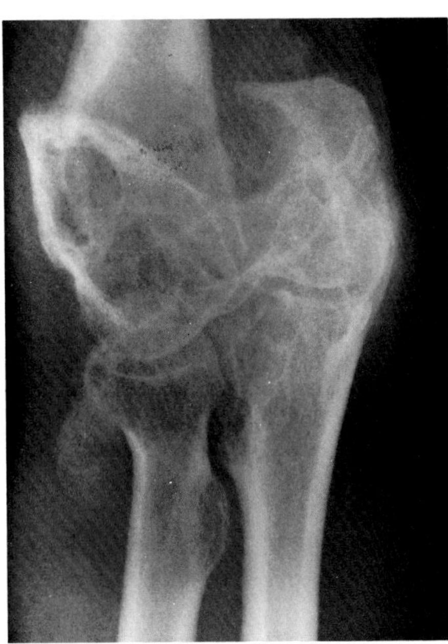

A

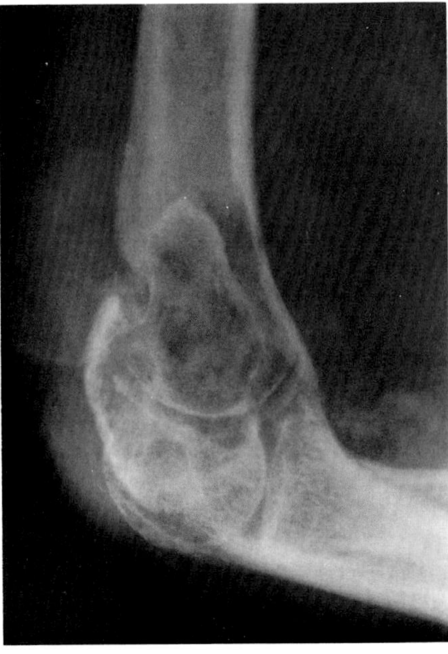

B

**Figure 64.26. Pigmented villonodular synovitis.** (A) Frontal and (B) lateral views of the elbow demonstrate a joint effusion with nodular soft tissue masses extending beyond the joint capsule. The soft tissue mass appears dense because of deposits of hemosiderin within it. Large bone erosions reflect a combination of pressure effect and direct invasion by the synovial growth.

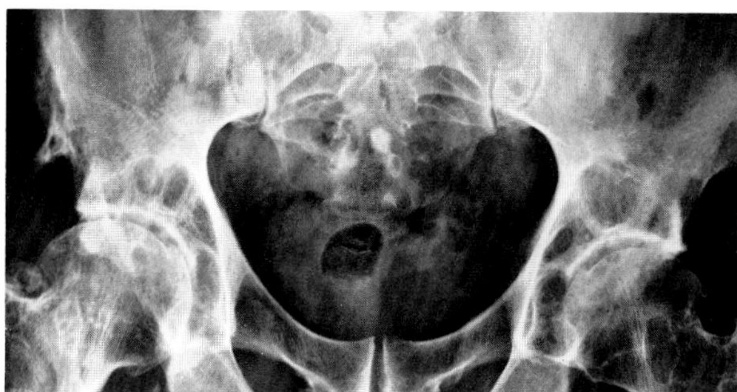

**Figure 64.27. Pigmented villonodular synovitis.** The synovial growth has produced marked bilateral cystic abnormalities involving the acetabulum and the femoral head and neck. (From Eisenberg and Hedgcock,[56] with permission from the publisher.)

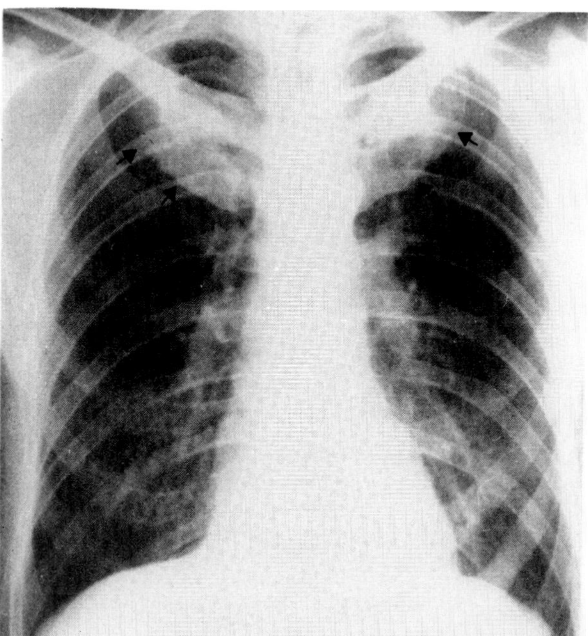

**Figure 64.28. Tietze's syndrome.** There is marked enlargement and increased density involving the entire left first rib and most of the right first rib (arrows). (From Skorneck,[41] with permission from the publisher.)

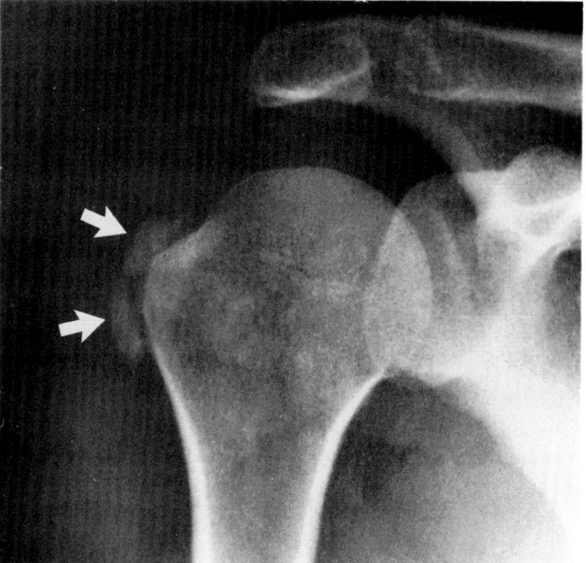

**Figure 64.29. Calcific tendinitis.** A frontal view of the shoulder demonstrates amorphous calcium deposits (arrows) in the supraspinatous tendon.

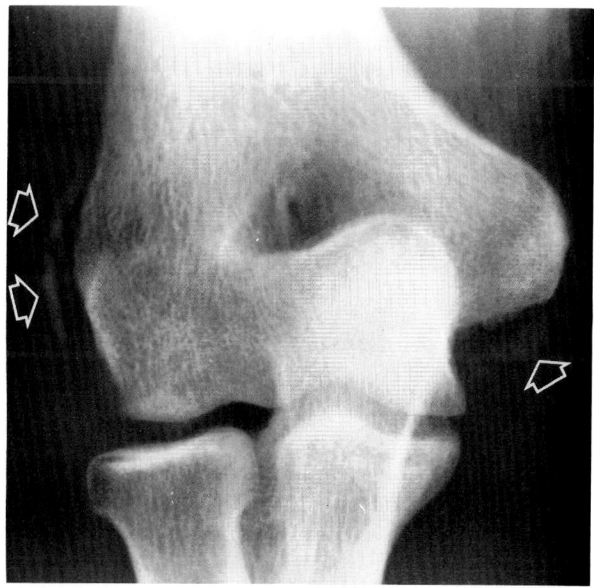

**Figure 64.30. Calcific tendinitis.** There is amorphous calcification in the region of the extensor and flexor tendons about the elbow (arrows). No radiographic evidence of arthritis is present. (From Brown and Forrester,[59] with permission from the publisher.)

deposits in the tendons are frequently associated with inflammation of an overlying bursa. Although the condition is often clinically termed *bursitis,* the calcification usually occurs within the tendon and not in the overlying bursa, though rupture of a mass of calcium into a bursa may occur. Calcific tendinitis most commonly involves the shoulder, and calcification may be demonstrated radiographically in about half the patients with persistent pain and disability in the shoulder region. However, calcification may also be detected in asymptomatic persons; conversely, severe clinical symptoms may occur without evidence of calcification. The elbow, knee, hip, and wrist may also be affected. Radiographically, calcific tendinitis appears as amorphous calcium deposits that most frequently occur in the supraspinatous tendon, where they are seen directly above the greater tuberosity of the humerus. The deposits vary greatly in size and shape, from thin curvilinear densities to large calcific masses.

## Rotator Cuff Tears

The rotator cuff of the shoulder is a musculotendinous structure composed of the teres minor, infraspinatus, supraspinatus, and subscapularis muscles. Rupture of the rotator cuff causes a communication between the glenohumeral (shoulder) joint and the subacromial bursa. At arthrography, opacification of the subacromial bursa is virtually pathognomonic of a rotator cuff tear.

Chronic rotator cuff tears can produce abnormalities on plain radiographs of the shoulders. These include narrowing of the normal distance between the inferior border of the acromion process and the humeral head, reversal of the normal acromial convexity on both internal and external rotation views, and cystic changes in the acromion process and around the humeral tuberosities.[42,43]

## Osteitis Condensans Ilii

Osteitis condensans ilii occurs almost exclusively in women during the childbearing period, usually following pregnancy. Although sometimes associated with chronic low back pain, the condition is usually asymptomatic and self-limiting and is rarely detectable in women past 50. Osteitis condensans ilii appears radiographically as a zone of dense sclerosis along the iliac side of the sacroiliac joint. It is usually bilateral and symmetric, though there may be some variation in density between the two sides. In contrast to ankylosing spondylitis, in osteitis condensans ilii the sacrum is normal and the sacroiliac joint space is preserved. Osteitis condensans ilii must also be differentiated from rheumatoid arthritis, in which sacroiliac changes are asymmetric and sclerosis rarely develops before extensive joint space destruction. Osteitis condensans ilii may represent a reaction to the increased stress to which the sacroiliac region is subjected during pregnancy and delivery. Indeed, a similar type of sclerotic reaction (osteitis pubis) may be seen in the pubic bones adjacent to the symphysis in women who have borne children.[44]

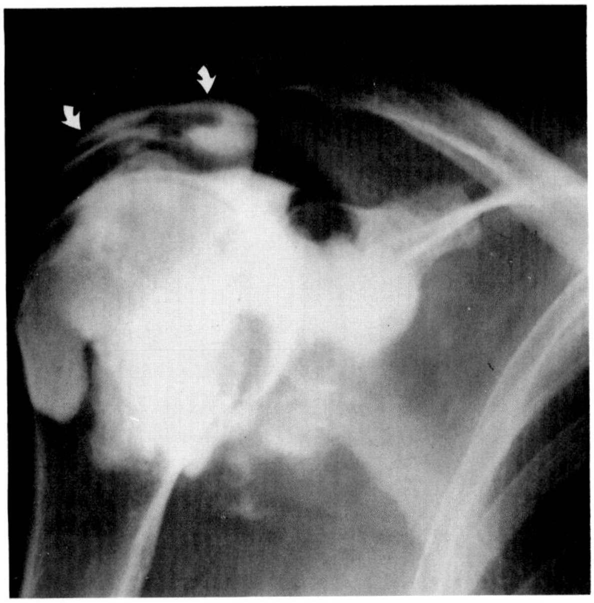

**Figure 64.31. Rotator cuff tear.** An arthrogram demonstrates contrast material within the subacromial bursa (arrows). The clear zone of radiolucency between the bursa and the joint capsule is occupied by the rotator cuff. (Courtesy of Robert H. Freiberger, M.D.)

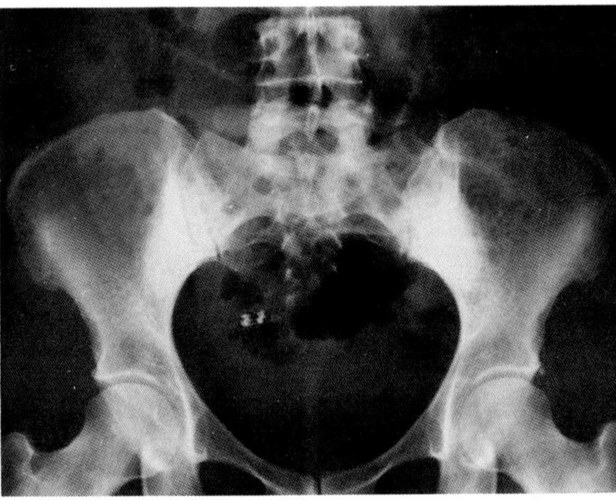

**Figure 64.32. Osteitis condensans ilii.** There is sharply demarcated sclerosis of the ilia adjacent to the sacroiliac joints. Note that the sacrum is not affected and that the margins of the sacroiliac joints are sharp and without destruction. The sclerosis that overlies the sacral wing is actually in the ilium, where it curves posteriorly behind the sacrum. (From Edeiken,[46] with permission from the publisher.)

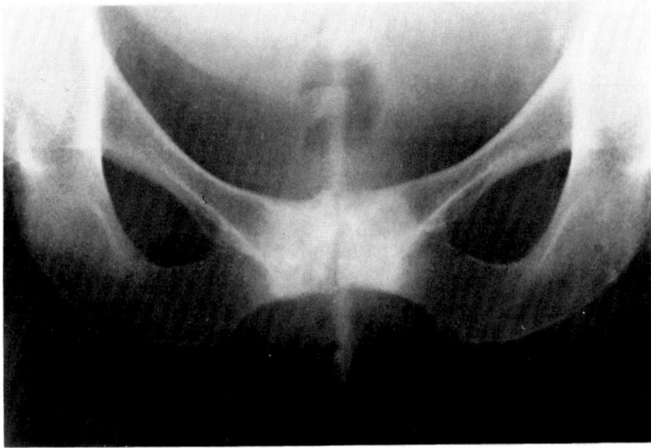

**Figure 64.33. Osteitis pubis.** There is dense sclerosis on either side of the symphysis pubis.

## Reflex Sympathetic Dystrophy Syndrome

The reflex sympathetic dystrophy syndrome (Sudeck's atrophy) is a symptom complex in which an affected extremity suffers partial loss of motor function, mild to severe vasomotor and trophic changes, bone demineralization, and mild to severe aching pain. In all instances, the degree of disability is out of proportion to the often rather trivial injury. The reflex sympathetic dystrophy syndrome is usually precipitated by trauma, myocardial infarction, infection, or neurologic disease (peripheral neuropathy, central nervous system abnormality, cervical osteoarthritis). Although the underlying pathogenic mechanism is unknown, the syndrome is most likely related to disordered function of the autonomic nervous system.

The reflex sympathetic dystrophy syndrome most commonly involves the hands and feet. The initial phase begins weeks or months after the precipitating event and consists of pain, swelling, and vasomotor disturbances. Radiographs may demonstrate periarticular and diffuse soft tissue swelling of the extremity, though this is much more easily recognized clinically. The major radiographic abnormality is a patchy or mottled osteoporosis that may be indistinguishable from the appearance caused by immobilization or paralysis. Although the osteoporosis involves all the bones of the affected extremity, the changes are most marked in the periarticular portions. Radionuclide bone scans may be of value in demonstrating increased uptake in these areas. As the disease progresses there is diffuse demineralization of the involved extremity and, occasionally, a moth-eaten appearance that suggests metastatic carcinoma or multiple myeloma. Magnification radiography shows endosteal scalloping and intracortical resorption, as identified by striations within the cortex. Linear resorption in the outer cortex may simulate periosteal new bone formation. During the acute stage, the mottled appearance of the bones and the severe loss of bone density is striking. As the process becomes more chronic, the mottled appearance is lost and the bones assume a ground-glass, uniform loss of density. In late stages, atrophy of the skin may lead to the development of contractures.

The term *shoulder-hand syndrome* refers to the pain and loss of range of motion in the shoulder that may develop in patients with reflex sympathetic dystrophy syndrome involving the upper extremity.[45,46]

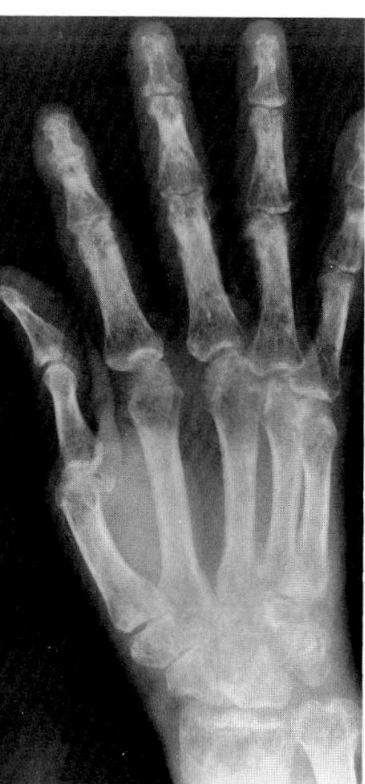

**Figure 64.34. Reflex sympathetic dystrophy syndrome.** Patchy osteoporosis predominantly affects the periarticular regions. Endosteal scalloping and intracortical striations are seen even without magnification films.

## Synovial Chondromatosis

In synovial chondromatosis, a hypertrophic synovial membrane produces multiple metaplastic growths of cartilage that are most often intra-articular but may occasionally involve bursae and tendon sheaths. The cartilaginous masses frequently calcify or even ossify in part, often becoming detached and lying free within the joint cavity. Synovial chondromatosis is usually monarticular and tends to occur in young adults or the middle-aged. The knee is most often involved, with the hip next in

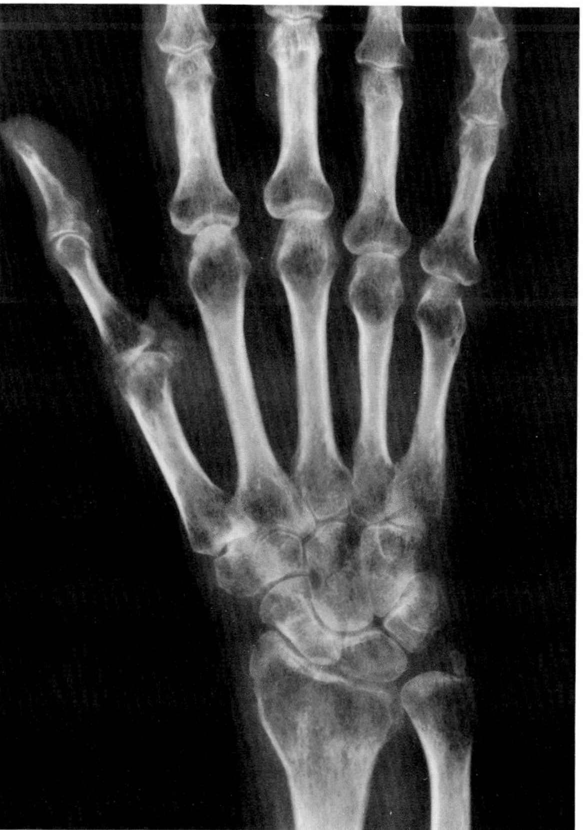

**Figure 64.35. Disuse osteoporosis.** Severe periarticular demineralization followed prolonged immobilization of the extremity.

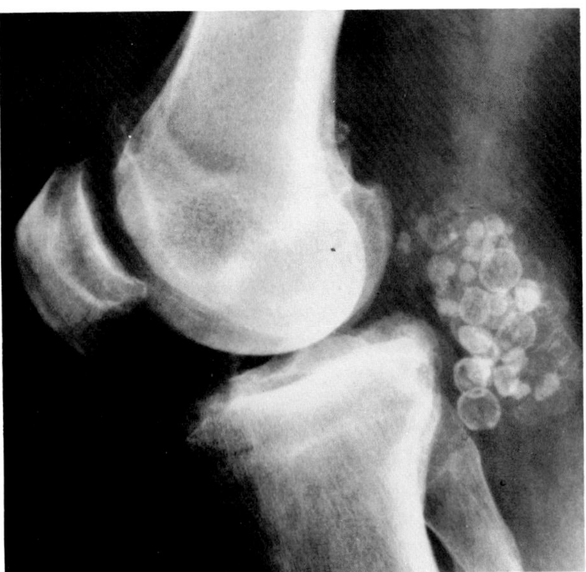

**Figure 64.36. Synovial chondromatosis of the knee.** Multiple calcified and ossified bodies lie free within the knee joint.

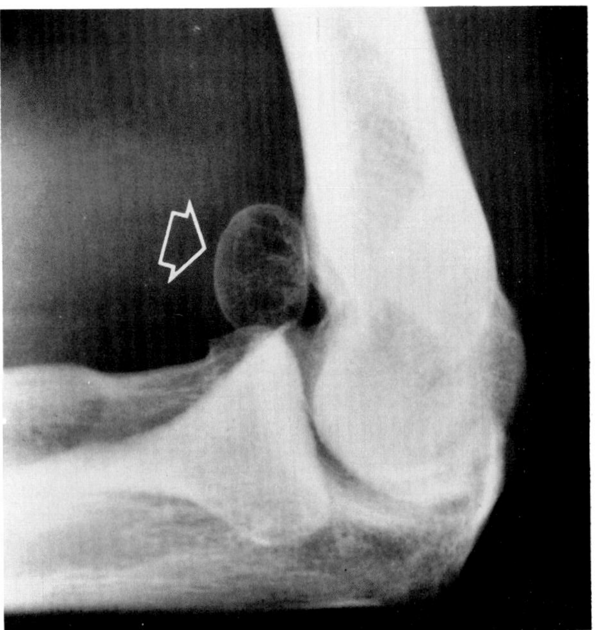

**Figure 64.37. Synovial chondromatosis of the elbow.** A single, large, free joint body (arrow) is seen within the elbow joint.

frequency; synovial chondromatosis has rarely been reported in the elbow, ankle, shoulder, and wrist. The involved joint may enlarge because of an effusion, and the intracapsular loose bodies may produce pain and limitation of joint motion.

Synovial chondromatosis appears radiographically as multiple calcified or ossified bodies in a single joint. The calcifications vary in size and are usually irregular, often with a laminated appearance. If the synovial chondromas do not calcify (about one-third of the cases), they cannot be detected on standard radiographs, and arthrography is necessary to demonstrate the cartilaginous bodies.[47]

## Synovioma (Synovial Sarcoma)

Synovioma is a malignant tumor that arises from a joint capsule, bursa, or tendon. It most frequently affects young adults. The tumor usually arises from synovial tissue in the vicinity of a large joint (para-articular soft tissues just beyond the capsule), rather than within the synovial lining of the joint itself. Although the knee is most often affected, the tumor may arise from a tendon sheath anywhere along a limb.

A synovioma appears radiographically as a well-defined, round or lobulated soft tissue mass adjacent to or near a joint. Amorphous punctate deposits or linear streaks of calcification frequently occur within the tumor. Although the underlying bone is initially normal, the expanding tumor may eventually invade the bone and destroy it. Unless the tumor arises within the joint capsule, joint effusion rarely occurs. A synovioma must be distinguished from pigmented villonodular synovitis, in which

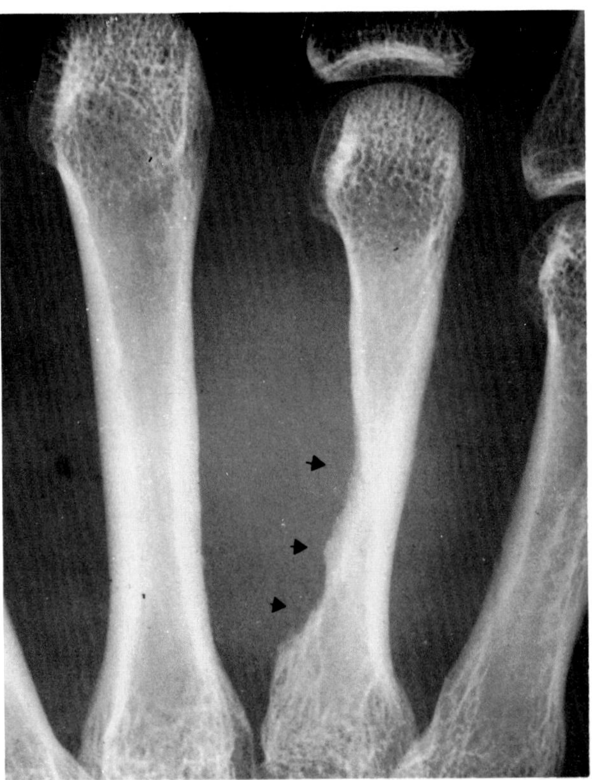

**Figure 64.38. Synovioma of the hand.** A soft tissue mass causes erosion of the third metacarpal (arrows). The slightly irregular margin and serrated configuration at the proximal aspect suggest an aggressive lesion. (From Cavanagh,[57] with permission from the publisher.)

joint effusion is common, calcification does not occur, and erosive bone changes usually involve the articular surfaces.[48,49]

## Calcinosis

Diffuse calcium deposits in subcutaneous fat, muscles, and tendons can develop in several disorders. In most cases the calcium deposits reflect a generalized disorder, such as dermatomyositis, scleroderma, Raynaud's phenomenon, chronic renal failure, or a hypercalcemic state (hyperparathyroidism, hypervitaminosis D, gout). Calcinosis in these conditions is discussed in the specific sections dealing with them.

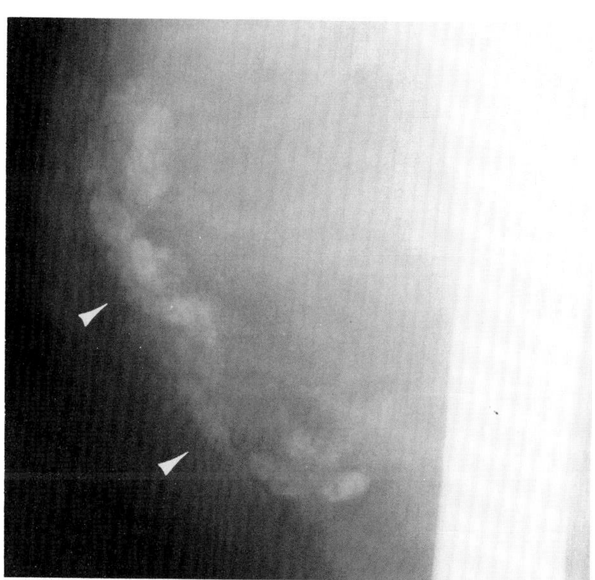

**Figure 64.39. Synovioma of the posteromedial thigh.** A lateral view shows a large soft tissue mass with extensive calcific deposits (arrows) within it. (From Cavanagh,[57] with permission from the publisher.)

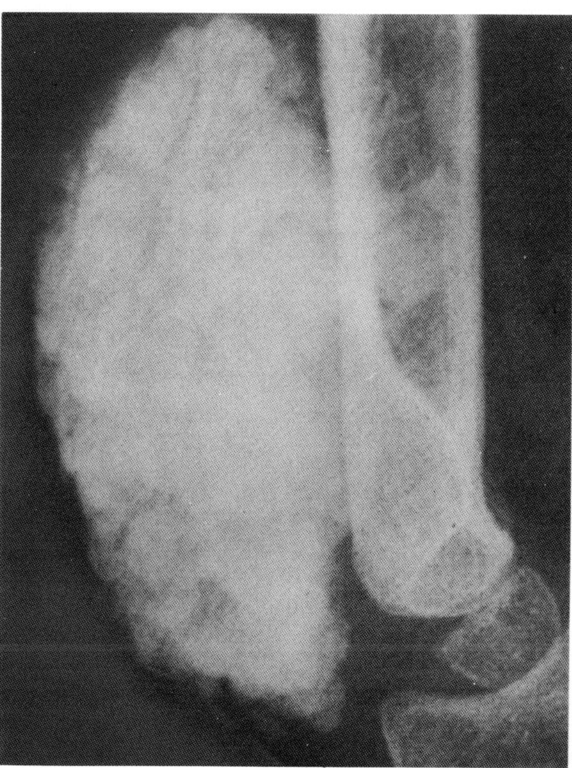

**Figure 64.40. Tumoral calcinosis.** There is a dense calcific mass about the elbow in a patient with secondary hyperparathyroidism due to chronic renal failure. (From Cockshott and Middlemiss,[58] with permission from the publisher.)

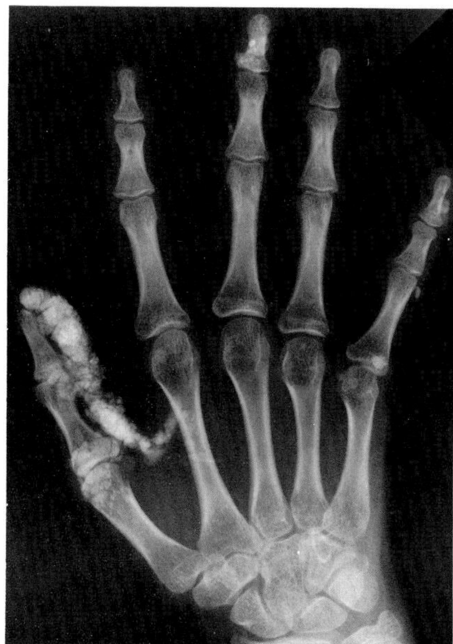

A

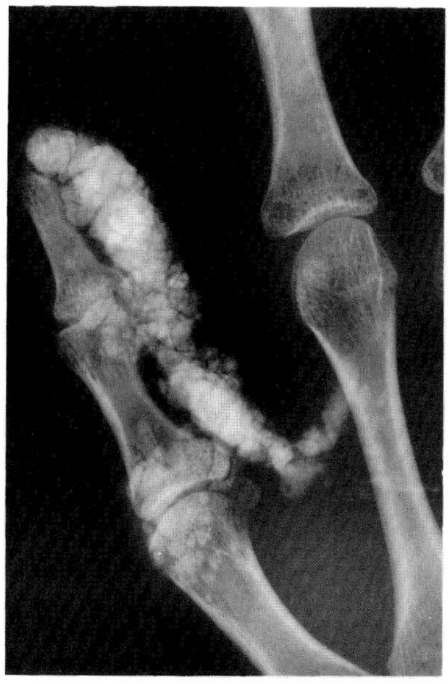

B

**Figure 64.41. Calcinosis circumscripta.** (A) A full view of the hand and (B) a coned view of the thumb demonstrate dense calcific deposits in the soft tissues on the ulnar aspect of the thumb. The patient was asymptomatic with no known connective tissue disorder. The normal distal phalangeal tufts virtually exclude scleroderma as the underlying cause for the calcification.

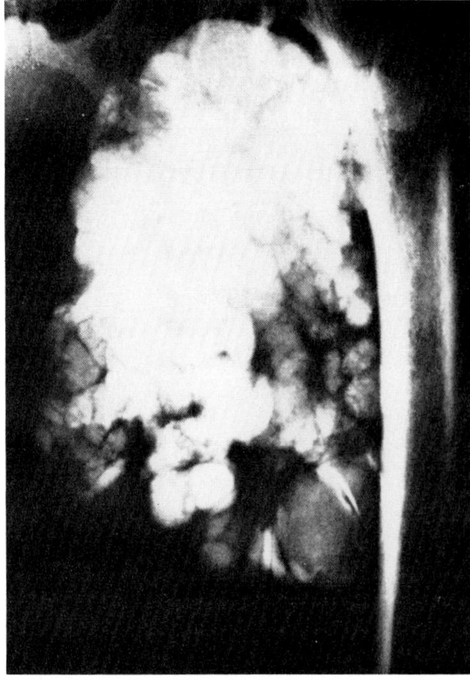

A

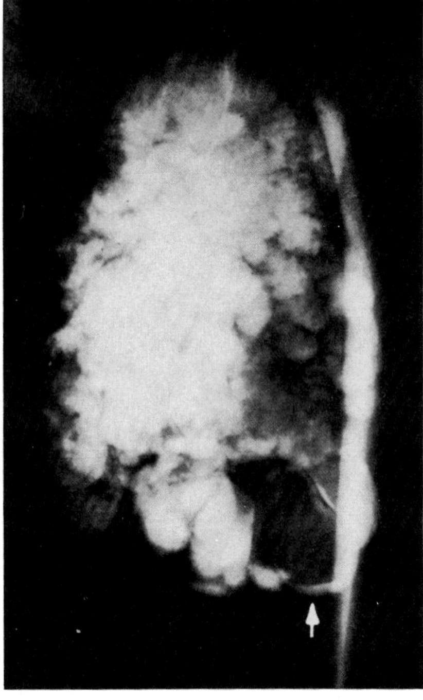

B

**Figure 64.42. Tumoral calcinosis.** (A) A supine view demonstrates a large, irregular calcific mass with some relatively lucent areas in the proximal thigh. (B) An upright view shows sedimentation in the liquid-filled cysts (arrow), with absence of sedimentation in the more amorphous gritty deposits. (From Hug and Guncaga,[52] with permission from the publisher.)

Calcinosis universalis is a disease of unknown cause in which calcium is initially deposited subcutaneously and later in deep connective tissues throughout the body. Frequently progressive, calcinosis universalis occurs in relatively young individuals and may lead to death. In calcinosis circumscripta, localized calcific deposits are found in the skin and subcutaneous tissues, especially in the hands and wrists.

Tumoral calcinosis refer to localized collections of calcium in periarticular soft tissues. The disorder involves young and otherwise healthy individuals, may affect single or multiple joints, and has a predilection for the hips, elbows, shoulders, ankles, and wrists. Although often painless, pain, swelling, and disability may be present. The condition begins as small calcified nodules that enlarge to form solid, lobulated tumors that are extremely dense and have rough, irregular borders. Pathologically, the calcific masses reflect honeycomb-like clusters of cysts in a dense fibrous capsule. Because the cysts are filled with a granular, pasty, or liquid material, on upright views there may be sedimentation of calcium phosphate crystals with resulting fluid-calcium levels.[50–53]

## REFERENCES


1. Kelly PJ, Martin WJ, Coventry MB: Bacterial (suppurative) arthritis in the adult. *J Bone Joint Surg* 52A:1595–1602, 1970.
2. Lloyd-Roberts GC: Suppurative arthritis of infancy. *J Bone Joint Surg* 42B:706–720, 1960.
3. Argen ST, Wilson CH, Wood P: Suppurative arthritis: Clinical features of 42 cases. *Arch Intern Med* 17:666–671, 1966.
4. Berney S, Goldstein M, Bishko F: Clinical and diagnostic features of tuberculous arthritis. *Am J Med* 53:36–41, 1972.
5. Hoaglund FT: Osteoarthritis. *Orthop Clin North Am* 2:3–18, 1971.
6. Forrester DM, Brown JC, Nesson JW: *The Radiology of Joint Disease,* Philadelphia, WB Saunders, 1978.
7. Solomon L: Patterns of osteoarthritis of the hip. *J Bone Joint Surg* 58B:176–183, 1976.
8. Hermodsson I: Roentgen appearances of arthritis of the hip. *Acta Radiol* 12:865–881, 1972.
9. Marmor L: Osteoarthritis of the knee. *JAMA* 218:213–215, 1971.
10. Martinez LO: Air in the esophagus as a sign of scleroderma (differential diagnosis with some entities). *J Can Assoc Radiol* 25:234–237, 1974.
11. Proto AV, Lane EJ: Air in the esophagus: A frequent radiographic finding. *AJR* 129:433–440, 1977.
12. Donner MW, Saba GP, Martinez CR: Diffuse disease of the esophagus: A practical approach. *Semin Roentgenol* 16:198–213, 1981.
13. Margulis AR, Koehler RE: Radiologic diagnosis of disordered esophageal motility: A unified physiologic approach. *Radiol Clin North Am* 14:429–439, 1976.
14. Mueller CF, Morehead R, Alter A, et al: Pneumatosis intestinalis in collagen disorders. *AJR* 114:300–305, 1972.
15. Queloz JM, Woloshin AJ: Sacculation of the small intestine in scleroderma. *Radiology* 105:513–515, 1972.
16. Horowitz AL, Meyers MA: The "hide-bound" small bowel of scleroderma: Characteristic mucosal fold pattern. *AJR* 119:332–334, 1973.
17. Poirier T, Rankin G: Gastrointestinal manifestations of progressive systemic sclerosis, based on a review of 364 cases. *Am J Gastroenterol* 58:330–344, 1972.
18. Farmer RG, Gifford RW, Hines EA: Prognostic significance of Raynaud's phenomenon and other clinical characteristics of systemic scleroderma. A study of 271 cases. *Circulation* 21:1088–1095, 1960.
19. Fraser RG, Pare JAP: *Diagnosis of Diseases of the Chest,* Philadelphia, WB Saunders, 1979.
20. Gondos B: Roentgen manifestations in progressive systemic sclerosis (diffuse scleroderma). *AJR* 84:235–247, 1960.

21. Winograd J, Schimmel DH, Palubinskas AJ: The spotted nephrogram of renal scleroderma. *AJR* 126:734–738, 1976.
22. Bassett LW, Blocka KLN, Furst DE: Skeletal findings in progressive systemic sclerosis (scleroderma). *AJR* 136:1121–1126, 1981.
23. Helms CA, Vogler JB, Simms DA, et al: CPPD crystal deposition disease or pseudogout. *RadioGraphics* 2:40–52, 1982.
24. Martel W, McCarter DK, Solsky MA, et al: Further observations on the arthropathy of calcium pyrophosphate crystal deposition disease. *Radiology* 141:1–15, 1981.
25. Resnick D, Niwayama G, Goergen TG, et al: Clinical, radiographic and pathologic abnormalities in calcium pyrophosphate dihydrate deposition disease (CPPD): Pseudogout. *Radiology* 122:1–15, 1977.
26. Ameri MR, Alebouyeh M, Donner MW: Hypertrophic osteoarthropathy in childhood malignancy. *AJR* 130:992–993, 1978.
27. Greenfield GB, Schorsch HA, Shkolnik A: The various roentgen appearances of pulmonary hypertrophic osteoarthropathy. *AJR* 101:927–931, 1967.
28. Hallis WC: Hypertrophic osteoarthropathy secondary to upper gastrointestinal neoplasm. *Ann Intern Med* 66:125–130, 1967.
29. Lazarus JH, Galloway JK: Pachydermoperiostosis. *AJR* 118: 308–313, 1973.
30. Banna M, Foster B: Roentgenologic features of acrodystrophic neuropathy. *AJR* 115:186–190, 1972.
31. Harrison RB: Charcot's joint: Two new observations. *AJR* 128:807–809, 1977.
32. Clouse ME, Gramm HF, Legg M, et al: Diabetic osteoarthropathy. *AJR* 121:22–34, 1974.
33. Resnick D, Niwayama G: Radiographic and pathologic features of spinal involvement in diffuse idiopathic skeletal hyperostosis (DISH). *Radiology* 119:559–568, 1976.
34. Littlejohn GO, Urowitz MB, Smythe HA, et al: Radiographic features of the hand in diffuse idiopathic skeletal hyperostosis (DISH). *Radiology* 140:623–629, 1981.
35. Alenghat JP, Hallett M, Kido DK: Spinal cord compression in diffuse idiopathic skeletal hyperostosis. *Radiology* 142:119–120, 1982.
36. Brodey PA: Multicentric reticulohistiocytosis: A rare cause of destructive polyarthritis. *Radiology* 114:327–328, 1975.
37. Barrow MV, Holubar K: Multicentric reticulohistiocytosis: A review of 33 patients. *Medicine* 48:287–305, 1969.
38. Gold RH, Metzger AL, Mirra JM, et al: Multicentric reticulohistiocytosis (lipoid dermatoarthritis). *AJR* 124:610–624, 1975.
39. Breimer VW, Freiberger RH: Bone lesions associated with villonodular synovitis. *AJR* 79:618–629, 1958.
40. Smith JH, Pugh DG: Roentgenographic aspects of articular synovitis. *AJR* 87:1146–1156, 1962.
41. Skorneck AB: Roentgen aspects of Tietze's syndrome: Painful hypertrophy of costal cartilage and bone—osteochondritis? *AJR* 83:748–755, 1960.
42. Kotzen LM: Roentgen diagnosis of rotator cuff tear. *AJR* 112:507–511, 1971.
43. Killoran PJ, Marcove RC, Freiberger RH: Shoulder arthrography. *AJR* 103:658–668, 1968.
44. Numguchi Y: Osteitis condensans ilii, including its resolution. *Radiology* 98:1–8, 1971.
45. Genant HK, Kozin F, Bekerman C, et al: The reflex sympathetic dystrophy syndrome: A comprehensive analysis using fine detail radiography, photon absorptiometry and bone and joint scintigraphy. *Radiology* 117:21–32, 1975.
46. Edeiken J: *Roentgen Diagnosis of Diseases of Bone*. Baltimore, Williams & Wilkins, 1981.
47. Prager RJ, Mall JC: Arthrographic diagnosis of synovial chondromatosis. *AJR* 127:334–346, 1976.
48. Horowitz AL, Resnick D, Watson RC: The roentgen features of synovial sarcomas. *Clin Radiol* 24:418–424, 1973.
49. Raben M, Calabrese A, Higenbotham L, et al: Malignant synovioma. *AJR* 93:145–153, 1965.
50. Greenfield GB: *Radiology of Bone Disease*. Philadelphia, JB Lippincott, 1980.
51. Yaghmai I, Mirbod P: Tumoral calcinosis. *AJR* 111:573–578, 1971.
52. Hug I, Guncaga J: Tumoral calcinosis with sedimentation sign. *Br J Radiol* 47:734–736, 1974.
53. Naidich TP, Siegelman SS: Para-articular soft tissue changes in systemic diseases. *Semin Roentgenol* 8:101–112, 1973.
54. Brown JC, Forrester DM: Back pain, in Eisenberg RL, Amberg JR (eds): *Critical Diagnostic Pathways in Radiology: An Algorithmic Approach*. Philadelphia, JB Lippincott, 1981, pp 263–278.
55. Eisenberg RL: *Gastrointestinal Radiology: A Pattern Approach*. Philadelphia, JB Lippincott, 1983.
56. Eisenberg RL, Hedgcock MW: Bilateral pigmented villonodular synovitis of the hip. *Br J Radiol* 51:916–917, 1978.
57. Cavanagh RC: Tumors of the soft tissues of the extremities. *Semin Roentgenol* 8:73–89, 1973.
58. Cockshott WP, Middlemiss P: *Clinical Radiology in the Tropics*. Edinburg, Churchill Livingstone, 1979.
59. Brown JC, Forrester DM: Arthritis, in Eisenberg RL, Amberg JR (eds): *Critical Diagnostic Pathways in Radiology: An Algorithmic Approach*. Philadelphia, JB Lippincott, 1981, pp 280–300.
60. Dalinka MK, Reginato AJ, Golden DA: Calcium deposition diseases. *Semin Roentgenol* 17:39–48, 1982.

## Chapter Sixty-Five
# Diseases of Striated Muscles

### DERMATOMYOSITIS AND POLYMYOSITIS

Polymyositis is an inflammatory disease of skeletal muscles in which a lymphocytic infiltration produces muscle fiber damage and degeneration. Dermatomyositis is a condition in which polymyositis is associated with skin inflammation and a characteristic rash. Adults with dermatomyositis have a relatively high incidence of underlying malignancy.

Dysfunction of the gastrointestinal musculature is most evident in the esophagus. A pattern of decreased peristalsis or esophageal atony, simulating scleroderma, is often produced. Because dermatomyositis and polymyositis involve striated muscle that is infrequently affected by scleroderma, there also may be disruption of the normal stripping wave in the upper third of the esophagus. Atony and dilatation of the stomach and small bowel can also occur.

Many patients with dermatomyositis or polymyositis have normal chest radiographs. In others, there may be a diffuse reticular or reticulonodular interstitial pattern that predominantly involves the lung bases. The radiographic findings usually respond dramatically to corticosteroid therapy. Unlike other connective tissue disorders that produce a similar radiographic appearance, pleural effusion is infrequent in dermatomyositis and polymyositis. Involvement of the muscles of respiration,

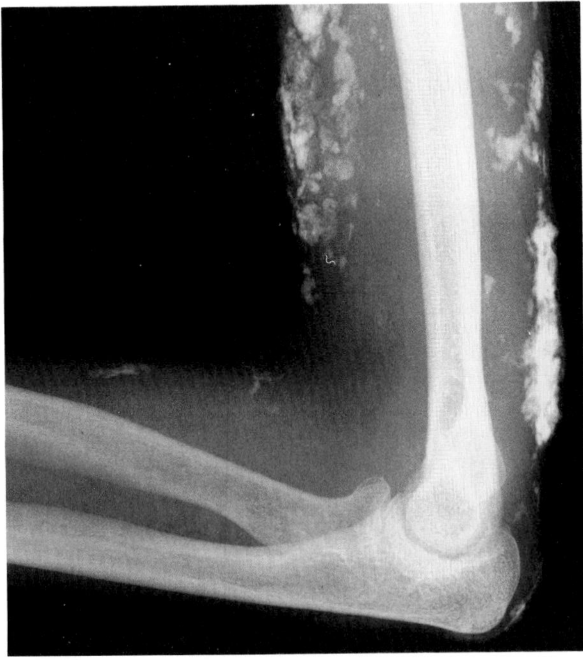

**Figure 65.1. Dermatomyositis.** There are extensive deposits of calcium in the soft tissues about the humerus and elbow, as well as a loss of the sharp demarcation between the muscles and the subcutaneous tissues.

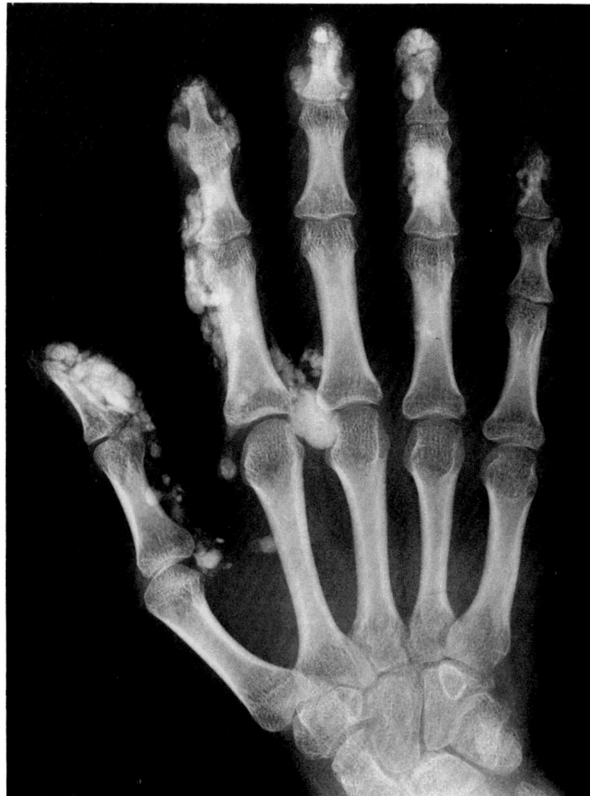

**Figure 65.2. Dermatomyositis.** Irregular calcific deposits involve all the fingers.

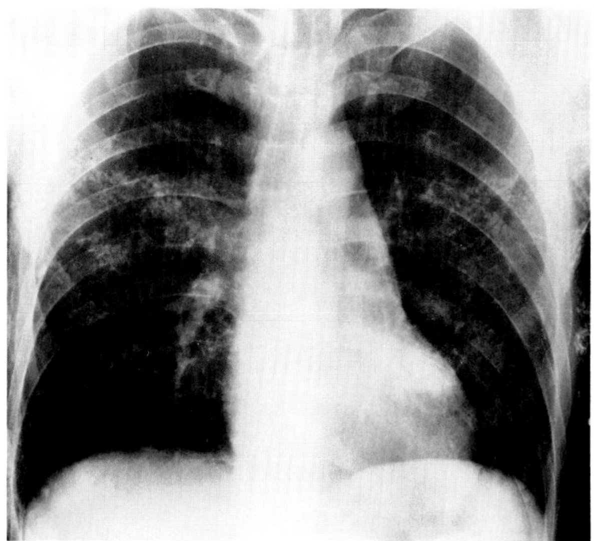

**Figure 65.3. Dermatomyositis.** Thin plaques of calcium in subcutaneous tissues and muscles over the thorax simulate an interstitial pneumonia.

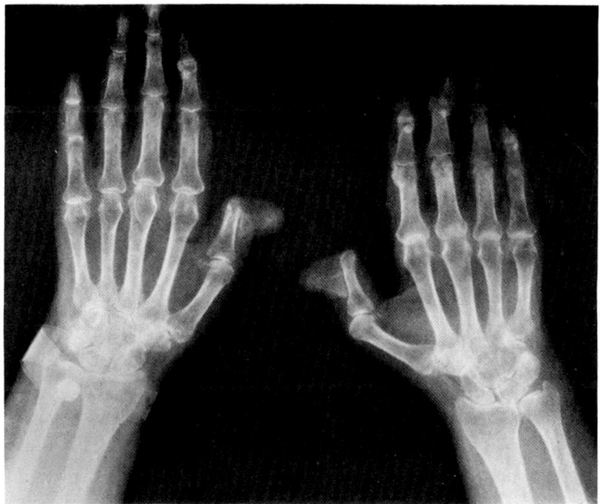

**Figure 65.4. Polymyositis.** A frontal radiograph of the hands and wrists demonstrates bilateral dislocations of the interphalangeal joints of the thumbs ("floppy thumb" sign). (From Bunch, O'Duffy, and McLeod,[4] with permission from the publisher.)

particularly the diaphragm, can cause diaphragmatic elevation with basilar atelectasis and a decrease in lung volume. Paralysis of the pharyngeal musculature can lead to patchy aspiration pneumonia.

The musculoskeletal changes are most severe in childhood dermatomyositis. Diffuse edema of the subcutaneous and muscular tissues leads to increased muscular bulk and density, poor delineation of the normally sharp subcutaneous-muscular interface, and edema of subcutaneous tissue septa. In chronic disease, there is decreased muscle bulk and joint contractures. A characteristic finding is extensive calcification in the muscles and subcutaneous tissues underlying the skin lesions. This calcification may appear as superficial or deep masses, as linear deposits, or as a lacy, reticular, subcutaneous deposition of calcium encasing the torso.[1–4]

## MUSCULAR DYSTROPHY

Replacement of muscle by fat in the muscular dystrophies results in a characteristic appearance on radiographs of the extremities. Although the total muscle mass is not decreased, the extensive accumulation of fat within the remaining muscle bundles produces a fine striated or striped appearance. In the pseudohypertrophic dystrophy of Duchenne, enlargement of specific muscle groups (calves, shoulder girdles) suggests muscular strength, though the patient is actually extremely weak. In wasting diseases, most of the muscle tissue is replaced by fat, so that the fascial sheath bounding the muscle stands out as a thin shadow of increased density as it is visualized on edge.

A characteristic sign of muscular dystrophy is unusual widening of the shaft of the fibula in its anteroposterior

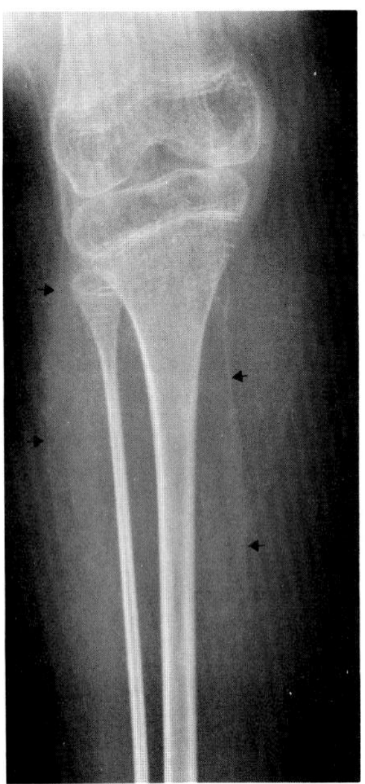
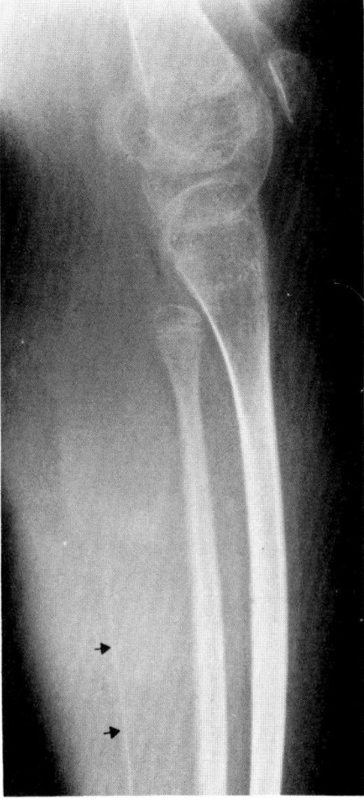

A B

**Figure 65.5. Muscular dystrophy.** (A) Frontal and (B) lateral views of the leg show increased lucency (arrows) representing fatty infiltration within muscle bundles. The fascial sheaths appear as thin shadows of increased density surrounded by fat. The bones of the lower leg are thin, especially the fibula on the frontal view.

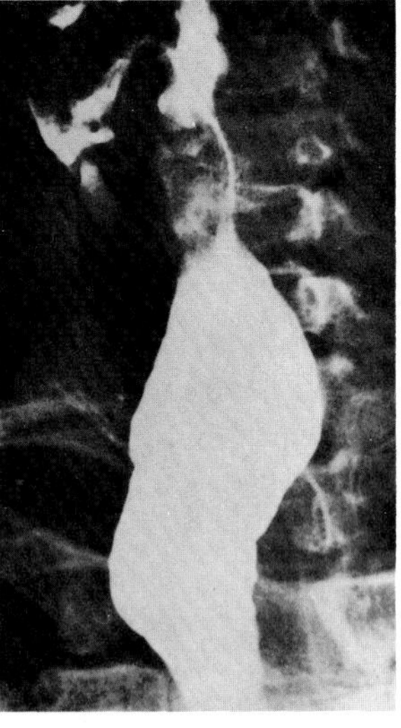

**Figure 65.6. Myotonic dystrophy.** A continuous column of barium extends from the hypopharynx into the cervical esophagus, caused by reflux across the level of the cricopharyngeus muscle. (From Seaman,[9] with permission from the publisher.)

diameter. Decreased muscular tone can lead to osteoporosis, bone atrophy with cortical thinning, scoliosis, and joint contractures.

An abnormal swallowing mechanism in muscular dystrophy can result in the failure to clear barium adequately from the valleculae and pyriform sinuses; this may lead to tracheal aspiration and nasal regurgitation.[5–7]

## MYOTONIC DYSTROPHY

Myotonic dystrophy is an inherited disease in which an anatomic abnormality of the motor end plate and striated muscle leads to atrophy and an inability of the contracted muscle to relax (myotonia). Associated findings include cataracts, frontal baldness and testicular atrophy in men, and a characteristic facial expression (myopathic facies).

In addition to severely disturbed pharyngeal peristalsis associated with the pooling of contrast material within the valleculae and pyriform sinuses, patients with myotonic dystrophy have a reduced or absent resting pressure of the cricopharyngeus muscle. Because a major function of this upper esophageal sphincter is preventing esophageal contents from refluxing into the pharynx, the diminished resting tone in patients with myotonic dystrophy permits easy regurgitation from the esophagus into the pharynx and leads to a high incidence of aspiration. Reflux across the cricopharyngeus results in the characteristic radiographic pattern of a continuous column of barium extending from the hypopharynx into the cervical esophagus, even during the resting phase when the patient is not swallowing. Concomitant smooth muscle atrophy can lead to dilatation and diminished peristalsis within affected segments of the gastrointestinal tract.

About half the patients with myotonic dystrophy demonstrate changes within the skull. These include a generalized increase in the width of the calvarium, a small sella turcica, and enlargement of the frontal sinus. Progressive atrophy of skeletal muscle is manifested radiographically by loss of muscle mass and a corresponding increase in fat and connective tissue deposition between the remaining muscle bundles, a nonspecific finding that can be seen in other forms of muscular dystrophy.

Pharyngeal dysfunction can lead to the development of patchy areas of aspiration pneumonia. Diaphragmatic involvement results in elevation of the diaphragm with basilar atelectasis and an increased incidence of infection. Rare involvement of the cardiac muscles may lead to generalized cardiac enlargement and congestive heart failure.[8,9]

## MYASTHENIA GRAVIS

Myasthenia gravis is a disorder of the neuromuscular junction characterized by weakness and undue fatigability on exercise. Weakness of the laryngeal and pharyngeal muscles may cause dysphagia, choking, aspiration of food, and nasal regurgitation. On an esophogram, peristalsis may appear normal on the initial swallow of barium. During repeated swallows, however, peristalsis in the upper esophagus becomes feeble or disappears completely. The pharynx is dilated and atonic, and barium collects in the pyriform sinuses and valleculae before slowly passing into the esophagus. Aspiration of contrast material into the trachea can usually be demonstrated; reflux into the nasal cavities can also occur.

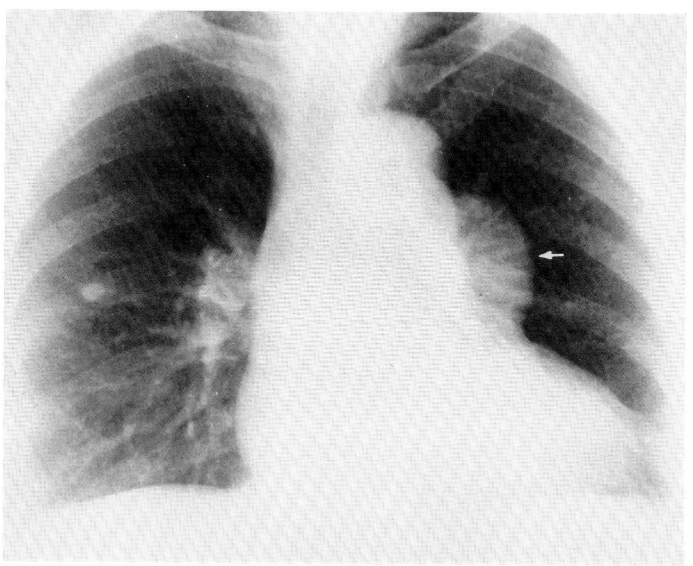

A

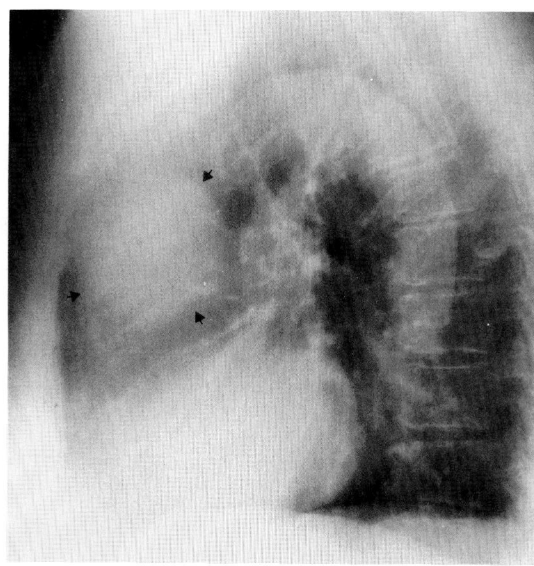

B

**Figure 65.7. Myasthenia gravis with thymoma.** (A) Frontal and (B) lateral views of the chest demonstrate a large mass in the anterior mediastinum (arrows).

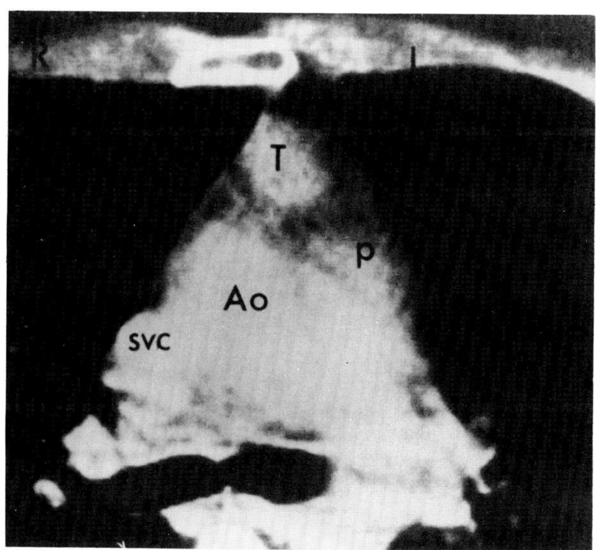


**Figure 65.8. Myasthenia gravis with thymoma.** A CT scan shows a thymoma (T) surrounded by fat. AO = aorta; P = pulmonary artery; svc = superior vena cava. (From Mink, Bein, Sukov, et al,[19] with permission from the publisher.)

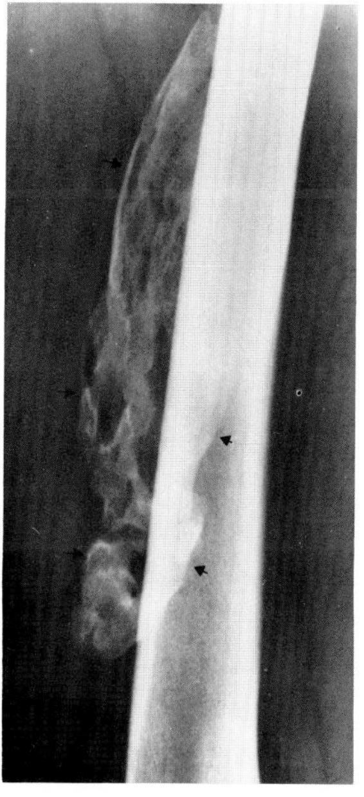

**Figure 65.9. Myositis ossificans circumscripta.** An ossified mass (arrows) lies parallel to and overlapping the shaft of the femur. The patient had sustained a severe muscle injury several years previously.

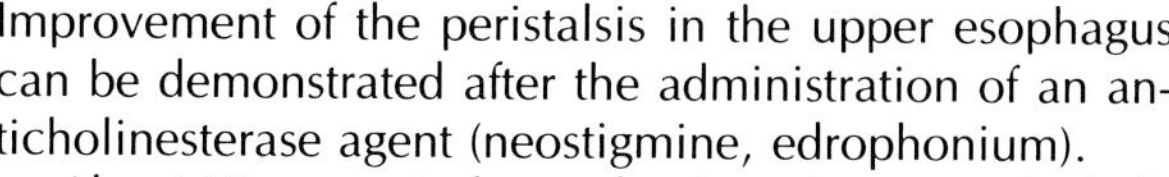

Improvement of the peristalsis in the upper esophagus can be demonstrated after the administration of an anticholinesterase agent (neostigmine, edrophonium).

About 10 percent of myasthenic patients, particularly older males, have a thymoma. Because this anterior mediastinal tumor may be impossible to detect on routine chest radiographs, computed tomography of the mediastinum is usually performed to screen for an underlying thymoma.[10–12,19]

## MYOSITIS OSSIFICANS

### Myositis Ossificans Circumscripta

Myositis ossificans circumscripta refers to the development of calcification or ossification within injured muscle that is usually related to acute or chronic trauma to the deep tissues of the extremities. Certain occupations and sports are prone to the development of myositis ossificans. Thus, heterotopic bone often arises in the adductor longus (rider's bone), brachialis (fencer's bone), and soleus (dancer's bone) muscles.

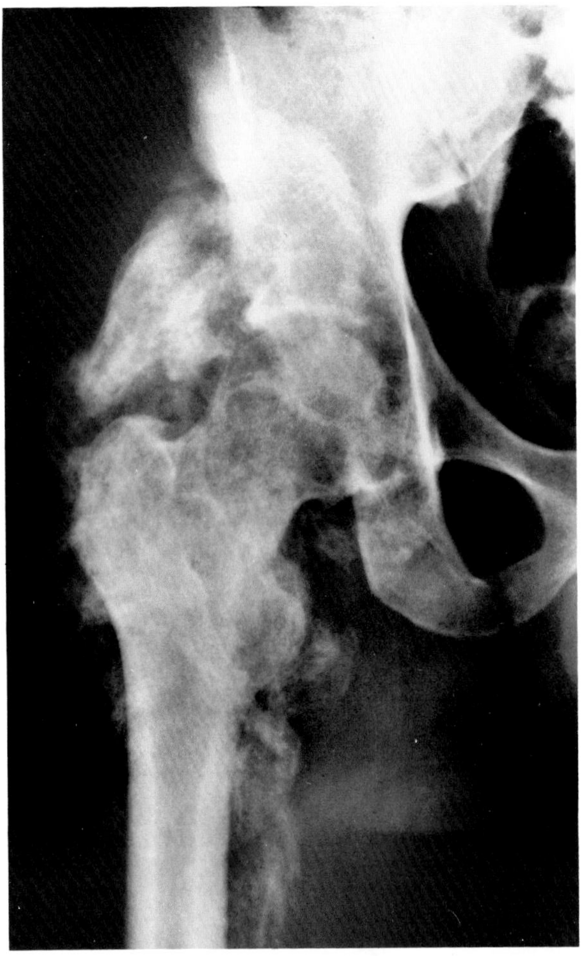

Figure 65.10. **Myositis ossificans associated with neurologic disorders.** There are diffuse osseous deposits in muscles, tendons, and ligaments about the hip in a patient with long-standing paralysis.

The earliest change is a soft tissue mass that develops soon after injury. After about 1 month, faint calcification appears as a hazy shadow of increased density. Over a period of several weeks the calcification becomes progressively denser and appears as a central lacy pattern of new bone formation surrounded by a sharply circumscribed peripheral cortex. The heterotopic calcification or ossification typically lies parallel to the shaft of the bone or along the axis of a muscle. Although the radiographic appearance may simulate parosteal sarcoma, myositis ossificans is completely separated from the bone by a radiolucent zone, unlike the malignant tumor, which is attached by a sessile base and a discontinuous radiolucent zone.[13–15]

## Myositis Ossificans Associated with Neurologic Disorders

Up to half of the patients with paraplegia demonstrate myositis ossificans in the paralyzed part. Osseous deposits occur in muscles, tendons, and ligaments. The heterotopic bone is most pronounced around large joints, especially the hip.[16]

## Generalized (Progressive) Myositis Ossificans

Generalized myositis ossificans is a rare congenital dysplasia characterized by an interstitial myositis or fibrositis that undergoes cartilaginous and osseous transformation. The first symptom is often a firm swelling in the paravertebral or cervical muscles with mild tenderness or discomfort. Within 6 to 12 months, palpable bony masses within the muscle can be detected radiographically. Thick columns and plates of bone eventually re-

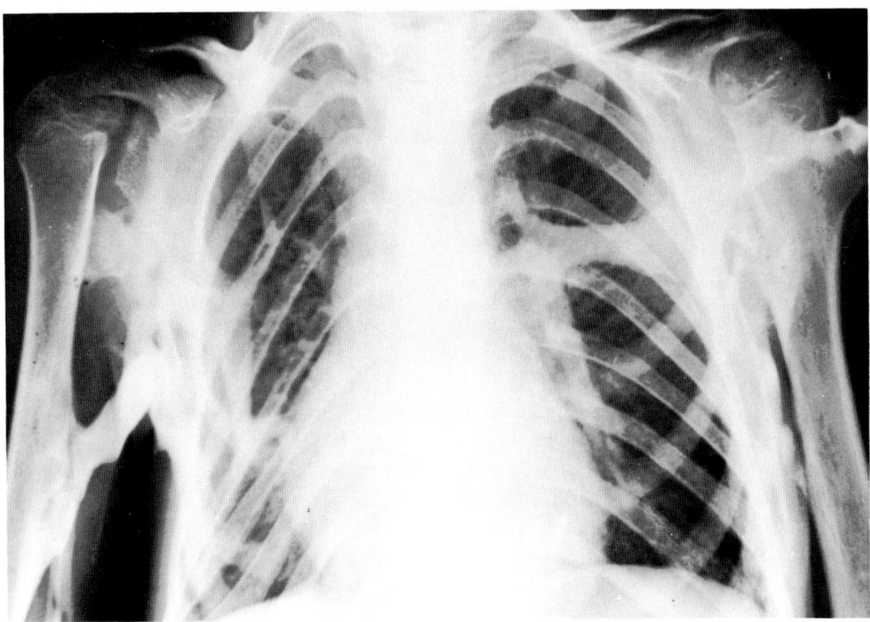

**Figure 65.11. Generalized (progressive) myositis ossificans.** A frontal view of the chest demonstrates extensive new bone formation in the soft tissues with severe limitation of arm motion. Note the exostosis of the left proximal humerus, due to blending of the ossific foci with the cortex of the bone. (From Greenfield,[20] with permission from the publisher.)

place tendons, fascia, and ligaments, causing such severe limitation of movement, contractures, and deformity that the patient becomes a virtual "stone man" and death ensues.

Patients with generalized myositis ossificans usually have a variety of congenital anomalies, most frequently hypoplasia of the great toes or thumbs.[17–18]

## REFERENCES


1. Ozonoff MB, Flynn FJ: Roentgenologic features of dermatomyositis of childhood. *AJR* 118:206–212, 1973.
2. Frazier AR, Miller RD: Interstitial pneumonitis in association with polymyositis and dermatomyositis. *Chest* 65:403–407, 1974.
3. Blane CE, White SJ, Braunstein EM, et al: Patterns of calcification in childhood dermatomyositis. *AJR* 142:397–400, 1984.
4. Bunch TW, O'Duffy JD, McLeod RA: Deforming arthritis of the hands in polymyositis. *Arthritis Rheum* 19:243–248, 1976.
5. Juhl JH: *Essentials of Roentgen Interpretation*, Philadelphia, Harper & Row, 1981.
6. Gay BB, Weems HS: Roentgen evaluation of disorders of muscle. *Semin Roentgenol* 8:25–36, 1973.
7. Simpson AJ, Khilnani MT: Gastrointestinal manifestations of the muscular dystrophies: A review of roentgen findings. *AJR* 125:948–955, 1975.
8. Krain S, Rabinowitz JG: The radiologic features of myotonic dystrophy with presentation of a new finding. *Clin Radiol* 22:462–465, 1971.
9. Seaman WB: Functional disorders of the pharyngo-esophageal junction: Achalasia and chalasia. *Radiol Clin North Am* 7:113–119, 1969.
10. Murray JP: Deglutition in myasthenia gravis. *Br J Radiol* 35:43–50, 1962.
11. Joseph WL, Murray JF, Mulder DG: Mediastinal tumors—problems in diagnosis and treatment. *Dis Chest* 50:150–160, 1966.
12. Baron RL, Lee JKT, Sagel SS, et al: Computed tomography of the abnormal thymus. *Radiology* 142:127–134, 1982.
13. Patterson DC: Myositis ossificans circumscripta. *J Bone Joint Surg* 52B:296–301, 1970.
14. Goldman AB: Myositis ossificans circumscripta: A benign lesion with a malignant differential diagnosis. *AJR* 126:32–40, 1976.
15. Norman A, Dorfman HD: Juxtacortical circumscribed myositis ossificans: Evaluation and radiographic features. *Radiology* 96:301–306, 1970.
16. Heilbrun N, Kuhn WF: Erosive bone lesions and soft tissue ossifications associated with spinal cord injuries (paraplegia). *Radiology* 48:480–483, 1947.
17. Fairbank HAT: Myositis ossificans progressiva. *J Bone Joint Surg* 32B:108–116, 1950.
18. Illingworth RS: Myositis ossificans progressiva. *Arch Dis Child* 46:264–268, 1971.
19. Mink JH, Bein ME, Sukov R, et al: Computed tomography of the anterior mediastinum in patients with myasthenia gravis and suspected thymoma. *AJR* 130:239–245, 1978.
20. Greenfield GB: *Radiology of Bone Disease*. Philadelphia, JB Lippincott, 1980.

# PART TWELVE
# DISORDERS OF THE NERVOUS SYSTEM

## Chapter Sixty-Six
# Diagnostic Methods in Neuroradiology

Recent technologic advances are dramatically changing the radiographic approach to the neurology patient. Uncomfortable and potentially dangerous examinations such as pneumoencephalography and arteriography are being replaced by modern computed tomography (CT). In the future, magnetic resonance imaging (MRI), also known as nuclear magnetic resonance (NMR), may replace many of the studies currently employed.

### PLAIN SKULL RADIOGRAPHS

Plain skull radiographs have long been used (and overused) in the diagnostic approach to the patient with head trauma, even though the presence or absence of a skull fracture does not correlate with intracranial abnormalities. Indeed, serious, treatable intracranial hematomas can be present without skull fractures. To prevent overutilization of skull films in trauma, the following indications have been established:

1. Unexplained focal neurologic signs
2. Unconsciousness (including the unarousable alcoholic)
3. Documented decreasing level of consciousness or progressive mental deterioration
4. History of previous craniotomy with shunt tube in place
5. Skull depression or subcutaneous foreign body palpable or identified by a probe through a laceration or puncture wound
6. Hemotympanum or fluid discharge from the ear
7. Discharge of cerebrospinal fluid from the nose
8. Ecchymosis over the mastoid process (Battle's sign)
9. Bilateral orbital ecchymosis (raccoon eyes)

Because the wide distribution of CT scanners has virtually eliminated the need for plain skull radiographs in patients with head trauma, these guidelines should now be considered to apply to the need for CT.

Plain skull radiographs have in the past been used in the evaluation of metastatic disease to the calvarium. Except in patients with suspected multiple myeloma, in whom false-negative radionuclide bone scans are common, isotopic scanning has replaced plain skull radiographs in the evaluation of metastatic disease.[1,2]

### PLAIN RADIOGRAPHS OF THE SPINE

The main indication for plain radiographs of the spine is acute trauma. Because plain radiographs are of limited value in assessing patients with low back pain, low back strain, or radicular symptoms, plain spine radiographs in such patients should be reserved for those who have

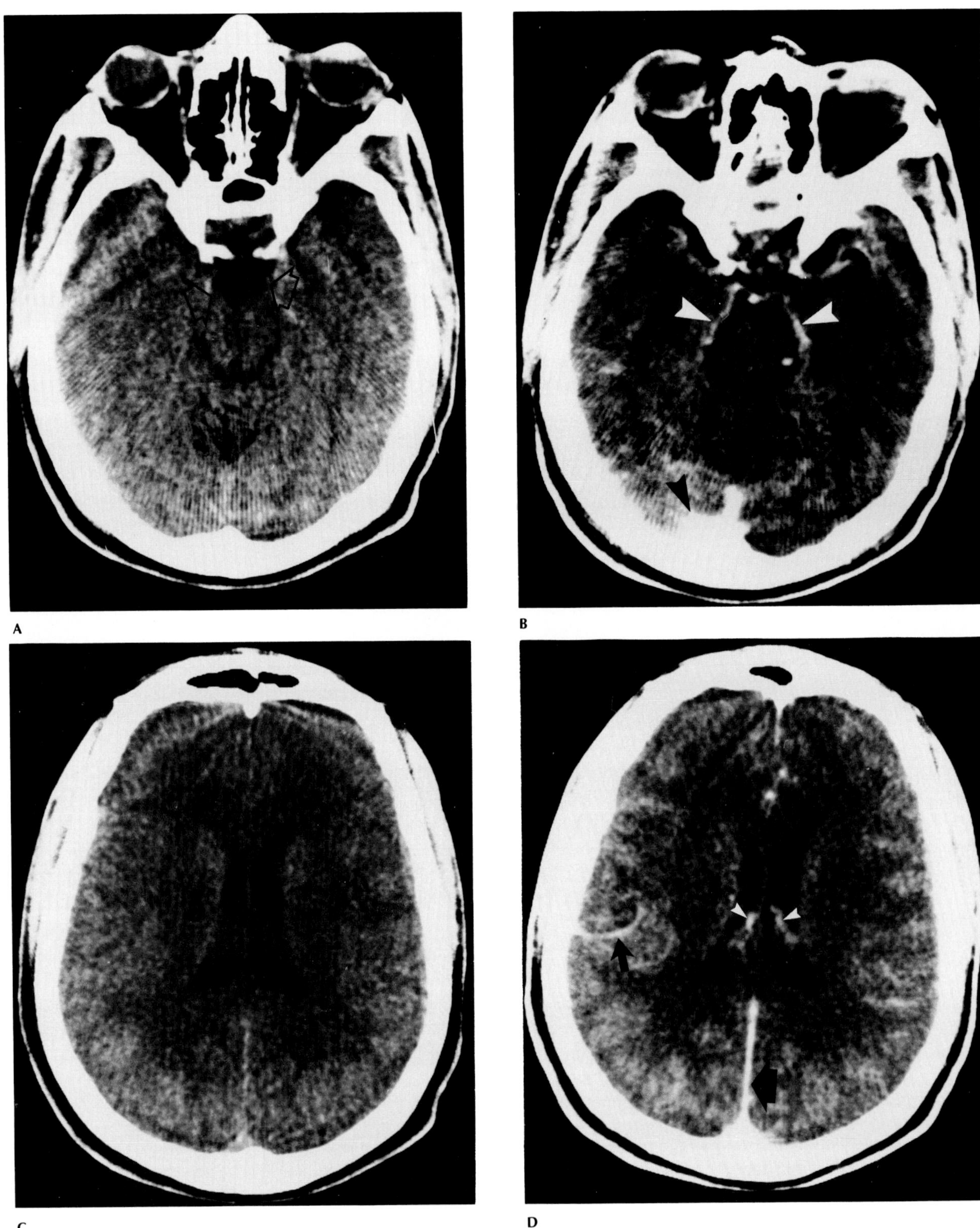


**Figure 66.1. Normal CT scan.** (A) Noncontrast and (B) contrast scans through the level of the suprasellar cistern. Before the intravenous injection of contrast material, the cerebrospinal fluid spaces anterior to the midbrain and suprasellar region are of low density (arrows). After contrast administration, there is enhancement of the normal vessels in the circle of Willis (white arrowheads) and the venous confluence (black arrowhead). (C) Noncontrast and (D) contrast scans through the lateral ventricles. After the administration of contrast material, there is marked enhancement of vascular structures such as the cortical veins (thin black arrow), falx (broad black arrow), and the choroid plexus (white arrows) of the lateral ventricles.

severe pain or who have symptoms that do not regress after appropriate conservative treatment. In patients with suspected metastatic bone disease, the radionuclide bone scan is the imaging modality of choice. If there is strong evidence that metastatic disease is causing radicular symptoms, myelography is indicated to outline a neoplastic lesion encroaching on the spinal canal.

## COMPUTED TOMOGRAPHY

CT is the examination of choice for suspected intracranial abnormalities. It has replaced most plain skull radiography and arteriography and has eliminated the need for pneumoencephalography. Spinal CT also is indicated in the evaluation of suspected disk herniation and spinal stenosis, as well as in trauma when plain radiographs of the spine are inadequate for diagnosis.

Patients with acute neurologic deficits (e.g., hemiparesis, aphasia) due to intracranial hemorrhage or stroke are scanned without intravenous contrast, which may be detrimental to such patients. Patients with slowly developing neurologic symptoms, which most likely reflect chronic processes such as tumors or demyelinating disease, are best evaluated by CT following the administration of intravenous contrast material, which defines vascular spaces (arteries, veins, and dural sinuses) and shows any disruption of the blood-brain barrier. As with all examinations involving the administration of contrast material, the patient must be carefully monitored to detect the rare instances of hypersensitivity reaction.[3–5]

## ANGIOGRAPHY

Angiography is used to evaluate such abnormalities of the vascular system as occluded or stenotic arteries, vascular malformations, aneurysms, and certain vascular tumors; however, CT scanning has supplanted angiography in localizing intracranial masses.

Depending on the clinical problem, angiography can be performed either by direct injection of contrast material into the artery in question or by using digital subtraction angiography (DSA), with injections of contrast material into the venous or arterial systems. Direct arterial injections are indicated when fine detail of blood vessels is necessary. However, if the angiogram is primarily required for the evaluation of areas of high-grade stenosis, as in the patient with a carotid bruit, intravenous DSA may be adequate.

Intra-arterial studies are used in the evaluation of patients with suspected saccular aneurysm and subarachnoid hemorrhage, vascular malformations, vasculitis, and certain vascular tumors (hemangioblastomas in the posterior fossa, meningiomas), and for the assessment of carotid vascular disease when the necessary equipment for intravenous DSA is unavailable. Intra-arterial angiography requires hospital admission and observation after the procedure. The use of new, smaller catheters and other techniques may allow outpatient intra-arterial angiography in the near future. Although serious complications of intra-arterial angiography are infrequent (less than 1 percent), this procedure should be performed only when it will affect patient management, not merely to document an untreatable condition.

Intravenous DSA is often the procedure of choice in evaluating patients with carotid bruits, transient ischemic attacks, postoperative endarterectomies, and pulsatile neck masses. It can be performed on an outpatient basis, because it requires only a venipuncture. Because intravenous DSA is very sensitive to motion, it may be impossible to perform this procedure in uncooperative patients. Dilution of the contrast bolus may preclude a successful examination in a patient with poor cardiac output or a cardiac arrhythmia.[6–7]

## PNEUMOENCEPHALOGRAPHY

Pneumoencephalography has been made obsolete by CT scanning. This extremely uncomfortable and time-consuming procedure involved a lumbar puncture and the injection of a large amount of air into the subarachnoid space. The air was then maneuvered into various portions of the ventricular system by altering the position of the patient.

## MYELOGRAPHY

Myelography is a procedure in which the introduction of radiopaque contrast material permits visualization of the subarachnoid space in the spine. Most myelography is now performed with water-soluble contrast material such as metrizamide (Amipaque), rather than the older, oil-based contrast agent iophendylate (Pantopaque). Water-soluble contrast material permits better visualization of the subarachnoid space, can be used with adjunctive CT scanning, and avoids arachnoiditis and other long-term complications of oil-based agents.

Myelography is indicated when the clinical history and the physical examination indicate a possible lesion in the spinal canal, spinal cord, intervertebral disk, or nerve roots. Initially, a puncture is made into the subarachnoid space at either the lumbar or cervical region. With water-soluble contrast, routine lumbar myelography for the evaluation of low back pain or radiculopathy involves a puncture at the L2–3 level, which avoids injury to the spinal cord (which usually ends at T12–L1) and is away from the L4–5 and S1 disk levels where herniated disks most frequently are located. Because of

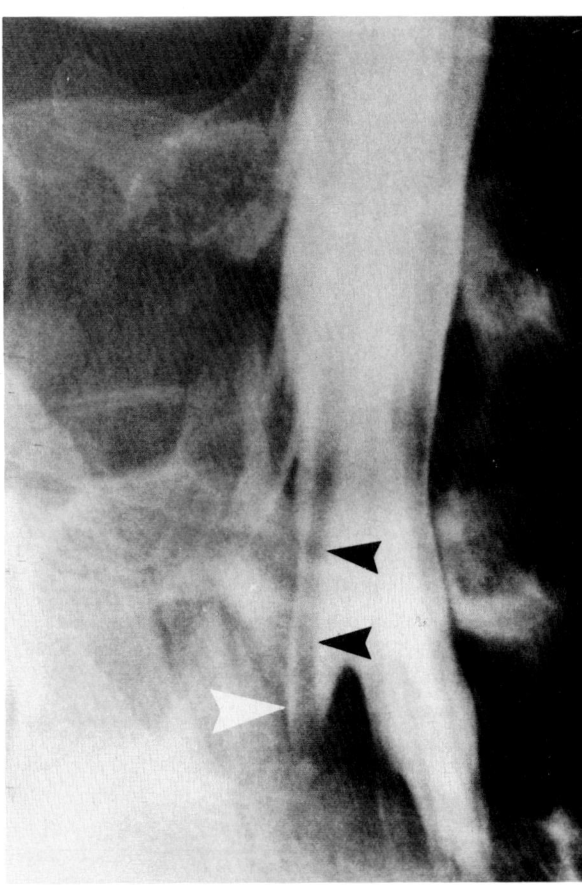

**Figure 66.2. Normal myelogram.** Radiopaque contrast material filling the subarachnoid space and nerve root sheaths (white arrow) appears white. The normal nerve roots (black arrows) displace the contrast-filled cerebrospinal fluid and appear black. Note that normal nerve roots cross the disk spaces.

the need to remove the contrast material at the termination of the study, myelography with oil-soluble agents requires a puncture at a lower lumbar level (L4–5) and leaving the needle in place during the entire examination. With either technique, a small amount of cerebrospinal fluid is routinely removed for laboratory examination of protein, glucose, and cell count.

Water-soluble myelographic contrast material tends to increase seizure activity and is thus relatively contraindicated in patients with a history of seizure disorders or alcoholism. Headaches, nausea, and occasionally vomiting are other common side effects. Oil-based myelography leads to a high incidence of postprocedure headaches and arachnoid adhesions, especially if there is blood in the cerebrospinal fluid. Because oil-based contrast material is slowly absorbed and is associated with a higher incidence of complications, its use is now usually reserved for the elderly or terminally ill or to preclude a second lumbar puncture in patients who require a follow-up examination for spinal cord compression.[8]

## RADIONUCLIDE BRAIN SCAN

Radionuclide brain scanning has been effectively replaced by computed tomography. This simple, noninvasive procedure depends on the detection of increased isotope uptake by lesions that disrupt the blood-brain barrier.

## ULTRASOUND

Ultrasound evaluation of the brain and spinal cord is limited because of the inability of this technique to penetrate bone. Its major use is in the evaluation of neonates who have patent fontanelles and patients with midline spinal defects (meningomyelocele), and as an intraoperative procedure after craniotomy for the localization of tumors and arteriovenous malformations.[9,10]

## MAGNETIC RESONANCE IMAGING

Magnetic resonance imaging (MRI), also known as nuclear magnetic resonance (NMR), obtains images of the body by utilizing magnetic fields and radio frequencies. Unlike conventional x-ray studies, MRI has greater contrast resolution and can directly image a lesion in multiple planes without the need for intravenous contrast or ionizing radiation. Although the usefulness and indications of MRI are still under investigation, it appears that this technique may eventually replace CT in the evaluation of many patients with neurologic disorders.

MRI is not affected by bone and gas and thus is especially useful in evaluating areas (e.g., posterior fossa) that have bone artifacts on CT. This technique also has the potential to evaluate blood vessels and blood flow and may eventually replace arteriography and other examinations of the extra- and intracranial vessels.

There are several important contraindications to the use of magnetic resonance imaging. Because cardiac pacemakers are designed to be switched on and off by magnets, patients with such devices cannot be studied by this technique. Some aneurysm clips may align with the strong magnetic fields and torque off cerebral aneurysms, causing rebleeding. Abdominal surgical clips are firmly adherent because of reactive fibrosis, so their presence is not a contraindication. Large metallic devices (total hip prostheses, total knee replacements) may cause degradation of the magnetic resonance image. In addition, although there was early concern about local heating effects, this has so far not proved to be a problem (at least with low field strength). Because iron-containing instruments (life-support equipment, respirators, oxygen tanks) are strongly attracted to the magnet and cannot be in the imaging room, patients requiring such equipment cannot be studied by MRI.[11–13]

## POSITRON EMISSION TOMOGRAPHY

Positron emission tomography (PET) is an experimental investigative technique that involves the intravenous injection of positron-emitting radionuclides of oxygen or F-18 deoxyglucose (FDG) followed by gamma camera tomography. The technique permits regional quantitation of oxygen uptake, blood flow, and glucose utilization within the brain. Initial studies have shown PET to be useful in patients with strokes, seizure disorders, and metabolic abnormalities. However, although PET scanning shows great promise in the biochemical analysis of brain functions, the high cost of the instrumentation and the requirement for a nearby cyclotron to produce the isotopes will probably restrict PET scanning to major medical centers and investigational use.[14]

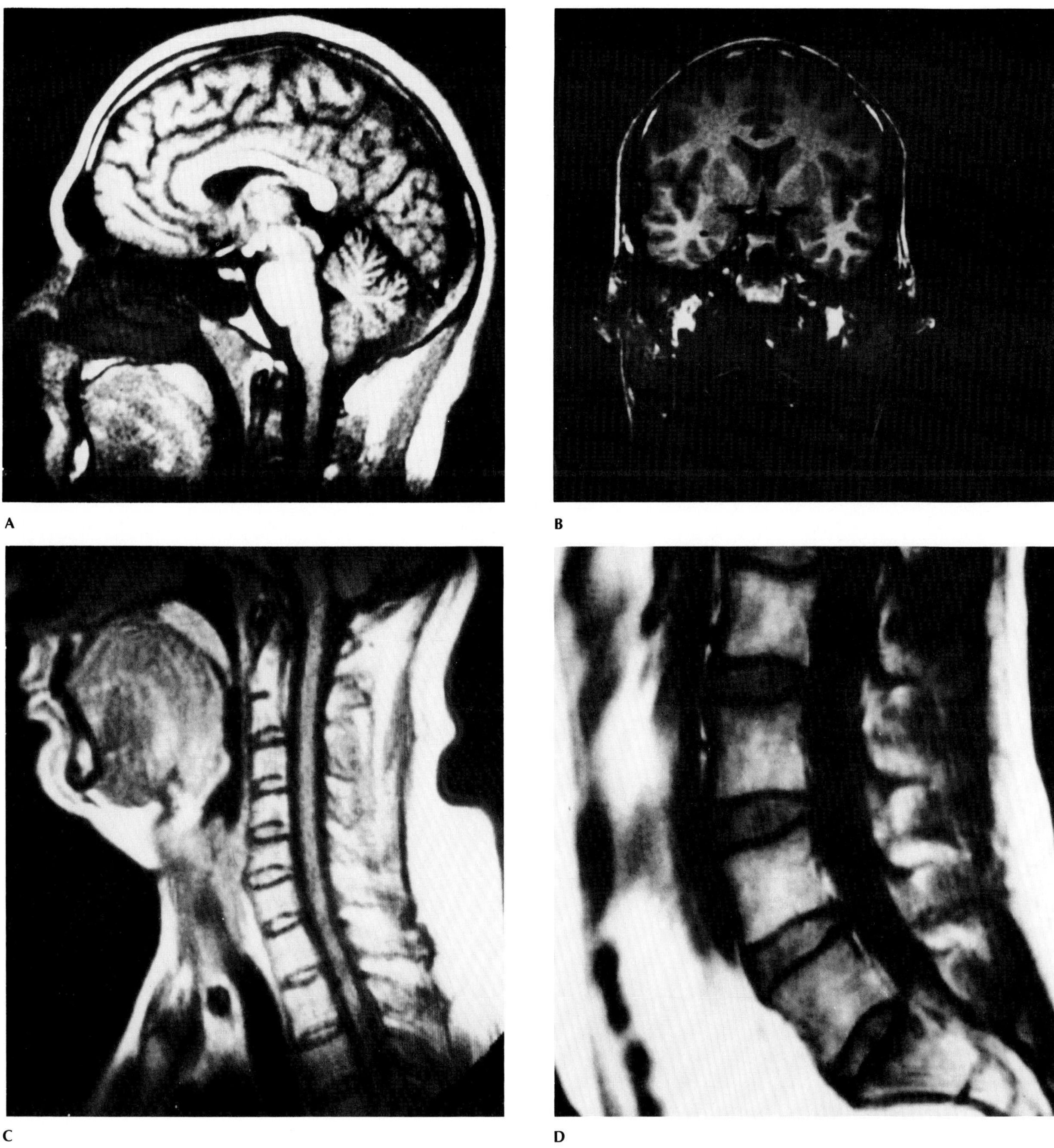


**Figure 66.3. Normal magnetic resonance imaging.** (A) Sagittal and (B) coronal views of the head. (C) In the cervical region, there is good delineation of the high-intensity spinal cord surrounded by the low-intensity cerbrospinal fluid spaces. Note the clear definition of the relation of the vertebral bodies and disks to the subarachnoid space. (D) Lumbosacral spine.

## REFERENCES


1. Bell RS, Loop JW: The utility and futility of radiographic skull examinations for trauma. *N Engl J Med* 284:236–239, 1971.
2. Masters SJ: Evaluation of head trauma: Efficacy of skull films. *AJNR* 1:329–335, 1980.
3. Lee SH, Rao KCVG (eds): *Cranial Computed Tomography*. New York, McGraw-Hill, 1983.
4. Mancuso AA, Hanafee WN: *Computed Tomography of the Head and Neck*. Baltimore, Williams & Wilkins, 1982.
5. Rosenberg RN, Heinz ER (eds): *The Clinical Neurosciences. IV. Neuroradiology*. New York, Churchill Livingstone, 1984.
6. Osborn AG: *Introduction to Cerebral Angiography*. Philadelphia, JB Lippincott, 1980.
7. Brody WR: *Digital Radiography*. New York, Raven Press, 1984.
8. Sackett JF, Strother CM: *New Techniques in Myelography*. Philadelphia, Harper & Row, 1979.
9. Babcock DS, Mack LA, Han BK: Congenital anomalies and other abnormalities of the brain. *Semin Ultrasound* 3:191–211, 1982.
10. Rubin JM, Dohrmann GJ: Intraoperative neurosurgical ultrasound in the localization and characterization of intracranial masses. *Radiology* 148:519–528, 1983.
11. Bydder GM, Steiner RE, Young IR, et al: Clinical NMR imaging of the brain. *AJR* 139:215–236, 1982.
12. Brant-Zawadzki M, Badami JP, Mills CM, et al: Primary intracranial tumor imaging: A comparison of magnetic resonance and CT. *Radiology* 150:435–440, 1984.
13. Han JS, Kaufman B, El Yousef SJ, et al: NMR imaging of the spine. *AJR* 141:1137–1145, 1983.
14. Ackerman RH: Clinical aspects of positron emission tomography (PET). *Radiol Clin North Am* 20:9–23, 1982.

## Chapter Sixty-Seven
# Cerebrovascular Diseases

The term *cerebrovascular disease* refers to any process that is caused by an abnormality of the blood vessels or blood supply to the brain. Pathologic processes causing cerebrovascular disease include abnormalities of the vessel wall, occlusion by thrombus or emboli, rupture of blood vessels with subsequent hemorrhage, and decreased cerebral blood flow due to lowered blood pressure or narrowed lumen caliber. Cerebrovascular diseases include arteriosclerosis, hypertensive hemorrhage, arteritis, aneurysms, and arteriovenous malformations.

The radiographic evaluation of cerebrovascular disease depends on the presenting symptoms and the most likely diagnosis. For ease of classification, patients with cerebrovascular disease can be divided into three categories: completed stroke; transient ischemic attacks (TIAs); and intracranial hemorrhage.

### STROKE SYNDROME

The term *stroke* denotes the sudden and dramatic development of a focal neurologic deficit, which may vary from dense hemiplegia and coma to only a trivial neurologic disorder. The specific neurologic defect depends on the arteries involved; this is discussed in great detail in Harrison's *Principles of Internal Medicine*. Strokes most commonly involve the circulation of the internal carotid arteries and present with symptoms that include acute hemiparesis and dysarthria.

The purpose of radiographic evaluation in the acute stroke patient is not to confirm the diagnosis of a stroke but to exclude other processes that can simulate the clinical findings (parenchymal hemorrhage, subdural hematoma). Although the abrupt onset of a stroke may permit differentiation from other conditions that have a more gradual onset of symptoms, patients with focal neurologic deficits of various causes may initially be found comatose, so that the history of gradual onset is not elicited. Obviously, it is essential to exclude an intracranial hemorrhage before considering the possibility of treating the stroke patient with anticoagulant therapy.

Noncontrast computed tomography (CT) is the examination of choice for the evaluation of the stroke patient. Intravenous contrast is contraindicated because it is a toxic substance that can cross the disrupted blood-brain barrier in the region of a cerebral infarct and lead to increased edema and a slower recovery for the patient. CT scans are normal in patients with small infarctions, such as lacunar infarcts, and in the early hours of large infarctions. The initial appearance of a cerebral infarction is a triangular or wedge-shaped area of hypodensity involving both the cortex and the underlying white matter down to the ventricular surface. The low density is

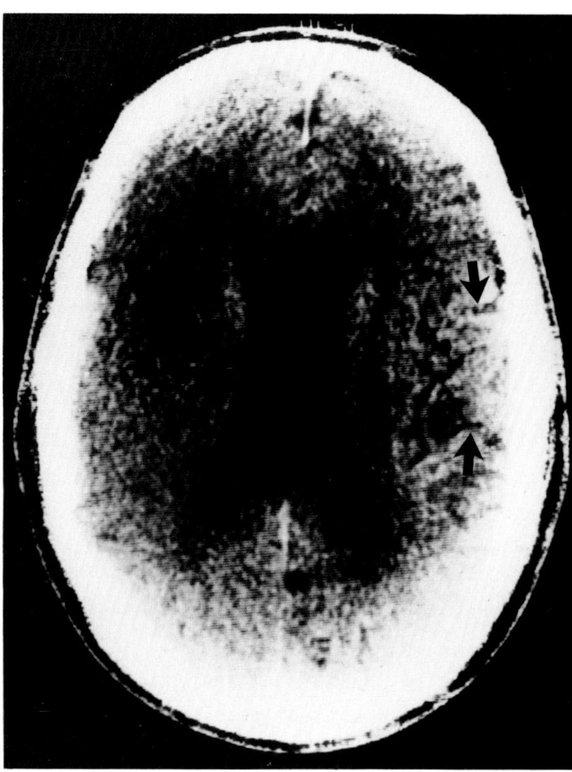

**Figure 67.1. Acute right middle cerebral artery infarct.** A CT scan obtained 20 hours after the onset of acute hemiparesis and aphasia shows obliteration of the normal sulci (black arrows) in the involved hemisphere. There is low density of the gray and white matter in the distribution of the right middle cerebral artery.

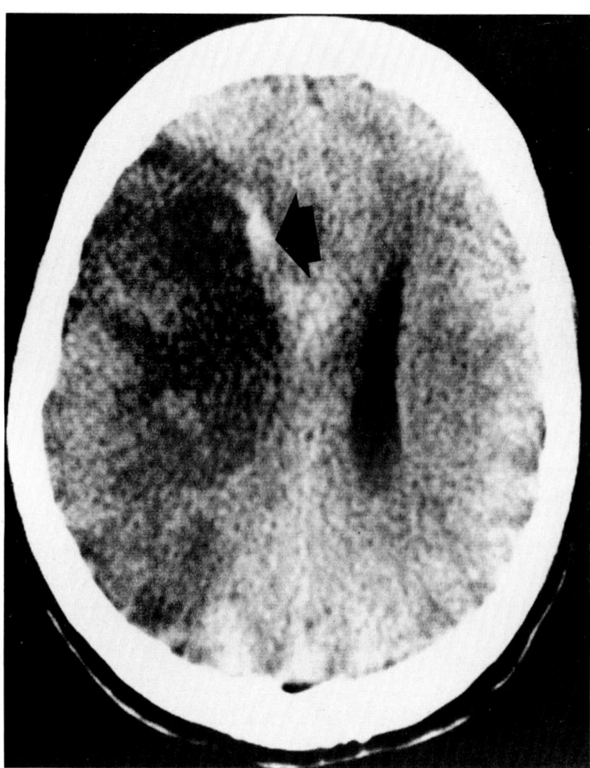

**Figure 67.2. Subacute right middle cerebral artery infarct.** A CT scan in a patient with acute left hemiplegia that developed 4 days previously shows an area of low density in the distribution of the right middle cerebral artery. Note the marked mass effect causing obliteration of the right lateral ventricle. A small portion of the infarct is hemorrhagic (black arrow).

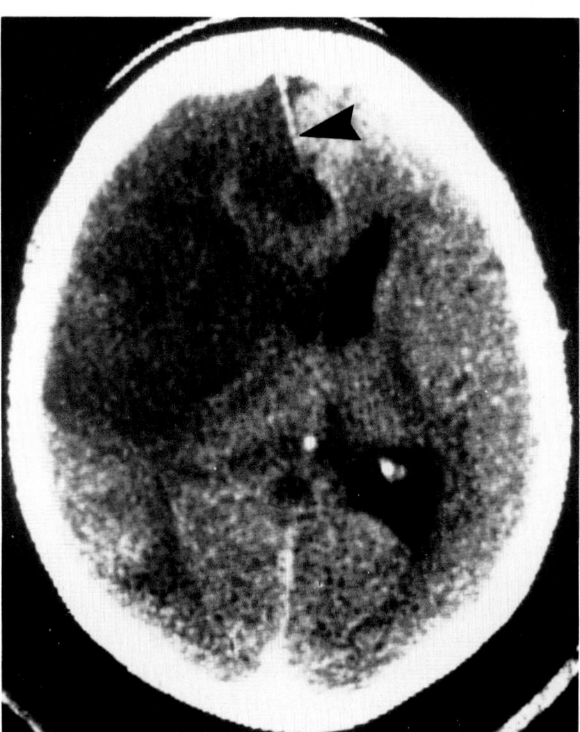

**Figure 67.3. Subacute right middle cerebral and anterior cerebral infarct.** A CT scan obtained 7 days after the onset of left hemiparesis shows an extensive low-density infarct involving the gray and white matter in the distribution of the right middle and anterior cerebral arteries. The marked mass effect causes shift of the ventricular system and falx (black arrowhead) from right to left.

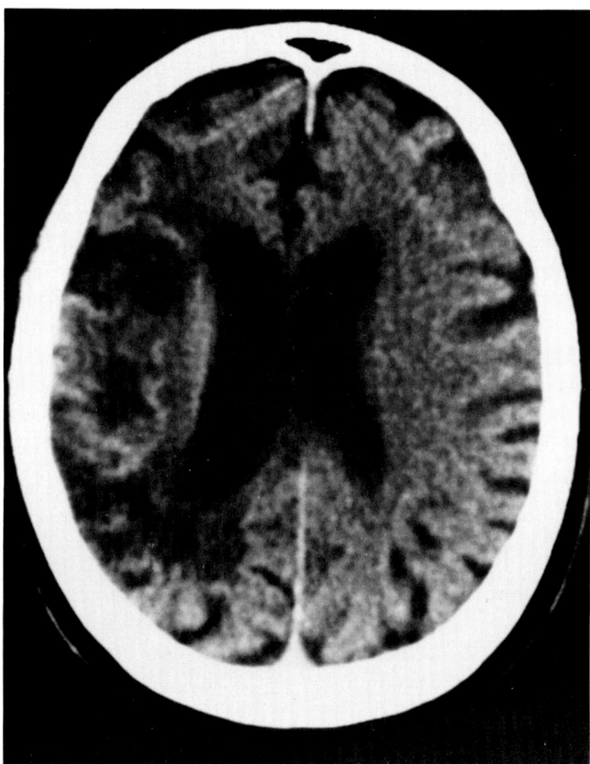

**Figure 67.4. Old right middle cerebral artery infarct.** A CT scan 1 year after infarction shows enlargement of the right lateral ventricle and hemispheric sulci.

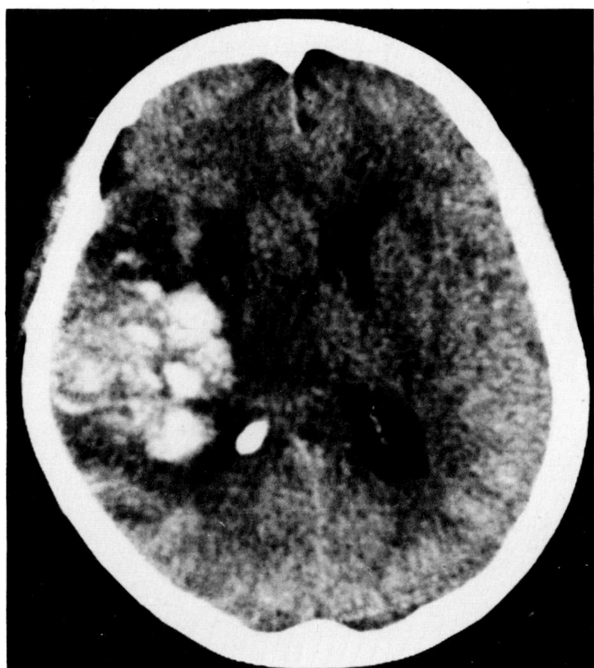

**Figure 67.5. Hemorrhagic cerebral infarct.** A CT scan obtained 4 days after the onset of acute left hemiplegia shows a mixed pattern of high-density hemorrhage and low-density edema in both gray and white matter. Involvement of both the gray and the white matter in a distribution corresponding to the right middle cerebral artery makes this appearance more likely to be a hemorrhagic infarct than an acute hypertensive hemorrhage.

confined to the vascular territory of the involved artery. Although little or no mass effect is evident during the first day, progressive edema produces a mass effect that is maximal 7 to 10 days after the acute event. As infarcts age, brain tissue atrophies and the adjacent sulci and ventricular system enlarge. In patients with classic stroke symptoms, follow-up CT scans are not indicated.

Arteriography is contraindicated in patients with an acute completed stroke. Since most surgeons postpone carotid endarterectomy until 6 weeks after a completed stroke, the arteriogram should be delayed until just prior to the anticipated surgery. Arteriography may be indicated in the patient who develops increasing neurologic deficits while under observation (stroke in evolution) if surgery is a therapeutic alternative. The purpose of arteriography in most stroke patients is to prevent further strokes by identifying treatable lesions in the extracranial arterial system.[1]

## TRANSIENT ISCHEMIC ATTACKS

Transient ischemic attacks (TIAs) present as focal neurologic deficits that completely resolve within 24 hours. They may result from emboli originating from the surface of an arteriosclerotic ulcerated plaque, which causes temporary occlusion of cerebral vessels, or from stenosis of an extracerebral artery, which leads to a reduction in critical perfusion. Because almost two-thirds of arteriosclerotic strokes are preceded by TIAs and the 5-year cumulative risk of stroke in patients with TIAs may be as high as 50 percent, accurate diagnosis and appropriate treatment (antiplatelet therapy, anticoagulation therapy, or carotid endarterectomy) are essential.

The most common location of surgically treatable arteriosclerotic disease causing TIAs is the region of the carotid bifurcation in the neck. Although the demonstration of carotid vascular disease by ultrasound is improving, angiography remains the best technique for evaluating the carotid arteries. Either intravenous or intra-arterial digital subtraction angiography or selective intra-arterial carotid angiography can be used to demonstrate TIA-producing stenotic or ulcerative lesions that may be amenable to surgical therapy. Angiographic evaluation of the aortic arch and vertebral arteries is infrequently indicated in evaluating patients with TIAs, both because surgery on the origins of the great vessels and posterior circulation is difficult and not established, and because of the higher morbidity from arch and vertebral angiograms reported in some studies.[2,3]

## INTRAPARENCHYMAL HEMORRHAGE

Aside from head trauma, the principal cause of intraparenchymal hemorrhage is hypertensive vascular disease. Rupture of a berry aneurysm or an arteriovenous malformation are less frequent causes. Hypertensive hemorrhages result in oval or circular collections that displace the surrounding brain and can cause a significant mass effect. Although they can occur at any location within the brain, hypertensive hemorrhages are most frequent in the basal ganglia, white matter, thalamus, cerebellar hemispheres, and pons. A frequent complication is rupture of the hemorrhage into the ventricular system or subarachnoid space. Intraparenchymal hemorrhages secondary to berry aneurysms usually are associated with subarachnoid hemorrhage and tend to develop in regions where these congenital vascular anomalies most commonly occur. These include the Sylvian fissure (middle cerebral artery) and the midline subfrontal area (anterior communicating artery). Arteriovenous malformations occur throughout the brain and tend to bleed into the white matter.

Patients with a suspected intraparenchymal hemorrhage should be evaluated with a noncontrast CT scan. A fresh hematoma appears as a homogeneously dense, well-defined lesion with a rounded to oval configuration. Hematomas produce ventricular compression and, when large, considerable midline shift and brain herniation. Steroid therapy, especially in nontraumatic hematomas,

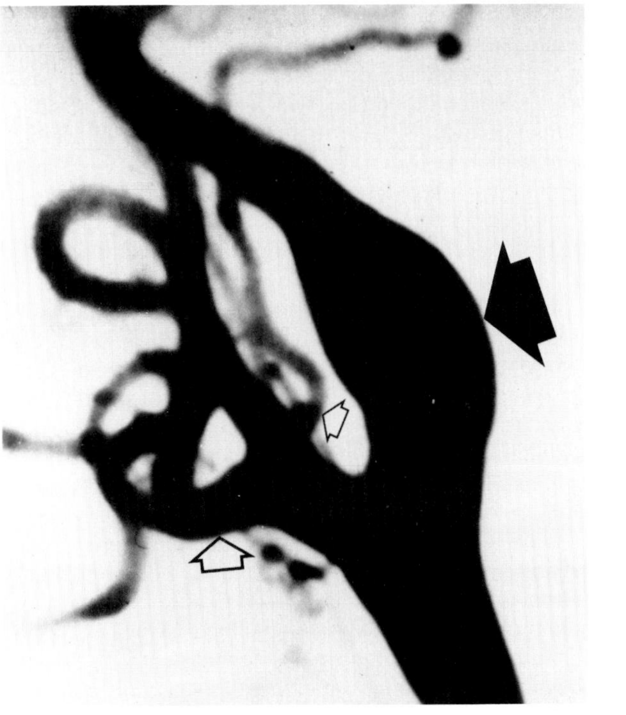
A

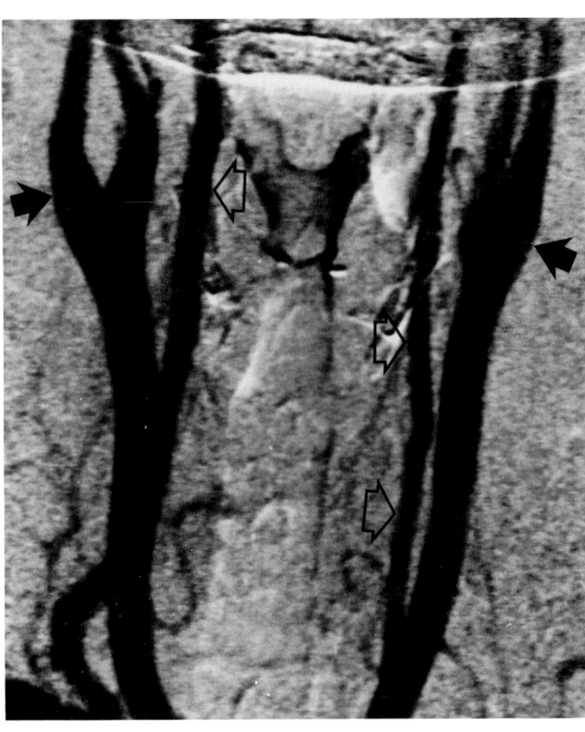
B

**Figure 67.6. Normal carotid artery bifurcation.** (A) A common carotid arteriogram shows the bulbous origin of the internal artery (black arrow) and multiple branches (open arrows) of the external carotid artery. (B) An intravenous digital subtraction angiogram shows the normal carotid bifurcation bilaterally (black arrows) and patent vertebral arteries (open arrows).

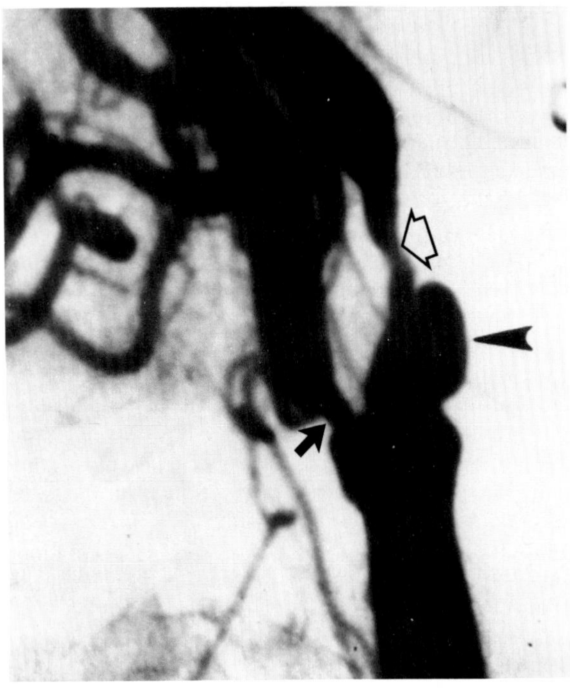
A

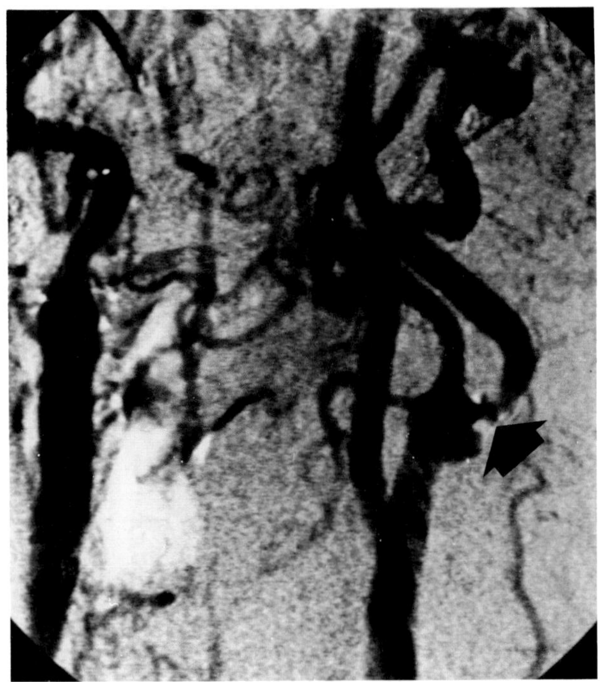
B

**Figure 67.7. Ulceration of the internal carotid artery.** (A) A common carotid arteriogram shows an ulcerated lesion (black arrowhead) at the origin of the internal carotid artery with severe stenosis of the internal carotid (open arrow) and external carotid (black arrow) arteries. (B) An intravenous digital subtraction angiogram shows an ulcerated lesion in the internal carotid artery (black arrows). This examination is diagnostic, even though the spatial resolution and detail are less than those of the arterial study.

usually controls the edema that produces much of the mass effect. A hematoma that is not homogeneously dense suggests hemorrhage occurring within a tumor, an inflammatory process, or an infarction. As the hematoma ages, its density becomes progressively less. After passing through an isodense stage, the hematoma becomes hypodense; by 6 months it appears as a well-defined, low-density region that is often considerably smaller than the original lesion. Contrast enhancement usually develops about the periphery of a hematoma after 7 to 10 days, the time at which radionuclide studies first become positive.

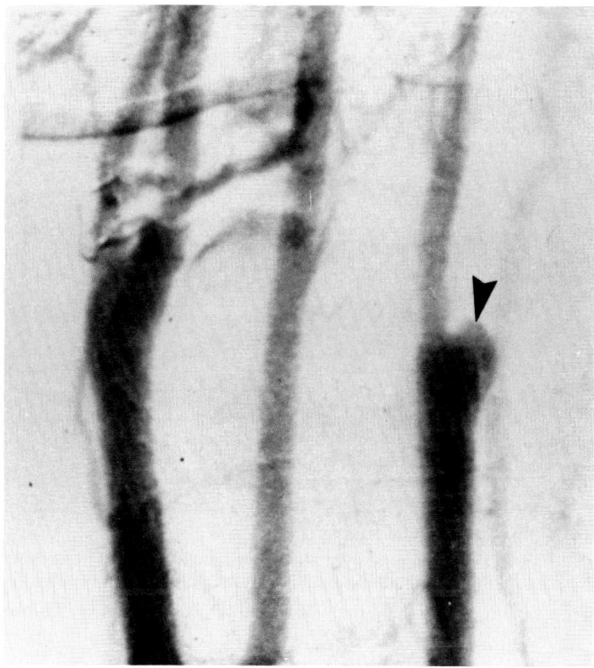

**Figure 67.8. Occluded internal carotid artery.** An intravenous digital subtraction angiogram shows occlusion of the left internal carotid artery at its origin (black arrow).

Arteriography is not indicated in patients with classic hypertensive hematomas. However, arteriography is of value in those patients in whom an arteriovenous malformation or aneurysm is suspected as the underlying cause of the hematoma.[4]

## SUBARACHNOID HEMORRHAGE

A major cause of subarachnoid hemorrhage is rupture of a berry aneurysm. Patients with this condition usually present with a generalized excruciating headache followed by unconsciousness. The radiographic procedure of choice is a noncontrast CT scan, which can demonstrate high-density blood in the subarachnoid spaces of the basal cisterns in more than 95 percent of the cases. Bleeding may extend into the brain parenchyma adjacent to the aneurysm. Contrast-enhanced CT scans are not indicated in subarachnoid hemorrhage, since the surgeon will not operate for a suspected aneursym without an angiogram, and the patient would thus be exposed to the risk of an excessive load of contrast material.

The timing of angiography in subarachnoid hemorrhage depends on the philosophy of the surgeon. Blood in the subarachnoid space is an irritant that causes vasospasm of the vessels of the circle of Willis and middle cerebral artery. This vasospasm, which can lead to ce-

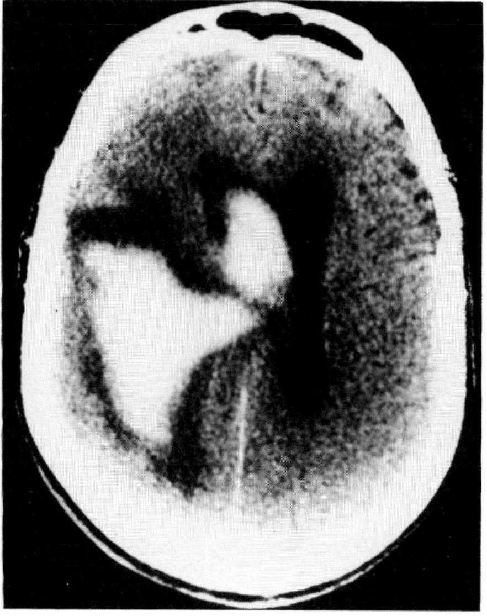

A

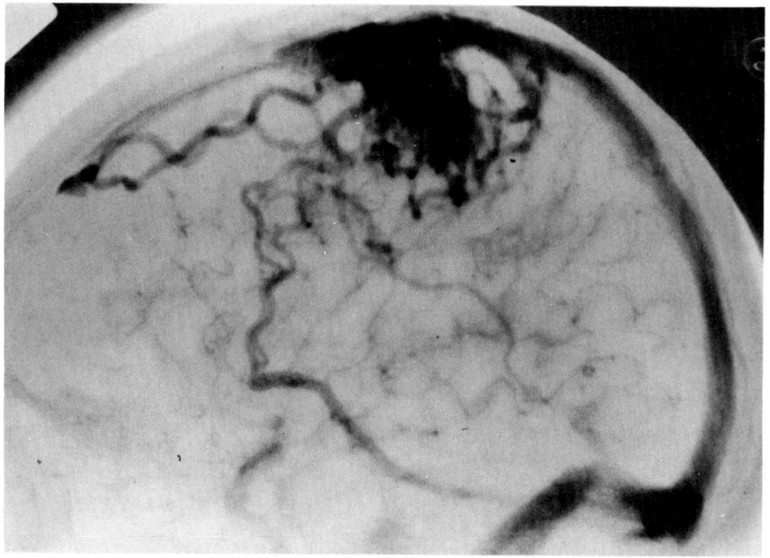

B

**Figure 67.9. Arteriovenous malformation with hemorrhage.** (A) A CT scan in a 34-year-old comatose man shows a high-density hemorrhage in the right parietal lobe that extends into the ventricular system. (B) A carotid arteriogram shows the dilated blood vessels constituting the arteriovenous malformation.

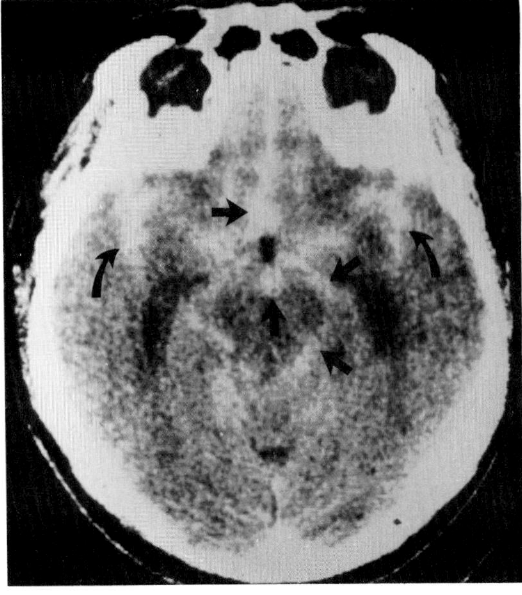

A

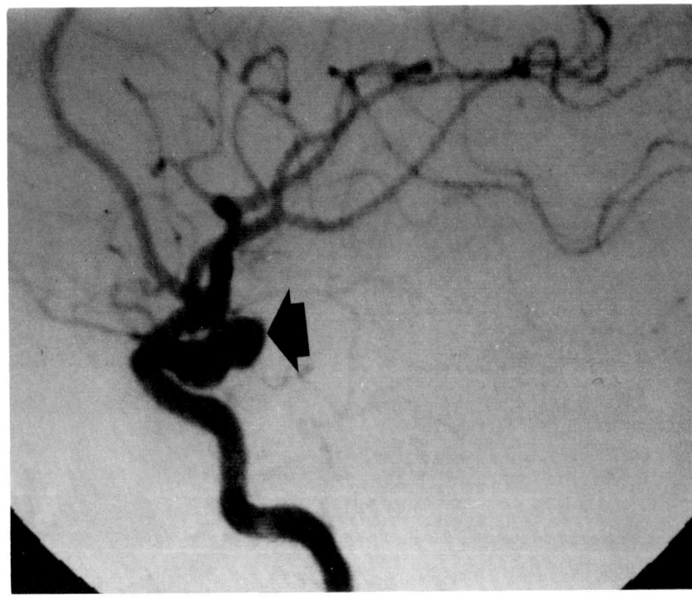

B

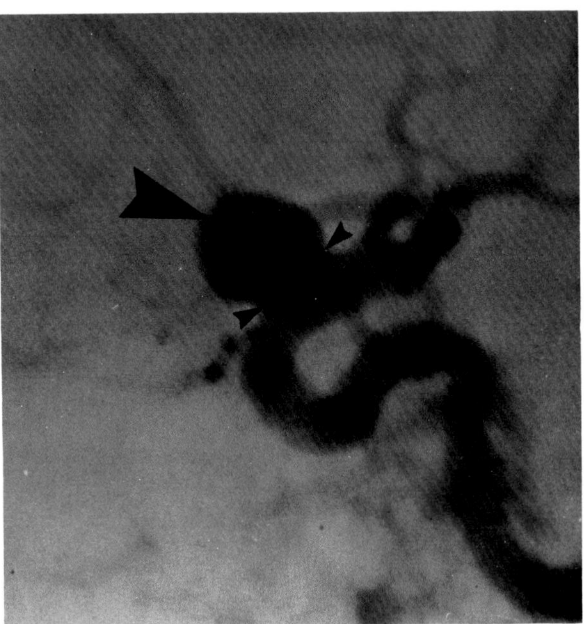

C

**Figure 67.10. Subarachnoid hemorrhage secondary to rupture of a berry aneurysm.** (A) A CT scan of an elderly comatose woman shows a subarachnoid hemorrhage in the basal cisterns (small black arrows) extending into the middle cerebral artery cistern (curved arrows) and Sylvian fissures (medium black arrows). There also is a parenchymal hemorrhage (large black arrow) in the left frontal lobe. (B) A left carotid arteriogram demonstrates a left supraclinoid aneurysm. The posterior bulge of the aneurysm (black arrow) represents the site of rupture. (C) In another patient with a subarachnoid hemorrhage, a right carotid arteriogram shows an aneurysm (large arrowhead) arising from the origin of the ophthalmic artery. In planning surgery, it is important to demonstrate the neck of the aneurysm (small arrowheads).

rebral ischemia and frank infarction, is greatest 3 to 14 days after the acute episode. If emergency surgery within the first 72 hours after the hemorrhage is planned, emergency selective angiography is indicated. If surgical intervention is to be delayed, angiography should be postponed until just prior to surgery. The most common locations for berry aneurysms are the origins of the posterior cerebral and anterior communicating arteries and the trifurcation of the middle cerebral artery. Because of the 20 percent incidence of multiple aneurysms, the angiographic procedure should include evaluation of the internal carotid and vertebral arteries bilaterally.[4,5]

## REFERENCES


1. Wall SD, Brant-Zawadzki M, Jeffrey RB, et al: High frequency CT findings within 24 hours after cerebral infarction. *AJNR* 138:307–311, 1982.
2. Goldstein SJ, Fried AM, Young B, et al: Limited usefulness of aortic arch angiography in the evaluation of carotid occlusive disease. *AJNR* 138:103–108, 1982.
3. Mistretta CA, Crummy AB, Strother CM: Digital angiography: A perspective. *Radiology* 139:273–276, 1981.
4. Scotti G, Ethier R, Melancon D, et al: Computed tomography in the evaluation of intracranial aneurysms and subarachnoid hemorrhage. *Radiology* 123:85–90, 1977.
5. Davis JM, Davis KR, Crowell RM: Subarachnoid hemorrhage secondary to ruptured intracranial aneurysm: Prognostic significance of cranial CT. *AJR* 134:711–715, 1980.

## Chapter Sixty-Eight

# Traumatic Diseases of the Brain

In the patient with head trauma, the purpose of radiographic imaging is to detect a surgically correctable hematoma. Emergency computed tomography (CT) has virtually replaced all other radiographic investigations in patients with suspected neurologic dysfunction secondary to head injury (see the section "Plain Skull Radiographs" in Chapter 66 for the indications for emergency CT). Because the presence or absence of a skull fracture does not correlate with intracranial abnormalities, plain radiographs of the skull are no longer indicated in the patient with head trauma.[1]

## EPIDURAL HEMATOMA

Epidural hematomas are caused by acute arterial bleeding and most commonly form over the parietotemporal convexity. Because of the high arterial pressure, epidural hematomas rapidly cause significant mass effect and acute neurologic symptoms. Because the dura is very adherent to the inner table of the skull, an epidural hematoma typically appears as a biconvex (lens-shaped), peripheral high-density lesion. There usually is a shift of the midline structures toward the opposite side unless a contralateral balancing hematoma is present.[2–4]

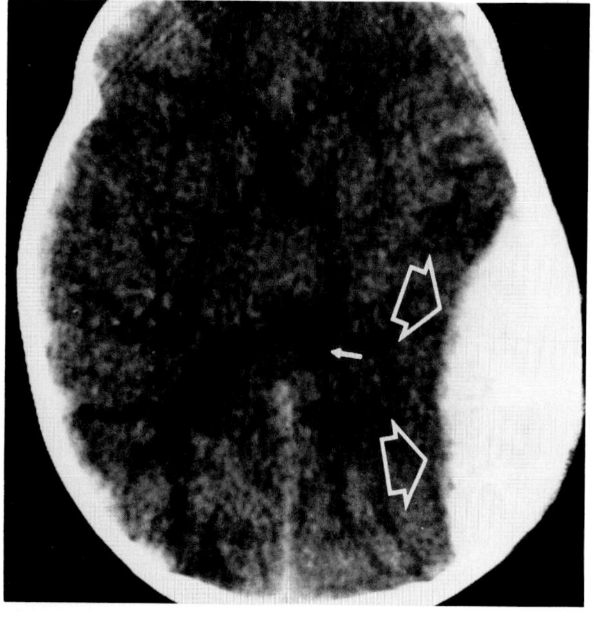

**Figure 68.1. Epidural hematoma.** A CT scan of a 4-year-old involved in a motor vehicle accident shows a characteristic lens-shaped epidural hematoma (open arrows). The substantial mass effect associated with the hematoma causes distortion of the lateral ventricle (closed arrow).

## SUBDURAL HEMATOMA

Subdural hematomas reflect venous bleeding, most commonly from ruptured veins between the dura and leptomeninges. Because of the low pressure of venous bleeding, patients with subdural hematomas tend to have a chronic labile course with symptoms of headache, agitation, confusion, drowsiness, and gradual neurologic deficits. An acute subdural hematoma typically appears as a peripheral zone of increased density that follows the surface of the brain and has a crescentic shape adjacent to the inner table of the skull. There usually is an associated mass effect with displacement of midline structures and obliteration of sulci over the affected hemisphere. Absence of midline displacement away from the side of a lesion suggests the not infrequent presence of bilateral subdural hematomas.

Serial CT scans demonstrate a gradual decrease in the attenuation value of a subdural lesion over a period of weeks. With absorption and lysis of the blood clot, the hematoma becomes isodense with normal brain tissue, and the lesion may be identified only because of its mass effect. At this stage, scanning after the administration of contrast material may be of value because of enhancement of the membrane around the subdural hematoma

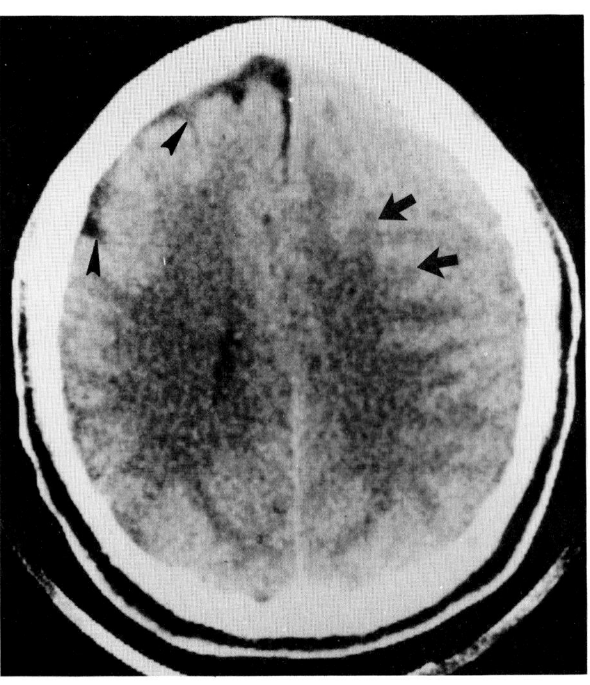

**Figure 68.3. Isodense left subdural hematoma.** A CT scan shows shift of the gray-white junction (black arrows) medially. There is loss of the normal sulci (black arrowheads) in the normal right hemisphere.

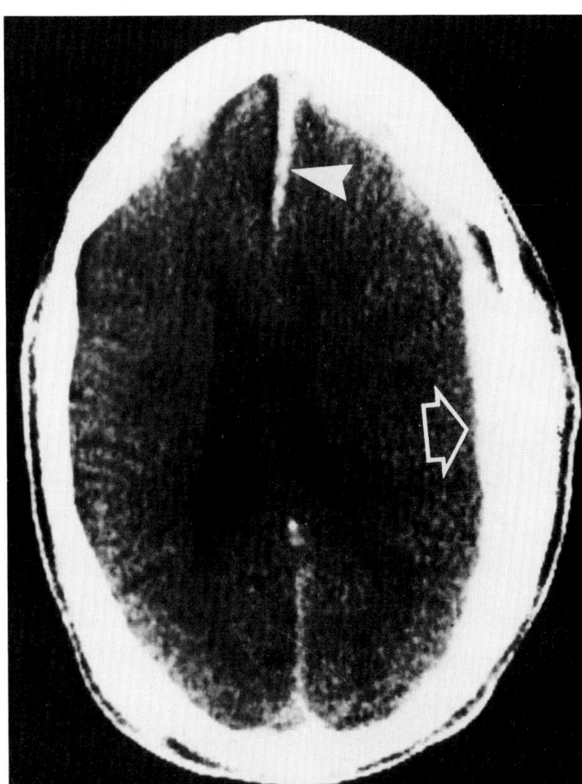

**Figure 68.2. Acute left subdural hematoma.** A CT scan shows a high-density, crescent-shaped subdural hematoma (open arrow) adjacent to the inner table of the skull. The hematoma extends into the interhemispheric fissure (white arrowhead).

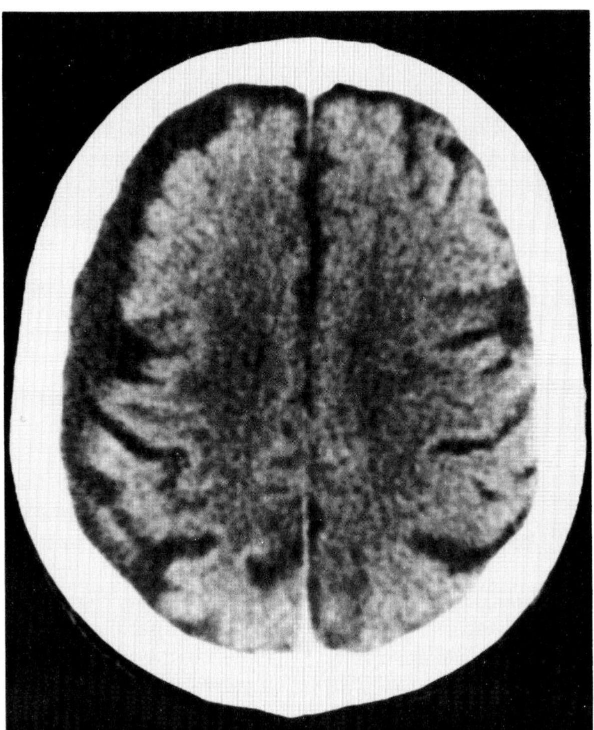

**Figure 68.4. Chronic right subdural hematoma.** A CT scan shows a crescent-shaped, low-density subdural hematoma of 1 month's duration in the right hemisphere of an elderly man who presented with mental obtundation and ataxia.

and identification of the cortical veins. A chronic subdural hematoma has an attenuation value similar to that of cerebrospinal fluid and substantially less than that of the underlying brain. At times, the small bridging veins associated with a chronic subdural hematoma may bleed and produce the difficult diagnostic problem of an acute subdural hematoma superimposed upon a chronic one.[3,4]

## CEREBRAL CONTUSION

Cerebral contusion reflects parenchymal damage caused by movement of the brain within the calvarium following blunt trauma to the skull. Contusions occur when the brain contacts rough skull surfaces such as the superior orbital roof and petrous ridges. They typically appear as low-density areas of edema and tissue necrosis, with or without nonhomogeneous high-density zones reflecting multiple small areas of hemorrhage. Because of a breakdown of the blood-brain barrier, contusions generally enhance with intravenous contrast material for weeks after the injury.[4,5]

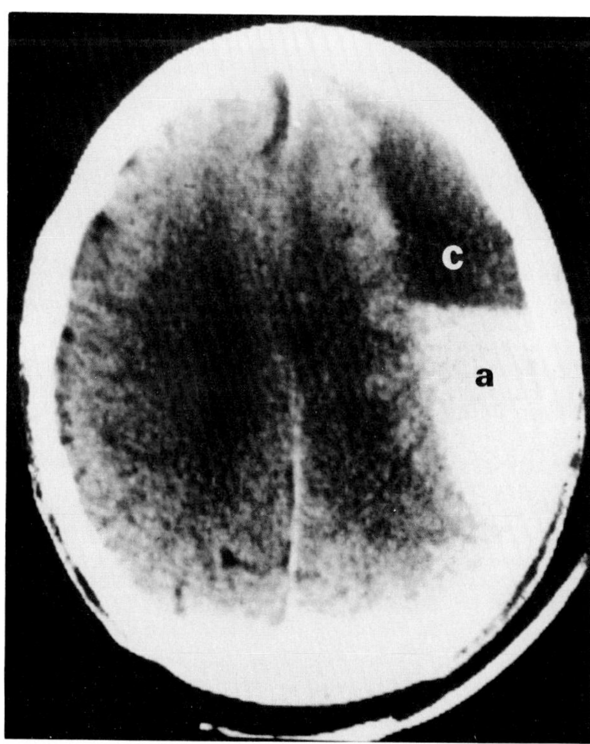


**Figure 68.5. Mixed acute and chronic left subdural hematoma.** A CT scan shows a high-density acute hemorrhage (a) layering in the dependent portion of the hematoma, with the lower-density chronic collection (c) situated anteriorly. Chronic subdural hematomas have bridging cortical veins which can bleed and cause a mixed acute and chronic subdural hematoma.

## INTRACEREBRAL HEMATOMA

Traumatic hemorrhage into the brain parenchyma can result from shearing forces to intraparenchymal arteries, which tend to occur at the junction of the gray and the white matter. Injury to the intima of intracranial vessels can cause the development of traumatic aneurysms, which can rupture. On CT scans, an intracerebral hematoma appears as a well-circumscribed, homogeneous, high-density region that usually is surrounded by areas of low density edema. As the blood components within the hematoma disintegrate, the lesion eventually becomes isodense with normal brain (usually 2 to 4 weeks after injury).

Although most intracerebral hematomas develop immediately after head injury, delayed hemorrhage is common. This is especially frequent after evacuation of acute subdural hematomas that are tamponading potential bleeding sites. Therefore, a repeat CT scan is often performed within 48 hours in patients who have undergone decompressive surgery.[4,6]

## CAROTID ARTERY INJURY

The extracerebral carotid arteries can be injured by penetrating trauma to the neck, as from gunshot wounds or stabbing. Arteriography can demonstrate laceration of the artery or intimal damage that may result in either

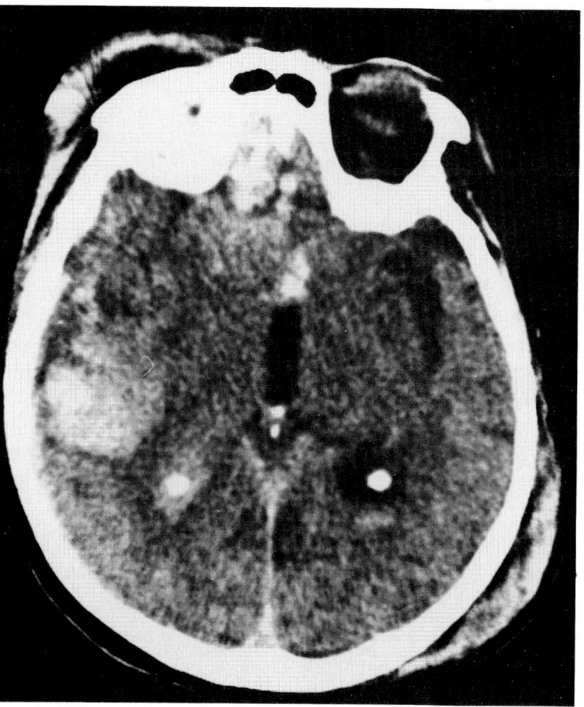

**Figure 68.6. Cerebral contusion.** A CT scan shows high-density lesions representing hemorrhagic contusions.

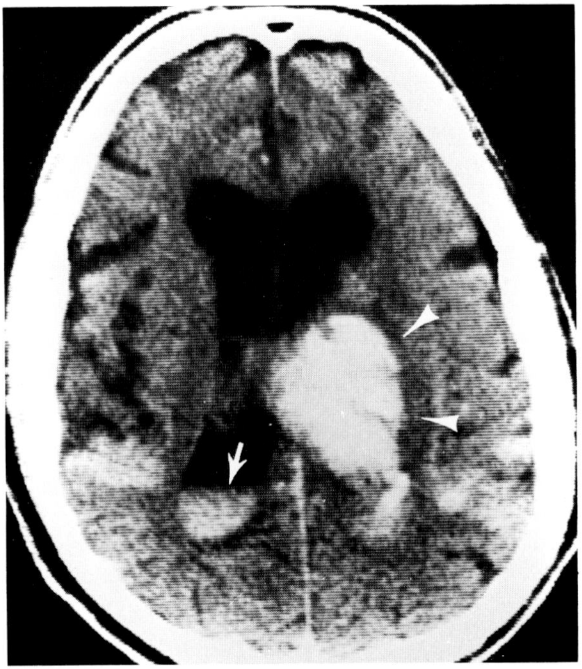

**Figure 68.7. Intracerebral hematoma.** A noncontrast CT scan shows a homogeneous, high-density area in the left thalamic region. A low-density area (arrowheads) adjacent to the hematoma represents associated ischemia and edema. The hematoma has entered the ventricular system, and a prominent CSF-blood-fluid level is seen in the dependent lateral ventricle (arrow). Such extension of blood into the ventricular system is an extremely poor prognostic sign. The mass effect caused by the hematoma has compressed the third ventricle and the foramen of Monro and resulted in obstructive enlargement of the lateral ventricles. (From Drayer,[6] with permission from the publisher.)

dissection or thrombotic occlusion. Hyperextension injuries from motor vehicle accidents can cause intimal damage to the carotid or vertebral arteries that may result in pseudoaneurysm formation. Traumatic arteriovenous fistulas usually arise between the internal carotid artery and the cavernous sinus. Carotid arteriography demonstrates opacification of the cavernous sinus during the arterial phase. Reverse flow from the cavernous sinus may rapidly opacify a markedly dilated ophthalmic vein. Placement of a detachable balloon catheter within the fistula under angiographic control may obviate surgical intervention.[7,8]

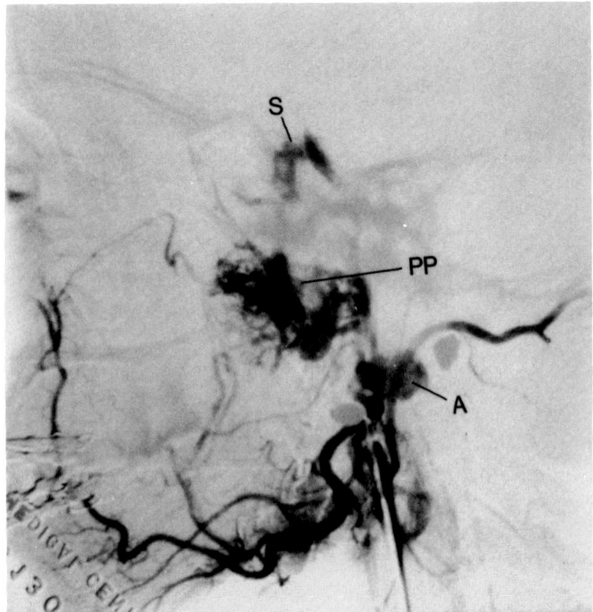


**Figure 68.8. Carotid-cavernous fistula.** A lateral view of an external carotid artery angiogram shows a pseudoaneurysm (A) and rapid filling of the cavernous sinus (S) and pterygoid plexus (PP). There is total interruption of the external carotid artery trunk beyond the takeoff of the facial artery.

## REFERENCES

1. Masters SJ: Evaluation of head trauma: Efficacy of skull films. *AJNR* 1:329–335, 1980.
2. Stone JL, Shaffer L, Ramsey RG, et al: Epidural hematomas of the posterior fossa. *Surg Neurol* 11:419–424, 1979.
3. Harris JH, Harris WH: *The Radiology of Emergency Medicine*. Baltimore, Williams & Wilkins, 1981.
4. Federle MP, Brant-Zawadski M: *Computed Tomography in the Evaluation of Trauma*. Baltimore, Williams & Wilkins, 1982.
5. Kaiser MC, Peterson H, Harwood-Nash DC, et al: CT for trauma to the base of the skull and spine in children. *Neuroradiology* 22:27–31, 1981.
6. Drayer BP: Diseases of the cerebral vascular system, in Rosenberg RN, Heinz ER (eds): *The Clinical Neurosciences. IV. Neuroradiology*. New York, Churchill Livingstone, 1984.
7. Debrun G, Lacour P, Caron JP, et al: Experimental approach to the treatment of carotid cavernous fistulas with an inflatable and isolated balloon. *Neuroradiology* 9:9–17, 1975.
8. Hamby WB: *Carotid Cavernous Fistula*. Springfield, Charles C Thomas, 1966.

## Chapter Sixty-Nine
# Neoplastic Diseases of the Brain

Intracranial neoplasms present clinically with seizure disorders or gradual neurologic deficits (deficient mentation, slow comprehension, weakness, headache). The specific clinical presentation and radiographic appearance depend on the location of the tumor and the site of the subsequent mass effect.

Computed tomography (CT) with contrast enhancement is the examination of choice in evaluating a patient for a suspected brain tumor. Although skull radiographs were used in the past to demonstrate tumoral calcification, bone erosion, and displacement of the calcified pineal gland, plain films are no longer indicated, since this information can be more effectively obtained on CT scans.

Prior to the advent of CT, cerebral arteriography was used to show such evidence of brain tumors as mass effect, contralateral displacement of midline arteries and veins, abnormal vessels with tumor staining, and early venous filling. At present, the major use of arteriography is to precisely delineate the arterial and venous anatomy prior to surgical therapy and to evaluate those cases in which a vascular anomaly is a strong consideration in the differential diagnosis of a tumor. Radionuclide brain scans have a relatively high rate of detection of cerebral tumors but are far less specific than CT.

### METASTATIC CARCINOMA

Carcinomas usually reach the brain by hematogenous spread. Infrequently, epithelial malignancies of the nasopharynx can spread into the cranial cavity through neural foramina or by direct invasion through bone. The most common neoplasms that metastasize to the brain arise in the lung and breast. Melanomas, colon carcinomas, and testicular and kidney tumors also cause brain metastases.

CT scans typically show brain metastases as multiple enhancing lesions of varying sizes with surrounding areas of low-density edema. On noncontrast scans, metastatic deposits may be hypodense, hyperdense, or similar in density to normal brain tissue, depending on such factors as cellular density, tumor neovascularity, and degree of necrosis. In general, metastases from lung, breast, kidney, and colon tend to be hypodense or isodense; hyperdense metastases often reflect hemorrhage or calcification within or adjacent to the tumor.

Single metastatic deposits in the brain may be indistinguishable from primary tumors. However, since primary brain neoplasms are unusual in older patients, single lesions in this population should suggest metastatic disease.[12]

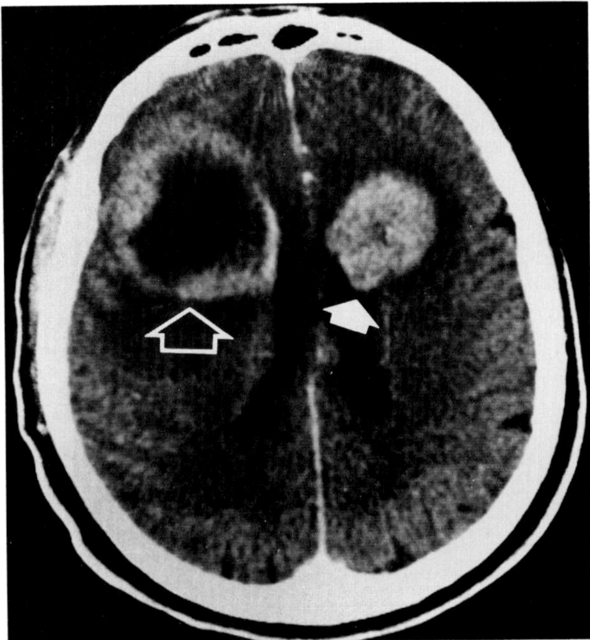

**Figure 69.1. Metastatic carcinoma.** A CT scan following the intravenous injection of contrast material shows enhancing metastases from squamous cell carcinoma of the lung. The metastases are either solid (solid arrow) or cavitating (open arrow).

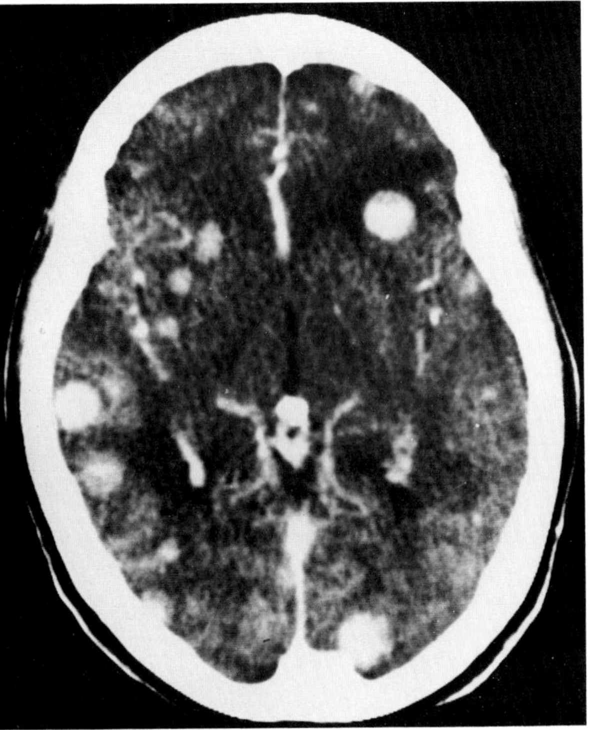

**Figure 69.2. Metastatic carcinoma.** A CT scan performed 2 years after mastectomy for carcinoma of the breast shows multiple enhancing masses of varying shapes consistent with hematogenous metastases.

## GLIOMAS

Gliomas are the most common primary brain tumors. Glioblastomas are highly malignant lesions that are predominantly cerebral in location, though similar tumors may occur in the brainstem, cerebellum, or spinal cord. Gliomas spread by direct extension and can cross from one cerebral hemisphere to the other though connecting white matter tracts such as the corpus callosum. They have a peak incidence in middle adult life and are infrequent in persons under the age of 30.

Astrocytomas are slowly growing tumors that have an infiltrative character and can form large cavities or pseudocysts. Favored sites are the cerebrum, cerebellum, thalamus, optic chiasm, and pons. Cerebral astrocytomas often undergo malignant degeneration and present as mixed astrocytomas and glioblastomas.

Less frequent types of gliomas are ependymoma, medulloblastoma, and oligodendrocytoma. Ependymomas most commonly arise from the walls of the fourth ventricle, especially in children, and from the lateral ventricles in adults. Medulloblastomas are rapidly growing tumors which develop in the posterior part of the vermis in children and, rarely, in the cerebellar hemisphere in adults. The tumor tends to spread throughout the subarachnoid space, with metastatic deposits occurring anywhere within the brain or spinal column. Oligodendrocytomas are slow-growing lesions that usually arise in the cerebrum and have a tendency to calcify.

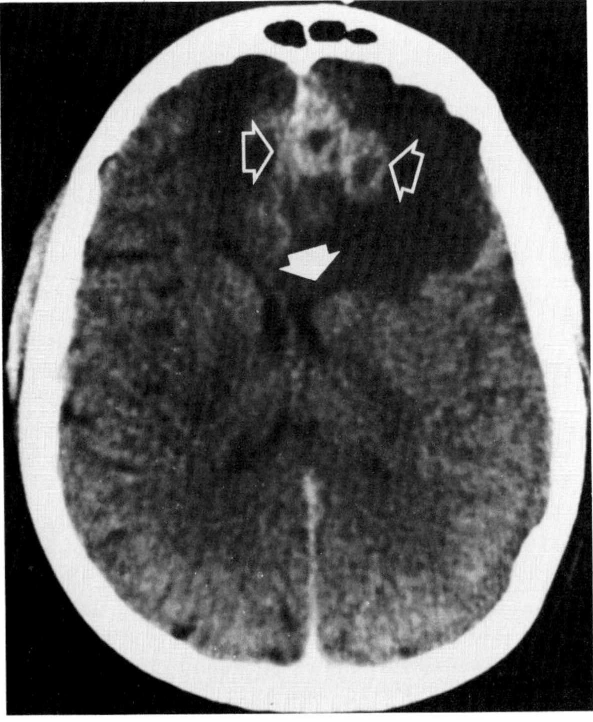

**Figure 69.3. Frontal glioblastoma.** A CT scan shows an irregular enhancing lesion (open arrows). The substantial mass effect of the tumor distorts the frontal horns (closed arrow).

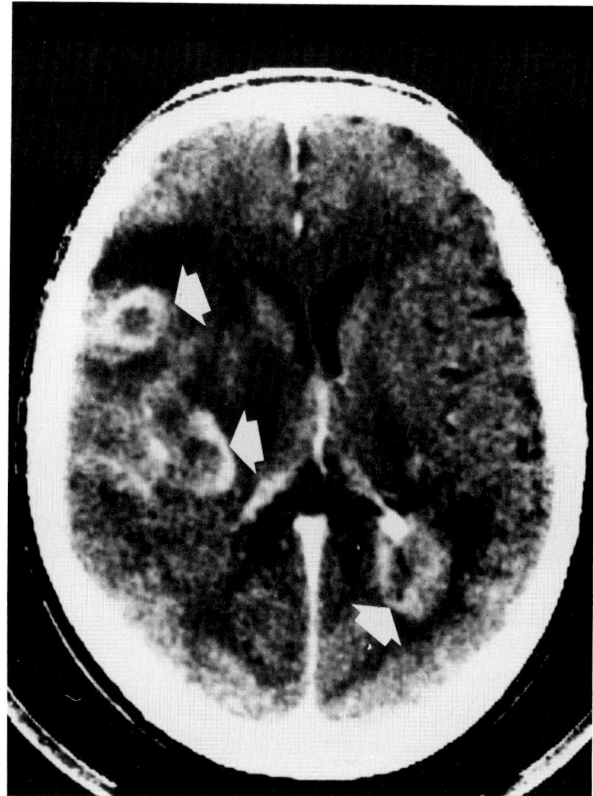

**Figure 69.4. Multicentric glioblastoma.** A CT scan shows bilateral irregular enhancing masses (solid arrows) with surrounding low-density edema.

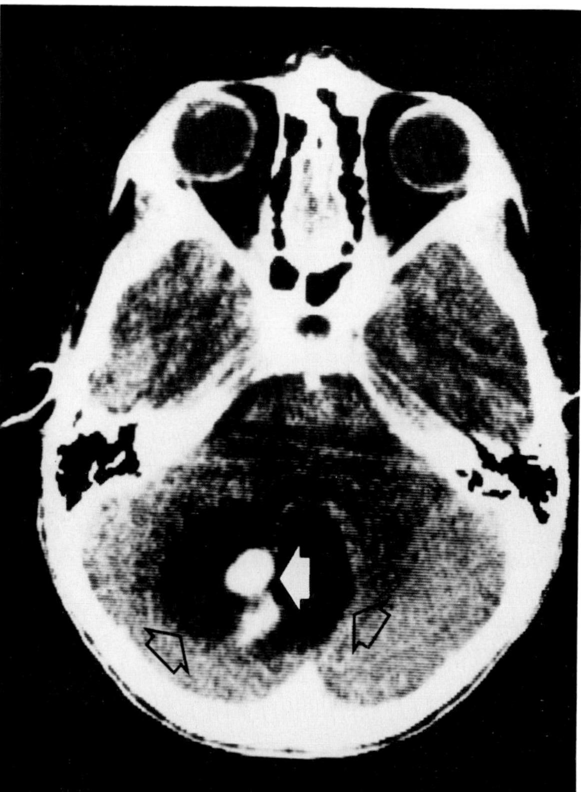

**Figure 69.5. Cystic astrocytoma.** A CT scan shows a cystic posterior fossa lesion (open black arrows) with a central nodule of enhancement (closed white arrow).

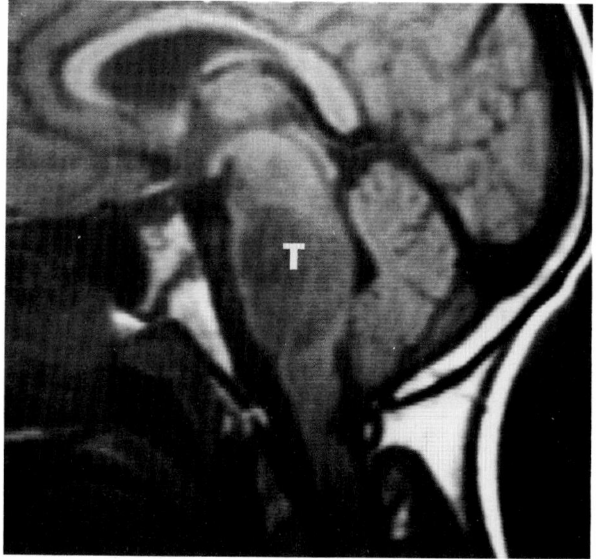


A

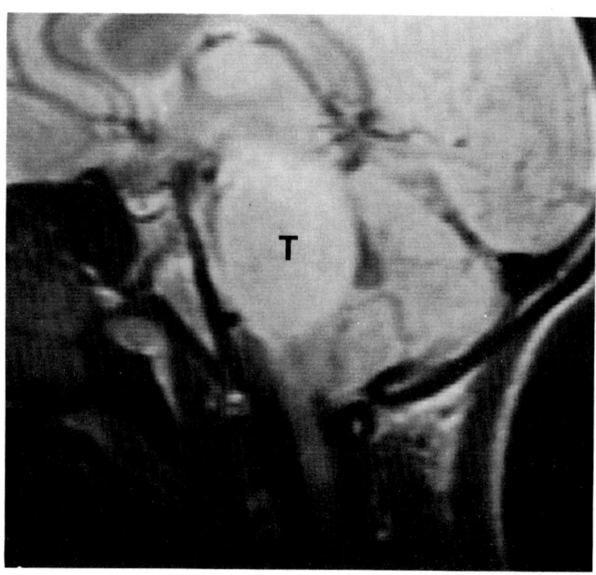


B

**Figure 69.6. Magnetic resonance imaging of a brainstem glioma.** Sagittal MRI scans show enlargement of the brainstem involving the pons and midbrain. Note that various imaging techniques can alter the characteristics of the tumor (T). With one imaging sequence (A), the tumor is gray (low intensity signal); with a different imaging sequence (B), the tumor now appears white (high-intensity signal).

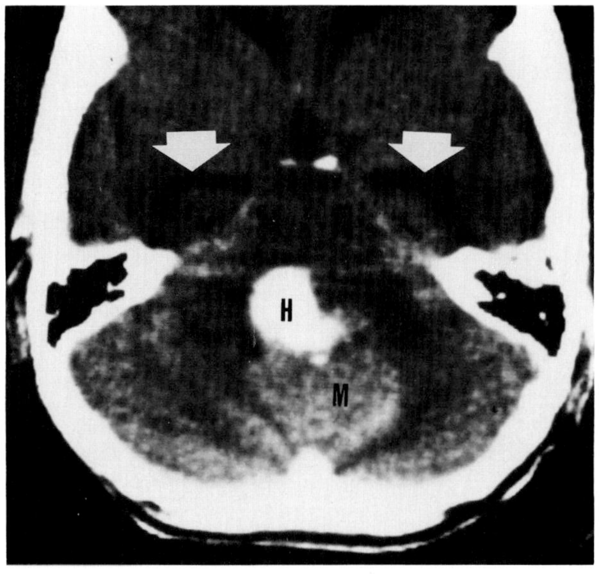


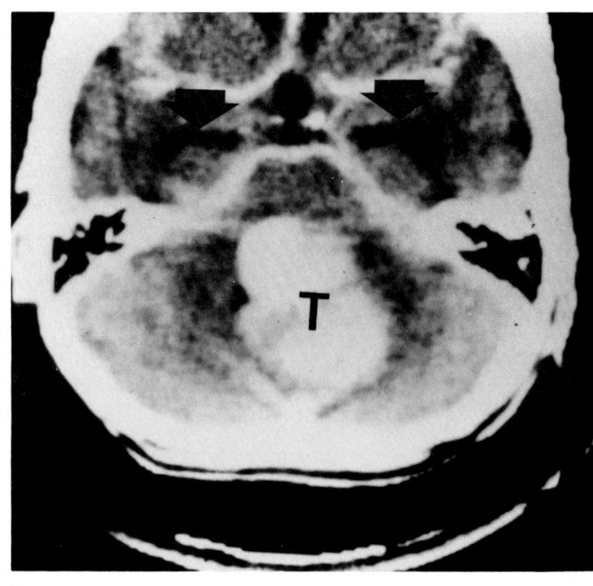


A B

**Figure 69.7. Medulloblastoma.** (A) A noncontrast CT scan in an 8-year-old girl shows the tumor as a mixed high-density (H) and medium-density (M) midline mass in the posterior fossa. (B) After the intravenous injection of contrast material, there is marked enhancement of the tumor (T). The arrows point to the dilated temporal horns representing hydrocephalus.

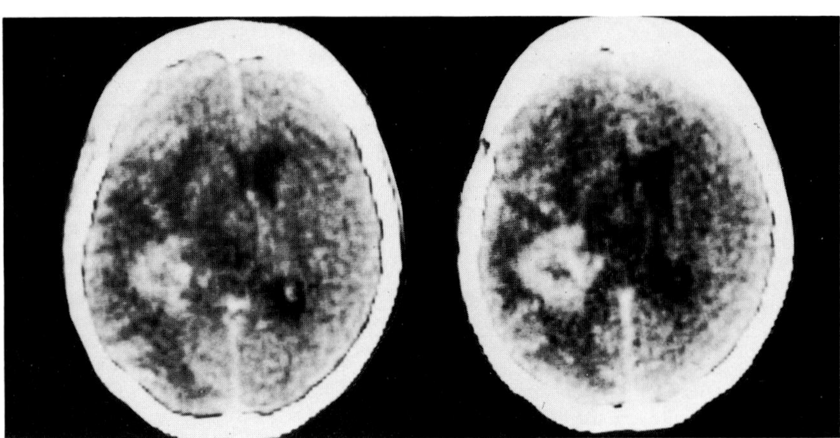

**Figure 69.8. Radiation necrosis.** The enhancing ring lesion with surrounding edema could represent either a primary or a metastatic tumor. At autopsy, the mass was found to represent postradiation necrosis with sarcomatous changes in a patient who had undergone surgery for a solitary metastasis. (From Rao and Williams,[12] with permission from the publisher.)

On noncontrast CT scans, gliomas most commonly present as single, nonhomogeneous masses. Low-grade astrocytomas tend to be low-density lesions; glioblastomas most frequently contain areas of both increased and decreased density, though a broad spectrum of CT appearances can occur. Perifocal edema often is seen in the subcortical white matter. Following the intravenous injection of contrast material, virtually all gliomas will enhance, with the most malignant lesions tending to enhance to the greatest degree. The most common pattern is an irregular ring of contrast enhancement, representing solid vascularized tumor, surrounding a central low-density area of necrosis. Contrast enhancement also can appear as patches of increased density distributed irregularly throughout a low-density lesion or as rounded nodules of increased density within the mass.[1,2]

## RADIATION NECROSIS

External radiation therapy, which often is combined with surgery in the treatment of brain tumors, may lead to the development of radiation necrosis of the brain 9 to 24 months after therapy. Radiation-induced necrosis of the brain may present on CT as a mass lesion that demonstrates contrast enhancement and is impossible to differentiate from recurrent or residual tumor.[3]

## MENINGIOMA

Meningioma is a benign tumor that arises from arachnoid lining cells and is attached to the dura. The most common sites of meningioma are the convexities of the cal-

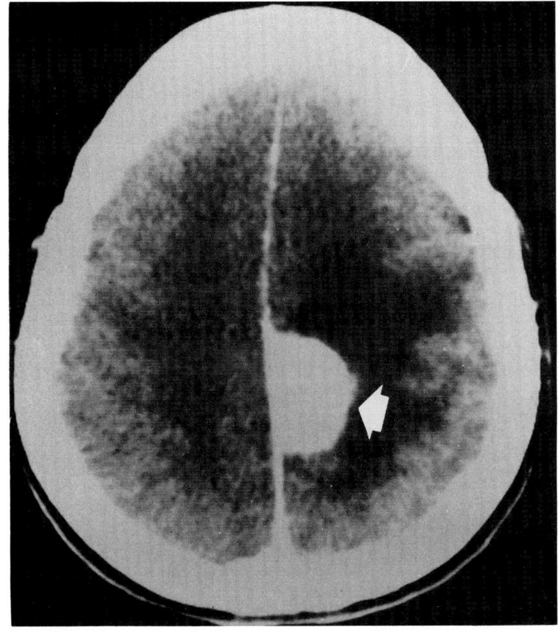

A

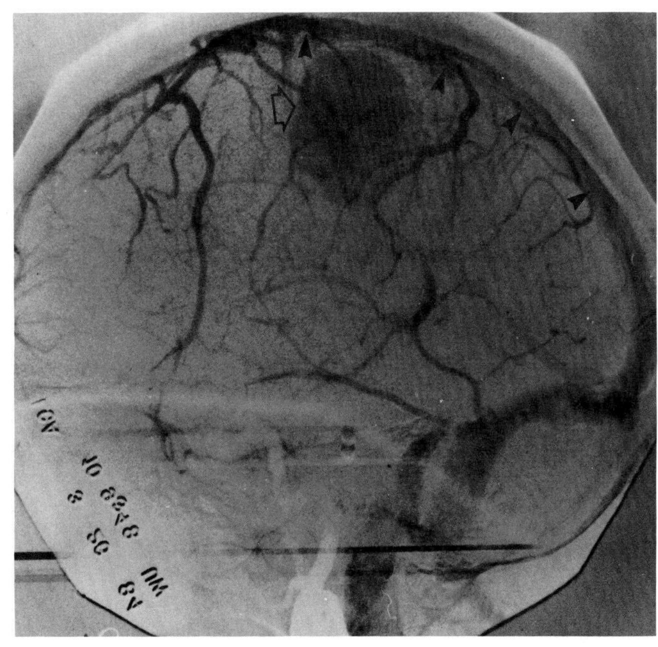

B

**Figure 69.9. Perifalcine meningioma.** (A) A CT scan following the intravenous injection of contrast material shows a uniformly enhancing mass (arrow) with surrounding low-density edema. (B) The venous phase of a carotid arteriogram shows the characteristic prominent vascular blush (open arrow) of the meningioma and the patent superior sagittal sinus (black arrowheads).

varium, olfactory groove, tuberculum sellae, parasagittal region, Sylvian fissure, cerebellopontine angle, and spinal canal.

CT typically shows a meningioma as a rounded, sharply delineated, isodense or hyperdense tumor in a juxtadural location. Calcification often is seen within the mass on noncontrast scans. Following the intravenous injection of contrast material, there is intense homogeneous enhancement, which reflects the highly vascular nature of the tumor.

Pronounced dilatation of meningeal and diploic vessels, which provide part of the blood supply to the tumor, may produce prominent grooves in the calvarium on plain films of the skull. Calvarial hyperostosis may develop because of invasion of the bone by tumor cells that stimulate osteoblastic activity. Dense calcification or granular psammomatous deposits may be seen within the tumor.

Arteriography can demonstrate the feeding arteries, which most commonly arise from both the internal and external carotid artery circulation. Preoperative embolization of the external carotid artery supply can improve operative hemostasis.[4]

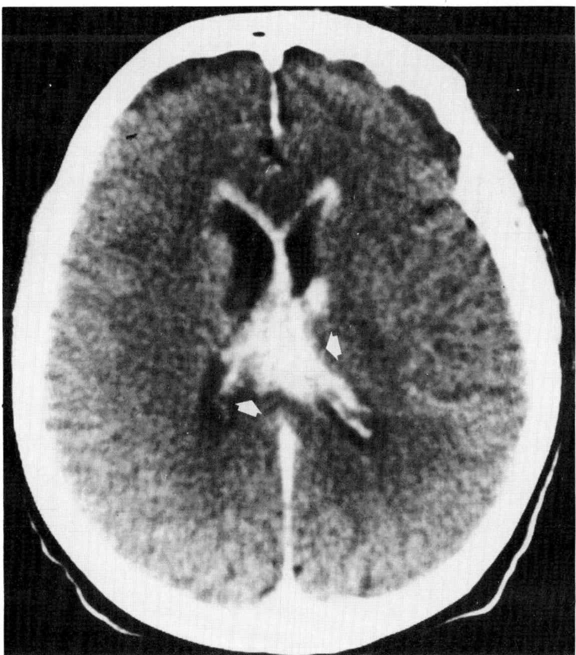

**Figure 69.10. Histiocytic lymphoma.** A CT scan shows homogeneously enhancing lesions (arrows) deep in the brain adjacent to the ventricular system.

## HISTIOCYTIC LYMPHOMA (RETICULUM CELL SARCOMA)

Cerebral lymphoma may involve the brain or meninges and present a clinical picture similar to that of glioblastoma. The tumor has a tendency to develop in immunosuppressed individuals, especially following renal transplantation. Cerebral lymphoma may be multifocal and usually arises in the periventricular and subcortical regions of the cerebrum and cerebellum.[5]

## HEMANGIOBLASTOMA

Hemangioblastomas are benign tumors of the posterior fossa that most frequently affect young and middle-aged adults. They have a high familial incidence and may be associated with retinal angiomas in von Hippel-Lindau disease. CT scans typically demonstrate a hemangioblastoma as a low-density cystic mass with single or multiple mural nodules that show homogeneous enhancement. Arteriography can be of value in delineating the large feeding artery and draining vein associated with the tumor.[6]

## PINEAL TUMORS

This topic is discussed in the section "Diseases of the Pineal Gland" in Chapter 4.

## COLLOID CYST OF THE THIRD VENTRICLE

Colloid cysts are papillomatous lesions situated in the anterior portion of the third ventricle. The tumor typically causes hydrocephalus by obstructing the foramen of

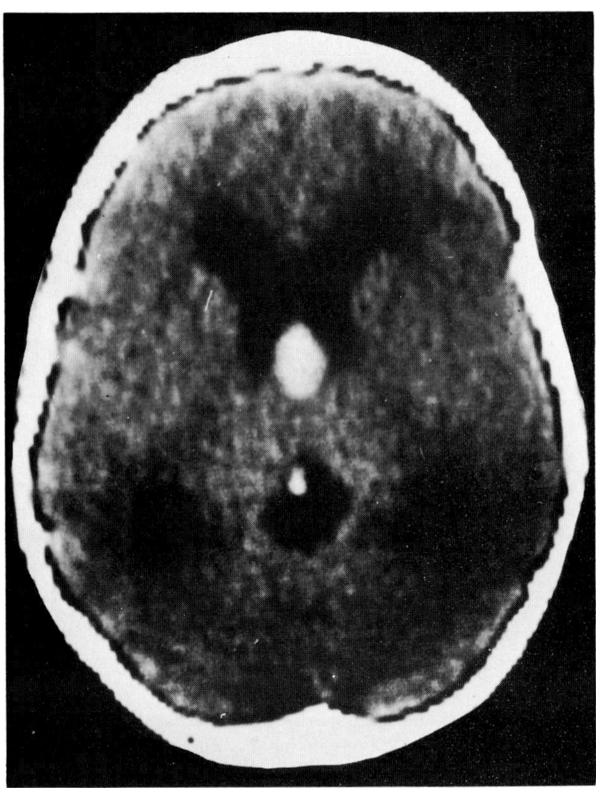

**Figure 69.12. Colloid cyst of the third ventricle.** A noncontrast CT scan shows an oval, hyperdense lesion in the region of the third ventricle with separation of the posteromedial sides of the frontal horns. Note the collapse of the posterior third ventricle. (From Lee and Rao,[11] with permission from the publisher.)

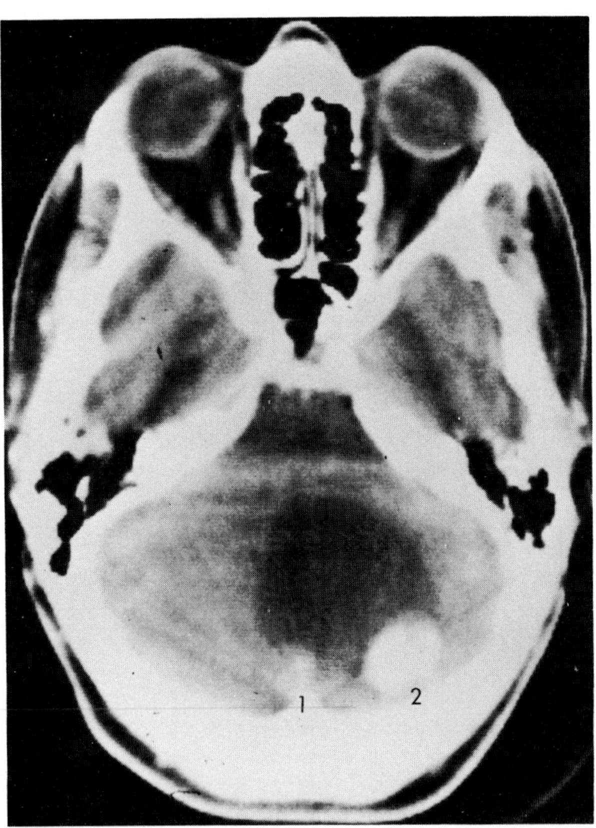


**Figure 69.11. Hemangioblastoma.** A CT scan shows a large, cystic posterior fossa tumor in the left cerebellum that contains two round, homogeneously enhancing tumor nodules (1,2). (From Lee and Rao,[11] with permission from the publisher.)

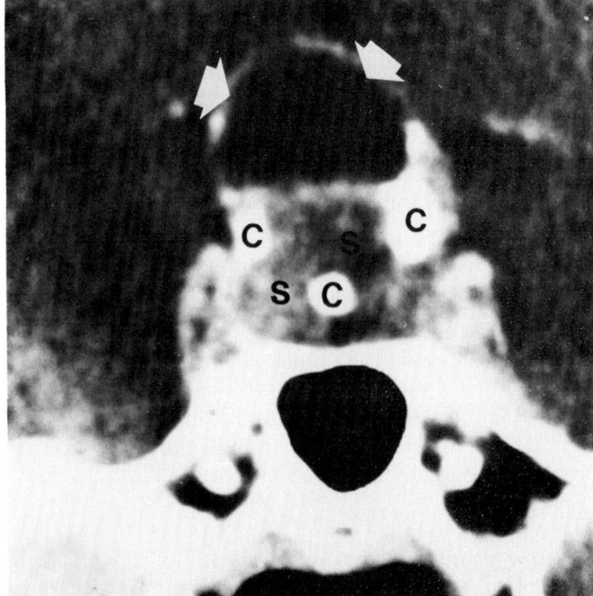


**Figure 69.13. Craniopharyngioma.** A coronal CT scan through the suprasellar cistern of a 14-year-old boy with progressive visual loss shows the mass to contain calcified portions (C), cystic areas (white arrows), and solid components (S).

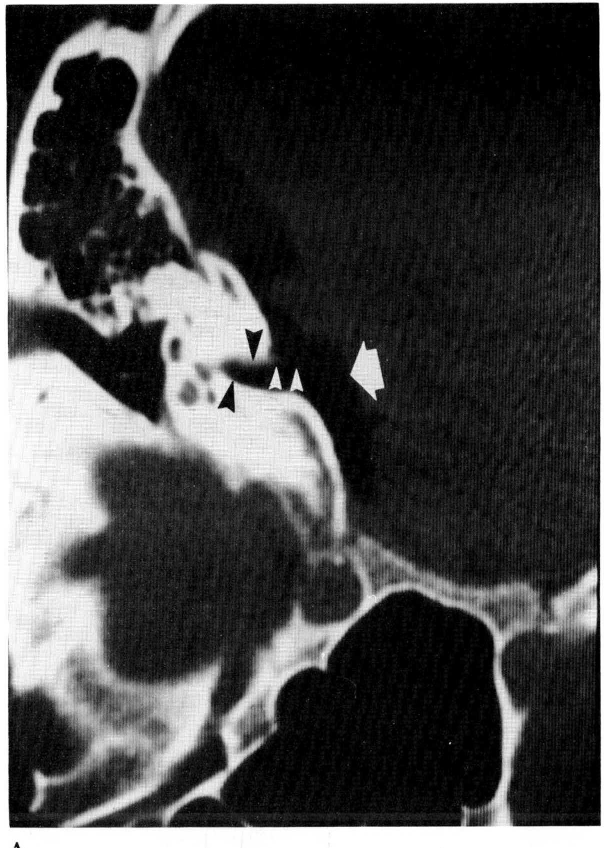

A

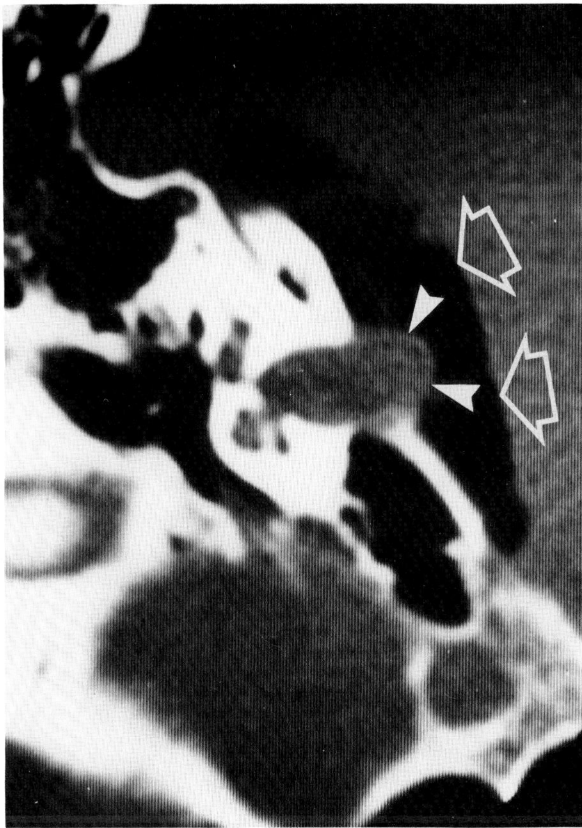

B

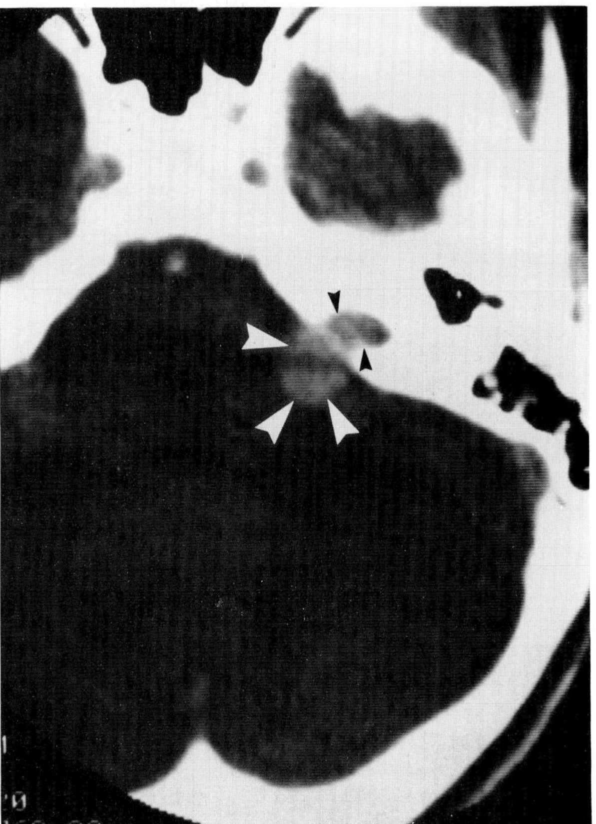

C

**Figure 69.14. Acoustic neuroma.** (A) Normal internal auditory canal air study. Air in the cerebellopontine cistern (large white arrow) and the internal auditory canal (black arrows) shows the eighth nerve (small white arrows) entering the canal. There is no evidence of tumor. (B) Intercanalicular acoustic neuroma. Air injected into the subarachnoid space fills the cerebellopontine angle cistern (open white arrows) and outlines the small tumor (white arrows). (C) In this patient, a CT scan shows the enhancing tumor (white arrows) that originates in the internal auditory canal and extends into the posterior fossa. Note the widening of the internal auditory canal (black arrows).

Monro. Noncontrast CT scans show a colloid cyst as a smooth, spherical or ovoid mass of homogeneously high density in the anterior part of the third ventricle.[7]

## CRANIOPHARYNGIOMA

Craniopharyngioma is a benign congenital, or "rest-cell," tumor with cystic and solid components. The lesion usually originates above the sella turcica, depressing the optic chiasm and extending up into the third ventricle. Less commonly, a craniopharyngioma lies within the sella, where it compresses the pituitary gland and may erode adjacent bony walls.

Most craniopharyngiomas have calcification that can be detected on plain skull films or CT scans. In cystic lesions, the shell-like calcification lies along the periphery of the tumor; in mixed or solid lesions, the calcification is nodular, amorphous, or cloudlike. CT scans clearly demonstrate the cystic and solid components of the mass. Following the administration of contrast material, there is variable enhancement depending on the type of calcification and the amount of cystic components within the tumor.[8]

## ACOUSTIC NEUROMA

Acoustic neuroma is a slowly growing benign tumor that may occur as a solitary lesion or as part of the syndrome of neurofibromatosis. The tumor arises from Schwann cells in the vestibular portion of the eighth cranial nerve. It usually originates in the internal auditory meatus and extends into the cerebellopontine angle cistern.

CT scans demonstrate enlargement and erosion of the internal auditory canal and a uniformly enhancing mass in the cerebellopontine angle. Small intercanalicular tumors confined to the internal auditory canal may cause bony changes or clinical findings suggesting an acoustic neuroma in the absence of CT evidence of a discrete mass. In such cases, repeat CT examination is required following the intrathecal administration of contrast material (metrizamide or air). Very large tumors may compress the fourth ventricle and lead to the development of hydrocephalus.[9]

## PITUITARY ADENOMA

This topic is discussed in the section "Pituitary Adenoma" in Chapter 4.

## CHORDOMA

Chordomas are tumors that arise from remnants of the notochord. Although any part of the vertebral column and base of the skull can be involved, the most common sites are the clivus and the lower lumbosacral region. The tumors are locally invasive but do not metastasize.

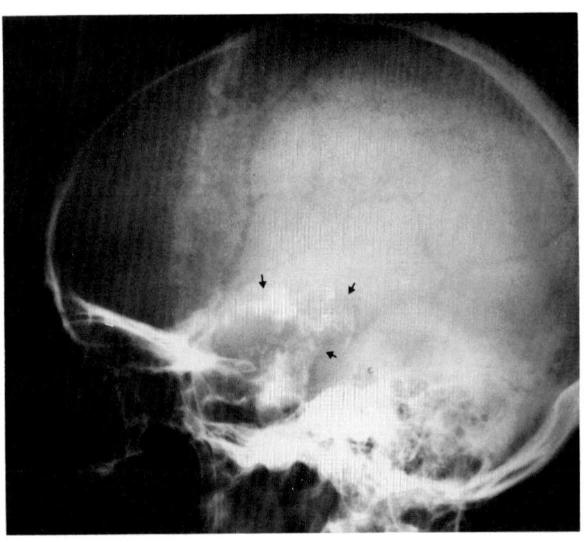

A

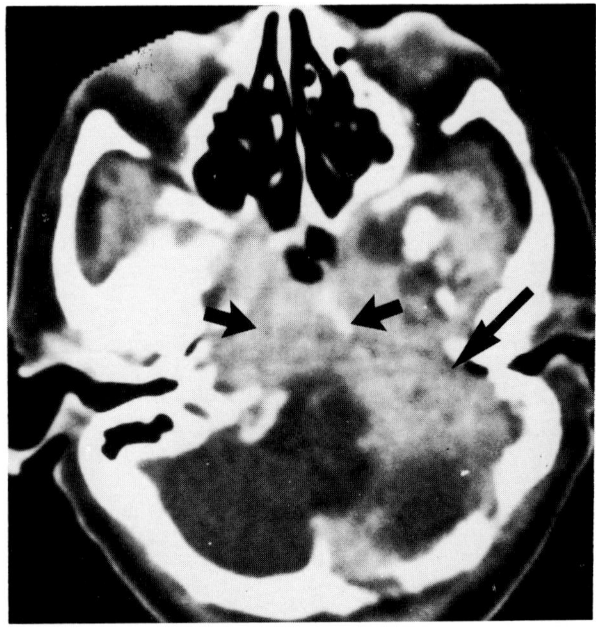

B

**Figure 69.15. Chordoma.** (A) A plain skull radiograph shows dense calcification (arrows) within a large soft tissue mass that has eroded the dorsum sellae and upper portion of the clivus. (B) In another patient, a CT scan shows an enhancing mass with destruction of the entire clivus (short arrows) and only small bone fragments remaining. The left petrous pyramid is also destroyed (long arrow). (B, from Levine, Kleefield, and Rao,[13] with permission from the publisher.)

Chordomas arising at the base of the skull produce the striking clinical picture of multiple cranial nerve palsies on one or both sides combined with a retropharyngeal mass and erosion of the clivus.

On plain radiographs, a chordoma tends to be a bulky mass causing ill-defined bone destruction or cortical expansion. Flocculent calcifications may develop within a large soft tissue mass. On CT scans, chordomas at the base of the skull tend to appear as lesions that are slightly denser than brain tissue and often demonstrate moderate contrast enhancement.[10]

## REFERENCES

1. Steinhoff H: CT in the diagnosis and differential diagnosis of glioblastomas. *Neuroradiology* 14:193–200, 1977.
2. Tans J, DeJongh IE: Computed tomography of supratentorial astrocytoma. *Clin Neurol Neurosurg* 80:156–168, 1978.
3. Mikhael MA: Radiation necrosis of the brain. *J Comput Assist Tomogr* 2:71–80, 1978.
4. Vassilouthis J, Ambrose J: Computerized tomography scanning appearance of intracranial meningiomas. *J Neurosurg* 50: 320–327, 1979.
5. Enzmann DR, Krikorian J, Norman D, et al: Computed tomography in primary reticulum cell sarcoma of the brain. *Radiology* 130:165–170, 1979.
6. Adair LB, Ropper AH, Davis JR: Cerebellar hemangioblastoma: CT, angiographic and clinical condition in seven cases. *J Comput Assist Tomogr* 2:281–294, 1978.
7. Ganti SR, Antunes JL, Louis KM, et al: CT in the diagnosis of colloid cysts of the third ventricle. *Radiology* 138:385–391, 1981.
8. Fitz CR, Wortzman G, Harwood-Nash DC, et al: Computed tomography in craniopharyngiomas. *Radiology* 127:687–691, 1978.
9. Kricheff II, Pinto RS, Bergeron RT, et al: Air-CT cisternography and canalography for small acoustic neuromas. *Am J Neuroradiol* 1:57–63, 1980.
10. Firoozina H, Pinto RS, Lin JP, et al: Chordoma: Radiologic evaluation of twenty cases. *AJR* 127:797–805, 1976.
11. Lee SH, Rao KCVG: Primary tumors in adults, in Lee SH, Rao KCVG (eds): *Cranial Computed Tomography*. New York, McGraw-Hill, 1983, pp 241–293.
12. Rao KCVG, Williams JP: Intracranial tumors: Metastatic, in Lee SH, Rao KCVG (eds): *Cranial Computed Tomography*. New York, McGraw-Hill, 1983, pp 345–370.
13. Levine HL, Kleefield J, Rao KCVG: The base of the skull, in Lee SH, Rao KCVG (eds): *Cranial Computed Tomography*. New York, McGraw-Hill, 1983, pp 371–460.

## Chapter Seventy
# Infections of the Central Nervous System

## PYOGENIC INFECTIONS

### Acute Bacterial Meningitis

Bacterial meningitis is most commonly due to *Hemophilus influenzae* in neonates and young children and to meningococci and pneumococci in adolescents and adults. Although the meninges initially demonstrate vascular congestion, edema, and minute hemorrhages, the underlying brain and its ependymal surfaces remain intact. Thus computed tomography (CT) scans are normal during most acute episodes of meningitis and remain normal if appropriate therapy is promptly instituted. If the infection extends to involve the subpial cortex of the brain and the ependymal lining of the ventricles, noncontrast scans demonstrate increased density in the basal cisterns, interhemispheric fissure, and choroid plexus. This appearance, which frequently simulates contrast enhancement, probably reflects a fibrinous or hemorrhagic exudate with a high protein level in the subarachnoid space of the interhemispheric fissure and basal cisterns. Diffuse brain swelling may symmetrically compress the lateral and third ventricles. Focal edema may produce localized areas of low density. CT is also of value in the early detection of such complications of acute meningitis as arterial or venous vasculitis or thrombosis with infarction, hydrocephalus due to adhesions or thickening of the arachnoid at the base of the brain, subdural effusion or empyema, encephalomalacia, and brain abscess.

Although acute bacterial meningitis is best evaluated by CT with bone-window settings, plain films of the sinuses and skull can demonstrate cranial osteomyelitis, paranasal sinusitis, or a skull fracture as the underlying cause of meningitis. Chest radiographs may show a silent area of pneumonia or a lung abscess.[1–3]

### Subdural Empyemas

Subdural empyema is a suppurative process in the cranial subdural space between the inner surface of the dura and the outer surface of the arachnoid. The most common cause of subdural empyema is the spread of infection from the frontal or ethmoid sinuses. Less frequently, subdural empyema may be secondary to mastoiditis, middle ear infection, purulent meningitis, penetrating wounds to the skull, craniectomy, or osteomyelitis of the skull. Subdural empyema is often bilateral and is associated with a high mortality rate, even if properly treated. The most common location of a subdural empyema is over the cerebral convexity; the base of the skull is usually spared.

CT is the procedure of choice in evaluating the patient with suspected subdural empyema. Noncontrast scans

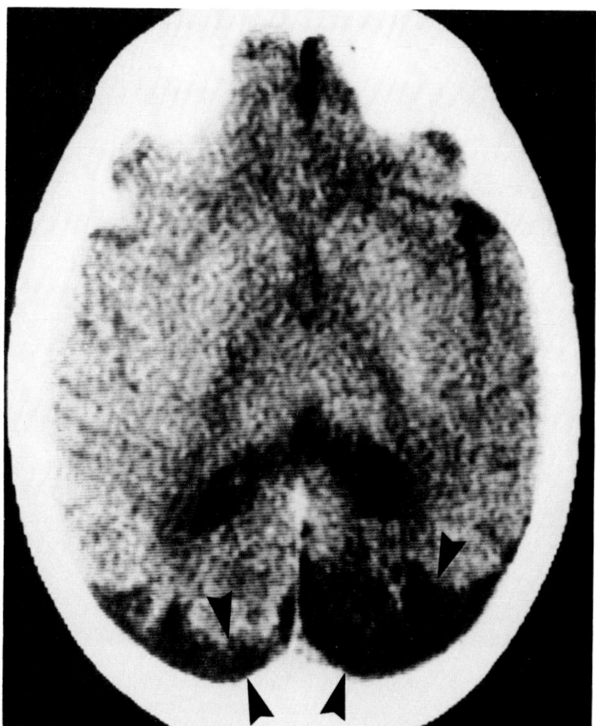

demonstrate a crescentic or lentiform, extra-axial hypodense collection (representing pus) adjacent to the inner border of the skull. There is compression and displacement of the ipsilateral ventricular structures. Following the intravenous administration of contrast material, a narrow zone of enhancement of relatively uniform thickness separates the hypodense extracerebral collection from the brain surface. CT can also demonstrate involvement of the adjacent parenchyma via retrograde thrombophlebitis with resultant infarction and/or abscess formation, signs associated with a poor prognosis.[4–6]

## Extradural Abscesses

Extradural abscess is almost invariably associated with osteomyelitis in a cranial bone originating from an infection in the ear or paranasal sinuses. The infectious

**Figure 70.1. Brain infarction secondary to childhood meningitis.** In this 12-year-old girl with cortical blindness due to previous meningitis at age 2, a CT scan shows bilateral occipital lobe infarctions (black arrowheads) as low-density lesions. Spasm of cerebral vessels due to meningitis led to the subsequent infarction.

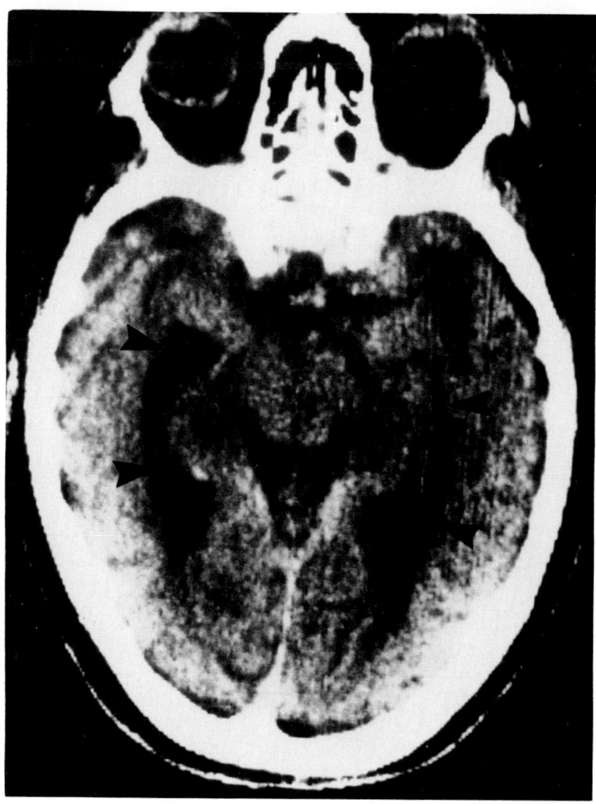

A

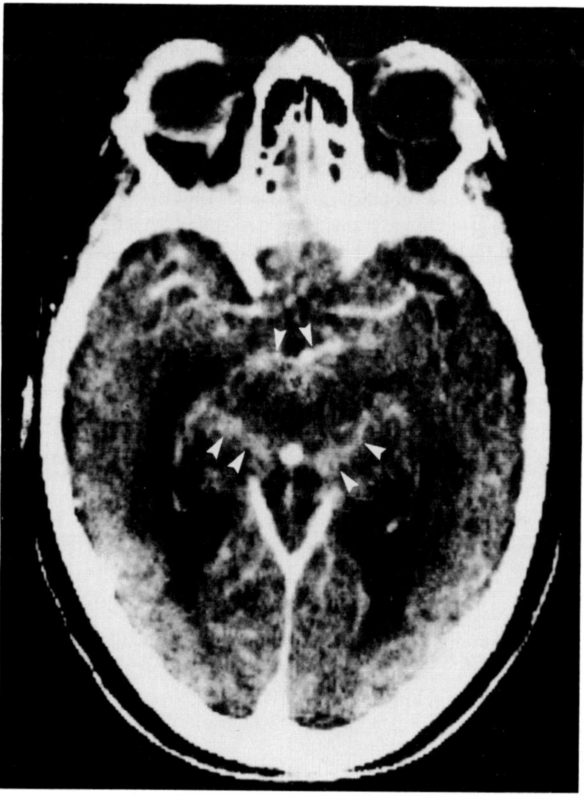

B

**Figure 70.2. Pneumococcal meningitis with hydrocephalus.** (A) A noncontrast CT scan shows dilatation of the temporal horns of the lateral ventricles (black arrows). (B) After the intravenous injection of contrast material, there is enhancement of the meninges in the basal cisterns (white arrowheads), reflecting the underlying inflammation due to meningitis. The hydrocephalus in meningitis is due to blockage of the normal flow of cerebrospinal fluid by inflammatory exudate at the level of the aqueduct and the basal cisterns.

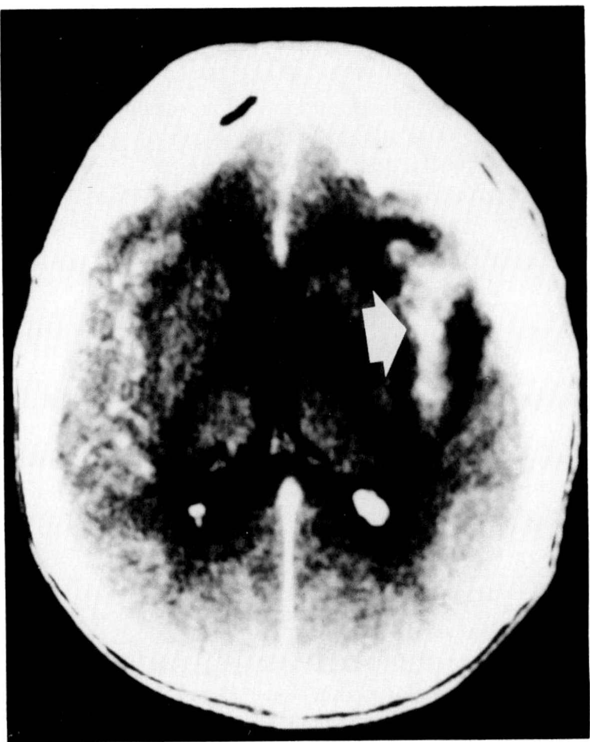

**Figure 70.3. Tuberculosis.** In this 50-year-old man with seizures, a CT scan following the intravenous injection of contrast material shows enhancement of the Sylvian fissure (white arrow). As in about half the patients with central nervous system tuberculosis, the patient's chest radiograph was normal.

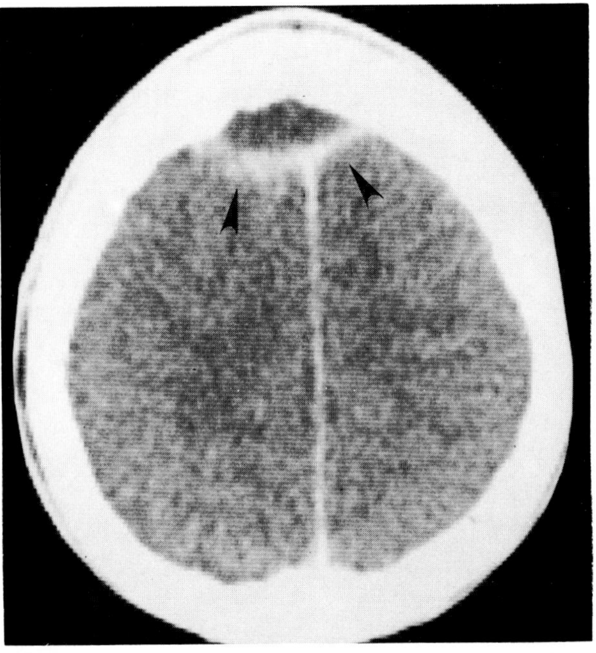

**Figure 70.4. Extradural abscess.** A CT scan following the intravenous injection of contrast material demonstrates a biconvex hypodense lesion with a contrast-enhanced dural margin (arrowheads), which crosses the falx and displaces the falx away from the inner table of the skull. (From Lee,[18] with permission from the publisher.)

process is localized outside the dural membrane and beneath the inner table of the skull. The frontal region is most frequently affected because of its close relation to the frontal sinuses and the ease with which the dura can be stripped from the bone.

Noncontrast CT scans demonstrate the epidural infection as a poorly defined area of low density adjacent to the inner table of the skull. An adjacent area of bone destruction or evidence of paranasal sinus or mastoid infection (fluid, soft tissue thickening) often can be demonstrated on CT or plain skull radiographs. Following the intravenous administration of contrast material, the inflamed dural membrane appears as a thickened zone of enhancement on the convex inner side of the lesion. If the collection lies in the midline, the attachment of the falx is displaced inward and separated from the adjacent skull, thus identifying its extradural location.[7,8]

## Brain Abscesses

Brain abscesses usually are secondary to chronic infections of the middle ear, paranasal sinuses, or mastoid air cells or to systemic infections (pneumonia, bacterial endocarditis, osteomyelitis). The most common organisms causing brain abscess are streptococci, most of which are anaerobic.

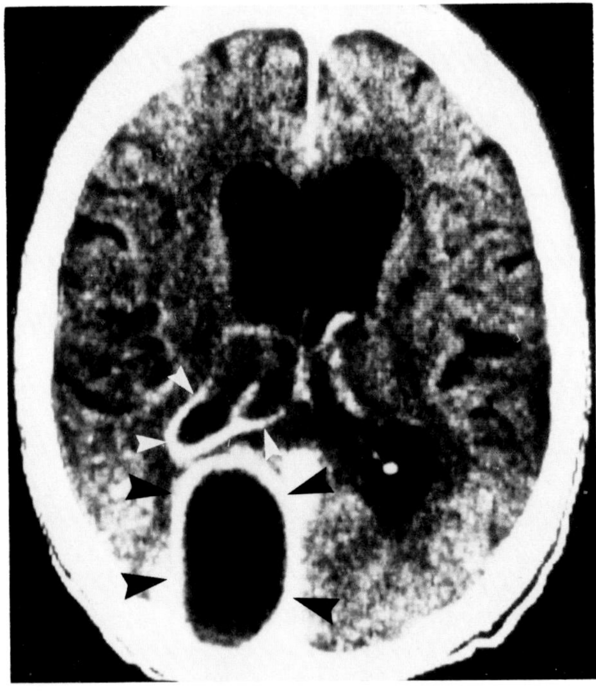

**Figure 70.5. Brain abscess with ventriculitis.** A CT scan following the intravenous injection of contrast material in a drug addict with lethargy and confusion demonstrates the abscess as a ring-enhancing lesion (black arrowheads) in the right occipital lobe. Extension of infection into the ventricular system has resulted in ventriculitis (white arrowheads).

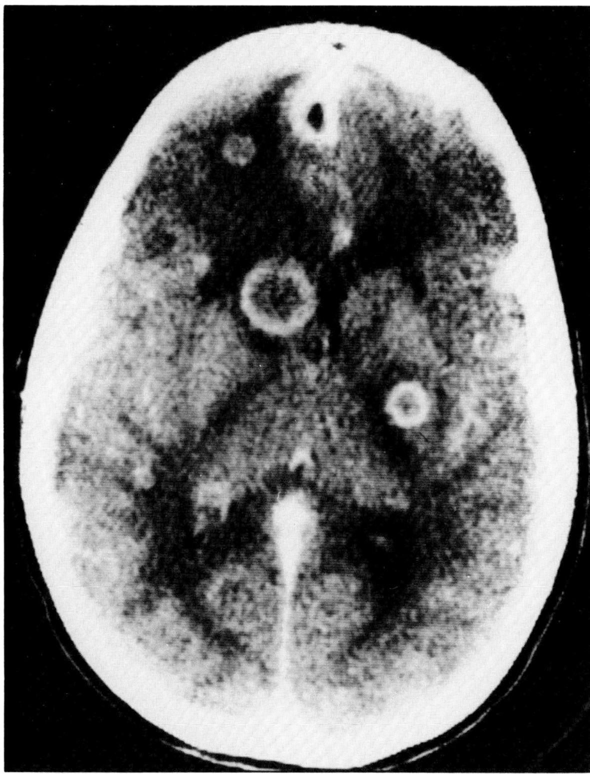

**Figure 70.6. Brain abscesses secondary to sinusitis.** A CT scan following the intravenous administration of contrast material in a 12-year-old girl with lethargy and headache shows multiple ring-enhancing lesions of varying size. The surrounding low-density edema is most prominent in the right frontal lobe. The multiple abscesses represented hematogenous spread of infection from underlying sinus disease.

In the early stage of focal cerebritis, CT shows an area of low density with poorly defined borders and a mass effect reflecting vascular congestion and edema. Further progression of the inflammatory process leads to cerebral softening that may undergo necrosis and liquefaction and result in an abscess. Noncontrast scans continue to demonstrate an ill-defined, low-density area with a mass effect on the ventricular system or midline structures. Following the intravenous administration of contrast material, an oval or circular peripheral ring of contrast enhancement outlines the abscess capsule. Although the wall is usually thin and of uniform thickness, an irregularly thick wall may be seen that mimics the wall of a malignant glioma. Multiple abscesses suggest the possibility of septic emboli from a systemic infection.

CT can be used to assess the results of therapy of brain abscess as well as to document complications. This modality can demonstrate the often fatal intraventricular rupture of an abscess, as well as the development of increased intracranial pressure that may lead to brain herniation.

Plain skull radiographs may show evidence of underlying sinusitis, mastoiditis, or osteomyelitis, though this is better evaluated by CT scanning with bone-window settings. Infection by gas-forming organisms occasionally produces an air-fluid level within the abscess cavity.

Because CT is far better than other modalities (radionuclide studies, arteriography) in assessing brain abscesses, if a CT scanner is unavailable, the patient should probably be transferred to another facility where CT can be performed.[9–12]

## VIRAL INFECTIONS

### Herpes Simplex (Type 1) Encephalitis

Herpetic encephalitis is an often-fatal, fulminant necrotizing encephalitis with petechial hemorrhages. The earliest and predominant CT finding in herpetic encephalitis is poorly marginated, hypodense areas in the temporal and frontal lobes, which are the characteristic sites of involvement in gross pathologic specimens. The low density probably represents a combination of tissue necrosis and focal brain edema. A mass effect is common and may present as a midline shift or as a focal mass compressing the ventricles or the Sylvian cisterns. Compromise of the blood-brain barrier in the area of rapid, progressive hemorrhagic necrosis results in a nonhomogeneous pattern of contrast enhancement that may be gyral or ringlike or may appear as linear streaks at the periphery of the low-density lesions. Follow-up scans typically demonstrate widespread low-density (encephalomalacia) involving the temporal and frontal lobes.

In addition to confirming the clinical diagnosis and excluding the presence of an abscess or tumor, CT is important in the evaluation of herpes simplex encephalitis to indicate the best site for biopsy. A definitive diagnosis of herpes infection, which is based on positive fluorescein antibody staining or culture of the virus from a brain biopsy, is essential before beginning treatment with adenine arabinoside, a chemotherapeutic agent that may be neurotoxic, mutagenic, and carcinogenic.[13,14]

### Progressive Multifocal Leukoencephalopathy

Progressive multifocal leukoencephalopathy (PML) is a rare condition that usually occurs in patients who have leukemia, lymphoma, carcinomatosis, or a variety of chronic diseases (tuberculosis, sarcoidosis, macroglobulinemia) or are receiving immunosuppressive therapy. Pathologically, PML produces multiple areas of demyelinization with relative sparing of neurons and little or no perivascular infiltration. CT demonstrates typical low-density foci that have a predilection for the subcortical white matter, especially in the parieto-occipital region. The lesions tend to be sharply marginated with a scalloped outer border. As with herpetic viral encephalitis, CT is important in selecting sites for biopsy to confirm the diagnosis.[15,16]

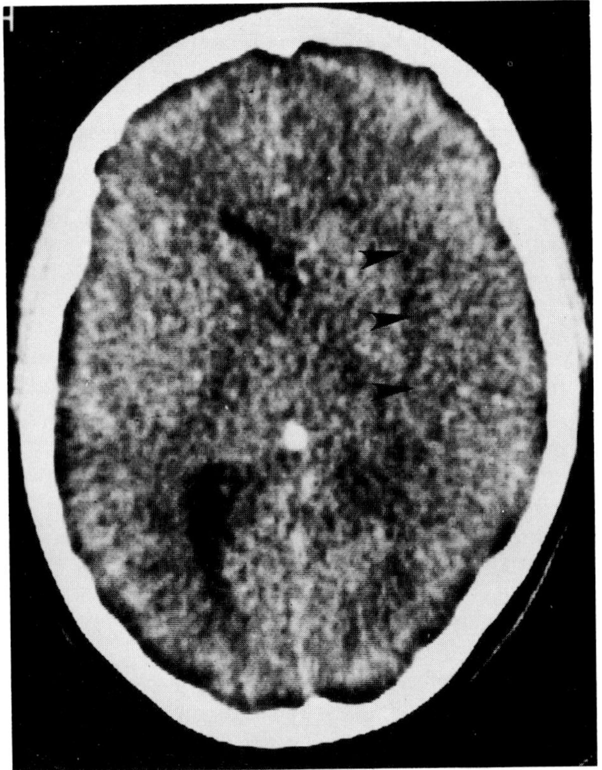
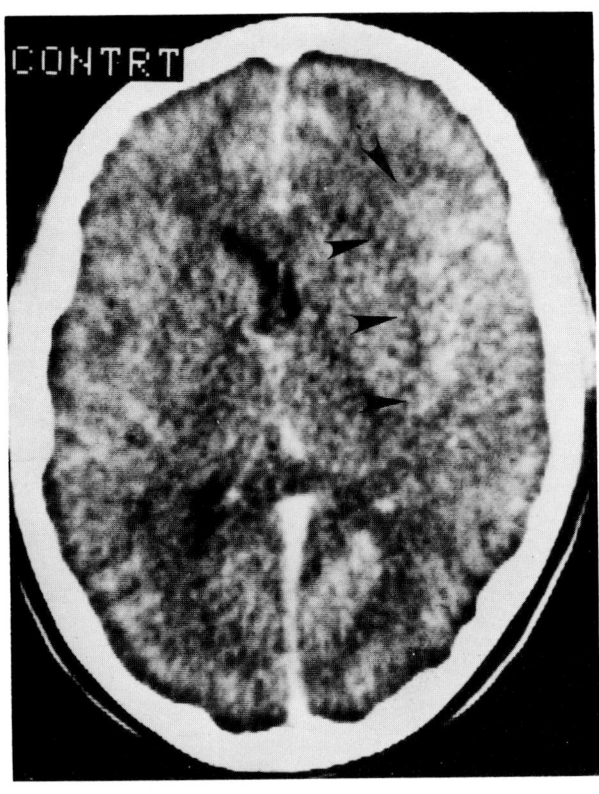


A B

**Figure 70.7. Herpes simplex (type I) encephalitis.** (A) A noncontrast CT scan shows a poorly defined, mixed-density area in the temporal lobe (arrowheads), with effacement of the ipsilateral ventricle. (B) Following the intravenous injection of contrast material, patchy areas of enhancement involve the temporal and frontal lobes (arrowheads). (From Lee,[18] with permission from the publisher.)

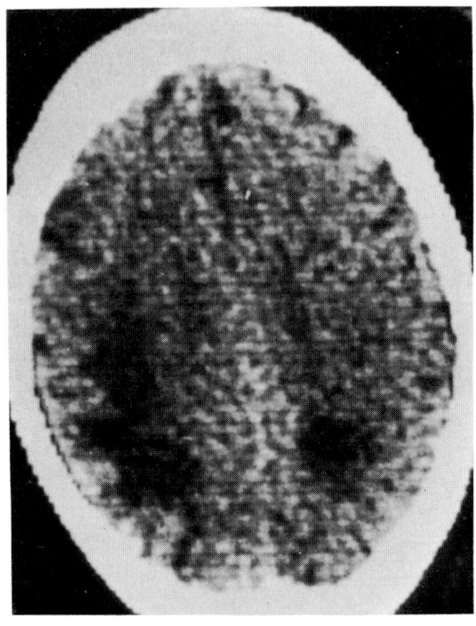

**Figure 70.8. Progressive multifocal leukoencephalopathy.** A noncontrast CT scan of an elderly man with chronic lymphocytic leukemia shows low-density zones in both cerebral hemispheres. Note the characteristic well-marginated medial aspect and scalloped outer border at the junction of the gray and the white matter. (From Carroll, Lane, and Norman, et al,[15] with permission from the publisher.)

## Creutzfeldt-Jakob Disease

Creutzfeldt-Jakob disease is an invariably fatal degenerative disease of the central nervous system that is caused by a slow virus and is characterized by rapidly progressive dementia and myoclonic seizures. The disease predominantly affects gray matter in the cerebrum (temporal and occipital lobes) and cerebellum. CT demonstrates cerebral atrophy with a rapid increase in ventricular size on serial studies.[17]

### REFERENCES


1. Zimmerman RA, Patel S, Bilaniuk LT: Demonstration of purulent bacterial intracranial infections by computed tomography. *AJR* 127:155–165, 1976.
2. Auh YH, Lee SH, Toglia JU: Excessively small ventricles on cranial CT: Clinical correlation in 75 patients. *J Comput Tomogr* 4:325–29, 1980.
3. Claveria LE, Du Boulay GH, Moseley IF: Intracranial infections: Investigations by CAT. *Neuroradiology* 12:59–71, 1976.
4. Kim SK, Weinberg PE, Magidson M: Angiographic features of subdural empyema. *Radiology* 118:621–625, 1976.
5. Zimmerman RD, Leeds NE, Danziger A: Subdural empyema: CT findings. *Radiology* 150:417–422, 1984.
6. Luken MG, Whelan MA: Recent diagnostic experience with subdural empyema. *J Neurosurg* 52:764–771, 1980.

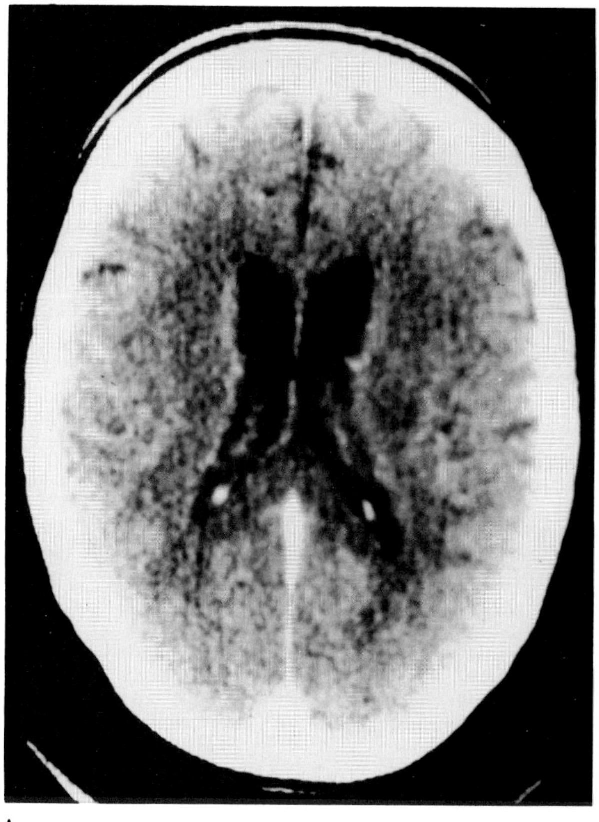
A

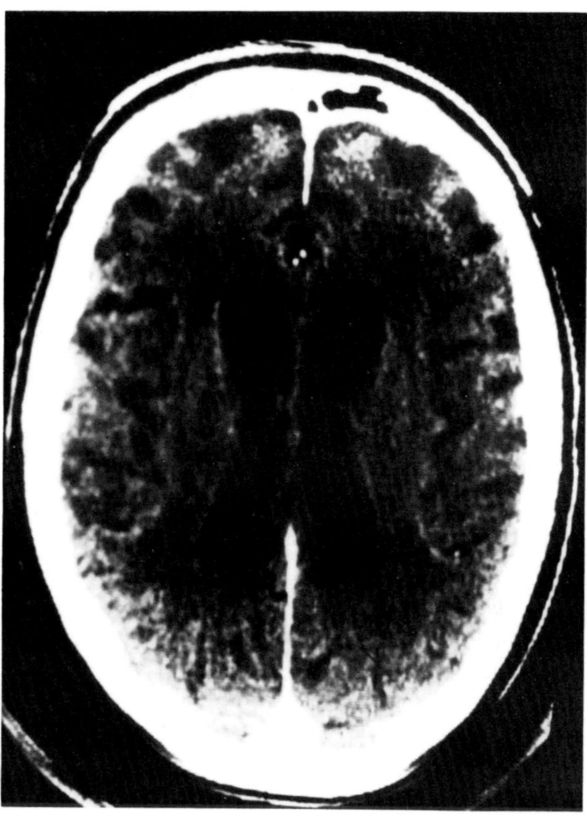
B

**Figure 70.9. Creutzfeldt-Jakob disease.** (A) An initial CT scan was normal in this 60-year-old man with unexplained progressive neurologic deficits leading to coma. (B) A CT scan obtained 4 months later shows a generalized loss of neurons with enlargement of the lateral ventricles and the development of prominent sulci.


7. Handel SF, Klein WC, Kim YU: Intracranial epidural abscess. *Radiology* 111:117–120, 1974.
8. Lott T, El Gammal T, Da Silva R, et al: Evaluation of the brain and epidural abscess by CT. *Radiology* 122:371–376, 1977.
9. Atinz ER, Cooper RD: Several early angiographic findings in brain abscess including "the ripple sign." *Radiology* 90: 735–739, 1968.
10. Nielsen H, Gyldenstadt C: CT in the diagnosis of cerebral abscess. *Neuroradiology* 12:207–217, 1977.
11. Stevens EA, Norman D, Kramer RA, et al: CT brain scanning in intraparenchymal pyogenic abscesses. *AJR* 130:111–114, 1978.
12. Robertheram EB, Kessler LA: Use of computerized tomography in nonsurgical management of brain abscess. *Arch Neurol* 36:25–26, 1979.
13. Leo JS, Weiner RL, Lin JP, et al: CT in herpes simplex encephalitis. *Surg Neurol* 10:313–317, 1978.
14. Davis JM, Davis KR, Kleinman GM et al: CT of herpes simplex encephalitis with clinicopathological correlation. *Radiology* 129:409–417, 1978.
15. Carroll BA, Lane B, Norman D, et al: Diagnosis of progressive multifocal leukoencephalopathy by computed tomography. *Radiology* 122:137–141, 1977.
16. Bosch EP, Cancilla PA, Cornell SH: Computerized tomography in progressive multifocal leukoencephalopathy. *Arch Neurol* 33:216, 1976.
17. Rao KCVG, Brennan TG, Garcia JH: Computed tomography in the diagnosis of Creutzfeldt-Jakob disease. *J Comput Assist Tomogr* 1:211–215, 1977.
18. Lee SH: Infectious diseases, in Lee SH, Rao KCVG (eds): *Cranial Computed Tomography*. New York, McGraw-Hill, 1983, pp. 505–546.

## Chapter Seventy-One

# Developmental and Other Congenital Abnormalities of the Nervous System

## MALFORMATION OF THE CRANIUM

By the age of 5 years, the brain (to which the skull always accommodates) approximates adult size and the sutures are so firmly closed that disease acquired later in life will have relatively little effect on the skull. Therefore, alterations in the size and shape of the head almost invariably have their origin during the intrauterine period or early childhood.

Enlargement of the head (macrocephaly) may reflect enlargement of the brain (either in normal individuals or in association with cerebral gigantism) or hydrocephalus. The latter can be easily identified by the demonstration of marked dilatation of the ventricular system and intracranial cerebrospinal fluid spaces. An abnormally small head (microcephaly) is related to a lack of brain growth or to a destructive lesion of the brain early in life. In these individuals the bones of the calvarium are often of increased thickness, and the facial bones, although of normal size, appear disproportionately large. In cerebral hemiatrophy or unilateral hypoplasia of the brain (Davidoff-Dyke syndrome), bony changes on the affected side compensate for the decreased brain volume. The petrous pyramid and the orbit on the atrophic side are elevated, and the vault is flattened and thickened. There may be unilateral enlargement of the frontal sinus and mastoid air cells, decreased convolutional markings, and deviation of the crista galli to the affected side.

An unusual shape of the head is usually due to premature closure of one or more cranial sutures (craniostenosis). The precise appearance of the skull depends on the location, degree, and age of onset of the premature closure. Restriction of the rapid growth of the normal brain and the resultant elevation of intracranial pressure produces a marked increase in the convolutional pattern of the cranial bones (beaten brass appearance).

Early fusion of the sagittal suture (dolichocephaly) causes the head to be long and narrow with a prominent brow and occiput. Premature closure of the coronal sutures produces a short anteroposterior diameter of the skull, with the head appearing more wide than long (brachycephaly). In plagiocephaly, premature fusion of a single coronal suture causes asymmetric flattening of the skull. The lesser sphenoid wing and the orbital roof are drawn upward so that the orbit has an elliptical appearance on the frontal view ("harlequin" eye). Premature closure of the lambdoid and coronal sutures, with the sagittal suture remaining open, permits the skull to expand in a vertical direction and results in a long verticobasal diameter and an extremely short anteroposterior diameter, with shallow orbits and bulging eyes (turricephaly). In Apert's syndrome, premature closure of the coronal sutures is associated with syndactyly of the hands and feet. In Crouzon's disease, bilateral coronal suture synostosis combined with hypertelorism and hypoplasia of the maxillas results in a typical froglike facial appearance.

The Dandy-Walker syndrome refers to congenital atresia of the foramina of Luschka and Magendie that causes a marked obstructive hydrocephalus in infants and children. The fourth ventricle is enlarged out of proportion to the remainder of the ventricular system. Indeed, the fourth ventricle is effectively replaced by a

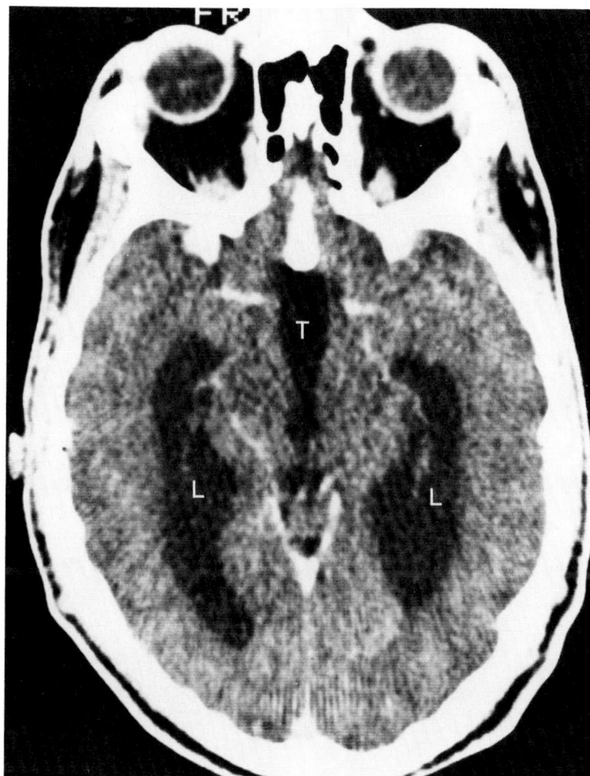


**Figure 71.1. Hydrocephalus.** A CT scan shows dilatation of the lateral (L) and third (T) ventricles due to aqueductal stenosis. The patient's symptoms of headaches and papilledema resolved after ventricular shunting.

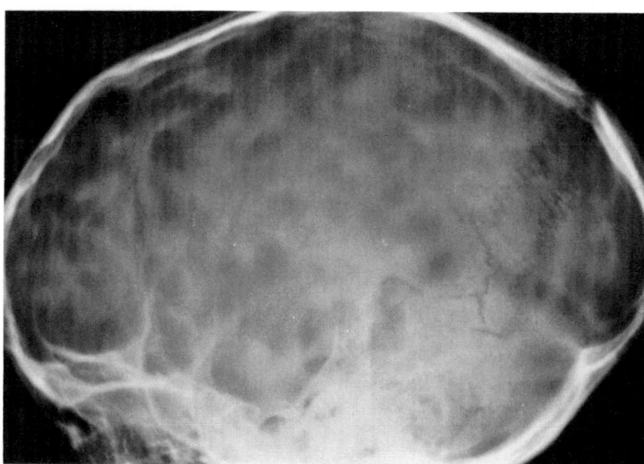

**Figure 71.3. Beaten brass appearance in craniostenosis.** A lateral view of the skull in a patient with premature closure of several sutures demonstrates prominence of the convolutional markings.

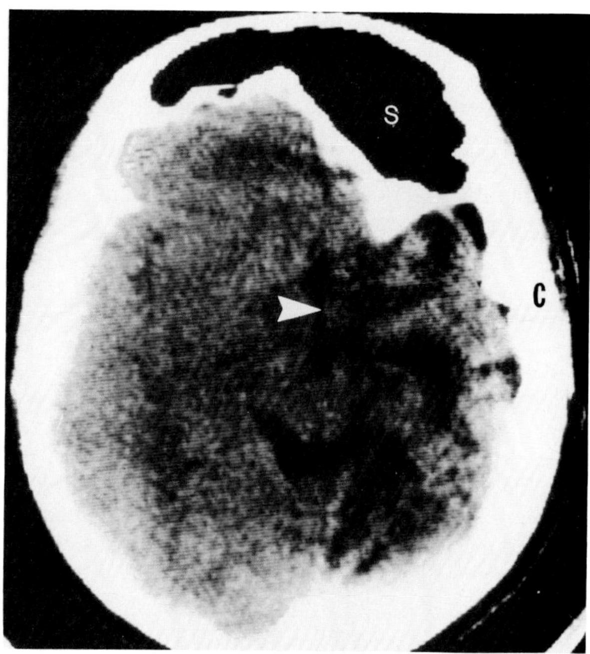


**Figure 71.2. Cerebral hemiatrophy (Davidoff-Dyke syndrome).** A CT scan of a 5-year-old boy who had intrauterine difficulties demonstrates extensive loss of brain volume in the left hemisphere. There is also enlargement of the left hemicalvarium (C), enlargement of the left frontal sinus (S), and a shift of midline structures such as the third ventricle (arrowhead) from right to left. The low density in the remainder of the hemisphere represents encephalomalacia.

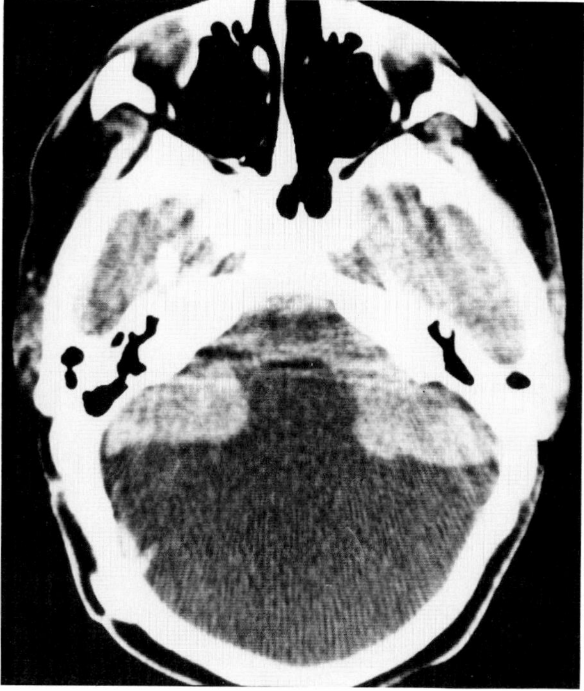

**Figure 71.4. Dandy-Walker syndrome.** A CT scan shows a huge low-density cyst that occupies most of the enlarged posterior fossa and represents an extension of the dilated fourth ventricle.

symmetrically enlarged midline cyst in the posterior fossa with hypoplasia of the inferior vermis. On computed tomography (CT), the Dandy-Walker cyst appears as a large, low-density mass occupying most of the posterior fossa. A thin rim of cerebellar tissue is seen anteriorly. The relative obstruction of the aqueduct can be demonstrated by introducing a small amount of dilute contrast material into the ventricles.[1–6]

## Basilar Impression

Basilar impression is a rare maldevelopment in which either the base of the skull is flattened or the occiput and upper cervical spine are invaginated into the posterior fossa. It may occur in association with anomalous development of the craniovertebral junction (occipitalization of the atlas) or be secondary to disease processes that produce a softening of the bones at the base of the skull (Paget's disease, rickets, osteomalacia, hyperparathyroidism, osteogenesis imperfecta). As a result of occipitalization, the odontoid process may compress the brainstem. In the more severe cases of basilar impression, the medial ends of the petrous pyramids are angled upward and the clivus is oriented more horizontally than normal (platybasia).

Basilar invagination is assessed on plain skull films by evaluating the relation of the tip of the odontoid process to two lines: Chamberlain's line (posterior margin of the hard palate to the posterior margin of the foramen magnum) and McGregor's line (posterosuperior margin of the hard palate to the lowermost point on the midline occipital curve).[7–9]

## Arnold-Chiari Malformation

In the Arnold-Chiari malformation, the medulla and inferior-posterior portions of the cerebellar hemispheres project caudally through the foramen magnum, often to the level of the second cervical vertebra. Other neurologic anomalies (spinal meningocele or myelomeningocele, basilar impression and platybasia, deformities of the cervical spine and cervico-occipital junction) are often associated. Occlusion of the foramina of Luschka and Magendie leads to obstructive hydrocephalus, which dominates the clinical picture in infants.

Plain skull radiographs demonstrate the typical appearance of long-standing increased intracranial pressure with prominence of the convolutional markings, enlargement of the sella, and thinning of the calvarial bones. In addition, the foramen magnum may be enlarged and the upper cervical vertebrae may show an expanded spinal canal due to the downward shift of the medulla and the cerebellar tonsils.

Myelography can demonstrate the downward displacement of the cerebellar tonsils, which project into and narrow the spinal subarachnoid space below the level of the foramen magnum. CT shows nonvisualization or pathologic downward displacement of the fourth ventricle, enlargement of the massa intermedia, dilatation of the lateral ventricles (representing hydrocephalus), hypoplasia of the falx and tentorium, and multiple associated bony abnormalies.[10–14]

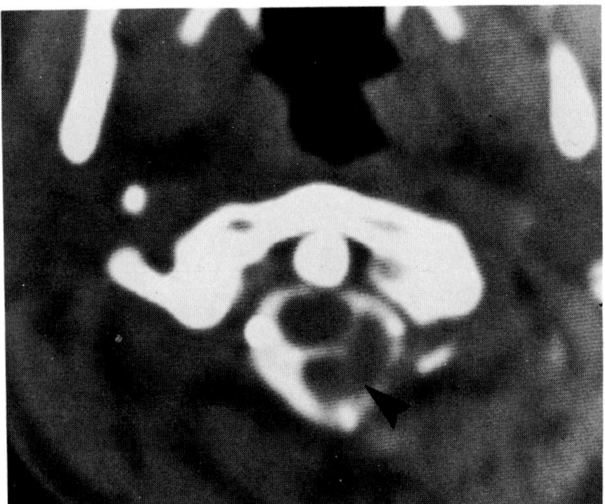

A

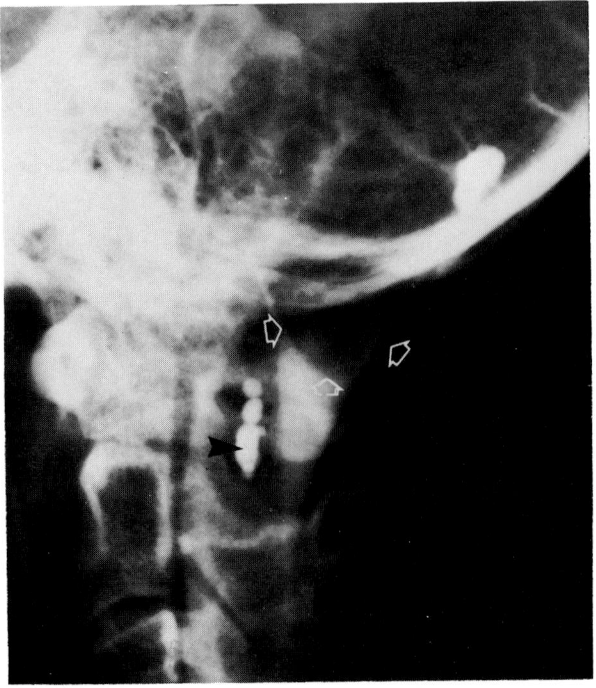

B

**Figure 71.5. Arnold-Chiari malformation.** (A) A close-up view of a CT scan following the administration of intrathecal metrizamide shows downward herniation of the cerebellar tonsil (arrowhead). (B) A lateral view from a metrizamide myelogram demonstrates the herniated cerebellar tonsil (open arrows). (From Rao and Harwood-Nash,[25] with permission from the publisher.)

## ABNORMALITIES OF THE SPINE

Congenital anomalies of the spine are frequent, particularly in the lumbosacral area. The most common anomaly is spina bifida occulta, an asymptomatic and insignificant cleft defect of the neural arch in the region of the spinous process that occurs in about 20 percent of the population. Spina bifida occulta tends to affect the lower lumbar and upper sacral spine but may occur in the lower cervical and upper thoracic regions. Large lumbar or cervical defects may be accompanied by herniation of the meninges (meningocele) or of the meninges and a portion of the spinal cord or nerve roots (meningomyelocele). These lesions can cause a neurogenic bladder and peripheral neuropathy and are associated with large bony defects, absence of the laminae, and an increased interpedicular distance. The meningocele itself presents as a soft tissue mass posterior to the spine. Myelography can demonstrate the presence of the spinal cord or nerve roots within the herniated sac.

Fusion of two or more vertebral bodies with absence of the intervening disk space produces block vertebrae. The posterior elements (laminae, spinous processes, articular facets) may also be fused. Maldevelopment and fusion of two or more cervical vertebrae may be part of the Klippel-Feil syndrome, which results in a patient with

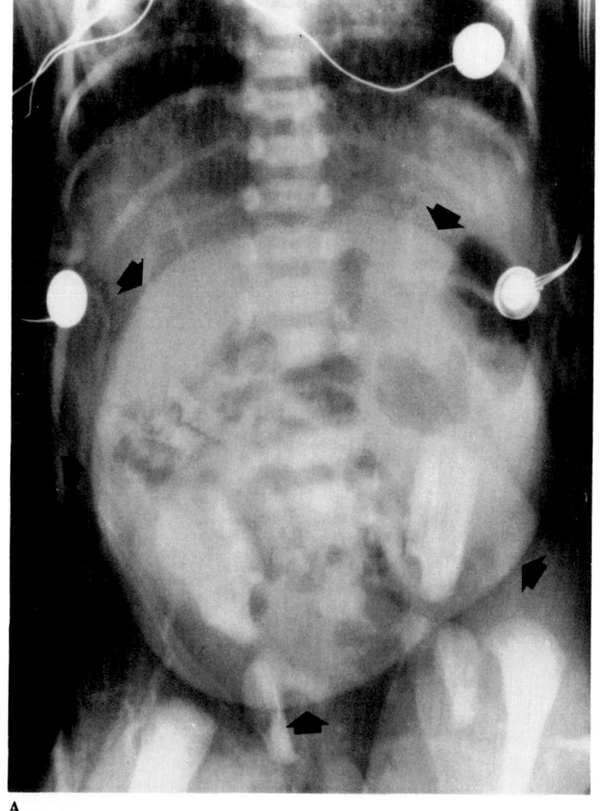

A

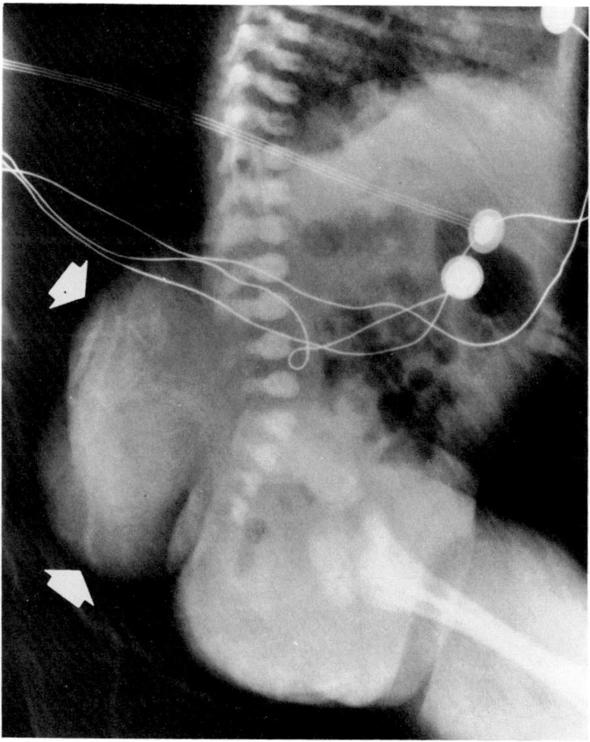

B

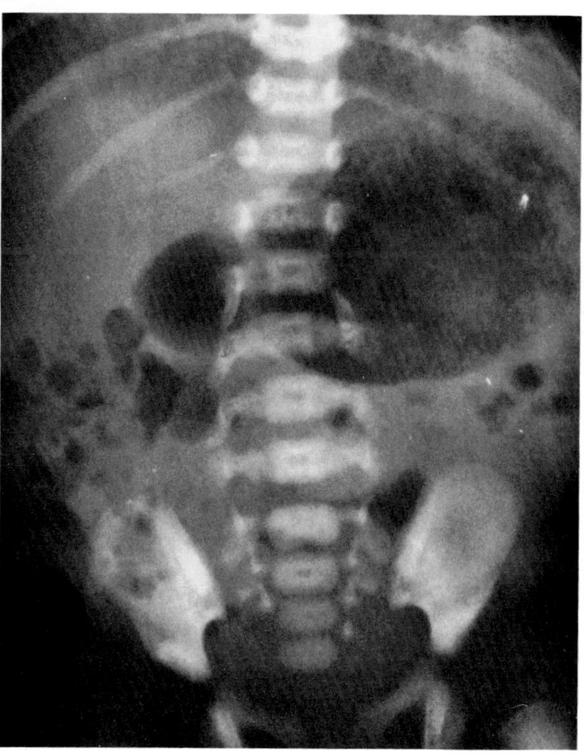

C

**Figure 71.6. Meningomyelocele.** (A) Frontal and (B) lateral views demonstrate a large soft tissue mass (arrows) situated posterior to the spine. Note the absence of the posterior elements in the lower lumbar and sacral regions. (C) A frontal view of the abdomen in another patient shows the markedly increased interpedicular distance of the lumbar vertebrae associated with a large meningocele.

a short neck of limited mobility. The height of the fused vertebral bodies is usually equal to the combined height of these vertebrae and their intervening disk spaces in the normal spine. Patients with the Klippel-Feil syndrome often have other developmental anomalies of the central nervous system, cardiovascular system, and genitourinary tract.[15–18]

## PHAKOMATOSES

The phakomatoses are a group of conditions in which neurologic abnormalities are combined with congenital defects of the skin, retina, and other organs.

### Neurofibromatosis (Von Recklinghausen's Disease)

Neurofibromatosis is an inherited condition in which spots of increased skin pigmentation are combined with multiple neurofibromas and bone dysplasias. The characteristic pigmented areas (café au lait spots) are irregular in shape, vary in size from a few millimeters to several centimeters, and are most prominent over the trunk, in the axilla, and about the pelvis.

A broad spectrum of radiographic abnormalities is seen in patients with neurofibromatosis. They may be caused either by erosions from an adjacent soft tissue neurofibroma or by a bone dysplasia in the absence of discrete neurofibromas. In the skeleton, dysplasia of vertebral bodies often causes a severe kyphoscoliosis. Posterior scalloping of one or more vertebral bodies may be due to dural ectasia. Pressure caused by neurofibromas of the intercostal nerves typically causes thinning and notching of the superior and inferior surfaces of multiple ribs ("twisted ribbon" ribs), an appearance that may sim-

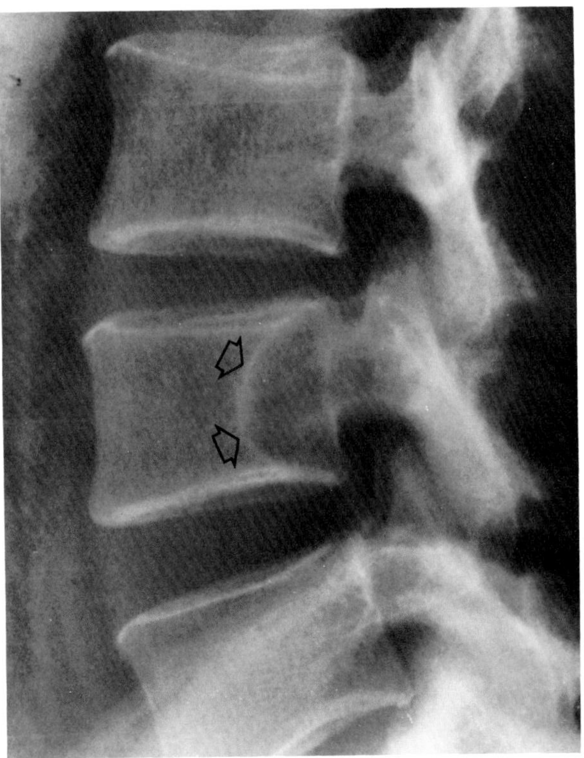

**Figure 71.7. Neurofibromatosis.** Well-circumscribed posterior scalloping (arrows) of a lumbar vertebral body.

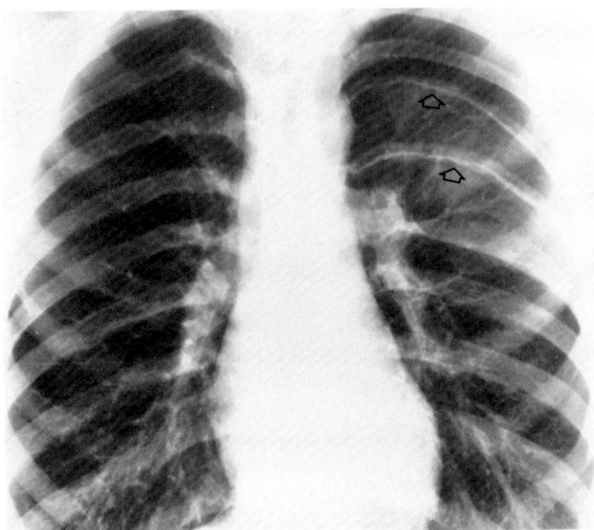

**Figure 71.8. Neurofibromatosis.** Thinning and notching of several of the upper left ribs (arrows).

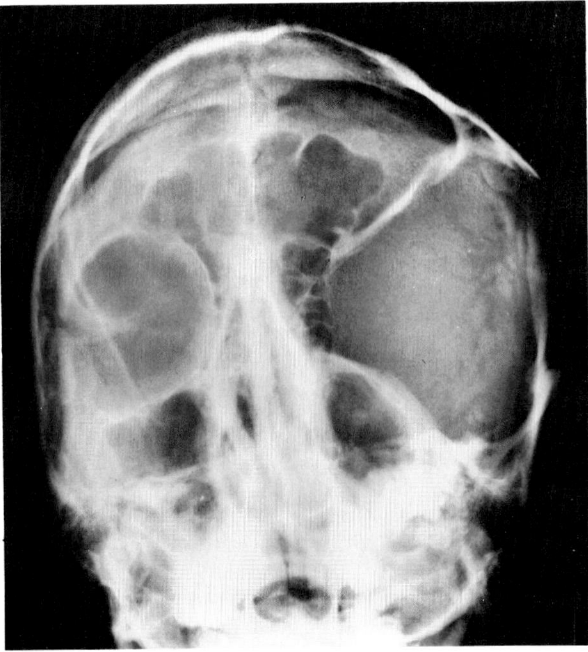

**Figure 71.9. Neurofibromatosis.** Severe orbital dysplasia with virtual absence of the posterolateral walls of the orbit.

ulate the effects of intercostal artery collaterals in coarctation of the aorta. Neurofibromas arising within bone may produce cystlike intramedullary lytic lesions. In some cases there may be localized enlargement of all or a part of one extremity (focal giantism), with the underlying bones appearing normal except for their increased size. A characteristic complication of neurofibromatosis is pseudoarthrosis, which reflects a fracture that failed to heal and may result in a bizarre deformity. Large collections of neurofibromatous masses may cause marked thickening of the soft tissues.

In the skull, a common anomaly in neurofibromatosis is orbital dysplasia, in which unilateral absence of a large part of the greater wing of the sphenoid and hypoplasia and elevation of the lesser wing result in a markedly widened superior orbital fissure. The temporal lobe may protrude through this large defect in the posterolateral wall of the orbit, resulting in pulsating exophthalmos.

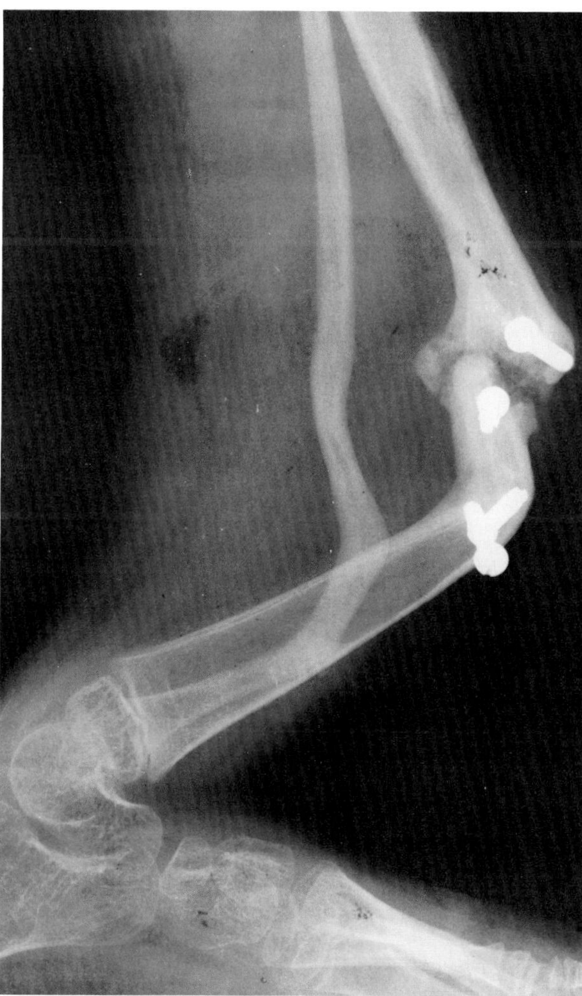

**Figure 71.10. Neurofibromatosis.** Post-traumatic pseudoarthrosis of the midshaft of the tibia. Note the severe disuse osteoporosis of the bones of the ankle and the ribbonlike shape of the lower fibula.

This malformation is a pure bone dysplasia and is associated with facial neurofibromas.

Other abnormalities less frequently associated with neurofibromatosis include tumors of the acoustic nerve, lateral intrathoracic meningoceles, dumbbell-shaped neurofibromas of spinal nerve roots (with intraspinal and extraspinal extension), absence or deformity of the clinoid processes, pulmonary interstitial fibrosis, pheochromocytomas and thyroid carcinomas, and coarctation of the abdominal aorta.[19,20]

## Tuberous Sclerosis (Bourneville's Disease)

Tuberous sclerosis is an inherited disorder manifested by the clinical triad of convulsive seizures, mental deficiency, and adenoma sebaceum. The disease is characterized by hamartomas that almost always involve the brain and may affect virtually any organ.

The most common site of intracranial hamartomas is the cerebrum, although the cerebellum, the medulla, and the spinal cord may be involved. These hamartomas vary in size and number. The vast majority lie adjacent to the cerebrospinal fluid pathways, predominantly as subependymal nodules in the lateral ventricles. On plain skull radiographs, the intracranial hamartomas appear as clusters of calcified nodules that primarily develop in the walls of the lateral ventricles. Even when uncalcified, the subependymal nodules can be detected on CT scans, in which they appear as rounded nodules of varying sizes projecting into the ventricular cavities. Large lesions may obstruct the ventricular foramina or the aqueduct of Sylvius, producing a pattern of obstructive hydrocephalus and radiographic signs of increased intracranial pressure.

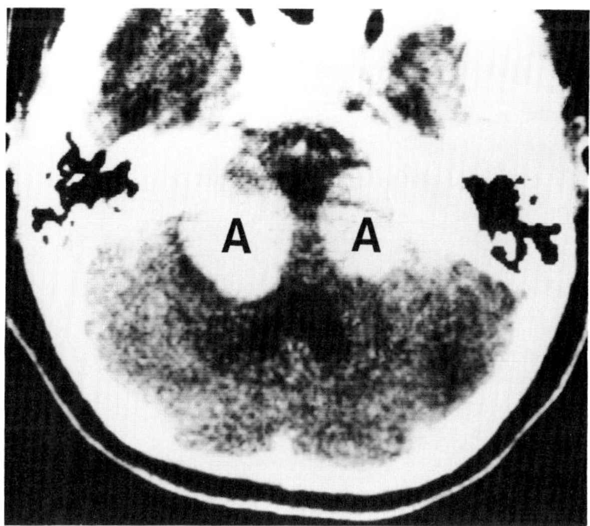


**Figure 71.11. Neurofibromatosis.** A CT scan shows bilateral acoustic neuromas (A) in a young girl with progressive bilateral sensorineural hearing loss.

The most common skeletal finding associated with tuberous sclerosis is the occurrence of dense sclerotic foci, which typically affect the bones of the cranial vault and the pedicles and posterior portions of the vertebral bodies. In the hands and feet, characteristic abnormalities are wavy periosteal new bone formation along the shafts of the metatarsals and metacarpals and cystlike changes in the phalanges.

Renal angiomyolipomas occur in about half the patients with lesions of tuberous sclerosis in the brain. These hamartomatous tumors are hypervascular masses that may contain sufficient fat to permit an accurate diagnosis on CT scans or even on plain abdominal radiographs. In patients with tuberous sclerosis, hamartomas are often multiple and may produce a bizarre appearance simulating adult polycystic kidney disease (see the section "Benign Renal Tumors" in Chapter 41).[21–24]

## Sturge-Weber Disease (Encephalotrigeminal Syndrome)

The Sturge-Weber syndrome typically consists of a port-wine nevus of the face (along the first branch of the trigeminal nerve) with leptomeningeal angiomatosis, mental retardation, seizure disorders, and hemiatrophy and hemiparesis. The characteristic radiographic appearance is that of undulating plaques of calcification within the brain cortex. The calcification appears to follow the cerebral convolutions and most often develops in the parietal area. Hemiatrophy leads to elevation of

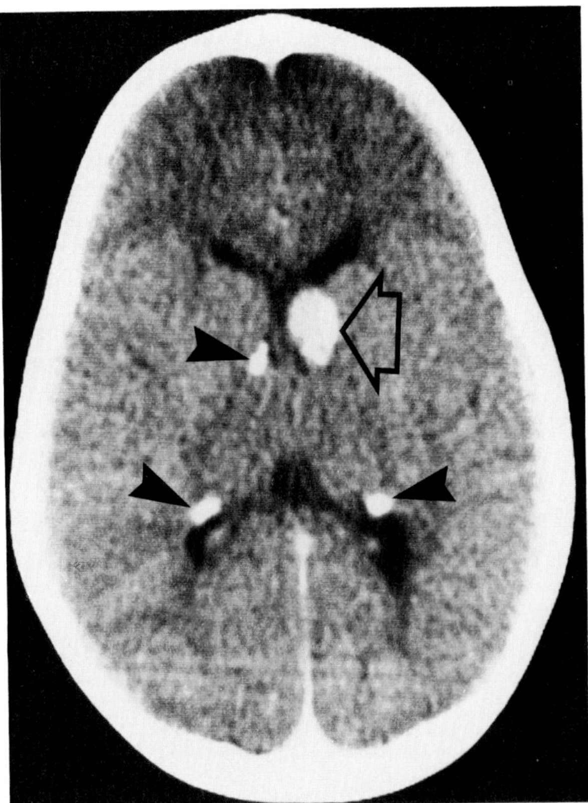

**Figure 71.12. Tuberous sclerosis.** A CT scan shows multiple calcified hamartomas (solid black arrowheads) lying along the ependymal surface of the ventricles. The open arrow points to a giant cell astrocytoma at the foramen of Monro.

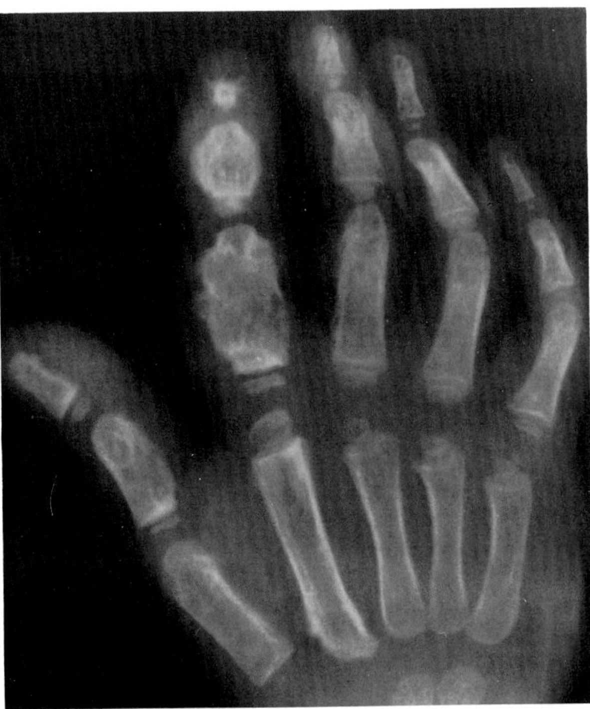

**Figure 71.13. Tuberous sclerosis.** There is wavy periosteal new bone formation about the proximal and middle phalanges of the second digit. Note the cystlike expansion of these bones. Periosteal new bone formation is also seen along the shaft of the second metacarpal.

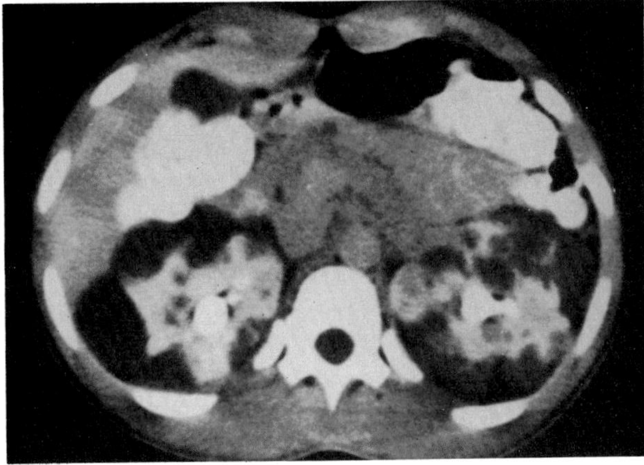

**Figure 71.14. Multiple renal hamartomas in tuberous sclerosis.** A CT scan shows innumerable low-density, cystlike masses bilaterally.

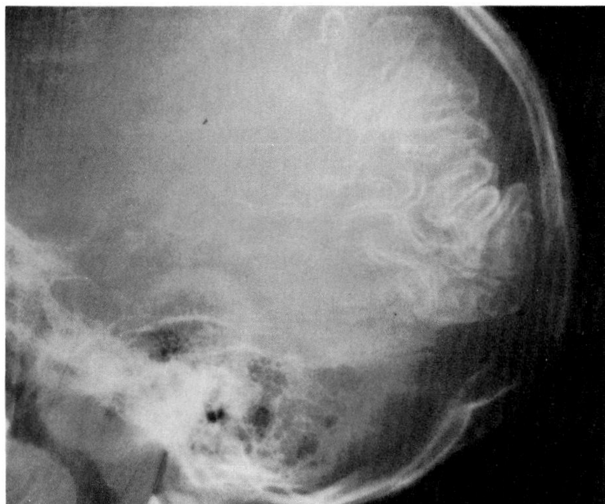

**Figure 71.15. Sturge-Weber disease.** A lateral view of the posterior portion of the skull demonstrates typical undulating plaques of cortical calcification, which appear to follow the cerebral convolutions.

the base of the skull and enlargement and increased aeration of the mastoid air cells on the side of the lesion.

CT usually shows that the calcification involves the periphery of the cerebral hemispheres and has a gyral pattern. Prominence of the ipsilateral sulci with mild to moderate enlargement of the lateral ventricle reflects some degree of cerebral atrophy.[25]

## Cerebelloretinal Hemangioblastomatosis (Von Hippel-Lindau Syndrome)

This syndrome consists of a vascular malformation of the retina, usually multiple capillary angiomas, and one or more slow-growing hemangioblastomas of the cerebellum. Patients with this syndrome often have a bizarre constellation of abnormalities, including angiomas and cysts of the liver, pancreas, and kidneys, renal tumors, and pheochromocytomas.[25]

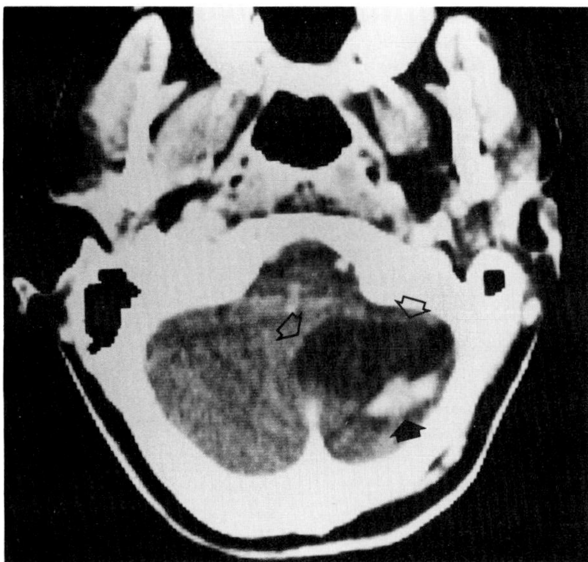

A

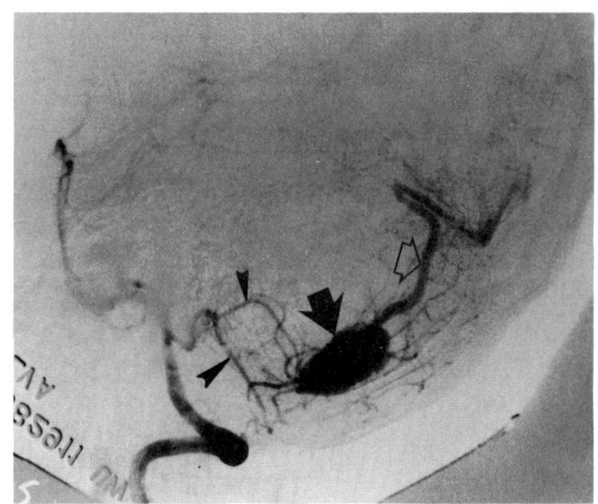

B

**Figure 71.16. Hemangioblastoma in von Hippel-Lindau syndrome.** (A) A CT scan shows a cystic lesion (open arrows) with an enhancing nodule (closed arrow) in the left cerebellar hemisphere. (B) A vertebral arteriogram shows the vascular nodule (solid arrow) of the tumor with multiple feeding arteries (black arrowheads) and a large draining vein (open arrow).

### REFERENCES


1. Tod PO, Yelland JDN: Craniostenosis. *Clin Radiol* 22:472–484, 1971.
2. Hope JW, Spitz EB, Slade HW: The early recognition of premature cranial synostosis. *Radiology* 65:183–193, 1955.
3. Schwarz E: The skull in skeletal dysplasias. *AJR* 89:928–937, 1963.
4. Seelenfreund M, Gartner S: Acrocephalosyndactyly (Apert's syndrome). *Arch Ophthal* 78:8–11, 1967.
5. Juhl JH, Wesenberg RL: Radiological findings in congenital and acquired occlusions of foramina of Magendie and Luschka. *Radiology* 86:801–813, 1966.
6. Naidich TP: Primary tumors and other masses of the cerebellum and fourth ventricle: Differential diagnosis by computed tomography. *Neuroradiology* 14:153–174, 1977.
7. McRae DL: Occipitalization of the atlas. *AJR* 70:23–46, 1953.
8. Peyton WT, Peterson HO: Congenital deformities in the region of the foramen magnum: Basilar impression. *Radiology* 38:131–137, 1942.
9. Lusted LB, Keats T: *Atlas of Roentgenographic Measurement.* Chicago, Year Book Medical Publishers, 1978.
10. Naidich TP, Pudlowski MR, Naidich JB, et al: Computed tomographic signs of Chiari II malformation: I. Skull and dural portions. *Radiology* 134:65–71, 1980.
11. Naidich JTP, Pudlowski MR, Naidich JB: Computed tomographic signs of the Chiari II malformation: II. Midbrain and cerebellum. *Radiology* 134:391–398, 1980.
12. Naidich TP, Pudlowski MR, Naidich JB: Computed tomographic signs of the Chiari II malformation: III. Ventricles and cisterns. *Radiology* 134:657–663, 1980.
13. Davies HW: Radiological changes associated with Arnold-Chiari malformation. *Br J Radiol* 40:262–269, 1967.
14. Kruyff E, Jeffs R: Skull abnormalities associated with the Arnold-Chiari malformation. *Acta Radiol (Diagn)* 5:9–15, 1966.

15. Brown MW, Templeton AW, Hodges FW: The incidence of acquired and congenital fusions in the cervical spine. *AJR* 92:1255–1259, 1964.
16. Shoul MI, Ritvo M: Clinical and roentgenological manifestations of Klippel-Feil syndrome. *AJR* 68:369–385, 1952.
17. Schwidde JT: Spina bifida: Survey of two hundred and twenty-five encephaloceles, meningoceles, and myelomeningoceles. *Am J Dis Child* 84:35–51, 1952.
18. Vasant C, Darab KD: Meningoceles and meningomyeloceles (ectopic spinal canal). *J Neurol Neurosurg Psychiatry* 33: 251–262, 1970.
19. Holt JF: Neurofibromatosis in children. *AJR* 130:615–639, 1978.
20. Hunt JC, Pugh DG: Skeletal lesions in neurofibromatosis. *Radiology* 76:1–20, 1961.
21. Hartman DS, Goldman SM, Friedman AC, et al: Angiomyolipoma: Ultrasonic-pathologic correlation. *Radiology* 139: 451–458, 1981.
22. Teplick JG: Tuberous sclerosis. *Radiology* 93:53–59, 1969.
23. Green GJ: The radiology of tuberous sclerosis. *Clin Radiol* 19:135–144, 1968.
24. Lee BCP, Gawler NJ: Tuberous sclerosis: Comparison of computed tomography and conventional neuroradiology. *Radiology* 127:403–407, 1978.
25. Rao KCVG, Harwood-Nash DC: Craniocerebral anomalies, in Lee FH, Rao KCVG (eds): *Cranial Computed Tomography*. New York, McGraw-Hill, 1983.

## Chapter Seventy-Two

# Miscellaneous Disorders of the Central Nervous System

## EPILEPSY AND CONVULSIVE DISORDERS

In the patient presenting with an acute seizure, the initial effort should be directed toward stabilizing the patient (securing adequate ventilation and perfusion) and stopping the seizure. Subsequently, a careful history, physical examination, and appropriate laboratory studies should be performed to exclude reversible chemical causes of seizures such as hypoglycemia, hypo- or hypernatremia, and hypo- or hypercalcemia. If the patient does not respond to routine anticonvulsive treatment, computed tomography (CT) may be indicated to search for possible surgical causes of an acute seizure disorder (e.g., subdural hematoma, intracranial hematoma). These conditions can be adequately assessed with a noncontrast CT scan; plain skull radiographs are not required.

Whenever possible, the radiographic evaluation of a patient with a seizure disorder should be done when the patient is clinically stable. The appropriate procedure is a contrast-enhanced CT scan to search for an unsuspected brain tumor or arteriovenous malformation. Since most seizure disorders are due to small areas of cortical brain injury secondary to trauma or infarction, CT scans are usually normal. If the CT scan is normal, no further radiographic examination is indicated.[1]

## DEMYELINATING DISEASES

### Multiple Sclerosis

Multiple sclerosis is a demyelinating disorder that presents as recurrent attacks of focal neurologic deficits that primarily involve the spinal cord, optic nerves, and central white matter of the brain. The disease has a peak incidence between 20 and 40 years of age and a clinical course characterized by multiple relapses and remissions.

The CT appearance of white matter disease in multiple sclerosis depends on whether the patient is in remission or in relapse. Old inactive disease typically shows multiple, rather well-defined areas of decreased attenuation situated in the deep white matter and the periventricular regions. Similar plaques of multiple sclerosis may be demonstrated in other white matter areas, such as the cerebellum and spinal cord. In the acute phase, CT following the intravenous administration of contrast material demonstrates a mixture of nonenhancing focal areas of decreased density (representing old areas of demyelination) and enhancing regions that represent active foci. Nonspecific enhancement patterns have been described as homogeneous, diffuse, peripheral, edge, ring, and central. In patients with prolonged disease, atrophy of

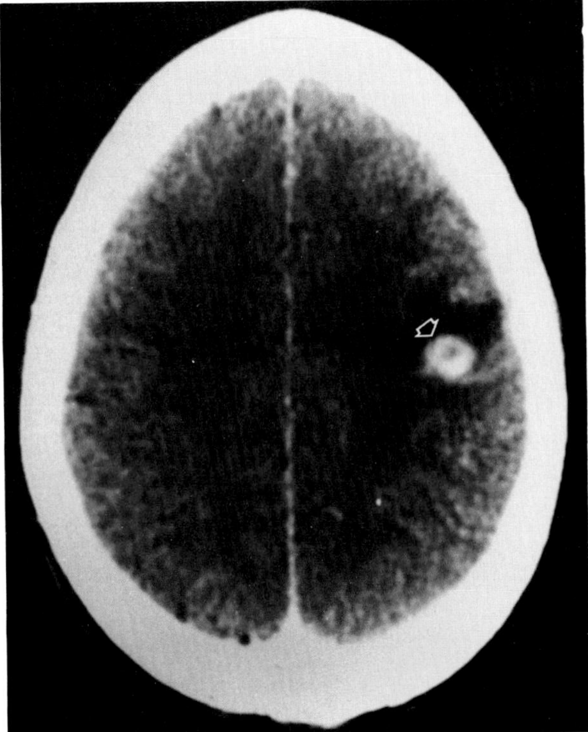

**Figure 72.1. Seizure disorder due to an astrocytoma.** A CT scan following the intravenous injection of contrast material in a 27-year-old woman with focal right arm seizures shows the enhancing left cerebral tumor surrounded by low-density edema (arrow). As in this patient, most seizure disorders can be evaluated with a contrast-enhanced CT scan.

the cerebral hemispheres and the cerebellum, especially the vermis, is common.

CT is also of value in excluding other entities, such as extra-axial tumors around the tentorium and the foramen magnum, that can cause neurologic findings simulating multiple sclerosis.[2–4]

## Acute Disseminated Encephalomyelitis

Acute disseminated encephalomyelitis is a widespread central nervous system inflammatory disorder characterized by the abrupt onset of symptoms and signs indicating damage that primarily involves the white matter of the brain or the spinal cord. The clinical course is varied; the disease may be fatal, completely reversible, or produce permanent neurologic disability. Major causes of acute disseminated encephalomyelitis include viral infections (measles, vaccinia, varicella), allergies (postvaccination), and nonspecific respiratory infections. A fulminating, nearly always fatal, form of acute disseminated encephalomyelitis characterized by multiple punctate intracerebral hemorrhages may reflect an immune complex disorder.

In the acute phase, CT demonstrates diffuse cerebral edema. In chronic disease, foci of demyelination develop in the white matter of the cerebrum and also may involve the cerebral cortex, cerebellum, basal ganglia, brainstem, and spinal cord.[1,5,13]

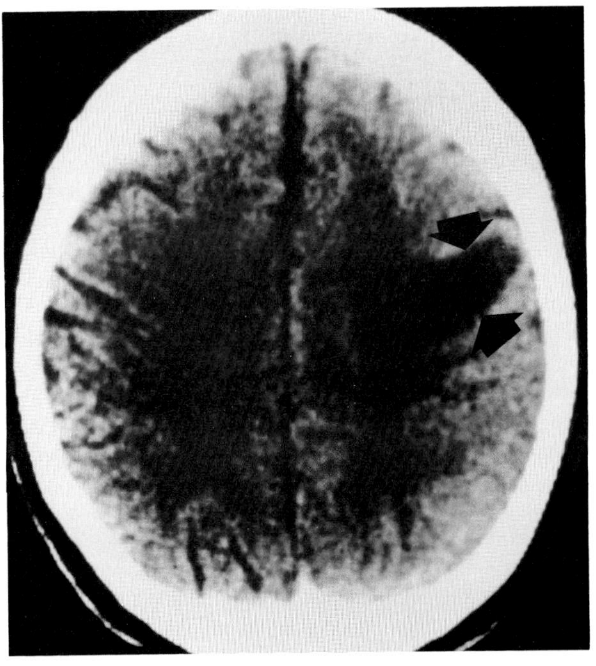

A

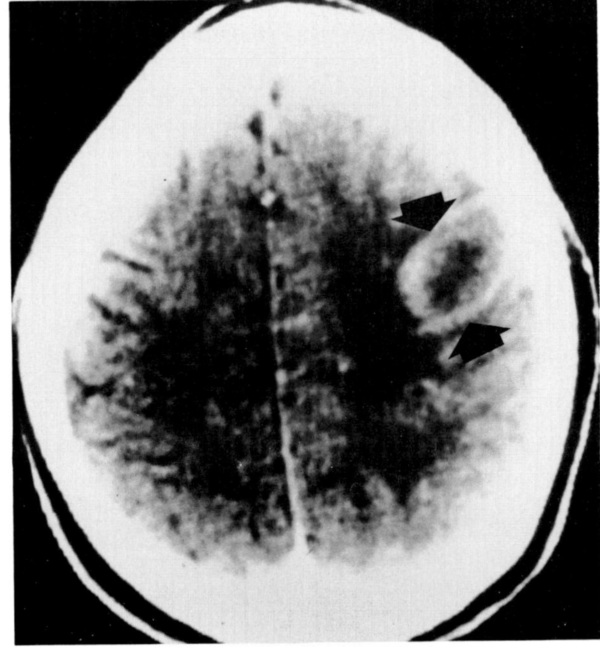

B

**Figure 72.2. Multiple sclerosis.** (A) A noncontrast CT scan shows a low-density plaque (black arrows) in the left centrum semiovale. (B) Following the intravenous injection of contrast material, there is a peripheral pattern of enhancement about the lesion (black arrows).

## DEGENERATIVE DISEASES

### Normal Aging

During normal aging, a gradual loss of neurons results in enlargement of the ventricular system and sulci. Demyelination, which is also a part of normal aging, leads to the development of low density in the periventricular regions.

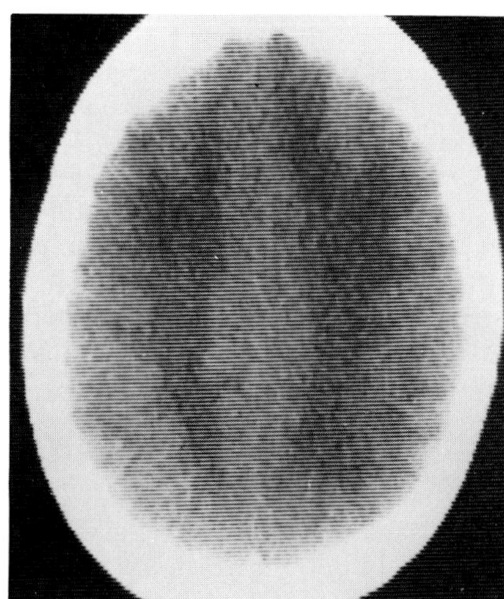

**Figure 72.3. Acute disseminated encephalomyelitis.** A CT scan shows diffuse low-density zones in both cerebral hemispheres. (From Fernandez and Kishore,[13] with permission from the publisher.)

### Alzheimer's Disease

Alzheimer's disease (presenile dementia) is a diffuse form of progressive cerebral atrophy that develops at an earlier age than the senile period. CT demonstrates nonspecific findings of cerebral atrophy, including symmetrically enlarged ventricles with prominence of the cortical sulci.[6]

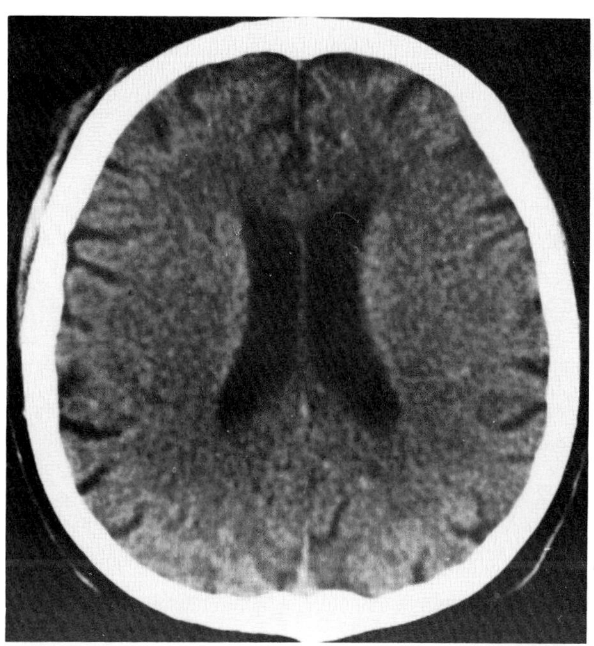

A

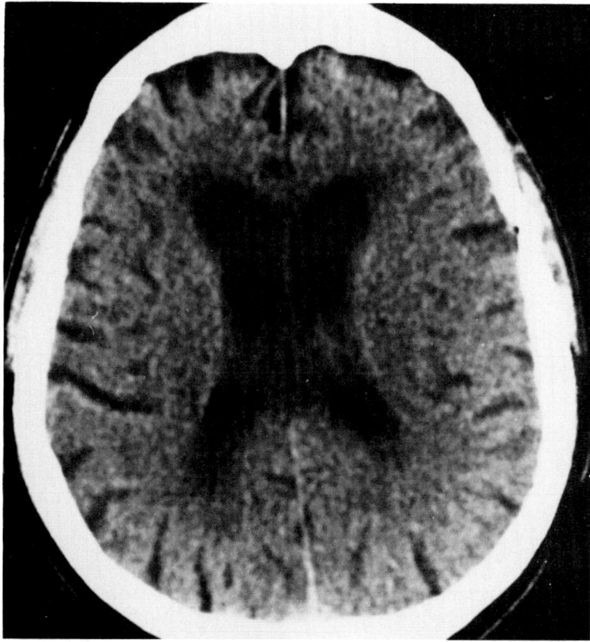

B

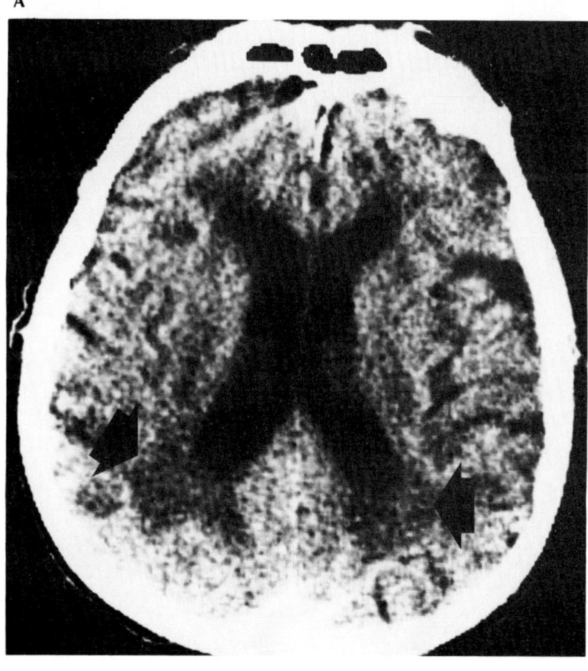

C

**Figure 72.4. Normal aging.** A series of CT scans obtained at the level of the lateral ventricles in (A) a 60-year-old, (B) a 70-year-old, and (C) an 85-year-old show progressive enlargement of the lateral ventricles and the development of prominent sulci. Figure C demonstrates the development of periventricular low-density areas (black arrows) that represent the normal demyelinating process of old age.

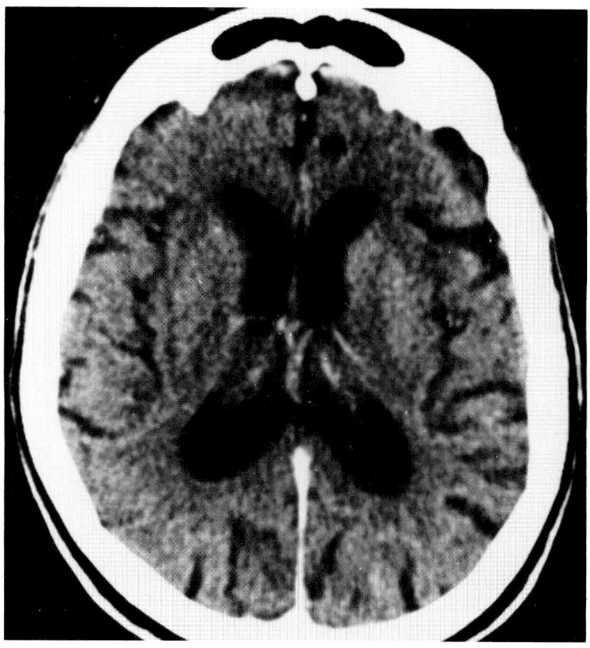

**Figure 72.5. Alzheimer's disease.** A noncontrast CT scan of a 56-year-old woman with progressive dementia shows enlargement of the ventricular system and sulci. Patients with this condition have a broad spectrum of CT findings, ranging from a normal examination to marked ventricular enlargement and prominence of the sulci.

## Pick's Disease

This rare disorder is characterized by lobar atrophy, in contrast to the diffuse atrophy of Alzheimer's disease. Because the atrophic process primarily involves the temporal and frontal lobes, dilatation of the frontal and temporal horns is especially marked. Substantial cortical atrophy and low density of white matter may be seen.[7]

## Huntington's Disease

Huntington's disease is an inherited condition (autosomal dominant) that predominantly involves males and presents in the early to middle adult years with dementia and typical choreiform movements. The pathologic hallmark of Huntington's disease is atrophy of the caudate nucleus and putamen, which produces the typical CT appearance of focal dilatation of the frontal horns and an increased bicaudate diameter of the ventricles. Generalized enlargement of the ventricles and dilatation of the cortical sulci can also occur.[14]

## Metachromatic Leukodystrophy

Metachromatic leukodystrophy is a genetically determined metabolic disorder in which a deficiency of the enzyme arylsulfatase A permits metachromatic sulfatides

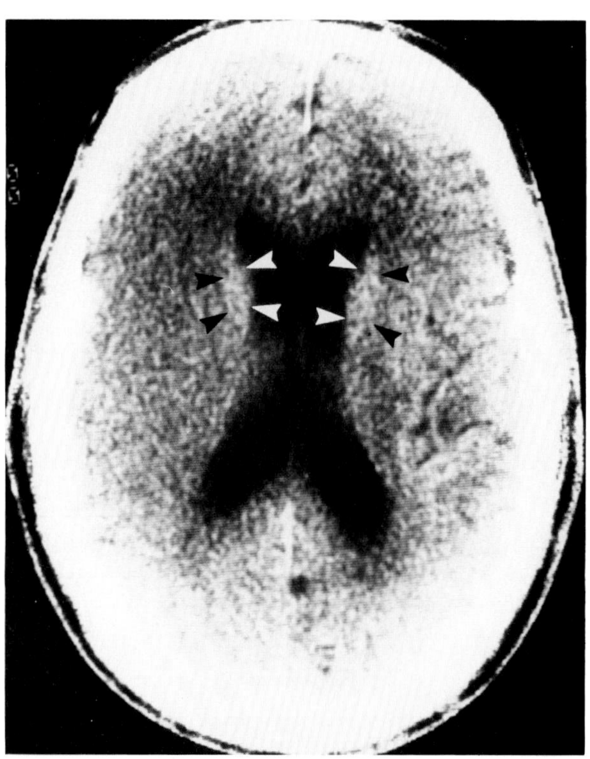

A

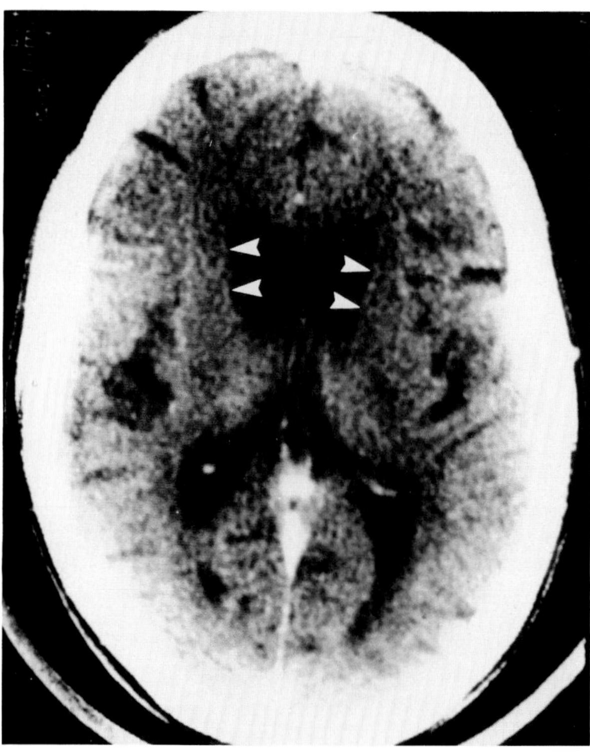

B

**Figure 72.6. Huntington's disease.** (A) A CT scan shows the normal heads of the caudate nucleus (black arrows) with a normal concavity of the frontal horns (white arrowheads). (B) In a patient with Huntington's disease, atrophy of the caudate nucleus causes a loss of the normal concavity (white arrowheads) of the frontal horns.

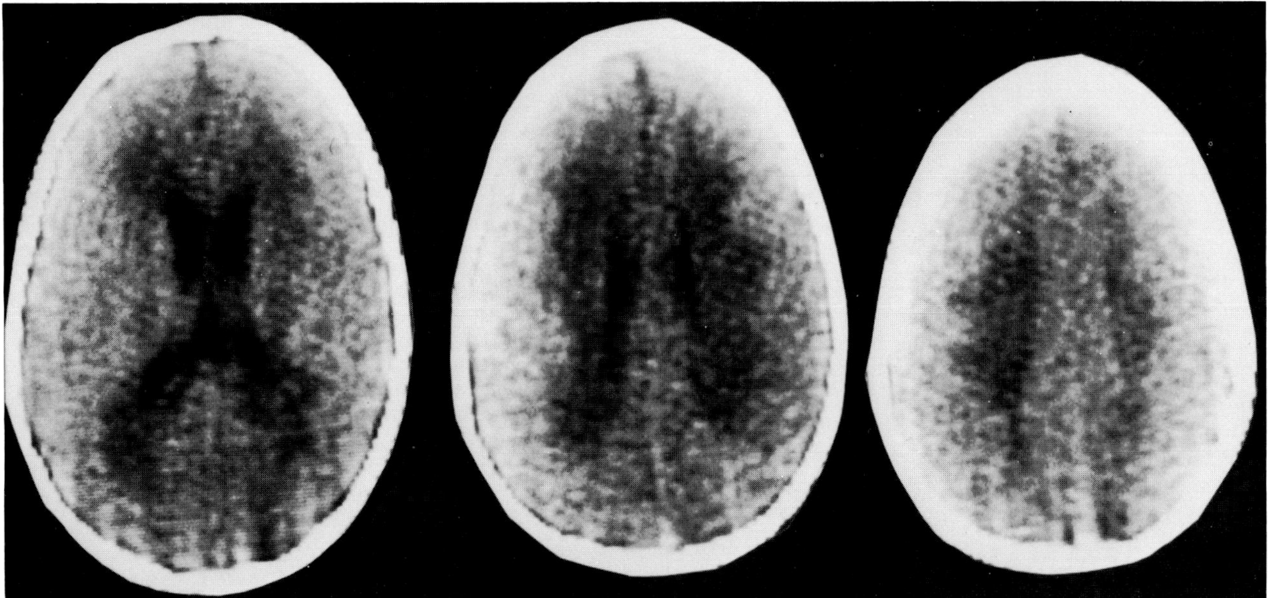

**Figure 72.7. Metachromatic leukodystrophy.** A CT scan of a 3-year-old girl shows hypodense zones in the white matter that do not conform to either a lobar or a vascular distribution. (From Buonanno, Ball, Laster, et al,[9] with permission from the publisher.)

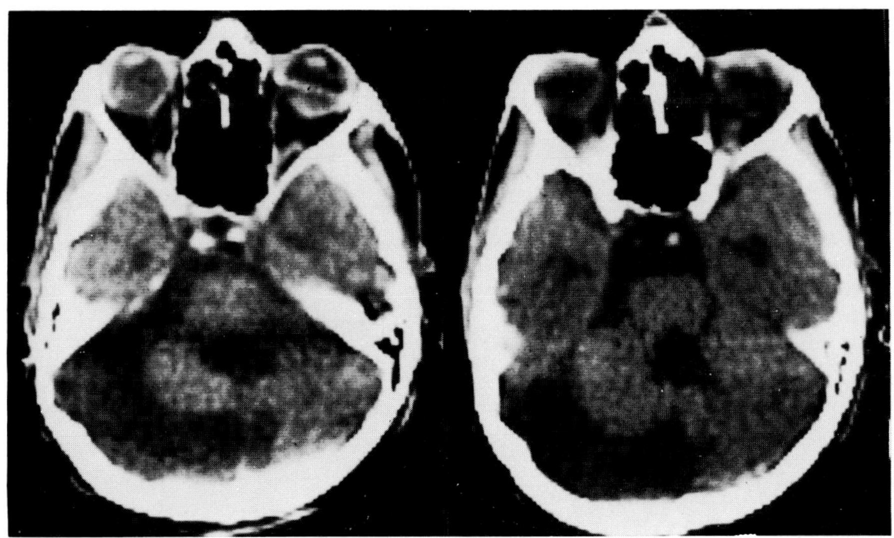

**Figure 72.8. Cerebellar atrophy secondary to vascular occlusion.** There is focal loss of cerebellar parenchyma on the right side, with resulting widening of the cerebellopontine cistern. The fourth ventricle is dilated. Absence of deformity or contralateral displacement of the dilated fourth ventricle excludes the possibility of a cerebellopontine angle mass. (From Ter Brugge and Rao,[14] with permission from the publisher.)

to deposit in various organs, especially the brain and kidneys. Although the disease is most often seen in infants and children, it may also occur in adults. Pathologic examination demonstrates diffuse widespread foci of disease in the centrum semiovale, cerebellum, and brainstem. The typical CT appearance is a diffuse, symmetric decrease in attenuation that predominantly involves the region of the centrum semiovale. However, the major role of CT in patients with metachromatic leukodystrophy is to exclude other causes for the clinical findings.

Deposition of the sulfatides in macrophages in the mucosa of the gallbladder leads to a progressive inability of the gallbladder to concentrate bile and, rarely, to the formation of single or multiple filling defects.[8,9]

## Paralysis Agitans (Parkinson's Disease)

Parkinson's disease, the most common of the extrapyramidal syndromes, is characterized by stooped posture, stiffness and slowness of movement, fixed facial expression, and involuntary rhythmic tremor of the limbs which subsides on active willed motion. A disorder of middle or late life, Parkinson's disease is very gradually progressive and exhibits a prolonged course.

CT in patients with Parkinson's disease often demonstrates cortical atrophy resulting from the degeneration of cells and fibers of the corpus striatum, globus pallidus, and substantia nigra. However, because Parkinson's disease is usually seen in older individuals, the ventricular enlargement and prominent cortical sulci found on CT scans may be indistinguishable from that due to the normal aging process. No significant correlation between the severity of the tremor and akinesia and the severity of the cerebral atrophy as shown by CT has been documented. Nevertheless, the atrophic process on CT is more marked in patients with Parkinson's disease than in persons in the same age group without the disease. Calcification of the basal ganglia is often seen in patients with Parkinson's disease, though it also can be seen in the normal population.

Gastrointestinal abnormalities may develop in patients with Parkinson's disease or reflect complications of levodopa therapy. Severe adynamic ileus, especially involving the colon, can produce massive dilatation of the bowel on plain abdominal radiographs. At times, a barium enema examination may be necessary to exclude the possibility of an obstructing lesion.[10,11]

## Cerebellar Atrophy

Isolated atrophy of the cerebellum may represent an inherited disorder, a degenerative disease, or the toxic effect of the prolonged use of such drugs as alcohol and diphenylhydantoin. Criteria of cerebellar atrophy on CT include enlargement of the cerebellar sulci (> 1 mm), cerebellopontine cisterns (> 1.5 mm), fourth ventricle (> 4 mm), and superior cerebellar cistern.[12,14]

## REFERENCES


1. Lee SH, Rao KCVG: *Cranial Computed Tomography*. New York, McGraw-Hill, 1983.
2. Wiezberg L: Contrast enhancement visualized by computerized tomography in acute multiple sclerosis. *J Comput Assist Tomogr* 5:293–300, 1981.
3. Cala LA, Mastaglia FL, Black JL: Computerized tomography of the brain and optic nerve in multiple sclerosis. *J Neurol* 36:411–426, 1978.
4. Oppenheimer DR: Demyelinating diseases, in Blackwood W, Corsellis JAN (eds): *Greenfield's Neuropathology*. Chicago, Year Book Medical Publishers, 1976.
5. Reik L: Disseminated vasculomyelinopathy on an immune complex disease. *Ann Neurol* 7:291–296, 1980.
6. Huckman MS: Computed tomography in the diagnosis of degenerative brain disease. *Radiol Clin North Am* 20:169–183, 1982.
7. McGeachi RE, Fleming JO, Sharer LR, et al: Diagnosis of Pick's disease by computed tomography. *J Comput Assist Tomogr* 3:113–115, 1979.
8. Kleiman P, Winchester P, Volberg F: Sulfatide cholecystosis. *Gastrointest Radiol* 1:99–100, 1976.
9. Buonanno FS, Ball MR, Laster DW, et al: Computed tomography in late-infantile metachromatic leukodystrophy. *Ann Neurol* 4:43–46, 1978.
10. Becker H, Schneider E, Hacker H, et al: Cerebral atrophy in Parkinson's disease—represented by CT. *Arch Psychiatr Nervenkr* 227:81–88, 1979.
11. Schneider E, Becker H, Fisher PA, et al: The course of brain atrophy in Parkinson's disease. *Arch Psychiatr Nervenkr* 227:89–95, 1979.
12. Allen JH, Martin JT, McLain LW: Computed tomography in cerebellar atrophic processes. *Radiology* 130:379–382, 1979.
13. Fernandez RE, Kishore PRS: White matter disease of the brain, in Lee SH, Rao KCVG (eds): *Cranial Computed Tomography*. New York, McGraw-Hill, 1983, pp 459–480.
14. Ter Brugge KG, Rao KCVG: Hydrocephalus and atrophy, in Lee SH, Rao KCVG (eds): *Cranial Computed Tomography*. New York, McGraw-Hill, 1973, pp 171–200.

## Chapter Seventy-Three

# Diseases of the Spine and Spinal Cord

### PROTRUSION OF INTERVERTEBRAL DISK (HERNIATED DISK)

Protrusion of a lumbar intervertebral disk is the major cause of severe acute, chronic, or recurrent low back and leg pain. Herniation of the nucleus pulposus most frequently occurs at the L4–5 and L5–S1 levels. In the cervical spine, most disk protrusions occur at the C6–7 or C5–6 levels; disk herniation in the thoracic spine predominantly occurs between T9 and T12.

Patients with symptoms suggestive of disk herniation should be treated conservatively with bed rest, muscle relaxants, and analgesics before being subjected to radiographic studies. Disk herniation can be evaluated by computed tomography (CT) or myelography. In general, the accuracy of CT is equal to or even greater than that of myelography. In addition, CT is more useful for determining the precise diagnosis of an epidural lesion or an associated paraspinal mass. This modality is most efficient when the level of clinical interest is accurately identified, as in patients with radiculopathy. When precise localization is impossible, or when a large segment of the spine must be evaluated (patients with myelopathy or polyradiculopathy), myelography is more efficient than CT. CT is more sensitive than myelography in lateral disk herniation or when the anterior epidural space is wide.

On CT scans, the normal intervertebral disk has a concave posterior border. The earliest abnormality is a minor bulging that causes the disk to develop a convex posterior border. Minimal disk bulges, especially when located centrally, may not affect the nerve roots and may thus be asymptomatic. Symptomatic compression of nerve roots usually is associated with disk herniations that are located laterally and posteriorly. Other CT findings suggesting protrusion of an intervertebral disk include displacement of epidural fat, a soft tissue density (representing the herniated nucleus pulposus) in the extradural space, deformity of the dural sac, and compression and displacement of the root sheath.

In the normal myelogram, radiopaque contrast material opacifies the subarachnoid space and fills the nerve root sheaths to the level of the pedicle. The nerve roots are well-delineated as filling defects that cross the level of the disk spaces. In the lumbar region, the nerve roots are numbered according to the pedicle under which they leave the thecal sac; for example, the L5 nerve root leaves the thecal space below the L5 pedicle. Nerve roots are usually affected by disease at the disk space just superior to the corresponding vertebral body. Thus, disk disease at the L4–5 interspace will affect the L5 nerve root.

At present, myelography is usually performed with a water-soluble agent (metrizamide) rather than an oil-based one (Pantopaque). The herniated disk typically produces a smooth extradural defect at the lateral or anterior portion of the contrast column at the level of the interver-

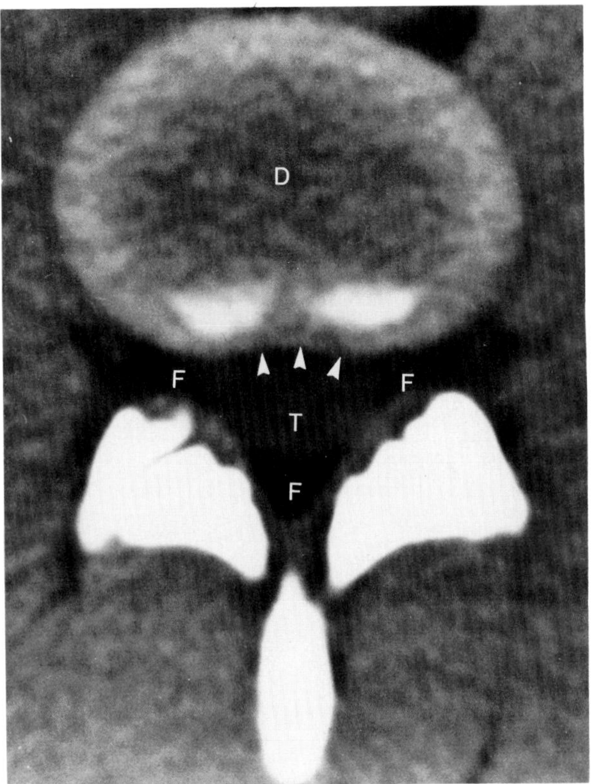


**Figure 73.1. CT of a normal lumbar disk.** The normal lumbar intervertebral disk (D) has a concave posterior border (arrowheads). Note the normal epidural fat (F) surrounding the thecal sac (T).

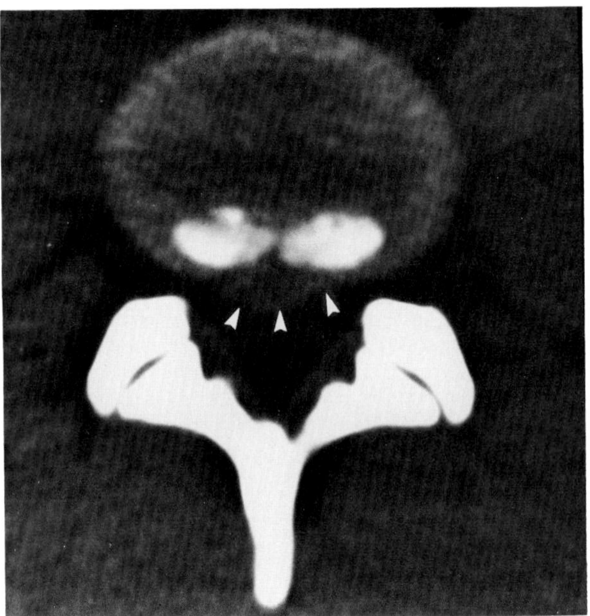

**Figure 73.2. Central bulging of an intervertebral disk.** A CT scan shows the convex posterior border of the disk (arrowheads). Note the preservation of the epidural fat.

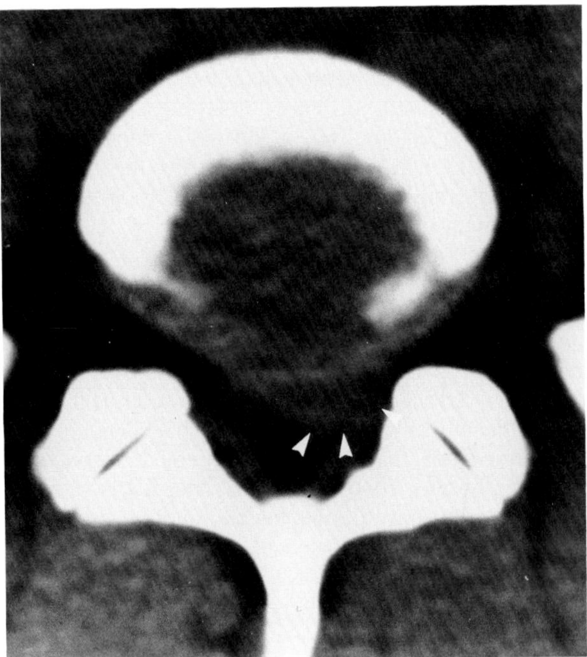

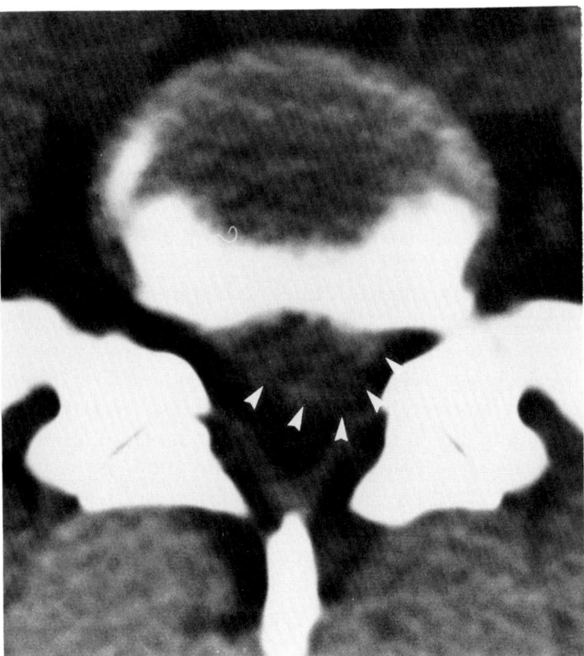

**Figure 73.3. Disk herniation at the L5–S1 level.** Two CT scans of a young man with a left S1 radiculopathy show herniation of the disk (arrowheads) to the left with obliteration of the epidural fat.

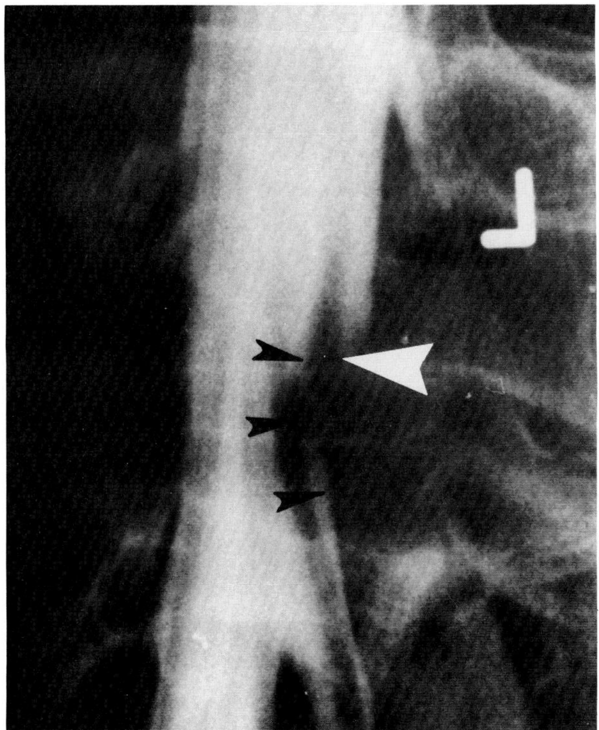

**Figure 73.4. Lumbar disk herniation.** A myelogram shows the herniated disk as an epidural lesion (black arrowheads) at the level of the intervertebral disk space. Compression by the disk causes amputation of a nerve root (white arrowhead).

tebral space. There often is amputation of the nerve root at the intervertebral disk space and incomplete filling of the nerve root sheath.[1–6]

## SPINAL STENOSIS

### Congenital (Developmental) Spinal Stenosis

Congenital spinal stenosis refers to a developmental abnormality of unknown cause in which a growth disturbance of the bony wall results in narrowing of the spinal canal at one or more vertebral levels. This anomaly is most common in the lower lumbar region, where it can cause compression of the cauda equina leading to back pain radiating into both lower extremities. The most consistent measurable abnormality in congenital stenosis is a midline sagittal diameter of the bony spinal canal of less than 10 mm. If the midline sagittal diameter is 10–13 mm (relative stenosis), slight degenerative changes may produce symptomatic stenosis.

Plain lumbar radiographs may be difficult to interpret because of obscuration of the landmarks by superimposed bony structures on both sides. CT is the procedure of choice, since it provides superb delineation of the borders of the bony vertebral canal in the axial projection and permits a precise measurement of the midsagittal diameter.[7]

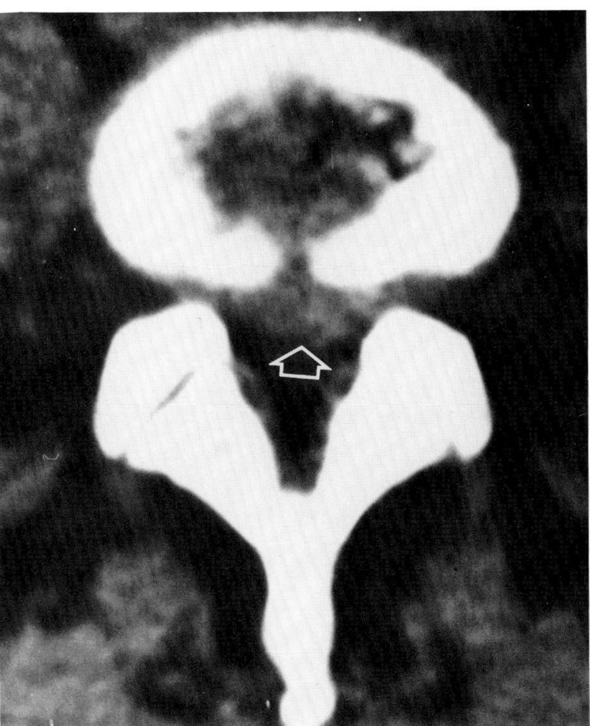

A

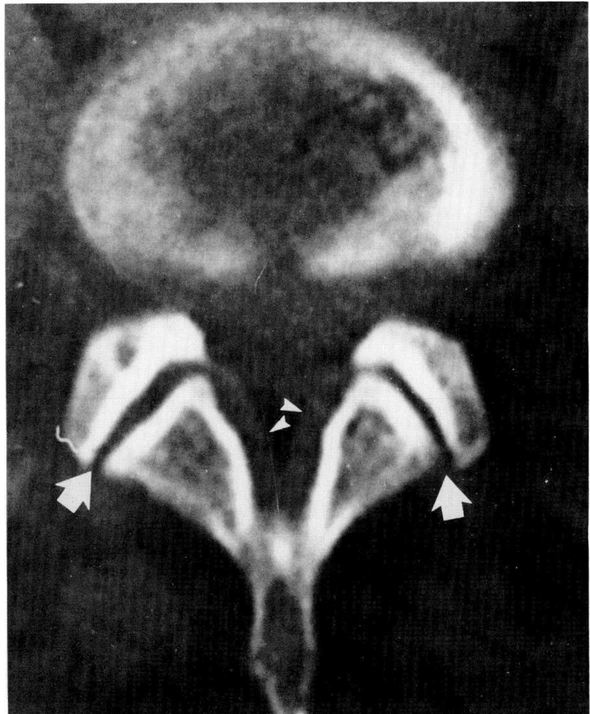

B

**Figure 73.5. Spinal stenosis.** (A,B) Two CT scans in an elderly woman with neurogenic claudication show obliteration of the subarachnoid space by the bulging disk (open arrow), ligamental hypertrophy (small arrowheads), and hypertrophy of the apophyseal joints (closed arrows). Note that the apophyseal joint spaces are best seen on figure B, which was obtained at a CT setting that optimizes bone detail.

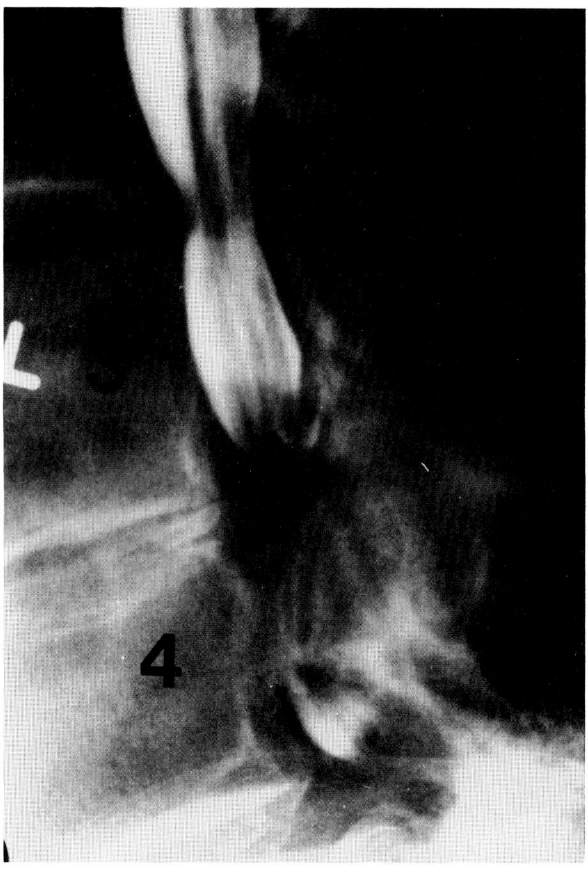


**Figure 73.6. Spinal stenosis.** A metrizamide myelogram in an elderly woman with neurogenic claudication and bilateral leg pain shows virtually complete block of the column of contrast material at the level of the L3–4 disk space. The severe obstruction was due to a combination of a bulging disk and apophyseal joint hypertrophy, both of which were better demonstrated by CT scans.

### Acquired (Degenerative) Spinal Stenosis

Acquired encroachment on the cauda equina and nerve roots may involve the spinal canal, the lateral recess (gutter), and the neural foramen. Central spinal stenosis is caused by hypertrophy of the inferior articular process and the lamina, which results in a decrease in the anteroposterior diameter of the bony spinal canal. This condition may be complicated by bulging of the posterior border of the disk or by associated osteophyte formation at the posterior border of the disk plates at the involved level. Stenosis of the lateral recess (gutter) is caused by hypertrophy of the superior articular process, which forms the posterior wall of the bony nerve root canal. This condition is of great clinical importance, since poor results from a laminectomy for decompression of the cauda equina may be due to a failure to appreciate the lateral extent of the stenosis and a subsequent failure to extend the decompression laterally to include the apophyseal joints. A severe type of lateral stenosis may be due to spondylolisthesis, in which forward slipping of the posterior elements of the vertebra above results in compression of the spinal nerve in the lateral recess against the posterior margin of the body of the vertebra below. Stenosis of the neural foramen may be caused by any process that causes compression on the nerve root in this region.

CT is the procedure of choice for demonstrating acquired spinal stenosis, since it provides superb delineation of the borders of the bony vertebral canal. Myelography in this condition typically shows an hourglass or a complete constriction of the contrast column and poor filling of nerve root sheaths.[8–10]

## INTRASPINAL MASSES

Intraspinal lesions can be divided into three groups according to the compartment which they occupy.

### Intramedullary Masses

Intramedullary masses arise within the spinal cord itself. The most common of the intramedullary tumors are the ependymomas, which may arise at any level but primarily involve the lower cord or the cauda equina. Astrocytomas and glioblastomas constitute most of the other intramedullary neoplasms. Syringomyelia and hydromyelia, which also widen the spinal cord, are discussed in the next section of this chapter.

On myelography, intradural masses typically produce a fusiform widening of the spinal cord with a symmetric narrowing of the surrounding contrast-filled subarachnoid space. Complete obstruction causes an abrupt concave termination of the contrast column, which appears similar in all radiographic projections. A large, expanding intraspinal mass may produce the characteristic plain film appearance of widening of the interpedicular distance and flattening of the medial margins of the pedicles. As with all intraspinal tumors, CT with dilute metrizamide in the subarachnoid space is slowly replacing myelography as the procedure of choice in assessing the patient with a suspected intramedullary lesion.[11–13]

### Extramedullary Intradural Masses

Extramedullary intradural masses arise ouside the spinal cord but within the dural sac. Most are benign tumors, primarily meningiomas and neurofibromas. On myelography, the spinal cord appears displaced and compressed against the opposite wall, causing narrowing of the contralateral subarachnoid space with widening of the space on the ipsilateral side. The presence of associated pressure erosions of adjacent vertebrae suggests a neurofibroma, which often extends extradurally.[18]

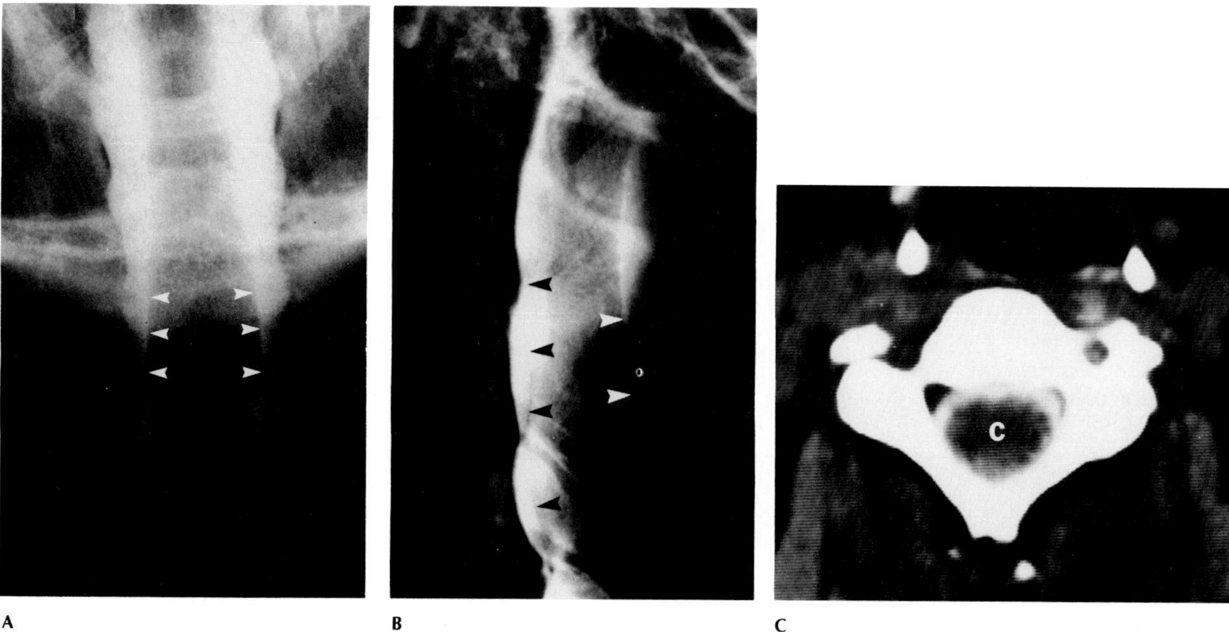


**Figure 73.7. Intramedullary mass (cervical cord astrocytoma).** (A) Frontal and (B) lateral views from a myelogram show enlargement of the cervical cord (arrowheads). It is important to demonstrate the cervical cord enlargement in both planes to exclude an extradural lesion, such as cervical spondylosis, that may simulate an intramedullary process. (C) A CT scan confirms the enlargement of the spinal cord. C = spinal cord.

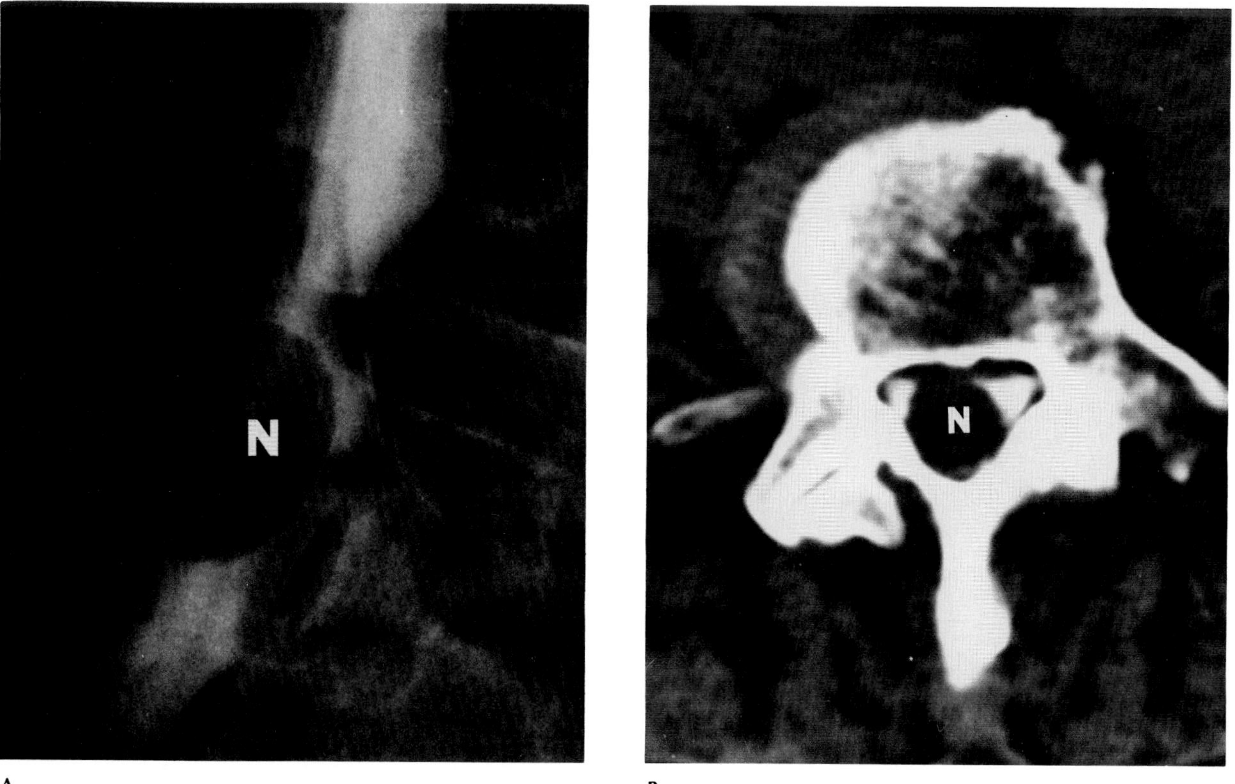


**Figure 73.8. Extramedullary intradural mass (lumbar neurofibroma).** (A) An oblique film from a myelogram shows a well-demarcated lesion (N) in the lower lumbar thecal sac. (B) A CT scan confirms that the neurofibroma (N) is intradural in location. Because the spinal cord ended at T12, the mass must represent an extramedullary intradural process.

### Extradural Masses

Extradural masses arise outside the dural sac and displace the entire thecal sac and its contents. These masses cause the distance between the lateral margin of the sac and the medial margin of the pedicle to be widened on the side of the tumor and narrowed on the opposite side. The margins of the sac remain smooth, unless the tumor has invaded through the dura to produce a concomitant intradural component. Large masses can compress the subarachnoid space on the side of the lesion and displace the spinal cord away from the mass.

The most common extradural lesion is disk herniation [see the section "Protrusion of Intervertebral Disk (Herniated Disk)" earlier in this chapter]. Most extradural tumors are malignant (metastatic carcinoma and lymphoma) and they often produce evidence of vertebral destruction that can be detected on plain films of the spine. In comparison with carcinomatous metastases, extradural defects due to lymphoma tend to be longer, smoother, and more often circumferential.

An extradural mass may surround the dura and produce a complete block or be eccentric and cause either a partial or a complete block.[18]

## SYRINGOMYELIA

Syringomyelia is an intraspinal cystic cavity that causes fusiform widening of the spinal cord which may extend over many segments. The syrinx is independent of, but may be connected with, the central canal. When the cavity is a distended central canal, the condition is termed *hydromyelia*. This congenital anomaly usually is associated with the Chiari type I malformation.

On positive-contrast myelography with the patient in a head-down position, syringomyelia and hydromyelia produce a pattern of fusiform widening of the spinal cord that is indistinguishable from that of an intramedullary tumor. In patients with hydromyelia who undergo air

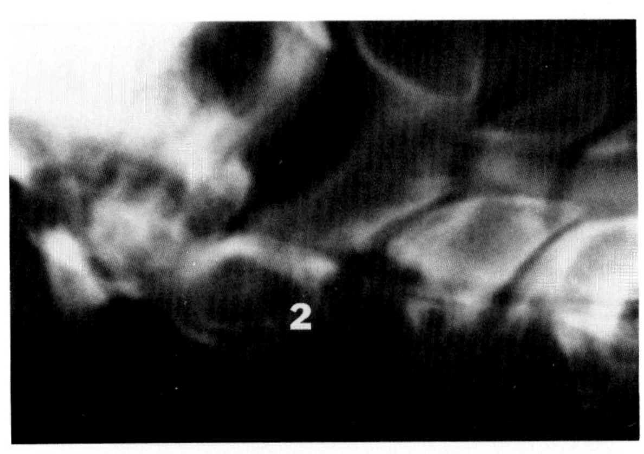


A

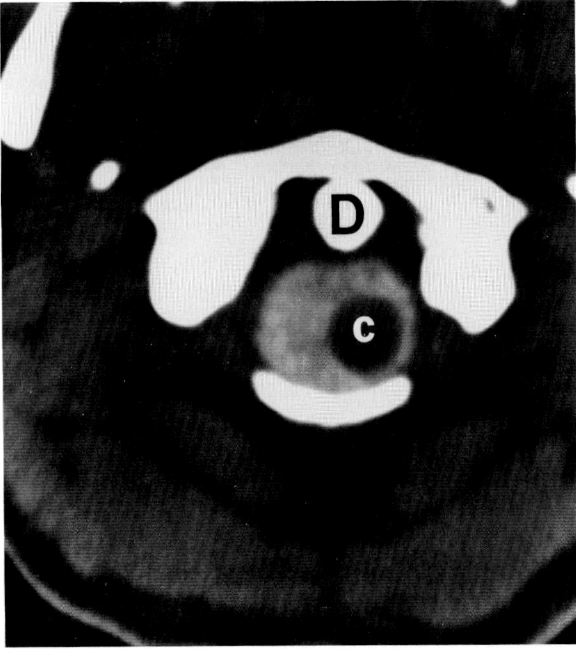


B

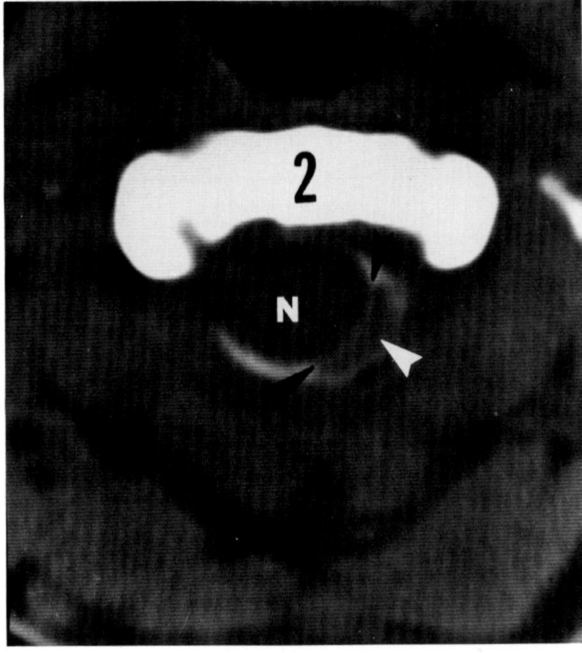


C

**Figure 73.9. Extramedullary intradural mass (cervical neurofibroma).** (A) A cervical myelogram performed via the lumbar approach shows incomplete filling of the subarachnoid space at the level of C2. (B) A delayed CT scan at the level of the dens (D) shows an abnormal position of the cervical spinal cord (C), which should be in the midline but is displaced to the left. (C) A CT scan at the level of C2 shows a distorted, small cervical cord (white arrowhead) displaced to the left by a large extramedullary intradural neurofibroma (N). Note that the mass and the spinal cord are separated by a small amount of white metrizamide (black arrowheads).

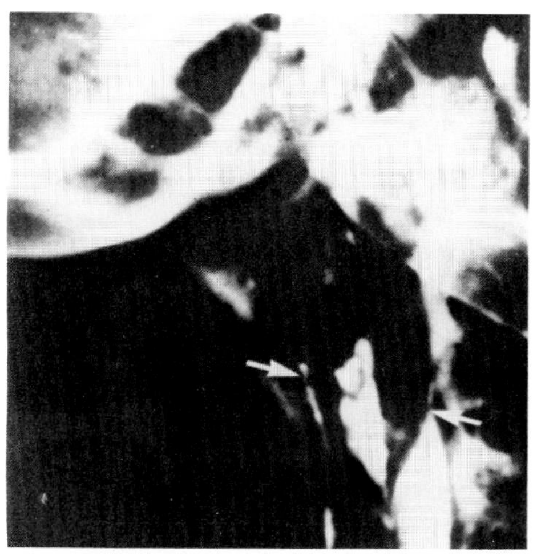


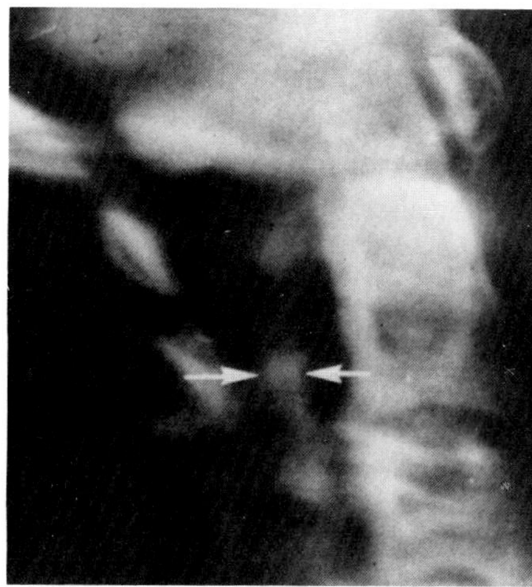


**Figure 73.10. Collapsing cord sign in hydromyelia.** (A) A lateral view from a myelogram performed with the patient in the head-down position shows the markedly increased anteroposterior diameter of the cervical cord (arrows). (B) A lateral tomogram of an air myelogram performed in the erect position shows a small cervical cord (arrows). (From Heinz, Schlesinger, and Potts,[14] with permission from the publisher.)

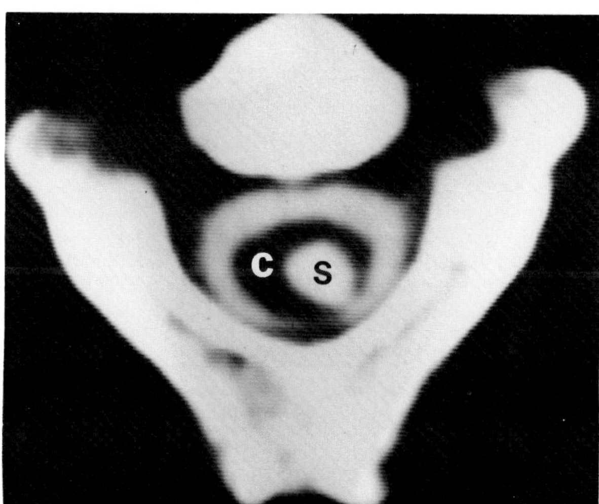


**Figure 73.11. Hydromyelia.** A CT scan after the subarachnoid injection of metrizamide shows enlargement of the cervical spinal cord (C) in a middle-aged woman with progressive bilateral upper extremity weakness. Note the high-density metrizamide filling the central cavity (S).

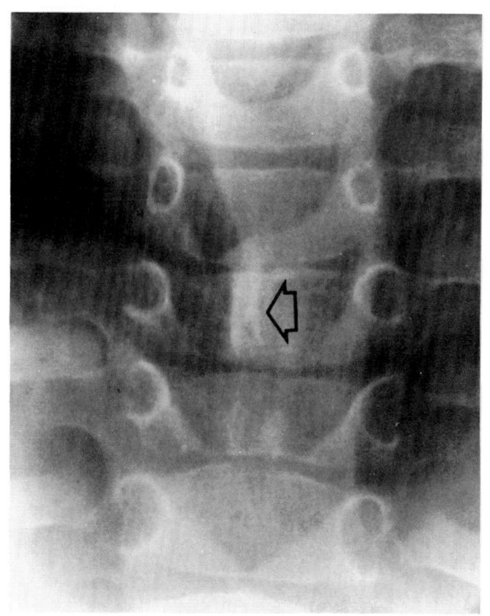


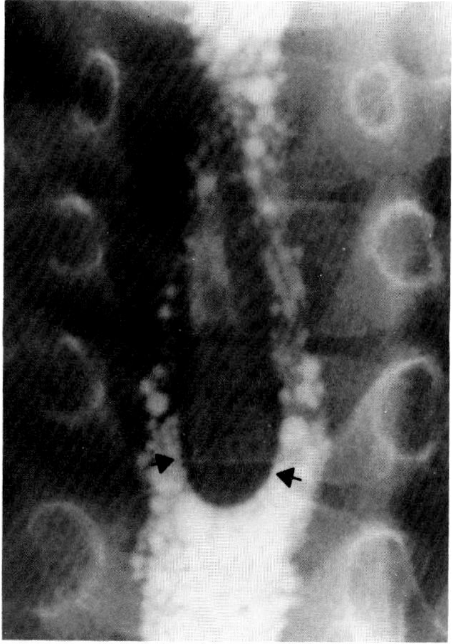


**Figure 73.12. Diastematomyelia.** (A) A plain radiograph of the spine shows fusiform widening of the spinal canal and an increase in the interpedicular distance extending over several segments. Note the pathognomonic ossified septum (arrow) lying in the midline of the neural canal. (B) A myelogram shows splitting of the contrast material around an oval midline defect (arrows), which represents the septum.

myelography performed in the semierect or erect position, fluid within the dilated central canal migrates caudally and permits the apparently enlarged cord to collapse. This collapsing cord sign indicates the presence of a cystic lesion rather than a solid intramedullary neoplasm.

CT can demonstrate the low-density fluid within a syrinx. Although some intramedullary gliomas also may have low density, these tumors tend to have a serrated border while the margin of a syrinx is smooth. A delayed scan obtained hours after a myelogram with water-soluble contrast material permits filling of the syrinx cavity either by a direct connection from the subarachnoid space or by diffusion of contrast material across the spinal cord. This technique gives better delineation of the lesion than does a CT scan performed without intrathecal contrast material.[14,15]

## DIASTEMATOMYELIA

Diastematomyelia is a rare malformation in which the spinal cord is split by a midline bony, cartilaginous, or fibrous spur extending posteriorly from a vertebral body. The condition most commonly occurs in the lower thoracic and upper lumbar regions and often is associated with a variety of skeletal and central nervous system anomalies.

Plain radiographs of the spine show fusiform widening of the spinal canal with an increase in the interpedicular distance that extends over several segments. If the septum dividing the cord is ossified, it may appear on frontal views as a pathognomonic vertical, thin bony plate lying in the midline of the neural canal. On myelography, the contrast material is split around a round or oval midline defect representing the septum. The septum and the two hemicords about it can be well demonstrated by CT.[16,17]

## REFERENCES


1. Gado MH, Hodges FJ, Patel JI: Spine, in Lee JKT, Sagel SS, Stanley RJ (eds): *Computed Body Tomography*. New York, Raven Press, 1983.
2. Meyer GA, Haughton VM, Williams AL: Diagnosis of herniated lumbar disk with computed tomography. *N Engl J Med* 301:1166–1167, 1979.
3. Williams AL, Haughton VM, Syvertsen A: Computed tomography in the diagnosis of herniated nucleus pulposus. *Radiology* 135:95–99, 1980.
4. Peterson HO, Kieffer SA: Radiology of intervertebral disk disease. *Semin Roentgenol* 7:260–271, 1972.
5. Sackett JF, Strother CM: *New Techniques in Myelography*. Philadelphia, Harper & Row, 1979.
6. Teplick JG, Haskins ME: CT and lumbar disc herniation. *Radiol Clin North Am* 21:259–272, 1983.
7. Verbiest H: The significance and principles of computerized axial tomography in idiopathic developmental stenosis of the bony lumbar vertebral canal. *Spine* 4:369–378, 1979.
8. McAfee PC, Ullrich CG, Yuan HA, et al: Computed tomography in degenerative spinal stenosis. *Clin Orthop* 161:221–234, 1981.
9. Postacchini F, Pezzeri G, Montanaro A, et al: Computerized tomography in lumbar stenosis. *J Bone Joint Surg* 62B:78–82, 1980.
10. Lancourt JE, Glenn WV, Viltse LL: Multiplanar computerized tomography in the normal spine and in the diagnosis of spinal stenosis. *Spine* 4:379–390, 1979.
11. Siebert CE, Barnes JE, Dreisbach JN, et al: Accurate CT measurement of the spinal cord using metrizamide. *AJNR* 2:75–78, 1981.
12. Aubin ML, Jardin C, Bar D, et al: Computerized tomography in three new cases of intraspinal tumor. *J Neuroradiol* 6:81–92, 1979.
13. Traub SP: Mass lesions in the spinal canal. *Semin Roentgenol* 7:240–248, 1972.
14. Heinz ER, Schlesinger EB, Potts DG: Radiologic signs of hydromyelia. *Radiology* 86:311–318, 1966.
15. Forbs WSC, Isherwood I: Computed tomography in syringomyelia and the associated Arnold-Chiari type I malformation. *Neuroradiology* 15:73–78, 1978.
16. Arrendondo F, Haughton VM, Hemmy DC, et al: Computed tomographic appearance of spinal cord in diastematomyelia. *Radiology* 136:685–688, 1980.
17. Hilal SK, Marton D, Pollack E: Diastematomyelia in children. *Radiology* 112:609–614, 1974.
18. Peterson HO, Kieffer SA: *Introduction to Neuroradiology*. Hagerstown, Harper & Row, 1972.

# PART THIRTEEN
# MISCELLANEOUS DISORDERS

## Chapter Seventy-Four

# Granulomatous Diseases of Unknown Etiology

### SARCOIDOSIS

Sarcoidosis is a multisystem granulomatous disease of unknown cause that is most often detected in young adults. Women are affected slightly more often than men, and the disease is far more prevalent among blacks than whites.

Ninety percent of the patients with sarcoidosis have radiographic evidence of thoracic involvement. Indeed, in most cases the presence of the disease is first identified on a screening chest radiograph of an asymptomatic individual.

Bilateral, symmetric hilar lymph node enlargement, with or without diffuse parenchymal disease, is the classic radiographic abnormality in sarcoidosis. There is usually also enlargement of the right paratracheal nodes, producing the typical 1-2-3 pattern. Conventional tomography frequently reveals additional enlargement of the left paratracheal nodes, which usually cannot be seen on routine frontal radiographs because they are obscured by the superimposed aorta and brachiocephalic vessels. Unilateral hilar enlargement, which is a common manifestation of primary tuberculosis or lymphoma, is rare in sarcoidosis.

Diffuse pulmonary disease develops in most patients with sarcoidosis. Although hilar and mediastinal adenopathy is often associated, there is often an inverse relationship between the degree of adenopathy and the extent of parenchymal disease, with the latter increasing while the adenopathy regresses. The most common appearance is a diffuse reticulonodular pattern that is widely distributed throughout both lungs. The alveolar pattern consists of ill-defined densities that may be discrete or may coalesce into large areas of segmental consolidation. This pattern resembles an acute inflammatory process and may contain an air bronchogram. Infrequently, large, dense, round lesions may simulate metastatic malignancy. Cavitation, atelectasis, and pleural effusion are rarely seen.

Although the pulmonary lesions usually regress spontaneously or after steroid therapy, irreversible pulmonary changes develop in up to 20 percent of the cases. Coarse scarring presents as irregular linear strands extending outward from the hilum toward the periphery, often associated with bleb or bulla formation. Severe fibrosis and emphysema can cause pulmonary hypertension and cor pulmonale.

The skeletal lesions in sarcoidosis primarily involve the small bones of the hands and feet. Perivascular granulomatous infiltration in the haversian canals causes destruction of the fine trabeculae, producing a mottled or lacelike, coarsely trabeculated pattern. Lytic destruction can produce sharply circumscribed, punched-out areas of lucency. Cortical thinning, expansion, or destruction

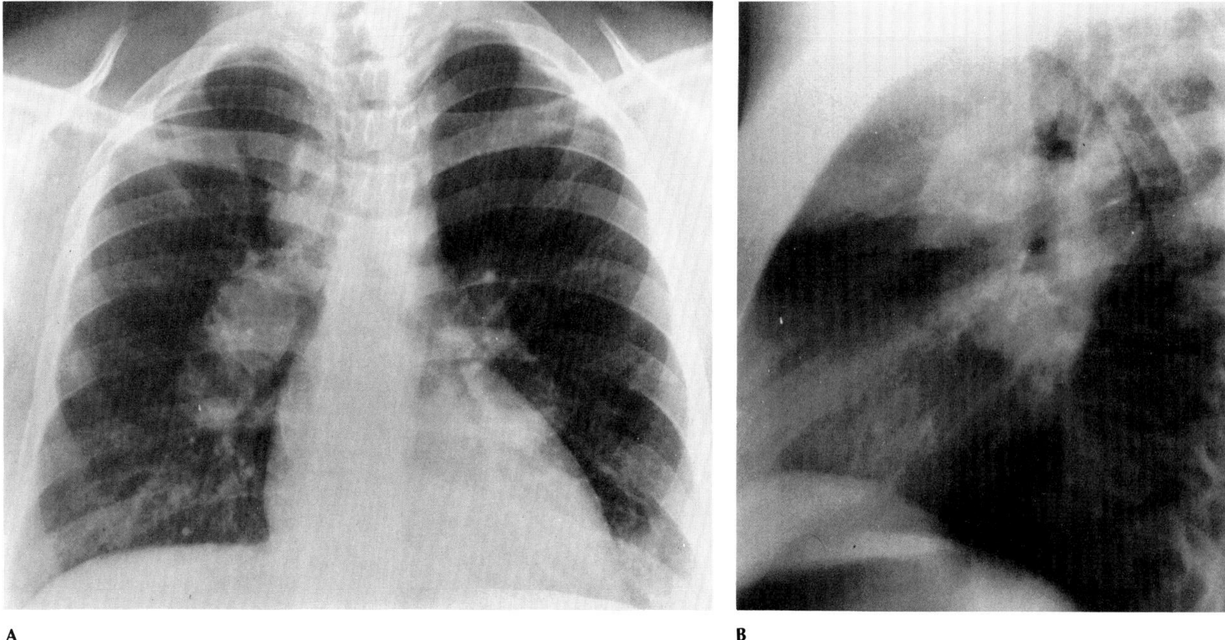


**Figure 74.1. Sarcoidosis.** (A) Frontal and (B) lateral views of the chest demonstrate enlargement of the right hilar, left hilar, and right paratracheal lymph nodes, producing the classic 1-2-3 pattern of adenopathy.

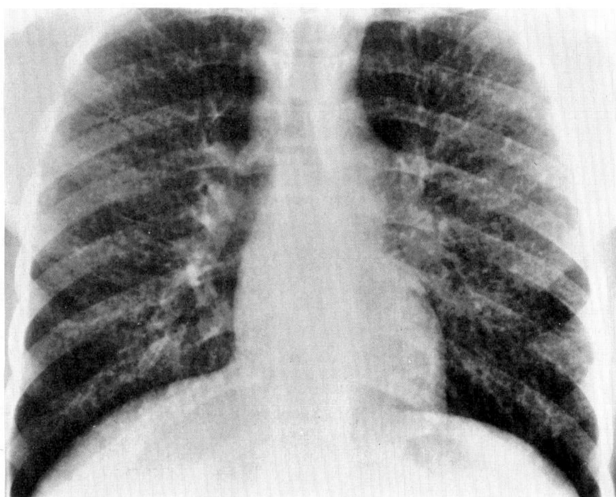

**Figure 74.2. Sarcoidosis.** A diffuse reticulonodular pattern is widely distributed throughout both lungs.

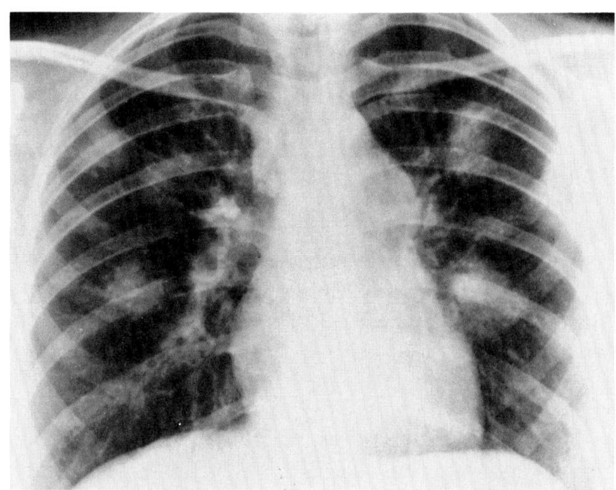

**Figure 74.3. Alveolar sarcoidosis.** Patchy, ill-defined areas of air space consolidation are scattered throughout both lungs.

may occur. Unusual manifestations of sarcoidosis include a destructive lesion in the vertebral body with a paravertebral soft tissue mass (simulating tuberculosis) and sclerotic changes in the skull, long bones, ribs, or pelvis.

Although sarcoidosis begins as a transient acute polyarthritis with periarticular soft tissue swelling in 15 percent of the patients, there are no chronic radiographic deformities. The liver and spleen are often involved, though there is rarely any radiographic evidence of enlargement of these organs. Granulomatous infiltration may occasionally enlarge one or more of the salivary glands (especially the parotid) and produce sialographic evidence of strictured ducts and parenchymal ectasia. Hypercalcemia, found in up to 10 percent of the patients with sarcoidosis, may cause nephrocalcinosis. Involvement of the heart may lead to congestive failure and recurrent pericardial effusion.

About 10 percent of the patients with sarcoidosis show evidence of stomach involvement on gastroscopic biopsy. Though usually asymptomatic, localized sarcoid granulomas can produce discrete mass defects. Diffuse

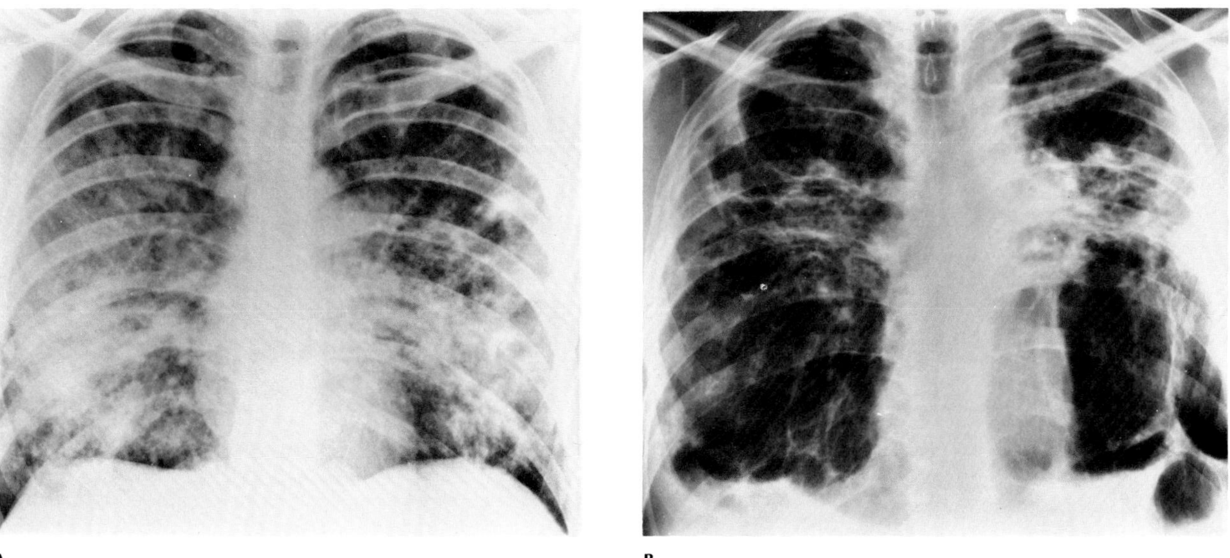


**Figure 74.4. Progression of pulmonary sarcoidosis.** (A) An initial film demonstrates diffuse reticulonodular and alveolar infiltrates. (B) In end-stage disease, there is severe fibrotic scarring, bleb formation, and emphysema.

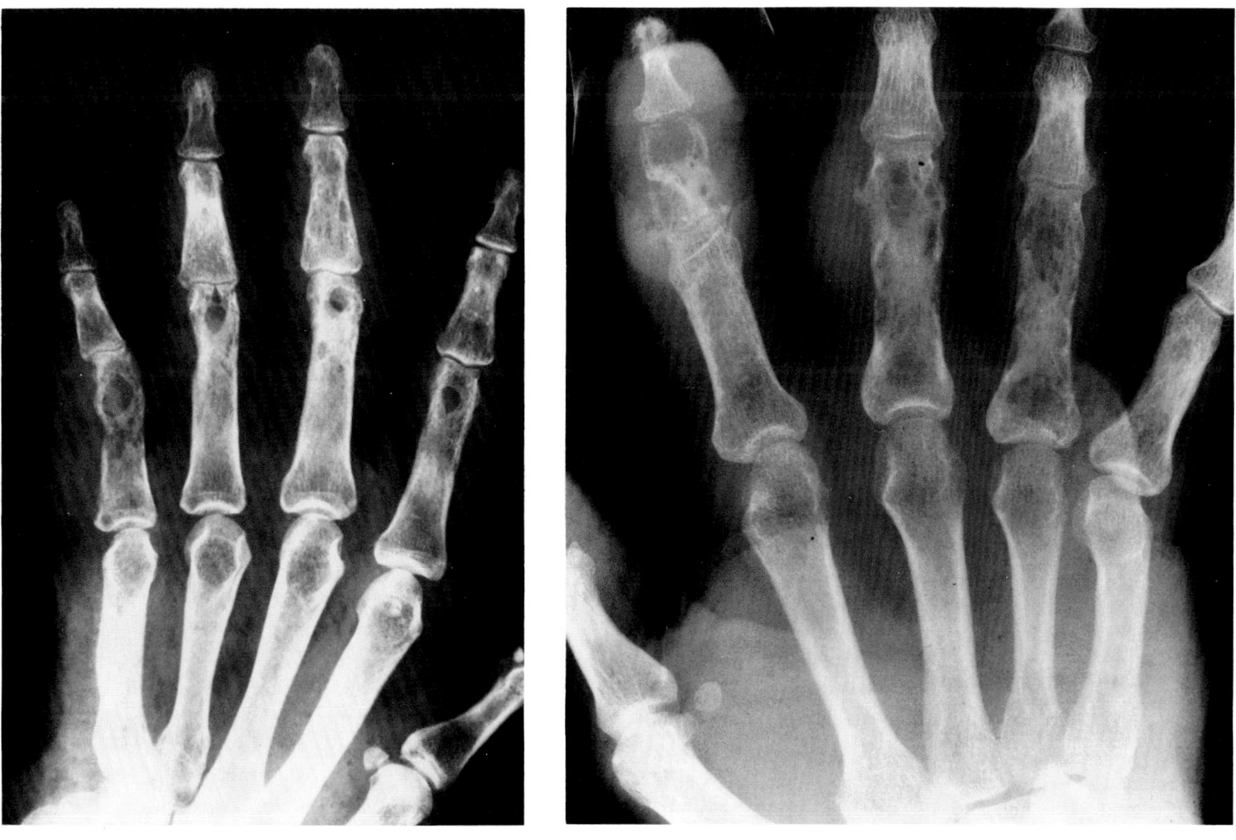


**Figure 74.5. Skeletal sarcoidosis.** (A) Multiple osteolytic lesions throughout the phalanges have a typical punched-out appearance. The apparent air density within the soft tissues is a photographic artifact. (B) The proximal phalanges of the third and fourth fingers demonstrate cortical thinning and a lacelike trabecular pattern. There is soft tissue swelling about the third proximal interphalangeal joint. Destructive changes involve the middle phalanx of the second finger.

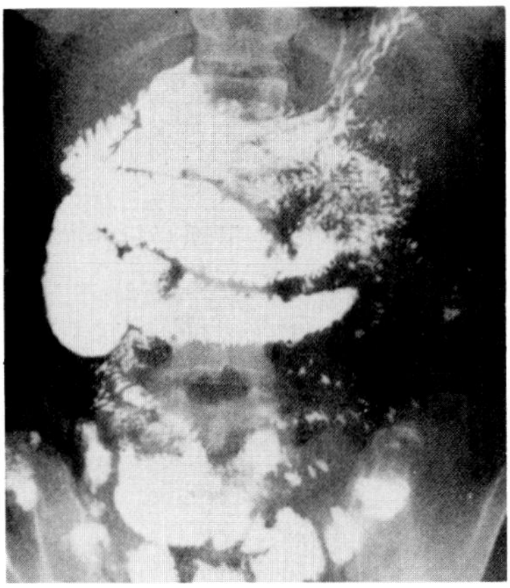
A

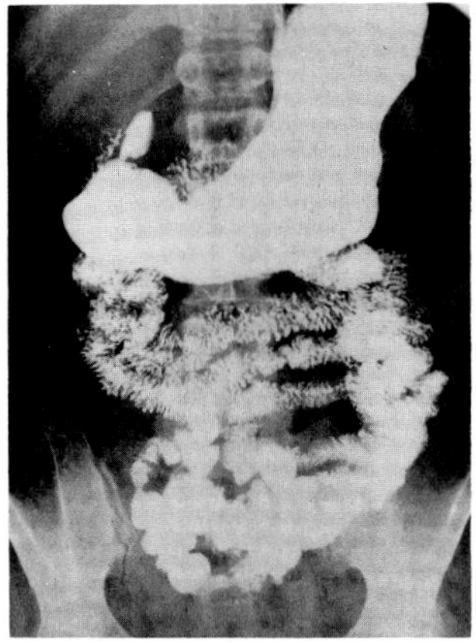
B

**Figure 74.6. Familial Mediterranean fever.** (A) During an attack of acute abdominal pain, a radiograph obtained 5 hours after the ingestion of barium shows dilatation of small bowel loops and a prolonged transit time. (B) During an asymptomatic interval, a film obtained ½ hour after the ingestion of barium shows a normal small bowel pattern and transit time. (From Eyler, Nixon, and Priest,[8] with permission from the publisher.)

lesions cause severe mural thickening and luminal narrowing, predominantly in the antrum, which mimic Crohn's disease and the radiographic pattern of linitis plastica (see the section "Granulomatous Infections of the Stomach" in Chapter 45). Ulcerations or erosions can lead to acute upper gastrointestinal hemorrhage.[1–6]

## FAMILIAL MEDITERRANEAN FEVER

Familial Mediterranean fever is an inherited disorder of unknown cause characterized by recurrent episodes of fever, peritonitis, and/or pleuritis. Arthritis and skin lesions are seen in some patients. The disease occurs predominantly in patients of Sephardic Jewish, Armenian, or Arabic ancestry. Amyloidosis develops in about 25 percent of the patients, and this complication usually leads to death.

Abdominal pain of varying severity occurs during acute attacks in almost all patients. Radiographic findings during acute episodes include adynamic ileus with dilatation of the large and the small bowel, edema of the wall of the small intestine, and a prolonged transit time through the gastrointestinal tract. A high incidence of gallbladder disease, with or without cholelithiasis, has been reported in men with this condition.

Acute pleuritic chest pain often develops, with or without abdominal symptoms. Chest radiographs obtained during acute episodes of pleuritis may demonstrate a transient pleural effusion that develops quickly and disappears rapidly with clinical improvement. Basilar atelectasis has also been described.

At least one episode of acute arthritis occurs in about 75 percent of the patients. The large joints of the lower extremities are the most frequently involved; the small joints of the hands or feet are rarely affected. Despite the occasional severity of the arthritis, complete functional recovery is the rule. The nonspecific radiographic findings are transient soft tissue swelling and osteoporosis; destructive changes are rare. Involvement of the sacroiliac joints may produce a pattern indistinguishable from the rheumatoid variants, with a loss of normal cortical definition, sclerosis with or without erosions, and even fusion.[7–11]

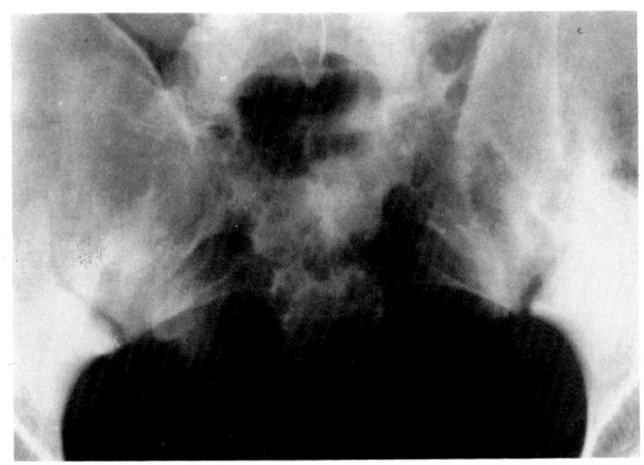

**Figure 74.7. Familial Mediterranean fever.** Asymmetric narrowing, erosive changes, and reactive sclerosis are seen about the sacroiliac joints bilaterally, producing an appearance indistinguishable from that of the rheumatoid variants.

## WEGENER'S GRANULOMATOSIS

Wegener's granulomatosis is characterized by a necrotizing vasculitis and a granulomatous inflammation that primarily affect the upper and lower respiratory tracts and the kidneys. The typical radiographic pattern in the lungs is that of rounded opacities that are usually sharply circumscribed and range in size from a few millimeters to 10 cm in diameter. The nodules are usually multiple and are generally bilateral and widely distributed. Thick-walled cavities with irregular, shaggy inner linings develop in about half the cases. Acute air space consolidation can occur. The radiographic abnormalities tend to clear rapidly with appropriate cytotoxic and immunosuppressive therapy.

The paranasal sinuses are affected in over 90 percent of the patients with Wegener's granulomatosis. Thickening of the mucous membrane, which occurs in early stages, may progress to destruction of the nasal bones, orbits, mastoids, and the base of the skull. Conventional or computed tomography is often of value in demonstrating the extent of the granulomatous process.

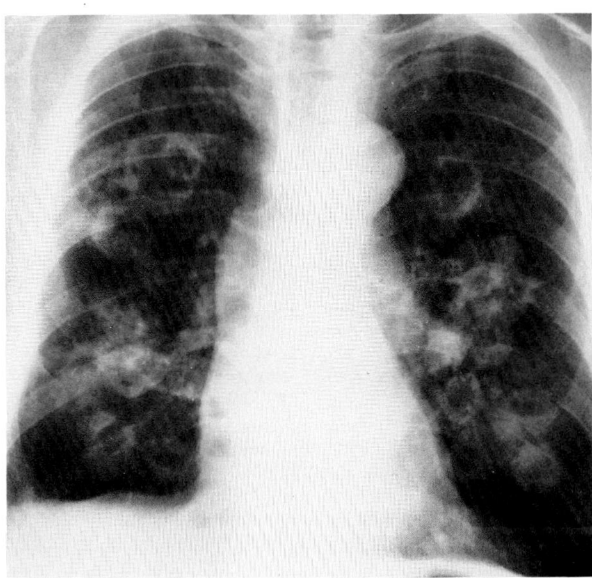

**Figure 74.8. Wegener's granulomatosis.** A plain chest radiograph demonstrates multiple rounded opacities in both lungs. Most of these lesions appear as thick-walled cavities with irregular, shaggy inner linings.

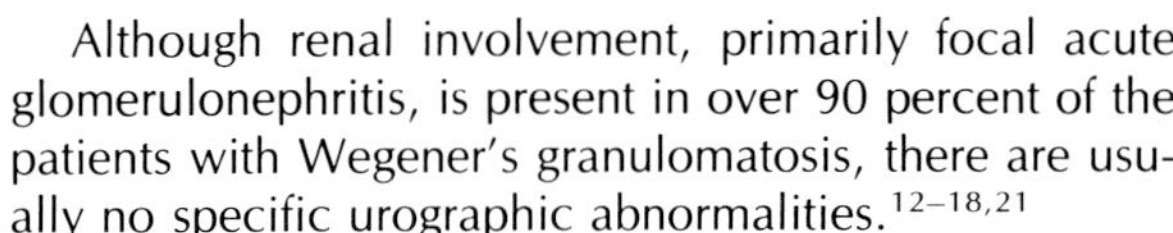

Although renal involvement, primarily focal acute glomerulonephritis, is present in over 90 percent of the patients with Wegener's granulomatosis, there are usually no specific urographic abnormalities.[12–18,21]

## MIDLINE GRANULOMA

Midline granuloma is an uncommon disease characterized by localized inflammation, destruction, and often mutilation of the tissues of the upper respiratory tract and face. Granulomatous masses in the nose and sinuses

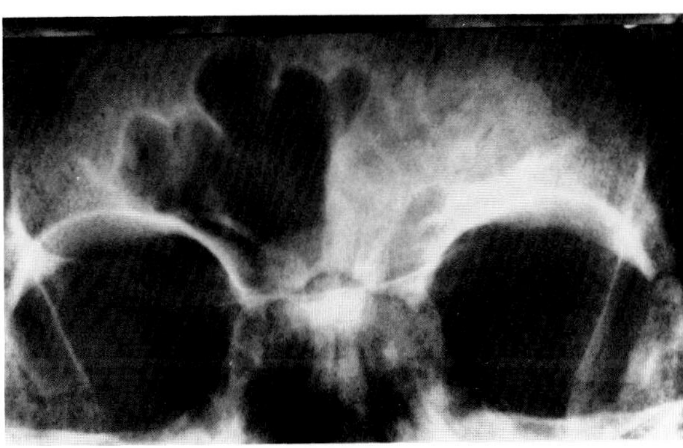

**Figure 74.9. Wegener's granulomatosis.** There is obliteration of the left frontal sinus, which has become densely opaque. (From Paling, Roberts, and Fauci,[21] with permission from the publisher.)

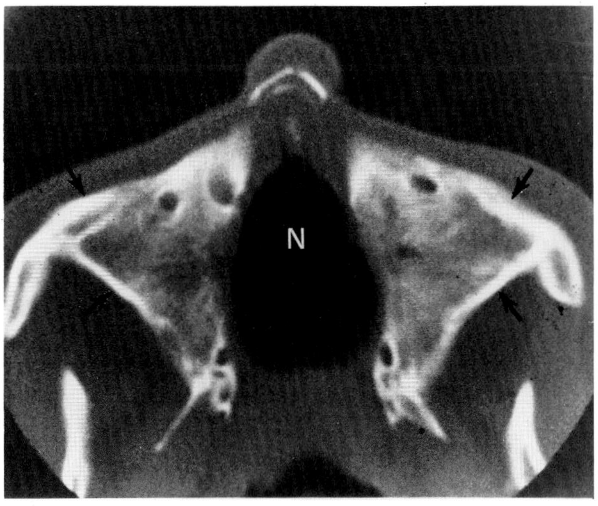


A

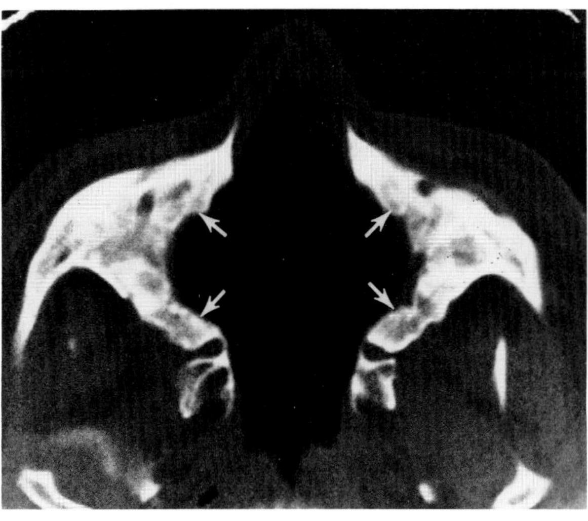

B

**Figure 74.10. Computed tomography scan of the maxillary sinuses in Wegener's granulomatosis.** (A) The anterior and infratemporal sinus walls (arrows) are only slightly sclerotic, but the sinus cavities are obliterated by trabeculated bone. Note the loss of the septum and turbinates within the nasal cavity (N). (B) The internal structures of the nasal cavity and the medial walls of the maxillary antra are destroyed. Gross thickening of the anterior and infratemporal walls of the sinuses (arrows) causes considerable, but incomplete, obliteration of the sinus cavities. (From Paling, Roberts, and Fauci,[21] with permission from the publisher.)

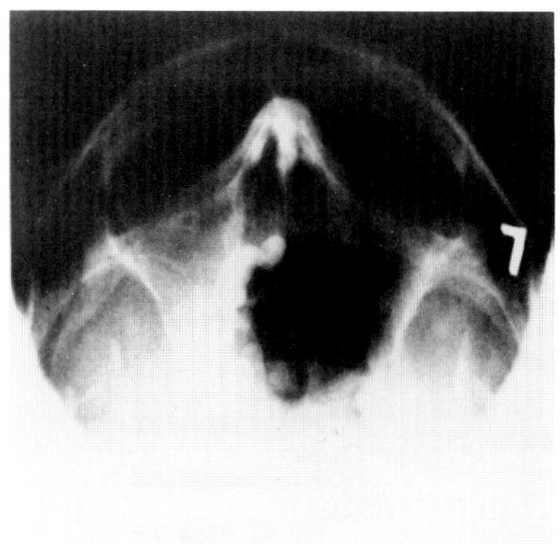
A

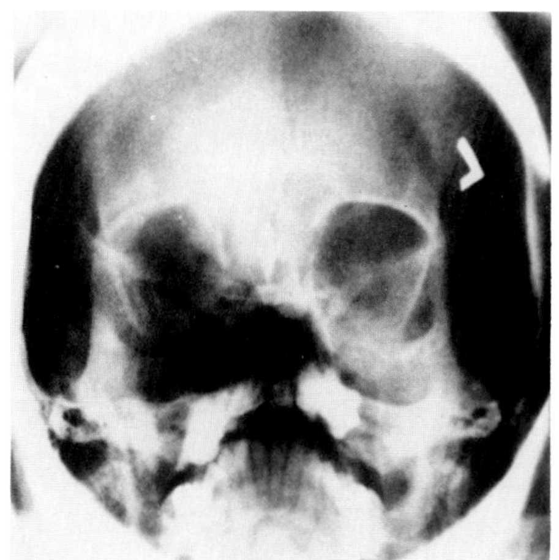
B

**Figure 74.11. Midline granuloma.** (A) A Waters' view of the sinuses shows extensive destruction of the palate and the left maxillary sinus with involvement of the floor of the left orbit. (B) A sinus film of another patient shows destruction of the nasal septum and the right maxillary sinus. (From Fauci, Johnson, and Wolff,[20] with permission from the publisher.)

cause radiographic clouding or complete obliteration of the air space. This progresses to extensive destruction of the nasal bone and the medial walls of the maxillary antra. The other paranasal sinuses are occasionally involved. Although the two entities are related, the characteristic pulmonary and renal lesions of Wegener's granulomatosis are absent in midline granuloma.[19,20]

## REFERENCES


1. Bonakdarpour A, Levy W, Aegerter EF: Osteosclerotic changes in sarcoidosis. *AJR* 113:646–649, 1971.
2. Kirks DR, McCormick VD, Greenspan RH: Pulmonary sarcoidosis. *AJR* 117:777–786, 1973.
3. Theros EG: RPC of the month from the AFIP. *Radiology* 92:1557–1561, 1969.
4. Holt JF, Owens WI: Osseous lesions in sarcoidosis. *Radiology* 53:11–30, 1949.
5. McLaughlin JS, Van Eck W, Thayer W, et al: Gastric sarcoidosis. *Ann Surg* 153:283–288, 1961.
6. Pfeiffer K: Boecks's disease of the salivary glands. *Radiology* 83:165–168, 1963.
7. Brodey PA, Wolff SM: Radiographic changes in the sacroiliac joints in familial Mediterranean fever. *Radiology* 114:331–333, 1975.
8. Eyler WR, Nixon RK, Priest RJ: Familial recurring polyserositis. *AJR* 84:262–268, 1960.
9. Wolff SM, Hathaway BE, Laster L: The gastrointestinal system in familial Mediterranean fever. *Arch Intern Med* 115:565–568, 1965.
10. Heller H, Gafni J, Michaeli D, et al: The arthritis of familial Mediterranean fever. *Arthritis Rheum* 9:1–17, 1966.
11. Shahin N, Sohar E, Dalith F: Roentgenologic findings in familial Mediterranean fever. *AJR* 84:269–274, 1960.
12. Landmann S, Burgener F: Pulmonary manifestations in Wegener's granulomatosis. *AJR* 122:750–757, 1975.
13. McGregor MBB, Sandler G: Wegener's granulomatosis. A clinical and radiological survey. *Br J Radiol* 37:430–439, 1964.
14. Raitt JW: Wegener's granulomatosis: Treatment with cytotoxic agents and adrenocorticoids. *Ann Intern Med* 74:344–356, 1971.
15. Gonzales L, Van Ordstrand HS: Wegener's granulomatosis. *Radiology* 107:295–298, 1973.
16. Hulfe R, Jung N: Wegener's granulomatosis in the region of the paranasal sinuses and skull base. *Fortschr Roentgenstr* 115:561–565, 1971.
17. Lynch ED, Herbert HL, Greenberg SD: Pulmonary cavitation in Wegener's granulomatosis. *AJR* 92:521–527, 1964.
18. Wolff SM: Wegener's granulomatosis. *Ann Intern Med* 81:513–525, 1974.
19. Fechner RE, Lamppin DW: Midline malignant reticulosis. *Arch Otolaryngol* 95:467–476, 1972.
20. Fauci AS, Johnson RE, Wolff SM: Radiation therapy of midline granuloma. *Ann Intern Med* 84:140–147, 1976.
21. Paling MR, Roberts RL, Fauci AS: Paranasal sinus obliteration in Wegener's granulomatosis. *Radiology* 144:539–543, 1982.

# Chapter Seventy-Five
# Diseases of the Spleen and Lymph Nodes

## ENLARGEMENT OF THE SPLEEN

Splenomegaly is associated with numerous conditions, including infections (subacute bacterial endocarditis, tuberculosis, infectious mononucleosis, malaria), connective tissue disorders, neoplastic hematologic disorders (lymphoma, leukemia, myeloproliferative syndromes), hemolytic anemia and hemoglobinopathies, and portal hypertension.

Plain abdominal radiographs can demonstrate the inferior border of the enlarged spleen well below the costal margin. An enlarged spleen can elevate the left hemidiaphragm, impress the greater curvature of the barium-filled stomach, and displace the entire stomach toward the midline. Splenomegaly also can cause downward displacement of the left kidney and the splenic flexure of the colon.

Computed tomography (CT) is of value when it is unclear whether a mass felt in the left upper quadrant represents an enlarged spleen or a separate abdominal mass. The enlarged spleen often loses the normal concavity of its visceral surface as the organ assumes a more globular shape. When splenomegaly is present, CT findings may suggest the cause of the splenic enlargement by demonstrating a tumor, abscess, or cyst. Associated abdominal lymph node enlargement suggests lymphoma; characteristic alterations in the size and shape of the liver and prominence of the venous structures in the splenic hilum and gastrohepatic ligament suggest that splenomegaly is secondary to cirrhosis and portal hypertension.

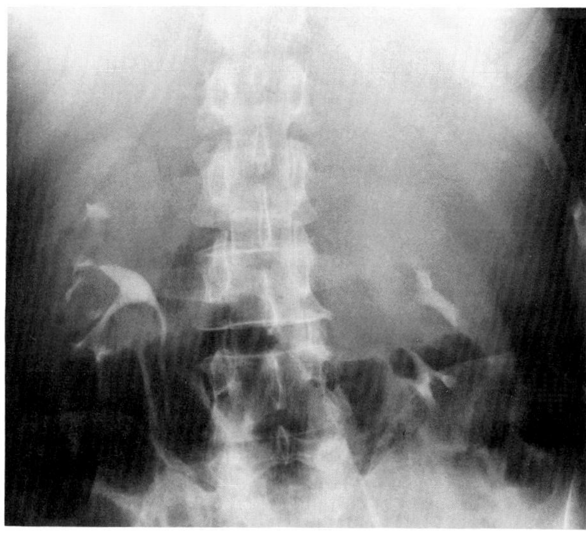

**Figure 75.1. Enlargement of the spleen.** An excretory urogram shows the enlarged spleen as a soft tissue mass filling the left upper quadrant and causing downward displacement of the left kidney.

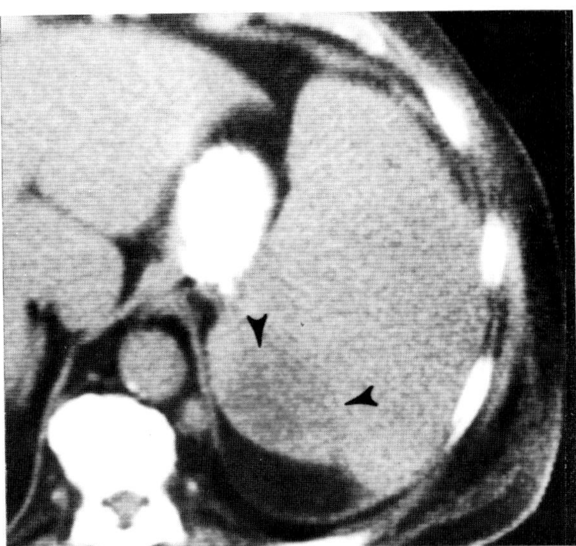

**Figure 75.2. Enlargement of the spleen due to lymphoma.** A CT scan demonstrates marked splenomegaly with a focal low-density lesion (arrowheads) posteriorly. (From Koehler,[1] with permission from the publisher.)

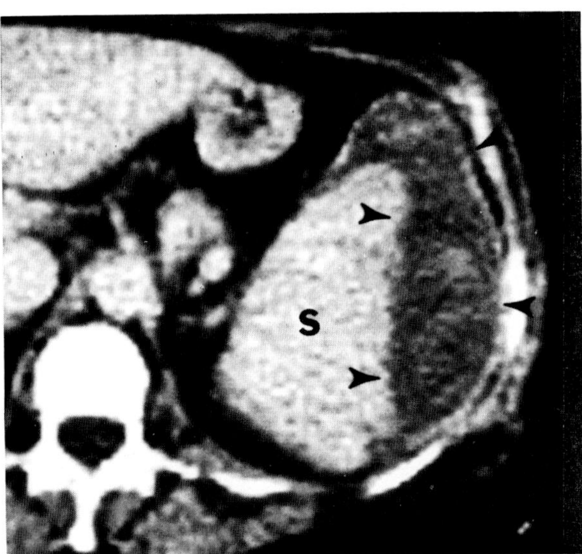


**Figure 75.4. Traumatic subcapsular hematoma.** A contrast-enhanced CT scan shows the hematoma as a large zone of decreased attenuation (arrowheads) that surrounds and flattens the lateral and anteromedial borders of the adjacent spleen (S). (From Koehler,[1] with permission from the publisher.)

Ultrasound and radionuclide scanning also can be used to assess splenomegaly. However, when the cause of splenomegaly is not known, CT is the preferred technique because it is more likely to demonstrate conditions that may be responsible for the splenic enlargement.[1,2]

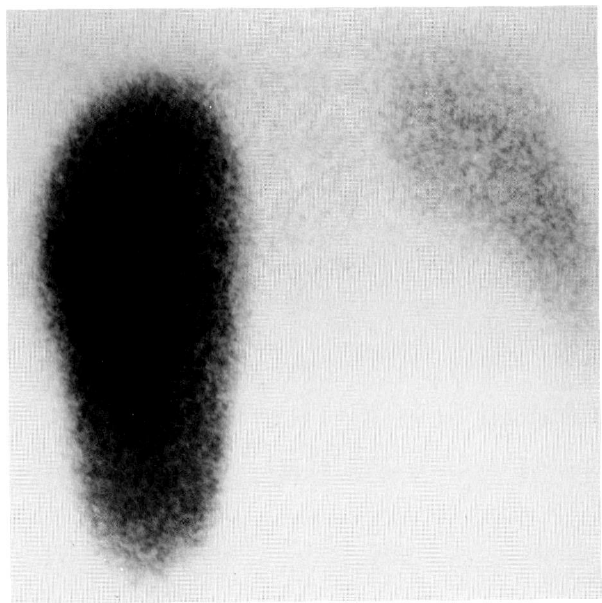

**Figure 75.3. Enlargement of the spleen.** A posterior view from a radionuclide liver-spleen scan demonstrates increased isotope uptake in the markedly enlarged spleen of a patient with cirrhosis and hypersplenism. Note the strikingly decreased uptake of isotope in the liver, a characteristic finding in patients with cirrhosis.

## SPLENIC RUPTURE

Rupture of the spleen is usually due to trauma. Infrequently, it may be a complication of palpation of a spleen enlarged by infection (especially infectious mononucleosis and sepsis) or leukemia. In many patients suffering traumatic rupture of the spleen, the severity of the clinical symptoms and the rapid loss of blood into the abdominal cavity require immediate surgery without radiographic investigation. However, in patients in whom bleeding stops temporarily or in whom there is slow bleeding over several days, radiographic studies may be of diagnostic value.

CT is the best imaging procedure for screening patients with blunt abdominal trauma for the presence of splenic injury. Indeed, its almost 100 percent sensitivity in detecting splenic injury has substantially decreased the need for abdominal arteriography and exploratory laparotomy. In addition, the noninvasive nature of CT is a clear advantage over splenic arteriography, and the ability of this technique to examine the abdomen simultaneously for signs of hepatic, renal, retroperitoneal, or other trauma is an advantage over radionuclide imaging.

Subcapsular hematomas appear on CT as crescentic collections of fluid that flatten or indent the lateral margin of the spleen. For the first day or two, the density of a subcapsular hematoma may be equal to or even greater than that of the splenic parenchyma; contrast enhancement may be required for detecting those hematomas that are isodense with the normal spleen. Over the next

10 days, the density of the fluid within the hematoma gradually decreases, becoming less than that of the normal spleen. Splenic lacerations, which may occur with or without an accompanying subcapsular hematoma, produce a CT appearance of splenic enlargement, an irregular cleft or defect in the splenic border, and free blood in the peritoneal cavity.

Many signs have been described as suggestive of splenic rupture on plain abdominal radiographs. These include prominence of mucosal folds on the greater curvature of the stomach (representing edema and hemorrhage in the gastrolienal ligament), gastric dilatation, displacement of the stomach to the right or downward, separation of intestinal loops by intraperitoneal fluid, and pleural reaction at the left base. However, these findings are nonspecific and of limited diagnostic value.

Prior to the advent of CT, splenic arteriography was the procedure of choice for demonstrating splenic rupture. Major positive findings include extravasation of contrast material into the splenic parenchyma; simul-

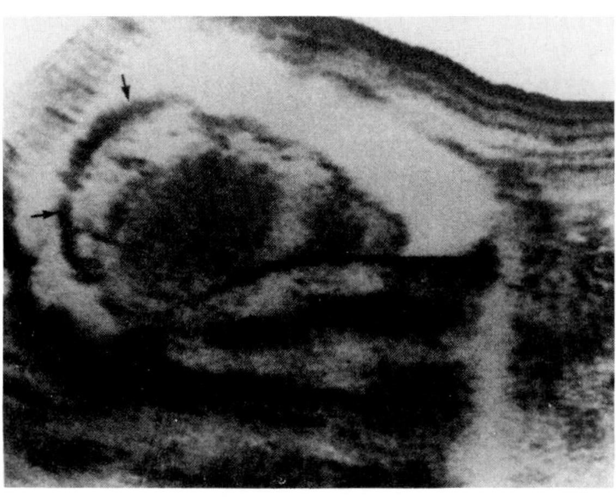
A

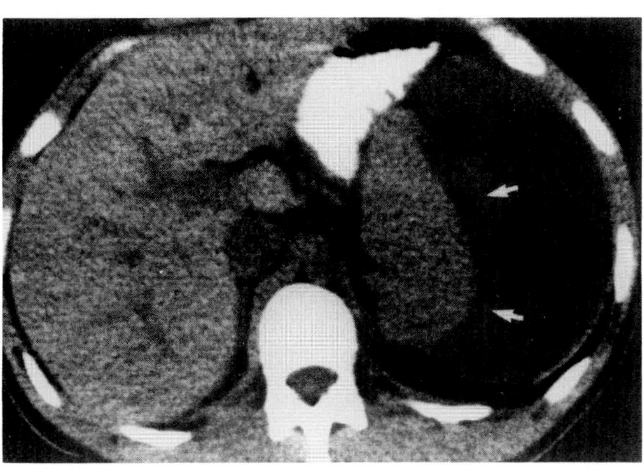
B

**Figure 75.5. Subcapsular hematoma from spontaneous splenic rupture in infectious mononucleosis.** (A) A longitudinal sonogram demonstrates a large fluid collection in the left upper quadrant surrounding the spleen. A prominent subcapsular fluid collection causes elevation of the splenic capsule (arrows). (B) A CT scan shows the subcapsular and pericapsular fluid collections. The arrows point to the faintly visible elevated splenic capsule. (From Johnson, Cooperberg, Boisvert, et al,[9] with permission from the publisher.)

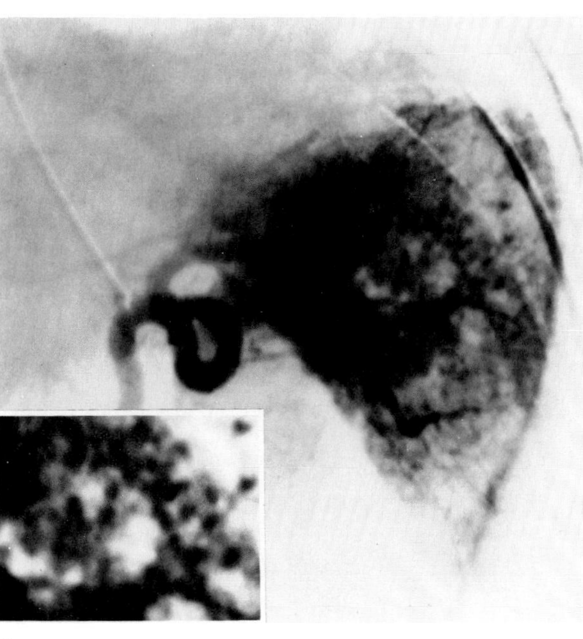
A

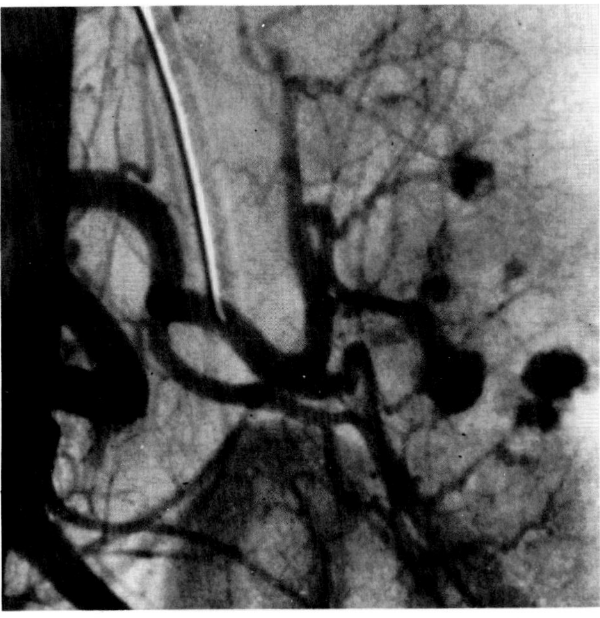
B

**Figure 75.6. Splenic rupture.** (A) A selective celiac arteriogram demonstrates multiple punctate areas (magnified in insert) that are most prominent at the midpole of the spleen. (B) An abdominal aortogram in another patient with a ruptured spleen demonstrates large globular areas of extravasation of contrast material. (From Kass and Fisher,[7] with permission from the publisher.)

taneous visualization of the splenic artery and vein, indicating rapid shunting by way of the sinusoids; amputation of a major splenic artery; and vascular defects in the splenogram of an enlarged spleen, indicating sites of rupture and hematoma.[1,3–11]

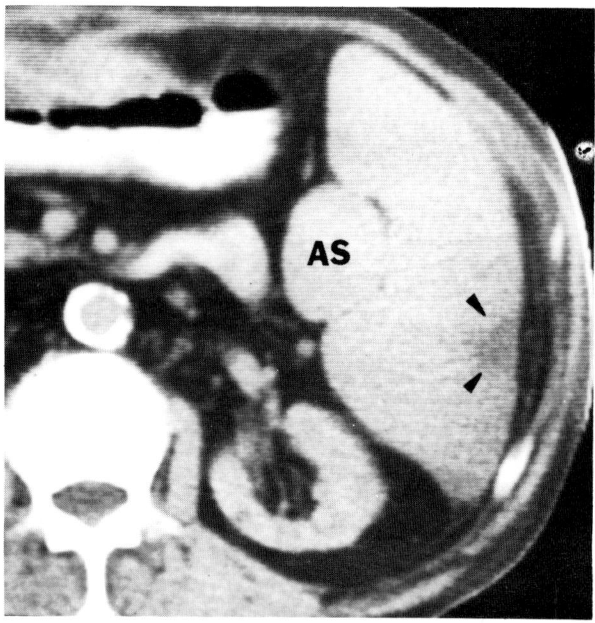


**Figure 75.7. Splenic infarct due to myeloid metaplasia.** A CT scan demonstrates the infarct as a peripheral, wedge-shaped, low-density region (arrowheads) in the posterolateral portion of a markedly enlarged spleen. Note the slight depression in the splenic contour over the infarct and the accessory spleen (AS). (From Koehler,[1] with permission from the publisher.)

## SPLENIC INFARCT

Splenic infarcts occur in patients with massive splenomegaly due to a myeloproliferative syndrome or in those with vascular occlusive phenomena associated with hemoglobinopathies such as sickle cell disease. CT demonstrates an infarction as a focal low-density region in the spleen. The defect often is wedge-shaped, with its base at the splenic capsule and its apex toward the hilum. Splenic infarcts may eventually calcify and be seen on plain abdominal radiographs.[1,8]

## SPLENIC ARTERY ANEURYSM

Calcification within the media of the splenic artery is extremely common and produces a characteristic tortuous, corkscrew appearance. When viewed end-on, splenic artery calcification appears as a thin-walled ring. A similar, though larger, circular pattern or bizarre configuration of calcification in the left upper quadrant can be due to a saccular aneurysm of the splenic artery. A definitive diagnosis of splenic artery aneurysm can be made by arteriography.[12]

## SPLENIC CYST

Splenic cysts may be congenital, parasitic (primarily echinococcal), or post-traumatic, reflecting the final stage in the evolution of a splenic hematoma. Splenic cysts are variable in size and, when large, can produce discrete soft tissue densities in the left upper quadrant and displace adjacent contrast-filled organs. Calcification most frequently occurs in echinococcal cysts, which are often

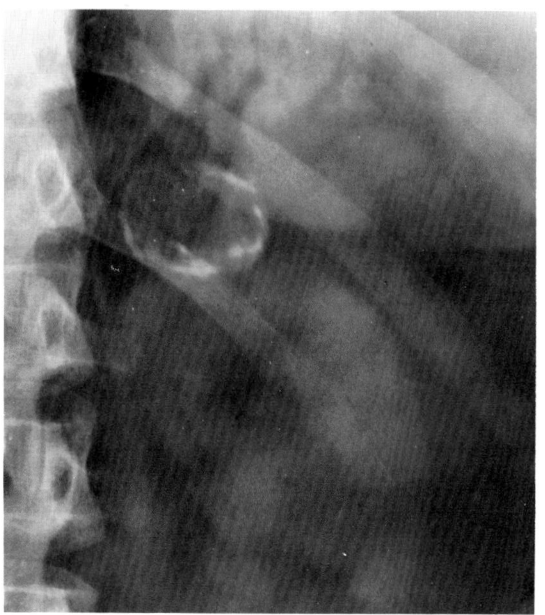

A

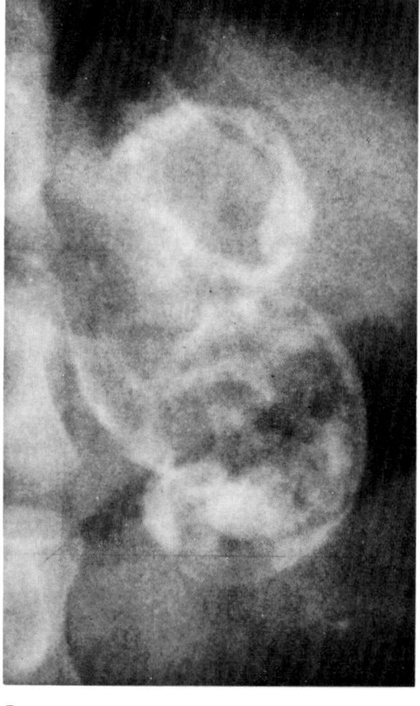

B

**Figure 75.8. Splenic artery aneurysm.** (A) Circular pattern of calcification. (B) Bizarre, lobulated calcification. (From Eisenberg,[19] with permission from the publisher.)

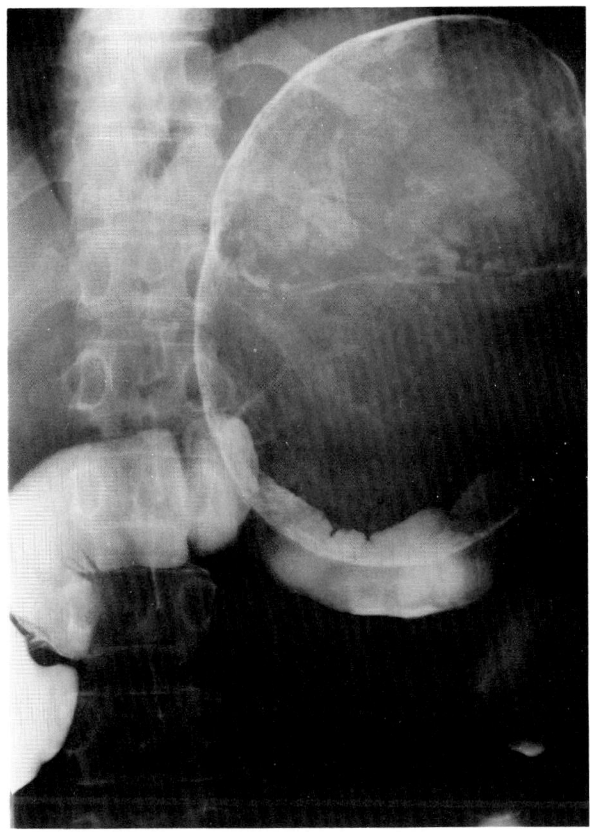

**Figure 75.9. Huge calcified splenic cyst.** (From Eisenberg,[19] with permission from the publisher.)

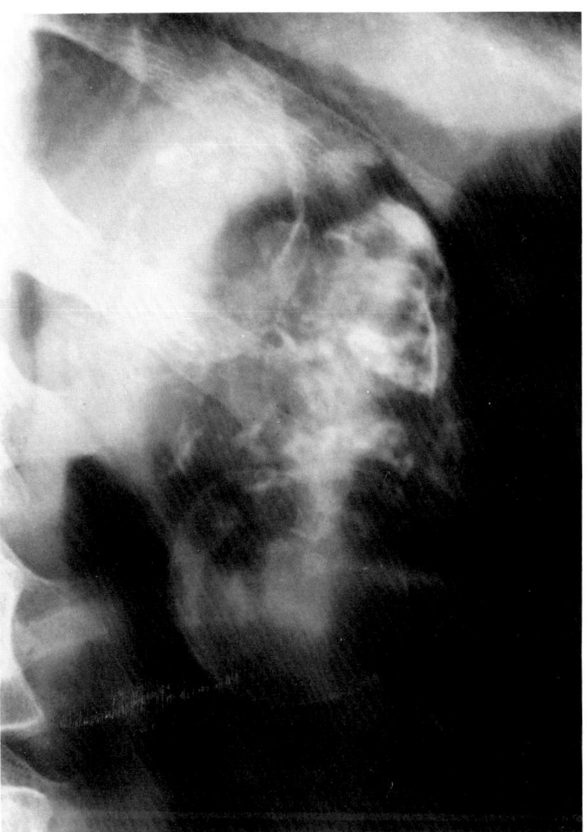

**Figure 75.10. Calcified echinococcal cyst of the spleen.** (From Eisenberg,[19] with permission from the publisher.)

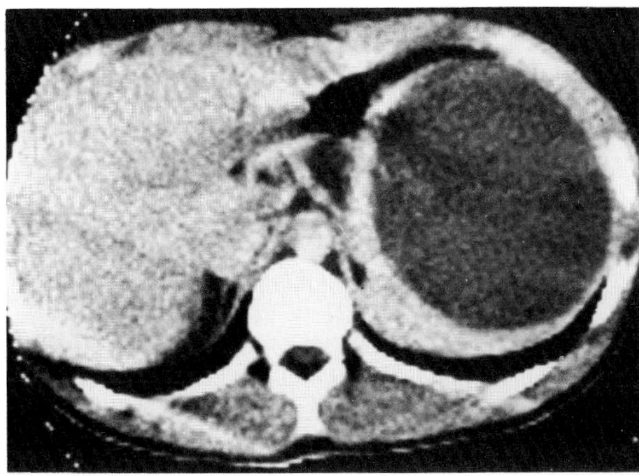

**Figure 75.11. Splenic cyst.** A CT scan demonstrates a large, low-density mass with pencil-sharp margins filling almost all of the spleen. (From Piekarski, Federle, Moss, et al,[8] with permission from the publisher.)

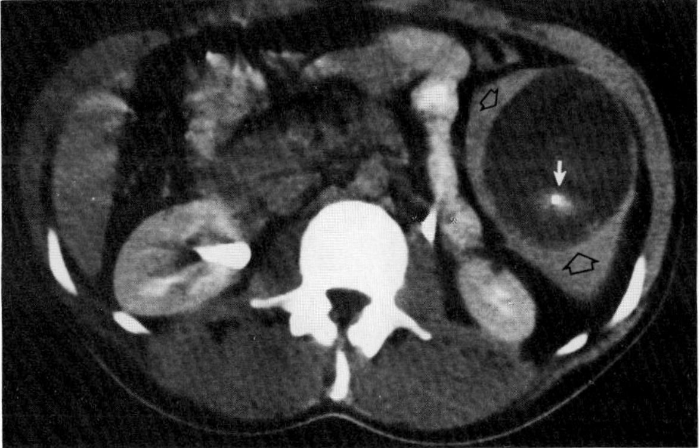

**Figure 75.12. Echinococcal cyst of the spleen.** A CT scan demonstrates a rounded, low-density intrasplenic mass with an area of intracyst calcification (solid arrow). The cyst has pencil-sharp margins and a rim (open arrows) that enhances after the injection of contrast material. (From Piekarski, Federle, Moss, et al,[8] with permission from the publisher.)

multiple and tend to have thicker and coarser rims of peripheral calcification than do congenital or post-traumatic cysts.

CT demonstrates splenic cysts as well-circumscribed, spherical lesions that are sharply demarcated from the adjacent splenic parenchyma and contain fluid of water density. Extensive calcification may be seen within the walls of echinococcal cysts. In some patients, ultrasound can be of value in confirming the cystic nature of a splenic mass suspected on CT.[1,8,13,14]

## CHRONIC CONGESTIVE SPLENOMEGALY (BANTI'S SYNDROME)

Chronic congestive splenomegaly is a syndrome characterized by enlargement of the spleen, pancytopenia, portal hypertension, and gastrointestinal bleeding due to varices. Although most commonly due to Laennec's cirrhosis, chronic congestive splenomegaly may be caused by thrombosis of the portal or splenic vein, intrahepatic portal obstruction due to schistosomiasis, cavernous transformation of the portal vein, or compression of the splenic vein by pancreatic tumor or fibrosis.

Plain films of the abdomen and contrast studies demonstrate the enlarged spleen as a prominent soft tissue mass displacing adjacent organs. Splenic and portal vein occlusion and the development of anastomotic venous collaterals (especially varices in the lower esophagus and gastric fundus) can be demonstrated during the venous phase of selective splenic or celiac arteriography or by splenoportography (direct percutaneous injection of contrast material into the spleen).

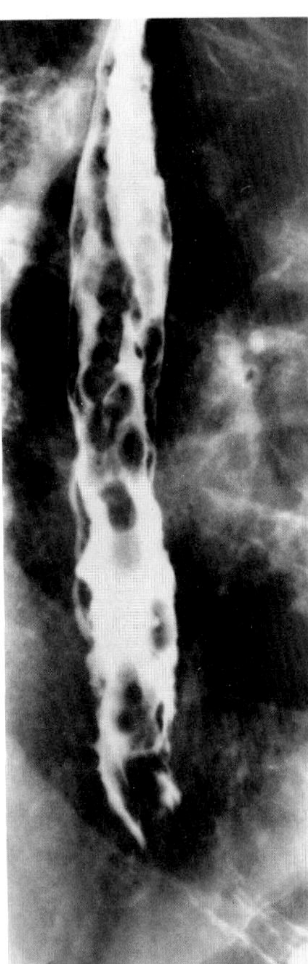

**Figure 75.13. Chronic congestive splenomegaly.** Portal hypertension has led to the development of esophageal varices, which appear as serpiginous submucosal masses representing dilated venous structures.

## HISTIOCYTOSES

The histiocytoses (eosinophilic granuloma, Hand-Schüller-Christian disease, Letterer-Siwe disease) are a group of disorders with similar pathologic lesions. Letterer-Siwe disease is a lymphomatous proliferation of poorly differentiated histiocytes which occurs in children under the age of 3 and has a rapidly fatal course. Hand-Schüller-Christian disease is considered to be a multifocal eosinophilic granuloma which, in addition to the bone involvement of unifocal eosinophilic granuloma, can appear in disseminated form, affecting multiple bones and producing pulmonary infiltrates.

Unifocal and multifocal eosinophilic granuloma produce osteolytic destructive lesions that may involve any portion of any bone, but predominantly affect the skull, pelvis, femurs, and spine. Although usually sharply defined, rapidly growing lesions may have indistinct, hazy borders. A characteristic finding is a peculiar beveled contour of the lesion with multiple undulating contours of the margin, which may produce a three-dimensional "hole-within-a-hole" effect. Periosteal reaction of a solid or laminated type may be local or extensive.

In the skull, eosinophilic granuloma typically begins as one or more small punched-out areas that originate in the diploic space and expand and perforate both the inner and outer tables. The calvarial defect may demonstrate a bony density in its center (button sequestrum).

Spine involvement typically begins as spotty destruction in the vertebral body that proceeds to collapse of the vertebra, which assumes the shape of a thin flat disk (vertebra plana).[15–18]

The pulmonary manifestations of the histiocytoses are discussed in the section "Diffuse Infiltrative Diseases" in Chapter 30.

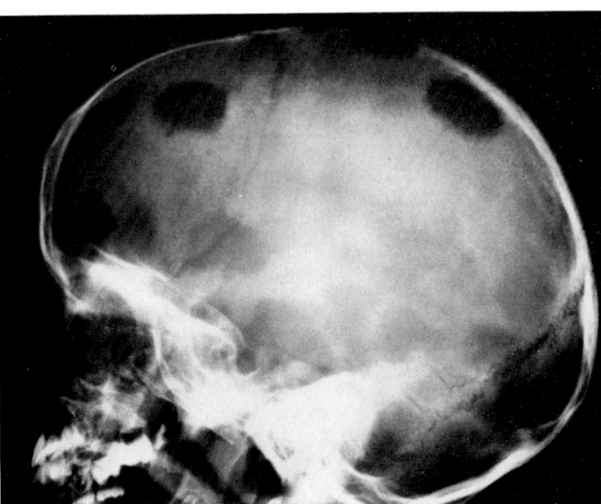

**Figure 75.14. Multifocal eosinophilic granuloma.** There are multiple punched-out lytic lesions in the skull.

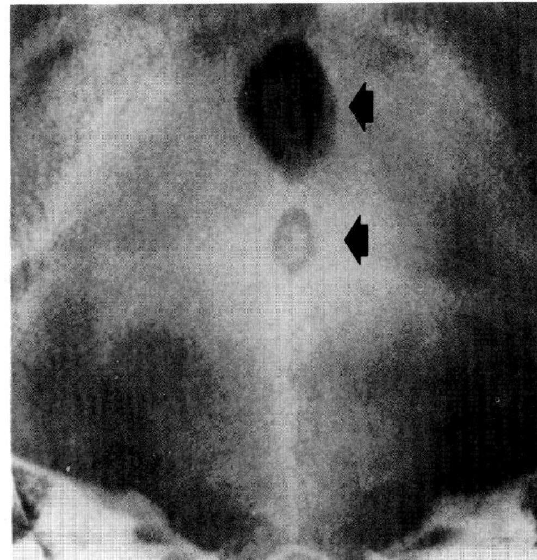

**Figure 75.15. Multifocal eosinophilic granuloma.** Central retained bone (button sequestrum) is seen in each of two midline parietal lesions (arrows). (From Sholkoff and Mainzer,[15] with permission from the publisher.)

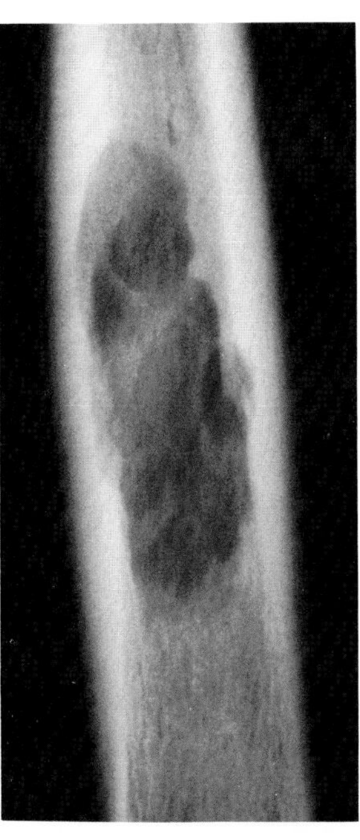

**Figure 75.16. Unifocal eosinophilic granuloma.** There is a bubbly osteolytic lesion in the femur, with scalloping of the endosteal margins and a thin layer of periosteal response.

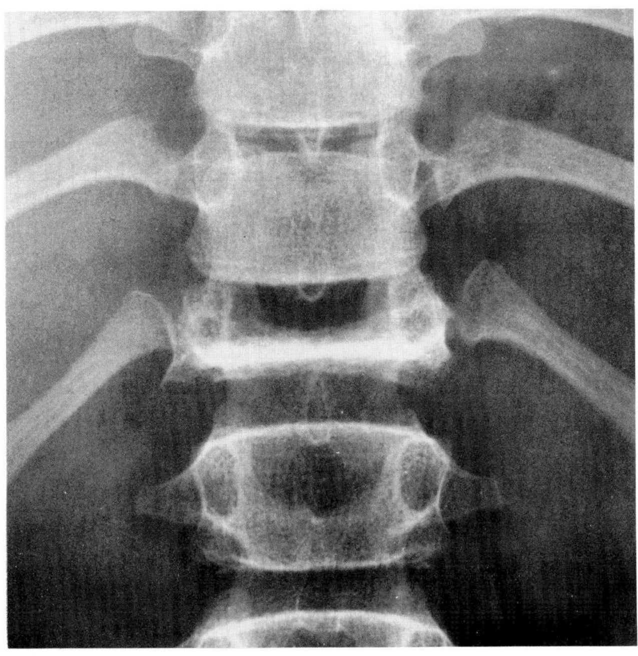

A

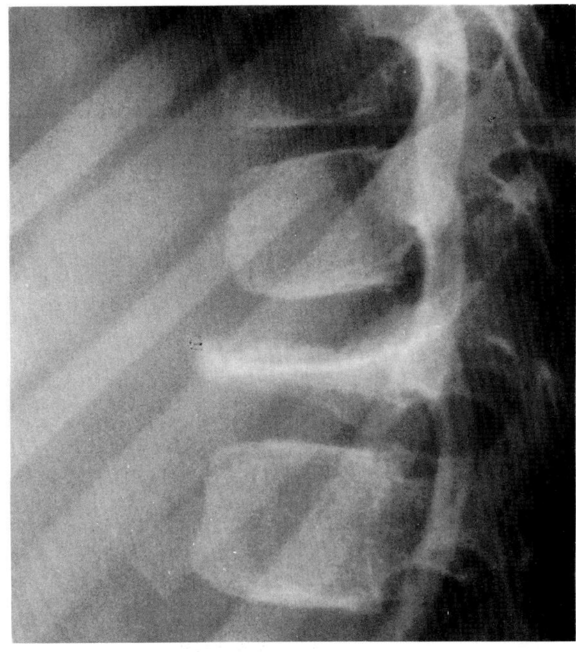

B

**Figure 75.17. Unifocal eosinophilic granuloma.** (A) Frontal and (B) lateral views of the spine show complete collapse with flattening of the T12 vertebral body (vertebra plana).

## REFERENCES


1. Koehler RE: Spleen, in Lee JKT, Sagel SS, Stanley RJ (eds): *Computed Body Tomography*. New York, Raven Press, 1983.
2. Juhl JH: *Essentials of Roentgen Interpretation*. Philadelphia, JB Lippincott, 1981.
3. Love L, Greenfield GB, Braun TW, et al: Arteriography of splenic trauma. *Radiology* 91:96–102, 1968.
4. Berk RN, Wholey MH: The application of splenic arteriography in the diagnosis of rupture of the spleen. *AJR* 104:662–666, 1968.
5. Berk RN: Changing concepts in the plain-film diagnosis of ruptured spleen. *J Canad Assoc Radiol* 21:67–74, 1970.
6. Jeffrey RB, Laing FC, Federle MP, et al: Computed tomography of splenic trauma. *Radiology* 141:729–732, 1981.
7. Kass JB, Fisher RG: The Seurat spleen. *AJR* 132:683–684, 1979.

8. Piekarski J, Federle MP, Moss AA, et al: Computed tomography of the spleen. *Radiology* 135:683–689, 1980.
9. Johnson MA, Cooperberg PL, Boisvert J, et al: Spontaneous splenic rupture in infectious mononucleosis. *AJR* 136:111–114, 1981.
10. Mall JC, Kaiser JA: CT diagnosis of splenic laceration. *AJR* 134:265–269, 1980.
11. Federle MP, Goldberg HI, Kaiser JA, et al: Evaluation of abdominal trauma by computed tomography. *Radiology* 138: 637–644, 1981.
12. Steinberg I, Finby N, Evans JA: Aneurysms of the splenic, hepatic and renal arteries. *AJR* 86:1108–1122, 1961.
13. Shirkoda A, McCartney WH, Staab EV, et al: Imaging of the spleen: A proposed algorithm. *AJR* 135:195–198, 1980.
14. Griscom NT, Harbreaves HK, Schwartz MZ, et al: Huge splenic cyst in a newborn: Comparison with ten cases in later childhood and adolescence. *AJR* 129:889–891, 1977.
15. Sholkoff SD, Mainzer S: Button sequestrum revisited. *Radiology* 100:649–652, 1971.
16. Ennis JT, Whitehouse G, Ross FGM: The radiology of bone changes in histiocytosis. *Clin Radiol* 24:212–220, 1973.
17. Lichtenstein L: Histiocytosis X. *J Bone Joint Surg* 46A:76–90, 1964.
18. Arcomano JP, Barnett JC, Wunderlich HO: Histiocytosis X. *AJR* 85:663–679, 1961.
19. Eisenberg RL: *Gastrointestinal Radiology: A Pattern Approach*. Philadelphia, JB Lippincott, 1983.

# Chapter Seventy-Six
# Diseases due to Environmental Hazards and Chemical and Physical Agents

## HEAVY METALS

### Lead

Lead poisoning results from the ingestion of lead-containing materials (e.g., in paint previously used for children's furniture and toys) or from the inhalation of lead fumes (e.g., burning storage batteries, solder, paint spraying). In adults, chronically ingested or inhaled lead is deposited in the bones, but does not appear radiographically because bone growth has ceased. In children, however, because lead and calcium are utilized interchangeably by bone, high concentrations of lead are deposited in the most rapidly growing portions of the skeleton, especially the metaphyses at the distal ends of the femur and radius and both ends of the tibia. This results in characteristic lead lines: dense transverse bands extending across the metaphyses of the long bones and along the margins of flat bones such as the iliac crest. Lead lines can be observed in growing bone about 3 months after the inhalation of lead and 6 months after the ingestion of the metal. After the intake of lead has ceased, normal bone forms on the epiphyseal side of the metaphysis and the lead line appears to migrate into the shaft, usually becoming wider and less dense before gradually disappearing. In patients with severe lead poisoning, a wide band of metaphyseal involvement can prevent normal remodeling and lead to residual deformity. Lead lines must be differentiated from the usual whiteness of the metaphyseal ends of tubular bones that is often seen in normal active children of less than 3 years of age.

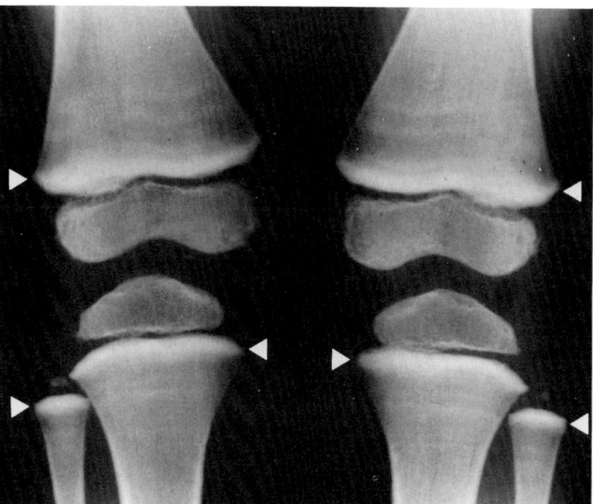

**Figure 76.1. Lead lines.** Dense transverse bands of sclerosis (arrowheads) extend across the metaphyses of the distal femurs and the proximal tibias and fibulas.

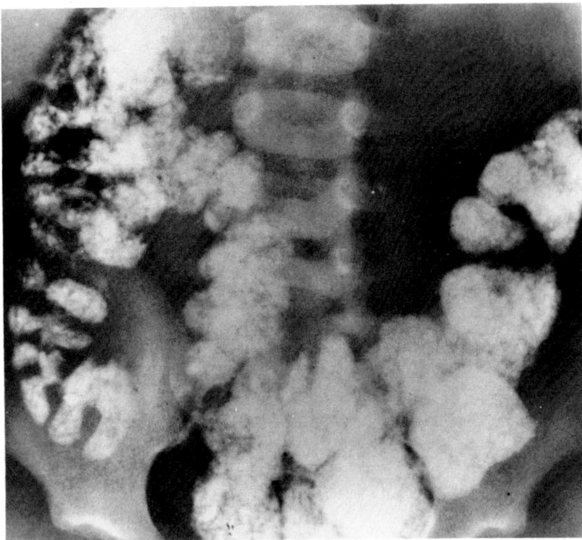

**Figure 76.2. Pica.** Ingested gravel and stones fill the colon of this child, who had received no contrast material. The mother complained that she heard a strange "plunking" noise whenever her child had a bowel movement. (From Eisenberg,[27] with permission from the publisher.)

In young children, irregular dense paint fragments (pica) in the gastrointestinal tract should suggest lead poisoning. Plain abdominal radiographs may demonstrate extensive mottled opacities suggesting intestinal barium even though no contrast material has been administered. Other radiographic manifestations of lead poisoning include persistent segmental or complete dilatation of the colon, calcification of the basal ganglia, and, in infants, widening of skull sutures because of increased intracranial pressure.[1,2]

## Bismuth and Phosphorus

Bismuth poisoning usually involves the long bones of syphilitic babies treated with bismuth or babies born to syphilitic women treated by injections of bismuth during pregnancy. Metallic phosphorus (yellow phosphorus) poisoning was formerly seen in children with rickets who were treated with phosphorized cod liver oil. The osseous manifestations of both these rare conditions are effectively identical to the radiographic changes seen in lead poisoning.

## Mercury

Acute mercury poisoning, usually caused by accidental ingestion, produces a severe ulcerative and necrotizing inflammation of the entire gastrointestinal tract. Secondary deposition in the kidneys causes destruction of tubular cells and may lead to fatal renal failure. Central nervous system involvement produces a broad spectrum of primary neurologic signs.

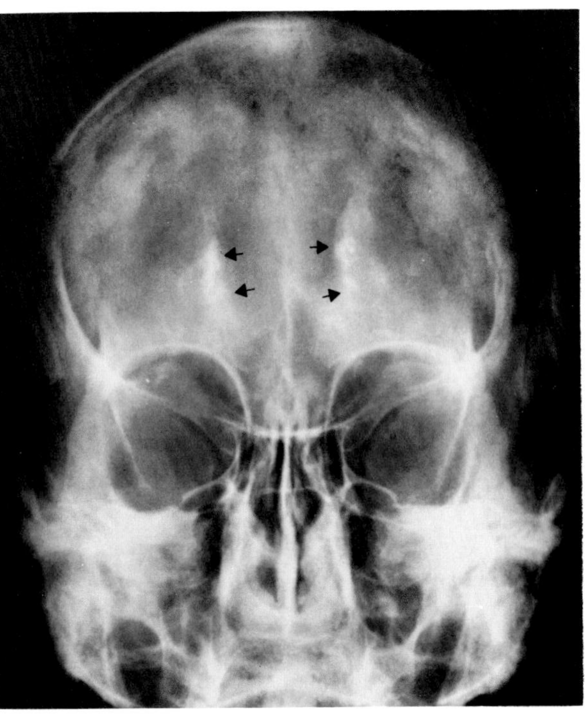

**Figure 76.3. Lead poisoning.** There is bilateral calcification of the basal ganglia (arrows).

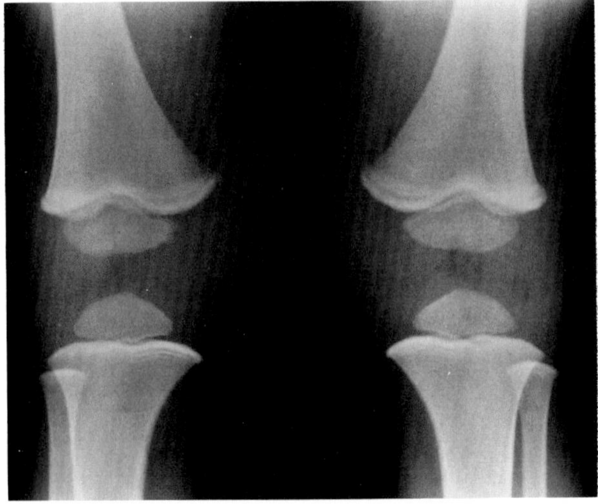

**Figure 76.4. Bismuth poisoning.** Transverse bands of metaphyseal density simulate lead lines.

Inhalation of mercury vapor, most often the result of an industrial accident, causes an acute necrotizing bronchiolitis and exudative pulmonary edema. Bronchiolar obstruction leads to a pattern of diffuse emphysema intermingled with patchy areas of hazy increased density that is more suggestive of pulmonary edema than of

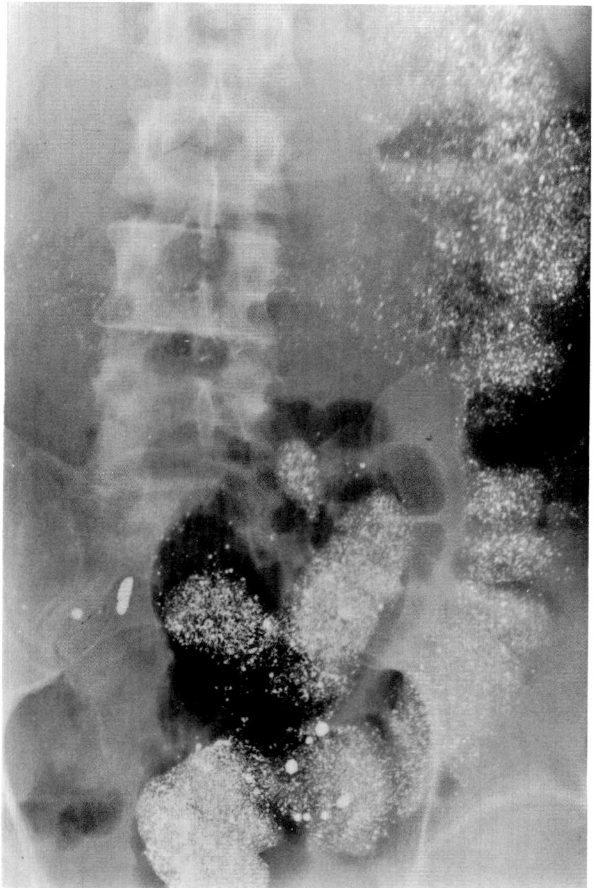

**Figure 76.5. Mercury in the bowel lumen.** The patient bit off the end of a thermometer and swallowed the mercury in it. (From Eisenberg,[27] with permission from the publisher.)

pneumonic consolidation. The hemidiaphragms are often depressed, and spontaneous pneumothorax frequently occurs. Interstitial fibrosis and secondary bacterial pneumonia can complicate the acute process.[3]

## Zinc

The fumes of zinc chloride, a substance used in smoke bombs, are extremely caustic to the mucous membranes. When inhaled, the fumes may cause severe damage to both the tracheobronchial tree and the lung parenchyma. Extensive pulmonary infiltrates and a pulmonary edema-like pattern may lead to the rapid development of diffuse interstitial fibrosis and an often fatal outcome.[4,5]

## Radium

Most victims of radium poisoning were workers who painted luminous dials on watches and clocks and ingested the radioactive luminous compound by licking their brushes to point them. Radium is metabolized in a manner similar to calcium and thus deposits in the skeleton. Long-term alpha radiation exposure causes severe tissue damage. Radiographically, radium poisoning causes ischemic necrosis and a pattern of irregular areas of destruction intermingled with zones of sclerosis. In radium poisoning, unlike other causes of ischemic necrosis, involvement of the articular cartilage causes joint space narrowing and secondary osteoarthritic changes. Pathologic fractures may occur, especially in the vertebrae.

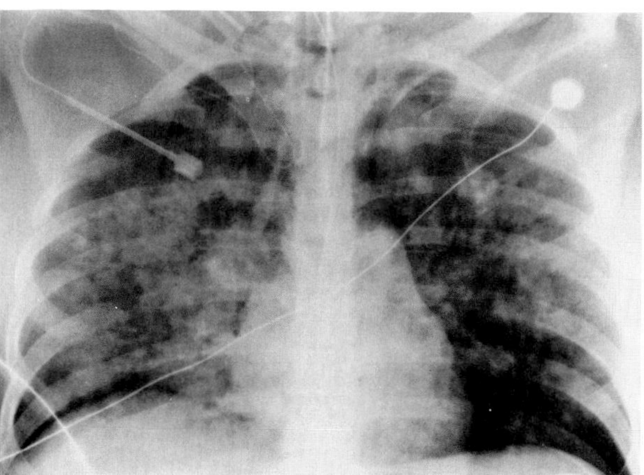

**Figure 76.6. Mercury poisoning.** Following the inhalation of mercury vapor, there is rapid development of an exudative pulmonary edema that is more marked on the right. The unusually sharp interface (with hyperlucency) between the right hemidiaphragm and the base of the right lung represents a pneumothorax in this patient, who is lying in the supine position. Note that the right subclavian line extends upward into the neck rather than downward into the superior vena cava or right atrium.

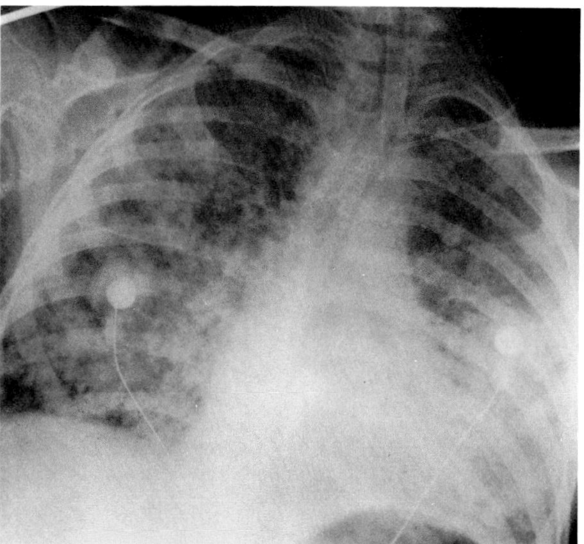

**Figure 76.7. Zinc chloride inhalation.** An extensive chemical pneumonia involving both lungs and simulating pulmonary edema developed in this army recruit following an exposure to smoke bombs during training maneuvers.

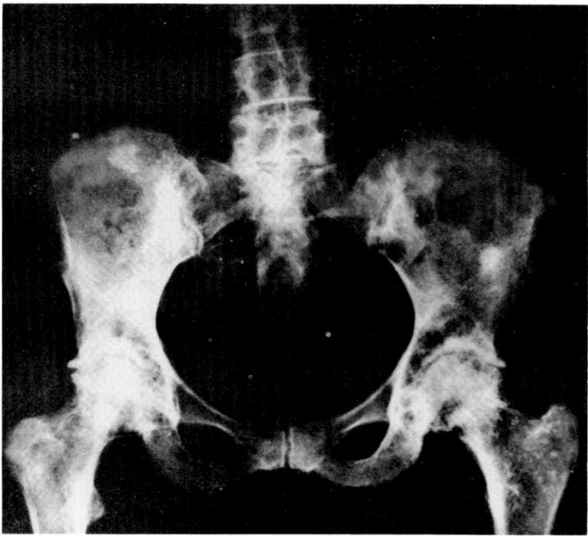

**Figure 76.8. Radium poisoning.** Multiple destructive lesions intermingled with zones of sclerosis are seen in the pelvis and proximal femurs. (From Looney, Hasterlik, Brues, et al,[8] with permission from the publisher.)

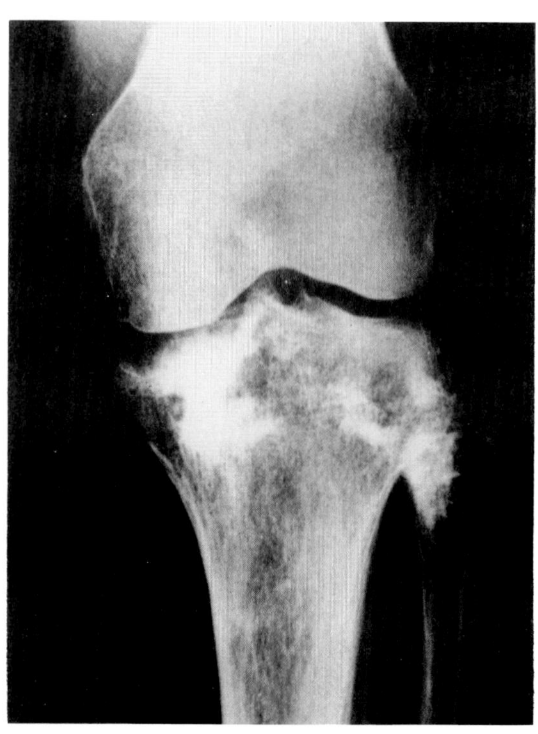

**Figure 76.9. Radium poisoning.** There is increased density in the proximal tibia and fibula. (From Looney, Hasterlik, Brues, et al,[8] with permission from the publisher.)

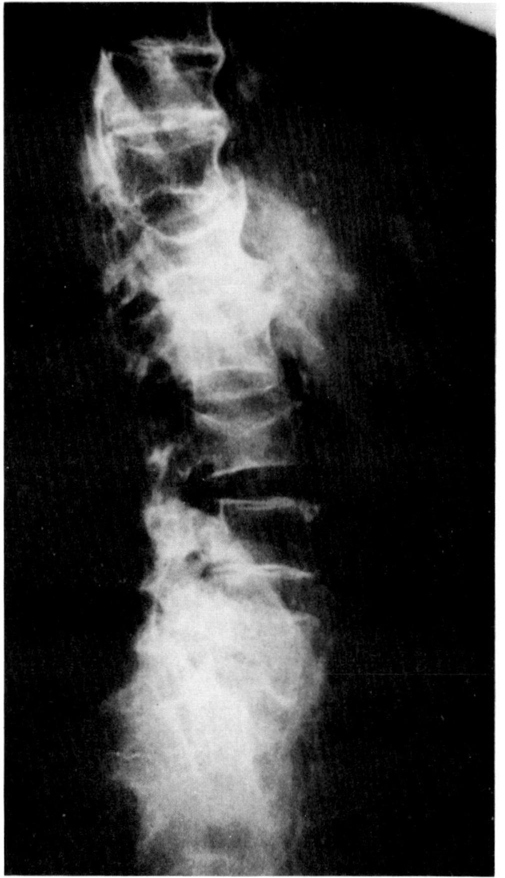

**Figure 76.10. Radium poisoning.** There is bone destruction with pathologic fractures of multiple vertebral bodies. (From Looney, Hasterlik, Brues, et al,[8] with permission from the publisher.)

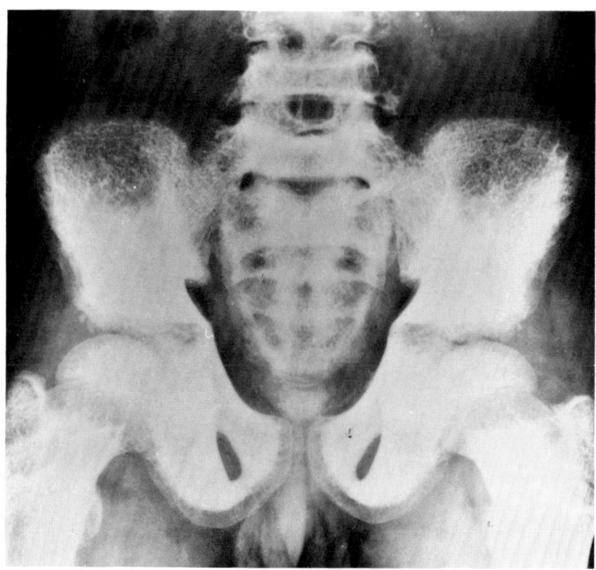

**Figure 76.11. Fluorosis.** Dense skeletal sclerosis with obliteration of individual trabeculae causes the pelvis and proximal femurs to appear chalky white.

Severe complications of radium poisoning include aplastic anemia and the development of radiation-induced tumors, such as carcinomas of the paranasal sinuses and the mastoids and sarcomas of bone.[6–8]

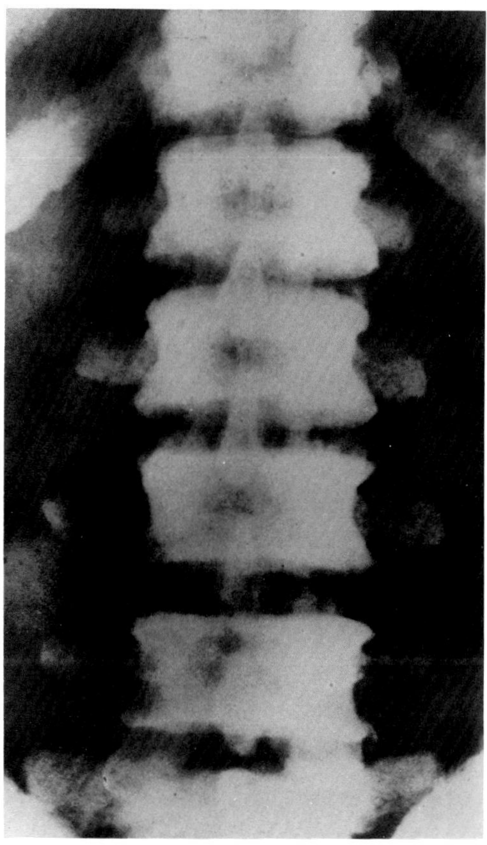

**Figure 76.12. Fluorosis.** Diffuse vertebral sclerosis. (From Cockshott and Middlemiss,[28] with permission from the publisher.)

## FLUOROSIS

Fluorine poisoning may be the result of drinking water with a high concentration of fluorides, industrial exposure (mining and conversion of phosphate rock into fertilizer; smelting of metals) or excessive therapeutic intake of fluoride (treatment of myeloma or Paget's disease).

Fluorosis typically causes dense skeletal sclerosis that is most prominent in the vertebrae and pelvis. Obliteration of individual trabeculae may cause the affected bones to appear chalky white. Although sclerosis of the long bones of the extremities is much less common, there usually is periosteal roughening and articular bone deposits that arise at sites of muscle and ligament attachment. Calcification of the interosseous membranes and the iliolumbar, sacrotuberous, and sacrospinous ligaments is a hallmark of fluorosis, though a similar pattern can represent a normal variant.

The associated findings of pelvic ligament calcification and irregular periosteal projections from the long bones should permit the differentiation of fluorosis from such other causes of generalized bone sclerosis as metastases, Paget's disease, myelofibrosis, and congenital osteopetrosis.[9,10]

## INTRAVENOUS DRUG ABUSE

Intravenous drug abuse, especially heroin addiction, is an ever-increasing problem. Narcotic overdose can produce a pattern of extensive pulmonary edema, which can lead to respiratory failure and death. The radiographic appearance clears rapidly in those patients who survive. A similar pulmonary edema pattern may occasionally develop after the oral administration of heroin or methadone.

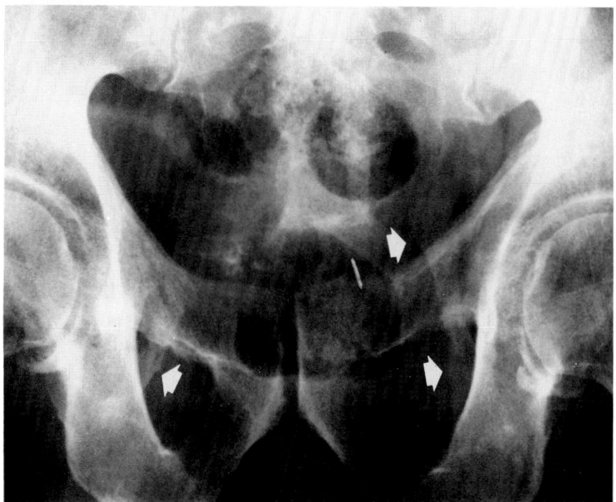

**Figure 76.13. Fluorosis.** Calcification of the sacrotuberous ligaments (arrows). (From Eisenberg,[27] with permission from the publisher.)

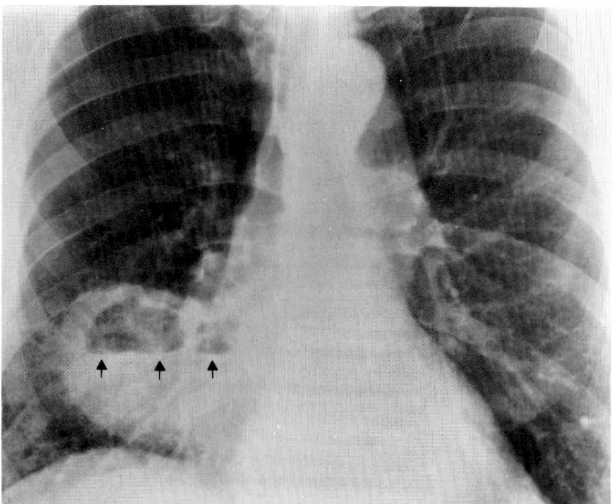

**Figure 76.14. Lung abscess secondary to intravenous drug abuse.** A frontal view of the chest shows a large right middle lobe abscess containing an air-fluid level (arrows).

Pulmonary infections, aspiration pneumonia, septic emboli, and lung abscesses are common complications of intravenous drug abuse. Bacterial and fungal endocarditis and mycotic aneurysms can also occur. Intra-arterial injection of drugs can result in the embolization of particulate material to peripheral vessels of the extremities, occasionally causing vascular obstruction. Hepatitis and osteomyelitis are other severe complications of intravenous drug abuse.[11,12]

## ELECTRICAL INJURIES

A heavy electric current passing through the body causes violent, tetanic muscular contractions that may result in fractures, dislocations, or torn tendons and ligaments. If the patient survives, late manifestations of electrical injury are determined by the type and extent of the immediate damage. Damage to blood vessels, with severe circulatory impairment, may result in ischemic necrosis of bone. This necrosis tends to develop away from the site where the current entered the patient, though along the path the current took through the body. Damage to peripheral nerves, with loss of sensation, may lead to trophic bone changes and a Charcot's joint. Soft tissue necrosis can cause subsequent contractures, joint dislocations, infection, and disuse atrophy. Occasionally, discrete areas of rarefaction persist as permanent cystic lucencies in the bone. Short, finely visible fracture lines and periosteal reaction may occur. Resorption of the phalangeal tufts (acro-osteolysis) may develop at the point of entry of the electrical current.[13,14]

## NEAR-DROWNING

Chest radiographs in patients saved from drowning usually demonstrate bilateral and symmetric changes indistinguishable from pulmonary edema. The severity appears to depend on the amount of water inhaled and may vary from fine perihilar alveolar infiltrates to almost complete opacification of both lungs. The pulmonary changes are secondary to hypoxic lung injury and usually clear completely in 7 to 10 days. Persistence of the infiltrates or an increase in their extent suggests a superimposed bacterial pneumonia.[15,16]

## FROSTBITE

Severe frostbite causes soft tissue swelling soon after the injury. Interstitial collections of gas imply superimposed infection, a poor prognostic sign that is usually an indication for amputation. Regardless of the severity of frostbite injury, no bone or joint changes are seen immediately. Bone demineralization, most marked where the soft tissue damage was greatest, is first seen from 4

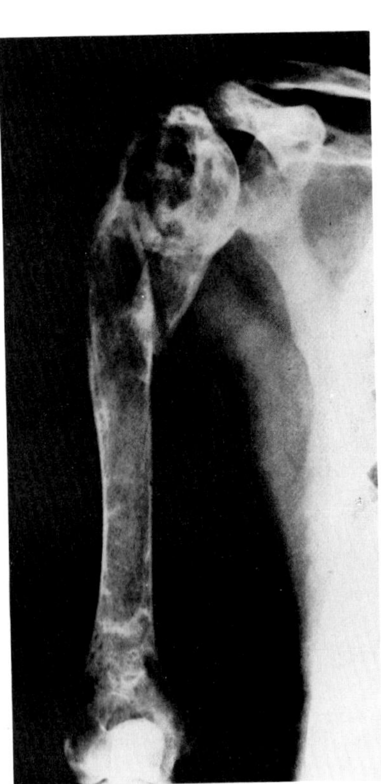

A

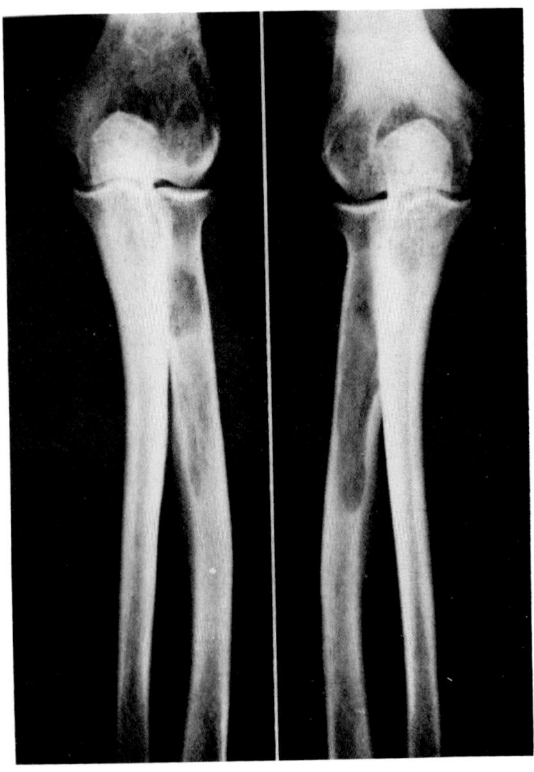

B

**Figure 76.15. Electrical injury.** (A) Comminuted fracture of the head and shaft of the humerus with a large bone fragment displaced medially. There is mottled decalcification of the humeral head. The cortex of the humerus is thin. The medullary cavity is widened, and discrete areas of rarefaction can be seen in the shaft and distal metaphyseal region. (B) A view of both forearms shows well-demarcated, elongated areas of rarefaction in the upper portion of the shafts of both radii. (From Brinn and Moseley,[14] with permission from the publisher.)

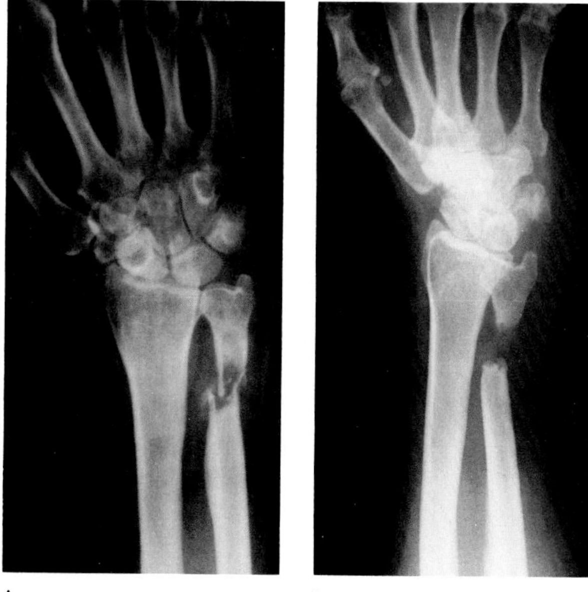

A B

**Figure 76.16. Electrical injury.** (A) A silent fracture of the ulna developed 15 months after injury. There is residual demineralization of the carpal bones and bases of the metacarpals. (B) Three and one-half years after the injury, there is a smooth loss of bone at the neck of the ulna and a quiescent, healed bony appearance. Note the rather striking reconstitution of the carpal bones with nearly normal joint spacing and mineralization. The interphalangeal joint of the thumb, which demonstrated bone and cartilage destruction a few months after the injury, now is smoothly fused. (From Barber,[13] with permission from the publisher.)

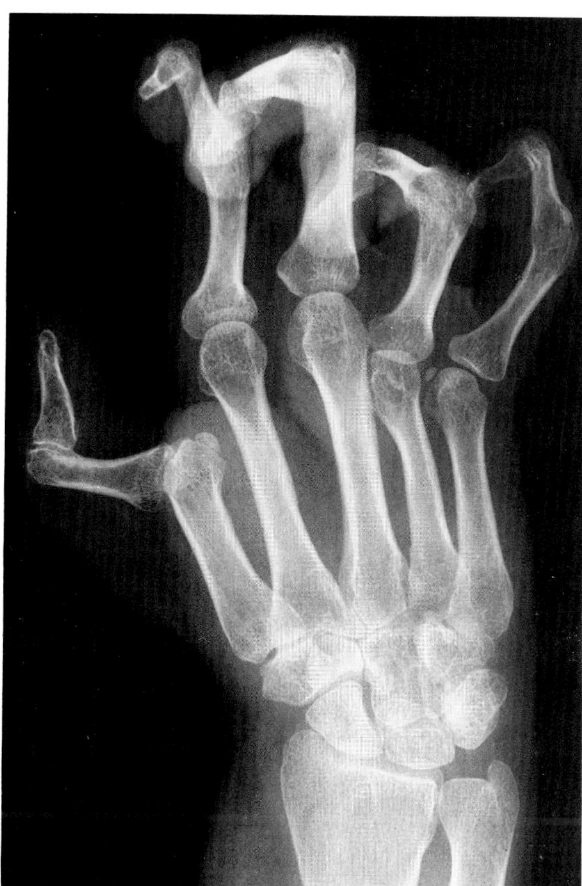

**Figure 76.17. Electrical burn scar.** There are contractures of all the digits without evidence of bone or cartilage destruction.

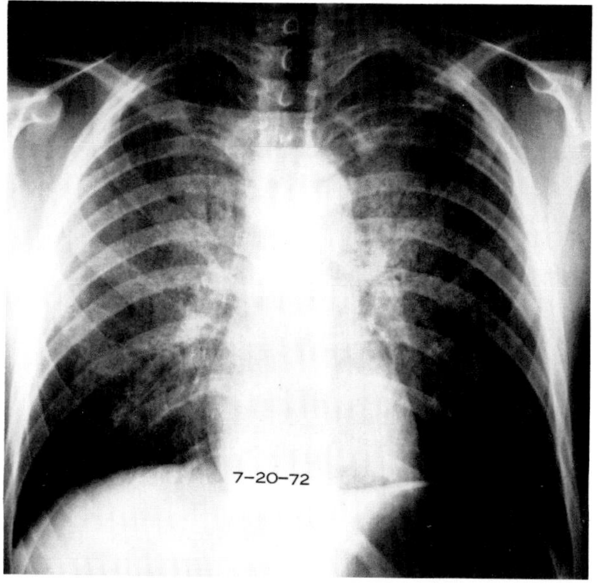


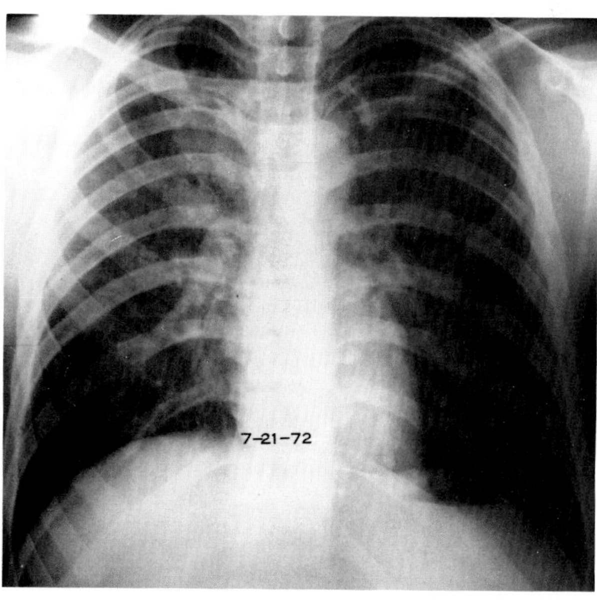


A B

**Figure 76.18. Near-drowning.** (A) An admission chest film shows fine, granular perihilar infiltrates with relative sparing of the apices, bases, and lateral portions of the lungs. Prominent air bronchograms are also evident. (B) A repeat study 18 hours later shows marked improvement. (From Hunter and Whitehouse,[15] with permission from the publisher.)

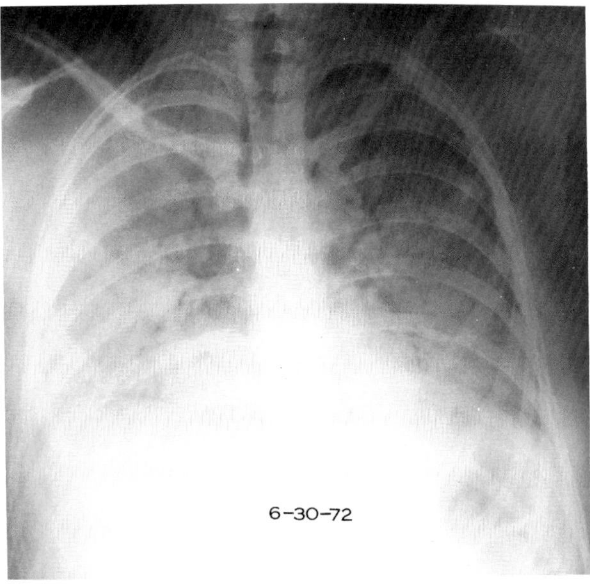


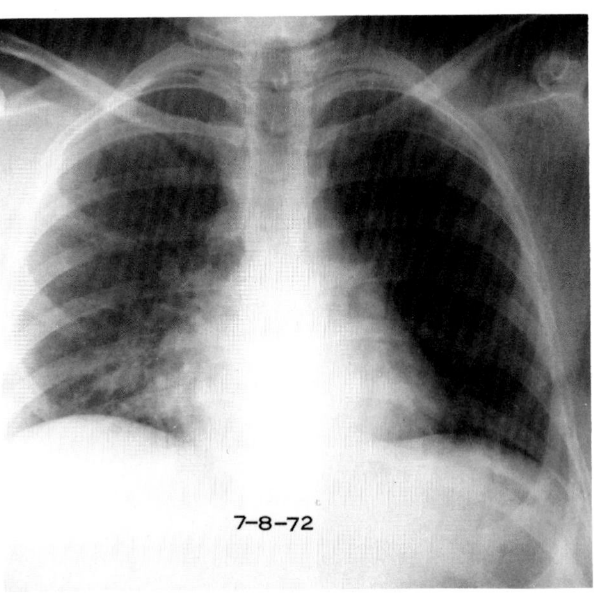


A B

**Figure 76.19. Near-drowning.** (A) An admission chest film demonstrates almost complete opacification of the lungs with prominent air bronchograms. (B) Eight days after admission, all the lung lesions have cleared except for a persistent right middle lobe infiltrate (bacterial pneumonia). (From Hunter and Whitehouse,[15] with permission from the publisher.)

to 10 weeks after the injury and may persist for many months. Bone and joint changes may occur 6 months to many years after the injury. The most common bone lesions are small, well-defined, punched-out periarticular defects about the distal interphalangeal, proximal interphalangeal, and metacarpophalangeal joints. Resorption of the terminal tufts of the distal phalanges is usually the result of superimposed infection or surgical intervention. There may be small areas of increased density in the phalangeal tufts, probably representing bone infarcts. Marginal spurring and joint space narrowing, indistinguishable from ordinary osteoarthritis, occasionally develop. Severe destruction of joint surfaces can result in ankylosis.

Frostbite injuries in the hand of a child may cause segmentation and premature fusion of the epiphyses, resulting in shortening or deformity of the involved phalanges.[17,18]

## RADIATION INJURY

Radiation injury is most commonly a complication of radiation therapy and may produce a broad spectrum of acute and chronic abnormalities. Significant radiographic changes can develop in the lungs, bones, gastrointestinal tract, and kidneys.

Radiation damage to the lung may be a complication of therapy for tumors of the breast, lung, esophagus, or mediastinum. Acute radiation pneumonitis, which is rarely detectable less than 1 month after the cessation of treatment, appears radiographically as a hazy, poorly defined increased density that is usually confined to the radiation port. This progresses to patchy areas of irregular consolidation, often associated with a considerable loss of volume. The late or chronic stage of radiation damage is characterized by extensive fibrosis, with the affected lung demonstrating a severe loss of volume and obliteration of all normal architectural markings. Radiation pneumonitis presents a challenging diagnostic dilemma. The acute pulmonary reaction must be differentiated from bacterial pneumonia; in chronic stages, the dense fibrotic strands that frequently extend from the hilum to the periphery may be difficult to distinguish from the lymphangitic spread of a malignant tumor.

In the skeleton, radiation changes are largely due to the effects on blood vessels supplying the bone. Initially, hyperemia causes local demineralization. More severe injury leads to coarsening and disorganization of the trabecular pattern, which results in disintegration of the bone and the radiographic changes of ischemic necrosis. Areas of dense sclerosis and focal lucency can produce a pagetoid appearance, especially in the pelvis. Pathologic fractures are common, are often painless, and rarely heal. Radiation-induced sarcoma is a rare complication of radiation therapy.

In children, radiation of the metaphyseal-epiphyseal area damages active centers of bone growth. This leads to the retardation or cessation of growth, which results in a shortening and narrowing of bones, a failure of proper tubulation, and other skeletal deformities. Radiation to half the growing spine may cause severe scoliosis.

Clinically significant radiation damage to the gas-

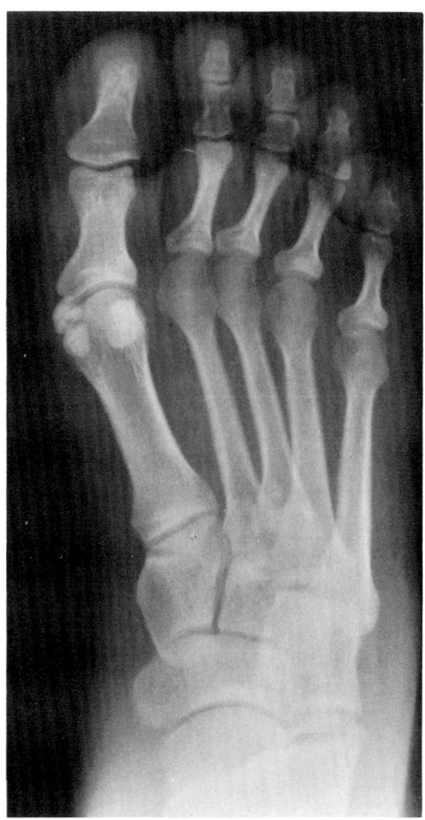

A

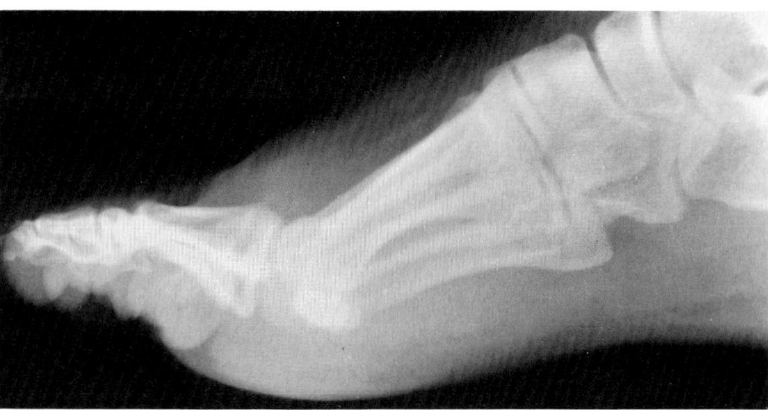

B

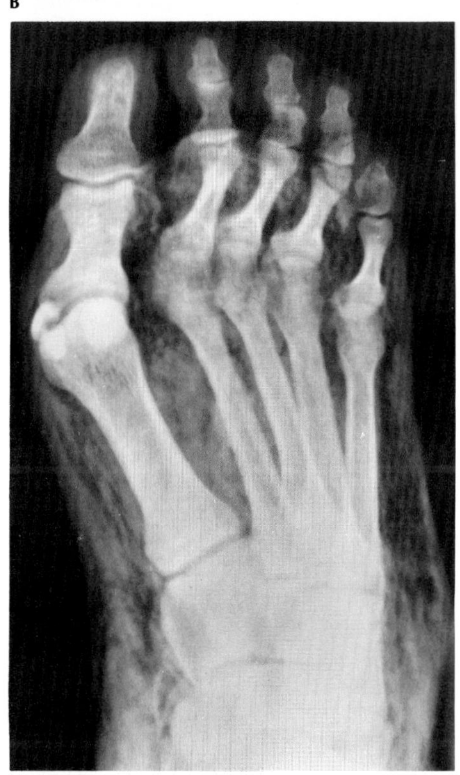

C

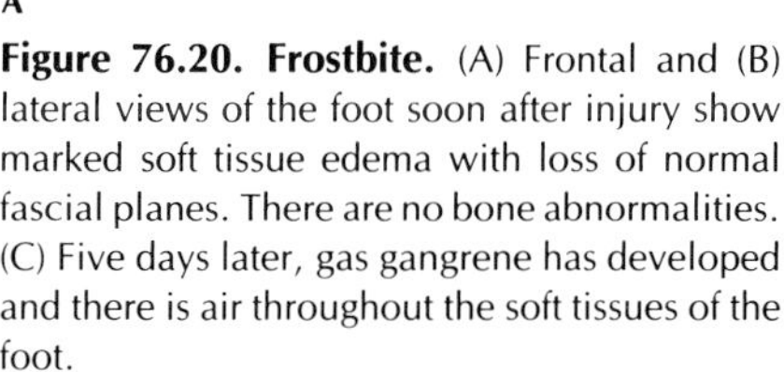

**Figure 76.20. Frostbite.** (A) Frontal and (B) lateral views of the foot soon after injury show marked soft tissue edema with loss of normal fascial planes. There are no bone abnormalities. (C) Five days later, gas gangrene has developed and there is air throughout the soft tissues of the foot.

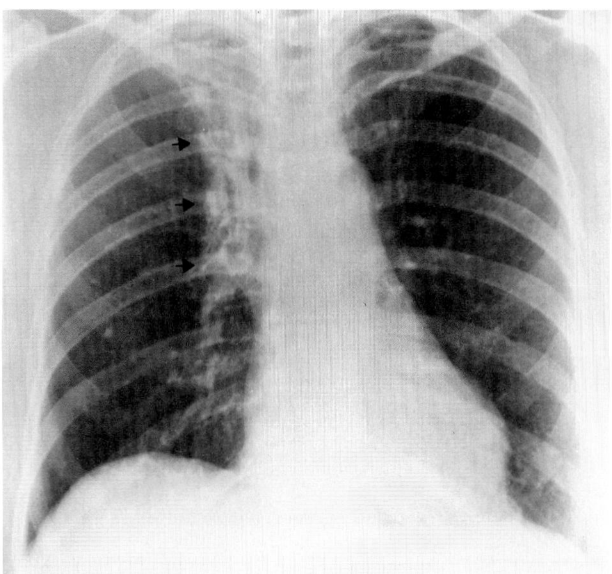

**Figure 76.21. Radiation pneumonitis.** Following postmastectomy radiation, a mass of fibrous tissue (arrows) extends from the right hilum to parallel the right border of the mediastinum. Note the absence of the right breast shadow.

trointestinal tract can follow radiotherapy to the abdomen. This radiation injury is probably secondary to an obliterative arteritis with vascular occlusion and bowel ischemia. Therefore, radiation effects are most common in patients with diabetes, arteriosclerosis, and hypertension, whose vessels are already compromised.

Radiation injury to the bowel primarily involves the colon. Although the small intestine is more radiosensitive than the colon, it is less susceptible to radiation injury from fixed ports because of its inherent mobility. Transient proctitis manifested as diarrhea, mucoid discharge, tenesmus, or crampy pain occurs in more than half the patients receiving pelvic irradiation for carcinoma of the cervix, endometrium, ovaries, bladder, or prostate. In patients with more severe acute radiation-induced colitis, barium enema examinations can demonstrate segmental changes of irregularity, spasm, and fine serrations of the bowel wall similar to the appearance of other ulcerating diseases of the colon. Discrete ulceration of the mucosa is frequent and can be superficial or penetrating. Diffuse fibrosis narrows the lumen and produces a characteristic long, smooth stricture that commonly involves the rectum and sigmoid colon and develops within 6 to 24 months of irradiation. A short, irregular radiation-induced stricture, especially if it has a relatively abrupt margin of transition, can closely resemble primary or metastatic malignancy. Fistulization to the bladder or to other bowel loops often occurs.

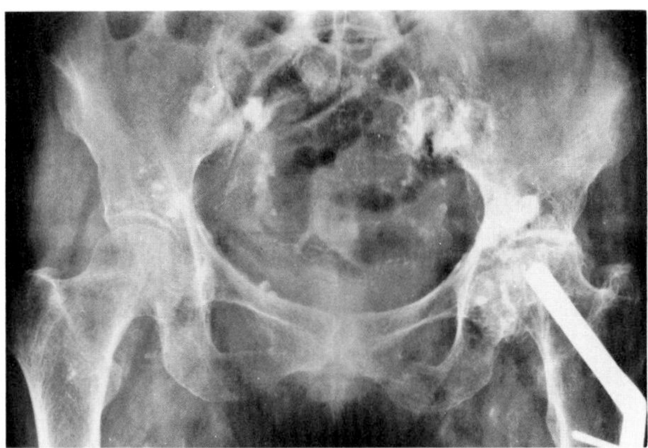

**Figure 76.22. Radiation necrosis of bone.** A frontal view of the pelvis shows patchy areas of dense sclerosis that developed following radiation therapy for carcinoma of the cervix. Flattening and sclerosis of the left femoral head reflects ischemic necrosis.

Small bowel changes associated with radiation injury include shallow mucosal ulcerations, thickened folds, nodular filling defects, and the separation of adjacent loops, which reflects thickening of the bowel wall from submucosal edema and fibrosis. Radiation injury to the stomach produces a diffuse gastritis, which can be associated with discrete gastric ulcers that often perforate or bleed. Healing with excessive fibrous scarring can lead to fixed luminal narrowing and mural rigidity, producing a linitis plastica pattern. Mediastinal radiation therapy can cause esophagitis, which usually appears as a benign-appearing stricture with tapered margins and relatively smooth mucosal surfaces.

Acute radiation nephritis can develop 6 to 12 months following radiation therapy (see the section "Radiation Nephritis" in Chapter 37). Excretory urography demonstrates a normal or slightly enlarged kidney with mild functional impairment. In chronic radiation nephritis, glomerular damage, tubular atrophy, arteritis, and in-

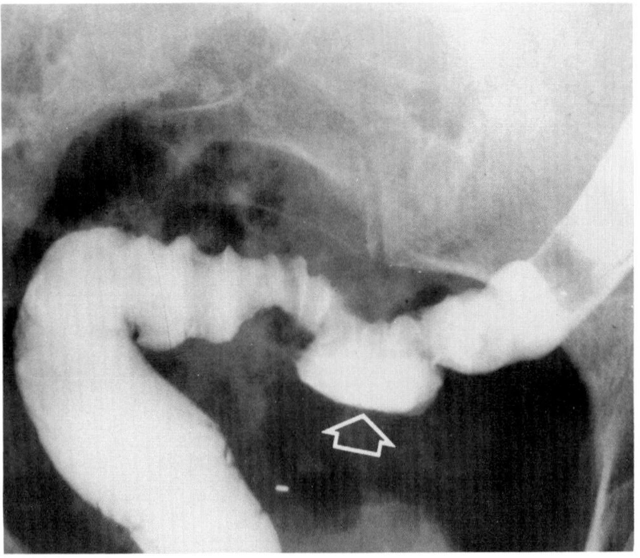

A

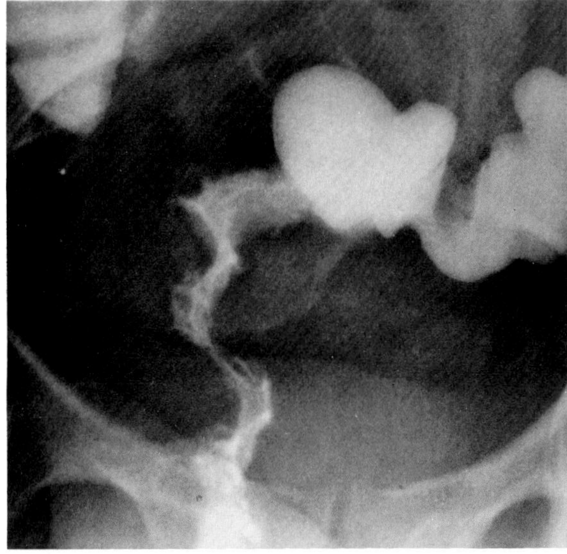

B

**Figure 76.23. Radiation injury to the colon.** (A) Large, discrete, penetrating ulcer (arrow) in the sigmoid colon. (B) Irregularity, spasm, and ulceration of the rectosigmoid. (A, from Rogers and Goldstein,[23] with permission from the publisher.)

**Figure 76.24. Radiation injury to the small bowel.** Thickening of the bowel wall and multiple nodular masses cause separation of small bowel loops in a patient who had received 7000 rad for treatment of metastatic carcinoma of the cervix.

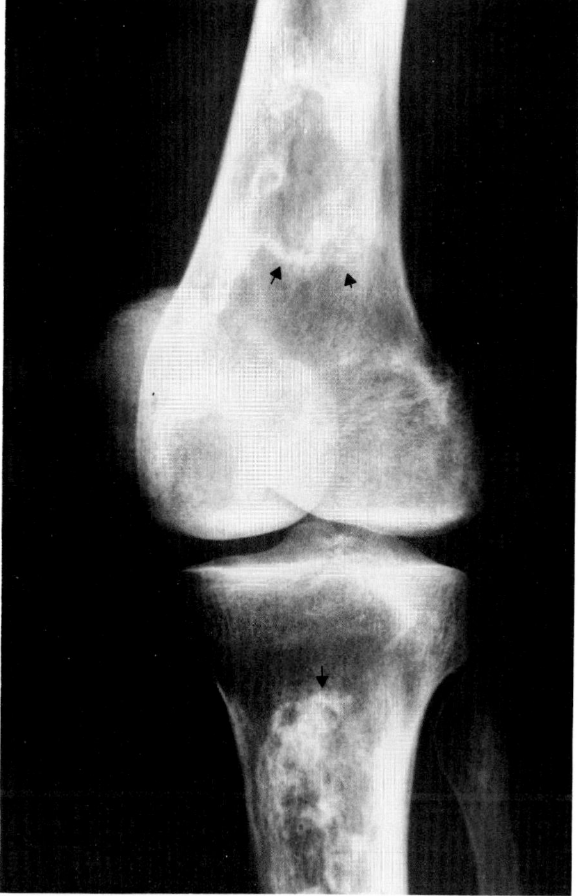

**Figure 76.25. Caisson disease.** Calcified bone infarcts (arrows) are seen in the distal femur and proximal tibia of a deep-sea diver.

terstitial fibrosis produce a small, poorly functioning kidney. This condition may progress to malignant hypertension unless the involved kidney is removed.[19–24]

## CAISSON DISEASE (DECOMPRESSION SICKNESS)

Caisson disease ("the bends") develops in persons working under increased atmospheric pressure who undergo too rapid decompression. Under high atmospheric pressure, an increased amount of inert nitrogen from the air is absorbed into the circulation. If the air pressure is released too rapidly, the nitrogen cannot stay in solution and is liberated in a gaseous state, with bubbles of nitrogen forming in the blood and tissues.

The typical radiographic bone changes in caisson disease often develop several months to years after exposure to hyperbaric conditions. In many cases the patients are asymptomatic. The lesions are typically extensive, multiple, and usually bilateral. The classic radiographic appearance is that of aseptic necrosis involving the articular ends of long bones (especially the femur and humerus) and irregularly calcified bone infarcts, which most commonly develop about the knee (see the section "Ischemic Necrosis" in Chapter 62).[25,26]

## REFERENCES

1. Leone AJ: On lead lines. *AJR* 103:165–167, 1968.
2. Pease CN, Newton GG: Metaphyseal dysplasia due to lead poisoning in children. *Radiology* 79:233–240, 1962.
3. Teng CT, Brennan JC: Acute mercury vapor poisoning. A report of four cases with radiographic and pathologic correlation. *Radiology* 73:354–361, 1959.
4. Milliken JA, Waugh D, Kadish ME: Acute interstitial fibrosis caused by a smokebomb. *Can Med Assoc J* 88:36–39, 1963.
5. Johnson FA, Stonehill RB: Chemical pneumonitis from inhalation of zinc chloride. *Dis Chest* 40:619–624, 1961.
6. Aub JC, Evans RD, Hempleman LH, et al: Late effects of internally deposited radioactive material in man. *Medicine* 31:221–229, 1952.
7. Martland HS: Occupational poisoning in manufacture of luminous watch dials. *JAMA* 92:466–467, 1929.
8. Looney WB, Hasterlik RJ, Brues AM, et al: Clinical investigation of chronic effects of radium salts administered therapeutically (1915–1931). *AJR* 74:1006–1037, 1955.

9. Stevenson CA, Watson R: Fluoride osteosclerosis. *AJR* 78:13–18, 1957.
10. Morris JW: Skeletal fluorosis among Indians of the American Southwest. *AJR* 94:608–614, 1965.
11. Morrison WJ, Wetherill S, Zyroff J: The acute pulmonary edema of heroin intoxication. *Radiology* 97:347–351, 1970.
12. Jaffe RB, Koschman EB: Intravenous drug abuse: Pulmonary, cardiac, and vascular complications. *AJR* 109:107–120, 1970.
13. Barber JW: Delayed bone and joint changes following electrical injury. *Radiology* 99:49–53, 1971.
14. Brinn LB, Moseley JE: Bone changes following electrical injury. *AJR* 97:682–686, 1966.
15. Hunter TB, Whitehouse WM: Fresh-water near-drowning: Radiological aspects. *Radiology* 112:51–56, 1974.
16. Rosenbaum HT, Thompson WL, Fuller RH: Radiographic pulmonary changes in near-drowning. *Radiology* 83:306–313, 1964.
17. Tishler JM: The soft-tissue and bone changes in frostbite injuries. *Radiology* 102:511–513, 1972.
18. Ellis R, Short JG, Simmonds BD: Unilateral osteoarthritis of the distal interphalangeal joints following frostbite. *Radiology* 93:857–858, 1969.
19. Dalinka MK, Edeiken J, Finkelstein JB: Complications of radiation therapy: Adult bone. *Semin Roentgenol* 9:29–40, 1974.
20. Bragg DG, Shidnia H, Chu FCH, et al: The clinical and radiographic aspects of radiation osteitis. *Radiology* 97:103–111, 1970.
21. Luxton RW: Radiation nephritis. *Lancet* 2:1221–1224, 1961.
22. Meyer JE: Radiography of the distal colon and rectum after irradiation of carcinoma of the cervix. *AJR* 136:691–699, 1981.
23. Rogers LF, Goldstein HM: Roentgen manifestations of radiation injury to the gastrointestinal tract. *Gastrointest Radiol* 2:281–291, 1977.
24. Smith JC: Radiation pneumonitis. A review. *Am Rev Resp Dis* 87:647–655, 1963.
25. Poppel MW, Robinson WT: The roentgen manifestation of caisson disease. *AJR* 76:74–81, 1956.
26. Mellen JR, Kindwal EP: Occupational aseptic necrosis of bone secondary to occupational exposure to compressed air. *AJR* 115:512–517, 1972.
27. Eisenberg RL: *Gastrointestinal Radiology: A Pattern Approach.* Philadelphia, JB Lippincott, 1983.
28. Cockshott WP, Middlemiss H: *Clinical Radiology in the Tropics.* Edinburgh, Churchill Livingstone, 1979.

## Chapter Seventy-Seven
# Cutaneous Diseases

## BENIGN MUCOUS MEMBRANE PEMPHIGOID

Benign mucous membrane pemphigoid, in contrast to other forms of pemphigus, particularly involves the mucous membranes of the mouth and conjunctiva, runs a chronic course, and tends to produce scarring. When the pharynx and the esophagus are affected, postinflammatory scarring leads to stricture formation that simulates an esophageal web or a malignant process.[1]

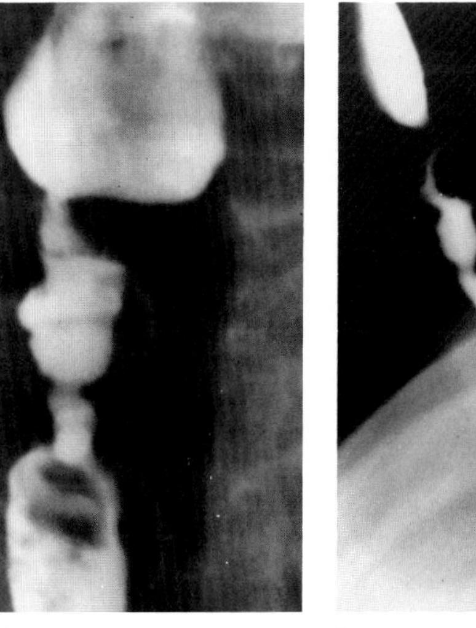
A

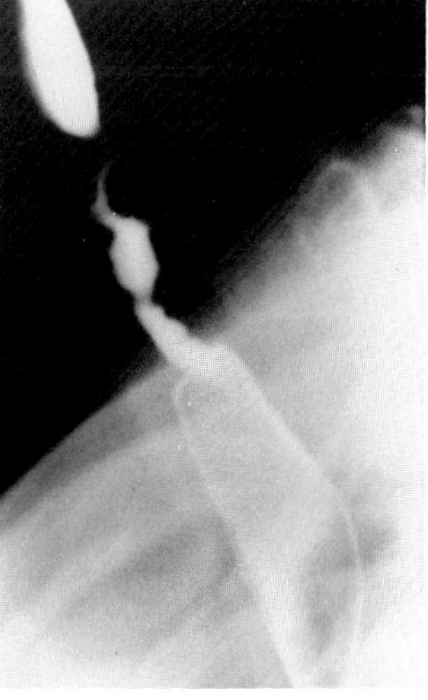
B

**Figure 77.1. Benign mucous membrane pemphigoid.** (A) Frontal and (B) lateral views of the upper esophagus demonstrate a long, irregular area of narrowing, suggestive of a malignant process, that was caused by postinflammatory scarring. (From Eisenberg,[9] with permission from the publisher.)

## EPIDERMOLYSIS BULLOSA

Epidermolysis bullosa is a rare hereditary disorder in which the skin blisters spontaneously or with injury. Subepidermal blisters characteristically affect the mucous membranes, giving rise to buccal contractures and feeding difficulties in infancy and childhood. Similar lesions in the esophagus can progress to stenotic webs. Epithelial bridging between apposing skin bullae leads to webbing between the fingers and flexion deformities that produce a clawlike hand. Severe scarring causes soft tissue atrophy and trophic changes of shortening and tapering of the distal phalanges.[2]

**Figure 77.2. Epidermolysis bullosa.** A stenotic web (arrow) in the upper esophagus resulted from the healing of subepidermal blisters involving the mucous membrane. (From Eisenberg,[9] with permission from the publisher.)

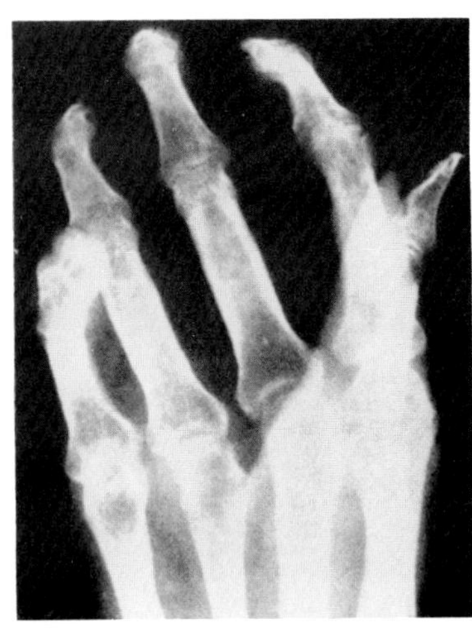

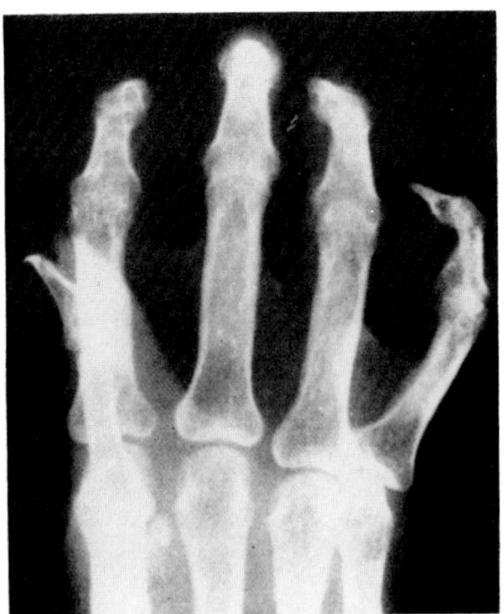

**Figure 77.3. Epidermolysis bullosa.** Frontal views of both hands show bilateral contracture deformities resulting in a claw hand. There is soft tissue webbing between the fingers. The terminal phalanges of the thumbs have a peculiar pointed, hooklike appearance. (From Brinn and Khilnani,[2] with permission from the publisher.)

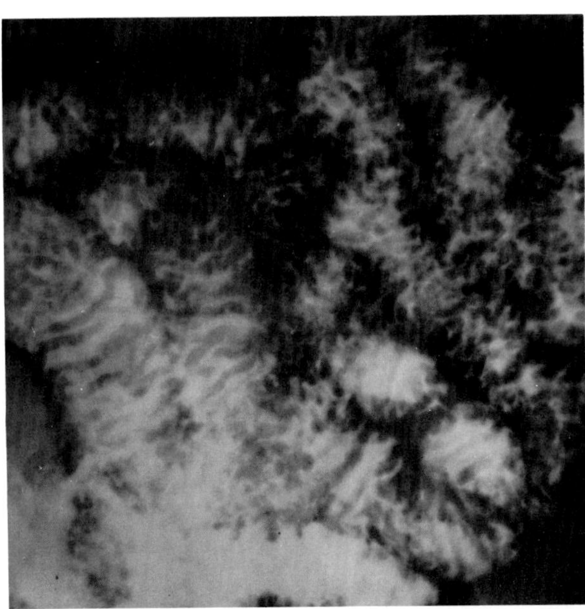

**Figure 77.4. Mastocytosis involving the small bowel.** Nodular filling defects are superimposed on a pattern of irregular thickened folds. (From Eisenberg,[9] with permission from the publisher.)

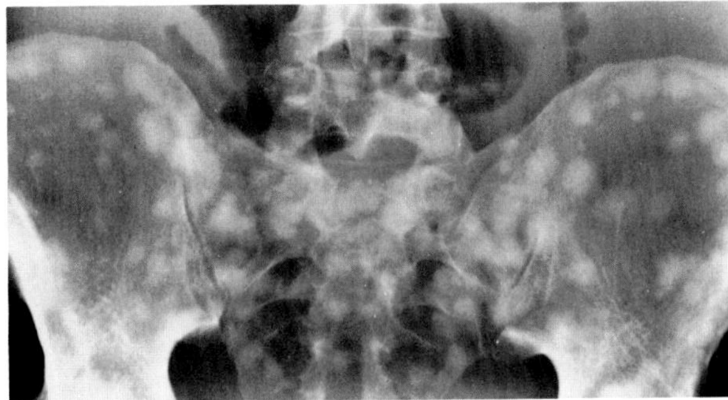

**Figure 77.5. Mastocytosis involving the skeleton.** Multiple scattered, well-defined sclerotic foci in the pelvis simulate blastic metastases.

## MASTOCYTOSIS

Systemic mastocytosis is characterized by mast cell proliferation in the reticuloendothelial system and the skin (urticaria pigmentosa). The episodic release of histamine from mast cells causes such symptoms as pruritus, flushing, tachycadia, asthma, and headaches. Nausea, vomiting, abdominal pain, and diarrhea are common gastrointestinal complaints. The incidence of peptic ulcers is high, presumably because of histamine-mediated acid secretion.

Lymphadenopathy and enlargement of the liver and the spleen are frequent in systemic mastocytosis. Because the lamina propria of the intestinal mucosa also is an important component of the reticuloendothelial system, mast cell infiltration of the small bowel is common. This produces the radiographic appearance of generalized irregular, distorted, thickened folds. Bulbous enlargement of small bowel villi can lead to a diffuse pattern of sandlike nodules superimposed on the irregularly thickened fold pattern.

Diffuse deposits of mast cells in the bone marrow cause both resorption and reactive sclerosis of bone, which produces a focal or generalized pattern of mixed osteolysis and osteosclerosis. Scattered, well-defined sclerotic foci can simulate blastic metastases; diffuse osteosclerosis may mimic myelofibrosis. In the skull, there may be stippling of the calvarium and thickening of the tables with obliteration of the diploic spaces.[3–6]

## ERYTHEMA NODOSUM

The erythema nodosum syndrome is characterized by multiple bilateral tender nodules, which appear primarily on the anterior aspect of the lower extremities and occasionally on the upper extremities and face. The nodules, which reflect a hypersensitivity vasculitis caused by a variety of conditions, can sometimes be seen on radiographs of the soft tissues.

Bilateral hilar adenopathy has been described in patients with erythema nodosum. However, this probably represents underlying sarcoidosis, one of the many conditions with which erythema nodosum is associated.[7]

## AINHUM

Ainhum is a tropical disease, usually affecting men, in which a chronic nonspecific inflammation of the deep fascial layers causes a deep soft tissue groove around the base of the fifth toe. Radiographically, this produces a sharply localized constriction of the soft tissues, which usually is located between the proximal and middle phalanges of the fifth digit. Resorption of the underlying bone produces eccentric tapering and often leads to autoamputation of the middle phalanx. The soft tissues and bony structures of the remaining toes remain normal.[8]

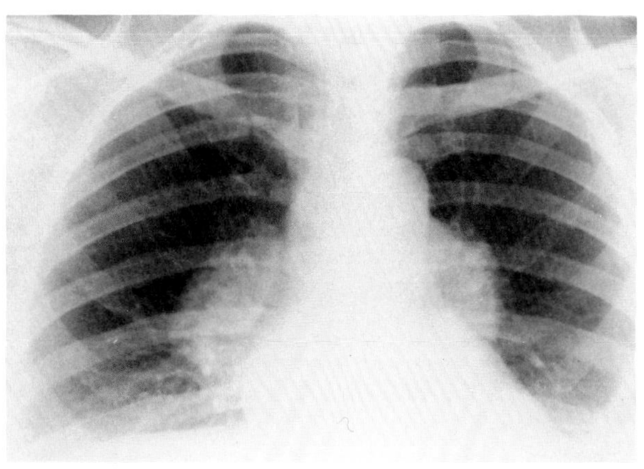

**Figure 77.6. Erythema nodosum.** A plain chest radiograph shows bilateral enlargement of hilar lymph nodes, which may reflect the patient's underlying sarcoidosis.

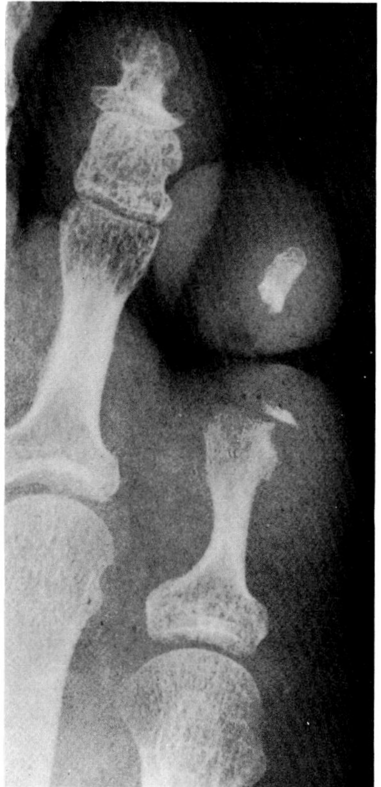

A

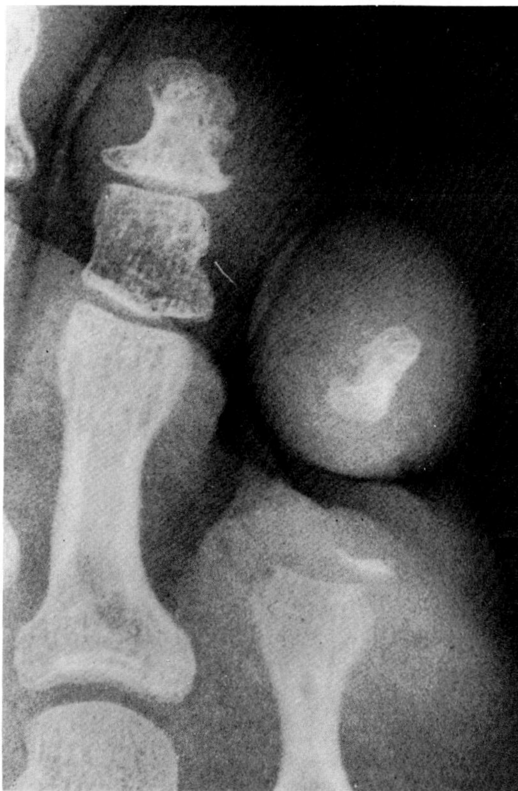

B

**Figure 77.7. Ainhum.** Two views in slightly different projections show the deep soft tissue groove around the affected fifth digit. There is virtual autoamputation with all but a fragment of the middle phalanx being absent. (From Fetterman, Hardy, and Lehrer,[8] with permission from the publisher.)

## REFERENCES


1. Agha FP, Raji MR: Esophageal involvement in pemphigoid: Clinical and roentgen manifestations. *Gastrointest Radiol* 7:109–112, 1982.
2. Brinn LB, Khilnani MT: Epidermolysis bullosa with characteristic hand deformities. *Radiology* 89:272–277, 1967.
3. Barer M: Mastocytosis with osseous lesions resembling metastatic malignant lesions in bone. *J Bone Joint Surg* 50A: 142–152, 1968.
4. Jensen WN, Lasser EC: Urticaria pigmentosa associated with widespread sclerosis of the spongiosa of bone. *Radiology* 71:826–832, 1958.
5. Clemett AR, Fishbone B, Levine RJ, et al: Gastrointestinal lesions in mastocytosis. *AJR* 103:405–412, 1968.
6. Robbins AH, Schimmel EM, Rao KC: Gastrointestinal mastocytosis. *AJR* 115:297–299, 1972.
7. Waisman M, Thomas MA: Bilateral pulmonary hilar lymphadenopathy in erythema nodosum. *Arch Dermatol* 82:754–757, 1960.
8. Fetterman LE, Hardy R, Lehrer H: The clinico-roentgenologic features of ainhum. *AJR* 100:512–522, 1967.
9. Eisenberg RL: *Gastrointestinal Radiology: A Pattern Approach*. Philadelphia, JB Lippincott, 1983.

**Chapter Seventy-Eight**

# Miscellaneous Congenital Disorders

## TRISOMY SYNDROMES

### Trisomy 13–15

In this rare syndrome, severe central nervous system defects, cleft palate, harelip, eye defects, and congenital heart disease are seen in small infants who do not thrive and usually die during the first year of life. The major radiographic abnormalities are confined to the skeleton and include polydactyly of the hands and feet, orbital hypotelorism and cleft palate, malformed ribs with asymmetry of the thorax, increased interpedicular distance in the cervical spine, and "rocker-bottom" feet. Visceral anomalies, in addition to congenital heart defects, include double vagina and uterus; diaphragmatic, umbilical, or inguinal hernias; and undescended testes.[1,2]

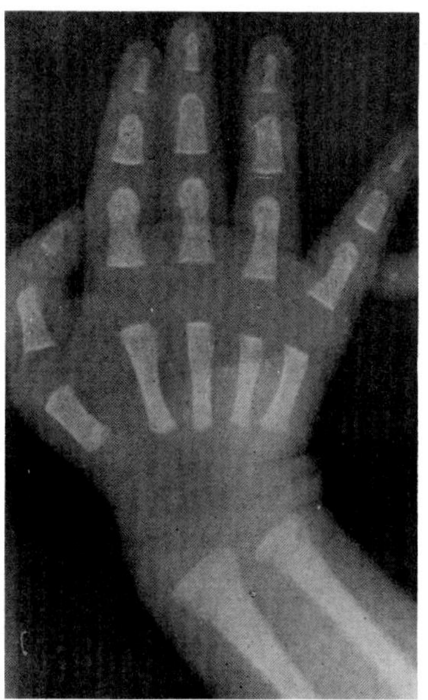

A

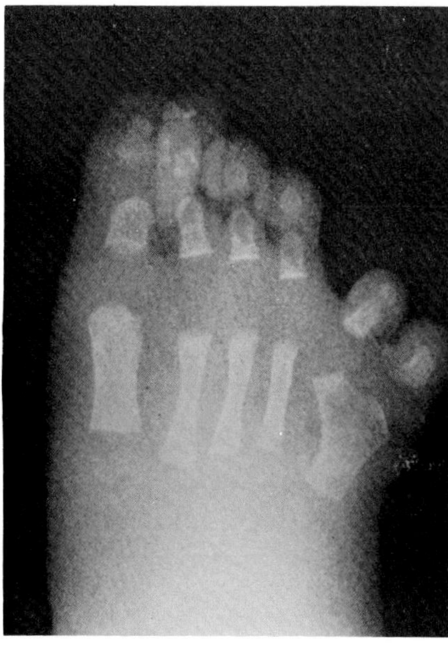

B

**Figure 78.1. Trisomy 13–15.** Polydactyly of (A) the hand and (B) the foot. (A, from Poznanski;[9] B, from Poznanski,[10] with permission from the publisher.)

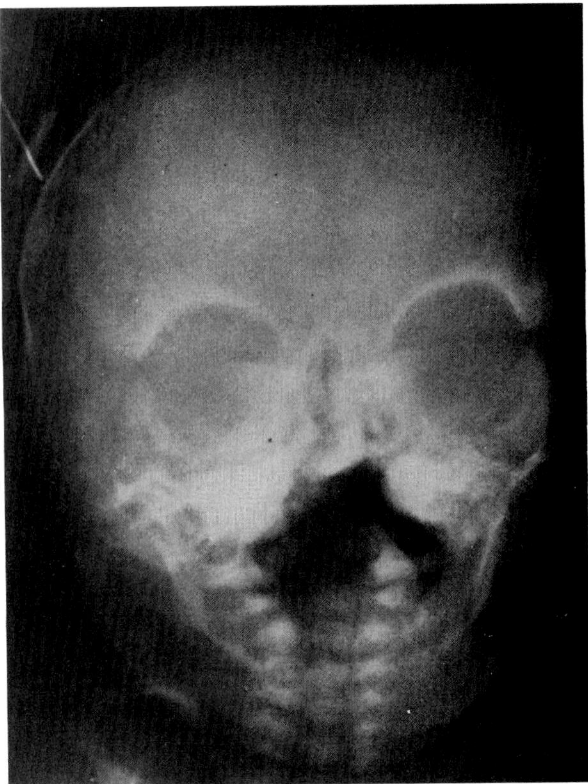

**Figure 78.2. Cleft palate in trisomy 13–15.** (From James, Belcort, Atkins, et al,[2] with permission from the publisher.)

### Trisomy 18

An extra chromosome 18 produces a clinical syndrome characterized by low-set, malformed ears, a small jaw and oral cavity (mandibular and maxillary hypoplasia), and severe somatic and mental retardation. Most infants with this condition die within the first few months. Radiographic findings include flexion deformities of the fingers and toes, ulnar deviation of the third, fourth, and fifth digits, thinning of the ribs, elongated and tapered clavicles, defective ossification of the sternum, and a narrow transverse diameter of the pelvis with vertical ilia and steep acetabular angles. Congenital heart disease, usually ventricular septal defect, is common and leads to cardiac enlargement with increased pulmonary vascularity.[1,3]

### Trisomy 21 (Down's Syndrome)

Down's syndrome, the most common of the trisomy disorders, is usually diagnosed from birth because of the characteristic clinical appearance—mental deficiency, short stature, muscular hypotonia, short neck, and typical facies. The major skeletal abnormality in infancy is in the pelvis, where there is a decrease in the acetabular and iliac angles with hypoplasia and marked lateral flaring of the iliac wings. Other common skeletal abnormalities include shortening of the middle phalanx of the fifth finger, squaring of the vertebral bodies (the superoinferior length becoming equal to or greater than the anteroposterior measurement), multiple ossification centers in the manubrium, the presence of only 11 ribs, and coxa vara. In the skull, the bones of the calverium are thin, the palate has a high short arch, the nasal sinuses are hypoplastic, and closure of the cranial sutures may be delayed.

Congenital heart disease, especially septal defects, occurs in about 40 percent of the patients with Down's syndrome. There is also a greater than normal incidence of duodenal obstruction (duodenal atresia or annular pancreas) and Hirschsprung's disease, as well as a substantially increased likelihood of developing acute leukemia.[4–6]

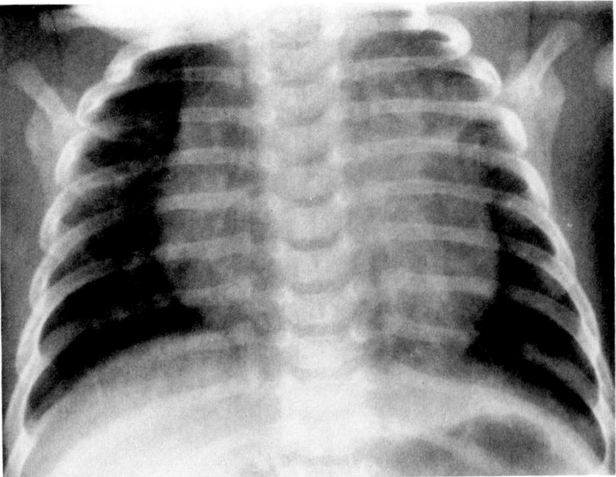

**Figure 78.3. Ventricular septal defect in trisomy 18.** The heart is enlarged with a globular configuration. Increased pulmonary vascularity reflects the left-to-right shunt.

## PROGERIA (HUTCHINSON-GILFORD SYNDROME)

Progeria is a nonhereditary congenital syndrome of dwarfism and premature senility developing in childhood. An affected infant appears normal at birth but soon develops the characteristic loss of subcutaneous fat, alopecia, and atrophy of muscles and skin that produce a wizened-old-man appearance. Premature arteriosclerosis in the coronary arteries and other vessels leads to death during late childhood or early adolescence.

Radiographic findings in progeria include hypoplastic facial bones, open cranial sutures and fontanelles, thin short clavicles, coxa valga, and progressive shortening

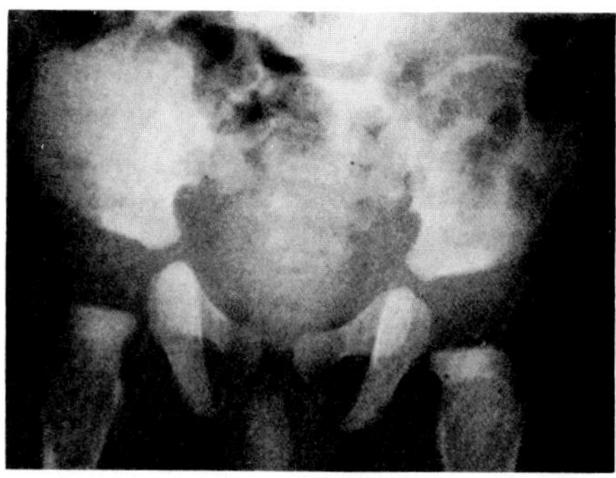

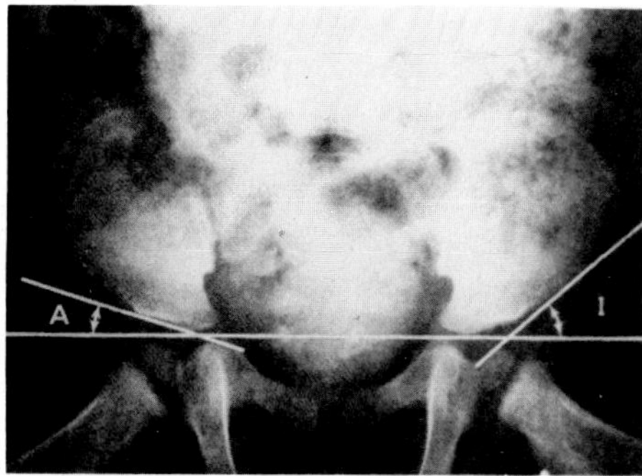


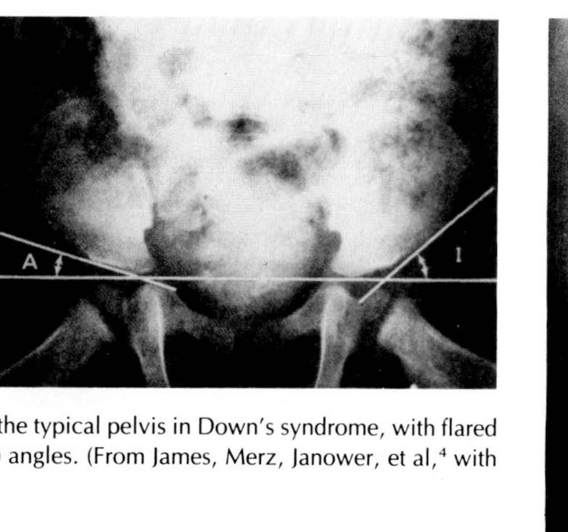

**Figure 78.4. Down's syndrome.** Two examples of the typical pelvis in Down's syndrome, with flared iliac wings and diminished acetabular (A) and iliac (I) angles. (From James, Merz, Janower, et al,[4] with permission from the publisher.)

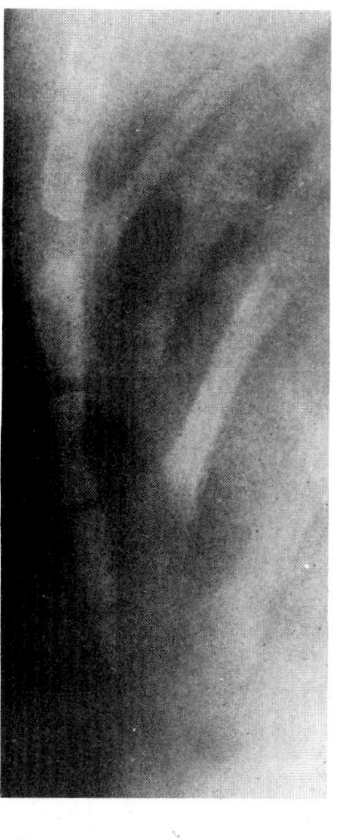

**Figure 78.5. Down's syndrome.** A lateral view of the sternum demonstrates two ossification centers of the manubrium. (From James, Merz, Janower, et al,[4] with permission from the publisher.)

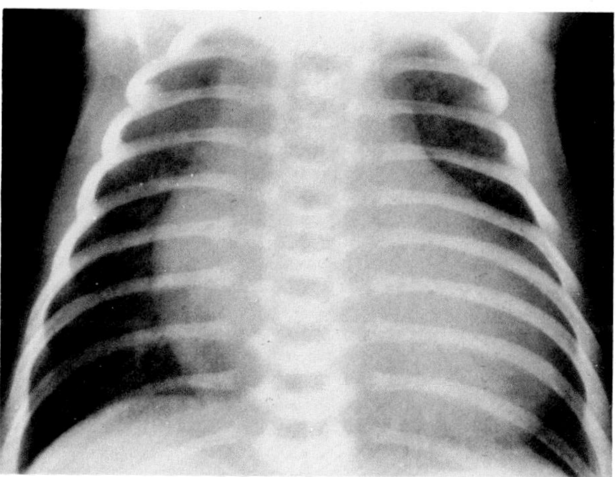

**Figure 78.6. Down's syndrome.** There is an endocardial cushion defect with globular enlargement of the heart and increased pulmonary vascularity.

and abrupt tapering (acro-osteolysis) of the terminal phalanges. The long bones are delicate and osteoporotic, though there is normal bone maturation. Coronary artery disease and hypertension lead to prominent cardiomegaly.[7]

## HOLT-ORAM SYNDROME

This syndrome is transmitted as an autosomal dominant and consists of congenital heart disease (most often atrial septal defect) and upper extremity malformations. The most common radiographic abnormalities are aplasia or hypoplasia of the thumb, a bifid navicular bone, inward curvature of the fifth digits, and hypoplasia of the radius and clavicle.[8]

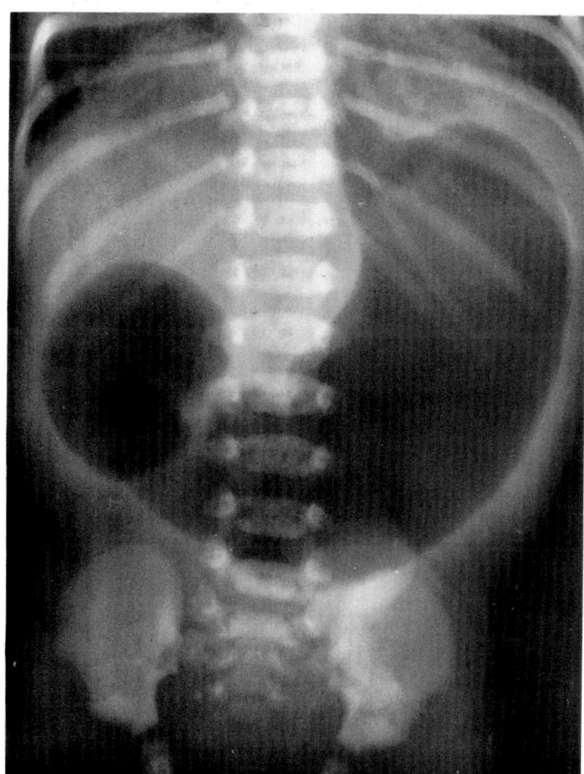

**Figure 78.7. Down's syndrome.** An annular pancreas with complete duodenal obstruction produces the double-bubble sign of gas in the stomach (left) and the duodenal bulb (right). (From Eisenberg,[11] with permission from the publisher.)

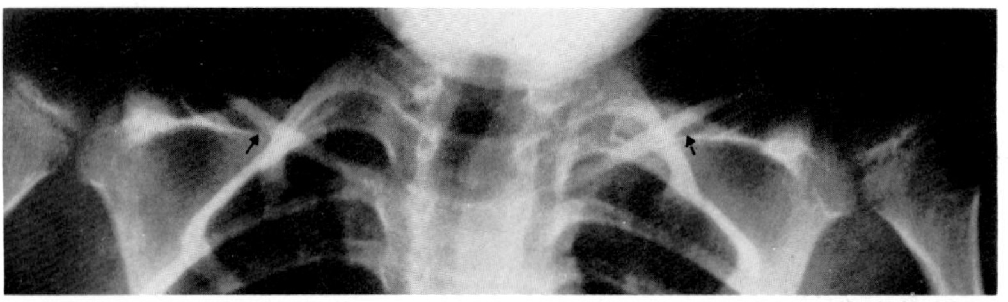

A

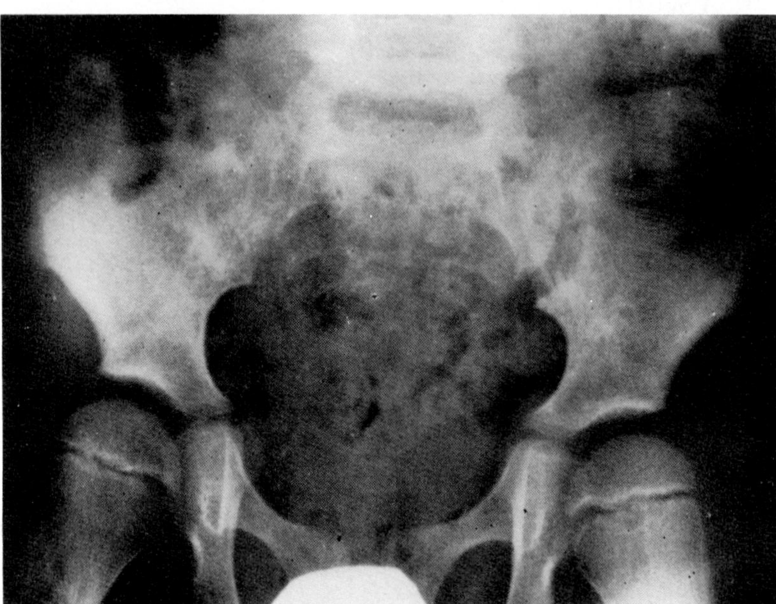

B

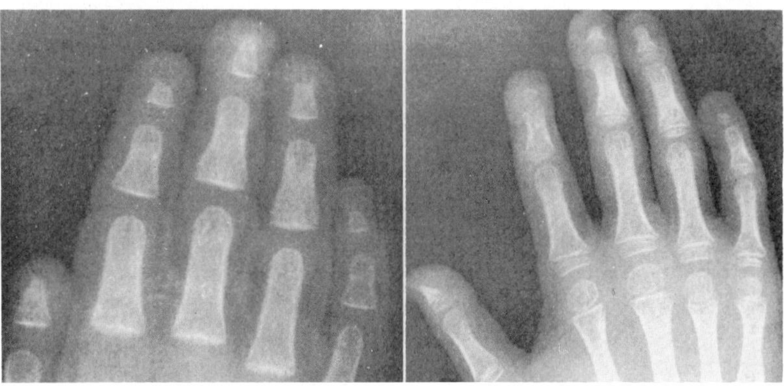

C

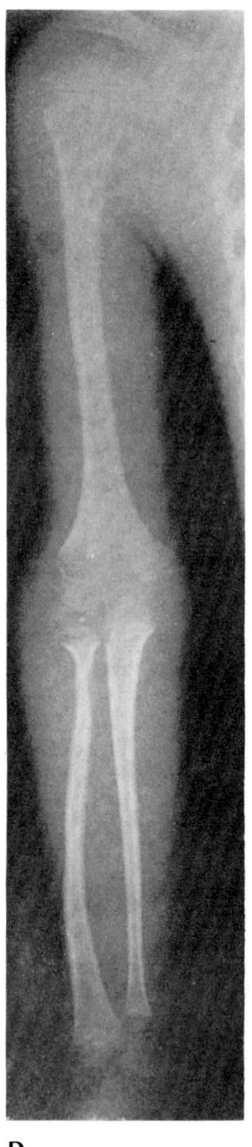

D

**Figure 78.8. Progeria.** (A) The clavicles are thin and dense (arrows). The lateral third of the left clavicle is absent. A small bone fragment represents the acromial portion of the right clavicle. (B) A radiograph of the pelvis shows bilateral coxa valga and a shallow impression of the lateral pelvic contour above the acetabulum. (C) Progressive resorption of the ungual tufts with preservation of soft tissues occurred over a 5-year period. (D) The ulna and radius are thin and osteoporotic. (From Margolin and Steinbach,[7] with permission from the publisher.)

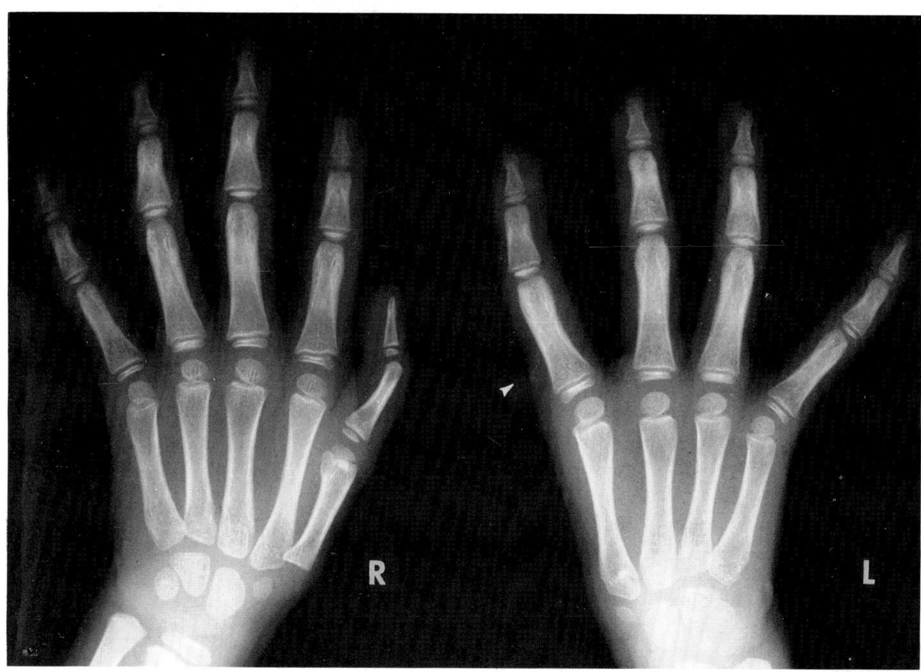


**Figure 78.9. Holt-Oram syndrome.** Radiographs of both hands show agenesis of the metacarpal and the phalanges of the left thumb. Note the minute soft tissue appendage arising from the lateral and proximal aspect of the left index finger (arrow). There is hypoplasia of the right thumb, the first metacarpal, and the thenar muscles. (From Chang,[8] with permission from the publisher.)

## REFERENCES

1. Juhl W: *Essentials of Roentgen Interpretation*, Philadelphia, Harper & Row, 1981.
2. James AE, Belcort CL, Atkins L, et al: Trisomy 13–15. *Radiology* 92:44–49, 1969.
3. James AE, Belcort CL, Atkins L, et al: Trisomy-18 syndrome. *Radiology* 92:37–43, 1969.
4. James AE, Merz P, Janower ML, et al: Radiologic features of the most common autosomal disorders. *Clin Radiol* 22:417–433, 1971.
5. Rabinowitz JG, Moseley JE: Lateral lumbar spine in Down's syndrome. *Radiology* 83:74–80, 1964.
6. Mortensson W, Hall B: Abnormal pelvis in newborn infants with Down's syndrome. *Acta Radiol* 12:847–851, 1972.
7. Margolin FR, Steinbach HL: Progeria. *AJR* 103:173–178, 1968.
8. Chang CH: Holt-Oram syndrome. *Radiology* 88:479–483, 1967.
9. Poznanski AK: The thumb in congenital malformation syndromes. *Radiology* 100:115–123, 1971.
10. Poznanski AK: Foot manifestations of the congenital malformation syndromes. *Semin Roentgenol* 5:354–365, 1970.
11. Eisenberg RL: *Gastrointestinal Radiology: A Pattern Approach*. Philadelphia, JB Lippincott, 1983.

**Appendix**

# Procedures in Diagnostic Imaging

## ARTERIOGRAPHY (ANGIOGRAPHY)

Arteriography is a procedure in which a rapid sequence of films is obtained following the injection of a bolus of contrast material through a catheter that is introduced percutaneously via either the femoral or the axillary artery. The tip of the catheter may be placed into the thoracic or the abdominal aorta or selectively into one of the major branches. Initial films show contrast material within the major arteries. Subsequent films demonstrate the filling of small arteries followed by the capillary phase (e.g., the nephrogram in renal arteriography); late films often show opacification of the venous drainage.

Arteriography can demonstrate intrinsic vascular abnormalities (atherosclerotic plaques, strictures, occlusions, malformations), the displacement of vessels by an adjacent mass, and abnormal vessels (neovascularity) due to tumor or inflammatory disease. In the extremities, arteriography not only demonstrates stenotic or occlusive lesions, but also can assess the degree of collateral circulation and the patency of distal vessels. These latter findings have a high predictive value for the success of a bypass graft.

Although various types of arteriography have different risks, the overall mortality rate is less than 1 percent; deaths are extremely rare in all but high-risk patients. Serious, permanent complications of arteriography have been reported to occur in about 0.7 percent of the patients; minor, transient complications (most related to the puncture site) are seen in fewer than 3 percent.

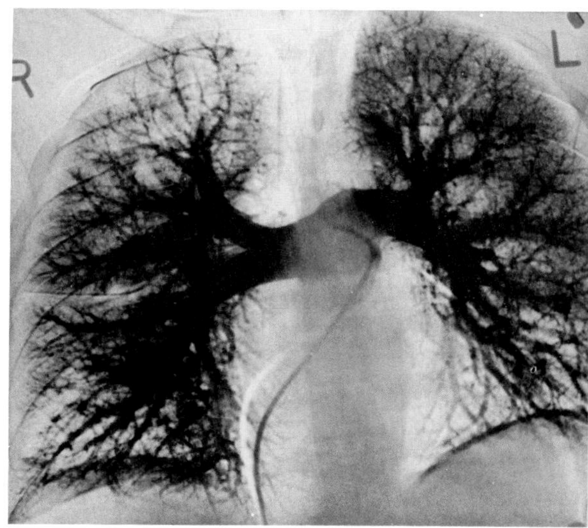


**Figure A.1. Normal pulmonary arteriogram.**

### Cardiac Catheterization (Angiocardiography)

This topic is discussed in the section "Cardiac Catheterization and Coronary Angiography" in Chapter 19.

### Cerebral Angiography

This topic is discussed in the section "Angiography" in Chapter 66.

### Coronary Angiography

This topic is discussed in the section "Cardiac Catheterization and Coronary Angiography" in Chapter 19.

### Digital Subtraction Angiography

This topic is discussed in the section "Angiography" in Chapter 66.

### Pulmonary Arteriography

Pulmonary arteriography is the most definitive and specific test for the diagnosis of pulmonary embolism. It is also of value in demonstrating congenital anomalies of the pulmonary vessels (the absence or hypoplasia of a major vessel, arteriovenous malformations) and was used to assess the operability of bronchogenic carcinoma before the advent of computed tomography.

Pulmonary arteriography is usually performed by percutaneous femoral or axillary catheterization. A catheter is threaded into the right atrium, through the tricuspid valve, and into the right ventricle and main pulmonary artery. The 1 to 2 percent morbidity of pulmonary arteriography includes patients who develop cardiac arrhythmias, cardiac arrest, or a reaction to contrast material. Complications relating to perforation or direct injury to vascular structures can be avoided with the use of an end- and side-hole pigtail catheter. Fatalities (< 0.5 percent) usually occur in patients with severe pulmonary hypertension, which may be primary or secondary to pulmonary emboli or congenital heart disease. Absolute contraindications to pulmonary arteriography are a history of severe anaphylactic reaction to iodinated contrast material and the presence of an acute myocardial infarction. Relative contraindications include marked ventricular irritability, left bundle branch block, and severe pulmonary hypertension.

## ARTHROGRAPHY

Arthrography refers to the direct injection of air or contrast material, or both, into the synovial cavity of a joint. Most often performed in the knee or shoulder, arthrography permits the visualization of enlarged joint spaces or communicating chambers, meniscal or ligamentous tears (in the knee), and synovial tumors or thickening.

## BARIUM ENEMA EXAMINATION

Opacification of the colon by a solid column of barium (single contrast) or by a combination of barium and air (double contrast) is the major technique for radiographically evaluating the colon. The double-contrast barium enema examination appears to have greater sensitivity in detecting polyps than the single-contrast technique, especially in the case of small colonic lesions. Nevertheless, reports of polyps that were missed on double-contrast studies but identified with single-contrast techniques indicate that the two procedures have complementary value.

A successful barium enema examination requires meticulous colon preparation. Several studies have reported that 10 to 20 percent of colon carcinomas are missed on the initial barium enema examination, primarily because poor preparation results in the tumor being overlooked or confused with fecal material. Proper cleansing of the colon (laxatives, dietary restrictions, cleansing

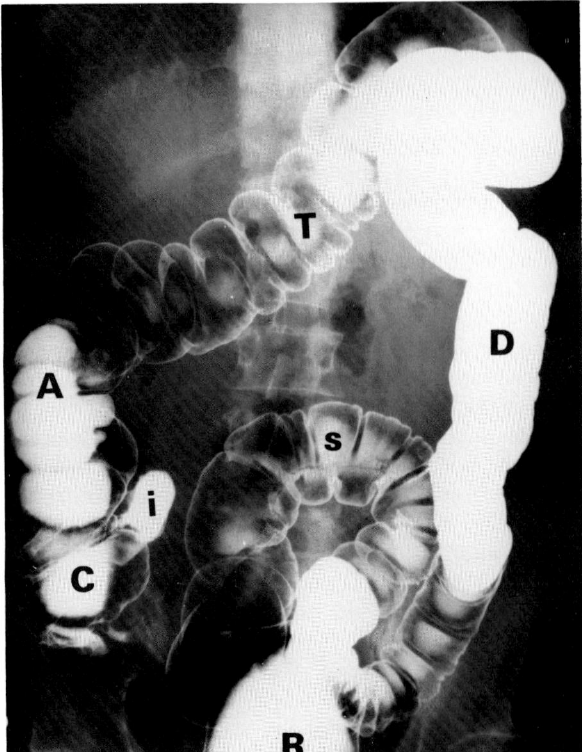


**Figure A.2. Normal barium enema examination.** C = cecum; A = ascending colon; T = transverse colon; D = descending colon; s = sigmoid colon; R = rectum; i = terminal ileum.

enemas in the radiology department) should improve these dismal statistics.

After the colon is filled with barium (or barium and air) under fluoroscopic control, a variable number of spot and overhead films are obtained. With the single-contrast technique, a postevacuation film is made to assess the mucosal pattern of the colon.

In addition to polyp detection, the barium enema examination can identify mucosal ulcerations, intramural masses (tumors, thumbprinting in ischemia or bleeding), and extrinsic impressions from adjacent lesions.

## BRONCHOGRAPHY

Bronchography is the study of the bronchial tree by the instillation of opaque contrast material into the desired bronchus or bronchi, usually under fluoroscopic control. This technique can demonstrate bronchiectasis; chronic bronchitis; bronchial obstruction, deformity, or displacement; communicating pulmonary cavities; and bronchopleural fistulas. In recent years, bronchography has been used much less frequently because of the availability of direct imaging and cytologic procedures (fiberoptic bronchoscopy with biopsy; brush biopsy; percutaneous biopsy).

## CAVOGRAPHY

Opacification of the superior vena cava can be obtained by the rapid injection of contrast material into an antecubital vein or through a subclavian vein catheter. The inferior vena cava is visualized by contrast injection through a catheter inserted into the femoral vein. Both studies permit the demonstration of narrowing, obstruction, or displacement of the venae cavae by extrinsic masses, extension of tumor, or progression of thrombi. The role of cavography has recently been reduced because of the ability of ultrasound and computed tomography to demonstrate lesions affecting these venous structures.

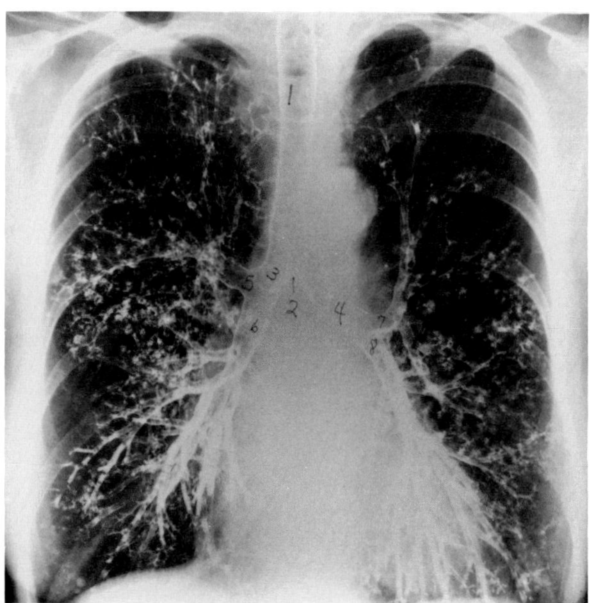


**Figure A.3. Normal bronchogram.** 1 = trachea; 2 = carina; 3 = right main bronchus; 4 = left main bronchus; 5 = right upper lobe bronchus; 6 = bronchus intermedius; 7 = left upper lobe bronchus; 8 = left lower lobe bronchus.

## CHOLANGIOGRAPHY

### Intravenous Cholangiography

Intravenous cholangiography (IVC) was formerly the mainstay of diagnosis in mild obstructive jaundice, especially in the postcholecystectomy patient. However, the advent of ultrasound, computed tomography, and thin-needle percutaneous transhepatic cholangiography has virtually eliminated the clinical application of intravenous cholangiography, since these new techniques

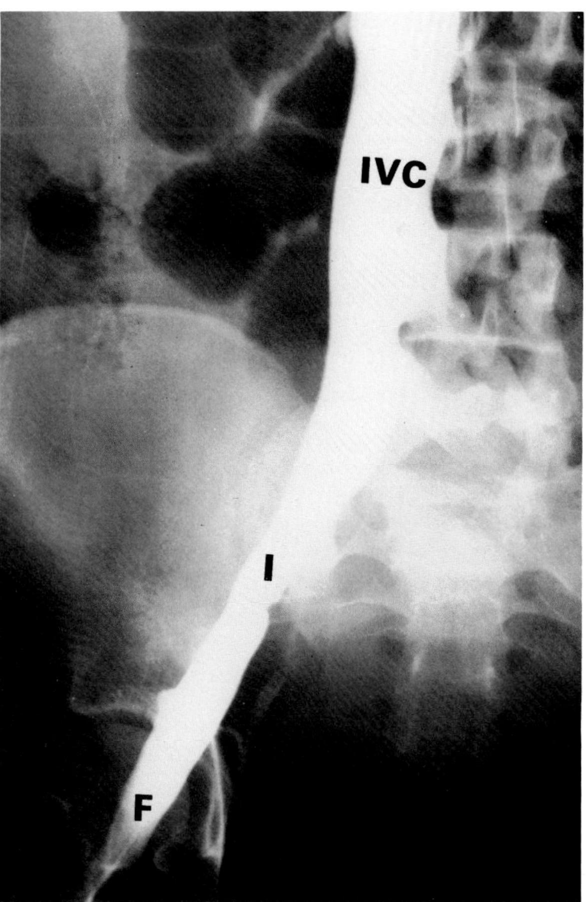


**Figure A.4. Normal venogram of the lower portion of the interior vena cava (IVC) and the right pelvic veins.** I = right common iliac vein; F = right femoral vein.

provide the same information without the morbidity (about 25 percent) and occasional death associated with the contrast material used for intravenous cholangiography.

## Percutaneous Transhepatic Cholangiography

Percutaneous transhepatic cholangiography (PTHC) performed with thin (23 gauge) flexible needles is the primary modality for localizing the proximal extent and character of obstructing lesions of the biliary system. After a needle is inserted percutaneously through the liver and into the bile duct, residual bile can be withdrawn and contrast material injected to opacify the hepatic and common ducts. In the past, when PTHC was performed with large, sheathed needles, it was considered too risky to perform unless surgical intervention was immediately available. Although the thin-needle technique decreases the chance of bile peritonitis or bleeding, fever or other signs of sepsis occur in 3.5 percent of the patients. A few individuals develop such complications as bile leakage with peritonitis, hemoperitoneum (usually not life-threatening), or pneumothorax from puncture of the pleural space. With a sufficient number of needle passes into the liver substance by an experienced physician, PTHC can demonstrate almost 100 percent of dilated biliary systems. Indeed, this technique can also demonstrate 50 to 75 percent of nondilated biliary systems. Some physicians even consider that the inability to demonstrate the biliary system after 10 to 15 passes implies that the patient is suffering from nonobstructive rather than obstructive jaundice.

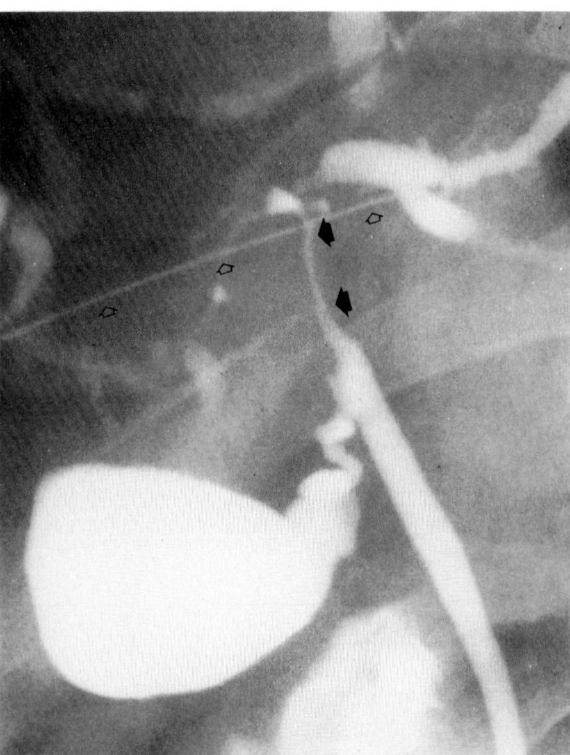

**Figure A.5. Percutaneous transhepatic cholangiogram.** The smooth stricture (closed arrows) extending from the carina to the junction of the cystic duct represents a cholangiocarcinoma. Areas of relative narrowing of the right and left hepatic ducts represent additional sites of tumor involvement. The open arrows point to the cholangiography needle.

## T-Tube Cholangiography

The injection of contrast material through the external opening of a drainage T tube that was placed in the common duct during surgery permits the direct opacification of the hepatic and common ducts. This study is primarily employed postoperatively to evaluate whether there is free passage of contrast material through the ampulla into the duodenum and to demonstrate any obstructive lesions or residual calculi in the common duct before removing the T tube.

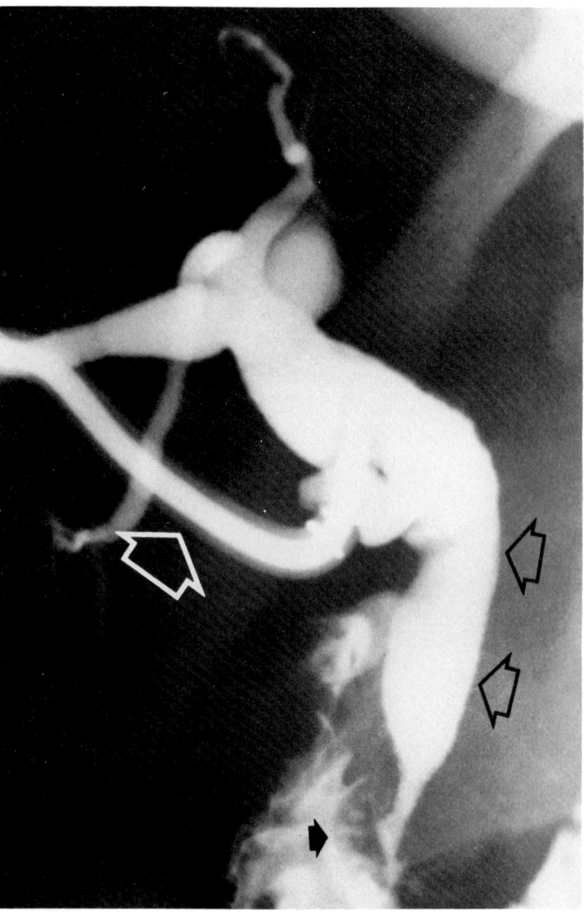

**Figure A.6. Normal T-tube cholangiogram.** The white arrow points to the T tube. The common bile duct (open black arrows) and the papilla (black arrow) are well seen, and there is prompt flow of contrast material into the duodenum.

## CHOLANGIOPANCREATOGRAPHY, ENDOSCOPIC RETROGRADE

Endoscopic retrograde cholangiopancreatography (ERCP) is a technique in which the biliary tract is cannulated via a flexible fiberoptic duodenoscope with its tip at the level of the ampulla. The instillation of contrast material permits visualization of the biliary tree and pancreatic duct. Unlike other procedures employed in patients with jaundice, ERCP permits endoscopic observation of the small bowel, visualization of the lower portion of a completely obstructed common bile duct, evaluation of a nondilated intrahepatic system (e.g., sclerosing cholangitis), and delineation of the pancreatic duct. It is the procedure of choice for the precise visualization of the biliary tract in patients with abnormal bleeding parameters and in those in whom percutaneous transhepatic cholangiography cannot be successfuly performed. A major disadvantage of ERCP is the requirement for an operator with extensive experience in biliary cannulation. The examination is time-consuming and expensive (personnel time and the significant cost of the endoscope itself). The most important disadvantage is the fairly high morbidity rate (10 to 15 percent), primarily pancreatitis and endoscopically induced complications.

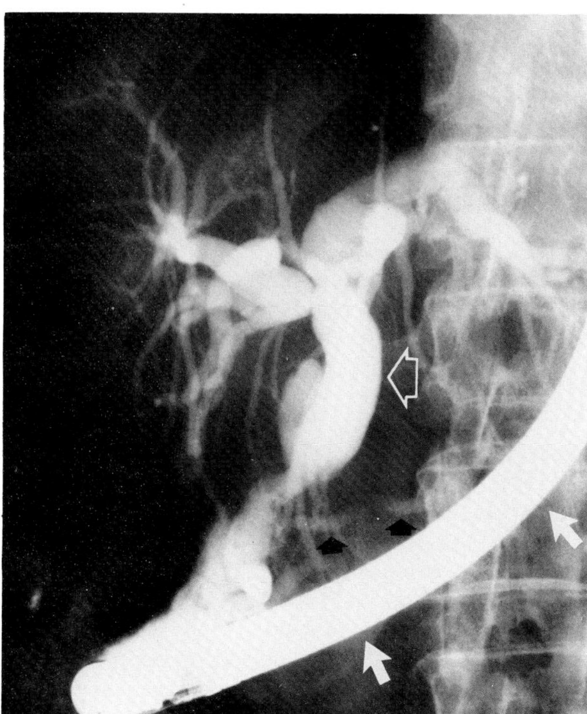

**Figure A.7. Endoscopic retrograde cholangiopancreatogram.** The open white arrow points to the common bile duct; the black arrows point to the pancreatic duct. The closed arrows indicate the duodenoscope.

## CHOLECYSTOGRAPHY, ORAL

Oral cholecystography has traditionally been the procedure of choice for diagnosing gallstones, the cholecystoses, and the rare tumorous causes of filling defects within the gallbladder. Oral cholecystographic contrast agents, whether fat-soluble (Telepaque) or water-soluble (Bilopaque), are absorbed from the proximal small intestine, flow through the portal venous system, are conjugated within the liver, and are excreted into the bile. With fat-soluble contrast agents, peak opacification of the gallbladder does not occur until 14 to 21 hours after ingestion; therefore, radiographs should be obtained during that period. With water-soluble agents, however, maximum radiographic opacification occurs 10 hours after ingestion. In addition to routine supine films, erect or decubitus views with a horizontal beam are essential for detecting small stones.

In addition to intrinsic gallbladder disease, there are many causes of nonvisualization of the gallbladder. Extrabiliary causes include the failure of the patient to ingest the contrast material and the failure of the contrast material to reach the absorptive surface of the small bowel (vomiting, diarrhea, proximal obstruction, malabsorption diseases). Liver disease with disturbed hepatocellular function can result in nonvisualization of the gallbladder either because of an inability of the patient to properly conjugate the oral cholecystographic contrast agent or because of insufficient production and flow of bile. Opacification of the gallbladder after the administration of oral contrast material rarely occurs if the serum bilirubin level is more than 3.5 mg/dl.

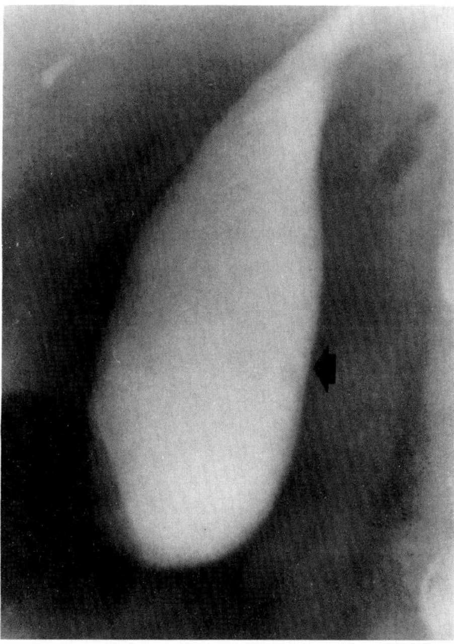

**Figure A.8. Oral cholecystogram.** There is a fibroadenoma (arrow) within the well-opacified gallbladder.

If extrabiliary and hepatocellular causes of nonvisualization can be excluded, the failure of the gallbladder to opacify after the administration of two doses of oral cholecystographic contrast material is highly reliable evidence of gallbladder disease. The major causes of nonvisualization intrinsic to the gallbladder are obstruction of the cystic duct or neck of the gallbladder, which prevents contrast-containing bile from reaching the gallbladder and opacifying the organ, and chronic cholecystitis, in which nonvisualization can be due to the obliteration of the gallbladder lumen by chronic inflammatory changes, the inability of the inflamed gallbladder mucosa to absorb and concentrate water, or the diffusion of contrast material through the diseased mucosa of the gallbladder into the bloodstream.

## COMPUTED TOMOGRAPHY

Computed tomography (CT) is a radiographic technique for producing cross-sectional tomographic images by first scanning a "slice" of tissue from multiple angles with a narrow x-ray beam, then calculating a relative linear attenuation coefficient for the various tissue elements in the section, and finally displaying the computer reconstruction as a gray-scale image on a television monitor. Unlike other imaging modalities (with the exception of the more recent magnetic resonance imaging), CT permits the radiographic differentiation among a variety of soft tissues. It is extremely sensitive to slight differences (1 percent) in tissue densities; for comparison, differences in tissue density of at least 5 percent are required for detection by conventional screen-film radiography. Thus, in the head CT can differentiate between blood clots, white matter and gray matter, cerebrospinal fluid, cerebral edema, and neoplastic processes. The CT number reflects the attenuation of a specific tissue relative to that of water, which is arbitrarily given a CT number of 0. The highest CT number is that of bone; the lowest is that of air. Fat has a CT number less than 0, while soft tissues have CT numbers higher than 0.

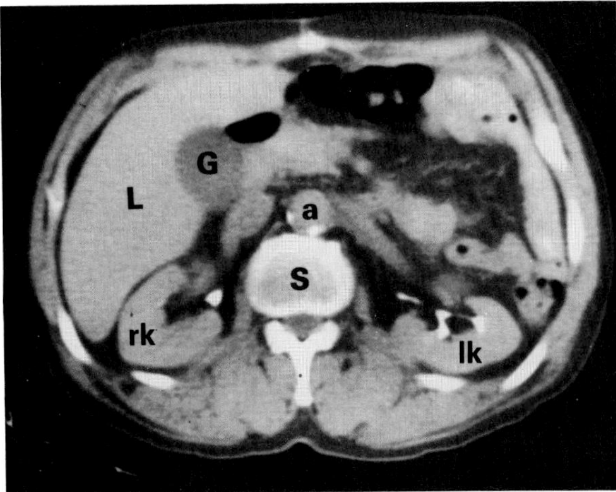


**Figure A.9. Normal CT scan of the lower abdomen.** a = aorta; G = gallbladder; L = liver; lk = left kidney; rk = right kidney; S = spine.

Technical improvements in CT instrumentation have greatly reduced the time required to produce a single slice, and this permits the CT evaluation of virtually any portion of the body. In most instances, some type of preliminary film is obtained (either a radiograph or a CT-generated image) for localization, the detection of potentially interfering high-density material (metallic clips, barium, electrodes), and correlation with the CT images. An overlying grid with numerical markers permits close correlation between the subsequent CT scans and the initial scout film.

The intravenous injection of iodinated contrast material is an integral part of many CT examinations. Scanning during or immediately following the administration of contrast material permits the differentiation of vascular from nonvascular solid structures. Differences in the degree and the time course of enhancement may permit the detection of neoplastic or infectious processes within normal parenchymal structures. Because of its relatively low CT number, fat can serve as a natural contrast material and can outline parenchymal organs. In patients with malignant lesions, the loss of adjacent fat is highly suspicious of tumor extension. For abdominal studies, especially those of the pancreas and retroperitoneum, dilute oral contrast material is frequently given to demonstrate the lumen of the gastrointestinal tract and permit the distinction between loops of bowel and solid abdominal structures.

## CYSTOGRAPHY

Cystography is a technique in which opaque contrast material is introduced into the urinary bladder through a urethral catheter. Cystography can demonstrate abnormalities in the shape or position of the bladder and can detect vesical diverticula and filling defects within the bladder due to tumors, blood clots, and calculi.

## CYSTOURETHROGRAPHY, VOIDING

Voiding cystourethrography is primarily used in the study of patients suspected of having lower urinary tract obstruction (bladder neck obstruction, urethral stricture, congenital urethral valves) or vesicoureteral reflux. Opaque contrast material is introduced into the bladder through a urethral catheter, and serial films or cine pictures are obtained before, during, and after urination.

## HYSTEROSALPINGOGRAPHY

This topic is discussed in the section "Female Infertility" in Chapter 4.

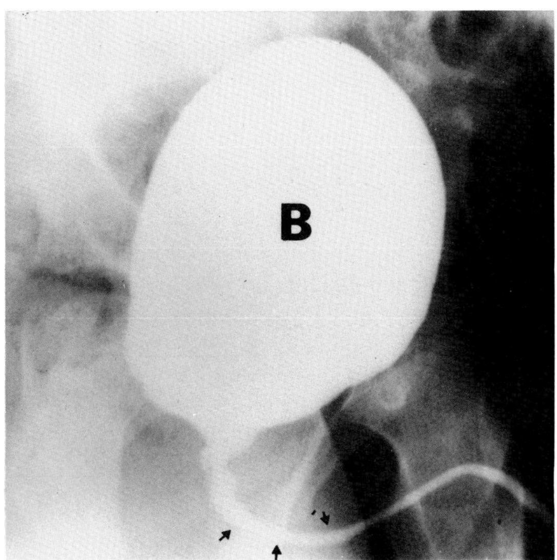


**Figure A.10. Normal voiding cystourethrogram in a young boy.** The arrows point to the urethra. B = bladder.

## LARYNGOGRAPHY

Laryngography, the study of the hypopharynx, the larynx, and the subglottic portion of the trachea, is a technique for demonstrating the size and location of a laryngeal tumor. It can supplement the results of direct laryngoscopy in determining possible subglottic extension of tumor and involvement of the pyriform sinus, and can assess the status of a lesion during and following radiation therapy. The procedure consists of coating the structures of the larynx with an oily opaque material after the administration of a topical anesthetic. Fluoroscopic spot films are then obtained in lateral and frontal projections during quiet respiration, phonating the sound *E*, the Valsalva maneuver, and the modified Valsalva maneuver with the mouth closed and the glottis open.

## LYMPHOGRAPHY

Lymphography is a technique in which the infusion of iodinated oil into a lymphatic vessel on the dorsum of the foot routinely opacifies the lymphatic channels of the lower extremity and those lymph nodes that lie adjacent to the ipsilateral external iliac and common iliac arteries and veins, the inferior vena cava, and the aorta approximately to the level of the renal vessels. Retroperitoneal lymph nodes cephalad to this level, or those situated more anteriorly or laterally, are not consistently opacified. Intra-abdominal lymph nodes, such as those in the hepatic and splenic hila, in the mesentery, and deep in the pelvis, are also generally not opacified by this technique.

Unlike other staging procedures for lymph nodes, lymphography possesses excellent spatial resolution and can detect "macroscopic" metastases within normal-sized lymph nodes. It also may suggest that large lymph nodes reflect benign reactive changes rather than the presence of tumor. Lymphography is also of value in demonstrating the blockage of lymphatic channels.

Lymphography is a minor surgical procedure that requires some degree of technical facility. Although the initial cost is relatively high, subsequent abdominal films are inexpensive and may provide useful information regarding the status of the residually opacified lymph nodes for as long as 18 months. This is of special value in assessing the effect of radiotherapy or chemotherapy on lymphatic metastases.

## MAGNETIC RESONANCE IMAGING

Magnetic resonance imaging (MRI), also called nuclear magnetic resonance (NMR), is a new modality for imaging proton density (and perhaps that of other nuclei) in the body and for obtaining dynamic studies of certain physiologic functions. Although the physics of MRI is beyond the scope of this book, the technique basically consists of inducing transitions between energy states by causing certain atoms to absorb and transfer energy. This is accomplished by directing a radio frequency pulse at a substance placed within a large magnetic field. Various measures of the time required for the material to return to a baseline energy state (relaxation time) can be translated by a complex computer algorithm to a visual image on a television monitor.

MRI images may be obtained from transverse, coronal, or sagittal slices or from any arbitrarily oriented slice. Because the radio frequency electromagnetic radiation can penetrate bone without significant attenuation, the underlying tissue can be clearly imaged. In the future, examinations using various chemical elements and high magnetic fields may permit the sophisticated biochemical analyses of tissues in vivo.

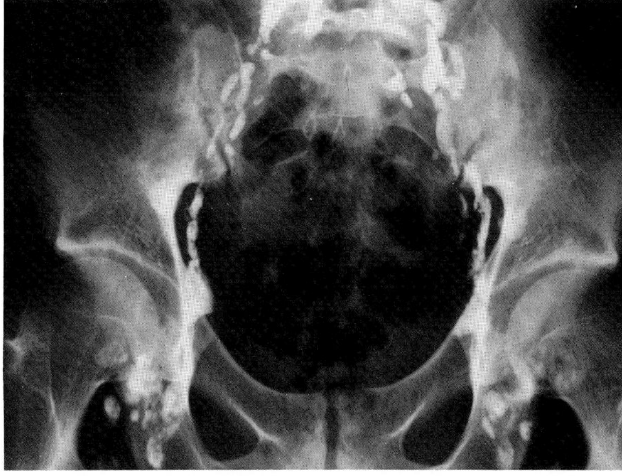

**Figure A.11. Normal pelvic lymphogram.** The opacified nodes lie along the iliac vessels and the aorta.

For a brief discussion of the use of MRI in neuroradiology and some contraindications to the procedure, see the section "Magnetic Resonance Imaging" in Chapter 66.

## MYELOGRAPHY

This topic is discussed in the section "Myelography" in Chapter 66.

## NEPHROSTOMY, PERCUTANEOUS

Percutaneous nephrostomy, performed under fluoroscopic or ultrasound guidance, was initially introduced for the emergency drainage of an obstructed upper urinary tract. More recently, this procedure has also been used in the treatment of sepsis aggravated or caused by ureteral obstruction; to assess the potential recovery of renal function of an obstructed kidney; and as a first step toward interventional techniques such as the dissolving or basketing of renal calculi, the dilatation of ureteral strictures, and the placement of antegrade stents in patients with ureteral fistulas.

The major contraindication to percutaneous nephrostomy is a severe bleeding diathesis. Short-term hematuria often develops in patients with normal clotting parameters, but this almost always subsides within 2 days without the need for further therapy. Complications of percutaneous nephrostomy include the development of gram-negative septicemia and/or ureteropelvic junction obstruction.

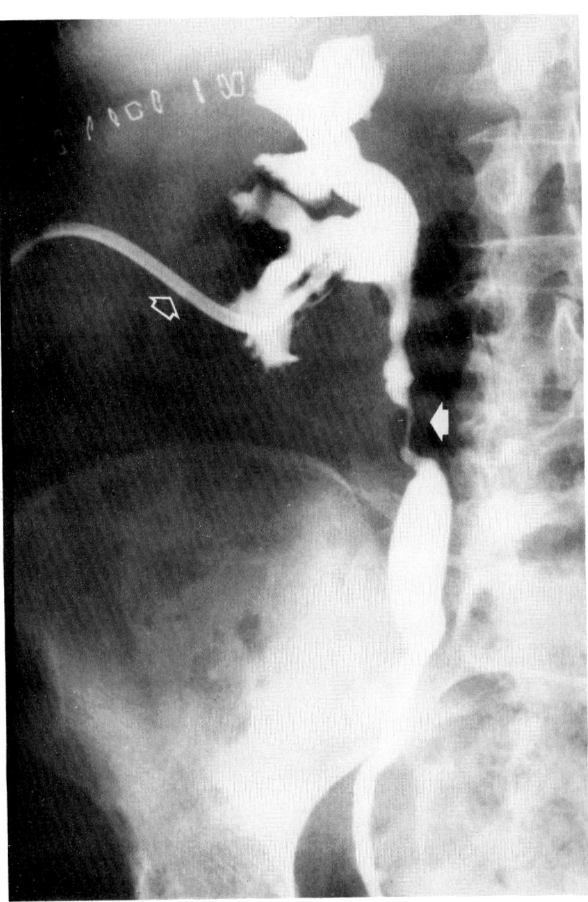

**Figure A.12. Percutaneous nephrostogram.** Antegrade filling of the right collecting system following the injection of contrast material through the nephrostogram catheter (open arrow) demonstrates an irregular narrowing of the proximal ureter (closed arrow) due to fibrosis secondary to prior stone removal.

## NUCLEAR MAGNETIC RESONANCE

This topic is discussed under the heading "Magnetic Resonance Imaging" in this section.

## PNEUMOENCEPHALOGRAPHY

This topic is discussed in the section "Pneumoencephalography" in Chapter 66.

## PYELOGRAPHY, INTRAVENOUS

This topic is discussed under the heading "Urography, Excretory" in this section.

## PYELOGRAPHY, RETROGRADE

In this technique, radiographs are obtained following the direct instillation of contrast material into the ureter or renal pelvis via a catheter inserted at cystoscopy. In the past, retrograde pyelography was frequently performed to better demonstrate a collecting system lesion seen on excretory urography or to show the pelvocalyceal systems and ureters in a patient with a completely nonfunctioning kidney. However, with improved contrast agents and the ability of ultrasound to diagnose hydronephrosis and ureteral obstruction, retrograde pyelography is now infrequently performed.

## RADIONUCLIDE EXAMINATIONS

Radionuclide examinations depend on the detection of gamma radiation emitted by radioactive tracers that have been introduced into the body, usually intravenously but occasionally into other sites such as the stomach, blad-

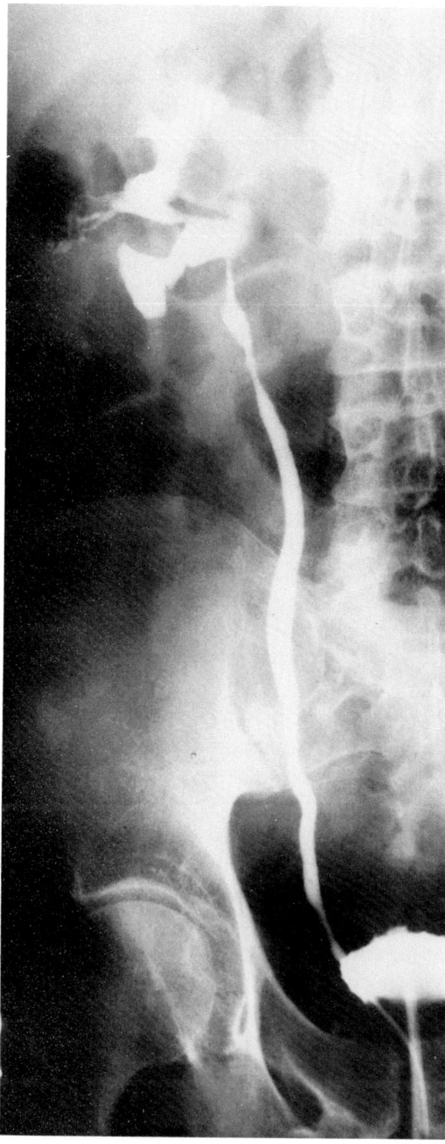

**Figure A.13. Retrograde pyelogram.** Contrast material injected into the right ureter permits visualization of a renal abscess cavity communicating with the pelvocalyceal system.

der, or subarachnoid space. A variety of metabolic and physical mechanisms cause the radioactive tracer to localize in specific anatomic sites. The results of radionuclide examinations can be displayed as static images that show the distribution of isotope activity within an organ or tissue, or as rapid-sequence images that demonstrate the transit of a radioactive bolus through a vessel or organ.

Although the resolution of radionuclide examinations is relatively low compared with those of standard radiography, radioisotope studies do provide a unique insight into the functional activity of various organ systems. Other advantages of radionuclide examinations are the lack of allergic or toxic reactions and the frequent high (compared with other modalities) sensitivity of radionuclide imaging in the early detection of disease states.

The most commonly used radionuclide is technetium 99m. This radioactive tracer is combined with DTPA for studies of the central nervous system and kidney, pyrophosphate for bone scans and infarct-avid imaging of the heart, macroaggregates or microspheres of albumin for lung scans, and sulfur colloid for studies of the liver and spleen. Other frequently used radionuclide compounds include gallium 67 citrate and indium 111–labeled red blood cells for abscess localization, iodine 131 and iodine 123 for thyroid scanning, thallium 201 to assess myocardial perfusion, and xenon 133 for inhalation lung scanning.

## SIALOGRAPHY

Sialography consists of filling the salivary ducts of the parotid or submaxillary glands with contrast material that is injected through a fine, thin-walled tube placed into a duct opening in the mouth. This technique can demonstrate calculi within the salivary ducts, ductal strictures, sialectasia, and inflammatory or neoplastic disease within the glands.

## SINUS TRACT INJECTION

In patients with a sinus tract extending to the skin or to an accessible mucous membrane, the introduction of oil- or water-soluble contrast material can demonstrate the site of origin and extent of the tract.

## SPLENOPORTOGRAPHY

Splenoportography refers to the visualization of the splenoportal venous system, and the liver parenchyma

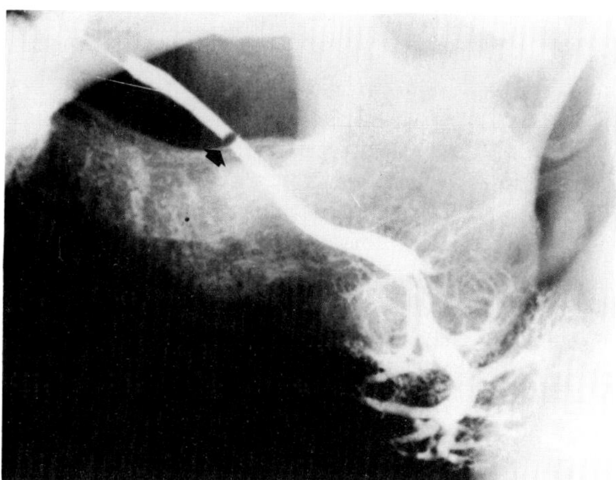

**Figure A.14. Normal submandibular gland sialogram.** The round lucency (arrow) within the duct represents an artifactual air bubble.

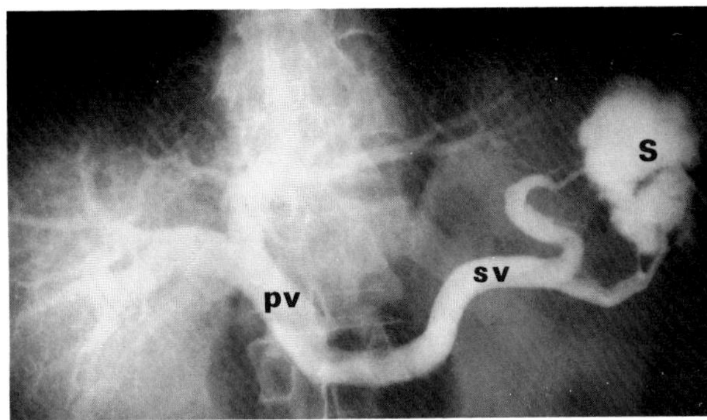


**Figure A.15. Normal splenoportogram.** S = spleen; sv = splenic vein; pv = portal vein.

to some extent, by the percutaneous injection of contrast material directly into the spleen. Although in the past this technique has been used to demonstrate obstruction of the splenic or portal veins and the presence of venous collaterals (varices), adequate visualization of the splenoportal system can generally be obtained during the venous phase of splenic or celiac axis arteriography (which are far safer and easier to perform).

## TOMOGRAPHY (CONVENTIONAL)

Tomography is a radiographic examination that permits the visualization of structures in a single plane of the body by using motion to blur out tissues above and below this level. This is accomplished by having the x-ray tube and cassette move simultaneously in opposite directions during the exposure. The stationary pivot point of this dual motion determines the level of the plane that is in focus. The amount of blurring depends upon the distance of the object or tissue from the level of the fulcrum. The thickness of the plane of tissue examined is determined by the distance traveled by the tube and film during the radiographic exposure. In addition to linear tomography, special tomographic units permit circular, elliptical, and hypocycloidal motion.

Tomography is used as a supplement to routine radiographic views whenever a lesion is obscured by overlying or underlying gas or other density. In the chest, tomography is of value in detecting cavities that are not visible on routine radiographs, calcification in small parenchymal nodules, and the integrity of the air-filled tracheobronchial tree. During excretory urography, tomography is often performed to eliminate confusing shadows due to overlying gas within the gastrointestinal tract as well as to provide improved detail of the renal margins and pelvocalyceal systems. In the skeleton, tomography can be employed to detect subtle fractures of the vertebrae and facial bones, to assess the degree of fracture healing, and to search for the lucent nidus within a suspected osteoid osteoma. Before the advent of computed tomography, plain tomography was the procedure of choice for detailing the tiny anatomic structures in the middle and inner ear that cannot be appreciated on conventional radiographs.

## ULTRASOUND

Ultrasound is a noninvasive imaging modality in which high-frequency sound waves are produced by the electric stimulation of a piezoelectric crystal, passed through the body, and attenuated (reduced in intensity) in relation to the acoustic properties of the tissues through which they travel. The crystal is mounted in a transducer, which also acts as a receiver to record echoes that are reflected back from the body whenever the sound wave strikes an interface between two tissues that have different acoustic impedances. A water-tissue interface produces strong reflections, while a solid tissue mass that contains only small differences in composition causes weak reflections. The display of the ultrasound image on a television monitor shows both the intensity level of the echoes and the position in the body from which they arose. Ultrasound images may be displayed as static gray-scale scans or as multiple images that permit movement to be viewed in "real time."

In general, fluid-filled structures have intense echoes at their borders, no internal echoes, and good through transmission of the sound waves. Solid structures produce internal echoes of variable intensity; since a solid structure tends to attenuate the sound waves, the posterior margin is less sharply defined than the anterior margin, and only a portion of the beam is transmitted.

The major advantage of ultrasound is its safety. To date, there is no evidence of any adverse effect on human tissues at the intensity level currently employed for diagnostic procedures. Therefore, ultrasound is the modality of choice for examinations of children and pregnant women, in whom there is a potential danger from the radiation exposure of other imaging studies. Ultrasound is used extensively to evaluate the intraperitoneal and retroperitoneal structures (with the exception of internal lesions of the gastrointestinal tract), to detect abdominal and pelvic abscesses, to diagnose obstruction of the biliary and urinary tracts, and to assess the patency of major blood vessels. Ultrasound may be used as a quick, inexpensive procedure to evaluate postoperative complications, though it may be difficult to perform in these patients because overlying dressings, retention sutures, drains, and open wounds may prevent the transducer from being in direct content with the skin.

The major limitation of ultrasound is the presence of acoustic barriers, such as air, bone, and barium. For example, air reflects essentially all of the ultrasound beam, so that structures beneath it cannot be imaged. This is a special problem when attempting to image the solid abdominal organs in a patient with adynamic ileus and is

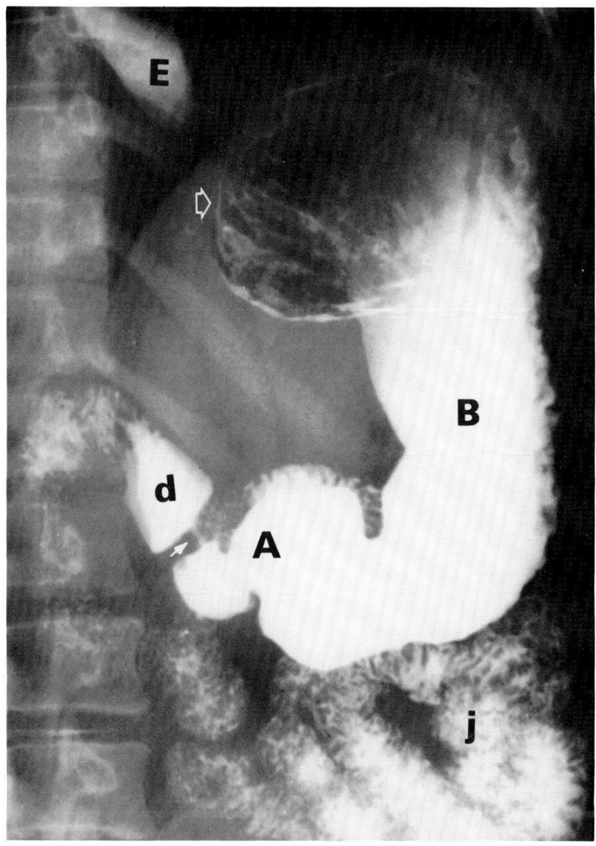


**Figure A.16. Normal upper gastrointestinal series.** E = distal esophagus; B = body of the stomach; A = gastric antrum; d = duodenal bulb; j = jejunum; open arrow = fundus of the stomach; closed arrow = pyloric channel.

the major factor precluding the ultrasound evaluation of the thorax. For an ultrasound examination of the pelvis, the patient is usually given large amounts of fluid to fill the bladder, thus displacing the air-filled bowel from the region of interest.

## UPPER GASTROINTESTINAL SERIES

The upper gastrointestinal series consists of a radiographic evaluation of the esophagus, stomach, duodenum, and proximal small bowel. It can be performed after the patient has ingested barium sulfate alone or barium combined with a gas-forming substance (double-contrast study). A variable number of fluoroscopic spot films and overhead views are obtained to search for disordered motility, hernias, ulcerations, and tumors or other causes of filling defects or narrowing. The identification of displacement or pressure deformities on the opacified esophagus, stomach, or small bowel indicates an extrinsic lesion outside the gastrointestinal tract.

The remainder of the small bowel can be assessed either by delayed films following the upper gastrointestinal series (small bowel follow-through) or by a specific study in which barium and air or another relatively low density contrast agent are injected directly into the small bowel through a tube positioned distal to the ligament of Treitz.

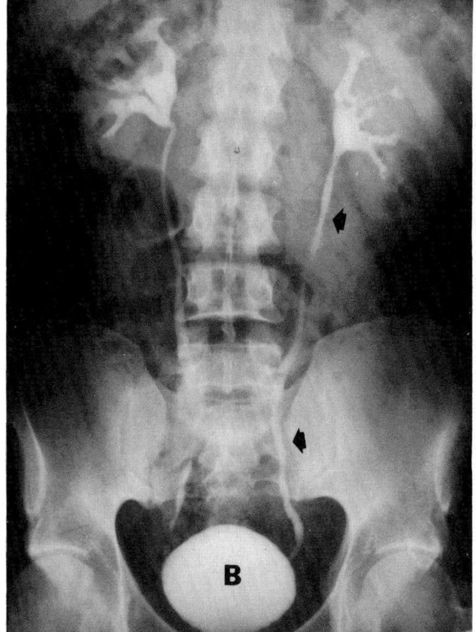


A

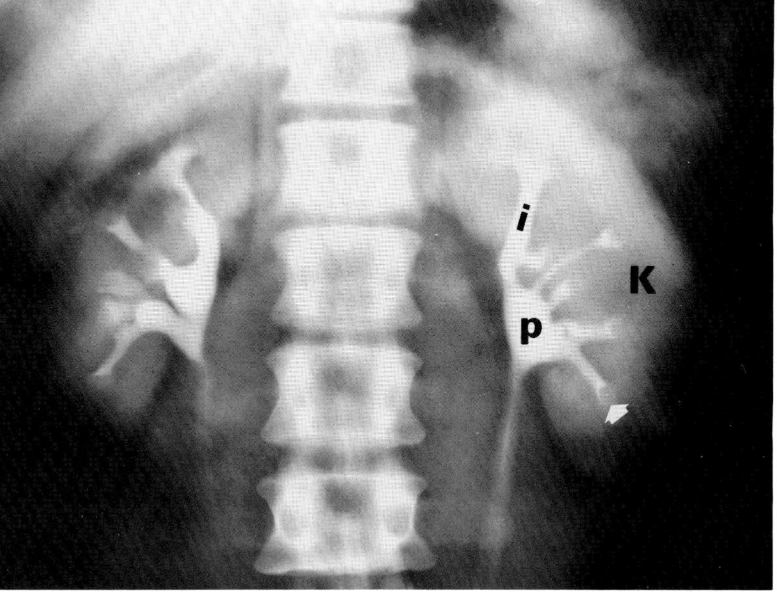


B

**Figure A.17. Normal excretory urogram (A) and normal nephrotomogram (B).** Note that the margins of the kidneys are visualized far better on the nephrotomogram. B = bladder; i = infundibulum; K = kidney parenchyma; p = renal pelvis; white arrow = renal calyx; black arrows = ureter.

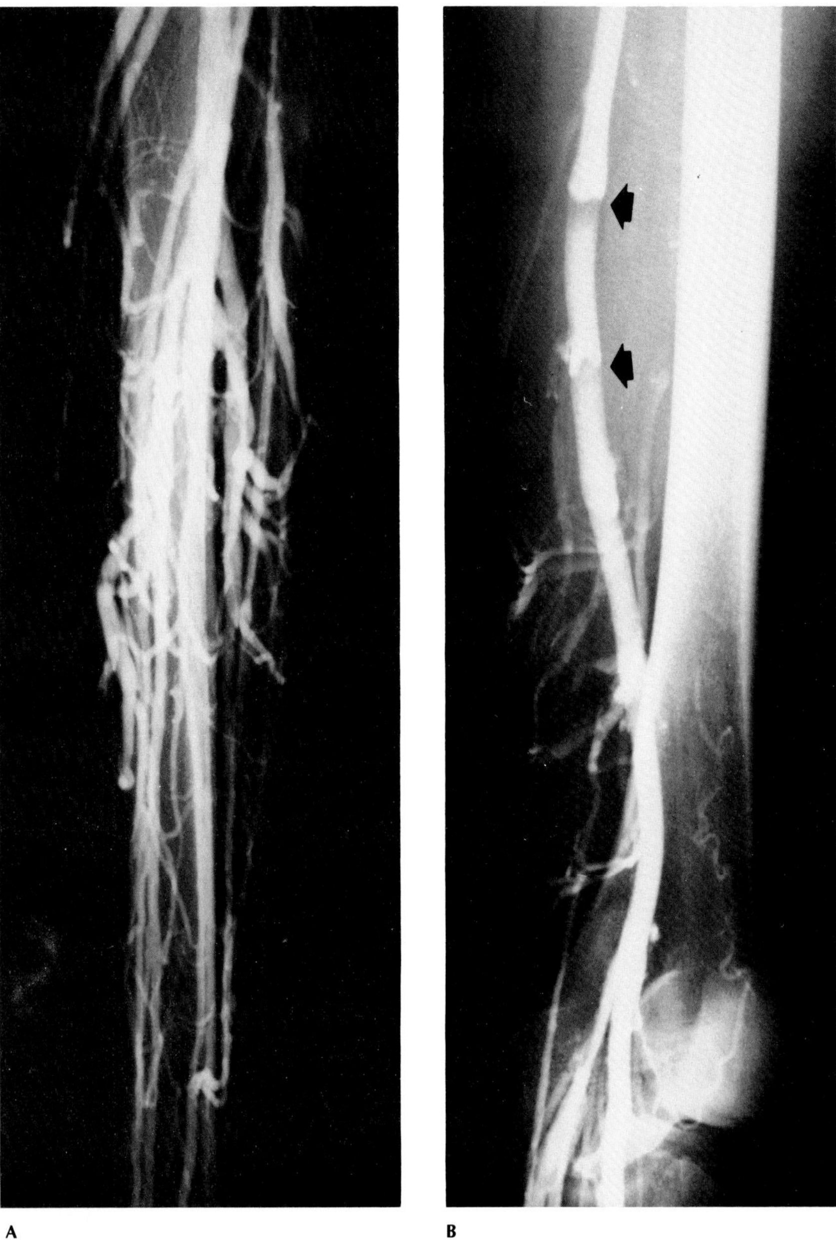


**Figure A.18. Peripheral venogram.** (A) Lower leg; (B) upper leg. Note the venous valves (arrows).

## UROGRAPHY, EXCRETORY

Excretory urography, also called intravenous pyelography (IVP), is an examination for demonstrating the size, shape, and position of the kidneys; intrinsic renal masses; and the pelvocalyceal system, ureters, and bladder. Following either the bolus injection or the drip infusion of contrast material, initial radiographs after 1 to 2 minutes demonstrate the nephrogram phase, in which there is opacification of the renal parenchyma. The pelvocalyceal system is opacified several minutes later, and delayed films show contrast material within the ureters and bladder. Tomography is often used to eliminate confusing shadows due to bowel gas; delayed films may be of value in demonstrating the site of obstruction in patients with ureteral calculi.

The excretory urogram is generally a safe procedure, with a mortality rate of approximately 1 in 50,000 examinations. There is a 1 to 2 percent incidence of morbidity (usually not severe), primarily urticaria, or nausea and vomiting.

## VENOGRAPHY, PERIPHERAL

Peripheral venography is performed primarily to assess the patency of the deep veins of the leg and to detect areas of deep venous thrombophlebitis or occlusion. The procedure consists of obtaining several radiographs during and following the slow injection of contrast material into one of the small veins on the dorsum of the foot or in the region of the ankle. In most cases, a tourniquet is used to occlude the superficial veins and thus enhance the opacification of the deep venous system in the leg and thigh.

The value of venography for the determination of deep venous thrombosis in the lower extremities is controversial. Because it may be impossible to fill all the veins in a normal extremity, it may be difficult to decide whether the failure to fill a given vein represents pathologic obstruction or merely a normal variant.

# Index

Page numbers in *italic* indicate illustrations.